Modern REFRIGERATION and AIR CONDITIONING

by

ANDREW D. ALTHOUSE, B.S., (M.E.), M.A.

Technical-Vocational Education Consultant
Life Member, American Society of
Heating, Refrigerating and Air Conditioning Engineers
Member, American Vocational Association

CARL H. TURNQUIST, B.S., (M.E.)., M.A.

Technical-Vocational Education Consultant
Associate Member, American Society of
Heating, Refrigerating and Air Conditioning Engineers
Member, Refrigeration Service Engineers Society
Member, Refrigerating Engineers and Technicians Association
Member, American Vocational Association

ALFRED F. BRACCIANO, B.S., M.Ed., Ed. Sp.

Technical-Vocational Education Consultant
Member, Refrigeration Service Engineers Society
Associate Member, American Society of
Heating, Refrigerating and Air Conditioning Engineers
Member, Air Conditioning Contractors of America
Member, American Vocational Association

South Holland, Illinois

THE GOODHEART-WILLCOX COMPANY, INC.

Publishers

Library of Congress Catalog Card Number 91-20208
International Standard Book Number 0-87006-915-2

2 3 4 5 6 7 8 9 10   92   96 95 94 93 92

The technological changes in the refrigeration and air conditioning industry in recent years have been very extensive. The accuracy, reliability, clarity, and reading level of this book have been achieved through the assistance of the following individuals:

Harold Boatman: General Service Technician, Michigan Consolidated Gas Company.

Daniel C. Bracciano, B.S.M.E.: Facilities Engineer, General Motors Corporation, Detroit, Michigan.

Robert H. Edgerton, B.S.M.E., M.S.M.E., Ph.D. Engineering: Professor of Engineering, Oakland University, Rochester, Michigan.

Robert M. Ottolini, B.M.E., M.S.M.E.: Engineering Manager, General Motors Corporation, Detroit, Michigan.

Jesse R. Riojas, B.S., Assoc. Climate Control Tech.: Instructor, Oakland Community College, Oakland Technical Center, Michigan.

Richard A. Romer, B.S.E.E., P.E.: Project Engineer, Rockwell International Corporation, Troy, Michigan.

Gloria M. Rossi, B.A., M.A., Ed. Sp.: Principal, Lapeer Community Schools, Lapeer, Michigan.

Connie Habermehl: Administrative Assistant, Associated Technical Authors, Rochester, Michigan.

Patricia I. Romer, Secretary, Associated Technical Authors, Rochester, Michigan.

Library of Congress Cataloging in Publication Data

Althouse, Andrew Daniel.
  Modern refrigeration and air conditioning / by Andrew D. Althouse, Carl H. Turnquist, Alfred F. Bracciano.
    p. cm.
  Includes index.
  ISBN 0-87006-915-2
  1. Refrigeration and refrigerating machinery.
2. Air conditioning. I. Turnquist, Carl Harold.
II. Bracciano, Alfred F. III. Title.
TP492.A43 1992
621.56--dc20   91-20208
                              CIP

INTRODUCTION

MODERN REFRIGERATION AND AIR CONDITIONING provides a thorough and authoritative course on the basic and advanced principles of refrigeration and air conditioning. It presents these important principles in an easily understood manner. It provides an excellent blend of theory with job-qualifying skill development.

MODERN REFRIGERATION AND AIR CONDITIONING also covers the practical application of refrigeration in all its branches: domestic, commercial, air conditioning, heat pumps, automotive air conditioning, thermoelectric, solar energy, absorption, special devices and applications. Up-to-date methods of installing, maintaining, diagnosing, and repairing refrigeration systems are explained.

Moreover, MODERN REFRIGERATION AND AIR CONDITIONING makes important use of four color photographs and line drawings, making principles and service information easy to understand. The unique color key used in this textbook shows different states and conditions of gases and liquids. Other colors indicate electrical, mechanical, and special components. Red is used to highlight safety, while blue is used to identify heads relating to service items. It is intended for use in refrigeration and air conditioning classes in high schools, technical schools, and community colleges. It may also be used in adult education classes and apprenticeship programs. It provides the foundation on which a solid, thorough knowledge of refrigeration and air conditioning may be based.

MODERN REFRIGERATION AND AIR CONDITIONING is written using both U.S. conventional units and SI metric units. The metric appears alongside the conventional unit throughout the textbook. Teachers and students may continue using that system with which they are most familiar.

MODERN REFRIGERATION AND AIR CONDITIONING is organized into parts, each covering a major area of knowledge. At the beginning of each chapter is a list of objectives identifying what will be learned by studying the chapter. Each chapter ends with questions that cover the content of the chapter and determine how well the user is meeting the objectives. A glossary of terms and an extensive collection of charts and technical characteristics are also provided for the user. A correlated Study Guide and Laboratory Manual provides lessons and assignments to help the student develop a solid background of principles and skills. The student is given a foundation for future studies.

MODERN REFRIGERATION AND AIR CONDITIONING is a valuable source for those interested in keeping abreast of the latest technological information. The rapid advances occurring in the field require a tremendous educational effort in order that the technician may remain knowledgeable. Ozone depletion has changed the types of refrigerants used, and new laws have altered the procedures that must be followed in removing refrigerants from systems. A knowledge of the recovery, recycling, and reclaiming of refrigerants is required.

MODERN REFRIGERATION AND AIR CONDITIONING provides an excellent basis of information for the technician in the areas of servicing and troubleshooting. The technician must have a thorough understanding of servicing and troubleshooting techniques in order to properly use the manufacturer's guides and remain current with today's technological changes.

Beginners and apprentices will find in MODERN REFRIGERATION AND AIR CONDITIONING an excellent aid for starting and pursuing a pleasant and profitable career. Experienced service technicians will find it a valuable guide and reference.

HOW TO USE THE COLOR KEY IN THIS BOOK

Colors are used throughout MODERN REFRIGERATION AND AIR CONDITION-ING to help show different states and conditions of gases and liquids. Other colors indicate electrical, mechanical, and special components. The following key shows what each color represents:

1. SYSTEM SCHEMATIC FOR REFRIGERATION/HEATING

Operating Medium

[DARK RED] —HIGH PRESSURE LIQUID

[LIGHT RED] —HIGH PRESSURE VAPOR

HOT

[DARK BLUE] —LOW PRESSURE LIQUID

[LIGHT BLUE] —LOW PRESSURE VAPOR

[DARK GREEN] —LOW PRESSURE WATER

COLD

[LIGHT GREEN] —LOW PRESSURE STEAM

Absorption Systems

[BROWN RED] —HIGH PRESSURE WATER AND AMMONIA
(WEAK SOLUTION—HOT)

[TURQUOISE] —LOW PRESSURE WATER AND AMMONIA
(STRONG SOLUTION—COLD)

Circulating Medium

COLD
[DARK GREEN]

LIQUID   AIR   VAPOR

HOT
[DARK RED]

WARM

Miscellaneous

FUEL
[DARK BROWN]

FLAME
[RED/YELLOW]

2. ELECTRICAL CONTROLS

[PURPLE] —COMPONENTS OF INTEREST

3. MECHANICAL

[BLUE GRAY] —HIGHLIGHT SYSTEM

[LIGHT ORANGE] —SPECIFIC COMPONENTS

4. SPECIAL COMPONENTS

[YELLOW] —HIGHLIGHT SYSTEMS

[DARK ORANGE] —SPECIFIC COMPONENTS

4

TABLE OF CONTENTS

1 FUNDAMENTALS OF REFRIGERATION 11

How mechanical refrigerator operates, 11
Heat and heat flow, 13
Cold, 13
Temperature measurement, 13
Thermometer scales, 14
Basic arithmetic, 15
Temperature conversion formulas, 16

Pressure, 19
Pascal's Law, 20
States of matter, 22
Specific gravity, 23
Force, work, power, 24
BTU, 25
First law, thermodynamics, 26
Latent heat, 27
Cooling capacity, 30

Energy units, 31
Mechanical refrigerating system, 32
Heat transfer principles, 33
Critical temperature/pressure, 34
Enthalpy, 34
Perfect gas equation, 35
Dalton's Law, 36
Abbreviations and symbols, 37
Safety, 37

2 REFRIGERATION TOOLS AND MATERIALS 39

Tubing, 39
Working with tubing, 41
Types of flares, 44
Annealing tubing, 46
Soldered/brazed fittings, 47
Pipe fittings and sizes, 52

Repairing threads, 53
Hand tools and gauges, 54
Instruments, 63
Gauge manifold, 65
Gauges, care and calibration, 66
Gaskets, 71

Fastening devices, 71, 72
Refrigerants, 72
Refrigerant oil, 73
Service valves, 73
Evacuating systems, 74
Safety, 74

3 BASIC REFRIGERATION SYSTEMS 75

Evaporative refrigeration, 76
Flooded systems, 76
External drive system, 77
High-side float system, 77
Automatic expansion valve control, 79
Thermostatically controlled

expansion valve, 80
Capillary tube control, 81
Multiple evaporator system, 81
Compound refrigerating systems, 82
Cascade refrigerating systems, 84
Modulating refrigeration cycle, 85

Drinking water cooler, 86
Expendable refrigerant systems, 88
Thermoelectric refrigeration, 88
Dry ice refrigeration, 90
Absorption systems, 91
Commercial systems, 95
Defrost systems, 97

4 COMPRESSION SYSTEMS AND COMPRESSORS 101

Laws of refrigeration, 101
Compression cycle, 102
Evaporator, 103
Accumulator, 104
Suction line, 104
Filter-driers, 105
Service valves, 105
Compressor, 105

Oil separator, 106
Condenser, 107
Liquid receiver, 108
Liquid line, 109
Controls, 110
Compressor types, 116
Compressor construction, 118
Motors, 135

Service valves, 135
Compressor cooling and lubrication, 136
Compressor volumetric efficiency, 137
Gaskets and seals, 139
Safety, 139

5 REFRIGERANT CONTROLS 141

Automatic expansion valves, (AEV), 141
Thermostatic expansion valve (TEV) principles, 144
TEV design, 147
Flash gas, 148
Superheat, 150

Sensing elements, 150
Thermal-electric expansion valve, 153
Pressure limiters, 153
TEV capacities, 156
Solenoid valves, 157
Equalizers, 158

Floats, 161
Capillary tubes, 162
Check valve, 165
Suction pressure valves, 165
Safety, 165

6 ELECTRICAL-MAGNETIC FUNDAMENTALS 169

Current, 169, 170
Types of electricity, 169, 170
Circuit fundamentals, 171
Meters, 174
Electrical values, 176
Power factor, 177
Resistors, 178
Connecting instruments, 180
Electrical materials, 181

Ohm's Law, 182
Electrical formulas, 183
Magnetism, 184
Capacitance, 188
Reactance, 190
Generators and motors, 190
Inductance, 193
Electronics, 194
Computers, 200

Electrical power, single- and
  three-phase, 201
Power circuits, 204
Transformer principles, 204
Relays, 208
Electrical codes, 209
Safety, 212

7 ELECTRIC MOTORS 213

Types and applications, 213
Speeds, 216
Start and run windings, 217
Starting current, 219
Connections, 219
Capacitor types, 221
Hermetic polyphase motors, 223

Terminals, hermetic motors, 225
Direct current and universal
  motors, 226
Horsepower, 227
Grounding and protective
  devices, 227
Motor temperature, 230

Standard data, 231
Fan motors, 233
Servicing, all types, 236
Servicing stuck hermetic
  compressor, 243
Safety, 244

8 ELECTRIC CIRCUITS AND CONTROLS 247

Complete wiring diagram, 247
Ladder diagram, 248
Fundamentals of control
  systems, 249
Temperature control principles, 251
Thermostatic control,
  construction, 253
Range adjustment, 253

Differential adjustment, 254
Controls for refrigeration,
  freezing, 258
Controls for air conditioning, 259
Motor safety controls, 265
Motor starting relays, 266
Servicing relays, checking and
  testing, 270

Defrosting controls, 271
Defrosting clocks, 277
Installing and servicing
  thermostats, 277
Temperature alarm system, 279
Grounding, 279
Safety, 279

9 REFRIGERANTS 281

Refrigerant identification, 281
Halide refrigerants, 283
Pressure-temperature curves, 282
Group one refrigerants, 283
Group two refrigerants, 292
Group three refrigerants, 294
Expendable refrigerants, 294

Cryogenic fluids, 295
Refrigerant cylinders, 295
Head pressures, 297
Refrigerator temperatures, 297
Using pressure-temperature
  tables, 298

Refrigerant applications, 298
Refrigeration oil, 301
Changing refrigerants, 301
New refrigerants, 301
Ozone protection, 301
Safety, 303

10 DOMESTIC REFRIGERATORS AND FREEZERS 305

Refrigerator storage, 305
Freezer storage, 305
Insulation, 306
Cabinets, 306
Mechanisms, manual defrost, 307
Electrical circuits, manual
  defrost, 308
Automatic defrost mechanisms, 313
Electrical circuits, automatic

defrost, 315
Mechanisms, automatic defrost, 315
Frost-free cabinets, 319
Frost-free circuits and
  mechanisms, 319
Solid-state controlled
  icemaker, 326
Chest-type freezers, cabinets,
  mechanisms, etc., 326

Upright freezers, cabinets,
  mechanisms, circuits, 331
Butter conditioner, 335
Cabinet hardware, 336
Gaskets, 337
Repairing finishes, 338
Thermometers, 339
Safety, 339

11 SERVICING AND INSTALLING SMALL HERMETIC SYSTEMS 341

Instruments, tools, and supplies, 341
Installing units, 342
Electrical supply, 342
Temperature-pressure
  conditions, 345
Trouble signals and
  troubleshooting, 346
External diagnosis and

servicing, 350
External cleaning, 353
Starting stuck compressors, 355
Internal diagnosis and service, 355
Gauge manifolds, 357
Hermetic service valves and
  adaptors, 359
Process tube and adaptors, 362

Piercing valves, 363
Core valves, 364
Leak detection and repair, 366
Charging a hermetic system, 369
Locating compressor faults, 374
Checking for restrictions, 374
Replacing filter-driers, 376
Hot gas defrost problems, 378

Dismantling system, 379
Installing a motor compressor, 381
Repairing condensers and
  evaporators, 382
Servicing capillary tubes, 382

Overhauling a hermetic
  system, 383
Repairing/rebuilding motor
  compressors, 384
Evacuating a system, 387

Charging/testing rebuilt
  systems, 393
Cycling times, 395
Shutting down units, 395
Safety, 395

12 COMMERCIAL SYSTEMS 397

Refrigerating mechanisms, 397
Commercial hermetic units, 401
Outdoor air-cooled condensing
  units, 406
Compressor types, 412
Condenser types, 415
Cooling towers, 418
Evaporative condensers, 421
Liquid receiver, 422
Evaporators, 422

Defrost systems, 436
Heat exchangers, 443
Motor controls and starters, 443
Ice maker controls, 448
Vending machine controls, 448
Defrost timers, 449
Valves, 454
Two-temperature valves, 454
Surge tanks, 458
Compressor protection

devices, 458
Oil control systems, 459
Water valves, 463
Refrigerant lines, 468
Mufflers, 469
Sight glasses, 469
Moisture indicators, 470
Filter-driers, 471
Engine-operated systems, 472
Safety, 473

13 COMMERCIAL SYSTEMS APPLICATIONS 475

Cabinets, construction and
  types, 475
Soda fountain, 484
Dispensing freezers, 486
Water cooler, 487
Modular refrigeration systems, 489

Automatic ice maker, 490
Vending machines, 491
Milk cooler/dispensers, 492
Industrial freezing of foods, 494
Truck refrigeration, 495

Railway car refrigeration, 497
Marine refrigeration, 499
Snow making, 499
Cryogenic applications, 500
Safety, 500

14 SERVICING AND INSTALLING COMMERCIAL SYSTEMS 501

Types of commercial
  installations, 501
Noncode installations, 501
Installing individual components,
  501, 503
Electrical connections, 505
Service valves, 507
Leak testing, 509
Evacuating and charging, 509
Code installation, 514
Welding and brazing equipment, 517
Testing Code installations, 518
Servicing commercial units, 519

Service equipment, 519
General instructions, 519
Removing compressor, 525
Overhauling the compressor, 526
Repairing condensers and
  receivers, 532, 537
Fixing water flow problems, 538
Evaporator service, 541
Installing, servicing expansion
  valves, 543
Installing, servicing two-
  temperature valves, 547
Removing moisture, 548

Servicing burnouts, 550
Servicing electrical circuits, 552
Servicing electric motors, 553
Troubleshooting and servicing
  hermetic compressors, 554
Servicing motor controls, 557
Servicing solenoid valves, 558
Servicing liquid line, 558
Servicing the suction line, 558
Locating troubles, 559
Refrigerant recovering and recycling,
  562
Safety, 562

15 COMMERCIAL SYSTEMS HEAT LOADS AND PIPING 565

Total heat load, 565, 573
Heat leakage variables, 565
K factor, 566
Air change heat load, 568
Product heat load, 568
Finding cabinet volume, 572
Computing heat loads, 576
Capacities of condensers and
  evaporators, 576
Locating and installing
  evaporators, 577
Types of evaporators, 580
Calculating evaporator area, 581
Evaporator design, 583
Water-cooling loads, 585

Ice cream freezing and storage
  loads, 586
Refrigeration cycle
  thermodynamics, 586
Pressure-heat diagrams and
  areas, 587, 588
Effective latent heat, 589
Saturated vapor, 589
Superheated vapor, 589
Specific heat, 589
Cascade system, 590
Two-stage compressors, 590
Bypass cycle, 590
Practical pressure-heat cycle, 591
Servicing, troubleshooting, 592

System capacity, 594
Compressor capacity, 594
Volumetric efficiency, 595
Coefficient of performance, 596
Calculating motor sizes, 596
Motor efficiency, 597
Condenser capacities, 597
Liquid receiver sizes, 598
Liquid line capacities, 601
Suction line capacities, 603
Discharge line piping, 608
Refrigerant control capacity, 608
Energy efficiency ratio, 609
Safety, 610

16 ABSORPTION SYSTEMS, PRINCIPLES AND APPLICATIONS................................**611**

Absorption systems types, 611
Principles of the absorption
  system, 612
Automatic defrosting, 616
Installing an absorption
  refrigerator, 616
Gas supply for continuous
  system, 618
Continuous system controls, 619

Pressure regulating valves, 620
Portable absorption
  refrigerators, 621
Residential absorption air
  conditioner cycle, 622
Construction, servicing residential
  absorption air conditioners, 624
Commercial absorption system
  cycle, 627

Absorption unit for air conditioning
  and heating, 628
Absorption system efficiency, 630
Installing, servicing absorption
  system refrigerators, 630
Servicing lithium bromide
  systems, 631
Safety, 631

17 SPECIAL REFRIGERATION SYSTEMS AND APPLICATIONS...........................**633**

Expendable refrigerant evaporator
  system, 633
Open cycle ammonia, 634
Cooling systems for liquefying
  gases, 639

Thermoelectric refrigeration, 640
Vortex tube, 640
Jet cooling systems, 642
Multistage systems, 643
Heat pipe, 645

Immersion freezing, 646
Cryogenic refrigeration, 646
Refrigerated containers, 646
Sterling cycle, 646
Safety, 647

18 FUNDAMENTALS OF AIR CONDITIONING.......................................**649**

Air and its properties, 649
Humidity, 650
Humidity measurement, 651, 653
Desiccants, 654
Humidity controls, 655
Psychrometric charts, 656
Vapor barriers, 659
Air velocity measurement, 659
Measuring velocity-pressure, 660
Draft measurement, 662

Ventilation, 662
Outdoor and indoor climate, 663
Temperature controls, 663
Sun heat load fundamentals, 663
Wind Chill Index, 664
Beaufort scale, 664
Comfort-Health index, 666
Degree days, 667
Types of air contaminants, 667
Pollutants, 668

Measuring filter efficiencies, 670
Air conditioning thermometers, 670
Manometer, 671
Barometers, 672
Heat insulation, 672
Types of heat exchange, 673
Noise and noise measurement, 673
Ecology and environment, 676
Safety, 677

19 BASIC AIR CONDITIONING SYSTEMS..**679**

Gravity warm air furnaces, 679
Fuel gas atmospheric burners, 679
Forced warm air furnaces, 679
Oil-fired hydronic heating
  system, 681
Forced warm air heating
  system, 683
Hydronic heating systems, 685
Room heating units and electrical

resistance, 686
Window or through-the-wall air
  conditioners, 687
Central air conditioner, complete
  system, 688
Absorption cycle, 689
Evaporative condenser, 690
Cooling towers, 693
Room humidifier and

dehumidifier, 694
Air-to-air heat pump, 695
Heat pumps, 697
Automobile air conditioning, 699
Steam jet cooling, 701
Vortex tube cooling, 702
Evaporative cooling, 703
Radiant heating, 705
Safety, 705

20 AIR CONDITIONING SYSTEMS HEATING AND HUMIDIFYING.......................**707**

Heating and humidifying
  systems, 707
Furnace design and
  construction, 710
Gas furnace operation, 711
High efficiency gas furnaces, 713
Furnace and chimney venting, 717
Fuel gases, 720
Gas burners, 720
Pilot lights and electric
  ignition, 722
Thermocouple circuits, 723
Gas controls, 724
Gas burner installation, 725

Servicing gas furnace, 727
Servicing warm-air heating systems
  and furnace inspection, 728
Hydronic heating systems, 728
Installing hydronic heating
  systems, 730
Servicing hydronic heating
  systems, 734
Preparing boiler system for heating
  season, 734
Steam heating systems, 734
Installing and servicing steam heating
  systems, 735
Fuel oils, 736

Oil furnaces, 737
Installing oil tank, 743
Servicing fuel oil burners, 746
Solid fuel heating, 747
Electric heat principles, 748
Radiant heat, 752
Installing and servicing heating
  coil, 754
Unit heaters, 754
Cogeneration, 755
Humidifiers, 755
Mobile home air conditioning, 758
Safety, 758

21 AIR CONDITIONING SYSTEMS, COOLING AND DEHUMIDIFYING 761

Atmosphere cooling
principles, 761
Cooling cycle, 762
Comfort cooling systems, 762
Cooling equipment, 762
Unit comfort coolers, 763

Installing, servicing window
units, 763
Installing, servicing console air
conditioners, 772
Remote comfort systems, 774
Recreational vehicle air

conditioning, 774
Evaporative cooling, 775
Storing cold for comfort
cooling, 775
Dehumidifying equipment, 775
Safety, 776

22 AIR CONDITIONING SYSTEMS—DISTRIBUTING, CLEANING, AND INSTRUMENTS 779

Conditioned air, 779
Weight of air, 779
Heat in air, 779
Ventilation requirements, 780
Noise, 781
Drafts, 783
Stratification of air, 783
Air ducts, types, sizes, 784
Calculating air volume, 793
Air circulation systems, 794
Room air movement, 795
Diffusers, grilles, and
registers, 796

Dampers, 796
Duct calculations, 799
Unit pressure drop system, 799
Total pressure drop system, 800
Return air ducts, 803
Balancing the system, 803
Special duct problems and
maintenance, 804
Fans, 804
Attic ventilation, 807
Air cleaning, 807
Electrostatic cleaning theory,
service, 810

Carbon filters, 814
Ultraviolet light, 814
Air curtains, 815
Instruments, 815
Smoke test, 819
Air volumes, 819
Visible airflow indicators, 820
Water analysis instruments, 820
Flow meters, 820
Electric meter, 820
Safety, 821

23 HEAT PUMPS AND COMPLETE AIR CONDITIONING SYSTEMS 823

Heat pump theory and
operation, 823
Heat pump efficiency, 826
Heat pump systems, 827
Supplemental resistance
heaters, 836
Installing heat pumps, 836
Servicing heat pumps, 836
Solar heat systems and heat
pumps, 839

Liquid solar collectors and heat
pumps, 839
Heat pump water heaters, 840
Complete through-the-wall air
conditioning systems, 841
Complete outside systems, 841
Installing and servicing rooftop
units, 841
Installing, inspecting, servicing
residential central air

conditioning systems, 852
Relative humidity control, 856
Two-duct systems, 857
Four-pipe systems, 857
Large comfort systems, 858
Total energy, 862
District heating and cooling
systems, 863
Safety, 863

24 SOLAR ENERGY 865

Electromagnetic energy, 865
Passive and active solar energy
systems, 866
Solar collectors, 867
Storage systems, 868
Closed water system heating
storage, 869
Warm air system heating
storage, 869

Angle of collector, 869
Collector surface heat
insulation, 869
Solar space heating
installations, 870
Solar domestic water heating, 870
Supplementary heat, 873
Heat pumps, 873
Solar energy cooling systems, 874

Converting solar energy to
electricity, 874
Solar cell construction, 874
Photovoltaic solar cell
applications, 875
Solar cell performance, 876
Safety, 876

25 AIR CONDITIONING AND HEATING CONTROL SYSTEMS 879

Controllers, control systems, 879
Thermostat types and
operation, 880
Heat anticipators, 884
Thermostat guard, 893
Servicing thermostats and
thermostat location, 893
Heating and air conditioning
controllers, 894
Relays, 894
Primary controls, 895

Sequential operating controls, 896
Limit controls, 896
Control circuits, 898
Control, types of, 899
Humidistats, 909
Penumatic systems, 911
Split-system controls, 914
Zone controls, 914
Total energy management
systems, 914
Energy consumption, 916

Direct digital control, 917
Localized controllers, 917
Remote controllers, 917
Centralized computerized
control, 918
System determination and
usage, 919
Servicing: Control system
diagnostics and repair, 922
Safety, 923

26 AIR CONDITIONING SYSTEMS HEAT LOADS 925

Types of heat loads, 925
Heat loads for heating or
  cooling, 925
Heat leakage, 926
Heat transfer rate, 927
U and R factors, 927, 931
Design temperatures, 929, 934
Building construction, 934
Roof design and construction, 935
Wall construction, 936

Basement heat loss, 936
Unheated spaces, 937
Infiltration, 937
Total heat load for heating, 938
Total heat gain for cooling, 938
Window heat load for cooling, 938
Sun heat load, 939
Heat lag, 943
Heat sources in buildings, 943
Insulation and vapor barriers, 944

Ponded roof, 944
Humidifier heat load, 945
Calculating unit air conditioner
  heat loads, 945
Electric heating insulation,
  ventilation, 946
Degree day method, 946
Commercial construction, 946
Energy conservation, 946
Safety, 947

27 AUTOMOBILE AIR CONDITIONING 949

Automotive air conditioner
  operation, 950
Operating conditions, 951
Cooling capacity, 952
Magnetic clutch, 952
Compressor protection and control
  switches, 953
Compressor types, 954
Compressor seals, 957
Belts, 957
Condensers, 959
Accumulator and receiver, 959

Refrigerant lines, 960
Evaporator and heater core, 961
Metering devices, 961
Suction pressure control valve, 962
Service valves, 964
Refrigerant, 964
Oil, 964
Air distribution, 964
Duct system, 965
Insulation, 965
Electrical circuits, 965
Blowers, 966

Manual control systems, 967
Electronic control systems, 967
Electronic control diagnostics, 967
Electronic control
  microprocessor, 967
Vacuum control systems, 968
Thermostat, 969
Truck air conditioning, 969
Servicing automobile air
  conditioners, 969
Servicing: Periodic maintenance, 975
Safety, 975

28 REFRIGERANT RECOVERY—RECYCLING—RECLAIMING 977

Chlorofluorocarbons (CFCs) and the
  ozone layer, 977
Physical properties of replacement
  refrigerants, 978
Recovering, recycling, reclaiming of
  refrigerants, 979
Refrigerant recovery equipment, 979
Vapor and liquid recovery

equipment, 980
Refrigerant recycling equipment, 981
Portable refrigerant management
  system, 982
Refrigerant reclaim procedure, 982
Identification of returnable
  cylinders, 984

Contractual agreements, 985
CFC recovery/recycle/reclaim safety
  standards, 984
Safety when removing refrigerant
  from system, 986
Mobile air conditioning, 986
Safety, 986

29 SERVICING AND TROUBLESHOOTING SIMPLIFIED 989

Servicing/troubleshooting, 989
Troubleshooting procedures, 989
Troubleshooting chart headings, 990
Owner's description of problem, 990
Checking possible cause, 990

Suggested remedy, 990
Troubleshooting charts, 991
Commercial system troubleshooting
  chart, 991
Domestic troubleshooting chart, 993

Ice flaker servicing, 995
Heating analysis guide, 998
Maintenance service contracts, 997
Safety, 1002

30 TECHNICAL CHARACTERISTICS 1005

Complete contents of tables, charts, and data listed on page 1004. Please refer to this page to find additional
technical information.

31 CAREER OPPORTUNITIES IN REFRIGERATION AND AIR CONDITIONING 1029

32 DICTIONARY OF TECHNICAL TERMS 1032

INDEX 1049

Chapter 1

FUNDAMENTALS OF REFRIGERATION

By studying this chapter, the technician will be able to:
- Describe the early development of refrigeration.
- Discuss the basic physical, chemical, and engineering principles applicable to refrigeration.
- Explain how cold preserves food.
- Define basic refrigeration terms.
- Explain principles of heat transfer.
- Compare Fahrenheit, Celsius, Kelvin, and Rankine temperature scales.
- Use temperature conversion formulas to convert from one temperature scale to another.
- Determine area and volume of cabinets.
- Explain the difference between psia (absolute pressure) and psig (gauge pressure).
- Describe the basic operation of a refrigerator.
- Discuss the differences between sensible heat, specific heat, and latent heat, and describe their applications.
- Explain physical laws which apply to refrigeration.
- Demonstrate and explain the relationship between SI metric and U.S. Conventional measurement.
- Recognize and use various symbols for SI metric units of measure.
- Make conversions between U.S. Conventional and SI metric systems of measurement.
- Calculate the enthalpy of water at various temperatures.

Users of this text who are unfamiliar with SI metrics have no cause for concern. Conventional measurements are carried alongside the metric. Reference is made to Chapter 30 as problems arise affecting metric usage.

1-1 DEVELOPMENT OF REFRIGERATION

Modern refrigeration has many applications. The first, and probably still the most important, is the preservation of food.

Most foods kept at room temperature spoil rapidly. This is due to the rapid growth of bacteria. At usual refrigeration temperatures of about 40 °F (4 °C), bacteria grow quite slowly. Food at this temperature will keep much longer. Refrigeration preserves food by keeping it cold.

Other important uses of refrigeration include air conditioning, beverage cooling, and humidity control. Many manufacturing processes also use refrigeration.

The refrigeration industry became important commercially during the 18th century. Early refrigeration was obtained by use of ice. Ice from lakes and ponds was cut and stored in the winter in insulated storerooms for summer use.

The use of natural ice required building insulated containers or iceboxes for stores, restaurants, and homes. These units appeared on a large scale during the 19th century.

Ice was first made artificially about 1820 as an experiment. Not until 1834 did artificial ice manufacturing become practical. Jacob Perkins, an American engineer, invented the apparatus which was the forerunner of our modern compression systems. In 1855 a German engineer produced the first absorption type of refrigerating mechanism, although Michael Faraday had discovered the principles for it in 1824.

Little artificial ice was produced until shortly after 1890. During 1890 a warm winter resulted in a shortage of natural ice. This helped start the mechanical ice-making industry.

Mechanical domestic refrigeration first appeared about 1910. J.M. Larsen produced a manually operated household machine in 1913. By 1918 Kelvinator produced the first automatic refrigerator for the American market. They sold 67 machines that year. Now millions of units are sold each year.

The first of the sealed or "hermetic" automatic refrigeration units was introduced by General Electric in 1928. It was named the Monitor Top.

Beginning with 1920, domestic refrigeration became one of our important industries. The Electrolux, which was an automatic domestic absorption unit, appeared in 1927. Automatic refrigeration units, for the comfort cooling part of air conditioning, appeared in 1927.

Fast freezing to preserve food for extended periods was developed about 1923. This marked the beginning of the modern frozen foods industry.

Mechanical refrigeration systems were first connected to heating plants to provide summer cooling in the late 1920's.

By 1940, practically all domestic units were of the hermetic type. Commercial units had also been successfully made and used. These units were capable of refrigerating large commercial food storage systems, comfort cooling of large auditoriums, and producing of low temperatures used in many commercial operations.

From a small, slow start in the late 1930's, air conditioning of automobiles has also grown rapidly. In 1935, Frederick McKinley Jones produced an automatic refrigeration system for long-haul trucks.

1-2 HOW A MECHANICAL REFRIGERATOR OPERATES

Removing heat from inside a refrigerator is somewhat like removing water from a leaking canoe. A sponge may be

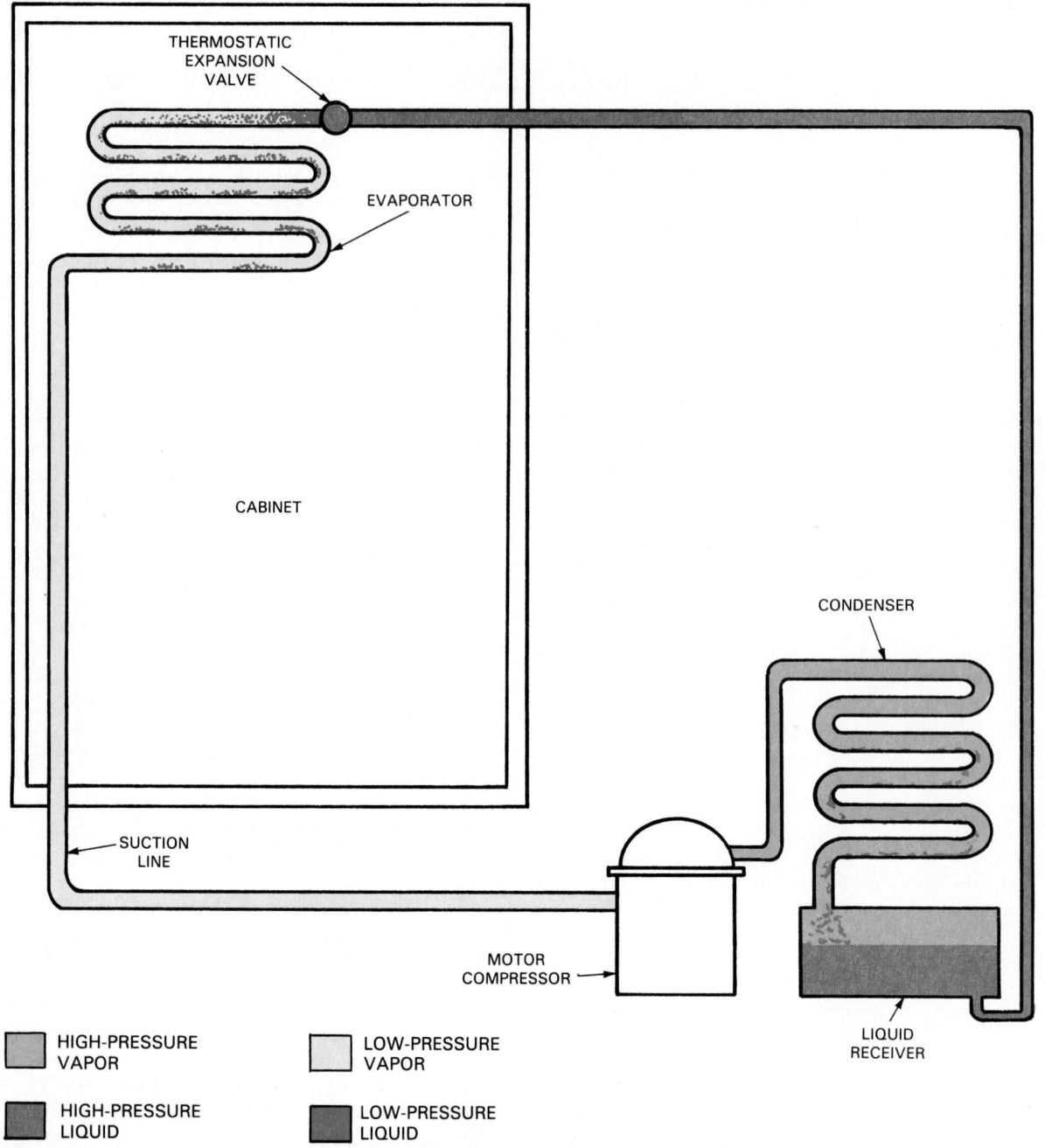

THERMOSTATIC
EXPANSION
VALVE

EVAPORATOR

CABINET

CONDENSER

SUCTION
LINE

MOTOR
COMPRESSOR

LIQUID
RECEIVER

| | HIGH-PRESSURE VAPOR | | LOW-PRESSURE VAPOR |
| | HIGH-PRESSURE LIQUID | | LOW-PRESSURE LIQUID |

Fig. 1-1. Elementary mechanical refrigerator. In operation, liquid refrigerant under high pressure (dark red) flows from liquid receiver to pressure reducing valve (refrigerant control) and into evaporator. Here pressure is greatly reduced (dark blue). Liquid refrigerant boils and absorbs heat from evaporator. Now a vapor, refrigerant (light blue) flows back to compressor and is compressed to high pressure (light red). Its temperature is greatly increased and, in the condenser, heat is transferred to surrounding air and the refrigerant cools, becoming liquid again. It flows back into liquid receiver and cooling cycle is repeated.

used to soak up the water. The sponge is held over the side, squeezed, and the water is released overboard. The operation may be repeated as often as necessary to transfer the water from the canoe into the lake.

In a refrigerator, heat instead of water is transferred. Inside the refrigerating mechanism, heat is absorbed, "soaked up," by evaporating the liquid refrigerant in the evaporator (cooling unit). This occurs as the refrigerant changes from a liquid to a vapor (gas), Fig. 1-1.

After the refrigerant has absorbed heat and has turned into a vapor, it is pumped into the condensing unit located outside the refrigerated space. The condenser works the

opposite of the evaporator. In the evaporator, liquid refrigerant enters one end and absorbs heat as it passes through the evaporator. By the time it reaches the end of the evaporator, it is all a vapor. As this vapor flows through the condenser under a high pressure and high temperature, it gives up its heat to the surrounding air. As it reaches the end of the condenser, the refrigerant, now cooled, has become a liquid again. We say that, in the condenser, the heat is "squeezed out." This cycle repeats until the desired temperature is reached.

Heat enters a refrigerator in many ways. It leaks through the insulated walls or enters when the door is opened. Still more heat is introduced when warm substances are placed in the refrigerator.

Heat is not destroyed to make the refrigerator cold. It is simply removed from the refrigerated space and released outside.

When presenting refrigeration system figures like Fig. 1-1, a common color coding system will be followed in this text:

Dark Red —High-Pressure Liquid ■

Light Red —High-Pressure Vapor ■

Dark Blue —Low-Pressure Liquid ■

Light Blue—Low-Pressure Vapor □

The paragraphs which follow will provide the technical foundation needed to understand the heat removal operation. This background is important for service and repair.

Service managers of refrigerating and air conditioning companies prefer service and installation technicians who are good mechanics. They also want employees who know the principles of mathematics and physics as these apply to refrigeration.

1-3 HEAT

Heat is a form of energy. It has a relationship to the atom, the smallest indivisible part of an element. (Indivisible means if one broke the atom down into more pieces it would no longer be that element.) All substances are made up of tiny atoms which combine to make molecules. All the atoms are in a state of rapid motion.

As the temperature of a substance increases, the atoms move more rapidly. As the temperature drops, they slow down. If all heat is removed from a substance (absolute zero), all molecular motion stops.

The U.S. conventional unit of heat is the British thermal unit (Btu). The metric unit of heat is the joule (J). If a substance is warmed, heat is added; if cooled, heat is removed.

The amount of heat in a substance is equal to the mass of the substance multiplied by its temperature. The amount of heat in a substance may greatly affect the nature of the substance. Adding heat causes most substances to expand. Removing heat causes them to contract.

Most substances change their physical state with the addition or removal of heat. For instance, water ice is a solid (under atmospheric pressure at a temperature below 0 °C). By adding heat to the ice, it will melt and become water (a liquid). Further addition of heat will cause the water to turn into a vapor (steam). The basic principle of operation of the compression type refrigeration cycle makes use of this principle in its operation.

1-4 HEAT FLOW

Heat always flows from a warmer to a cooler substance. What happens is that the faster moving atoms give up some of their energy to slower moving atoms. Therefore, the faster atom slows down a little and the slower one moves a little faster.

Heat causes some solids to become liquids or gases, or liquids to become gases. Cooling will reverse the process. This happens because the atoms making up the molecules of these substances act in a different way to temperature. Instead of moving faster or slower, one or more of the atoms in the molecule shift their positions.

1-5 COLD

Cold means low temperature or lack of heat. Cold is the result of removing heat. A refrigerator produces "cold" by drawing heat from the inside of the refrigerator cabinet.

The refrigerator does not destroy the heat, but pumps it from the inside of the cabinet to the outside. Heat always travels from a substance at a higher temperature to a substance at a lower temperature (second law of thermodynamics—see Chapter 30). Heat cannot travel spontaneously from a cold body to a hot body.

1-6 COLD PRESERVES FOOD

Spoiling of food is actually the growth of bacteria in it. As the molecules move slowly, they have an important effect on the bacteria present in most foods. Slowing movement by cooling the molecules makes all organisms more sluggish. Cold, or low temperature, slows up the growth of these bacteria. Foods, thus, do not spoil as fast. If the bacteria can be kept from increasing, the food will be edible longer.

Most foods contain a considerable amount of water. Food, therefore, must be kept slightly above freezing temperatures (32 °F, 0 °C).

If food is frozen slowly at or near the freezing point of water, the ice crystals formed are large, and their growth breaks down the food tissues. When defrosted, it spoils rapidly; appearance and taste are ruined.

Fast freezing at very low temperatures, O to −15 °F (−18 to −26 °C), forms small crystals which do not injure the food tissues. Food freezers are maintained at or below 0 °F (−18 °C). Food placed in them will freeze quickly.

Keep in mind the difference between refrigerating and freezing. The correct refrigerating temperature for fresh food is 35 °F (1.7 °C) to 45 °F (7.3 °C). To make ice, a temperature lower than 32 °F (0 °C) is needed.

1-7 TEMPERATURE AND TEMPERATURE MEASUREMENT

Temperature measures the heat intensity or heat level of a substance. Temperature alone does not give the amount of heat in a substance. It indicates the degree of warmth, or how hot or cold the substance or body is.

In the molecular theory of heat, temperature indicates the speed of motion of the molecule. It is important not to use the words "heat" and "temperature" carelessly.

Temperature measures the speed of motion of the atom. Heat is the speed of motion of the atom multiplied by the number of atoms (mass) so affected.

For example, a small copper dish weighing a few grams, heated to 1340 °F (727 °C) does not contain as much heat as 5 kilograms of copper heated to 284 °F (140 °C). However, its heat level is higher. Its intensity of heat is greater.

The U.S. conventional unit of temperature is the degree Fahrenheit. The SI unit of temperature is the kelvin (K). The temperature intervals (space between degrees) on the Kelvin scale are the same as Celsius.

Temperature is measured with a thermometer, usually through uniform expansion of a liquid in a sealed glass tube. There is a bulb at the bottom of the tube and a quantity of liquid (mercury or alcohol) inside.

The glass does not expand or contract as much as the

liquid during a temperature change. The liquid will rise and fall in the tube as the temperature changes. The tube is calibrated or marked off in degrees using the desired temperature scale. Fig. 1-2 shows a glass stem thermometer used in refrigeration and air conditioning work.

Some thermometers use metal to measure temperature. Metal will expand and contract as temperature rises and falls. This moves an indicator up and down the scale.

Thermometers have also been developed which indicate temperature by measurement of a very small electric voltage generated in a thermocouple. (See Chapters 6 and 18.) These instruments are called potentiometers. They are especially useful in the measurement of high temperatures above the useful range of a glass stem thermometer.

A radiometer is a thermometer which detects infrared rays released by a substance. This thermometer is very easy to use. No contact is needed with the substance whose temperature is to be measured. (See Fig. 30-1.)

Thermistors may also be used to measure temperatures. (See Chapter 6.)

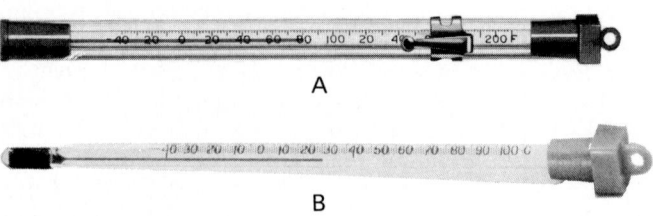

Fig. 1-2. Glass stem thermometers are used in refrigeration work. A—Fahrenheit thermometer has range from −40 to 210 °F. B—Celsius thermometer has range from −40 to 100 °C. (Marsh Instrument Co.)

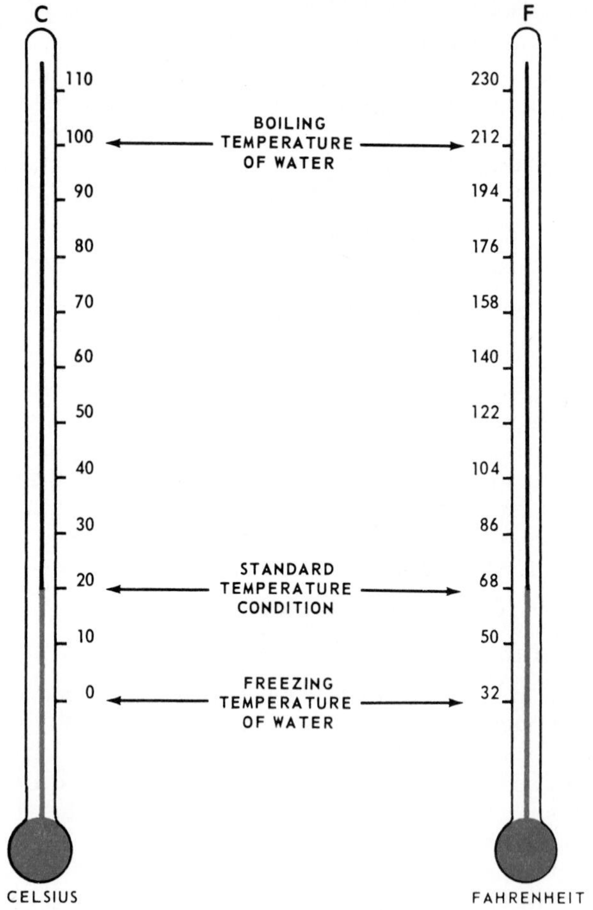

Fig. 1-3. A comparison of Celsius and Fahrenheit thermometer scales.

1-8 THERMOMETER SCALES, FAHRENHEIT AND CELSIUS

The two most common thermometer scales are the Fahrenheit and the Celsius, sometimes called the Centigrade, scale. The Celsius scale is named in honor of Anders Celsius, the Swedish astronomer who recommended the new system.

Two temperatures determine the calibration of a thermometer:
1. The temperature of melting ice.
2. The temperature of boiling water.
(Both must be at a pressure of 1 atmosphere or at sea level.)

On the Fahrenheit thermometer, the temperature of melting ice or the freezing temperature of water is 32 °F. The temperature of boiling water is 212 °F. This provides 180 spaces or degrees between the freezing and boiling temperatures.

On the Celsius thermometer, the temperature of melting ice or the freezing temperature of water is 0 °C. The temperature of boiling water is 100 °C. There are 100 spaces or degrees on the scale between freezing and boiling.

For a comparison of the Fahrenheit and Celsius scales, see Fig. 1-3. (Also, see Chapter 30.)

The freezing and boiling points are based on freezing and boiling temperatures of water at standard atmospheric pressure. Effects of pressure on these temperatures is explained in Para. 1-34 and Para. 1-35.

1-9 ABSOLUTE TEMPERATURE SCALES, KELVIN AND RANKINE

Absolute zero is that temperature where molecular motion stops. It is the lowest temperature possible. There is no more heat in the substance at this point.

Two absolute temperature scales are used with very low temperature work such as cryogenics. (See Para. 1-56.) These two scales are the Rankine (Fahrenheit Absolute) scale and the Kelvin (Celsius Absolute) scale.

The Rankine scale uses the same divisions as the Fahrenheit scale. Zero on this scale (0 °R) is located 460 degrees below 0 °F.

The Kelvin scale uses the same divisions as the Celsius scale. However, zero on the Kelvin scale (0 K) is 273 degrees below 0 °C.

Fig. 1-4 compares the Celsius, Fahrenheit, Kelvin, and Rankine thermometer scales.

Problem: What is the temperature at which water freezes and boils using the Kelvin scale?

Solution, Freezing Point: Water freezes at 0 °C. The Kelvin scale zero is 273 degrees below 0 °C. The freezing temperature of water is, therefore, 273 degrees above zero kelvin (K), or 273 kelvin. The freezing temperature is 273 K.

Solution, Boiling Point: Water boils at 100 degrees above 0 °C. The boiling point of water on the Kelvin scale will be: 100 + 273 = 373 K. Therefore, the boiling point is 373 K.

1-10 BASIC ARITHMETIC

+ means plus or add.
  Example: 4 + 4 = 8
= means equal to or of the same value.
− means minus, subtract or take away.
  Example: 4 − 3 = 1
× means multiply by, or times.
  Example: 4 × 5 = 20
÷ means divide by.
  Example: 12 ÷ 2 = 6
• means multiply by, or times.
( ) means parenthesis; do the arithmetic inside the parenthesis first.
  Example: (7 − 3) + 2 = (4) + 2 = 6
  Some calculations use parentheses instead of a multiplication sign.
  Example: (4) (5) = 20
( )^2 means that the number inside the parenthesis is to be multiplied by itself or squared.
  Example: (4)^2 = 4 × 4 = 16
( )^3 means that the number inside the parenthesis is to be multiplied by itself twice, or cubed.
  Example: (4)^3 = 4 × 4 × 4 = 64
$\frac{a}{b}$ means that the top number, ''a,'' is to be divided by the bottom number, ''b.''
  Example:
  If ''a'' = 6, and ''b'' = 2, $\frac{a}{b} = \frac{6}{2} = 6 ÷ 2 = 3$

Δ (delta) means a difference
  Example:
  Δ T = temperature difference,
  for instance, 0 °C to 40 °C.

Most calculations include the use of basic units. Basic units are expressed in digits. In the statement, 7 × 8 = 56, 7 and 8 are digits; 56 is made up of two digits, 5 and 6. In the metric system, multiples of digits are on the basis of 10. For example: The digit 1, if divided by 10, would be 0.1; each subsequent division of 10 would result in 0.01, 0.001, and the like. The prefix (name) for these follow. The digit 1, if multiplied by 10, would be 10; each subsequent multiplication by 10 would result in 100, 1 000, 10 000, 100 000, and the like. *Each level of multiplication or division has a name:*

Symbol	Prefix	Quantity	Pronunciation
M	mega	= 1 000 000	like megaphone
k	kilo	= 1 000	kill'-oh
h	hecto	= 100	heck'-toe
da	deka	= 10	deck'-uh
basic unit		= 1	
d	deci	= 0.1	dess'-ih
c	centi	= 0.01	sen'-tih
m	milli	= 0.001	like military
μ	micro	= 0.000 001	my'-crow

In many calculations, it is difficult to work with numbers using many zeros either ahead of or behind the decimal point. A special number, called a ''powers of ten,'' may be used to express these types of numbers.

''Power of 10'' means that the number 10 is multiplied by itself the desired number of times to obtain the required number of zeros. The number of times the ten is to be

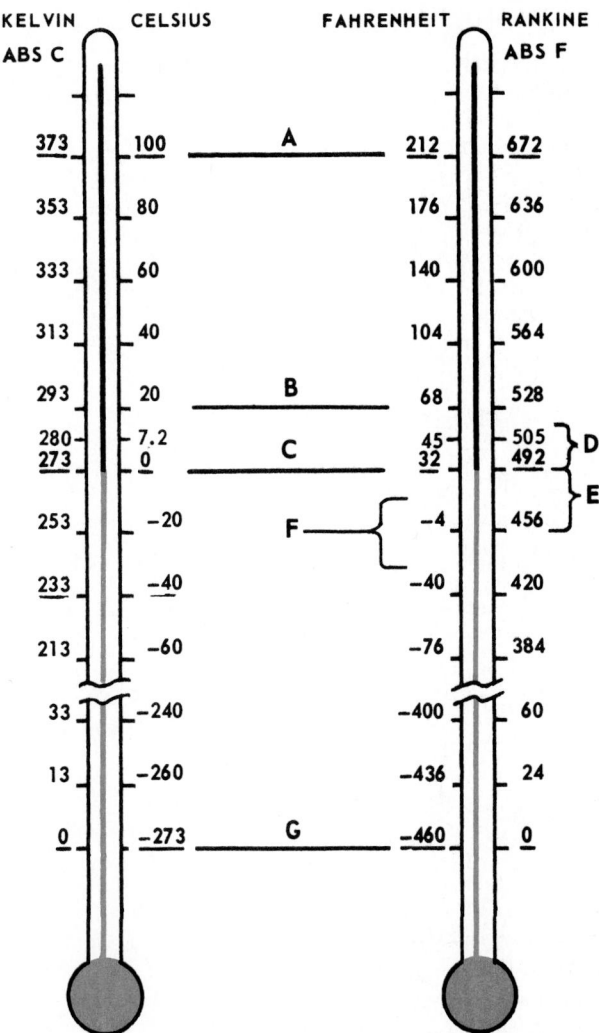

Fig. 1-4. Celsius, Fahrenheit, Kelvin, and Rankine thermometer scales are compared. A—Boiling temperature of water. B—Standard conditions temperature. C—Freezing temperature of water. D—Temperature range for fresh food storage. E—Range of evaporator temperatures for food. F—Temperature range for frozen food storage. G—Absolute zero.

multiplied by itself is shown by the small number above and to the right of the number 10. This number is also called the ''exponent.'' It works as follows:

For numbers larger than one:
  1 000 = 10^3 or (10 × 10 × 10)
  100 = 10^2 or (10 × 10)
  10 = 10^1 or (10)
For numbers less than one:
  0.1 = 10^{-1} or (0.10)
  0.01 = 10^{-2} or (0.10 × 0.10)
  0.001 = 10^{-3} or (0.10 × 0.10 × 0.10)

ROUNDING NUMBERS

In refrigeration calculations, it is not usually necessary to use fractions or decimals of a unit. When the decimal is less than 5, round to the number, ignoring the decimal. When the decimal is 5 or over, round to the next larger number. For instance, 35.5 becomes 36. If a problem has been carried two or more decimal places and less accuracy is required, it is acceptable to round such numbers to a single decimal. For instance: 3.52 may be rounded to 4.

1-11 TEMPERATURE CONVERSION FORMULAS

It is sometimes necessary to convert a temperature from one scale to another. Formulas have been developed for this:

°C = temperature in degrees Celsius
°F = temperature in degrees Fahrenheit
 K = temperature in degrees Kelvin
°R = temperature in degrees Rankine

1. To convert Celsius degrees to Fahrenheit degrees:
   Formula:

   $$\text{Temp. to: } °F = (\frac{180}{100} \times \text{Temp. } °C) + 32 \text{ or}$$

   $$°F = (\frac{9}{5} \times °C) + 32$$

   Example: Convert 75 °C into Fahrenheit.
   Solution:

   $$°F = (\frac{9}{5} \times 75) + 32 = (135) + 32 = 167 °F$$

2. To convert Fahrenheit degrees into Celsius degrees:
   Formula:

   $$\text{Temp. } °C = \frac{100}{180} \times (\text{Temp. } °F - 32) \text{ or}$$

   $$°C = \frac{5}{9} \times (°F - 32)$$

   Example: Convert 212 °F into °C.
   Solution:

   $$°C = \frac{5}{9} \times (212 - 32) = \frac{5}{9} \times (180) = 100 °C$$

3. To convert Fahrenheit degrees into Fahrenheit Absolute (Rankine) degrees:
   Formula: Temp. F_A (°R) = °F + 460
   Example: Convert 40 °F to F_A (°R)
   Solution:
   $°F_A$ (°R) = 40 + 460 = 500 degrees F_A (°R)

4. To convert Rankine degrees to Fahrenheit degrees:
   Formula: Temp. °F = °R − 460
   Example: Convert 180 °R to °F
   Solution: °F = 180 − 460
   $\qquad\qquad$ °F = −280 °F

5. To convert Celsius degrees to Kelvin:
   Formula: Temp. K = °C + 273
   Example: Convert − 10 °C to K
   Solution: K = −10 + 273 = 263 K

6. To convert Kelvin to Celsius:
   Formula: Temp. °C = K − 273
   Example: Convert 400 K to °C
   Solution: °C = 400 − 273
   $\qquad\qquad$ °C = 127 °C

7. To convert Rankine to Kelvin or Kelvin to Rankine, use the same ratios as for converting Fahrenheit to Celsius and for converting Celsius to Fahrenheit.

Temperature Difference Calculations:

Calculations which require converting Fahrenheit temperature difference to Celsius temperature difference and Celsius temperature difference to Fahrenheit temperature difference may be computed as follows:

$$°C \text{ temp. diff.} = \frac{5}{9} °F \text{ temp. diff.}$$

$$°F \text{ temp. diff.} = \frac{9}{5} °C \text{ temp. diff.}$$

Examples:
   Fahrenheit to Celsius:
   When the outside temperature is 10 °F and the inside temperature is 75 °F, the temperature difference is 65 °F. What is the temperature difference in °C?

   $$°C \text{ (temp. diff.)} = \frac{5}{9} \times 65 = 36 °C$$

   Celsius to Fahrenheit:
   When the outside temperature is 10 °C and the inside temperature is 26 °C, the temperature difference is 16 °C. What is the temperature difference in Fahrenheit?

   $$°F \text{ (temp. diff.)} = \frac{9}{5} \times 16 = 28.8 \text{ or } 29 °F$$

Throughout this text temperatures are given in both Fahrenheit and Celsius. Most of the Fahrenheit temperatures are rounded to whole numbers and the equivalent Celsius temperature is shown rounded to the nearest possible whole

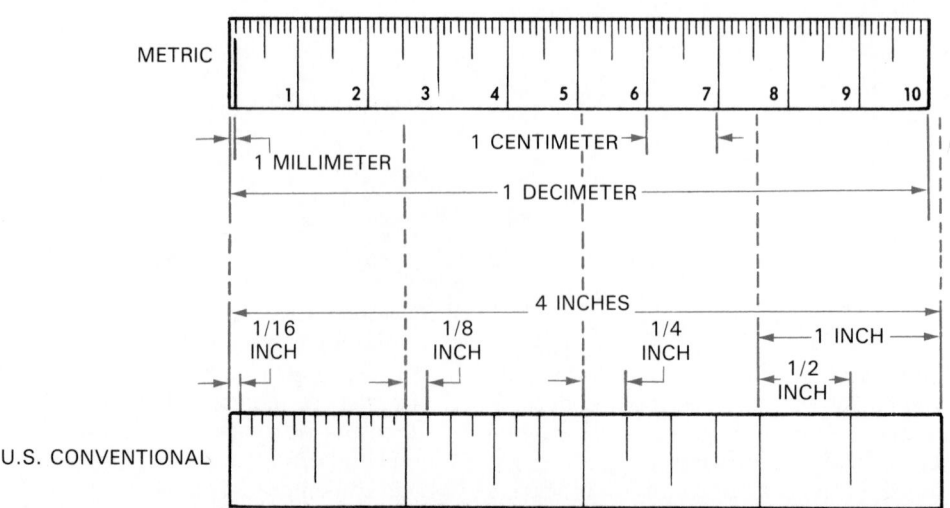

Fig. 1-5. U.S. Conventional and Metric Units of linear measurement are compared.

number. Where the Celsius temperature ends in 0.5 or more, the next higher temperature is used. If the Celsius temperature ends in a 0.4 or less, the next lower Celsius temperature is chosen.

Example: 40°F is equivalent to 4.4°C. This number is rounded off to 4°C.

Another example is 0°F. Carried to one decimal place, it equals −17.8°C. This is rounded to −18°C.

1-12 DIMENSIONS

Dimensions are measurements which are necessary in determining lengths, areas, and volumes.

LINEAR MEASUREMENT (Length)
Linear measurement considers only one dimension. Finding the length of a piece of copper tubing is an example.

U.S. Conventional Units-Decimals and Fractions of an Inch:

Measurement	How to Express the Measurement
0.001 in.	one thousandth of an inch
0.01 in.	one hundredth of an inch
0.1 in.	one tenth of an inch
1/64 in.	one sixty-fourth of an inch
1/32 in.	one thirty-second of an inch
1/16 in.	one sixteenth of an inch
1/8 in.	one eighth of an inch
1/4 in.	one fourth of an inch
1/2 in.	one half of an inch

Sometimes the symbol ('') indicates inches; for example, 6''. Occasionally the symbol (') indicates feet. The following is an example: 6'.

Units of Conventional Linear Measurement:

    12 inches = 1 foot
    3 feet = 1 yard
    5280 feet = 1 statute mile
    6080 feet = 1 nautical mile

Metric Units and U.S. Conventional Unit Equivalents:

    1 millimeter (mm) = 0.039 in.
    10 mm = 1 centimeter (cm) = 0.394 in.
    10 cm = 1 decimeter (dm) = 3.937 in.
    10 dm = 1 meter (m) = 100 cm = 39.37 in. = 3.28 ft.
    1 000 m = 1 kilometer (km) = 3280.8 ft.
    2.54 cm = 1 in.

Some linear metric units useful to the service technician are shown in Fig. 1-5. In measuring very tiny particles, the micron (μ) unit has been most used. The micron is one-thousandth of a millimeter (mm).

AREA MEASUREMENT
The measurement of area involves the measurement of two-dimensional space.

The area of an object is found by multiplying its length by its width (L × W). For example, the width of a table top is 90 cm and the length of the table is 150 cm. The area of the table top is 90 × 150 = 13 500 cm².

Some special formulas must be used when finding the area of certain objects. For example, the area of a circle is found by using the formula $A = \pi r^2$. The symbol, π, is always 3.1416, and r is the radius of a circle. It is equal to 1/2 the diameter. Therefore, this formula may also be

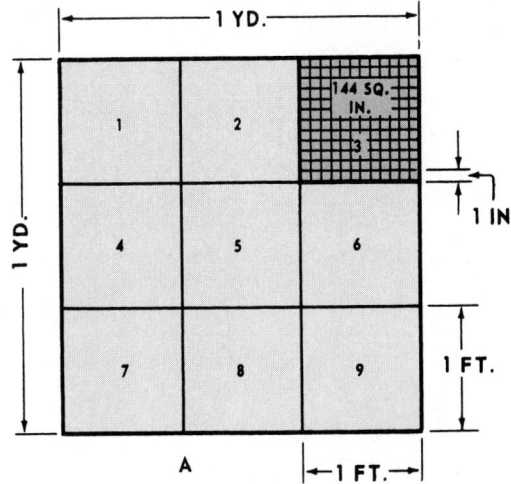

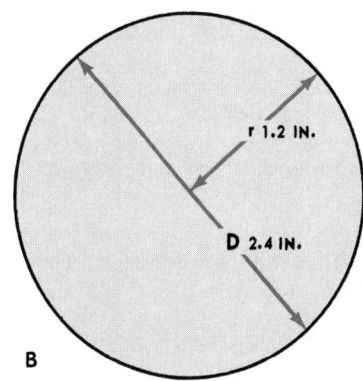

Fig. 1-6. Calculating standard areas in inches and feet. A—Area of rectangle is calculated by multiplying width by length. Remember, 144 sq. in. equal 1 sq. ft. and 9 sq. ft. equal 1 sq. yd. B—Area of a circle is calculated using the formula πr^2. Value of π is 3.1416. If diameter (D) of circle is 2.4 in., the radius (r), which is half the diameter, is 1.2 in.: $r^2 = r \times r = 1.2 \times 1.2 = 1.44$. Area of circle is 3.1416 × 1.44 = 4.52 in.².

expressed as:

$$A = \pi \frac{D^2}{4}$$

U.S. Conventional Units:
    square inches (sq. in.)
    144 sq. in. = 1 square foot
    9 sq. ft. = 1 square yard (sq. yd.)

These units are shown in view A of Fig. 1-6. The area of a circle is shown in view B, Fig. 1-6.

Metric Units:
    1 square centimeter (cm² or sq. cm) = 0.155 square inch
    1 square decimeter (dm² or sq. dm) = 10 cm × 10 cm = 100 cm² = 15.5 sq. in.
    1 square meter (m² or sq. m) = 1550 sq. in. = 10 dm × 10 dm = 100 square decimeters (dm²) = 10.76 sq. ft.

These units are shown in view A, Fig. 1-7. Refer back to Fig. 1-6 for the area of a circle.

VOLUME MEASUREMENT
The measurement of volume involves the measurement of three-dimensional space (cubic).

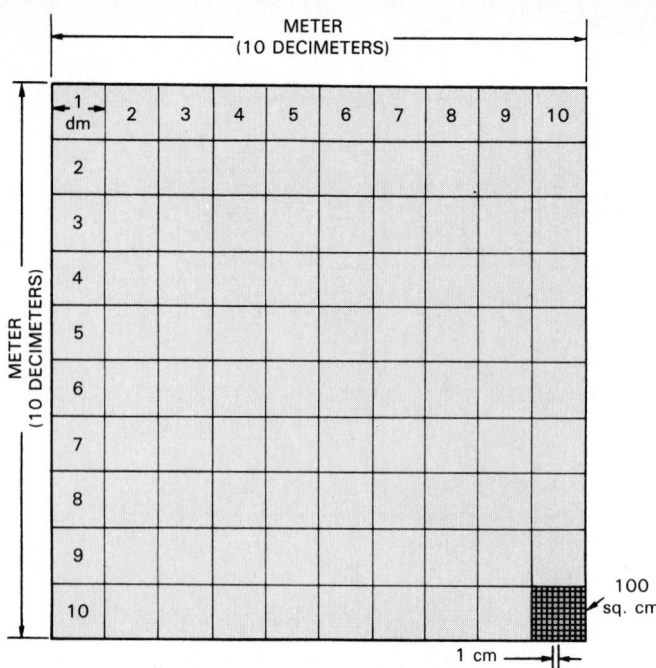

Fig. 1-7. Calculations of standard areas using metric units. Area of a rectangle is calculated by multiplying width by length. Area of circle is calculated with same formula as in Fig. 1-6, view B.

The volume of an object is determined by multiplying the width by the length by the height. An example is finding the volume of a cube (width × length × height, or W × L × H).

Once again, some special formulas must be used when finding the volume of certain objects. The volume of a cylinder, for example, is determined by multiplying the area of one end ($\pi\, r^2$) by the length (L) of the cylinder.

U.S. Conventional Units:
cubic inches (cu. in.)
cubic feet (cu. ft.)
cubic yards (cu. yd.)

1728 cu. in. = 1 cu. ft.
27 cu. ft. = 1 cu. yd.
1 cu. ft. = 7.48 gal.

These units are shown in view A, Fig. 1-8. The volume of a cylinder is shown in view B, Fig. 1-8.

Metric Units:
1 liter (L) = 1 000 cubic centimeters (cm³)
= 1.05 quarts (qt.)
= 61 cu. in. = 0.035 cu. ft.
1 000 cubic centimeters (cm³) = 1 cubic
decimeter (dm³)
1 cubic meter (m³) = 1.3 cu. yd.

These are shown in view A, Fig. 1-9. The volume of a cylinder is shown in view B, Fig. 1-9.

1-13 ANGULAR MEASUREMENT

Circles and arcs of a circle are measured in degrees. A complete circle has 360 degrees. See Fig. 1-10. A degree is further divided into minutes. Sixty (60) minutes equal one degree.

Minutes are further divided into seconds. Sixty (60) seconds equal one minute. These are the same names—minutes and seconds—that are used in measuring time. In angular measurement, they have a different meaning.

The angle does not depend on the size of the circle or the length of the diameter or radius. A part of a circle is called an arc. It is formed by two lines going out from the center of the circle and cutting across the circumference. It can be seen that, for a given central angle, the larger the circle, the longer the arc. The arc of the circle which includes a 90-degree central angle is called a 90-degree arc.

1-14 WEIGHT AND MASS

The amount of a substance is commonly related to how much it weighs. Food and metals, for example, are sold on the basis of their weight. The U.S. conventional units of weight are the grain, ounce, pound, and ton. This weight is often expressed as the force an object exerts on a scale.

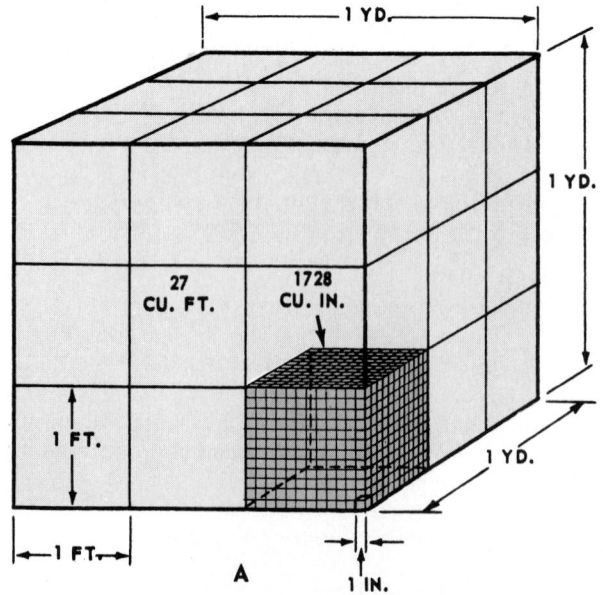

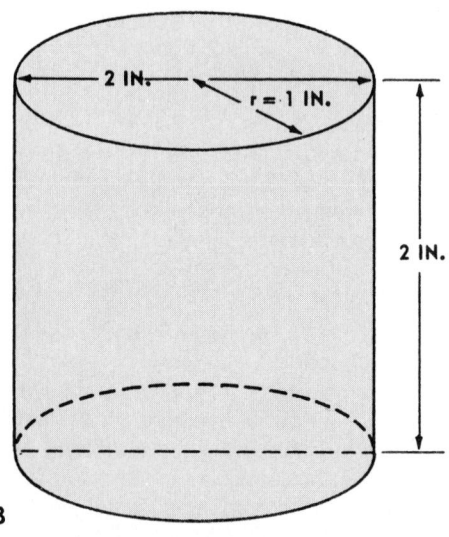

Fig. 1-8. Calculation of standard volumes in inches and feet. A—Volume is calculated by multiplying width by length by height. B—Volume of cylinder is worked out by multiplying area of end by length. (Cylinder dimensions are usually expressed in inches and decimals of an inch.)

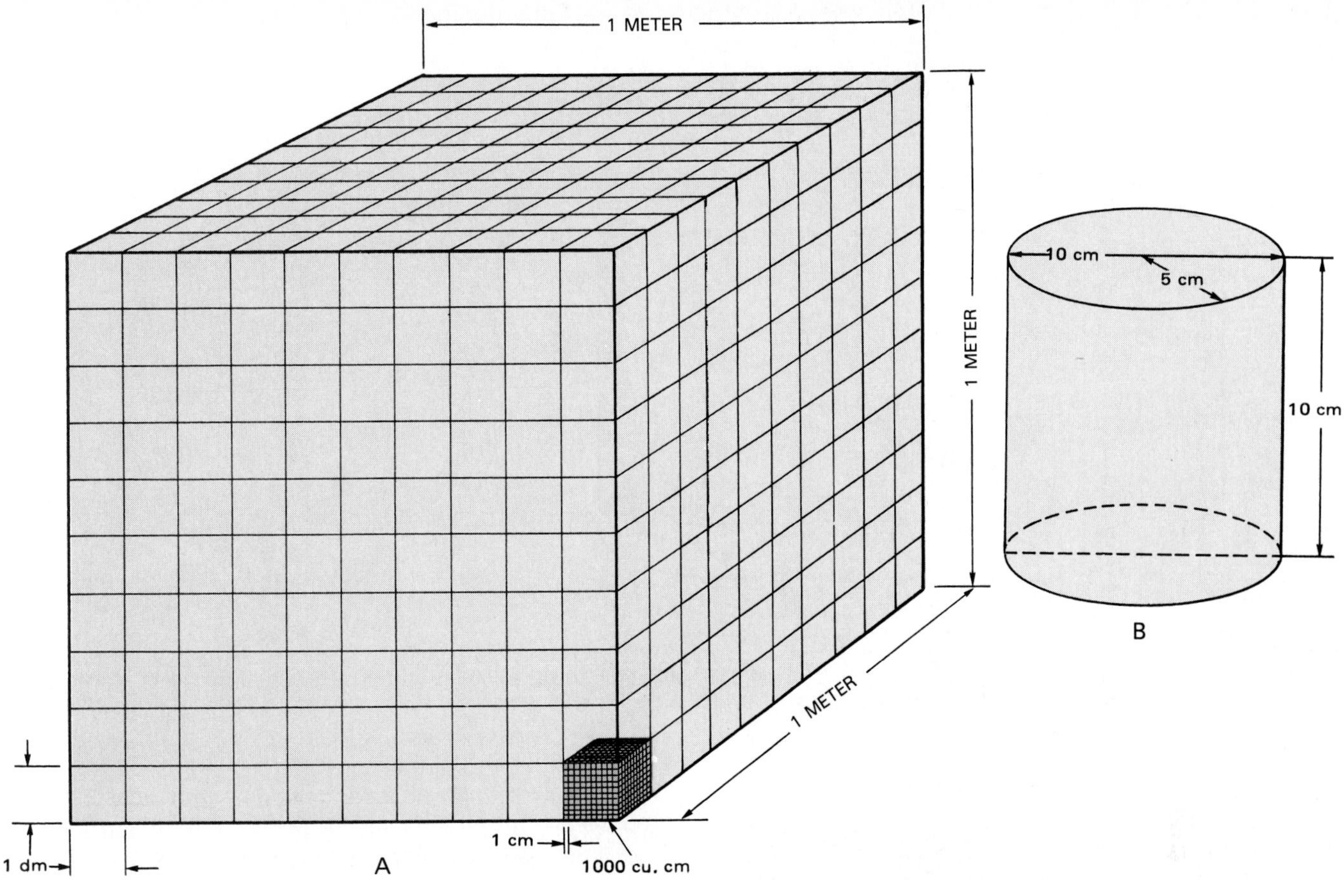

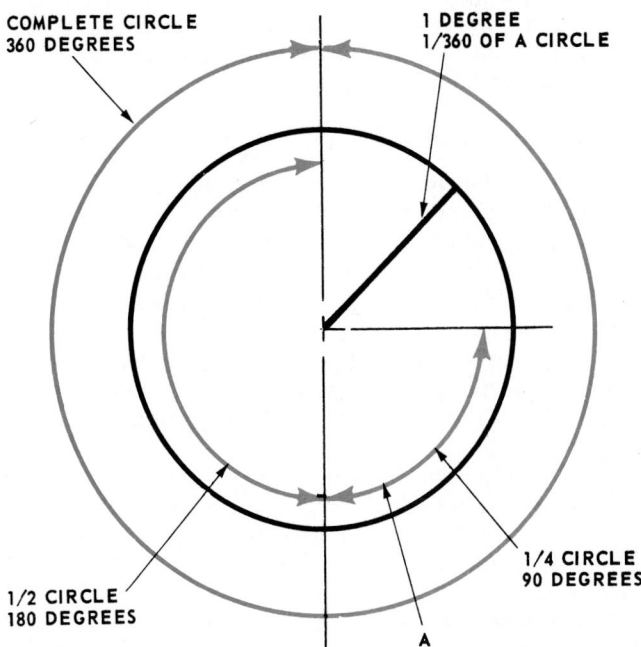

Fig. 1-9. Calculation of standard volumes using metric units. A—Volume is calculated by multiplying width by length by height. B—Volume of cylinder is calculated by multiplying area of end by length.

Gravitational force exerted on an object by the earth is expressed as its weight.

This attractive force (force of gravity) will accelerate an object if the object is released and allowed to fall. The same quantity of material will accelerate at a different rate depending upon the distance it is from the earth. To express the fact that it is the same quantity of material, even if the force of attraction upon it is less, one defines the quantity of material as its mass.

If an object weighs 1 lb., then at the earth's surface—where the gravitational force will accelerate it at 32.2 ft./sec.²—it is said to have a mass of 1 lb. Under these conditions, lb_f represents the weight, and lb_m represents the mass of the object.

In SI metric measurement units, the mass is measured in kilograms (kg). A kilogram mass is equivalent to 2.2 lb_m. In SI units, the unit of force is called the newton (N). A newton represents the force exerted on an object having a mass of 1 kilogram, where the gravitational acceleration is 1 m/sec². A 1-kilogram mass at the earth's surface will exert a force of 9.8 newtons, because in SI units the gravitational acceleration is 9.8 m/sec².

A 1-lb. mass will then have a mass of $\frac{1}{2.2} = 0.455$ kilograms, but will exert a force of $\frac{9.8}{2.2} = 4.455$ newtons on a weighing scale.

1-15 PRESSURE

Pressure is the force per unit area, and it is expressed in pounds per square inch (psi). It is also expressed in pascals (Pa) or kilopascals (kPa).

The normal pressure of the atmosphere at sea level is 14.7 pounds per square inch (psi) or 101.3 kPa. In engineering practice, this is usually rounded to 15 psi or 100 kPa.

Fig. 1-10. Angular measurement. Complete circle consists of 360 degrees. Half circle equals 180 degrees and 1/4 circle equals 90 degrees. ''A'' indicates 90 degree arc.

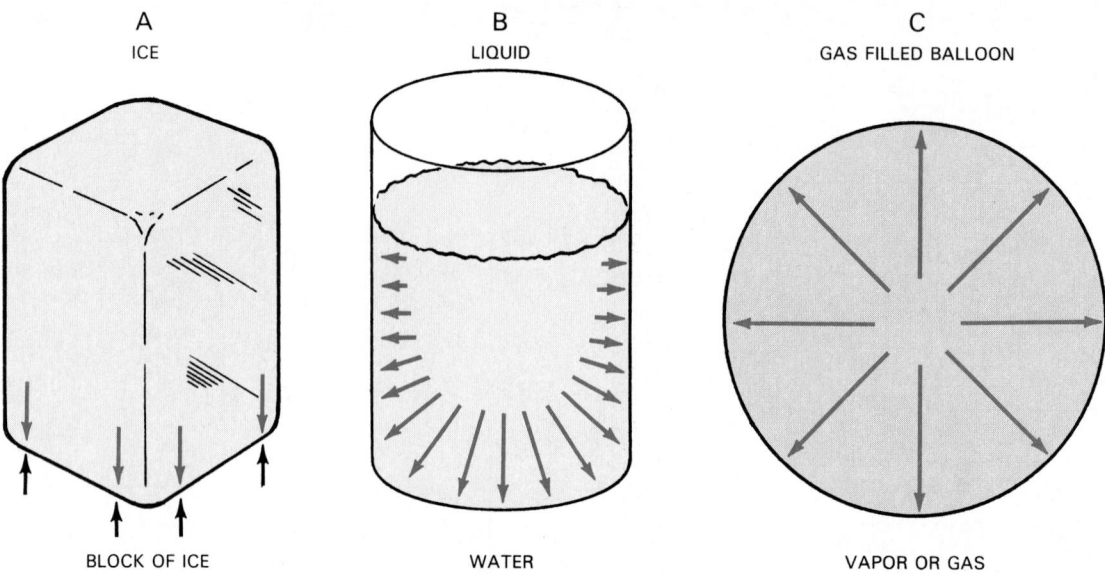

Fig. 1-11. Three states of substance such as water. A—Solid (ice). Mass of ice exerts downward force only. B—Liquid state. Water exerts pressure on container both vertically and horizontally. C—Vapor or gas in rubber balloon exerts pressure uniformly in all directions.

Operation of a refrigerating system depends mainly on pressure differences in the system. Substances always push on the surfaces supporting or containing them. A block of ice (a solid) exerts a pressure on its support. If the support were removed, the block would fall to another supporting level.

A liquid exerts a pressure on the sides and bottom of its container. A gas exerts a pressure on all surfaces of its container. Fig. 1-11 illustrates these types of pressures.

Using the U.S. conventional system, a solid weight of 1 pound with a bottom surface area of 1 inch square would exert a pressure of 1 pound (lb.) per square inch (1 psi) upon a flat surface.

A liquid in a container maintains an increasing pressure on the sides and bottom as the liquid depth increases. The pressure of gas in the container will depend on the quantity of the gas and the temperature.

1-16 PASCAL'S LAW

To honor Pascal, the SI metric system uses the term "pascal" as a unit of pressure. *A pascal is a newton per square meter (N/m²).*

A newton is the metric unit of force. *One newton is equal to the mass of 1 kilogram being accelerated at a rate of 1 meter per second per second.*

Pascal's Law states that: *Pressure applied upon a confined fluid is transmitted equally in all directions.* It is the basis of operation of most hydraulic and pneumatic systems.

Fig. 1-12 illustrates Pascal's Law. It shows a fluid-filled cylinder. A piston having a cross-sectional area of 645 mm² (one square inch) is fitted into a small cylinder connected to the larger cylinder. A force of 85 psig or 100 psia (690 kPa) is applied to the piston in the small cylinder. The pressure gauges show the pressure being transmitted equally in all directions.

One psia equals 6 894.8 pascals (Pa). This can be rounded to 6.9 kPa. One psig equals 15.7 psia. Then 15.7 psia is converted to the metric unit as 108 kPa.

The refrigeration engineer must deal with pressures both above and below atmospheric pressure. Metric gauges are calibrated so that zero on the gauge means that there is no pressure at all—not even atmospheric pressure. No

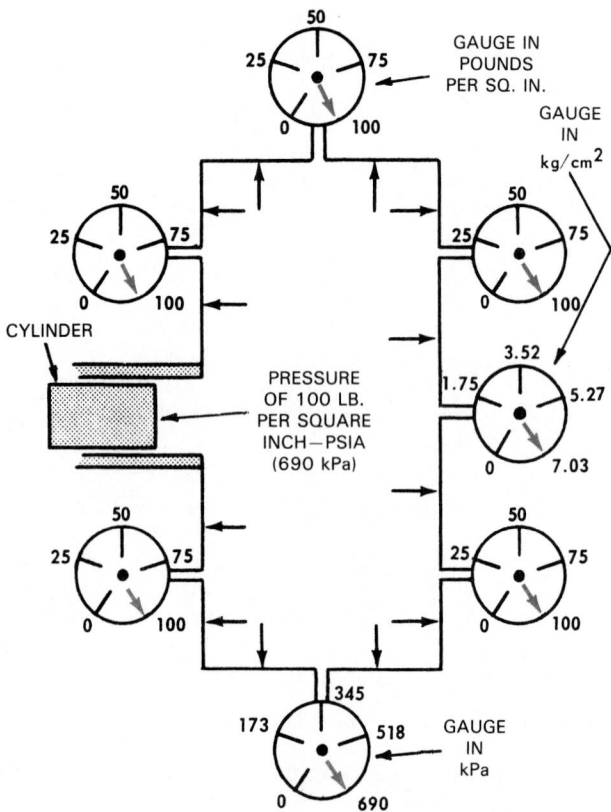

Fig. 1-12. Illustration of Pascal's Law. Pressure of 100 psia (690 kPa) is pressing against piston having head area of 1 sq. in. All gauges have same reading. Bottom gauge, calibrated in kilopascals, reads 690 kPa (100 psia). Fig. 1-17 is a reference which summarizes data about pressure scales.

negative pressures are used. A pressure of 5 kPa, then, is the same as 0.75 pounds per square inch (psi). Expressed in inches of mercury vacuum, it would be about the same as 28.5 inches of Hg vacuum.

In the U.S. conventional units, pressures above atmospheric pressure are measured in pounds per square inch (psi). Pressures below atmospheric are measured in inches of mercury (in. Hg) column. See Para. 1-17.

Some older references may show instruments calibrated in atmospheres, bars, and torrs. The bar is equal to one atmosphere; a millibar (mb) is equal to 0.001 of a bar. An atmosphere is approximately 14.7 pounds per square inch (psi). Gauges calibrated in atmospheres are marked 1, 2, 3, 4, and up. The numbers represent the number of atmospheres. Arbitrarily, the atmospheric gauges have been calibrated in atmospheres of 15 pounds per square inch (psi). One torr is 1/760 of an atmosphere.

1-17 PRESSURES—ATMOSPHERIC, GAUGE, AND ABSOLUTE

Atmospheric pressures are expressed in pounds per unit of area or in inches of liquid column height. The most popular instruments (gauges) are those that register in pounds per square inch *above* the atmospheric pressure (psig or psi).

A reading of 0 psi on the gauge is equal to the atmospheric pressure, which is about 14.7 psi (15 pounds per square inch is often used for working out problems).

The absolute pressure scale registers zero at a pressure which cannot be further reduced. A perfect vacuum is 0 pounds per square inch absolute (0 psia).

Pressure may also be indicated in other ways: 1. Inches of mercury (in. Hg). 2. Feet or inches of water column. These gauges may be calibrated either above atmospheric pressure or absolute pressure, depending on the construction. A mercury gauge is often used for measuring below atmospheric pressure.

The barometer in Fig. 1-13 is a mercury gauge. With a vacuum at the closed top of the tube, the atmospheric pressure will support a mercury column of 29.92 in. Hg in height at sea level under standard conditions.

A unit of measure which has been used for reading high vacuums (pressure close to an absolute vacuum) is the *torr*. One torr is equal to a pressure of 1 mm of mercury (mm Hg, 0 °C). It is named after the man who invented the mercury barometer. The unit torr may be expressed in fractions of an atmosphere. One torr = 1/760 of an atmosphere. A pressure of one torr is almost a perfect vacuum.

In solving most pressure and volume problems, it is necessary to use absolute pressures (psia). *Absolute pressure is gauge pressure plus atmospheric pressure.*

Example: Calculate absolute pressure when the pressure gauge reading is 21 psi (psi always indicates gauge pressure, and psia indicates absolute pressure).

Solution: absolute pressure equals gauge pressure plus atmospheric pressure:

psi + 15 = psia
21 + 15 = 36 psia

Air pressure or a vacuum can be measured with a column of water. To equal 29.92 in Hg, it would be about 34 feet high. The height is greater because water is so much lighter than mercury.

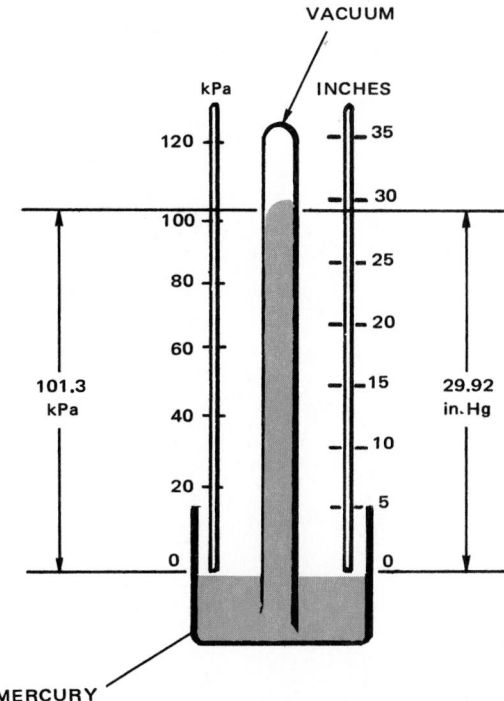

Fig. 1-13. Mercury barometer used for measuring atmospheric pressure. It consists of glass tube sealed at one end and open at other end. Fill tube with mercury. Then, sealing open end, invert it in container of mercury. When seal is removed, mercury will drop to level corresponding with atmospheric pressure. Glass tube should be about 34 in. (86.4 cm) long.

Since the service engineer must often test both pressures and vacuums in the same system, pressure gauges are made which will measure both. They are called compound gauges.

Compound gauges have two or more pressure scales. One measures pressure below atmospheric pressure. The other measures pressures above atmospheric pressure. Fig. 1-14 illustrates such a gauge.

It is also frequently necessary to convert inches of mercury into pounds per square inch absolute or to convert pounds per square inch absolute into inches of mercury. Formulas are available for making an accurate conversion.

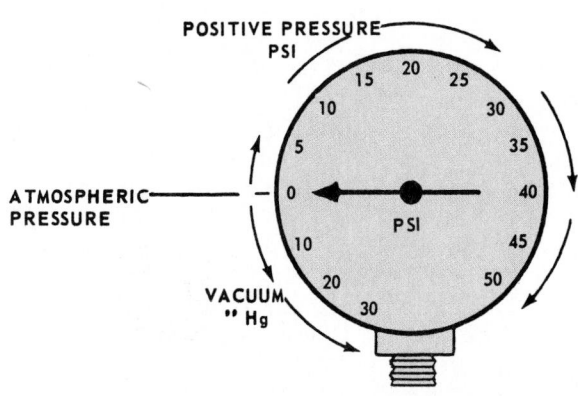

Fig. 1-14. Compound gauge measures both pressure above atmospheric in psi and pressure below atmospheric using units of in. Hg. Zero on this scale is atmospheric pressure.

INCHES of Hg	mm of Hg	psia	FT. of WATER
30		15	
(29.92)	760	(14.7)	33.40
29		14.5	
28	711	14	32.2
27		13.5	
26	660	13	29.9
25		12.5	
24	610	12	27.6
23		11.5	
22	559	11	25.3
21		10.5	
20	508	10	23.0
19		9.5	
18	457	9	20.7
17		8.5	
16	408	8	18.4
15		7.5	
14	356	7	16.1
13		6.5	
12	305	6	13.8
11		5.5	
10	254	5	11.5
9		4.5	
8	203	4	9.2
7		3.5	
6	152	3	6.9
5		2.5	
4	102	2	4.6
3		1.5	
2	51	1	2.3
1		0.5	
0	0	0	0

Fig. 1-15. Chart converts inches of mercury (in. Hg) into pounds per square inch absolute (psia).

Roughly 2 in. Hg equals 1 psi. A chart, Fig. 1-15, makes converting easy.

Water columns are usually designed for measuring small pressures above or below atmospheric pressure. For example, they can be used for pressures in air ducts, gas lines, and the like. A water column 2.3 feet high (or about 28 in.) equals 1 psi.

These pressure measuring devices are called manometers. They are calibrated in inches of water column. Fig. 1-16 shows types of water manometers.

In some high-pressure refrigerating machines, pressure gauges are calibrated in atmospheres. One atmosphere corresponds to about 15 pounds per square inch (psi), two atmospheres to 30 psi, three atmospheres to 45 psi, four atmospheres to 60 psi, and so on.

SI UNITS USED

Fig. 1-17 uses SI units. Compare the scales.

In SI units, atmospheric pressures are expressed in kPa (kilopascals). Normal atmospheric pressure is 101.3 kPa. But, for practical engineering purposes, gauges are often calibrated at 100 kPa for atmospheric pressure.

Pressures lower than atmospheric are called partial vacuums. Zero on the absolute pressure scale is at a pressure which cannot be further reduced. A perfect vacuum is 0 Pa (pascals). The pascal, rather than the kilopascal, is used for measuring high vacuums (pressures close to an absolute vacuum).

Fig. 1-18 illustrates a pressure gauge used in refrigeration work. It is calibrated in kilopascals rather than in psi.

1-18 THE THREE PHYSICAL STATES

Substances exist in three states, depending on their temperature, pressure, and heat content. For example, water at atmospheric pressure is a solid at temperatures below 32 °F (0 °C), and a liquid from 32 °F (0 °C) to 212 °F (100 °C). At 212 °F (100 °C) and above it is a vapor (gas).

Water is shown in its three states in Fig. 1-11. In this example, the physical state is controlled both by temperature and pressure. The temperatures just given apply only when atmospheric pressure is at 14.7 pounds per square inch (psi) or 101.3 kilopascals (kPa).

1-19 SOLIDS

A solid is any physical substance which keeps its shape even when not contained. It consists of billions of molecules, all exactly the same size, mass, and shape. These stay in the same relative position to each other. Yet, they are in a condition of rapid motion or vibration. The rate

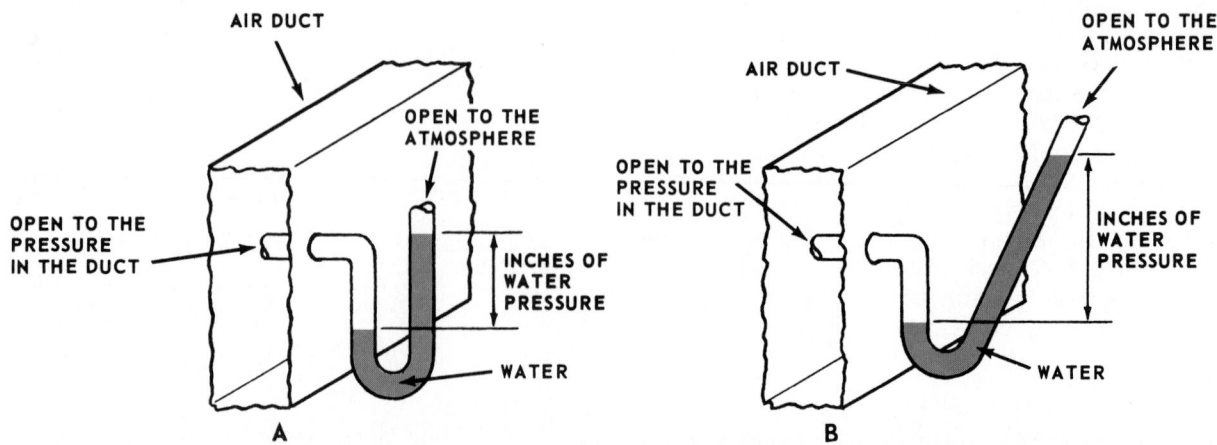

Fig. 1-16. Types of water manometer. This manometer is used to measure low pressure in air ducts and gas lines. Pressure is indicated in inches of water. It is measured by difference in level between surface of water in two branches of tube. B—For easier reading, the end open to atmosphere is often placed at a low angle. Red dye in water makes gauges easier to read.

	POUNDS PER SQUARE INCH		INCHES MERCURY VACUUM	cm Hg	kPa
	ABSOLUTE psia	GAUGE psig or psi	in. Hg.		
Positive Pressure	105	90			725
	90	75			621
	75	60			518
	60	45			414
	45	30			311
	30	15			207
Atmospheric Pressure	14.7	0	0	0	101.3
Negative Pressure or Vacuum	10	−5	10	25.4	69
	5	−10	20	50.8	35
	0	−15	29.92	76.0	0

Fig. 1-17. Table compares various pressure scales.

of vibration will depend upon the temperature. The lower the temperature, the slower the molecules vibrate; the higher the temperature, the faster the vibration.

The molecules are strongly attracted to each other. Considerable force is necessary to separate them.

A solid must always be supported by an upward force or it will fall. See view A, Fig. 1-11.

1-20 LIQUIDS

A liquid is any physical substance which will freely take on the shape of its container (view B, Fig. 1-11), yet its molecules strongly attract each other.

Think of the molecules as swimming among their fellow molecules without ever leaving them. The higher the temperature, the faster the molecules swim. Warmer molecules will move upwards toward the top of the container. This is because they take up more space by their rapid movement. They become lighter than colder molecules.

1-21 GASES

A gas is any physical substance which must be enclosed in a sealed container to prevent its escape into the atmosphere.

The molecules, having little or no attraction for each other, travel (fly) in a straight line. They bounce off each other, off molecules of other substances, or off the container walls. They have little or no attraction for any other substance. The pressures shown in the gas-filled balloon in view C, Fig. 1-11, illustrate how gases behave.

Any substance can be made to exist as a solid, a liquid, or a gas. Any molecule can be made to vibrate, swim, or fly. It depends on two things: temperature and pressure. To understand this change of state, one must study temperature and pressure relationships.

1-22 DENSITY

Some substances are heavier than others. Comparative weights of gases, liquids, and solids may be shown by either density or specific gravity (see Para. 1-24).

Density is a substance's mass per unit of volume.

Density is expressed as pounds per cubic foot (lb./ft.³) or kilograms per cubic meter (kg/m³).

1-23 SPECIFIC VOLUME

When comparing densities of gases, it is common to express the densities in specific volumes. *Specific volume is the volume of one pound of a gas at standard conditions.*

Standard conditions are 68°F at 29.92 in. of mercury column pressure. The volume of 1 lb. of dry, clean air at standard atmospheric conditions is 13.454 cu. ft. By comparison, 1 lb. of hydrogen occupies 178.9 cu. ft., and 1 lb. of the refrigerant, ammonia (R-717), occupies 21 cu. ft. Carbon dioxide (R-744) only occupies 8.15 cu. ft.

In SI terms, specific volume is the volume of one kilogram of a gas at standard conditions. Standard conditions are 20°C at 101.3 kPa pressure. The volume of 1 kg of dry, clean air at standard atmospheric conditions is 0.840 m³. By comparison, 1 kg of hydrogen occupies 11.17 m³, and 1 kg of the refrigerant ammonia (R-717) occupies 1.311 m³. Carbon dioxide (R-744) only occupies 0.509 m³.

If 1 kg of gas occupies a greater space than air, it is called a light gas. If it occupies less space than air, it is classified as a heavy gas. The specific volume is $\frac{1}{density}$.

Equivalents:
1 lb./ft.³ = 16 kg/m³
1 kg/m³ = 0.0625 lb./ft.³

1-24 SPECIFIC GRAVITY (RELATIVE DENSITY)

Specific gravity (sp. gr.), which is sometimes called relative density, is the ratio of the mass of a certain volume of a liquid or a solid as compared to the mass of an equal volume of water.

Water is given a specific gravity of one. Objects which float on water have a specific gravity of less than one. Objects which sink in water have a specific gravity greater than one.

Mixtures of salt and water (brine) have a specific gravity greater than one. A calcium chloride brine adjusted to freeze at 0°F (−18°C) will have a specific gravity of 1.18. See Chapter 30 for a table of brine densities and freezing temperatures.

The relative density of gases is defined as the ratio of the mass of a certain volume of a gas as compared to the mass of an equal volume of hydrogen. The readings are taken at 68°F and 29.92 in. Hg pressure.

1-25 FORCE

Force applied to a body at rest causes it to move. The unit of force is the pound force (lbf) or newton (N). *The pound force is that force which, applied to a one-pound mass, will result in an acceleration of 32.173 ft/s².*

At the surface of the earth, where the acceleration of gravity is 32.173 ft/s², a 1-lb. mass weighs 1-lb. force (it exerts 1-lb. force on the surface upon which it rests). If the object of 1-lb. mass were on the moon, where the gravity is about 1/6 that on earth, the weight would be 1/6 lbf.

To determine the force on the head of a piston 10 square inches (sq. in.) in area and under a pressure of 25 psi, use the following formula:

$$F = A \times P$$

in which:

F = force
A = area of the piston head (10 sq. in.)
P = pressure (25 psi)

To solve the problem:

$$F = 10 \times 25$$
$$F = 250 \text{ pounds (lb.)}$$

In SI units, the newton is that force which, when applied to a body having a mass of one kilogram, gives it an acceleration of one meter per second per second. Force may also be called accumulated pressure.

One newton equals one kilogram times one meter divided by seconds squared: 1 kg × 1 m/s² or 1 kg m/s² = 1 N. (This unit of force is similar to the pound force in the U.S. conventional system.)

Examples: Determine the force on the head of a piston 645 mm² in area and under a pressure of 0.172 MPa. Use the following formula:

$$F = A \times P$$

in which:

F = force
A = area of the piston head (645 mm²)
P = pressure (0.172 MPa)

Solution:

$$F = 645 \times 0.172$$
$$F = 111 \text{ newtons (N)}$$

A pascal is the unit of pressure. The newton is the total force, which equals the unit pressure times the area.

Equivalents:

1N = 0.225 lb. force
1 lb. force = 4.45 N

1-26 WORK AND ENERGY

Work (W) is force (F) multiplied by the distance (D) through which it travels.

The unit of work is called the foot-pound. One foot-pound is the amount of work done in lifting a 1 lb. weight a vertical distance of 1 ft. Work is sometimes expressed in inch-pounds. At such times, the distance through which the force acts is measured in inches.

Example: Calculate the work when lifting a weight of 2000 lb. a vertical distance of 10 ft.

Work = Force × Distance
W = F × D (F = 2000 lb.; D = 10 ft.)
W = 2000 × 10 = 20,000 foot-pounds (ft.-lb.) or, expressed in inch units,
W = 2000 × 10 × 12 = 240,000 inch-pounds (in.-lb.)

The unit of work is called the joule (J) in SI. The joule is the amount of work done by a force of one newton moving its point of application a distance of one meter.

Example: The propeller on a boat pushes the boat through the water with a force of 200 newtons. If the boat travels 10 km, how much work is done?

Solution:

10 km = 10 000 m
Work = F × D = 200 × 10 000
        = 2 × 10² × 1 × 10⁴
Work = 2 × 10⁶ N·m = 2 MJ

Energy is the capacity or ability to do work.

The electric motor supplies the energy to drive the refrigerator compressor.

There are three kinds of energy:

1. Potential energy is stored energy. Examples are: water behind a dam, electrical energy in a battery, and weight which can fall or drop.
2. Kinetic energy is energy doing work. Examples are: water flowing over a dam, a battery lighting a bulb, and a falling weight.
3. Heat energy (see Para. 1-3).

The work formula is expressed as W = F × D, or newtons times meters (N·m). 1 J = 1 N × 1 m = 1 N·m.

Equivalents:

1 ft. lb. force = 1.356 J = 1.356 N·m
1 J = 1 N·m = 0.737 ft. lb.

1-27 POWER

Power is the time rate of doing work. The unit of mechanical power is the horsepower. One horsepower (hp) is the equivalent of 33,000 foot-pounds (ft. lb.) of work per minute. If a 2000-lb. weight is lifted 10 ft. in two minutes, the power required would be:

$$\text{Horsepower} = \frac{\text{weight in pounds} \times \text{distance in feet}}{\text{time in minutes} \times 33,000}$$

$$\text{Horsepower} = \frac{2000 \times 10}{2 \times 33,000} = \frac{20,000}{66,000} = 0.3 \text{ hp}$$

In SI units, power is expressed in watts. A watt is the force of one newton moving through a distance of one meter in one second.

The common unit of mechanical power is the kilowatt (kW). A kilowatt is equal to 1 000 watts. The formula for power is force times distance divided by time. It is expressed in watts (W). 1 W = 1 joule per second = 1 J/s (see Para. 1-26).

Example: What is the power required to lift a mass of 100 kilograms at the rate of 10 meters per second?

Solution:

$$\text{Power} = \frac{\text{Force} \times \text{Distance}}{\text{Time}} = \frac{\text{newtons} \times \text{meters}}{\text{seconds}}$$

Force = 100 × 9.8 newtons
Distance = 10 meters
Time = 1 second

$$\text{Power} = \frac{100 \times 9.8 \times 10}{1}$$

$$= 9\ 800 \text{ W} = 9.8 \text{ kW}$$

Equivalents:

1 hp = 746 W
1 W = 0.0013 hp

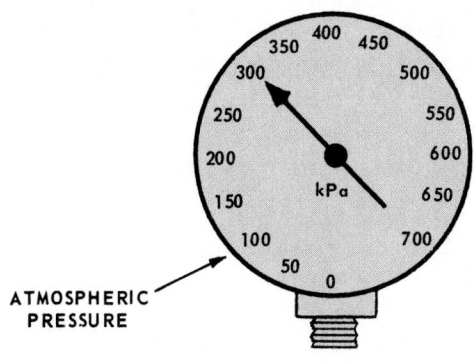

Fig. 1-18. Pressure gauge calibrated in kilopascals. Pressures from 0 to 100 are partial vacuums. Atmospheric pressure is set at 100 kPa.

1-28 UNIT OF HEAT

The unit of heat is the British thermal unit (Btu). The Btu is the amount of heat required to raise the temperature of one pound of water one degree Fahrenheit. (See Fig. 1-19, Part A.)

Whether a substance such as water is cooled or heated, the heat calculation is made in the same way. The temperature difference multiplied by the number of pounds of water gives the number of Btu. Where large heat loads are involved, the unit therm, which equals 100,000 Btu, is often used.

Example: Calculate the amount of heat required to raise the temperature of 62.4 lb. (1 cu. ft.) of water from 40°F to 80°F:

Btu = wt. in lb. × temperature change in degrees Fahrenheit
  = 62.4 lb. × (80 − 40)
  = 62.4 × 40
  = 2496 Btu

Conversely, if a substance is cooled, heat is removed.
Example: The amount of heat removed to cool 50 lb. of water from 80°F to 35°F:

Btu = wt. in lb. × temperature change in degrees Fahrenheit
  = 50 lb. × (80 − 35)
  = 50 × 45
  = 2250 Btu

In SI, the unit of heat is the joule (J). A joule is a very small unit of heat. For refrigeration work, the *kilojoule (kJ),* 1 000 joules, is used. The amount of heat required to raise the temperature of 1 kg of water 1 °C is equal to 4.187 kJ. (See Fig. 1-19, view B.) Conversely, the amount of heat removed to lower the temperature of 1 kg of water 1 °C is also equal to 4.187 kJ.

The mass in kilograms multiplied by the degrees Celsius temperature difference multiplied by 4.187 kJ equals the amount of heat added or subtracted.

Example: Find the amount of heat required to raise the temperature of 1 kg (approximately 1 liter) of water from 4 °C to 27 °C:

kJ = 4.187 × mass in kilograms × temperature change in degrees Celsius
  = 4.187 × 1 × (27 − 4)
  = 4.187 × 1 × 23
  = 96.301 kJ

Conversely, if a substance is cooled, heat is removed:
Example: The amount of heat removed to cool 19 kg of water from 27 °C to 1 °C:

kJ = 4.187 × mass in kilograms × temperature change in degrees Celsius
  = 4.187 × 19 × (27 − 1)
  = 4.187 × 19 × 26
  = 2 068.378 kJ

Another metric unit, the calorie, is the amount of heat required to raise the temperature of one gram of water one

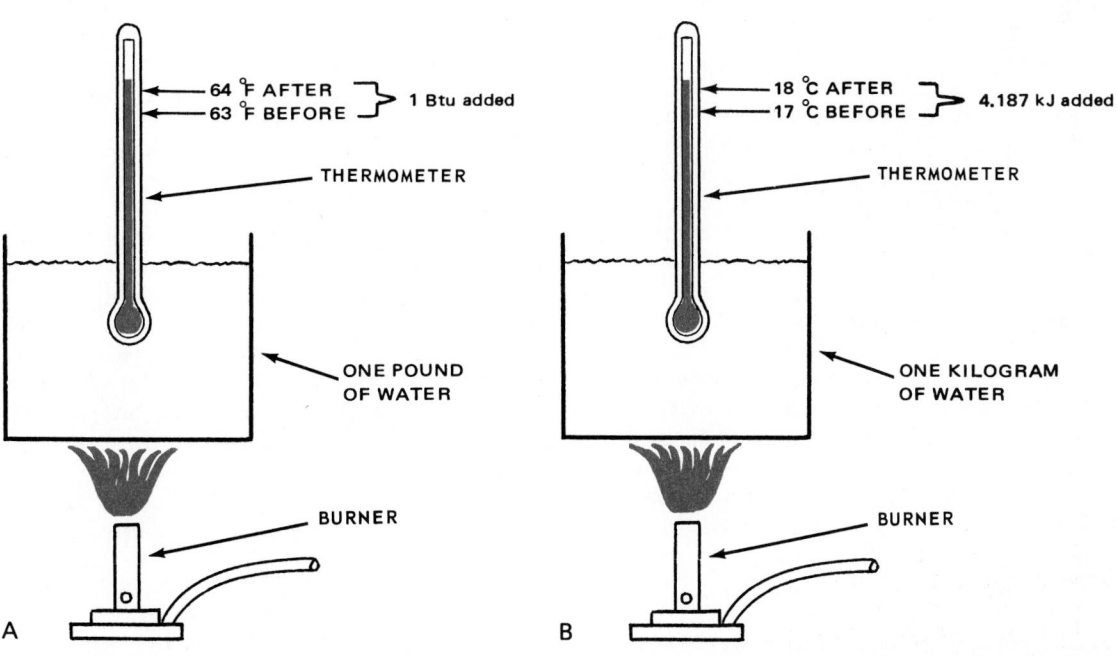

Fig. 1-19. Experiment in heating. A—Raising temperature of one pound of water from 63 to 64° F requires one British thermal unit of heat. B—It takes 4.187 kJ of heat to raise temperature of 1 kg of water from 17 to 18 °C. Note: An accurate definition of the calorie uses 4.1840, not 4.187.

MATERIAL	SPECIFIC HEAT CAPACITY	
	Btu/lb./F.	kJ/kg·K
Wood	0.327	1.369
Water	1.0	4.187
Ice	0.504	2.110
Iron	0.129	0.540
Mercury	0.0333	0.139
Alcohol	0.615	2.575
Copper	0.095	0.398
Sulphur	0.177	0.741
Glass	0.187	0.783
Graphite	0.200	0.837
Brick	0.200	0.837
Glycerine	0.576	2.412
R-717 (Liquid ammonia at 40 °F)	1.1	4.606
R-744 (Carbon dioxide at 40 °F)	0.6	2.512
R-502	0.255	1.068
Salt Brine 20%	0.85	3.559
R-12	0.213	0.892
R-22	0.26	1.089

The above values may be used for computations which involve no "change of state." If a change of state is involved, the specific heat for each state of the substance must be used.

Fig. 1-20. Table of specific heat capacity values for some substances.

degree Celsius. But the calorie is such a small unit that it is no longer used. Most calculations in engineering science are made using the kilocalorie which equals 1 000 calories.
Example: Find the amount of heat required to raise the temperature of 150 grams of water from 10 °C to 90 °C:
Calorie = mass in grams
×temperature change in degrees Celsius
= 150 grams × (90 − 10)
= 150 × 80
= 12 000 calories = 12 kilocalories
Equivalents:
1 kJ = 239 cal = 0.948 Btu
1 cal = 0.004 kJ = 0.004 Btu
1 Btu = 1.055 kJ = 252 cal = 0.252 kilocalories

1-29 FIRST LAW OF THERMODYNAMICS

The First Law of Thermodynamics states that "heat and mechanical energy are mutually convertible."

In Para. 1-3, the Btu is defined as the unit of heat. In Para. 1-26, the unit of work is defined as the ft.-lb. Since work is convertible to heat, the conversion factor from Btu to ft.-lbs. is used:
1 Btu = 778 ft.-lbs.
Example: Change 10 Btu to ft.-lbs.
10 Btu = 10 × 778 = 7780 ft.-lbs.
Similarly, in SI units, heat and mechanical energy are mutually convertible. In Para. 1-3, the joule is defined as the unit of heat. In Para. 1-26, the unit of work is also the joule, or newton-meter.
Example: 100 joules of work equal 100 joules of heat, or 100 newton-meters.

1-30 SENSIBLE HEAT

Heat which causes a change in temperature in a substance is called sensible heat.

If a substance is heated (heat added) and the temperature rises as the heat is added, the increase in heat is called sen-

sible heat. Likewise, heat may be removed from a substance (heat subtracted). If the temperature falls, the heat removed is, again, sensible heat.

1-31 SPECIFIC HEAT CAPACITY

The specific heat capacity of a substance is the amount of heat added or released to change the temperature of one pound of the substance one degree Fahrenheit.

The sensible heat required to cause a temperature change in substances varies with the kind and amount of substance. The specific heat capacity of water is 1.0 Btu/lb. °F. Different substances require different amounts of heat per unit mass to cause these changes of temperature. The specific heat capacity of several common substances is shown in Fig. 1-20.

Para. 1-28 shows how to calculate the heat (Btu) added or removed from water.

The amount of heat necessary to cause a desired change of temperature is calculated in the following manner: Multiply the mass of the substance by the specific heat capacity and by the temperature change (provided there is no change of state).

To calculate the amount of heat added or removed from substances, use the following formula:
Amount of heat added or removed in Btu = wt. of substance in lb. × sp. ht. capacity × temperature change in °F, or Btu = wt. × sp. ht. capacity × °F change.
Example: The amount of Btu which must be removed to cool 40 lb. of 20 percent salt brine (see Fig. 1-20) from 60 °F to 20 °F:
Solution:
Btu = wt. × sp. ht. capacity × °F change
Btu = 40 lb. × 0.85 sp. ht. capacity × (60 °F − 20 °F)
Btu = 40 × 0.85 × 40
Btu = 1360
In SI, the specific heat capacity of a substance is the amount of heat that must be added or released to change the temperature of one kilogram of the substance one degree kelvin (K). The specific heat capacity of water is 4.187 kJ/kg·K. See Fig. 1-20 for the specific heat capacity of several common substances. Para. 1-28 shows how to calculate the heat in kilojoules (kJ) added or removed from water.

The specific heat capacity unit is expressed as joules per kilogram kelvin. The expression is J/kg·K. Amount of heat added or removed in kJ = mass of substance × specific heat capacity × temperature change in kelvins, or
kJ = mass × sp. ht. capacity × K change
Example: Find the amount of heat, in kJ, which must be removed to cool 15 kg of 20 percent salt brine (see Fig. 1-20) from 16 °C to 7 °C:
Solution:
kJ = mass × sp. ht. capacity × K change
kJ = 15 kg × 3.559 sp. ht. capacity × (16 °C − 7 °C)
kJ = 15 × 3.559 × 9
kJ = 480.5
Specific Heat Capacity Equivalents:
1 cal/g· °C = 4.187 J/g·K
1 Btu/lb. °F = 4.187 kJ/kg·K
(kilojoule per kilogram kelvin)
1 kJ/kg·K = 0.2388 Btu/lb. °F

1-32 LATENT HEAT

Heat which brings about a change of state with no change in temperature is called latent (hidden) heat.

All pure substances are able to change their state. Solids become liquids, liquids become gas. These changes of state occur at the same temperature and pressure combinations for any given substance. It takes the addition of heat or the removal of heat to produce these changes.

In Fig. 1-21, note that considerable heat (144 Btu/lb., 335 kJ/kg) was added between points B and C. Even so, the temperature did not change. This heat was required to change the ice to water. This heat is called "latent heat of melting" during a heating operation, or "latent heat of fusion" during a cooling operation.

Likewise, between points D and E, 970 Btu/lb. (2 257 kJ/kg) were added and the temperature did not change. This heat was required to change the water to steam. This heat is called "latent heat of vaporization." When cooling the steam to water, the latent heat removed is called the "latent heat of condensation."

There are two latent heats for each substance, solid to liquid (melting and freezing) and liquid to gaseous (vaporizing and condensing). Fig. 1-22 shows the latent heat for water and several common refrigerants.

The latent heat of ice is 144 Btu/lb. (335 kJ/kg). The latent heat of vaporization of water (at 212 °F, 100 °C) = 970 Btu/lb. (2 257 kJ/kg).

The addition of heat to a solid increases the vibration of the molecules until they separate at the change-of-state point. In the liquid form, the molecules are only weakly attracted to other molecules and are free to move around. At the transition from a solid to a liquid, some molecules are attached in a solid form while others are weakly attracted in a liquid form. When all the solid attachments are broken, then further heating causes the molecules in liquid form to move around faster.

It requires as much energy to change the molecular at-tachment of a block of ice as it does to raise the temperature of the same amount of liquid from 32 °F to 176 °F. In SI, it requires as much heat to change 1 kg of ice to 1 kg of water as it does to raise the temperature of that same kg of water from 0 °C to 80 °C.

All of the basic operations of the compression refrigeration cycle are based upon these two heats, SENSIBLE and LATENT.

Equivalents:

1 kJ/kg   = 0.4299 Btu/lb.
1 Btu/lb. = 2.326 kJ/kg
1 kJ/kg   = 0.2388 kcal/kg
1 kcal/kg = 4.187 kJ/kg
1 kJ/kg   = 0.2388 cal/g
1 cal/g   = 4.187 kJ/kg

1-33 APPLICATION OF LATENT HEAT

In refrigeration work, the physics of latent heat is especially important. Applications of this principle give the cold or freezing temperature desired.

As ice melts, its temperature remains constant. Nevertheless, it absorbs a considerable amount of heat in changing from ice to water. To melt 1 lb. of ice, 144 Btu are required (288,000 Btu are required to melt 1 ton of ice). To melt one kg of ice, 335 kJ of heat are required.

When a substance passes from a liquid to a vapor, as in a mechanical refrigerator, its ability to absorb heat is very high. This principle is useful in the operation of the refrigerator.

The temperature at which a substance changes its state depends on pressure. The higher the pressure, the higher the temperature needed to bring about the change. Conversely, if the pressure is lowered, the temperature at which the change of state will take place is lowered. This principle is shown in Fig. 1-23.

A liquid under low pressure will boil at a lower temperature. If the vapor resulting from this boiling is then

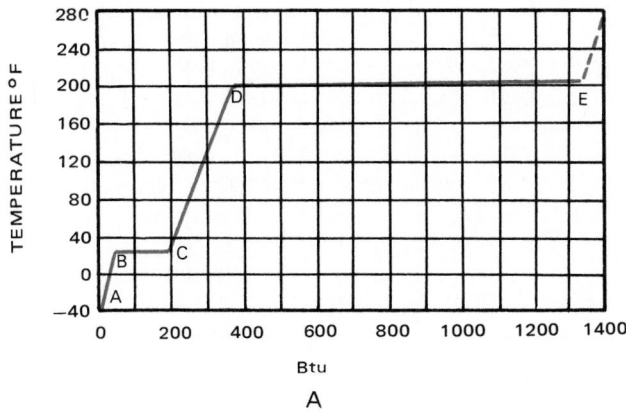

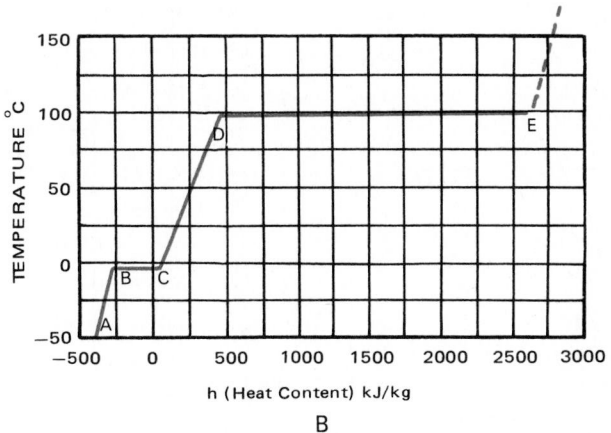

A   B

Fig. 1-21. A—Temperature-heat diagram for one pound of water at atmospheric pressure, heated from −40 °F through complete vaporization. From A to B, 36.3 Btu were added to heat ice from −40 °F to 32 °F (−40 to 32 °F = 72 °F temperature change). (Btu = 1 lb. × 0.504 × 72 = 36.288). From B to C, 144 Btu were added to melt ice. The temperature did not change from B to C. From C to D, 180 Btu were added to heat water from 32 °F to 212 °F. From D to E, 970 Btu were added to vaporize water. Note that temperature did not change from D to E. B—Temperature-heat content diagram for one kilogram of water at atmospheric pressure (100 kPa) heated from −50 °C through complete vaporization. From A to B, 100 kJ of heat are added to increase ice temperature from −50 °C to 0 °C. This is 2 kJ/kg °C × 50 °C = 100 kJ/kg. From B to C, 335 kJ are added to melt ice without changing its temperature. From C to D, 420 kJ were added to heat water to boiling point (4.2 kJ/kg °C × 100 °C = 420 kJ). From D to E, 2 260 kJ were added to convert water to steam without changing temperature. More heat increases temperature of steam as shown in dotted line.

MATERIAL	FREEZING OR MELTING Btu/lb.	LATENT HEAT OF VAPORIZATION OR CONDENSATION Btu/lb.	FREEZING OR MELTING kJ/kg	LATENT HEAT OF VAPORIZATION OR CONDENSATION kJ/kg
Water	144	970.4 at 212 °F	335	2257 at 100 °C
R-717 (Ammonia)	—	565.0 at 5 °F	—	1314 at −15 °C
R-502	—	68.96 at 5 °F	—	160 at −15 °C
R-40 (Methyl chloride)	—	178.5 at 5 °F	—	415 at −15 °C
R-12	—	68.2 at 5 °F	—	159 at −15 °C
R-22	—	93.2 at 5 °F	—	217 at −15 °C

Fig. 1-22. Table of latent heat of vaporization of water and some common refrigerants. Latent heat of fusion is only given for water, as refrigerants do not freeze at temperatures commonly handled by refrigeration service engineer.

compressed, it will condense back into a liquid at a higher temperature.

Every substance has a different latent heat value because each substance has a different molecular structure. Latent heat temperatures for water and the more common refrigerants are shown in Fig. 1-22. See Chapter 9 for more information concerning refrigerants.

In a modern refrigerator, freezer, or air conditioner, liquid refrigerant is piped under pressure to the evaporator. In the evaporator, the pressure is greatly reduced. The refrigerant boils (vaporizes), absorbing heat from the evaporator. This produces a low temperature and cools the evaporator.

The compressor pumps this vaporized refrigerant out of the evaporator and compresses (squeezes) it into the condenser. Here the heat that was absorbed in the evaporator is "squeezed out," released, to the surrounding atmosphere.

Having lost this heat of vaporization, the refrigerant becomes a liquid again. The cycle is then repeated.

1-34 EFFECT OF PRESSURE ON EVAPORATING TEMPERATURES

The evaporating (boiling) temperature for any liquid is controlled by the pressure placed upon it. Water at atmospheric pressure (15 psia or 100 kPa) normally boils at 212 °F (100 °C). If the pressure is increased to 45 psia (311 kPa), the boiling temperature is raised to 271 °F (133 °C). If the pressure is lowered to 7 psia (48 kPa), the boiling temperature will be lowered to 176 °F (80 °C), as shown in Fig. 1-23.

Mechanical and absorption refrigerators use the effect of reduced pressure to lower the boiling temperature. Consider

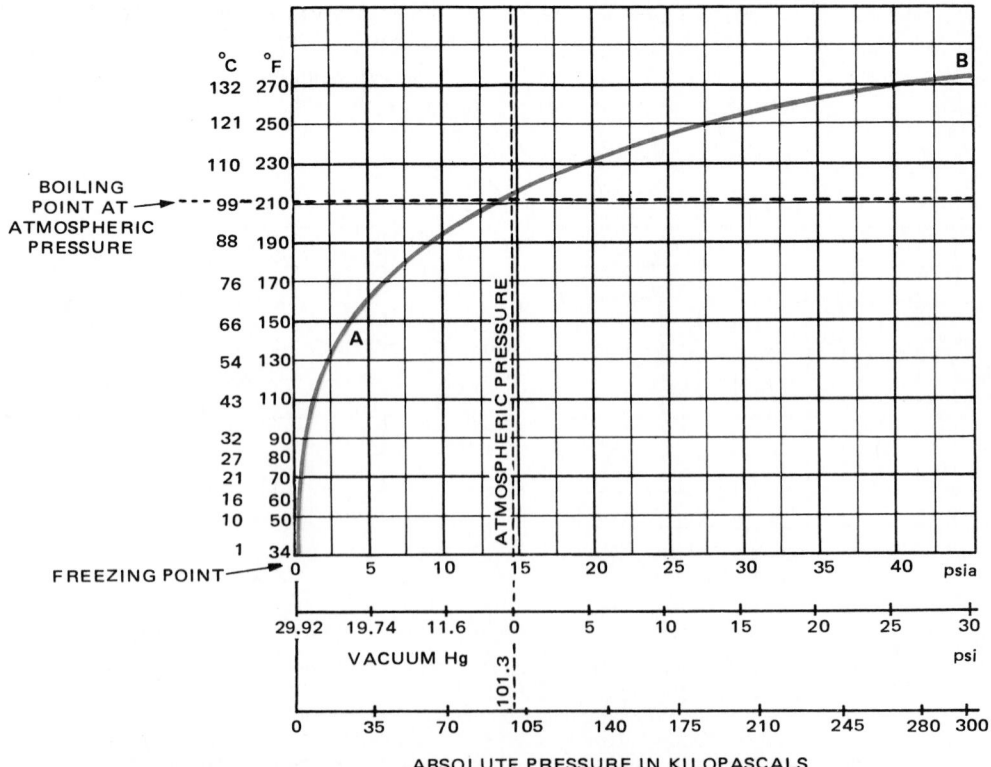

Fig. 1-23. Temperature-pressure curve for water. At atmospheric pressure, water boils at 212 °F (100 °C). At point A, with pressure of 24″ Hg (20 kPa), water boils at 142 °F (62 °C). Increasing pressure above atmospheric raises boiling temperature. At B, which is at a pressure of 45 psia (311 kPa), boiling temperature is 271 °F (133 °C).

SUBSTANCE	EVAPORATING TEMPERATURE IN °C AT:		
	60 kPa PRESSURE	ATMOSPHERIC PRESSURE 100 kPa	200 kPa PRESSURE
Water	89	100	122
R-12	−41	−29	−10
R-717 (Ammonia)	−40	−38	−18

Fig. 1-24. Effect of pressure is shown on evaporating temperatures of three substances.

the refrigerant, R-12. It boils under atmospheric pressure (15 psia or 100 kPa) at −20 °F (−29 °C). If the pressure is lowered to 9 psia (62 kPa), the boiling temperature is −42 °F (−41 °C). A refrigerator can then cool to −42 °F (−41 °C), if the pressure on the evaporator is lowered to 9 psia (62 kPa). But, if the evaporator were to operate at atmospheric pressure, the lowest temperature possible with R-12 would be −20 °F (−29 °C). Fig. 1-24 shows the effect of pressure change on the boiling temperature of three substances used in refrigeration work.

1-35 EFFECT OF PRESSURE ON FREEZING TEMPERATURE

The temperature at which water freezes is affected by the pressure on the surface of the water. Increasing the pressure lowers the freezing temperature. Decreasing the pressure raises the freezing temperature. Fig. 1-25 shows this relationship.

1-36 REFRIGERATING EFFECT OF ICE

Ice is still important to the refrigeration industry. As stated before, ice changes to water at 32 °F (0 °C) and atmospheric pressure. Heat absorption to produce this change

is 144 Btu/lb. (335 kJ/kg).

The specific heat equation for changing ice to water is:
Heat = wt. of ice × sp. ht. capacity of ice × temperature change.
Heat will be in Btu.
Wt. will be in pounds.
Sp. ht. will be in Btu/lb.
The specific heat of water is 1 Btu/lb.
The specific heat of ice is 0.50 Btu/lb.
The latent heat of fusion of ice is 144 Btu/lb.

Example: How many Btu will be absorbed in changing 25 lb. of ice at 5 °F to water at 40 °F?
Solution (in steps):
1. Raise the temperature of ice from 5 °F to 32 °F:
   Btu = wt. of ice × sp. ht. capacity of ice × temperature change
   = 25 × 0.50 × (32 − 5)
   = 25 × 0.50 × 27
   = 337.5 Btu
2. To melt the ice at 32 °F:
   Btu = wt. of ice × latent heat of fusion of ice
   = 25 × 144
   = 3600 Btu
3. To warm the water from 32 °F to 40 °F:
   Btu = wt. of water × sp. ht. capacity of water × temperature change
   = 25 × 1 × (40 − 32)
   = 25 × 1 × 8
   = 200 Btu
4. Total heat required to change 25 lb. of ice at 5 °F to water at 40 °F:
   = 337.5 + 3600 + 200 = 4137.5 Btu

In SI, the specific heat capacity of ice = 2.11 kJ/kg·K. Its heat absorption ability, when changing from a temperature below 0 °C to 0 °C, is 2.11 kJ/kg per degree

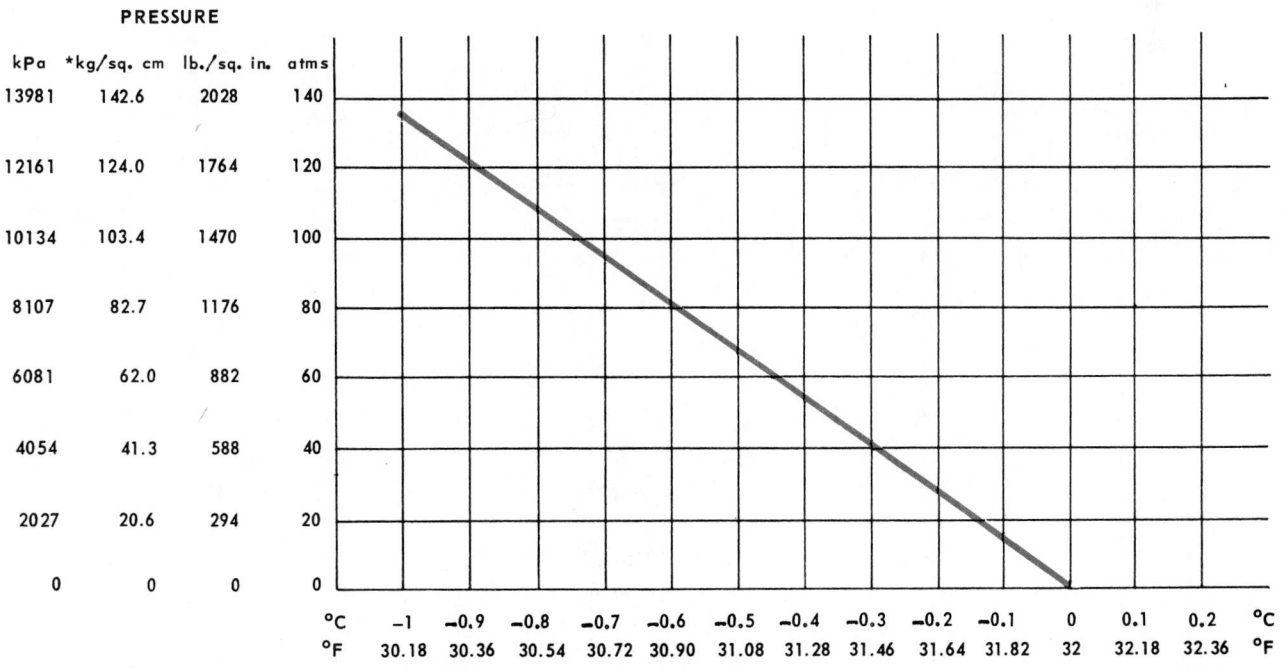

Fig. 1-25. Chart shows effect of pressure on freezing temperature of water.

change. The latent heat of fusion (melting) of ice = 335 kJ/kg. The specific heat of water = 4.19 kJ/kg·K.

Example: How many kJ will be absorbed in changing 93 kg of ice at −20 °C to water at 4 °C?

Solution (in three steps):

1. To find heat needed to bring ice from −20 °C to 0 °C:
   kJ = mass of ice × sp. ht. capacity of ice × temperature change
   = 93 × 2.11 × 20 = 3 924.5 kJ

2. To find heat needed to melt ice at 0 °C:
   kJ = mass of ice × latent heat of fusion of ice
   = 93 × 335 = 31 155.0 kJ

3. To find heat needed to raise water temperature from 0 °C to 4 °C:
   kJ = mass of water × sp. ht. capacity of water × temperature change
   = 93 × 4.19 × (4 −0)
   = 93 × 4.19 × 4 = 1 558.7 kJ

Total heat required to change 93 kg of ice at −20 °C to water at 4 °C
= 3 924.6 kJ + 31 155.0 kJ + 1 558.7 kJ
= 36 638.3 kJ

Equivalents:
1 kJ/kg·K = 0.239 Btu/lb. °F
1 Btu/lb. °F = 4.187 kJ/kg·K

1-37 ICE AND SALT MIXTURES

Refrigerating by ice alone will not provide temperatures below 32 °F (0 °C). Therefore, to get the lower temperatures required in some instances, ice and salt mixtures are used. These mixtures, ice and salt (sodium chloride or NaCl) and ice and calcium chloride (CaCl), lower the melting temperature of ice. An ice and salt mixture may be made which will melt at 0 °F (−18 °C).

The reason that a solution of water and salt freezes at a lower temperature is that more energy must be removed from the solution before ice will form. This phenomenon is called "freezing point depression."

1-38 TON OF REFRIGERATION EFFECT

The cooling capacity of older refrigeration units is often indicated in "tons of refrigeration." *A ton of refrigeration represents the rate of cooling produced when a ton (2000 lb.) of ice melts during one 24-hour day.* The ice is assumed to be a solid at 32 °F (0 °C) initially and becomes water at 32 °F (0 °C). The energy absorbed by the ice is the latent heat of ice times the total weight.

Ratings of refrigeration or air conditioning machines in Btu/hr may be found. A 12,000 Btu/hr cooling capacity is equivalent to 1 ton of refrigeration.

The Btu equivalent of 1 ton of refrigeration is 288,000 Btu per 24 hours. This may be calculated by multiplying the weight of ice (2000 lb.) by the latent heat of fusion (melting) of ice (144 Btu/lb.).

One ton of refrigeration effect =
2000 lb./24 hours × 144 =
288,000 Btu/24 hours

A refrigerating or air conditioning mechanism capable of absorbing heat is usually rated in tons per 24 hours by its heat absorbing ability (HA) in Btu divided by 288,000.

T = tons of refrigeration effect
HA = heat absorbing ability

$$T = \frac{HA}{288{,}000} = \text{tons of refrigeration effect}$$

Example: The heat absorbing ability of a refrigerator unit is rated at 1,440,000 Btu per 24 hours. What is its ton rating?

Solution:

$$T = \frac{1{,}440{,}000}{288{,}000} = 5 \text{ tons of refrigeration effect}$$

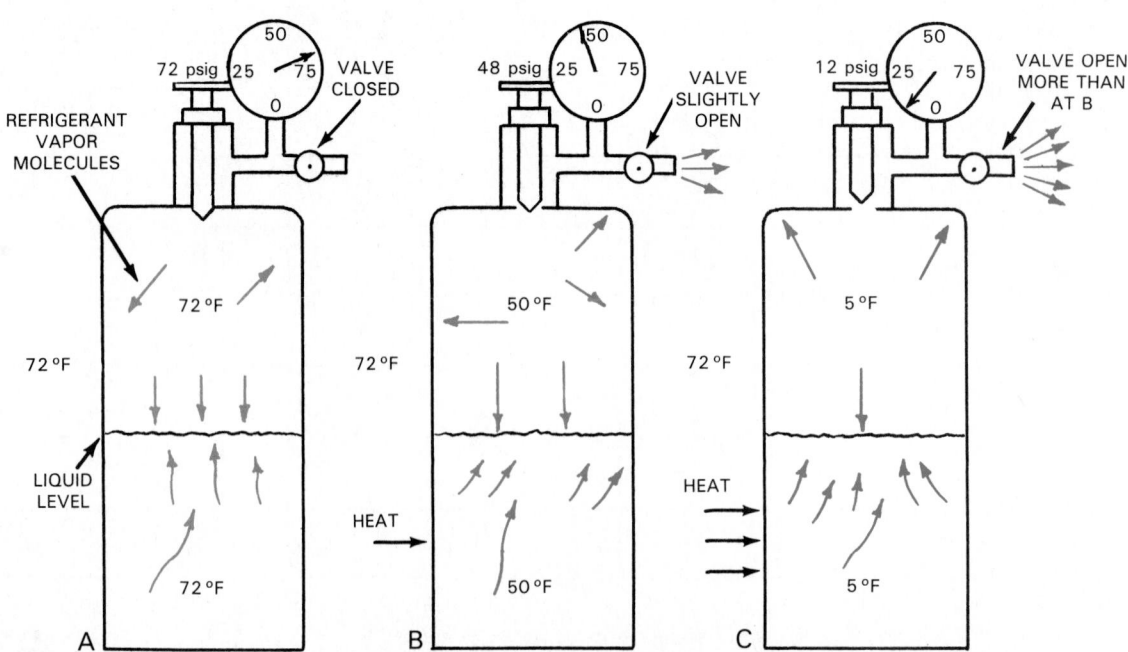

Fig. 1-26. Cooling effect of different pressures operating on surface of liquid refrigerant R-12. A—Cylinder of refrigerant with valve closed. Refrigerant is in state of equilibrium. B—Cylinder of refrigerant with valve slightly opened and refrigerant vapor beginning to leave cylinder. C—Cylinder of refrigerant with valve open wider. Large amount of refrigerant vapor is flowing from cylinder.

Example: What will be the "ton" rating of a refrigerating mechanism capable of absorbing 1,728,000 Btu in 24 hours?

Solution:

$$T = \frac{1,728,000}{288,000} = 6 \text{ tons of refrigeration effect}$$

Example: What is the Btu heat absorbing capacity of a 1/2-ton refrigerating system?

Solution:

1/2 × 288,000 = 144,000 Btu per day, or
                        6000 Btu per hour

Most room air conditioners are rated on their heat-absorbing ability in Btu per hour. A 1-ton machine is rated:

$$\frac{288,000}{24} = 12,000 \text{ Btu per hour.}$$

The SI metric system has no unit which can be compared with the "ton of refrigeration."

1 ton = approximately 907 kg
latent heat = 335 kJ/kg
energy absorbed = latent heat × weight
                = 335 kJ/kg × 907 kg
                = 303 845 kJ

The melting of this ice in one day has a cooling or refrigeration capacity of 303 845 kJ.

To convert the rating to kilowatts:

1kW = 1 kJ/sec
1 ton refrigeration capacity
    = 303 845 ÷ (24 × 3 600 sec)
    = 303 845 ÷ 86 400 sec
    = 3.52 kJ/sec
    = 3.52 kW

A refrigerator which produces an equivalent cooling rate of this ice melting will be rated as a 1-ton unit.

Equivalents:

1 kW = 3415 Btu/hr
1 Btu/hr = 0.3 W

1-39 AMBIENT TEMPERATURE

Ambient temperature is the temperature of the air surrounding a motor, a control mechanism, or any other device. A motor, for example, may be guaranteed to deliver its full power when its temperature does not go higher than 72 °F (40 °C) above the ambient temperature. This means that the temperature of the motor must not be 72 °F (40 °C) warmer than the surrounding air if the motor is to maintain its operating efficiency.

Example: If room temperature is 72 °F (22 °C) and motor temperature rise is 72 °F (40 °C), the motor temperature should not go over 144 °F (72 + 72) or 62 °C (22 + 40).

Ambient temperature is not usually constant. It may change day by day and hour by hour, depending on usage of the space, sunshine, and many other factors.

1-40 HEAT OF COMPRESSION

As a gas is compressed, its temperature rises. This is due to the work (energy) added to the gas by the compressor. The energy added is often termed "heat of compression."

The temperature of the vaporized (gas) refrigerant returning to the compressor from the evaporator will probably be at, or slightly below, the room temperature. This same vapor, as it leaves the compressor and enters the condenser, will be at a much higher temperature.

The compression of vaporized refrigerant in a refrigerator compressor is considered to be near a state of adiabatic compression. Adiabatic compression is a process in which a gas is compressed without losing heat to the surroundings.

Because the compressed vapor in the condenser is now warmer than the temperature of the surrounding air, the heat will be rapidly transferred through the condenser walls to the surrounding air or condenser cooling water. (See Second Law of Thermodynamics, Chapter 30.)

1-41 ENERGY UNITS

In refrigeration work, three common, related forms of energy must be considered: mechanical, electrical, and heat. The study of refrigeration deals mainly with heat energy, but the heat energy is usually produced by a combination of electrical and mechanical energy.

In a compression refrigerating unit, electrical energy flows into an electric motor, and this electrical energy is turned into mechanical energy. This mechanical energy is used to turn a compressor. The compressor, in turn, compresses the vapor to a high pressure and high temperature, transforming mechanical energy into heat energy.

Various units are used for measuring mechanical, heat, and electrical energy. Energy conversion units are expressed as follows:

Mechanical to heat,   1 hp = 2546 British thermal
                                   units (Btu)/hr
                     778 ft. lb. = 1 Btu
Mechanical to electrical, 1 hp = 746 watts (W)
Electrical to mechanical, 746 watts = 1 hp
Electrical to heat,   1 watt (1 joule/sec)
                           = 3.412 Btu/hr
                     1 kilowatt (kW)
                         = 3412 Btu/hr
Heat to mechanical,   1 Btu/hr = 0.000393 hp
Heat to electrical,   1 Btu/hr = 0.293 watts

These conversion units are used in calculating loads and determining the capacity of equipment required for specific refrigeration applications.

In SI, the unit for measuring energy in all three of these forms is the joule. The kilowatt hour is widely used, however, as a measure of electric energy.

Equivalents:

1 J = 0.7376 ft.-lb.
1 ft.-lb. = 1.3558 J
1 W = 0.7376 ft. lb./s
1 ft.-lb./s = 1.3558 W
1 kW = 1.34 hp = 3412 Btu/h
1 hp = 0.746 kW

1-42 REFRIGERANT

In refrigerating systems, fluids which absorb heat inside the cabinet and release it outside are called refrigerants.

These fluids, in their liquid form, under a reduced pressure, absorb heat in the evaporator and, in absorbing heat, change to a vapor. In their vapor form, the fluids are taken into the compressor where the temperature and pressure are increased. This allows the heat that was absorbed in the evaporator to be released (squeezed out) in the condenser, and the refrigerant is then returned to a liquid form for another cycle.

Chapter 9 describes the refrigerants most used. Their technical characteristics, likewise, are explained.

1-43 ELEMENTARY REFRIGERATOR

In Fig. 1-26, a refrigerant cylinder, A, is shown with the valve closed. The pressure inside is 72.7 psig (87.4 psia, 603 kPa) and the temperature is 71.5 °F. All conditions inside the cylinder are balanced. The number of molecules leaving the vapor state, the number diving back into the liquid, and the liquid molecules flying out of the liquid into the vapor state are equal.

In cylinder B, the valve has been opened slightly and some of the vapor is escaping. The results are twofold:

1. The pressure over the liquid refrigerant in the cylinder is reduced to 47.4 psig (62.1 psia). This causes change. There is more liquid changing to a vapor than there is vapor changing back into a liquid.
2. With more liquid turning into a vapor than vapor returning to a liquid, heat is absorbed. The liquid refrigerant and cylinder will be cooled. The temperature of the refrigerant and cylinder is now 50 °F. Some heat from the surrounding area, which is at 72 °F, will now flow into the cylinder and the refrigerant.

In cylinder C, the valve has been opened more than in B. Refrigerant vapor now flows out more rapidly. This brings about a still lower pressure on the liquid refrigerant and a more rapid evaporation of the refrigerant.

The increase in the rate of evaporation lowers the temperature of the refrigerant and the cylinder still more. The 72 °F air surrounding the cylinder is thus made to give up its heat to the colder cylinder more rapidly.

In cylinder A, there is a state of equilibrium (balance), with all temperatures and pressures in balance.

In cylinder B, there is a slight imbalance due to the vapor escaping through the valve. If this condition were to continue for a considerable time, a condition of balance would again prevail. But, in this new condition, its balance would not be static as in A. Rather, it would be a balance between the rate of heat flow into the cylinder, the evaporation of refrigerant, and the flow of refrigerant vapor out of the cylinder valve. In this condition of balance, the refrigerant is cooling the cylinder and its surroundings.

As long as the valve is open and vapor can escape, the temperature will be lower, because more liquid molecules are becoming vapor molecules than vapor molecules are returning into the liquid.

This vapor bombardment is called vapor pressure. If pressure can be reduced, the temperature of the liquid can be reduced since evaporation will be increased.

If the vapor molecules can be removed fast enough, vapor pressure may be low enough to create refrigerant boiling temperatures in the refrigerating range (low temperature). Vapor molecules are usually removed with a compressor or by using chemicals which absorb the molecules.

The operation of the mechanical refrigerator is based on the heat absorption property of a fluid passing from the liquid to the vapor state. By placing a cylinder of refrigerant into a box and venting the vapor to the outside, one would have a heat absorber in the box, as shown in Fig. 1-27.

The liquid can boil only at or above its evaporation temperature, say 20 °F (−7 °C). This liquid will remain at this temperature until it has completely evaporated.

Additional heat would only cause more rapid evaporation

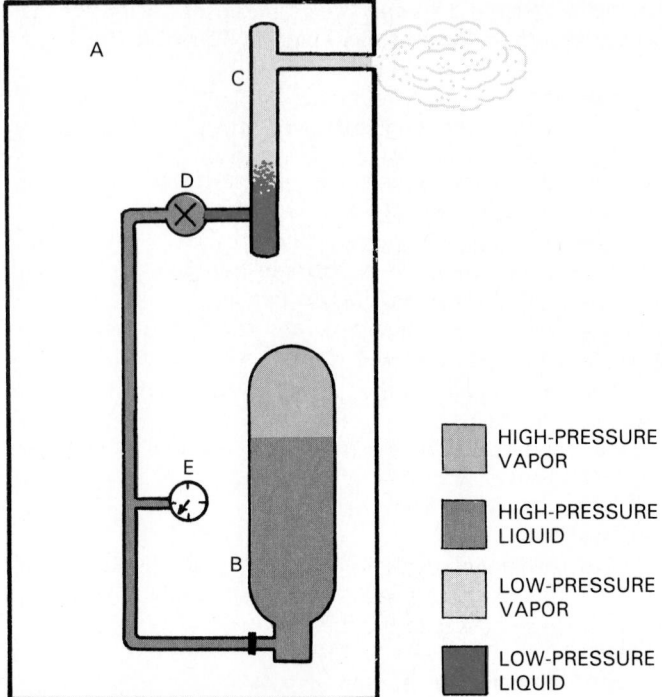

Fig. 1-27. Elementary refrigerator. A—Refrigerated space. B—Cylinder of liquid refrigerant. C—Evaporator. D—Control valve. E—Pressure gauge.

HIGH-PRESSURE VAPOR

HIGH-PRESSURE LIQUID

LOW-PRESSURE VAPOR

LOW-PRESSURE LIQUID

of the liquid. Pressure, of course, must remain constant.

Since the liquid is at this low temperature, there is a transfer of heat to it from the surrounding objects. This heat increases the evaporation, and the heat itself is carried away as the vapor passes off. Thus, the fluid changing its state to a vapor gets the energy (or heat) for doing this from the objects surrounding it. That same heat is removed with the vapor to the outside of the box.

This type of refrigerating system works nicely. But it is expensive because the refrigerant fluid is lost. Some mobile refrigerating units (trucks) use this method. The refrigerant is usually a liquid nitrogen which is relatively inexpensive. It is called "expendable refrigerant" refrigeration.

1-44 MECHANICAL REFRIGERATING SYSTEM

In a mechanical refrigerating system, the vaporized refrigerant is captured, compressed, and cooled to a liquid state again. This is done so that it can be reused, as shown in Fig. 1-28.

To follow the refrigerant cycle, begin with the refrigerant in the receiver B. The refrigerant is R-12.

The refrigerant at B is under a pressure corresponding to the room temperature of 72 °F (22 °C). For R-12, this pressure will be approximately 74 psig (610 kPa) when the unit is idle, but higher when the unit is running.

At the refrigerant control, C, this pressure is reduced to provide low-pressure, low-temperature evaporation in the evaporator (cooling coil). Since the box or inside temperature is to be held at 35 °F (1.5 °C), the pressure in the evaporator, D, must be held at or below 30 psig (310 kPa).

The purpose of the refrigerant control is to allow refrigerant to flow from the liquid receiver (high side) into the evaporator (low side). At the same time, it must maintain the pressure difference between the high-pressure side (high side) and the low-pressure side (low side).

In the evaporator, D, the liquid refrigerant is now under a much reduced pressure, and it will evaporate or boil very rapidly. This, in turn, cools the evaporator. The compressor, E, creates a low pressure and draws (sucks) the evaporated refrigerant vapor from the evaporator. Then it compresses it back to the high side.

From the compressor, the high-temperature (see heat of compression, Para. 1-40), high pressure vaporized refrigerant flows into the condenser, F. The temperature of the vapor, as it enters the condenser, will be several degrees warmer than the room temperature. This difference causes a very rapid transfer of the heat from the condenser to the surrounding air.

The vapor, as it flows through the condenser, cools and loses its heat of vaporization. It returns to the liquid state. As a liquid, it flows from the condenser back into the liquid receiver, B.

This refrigeration cycle is repeated over and over until the desired temperature is reached. Then a thermostat opens the electrical circuit to the driving motor. The compressor stops.

The temperature of the evaporator determines the pressure at which the refrigerant is evaporated. The amount of heat removed depends on the amount (mass) of refrigerant vaporized.

1-45 HEAT TRANSFER

Heat may be transferred or moved from one body to another by one of three methods: radiation, conduction, or convection. Some systems of heat transfer use a combination of these three methods.

1-46 RADIATION

Radiation is the transfer of heat by heat rays. The earth receives heat from the sun by radiation. Light rays from the sun turn into heat as they strike opaque or translucent materials which will absorb some or all of the rays. (Opaque means light cannot shine through; translucent means light can go through but one cannot see through, Chapter 24.)

Air is heated very little as light rays pass through it. Likewise, a glass pane absorbs little heat as rays pass through it.

Sunlight generates more heat when its rays strike dark-colored objects than when they strike light-colored or polished surfaces. This is because light-colored and polished objects reflect the rays. They are not absorbed and changed into heat. Rough, dark-colored surfaces will radiate more heat than will light-colored or polished surfaces (because they get hotter).

Any heated surface loses heat to cooler surrounding space or surfaces through radiation.

Likewise, a cold surface will absorb radiated heat. Some space heating systems depend on radiant heating sources located in the ceilings, walls, or floors to heat a room.

1-47 CONDUCTION

Conduction is the flow of heat between parts of a substance by molecular vibrations. The flow can also be from one substance to another substance in direct contact.

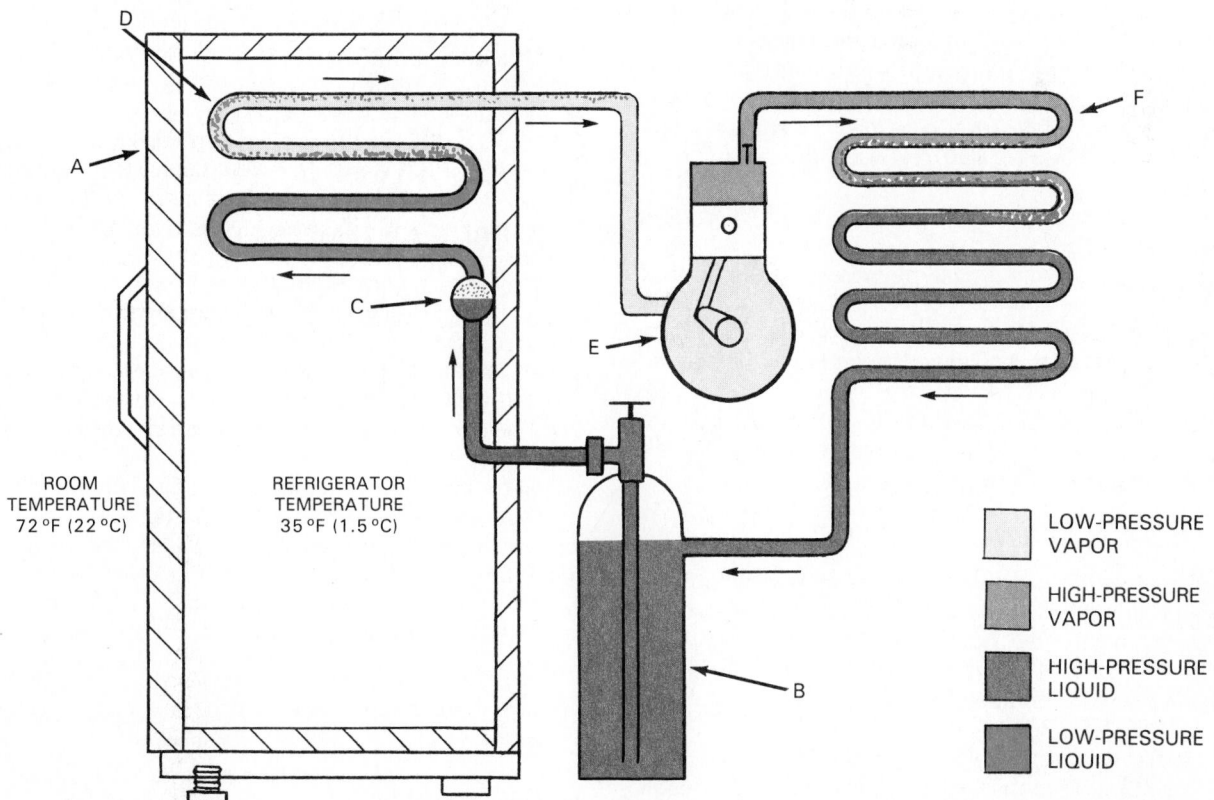

ROOM TEMPERATURE 72 °F (22 °C)

REFRIGERATOR TEMPERATURE 35 °F (1.5 °C)

LOW-PRESSURE VAPOR

HIGH-PRESSURE VAPOR

HIGH-PRESSURE LIQUID

LOW-PRESSURE LIQUID

Fig. 1-28. Elementary mechanical refrigerator. A—Cabinet. B—Liquid refrigerant receiver. C—Refrigerant control. D—Evaporator. E—Motor-driven compressor. F—Condenser. Refrigerant is recycled, as needed, to get desired temperature.

A piece of iron with one end in a fire will soon become warm from end to end. This is an example of the transfer of heat by conduction. The heat travels through the iron, using the metal as the conducting medium.

Substances differ in their ability to conduct heat. In general, substances which are good conductors of electricity are also good conductors of heat (Wiedmann-Frank Law).

Substances which conduct heat poorly are called insulators. Such substances are used to insulate refrigerators, homes, and any structure that is to be maintained at a temperature difference from its surroundings.

1-48 CONVECTION

Convection is the movement of heat from one place to another by way of fluid or air. For example, heated air moves from a furnace into the rooms of a house. It releases its heat to the rooms. Then cooled air returns through cold air ducts to receive another supply of heat.

The same method may be used to cool a space. Unwanted heat is collected and discharged outside the space.

1-49 CONTROL OF HEAT FLOW

The flow of heat by radiation, conduction, and convection can be controlled. The transfer of heat by each can be increased or cut back according to need.

The transfer of heat by radiation may be improved by making the radiating surfaces out of a material known to be a good radiator of heat. A color known to be a good radiator can also be used. Radiation may also be improved by making the receiving surfaces of a material or color known to be a good absorber (or poor reflector) of radiated heat. Radiation may be reduced by reversing this application.

Dark materials or colors absorb and radiate readily. Light colored or shiny materials have opposite properties.

Conduction may be improved by providing large conducting surfaces and good conducting materials, such as copper, aluminum, and iron. Cork, foam plastics, mineral wool, and many other similar materials are poor conductors of heat.

Poor conductors of heat are often referred to as heat insulators (insulation).

Convection may be improved by increasing the flow of the conveying medium. Forced-air circulation heating systems are an example. A blower speeds up airflow. Conversely, convection can be slowed by retarding the circulation of air.

Heat transfer is also controlled (affected) by the temperature difference (T.D.). The greater the temperature difference, the greater the heat flow.

1-50 BRINE AND SWEET WATER

Some refrigeration and air conditioning applications require that the water be kept from freezing at temperatures considerably below the normal freezing temperature of 32 °F (0 °C). Other applications require that water at atmospheric pressure be kept from boiling at temperatures above 212 °F (100 °C).

Salt, sodium chloride (NaCl), or calcium chloride (CaCl), added to water, raises the temperature at which the water will boil while lowering the temperature at which it will freeze. See Chapter 30 for tables of brine solutions with a freezing point and specific gravity for each.

There are some refrigerating and air conditioning installations which use tap water without any salt or other substance added. This is referred to as "sweet water."

1-51 DRY ICE

Solid carbon dioxide (CO_2) is sometimes used for refrigeration. A white crystalline (like crystals) substance, it is formed when liquid carbon dioxide is allowed to escape into a snow chamber (heat-insulated box).

The heat for vaporizing the liquid is drawn from the interior of the chamber so that a very low temperature (−108 °F, −78 °C) is formed. As a result, quantities of the carbon dioxide solidify.

This solid is pressed into various shapes and sizes and sold for refrigeration purposes. It is given such names as dry ice, zero ice, and so forth. It remains at a temperature of −108 °F (−78 °C) while in a solid state at atmospheric pressure.

Dry ice does not turn into a liquid as it melts. It goes directly from the solid to the vapor state. This is called "sublimation." It has some desirable characteristics. It does not wet the surfaces that it touches, and the vapor given off is a preservative. The low temperature maintained permits handling frozen foods without using a heavily insulated container.

The latent heat of sublimation is 248 Btu/lb. (577 kJ/kg). The heat absorbed by the vapor in passing from −108 °F (−78 °C) to 32 °F (0 °C) is approximately 27 Btu/lb. (63 kJ/kg). This, added to the latent heat of sublimation, makes a total heat-absorbing capacity of 275 Btu/lb. (640 kJ/kg).

Dry ice has a greater heat-absorbing capability than does water ice. It is generally more expensive than water ice. Equivalents:

1 kJ/kg = 0.4299 Btu/lb.
1 Btu/lb. = 2.326 kJ/kg

Never place dry ice in a sealed container! At ordinary temperatures, the dry ice will sublime (turn into a vapor). The resulting pressure may cause the container to explode. Avoid touching dry ice. It will instantly freeze the skin.

1-52 CRITICAL TEMPERATURE

The critical temperature of a substance is the highest temperature at which the substance may be liquefied, regardless of the pressure applied upon it. Chapter 30 lists critical temperatures for common refrigerants.

The condensing temperature for a refrigerant must be kept below its critical temperature. Otherwise, the refrigerator will not operate. Carbon dioxide (R-744) has a critical temperature of 87.8 °F (31 °C). This refrigerant is not used in air-cooled compression systems because the condensing temperature would usually be above this temperature.

1-53 CRITICAL PRESSURE

The critical pressure of a substance is the minimum pressure necessary to liquefy a gas that is at its critical temperature. Less pressure will not liquefy it.

1-54 ENTHALPY

Enthalpy is the measure of the energy content of a substance. The amount of enthalpy is determined by both the temperature and the pressure of the substance.

Enthalpy is all the heat in one pound of a substance calculated from an accepted reference temperature, 32 °F. This reference temperature can be used for water and water vapor calculations. For refrigerant calculations, the accepted reference temperature is −40 °F. See Fig. 1-21, view A.

Example: What is the enthalpy of 1 lb. of water at 212 °F, assuming 0 enthalpy at 32 °F?

Solution:

Enthalpy at 32 °F = 0

The specific heat of water is C_p = 1 Btu/lb./°F

Heat to raise the temperature of 1 lb. of water from 32 °F to 212 °F:

212 − 32 = 180 °F

H (enthalpy) = m (mass of water) × C_p (specific heat of water) × 180 = 1 × 1 × 180 = 180 Btu

Total enthalpy at 212 °F = 180 Btu

In SI, the zero enthalpies for water, refrigerants, and air are also taken at some convenient temperature (reference temperature or T_r) and pressure, as shown:

1. For water, zero enthalpy is at 0 °C and 100 kPa.
2. For refrigerants, −40 °C and 100 kPa.
3. For air, 25 °C and 100 kPa.

The enthalpy is measured in joules (J) or kilojoules (kJ).

Example: What is the total enthalpy of 5 kilograms of water at 80 °C?

Solution: The enthalpy (H) of 5 kilograms of water at 80 °C is calculated using the following formula:

H = mass (M) × specific heat (C_p) × the temperature change (T − T_r)
= kJ

H = M × C_p × (T − T_r)

M = mass of water = 5 kg

C_p = the specific heat of water = 4.19 kJ/kg·K (Para. 1-31)

T = 80 °C temperature

T_r = 0 °C, the reference temperature for water

H = 5 × 4.19 × 80 = 1 676 kJ

Fig. 1-21, view B, shows the relationship of enthalpy to temperature.

1-55 SPECIFIC ENTHALPY

Specific enthalpy is enthalpy per unit mass. It is measured in Btu per pound (J/kg). Tables of the enthalpy of substances and pressure-enthalpy diagrams, such as Fig. 1-21, use specific enthalpy.

Example: If 100 lb. of a substance absorbs 2000 Btu of energy when heated from the reference state of 0 Btu/lb., what is the specific enthalpy?

Solution: Specific enthalpy = enthalpy absorbed divided by the mass:

$$h = \frac{H}{M} = \frac{2000 \text{ Btu}}{100 \text{ lbs.}} = 20 \text{ Btu/lb.}$$

1-56 CRYOGENICS

Cryogenics refers to the use of, or the creating of, temperatures in the range of 116 °K down to 0 °K (−251 °F down to −460 °F, or −157 °C down to −273 °C).

The frequent use of liquid helium, nitrogen, and liquid hydrogen in refrigeration has increased the common use of the term cryogenics. The same term is applied to the low-

BOILING TEMPERATURE AT ATMOSPHERIC PRESSURE

FLUID	FAHREN-HEIT °F	RANKINE R (Absolute F Scale)	CELSIUS °C	KELVIN K (Absolute °C Scale)
Water	212	672	100	373
R-12 Refrigerant	-22	438	-30	243
R-22 Refrigerant	-41	419	-41	230
R-744 Refrigerant Carbon Dioxide	-109	351	-78	195
R-1150 Refrigerant Ethylene	-135	325	-93	180
Beginning of the Cryogenic Range	-250	210	-157	116
Methane	-258	202	-161	112
Oxygen	-297	163	-183	90
Air	-313	147	-192	81
Nitrogen	-320	140	-196	77
Neon	-411	49	-246	27
Hydrogen	-423	37	-253	20
Helium	-452	8	-270	3
Absolute Zero	-460	0	-273	0

Fig. 1-29. Boiling temperatures of some common refrigerants and some other fluids at atmospheric pressure. Note difference between boiling points of some commonly used refrigerants and boiling points of fluids in the cryogenic range.

temperature liquefaction of gases, handling and storage, insulation of containers, instrumentation, and techniques used in such work.

Fig. 1-29 shows the evaporating temperature at atmospheric pressure of some of the common cryogenic fluids. It also indicates the cryogenic range of temperatures. These temperatures are in kelvin (K).

1-57 PERFECT GAS EQUATION

If a quantity of gas is enclosed in a tight container, the relationship between pressure, temperature, and volume may be expressed by the formula: PV = MRT

P = Pressure in pounds per square foot absolute

V = Volume in cubic feet

M = Mass of gas in pounds

R = Gas constant (R will differ for different gases). (Fig. 1-30 gives the value of R for some common substances.)

T = Absolute temperature R

MATERIAL	GAS CONSTANT (R) J/kg·K	ft.-lb. lb. R	SPECIFIC HEAT kJ/kg·K C_p	C_v
Air	288.68	53.34	1.00	0.71
R-717 (Ammonia)	666.98	123.24	2.13	1.46
R-744 (Carbon Dioxide)	210.10	38.82	0.92	0.71
Ether	125.08	23.11	2.01	1.88
Oxygen	262.76	48.55	0.92	0.67
Alcohol	224.87	41.55	1.88	1.67
Water Vapor	450.45	83.23	2.03	1.55

Fig. 1-30. Table lists gas values (constants) for some substances used in refrigeration work.

Example: What will be the volume of 2 lb. of carbon dioxide at 240 °F when the pressure is 185 psi?

Formula: $PV = MRT$

$P = (185 + 15) = 200$ psia $= 200 \times 144$
$\quad = 28{,}800$ lb. per sq. ft. absolute

$M = 2$ lb.

$R = 38.82$ ft.-lb./lb. °R*

$T = (240 + 460) = 700$ °R

$V = \dfrac{MRT}{P}$

$V = \dfrac{2 \times 38.82 \times 700}{28{,}800}$

$V = \dfrac{54{,}348}{28{,}800} = 1.89$

$V = 1.89$ cu. ft.

The equation in SI may be expressed: $PV = MRT$

$P = $ the pressure in pascals (Pa)

$V = $ the container volume in cubic meters (m³)

$M = $ the mass of the gas in the container in kilograms (kg)

$R = $ the gas constant which has a value depending on the gas properties. Fig. 1-30 gives values of R for some substances used in refrigeration work.

$T = $ the absolute temperature, kelvin (K) which is: $(273 + T°C)$

The equation shows that if the container of gas is heated so that the temperature increases, then the pressure will also rise. Cooling a container will reduce both the temperature and the pressure.

Example: If 0.2 kg of air at a pressure of 1.00 kPa is contained in a volume of 50 m³, what is the temperature?

Solution: Solving for T (absolute temperature):

$T = \dfrac{PV}{MR}$

$\quad P = 1.00$ kPa $= 1\,000$ newtons/m²

$\quad V = 50$ m³

$\quad R = 288.68$ J/kg·K (from Fig. 1-30)

$\quad M = 0.2$ kg

$\quad T = \dfrac{1\,000 \times 50}{0.2 \times 288.68} = 866$ K

The temperature in Celsius is then:

$T°C = T K - 273 = 866 - 273 = 593$ °C

Note: The temperature used in the perfect gas equation must always be in kelvin.

Example: If the air in the container is heated until the temperature reaches 1 000 °C, find the new pressure:

Solution: $T = 1\,000 + 273 = 1\,273$ K

$\quad P = \dfrac{MRT}{V} = \dfrac{0.2 \times 288.68 \times 1\,273}{50}$

$\quad P = 1\,470$ Pa $= 1.47$ kPa

1-58 DALTON'S LAW

Dalton's Law of partial pressures is the foundation of the principle of operation of one of the absorption type refrigerating systems. The law states:

The total pressure of a confined mixture of gases is the sum of the pressures of each of the gases in the mixture.

The total pressure of the air in a compressed air cylinder is the sum of the oxygen gas, the nitrogen gas, the carbon dioxide gas, and the water vapor pressure.

The law further explains that each gas behaves as if it occupies the space alone. To illustrate, the Electrolux

refrigerator uses two gases, ammonia and hydrogen. The ammonia, at room temperature, is absorbed by the water in the closed system.

Heating this solution drives out the ammonia. (The hydrogen is not absorbed by the water and remains as a gas.) Due to the pressure it is under, the ammonia condenses into a liquid in the condenser. The pressure is uniform throughout the system. Total pressure in the system is the sum of the vapor pressure of the ammonia plus the hydrogen pressure. When the pressure of the ammonia vapor is below the pressure corresponding to the vapor pressure for ammonia alone, the ammonia continues to evaporate as it tries to reach a vapor pressure corresponding to the temperature in the absorber.

1-59 EVAPORATOR

In this text, the word evaporator is used to indicate that part of the refrigerator mechanism where the liquid refrigerant boils or evaporates and absorbs heat. In some trade literature, the word cooling coil is used to indicate the part in which such cooling takes place. The correct technical term is evaporator.

Coils cooled by brine or any fluid which does not evaporate in the coil may properly be called cooling coils.

1-60 VAPOR—GAS

The word vapor in this text indicates refrigerant which has become heated, usually in the evaporator, and has changed to a vapor or gaseous state. In some trade and service literature, refrigerant in this state is called a gas. The correct technical term is vapor.

1-61 SATURATED VAPOR

The term "saturated vapor" identifies a condition of balance on an enclosed quantity of a vaporized fluid. The balance is such that some condensate (liquid) will be produced if there is even the slightest lowering of the temperature or increase in pressure.

There is usually some of the substance present in liquid form when the vapor is saturated. In a saturated condition, all of the substance has been vaporized that can be vaporized under the existing conditions of pressure and temperature.

1-62 HUMIDITY—RELATIVE HUMIDITY

The word humidity, as used in connection with refrigeration, air conditioning, and weather information, refers to water vapor or moisture in the air. Air absorbs moisture.

The amount depends on the pressure and temperature of the air. The higher the temperature of the air, the more moisture it will absorb. The higher the pressure of the air, the smaller the amount of moisture it will absorb.

Relative humidity is the amount of moisture carried in a sample of air, compared to the total amount which it can absorb at the stated pressure and temperature. Relative humidity is covered in Chapter 18.

A relative humidity of 50 percent indicates that the air has 50 percent as much moisture as it will hold at that particular temperature and pressure. See Chapter 30 for tables of moisture-holding ability for air at various temperatures and pressures, with volume and heat data.

*The unit "ft.-lb." is used instead of Btu. Btu is not compatible with feet, pounds, and °R, just as cal are not compatible with N·m = Joule, kg, and °K.

1-63 REVIEW OF ABBREVIATIONS AND SYMBOLS

The following is a review of the various abbreviations and symbols studied so far in this chapter. This review includes both the U.S. conventional units and the SI metric units.

U.S. Conventional Units:

- Btu = British thermal unit
- Btu/h = British thermal units per hour
- °F = degrees Fahrenheit
- F_A = degrees Fahrenheit absolute
- °R = degrees Rankine = degrees absolute F
- lb. = pound
- psi = pounds per square inch = lb. per sq. in.
- psia = pounds per square inch absolute = psi + atmospheric pressure
- in. = inches = i = ''
- ft. = foot or feet = f = '
- sq. in. = square inch = in^2
- sq. ft. = square foot = ft^2
- cu. in. = cubic inch = in^3
- cu. ft. = cubic foot = ft^3
- ft.-lb. = foot-pound
- ton = ton of refrigeration effect
- lb./cft. = pounds per cubic foot
- in. Hg = inches of mercury vacuum = ''Hg
- hp = horsepower
- qt. = quart
- gr = grain

SI Metric Units:

- °C = degrees Celsius
- K = kelvin
- mm = millimeter
- cm = centimeter
- cm^2 = centimeter squared
- cm^3 = centimeter cubed
- dm = decimeter
- dm^2 = decimeter squared
- dm^3 = decimeter cubed
- m = meter
- m^2 = meter squared
- m^3 = meter cubed
- L = liter
- g = gram
- kg = kilogram
- J = joule
- kJ = kilojoule
- N = newton
- Pa = pascal
- kPa = kilopascal
- W = watt
- kW = kilowatt
- MW = megawatt

Miscellaneous Abbreviations:

- P = pressure
- h = hours
- sec = seconds
- T_r = reference temperature
- Δ = difference
- C_p = specific heat (sp. ht.)
- h = enthalpy per unit mass
- H = total enthalpy
- D = diameter
- r = radius of circle
- A = area
- π = 3.1416 (a constant used in determining the area of a circle)
- V = volume
- R = gas constant
- ∞ = infinity
- M = mass

1-64 REVIEW OF SAFETY

The term "safety," as applied to any refrigeration or air conditioning activity, may have three different applications. It may apply to:

1. *Safety of the operator.* When refrigeration and air conditioning equipment is properly handled, there is relatively little danger to the operator.

   Always pull on a wrench (instead of pushing), to prevent possible slippage of the wrench which could cause rounded corners on nuts and bolts and possible injury to hands. A hoist is recommended for lifting anything weighing over 35 lb. (13 kg).

   Always use leg muscles when lifting objects, never the back muscles. Make certain there is no oil or water on the floor. Always wear safety goggles when working with refrigerants.

   Most refrigerating mechanisms are electrically driven and controlled. When working on electrical circuits, make sure that the circuit is disconnected from the power source. This can usually be accomplished by opening the switch at the power panel. Never work on "hot" electrical circuits.

2. *Safety of the equipment.* Many parts of refrigeration and air conditioning equipment are quite fragile. Parts may be ruined by overtightening nuts and bolts, not tightening them in the correct order, or using the wrong size wrench. Make certain that all connections are tightened before operating a compressor. Before operating open compressors, be sure the flywheel and pulley are in alignment and that guards are in place.

3. *Safety of the contents.* Safety of the contents of the refrigerated space depends entirely on the accuracy and care given the installation and adjustment of the various parts of the system.

   Throughout this text, tables are provided which list the proper operating temperatures for various types of refrigerating space. These operating temperatures must be observed if the unit is to provide safe conditions for the refrigerated or air conditioned space.

It is advisable to observe these three points during the service work explained in the chapters following.

Each will have a "Review of Safety" as a reminder of the hazards which may be present when working with the equipment and supplies.

There is no exception to the rule that "The safe way is the right way."

1-65 TEST YOUR KNOWLEDGE

Fundamentals:
1. What is dry ice?
2. Given two objects of the same size, one chrome plated and one painted black, which one will absorb more radiant heat?
3. Name a condition which illustrates the principle of convection.

4. Will glass or copper conduct heat the most rapidly?
5. Express one millimeter in decimals of a centimeter.
6. How many cubic centimeters in a liter?
7. How many centimeters are equal to one inch?
8. What is the piston displacement of:
    a. Compressor with 2-in. bore, 3-in. stroke?
    b. Compressor with 30-mm bore, 45-mm stroke?
9. How many 90° arcs fit in a complete circle?

Pressure:
10. What determines the temperature at which a refrigerant will vaporize?
11. Should dry ice ever be put in a sealed container? Why?
12. What is the U.S. conventional absolute pressure equivalent in pounds per square inch (psia) of 8 in. of mercury vacuum (8″ Hg)?
13. a. What is the pressure difference in pounds per square inch between 6 in. of mercury vacuum (6″ Hg) and 8 pounds per square inch gauge (psig)?
    b. What is the pressure difference between 3 kPa and 35 kPa?
14. If the head pressure is 85 lb. per square inch gauge (psig), what is the total force on one face of a circular disk with a 5 in. diameter?
15. a. Express standard atmospheric pressure in pounds per square feet.
    b. What is standard atmospheric pressure in SI units?

Temperature:
16. Should refrigerants be operated at temperatures above or below their critical temperature?
17. What is the temperature desired in a domestic cabinet:
    a. in Fahrenheit?
    b. in Celsius?

18. A substance has a temperature of 78 °F. What would be the temperature in °C?
19. A substance has a temperature of 20 °C. What would be the temperature in °F?
20. A substance has a temperature of 5 °F. (a) What would be the temperature in °C? (b) What would be the temperature in R?
21. A substance has a temperature of 432 °F.
    a. What would be the temperature in °C?
    b. Give the temperature in K.
22. A substance has a temperature of 14 °R. Give its temperature in K.

Heat:
23. What is the unit of heat? What is the unit of heat in SI units?
24. a. How many Btu will be required to change 5 lb. of ice at 32 °F into water at 82 °F?
    b. How much heat will be required to change 3 kg of ice at 0 °C into water at 29 °C?
25. How much heat will 1 lb. of ice at 32 °F absorb in changing to water at 32 °F?
26. How much heat will 1 kg of ice at 0 °C absorb in changing to water at 0 °C?
27. Calculate the number of British thermal units (Btu) that would be required to convert 1 lb. of ice at 0 °F to steam at a temperature of 212 °F.
28. Calculate the number of joules required to convert 1 kg of ice at 0 °C to steam at 100 °C.
29. What is the enthalpy of 1 kilogram of water if it is at a temperature of 70 °C?
30. What is equivalent of 10 ft.-lb. of work in SI metric units?

Chapter 2

REFRIGERATION TOOLS AND MATERIALS

By studying this chapter, the technician will be able to:
☐ List and discuss the various types of tubing used in refrigeration work.
☐ Cut and fit tubing using approved methods.
☐ Demonstrate soldering and brazing techniques.
☐ Repair cracks and leaks in evaporators.
☐ Select the proper tools for servicing and maintaining domestic refrigerators.
☐ Explain how to use various hand tools.
☐ Discuss the procedures for threading steel pipe.
☐ Identify thread types.
☐ Identify different types of screw fasteners.
☐ Demonstrate standard procedures for basic mechanical service and repair operations.
☐ Explain how to maintain and calibrate gauges.
☐ Compare cleaning methods and use of solvents.
☐ Explain the use of vacuum and compound gauges.
☐ Define various types of service valves.
☐ Discuss the importance of oil in the refrigerating system.
☐ Define purging and explain how it is done.
☐ Discuss the evacuation of a system.
☐ Follow approved safety procedures.

2-1 TUBING

Most tubing used in refrigeration and air conditioning is made of copper. However, some aluminum, steel, stainless steel, and plastic tubing is being used.

Instructions in this chapter will deal mainly with copper tubing. All tubing used in air conditioning and refrigeration work is carefully processed to be sure that it is clean and dry inside. The service technician must keep the ends sealed to be sure that it remains clean and dry in handling.

Most copper tubing used in air conditioning and refrigeration work is known as Air Conditioning and Refrigeration (ACR) tubing. This designation means that the tubing is intended for air conditioning and refrigeration. Furthermore, it has been processed to give the desired characteristics.

ACR tubing is usually charged with gaseous nitrogen to keep it clean and dry until it is used. Nitrogen should be fed through it during brazing and soldering operations. Take care, it is dangerous to use! See Chapter 11.

This will eliminate the danger of oxidation inside the tube. All tubing should have the ends plugged immediately after cutting a length from the piece.

Copper tubing is available in soft and hard types. Both are available in two wall thicknesses, K and L. Type K is a heavy wall, type L is medium thick. Most ACR tubing used

at present is the L thickness. Soft copper tubing is supplied in 25- and 50-foot rolls.

Another type of copper tubing used in heating and plumbing is called "nominal size."

2-2 SOFT COPPER TUBING

Soft copper tubing is used in domestic work and in some commercial refrigeration and air conditioning work. It is annealed (heated and then allowed to cool). This makes it flexible, therefore easy to bend and flare. Being easily bent, this tubing must be supported by clamps or suitable brackets. Soft copper tubing is most often used in connection with flared fittings (Society of Automotive Engineers (SAE) standards) and soft soldering fittings. It is sold in rolls 25+, 50+, and 100 feet long. Sizes most commonly used are 3/16, 1/4, 5/16, 3/8, 7/16, 1/2, 9/16, 5/8, and 3/4 in. outside diameter (OD). Wall thickness is usually specified in thousandths of an inch. Fig. 2-1 is a table of common copper tube diameters and thicknesses.

OUTSIDE DIAMETER	WALL THICKNESS
1/4	.030
3/8	.032
1/2	.032
5/8	.035
3/4	.035
7/8	.045
1 1/8	.050
1 3/8	.055

Fig. 2-1. Copper tube sizes used in refrigeration work. Both soft and hard drawn sizes are the same as the measurements listed in table. OD size for this tubing is the actual outside diameter of tube.

Soft copper tubing may be worked to give it certain properties. It can be hardened by repeated bending or hammering. This is called work hardening. It can be softened by annealing, as explained earlier.

Tubing must be installed in such a way that there is no strain on it when the job is completed. Horizontal loops may be used to keep vibration from crystallizing the copper, making it crack or break.

2-3 HARD-DRAWN COPPER TUBING

Hard-drawn copper tubing is used in commercial refrigeration and air conditioning applications. Being hard and stiff, it needs few clamps or supports, particularly in larger diameters.

Hard tubing should not be bent. Use straight lengths and fittings to form necessary tubing connections.

Hard-drawn refrigeration tubing joints should be brazed. Soft solder should be used only on water lines. Hard-drawn tubing is supplied in 20 ft. lengths. It is available in the same diameters and thicknesses as soft copper tubing.

2-4 STEEL TUBING

Some thin-wall steel tubing is used in refrigeration and air conditioning work, sizes being practically the same as for copper tubing. Connections may be made on steel tubing by using either flared joints or brazed joints.

Copper or brass tubing should not be used with refrigerant R-717 (ammonia). Use steel tubing. There is a chance of chemical reaction (corrosion) between ammonia and copper.

Two types of steel tubing are in common use. One type has a double lap brazed construction using SAE 1008 mild steel. The other is butt welded, using the same type steel.

2-5 STAINLESS STEEL TUBING

Stainless steel tubing comes in the usual refrigeration tube sizes. The most common sizes are listed in the table, Fig. 2-2. Stainless steel is strong and very resistant to corrosion. It may be easily connected to fittings by flaring or brazing.

Stainless steel tubing No. 304 is most used. This is a low carbon (C) nickel (Ni) Chromium (Cr) stainless steel. It is often required in food processing, ice cream manufacture, milk handling systems, and the like.

2-6 METRIC TUBE SIZES

Metric sized tubing is used in some refrigeration and air conditioning systems. The standard sizes are: 6, 8, 10, 12, 14, and 15 millimeters (mm) OD.

2-7 PLASTIC TUBING

Polyethylene is one of the most common substances used in the manufacture of plastic tubing. Sizes and suggestions for its use are shown in Fig. 2-3.

The usual safe temperature range is from −100 to +175°F (−73 to 79°C). Therefore, never use this tubing in installations where fluid temperature goes beyond these limits.

In general, polyethylene tubing is not used in the

OUTSIDE DIAMETER	WALL THICKNESS	BURST PRESSURE psi	MINIMUM BEND RADIUS
1/8"	.020	500	1/2"
3/16"	.030	500	1/2"
1/4"	.040	400	1"
5/16"	.062	600	1 1/8"
3/8"	.062	350	1 1/4"
1/2"	.062	250	2 1/2"

Fig. 2-3. Plastic tube specifications. Note that there are four different thicknesses used in this size range of plastic tubing. (Imperial Eastman, Imperial Division)

Fig. 2-4. Thermoplastic hose used in refrigeration systems. A—Inner nylon tube. B—Yarn reinforcement. C—Polyethylene cover.

refrigerating cycle mechanisms, but for cold water lines, water-cooled condensers, and the like. Being very easy to use, polyethylene tubing may be cut with a knife. It may also be easily bent.

Special fittings are available for connecting polyethylene tubing to refrigeration and air conditioning mechanisms.

2-8 FLEXIBLE TUBING (HOSE)

In many refrigeration and air conditioning applications, the liquid lines and suction lines must be flexible, Fig. 2-4. This is particularly true in many commercial and industrial refrigeration and air conditioning applications.

Air conditioning equipment on motor vehicles requires flexible tubing. Hose for this purpose is usually made from a variety of special materials. Such materials do not age, thereby remaining flexible. These materials allow very low leakage through the hose wall and are easy to attach to fittings.

2-9 FLEXIBLE HOSE FITTINGS

There are various types of flexible hose fittings available. See the descriptions listed in Fig. 2-5.

	OUTSIDE DIAMETER							WALL THICKNESS
FRACTIONS	1/4	3/8	1/2	5/8	3/4	1	1 1/4	(All of the stainless steel tubing is available in various wall thicknesses (BWG)* — 31 to 20 gage.)
DECIMALS	.250	.375	.500	.625	.750	1.000	1.250	
MILLIMETERS	6.35	9.52	12.7	15.87	19.05	25.40	31.75	

*— Birmingham Wire Gage

Fig. 2-2. Stainless steel tubing sizes are given in fractions, decimals, and millimeters.

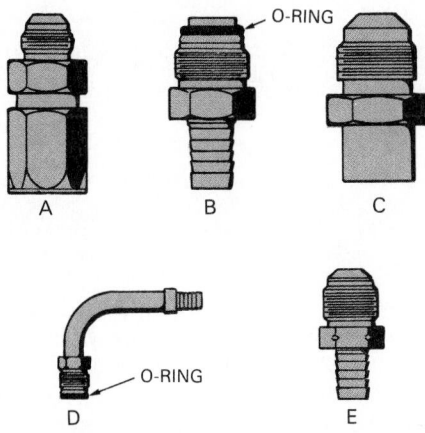

Fig. 2-5. Assorted nylon fittings suitable for use with refrigerant hose: A—Coupling, straight male 45° flare, screw-on reusable. B—Coupling, straight male push-on barb type reusable, with O-ring seal. C—Coupling, straight male 45° flare permanent (crimped-on and not reusable). D—Coupling, 90° male push-on barb type reusable, with O-ring seal. E—Coupling, straight male 45° flare push-on.

NOMINAL SIZE INCHES	TYPE	OD INCHES		WALL THICKNESS INCHES
1/4	K	0.375	3/8	0.035
	L	0.375	3/8	0.030
3/8	K	0.500	1/2	0.049
	L	0.500	1/2	0.035
1/2	K	0.625	5/8	0.049
	L	0.625	5/8	0.040
5/8	K	0.750	3/4	0.049
	L	0.750	3/4	0.042
3/4	K	0.875	7/8	0.065
	L	0.875	7/8	0.045
1	K	1.125	1 1/8	0.065
	L	1.125	1 1/8	0.050

Fig. 2-6. Nominal size copper tubing. Type K—heavy wall is available in hard and soft temper. Type L—medium wall is available in hard and soft temper. Type K is used where corrosion conditions are severe. Type L is used where conditions may be considered normal. OD sizes indicated by dimension are 1/8 in. (.125) larger than nominal size.

These fittings are usually made of brass; however, nylon fittings are sometimes used. Synthetic rubber O-rings are put on some of these fittings to provide a better seal.

The attachment end of these fittings conforms to the standard SAE fittings specifications.

2-10 NOMINAL SIZE COPPER TUBING

Nominal size copper tubing is a type used on water lines, drains, and in other applications. Nominal size copper tubing is never used with refrigerants. It is available in both soft- and hard-drawn grades. The table, Fig. 2-6, shows commonly used sizes and wall thicknesses of this tubing.

The classification (type) listed after nominal size is used when referring to copper tubing for heater lines, drains, and so forth.

Copper tubing used for such applications is often referred to by its nominal size. If the service technician measures the OD, he will notice that the OD is 1/8 in. larger than that listed under nominal size.

When purchasing fittings for this tubing, it is important that the fitting size be the same as the size tubing purchased. One should order all the tubing, valves, and fittings by nominal size or order all by OD, to avoid problems.

2-11 CUTTING TUBING

To cut tubing, use either a hacksaw or a tube cutter. The tube cutter is usually used on smaller, annealed (soft) copper tubing. The hacksaw is preferred for cutting larger, hard copper tubing. Fig. 2-7 illustrates a wheel type cutter. The tubing should be straight and cut squarely (90 deg.) to eliminate an off-center flare. After the tubing has been cut, its ends must be scraped or reamed with a pointed tool to remove any sharp burrs on the end of the tubing. Most tube cutters have a reamer.

If a saw is used, a wave set blade of 32 teeth per inch will do the best job. (See Para. 2-68.)

It is important that no filings or chips of any kind enter the tubing.

When cutting tubing with a hacksaw, hold the tubing in such a way that chips will not fall into the section that is to be used. Fig. 2-8 illustrates a sawing fixture.

If soft tubing is used, pinching the end of the tube that is not going to be used eliminates the danger of chips entering the tubing. It also seals the tubing against moisture and protects it until used. Again in hard copper tubing installation, the tubing ends of the part not being used should be capped or plugged.

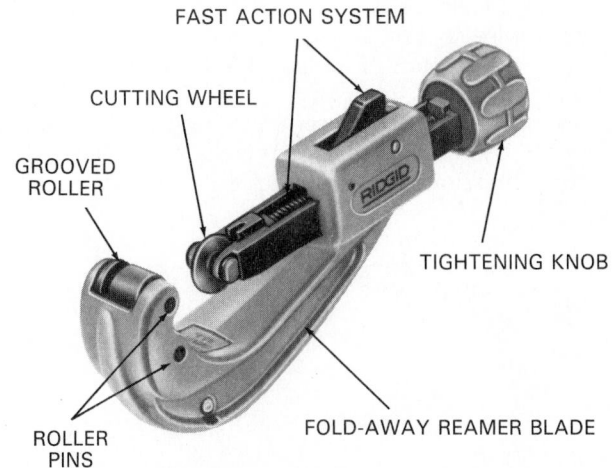

Fig. 2-7. A tube cutter. Note attached reamer which is used to remove burrs from inside of tube after cutting. Grooves in the rollers allow cutter to be used to remove flare from tube with little tubing waste. (Ridge Tool Company)

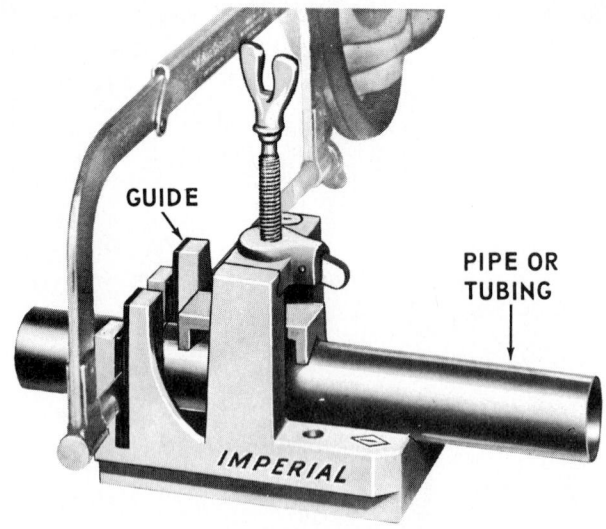

Fig. 2-8. Sawing fixture to ensure square and accurate cuts when using a hand hacksaw to cut tubing. This method is recommended for cutting hard drawn and steel tubing.
(Imperial Eastman, Imperial Division)

To provide a full-wall thickness at the end of the tubing, many service technicians file the end of the tubing with a smooth or medium cut mill file. (See Para. 2-67.)

2-12 BENDING TUBING

It takes practice to become good at bending tubing. For the smaller sizes used in domestic models, it is not necessary to use special bending tools. However, a much neater job and a much more satisfactory one is possible with such tools.

The tubing should be bent so that it does not place any strain on the fittings after it is installed. The tubing, at the bend, should not be reduced in cross-sectional area (kinked). Be very careful when bending the tubing to keep it round. Do not allow it to flatten or buckle. The minimum radius for a tubing bend is between five and 10 times the diameter of the tubing as shown in Fig. 2-9.

Tubes should be bent quite slowly and carefully. It is always wise to use as large a radius as one can. This reduces the amount of flattening. It is also easier to bend a large radius.

Do not try to make the complete bend in one operation, but bend the tubing gradually. There is less danger that the sudden stress will break or buckle the tubing.

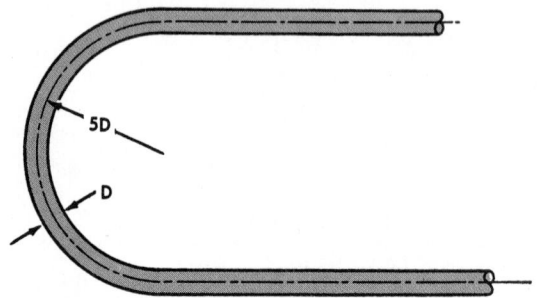

Fig. 2-9. Minimum safe bending radius for bending tubing. D is the outside diameter of tube being bent.

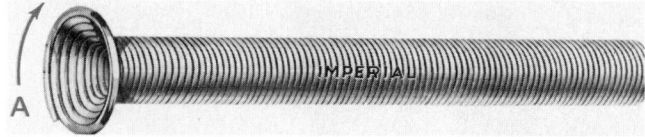

Fig. 2-10. Tube bending spring may be fitted either outside or inside copper tube while bending tube. Bending spring reduces danger of flattening tube while it is being bent. A—Twist to remove spring.
(Imperial Eastman, Imperial Division)

An inexpensive tool called a bending spring is illustrated in Fig. 2-10. It may be easily carried in a kit. These are available in a variety of sizes and are made for both external and internal use. The spring bender may be used internally for making bends near the end of the tubing or even after the tubing has been flared.

An external bending spring for 1/4 in. OD tubing may be used as an internal bending spring for 1/2 in. OD tubing. Use the spring bender externally in the middle of long lengths of tubing.

Bending springs tend to bind on the tubing after the bend. They may be easily removed by twisting the spring. This causes the portion on the outside of the bend to expand, causing the part of the spring on the inside to contract.

If a bend is to be made near the flare and an external spring is to be used, bend the tubing before making the flare. An internal spring can be used either before or after the flaring operation.

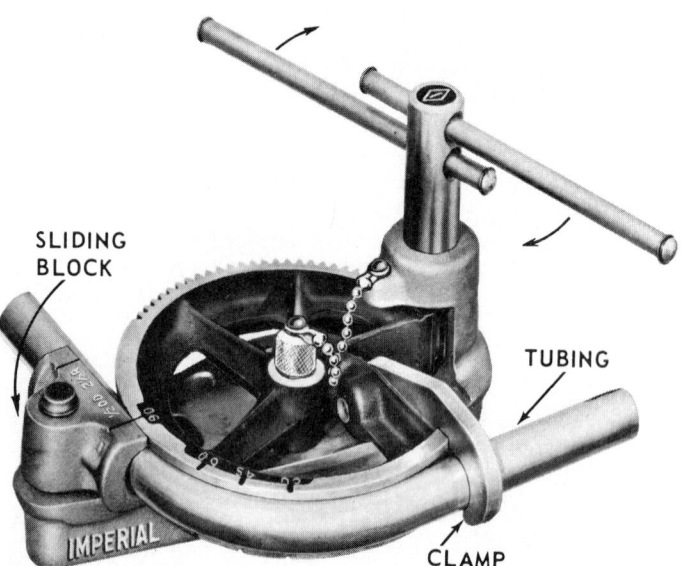

Fig. 2-11. Tube bender which will produce accurate bends and will reduce danger of flattening or buckling tube while it is being bent.
(Imperial Eastman, Imperial Division)

Other tools are available for bending operations. A gear-type bending tool is shown in Fig. 2-11.

A lever-type bender for accurate bending to within 1/32 in. for soft tubing is shown in Fig. 2-12. It can be purchased in five different sizes to match the diameter of the tube to be bent. Always use a bending tool when bending steel tubing. Fig. 2-13 shows some practice bends on tubing.

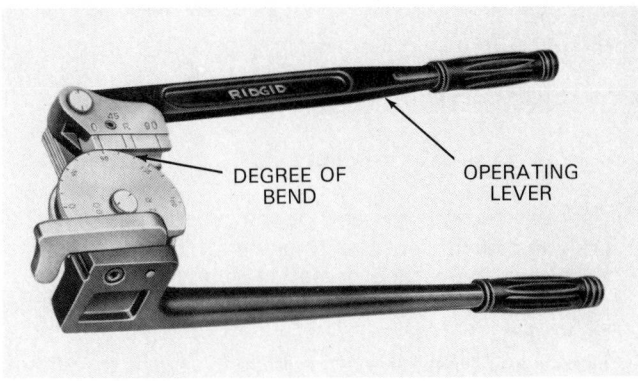

Fig. 2-12. Lever type tube bending tool. As shown, tool is making 90° bend.   (Ridge Tool Company)

2-13 CONNECTING TUBING

Since tubing walls are too thin for threading, other methods of joining tubing to tubing and tubing to fittings must be used. The three common methods are:
1. Flared connections.
2. Soldered connections.
3. Brazed connections.

2-14 FLARED CONNECTIONS

When connecting tubing to fittings, it is common practice to flare the end of the tube and to use fittings designed

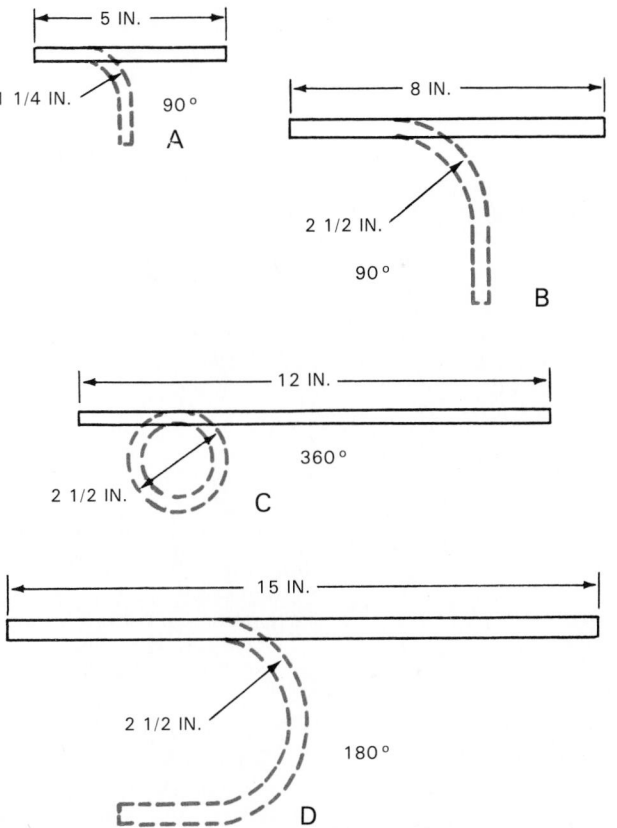

Fig. 2-13. Some practice bends on tubing. A—90° bend on 1/4 in. tubing. B—90° bend on 1/2 in. tubing. C—360° bend on 1/4 in. tubing. D—180° bend on 1/2 in. tubing.

to grip the flare for a vapor-tight seal. Special tools are used for making flares.

Fig. 2-14 illustrates how a flare is used to form a leakproof joint between a tube and a fitting. It also shows some incorrectly made flares.

Some flares are made which use a single thickness of

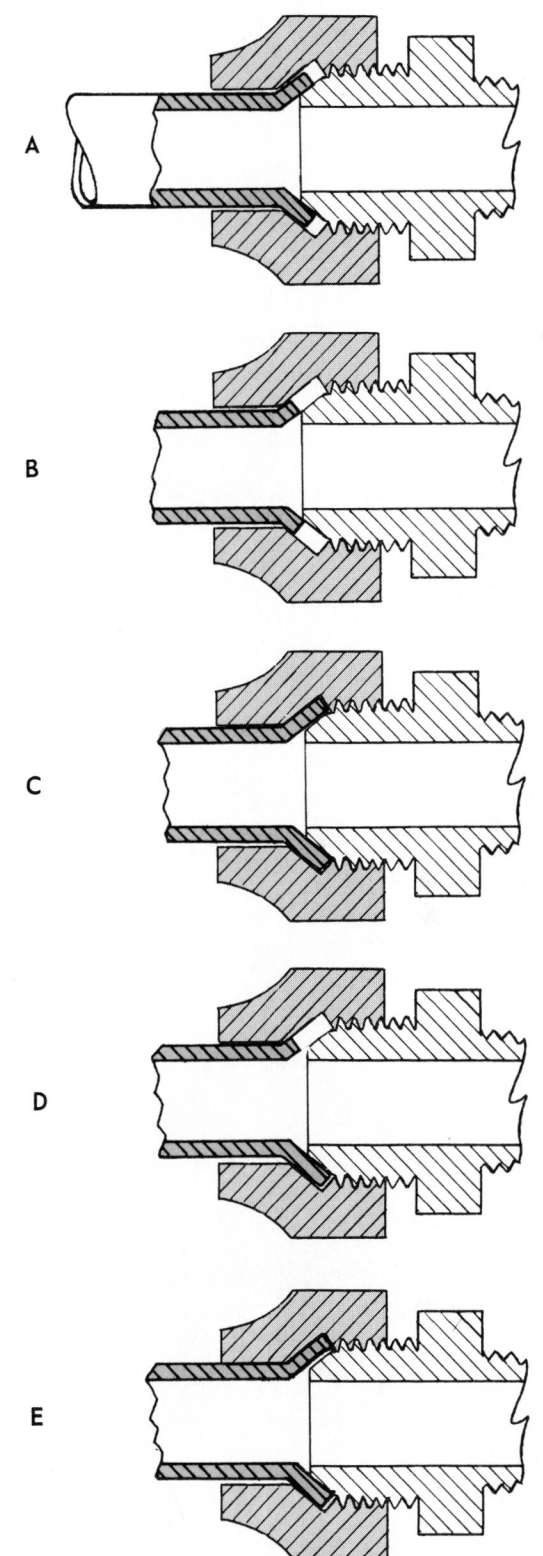

Fig. 2-14. Flared fittings. A—Correctly made flare. B—Flare too small. C—Flare too large. D—Flare is uneven. E—Flare has burrs on edge.

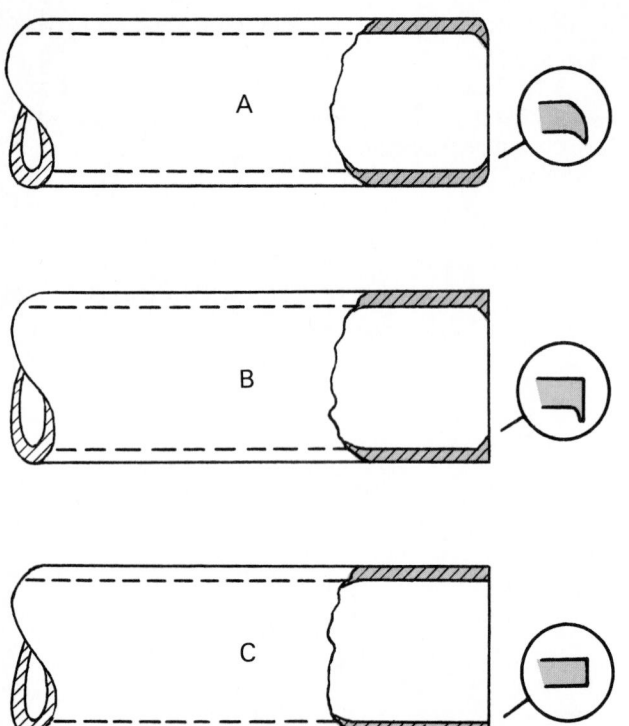

Fig. 2-15. End of tube must be carefully prepared before flaring. A—Tube after being cut. B—Tube after being squared with file. C—Tube filed, reamed, and ready for flaring.

the tube. Other flares are made with a double thickness of metal in the flare surface. Called "double flares," they are stronger and usually cause less trouble if properly made.

Most flares are made at a 45° angle to the tube. Flares on steel tubing, however, are usually made at a 37° angle.

This is because steel tubing is not as easily flared as copper tubing.

2-15 SINGLE THICKNESS FLARE

To make a flare of the correct size using a flaring block, do the following:
1. Carefully prepare the end of the tube for flaring. The end must be straight and square with the tube, and the burr from the cutting operation removed by reaming. Fig. 2-15 shows the steps necessary to prepare a tube for flaring.
2. First, use a 10-in. smooth mill file to square the end of the tube. Use great care that no filing enter the tubing. Next, use a burring reamer to remove the slight burr remaining after the cut-off operation.
3. A flaring tool which may be used to make a single thickness flare is shown at view A in Fig. 2-16. A flaring tool suited for flaring either fractional size or millimeter size tubing is shown at view B in Fig. 2-16.
4. Place the flare nut on the tubing with the open end toward the end of the tubing. Insert the tube in the flaring tool so that it extends above the surface of the block as shown at view A in Fig. 2-17. This allows enough metal to form a full flare.
5. If the tube extends above the block more than the amount shown, the flare will be too large in diameter and the flare nut will not fit over it. If the tube does not extend above the block, the flare will be too small, and it may be squeezed out of the fitting as the flare nut is tightened. View B in Fig. 2-17 shows appearance of completed flare.
6. To form the flare, first put a drop of refrigerant oil on the flaring tool spinner where it will contact the tubing. Tighten the spinner against the tube end one-half turn

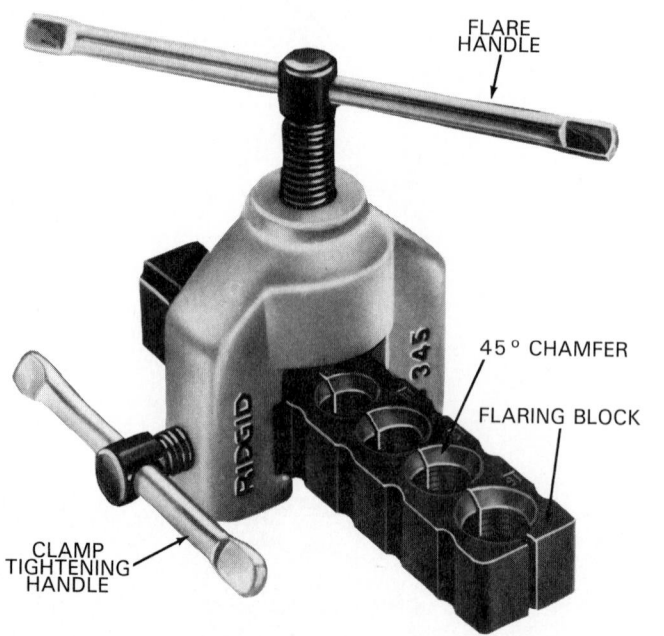

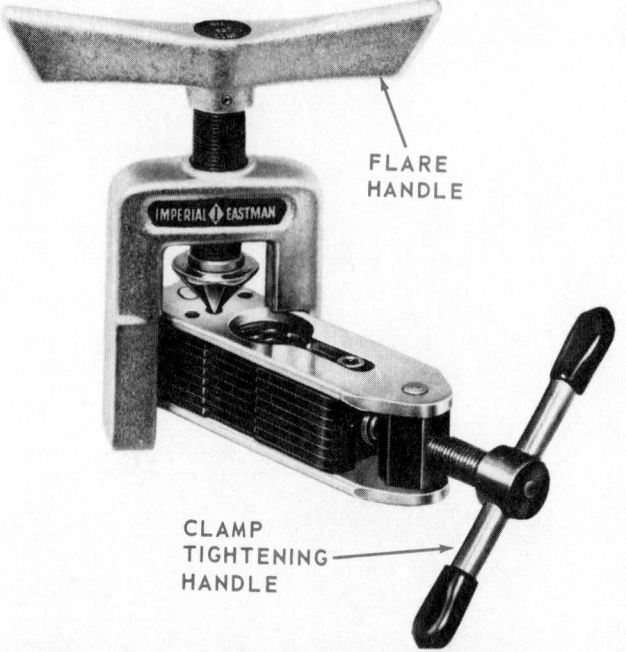

Fig. 2-16. Flaring tools. Left. Popular style used for making single thickness flares on refrigeration tubing. Flaring block is split, making it easy to insert the clamp tubing in place for flaring. Note 45° chamfer in block, which gives the flare its correct shape. (Ridge Tool Co.) Right. Flaring tool having an adjustable tube-holding mechanism which permits flaring tubing 3/16 to 5/8 in. OD and 5 to 16 mm OD. (Imperial Eastman, Imperial Division)

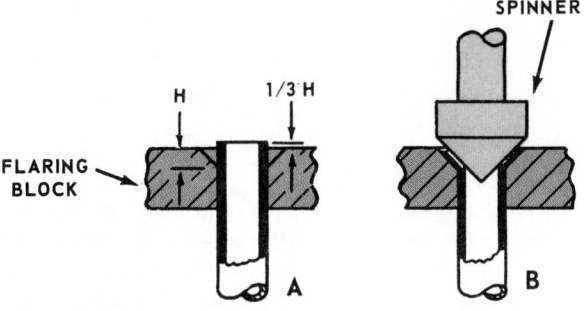

Fig. 2-17. Tubing to be flared should extend slightly above flaring block to allow enough metal to form satisfactory flare. Amount to allow is about a third of the height of flare. A—Proper position of tube in flaring tool before flaring. B—Completed flare.

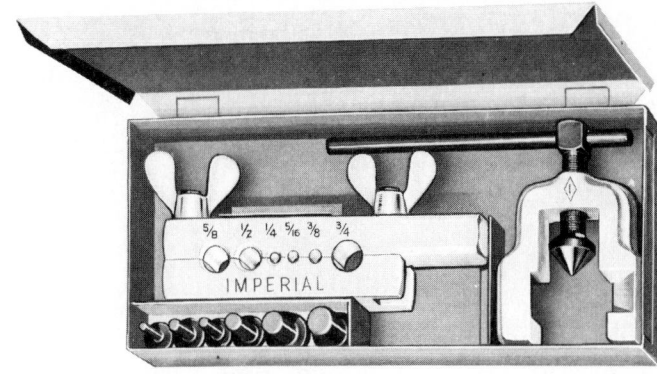

DOUBLE FLARE ADAPTORS

Fig. 2-19. A flaring tool which, with necessary adaptors, is capable of producing either single or double flares. (Imperial Eastman, Imperial Division)

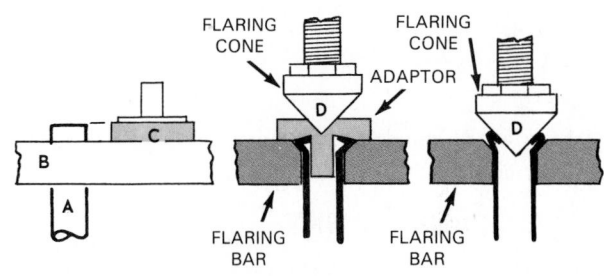

Fig. 2-20. Correct procedure for forming double flare using adaptors with combination single flare-double flare, flaring tool. A—Tubing. B—Block. C—Adaptor. D—Flaring cone.

and back it off one-quarter turn. Advance it three-quarters of a turn and again back it off one-quarter turn. Repeat the forward movement and backing off until the flare is formed.

Some service technicians make the flare using one continuous motion of the flaring tool; that is, without a back-and-forth motion. It is believed by some that the constant turning of the tool, without back turning, may work harden the tubing and make it more likely to split.

Other technicians like to use a flare which is not completely formed—about seven-eighths complete. They depend on the tightening of the flare nut on the flare to complete it.

Do not tighten up the spinning tool too much because this will thin the wall of the tubing at the flare and weaken it.

Always place the flare nut on the tube in the proper position before the flare is made. It cannot, in most cases, be installed on the tube after it has been flared.

2-16 DOUBLE THICKNESS FLARE

Double thickness flares are formed with special tools. Fig. 2-18 illustrates a cross section through a simple block-and-punch type of tool used to make a double flare. The correct shape of the double flare is shown in a final operation in this figure. Some flaring tools are fitted with adaptors which makes it possible to form either single or double flares with the same tool, Fig. 2-19.

Fig. 2-20 shows the steps for making a double flare (using the tool shown in Fig. 2-19).

Double thickness flares are recommended for only the larger size tubing, 5/16 in. and over. Such flares are not

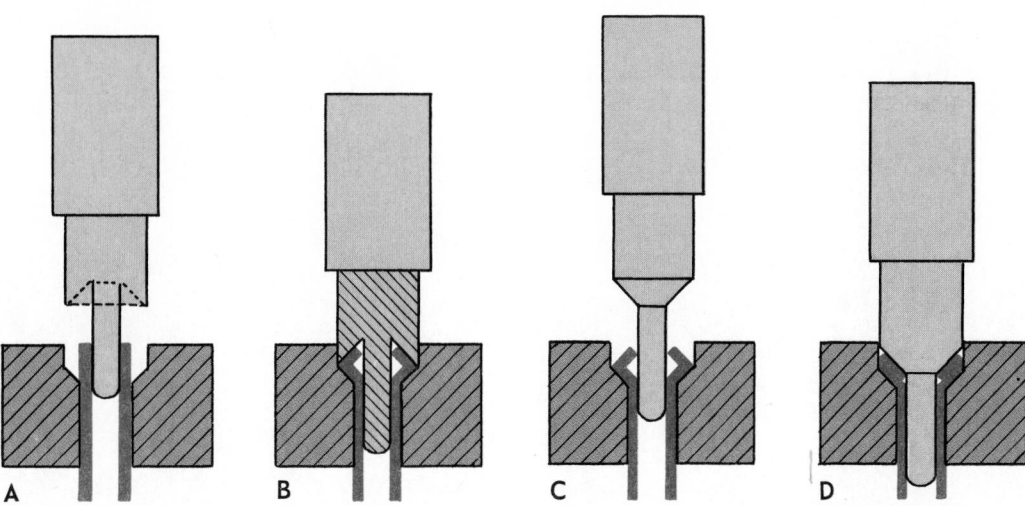

Fig. 2-18. Simple block and punch tool for forming double flares on copper tubing. A—Tube is clamped in body of flaring block. B—First female punch bends end of tube inward. C—Male punch is inserted in partially flared tube. D—In operation, male punch folds end of tube downward to form double thickness and expand flare into final form.

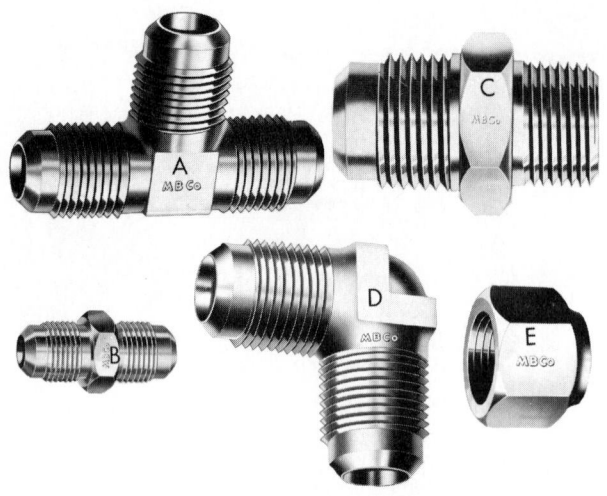

Fig. 2-21. Some of the more common flared type fittings used in refrigeration and air conditioning work. A—Flared Tee fitting, male flare x male flare. B—Flared Union coupling, male flare x male flare. C—Flared Half Union coupling, male flare x male pipe. D—Flared 90° Elbow, male flare x male flare. E—Flare Nut. (Mueller Brass Co.)

easily formed on smaller tubing. The double flare makes a stronger joint than a single flare.

2-17 ANNEALING TUBING

If a flare splits when being made, it may be due to the age of the tubing. Old tubing becomes brittle after a period of use and cannot be flared satisfactorily.

To remedy, anneal the tubing by heating to a dull, cherry red or blue color and allow it to cool. Pounding, rough handling, or bending the tubing tends to work harden it. Hard drawn tubing cannot be bent or flared unless annealed.

2-18 FLARED TUBE FITTINGS

To attach a fitting to soft copper tubing, a flared type connection is generally used. There are many different fitting designs on the market. However, the accepted standard for refrigeration is a forged fitting, using either pipe thread or Society of Automotive Engineers (SAE) National Fine Thread. See Fig. 2-21.

The fittings are usually made of drop-forged brass. They are accurately machined to form the National Fine (NF) threads, the National Pipe (NP) threads, the hexagonal shapes for wrench attachment, and the 45° flare for fitting against the tubing flare. These threaded fittings must be carefully handled to prevent damage to them.

All fitting sizes are based on the tubing size. For example, a 1/4 in. flare nut attaches 1/4 in. tubing to a flared fitting even though it has 7/16 in. NF internal threads and uses a wrench with a 3/4 in. opening to turn it.

Fig. 2-22 is a table of common flared fitting sizes.

Catalog listings of tube fittings usually provide a code number to indicate the size. The code number 3 indicates that the fitting fits 3/16 in. tubing. Code number 4 indicates that it fits 1/4 in. (4/16). Code number 8 fits 1/2 in. tubing (8/16).

Some tubing fittings have pipe threads on one end. Pipe threads taper 1/16 in. in diameter for every inch in length.

With more plastic tubing being used, it has become

REFRIGERATION FITTINGS (FLARED TYPE)
Sizes are based on the Outside Diameter of the Tubing

Name and Description	1/4	5/16	3/8	7/16	1/2	5/8
Nut Forged	X	X	X	X	X	X
Union (Threads same size)	X	X	X	X	X	X
Half Union (1/8 Pipe)	X	X				
Half Union (1/4 Pipe)	X	X	X	X		
Half Union (3/8 Pipe)					X	
Half Union (1/2 Pipe)						X
Elbow	X	X	X	X	X	X
Elbow (One 1/8 Pipe)	X	X				
Elbow (One 1/4 Pipe)	X	X	X	X		
Elbow (One 3/8 Pipe)					X	
Elbow (One 1/2 Pipe)						X
Tee (Threads same size)	X	X	X	X	X	X
Tee (One 1/8 Pipe)	X	X				
Tee (One 1/4 Pipe)			X	X		
Tee (One 3/8 Pipe)					X	
Tee (One 1/2 Pipe)						X
Cross	X	X	X	X	X	X
Flared Tube Sealing Plug	X	X	X	X	X	X
Flared Tube Sealing Cap		X	X	X	X	X
Flared Tube Copper Seal Cap	X	X	X	X	X	X
Union (Reducing)	5/16-1/4	3/8-1/4	1/2-1/4	1/2-3/8		
Elbow (Reducing)	5/16-1/4	3/8-1/4	1/2-1/4	1/2-3/8	5/8-1/2	
Tee (Reducing)	5/16-1/4	3/8-1/4	1/2-1/4	1/2-3/8	5/8-1/2	

Fig. 2-22. Some of the more popular flared type copper tube fittings used by the refrigeration and air conditioning service engineer. The reducing fittings are used for connecting tubing of different sizes.

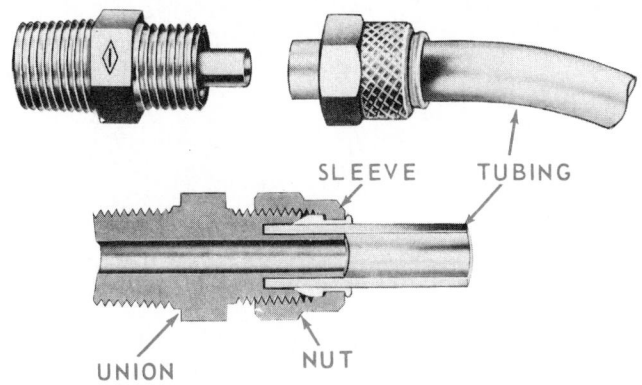

Fig. 2-23. A compression type fitting used with polyethylene tubing. Caution: Polyethylene is a soft substance and very little tightening is needed. While most fittings are made with flats for wrench tightening, most polyethylene installations require little more than ''finger tightness.'' (Imperial Eastman, Imperial Division)

necessary to provide special fittings. Brass, aluminum, and polyethylene materials are commonly used. Fitting connections made on plastic tubing are not flared. Rather, a compression type fitting, as shown in Fig. 2-23, is used.

2-19 METRIC SIZE TUBE FITTINGS

Metric size tubing, as described in Para. 2-6, requires metric size fittings. These are made of the same materials and in the same general styles and shapes as U.S. conventional size fittings and are used in the same way. The service technician must be careful that U.S. conventional size fittings are not mixed with metric size fittings.

2-20 SOLDERED OR BRAZED TUBING FITTINGS

In modern practice, most tube and fitting connections are made by either soldering or brazing. Soldered joints are used for water pipes and drains. Brazed joints are used for refrigerant pipes and tubing. The difference between soldering and brazing is the temperature at which solder flows.

Soldered joints use a capillary action to draw molten solder into the area between the fitting and the tube. The selection of a solder is based upon two factors: operating pressure and temperature of the line. A 50/50 tin-lead solder is appropriate for moderate pressures and temperatures. It is a mixture of one-half tin and one-half lead. It melts at 360°F (182°C) and is fluid at 415°F (213°C). For higher pressures or greater joint strength, a 95/5 tin-antimony solder is used. This mixture contains 95% tin and 5% antimony and is harder than a 50/50 solder. A 95/5 solder melts at 450°F (232°C) and is fully liquid at 465°F (241°C).

In brazing, brazing filler metals are used to produce a stronger bond. They also have the advantage of joining similar and dissimilar metals at low brazing temperatures. Brazing filler metals melt at temperatures in the range of 1000°F (538°C) and 1500°F (816°C). Brazing filler metals used for joining copper tubing are of two categories: alloys containing 30 to 60% silver, and copper alloys which contain phosphorus. These two classes vary in melting, flowing, and fluxing characteristics. Strength of a brazed copper joint is not as dependent upon choice of filler metals, but depends upon proper clearance between the tube and the socket of the fitting.

2-21 SOLDERING

Soldering is a process of applying molten (melted) metal to metals that are heated but are not molten. It is an adhesion process. (In adhesion, one part is bonded to or is stuck to another by a third material.) The molten solder flows into the pores of the surface of the metals being joined. As the solder solidifies (hardens), a good bond is obtained.

A fitting soldered to a tube is shown in Fig. 2-24. A good sweat joint begins with first cleaning the parts to be joined, then fluxing and assembling them. The assembly is then heated. As soon as the joint reaches the flowing temperature of the solder, solder is added to the joint and flows into it. After the solder cools, it will seal and connect the surfaces. The step-by-step procedure for making a sweat joint is shown in Fig. 2-25.

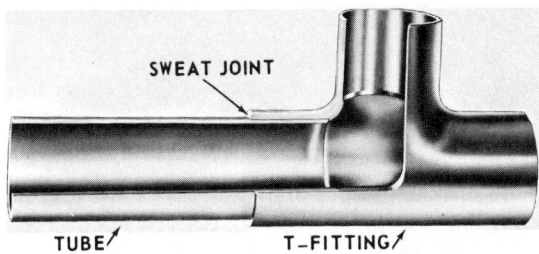

Fig. 2-24. A cross section of tee fitting soldered to hard ACR tube.

When assembling swaged (shaped) tube-to-tube joints or tubing to a fitting, thoroughly clean the mating parts. Next, apply flux to the outside of the tube. Insert the tube into the fitting 1/16 to 1/8 in. Rotate one of the pieces to spread the flux evenly over both the internal and external surface. See Fig. 2-26. Using the rotation procedure will eliminate any possibility of flux entering the system. Then apply the necessary heat for soldering or brazing.

Avoid swaging steel tubing (shaping by hammering). It is harder (less ductile) and may crack or split. Sometimes the process tube of a hermetic motor compressor is made of steel. Many clean the tubing before cutting it, to reduce the amount of dirt that may get into the system. One tubing should extend into the other the same distance as the diameter of the larger tubing (i.e., 1/4 in. into 5/16 in. should overlap 5/16 in.)

Fig. 2-27 illustrates some common fittings, which may be either soldered or brazed to tubing. All brass and copper parts may be easily soldered.

To solder:
1. The surfaces to be soldered must be very clean.
2. A good clean flux must be used.
3. A good source of clean heat must be on hand.
4. The parts being soldered must be firmly supported during the soldering operation.

Flux does not clean the metal. It keeps the metal clean once soil has been removed by filing, scraping, sanding, and using steel wool or wire brushes. Surfaces being soldered must be free of grease, dirt, and oxides.

A 50/50 alloy of tin and lead solder is usually satisfactory for soft soldering. Solders containing as much as 95 percent tin are now being recommended for soldered joints subjected to very low temperatures.

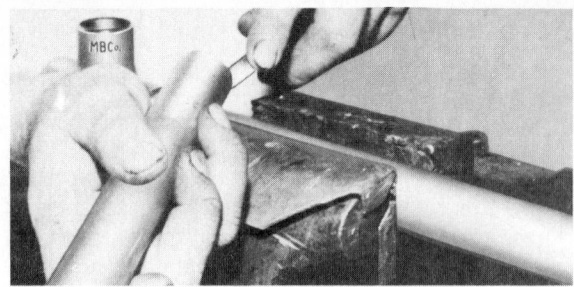

Step 1. Cut tube to length and remove burr with file or scraper.

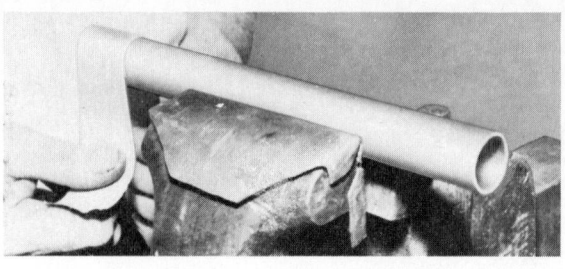

Step 2. Clean outside of tube with clean sandpaper or sandcloth.

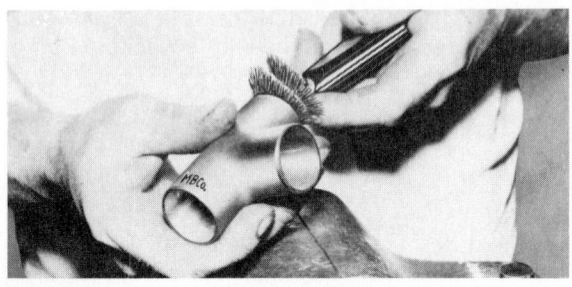

Step 3. Clean inside of fitting with a clean wire brush, sandcloth or sandpaper. Do not use emery cloth.

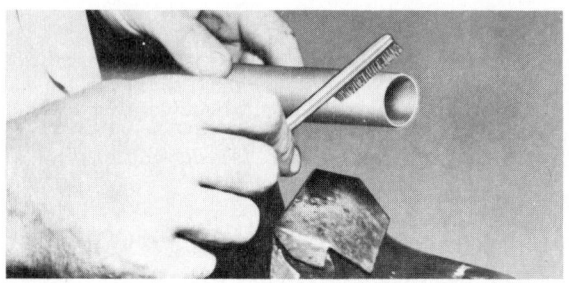

Step 4. Apply flux thoroughly to outside of tube — assemble tube and fitting.

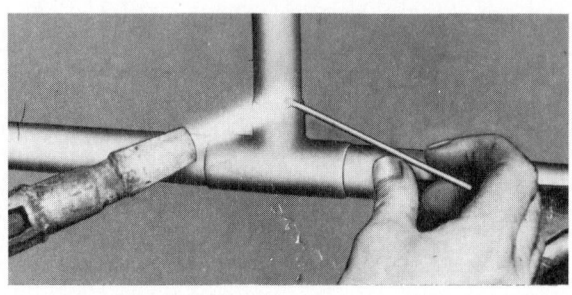

Step 5. Apply heat with torch. When solder melts upon contact with heated fitting, the proper temperature for soldering has been reached. Remove flame and feed solder to the joint at one or two points until a ring of solder appears at the end of the fitting.

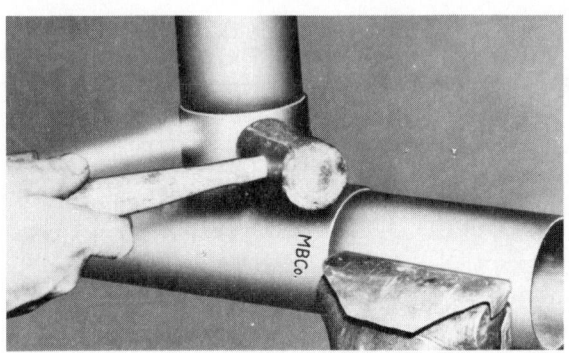

Step 6. Tap larger sized fittings with mallet while soldering, to break surface tension and to distribute solder evenly in joint.

Fig. 2-25. Recommended step-by-step procedures to follow when soldering tubing.   (Mueller Brass Co.)

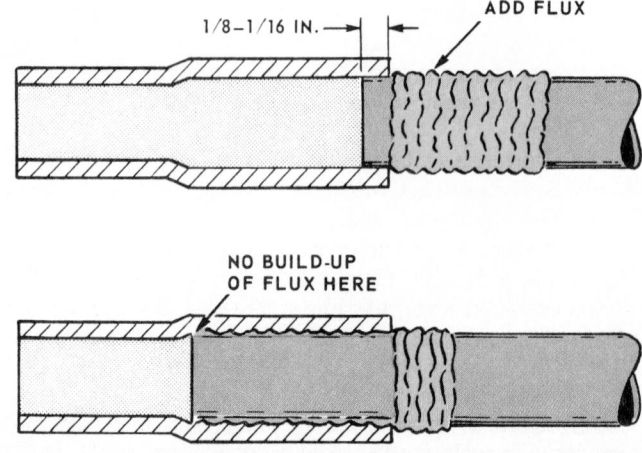

Fig. 2-26. Brazing flux may be a source of corrosion in a system. Apply flux to joints as above so that it will not get into system.

Fig. 2-27. Common fittings which may be either soldered or brazed to tubing. A—Coupling with rolled stop, sweat x sweat. B—Tee, sweat x sweat x sweat. C—90° elbow, sweat x sweat. D—Adaptor, sweat x male pipe thread (mpt).   (Mueller Brass Co.)

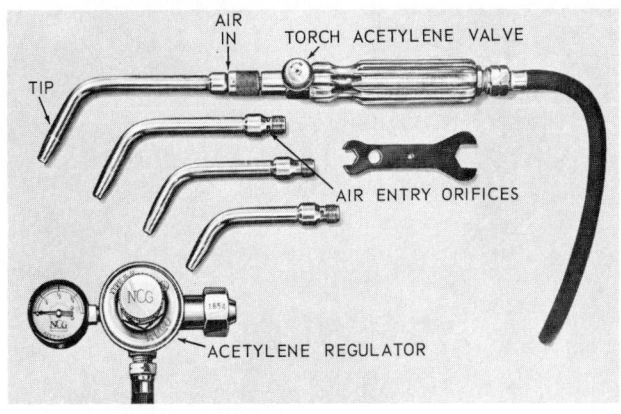

Fig. 2-28. Air-acetylene torch suitable for both soldering and brazing tubing. These torches use the small portable acetylene cylinders. Flame is very clean and hot and torch is easy to use.

A portable air-acetylene torch, Fig. 2-28, is a practical tool for heating surfaces to be soldered. Acetylene is the fuel recommended for this type of torch. However, liquid propane may be used.

When joining tubing and tubing fittings by soldering, thoroughly clean the surfaces to be soldered. Rolls of clean abrasive paper or abrasive cloth are good cleaning materials. Brushes may also be used to clean the inside and outside.

Apply flux thoroughly to outside of tube. Flux for this type of work should have no corrosive properties. Acid flux should not be used. It tends to corrode fittings, making them unsightly and difficult to work on later.

An important fundamental of good soldering is that the metal being joined be hot enough to melt the solder. This is the only way the solder will go into the pores of the metal.

Solder in wire form is usually used because of the difficulty in getting at the soldering surfaces.

While soldering, it helps to "wipe" the surfaces after putting some solder on. Use a clean cloth, a brush, or the solder wire itself. This action will remove any dirt and will help coat the surface.

Apply the heat to the metal to be soldered; then touch the solder to the metal. Do not overheat. Keep testing the

metal with the solder wire. Heat only until solder flows.

If the parts to be soldered are of the correct temperature and have been cleaned and fluxed, the solder will flow over the surface quickly. Remember not to heat the solder directly with the torch.

Fig. 2-29 illustrates a tube soldering practice assembly. This assembly can then be connected to an R-12 cylinder or compressed air line, and the quality of the soldered joint determined by checking for leaks.

Never use oxygen when testing for leaks. Any oil in contact with oxygen under pressure will form an explosive mixture.

2-22 BRAZING

One of the best methods of making leakproof connections while providing maximum strength is to braze the joints. These joints are very strong and will stand up under the most extreme temperature conditions.

Oxyacetylene brazing equipment is used to achieve maximum strength and a leakproof joint. Oxyacetylene is the introduction of pure oxygen to acetylene. This combined mixture produces a very hot flame.

Correct use of oxyacetylene depends upon the technician constantly metering the flow of oxygen and acetylene. The oxygen tank and the acetylene tank have pressure regulators and a set of gauges. One gauge registers tank pressure, the other displays pressure of the torch.

Acetylene is a highly flammable gas, especially when mixed with oxygen. Therefore, safety glasses should always be worn when brazing. Never point the torch (lit or unlit) toward an open flame or source of sparks. Light the torch only with a sparker—do not use matches. The acetylene valve adjusts the needed flame size. *Slowly* turn the oxygen valve to obtain type of flame required. A "neutral flame" has a blue cone with a bit of reddish purple at the tip and is most efficient in brazing.

Fig. 2-30 illustrates an oxyacetylene outfit. Brazing can be done easily if the correct procedures are followed:
1. Degrease parts and clean the joints thoroughly.
2. Fit the joints closely and support all parts.
3. Apply the clean flux recommended for the brazing alloy. Follow the manufacturer's instructions.
4. Heat evenly to recommended temperature. Keep the torch moving constantly in a "figure-eight" motion.
5. Apply brazing alloy to the heated parts. Do not heat (melt) the brazing alloy with the torch.
6. Cool the joint.
7. Clean the joint thoroughly, using warm water and a brush. Be sure all flux has been removed.

An oxyacetylene torch is an excellent heat source for brazing. However, one must have training in its safe use. Be sure to use flashback arrestors at both the acetylene and oxygen regulators.

There are various brazing alloys on the market. Most have a 35 to 45 percent silver content. This material usually melts at 1120 °F (604 °C) and flows at 1145 °F (618 °C). Contact the local welding supply house or air conditioning and refrigeration wholesaler for suitable brazing supplies.

CAUTION: Carefully check the specifications of the brazing alloy used. If it contains any amount of cadmium, be SURE that the work space is well ventilated and that none of the fumes are inhaled or come in contact with the eyes or skin. Cadmium fumes are very poisonous.

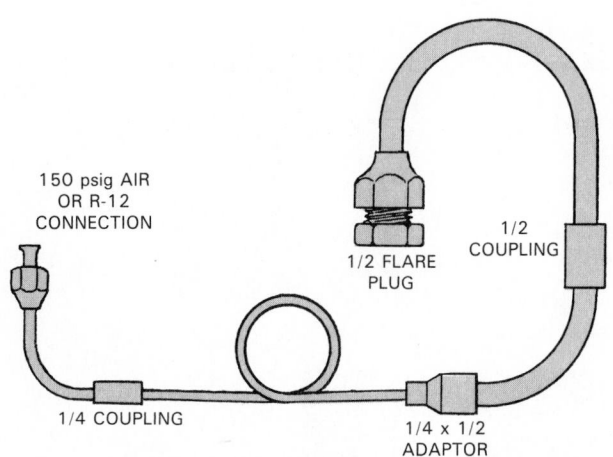

150 psig AIR OR R-12 CONNECTION

1/2 FLARE PLUG

1/2 COUPLING

1/4 COUPLING

1/4 x 1/2 ADAPTOR

Fig. 2-29. Practice piece used to develop skill in copper tube soldering. Completed assembly should be pressurized as indicated and all of the joints tested for leaks. Use soapsuds.

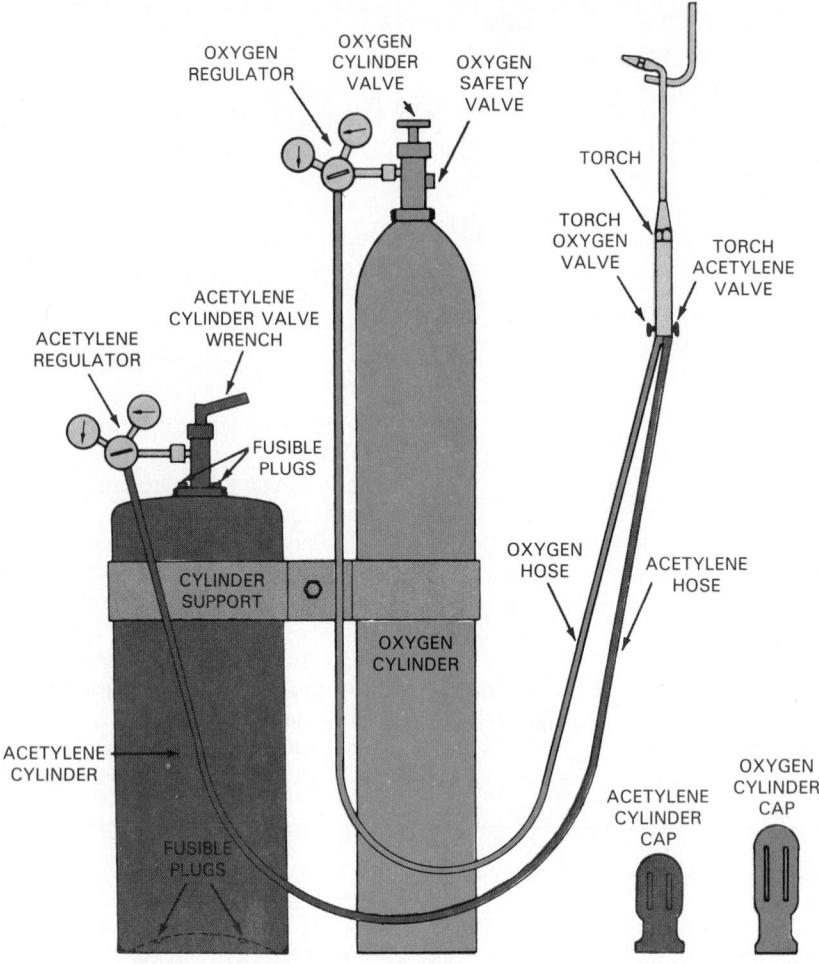

Fig. 2-30. An oxyacetylene welding outfit.

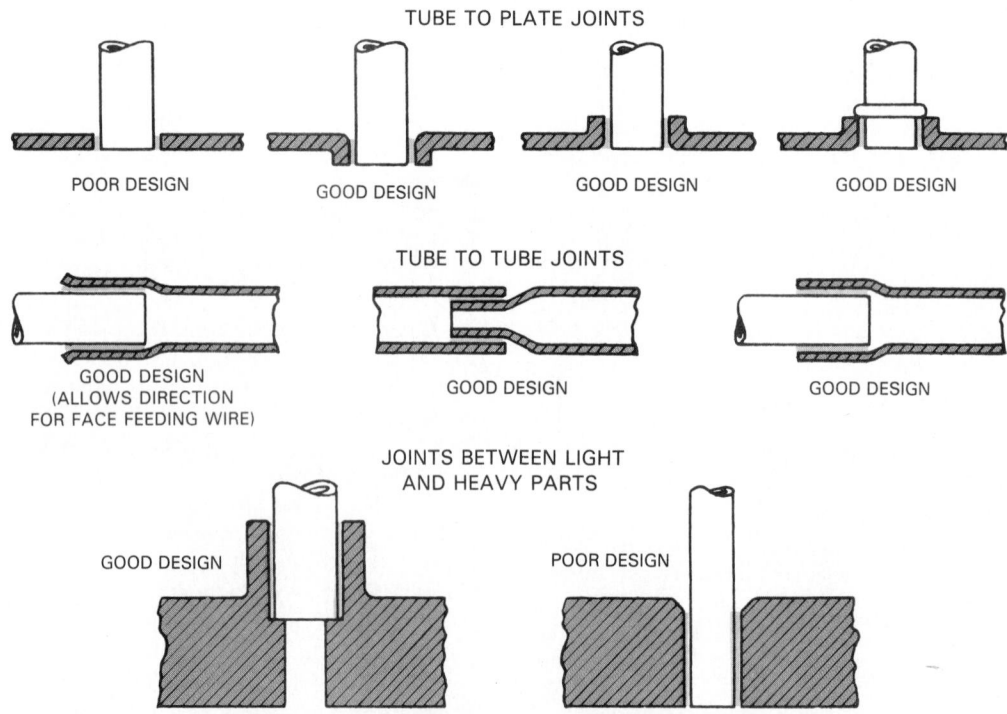

TUBE TO PLATE JOINTS

POOR DESIGN   GOOD DESIGN   GOOD DESIGN   GOOD DESIGN

TUBE TO TUBE JOINTS

GOOD DESIGN
(ALLOWS DIRECTION
FOR FACE FEEDING WIRE)   GOOD DESIGN   GOOD DESIGN

JOINTS BETWEEN LIGHT
AND HEAVY PARTS

GOOD DESIGN   POOR DESIGN

Fig. 2-31. Suggestions for making joints to be brazed. Actual thickness of brazing is exaggerated to show its application. (Lucas-Milhaupt, Inc., A Handy & Harman Company)

The part to be brazed must be carefully cleaned and fitted accurately. Dirt must be removed from any external surface. Use a fine grade of stainless steel wool for cleaning the exterior. Internal surfaces can be cleaned with stainless steel wire brushes or stainless steel wool rolled on a rod.

The parts must have contacting surfaces of sufficient area, such as a tube sliding into a fitting (not a drive fit), Fig. 2-31. The contacting surfaces need not be very large (three times smallest section.).

If the parts are dented or are out of round, these faults must be corrected before the brazing is done. It is important to support all the parts securely during the operation so that the parts will not move.

Make sure that no flux enters the system during the brazing operation, as it cannot be easily removed. Avoid overfluxing by applying the flux to the surface that is to slide into the part, Fig. 2-26. The excess flux will then stay on the outside.

All air must be removed from the tubing being brazed. This can be done best by purging the tubing with either carbon dioxide or nitrogen, as shown in Fig. 2-32. If there is any oil inside the tubing or part, the heat of the torch may cause this oil to vaporize. Oil vapor mixed with air will explode if ignited. Using a nonflammable gas such as carbon dioxide or nitrogen will eliminate this hazard. HAZARD—See Chapter 11.

CAUTION: NEVER USE A REFRIGERANT, OXYGEN, OR COMPRESSED AIR.

Heating of the joint must be done carefully. The flux behavior is a good indication of the temperature of the joint as the heating progresses.
1. Keep the joint covered with the flame all during the operation to prevent air getting to it.
2. The flux will dry out; the moisture (water) will boil off at 212 °F (100 °C), then the flux will turn milky in color.
3. Next, it will bubble at about 600 °F (316 °C).
4. At 800 °F (427 °C), the flux lies on the surface and has a milky appearance.
5. Following this, it will turn into a clear liquid at about 1100 °F (593 °C). This point is just short of the brazing temperature.

The alloy itself melts at 1120 °F (604 °C) and flows at 1145 °F (618 °C). A torch tip several sizes larger than the one used for soldering should be used. Be sure to heat BOTH pieces which are to have the alloy adhered to them.

The proper brazing temperature will be indicated by the color of the secondary flame. The flame will start to show a green shade as the brazing temperature is reached. For silver brazing, a clear flux and/or a green flame show the proper temperature.

When heating a copper-to-steel joint, heat the copper first (it takes more heat because it carries it away faster). Put some flux on the brazing rod to help it flow quicker.

When cutting capillary tubing, notch all around with a three-corner file. Break tubing by bending back and forth (small bends). The tubing ID will then remain full size. A tube cutter will reduce inside diameter.

Do not clean the end of the tube, or brazing material may run to the end and close or partially close the hole (ID) of the capillary.

When brazing, the torch is never held in one spot, but must be moved around the entire area to be brazed. Never hold the heat in one area but keep the torch moving. Many

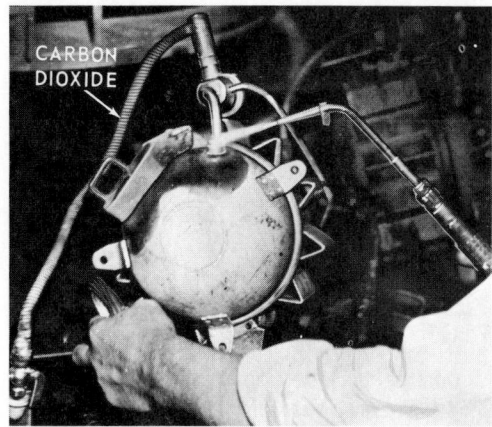

Fig. 2-32. A fitting being brazed to compressor dome. Note tube carrying low-pressure dioxide or nitrogen through fitting and into compressor during brazing to prevent fire or explosion.

service technicians prefer to move the torch in a "figure-eight" motion. Larger torch tip sizes are recommended to allow a soft flame and large quantity of heat without excess pressure or "blow." A slight feather on the inner cone is good. See Fig. 2-33.

2-23 CLEANING THE BRAZED JOINT

Thoroughly wash with water and scrub the outside of the completed brazed joint. This is always necessary. Flux left on the metals will tend to corrode them or may temporarily stop a leak which will only show up later.

The joint may be cooled quickly or slowly. Cooling with water is allowable. The same water may be used to wash the joint.

Visual inspection will quickly reveal any places where the alloy did not adhere. It is advisable to watch for poor adhesion (dark cup-shaped areas) and make any corrections during the brazing operation.

Fig. 2-33. Brazing copper tubing connection. See text for suggestions on flame movement. (Kramer Trenton, Co., BRAZING BOOK)

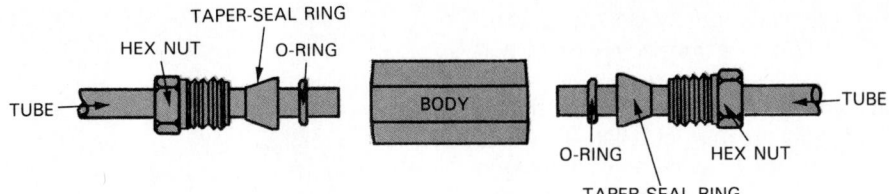

Fig. 2-34. Screw-on mechanical fitting. Note the hex nut, tapered-seal ring, and O-ring. (Watsco Components, Inc.)

2-24 TUBE COUPLINGS

Tube couplings may be used to join aluminum tubes to copper tubes. This requires a process unlike that of joining copper to copper. There are a variety of methods available for joining aluminum to copper. These include: screw-on mechanical fittings, flared and compression fittings, and epoxy resin and adhesive kits.

Fig. 2-34 illustrates a screw-on mechanical fitting. In the adhesive kit is a tube coupling that, when heated with a propane torch, shrinks in approximately 20 seconds to form a copper-aluminum joint, Fig. 2-35.

2-25 SWAGING COPPER TUBING

Swaging, Fig. 2-36, permits two pieces of soft copper tubing of the same diameter to be joined together without the use of fittings.

Swaging of copper tubing is often done. It is more convenient to solder one joint than to make two flared connections.

The length of overlap of the two pieces of tubing is important. As a general rule, the length of the overlap should equal the outside diameter (OD) of the tubing.

Two types of swaging tools are commonly used: the punch type and the lever type. In both cases, different tool sizes are available for use with the many sizes of copper tubing.

When using the punch type swaging tool, the copper tubing is inserted into the correct hole size in the anvil block.

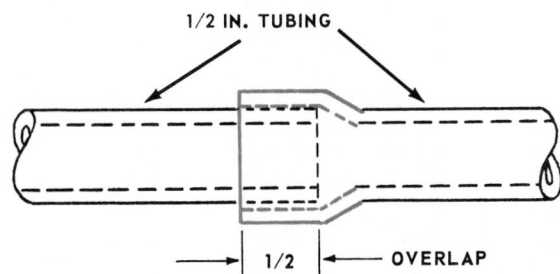

Fig. 2-36. Two pieces of soft copper tubing assembled and ready for soldering or brazing to make joint. Note that pieces were of same diameter before one was swaged.

Then a punch is inserted into the copper tubing and hammered down until it has entered the tubing the desired distance. See Fig. 2-37, view A.

When using the lever type tool, the tubing is placed over the expander. Squeezing the lever expands the tube to the proper size, Fig. 2-37, view B. Fig. 2-37, view C, illustrates the end of the tubing expanded and the pieces fitted together, ready for soldering.

2-26 TUBE CONSTRICTOR

In many cases a service technician must make a soldered or brazed joint between two tubes where one tube fits rather loosely inside the other. Good practice demands that the tubes be sized as close as .003 in. to each other. Fig. 2-38 shows a special tool used to constrict the outer tube until it fits the OD of the inner tube. By using this tool, the service technician can easily solder or braze the joint without danger of leaks or of flux getting into the system.

2-27 PIPE FITTINGS AND SIZES

Air conditioning and refrigeration installations make wide use of pipe fittings and pipe threads (National Pipe Threads or NP). Taper pipe threads are specially formed V-threads made on a conical spiral. This causes the threads to seal as the fitting is tightened. Pipe threads taper 1/16 in. in diameter for every inch of length. If the threads were not tapered, a gasket or an American Standards Association (ASA) machined shoulder would be necessary to provide a leakproof joint.

Besides being tapered (or in a conical spiral), pipe threads are different from fine-thread series National Fine (NF) and the coarse-thread series National Coarse (NC). NF and NC sizes are based on outside diameter. Pipe thread sizes are based on flow diameter, or roughly the diameter of the hole in the pipe (inside diameter or ID). Fig. 2-39 illustrates a male thread on a 1/2 in. pipe. The external threads are cut with a pipe die. The die is turned by a standard die stock,

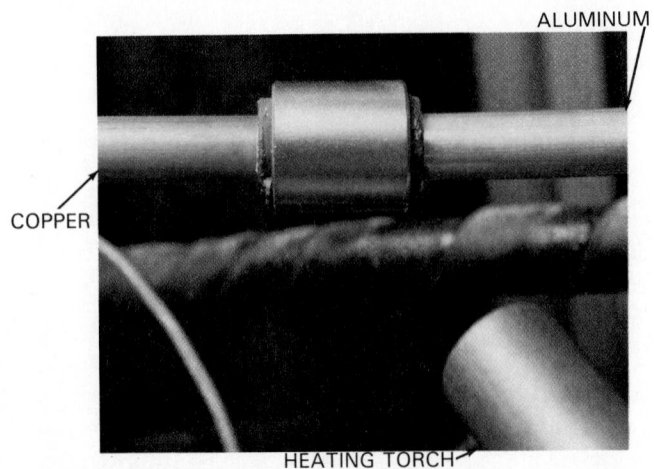

Fig. 2-35. Copper and aluminum tubing being joined with adhesive kit and heat from torch.

52

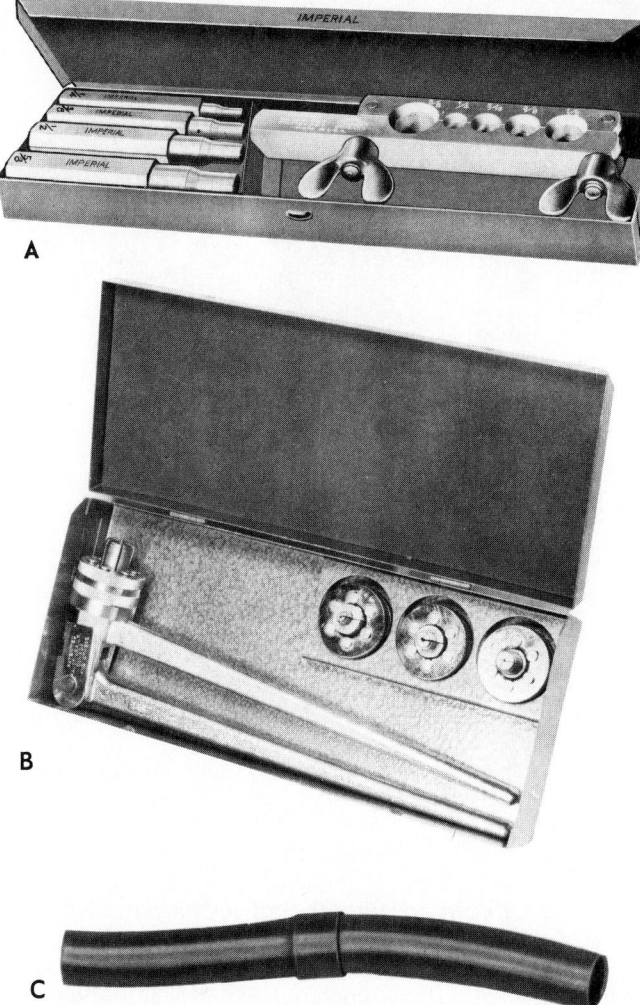

A

B

C

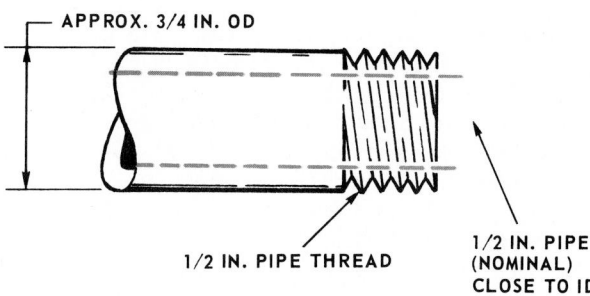

Fig. 2-39. A male thread on a 1/2 in. pipe.

Fig.2-37. Swaging tools. A—Punch type swaging tool with various size punches and anvil for swaging tubing from 1/4 to 1/2 in. diameter. (Imperial Eastman, Imperial Division) B—Lever type (swaging tool) tube expander. This expander has various sized adaptors to fit various sized tubes. C—Tube expanded and fitted to another ready for soldering.

a ratchet die stock, or power driven die stock. In Fig. 2-40, the pipe thread tap (for cutting the female or internal threads) is turned with a tap wrench. The male threaded pipe should be turned into the female fitting five threads.

Pipe fittings are supplied with the threads already cut. The most common fittings are the coupling, the reducing coupling, the union, the nipple, the 90 deg. elbow, the reducing elbow, the 45 deg. elbow, and the street ell. A street ell is usually a 90 deg. fitting having a male thread on one end and a female thread on the other.

The threads are made self-sealing by the pressing together of the sharp V-threads as they are assembled. Various commercial compounds are available to help seal these threads. Brushed on pipe threads before assembly, the compound will make a strong, leakproof joint.

Special thread cutting compound should be used when cutting pipe threads. The taps and dies must be kept clean and sharp.

Fig. 2-40. Pipe tap. Note taper of threads. (TRW Greenfield Tap & Die Div.)

2-28 REPAIRING THREADS

Occasionally the threads of a fitting or fixture—especially pipe threads—may become so worn that they are no longer leakproof. A very convenient and rapid remedy for this condition is to coat the threads with solder and then remove the excess solder by sharply rapping the fitting while it is still hot. The threads will then be coated with a thin film of solder. This will usually remedy the trouble.

2-29 EPOXY RESIN

Epoxy resin may be used to repair cracks and leaks in evaporators and joints. Such resins have good adhesion (sticking) qualities when used with steel, copper, wood and many types of plastics. Epoxy adhesives are available from many manufacturers or refrigeration supply wholesalers.

The most desirable type is the two-part system. This consists of an epoxy resin and a hardener. Combinations of

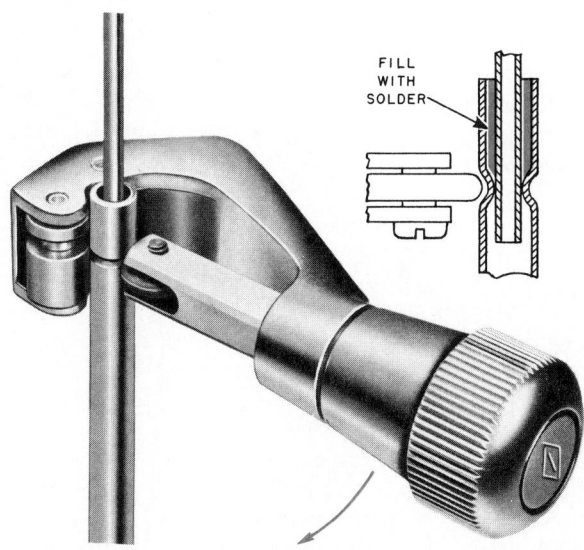

Fig. 2-38. Combination tube cutter and constrictor. Different wheel is used for cutting than for constricting. Insert shows two tubes joined by soldering or brazing. Outer tube has been constricted before soldering. (Imperial Eastman, Imperial Division)

these two will harden at room temperature. The one-part adhesive must be heated in order to cause it to harden.

The two-part system consists of two jars or tubes of a paste-like substance in contrasting colors.

When repairing with epoxy cement, it is necessary to first determine whether a patch is necessary or whether the hole can be repaired by placing the mixed epoxy over the hole or leak. Small leaks or holes up to 1/16 in. in diameter can often be successfully sealed by placing the mixed epoxy over the leak and allowing it to cure. The same procedure is recommended for tubing cracks. For larger holes, a patch of the same type of tubing material is recommended.

The refrigeration service technician is cautioned to purchase the epoxy compound from a refrigeration wholesaler because some epoxies available elsewhere are not compatible with refrigerants R-12 and R-22. Furthermore, the shelf life of most epoxy resins is about six months.

The service technician should be careful when using epoxy compounds because they contain chemicals which may irritate the skin. Long contact with the skin should be avoided. (In case of contact, remove the epoxy and clean the skin with rubbing alcohol followed by a thorough washing with soap and water.)

The procedure for using epoxy compound is as follows:
1. Obtain a kit having a two-part epoxy, Fig. 2-41.
2. Clean the surface or surfaces which are to be bonded by sanding with clean, coarse sandpaper or scrubbing with clean steel wool.
3. Clean surface with recommended solvents such as methyl ethyl ketone, toluene, acetone, or a similar industrial solvent.
4. Mix epoxy on a clean surface such as a piece of cardboard. The two parts of epoxy compounds are mixed together (equal parts of each) until all color streaks have been eliminated and the substance has a uniform color.
5. Apply the epoxy mixture to the surface if it is a small hole or apply to mating surfaces if a patch is to be used as shown in Fig. 2-42. Epoxy compounds should be used immediately after mixing, as chemical hardening, or setting starts immediately.

2-30 HAND TOOLS

The refrigeration and air conditioning service technician performs work chiefly with hand tools. To be successful, the technician must pick good ones, take good care of them, and be skilled in their use. Most service failures can be traced to poor hand tool skills.

The refrigerating mechanism, in comparison to an automobile engine, is relatively light. It can easily be damaged by abuse. Great care is necessary to avoid damaging the units.

Fig. 2-43 shows an assortment of basic hand tools needed by the service technician.

The following paragraphs provide useful suggestions for the selection, care, and use of hand tools.

2-31 WRENCHES

Most refrigeration and air conditioning installation and service work requires the use of various types of wrenches. Many fasteners and parts are copper or brass and therefore are rather soft. Never use pliers on parts designed to be handled with wrenches.

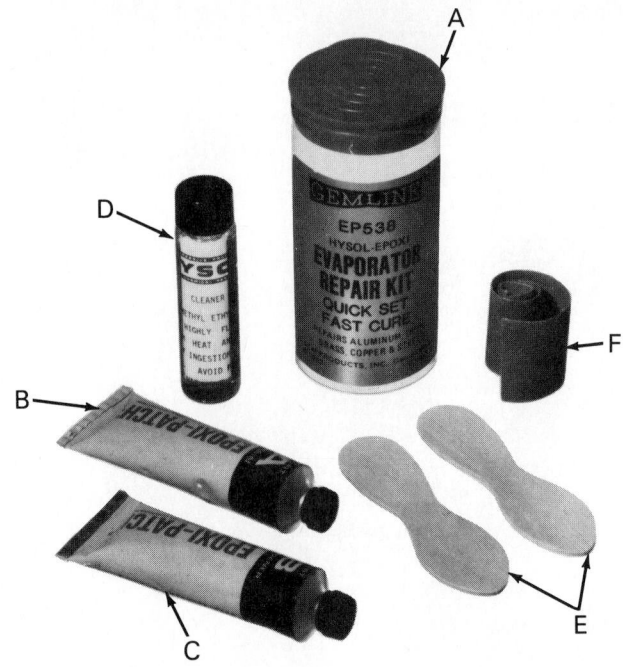

Fig. 2-41. Evaporator repair kit is shown at A. Contents include: B—Epoxy resin. C—Hardener. D—Vial of methyl ethyl ketone. E—Mixing spatulas. F—Sandpaper. (Gem Products, Inc.)

A service technician needs wrenches of several types and sizes. Wrenches should be made of good alloy steel, properly heat treated, accurately machined, and ground to fit the assembly devices. The wrench must fit the nut or bolt head accurately and it must fit as much of the hexagon as possible. For these reasons, the wrench types are listed in the order preferred.
1. Socket wrenches.
2. Box wrenches.
3. Open end wrenches.
4. Adjustable wrenches.

A loose or worn wrench may slip and spoil the corners on nuts. Proper servicing then becomes impossible without replacing the part.

It is important to use wrenches properly so that they fit completely on the nut or bolt. Always pull on a wrench rather than push on it. Otherwise, sudden loosening may cause a serious hand injury.

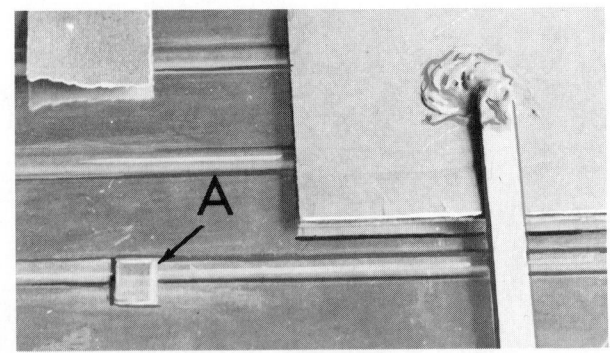

Fig. 2-42. Epoxy compound is used to join the patch material to the injured part. A—Completed patch.

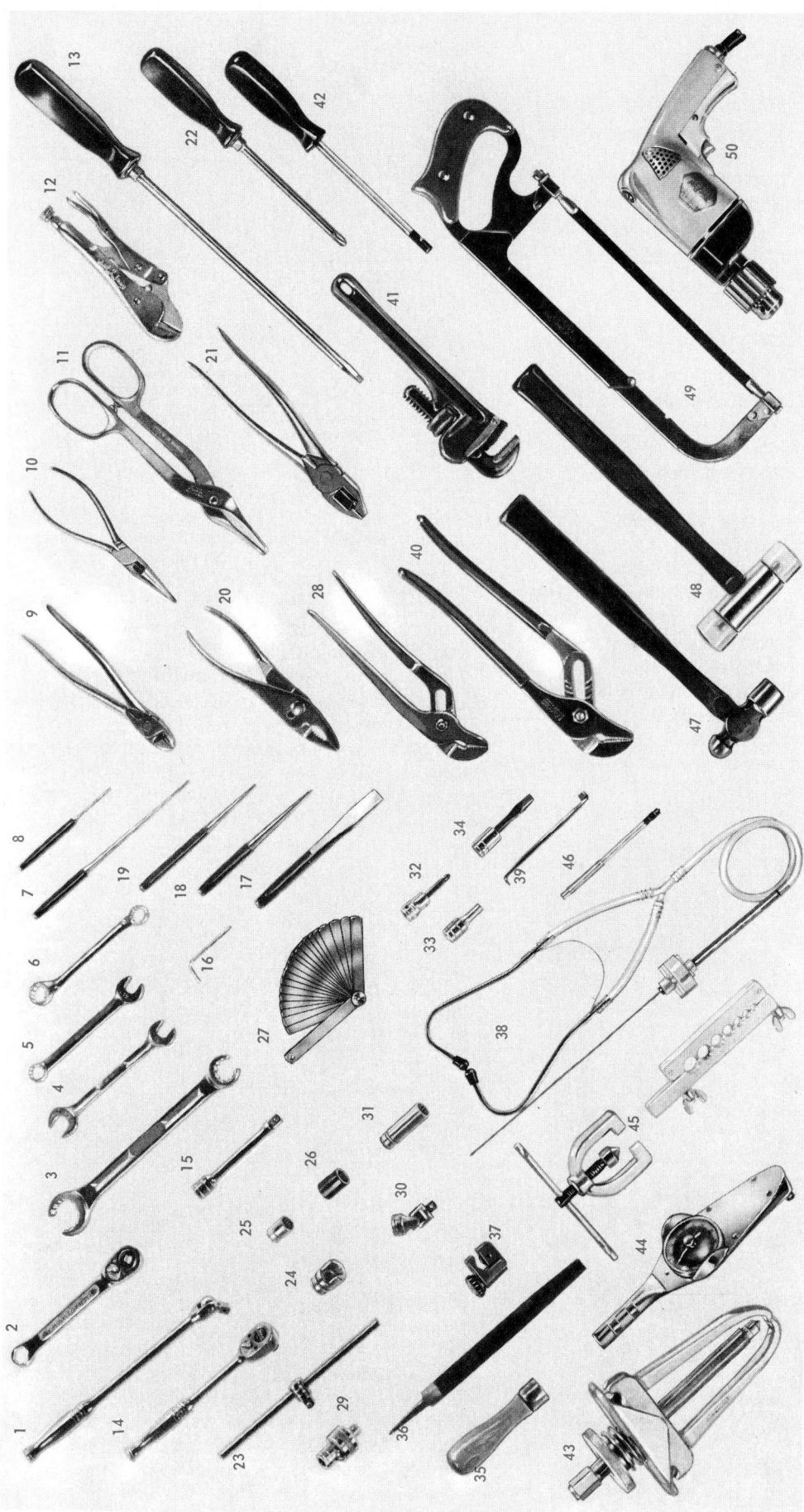

Fig. 2-43. Basic hand tool assortment for refrigeration service. 1—Nut spinner. 2—Wrench, open end. 5—Wrench, double hex, combination. 6—Wrench, box socket—double hex offset. 7—Punch, taper. 8—Punch, pin. 9—Pliers, diagonal cutting. 10—Pliers, needle-nose. 11—Snip, tinner's. 12—Pliers, pinch-off. 13—Screwdriver, standard tip. 14—Handle ratchet. 15—Extension, 4-in. 16—Wrench, hex head (Allen). 17—Chisel, cold. 18—Punch, center. 19—Punch, starter. 20—Pliers, gripping—slip joint. 21—Pliers, lineman. 22—Screwdriver, Phillips. 23—Handle, sliding bar. 24—Socket, Weatherhead. 25—Socket, double hex. 26—Socket, magnetic. 27—Gage, feeler. 28—Pliers, interlocking joint. 29—Adaptor. 30—Universal joint. 31—Socket, double hex deep. 32—Socket, Phillips screwdriver. 33—Socket, clutch screwdriver. 34—Socket, standard screwdriver. 35—File handle. 36—File, half round. 37—Cutter, tube. 38—Stethoscope, mechanic's. 39—Screwdriver, offset. 40—Pliers, large, interlocking joint. 41—Wrench, pipe. 42—Screwdriver, clutch. 43—Puller, two-jaw. 44—Torque wrench, English-metric. 45—Flaring tool. 46—Screw starter. 47—Hammer, ball peen. 48—Hammer, plastic tip. 49—Hacksaw, hand. 50—Drill, electric. (Snap-on Tools Corp.)

The sockets should be inserted all the way on the nut or bolt head.

Fig. 2-51 shows the proper direction to pull on an adjustable wrench.

Avoid pounding on a wrench to obtain greater turning movement or torque. Avoid using a length of pipe or another wrench for more turning force or torque.

A tight bolt or nut may be loosened safely by soaking the threads with a penetrating oil or by heating the nut or bolt. Some service technicians tap a nut or bolt lightly with a hammer. Any of these methods can be used to loosen corroded threads.

2-32 SOCKET WRENCHES

If the nut or bolt head has enough room around it, the six- or twelve-point socket is the best wrench to use. These sockets are usually made of chromium-vanadium steel and are turned by handles that have a 1/4-, 3/8-, or 1/2-in. square drive, Fig. 2-44. The handles come in a variety of designs as shown in the illustration.

A variation of the socket wrench is the nut driver. A nut driver is a small direct-drive socket wrench. It has a plastic handle that can be used with assorted drive sockets.

Sockets are now available which will hold the nut or cap screw firmly while it is being aligned and threading started. This is especially important where the nut or cap screw will cause problems if it drops out of the socket.

Box end and socket wrenches are more usable if they are double broached (12-point). Fig. 2-45 illustrates both the 6-point and the 12-point box end wrench. The 12-point type wrench is easier to use if the handle must be operated

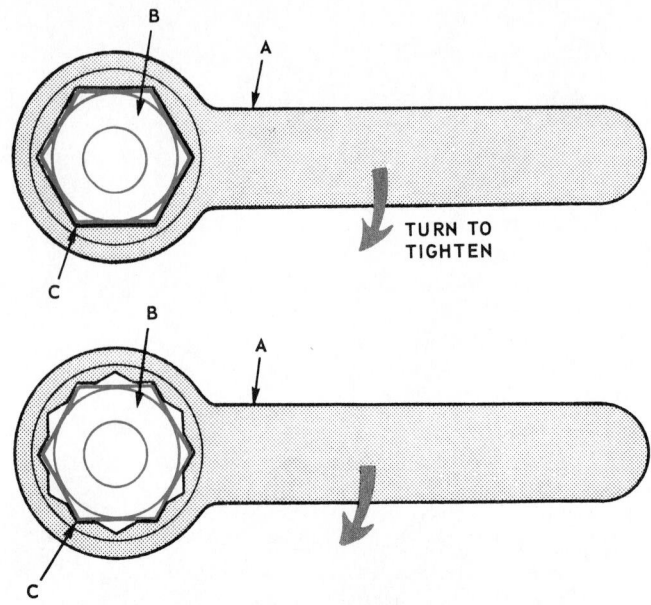

Fig. 2-45. Box end socket wrenches as they appear fitting over hex nuts. Upper wrench is a 6-point box end. Lower is a 12-point box end. A—Wrench handle. B—Nut. C—Contact points (6 contact points).

in a small or restricted space. The 6-point socket is best for worn hex nuts or bolts.

Metric size nuts and bolts require metric size wrenches. Fig. 2-46 shows a set of metric 6-point sockets commonly used when working with metric size nuts and bolts. The size marked on the socket corresponds to the diameter of the cap screw or bolt and is not the distance across the flats, as is the general practice with common fractional-inch wrenches.

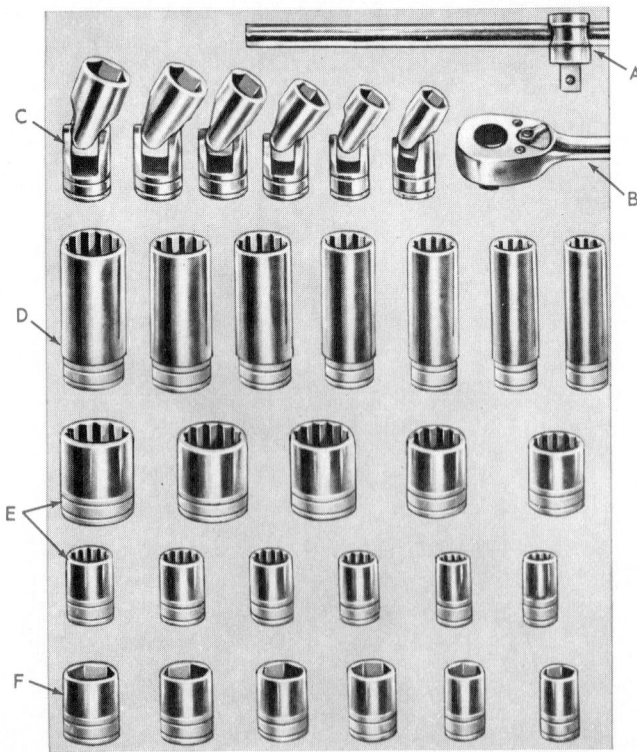

Fig. 2-44. Typical socket wrenches and handles. A—T-handle. B—Ratchet handle. C—Swivel 6-point sockets. D—12-point deep sockets. E—12-point sockets. F—6-point sockets.

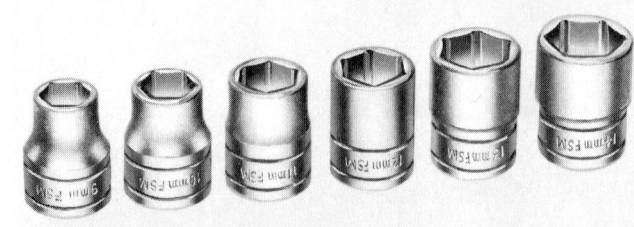

Fig. 2-46 A set of metric size sockets. Note that the size marked on each socket corresponds to the diameter in millimeters (mm) of the bolt or capscrew. Black rings indicate metric sockets. (Snap-on Tools Corp.)

2-33 BOX WRENCHES

When the nuts or bolt heads are in close spaces where one cannot use a socket wrench, the box wrench is satisfactory. Box wrenches are usually 12-point and provide a powerful noninjuring grip on the nut or bolt, Fig. 2-47.

Box wrenches may be either straight, offset, or double offset. Most box wrenches are double ended. Both ends may be of the same size with one end offset, or they may be of the same pattern and different sizes.

The refrigeration service technician will find that, for standard bolts and nuts, the table in Fig. 2-48 will give the size

Fig. 2-47. An alloy steel box wrench with 12-point or double hex ends. Ends are offset (double offset) to provide gripping or swinging clearance above mechanism. Socket wrenches are safest; box wrenches are next safest. Box wrenches are less likely to slip than open end wrenches. (Duro/Indestro, Duro Metal Products Co.)

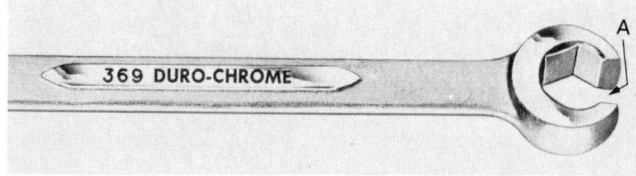

Fig. 2-49. Flare nut wrench used when turning SAE flare nuts. A—Note opening for tubing. (Duro/Indestro, Duro Metal Products Co.)

of the wrench across the flats for the sizes of bolts and nuts in most common use. Below 1/2-in. bolt size, the wrench size is 3/16 in. larger than the bolt size. A 1/4-in. bolt uses 4/16 + 3/16 = 7/16-in. wrench size.

At 1/2-in. bolt size and larger, the wrench size is 1/4 in. larger than the bolt's size. For example, on a 5/8-in. bolt: 5/8 + 1/4 = 5/8 + 2/8 = 7/8-in. wrench size.

The size of the wrench opening across the flats is marked at each wrench opening. Box wrenches having both flat and 15° handles are necessary for a complete tool kit.

NOMINAL BOLT SIZE	WRENCH SIZE HEAD AND NUT WIDTH ACROSS FLATS	
1/4	7/16	
5/16	1/2	WRENCH IS 3/16" LARGER THAN THE BOLT
3/8	9/16	
7/16	5/8	
1/2	3/4	
9/16	13/16	WRENCH IS 1/4" LARGER THAN THE BOLT
5/8	7/8	
3/4	1	

Fig. 2-48. Table of wrench openings for standard bolt heads and nuts.

2-34 FLARE NUT WRENCHES

Flare nuts used on SAE flared connections require special wrenches. Since the nut is on a fitting connected to tubing, the common box wrench cannot be used. Special flare nut wrenches have been developed for this purpose. An opening permits the box end to slip over the tubing. Note the special wrench shown in Fig. 2-49.

Since the flare nut wrench is often used in limited space, special wrenches have been devised. Fig. 2-50 shows a strong, easy-to-operate opening type flare nut wrench.

Forged flare nut sizes are an SAE standard used in automotive, marine, and refrigeration service. See Chapter 30 for a table of flare nut wrench sizes.

2-35 OPEN END WRENCHES

Open end wrenches can slide on the nut or bolt head from the side and are used in close spaces on unions and other places where the socket wrench and box wrench cannot be used on the assembly device.

End wrenches should not be used for refrigeration work when the jaws are spread or have burrs. End wrenches used in servicing work should have a thick jaw. Care must be taken when using thin wrench jaws as they have a tendency to bite into soft brass and copper parts.

Popular sizes for open end wrenches are:
1. The 7/16 in. across flats often needed for 1/4-in. screws and bolts.
2. The 1/2 in. across flats for 5/16-in. NC and NF cap screws commonly used on compressors and expansion valves.
3. The 3/4 in. across flats used for 1/4-in. flare nuts.
4. The 1 in. across flats which fit the 1/2-in. flare nuts.

A typical open end wrench, No. 4, is shown in the assortment making up Fig. 2-43.

Another very popular wrench used in refrigeration work is the combination open end and box socket. Both ends are the same size. This wrench is illustrated in Fig. 2-43, No. 5.

2-36 ADJUSTABLE WRENCHES

Wrenches with adjustable jaws, Fig. 2-51, are necessary in the tool kit because of the odd size nuts and bolts often found in this work. Adjustable wrenches must be kept in good repair. If the wrench does not fit tightly, it may slip and result in a ruined wrench, bruised hand, and a ruined nut or bolt head.

The direction of the forces operating on the jaws of the adjustable wrench should be in the direction which will give solid support against both the nut and the body of the wrench, Fig. 2-51.

2-37 PIPE WRENCHES

The pipe wrench is designed to grip pipes, studs, and other cylindrical (round) surfaces. The greater the torque on

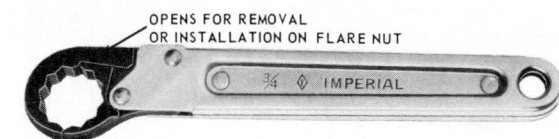

Fig. 2-50. Special type of flare nut wrench opens to pass over tubing and closes on flare nut to give positive contact. (Imperial Eastman, Imperial Division)

Fig. 2-51. A popular type of adjustable wrench. Handle should be pulled as shown by direction of arrow on handle. Note that wrench is adjusted to fit nut tightly. The red arrows show the pressure of the wrench against the corners of the nut. Turning wrench in the direction shown tends to press movable jaw against wrench body.

the wrench handle, the tighter the wrench will grip the object. Pipe wrenches should not be used on nuts or bolt heads. The typical pipe wrench is pictured in Fig. 2-43, item number 41.

An internal type pipe wrench, Fig. 2-52, may be used for installing pipe, nipples, or fittings.

A chain wrench, Fig. 2-53, is another type of adjustable pipe wrench. The chain wrench is often used in confined areas or on square, round, or irregular shapes.

2-38 HEX KEY WRENCHES

Hex key wrenches are constructed of alloy steel with a hexagonal (six-point) tip. Common variations of hex keys are fold-up tools with many key sizes in one handle, Fig. 2-54, view A. Also, individual L-keys and T-handle hex keys, Fig. 2-54, view D, are frequently used for long-reach operations, such as set screws on pulleys.

Another type of wrench similar to the hex wrench is the TORX®, which is star-like in appearance, Fig. 2-54, view C. This shape allows more metal-to-metal contact, resulting in increased transfer of torque.

2-39 SERVICE VALVE WRENCHES

Service valve stems are usually constructed with a square end milled on the valve shaft. A special wrench is needed to turn them, Fig. 2-55. This tool, called a service valve wrench, usually has a ratchet and a fixed end for this kind of work.

When "cracking" valves, the fixed end only should be used. "Cracking" is the slight opening required to cause the valve needle or plunger to leave its seat, but allow only a very little flow of refrigerant.

Using the fixed end of the wrench gives the service technician quick control of the slight opening and closing

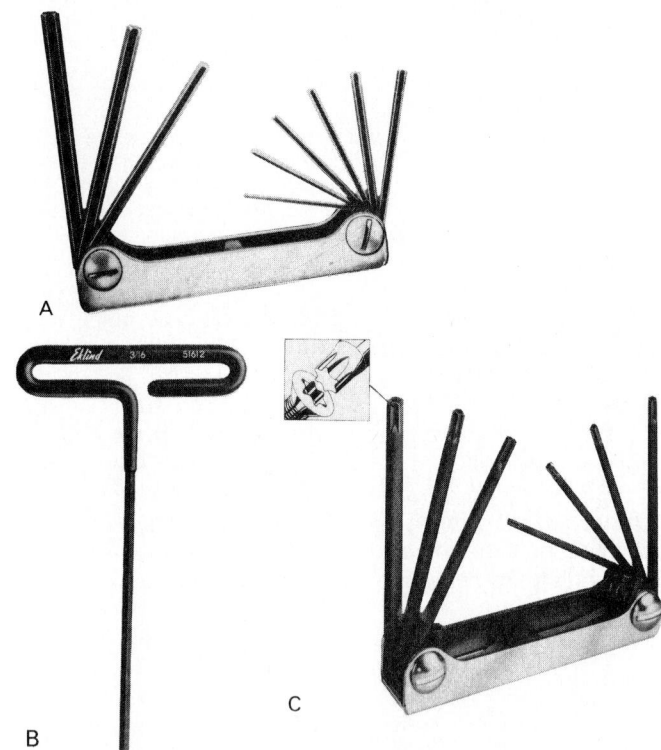

Fig. 2-54. Fold-up wrenches. A—Hex key. B—T-handle hex key. C—Torx key wrench set used on recessed screws. (Eklind Tool Co.)

of a valve. When rapidly opening and closing valves, the ratchet end may be used.

Some service valve wrenches have a reversible ratchet which enables the operator to reverse the turning without removing the wrench from the stem. Fig. 2-56 is a photograph of such a wrench.

2-40 SERVICE VALVE WRENCH ADAPTORS

Many manufacturers use valve stems other than the 1/4 in. square. Some valve stems are made so that the milled end is inside the valve body. This requires a good socket wrench to turn it.

To accommodate these valves, adaptors are available in various sizes. The male, or drive part of the socket, is usually 1/4 in. square. There are a few which use a larger drive (9/32 in.). The socket or opening which fits the valve stem comes in five sizes: 3/16 in., 7/32 in., 1/4 in., 5/16 in., and 3/8 in. These sockets usually have eight points to simplify their use.

Most valve stems have internal packing gland nuts, and special sockets must be used on these. It is best to use sockets with ball bearing grippers. There is less chance of losing tools when working in difficult positions. Fig. 2-57 illustrates a set of these special tools. Also included are sockets for packing gland fittings.

2-41 TORQUE WRENCHES

All materials are elastic (will stretch, compress, and twist). Even cast iron and hardened steels used in the construction of compressors are elastic up to a point. When tightening bolts, nuts, and other attachments on compressor parts and assemblies, it is important that the amount of tightness

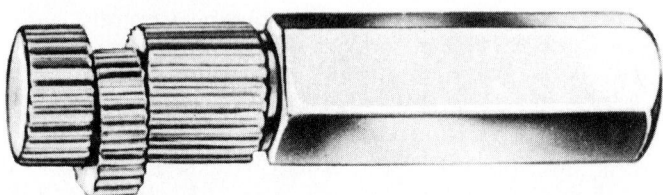

Fig. 2-52. Internal type of pipe wrench. It grips the pipe from the inside. (Snap-on Tools Corp.)

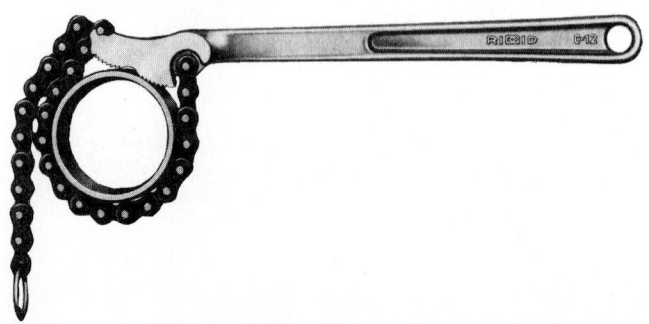

Fig. 2-53. A light-duty chain wrench, for use in close quarters. (Ridge Tool Co.)

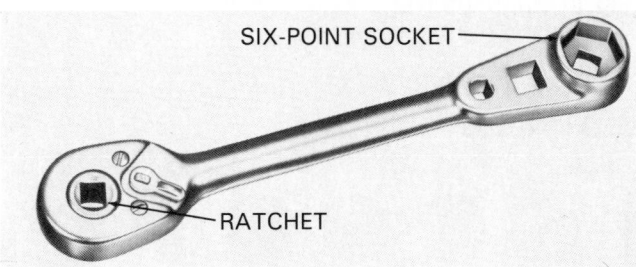

Fig. 2-55. Refrigeration service valve wrench. Fixed end is for "cracking" valves. Ratchet end is for rapid valve stem operation. Left end has 1/4 in. square drive for use with valve stem and packing gland nut sockets. Other openings are 3/16, 1/4, and 5/16 in. square. The 6-point socket fits 3/8 in. nuts.
(Duro/Indestro, Duro Metal Products Co.)

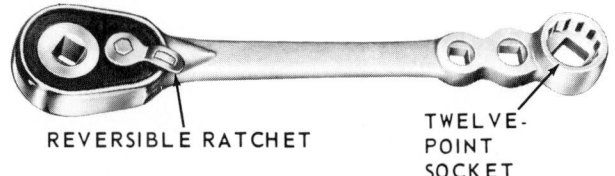

Fig. 2-56. Reversible ratchet wrench. Fixed end has openings for three valve stem sizes, 5/16, 1/4, and 3/16. The 12-point opening is for 5/16 in. nuts and bolts (1/2 in. across the flats).
(Duro/Indestro, Duro Metal Products Co.)

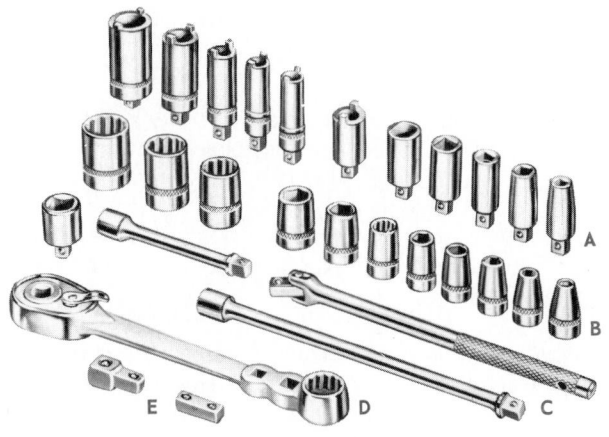

Fig. 2-57. Special service valve socket set. A—Packing gland sockets and valve stem sockets. B—Variety of 6-point and 12-point sockets. C—Handles. D—Ratchet wrench. E—Adaptors.

be measured to avoid warpage or other damage to parts. To measure the amount of tightness, a torque wrench is used, Fig. 2-58. Torque is twisting force.

Torque wrenches are usually wrench handles only and are made to be used with sockets of different sizes. The handle is equipped with a dial or pointer which measures the foot-pounds or inch-pounds of torque.

One can work out the torque by multiplying the length of the handle in feet by the pull in pounds applied to the handle (foot-pounds). A wrench handle one foot long and pulled by a spring scale reading 50 pounds will produce a torque of 50 foot-pounds.

(Technically, foot-pounds is the wrong term. The correct term should be pounds-feet. The foot-pound is a unit of work. However, popular usage has made the term foot-pound acceptable for the measurement of torque.)

Inch-pounds are calculated by multiplying the length of the handle in inches by the pull on the handle in pounds.

The manufacturers of equipment such as automobiles, airplanes, refrigerating equipment, and the like, are able to determine the proper torque (twist) that should be applied to the fasteners on their various mechanisms. The recommended torque for the many parts of refrigerating mechanisms are specified as part of the service data. Factory service manuals include this data.

To use a torque wrench, the operator fits the proper size of socket into it. The wrench is then applied to the nut and the handle of the wrench pulled until the indicator shows that the required torque has been applied. At that torque the nut is at the right tightness recommended by the manufacturer.

2-42 HAMMERS

A hammer is a necessity in the refrigeration shop and the 12- or 16-ounce ball peen hammer is a useful tool. See Fig. 2-43, Part 47. It is important that the hammer head be firmly fastened to the handle and that the handle be in good condition.

To use the hammer, grasp the handle about two-thirds of the way back from the head. For light accurate blows, hold the hammer with the index finger on the top of the handle and use wrist action. For heavy blows, hold the hammer with fingers around the handle and use elbow muscles.

A carpenter's claw hammer may also be needed for mounting pipe supports and fastening sheet metal to wood.

2-43 MALLETS

In refrigeration and air conditioning service work, the mallet is often needed to drive parts into place or to separate them without injury to their surfaces. For such work, a 1 1/2-lb. to 2-lb. mallet is desirable. A rawhide, rubber, wood, plastic, or lead mallet should be used. A mallet is shown in Fig. 2-43, Part 48.

2-44 PLIERS

Pliers are universal tools. Many different types are available.
1. Gas pliers are slip joint combination pliers, handy for general use. However, they should not be used on nuts, bolts, or fittings. They could slip and injure the surface. See Fig. 2-43, Part 20.
2. Cutting pliers are mostly used when working on the wiring of the refrigerator. One type, called the lineman's pliers, is a powerful cutting and gripping tool. Another type called "diagonal" pliers is used to cut in close quarters. Refer to Fig. 2-43, Part 9.

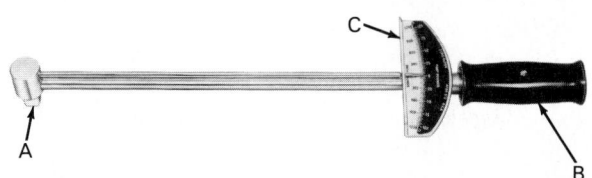

Fig. 2-58. Torque wrench used to measure the amount of tightness of nuts and screws. This wrench is made to be used with standard sockets. A—Socket drive. B—Handle. C—Torque scale.
(Duro/Indestro, Duro Metal Products Co.)

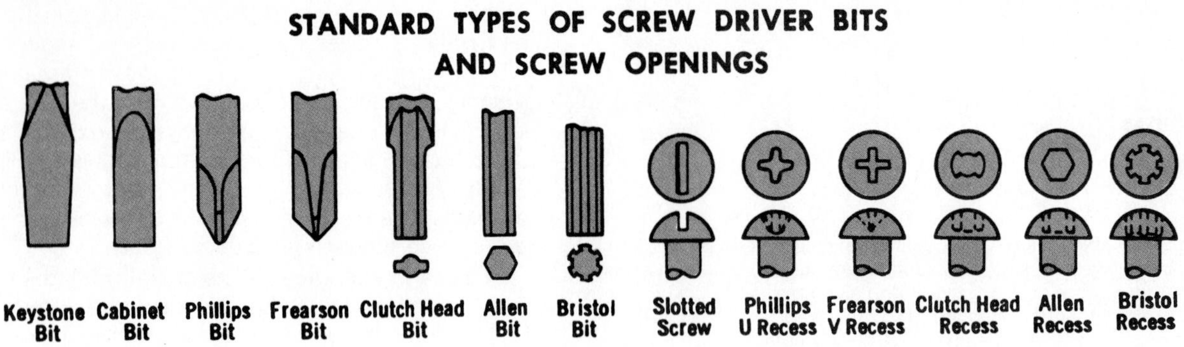

STANDARD TYPES OF SCREW DRIVER BITS
AND SCREW OPENINGS

| Keystone Bit | Cabinet Bit | Phillips Bit | Frearson Bit | Clutch Head Bit | Allen Bit | Bristol Bit | Slotted Screw | Phillips U Recess | Frearson V Recess | Clutch Head Recess | Allen Recess | Bristol Recess |

Fig. 2-59. Several types of screwdrivers. Flat bladed Keystone and Cabinet bits and the Phillips bit are most popular. (Klein Tools, Inc.)

3. Nut pliers are used to good advantage on some jobs. Jaws are parallel and some have an adjustable cam action that locks the jaws on the nut or bolt.

   In general, it is not considered good practice to use pliers to hold bolts or nuts. However, on a job such as holding the head of a bolt while turning the nut with a wrench, the use of nut pliers is permissible.

4. Slim nose pliers and duckbill pliers are frequently used in hard-to-reach places. See Fig. 2-43, Part 10. It is sometimes called a needle-nose pliers.

5. Round nose pliers are used to shape wire into loops and to bend sheet metal edges.

Pliers are made of alloy steel, usually with manganese, although some are chrome-vanadium steel. When of top quality they are usually drop forged. Only a pair of pliers with insulated handles should be used when working on electrical parts.

2-45 SCREWDRIVERS

Screwdrivers are widely used in refrigeration service work both for installation and for shop work. A complete set will be found very necessary. The length of a screwdriver is measured from the blade tip to the handle. Handles are not measured. The recommended average sizes are 2 1/2, 4, 6, and 8 in.

The types of screwdrivers are named for the shape of the end of the blade or bit. See Fig. 2-59. Most popular is the straight blade, slot blade, or regular screwdriver. The screwdriver bit should fit the screw slot snugly while the blade should be wide enough to fill the screw slot end-to-end. Also see Fig. 2-43, Part 13.

The Phillips type has a tip which fits a recessed cross in the head of the screw. Phillips screwdrivers are available in four sizes: the 3 in. size for No. 4 and smaller screws; the 4 in. size for No. 5 to No. 9 screws; the 5 in. size for No. 10 to No. 16 screws and the 8 in. size for No. 18 screws and larger.

Stubby (short) screwdrivers are available for working in small spaces. Some screwdrivers may be equipped with a clip that holds screws while starting them. Better quality screwdrivers have strong handles firmly bonded to the blade. Plastic handles are popular.

An offset screwdriver is necessary in refrigeration work. There are many places where it alone can be used.

Never use a hammer to pound on a screwdriver. If a screwdriver is needed for heavy service, use one with a solid steel handle.

2-46 STAMPS

It is good practice for refrigeration service technicians to stamp their names and the date on units sold or serviced. This often eliminates arguments. Many companies have a code system allowing their own employees to interpret the information stamped on the unit.

This stamping is done with hardened steel stamps which can be purchased in a variety of sizes, letters, figures, or symbols. One-eighth in. letters are a popular size. Such stamps are not suitable for use on hardened materials and tools, however. An electric marking tool can be used in these cases.

2-47 VISES

Sturdy machinist's vises are necessary in the shop. They are particularly convenient for holding parts during drilling, filing, or assembling. Also useful is a pipe vise.

One vise should be large enough to hold most compressor bodies. A special pipe vise that has a hacksaw blade slot for accurate cutting of pipe and tubing is useful for a large service shop.

Always use soft jaws made of sheet copper or aluminum when clamping a part which must not be marred. These are available as inserts which fit over regular vise jaws.

2-48 TWIST DRILLS

Twist drills, frequently used for installation and repair work, are available for working metal, wood, plastic, and (with special designs) masonry. Twist drills may be turned by drill presses, portable electric drills, or hand braces.

Those intended for working metal come in three different set sizes. Identification systems for sizes include fractional numbers, whole numbers, and letters. Sizes go by the bit diameter.

Usually, twist drills are of the straight shank type. This means that the section gripped by a three-jaw chuck is straight and perfectly round (cylindrical) in shape. See Fig. 2-60. Split joint twist drills are often used with portable electric drills because they penetrate many metals easily.

The shank carries a stamped identification giving the kind and size of the drill. Depending on quality and use, twist drills may be made from either high carbon steel or alloy steel (HSS) for high speed use.

Fractional sizes come in sets usually beginning with 1/16 in. and going up to 1/2 in. in steps of 1/64 in. Larger fractional sizes are also available.

Numbered sets begin with No. 1 and go up through No. 80 (.228-.0135 in.). However, most commonly used are No. 1 through No. 60. The higher the number, the smaller the drill.

Letter size twist drills are larger than 1/4 in. in diameter and vary from .234 for the "A" to .413 for the "Z" drill.

Note that the numbered twist drills cover a range of sizes approximately .013 through 1/4 in. Letter sizes range from approximately 1/4 in. to nearly 1/2 in. These two twist drill sets are often used as tap drills in making holes for inside threads. They provide a greater range of sizes than the fractional twist drills. For a table of various drill diameters, see Chapter 30.

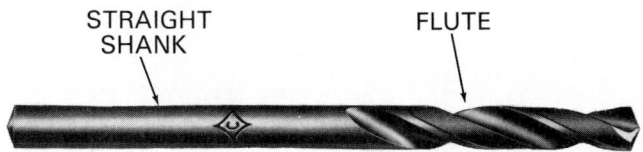

Fig. 2-60. Straight shank twist drill for use on metal. (Cleveland Twist Drill Co.)

Speed of drilling depends upon the type of material being drilled and the diameter of the hole. In general, the smaller the twist drill, the faster it should be turned.

Most twist drills have two cutting edges or "lips." These edges must be sharp, equal in length, and must have clearance and a rake angle. See Fig. 2-61.

Twist drills have flutes which remove chips from the hole. Most flutes are spiraled at an angle which automatically provides a rake angle for the cutting edges.

Always be sure the drill is cutting when it is being used. If the cutting edges are just rubbing against the stock, they will quickly heat up, destroying the hardness of the drill.

To insure that the drill forms the correct size hole, both cutting lips must be exactly the same length and angle. See

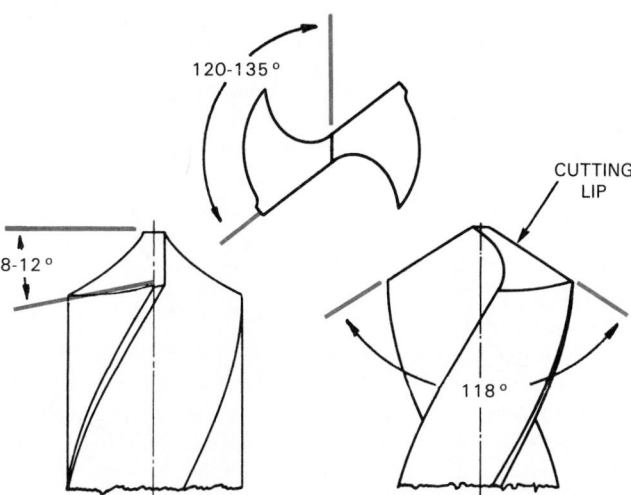

Fig. 2-61. Correctly ground twist drill point for steel. Clearance angle shown—8-12°—is rake angle for mild steel and cast iron for drills in 1/2 in. range. As diameters are reduced, clearance angles increase. A 1/16 in. diameter twist drill should have a clearance angle of about 20°. (Cleveland Twist Drill Co.)

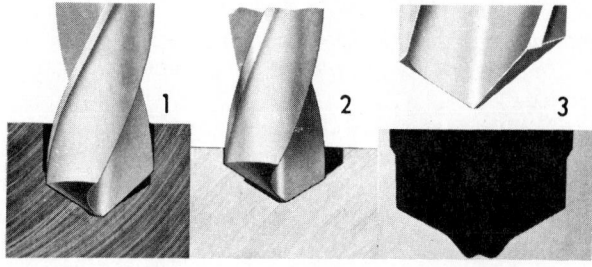

Fig. 2-62. An illustration of what happens when a drill is incorrectly sharpened. 1—Lips are equal in length but at different angles. 2—Lips are at equal angles but are of different lengths. 3—Lips are at different angles and at different lengths.

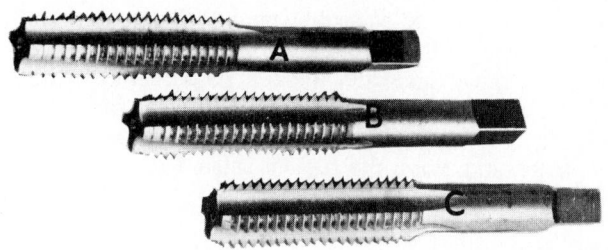

Fig. 2-63. Set of taps for cutting 1/2 in. NC threads. A—Taper tap. B—Plug tap. C—Bottoming tap.

Fig. 2-62. If one lip is longer, the hole being drilled will be oversize; or, if one lip has a smaller angle, it will do all the cutting and soon grow dull.

Always wear safety glasses to protect eyes from flying chips when using either a drill press or portable drill.

Electric drills should be grounded for safety. The metal frame of the drill should be electrically connected to a good ground (water pipe or a ground rod). Most electric drills are equipped with a three-prong grounded plug. If the circuit to which the drill is connected is not provided with a three-prong grounded socket, a grounded adaptor should be used.

2-49 TAPS

Many assemblies of metal parts are fastened with machine screws or tap screws threaded into tapped holes.

A tap is used for making threads inside a hole. Taps are of hard alloy steel accurately made with clearance pockets provided for chips. The threads are made with small clearance to provide good cutting.

There are taps for every size or diameter thread and also for each kind of thread—National Fine (NF), National Coarse (NC), American Standard Taper Pipe Thread (ASA), or metric. Taps are of three types: taper, plug, and bottoming. Most common is the plug type. See Fig. 2-63.

Taper taps are used for starting a cut and for tapping thin pieces in which the tapped hole goes all the way through. Plug types are used to do most of the cutting in blind holes, while bottoming taps are used to cut full threads to the bottom of a blind hole.

The shank of the tap is ground to a square at the end, and a tap wrench is used to turn it. Power tools may also be used for driving. However, a special tap-driving accessory must also be used.

Since tapping is basically a cutting operation, the general rules for cutting metals apply. Most taps have four cutting edges for each thread. These edges must be kept sharp. They must have a ground cutting face and cutting clearance.

Always use a special thread cutting lubricant when doing any kind of threading. The single exception is gray cast iron, which contains enough graphite to provide necessary lubrication. Thread cutting lubricants, if applied generously, also serve as a coolant.

2-50 TAP-DRILL SIZES

It is very important that the hole to be tapped is first drilled to the correct size. If the hole is oversize, threads will not be full size. If the hole is undersize, the tap must remove too much metal and will probably break. See Fig. 2-64.

The tap drill should be slightly larger in diameter than the root diameter of the threads for which the hole is being drilled. Generally speaking, threads 75 percent of full size are considered satisfactory. Always refer to tap-drill size tables for the correct size drill. For most refrigeration and air conditioning work, the tap-drill table, Fig. 2-65, will be satisfactory.

2-51 DIES

Dies cut external threads on round stock. The threads match those cut by the tap. Because a tap is nonadjustable, dies can usually be adjusted to permit careful matching of the threads. Dies are held and turned by a diestock. A diestock is shown in Fig. 2-66.

As with taps, there are dies for each type of thread and size. Because they are cutting tools, they, like taps, must

TAP	TAP DRILL	TAP	TAP DRILL
4/36	No. 43	14-20	No. 9
	No. 44		No. 10
	No. 45		No. 11
4-40	3/32	14-24	No. 6
	No. 43		No. 7
	No. 44		No. 8
4-48	No. 41	1/4-20	No. 5
	No. 42		No. 6
5-40	No. 37		13/64
	No. 38		No. 7
	No. 39		No. 8
5-44	No. 36	1/4-28	7/32
	No. 37		No. 3
	No. 38	5/16-18	17/64
6-32	No. 33		G
	No. 34		F
	7/64	5/64-24	J
	No. 36		I
6-40	No. 32	3/8-16	O
	No. 33		5/16
8-32	No. 29	3/8-24	R
8-36	No. 28		Q
	No. 29	7/16-14	3/8
10-24	No. 24		U
	No. 25	7/16-20	25/64
	No. 26		W
10-32	No. 19	1/2-13	27/64
	No. 20	1/2-20	29/64
	No. 21	9/16-12	31/64
	No. 22	9/16-18	33/64
12-24	No. 15	5/8-11	17/32
	No. 16	5/8-18	37/64
	No. 17	3/4-10	21/32
12-28	3/16	3/4-16	11/16
	No. 13		
	No. 14	A	B
	No. 15		

Fig. 2-65. Tap drill sizes recommended for common tapping operations. Note that for certain sizes, the tap drill may be a fractional size, a number size, or a letter size drill bit. A—Outside diameter. B—Number of threads per inch.

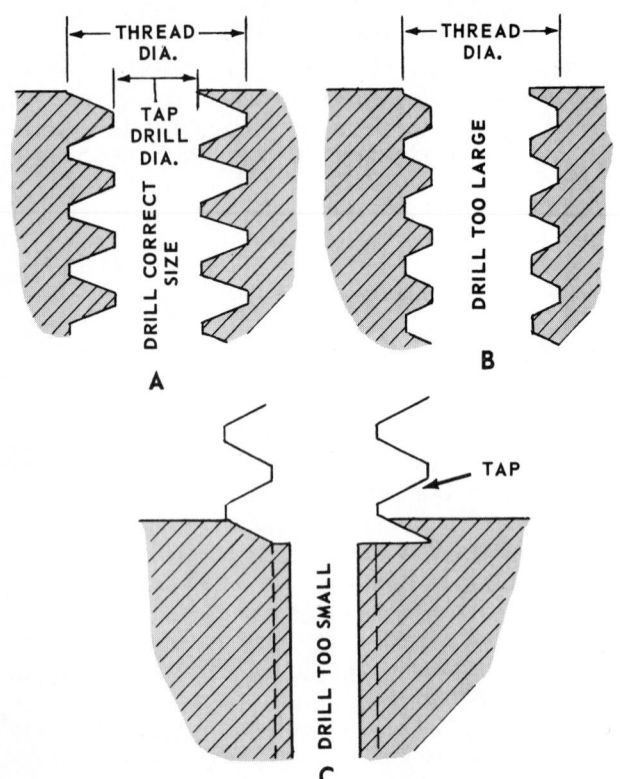

Fig. 2-64. Hole size is important when tapping threads in metal. A—Tap drill correct size, correct thread depth. B—Tap drill too large, threads not full depth. C—Tap drill too small, tap likely to break.

be made of tool steel. They, too, must be carefully shaped to cut correctly.

Special precautions should be taken to start threading very straight. Guides are available for this purpose. The small guide is located inside the threading die.

Round stock must also be accurately sized. Even if the stock is only a few thousandths of an inch oversize, it might break the die.

First adjust the die to full open position to make the initial cut. Then adjust the die to cut deeper until the threads match the tapped thread.

Always advance the die a quarter to half turn and then back off by reversing the direction of the diestock when using either the tap or die.

Taps and dies may also be used to clean threads that are corroded or damaged.

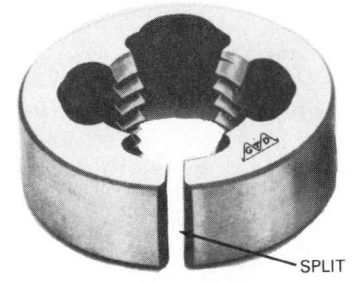

Fig. 2-66. Adjustable diestock has three screws. Two hold die and apply contracting pressure. Third expands die at split. (TRW Greenfield Tap & Die Div.)

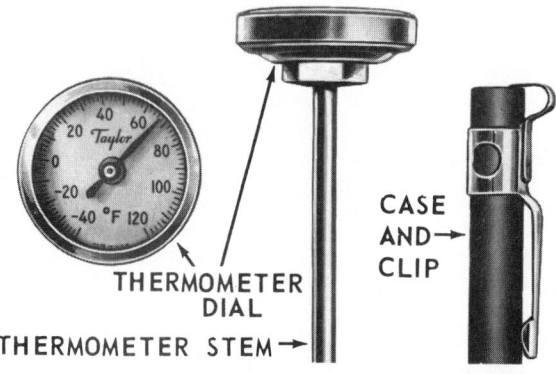

Fig. 2-67. Dial-stem thermometer is calibrated in 2-degree increments from −40 °F to + 120 °F (−40 °C to 49 °C). This is the temperature range most used by technicians in refrigeration and air conditioning work. (Taylor Scientific, Consumer Instruments, Sybron Corp.)

2-52 INSTRUMENTS AND GAUGES

The refrigeration service technician must use instruments and gauges to determine conditions (pressure and temperature) inside the operating mechanism. The most common instruments are thermometers, pressure gauges, and vacuum gauges. Later chapters will treat special instruments such as recording thermometers, hygrometers, ammeters, voltmeters, ohmmeters, and others.

Instruments must be kept in good condition and carefully handled if they are to remain accurate. If accuracy is doubtful, the instrument should be sent to a repair company for testing and calibration.

2-53 THERMOMETERS

The thermometer records the temperature of the evaporator, refrigerator cabinet, liquid line, suction line, and condensing unit. An ice and water bath may be used to determine a thermometer's accuracy. When in this solution, the thermometer should check within 1 deg. F of 32 °F (within 1 deg. C of 0 °C).

Many sizes and types of thermometers have been developed for the refrigeration service and installation technician. A popular type has the glass stem mounted in a metal case and is fitted with a pocket clip, as in Fig. 1-2.

Glass-stem thermometers usually read from − 30 °F to 120 °F (−35 °C to 49 °C), in 2 deg. increments (marked spaces). The tube may contain either mercury or a red fluid. The mercury-filled thermometer is faster but more difficult to read. Some thermometers have a special magnifying front built into the glass to enlarge the liquid line for easier reading.

There are numerous thermometers that are popular and easy to use. The dial-stem thermometer shown in Fig. 2-67 may be operated either by a bimetal strip or by a bellows charged with a volatile (vaporizes readily) fluid. The temperature range for this instrument varies, but is usually from − 40 °F to 120 °F (−40 °C to 49 °C) in 2 deg. increments. When using any kind of thermometer, never expose it to temperatures beyond the

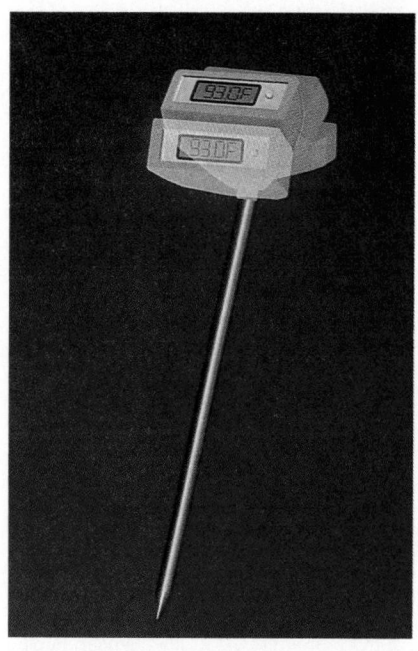

Fig. 2-68. A digital-type thermometer with swivel head for greater flexibility. (PSG Industries, Inc.)

limits of the scale. To do so may ruin the instrument.

A battery-powered digital thermometer with dial/head swivel is shown in Fig. 2-68. The temperature range for this instrument also varies, but may be from − 40 °F to 230 °F (−40 °C to 110 °C).

Dial thermometers are very useful for taking pipe temperatures. Fig. 2-69 illustrates two of these thermometers and shows how the instrument may be clamped to pipes or tubes to check temperatures.

Two other types of thermometers, the thermocouple and the thermistor, are operated by electrical current. The fundamentals of these are described in Chapter 6. Additional information regarding thermometers and their applications may be found in Para. 18-41. Fig. 2-70 illustrates a hand-held digital thermometer used for troubleshooting systems where the knowledge of the specific temperature of the evaporator or condenser is necessary.

Fig. 2-71 shows a maximum and minimum thermometer. This type is particularly useful when attached to a system which is unattended for a few cycles.

Occasionally, fluid in the liquid column of the glass-stem thermometer may separate. To make the column solid again, try cooling the bulb by spraying a small quantity of liquid refrigerant R-12 on it. The column will shrink into the bulb; and, when it re-expands, the break should have disappeared.

CAUTION: If the mercury is frozen into a solid, the thermometer will break.

Another way to re-connect the column is to heat the thermometer as in Fig. 2-72. Heat the thermometer very, very slowly. Do not allow fluid to reach top end of stem. If overheated, thermometer will burst. Wear goggles.

2-54 PRESSURE GAUGES

Pressure gauges are used by the technician to help determine what is happening inside the system.

Gauges use a Bourdon tube as the operating element. The Bourdon tube is a flattened metal tube (usually copper alloy) sealed at one end, curved and soldered to the gauge fitting at the other end. Fig. 2-73 shows the typical construction of a Bourdon tube.

A pressure rise in a Bourdon tube makes it tend to straighten. This movement will pull on the link, which will turn the gear sector counterclockwise. The pointer shaft will then turn clockwise to move the needle.

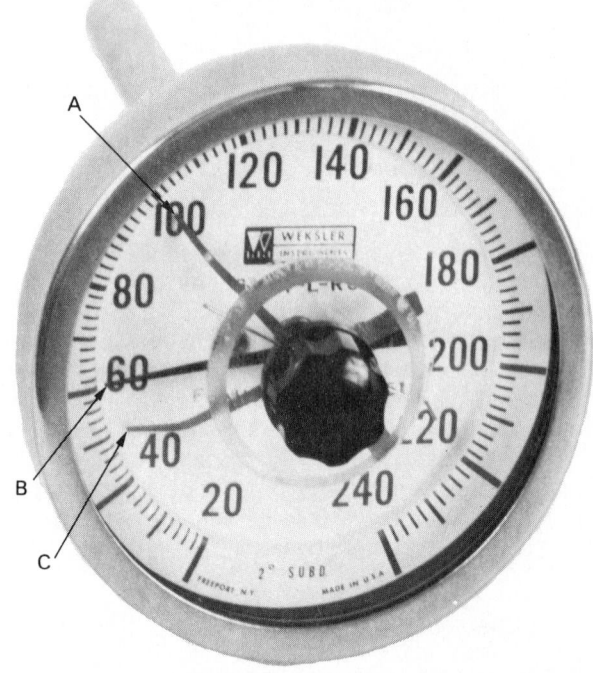

Fig. 2-71. Maximum and minimum thermometer. A—Hand indicating highest temperature reached. B—Hand indicating present temperature. C—Hand registering lowest temperature reached. (Weksler Instruments Corp.)

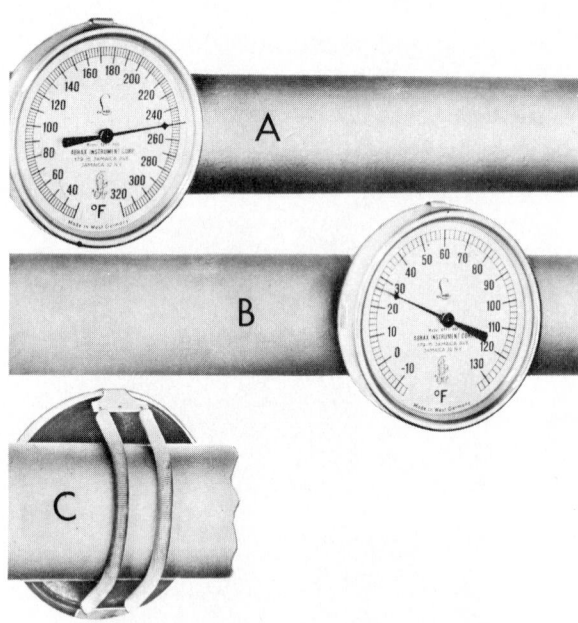

Fig. 2-69. Dial thermometers are easily clamped to pipes to indicate temperature of pipe. A—Shows calibration from 40 °F to 320 °F by 2-degree increments. Temperature of 250 °F indicates that this thermometer is attached to steam or hot water pipe. B—Calibration is from −10 °F to 130 °F by 2-degree increments. Temperature (26 °F) indicates thermometer is attached to evaporator suction line. C—Spring arrangement for attaching thermometer to pipe.

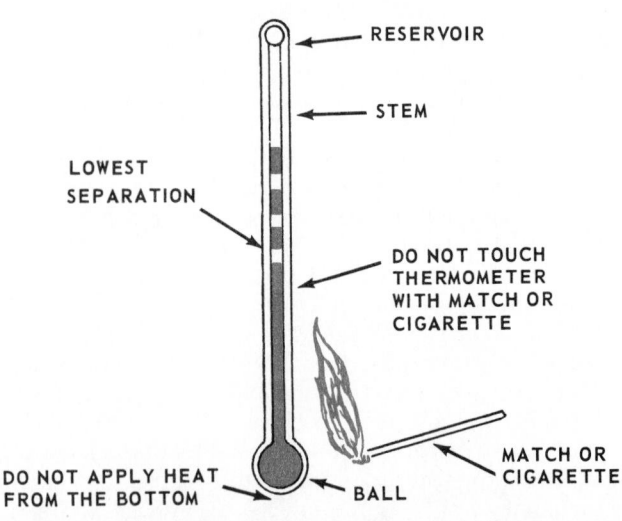

Fig. 2-72. Using heat to connect a break in the liquid column of a glass stem thermometer. Be careful. Wear goggles. Avoid putting match below bulb. Keep flame moving to avoid hot spots. (White-Rodgers Div., Emerson Electric Co.)

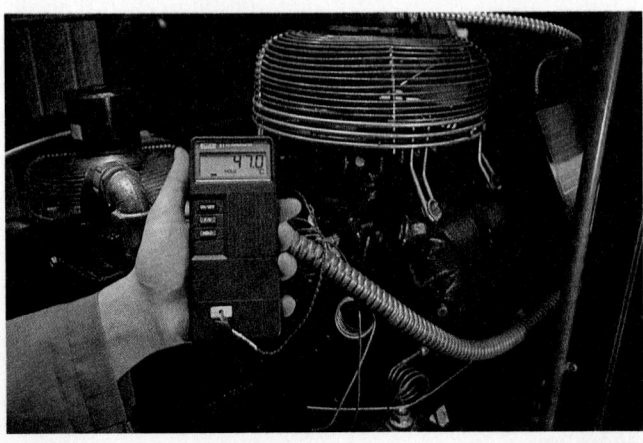

Fig. 2-70. A hand-held thermometer indicating the condensing temperature. (John Fluke Mfg. Co., Inc.)

Some gauges have a retarder to permit accurate readings in the usual operating range. The retarder uses an extra spring at pressures above normal. These gauges are easily recognized by the change in graduations at the higher readings of the positive pressure scale.

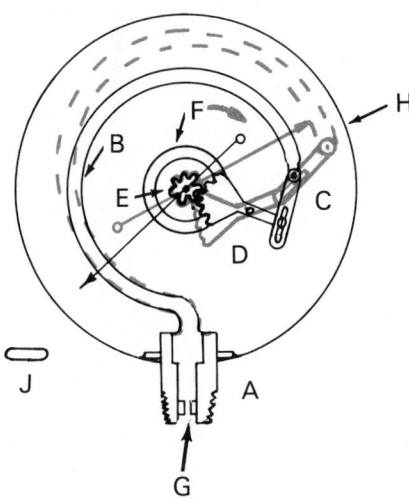

Fig. 2-73. Internal construction of pressure gauge. A—Adaptor fitting, usually 1/8 in. pipe thread. B—Bourdon tube. C—Link. D—Gear sector. E—Pointer shaft gear. F—Calibrating spring. G—Restrictor. H—Case. J—Cross-section of Bourdon tube. Red outline at top indicates how pressure in Bourdon tube causes it to straighten and operate gauge.

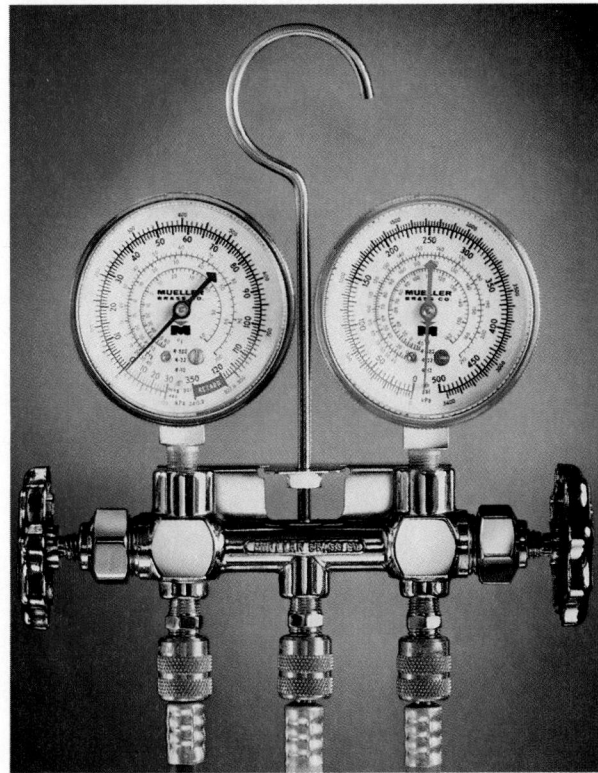

Fig. 2-74. A testing manifold. The compound (suction) gauge (blue) is mounted on the left. Its hose leads to the equipment suction service valve. The high pressure (discharge) gauge (red) is mounted on the right. Its hose leads to the discharge service valve or liquid line. (Mueller Brass Co.)

Most popular gauges have a 2 1/2 in. dial and are connected into the refrigerating system with a 1/8 in. male pipe thread. Some gauges, however, have a 1/8 in. female pipe thread.

It is advisable to use a 1/8 in. pipe nipple on any gauge, as the continual installing and removing quickly wears the threads. Maximum pressure at which a gauge should be continuously used should be no greater than 75 percent of the full-scale range.

Three types of pressure gauges are used in refrigeration service work: high-pressure, vacuum, and compound gauge.

2-55 GAUGE MANIFOLDS

A gauge manifold includes both a high-side gauge and a low-side (vacuum) gauge. It allows the service technician to check operating pressures, add or remove refrigerant, add oil, and perform other necessary operations.

The gauge manifold illustrated in Fig. 2-74 is color coded. Manufacturers often color code the exterior of the gauge—low-side blue, and high-side red. Low-side hoses are color coded blue, and high-side hoses are red.

2-56 HIGH-PRESSURE GAUGES

The high-pressure gauge has a single continuous scale, usually calibrated (marked off) to read from 0 to 500 psi. The scale is usually in either 2-lb. or 5-lb. increments and is usually connected into the high-pressure side of the refrigerating mechanism. Fig. 2-75 shows typical gauges.

The gauge shown in view B of Fig. 2-75 is a two-way manifold with liquid crystal display (LCD) and is operated through the use of a battery. The low-side gauge indicates either a vacuum or pressure reading from 0 to 29.9 inches of mercury or 0 to −99.9 psi pressure. The high-side gauge reads up through a maximum of 999 psi.

2-57 VACUUM GAUGES

The vacuum gauge measures lower-than-atmosphere pressure. It will have one of four calibrations: inches of mercury (Hg); pounds per square inch absolute (psia); millimeters of mercury (mmHg); or, for very high vacuum, torrs or microns. The micron is explained in Chapter 11.

Generally, the mercury barometer, Fig. 1-13, measures vacuum in the normal ranges of refrigeration work. For measurement of very high vacuums, a special instrument, the McLeod gauge, Fig. 2-76, is usually used. Such instruments are calibrated in millimeters of mercury (torr). See Para. 1-17 for definition of a torr. The vacuum calibration on the compound gauge (inches of Hg) is most used in refrigeration work for measuring pressures below atmospheric.

For very high vacuums, the thermocouple gauge, Fig. 11-105, should be used. It is accurate between 1 and 1 000 microns. A mercury manometer is accurate from 1 000 microns and above.

2-58 COMPOUND GAUGES

The compound gauge, Fig. 2-77, measures both pressure and vacuum. It is usually calibrated from 0 to 30″ Hg and from 0 to 200 psi.

Some compound gauges have scales calibrated according to the evaporating temperature of various refrigerants,

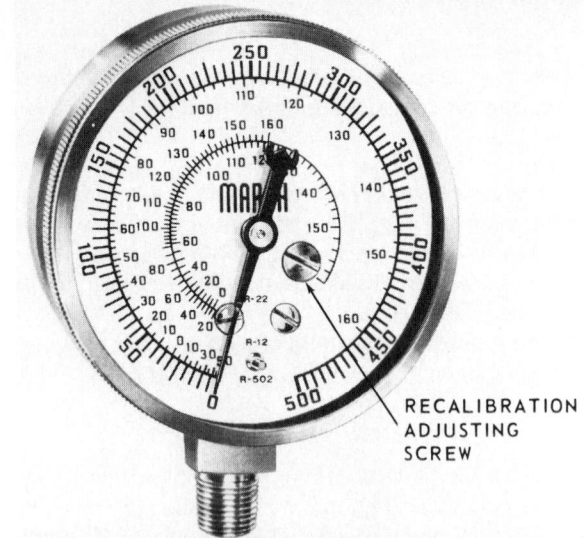

A

RECALIBRATION
ADJUSTING
SCREW

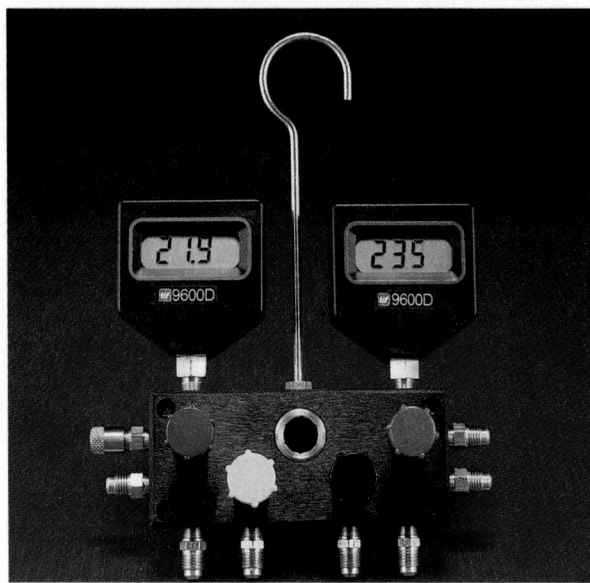

B

Fig. 2-75. High-pressure gauges. A—Gauge calibrated in U.S. conventional units. Note temperature-pressure scales for refrigerants R-502, R-12, and R-22. (Marsh Instrument Co.) B—Two-way digital manifold gauge set with LCD. (TIF Instruments, Inc.)

liquid, which tends to prevent rapid fluctuations in the instrument.

Gauges that are used in refrigeration work must be accurate. When it is time for a periodic re-calibration of gauges, instruments are available to do this accurately. Any shop that uses a large number of gauges or is remotely situated should have one. Such instruments are usually constructed using "dead weights" for calibrations above atmospheric pressure and a mercury column indicator for pressures below atmospheric or vacuum. Gauges should be checked over their full operating range or scale.

When checking gauge accuracy, it is important to remember that calibrating equipment is made to show a "O" reading at sea level. A gauge calibrated on equipment so adjusted will not be accurate at either above or below sea level.

Fig. 2-78 shows change in atmospheric pressure with altitude. To make sure that either a pressure gauge or a vacuum gauge is calibrated for elevation:

1. For a pressure gauge, set the needle at 0 when the gauge is unconnected.
2. For a vacuum gauge, the needle should rest at 0 when the instrument is not connected to any vacuum source.
3. A pressure gauge or vacuum gauge should be adjusted to 0 for the location at which the gauge is being used.

such as R-12, R-22, R-502 and the like. With this extra scale it unnecessary to refer to pressure-temperature tables and curves for common refrigerants in order to check for correct pressure-temperature relationships.

A 30-inch Hg—0-200 psi gauge should be used when connecting to a refrigerating mechanism in which the high pressure may back up through the compressor or balance through the refrigerant control while the compressor is stopped. Never use compound gauges continuously on the high pressure side of the system.

2-59 CARE AND CALIBRATION OF GAUGES

Rapidly changing fluctuating pressures quickly destroy gauge accuracy. A sudden release of high pressure (300 psi) into a gauge also may injure it. If it is necessary to connect a gauge into a rapidly fluctuating pressure condition, it should be attached through a connector having a very tiny bore. This will help to dampen (choke) the pressure fluctuations entering the gauge. Some gauges are filled with

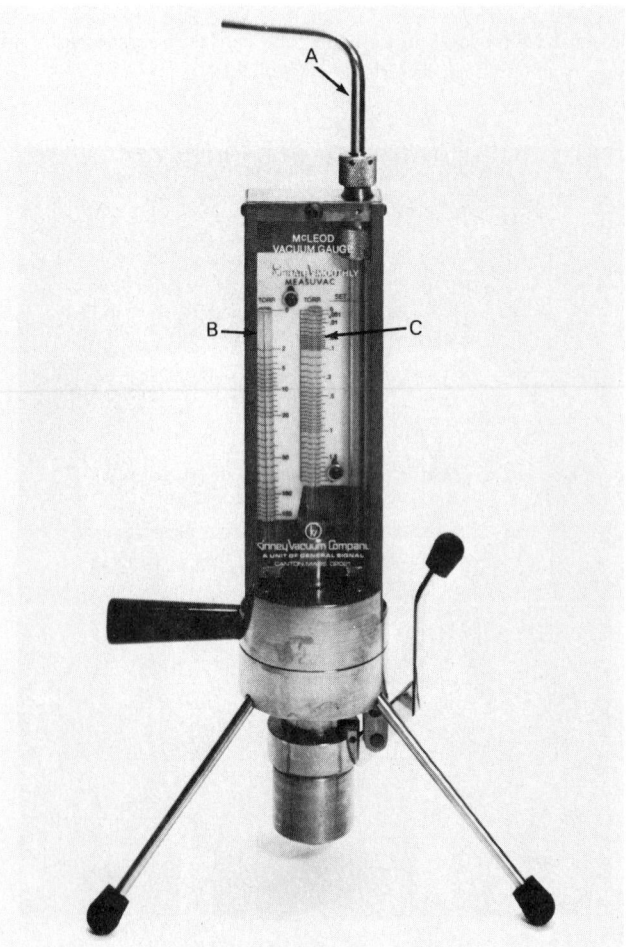

Fig. 2-76. A very high vacuum gauge using the McLeod principle. It can measure from 150 torr (150 mm) to 1 millitorr. A—Connecting tube. B—0-150 mm Hg scale. C—0-20 mm Hg scale. (Kinney Vacuum Co.)

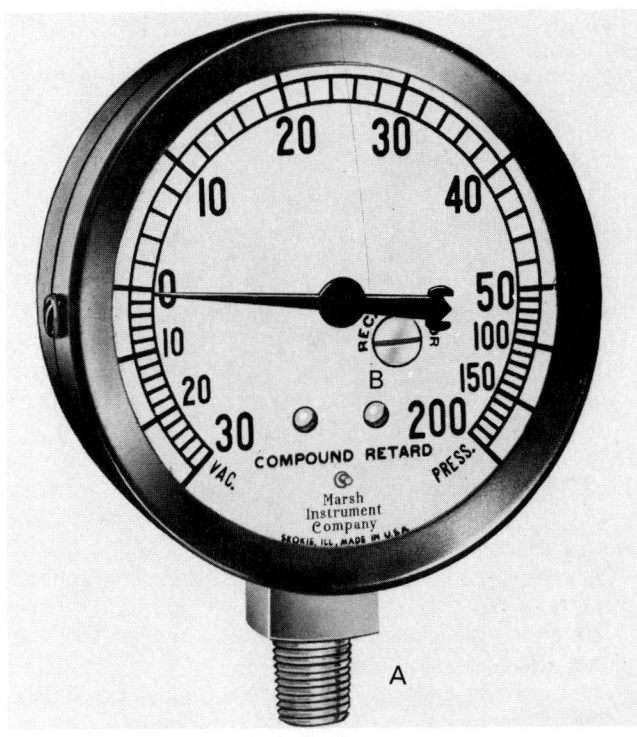

Fig. 2-77. Compound gauge with scale of 30'' Hg vacuum to 0 psi to 200 psi. Note that scale is greatly shortened (retarded) between 50 psi and 200 psi. This is accomplished by use of retarder spring. A—1/8 in. npt connection. B—Calibration adjustment. (Marsh Instrument Co.)

ELEVATION	PRESSURE
Sea Level	14.7 psi
2,000 ft.	13.7 psi
4,000 ft.	12.9 psi
5,000 ft.	12.2 psi

Fig. 2-78. Atmospheric pressure change with altitude.

The refrigerating machine pressures recorded by the gauge will then be accurate enough for the service technician to use.

A compound gauge is accurate to about 28-in. vacuum; a mercury manometer to 1 mm of mercury (1 000 microns).

There are many dial scales in use; some common ones are shown in Fig. 2-79. One must be careful when reading a pressure gauge to use the correct scale spacings and values.

To speed installation of gauges, one may use a quick coupler system, as in Fig. 2-80.

2-60 MEASURING RULES AND TAPES

A 9 in. or 12 in. steel rule is frequently needed when overhauling or installing units. The rule should be graduated in increments of 1/32 in. and, if possible, should be stainless steel to avoid rusting. Numerals and graduations should be clearly visible. Installation workers will find a 6 ft. flexible steel tape useful when laying out a job.

2-61 MICROMETERS

Often, the service technician must inspect, check sizes, dimensions, and make accurate measurements to a few thousandths of an inch or hundredths of a millimeter. The common micrometer is most used for this purpose.

Micrometers are available in both English units and in metric units. The English micrometer is calibrated in inches and decimals of an inch; the metric in millimeters and decimals of a millimeter.

ENGLISH MICROMETERS

Fig. 2-81 illustrates a 1 inch micrometer caliper. When reading it, be careful to observe the micrometer calibration range. Note that a 1-inch micrometer, Fig. 2-81, has a range of 0 to 1 in. A 2-inch micrometer has a range of 1 to 2 in., and so on. If a 2-inch micrometer is being used, 1 in. must be added to the reading. If it is a 3-inch micrometer, 2 in. must be added to the micrometer reading to compensate for the two-inch gap left uncalibrated on the micrometer.

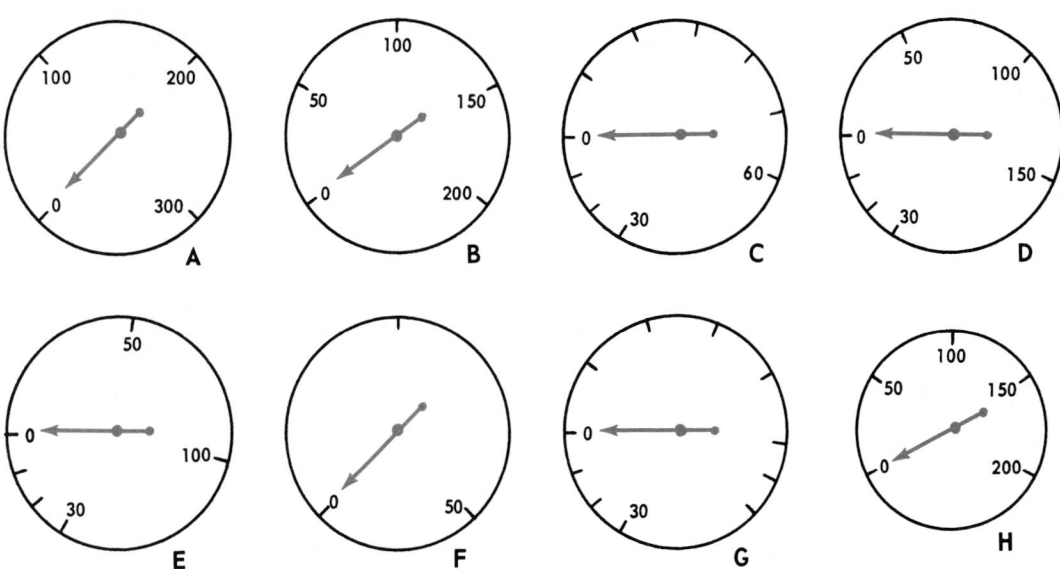

Fig. 2-79. Some common pressure gauge dials. A, B, F, and H are pressure gauges. C, D, E, and G are compound gauges.

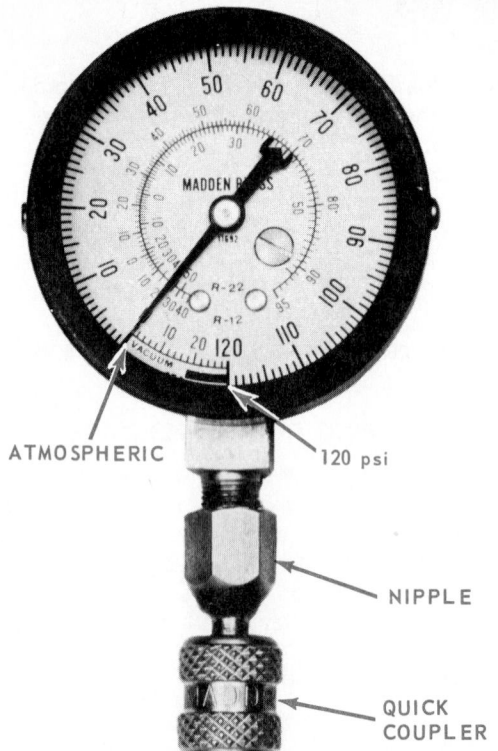

Fig. 2-80. Compound refrigeration gauge shows evaporating temperature for refrigerant R-12 and R-22 corresponding to gauge pressure. Gauge reads from 30 in. vacuum to 0 psig to 120 psig. On these gauges, the 30 in. vacuum and the 120 psig reading are on the same spot on dial. However, hand must travel a complete turn clockwise from the 30 in. to reach the 120 psig reading. (Robinair Division, SPX Corporation)

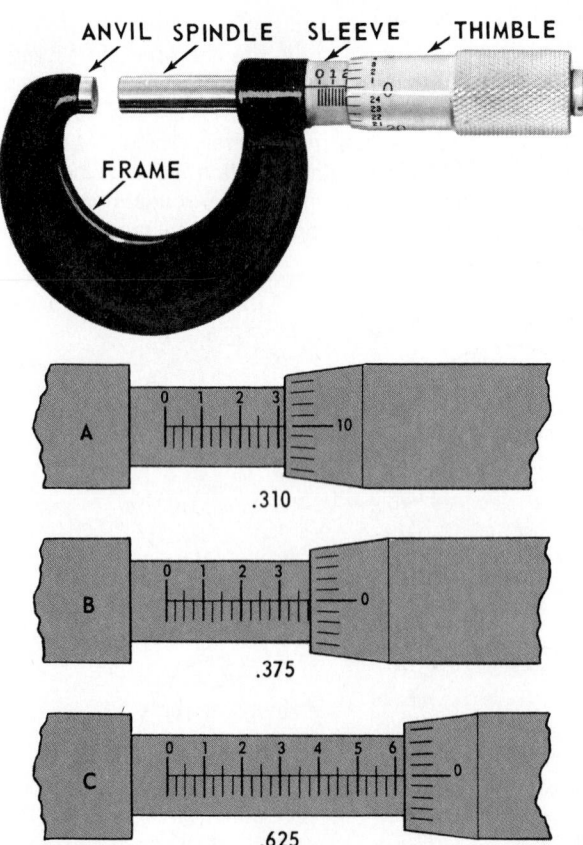

Fig. 2-81. Reading a micrometer caliper. Can you tell why the reading in the photograph is .200? Note drawings A, B, and C to see how readings are made. Answers are in thousandths of an inch.

The following points may help the beginner learn to read the micrometer:

1. The divisions on the sleeve are tenths of an inch.
2. The four divisions between the tenths markers are 1/40 in. or .025 (twenty-five thousandths of an inch).
3. The thimble (right hand end of the micrometer) carries the spindle and is threaded into the micrometer body on a 40-thread-per-inch screw. One turn of the thimble moves the spindle 25 thousandths (.025) of a inch.
4. There are 25 graduations on the thimble. Turning the thimble one graduation moves the spindle 1/1000 (.001) of an inch.
5. The micrometer, Fig. 2-81, reads .200 inch. The inserts (portions of a micrometer) read as noted in the caption.

METRIC MICROMETERS

Fig. 2-82 shows a metric micrometer. The calibration range is 0 to 25 millimeters. Twenty-five millimeters are almost one inch (1 inch = 25.4 mm).

The following points may help the beginner learn to read the metric scale:

1. The metric micrometer is calibrated in millimeters and decimals of a millimeter.
2. There are two calibrations on the sleeve (see Fig. 2-82). The lower calibration (A scale) is in millimeters. The upper calibration (B scale) is half (0.5) of a millimeter. The lines of B are halfway between those of A.
3. The thimble (C scale) is calibrated in hundredths of a millimeter. There are 50 graduations on the thimble.
4. The thread pitch on the metric micrometer spindle is 2 threads per millimeter. This means that the spindle moves a half (0.5) millimeter for each turn.
5. There are 50 divisions on the thimble. Since the thimble moves a half millimeter in one turn, one division on the thimble (1/50) moves the spindle 1/2 of 1/50 of a millimeter. This equals 1/100 (0.01) of a millimeter.
6. When reading a metric micrometer, it is not necessary to observe the micrometer calibration range. The lower calibration on the sleeve starts with a millimeter reading corresponding to the gap between the anvil and the spindle. As an example, the 0 to 25 mm range starts with 0 on the sleeve; the 25 to 50 mm range starts with 25 on the sleeve; and the 50 to 75 mm range starts with the number 50 on the sleeve.

2-62 REFRIGERATION SUPPLIES

Space prevents giving specifications for all refrigeration supplies. However, a few of the basic items will be named and some information given about their specifications, handling, and use.

2-63 ABRASIVES

Surfaces may be cleaned, smoothed, or made to accurate size with abrasives. Abrasives are sand-like grinding particles attached to paper or cloth by glue or other adhesives. Various abrasive materials are used.

Sandpaper was the only abrasive for many years. It is still excellent for wood finishing or where a dry surface is wanted. Today, emery, aluminum oxide, and silicon carbide are also commonly used.

Each abrasive has several grades or variations in coarseness. Emery cloth: 0000 (finest), 000 (extra fine),

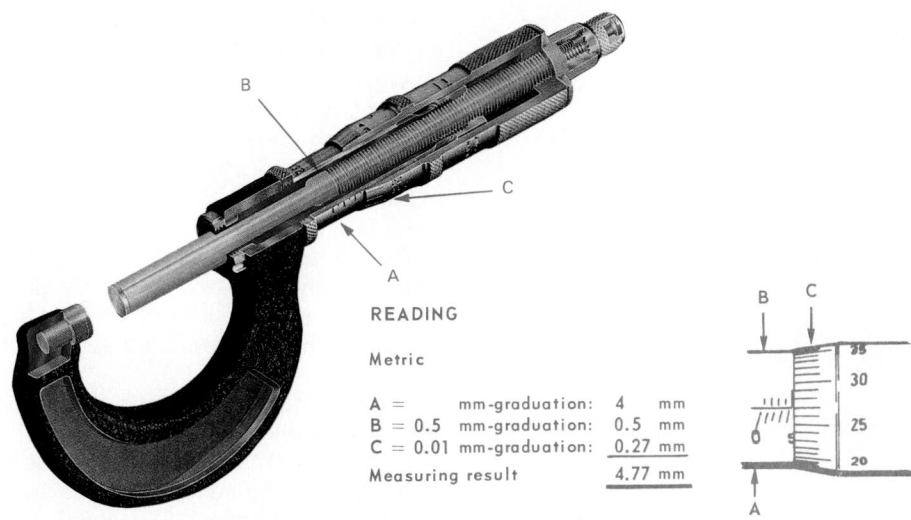

READING				
Metric				
A =	mm-graduation:	4	mm	
B = 0.5	mm-graduation:	0.5	mm	
C = 0.01	mm-graduation:	0.27	mm	
Measuring result		4.77	mm	

Fig. 2-82. A 0-25 mm micrometer. The reading as shown in the insert is 4.77 mm.   (S-T Industries, Inc.)

OO (very fine), O (fine), 1/2 (medium fine), and 1 (medium).

Silicon carbide: 500 (finest), 360 (very fine), 320 (fine), 220 (medium fine), and 180 (medium).

Aluminum oxides: 320 (extra fine), 240 (fine), 150 (medium fine), and 100 (medium).

These abrasives come in 9- by 11-in. sheets or in rolls usually 1 in. wide.

Sheet abrasives, whether paper or cloth, should be backed by a block of wood, metal, felt, or rubber. Special sanding blocks may also be used. Always use clean abrasive paper.

2-64 BRUSHES

A clean steel wire brush is an excellent tool to prepare copper and steel surfaces for welding or brazing. These brushes can be bought in a variety of shapes and sizes. They should have fine steel wire bristles, thickly set, and the handle should be comfortable. Special cylindrical brushes, available in all sizes, are good for cleaning outside and inside surfaces of tubing and fittings. Fig. 2-83 shows a cylindrically shaped wire brush.

2-65 CLEANING SOLVENTS

Any refrigeration mechanism must be thoroughly cleaned before and after repair. Many methods have been used. Some do not do a thorough job; some are dangerous.

Any method must remove oil, grease, and sludge. In refrigeration and air conditioning applications, the cleaning method must remove moisture, or at least it should not add moisture. It must not injure the parts nor harm the user.

There are several cleaning methods:

1. Steam Cleaning. If the parts are exposed to hot water or steam, the grease will usually become fluid and float off the surface. However, steam and hot water will burn the service technician if they are carelessly used.
2. Caustic Solution Cleaning. When an alkaline cleaner is dissolved in hot water, the mixture will remove grease and oil. This solution must be carefully used or the user may suffer burns or eye injury.
3. Oleum or Mineral Spirits. This petroleum product is popular for cleaning. It has a flash point of approximately 140 °F (60 °C), kerosene has a flash point of 130 °F (54 °C). It cleans well and leaves a smudge-free surface. However, it presents a fire hazard and should always be used in small amounts. It should be contained in self-closing tanks. The area should be exhaust ventilated (have a hood and an explosion-proof exhaust fan).
4. The refrigerant R-11 is a good cleaning solvent for flushing systems contaminated by hermetic motor burnout. R-11 has a high boiling point, 74.8 °F (24 °C). It is nontoxic and nonflammable. It leaves no noncondensable residue and has no reaction with the electrical insulation. R-11 refrigerant can be filtered and reused.

Fig. 2-83. Wire brush used for cleaning inside surfaces of fittings before soldering or brazing.   (Schaefer Brush Mfg. Co., Inc.)

5. Alcohol is also a good cleaning fluid. However, it is both flammable and toxic. Special precautions—such as excellent ventilation, no open flames, use in small amounts—must be taken.
6. Vapor degreasing is a system using a cleaning fluid contained in tank. The fluid is warmed, filling the upper part of the tank with vapors of the cleaner. Any parts suspended in this cleaning vapor are quickly and thoroughly cleaned. Such a tank must be specially vented.
7. There are several patented cleaning fluids available. When these are used, one should read and carefully

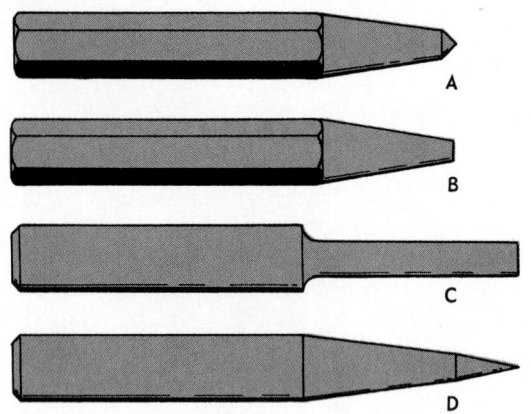

Fig. 2-84. Punches most used in refrigeration and air conditioning work. A—Center punch. B—Drift punch. C—Pin punch. D—Prick punch.

follow the manufacturer's instructions.

8. Carbon tetrachloride. Carbon tetrachloride should never be used in cleaning refrigeration or air conditioning mechanisms. It is toxic to the respiratory system and to the skin.

9. Never use gasoline for cleaning. It has a low flash point, its fumes are heavy and may travel far to ignition sources causing an explosion or flash fire.

10. Do not use a propane or LP cylinder to clean parts. The liquid or gas will quickly evaporate and may burn or explode.

Refer to Chapter 30 for more information on cleaning.

2-66 COLD CHISELS AND PUNCHES

A cold chisel is used on various jobs. As an example, one may find the assembly devices on evaporators corroded and a chisel is needed to remove the nut or screw.

A 3/4-in. flat cold chisel is a popular size. Be sure to keep the head (hammering end) of the chisel free from "mushrooming." Flying pieces from a mushroomed head may cause injuries. Always wear goggles or head shield. See Fig. 2-43, Part 17.

Four common types of punches are the center punch, drift punch, pin punch, and prick punch. See Fig. 2-43, Parts 7, 8, 18, and 19.

The center punch has a 60-degree to 90-degree point and is used for center punching the location of a hole to be drilled. A heavy blow on the punch makes a depression in which the point of the drill may be started. Center punches may also be used to make alignment marks on refrigeration parts before dismantling.

A drift punch is used to drive out keys and to line up holes in mating surfaces. The punch tapers from its flat point to the stock diameter.

The pin punch is used for driving retainer pins in or out. The blunt end is called the bill. Pin punches are measured in overall length, by diameter of the stock, and by diameter of the bill. Bill diameters are available in sizes from 3/32 to 5/16 in.

The prick punch has a long, sharp point and is used only for layout work. Punches are available in various lengths and are usually made of carefully heat-treated chrome-alloy steel. The cutting edge or point is hard, while the head is tough and shatterproof. Always grind away any mushroom head that forms. A fairly heavy 6 in. punch will be the most satisfactory.

Fig. 2-84 illustrates the shapes of various punches.

2-67 FILES

Various sizes and types of files are needed for cleaning metal surfaces and shaping metal parts. They are classified according to tooth size, shape, and the number of directions the teeth are cut on the file.

Files are either single or double cut, Fig. 2-85. The single cut is used for finishing surfaces and double cut for fast metal removal.

Files come in different lengths: 4, 6, 8, 10, 12 in., and so on. The longer, the coarser the teeth. The size of the teeth varies from dead smooth, smooth, second cut, bastard, to coarse. A second cut 6 in. file has finer teeth than one that is a second cut 12 in.

File shapes are many: rectangular, half round, round, triangular, square, wedge shape, and so on.

One oddity is that there are three types of rectangular cross-section files: mill, hand, and flat. The mill file has only single cut teeth; the mill file is uniform in thickness but tapers slightly in width. The hand and flat files have double cut teeth, but the hand file has one edge that has no teeth, called a safe edge. The edges are parallel but the thickness

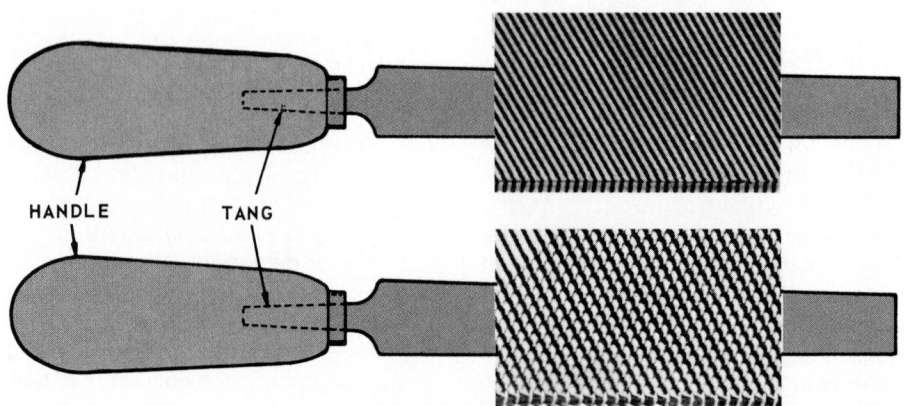

HANDLE   TANG

Fig. 2-85. Hand files may be either single cut (upper) or double cut (lower). Files should always have handles to avoid hand injuries. (The Cooper Group, Nicholson)

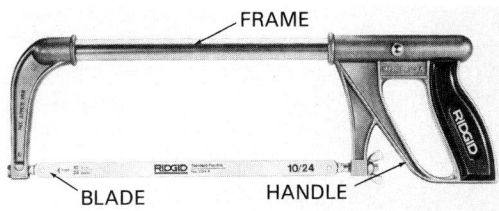

Fig. 2-86. Hand hacksaw must have rigid frame to hold blade in proper tension. (Ridge Tool Company)

varies slightly. The flat file has teeth on all four surfaces. See Fig. 2-43, Parts 35 and 36.

Use file brushes and file cards to clean file teeth which quickly become filled with metal. If clogging material is not removed, the files become useless. Do not use a file card for any other purpose. The bristles may become clogged with dirt.

2-68 HACKSAWS

A hand hacksaw is used for cutting tubing and other installation and maintenance work requiring metal cutting.

Fig. 2-86 illustrates a popular saw with a rigid frame and a 10 in. blade. Blades are available with different numbers of teeth per inch. Blades with 14 teeth per inch are used for soft metal and wide cuts; 18 teeth per inch for medium soft metals; 24 teeth per inch for general work; and 32 teeth per inch for thin metal, tubing, or hard metal.

A hacksaw blade should not be stroked over 60 strokes per minute. Most blades are made of high carbon steel, and the cutting edges (points) are very sharp and very small. Too rapid use will cause these points to overheat and lose their temper.

Always lift the blade slightly on the back stroke. This helps to keep the cutting edges sharp. If the blade is not lifted, particles of metal may roll between the work and the cutting edge of the blade, dulling the teeth.

With most blades, the teeth are hardened while the back of the blade is soft and flexible. Some higher quality blades may be made of tungsten or molybdenum steel alloy. Fig. 2-87 illustrates a bimetal hand hacksaw blade. It has a high-speed steel cutting edge and a die steel flexible backing.

Special hacksaw frames are available for working in small holes. There is also a stub hacksaw blade and an adaptor drive to fit electric drills.

2-69 FASTENING DEVICES

Many items are made in one piece today that were considered impossible to fabricate a short time ago. However, some mechanisms must still be made of several pieces and then assembled. If there is motion in the mechanism—such

as a piston in a cylinder—the apparatus must be made of two or more pieces.

Various techniques have been developed to fasten pieces together. In metal work, soldering, brazing, welding, crimping, rivets, bolts, machine screws, pins, spring fasteners, and force fits have been used with success.

The type of assembly device used depends first on the kind and condition of the metal, and secondly on how frequently the pieces must be dismantled.

If the parts are to be put together permanently, riveting, welding, soldering, and brazing are popular assembly methods.

If the parts must be dismantled for frequent repair or service, assembly devices must be used that can be easily removed without injuring the parts. Nuts and bolts, cap screws, machine screws, and set screws are then used. Fig. 2-88 shows an assortment of fastening devices. In the metric system, the diameter of fastening sizes is specified in millimeters.

2-70 MACHINE SCREWS, BOLTS, AND CAP SCREWS

Many small parts are assembled using specially threaded devices called machine screws. Machine screws may be made of steel, stainless steel, brass, Monel metal, or other materials. They are available in a variety of head shapes. Various methods have been used to turn these screws, see Figs. 2-54, 2-59, and 2-88. Some less commonly used screw heads are shown in Fig. 2-89.

Machine screws come in various diameters. Eight are in the numbered sizes, three in the fractional sizes. Each size may have either fine or coarse threads; the larger the number, the larger the diameter. A table of machine screw sizes and threads is given in Fig. 2-90.

In general, bolts and cap screws are used in sizes 1/4 in. and larger. The length of threads on a bolt is usually two times the bolt diameter. Most bolts are provided with nuts.

The threading on a cap screw is usually longer than the threading on a bolt and sometimes extends up to the head of the cap screw. Cap screws are threaded into a part of the mechanism and do not require a nut.

Diameters of metric bolts, nuts, and screws, as well as the thread pitches, are in millimeters. For example: 10 mm diameter with a 1.50 mm pitch.

The wrench size is the same as the size given for the diameter of the bolt. A 10 mm wrench fits a 10 mm bolt.

The metric system uses:
1. Extra fine thread screws.
2. Fine thread screws.
3. Average thread screws.
4. Coarse thread screws.

The popular diameters and thread pitch sizes are shown in Fig. 2-91.

2-71 GASKETS

Most mating surfaces are somewhat rough. To make a leakproof joint, gaskets are used between surfaces. Gaskets, being soft, seal joints between assembled parts. They keep the refrigerant from leaking out, prevent oil leakage, and keep air from leaking into the system. They are used between the valve plate and the compressor body, between the service valve and the compressor body, and

Fig. 2-87. Hand hacksaw blade. Three basic types are: 1—Standard high-speed blades for general use; 2—High-speed flexible blades made of molybdenum steel; 3—Bi-metal blades with high-speed steel teeth and flexible die steel backing. The type of blade, its length, and number of teeth per inch are indicated on the blade. (Ridge Tool Company)

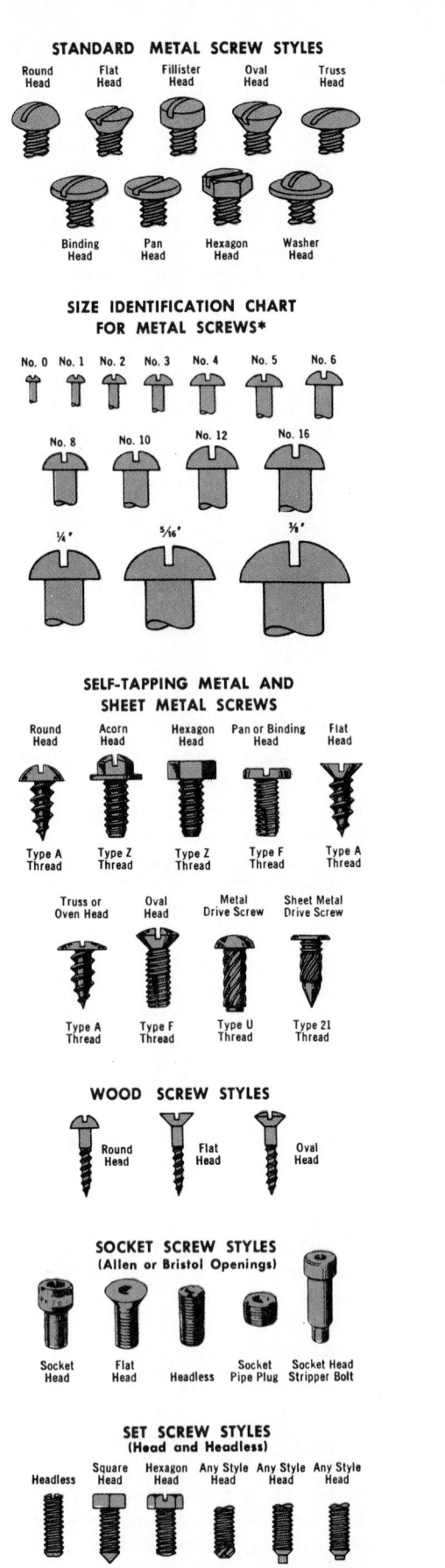

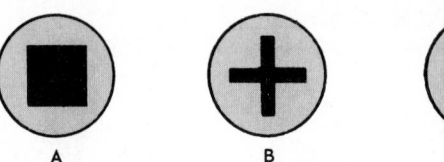

Fig. 2-89. Less commonly used screw heads. A—Square countersunk head (Scrulox). B—Reed Prince head (similar to Phillips). C-Spline.

between the valve plate and the compressor head. Gaskets are also used on the crankcase and at the crankshaft seal on open or external drive units.

Metal is the most common gasket material. Lead is popular, being soft and noncorrosive. Aluminum has also been used. Composition gaskets made of plastic impregnated paper are also popular.

Gaskets must not restrict the openings. They must not lose their compressibility. Replacement gaskets must not be thicker than the original gaskets.

The surfaces of parts that contact the gasket must be free of burrs, bruises, and foreign matter.

MACHINE SCREW NUMBER	DIAMETER IN.	THREADS PER INCH	
		COARSE	FINE
2	.086	56	64
3	.099	48	56
4	.112	40	48
5	.125	40	44
6	.138	32	40
8	.164	32	36
10	.190	24	32
12	.216	24	28
1/4	.250	20	28
5/16	.3125	18	24
3/8	.375	16	24

Fig. 2-90. Common machine screw sizes.

2-72 REFRIGERANTS

The refrigeration service technician is required to handle and charge refrigerants into refrigerating mechanisms. Several different refrigerants are carefully described in Chapter 9 along with safe handling methods.

Refrigerants must be kept DRY and CLEAN. Remember, all exposed surfaces absorb moisture if left in the open. If a compressor is torn down, overhauled, and reassembled, it must first be completely dried before it can be charged with refrigerant. See Chapter 11 for detailed instructions.

Refrigerants are stored and handled by the service technician using portable refrigerant cylinders. Refrigerants are identified by a cylinder color code. See Chapter 9 for a table of color codes for common refrigerants.

Cylinders for different refrigerants must not be interchanged. Refrigerants should always be stored in the cylinder specified (color code).

Never fill refrigerant cylinders over 85 percent of capacity. With a temperature increase, hydrostatic pressure may burst the overfilled cylinder.

Fig. 2-88. Fasteners must be carefully used and driven with proper tools. (Klein Tools, Inc.)

2-73 REFRIGERATION OIL

In the mechanical refrigerating system, moving parts must be lubricated with oil for long life and efficient performance.

Refrigerant oil is a specially prepared mineral oil. Refining steps are taken to remove excess wax, moisture, sulphur, and other impurities. Most refrigeration oils have a foaming inhibitor added.

Use oils that have a low pour point. This will avoid wax separation at the lowest temperature in the system. Wax could clog the refrigerant control orifice.

Because of the low temperatures at which they operate, food freezer and frozen food units need oils with extra low pour points and a very low wax content. Nor can the oil have any hydro-carbons that may collect on the compressor valves or other parts.

The viscosity of the oil must be accurately determined for the temperature ranges to which the refrigerating system may be exposed. Three viscosities are available—150, 300, and 500. Most automatic refrigerating systems use 300.

When possible, follow the manufacturer's recommendations. See Chapter 30.

It is highly important that refrigerant oil be kept in sealed containers, that it be transferred in chemically clean containers and lines, and that it not be left exposed to air where it will absorb moisture. When refilling a refrigerant, always use new oil.

Refrigeration oils come in one or five gallon cans and in barrels. It is advisable to purchase it in small sealed containers, holding only enough for each separate service operation.

Unused oil allowed to remain in the container or oil transferred from one container to another will pick up some moisture and perhaps even dirt. Special pressure pumps may be used to pump oil into the low side of the system.

The pour point of any oil is the temperature at which it starts to flow. The price of refrigeration oil varies with the grade. A low pour point oil is more expensive.

Domestic machines with refrigerant temperatures as low as 0 °F to 5 °F (−18 °C to −15 °C) need oil with a pour point of −20 °F (−29 °C). For food freezers with refrigerant temperatures as low as −50 °F (−46 °C), a pour point of −60 °F (−51 °C) is desirable. Always seal an oil container after having drawn oil from it.

DIAMETER mm	PITCH mm	WRENCH SIZE mm
3	0.60	3
4	0.70	4
4	0.75	4
5	0.80	5
5	0.90	5
6	1.00	6
7	1.00	7
8	1.00	8
8	1.25	8
9	1.00	9
9	1.25	9
10	1.25	
10	1.50	
11	1.50	
12	1.25	
12	1.50	
12	1.75	

Fig. 2-91. Table of metric screw information.

2-74 SERVICE VALVES

Service technicians must be familiar with manual valves in refrigerating systems. These valves enable them to seal off parts of the system while installing gauges, recharging, or discharging a system.

Several kinds of manual or hand valves are used. Such valves may have handwheels on their stems, but most are so made that a valve wrench is needed to turn them. Valve stems are made of steel or brass while the body of the valve is usually made of drop-forged brass. Packing is installed around the valve stem and a packing adjusting nut keeps the joint from leaking.

In common use are the one-way and two-way service valves. "One-way" means there is only one opening which can be either opened or closed. The "two-way" valve has two openings. One may be open while the other is closed, or both may be open.

The two-way valve usually closes or shuts off the refrigerant flow in the system when the stem is turned all the way in (clockwise). It shuts off the charging, discharging, or gauge opening when the valve stem is turned all the way out (counterclockwise). When the valve stem is turned part way, both of the openings allow the fluid (refrigerant) to flow through, Fig. 2-92.

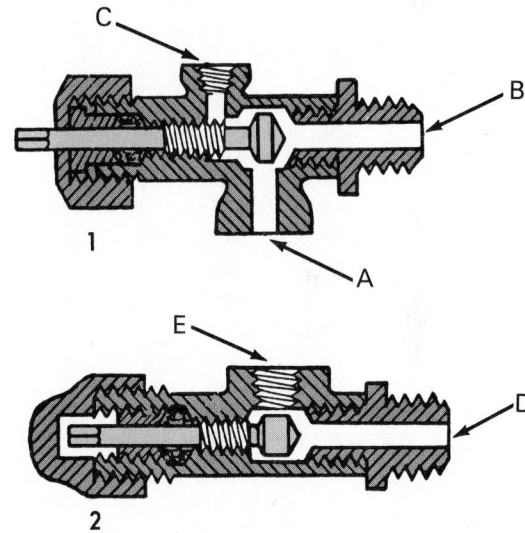

Fig. 2-92. Two typical service valve designs. No. 1 is a two-way valve. A—Opening to compressor. B—Opening to the refrigerant line. C—Service opening. No. 2 is a one-way valve. D—Opening to the liquid line. E—Opening to the liquid receiver. Valve No. 1 has an open valve stem. Valve No. 2 has a cap over the valve stem.

The tubing or pipe is fastened to valves by flare connection or by brazing. The valve may also be attached to the refrigerating mechanism by either pipe threads or by bolted flanges.

It is good practice to open a valve by first "cracking" it (opening it 1/16 or 1/8 turn). This prevents a shock pressure rush which may injure mechanisms, gauges, flush oil in abnormal amounts, or injure the service technician.

Be sure the valve stem is clean before turning it in (clockwise). A scarred or dirty valve stem will ruin the valve packing. See Chapter 14.

2-75 PURGING

Purging is a term used to describe the process of removing unwanted air, vapors, dirt, or moisture from the system. A neutral gas or the recommended refrigerant is allowed to flow through the refrigerator part or tubing, forcing unwanted air and vapors out.

2-76 EVACUATING

A refrigerating system must contain only the refrigerant in either liquid or vapor state along with dry oil. All other vapors, gases, and fluids must be removed.

These substances can be removed best by connecting the system to a vacuum pump and allowing the pump to run continuously for some time while a deep vacuum is drawn on the system.

It is sometimes necessary to warm the parts to 120 °F (49 °C) while under a high vacuum, in order to remove all unwanted moisture. Heat the parts using warm air, heat lamps, or warm water.

Never use a torch!

2-77 A REVIEW OF SAFETY IN HANDLING REFRIGERATION SUPPLIES AND EQUIPMENT

Before working with tools, always review the safety steps. The following are only a few of the items to remember:

1. Tubing should be bent in as large a radius as possible.
2. Epoxy bonding materials may irritate the skin or many membranes of the user.
3. "Mushroom" heads should be ground from chisels and punches, as these particles may fly when struck with the hammer, causing serious injury to the operator or a bystander.
4. Files should never be used without handles. The tang may injure the hands.
5. Wear goggles when drilling. Sometimes chips may fly. Eyes should therefore always be protected.
6. Emery cloth should not be used to clean tubing preparatory to soldering. It may leave an oily deposit. Also, the grit is hard and could cause considerable damage if allowed to enter the refrigerating mechanism.
7. When pressure testing for leaks in tubing circuits, carefully and safely use low or medium pressure carbon dioxide or nitrogen. Never use oxygen. CAUTION—see Chapter 11.
8. Carbon tetrachloride should not be used for any purpose as it is toxic and harmful to the skin.

9. Brazing materials sometimes contain cadmium. Fumes from heated cadmium are very poisonous. Be sure that the work space is well ventilated. If at all possible, use brazing alloys which DO NOT contain cadmium.
10. It is recommended that refrigerant cylinders never be filled above 85 percent of their capacity. If overfilled, hydrostatic pressure may cause them to burst.
11. Wrenches used on refrigeration line fittings should always fit snugly. Poorly fitted wrenches will ruin nuts and bolt heads. Also, the wrench may slip and cause an injury to the technician.
12. Always "crack" service valves and cylinder valves before opening. This gives quick control of the flow of gases if there is any danger.
13. Moisture is always a hazard to refrigerating mechanisms. Keep everything connected with a refrigerating mechanism thoroughly dry.

2-78 TEST YOUR KNOWLEDGE

1. List the fittings to be used when connecting a compound gauge with 1/8-in. pipe male thread to a service valve having 1/4-in. pipe female thread.
2. When cutting tubing with a hacksaw, what two precautions must be taken?
3. What is the most common type of tubing used in refrigeration applications?
4. What is the thickness of refrigeration tubing? What is the inside diameter of 1/4-in. tubing?
5. What may cause copper tubing to harden?
6. Why do some service technicians file the ends of copper tubing?
7. How may an external bending spring be easily removed from the tubing after making a bend?
8. Is it enough to clean metal for soldering with flux only? Why?
9. When soldering, what temperature must the soldered metal be?
10. Why must a joint be cleaned after brazing?
11. How may one tell when the correct brazing temperature is reached?
12. What type of wrench should be used on a valve stem?
13. Should one push or pull a wrench?
14. Describe an easy way to check the accuracy of a thermometer.
15. What is the purpose of a compound gauge?
16. Why must refrigerant oil be practically wax free?
17. What is a very important precaution one must take when filling refrigerant cylinders?
18. What is a double cut file?
19. Name five types of pliers.
20. Why are torque wrenches used?

Chapter 3

BASIC REFRIGERATION SYSTEMS

By studying this chapter, the technician will be able to:

☐ Explain the operation of a simple ice refrigerator.

☐ Explain how evaporation provides a cooling effect.

☐ Name the basic mechanical refrigeration systems.

☐ Explain various applications for mechanical refrigeration systems.

☐ Describe the operation of various mechanical refrigeration systems.

☐ Compare compression and absorption type systems.

☐ Discuss refrigeration systems using icemakers and water coolers.

☐ Explain how a system using an expendable type of refrigerant works.

☐ Discuss and compare domestic and commercial refrigeration systems.

☐ Explain the operation of thermoelectric refrigeration.

☐ Compare the differences between hot gas and electric defrost systems.

3-1 ICE REFRIGERATION

For years, ice (frozen water) was the only refrigerating means available. It is still used in many refrigerating applications. The usual ice refrigerator, Fig. 3-1, is an insulated cabinet equipped with a tray or tank at the top for holding blocks or pieces of ice (see blue).

Shelves for food are located below the ice compartment. Cold air (see light blue), flows downward from the ice compartment and cools the food on the shelves below. The air becomes warmer and rises from the bottom of the cabinet (see light red), up the sides and back of the cabinet, flows over the ice, cools, and again flows down over the shelves.

Ice refrigeration has the advantage of maintaining the interior of the cabinet at a fairly high humidity. Food stored in this type refrigerator does not dry out rapidly.

Until the development of the mechanical refrigerator, natural ice refrigeration was quite widely used. Since then, artificial ice has been manufactured for refrigeration. Temperatures inside an ice refrigerator are controlled by the flow of air over the ice and through the cabinet. Temperatures will usually range between 40° and 50°F.

When it is necessary to use ice for cooling temperatures below 32°F (0°C), ice and salt mixtures may be used. Temperatures down to 0°F (−18°C) may be easy to get with ice and salt mixtures. See Chapter 30 for a table of ice and salt mixtures.

3-2 EVAPORATIVE REFRIGERATION (DESERT BAG)

When a fluid evaporates, heat is absorbed. Evaporation of water is an example. This is why humans and animals perspire. Evaporation of moisture from the skin surface helps to keep one cool.

Another common application of this principle is the "desert bag" used to keep drinking water cool. This bag, Fig. 3-2, made of a tightly woven fabric, is filled with drinking water. Since the bag is not waterproof, some water seeps through. Thus, the outside surface of the bag remains

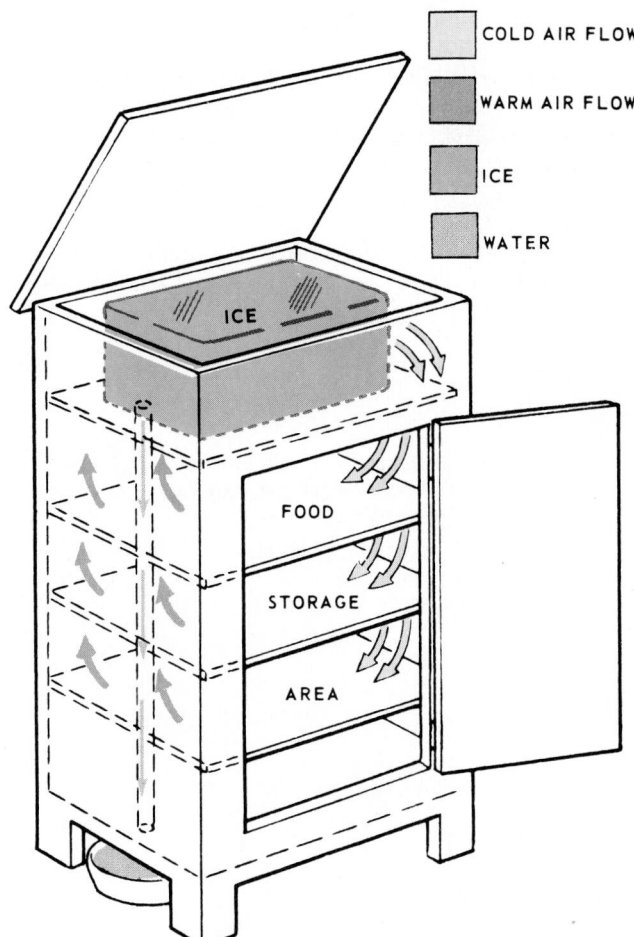

COLD AIR FLOW

WARM AIR FLOW

ICE

WATER

ICE

FOOD

STORAGE

AREA

Fig. 3-1. Basic design and operation of ice refrigerator.

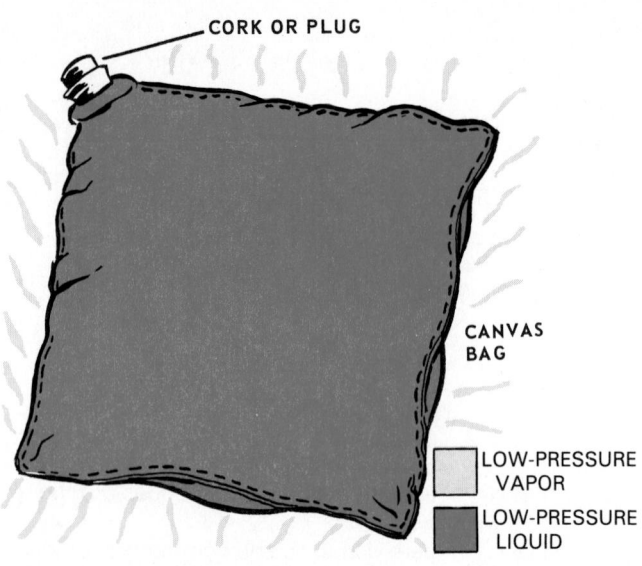

CORK OR PLUG

CANVAS BAG

LOW-PRESSURE VAPOR

LOW-PRESSURE LIQUID

Fig. 3-2. The "desert bag" is an example of cooling by evaporative refrigeration.

moist. Under desert conditions, which are usually both hot and dry, moisture on the surface of the bag evaporates rapidly.

A large part of the heat necessary to cause this evaporation comes from the bag and the water in it. This heat removal cools the drinking water inside the canvas, keeping it at a temperature several degrees below the temperature of the surrounding air.

3-3 EVAPORATIVE REFRIGERATION (SNOW-MAKING)

Another common application of water evaporation refrigeration is the method of making artificial snow for ski slopes. This device (snow machine) consists of a water nozzle into which a high-pressure jet of air is inserted. Water

(see dark green) flows from the nozzle, as shown in Fig. 3-3, and the air (see green stripe), under high pressure, causes the water to break up into tiny droplets (almost a fog). If the surrounding air temperature is near freezing or below freezing, the droplets of water tend to evaporate and rapidly cool to the point where tiny drops of ice are formed. Using this method, artificial snow can be made when the temperature of the surrounding air is 32 °F (0 °C) or lower.

If the humidity is low, artificial snow can be made when the temperature is as high as 34 °F (1 °C). This is possible because of the rapid evaporation and evaporative cooling caused by the low humidity.

Evaporative condensers, often used in connection with air conditioners, are another example of evaporative cooling. See Chapter 12. The evaporation of water helps cool the condenser.

3-4 COMPRESSION SYSTEM USING LOW-SIDE FLOAT REFRIGERANT CONTROL

The "low-side float" refrigerant control system was very popular in many of the early refrigerating mechanisms. It is also known as a flooded system.

Fig. 3-4 is a schematic diagram of this system. The liquid refrigerant flows from the liquid receiver through the liquid line, up to the low-side float needle. The evaporator in this system consists of a finned tank (evaporator). It contains a float and needle control, which maintains a constant level of liquid refrigerant under a low-side pressure.

This refrigerant, since it is a liquid on the low side, is at a low temperature. The cold liquid refrigerant will absorb much heat on both the on cycle and the off cycle.

Vaporized refrigerant moves through the suction (vapor) line to the compressor, where it is compressed to a high pressure and discharged into the condenser. It is cooled by the condenser, returns to a liquid and flows into the liquid receiver. The operation continues until the desired low temperature is reached.

The pressure on the low side in a flooded system such as this will vary with the temperature. The higher the

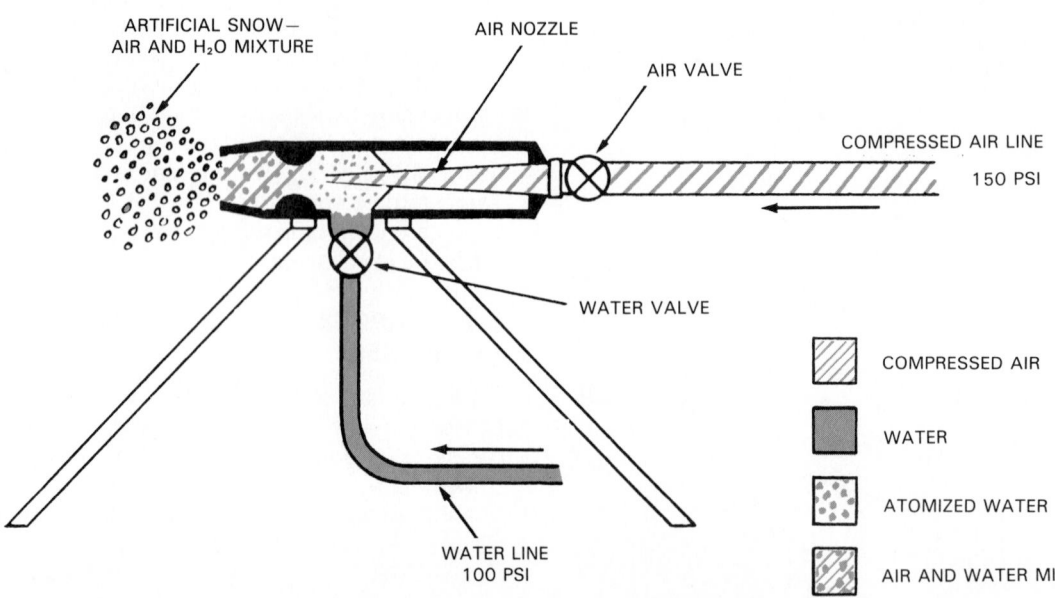

ARTIFICIAL SNOW— AIR AND H₂O MIXTURE

AIR NOZZLE

AIR VALVE

COMPRESSED AIR LINE 150 PSI

WATER VALVE

WATER LINE 100 PSI

COMPRESSED AIR

WATER

ATOMIZED WATER

AIR AND WATER MIX

Fig. 3-3. Water-compressed air nozzle is used for making artificial snow.

temperature, the higher the low-side pressure.

The system shown in Fig. 3-4 uses a pressure motor control. A spring-loaded pressure sensitive device on the suction line, or on the evaporator, activates a motor control switch. As the motor drives the compressor, the pressure

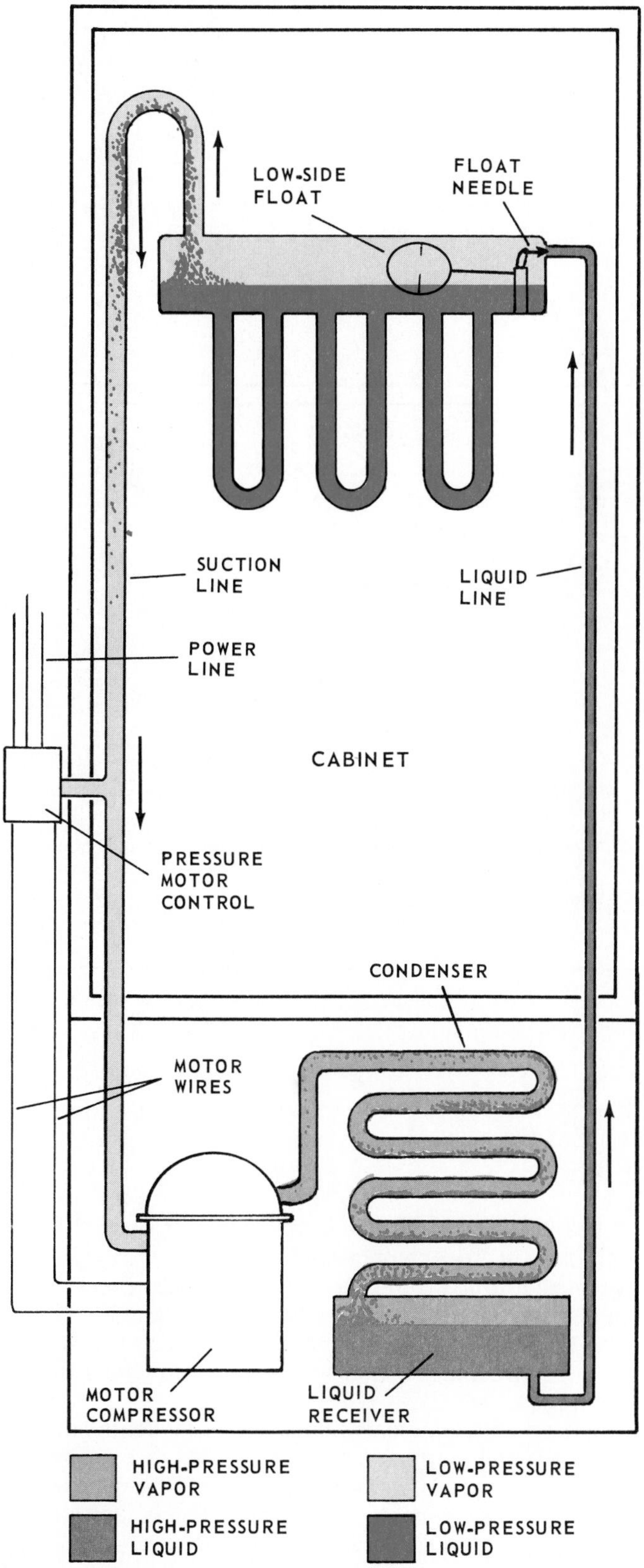

and temperature in the evaporator will be reduced. At a given pressure setting, the motor compressor will stop.

The cycle will be repeated as soon as the pressure in the evaporator rises to a level corresponding to the refrigerant temperature at which the motor compressor is to start again.

The cabinet temperature may be controlled by the temperature control switch. In this case, the temperature sensitive element may be clamped to the fins on the evaporator.

This refrigerating cycle is often used on drinking fountains and other installations, where a constant temperature is desired.

Since the pressures do not balance on the off cycle, it is necessary to use a motor which will start under a load.

Such a system requires a rather large refrigerant charge, as there is liquid refrigerant in both the liquid receiver and in the evaporator.

All flooded systems are quite efficient since cold liquid refrigerant wets the evaporator surfaces, providing excellent heat transfer. These systems are easy to service.

The float needle and seat must be kept in good condition to avoid possible flooding of the low side.

3-5 OPEN—EXTERNAL DRIVE— REFRIGERATING SYSTEM

In this system, the compressor is usually belt-driven from an electric motor. The speed of the compressor is usually considerably less than the speed of the motor. This is done by using a small pulley on the motor shaft and a larger pulley (flywheel) on the compressor shaft. Most early refrigerating systems were like this. Fig. 3-5 illustrates an open system.

The liquid refrigerant (see dark red) flows through the thermostatic expansion valve into the evaporator where it is under low pressure. It boils, vaporizes, and absorbs heat in the evaporator.

When the compressor is running, the vaporized refrigerant (see light blue) is drawn through the suction line to the compressor. It is compressed to a high pressure (see light red) before being discharged into the condenser.

In the condenser, the vapor gives up its latent heat of vaporization, is cooled, and returns to a liquid (see dark red). From here the cycle is repeated.

A thermostatic bulb motor control is shown. The starting mechanism on external drive (open) system motors is usually built into the motor.

This system requires a crankshaft seal on the compressor. The motor and the compressor drive are at atmospheric pressure. The pressure inside the crankcase will vary depending upon the refrigerant used and the temperature. Sometimes it may be considerably above atmospheric pressure; at other times, it may be below. Refrigerant vapor cannot be allowed to flow out or air to flow into the crankcase. Either would quickly ruin the operation.

3-6 COMPRESSION SYSTEM USING HIGH-SIDE FLOAT REFRIGERANT CONTROL

The high-side float system is a flooded system. The evaporator is always filled with liquid refrigerant.

Fig. 3-6 is a schematic diagram of this system. As the compressor runs, refrigerant from the condenser flows into the high-side float mechanism.

Fig. 3-4. Compression system using low-side float refrigerant control.

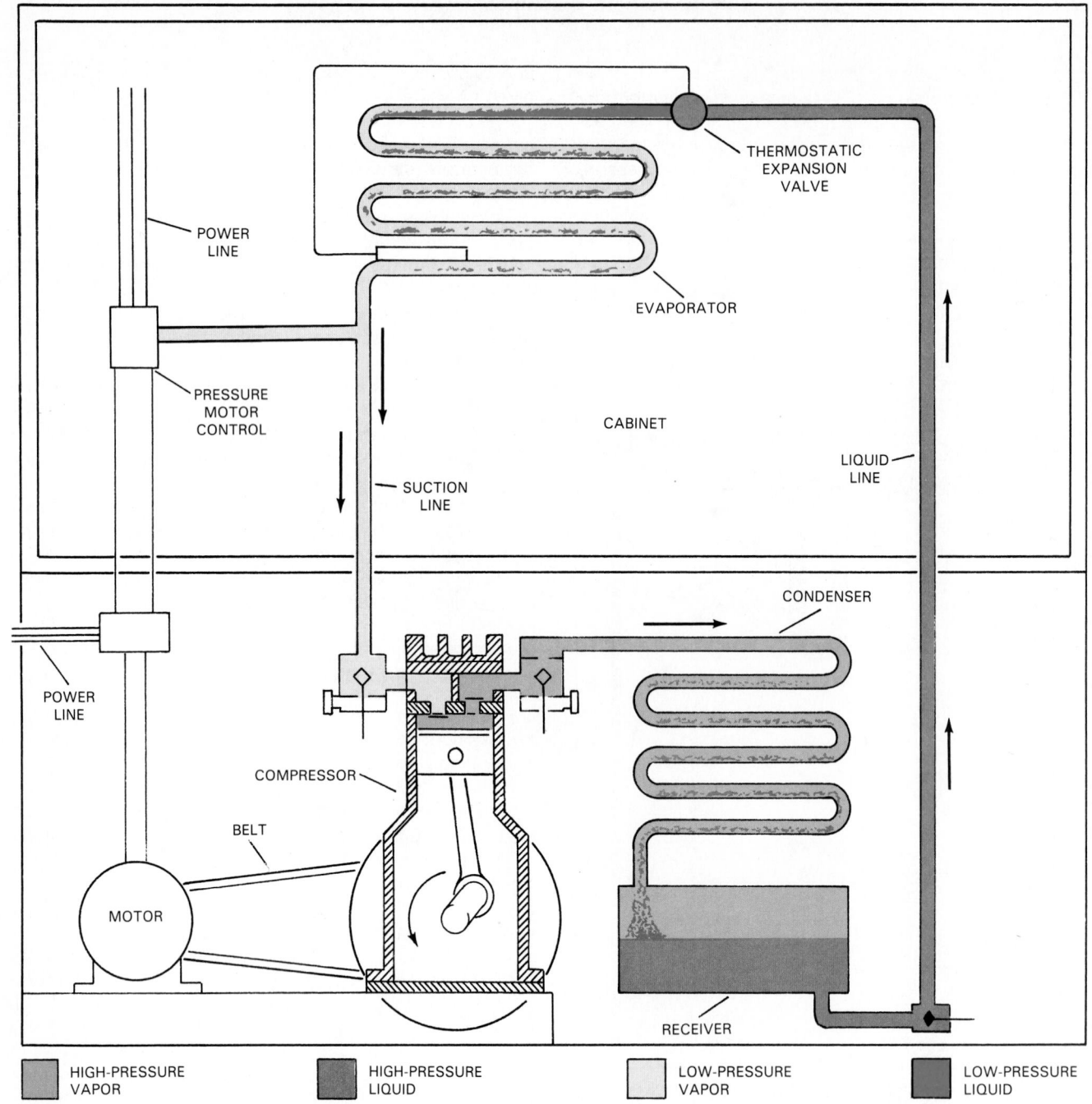

| HIGH-PRESSURE VAPOR | HIGH-PRESSURE LIQUID | LOW-PRESSURE VAPOR | LOW-PRESSURE LIQUID |

Fig. 3-5. Compression system using external drive (open) type compressor. Crankshaft seal is required at the place where crankshaft extends through crankcase of compressor.

As soon as enough liquid refrigerant has entered the high-side float mechanism, it will raise the float ball. The refrigerant will then begin to flow through the control to the evaporator. Since the evaporator is under low pressure, the tubing connecting the high-side float and the evaporator should be insulated. A capillary tube refrigerant line is frequently used.

If a different size line is used, it should have a weight valve at the evaporator to prevent the refrigerant from evaporating in the connecting line. Fig. 3-6 shows a weight valve in the connecting line.

Refrigerant entering the evaporator is under low pressure (see dark blue). It will rapidly evaporate (boil) and absorb heat from the evaporator.

The vapor (see light blue) then flows through the suction line to the compressor, where it is compressed (squeezed) to the high-side pressure (see light red). In the condenser, the heat absorbed in the evaporator is removed and the refrigerant is returned to the liquid state (see dark red). It flows into the high-side mechanism where the cycle is repeated.

Either a temperature or a pressure motor control may be used on this refrigeration cycle. Fig. 3-6 shows a temperature motor control located in the refrigerated space.

This system is most used in commercial applications where high operating efficiency is desired.

It is easy to service, but the amount of refrigerant charged into the system must be very accurately measured.

3-7 COMPRESSION SYSTEM USING AUTOMATIC EXPANSION VALVE REFRIGERANT CONTROL (AEV)

The operation of an automatic expansion valve refrigerant control (AEV) refrigerating mechanism is shown in Fig. 3-7.

Compressor, motor, and condenser (condensing unit) are in the base of the cabinet. Liquid refrigerant (see dark red) flows from the liquid receiver through the liquid line, through the filter to the automatic expansion valve.

The automatic expansion valve is designed so that no liquid refrigerant will flow through it unless the pressure in the

evaporator is reduced by the running of the compressor. As the compressor runs and liquid refrigerant flows through the automatic expansion valve, it is sprayed into the evaporator (see dark blue). Here, due to low pressure, it

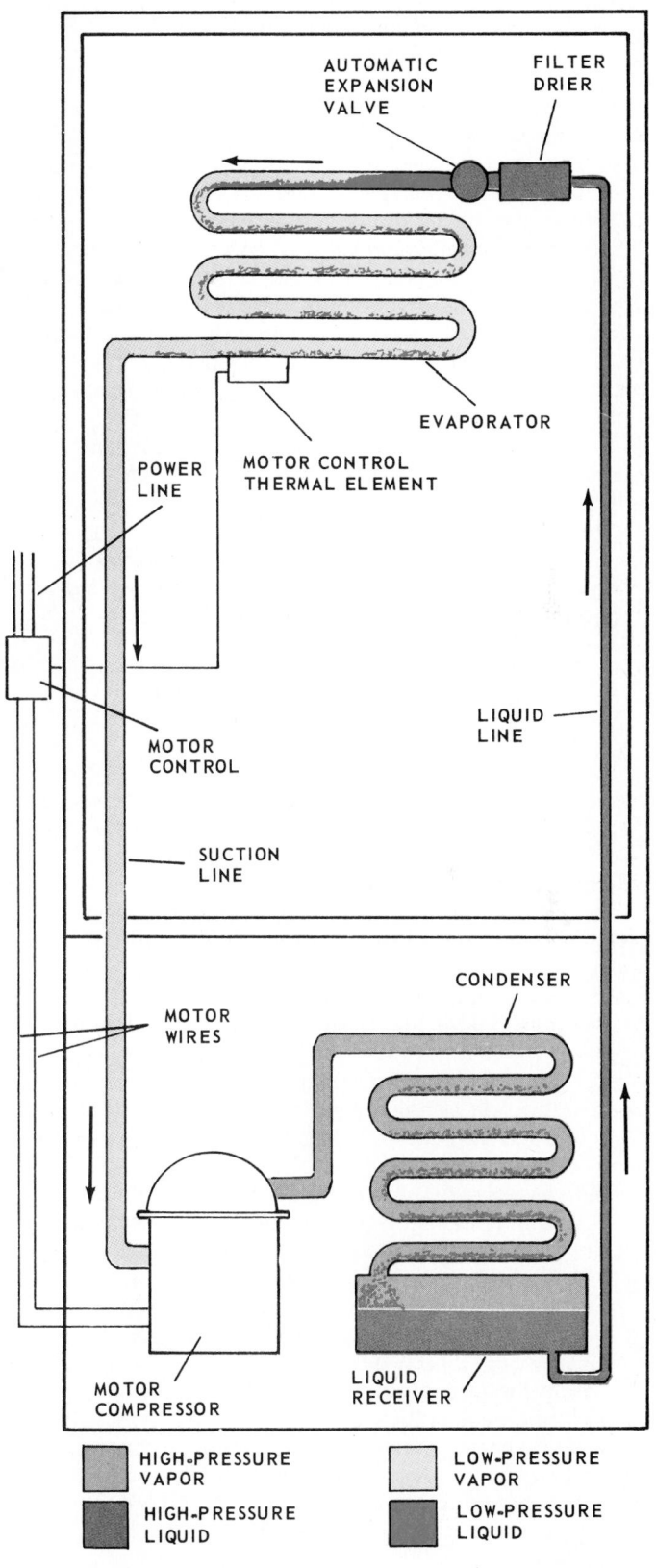

HIGH-PRESSURE VAPOR		LOW-PRESSURE VAPOR
HIGH-PRESSURE LIQUID		LOW-PRESSURE LIQUID

Fig. 3-7. Compression system using automatic expansion valve refrigerant control (AEV).

Fig. 3-6. Compression system using high-side float refrigerant control.

HIGH-PRESSURE VAPOR		LOW-PRESSURE VAPOR
HIGH-PRESSURE LIQUID		LOW-PRESSURE LIQUID

boils rapidly and absorbs heat. This vaporized refrigerant (see light blue) moves back to the compressor through the suction line.

In the compressor, it is compressed to the high-side pressure as vapor (see light red). While flowing through the condenser, it is cooled, giving up the heat that it absorbed in the evaporator and returning to a liquid (see dark red). It then flows into the liquid receiver ready to repeat the cycle.

The motor control thermal element is clamped to the end of the evaporator at the beginning of the suction line. After the evaporator is cooled to its proper temperature, the control bulb pressure causes the motor control to turn off the current to the driving motor. The compressor is stopped.

The operating characteristics of this system are quite satisfactory. The refrigerant oil is circulated without trouble. The temperature control limits can also be kept quite close.

This type of refrigeration cycle is used widely in small commercial applications. Because the pressures do not balance on the off cycle, the motor compressor must start while under load.

If the needle or seat in the expansion valve is faulty and refrigerant leaks through the valve on the off cycle, liquid refrigerant may flow into the suction line. When the compressor starts, this will be indicated by frosting of the suction line. If the trouble is severe, it may result in liquid refrigerant entering the compressor through the suction line. This may cause the compressor to knock severely.

3-8 COMPRESSION SYSTEM USING THERMOSTATICALLY CONTROLLED EXPANSION VALVE (TEV)

A schematic diagram of a thermostatically controlled expansion valve (TEV) refrigeration cycle is shown in Fig. 3-8. The liquid refrigerant (see dark red) flows from the liquid receiver through the liquid line to the filter-drier and to the thermostatic expansion valve.

The operation of the thermostatic expansion valve is controlled by both the temperature of the TEV control bulb and the pressure in the evaporator. The temperature of the TEV control bulb must be higher than the evaporator refrigerant temperature before the valve will open. The amount of opening will be governed by the temperature of the evaporator. If the evaporator is quite warm, the needle will open quite wide allowing a rapid flow of liquid (see dark blue) into the evaporator. In this way, cooling is speeded up. As the temperature of the evaporator drops, the TEV needle valve will cut down the flow of refrigerant.

Vaporized refrigerant (see light blue) from the evaporator moves back into the compressor where it is compressed back to the high-side pressure (see light red). As it flows through the condenser, it gives up the heat absorbed in the evaporator. Now cooled, the refrigerant is condensed to a liquid (see dark red) and flows back into the liquid receiver. The refrigerating cycle is repeated.

When the evaporator reaches the desired temperature, the motor control will turn off current to the motor to stop the compressor. When this happens, the TEV needle valve will close, allowing no more refrigerant to flow through it until the compressor again lowers the pressure in the evaporator.

This system is used on large commercial refrigerators as well as on many air conditioning applications.

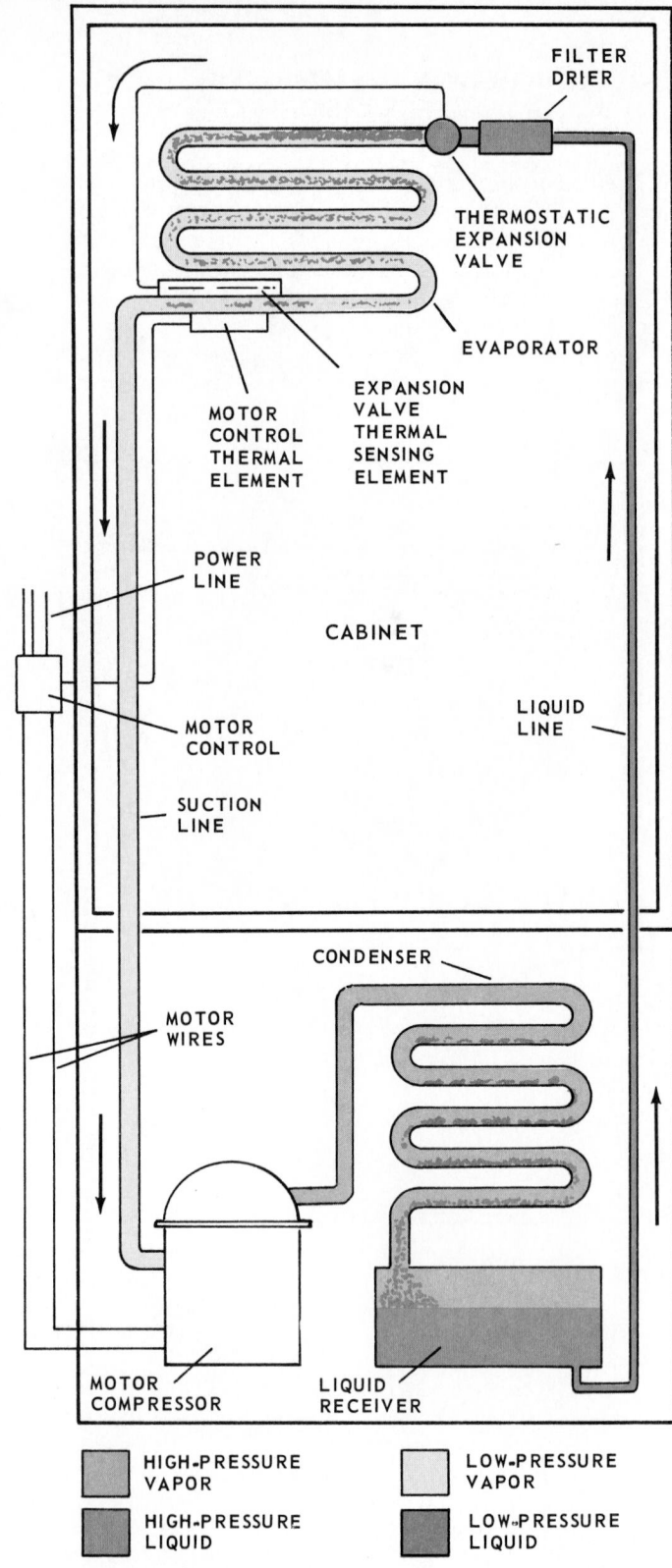

Fig. 3-8. Compression system using thermostatically controlled expansion valve (TEV).

Since pressures do not balance on the off cycle, it is necessary to provide a motor compressor which will start under load.

The TEV control remains closed unless the evaporator is under reduced pressure and the temperature is above normal. A leaking valve will usually be indicated by a frosted or sweating suction line.

3-9 COMPRESSION SYSTEM USING CAPILLARY TUBE REFRIGERANT CONTROL

The "capillary tube" system, Fig. 3-9, is one of the most popular of the compression type systems. Liquid refrigerant (see dark red) flows from the condenser up through the liquid line to the filter (which may also be a drier). From the filter, refrigerant flows through the capillary tube refrigerant control into the evaporator. The pressure of the liquid refrigerant, as it enters the capillary tube at the filter end, is at a high pressure (see dark red). This is the high-pressure side. The pressure in the evaporator is low.

The design of the capillary tube is such that it maintains a pressure difference while the compressor is operating. The compressor maintains a low pressure in the evaporator and the refrigerant boils, rapidly absorbing heat. The vaporized refrigerant (see light blue) moves through the suction line back to the compressor. Here it is compressed to a high pressure and discharged into the condenser (see light red). It is cooled in the condenser and returns to a liquid (see dark red) and again flows into the liquid line.

This operation continues until the thermal element has been cooled to a preset low temperature. When that temperature is reached, the thermal element operates the motor control mechanism and turns off power to the motor. The refrigeration cycle stops. It will remain stopped until the thermal element warms up and the thermal bulb pressure closes the motor control contacts to again operate the compressor.

On the off cycle, the capillary tube allows the pressures to balance between the high and low sides. It is not usually necessary, then, to use a motor with a high starting torque.

This system is commonly used in household refrigerators, freezers, air conditioners, dehumidifiers, and many small commercial applications. This type of cycle is quite satisfactory for most refrigerating applications.

3-10 MULTIPLE EVAPORATOR SYSTEM

Some commercial refrigerating systems have one condensing unit connected to two or more evaporators. Liquid refrigerant (see dark red) flows through the thermostatic expansion valves to the evaporators. The evaporators may have the same evaporator temperatures or they may evaporate the refrigerant at different temperatures.

If the evaporator temperatures are the same, the system uses only a low-side float or the thermostatic expansion valve to control the refrigerant.

If two or more evaporating temperatures are desired (a frozen foods temperature and a water cooling temperature for example), a device must be used to keep one of the evaporators at a higher low-side pressure. Look at schematic, Fig. 3-10. A two-temperature valve in the suction line (upper left) keeps the low-side pressure refrigerant liquid (see dark blue) and vapor (see light blue) in evaporator B at a higher pressure than at evaporator A. The evaporator temperature is governed by the evaporating pressure. The lower the pressure, the lower the temperature.

A check valve in the suction line coming from the colder evaporator, A, prevents the warmer, higher pressure low-side vapor (see light blue) from moving into the colder evaporator, A, during the off cycle.

The vaporized refrigerant (see light blue) is returned to

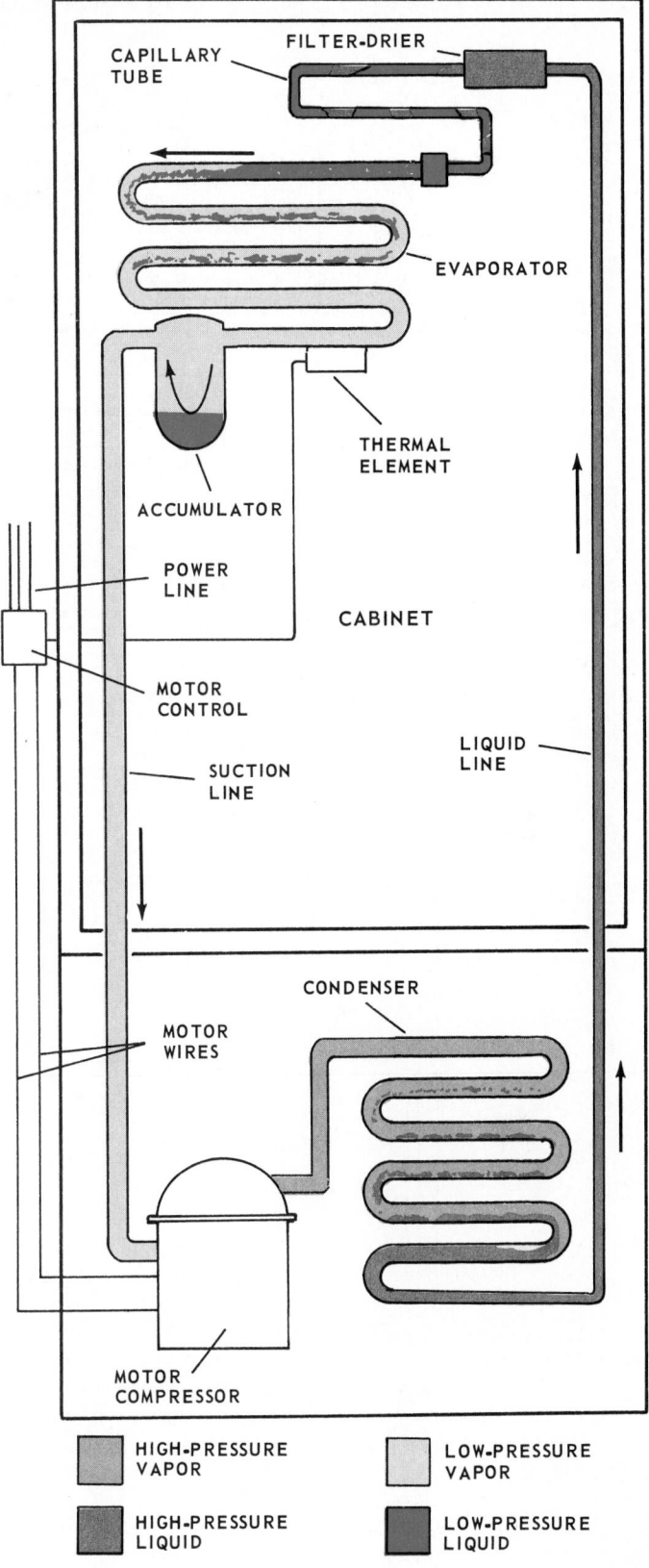

HIGH-PRESSURE VAPOR   LOW-PRESSURE VAPOR

HIGH-PRESSURE LIQUID   LOW-PRESSURE LIQUID

Fig. 3-9. Compression system using capillary tube type refrigerant control.

the motor compressor. It is compressed to a high-pressure and high-temperature vapor (see light red). This vapor is cooled in the condenser, becoming a high-pressure liquid (see dark red) to be stored in the receiver until needed.

Note the filter-drier on the liquid line. It keeps the refrigerant clean and dry.

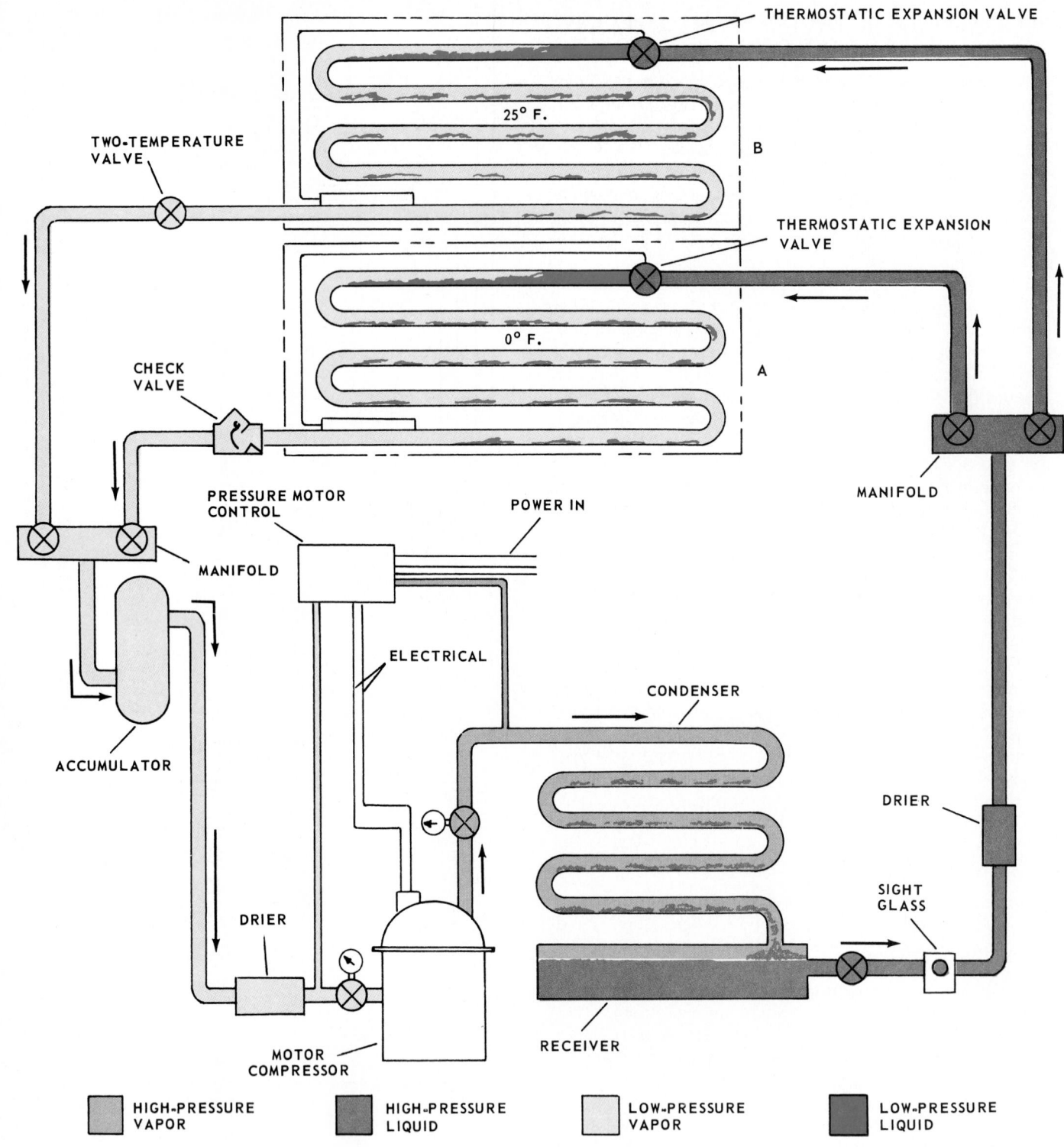

THERMOSTATIC EXPANSION VALVE

TWO-TEMPERATURE VALVE

25° F.

B

THERMOSTATIC EXPANSION VALVE

CHECK VALVE

0° F.

A

MANIFOLD

PRESSURE MOTOR CONTROL

POWER IN

MANIFOLD

ELECTRICAL

CONDENSER

ACCUMULATOR

DRIER

DRIER

SIGHT GLASS

MOTOR COMPRESSOR

RECEIVER

| | HIGH-PRESSURE VAPOR | | HIGH-PRESSURE LIQUID | | LOW-PRESSURE VAPOR | | LOW-PRESSURE LIQUID |

Fig. 3-10. A multiple evaporator system. Evaporator A operates at 0 °F (−18 °C). Evaporator B operates at 25 °F (−4 °C).

A liquid indicator (sight glass) is often included in the liquid line. The service technician may then check to see if there is enough refrigerant in the system. Bubbles will indicate a refrigerant shortage. This system, as shown, uses a pressure motor control. The operating pressure is taken from the low side of the system.

A line from the high-pressure side also enters the motor control. This operates a safety device which stops the motor if the condensing pressure (high side) goes too high.

Multiple (two or more) evaporator refrigeration systems are commonly used in commercial refrigeration applications.

3-11 COMPOUND REFRIGERATING SYSTEMS

In compound refrigerating systems, two or more compressors are connected in series, Fig. 3-11. In this illustration, compressor No. 1 discharges into the intake side of compressor No. 2. Compressor No. 2 then discharges into

Basic Refrigeration Systems

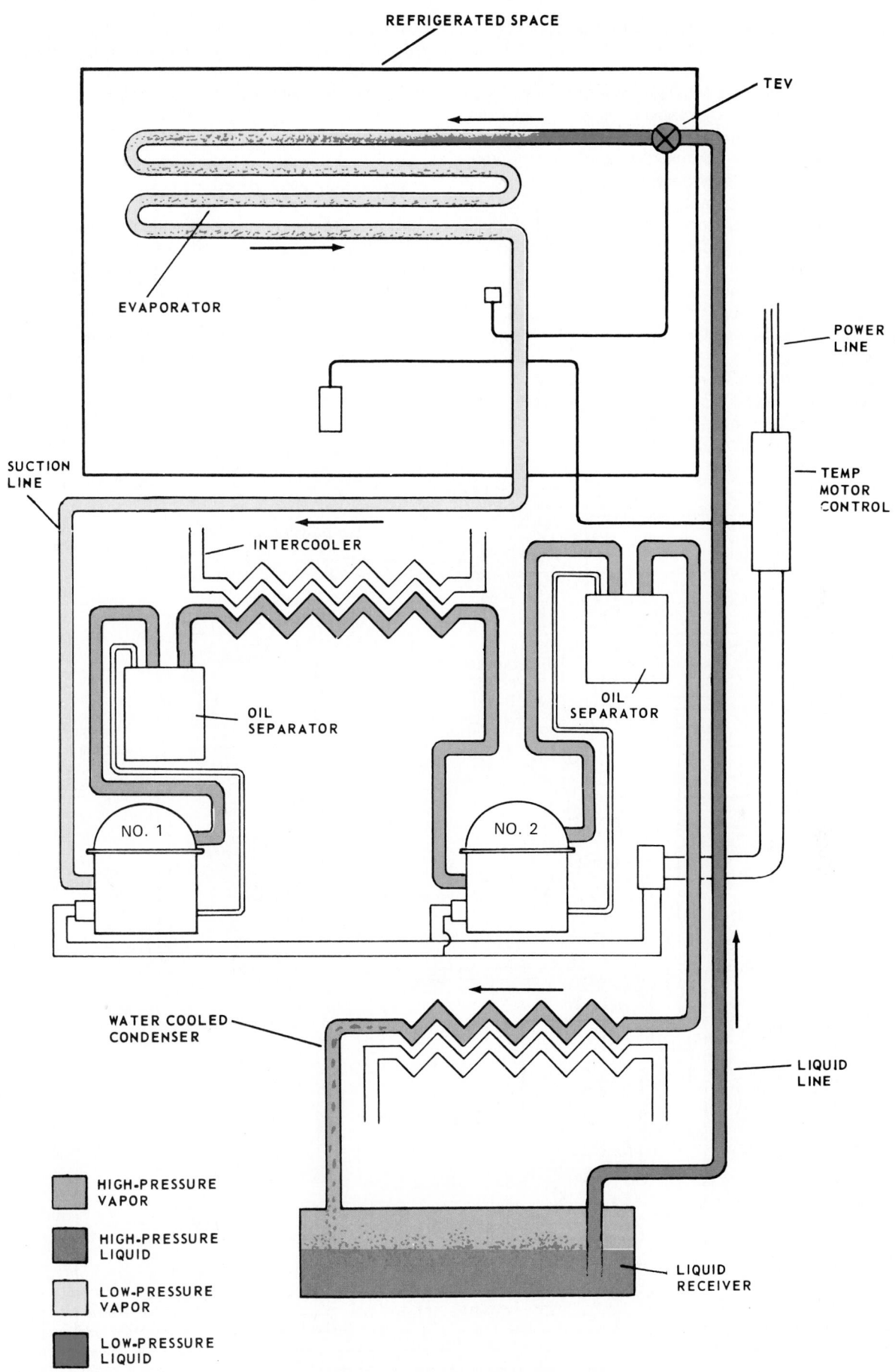

Fig. 3-11. Compound refrigerating system.

the condenser (see light red). Here the vapor condenses. Then the liquid refrigerant (see dark red) flows into the liquid receiver.

From the liquid receiver the liquid refrigerant (see dark red) flows up to the thermostatic expansion valve and into the evaporator. In the evaporator (see dark blue) the refrigerant boils and absorbs heat (see light blue). From the evaporator the vaporized refrigerant flows back to compressor No. 1. From here the cycle is repeated.

Such a compound system increases capacity when pulling down to such a low pressure (low temperature) that one compressor cannot do it well.

Refrigerant vapor is not condensed between compressors. An intercooler lowers the vapor temperature. This type of installation usually requires an oil separator for each compressor.

A single-temperature motor control operates all motors and a thermostatic expansion valve controls the liquid refrigerant flow into the evaporator.

Since the pressures do not balance on the off cycle, motors capable of starting under load are required.

Compound installations usually operate under rather heavy service requirements. Condensers and refrigerant must be kept clean. Compressor valves must be kept in good condition.

3-12 CASCADE REFRIGERATING SYSTEMS

In a cascade refrigerating installation, two or more refrigerating systems are connected as shown in Fig. 3-12. Both systems operate at the same time. System A (on the right) has its evaporator, A, (heat absorbing part) arranged to cool the condenser B for the system B. The evaporator for system B supplies the cooling effect desired. Each system has a thermostatic expansion valve (TEV) for refrigerant control.

The low-pressure liquid (see dark blue) of system A cools the high-pressure vapor (see light red) of system B.

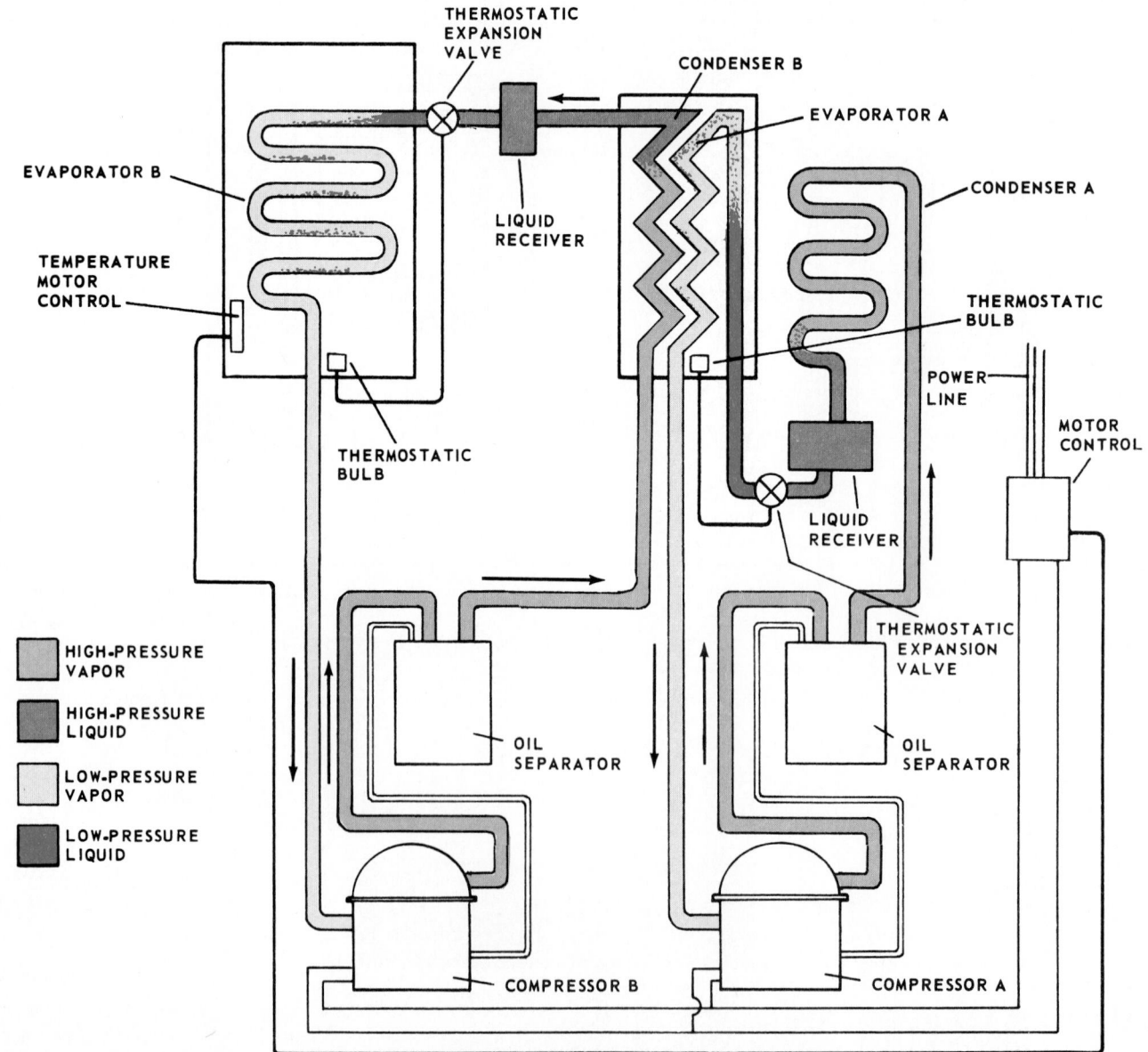

Fig. 3-12. Cascade refrigerating system.

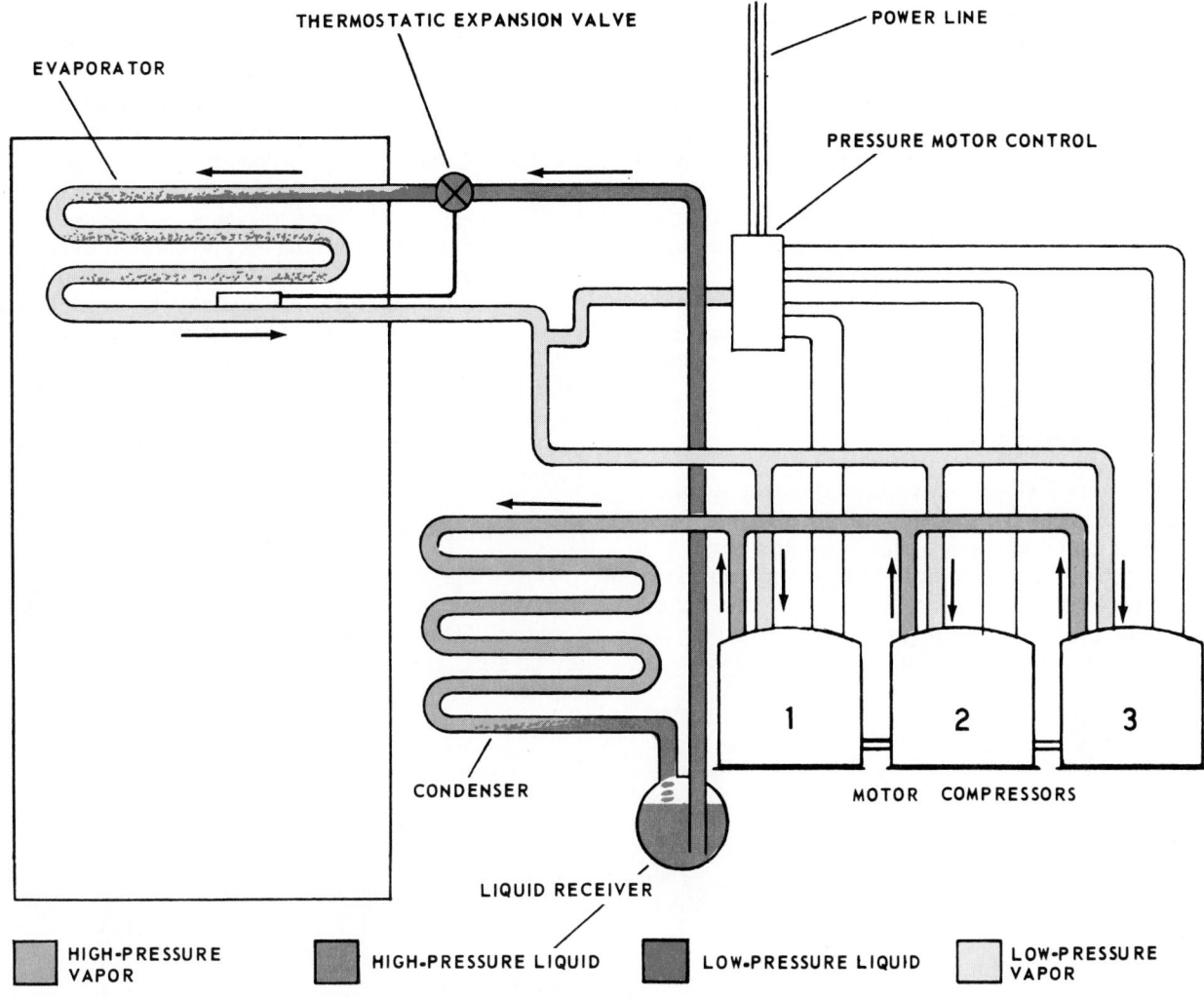

Fig. 3-13. Modulating refrigeration cycle mechanism, which uses three motor compressors: Pressure motor control is arranged to operate one or more compressors as needed.

Cascade systems are often used in industrial processes where objects must be cooled to temperatures below −50 °F (−46 °C).

One motor control is used for both motors. It is connected to a temperature sensing bulb on evaporator B.

Motors used on cascade systems must be capable of starting under load. With the use of thermostatic expansion valves, the pressures do not balance on the off cycle.

The condenser-evaporator is usually of the shell-and-tube flooded evaporator type.

Since these systems operate at very low temperatures, the refrigerant must be very dry. Otherwise, any moisture would condense at the needle-seat of the TEV and stop the flow of refrigerant. System B must have special refrigerant oil (wax-free, moisture-free, and flowable at extra low temperatures).

Oil separators should be installed in the compressor-to-condenser lines on both of these condensing units to help keep the oil in the compressors.

3-13 MODULATING REFRIGERATION CYCLE

In most refrigeration installations, the cooling or refrigerating capacity is enough to maintain the desired

temperature under the heaviest load. This temperature is maintained by the motor control. It starts the motor compressor when cooling (or heat removal) is required and shuts if off as soon as the desired temperature is reached.

However, if the heat load is light, this single system may be over-capacity for the job. The operating expense is greater than it would be if the machine capacity more nearly matched the needed load. The system also tends to cool too fast and it operates on and off too quickly.

A modulating (varying capacity) system has been developed to fit the machine capacity more closely to the needed heat load. This is sometimes done by using two or more compressors connected in parallel. Each compressor is operated by a motor control.

During operation, if the heat load increases and the temperature starts to rise, one compressor will continue to run. But if the temperature keeps on rising, the second compressor will start to operate. Additional compressors may cut in until enough capacity is obtained.

Fig. 3-13 illustrates a typical cycle diagram for a modulated installation. This installation has three compressors. A pressure control connected to the suction lines operates the motors. The control contains a special switching device which rotates the service of the various compressors. Thus, each compressor will be used about the same amount of time.

The modulating cycle maintains uniform temperatures and operates economically.

Any conventional refrigerant control can be used. However, the thermostatic expansion valve is most common.

The same condenser and liquid receiver may be used by all the compressors, or each may have its own. The same evaporator is connected to all the compressors.

A modulating system may use a multiple cylinder compressor, each cylinder being equipped with an unloader device. Variable speed motors are also used to provide a modulated refrigeration capacity.

3-14 ICE MAKER

Ice makers use various types of refrigerating systems. The simple unit in Fig. 3-14 operates as follows: the motor compressor and condenser are usually located in the bottom of the cabinet. Liquid refrigerant (see dark red) flows from the bottom of the condenser up through a filter-drier. It enters the evaporator through a capillary tube. The evaporator surrounds the inverted (upside-down) ice cube molds.

From the evaporator, the refrigerant vapor (see light blue) flows into an accumulator. This is a type of container which has a coil from the liquid refrigerant line in it or around it. Such an arrangement serves as a heat exchanger. The refrigerant vapor (see light blue) is drawn from the accumulator back to the compressor. Here it is compressed up to the high-side pressure (see light red) and is forced into the condenser. From here the cycle is repeated.

The mechanism which makes and handles the ice is also shown. Cold water is sprayed into the inverted ice cube molds. The temperature of the molds is very low. Water striking the molds freezes to the mold surface and gradually builds up until complete ice cubes are formed. Then the refrigerating cycle is stopped. Now an electric heating unit heats the ice cube molds until the cubes fall out and slide down a chute into the ice cube bin. Most surfaces in contact with water and ice are stainless steel for cleanliness.

3-15 DRINKING WATER COOLER

The water cooler is a special use of a refrigerating mechanism. It is used to cool water "on tap" at a drinking fountain. The usual hermetic (airtight) compression refrigerating system is used. The refrigerant control is a capillary tube. The cycle is shown in Fig. 3-15.

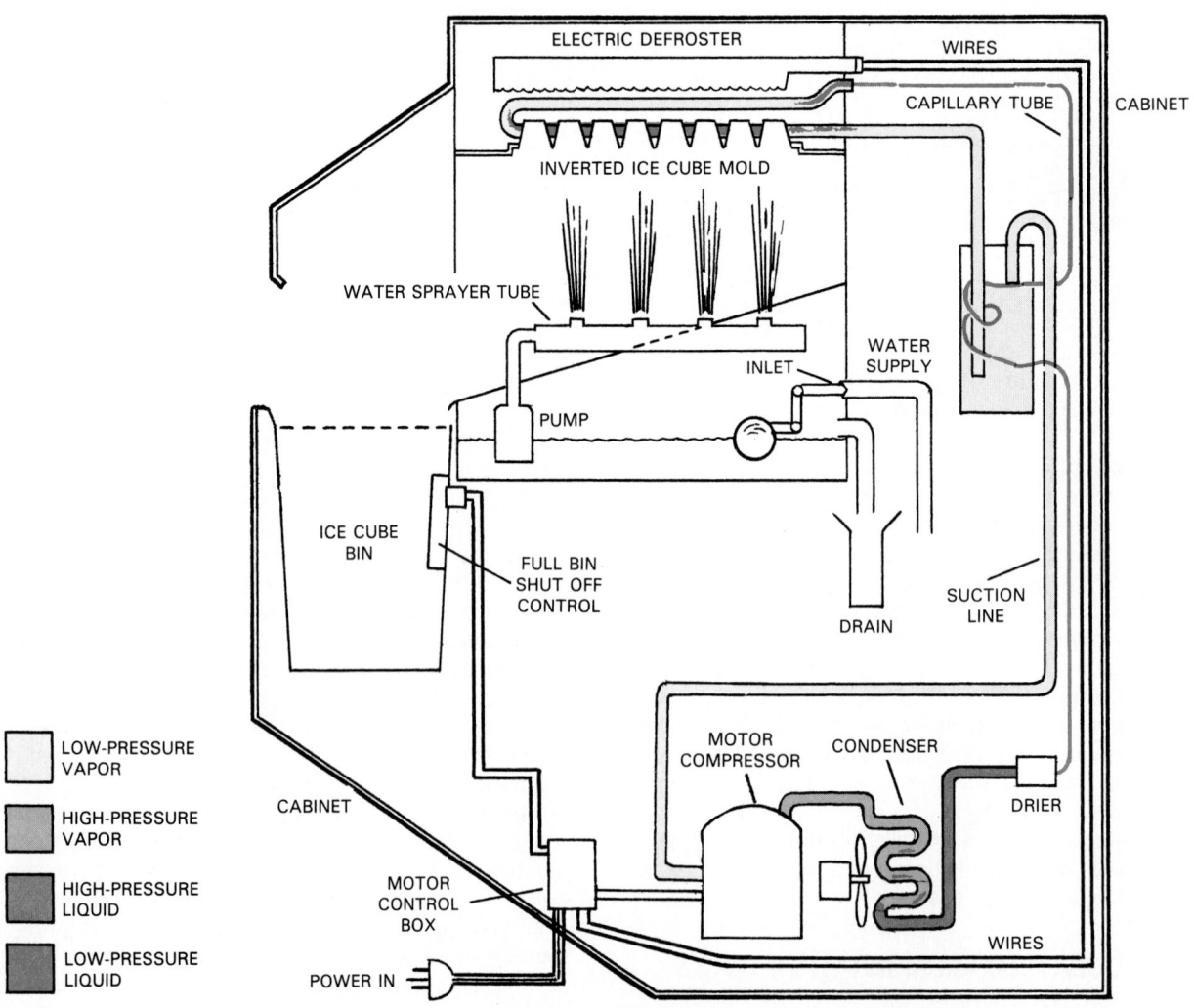

Fig. 3-14. Ice maker. Water is sprayed into ice cube molds to produce clear ice cubes.

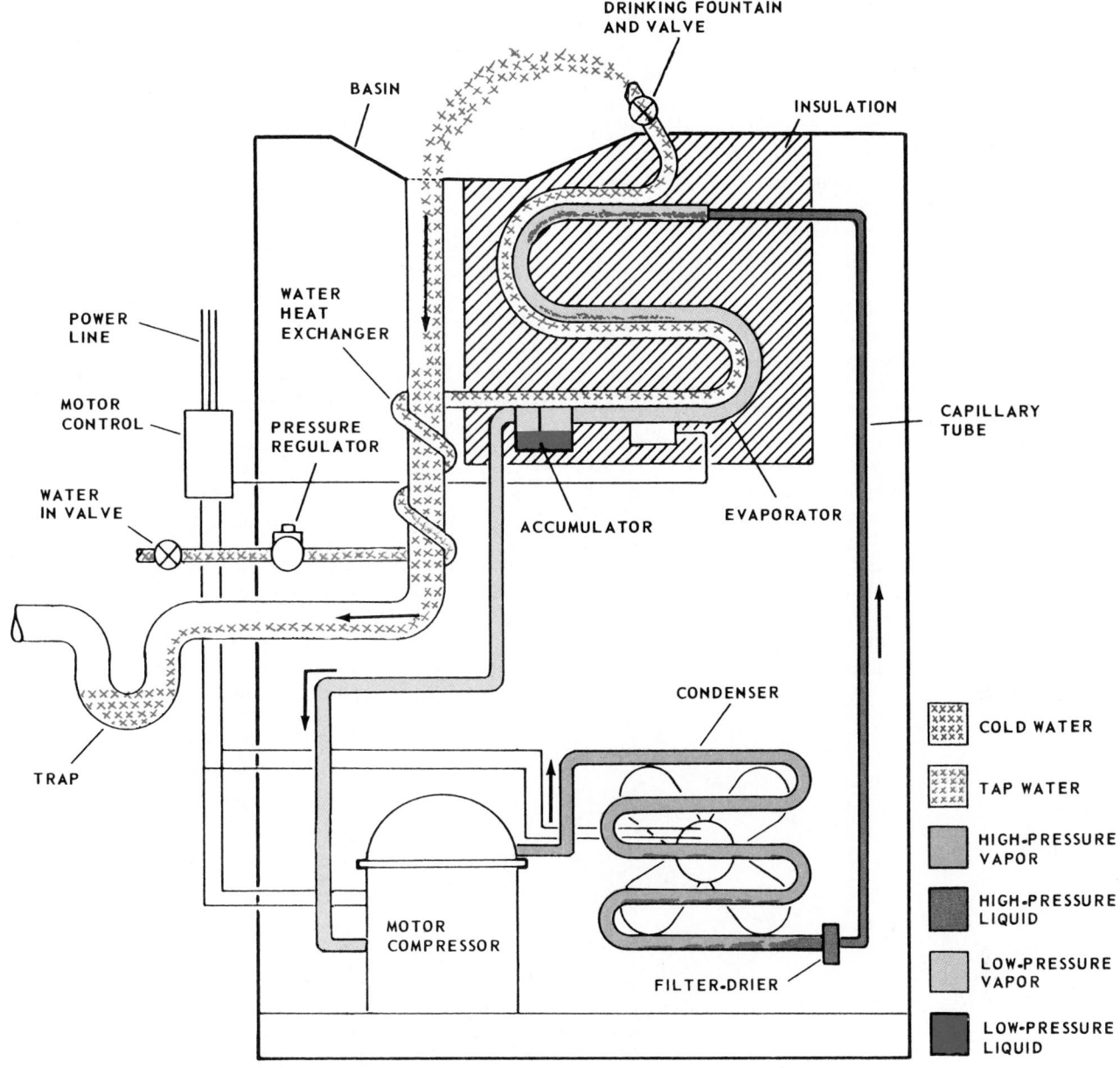

Fig. 3-15. A drinking fountain cooled by a compression system refrigerating mechanism.

Liquid refrigerant flows from the bottom of the condenser through the liquid line, into a filter-drier (see dark red) and into the capillary tube. As it flows into the evaporator, it vaporizes and absorbs heat from the evaporator surface (see light blue). The evaporator is either adjacent to or surrounds the drinking water coil or water cooling tank.

From the evaporator, the refrigerant vapor goes into an accumulator in the suction line. (The accumulator stops any liquid refrigerant from flowing into the suction line and on into the motor compressor.)

From the accumulator, the vapor is drawn into the motor compressor where it is pumped into the condenser (see light red). Here the heat picked up in the evaporator is released. Meanwhile, the refrigerant returns to a liquid and collects in the bottom of the condenser. From here, the cycle is repeated.

Since the demand on a drinking fountain is very irregular, it is necessary that it have some hold-over capacity. Still it must not over-cool the water. The necessary capacity is provided by using either an insulated storage tank or large cooling surfaces in the evaporator.

To increase the mechanism's efficiency, the waste water flows down a tube alongside or attached to the fresh water inlet. In this way, the warmer fresh water (water-in) is cooled, to some extent, by the cooler waste water leaving the fountain.

A water pressure regulator adjusts the water flow. The condensing unit is air cooled. However, to make sure the fountain can deliver enough cold water under heavy demand, a condenser fan is used to increase the condenser capacity. The fan is connected into the electrical circuit and runs whenever the condensing unit is running.

A thermostat with the control bulb attached to the water dispensing tube maintains the desired drinking water temperature in the fountain. Water leaving the fountain should be at approximately 50 °F (10 °C).

3-16 EXPENDABLE REFRIGERANT REFRIGERATION SYSTEM

This simple system, sometimes called chemical refrigeration or open cycle refrigeration, is becoming increasingly popular. It is used on trucks and other vehicles in the transportation and storage of refrigerated or frozen foods. Basically it is a heavily insulated space which is cooled either by being surrounded by tubes carrying evaporating liquid nitrogen or by spraying liquid nitrogen directly into the space to be cooled.

Fig. 3-16 illustrates the spray system. The liquid nitrogen (see dark red), supplied from a cylinder inside the refrigerated space, is kept under pressure (200 psi). Dark blue indicates low-pressure liquid refrigerant.

Although the pressurized cylinder is insulated, an automatic pressure relief valve will open as a safety measure and allows the nitrogen vapor to escape should pressure exceed the relief valve setting. Heat surrounding the cylinder will sometimes make the vapor pressure rise above the automatic pressure release setting. Cold nitrogen vapor, released by the automatic pressure release valve, is discharged into the refrigerated space, or into the refrigerating tubes, depending on the system being used.

A temperature sensing element, control box, and liquid control valve, control the flow of liquid nitrogen from the nozzles. They maintain the desired temperatures inside the refrigerated space.

Liquid nitrogen (see dark red) vaporizes (boils and turns into a gas) at a temperature of $-320\,°F$ ($-196\,°C$) at atmospheric pressure (Fig. 1-29). This type system is excellent for shipping frozen foods. Temperatures may be kept as low as desired—usually about $-20\,°F$ ($-29\,°C$).

Simple construction such as this demands little attention other than to replace or recharge the nitrogen storage cylinder. Another advantage is its ability to operate without a power source. Safety devices in spaces refrigerated by liquid nitrogen shut off the flow of nitrogen when one opens a door to the space. See Chapter 17.

Another form of expendable refrigeration system is natural gas shipped in liquid form in large tanker ships. Natural gas, liquid when under pressure, will evaporate. Some of the gas is allowed to evaporate. This evaporative cooling maintains the remaining natural gas in liquid form. The evaporated natural gas is then ducted to the tanker engines where it is burned to provide power to drive the tanker.

3-17 THERMOELECTRIC REFRIGERATION

The physical principle (Peltier effect), upon which thermoelectric refrigeration is based, has been known since 1834. This system of transferring heat energy from one place to another uses electrons instead of a refrigerant.

Fig. 3-17A represents a simple thermoelectric couple. The couple moves heat from the inside of an insulated space

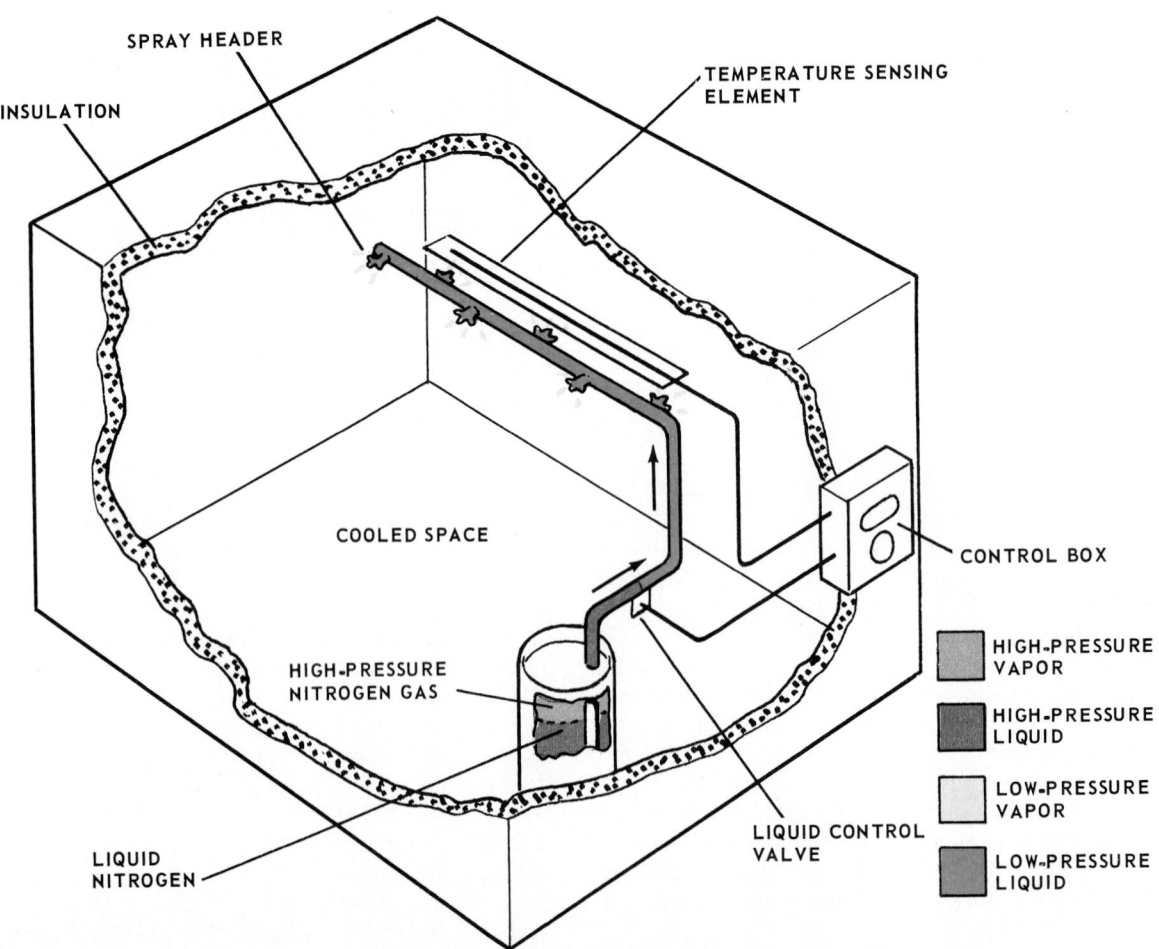

Fig. 3-16. Expendable refrigerating system.

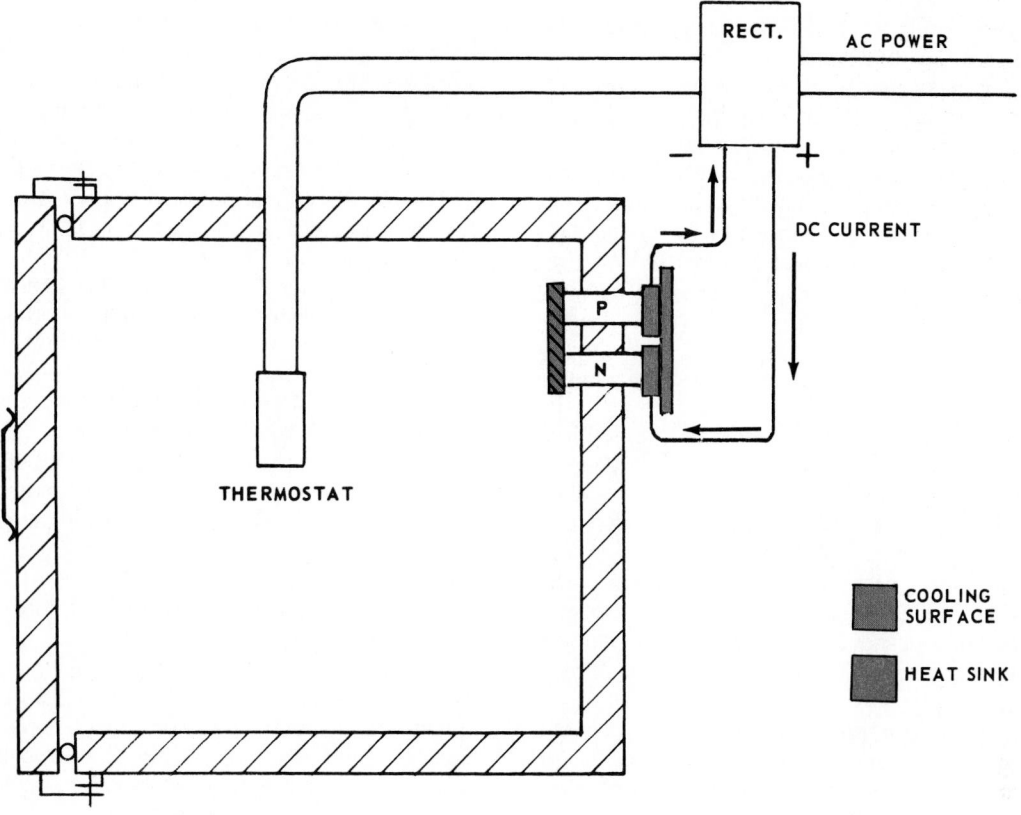

Fig. 3-17A. Diagram of simple thermoelectric couple, used for refrigerating an insulated space. Heat absorbed by thermoelectric couple is released to outside by fins attached to heat radiating surface (heat sink).

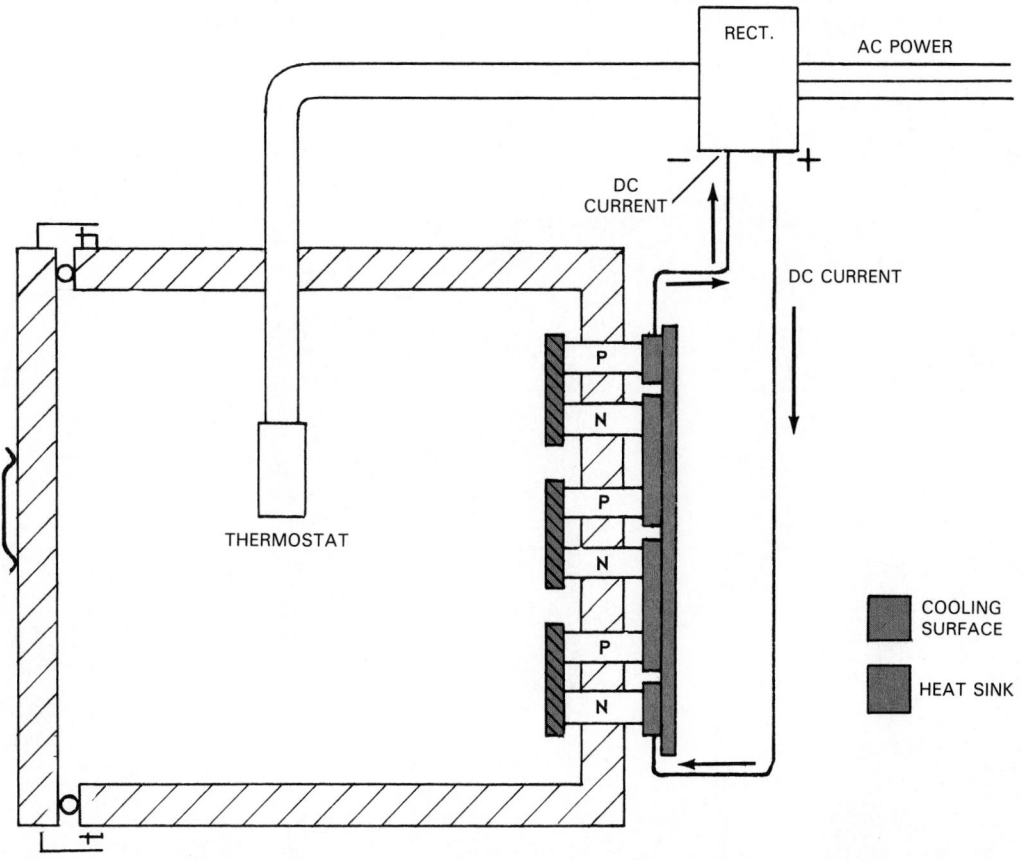

Fig. 3-17B. Thermoelectric module cooling device. Three couples are connected in series to increase heat absorbing effect. Electrons flow into N-type section. See Fig. 6-75.

to a heat exchanger on the outside. Electrons, rather than refrigerants, carry away the heat.

Fins on the evaporator (see dark blue) increase the heat flow. Fins on the outside of the heat exchanger (see dark red) help give off the heat to the surrounding air.

Semiconductors are materials that conduct electricity, but not as well as typical metals. They may be made from elements such as silicon or germanium or a combination of elements. Semiconductors may be processed so that some, called N-type semiconductors, conduct electricity by the flow of negatively charged particles (usually electrons). Others, called P-type semiconductors, conduct electricity by the flow of positively charged particles (often called "holes" or electron holes).

When current is forced to flow from an N material into a P material, as shown in Fig. 3-17A, the junction where N and P are connected absorbs heat. The opposite ends become hot and give off heat. This is the Peltier effect. A single junction produces only a small cooling effect, so that several N-P paired junctions are connected in series to produce significant cooling. See Fig. 3-17B. Groups of modules may be connected together in parallel, to increase the capacity still further. (See Chapter 6 for series-parallel connections.)

A thermostat inside the refrigerated space controls the current flow through the transformer-rectifier, which supplies a controlled dc current to the modules. In this way, the temperature inside the refrigerator is controlled.

There are no moving parts in this refrigerator. Aside from the construction of the modules, it is quite simple. Thermal efficiency is low. That is, the amount of refrigerating effect obtained for the electrical energy spent is less than with a conventional compressor type refrigeration system.

By reversing the direction of the flow of current through a thermoelectric device, the hot and cold surfaces will be reversed. Thus, the same device can be used for both heating and cooling an insulated space.

One application of this thermoelectric device has been in the air conditioning and heating of nuclear submarines. The thermoelectric device is also used extensively to control temperatures in electronic equipment (computers, aerospace devices, and so forth).

Refer to Chapter 17 for further technical information concerning thermoelectric refrigeration and air conditioning devices.

3-18 DRY ICE REFRIGERATION

Dry ice is solid carbon dioxide. It may be pressed into various sizes and shapes, blocks, or slabs. As it absorbs heat, it changes directly from a solid to a vapor. It does not go through the liquid state. This change, from solid to vapor, is called sublimation. At atmospheric pressure, solid carbon dioxide vaporizes at $-109\,°F$ ($-78\,°C$).

Fig. 3-18A illustrates a common method of using dry ice as a frozen food refrigerating device.

Dry ice (see dark blue) is usually packed with frozen food cartons either beside or on top of the food packages. Carbon dioxide, as it changes to a vapor, keeps the food frozen. The dry vapor tends to replace the atmospheric air in the container or cabinet which helps to preserve the food.

A device has been developed which uses dry ice for refrigerating materials carried on aircraft. See Fig. 3-18B. A closed refrigerating circuit, containing a common refrigerant, is connected to an evaporator in the space to be refrigerated and a condenser located in an insulated bin. The bin holds dry ice pellets.

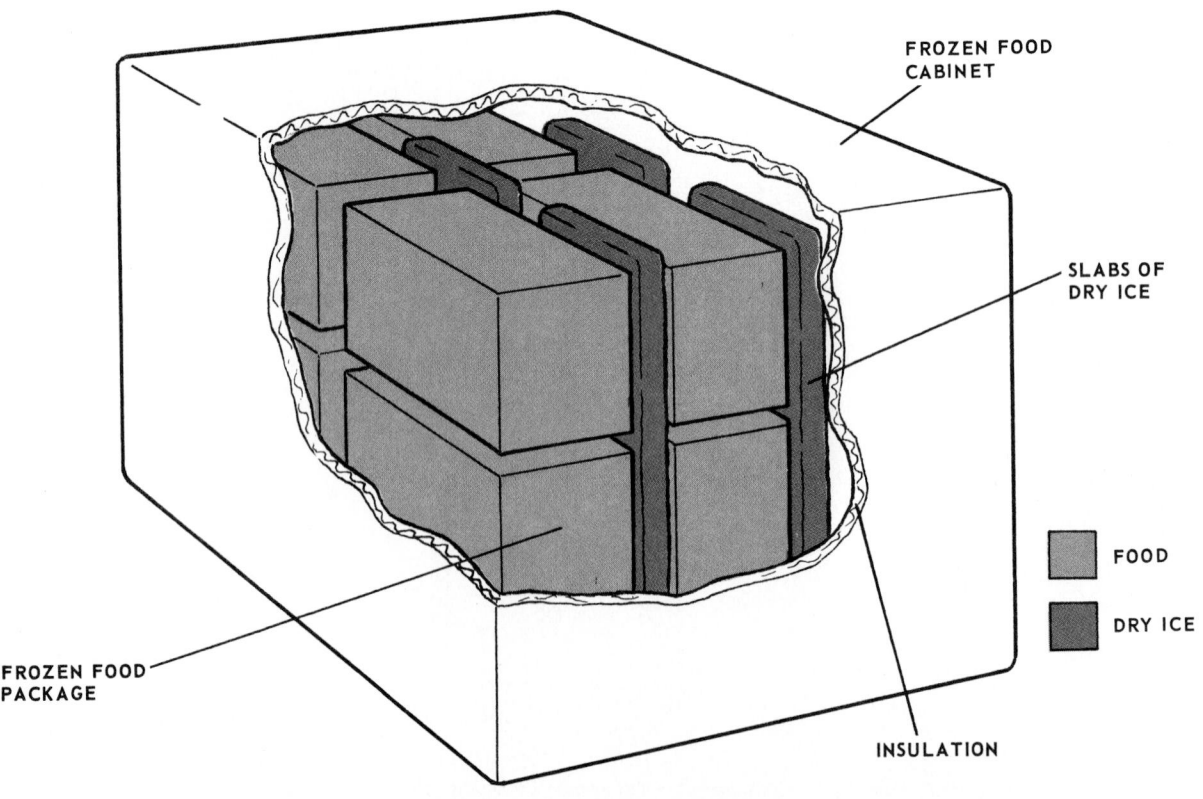

FROZEN FOOD CABINET

SLABS OF DRY ICE

FOOD

DRY ICE

FROZEN FOOD PACKAGE

INSULATION

Fig. 3-18A. Dry ice frozen food container.

In operation, the very low temperature in the condenser (−109 °F or −78 °C) causes refrigerant vapor entering the condenser to condense quickly to a liquid. This liquid flows by gravity into the evaporator. There, it absorbs heat as it vaporizes and flows upward into the condenser. From here the cycle is repeated.

A thermostatically operated control valve, located in the liquid line, controls the flow of refrigerant into the evaporator. This device is illustrated in Fig. 3-18B.

Dry ice is usually stored in heavily insulated cabinets. Never handle it with bare hands. It will cause instant freeze burns. Always wear heavy gloves.

3-19 AN INTERMITTENT ABSORPTION SYSTEM

The intermittent absorption system uses a generator charged with water and ammonia. A heat source, usually a kerosene flame, heats this solution in the generator. The ammonia becomes vaporized and is driven off.

A condenser, at the top of the system, condenses the ammonia vapor into a liquid. The liquid flows, by gravity, into the liquid receiver and then into the evaporator. During the generating cycle, as explained above, little or no refrigerating effect is taking place. As the system cools, pressure drops, causing the liquid ammonia in the evaporator to boil and absorb heat. The cycle is completed when vaporized ammonia is reabsorbed in the generator.

Fig. 3-19A diagrams the generating cycle. In operation, the kerosene burner tank is filled with just enough kerosene for one cycle. This usually is once a day. The burner is filled and lighted. It heats the water and ammonia mixture (see reddish-brown) in the generator. The ammonia vapor (see light red) is driven off through the tube, A, up to the condenser, C, where the ammonia gas is cooled and condensed

to liquid ammonia (see dark red). The liquid flows into the receiver.

When the kerosene has all been burned (usually from 20 to 40 minutes), the generating cycle ends. The refrigeration cycle now begins. See Fig. 3-19B.

The pressure in the system drops as the water cools and absorbs ammonia vapor. Liquid ammonia (see dark blue) flows into the evaporator, begins to evaporate, and cools it. Evaporated ammonia (see light blue) flows back through the tube, B, and is again absorbed by the water in the generator. Refrigeration continues, usually until the next firing of the kerosene burner.

This type of refrigerating system is quite simple. The piping is welded steel, since the pressures on the generating cycle are quite high. The refrigerating ability is quite good. Kerosene flame heated absorption refrigerators are popular in areas where electric power is not available.

3-20 A CONTINUOUS CYCLE ABSORPTION SYSTEM

The continuous absorption type of cooling unit is operated by the application of a limited amount of heat furnished by gas, electricity, or kerosene. No moving parts are employed. The operation of the refrigerating mechanism is based on Dalton's Law. See Para. 1-58.

The unit consists of four main parts—the boiler, condenser, evaporator, and absorber. See Fig. 3-20.

When the unit operates on kerosene or gas, the heat is supplied by a burner which is fitted underneath the central tube (A). When the unit operates on electricity, the heat is supplied by an element inserted in the pocket (B).

The unit charge consists of a quantity of ammonia, water, and hydrogen at a sufficient pressure to condense ammonia at room temperature. When heat is supplied

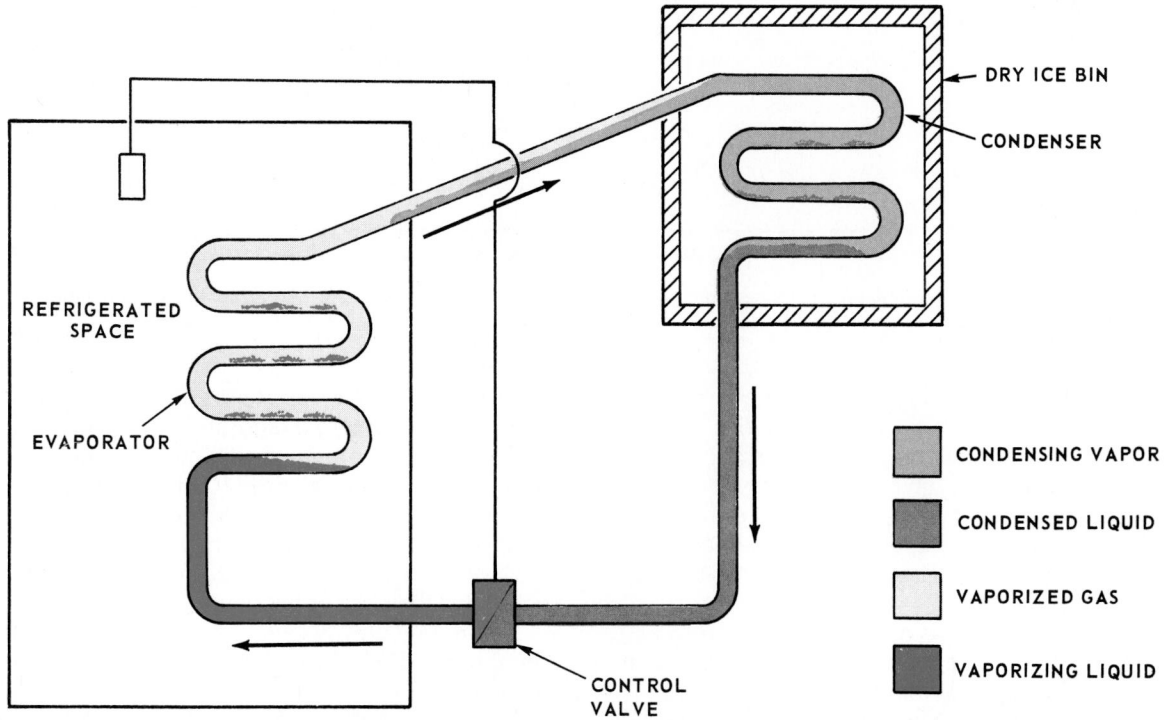

DRY ICE BIN

CONDENSER

REFRIGERATED SPACE

EVAPORATOR

CONTROL VALVE

☐ CONDENSING VAPOR

☐ CONDENSED LIQUID

☐ VAPORIZED GAS

☐ VAPORIZING LIQUID

Fig. 3-18B. Dry ice refrigerator does not require a compressor.

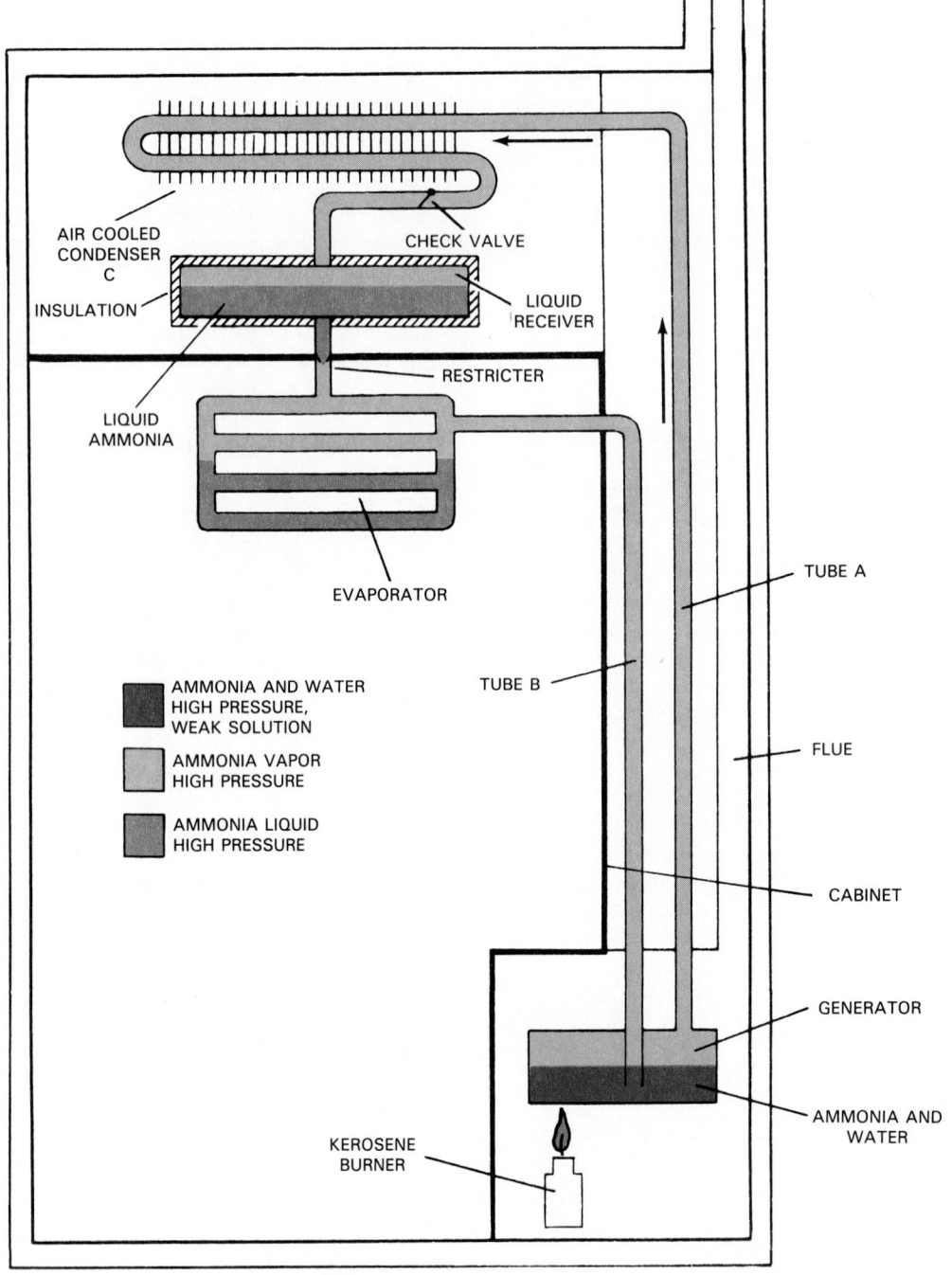

Fig. 3-19A. Intermittent absorption system during generating cycle. System is under high or condensing pressure.

to the boiler system, bubbles of ammonia gas are produced which rise and carry with them quantities of weak ammonia solution through the syphon pump (C). This weak solution passes into tube (D), while the ammonia vapor passes into the vapor pipe (E) and on to the water separator. Here any water vapor is condensed and runs back into the boiler system, leaving the dry ammonia vapor to pass to the condenser.

Air circulating over the fins of the condenser removes heat from the ammonia vapor, and it condenses into liquid ammonia. It then flows into the evaporator.

The evaporator is supplied with hydrogen. The hydrogen passes across the surface of the ammonia and

lowers the ammonia vapor pressure sufficiently to allow the liquid ammonia to evaporate. The evaporation of the ammonia extracts heat from the evaporator which, in turn, extracts heat from the food storage space, lowering the temperature inside the refrigerator.

The mixture of ammonia and hydrogen vapor passes from the evaporator to the absorber. Entering the upper portion of the absorber is a continuous trickle of weak ammonia solution fed by gravity from the tube (D). This weak solution, flowing down through the absorber, comes into contact with the mixed ammonia and hydrogen gases. This readily absorbs the ammonia from the mixture, leaving the hydrogen free to rise through the absorber coil and to return

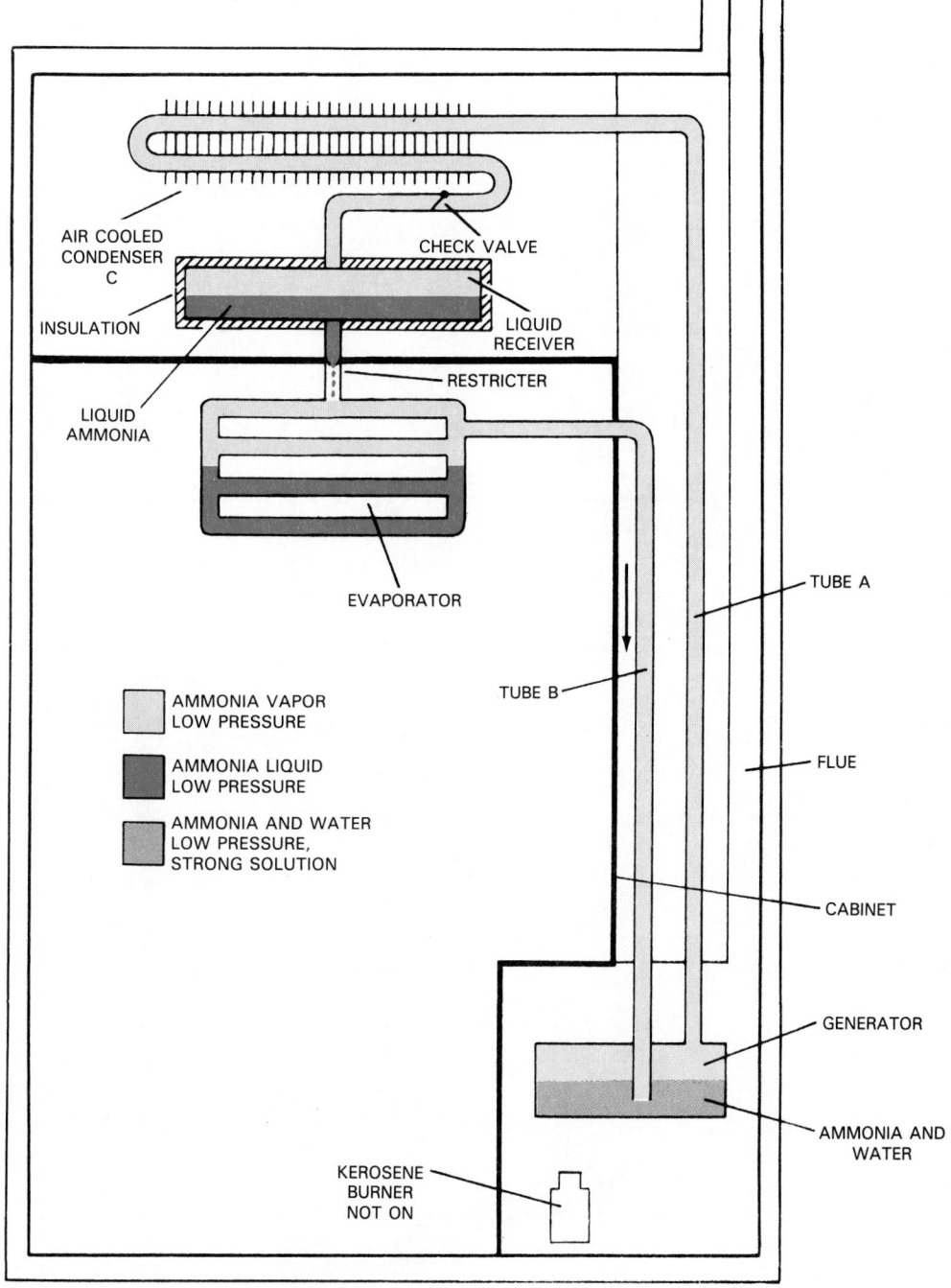

Fig. 3-19B. Intermittent absorption system during refrigerating cycle. System is under low or refrigerating pressure.

to the evaporator. The hydrogen thus circulates continuously between the absorber and the evaporator.

The strong ammonia solution produced in the absorber flows down to the absorber vessel and on to the boiler system, thus completing the full cycle of operation.

This cycle operates continuously as long as the boiler is heated. A thermostat which controls the heat source regulates the temperature of the refrigerated space.

Since the refrigerant is ammonia, it can produce quite low temperatures. Most systems require electrical devices, so both gas and electricity must be supplied. With the exception of the thermostatic controls and (in some cases) fans, there are no moving parts.

This refrigerating device is widely used in domestic refrigerators, recreation vehicles, and in year-around air conditioning of both homes and larger buildings.

Service is usually quite simple. The burner and stack must be kept clean. The refrigerator should be carefully leveled before being placed in operation.

Modern absorption systems are illustrated in Chapter 16.

3-21 SOLID ABSORBENT REFRIGERATION

Various kinds of solid absorbent refrigerators have been developed. All have depended on the original Faraday experiment.

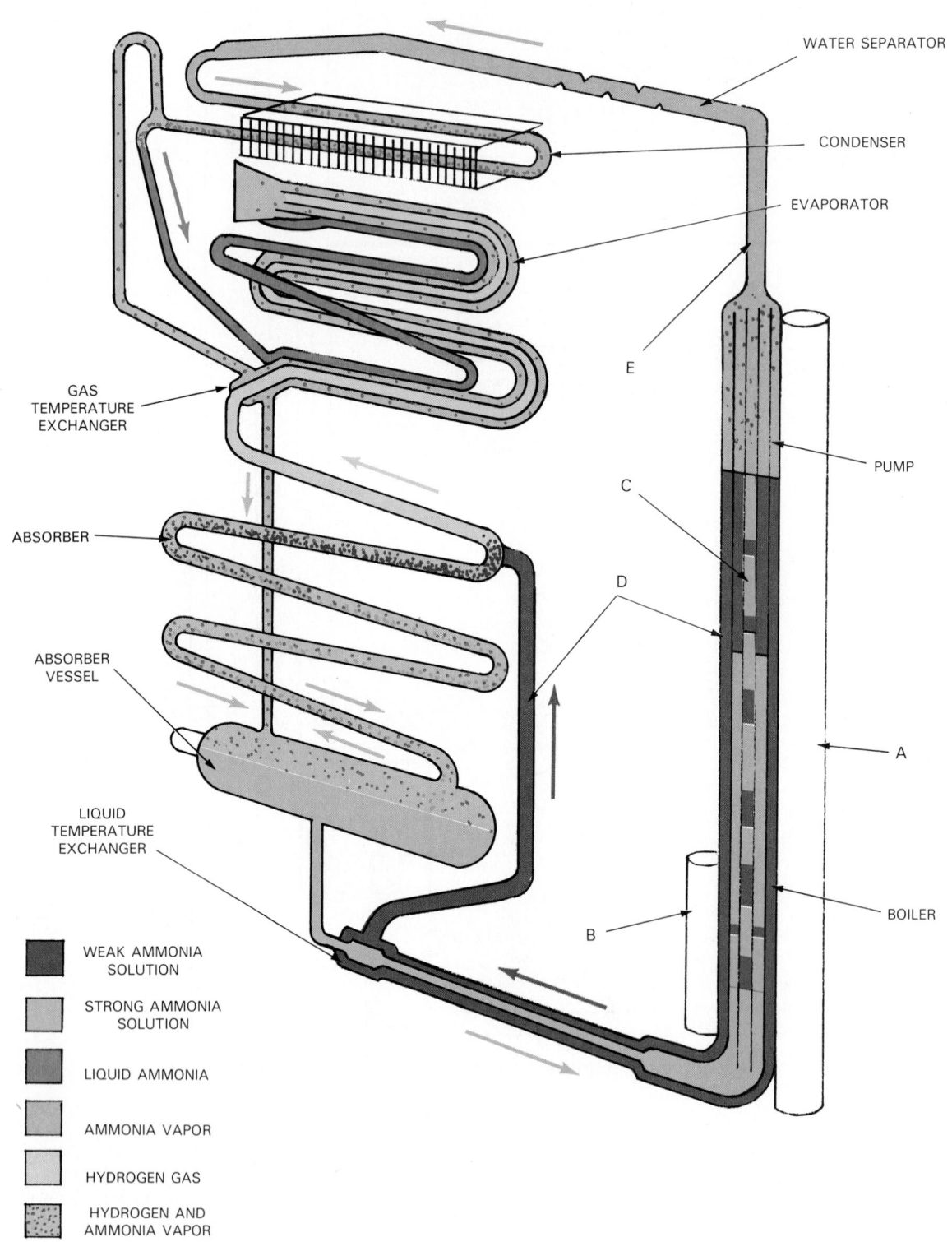

WATER SEPARATOR

CONDENSER

EVAPORATOR

E

PUMP

C

D

A

B

BOILER

GAS
TEMPERATURE
EXCHANGER

ABSORBER

ABSORBER
VESSEL

LIQUID
TEMPERATURE
EXCHANGER

WEAK AMMONIA
SOLUTION

STRONG AMMONIA
SOLUTION

LIQUID AMMONIA

AMMONIA VAPOR

HYDROGEN GAS

HYDROGEN AND
AMMONIA VAPOR

Fig. 3-20. Continuous cycle absorption system. (Electrolux AB)

In 1824, Michael Faraday tried to liquefy certain "fixed" gases—gases which certain scientists believed could exist only in vapor form. Among them was ammonia, then regarded as a "fixed" gas.

Faraday knew that silver chloride, a white powder, could absorb large amounts of ammonia vapor. He exposed silver chloride to dry ammonia vapor.

When the powder had absorbed all of the vapor it would take, he sealed the ammonia-silver chloride compound in a test tube which was shaped like an inverted "V". (See

Fig. 3-21, view A.)

He then heated the end of the tube containing the powder (see dark red), and at the same time cooled the opposite end with water. The heat released ammonia vapor. Drops of a colorless liquid soon began to appear in the cool end of the tube. It was liquid ammonia.

Faraday continued the heating process until he had enough liquid ammonia for his purpose. Then, he took away the heat, removed the cooling water, and watched the newly discovered substance.

Moments later, Faraday saw something unusual. The liquid ammonia, instead of remaining quietly in the sealed test tube, began to bubble and then to boil violently, Fig. 3-21, view B. The liquid was rapidly changing back into a vapor and the vapor was being reabsorbed by the powder.

Touching the end of the tube containing the boiling liquid, Faraday found it intensely cold. Ammonia, in changing from liquid to vapor form, had removed heat. It took this heat from the nearest thing at hand—the test tube itself.

At one time, many refrigerating cycles used this principle. These are not in common use at present. However, cooling mechanisms have been developed on this principle. They use water as the refrigerant and lithium bromide or lithium chloride as the absorbent. See Chapter 16.

3-22 SOPHISTICATED COMMERCIAL SYSTEMS

Up to now, the compression type refrigerating systems described have been quite simple. Different conditions and refrigeration requirements require accessory (add-on) devices. Pressure regulators, vibration dampeners, crankcase heaters, and other mechanisms make the refrigerating systems more efficient and safer. Fig. 3-22 illustrates a small commercial type refrigerating system using a variety of accessory devices.

Beginning with the evaporator in the top (warmer) cabinet, refrigerant evaporates (see light blue) and flows back through the suction line toward the motor compressor. The suction line then enters an evaporator pressure regulator. From there it leads to the suction line accumulator. Any liquid refrigerant which may come from the evaporator will stay here and evaporate, preventing it from slugging into or entering the compressor.

The vapor then goes through a suction line filter-drier, which traps any moisture or solid impurities.

A compressor pressure regulator protects the compressor from excessive low-side pressures.

The suction line vapor then enters a vibration dampener. This is a flexible connection on the low side between the motor compressor and the suction line. The sensing element for the motor compressor control is also attached to the suction line.

A service valve is located at the entrance to the low side of the motor compressor. Along with the service valve on the high side, this valve makes servicing the motor compressor easy.

A crankcase heater keeps refrigerant from liquefying in the motor compressor during the off cycle when the unit is operating in a cold space.

From the high-side service valve, the compressed vapor (see light red) enters an oil separator. The oil which the separator has removed from the high-pressure refrigerant is returned to the compressor crankcase.

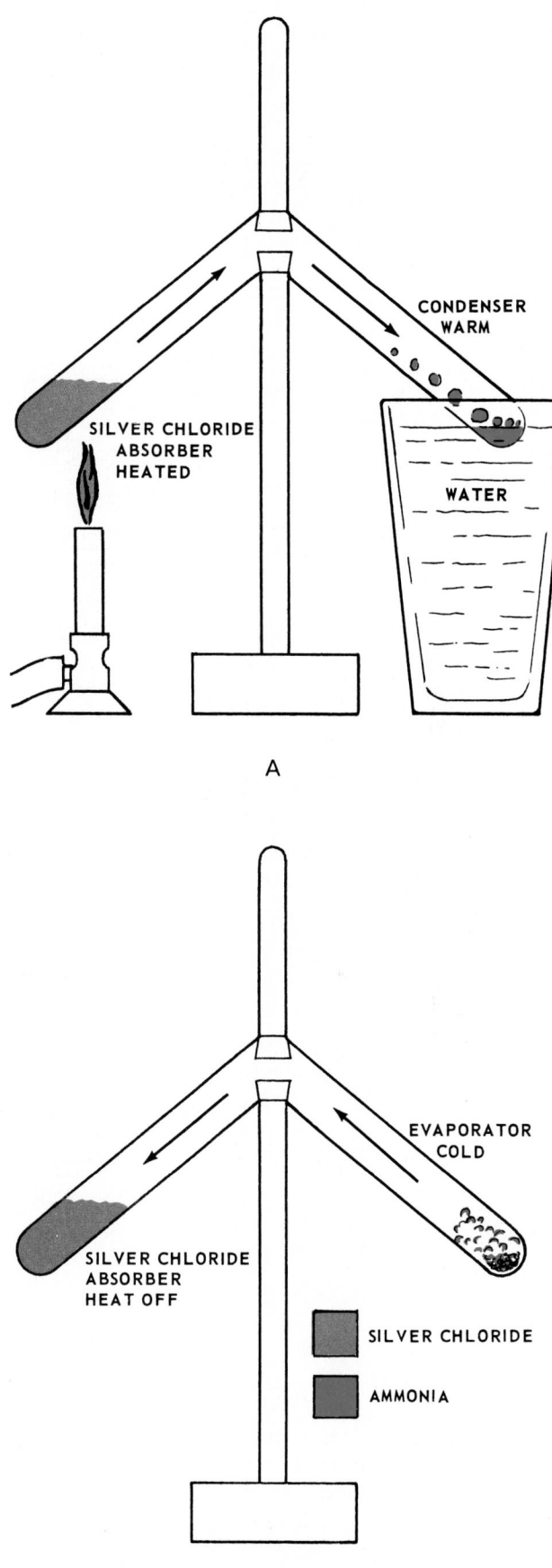

Fig. 3-21. Solid absorbent refrigerator principle in experiment done by Michael Faraday.

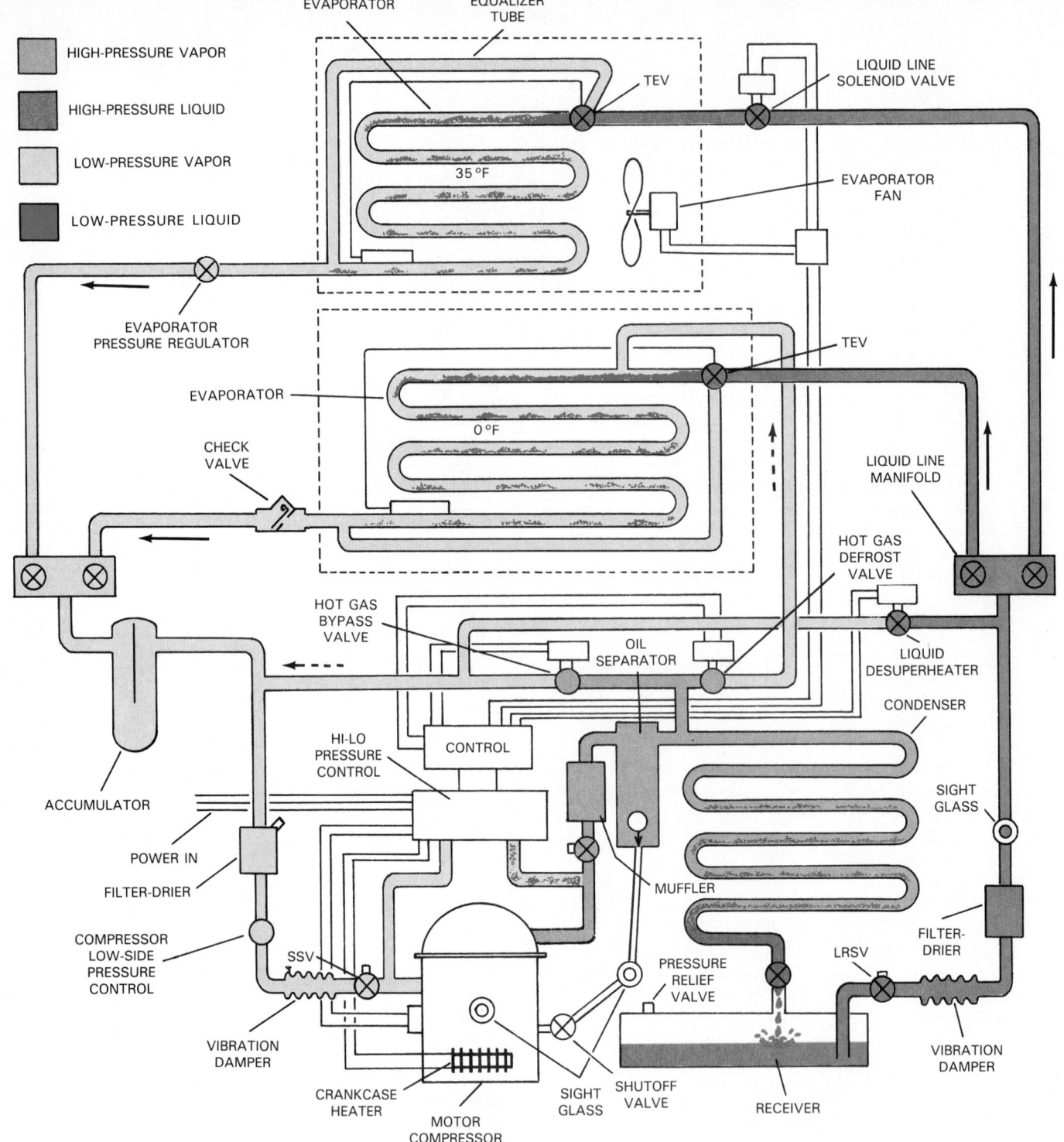

Legend:
- HIGH-PRESSURE VAPOR
- HIGH-PRESSURE LIQUID
- LOW-PRESSURE VAPOR
- LOW-PRESSURE LIQUID

EVAPORATOR

EQUALIZER TUBE

LIQUID LINE SOLENOID VALVE

TEV

EVAPORATOR FAN

35°F

EVAPORATOR PRESSURE REGULATOR

EVAPORATOR

TEV

0°F

LIQUID LINE MANIFOLD

CHECK VALVE

HOT GAS DEFROST VALVE

HOT GAS BYPASS VALVE

OIL SEPARATOR

LIQUID DESUPERHEATER

CONDENSER

SIGHT GLASS

ACCUMULATOR

HI-LO PRESSURE CONTROL

CONTROL

MUFFLER

POWER IN

FILTER-DRIER

COMPRESSOR LOW-SIDE PRESSURE CONTROL

SSV

PRESSURE RELIEF VALVE

LRSV

FILTER-DRIER

VIBRATION DAMPER

CRANKCASE HEATER

MOTOR COMPRESSOR

SIGHT GLASS

SHUTOFF VALVE

RECEIVER

VIBRATION DAMPER

Fig. 3-22. Accessory parts on this commercial system make unit work better. It is also easier to service.

From the oil separator, the high-pressure vapor enters the condenser. The condenser has a head pressure control. When head pressure gets too high, it shuts off the system.

A service valve is placed in the line between the condenser and the liquid receiver. The liquid receiver is a reservoir for liquid refrigerant (see dark red).

Another service valve is located at the outlet of the liquid receiver. This makes it possible to remove the receiver from the system or store the refrigerant in the receiver during service operations.

The line leaving the liquid receiver is also fitted with a vibration dampener. This flexible tube stops carry-over of vibration to other parts of the system.

As the liquid moves on, it passes through a filter-drier which helps keep the refrigerant clean and dry. Then a moisture and liquid indicator allows visual inspection to see if enough refrigerant is flowing. At the same time, its color indicates presence of any moisture in the refrigerant.

Sometimes the motor compressor overheats. This can occur when the suction gas temperature is too high. The liquid desuperheater valve permits light injections of refrigerant to flow into the low side of the system, so that the suction gas is cooled immediately.

A manifold with hand valves allows the liquid to pass into one of the two evaporators. Then the refrigerant flows through a solenoid valve to the top evaporator. This makes it possible to automatically control the flow into the evaporator.

From here, the liquid refrigerant flows into a thermostatic expansion valve. This valve regulates the rate of flow (see dark blue) depending upon the temperature and pressure of the refrigerant as it leaves the evaporator.

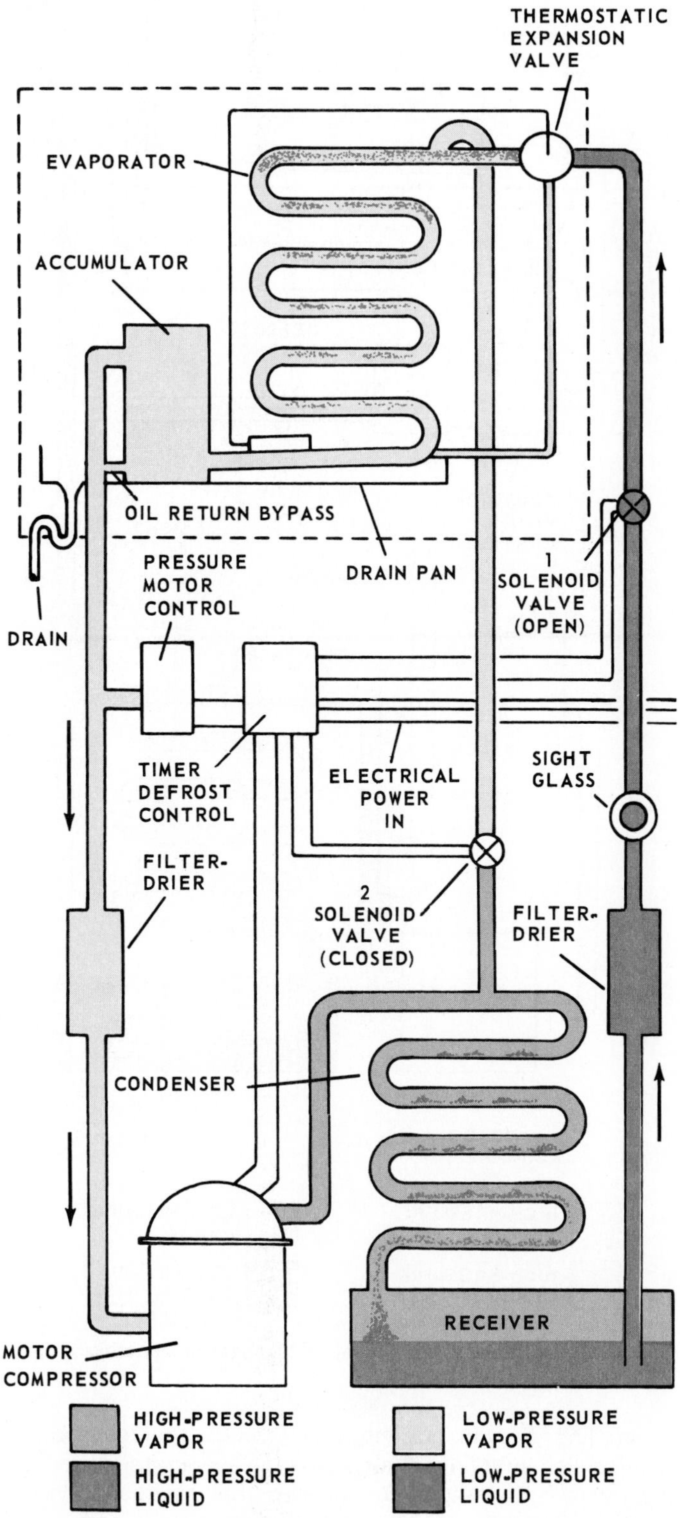

The second evaporator is fitted with an electrically operated defrost control device. When opened, it allows hot compressed vapor to enter the evaporator and flow back to the compressor without going through the expansion valve. This heats the evaporator quickly and any frost accumulation on it will be quickly melted.

A hot gas bypass solenoid valve is used to allow hot refrigerant vapor to enter the suction line in case the suction line gets too cold and may allow liquid refrigerant to enter the compressor.

This brief description of some of the accessories and their purpose may help in understanding the commercial systems as described later in the text.

3-23 HOT GAS DEFROST

In the hot gas defrost system, a timing mechanism directs hot high-pressure vapor (see light red) through the evaporator to remove frost and ice. Fig. 3-23A shows how it operates during the refrigerating cycle; Fig. 3-23B illustrates defrost cycle operation.

Two solenoid valves in the refrigerant circuit control the system to provide either the refrigerating cycle or the defrost cycle. During the refrigerating cycle, as in Fig. 3-23A, solenoid valve No. 1 is open. The refrigerator is operating normally for a refrigerator using a thermostatically controlled expansion valve.

The liquid refrigerant (see dark red) flows from the liquid receiver up through the liquid line, through solenoid valve No. 1, through the thermostatic expansion valve, and into the evaporator. It evaporates under low pressure, absorbs heat, and returns as a vapor (see light blue) first through the accumulator and then through the suction line back to the compressor. From the compressor, the hot compressed vapor (see light red) is forced into the condenser. Its heat is removed and it is condensed back into a liquid.

During the defrost cycle, Fig. 3-23B, solenoid valve No. 1 is closed. Solenoid valve No. 2 is open. Since solenoid valve No. 1 is closed, no liquid refrigerant is flowing through the thermostatic expansion valve into the evaporator. Since solenoid valve No. 2 is open, the hot compressed refrigerant vapor (see light red) flows through it directly into the evaporator. It passes through the evaporator, through the accumulator, and back through the suction line to the suction side of the hermetic (airtight) compressor. The hot vapor (see light red), as it passes through the evaporator, melts the ice from the evaporator surface.

The vaporized refrigerant (see light blue) picks up some heat passing through the suction line. The heat of compression, as it passes through the compressor, raises the temperature to such a level that it continues to heat the evaporator and remove the frost. Little or no condensation of vaporized refrigerant takes place during this cycle.

3-24 ELECTRIC DEFROST

Electric heating elements placed alongside the evaporator surfaces heat up to melt the frost and ice buildup from the evaporator. A timer or control mechanism operates the heater during the time that the refrigerating mechanism is on the off cycle. Fig. 3-24A shows the refrigerating cycle; Fig. 3-24B shows the defrost cycle.

In Fig. 3-24A, the electric heating mechanism is in the refrigerating part of the cycle. Liquid refrigerant is

Fig. 3-23A. Normal refrigerating cycle of hot gas defrost system.

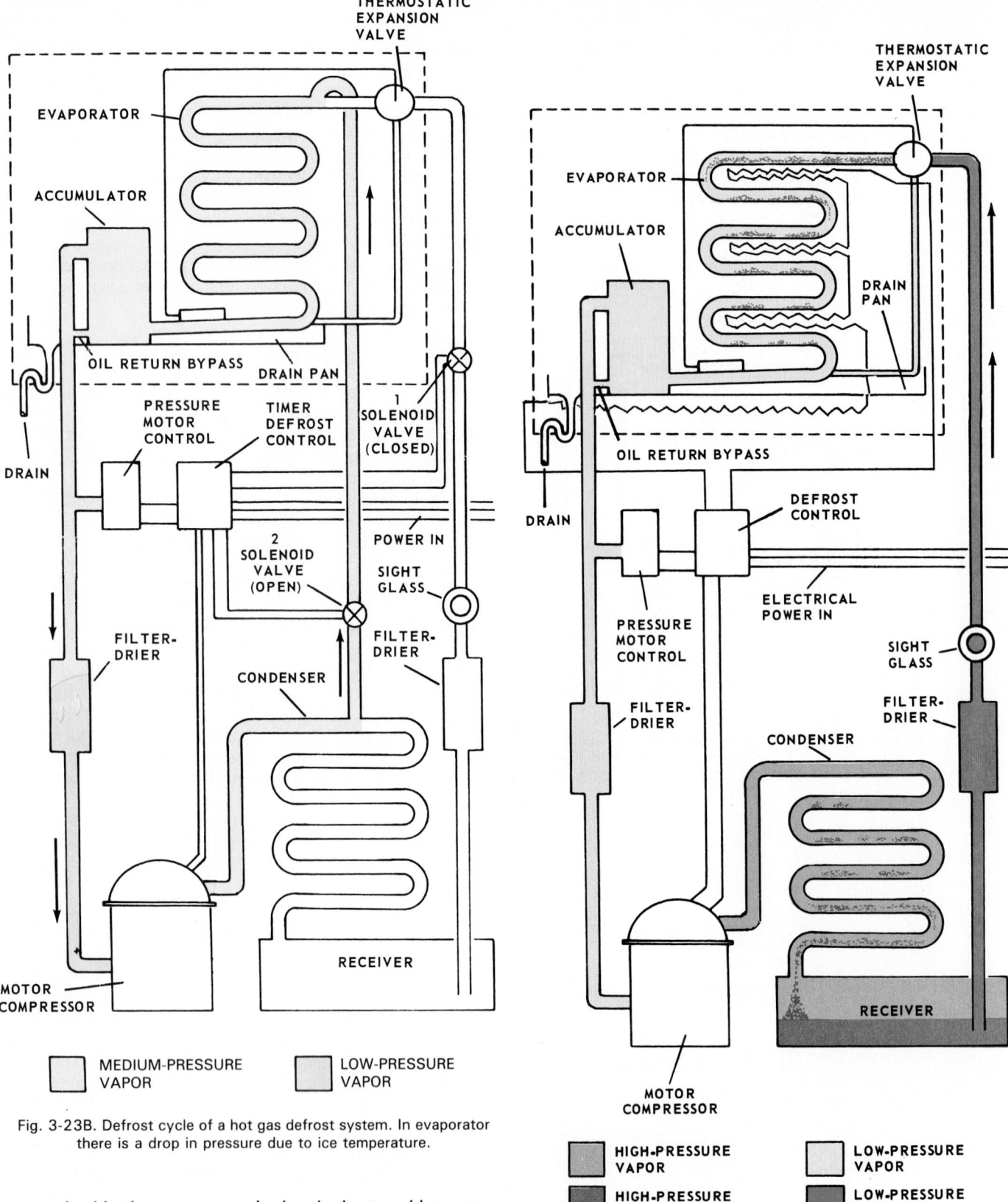

THERMOSTATIC EXPANSION VALVE

EVAPORATOR

ACCUMULATOR

OIL RETURN BYPASS

DRAIN PAN

DRAIN

PRESSURE MOTOR CONTROL

TIMER DEFROST CONTROL

1 SOLENOID VALVE (CLOSED)

2 SOLENOID VALVE (OPEN)

POWER IN

SIGHT GLASS

FILTER-DRIER

FILTER-DRIER

CONDENSER

MOTOR COMPRESSOR

RECEIVER

☐ MEDIUM-PRESSURE VAPOR   ☐ LOW-PRESSURE VAPOR

Fig. 3-23B. Defrost cycle of a hot gas defrost system. In evaporator there is a drop in pressure due to ice temperature.

THERMOSTATIC EXPANSION VALVE

EVAPORATOR

ACCUMULATOR

DRAIN PAN

OIL RETURN BYPASS

DRAIN

DEFROST CONTROL

PRESSURE MOTOR CONTROL

ELECTRICAL POWER IN

FILTER-DRIER

SIGHT GLASS

FILTER-DRIER

CONDENSER

MOTOR COMPRESSOR

RECEIVER

☐ HIGH-PRESSURE VAPOR   ☐ LOW-PRESSURE VAPOR

☐ HIGH-PRESSURE LIQUID   ☐ LOW-PRESSURE LIQUID

Fig. 3-24A. An electric defrost system during the refrigerating cycle.

vaporized in the evaporator. It absorbs heat and becomes a vapor. While in the evaporator, it flows through an accumulator and passes on to the suction line back to the compressor. In the compressor, it is compressed to a high pressure, high temperature and flows into the condenser. Here the heat of vaporization is removed and the refrigerant returns to a liquid flowing into the liquid receiver. From here the cycle is repeated.

In Fig. 3-24B, the same system is in the defrost cycle. The compressor is stopped and the defrost control mech-

anism lets electric current flow through the resistance heating elements (in red) alongside the evaporator surface. Heat warms the evaporator surfaces until the frost and ice are melted and the moisture empties into a drain pan. The drain tubes go to the building drain.

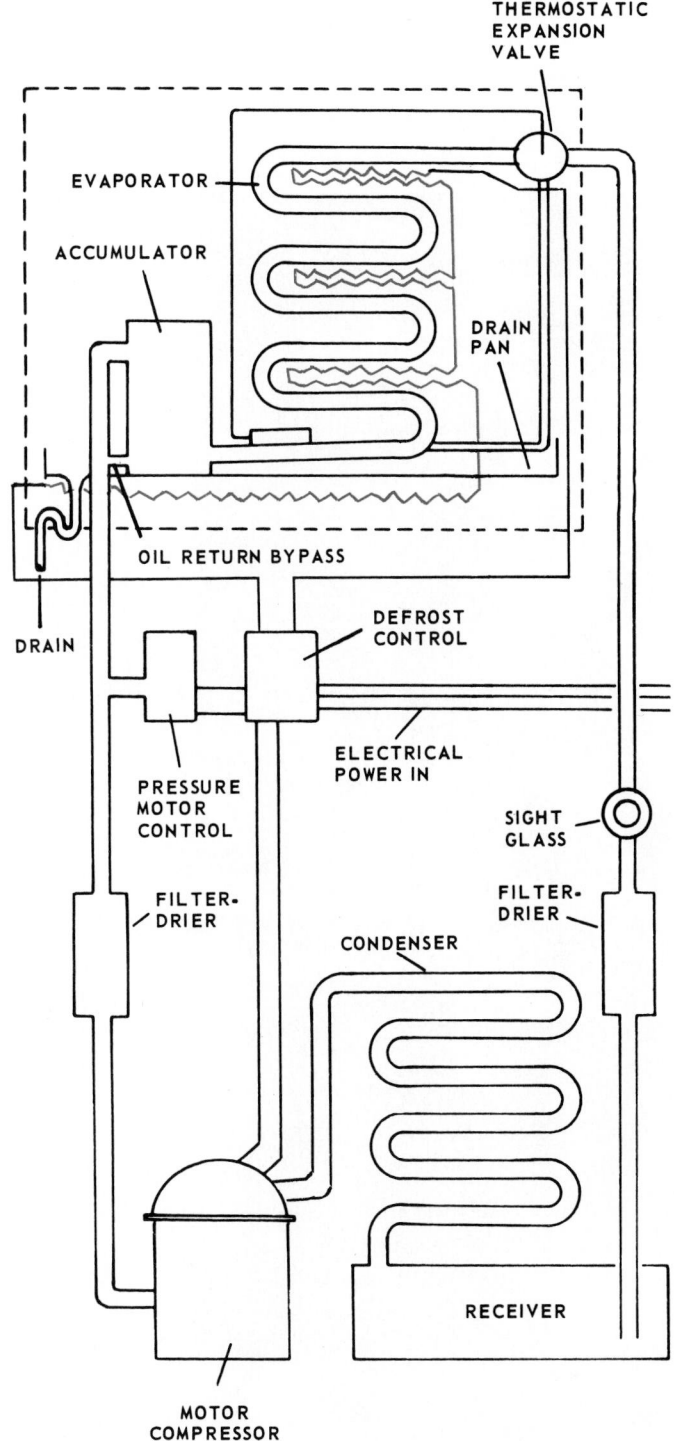

THERMOSTATIC
EXPANSION
VALVE

EVAPORATOR

ACCUMULATOR

DRAIN
PAN

OIL RETURN BYPASS

DRAIN

DEFROST
CONTROL

ELECTRICAL
POWER IN

PRESSURE
MOTOR
CONTROL

SIGHT
GLASS

FILTER-
DRIER

FILTER-
DRIER

CONDENSER

RECEIVER

MOTOR
COMPRESSOR

Fig. 3-24B. An electric defrost system during the defrost cycle. Note that the refrigerating unit is not running; therefore there is no refrigerant flow.

The operation of the resistance units is usually timed to control both the frequency and the duration of the electric heating. This timing provides for adequate frost removal and the system is helped to operate efficiently with little or no frost on the evaporator surfaces.

3-25 TEST YOUR KNOWLEDGE

1. List some advantages of the old icebox type of refrigeration.
2. What is the principle upon which the "desert bag" water cooling device operates?
3. What are some of the advantages of the capillary tube type refrigerant control?
4. What causes the needle valve to open in an automatic expansion valve refrigerant control?
5. What causes the needle valve to open in the thermostatically controlled expansion valve?
6. What controls the cabinet temperature in a system which uses a thermostatically controlled expansion valve?
7. What substances are most used in an intermittent absorber-type absorption refrigerator?
8. What did Faraday discover in his famous experiment?
9. What is the chief advantage of a compound system?
10. What is the chief advantage of a cascade system?
11. In a thermoelectric refrigeration system, how is heat carried out of the space to be refrigerated?
12. How is modulation (varying capacity) in refrigeration accomplished?
13. Why is it sometimes necessary to use a multiple evaporator system?
14. What is the basic principle behind the operation of most ice cube makers?
15. Name some advantages of the low-side float type of refrigerant control.
16. Is amount of refrigerant charged into system very important in a high-side float type refrigerating system?
17. What principle is used in the continuous type absorption refrigerator?
18. Where is "expendable refrigerant" cooling most used?
19. What is the chief disadvantage of thermoelectric refrigeration?
20. What are some of the advantages of dry ice refrigeration?
21. In what kind of refrigerating mechanism are semiconductors used?
22. What is the refrigerant most commonly used in "expendable refrigerant" cooling systems?
23. In "cascade" systems, are the refrigerating temperatures usually above or below 0 °F (− 18 °C)?
24. What is another name for an "open" refrigerating system?
25. Why is hydrogen used in some absorption systems?

Navstar satellite, designed to withstand the extremes of heat and cold in space, being tested in a Mark I chamber. The chamber was fitted with 500 lamps, each 1,000 watts bright, to simulate solar heat. The chamber walls are cooled with liquid nitrogen to simulate space temperatures as low as −300 °F (−184 °C). (Arnold Engineering Development Center, Arnold Air Force Base, TN)

Chapter 4

COMPRESSION SYSTEMS AND COMPRESSORS

After studying this chapter, the technician will be able to:
☐ State the five thermal laws relating to refrigeration.
☐ Explain the compression cycle for a domestic refrigerator.
☐ List the components of a refrigeration compression system.
☐ Explain the operation of each component of a compression system.
☐ Trace the flow of refrigerant through a complete refrigeration system.
☐ Name the two types of motor control and discuss their operation and purpose.
☐ Describe the five principal types of refrigerant controls and their operation.
☐ Name four different types of compressors.
☐ Explain how compressors operate.
☐ Identify the internal parts of a compressor.

4-1 LAWS OF REFRIGERATION

All refrigerating systems depend on five thermal laws:
1. Fluids absorb heat while changing from a liquid state to a vapor state and give up heat in changing from a vapor to a liquid.
2. The temperature at which a change of state occurs is constant during the change provided the pressure remains constant.
3. Heat flows only from a body which is at a higher temperature to a body which is at a lower temperature (hot to cold).
4. Metallic parts of the evaporating and condensing units use metals which have a high heat conductivity (copper, brass, aluminum).
5. Heat energy and other forms of energy are interchangeable. For example, electricity may be converted to heat; heat to electrical energy; and heat to mechanical energy.

4-2 COMPRESSION CYCLE

The compression cycle is so named because it is the compressor which changes the refrigerant vapor from low pressure to high pressure. This pumping causes the transfer of heat energy from the inside of the cabinet to the outside. Since the compression machine transfers heat from one place to another, it may also be called a heat pump.

A refrigerating system consists principally of a high-pressure side and a low-pressure side, Fig. 4-1.

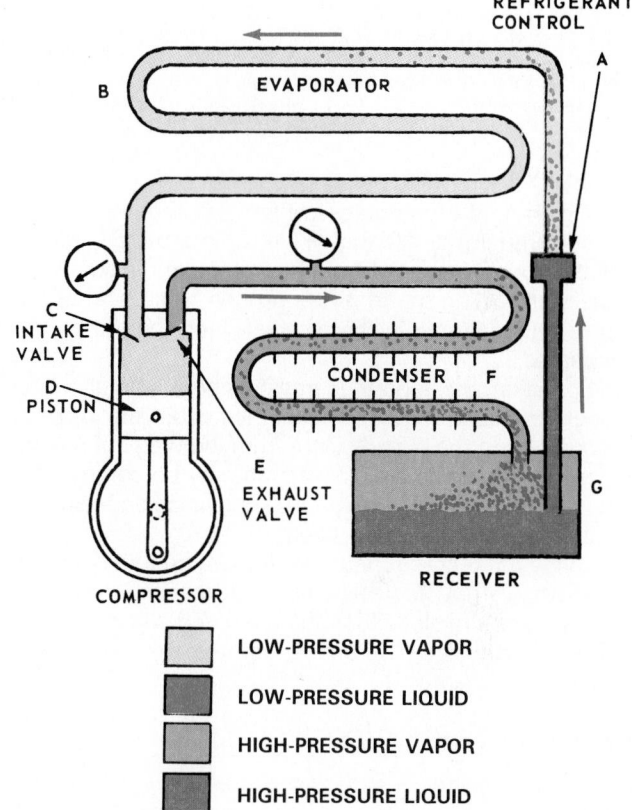

LOW-PRESSURE VAPOR

LOW-PRESSURE LIQUID

HIGH-PRESSURE VAPOR

HIGH-PRESSURE LIQUID

Fig. 4-1. Compression cycle showing the two pressure conditions. Low-pressure side extends from refrigerant control, A, through evaporator, B, to the compressor intake valve, C. High-pressure side begins in the cylinder above the piston, D, on the compression stroke. It extends from exhaust valve, E, through condenser, F, liquid receiver, G, and liquid line, to refrigerant control, A.

A refrigeration cycle follows these steps: From the liquid receiver, G, liquid refrigerant, at a high pressure, flows through the refrigerant control, A, (a pressure reducer). It moves into the evaporator, B. The evaporator is under a low pressure. Here the liquid refrigerant vaporizes (boils) and absorbs heat.

The vapor then flows into the compressor through the intake valve, C, back into the compressor cylinder. The piston, D, on the compression stroke, squeezes the vapor into a small space with an increase in temperature. Fig. 4-2 illustrates this principle.

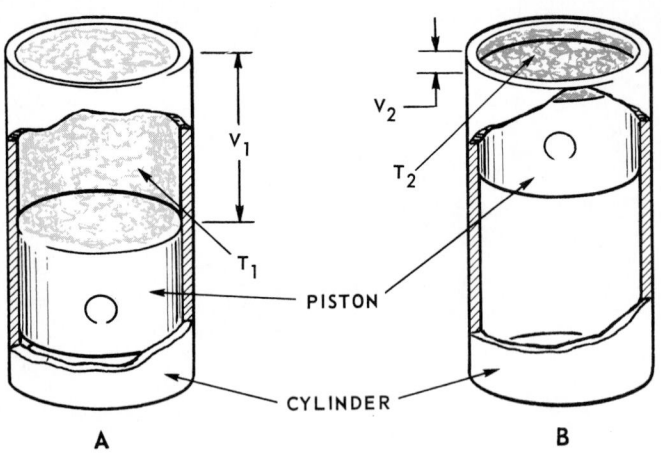

Fig. 4-2. Heat of the vapor compressed into a small space raises vapor temperature greatly. A—Volume V_1 = volume of vapor at the end of the intake stroke = 8 cu. in. (131 cm³). T_1 = temperature of vapor at end of intake stroke = 50 °F (10 °C). B—Volume V_2 = the volume of the vapor at the end of compression stroke = 1/2 cu. in. (8.2 cm³). T_2 = the temperature of the vapor at the end of the compression stroke = 250 °F (121 °C).

In Fig. 4-1, the compressed high-temperature vapor is pushed through the exhaust valve, E, into the condenser, F. In the condenser, heat from the refrigerant is passed on to the surrounding air. In giving up this heat, it returns to a liquid and is stored in the receiver. From here the cycle is repeated.

In operation, the apparatus transfers heat from one place to another place. That is, it takes heat from the inside of a refrigerator to the outside air or from the water of a water cooler to the outside air. This action may be compared to using a sponge to pick up water in one place and releasing it in another by squeezing it.

To have a transfer of heat, there must be a temperature difference. To get the temperature difference, there must be a low-pressure side (heat absorber) and a high-pressure side (heat dissipator). Various compression cycles are illustrated in Chapter 3.

4-3 OPERATION OF COMPRESSION CYCLE

Fig. 4-3 illustrates a typical compression cycle as used in a domestic refrigerator. It has certain necessary parts which will be explained.

In any compression refrigeration system, there are two different pressure conditions. One is called the low side and the other the high side. The evaporator is in the low side. Heat is absorbed in the low side.

The accumulator, suction line, and entrance to the compressor suction valve are also on the low side.

The condenser is in the high side, where the heat is released from the refrigerant. The compressor exhaust valve, liquid receiver (if used), liquid line filter-drier, liquid line, and the refrigerant control are also on the high side.

A thermostat maintains correct operating temperature by controlling the motor electrical circuit.

4-4 TEMPERATURE AND PRESSURE CONDITIONS IN THE COMPRESSION CYCLE

When the compressor starts, it moves molecules of refrigerant from the low-pressure side to the high-pressure side without much difficulty. After this change, the molecules are not moving about much faster. See Fig. 4-4, view A. These molecules of refrigerant enter the condenser from the compressor through opening at 1. The temperatures are the same (70 °F) inside and out.

Pressure is the sum of the bombarding molecules and temperature is the speed of molecular motion. (How fast the molecules move to and fro.)

It is necessary then to speed up the molecules. This will raise their temperature to a point at which they can give up heat to surrounding cooling surfaces (air and water).

The longer the compressor runs, the more vapor molecules it squeezes into the condenser. With each stroke, the pressure and temperature increase since more molecules are hitting the sides of the container. The compressor piston, pushing the vapor molecules against the higher pressure, hits them harder, speeding them up (increasing temperature).

During compression, the pressure increases (due to Boyle's Law, Chapter 30). At the same time, the temperature increases (explained by Charles' Law, Chapter

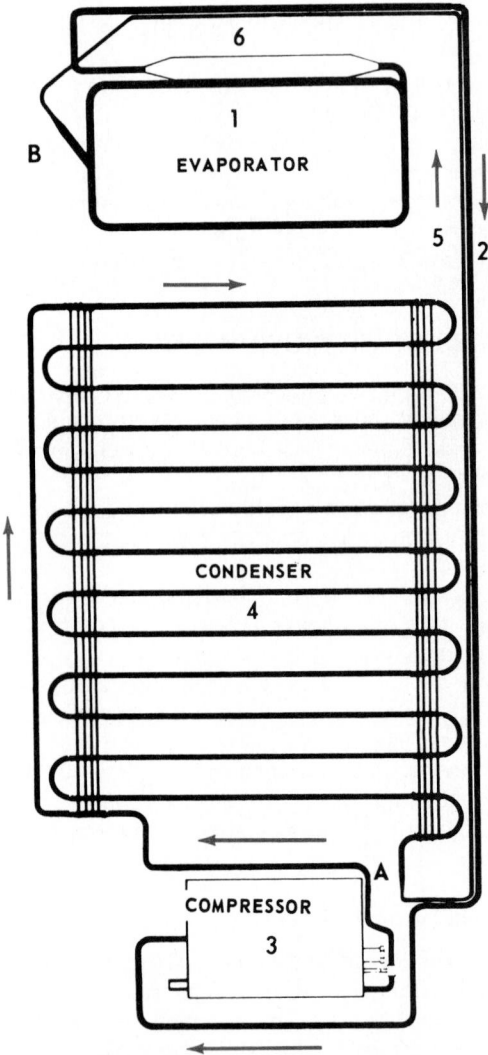

Fig. 4-3. Compression cycle showing the flow of refrigerant. 1—Evaporator. 2—Suction line. 3—Compressor. 4—Condenser. 5—Capillary tube, A to B. 6—Accumulator. (Hotpoint Div., General Electric Co.)

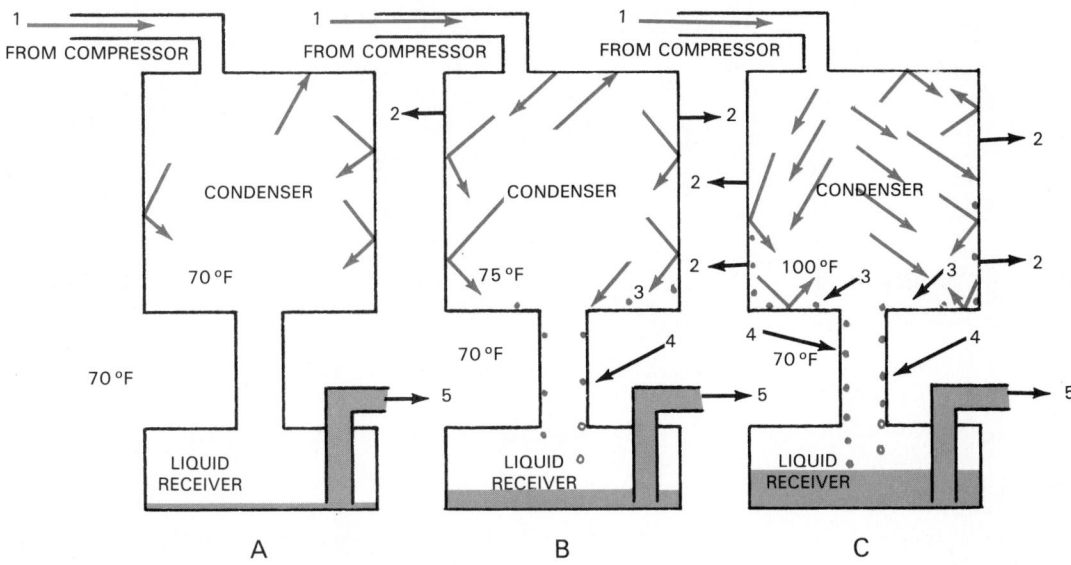

Fig. 4-4. Refrigerant condition as it changes from vapor to liquid in condenser. At A, condensing unit is just starting; B, unit has been operating long enough to condense some refrigerant vapor. At C, unit is in state of equilibrium (balance), heat is being removed and vaporized refrigerant is being condensed at same rate it is being pumped into condenser. A, B, C—1. Vapor enters under pressure. B—2. Heat moving from condenser (small amount). B—3. Vapor losing heat and condensing to liquid (small amount). B—4. Condensed refrigerant enters liquid receiver (small amount). C—2. Heat moving from condenser (large amount). C—3. Condensed refrigerant droplets (large amount). C—4. Condensed refrigerant flowing into liquid receiver (larger amount). Refrigerant liquid line to refrigerant control is shown at 5.

30) until the vapor temperature is higher than the temperature of the condenser cooling medium.

The higher temperature, as shown at B, Fig. 4-4, causes a flow of heat to the surrounding metal and air. Heat now moves from the vaporized refrigerant as shown at B2 to the cooling medium.

This cooling continues until enough heat loss makes some vapor molecules become liquid molecules as in B3. As these collect, they flow into the liquid receiver. See B4.

The temperature and pressure will continue to rise until a balance is reached. That is, just as many vapor molecules condense into a liquid as the compressor pumps into the condenser, Fig. 4-4C.

If anything changes this balance, the condensing pressure and temperature will adjust accordingly. For example, if the room were to get warmer, the pressure and temperature will rise again until just as many vapor molecules are condensing as are being pumped into the condenser.

After condensing (liquefying), the refrigerant is stored in the liquid receiver until needed. When needed, it passes through the high-pressure liquid line, 5, to the refrigerant control. Here refrigerant pressure is reduced to allow evaporation of the liquid at a low temperature.

The evaporating liquid absorbs much heat and so furnishes refrigeration. The increase in its volume, as it evaporates, pushes the vaporized refrigerant through the "suction" line. It comes, finally, to the compressor intake, where pressure is greatly reduced. It then passes through the intake valve of the compressor into the cylinder where a new cycle begins.

4-5 EVAPORATOR

The liquid refrigerant entering the evaporator from the refrigerant flow control is suddenly under low pressure. This makes it vaporize (boil) and absorb heat. The vapors move on into the suction line. If all of the liquid refrigerant has

not vaporized in the evaporator, there is usually a cylinder (accumulator) to prevent liquid refrigerant from flowing into the suction line.

Evaporators are mainly of two types, a dry system and the flooded system. Refrigerant is fed into the dry system evaporator only as fast as is needed to maintain the temperature wanted. In the flooded system, the evaporator is always filled with liquid refrigerant. The type of refrigerant control used determines the type of evaporator to be used.

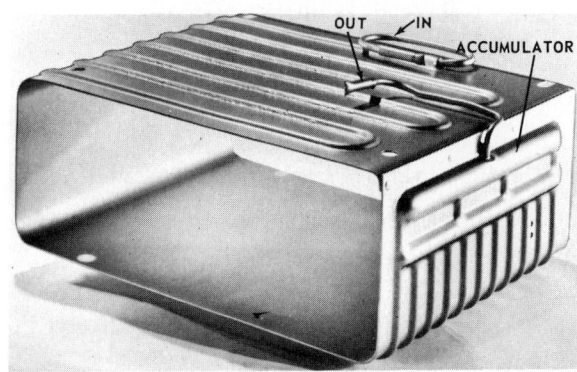

Fig. 4-5. Shell type evaporator used with capillary tube or high-side float type refrigerant control.

Evaporators are made in four different styles:
1. Shell type, Fig. 4-5.
2. Shelf type, Fig. 4-6.
3. Wall type, used in chest type freezer, Fig. 4-7.
4. Fin tube type with forced circulation. This type of evaporator is most used with frost-free construction. (See views A and B of Fig. 4-8.)

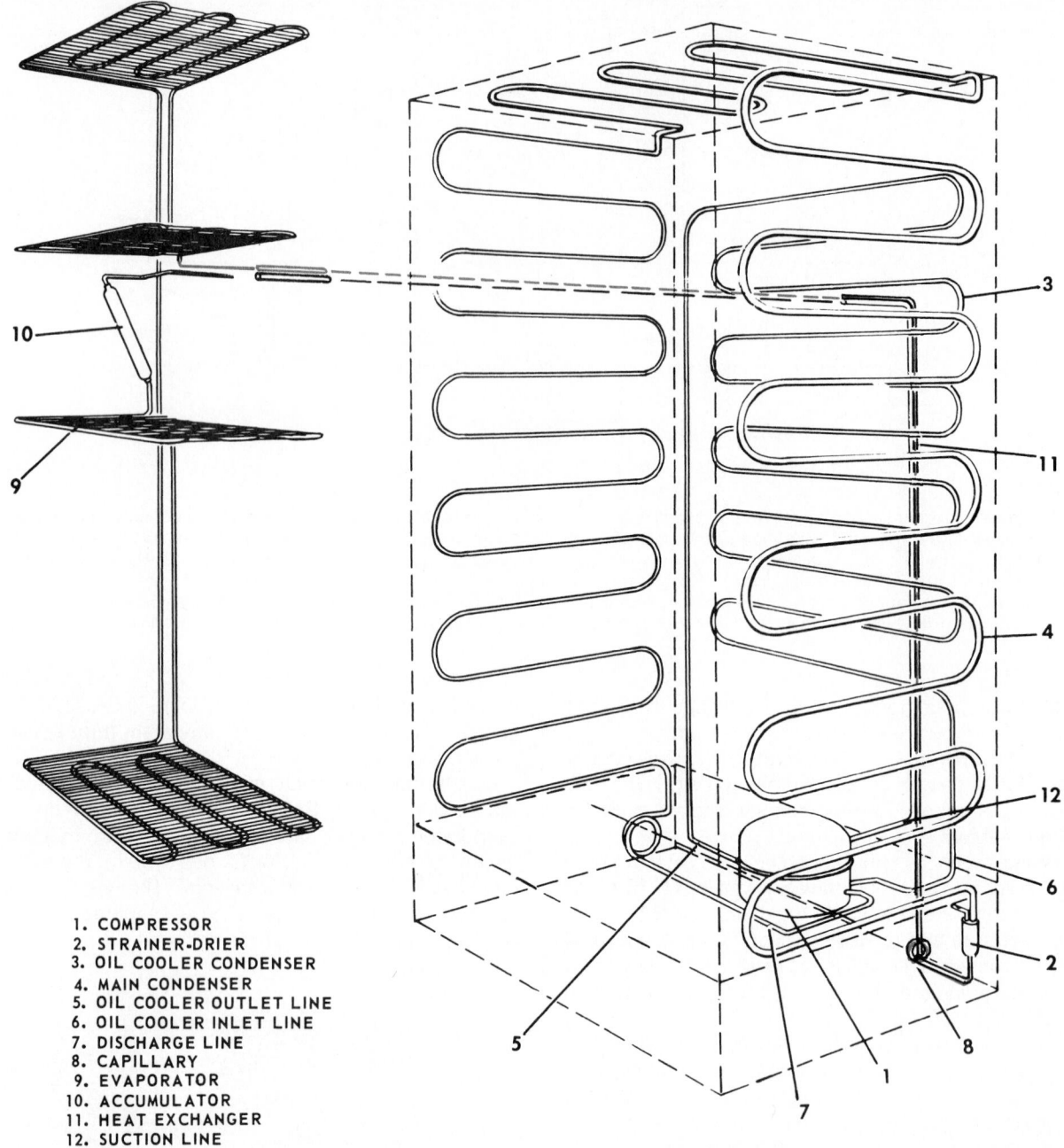

1. COMPRESSOR
2. STRAINER-DRIER
3. OIL COOLER CONDENSER
4. MAIN CONDENSER
5. OIL COOLER OUTLET LINE
6. OIL COOLER INLET LINE
7. DISCHARGE LINE
8. CAPILLARY
9. EVAPORATOR
10. ACCUMULATOR
11. HEAT EXCHANGER
12. SUCTION LINE

Fig. 4-6. Shelf type evaporator 9. This shows evaporator as it forms the shelf in upright freezer. Accumulator 10 is located at outlet of evaporator. This is a small reservoir to catch refrigerant not needed in evaporator.

Frost-free refrigerators usually need a fan or fans. They circulate the air over the evaporator and distribute cold air throughout the cabinet.

There are many types of commercial system evaporators. Air cooling and liquid cooling are two of the basic designs. They are constructed of plain or finned tubing or have a flat plate design.

Details of commercial system evaporators are explained in Chapter 12.

4-6 ACCUMULATOR

The accumulator is a safety device to prevent liquid refrigerant from flowing into the suction line and into the compressor. Liquid refrigerant, if it were to flow into the

compressor, could cause considerable knocking and damage to the compressor.

A typical accumulator, Figs. 4-5 and 4-6, has the outlet at the top. Any liquid refrigerant that flows into the accumulator will be evaporated. Then vapor only will flow into the suction line. Since the accumulator is located inside the cabinet, it also provides some refrigeration.

Not all refrigerating systems have accumulators. Commercial system accumulators are explained in Chapter 12.

4-7 SUCTION LINE

The suction line carries the refrigerant vapor from the evaporator to the compressor. The line must be large enough to carry the vaporized refrigerant with minimal flow

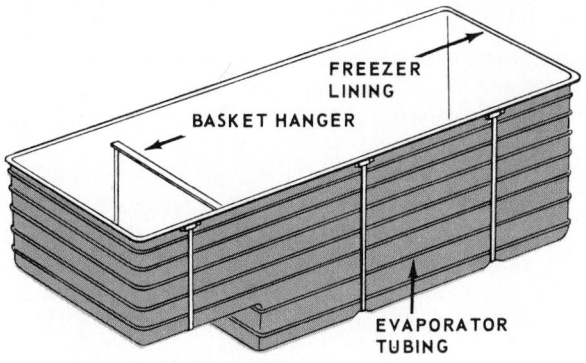

Fig. 4-7. Wall type evaporator. Note that evaporator tubing is attached to lining of freezing cabinet. This arrangement provides smooth inside surface with uniform cooling throughout cabinet.

resistance. It should slope from the evaporator or accumulator down to the compressor. Otherwise, pockets of oil collect.

In many units, the liquid line is in contact with all or part of the length of the suction line. This cools the liquid

refrigerant, helping to reduce flash gas in the evaporator. It also adds some superheat to the refrigerant vapor entering the compressor. See Para. 5-10 for an explanation of superheat.

4-8 LOW-SIDE FILTER-DRIER

Some systems include a low-side filter-drier at the compressor end of the suction line. These may be a part of the original system or may be placed in the system, for a short time, to clean it up. Fig. 4-9 is a typical suction line filter-drier. The filter-drier used in the suction line should offer little resistance to the flow of the vaporized refrigerant. This is desired because the pressure difference between the pressure in the evaporator and the inlet to the compressor should be small.

4-9 COMPRESSOR LOW-SIDE OR SUCTION SERVICE VALVE

Many systems have some means that allow the service technician to connect gauges to the system, check pressures, and add or take out refrigerant or oil.

AIR FLOW

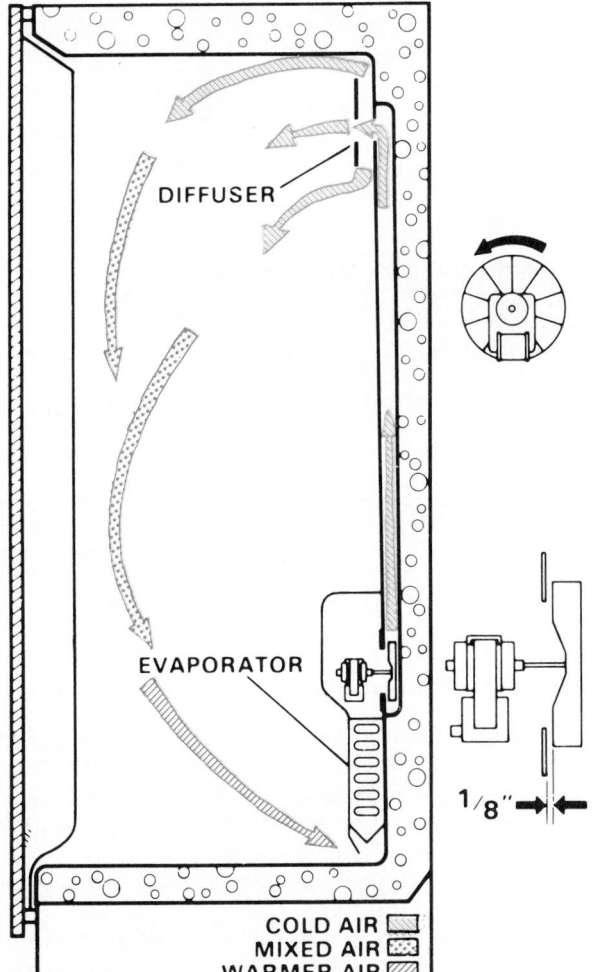

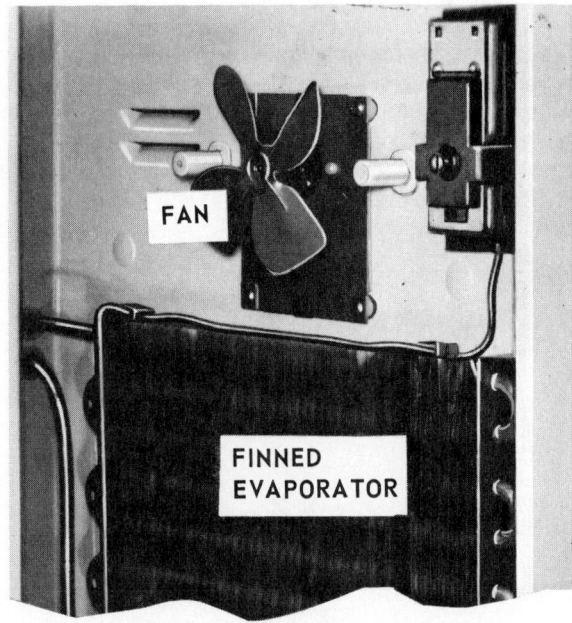

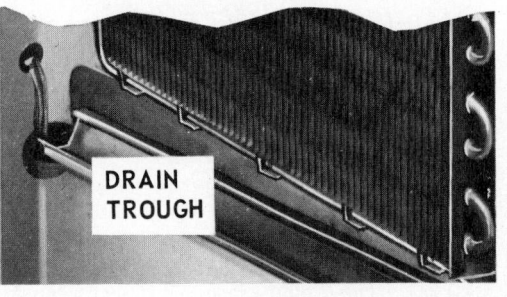

Fig. 4-8. Forced convection evaporator. A—A forced circulation evaporator is used in this upright freezer cabinet. Door switch stops fan when cabinet door is opened. B—Fin type evaporator. Note the trough to collect and carry away the defrost moisture which drains from the evaporator during the defrost part of the cycle. (Frigidaire Company)

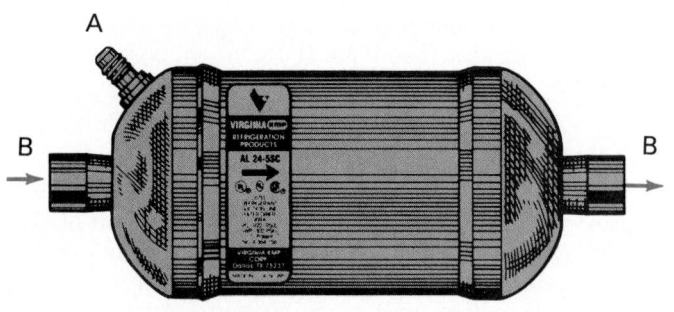

Fig. 4-9. Suction line filter-drier. Direction of refrigerant vapor flow is indicated. A—Service connection. B—Tubing connection. (Virginia KMP Corp.)

A typical compressor suction service valve is pictured in Fig. 4-10. This valve is connected to the compressor at the compressor inlet union. The suction line from the evaporator is attached at the low-side inlet. Sealing caps protect the charging and gauge opening port and the valve stem when the valve is not in use.

More recent domestic models do not have service valves. The service technician must use a saddle valve, Chapter 11.

4-10 COMPRESSOR

The refrigeration compressor is a motor-driven device which moves the heat-laden vapor refrigerant from the evaporator and compresses (squeezes) it into a small volume and to a high temperature. The various types of pumping mechanisms (compressors) used are explained later in this chapter.

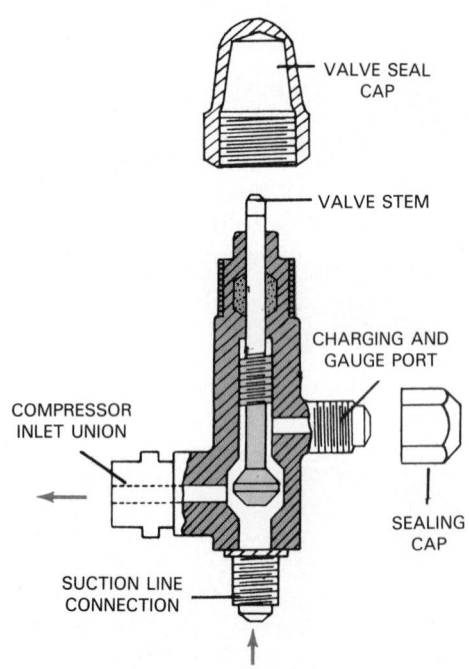

Fig. 4-10. Compressor low-side or suction service valve. If valve stem is turned all the way in, it closes off the connection from the compressor to the suction line. In this position, if valve is removed from compressor, suction line remains sealed. If valve stem is turned out as far as possible, it closes off the opening called the "charging and gauge port." This makes it possible to install compound gauge or charging line.

4-11 COMPRESSOR HIGH-SIDE SERVICE VALVE

The compressor high-side service valve provides a shutoff between the compressor and the condenser. It also provides an opening for a high-pressure gauge or a gauge manifold.

With the valve closed (all the way in), it is possible to disconnect the compressor from the condenser without any leakage of refrigerant from the condenser. When the valve stem is all the way out, the opening for the gauge is closed.

Fig. 4-11 illustrates a cross section of the service valve. It is not used on all refrigerating systems.

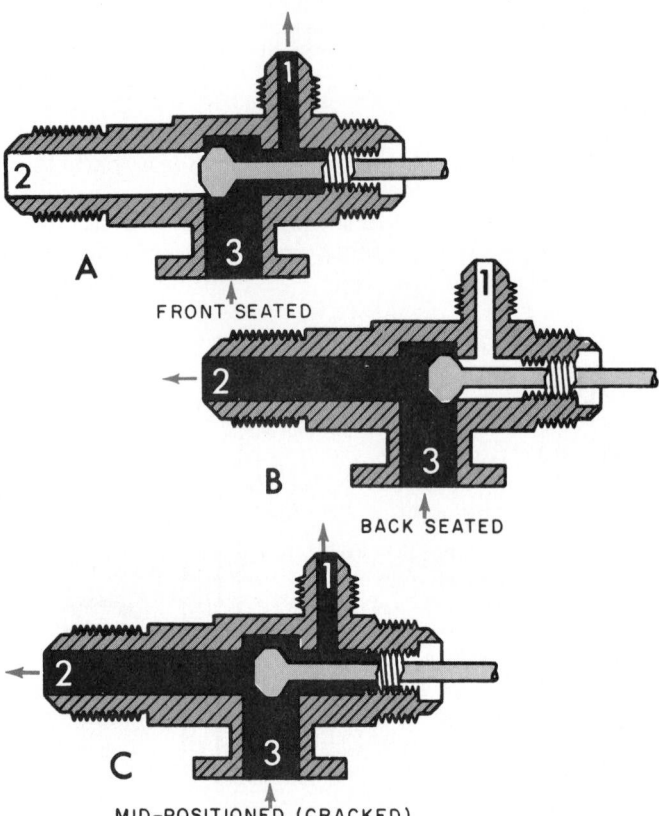

Fig. 4-11. A compressor high-side service valve. If valve is turned all the way in, as in view A, it shuts off connection between compressor, 3, and condenser, 2. If valve is turned all the way out, as in view B, it closes off connection to gauge port, 1. At mid-position, view C, all passages are open. (Murray Corp.)

4-12 OIL SEPARATOR

Refrigeration compressors get their lubrication from a small amount of special lubricating oil placed inside the compressor crankcase or housing. This oil is circulated to various compressor parts. In a hermetic (airtight) system, this oil also lubricates the motor bearings.

When the compressor operates, small amounts of oil will be pumped out with the hot compressed vapor. A small amount of oil throughout the system does no harm. However, too much oil in such parts as the condenser, refrigerant flow controls, evaporator, and filters interferes with their operation.

It is possible to separate the oil from the hot compressed vapor. This involves placing an oil separator between the

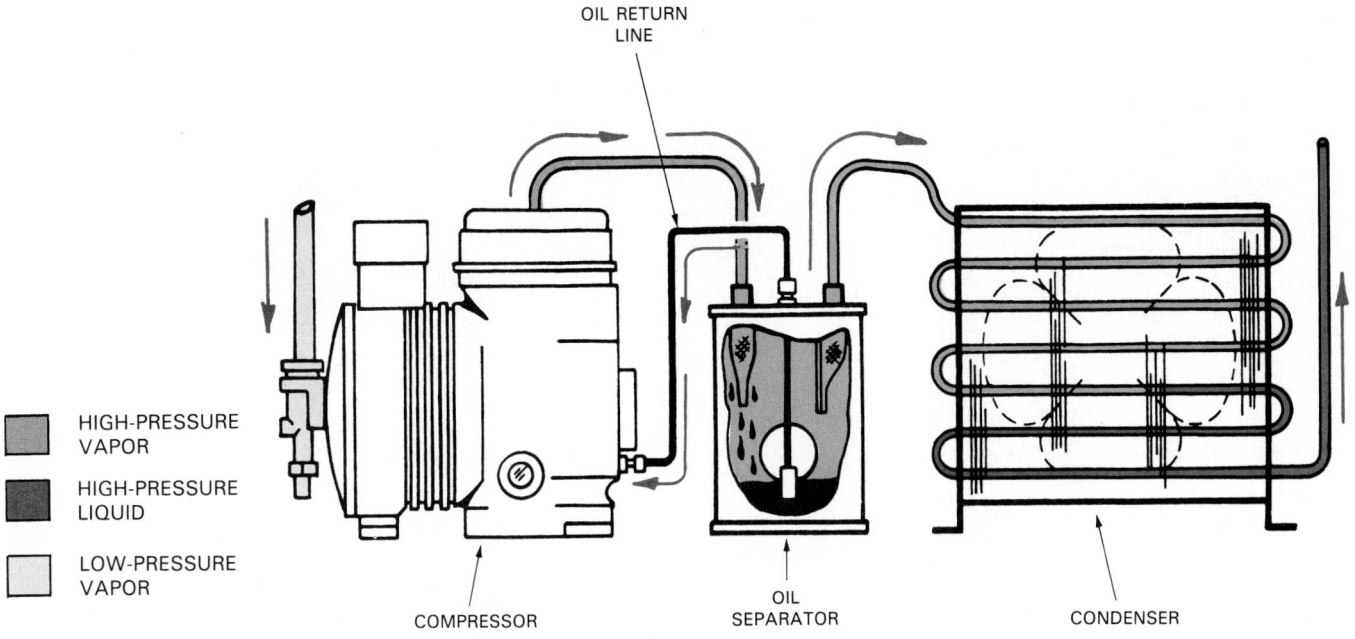

OIL RETURN
LINE

HIGH-PRESSURE
VAPOR

HIGH-PRESSURE
LIQUID

LOW-PRESSURE
VAPOR

COMPRESSOR

OIL
SEPARATOR

CONDENSER

Fig. 4-12. An oil separator located in the discharge line. Note flow of refrigerant and oil. (A C and R Components, Inc.)

compressor exhaust and the condenser. The location and operation of such a separator is shown in Fig. 4-12. The separator is enlarged in the illustration to help show details.

The oil separator is a tank or cylinder with a series of baffles or screens which collect the oil. The oil, separated from the hot, compressed vapors, drops to the bottom of the separator.

A float arrangement controls a needle valve which opens an oil return line to the compressor crankcase. When the oil level is high enough, the float rises and opens the needle valve. This oil returns quickly to the compressor crankcase, as the pressure in the separator is considerably higher than the pressure in the compressor crankcase.

Oil separators are quite efficient. Very little oil passes on into the system. They are most commonly used in large commercial installations.

4-13 CONDENSER

As mentioned in Para. 4-2, the condenser in the refrigeration cycle removes the condensation heat from the refrigerant vapor. This heat is picked up in the evaporator. Domestic refrigerators commonly use the four following types of condensers (see Fig. 4-13):
1. Finned-static (natural convection).
2. Finned-forced convection.
3. Wire-static.
4. Plate-static.

Illustrated in view A, Fig. 4-13, is a common finned type static condenser. Static means that air circulation through the condenser tubing and fins is by natural convection; that is, warm air tends to rise. As the air in contact with the fins and tubes becomes heated, it rises and cooler air takes its place. The tubes and fins are usually made of copper or steel.

View B, Fig. 4-13, is a forced convection fin type condenser. Whenever the compressor is operating, the motor-driven fan forces air through the condenser.

Shown in view C is a wire type condenser which uses

small metal wires brazed or spotwelded to the condenser tubing. It is usually of the static type.

The plate type condenser is pictured in view D. In this type, the condenser tubes are soldered or brazed to a flat metal surface. This is a very common type of condenser construction. It is used on many chest-type freezers. The condenser tubes are attached to the inside (insulation side) of the freezer's outer shell. This type of condenser is very easy to keep clean. It is only necessary to wipe off the surface of the cabinet shell. To get proper removal of heat from the refrigerant vapor, always keep the condenser area clean.

Commercial systems use three types of condensers:
1. Finned-static, air-cooled.
2. Finned-forced convection, air-cooled.
3. Water-cooled, tube-in-a-tube, and shell.

The finned-static and the finned-forced convection condensers are built much the same as the domestic condensers, except that they are larger. Forced convection finned condensers are used on many commercial refrigeration installations. Some applications of these are shown in Chapter 12.

Water-cooled condensers usually consist of two tubes, one within the other. Water circulates through the inside tube. Hot compressed vapor circulates through the space between the tubes. These condensers are usually called tube-within-a-tube condensers. They are very efficient.

With the rapid development of commercial refrigeration and air conditioning, many communities have difficulty in supplying enough water for water-cooled condensers. Thus, more and more forced convection air-cooled condensers and "cooling towers" are being used in connection with water-cooled condensers.

See Chapter 12 for illustrations of cooling tower applications and water-cooled condensers. Fig. 4-14 shows a large, three-fan, roof-mounted air-cooled condenser installation. Diagram of a water-cooled condenser is shown in the bottom view.

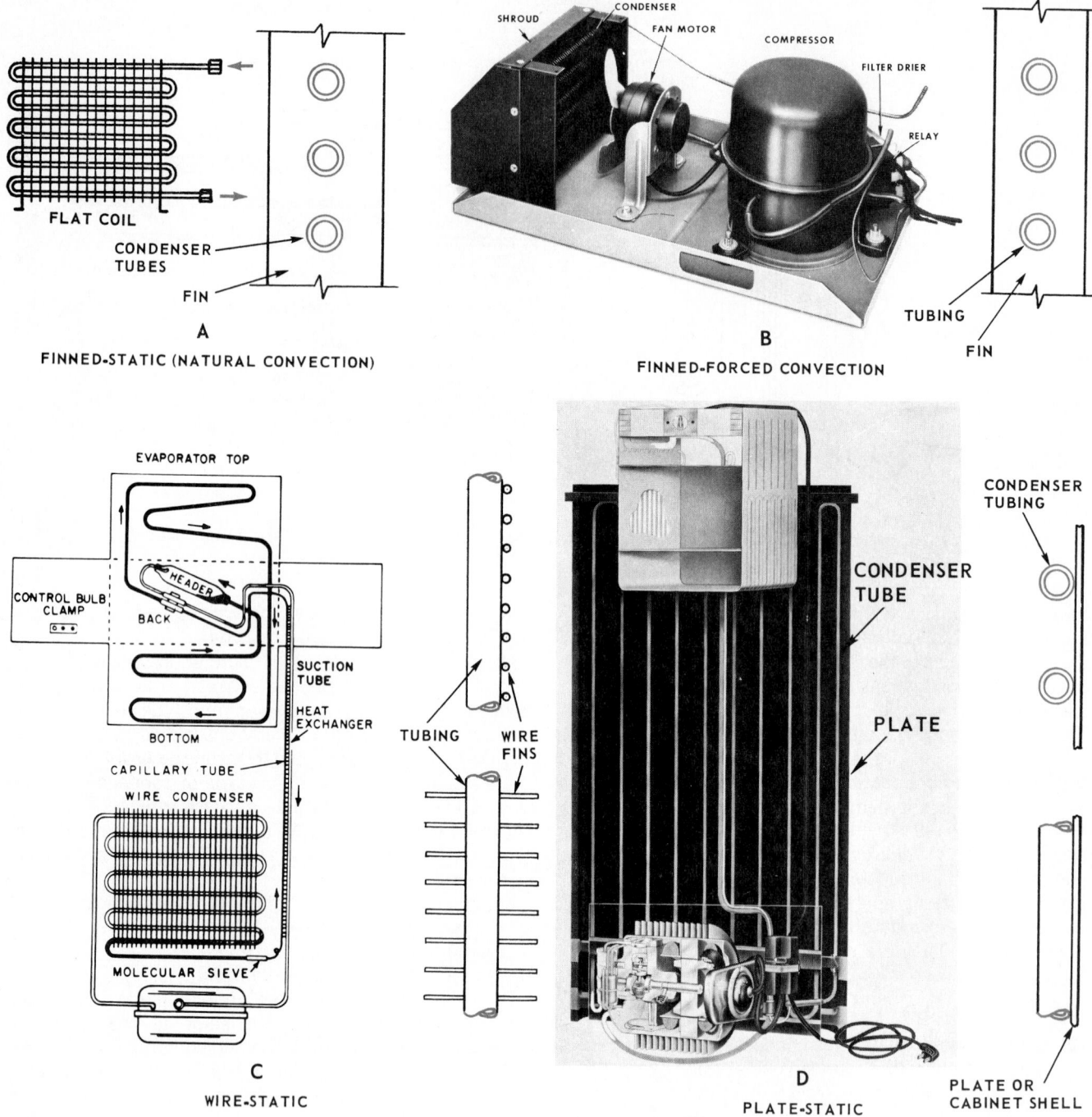

Fig. 4-13. Common domestic type condensers.

4-14 LIQUID RECEIVER

The liquid receiver is a storage tank for liquid refrigerant. When a refrigerating mechanism has one, the refrigerant is usually pumped out of the various parts and stored in it during servicing. Its use makes the quantity of refrigerant in a system less critical.

Occasionally one may find a liquid receiver built into the bottom of the condenser. Most receivers have service valves. See Fig. 4-15. A fine copper mesh in the outlet prevents dirt from entering the refrigerant control valves.

Liquid receivers are used on most systems with the low-side float type or the expansion valve type refrigerant con-

trol. Capillary tube systems do not use them because all the liquid refrigerant is stored in the evaporator during the off part of the cycle. Greater use of hermetic systems and capillary tube refrigerant controls has removed the need for liquid receivers in domestic systems and many small commercial units.

On larger commercial systems, the receiver provides enough reserve liquid refrigerant to insure that the liquid line refrigerant is subcooled and free of flash gas. The receiver must provide enough room for refrigerant during automatic pumpdowns (for defrost purposes and when some of the evaporators are not in use).

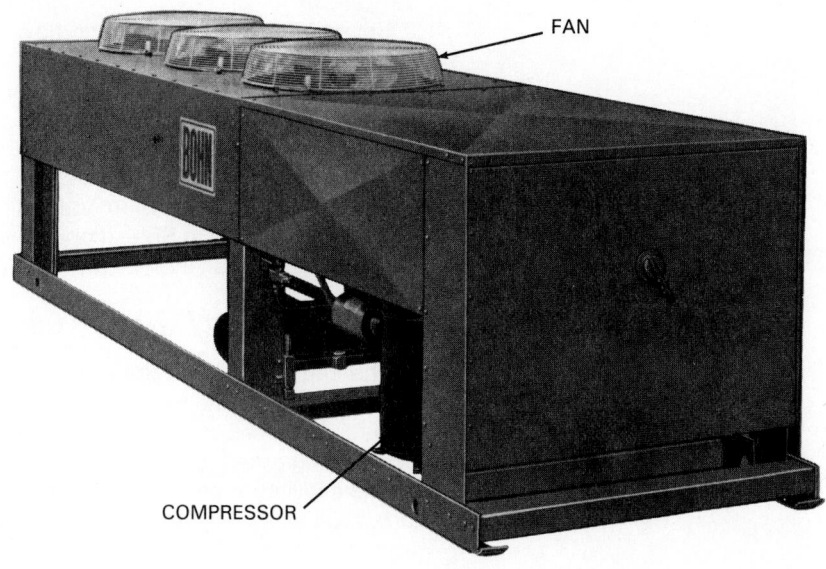

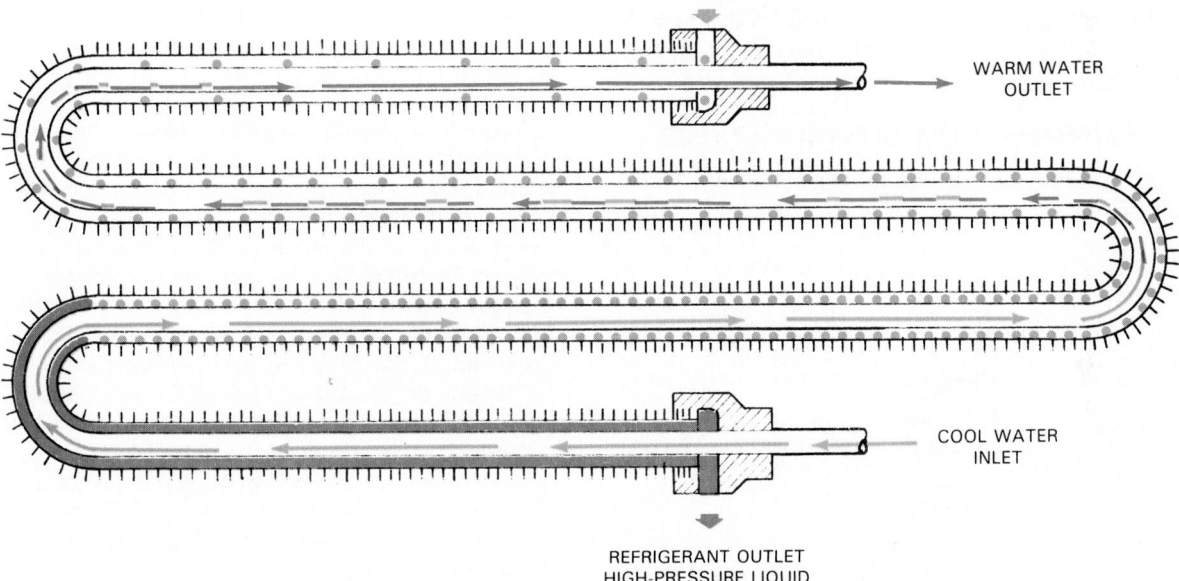

Fig. 4-14. Typical commercial refrigeration condensers. Top. Large air-cooled condenser which uses motor driven fans to increase airflow over condenser surfaces. (Bohn Refrigeration Products, a unit of Heatcraft, Inc.) Bottom. Water-cooled condenser flow diagram. Refrigerant flows through condenser in opposite direction of water.

Some systems which have an outdoor air-cooled condenser need room in the receiver for extra refrigerant. Without extra room, liquid partly fills the condenser when the head pressure is too low to move the liquid through.

4-15 LIQUID LINE FILTER-DRIER

It is common practice to install a filter-drier in the liquid line. This tank-like accessory keeps moisture, dirt, metal, and chips from entering the refrigerant flow control. What is more, the drying element in the filter removes moisture which might otherwise freeze in the refrigerant flow control.

Moisture is also harmful when mixed with oil in a system. It forms sludges and acids. It is especially harmful to hermetic units. A liquid line filter-drier is shown in Fig. 4-16.

Some filter-driers are equipped with a sight glass which will indicate refrigerant level. Many sight glasses also have a chemical which will change color when the system has moisture in it.

4-16 LIQUID LINE

While copper tubing is commonly used to carry the liquid refrigerant from the condenser to the evaporator, domestic units often use steel. These lines are mounted in back of the refrigerator cabinet or are hidden behind the breaker strip at the refrigerator door jamb (frame).

The lines are soldered or brazed to fittings. It is important

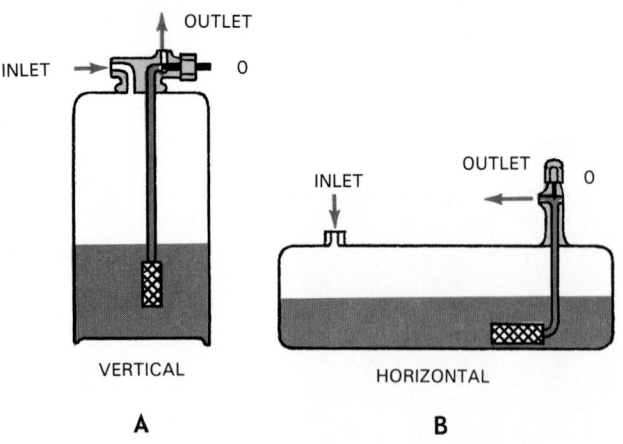

Fig. 4-15. Two common types of liquid receivers. A—Vertical liquid receiver. B—Horizontal liquid receiver. Note liquid line service valve at O. It allows easy charging of refrigerant.

to avoid pinching or buckling these lines. Support them to prevent wear or breakage from vibration.

Refrigerant lines in commercial units may be connected by soldering, brazing, or by flared fittings.

In many installations, the liquid line runs parallel to and in contact with the suction line. The reason for this is explained in Para. 4-7.

4-17 REFRIGERANT FLOW CONTROL—TYPES

The refrigerant flow control has two jobs. It allows liquid refrigerant to enter the evaporator and, at the same time,

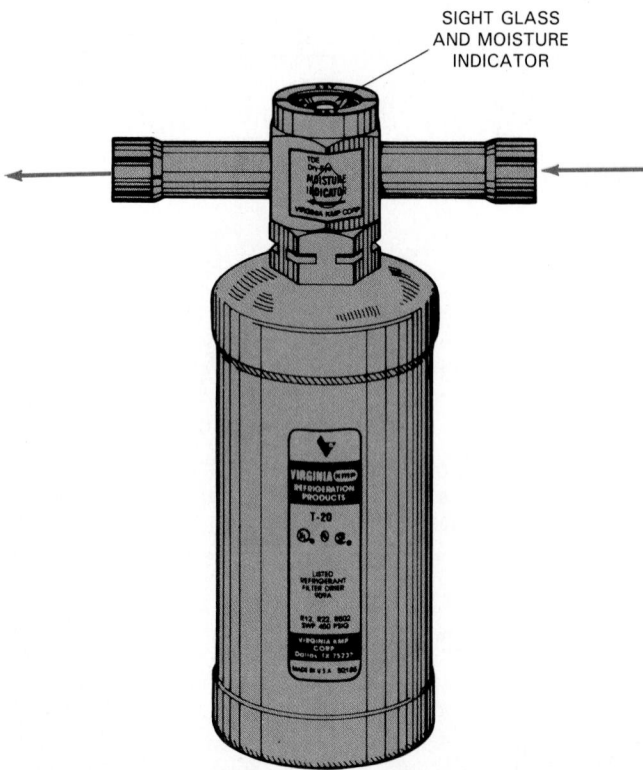

Fig. 4-16. A liquid line filter-drier with sight glass indicator. (Virginia KMP Corp.)

maintains the required evaporating pressure in the evaporator.

Several types of refrigerant flow controls are used in modern refrigerating mechanisms. These controls and their characteristics are explained in Chapter 5.

There are five principal types:
1. Capillary Tube (Cap.)—(dry system).
2. Automatic Expansion Valve (AEV)—(dry system).
3. Thermostatic Expansion Valve (TEV)—(dry system).
4. Low-Side Float (LSF)—(flooded system).
5. High-Side Float (HSF)—(flooded system).

4-18 CAPILLARY (CAP.) TUBE

The capillary tube is the most common refrigerant flow control for domestic refrigerators, freezers, room air conditioners, and small commercial installations. A typical installation is diagramed in Fig. 4-17.

The capillary tube is a long length of small diameter tubing. It reduces pressure, by reducing the flow of refrigerant through its length.

Fig. 4-18 is another drawing of a capillary tube refrigerant control. The tube's inside diameter may vary, depending upon the refrigerant, the capacity of the unit, and the length of the line.

The tube is placed between the liquid line and the evaporator. Just enough liquid passes through it to make up for the amount that is vaporized in the evaporator as the compressor operates.

It reduces the liquid refrigerant from its high pressure to its evaporating pressure. There is no change in the liquid except a slight drop in pressure for about the first two-thirds of the length of the capillary tube.

Then some of the liquid starts to change to vapor. By the time the refrigerant reaches the end of the tube, from 10 to 20 percent of it has vaporized. The increased volume of the vapor causes most of the pressure drop to take place in the end of the tube nearest the liquid line.

A recent development in capillary tube design uses a larger and longer tube (20 to 30 ft.). Being larger in diameter, it is less likely to become plugged.

The capillary tube refrigerant control does not use a check valve or a direction control valve. The high and low pressures equalize during the off part of the cycle. This permits easier starting, since the compressor starts with equal pressures on the high and low side. The system must not have an overcharge of refrigerant, as the extra refrigerant would tend to fill the evaporator too full. An overcharge will be indicated by severe frosting of the suction line when the motor starts.

4-19 AUTOMATIC EXPANSION VALVE (AEV)

One of the dry systems uses an automatic expansion valve (AEV) as refrigerant control, Fig. 4-19. As pressure drops on the low side, the expansion valve opens and liquid refrigerant flows into the evaporator. It absorbs heat then, while evaporating under low pressure. The valve maintains constant pressure in the evaporator when the system is running. This system operates independently of the amount of refrigerant in the system.

The AEV is, therefore, one of the division points between the high-pressure and low-pressure sides of the system. See Chapter 5 for a detailed explanation of expansion valves.

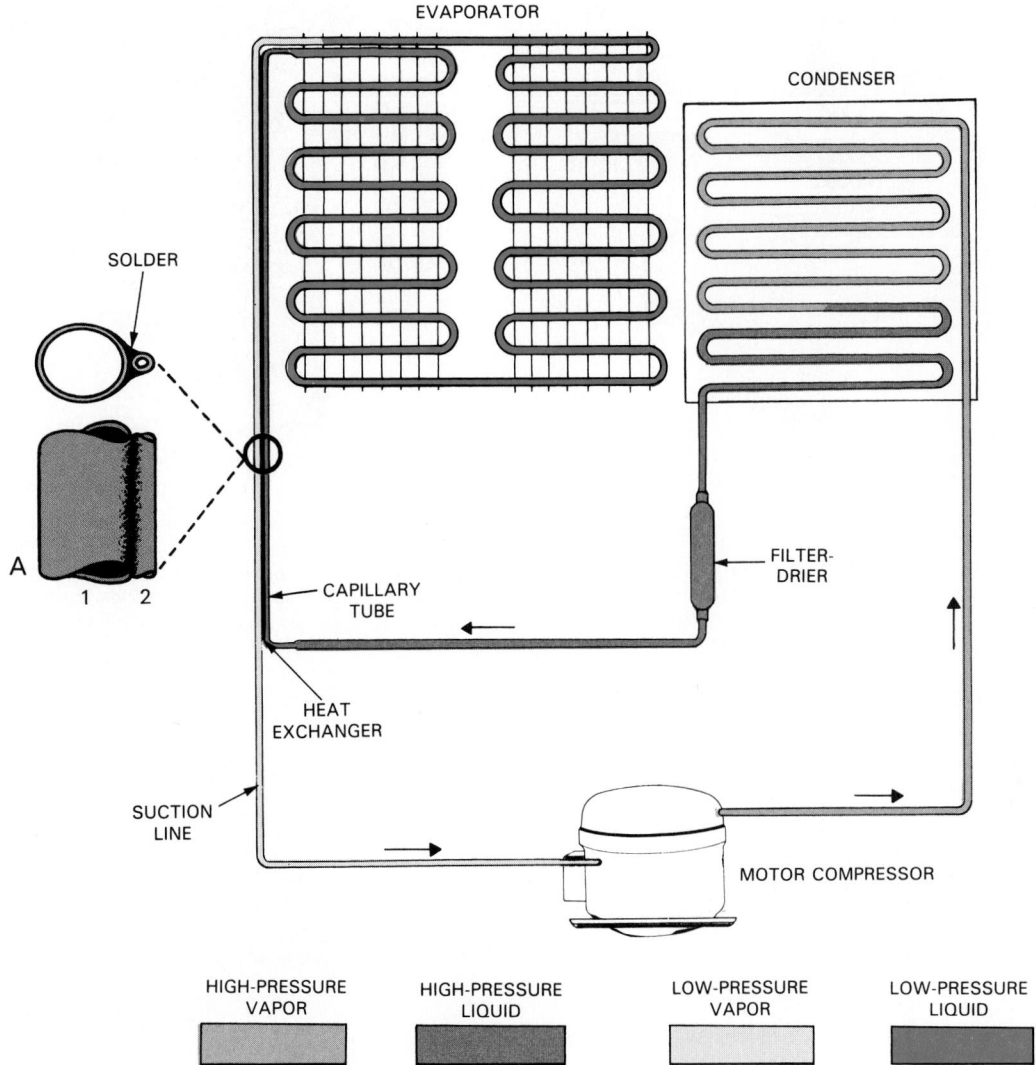

EVAPORATOR

CONDENSER

SOLDER

A

1   2

CAPILLARY
TUBE

FILTER-
DRIER

HEAT
EXCHANGER

SUCTION
LINE

MOTOR COMPRESSOR

HIGH-PRESSURE VAPOR	HIGH-PRESSURE LIQUID	LOW-PRESSURE VAPOR	LOW-PRESSURE LIQUID

Fig. 4-17. Refrigerating system using capillary tube refrigerant control. Filter-drier is located in liquid line ahead of connection to capillary tube. Most of the capillary tube is fastened to suction line, which provides heat exchange. A—Enlarged cross section of suction line, 1, and capillary tube, 2, showing how they are soldered or brazed together.   (Frigidaire Company)

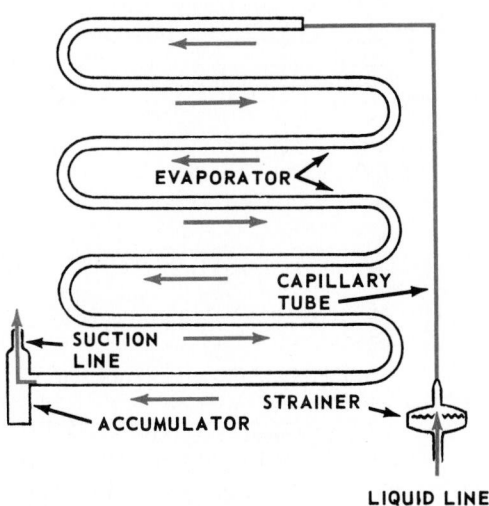

EVAPORATOR

CAPILLARY
TUBE

SUCTION
LINE

STRAINER

ACCUMULATOR

LIQUID LINE

Fig. 4-18. Capillary tube refrigerant control. A strainer is located in liquid line at entrance to capillary tube. Note accumulator. This serves as receiver for any liquid refrigerant overflow.

The automatic expansion valve, Fig. 4-20, may be adjusted to the correct evaporator pressure. Turning the adjustment clockwise will increase the rate of flow and, therefore, increase the low-side pressure.

The rate of refrigerant flow through the automatic expansion valve is controlled by the pressure in the evaporator. No refrigerant will flow through the valve unless the compressor is running and the evaporator is under a low pressure.

Remember, lowering the pressure in the evaporator lowers the temperature at which the refrigerant evaporates.

This valve may be used only with the temperature-operated motor control.

4-20 THERMOSTATIC EXPANSION VALVE (TEV)

Many units, especially commercial ones, are equipped with a temperature controlled expansion valve called a thermostatic expansion valve (TEV). Fig. 4-21 shows a cross section of a thermostatic expansion valve.

This valve has a temperature sensing bulb mounted on

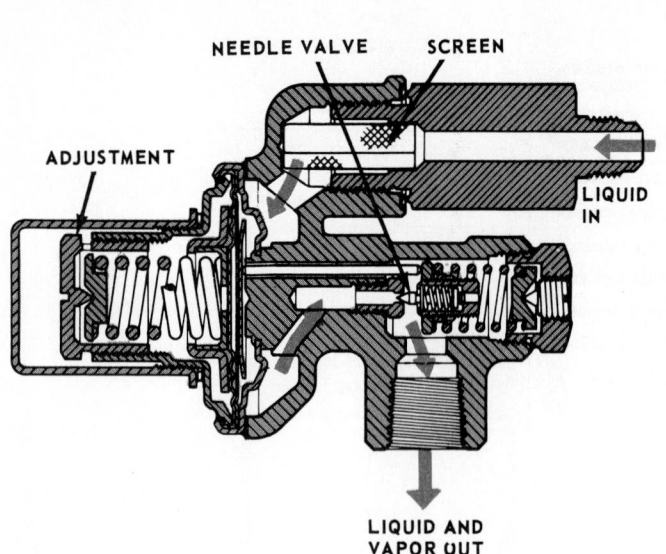

Fig. 4-19. Cross section of an automatic expansion valve shows flow of refrigerant through valve. This valve is designed to control the flow of liquid into evaporator. It also maintains constant low pressure in evaporator while compressor is running.

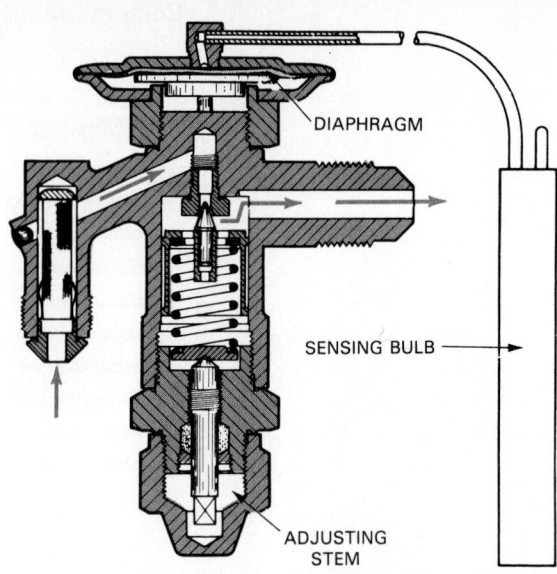

Fig. 4-21. Diaphragm type thermostatic expansion valve. Sensing bulb pressure operates on upper surface of diaphragm. As sensing bulb temperature increases, pressure on top of diaphragm increases and tends to open valve, allowing liquid refrigerant to enter evaporator. Note screen at liquid line connection and direction of refrigerant flow through valve. (Sporlan Valve Co.)

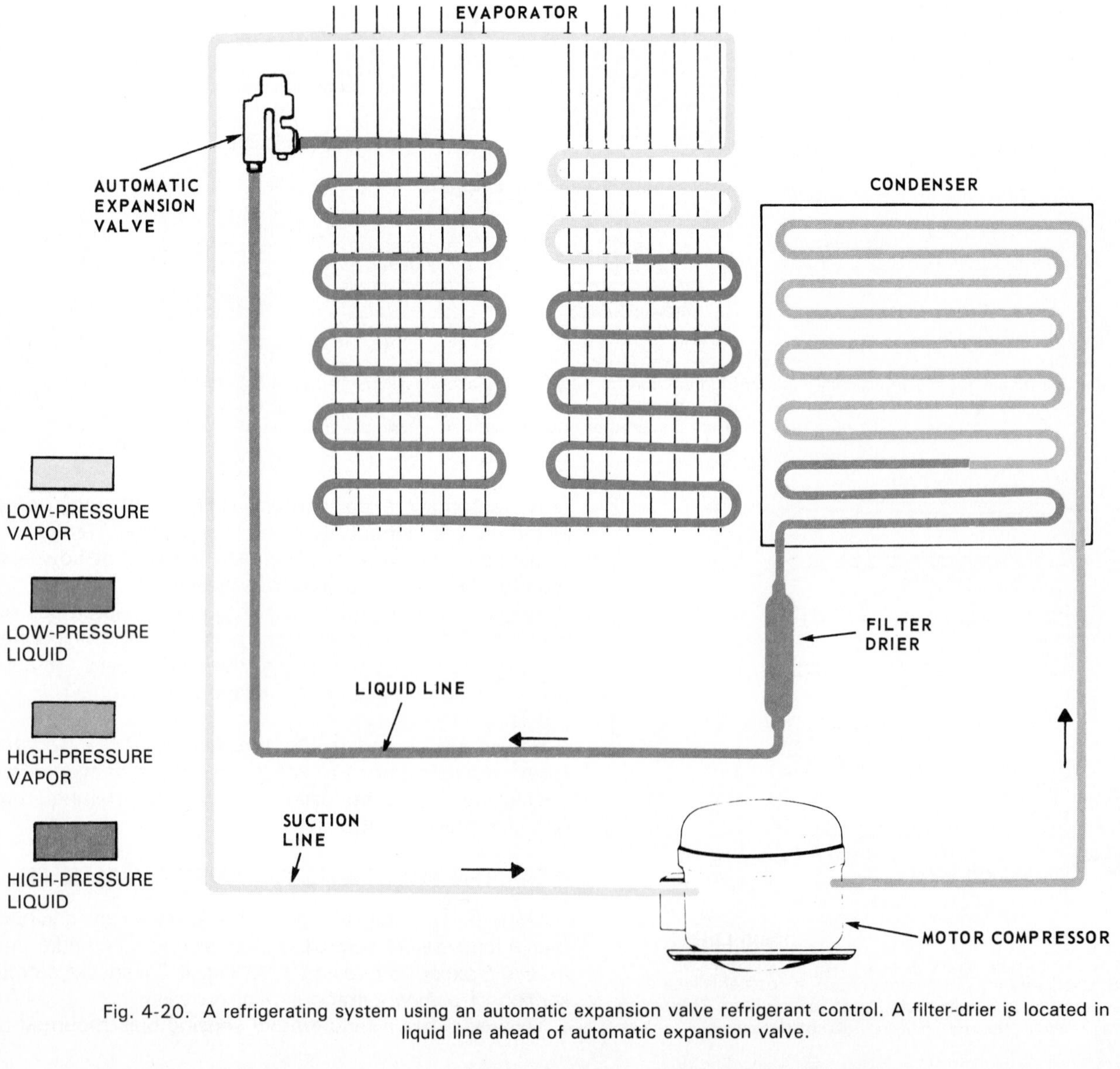

EVAPORATOR

AUTOMATIC
EXPANSION
VALVE

CONDENSER

LOW-PRESSURE
VAPOR

LOW-PRESSURE
LIQUID

HIGH-PRESSURE
VAPOR

HIGH-PRESSURE
LIQUID

FILTER
DRIER

LIQUID LINE

SUCTION
LINE

MOTOR COMPRESSOR

Fig. 4-20. A refrigerating system using an automatic expansion valve refrigerant control. A filter-drier is located in liquid line ahead of automatic expansion valve.

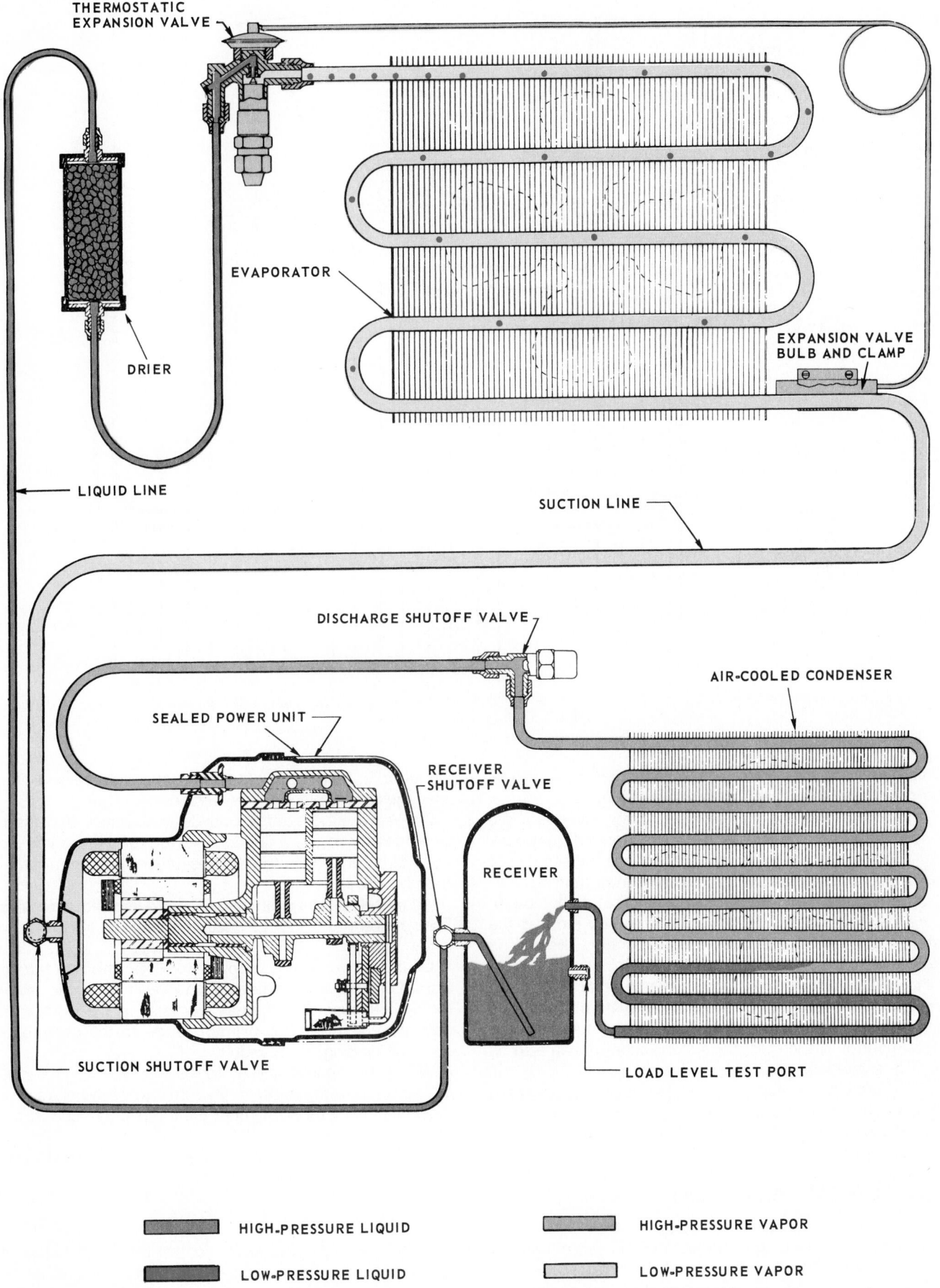

THERMOSTATIC
EXPANSION VALVE

DRIER

EVAPORATOR

EXPANSION VALVE
BULB AND CLAMP

LIQUID LINE

SUCTION LINE

DISCHARGE SHUTOFF VALVE

AIR-COOLED CONDENSER

SEALED POWER UNIT

RECEIVER
SHUTOFF VALVE

RECEIVER

SUCTION SHUTOFF VALVE

LOAD LEVEL TEST PORT

HIGH-PRESSURE LIQUID

HIGH-PRESSURE VAPOR

LOW-PRESSURE LIQUID

LOW-PRESSURE VAPOR

Fig. 4-22. Typical thermostatic expansion valve system. Note the two-cylinder hermetic compressor.

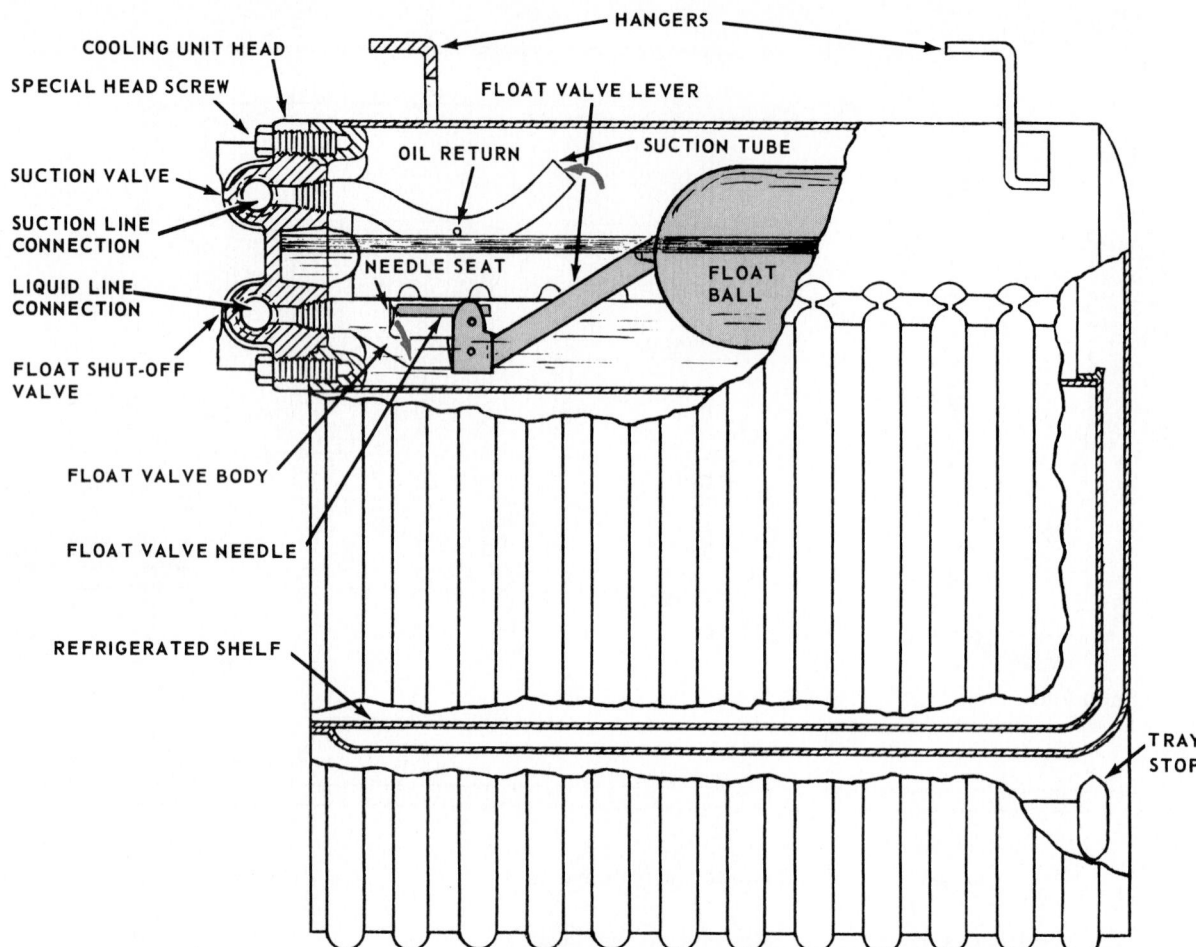

Fig. 4-23. Low-side float refrigerant control. Note suction line and liquid line connections. Float and needle mechanism maintain constant level of liquid refrigerant in evaporator.

the outlet of the evaporator. The bulb temperature controls the opening of the thermostat valve needle.

Addition of this thermal element to the valve lets the evaporator fill more quickly and permits more efficient cooling. The thermostatic expansion valve keeps the evaporator full of liquid refrigerant when the system is running. See Chapter 5 for a more detailed explanation of its operation.

As the evaporator becomes colder, the TEV reduces the rate of flow of refrigerant into the evaporator. There is no flow at all unless the compressor is running.

Fig. 4-22 demonstrates how a thermostatic expansion valve is connected into a refrigerating system. The valve may be used with either a pressure or temperature operated motor control. A thermostatic expansion valve can also be used in a multiple evaporator system.

In a recent development, a thermoelectric element replaces the temperature sensing element in the thermostatic expansion valve. More will be said about this valve in Para. 5-15.

4-21 LOW-SIDE FLOAT (LSF)

This control is used on a flooded system where the evaporator is flooded with refrigerant and the refrigerant level is controlled by a float valve. A float is used for control. As the refrigerant evaporates, the liquid level falls. This lowers a float and, in turn, opens the needle valve con-

nected to it. More liquid enters from the high-pressure liquid line, taking the place of the evaporated liquid.

Fig. 4-23 suggests the outside appearance and the inside construction of a flooded evaporator using a low-side float. Either a temperature or pressure motor control switch may be used.

A low-side float system usually has a large liquid receiver. The receiver must be large enough to store all the refrigerant in the system.

Oil picked up by the vapor is normally returned through a small opening at a predetermined level in the suction return tubing. Because of the small diameter of the hole, if the unit is not level, the oil will not return to the compressor and "oil binding" may result.

When this occurs, the oil forms a layer on the surface of the liquid refrigerant. It prevents the refrigerant from evaporating at a rapid rate or at the temperature corresponding to the pressure.

The low-side float refrigerant control can be used in multiple evaporator systems.

4-22 HIGH-SIDE FLOAT (HSF)

A float located in the liquid receiver tank or in a chamber in the high-pressure side makes this system operate. When enough liquefied refrigerant has collected in the float chamber, the float will rise enough to open the needle valve.

Liquid flows into the low-pressure side or evaporator. The float controls the level of liquid refrigerant on the high-pressure side.

The amount of refrigerant in the system must be carefully measured if the evaporator is to receive the correct amount and if the system is to operate correctly.

Extra refrigerant will overcharge the evaporator and cause frosting of the suction line.

Refer to Fig. 4-24 for an illustration of a high-side float mechanism. A more detailed description is given in Chapter 5. This refrigerant flow control can be used with either a pressure motor control or a thermostatic motor control.

4-23 MOTOR CONTROL

Practically all automatic electric refrigerators are designed with more cooling capacity than needed. Therefore, under normal use, they do not run all of the time. To get correct refrigeration temperature, the motor must be turned off when the desired low temperature is reached and turned on again when the evaporator has warmed to a certain temperature. Two principal types of motor controls are used to turn the motors on and off:

1. Temperature motor control (thermostatic).
2. Pressure motor control (low-side pressure).

The temperature control is the most popular, especially on small installations.

The thermostatic temperature control, Fig. 4-25, has a sensing bulb connected by a capillary tube to a diaphragm or bellows. This element is charged with a volatile fluid which expands to increase the pressure as the bulb

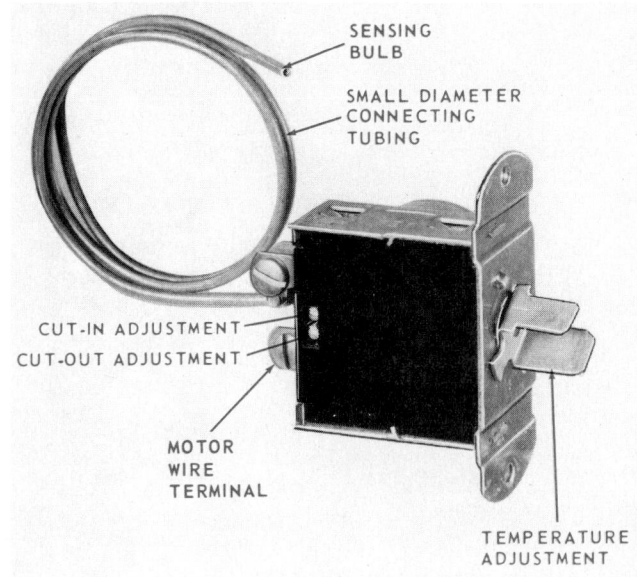

Fig. 4-25. A temperature motor control. Note temperature adjustment. (Eaton Corp.)

becomes warmer and will contract again to decrease the pressure as the bulb cools.

As bulb pressure increases, the diaphragm moves. Since it is connected to a toggle or snap action switch, it will turn this switch on (close the circuit). Then, as the bulb cools and the diaphragm or bellows moves the other way, the toggle switch will move (to open the circuit).

These controls have adjustments that permit differences in operating temperatures. Many controls have a manual switch to permit shutting off or turning on of the system as desired.

They also may include an overload protector which will open the switch if the unit draws too much current.

Thermostats may also be electrically connected to timers for automatic defrosting of the evaporator.

Many commercial units use a pressure type motor control. It opens the circuit when the pressure drops enough and closes the circuit when the pressure has risen enough. A pressure motor control may be used with the low-side float, the TEV, and the high-side float refrigerant control systems.

It should be kept in mind that the pressure of the vapor in the low-pressure side varies with the temperature. Therefore, the pressure may indicate temperature. This permits the use of pressures to control the stopping and starting of the motor, and thus controls the temperature of the cabinet. The details concerning the operation of these controls is explained fully in Chapter 8.

4-24 EXTERNAL DRIVE COMPRESSORS

The purpose of a compressor is described in Para. 4-10.

An external drive (open) compressor is bolted together. Its crankshaft extends through the crankcase. The crankshaft is driven by a flywheel (pulley) and belt, or it can be driven directly by an electric motor.

Fig. 4-26 illustrates a cross section through an open compressor. This is a four-cylinder V-type. An eccentric type crankshaft is used. The pistons are fitted with rings.

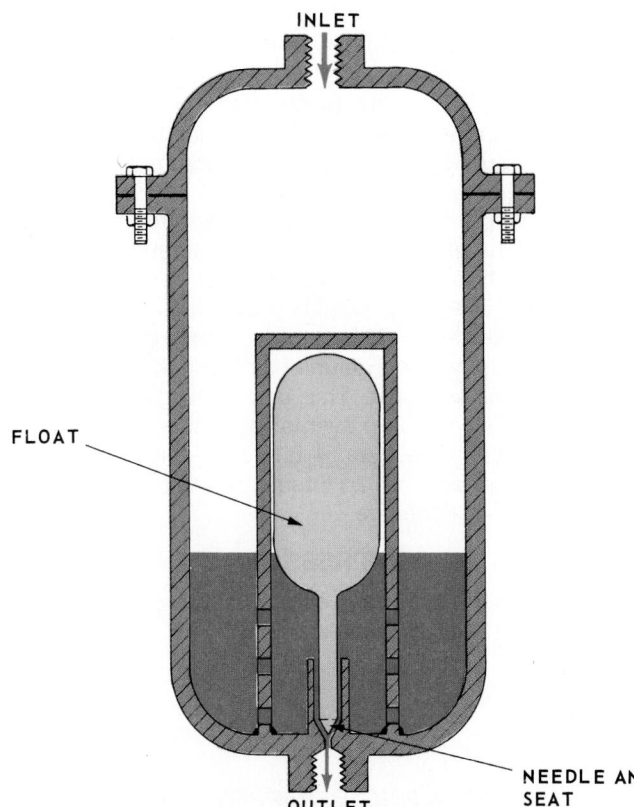

Fig. 4-24. High-side float allows liquid refrigerant to flow to the evaporator only when enough refrigerant collects to raise the float and open the needle valve.

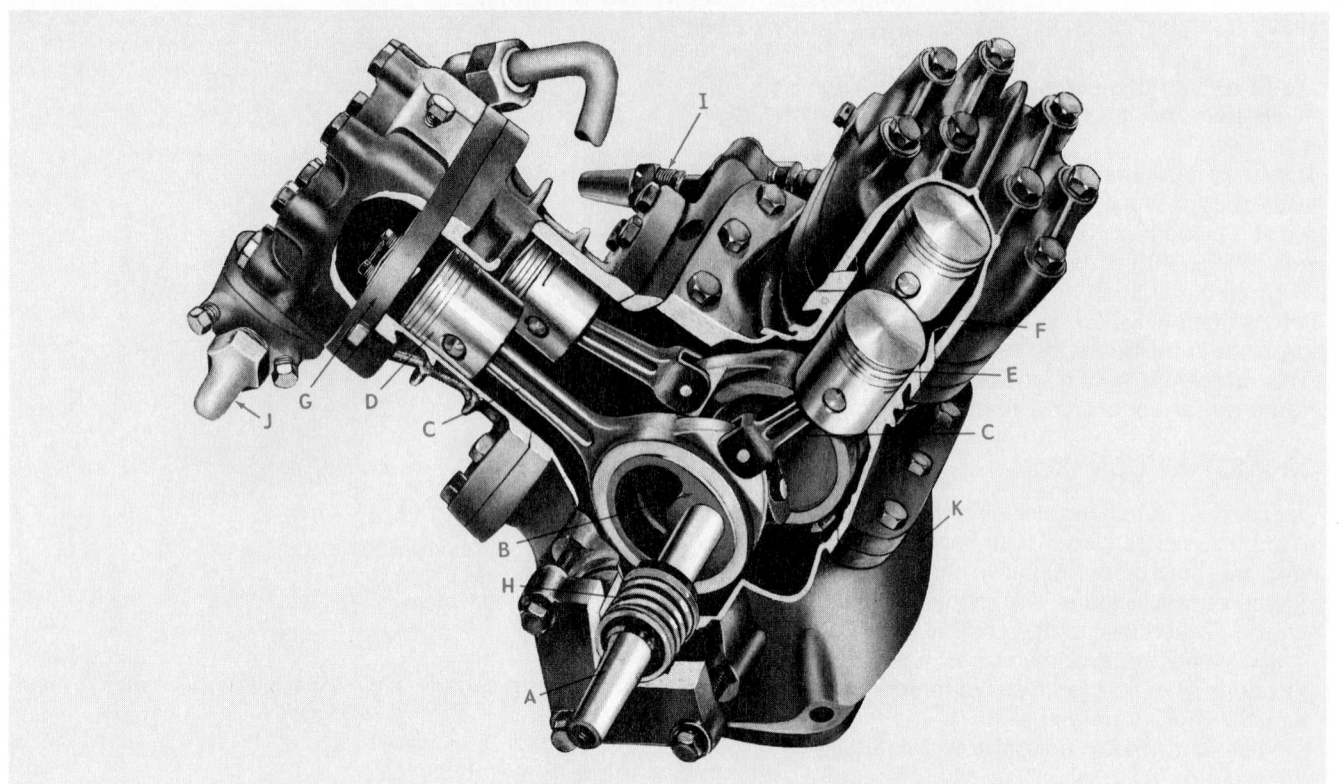

Fig. 4-26. A four-cylinder external drive V-type compressor, air cooled. A—Crankshaft. B—Eccentric. C—Connecting rod. D—Piston. E—Piston rings. F—Cylinder. G—Valve plate. H—Crankshaft seal. I—Suction service valve. J—Exhaust service valve. K—Crankcase. (Frick Co.)

A master connecting rod is mounted on each eccentric and is connected to a piston in one bank of the "V". The connecting rod which is attached to the piston in the other bank is connected with a pin through a flange on the master connecting rod. These connecting rods are somewhat shorter than the master connecting rod and are called articulated connecting rods.

A crankshaft seal is required where the crankshaft comes through the crankcase.

4-25 HERMETIC COMPRESSORS

The motor in a hermetic compressor is sealed inside a dome or housing with the compressor and is directly connected to the compressor. A crankshaft seal is not needed.

A motor rotor is usually press fit onto the compressor crankshaft. Some motor compressors are made with the motor at the top, while others have the motor at the bottom and the compressor at the top.

The unit is usually spring mounted inside the hermetic dome. This prevents most of the compressor vibration from being felt outside of the dome.

The exhaust and suction lines inside the dome are made flexible. A connection through the dome provides means of fastening the compressor lines to the remainder of the system. The electrical connections to the motor pass through the dome by means of an insulated leakproof seal.

To lubricate the compressor, the return suction gas is fed into a hollow disk mounted on the motor compressor shaft. Centrifugal force throws the oil and a liquid refrigerant to the outer rim of the disk and flows over the motor windings. (Centrifugal force action rotates things to pull spinning particles away from the center of the rotation.) Only the vapor refrigerant remains at the center and is drawn into the cylinders of the compressor.

A hermetic motor compressor usually requires an outside electrical relay starting mechanism. Fig. 4-27 shows a section through such a motor.

Some motor compressors are two-speed. These are popular in large systems and in air conditioning where heat loads change.

Fig. 4-28 illustrates a hermetically sealed compressor with an internal and external steel shell combined into a single housing. The suction gas goes through the motor area, in effect cooling the motor. The unit has an internal accumulator that prevents liquid from returning to the cylinder area. The discharge line is coiled to the unit and keeps the oil warm enough to evaporate any liquid refrigerant that may have returned. The piston head is sculptured and has a circular valve plate design. This provides a balanced inlet and outlet flow of the vapor. The unit has rotolock connections which allow for servicing. This unit contains an internal discharge muffler which prevents excessive vapor pulsation and vibration. See Para. 4-51 for additional information. The system also has an internal overload that senses temperature and amperage.

4-26 TYPES OF COMPRESSORS

There are five basic types of compressors in use:
1. Reciprocating (piston-cylinder).
2. Rotary.
3. Screw-type.
4. Centrifugal.
5. Scroll.

4-27 RECIPROCATING COMPRESSORS

The original energy source is usually an electric motor. Its rotary motion must be changed to reciprocating motion. This change is usually made by a crank and a rod connect-

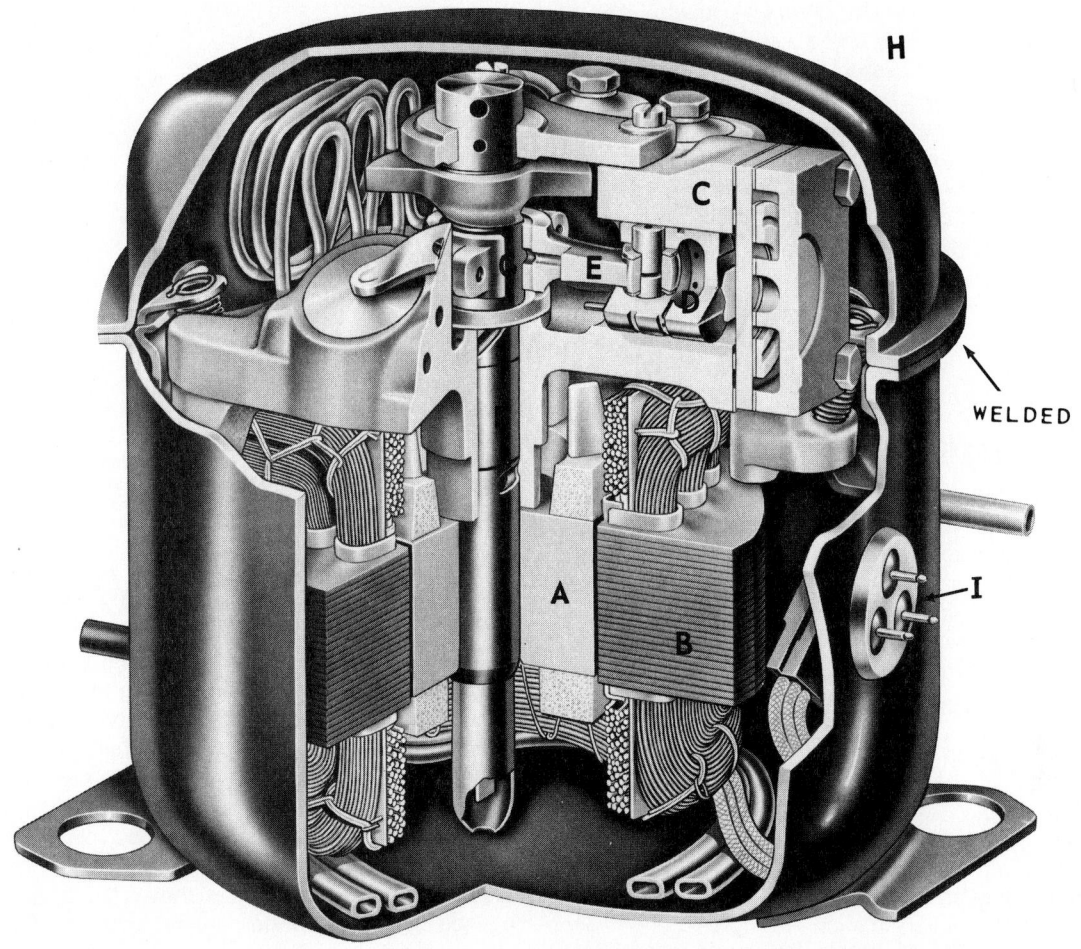

Fig. 4-27. Reciprocating hermetic compressor. Compressor is at top and motor at bottom. Assembly is mounted on springs inside dome. A—Motor rotor. B—Motor stator. C—Compressor cylinder. D—Compressor piston. E—Connecting rod. F—Crankshaft. G—Crank throw. H—Compressor shell. I—Glass sealed electrical connections through compressor shell. (Tecumseh Products Co.)

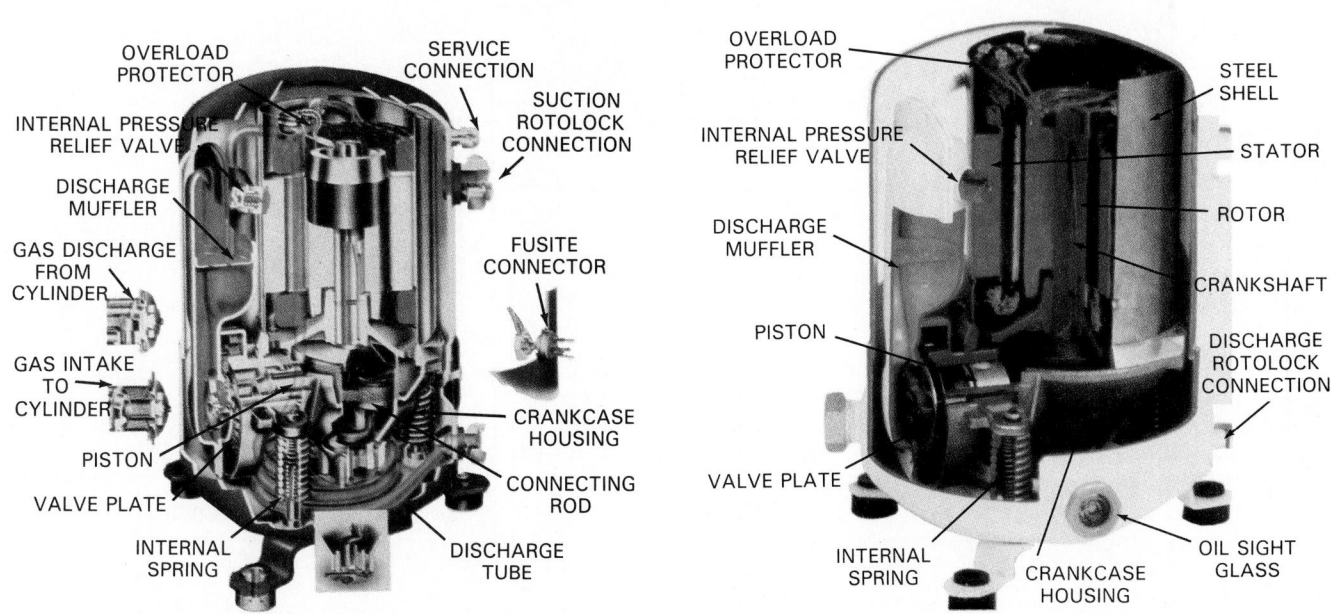

Fig. 4-28. Hermetic compressor steel shell with a winding that is sealed and running gear enclosed, allowing the compressor to take continuous liquid slugging. (Maneurop Inc.)

ing the crank to the piston. The complete mechanism is housed in a leakproof container called a crankcase. It is very efficient. Its construction resembles, in many ways, that of the automobile engine.

Basically, this compressor is a cylinder and a piston. Fig. 4-29 shows the principle of operation of a reciprocating compressor. In illustration No. 1, the piston, B, has moved downward in the cylinder, A, and has moved refrigerant vapor from the suction line, C, through the intake valve, E, and into the cylinder space, G. In illustration No. 2, the piston has moved upward and has compressed the vaporized refrigerant into a much smaller space (clearance space) marked H, and has pushed the compressed vapor through the valve, F, into the condenser.

4-28 PISTON CYLINDER CRANK ARRANGEMENTS

In compressors having more than one cylinder (multicylinder), the crankshaft and cylinders are usually arranged to make the compressor as compact as possible. At the same time, it should provide more pumping capacity for each revolution of the crankshaft. Most two-cylinder compressors use a side-by-side arrangement of the cylinders and a 180° crankshaft. While one piston is at the top of the stroke, the other piston is at the bottom. Other two-cylinder compressors have two cylinders at a 90°V. See Fig. 4-26. With this type of cylinder arrangement, a single throw crank is used.

Fig. 4-30 illustrates some common cylinder and crankshaft arrangements.

4-29 CYLINDERS

Compressor cylinders for external drive compressors are usually made of cast iron. The cast iron must be dense

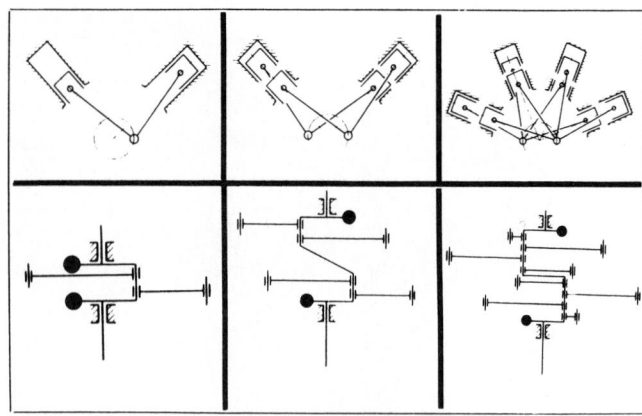

Fig. 4-30. Piston, cylinder, and crankshaft arrangements for two-, four-, and eight-cylinder compressors.

enough to prevent the seepage of refrigerant through it. Some nickel is usually added to give the casting this density.

Small compressors usually have fins cast with the cylinders to provide better air cooling. Larger compressors may have water jackets surrounding the cylinders for cooling. Some compressors are built with cylinder liners or sleeves which may be replaced when worn.

Usually, the crankcase is part of the same casting as the cylinder. This practice cuts down the number of joints that might leak. It also permits close alignment between crankshaft main bearings and cylinder. The main bearings are ball type. Construction is shown in Fig. 4-31. The side-by-side cylinder arrangement is commonly used on open compressors.

This compressor is designed for use on vehicle air condi-

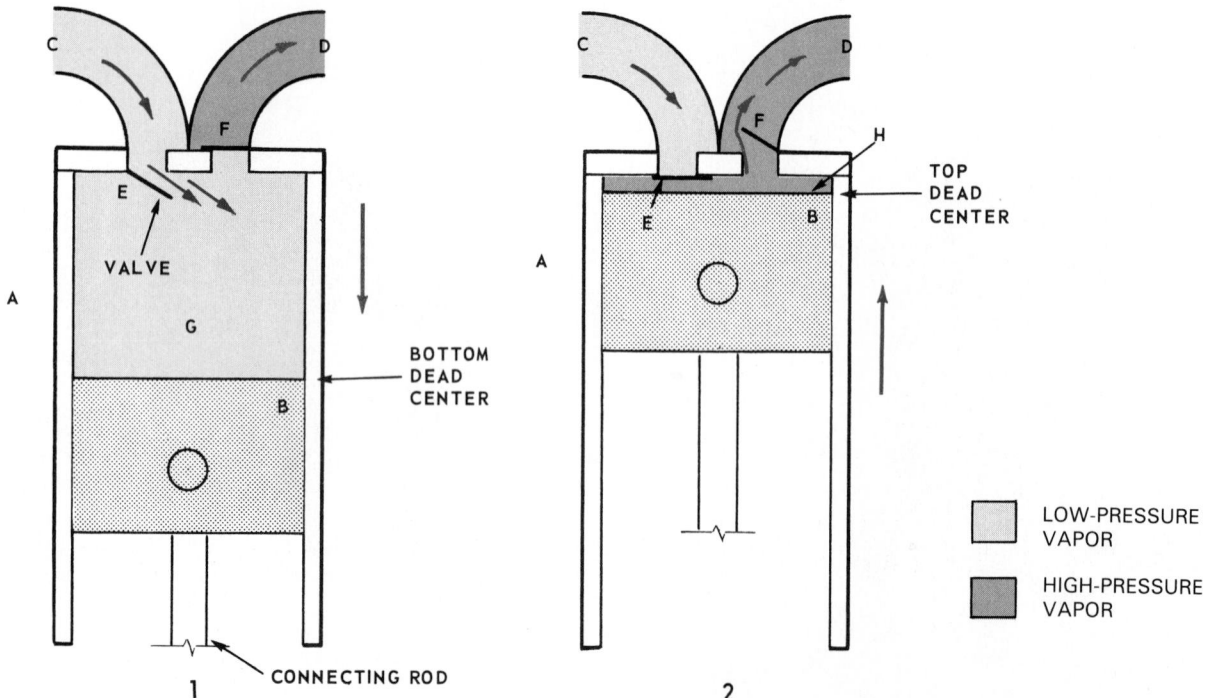

Fig. 4-29. Basic construction of reciprocating compressor. A—Cylinder. B—Piston. C—Intake port from suction line. D—Exhaust port to condenser. E—Intake valve. F—Exhaust valve. G—Piston displacement indicates volume of vapor drawn into cylinder on intake stroke. H—Clearance space at end of compression stroke. Left-hand illustration shows intake stroke; right-hand the exhaust stroke.

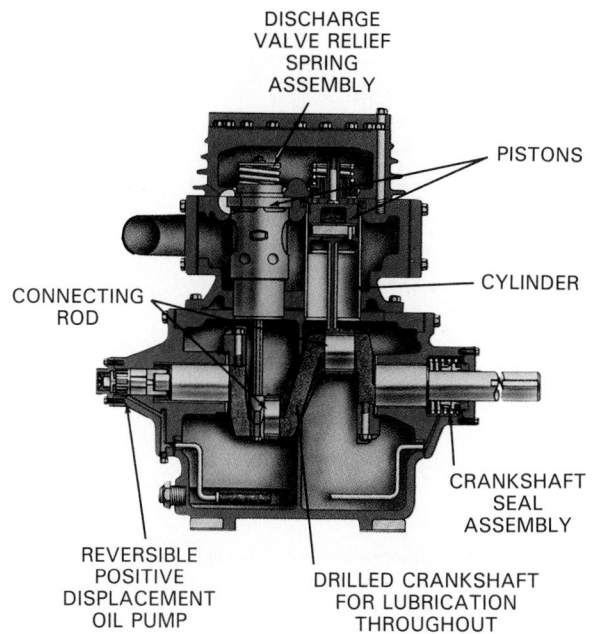

Fig. 4-31. Cutaway view of small, external drive, two-cylinder reciprocating compressor. The body is a casting using lightweight alloy. Cast-iron cylinder liners are permanently cast into crankcase body. (Grasso, Inc.)

tioning. However, the same design may be used in other air conditioning applications.

Hermetic (sealed) compressors usually have cast iron cylinders. Some may be of aluminum or other materials. (Typical of hermetic compressor cylinders is C in Fig. 4-27.) Another type of hermetic compressor is pictured in Fig. 4-32. This is a bolted type hermetic and can be dismantled easily for servicing.

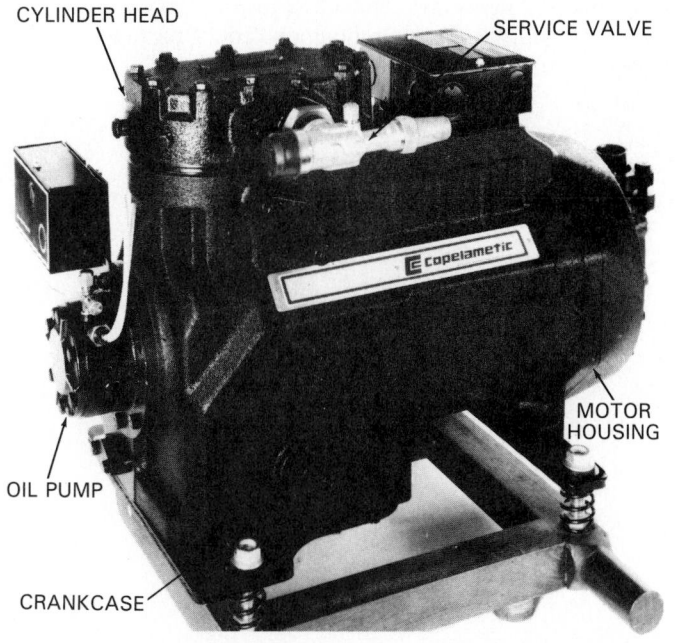

Fig. 4-32. Bolted type hermetic motor compressor. Motor is at right; compressor is at left. (Copeland Corp.)

4-30 PISTONS AND PISTON RINGS

Pistons used in external drive compressors are usually made of cast iron, while in small high-speed hermetic compressors they are of die-cast aluminum. Smaller sizes do not have piston rings.

Since the temperature of pistons seldom goes higher than 250°F (121°C), there is not much expansion of either piston or cylinder. Pistons may be fitted with as little as .0002 in. (.0051 mm) clearance for each inch in diameter.

The smaller pistons have oil grooves cut in them. Fig. 4-33 illustrates a common piston connecting rod assembly. Fig. 4-34 illustrates a commercial type piston and connecting rod assembly. This one is fitted with piston rings.

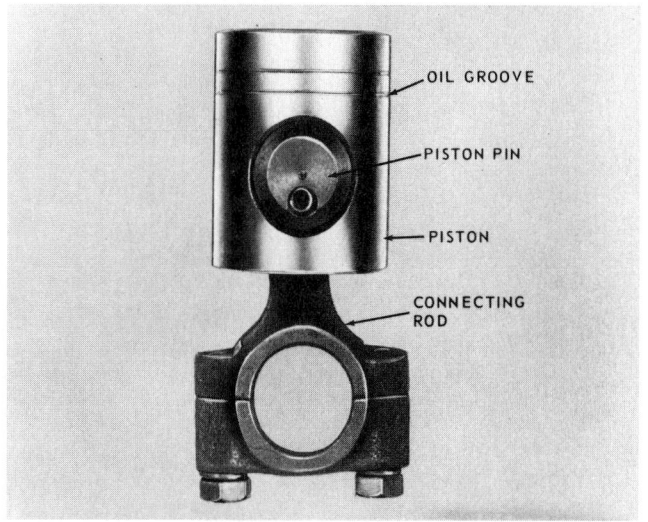

Fig. 4-33. Piston and connecting rod assembly. Note oil grooves cut in piston. (General Electric Co.)

There are two types of piston rings. The upper ring or rings are known as compression rings and the lower is designed to control the oil "flow" past the piston. It is an oil ring.

Piston rings are usually made of cast iron. Some bronze rings have been used, however. Rings should be fitted to the groove as closely as possible and still allow movement. A ring is a complete circle with a gap in it. A 45° tapered or angled ring gap permits the ring to push out against the cylinder wall. This gap should be about .001 in. (.0254 mm) for each inch of piston diameter.

Piston pins are made of case hardened high carbon steel accurately ground to size. They are hollow to reduce weight.

Piston pins are usually of the full floating type. This means the pin is free to turn in both the connecting rod bushing and the piston bushings.

The piston is designed to come as close as possible to the cylinder head without touching it. This is to press as much of the vapor into the high-pressure side as possible.

When the piston is at top dead center of its stroke, the clearance between the piston and cylinder head is approximately .010 in. to .020 in. (.254 mm to .508 mm). The volume of space created is called clearance space. (It is illustrated at H in the right half of Fig. 4-29.)

There is a valve plate under the cylinder head with both

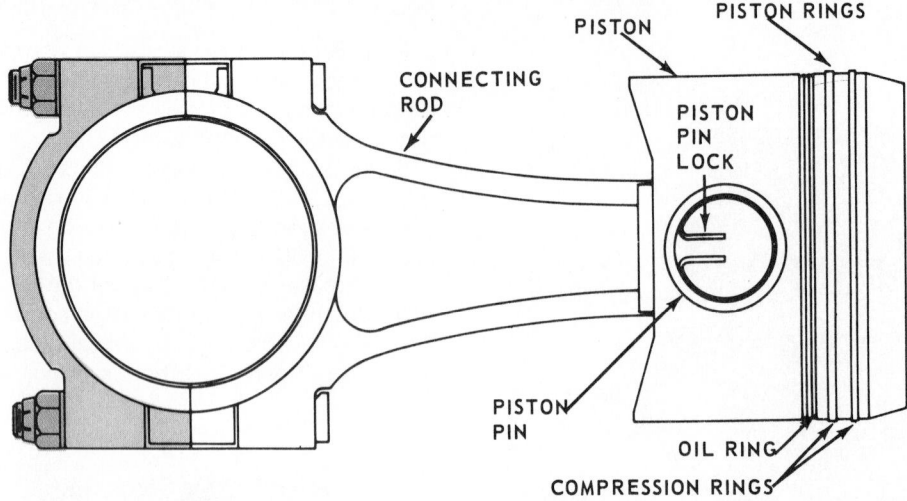

NOTE: PISTON, ROD, AND PIN ARE A MATCHED SET.

Fig. 4-34. Compressor piston and connecting rod assembly. Note how connecting rod's lower (left) end is split and then bolted together to provide bearing to fit crankshaft journal.

the intake and exhaust valve located in it, Fig. 4-35.

In hermetic systems, the construction of the pistons and rings, if used, are much the same as those used in external drive compressors. However, since the hermetic compressors usually run at a higher speed than external drive compressors, the pistons are smaller in diameter and are made as light as possible.

Fig. 4-36 illustrates a hermetic compressor. It uses a Scotch yoke piston arrangement. Fig. 4-37 shows a hermetic motor compressor with four opposing pistons on a flat plane.

4-31 CONNECTING RODS

The connecting rod attaches the piston to the crankshaft. A conventional connecting rod is shown in Fig. 4-34.

Connecting rods for external drive compressors are usually made of drop forged steel. Crankshafts with a throw use a connecting rod having a split lower end that clamps around the crankshaft journal. The rod bearing must be fitted to a clearance of about .001 in. (.0254 mm). It is, therefore, important that the bolts be carefully tightened (torqued).

Fig. 4-38 shows four connecting rods mounted on one crankshaft journal. This arrangement could be used in a radial or a V-type compressor.

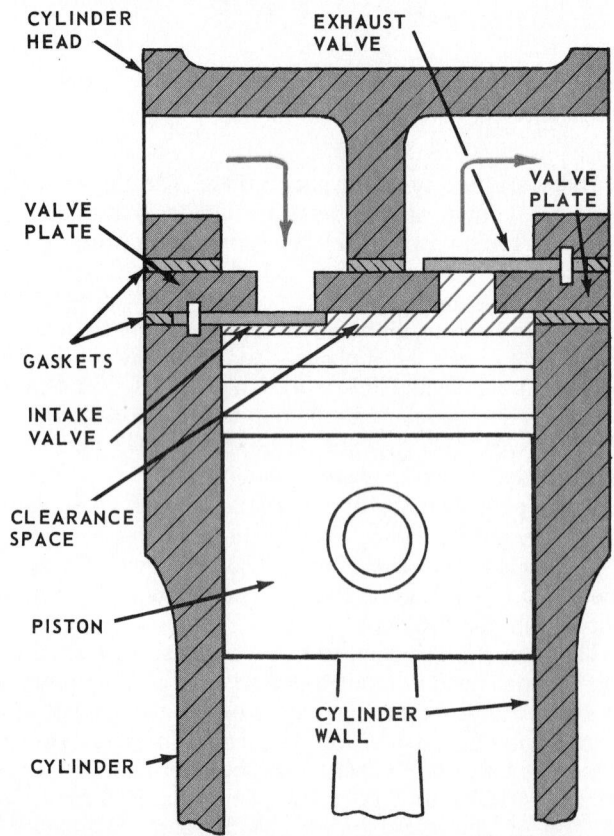

Fig. 4-35. Cross section through compressor cylinder showing cylinder, piston, valve plate, valves, gaskets, and cylinder head.

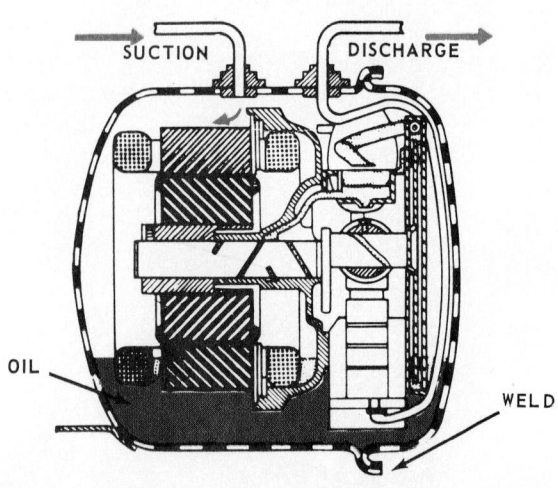

Fig. 4-36. Single cylinder hermetic compressor. Cool refrigerant vapor from suction line flows over motor windings to aid in cooling motor. Compressor uses Scotch yoke piston crank mechanism. Compressor is inverted and a horizontal motor shaft is used.

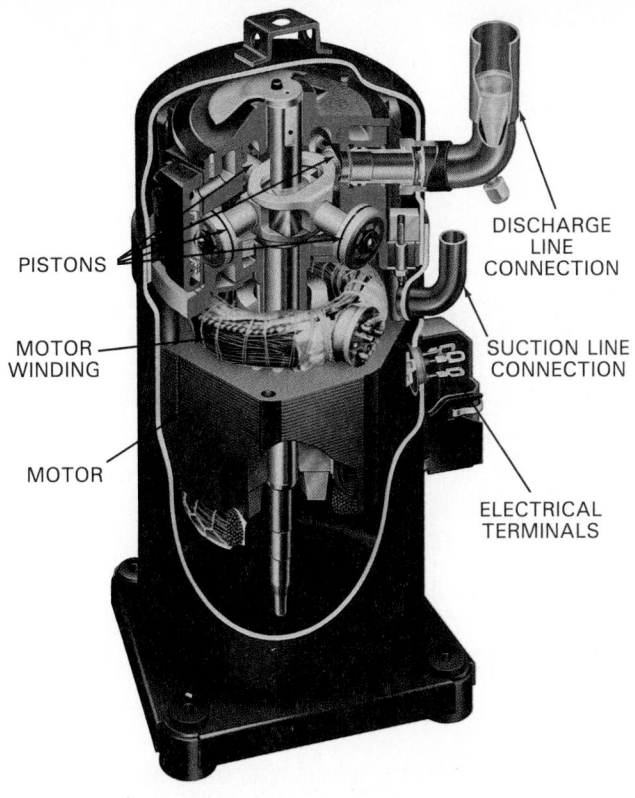

Fig. 4-37. Four opposing pistons—"Quadro-Flex." The four opposing pistons on a flat plane reduce the vibration of the unit and, therefore, it can be used for rooftop heat pumps. (Tecumseh Products Company)

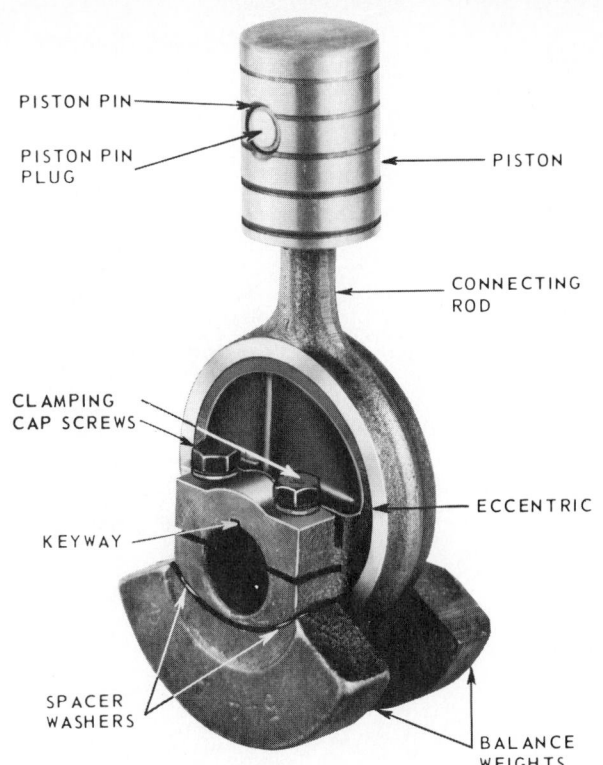

Fig. 4-39. Eccentric crankshaft assembly. Note clamping cap screws and balance weights.

The eccentric type connecting rod usually has a cast iron bearing surface. The crank throw end is a solid ring and must be mounted on the eccentric before the crankshaft proper is assembled to the eccentric. The construction is shown in Fig. 4-39.

A small piston and connecting rod assembly, used in a hermetic unit, is shown in Fig. 4-33. The connecting rod is attached rigidly to a large piston pin by means of a locking pin and spring. The dismantled unit is illustrated in Fig. 4-40.

4-32 CYLINDER HEAD

Cylinder heads for both external drive and hermetic compressors are usually made of cast iron. The head serves as a pressure plate to support and hold the valves and valve plate in position. It also provides the vapor passages into

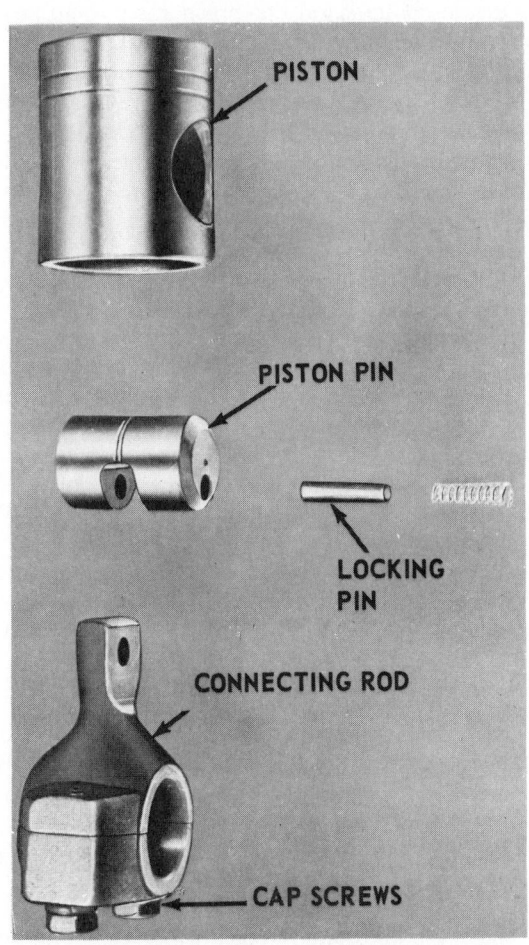

Fig. 4-40. A piston and connecting rod assembly dismantled. This is the same assembly shown in Fig. 4-33. Note small pin which locks connecting rod to large piston pin. Also note the two cap screws used to attach cap to crankshaft end of connecting rod.

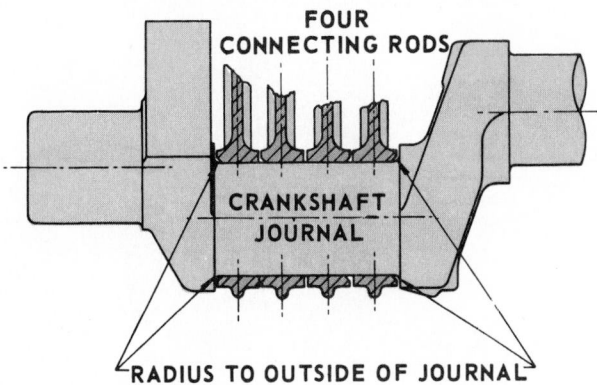

Fig. 4-38. Four connecting rods mounted on one crank journal. This arrangement could be used on radial type compressor.

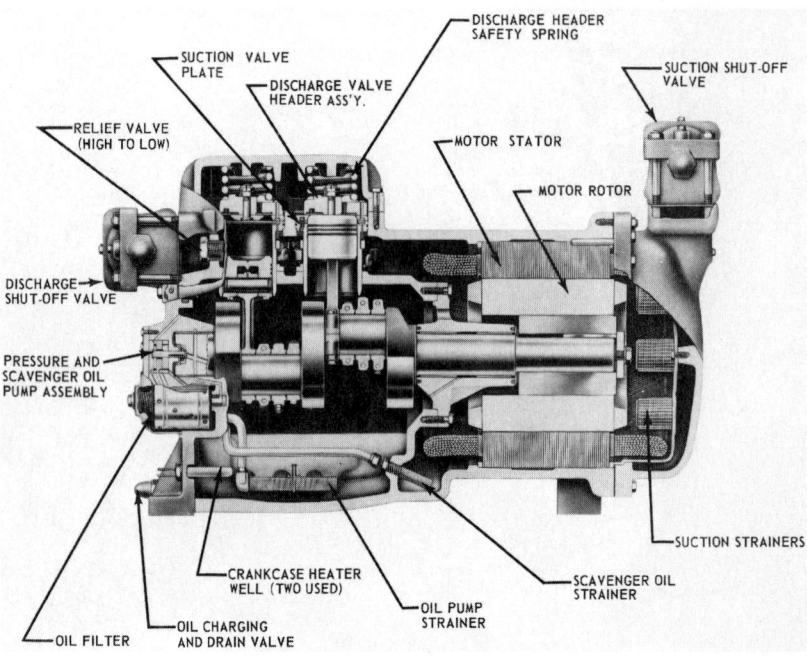

Fig. 4-41. A serviceable hermetic reciprocating type commercial compressor. A bolted type, it has four banks of two cylinders each (four connecting rods on each crankthrow).

and out of the compressor. The pressures of compression may amount to as much as 300 psi (2 170 kPa) depending upon the kind of refrigerant used. The valve plate must, therefore, have good support so that there will be no leakage at the gaskets on either side of the valve.

In some hermetic systems, the entire compressor housing is inside a dome. The entire space within the dome is open to the suction line. Consequently, the whole dome is under low-side pressure. In such systems, no intake manifold is required—merely an opening into the intake valve or valves.

The cylinder head is usually attached to the cylinder with cap screws. Fig. 4-41 is a cutaway view of a commercial multi-cylinder reciprocating type hermetic compressor.

The suction line connects to the shutoff valve on the right end. The exhaust, which connects to the condenser, is connected to the discharge shutoff valve at the left end.

Note the location of the crankcase heater (see Para. 4-60). Note, also, the discharge header safety spring. This relieves the pressure in the cylinder should it exceed a safe working limit (due to slugging of liquid refrigerant or refrigerant oil).

4-33 VALVES AND VALVE PLATES

The usual valve assembly consists of a valve plate, an intake valve, an exhaust valve, and the valve retainers. Refer to Fig. 4-42.

Valve plates are sometimes made of cast iron, but hardened steel is also used, as plates can be thinner with longer wearing valve seats.

Compressor valves are usually made of high carbon alloy steel. They are heat treated to give them the properties of spring steel and ground to a perfectly flat surface.

The intake valve is usually kept in place by small pins or the clamping action between the compressor head and valve plate. Exhaust valves may be clamped the same way.

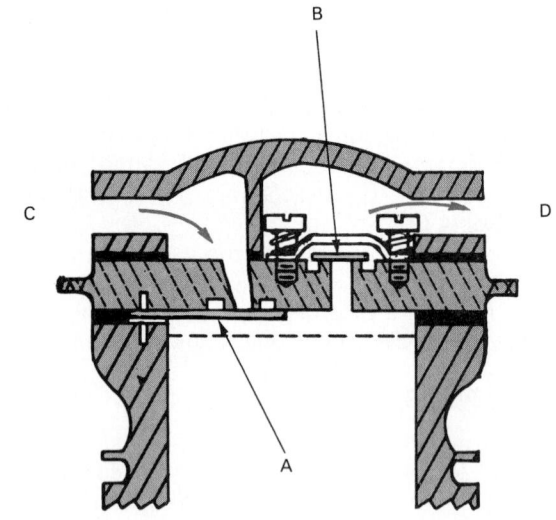

Fig. 4-42. Typical compressor valve plate. A—Intake valve. B—Exhaust valve. Heavy springs on exhaust valve cage permit a greater valve lift to protect compressor in case of severe liquid refrigerant or oil pumping.

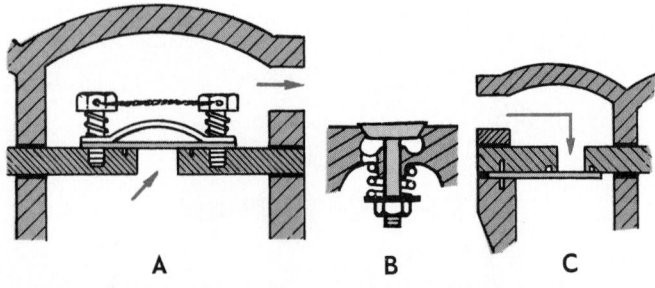

Fig. 4-43. Some typical compressor valve designs. A—Reed valve, spring closed. B—Poppet valve, spring closed. It is used on some large compressors. C—Reed valve. Pressure difference keeps valve closed.

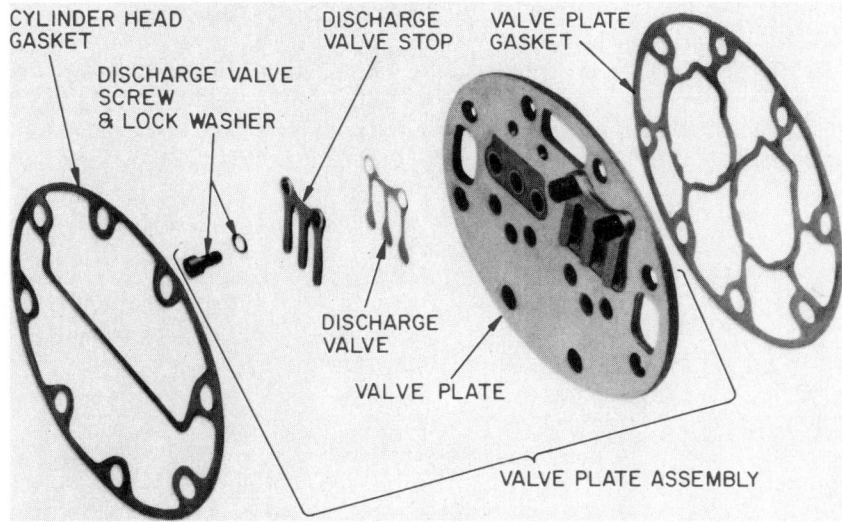

Fig. 4-44. Centrifugal valve plate assembly.
(Carrier Corp., Subsidiary of United Technologies Corp.)

Some different valve designs are displayed in Fig. 4-43. Fig. 4-44 shows a typical valve plate assembly.

The valve disks or reeds must be perfectly flat. A defect of only .0001 in. (.00254 mm) will cause valve to leak.

Of the two valves, the intake gives the least trouble. This is because it is constantly lubricated by oil circulating with the cool refrigerant vapors. Also, it operates at a relatively cool temperature.

The exhaust valve must be fitted with special care. It operates at high temperatures and must be leakproof against a relatively high pressure difference.

Because of the high vapor pressures and the high temperatures, there is a tendency for the heavy ends (heavy molecules of hydrocarbon oils) to collect on the valve and valve seat as carbon.

The valves open about .010 in. (.254 mm). If the movement is more, a valve noise develops. If the movement is too little, not enough vapor can move past the valve.

In small high-speed hermetic compressors, the intake valves are made very light and as large as possible. Greater amounts of refrigerant vapor are thus allowed to enter the cylinder during the very small fraction of a second that the intake valve is open.

4-34 CRANKSHAFT SEAL

Refrigerating systems that use an external motor (open type) to drive the compressor need a leakproof joint where the crankshaft comes out of the compressor crankcase. This is absolutely necessary, as the pressures vary greatly in the crankcase.

This joint requires seals that are carefully designed and installed, for it is a place where the shaft rotates part of the time and then is at rest part of the time.

All seals use two rubbing surfaces. One surface turns with the crankshaft and is sealed to the shaft with an O-ring of synthetic material. The other surface is stationary and mounted on the housing with leakproof gaskets. The surface materials (accurate to .000001 in. [.0000254 mm] and optically flat) are made of these different combinations: hardened steel and bronze, ceramics and carbon. The two

rubbing surfaces must be lubricated or they will wear and start to leak.

Teflon is often used as a gasket material on automobile air conditioning compressors. The crankshaft seal must operate at a high temperature. It is usually made of carbon and ceramic.

Fig. 4-45 illustrates some popular crankshaft seals.

4-35 COMPRESSOR DRIVE (EXTERNAL DRIVE)

External drive compressors are usually driven by a V-belt. Most are driven at less than the motor speed. This means that the motor belt pulley will be smaller than the compressor pulley. The V-belt provides a quiet, efficient drive.

In large capacity installations, more than one belt may be used in order to transmit the required horsepower. Fig. 4-46 illustrates a two-belt external drive system.

Pulleys must be in perfect alignment and proper tension must be provided on the belt. Pulley shafts (motor and compressor) must be exactly parallel to each other. Most V-belt pulleys are made of cast iron but some are built up from stamped steel parts.

The diameter of the motor drive pulley and the compressor flywheel will govern the speed of the compressor.

4-36 CRANKSHAFT

Reciprocating type compressors must use some means to change the rotary motion of the motor into reciprocating motion in the compressor. The crank throw-connecting rod-piston combination is most frequently used. The crankshaft in these designs is usually made of forged or cast steel. Refer to Fig. 4-47.

Some compressors use an eccentric fastened to a straight shaft in place of the conventional crankshaft (eccentric shaft). (An eccentric is a section of a shaft which is larger and has a different center than the shaft.) This construction is used to reduce vibration and to remove the need for connecting rod caps and bolts. See Figs. 4-48 and 4-39.

The crankshaft main bearings support the crank. They also must carry any end load. Crankshaft and connecting rod bearings are fitted with great accuracy. Clearance for lubrication is usually .001 in. (.0254 mm). In external drive compressors, the drive pulley is usually attached to the crankshaft with a standard taper, a key, and a nut-lock washer combination.

Many hermetic systems use the crank throw-crankshaft-connecting rod-piston arrangement, Fig. 4-27.

4-37 SCOTCH YOKE

The Scotch yoke mechanism is shown in Figs. 4-49 and 4-50. There is no connecting rod. The cylinder and piston are both quite long, and even at the lower end of the stroke the piston is still guided by the cylinder wall. The crankshaft

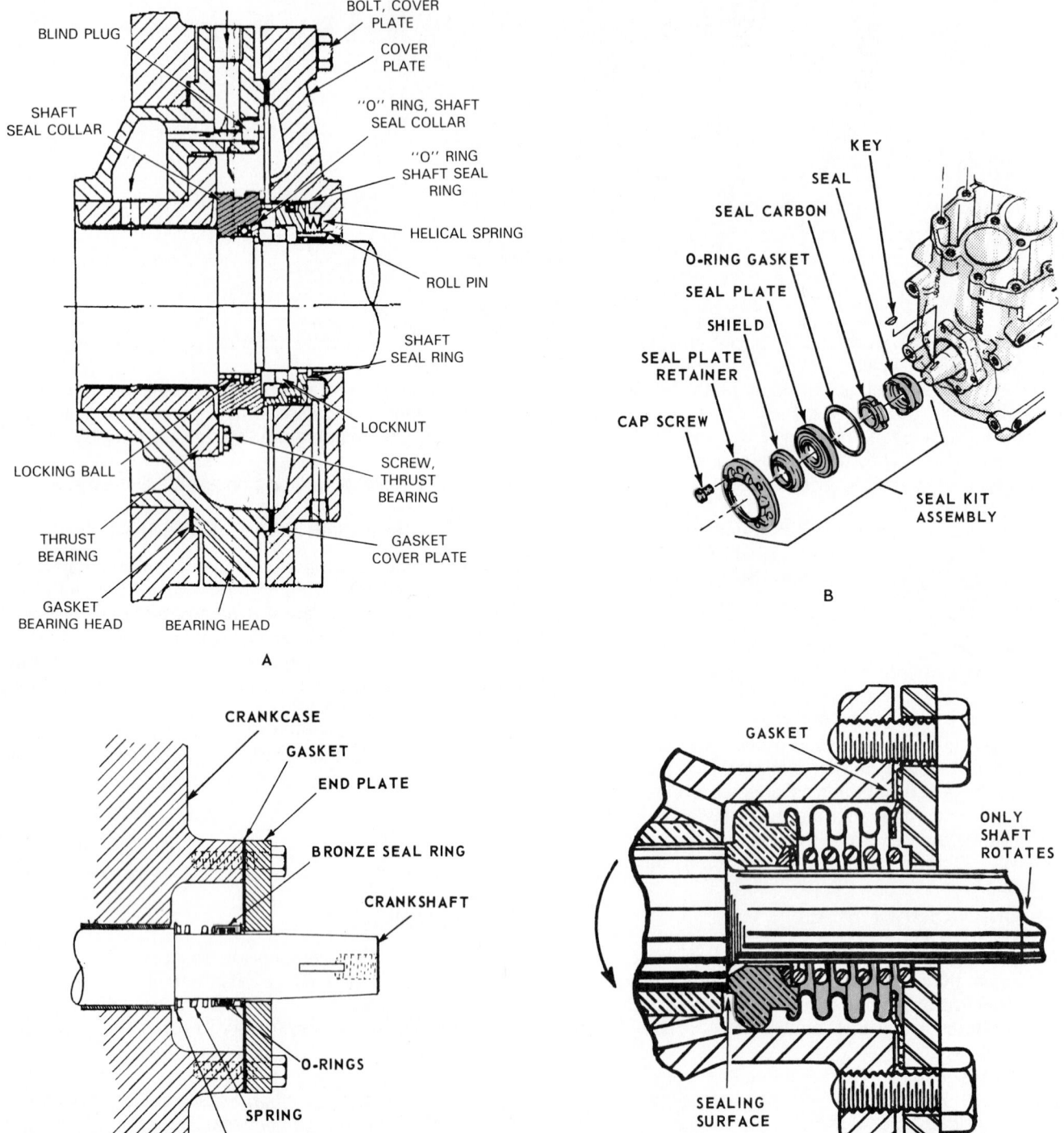

Fig. 4-45. Crankshaft seal construction for external drive compressors. A—Seal used in commercial compressors. (Mycom Corp.) B—Seal used with an automobile air conditioning compressor. (Ford Division) C—Replacement seal. (Chicago Valve Plate & Seal Co.) D—Bellows type seal.

Fig. 4-46. External drive compressor with two-belt drive. Service technician is checking for proper belt tension.   (The Gates Rubber Co.)

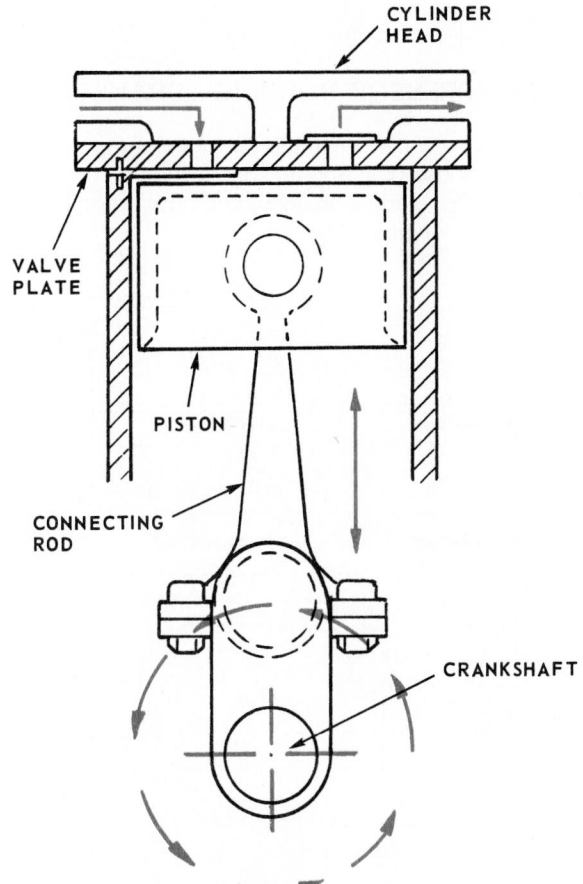

Fig. 4-47. Crank throw type crankshaft. As crankshaft revolves, piston reciprocates (moves up and down). Piston pin oscillates (swings back and forth) as it reciprocates with the piston. Lower end of the connecting rod rotates with crankshaft.

pin, also called the crankthrow, connects to the lower end of the piston by means of a floating bearing. The Scotch yoke is popular in small high speed compressors.

4-38 SWASH PLATE

A popular type of reciprocating compressor used on many automobile air conditioning systems is known as a "swash" plate or "wobble" plate compressor. No connecting rod is used in this type of compressor. The cylinder and pistons are mounted as in Fig. 4-51.

As the shaft revolves, the swash plate causes the pistons to reciprocate in the cylinders. Usually the swash plate

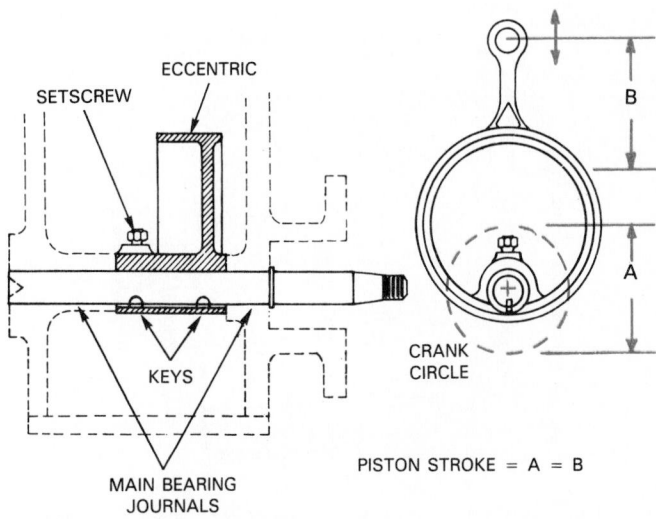

Fig. 4-48. Eccentric type crank mechanism. Note that eccentric is attached to shaft with key and setscrew.

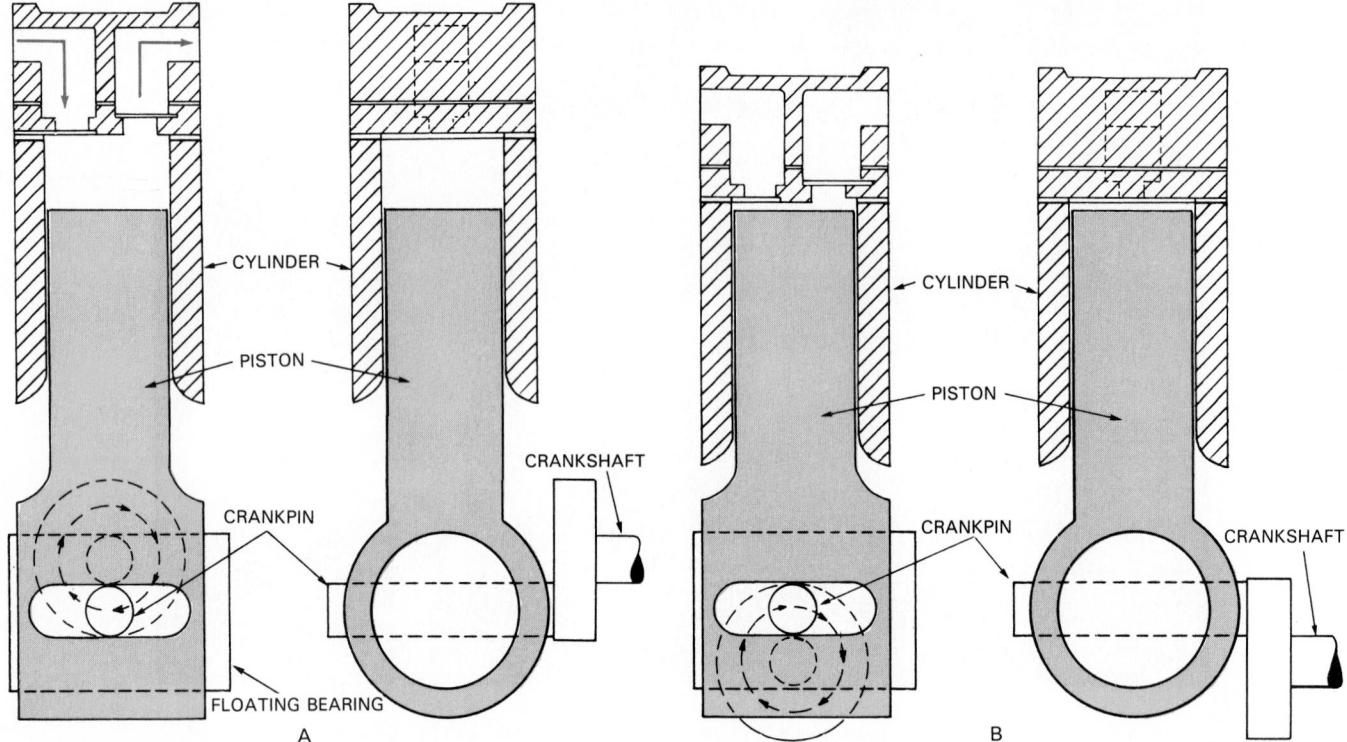

Fig. 4-49. Scotch yoke mechanism used to connect piston to crankshaft. No connecting rod is used. Piston extends to yoke mechanism and compressor cylinder serves as a guide. A—Shows piston at bottom of stroke (end of intake stroke). B—Shows piston at top of stroke (end of exhaust stroke).

compressor has three or more cylinders arranged in a circle around the drive shaft.

Since the compressor is double acting—that is, compression takes place at each end of the stroke—a three-cylinder compressor will give a pumping action like a six-cylinder conventional compressor of the same cylinder and stroke dimensions. This is an external drive compressor. It requires a seal where the drive shaft passes through the compressor housing.

4-39 COMPRESSOR HOUSING—CRANKCASE

In both the external drive and hermetic compressor, the compressor housing supports the cylinders, crankshaft, valves, oil pump, lubrication lines, and both refrigerant inlet and exhaust openings.

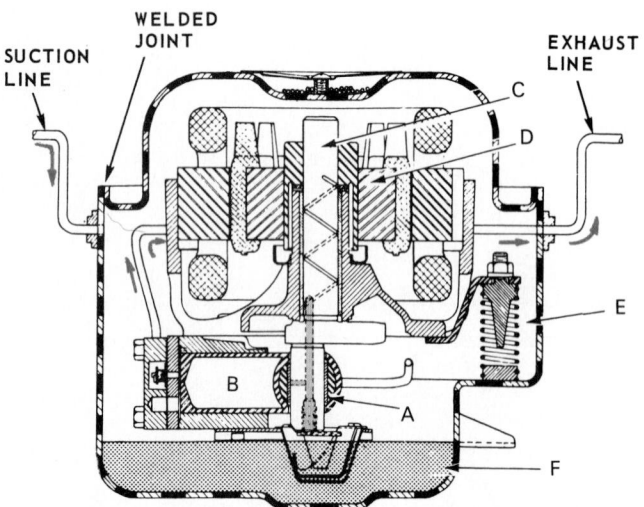

Fig. 4-50. Hermetic compressor using Scotch yoke mechanism. A—Crank throw and Scotch yoke. B—Hollow piston. C—Combined motor shaft and crankshaft. D—Crankshaft thrust bearing. E—Internal mounting spring. F—Oil reservoir.

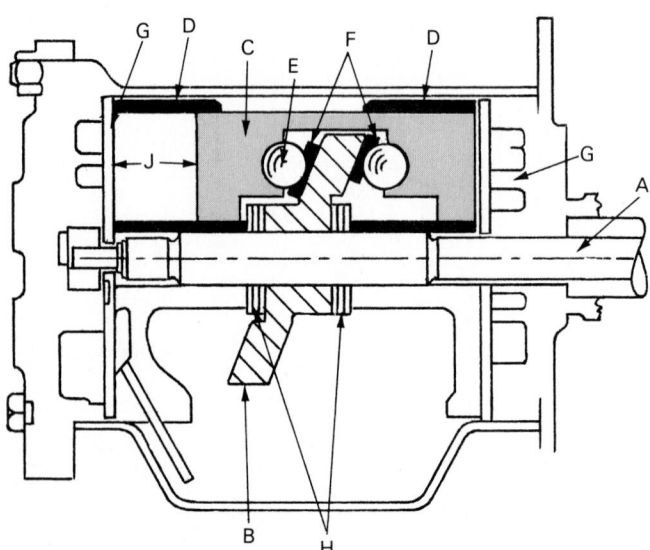

Fig. 4-51. Cross section through a "swash" plate type of reciprocating compressor. A—Drive shaft. B—Swash plate. C—Piston. D—Cylinder wall. E—Drive ball. F—Ball shoe. G—Valve plate (valve not shown). H—Thrust bearing. J—Piston stroke. As drive shaft and swash plate revolve, double end piston is moved back and forth in cylinder.

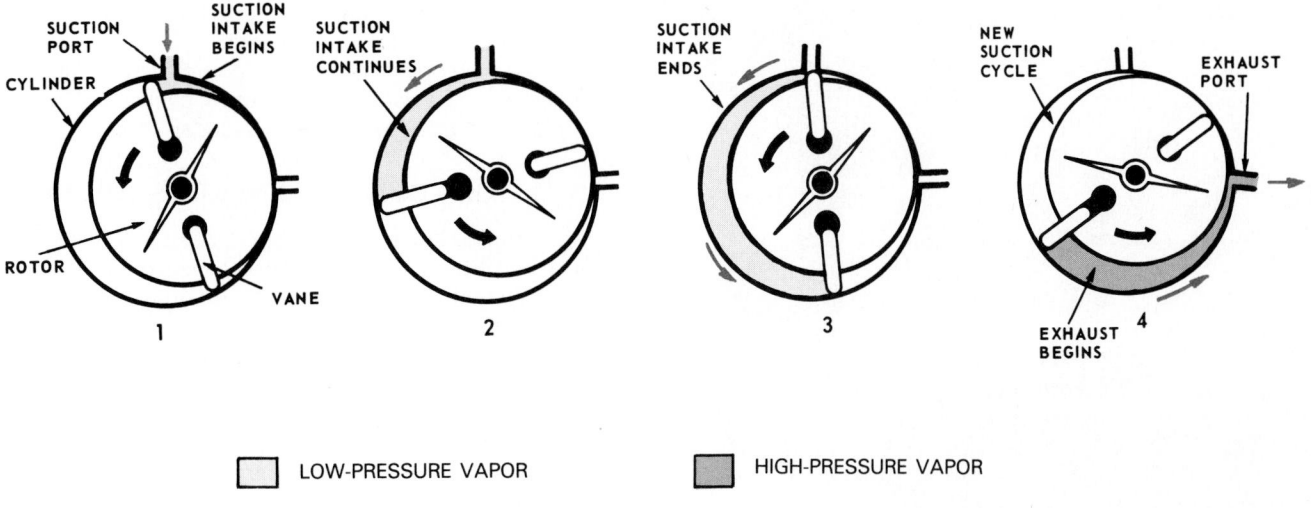

LOW-PRESSURE VAPOR   HIGH-PRESSURE VAPOR

Fig. 4-52. A rotary blade compressor. Black arrows indicate direction of rotation of rotor. Red arrows indicate refrigerant vapor flow.

In the case of hermetic systems, the housing also supports and aligns the driving motor. Typical compressor housing designs are shown in Figs. 4-32 and 4-50.

Some hermetic designs are bolted together and are provided with service valves. These are called serviceable hermetics, Fig. 4-32. Many hermetic housings, particularly in the smaller sizes, are welded together. Refer to Figs. 4-27 and 4-36.

4-40 INTAKE AND EXHAUST PORTS

Conventional external drive compressors provide inlet and exhaust ports as part of the cylinder head. These ports are usually fitted with service valves. See Fig. 4-26.

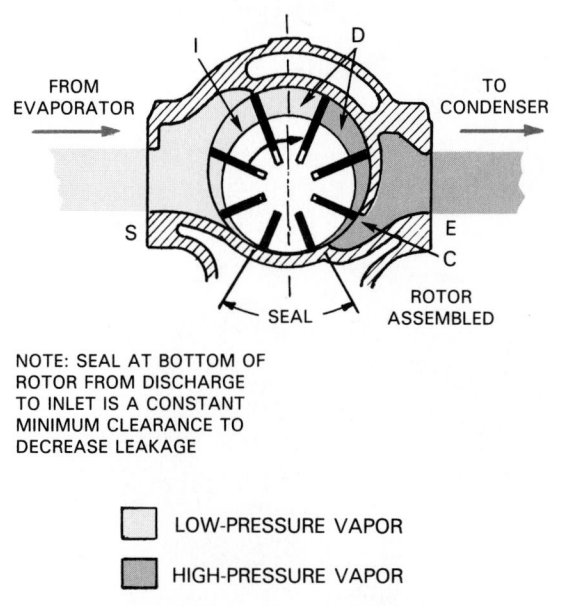

NOTE: SEAL AT BOTTOM OF ROTOR FROM DISCHARGE TO INLET IS A CONSTANT MINIMUM CLEARANCE TO DECREASE LEAKAGE

LOW-PRESSURE VAPOR

HIGH-PRESSURE VAPOR

Fig. 4-53. Eight-blade rotary compressor with rotating blades. Black arrow indicates direction of rotation. Red arrows show direction of vapor flow. C—Exhaust port. D—Displacement. E—Exhaust. I—Inlet port. S—Suction. Inlet port is much larger than exhaust port. Large inlet port is needed to collect enough refrigerant vapor from the sparse low pressure side (light blue).

Some hermetic compressors also have service valves. In the small hermetic compressors in which the motor-compressor mechanism is enclosed in a welded dome, the inlet and exhaust lines go directly from the inlet and exhaust port of the compressor through the compressor dome and are not generally supplied with service valves. See the line connections in Fig. 4-50.

4-41 ROTARY COMPRESSOR

There are two basic types of rotary compressors. One has blades (vanes) that rotate with the shaft. The other has a blade which remains stationary and is part of the housing assembly. In both types, the blade itself slides to provide a continuous seal for the refrigerant vapor. Fig. 4-52 shows a typical rotating two-blade compressor. The low-pressure vapor from the suction line is drawn into the opening and fills the space behind the blade as it revolves. As the blades revolve, trapped vapor in the space ahead of the blade is compressed until it can be pushed into the exhaust line to the condenser.

A commercial rotary blade compressor, using eight blades, is pictured in Fig. 4-53. The basic operation of the eight-blade compressor is the same as the two-blade.

Fig. 4-54 illustrates a section through an eight-blade rotary blade compressor. This is an external drive compressor. The shaft seal is shown at the right end.

Fig. 4-55 shows the cylinder with the intake and exhaust ports in Part A. The relative positions of the rotor and cylinder are shown in Part B.

Rotating vane compressors are frequently used as the ''booster'' compressor in cascade systems. This is the name commonly given to the first compressor in a cascade system.

These compressors have three advantages:
1. They provide a large size opening into the suction line.
2. They provide large inlet port openings.
3. They have a very small clearance volume.

Since the low-side pressure may be quite low, this means that the low-side vapor will be drawn into the compressor under a very small pressure difference. Because these compressors provide a large opening into the compressor from

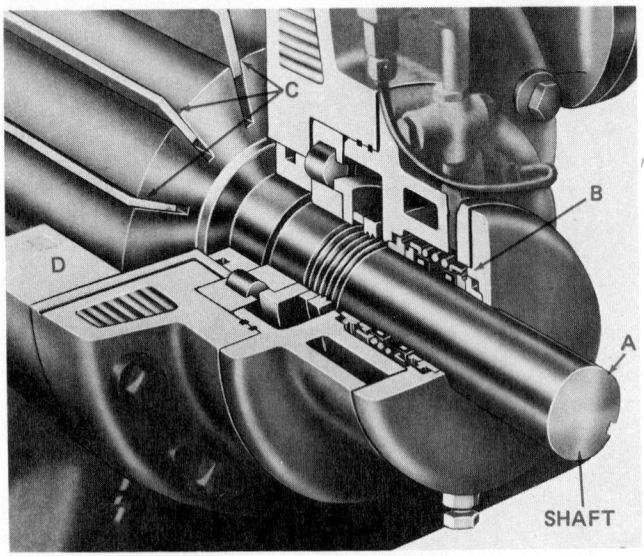

Fig. 4-54. Section through rotating vane compressor. This is an external drive compressor. Note shaft seal at the right end. A—Shaft. B—Shaft seal. C—Blades. D—Housing.

the low side, more vapor will be drawn in on the intake stroke. The small clearance space provided in these compressors means that all the vapor drawn in on the intake stroke will be pushed out on the exhaust stroke. This increases the compressor efficiency. The cascade system is explained in Chapter 17.

Fig. 4-56 represents a stationary blade (often called a divider block) rotary compressor. An eccentric shaft rotates an impeller in a cylinder. This impeller constantly rubs against the outer wall of the cylinder.

As the impeller (or roller) revolves, the blade traps quantities of vapor. The vapor is compressed into a smaller and smaller space, building up the pressure and temperature. Finally the vapor is forced through the exhaust port into the high-pressure side of the system (condenser).

The compression action on one quantity of vapor takes place at the same time another quantity of vapor is filling the cylinder on the intake stroke. All of the parts must be fitted to extremely close tolerances and clearances. The dimensions are so accurate and the surfaces so smooth that no gaskets are needed in the compressor assembly.

Fig. 4-57 shows a hermetic rotary compressor using a stationary blade (dividing block).

In rotary compressors, check valves are usually used in the suction line to prevent the high-pressure vapor and compressor oil from flowing back into the evaporator.

4-42 ROTARY CYLINDER CONSTRUCTION

Rotary cylinders are usually made of cast iron. Each is accurately machined, honed, and lapped on the inner surface and on the ends.

All cylinders have intake and exhaust ports; some models have oil passages for lubrication. Cylinders are usually mounted on an end plate that is part of the main crankcase of the compressor. Refrigerant passages continue into the end plate.

The exhaust valve reed is mounted on the exhaust port outlet of the compressor as close to the compression

chamber as possible. Four or more bolts hold the cylinder to the main part of the compressor.

There are also one or more steel dowel pins to help align the cylinder on the back plate. Another accurately finished plate seals the other end of the cylinder.

4-43 ROTOR—COMPRESSOR CONSTRUCTION

In the rotating blade compressor, the rotor is a fixed part of the shaft. The rotor length must be accurate to .0005 in. (.0127 mm). Usually the slots for the blades are on a radius to the center of the shaft. To lower the starting load, one company puts the slots at an angle to prevent the blades from touching the cylinder until the compressor nears its operating speed.

In the stationary blade compressor, the rotor, sometimes called the impeller, accurately fits the eccentric, a fixed part of the shaft.

Fig. 4-54 illustrates a popular type of rotor construction used on external drive commercial compressors. Fig. 4-58

Fig. 4-55. Cylinder and rotor from Fig. 4-54 in detail. A—The inside of the cylinder showing port openings. Note that intake ports are longer than exhaust ports. B—Relative position of rotor and blades inside cylinder.

illustrates a stationary blade rotary compressor. This is a hermetic compressor used both in refrigeration and air conditioning units.

4-44 BLADE (VANE) CONSTRUCTION

Rotating blade compressors use two or more blades. Fig. 4-55, Part B, shows a rotor with eight of them. These blades may be made of materials like cast iron, steel, aluminum, or carbon.

The efficiency of the compressor depends to a great extent on the condition of the edge of the blade where it rubs on the cylinder. The blade must be very accurately ground to fit the slots, the ends of the cylinder, and other contact surfaces with the cylinder.

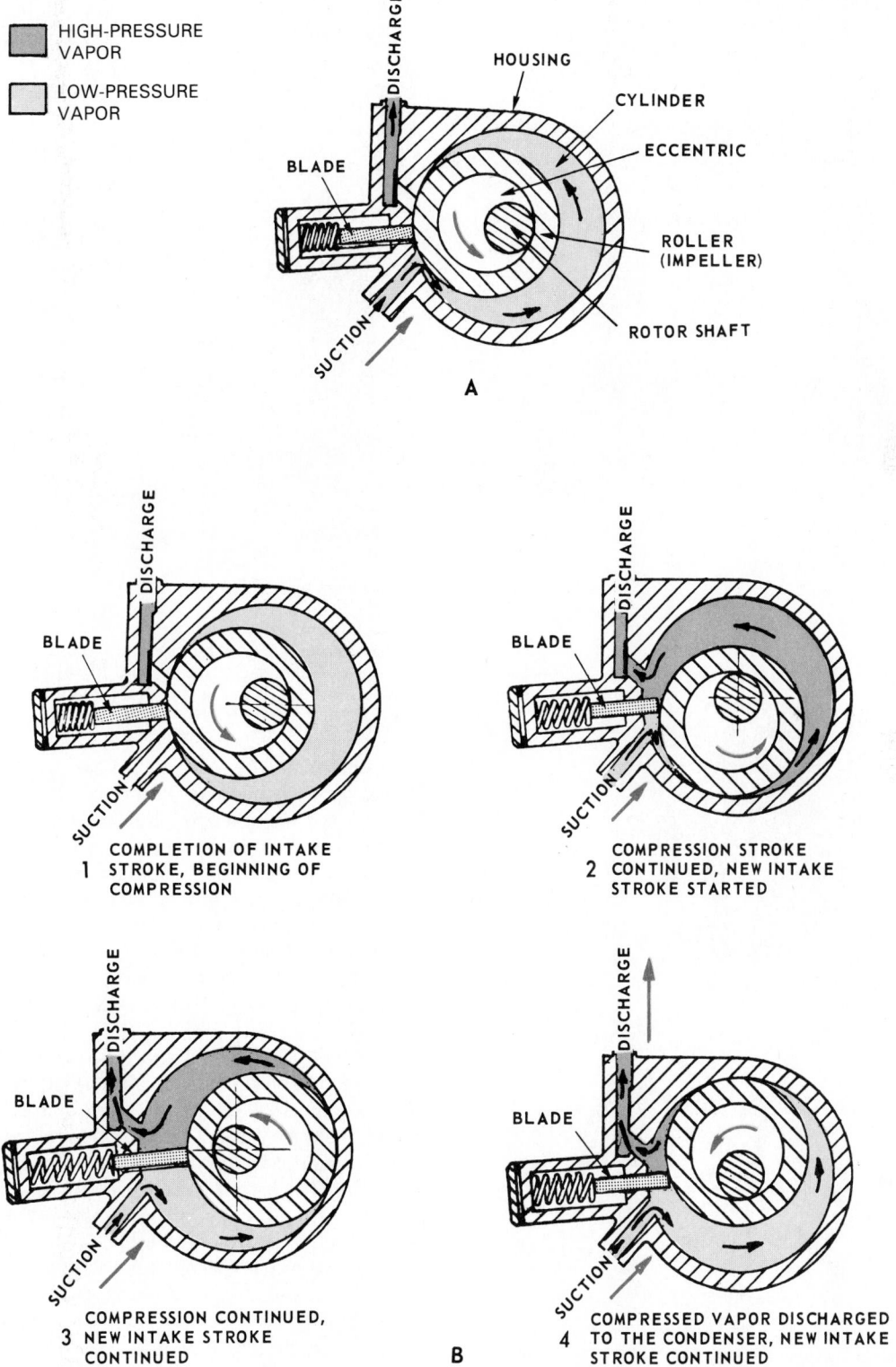

HIGH-PRESSURE VAPOR

LOW-PRESSURE VAPOR

DISCHARGE

HOUSING

CYLINDER

ECCENTRIC

BLADE

ROLLER (IMPELLER)

SUCTION

ROTOR SHAFT

A

BLADE

DISCHARGE

SUCTION

1 COMPLETION OF INTAKE STROKE, BEGINNING OF COMPRESSION

BLADE

DISCHARGE

SUCTION

2 COMPRESSION STROKE CONTINUED, NEW INTAKE STROKE STARTED

BLADE

DISCHARGE

SUCTION

3 COMPRESSION CONTINUED, NEW INTAKE STROKE CONTINUED

BLADE

DISCHARGE

SUCTION

4 COMPRESSED VAPOR DISCHARGED TO THE CONDENSER, NEW INTAKE STROKE CONTINUED

B

Fig. 4-56. Rotary compressor. Stationary blade or divider block is in contact with a roller (impeller). A—Identification of parts. B—Operation.

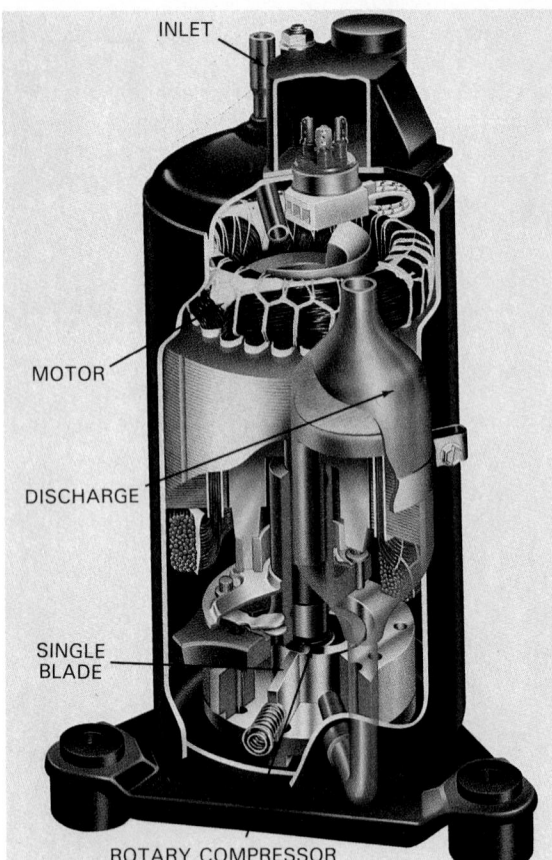

Fig. 4-57. Hermetic rotary single stationary-blade type compressor. (Tecumseh Products Company)

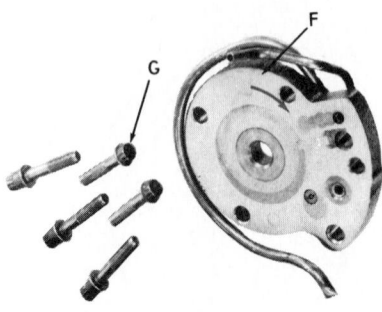

Fig. 4-58. A stationary blade rotary hermetic compressor. A—Cylinder. B—Eccentric. C—Roller. D—Single blade. E—Rotor bearing. F—Cylinder end plate. G—End plate attaching cap screw. H—Suction line from evaporator. I—Exhaust to the condenser. J—Motor winding.

4-45 SCREW-TYPE COMPRESSOR

The screw-type compressor uses a pair of special helical rotors. They trap and compress air as they revolve in an accurately machined compressor cylinder. These compressors are available in either external drive or hermetic construction. They are used in large systems (20 ton capacity and up).

Fig. 4-59 illustrates a cross section. The two rotors are not the same shape. One is male, the other female. The male rotor, A, is driven by the motor. It has four lobes. The female rotor, B, meshes with and is driven by the male rotor. It has six interlobe spaces. The cylinder, C, encloses both rotors.

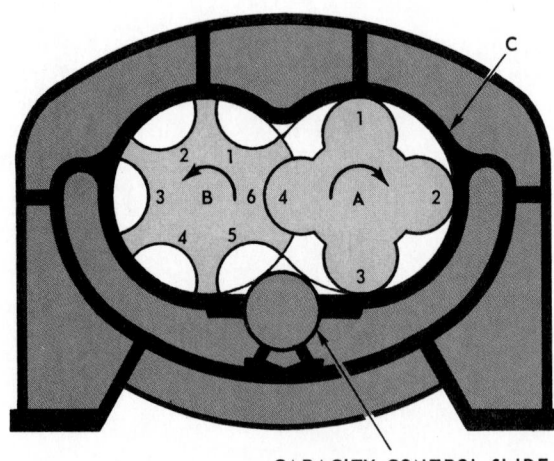

CAPACITY CONTROL SLIDE

Fig. 4-59. Cross section of screw-type compressor. A—Male rotor. B—Female rotor. C—Cylinder. Vaporized refrigerant enters at one end and exhausts at other end. (Stal Refrigeration AB)

In operation, the refrigerant vapor is drawn in as shown in Fig. 4-60. The intake (low-pressure vapor) enters at one end of the compressor and is discharged (compressed vapor) at the opposite end.

Since there are four lobes on the male rotor and six on the female rotor, the male rotor will revolve more rapidly than the female rotor. The rotors are helixes so the pumping action will be a continuous action rather than pulsating as with a reciprocating compressor. In the absence of this reciprocating motion, there is very little vibration during operation.

Fig. 4-61 illustrates a cutaway view of an external drive screw-type compressor. It is powered by an external electric motor, which drives the shaft marked "E." The motor drives the male rotor. Two matched helical rotors, male and female, trap and compress the refrigerant as they turn together. The rotor marked "H" in the illustration is an extension of the drive shaft. The other rotor is made to turn by the action of the male rotor. A control device mounted outside the housing regulates the capacity of the unit. The device can be seen at the left, marked "J."

A hermetic screw-type compressor is shown in Fig. 4-62. Its capacity control device is mounted inside the housing. Its inlet port is located at right angles to the rotors. Outlet port is through the motor housing.

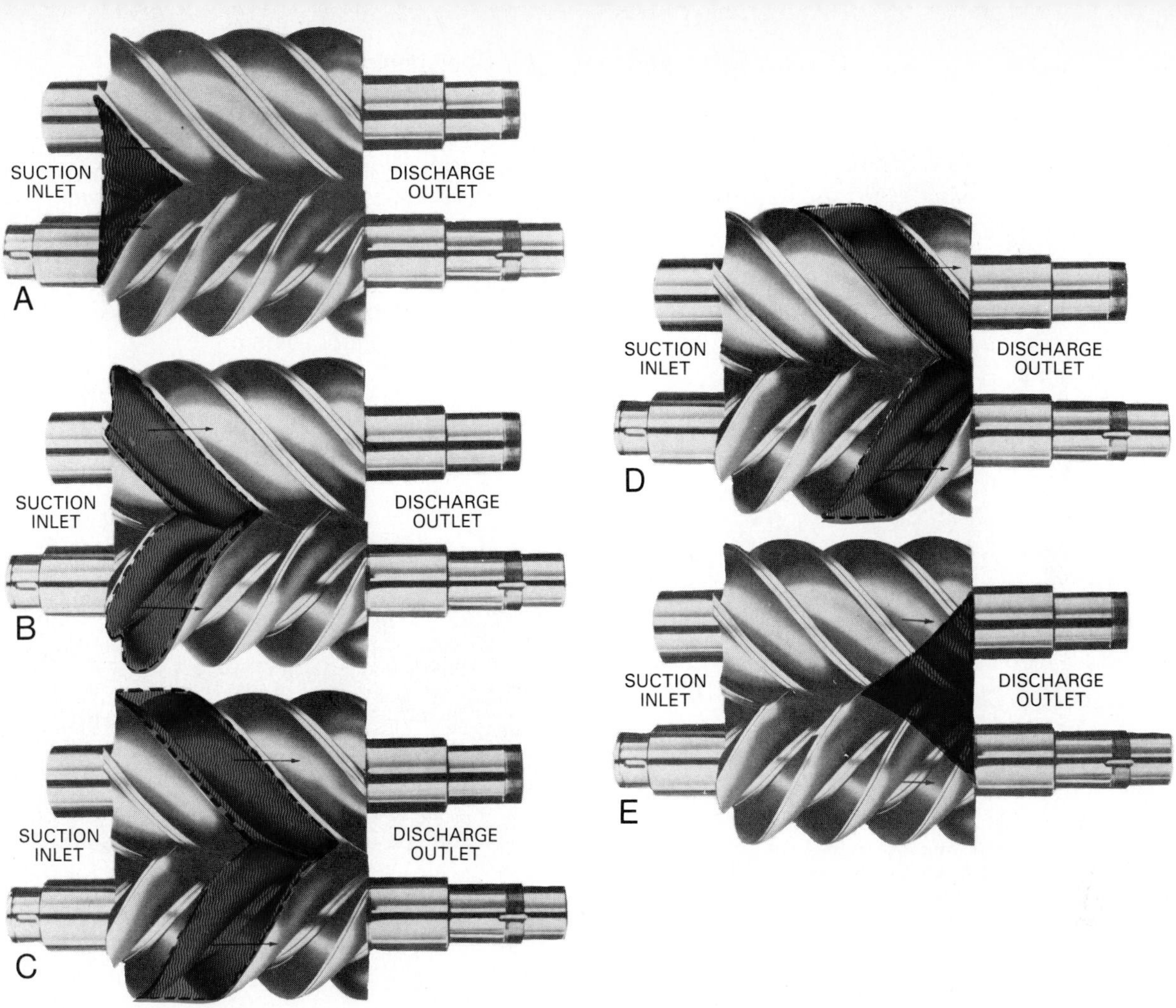

Fig. 4-60. Basic operation of screw-type compressor. Revolving rotor compresses vapor. A—Compressor interlobe spaces being filled. B—Beginning of compression. C—Full compression of trapped vapor. D—Beginning of discharge of compressed vapor. E—Compressed vapor fully discharged from interlobe spaces.   (Dunham-Bush, Inc.)

Fig. 4-61. Screw-type compressor which uses matched set of helical rotors. A—Suction inlet. B—Female idler rotor. C—Female section of compressor cylinder. D—Shaft seal. E—Drive shaft. F—Discharge outlet. G—Male section of compressor cylinder. H—Motor driven male rotor. J—Capacity control.   (Dunham-Bush, Inc.)

Fig. 4-63 illustrates a 3600-rpm single screw-type compressor which utilizes one main rotor that meshes with two diametrically opposed star-shaped gate rotors. The main rotor contains six grooves and has straight roller bearings at the shaft ends. Two capacity control slide valves, one on each side, help to determine the capacity control. Fig. 4-64 is a complete single screw compressor unit using a microprocessor control system.

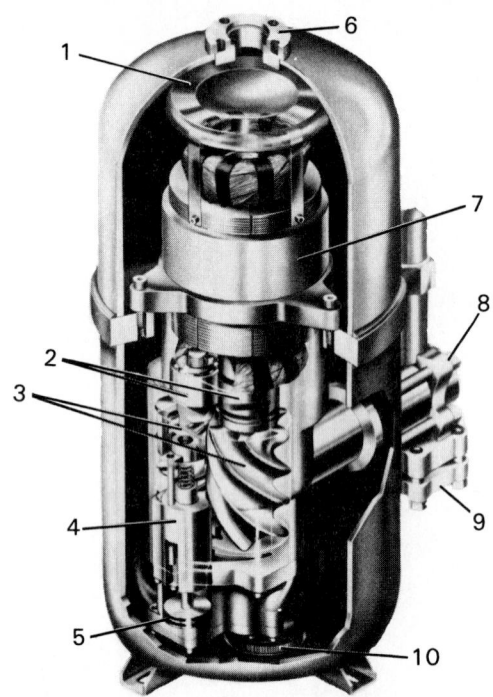

Fig. 4-62. Screw-type hermetic motor compressor in cutaway view. Note helical rotors and capacity control slide valve. 1—Oil separation plate. 2—Inlet main bearings. 3—Rotors. 4—Slide valve. 5—Unloader piston. 6—Discharge port. 7—Motor frame. 8—Suction service valve. 9—Suction port. 10—Oil strainer. (Dunham-Bush, Inc.)

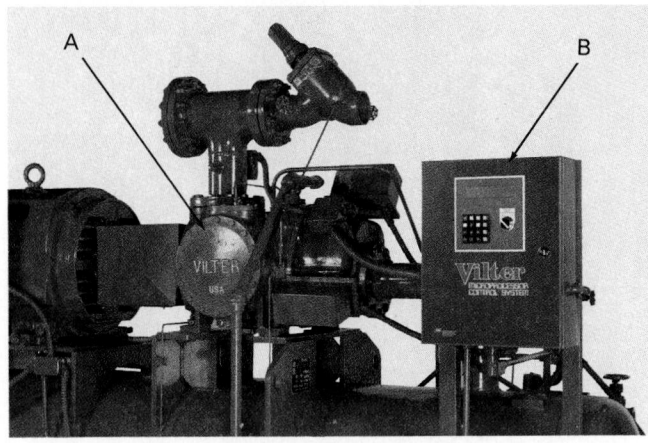

Fig. 4-64. A commercial single screw-type compressor system. A—Single screw-type compressor. B—Microprocessor control system. (Vilter Manufacturing Corporation)

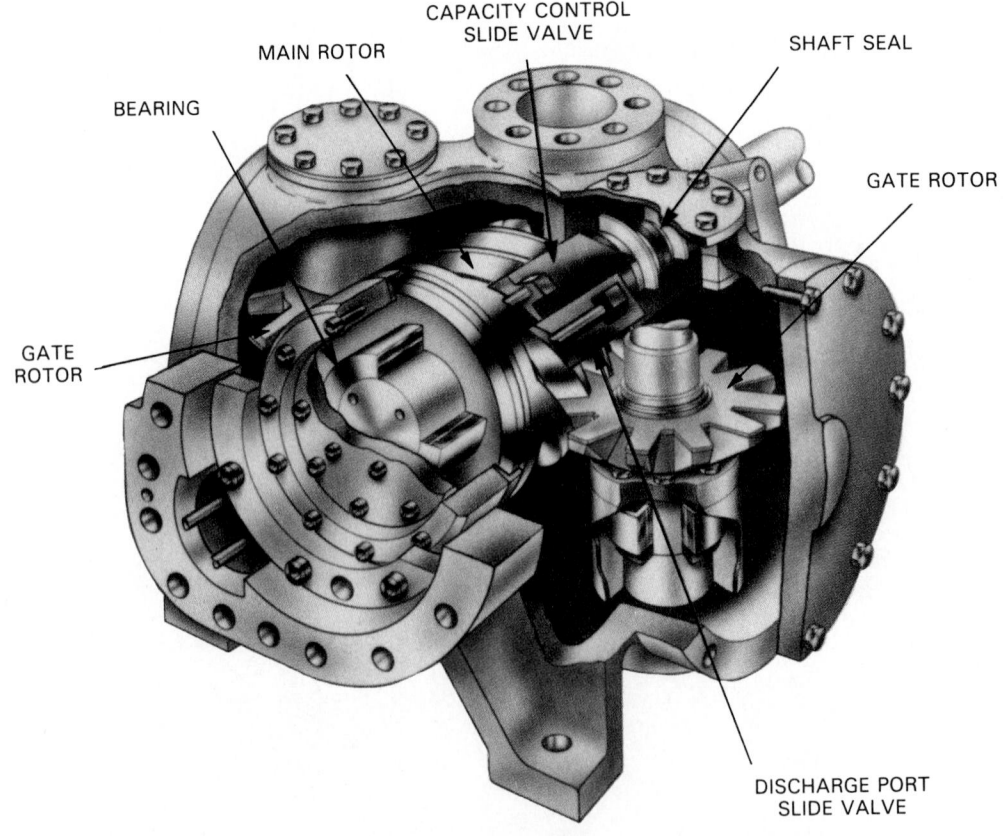

Fig. 4-63. Single screw compressor. Note the location of the main rotor to the two gate rotors. (Vilter Manufacturing Corporation)

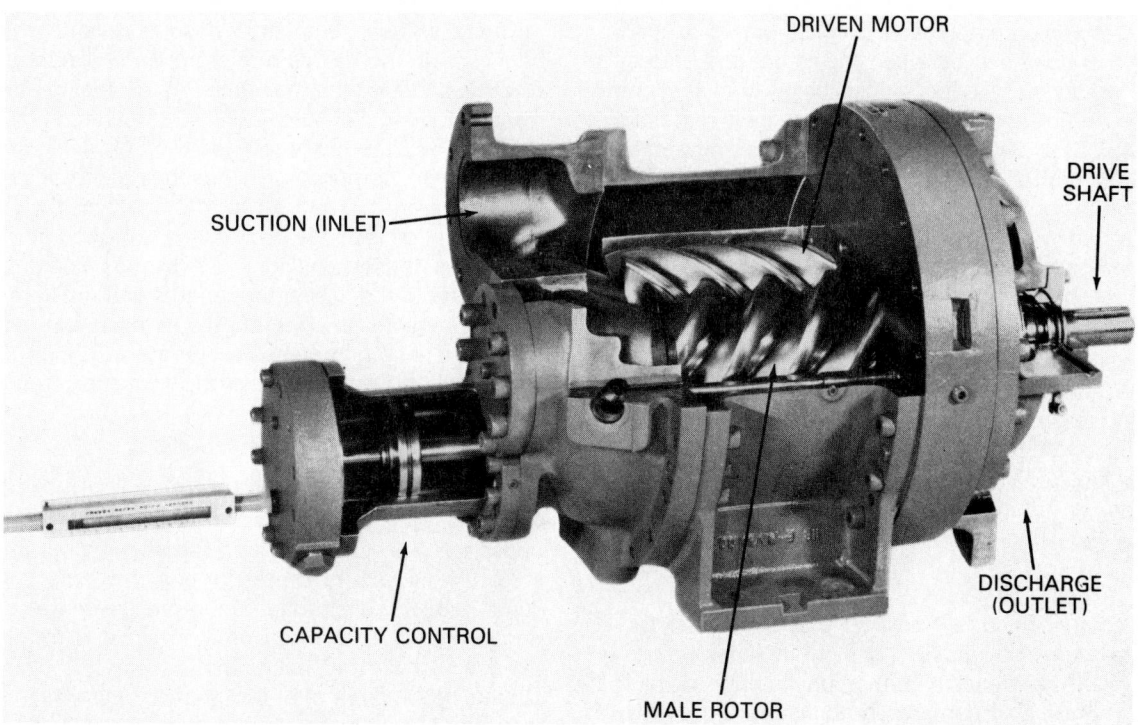

Fig. 4-65. Matched set of helical rotors in screw-type compressor. Male rotor drives female rotor. Refrigerant vapor is compressed as it flows from left to right through screw compressor. (Dunham-Bush, Inc.)

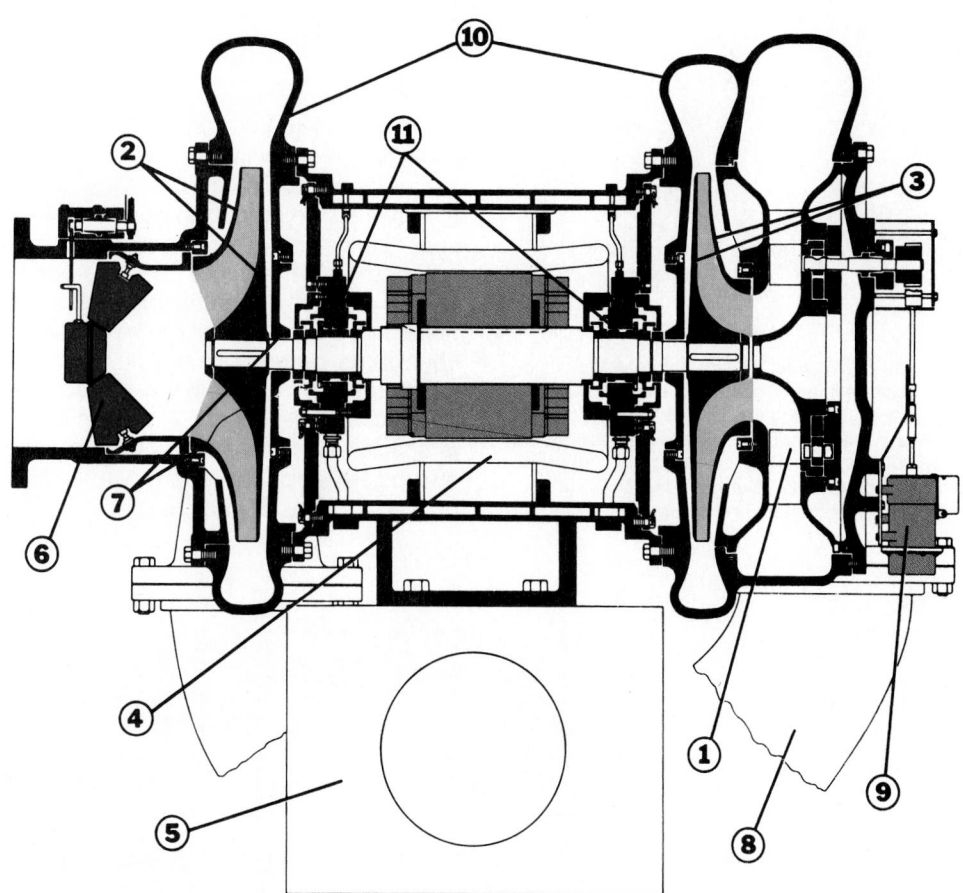

Fig. 4-66. Two-stage centrifugal compressor. 1—Second stage variable inlet guide vane. 2—First stage impeller. 3—Second stage impeller. 4—Water-cooled motor. 5—Base, oil tank, and lubricating oil pump assembly. 6—First stage guide vanes and capacity control. 7—Labyrinth seal. 8—Cross-over connection. 9—Guide vane actuator. 10—Volute casing. 11—Pressure lubricated sleeve bearing. Note that discharge opening is not shown.

Many screw-type compressors operate with oil injection. This seals the clearance between the rotors and between the rotors and the cylinder. It also helps cool the compressor. The efficiency of these compressors is quite high. They may be used with most of the common refrigerants.

Fig. 4-65 illustrates a pair of screw-type rotors in operating position. Since the screw-type compressor operates at fairly high speed, adequate bearings must be provided for good rotor bearing life.

4-46 CENTRIFUGAL COMPRESSORS

Centrifugal compressors are used successfully in large refrigerating systems. In this type compressor, vapor, as it is moved rapidly in a circular path, moves outward. This action is called centrifugal force, but the correct term is centripetal force.

The vapor is fed into a housing near the center of the compressor. A disk with radial blades (impellers) spins rapidly in this housing, forcing vapor against the outer diameter.

The pressure gained is small, so that several of these compressor wheels or impellers are put in series to create greater pressure difference and to pump a sufficient volume of vapor. This type of compressor looks like a steam turbine or an axial flow air compressor for a gas turbine engine.

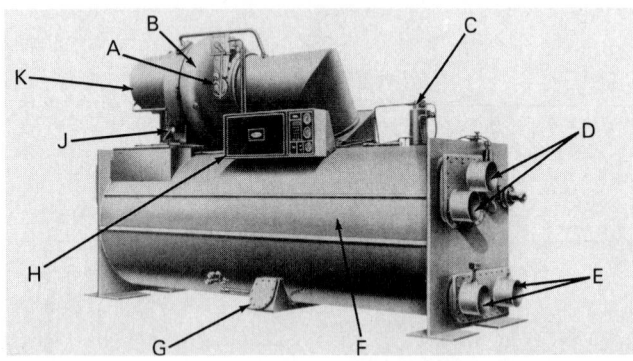

Fig. 4-67. A hermetic centrifugal water chiller, single-stage, using R-11, with microprocessor controls. A—Guide vane motor and linkage. B—Compressor. C—Purge unit. D—Condenser water connectors and rupture disk. E—Cooler water connectors. F—Condenser subcooler. G—Refrigerant liquid control valve. H—Microprocessor control panel. J—Oil sump thermometer. K—Hermetic motor.
(Carrier Corp., Subsidiary of United Technologies Corp.)

The centrifugal compressor has the advantage of simplicity. There are no valves or pistons and cylinders. The only wearing parts are the main bearings. Pumping efficiency increases with speed, so compressors operate at high speeds.

Fig. 4-66 is a cross-section through a two-stage centrifugal type compressor. The driving motor is mounted between stages. The inlet is at the left on the illustration. The discharge is in the back at the right end of the illustration and is not shown. Fig. 4-67 pictures a centrifugal compressor mounted in a refrigerating system. Fig. 4-68, right view, shows a section through a hermetic centrifugal compressor. These compressors operate at a high speed and are usually driven by an electric motor or steam turbine.

4-47 SCROLL COMPRESSOR

The scroll compressor uses two offset spiral disks to compress the refrigerant vapor. See part "a" of Fig. 4-69A. The upper scroll is stationary, and the lower scroll is the driven scroll. The intake of the vapor is at the outer edge of the scroll, and the discharge occurs at the center of the stationary scroll. The driven orbiting scroll is rotated around the fixed scroll in an orbiting motion, part "b" of Fig. 4-69A. During this movement, the suction vapor is trapped between the two scrolls. As the driven scroll rotates, it compresses the refrigerant vapor through the discharge port. The scroll compressor, Fig. 4-69B, has fewer moving parts than reciprocating compressors and has less torque variation. This results in very smooth and quiet operation.

4-48 STATOR AND ROTOR CONSTRUCTION

The stator or casing of a centrifugal compressor is usually made of cast iron. It has a changing radius inside to adapt itself to the vapor pickup by the impellers.

The casing (cylinder) also holds the main bearings, and the oil pressure producing pump, as well as the refrigerant vapor intake and exhaust ports. It also holds the shaft seal where the shaft extends or sticks out from the casing for the power drive, when an external motor is used. Both the first stage and second stage have adjustable inlet vanes to control the capacity of the pump.

The rotor or impeller in a centrifugal compressor is keyed to the compressor shaft. It is made of cast iron or steel and is specially designed to move the vapors without going above gas velocity limits and without having vapor-trapping pockets. A typical rotor is shown in Fig. 4-68, left view.

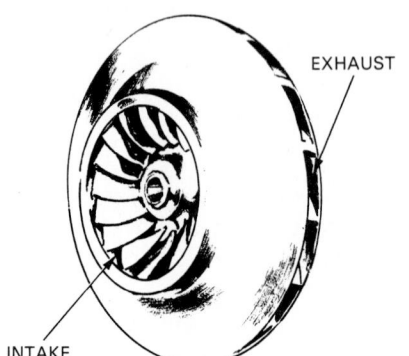

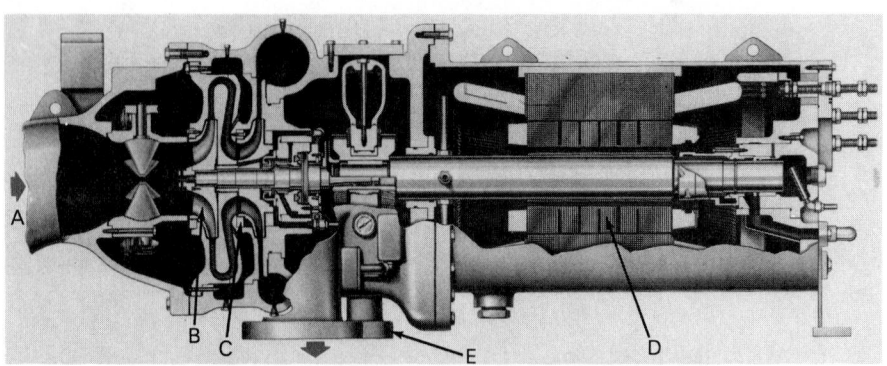

Fig. 4-68. Hermetic centrifugal compressor. The impeller is shown at left above. Major components of the compressor are: A—Intake. B—First stage impeller. C—Second stage impeller. D—Heremtic motor. E—Exhaust. (Carrier Corporation, Subsidiary of United Technologies Corporation)

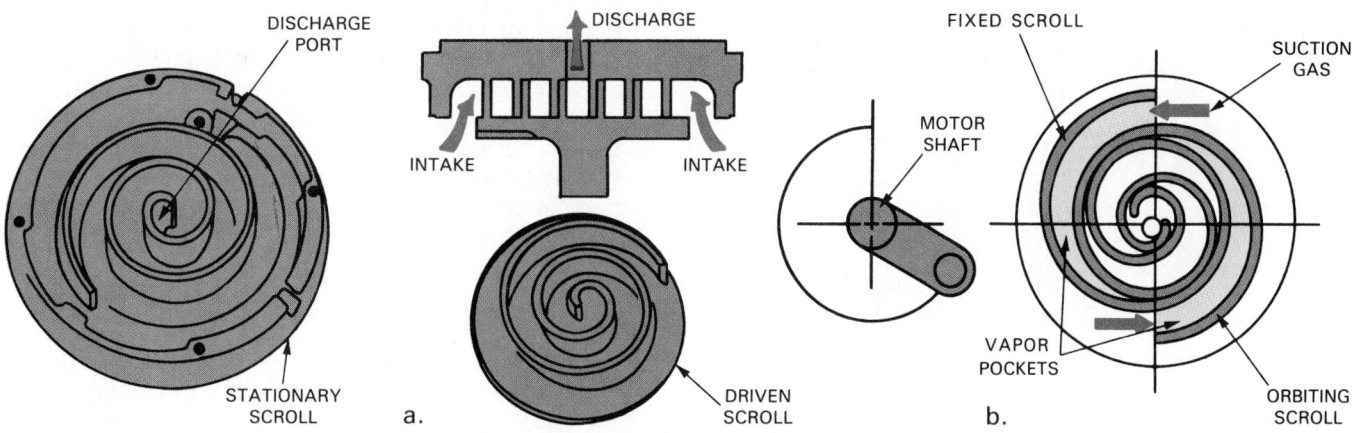

Fig. 4-69A. Scroll compressor design. In part "a," two scrolls are used to produce a vapor compression. The upper scroll is stationary and the lower scroll is driven. Note intake and discharge ports. Note in part "b" how the rotation of motor shaft causes orbiting scroll to orbit—not rotate—about the shaft center. (The Trane Co.)

4-49 MOTORS

Two different types of motors are used for driving refrigeration compressors.

External drive compressors use conventional motors and usually drive the compressor with one or more V-belts or by direct drive. A sectional view of the kind of motor used on small external drive compressors is shown in Fig. 4-70.

In hermetic compressors, the motor is mounted under the same dome as the compressor. These motors are lubricated by the oil carried in the refrigerant.

Hermetic motors do not use brushes or open points inside the dome. Arcing would cause pollution in both the oil and the refrigerant. This would lead to an electrical burnout.

Fig. 4-70. Motor used on external drive compressor. A—Motor shaft. B—Motor rotor. C—Motor stator. D—Starting mechanism. E—Motor bearing. F—Motor frame. (A.O. Smith Electrical Products Co.)

A special electrical starting device is located outside the dome. Fig. 4-71 shows the three main parts of a motor used in a hermetic refrigeration compressor.

See Chapter 7 for information concerning motors. Chapter 8 has more information about starting devices and motor controls.

4-50 SERVICE VALVES

Service valves are shown in Figs. 4-10 and 4-11.

Many small hermetic systems do not have service valves of any type. To attach gauges and service manifolds to these systems, refrigerant lines must be tapped. Special tapping valves are available. These are clamped to a tube so that the valves pierce the tube and, at the same time, provide for necessary gauge and service connections.

4-51 MUFFLERS

Most hermetic units and many external drive systems have noise reducing devices called mufflers on both the

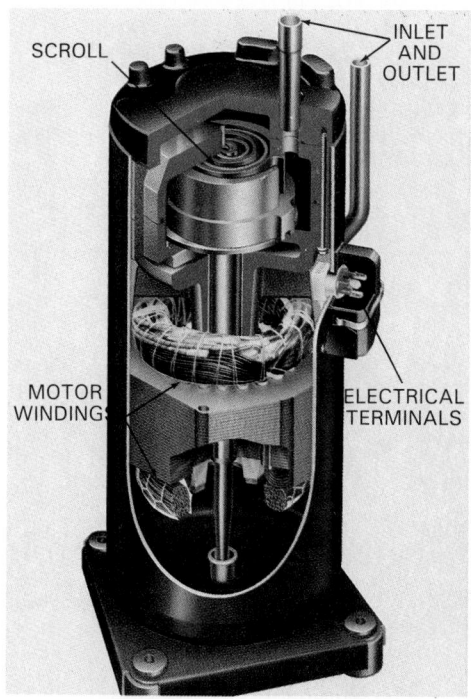

Fig. 4-69B. Scroll compressor for use on domestic room air conditioners. (Tecumseh Products Company)

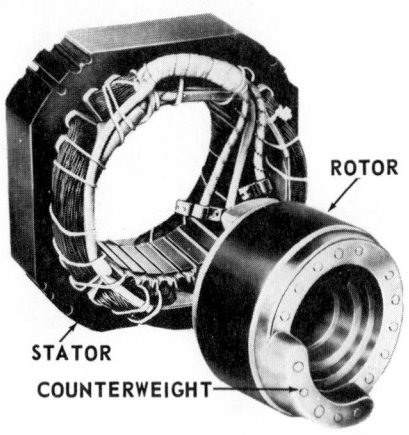

Fig. 4-71. Hermetic unit motor stator and rotor. Rotor is mounted directly on compressor crankshaft. Note counterweight, which balances weight of crank, connecting rod, and piston.

intake and the exhaust openings of compressors. Mufflers reduce the sharp gasping sound on the intake stroke and the even sharper puff of the exhaust. These mufflers are brazed cylinders with baffle plates mounted inside. Fig. 4-72 shows a unit equipped with a suction "intake" muffler.

In one design, suction and exhaust mufflers are connected directly to the compressor cylinder head with lines as in Fig. 4-73.

4-52 COMPRESSOR COOLING

The temperature of the compressor is greatly affected by the heat of compression, which raises the temperature of the vapor as it is "squeezed" and forced into the condenser. Friction (rubbing) between moving parts also adds to compressor temperature. This heat must be removed to prevent loss of efficiency of the pump and to maintain the lubricating qualities of the oil.

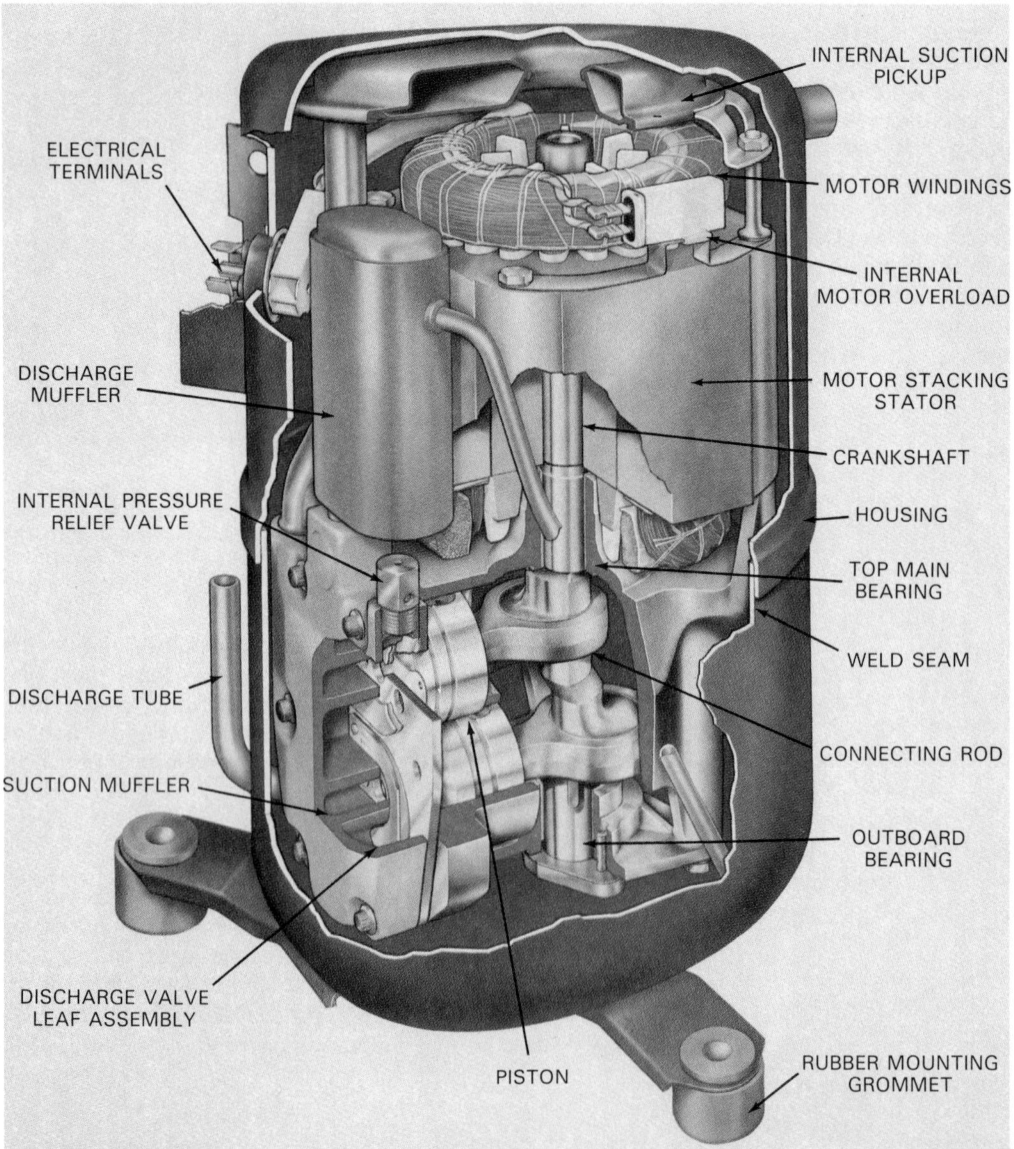

INTERNAL SUCTION PICKUP

MOTOR WINDINGS

INTERNAL MOTOR OVERLOAD

ELECTRICAL TERMINALS

MOTOR STACKING STATOR

DISCHARGE MUFFLER

CRANKSHAFT

HOUSING

INTERNAL PRESSURE RELIEF VALVE

TOP MAIN BEARING

WELD SEAM

DISCHARGE TUBE

CONNECTING ROD

SUCTION MUFFLER

OUTBOARD BEARING

DISCHARGE VALVE LEAF ASSEMBLY

PISTON

RUBBER MOUNTING GROMMET

Fig. 4-72. A two-cylinder hermetic compressor suitable for use either in a commercial application or as part of a residence air conditioning system. Note suction and discharge mufflers. (Tecumseh Products Company)

The oil that circulates in the compressor removes much of the heat from motor and compressor. As it flows over the heated surfaces, it picks up and carries this heat to cooler surfaces.

Many compressors and many domes have metal fins on their outer surfaces to help carry this heat away. Some units even use a motor driven fan to force cooling air over the compressor.

When a water-cooled condenser is used, this water is often also used to cool the compressor or dome.

Motors are often cooled by passing the suction vapors and return oil over the windings. Some units circulate the crankcase oil through an air-cooled coil. The cooled oil then helps to cool the motor compressor. Some larger hermetic units are water cooled.

4-53 LUBRICATION

Lubricating oils have been developed especially for reciprocating and rotary refrigeration compressors. Usually, these are mineral oils, completely dehydrated, wax-free, and nonfoaming. They have a viscosity found to be best for the refrigerant and for the refrigeration temperatures.

Some refrigerating oil contains additives to improve lubricating qualities. The additives may also improve the oil's viscosity properties (ability to flow at given temperatures).

Reciprocating compressors may be lubricated either by a splash or by a pressure (force feed) system.

In the splash system, the crankcase is filled with the correct oil up to the bottom of the main bearings or to the middle of the crankshaft main bearings. At each crankshaft revolution, the crank throw, or the eccentric, dips into the oil and splashes it around the inside of the compressor. Oil is thrown onto cylinder walls and piston pin bushings and into small openings where it can drain into the main bearings.

This is an excellent system for normal use in small compressors. Some compressor connecting rods have little dips or scoops attached to the lower ends. These scoop up small amounts of oil and sling it around to other parts.

Clearances between the moving parts must be less in this type system. Noisy bearings will occur at smaller clearances than in the pressure system, because there is no oil under pressure to cushion the bearing surface.

The force feed or pressure system uses a small oil pump to force oil to the main bearings, lower connecting rod bearings, and in some cases, piston pins. It is a more expensive system. It needs a pump, and the crankshaft and connecting rod must have oil passages drilled in them.

With the pressure system, the compressor gets better protection from the oil. It will also run more quietly even though it has greater bearing clearances.

The oil pump is usually mounted on one end of the crankshaft. Whenever used, an overload relief valve must be built into the pump to protect it and the system against oil pressures that are too high.

Fig. 4-74 shows the oil path in a pressure type lubrication system. The oil pump delivers oil, under pressure, to all bearing surfaces.

Larger pressure-lubricated compressors sometimes use pressure-controlled electric switches which will stop the unit if the oil pressure drops too low. It is sometimes necessary to use some kind of an unloading device which enables the

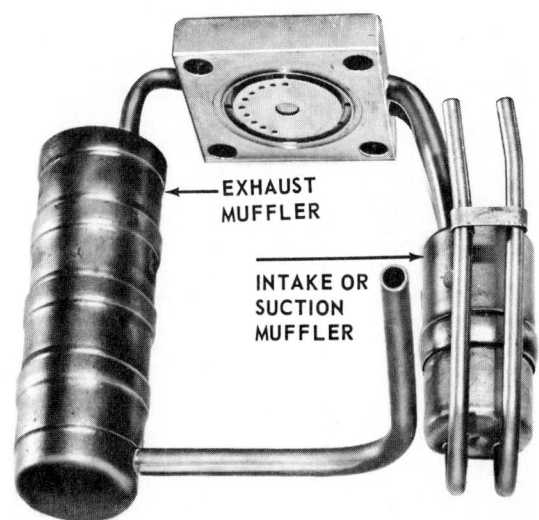

Fig. 4-73. Hermetic compressor cylinder head. Mufflers are attached to suction and exhaust openings.

compressor to start easily with no vapor pressure load in the cylinder.

In a rotary compressor, it is best to have a constant film of oil on the cylinder, roller, and blade surfaces. When the compressor operates, the oil feeds through the main bearings into the cylinder. The cylinder is located so the oil level half covers the main bearings.

In larger units and even in some of the smaller units, a forced feed lubrication system is used. Some units use a separate oil pump, but some use the pumping action of the blades moving in and out of their slots.

4-54 COMPRESSOR VOLUMETRIC EFFICIENCY

Volumetric efficiency of a compressor is the actual volume of refrigerant gas pumped, divided by the calculated volume. A compressor may be designed to pump 10 cu.in. of vapor each revolution or stroke (this is called the piston displacement).

If it pumps only 6 cu. in. each revolution, the volumetric efficiency of the pump is 60 percent (6/10):

$$\frac{6 \text{ cu. in.}}{10 \text{ cu. in.}} = .6:$$

.6 × 100 (to change to percentage)
= 60 percent.

Metric: $\frac{98 \text{ cm}^3}{164 \text{ cm}^3} = .6$. Thus, .6 × 100 = 60 percent.

For efficient operation, the volumetric efficiency must be as high as possible. Several things affect this efficiency.

First, if the head pressure (the pressure the compressor must pump against) increases, the amount pumped per stroke will decrease. This is because the compressed vapor in the clearance space will expand on the intake stroke, and fresh vapor cannot move into the cylinder until the pressure in the cylinder is lower than the pressure in the suction line. The higher the compressed pressure, the greater the compressed vapor in the clearance space will expand.

Second, if the low-side pressure decreases, it is more difficult for the vapors to fill the cylinder, and the amount pumped per stroke will decrease.

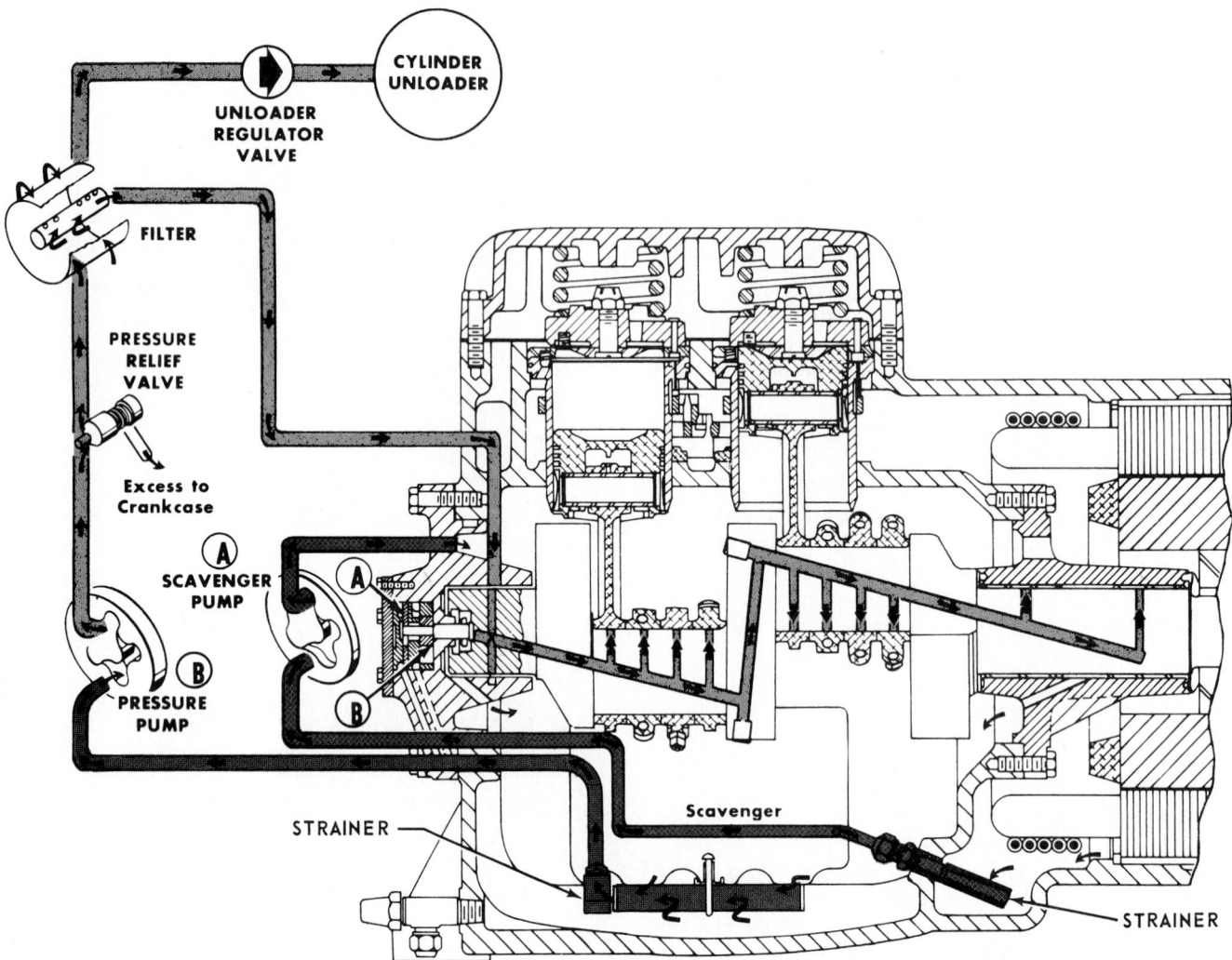

Fig. 4-74. A pressure lubricated hermetic radial multi-cylinder reciprocating refrigerator compressor. Scavenger pump returns oil from motor end of the motor compressor back to the lower portion of the crankcase at the cylinder, which serves as an oil sump to store the oil charge. Note pressure relief valve in pressure line. Cylinder unloader is operated by the oil pressure.

Third, if the clearance pocket is enlarged, the amount pumped per stroke will decrease. The clearance space is the space left in the cylinder when the roller or piston is at the end of its pumping stroke as shown in Fig. 4-35.

The efficiency of a compressor also depends on the size of the valve openings. If the intake valve reduces the flow of low-side vapor into the cylinder, the cylinder will not be filled, and the efficiency of the compressor will be lowered. Also, if the exhaust valves stick or if the line from the compressor to the condenser is pinched, this extra pressure in the cylinder will cut down the compressor's pumping efficiency.

4-55 COMPRESSION RATIO

Compression ratio is the ratio of the total volume of the cylinder to the clearance space (that space which remains at the end of the compression cycle). In a refrigeration system, this is the relationship of the absolute pressure of the high side to the absolute pressure of the low side. Ratios vary up to 10 to 1 for single-stage compressors. If the ratio is higher, two-stage compressors must be used. This factor is discussed in Chapter 30.

4-56 CHECK VALVES

Check valves, Fig. 4-75, allow fluid flow in only one direction in a system. Sometimes they are put in refrigerant suction lines. They prevent refrigerant vapor, oil, or even liquid refrigerant from backing up into the evaporator or other devices where it might condense or lodge. Check valves are often used in multiple evaporator installations and in heat pump installations.

4-57 UNLOADER

To make it easier to start the compressor, some installations provide an unloader. This unloader temporarily reduces the high-side pressure at the cylinder head while the compressor is starting. The unloader may be operated mechanically, electrically, hydraulically, or by a solenoid valve. A system using an unloader is shown in Fig. 4-74.

On small systems which use a capillary tube refrigerant control, the balancing of the low and high pressures when the compressor is off serves as an unloader. Unloader mechanisms may also be used to vary the pumping capacity in case of a changing heat load, such as in an air conditioner.

4-58 GASKETS

On external drive compressors, the joining surfaces between bolted parts, such as cylinder heads, valve plates,

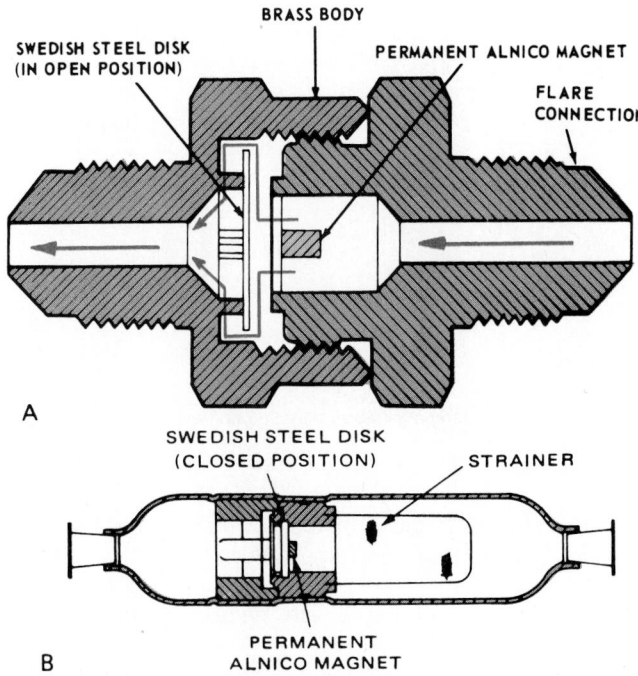

Fig. 4-75. Check valves in which a permanent magnet provides a slight pull. This tends to keep the valves closed. A—Flare type connector in open position. B—Solder type connector in closed position. (Watsco Components, Inc.)

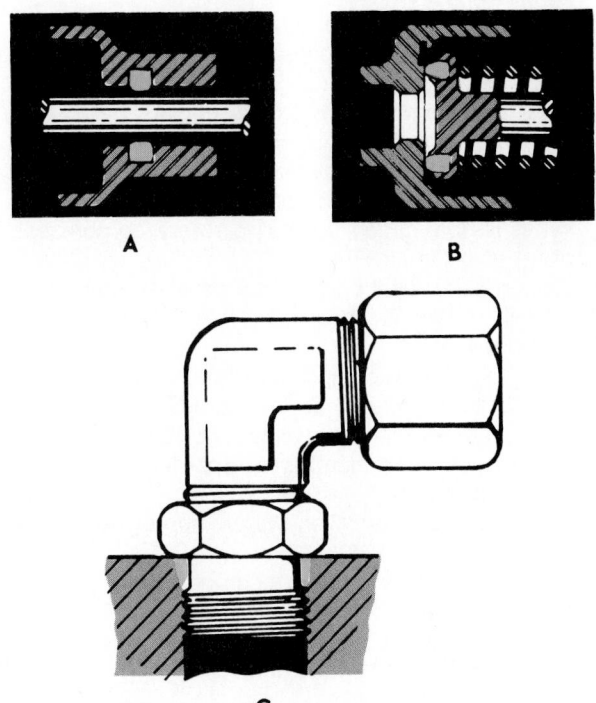

Fig. 4-76. Typical O-ring installations. A—An O-ring installed as seal between a shaft and its housing. B—O-ring installed to serve as a seat seal. C—An O-ring installed as a pipe fitting assembly seal. (Parker Seal Co.)

crankcase openings and the like, are usually sealed with gaskets.

The gaskets may be made of either special paper, synthetic material, or lead. Some are made of a plastic substance. Gaskets must be completely free from moisture before use.

4-59 O-RINGS

An O-ring is commonly used as a sealing device, particularly in assembling parts where there may be some motion between the assembled parts. Fig. 4-76 illustrates three typical O-ring installations.

The materials used in making O-rings will depend on various factors such as temperature, pressure, fluids to be controlled, and useful life required. They are usually made of fluid-resistant elastomer compounds.

4-60 CRANKCASE HEATER

In many condensing unit applications, it is necessary to heat the compressor crankcase to evaporate the liquid refrigerant trapped in the oil. Most large compressors used in commercial applications are fitted with a crankcase heater at the time the compressor is manufactured—especially if the compressor may be exposed to cold temperatures (outdoor units). These crankcase heaters may be operated during the off cycle or they may be thermostatically controlled.

For smaller installations which do not usually require a built-in crankcase heater, an accessory heater may be purchased and attached to the crankcase. Crankcase heaters are generally required on remote installations in which the compressor may at times operate at an ambient temperature which is lower than the evaporator temperature.

A compressor with a built-in crankcase heater can be seen in Fig. 4-41. Fig. 4-77 shows a crankcase heater which may be attached to the housing of a hermetic compressor.

4-61 REVIEW OF SAFETY

The service technician should always be alert to avoid service procedures which may:
1. Be a hazard to the technician.
2. Be injurious to the equipment.
3. Cause refrigeration to fail.

Normal servicing of refrigerator mechanisms is not considered hazardous. There are, however, recommended procedures which should be followed to be sure that the service operations are performed under the safest possible conditions. If the service technician knows and understands the construction and operation of all the parts in a refrigerating system, it will help in servicing it safely.

The technician should always wear goggles when working on refrigerating systems.

As an installation or servicing technician, one should always handle the parts with care, keeping every dismantled part clean and dry.

It is the little things that count most in servicing refrigerator mechanisms. Care in tightening a tube connection, installing a gasket, replacing an electrical terminal, and soldering a fitting often determines whether the entire job will be safe and satisfactory.

Even a slight amount of moisture allowed to enter a refrigerating mechanism may cause the system to fail either

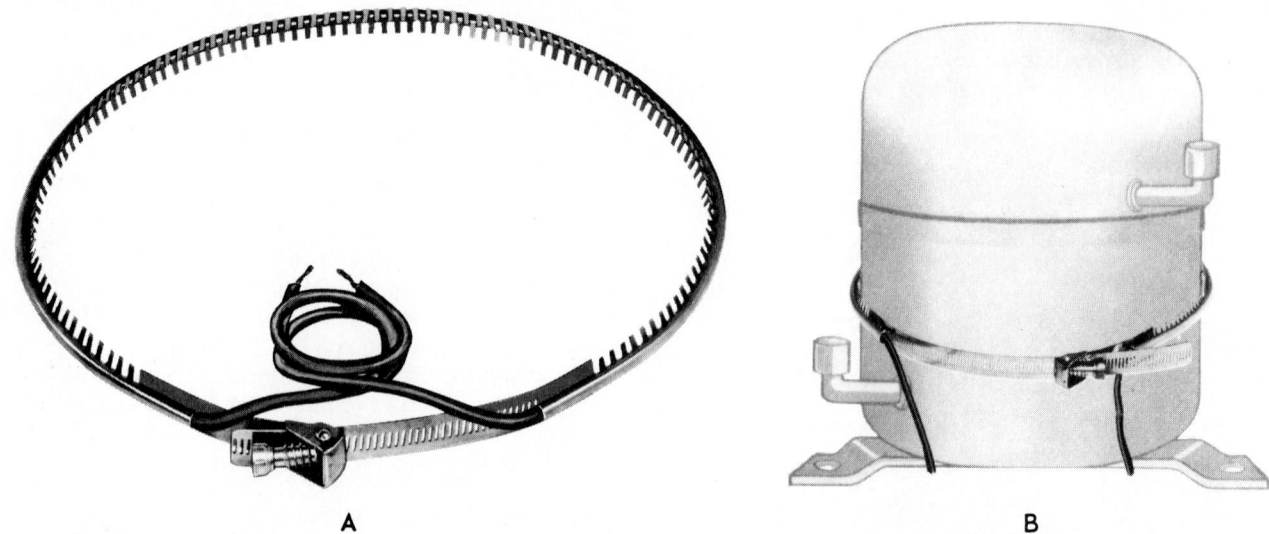

Fig. 4-77. Crankcase heater. A—Accessory type electrically operated crankcase heater. B—Crankcase heater installed on hermetic compressor. (Tutco)

through the formation of ice in the refrigerant control device or the formation of a "sludge" in combination with the refrigerant oil.

One important purpose of refrigeration is to preserve food. If the refrigerator does not maintain the desired temperatures, food may spoil. The service technician must make the necessary settings and adjustments correctly.

4-62 TEST YOUR KNOWLEDGE

1. Name the eight most important parts found in all compression cycle refrigerators.
2. Name the main parts commonly located in the low-pressure side.
3. Name the main parts commonly located in the high-pressure side.
4. Name four types of compressors.
5. What type of compressor does not use a crankshaft seal?
6. Why does the temperature rise as a gas is being compressed?
7. Why are check valves used in the suction line on some rotary compressors?
8. Why should condensers be cleaned?
9. Name four different styles of evaporators.
10. What control determines the refrigerator cabinet temperature?
11. Name five types of refrigerant controls.
12. How much clearance is allowed between the piston and the cylinder on small compressors?
13. How much lift is allowed the intake and exhaust valves?
14. What does a low-side float control?
15. Why are expansion valves (AEV) usually adjustable?
16. What is a high-side float control?
17. What is the purpose of the crankshaft seal?
18. How does a capillary tube reduce the pressure?
19. When does the condensing pressure stop rising?
20. What basic conditions are necessary to produce refrigeration?
21. What type of reciprocating compressor is often used in automobile air conditioning systems?
22. What is the purpose of a compressor?
23. What is a full floating piston pin?
24. Why must compressor valves be light weight?
25. How is the crankshaft joint sealed where it leaves the compressor body?
26. Of what materials are compressor pistons made?
27. Why must the clearance pocket volume be kept as small as possible?
28. How are compressor cylinders cooled?
29. How are hermetic motors usually cooled?
30. What is the purpose of the stationary blade in a stationary blade rotary compressor?
31. What is sometimes used in a rotary compressor in place of an intake valve?
32. Does a centrifugal compressor have exhaust valves?
33. Explain what is meant by the term volumetric efficiency.
34. On external drive compressors, should the crankshaft seal rubbing surfaces be lubricated?
35. Are some compressors water-cooled?
36. What is the least number of impellers a centrifugal compressor may have?
37. Why does a compression system sometimes use two mufflers?
38. How many openings does a compressor service valve have?
39. What does an oil separator do?
40. Why is a liquid line filter-drier used?

Chapter 5

REFRIGERANT CONTROLS

After studying this chapter the technician will be able to:
☐ Explain the purpose and operation of refrigerant control devices.
☐ Name the six main types of controls and explain their operation.
☐ Define terms related to refrigerant control operations.
☐ Compare the various charging elements used on refrigerant controls.
☐ Determine the proper size capillary tube to be used for specific applications.
☐ Explain the operation of special refrigerant controls.
☐ Define the purpose and function of three types of solenoid valves.

A refrigerant control is the device used in a refrigeration system to change the pressure of the refrigerant.

Refrigerant in the evaporator must be at a low pressure so it will evaporate at a low temperature. The liquid refrigerant in the condensing unit is at a relatively high pressure. So that the refrigerating unit may operate automatically, an automatic refrigerant flow control must be placed in the circuit between the liquid line and the evaporator. This control reduces the high pressure in the liquid line to the low pressure in the evaporator. There are six main types of automatic refrigerant flow controls:
A. Automatic Expansion Valve (AEV or AXV).
B. Thermostatic Expansion Valve (TEV or TXV).
C. Thermal-Electric Expansion Valve (THEXV).
D. Low-Pressure Side Float (LSF).
E. High-Pressure Side Float (HSF).
F. Capillary Tube (Cap. Tube).

The basic systems using these controls are described in Chapter 3 and Chapter 4.

5-1 COMPRESSION SYSTEM REFRIGERANT CONTROLS

Modern refrigeration systems are automatic in operation. Devices have been developed for controlling refrigerant flow into the evaporator and also for controlling the electric motor which drives the mechanism.

The refrigerant controls may be divided into three principal classes:
1. Control based on pressure changes.
2. Control based on temperature changes.
3. Control based on volume or quantity changes.

A combination of controls may also be used.

Automatic refrigerant and motor controls are needed that will maintain the temperature in the refrigerated space within specific limits. Automatic motor controls are explained in Chapter 8.

Popular applications and temperature ranges include:
1. For fresh food storage, temperatures usually are maintained at 35 to 45 °F (2 to 7 °C).
2. For frozen food storage, temperatures usually are between -10 °F and 0 °F (-23 and -18 °C).
3. For comfort cooling, temperatures are usually maintained not to exceed 10 to 12 °F (5.6 to 6.6 °C) below the ambient temperature.

5-2 AUTOMATIC EXPANSION VALVE PRINCIPLES

An automatic expansion valve (AEV or AXV) or pressure controlled expansion valve is a refrigerant control operated by low-side pressure. It throttles the liquid refrigerant in the liquid line down to a constant pressure on the low-pressure side while the compressor is running.

The valve acts like a spray nozzle. While the compressor is running, the liquid refrigerant is sprayed into the evaporator. A system using an automatic expansion valve is sometimes called a "dry" system. This is because the evaporator is never filled with liquid refrigerant, but with a mist or fog.

5-3 AUTOMATIC EXPANSION VALVE DESIGN

The diagram in Fig. 5-1 is of a flexible bellows linked up to a needle valve with evaporator pressure, P_2, on the inside and atmospheric or confined gas pressure, P_1, on the outside. Spring force, F_1, tends to open the valve, while spring force, F_2, tends to close the valve.

From the illustration it may be seen that, as the pressure in the evaporator decreases, the difference in pressures will force the bellows toward the valve body. Being attached to the needle, it will open the needle valve and some liquid refrigerant will spray into the evaporator. As the refrigerant evaporates at a constant low pressure, it keeps the evaporator and cabinet temperature within design limits.

The expansion valve opens only when the evaporator pressure drops. Pressure drop occurs only when the compressor is running. The expansion valve, however, will not flood the low side when the compressor is running. As soon as the evaporating refrigerant, liquid, spray, and vapor reach

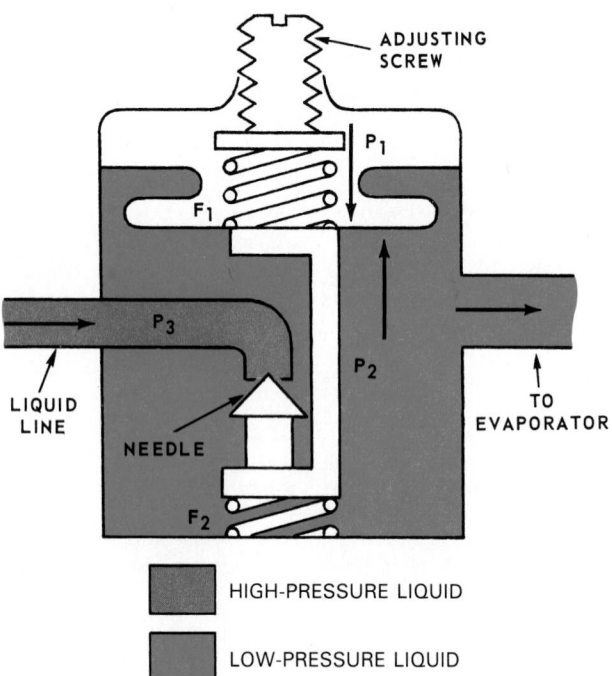

Fig. 5-1. Automatic expansion valve shows various pressures inside valve which cause it to operate. P_1—Atmospheric pressure. P_2—Suction or evaporator pressure. P_3—Liquid line pressure. F_1—Adjustable spring. F_2—Nonadjustable pressure spring.

5-4 BELLOWS TYPE AUTOMATIC EXPANSION VALVE

A bellows type automatic expansion valve, Fig. 5-3, usually has valve seats softer than the needles to eliminate, as much as possible, the wearing of a shoulder on the needle. Usually, needles are Stellite and seats Monel metal. The spring, C, is attached at both ends. It can be adjusted for either pressure or tension. This eliminates the need to have a spring inside the refrigerant space. The needle, L, is mounted in a ball and socket to insure full seating by allowing the needle to align with the seat, M. The bellows, D, made of special brass, is flexible and is soldered to both the body and the disk. This bellows is made of metal .005 to .010 in. thick.

The liquid line is usually 1/4 in. outside diameter (OD) and is fastened to the valve, H, by a special nut permitting easy removal of the screen. Note how the valve is sealed at O and B to insure that moisture will not enter. If present, it could freeze on the bellows and interfere with the accurate operation of the valve.

The valve may be attached to the evaporator either by

the end of the evaporator, the motor control (sensing bulb) attached to the suction line will cool. This opens the switch and stops the motor. The low-side pressure will then immediately build up enough to close the expansion valve.

These valves are adjustable to permit opening of the needle valve over a wide range of pressures. Expansion valves must be adjusted for atmospheric pressure, P_1, which affects their operation. High altitudes will cause a decrease in atmospheric pressure. The adjusting screw must be turned in to make up for the lower atmospheric pressure. Various refrigerants have different expansion valve settings. Their evaporating pressures are not the same.

Automatic expansion valves are of many different designs. The flexible part is either a diaphragm, Fig. 5-2, or a bellows. Usually it is made of phosphor bronze soldered or brazed to the valve body. Flexible elements must move in and out time after time without losing flexibility. The valve body is usually drop-forged brass, but sometimes it is cast. It must be seepage (leak) proof. The liquid inlet has either a soldered connection, a standard flange, a flared connection, or a pipe thread. It usually has a screen designed for easy removal. The screen is made of brass or stainless steel wire 60 to 100 mesh. (This means 60 to 100 openings per square inch.)

It is important to remember that the liquid refrigerant flowing past the expansion valve needle is the same weight as the vapor (gas) pumped by the compressor. Valve capacity should equal pump capacity.

Use a "one-ton valve" with a one-ton capacity condensing unit. An under-capacity valve tends to "starve" an evaporator (too little refrigerant gets through). An over-capacity valve will tend to allow too much refrigerant into the evaporator when the valve opens. This may cause sweat backs or frost backs on the suction line.

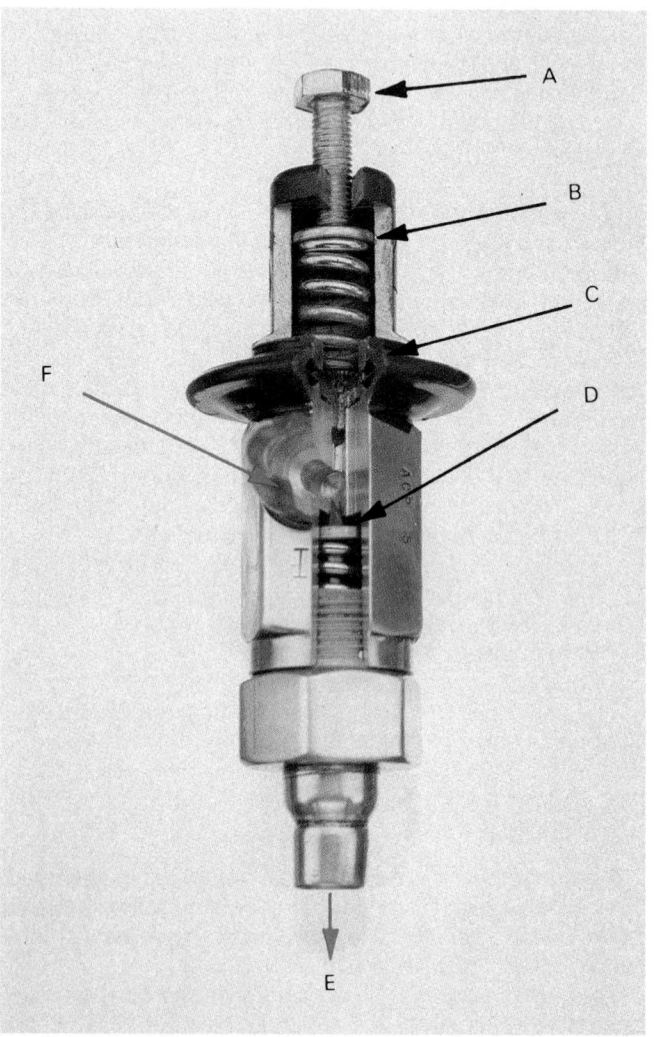

Fig. 5-2. Diaphragm type automatic expansion valve. A—Valve adjustment. B—Adjustment spring. C—Control diaphragm. D—Refrigerant needle and seat. E—Refrigerant outlet leading to evaporator. F—Refrigerant inlet.   (Alco Controls Division, Emerson Electric Company)

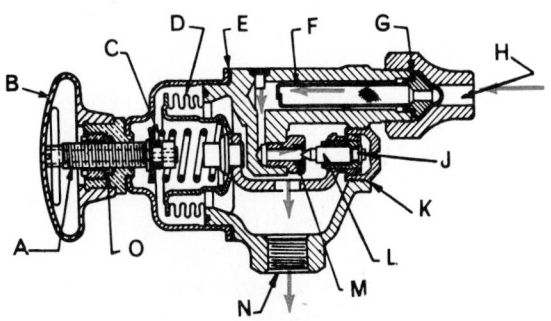

Fig. 5-3. Bellows type expansion valve. A—Adjusting screw. B—Rubber cap. C—Adjustment spring. D—Bellows. E—Gasket. F—Screen. G—Shipping plug. H—Liquid refrigerant inlet. J—Needle shoulder. K—Seating plug. L—Needle. M—Needle valve seat. N—Refrigerant outlet. O—Packing gland. Arrows show direction of refrigerant flow through valve.

means of a threaded fitting or by a two-bolt flange. Flanges sealed with lead gaskets are usually preferred.

The bellows type automatic expansion valve is used chiefly on domestic air conditioning units, vending machines, and as a replacement for capillary tubes.

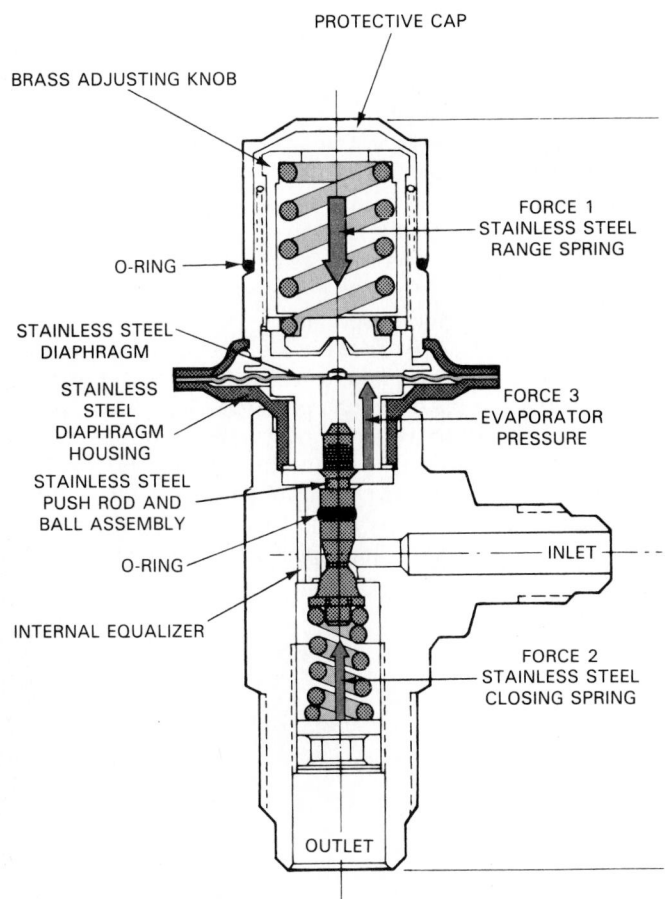

Fig. 5-4. Cross section of a diaphragm type automatic expansion valve, showing three forces that control operation of valve. Force 1, adjustable range opening, moves diaphragm. Force 2 moves push rod and ball assembly. Force 3 is evaporator pressure. Valve is designed to control flow of refrigerant to evaporator, maintaining a constant evaporator pressure.
(Refrigeration & Air Conditioning Div., Parker Hannifin Corp.)

5-5 DIAPHRAGM TYPE AUTOMATIC EXPANSION VALVE

A diaphragm type automatic expansion valve has stops to prevent too great a movement of the diaphragm. See Fig. 5-4. Note that the diaphragm has concentric corrugations (ripples) to improve its flexibility. The diaphragm separates the atmospheric pressure and the system pressure. Three basic forces control the operation of the valve. Force 1 is the adjustable spring. This moves the diaphragm down, opening the valve. Force 2 is a spring beneath the diaphragm which moves the push rod and ball assembly up, closing the valve. Force 3 is the outlet pressure underneath the diaphragm. This is the pressure that is controlled when the spring force, F_1, is equal to sum of the forces F_2 and F_3. See Fig. 5-4.

Fig. 5-5 is an external view of an automatic expansion valve. The word IN is marked on the inlet port to insure proper installation (not shown in Fig. 5-5). Its capacity, using a particular refrigerant, is often marked on the cap.

Another diaphragm type expansion valve design is shown in Fig. 5-6. The diaphragm movement is limited by the body and the cap. Threads fasten the diaphragm assembly to the body of the valve. The cap or cover plate, protecting the pressure adjustment, is tightly fitted over the entire assembly. It can be removed to adjust the valve.

The diaphragm has a disk on its low-pressure side which presses on a pin that moves the ball valve away from the seat. When the low-side pressure increases, the diaphragm

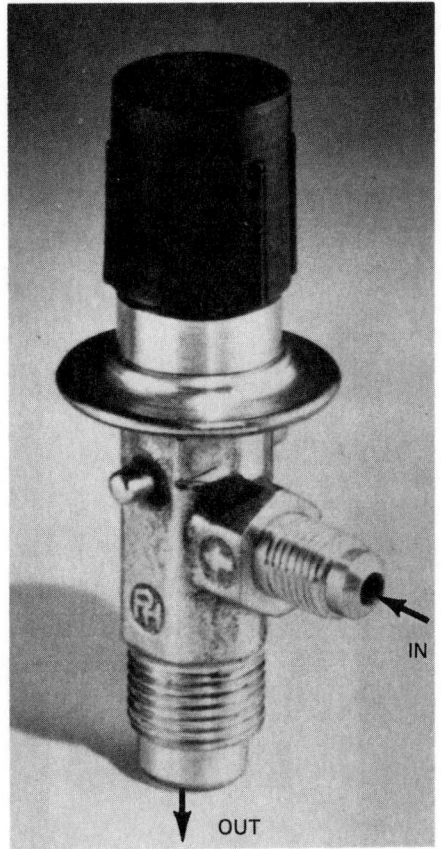

Fig. 5-5. Diaphragm type automatic expansion valve. Note refrigerant flow direction. (Refrigeration & Air Conditioning Div., Parker Hannifin Corp.)

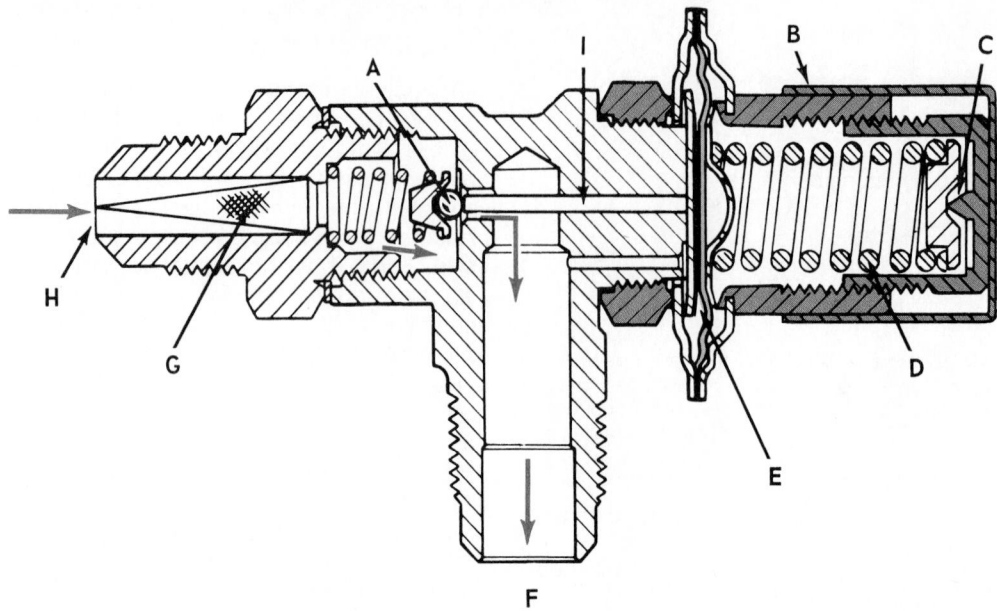

Fig. 5-6. Parts and operation of an automatic expansion valve. A—Valve (ball) and seat. B—Metal cap. C—Adjusting screw. D—Adjusting spring. E—Diaphragm. F—Outlet to evaporator (low-pressure). G—Screen. H—Liquid refrigerant inlet (high-pressure). I—Valve opening pin. Red arrows indicate direction of flow of refrigerant through expansion valve.

moves against the adjustment spring. This allows the spring at the ball valve to push the ball valve against the seat. The inlet is a 1/4-in. male flare connection. The outlet is a 3/8-in. male flare.

5-6 AUTOMATIC EXPANSION VALVE BYPASS

Many motor-compressor units are designed to start under a low load (torque) condition, such as when the low-side and high-side pressures are equal (balanced). The equal pressures allow the compressor to start without pushing against a high pressure. The motor will therefore require less starting torque. Capillary tube systems have balanced pressures during the off cycle.

Automatic expansion valves seal the refrigerant orifice during the off part of the cycle. To balance pressures, an opening is designed into the valve to permit refrigerant to pass through. A typical way to do this is to make a V-shaped slot in the valve seat as shown in Fig. 5-7. The bypass or bleeder openings are so small that they do not interfere with the operation of the valve when the compressor is running.

When using this type of expansion valve, the right amount of refrigerant charge must be used. The evaporator outlet must have an accumulator; otherwise, liquid refrigerant may travel down the suction line and cause sweating or frost on the suction line; also, dangerous liquid refrigerant slugging in the compressor may occur.

5-7 THERMOSTATIC EXPANSION VALVE (TEV) PRINCIPLES

Thermostatic expansion valves are of two basic types:
1. Sensing bulb.
2. Thermal-electric.
The sensing bulb type is further divided into four types:
1. Liquid charged.

2. Gas charged.
3. Liquid cross charged.
4. Gas cross charged.

In the liquid charged and the gas charged elements, the refrigerant is the same as is used in the system. Cross charged means that the fluid in the sensing bulb is different than the refrigerant in the system.

In the automatic expansion valve, the refrigerant flow (through the valve and into the evaporator) is controlled only by the pressure in the evaporator.

In the thermostatic expansion valve, the flow through the valve and into the evaporator is controlled by both the low-side pressure and the temperature of the evaporator outlet. The valve provides a rather high rate of flow if the evaporator is quite empty (warm). It slows up the flow as the evaporator fills (cools) with refrigerant.

The sensing bulb type thermostatic expansion valve is operated by the accumulated pressure difference or force

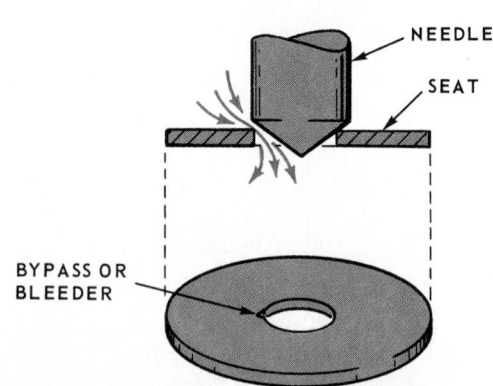

Fig. 5-7. Needle valve and seat used on bypass or "bleeder" type automatic expansion valve. Bypass permits pressures to balance during "off" part of cycle.

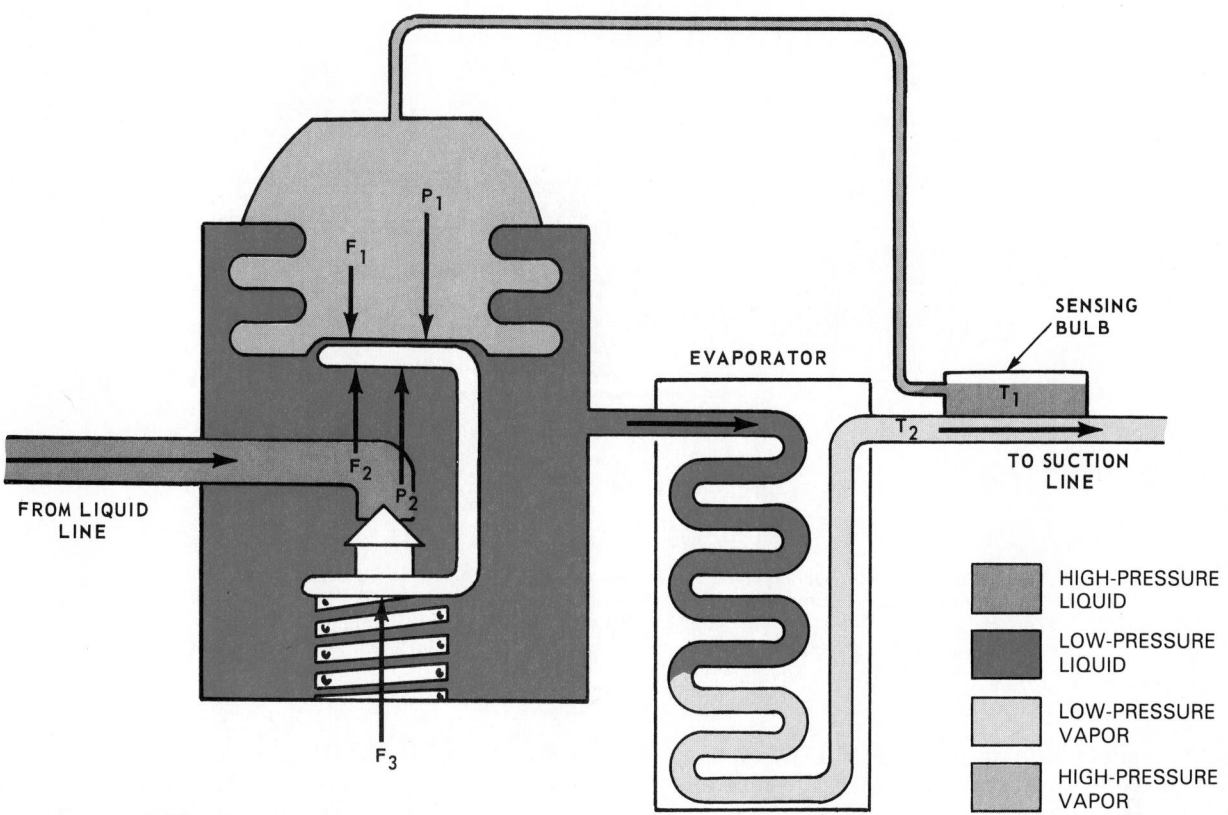

Fig. 5-8. Thermostatic expansion valve showing various pressures and temperatures within valve which cause it to operate. F_1—Sensing bulb pressure (force) tending to open valve. F_2—Low-side pressure (force) tending to close valve. F_3—Spring force tending to close valve. P_1—Sensing bulb pressure tending to open valve. P_2—Suction pressure (low side) tending to close valve. T_1—Sensing bulb temperature. T_2—Evaporator refrigerant temperature (low side). Valve opens when F_1 is greater than combined force of F_2 and F_3. Valve closes when combined F_2 and F_3 forces are greater than F_1.

difference between the sensing bulb bellows and the valve low-side pressure. See Fig. 5-8.

With the unit running, the refrigerant, T_1, in the expansion valve sensing bulb is usually about 10 °F (5.6 °C) warmer than the refrigerant in the evaporator, T_2. This temperature difference produces the different pressures and therefore the different forces.

This means that the unit pressure in the sensing bulb, P_1, is greater than the unit pressure, P_2, in the evaporator. This temperature difference is often described as the superheat of the bulb over the refrigerant temperature inside the evaporator. It should be noted that, as the temperatures increase or decrease, the pressure will also increase or decrease.

When the compressor stops, the low-side pressure and the sensing bulb pressure tend to equalize. The total expansion valve internal force, F_2 plus F_3, overpowers the sensing bulb force, F_1, and the needle is forced firmly into its seat. Refrigerant flow stops. The needle will stay closed until the sensing bulb force overcomes the low-side force.

This valve opening action should only happen when the unit is running. If the valve is adjusted correctly, the closing of the valve while the compressor is idle prevents the flooding of the low side with liquid refrigerant.

The thermostatic expansion valve does not regulate the low-side pressure, but rather controls the filling of the evaporator with refrigerant. The pumping action of the compressor establishes the low-side pressure. Fig. 5-9 illustrates a bellows type thermostatic expansion valve.

The adjustment enables one to set the valves so the needle can seat itself sooner while the unit is running. The needle will then close even though there is a greater temperature difference (about 15 °F or 8.3 °C) between the refrigerant in the evaporator and that in the sensing bulb, as shown in Fig. 5-10. The evaporator liquid refrigerant will not reach the sensing bulb location in this case. The temperature of only the low-pressure vapor will be cold enough to reduce the sensing bulb temperature (and therefore the pressure) to the closing point. The needle will close before the evaporator becomes full of liquid refrigerant droplets and the evaporator will be "starved."

If the adjustment is turned in too much the other way (one or two revolutions) to move the needle away from the seat, the temperature of the sensing bulb refrigerant will become closer to the temperature of the evaporator refrigerant 5 to 7 °F (2.8 to 3.9 °C). See Fig. 5-11.

The evaporator must now become more than full of liquid refrigerant droplets to bring the temperature (pressure) difference down to this value. The evaporator will be completely flooded. Some liquid droplets may even go into the suction line causing a sweating or frosting of the suction line and may harm the compressor (slugging). This adjustment is sensitive and should never be turned more than one-quarter of a turn each 10 to 15 minutes while the unit is operating.

Some thermostatic expansion valves use diaphragms instead of bellows. In either design, the valve is closed when the unit is not running.

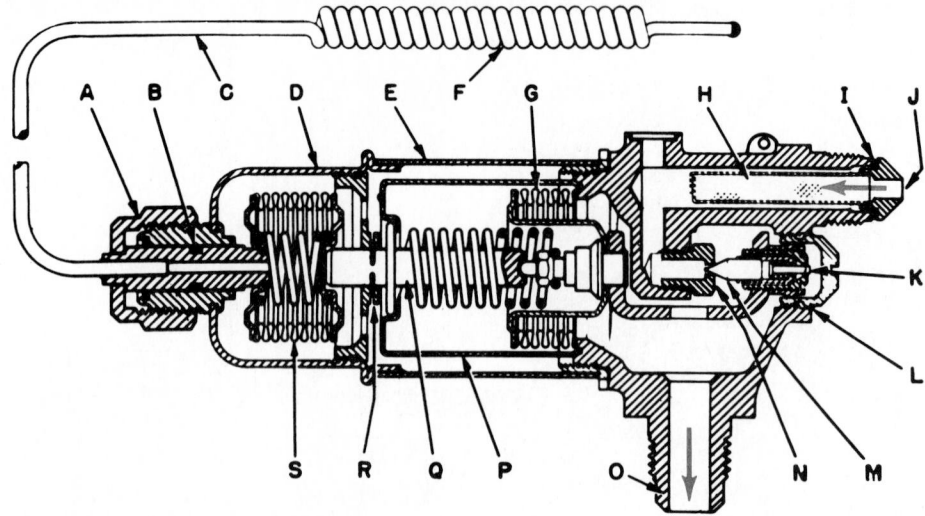

Fig. 5-9. Thermostatic expansion valve. A—Adjusting nut. B—Seal ring. C—Capillary tube. D—Bellows housing. E—Housing spacer. F—Temperature sensing bulb. G—Body bellows. H—Screen. I—Gasket. J—Refrigerant inlet. K—Needle pin. L—Sealed fitting. M—Needle. N—Seat. O—Evaporator connection. P—Inner spacer. Q—Spacer rod. R—Snap ring. S—Thermal bellows.

Chapter 3 describes the basic principles of a TEV system.

See Chapter 14 for instructions on installing and servicing thermostatic expansion valves.

In some large refrigeration installations (50 tons and over), a pilot-controlled thermostatic expansion valve may be used. In these installations, a conventional thermostatic expansion valve is mounted on a large auxiliary valve body. The auxiliary valve provides a larger pressure operated needle and orifice. The conventional thermostatic refrigerant control (pilot) regulates the pressure which operates the large refrigerant orifice control. Fig. 5-12 illustrates one type of pilot-controlled thermostatic expansion valve.

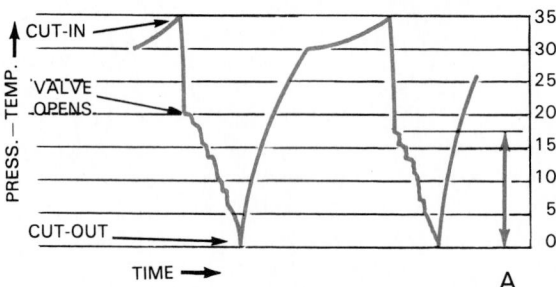

Fig. 5-10. A thermostatic expansion valve low-side pressure-time cycle diagram using a high superheat adjustment. A—Pressure drop on low side between opening of valve and cut-out point.

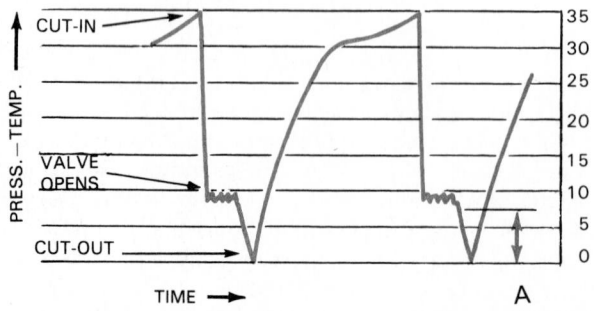

Fig. 5-11. A thermostatic expansion valve low-side pressure-time cycle diagram using a low superheat adjustment. A—Pressure drop on low side between opening of valve and cut-out point. Note: Compare with A, in Fig. 5-10, to understand how superheat adjustment controls evaporator (low side) pressure.

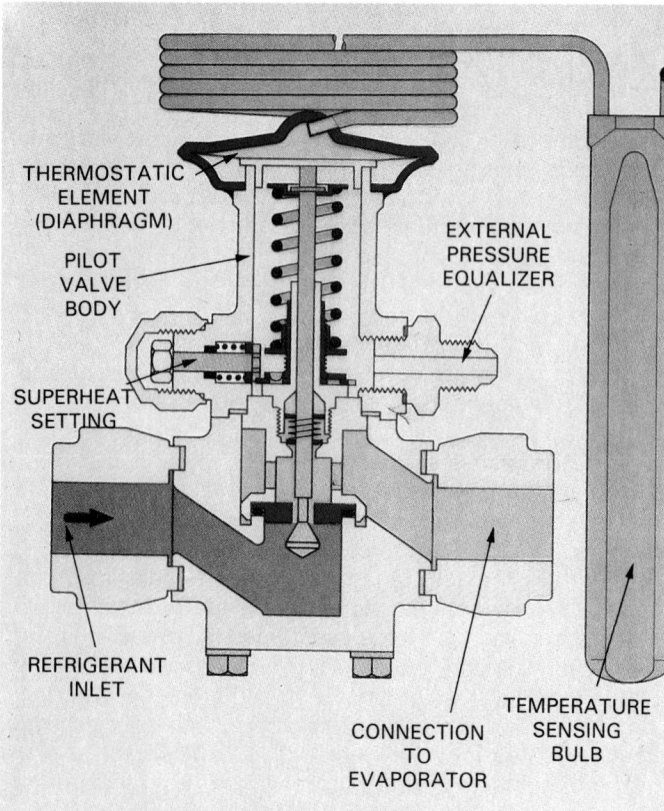

Fig. 5-12. A pilot-controlled thermostatic expansion valve. (Danfoss Inc.)

5-8 THERMOSTATIC EXPANSION VALVE DESIGN

Thermostatic expansion valves are usually used on multiple evaporator systems. However, the low-side float may also be used on multiple systems. It is possible in a multiple system using thermostatic expansion valves to provide a variety of temperatures in the various cabinets. This valve is also commonly used on air conditioning systems.

One must choose the correct sensing element charge and the correct size valve for each installation.

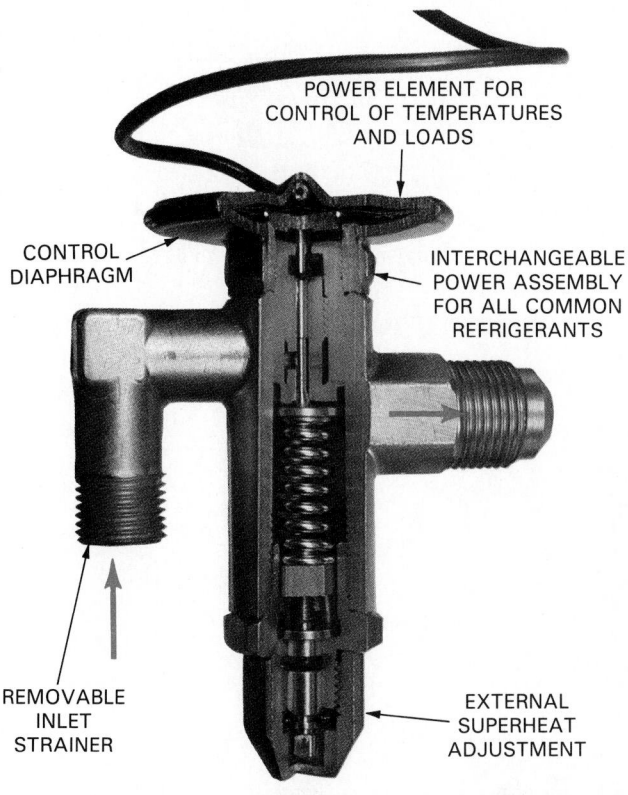

Fig. 5-14. Diaphragm type thermostatic expansion valve using flared inlet and outlet connections. Turn adjustment screw in to "starve" evaporator. Turn it out to flood evaporator.
(Alco Controls Division, Emerson Electric Company)

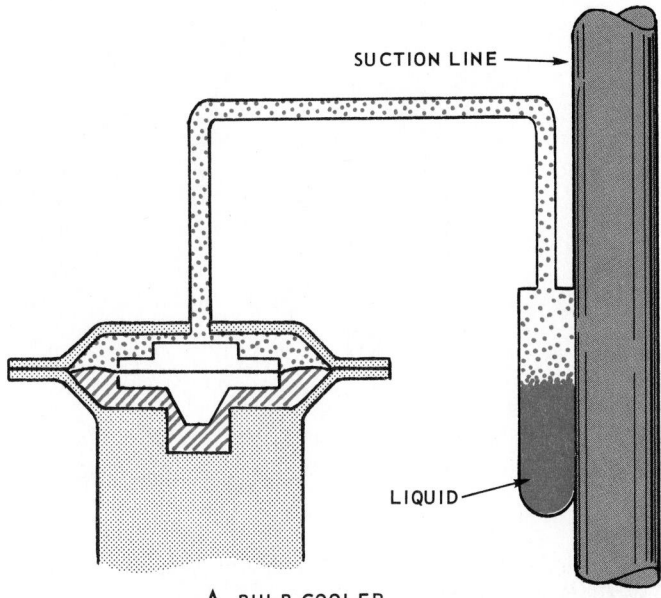

A BULB COOLER

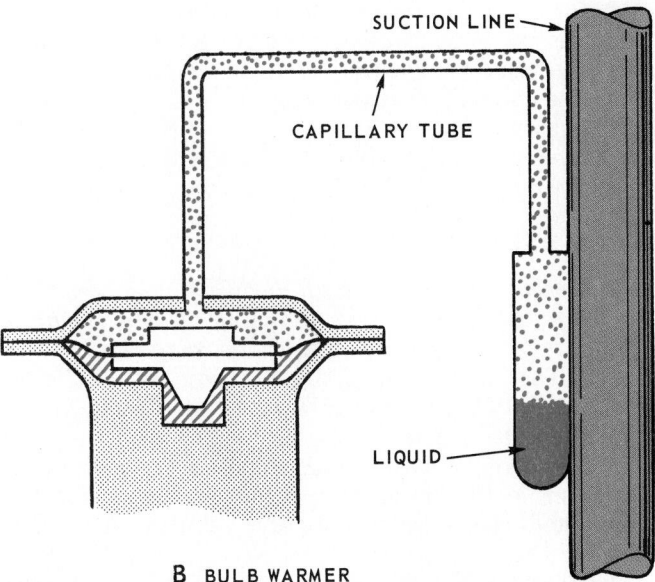

B BULB WARMER

Fig. 5-13. Effect of temperature on sensing bulb of thermostatic expansion valve. A—Sensing bulb is cold, pressure is low, and a considerable quantity of control fluid is shown as a condensed liquid in bulb (solid red). B—Sensing bulb is warmer and some of control fluid has evaporated (red dots) and is exerting pressure on expansion valve diaphragm, which will cause it to admit more refrigerant into evaporator. Note that for accurate control, there is enough control fluid in sensing bulb to insure control fluid in sensing bulb at all times. Red cross-hatching indicates suction line pressure.

The thermostatic expansion valve consists of a brass body into which the liquid line and evaporator line are connected. The needle and seat are inside the body. The needle is joined to a flexible metal bellows or diaphragm. This bellows, in turn, is made to move by a rod connected at the other end to a sealed bellows or diaphragm (power element) which is joined to the sensing bulb by means of a capillary tube.

Fig. 5-13 shows the refrigerant behavior in the sensing element. Each manufacturer has a code for identifying the fluid that charges the sensing element. Some use letters; others use colors or numbers to identify the charge. Some valves are marked with the refrigerant number. Some valves intended for use with refrigerant R-12 are color coded yellow.

The valve is sealed to prevent moisture seeping into any part. A strainer (screen) is always located between the liquid line connection and the orifice to keep dirt away from the needle and seat. See Fig. 5-14.

Some large air conditioning systems may use as many as six thermostatic expansion valves on one evaporator. In this way it is possible:

1. To maintain constant pressures and temperatures.
2. To make sure that all of the evaporator has a full charge of refrigerant.
3. To reduce pressure drop through the evaporator.

Fig. 5-15 is an exploded view of a thermostatic expansion valve. The body is usually made of brass.

Note that the inlet flare surface is mounted on the strainer. The pin or needle is usually made of Stellite,

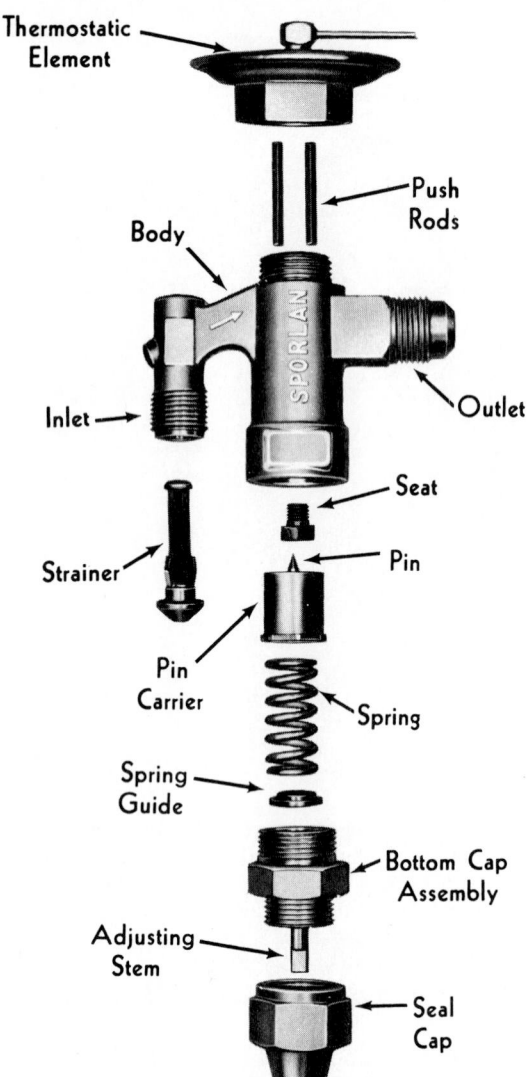

Fig. 5-15. Parts of a thermostatic expansion valve. In this valve, thermostatic element is threaded on body of valve.
(Sporlan Valve Co.)

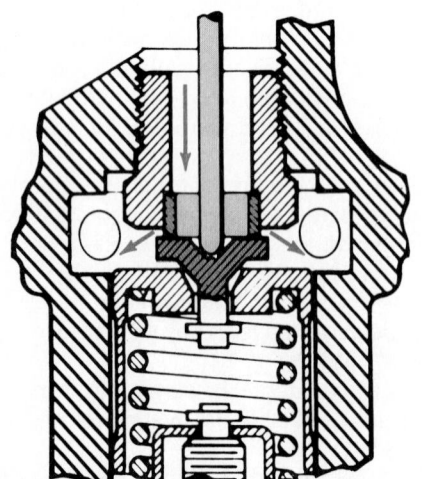

Fig. 5-16. This cutaway shows a thermostatic valve orifice that has been designed for large capacity. Note that both valve and valve seat are formed by flat surfaces. Arrows indicate direction of flow of refrigerant through valve seat mechanism.

Hastelloy, or stainless steel. The needles are usually sharp pointed cones, but spherical valves (balls) and flat orifice closers may also be used. The cone needle is popular for small capacity valves, while the ball type or the flat type is used in larger capacity valves.

Fig. 5-16 illustrates a large capacity flat valve seat. It is always good practice to place a filter-drier in the liquid line immediately ahead of the thermostatic expansion valve.

Fig. 5-17 shows the outside of a diaphragm type TEV with a ball type valve. Fig. 5-18 shows a cutaway of same.

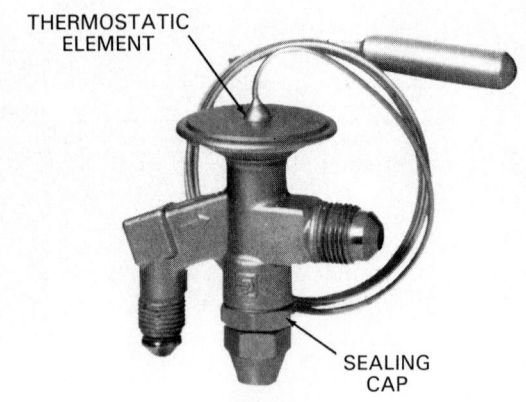

Fig. 5-17. Exterior view of thermostatic expansion valve. Valve may be adjusted after removing sealing cap.
(Refrigeration & Air Conditioning Div., Parker Hannifin Corp.)

Another type thermostatic expansion valve is shown in Fig. 5-19. It is the single diaphragm type designed particularly for service in air conditioning applications. It is equipped with a bleed valve for rapid pressure balancing (RPB). The rapid pressure bleed mechanism is shown at B in the illustration. The bleed mechanism works only on the off cycle. When the compressor starts up again, the secondary bleed port closes and the valve operates in a normal manner. Rods carry the diaphragm action to the needle. The liquid inlet is on the left; the evaporator connection on the right.

5-9 FLASH GAS

The term ''flash gas'' is used to indicate that portion of the refrigerant which evaporates instantly (flashes) and turns into a vapor as it passes through the refrigerant control orifice. The instant vaporizing of some of the liquid refrigerant (flash gas) cools the rest of the liquid to the evaporating temperature.

The amount of flash gas depends on temperature of refrigerant in the liquid line and the pressure inside the evaporator. Flash gas reduces the valve capacity. See Chapter 15.

One method used to reduce ''flash gas'' is to clamp the liquid line to the suction line. This is often called a heat exchanger.

Since the liquid coming from the condenser is often quite warm and the vapor coming from the evaporator is quite cold, clamping the two lines together causes a heat transfer from the liquid line to the suction line.

Cooling the liquid in the liquid line decreases the ''flash gas''. Raising the temperature of the vapor in the suction

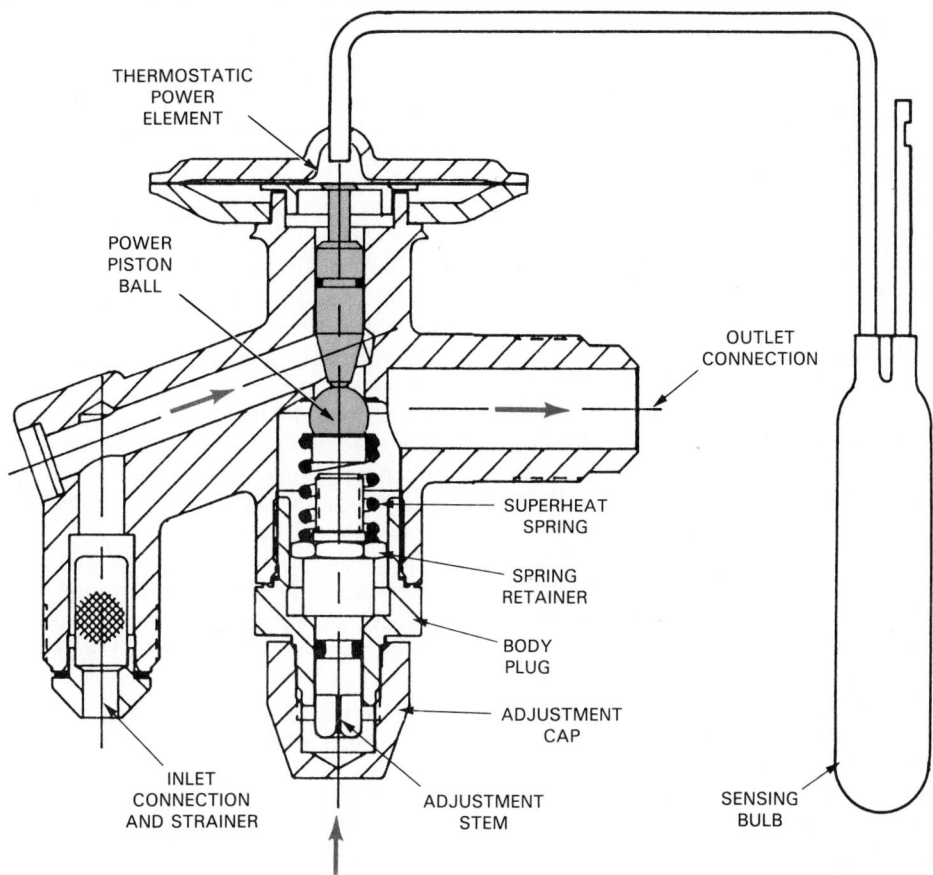

THERMOSTATIC
POWER
ELEMENT

POWER
PISTON
BALL

OUTLET
CONNECTION

SUPERHEAT
SPRING

SPRING
RETAINER

BODY
PLUG

ADJUSTMENT
CAP

INLET
CONNECTION
AND STRAINER

ADJUSTMENT
STEM

SENSING
BULB

Fig. 5-18. Diaphragm type thermostatic expansion valve. Note that a ball is used as the valve, replacing needle. Direction of refrigerant flow through expansion valve is marked by arrows. (Refrigeration & Air Conditioning Division, Parker Hannifin Corporation)

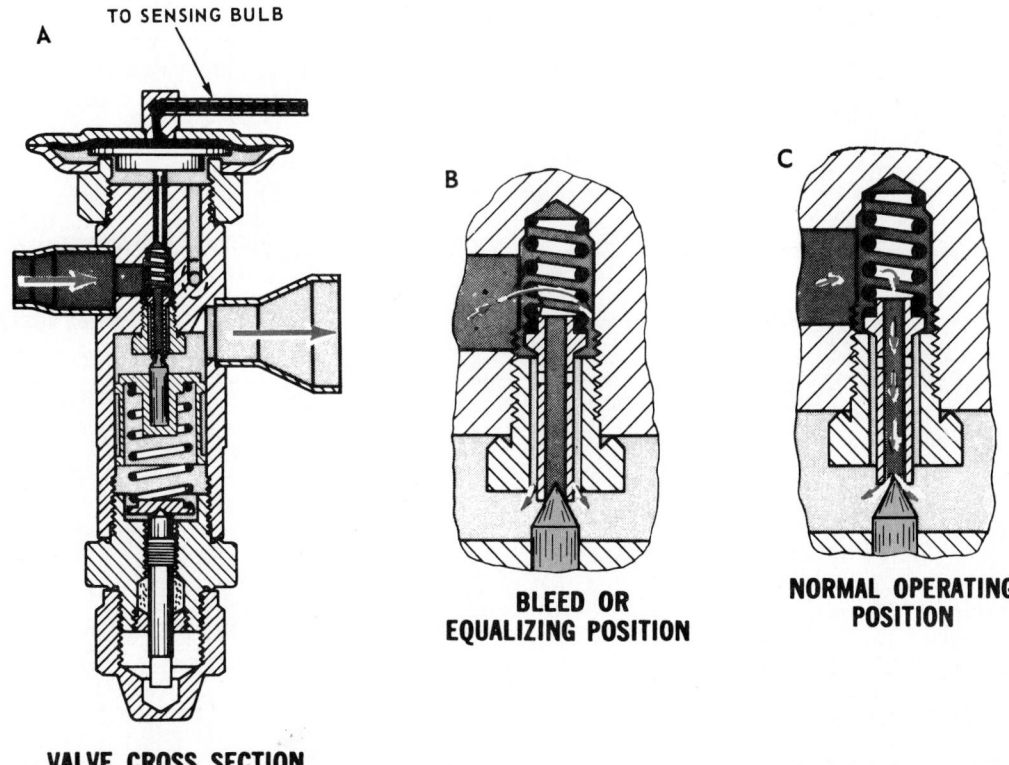

A TO SENSING BULB

B

C

BLEED OR
EQUALIZING POSITION

NORMAL OPERATING
POSITION

VALVE CROSS SECTION

Fig. 5-19. Diaphragm type thermostatic expansion valve. Sensing bulb pressure operates on top surface of diaphragm. As sensing bulb temperature increases, pressure on top of diaphragm tends to open valve, allowing refrigerant to enter evaporator. Note direction of refrigerant flow through valve. A—Cross section through entire valve. B—Bleed valve shown in normal operation. C—Needle valve in normal operating position. (Sporlan Valve Co.)

line decreases the possibility of any liquid refrigerant entering the compressor.

5-10 SUPERHEAT

The term "superheat," as used with thermostatic expansion valves, refers to the difference in temperature between the vapor in the low side and in the sensing bulb. A system adjusted to operate at a normal 10 °F (5.6 °C) superheat is shown in Fig. 5-20.

Increasing the superheat tends to "starve" the evaporator. *"Starving" the evaporator means that only part of the evaporator is filled with drops of liquid refrigerant.* Fig. 5-21 illustrates a superheat setting of 15 °F (8.3 °C). At this setting, the evaporator is said to be "starved."

The amount of superheat in the suction line will be governed considerably by the type of refrigerant control used and its adjustment.

The best superheat setting for an evaporator is the point at which the temperature of the thermal bulb of the thermostatic expansion valve changes the least when the system is running. This setting is called the MSS point or setting. It means Minimum Stable Signal. This setting is a result of the evaporator flow behavior as well as the behavior of the thermostatic expansion valve.

For example, if a valve and evaporator combination behaves as follows when adjusted:

At 12 °F superheat, bulb temperature
   changes from 14 to 10 °F.
At 10 °F superheat, bulb temperature
   changes from 11 to 9 °F.
At 8 °F superheat, bulb temperature
   changes from 8 1/2 to 7 1/2 °F.
At 6 °F superheat, bulb temperature
   changes from 8 to 4 °F.

The least change (most stable condition) is at 8 °F superheat setting.

5-11 LIQUID CHARGED SENSING ELEMENT

The liquid charged sensing bulb is charged with the same refrigerant as is used in the system. Thus, the valve is able to maintain a constant superheat setting even though the low-side pressures and temperatures change as shown by the graph in Fig. 5-22.

In the liquid charged sensing bulb, the quantity of fluid is sufficient so there is always some liquid in the bulb regardless of its temperature.

The sensing element will always control thermostatic valve operation, even if the temperature of the valve body is lower than the temperature of the sensing element. These elements are designed for a temperature range of from approximately −20 to 40 °F (−28.9 to 4.4 °C).

As the temperature of the evaporator drops, the amount of superheat increases.

This valve may cause some evaporator flooding when pulling down from normal ambient temperatures.

Air conditioners usually use this type of expansion valve. However, normal refrigerating systems may use them too.

5-12 LIQUID CROSS CHARGED SENSING ELEMENT

The liquid cross charged sensing bulb uses a liquid different from the refrigerant in the system. It may use a mixture of fluids to give the desired operating characteristics.

The amount of charge is such that some liquid will remain in the sensing element under all temperature conditions.

The sensing element will always control the valve operation even if the valve body temperature is lower than the temperature of the sensing element.

The valve closes quickly when the compressor stops because the evaporator pressure increases more rapidly than the sensing bulb pressure as the evaporator warms up.

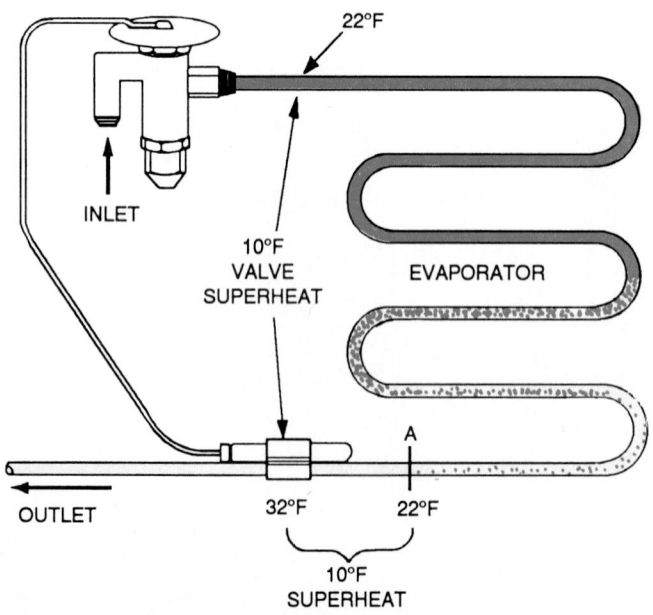

Fig. 5-20. A thermostatic expansion valve adjusted to give a normal 10 °F (5.6 °C) superheat. Liquid refrigerant will reach point A before valve controlling refrigerant closes.
(Alco Controls Division, Emerson Electric Co.)

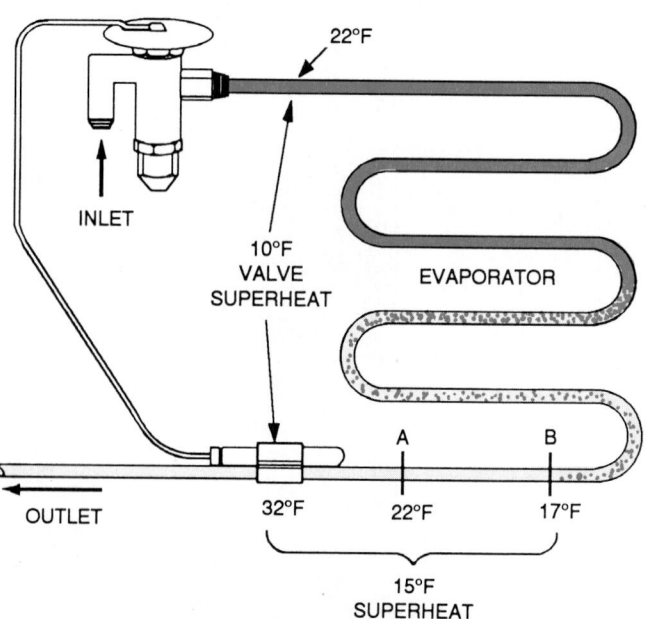

Fig. 5-21. A "starved" evaporator results when there is a pressure drop. Liquid reaches point B at 17 °F (−8 °C) and vapor warms to 22 °F (−6 °C). Valve closes at this point with too little liquid in evaporator.
(Alco Controls Division, Emerson Electric Co.)

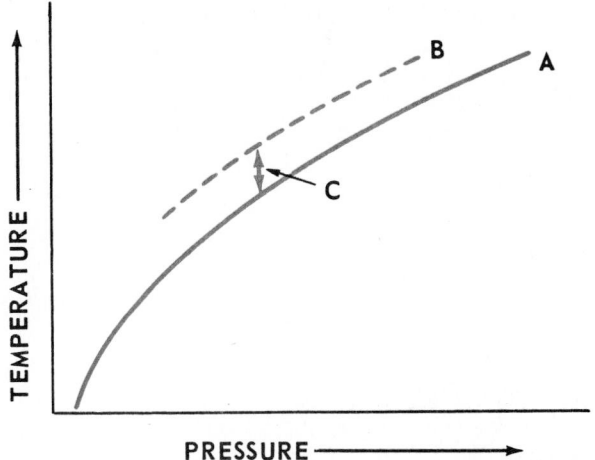

Fig. 5-22. A liquid charged thermostatic expansion valve superheat. A—Vapor pressure curve of refrigerant in system. B—Vapor pressure curve of charge in sensing bulb. C—Common value of superheat normally is 10 °F (5.6 °C).

The load on the compressor is reduced at start-up. As the suction pressure is reduced, the superheat is reduced, thus utilizing maximum evaporator surface.

"Hunting" is reduced because the flatter pressure-temperature curve of the sensing element pressure makes the valve more responsive to changes in suction pressure than to changes in sensing bulb temperature.

The liquid cross charged elements are designed for a temperature range of from −40 to 40 °F (−40 to 4.4 °C). These valves are usually used for either commercial low-temperature applications or with extremely low-temperature systems.

Fig. 5-23 shows the operating characteristics of the liquid cross charged element in graph form.

Fig. 5-24 shows the difference in the superheat curve of the cross charged element as compared to a normal charged element.

5-13 GAS (VAPOR) CHARGED SENSING ELEMENT

The gas charged sensing bulb is charged with the same refrigerant used in the system. The amount of charge is such that, at a predetermined temperature, all the liquid has vaporized.

Increasing the temperature above this point does not cause an increase in element pressure. Consequently the expansion valve does not open more with an increase in the cabinet temperature.

However, if the valve body becomes colder than the sensing bulb, the vaporized control fluid will condense in body of valve and control is lost. The valve closes.

For example, if just enough control fluid is put in the element to produce a maximum pressure of 40 psig (377 kPa), the element pressure will never exceed this pressure no matter how warm the bulb is. When the low side exceeds this pressure, the valve will not open. See Fig. 5-25. Thus, low-side pressure will not have to operate above 40 psi.

Vapor charged elements are designed for a temperature range of from 30 to 60 °F (−1.1 to 15.6 °C).

5-14 GAS (VAPOR) CROSS CHARGED SENSING ELEMENT — ADSORPTION

The gas cross charged sensing bulb is charged with a liquid different than the refrigerant in the system. The amount of charge is such that, at the desired temperature,

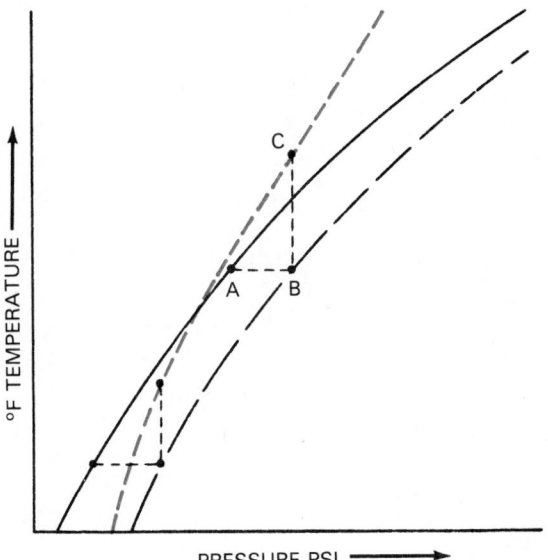

Fig. 5-23. A graph of action of a liquid cross charged element designed for application within a specific temperature range. It provides a rapid pull-down and is used for normal refrigeration. Superheat setting at top end is wide to prevent flooding of unit. A—Given evaporator pressure and corresponding saturation temperature. B—Evaporator pressure plus equivalent superheat spring pressure. C—Corresponding sensing bulb and power assembly pressure. B—C is superheat setting for this evaporator pressure and particular superheat spring setting.

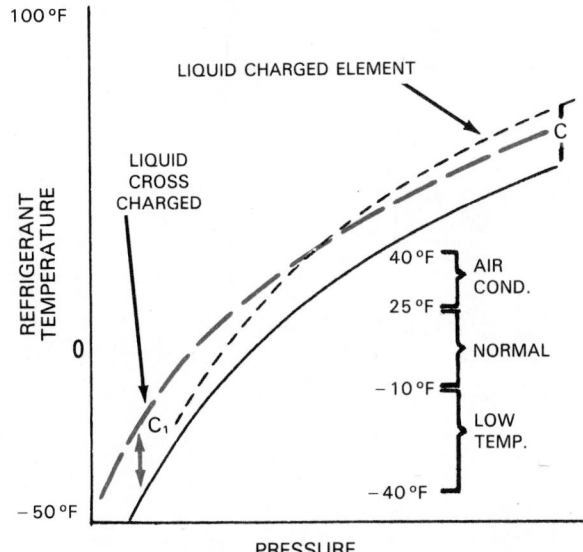

Fig. 5-24. Graph showing constant superheat of liquid cross charged power element designed to be used for all three applications—low temperature, normal, and air conditioning—as compared to a liquid charged power element during wide temperature ranges. Red dotted line shows degrees of superheat for a liquid cross charged thermostatic power element. Black dotted line shows degrees of superheat for liquid charge. The C liquid charge changes as temperature drops, but the C_1 liquid cross charge remains constant as temperature drops.

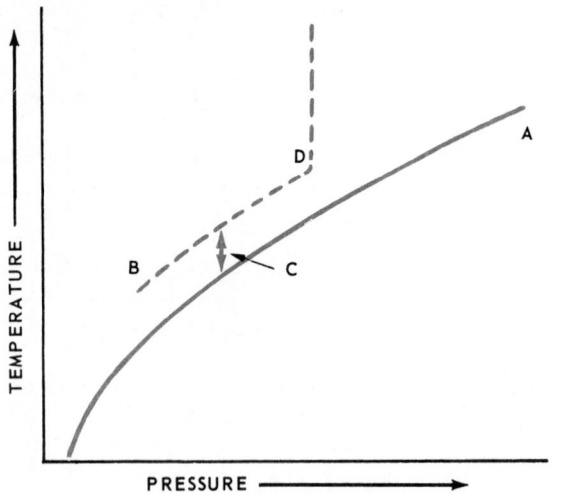

Fig. 5-25. A gas (vapor) charged thermostatic expansion valve superheat. A—Vapor pressure curve of refrigerant in system. B—Vapor pressure curve of charge in sensing bulb. C—Superheat. D—Point at which all fluid in sensing bulb becomes vaporized.

ing element contains two substances. One is a noncondensing gas, such as carbon dioxide, which provides the pressure in the element. The other is a solid such as carbon, silica gel, or charcoal.

These substances have the ability to adsorb gas depending upon their temperature. *Adsorption means the adhesion of a layer of gas one molecule thick over the surface of a solid substance.* There is no chemical combination between the gas and the solid substance (adsorber).

The ability of a substance to adsorb gas depends upon the temperature of the substance. Substances more readily adsorb gas at low temperatures. As the sensing element warms, the pressure in the element will increase due to release of the adsorbed gas. As the sensing element cools, its pressure will decrease due to the adsorption of gas back to the adsorbing substance. This pressure change is used to control the refrigerant needle valve opening in the thermostatic expansion valve.

The appearance and general construction of these thermostatic expansion valves is the same as with the usual gas cross charge. The only difference is in the gas and substances contained in the sensing element to control the sensing element pressures.

These thermostatic expansion valves have the advantage of a pressure-temperature lag in their operation. They have very wide temperature applications and can be used on any type refrigerating or air conditioning system. Their range is sufficient to cover almost any refrigerating application. The gas in the sensing element is a noncondensing gas and remains a gas throughout the operating range of the valve.

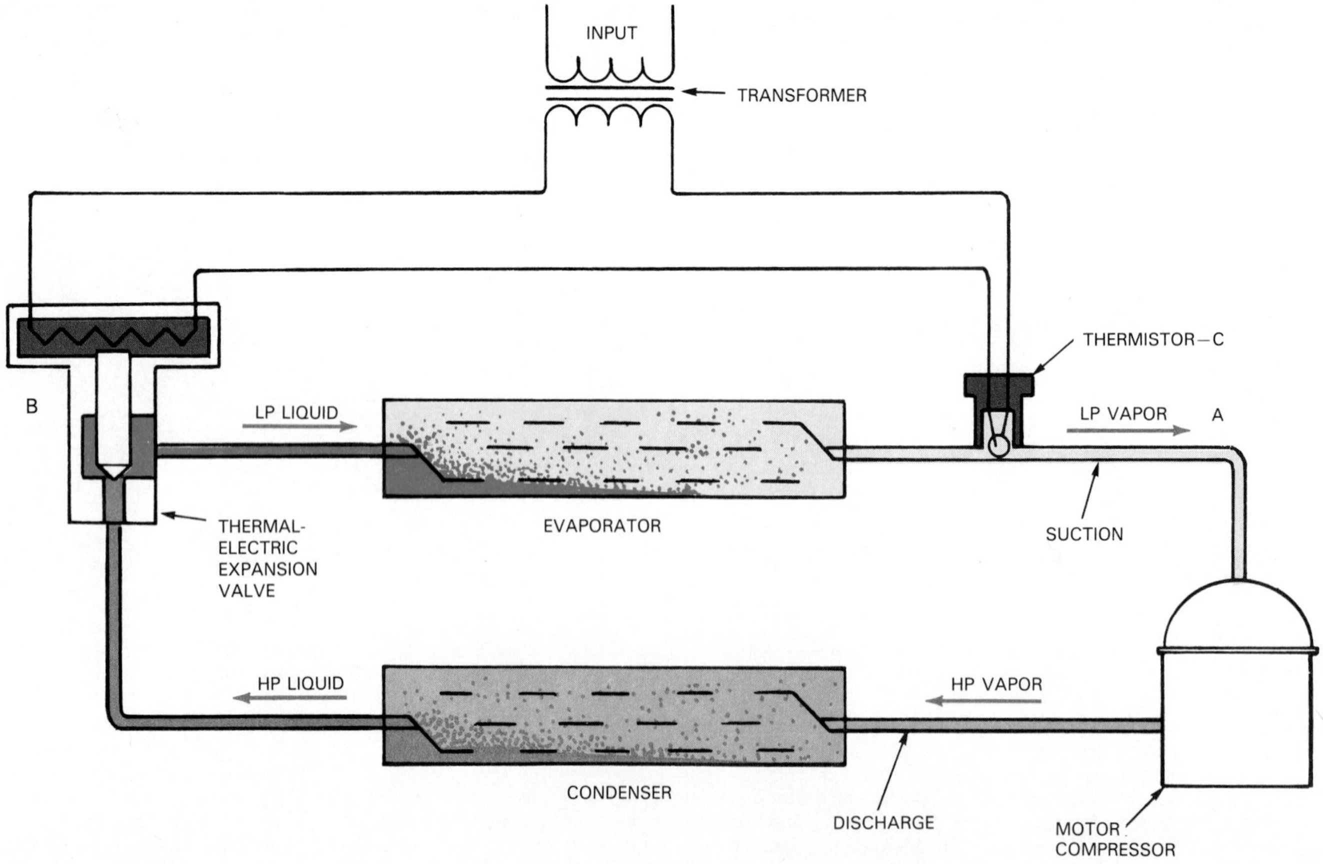

all the liquid has vaporized. Increasing the temperature above this point does not cause a usable increase in element pressure. The superheat characteristics are like the liquid cross charged element.

Some types of gas cross charged sensing elements depend upon a different principle than the one explained above. In these thermostatic expansion valves, the sens-

Fig. 5-26. A thermal-electric expansion valve installation.

5-15 THERMAL-ELECTRIC (SOLID STATE) EXPANSION VALVE

The thermal-electric controlled expansion valve depends upon the use of thermistors (see Chapter 6), directly exposed to the refrigerant in the suction line, to control the expansion valve needle opening. It does not use a pressure element as in the thermostatic expansion valve.

The resistance to electrical flow in the thermistor changes with its temperature. Increasing temperature reduces resistance. Therefore, with a given voltage, it increases the current rate of flow. This increased current flow heats the bimetal in the valve body and makes the bimetal bend, opening the valve.

Fig. 5-26 illustrates a typical thermal-electric expansion valve installation. The thermistor, C, is placed in immediate contact with the refrigerant vapor inside the suction line from the evaporator.

A low voltage transformer is the power source. This is connected to the expansion valve control mechanism at B. The transformer is in series with the thermistor and electric device at B so that increasing current flow through the thermistor increases the opening of the expansion valve and, therefore, increases the rate of flow of the refrigerant into the evaporator.

Increasing the current causes the valve needle to open, while a decrease closes the valve. Thereby the refrigerant flow is controlled. The thermal-electric expansion valve is not dependent on the pressure in the evaporator. It restricts the flow of refrigerant and controls the suction line superheat in order to prevent flooding of the compressor.

Fig. 5-27 shows a cross section of such a valve. This illustrates, in some detail, the electrical connections and the mechanisms which control the operation of this expansion valve. A complete thermal-electric expansion valve and thermistor ready for installation is shown in Fig. 5-28.

Off-cycle operation is possible in one of two ways. In one case, the thermal-electric expansion valve may be electrically connected in parallel to the operating system. During the off cycle, the thermistor will become warmed and the thermal-electric expansion valve will remain open. This unloading is similar to the pressure-balancing effect when a capillary tube refrigerant control is used.

In the second case, it may be electrically connected into the motor circuit (interlocked) in such a way that it is only energized when the compressor is running. With this type of connection, the valve will be closed on the off cycle.

5-16 PRESSURE LIMITERS

Sometimes a pressure-limiting expansion valve is used to prevent overloading the condensing unit. *Pressure limiters are designed for systems in which the evaporator pressure must not exceed a safe operating limit.*

One of these devices is placed between the sensing element mechanism and the needle mechanism. It is a second diaphragm and a spring located beneath the control diaphragm and is designed to collapse at a certain force. Thus, if this element is designed to collapse at 40 psig (377 kPa), the needle will close if the low-side pressure ever exceeds this amount. It does not matter what the evaporator temperature-pressure is. These valves offer rapid pull-down on start-up and maintain the evaporating pressure within the prescribed safe limits. See Fig. 5-29.

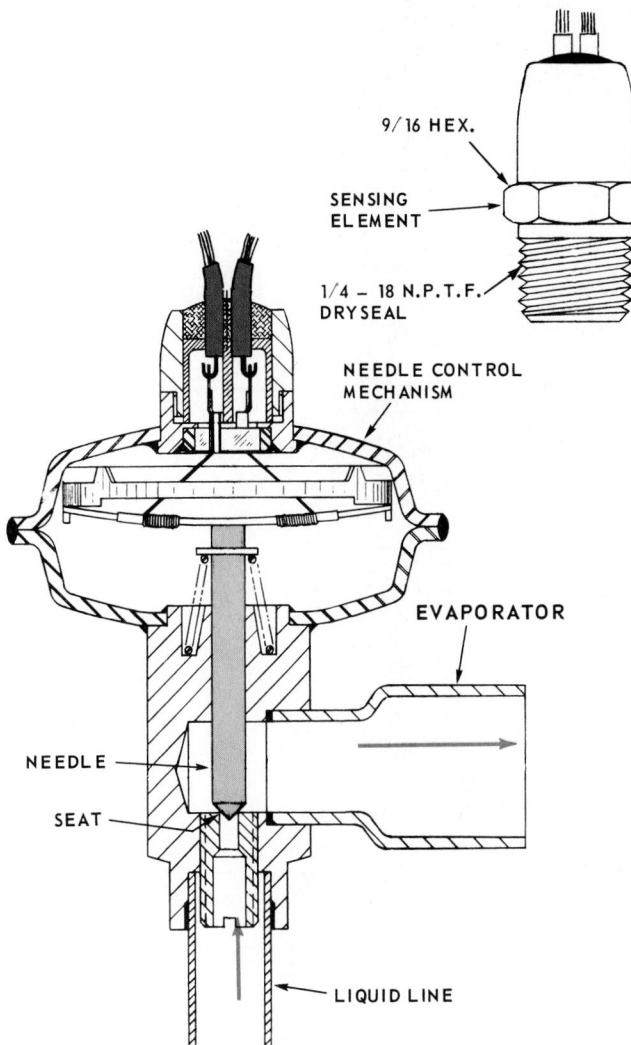

Fig. 5-27. A thermal-electric expansion valve.

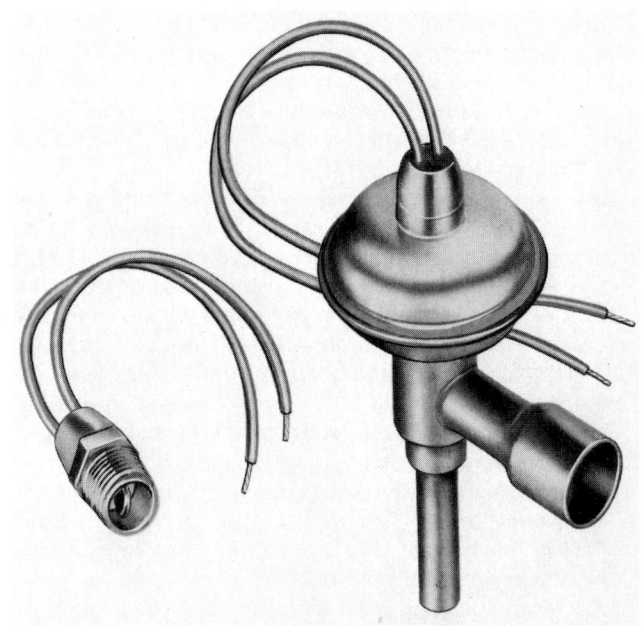

Fig. 5-28. A thermal-electric expansion valve. A—Control mechanism. B—Liquid line. C—Connection to evaporator. D—Thermistor.

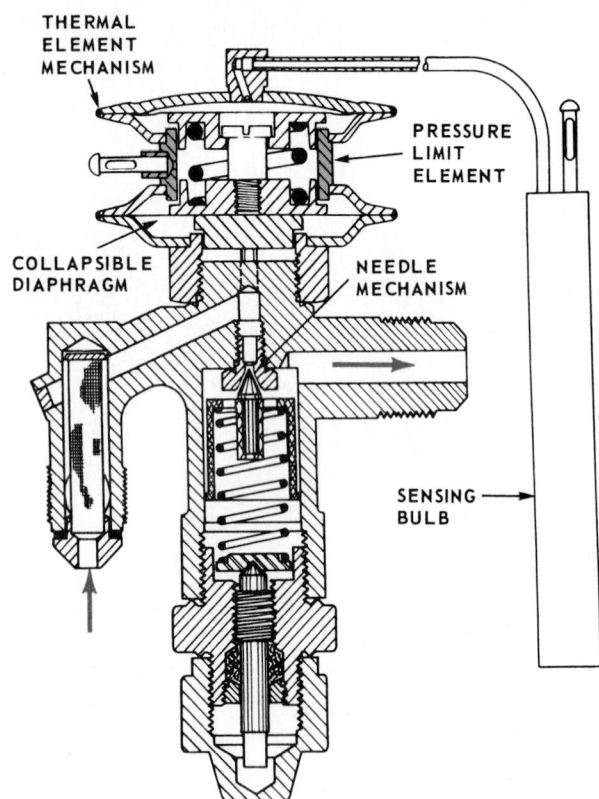

Fig. 5-29. Thermostatic expansion valve with mechanical pressure limit element. Element limits pressure by means of two diaphragms and a spring. Whenever suction pressure gets near motor overload point, spring between two diaphragms compresses and valve reduces flow of refrigerant to evaporator.

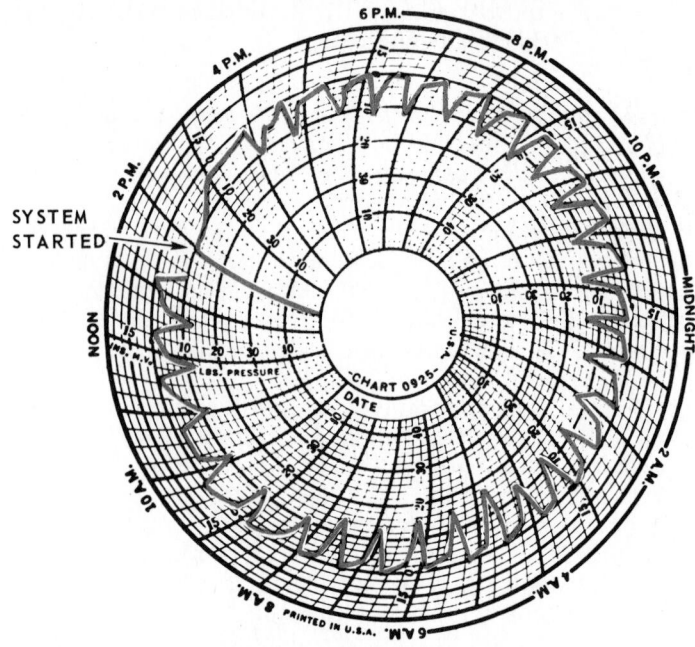

Fig. 5-30. A 24-hour record of pressure-time relationship for a food freezer as it is cycled, beginning with a warm condition at 2 p.m. Freezer is equipped with pressure-limiting thermostatic expansion valve and thermostatic motor control. Note quick reduction of pressure until valve opened at 2 p.m. (Sporlan Valve Co.)

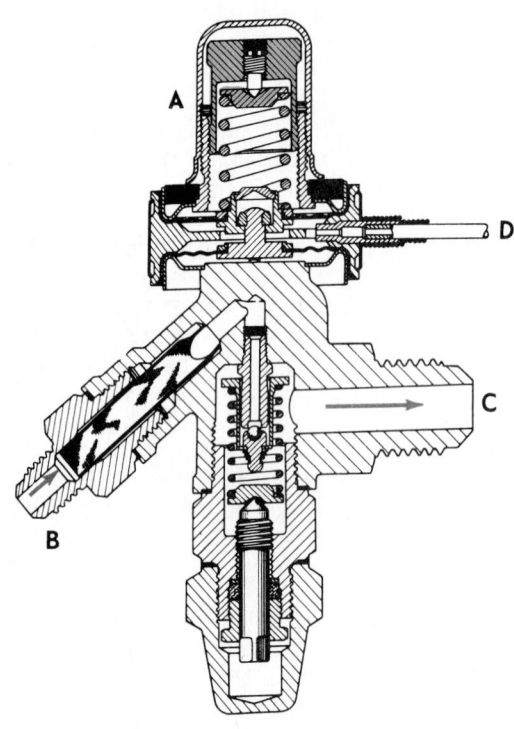

Fig. 5-31. Thermostatic expansion valve with adjustable pressure limiter device. A—Pressure limiter adjustment. B—Liquid refrigerant inlet. C—Evaporator connection. D-Sensing bulb connection.

A gas charged collapsible element can also be used to provide a limit to the pressure which will open the valve. When the pressure in the low side exceeds a certain set value, the diaphragm will collapse. The gas used is non-condensable and obeys Charles's and Boyle's Laws.

An example of how the pressure limiting device prevents a long running time at excessive low-side pressures when the refrigerator is warm is shown in Fig. 5-30. The cycle record shows a pressure drop from over 50 psig to 10 psig (446 kPa to 170 kPa) in just a few minutes. The unit then runs for two hours before it shuts off.

Still another type of pressure limiter thermostatic expansion valve is shown in Fig. 5-31. This valve has an adjustable pressure limiter, A. Above a certain pressure setting at A, the spring above the diaphragm will compress instead of the valve needle being opened. The liquid inlet is at B, and the evaporator connection is at C. The capillary tube and sensing bulb are not shown, but the capillary tube connection is at D.

Frequently in a refrigeration device it is necessary to have two totally different pressure levels controlling a given valve. *A control valve with two pilot pressure regulators is called a "dual pressure regulator."* This type of system uses a switching mechanism to allow for the selection of either high- or low-pressure control.

Maximum operating pressure (MOP) or pressure limit can also be achieved by the use of a specially designed thermostatic charge. The MOP thermostatic charge is a modification of the conventional limited liquid thermostatic

charge. This type of thermostatic charge allows a predetermined valve bulb temperature and corresponding bulb pressure (opening pressure) to be reached and the valve throttles or closes. Any increase in bulb temperature above the predetermined value results in little or no increase in bulb pressure. The MOP setting is comparable to the pressure limit setting of the mechanical pressure limit valve.

MOP thermostatic charges have two distinct differences over conventional thermostatic charges. The first is that MOP thermostatic expansion valves close tightly during off-cycle. As the evaporator warms up in the off-cycle, the point of maximum bulb pressure is reached. Since any increase in bulb temperature results in no increase in bulb pressure (opening pressure), the evaporator pressure (closing pressure) continues to rise and, assisted by the spring pressure (closing pressure), closes the valve tightly.

The second difference is that thermostatic expansion valves remain closed during pull-down. Although temperatures and pressures are relatively high in the evaporator during pull-down, the valve remains closed until the evaporating temperature is reduced below the MOP of the thermostatic charge. This permits rapid pull-down, avoiding floodback and overloading of compressor motor.

Thermostatic charges using a maximum operating pressure (MOP) require the diaphragm and capillary tubing to be kept at a temperature warmer than the bulb during the operating cycle. This is necessary so the valve will be controlled by the bulb. If the diaphragm case is allowed to become colder than the bulb, the thermostatic charge may migrate to the diaphragm case and control from the bulb will be lost. The valve will then close or throttle.

Maximum operating pressure thermostatic charges are used with comfort cooling systems and indoor and outdoor coils of heat pumps.

5-17 SENSING BULB MOUNTING

The location and the actual mounting of the sensing bulb is very important. It must be in good thermal contact with the evaporator outlet, Fig. 5-32. The bulb should be mounted on the top of the suction line so the liquid in the bulb is close to the suction line as shown at A in the illustration. If it is necessary to mount the bulb on a vertical suction line, the capillary tube of the bulb should always enter from the top of the bulb as shown at B, never from the bottom!

To keep the bulb from being affected by the air or liquid being cooled, it should be wrapped in insulation, as in C. Special insulation forms are available or plastic tape can be used so that only suction line temperature affects the bulb.

Copper straps and nonrusting machine screws and nuts should be used to fasten the bulb to the suction line. The bulb must have excellent thermal contact with the suction line. The connection must be clean and tight. One should clean both the suction line and the bulb with steel wool before assembling.

The suction line carries chiefly vaporized refrigerant. However, there will be some droplets of liquid refrigerant and some oil.

Fig. 5-33 illustrates conditions inside the suction line, particularly on installations which require the line to be a large diameter. At A, refrigerant vapor and some droplets of liquid refrigerant are flowing through a rather large diameter suction line. Due to the large diameter, the velocity of the vaporized refrigerant at times will be quite slow. The droplets of liquid refrigerant and oil will settle on the bottom of the line. The suction area at B is smaller. As a result, the velocity of the vaporized refrigerant will be higher than at A. This means less separating of the oil and liquid

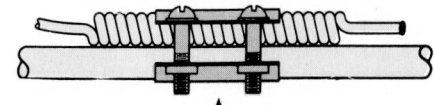

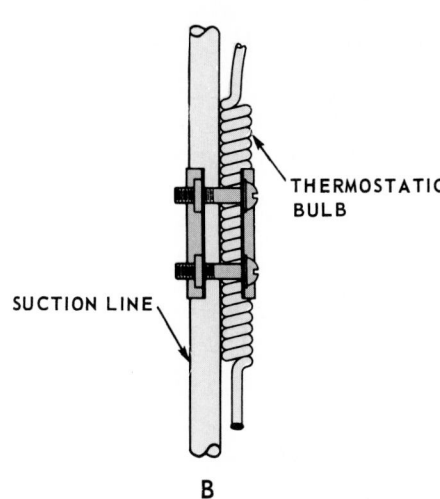

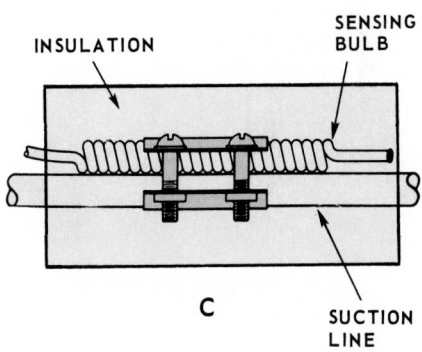

Fig. 5-32. Correct way to attach thermostatic expansion valve sensing bulb to suction line. A—Thermostatic sensing bulb mounted in horizontal position. B—Thermostatic sensing bulb mounted in vertical position. C—Insulated sensing bulb installation. Note that thermal bulb is best mounted in horizontal position and on top of suction line.

refrigerant from the flowing vapor. The inside of the tube will be rather uniformly coated with oil. The suction line at C is shown in vertical position. In this position, there will be no separation of the droplets of refrigerant from the vapor. However, the oil will rather uniformly coat the inside of the suction line.

The temperature of the vaporized refrigerant and the droplets of liquid refrigerant will be a few degrees colder than the surface of the suction line. This is because of the insulating quality of the oil which coats the inside of the suction line.

The sensing bulb on large suction lines should be located near the underside of the line rather than on top. This is better because droplets of liquid refrigerant tend to separate from the flowing vapors. Also, as mentioned before, oil has an insulating effect. The recommended bulb position (on large suction lines) is shown at D, Fig. 5-33. Some sensing bulbs are crimped or creased lengthwise to provide double contact and to help align the sensing bulb with the surface of the suction line.

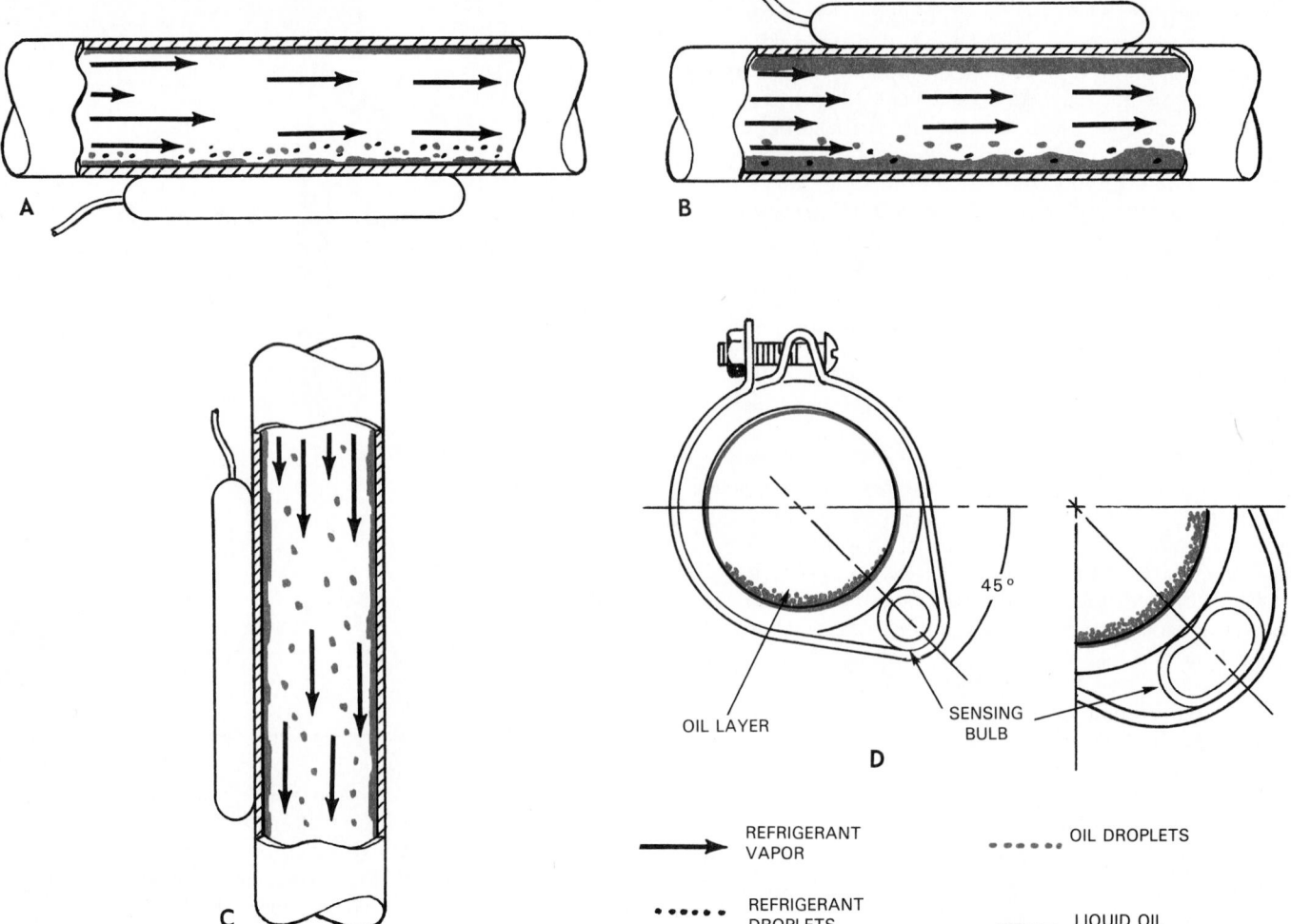

Fig. 5-33. Flow conditions in suction line. A—Horizontal suction line (large). B—Horizontal suction line (small). C—Vertical suction line. D—Sensing bulb on large diameter suction line. Oil is shown in red; refrigerant vapor by black arrows; refrigerant droplets by black dots.

5-18 THERMOSTATIC EXPANSION VALVE CAPACITIES

The capacity of a thermostatic expansion valve (TEV) varies according to:
1. Orifice size.
2. Pressure difference between the high side and the low side.
3. The temperature and condition of the refrigerant in the liquid line. The amount of flash gas will increase with a rise of liquid line temperature.

The capacity of most thermostatic expansion valves may be selected from the size of the orifice and needle assembly. The same body may be used for many capacities.

The larger the orifice, the more liquid refrigerant that can be fed into the evaporator for each unit of time.

Valves are rated in tons of refrigeration. However, the same orifice usually has three different tonnage capacities. This range of capacity depends upon the difference in pressure between the high side and the low side.

Increasing this pressure difference will increase the rate of refrigerant flow. Therefore, if the valve is used on an R-12 refrigerant system, a valve that has a 1/2 ton (.455 metric ton) rating at 13 psi (193 kPa) pressure on the low side will have a 3/4 ton to 1 ton (.72 to .98 metric ton) capacity

at 5 in. (12.7 cm) Hg vacuum on a frozen food unit. The same valve has only a 1/3 ton (.3 metric ton) capacity at a low-side pressure of 40 psi (380 kPa) on an air conditioner.

In the first case, there is a 130 − 13 = 117 psi (807 kPa) pressure difference, assuming a 130 psi (1 000 kPa) head pressure. In the second case, it is a matter of 130 plus 2 1/2 psi (5 in. of vacuum = 2 1/2 psi = 84 kPa) = 132 1/2 psi (914 kPa) pressure difference. In the last case, it is 130 − 40 = 90 psi (621 kPa) pressure difference.

It is important to use a valve of the correct capacity. If the valve orifice is undersize, the evaporator will be starved regardless of the superheat setting. The full capacity of the evaporator cannot be reached.

If the orifice is oversize, the valve will "hunt" or surge. When the valve opens, too much refrigerant will pass into the evaporator and the suction line will sweat or frost before the thermal element can close the valve. If one tries to correct this condition by increasing the superheat setting, the evaporator will be starved much of the time.

5-19 SOLENOID VALVE PRINCIPLES

A solenoid valve is used in many refrigerating applications to automatically close off or open refrigerant circuits to get

156

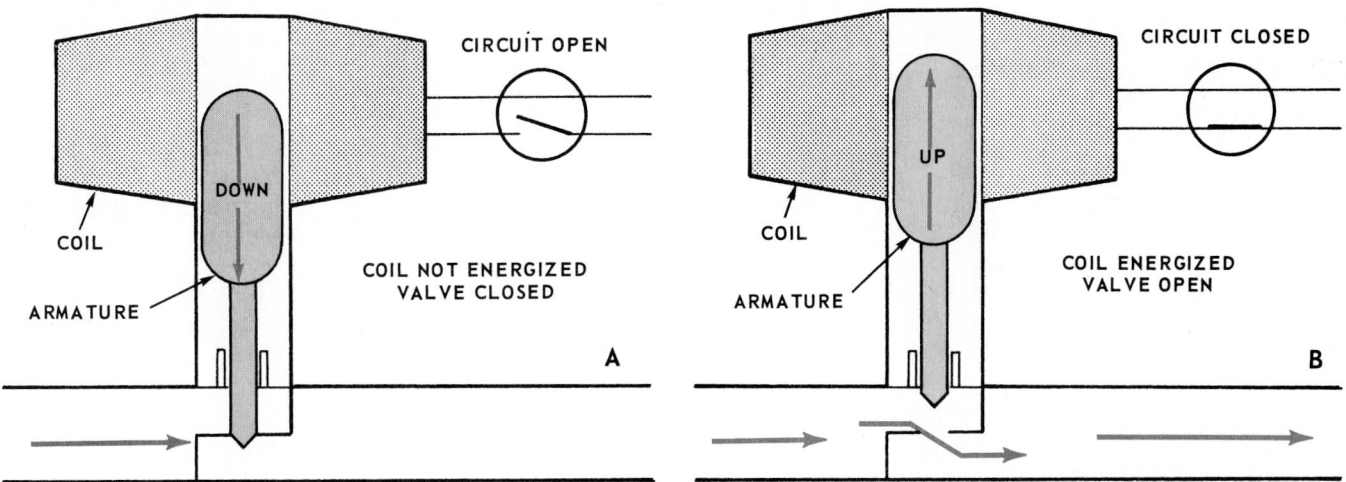

Fig. 5-34. Solenoid valve. A—Circuit is open, coil is not magnetized and armature, due to its weight, drops and closes valve. B—Circuit is closed, coil is magnetized and armature, due to magnetic field, is pulled up and opens valve.

the desired refrigerating effect. It is easily installed and uses only simple electrical control circuits.

A solenoid valve is simply an electromagnet with a movable core or center. The basic construction of an electrical solenoid valve includes a movable armature. This is made of an iron alloy and is attached to the valve needle. This element is sealed into the valve body so the armature can raise and lower the valve needle. A coil is wound around the valve housing which contains the armature.

The basic construction of a solenoid valve is shown in Fig. 5-34. As the coil is energized, the armature, which is magnetic, moves upward toward the center of the coil, opening the valve. When the circuit is opened, the coil is de-energized and, therefore, demagnetized. The spring and weight of the armature force the valve against the valve seat.

5-20 TYPES OF SOLENOID VALVES

Three types of solenoid valves are in common use:
1. The two-way valve which controls the flow of refrigerant through a single line. See Fig. 5-35.
2. The three-way valve with an inlet which is common to either of two opposite openings. It controls refrigerant flow in two different lines. See valve shown in Fig. 5-36.
3. The four-way reversing valve which is used often on heat pumps. These are illustrated in Fig. 23-16.

Solenoid valves may be turned on by means of a thermostat. These valves are used to control the temperature of a refrigerator or a room.

Three-way solenoid valves are used mainly on commercial refrigerating units. They may be used to control two separate refrigerant circuits for defrosting, two-temperature evaporators, and so on.

In the three-way solenoid valve shown in Fig. 5-36, opening A is the common opening and is never closed. As the electromagnet is de-energized, the weight of the solenoid plunger assembly and the force of the upper spring hold the valve firmly against the lower seat. This action closes port B to the common port and opens port C.

When the electrical circuit is closed, the solenoid becomes energized. The piston with the two valves attached to it will move in the opposite direction. This action opens port

B in Fig. 5-36 to common port A and, thereby, closes port C to the common port.

Four-way solenoid valves are often called reversing valves. See Chapter 23. Four-way solenoid valves are used chiefly on heat pumps to control the cycle for either heating or cooling, as needed.

When the solenoid is de-energized, the valve stem closes several ports and opens others to reverse the flow of the refrigerant to the condenser and evaporator. The heat pump then becomes a cooling system.

When the solenoid valve becomes energized, the four-way valve stem is drawn upward. The heat pump system becomes a heating system again.

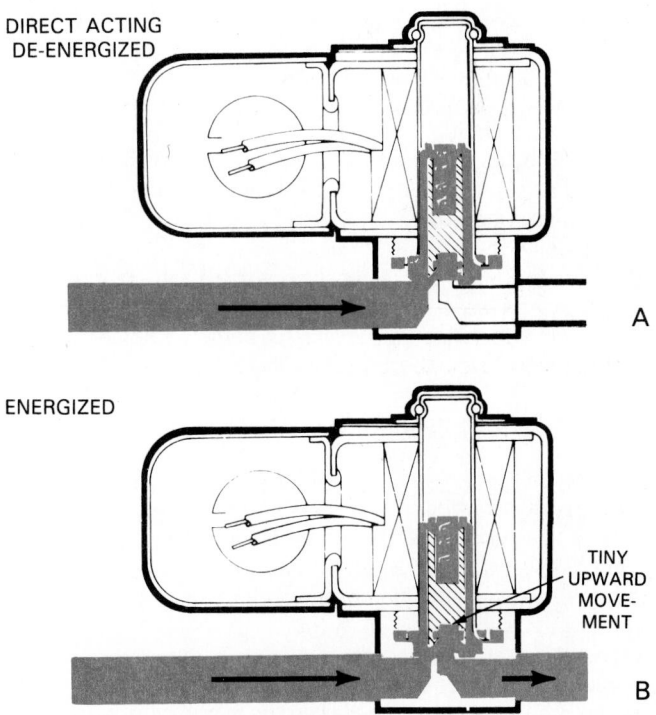

Fig. 5-35. Two-way solenoid valve. A— De-energized, closed position. B—Energized, open position.   (Alco Controls Div., Emerson Electric Co.)

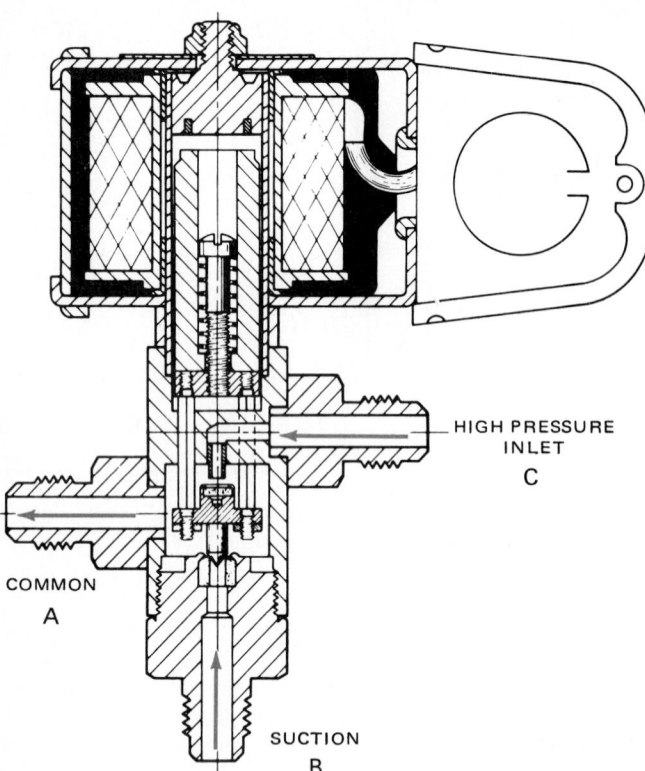

Fig. 5-36. Three-way solenoid valve used to close thermostatic expansion valve during "off" part of cycle. When compressor is running, openings marked "suction" and "common" are connected. When compressor is off, "high pressure inlet" and "common" are connected. (Sporlan Valve Co.)

For large commercial applications, it is desirable to use a pilot-operated solenoid valve. In these valves, the solenoid operates a pilot mechanism and the pressures in the lines concerned operate upon a piston arrangement. This causes the opening and closing of the main or large control valve.

Fig. 5-37 is a cutaway of a pilot-operated solenoid valve. When the solenoid, A, is energized, the plunger, B, will be pulled from its seat and the pressure in D will leave the cylinder and cause piston, E, to move up. The movement of the piston controls the opening and closing of the control valve, F.

When the solenoid valve is de-energized, plunger, B, returns to its seat. Pressure from G goes through small opening H and builds up pressure in control cylinder D. The spring then closes the control valve, F.

5-21 EQUALIZERS

An equalizer is a small tube—usually 1/4 in. OD—which joins the suction line at the outlet of the evaporator. The other end opens beneath the expansion valve diaphragm. *The equalizer compensates for any pressure drop through the evaporator while the compressor is running.* An equalizer tube is shown in Fig. 5-38.

There is always some pressure drop through the evaporator. It is recommended that an equalizer device be used if the pressure drop between the inlet of the evaporator and the outlet is more than 4 psi (28 kPa). In operating the control valve, the equalizing tube provides the same pressure as is in the suction line at the sensing bulb loca-

tion. This equalizing of pressure will permit accurate superheating adjustments. Pressure drop in the evaporator always tends to increase the superheat effect and tends to starve the evaporator.

The thermostatic expansion valve may tend to open intermittently (opening and closing frequently) during the off cycle. This may be caused by temperature fluctuations (changes) which occur during cabinet opening and closing. To prevent this, one may use the high-side pressure on the valve to force it closed during the off part of the cycle. A solenoid valve may be used to control this pressure.

A special solenoid valve connected into the equalizer tube is shown in Fig. 5-39. Note valve construction in Fig. 5-36.

The electrical circuit to the solenoid valve is opened at the time the motor-compressor circuit opens. On opening the circuit, the solenoid core falls and closes the equalizer tube to the suction line. The high-pressure refrigerant enters the top of the solenoid valve and passes upward through the equalizer tube. This forces the thermostatic diaphragm up, closing the valve.

The thermostatic expansion valve, Fig. 5-40, uses an unusual adjustment. The equalizer tube connects into the valve at connection K.

Fig. 5-41 is a large capacity thermostatic expansion valve. It has flanged refrigerant line connections bolted together and gasketed. Instead of a needle and seat, it has

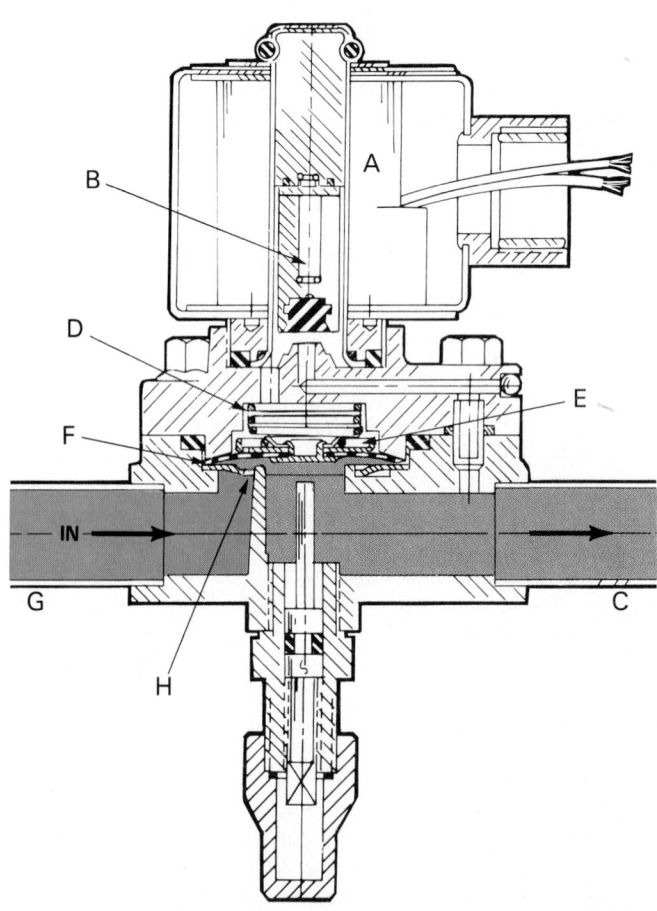

Fig. 5-37. A pilot-operated solenoid valve in open position. A—Solenoid. B—Solenoid plunger. C—Pressure outlet. D—Control cylinder. E—Control piston. F—Control valve. G—Pressure inlet. H—Small opening. (Alco Controls Div., Emerson Electric Co.)

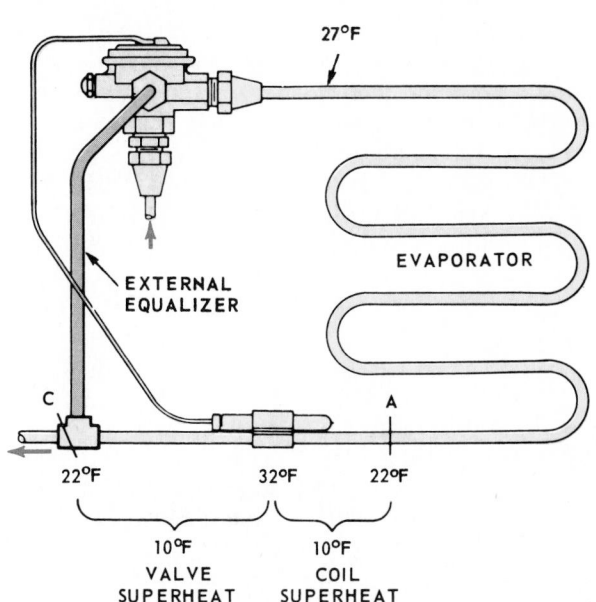

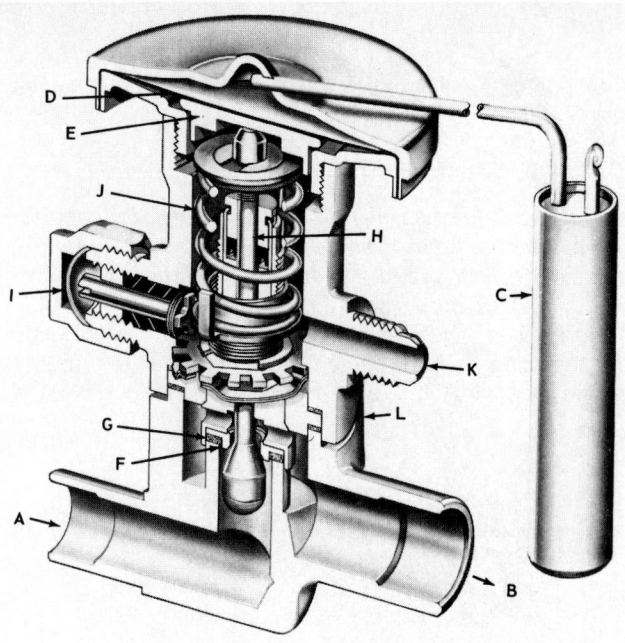

Fig. 5-38. A thermostatic expansion valve fitted with an equalizer tube. Equalizer connects suction line pressure at sensing bulb to under-side of valve bellows or diaphragm (low-pressure side). This enables low-side pressure operating valve to be same as pressure at sensing bulb. This compensates for any pressure drop through evaporator while compressor is running.

Fig. 5-40. A thermostatic expansion valve which uses an equalizer tube. A—Liquid line connection. B—Suction line connection. C—Sensing bulb. D—Diaphragm. E—Buffer plate. F—Valve pin. G—Valve seat. H—Push rod. I—Superheat adjustment. J—Superheat spring. K—Equalizer tube connection. L—Valve body.
(Alco Controls Div., Emerson Electric Co.)

a flat (disk) valve surface and seat, B. An equalizer tube connection is shown in the upper right part of the body, A.

5-22 SPECIAL THERMOSTATIC EXPANSION VALVES

Many different thermostatic expansion valve designs are available. One type uses a separate six-circuit refrigerant

distributor connected to the expansion valve. See Fig. 5-42.

This design is used to reduce the pressure drop in a large evaporator by providing several parallel refrigerant paths through the evaporator. It is popular for air conditioning applications. Careful engineering is needed, as each tube must receive an equal amount of refrigerant.

A diagram of a special thermostatic expansion valve, which provides multiple connections to the evaporator, is

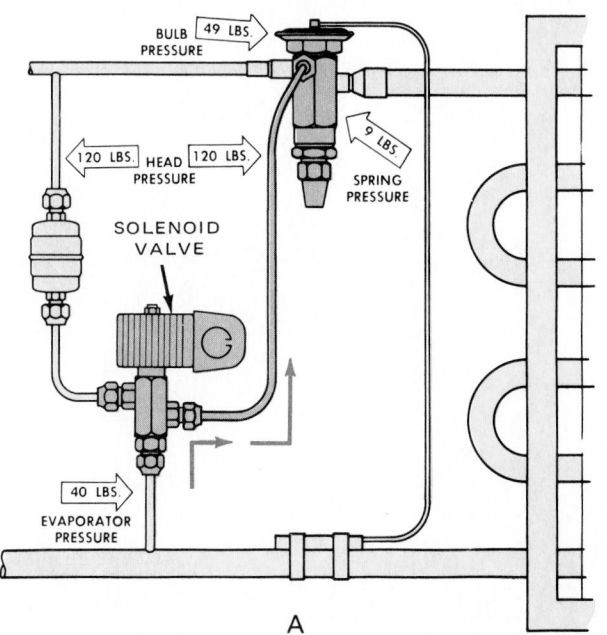

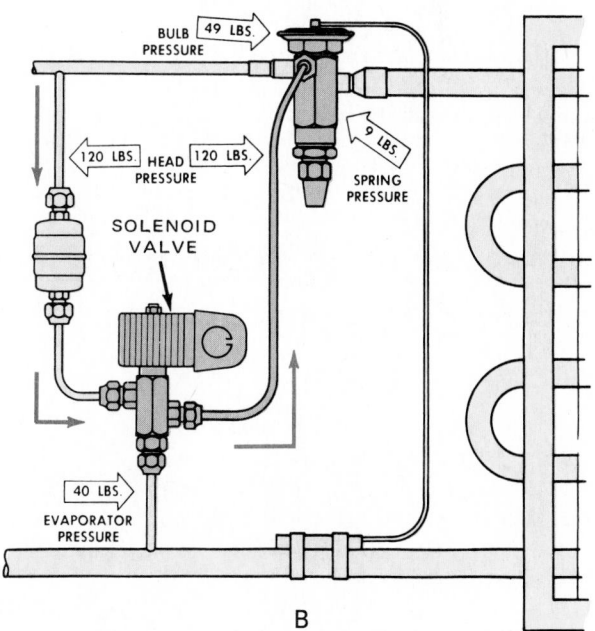

Fig. 5-39. A three-way solenoid valve is used to keep thermostatic expansion valve tightly closed during "off" cycle. A—When motor compressor is on the "on" part of cycle, valve is closed. Pressure in suction line is transmitted through equalizer tube to thermostatic expansion valve. B—With motor compressor on the "off" part of cycle, valve opens so high pressure can flow up equalizer tube to close expansion valve. (Sporlan Valve Co.)

shown in Fig. 5-43. Fig. 5-44 shows a distributor used on large capacity evaporators.

5-23 HUNTING

The term "hunting," applied to any type of mechanism, means that the mechanism is first going too far in one direction, then returning too far in the other direction. Sometimes, this is also called "surging."

Hunting, in refrigerating systems, is a term used to identify the changes in refrigerant flow through the refrigerant control operation. There is always some hunting while the

system is operating. If the valves hunt, they will alternately open up too wide, allowing too much refrigerant to flow into the evaporator, then close down too far, not allowing enough refrigerant into the evaporator. This effect is suggested by dashed lines in Fig. 5-45.

The less hunting, the more effective the system will be. During the interval that the valve is hunting too much, it is not providing a uniform amount of refrigerant to the

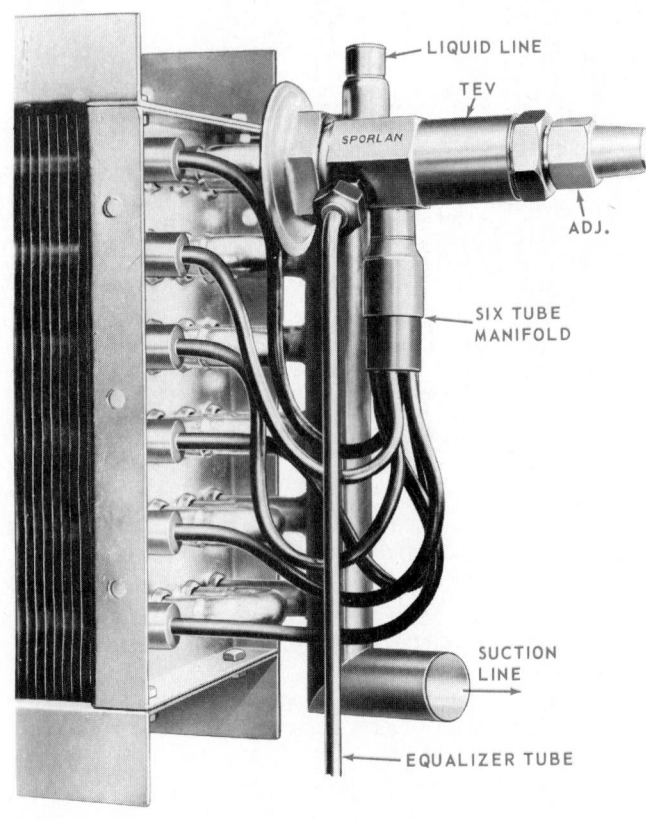

Fig. 5-42. Special thermostatic expansion valve installation for air conditioning applications. Refrigerant distributor is brazed onto outlet connection of valve. This valve uses an equalizer tube which connects to suction line and enters under valve diaphragm. (Sporlan Valve Co.)

Fig. 5-41. Large capacity thermostatic expansion valve shown is fitted with an equalizer connection. A—Equalizer tube opening. B—Valve seating mechanism. (Sporlan Valve Co.)

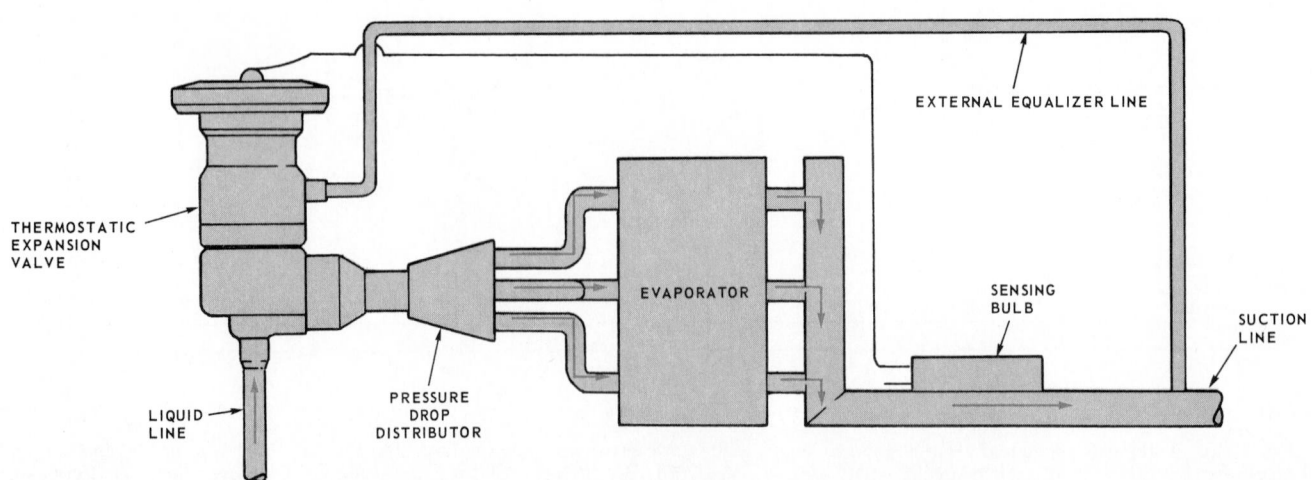

Fig. 5-43. A thermostatic expansion valve which provides multiple connections to evaporator. (Alco Controls Div., Emerson Electric Co.)

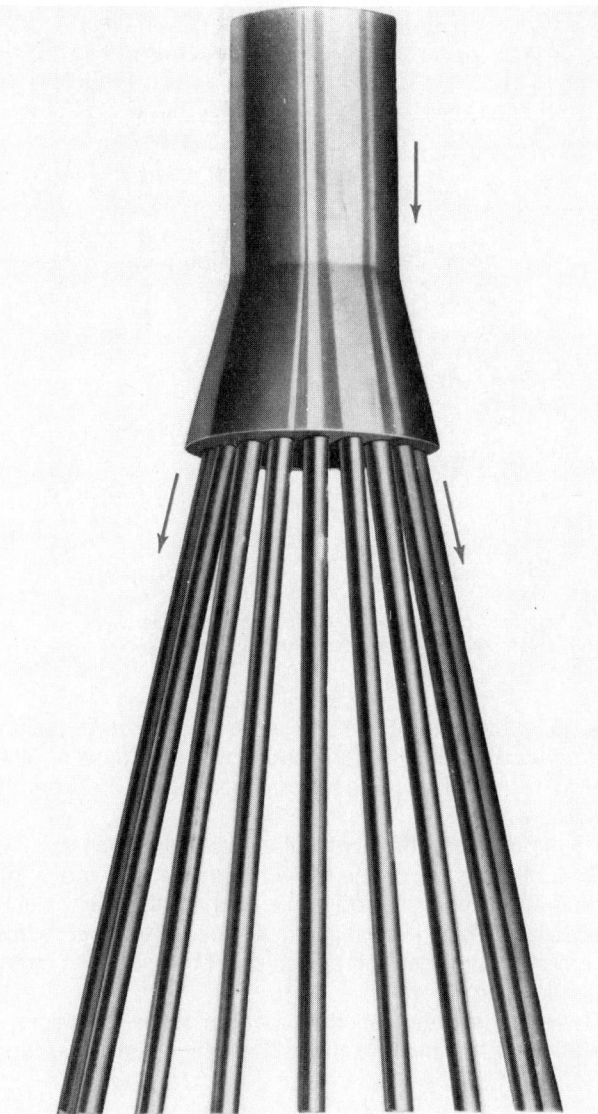

Fig. 5-44. A distributor tube. Thermostatic expansion valve supplies refrigerant at top. In turn, distributor tube supplies equal quantities of refrigerant to parallel paths through evaporator (bottom). (Alco Controls Div., Emerson Electric Co.)

evaporator. It may even allow liquid refrigerant to reach the compressor and cause compressor damage. In some cases, this condition may be caused by a valve that is too large for the system.

Each thermostatic expansion valve and evaporator combination has to be adjusted. The superheat adjustment should be used to reduce surging and hunting to a minimum but still permit full evaporator use.

5-24 LOW-SIDE FLOAT

The low-side float is an efficient, yet simple, refrigerant control. Its job is to maintain a constant level of liquid refrigerant in the evaporator. The system is an efficient heat transfer device. Heat moves easily from the evaporator to the liquid refrigerant.

The float itself may be a sealed ball, a cylinder, or an open pan. It is connected by levers to a needle which closes an orifice when the liquid level reaches the correct height. The

needle opens the orifice when some of the refrigerant evaporates and the liquid level drops.

The low-side float has the possible disadvantage of "oil binding." Because the mechanism will float in either oil or liquid refrigerant, special provisions must be made to return any excess oil to the compressor. This is sometimes done by the use of a wick or a small bypass at the surface of the liquid refrigerant.

Such a system uses a liquid receiver. Extra refrigerant is stored in the liquid receiver.

The basic principles of the low-side float system are explained in Chapter 3.

The low-side float may be connected either to a needle or a ball valve. It is calibrated so that the valve will close when the float is at the proper level—that is, when there is a certain level of liquid refrigerant in the evaporator.

The suction tube on these evaporators extends to the float chamber and, with pan-type floats, extends to the bottom of the pan. See Fig. 5-46. This design insures a more positive oil return than the ball type float.

Low-side float systems are used in large industrial systems and in some water cooling systems.

Either a pressure-operated motor control or a thermostatic motor control may be used with a low-side float system.

Either the thermostatic expansion valve (TEV) or the low-side float may be used in multiple evaporator systems. This low-side float control is also used for controlling water level in cooling towers, evaporative condensers, and in humidifiers.

5-25 HIGH-SIDE FLOAT

A high-side float refrigerant control is somewhat like the low-side float mechanism. However, it is located in the high-pressure side.

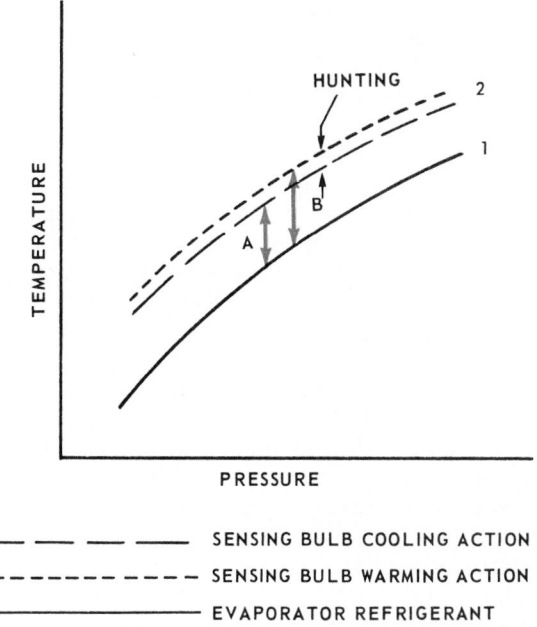

Fig. 5-45. Thermostatic expansion valve superheat. 1—Vapor pressure curve of refrigerant in system. 2—Pressure-temperature curve of liquid in sensing bulb. A—Superheat during warming of bulb. B—Superheat during cooling of bulb. The difference between the two lengths for A and B is called "hunting." It is mainly the result of weight of valve parts and bending resistance of the bellows or diaphragm.

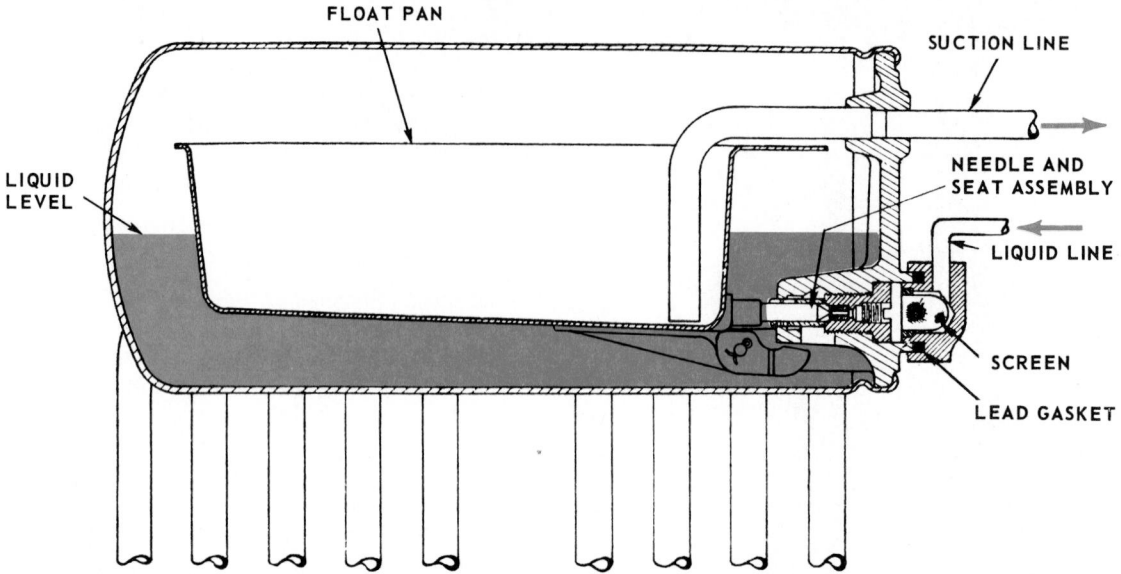

Fig. 5-46. Low-pressure side float refrigerant control. A bucket or pan type float is used in this refrigerant level control. Suction line dips to bottom of open float in order to remove oil which might otherwise accumulate in open pan.

When the compressor is running, the condensed refrigerant from the condenser flows directly to the high-side float chamber. No liquid receiver is used. As the liquid refrigerant level rises, the float inside the chamber opens a valve, allowing the liquid refrigerant to flow into the evaporator.

In this high-side float type of refrigerant control, the liquid refrigerant is stored in the evaporator, which makes it a flooded evaporator. It is designed accordingly.

As the liquid level falls in the float chamber, the float will move down and close the valve opening into the evaporator. In this way the pressure difference between the high side and the low side is maintained. *The high-side float maintains a constant level of liquid refrigerant on the high-pressure side.*

Floats are made of either copper or steel. In the hermetic units, steel is usually used. Such units cannot use a liquid receiver unless the float is located within it.

High-side float controls do not have as much difficulty with oil binding as the low-side float control. This is because, at the higher pressure, the oil is dissolved more readily in the liquid refrigerant and circulates with it. However, the evaporator used in connection with a high-side float control must be equipped with a special oil return or oil binding will occur.

The float may be connected directly to the needle, or it may operate the needle with a simple lever, as shown in Fig. 5-47. The needles and seats are made of long-wearing alloys such as stainless steel or hard-surfaced alloys.

Some high-side float systems use a weight check valve when the float chamber is located a long way from the evaporator. Chapter 3 explains how this type system operates.

Either a thermostatic or pressure motor control may be used with a high-side float refrigerant control.

5-26 CAPILLARY TUBE

*The capillary tube type of refrigerant control is simply a length of seamless tubing with a small and accurate in-*side diameter. It acts as a constant throttle on the refrigerant. *It usually is equipped with a fine filter or filter-drier at its inlet. This removes any moisture or dirt from the refrigerant.*

The amount of refrigerant in the system must be carefully calibrated, since all of the liquid refrigerant will move into the low side during the off-cycle as the pressures balance or equalize. Too much refrigerant will cause the unit to frost back on the low side. This control must be used with a thermostatic motor control.

There are several theories concerning the principle of operation of the capillary tube. The tube resists fluid flow.

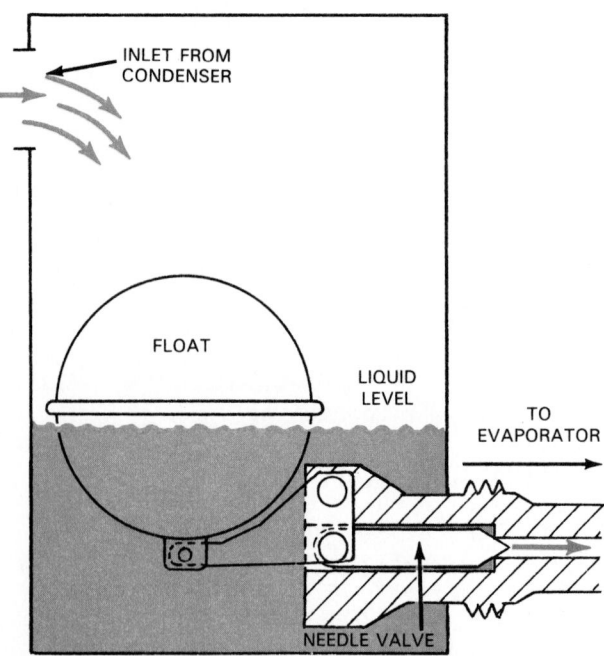

Fig. 5-47. High-pressure side float refrigerant control mechanism. Liquid refrigerant flowing in from condenser will cause float to rise and open needle valve. Then liquid refrigerant flows into evaporator.

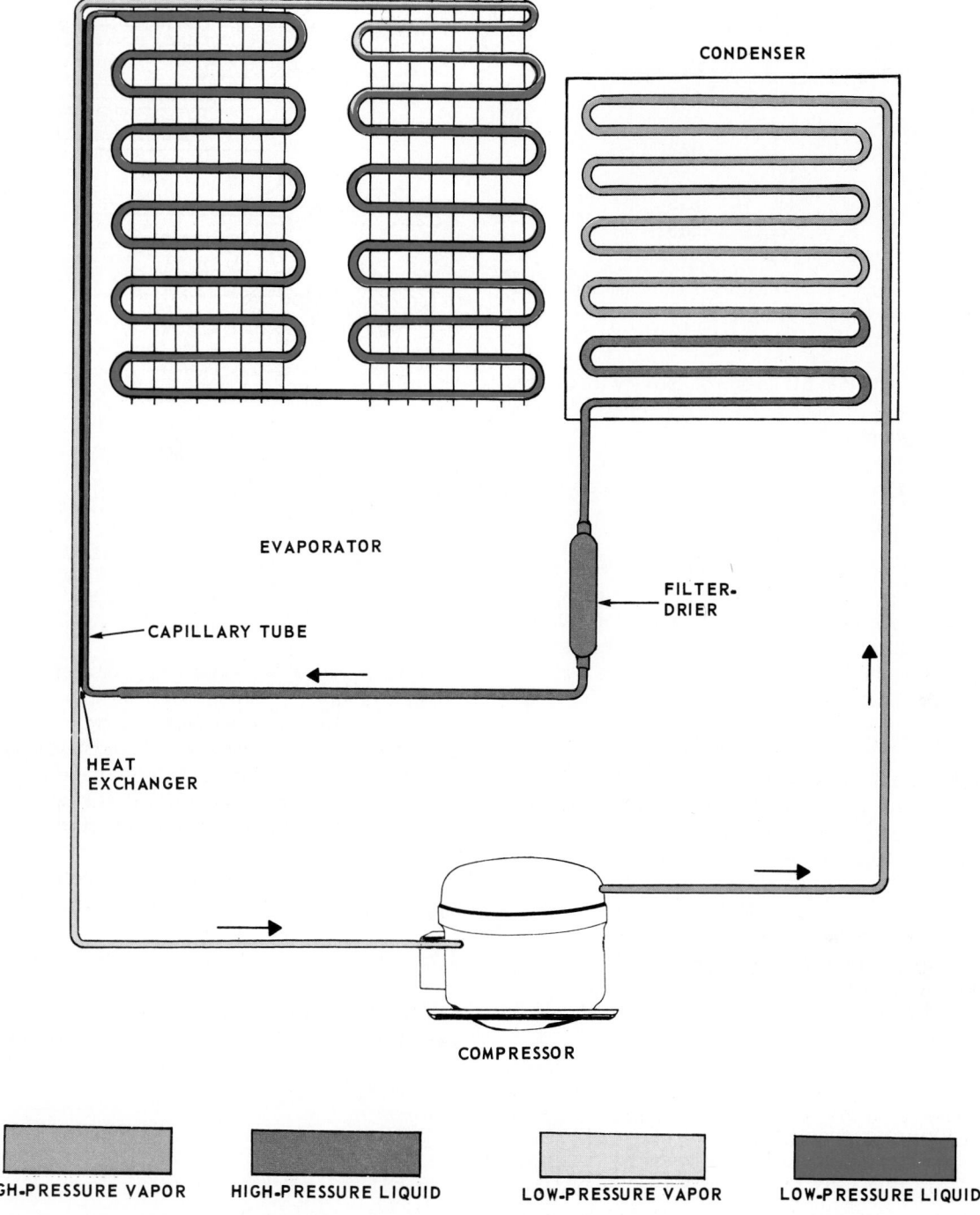

CONDENSER

EVAPORATOR

FILTER-DRIER

CAPILLARY TUBE

HEAT EXCHANGER

COMPRESSOR

| HIGH-PRESSURE VAPOR | HIGH-PRESSURE LIQUID | LOW-PRESSURE VAPOR | LOW-PRESSURE LIQUID |

Fig. 5-48. A capillary tube refrigerant control. Note heat exchanger which cools capillary tube by transferring heat from capillary tube to suction line. (Frigidaire Company)

Pressure goes down as the small liquid flow moves through the tube—until the liquid starts to evaporate in the tube. This vapor formation provides a sudden pressure and temperature drop in approximately the last quarter of the length of the tube. The refrigerant finally is cooled to evaporator temperature and its pressure is reduced to evaporator pressure.

This vapor formation in the capillary tube is called "vapor lock." The design of the capillary tube depends on four variables:
1. Tube length.
2. Inside diameter.

3. Tightness of tube windings.
4. Temperature of tubing.

A typical capillary tube refrigerant control is shown in Fig. 5-48. Most domestic refrigerators use this type of control.

The capillary tube refrigerant control has no moving parts. Therefore, it has several advantages. First, there are no parts to wear or stick. Second, the pressures balance in the system when the unit stops. This condition places a minimum starting load on the motor.

The capillary tube may be coiled for part of its length. It is usually attached to the suction line, permitting suction line vapor to cool the liquid refrigerant in the capillary tube.

CAPILLARY TUBE DIAMETERS

Outside Diameter	Inside Diameter
.083	.031
.094	.036
.109	.042
.114	.049
.120	.055
.130	.065

Fig. 5-49. Some common capillary tube diameters. Other application tables are listed in Chapters 11 and 21.

A recent development in the design of capillary tubes uses a larger diameter and a longer tube. The larger diameter tube is less likely to become plugged with dirt, ice, or wax than the smaller diameter tube. These larger diameter capillary tubes are most used on air conditioning applications.

5-27 CAPILLARY TUBE CAPACITIES

Popular capillary tube sizes are shown in Fig. 5-49. Tubing .114 OD by .049 ID is often used for refrigerant R-12 in domestic units. This is suitable for average temperatures and frozen food temperatures, dependent on the power of the unit and the use. Approximate sizes for capillary tube installations for R-12 are shown in Fig. 5-50.

5-28 CAPILLARY TUBE FITTINGS

The capillary tube connections are most often brazed at both the condenser and evaporator end. In other applications, the capillary tube may be attached to the evaporator and condenser or drier with fittings that are leakproof and vibration-proof.

Fig. 5-51 illustrates three popular ways to make these connections. Detail 1 shows the use of a special nut that squeezes against both the capillary tube and the fitting. Because the nose section is deformed as the nut is tightened, the nut should always be replaced when the capillary tube is serviced.

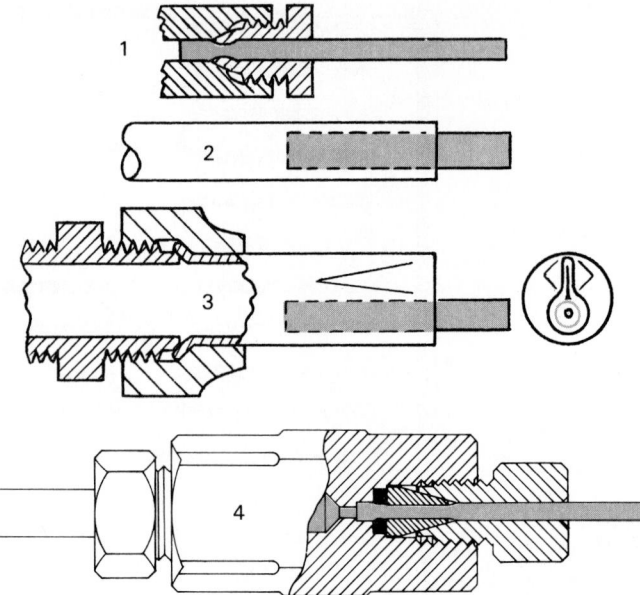

Fig. 5-51. Some typical capillary tube connections. A standard flared fitting connected to a special capillary tube fitting is shown in View 4. (Watsco Components, Inc.)

Detail 2 shows the capillary tube brazed to a 1/4 OD soft copper tubing. The larger tube may then be connected to the system by the usual flared fitting.

Detail 3 shows a larger tube brazed to the capillary tube. The larger tube connection is then made with a flared fitting. Mechanical connections are often substituted during overhaul using special capillary tube fittings.

Detail 4 shows a standard flared fitting connected to a special capillary tube fitting. The diameter of the capillary tube determines the size of the fitting.

5-29 COMPARING REFRIGERANT CONTROLS

The pressure-time diagram for one operating cycle for any particular refrigerant control will show its operating

HORSEPOWER	TEMP.	REQUIRED LENGTH OF CAPILLARY TUBING IN FEET						
		.031 ID	.036 ID	.040 ID	.042 ID	.049 ID	.055 ID	.065 ID
1/8	H	1.1	2.2	3.5	4.5	9	15	
	M	4	8	13	16	32	56	
	L	9	18	29	36	72	126	
1/5	H						10	
	M	2.2	4.4	7.0	9	18	31	
	L	5.2	10.5	17	21	42	73	
1/4	H						5	
	M	1.1	2.2	3.5	4.5	9	15	
	L							7.5
1/3	M						9.5	
	L	1.75	3.5	5.6	7	14	2.5	

Fig. 5-50. Capillary tube sizes and applications. Temperatures are shown as: high, "H"; medium, "M"; or low, "L." Required length of tube in feet is indicated for each tube inside diameter (ID). Length in feet is for R-12 refrigerant.

characteristics for one running cycle of the refrigerator. The pressure-time diagram will vary with the refrigerant used. Fig. 5-52 shows the pressure-time characteristics for six different refrigerant controls using refrigerant R-12.

5-30 CHECK VALVE

Check valves allow fluid to flow through them in only one direction. Rotary and gear compressors have check valves in the suction line. This is to prevent the high-pressure vapor and the refrigerant oil from backing up into the evaporator during the off part of the cycle.

A typical check valve is shown in Chapter 4, Fig. 4-75. These check valves may use either a disk or a solid ball in their construction. They may also operate differently. Some use a spring or magnet to keep the valve against the seat. Others are mounted so that the weight of the valve keeps it against its seat.

Multiple systems have evaporators which operate at different temperatures. They use check valves to keep the refrigerant vapors in warmer evaporators from backing up into the colder evaporators. This is explained in Chapter 12.

5-31 SUCTION PRESSURE VALVES

Many systems use bellows or diaphragm-operated suction pressure regulating valves on the low side of the system. These valves are required on multiple systems in which the evaporators operate at different temperatures.

Some of these valves are used to keep the compressor suction pressure at a safe level to avoid overloading the compressor. A modified suction pressure valve is required on most automobile air conditioning systems because the compressor is operating at various speeds. The valve is needed to maintain a fairly constant temperature in the evaporator.

Suction pressure valves may be controlled from the evaporator pressure. These valves control the evaporator temperature. Some operate from crankcase pressure, working to keep the compressor from being overloaded.

As a crankcase pressure regulator, the valve modulates (adjusts) suction pressure to the compressor intake to provide overload protection for the condensing unit motor. Fig. 5-53 illustrates a low-pressure side pressure regulator installation.

Additional information concerning suction pressure valves is given in Chapter 12.

5-32 REVIEW OF SAFETY

Perhaps the most common damaging condition of refrigeration operation is allowing liquid refrigerant to enter the compressor. This is called "refrigerant slugging." Liquid refrigerant is not compressible. The piston striking the liquid refrigerant in the cylinder will cause knocking and considerable overloading of the piston, connecting rod, bearings, and compressor valves. Most compressors provide some kind of a safety overload device in the valve arrangement to attempt to protect the compressor under severe slugging conditions.

A 24-hour pressure time recorder attached to any new installation is an excellent safety device to make sure that the system is operating within safe pressure limits. A 24-hour temperature recorder may also be used for this purpose.

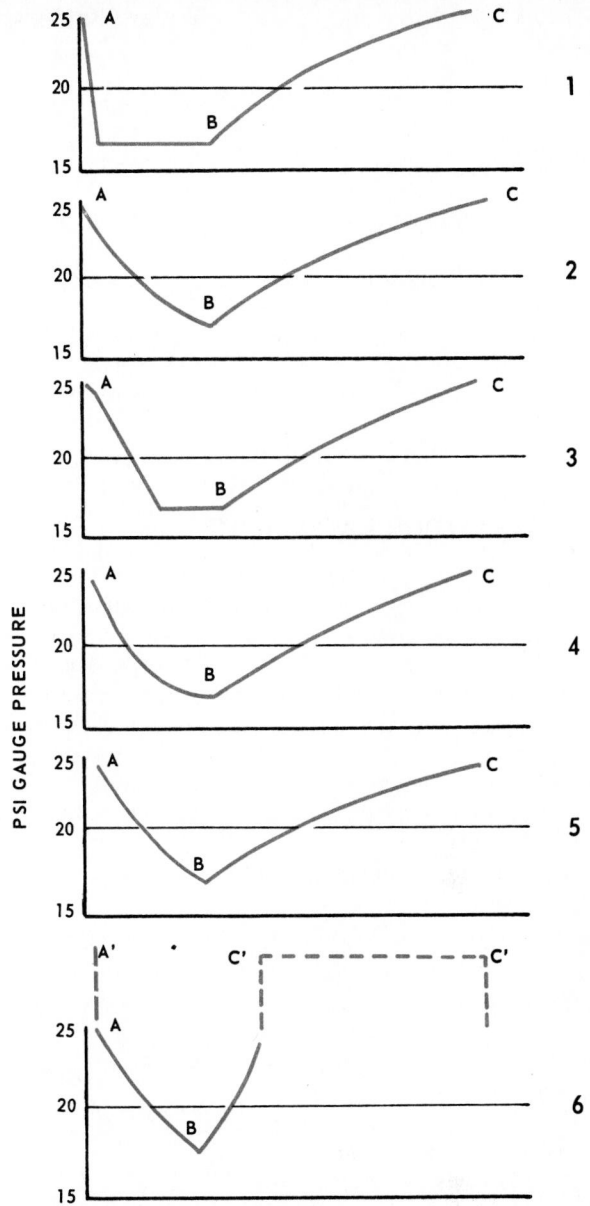

Fig. 5-52. A comparison of low-side pressure-time cycles for various refrigerant controls using R-12 refrigerant. 1—Automatic expansion valve. 2—Thermostatic expansion valve. 3—Thermal-electric expansion valve. 4—Low-side float. 5—High-side float. 6—Capillary tube. Note that pressure goes up until it balances with high-side pressure. Point A—Cut-in pressure. Point B—Compressor stops. Point C—Cycle repeats. Pressures at point A' and point C' are pressures at cut-in.

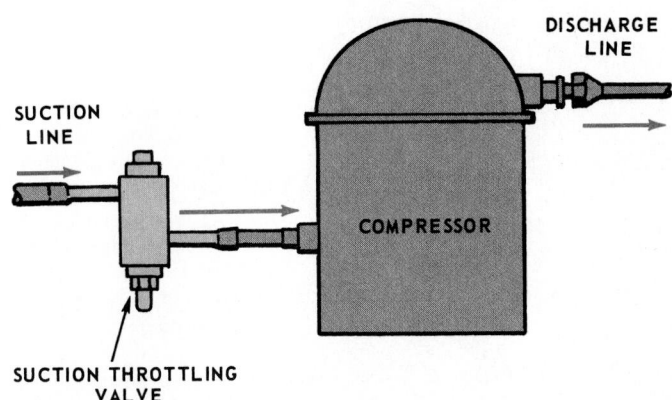

Fig. 5-53. A low-side pressure control valve located in suction line.

Keep the floor clear of debris—oil, water, or other slippery material.

Wear safety goggles when working on refrigerating systems.

Avoid lifting objects which weigh over 35 pounds (15.9 kg). Use leg muscles when lifting.

Always have good ventilation and good lighting when working on a refrigerating system.

In order to avoid the possibility of electrical shock, make sure that all electrical circuits are well insulated.

All metal parts of refrigerating mechanisms should be grounded.

When removing a valve from a system, always use two wrenches to avoid twisting the valve or the tubing.

5-33 TEST YOUR KNOWLEDGE

1. How many pressures or forces influence the needle movement of an automatic expansion valve needle?
2. Why are expansion valves adjustable?
3. What is the purpose of the automatic expansion valve?
4. What is thermostatic expansion valve superheat?
5. What is meant by a gas charged element used in connection with a thermostatic expansion valve?
6. What mesh screens are used in refrigerant controls?
7. What is the most common expansion valve body material?
8. How is the liquid line usually attached to the expansion valve?
9. Why does the suction line extend down into the pan type low-side float?
10. High-side floats are usually made of what materials?
11. What is the purpose of an automatic refrigerant control?
12. What is meant by the term "pilot-operated" solenoid valve?
13. How does a capillary tube operate?
14. Does a capillary tube system usually use a liquid receiver? Why?
15. Does a capillary tube need a filter or a screen at its inlet?
16. What type of system needs a check valve?
17. What energy opens a solenoid valve?
18. Why is a thermostatic expansion valve equalizer tube used?
19. What may be the trouble if a capillary tube unit frosts down the suction line?
20. What is meant by a four-way solenoid valve?
21. How is a capillary tube usually fastened to a 3/8 OD copper tubing?
22. What is the most popular TEV superheat setting?
23. Does a system using an automatic expansion valve have a "dry" or a "flooded" evaporator?
24. What is a possible disadvantage of a low-side float?
25. If the bulb temperature of a MOP pressure limiter continues to increase, what happens to the bulb pressure?

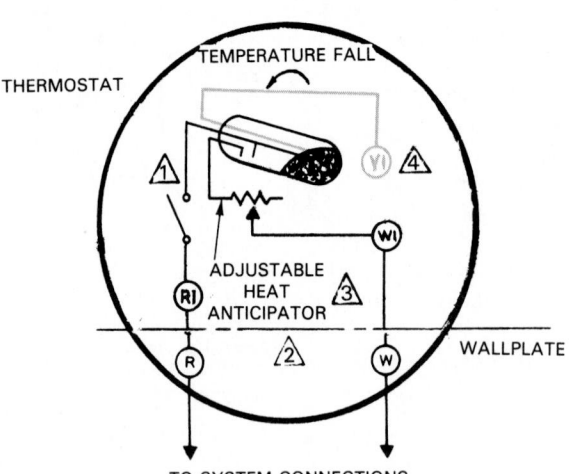

 Model with positive off switch opens the thermostat circuit when set point dial is moved to the off position.

 Make system wiring connections to terminals on wallplate.

 R1, W1 terminals on thermostat are directly connected to R, W terminals on wallplate when thermostat is mounted on wallplate.

 R1, Y1, terminals are thermostat connection points for cooling unit.

Operation of thermostat switch. Tilting of glass vial lets drop of mercury make contact. "Heat anticipator" helps avoid overshooting desired temperature setting: a resistor adds a little heat to bimetal strip.
(Honeywell)

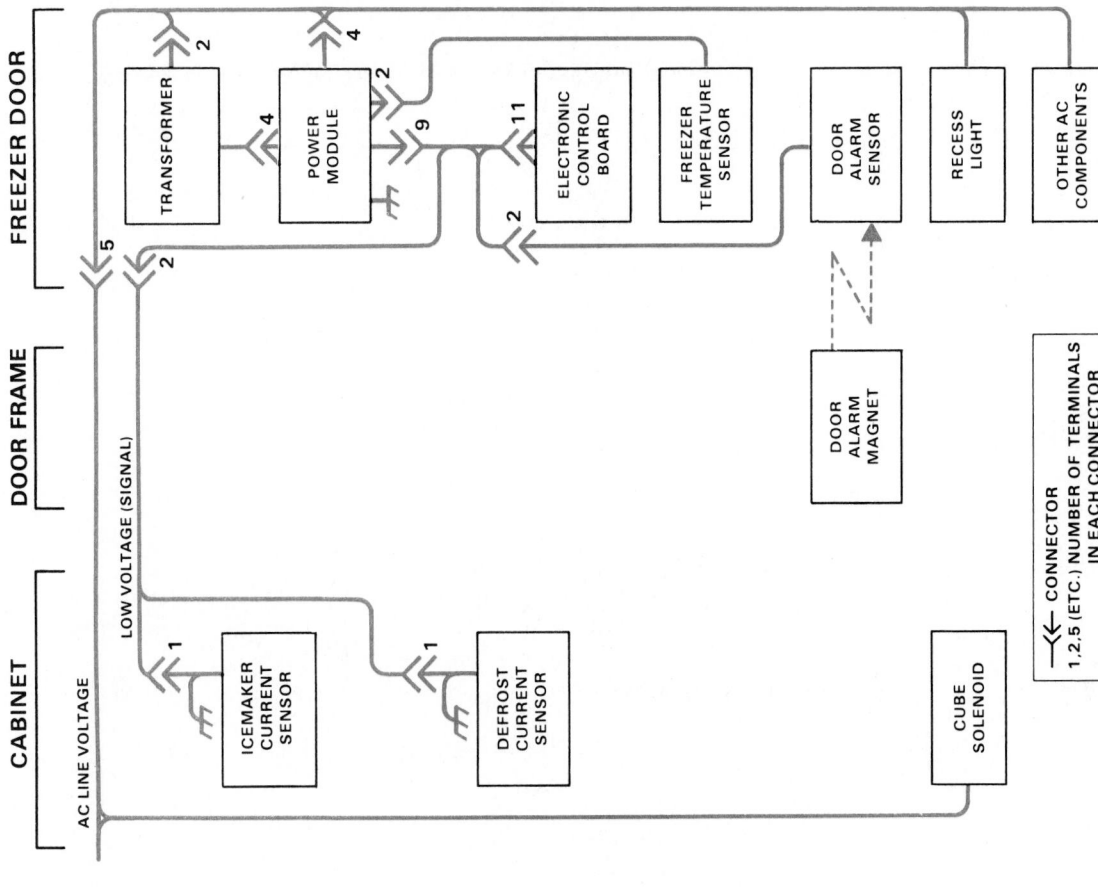

Electronic block diagram. The individual components of the electronic system and the interrelationship of line voltage components controlled are illustrated. (General Electric Co.)

ELECTRONIC CONTROL BOARD CONNECTOR

VOLTAGE MEASUREMENTS — (CONNECTOR UNPLUGGED)

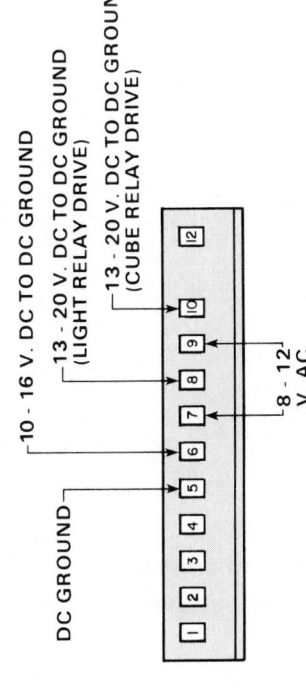

RESISTANCE MEASUREMENTS — (CONNECTOR UNPLUGGED)

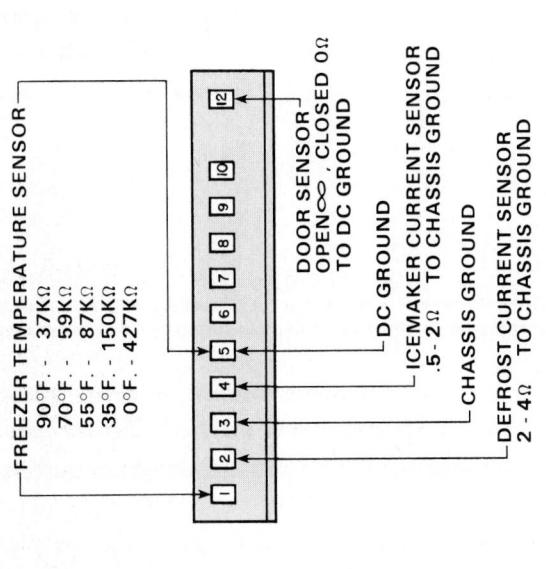

Electronic control board connector. This is part of an electronic control console monitor on a domestic refrigerator. It is used to diagnose various voltage and resistance measurements of electronic systems, such as those for an icemaker or for defrost. (General Electric Co.)

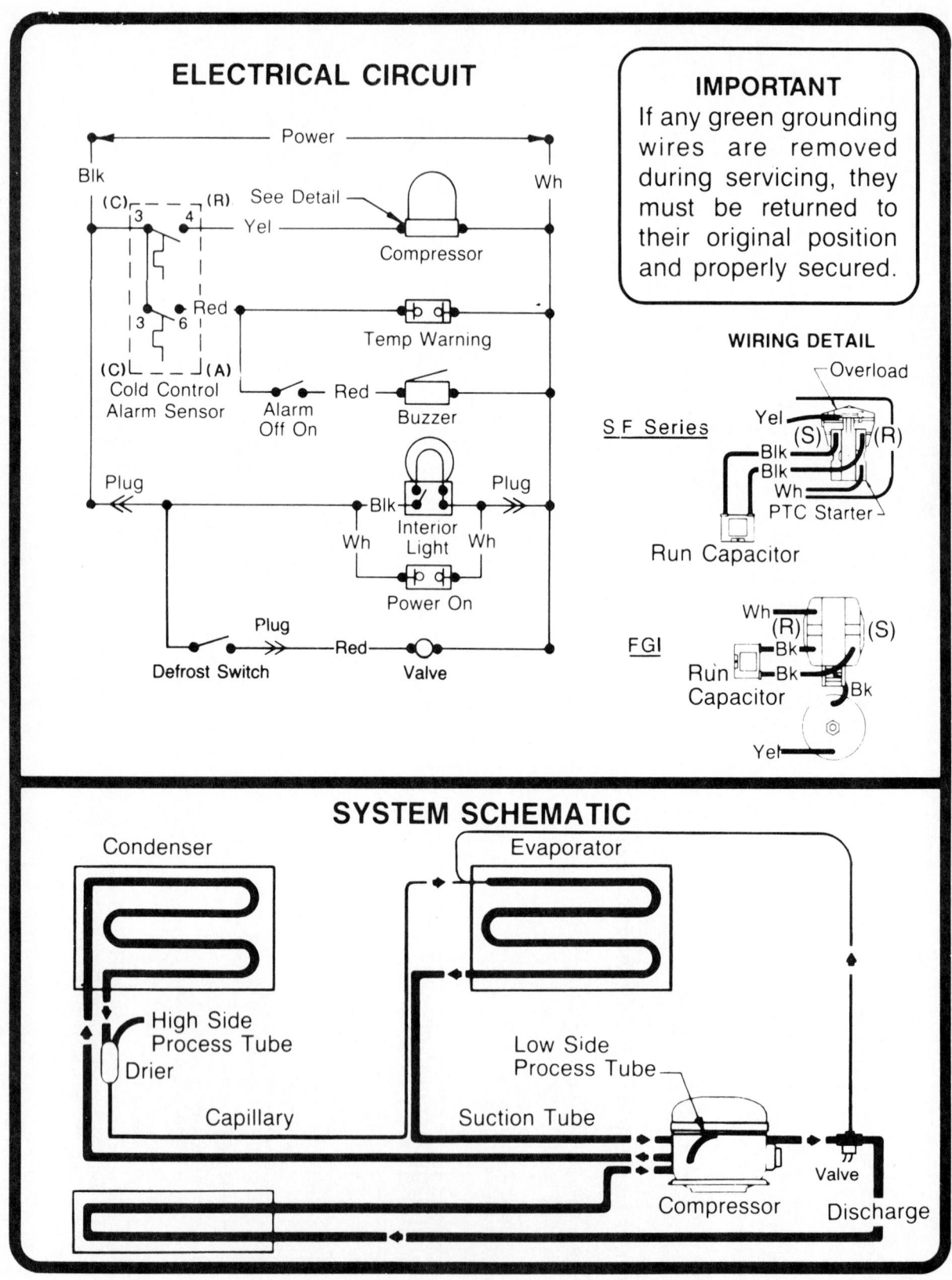

Fig. 6-1. Typical electrical circuit diagram and system schematic for a freezer. Note the temperature warning system.　(WCI Freezer Division)

Chapter 6

ELECTRICAL-MAGNETIC FUNDAMENTALS

After studying this chapter, the technician will be able to:
☐ Define the terms, "electricity" and "electronics."
☐ Distinguish between types of electricity—static and current.
☐ Explain the difference between direct and alternating current.
☐ Define electrical and electronic terms.
☐ Describe the difference between parallel circuits and series circuits.
☐ Discuss the basic theory of electric motors and related devices.
☐ Use various electrical formulas to solve problems.
☐ Explain the use of computers in refrigeration controls.
☐ Explain the components of an electrical circuit.
☐ Identify and use the proper electrical symbols.

There may be some confusion in the understanding of the terms "electricity" and "electronics." Electronics is the branch of physics covering the behavior and effects of electron flow in vacuum tubes, gases, and semiconductors. Electricity is the branch of physics concerning the behavior of a natural phenomenon known only by its effects: electric charge, electric current, electric field, and electromagnetism. Electricity usually refers to power generation, transmission, and use. Electronics usually involves the use of low power to control high power circuits.

6-1 BASIC CURRENT ELECTRICITY

Electricity is usually supplied to homes and industries by electric utility companies. The supply from the utility company to the home ends in a control panel. This panel contains fuses or circuit breakers for each of the circuits.

The electricity supplied to most homes is at 120 or 240 volts (V) alternating current. Most lighting and appliances use 120 volts (V). Electric stoves, water heaters, air conditioners, and refrigerating systems (the larger sizes) uses 240 volts (V).

Service technicians should understand the fundamentals of electricity. This will help them service the electrical system of a refrigerator or air conditioner. See Fig. 6-1.

Terms such as "voltage," "direct current," "cycles," and "Hz" are explained in later paragraphs.

6-2 TYPES OF ELECTRICITY

Refrigeration and air conditioning is concerned with two common types of electricity. These are static electricity and current electricity. *Static electricity is often defined as electricity at rest. Current electricity is electricity flowing through conductors (wires).*

Lightning is created through the discharge of static electricity. Under certain conditions, materials such as sheets of paper and clothing may become charged with static electricity. That is why they sometimes cling together. Static electricity is often produced by friction.

Current electricity is commonly used in homes and in industries to drive motors, weld, and start engines. It is usually produced by coiled wires moving in an electromagnetic field.

6-3 GENERATING ELECTRICITY

The electricity used in the refrigeration and air conditioning industry is usually generated by electromechanical equipment (a generator). This may be either direct or alternating current.

Electricity also may be generated chemically. Dry cells used in flashlights are an example. The voltage of a dry cell is approximately 1.5 volts (V). Chemically generated electricity is always direct current. The chemicals combine or react to generate electricity and, in the process, the chemicals are used up. The cell is discharged.

The automobile storage battery does not generate electricity, but merely stores it. Electricity from the vehicle's generator flows into the storage battery, causing a reversible chemical action between the electrolyte (acid solution) and the battery plates (lead or lead oxide). The electrical energy has been stored and can be used later.

Electricity can be caused to flow from other energy forms, such as heat energy, friction, mechanical energy, light, chemistry, and magnetism. Any method which produces a movement of free electrons causes an electrical potential (pressure), and free electrons will flow if a conductor (wire) is present.

6-4 ELECTROSTATIC ELECTRICITY

There are two kinds of electrostatic charges—positive and negative. Substances charged with the same kind of electrostatic electricity tend to push apart or repel each other. Two substances, one with a positive charge and the other with a negative charge, tend to be attracted to one another.

A common electronic device used to indicate electrostatic charge is an electrometer. Another is the electroscope. A simple electroscope consists of two small strips of aluminum foil attached to a metal rod. The rod is extended through the top of a stoppered glass bottle. When this rod contacts a body having an electrostatic charge, the two strips of aluminum immediately move away from each other. They have both become charged with the same kind of electrical charges. They are both negatively charged or both positively charged.

Placing like poles of two magnets close to each other demonstrates the same reaction. The magnets act as though a strong invisible spring had come between them, preventing contact.

A common example of electrostatic generation occurs when one walks across a carpet. This charges the entire body with static electricity. If one then touches a filing cabinet, a faucet, or a doorknob, this charge will be quickly dissipated (spent). There may be a visible and easily heard spark discharge.

In another example, silk rubbed over glass produces a positive charge of static electricity on the surface of the glass. Plastic rubbed with wool or fur will generate a negative charge.

The refrigeration service technician is concerned with static electricity in two applications. It applies to condensers used with capacitor type electric motors. *The term "capacitor" means practically the same as "condenser." Capacitance, listed in farads, is the unit of measurement that describes the capacity of a capacitor for storing electrostatic charge.* See Para. 6-46. Electrostatic electricity is also used in electrostatic air cleaners.

6-5 CURRENT ELECTRICITY

Current electricity is the movement of electrons along an electrical conductor. For example, pushing the button on a doorbell closes (completes) the circuit and causes electrons to flow through the circuit. This electrical flow rings the bell.

There are two common types of electric current:
1. Direct current (dc).
2. Alternating current (ac).

Direct current is the continuous flow of electrons in the same direction. It is the type of current used in automobile starting, lighting, and ignition. Direct current is used in most solid-state circuits.

Direct current is used on cordless electric appliances such as toothbrushes, electric shavers, and drills. It is also used extensively in electronics. Direct current is needed for battery charging since batteries produce only direct current.

Alternating current is the flow of electrons along a conductor first in one direction, then in the other. It is the type of current used for most power and light applications.

6-6 DIRECT CURRENT

Direct current (dc) is electron flow along a conductor in one direction. It is the type of current produced by batteries. Vehicles—except some larger buses and earth-moving equipment—operate on a 12 V direct current circuit supplied by the storage battery.

Some power companies still produce and sell direct current. However, its use is fast disappearing for power and lighting. At present its chief uses are in electronics, elevator service, electric welding, and automobiles.

Generally, in both elevator operation and electric welding, the direct current is generated at the site. It is made by using an alternating current rectifier, by driving a dc generator with an ac motor, or by driving a dc generator with a gasoline or diesel engine.

With the development of new lightweight storage cells and rectifiers, many small appliances are using "cordless" dc power. Motors in some appliances operate on either ac or dc. These units are called universal motors.

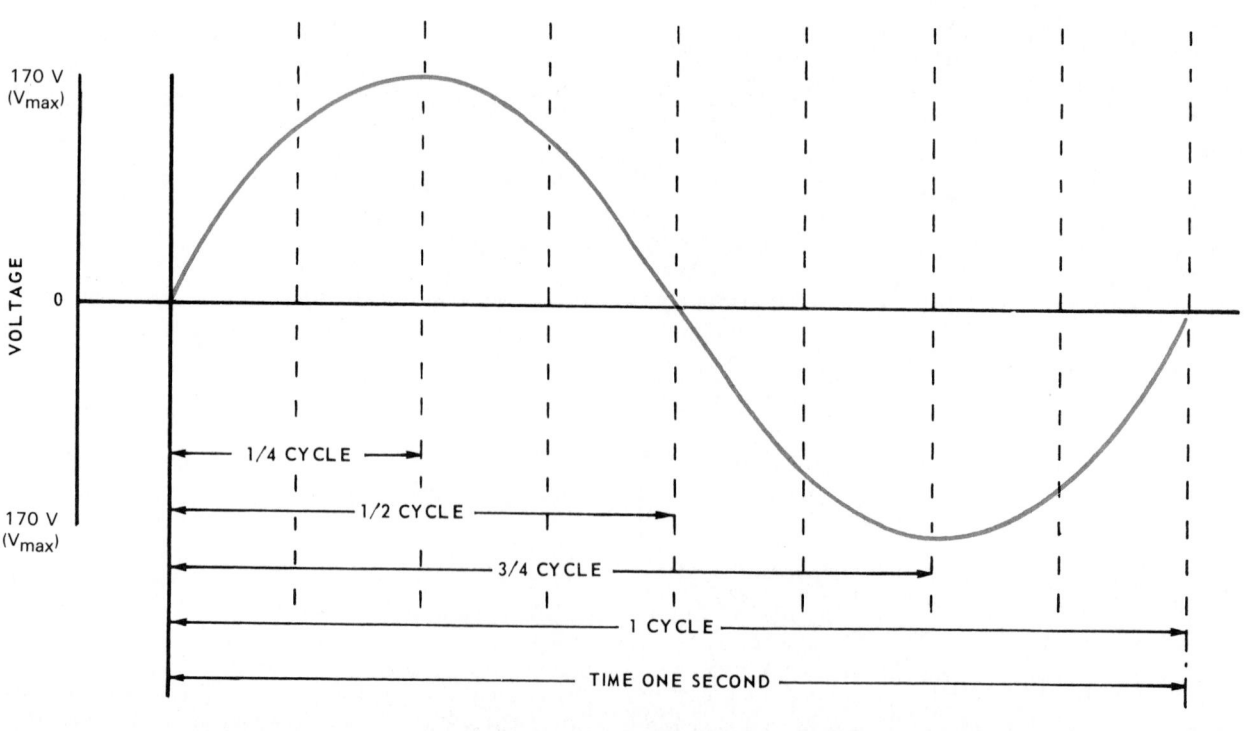

Fig. 6-2. Red line represents one complete cycle of alternating current flow per second in a typical household (120 V) circuit. Most ac is at rate of 60 cycles per second (60 Hz). Note that voltage varies from 0 to V_{max} twice each cycle.

6-7 ALTERNATING CURRENT

Alternating current (ac) is electron flow along a conductor, first in one direction, then in the other. Because the current alternates its direction of flow, it is called alternating current.

A cycle diagram for an alternating current is shown in Fig. 6-2. In this illustration, the voltage starts at zero. It reaches its maximum (V_{max}) at a quarter of the cycle. At half cycle, it is again at zero. The voltage reaches its maximum in the opposite direction at three-quarters of the cycle, and is again zero at the end.

This diagram pictures the ideal alternating current voltage cycle. Since most alternating current is 60 cycle, it means that this pattern is being repeated in the alternating current circuit at the rate of 60 times per second. The peak voltage (V_{max}) depends upon the voltage supplied by the power source.

Due to this alternating nature of a typical ac circuit, most references to voltage, current, and power refer to the root mean square (rms) values.

The rms value is an average which is equal to the maximum value times a constant. For standard ac this constant is .707.

Typical household voltage:

$$\begin{array}{ccc} V_{rms} & V_{max} & \text{Constant} \\ 120\text{ V} = & 170\text{ V} \times & .707 \end{array}$$

Many ac meters measure the ac voltage, current, and power, and indicate the root mean square (rms) value. Chapter 30 provides additional information on root mean square values.

Hertz (Hz) is the accepted word for electrical cycles. Household electricity in most countries is 60 Hz.

6-8 CIRCUIT FUNDAMENTALS

In order to understand an electrical circuit's operation, compare it to a water system. Just like water in pipes, the wires must be large enough to carry the current (amount).

There is always a loss of pressure (volts) in a wire when electricity flows along it. This action is due to resistance in the wire. To relate this to water pressure, view 1 in Fig. 6-3 shows a water pipe with the inlet valve open and the outlet valve closed. Pressure produced by the water pumps is the same throughout the system when the water is not flowing. When the outlet valve is opened, water flows and the pressure drops.

Notice in view 2 in Fig. 6-3 that the pressure drops a little for increased distance from the pump. This pressure drop indicates energy lost pushing the water that distance. Pressure drop between gauges C and D is greatest because additional effort is needed to push water around the three sharpest bends of the four total bends.

In Fig. 6-4 a similar electrical system setup is shown (drawing is in symbols). In both views, a wire extends from switch No. 1 to the resistance, light bulb, or motor at M. In view 1, switch No. 1 is closed, but switch No. 2 is open. The potential (voltage) is 120 V up to switch No. 2. However, no electricity is flowing (open circuit). Therefore, there is no voltage drop along the line. Voltmeters A, B, C, and D show no voltage (no pressure difference) between their lead connections to the main line. However, the voltmeters at E, F, and G show the full 120 V, because

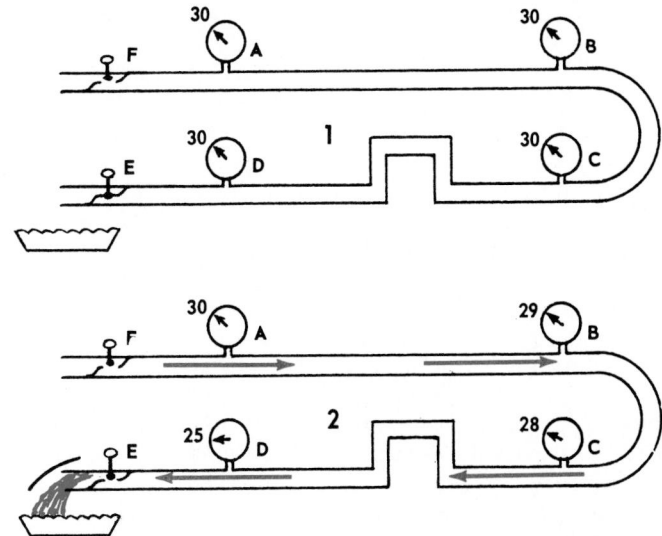

Fig. 6-3. Water flow compared to electrical flow. 1 — Valve E closed, valve F open. No water flow, all gauges read the same. 2 — Valve E open, valve F open, water flowing. All gauges show a pressure drop. Most of drop is between C and D because of bend in pipe.

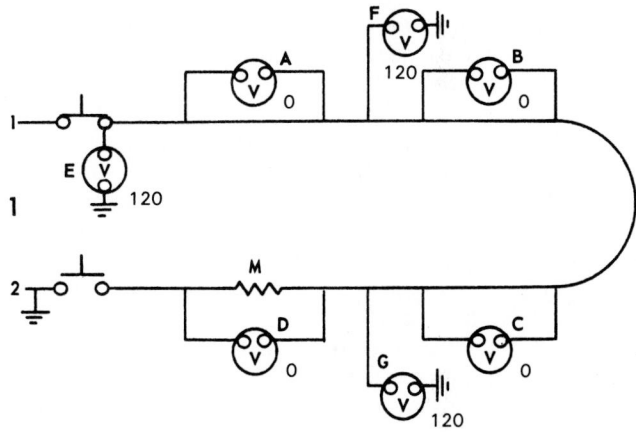

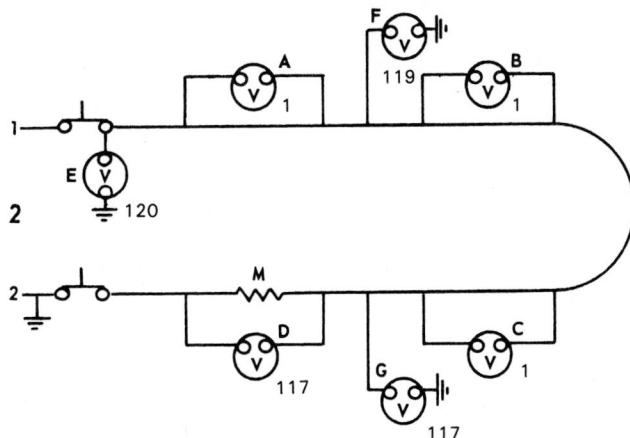

Fig. 6-4. Voltmeter tests of electricity flow show how potential (pressure), resistance, and potential drop operate in a system. View 1 — Electricity not flowing, with switch No. 1 closed, switch No. 2 open. Pressure to ground is 120 V, but other pressure drops are zero. View 2 — Electricity is flowing, with switch No. 1 and No. 2 closed. Voltage drop at A, B, and C is 1 V each, and resistance at D has a 117 V drop.

there *is* an electric pressure difference between the line and the ground.

View 2 in Fig. 6-4 shows switch No. 2 closed and current flowing through the circuit (closed circuit). Note that now there is a small voltage drop at A. This drop takes place in any line in which current is flowing. If the line is large enough to carry the current flow, the voltage drop will be very small (.001 to .0001 V). If the line is too small, voltage drop will be greater. Usually, an undersize wire may be detected by the fact that it gets warmer than usual while current is flowing.

Note in Fig. 6-4 that the voltage difference between the line and the ground is less at points F and G in view 2 than in view 1 because of voltage drop in the line up to these measuring points. Also note that at D the voltage drop is 117 V, which is the remaining pressure difference between the hot wire and ground.

If voltage drop was greater than shown at A, B, and C of view 2 in Fig. 6-4, the voltage at D would be less. This is one serious cause of motor trouble because voltage drop reduces motor voltage. If a motor is designed to operate at 120 V and there is a large amount of voltage drop in the circuit, the motor will lose speed. The rotor will start slipping relative to the magnetic field in the stator (stationary windings). The rotor will slow down too far below its synchronous speed. This causes the magnetic fields to grow large at the wrong time, and the motor will heat up. It may even burn up.

6-9 CIRCUITS AND CIRCUIT SYMBOLS

An electrical circuit is a complete path (or paths) for the electrons to follow. (See Chapter 30.) It might consist of a battery, switch, lamp, and conductors (wires). See Fig. 6-5. When a conductor is used to connect the battery to the lamp, the lamp to the switch, and the switch to the battery, an electrical circuit is made. If the switch is closed, the electron path is complete and the lamp will light. This shows that current is flowing out of the battery along one conductor, Fig. 6-5, and returns along the other conductor.

There are two terms that should be understood: open circuit and closed circuit. *An open circuit means that an electrical switch or other device is open or disconnected and current cannot flow. A closed circuit means that an electrical switch or other device is closed or connected, per-*

mitting the electrons to leave from the original source and return to it. This creates a closed path for electron flow. A closed circuit may also be called a continuous circuit.

Fig. 6-6 illustrates three types of circuit troubles:

No. 1 shows an open (disconnected) circuit, meaning it has an open switch or a broken conductor.

No. 2 shows a short circuit. The electrons have taken a short cut back to their source. For example, if a small conductor is placed across the terminals of a lamp, most of the electrons will flow along the path of least resistance (the new conductor), and the light will go out. Because of low resistance, too many electrons may flow, and the wires may overheat. Amperage flow will greatly increase (because of a decrease in resistance), and trouble usually occurs. Another more common example of a "short" is the touching of two adjacent conductors.

No. 3 shows a ground condition. This can occur when a conductor touches the metal structure of a device. For example, a bare conductor may touch the metal frame of the lamp. The metal frame becomes "hot."

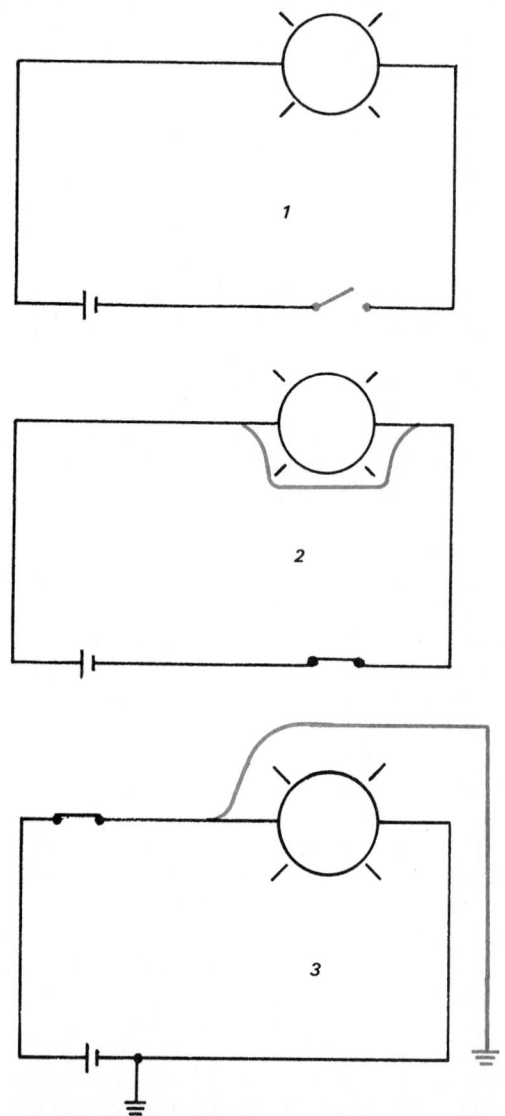

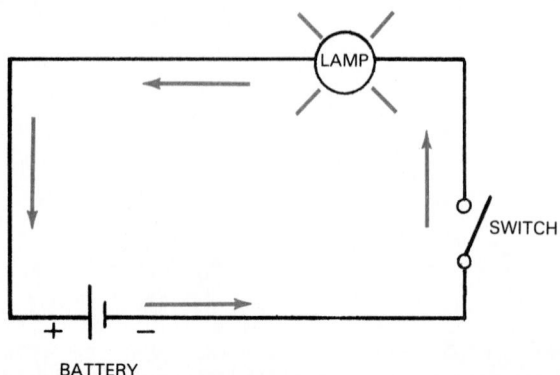

Fig. 6-5. Simple electrical circuit using a one-cell battery, three conductors, a switch, and one lamp. Arrows show direction of electron (current travel) from − to +.

Fig. 6-6. Three common electrical circuit troubles. 1—Open circuit (open switch shown here or a broken conductor). 2—Shorted circuit. 3—Grounded circuit.

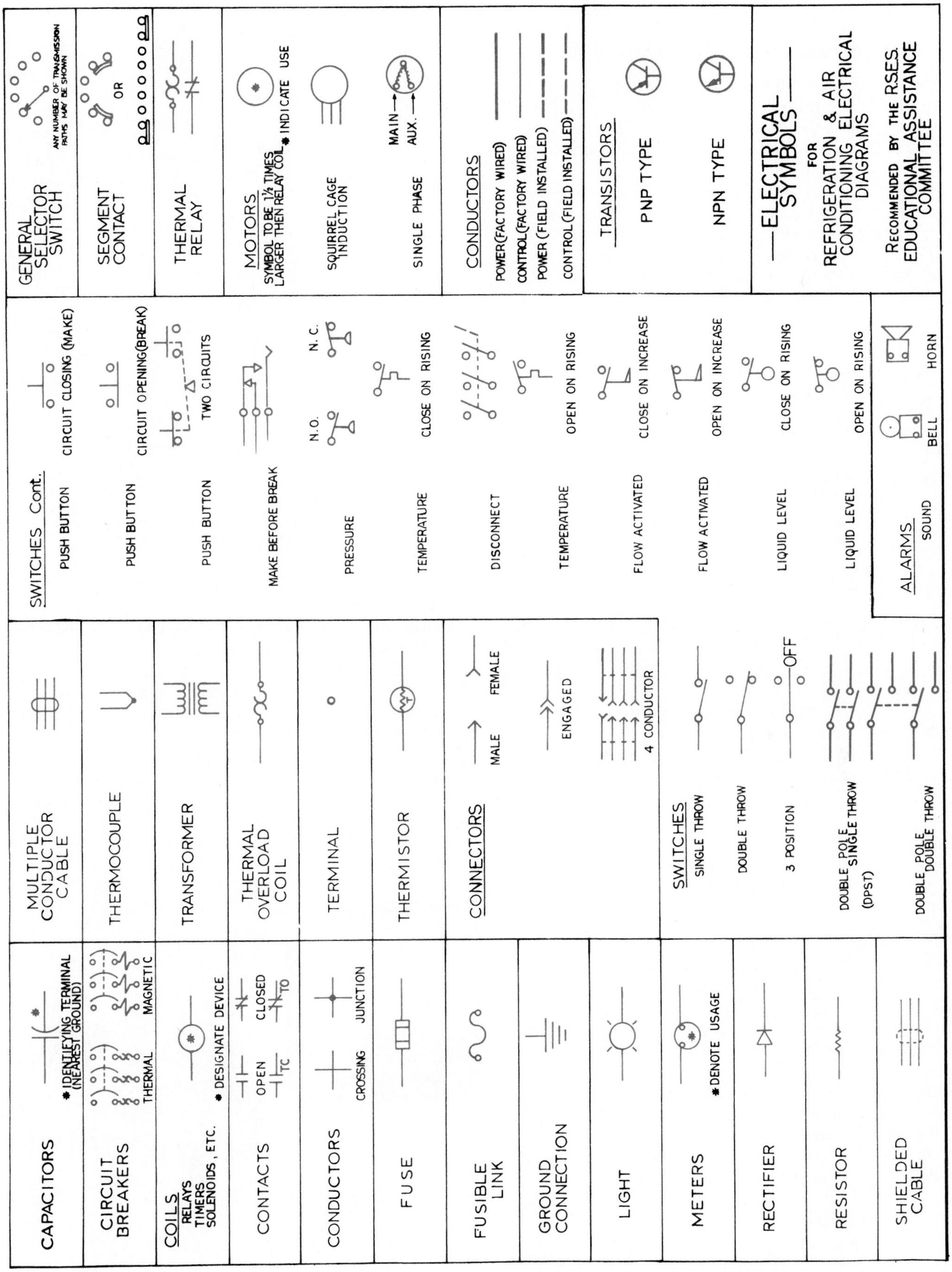

CAPACITORS

* IDENTIFYING TERMINAL (NEAREST GROUND)

CIRCUIT BREAKERS

THERMAL   MAGNETIC

COILS
RELAYS
TIMERS
SOLENOIDS, ETC.

* DESIGNATE DEVICE

CONTACTS

OPEN   CLOSED

CONDUCTORS

CROSSING   JUNCTION

FUSE

FUSIBLE LINK

GROUND CONNECTION

LIGHT

METERS

* DENOTE USAGE

RECTIFIER

RESISTOR

SHIELDED CABLE

MULTIPLE CONDUCTOR CABLE

THERMOCOUPLE

TRANSFORMER

THERMAL OVERLOAD COIL

TERMINAL

THERMISTOR

CONNECTORS

MALE   FEMALE

ENGAGED

4 CONDUCTOR

SWITCHES
SINGLE THROW

DOUBLE THROW

3 POSITION   OFF

DOUBLE POLE SINGLE THROW (DPST)

DOUBLE POLE DOUBLE THROW

GENERAL SELECTOR SWITCH
ANY NUMBER OF TRANSMISSION PATHS MAY BE SHOWN

SEGMENT CONTACT
OR

THERMAL RELAY

MOTORS
SYMBOL TO BE 1½ TIMES LARGER THEN RELAY COIL * INDICATE USE

SQUIRREL CAGE INDUCTION

MAIN   AUX.

SINGLE PHASE

CONDUCTORS
POWER (FACTORY WIRED)

CONTROL (FACTORY WIRED)

POWER (FIELD INSTALLED)

CONTROL (FIELD INSTALLED)

TRANSISTORS
PNP TYPE

NPN TYPE

— ELECTRICAL SYMBOLS —
FOR
REFRIGERATION & AIR CONDITIONING ELECTRICAL DIAGRAMS

RECOMMENDED BY THE R.S.E.S. EDUCATIONAL ASSISTANCE COMMITTEE

SWITCHES Cont.
PUSH BUTTON   CIRCUIT CLOSING (MAKE)

PUSH BUTTON   CIRCUIT OPENING (BREAK)

PUSH BUTTON   TWO CIRCUITS

MAKE BEFORE BREAK

N.O.   N.C.

PRESSURE

TEMPERATURE   CLOSE ON RISING

DISCONNECT

TEMPERATURE   OPEN ON RISING

FLOW ACTIVATED   CLOSE ON INCREASE

FLOW ACTIVATED   OPEN ON INCREASE

LIQUID LEVEL   CLOSE ON RISING

LIQUID LEVEL   OPEN ON RISING

ALARMS
SOUND

BELL   HORN

Fig. 6-7. Electrical symbols commonly used in wiring diagrams.

Electrical wiring diagrams use symbols for many of the electrical parts. Fig. 6-7 illustrates standard electrical symbols.

6-10 ELECTROMOTIVE FORCE (EMF)—VOLT

The term "electromotive force," or emf, is used to indicate electrical pressure or voltage which causes current to flow. One volt is the electromotive force (emf) required to send one ampere through a resistance of one ohm. The abbreviation for volt (emf) is E.

A volt is the unit of electrical pressure and is similar to pressure used to make gases or liquids flow. Water pressure in psi (pounds per square inch) or kPa causes the water to flow through pipes, hoses, nozzles, and the like. Increasing the water pressure increases the water flow, Fig. 6-8.

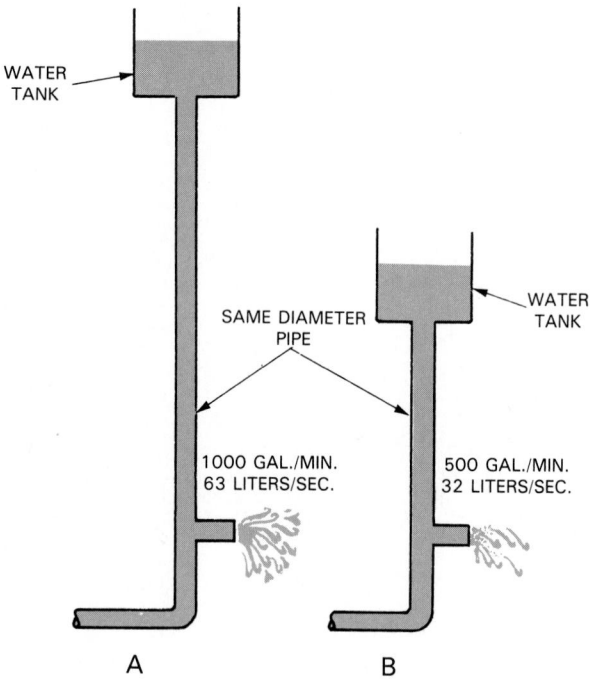

Fig. 6-8. Rate of flow depends upon pressure. A—Higher tank of water provides greater pressure, greater flow. B—Lower tank provides lower pressure, lesser flow.

In an electrical circuit, increasing the voltage increases the current flow. This is indicated by the increase in the amount of light coming from the lamp. The effect of increasing the electrical pressure—voltage (emf)—on an electrical circuit is shown in Fig. 6-9. Note the increase in the amount of light coming from the lamp in the circuit with a two-cell battery.

Voltage in a direct current circuit is measured with the use of a direct current voltmeter.

To measure the voltage in an alternating current circuit, such as a home lighting circuit, an alternating current voltmeter must be used.

6-11 VOLTMETER

Volts are units of electromotive force. Volts are pressure or force. Like charges (two electrons) repel each other; this is a force (volts). Unlike charges (electron and proton) attract

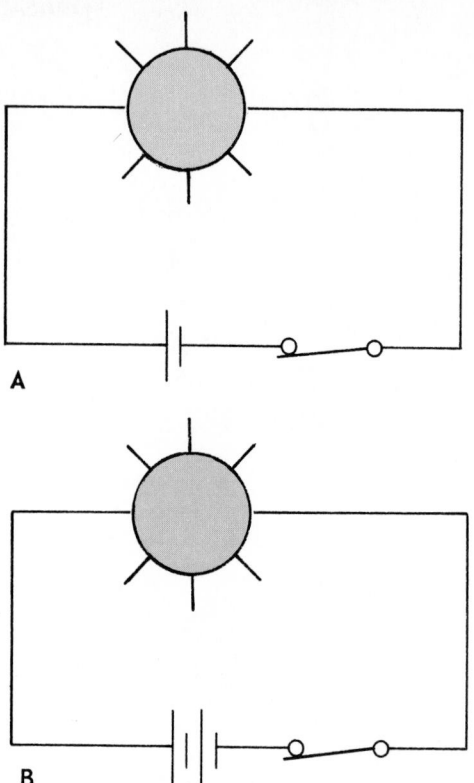

Fig. 6-9. A—Circuit with one-cell battery (1 1/2 V). Bulb lights, but is dim. B—Same circuit with two one-cell batteries (3 V). Light is much brighter (more current flowing).

each other; this is force.

Instruments called voltmeters have been developed to measure the electromotive force.

The force (electromotive force) needed to move one ampere through a one-ohm resistance is called a volt. One kilovolt (kV) is 1000 volts. A millivolt (mV) is 1/1000 (.001) of a volt. A microvolt (µV) is 1/1,000,000 (.000 001) volt.

There are five types of electromechanical voltmeters for measuring electromotive force. The first four use forces made by current flow and magnetism. The fifth uses electrostatic forces:

1. Permanent magnet-moving coil (D'Arsonval), Fig. 6-10.

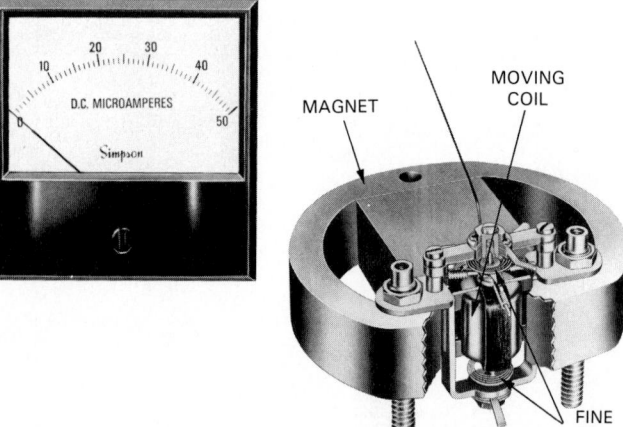

Fig. 6-10. Left. A microampere instrument used for testing dc circuits. Right. The permanent magnet and moving coil movement. (Simpson Electric Company)

2. Electrodynamic (dynamotor movement), Fig. 6-11.
3. Moving vane (iron), Fig. 6-12.
4. Moving magnet (polarized iron or iron vane), Fig. 6-13.
5. Moving plate (the electrostatic type), Fig. 6-14.

A sixth type of voltmeter is a digital voltmeter. Most new meters use electronic circuitry instead of electromagnetic effects. See Fig. 6-15. They have several advantages over the electromagnetic effect devices:
1. No moving mechanical parts.
2. Easy readability.
3. Smaller size.

These devices use solid-state semiconductors which

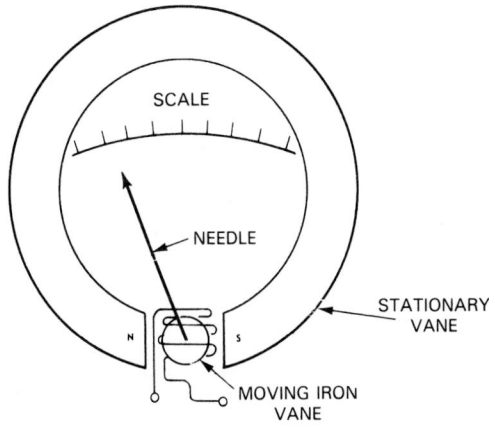

Fig. 6-13. A moving magnet type voltmeter is designed for use with dc.

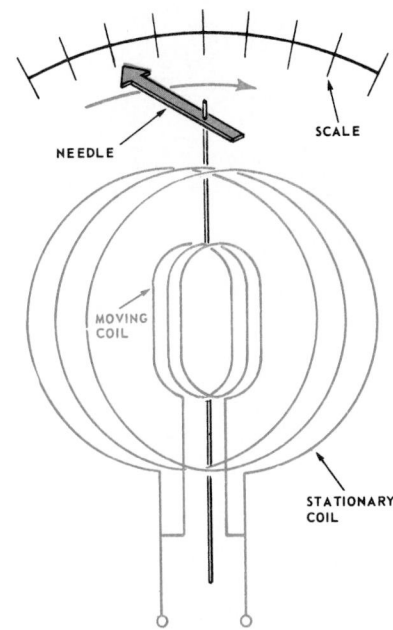

Fig. 6-11. Electrodynamic instrument. Two coils create magnetism, and inner coil movement is related to electron flow. Used with dc, or ac to 200 Hz. Single coil type is used for voltage, current, or power. Double coil type is used for power, motor phase angle, frequency, and capacitance.

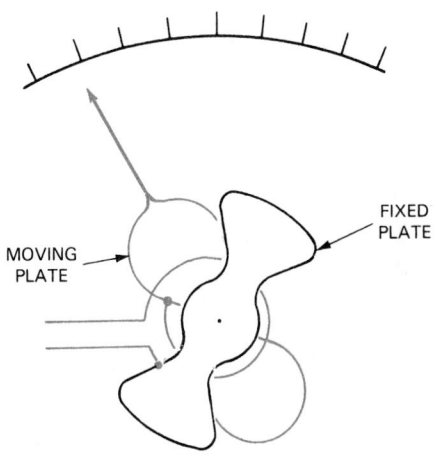

Fig. 6-14. Electrostatic type meter. Potential difference exists between the moving plate and the fixed plate. Used for testing voltage in dc or ac circuits over 10 V.

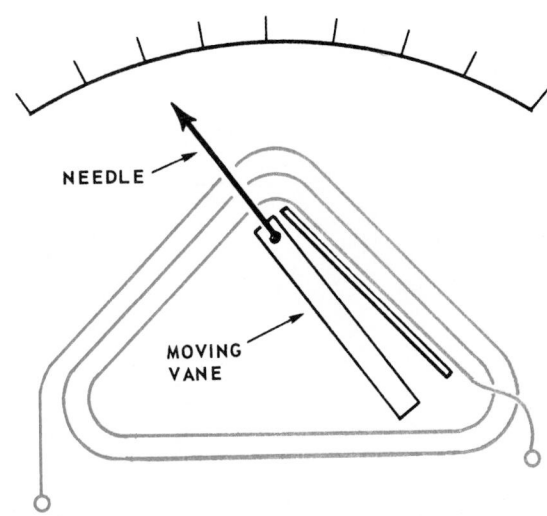

Fig. 6-12. This moving iron or moving vane instrument is used for dc, or ac to 125 Hz.

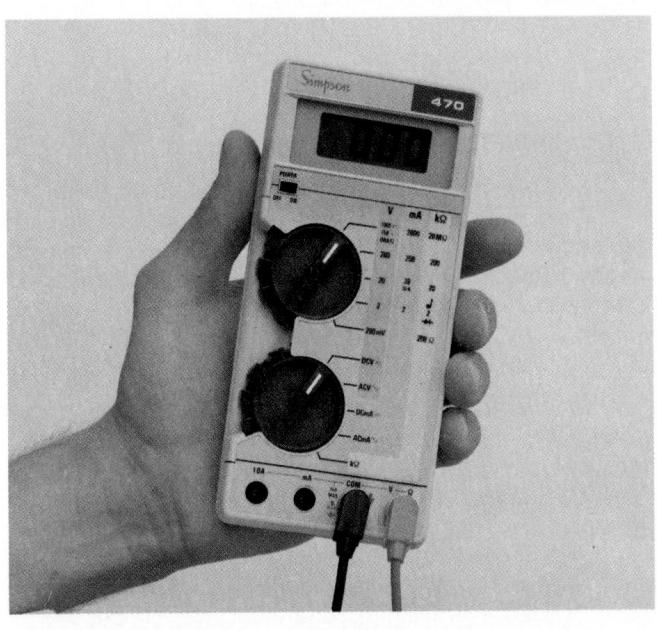

Fig. 6-15. Hand-held digital multimeter. This device measures current, voltage, and resistance, using electronic circuitry.
(Simpson Electric Co.)

withstand shock and vibration better than delicate meter movements. They require no lubrication or compensation for the position of the instrument. The display is a number (digit) instead of a pointer, which allows rapid reading.

Some designs offer auto-ranging. This means that one does not have to select the best scale on the instrument for the voltage to be measured. The voltmeter automatically senses the voltage level and selects the proper scale. These devices are also smaller in size and more portable. Electrical power requirements for electronic meters are lower than those for electromagnetic meters.

Voltmeters are always connected in parallel with the circuit. The method of connecting them is shown in Fig. 6-16.

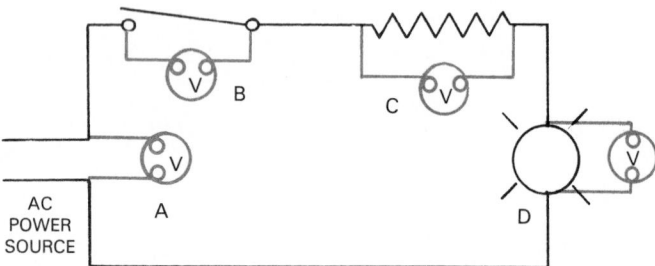

Fig. 6-16. Correct connections for testing with voltmeter. Meter will show: A—115 to 120 V if power is on. B—Whether switch is open or closed (zero reading if closed, 120 V if open). C—Voltage drop across resistor. D—Voltage drop across lamp (zero reading if shorted, 120 V if broken or open). Voltages of B + C + D should equal A when switch is closed.

6-12 COULOMBS

A coulomb is a count of the number of electrons passing a given point on a conductor. It is the *amount* of electricity passing a given point in a conductor in one second when the rate of current flow is constant at one ampere.

The number of electrons in a coulomb equals 6.24×10^{18}, or 6,240,000,000,000,000,000. A one-coulomb flow per second equals one ampere. This is like the rate of water flow in gallons per minute (gpm).

6-13 AMPERE

The ampere measures the rate of flow of current. It does not measure electrons. It has a one-to-one relationship with coulombs. Ten amperes flowing past a point in one second equals 10 coulombs. "Current" usually means rate of flow. It is measured with an ammeter.

6-14 AMMETER

An ammeter is an instrument that measures the rate of current flow in amperes. There are two types. One measures direct current flow; the other, alternating current. Most refrigeration and air conditioning work requires an ac ammeter. See Fig. 6-17. The operation of the ammeter in Fig. 6-17 depends upon the magnetic effect of current in a conductor. The magnetism causes a current flow at the base of the tongs (clamp) surrounding the conductor.

The tongs can be opened and slipped around a conductor. Electricity generated in the tongs by the magnetism moves the indicator to give the ampere flow reading.

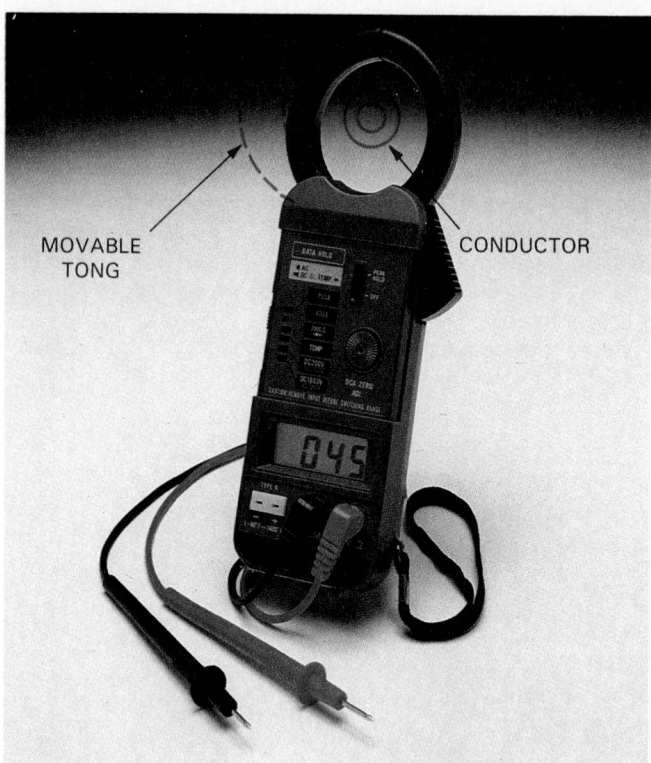

Fig. 6-17. Digital clampmeter with six functions and red and black test leads. Meter measures ac and dc voltage and temperature. (The Dickson Company)

Only one wire of the circuit should be placed in the jaws of the meter. This type of ammeter can only be used on an alternating current circuit. See Fig. 6-18.

If the reading from a "clip on" (tong-type) ammeter is too low to be read accurately, wrap the wire once around one of the jaws to double the reading (wire goes between jaws twice). Divide this reading by two to get the correct reading. Some of these meters have "multipliers" available as accessories.

The method of connecting the two types of ammeters is shown in Fig. 6-19. CAUTION: Ammeters should always be connected in series with a circuit. (For series circuit, see Para. 6-30.) If an ammeter is accidentally connected in parallel, it will be burned up.

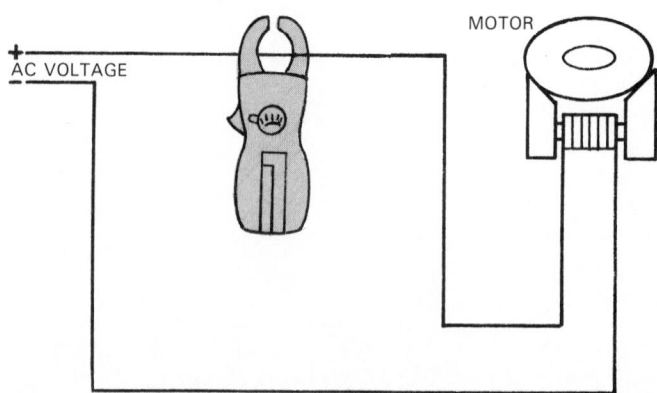

Fig. 6-18. Tong type ammeter usage.

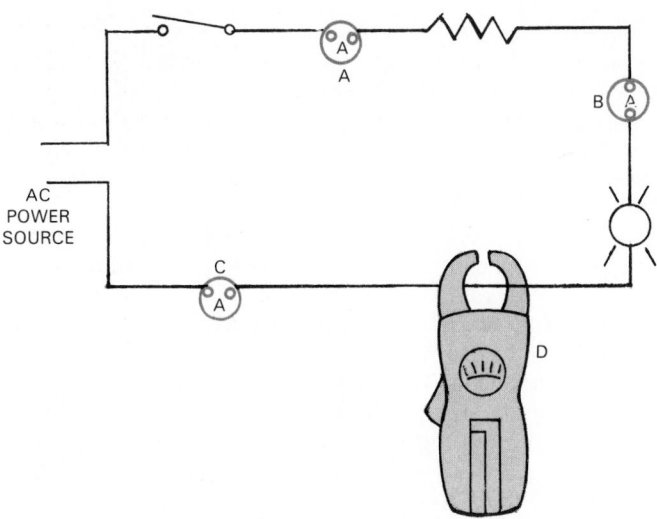

Fig. 6-19. Ammeter connections. Ammeters A, B, C, and D should all read the same: Zero when switch is open, 5 A when switch is closed. D is special induction type ammeter described in Fig. 6-17.

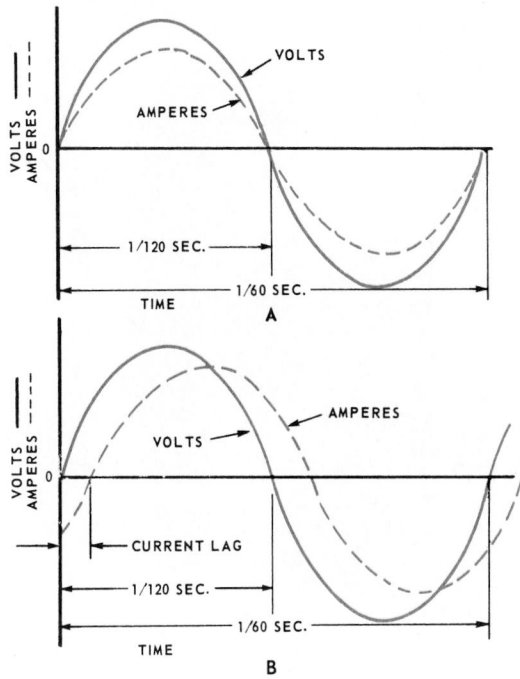

Fig. 6-20. Alternating current cycles. A—Current and voltage in phase. B—Current and voltage not in phase.

6-15 WATTS

Power is the time rate of doing work. When amperes flow (coulombs per second) at a certain pressure (emf or volts), this is power.

Electrical power is measured in watts (W) and kilowatts (kW). See Para. 6-70. When computing electrical horsepower (hp), 746 W or 0.746 kW equal one horsepower.

In direct current circuits, the wattage may be obtained by multiplying the current in amperes, intensity, by the emf (volts), $I \times E = W$.

In alternating current circuits, the wattage may be less than the amount obtained by multiplying the amperes and volts together. This is because the current in an ac circuit may not be exactly in phase with the voltage.

6-16 POWER FACTOR

The power factor represents that fraction of the total possible power that can be generated in a circuit. See Para. 6-71 for additional information.

Current usually lags behind the emf, depending on the type of circuit. That is, when the emf is at 50 percent of its maximum, the current flow may be 35 or 40 percent of its maximum. Therefore, the wattage or power of an ac circuit must be obtained by multiplying the voltage times the current times the power factor ($E \times I \times PF = watts$).

View A in Fig. 6-20 shows an alternating current cycle in which the voltage and the current are in phase. The current is at its maximum at the same time that the voltage is at its maximum. The voltmeter reads 120 V and an ammeter connected into the circuit reads 10 A. The product of the voltage times the current equals 1200 W. A wattmeter connected to this circuit will also read 1200 W. This indicates a power factor of 100 percent.

View B in Fig. 6-20 shows an alternating current cycle out of phase. The current is lagging behind the voltage one-eighth of a cycle. A voltmeter connected to the circuit reads 120 V and an ammeter connected to the circuit reads 10 A. The product of the voltmeter reading multiplied by

the ammeter reading will be 1200. However, a wattmeter applied to this circuit will read 1000 W. The power factor is the ratio of the wattmeter reading over the calculated wattage, which is the current multiplied by the voltage.

Power Factor =

$$PF = \frac{wattmeter\ reading}{ammeter\ reading \times voltmeter\ reading}$$

Example: wattmeter reading 1000
voltmeter reading 120
ammeter reading 10

$$PF = \frac{1000}{120 \times 10} = \frac{1000}{1200} = .833$$

To change to percentage: $.833 \times 100 = 83.3$ percent

To calculate the wattage, multiply the voltmeter reading by the ammeter reading. The answer must be multiplied by .83 (the power factor), to give the correct wattage:

$120 \times 10 \times .83 = 1000$.

Various things in the electrical circuit affect the power factor. If the electrical load is entirely a resistance load, such as electric heating and the like, this is called a noninductive load. It does not seriously affect the power factor.

Most electric motors, transformer welders and the like, do affect the power factor. These are called inductive loads. It takes time for current to build up in any coil. This is due to the counter emf generated in the coil. This lag affects the power factor. See Para. 6-51, which discusses the effects of counter electromotive force.

6-17 WATTMETER

A wattmeter is an instrument used to measure the wattage consumed by an electric motor or an electrical device.

It is connected in series with the circuit being measured. (See Para. 6-30.) Fig. 6-21 illustrates a portable wattmeter

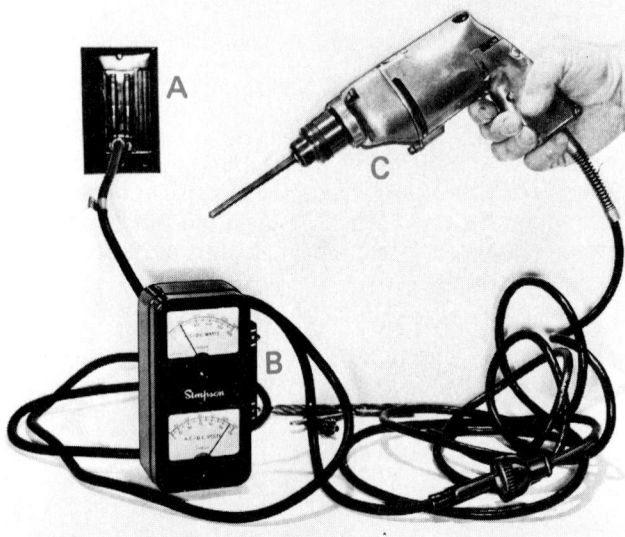

Fig. 6-21. Top. Combination voltmeter and wattmeter. Bottom. Volt-wattmeter being used to test an electric portable drill. A—Power source. B—Volt-wattmeter. C—Electric load. (Simpson Electric Company)

and shows how it is connected to measure wattage drawn by an electric drill. This is a combination instrument and may be used for measuring volts or watts.

A wattmeter indicates the true wattage in a circuit. It automatically adjusts for the power factor. This is done by two coils in the wattmeter—a voltage coil and a current coil. The influence of the two coils drives the wattmeter. If the voltage and current are out of phase, the influence of the two coils will be reduced. This is how the wattmeter reading adjusts for the power factor.

6-18 POWER FACTOR METER

A power factor meter is an instrument that is used to provide a direct reading of the power factor in an electrical circuit. A direct-reading meter saves time.

This meter first measures the total available wattage in a circuit by finding the product of the measured voltage times the measured current. It then determines the true wattage using the same method described in Para. 6-17. Lastly, the ratio of the true wattage to the available wattage is determined and then displayed. This value is the power factor. See Fig. 6-22.

A power factor meter can be used to assist in increasing the effectiveness of a circuit. For example, assume a circuit has a measured power factor of .80 (80% of the total possible wattage is being used). As mentioned before, a value similar to this is quite common in an inductive (motor, transformer) circuit. The proper addition of a capacitive load would result in an increase in true wattage. This increase would then bring the power factor closer to 1.00.

6-19 RESISTANCE—RESISTORS

Most electrical conductors are made of metal. No material is a perfect conductor of electricity, but some metals are better conductors than others. Silver, copper, and aluminum are very good. Iron, steel, and carbon will also conduct electricity, but their resistance is quite high. Carbon is sometimes used in electrical circuits, but it is not a very good conductor.

Extremely poor conductors are called resistors or resistances. They have no free electrons or, at best, very few free electrons in the atom. It is difficult for these free electrons to travel through and around the other atoms. The word impedance is often used in place of the word resistance when a part of an ac circuit or a complete ac circuit resists the flow of free electrons.

The total resistance in a series circuit (Para. 6-30) is the sum of the resistance of each part of the circuit.

The harder it is for the free electrons to move, the greater the heat generated in the conductor. This heating action is shown best in iron, steel, and steel alloys used for elec-

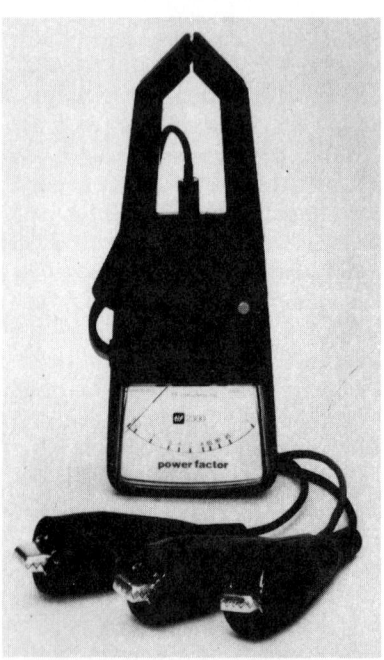

Fig. 6-22. Power factor meters determine the effect of individual loads and systems. (TIF Instruments, Inc.)

tric heating purposes. See Chapter 30 for a more complete discussion of the resistance values of conductors, semiconductors, and nonconductors.

All materials have some resistance to the flow of electricity. The resistance of electrical conductors usually increases with an increase in temperature or in the length of the conductor. Also, the resistance increases as the diameter (thickness) of the conductor decreases. In the case of some semiconductors, increasing the temperature increases their ability to conduct electricity.

Components designed to offer specific levels of resistance in a circuit are called resistors. See Fig. 6-23. These resistors usually have a series of color bands which represent both the amount and accuracy of resistance. See Chapter 30 for a brief summary of these color bands.

6-20 OHMS

Electrical resistance is measured in ohms. An ohm is the amount of resistance in an electrical circuit which allows an emf of 1 volt to cause 1 ampere to flow through the circuit.

The resistance in a conductor depends upon four things:
1. Material used.
2. Diameter or size (thickness) of conductor.
3. Length of conductor.
4. Temperature of conductor.

6-21 OHMMETER

Ohmmeters are used to check circuits for resistance, open circuits, and grounds. Power must be off when the ohmmeter is being used. The instrument may be ruined otherwise. The ohmmeter, Fig. 6-24, has a small battery or a magneto as a power source. Fig. 6-25 shows the correct

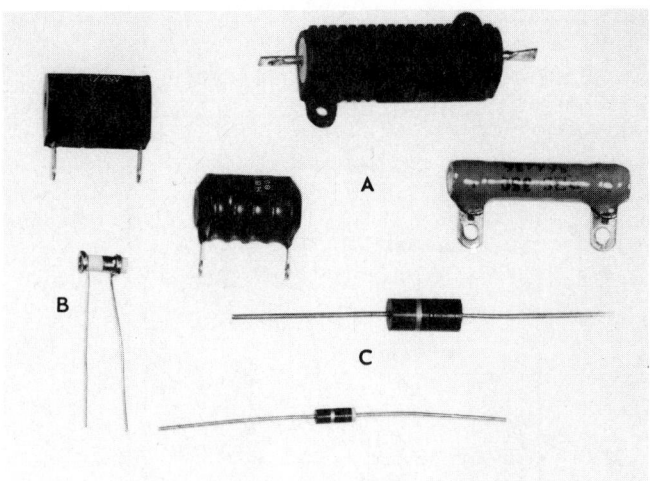

Fig. 6-23. Typical resistors. A—Wirewound. B—Thin film. C—Composition.

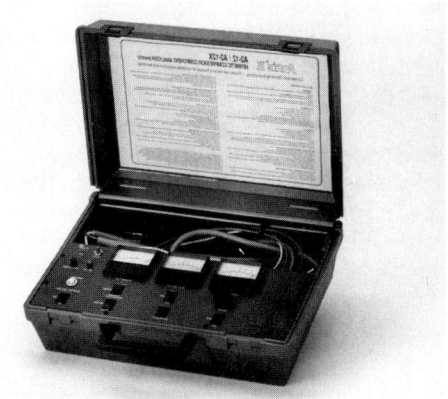

Fig. 6-24. Combination ammeter-voltmeter-ohmmeter. Note test leads and power cord. (Imperial Eastman, Imperial Division)

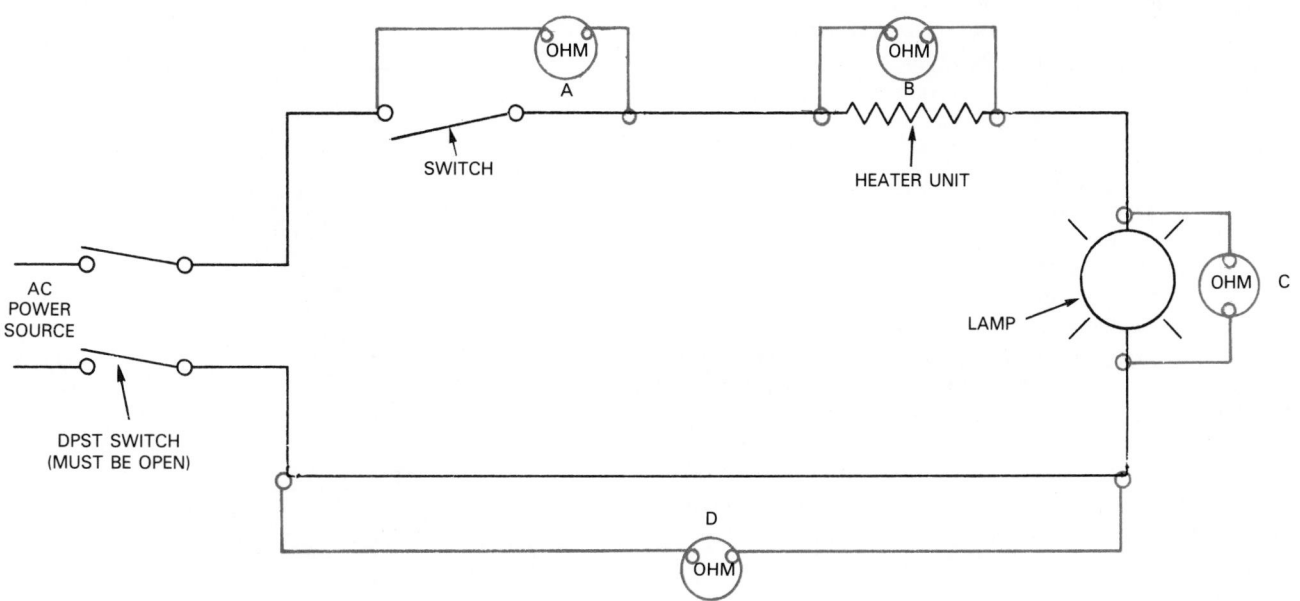

Fig. 6-25. Ohmmeter connections. There must be no power in circuits. A—Testing resistance across a switch in open, then closed position. With switch closed, resistance indicated on meter indicates a poor connection somewhere in switch. B—Testing resistance (ohms) across heater unit. C—Testing resistance across a lamp. D—Measuring continuity of conductor. If meter reads zero, conductor is not broken and connections are good. If meter reads infinity (maximum resistance), conductor or a connection is broken.

method of connecting an ohmmeter into a circuit to measure the resistance.

6-22 INSTRUMENT CONNECTING AND HANDLING

Basic electrical instruments may be used to measure either volts, amperes, or ohms. The difference is in the external wiring of the instrument. The scale and use of instruments vary depending upon the wiring inside of the instrument. Any instrument needs only a very few electrons to activate the pointer mechanism.

The voltmeter has a high internal resistance. The ammeter is designed to bypass (shunt around) most of the current outside of the instrument. The ohmmeter allows only a few electrons in the circuit.

The mechanisms for each type of meter vary widely. These instruments must be delicate enough to read accurately. Therefore, they should not be dropped, used above their maximum reading, or carelessly handled or stored.

The voltmeter is connected in parallel with the circuit. A high value resistor in the instrument coil circuit will allow a maximum of 5/1,000,000 of an ampere, or 5 microamperes, to flow into the moving coil through the springs.

Be careful! If a 120 V voltmeter is used to measure a 240 V circuit, the meter will be ruined. (Double the milliamperes will flow through the meter.)

A low resistance shunt is placed across the terminals of an ammeter. The instrument is placed in series with the circuit when current is being measured.

An ohmmeter is connected in parallel with the circuit while a known voltage is appied (usually a low voltage battery inside the meter housing).

6-23 SHUNT

As indicated in Para. 6-22, internal construction of many electrical meters is very much alike. Methods of connecting into the circuit, and accessory devices used, enable the same basic instrument to be used as either an ammeter, voltmeter, or ohmmeter. See Fig. 6-26.

Voltmeters are connected across or in parallel with the circuit in which the voltage is being measured. Ammeters are connected in series with the circuit. This would indicate that all of the current being measured by the ammeter must go into the meter. This is not true. A shunt (a second circuit path) is placed in parallel with the instrument, and the greater portion of the current goes through the shunt rather than through the meter.

Several different shunts can be used with the same meter. Shunts are calibrated for the current range in which each is to be used. Sometimes shunts are built into the meter. In such cases, the electrical connection to the meter should be made to terminals within the range of the meter.

If in doubt about the voltage or amperage of a circuit, connect first to the upper range of the instrument. If the reading is within the lower range, a more accurate reading may be made by making a connection which uses almost the full scale of the instrument. See Fig. 6-27.

6-24 ELECTRICAL MATERIALS

There are three physical materials which are used in electrical and electronic systems:

1. Conductors (metals, such as silver, copper, and aluminum).

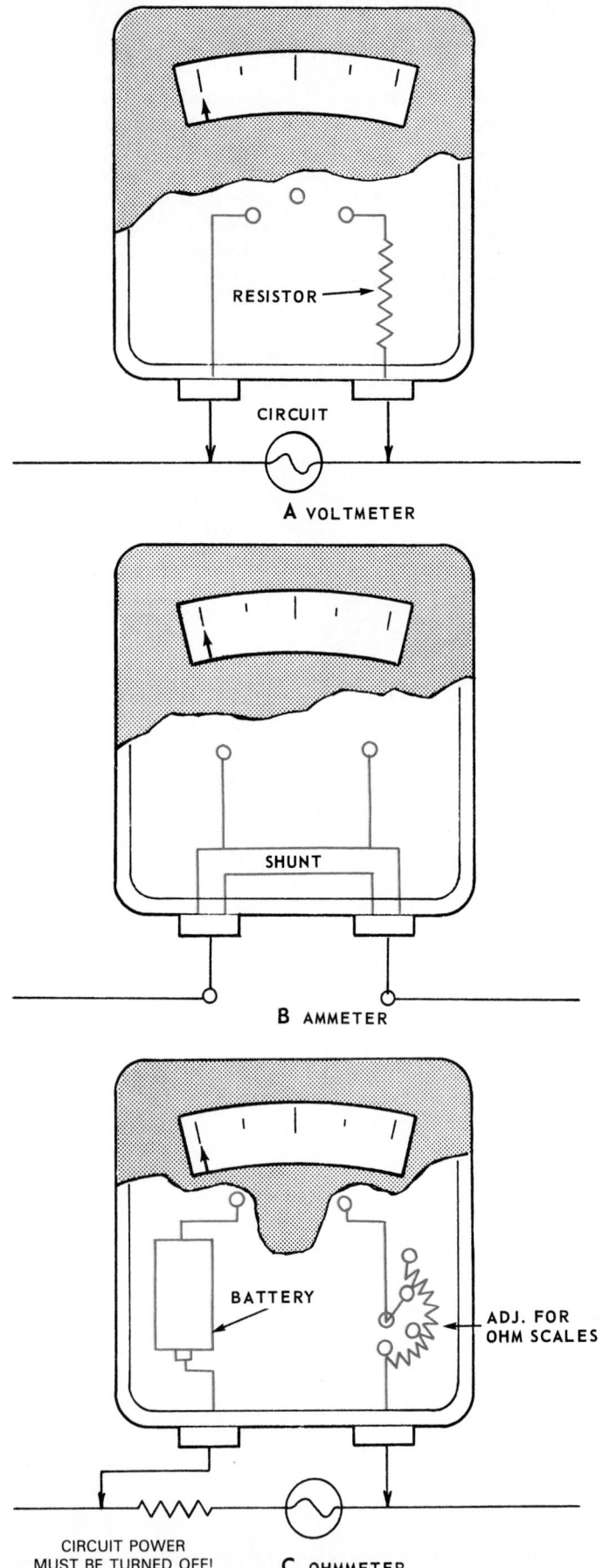

Fig. 6-26. How the basic meter movement is usually wired inside the voltmeter, ammeter, and ohmmeter.

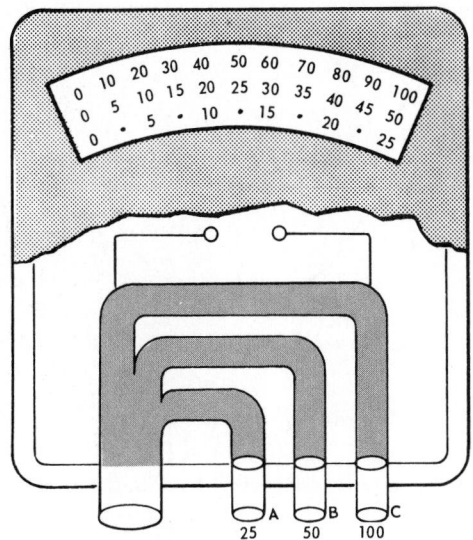

Fig. 6-27. Ammeter having built-in shunts. A—0 to 25 A. B—0 to 50 A. C—0 to 100 A.

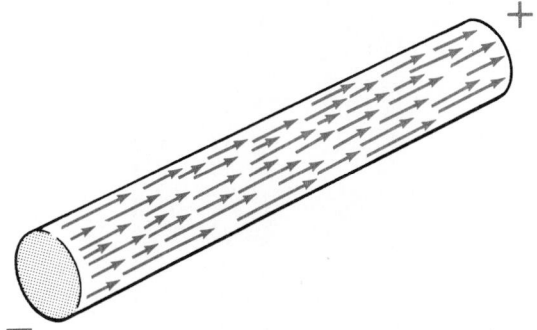

Fig. 6-28. A wire with free electrons traveling from negative (−) to positive (+).

2. Semiconductors (metal oxides or metal compounds).
3. Nonconductors or insulators (nonmetals such as glass, wood, paper, and mica).

Any of these three may exist in any of the three forms of matter, but most are solids rather than liquids or gases.

6-25 CONDUCTORS

A conductor has atoms with free electrons in its structure. Any electromotive force (pressure) will cause these electrons to travel from one atom to another. This moves the electrical energy through the material. In a wire, for example, energy moves from one end to the other, Fig. 6-28. Note that electron movement is from negative (−) to positive (+).

Each type of atom allows the free electrons to move with varying ease. In this respect, gold and silver atoms are excellent; copper, mercury, and aluminum atoms are very good, also. Most good conductors are metals, yet the movement of free electrons in iron is somewhat more difficult.

Conductivity of metal is usually expressed in ohms per circular mil foot at standard temperature. Wire with 1 circular mil cross-sectional area has a diameter of 0.001 in. The standard temperature used when measuring the conductivity is 68 °F (20 °C).

Wires (solid conductors) are popular for carrying electricity from one electrical device to another. Most electricians call wires "electrical leads" (pronounced like "lead the way").

6-26 SEMICONDUCTORS

Conductors such as metals conduct heat and electricity readily. Insulators conduct heat and electricity very poorly. In between these materials are semiconductors. Semiconductors are ordinary insulators which, under certain conditions, can be made to conduct electricity easily. These materials are the basis of the present electronics industry. The term "solid-state electronics" refers to electronic devices which are made up of semiconductor elements.

Common devices which are semiconductors are transistors, diodes, and photocells. Many modern motor con-

trols consist of silicon controlled rectifiers (SCRs), which are semi-conductive switching devices.

The conductivity of a semiconductor can be controlled by an electrical signal, light intensity, pressure, temperature, and other signaling devices. Semiconductors can, therefore, serve as relays and switches.

The ac current produced by an alternator is converted to dc current by semiconductor diodes. These are the sensitive devices which can be damaged if one starts a car with the wrong jumper connection between batteries.

Photocells used on automatic door openers are semiconductor switching devices activated by light.

In troubleshooting of motors which fail to operate, the most likely cause of motor control failure is a burned-out semiconductor diode or SCR. See Para. 6-55 for some semiconductor applications.

6-27 NONCONDUCTORS (INSULATORS)

Nonconductors resist electron flow. The atoms have virtually no free electrons. A perfect vacuum is also a nonconductor.

Nonconductors are as useful in electrical systems as conductors or semiconductors. There are many parts of the system in which electrical flow must be stopped (where insulation is needed).

Some common nonconductors are quartz, ceramics, mica, glass, and organic substances (rubber, wood, paper, and plastics). The resistance values for nonconductors (insulators) range between 10^9 to 10^{18} ohms.

6-28 INSULATION TESTERS

Poor insulation can cause equipment breakdowns and may also create a dangerous shock hazard. *An insulation tester is a meter used to detect leaks or possible areas of failure along nonconductors or insulators.* See Fig. 6-29.

Insulation testers may be used in two ways. They can be applied to a "live" circuit, where one lead is connected to ground and the probe is run along the insulation to see if any voltage reading (leak) occurs. This would indicate a break in the insulation. They can also be used on a circuit where power has been disconnected. In this case, one lead is connected to the conductor and the probe is run along the insulation. This reading will indicate the high resistances (megohms) that exist. A quick drop in this reading would identify either a break in the insulation or an area of heavy wear where failure is likely to occur.

Fig. 6-29. Insulation tester measures volts, ohms, and megohms. Measurement can be taken through insulation. (A.W. Sperry Instruments, Inc.)

6-29 OHM'S LAW

The relationship between the volt, the ampere, and the ohm is known as Ohm's Law.

If a conductor 1 ft. long will allow 1 A to flow with an emf of 1 V, this same conductor will allow 2 A to flow if the emf is 2 V. Also, if the conductor is 2 ft. long and the emf is 1 V, only 1/2 A will flow. Therefore, it is evident that emf is the product of the current intensity (amperes) multiplied by the resistance (ohms); or developing a formula, begin with these symbols:

E = Electromotive force (emf) in volts.
I = Intensity of current in amperes.
R = Resistance in ohms.

English System:

$$\frac{Emf}{(Volts)} = \frac{Intensity}{(Amperes)} \times \frac{Resistance}{(Ohms)}$$

$$E = IR$$

and therefore $I = \frac{E}{R}$

or $R = \frac{E}{I}$

Fig. 6-30 shows the relationship of I, E, and R.

SI Metric System:

Voltage = Amperes × Ohms

V = A × Ω (This particular symbol is the Greek letter omega. It is the symbol used for ohms.)

From this basic law, one may learn that, if the resistance stays constant, the current can only be increased by increasing the emf. Also, if the resistance in a circuit becomes low, the amperage will become high.

For example, if a 240 W lamp draws 2 A at 120 V, its resistance is as follows:

$$R = \frac{E}{I}$$

$$R = \frac{120\ V}{2\ A}$$

R = 60 ohm (Ω)

The formula E = I × R also means that if one doubles the emf in a circuit (the resistance remaining the same), the

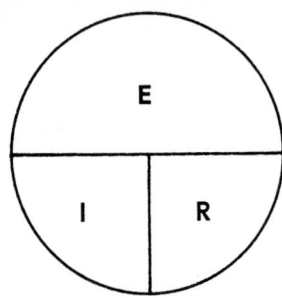

If solving for **E**, cover **E** with finger and the circle shows

E = I x R   **E** =

If solving for **R**, cover **R** with finger and the circle shows

$$R = \frac{E}{I}$$ R =

If solving for **I**, cover **I** with finger and the circle shows

$$I = \frac{E}{R}$$ I =

Fig. 6-30. Ohm's Law equations shown in graphic form.

current flow will double and may cause trouble. An example would be to connect a 120 V motor into a 240 V power outlet. If the fuse does not "blow," the motor windings will become too hot carrying this excessive current. The insulation on these wires may be destroyed and the motor may be ruined.

If the resistance is increased (putting in a smaller diameter conductor or having a dirty or loose connection) and the emf stays the same, the current will decrease. This causes a loss of power and the wire will heat or the poor connection will become very warm (and may even cause a fire).

6-30 CIRCUITS, SERIES

A circuit which has only a single path for current flow is called a series circuit. See Fig. 6-31.

In Fig. 6-31, each resistance has the same current flowing through it. In a series circuit, all the resistances are added together to determine the total resistance. In Fig. 6-31, the total voltage at the power supply is equal to the sum of the voltages across the three resistances.

If one of the objects in a series circuit—switch, light bulb, resistor, etc.—does not allow current to flow, the circuit will not operate.

6-31 CIRCUITS, PARALLEL

A circuit which allows the current to flow along either one of two or more conductors or electrical paths at the same time is called a parallel circuit. See Fig. 6-32.

In Fig. 6-32, the sum of the current flowing through the lamp and the motor equals the total input current. The current flow is based on the resistance of the conductors it flows through. If the lamp has 1/4 of the resistance of the motor, 4/5 of the current will flow through the lamp and

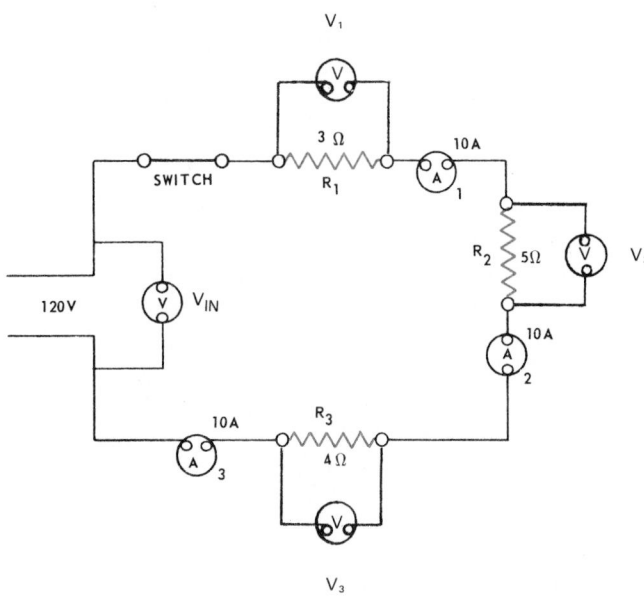

$$V_{IN} = V_1 + V_2 + V_3$$
$$R_{TOTAL} = R_1 + R_2 + R_3$$
$$I = V_{IN}/R_{TOTAL}$$

Fig. 6-31. A series circuit with three resistances. R_1 = 3 ohms, R_2 = 5 ohms, and R_3 = 4 ohms. Total resistance is 12 ohms. All three ammeters will read the same, 10 A. V_1 = 30 V, V_2 = 50 V, and V_3 = 40 V. $V_{IN} = V_1 + V_2 = V_3$ = 120 V.

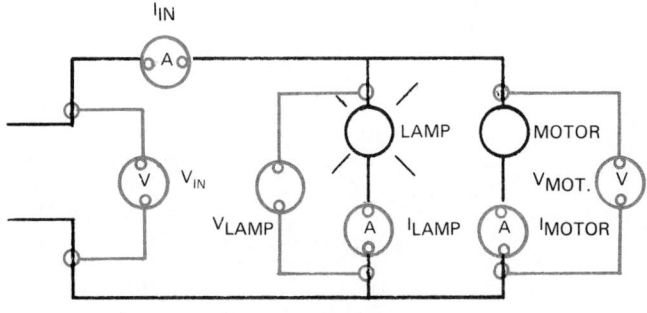

$$I_{IN} = I_{LAMP} + I_{MOTOR}$$
$$V_{IN} = V_{LAMP} = V_{MOTOR}$$

Fig. 6-32. A parallel circuit showing the incoming current and the current draw. I_{IN} represents the incoming current. I_{LAMP} equals the current through the lamp. I_{MOTOR} is the current through the motor. $I_{IN} = I_{LAMP} + I_{MOTOR}$. $V_{IN} = V_{LAMP} = V_{MOTOR}$.

1/5 through the motor.

The voltage is the same across each of the parallel paths in the circuit.

6-32 CIRCUITS, SERIES—PARALLEL

A combination of series and parallel circuits is sometimes used. In some automatic defrost systems, two or more heating elements are connected in parallel. These heating elements in turn are connected into the system when the compressor motor is stopped. A timed switch controls the heating elements. This switch is in series with the heating elements. See Fig. 6-33.

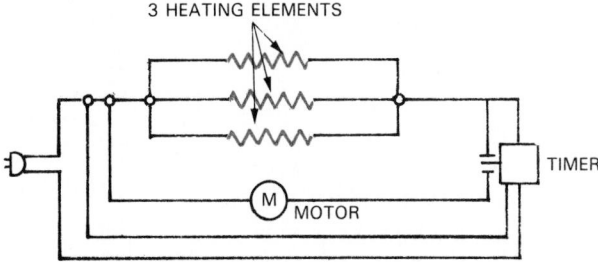

Fig. 6-33. Series parallel circuit. Timer controls three heating elements (resistances). Current through heating element section divides three ways. Some current goes through each of the three heating elements.

6-33 VOLTAGE DROP (IR)

The sum of the voltage drops in a circuit is always equal to the voltage applied. The voltage drop across any part of a circuit is equal to the current (amperes) multiplied by the resistance (ohms) across that part of the circuit.

In Fig. 6-34, for example, the voltmeter indicates an applied voltage of 120 V in a typical refrigeration circuit. The alternating current ammeter indicates a current draw of 5 A.

Equivalent resistance of circuit wiring is 0.5 Ω.	5 × .5 =	2.5
Equivalent resistance of thermostat is 0.5 Ω.	5 × .5 =	2.5
Equivalent resistance of starting relay is 1.0 Ω.	5 × 1 =	5.0
Equivalent resistance of motor compressor is 22.0 Ω.	5 × 22 =	110.0
Therefore, total voltage drop is		120.0 V

There is always some electrical resistance across any electrical control, relay, or circuit wiring. A very sensitive instrument is needed to measure this low resistance.

6-34 POWER LOSS (I²R)

The power loss in a circuit due to resistance is equal to the square of the current multiplied by the resistance. This is usually expressed as I^2R loss = $I^2 × R$ = Power Loss.

Referring to Fig. 6-34, the power applied is at 120 V. The total current in the circuit is 5 A; the total resistance in the circuit is 24 Ω.

$$5^2 = 25$$
$$25 × 24 = 600 \text{ W power loss.}$$

Power applied is equal to voltage times amperage:
$$5 \text{ A} × 120 \text{ V} = 600 \text{ W.}$$

Power loss results in the generation of heat. One watt converted into heat equals:

3.4144 Btu/hr. = 860 calories/hr. (1 Btu = 252 calories = 0.252 kg-calories).

6-35 ALTERNATING CURRENT CYCLES (HERTZ or Hz)

If electricity flows for 1/120 second in one direction and then for 1/120 second in the other direction, this is called one cycle. Most alternating current is generated in 60 cycles (Hertz). This means that in one second, the electricity flows 60 times one way and 60 times the other, or for 1/120 of a second it flows in one direction and then for 1/120 of a second in the other direction. The current, therefore,

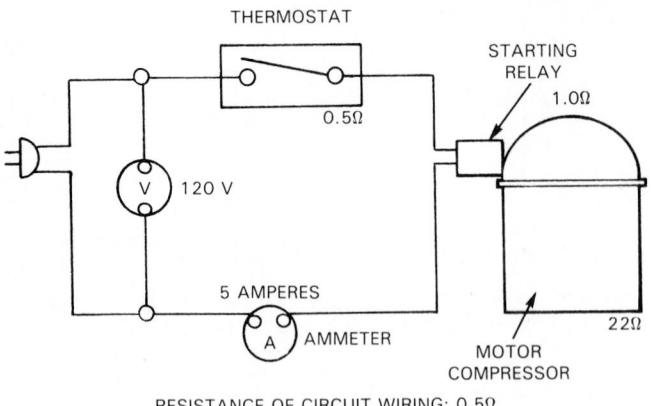

Fig. 6-34. Voltage drop in an electrical circuit is equal to voltage applied.

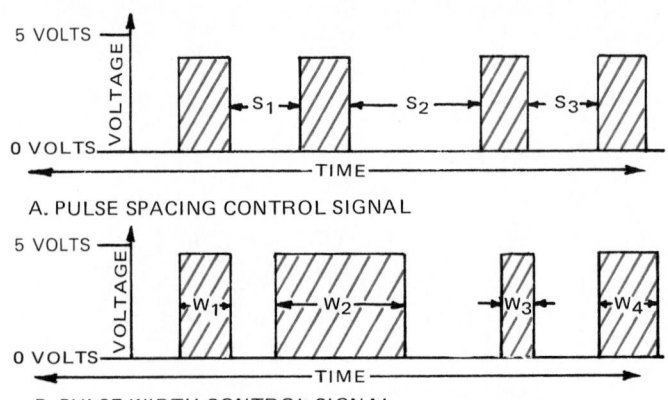

Fig. 6-36. Digital control signals used in pulse wave applications. A—Control using variable spacing between pulses, (S). B—Control using variable pulse width, (W).

makes a complete cycle (Hertz) in 1/60 of a second.

This reversal of the current flow can be shown by a graph, as shown in Fig. 6-35.

This drawing shows a sine wave of 60 cycles (Hertz or Hz) alternating current. In some industries, 180 cycle (Hertz or Hz) current is generated to operate hand tools such as wrenches, drills, and screwdrivers. These high cycle tools do not draw heavy currents and do not overheat if stalled.

6-36 PULSE WAVE AND DIGITAL CONTROL SIGNALS

In computer or digital control applications, a second type of alternating current is used. This is pulse wave electronics. The signals in these applications are electrical pulses, as shown in Fig. 6-36. The control is obtained by the spacing of the pulses and the width of the pulses. Most control systems using computers have 5-volt pulses. If they are used in motor control, the voltage is amplified to the voltage required by the motor.

6-37 MAGNETISM

Operation of the magnetic compass depends on the fact that the earth is a magnet. The north magnetic pole is near the north geographic pole. The compass needle is free to turn and "line up" with the earth's magnetic field. See Fig.

6-37. It is more correct to say there is a south magnetic pole at the earth's north geographic pole. The north pole of the compass is attracted to the south magnetic pole. Therefore, it points north.

All magnets have a north pole and a south pole. Like poles repel each other (try to move apart) and unlike poles attract (pull toward each other). Fig. 6-38 illustrates several shapes of magnets. The attraction and repulsion of magnetic poles is shown in Fig. 6-39.

There are lines of magnetic force connecting the north and south poles of a magnet. These lines of force are called flux. The space in which a magnetic force is operating is called a magnetic field. Magnetic flux will flow through any substance. It is not stopped by glass, mica, wood, air, or any other usual electrical insulating substances. Some substances, particularly soft iron, are better conductors of magnetic flux than other substances. This is the reason that certain parts of electric motors and generators are made

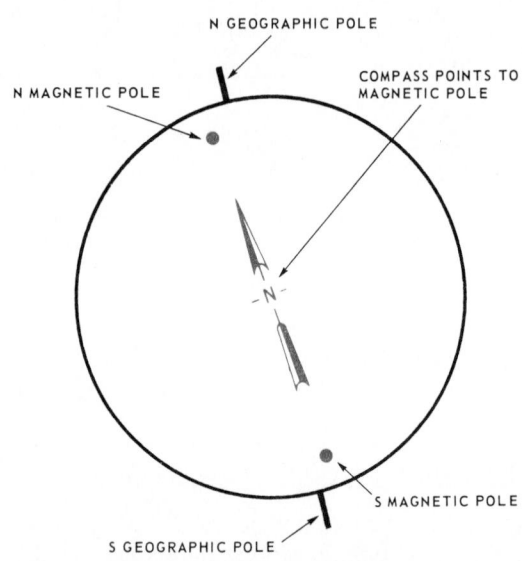

Fig. 6-37. A magnetic compass is a bar of magnetized steel mounted to turn freely on a vertical axis. It points in direction of earth's magnetic lines of force. It is more correct to say there is a south magnetic pole at the earth's north geographic pole which attracts the north pole of compass magnet.

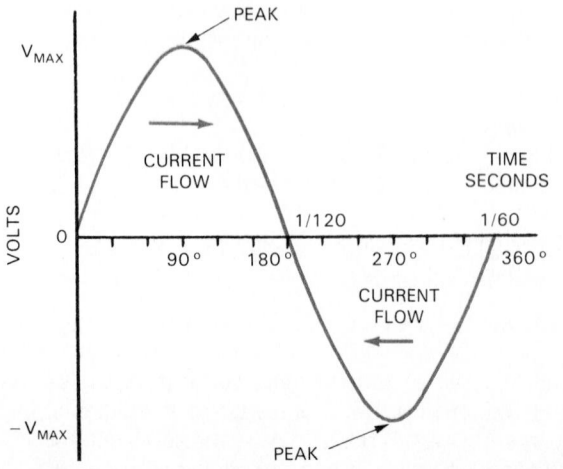

Fig. 6-35. A graph showing flow of 60 Hz alternating current (ac).

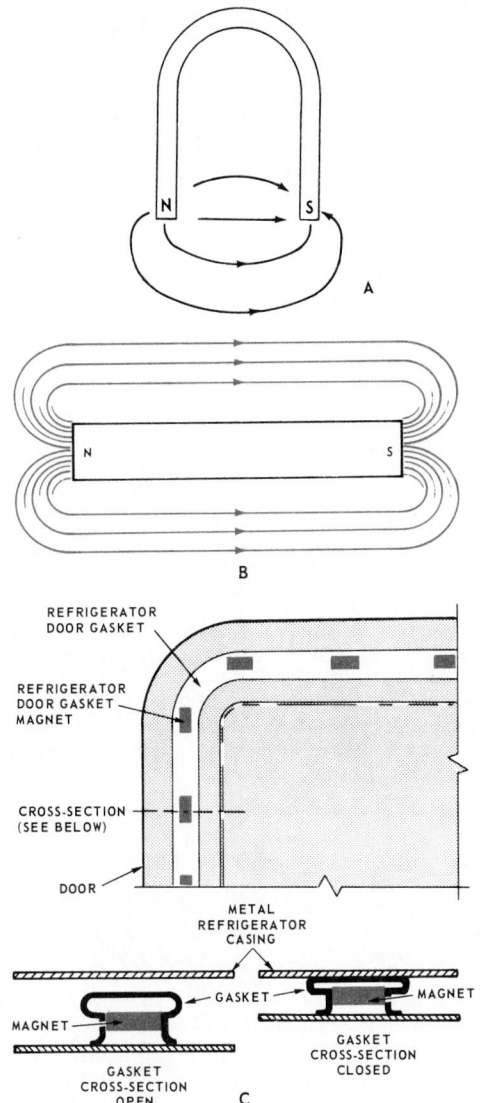

Fig. 6-38. Various shapes of magnets. A—Horseshoe shaped magnet. B—Bar magnet. C—Refrigerator door gasket magnet.

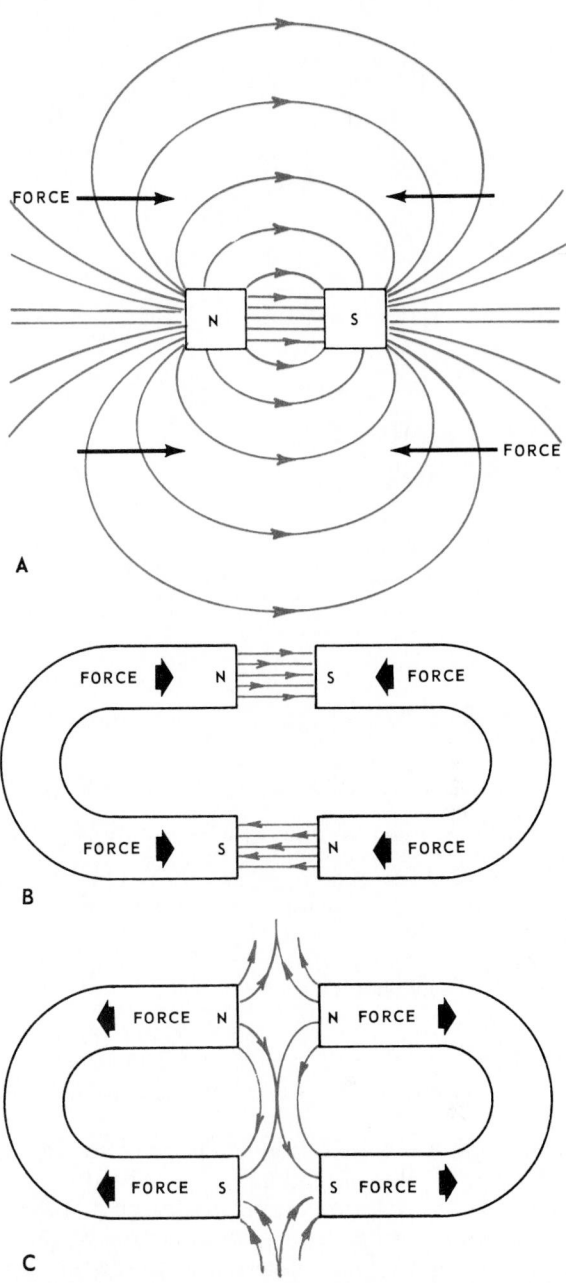

Fig. 6-39. Attraction and repulsion of magnetic poles. A—Magnetic flux around poles of a horseshoe magnet, shown end on. B—Magnetic flux provides a force which tends to pull unlike poles of magnets together (attract). C—Magnetic flux provides a force which tends to push like poles away from each other (repel).

of soft iron.

Instruments may be shielded from a magnetic field by encasing them in a ring or case of soft iron. The magnetic field passes around the instrument due to the fact that the soft iron is a good conductor of magnetic flux. Some non-magnetic watches use this method.

6-38 PERMANENT MAGNETISM

Permanent magnets are usually made of hardened steel. Once magnetized, they remain magnetized. There are some patented alloys of iron, aluminum, nickel, and cobalt which make strong and super-permanent magnets.

Magnetic lines (flux) are like stretched rubber bands. They tend to become as short as possible. This shortening force has many industrial applications.

Permanent magnets are used in some controls to provide a snap action for electrical contacts. They are also used in small control motors. In these motors, either the stator or rotor has permanent magnets to provide precision movement of a motor rotor (dc, servo motors).

6-39 INDUCED MAGNETISM

A property of magnetism is that it may produce magnetism in some other metals near it.

In Fig. 6-40, the permanent magnet is surrounded by its magnetic field, but the soft (low carbon), mild steel piece has no magnetism. If the mild steel piece is placed near the permanent magnet as in view B, or is touching it, as in view C, it too becomes a magnet.

Any material capable of being magnetized will become a magnet if it is placed in a magnetic field. This is called induced magnetism. Since the bar is soft iron, it will lose its magnetism when it is removed from the magnetic field.

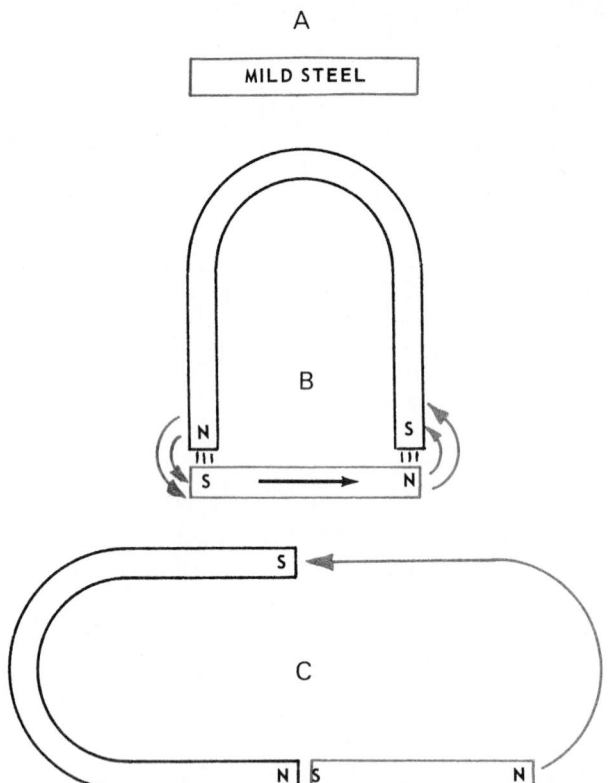

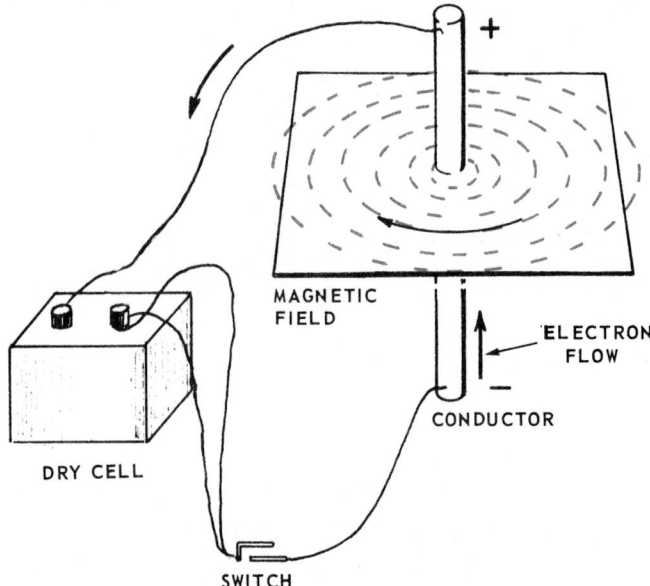

Fig. 6-41. Magnetic field encircles current flowing in a straight conductor. To demonstrate, conductor is passed vertically through the center of a sheet of cardboard. Iron filings are sprinkled over surface of cardboard. When ends of vertical conductor are connected to a dry cell, iron filings form pattern of circles around conductor. Changing the placement of cardboard will not change pattern of magnetic field.

Fig. 6-40. Induced magnetism. A—Nonmagnetized soft iron bar. B—Bar has become a magnet by being placed in a magnetic field. C—Bar is again magnetized by being placed against one pole of permanent magnet.

6-40 ELECTROMAGNETISM

If a current of electricity is passed through a conductor, the conductor becomes surrounded by a magnetic field, Fig. 6-41. If the current is turned off, the magnetic field will disappear.

If the conductor is wound around a piece of soft iron (a good magnetic conductor) and current is passed through the conductor, the soft iron becomes a magnet. Turning off the current (opening the circuit) stops the magnetic effect. The magnetic effect is called electromagnetism, and magnets formed in this manner are called electromagnets, Fig. 6-42. The iron part is called the core; the current-carrying conductor is called the winding.

The characteristics of a magnetic field surrounding a current-carrying conductor (wire) are shown at views A and B in Fig. 6-43. One very important characteristic is that the magnetic field travels around the wire in a specific direction, depending on the direction of electron flow in the wire. This, in turn, determines which end of the electromagnet is the north pole. To determine the direction of magnetic field travel around a wire, use the left hand rule as illustrated in A in Fig. 6-43.

Left hand rule: If one wraps the left hand fingers around a wire with the thumb pointing in the direction of electron flow (− to +), the fingers will point in the direction of the magnetic field travel.

The strength of an electromagnet is determined by the number of turns in the winding around the core and the number of amperes flowing through the winding. This strength is indicated by the term "ampere-turns." This is

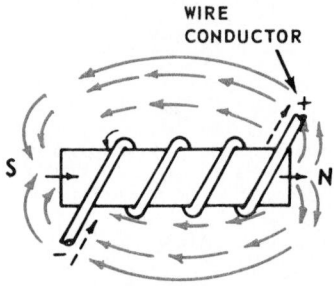

Fig. 6-42. Simple electromagnet has several turns of conductor (wire) placed on soft iron core. Individual circular magnetic fields are combined in core to form one magnet.

worked out by multiplying the number of turns in the winding by the amperes flowing through the winding.

Electromagnets are used in motors, relays, and solenoids, and in many other electromagnetic applications. Fig. 6-44 shows magnetic flow through the field poles and rotor of a four-pole motor. Note that the outer part of the stator to which the pole shoes are attached carries the same magnetic field as the pole shoes. This is the reason why the iron in the motor stator needs to be so thick and heavy.

6-41 MAGNETIC FIELD STRENGTH

Magnetic field strength depends upon the density of magnetic lines of force. As the lines of force become less dense away from the magnet, the strength of the magnetic field rapidly decreases. To illustrate this effect, magnets placed 1/2 in. from each other may be pulled together with a force of 20 lb. At a distance of 1 in., the pull would be only 5 lb. The magnetic field density is measured in units of Gauss (pronounced like "house").

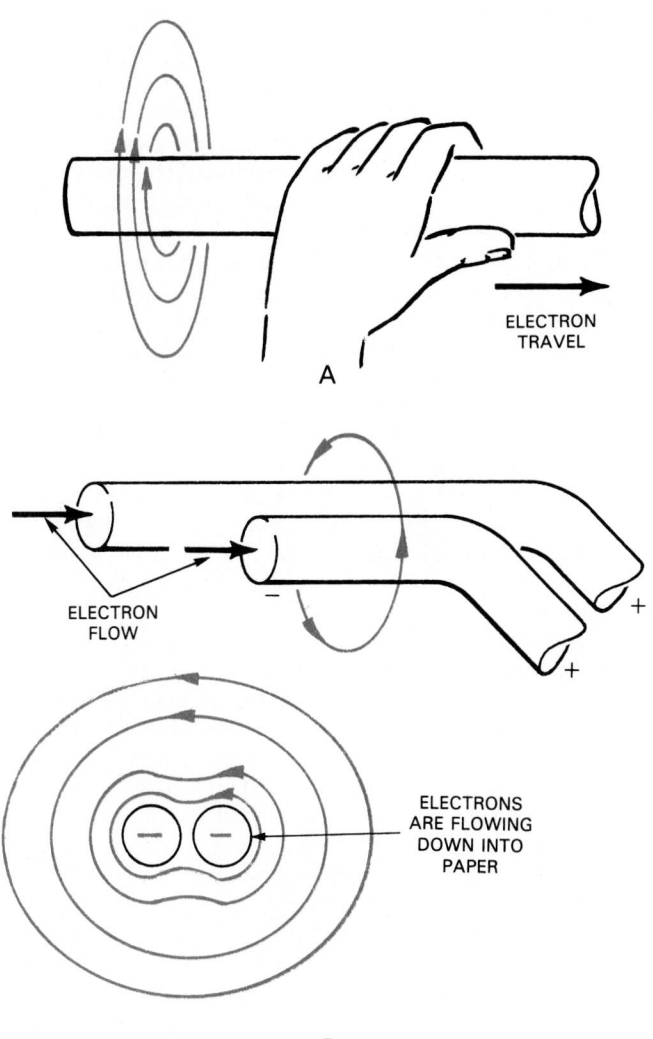

ELECTRON
TRAVEL

A

ELECTRON
FLOW

ELECTRONS
ARE FLOWING
DOWN INTO
PAPER

B

Fig. 6-43. Magnetic fields. A—Around a single wire. B—Around groups of wires. Note that left-hand rule establishes direction of magnetic field arrows.

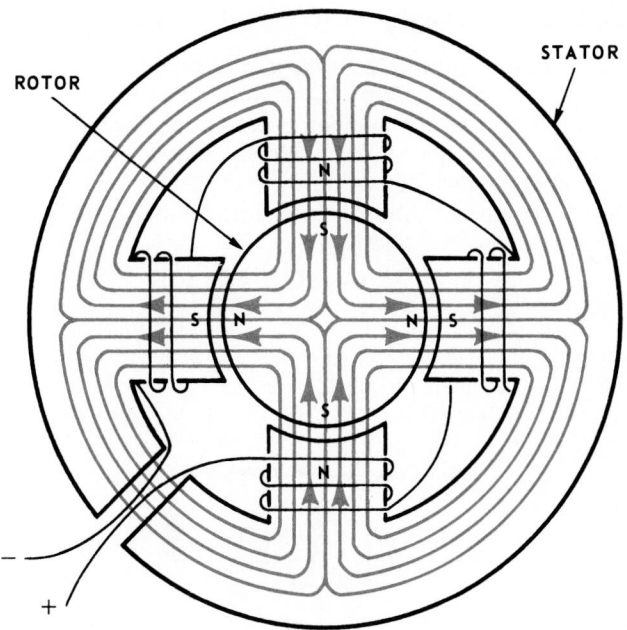

ROTOR

STATOR

Fig. 6-44. Magnetic lines of force flowing through rotor and stator on a four-pole motor.

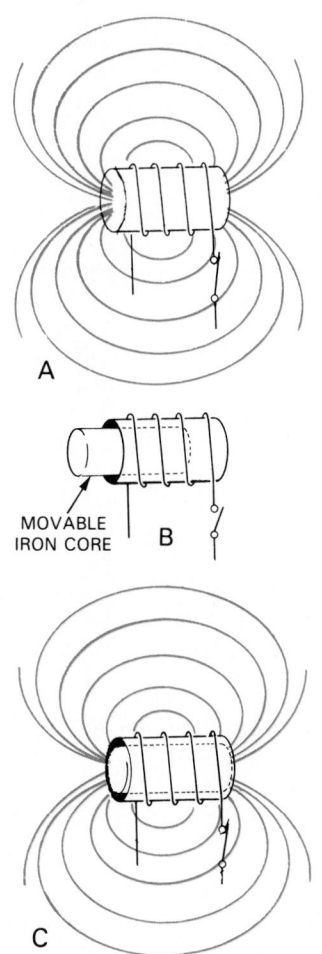

A

MOVABLE
IRON CORE

B

C

Fig. 6-45. A—Magnetic field is created if a solenoid is nonmagnetic material and current is flowing through windings. B—Solenoid fitted with an iron core with no electric current flowing. C—With current flowing, iron core is drawn into solenoid winding.

6-42 SOLENOID

If a coil is wound on a nonmagnetic substance (paper or plastic) and the coil is made to carry a current of electricity, the coil will become a magnet. See view A in Fig. 6-45. However, the magnetic effect will not be as great as it would be if an iron core were placed inside the coil, as illustrated in view B.

If this electromagnet is fitted with a movable iron core and the coil is made to carry a current of electricity, the soft iron core will be quickly drawn into the coil as shown at view C in Fig. 6-45. This is because magnetic lines of force are like stretched rubber bands. They like to be as short as possible.

The coil shown at views A, B, and C are commonly known as solenoids. The solenoid principle is used in many places in refrigeration and air conditioning work. They may be used to open and close valves, operate dampers, and make other desired movements in a mechanism.

A solenoid may be used on either alternating or direct current, since the lines of magnetic force tend to become as short as possible in either case.

6-43 POLARITY OF ELECTROMAGNETS

The polarity of an electromagnet is determined by the direction of flow of current (or electrons) through the winding on the core of the magnet. If the electromagnet is held in the left hand, Fig. 6-46, with the fingers grasping the magnet in the direction of flow of the electric current, the thumb will point to the north pole of the magnet. This is known as the left-hand rule.

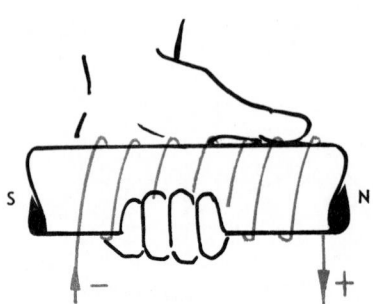

Fig. 6-46. Electromagnetic polarity. Use left-hand rule to determine which end of magnet is north pole.

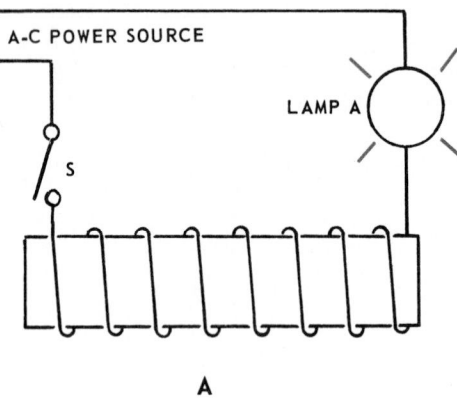

A

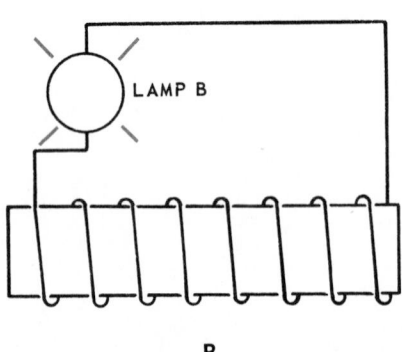

B

Fig. 6-47. Electromagnetic induction. Coils A and B are wound on soft iron cores. Coil A is connected through a switch to an alternating current source. Lamp in circuit controls amount of current flowing through coil A. Coil B is wound on a core adjacent to coil A, but they are not connected electrically. A low voltage lamp is connected to terminals of coil B. When switch to coil A is closed, both lamps will light, indicating that a current is being generated in coil B by electromagnetic induction.

6-44 ELECTROMAGNETIC INDUCTION

A magnet may cause another substance to become a magnet. This is shown in Fig. 6-40. Fig. 6-42 illustrates how the flow of electricity may produce magnetism. Electromagnetic induction uses both of these principles. It is the basis of operation of the induction motor which is most used for driving refrigeration and air conditioning compressors.

The principle of electromagnetic induction is illustrated in Fig. 6-47. Views A and B are separate circuits. The leads to the coil in view A are connected to an alternating current supply as shown. When switch S is closed, the lamps in views A and B will light. This is because the alternating field around the coil in view A induces a magnetic field around the coil in view B. As the magnetic flux builds up, then reverses, the coils of wire around coil B have an emf generated in them. The building up and collapsing of the magnetic field produces the effect of wires cutting across a magnetic field. This is the basic principle of the electric generator explained in Para. 6-48. This also explains the basic operation of the electric transformer, as described in Para. 6-75.

Magnetic induction is used in electric motors. The rotor is made of mild steel. Magnetism in the rotor is induced by magnetism created in the stator magnets or field windings.

If a current travels in a coil of wire, it will create a magnetic field. If there is a wire or another coil close enough to the first coil, the magnetic field cutting across these wires will create an emf in this wire or in this coil.

Remember: In order to create magnetic induction, either:
1. The magnetic field must be changing.
2. The magnet must be moving.
3. The wires must be moving.
4. Any one of these conditions or a combination of the conditions must be taking place.

6-45 PERMEABILITY—RELUCTANCE

Some substances, particularly soft iron, are better conductors of magnetic flux than other substances. The property of the material which determines its flux density under a magnetic field is called its magnetic permeability. It indicates the ease with which a material may be magnetized. Air is considered to have a permeability of one. Soft iron has a much greater permeability than air.

Reluctance is the resistance offered to the passage of magnetic lines of force through a substance.

Fig. 6-48 shows some fundamentals of magnetic behavior when glass or iron is placed near a magnet.

6-46 CAPACITANCE—CAPACITORS

Capacitance may be defined as a system of conductors and insulators which permit the storage of electricity (free electrons). This ability is indicated by the letter C. In referring to the capacities of a series or group of capacitors, they are usually designated as C_1, C_2, C_3, and so on.

The unit of capacity is the farad. The symbol for the farad is the capital letter F. A farad may be defined as a charge of one coulomb on the capacitor surface with a potential difference of one volt between the plates.

A farad is a rather large unit of capacity. Most capacitors (condensers) used in the refrigeration industry are rated in microfarads. A microfarad is one-millionth (.000 001) of a farad, or 10^{-6}. The symbol for the microfarad is (μf).

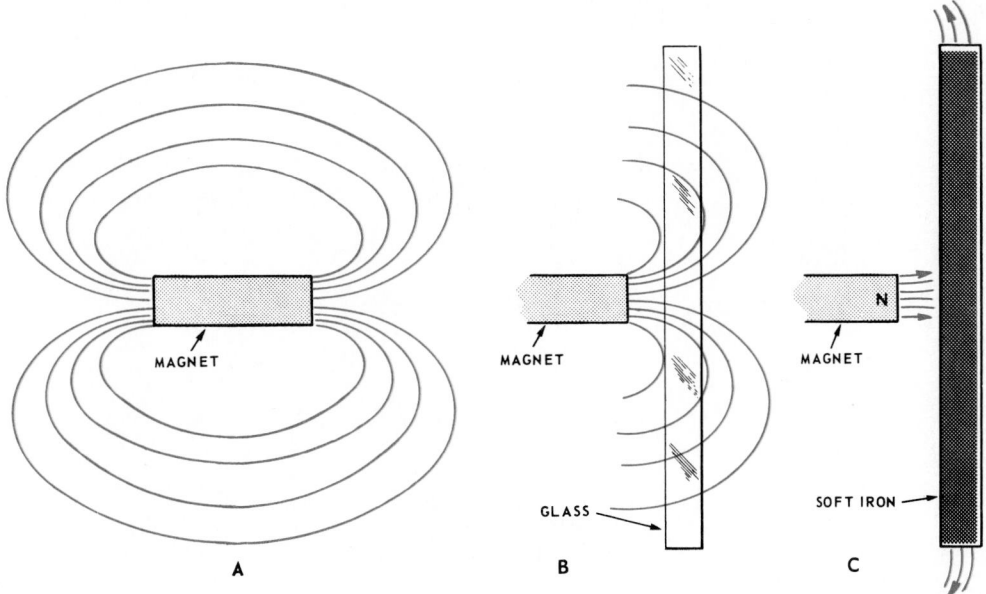

Fig. 6-48. Fundamentals of magnetic behavior A—Typical magnetic field surrounding a magnet. B—Plate of glass has no effect on shape of magnetic field. C—Effect of a soft iron plate placed in a magnetic field.

The Greek letter μ (mu) is used instead of the English m.

Any device that has capacitance (stores free electrons) is a capacitor. Large capacitors are large metal surfaces such as aluminum foil separated by insulating material (dielectric). See Fig. 6-49.

Capacitors are classified by the materials used for the dielectric. They include air, mica, paper, oil-filled, ceramic, and electrolytic. Fig. 6-50 shows several types of capacitors.

The capacitor, in series with the load, changes the sine wave and makes the current wave (lead) the voltage curve of an ac circuit.

Capacitors are used to help start motors, increase their efficiency, and improve the power factor.

The capacity value of capacitors in series may be expressed by the formula:

$$\frac{1}{\mu F} = \frac{1}{C_1} + \frac{1}{C_2}$$

μF = net capacitance (effective value)

C_1 = capacity of capacitor No. 1 in microfarads

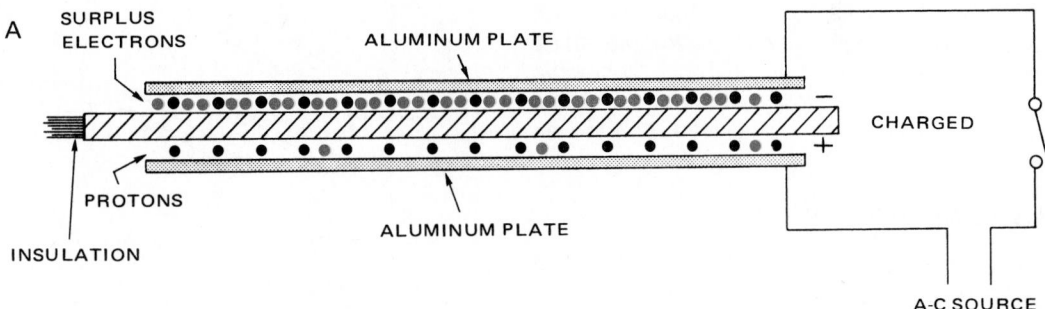

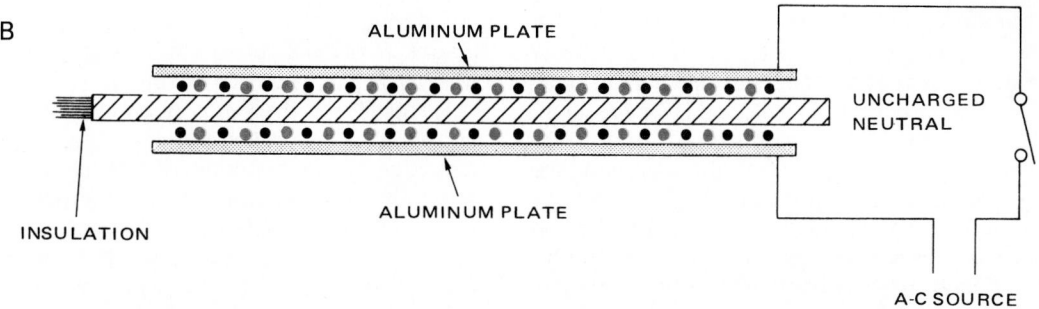

Fig. 6-49. Capacitor construction. A—Charged condition. B—Discharged condition.

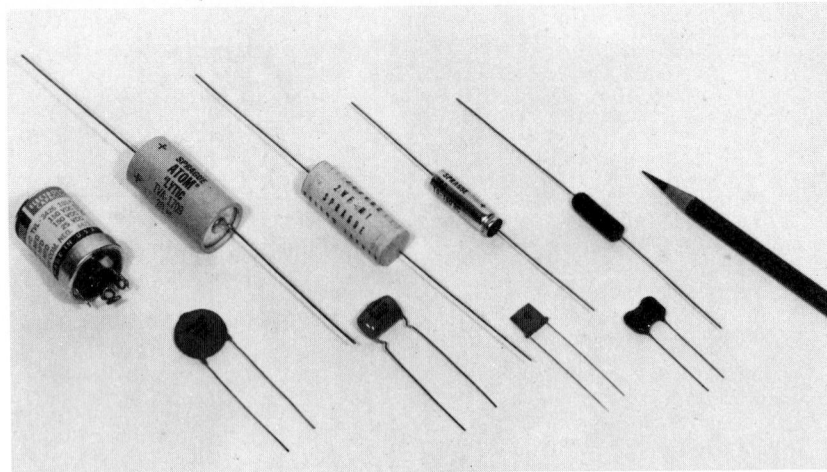

Fig. 6-50. Photo of some typical capacitors, including mica, paper, and electrolytic. (Sprague Electric Co.)

C_2 = capacity of capacitor No. 2 in microfarads

The capacity of capacitors in parallel may be expressed by the formula:

$$\mu F = C_1 + C_2$$

or by simply adding together the values of all the capacitors connected in parallel.

Large capacitors in a machine can store up dangerously high voltages. Before handling or replacing them, drain off the charge. A 20,000 ohm (20KΩ), 2 watt resistor may be used. A high voltage capacitor may store electricity at 600 V in a three-phase electrical motor circuit. This is dangerous.

6-47 REACTANCE

Reactance is the opposition to an alternating current flow in a circuit.

There are two kinds of reactance: capacitive reactance and inductive reactance. Capacitive reactance is caused by the capacity or condenser effect in the circuit. Inductive reactance is caused by the generation of counter-emf in a circuit. It is usually produced by a wire coil or an electromagnet.

6-48 ELECTRICAL GENERATOR

If a conductor is moved across a magnetic field, an electrical potential (emf) will be generated in the conductor. This is illustrated at views A and B in Fig. 6-51.

The loop in view A is not cutting across the lines of magnetic force—it is running parallel to them. No emf will be generated in the loop during the interval that it is in this position. View B shows the loop in a position so that the two sides are cutting across the magnetic field. In this position, an emf will be generated in both sides of the loop.

If the two ends of the loop are connected into an electrical circuit, no current will flow when the loop is in the position shown in view A. As the loop starts to revolve, it will cut across the magnetic field, and current will begin to flow. When the loop has revolved 90°, view B, current flow will reach its maximum. Once past this point, current flow will diminish until, after another 90°, no current will be generated. Then, as the loop continues to revolve, it will

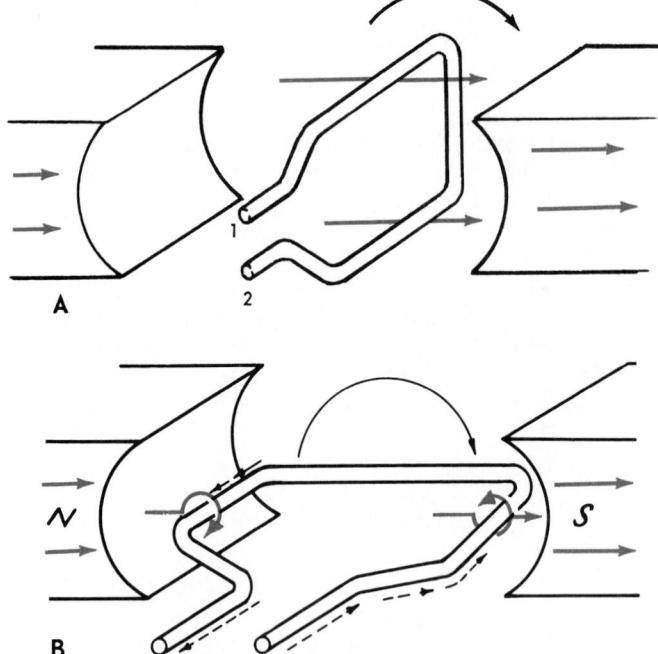

Fig. 6-51. Generation of current. Conductor loop has two sides, 1 and 2, and it is revolving clockwise. A—In this position, wires are traveling nearly parallel to lines of force and no emf is being generated. B—Loop has revolved 90° and current is being generated.

cut through the magnetic field at opposite poles of the magnet. This will cause the current generated in the loop to flow in the opposite direction.

Relating this theory to generator operation, Fig. 6-52 shows the voltage and current changes in the alternating circuit as the generator conductor (armature) revolves. The wires of the generator first cut the magnetic field in one direction, then in the other to generate an alternate (or opposite) flow. This happens each revolution of the armature as the conductors pass the magnetic poles and generate alternating current (ac). However, commutators and brushes on direct current generators rectify (correct)

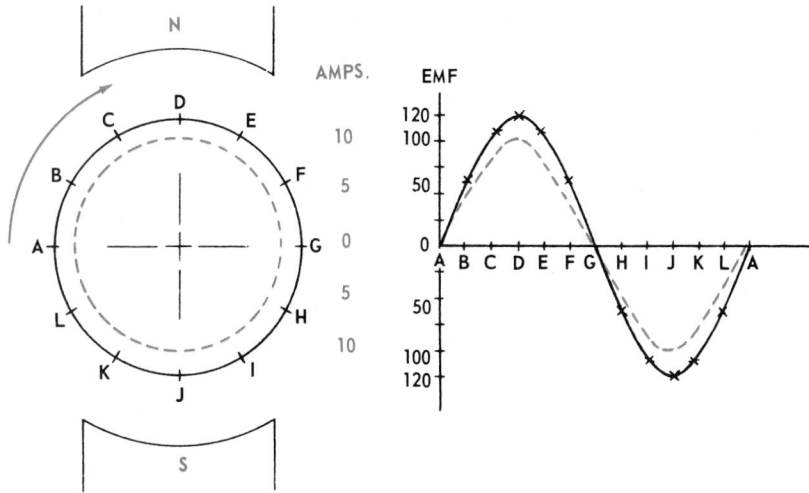

Fig. 6-52. Current (amps) and voltage (emf) changes in an alternating current circuit are graphed for one revolution of armature.

the current leaving the generator so that it flows in one direction only.

As current flows from the generator, a magnetic field surrounds the conductor in which the current is generated. This field is such that it opposes the movement of the conductor across the generator field. (See Fig. 6-56.)

This is the reason that it takes a great deal of power to drive a generator while it is producing a current. The greater the current produced, the greater the power required to drive the generator.

This principle is stated in Lenz's Law: "The magnetic effect surrounding the conductor in which a current is induced opposes the movement by which the current is induced."

6-49 COMMUTATORS

A conductor moving across a magnetic field will have an emf generated in it. In Fig. 6-52, the generated current in the wires of an armature will flow in one direction, then in the other as the conductors move from one field pole to the other. Alternating current is being produced.

In order to produce direct current, it is necessary to use a commutator and brushes in the generator. Fig. 6-53 illustrates an elementary type of armature fitted with a commutator and brushes.

The commutator and brushes provide a movable electric contact between the rotating armature wires and the stationary electrical device. The movable contact surface on the rotating armature is called the commutator, and the parts that contact the rotating parts of the commutator are called the brushes.

As the armature revolves, the commutator contacts are made in such a way that brush A always carries current into the commutator in a negative to positive direction. Brush B always carries current from the commutator to the charging circuit. Therefore, direct current generators are always fitted with commutators and brushes.

The starting mechanism on some ac motors with wound motor armatures also is fitted up with commutators and brushes. On some motors, these commutators and brushes are used only during the interval that the motor is starting. Universal motors (both ac and dc operation) use the com-

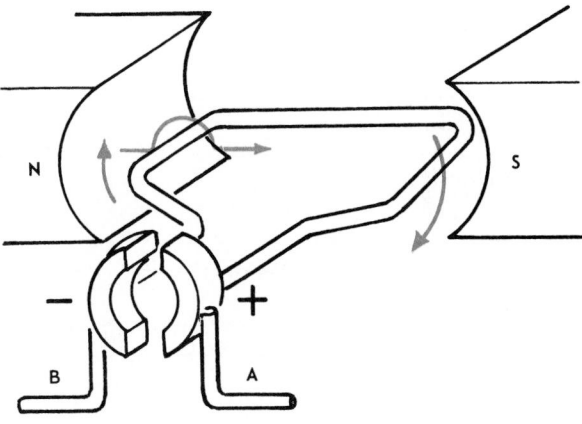

Fig. 6-53. A generator commutator. Brushes A and B contact ends of generating loop. Current from brushes always flows in same direction.

mutators and brushes whenever the motor is in use. Motors of this type are used in electric drills and mixers.

6-50 THE ELEMENTARY ELECTRIC MOTOR

Electrical energy is changed to mechanical energy in an electric motor. First, electrical energy becomes magnetism. Magnetism may then be used to cause motion.

Because like poles repel (N repels N and S repels S), and because unlike poles attract (N attracts S and S attracts N), motion can be produced by putting one magnet on a shaft and mounting another in a fixed position as shown in Fig. 6-54.

The bar magnet on the shaft will turn until its S pole is near the fixed N pole. This movement places the shaft bar magnet's N pole near the fixed S pole. (The fixed magnets, which are stationary, are called the stator. The other magnets, which rotate, are called the rotor or armature.)

The rotor will stay in the vertical position until the magnetism of the stator or rotor is reversed. Then the rotor will rotate another half turn.

The magnetism may be reversed by using electromagnets instead of permanent magnets, Fig. 6-55. With this setup, when the rotor reaches the vertical position, the alternating current reverses (due to its cycling) and stator polarity reverses. The rotor S pole is now near the S pole of the stator, and they will repel each other. The rotor will now turn or revolve one half a revolution or 180 deg.

The direction of movement of the armature (or rotor) depends on polarity, as shown at A in Fig. 6-56. The direction of movement of a current carrying conductor in a magnetic field may be determined by what is popularly called the "left hand motor rule." To use this rule, place the thumb and first two fingers of the left hand at right angles to each other with the index finger pointed in the direction of the magnetic force. Point the middle finger in the direction of the current flow in the conductor. The thumb is then pointed in the direction of the force or movement of the conductor. This is shown at view B in Fig. 6-56.

If the rotor turns a half revolution during one half of 60 Hz (cycles), then it turns a half turn during 1/120 of a second. It will then turn 60 revolutions per second or 60 × 60 seconds or 3600 revolutions in one minute. The two-pole motor (3600 rpm) has become a popular hermetic motor for refrigerating and air conditioning units. Its construction is shown in Fig. 6-57.

If four poles are used in the stator, the motor will turn only 1800 rpm. That is, the rotor will only turn one fourth revolution in 1/120 of a second, as shown in Fig. 6-58. Most open motors are of this design and some hermetic motors also use the four-pole design principle.

Speeds of 3600 rpm and 1800 rpm are called synchronous speeds. Under actual conditions, the 3600 rpm motor usually operates at approximately 3450 rpm, while the 1800 rpm motor operates at approximately 1750 rpm. This reduction in speed is due to a slight magnetic slippage, depending on the load. An overloaded motor will slow down slightly.

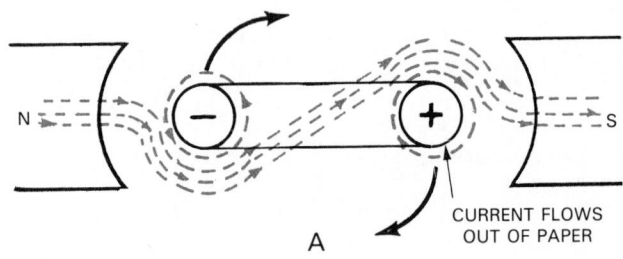

A

CURRENT FLOWS OUT OF PAPER

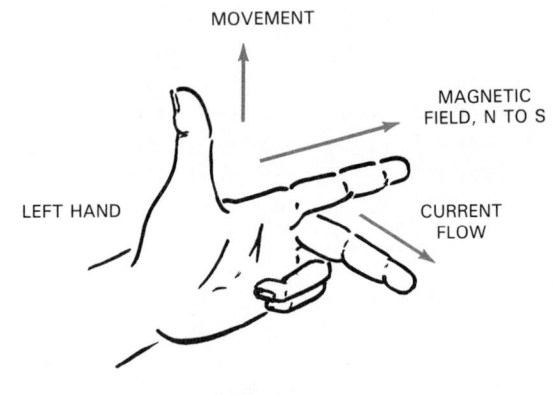

MOVEMENT

MAGNETIC FIELD, N TO S

CURRENT FLOW

LEFT HAND

B

Fig. 6-56. Left-hand motor role. A—Magnetic lines of force from north pole to south pole merge with magnetic field around conductor. The combined fields cause movement of rotor. B—Position fingers of left hand as shown, and thumb will point in direction of movement.

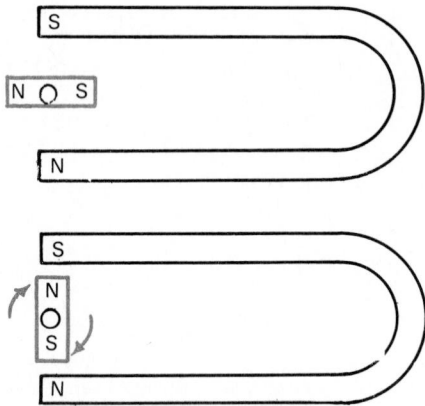

Fig. 6-54. Magnet mounted on an axis will turn when put in another magnetic field. Note direction of movement of pivoting magnet.

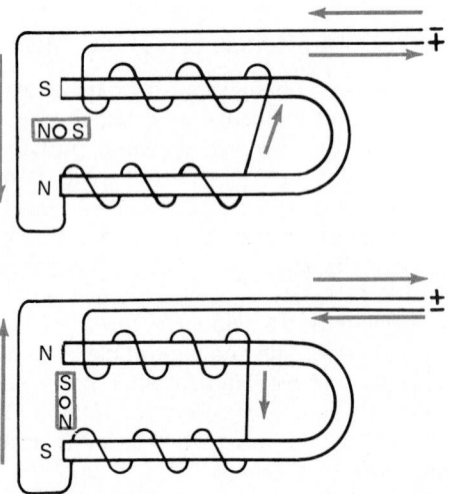

Fig. 6-55. Elementary electric motor. By reversing flow of current through field (stator) winding, its magnetic polarity will be reversed. This causes rotor magnet to turn on its axis.

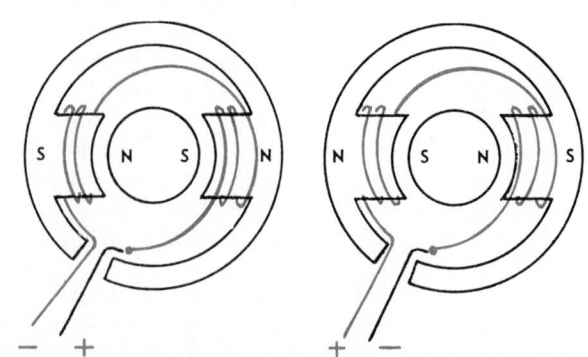

Fig. 6-57. Two-pole stator motor. Rotor will make a half turn with each half of current cycle.

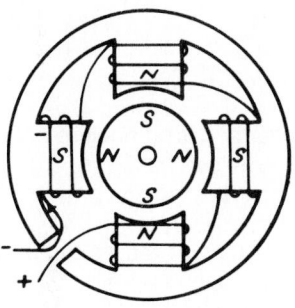

$\frac{1}{2}$ CYCLE

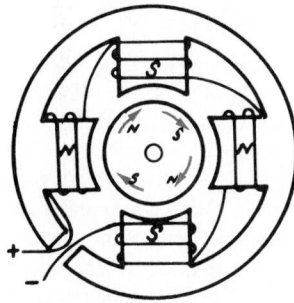

OTHER $\frac{1}{2}$ CYCLE

Fig. 6-58. Four-pole stator. Armature makes one half turn as current completes one cycle in field windings.

The stator of an induction motor produces a rotating magnetic field. The rotor cannot keep up with the field.

The electric motor shown in Fig. 6-59 is an open motor of the capacitor-start type. It can be used on external drive refrigeration compressors, pumps, and the like. See Chapter 7 for further information on various types of motors used on refrigeration and air conditioning machines.

Fig. 6-59. Section through motor suitable for use on external drive refrigeration and air conditioning installations. A—Sleeve bearing. B—Shaft extension end. C—Ball bearing assembly at other end of shaft. (Magnetek)

6-51 COUNTER EMF

A running motor develops an emf in the rotor bars or windings. This is called a counter emf. It opposes the applied emf which is driving the motor.

Counter emf depends upon the speed of the rotor. Under no-load, synchronous speed, the counter emf practically balances the applied emf. As the load increases and the speed decreases, the counter emf drops. As a result, the applied emf sends more current through the windings, tending to maintain the speed constant.

If the motor is slowed down considerably by a heavy load, the current supplied will increase greatly due to the reduced counter emf. The motor then will overheat. Under continuous overload, it is likely to burn out.

If the rotor is locked so it cannot turn and the current is applied, the current will be very high. A motor will quickly burn out under this "locked rotor condition." Counter emf also occurs in all coils and electromagnets.

6-52 INDUCTANCE

A magnet, formed by winding a coil of wire (conductor) on a soft iron core, will become magnetized if an electrical current is passed through the conductor. The entire coil is surrounded and saturated by magnetic lines of force (flux).

As the current is switched on in such a coil, the magnetism is not built up instantly. There is a delay of perhaps a few hundredths of a second during which time the current continues to increase until it reaches its full value. This value depends on the resistance and emf of the circuit. Likewise, when the switch is opened and the current is turned off, it does not stop flowing instantly. The magnetic lines build up as the switch is turned on and collapse as the switch is turned off.

There is a tendency to generate within the coil an electromotive force which counteracts the change in the current flow. This counteracting force is the counter electromotive force or counter emf (discussed in Para. 6-51).

This principle of inducing a voltage in a coil due to the change in the rate of flow of current in the coil is called inductance. It acts much like a flywheel on a piece of machinery. The flywheel requires power to give it a rotating motion and, likewise, it gives up power if it is forced to stop.

6-53 INDUCTORS

Induced magnetism is useful in electric motors. By putting electrical windings on the field poles, magnetism can be created or induced in the rotor. There is a slight time delay in this induction. Therefore, if the rotor has once started to turn, it will continue to do so because the field pole (stator) always has the opposite polarity.

During the time the field pole changes its polarity, it induces an opposite pole in the rotor. This repels the rotor pole toward the next stator pole, which has become an opposite pole in the meantime. The rotor must turn the full distance between the two poles before the induced magnetism can be changed or the motor will not keep running. Consider that it takes only 1/120 of a second to change the magnetism from a full strength N pole to a full strength S pole.

This principle is used in the design of the split-phase motor. There are two windings in this motor. One is a starting winding and the other a running winding. The starting winding is a smaller diameter wire than the running winding. However, it has a greater number of turns. As a result, its magnetic inductance will be greater than for the running winding. This means that the starting winding is always behind the running winding in both building up and stopping its magnetic field. This type of inductance is generally known as self inductance.

Mutual inductance, on the other hand, is another type of flow of electricity and a magnetic field in another coil that is close to it. The principle of mutual induction is used in all induction type motors.

In induction motors, the magnetic effect of the current flowing in the field windings induces the current in conductors on the rotor. The magnetic effect of this induced rotor current causes the rotor to revolve. This characteristic of the flow of current, in an electromagnet, to resist the flow when the current is turned on and to resist the stopping of the flow when the current is turned off, depends on Lenz's Law. The law states that the polarity of an induced voltage is such that it opposes the motion of the flux inducing it. (See Para. 6-48.)

6-54 ELECTRONICS

Electronics concerns electron flow through gases, vacuums, and semiconductors. Electronics was developed with the discovery of vacuum tubes and gas-filled tubes.

It was found that electrons would flow from a heated element in a tube to another element if a potential difference existed between the two elements.

Fig. 6-60 shows a simple circuit of this type. The first radios used the vacuum tube to control and amplify the radio signals. These tubes were fragile and relatively large. The invention of the solid state semiconductor as a replacement for tubes has greatly enlarged the application of electronics.

In the following paragraphs, some electronics will be explained. This includes parts such as:

1. Semiconductors.
2. Diodes—diacs.
3. Rectifiers.
4. Silicon controlled rectifiers—triacs.
5. Transistors.

6. Sensors.
7. Thermistors.
8. Amplifiers.
9. Transducers.
10. Thermocouples and thermoelectric devices.
11. Photoelectric devices.
12. Integrated circuits.
13. Integrated circuit boards.

With the development of rather complicated automatic controls used on refrigerating and air conditioning appliances, more and more electronic devices are being used. It will be necessary to understand these devices in order to understand the circuits in which they are used.

6-55 SEMICONDUCTOR APPLICATIONS

Solid-state electronics devices make up most of the modern electrical control system. They are important in computer control systems and in devices which must withstand hard use.

Semiconductors are of two general types:

1. Intrinsic semiconductors, which are pure substances like silicon and germanium or combined substances like lead sulfide. These are particularly useful as thermometers and as other temperature sensing devices.
2. Extrinsic semiconductors, which are combinations of intrinsic semiconductors with very small "impurities." These semiconductors are very sensitive to electrical forces and are the basic materials used in electronics. Thermoelectric refrigerators (discussed in Para. 3-17) use them to produce cooling.

6-56 DIODES, DIACS

A solid-state diode is a solid wafer or capsule, composed of two materials, that allow the electrons to flow through in one direction only, Fig. 6-61. The diode acts as an electron flow check valve.

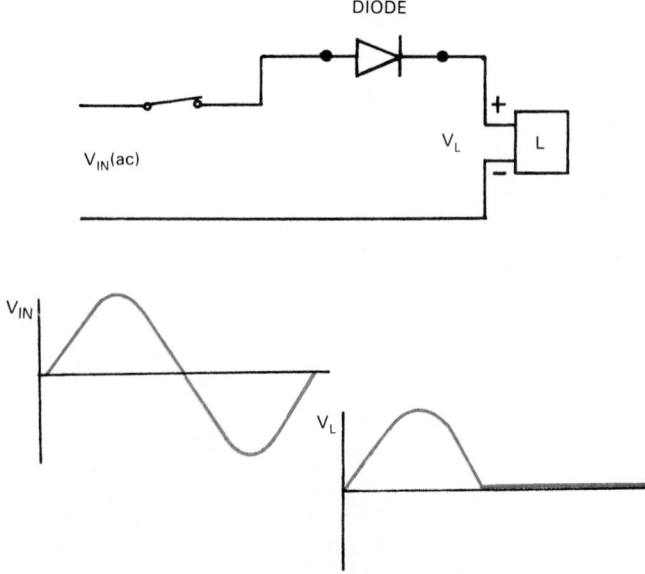

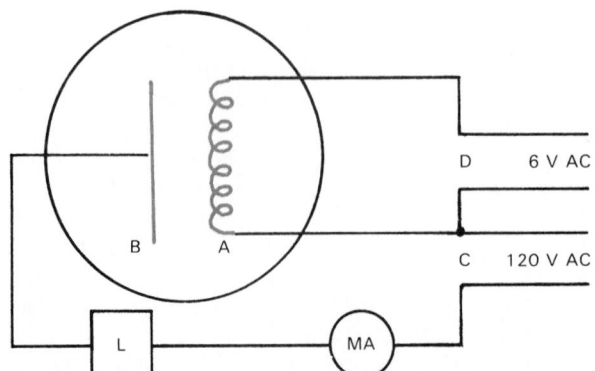

Fig. 6-60. Diagram of vacuum tube rectifier. A—Heating element from which electrons will flow. B—Plate to which electrons flow from A. C—120 V ac power source. D—6 V ac to operate heating element. L—Direct current load. This might be a storage cell being charged. MA—Milliammeter, which shows rate of dc flow.

Fig. 6-61. When an alternating voltage is applied to a diode (V_{IN}), the diode allows current flow only when the voltage is positive. As a result, voltage at load (V_L) reflects only the positive component of input voltage.

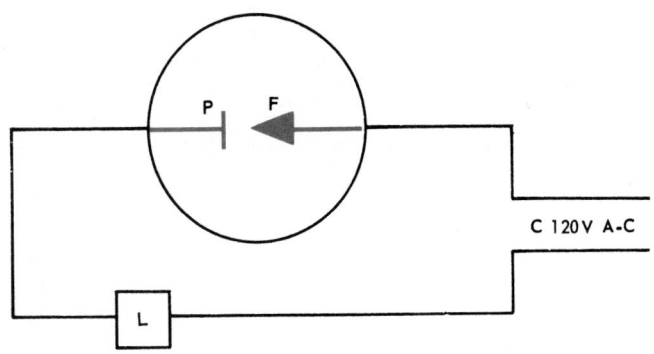

Fig. 6-62. Diagram of a vacuum tube serving as an alternating current rectifier or diode. Note that electrons will only flow from pointed filament F to plate P. L is direct current load, which may be a storage battery being charged.

Vacuum tubes can also serve as diodes, Fig. 6-62. Electrons will flow from the pointed filament to the plate on half the cycle. During the other half cycle, the electrons will not flow from the plate to the filament.

A diac (ac diode) is similar to a diode; however, it allows current to flow in both directions. The diac will not conduct current until a ''preset'' voltage is exceeded. See Fig. 6-63. Schematically it operates similar to two diodes in parallel. See Fig. 6-64. The diac is used in ac circuits where both halves of the sinusoidal voltage are required. They are used often as part of the switching circuit in motor controls (see Chapter 8).

6-57 RECTIFIERS

Rectifiers are electronic valves which permit the flow of current through them in one direction only. These devices change alternating current to an output of direct current.

A simple rectifier uses only one half of the sine wave. To use both halves of the sine wave and still produce only direct current, one must use four diodes. See Fig. 6-65.

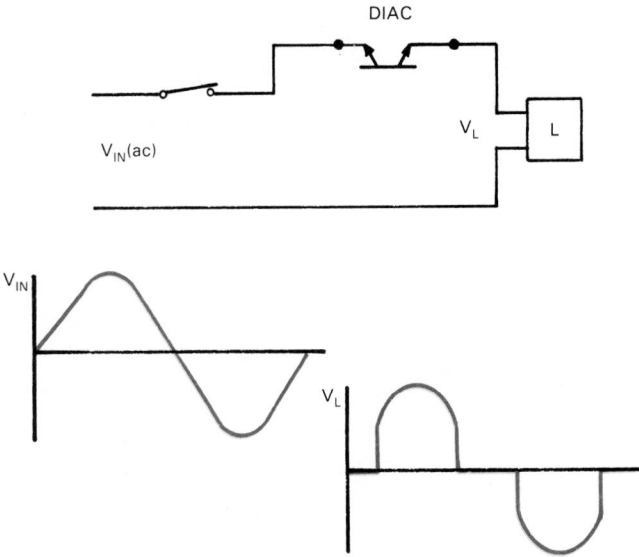

Fig. 6-63. When an alternating voltage (V_{IN}) is applied to a diac, the diac allows current to flow after a preset voltage level. Result is voltage at load (V_L) as shown.

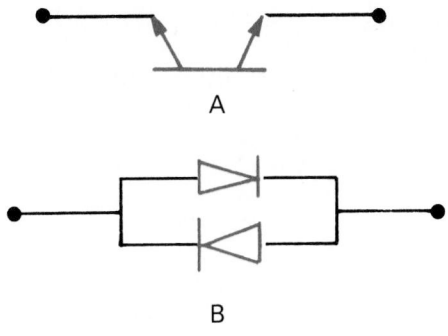

Fig. 6-64. Schematic representation of a diac. A—Standard schematic. B—Dual diode representation.

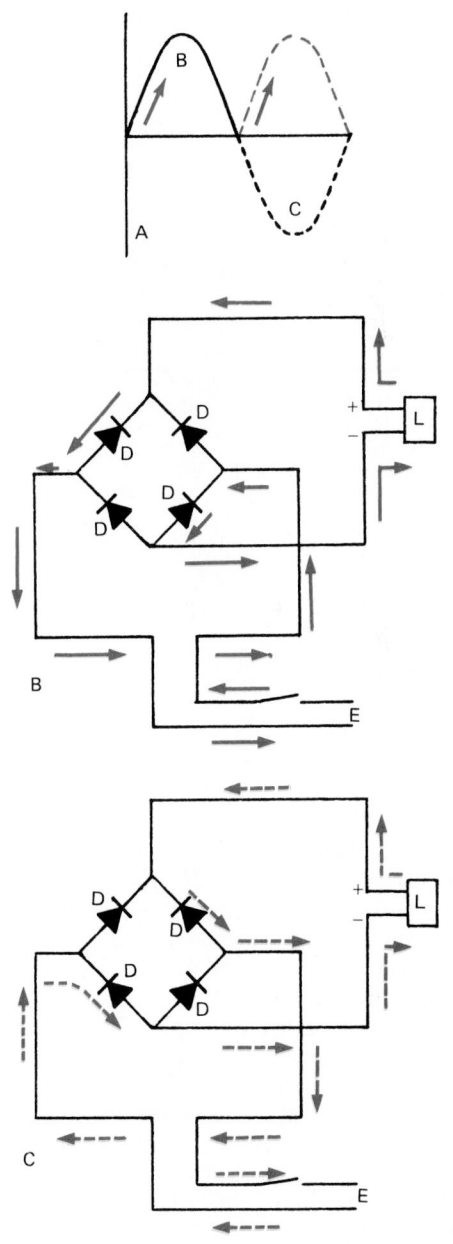

Fig. 6-65. Circuit for full wave rectifier. A—Note how two halves of an ac cycle are made to provide dc during all parts of cycle. B—Solid red arrows show current flow for one-half wave. C—Dashed red arrows show current flow for other half wave. D—Four diodes are needed. E—60 cycle ac power supply. L—The dc load.

6-58 SILICON CONTROLLED RECTIFIERS, TRIACS

Among semiconductors using silicon are diodes and silicon controlled rectifiers (SCRs). The SCR has three connections as shown in view A of Fig. 6-66. It conducts current from A to C when both the voltage at A is greater than at C, and a preset voltage has been applied at B. If these conditions are not met, the device is "off" and not conducting current.

SCRs are used extensively in electrical motor controls. They are also the basic elements used to convert ac voltage to dc voltage in inverter devices. SCR devices are now used in many applications where relays were formerly used for switching.

A triac, Fig. 6-66, view B, is similar to an SCR. However, it can conduct current in both directions, from A to C and C to A, when a preset voltage is applied at B.

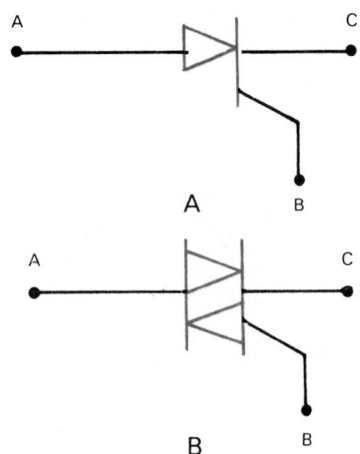

Fig. 6-66. Schematics of semiconductors. A—Silicon controlled rectifier. B—Triac.

6-59 INVERTER

Electrical energy stored in a battery is available as direct current (dc) energy. The voltage supplied by the battery is a steady voltage, gradually decreasing with time as the charge is drained from the battery. An electric motor powered by a battery must be a dc motor.

Since dc motors are heavier and usually more expensive than ac motors, it is often advantageous to change the voltage from the battery so that an ac motor can be used. The device to do this is called an inverter. The device does the opposite of the rectifier discussed in Para. 6-57, which converts ac power to dc power. Fig. 6-67 indicates the inverter function, converting direct current (dc) to alternating current (ac).

Older electrical systems used a combination of a dc motor connected to an ac generator to do this inverting. Newer solid-state electronic devices do this without any mechanically moving parts. The basic elements used in a solid-state inverter are:
1. A crystal which oscillates at the frequency of the ac power required.
2. A switching circuit using silicon controlled rectifiers (SCRs) to switch the dc power on and off.

A simple inverter, using a set of standard diodes, pro-

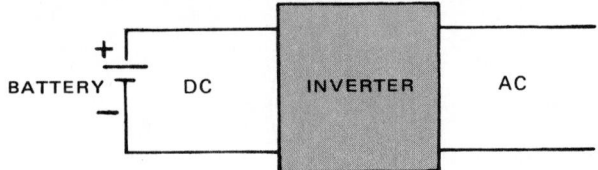

Fig. 6-67. Inverter converts direct current (dc) supplied by battery to alternating current (ac).

duces a square wave output.

Most ac motors and controls are designed to operate only with alternating (ac) power, similar to that provided by the power company. These devices will operate with a square wave but not as efficiently, and their lifetimes will usually be reduced.

An inverter is usually required in solar electric energy systems, since the output of solar cells is dc power.

6-60 TRANSISTORS

A transistor is a three-layer sandwich of two different components which consist chiefly of silicon semiconductor material. Electrically, the three wafers are connected as shown in Fig. 6-68.

The materials are labeled for their properties. P is for positive, meaning a lack of electrons (it has "holes" ready to receive electrons). N is for negative, meaning the material has a surplus of electrons.

Three conductors are connected to the transistor. One attaches to the base or middle wafer, one connects to one of the outer wafers, called the emitter. The third connects to the collector. The outer two wafers are of the same material. The base material (middle wafer) is different.

A small electron flow from the base to the emitter will control a large electron flow from the emitter to the collector. The device, therefore, acts as a valve and as a relay. An electron signal circuit from base to emitter may control a collector electron flow as much as 1000 times larger than base to emitter.

All three parts are semiconductors having added substances to give the desired characteristics. View A in Fig. 6-68 shows, in a schematic way, the basic construction of a transistor. The two basic types of transistors are shown in Fig. 6-69.

A transistor connected as an amplifier is shown in Fig.

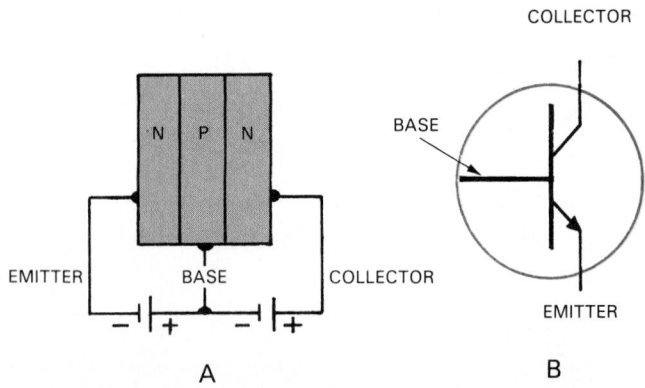

Fig. 6-68. A—Transistor in a circuit. B—Symbol for a transistor.

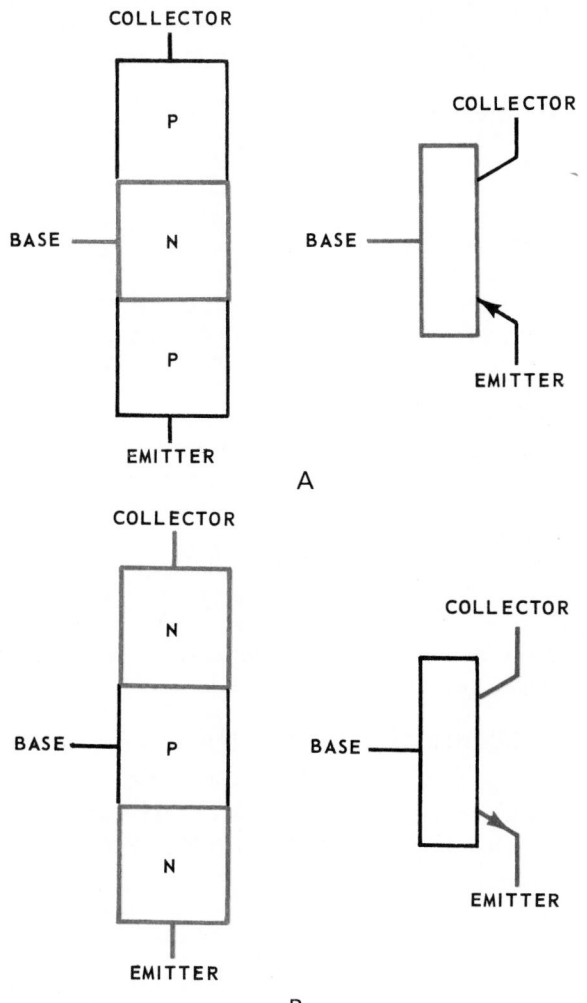

Fig. 6-69. Two basic types of transistors. A—PNP transistor. B—NPN transistor. In A, current flow is controlled by negative charge carried in base. In B, current flow is controlled by positive charge in base.

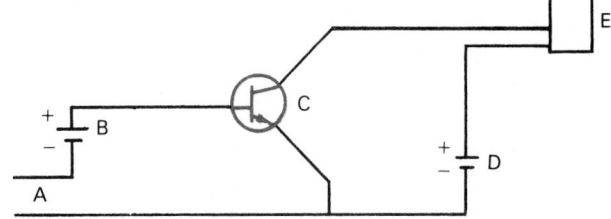

Fig. 6-70. Circuit diagram showing transistor used in amplifier circuit. A—Signal (current) to be amplified enters here. B—Battery. C—Transistor. D—Battery. E—Load, which receives amplified current.

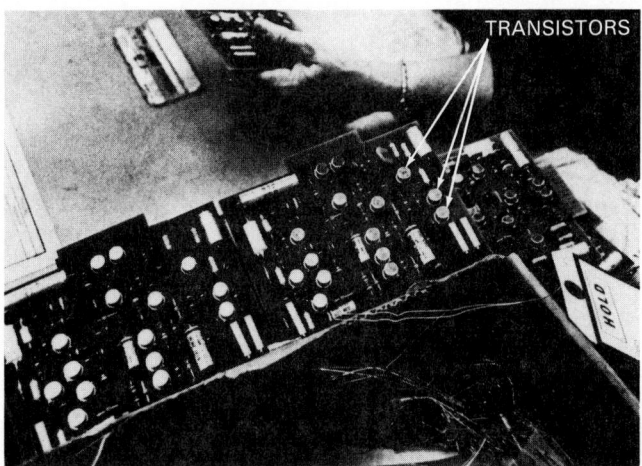

Fig. 6-71. Circuit board with transistors and other solid-state devices. (Acme Electric Corp.)

6-70. The low energy signal enters the circuit at the left. The signal is amplified by the action of the transistor and energy supplied by the batteries. The amplified signal (at the load) is shown leaving the circuit at right.

Note the transistors in the circuit board, Fig. 6-71. Each transistor amplifies one signal. A unit may control parts of a refrigeration system.

6-61 SENSORS

In many electronic circuits, the control signal is triggered by a sensor. The sensor is a solid-state semiconductor material which controls electron flow as its temperature or pressure changes. Pressure sensors can respond to pressure changes in the refrigerating system.

A typical, completely automatic automotive air conditioner uses three sensors in its circuitry:
1. For outside (ambient) temperature.
2. For in-the-car temperature.
3. For the air discharge duct temperature.

6-62 THERMISTORS

A thermistor is a solid-state semiconductor which allows fewer electrons to flow through it as the material's

temperature increases. Most thermistors are made of lithium chloride or doped barium titanate.

The resistance changes about 3 percent for each degree F change (6 percent for 1 °C). In some circuits, the thermistor is used instead of a bimetal strip or temperature-sensitive power element.

The thermistor is used in three ways:
1. A temperature-operated electric circuit control.
2. To measure temperatures.
3. To stop the electric power flow to a motor if the motor windings' temperature increases to the danger point.

In a typical thermistor circuit, Fig. 6-72, the thermistor temperature sensor, C, is used to control the temperature of a room or conditioned space which is heated by a 120 V ac electric resistance heater.

The temperature control, A, is set to the desired room temperature. If the room is below this desired temperature, the sensor, C, will change current flow to transistors D and E. Here the changes are amplified and the thermal relay heater, I, will cause the contact points at J to close. This action brings the electric space heater, K, into operation.

As soon as the desired temperature is reached, the sensor, C, will cause the current to the heater coil to be switched off. Then contact points at J will open and stop the flow of current to the space heater. These devices are very sensitive and will maintain the space temperature within a fraction of a degree.

There is a special thermistor which increases its resistance as the temperature rises. However, it changes from low resistance to high resistance within two degrees. It can,

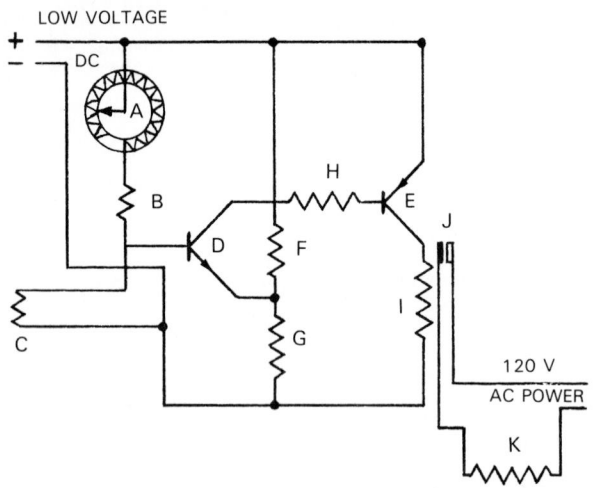

Fig. 6-72. Typical thermistor circuit. A—Temperature control knob, which controls variable resistance. B—Fixed resistance. C—Thermistor temperature sensor. D and E—Transistors. F and G—Bias resistors. H—Current-limiting resistor. I—Thermal relay heater. J—Thermal relay, K—120 V ac electric space heater. Sensor will maintain temperature of heated space within close limits.

therefore, be used as a switch. At present, these operate between 150 °F (65 °C) and 356 °F (180 °C).

The thermistor may be used to control crankcase heaters (shuts off current when oil temperature reaches design conditions). It also may be used to sequence heating systems. The number of heaters in operation would increase or decrease as heating need increased or decreased.

The thermistor can do the same for a cooling system. It may be used to control a defrosting system on an ice cube maker release circuit. It is small enough to be put in motor windings to protect them from too much heat.

6-63 AMPLIFIERS

An amplifier is an electronic device that, upon receiving a small input signal, will increase it to produce a larger output signal. For example, the signal from a phonograph record is not large enough to be heard. It must be amplified.

Amplifiers are often used in control systems when a small signal from a sensor needs to be increased to a level which is high enough to control another device. See Fig. 6-73.

Amplifiers can also be used to determine the difference between a changing input voltage and a constant base voltage. These are called differential amplifiers.

6-64 TRANSDUCERS

The term, transducer, is used to identify a great variety of devices which are sensitive to changes in the intensity of some form of energy. The transducer then responds by controlling the intensity of some other form of energy.

Transducers may be operated by pressure, temperature, fluid flow, vibration, electrical potential, and others.

A small varying current flow through a transducer may be amplified by an amplifier. The amplified current may then operate a control circuit.

An application of a transducer is shown in Fig. 6-74. In this application, a pressure transducer is connected to a pipe, A, which is carrying fluid under pressure. The transducer, B, will change pressure variation into electric current variation.

In the amplifier, C, electric current variations are amplified and connected to D. This is a relaying device which may translate what was a weak pipe pressure variation into E. E may be a recorder, pressure gauge, signal light, oscilloscope, or some other signal indicating device.

6-65 THERMOCOUPLE AND THERMOELECTRIC

Thermocouples may be used to measure temperatures or to operate controls. Their principle of operation depends on the fact that if two dissimilar (different) metals are con-

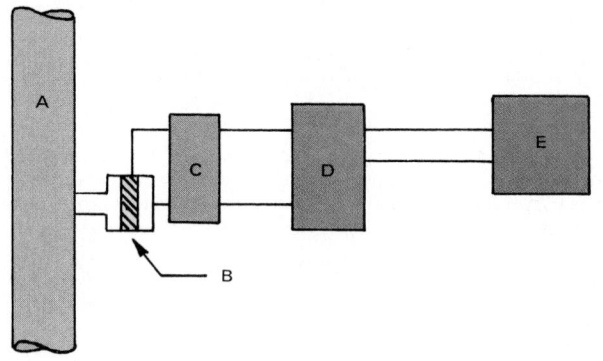

Fig. 6-74. This transducer application indicates pressure in a pipe. A—Water pipe. B—Transducer. C—Amplifier. D—Relay. E—Indicator or pressure control.

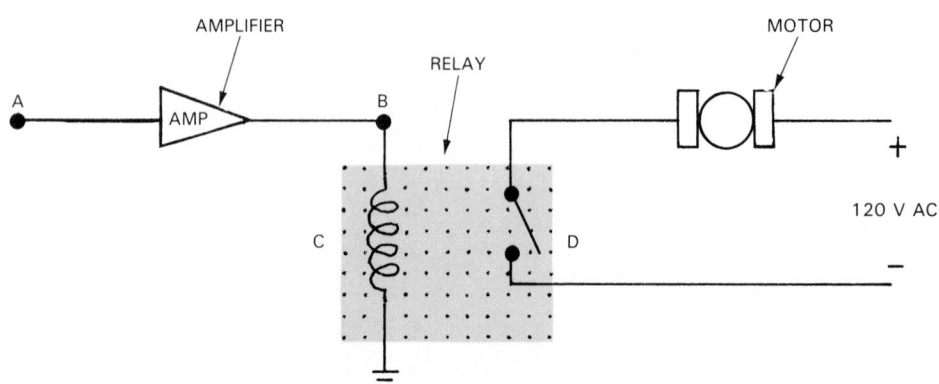

Fig. 6-73. Typical amplifier circuit. Small signal received at A is amplified to produce a voltage at B. Voltage at B is strong enough to energize coil C and close relay switch D. Relay then turns motor on.

nected together and the point of connection heated, an emf (voltage) difference will be developed across the other ends of the two metals. Copper and iron may be used as the two different metals. However, other thermocouples have been developed using tungsten and rhenium as well as other materials.

If a thermocouple is connected to a sensitive voltmeter, the voltage indicated will vary with the temperature of the junction. This principle is the basis of operation of the thermocouple thermometer. The instrument is calibrated to either Fahrenheit or Celsius.

If two or more such junctions are connected in series and connected to a sensitive voltmeter, the voltage indicated will vary directly with the number of the junctions. A multiple thermocouple installation can generate as much as 500 mV. It can, therefore, operate special solenoids which will operate valves on a gas furnace without the need of connecting the system to an electric utility. It may also be used as a safety device to shut off a gas supply if a pilot light is extinguished.

The electrical efficiency of a thermocouple is quite low. It is not an efficient way to generate electricity.

Typical thermocouple applications are shown in Fig. 6-75. In 1820, the German physicist, Thomas J. Seebeck, discovered that when a closed circuit is made through two different metals in contact with each other, an electric current will flow in the circuit when heat is applied to one of the junctions.

In 1834, Jean Peltier discovered that, if direct current is passed through a junction of two dissimilar metals, the junction becomes either hot or cold, depending on the direction of the current flow.

Emil Lenz, in 1837, showed the importance of both Peltier's and Seebeck's discovery. He placed a drop of water on the junction of two dissimilar metals. When current passed in one direction, the drop of water froze. When the current was reversed, the ice melted and the water was warmed.

The principle of the thermocouple, therefore, can be used either as a temperature measuring instrument, a current generator for some sensitive controls, or it may be used for its refrigerating effect. Fig. 6-75 shows these three applications of the Seebeck, Peltier, and Lenz discoveries.

View 1 of Fig. 6-75 shows the thermocouple being used as a temperature measuring instrument. View 2 shows the current-generating effect. This current is used to control a solenoid valve. View 3 illustrates the effect when the thermocouple junction is used for cooling. View 4 illustrates the effect when the thermocouple is used for heating.

6-66 PHOTOELECTRICITY

Photoelectric devices are of three particular types:
1. Photoconductor.
2. Photovoltaic.
3. Photoemissive.

Photoconductors are semiconductor devices which increase their conductivity of electricity when they are illuminated. They are used in electric eye devices and in infrared camera devices.

Photovoltaic devices include solar cells. These are semiconductor devices which produce electrical energy when they absorb light. They are used in solar energy conversion and in light meters for photography.

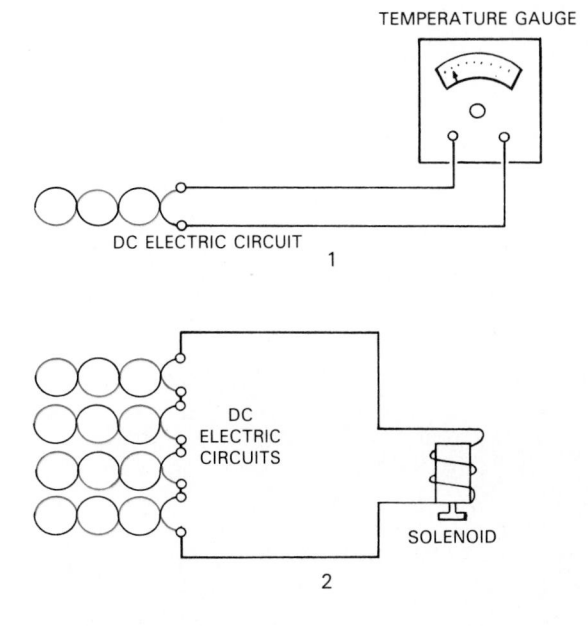

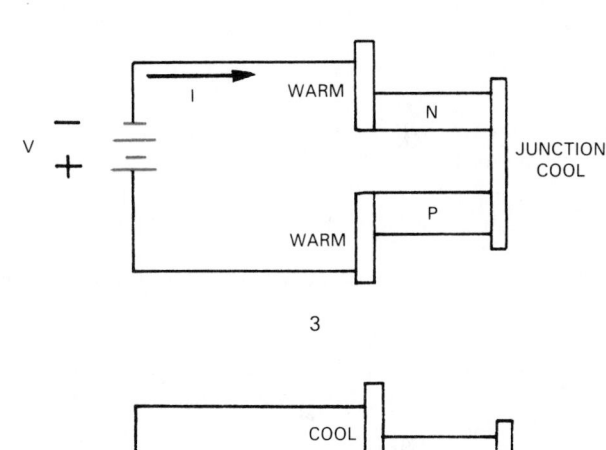

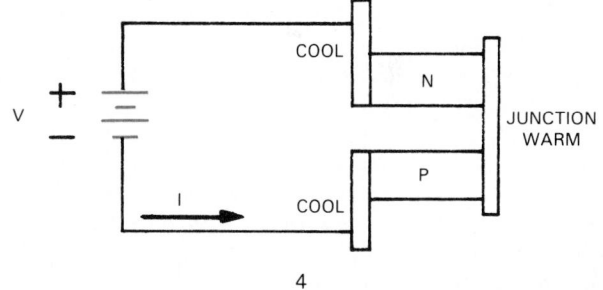

Fig. 6-75. Three basic uses of thermocouple principle. 1—Temperature measurement (used for −300 to 1700°F temperatures). 2—Generating dc electricity to operate solenoid. Solenoid then operates electric circuit for gas valve. 3—When current enters N-type material, junction cools. 4—When current enters P-type material, junction warms.

Photoemissive devices give off light when electrical energy is added. The semiconductor in a "light emitting diode" gives off the light which shows the numbers in a calculator or wristwatch. Other light emitting devices include fluorescent lights and lasers.

6-67 INTEGRATED CIRCUITS

The electronics industry has developed ways to integrate many previously separate semiconductor circuit components into small self-contained devices.

The device that incorporates multiple transistors and other semiconductor devices in a single small component is known as an integrated circuit chip. Fig. 6-76 shows typical "chips" of this type.

Fig. 6-76. Circuit board using seven integrated chips. Board controls automatic start/stop system for trailer refrigeration. (Carrier Transicold Division, Carrier Corporation)

An integrated circuit chip is usually constructed as follows:
1. The proper base material is selected. This material is in the form of semiconductor layers similar to those found in a transistor.
2. A circuit is designed and laid out on the chip material.
3. The circuit is "burned" into the material, usually using lasers or acid.
4. Input and output locations are identified and attached to metal connectors (legs) on the chip.
5. The chip is then tested and packaged.

Utilizing the integrated circuit technique, a single component containing many circuits can then be designed. This device is commonly known as a microprocessor. Microprocessors are capable of accepting information, storing it, and reacting in some preset way. See Fig. 6-77.

The microprocessor is the "brains" of many of the electronic devices used in modern HVAC (heating, ventilating, air conditioning) systems. A typical programmable thermostat or the electronic controls on a modern refrigeration system often utilize a microprocessor.

6-68 PRINTED CIRCUIT BOARDS

A printed circuit board is a support for electronic circuits. Electronic components are combined with resistors, capacitors, and other electrical devices to form the various circuits on the board. A given circuit board is usually related to a specific function in a device. A series of individual circuit boards are then used like building blocks to construct the electrical system for that device.

Circuit boards help to simplify the servicing of electrical systems. See Fig. 6-78. Usually when a specific electrical function does not work in a device, replacement of the cir-

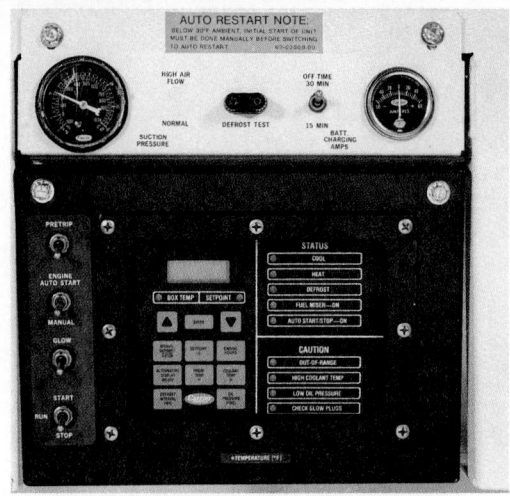

Fig. 6-77. Microprocessor control panel used for large commercial trailer refrigeration units. (Carrier Transicold Division, Carrier Corporation)

Fig. 6-78. Checking a printed circuit board. (Simpson Electric Co.)

cuit board relating to that function is recommended. The failed circuit board can then be repaired at a later time or discarded.

6-69 COMPUTERS

Computers have assumed an important role in the HVAC industry. They are being used as an integral part of many electronic controls. From the increasingly popular programmable home thermostats, Fig. 6-79, to the more complicated computerized industrial control systems, Fig. 6-80, they are helping to revolutionize the industry. In most cases, a computer supported HVAC system can even provide a diagnostic analysis for that system.

A computer is any device constructed to provide specific outputs based on input data. The electronic components reviewed in this book can be assembled to do that. Most modern computers have microprocessors as their "thinking" component. These microprocessors are then combined with input devices (keyboards, transducers, etc.), output devices (visual screen—CRT, LEDs, etc.), and data storage to make up the total computer system.

Fig. 6-79. Programmable home thermostat. Unit is designed to control a heat pump. (Harper-Wyman Co.)

Fig. 6-80. Computerized industrial control system. (McQuay, SnyderGeneral Corp.)

Computers use the on-off characteristics of integrated circuitry (i.e., the one-way current flow of a diode or the relay action of a transistor). This information can be *programmed* as a series of 1s (on) and 0s (off). All computer devices from the large computers used to control corporate HVAC systems to a hand held calculator use the on-off principle of storing and reacting to information.

Using 1s and 0s for all programming of a computer can be very cumbersome. Additional computer languages have been developed to make this process easier. These languages try to make programming more closely related to actual commands or instructions. Some of the more common languages include BASIC, FORTRAN, and COBOL.

The electronic and computer industries will continue to have a growing effect on future HVAC technology.

6-70 ELECTRICAL POWER

Electrical power is measured in watts (W), kilowatts (kW), and megawatts (MW). *A watt is the rate at which energy is produced by a current of one ampere flowing under an electrical potential of one volt.* A simple example is the rate at which heat is given off by a wire connected to a 1-volt battery when the current in the wire is 1 ampere.

This can be expressed mathematically:

Power (watts) =
    current (amps) × electrical potential (volts)
$P = I \times V$

Example: What is the power used by an electric motor that draws a current of 20 amperes (A) from a 120-volt (V) power source?

Solution:
$P = 20 \times 120 = 2400 \text{ W} = 2.4 \text{ kW}$

The electrical potential is sometimes called the electromotive force (emf).

In Para. 6-34, the power loss was indicated as I^2R. If the load on a circuit is only a resistance load, then the power is lost as heat. The power loss can then be calculated as either I^2R or $I \times V$, where $V = I \times R$, from Para. 6-33.

A simple electrical circuit which represents a speed control for a dc fan motor is shown in Fig. 6-81. When the switch is at A, the fan is off, and no current flows. When the switch is at C, the fan is on high speed and the power used is the current I_C times the motor resistance R_M.

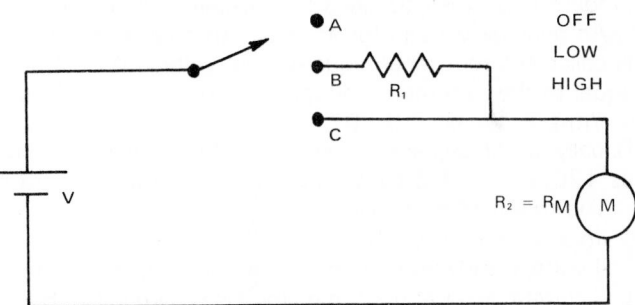

Fig. 6-81. Example of two-speed fan motor circuit. A—Off. B—Low speed. C—High speed.

$$P_C = I_C^2 R_M$$

The current $I_C = \dfrac{V}{R_M}$

Then $P_C = V I_C = \dfrac{V^2}{R_M}$

If the switch is moved to B, the fan is on low speed. The power is then:

$$P_B = I_B^2 (R_1 + R_M)$$

The current is $I_B = \dfrac{V}{R_1 + R_M}$

Then, $P_B = V I_B$

$$P_B = \frac{V^2}{R_1 + R_M}$$

The power (P_B) used with the resistance R_1 in the circuit is less than the power (P_C) used without the resistance. With the switch on low speed, the power used by the fan motor is lower.

6-71 ELECTRICAL EFFICIENCY—POWER FACTOR

If voltage and current vary within the cycle as shown in Fig. 6-82, the power must be calculated differently. The fact that the current and voltage are not varying together

circuit, due to its length, conductor size, and ampere flow load, has only 197.6 V (208 V × 95 percent = 208 V × .95 = 197.6 V), it is at the very lowest usable voltage. If the voltage goes below this, the motor may work poorly or the windings may burn out.

A serious situation may occur if a 240 V motor compressor should happen to be connected to a 208 V line (240 × 95 percent = 240 × .95 = 228 V). A voltage of 228 V is the lowest voltage on which a 240 V motor will work satisfactorily. The 208 V circuit is much below the 228 V circuit desired. The 240 V motor will operate poorly, the overload protection will operate, or the motor may fail.

A 230 V motor will operate down to 218.5 V (230 × 95 percent = 230 V × .95 = 218.5 V). Therefore, a 208 V circuit is dangerous if used with other than 208 V motors.

A 220 V motor will operate down to 209 V (220 × 95 percent = 220 V × .95 = 209 V) and this motor may operate on 208 V, but it is a borderline case and the efficiency will be low.

However, motors can use a voltage slightly over their rating. In fact, a 208 V motor connected to a 220 V line will operate very well, will start faster, and will give more power to the compressor. A 10 percent over-normal voltage will let a motor carry a 20 percent overload. However, a 208 V motor on a 240 V line will be much noisier—a handicap in air conditioning.

A 220 V motor operates very well on a 240 V line. A 220 V motor can operate on a 208 V line, but will have lower torque. These are single-phase motors. When only single-phase is available and more than 3 horsepower is needed, two or more motor compressors should be used in the system.

Fig. 6-92 is a table of changes in a motor's operating characteristics as the input voltage changes.

For example, in a home air conditioner, the power circuit is usually 240 V, one-phase, three wire. This means there is 240 V between the two hot conductors and there is 120 V between the hot conductors and the ground conductor, as shown in Fig. 6-93.

These voltages should be checked (verified) with a voltmeter. Do not trust a test light and guess at its bulb brilliance.

When studying a building wiring system to decide on the circuit necessary for a motor compressor, the wire sizes should be checked for capacity. All electrical appliances should be on at the same time. Also, the average load per day should be checked. A recording ammeter may be used for this purpose if the utility does not already have the data. A demand meter will usually be put in by the electrical utility if requested.

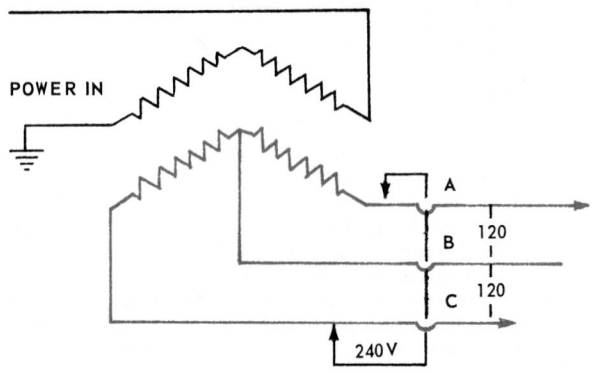

Fig. 6-93. Schematic circuit diagram of transformer used to serve average home. Note that 120 V are available between A and B, and between B and C. B is a ground. Also, 240 V are available between lines A and C.

INPUT VOLTAGE VARIATIONS VERSUS MOTOR CHARACTERISTICS					
INPUT VOLTAGE	CURRENT	TORQUE	TEMPERATURE	CYCLE SLIP	EFFICIENCY
+ 10%	− 7%	+ 21%	−3 deg. C. = (5.5 deg. F.)	17% decrease	1% increase
−10%	+ 11%	−15%	+7 deg. C. = (12.5 deg. F.)	23% increase	2% decrease

Fig. 6-92. Effect of input voltage on operating characteristics of motor.

A 240 V motor compressor works with the same efficiency as a 120 V motor compressor. The notion that a 240 V motor uses less kilowatt hours to do the same amount of work is false. For example, a 120 V motor uses 5 A to create 600 W (120 V × 5 A = 600 W) or .600 kW. To provide .600 kW, a 240 V circuit must carry 2.5 A, but the electrical cost is the same. The only advantage is that the 240 V unit may use smaller size conductors from the meter box to the unit. There will be less voltage drop between the meter box and the motor.

Some motors are labeled 208-220 V to indicate that they may be used with either voltage. However, these motors are sensitive to voltages below 208 V, and voltages of 195 V or lower must not be used.

6-77 LINE VOLTAGE TRANSFORMER

Improper line voltage may cause refrigeration and air conditioning motors to burn out. Line voltages may vary as much as 5 to 10 percent. Equipment designed to operate on 120 V plus or minus 10 V may not operate well where the voltage drops to 100 V or lower.

When equipment designed for 120 V service is used on 240 V lines, a line voltage transformer is used to overcome the problem of voltage fluctuation. The line voltage transformer is designed to increase and decrease the input voltage to the motor as needed.

Fig. 6-94 shows a basic wiring diagram of a "boost-and-buck" voltage transformer. It is being used to correct the

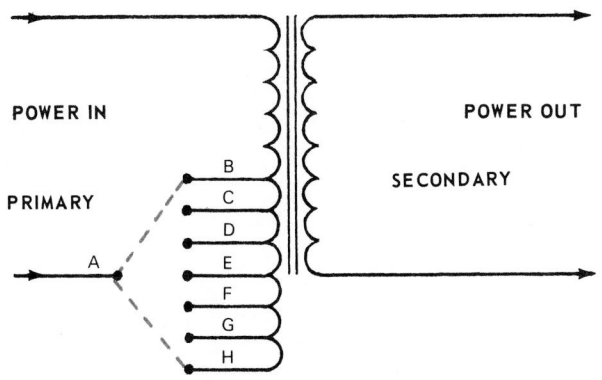

Fig. 6-94. Wiring diagram of transformer designed to step up or step down voltages to provide correct voltage for driving motor. Connecting terminal A to B, C, or D will step up power-out voltage. Connecting A to F, G, or H will step down power-out voltage. Connecting A to E will not change voltage, but will match load to source and will limit power surges. A transformer like this used as an isolation transformer can improve safety against electrical shock.

line voltage to 120 V. Fig. 6-95 shows a wiring diagram of a transformer designed to change a 208 V circuit to 240 V. Some line voltage transformers have one coil. Part of the secondary coil is shared with the primary coil. This is usually called an autotransformer, view B in Fig. 6-95.

A voltage of 208 V is quite common where the main electrical load is lighting. A line transformer will be required if 120 V or 240 V motors are to be used on the 208 V current supply. This also is sometimes called an autotransformer. See Fig. 6-96.

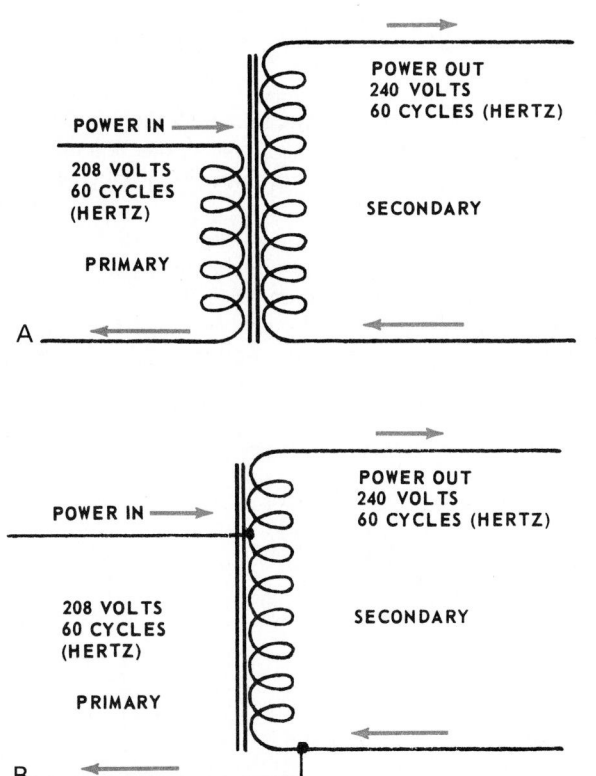

Fig. 6-95. Wiring diagram of voltage transformer used to step up 208 V to 240 V. B—An autotransformer used to step up 208 V to 240 V. Note that part of the secondary coil is shared with the primary coil.

Fig. 6-96. Line transformer used to transform 208 V ac to 240 V ac. (Acme Electric Corp.)

A 120 V transformer output can be produced through the use of various connections on the input side ranging from 95 to 260 V. The service technician must first determine the required voltage from the motor identification plate and specifications.

A voltmeter is used to measure the voltage and ampere flow at the motor compressor when the unit is running. A transformer is then selected which will carry the current. It is adjusted to raise or lower the voltage as needed.

Transformers are rated in kVA or VA output. The abbreviation, kVA, means "kilovolt amperes." VA means "volt amperes." To determine kVA, one must multiply the output voltage by the amperes.

$$kVA = \frac{Volts \times Amperes}{1000}$$

There is often confusion between "volt amperes" and watts. Likewise, there is confusion between "kilovolt amperes" and kilowatts. The difference is explained as follows. Suppose a 10 kVA transformer is supplying power to a load which is partially inductive. The load voltage multiplied by the load current is 10 kVA. Because of the inductance of the load, the power factor is 0.8. The quantity of kilowatts supplied by the transformer is calculated as follows:

kW = kVA × Power Factor
kW = 10 kVA × 0.8 = 8 kW

Only 8 kW are being supplied to the load for producing heat or useful mechanical motion. Although the transformer is rated at 10 kVA, it can only supply a real power of 8 kW. The transformer will supply 10 kW to a purely resistive load, such as an electric space heater. Transformers that drive inductive or capacitive loads must always be larger and heavier than transformers that drive resistive loads. Note that a similar reasoning process applies when talking about "volt amperes" and watts.

The total load of the circuits to be fed by the transformer must not be more than the transformer output. The transformer is connected into the circuit between the electrical service and the motor. The service technician should check the completed installation for both ampere flow and voltage. If these values are different than the motor ratings, further adjustments will be necessary.

6-78 THREE-PHASE, FOUR-WIRE TRANSFORMER

Another transformer design used by utilities is the Wye type shown in Fig. 6-97. This system uses one transformer, connected as shown. Note than a 208 V circuit is also possible from this system. In fact, any two of the three conductors produces 208 V. This is made possible by the angle of the two secondary windings.

Some utilities recommend a 277-480 V system. It is the least critical and therefore is good for commercial-industrial use. This system is shown in Fig. 6-98.

Utility circuits may show as much as 212 V at the meter on a 208 V system, or as much as 250 V on a 240 V system. However, because an increase in voltage is as much a disadvantage to light bulb life as it is an advantage for a motor compressor, the utilities keep their circuit voltages as close to the required value as possible.

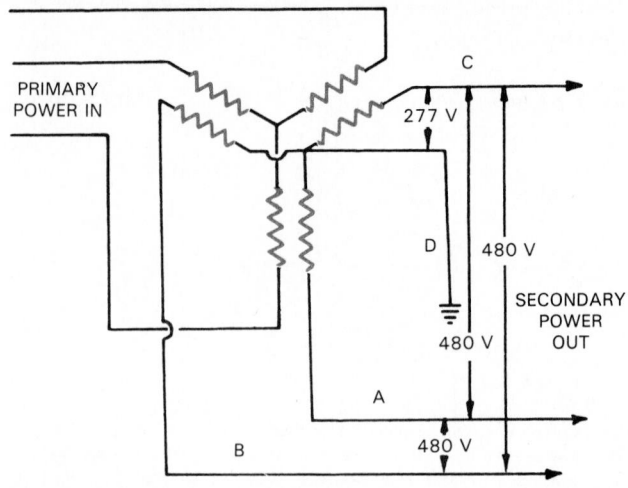

Fig. 6-98. Schematic of Wye-type transformer used in serving industrial and commercial applications. Following voltages are available from this transformer. Between C and D, 277 V. Between A and B, 480 V. Between A and C, 480 V. Between B and C, 480 V.

A 277 V circuit is used for some refrigeration or air conditioning units of 15,000 to 35,000 Btu/hour capacity. It may be obtained by an angular tap from a Wye type transformer, when one lead is from any of the three legs and the other lead is from the center of the Wye, Fig. 6-98. The secondary voltage between A and B and between A and C is 480 V.

The Closed Delta system, Fig. 6-99, is sometimes found in large industries. The transformer is not grounded. The voltage between any two terminals in Fig. 6-99 is 480 V.

6-79 RELAYS

Various types of relays are used in electrical circuits on both refrigerating and air conditioning mechanisms. Fan relays, motor starting relays, and main power line relays are very popular in air conditioning and refrigerating electrical systems.

These relays usually have an enclosed switch operated by an electromagnet. Relays are designed to protect motors by disconnecting them from the line when they become overheated or overloaded. The amount of current used to operate a relay is very small. The contact points in the relay

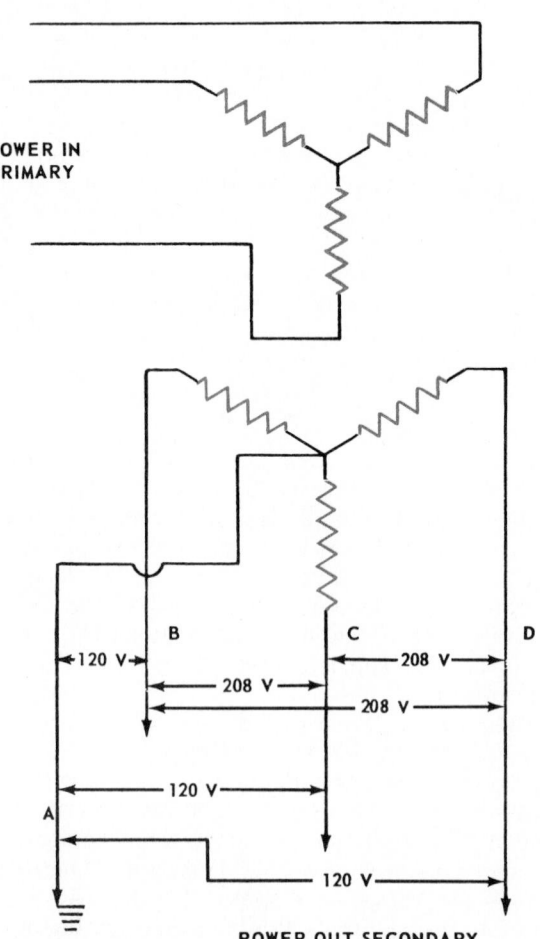

Fig. 6-97. Wye-type transformer. Note that center of Wye is grounded in secondary circuit. Choice of emf may be obtained from this transformer by taking power from pairs of terminals as follows: A and B — 120 V. A and C — 120 V. A and D — 120 V. B and C — 208 V. C and D — 208 V. B and D — 208 V.

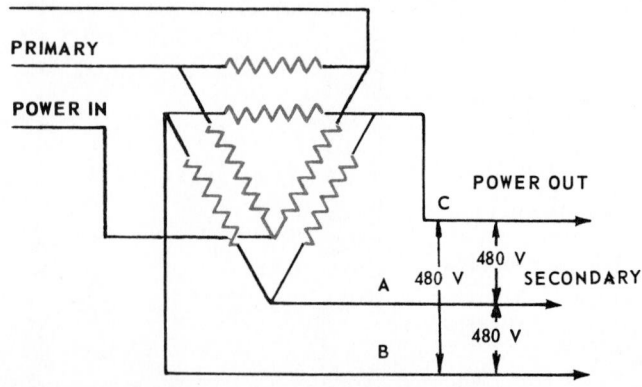

Fig. 6-99. Diagram of Closed Delta type transformer which will deliver 480 V between either A and B or A and C.

then operate circuits which can carry quite heavy currents. Fig. 6-100 shows the electrical circuit for a typical relay.

Electronic relays are also now in use. These relays are usually self-enclosed devices with connections for *normally open (NO), normally closed (NC),* and input/controlled voltage.

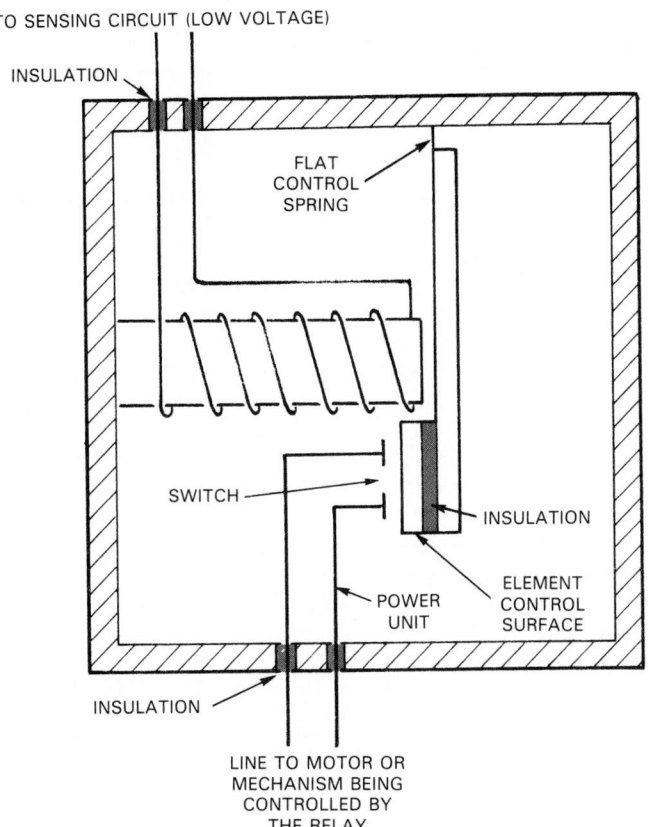

TO SENSING CIRCUIT (LOW VOLTAGE)

INSULATION

FLAT CONTROL SPRING

SWITCH

INSULATION

ELEMENT CONTROL SURFACE

POWER UNIT

INSULATION

LINE TO MOTOR OR MECHANISM BEING CONTROLLED BY THE RELAY

Fig. 6-100. Electrical relay. Low voltage thermostat or sensing instrument controls operation of relay. Relay in turn switches power circuit on and off. Power circuit is normally off (called normally open, NO). Power circuit is on only when relay low voltage circuit is energized. Some relays are normally closed, NC.

6-80 ELECTRICAL CODES

The National Electrical Code establishes rules and regulations covering materials and methods used in installing electrical systems. In addition, many cities and communities have supplementary local codes. All electrical installations should be made in conformity with national and local codes.

The National Electrical Code provides for a type of wiring called Class 2. These circuits are usually used for controlling relays, bells, signal systems, communications, and the like. The current supply is usually from a small transformer having two windings. These transformers have a capacity of about 100 volt amperes.

Refrigeration and air conditioning service technicians are permitted to make Class 2 connections and installations. Under this Code, the current from the secondary winding of the transformer is limited according to the voltage: below 15 V, up to 5 A; 15 to 30 V, up to 3 A; 30 to 60 V, up to 1.5 A; above 60 V, not over 1 A.

Underwriters Laboratory tests and approves electrical components such as switches, extension cords, and relays.

6-81 CIRCUIT PROTECTION

Previous paragraphs explain the heating and magnetic effect of current flowing through electrical circuits. Instruments and appliances may be seriously injured or ruined by either overheating or, in the case of instruments, too much magnetism.

Four types of safety devices are commonly used to avoid the possibility of an accidental current surge damaging either the instruments or the appliances.

The most common protection is the fuse. See Fig. 6-101. This is a soft metal conductor wire which may be placed in series with the circuit with either a plug or a cartridge arrangement. The size, length, and material in the fuse are such that its resistance will cause enough heat to melt the fuse. Therefore, the fuse will open the circuit if the current exceeds the fuse rating.

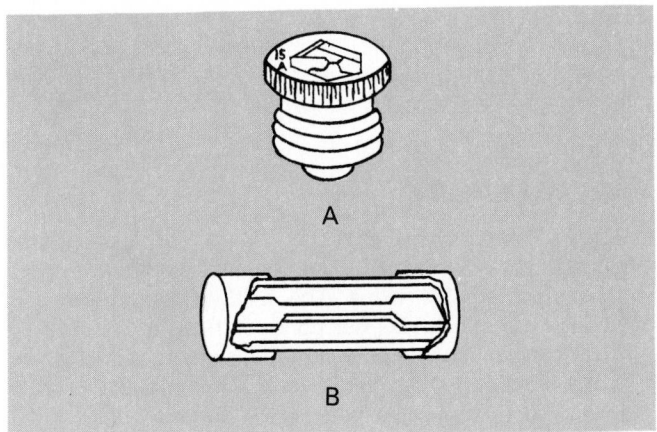

A

B

Fig. 6-101. Standard types of fuses. A—Household. B—Single element cartridge.

Fuses are rated in amperes. Those used in refrigeration and air conditioning circuits are usually designed to carry 5, 10, 15, 20, or 30 A. Fuses used in electronics circuits are available from 1/500 A to 2 A. Some electronic fuses, however, are available with higher ampere ratings.

The circuit breaker is another protective device. See Fig. 6-102. Current flowing through the circuit being protected passes through a solenoid in the breaker. The magnetic effect of the solenoid current will cause the solenoid to trip a spring-loaded switch if the current in the circuit exceeds a predetermined level (amperage).

Another common circuit breaker device has a bimetal breaker. The bimetal strip is usually connected in series with the breaker point and a resistance heater. The current flowing in the circuit will cause the resistance heater to heat the bimetal strip, causing it to bend. The heating effect will depend upon the current flow—the more current, the more heat. The instrument is calibrated to open the points when the current flow is greater than safely allowed.

The fourth type of circuit protection makes use of the thermistor. Thermistors have been developed for circuit protection which either regulate (modulate) the current flow

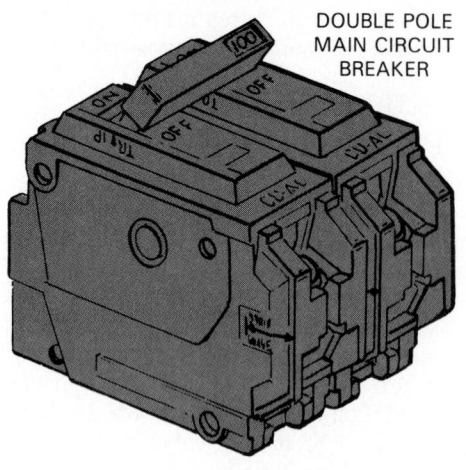

DOUBLE POLE
MAIN CIRCUIT
BREAKER

Fig. 6-102. A circuit breaker will automatically switch to "off" during an overload. (General Electric Co., Wiring Devices Dept.)

or which may, in some cases, cause the current flow to be reduced to a safe value.

CAUTION: Never connect alternating current appliances or instruments into direct current circuits. Never connect direct current appliances or instruments into alternating circuits.

6-82 WIRE SIZES

The current-carrying capacity of a solid conductor depends on its diameter. Larger wires may carry a heavier current than smaller wires. The following information is based on the use of copper conductors (wires).

The electrical supply wire to outlet receptacles must be of adequate size. Fifteen- and twenty-ampere outlets should be supplied with No. 12 wire, 30 A outlets with No. 10 wire, and 40 A outlets with No. 8 stranded wire.

Wire sizes are measured in circular mils. A mil is the area of a circle 1/1000 (.001) in. in diameter. Fig. 6-103 is a table of the more common conductor sizes.

WIRE SIZE AWG*	CONDUCTOR DIAMETER IN INCH DECIMALS	CIRCULAR MILS	OHMS PER 1000 FT. AT 68°F
14	.0641	4,110	2.52
13	.0720	5,180	2.00
12	.0808	6,530	1.59
11	.0907	8,230	1.26
10	.1019	10,380	1.00
9	.1144	13,090	.7925
8	.1285	16,510	.6281
7	.1443	20,820	.4981
6	.1620	26,240	.3952
5	.1819	33,090	.3134
4	.2043	41,740	.2485
3	.2294	52,620	.1971
2	.2576	66,360	.1563
1	.2893	83,690	.1239
0	.3249	105,600	.09825

* American Wire Gauge

Fig. 6-103. Wire size data for rubber or thermoplastic covered conductors.

The cross-sectional area of a No. 10 wire is about 10,000 circular mils. Its resistance is about 1.0 ohm per 1000 ft. A No. 7 wire has approximately double the circular mil area of a No. 10 wire, and half the resistance (approximately 0.5 ohm per 1000 ft.). A No. 4 wire has a circular mil area of approximately 40,000 and its resistance is approximately .25 ohms per 1000 ft. This indicates that the circular mil area doubles every third size as the wire becomes larger and the resistance is cut in half. Using this data, the electrician can easily calculate the wire size and resistance for commonly used conductors.

6-83 ATTACHMENT PLUG CONFIGURATIONS (TERMINALS)

It is sometimes necessary to connect electrical devices using flexible leads and attachment plugs. Since most electrical devices are designed for a particular power supply specification, it is important that connections to the power supply conform to the electrical specifications for the equipment.

For instance, if an appliance designed for 120 V were to be connected into a 240 V circuit, the appliance would very quickly "burn out." Likewise, if an appliance has protection for only up to 15 A and is connected into a circuit of 30 A capacity, it could be burned out or the safety device could be ruined.

The National Electrical Manufacturers Association (NEMA) has established attachment plug configurations (terminals). Fig. 6-104 shows some common receptacle and plug configurations for the most commonly used refrigeration and air conditioning equipment.

6-84 SWITCHES

Any device used to open or close an electrical circuit is called a switch. Switches are made of many materials and

		15 AMPERE		20 AMPERE	
		RECEPTACLE	PLUG	RECEPTACLE	PLUG
2-POLE 3-WIRE GROUNDING	125V				
	250V				
	277V				
3-POLE 3-WIRE	3-PHASE 250V				
3-POLE 4-WIRE GROUNDING	3-PHASE 250V				
4-POLE 4-WIRE	3-PHASE 120/208V				

Fig. 6-104. NEMA configurations are shown for general purpose nonlocking plugs and receptacles. Blade spacing and configuration vary with specified amperage.

there are numerous designs. Some are manually operated, while some are operated automatically.

Fig. 6-105 shows the basic types of switches. All are open contact point. Other types, such as knife blade contact and mercury-in-a-tube contact, use the same basic arrangements.

Normally closed (NC) switches have the contacts touching when there is no power in the circuit.

Normally open (NO) switches have the contacts separated when there is no power in the circuit.

6-85 CIRCUIT TESTING INSTRUMENTS

A test light, Fig. 6-106, is a tool which makes it easy to test electrical circuits without causing damage. The test light with the two prongs may be used when the device being tested is connected to electrical power.

For example, if the refrigerator, air conditioner, or heating system is plugged in and will not start, this test light (about 25 to 50 W) may be used to determine if power is coming to the wall outlet. The lighting of the light will indicate power is available up to the wire probes. If the wall outlet has power, then the open circuit is in the refrigerator wiring.

The test light at view B in Fig. 6-106 may be used to check electrical devices not connected to power. This test

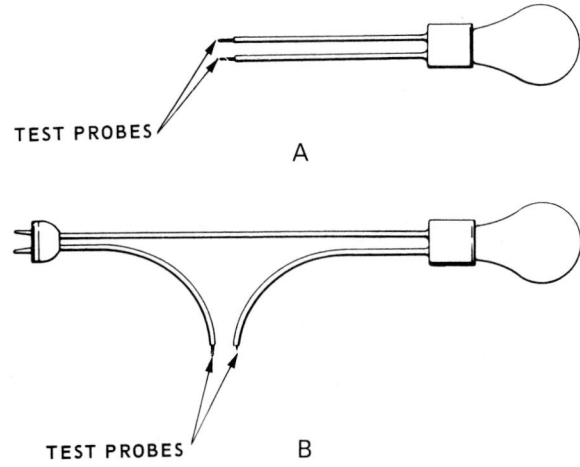

Fig. 6-106. Two types of test lights. A—Light bulb with two probes is used to test circuits which have power on. B—Test light connected to electric power source is used to check circuits not connected to power source.

light should be connected to a power source. If the bulb lights when the probes are touched together, the test light is functioning.

To use this light, put one probe on one end of a wire and the other probe on whatever that wire is supposed to connect to. If the bulb lights, the circuit is continuous (there is continuity).

Fig. 6-107 shows a special test light used to test a three-phase circuit. It may be used to determine which leads of a three-phase circuit connect to the three-phase power lines. If the tester glows, the rotation is 1, 2, 3. Reverse any of the two leads and the tester should not glow. If the tester glows when connected in any order, one of the three-phase circuits is open.

These test lights should not be used on solid state circuits. They may damage diodes, transistors, or other parts.

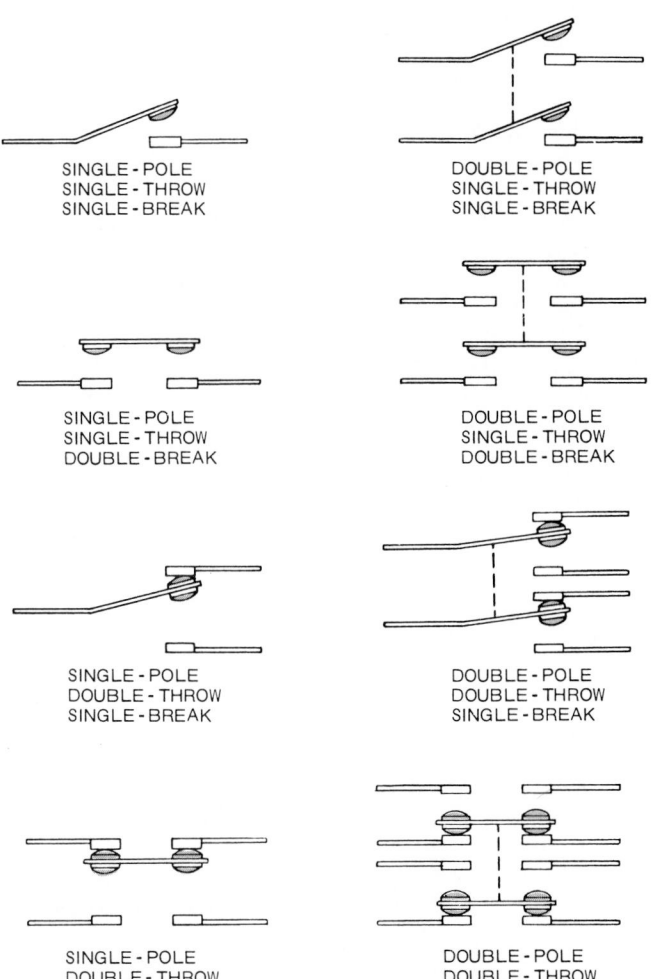

SINGLE - POLE
SINGLE - THROW
SINGLE - BREAK

DOUBLE - POLE
SINGLE - THROW
SINGLE - BREAK

SINGLE - POLE
SINGLE - THROW
DOUBLE - BREAK

DOUBLE - POLE
SINGLE - THROW
DOUBLE - BREAK

SINGLE - POLE
DOUBLE - THROW
SINGLE - BREAK

DOUBLE - POLE
DOUBLE - THROW
SINGLE - BREAK

SINGLE - POLE
DOUBLE - THROW
DOUBLE - BREAK

DOUBLE - POLE
DOUBLE - THROW
DOUBLE - BREAK

Fig. 6-105. Types of switches. Orange contacts indicate movable breaker points. (Micro Switch, Div. of Honeywell, Inc.)

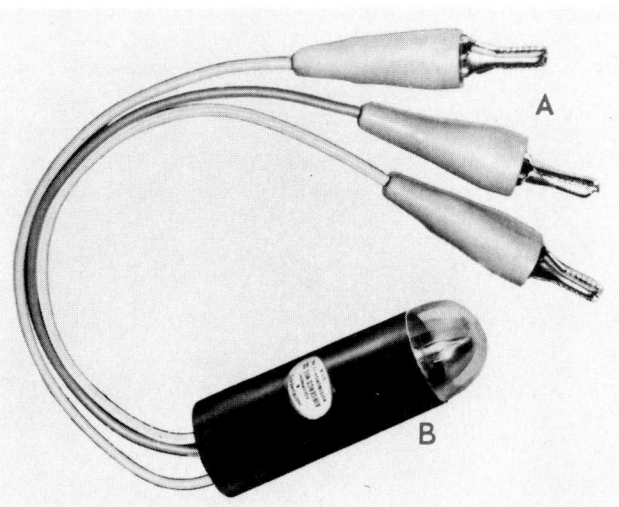

Fig. 6-107. Test light for testing a three-phase circuit. Connect three leads to three-phase terminals. If tester bulb glows, rotation is 1, 2, 3. If it does not glow, reverse pairs of leads until it does glow. If tester glows regardless of connection order, one circuit is open. A—Terminals. B—Bulb. (Airserco Mfg. Co., Inc.)

6-86 REVIEW OF SAFETY

There are two different factors to be considered in reviewing safety in this chapter.
1. Safety to the operator.
2. Safety to the equipment.

One must have great respect for electricity. It only requires 25 mA to seriously injure or kill a person. Voltages as low as 30 V can force this much current through vital parts of a person (the heart and brain are the most delicate). Never handle live circuits when in contact with pipes, other wires, damp floors, damp ground, or water.

REMEMBER: Electricity cannot be seen. One can only see its effect in light, heat, magnetism, and so on. It is too late to learn that a circuit is live after one has touched it. Use instruments to find out if the wires or equipment are safe before starting to work.

Always use insulated tools (or dry rubber gloves) when working on circuits of 30 V or higher. Even a large battery can injure one if a tool, a wrist watch, or a ring should short circuit the battery. The metal will become hot enough to cause severe burns.

Before working on household, commercial, or industrial electrical circuits, disconnect the power. Make sure no one can turn the power on while someone is working on the circuit (use signs, locks, or other devices).

Instruments and equipment used in refrigeration and air conditioning work, while durable, are quite sensitive to abuse. When connecting an electrical instrument into a circuit, make sure that the instrument and its setting are within the voltage and current range which may be applied to the instrument. An instrument adjusted to measure a voltage within a 150 V range will be ruined if connected into a 440 V circuit.

In making capacitor tests, the capacitor should not be left in the circuit any length of time. Usually a second or two is sufficient. A longer period may destroy the capacitor.

6-87 TEST YOUR KNOWLEDGE

1. Name the unit of electromotive force.
2. What measurement of electricity is similar to "gallons per hour" of water?
3. Why is the statement "dc current" incorrect?
4. What is the unit of electrical power?
5. In ac, does the power equal the emf (volts) times the current (amperes)?
6. In Ohm's Law, the letter I stands for what word?
7. Write the three formulas of Ohm's Law.
8. How does a transistor's operation compare to a relay?
9. Describe an electrical circuit.
10. Where is a thermistor used?
11. What is voltage drop?
12. What is electrical continuity?
13. How long does current flow in one direction when 60 Hertz (cycle) current is used?
14. Should a voltmeter be connected in series or in parallel with the circuit being tested?
15. Should an ammeter be connected in series or in parallel with the circuit being tested?
16. In order to test the resistance in a circuit, an ohmmeter is used. Should the ohmmeter be connected to the circuit while the power is being applied to the circuit or only when the circuit is turned off?
17. Is a shunt placed in parallel or in series with an instrument?
18. Which has the least electrical resistance, a conductor or a semiconductor?
19. What is the formula for voltage drop?
20. What is the formula for power loss?
21. What is the integrated circuit device used as the "brains" in many HVAC programmable controls?
22. What does an amplifier do to a small input signal?
23. In a domestic refrigerator, is the cabinet light in series or in parallel with the cabinet light switch?
24. When using a voltmeter to measure an ac voltage, is the maximum value or an average value being displayed?
25. What is the unit of capacitance?
26. Is a counter emf generated in all running motors?
27. Does the electrical resistance increase or decrease in a thermistor as its temperature is increased?
28. What is the purpose of a transformer?
29. Which is the most popular polyphase circuit—two-phase, three-phase, or four-phase?
30. Computers use what characteristic of integrated circuits as the basis for their performance?

Chapter 7

ELECTRIC MOTORS

After studying this chapter, the technician will be able to:
☐List several types of electric motors.
☐Explain the operating principles of various types of electric motors.
☐List and describe devices that protect motors from overloads and overheating.
☐Demonstrate service procedures required for several types of motors.
☐List safety procedures for servicing electric motors.
☐Use various electrical testing instruments to check motor windings, shorts, and grounds.

7-1 ELECTRIC MOTOR APPLICATIONS

Refrigerating systems operate either with open or sealed-in (hermetic) motors. The open motor drives the compressor directly off the shaft or by means of a belt. The sealed-in or hermetic type is built inside the compressor dome. It usually drives the compressor directly. Open electric motors are also often used to drive many accessory devices.

Motors may be grouped into four general classifications according to use:
1. To drive compressors.
   A. Open belt drive (external drive).
   B. Hermetic direct drive.
2. To drive fans for:
   A. Condensers.
   B. Evaporators.
   C. Air circulation.
   D. Induced draft.
   E. Forced draft.
3. To drive pumps.
   A. Condensate pumps.
   B. Chilled water pumps.
   C. Condenser water pumps.
   D. Ice making machine water pumps.
   E. Oil pumps.
4. To drive miscellaneous devices.
   A. Vending machines.
   B. Automatic ice cube makers.

7-2 THE MOTOR STRUCTURE

All motors have a similar basic construction.
Each has two main parts:
1. The stator.
2. The rotor.

The stator may also be known as the frame. This frame is usually cylindrical in shape. The field poles with field windings on them are part of the stator. The identification plate is also mounted on the stator.

The rotor is mounted on a shaft which has two journal bearings, one at each end.

The frame has end bells or plates attached to it. These hold the bearings. When the rotor shaft journals are mounted in the bearings, the bells support the rotor.

The bearings are accurately machined to provide the proper amount of end play for the rotor. There is a clearance of .001 to .002 in. between the motor shaft and the bearing. In hermetic units built into the compressor dome, the compressor bearings may also serve as rotor bearings.

The windings are insulated copper wire. This insulation is usually a polyester material. It is resistant to moisture and has considerable dielectric and mechanical strength.

7-3 TYPES OF ELECTRIC MOTORS

Both alternating current (ac) and direct current (dc) may be used to operate electric motors. Alternating current is most commonly used, but direct current motors are found in areas supplied with direct current only.

Following are the alternating current motor types used to drive compressors. Those listed in red are most common and will be explained in greater detail:
1. Basic types:
   A. Single-phase.
   B. Two-, three-, and four-phase (polyphase).
2. The open motor. These are used on open, belt-driven compressors. Types include:
   A. Repulsion-start, induction-run.
   B. Capacitor-start, induction-run.
   C. Capacitor-start, capacitor-run.
   D. Capacitor-run.
   E. Permanent split capacitor.
   F. Induction polyphase.
3. Hermetic motors. These types are used on sealed systems in which the motor and compressor are enclosed in a common housing or dome:
   A. Capacitor-start, induction-run.
   B. Capacitor-start, capacitor-run.
   C. Capacitor-run.
   D. Permanent split capacitor.
   E. Induction two-phase and polyphase.

213

4. The condenser and evaporator fan motor type:
   A. Split-phase.
   B. Shaded-pole.
   C. Capacitor.
   D. Permanent split capacitor.

Today, the capacitor motor is being used on most applications. It is one of the most popular for single-phase hermetic machines.

7-4 EXTERNAL DRIVE MOTORS

Three main types of motors are used to drive external drive compressors:
1. Repulsion-start, induction-run motor, Fig. 7-1.
2. Capacitor-start, induction-run motor, Fig. 7-2.
3. Three-phase motors.

V-belts connect the motors to the compressor or a direct connection is made with a coupling. The speed reduction for belt drives is usually about three to one. This means that the compressor flywheel diameter is three times larger than the motor pulley diameter. Para. 7-5 through 7-12 explain features common to most external drive motors. Fig. 7-3 shows the parts of a capacitor type motor.

7-5 INDUCTION MOTORS

Induction motors have no windings on the armature (rotor). However, the rotor does have copper bars or other conducting material. These are in its outer surface and lie about parallel to the motor shaft. When current is induced (made to flow) in them, turning power or torque is produced by the magnetism that is created.

One or more field windings are mounted in the stator. Alternating current, passing through these windings, creates a changing magnetic field. This magnetism passes through the rotor, inducing (building up) current in the rotor bars.

The effect of the induced current is to create an opposite magnetic field in the rotor. The opposing magnetism causes the rotor to revolve as it tries to keep up with the changing field polarities in the field windings.

All hermetic machines use induction motors. Most of these motors use two field windings:
1. A starting winding.
2. A running winding.

In hermetic motors, it is often necessary to use a starting relay. This is always located outside of the motor compressor dome. This is because electrical contacts would quickly fail from the oil and refrigerant mist inside the dome.

Para. 8-25 describes the various types of starting relays and how they are used.

The starting relay temporarily connects the starting winding of the motor to the power circuit. When the motor reaches about 75 percent of operating speed, the relay opens the circuit. The starting winding is then disconnected from the power line.

7-6 SPLIT-PHASE INDUCTION MOTOR

The split-phase induction motor is used for most fractional horsepower appliance operation requirements. Two stator windings are used, one for starting and one for running.

The running windings are made into two or four coils in series depending on the motor speed desired. These windings are energized (have current flowing in them) during the whole time the motor is running.

Fig. 7-1. Section view of repulsion-start induction-run 1/2 hp motor. Note that this motor has wound rotor and centrifugal mechanism. Centrifugal mechanism shorts the rotor winding and raises brushes off commutator as soon as motor attains about 75 percent of running speed. A—Centrifugal mechanism. B—Commutator shorting segments. C—Brushes and brush holder. Wick oilers are used at bearings.

The starting windings have the same number of coils usually rotated several degrees from the running windings. See Para. 1-13 for information about degrees of arc measurement.

A smaller gage of wire is used in the starting winding. However, it has more turns than the running winding. Counter emf in the greater number of turns causes the current to build up more slowly in this winding. Therefore, the magnetic effect will be several electrical degrees behind the running winding. This will create torque on the rotor to make it start turning in the correct direction. The starting windings

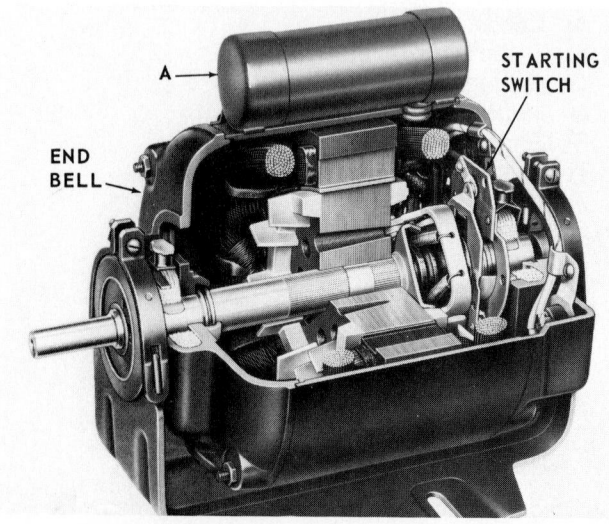

Fig. 7-2. Capacitor-start induction-run motor. Motor has rubber mounts at each end which provide flexibility. A—Capacitor is mounted in housing on top of motor. Wick oiling is used.

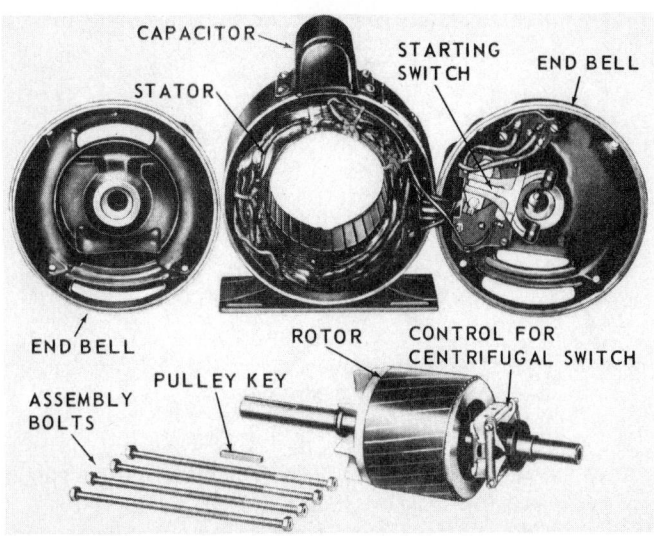

Fig. 7-3. Parts of a capacitor type motor used on open compressor. (Emerson Electric Co.)

are disconnected when the motor reaches approximately 75 percent of its running speed.

There are no windings on the rotor, but magnetism builds up around the rotor bars. Some rotors have heavy copper bars fitted into slots in the laminated iron. The ends of these copper bars are braze-welded to heavy copper rings at each end. This completes the induced (by magnetism) electrical circuit. Such an arrangement is often called a squirrel-cage winding.

The term "electrical degrees" means that the maximum magnetic effect on the rotor is a few degrees away from (behind) the magnetic effect in the stator.

7-7 REPULSION-START, INDUCTION-RUN MOTOR

A repulsion-start, induction-run motor has an electrical winding on the rotor for starting purposes. Typical applications are external-drive refrigerators and air compressors. In many cases, they have been replaced by capacitor-type motors.

A special winding in the armature gives it a high starting torque. The motor starts as a repulsion motor, using brushes against a commutator in the armature winding circuit. This increases the induced electrical flow in the armature and produces more magnetic power. As soon as it reaches a certain speed, the armature windings are shorted. Then the brushes are usually lifted from the commutator and the motor operates as an induction motor, Fig. 7-4. Para. 6-49 explains the operation of commutators and brushes.

7-8 CAPACITOR-START, INDUCTION-RUN MOTOR

The capacitor-start, induction-run motor uses a capacitor in the starting winding. (Recall that a capacitor is a device for storing electrical energy.) One type, Fig. 7-5, becomes a two-phase motor when starting. More torque is provided that way. Then it becomes a single-phase motor at about 75 percent of its full speed. The capacitor-start, induction-run motor is a popular type.

During starting, the capacitor changes the phase angle of the current in the starting winding to produce two-phase electrical characteristics.

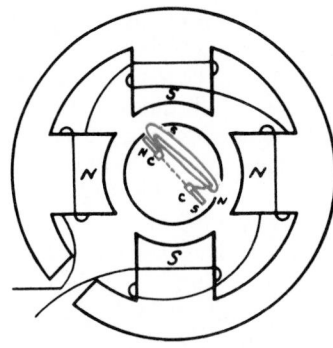

Fig. 7-4. Schematic sketch of repulsion-start induction-run motor. C—Commutator. Brushes that contact commutator are grounded and complete circuit between two commutator bars.

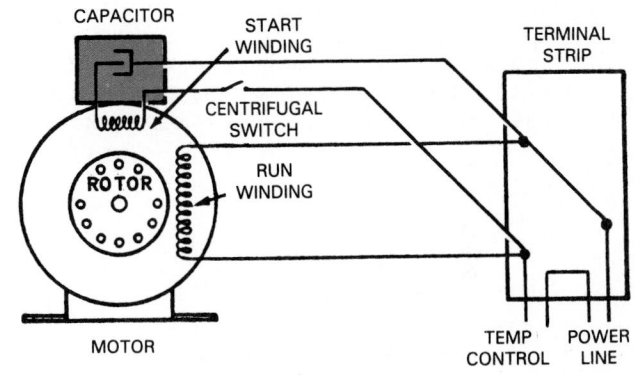

Fig. 7-5. Wiring diagram of capacitor-start induction-run motor. Note that capacitor is in series with starting field winding as motor starts. Rotor-operated centrifugal switch opens this winding as soon as motor reaches about 75 percent of full speed.

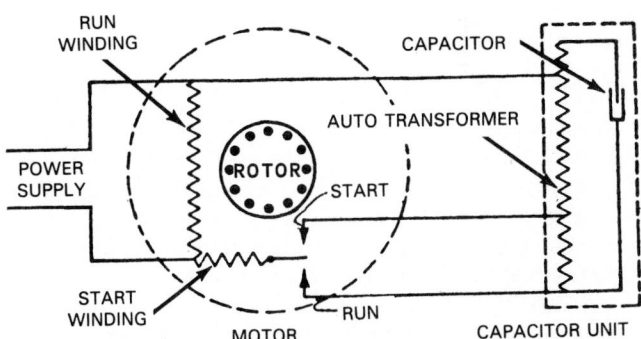

FOR COUNTERCLOCKWISE ROTATION CONNECT AS SHOWN

Fig. 7-6. Wiring diagram of combination capacitor and autotransformer motor. Note double-throw single-pole switch for changing transformer output when centrifugal switch moves from start to run. (Fedders North America)

The capacitor is usually placed on top of the motor in a metal or plastic cylinder. The capacitor is connected to a centrifugal switch built into the motor and to a starting winding in the stator. See Fig. 7-6. (A centrifugal switch is one which whirls around with the shaft and weights are moved by the whirling motion to open the contacts.)

Operation is quite simple. When the motor starts the centrifugal switch (closed when the motor is not running) causes the current to pass through both the starting and the running winding. The starting winding is connected in series with the capacitor. This capacitor puts the electrical surges in the starting winding out of step or out of phase with those of the running winding. The motor then acts as a temporary two-phase motor and has a very high starting torque.

At about 75 percent of the motor's rated speed, the centrifugal switch opens and disconnects the starting winding. The unit, however, continues to run as an induction motor.

The capacitor has two terminals. One connects to one leg of the power line and the other to the starting winding terminal.

The simplest method for producing greater torque is to change the single-phase motor into a two-phase during starting and/or running. A capacitor is used.

There are two types of capacitors:
1. Starting capacitor.
2. Running capacitor.

The starting capacitor is usually a dry type. It is constructed of two sheets of conductor metal separated by an insulator, as shown in Fig. 7-7. Typical of capacitors is the one shown at the top of Fig. 7-2.

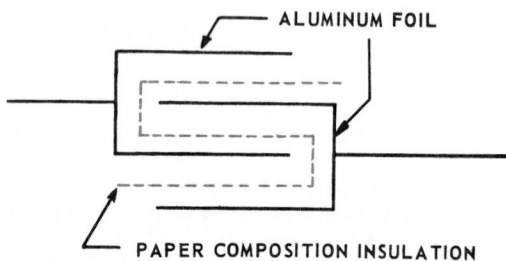

Fig. 7-7. Capacitor construction. Two layers of metal foil are separated by special insulating paper. The two sheets of foil are connected to the two terminals of the unit. Capacitors used on motors are rolled or folded into compact package and installed in metal housing.

A capacitor placed in an alternating current line is charged during the buildup of the voltage and current. The surge of power in the line causes this buildup. Then, during the decrease in current flow in the power line, as the ac reverses, the capacitor discharges. This causes another power surge in the motor starting windings, Fig. 7-8.

The running capacitor operates in the same way, except that it keeps on operating when the motor is running. This capacitor is usually smaller and is made to provide better heat removal. The insulation, plates, and terminals of capacitors are designed to last.

7-9 MOTOR SPEEDS

Induction motors—including split-phase, repulsion-start, and capacitor-start—are made for use on 25, 50, or 60 cycle (Hz) current. *The speed of a motor depends on the cycles (Hz) and number of field poles. Motor speeds are calculated from the synchronous speed.*

Synchronous speed is related to rotating magnetic fields. The stator windings of a motor produce a rotating magnetic field. (The field advances one pole for every 1/2 cycle of current.) If the rotor can keep up with the rotating field, the

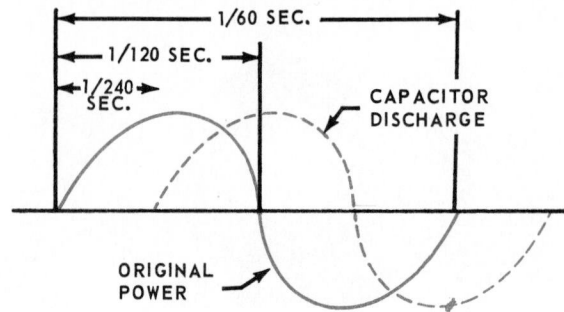

Fig. 7-8. Cycle diagram showing effect of capacitor in series with power flow in single-phase circuit.

motor runs at synchronous speed. The synchronous speed for two- and four-pole motors is shown in Fig. 7-9.

Motors do not operate exactly at synchronous speed. They are operating under some load and there is some magnetic slippage. Thus, motors are not rated at the synchronous speed, but rather at a speed which corresponds with a normal load. Fig. 7-9 shows operating speed for two- and four-pole motors operating at 60, 50, and 25 Hz.

There are hermetic motor compressors which may be made to operate as either two-pole or four-pole motors. Such an arrangement makes it possible to operate the motor compressor either at 3600 rpm or 1800 rpm.

The residential hermetic motor compressor, Fig. 7-10, may be wired to operate at a low speed or a high speed.

In Fig. 7-11, view A, the run windings are connected in series. Leads 1 and 2 are connected to terminals 1 and 7. These windings form a four-pole motor operating at a low speed of 1800 rpm. In Fig. 7-11, view B, the unit operates at a high speed when the run windings are connected in parallel, L1 and L2 to terminals 1 and 2. This winding forms a two-pole motor operating at 3600 rpm.

A two-speed three-phase motor compressor is shown in Fig. 7-12. The motor operates as a two-pole motor at high speed and as a four-pole at low speed. The circuit diagram for a two-speed motor compressor is shown in Fig. 7-13. Solid-state circuits control the compressor speeds.

Motors designed and constructed to operate on a certain Hz will not operate at a different Hz. This is because the number of turns of wire on the field poles and the amount of iron in the magnetic circuit is different for each Hz. Some motors have field connections in either series or parallel, allowing either 120 V or 240 V operation. On 120 V, the field winding would be connected in parallel. Operating on 240 V, the field winding would be connected in series.

It is possible to build motors having six, eight, or more poles. However, such motors are not in common use.

	MOTOR SPEED – RPM					
NO. OF POLES	60 Hz		50 Hz		25 Hz	
	SYN.	OP.	SYN.	OP.	SYN.	OP.
2	3600	3450	3000	2850	1500	1450
4	1800	1750	1500	1450	750	700

Fig. 7-9. Synchronous and operational speed for two- and four-pole motors. Note that operational (op.) speed is approximate.

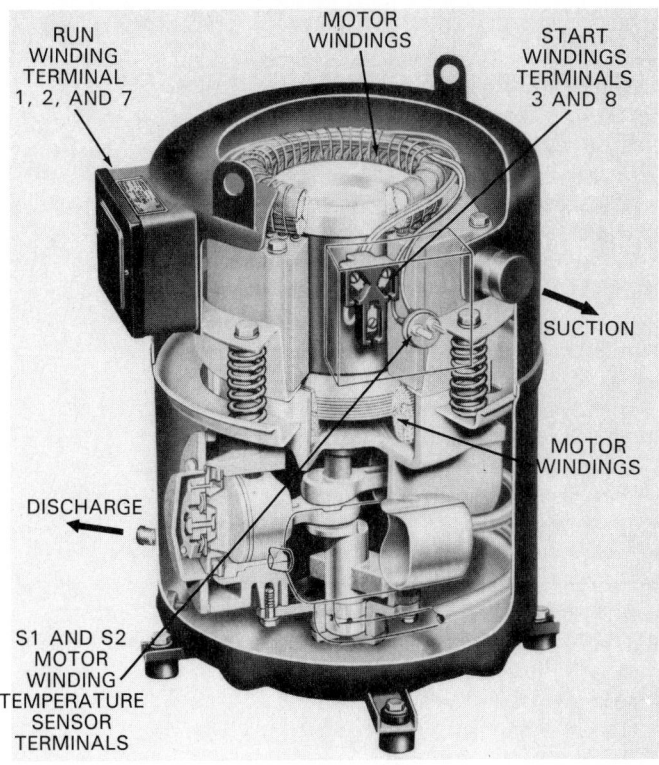

Fig. 7-10. Residential two-speed hermetic motor compressor. It uses single-phase electrical power. (Lennox International, Inc.)

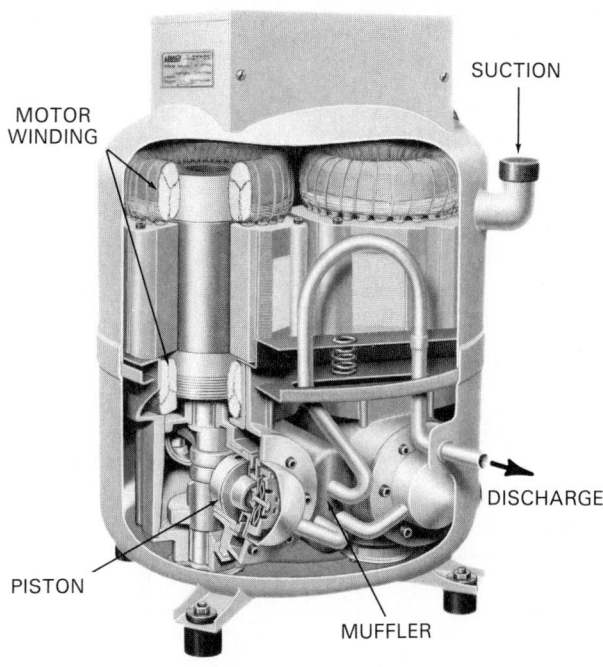

Fig. 7-12. Commercial two-speed hermetic motor compressor. It uses three-phase electrical power. Note muffler in discharge line. Muffler reduces noise and stabilizes discharge pressure. (Lennox International, Inc.)

7-10 STARTING AND RUNNING WINDINGS

As mentioned in previous paragraphs, most motors have a running winding and a starting winding. These windings are mounted on the stator.

During starting, current goes through both windings. When the motor reaches 60 to 75 percent of its running speed, the starting winding circuit is opened and the motor operates on the running winding only. The 60 percent value is for 6- and 12-pole motors. *A starting winding is found on split-phase motors and on all types of capacitor motors.*

This winding has the same number of coils as the running winding. However, it has smaller diameter wire and a greater number of turns. Its induction action, therefore, splits the phase, creating a rotating magnetic field.

Electricity passes through the running winding all the time the motor is in operation. Heavier wire is used than in the starting winding. It is installed on the field poles on both the two-pole and the four-pole motors.

Fig. 7-14 shows the electrical circuits of a split-phase electric motor while the motor is starting. When the current goes into the R terminal (during one-half of the cycle),

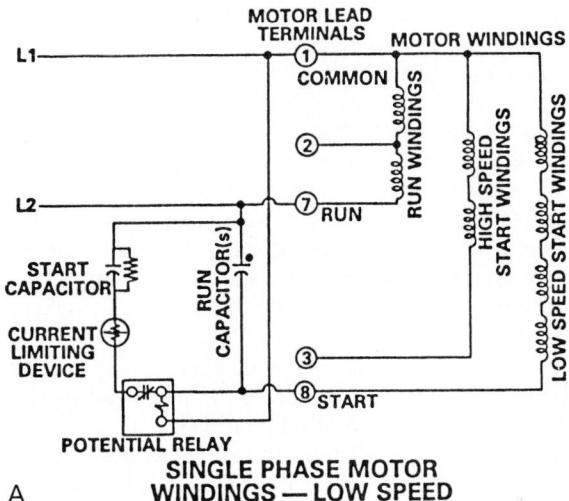

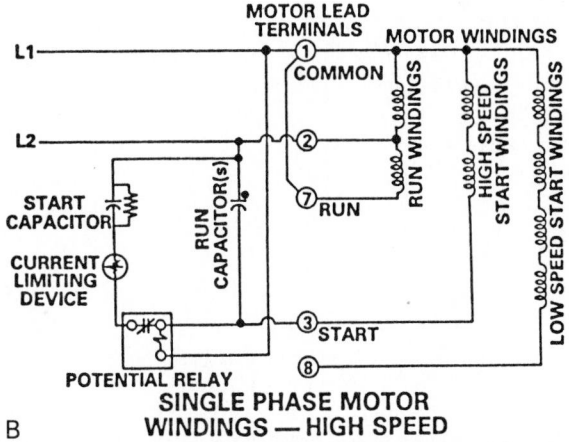

Fig. 7-11. Residential hermetic motor compressor winding connections. Connections are shown for low speed, A, and high speed, B. (Lennox International, Inc.)

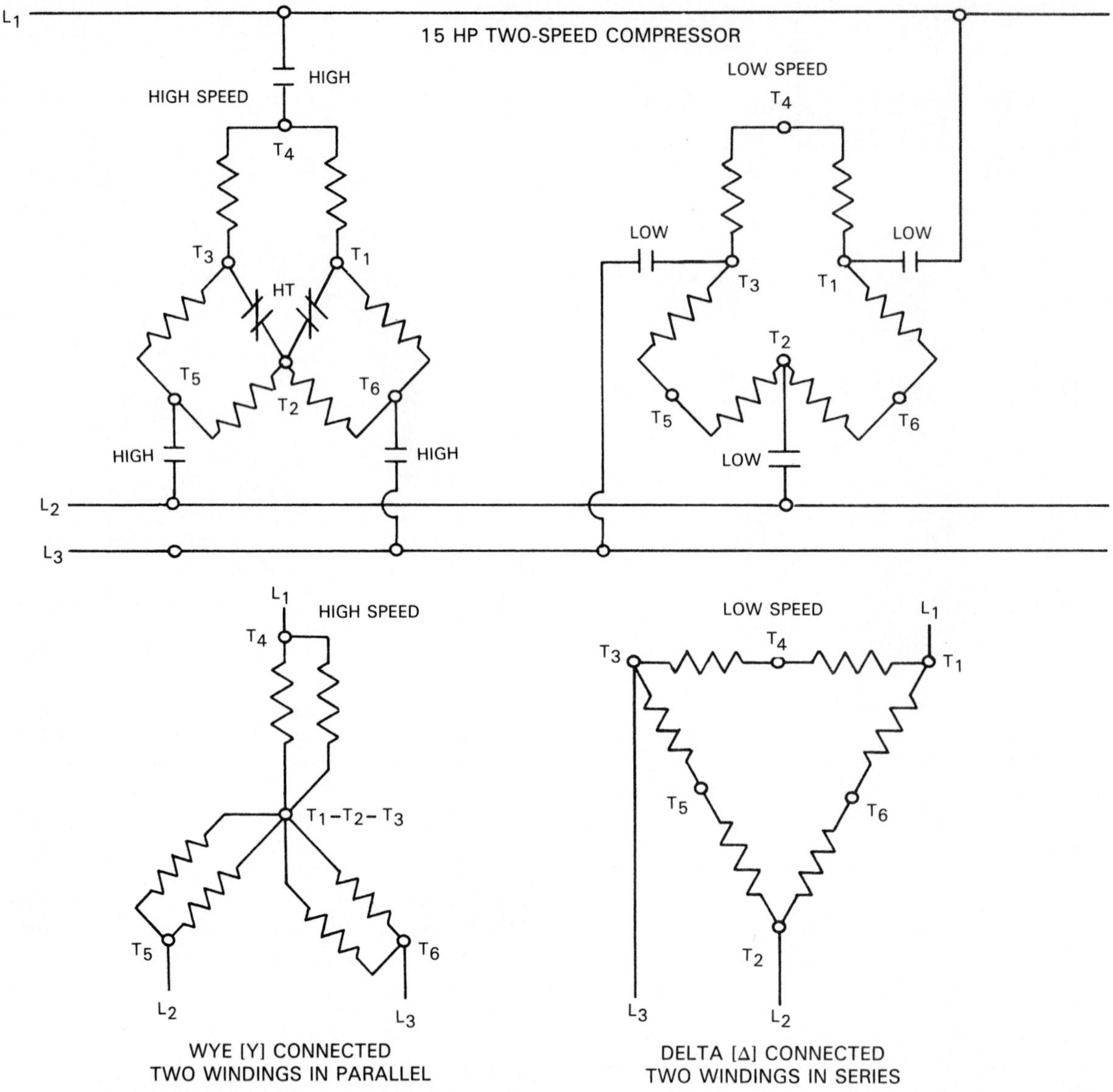

15 HP TWO-SPEED COMPRESSOR

HIGH SPEED

LOW SPEED

WYE [Y] CONNECTED
TWO WINDINGS IN PARALLEL

DELTA [Δ] CONNECTED
TWO WINDINGS IN SERIES

Fig. 7-13. Circuit diagram of two-speed, three-phase hermetic motor compressor. (Lennox International, Inc.)

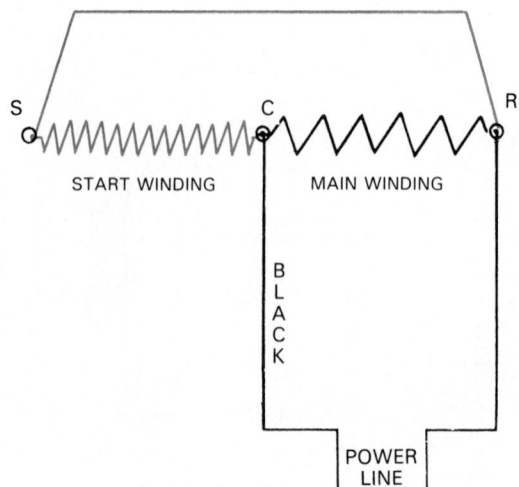

Fig. 7-14. Circuit diagram of split-phase motor. C—Common terminal for both windings. C to R—Main or running winding. C to S—Starting winding. R—Running winding terminal. S—Starting winding terminal. (Copeland Corporation)

the electrons separate. Most will go through the running winding but some will go through the starting winding. Magnetism builds up faster in the running winding. Thus, when the magnetism is created a few thousandths of a second later in the starting winding, a turning force is exerted upon the rotor.

When current enters terminal R during the other half cycle, most will go through the running winding. Some will go through the starting winding. This action reverses the polarity of the electromagnets and, again, there is a delay of magnetic buildup in the starting winding. The rotor is attracted and repulsed in the same direction of rotation as in the first half of the cycle.

To change the direction of rotation on some of these types of motors, disconnect the starting winding leads from the two terminals and reverse them. That is, put the old S connection (lead) on terminal R and put the old R connection (lead) on terminal S. Reversing the two main leads will not reverse the rotation of the motor. (With start windings disconnected, one can start the motor manually in either direction.)

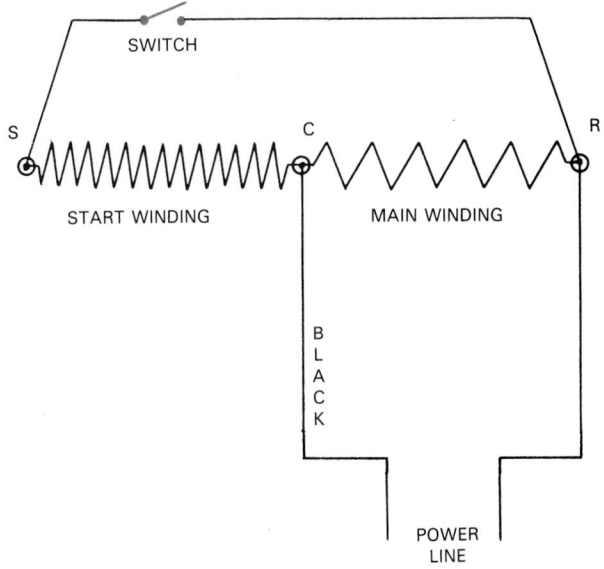

Fig. 7-15. Circuit diagram of split-phase motor using switch in starting winding circuit. Switch is operated by rotor speed or relay and is closed as motor starts. Switch will open when motor reaches nearly 75 percent of running speed. (Copeland Corp.)

If the starting winding is left in the circuit, it may overheat. Therefore, a switch is mounted in the starting windings circuit. See Fig. 7-15. After a motor reaches approximately 75 percent of the motor's rated speed, the switch disconnects the winding. It is open while the motor runs.

7-11 STARTING CURRENT

The instant the starting circuit is closed on a motor, there is little or no counter emf. Therefore, the amount of current starting to flow will be determined by the resistance in the circuit. At the instant of starting, the current flow will be quite high (two to four times the running current).

This initial high flow of amperage is called "locked rotor amperage." During this high amperage flow, the voltage drops to its lowest reading.

Frequent starting of a motor causes these high current flows to happen too often. The motor will overheat—especially those of 1/2 hp or more.

Fuses and circuit breakers must be designed with this fact in mind. If the protective devices are set to allow only running current, they will open at the instant of start. The protective device must have delayed action—usually a heating strip and a bimetal switch.

7-12 MOTOR CONNECTIONS

Except for those that are reversible or of some special design, motors usually come with two or four leads to connect the motor to one of two different voltages.

Provided the correct voltage is maintained at the motor terminals, 120 V, 240 V, or other corresponding voltages are equally good. These motors may be used on circuits having voltages of 105 to 125 or 210 to 250 volts.

Most electrical power companies are now providing 120 V and 240 V electrical power. These voltages are safe to use on motors labeled 110-220 V or 115-230 V.

It is recommended that, when possible, all motors be connected to 240 V service. If the wires are of good size, this will result in a good voltage at the motor terminals, especially if the motor is heavily loaded.

Most repulsion-start induction-run motors use four motor leads. These leads are connected to field windings, two leads to a winding. The windings are connected in parallel if the voltage to be used is 120 V. The leads are connected in series if the voltage used is 240 V. Since the power is the product of the emf in volts multiplied by the amperage, the amperage flow for any specified power using 240 V will be half the current flow if using 120 V.

Fan motors and similar units are usually wired either for 120 V or 240 V only. Only two leads are connected to the field windings. Air conditioning evaporator fans often have two or three speeds. These motors have three to four leads.

One must be careful to avoid connecting a 240 V motor to a 120 V circuit. If this is done, the motor will overheat and "burn out." The maintenance of proper voltage at the motor terminals is absolutely necessary.

The power that an alternating current motor will develop varies directly to the square of the voltage at the motor terminals. Low voltage will also result in a rapid drop in the horsepower capacity of any motor.

See Chapter 30 for information on the wire size to be used. Loose connections cause excessive voltage drop and are a fire hazard. All connections should be made with solder or with Underwriters approved pressure connectors.

A 120 V motor supplied with current at 100 V will develop only about 75 percent as much power as it would if supplied with 120 V current. Therefore, to do the job and to develop its full horsepower, it must draw more current. This increase in current may overheat the "power in" leads and/or the motor windings, causing insulation failure.

Fig. 7-16 shows a method of making connections for single-phase motors. Note how the direction of rotation may be changed. Special motors requiring other connections are shipped with diagrams showing proper connections.

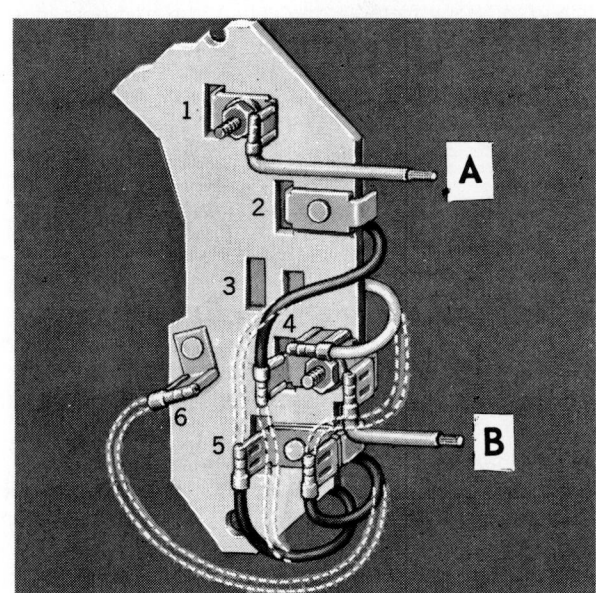

Fig. 7-16. How to make motor leads give desired direction of rotation and connect motor to voltage supplied. Leads for power input are shown at A and B. Wiring diagram inside motor cover plate indicates correct connection for each voltage and direction of rotation. Dotted lines show method of reversing. (Magnetek)

A 120-240 V motor is connected to a 120 V line through an across-the-line switch. This switch is usually operated by the cabinet temperature, although low-side pressure may also be used. The same motor control may be used to operate a 120-240 V motor installed on either the 120 V or 240 V circuit.

The field winding leads come out of the motor frame through an insulated opening. The wire is normally protected with rubber, fabric, or plastic grommet. A small metal box is usually mounted over the leads to protect the electrical connections.

7-13 HERMETIC SYSTEM MOTORS

When the motor and the compressor are placed inside a dome or housing, it is called a hermetic unit. In such systems, the motor must drive the compressor directly. This requires good, leakproof electrical connections. The motor must have good power characteristics and must be of the induction type.

Motors using rotor windings requiring either brushes or slip rings cannot be used. Development of the split-phase motor and the capacitor motor made the hermetic motor possible.

The original hermetic motors were four-pole operating at approximately 1750 rpm.

To calculate the motor speed, the formula is:

$N = \dfrac{120f}{P}$ where N = rpm

f = frequency (cycle)

P = number of poles

$N = \dfrac{120 \times 60}{4} = 30 \times 60 = 1800$ revolutions per minute (rpm)

The 120 in the formula is the 60 seconds per minute × 2 (the magnetism or polarity changes 2 times per cycle).

Using the above formula, the speed of a two-pole motor will be:

$N = \dfrac{120 \times f}{P}$

$N = \dfrac{120 \times 60}{2} = 60 \times 60 = 3600$ rpm synchronous speed or about 3400 rpm actual speed.

The reason the motor is almost twice as fast is that the motor rotor has to travel one-half revolution during one-half cycle instead of one-fourth revolution as with the four-pole motor. These two-pole motors are about two-thirds the size of the four-pole motors of the same power.

Hermetic motors present some problems not found with open motors:

1. Special cooling provisions must be made.
2. Wiring insulation must be resistant to oil and chemicals in the refrigerant (particularly in the presence of moisture and/or high temperatures).
3. Manufacturing standards must provide exact alignment of the stator, rotor, and compressor.
4. Electrical connections through the dome must be electrically perfect and leakproof.

The motors may be cooled by several methods. One successful method is to press the stator into the dome and then put cooling fins on the dome.

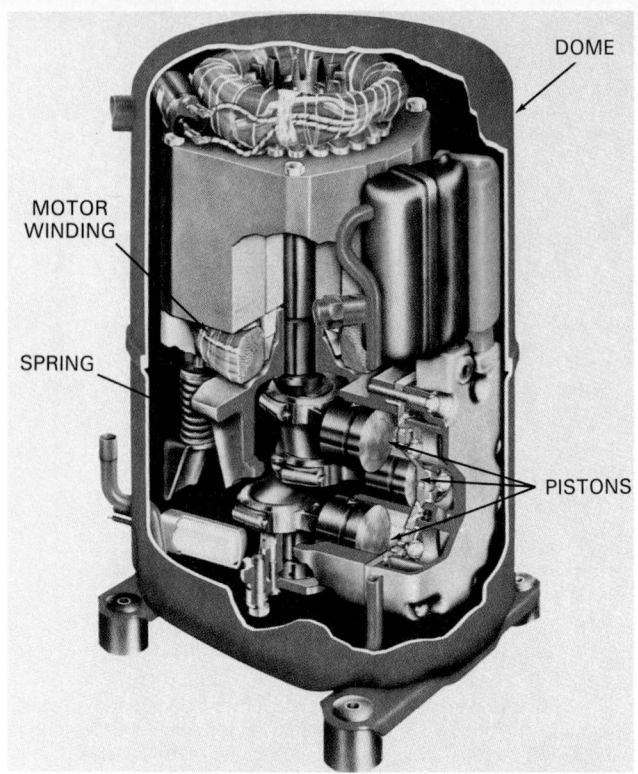

Fig. 7-17. Hermetic motor compressor unit. Note that compressor is mounted on springs inside dome. Muffler helps prevent liquid refrigerant from filling tubing on top of cylinder valve. (Tecumseh Products Co.)

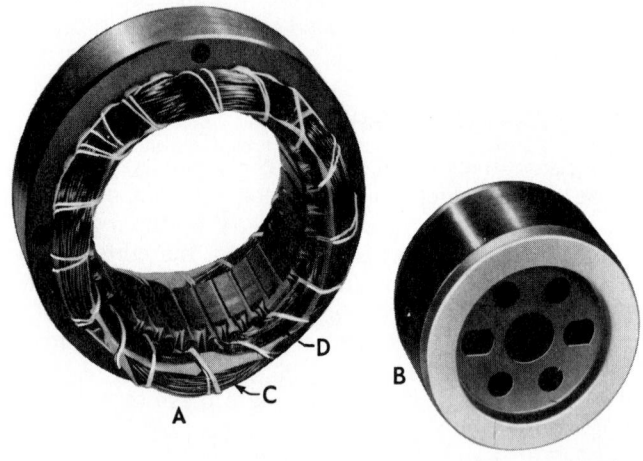

Fig. 7-18. Rotor and stator from a 1/4 hp split-phase hermetic motor. A—Stator. B—Rotor. C—Running winding. D—Starting winding.

Another method uses a water coil to cool the motor windings while the unit is running. Some designs pass partly cooled condenser refrigerant around the motor housing to help cool the electric motor. Most systems pull the cool return suction line vapor and oil over the motor windings.

Most manufacturers have developed special synthetic wire coatings (usually synthetic enamels) which have good insulating qualities while being safe to use with most of the popular refrigerants. The air gap between the rotor and the stator is only a few thousandths of an inch. It must be equal on all sides; otherwise, the motor may hum. Note, in particular, that many motor compressors are mounted on springs inside the dome.

Fig. 7-17 shows a hermetic motor compressor unit. The stator and rotor of a hermetic motor are shown in Fig. 7-18.

7-14 HERMETIC MOTOR ELECTRICAL CHARACTERISTICS

The design characteristics of electric motors used in hermetic machines will depend on whether the unit starts under load, no-load, or a balanced pressure condition.

The basic operation of the motor is the same as for open-type motors. All torque is developed by induction only.

The higher starting torque (turning effort) needed for those units which start under load requires the use of larger conductors in the starting circuit. Usually, manufacturers try to provide starting power equal to twice the running power. In other words, a 1/6 hp motor is designed to produce 1/3 hp during starting. Various methods are used to shut off the special starting devices after the motor has reached full speed.

Fig. 7-19 shows the external circuit of a hermetic motor, including a solid state positive temperature coefficient (sometimes referred to as a PTC) resistor that decreases the resistance as the temperature is increased; a thermally-operated overload protector; and a run capacitor.

7-15 HERMETIC MOTOR TYPES

Hermetic motors are either single-phase or polyphase. Four types of single-phase induction motors are used:
1. Split-phase (SP).
2. Capacitor-start, induction-run (CSIR).
3. Capacitor-start, capacitor-run (CSCR).
4. Permanent-split capacitor (PSC).

7-16 HERMETIC SPLIT-PHASE INDUCTION MOTOR

Split-phase induction is the basic type motor for small hermetic condensing units.

The principle of operation is simple. There are two windings—one for running and one for starting. Since starting torque is low, such motors must be used on systems with low starting load.

The split-phase motor is very popular on systems which use the capillary tube refrigerant control, since pressures balance on the off cycle. Thus, the compressor is not required to start under a load.

These motors may also be used where the system provides an electrical, mechanical, or hydraulic pressure unloading device. In these installations, any type refrigerant control may be used.

A split-phase motor used in hermetic motor compressor systems must have some type of outside starting relay. This may be a thermal, current, or potential type relay. See Chapter 8 for information on these relays.

A schematic wiring diagram for a hermetic split-phase induction motor with a current type starting relay is shown in Fig. 7-20. Fig. 7-21 shows the same motor using a potential type relay. These motors are most used on small units—1/10, 1/6, to 1/3 hp.

7-17 MOTOR CAPACITORS

The capacitor in an alternating current circuit basically changes a single-phase electrical flow to a two-phase electrical flow. The capacitor is usable in both the starting winding and the running winding.

If used in the starting winding only, it is called a start capacitor. Being used only for a few seconds at a time, it should have no overheating problems. However, if a motor

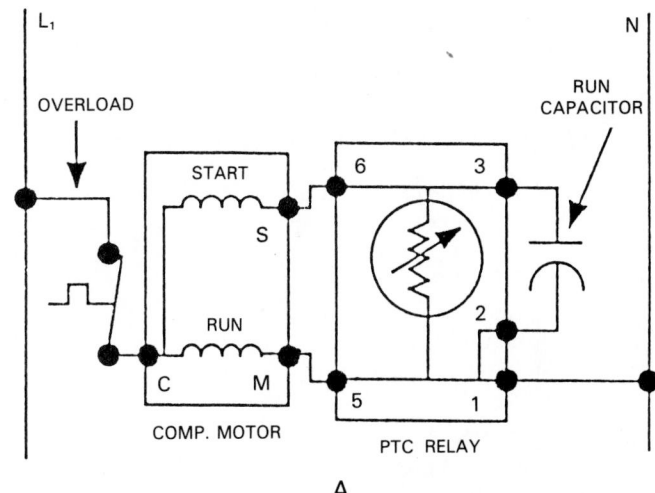

A

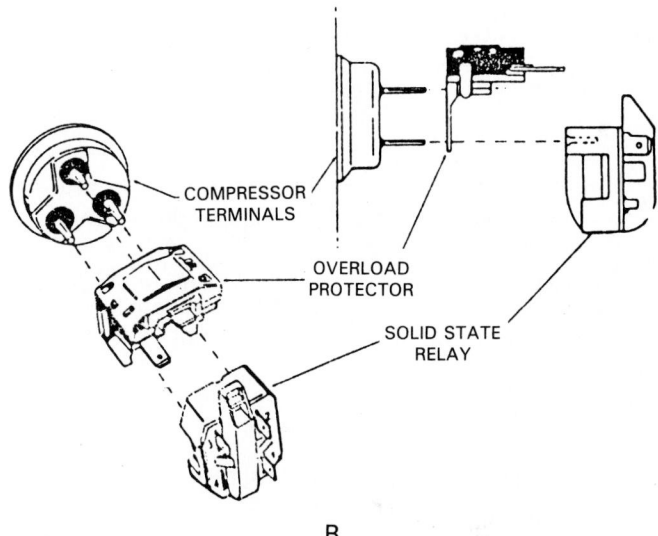

B

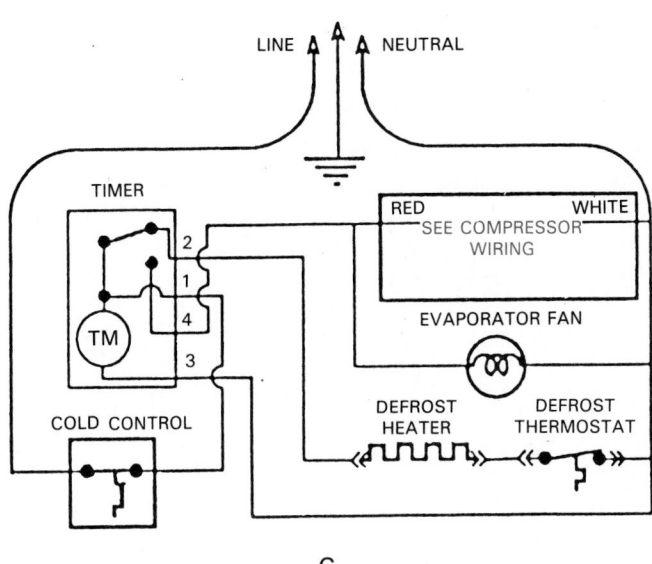

C

Fig. 7-19. External circuit of a hermetic motor. A—Compressor electrical circuit. B—Compressor electrical components. C—External circuit used with high-efficiency compressor. (Frigidaire Company)

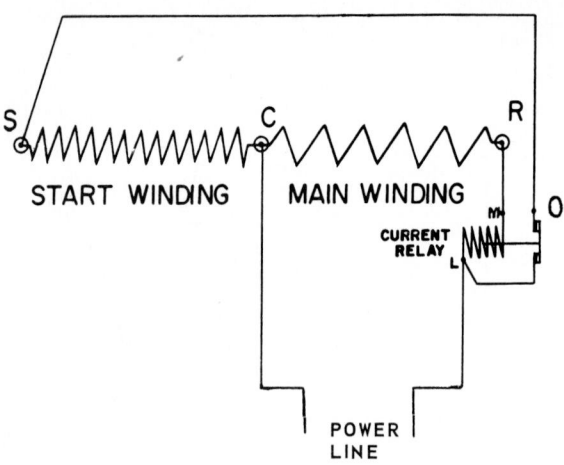

Fig. 7-20. Wiring diagram of hermetic split-phase induction motor. With first burst of current from power line, current relay control switch O is closed and start winding is cut in. As motor begins to start, current flowing through relay control winding and main winding keeps switch O closed. As motor reaches about 3/4 of running speed, current through main winding and relay coil is reduced. This opens switch O to cut out start winding, and motor operates on main winding. C—Common terminal. R—Running terminal. S—Starting terminal. (Copeland Corp.)

is started too often or if the starting winding is used longer than it is designed for, the capacitor insulation will overheat and the capacitor may fail.

If the capacitor is used for the running winding, it is carefully designed to discard any heat generated in it during operation. *Never use a starting capacitor in the running circuit.*

Always use the correct voltage and the correct microfarad capacity when replacing a capacitor.

7-18 HERMETIC CAPACITOR-START, INDUCTION-RUN MOTOR

This is a very popular hermetic motor for refrigerating units. It has a good starting torque which is obtained by

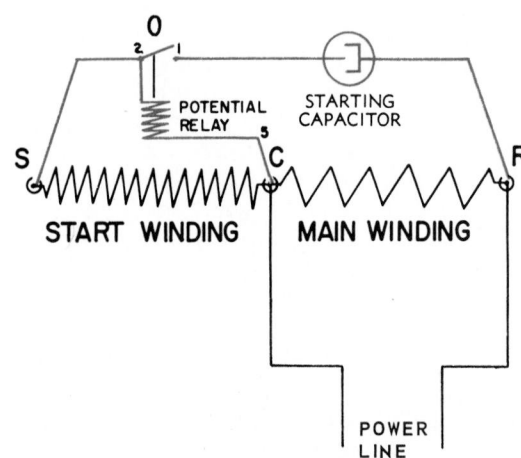

placing a capacitor in series with the starting winding. It can use any one of several starting relays: current relay, potential relay, or a hot wire (thermal) relay.

Fig. 7-22 is a schematic wiring diagram for such a motor using a current type starting relay. A similar installation using a potential type relay is shown in Fig. 7-23. The starting capacitor circuit is kept open during the running cycle by the induced emf generated in the starting winding (across terminals S and C). Induced voltage pushes enough current through the potential relay coil to produce a magnetic effect. The magnetic force keeps the relay points open.

See Chapter 8 for detailed information concerning construction and operation of the different types of motor controls.

7-19 HERMETIC CAPACITOR-START, CAPACITOR-RUN MOTOR

The capacitor-start, capacitor-run motor generally uses two capacitors. Both are in the starting winding circuit but only the start capacitor is controlled by the relay switch.

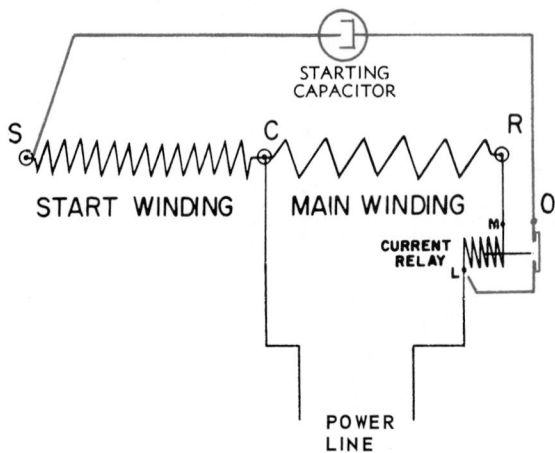

Fig. 7-22. Current type relay is shown in starting position (points closed). C—Common terminal. R—Running terminal. S—Starting terminal. O—Relay control switch. (Copeland Corporation)

Fig. 7-21. Wiring diagram of hermetic split-phase induction motor with a potential type relay in running position. Relay control winding develops current from generation of electricity by starting winding. When enough magnetic effect is produced, it will open switch O just after motor starts. C—Common terminal. R—Running terminal. S—Starting terminal. O—Relay control switch. (Copeland Corporation)

Fig. 7-23. Capacitor-start induction-run motor with potential type relay in running position (points open). C—Common terminal. R—Main or running terminal. S—Starting terminal. O—Relay control switch. (Copeland Corp.)

When the motor is started, the capacitors turn the motor power surges into two-phase power and produce a high starting torque. After the motor reaches two-thirds or three-fourths of its rated speed, the relay opens the circuit to the starting capacitor. The running capacitor is left in the circuit.

This action produces a two-phase motor that is very efficient. The power factor is improved. Larger hermetic units in commercial systems use this type of motor.

Fig. 7-24 shows the wiring diagram. Note that the running capacitor is in series with the starting winding. In Fig. 7-25, the motor wiring circuit shows two starting capacitors and two running capacitors. These are connected in series to increase the voltage capacity. (Two 120 V capacitors in series can be used in a 240 V circuit.)

7-20 PERMANENT SPLIT CAPACITOR MOTOR (PSC)

The permanent split capacitor motor is popular for air conditioning systems. It does not use a relay. Current flows through both the running winding and the starting winding when the power is on. (See diagram, Fig. 7-26.) A running capacitor is connected between the running (R) and starting (S) terminals and is in series with the starting winding.

Such motors are sensitive to line voltage. A 5 to 10 percent drop will cause starting difficulty and overheating. To prevent damage, thermal protectors will open the circuit.

Starting torque is low. Thus, if it tries to start when the system's pressures are not balanced (equal), the motor will overheat and thermal protectors will open the circuit.

Chapter 21 explains how to change a PSC into a capacitor-start, capacitor-run motor in an air conditioner.

7-21 HERMETIC POLYPHASE MOTOR

Large hermetic compressors usually are driven by three-phase motors. The surges of current travel in these motors are closer together than with single-phase current supply. Therefore, they are more efficient power sources. The sine curves for a two-phase motor are shown in Fig. 7-27, while

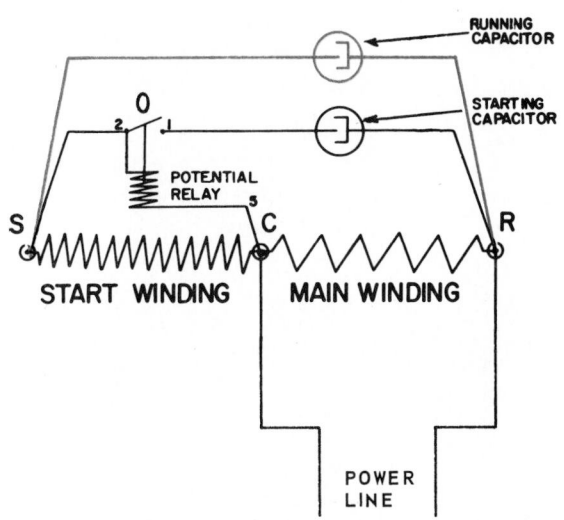

Fig. 7-24. Schematic wiring diagram of capacitor-start capacitor-run motor using potential relay. Relay is in running position (points open). C—Common terminal. R—Main or running terminal. S—Starting terminal. O—Relay control switch. (Copeland Corporation)

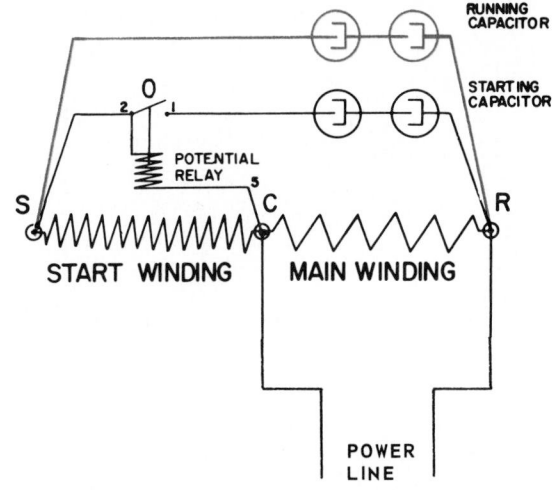

Fig. 7-25. Schematic wiring diagram of hermetic motor. This circuit has potential relay in running position. C—Common terminal. R—Running terminal. S—Starting terminal. O—Relay control switch. (Copeland Corporation)

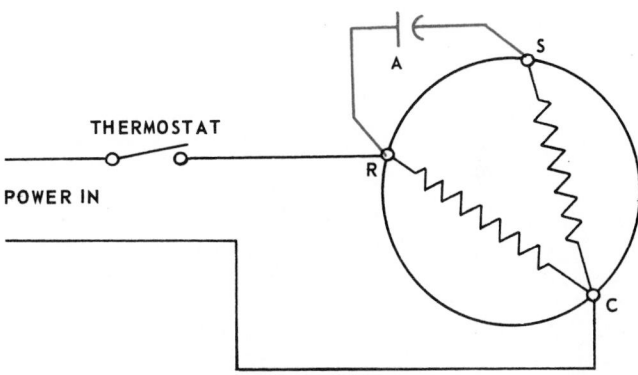

Fig. 7-26. A PSC (permanent split capacitor) motor. A—Running capacitor.

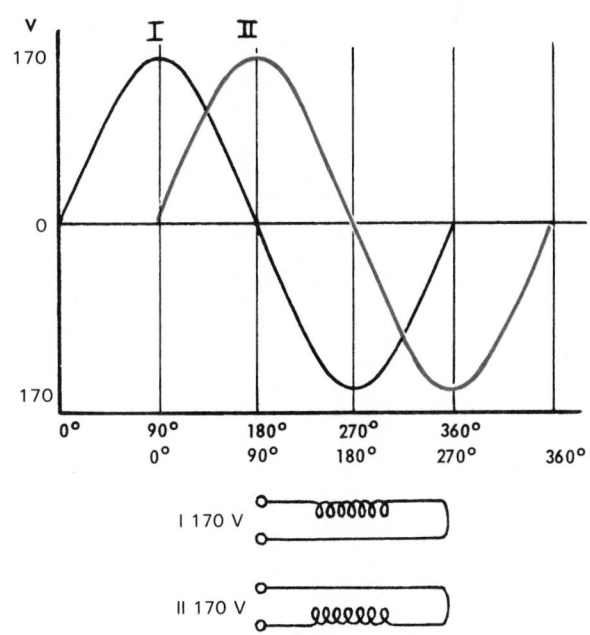

Fig. 7-27. Two voltage sine curves for two-phase power circuit. Note that there are two separate circuits. They are 90 deg. out of phase.

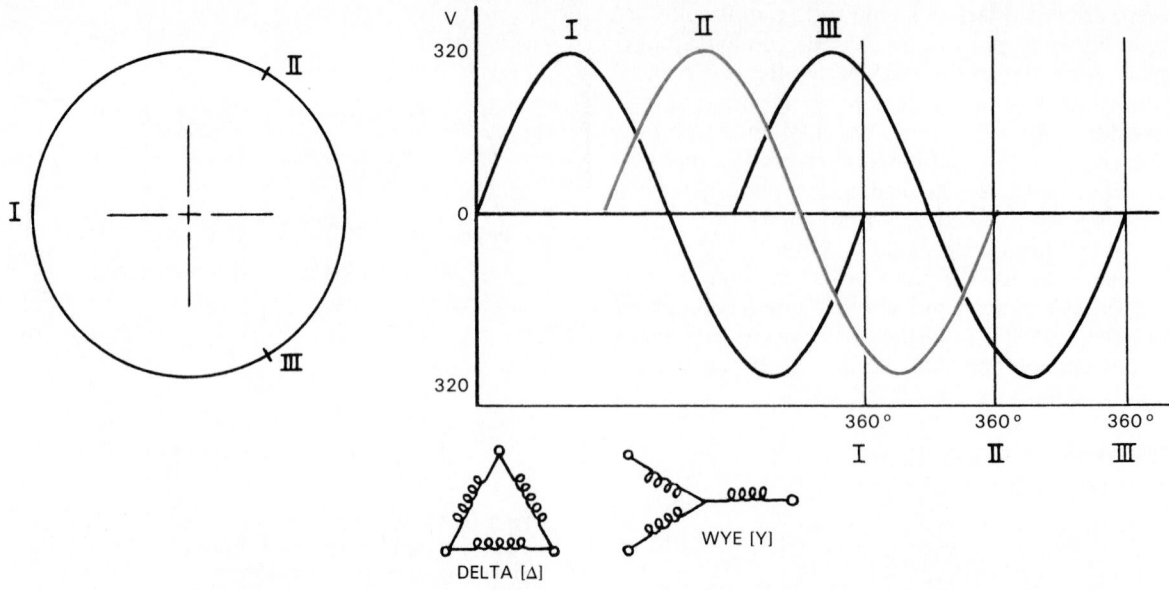

Fig. 7-28. Three voltage sine curves of three-phase circuit. System can be wired for either Delta or Wye design.

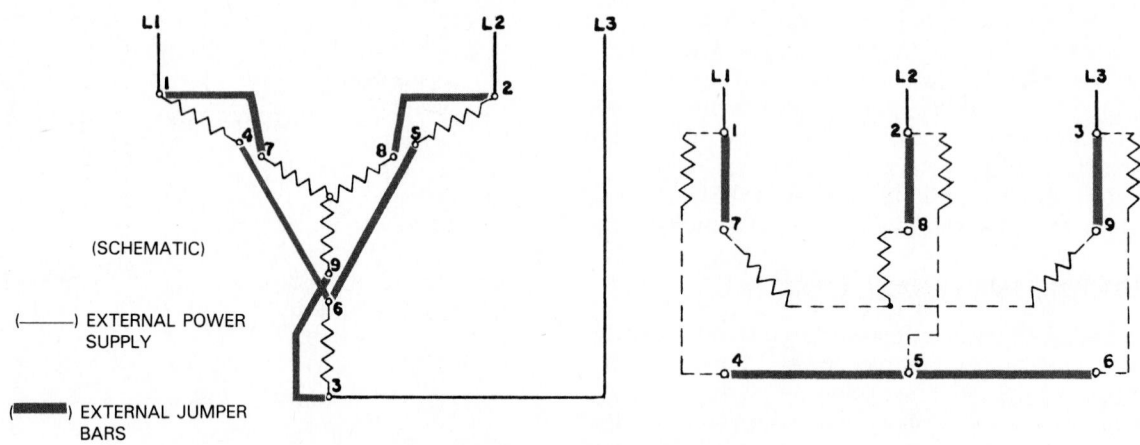

NINE TERMINAL DUAL VOLTAGE MOTOR
(PARALLEL CONNECTED FOR 220 VOLTS ACROSS-THE-LINE)

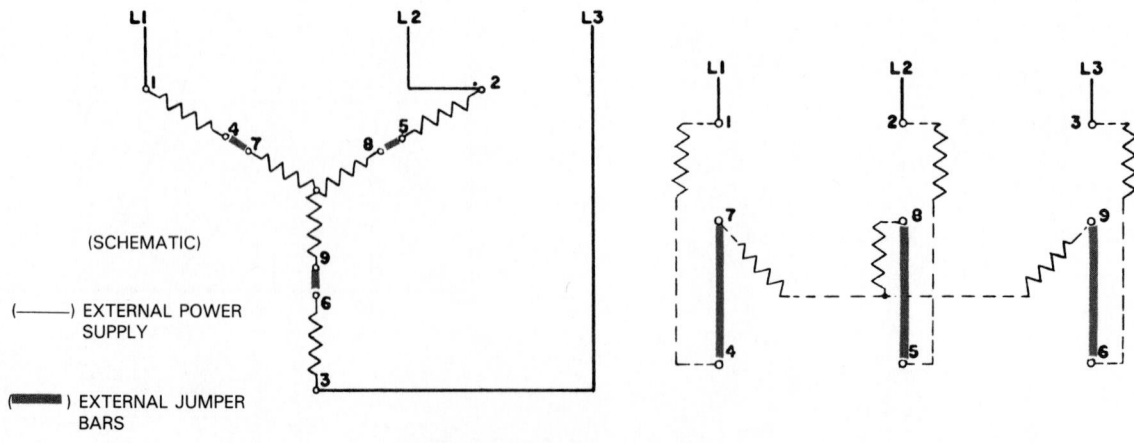

NINE TERMINAL VOLTAGE MOTOR
(SERIES CONNECTED FOR 440 VOLTS)

Fig. 7-29. Schematic wiring diagram showing circuits and connections for three-phase hermetic motor. Top. Circuit as connected for 220 V. Bottom. Circuit as connected for 440 V. Note that L_1, L_2, and L_3 are the three-phase line connections. Numbers 1-2-3, 4-5-6, 7-8-9 are the connections to the motor windings. Each motor winding coil is designed for 220 V; example, the coil between terminals 1 and 4 is rated for 220 V.

the sine curves for a three-phase motor are shown in Fig. 7-28. These motors are usually of the 220 V or 440 V type. Since the dome terminal block has nine terminals, the service technician may wire the motor for either 220 V or 440 V. (See Fig. 7-29.) Some of these motors use a 550 V supply.

Three-phase motors are available from 1/2 hp up. The building in which the unit is to be placed must be wired for three-phase service. Very few residences have three-phase electrical power, but most industries and some commercial buildings do.

The higher voltages are very dangerous. Disconnect the power and lock the switch open (use an actual lock) before starting to service the machines.

Three-phase motors use contactors or motor starters. They do not have the usual starting relays.

Fig. 7-30 shows a three-phase motor wiring circuit with its starting and protector circuit. It is always best to have an electrical journeyman do the electrical work on these units. Since each unit may have certain differences, when servicing the system, it is important to get the wiring diagram for the unit from the manufacturer.

Operating characteristics of two types of polyphase motors are shown in Fig. 7-31.

Occasionally, a three-phase motor may blow a fuse or open a circuit breaker on one phase only. The motor will attempt to operate on the remaining two phases. If there is much load on the motor, it will quickly overheat and may burn out. This is because the two windings must carry all the load so each one will need 1 1/2 times the current.

Phase loss monitors are sometimes used to shut down a motor to prevent it from being damaged.

Each phase of a three-phase motor must be tested individually using a voltmeter. There will be about 50 volts difference between the open line and one of the other lines. The circuit having the "blown" fuse will indicate below-normal voltage. The direction of rotation of a three-phase motor may be reversed by changing any two of the power leads to the motor.

7-22 HERMETIC MOTOR TERMINALS

The electrical terminals which carry the circuit through the dome must be electrically insulated from the dome or housing. They must also be leakproof.

Most motor terminals are fused to glass while the glass, in turn, is fused to a metal disk. See Fig. 7-32. This assembly may be welded to the hermetic dome or housing. The terminal must be leakproof after thousands of heating and cooling, expansion and contraction cycles. Furthermore, it must have a high insulating value.

A fused glass multiple terminal installation is shown in Fig. 7-33. The wire terminals are spring clips that tightly grip the hermetic terminals.

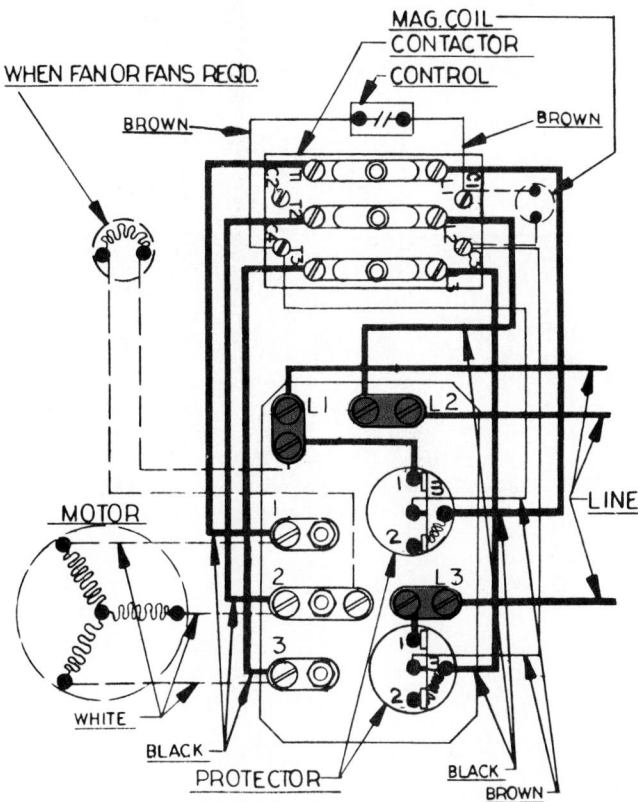

EXTERNAL INHERENT PROTECTION
(2) 3 TERMINAL PROTECTORS & CONTACTOR

Fig. 7-30. Three-phase hermetic motor circuit. L₁, L₂, and L₃ are three-phase line connections. Overload protector is shown in L₁ and L₃ circuits and operates magnetic coil of starter switch (top part of drawing). Special magnetic type starter is required. Automatic temperature control would be connected to control at top of illustration. (Copeland Corp.)

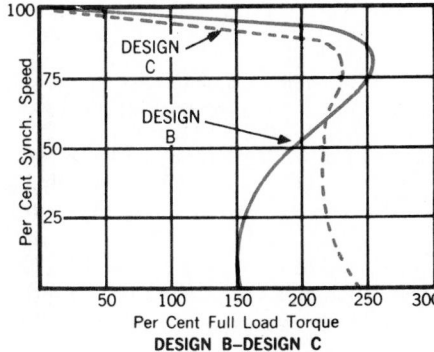

DESIGN B–DESIGN C

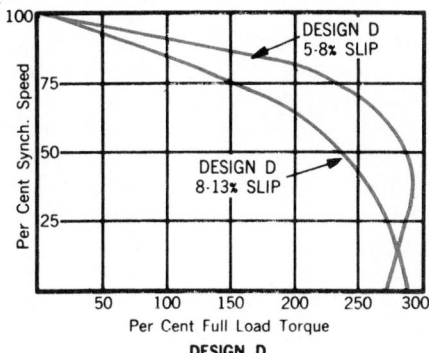

DESIGN D

Fig. 7-31. Speed/torque curves compared for two polyphase motor designs. Note that Design B and C motors provide starting torque of 150 and 250 percent of full load torque. Design D motors provide starting torque of 260 to 280 percent. However, speed of Design D motors will fluctuate more than Design B and C motors as load changes. (Magnetek)

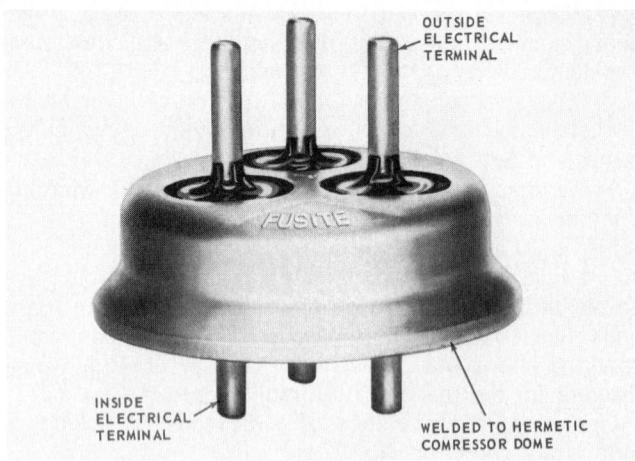

Fig. 7-32. Metal electric terminals which may be welded to hermetic compressor dome. (Fusite Division, Emerson Electric Co.)

Fig. 7-33. Cross-sectional view of metal-glass fused hermetic electrical terminal installed. Wires to relay are connected to terminals with spring-loaded clips. Note metal structure around terminals to protect them from abuse.

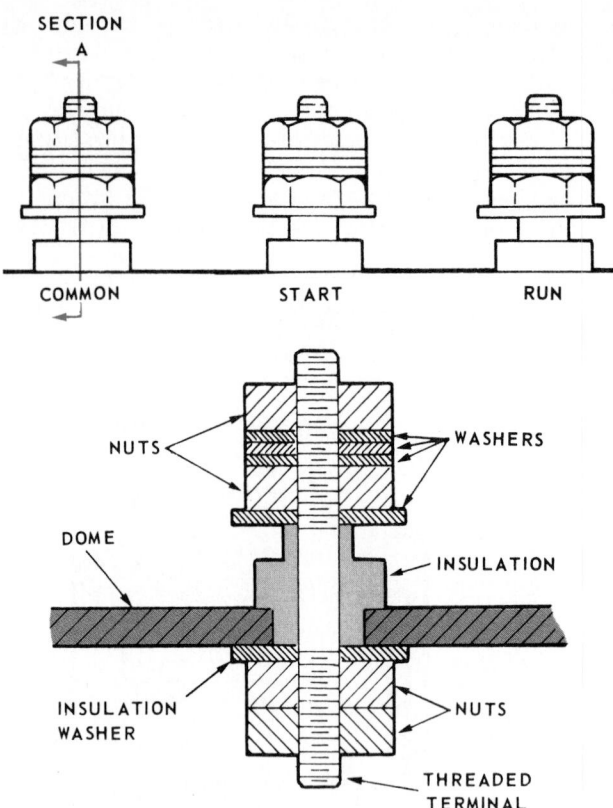

Fig. 7-34. Built-up hermetic motor terminals. A—Cross-section of one of the terminals, showing insulation.

Some compressors use what is known as built-up terminals. These are attached to the compressor dome, as in Fig. 7-34.

Replacement terminals are used by many service technicians. Gaskets of synthetic material are used to make a leakproof joint.

7-23 DIRECT CURRENT AND UNIVERSAL MOTORS

Areas with dc power must use direct current motors in refrigerators. These motors are compound wound and have a mechanical likeness to both the capacitor- and repulsion-start induction type motors. However, the electrical circuits are quite different. A circuit diagram is shown in Fig. 7-35.

Compound wound motors have two types of field windings; one is in parallel while the other is in series with the armature winding. Because dc is used, the field poles always have the same magnetic polarity. Also since a dc current is going through the armature coil, the magnetic polarity of the armature will remain constant. It is positioned to cause a turning effect or torque in the armature. The series field helps to keep the motor speed constant.

The series winding strength builds up to increase the rpm if the motor tends to slow down or weakens to reduce the

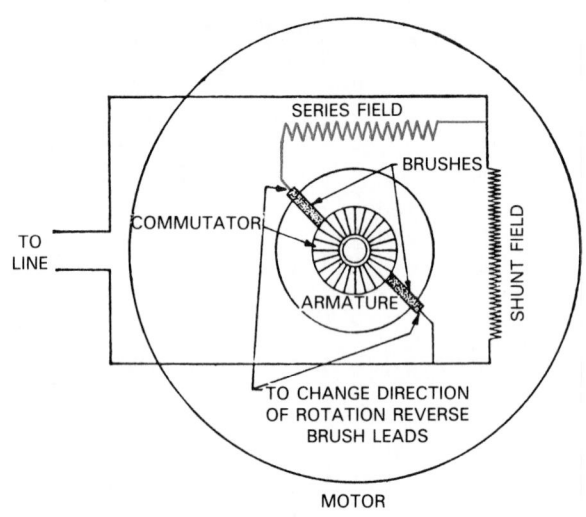

Fig. 7-35. Wiring diagram of compound wound direct current motor. Note that current going through armature must pass through series field. Shunt field winding is in parallel with armature/series field winding circuit. Series winding gives motor high starting torque. (Fedders North America)

rpm if the motor tends to speed up. One has only to reverse the brush leads to reverse the direction of rotation of these motors. (It reverses the armature magnetic polarity.)

The possibilities for wear are slightly greater in the direct current motors than in the others because of the armature design. The brushes are in constant contact with the commutator, giving rise to troubles such as dirty commutator,

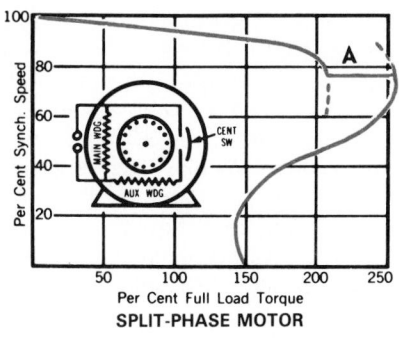
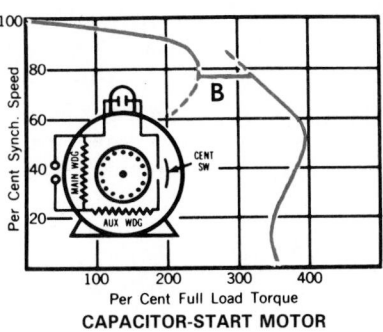
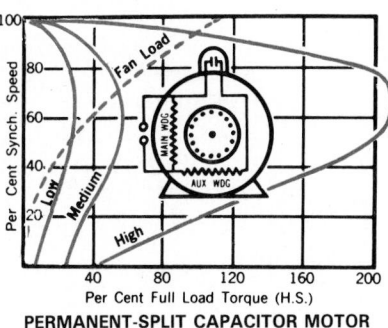

SPLIT-PHASE MOTOR   CAPACITOR-START MOTOR   PERMANENT-SPLIT CAPACITOR MOTOR

Fig. 7-36. Torque and speed characteristics of three types of single-phase fractional horsepower motors: Split-phase, left. Starting winding is disconnected by rotor-operated centrifugal switch at 75 percent of synchronous speed as shown at A. This type gives fair starting torque. It is available in horsepower ratings up to 3/4 hp and for either 120 V or 240 V. Capacitor-start, center. Starting winding is disconnected by rotor-operated centrifugal switch at about 75 percent of synchronous speed as shown at B. Higher starting torque is obtained by adding capacitor in starting circuit. Available up to 3 hp and for either 120 V or 240 V. Permanent-split capacitor, right. Capacitor stays in auxiliary winding whenever motor is running. Its use is limited to easy-to-start loads. PSC type may be used as multi-speed motor using simiple and inexpensive controls (see high, medium, low, fan in chart). Used in low horsepower applications. Available for either 120 V or 240 V. (Magnetek)

worn brushes, high mica insulation between the bars, shorted armature, and squeaky brushes.

Direct current may be used only on open type systems.

7-24 MOTOR HORSEPOWER AND MOTOR CHARACTERISTICS

Energy, work, and power are explained in Chapter 1. Motors are rated by horsepower. One horsepower is equal to lifting 33,000 ft. lb. per minute or 550 ft. lb. per second. At 100 percent efficiency, 746 W equals one hp. Refrigeration motors of 1/100 hp to several hundred hp are in use. Motors as small as 1/20 hp have been used to drive compressors.

In refrigeration, a motor's torque is as important as its horsepower. Torque is the force of a twisting or turning action (such as turning the crankshaft of a compressor).

It is important to know the properties of single-phase motors, the most popular of all motors. Fig. 7-36 shows the operating curves of the various types based on motor speed and percent of full load torque.

The table in Fig. 7-37 gives the average full load amperage and locked rotor amperage for various size of ac motors.

7-25 ELECTRIC MOTOR GROUNDING

Electric motors used on refrigeration and air conditioning systems must be grounded. Grounding of electric motors and other refrigerator parts and appliances is explained in some detail in Para. 6-72.

In all refrigeration and electrical circuits, the ground wire is green. This is never used as a current carrying conductor. Its main purpose is to provide protection to the operator in the event of an accidental ground in one of the mechanisms.

7-26 MOTOR PROTECTION

The most common causes of motor failure are overloads and overheating. An overload condition is produced when current to the motor exceeds the motor's normal operating current flow. An overload condition may result in melted conductors or burned insulation on the motor's conductors.

Considerable damage may also result if the compressor and/or motor overheat. This overheating may occur without the current draw becoming excessive.

It is important to protect a motor from both current overloads and overheating. Therefore, it is necessary to use both current and heat-sensitive devices which will open the circuit before damage is caused to the motor. The following devices are used for motor protection:
1. Fuses and circuit breakers.
2. Bimetal switches.
3. Electronic thermistors.

Each of these types of motor protection devices is described in detail in the following paragraphs.

7-27 MOTOR PROTECTION—FUSES AND CIRCUIT BREAKERS

Fuses and circuit breakers are often used to protect motors from burning out due to current overloads. Fuses and circuit breakers are usually located outside the motor.

Fuses conduct electrical current normally when operating below their maximum rating. (Fig. 7-38 shows the full-load current draw in amperes for various sizes of ac and dc motors, and the fuse ratings.)

When an overcurrent condition exists that exceeds the fuse's maximum rating, heat builds up inside the fuse,

	AMPERAGE			
	120 VOLTS		240 VOLTS	
HP	FULL LOAD	LOCKED ROTOR	FULL LOAD	LOCKED ROTOR
1/6	4.4	26.4	2.2	13.2
1/4	5.8	34.8	2.9	17.4
1/3	7.2	43.2	3.6	21.6
1/2	9.8	58.8	4.9	29.4
3/4	13.8	82.8	6.9	41.4
1	16.0	96.0	8.0	48.0
1 1/2	20.0	120.0	10.0	60.0
2	24.0	144.0	12.0	72.0
3	34.0	204.0	17.0	102.0

Fig. 7-37. Full load and locked rotor amperage for single-phase ac motors.

HP	AC Motors Single-Phase Split-Phase or Capacitor		DC Motors Compound Wound	
	120V	240V	120V	240V
1/6	4.4	2.2	- - -	- - -
1/4	5.8	2.9	2.9	1.5
1/3	7.2	3.6	3.6	1.8
1/2	9.8	4.9	5.2	2.6
3/4	13.8	6.9	7.4	3.7
1	16.0	8.0	9.4	4.7
1 1/2	20.0	10.0	13.2	6.6

Fig. 7-38. Maximum fuse ratings for motor running protection.

causing the conduction element inside to melt. After the element melts and opens the circuit, electrical current can no longer flow to the motor. A fuse with a melted element is called a blown fuse. A circuit using this type of protection is shown in Fig. 7-39.

There are four basic types of fuses:
1. Fast-acting.
2. Time-delay.
3. Multipurpose.
4. Current-limiting.

The starting current of a motor can be from two to six times the running current of the motor. Fast-acting fuses blow immediately after the maximum rating of the fuse is exceeded. If a fast-acting fuse is used on a motor with a high starting current, it will blow before the motor can start running.

Time-delay fuses will not blow unless an overload condition exists for an extended period of time, typically 10 seconds. The time delay is usually required when a motor has high starting currents.

The time-delay has a disadvantage to the fast-acting fuse if an extremely high current overload occurs. The motor could be damaged from the high current before the time-delay is over. The fast-acting fuse does not have the time-delay and can shut the motor off before damage may occur.

The multipurpose fuse has the advantages of both the fast-acting and time-delay fuses. The multipurpose fuse will not blow during small overloads that last for only short periods of time, such as when the motor is starting. If an extremely high overload occurs, over 500 percent maximum current rating, the fuse will blow immediately. The multipurpose fuse provides good motor protection from both long-term small overloads and short-term large overloads.

The current-limiting fuse will never blow regardless of conditions. It prevents the electrical current to the motor from exceeding the rated current.

If a fuse continues to blow, check to be certain that it is the proper size rating for the application. If the fuse is the correct size, there could be another problem.

Many homes and businesses use circuit breakers rather than fuses. This is an automatic switch which will open a circuit if the current draw is too great.

Circuit breakers are usually rated the same as fuses. An "opened" circuit breaker must be manually "reset." As with fuses, a circuit which is continually opening the breaker should be carefully examined. If the breaker has sufficient capacity, there may be a short or other trouble in the circuit.

Fuses and circuit breakers are not necessarily interchangeable. The UL (Underwriters Laboratories) nameplate on an HVAC device may indicate the type of overcurrent protection required by the National Electrical Code.

7-28 MOTOR PROTECTION—BIMETAL SWITCHES

Bimetal overload devices are now the most common safety device and are located at important places in the unit. If these parts overheat for any reason, the circuit will be opened by the bimetal snap switches.

All these devices, however, will only stop the mechanism if the current load is too high. If the motor should overheat from other troubles, the unit may still run and cause damage. Other sources of excessive heat may be: high exhaust temperatures, poor air circulation, poor refrigerant circulation, and friction.

Present refrigerating systems have safety devices installed which open the electrical circuit if the motor draws too much current, overheats for any reason, or if the compressor becomes too hot.

Fig. 7-40 illustrates the action of a bimetal device. The device opens the circuit if the bimetal disk should reach a

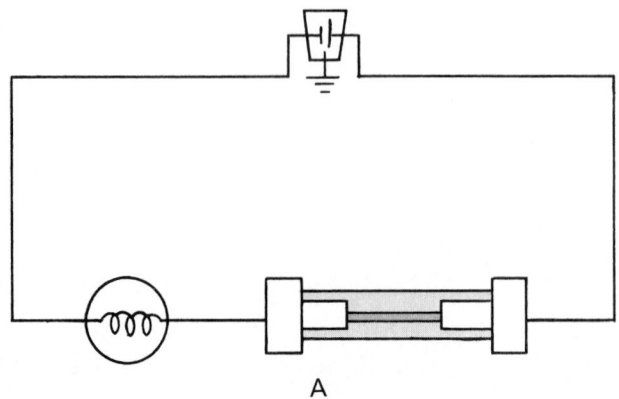

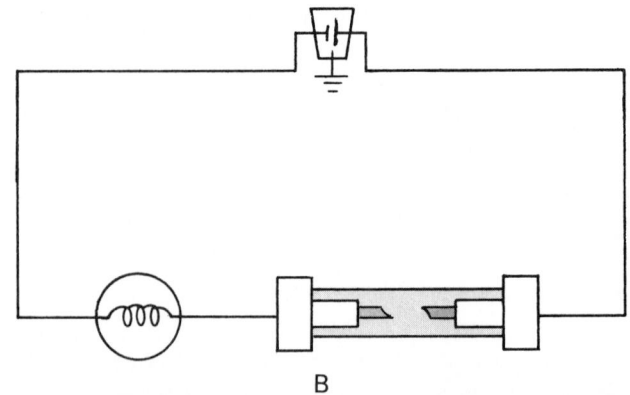

Fig. 7-39. One way to get motor protection. A—During normal conditions, fuse conducts current and motor is "ON." B—After overload condition, fuse can no longer conduct current, due to burnout, and motor is "OFF."

temperature that will cause it to snap in the other direction. Fig. 7-41 shows the construction of a motor protector.

Overheating could occur without the current draw becoming excessive. It is, therefore, necessary to use heat sensitive overload devices which will open the circuit before heat can cause damage. These controls are temperature operated and receive their heat from the motor, compressor, and the current draw of the motor. Fig. 7-42 shows an overload protector located on the motor compressor dome or housing. An external overload protector used on polyphase motors is shown in Fig. 7-43.

The action of a bimetal snap action current-actuated and heat-actuated overload protector as used on an open type motor is shown in Fig. 7-44.

Some of the temperature sensitive motor protectors are quite compact and may be easily installed inside the motor winding. Such a protector is described in Para. 7-29.

7-29 MOTOR INTERNAL OVERLOAD AND OVERHEATING PROTECTION

Internal overload motor protectors are mainly used on hermetic motors. They were developed, along with the two-pole motor, for large units. These are another example of a bimetal protection device.

A motor may overheat if there is too little refrigerant flow.

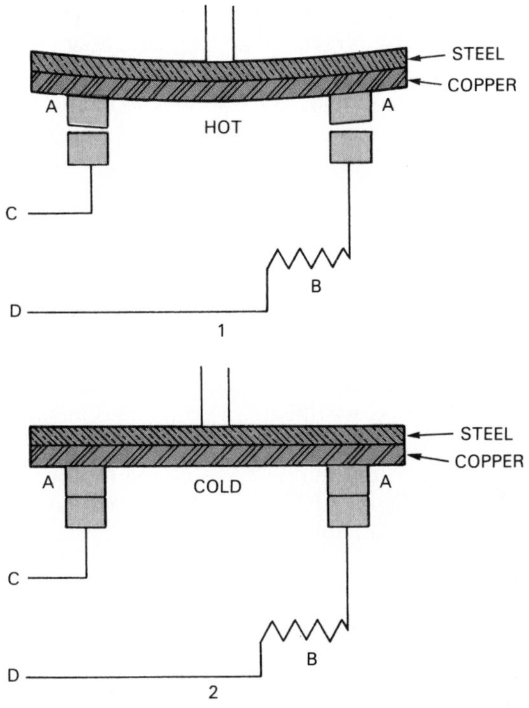

Fig. 7-40. Snap action type bimetal motor overload protector. High speed of snap action motion avoids burning of points due to arcing. A—Contact points. B—Heater coil. C—Motor winding connection. D—Power source connection. 1—Circuit open because of increased temperature or voltage. 2—Points closed in normal operation.

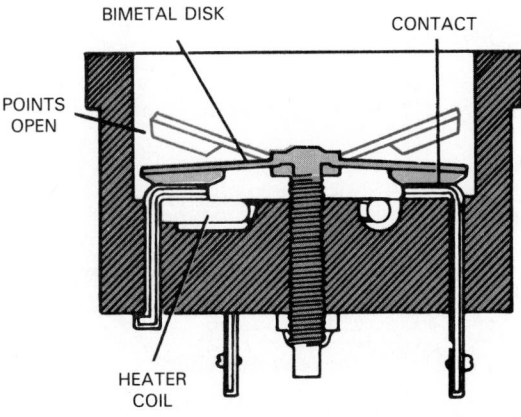

Fig. 7-41. Motor overload protector. Excess heat will cause bimetal disk to snap (oil-can effect) and open contact points. (Texas Instruments, Inc.)

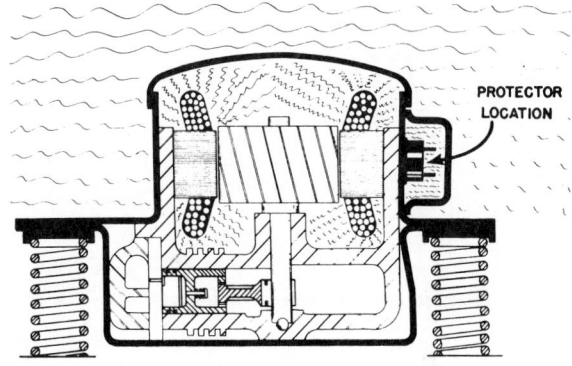

Fig. 7-42. Suitable location for motor compressor overload or excess temperature protector.

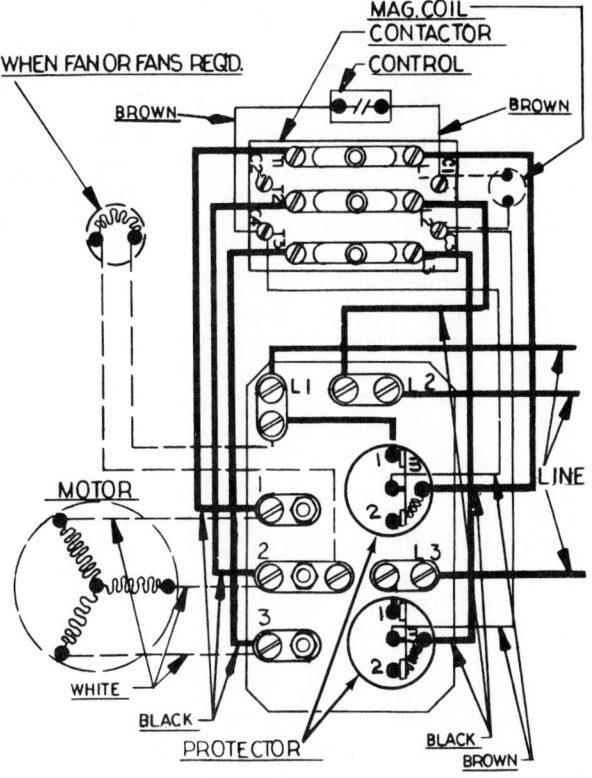

Fig. 7-43. Wiring diagram for three-phase hermetic motor which uses two external motor protectors. (Copeland Corp.)

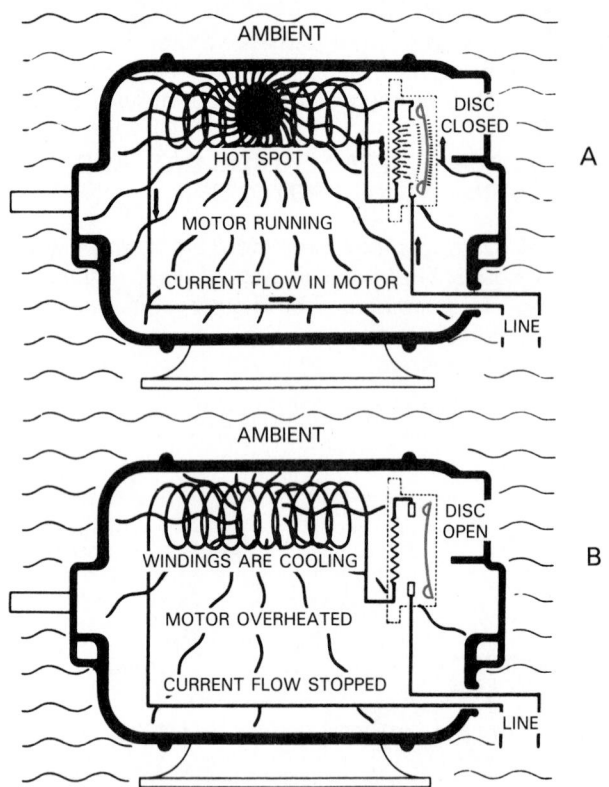

Fig. 7-44. Bimetal overload protector installed inside open type electric motor. A—Current flowing, motor running. B—Current cut off, motor stopped.

(The refrigerant gas cools the motor.) Or, it may overheat if the unit has to start too soon after shutting off. Too much current draw may be caused by a stuck or locked rotor or compressor.

In most refrigeration and air conditioning equipment, the motor compressor unit is designed to start under a condition of balanced pressures. There is danger of overheating the motor if it has to start against a high head pressure. If there is an increased starting load, the internal overload protector will open the motor circuit and protect the motor from such abuse.

The units normally operate at 125 °F (52 °C). When the temperature reaches 200 to 250 °F (93 to 121 °C), the protection device will open the circuit and stop the motor. It will then close at about 150 to 175 °F (66 to 79 °C).

Do not tap on the controls in an attempt to operate the points. The points may vibrate and arc, causing them to burn out quickly.

One type of internal motor overload protector is the bimetal disk.

In the bimetal type, the contact points are on a bimetal strip and are normally in a closed position as in Fig. 7-45. When there is too high a temperature rise, the points will open the circuit. When the temperature in the disk decreases enough, the strip will return to its normal position and the contact points will close.

The internal overload protector is placed inside the hermetically sealed compressors, directly on or in the windings. The internal overload protector will open if there is excessive current draw, excessive temperature, or both. Loss of refrigerant, a restriction in the system, or low suc-

tion pressure could lead to a motor burnout if the overload protector were not installed.

Depending on ambient temperature conditions, it may be an hour to two hours after the protector opens the circuit before it will close. Cooling the dome with forced air, dry ice, or carbon dioxide spray will speed up the cooling process.

The leads to these protectors must never be shorted, because even a few moments of operating a unit without this protector may burn out the motor. These protectors cannot be taken out of the circuit. They are the best possible protection for a hermetic compressor. Motors having this protection are usually labeled "Internal Overload Protected."

In Wye type three-phase motors, the internal overload is placed at the common terminal of the three windings and it will open all three circuits when its points open, Fig. 7-46.

These internal protectors are very reliable. Cases of failure are almost unknown. Fig. 7-47 shows an instrument used to check the three circuits of a three-phase system. An instrument is used to check if one lead of a three-phase system is open.

7-30 MOTOR PROTECTION— ELECTRONIC THERMISTOR

Thermistors that have a positive temperature coefficient, PTC, are used for motor protection. As temperature increases, the resistance of the PTC type thermistor also increases. The PTC thermistor is connected in series with the copper windings of a motor. It prevents current from conducting when the temperature of the motor increases beyond a safe value. After the motor and thermistor cool back to a safe value, current can begin to conduct and the motor will start up.

Another type of electronic thermistor is one which has a negative temperature coefficient, NTC. The thermistor is placed in a capsule within the motor. As the temperature increases, the resistance of the NTC type thermistor decreases. If the temperature rises to about 200 °F (93 °C), the increased current flow through the thermistor will operate a relay circuit and open the circuit. This shuts the motor off. When the temperature falls back to a safe value, the current decreases below the amount required to hold the relay open. The relay will then close so the motor can run again.

7-31 MOTOR TEMPERATURE

The temperature of the hottest part of motor should not be more than 72 °F (40 °C) over the room temperature. This

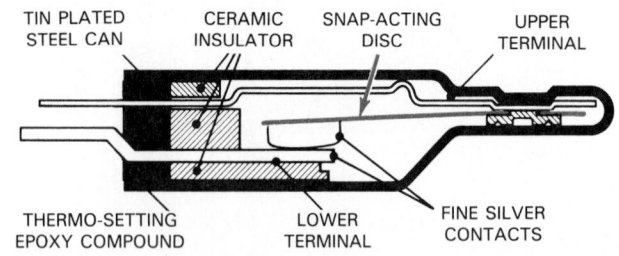

Fig. 7-45. Compact motor protector designed to be fitted into motor windings.

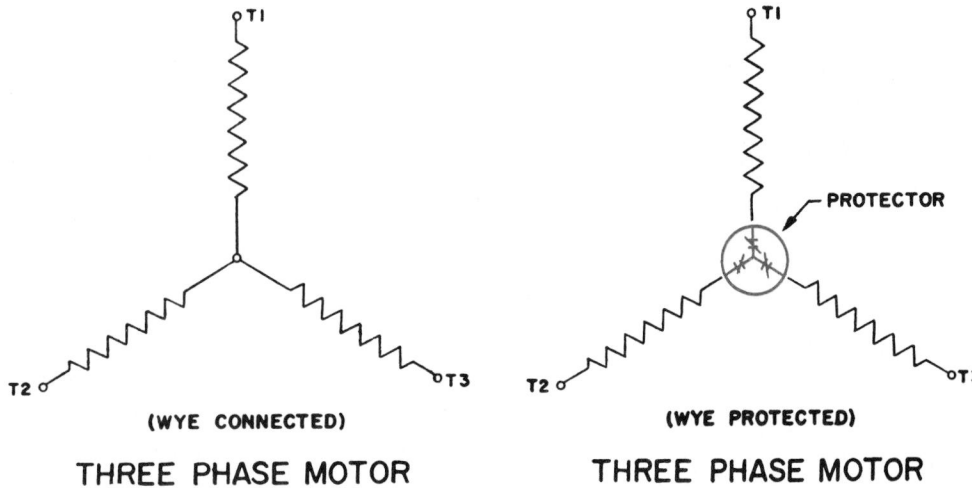

Fig. 7-46. Method of connecting motor protector in three-phase Y type motor. (Copeland Corp.)

means an average maximum temperature of approximately 150 °F (66 °C).

This temperature is difficult to measure except in a laboratory; so, it is better to check the ambient (surrounding) temperature to be sure it is not too high. Check the motor for cleanliness and airflow. Then check the current draw of the motor. If draw exceeds the rating on the motor identification plate or in the motor manual specifications, the motor will overheat. Current overdraw may be due to overloading the motor or to a shorting in the motor windings.

Always measure motor temperature with a thermometer. A motor with too high a frame or stator temperature may have motor winding temperatures so high that the insulation on the wires may fail.

The thermal overload protector for motors usually opens the circuit when the temperature reaches 200 °F (93 °C) and closes the circuit when the temperature drops to about 150 °F (66 °C).

Motors depend on rather cool ambient air for cooling. If this air is too warm or if the airflow is restricted, the motor will overheat.

7-32 STANDARD MOTOR DATA

The data on the motor identification plate usually gives the following information:
1. Required voltage (emf) supply.
2. Hertz (cycle).
3. Running current (amperes).
4. Locked current draw. The locked current draw indicates the condition of the internal circuit when the rotor is locked so it cannot turn. The locked current draw is also the starting current draw.
5. Temperature rise. The temperature rise is usually specified in degrees Celsius.
6. Seasonal energy efficiency ratio: SEER (Chapter 15).

Compressor speed on external drive systems can be controlled by using either two- or four-pole motors and by changing pulley sizes. Direct connected compressors must operate at motor speed. If a four-pole motor is used to replace a two-pole motor, a compressor of greater displacement must be used with the slower rpm motor. Because

the unit is running at half its former speed, the compressor displacement per stroke must be double.

The size of the conductor used in the refrigeration mechanism is very important. If too small or too long, it will heat up and eventually cause a fire. Long circuits also add an unnecessary resistance to the flow of electricity. This causes excessive voltage drop. If the voltage to the motor is less than 90 percent of the rating of the motor, there is danger of the motor being overheated and ruined. A table of wire sizes recommended for 120 V circuits is given in Fig. 7-48.

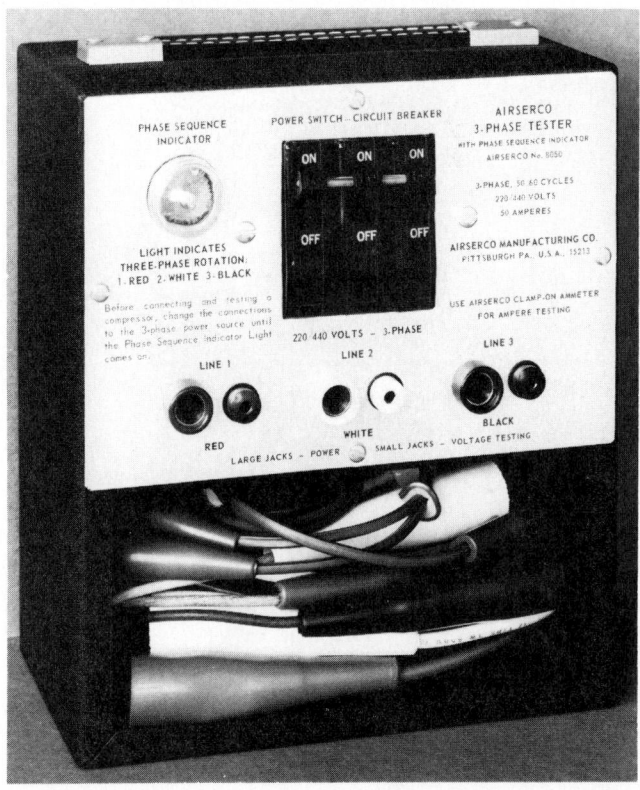

Fig. 7-47. Instrument which may be used to test all three circuits of a three-phase system. (Airserco Mfg. Co., Inc.)

Wire No.	Diameter of Wire in Inches	Ampere Capacity Plastic Insulation
18	.040	5
16	.051	10
14	.064	20
12	.081	25
10	.102	30
8	.128	50
6	.162	70
4	.204	90
2	.258	125

Fig. 7-48. Recommended wire sizes for various ampere capacity circuits. This table is calculated on the basis that wire is used for 120 V circuit.

The efficiency of small motors is only 50 to 60 percent because of clearances and efficiencies of the winding. Therefore, they consume nearly twice as much current as they should, compared to larger motors such as 1/2 hp and over. A 1/6 hp motor, which should theoretically use only 124 W or 1 1/4 A, will be found to need approximately 2 1/2 to 3 A (280 to 400 W) to develop the 1/6 hp.

When making electrical connections to a domestic refrigerator, the thermostat should be connected into the hot wire of the circuit. This wire has black insulation.

The other wire is called the common wire of the circuit. It has white insulation. It should be run directly to the motor.

The white wire (common) carries the same amount of current as is carried in the black wire.

A system of green grounding wires grounds all of the mechanisms in a refrigerator or air conditioner. This ground is *not* a current carrying wire. It is for safety only and is used to avoid the possibility of an electric shock should a short circuit or a ground occur in the electrical system.

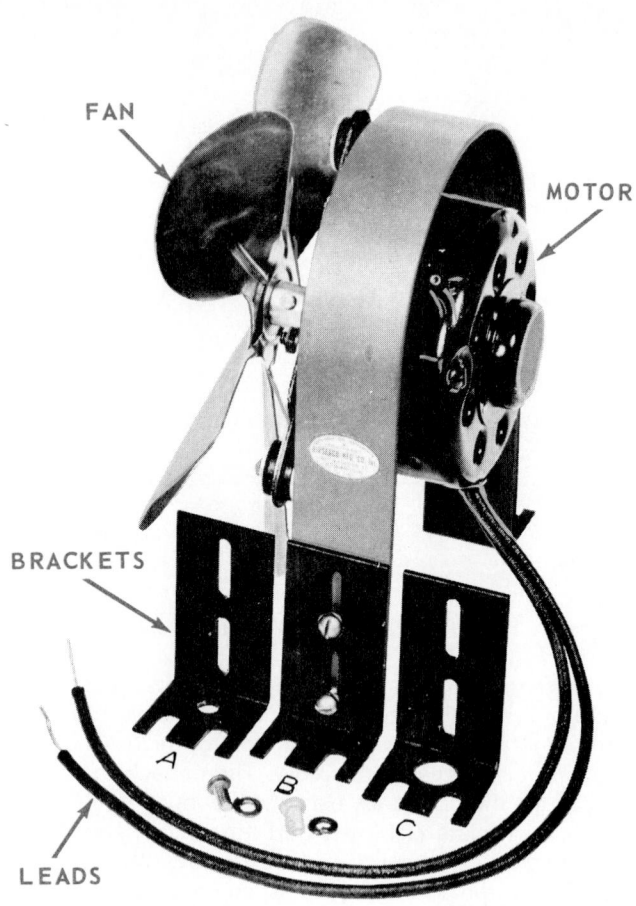

Fig. 7-50. Replacement motor and fan for refrigerant condenser. Motor unit has universal mounting brackets (A, B, C) which may be used with a variety of condensing unit designs.

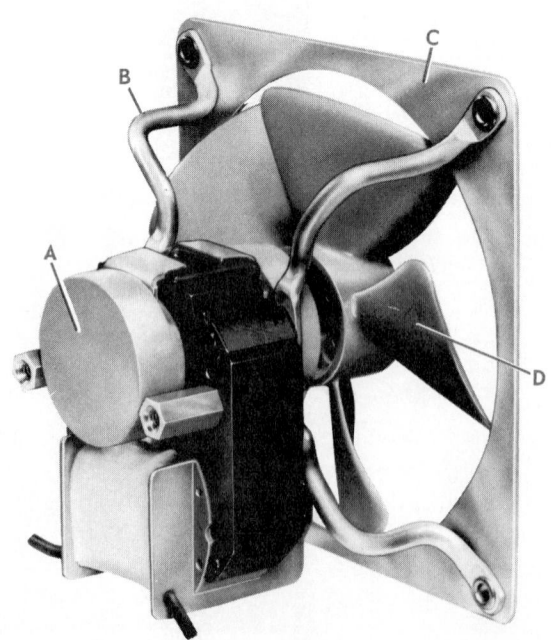

Fig. 7-49. Shaded-pole motor and fan. A—Motor. B—Motor support brackets. C—Mounting plate. D—Molded fan. (General Electric Co.)

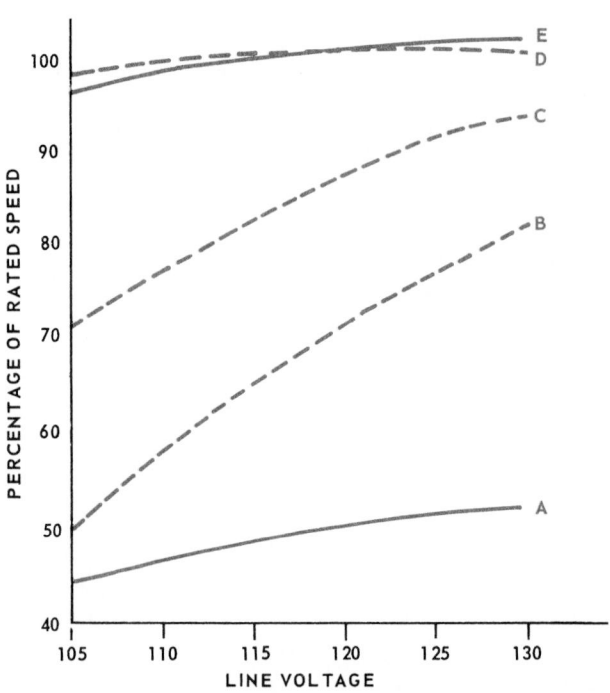

Fig. 7-51. How voltage affects fan speed for six- and twelve-pole motors. A—Twelve-pole. B—Low speed for a high-efficiency twelve-pole motor. C—Medium speed, twelve-pole. D—High speed, twelve-pole. E—Six-pole.

7-33 FAN MOTORS

Many hermetic units use motor-driven fans to:

1. Force condenser cooling air through the ducts and over the condenser and condensing units.
2. Circulate air in refrigerated parts of domestic and commercial units. See Fig. 7-49.

To create efficient air movement, the fan and condenser are in a housing of sheet metal or plastic. The fans are carefully balanced and run almost noiselessly. They are usually attached to the motor shaft with Allen setscrews.

Some of these motors have sealed bearings (bushings) and require no oiling. Others need oiling (SAE 10 or 20) amounting to one drop per bearing each six months. A few motors on the market have only one bearing. The motors are usually attached to brackets and are mounted in rubber. Fig. 7-50 shows a replacement condenser fan.

Generally, the condenser fan motor leads are connected to the common terminal and the running winding terminal of the compressor motor. This connection puts the fan motor in parallel with the compressor motor and allows it to be controlled by the thermostat. The safety overload cutout is also put in the circuit ahead of the fan so that it will also cut out the fan motor.

Some fan motors have their own thermal safety controls and many are of the two- or three-speed type. The variation in speed may be obtained by using extra poles in the stator or by using a solid state control.

The speed of a fan motor is quite sensitive to the applied voltage. As the voltage drops, so will the fan speed. Fig. 7-51 shows the relationship between the voltage and fan speed.

Fig. 7-52 is a schematic of some of the common fan motor circuits. One-, two-, and three-speed motor circuits

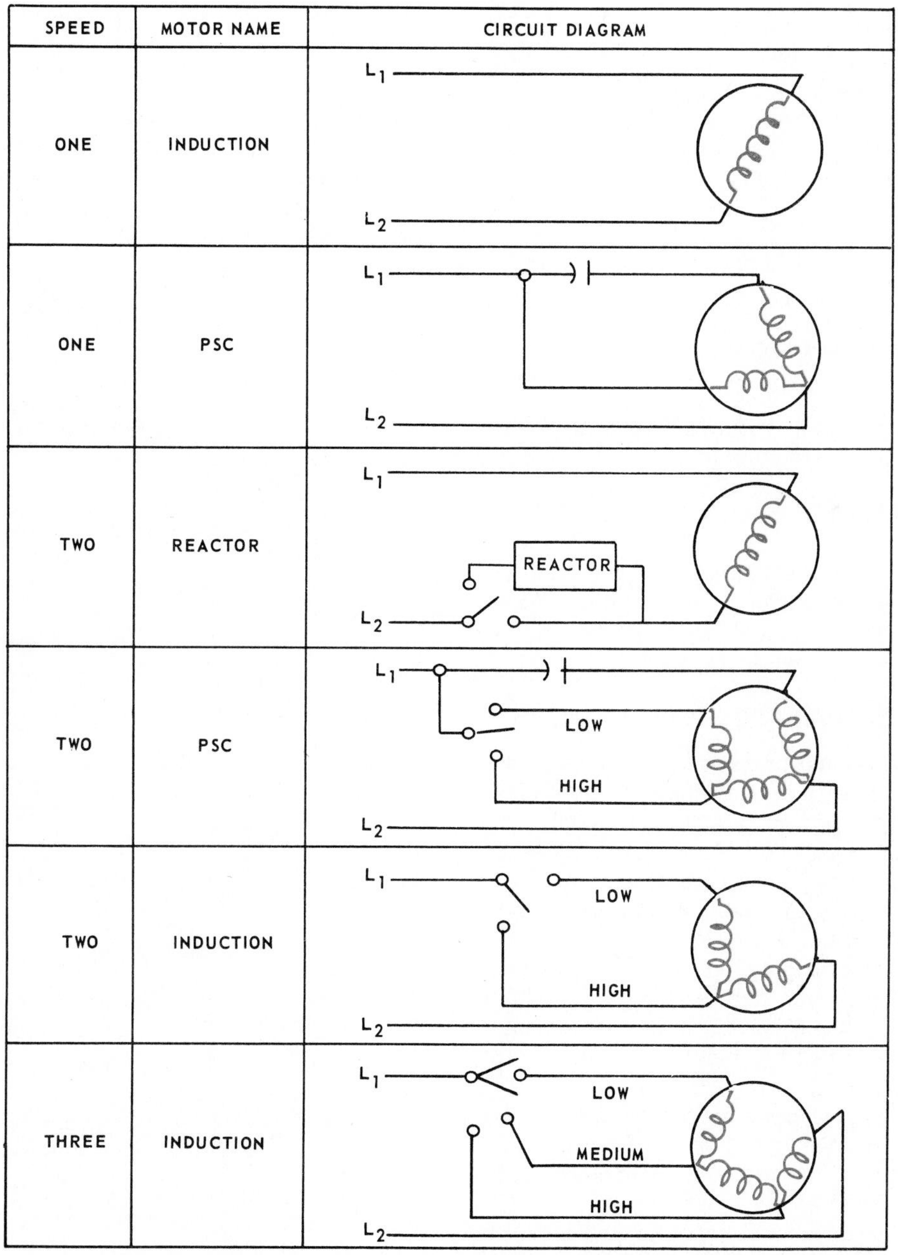
SPEED
MOTOR NAME
CIRCUIT DIAGRAM
ONE
INDUCTION
ONE
PSC
TWO
REACTOR
REACTOR
TWO
PSC
LOW
HIGH
TWO
INDUCTION
LOW
HIGH
THREE
INDUCTION
LOW
MEDIUM
HIGH
L1
L2

Fig. 7-52. Schematic of some common fan motor circuits.

are shown. Refer to Para. 7-48 for information on voltage drop tests, troubleshooting, and how to service fan motors.

7-34 SHADED-POLE MOTORS

The construction of a shaded-pole motor is somewhat different than the construction of the motors previously described. See Fig. 7-53. The shaded-pole produces a moving magnetic field perpendicular to the field pole.

Approximately half of each pole face has a small copper plate insert, A, with a small winding. This insert slows down the buildup of the magnetic field through the copper plate just enough to cause a magnetic motion toward the copper plate.

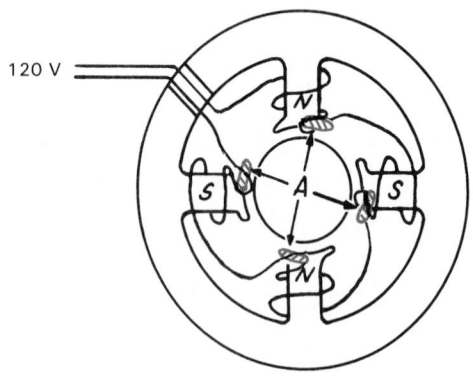

Fig. 7-53. Shaded-pole fan motor. S—South polarity. N—North polarity. A—Shaded-pole plate (copper).

This action produces a lag for induced magnetism in the rotor (opposite magnetism). The rotor turns as it is attracted by the magnetism. Movement of the rotor will continue as the alternating current changes the polarity of the poles and the rotor.

Although this design has less starting torque than other type motors, it is very successful in small motors 1/6 to 1/100 hp. Fig. 7-54 shows a double-shaft fan motor. The end bell of this motor is shown in Fig. 7-55.

7-35 ELECTRONIC VARIABLE SPEED MOTORS

A method employing transistor switching instead of the brush or commutator has been developed for low horsepower motors. Brushless motors operate with silicon rectifiers, transistors, and special circuitry. The advantages of the transistorized motor are its high speed, compactness, performance, durability, variable speed, and the elimination of sparking and brush noise.

Speed is changed by adding a small variable resistance (potentiometer) which will vary the resistance within the circuitry of the control. The direction of the motor's rotation can be quickly reversed by manipulating the motor control's switching devices.

Fig. 7-56 is the wiring diagram for a solid state motor speed control. The control reacts to signals from the unit-mounted sensor. Fig. 7-57 illustrates the same unit complete with case and controls.

The diagram in Fig. 7-58 is for yet another control which is commonly used to regulate motor speed. The internal schematic diagram of the control is shown in Fig. 7-59.

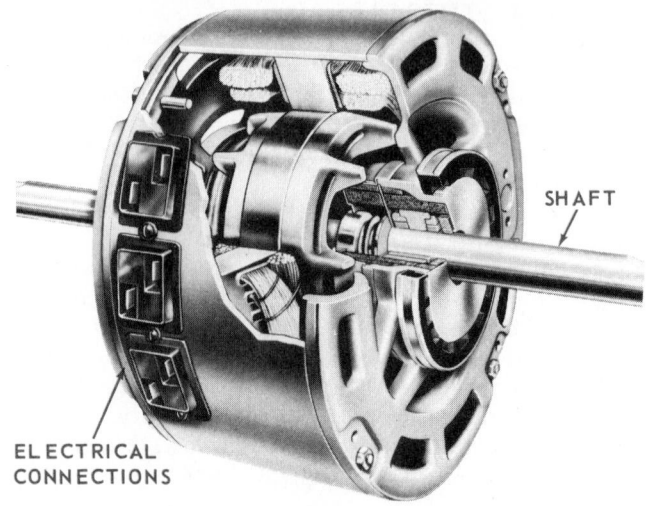

Fig. 7-54. Low starting torque shaded-pole, double-shaft fan motor. Three terminals for electrical connections provide a variety of fan speeds. (General Electric Co.)

The circuit operates as follows:

Resistances R_2, R_3, R_4, and capacitor C_2 form an R-C (resistance-capacitance) charging network. The effective resistance of R_2, R_3, and R_4 charges the capacitor, C_2, up to a voltage that causes diac, D_1, to conduct an electric current and fire triac, Q_1.

When Q_1 is fired, it conducts until the current through it drops below the holding current (25 milliamperes) of the

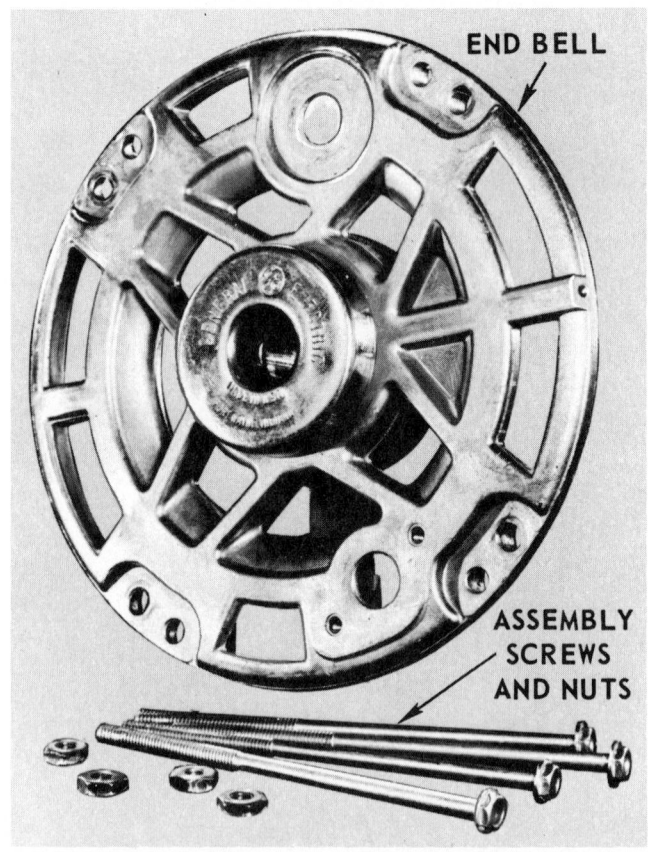

Fig. 7-55. Aluminum end bell for fan motor. Note assembly screws and nuts.

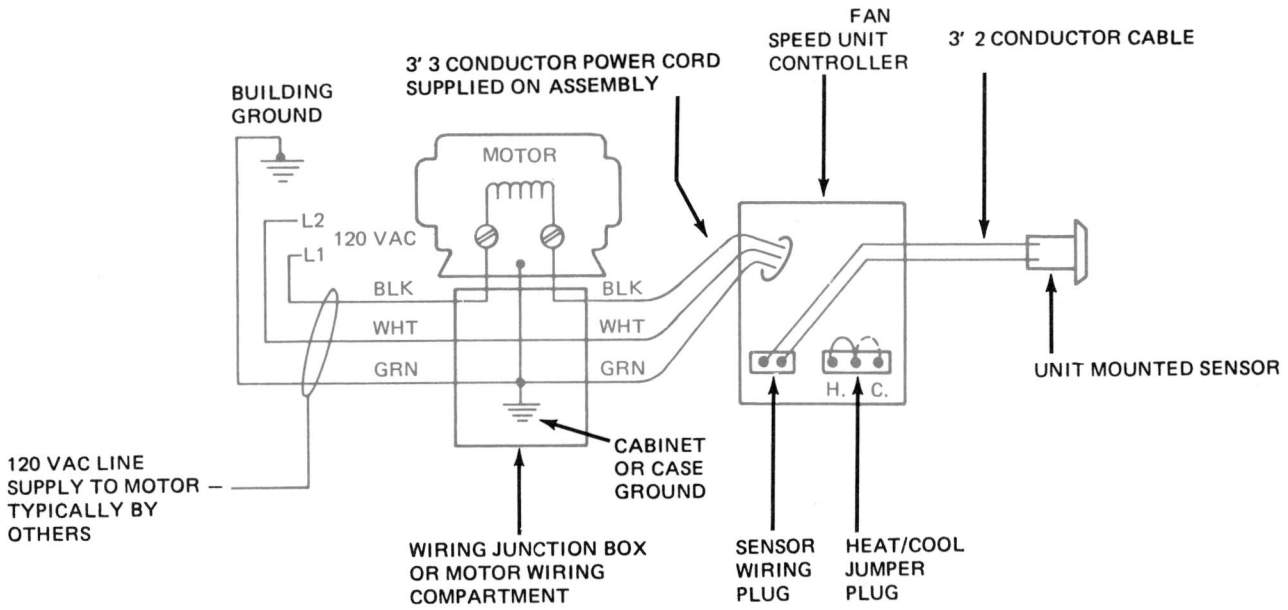

Fig. 7-56. Installation wiring diagram for solid state motor control. (Barber-Colman Co.)

triac. At that point, the triac turns off. The control's chargeup, fire, conduct, and turn-off process repeats every half cycle of alternating current power.

The speed of the motor is controlled by varying the resistance of R_4. The variable resistor is operated by the knob on the front of the control.

The lower the resistance of R_4, the faster the motor will turn. The minimum speed of the motor can be set by adjusting the resistance of trimmer, R_3.

R_1, C_1, and L_1 provide suppression of radio frequency interference caused by the fast switching of Q_1. This circuit can be used with permanent split capacitor, shaded-pole, and universal motors.

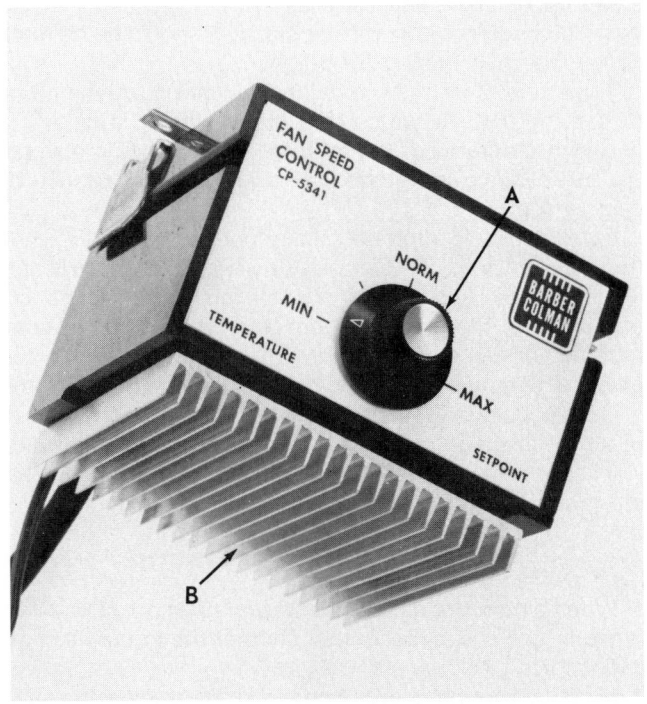

Fig. 7-57. Solid state fan speed control. A—Adjustment. B—Heat sink.

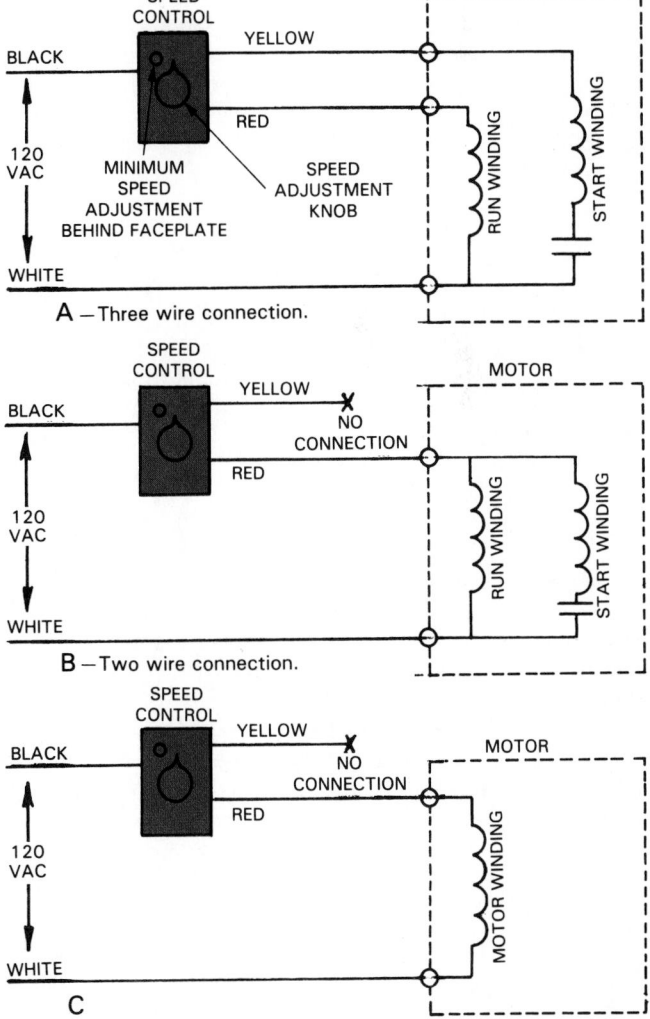

Fig. 7-58. A and B—Connecting an electronic variable speed control to a permanent split capacitor motor. C—Same control connected to a shaded-pole motor. (Lutron Electronics Co., Inc.)

235

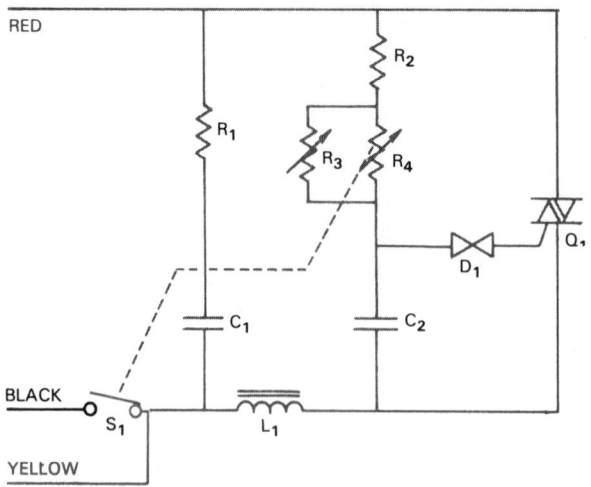

Fig. 7-59. The electronic circuit for a variable speed control used with permanent split capacitor (PSC) motors, shaded-pole motors, and universal motors. Refer to Fig. 7-58. (Lutron Electronics Co., Inc.)

7-36 SERVICING ELECTRIC MOTORS

The maintenance, troubleshooting, removal, repair, and installation of electric motors as well as their accessories is a major portion of a service technician's job.

Such persons must know how to use instruments and be knowledgeable about electricity to accurately determine the trouble. The following paragraphs describe these operations:
1. General service.
2. Open motors.
3. Hermetic motors.
4. Fan motors.

It is important that a solid base be provided for the installation of any motor and that it be bolted down securely. The armature shaft should be level for a horizontal motor and exactly vertical (plumb) for a vertical motor.

If a motor is severely damaged and must be replaced, the following information must be taken:
1. Type of motor.
2. Operating line voltage.
3. Maximum current draw.
4. Direction of rotation and speed.
5. Mounting hardware.

Use the manufacturer's nameplate to get all the required motor specifications. Sometimes a replacement motor from another manufacturer can be found by using cross-reference charts.

Make certain that the replacement motor has all the same specifications as the original motor. The replacement motor may have a maximum current draw 10 percent larger, but never less than, the motor being replaced.

7-37 SERVICE: WATT READING TO DETERMINE MOTOR TROUBLES

One way of learning the condition of a motor compressor unit is to observe the wattage consumption (watts drawn) of the unit.

The meter will give two different wattage readings:
1. Combined starting and running winding reading (only 1 to 1 1/2 seconds long).

2. The running winding reading during the time the unit is running.

Note: The wattmeter needle will overswing slightly at the instant that the motor compressor starts. Since the combined starting and running winding watt reading is for only a fraction of a second, the service technician must allow for the overswing.

When the thermostat contacts close, the wattmeter indicator will swing to the right; then it will quickly move back to the combined reading. In a few seconds, the pointer will fall to the running winding reading only.

If the starting winding circuit is open, the wattmeter pointer will swing to the right and then move back to the running winding value only. This action indicates a bad relay or starting winding. The overload safety cutout should open the circuit in two or three seconds.

Approximate watt readings for small hermetic motors are shown in Fig. 7-60.

MOTOR HP	WATTS AT 120 VOLTS		STARTING OR LOCKED
	RUNNING		
	AT 70°F	AT 110°F	
1/16	66	100	375
1/9	117	160	740
1/8	108	163	743
1/7	160	218	970
1/5	242	295	1450
1/4	235	320	1250

Fig. 7-60. Approximate watt readings for small hermetic motor compressors. Indicated temperatures are ambient.

7-38 RADIO AND TV INTERFERENCE

Some conventional refrigerators cause a slight amount of radio or TV interference. This interference will usually amount to a slight snap or click in the radio or TV at the instant the refrigerator stops or starts. It should be no more noticeable than turning off a light.

This interference may be reduced by grounding the frame of the motor to a water pipe or by placing a capacitor between the frame of the motor and a ground. In general, the interference is not disagreeable and may usually be disregarded.

Excessive radio interference of a continuing nature when the refrigerator motor operates, or when it starts, indicates a loose electrical connection or some fault in the mechanism of the motor. These include worn brushes, badly pitted commutator, or loose connections. The particular trouble can be easily determined by a careful examination of the motor.

Occasionally a static charge will be built upon the belt of a belt-driven compressor. The discharge of this build-up will cause radio interference. If the motor and compressor are grounded together, this noise will be eliminated.

7-39 SERVICE: TESTING CAPACITORS

When a motor or motor compressor does not start or run properly, there is a good possibility that the trouble is in the capacitor.

Most motors have only one—a starting capacitor—but some have two or more. In such cases, one might be a starting capacitor and the other a running capacitor. Some motors have two or more capacitors connected in parallel for additional capacitance or in series for additional voltage.

The starting capacitor is connected in series with the starting winding. It is usually wired into the circuit between the relay and the starting winding terminal of the motor (off-and-on operation).

The running capacitor is also in the starting winding circuit but stays in operation while the unit runs (continuous operation).

There are two types of capacitors:
1. The dry capacitor (for intermittent operation).
2. The electrolytic capacitor (for continuous operation). Both types may be tested in the same way. (The dry capacitor is described in Para. 7-8.)

The simplest capacitor test is to substitute a good capacitor for the one being tested. If the motor operates, the old capacitor is faulty. The replacement capacitor should be the same capacity as the old. If one must be used of a different capacity, it should be 10 percent over capacity rather than under.

Fig. 7-61 illustrates a commercial model of a capacitor tester. This type of tester is used for checking capacitor shorts, leaks, grounds, and open circuits. The capacitor tester will indicate to the service technician whether the capacitor is the problem in the circuit and should be replaced.

IMPORTANT: Never place fingers across the terminals of a capacitor. When discharging a capacitor, place it in a protective case, then discharge through a resistor connected between the terminals.

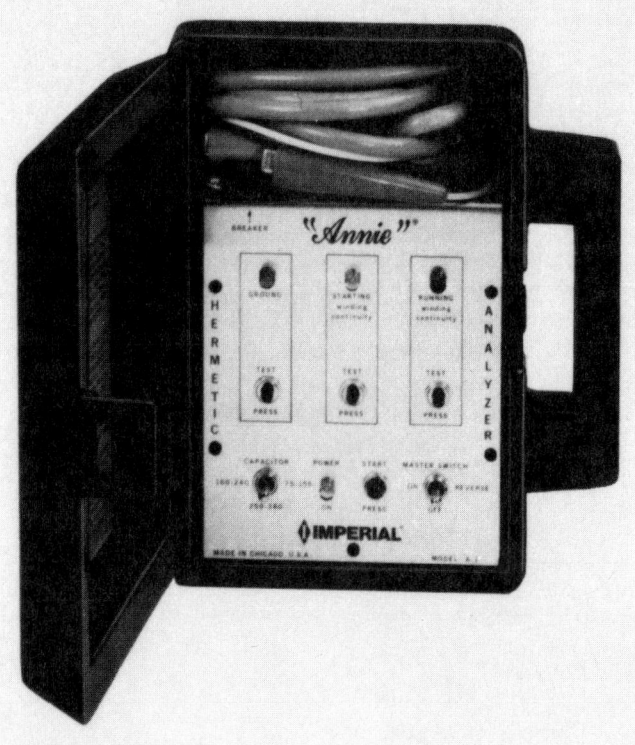

Fig. 7-61. Capacitor tester and analyzer used to detect hermetic motor troubles. It may also be used to reverse the motor. (Imperial Eastman, Imperial Division)

7-40 SERVICING CAPACITORS

Some capacitors have resistors connected across the terminals. This resistor, visible at the capacitor terminal, slowly bleeds the capacitor of its charge to lessen the arcing of the motor control points. Such arcing may occur if the unit cycles frequently.

When testing such a capacitor, remove the resistor from one terminal. To discharge a capacitor, use a 20,000 ohm, 2 W resistor in the circuit. Avoid shorting the terminals. The sudden discharge may rupture the thin metal foil in the capacitor.

Capacitor size must be accurately suited to the motor and the motor load. It is general practice to permit up to 10 percent over capacity. For example, a 110 μf can be used for a 100 μf capacitor, but an undersized capacitor should never be used. If at all possible, use an exact replacement. The make, model, and model number are usually placed on each capacitor.

If this information is unavailable, or if an emergency capacitor must be used temporarily, there are several ways to determine the proper size. One way is to look at the capacitor size used in a similar refrigerating unit by the same manufacturer or look in a service manual for a similar unit by any manufacturer.

A better method is to use a specially designed capacitor selector unit. This selector has a variable capacitance and increasing amounts of microfarads are put in the circuit (in series) until the correct voltage reading is reached. The capacitance registered on the selector indicates the capacitor size that should be put in the circuit.

Some of the capacitors have mechanical connectors (machine screws) while some have solder-type leads. It is necessary to use a small electric soldering copper or soldering gun to connect the second type.

Some running capacitors that used polychlorinated biphenyl dielectric fluid (PCB) are still in use. This fluid is dangerous. Do not open the shell of this capacitor. If the shell is accidentally pierced or broken, be very cautious not to touch the fluid or breathe its fumes. Dispose of capacitors containing this dielectric fluid according to local environmental rules. The manufacturer can help.

7-41 SERVICING EXTERNAL DRIVE MOTORS

Troubles found in open type motors are few and they can be classified as:
1. Electrical troubles.
2. Mechanical troubles.

The electrical troubles found in electric motors may be:
1. An open circuit, short circuit, or ground may occur in the field windings. In these cases, replacing the motor is recommended.
2. Frequent starting of the motor may result in overheating. Overheating the capacitor may cause the switch and the insulation to fail. If the motor will not start until the pulley is spun but has the characteristic ac hum, it is a sign that the capacitor or the centrifugal switch points have failed. It is easy to replace the old capacitor with a good one of the same capacity.

If the motor still will not start, the trouble is probably in the centrifugal switch. If the points are dirty, pitted, or dark from overheating, do not try to repair them. Filing or sanding does little good, as the contact material is

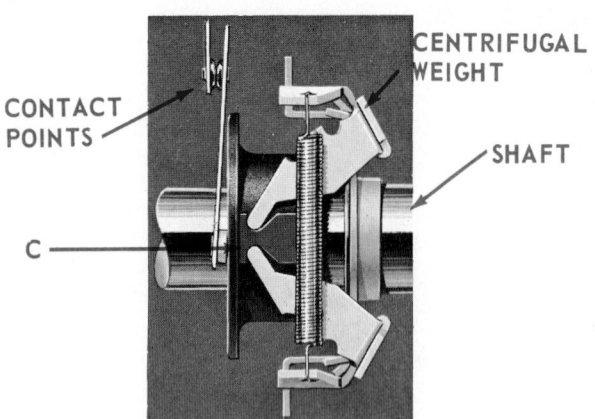

Fig. 7-62. Centrifugal switch mechanism used on ''open type'' compressor motor. Centrifugal weights and spool, C, revolve with motor shaft (rotor). At about 75 percent of full operating speed, centrifugal weights cause spool, C, to move to right. This allows contact points to open, breaking the starting winding circuit.   (Magnetek)

worn away. Repaired points usually last just a few hours and a call back may have to be made. It is best to replace them.

Mechanical troubles are almost the same in all open type motors:

1. There is a possibility that the centrifugal switch used for connecting and disconnecting the capacitor and/or the starting winding may become worn. Replacement of the switch is necessary, in such cases. Fig. 7-62 shows a centrifugal switch for a capacitor-start type electric motor.

2. Other troubles may include bearing wear, end play, excessive vibration, misalignment of the motor with the compressor, and improper air gap between the rotor and stator.

Information on the repair and testing of motors is covered in Chapter 14.

7-42 SERVICE: MOTOR LUBRICATION

Open motors may be lubricated in various ways. It depends on the type of bearing used and the position of the motor. Open motors using bronze bushings, plain or sleeve, may be lubricated in two different ways:

1. Wick system.
2. Slip ring system.

The wick system uses a well or reservoir in the end bell. A wick (cotton or wool yarn) carries oil from this well to the bushing and shaft. This system allows long intervals between servicing bearings and prevents the bearing getting too much oil. This type of lubrication is shown in Fig. 7-63.

Motors with this lubrication system have the cotton or wool yarn saturated with oil when shipped from the factory. However, before starting the motor, the oil wells should be filled by adding the amount of oil designated or until oil appears in the lower oil level cup.

If the bearing is to be removed from the shaft or if the bushing is to be removed from the end bell, the yarn should be lifted clear of the bearing to prevent the yarn being forced between the shaft and the bearing upon replacement.

When replacing the yarn, pack equal amounts on each side of the bearing, and over the slot of the bearing so the spring on the oil well cover will push the yarn down on the

shaft. Wick lubricated bearings should be oiled with one or two drops every six months.

Some larger refrigeration motors use the ring lubricating method. A brass ring rests on the motor shaft through a slot in the top of the bearing. The ring is large enough to dip into the oil pocket below. As the motor shaft turns, the ring turns slowly and the wet portion lubricates the bearing.

Be sure to check the rings when working on these motors. Use a medium viscosity nondetergent oil such as SAE 20 or SAE 30 (200 to 300 viscosity).

Some motors use ball bearings, as in Fig. 7-64. These bearings are grease lubricated. Most are sealed and do not need any lubrication service. Some, however, come with grease cups and can be lubricated with a grease gun.

These motors, when new, are supplied with enough grease in the bearings to lubricate them for a number of months. A small amount of grease should be added every two or three months. Use a high grade ''medium'' grease on fully enclosed motors. Too much grease may cause the bearings to overheat.

The life of bearings depends, to a considerable extent, on cleanliness. Use only clean grease and keep all dirt out of bearing. Clean all fittings before using grease gun.

Most greases and oil oxidize and will collect dirt while in use. When a motor is reconditioned, the old lubricant must be discarded, the lubricated portions thoroughly cleaned, and new lubricant used.

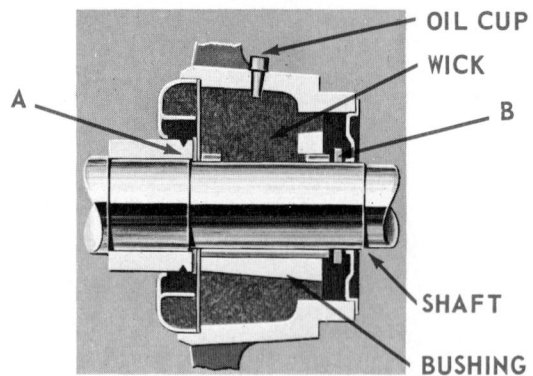

Fig. 7-63. Motor end bearing which uses wick oiler. Oil slingers A and B prevent oil from leaking into motor or out along motor shaft.

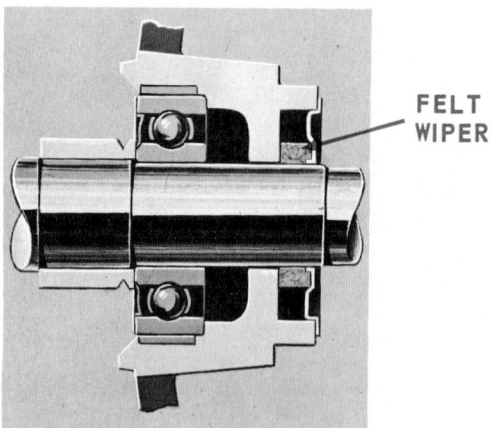

Fig. 7-64. Motor shaft mounted on grease-lubricated ball bearings. Felt wiper keeps out dirt and dust.

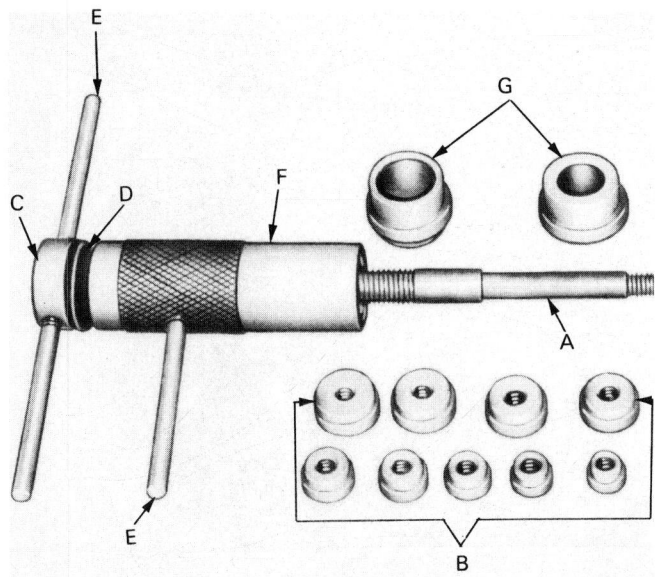

Fig. 7-65. Bushing and bearing tool kit for removing and replacing motor bearings or bushings. A—Tool shaft. B—Engaging taps. C—Tension nut. D—Thrust washer. E—Tension nut handle. F—Bearing tool housing. G—Adaptor sleeves. (Grainger, Div. of W.W. Grainger, Inc.)

Another method of lubrication presently used is the oilless bushing. In this arrangement, the shaft passes through a sintered (porous bronze) bushing which has been impregnated with oil at the factory. The total tolerance between the shaft and bushing is less than .001 in. and may have a tolerance of as little as .0003 in. As the tolerance between the bushing and shaft increases due to wear, the motor will become noisy.

The oilless bushing is considered to be permanently lubricated. It is often used on fan motors and on other low horsepower applications.

Fan motors may become very cold when idle. The bearing oil may become very thick and the motor (usually low torque) will start with difficulty. It may even burn out or activate the overload switch. Be sure to use oil that will remain quite fluid at the temperatures (0 °F [−18 °C] to −40 °F [−40 °C]) the bearings may reach during idle time.

7-43 SERVICING MOTOR BEARINGS

In any service to motors, bearings should be checked to see if they are worn. If the rotor is hitting the stator, bearings are worn out and must be replaced.

Clearance between rotor and stator varies from .015 in. to .030 in., depending on the size of the motor. This clearance should be the same all the way around the rotor. A heavy rumbling sound at starting usually indicates that the bearing is badly worn, even though the rotor may not be touching the field poles.

Bearings are usually made of phosphor bronze and are pressed into the end brackets or end bells. Sometimes they are locked in by a pin pressed through the bearing housing and into the bearing. The bearing must always be pressed inward to remove it.

Care should be taken not to put an out-of-line force on the end bell when pressing bearings out. This would probably crack the end bell. A special tool, Fig. 7-65, can be used to remove or install bushings or sleeve bearings.

After the new bearing is pressed into the bracket, the bearing must be reamed. It is best to ream the two in-line with adjustable reamers.

The surface of the shaft in contact with the bearing must be perfectly smooth. A scored shaft may be repaired in a lathe using a grinder mounted on the tool post.

If a bearing is overheating, any one of the following may be the cause:
1. Oil too heavy.
2. Oil too thin. Select a good grade of mineral lubrication oil which is not greatly affected by a change in temperature and which does not foam or bubble too freely.
3. Dirt or grit in the oil.
4. Belt too tight.
5. Pulley hub rubbing against the bearing.
6. Motor not properly lined up, causing the armature shaft shoulder to pull on or be pushed against one bearing.

7-44 CLEANING MOTORS

While in service, the motor should be cleaned regularly. Dust and lint in the motor will prevent proper air circulation. Compressed air or a hand bellows should be used frequently to blow dirt out of the motor.

Any oil which may overflow from the bearings should be wiped off. A little attention will result in efficient operation with the motor giving good service for many years.

If the motor must be dismantled, all parts should be carefully cleaned before being worked on or reassembled. Cleaning fluids must be used that are not harmful to the electrical insulation material or to the technician's health. There are many cleaning fluids on the market. Be sure to check the one being used for safety.

All cleaning fluids should be used in well ventilated and fireproof surroundings. Use only enough cleaning fluid for the job. Too much may be dangerous. Always avoid using carbon tetrachloride as the fumes are very toxic and can be fatal.

Motor interiors are well designed but rough handling may harm them. Fig. 7-66 illustrates typical internal wiring construction.

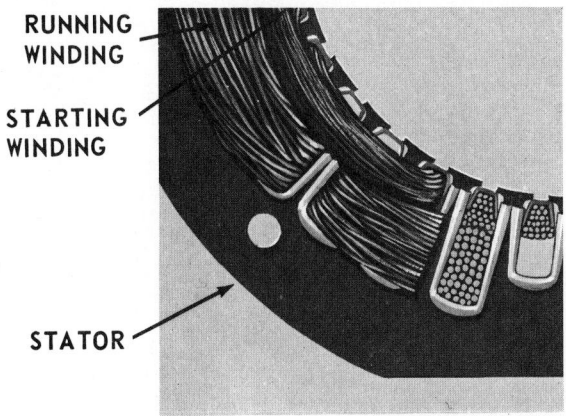

Fig. 7-66. Section through stator windings showing method of installing, insulating, and retaining windings. Wire insulation is heavy, modified polyester varnish. Slots are lined, and "cuffed" ends prevent grounds where coils cross corners of core. Slots are closed with shaped sticks of heavy electrical paper.

7-45 PULLEYS

Motor shaft pulleys or sheaves are available in many sizes and types of construction. Some are made of cast iron and some of steel stampings. They come in various shaft sizes and diameters.

The most popular shaft sizes for fractional horsepower motors are 1/2 in., 5/8 in., and 3/4 in. diameter. Practically all pulleys have a keyway and a setscrew.

Pulley diameters vary from 3 in. to 38 in.

The V-pulleys come in two popular widths. The A width is for belts up to 1/2 in. width, while the B width fits belts 1/2 in. to 21/32 in. wide.

Multiple-groove pulleys are available for units that use two or more belts to drive the flywheels. Some air conditioning units use step pulleys for driving the air movement fan. By changing the belt from one groove to another, the speed of the fan can be changed.

Special variable-pitch pulleys are also available. These are made with half of the pulley threaded on the hub of the other half, as shown in Fig. 7-67. A setscrew locks the variable half in place when it is properly adjusted.

By turning the variable half, the V-groove can be widened to let the belt ride closer to hub. This reduces speed of a driven flywheel or fan. The speed of the driven unit can be varied by as much as 30 percent using these pulleys. Fig. 7-68 illustrates a double-groove, variable-pitch pulley.

Bushings are available to adapt large bore pulleys to small shafts. For example, a bushing can reduce a 3/4 in. bore to a 1/2 in. bore.

7-46 BELTS

The V-belt is the most popular way to drive the open compressor and large fans. These belts are made in layers of rubber, fabric, and cord. Some belts are a mixture of natural and synthetic rubber. Fig. 7-69 illustrates the most commonly used belts for refrigeration and air conditioning—

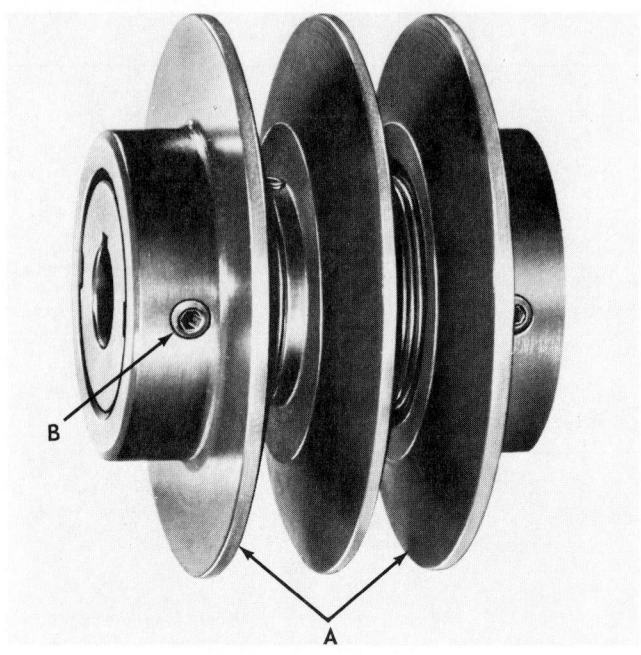

Fig. 7-68. Adjustable pulley with double V groove. Both grooves are adjustable and must be evenly spaced or one belt will take all the load. A—Adjustable pulley halves. B—Setscrew. (Maurey Mfg. Corp.)

the standard multiple-cord industrial system and the narrow industrial system.

The belts are made in many lengths from 15 in. outside length to 660 in. outside length. (Outside length is the

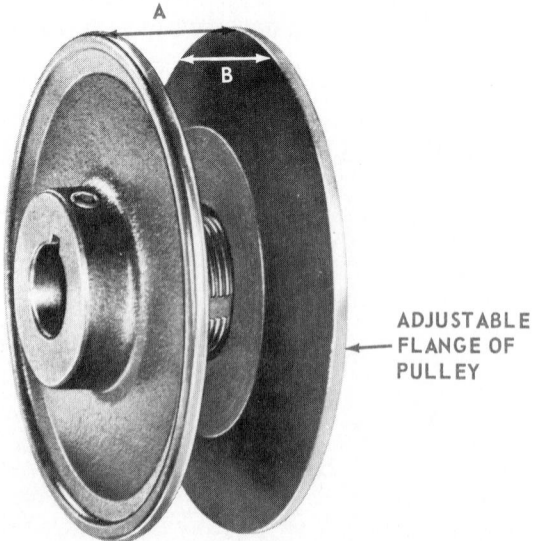

Fig. 7-67. Adjustable V-pulley is used to change speed of belt-driven appliances. A—Moving the adjustable flange away from other flange widens the V and gives the effect of using smaller diameter pulley. B—Narrowing V by moving flange in gives effect of larger diameter pulley. (Maurey Mfg. Corp.)

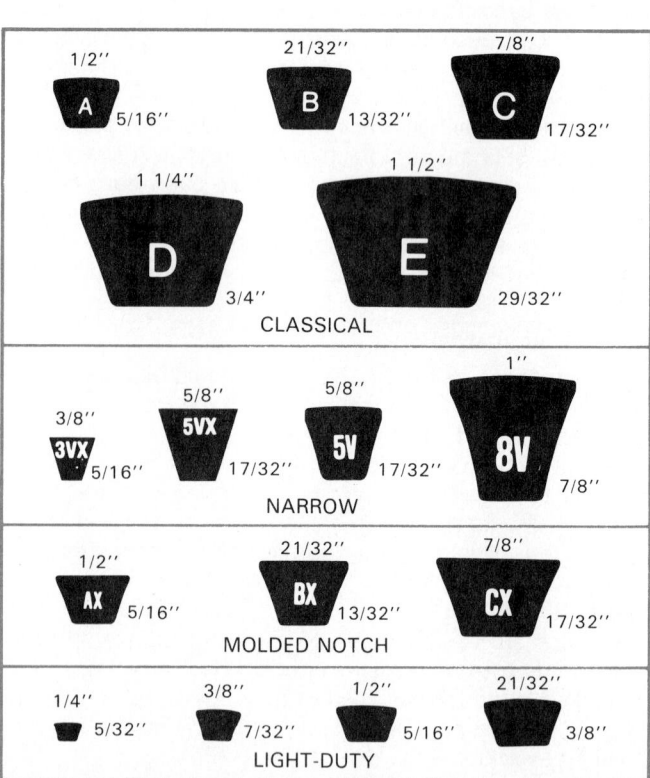

Fig. 7-69. Above are industry accepted cross sections and dimensions for (from the top): classical, narrow, molded notch and light-duty industrial belts. (The Gates Rubber Company)

distance around the outside of the belt.) A steel tape, cloth tape, or a special belt-measuring fixture can be used to quickly determine this length.

Most belts fall into one of four standard widths. The industry designation for these widths are: Classical, Narrow, Notched, and Light-Duty (see Fig. 7-69). Measurement is made at the belt's greatest width.

When the motor is belted to the driven machine, both shafts must be parallel to make the belt ride properly on the pulleys.

The development of new rubber and new cording design and material permits the use of belts with a small cross section. When installing belts, be careful to adjust them for proper tension and alignment. They should be snug but not tight. The correct tension is the lowest tension at which belts will not slip when the drive is under full load. On modern power transmission equipment, tensions are higher and are more critical. Therefore, the old hand or finger deflection method is not recommended. A commercial tension gauge will provide optimum performance.

The compressor flywheel and the motor pulley must be in line with each other in two different ways to give long life to the belt and to the electric motor. First, the centerline of the compressor must be parallel with the centerline of the electric motor shaft. Secondly, the pulley grooves must be in line with each other.

A poorly aligned belt will shorten the life of the motor because the motor is not designed to stand an excessive end load. A noisy, poorly operating motor may be the result. Fig. 7-70 shows a tool which may be used in adjusting and aligning belt drives.

Automobile air conditioner belts are designed with exacting standards to transmit power to the compressor. It is very important to follow factory instructions when adjusting these belts. See Chapter 27.

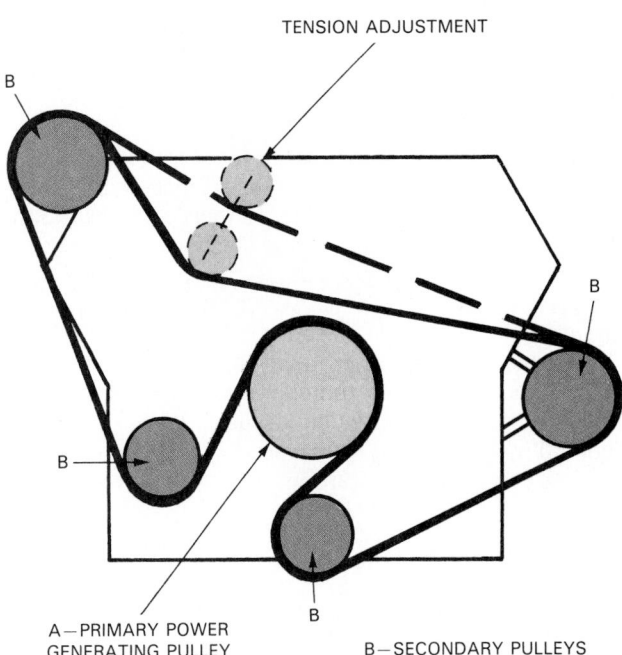

Fig. 7-71. Typical use of a serpentine belt, which transmits power from primary power-generating pulley, A, to secondary pulleys, B. Notice the tension adjustment.

Fig. 7-71 illustrates how a serpentine belt is able to transmit power from a single primary pulley to numerous secondary pulleys. Note the self-adjusting tension mechanism to maintain the proper tension in belt. Many automobiles are equipped with serpentine belts to drive the air conditioner's compressor and other devices.

7-47 MOTOR TESTING STAND

Worn electric motors are frequently taken to an electric motor repair shop for overhaul. However, some refrigeration service companies prefer to repair as many of these motors as possible. When rewinding is necessary, this should be done in a shop equipped for such work.

In each case, the motor should be carefully tested to determine whether it needs rewinding or simple repair. The technician must decide by studying operating characteristics. To do this, a torque testing stand should be used. The term "dynamometer" refers to such a test stand.

A mechanical dynamometer consists of a stand and a group of pulleys with equal outside diameters but with shaft diameters to fit various size motor shafts. The faces of these pulleys are smooth and flat. A torque arm lined with automobile brake lining is arranged to fit the pulleys. The torque arm is exactly one foot long between its point of support on the scales and the center of the motor pulley. A spring-loaded adjustment mechanism is arranged on the friction surfaces of the torque arm to enable various frictional forces to be produced between the torque arm and the motor pulley.

The end of the torque arm rests either on a spring scale or, preferably, on a platform scale. The arm should be balanced to remove any prior loading of the scale.

At the time the readings are taken, the torque arm must be level. The motor torque can then be very accurately checked for stall condition and full-speed load.

Fig. 7-70. Belt adjusting and aligning tool. Tool is inserted between pulleys to tension belt. A—Pulley. B—End with flywheel. C—Tool.

The torque obtained in this way may then be read directly from the scales in pound-feet. This data, when checked with the manufacturer's torque ratings, will show the condition of the motor.

Switch and brush mechanisms of these motors may also be checked on this stand. Ammeters, voltmeters, and wattmeters will determine electrical characteristics of the motor.

Temperature rise of the bearings should be carefully noted. A thermometer placed in the bearing oil reservoir is recommended for this purpose. Temperature rise should not be more than 72 °F (40 °C) above the room temperature.

Electric dynamometers are available for measuring the power output of electric motors. They can also run the motor to determine its friction losses.

Electric meters should be installed in a panel in such a way that push switches can put an ammeter, voltmeter, or wattmeter into the circuit.

7-48 SERVICING FAN MOTORS

The most common fan motor troubles are:
A. Loose connections.
B. Dry bearings.
C. Worn bearings.
D. Burned-out motor.
E. Loose fan.
F. Out-of-balance fan.
G. Fan blades touching the housing.

Loose or dirty connections will cause too much voltage drop at the motor and the fan motor will lose speed, hum loudly, and overheat. A sensitive voltmeter or an ohmmeter will quickly locate the faulty connection. Do not rely on visual inspection.

A dry bearing will cause the same symptoms, but this condition will last only a short time before the bearings will either seize (bind) or become badly worn.

Occasionally, the end play of the rotor becomes excessive (see D in Fig. 7-72) and causes the rotor to shift back and forth. As it shifts, it produces a distinct knock. Occasionally, the bearing (bushing) inserts in a reconditioned motor are out of position. This may force the rotor out of the magnetic center along its shaft.

When the motor is running, it should float between the extremes of its end play. One may check this by lightly touching the end of the rotor shaft with a wooden stick as the motor is running. It should move back and forth and then settle in between the extremes of the end play.

If the rotor cannot assume its magnetic center, it will hum excessively and heat. When running, the heaviest magnetic flow from the stator tries to line up with the heaviest magnetic flow from the rotor, B. This aligning must take place with end play clearance at D and D^1. The total clearance is usually about .030 or 1/32 in. Fig. 7-73 shows the bushing and thrust bearing washer on a fan motor.

A rattle in the fan motor may sometimes be nothing more than a loose fan on the motor shaft. This noise can be remedied by tightening the setscrew that fastens the fan hub to the shaft. Smaller fans have either a round shaft or a flat spot milled on the shaft.

If the fan is abused, the blades may be forced out of position and one or more blades may vibrate. The easiest repair is to replace the fan. Any attempt to rebalance the blades is difficult unless special static and dynamic balancers are available.

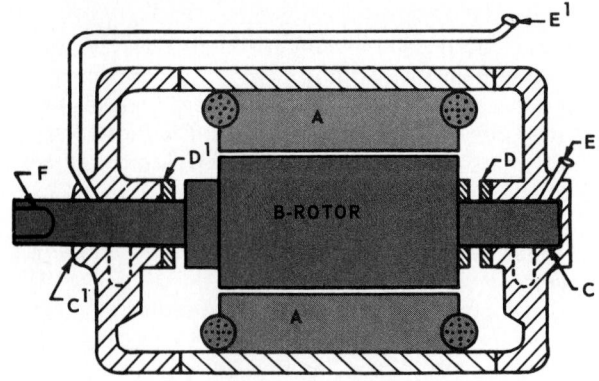

Fig. 7-72. Rotor running in its magnetic center. A—Stator. B—Rotor. C—Bearings. D—End play bearings. E—Oil cups. F—Pulley setscrew contact surface. E and E^1—Oilers for bearings.

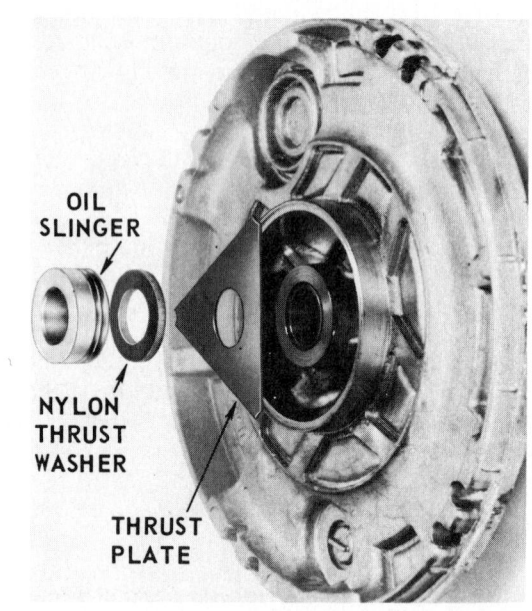

Fig. 7-73. Fan motor bearing and thrust bearing washer. (General Electric Co.)

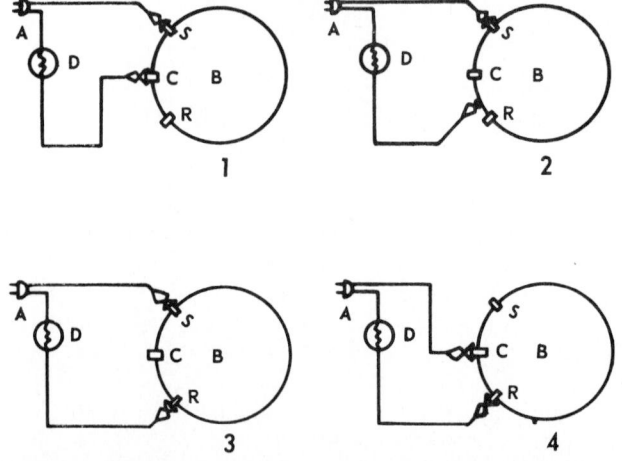

Fig. 7-74. Method of testing motor electrical circuits. A—Power circuit plug, 120 V ac. B—Motor being tested. C—Motor common terminal. D—25 W electric light bulb. S—Starting winding terminal. R—Running winding terminal.

If the fan blades touch the fan housing, the motor may be out of line or the shroud or housing may be bent. The contact spot is usually easily detected and remedied by moving the fan on the shaft or moving the shroud or housing. Do not bend the fan blades, as this will cause the unit to vibrate.

7-49 SERVICING AND REPAIRING HERMETIC MOTORS

Servicing hermetic motors involves two major operations:
1. External servicing.
2. Internal servicing.

Most hermetic motor troubles are external—either in the wiring or in the motor control devices. It is important to find out exactly where the electrical troubles are before deciding whether the motor is at fault.

Furthermore, it is essential that any outside trouble be remedied as soon as possible, because eventually it may cause the motor to fail. The proper steps for internal servicing of motors are explained in Chapter 11.

To help in checking electrical circuits, many companies list the volt ampere (VA) values for their equipment. This data does away with the need for a wattmeter reading. The Power Factor need not be considered or calculated.

For example, a manufacturer may list the motor as a 120 V single-phase motor with 2.9 VA locked rotor (starting load) and 1.1 VA full load. The service technician can then check the motor with a combination volt-ammeter to find out if the motor is operating correctly.

7-50 SERVICING AND EXTERNAL TESTING OF HERMETIC MOTORS

The condition of the refrigeration motor can be readily and easily determined with the use of instruments, without the necessity of opening the unit. This may be done by disconnecting the wires to the motor terminals and then testing the motor independent of its outside electrical connections, Fig. 7-74. Point A is the plug, B is the hermetic motor, and D is the light bulb (25 W). The starting winding terminal is S, the common terminal is at C, and the running or main winding terminal is at R.

In view 1, the starting winding is being checked for continuity. If the bulb, D, lights, it means the current is flowing through the starting winding from S to C. In this case, there is no open spot in the windings.

In view 2 of Fig. 7-74, the windings are being checked for a ground. Grounding means that some part of the internal wiring or the terminals are touching or making electrical contact with the metal parts of the unit.

If the bulb, D, lights when one of the electrical leads is touched to any of the terminals, and when the other lead or clip is touched to the clean or bare metal body of the dome, it means that electricity is flowing along the internal wires and through a grounded wire into the metalwork. Be sure the terminals are clean and dry during the test. They may be temporarily grounded by dirt.

Insulation on the windings should show no breaks. There should be infinite resistance between the terminal and the casing (dome) at 1000 to 1500 V. The instrument in Fig. 7-75 checks motor windings at high voltages. One must be extremely cautious not to touch the parts when power from the test instrument is on.

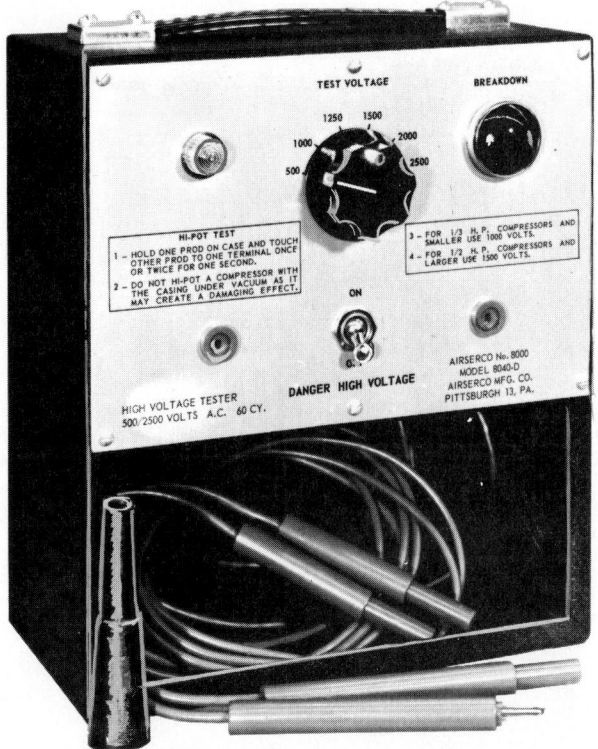

Fig. 7-75. High voltage test instrument checks motor windings for shorts or grounds. Use 1000 V for units up to 1/3 hp and 1500 V for units 1/2 hp and higher. Duration of test should not be longer than one second. (Airserco Mfg. Co., Inc.)

In view 3 of Fig. 7-74, the continuity of both the running and starting windings is being checked. In view 4, the running winding only is being checked for continuity.

Sometimes—especially if the unit has been overheated—motor windings short out without a ground being formed. Any shorting of the motor windings will increase the current draw, decrease the power, and overheat the unit. A shorted unit can sometimes be detected by an interruption in the steady hum of the motor when it is running. This is a noticeable beat added to the steady hum of the motor. To check for this short, one can roughly determine its existence by the test light, D, as shown in all views of Fig. 7-74. The test light will be brighter than normal if some of the windings are shorted.

A better way to check is to use an ohmmeter and determine the resistance of the coils. As models are checked, record the data. Many services technicians use an ohmmeter to check for continuity, shorts, and grounds. An ohmmeter is more accurate than a test light. Repeated tests have shown that the approximate resistance of domestic unit windings are as shown in Fig. 7-76.

7-51 SERVICING A STUCK HERMETIC MOTOR COMPRESSOR

Occasionally, a unit will not start even though all the electrical tests indicate that it is in good condition. This condition may have several causes. The unit may have been idle for a considerable time or a particle of dirt has gotten into it. On the other hand, some electrolytic plating may have taken place. A more than normal amount of liquid

	Ohmmeter Readings	
Running HP	Running Winding	Starting Winding
1/8	4.7	18
1/6	2.7	17
1/5	2.3	14
1/4	1.7	17

Fig. 7-76. Approximate ohmmeter readings for a typical fractional horsepower, single-phase motor.

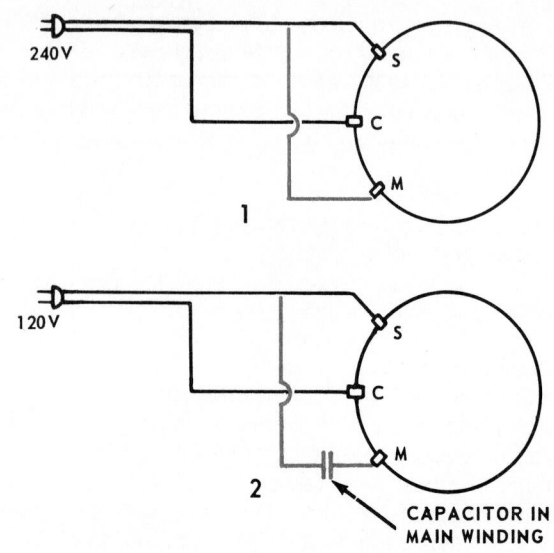

Fig. 7-78. Methods which may start a stuck compressor. 1—Use voltage above normal. 2—Use capacitor in series with running winding. CAUTION: Close the circuit for only one or two seconds at a time.

refrigerant in the compressor may also bind the unit. Three methods are recommended to break loose a stuck unit:

1. Connect the power line directly to the motor connections, eliminating the starting relay, as in Fig. 7-77.
2. Use above-normal voltage, such as 240 V on a 120 V circuit, to break it loose, Fig. 7-78, view 1. This method can only be used for a very short period of time.
3. Reverse the unit to make it run backward. This reversal of rotation may be done by putting a capacitor in series with the running winding, as illustrated in Fig. 7-78, view 2.

It is important to be continually on the alert when working with high voltage circuits. Discharge capacitors before handling them.

7-52 REVIEW OF SAFETY

Electrical hazards can be considered in two parts:
1. Electricity as a source of ignition to start fires.
2. Electrical shock.

When electricity is passed through a part of an animal or human, it causes muscle spasms. If it passes across the heart or brain, it can be fatal. If enough current (amperes) is present, the electric current can actively overheat the body, cause burns and high temperature which may result in permanent body damage. As little as .25 mA can kill a human being!

Disconnect the electrical power source before any repair or service to electrical parts. Lock the switches open to prevent someone from closing them during installation or ser-

ELECTRIC MOTORS

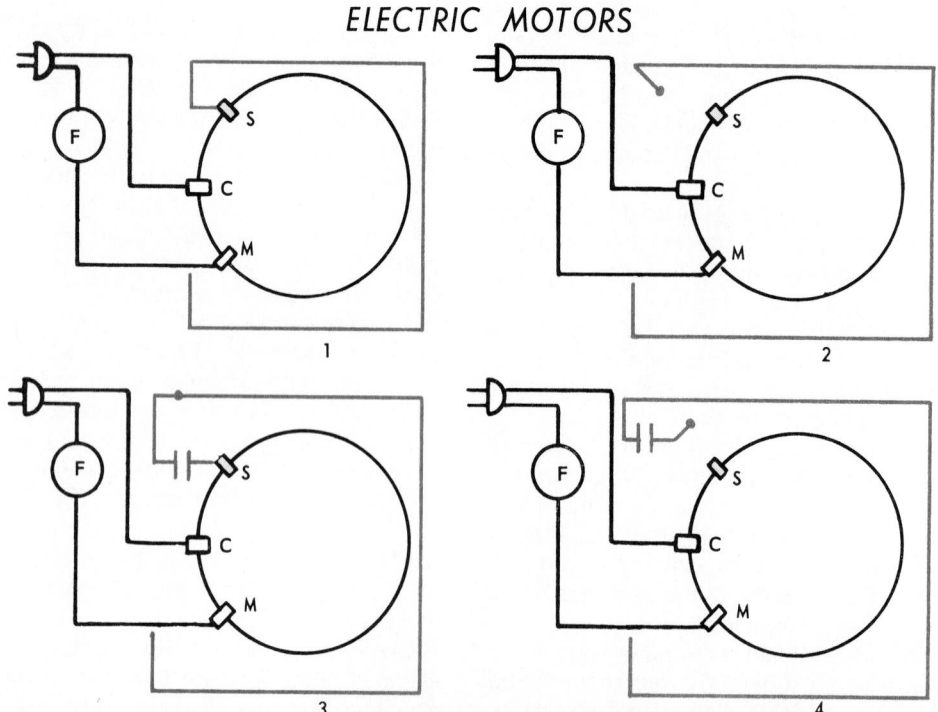

Fig. 7-77. Some methods which may be used to break loose a stuck hermetic compressor. Views 1 and 2—Without a capacitor in the starting circuit, touch S terminal for one or two seconds maximum. Then open circuit as shown in view 2. View 3 and 4—With a capacitor in the starting circuit, touch S terminal for one or two seconds maximum, release as in view 4.

vice operations. The switch should also be tagged to warn other people.

Always discharge a capacitor before touching its terminals. If it is charged, it may discharge 200 to 500 V into the body.

Persons subjected to electrical shock should be made to lie down. Keep them warm. If they are unconscious, give artificial respiration. Always call a physician if someone has suffered severe electrical shock.

Avoid touching moving belts. One may be seriously injured if the fingers or hand are caught between the belt and pulley or flywheel. Revolving fans can badly mutilate the fingers and hand.

7-53 TEST YOUR KNOWLEDGE

1. List the most common types of motors used on hermetically sealed domestic units.
2. How many field windings are used in a 120-240 V repulsion-start, induction motor?
3. Why must compressor, fan, and pump motors be cleaned regularly?
4. Why do some capacitors fail?
5. Why do some capacitors have resistors?
6. What is the running winding wattage of a 1/8 hp hermetic motor?
7. Should a replacement capacitor be undersize?
8. What type of belt is used to drive an automotive compressor?
9. How often should external drive motor bearings or bushings be oiled?
10. What kind of electric motors are usually used in hermetic refrigerators?
11. What are the common voltages used on direct current motors?
12. How can radio interference caused by the refrigerator motor be reduced?
13. How many terminals does a domestic hermetic motor usually have?
14. How many windings does a single-phase hermetic motor stator have?
15. Why are some electric motors called split-phase motors?
16. What material is used in the construction of replacement terminals?
17. Why are hermetic motor field windings insulated in a special way?
18. What is a capacitor?
19. Is the starting capacitor connected in series with the starting winding?
20. What usually happens in the starting winding circuit when the motor reaches the correct speed?
21. What type motors are used to power the fans used on hermetic systems?
22. How is the fan motor connected electrically to the compressor motor?
23. How may a shorted capacitor be detected?
24. What type of electric motor uses a start winding?
25. How is the wiring marked in many wiring harnesses for easy tracing?
26. What is the magnetic center of a motor?
27. Why do some systems use two running capacitors connected in series?
28. What information is usually recorded on the motor identification plate?
29. What is the color of the grounding wire?
30. What can be done to free a stuck compressor motor?

This solid-state logic module is part of a preprogrammed computer used on a heat pump. Technicians carefully inspect the modules for quality.
(York International Corp., Central Environmental Systems)

Surface-mounted thermistors save space. (Dale Electronics, Inc.)

Chapter 8

ELECTRIC CIRCUITS AND CONTROLS

After studying this chapter, the technician will be able to:
☐ Identify the common types of electrical circuit controls used in refrigeration and air conditioning systems.
☐ Discuss different refrigeration circuits.
☐ Differentiate between standard electrical circuit diagrams and ladder diagrams.
☐ Read and understand electrical circuitry.
☐ Adjust various refrigeration controls.
☐ Demonstrate how to test and service thermostats.
☐ Identify safety devices used in an electrical circuit.
☐ Work safely around electrical circuits and controls.

8-1 ELECTRICAL CIRCUITS—COMPLETE WIRING DIAGRAM

Electrical circuits contain a power source, a switch, and an electrical load. The power source is usually the building's electrical power. The electrical load is most often a motor.

Fig. 8-1 illustrates an electrical circuit with two switches. Switch 1 and 2 must be in a closed position for the motor to be "on." If Switch 1 or 2 is in an open position, the circuit will not be completed and the motor will be "off." This is called a series circuit.

A parallel circuit is shown in Fig. 8-2. Switch 1 and 2 must be in an open position for the motor to be "off." If Switch 1 or 2 or both are in the closed position, the circuit will be complete and the motor will be "on."

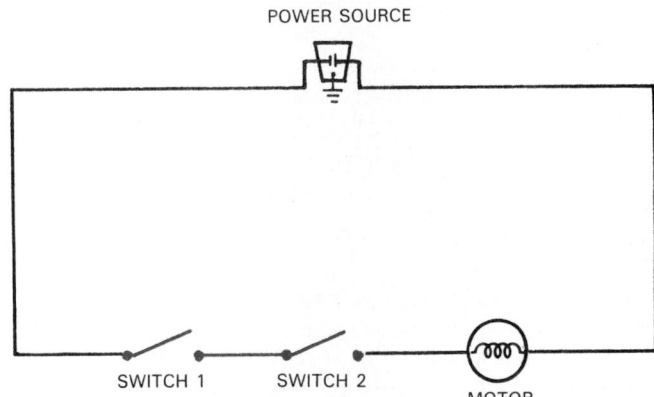

Fig. 8-1. A basic electrical circuit, connected in series. Switches 1 and 2 are open and the motor will not operate.

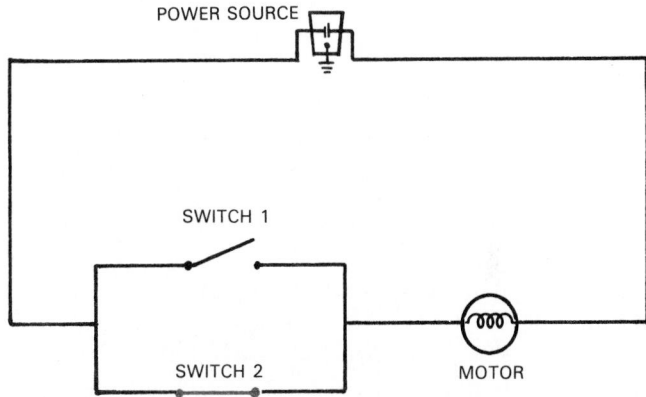

Fig. 8-2. A parallel circuit. Note that electrical Switch 1 is open and Switch 2 is closed. The motor will operate.

Complete wiring diagrams of an electrical circuit show the connections and/or actual location of individual devices that form the system. A complete electrical circuit diagram for a modern domestic refrigerator is shown in Fig. 8-3. To the beginner, this may appear to be a rather complicated diagram. However, the various circuits and controls may be fairly easily broken down into several individual circuits. The parts should first be identified:

1. Temperature control.
2. Freezer compartment fan (or evaporator fan).
3. Center rail heater (mullion).
4. Motor compressor.
5. Motor control (temperature control in Fig. 8-3).
6. Starting relay.
7. Cabinet light.
8. Cabinet light switch.
9. Ice maker.
10. Water valve.
11. Defrost heater.
12. Drain heater.
13. Defrost control.
14. Defrost terminating thermostat.
15. Cabinet ground.
16. Starting capacitor.

Other devices sometimes found in these circuits include:
1. Condenser fan motor.
2. Evaporator fan.

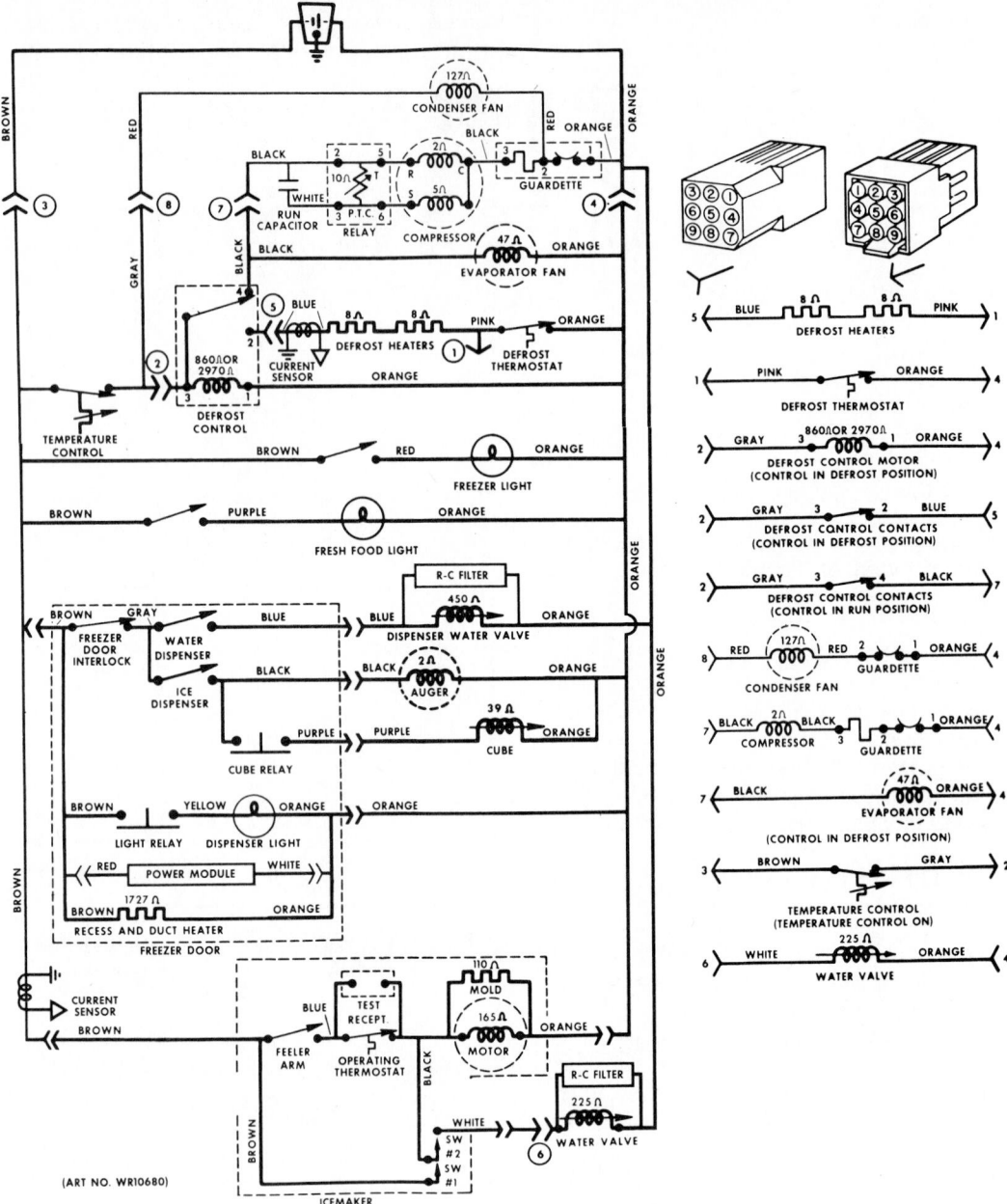

Fig. 8-3. Typical wiring diagram for domestic refrigerator. In addition to temperature control, this circuit includes defrost heater, defrost thermostat, ice maker, water dispenser, and ice dispenser. Service cord attachment plug contains a ground terminal. (General Electric Co.)

3. Condenser fan.
4. Ultra-violet ray lamp.
5. Butter warmer.
6. Humidity controls (see Chapter 20).

Electrical power to the unit is obtained through an insulated electrical cord. Usually this is a No. 18 stranded wire for domestic refrigerators. Sizes up to No. 12 are used for the heavier duty air conditioners. Additional information concerning wire sizes is provided in Para. 6-82 and Fig. 7-48. These cords are insulated to withstand at least twice their normal voltage. All power cords, for refrigeration and air conditioning units, are of the three-wire (one green ground wire) type. The cord is usually connected to a junction box mounted in the condensing unit compartment.

In all cases, the green wire is the ground. It is never to be used as a current-carrying conductor. There is no accepted color code for the various electrical circuits of a

domestic refrigerator except the green for ground. Standard extension cord plug configurations are explained in Para. 6-83. Various refrigeration circuits will be identified and described in some detail later in this chapter.

8-2 ELECTRICAL CIRCUITS—LADDER DIAGRAM

It may be difficult to determine the correct order of operation in detail using a standard electrical circuit diagram. Therefore, equivalent ladder diagrams are drawn to aid the service technician when troubleshooting a system. Ladder diagrams identify the order of operation of specific devices, not their actual location. The vertical lines identify the power supply and the horizontal lines contain the various switches and loads located in the circuit, Fig. 8-4.

Most electrical systems are a combination of parallel and series circuits switching on and off electrical loads.

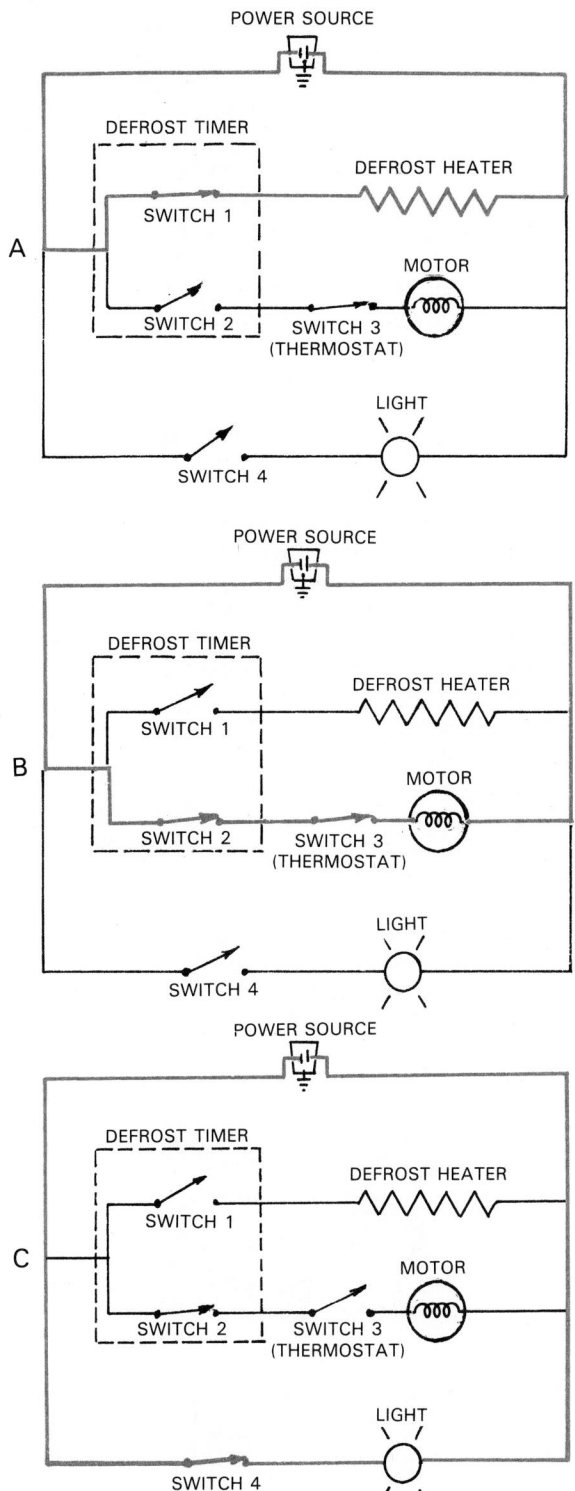

Fig. 8-4. Ladder diagram for a domestic refrigerator. A—Defrost heater is turned on by Switch 1, and compressor motor is turned off by Switch 2. B—Compressor motor is on when Switches 2 and 3 are closed. C—Light is on when Switch 4 is closed.

8-3 CONTROL SYSTEMS—FUNDAMENTALS

Refrigeration and air conditioning systems control conditions within a specified area. The devices that maintain the desired conditions are known as operating controls. The controls that keep the equipment operating at proper levels,

preventing damage to the system, and avoiding injury to people, are known as safety or limit controls. Refrigeration and air conditioning control systems are often divided into the following three categories:

1. Conditioned area: The area in which the temperature, pressure, and humidity are being controlled.
2. Controlling instrument: Instrument that is responsive to changes. This is accomplished through the use of sensing devices, thermostats, motor controls, pressurestats, humidistats, and air distribution controls.
3. Operating device: The mechanism that directly affects the actual conditions and is regulated by the controlling instrument. Examples of this are valves, dampers, fans, and compressors.

A control system that is constantly correcting the condition is called a closed-loop, or feedback, control system.

In the operation of a typical refrigerator, the thermostat is the controlling instrument, the compressor is the operating device, and the space inside the refrigerator is the conditioned area. As the temperature in the space becomes too warm, the thermostat turns on the compressor. The compressor then circulates the refrigerant so that it can cool the space. After the space reaches the desired temperature, the controlling instrument turns off the compressor. The cycle is repeated as the controlled space continues to warm up and cool down.

8-4 TYPES OF CONTROLLING INSTRUMENT ACTION

There are five basic types of controlling instrument action:
1. Two-position "on-off."
2. Timed "on-off."
3. Variable.
4. Proportional.
5. Proportional with automatic reset.
A description of each type of action is as follows:
1. Two-position on-off control action is the most common type of action. The control either turns the operating device "on" or "off." A typical example of this type of action is the operation of a refrigerator. The thermostat turns the compressor "on" to cool the refrigerated space when it gets too warm, and turns it "off" after the space reaches its desired temperature. The following terms define the operating characteristics of a two-position on-off control device:
   a. Range represents the temperature or pressure at which a control is set to operate. Adjusting the range (Para. 8-10) changes the set point level.

For example:

	Temp. Thermostat On	Temp. Thermostat Off
Original setting	72 °F	77 °F
Setting after + 3 °F range adjust	75 °F	80 °F

   b. Desired differential is the difference between the high and low temperature or pressure set points. Adjusting the differential increases or decreases this difference (Para. 8-12).

For example:

	Temp. Thermostat On	Temp. Thermostat Off	
Original setting	72 °F	77 °	(5 °F diff.)
Setting after additional +3 diff. adjust	72 °F	80 °F	(8 °F diff.)

c. The operating differential is the actual temperature or pressure difference which occurs in the conditioned area. If the desired differential is less than the operating differential, the system lags. If the desired differential is greater than the operating differential, the system leads. If the operating differential is low, the system will cycle frequently. If the operating differential is too great, the system may not be keeping the conditioned area within the desired range. Fig. 8-5 shows a two-position on-off control. When the damper is open, A, the area is heated and the temperature rises. When the damper is closed, B, no warm air can flow to the area and the temperature falls.

2. Timed on-off control: This is the action that is required when the operating differential is too great. One method is to have an anticipator built into an on-off control so that it can turn the operating device on or off before it normally would. This reduces the operating differential. A heating thermostat with this type of action would contain a small internal heater as the anticipator. The small heater warms the thermostat faster than the room air does. This enables the heating thermostat to shut off the furnace sooner than it otherwise would, and prevents the room from becoming too warm. In a cooling system, heat is added to the thermostat during the off cycle (cooling anticipation). The control turns on before the room gets warmer than it should. Turning on for cooling is the opposite of turning off a furnace with heat anticipation. Para. 25-8 provides additional information on heat and cooling anticipators.

One type of timed on-off control uses a clock that opens and closes electrical contacts. For example, the compressor may be on for 5 minutes, then off the 20 minutes. It will again be on for 5 minutes, off for 20, and so on. There may be no problem with overcooling the cabinet for the following reason. Most of the cooling happens in the first 1 1/2 minutes of the cooling

period, and less heat is transferred out of the cabinet as the cabinet becomes cooler. The cooling system tends to regulate itself, since the cabinet will not become cooler than a certain limiting cold temperature.

3. Variable control action is as follows: a control moves gradually from one position to another (for example, a variable position damper). When the desired temperature is within the control range, there is no signal transmitted. The damper is left in a partially open or closed position until another signal is received from the control. Variable control action is illustrated in Fig. 8-6. In view A, the damper moves gradually from position 1 to 2, allowing more warm air to enter the area and raising the temperature above the lower limit. When the temperature reaches the proper range, the thermostat signals the damper to stop opening and remain in position 2. In view B, the damper will remain in position 2 while the temperature varies within the upper and lower limits of the range. When the temperature exceeds the upper limit of the range, the thermostat will signal the damper to close gradually, view C. It will stop closing, 3, when the temperature falls below the upper limit. This cycle is repeated whenever a limit is exceeded.

4. Proportional control action: The output signal of the control is not an on-off signal, but one which varies in strength. This variation is dependent upon the amount of change needed by the actual condition. The operating device is adjusted in proportion to the control signal's strength. Proportional control action is much more sensitive to changes than on-off control or variable control action. The following terms are used when referring to proportional control action devices:

a. Set point is the setting placed on the control to maintain the desired conditions.

b. Control point is the condition that is being maintained.

c. Offset is the deviation between the set point and the control point. Offset is sometimes called error.

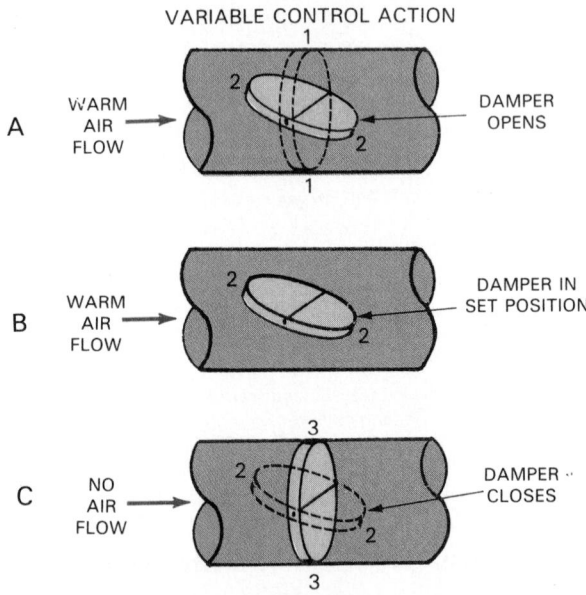

TWO POSITION CONTROL ACTION

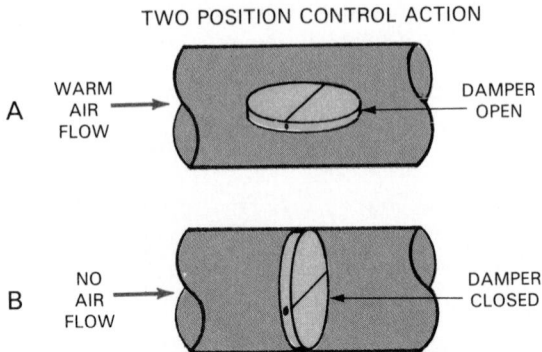

Fig. 8-5. The action of a two-position on-off control. The area is heated when the damper is open. A—Damper is open, allowing air to flow. B—Damper is closed, stopping air flow.

Fig. 8-6. Action of a variable control damper. A—Damper opens until the temperature is within desired range. B—Damper remains in position as long as temperature stays within desired range. C—Damper closes until temperature is within desired range.

PROPORTIONAL CONTROL ACTION

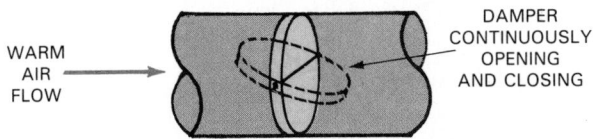

Fig. 8-7. Action of a proportional control. Damper opens and closes continuously to keep temperature constant.

The damper in Fig. 8-7 is continuously changing its position. As the temperature changes, the damper readjusts to maintain the desired temperature.

5. Proportional control action with automatic reset: This is the most advanced method of maintaining a desired condition. The control point (actual pressure, temperature, etc.) is automatically adjusted to the set point (desired pressure, temperature, etc.) when the need occurs. Sometimes the control is designed to foresee an expected condition. The control then takes action very early before trouble starts.

The specific type of controlling instrument to use in a refrigeration or air conditioning system depends upon the specific application and the desired result.

8-5 TEMPERATURE CONTROL PRINCIPLES— SENSING BULB

There are three types of mechanisms used in motor control thermostats:
1. Sensing bulb.
2. Bimetal.
3. Solid state (usually with a thermistor).

The vapor pressure thermostat with sensing bulb is most often used on refrigeration systems. See Fig. 8-8. The operation of the sensing bulb depends upon the vapor pressure of a volatile fluid. (A volatile fluid is one that becomes vapor at low temperatures.) The vapor pressure of the volatile fluid acts on a flexible bellows or diaphragm to control the mechanism in the thermostat.

The volatile liquid is located in the sensing bulb and the bulb is in contact with the refrigerator evaporator. A

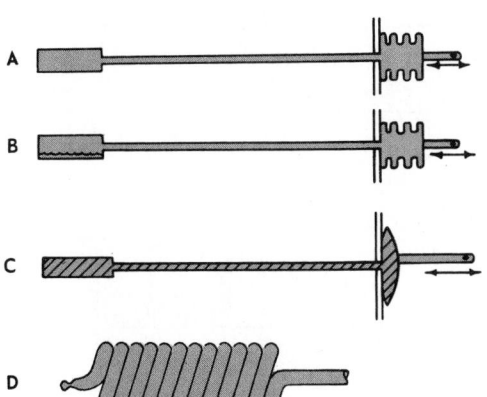

Fig. 8-8. Methods used to obtain motion from temperature changes. A—Gas-charged temperature response bellows. B—Vapor pressure temperature response bellows. C—Liquid charge temperature response diaphragm. D—Capillary tube coil used as a bulb.

capillary tube connects the temperature sensing bulb to the operating mechanism in the thermostat. This type of thermostatic element is illustrated in Fig. 8-9.

8-6 TEMPERATURE CONTROL PRINCIPLES— BIMETAL

Bimetal strip temperature control devices consist of two different metals that are bonded together. The two metals commonly used are copper and steel. Copper has a greater coefficient of expansion than steel; it expands more with an increase in temperature than steel. This causes the bimetal strip to bend as temperature rises. Bending action of the strip opens and closes the contact points in the electrical circuit. The strip bends as shown in Fig. 8-10.

Another common type of switch is the mercury switch mounted on one end of a coiled bimetal strip. This type of thermostat is most commonly used in air conditioning and heating thermostats.

Fig. 8-11 illustrates a bimetal coil heating thermostat. As the air surrounding the coil gets warmer, the coil expands. This tilts the bulb upward, resulting in the movement of the mercury. The contacts are no longer closed and the unit is "off." During the cooling cycle, the opposite occurs.

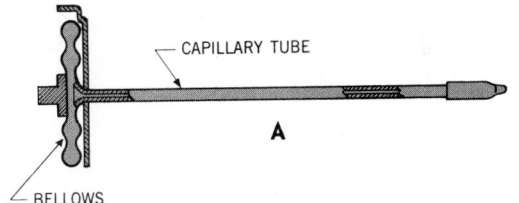

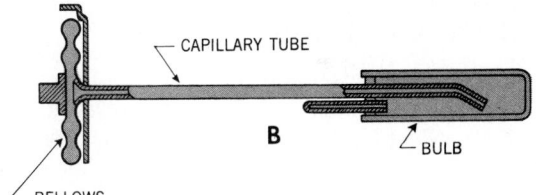

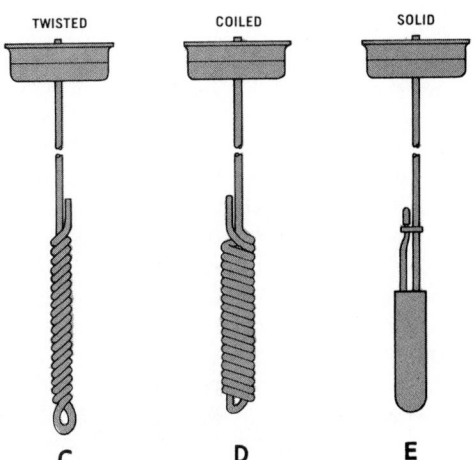

Fig. 8-9. Several thermal element designs. A—Capillary tube sensitive element. B—Bulb sensitive element. C—Twisted sensitive element. D—Coiled sensitive element. E—Solid (bulb) sensitive element.
(Ranco North America)

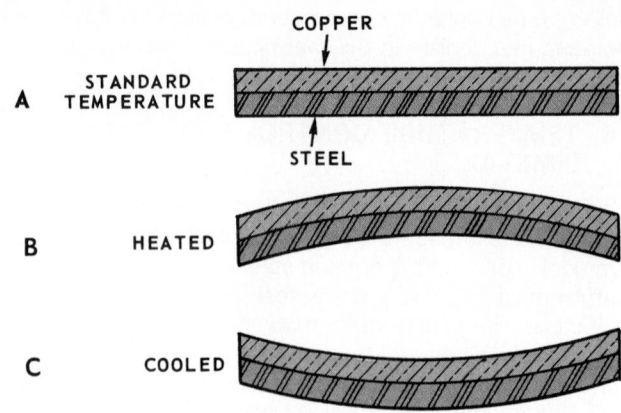

Fig. 8-10. A bimetal strip bends or warps with temperature change. A—Strip at controlled temperature. B—Strip warmed above control temperature. C—Strip cooled below control temperature.

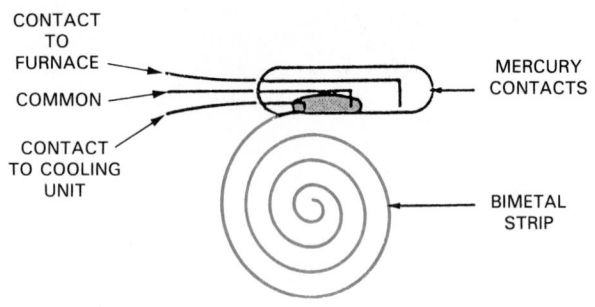

Fig. 8-11. Mercury bimetal type thermostat.

8-7 TEMPERATURE CONTROL PRINCIPLES— ELECTRONIC (SOLID-STATE)

Electronic controls have advantages over other controls because of their compactness, increased reliability, rapid response times, and lack of moving parts. Electronic controls can be identified by their low operating voltages. Most operate at 5 to 15 volts, provided by a step-down transformer and rectifying circuit.

The sensing device in an electronic temperature control is usually a thermistor. See Para. 6-62. The resistance varies as temperature varies. If the resistance of the thermistor goes down as temperature increases, it has a negative temperature coefficient (NTC). There are also thermistors with a positive temperature coefficient (PTC). Their resistance increases as temperature increases. PTC thermistors are used in motor safety controls.

Changes in a thermistor's resistance are detected by an electronic circuit called a wheatstone bridge. As the resistance changes due to a change in temperature, the resistances in the bridge become unbalanced. This causes the output voltage of the bridge to change. A simple form of this type of wheatstone bridge is shown in Fig. 8-12. The output voltage can be amplified to signal the control device. The signal to the control device indicates the action that needs to be taken.

Electronic thermistors have the ability to accurately sense extremely high and low temperatures. Without these types of devices, it would be difficult to measure temperatures in situations where conventional methods do not work.

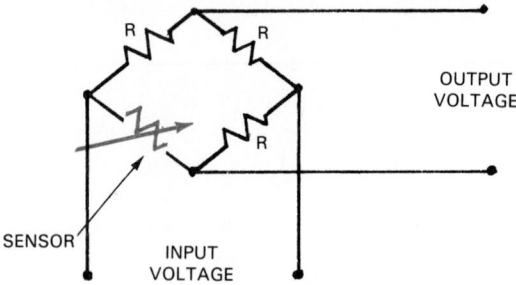

Fig. 8-12. Wheatstone bridge circuit principles. Wheatstone bridge measures changes in sensor resistance.

8-8 THERMOSTATIC CONTROL MECHANISM CONSTRUCTION—VAPOR PRESSURE

Many mechanisms are used in vapor pressure type thermostats. The action of the contact points needs to be quite rapid. If contact points are allowed to open very slowly, there will be some electric arcing as the current tends to jump across the tiny gap. This arcing action would very quickly burn the contact points and ruin their ability to make a good electrical contact.

There are two ways to get this snap action. One is through a snap action toggle mechanism. The other is with a permanent magnet. These two mechanisms are shown in Fig. 8-13.

The toggle mechanism is shown in view 1. In this mechanism, the fulcrum points at A and B are under some pressure. This tends to push them together. As the bellows expand, due to warming of the sensing bulb, the toggle

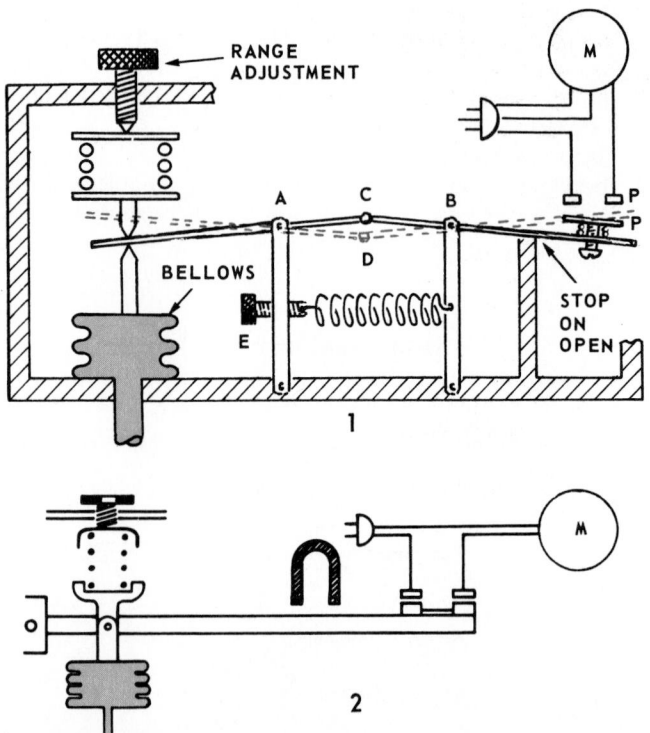

Fig. 8-13. Construction detail drawings show how bellows are connected to snap action motor control switch. 1—Toggle snap action. 2—Magnet snap action.

point, C, will move down. The instant it passes the center point, it will snap into position, D, closing contact points, P.

As the sensing bulb cools and the bellows contacts (shrinks), the toggle will be moved toward point C. As soon as it passes the center point, it snaps open very quickly. An adjustment, at E, controls the pressure tending to move points A and B together. Increasing this pressure will lengthen the running time. This toggle arrangement gives good snap action.

The system with the magnet snap action is shown at 2 in Fig. 8-13. The bar carrying the contact points is made of a magnetic metal like iron.

The permanent magnet tries to draw this material toward it. Magnetic lines of force always try to be as short as possible. As the pressure in the bellows tends to close the points, the magnetic effect increases as the iron bar approaches the magnet. This causes a snap action which closes the points quickly.

Likewise, as the sensing element cools and the bellows contracts, it will take force to pull the points apart. However, the magnetic pull decreases rapidly as soon as the points separate, which allows a quick opening of the points.

With this type of snap action, the running interval can be shortened by moving the magnet away from the iron bar. Or, it can be lengthened by bringing the magnet closer to the iron bar. The magnet should never touch the iron bar.

A typical vapor pressure type motor control is shown in Fig. 8-14. A magnetic type snap action control is shown. The sensing bulb is at the end of the coiled capillary tube. It has a range control which the owner may use to adjust the cabinet temperature.

8-9 THERMOSTATIC CONTROL MECHANISM CONSTRUCTION—BIMETAL

How the bimetal thermostat operates is explained in Para. 8-6. These thermostats are used, to some extent, in refrigeration but not as often as in air conditioning.

The construction of a bimetal thermostat is shown in Fig.

8-15. These thermostats use bimetal discs for fast snap action. The bimetal control usually consists of a dished disc. Its construction is such that it is dished in one direction when it is cold; and, as it warms, it suddenly snaps into a dish position in the other direction. Snap action discs are also used on pilot lights and gas burner controls.

The bimetal strip serves two purposes: one, it provides adjustable calibration; and, two, it provides a small effective temperature differential.

8-10 RANGE ADJUSTMENT

Range adjustment provides for the correct minimum and maximum temperature or pressure in an automatically operated system. Examples: the range adjustment will keep a refrigerator between certain temperatures, an air conditioned room between certain temperatures, and an air compressor between certain pressures.

It is very difficult to keep any device at one particular temperature and/or pressure, such as 34.4 °F (1.3 °C) or 150.2 psig, which is 164.9 psia (1 136 kPa). That is why a range adjustment is used.

A range adjustment is shown in Fig. 8-16. The unit cuts in at 25 °F and cuts out at 15 °F on the evaporator temperature. To make the cabinet operate at a warmer temperature, the range adjustment may be changed so that the cutout becomes 16 °F and the cut-in 26 °F, as shown in A of Fig. 8-16.

Note that the temperatures are higher but the temperature distance between the two has not been changed. This distance is still 10 °F. The new settings will only slightly affect the running time of the unit. There will be a slight decrease in compressor motor current instead of a decrease in running time. This is because the condition desired is not as cold and, therefore, the compressor will not have to do as much work.

The temperature of the refrigerator may be lowered a degree by making the range adjustment cut in at 24 °F and cut out at 14 °F, as shown in B of Fig. 8-16.

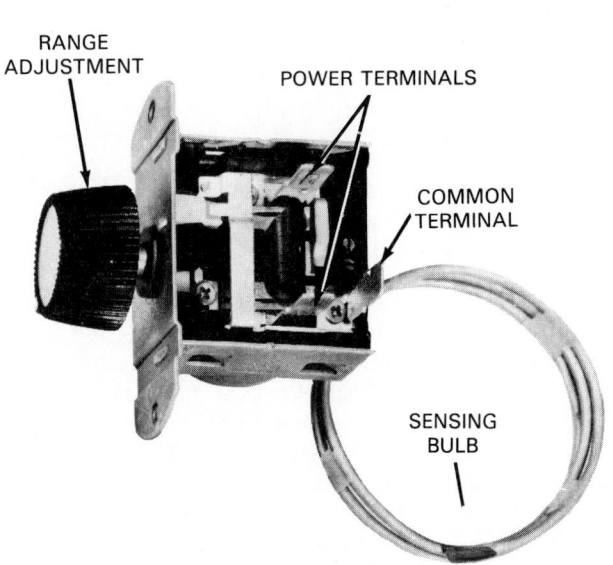

Fig. 8-14. Internal construction of vapor pressure type thermostatic motor control. (Eaton Corp., Controls Div.)

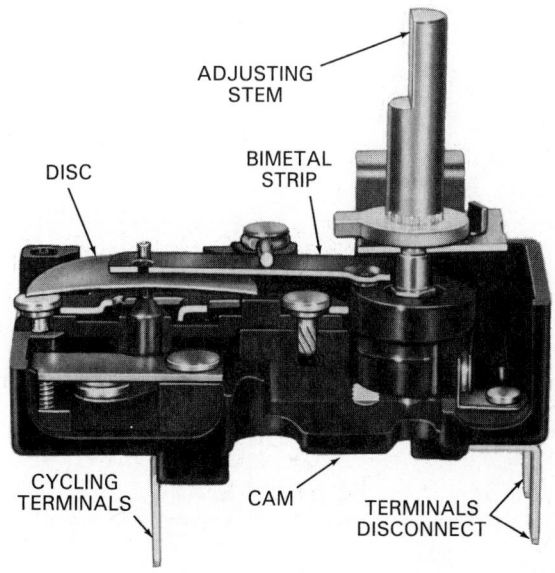

Fig. 8-15. Construction of thermostatic motor control. Note location of bimetal strip and disc. This type of unit is used for control in refrigeration, electric heat, and air conditioning. (Therm-O-Disc, Inc.)

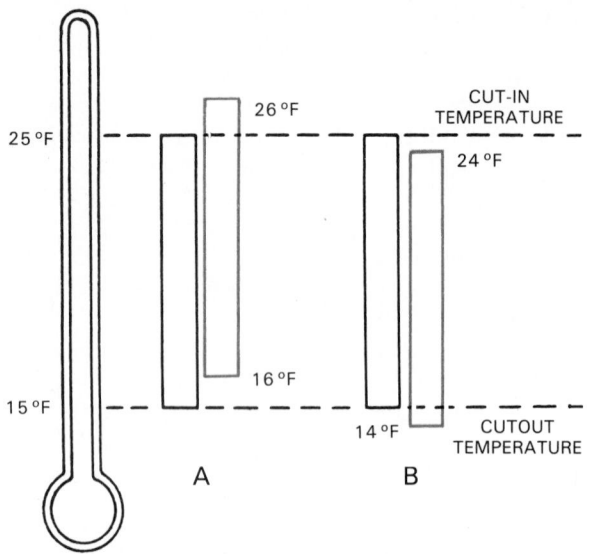

Fig. 8-16. Typical range adjustments Basic range is shown in black.

8-11 RANGE ADJUSTMENT MECHANISMS

The range adjustment is easily recognized. It is an adjustable force pressing directly upon the bellows or diaphragm which operates the switch. This force is always being exerted on the bellows whether the switch is in either the cutout or the cut-in position.

The adjustable force may be an adjustable weight that always presses against the bellows. More commonly, it is a spiral spring with an adjustable screw. Turning the screw changes the pressure or the tension of the spring. The spring may press or pull either directly on the bellows or diaphragm or on a lever attached to the bellows. See Fig. 8-17.

Most of the range adjusting screws have a calibrated dial or a pointer connected to them. It will indicate the direction the screw should be turned to provide a warmer or colder setting.

8-12 DIFFERENTIAL ADJUSTMENT

The differential adjustment is built into the temperature control mechanism. It is not adjusted or changed by the

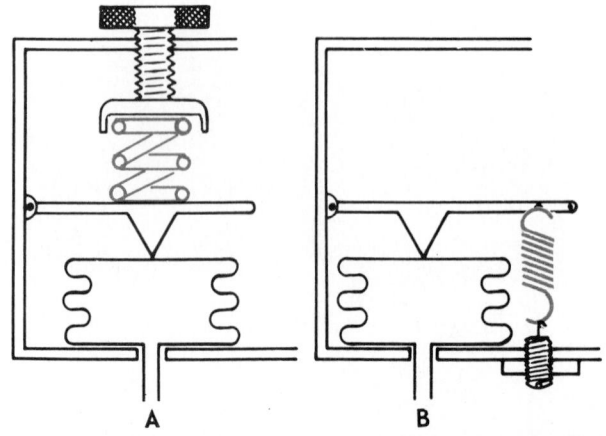

Fig. 8-17. Adjusting range settings of motor control. A—Compression spring range adjustment. B—Tension spring range adjustment.

operator when making a temperature selection by use of the control knob. The differential adjustment should be made by a service technician who understands the working of the differential adjustment mechanism.

The differential adjustment controls the temperature difference between the cutout and the cut-in settings.

If the evaporator is set to cut in at 25 °F and cut out at 15 °F, the difference, or differential, is 10 °F. Whenever the differential (the distance between the settings) is changed, the range is also changed. If only the range is adjusted, the differential will not be affected.

The thermostat differential for capillary tube systems must be large enough to allow the pressures to equalize in the system. This increases the off cycle, but the change should not be too much. If it is, the fixture temperature will rise too high. A differential of about 14 °F (8 °C) is common for fresh food refrigerators and about 11 °F (6 °C) for freezers.

Fig. 8-18 shows three types of differential adjustments:
1. Cut-in type. The cut-in point may be moved without changing the cutout as shown at 1 and 2 in view A.
2. Cutout type. The cutout point may be moved without changing the cut-in setting as at 1 and 2 in view B.
3. Double type. The cut-in and cutout settings may be brought closer together or moved farther apart, as shown at 1 and 2 in view C.

A person may learn to recognize cut-in and cutout differential adjustments by the way that they affect the operation of the switch mechanism. In the case of cut-in type, the effect is seen only when the switch is in the cutout position, ready to snap back into cut-in; in the case of cutout type, only when the switch is in the cut-in position, ready to snap to cutout position.

The third type is usually an adjustable arrangement which affects the effort of the toggle to snap off and on. This adjustment affects both the cut-in and the cutout. They may be adjusted farther apart or closer together.

To get a certain cabinet temperature, it is usually necessary to adjust both the range and the differential.

To keep the same average cabinet temperature, the control may be adjusted to 14 ° and 25 °F, shown in A_1 of Fig. 8-19. With this setting, the compressor will run longer but will not cycle as often.

If this control range adjustment is set to cut in at 25 °F and cut out at 13 °F, as at A_2 in Fig. 8-19, the compressor will run longer than normal. The cabinet temperature also will be lower than normal.

By setting the control to cut in at 27 °F and cut out at 15 °F, as shown in A_3 in Fig. 8-19, the compressor will run less and the cabinet temperature will be higher.

In each of the above examples, the differential has been 12 °F, but the range has changed.

If the control is adjusted to cut in at 24 °F and cut out at 16 °F, as shown at B_1 in Fig. 8-19, the cabinet temperature will be normal. However, both the on and off cycles will be shorter and the cabinet temperature will vary less than normal. The differential is now 8 °F, although the range is the same setting as one which would produce the scale reading in black.

With the control adjusted to cut in at 25 °F and to cut out at 17 °F, as shown in B_2 of Fig. 8-19, the cabinet temperature will be warmer than normal. The on and off cycles will be shorter than normal. Cabinet temperature will

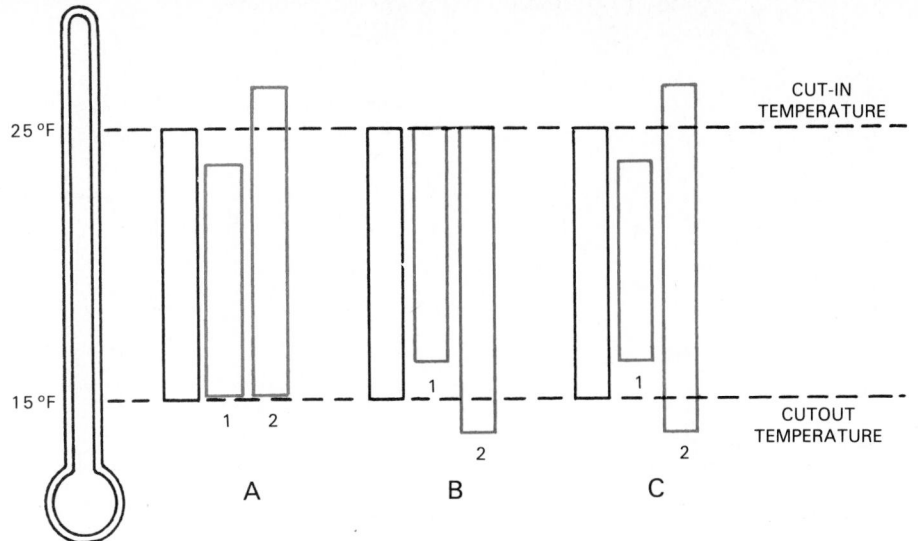

Fig. 8-18. Effect of different types of differential control adjustments. A—Cut-in setting type. B—Cutout setting type. C—Cut-in and cutout type. In each illustration, normal setting is shown in black.

vary less than normal. Note that the range has been raised compared to the black scale.

With the control adjusted to cut in at 23 °F and cut out at 15 °F, as shown at B_3 in Fig. 8-19, cabinet temperature will be below normal. But it will vary less than normal. The on and off cycles will be shorter than normal. The range is lower than for black scale.

For a slightly lower cabinet temperature with little or no increase in running time, the cutout type differential may be adjusted to cut out at 14 °F, as in A_1 of Fig. 8-20.

To decrease running time, compared to A_1 and keep the same running time as for black scale, the range adjustment may be set as shown in A_2 of Fig. 8-20. A cutout differential adjustment, as shown in B_1 of Fig. 8-20, will give a slightly warmer cabinet temperature and a shorter cycle interval. Also, the cabinet temperature will vary less than for black scale.

A range adjustment, as shown at B_2 of Fig. 8-20, will

provide a normal cabinet temperature, which varies less than normal. The cycling interval will be shorter than normal.

A typical control adjustment follows:

1. To increase the cycle time when a control with a cut-in differential is used, (a) adjust the differential from a 25 °F cut-in to a 27 °F cut-in, as shown at A_1 of Fig. 8-21; (b) adjust the range to move the settings to 26 °F cut-in, and 14 °F cutout, as shown at A_2 of Fig. 8-21.
2. To decrease the cycling time (interval) when a control with a cut-in differential is used, (a) adjust the cut-in differential to cut in at 23 °F, as shown at B_1 of Fig. 8-21; (b) adjust the range to cut in at 24 °F and cut out at 16 °F.
3. To decrease the running time with a cutout differential, (a) adjust the cutout differential to cut out at 17 °F, as shown at C_1, of Fig. 8-21: (b) adjust the range to

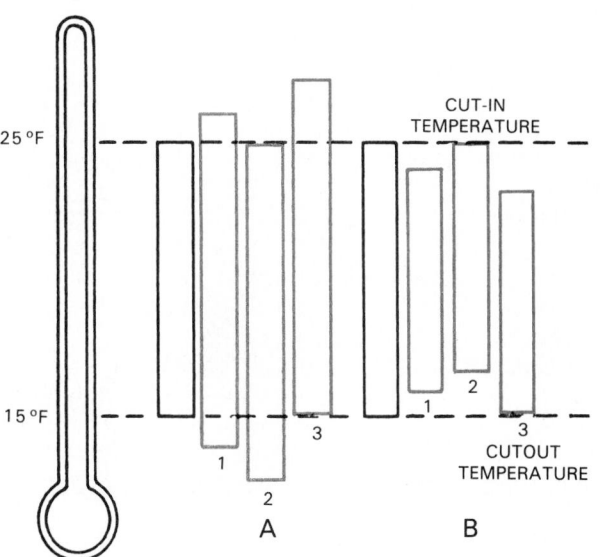

Fig. 8-19. Cut-in, cutout, and differential adjustment. A_1—Differential increased. A_2—Range lowered. A_3—Range raised. B_1—Differential decreased. B_2—Range raised. B_3—Range lowered.

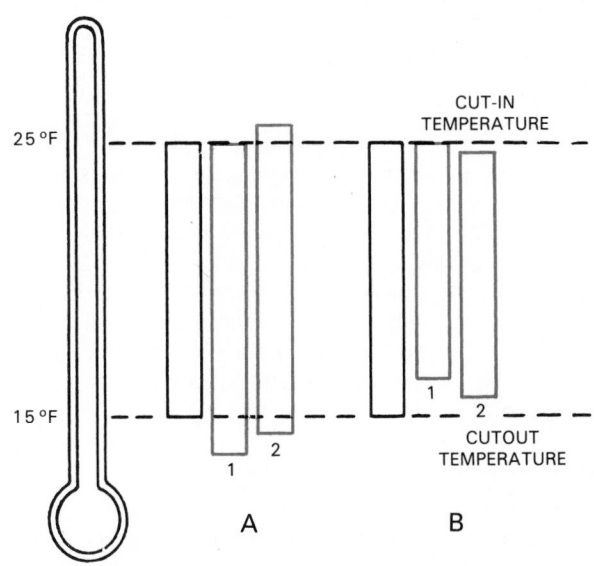

Fig. 8-20. Cutout differential adjustments. A_1—Differential lowered from 15 °F to 14 °F cutout. A_2—Range is then adjusted to 14.5 °F cutout and 26.5 °F cut-in. B_1—Differential is changed from 15 °F to 16 °F cutout. B_2—Range is then adjusted to 15.5 °F cutout and 24.5 °F cut-in.

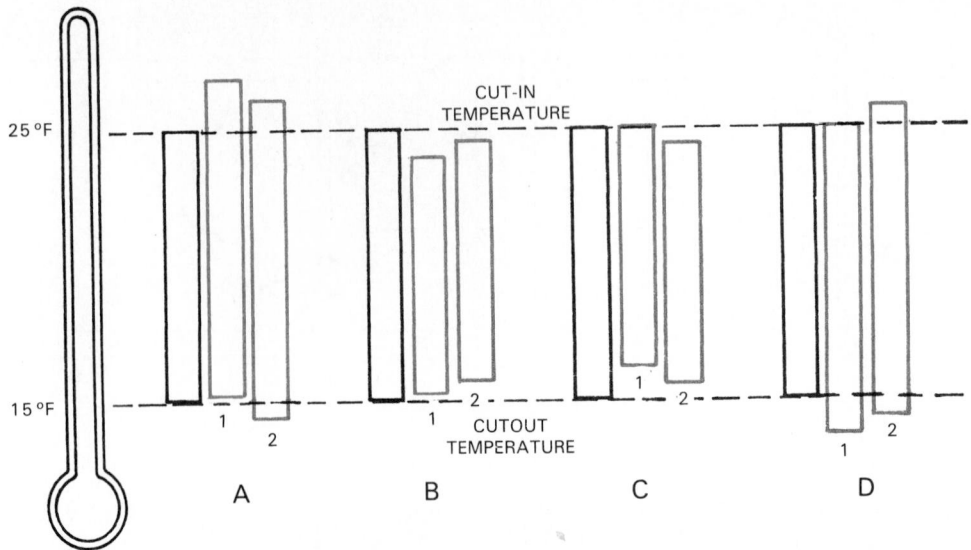

Fig. 8-21. Effect of a cut-in differential adjustment. A_1—Cut-in differential raised. A_2—Range lowered. B_1—Cut-in differential lowered. B_2—Range raised. Effect of cutout differential adjustment. C_1—Cutout differential raised. C_2—Range lowered. D_1—Cutout differential lowered. D_2—Range raised.

cut out at 16 °F and cut in at 24 °F, as shown at C_2 of Fig. 8-21.

4. To increase the running time with a cutout differential, (a) adjust the cut-out differential to cut out at 13 °F, as shown at D_1; (b) adjust the range to cut out at 14 °F and cut in at 26 °F, as shown at D_2 of Fig. 8-21.

8-13 DIFFERENTIAL ADJUSTMENT MECHANISMS

Two types of differential adjustments have a spring that affects the movement of the bellows, either just before the contact points cut out, or cut in, but not both. This limited action is controlled by a stop on the spring as shown on the right in each unit in Fig. 8-22. View A is a cut-in differential, view B is a cutout differential, and view C a cut-in differential.

A third type of differential adjustment affects both the cutout and cut-in. A spring makes it easier for the points to open and close, or more difficult for the control to open or close electrical contact points. (See view D, Fig. 8-22.)

Fig. 8-23 shows a mechanism with a cut-in differential. Fig. 8-24 shows a control with a cut-in calibration screw and lever, a cutout calibration screw and lever, and a range adjustment screw.

Some controls have a small heater unit. The heat from this unit keeps the bellows and diaphragm from becoming too cold. A too-cold thermostat body will not cut in as soon as it should, and may cause erratic cabinet temperatures.

A complete control is shown in Fig. 8-25.

8-14 ADJUSTING CONTROLS

It is advisable to adjust controls as the manufacturer specifies. Some controls are marked "cut-in adjusting screw" and "cutout adjusting screw." Others mark the adjustment "range."

1. Some controls have a cut-in differential with the control marked "step one, cutout adjustment" for one adjusting screw, and "step two, cut-in adjustment" for

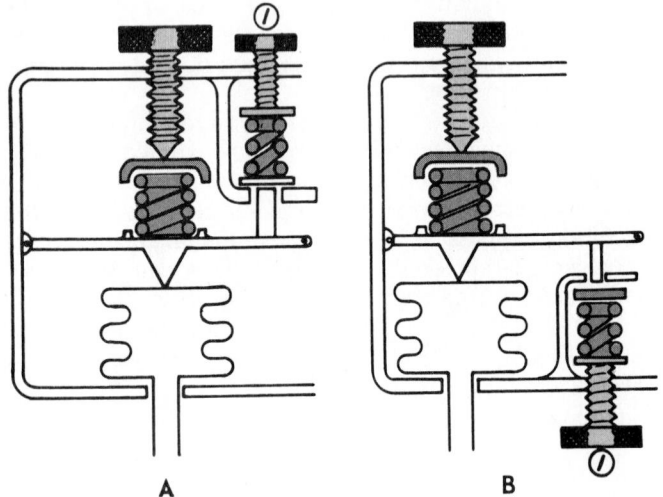

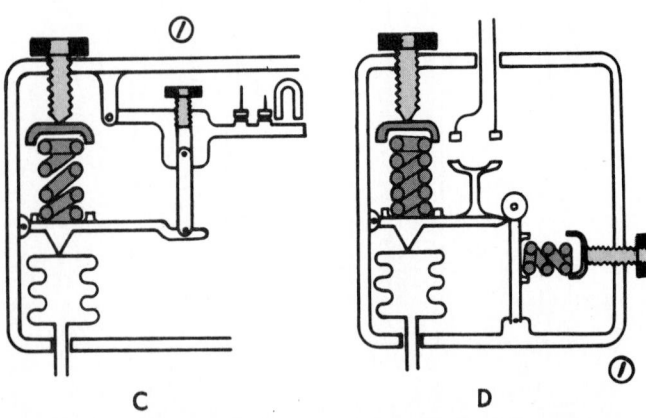

Fig. 8-22. Various types of differential adjustment mechanisms. A—Cut-in type. B—Cutout type. C—Cut-in type using a slot and magnet. D—Double type. 1—Differential adjustment.

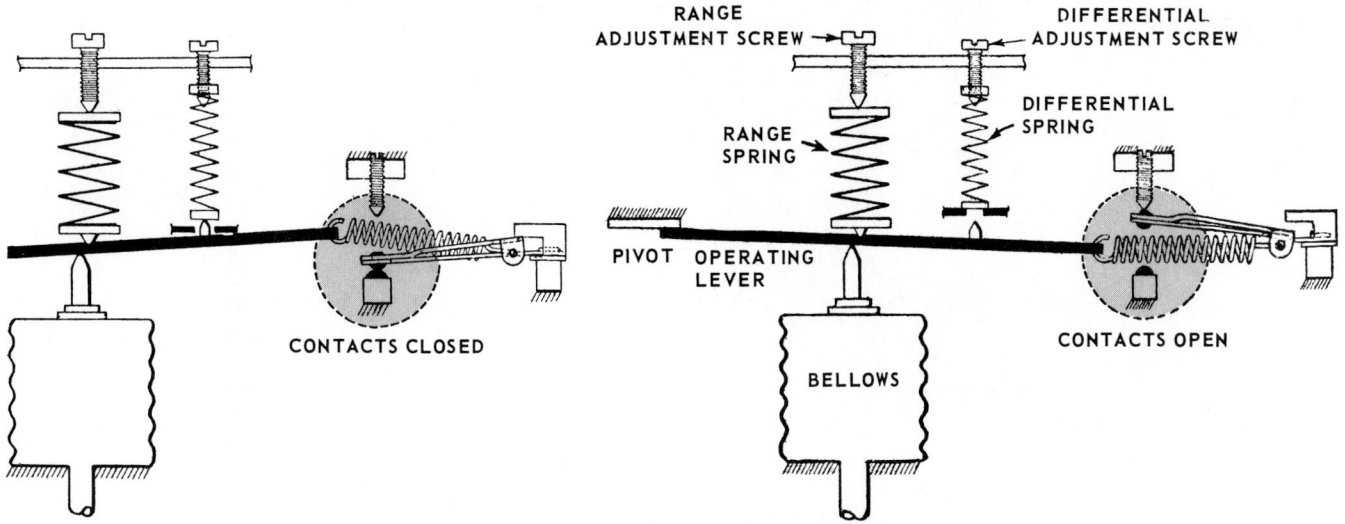

Fig. 8-23. Toggle switch mechanism showing range adjustment and a cut-in type differential adjustment.

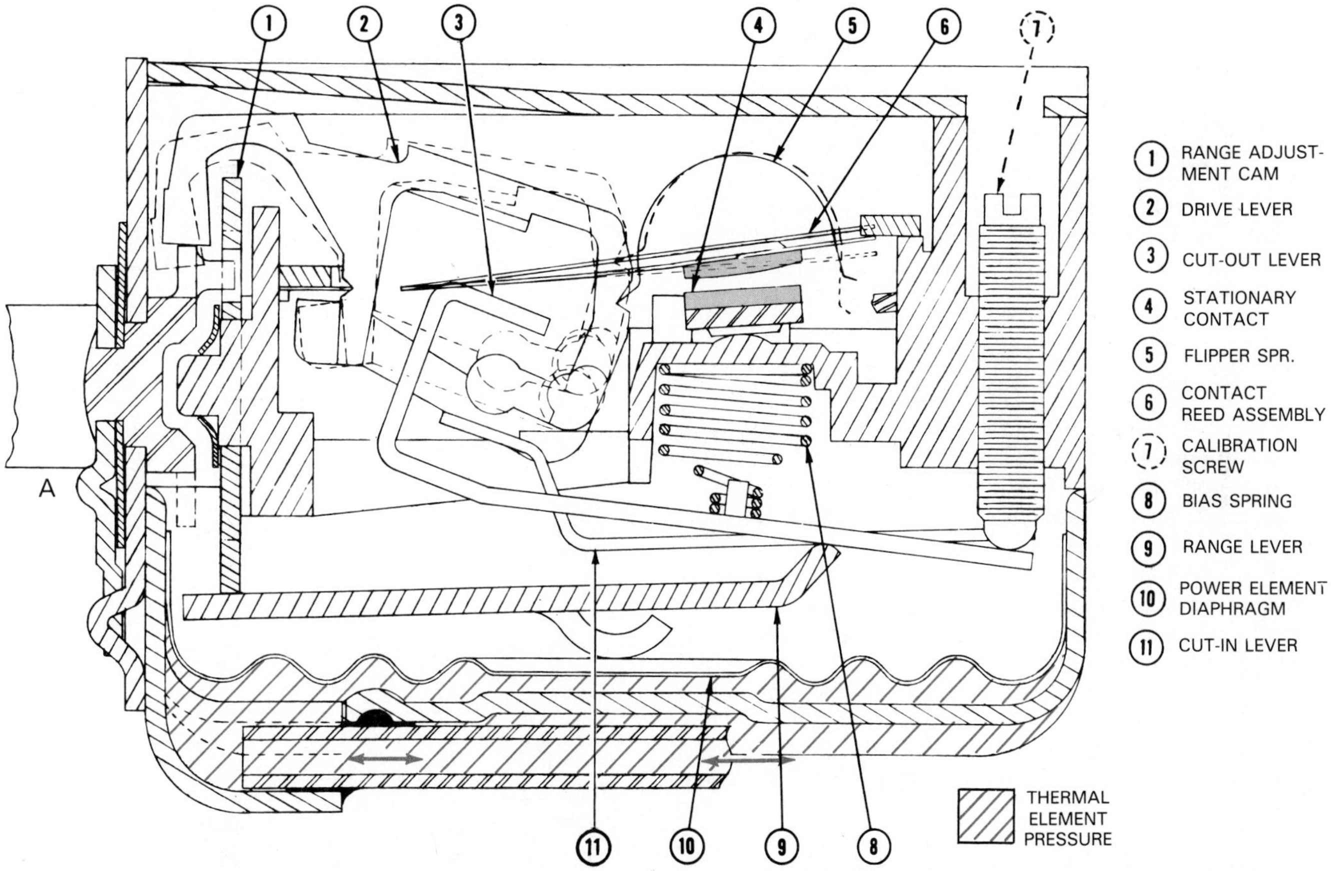

Fig. 8-24. Operating mechanisms and adjustments of a thermostat. 1—Range adjustment (cam). 7—Cutout differential adjustment. A—Range adjustment knob. Contact points are shown in orange.

a second adjusting screw. In such cases, the cut-in screw is the differential adjustment and the cutout screw is the range.

2. If the control has a cutout differential, the opposite applies. The cut-in will be the range and the cutout, the differential.

3. Some controls have a differential adjustment which con-

trols the distance between the cutout and the cut-in. The first setting should be made by using the differential adjustment to get the correct distance between the settings. Then the range adjustment is turned to obtain the correct settings.

The adjustment used by the owner is usually a limited range adjustment. However, some models allow the owner

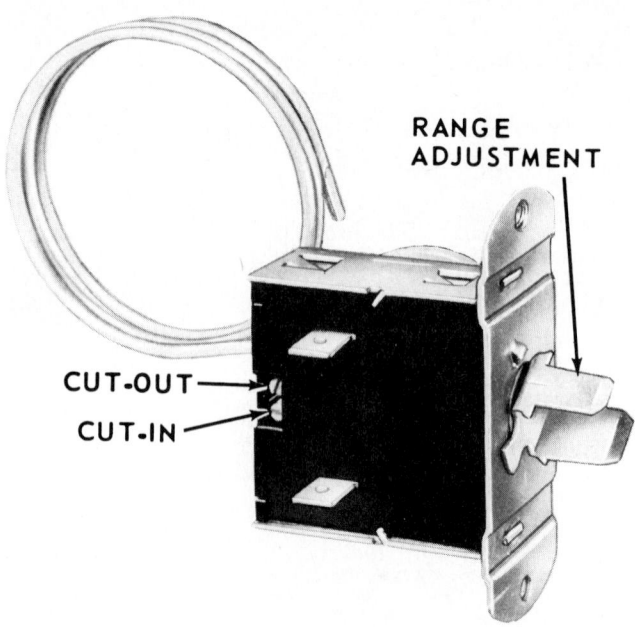

Fig. 8-25. Domestic cabinet thermostat. Range adjustment used by the owner is at right. Note cut-in and cutout adjustment at left. (Eaton Corp., Controls Div.)

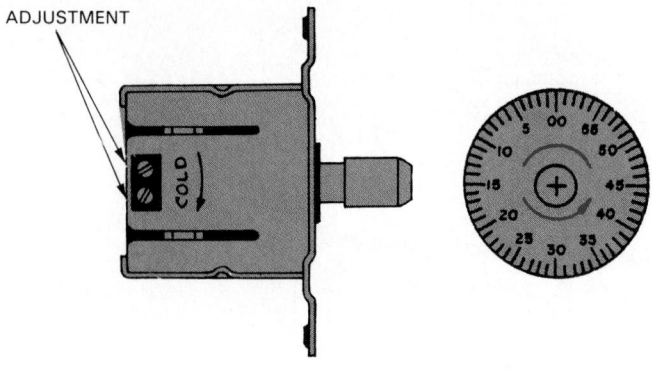

ALTITUDE CORRECTION BOTH "CUT-IN" AND "CUTOUT" SCREWS MUST BE ADJUSTED

ALTITUDE IN FEET	COUNTER CLOCKWISE TURNS
2,000	7/60
3,000	13/60
4,000	19/60
5,000	25/60
6,000	31/60
7,000	37/60
8,000	43/60
9,000	49/60
10,000	55/60

THIS SCALE MAY BE USED AS A GUIDE FOR MEASURING DEGREES OF ROTATION REQUIRED FOR ALTITUDE CORRECTION. THE ARROWS INDICATE DIRECTION OF SCREW ROTATION.

Fig. 8-26. Thermostat equipped with altitude adjustment. Table indicates correct setting. (Amana Refrigeration, Inc.)

to adjust only the cutout setting. This design insures a safe cut-in temperature at all times.

Contact points may chatter as they open or close. Operate control to check for this condition. If there is any visual indication of pitting or burning of points, replace the entire control.

8-15 EFFECT OF ALTITUDE ON REFRIGERATOR TEMPERATURES

Refrigeration systems that contain a sensing bulb type thermostat calibrated at sea level may be too cold at elevations above 5000 ft. This is a result of the decreased pressure at higher elevations. Bimetal and electronic thermostats are not affected by different altitudes.

At high altitudes, the atmospheric pressure drops. Then pressure on the control diaphragm or bellows is lowered enough to affect the settings. The altitude adjustment and range control pressure for the bellows or diaphragm should, therefore, be increased to make up for the lower atmospheric pressure.

Fig. 8-26 shows such a control and its altitude adjustment table. Note that the dial is divided into 60 equal calibrations.

Fig. 8-27 shows two controls and a table of adjustments used to correct each for various altitudes.

8-16 REFRIGERATOR AND FREEZER CONTROLS

Automatic refrigeration and air conditioning is designed to provide correct temperatures with the least attention. To produce these temperatures under all conditions, a refrigerating unit must have more capacity than is needed under average conditions. This unit would over-refrigerate or overheat if it ran all the time.

In domestic refrigeration, for the northern temperate latitudes, the unit will run 35 to 40 percent of the time.

In semitropic latitudes, it will run about 50 percent of the time. Domestic refrigerators usually run 5 to 10 minutes and are idle 10 to 20 minutes.

Most refrigerator manufacturers design their units to operate only 8 to 14 hours out of each 24. This is about 40 percent of the time on the average.

This 14-hour operating time is based on average use of the cabinet. The ambient (surrounding) temperature also affects the running time. A refrigerator in a room at 95 °F (35 °C) willl run longer than the same refrigerator operating in a room at 75 °F (24 °C).

Frost-free refrigerators and some automatic defrost refrigerators may run a little longer or more often than the conventional older style machine. Defrosting energy adds to the heat load.

Domestic refrigerator cabinets usually have a temperature range between 35 °F (2 °C) and 45 °F (7 °C). The adjustment on the motor control allows the owner to select the desired temperature within this range.

Food freezers are usually designed to operate from about 5 °F (−15 °C) to −30 °F (−34 °C). The motor control must be designed for food freezer use. Most of these controls provide for an adjustment which the owner can make in selecting the temperature range desired.

The principle of operation of the motor control used on food freezers is exactly the same as that used on domestic refrigerators.

ALTITUDE ADJUSTMENT SCREW

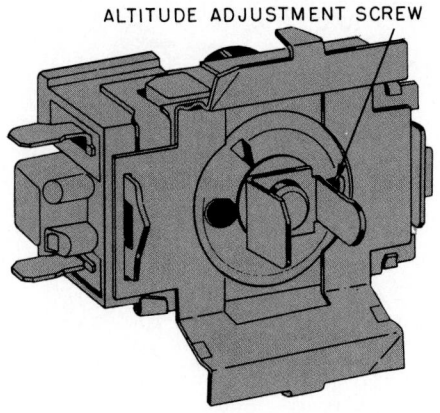

CUT-OUT ADJUSTMENT SCREW
CUT-IN ADJUSTMENT SCREW

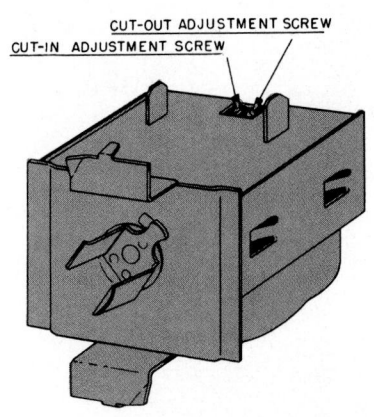

Altitude Above Sea Level - Feet	Constant Cut-In	ALTITUDE ADJUSTMENT		
	Altitude Screw Adjustment (Turns Clockwise)	Both Cut-In And Cut-Out Screws Must Be Adjusted		
		Altitude Above Sea Level - Feet	Constant Cut-In	
			Turns Counter-Clockwise	
			Cut-In Screw	Cut-Out Screw
1000	No Change	2000	1/8 CCW	1/16 CCW
2000	1/16	3000	7/32	1/8
3000	1/8	4000	5/16	5/32
4000	3/16	5000	13/32	7/32
5000	1/4	6000	1/2	1/4
6000	5/16	7000	19/32	5/16
7000	3/8	8000	11/16	3/8
8000	3/8	9000	13/16	13/32
9000	-	10000	15/16	7/16
	Variable Cut-In	Altitude Above Sea Level - Feet	Variable Cut-In	
			Turns Counter-Clockwise	
Altitude Above Sea Level - Feet	Range Screw Adjustment (Turns Clockwise)		Cut-In Screw	Cut-Out Screw
		2000	1/16 CCW	1/16 CCW
1000	3/32	3000	1/8	1/8
2000	3/16	4000	5/32	5/32
3000	7/32	5000	3/16	3/16
4000	1/4	6000	1/4	1/4
5000	3/8	7000	5/16	5/16
6000	7/16	8000	3/8	3/8
7000	15/32	9000	13/32	13/32
8000	1/2	10000	7/16	7/16
9000	9/16			

Fig. 8-27. Two domestic type thermostats are equipped with altitude adjustments. Left—General Electric. Right—Eaton Corp.

8-17 COMFORT COOLING AND AIR CONDITIONING CONTROLS

Comfort cooling air conditioners have two standard controls:
1. Thermostat, to control the temperature.
2. Defrost control, to eliminate icing of the evaporator.

The thermostat sensing element is usually located in the return air duct of the air conditioner. When this incoming air cools enough, the thermostat will stop the unit.

A two stage (two-switch) motor control thermostat is shown in Fig. 8-28. The sensing element reacts to the return air temperature. A rise in temperature will first cause the fan switch to cut in. A further rise in temperature will cause the compressor switch to cut in. As the room temperature drops, this operation will be reversed (the fan is the last to turn off). Fig. 8-29 shows a schematic which helps explain the operation of this control.

A control knob provides an adjustment which the owner may use in setting the range of the air conditioner. If the room is too cold, adjustment may be made to reduce the running time of the air conditioner.

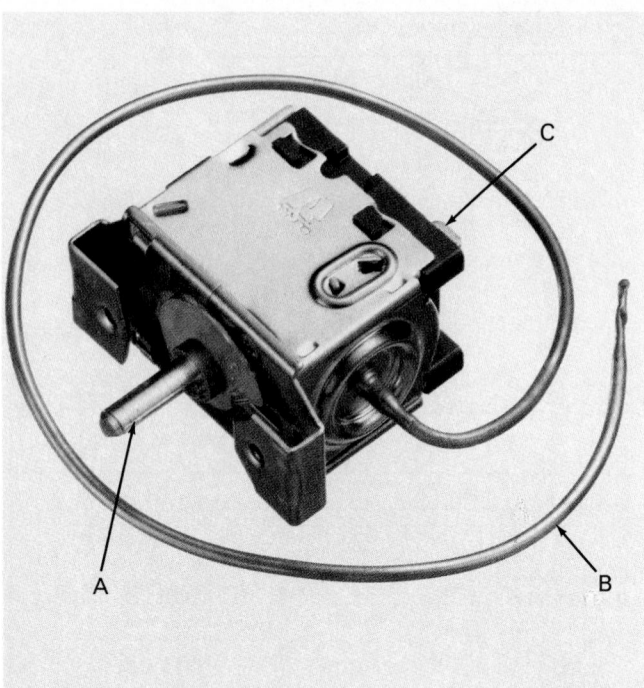

Fig. 8-28. A comfort cooling air conditioner motor control thermostat. It controls operation of fan and compressor. A—Range adjustment. B—Sensing bulb (located in return air duct). C—Electrical terminals. (Ranco North America)

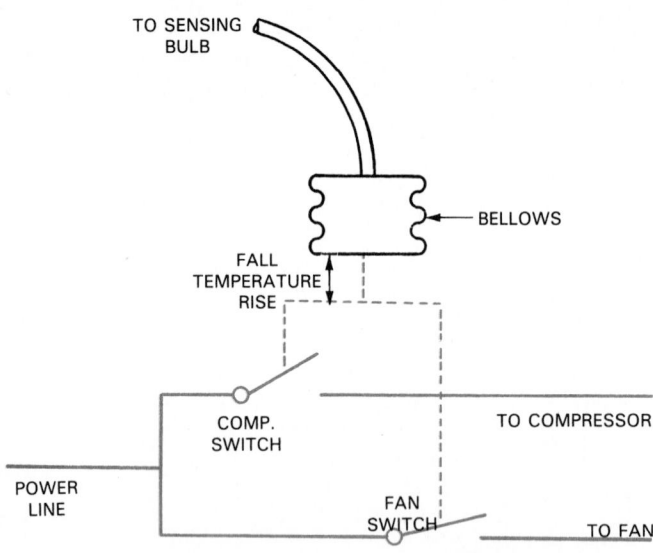

Fig. 8-29. Schematic showing how room air conditioner thermostat works. (Ranco North America)

The defrost control, Fig. 8-30, prevents ice formation on the air conditioner evaporator. This control opens the circuit if the evaporator temperature is close to freezing. It has a factory adjustment only, and should not require adjusting on the job.

This control, Fig. 8-31, is an SPST (single pole, single throw) control if used for defrost only (provides defrost by keeping compressor off for sufficient time after sensor shows ice formation temperature). When defrost heating

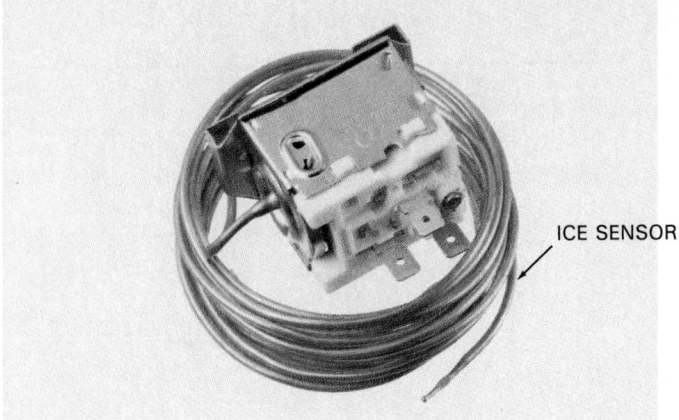

Fig. 8-30. Air conditioner control unit used to pervent formation of ice on the evaporator. (Ranco North America)

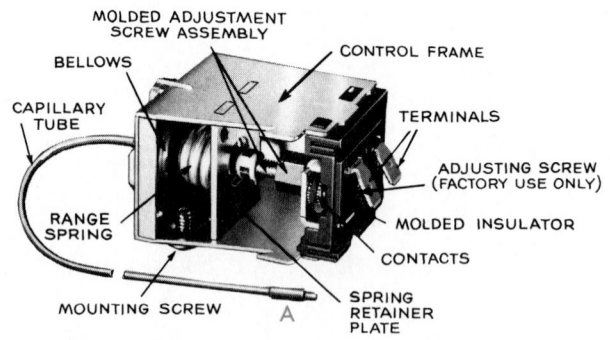

Fig. 8-31. Mechanism of air conditioner defrost control. In operation, control bulb, A, is mounted on evaporator.

units are used on the off cycle, it is an SPDT (single pole, double throw) control.

The same controls are used in refrigerators and heat pumps. Fig. 8-32 is a schematic diagram of the control which shows how it is connected into the air conditioning electrical and refrigerating circuits. At A in Fig. 8-32, it is shown without a defrost heater and view B shows a circuit with a defrost heater.

8-18 CENTRAL AIR CONDITIONING CONTROLS

Central air conditioning implies that the system can provide heating, cooling, humidification, dehumidification, and, in many systems, electrostatic air cleaning. Each of these systems has electrical controls.

The heating controls, which usually operate from a room thermostat, govern the running time of the heating device. The thermostat, then, must make suitable electrical contact and operate the heating mechanism until the desired room temperature is reached. Then, the electrical circuit is opened and the heating device stopped. Details concerning installation and operation are given in Chapters 19, 20, 21, and 22.

Cooling is usually controlled by a room thermostat. This thermostat may be built in with the heating thermostat, or it may be a separate instrument. A refrigerating mechanism usually provides cooling. The room thermostat turns on the refrigerating mechanism when the temperature of the room

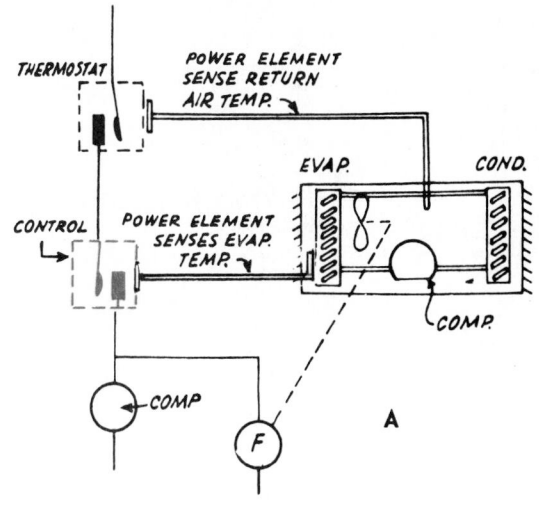

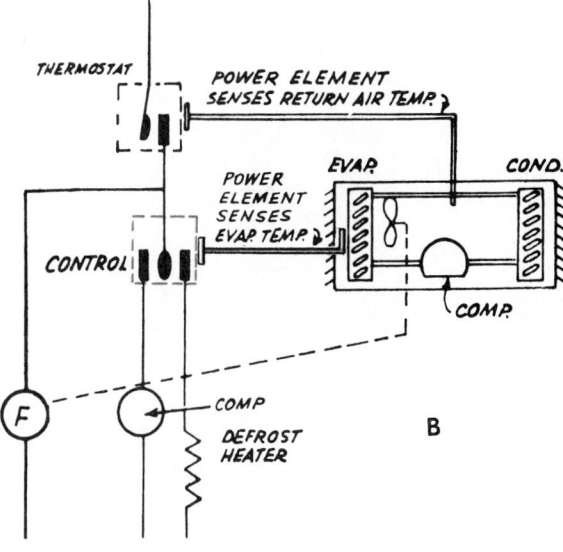

Fig. 8-32. Schematic diagram of air conditioning defrost control. A—Control without defrost heater. B—Control with defrost heater.

goes above a certain temperature. It shuts off the refrigerating mechanism when the temperature of the room has been brought down to the desired temperature.

Automatic humidification controls usually govern the flow of water or steam into a humidifier. If the air is too dry, additional moisture is provided. The humidistat is usually located in the same room as the temperature controls.

In some central air conditioning systems, the humidistat will cause the refrigerating mechanism to operate if the air is too moist. The evaporator is located in the plenum chamber of the central air conditioning system. The cold surfaces of the evaporator will condense out moisture, thus drying the air being circulated.

Electrostatic air cleaners usually operate at the same time as either the heating or cooling mechanisms. Air being drawn through the air conditioner is "scrubbed" while passing through the electrostatic air cleaner.

Requirements of air conditioning systems using a heat pump are described as follows:

1. Air conditioning systems which use a heat pump provide a motor control to turn the motor compressor on and off. Usually, the same motor compressors are used for both heating and cooling. Their operation is controlled by the room thermostat.

2. During summer cooling, this system works without supplementary controls. During winter heating, however, more controls are required.

3. In addition to the motor controls used on the heat pump, an electric resistance heater is usually required at the evaporator. This heater will operate if the evaporator surface starts to ice over. (Ice would keep air from flowing through it.)

4. During certain weather conditions when it is not possible to extract sufficient heat from the outside air, some systems use an electric heating unit. The operation of this auxiliary electric heating unit is automatically controlled by the temperature of the condenser.

8-19 WATER COOLER CONTROLS

Drinking fountains are cooled with small refrigerating mechanisms. The evaporator is placed so that the water flowing to the outlet faucet is cooled to a temperature of about 50 °F (10 °C).

These water coolers usually provide storage cylinders where water is maintained at a lower temperature. In some cases, ice is formed. The motor control is operated from the accumulator, which operates to keep a "cold reserve" available to supply needed water to the faucet. Details of some of these water coolers are shown in Chapter 13.

8-20 ICE MAKER CONTROLS

Many domestic refrigerators now have an automatic ice maker. A small fresh water line is connected to the refrigerator. A combination of solenoid valve, timer, evaporator, defroster, and bin level control lets the refrigerator keep a storage bin full of ice cubes at all times. Fig. 8-33 shows the construction of a typical unit.

The electrical wiring diagram, Fig. 8-34, shows an ice maker connected to a refrigerator mechanism. A control circuit works with a temperature sensing device called a sensor. This senses the amount of time required to freeze the ice cubes and start the harvest cycle. A commutator and

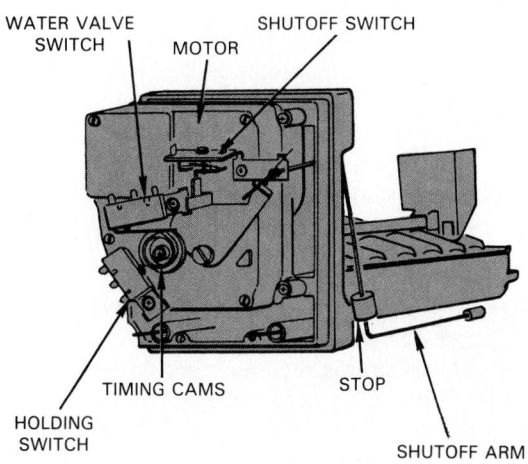

Fig. 8-33. One type of ice cube maker used on domestic refrigerators. Note the shutoff arm which stops the cycling when cube bin is full.

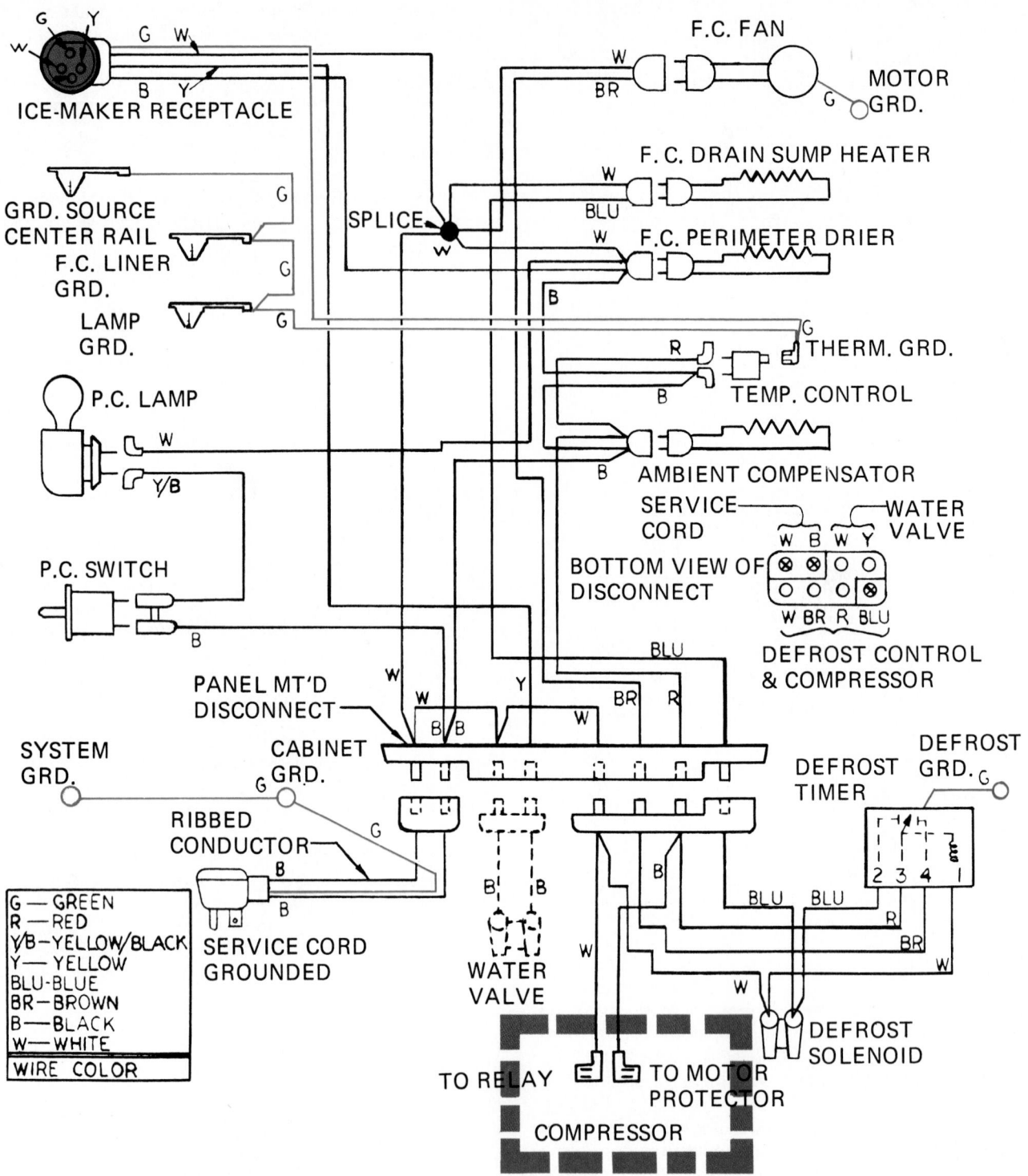

Fig. 8-34. Wiring diagram of domestic refrigerator, including ice cube maker. Note that green is the ground wire. Terminals at bottom center go to compressor and relay.

brush assembly moves the mechanism through the refill, freezing, and harvest operation.

Fig. 8-35A illustrates the important parts of a common ice maker. The step-by-step operation of this ice maker is shown in the sequence of drawings presented in Fig. 8-35B through Fig. 8-35H.

The filling time in Fig. 8-35B is controlled by the motor-driven commutator. The water valve solenoid is energized and water flows into the ice tray for 7.5 seconds.

During the freezing cycle, Fig. 8-35C, the temperature in the freezing compartment is at 8 °F or lower.

Start of the harvest cycle is shown at Fig. 8-35D. A "cold" cut-in temperature thermostat (T_1), and high cutout temperature thermostat (T_2), are part of the temperature sensor. They are subject to the same temperature and conditions as the ice tray.

The water and temperature sensor are gradually cooled at the same rate to 11 °F (plus or minus 3 deg.). The cold

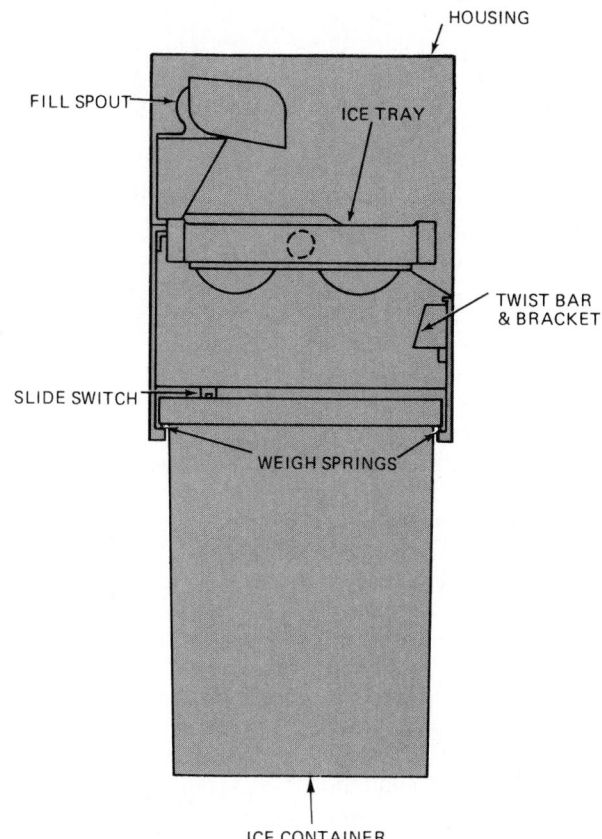

Fig. 8-35A. Parts of a common ice maker.

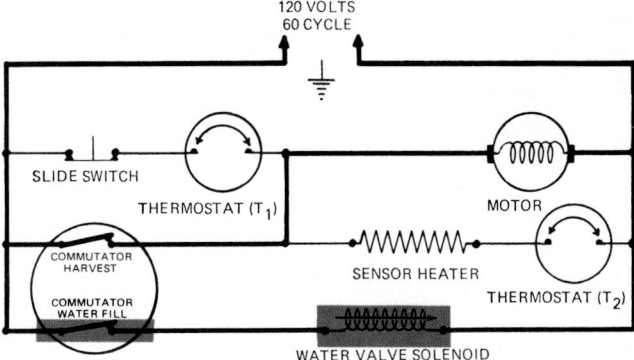

Fig. 8-35B. Electrical circuit at time ice tray is being filled with water.

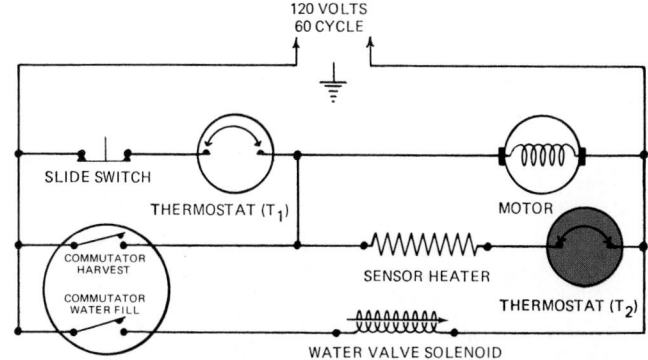

Fig. 8-35C. Electrical circuit during freezing cycle. Note that thermostat (T₂) is closed.

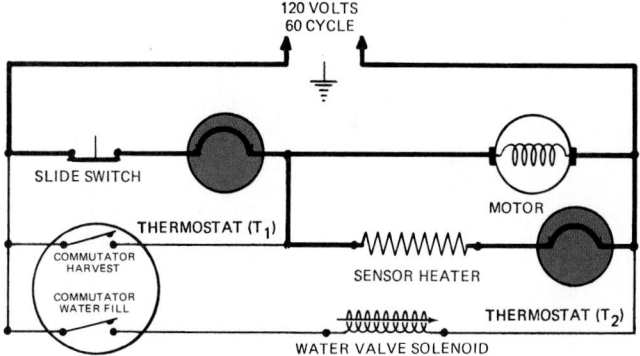

Fig. 8-35D. Beginning the harvest cycle. Both thermostats are closed. One circuit path for motor and sensor heater is provided.

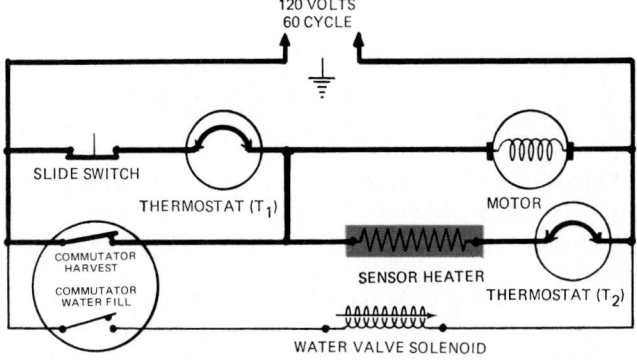

Fig. 8-35E. After 35 deg. tray rotation, sensor heater is still energized, and uses second circuit path.

cut-in thermostat closes to start a harvest cycle. The drive motor starts and rotates the ice try, at a speed of 1/4 rpm. A 22 W heater in the temperature sensor is energized and heats the temperature sensor.

The harvest cycle continues in Fig. 8-35E. After a 35 deg. tray rotation, a parallel circuit to the motor and temperature sensor is made through the commutator.

In Fig. 8-35F, the harvest cycle is still in operation. After a 90 deg. tray rotation, the "cold" cut-in temperature thermostat opens. Both the motor and sensor heater remain energized. The sensor temperature rises to 35 °F.

At Fig. 8-35G, the tray has rotated 120 deg. The temperature rises to 59 °F and the thermostat (T₂) opens the temperature sensor heater circuit.

The harvest cycle is complete in Fig. 8-35H, after 180 deg. of tray rotation. The rear bracket of the ice tray comes to rest on top of the twist bar. For approximately 18 seconds, the rear portion of the ice tray is stopped, while the front portion continues to rotate. This twisting action loosens the ice cubes. The cam action to the tray front bracket moves the twist bar forward, releasing the rear tray bracket. This dumps the ice cubes into the container.

Many types of commercial ice cube makers are described in Chapters 12 and 13. These systems have normal units except for the evaporator and cycling thermostat. In each case, ice making continues until the ice cube storage bin has enough cubes stored. A lever-operated control stops the process when the bin is full by opening a switch. It closes the switch again when the level drops to about one-fourth to one-third full.

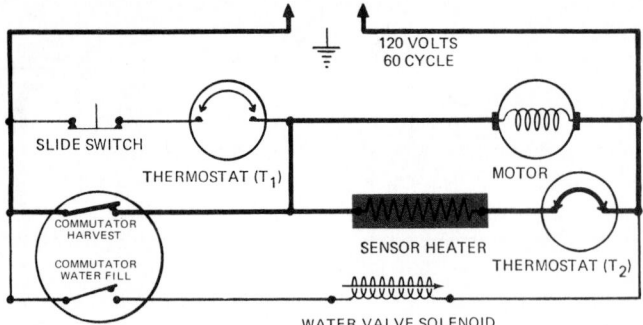

Fig. 8-35F. After 90 deg. tray rotation, sensor heater is still energized. Temperature rises to 35 °F.

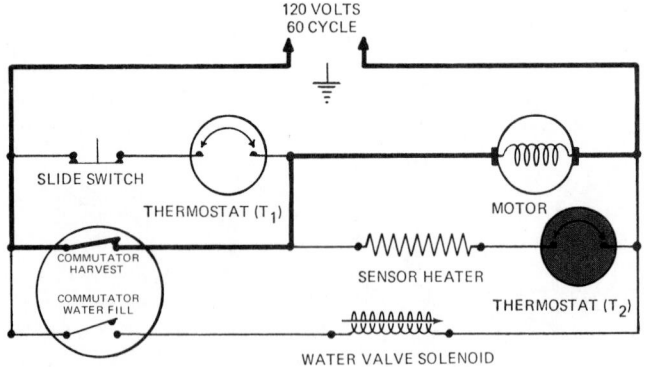

Fig. 8-35G. After 120 deg. tray rotation, thermostat (T_2) opens, breaks circuit to sensor heater. Temperature is now 59 °F.

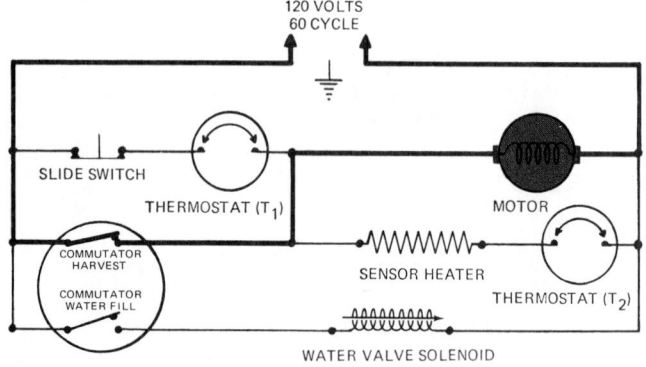

Fig. 8-35H. Completion of harvest cycle after 180 deg. tray rotation. Ice is dumped into container.

8-21 REMOTE TEMPERATURE SENSING ELEMENT REQUIREMENTS

Remote temperature sensing elements are available in several designs. One major difference in type is based on the pressure in the element. There are two common designs:

1. Above atmospheric pressure in the element (used for controlling refrigeration temperatures).
2. Below atmospheric pressure in the element (used for controlling heating units).

The above-atmospheric-pressure element is used for con-

trols which close the electrical circuit on temperature rise. If the element loses its charge, the unit is unable to start. This is known as a "fail safe" action. This element is used where continuous running would be harmful such as in refrigerators and comfort cooling units.

The below-atmospheric-pressure element is used for controls which open the electrical circuit on temperature rise. It is found on electric heating and electric defrost units. If the element loses its charge, the bellows or diaphragm will be unable to contract due to "loss of vacuum," and the points will open (points are designed to close on temperature drop and open on temperature rise). As in units above atmospheric pressure,. this is a "fail safe" device that prevents an electric heating coil from overheating.

The temperature of the mechanism at the point of contact with the sensing element, in most cases, controls the motor operation. This sensing element may be used to stop the motor if the condensing temperature rises too high. Sensing elements may also be used as safety devices to control the motor operation. Commercial installations, in particular, may use them to stop the motor when:

1. The flow of water through the condensing unit stops.
2. The oil pressure in the compressor is too low.
3. The head pressure is too high.
4. The desired low temperature in the cabinet is reached.

Many motor controls may satisfy several of these requirements. This means that there may be more than one capillary tube entering the motor control.

8-22 PRESSURE MOTOR CONTROLS

A low pressure must be maintained in the evaporator to permit evaporation of the refrigerant at a low temperature. Therefore, automatic control of the motor may be based on pressure differences in the evaporator. This control is used on commercial systems. A bellows-operated low pressure control is shown in Fig. 8-36.

This is how it operates: as the evaporator warms, the low-side pressure increases, the bellows expands, the switch is closed, and the motor starts. When the pressure (and temperature) becomes low enough, the bellows assembly contracts, contacts open, and the motor automatically shuts off.

A pressure motor control with cutout and cut-in adjustments is shown in Fig. 8-37. The cutout and cut-in controls are the only ones needed to have full control over both the differential and the range.

The electrical switch may be of either the mercury bulb or the open-contact point type.

The range adjustment will lower both the cut-in and the cutout an equal distance if the screw is moved out (counterclockwise). Cutout and cut-in pressure will be raised if turned in (clockwise). The spring is under compression and presses on the bellows at all times.

The differential adjustment will raise the cutout pressure when turned to the right (clockwise). This is the cutout type differential adjustment. If the spring tension is increased by turning the screw clockwise, it is harder for the bellows to reach its cutout setting, but the spring has no effect on the cut-in setting.

Some models of the pressure control are also equipped to act as a safety control. A bellows construction with a pressure tap to the high-pressure side of the compressor is used. If the compression pressure or head pressure should

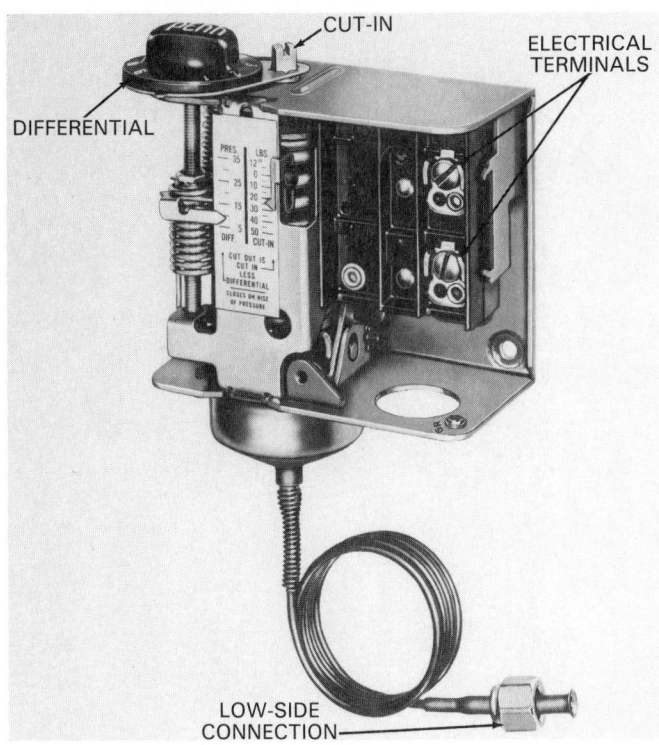

Fig. 8-36. Pressure type motor control with cover removed. Cutout pressure is cut-in minus differential setting. (Johnson Controls, Inc.)

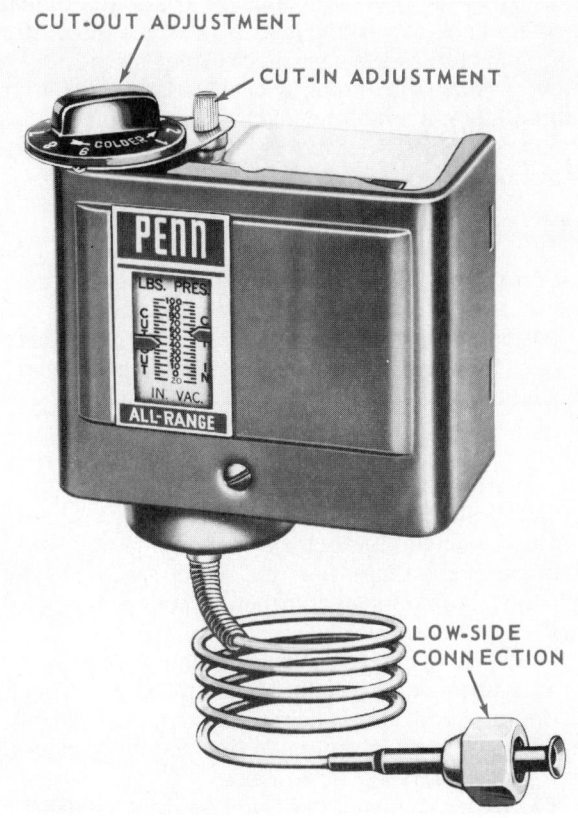

Fig. 8-37. Typical pressure motor control. It is possible to adjust cut-in and cutout pressure on this motor control. (Johnson Controls, Inc.)

become too high, the bellows will expand. This movement will open the switch and stop the motor.

Such a control is a safety device for the motor. It is especially necessary when a water-cooled unit is used. See Chapter 12 for more information on commercial controls.

The low-pressure type of control is easy to adjust on the job. It can also be easily adjusted with a vacuum pump and a compound gauge after it has been removed from the unit.

The adjustment of these controls is important, because satisfactory operation of the unit depends, to a great extent, on their proper operation.

8-23 MOTOR SAFETY CONTROLS

Several safety mechanisms are used in the electrical circuit to make sure that the motor compressor is protected. The three most common safety controls are:
1. Head pressure safety cutout.
2. Low pressure safety cutout.
3. Oil pressure safety cutout.

One of the most harmful things that can happen to a hermetic system is to have high head (condensing) pressures. These high pressures raise the temperature of the vapor and oil moving past the compressor exhaust valve to a point which may cause oil and refrigerant breakdown.

This condition is worsened if a little moisture and dirt are present. Carbon, acids, and sludges may be formed.

It is important to stop the system before these dangerous temperatures are reached. A high pressure safety cutout is often used for this purpose. If pressure exceeds a certain amount, the current to the motor will be shut off and the motor will be stopped.

Several conditions might cause this control to operate:
1. Lack of proper air circulation through the condenser.
2. Lack of flow of water through a water-cooled condenser.
3. A greatly increased refrigeration load of some kind. (See Chapter 12.)

A low-pressure switch can be used to cycle a refrigerating system using a TEV (thermostatic expansion valve). A low-pressure switch can also be used as a safety device. The cooling of a motor compressor depends on the amount and temperature of the suction vapor. If this vapor pressure is too low (system low on refrigerant or flow restricted), the motor compressor may overheat and burn out. The low-pressure safety control will stop the motor compressor before it is damaged.

The schematic wiring diagram for a low-side or suction pressure control is shown in Fig. 8-38.

Some larger units use a safety control device connected to the compressor lubrication system. These shut off the unit if the oil pressure decreases or falls below a predetermined safe pressure above the low-side pressure. The construction is similar to a low-pressure control but usually with a fixed or nonadjustable differential. This control has one connection to the oil pump and one connection to the low-pressure side. See Chapter 12.

8-24 OVERLOAD PROTECTION

All refrigeration and air conditioning units ought to be connected to separate circuits from the control panel. This applies to both domestic units and commercial units. The fuse or circuit breaker in the individual circuit should have enough

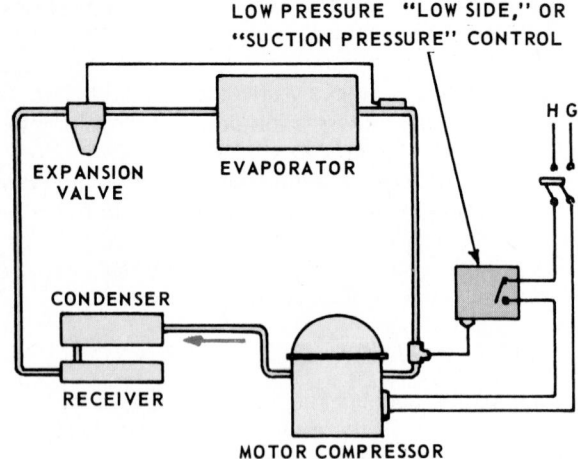

Fig. 8-38. Schematic of a low-pressure motor control installed in a refrigerating system.

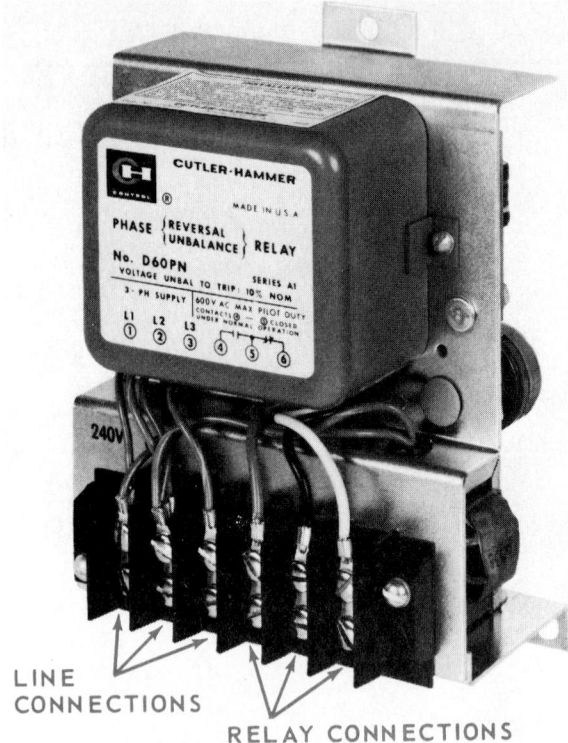

Fig. 8-39. Three-phase protection system with cover removed. (Eaton Corp., Controls Div.)

capacity to provide a continuous flow of current under normal operating conditions. But they should open the circuit in the event of continuous overload of over 25 percent.

At the instant of starting, all motors draw an overload of current. This may amount to 600 percent. However, this is for a very short time and the circuit breaker or fuse should not open the circuit during this short time.

High horsepower motors usually incorporate a starting device in the electrical circuit. This starter does not throw the motor directly on the line. It brings into use a resistance or induction unit which restricts the flow of current at the instant of starting but allows an increase later as the motor speed increases.

All starting relays have some type of overload protection. The most popular type is a bimetal control in series with the power supply to the motor. A resistance heating unit, also in series with the power supply, is alongside the bimetal control. This resistance unit will heat up if the motor is overloaded. The bimetal safety device, reacting to the heat of the resistance unit, will bend. The points at the end of the bimetal strip will open and stop the motor. The motor will not restart until the safety device cools down.

Three-phase circuits and motor compressors are often used in sizes above 1 hp. The three voltages of the three phases should be kept within 10 percent of each other. This is to prevent damage to the motor and to prevent motor reversing.

Controls are used to open the circuit if the voltage in any of the three circuits changes over 10 percent, or if the motor reverses. Some controls will close the circuit again when normal conditions are reached. These controls usually act from about 0.1 of a second up to two seconds (if desired).

These protective devices are basically voltage-controlled relays. Transformers are used to step the voltage down for each potential relay. The stepped-down voltage is then rectified and is fed to the solenoid coil. The same voltage is also fed to the control of a transistor. If the voltage varies too much, the transistor will bypass the solenoid, causing the circuit to open. Fig. 8-39 shows a three-phase circuit protector.

Thermistors with a positive temperature coefficient (PTC) are also used in electronic motor safety controls. A PTC type thermistor is connected in series with the motor windings. When an overload condition exists, the temperature increases and causes the resistance of the thermistor to go up. This increase in resistance limits the current to the motor. After the motor and thermistor cool to a safe temperature, the motor can draw current to start up again. Refer to Para. 7-26 for further information concerning motor protection.

8-25 MOTOR STARTING RELAYS

Some motor controls for hermetic systems are different from those used on external drive systems.

Starting relays are found on the outside of hermetic compressor systems. These relays are usually one of the following types:

1. Current (magnetic).
2. Potential (magnetic).
3. Thermal.
4. Solid state electronic.

The relay permits electricity to flow through the starting winding of the motor until the motor reaches about two-thirds of its rated speed. It then disconnects or opens the starting winding circuit.

The starting winding should be energized only for three or four seconds at a time. If current flows through it for a longer period, the winding may overheat. Many relays have current and/or thermal protection devices to prevent the starting winding from abuse.

To operate correctly, these relays must be the right size for the motor. When replacing one, be sure it has the same electrical specifications as the original. It is impossible to use open electrical contacts inside a sealed system.

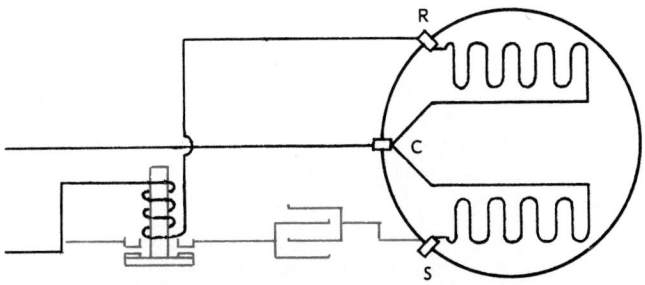

Fig. 8-40. Current type relay schematic. Relay is shown in open position. R—Running winding terminal. S—Starting winding terminal. C—Common terminal.

8-26 CURRENT (MAGNETIC) RELAY

Current relays are usually found on low torque, smaller horsepower motors. The current (magnetic type) relay uses the electrical characteristics of the motor to operate it.

As the rotor picks up speed, the magnetic fields build up and collapse in the motor. This produces a bucking, counter electromotive force (emf), or voltage on the running winding. The running winding consumes more current when the rotor is not running, or is turning slowly, than it does when it reaches full speed. Current-operated relay switches, used to close and open the starting winding, operate on the change in current flow of the running winding, as it goes from a start condition to run.

The magnetic relay is an electromagnet much like a solenoid valve. Either a weight or a spring holds the starting winding contact points open when the system is idle. Fig. 8-40 is a schematic of a weight-operated unit. When the motor control (thermostat or pressurestat) contacts close, the high current flows into the running winding. The magnetic current relay coil is then heavily magnetized. It lifts the weight or overcomes the spring pressure and closes the contacts.

This action closes the starting winding circuit and the motor will quickly accelerate (speed up) to two-thirds or three-fourths of the rated speed. As it does so, the amperage draw of the running winding of the motor decreases. This decreases the magnetic strength of the magnetic current relay enough to allow the weight or the spring to open the points. Fig. 8-41 shows a magnetic current relay in the closed position (starting) and also in the open running position.

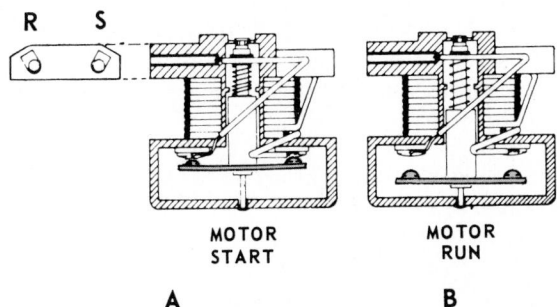

Fig. 8-41. Current (magnetic) relay. A—Relay is in motor starting position. B—Relay is in motor running position. (WCI Freezer Division)

Current relays are sometimes called amperage relays, since it is the ampere draw on the circuit that operates the relay. One type of magnetic current control uses a rotary solenoid. This type can be mounted in any position. The weight type must be mounted level.

One type of weight-operated magnetic current relay is shown in Fig. 8-42. A spring-operated type is shown in Fig. 8-43. A method of mounting a current-starting relay is shown in Fig. 8-44.

These relays are available in a number of capacities. The difference between closing amperage and opening amperage settings is small. This small difference in current flow will close the starting circuit and then open it when the motor reaches approximately three-fourths speed. Fig. 8-45 is a circuit using a current relay.

The utility companies sometimes reduce line voltage to

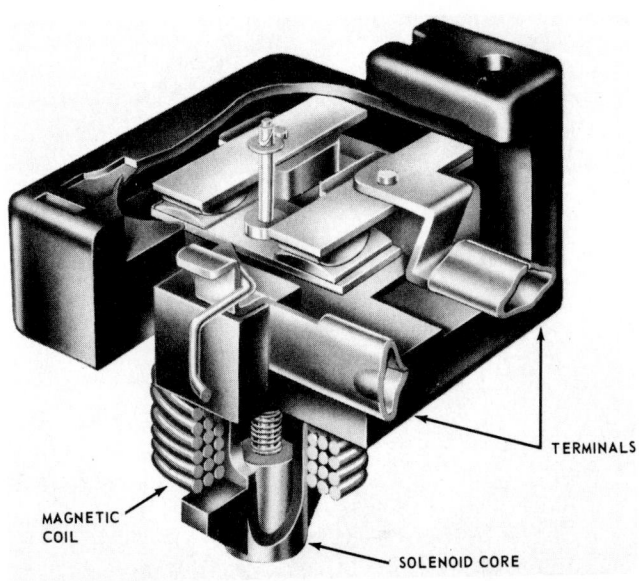

Fig. 8-42. Current type magnetic relay. Plastic housing contains solenoid which has movable center. Heavy current draw when starting raises center plunger and closes starting winding circuit. (Tecumseh Products Co.)

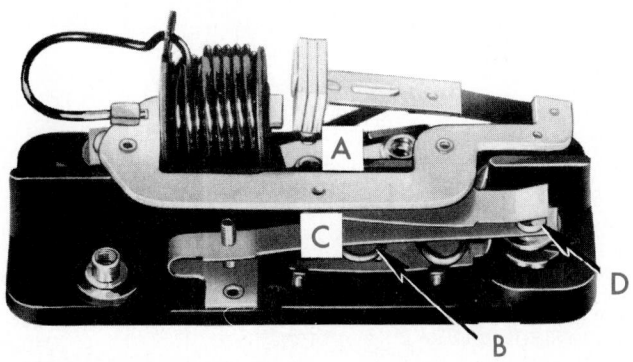

Fig. 8-43. Current type magnetic starting relay with overload safety switch. Starting winding points are kept open by means of cantilever spring at A. At instant of starting, magnetism pulls spring down and closes starting points at D. If current draw is too great, resistance wire at B will heat and cause bimetal strip at C to bend and open circuit at D.

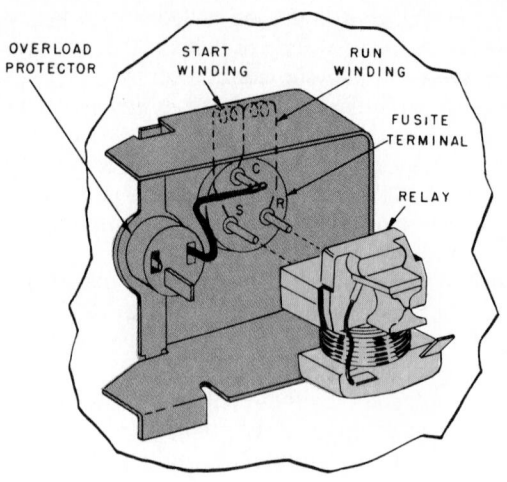

Fig. 8-44. Starting relay mounted on compressor housing. (Frigidaire Company)

Fig. 8-46. Potential starting relay. Weight, A, closes points, C, during off cycle. On starting, increasing voltage into coil, B, will pull contact points apart and stop current through starting winding.

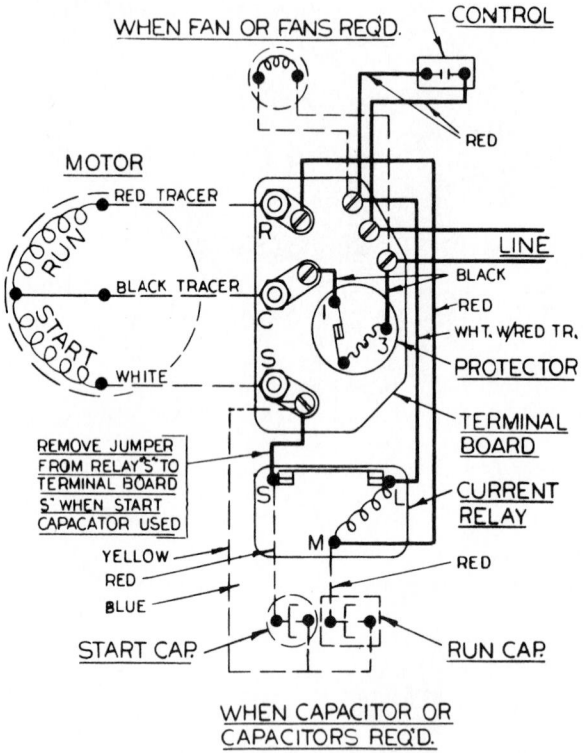

Fig. 8-45. Complete wiring diagram of system using current relay.

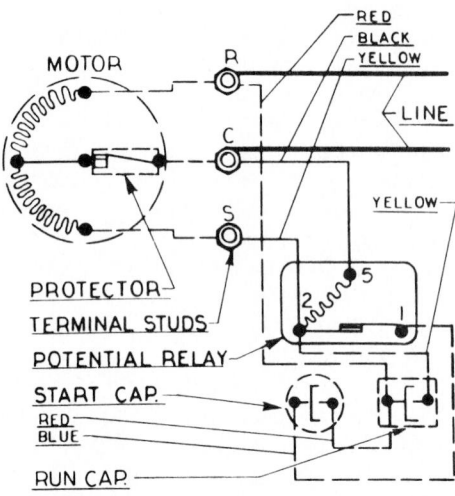

Fig. 8-47. Wiring diagram for potential (voltage) type magnetic starting relay. Note that starting capacitor circuit is opened when relay contacts open, but running capacitor is still connected across starting and running winding in series. (Copeland Corp.)

save current. The current-type starting relay gives some protection to the motor under these conditions. However, there is still some danger of a motor "burnout."

8-27 POTENTIAL (MAGNETIC) RELAY

Potential relays, sometimes called voltage relays, are usually used with high torque, capacitor-start motors. They look somewhat like the amperage relay. However, the operation of these relays is based on the increase in the

voltage as the unit approaches and reaches its rated speed. Fig. 8-46 is a potential magnetic relay.

The contact points remain closed during the off part of the cycle. This feature is its biggest advantage. If the points are closed as the thermostat closes the power circuit, there will be no arcing of the relay points. This quite often occurs with the current relay.

As the motor speed increases, higher voltage from the starting winding creates more magnetism in the relay coil, pulling the contact points apart, opening the starting circuit. The relay coil is connected across the starting winding. The relay coil is made of small wire, so very little current passes through it. This minimizes the heating of the coil and core.

Resistance of the relay coil to voltage must be high enough to prevent the contact points from opening before the motor reaches 80 to 90 percent of its full speed. But such resistance must be low enough to positively open the points and remove the starting winding from the circuit at the right time. If not, the motor will overheat.

Fig. 8-47 is a circuit diagram of a unit with a potential relay. The relay itself is shown in Fig. 8-48. The wiring diagram for its installation is shown in Fig. 8-49.

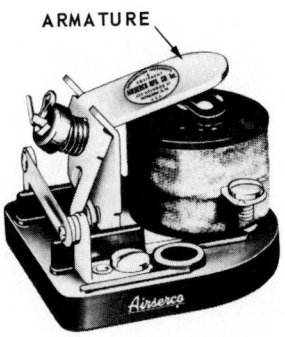

Fig. 8-48. Potential type relay. As armature lever is pulled down by electromagnet, lever at left will open starting circuit points.

Fig. 8-50. Thermal starting relay using two bimetal strips. A—Heating wire. B—Running winding bimetal. C—Starting winding bimetal. P—Power wire connection. R—Running winding connection. S—Starting winding connection. Above, starting circuit is open, unit is operating on running winding only. Below, both circuits are open and have been drawing too much current.

8-28 THERMAL RELAY

There are two types of thermal relays. One type uses two bimetal strips to control the contact points. The other controls the contact points through a resistance wire which is under tension.

In the first type, one strip controls the starting winding and the other the running windings. See Fig. 8-50.

When cold, both sets of contact points are closed. A resistance wire is mounted near the bimetal strips. It is in series with both the starting and the running winding. It is the right size and distance from the bimetal strip so that its heat opens the starting winding contact points as soon as the motor reaches its proper operating speed.

This control also serves as a safety cutout. If the motor should use too much current, the resistance wire will heat the bimetal strip. The heated bimetal strip will open the contact points and stop the motor.

In the second type, the resistance wire is attached in series with both the starting and running windings. The tension of this wire, when cold, keeps both sets of contact points closed. See Fig. 8-51.

While the current passes through, the resistance wire is heated and expands or stretches. At a predetermined setting, the stretched wire opens the starting winding contacts. This control also serves as a safety cutout. If the motor should use too much current, the wire will stretch enough to open the running winding contact points.

The complete wiring diagram for a refrigerator using a thermal relay is shown in Fig. 8-52.

8-29 SOLID-STATE ELECTRONIC RELAYS

Relays using solid-state transistors, diodes, silicon controlled rectifiers, diacs, and triacs are now being used to control starting of hermetic motors. Changes in voltage in the motor, as it starts and then gathers speed, are used to open the starting winding circuit at the correct time. These relays are not as sensitive to the size of the motor

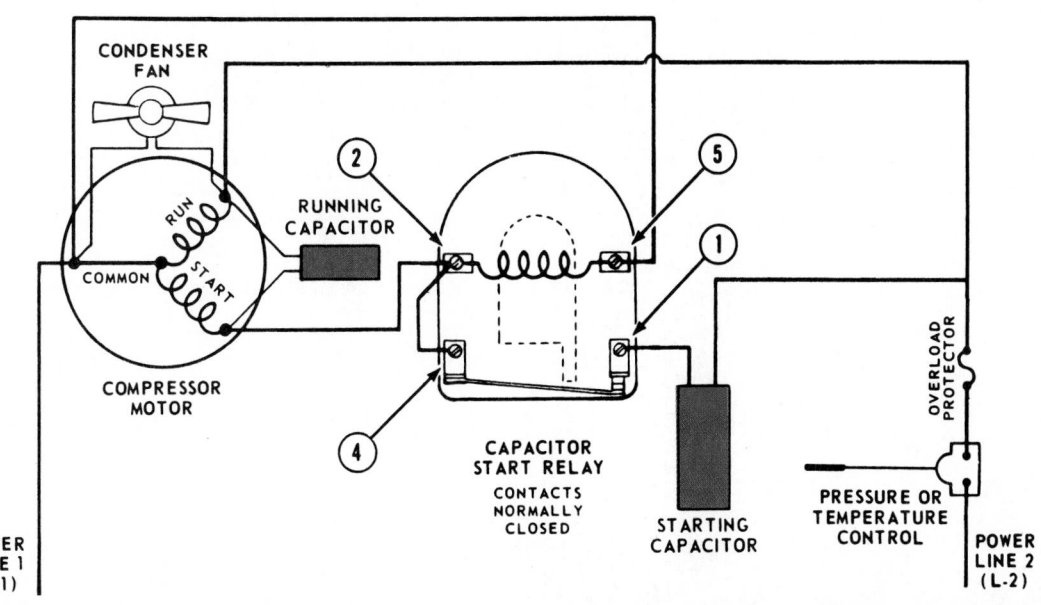

Fig. 8-49. Wiring diagram for potential relay shown in Fig. 8-48. Note that this is a capacitor-start, capacitor-run motor.

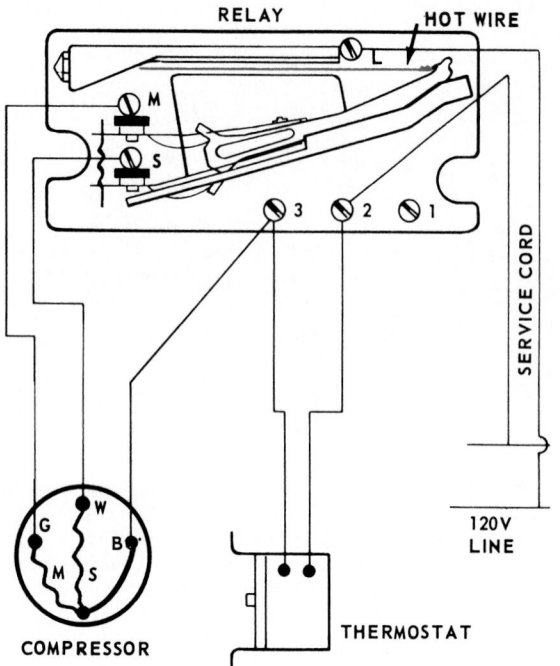

Fig. 8-51. Circuit diagram for thermal (hot wire) relay. Current draw will cause "hot wire" to heat and stretch. Slight heating will cause starting points at S to open. Further heating by too-heavy current will cause points at M to open and stop unit.

as other relays. The same solid-state unit can be used for motors varying from 1/12 to 1/3 hp.

8-30 SERVICING: CHECKING AND TESTING RELAYS

In general, relays should be replaced, not repaired. The service technician's job is mainly to determine if the relay is defective. If so, it is replaced with an exact duplicate. Fig. 8-53 shows the wiring connections to three types:

A. Klixon magnetic.
B. General Electric magnetic.
C. Delco hot wire.

The wire size, the contact point area, and the spring tension or weight plus the air gaps, must be accurately set for each unit. A slight difference in weight or spring tension, for example, might result in 100 motor revolutions difference.

The most effective way to determine if the relay is causing the trouble is to check the other parts of the circuit first. Check the motor, the capacitor, the overload cutout and the thermostat. Only if these parts test all right, should the relay be replaced.

A tester in Fig. 8-54 will check out either a current relay or potential relay. It will also check the relay for line voltage. The procedure is simple and quickly done. Connect the numbered leads from the numbered jacks on the tester to

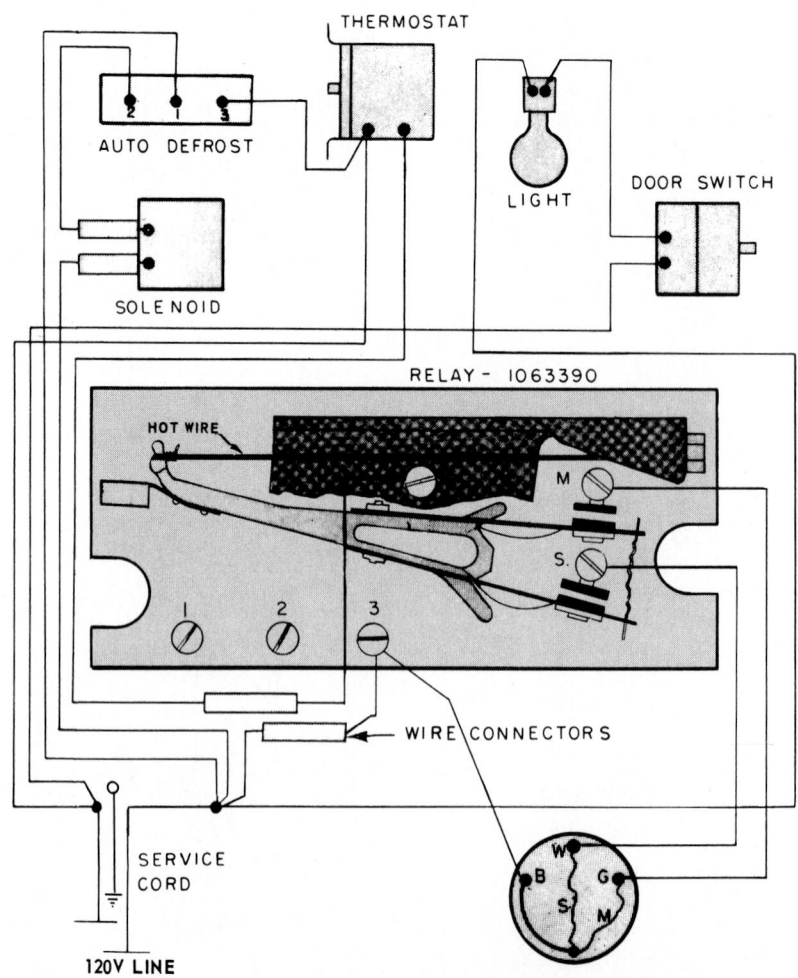

Fig. 8-52. Electrical circuit for refrigerator using thermal relay.

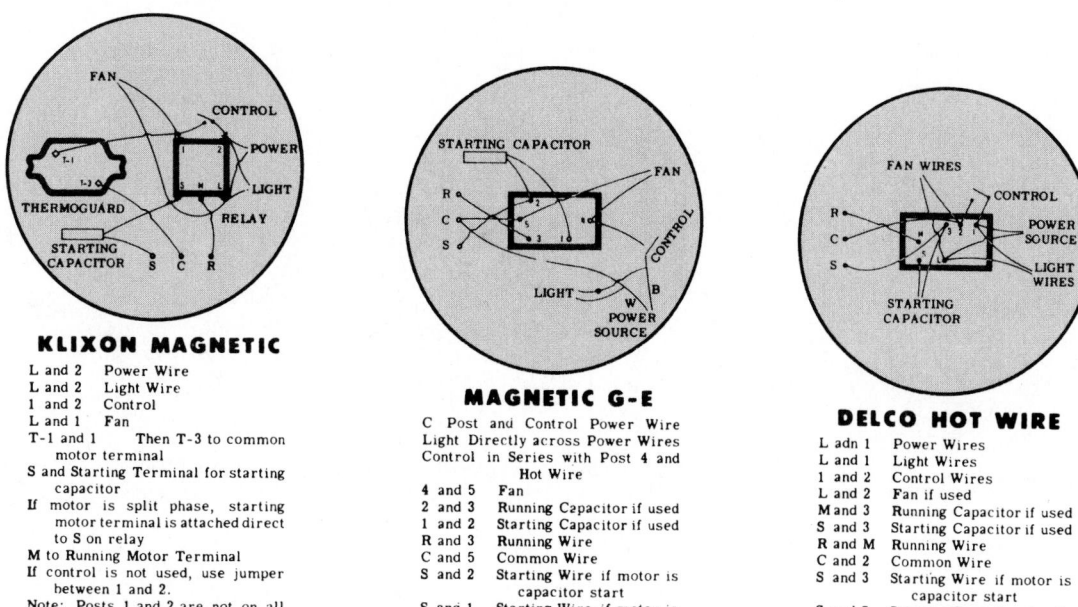

KLIXON MAGNETIC

L and 2   Power Wire
L and 2   Light Wire
1 and 2   Control
L and 1   Fan
T-1 and 1   Then T-3 to common
      motor terminal
S and Starting Terminal for starting
      capacitor
If motor is split phase, starting
      motor terminal is attached direct
      to S on relay
M to Running Motor Terminal
If control is not used, use jumper
      between 1 and 2.
Note: Posts 1 and 2 are not on all
      Klixons, but all are interchangeable

A

MAGNETIC G-E

C Post and Control Power Wire
Light Directly across Power Wires
Control in Series with Post 4 and
       Hot Wire
4 and 5   Fan
2 and 3   Running Capacitor if used
1 and 2   Starting Capacitor if used
R and 3   Running Wire
C and 5   Common Wire
S and 2   Starting Wire if motor is
      capacitor start
S and 1   Starting Wire if motor is
      split phase

B

DELCO HOT WIRE

L adn 1   Power Wires
L and 1   Light Wires
1 and 2   Control Wires
L and 2   Fan if used
M and 3   Running Capacitor if used
S and 3   Starting Capacitor if used
R and M   Running Wire
C and 2   Common Wire
S and 3   Starting Wire if motor is
      capacitor start
S and S   Starting Wire if motor is
      split phase

C

Fig. 8-53. Circuit diagram for three starting relays. A—Current type relay. B—Potential (voltage) type relay. C—Hot wire type relay.

the same numbered terminals on the relay. Test on the 120 V switch before using the 208-240 V switch.

Keep the relay cover in place; never discard it. Dust collecting on the contact points will quickly cause them to burn. This causes excessive voltage drop across the points and a poorly working control.

The weight-type amperage relay must be mounted in a straight up-and-down position. Otherwise, the plunger will rub and bind against the sides of the relay body. The relay should open in about three seconds if it is working correctly.

When replacing the relay, it is important to disconnect

the power supply. Label each wire as it is disconnected and use the correct size screwdriver. The terminals are usually numbered so a tag or clip on each wire with the corresponding number makes it easier to connect the new relay. Masking tape and a marking pencil are useful for such labeling.

If an exact potential relay replacement is not available, use one of the lower rather than a higher voltage rating (90 percent of rating). Because capacitors can discharge at 300 V or higher, potential relay points may be burned (fused) by this discharge if the unit short cycles. To eliminate this trouble, use capacitors equipped with resistors across the capacitor terminals or use a time delay switch to prevent the short cycling.

Frequently a unit will short cycle because the thermostat is exposed to vibration (on a shaky wall or a stairs). Be sure to mount the thermostat to something solid. Avoid tapping a relay to check it. Such tapping may cause the points to touch. This brief contact may ruin the points and damage the motor. The relay must function correctly without being tapped or it should be replaced.

8-31 AUTOMATIC DEFROST CONTROLS

A number of refrigerators have a standard temperature section and a frozen foods section in the cabinet. These dual-purpose cabinets need a special series of motor controls. First, the controls must give correct temperatures in both sections. Second, the controls must provide completely automatic defrost.

One type of control is shown in Fig. 8-55. The wiring diagrams for it are shown in Fig. 8-56.

There are four basic means of controlling the defrosting interval:

1. An electric timer which defrosts the unit at certain time intervals.
2. A device which defrosts the units based on the number of times the refrigerator door is opened.
3. A clock which runs only when the unit is running. It

Fig. 8-54. Motor starting relay tester will test both current and potential type relays. May be used on both 120 and 240 V circuits. (Airserco Mfg. Co., Inc.)

271

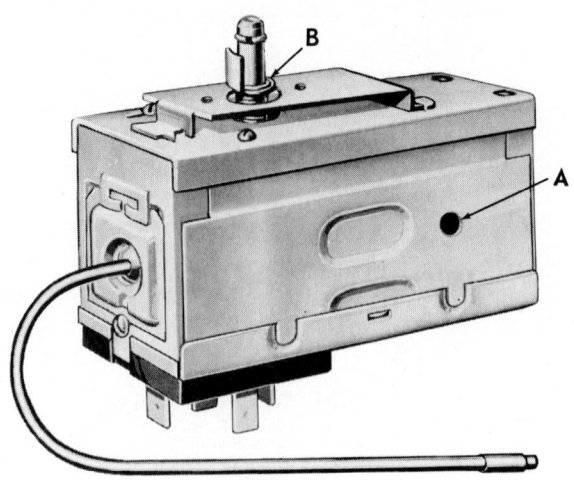

Fig. 8-55. Fully automatic defrost control. A—Electric timer motor. B—Range adjustment.

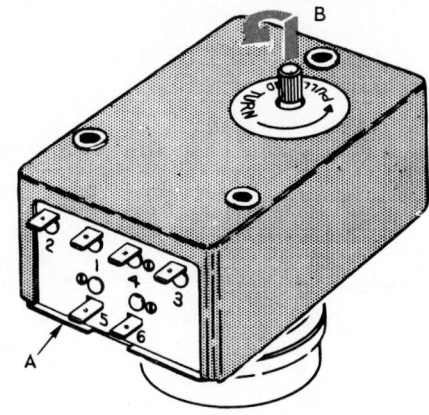

Fig. 8-57. Defrost timer used to operate defrost cycle in three-step nofrost refrigerator. A—Terminals. B—Adjustment.

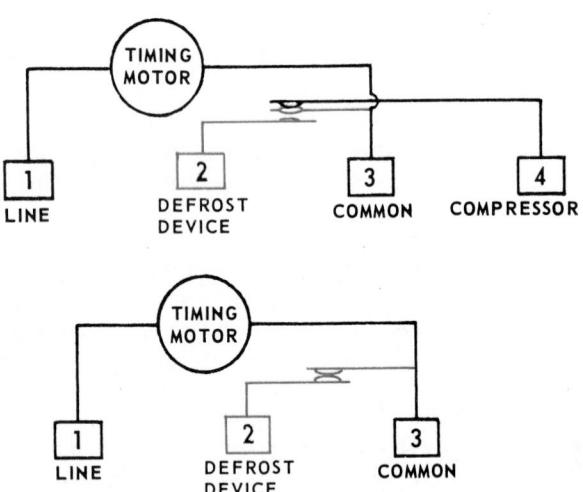

Fig. 8-56. Wiring diagram of automatic defrost control. Control terminals are labeled. Above, electric defrost system breaks motor circuit during defrost time. Lower diagram is for hot gas defrost system, which has compressor running during defrost cycle.
(Ranco North America)

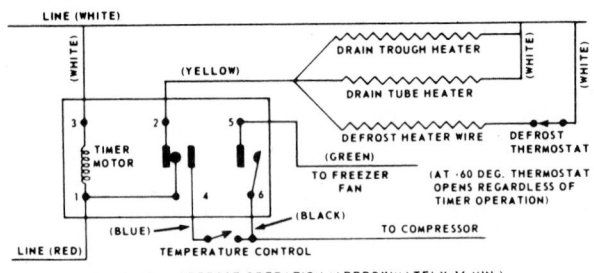

FIRST STEP

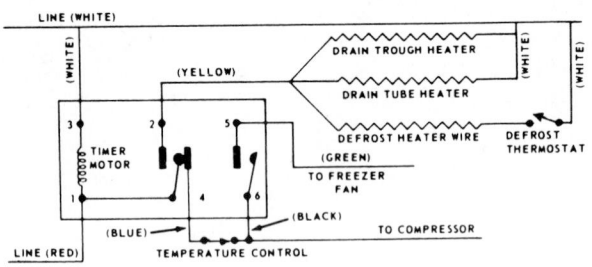

SECOND STEP

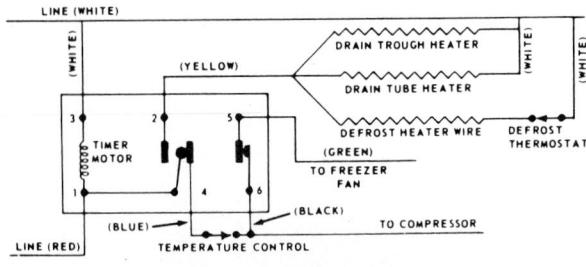

THIRD STEP

keeps the motor off to defrost after a predetermined number of hours of running time.

4. A nofrost system which uses forced convection evaporators and defrosts these evaporators during each off-portion of the operating cycle. Either hot gas or electric heating elements are used.

For example, a defrost system which uses a timer to operate the defrost cycle every 12 hours, works this way:

1. It shuts off the compressor and the evaporator fans and starts the electric heaters. The heaters will be on for about 15 minutes.
2. Then it shuts off the electric heaters and starts the compressor.
3. Evaporator fans start after compressor has run about four minutes, and unit returns to normal operations.

Fig. 8-57 shows such a timer, while Fig. 8-58 shows its electrical circuits. A thermostat controlling the on-and-off circuit for the electric heater elements is shown in Fig. 8-59.

Fig. 8-58. Electric circuits used in three-step defrost method. First step—timer stops compressor and freezer fan, closes circuit to three heaters. Second step—Timer stops heater and starts compressor, but freezer fan does not start. Third step—Freezer fan circuit closed.

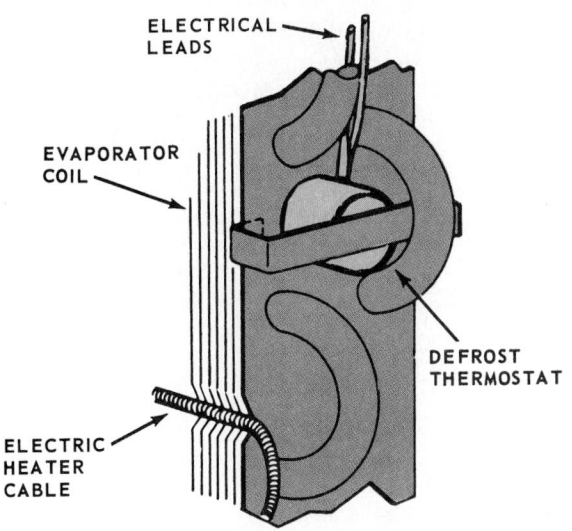

Fig. 8-59. Bimetal defrost thermostat. Control closes at 20°F and opens at 50°F, during defrost.

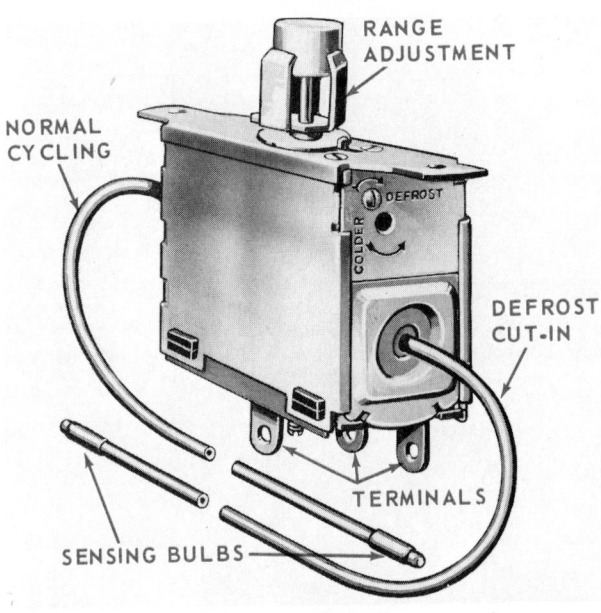

Fig. 8-61. Combination temperature and defrost control.

One control has a timer-operated cam, as shown in Fig. 8-60. This returns the refrigerating circuit to normal operation after a certain time interval even though the thermostat may not call for cooling. This device prevents too long a defrost interval and also serves as a safety device.

8-32 SEMIAUTOMATIC DEFROST CONTROLS

Some domestic refrigerators use semiautomatic defrosting controls. These devices do two things:
1. Defrost the unit when the owner presses the button.
2. Return to regular operation automatically after the unit has defrosted.

A second system raises the range a fixed amount when a button is pressed. The evaporators will run at a temperature warm enough to permit defrosting but will still give satisfactory refrigeration. Pulling on the button returns the system to regular operation.

Refrigerators with manual defrost frequently use a double capillary tube control, Fig. 8-61. This control has a power element for normal cycling and another element as a cut-in control for the defrost. Defrosting starts when the control knob is pushed in. This movement opens the motor circuit and closes the circuit to either a defrost solenoid (hot gas) or to electric heater elements.

When the coils are defrosted, the defrost capillary tube will create enough bellows pressure to open the defrost circuit and, once again, return the control connections to the motor circuit.

A wiring diagram of how this type of control is connected to a hot gas defrost system is shown in Fig. 8-62. The circuits used for an electric defrost system are shown in Fig. 8-63.

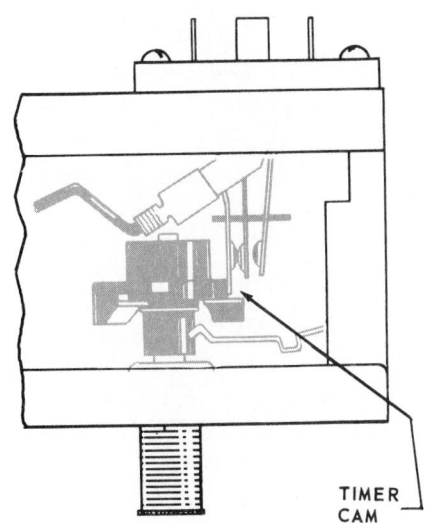

Fig. 8-60. Timer-operated cam which returns unit to normal cycling after about 45 minutes.

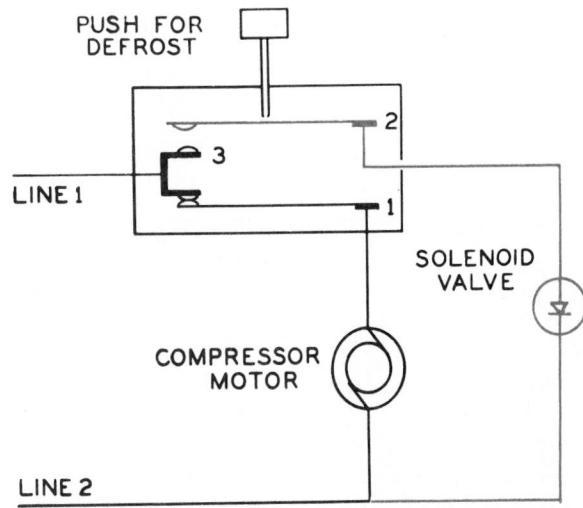

Fig. 8-62. Wiring diagram for "hot gas" defrosting device which uses combination control. Control is shown in refrigerating position. Defrost system has solenoid valve that controls hot gas flow.

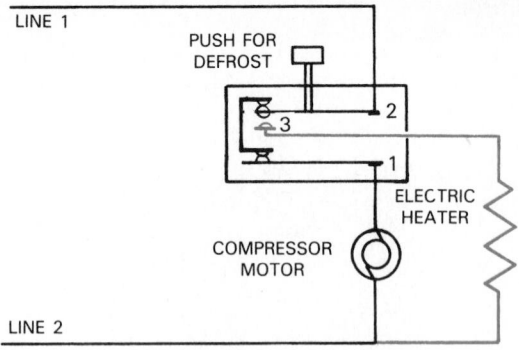

Fig. 8-63. Wiring diagram for electric heater defrost system using combination control. Control is shown in refrigerating position. Heater defrost circuit is shown in red.

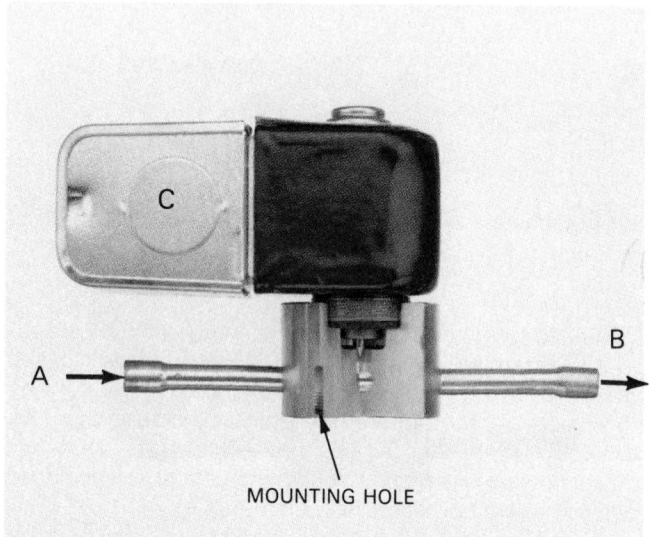

Fig. 8-64. Solenoid valve which may be used with thermostats for either hot gas defrosting or secondary system control. A—Inlet. B—Outlet. C—Solenoid. (Alco Controls Division, Emerson Electric Company)

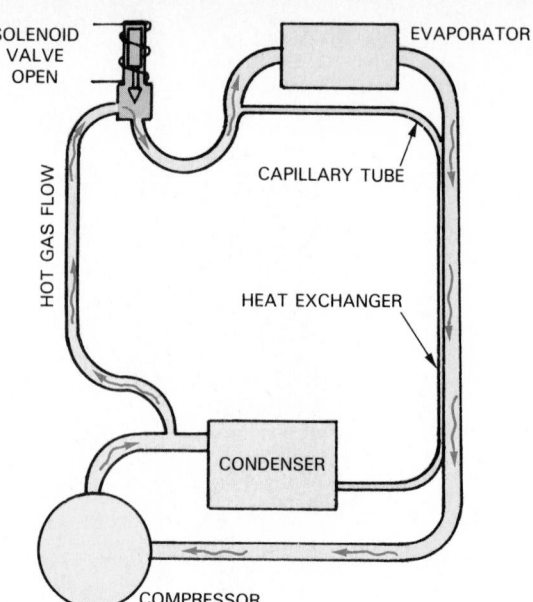

Fig. 8-65. Hot gas defrost system using solenoid valve. Illustration shows valve open and hot gas passing from compressor to evaporator.

8-33 HOT GAS DEFROST CONTROLS

The hot gas method of defrosting uses a solenoid valve to open and close the bypass from the compressor discharge to the evaporator. Fig. 8-64 shows the valve equipped with connector lines. It is similar to the solenoid valves used to control refrigerant flow in secondary systems of two-temperature refrigerators. The valve must be mounted in a vertical position to work correctly.

Fig. 8-65 shows the solenoid valve installed in a hot gas defrost system, while Fig. 8-66 is an installation in a bypass refrigeration system.

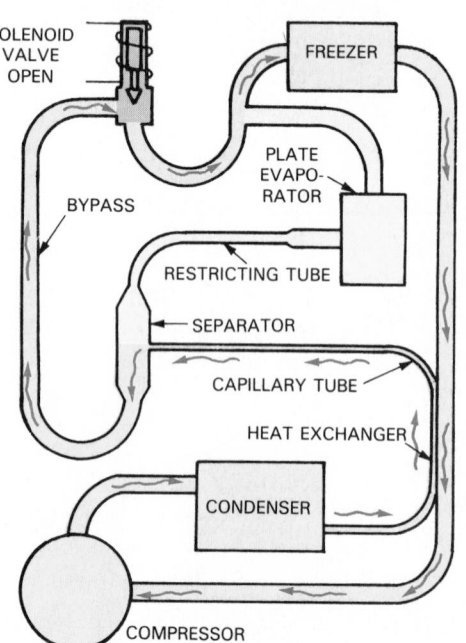

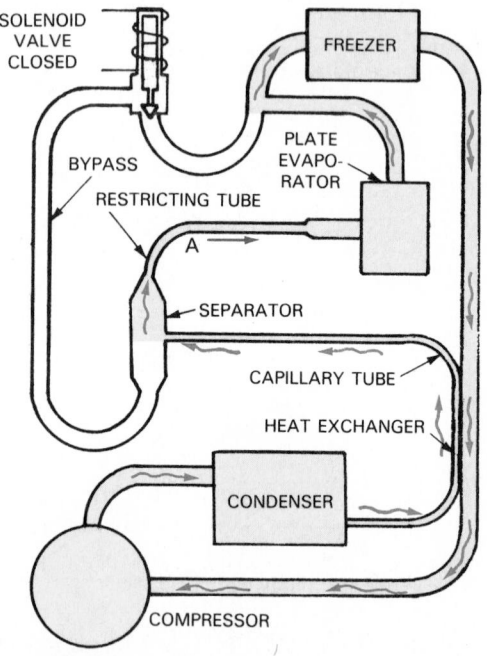

Fig. 8-66. Bypass system using solenoid valve. Left. With valve open as shown, main refrigerating effect is in freezing compartment. Right. With valve closed, plate evaporator and freezer evaporator will be refrigerated (low pressure liquid flows as shown by arrow at A).

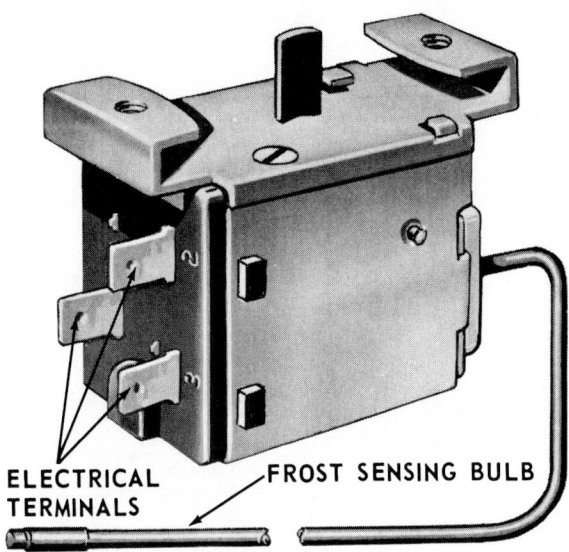

ELECTRICAL TERMINALS   FROST SENSING BULB

Fig. 8-67. Defrost control used in addition to main temperature control. (Ranco North America)

Some refrigerators use two separate controls. One control is the regular thermostat while the other control is the defrost control.

A single control is shown in Fig. 8-67. It can be used with either the hot gas or electric defrost systems. The control is designed with a vacuum in the sensitive element. If the bellows loses its charge, the pressure will rise and open the circuit (a fail safe control).

8-34 ICE BANK CONTROLS

Many types of coolers, vending machines, and other medium temperature refrigeration systems use ice banks to provide reserve cooling capacity. During periods when the system cooling demand is low, a bank of ice is formed around the evaporator, Fig. 8-68. When demand increases,

the cooling offered by the ice assists the refrigeration system.

A typical system of this type is found in some drink dispensers. Ice is built up during evening or low activity hours, and is then used for cooling beverages during high activity periods. Another use for this type of refrigeration would be in an air conditioning system that is used on a periodic basis, such as in a church building. In this example, ice would be built up during the week and then used to cool the building on the weekend.

A control is needed to limit the amount of ice that is formed. A conventional temperature control cannot distinguish between ice and water, since both can be present at 32 °F (0 °C). An ice bank control has been developed, however, which can determine the presence of ice by sensing a pressure change.

This type of control uses a sensing bulb which consists of two compartments divided by a membrane. See Fig. 8-69. One side contains water, which, when frozen, expands and flexes the membrane. The other side contains a liquid which transmits the membrane movement up to the head of the sensing bulb which, in turn, operates a control switch. This switch turns off the compressor, resulting in the melting of some of the ice. The cycle for making and detecting ice can then begin again.

Fig. 8-70 shows how an ice bank control is mounted to control the size of the ice bank. Ice forms approximately 1/8 in. beyond the outermost mounting of the sensing bulb before the compressor is turned off.

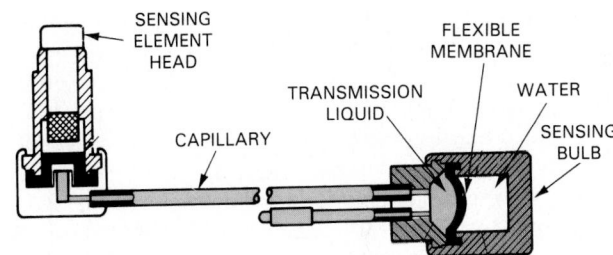

Fig. 8-69. Sensing element bulb. Note transmission liquid in capillary. (Ranco North America)

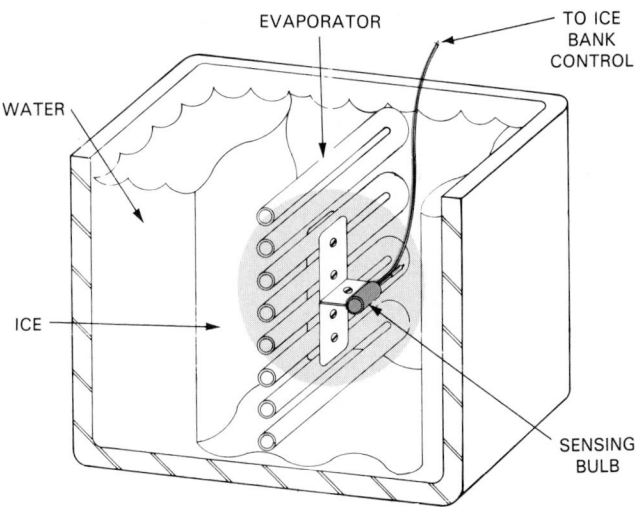

Fig. 8-68. Electric ice bank control. Bulb senses temperature of ice, which can be below 32 °F. If temperature is below 32 °F, ice bank control turns compressor off. If temperature is at 32 °F, bulb senses the pressure increase due to ice crystals forming in bulb. Again, control turns compressor off. (Ranco North America)

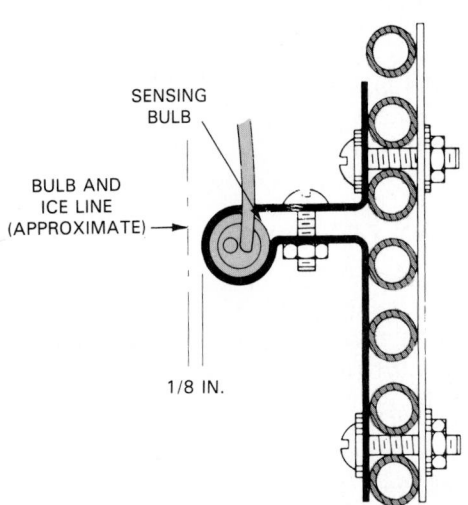

Fig. 8-70. Cross section of evaporator with sensing bulb mounted. (Ranco North America)

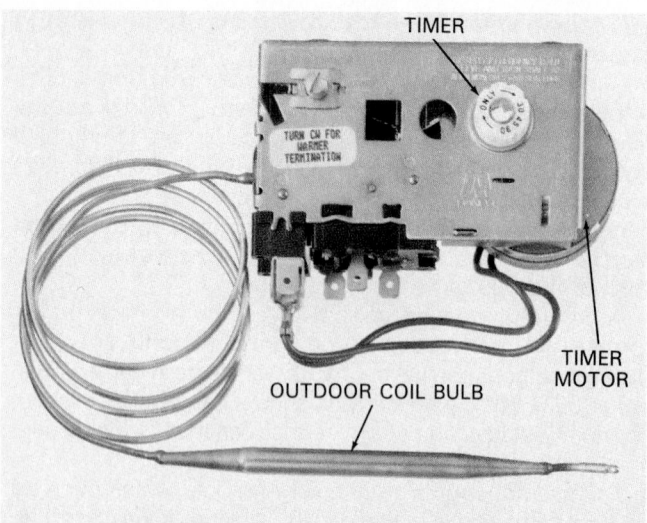

Fig. 8-71. Heat pump de-ice control. Control reverses heat pump cycle and defrosts outdoor coil each 30 to 90 minutes if outdoor coil is at 26°F or below. (Ranco North America)

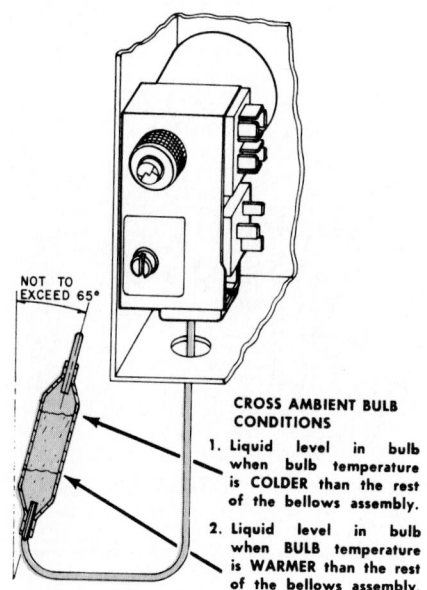

CROSS AMBIENT BULB CONDITIONS

1. Liquid level in bulb when bulb temperature is COLDER than the rest of the bellows assembly.
2. Liquid level in bulb when BULB temperature is WARMER than the rest of the bellows assembly.

Fig. 8-72. De-icer sensitive bulb operation. Note how temperature sensitive bulb is positioned.

8-35 DE-ICE CONTROLS

Air-to-air heat pumps which have an outdoor coil sometimes have an icing problem. Ice accumulates on the outdoor coil during the cold season when the heat pump is used as a heating unit. This ice reduces the heat flow into the outdoor coil and it also tends to block the airflow through the coil.

A special de-ice control is used to prevent ice accumulation (buildup). The heat pump de-ice control, shown in Fig. 8-71, combines the actions of a timer and a thermostat. The timer mechanism will start a defrost cycle unless the thermostat has reacted to a certain temperature level at the outside air coil.

The timer is adjustable for a coil defrost cycle of 30, 45, or 90 minutes. The sensitive bulb is cross charged (see Para. 5-12) to insure constant bulb control. This control permits the defrost cycle only if the coil is at 26°F (−3°C) or below. If above this temperature, the defrost cycle is skipped until the next defrost interval. The sensitive bulb is usually mounted at the place where the ice last melts from the coil.

Fig. 8-72 shows the sensitive bulb details. The temperature cut-in adjustment, the dimensions and the wiring diagram are shown in Fig. 8-73. Note that the defrost cycle reverses the heat pump and the outdoor coil temporarily acts as a condenser during the defrost cycle.

8-36 HUMIDITY CONTROLS

Humidity controls are used to control humidifiers and dehumidifiers. As humidity changes, the resistance value of the sensor in an electronic humidistat varies accordingly. The change in resistance is detected by a wheatstone bridge type circuit, as shown in Fig. 8-12.

In humidifiers, the humidistat closes the circuit when humidity drops or decreases, see Fig. 8-74. Chapter 20 explains the design and operation of humidifers.

On a dehumidifier, the humidistat closes the circuit when the humidity increases, see Fig. 8-75. Chapter 21 explains dehumidifiers in more detail.

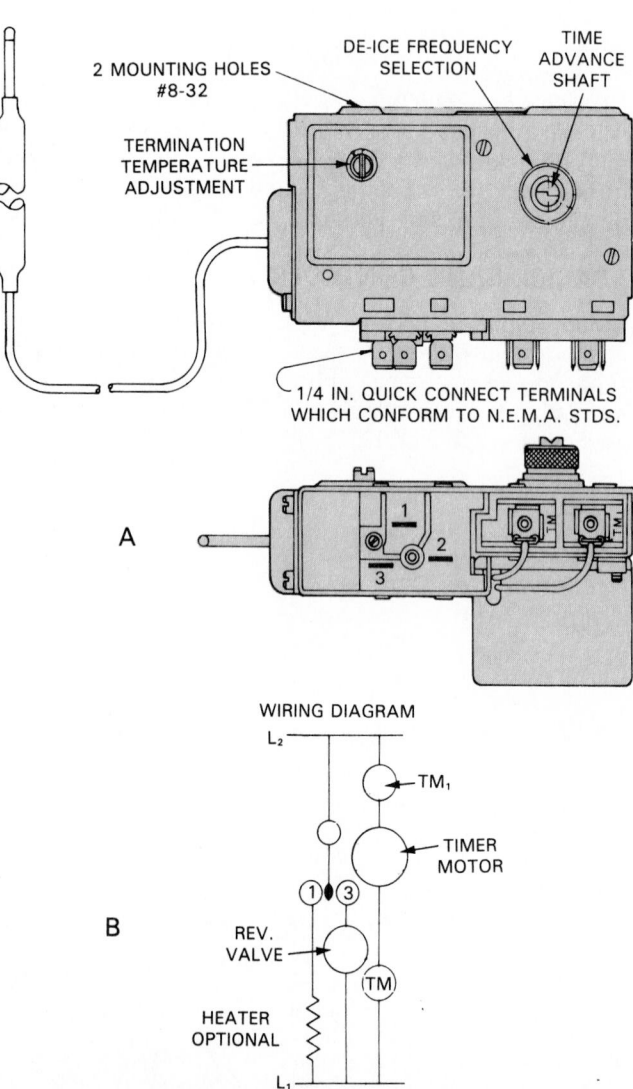

Fig. 8-73. De-icer control. Mechanisms of control and its adjustments are shown at top. Wiring diagram is shown at bottom.

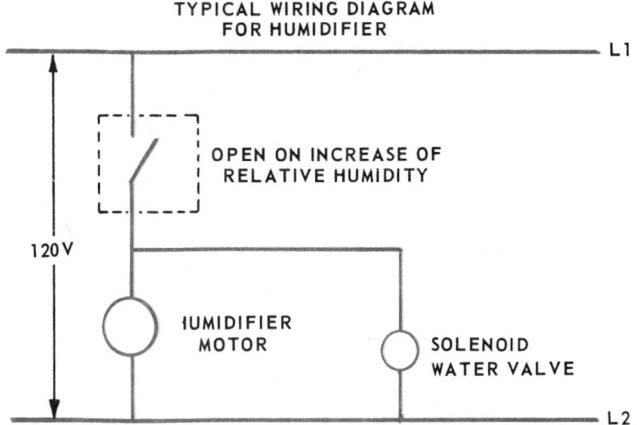

Fig. 8-74. Wiring diagram of humidifier circuit. Control is adjustable to maintain relative humidity between 20 and 80 percent.

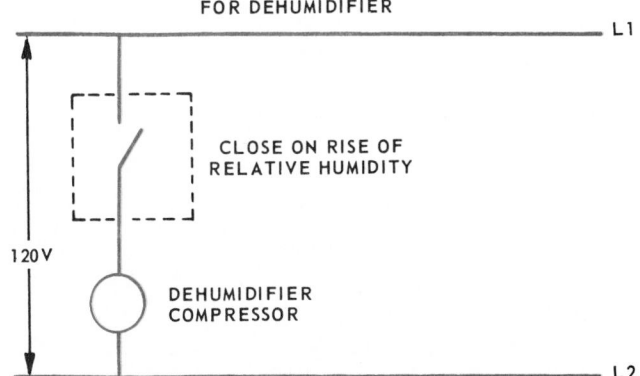

Fig. 8-75. Wiring diagram of dehumidifier circuit. Control is adjustable between 20 and 90 percent relative humidity.

8-37 DEFROSTING CLOCKS

To ease the burden of the user and to permit more efficient operation of the refrigerating unit, many companies have made defrosting clocks standard equipment on their domestic refrigerators (Para. 8-31).

Defrosting clocks, Fig. 8-76, are available for any make of refrigerator. They are also designed for shut-down and defrost of commercial beverage coolers and dispensers. Also called defrost timers, such clocks may be used on both hot gas defrost systems and, most easily, with electric heating defrost systems.

8-38 INSTALLING THERMOSTATS

Thermostats must be correctly installed or the system will not operate accurately or regularly.

The electrical connections must be clean and tight. Wrapping stranded wires around a terminal screw does not make a good or permanent connection. Strands of wire may work loose and ground the wire or short the terminals. Use only U-shaped (spade) or O-shaped (ring) wire terminals. The connections should be cleaned with clean steel wool before installing wires.

Many types of terminals have been developed to do this job. The terminal must be large enough to carry the cur-

Fig. 8-76. Defrosting clock may be wired into power circuit. Clock may be set to defrost the refrigerator or commercial unit at any selected time. Clock has 24-hour dial. Time and length of defrost may be selected. (Paragon Electric Co., Inc.)

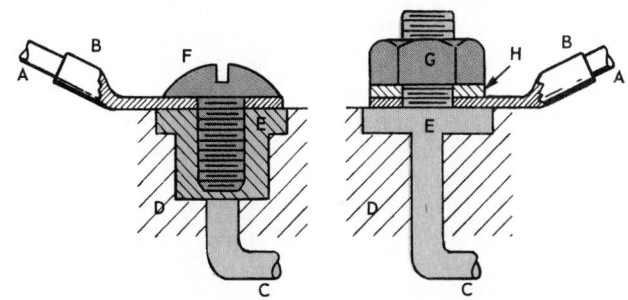

Fig. 8-77. Two types of solderless electrical terminal assemblies. Screw type is at left, nut type at right. A—Wire (lead). B—Wire terminal. C—Wire (lead) to inside of control. D—Insulation. E—Insulated terminal block. F—Machine screw (round head). G—Nut. H—Washer.

rent used in the wire (lead).

Some terminals are connected to posts which either have screws or threaded studs with a nut and washer. See Fig. 8-77. The wire (lead) terminal may be one of several types. Some are designed for easy removal. See Fig. 8-78.

To help service technicians remove and replace wire leads quicker, quick-disconnect terminals have been developed and are becoming very popular. These are made in many designs. They must be clean and tight.

The sensing bulb of the thermostat must be very carefully mounted. It should be attached tightly to the evaporator or the tubing. The sensing bulb and the place to which it is clamped should be cleaned with clean steel wool before assembly.

The best place for attaching the thermostat sensing bulb is on the last one-third of the evaporator.

When installing the control, be careful not to bend the

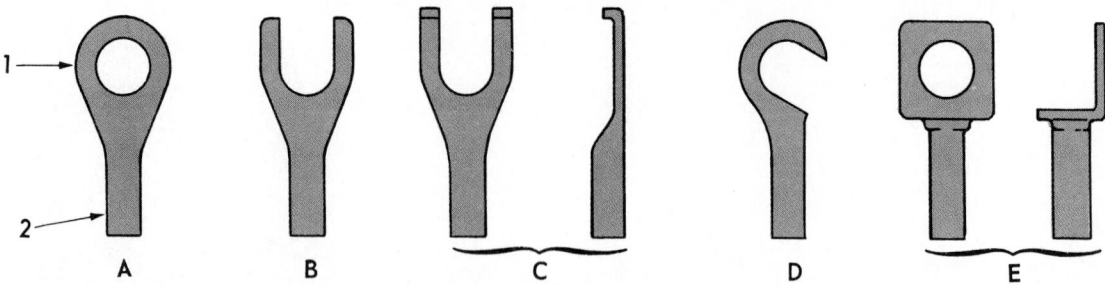

Fig. 8-78. Several types of terminals which are used to connect leads (wires) to terminal posts. A—Ring. B—Spade. C—Flanged spade. D—Hook. E—Flag. 1—Tongue of terminal. 2—Barrel of terminal.

capillary tube back and forth. This small copper tube will work harden and may break. Also, be sure the capillary tube does not touch any part of the evaporator. If any of the capillary tube is coiled, tape the coil to prevent vibration. If any of the tubing rubs against any part, it may wear through or work harden at that spot and crack.

8-39 SERVICING AND TESTING THERMOSTAT

A mixture of crushed ice and water may be used to check and adjust a refrigerator thermostat. With these materials, it is easy to determine the operation of the thermostat at 32 °F (0 °C), since this is the temperature of melting ice.

To use this method, place the thermostat control bulb in the ice and water mixture. It is a good idea to use a thermometer when making this test. With the control bulb in the ice and water mixture, set the control dial temperature reading to 32 °F (0 °C). The control contact points should be open. After a few minutes, lift the control bulb from the mixture; and, as the control bulb warms, the points should again close. Use a test cord with a small light bulb to check the opening and closing of the contact points.

To check further, place the control bulb in water and adjust the water temperature to 45 °F (7 °C). Set the thermostat control to 45 °F. At this setting, the points should be open (they begin to open below 45.5 or 46 °F).

Lift the control bulb out of the water, let it warm up for a few minutes and the points should close. If the points do not open and close properly, the thermostat should be replaced.

Freezer cabinet temperatures are usually in the 0 °F to −20 °F range. Ice and water mixtures cannot be used in setting these thermostats. A thermostatic control tester which uses the refrigerant R-12 or R-22 is recommended. Fig. 8-79 illustrates one of these instruments.

In operation, the freezer thermostat control bulb is placed in the tester and refrigerant is allowed to flow into the control pressure chamber. By adjusting the refrigerant pressure, it is possible to obtain any desired temperature. Freezer thermostats have a predetermined cut-in and cutout temperature much the same as refrigerators. However, the freezer temperatures are much lower. Also, the differential is usually a little greater, 10 to 12 degrees.

To test a freezer thermostat, with the freezer plugged in and operating, place the control bulb in the control tester. Adjust the refrigerant flow and pressure until the desired cutout temperature is indicated on the thermometer. If the freezer control cuts out before the desired cutout temperature is reached, the control may be adjusted to cut out at a lower temperature.

Leave the freezer connected and open the cabinet door. Adjust the control tester to the desired cut-in temperature. If the thermostat does not cut in at the desired temperature, an adjustment should be made to the thermostat. If it is not possible to get a satisfactory thermostat adjustment, the thermostat should be replaced.

To check completely the operation of the freezer thermostat, it is recommended that a recording thermometer be used to chart the temperature and time for the freezer over a 24-hour period.

Several kinds of trouble may be encountered in thermostats:

1. Corrosion occurring at the contact points causing a poor electrical circuit. Use a test light to check electrical circuits. Once contact points become worn and cause trouble, it is best to replace them. Any repair would only be temporary, and a repeat call (call back) would soon be necessary.

   Contact points must be clean. Corroded or pitted points (best detected by using an ohmmeter) usually should not be repaired. The control should be replaced. In an emergency, the points can be cleaned with a small, clean file. A single-cut or mill file is best, but even a clean nail file may do a temporary job.

   Thermostats must have good electrical connections. They must also be adjusted to correct temperatures. The sensitive element must accurately sense the temperature of the evaporator or the cabinet temperature.

   Electrical connections must be clean and tight. Only metal terminals should be used. Clean the terminals, the terminal jacks, or the screw posts with clean steel wool.

2. The overload protection devices may also have dirty contact points, and low current flow may result.

3. The power element and bellows may lose its charge. This may be detected by a simple check. If the charge is lost, the bellows are very easily compressed. If the

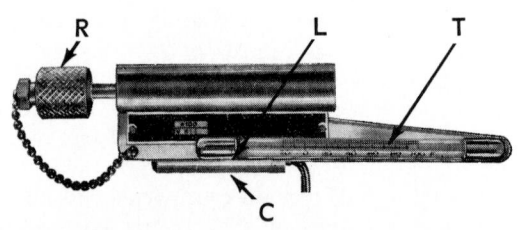

Fig. 8-79. Thermostat testing and adjusting instrument. T—Accurate thermometer. C—Clamp for holding control bulb. R—Fitting for attaching refrigerant cylinder. L—Shallow trough for liquid refrigerant.

bellows were charged, the pressure inside would probably be around 75 psi (517 kPa) or more, and finger pressure would not affect it. In the event of leakage, replace this part of the control or the complete control.

4. Frequently the control bulb is not attached tightly to the evaporator, requiring a great change of temperature range before the motor will cut in and cut out.

The power element must be firmly clamped to the evaporator. Good thermal contact must be obtained between the thermal bulb and the evaporator. Many evaporators have metal sockets into which the thermal elements are inserted. Clean the contact surfaces.

To adjust to the correct temperature settings, mount a thermometer at the sensing bulb, then cycle the unit. It is difficult to obtain accurate settings until the unit cycles several times. Time may be saved by using a thermal bath or a thermostat adjusting tool.

The method of checking and adjusting thermostats is explained in Chapter 11.

8-40 TEMPERATURE ALARM SYSTEM

In some installations, such as food freezers, an electrical alarm system will be sounded if the temperature in the cabinet rises above an upper safe limit. These systems sometimes operate from the electrical circuit powering the compressor. Others are provided with a dry cell arrangement. With the dry cell arrangement, if the dry cells are in good condition, the alarm will work even though the electrical power supply may fail or be interrupted.

8-41 LOW-SIDE PRESSURE LIMITER

A popular refrigerator and air conditioning safety device incorporates a low-side pressure limiter. The condensing unit cannot be overloaded if the low-side pressure is maintained at a low enough limit.

Low-side pressure limiters consist of a pressure sensitive element such as a diaphragm or bellows placed in series with a condensing unit circuit. The electrical circuit will be open if the low-side pressure is higher than a desired limit. Some low-side pressure limiters operate through a relay. The low-side pressure operates the relay, and the relay, in turn, controls the electrical circuit.

8-42 GROUNDING

See Para. 6-72 which explains the requirements for grounding the various parts of refrigerator mechanisms. To avoid any possibility of electrical shock, refrigerators should always be properly grounded. Fig. 8-80 illustrates a typical refrigerator mechanism and the required grounds.

8-43 REVIEW OF SAFETY

Anything in motion, anything which holds back a pressure, anything which can conduct electricity or heat, anything which is rough or sharp, and anything which can be dropped is a potential safety hazard.

Most accidents are the result of carelessness. When one is concentrating on a job, or on getting a job done, one tends to, momentarily, neglect safety.

Therefore, service technicians must train themselves to do things safely. They must study the job for its safety problems and their solutions before starting the job.

Think about the safety aspects before each step of a job.

Always disconnect the electrical power and make sure no one can turn it on while working on the electrical parts of a refrigerator or air conditioning system. Replace worn electrical wires or wires which have brittle insulation (brittle insulation cracks when the wire is bent into a loop).

Use only screwdrivers with insulated handles (wood or plastic). Use only wrenches and pliers with insulated handles. This habit is double insurance against shocks.

If one must work in a damp or wet room—stand on a dry, insulated platform.

Always use instruments to check a circuit to see if it is electrically charged before handling wires, terminals,

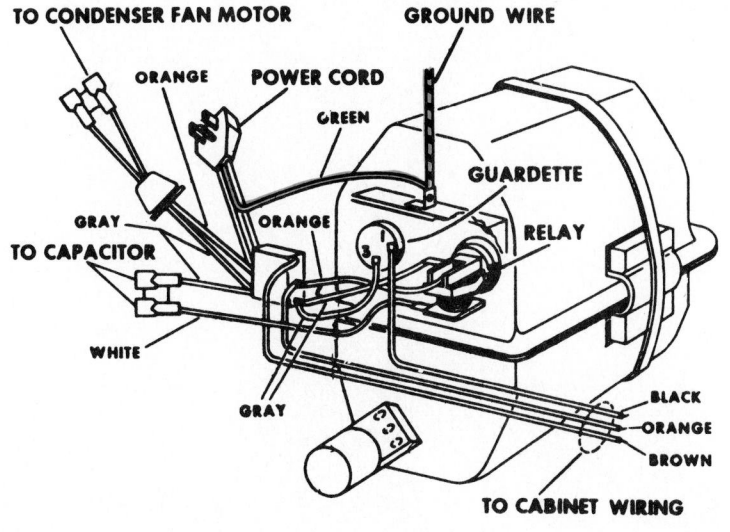

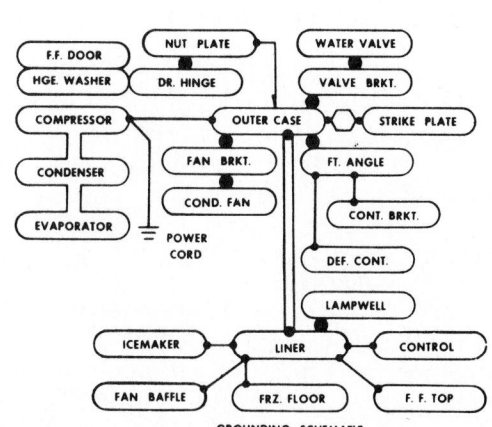

Fig. 8-80. All parts that could cause electrical shock are grounded. Note green ground wire of power cord and ground wire connected to motor compressor. (General Electric Co.)

capacitors, or other parts.

The electrical resistance of the human body is too low to provide any great protection against the flow of electrical current. Electrical resistance is lowered further if the skin on the body is wet. The technician must be certain to be dry in order to help reduce the possibility of electrical shock. Rubber gloves and rubber-soled shoes or rubber boots or rubbers provide considerable added protection.

Most of the resistance to electrical flow is on the skin or surface of the skin. The human body is made up of many fluids which have fairly high electrical conductivity. When a service technician is working in a warm place where the body is wet with perspiration, there is much more chance of receiving an electrical shock than if the body were dry. Many technicians carry a 3-ft. by 3-ft. rubber mat to stand on when doing electrical service work.

Always short circuit any capacitors in the circuits before working on them. Use a 100 kΩ resistor across the capacitor terminals. This will discharge the capacitor.

Water pipes, radiators, heating plants, and the like are grounded. If the "live" wire (black) comes in contact with one of these grounds, either a circuit breaker or a fuse will be blown. A person touching both a live wire and a ground will receive a very severe shock.

8-44 TEST YOUR KNOWLEDGE

1. Can a motor control be operated by low-side pressure? High-pressure? Both?
2. In what ways are pressure and thermostatic types of motor controls essentially similar?
3. What two purposes does a motor control serve?
4. How many types of differential adjustments are there?
5. With what are the thermostatic motor control bulbs charged?
6. Why are starting relays used in connection with most hermetic motor controls?
7. What does the range adjustment control?
8. What does the differential adjustment control?
9. What devices are built into motor controls to protect the motor from using too much current?
10. Will turning the cut-in differential adjustment affect the cutout temperature?
11. Will turning the double type differential adjustment affect the cutout temperature?
12. What is the purpose of a timer on a domestic refrigerator?
13. What is the purpose of having ladder diagrams for electrical circuits?
14. How may a thermostat be checked quickly?
15. How does a bimetal strip respond to a temperature change?
16. What are the differences between two-position on-off and proportional controlling instrument action?
17. When a service technician is testing a refrigerator thermostat, how is a 32 °F (0 °C) temperature accurately generated for the control bulb?
18. Within what temperature range are most domestic refrigerators designed to operate?
19. Are current relay contact points open or closed during the off part of the cycle?
20. Why do electrical wires have different colored insulations?
21. Is the current flow more when the motor is starting or when it is running?
22. Is the voltage drop more when the motor is starting or when it is running?
23. How does altitude affect a diaphragm or bellows type thermostat?
24. Why are some thermostat bodies heated?
25. Are potential relay contact points open or closed during the off cycle?
26. Are current relay contact points open or closed when the system is running?
27. What size wires are most commonly used on domestic refrigerators?
28. What type system uses a de-ice control?
29. Upon what is the operation of a magnetic relay based?
30. Name four common starting relays.

Chapter 9

REFRIGERANTS

After studying this chapter, the technician will be able to:

☐ Correctly identify and classify common refrigerants by their numbers.

☐ List the necessary properties of refrigerants.

☐ Read a pressure-temperature curve and identify the proper refrigerant.

☐ Demonstrate ability to read pressure-enthalpy diagrams.

☐ Discuss properties of different refrigerants and their applications in a system.

☐ Demonstrate handling of refrigerant cylinders and identify color code.

☐ Identify the safety procedures for using refrigerant cylinders.

9-1 REQUIREMENTS FOR REFRIGERANTS

Fluid used as a refrigerant should have certain properties:

1. It should be nontoxic (not harmful if inhaled or spilled on the skin) and nonpoisonous.
2. It should be nonexplosive.
3. It should be noncorrosive.
4. It must be nonflammable.
5. Leaks should be easy to detect.
6. Leaks should be easy to locate.
7. It should operate under low pressure (low boiling-point).
8. It should be a stable gas.
9. Refrigerator or compressor parts moving in the fluid should be easy to lubricate.

10. It should have a high liquid volume per pound to provide durable refrigerant controls.
11. It should have a high latent heat per pound to produce good cooling effect per pound of vapor pumped.
12. It should have a low vapor volume per pound. This will reduce compressor displacement needed.
13. The pressure difference between evaporating pressure and condensing pressure should be as little as possible to increase pumping efficiency.

It is desirable to keep normal pressures in the refrigerator as close to atmospheric pressure as possible, because excessive differences may cause leaks, overwork the compressor, and decrease the efficiency of the valves.

The standard comparison of refrigerants, as used in the refrigeration industry, is based on an evaporating temperature of 5°F (−15°C) and a condensing temperature of 86°F (30°C). In this chapter, each refrigerant discussed is compared on this basis.

9-2 IDENTIFYING REFRIGERANTS BY NUMBER

Refrigerants are identified by number. The number follows the letter R, which means refrigerant. The identifying system of numbering has been standardized by the American Society of Heating, Refrigerating and Air Conditioning Engineers (ASHRAE). One should become familiar with refrigerant numbers, as well as with the names.

Some refrigerants in common use are shown in Fig. 9-1. See Chapter 28 for a more complete list.

REFRIGERANT NO.	NAME AND CHEMICAL FORMULA
R-11	Trichloromonofluoromethane CCl_3F
R-12	Dichlorodifluoromethane CCl_2F_2
R-22	Monochlorodifluoromethane $CHClF_2$
R-123	Dichlorotrifluoroethane $CHCl_2CF_3$
R-134a	Tetrafluoroethane CF_3CH_2F
R-500	Azeotropic mixture of 73.8% (of R-12) and 26.2% (of R-152a)
R-502	Azeotropic mixture of 48.8% (of R-22) and 51.2% (of R-115)
R-503	Azeotropic mixture of 40.1% (of R-23) and 59.9% (of R-13)
R-717	Ammonia NH_3

Fig. 9-1. The most commonly used refrigerants.

281

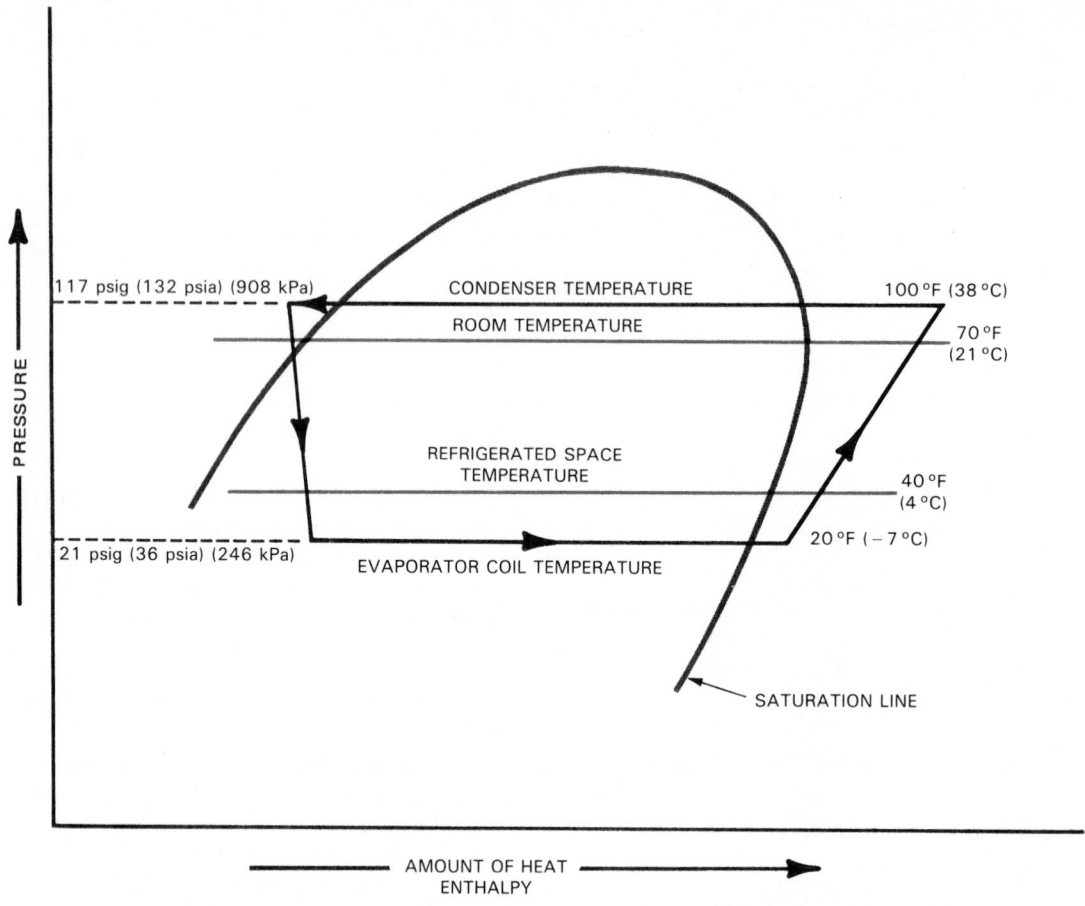

Fig. 9-2. Typical refrigeration cycle with temperatures of one evaporator and condenser shown in comparison to room and refrigerated space temperatures.

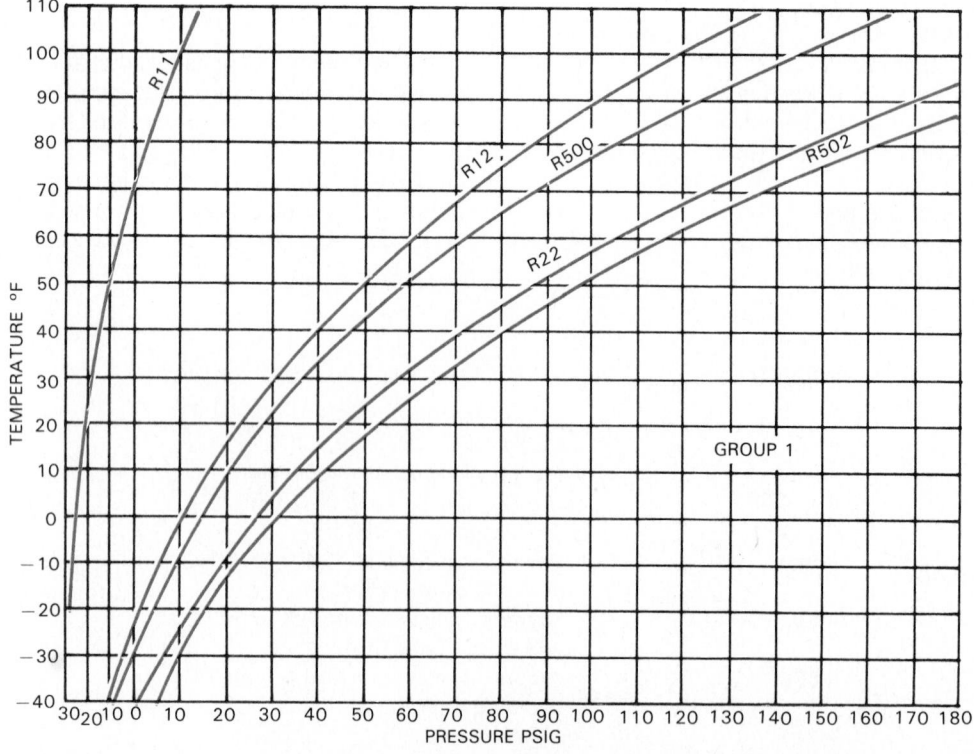

Fig. 9-3. Pressure-temperature curves for popular Group One refrigerants. R-11 is called a low-pressure refrigerant and R-502 a medium-pressure refrigerant.

9-3 STANDARD EVAPORATOR AND CONDENSER TEMPERATURES

In domestic refrigerators, the service technician usually adjusts the controls using a temperature of 5 °F (−15 °C) in the evaporator. Also, the controls are adjusted to give a standard condensing temperature of 86 °F (30 °C). Cycle with 80 °F temperature difference is shown in Fig. 9-2.

9-4 USE OF PRESSURE-TEMPERATURE CURVES

The pressure-temperature curves in Fig. 9-3 (and in Fig. 9-16) show the refrigerants in a state of equilibrium. (See Fig. 1-26, view A.) Vertical scale is temperature in Fahrenheit. Horizontal scale is pressure in psig.

To find the pressure of the refrigerant at any particular temperature, read horizontally from the temperature reading until the curve of the particular refrigerant is reached, then move directly down to the pressure reading.

For example, the vapor-pressure of R-12 refrigerant at a temperature of 100 °F (38 °C) is 117 psig or 132 psia (909 kPa). The temperature is always the temperature of the refrigerant. The same curve may be used for determining both the condensing and evaporating temperatures and pressures. The condensing values for both temperature and pressure are higher than the equilibrium state.

When using this chart, keep several things in mind:

1. The temperature of the refrigerant in the evaporator is about 8 °F to 12 °F (4 °C to 7 °C) colder than the evaporator when the compressor is running.
2. The temperature of the refrigerant in the evaporator is the same as the evaporator temperature when the compressor is not running.
3. The temperature of the refrigerant in an air-cooled condenser is approximately 30 °F to 35 °F (17 °C to 19 °C) warmer than the room temperature.
4. The temperature of the refrigerant in a water-cooled condenser is approximately 20 °F (11 °C) warmer than the water temperature at the drain outlet.
5. The temperature of the refrigerant in the condenser will be about the same as that of the cooling medium after the unit has been shut off for 15 to 30 minutes.

9-5 HALIDE REFRIGERANTS

Many refrigerants are made up of compounds containing one or more of the halogens. Halogen substances are fluorine, chlorine, iodine, and bromine.

Halogen compounds produce a bright green color when exposed to a hot copper surface, such as a torch flame. It is possible to use the halide torch to find a refrigerant leak. See Chapter 11 on using the halide torch.

In the presence of a flame or incandescent (bright red) electric heating element, halogen compounds may break down into their elements. A sharp, pungent odor develops when the compound decomposes.

This is the basis of the halide leak detector. These elements, in large quantities, may be very harmful to human tissue, particularly the respiratory system.

Leaking refrigerants which do not contain halogen cannot be detected with a halide torch.

Avoid releasing any sizable quantities of these halogen refrigerants where there are flames and/or incandescent (bright red) electric heating elements. None of the halogen refrigerants are flammable or explosive.

Halide refrigerants are heavier than air and will settle to the lowest place in a room or container.

9-6 GROUPING AND CLASSIFICATION OF REFRIGERANTS

Refrigerants have been catalogued by two different national organizations. They are: The National Refrigeration Safety Code (NRSC), groups one through three, and The National Board of Fire Underwriters (NBFU), class 1-6.

The National Refrigeration Safety Code (NRSC) catalogs all the refrigerants into three groups:

Group One—Safest of the refrigerants.
Group Two—Toxic and somewhat flammable refrigerants.
Group Three—Flammable refrigerants.

The National Board of Fire Underwriters (NBFU) classifies refrigerants mainly on their degree of toxicity. (Toxic means poisonous or injurious to health.) There are six classifications in this scale. Class One is the most toxic, while Class Six is the least toxic. Fig. 9-4 shows the grouping and classification of some common refrigerants.

REFRIGERANT	NRSC GROUP	NBFU CLASSIFICATION
R−11	1	6
R−12	1	6
R−22	1	5
R−500	1	6
R−502	1	6
R−503	1	6
R−744	1	5
R−717	2	2

Fig. 9-4. Popular refrigerants listed by the National Refrigeration Safety Code and National Board of Fire Underwriters.

9-7 GROUP ONE REFRIGERANTS

The important characteristics of the most-used Group One refrigerants are explained at length in Para. 9-8 through 9-14. (The pressure-temperature curves for five common Group One refrigerants are in Fig. 9-3.)

The refrigerants in this group may be used in the greatest quantities in any installation. The allowable quantities are specified by the "American Standard Safety Code for Mechanical Refrigeration." The amounts are:

1. Up to 20 lb. in hospital kitchens.
2. Up to 50 lb. (indirect system) in public assemblies.
3. Up to 50 lb. in residential use if precautions are taken.
4. Up to 20 lb. in residential air conditioning systems.

Some refrigerants in Group One are:

R-11   Trichloromonofluoromethane CCl_3F
R-12   Dichlorodifluoromethane CCl_2F_2
R-22   Monochlorodifluoromethane $CHClF_2$
R-500   73.8 percent R-12 and 26.2 percent R-152a
R-502   48.8 percent R-22 and 51.2 percent R-115
R-503   40.1 percent R-23 and 59.9 percent R-13
R-744   Carbon Dioxide CO_2

Halogenated chlorofluorocarbons in Group One refrigerants (CFC 11, 12, 113, 114, and 115) are under Environmental Protection Agency (EPA) investigation. There is some evidence that these chemicals damage the earth's protective ozone layer. Companies supplying these refrigerants are

	PRESSURE		VOLUME VAPOR	DENSITY LIQUID	HEAT CONTENT BTU/LB.	
Temp °F	Psia	Psig	Cu. Ft./Lb.	Lb./Cu. Ft.	Liquid	Vapor
−150	0.154	29.61*	178.65	104.36	−22.70	60.8
−125	0.516	28.67*	57.28	102.29	−17.59	63.5
−100	1.428	27.01*	22.16	100.15	−12.47	66.2
− 75	3.388	23.02*	9.92	97.93	− 7.31	69.0
− 50	7.117	15.43*	4.97	95.62	− 2.10	71.8
− 25	13.556	2.32*	2.73	93.20	3.17	74.56
− 15	17.141	2.45	2.19	92.20	5.30	75.65
− 10	19.189	4.49	1.97	91.69	6.37	76.2
− 5	21.422	6.73	1.78	91.18	7.44	76.73
0	23.849	9.15	1.61	90.66	8.52	77.27
5	26.483	11.79	1.46	90.14	9.60	77.80
10	29.335	14.64	1.32	89.61	10.68	78.335
25	39.310	24.61	1.00	87.98	13.96	79.9
50	61.394	46.70	0.66	85.14	19.51	82.43
75	91.682	76.99	0.44	82.09	25.20	84.82
86	108.04	93.34	0.38	80.67	27.77	85.82
100	131.86	117.16	0.31	78.79	31.10	87.03
125	183.76	169.06	0.22	75.15	37.28	88.97
150	249.31	234.61	0.16	71.04	43.85	90.53
175	330.64	315.94	0.11	66.20	51.03	91.48
200	430.09	415.39	0.08	60.03	59.20	91.28

* Inches of mercury below one atmosphere.

Fig. 9-5. Properties of liquid and saturated vapor of refrigerant R-12. Note pressures corresponding to standard evaporating temperature of 5 °F (−15 °C) and condensing temperature of 86 °F (30 °C).
(Du Pont Company)

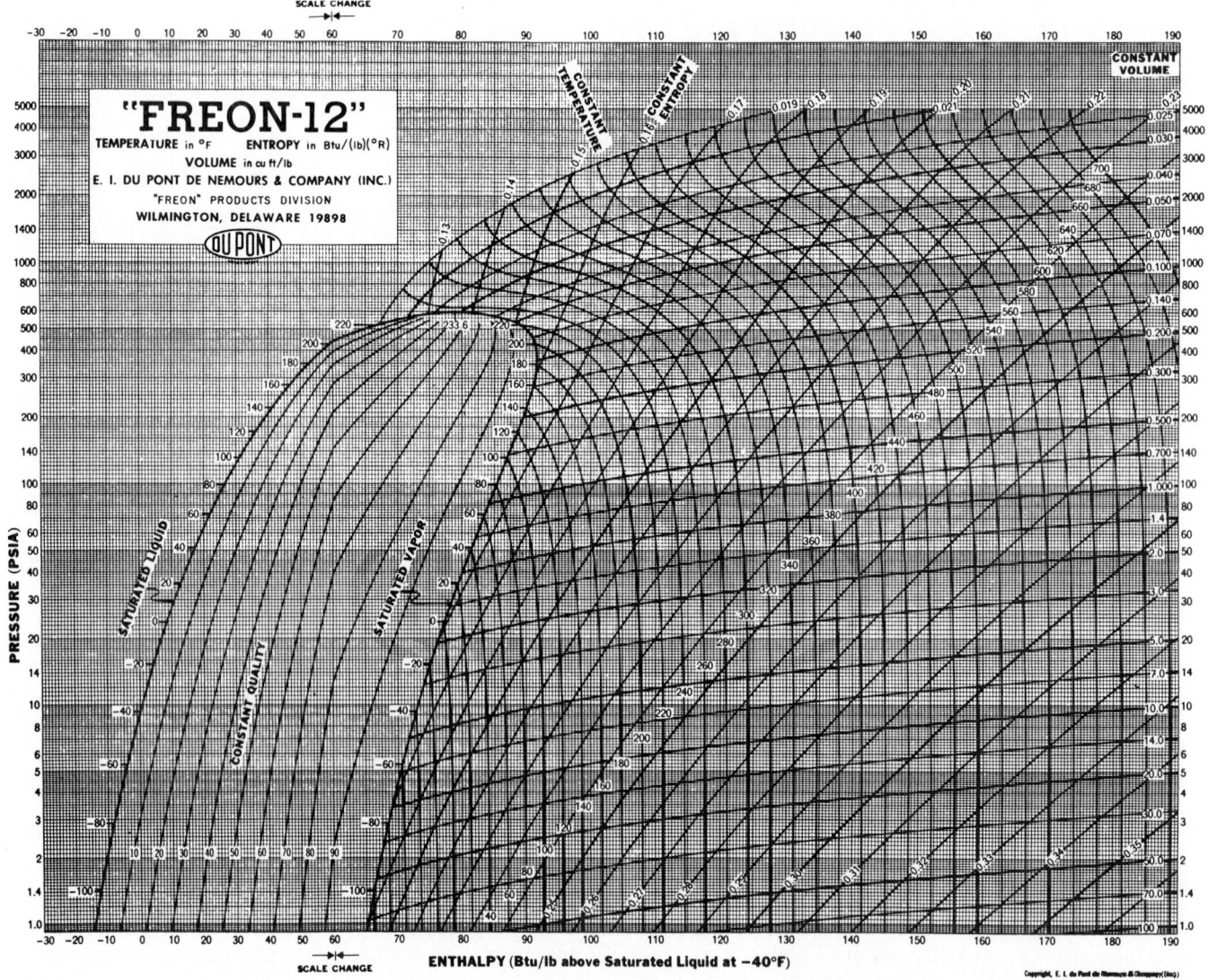

Fig. 9-6. Pressure-enthalpy diagram for refrigerant R-12. Note that 0 of enthalpy scale is taken at −40 °F (−40 °C). (Du Pont Company)

284

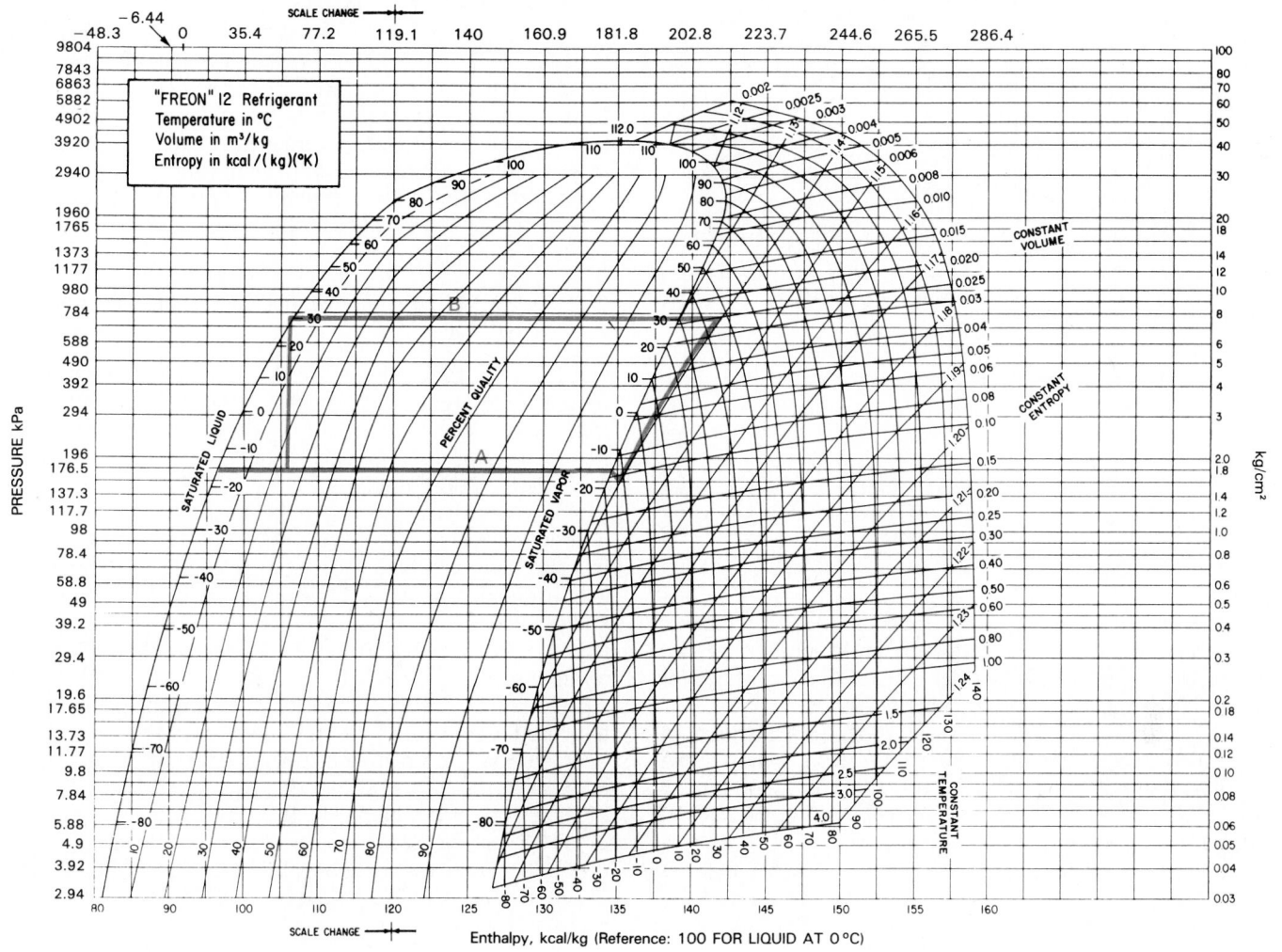

Fig. 9-7. Pressure-heat diagram for R-12 expressed in metric units. The standard refrigerating cycle of evaporating temperature is shown at A and condensing temperature at B. (Adapted from Du Pont Company)

finding alternatives (which could result in increased refrigerant costs, reductions in refrigerating efficiencies, and the use of new refrigerants). See Para. 9-37 for more information.

9-8 R-12 DICHLORODIFLUOROMETHANE (CCl₂F₂)

R-12 is a very popular refrigerant. It is a colorless, almost odorless liquid with a boiling point of −21.7 °F (−29 °C) at atmospheric pressure. It is nontoxic, noncorrosive, non-irritating, and nonflammable.

Chemically, it is inert at ordinary temperatures, and thermally stable to above 800 °F (427 °C). This temperature is well above the safe operating temperatures of most refrigerating mechanism materials and lubricants. A table of properties of R-12 is shown in Fig. 9-5.

R-12 has a relatively low latent heat value. In the smaller refrigerating machines, this is an advantage. The large amount of refrigerant circulated will permit the use of less sensitive and more positive operating and regulating mechanisms. It is used in reciprocating, rotary, and large centrifugal compressors. It operates at a low but positive head and back pressure and with a good volumetric efficiency.

R-12 has a pressure of 26.5 psia or 11.8 psig (183 kPa) at 5 °F (−15 °C), and a pressure of 108.0 psia or 93.3 psig (745 kPa) at 86 °F (30 °C). The latent heat of R-12 at 5 °F (−15 °C) is 68.2 Btu/lb. (159 J/g). Latent heat is difference of last two columns of table in Fig. 9-5.

An R-12 leak may be detected by several means:
1. A soap solution.
2. A halide lamp.
3. Colored oil added to the system.
4. An electronic leak detector.

See Chapter 11 about the use of leak detectors.

Water is only slightly soluble in R-12. At 0 °F (−18 °C), it will only hold six parts per million by weight. The solution formed is only very slightly corrosive to any of the common metals used in refrigerator construction. The addition of mineral oil to the refrigerant has no effect upon the corrosive action, except to lessen the amount of discoloration caused by the free water.

R-12 is more critical as to its moisture content when compared to R-22 and R-502. R-12 is soluble in oil down to −90 °F (−68 °C). This helps the oil flow in very cold evaporators. The oil will begin to separate at this temperature and, because it is lighter than the refrigerant, will collect on the surface of the liquid refrigerant.

The pressure-heat (enthalpy) diagram for this refrigerant

285

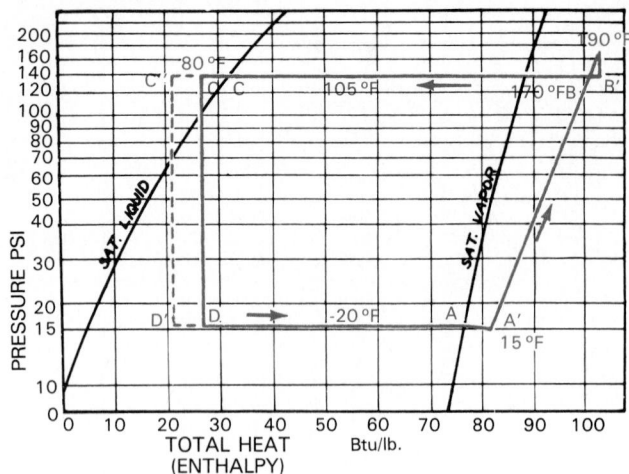

Fig. 9-8. Pressure-heat diagram for freezer application using R-12 as refrigerant. Heat is absorbed in evaporator from D to A. Vapors are compressed by compressor A' to B'. Heat is given off in the condenser B' to C'. Pressure drops from C' to D as refrigerant passes through refrigerant control without change in heat content.

is shown in Fig. 9-6. See Para. 1-54 for an explanation of enthalpy. Metric unit pressure-enthalpy diagram for R-12 is shown in Fig. 9-7. It is safe to use 30 lb. of R-12 for each 1000 cu. ft. of air conditioned space.

A typical R-12 cycle for a frozen foods unit is shown in Fig. 9-8. The dotted line labeled C'' to D' shows the increased refrigerating effect when the liquid is sub-cooled in the condenser or liquid line. This sub-cooling uses a low ambient temperature or a heat exchanger.

R-12 is available in a variety of cylinder sizes. The cylinder code color is white. This is an important fact to remember when purchasing or using refrigerants. See Para. 9-27 for

more color code information.

A patented refrigerant, Genetron 12/31, is available. In some installations it may be used as a replacement for R-12. It is a mixture of R-12 and R-31. R-31 is monochloromonofluoromethane. The chemical composition is CCl_2F_2 (78 percent) and CH_2CIF (22 percent). Its latent heat of vaporization is slightly higher than for R-12. Its head pressure is also a little higher than for R-12. Critical temperature is 244 °F (118 °C).

9-9 R-22 MONOCHLORODIFLUOROMETHANE (CHClF₂)

R-22 is a synthetic (made by humans; not found in nature) refrigerant developed for refrigeration installations that need a low evaporating temperature. (See Fig. 9-9.) It is referred to as "monochlorodifluoromethane" and also as "chlorodifluoromethane."

One application is in fast freezing units which maintain a temperature of − 20 °F to − 40 °F (− 29 to − 40 °C). It has also been successfully used in air conditioning units and in household refrigerators. It is used with both reciprocating and centrifugal compressors. It is not necessary to use R-22 at below-atmosphere pressures in order to obtain these low temperatures.

R-22 is stable and is nontoxic, noncorrosive, nonirritating, and nonflammable. R-22 has a boiling point of − 41 °F (− 41 °C) at atmospheric pressure. The normal head pressure at 86 °F (30 °C) is 173 psia or 158 psig (1 190 kPa), as shown in table in Fig. 9-9. The evaporator pressure or R-22 is 43 psia or 28 psig (296 kPa) at 5 °F (− 15 °C). The latent heat of R-22 at 5 °F (− 15 °C) is 93.2 Btu/lb. (217 J/g). Latent heat is difference of last two columns in Fig. 9-9.

	PRESSURE		VOLUME VAPOR	DENSITY LIQUID	HEAT CONTENT BTU/LB.	
Temp °F	Psia	Psig	Cu. Ft./Lb.	Lb./Cu. Ft.	Liquid	Vapor
−150	0.272	29.37*	141.23	98.24	−25.97	87.52
−125	0.886	28.12*	46.69	96.04	−20.33	90.43
−100	2.398	25.04*	18.43	93.77	−14.56	93.37
− 75	5.610	18.50*	8.36	91.43	− 8.64	96.29
− 50	11.674	6.15*	4.22	89.00	− 2.51	99.14
− 25	22.086	7.39	2.33	86.48	3.83	101.88
− 15	27.865	13.17	1.87	85.43	6.44	102.94
− 10	31.162	16.47	1.68	84.90	7.75	103.46
− 5	34.754	20.06	1.52	84.37	9.08	103.96
0	38.657	23.96	1.37	83.83	10.41	104.47
5	42.888	28.19	1.24	83.28	11.75	104.96
10	47.464	32.77	1.13	82.72	13.10	105.44
25	63.450	48.75	0.86	81.02	17.22	106.84
50	98.727	84.03	0.56	78.03	24.28	108.95
75	146.91	132.22	0.37	74.80	31.61	110.74
86	172.87	158.17	0.32	73.28	34.93	111.40
100	210.60	195.91	0.26	71.24	39.27	112.11
125	292.62	277.92	0.18	67.20	47.37	112.88
150	396.19	381.50	0.12	62.40	56.14	112.73

*Inches of mercury below one atmosphere.

Fig. 9-9. Properties of liquid and saturated vapor of refrigerant R-22. Note pressures corresponding to standard evaporating temperature of 5 °F (− 15 °C) and condensing temperature of 86 °F (30 °C).

Water mixes better with R-22 than R-12 by a ratio of 3 to 1 or 19.5 ppm by weight (ppm means parts per million). Water must be kept at a minimum, and desiccants (driers) should be used to remove most of the moisture.

Because of the ability of water to mix with R-22, larger amounts of desiccant are needed to dry the refrigerant. R-22 has good solubility in oil. This solubility remains high down to about 16°F (−9°C). The oil will remain fluid enough to flow down the suction line at temperatures down to −40°F (−40°C). However, around this temperature or not far below it, the oil will begin to separate from the refrigerant.

Because oil is lighter, it will collect on the surface of the liquid refrigerant. Leaks may be detected with a soap solution, a halide torch, or with an electronic leak detector. Some of the properties of R-22 are shown in Fig. 9-10. Metric unit pressure-enthalpy diagram for R-22 is shown in Fig. 9-11.

The cylinder code color of R-22 is green.

9-10 R-11 TRICHLOROMONOFLUOROMETHANE (CCl₃F)

R-11 is a synthetic chemical product which can be used as a refrigerant. It is stable, nonflammable, and nontoxic. This means it will not burn and is not a poison. Considered to be a low-pressure refrigerant, it has a low-side pressure of 24 in. vacuum (610 mmHg.) at 5°F (−15°C), and a high-side pressure of 18.3 psia or 3.6 psig (126 kPa) at 86°F (30°C). The latent heat at 5°F (−15°C) is 84.0 Btu/lb. (195 J/g).

This refrigerant is extensively used in large centrifugal compressor systems. As much as 35 lb. of this refrigerant may be used for each 1000 cu. ft. (28.3 m³) of air conditioned space. (This would be a room about 10 ft. (3 m) by 12.5 ft. (3.8 m) by 8 ft. (2.4 m).) Leaks may be detected by using a soap solution, a halide torch, or by using an electronic detector.

R-11 is often used by service technicians as a flushing agent for cleaning the internal parts of a refrigerator com-

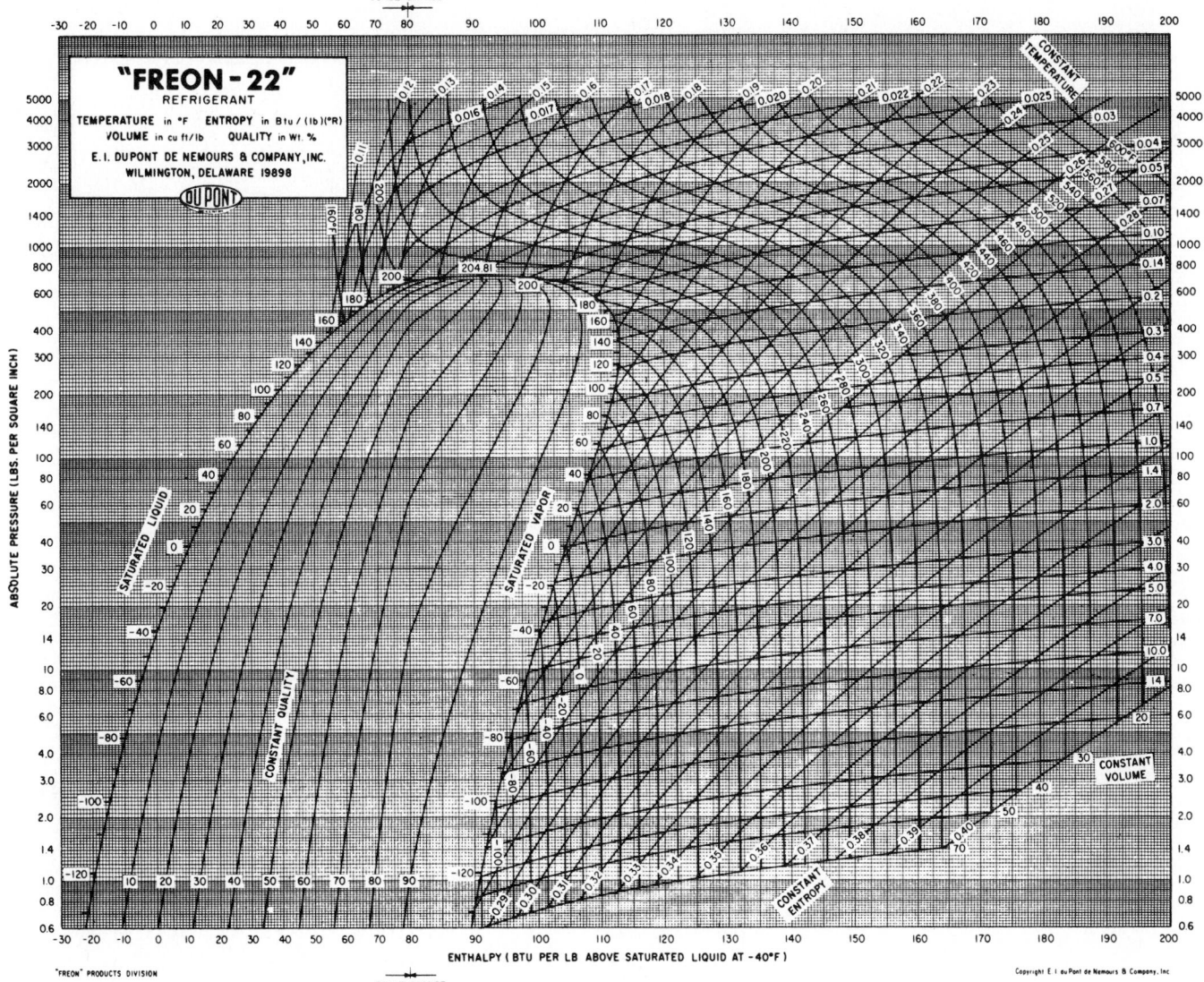

Fig. 9-10. Pressure-enthalpy diagram for refrigerant R-22. Note that 0 of enthalpy scale is taken at −40°F (−40°C). Also note that heat (enthalpy) scale changes at 80 Btu per lb. to help technician read superheat values more easily.   (Du Pont Company)

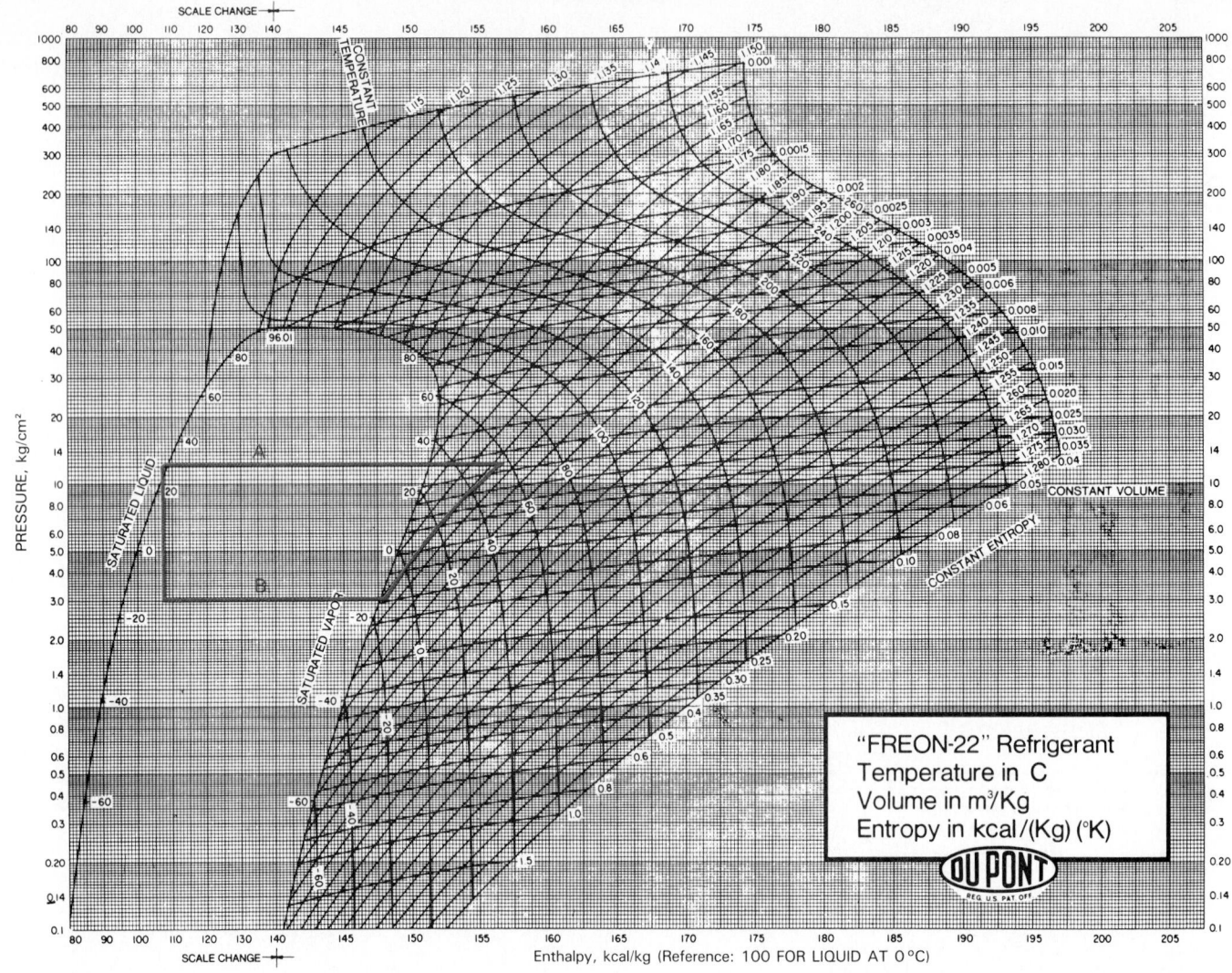

Fig. 9-11. Pressure heat diagram for R-22 expressed in metric units. The standard refrigerating cycle of evaporating temperature is shown at A and condensing temperature is shown at B. (Du Pont Company)

Fig. 9-12. Properties of liquid and saturated vapor of refrigerant R-500. Note pressures corresponding to evaporating temperature of 5 °F (− 15 °C) and condensing temperature of 86 °F (30 °C).
(Allied Signal, Inc.)

Temp °F	PRESSURE		VOLUME VAPOR	DENSITY LIQUID	HEAT CONTENT BTU/LB.	
	Psia	Psig	Cu. Ft./Lb.	Lb./Cu. Ft.	Liquid	Vapor
−40	10.95	7.62*	4.0	84.28	0.00	87.74
−30	14.10	1.22*	3.15	83.35	2.38	89.04
−20	17.92	3.23	2.52	82.40	4.79	90.31
−10	22.52	7.82	2.03	81.44	7.22	91.57
0	27.98	13.3	1.66	80.46	9.71	92.81
5	31.07	16.4	1.501	79.96	10.96	93.42
10	34.43	19.7	1.36	79.46	12.23	94.03
20	41.96	27.3	1.13	78.45	14.79	95.22
30	50.70	36.0	0.94	77.41	17.40	96.39
40	60.75	46.1	0.79	76.34	20.05	97.53
50	72.26	57.6	0.67	75.26	22.75	98.64
60	85.33	70.6	0.57	74.14	25.48	99.71
70	100.1	85.4	0.48	72.98	28.28	100.75
80	116.7	102.0	0.42	71.80	31.12	101.75
86	127.6	113.0	0.38	71.06	32.85	102.33
90	135.3	121.0	0.36	70.56	34.01	102.70
100	155.9	141.0	0.31	69.28	36.97	103.60
110	178.8	164.0	0.27	67.95	40.00	104.44
120	204.1	189.0	0.23	66.55	43.10	105.22
130	231.9	217.0	0.20	65.08	46.29	105.91
140	262.4	248.0	0.17	63.51	49.58	106.51

* Inches of mercury vacuum.

288

pressor when overhauling systems. It is useful after a system has had a motor burnout or after it has had a great deal of moisture in the system. By flushing moisture from the system with R-11, evacuation time is shortened. Other refrigerants are being considered as replacements for R-11, in keeping with the E.P.A. rulings. See Para. 9-37 and Chapter 28.

The cylinder code color of R-11 is orange.

9-11 AZEOTROPIC MIXTURES

Azeotropic refrigerants are liquid mixtures of refrigerants which exhibit a constant maximum and minimum boiling point. However, these mixtures act as a single refrigerant. Three commonly used azeotropic refrigerants are:

R-500—composed of 73.8 percent R-12 and 26.2 percent R-152a.

R-502—composed of 48.8 percent R-22 and 51.2 percent R-115.

R-503—composed of 40.1 percent R-23 and 59.9 percent R-13.

These are patented refrigerants and the manufacturing process is rather complicated. Service technicians should never attempt to make their own mixtures.

Properties of these refrigerants are explained in the following paragraphs.

Azeotropic mixtures are chiefly used with reciprocating compressors.

See Para. 9-37 for further information concerning the use of R-502 and the depletion of the earth's ozone layer.

9-12 R-500 REFRIGERANT (R-152a + R-12) (CCl₂F₂/CH₃CHF₂)

Refrigerant R-500 is an azeotropic mixture of 26.2 percent R-152a and 73.8 percent R-12. It is used in both industrial and commercial applications, but only in systems with reciprocating compressors. It has a fairly constant vapor-pressure temperature curve which is different from the vaporizing curves for either R-152a or R-12.

R-500 offers about 20 percent greater refrigerating capacity than R-12 for the same size motor when used for the same purpose. The evaporator pressure of R-500 is 31.1 psia or 16.4 psig (214 kPa) at 5 °F (−15 °C). It has a boiling point, at atmospheric pressure, of −28 °F (−33 °C). Its condensing pressure is 128 psia or 113 psig (879 kPa) at 86 °F (30 °C). Its latent heat at 5 °F (−15 °C) is 82.5 Btu/lb. (192 J/g), as shown in table in Fig. 9-12.

R-500 can be used whenever a higher capacity than that obtained with R-12 is needed. There is little change in condensing temperatures, as shown in Fig. 9-13. R-500 is also recommended where electrical service varies from 60 cycle to 50 cycle (Hz).

The solubility (mixing with or going into solution) of water in R-500 is highly critical. R-500 has fairly high solubility with oil. Use a halide leak detector, an electronic leak detector, soap solution, or a colored tracing agent to detect leaks.

Servicing refrigerators using this refrigerant does not present any unusual problem. Water is quite soluble in this refrigerant. It is necessary to keep moisture out of the system by careful dehydration and by using driers.

The cylinder code color of R-500 is yellow.

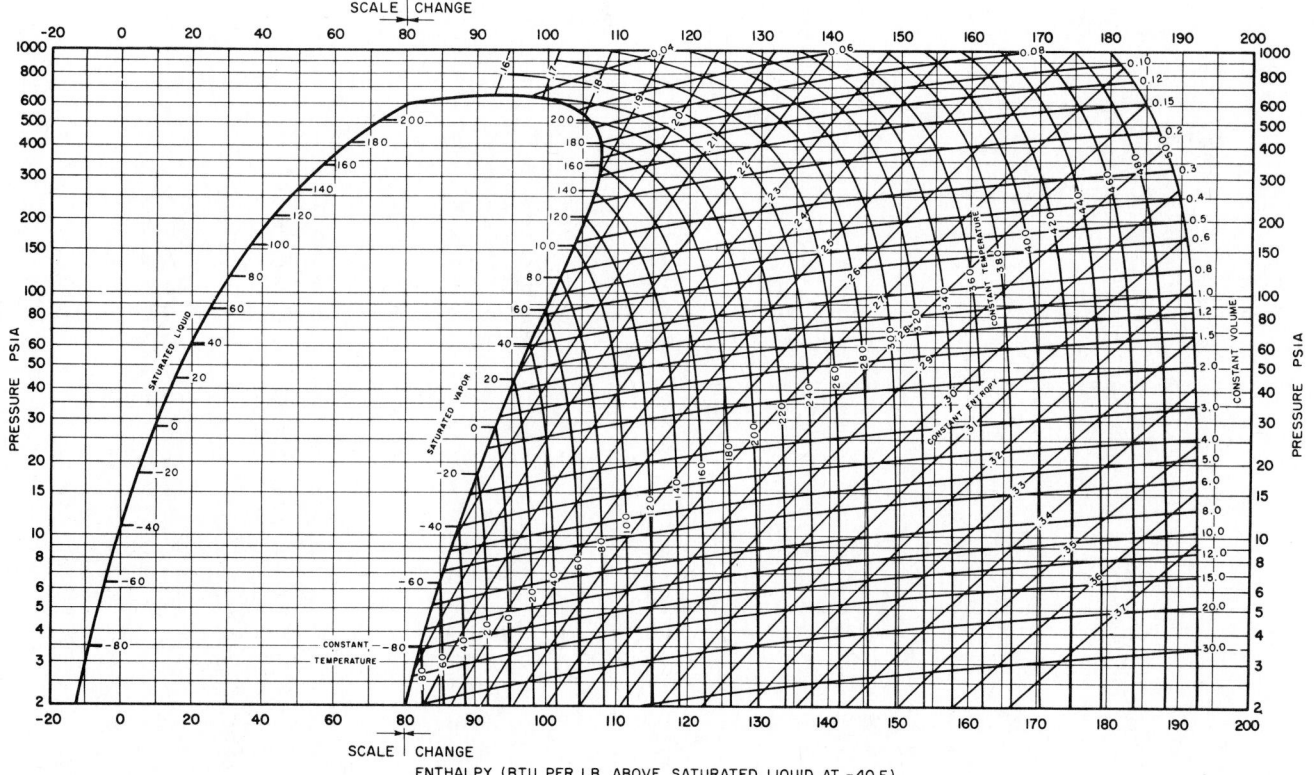

Fig. 9-13. Pressure-enthalpy diagram for refrigerant R-500. Note that 0 of enthalpy scale is taken at −40 °F (−40 °C). Also note that enthalpy scale changes at 80 Btu per lb. to help technician read superheat values.

9-13 R-502 REFRIGERANT (R-22 + R-115) (CHClF₂/CClF₂CF₃)

Refrigerant R-502 is an azeotropic mixture of 48.8 percent R-22 and 51.2 percent R-115. It has been used since 1961. It is a nonflammable, noncorrosive, practically nontoxic liquid. A good refrigerant for obtaining medium and low temperatures, it is suitable where temperatures from 0 to $-60\,°F$ (-18 to $-51\,°C$) are needed. It is often used in frozen food lockers, frozen food processing plants, frozen food display cases, and in storage units for frozen foods and ice cream. It is only used with reciprocating compressors.

Its boiling point is $-50\,°F$ ($-46\,°C$) at atmospheric pressure. The condensing pressure is 191 psia or 177 psig (1 320 kPa) at 86 °F (30 °C). Its evaporating pressure at 5 °F ($-15\,°C$) is 50.6 psia or 35.9 psig (348 kPa). Its latent heat at $-20\,°F$ ($-29\,°C$) is 70.8 Btu/lb. (165 J/g) as shown in the table of properties in Fig. 9-14. The latent heat of R-502 at 5 °F ($-15\,°C$) is 67.3 Btu/lb. (157 J/g).

Refrigerant R-502 combines many of the good properties of both R-12 and R-22. It gives a machine the approximate capacity of R-22 with just about the condensing temperature of a system using R-12. A pressure-enthalpy diagram of the refrigerant is shown in Fig. 9-15.

When this refrigerant is used at a condensing temperature of 30 °F ($-1\,°C$), as is common for the frozen food applications mentioned above, the low condensing pressure and temperature increases the life of the compressor valves and other parts. Better lubrication is possible because of the increased viscosity of the oil at the lower condensing temperature. Because of the lower condensing pressures,

it is possible to eliminate liquid injection to cool the compressor. This is often necessary with R-22.

R-502 has all the qualities found in the other halogenated (fluorocarbon) refrigerants. It is nontoxic, nonflammable, nonirritating, stable, and noncorrosive. Leaks are detected with soap solution, halide torch, or electronic leak detector.

R-502 will hold 1.5 times more moisture at 0 °F ($-18\,°C$) than R-12 (12.0 ppm by wt.). R-502 has fair solubility in oil above 180 °F (82 °C). Below this temperature, the oil tries to separate and tends to collect on the surface of liquid refrigerant. However, oil is carried back to the compressor at temperatures down to $-40\,°F$ ($-40\,°C$). Special devices are sometimes used to return the oil to the compressor.

The cylinder code color of R-502 is orchid.

9-14 R-503 REFRIGERANT (R-23 + R-13) (CHF₃/CClF₃)

Refrigerant R-503 is an azeotropic mixture of 40.1 percent R-23 and 59.9 percent R-13.

This is a nonflammable, noncorrosive, practically nontoxic liquid classified under Group 6 in the Underwriter's Laboratory Classification Scale. See Fig. 9-4.

Its boiling temperature at atmospheric pressure is $-126\,°F$ ($-88\,°C$). This is lower than either R-23 or R-13.

Its evaporating pressure at 5 °F ($-15\,°C$) is 266 psia or 252 psig (1 830 kPa). Its critical temperature is 67 °F (20 °C), and its critical pressure is 607 psia or 592 psig (4 180 kPa).

This is a low temperature refrigerant and good for use in the low section of cascade systems which require

| Temp °F | PRESSURE | | VOLUME VAPOR | DENSITY LIQUID | HEAT CONTENT BTU/LB. | |
	Psia	Psig	Cu. Ft./Lb.	Lb./Cu. Ft.	Liquid	Vapor
−100	3.261	23.281*	10.461	97.857	−12.548	65.885
− 75	7.281	15.097*	4.959	95.234	− 7.597	68.919
− 50	14.602	0.190*	2.596	92.513	− 2.251	71.928
− 25	26.817	12.121	1.465	89.673	3.496	74.866
− 20	30.006	15.310	1.317	89.088	4.693	75.442
− 15	33.480	18.784	1.187	88.496	5.905	76.012
− 10	37.256	22.560	1.073	87.898	7.133	76.577
− 5	41.349	26.653	0.973	87.293	8.376	77.137
0	45.775	31.079	0.881	86.681	9.633	77.690
5	50.553	35.857	0.801	86.062	10.906	78.237
10	55.697	41.001	0.731	85.434	12.193	78.777
15	61.225	46.529	0.666	84.797	13.494	79.310
20	67.155	52.459	0.612	84.152	14.809	79.836
25	73.503	58.807	0.557	83.497	16.138	80.353
50	112.12	97.42	0.367	80.058	22.977	82.800
75	163.81	149.11	0.248	76.269	30.122	84.958
86	191.28	176.59	0.210	74.453	33.359	85.789
100	230.89	216.19	0.171	71.967	37.563	86.711
125	316.04	301.35	0.118	66.838	45.361	87.834
150	423.06	408.35	0.079	60.092	53.850	87.757
160	473.38	458.69	0.066	56.429	57.732	87.013

* Inches of mercury below one atmosphere.

Fig. 9-14. Properties of liquid and saturated vapor of refrigerant R-502. Note pressures corresponding to standard evaporating temperature of 5 °F ($-15\,°C$) and condensing temperature of 86 °F (30 °C). (Du Pont Company)

"FREON" 502 REFRIGERANT PRESSURE-ENTHALPY DIAGRAM
TEMPERATURE in °F, ENTROPY Btu/(lb)(°R), VOLUME in cu ft/lb

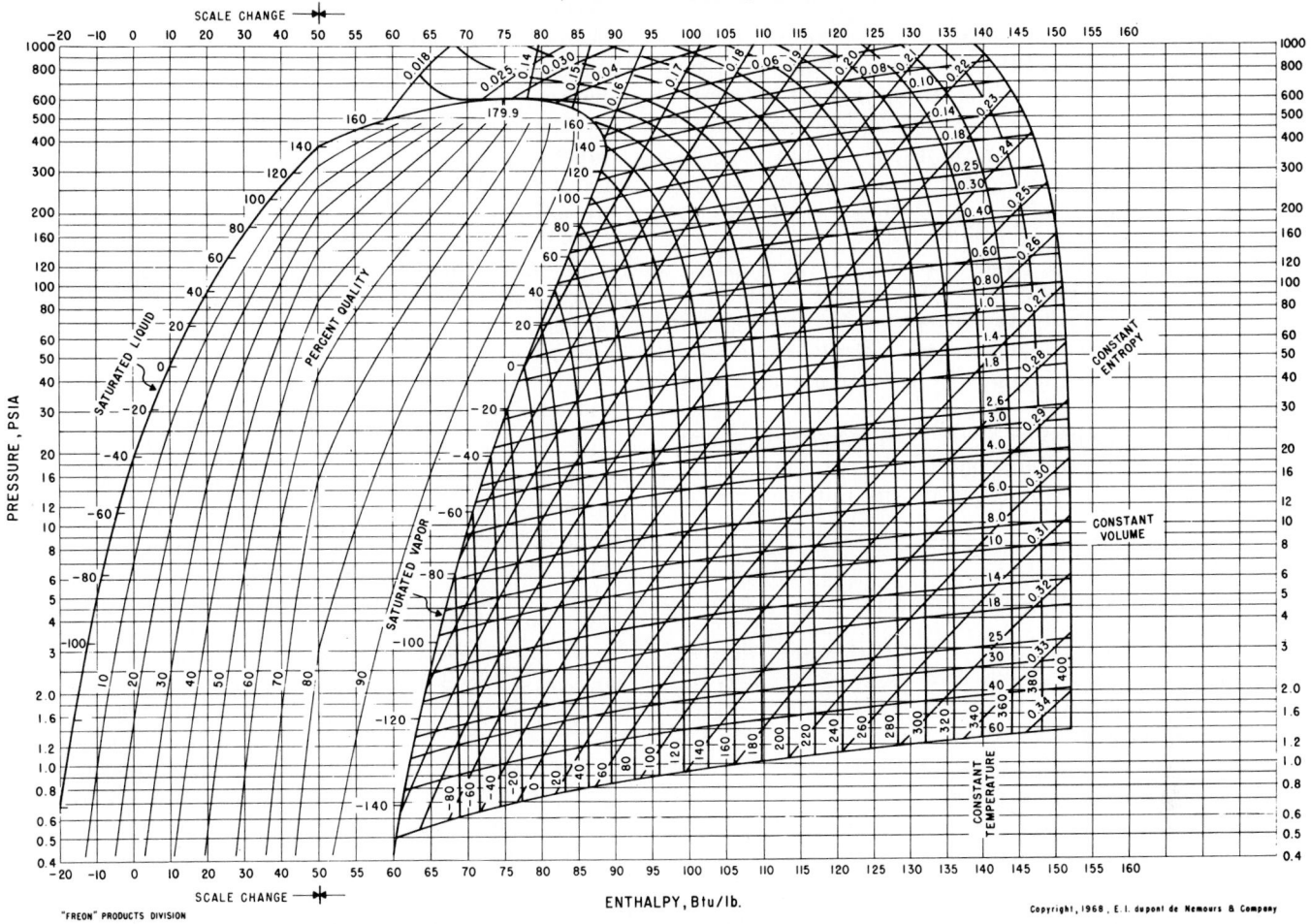

Fig. 9-15. Pressure-enthalpy diagram for refrigerant R-502. Note that 0 of enthalpy scale is at −40 °F (−40 °C). Also note that scale changes at 50 Btu per lb. above saturated liquid at −40 °F (−40 °C) to permit easier reading of superheat values.
(Du Pont Company)

Fig. 9-16. Properties of liquid and saturated vapor of refrigerant R-503. Note that to operate above 0 °F (−18 °C), condenser pressure in excess of 231 psig (1 690 kPa) is required. Note pressure corresponding to condensing temperature of 5 °F (−15 °C). R-503 is used in cascade system with evaporating temperature as low as −50 °F (−46 °C). Also note that 0 of liquid heat content is taken at −40 °F (−40 °C).
(Allied Signal, Inc.)

Temp °F	PRESSURE		VOLUME VAPOR	DENSITY LIQUID	HEAT CONTENT BTU/LB.	
	Psia	Psig	Cu. Ft./Lb.	Lb./Cu. Ft.	Liquid	Vapor
−140	9.234	11.1 *	4.123	93.49	−26.45	52.88
−130	12.98	3.49*	2.998	92.47	−23.93	53.84
−120	17.83	3.13	2.227	91.39	−21.36	54.77
−110	23.98	9.28	1.685	90.25	−18.77	55.66
−100	31.64	16.9	1.296	89.05	−16.16	56.52
− 90	41.05	26.3	1.012	87.78	−13.53	57.35
− 80	52.42	37.7	0.8008	86.44	−10.87	58.13
− 70	66.00	51.3	0.6409	85.02	− 8.19	58.86
− 60	82.05	67.4	0.5182	83.52	− 5.49	59.54
− 50	100.8	86.1	0.4227	81.93	− 2.76	60.16
− 40	122.6	108	0.3474	80.25	0.00	60.72
− 30	147.6	133	0.2872	78.46	2.81	61.20
− 20	176.2	161	0.2387	76.56	5.66	61.60
− 10	208.6	194	0.1991	74.52	8.59	61.89
0	245.3	231.0	0.1664	72.33	11.60	62.05
5	265.9	251.5	0.15275	70.99	13.17	62.04
10	286.4	272.0	0.1391	69.65	14.74	62.04
20	332.6	318	0.1160	67.35	18.05	61.82
30	384.1	369	0.0962	64.45	21.60	61.31
40	440.6	426	0.0793	61.12	26.03	60.45
50	503.3	489	0.0640	57.09	26.69	58.95
60	574.8	560	0.0485	51.40	34.32	55.77

* Inches of mercury vacuum.

temperatures in the −100°F to −125°F range (−73° to −87°C). Properties of R-503 are shown in Fig. 9-16. A pressure-enthalpy diagram for this refrigerant is shown in Fig. 9-17.

The latent heat of vaporization at atmospheric pressure (−127 °F or −88°C) is 77.2 Btu/lb. (180 J/g). The latent heat at 5°F (−15°C) is 48.9 Btu/lb. (114 J/g).

Leaks in R-503 systems may be detected with the use of soap solution, a halide torch, or an electronic leak detector. This refrigerant will hold more moisture than some other low temperature refrigerants. All low temperature applications must have extreme dryness. Any moisture not in solution with the refrigerant is likely to form ice at the refrigerant control devices.

Oil does not circulate well at low temperatures. Cascade systems and other low-temperature units are usually fitted with oil separators and other devices for returning the oil to the compressor.

The code color for R-503 cylinders is aquamarine.

9-15 GROUP TWO REFRIGERANTS

The Group Two refrigerants are toxic. They are irritating to breathe and may be slightly flammable. Methyl chloride is quite toxic. Refrigerants in this group include:

R-717   Ammonia NH_3
R-40   Methyl Chloride CH_3Cl

R-764   Sulphur Dioxide SO_2

Pressure-temperature curves for some Group 2 refrigerants are shown in Fig. 9-18.

R-717 was one of the first refrigerants used. However, with the exception of the absorption refrigerators, it is now used only in large industrial installations.

R-764 and R-40 are seldom used today. At one time, R-764 was the refrigerant most used in domestic refrigerators. However, there are sulphur dioxide (R-764) and methyl chloride (R-40) charged units still in use.

Additional information about Group Two refrigerants can be found in Chapter 30.

9-16 R-717 AMMONIA (NH_3)

R-717 is commonly used in industrial systems. It is a chemical compound of nitrogen and hydrogen (NH_3) and under ordinary conditions is a colorless gas. Its boiling temperature at atmospheric pressure is −28°F (−33°C) and its melting point from the solid is −108°F (−78°C).

The low boiling point makes it possible to have refrigeration at temperatures considerably below zero without using pressures below atmospheric in the evaporator. Its latent heat is 565 Btu/lb. (1 310 J/g) at 5°F (−15°C). Thus, large refrigerating effects are possible with relatively small-sized machinery. Condensers for R-717 are usually of the water-cooled type, although air-cooled condensers are be-

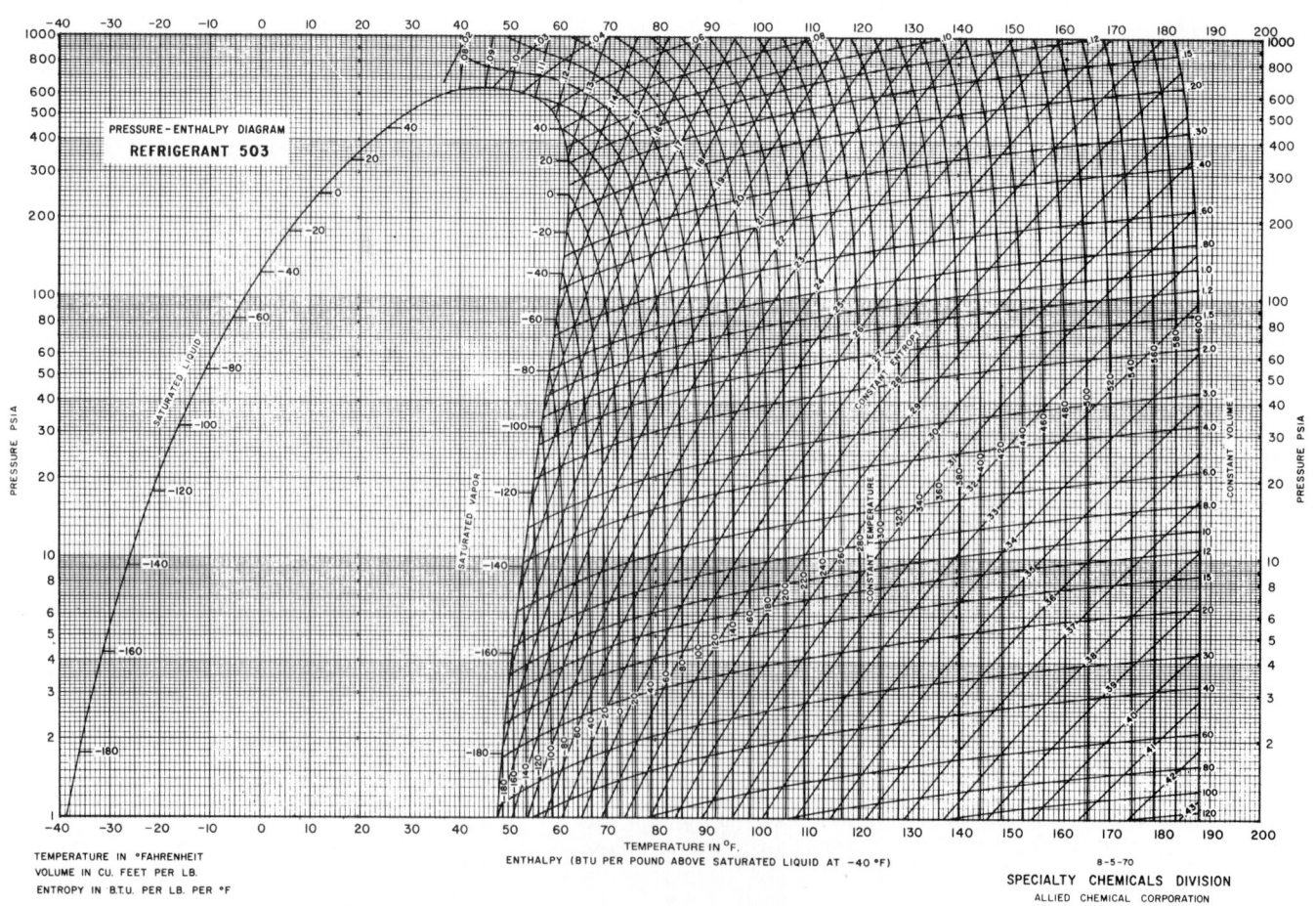

Fig. 9-17. Pressure-enthalpy diagram for refrigerant R-503.   (Allied Signal, Inc.)

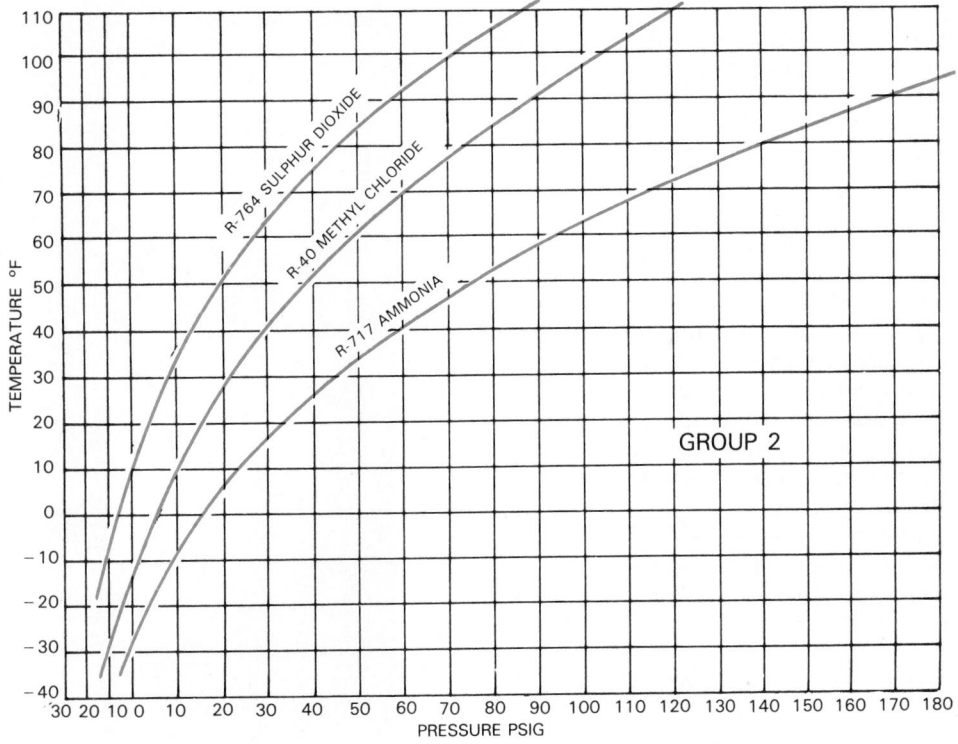

Fig. 9-18. Saturation temperature-pressure curves for some Group 2 refrigerants.

ing developed. The evaporator pressure at 5 °F (−15 °C) is 34.3 psia or 19.6 psig (236 kPa). The condenser pressure is 169 psia or 155 psig (1 170 kPa) at 86 °F (30 °C) as shown in Fig. 9-19.

R-717 is somewhat flammable and, with the proper proportions of air, will form an explosive mixture. Accidents from this source, however, are rare.

While not classed as poisonous, its effect on the respiratory system is so violent that only very small quantities of it can be breathed safely. About .35 volumes per 100 volumes of air is the strongest concentration one can bear for any length of time. Because of its pronounced and distinguishable odor, it is easily detected in the air.

At 3 to 5 ppm, ammonia is identified by smell. At 15

Temp °F	PRESSURE		VOLUME VAPOR	DENSITY LIQUID	HEAT CONTENT BTU/LB.		LATENT HEAT BTU/LB.
	Psia	Psig	Cu. Ft./Lb.	Lb./Cu. Ft.	Liquid	Vapor	
−100	1.24	27.4*	182.4	45.52	−63.3	572.5	635.8
− 75	3.29	23.2*	72.81	44.52	−37.0	583.3	620.3
− 50	7.67	14.3*	33.08	43.49	−10.6	593.7	604.3
− 35	12.05	5.4*	21.68	42.86	5.3	599.5	594.2
− 25	15.98	1.3	16.66	42.44	16.0	603.2	587.2
− 20	18.30	3.6	14.68	42.22	21.4	605.0	583.6
− 15	20.88	6.2	12.97	42.00	26.7	606.7	580.0
− 10	23.74	9.0	11.50	41.78	32.1	608.5	576.4
− 5	26.92	12.2	10.23	41.56	37.5	610.1	572.6
0	30.42	15.7	9.12	41.34	42.9	611.8	568.9
5	34.27	19.6	8.15	41.11	48.3	613.3	565.0
10	38.51	23.8	7.30	40.89	53.8	614.9	561.1
15	43.14	28.4	6.56	40.66	59.2	616.3	557.1
20	48.21	33.5	5.91	40.43	64.7	617.8	553.1
25	53.73	39.0	5.33	40.20	70.2	619.1	548.9
35	66.26	51.6	4.37	39.72	81.2	621.7	540.5
50	89.19	74.5	3.29	39.00	97.9	625.2	527.3
75	140.5	125.8	2.13	37.74	126.2	629.9	503.7
86	169.2	154.5	1.77	37.16	138.9	631.5	492.6
100	211.9	197.2	1.42	36.40	155.2	633.0	477.8
125	307.8	293.1	0.97	34.96	185.1	634.0	448.9

*Inches of mercury below one atmosphere.

Fig. 9-19. Properties of refrigerant R-717 (ammonia). Note high latent heat. Latent heat is difference of columns 6 and 7.

ppm, the odor is quite irritating. At 30 ppm, the service technician will need a respirator. Exposure of 5 minutes to 50 ppm is the maximum allowed by OSHA. It becomes a hazard to life at 5000 ppm, and is flammable at 150,000 to 270,000 ppm.

Always stand to one side when operating an ammonia valve. A small stem leak may burn, damage the eyes, and cause almost instant loss of consciousness. Wear a tight fitting mask. R-717 leaks may be quickly and easily detected by the white smoke-like fumes that it forms in the presence of a sulphur candle or sulphur spray vapor.

R-717 attacks copper and bronze in the presence of a little moisture but does not corrode iron or steel. It presents no special problems in connection with lubrication unless extreme temperatures are encountered. R-717 is lighter than oil and no problems are encountered in dealing with the separation of the two. Excess oil in the evaporator may be removed by opening a valve in the bottom of the evaporator. The solubility of oil in liquid R-717 is only 20 ppm at 5 °F (− 15 °C) and only 125 ppm at 86 °F (30 °C). R-717 vapor is extremely soluble in water. It is used in large compression machines using reciprocating compressors and in many absorption type systems.

The basic properties of R-717 are shown in Chapter 30.

A recognized safety code for the use and handling of ammonia in refrigerating systems is supplied by The International Institute of Ammonia Refrigeration.

9-17 GROUP THREE REFRIGERANTS

Group Three refrigerants may form a combustible mixture when mixed with air. The more common refrigerants in this group are:

R-600	Butane	C_4H_{10}
R-170	Ethane	C_2H_6
R-290	Propane	C_3H_3

These refrigerants are no longer commonly used. Their characteristics are not covered in detail in this chapter. See Chapter 30 for further information.

9-18 EXPENDABLE REFRIGERANTS

Expendable refrigerants are used to cool a substance or evaporator and the refrigerant is then released to the atmosphere. It is not collected and recondensed, as is the case with the usual compression system. It is used only once. Systems using expendable refrigerants are sometimes referred to as chemical refrigeration or open cycle refrigeration. Refrigerants of this type have a low boiling temperature.

The most common expendable refrigerants are:
1. Liquid nitrogen (R-728); boiling temperature at atmospheric pressure, − 320 °F (− 196 °C).
2. Liquid helium (R-704); boiling temperature at atmospheric pressure, − 452 °F (− 269 °C).
3. Carbon dioxide (R-744); boiling temperature at atmospheric pressure, − 109 °F (− 78 °C) either in the solid or liquid state.

9-19 WATER AS A REFRIGERANT

Water is never used in the compression cycle refrigerating mechanism. However, it is the refrigerant for steam jet refrigeration used in connection with air conditioning

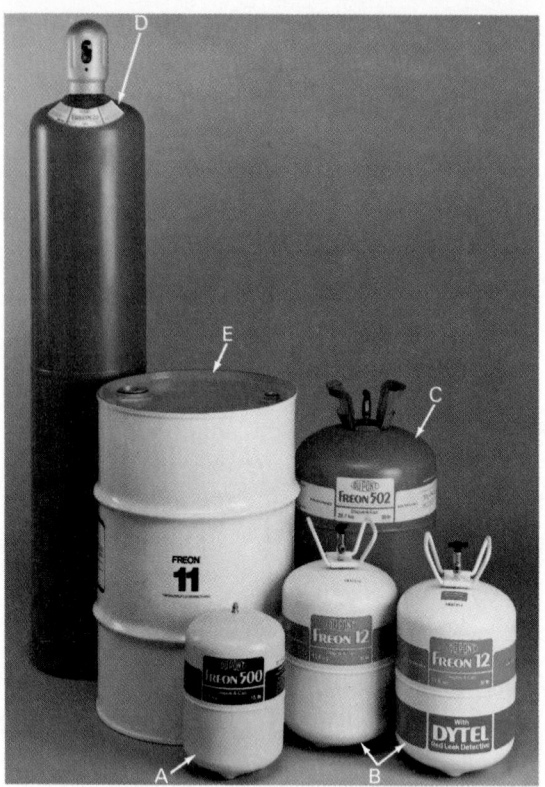

Fig. 9-20. Common refrigerant storage cylinders. A, B, C—Disposable cans: A — 1 5-lb., B — 30 lb., C — 50 lb. D — Returnable and refillable container. E — Drum.   (Du Pont Company)

systems (see Chapter 19). At atmospheric pressure, water boils at 212 °F (100 °C). One pound of water absorbs 970 Btu in changing from a liquid at 212 °F to a vapor (steam) at 212 °F. One kilogram of water absorbs 2 260 kJ in changing from liquid to 100 °C to a vapor at 100 °C. The usual temperature range in using water as a refrigerant is above 45 °F (7 °C). Water, in changing from a liquid to a vapor, absorbs a considerable amount of heat.

The heat absorbed can be stated in a more familiar way, using the concept of latent heat. The latent heat of water at 212 °F (100 °C) is 970 Btu/lb. (2 260 kJ/kg or 2 260 J/g). Compare this with the latent heats for standard refrigerants to see how effective water can be.

The volume of vapor formed is large. At 45 °F (7 °C), one pound of water will turn into 2040 cu. ft. (57.7 m³) of vapor. Water vaporizing at 29.6 in. Hg. vacuum, or at 0.15 psia (7 800 microns), will produce a refrigeration temperature of 45 °F (7 °C).

9-20 FOOD FREEZANTS

When processing frozen foods, it is best to complete the freezing operation in the shortest possible time. Many commercial food freezing companies submerge the food to be frozen in a liquid refrigerant. The United States Department of Agriculture has approved certain refrigerants for this purpose. They are of a high purity and are designated by the name "Food Freezants."

This method is very rapid, since the heat transfer from the food to be frozen to the liquid is much faster than the heat transfer would be if the food were surrounded by air at the same temperature. The refrigerant used in this process does not affect the wholesomeness of the food.

See Chapter 13 for description of the equipment required for the use of Food Freezants.

9-21 CRYOGENIC FLUIDS

Use of cryogenic fluids is becoming quite general in modern industry. They range in temperature from $-250\,°F$ $(-157\,°C)$ to absolute 0 $(-459.69\,°F$ or $-273\,°C)$. This is called the cryogenic range. Chapter 30 has a listing of these temperature ranges using various temperature scales.

Such low temperatures may be easily reached by evaporating cryogenic fluids. Common cryogenic fluids are:

R-702	Hydrogen
R-704	Helium
R-720	Neon
R-728	Nitrogen
R-729	Air
R-732	Oxygen
R-740	Argon

Fig. 1-29 lists the boiling temperature of cryogenic fluids.

Containers must be of special materials able to withstand extremely low temperatures without losing their strength. Insulation is very heavy as the temperature of the fluids inside is very low.

Small containers are of a thermos bottle type construction. Pressures are kept at a relatively low level which corresponds to the vapor pressure of the fluid.

No attempt is made to seal the fluid in a pressure-tight container. As a result, some of the fluid is continually boiling. This maintains the rest of the fluid at a very low temperature. The vapor from the boiling fluid is allowed to escape.

In general, these fluids for low temperature application are expendable. This means that they are used but once and the vapor is vented to the atmosphere.

There are certain cautions which must be observed by anyone handling these fluids:

Do not attempt to use any of these fluids in any container or mechanism which was not designed for its use.

An additional caution: Cryogenic fluids must never be allowed to touch the skin. Such a contact would result in immediate freezing of the flesh. Persons handling cryogenic fluid must have their entire body protected by suitable clothing, helmets, gloves, and the like.

See Para. 1-56 and Fig. 1-29, also Chapter 30, for further information on cryogenic fluids and temperatures.

9-22 REFRIGERANT CYLINDERS

There are three types of refrigerant cylinders:
1. Storage cylinder.
2. Returnable service cylinder.
3. Disposable (throw-away) cylinder.

Cylinders are made of steel or aluminum. The larger ones usually have a fusible plug safety device threaded into the concave bottom as a protection against overheating or excessive pressures. A valve at the top provides a connection for charging or discharging service cylinders.

Regulations are prescribed by the Department of Transportation (DOT) to insure the safety of those working with cylinders containing refrigerants.

The DOT regulation requires that cylinders which have contained a corrosive refrigerant must be checked every five years. Cylinders containing noncorrosive refrigerants must be checked every 10 years. All cylinders over a 4 1/2 in. diameter and 12 in. long must contain some type of pressure release protective device. This can be a fusible plug or a spring-operated relief valve, for example.

Never recharge a disposable service cylinder. It may explode.

9-23 STORAGE CYLINDERS

It is cheaper to purchase refrigerants in 100 and 150 lb. cylinders. These become storage cylinders which are frequently positioned upside down with the valve at the bottom. This makes charging service cylinders much easier.

The transferring of refrigerant from the large cylinder to the smaller service cylinder should be done carefully and a record kept of the quantity of refrigerant removed. To the total figure, add three percent to account for vapor losses.

The storage cylinders should be dated and stamped with a DOT stamp (Department of Transportation). No cylinder should be used beyond six years from the date on it. Refrigerant manufacturers request that all cylinders be returned to them every six months, or more often, so that the valve fittings and the complete cylinder may be carefully checked. This service helps to assure safe cylinders.

Storage cylinders are fitted with a valve and usually a protective cap, which may be screwed over the valve for shipment. This is shown in Fig. 9-20.

The cylinder valves, usually of the packed one-way type, should receive the same care as the service valves of a refrigeration system. The packing nut should be kept tight unless the valve is being used. The refrigerant opening should be sealed with a plug or cap when not in use.

It is best to use a hoist to lift and move cylinders which weigh over 35 pounds (16 kg).

9-24 SERVICE CYLINDERS

The service technician carries small returnable service cylinders, 4- to 25-lb., which are used to charge refrigerating systems. The cylinder valve is usually fitted with 1/4 in. male flare. The service cylinders are usually filled from the storage cylinders located at the shop.

CAUTION: Never completely fill a refrigerant cylinder with liquid refrigerant. Allow space for expansion. Liquid refrigerant expands with an increase in temperature. A cylinder completely filled with cold or cool refrigerant will burst if allowed to warm up. The safe limit is 80 percent full.

Service cylinders should be weighed before and after filling. In this way, the amount of refrigerant in the cylinder may be readily determined. During charging, the service cylinder should be placed on an accurate scale. Only the specified weight of refrigerant should be charged into it.

9-25 RETURNABLE SERVICE CYLINDERS

Most refrigeration supply houses provide service cylinders on an exchange basis. Empty cylinders are returned and full ones are provided as a replacement. The supply house then arranges to refill the empty service cylinder.

Fig. 9-20 illustrates several common refrigerant storage cylinders. Some refrigerant service cylinders are fitted with a carrying handle. See Fig. 9-20B and C.

A new liquid-vapor valve, Fig. 9-21, is available on standard size cylinders. This valve enables the service technician to charge a system in the usual manner from a standard valve fitting either as a vapor or as a liquid without inverting the cylinder.

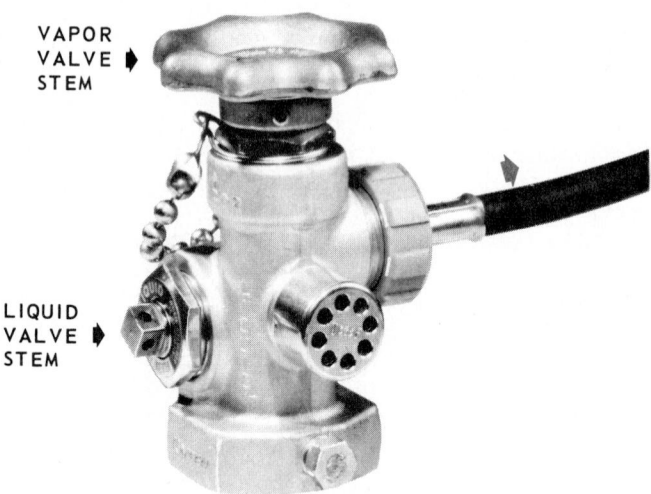

Fig. 9-21. Refrigerant storage cylinder has valve that uses two valve stems. Refrigerant may be drawn from cylinder either as a liquid or vapor by using correct valve stem.

To transfer refrigerant either as a liquid or vapor, use one or the other of the two valve stems. The vapor valve stem is located at the top of the valve and is marked "vapor." The liquid valve stem is located on the side of the valve and is marked "liquid." It is attached to a tube which extends inside to the bottom of the cylinder. The valve has only one valve stem wheel to prevent opening both stems

Fig. 9-23. Disposable refrigerant cylinder. A—Handle and safety guard for refrigerant valve. B—Refrigerant valve.
(Du Pont Company)

at the same time. A standard hood cap is placed over the valve when it is not in use. Fig. 9-22 is a cross section of the liquid-vapor valve.

The reclaiming of refrigerants has increased the use of returnable cylinders. See Chapter 28.

9-26 DISPOSABLE CYLINDERS

Many popular refrigerants are available in small quantities—from a few ounces up to 50 lb.—in "throwaway" (disposable) cylinders. These containers are easy to handle and they eliminate the problem of refilling. Fig. 9-23 illustrates a popular disposable cylinder.

Most disposable cylinders are fitted with relief valves. Usually these are located in the valve body. Some "throwaway" refrigerant containers are sealed cans. The top is made in such a way that a special service valve can be tightly clamped to the top of the can. This valve, when clamped

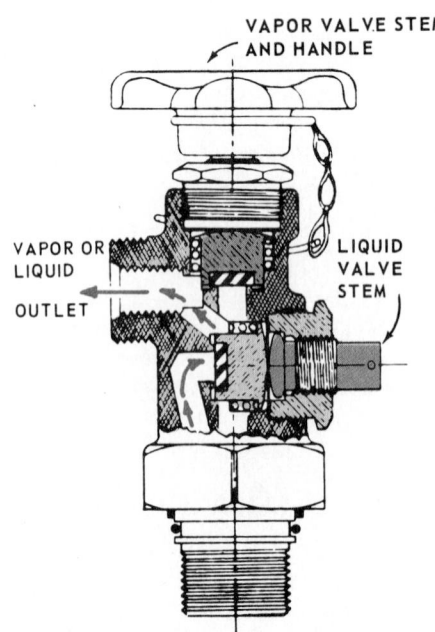

Fig. 9-22. Cross section of a refrigerant cylinder valve which allows either refrigerant liquid or refrigerant vapor to leave the cylinder.

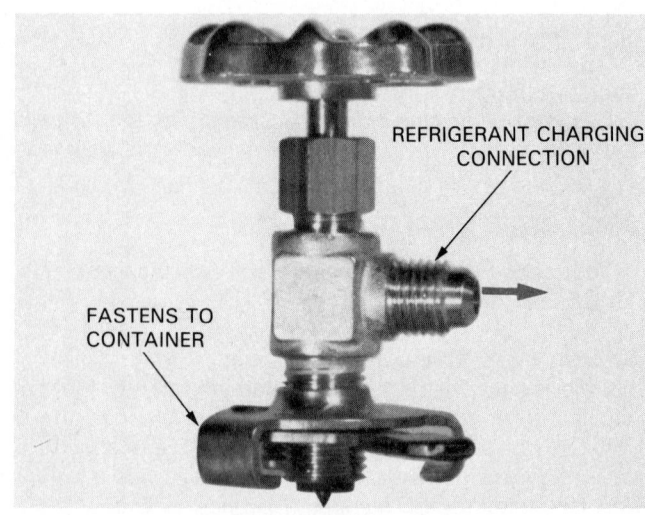

Fig. 9-24. Hand valve which may be attached to throw-away refrigerant containers. Valve clamps to top of refrigerant container.

on the can, can be made to puncture it and provide a means of drawing refrigerant from the can. Fig. 9-24 illustrates such a valve.

CAUTION: Disposable cylinders should not be recharged nor used to store refrigerant removed from a system.

9-27 COLOR CODE FOR REFRIGERANT CYLINDERS

Often, cylinders used for transporting refrigerants are color coded to permit easy identification of the refrigerants in the cylinders. This practice helps to prevent accidental mixing of refrigerants within a system.

However, one must always read the label and identify the refrigerant before using the cylinder. The color code shown is not a requirement for all manufacturers. Popular refrigerants, with their R-number and cylinder color code, are shown in Fig. 9-25.

NUMBER	REFRIGERANT NAME	COLOR
R-11	Trichloromonofluoromethane	Orange
R-12	Dichlorodifluoromethane	White
R-13	Monochlorotrifluoromethane	Pale Blue
R-22	Monochlorodifluoromethane	Green
R-113	Trichlorotrifluoroethane	Purple
R-114	Dichlorotetrafluoroethane	Dark Blue
R-123	Dichlorotrifluoroethane	Light (silver) Gray
R-134a	Tetrafluoroethane	Light (sky) Blue
R-500	Refrigerant 152A/12	Yellow
R-502	Refrigerants 22/115 (48.8% R-22/51.2% R-115)	Orchid
R-503	Refrigerants 23/13 (40.1% R-23/59.9% R-13)	Aquamarine
R-717	Ammonia	Silver

Fig. 9-25. Cylinder color code for some common refrigerants.

9-28 HEAD PRESSURES (HIGH SIDE)

Pressures will vary with refrigerants. In air-cooled condensers, the head pressure should correspond to a temperature between 30 °F (17 °C) and 35 °F (19 °C) higher than the ambient (surrounding) temperature or the temperature of the air passing over the condenser.

In a water-cooled condenser, the head pressure should correspond to a temperature 15 °F (8 °C) to 20 °F (11 °C) above the exhaust temperature of the water. Using this information, the correct head pressure for common refrigerants may be found by referring to Figs. 9-3 and 9-18. Read across the chart.

In all cases, the condensing temperature will rise until the heat loss from the condenser equals the heat input into the condenser. If condensing pressure is too high, the compressor has to work too hard. Too much vapor will be left in the compressor clearance pocket. This lowers its volumetric efficiency. The temperature of the exhaust vapor will be too high and may cause oil deterioration. Usual causes of above-normal head pressures are listed below:

1. A noncondensable vapor or gas trapped in the condenser—air, for example. Due to Dalton's Law, the head pressure will be the sum of the refrigerant vapor pressure plus the air pressure.

2. An overcharge of refrigerant in systems using a low-side float, an expansion valve, or a thermostatic expansion valve. Some of the heat-radiating space in the condenser will fill with liquid refrigerant and reduce the condenser's heat-radiating ability.

3. Either the inside or outside of the condenser is dirty. This dirt will act as an insulator, lowering the heat-radiating capacity of the condenser. Then, the condenser temperature will rise.

4. If the air movement or the water movement through the condenser is reduced by blocked passages or poor water flow, there will not be enough heat-removing material to remove the heat from the condenser.

5. A restriction in the system, for example, a clogged capillary tube or a stuck refrigerant control, may temporarily cause a high head pressure.

6. If the low-side pressure is above normal, the head pressure will be higher than normal.

9-29 REFRIGERATOR TEMPERATURES

The low-side pressure in a refrigerating system determines the temperature in the evaporator.

One must first determine the temperature that is wanted in the cabinet or fixture, then adjust the motor control until this temperature is maintained. However, there are many cases where both a certain evaporator temperature and a cabinet temperature relationship should exist.

Cabinet temperatures are fairly standard. Fig. 9-26 shows recommended fixture (cabinet) temperatures for some common fixtures (cabinets).

FIXTURE (Cabinet)	TEMP. °F	TEMP. °C
Back Bar	37—40	3—4
Beverage Cooler	37—40	3—4
Beverage Precooler	35—40	2—4
Candy Case (Display)	60—65	16—18
Candy Case (Storage)	58—65	15—18
Dairy Display Case	36—39	2—3
Double Display Case	36—39	2—3
Delicatessen Case	36—40	2—4
Dough Retarding Refrigerator	34—38	1—3
Florist Display Refrigerator	40—50	4—10
Florist Storage Case	38—45	3—7
Frozen Food Cabinet (Closed)	−10 to −5	−23 to −21
Frozen Food Cabinet (Open)	−7 to −2	−22 to −19
Grocery Refrigerator	35—40	2—4
Retail Market Cooler	34—39	1—3
Pastry Display Case	45—50	7—10
Restaurant Service Refrigerator	36—40	2—4
Restaurant Storage Cooling	35—39	2—3
Top Display Case (Closed)	35—42	2—6
Vegetable Display Refrigerator (Closed)	38—42	3—6
(Open)	38—42	3—6

Fig. 9-26. Recommended fixture (cabinet) temperatures.

The recommended temperature for various applications is shown in Fig. 9-27.

It is necessary to have the correct size evaporator for the temperature desired. If the evaporator is too large, temperature will be above normal. If the evaporator is undersized, temperature will be below normal.

The evaporator will have a lower temperature than the fixture temperature (a temperature difference is needed for heat flow).

APPLICATION	TEMP. °F	TEMP. °C
Service	34–38	1–3
Meats	30–34	−1 to 1
Bananas	60–65	16–18
Fresh Meats	28–32	−2 to 0
Aging Room	30–34	−1 to 1
Chill Room	35–39	2–3
Curing Room	32–36	0–2
Freezer Room	−15	−26
Poultry	30–34	−1 to 1
Vegetables, Fresh	36–42	2–6
Ice Cream Hardening	−25	−32
Ice Cream Storage	−20 to −10	−29 to −23
Plants and Flowers	38–50	3–10
Fur Storage	33–37	0–3
Locker Room	−5 to 0	−21 to −18

Fig. 9-27. Recommended temperatures for various refrigeration applications.

Normally, the refrigerant will be 10 °F (6 °C) colder than the evaporator temperature when the unit is running. The refrigerant and the evaporator will become the same temperature during the off cycle.

The evaporator surface temperature depends on its size and the rate at which heat is being removed from the fixture.

The temperature of a typical frosting type evaporator (domestic type) will vary from 0 °F to 25 °F (−18 °C to −4 °C), and the refrigerant temperature will be about 10 °F (6 °C) lower than this or in the range of −10 °F to 15 °F (−23 °C to −9 °C) while the unit is running.

The table in Fig. 9-28 gives the pressure corresponding to the evaporating temperatures and condensing temperatures for five popular refrigerants. The recommended average low-side pressure is shown under a blue tint opposite ''Ev.'' The recommended high-side pressure is shown with blue opposite ''WC'' for water-cooled systems and opposite ''AC'' for air-cooled systems.

To change psig to the metric kPa, first change psig to psia by adding 14.7 to the psig value. Then divide the psia value by 14.7 and multiply by 101.3. Most technicians just multiply psia by 6.9.

Example: R-12 at 5 °F (−15 °C) has a pressure of 11.8 psig. The pressure in psia is: 11.8 + 14.7 = 26.5 psia. The metric pressure is:

$$\frac{26.5}{14.7} \times 101.3 = 182.6 = 183 \text{ kPa}$$

Using 6.9 directly: 26.5 × 6.9 = 182.9 = 183 kPa.

9-30 USE OF PRESSURE-TEMPERATURE TABLES

The pressure-temperature relationship of the refrigerants under saturated conditions (holding as much vapor as they can) can be shown in tables.

The table can also be used to show the volume of one pound of the vapor at that temperature, as well as the latent heat, the specific heat, and the density of the liquid.

Tables of these values for some popular refrigerants are shown in Figs. 9-5, 9-9, 9-12, 9-14, 9-16, and 9-19. These values are of great value to a service technician and an engineer.

To use the tables, find in the left-hand, vertical column, the temperature being investigated. Then move across to the other columns horizontally to determine the pressure.

9-31 REFRIGERANT APPLICATIONS

Some popular refrigerant applications are shown in Fig. 9-29. One type of refrigerant may be used in a number of applications. Some refrigerant applications recommended for different types of compressors are shown in Fig. 9-30.

The selection of the type of refrigerant to be used in a given system is determined by the manufacturer. Several items are considered in the selection of the refrigerant:
1. The capacity, governed by the refrigerant boiling point.

TEMPER-ATURE °F	REFRIGERANT—CODE				
	12-F	22-V	502-R	717-A	500-D
−60	19.0	11.9	7.2	18.7	17.0
−55	17.3	9.2	3.9	16.7	15.0
−50	15.4	6.1	0.2	14.4	12.8
−45	13.3	2.7	1.9	11.8	10.4
−40	11.0	0.6	4.1	8.8	7.6
−35	8.4	2.6	6.5	5.5	4.6
−30	5.5	4.9	9.2	1.7	1.2
−25	2.3	7.5	12.1	1.2	1.2
−20	0.6	10.2	15.3	3.5	3.2
−18	1.3	11.4	16.7	4.5	4.1
−16	2.1	12.6	18.1	5.6	5.0
−14	2.8	13.9	19.5	6.7	5.9
−12	3.7	15.2	21.0	7.8	6.8
−10	4.5	16.5	22.6	9.0	7.8
−8	5.4	17.9	24.2	10.2	8.8
−6	6.3	19.4	25.8	11.5	9.9
−4	7.2	20.9	27.5	12.8	11.0
−2	8.2	22.4	29.3	14.2	12.1
0	9.2	24.0	31.1	15.6	13.3
1	9.7	24.8	32.0	16.4	13.9
2	10.2	25.7	32.9	17.1	14.5
3	10.7	26.5	33.9	17.9	15.1
4	11.3	27.4	34.9	18.7	15.7
Ev 5	11.8	28.3	35.9	19.5	16.4
6	12.4	29.1	36.9	20.3	17.0
7	12.9	30.0	37.9	21.1	17.7
8	13.5	31.0	38.9	22.0	18.3
9	14.1	31.9	39.9	22.8	19.0
10	14.7	32.8	41.0	23.7	19.7
11	15.3	33.8	42.1	24.6	20.4

TEMPER-ATURE °F	REFRIGERANT—CODE				
	12-F	22-V	502-R	717-A	500-D
12	15.9	34.8	43.2	25.5	21.1
13	16.5	35.8	44.3	26.4	21.9
14	17.1	36.8	45.4	27.4	22.6
15	17.7	37.8	46.5	28.3	23.3
16	18.4	38.8	47.7	29.3	24.1
17	19.0	39.9	48.9	30.3	24.9
18	19.7	40.9	50.0	31.3	25.6
19	20.4	42.0	51.2	32.4	26.4
20	21.1	43.1	52.5	33.4	27.2
21	21.8	44.2	53.7	34.5	28.0
22	22.5	45.3	54.9	35.5	28.9
23	23.2	46.5	56.2	36.6	29.7
24	23.9	47.6	57.5	37.7	30.6
25	24.6	48.8	58.8	38.8	31.4
26	25.4	50.0	60.1	40.0	32.3
27	26.2	51.2	61.5	41.2	33.2
28	26.9	52.4	62.8	42.4	34.1
29	27.7	53.7	64.2	43.7	35.0
30	28.5	54.9	65.6	44.9	36.0
31	29.3	56.2	67.0	46.1	36.9
32	30.1	57.5	68.4	47.4	37.9
33	30.9	58.8	69.9	48.7	38.8
34	31.8	60.2	71.3	50.0	39.8
35	32.6	61.5	72.8	51.4	40.8
36	33.5	62.9	74.3	52.7	41.8
37	34.3	64.3	75.9	54.1	42.9
38	35.2	65.7	77.4	55.5	43.9
39	36.1	67.1	79.0	57.0	45.0
40	37.0	68.6	80.5	58.4	46.0
41	37.9	70.0	82.1	59.2	47.1

TEMPER-ATURE °F	REFRIGERANT—CODE				
	12-F	22-V	502-R	717-A	500-D
42	38.9	71.5	83.8	61.4	48.2
43	39.8	73.0	85.4	62.9	49.3
44	40.8	74.5	87.0	64.5	50.4
45	41.7	76.1	88.7	66.1	51.6
46	42.7	77.6	90.4	67.6	52.7
47	43.7	79.2	92.1	69.3	54.0
48	44.7	80.8	93.9	70.9	55.1
49	45.7	82.4	95.6	72.6	56.3
50	46.7	84.1	97.4	74.3	57.5
55	52.1	92.6	106.6	83.2	63.9
60	57.8	101.6	116.4	92.6	70.6
65	63.8	111.3	126.7	102.8	77.7
70	70.2	121.5	137.6	113.8	85.3
75	77.0	132.2	149.1	125.5	93.4
80	84.2	143.7	161.2	138.0	101.9
WC 85	91.7	155.7	174.0	151.4	110.9
90	99.7	168.4	187.4	165.5	120.5
95	108.1	181.9	201.4	180.6	130.5
100	117.0	196.0	216.2	196.7	141.1
AC 105	126.4	210.8	231.7	213.9	152.2
110	136.2	226.4	247.9	231.8	164.0
115	146.5	242.8	264.9	251.0	176.3
120	157.3	260.0	282.7	271.1	189.2
125	168.6	278.1	301.4	292.5	202.8
130	180.5	297.0	320.8	314.9	217.0
135	192.9	316.8	341.2	338.8	231.8
140	205.9	337.5	362.6	363.5	247.4
145	219.5	359.1	385.0	390.2	263.7
150	233.7	381.7	408.4	417.4	280.7
155	248.6	405.4	432.9	447.0	298.5

Vacuum-Inches of Mercury Italic Figures

Pressure-Pounds Per Square Inch Gage Bold Figures

Fig. 9-28. Temperature-pressure chart which may be used to determine operating pressures for various temperatures and for various refrigerants. Ev—Average low-side pressures. WC—Average water-cooled system head pressures. AC—Average air-cooled system head pressures. (Sporlan Valve Co.)

BOILING POINT AT ATMOSPHERIC PRESSURE: °F (°C)	REFRIGERANT	APPLICATION
		HIGH TEMPERATURE
118 °F (48 °C)	R-113	Used in low capacity centrifugal packaged units for commercial and industrial air conditioning and chilling. Also used for waste heat recovery in Organic Rankine Cycle engines. Operates at very low pressures and high gas volumes.
75 °F (24 °C)	R-11	Used in centrifugal packaged units at somewhat higher system pressure and capacity than R-113. Also used as a secondary coolant in low temperature systems and for waste heat recovery.
82 °F (28 °C)	R-123	Replacement for many R-11 applications.
39 °F (4 °C)	R-114	Used in high capacity multi-stage centrifugal and rotary systems operating at intermediate pressure and displacement; to improve heat transfer in solar water heaters; and to reduce evaporator temperature in process chillers.
		MEDIUM TEMPERATURE
−22 °F (−30 °C)	R-12	Most widely used refrigerant for air conditioning and refrigeration. It is the principal refrigerant for automotive air conditioning.
−16 °F (−27 °C)	R-134a	Replacement for many R-12 applications.
−28 °F (−33 °C)	R-500	Used in place of R-12 in increase capacity at the same compressor displacement. It is an azeotrope of R-12 and FC-152a (73.8/26.2% by weight).
−41 °F (−41 °C)	R-22	Used in air conditioners and heat pumps for residential and commercial applications and in refrigeration systems. Operates at higher pressures and lower compressor displacement than R-12.
−50 °F (−46 °C)	R-502	Used in supermarket freezers and refrigerated cases. Operates at lower compressor discharge temperatures than R-22, it provides lower compression ratios and discharge temperatures and higher capacity.
−72 °F (−58 °C)	R-13B1	Used in medium to low-temperature applications with one or two stages of compression. Compared to R-22, it provides lower compression ratios and discharge temperatures and higher capacity.
		LOW TEMPERATURE
−109 °F (−78 °C)	R-116	Used in specialty low temperature applications.
−115 °F (−82 °C)	R-13	Used to produce evaporator temperatures as low as −100 °F (−73 °C) in the low temperature stage of cascade refrigeration systems that employ other FREON refrigerants in the high temperature cycle.
−128 °F (−89 °C)	R-503	Used to improve compressor capacity and low temperature capability in the second stage of cascade systems that employ R-502, R-12, or R-22 in the first stage. It is an azeotrope of R-23 and R-13 (40.1 / 59.9% by weight).
−198 °F (−128 °C)	R-14	Used with reciprocating compressors to produce evaporator temperatures down to −200 °F (−129 °C) in the third stage of triple cascade systems.

Fig. 9-29. Some popular refrigerant applications.

REFRIGERANT	COMPRESSOR TYPE	APPLICATION
R-11	Centrifugal	Large air conditioning systems ranging from 200 to 2000 tons in capacity. Refrigerating systems for industrial process water and brine.
R-123	Centrifugal	Replacement for many R-11 applications.
R-12	Reciprocating Centrifugal Rotary	Large air conditioning and refrigeration systems. Small household refrigerators including frozen food and ice cream cabinets, food locker plants, water coolers, room and window air conditioners and others. Principal refrigerant in automobile air conditioning.
R-134a	''	Replacement for many R-12 applications.
R-22	Reciprocating Centrifugal Rotary	Residential and commercial air conditioning. Food-freezing plants, frozen-food storage and display cases and many other medium and low-temperature applications.
R-500	Reciprocating Centrifugal	Small home and commercial air conditioning equipment. Household refrigeration and commercial chillers.
R-502	Reciprocating	Frozen food and ice cream display cases, warehouses and food freezing plants. Medium-temperature display cases, truck refrigeration, and heat pumps.
R-503	Reciprocating	Low-temperature systems down to about −130 °F (−90 °C).
R-13	Reciprocating	Low-temperature systems down to about −130 °F (−90 °C) in cascade systems.
R-113	Centrifugal	Small to medium air conditioning systems. Industrial cooling, food freezing, and storage.

Fig. 9-30. Refrigerants suitable for different applications of reciprocating, centrifugal, and rotary compressors.

2. The volume of the vapor pumped to provide the necessary refrigeration.
3. The latent heat of the refrigerant.
4. The operating temperatures required.
5. The size of the equipment.

9-32 REFRIGERATION OIL

Oil circulates through the system with the refrigerant. Oil gives lubrication and cools the compressor's moving parts.

Refrigerant oil must have certain properties, because it is mixed with the refrigerant. The oil comes in direct contact with hot motor windings in hermetic units. Thus, it must be able to withstand extreme temperatures and be harmless to refrigerants and equipment.

Oil in the refrigeration system is cooled to low temperatures. Yet it must be able to withstand high temperatures at the compressor. It must remain fluid in all parts of the system.

The fluidity of an oil-refrigerant mixture is determined by the refrigerant used, the temperature, properties of the oil, its solubility in refrigerant, and the solubility of refrigerant in the oil (to keep oil fluid at low temperatures).

The properties of a good refrigerant oil are:
1. Low wax content. Separation of wax from the refrigerant oil mixture may plug refrigerant control orifices (openings).
2. Good thermal stability. It should not form hard carbon deposits at hot spots in the compressor (such as valves or discharge ports).
3. Good chemical stability. There should be little or no chemical reaction with the refrigerant or materials normally found in a system.
4. Low pour point. Ability of the oil to remain in a fluid state at the lowest temperature in the system.
5. Low viscosity. This is the ability of the oil to maintain good oiling properties at high temperatures and good fluidity at low temperatures; to provide a good lubricating film at all times.

SERVICE CONDITION	REFRIGERANT	VISCOSITY
Compressor Temperature Normal	All	150 150/additives
High	Halogen Ammonia	150/additives 300 300/additives
Evaporator Temperature Above 0°F (−18°C)	Halogen	150 150/additives
	Ammonia	300
0°F to −40°F (−18°C to −40°C)	Halogen	150 150/additives
	Ammonia	150 150/additives
Below −40°F (−40°C)	Halogen	150 150/additives
	Ammonia	150 150/additives
Automotive Compressors	Halogen	500

Fig. 9-31. Refrigerant oils should be chosen according to compressor temperature, evaporator temperature, and kind of refrigerant used.

Looking to improve the performance of the oil, many manufacturers add chemicals which are designed to inhibit (slow down or stop) the formation of sludge or foaming. (Oil which contains moisture or air will form sludge or varnish and may cause damage to the unit.)

Oil removed from a system should be clear. Discoloration means that it is impure. When this has happened, new driers and filters should be placed in the system to keep the new oil clean.

Another indicator of contaminated oil is odor. Dirty oil from a hermetic system may be acidic and will burn the hands. For additional information on oil, see Chapters 2, 11, and 14.

Only oil recommended by manufacturers of the equipment should be used. When the manufacturer's recommendation cannot be found, the viscosity of oil indicated in Fig. 9-31 may be used for most applications. Refrigerant oil containers must always be kept tightly sealed. Oil exposed to the atmosphere will absorb moisture.

9-33 MOISTURE IN REFRIGERANT

Moisture (water) in a system may freeze at the refrigerant control and clog or partly clog it. Moisture and some refrigerants in the presence of high compressor temperatures may cause the refrigerant to break down and may form harmful acids. It may cause rusting, corrosion, or oil sludging, which could result in a motor burnout in hermetic systems.

Refrigerants must be kept in sealed containers and must be kept completely dry. Methods of drying refrigerants are explained in Chapters 11, 14, and 30.

Moisture indicators placed in the liquid line will help the service technician determine if moisture is present in the refrigerant. The indicator is a chemical which changes color depending upon the amount of moisture present. A sight glass or "window" in the liquid line helps one see the change in color.

It may be impossible to remove all moisture from a refrigerant. However, the amount of moisture must be kept very low. The maximum amount of moisture which may be allowed in a system varies with the kind of refrigerant and the low-side temperature.

Most refrigerant manufacturers supply refrigerants that are dry. The moisture content never exceeds five parts per million (ppm). Liquid refrigerants can hold more moisture (in solution) as the low-side temperature rises.

This enables the refrigerant to circulate without danger of the moisture separating from the refrigerant and freezing or forming harmful compounds. For example, R-12 is safe to use at 20°F (−7°C) with 17 ppm, at 0°F (−18°C) with 8.3 ppm, at −20°F (−29°C) with 3.8 ppm, and at −40°F (−40°C) with only 1.7 ppm.

A table of safe moisture content for certain refrigerants is shown in Fig. 9-32. Any amount of water at or above the value of the "wet color" will be harmful to the system. The service technician must depend on the moisture indicator to determine the amount of moisture in the system.

If the moisture indicator shows a "wet" color, a new drier should be placed in the line and the system operated until the moisture indicator indicates a "dry" color. It may sometimes be necessary to replace the drier several times to dry up the system.

When servicing, avoid exposing cold internal parts of the system to air. Moisture from the air will condense on the parts and get inside the system. Warm the parts to room temperature with a heat lamp before opening the system.

REFRIGERANT	DRY COLOR	WET COLOR
12	Below 5	Above 15
22	Below 30	Above 100
502	Below 15	Above 50

Fig. 9-32. Safe and unsafe moisture content for three types of refrigerants. The water concentration is shown in parts per million at 75 °F (24 °C). Dry color column shows allowable concentration. Wet color column shows concentrations that will cause trouble. When these quantities are present, water will start freezing in low side and may block automatic expansion valve.

9-34 REFRIGERANT IDENTIFICATION

Basically, refrigerants are identified by the use of a pressure-temperature chart. Identifying them by smell or color is very difficult. The two exceptions are R-764 which is sulphur dioxide and R-717 which is ammonia. Sniffing some refrigerants can even cause death.

The Airserco Refrig-I-Dent, Fig. 9-33, may be used to help identify refrigerants. With this instrument, the service technician may test and identify an unknown refrigerant either from an operating system or from a refrigerant cylinder. It combines a thermometer and pressure gauge attached to a special manifold. A pressure-temperature chart is also supplied.

The refrigerant sample taken from the system or cylinder is trapped in the manifold of the instrument. The pressure and temperature registered by the instrument is then compared with the pressure-temperature chart. The "R" number of the refrigerant is determined.

9-35 CHANGING REFRIGERANTS

It is not good practice to change the kind of refrigerant in a unit. Engineering features designed into the unit to make it efficient may be wasted.

If for some reason the refrigerant must be changed, it is best to use an expansion valve designed for the new refrigerant. If the unit uses a capillary tube, and the density or pressure differences are changed, the tube should be replaced to match the new refrigerant.

Be sure to clean the mechanism thoroughly and replace the oil. A new filter-drier should be installed when a system is changed from one refrigerant to another.

9-36 AMOUNT OF REFRIGERANT REQUIRED IN A SYSTEM

The amount of refrigerant that should be used varies with the type of system. The low-side float, automatic expansion valve, and thermostatic expansion valve systems are not sensitive to the amount of refrigerant in the system. High-side float systems and capillary tube systems are very sensitive to the amount of refrigerant charge.

Fig. 9-33. This instrument helps identify the refrigerant used in a system.

A sure way to determine if the system has sufficient refrigerant is to put a sight glass in the liquid line. The appearance of vapor bubbles in this glass is a sign that the system is short of refrigerant. It should be charged until the vapor bubbles disappear.

Methods of charging refrigerant into different types of systems are explained in Chapters 11 and 14.

9-37 NEW REFRIGERANTS AND OZONE PROTECTION

A ruling by the Environmental Protection Agency (EPA), Title VI of Clean Air Act Amendments of 1990, states fully halogenated refrigerants (CFCs) will be phased out. It also calls for the phase-out of HCFCs by the year 2030.

Currently HCFC-22 is the principle refrigerant in unitary air conditioners, and is also used in non-industrial heat pumps and positive displacement chillers. It will likely remain in use for some years to come. CFC-11 and CFC-12, primarily used in chillers, residential and automotive refrigeration, may be replaced by HFC-134a, tetrafluoroethane, and HCFC-123, dichlorotrifluoroethane. Additional future replacement refrigerants include R-124, chlorotetrafluoroethane, replacing R-114, used in marine chillers; and R-125, pentafluoroethane, replacing R-502, used in stores and supermarkets.

The thermodynamic properties of R-12 and R-134a are very similar. The major difference is that 134a has no harmful influence on the ozone layer of the earth's atmosphere. The ozone depletion potential of HFC-134a is nil because it contains no chlorine. Therefore, it is considered completely safe concerning the ozone depletion problem.

However, new refrigerants cannot simply be dropped into a system designed to use CFCs. If the replacement

Temp. °F	Pressure		Volume cu.ft./lb.	Density lb./cu.ft.	Heat Content (Enthalpy) Btu/lb.	
	Absolute	Gauge	Vapor	Liquid	Liquid	Vapor
−150	0.07107	29.776*	457.0719	102.344	−32.781	80.212
−125	0.28333	29.344*	123.4418	99.641	−25.383	83.716
−100	0.89915	28.090*	41.5241	96.891	−17.939	87.245
−75	2.3866	25.062*	16.5646	94.087	−10.472	90.760
−50	5.4966	18.730*	7.5560	91.220	−2.995	94.248
−25	11.2964	6.9214*	3.8338	88.278	4.503	97.721
−15	14.6686	0.0555*	2.9960	87.078	7.518	99.109
−10	16.6293	1.9334	2.6610	86.472	9.030	99.804
−5	18.7906	4.0947	2.3705	85.862	10.546	100.499
0	21.1665	6.4706	2.1177	85.248	12.067	101.195
5	23.7710	9.0751	1.8969	84.630	13.593	101.891
10	26.619	11.923	1.7036	84.007	15.125	102.587
25	36.773	22.078	1.25178	82.110	19.763	104.677
50	60.032	45.335	0.78067	78.836	27.666	108.149
75	93.080	78.384	0.50743	75.387	35.851	111.553
86	111.321	96.626	0.42412	73.799	39.560	113.004
100	138.28	123.58	0.33993	71.701	44.393	114.782
125	198.27	183.57	0.23204	67.679	53.385	117.660
150	276.12	261.42	0.15914	63.126	62.989	119.879
175	375.69	360.99	0.10705	57.601	73.581	120.788
200	502.54	487.85	0.06542	49.439	86.528	118.155

*Inches of mercury below one atmosphere.

Fig. 9-34. Properties of liquid and saturated vapor of refrigerant 134a. Note pressures corresponding to standard evaporating temperature of 5 °F (−15 °C) and condensing temperature of 86 °F (30 °C). (ICI Americas, Inc.)

refrigerant is not identical to the original, the CFC system may be inefficient or unable to operate properly. A key determining factor is the compatibility of the new refrigerant with the materials of construction in the CFC system. As an example, it may be necessary to replace the elastomers or the plastic components throughout a system when new

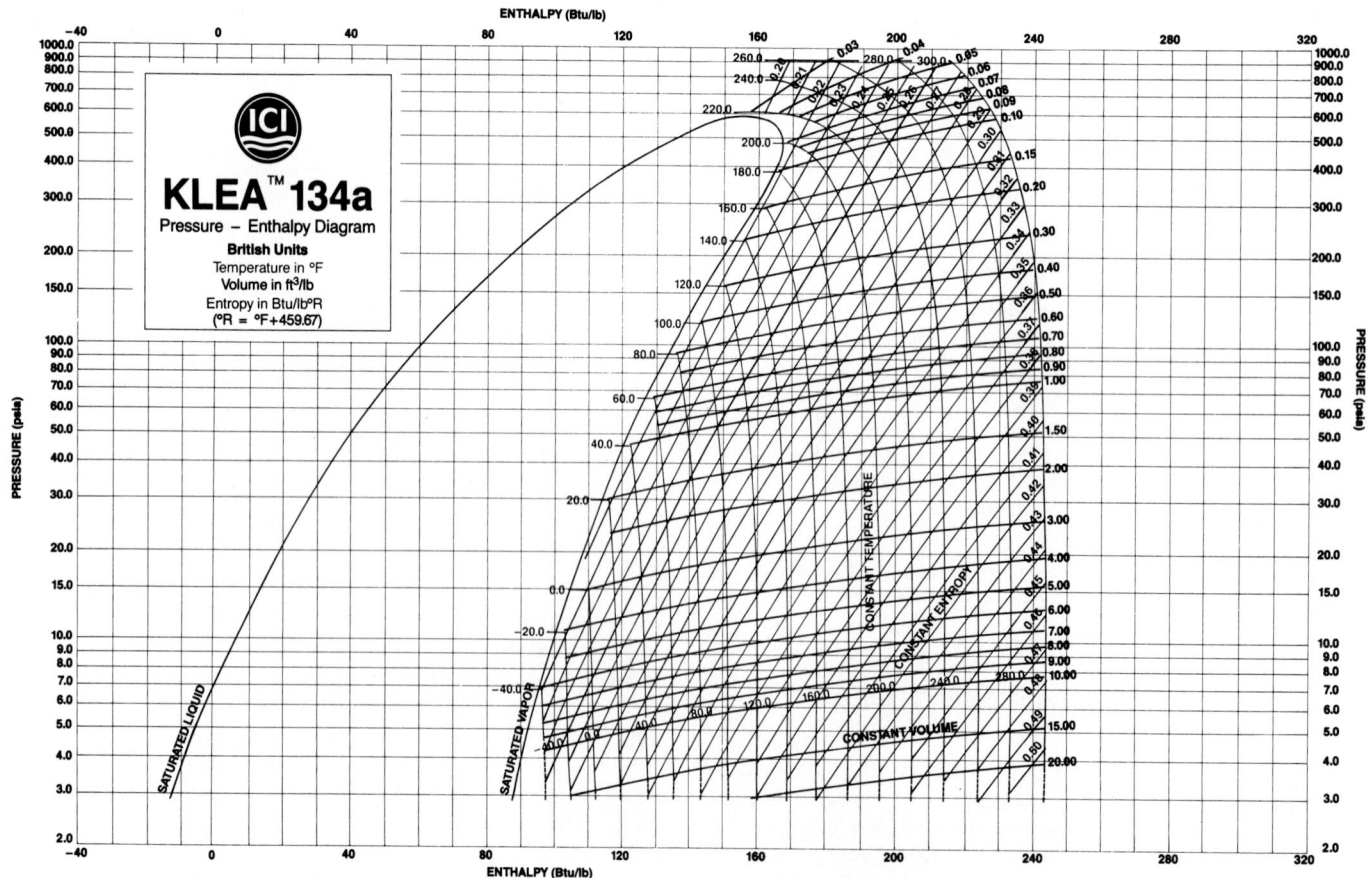

Fig. 9-35. Pressure-enthalpy diagram for KLEA™ 134a. (ICI Americas, Inc.)

refrigerants are used. The replacement of oil is also a concern. Therefore, the service technician cannot simply place a new refrigerant into the system without checking with the system's manufacturer.

Refrigerant 134a is not compatible with the present mineral-based refrigerant oils and lubricants used for R-12. Specific oils will be identified for use with it. Numerous design changes have been developed and will be implemented for use, such as condenser and evaporator sizing (possible 30% increase), desiccant type (from silicone gel to molecular sieve), hose design (134a will require smaller hoses), and control pressure regulations (30% increase). Many component parts are being manufactured and are available to use and replacement.

R-134a is presently being used as a standard refrigerant in vehicular air conditioning. It has been identified as a substitute for a wide range of air conditioning and refrigeration systems in residential, commercial, and industrial applications. The properties of the liquid and saturated vapor of R-134a are shown in Fig. 9-34.

Pressure-heat enthalpy diagram for 134a is shown in Fig. 9-35. See Para. 1-54 for an explanation of enthalpy. The color code for R-134a cylinders is light blue.

HCFC-123, Fig. 9-36, is being considered as a replacement for CFC-11. It is similar to R-11 in that it is a low-boiling-point nonflammable liquid with low chemical reactivity. R-11 has been used extensively as a refrigerant in centrifugal chillers and in foam-blowing applications, mainly for polyurethanes.

Molecular Formula	$CHCl_2CF_3$
Molecular Weight	152.91
Normal Boiling Point, (°F)	82.2
Vapor Pressure[a], (psia)	13.24
Liquid Density[a], (lb./cu.ft.)	91.29
Vapor Thermal Conductivity[a] (Btu in/hr ft² °F)	0.0722
Liquid Viscosity[a], (cP)	0.481
Solubility in H_2O[a], (ppm)	2100.
H_2O Solubility[a], (ppm)	662.
Enthalpy of Vaporization (Btu/mol)	24.93
Vapor Flammability[a,b,d]	
Lower limit (vol. %)	none
Upper limit (vol. %)	none
Flash Point[c,d], (°F)	none

[a]Measured at 77 °F.

[b]Flame limits determined using ASTM E 681 with electrically activated match ignition.

[c]No flash point via open cup (ASTM D 1310-67) or closed cup (ASTM D 56-82).

[d]Flammability properties of these materials are not intended to reflect the fire hazards of any resultant cellular or foamed plastic products.

Fig. 9-36. Properties of the liquid and saturated vapor of Refrigerant 123. (Allied Signal, Inc.)

Figs. 9-37 and 9-38 give the pressure corresponding to evaporating temperatures and condensing temperatures for some popular refrigerants and possible replacements.

The laws forbidding release of CFCs to the atmosphere have resulted in the development of procedures and equipment for recovery, recycling, and reclamation of current refrigerants. Information concerning these procedures and equipment can be found in Chapters 14 and 28.

VAPOR PRESSURES

TEMP °F	113	141b	123	11	114	134a	12	500	22	502	13	503
−150.0							29.6	29.5	29.4	29.1	20.9	16.9
−140.0						29.6	29.4	29.2	29.1	28.5	16.8	11.1
−130.0						29.4	29.1	28.8	28.5	27.8	11.5	3.5
−120.0						29.1	28.6	28.3	27.7	26.7	4.5	3.1
−110.0					29.7	28.7	27.9	27.5	26.6	25.3	2.1	9.3
−100.0					29.5	28.0	27.0	26.9	25.1	23.3	7.6	16.9
−90.0				29.7	29.3	27.1	25.8	24.9	23.0	20.6	14.3	26.3
−80.0			29.7	29.6	29.0	25.7	24.1	22.9	20.2	17.2	22.5	37.7
−70.0		29.7	29.6	29.4	28.6	24.0	21.9	20.3	16.6	12.7	32.3	51.3
−60.0		29.5	29.4	29.2	28.0	21.6	19.0	17.0	11.9	7.2	43.9	67.3
−50.0	29.6	29.3	29.2	28.9	27.1	18.6	15.4	12.8	6.1	0.2	57.6	86.1
−40.0	29.5	29.0	28.8	28.4	26.1	14.7	11.0	7.6	0.6	4.1	73.3	107.8
−35.0	29.4	28.8	28.6	28.1	25.4	12.3	8.4	4.6	2.6	6.5	82.2	119.9
−30.0	29.3	28.6	28.3	27.8	24.7	9.7	5.5	1.2	4.9	9.2	91.6	132.8
−25.0	29.2	28.3	28.1	27.4	23.8	6.8	2.3	1.2	7.5	12.1	101.7	146.7
−20.0	29.0	28.1	27.7	27.0	22.9	3.6	0.6	3.2	10.2	15.3	112.5	161.4
−15.0	28.8	27.7	27.3	26.6	21.8	0.0	2.5	5.4	13.2	18.8	123.9	177.1
−10.0	28.7	27.3	26.9	26.0	20.6	2.0	4.5	7.8	16.5	22.6	136.1	193.9
−5.0	28.4	26.9	26.4	25.4	19.3	4.1	6.7	10.4	20.1	26.7	149.1	211.6
0.0	28.2	26.4	25.8	24.7	17.8	6.5	9.2	13.3	24.0	31.1	162.9	230.5
5.0	27.9	25.8	25.2	23.9	16.2	9.1	11.8	16.4	28.3	35.9	177.4	250.5
10.0	27.5	25.2	24.5	23.1	14.4	12.0	14.7	19.7	37.8	46.5	209.1	294.1
15.0	27.2	24.5	23.7	22.1	12.4	15.1	17.7	23.3	37.8	46.5	209.1	294.1
20.0	26.7	23.7	22.8	21.1	10.2	18.4	21.1	27.2	43.1	52.5	226.3	317.8
25.0	26.3	22.8	21.8	19.9	7.8	22.1	24.6	31.4	48.8	58.8	244.4	342.8
30.0	25.7	21.8	20.7	18.6	5.1	26.1	28.5	36.0	54.9	65.6	263.5	369.3
35.0	25.1	20.7	19.5	17.1	2.2	30.4	32.6	40.8	61.5	72.8	283.6	397.2
40.0	24.4	19.5	18.1	15.6	0.4	35.0	37.0	46.0	68.5	80.5	304.8	426.6
45.0	23.7	18.1	16.6	13.8	2.1	40.0	41.7	51.6	76.1	88.7	327.1	457.5
50.0	22.9	16.7	15.0	12.0	3.9	45.4	46.7	57.5	84.1	97.4	350.4	490.2
55.0	21.9	15.1	13.1	9.9	5.9	51.2	52.1	63.8	92.6	106.6	375.0	524.5
60.0	20.9	13.4	11.2	7.7	8.0	57.4	57.8	70.6	101.6	116.4	400.9	560.7
65.0	19.8	11.5	9.0	5.2	10.3	64.0	63.8	77.7	111.3	126.7	428.1	598.7
70.0	18.6	9.4	6.6	2.6	12.7	71.1	70.2	85.3	121.4	137.8	456.8	
75.0	17.3	7.2	4.1	0.1	15.3	78.6	77.0	93.4	132.2	149.1	487.2	
80.0	15.8	4.8	1.3	1.6	18.2	86.7	84.2	101.9	143.7	161.2	519.4	
85.0	14.2	2.3	0.9	3.3	21.2	95.2	91.7	110.9	155.7	174.0		
90.0	12.5	0.2	2.5	5.0	24.4	104.3	99.7	120.5	168.4	187.4		
95.0	10.6	1.7	4.2	6.9	27.8	113.9	108.2	130.5	181.8	201.4		
100.0	8.6	3.2	6.1	8.9	31.4	124.1	117.0	141.1	196.0	216.2		
105.0	6.4	4.8	8.1	11.1	35.3	134.9	126.4	152.2	210.8	231.7		
110.0	4.0	6.6	10.2	13.4	39.4	146.3	136.2	163.9	226.4	247.9		
115.0	1.4	8.4	12.6	15.9	43.8	158.4	146.5	176.3	242.8	264.9		
120.0	0.7	10.4	15.0	18.5	48.4	171.1	157.3	189.2	260.0	282.7		
125.0	2.1	12.4	17.7	21.3	53.3	184.5	168.6	202.7	278.1	301.3		
130.0	3.7	14.6	20.5	24.3	58.4	198.7	180.5	216.9	297.0	320.6		
135.0	5.3	16.9	23.5	27.4	63.9	213.5	192.9	231.8	316.7	341.2		
140.0	7.1	19.3	26.7	30.8	69.6	229.2	205.9	247.4	337.4	362.6		
145.0	9.0	21.8	30.2	34.3	75.6	245.6	219.5	263.7	359.1	384.9		
150.0	11.1	24.4	33.8	38.1	82.0	262.8	233.7	280.7	381.7	408.4		

Fig. 9-37. Temperature-pressure chart may be used to determine operating pressures for various refrigerants and temperatures. Vapor pressures are shown as psig. (Allied Signal, Inc.)

9-38 REVIEW OF SAFETY

When a leak is suspected, make certain that the room is thoroughly ventilated before starting to work on the unit. Always check for recommended operating pressures for each refrigerant and install gauges to find the pressures in the system.

Check refrigerant R-number before charging to avoid mixing refrigerants. If the refrigerant is a fluorocarbon, make certain there are no lighted flames near a system that is suspected of having a bad leak. The refrigerant may break down and produce dangerous gases. Always check the ICC cylinder stamp to make sure it is a safe cylinder.

Wear goggles and gloves at all times, especially when charging or discharging. These will protect the eyes, skin, and hands in case of a sudden leak.

Always charge refrigerant vapor into the low side of the system. Liquid refrigerant entering a compressor may injure the compressor and may cause the unit to burst.

Liquid refrigerant on the skin may freeze the skin surface and cause "frostbite." If this should happen, quickly wash away the refrigerant with water. Treat the damaged surface for "frostbite."

Accidents involving refrigerants should be immediately referred to a doctor.

Make sure that a refrigerant service cylinder is never completely filled with liquid refrigerant. If a service cylinder is completely filled with liquid refrigerant and then allowed to

TEMP (°F)	VAPOR PRESSURES	
	HCFC 124	HCFC 125
-100	29.2*	24.4*
-90	28.8*	21.7*
-80	28.2*	18.1*
-70	27.4*	13.3*
-60	26.3*	7.1*
-50	24.8*	0.3
-40	22.8*	4.9
-30	20.2*	10.6
-20	16.9*	17.4
-10	12.7*	25.6
0	7.6*	35.1
10	1.4*	46.3
20	3.0	59.2
30	7.5	74.1
40	12.7	91.2
50	18.8	110.6
60	25.9	132.8
70	34.1	157.8
80	43.5	186.0
90	54.1	217.5
100	66.2	252.7
110	79.7	291.6
120	94.9	334.3
130	111.7	380.3
140	130.4	430.2
150	151.0	482.1
160	173.6	
170	198.4	
180	225.6	
190	255.1	
200	287.3	
210	322.1	
220	359.9	
230	400.6	
240	444.5	
250	491.8	

Vapor pressure are psig, except (*) which are inches of mercury vacuum.

Fig. 9-38. Temperature pressure chart for some additional refrigerants.

become warm, the hydrostatic pressure inside the cylinder will burst it.

Refrigerant cylinders should always be stored in a cool, dry place. Carefully assemble fittings and tubing to cylinders. Stripped threads are dangerous and costly. Never use refrigerant cylinders for supports or for rollers.

Refrigerant cylinders should be used only for storing the refrigerant marked on the label. Using the cylinder for compressed air is very dangerous. The cylinder may explode when exposed to these high pressures.

Refrigerants R-717 and R-764 are very irritating to the eyes and lungs. The service technician must always avoid exposure to these refrigerants.

Refrigerant oil contained in a hermetic compressor which has had a burnout may be very acidic. This oil should never be allowed to touch the skin. It may cause an acid burn.

If moisture is allowed to enter a refrigerating system, it is likely to cause considerable damage to the system. All parts of refrigerating mechanisms must be kept dry at all times. Containers of oil must always be kept tightly sealed to avoid the possibility of absorbing moisture from the air.

Many refrigerants have no disagreeable odor and it might be possible to work in an area without being aware that there is a considerable amount of refrigerant vapor present. Also, many refrigerants are heavier than air and will replace the air in a room. This can be very dangerous because if the air one breathes does not contain at least 19.1 percent oxygen, the person breathing the mixture will lose consciousness. Instruments are available to warn the person if the oxygen content of the air is below a safe limit. Make sure that every worker in your group knows how to read the instruments which warn of a poor oxygen level. Practice some of the steps which should be taken if a warning is given.

Sniffing of a refrigerant can cause death.

9-39 TEST YOUR KNOWLEDGE

1. What does "halide" mean?
2. How may one test for R-502 leaks?
3. What is a common head pressure for air-cooled R-12 refrigerating systems?
4. What is the method used to locate R-22 leaks?
5. What are the low-side pressures at 5 °F (−15 °C) for R-22 and R-12: In U.S. Conventional units? In metric?
6. What does toxic mean?
7. What oil properties are required with R-12?
8. What is the pressure of R-12 at 105 °F (41 °C) in U.S. conventional units? In metric units?
9. What is the evaporating temperature of R-12 under a pressure of 20 pounds per square inch gauge?
10. What is the pressure of R-22 at 0 °F (−18 °C): In U.S. conventional units? In metric units?
11. Is it advisable to substitute refrigerants in a system?
12. How may one determine the refrigerant temperature in an air-cooled condenser?
13. Name the refrigerants that may be tested for leaks with the halide torch.
14. What will be the condenser temperature and high-side pressure (gauge) in an air-cooled condenser using R-502 while the condensing unit is running, if the room temperature is 75 °F (24 °C)?
15. What is the average head pressure for an air-cooled R-22 system: In U.S. conventional units? In metric?
16. Is the refrigerant temperature in the evaporator the same temperature as the evaporator when the unit is running?
17. What effect does air in the system have on the head pressure?
18. Name two refrigerants that may not be tested with the halide torch.
19. How does one determine the correct head pressure of a water-cooled condenser?
20. What is the meaning of Group 1, Group 2, and Group 3 refrigerants?
21. May carbon dioxide be used as a refrigerant?
22. How many divisions are there in the National Board of Fire Underwriters classification of refrigerants?
23. Which number in the above classification is the least toxic?
24. Must an automatic expansion valve system have an exact amount of refrigerant charge?
25. What is an indicator of moisture (water) in a refrigerant system?
26. Can an R-22 system safely have more moisture in it than an R-12 system?
27. Why is it dangerous to fill a refrigerant cylinder completely full of liquid refrigerant?
28. How will the head pressure be affected by a covering of lint on the air-cooled condenser?
29. What are soap solutions used for in relation to refrigeration service?
30. Why is it necessary to be cautious when handling contaminated refrigerant oil?

Chapter 10

DOMESTIC REFRIGERATORS AND FREEZERS

After studying this chapter, the technician will be able to:

☐ Discuss the construction of domestic refrigerators and freezers.

☐ Describe the mechanisms and cabinets for different types of refrigerators and freezers.

☐ Compare standard circuit diagrams and ladder diagrams of a system.

☐ Discuss differences in circuits for manual defrost and automatic defrost systems.

☐ Demonstrate how to repair damaged cabinet finishes.

☐ Identify the various types of condensing units and evaporators used in domestic systems.

A modern domestic fresh food refrigerator or food freezer consists primarily of three parts:

1. The cabinet.
2. The mechanism (condensing unit and evaporator).
3. The electrical circuit.

The cabinet contains and supports the evaporator and condensing unit. It also supplies shelving and storage space for the foods or beverages.

In the evaporator, the liquid refrigerant expands and becomes a vapor. This vapor absorbs heat from the foods or beverages in the cabinet.

The condensing unit removes the heat absorbed in the evaporator. The liquid refrigerant then returns to the evaporator to repeat the refrigerating cycle. It is suggested that one carefully review Chapter 4 before continuing study of this chapter.

10-1 PRESERVING FOODS BY REFRIGERATION AND FREEZING

Foods (vegetables and fruits) last longer when kept at temperatures just above freezing. The lower temperatures slow down oxidation of the food, reduce the multiplication of the bacteria in the cells and fibers, and reduce the evaporation (loss of fluid) from the food.

10-2 HOW COLD PRESERVES FOOD

Food has cells, enzymes, colloids, water, and a few microorganisms. If not kept cold, food will spoil.

Enzymes, which cause food spoilage, are controlled by low temperatures. Research has discovered that to preserve some foods for long periods (a year or more) the temperature must be well below 0 °F (-18 °C). For best results, it should be -20 °F (-29 °C).

Enzymes are tiny particles of matter that exist in food substances. They are not destroyed by fast freezing but their increase is slowed down by the low temperatures. They seem to stimulate organic change. They are destroyed by pasteurization.

Colloids are found in flesh foods. They are tiny cells in meats, fish, and poultry. If they are abused in any way, such as cell disruption (breaking), the food quickly becomes rancid (spoiled). Colloids seem to be cell containers or capsules. If the container is broken, the food rapidly deteriorates.

Meat, poultry, and fish have important colloidal (miniature cell) changes that can be slowed by using low temperatures.

Water in food forms ice crystals when frozen. Fast freezing produces small ice crystals and is less damaging to food. Slow freezing allows time for larger crystal growth. The larger the ice crystals, the more the food cell walls are damaged.

10-3 STORAGE OF FRESH FOODS IN THE REFRIGERATOR

The air in a fresh food refrigerator is always quite dry. What moisture there is in the refrigerator tends to collect and condense on the evaporator surfaces. Therefore, food containers should be covered and as airtight as possible to keep food moist.

The temperature inside the fresh food cabinet should be kept at 35° to 45°F (2° to 7°C).

Most fresh foods may be kept from three days to a week at the above temperatures. Unfrozen meat and fish should be stored at as close to 32 °F (0 °C) as possible. The recommended storage temperatures for various foods are listed in Chapter 15.

Fruits and vegetables should be cleaned and prepared for the table before being refrigerated.

10-4 STORAGE OF FROZEN FOOD IN THE FREEZER

The air in a food freezer, as in a refrigerator, is very dry. Any moisture in the air of the freezer quickly condenses on the evaporator surfaces. It is very important, therefore, that all frozen foods be packaged in moistureproof containers.

When packaging food for the freezer, as much air as possible should be removed from the packaging.

Frozen food packages must be tightly sealed. Ordinary paper is too porous for freezer use. If not properly packaged, frozen food will develop "freezer burn."

Freezer burn is indicated by a change in color of the food. Food value is not affected but there is a change in color and outside appearance.

Fig. 10-1. Fresh food compact refrigerator. A—Evaporator. B—Thermostat control knob. C—Vegetable crisper. (Whirlpool Corporation)

Most frozen foods, if kept at temperatures of 0°F to −10°F (−18°C to −23°C), may be kept for several weeks. Food to be kept for a year or more should be frozen at −20°F (−29°C) or lower. Some frozen foods keep better than others; beef keeps better than pork. Commercial systems for food freezing are explained in Chapter 13.

10-5 REFRIGERATOR AND FREEZER INSULATION

Insulation lines the walls of the refrigerator and freezer cabinet to keep heat from leaking through.

The insulation materials most used in household refrigerators and freezers are urethane foam or fiberglass. Other insulating materials are used in some commercial and industrial systems.

Tables giving the insulating properties of various materials are shown in Chapter 30.

10-6 REFRIGERATOR, FRESH FOOD, SINGLE DOOR, MANUAL DEFROST

The simple fresh-food refrigerator consists essentially of an evaporator placed either across the top or in one of the upper corners of the cabinet. Fig. 10-1 is typical. The evaporator is across the top of the cabinet and has a place to store frozen food for a short period of time. The condensing unit is in the bottom of the cabinet.

Some additional food storage is provided in the door. A vegetable crisper drawer is located below the bottom shelf, a butter conditioner in the door. This conditioner has a door, closing it off from the refrigerated compartment, to keep the butter at a slightly higher temperature than the rest of the food in the cabinet.

10-7 CABINETS (Refrigerator, Fresh Food, Single Door, Manual Defrost)

Refrigerator cabinets are made of pressed steel. The seams are welded. The outside shell must be smooth and vapor-proof. The inside shell provides a surface for the interior finish of the cabinet. It also provides brackets for mounting shelves, lights, thermostats, temperature controls, and the like.

The insulation is installed between the outer and the inside shell. Urethane foam, when used, is expanded in-place and, therefore, fits with no crevices or open places. The hinge arrangement is usually a part of the outside shell and door assembly.

In the simple fresh-food refrigerator, the cabinet provides a space for the evaporator along the top or upper corner. The cold air from the evaporator flows by natural circulation through the refrigerated space. The shelves are usually constructed so that air can circulate freely past the ends and sides. In such an installation, there is no need to use a fan. The crisper for fresh vegetables is usually located in the bottom shelf of the refrigerator. It generally has a cover in order to maintain fairly high humidity around the vegetables.

In most cases, the light switch is located at the hinge side of the door. The light is turned on and off as the door is opened and closed.

To reduce the heat flow from the outer shell to the inner shell at the door opening, a connecting trim of special plastic is usually used. This plastic is sometimes called a "cold ban," being a poor conductor of heat. This trim is usually attached to the refrigerator liner and shell with trim clips. The finish on fresh food refrigerators—both outside and inside—is usually a good grade of baked-on enamel. Porcelain enamel is found on steel cabinet liners.

10-8 MECHANISMS
(Refrigerator, Fresh Food, Single Door, Manual Defrost)

The typical mechanism in a simple fresh food refrigerator consists of a hermetic compressor placed in the cabinet base. The condenser is either at the bottom or at the back of the cabinet. An evaporator is placed inside the cabinet at the top. A typical mechanism is illustrated in Fig. 10-2.

In this illustration, the parts are shown in diagram fashion.

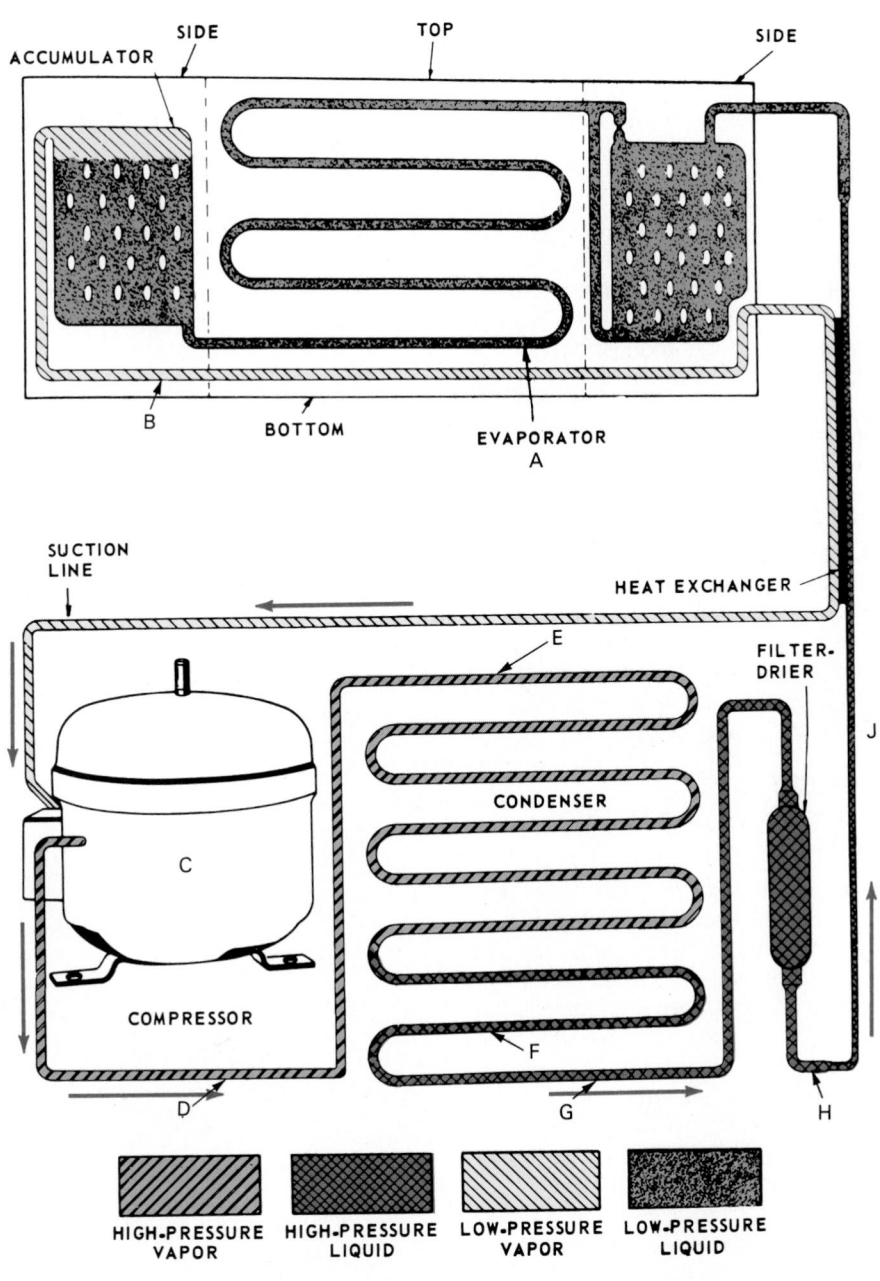

Fig. 10-2. Cycle diagram for fresh food refrigerator.

The cycle operation is as follows:

1. The liquid refrigerant (usually R-12) enters the evaporator, A.
2. As the refrigerant boils and absorbs heat in the evaporator, the vapor is drawn through the suction line, B, back to the compressor, C.
3. In the compressor, it is compressed to a high pressure. Its temperature is increased and the compressed vapor flows through the high pressure vapor line, D, into the condenser, E. The condenser, in this case, is a vertical natural draft, wire-tube type.
4. In the condenser, the high-pressure high-temperature vapor, F, gives up its heat to the surrounding air and the vapor is condensed back to a liquid. The liquid is shown at G in the bottom of the condenser.
5. Liquid refrigerant then flows through the filter-drier and enters the capillary tube at point H. The capillary tube refrigerant control, J, is attached to the suction line at the heat exchanger.
6. The warm refrigerant passing through the capillary tube gives up some of its heat to the cold suction line vapor. This increases the heat absorbing ability of the liquid refrigerant slightly and increases the superheat of the vapor entering the compressor.
7. The low-pressure liquid now enters the evaporator, A, and the cycle is repeated.

This is the simplest type of automatic domestic refrigerator. These refrigerators are manually defrosted. As frost builds up on the evaporator it will be necessary, from time to time, to remove it. The ice accumulation on the evaporator greatly reduces the refrigeration effect.

There are two common methods for manually defrosting these refrigerators:

1. The refrigerator is turned off and allowed to remain off overnight. A drip pan must be used to catch the condensation that comes from defrosting the refrigerator.
2. Turn off the refrigerator and place a pan of hot water in or near the evaporator. This will quickly remove the frost and the refrigerator can be returned to normal service in a few minutes. *Never use a metal scraper on an evaporator. There is danger of puncturing it.*

Evaporator surfaces and inside surfaces of the refrigerator ought to be cleaned each time the mechanism is defrosted. A solution of baking soda and water is best.

10-9 ELECTRICAL CIRCUITS
(Refrigerator, Fresh Food, Single Door, Manual Defrost)

The electrical supply comes through the grounded extension cord and plug, Fig. 10-3. The electrical circuit then supplies current to the panel disconnect. Two separate circuits lead away from this panel. One circuit supplies current to the cabinet light. The light is controlled by the cabinet switch. The cabinet switch is in series with the light. The light, then, comes on when the refrigerator door is opened and is turned off when the door is closed.

The second circuit brings current to the motor compressor. The thermostat in Fig. 10-3 is in series with this circuit and controls the running of the compressor.

When the temperature in the cabinet rises above a determined point, the thermostat completes the circuit through the motor. The compressor runs and the refrigeration cycle goes into operation. As soon as the temperature inside

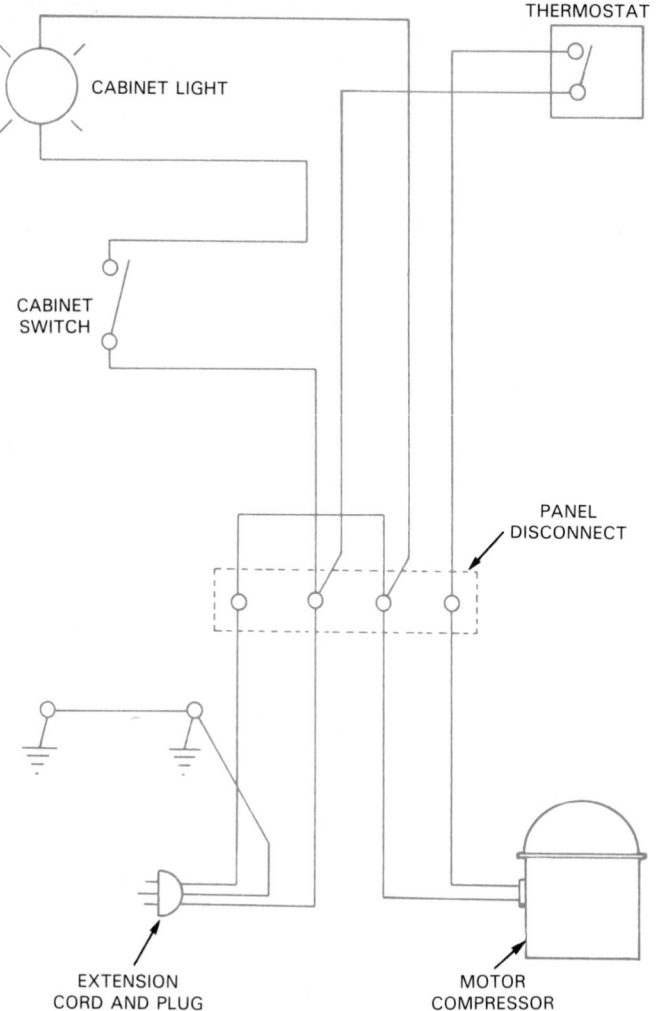

Fig. 10-3. Wiring diagram of fresh food refrigerator.

the cabinet has been brought down to the minimum desired temperature, the thermostat turns off (opens) the current and the motor compressor stops.

A motor control thermostat is shown in Fig. 10-4. The thermostat control knob is attached to the control adjust-

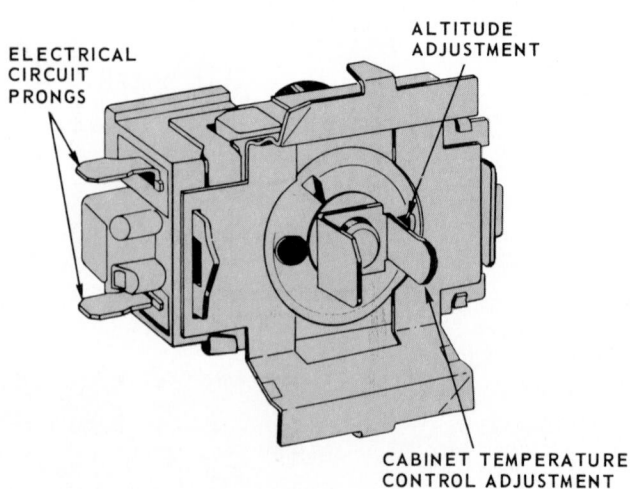

Fig. 10-4. Thermostat for fresh food refrigerator.

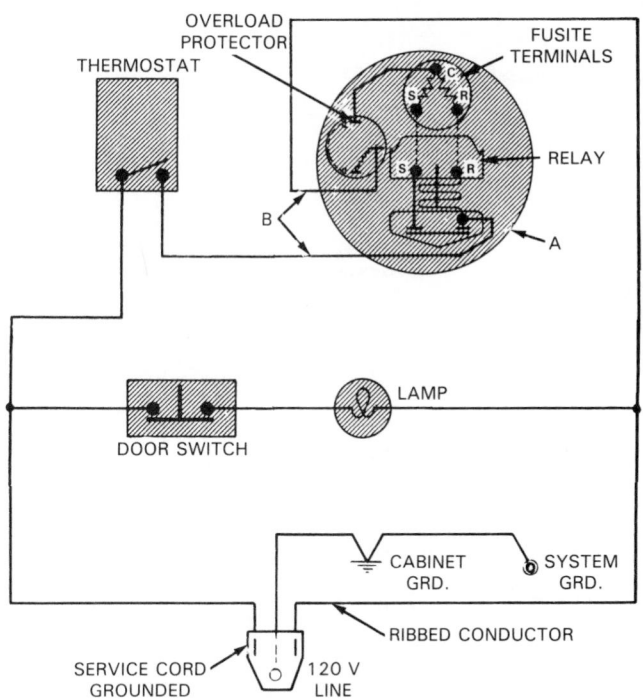

Fig. 10-5. Motor control relay for fresh food refrigerator. Relay is of the current type, using running winding current. Diagram shows electrical connections to thermostat and motor control relay. This motor compressor relay has an overload protector. A—Starting relay. B—Power supply leads. S—Starting winding terminal. R—Running winding terminal. C—Common terminal.

ment. The altitude adjustment is turned clockwise as the altitude increases. Refer to Chapter 8 for instructions on how to adjust these thermostats.

Most hermetic refrigerators use a starting relay which is usually mounted on the body of the motor compressor. These starting relays also provide overload protection for the motor.

The overload protector contains a resistor wired in series with the running current. Should the current draw be too great (overload), the resistor will heat up and cause a bimetal contact to break the circuit.

Fig. 10-5 illustrates the electrical circuits for this refrigerator. This starting relay is shown at A. It connects both the starting winding and the running winding to the power circuit. Then it disconnects the starting winding from the power circuit as soon as the compressor motor has reached about 75 percent of running speed.

10-10 REFRIGERATOR-FREEZER, MANUAL DEFROST

This unit, Fig. 10-6, consists essentially of two refrigerated spaces. It has a frozen food compartment across the top of the cabinet where temperature is kept at approximately 0 °F (−18 °C). A fresh food compartment below the frozen food compartment maintains a temperature of about 35° to 45 °F (2° to 7 °C).

Each of these compartments has a separate door. The condensing unit is usually in the bottom of the cabinet with the condenser either in the bottom or at the back.

These refrigerators provide shelves in both compartments. A butter conditioner is usually located in the door of the

fresh food compartment. The doors of both compartments are often fitted with narrow shelves for storage of small containers.

10-11 CABINETS
(Refrigerator-Freezer, Manual Defrost)

Cabinet construction for this refrigerator is much the same as for the simple fresh food refrigerator. Insulation of the refrigerator with a frozen food compartment must be thicker than that of the fresh food refrigerator. This added insulation maintains the lower temperatures necessary in the frozen food compartment. A separate freezer door is provided.

The motor control, temperature control, light switch, shelf support, and crisper are the same as those on the fresh food refrigerator. The finish is usually a good grade of lacquer, although some cabinets have porcelain finished interiors.

10-12 MECHANISMS
(Refrigerator-Freezer, Manual Defrost)

Refrigerators with a frozen food compartment have a hermetic compressor in the base of the cabinet. The condenser is either at the bottom or the back of the cabinet. The liquid refrigerant flows from the capillary tube into the evaporator in the freezing compartment.

A typical mechanism is illustrated in Fig. 10-7. The refrigerant charge is usually sufficient to keep the freezer evaporator, A, filled. It also has enough for spill-over from

Fig. 10-6. Two-door model refrigerator-freezer with freezer at top. Note evaporator in upper part of lower compartment.
(Whirlpool Corporation)

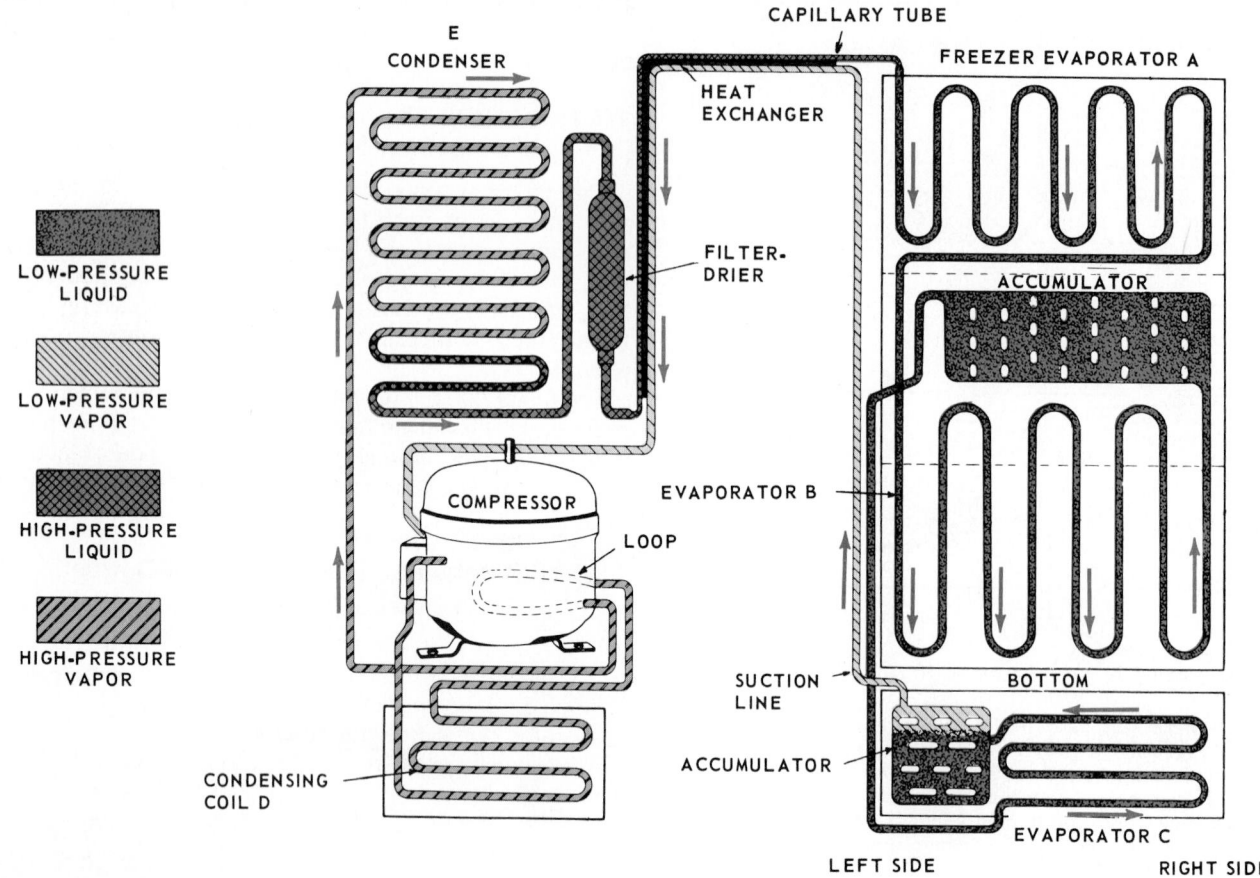

Fig. 10-7. Cycle diagram, refrigerator with frozen food compartment, manual defrost.

the freezer evaporator into the fresh food compartment evaporator, B. Usually located at the back of the fresh food compartment, this evaporator has a rather large accumulator. Any possible spill-over from this enters a third evaporator, C, usually located at the side of the fresh food compartment. This third evaporator is also fitted with an accumulator. This assures that all the refrigerant is evaporated before the vapor is allowed to enter the suction line.

From this third coil accumulator, the vaporized refrigerant is drawn back to the suction line into the compressor. The vapor is compressed and pumped first into a small condensing coil D. From here the high-pressure vapor is pumped through a loop in the base of the compressor which serves as an oil cooler. From here the compressed vapor flows into condenser, E, at the bottom or at the back of the refrigerator.

Here the heat of the vapor is radiated to the surrounding air and the refrigerant is condensed back to a liquid. The liquid flows from the bottom of the condenser through the filter-drier and into the capillary tube attached to the suction line. The capillary tube controls the refrigerant flow into the freezer evaporator A. The cycle is then repeated.

This refrigerator is manually defrosted.

10-13 ELECTRICAL CIRCUITS
(Refrigerator-Freezer, Manual Defrost)

The electrical supply is fed through the grounded extension cord and plug at 1 in Fig. 10-8. The electrical circuit supplies current to the panel disconnect at 2.

From the panel mounted disconnect, two separate circuits are provided. One circuit supplies current to the cabinet light at 3. It is controlled by cabinet switch 4. The cabinet switch is in series with the light. The light comes on or goes off as the refrigerator door is opened and closed.

The second circuit goes to the motor compressor. The thermostat at 5 is in series with the circuit and controls the running of the compressor at 6.

When the temperature in the cabinet climbs to a determined temperature, the thermostat completes the circuit through the motor. The compressor runs and the refrigeration cycle comes into operation. As soon as the temperature inside the cabinet has been brought down to the minimum desired temperature, the thermostat turns off the current and the motor compressor stops.

Two additional electrical devices are usually used on this type of refrigerator. One is an electrical resistance heat wire, also called a perimeter drier. It is shown at 7 and is located in the trim of the freezer door.

This heater operates all the time. It provides enough "warming effect" to stop condensation on the exterior of the cabinet and around the freezing compartment door.

The second electrical device is an ambient compensator shown at 8. This is also an electrical resistance heat wire. The ambient compensator draws 15 to 20 VA (formerly watts). It is attached to the insulation side of the fresh food compartment thermostat. It is energized only on the "off" cycle (thermostat contacts open).

The purpose of the ambient compensator is to provide a continuous small heat flow into the fresh food compart-

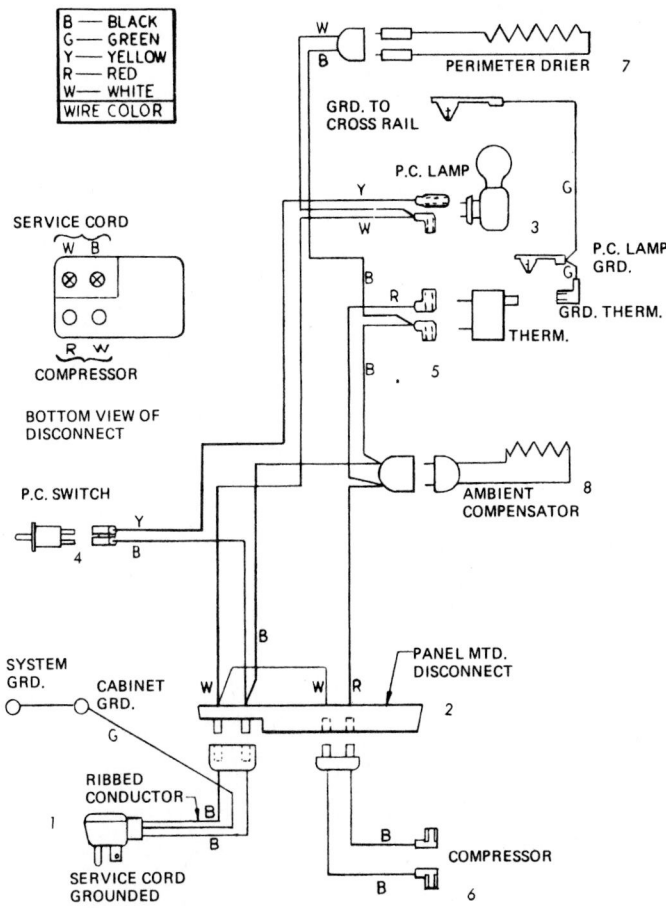

WIRE COLOR	
B	BLACK
G	GREEN
Y	YELLOW
R	RED
W	WHITE

Fig. 10-8. Wiring diagram, refrigerator with frozen food compartment, manual defrost.

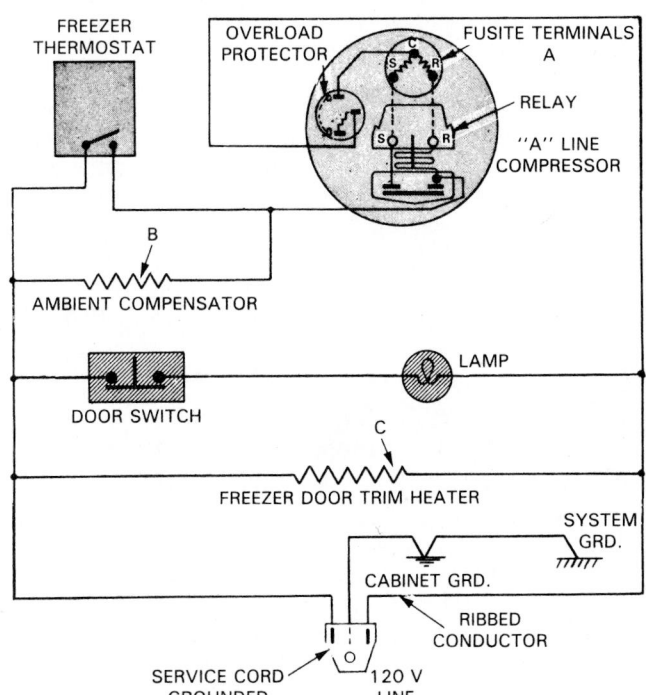

Fig. 10-9. Motor control relay, refrigerator with frozen food compartment, manual defrost. This ladder diagram illustrates the electrical connections to thermostat and motor control relay. This motor compressor relay is provided with an overload protector.

ment. This will cause the refrigerator to cycle in the event the ambient temperature of the room in which the refrigerator is located is lower than the normal thermostat setting.

Fig. 10-9 illustrates the electrical circuits for this refrigerator. The area near the letter A shows the starting relay. The starting relay connects both the starting winding and the running winding to the power circuit. It disconnects the starting winding as soon as the compressor motor has reached approximately 75 percent of running speed.

This starting relay also provides overload protection for the motor. The overload protector contains a resistor in series with the running current. In the event the current draw is too great (overload), the resistor will heat up and cause a bimetal contactor to break the circuit. The ambient compensator is shown at B. The freezer door trim heater is shown at C.

10-14 REFRIGERATOR-FREEZER, AUTOMATIC DEFROST

All air contains some moisture. When air comes in contact with an evaporator surface which is below the freezing temperature, the moisture will condense and form ice on the evaporator.

As mentioned in the earlier paragraphs, it is frequently necessary to defrost the evaporator in order to maintain good refrigerating efficiency. This applies to the evaporator for both the fresh food and the frozen food compartments.

Fig. 10-10. Refrigerator with frozen food compartment, automatic defrost. Evaporator in frozen food compartment serves as fast-freezing shelf. Fresh food cabinet provides butter conditioner, fresh meat storage, and vegetable crisper. (Whirlpool Corporation)

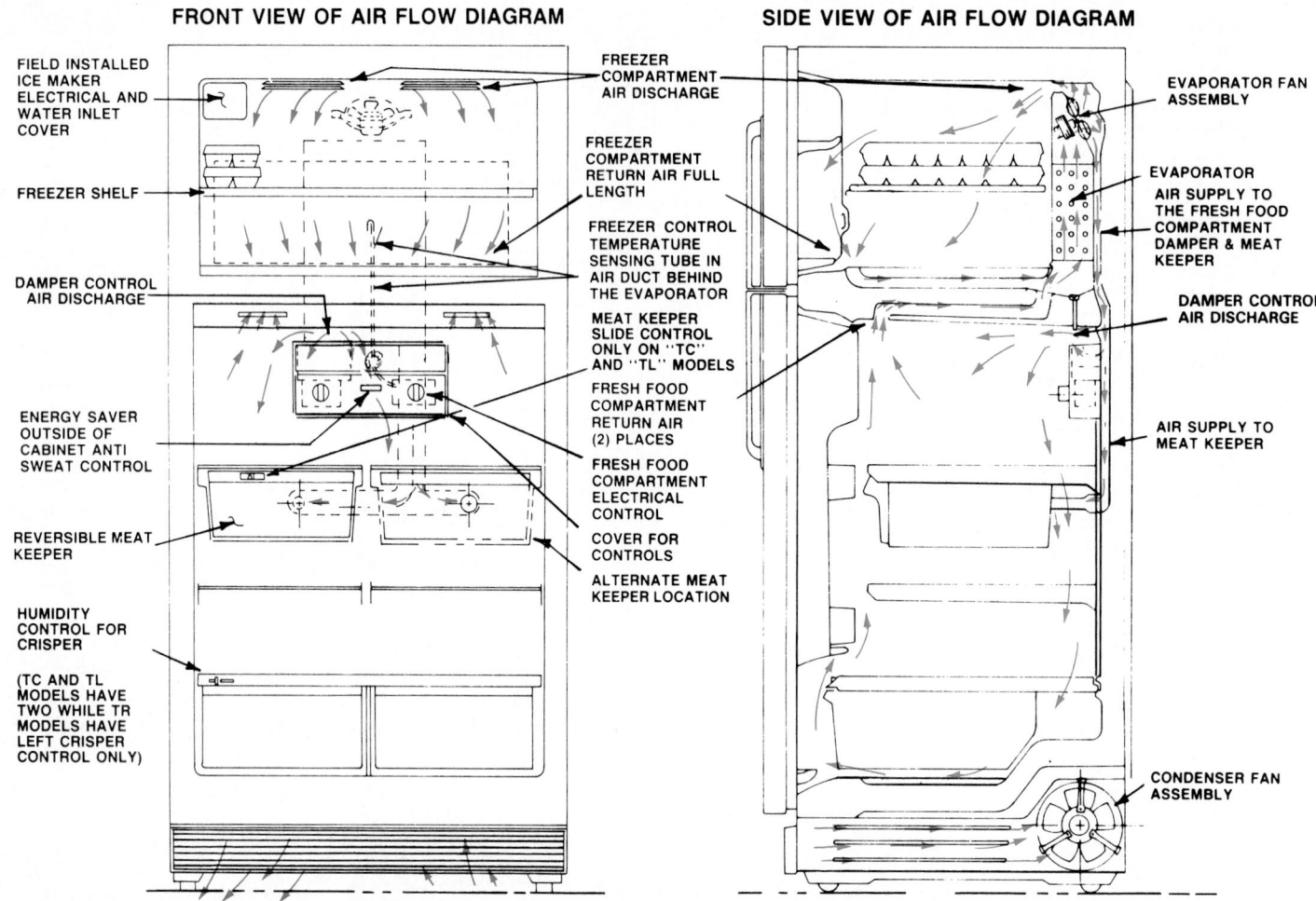

FRONT VIEW OF AIR FLOW DIAGRAM

SIDE VIEW OF AIR FLOW DIAGRAM

FIELD INSTALLED ICE MAKER ELECTRICAL AND WATER INLET COVER

FREEZER COMPARTMENT AIR DISCHARGE

EVAPORATOR FAN ASSEMBLY

FREEZER SHELF

FREEZER COMPARTMENT RETURN AIR FULL LENGTH

EVAPORATOR AIR SUPPLY TO THE FRESH FOOD COMPARTMENT DAMPER & MEAT KEEPER

DAMPER CONTROL AIR DISCHARGE

FREEZER CONTROL TEMPERATURE SENSING TUBE IN AIR DUCT BEHIND THE EVAPORATOR

DAMPER CONTROL AIR DISCHARGE

MEAT KEEPER SLIDE CONTROL ONLY ON "TC" AND "TL" MODELS

ENERGY SAVER OUTSIDE OF CABINET ANTI SWEAT CONTROL

FRESH FOOD COMPARTMENT RETURN AIR (2) PLACES

AIR SUPPLY TO MEAT KEEPER

FRESH FOOD COMPARTMENT ELECTRICAL CONTROL

REVERSIBLE MEAT KEEPER

COVER FOR CONTROLS

HUMIDITY CONTROL FOR CRISPER

ALTERNATE MEAT KEEPER LOCATION

(TC AND TL MODELS HAVE TWO WHILE TR MODELS HAVE LEFT CRISPER CONTROL ONLY)

CONDENSER FAN ASSEMBLY

ENERGY SAVING 32 IN. TOP MOUNT REFRIGERATOR AIR FLOW

Fig. 10-11. Air circulation in refrigerator with frozen food compartment. Evaporator fan forces circulation of cooled air through various compartments. Dampers control air duct openings. Therefore, they control cabinet temperatures. (Amana Refrigeration, Inc.)

The owner considers it a chore to manually defrost the refrigerator. As a result, many refrigerators provide a system for automatic defrosting. There are two basic systems used in automatic defrosting. The hot gas system, through the use of solenoid valves, uses the heat in the vapor from the compressor discharge line and the condenser to defrost the evaporator.

The other system uses electric heaters to melt the ice on the evaporator surface. A paragraph in this chapter will be devoted to each of these systems.

Fig. 10-10 illustrates a typical refrigerator with a frozen food compartment. This refrigerator uses an automatic defrost. As is the case with most domestic refrigerators, the condensing unit is mounted in the base of the cabinet, with the condenser either in the bottom or at the back of the cabinet. Shelving is provided in both the fresh food compartment and the frozen food compartment.

Some refrigerators operate on what is called "frost free" or "no frost" cycle. In these refrigerators, the evaporator is located outside the refrigerated compartment. On the running part of the cycle, air is drawn over this evaporator and is forced into the freezing and refrigerator compartment by the use of a motor driven fan. On the off part of the cycle, these evaporators automatically defrost.

Such refrigerators may use a single evaporator for both

the frozen food compartment and the fresh food compartment or a separate evaporator may be used in each compartment. The condensation from the evaporator which melts off during the off cycle is carried to an evaporating pan or collecting surface directly over the compressor and condenser. The heat from the compressor evaporates this moisture and it is returned to the room's atmosphere. There is never any visible frost accumulation in this type of frost control.

10-15 CABINETS
(Refrigerator-Freezer, Electric Automatic Defrost)

Automatic defrosting requires some important differences in the cabinet construction. Since the condensation which collects on the evaporator must be melted from time to time, different methods are used for disposing of this condensation. Usually, it is collected on a plate or surface just over the motor compressor. This surface, or plate, is heated by the motor compressor and by heat from the condenser. The moisture then reevaporates and goes back into the room. Provisions must be made in the cabinet for tubing to conduct this moisture to the top of the motor compressor.

Some refrigerators use electric defrost. Provisions must

be made in the cabinet for housing electric heaters and their controls.

Fig. 10-11 shows a refrigerator with a frozen food compartment that provides other temperatures besides the temperature of the food freezer and the temperature of the fresh food compartment. The air circulation provides lower temperatures to areas reached first and provides higher temperatures to those areas reached last. The evaporator is beneath the fast-freezing shelf at the bottom of the freezing compartment.

All of the refrigerating effect comes from this evaporator. An air circulation system, consisting of a fan, ducts, and a damper at the back of the cabinet, provides the necessary airflow to give the temperature desired in each compartment. The condensing unit is located in the bottom of the cabinet.

10-16 MECHANISMS
(Refrigerator-Freezer, Electric Automatic Defrost)

Fig. 10-12 is typical of a mechanism which provides for more than two temperatures in the refrigerator with a frozen food compartment.
EVAPORATOR
The evaporator is located at the back of the shelf which separates the freezing compartment from the fresh food compartment. The refrigerant used is R-12. The evaporation of refrigerant in the evaporator provides all of the heat absorption (cooling) required in the cabinet. Usually, a motor-driven fan forces air over the evaporator surface and through the various ducts to provide all the necessary refrigerator temperatures for the compartments.
MOTOR COMPRESSOR
The suction line from the evaporator extends down the wall of the cabinet to the inlet side of the hermetic motor compressor, which is located in the base of the cabinet.
CONDENSER
The condenser is a wire and tube type. Forced air circulation is provided by a motor and fan located at the back of the compartment containing the compressor and the condenser.
CAPILLARY TUBE
As the refrigerant is condensed in the condenser, it flows through a high-side filter-drier into a capillary tube attached to a section of the suction line. This provides a heat exchange between the capillary tube and the suction line. The refrigerant from the capillary tube flows into the evaporator and the cooling cycle is completed.
FEATURES
Only the fresh food compartment has a light. It is operated by a switch controlled by the door. The butter conditioner temperature is slightly above the cabinet temperature. There are control dampers for the refrigerator

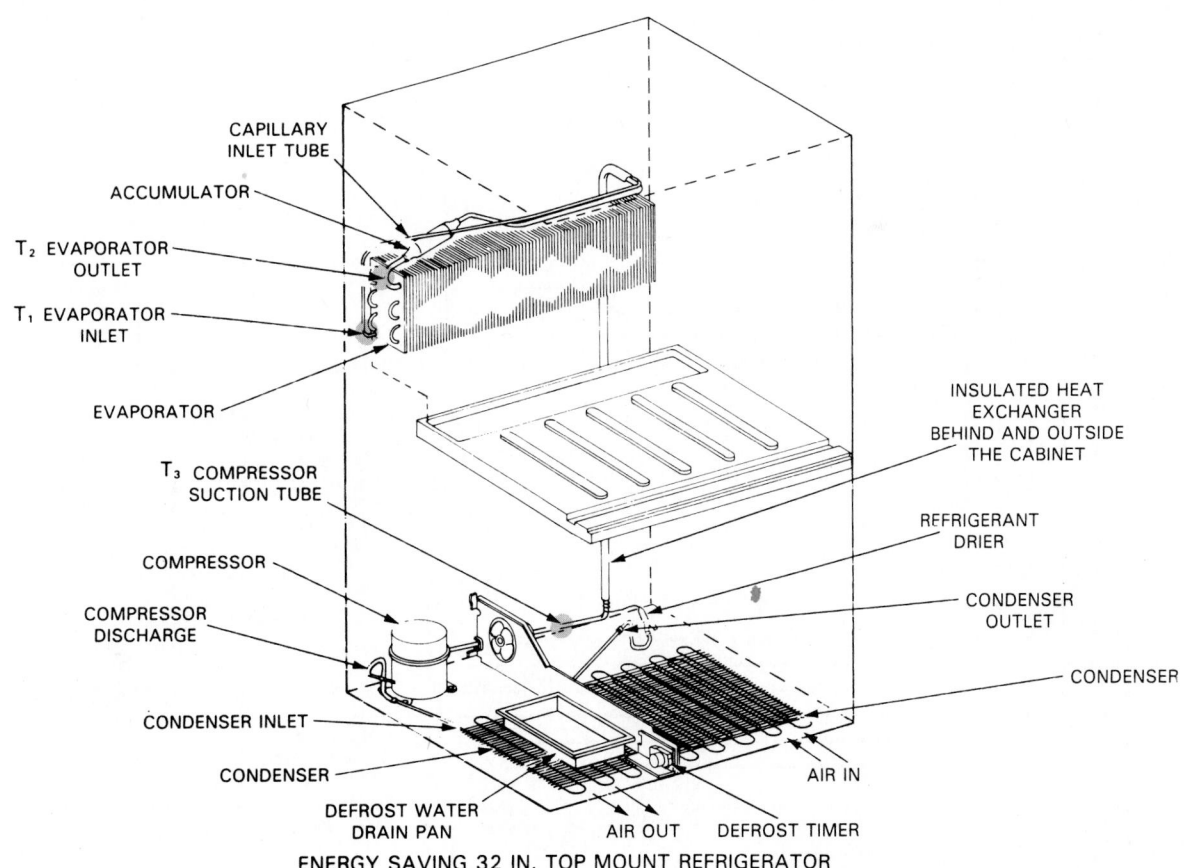

CAPILLARY INLET TUBE

ACCUMULATOR

T_2 EVAPORATOR OUTLET

T_1 EVAPORATOR INLET

EVAPORATOR

T_3 COMPRESSOR SUCTION TUBE

COMPRESSOR

COMPRESSOR DISCHARGE

CONDENSER INLET

CONDENSER

DEFROST WATER DRAIN PAN

AIR OUT

DEFROST TIMER

AIR IN

CONDENSER

CONDENSER OUTLET

REFRIGERANT DRIER

INSULATED HEAT EXCHANGER BEHIND AND OUTSIDE THE CABINET

ENERGY SAVING 32 IN. TOP MOUNT REFRIGERATOR

Fig. 10-12. Cycle diagram for refrigerator with frozen food compartment and automatic electrical defrost. Correct operation may be checked by determining temperatures at test points, T_1, T_2, and T_3. Recommended operating temperatures are: $T_1 = -15$ to $-13\,°F$ (-26 to $-25\,°C$), $T_2 = -15$ to $-14\,°F$ ($-26\,°C$), $T_3 = 80$ to $103\,°F$ (27 to $39\,°C$). Temperature T_3 is high because capillary tube is in contact with suction line to transfer excess condenser heat to suction line. This increases superheat at compressor inlet. (Amana Refrigeration, Inc.)

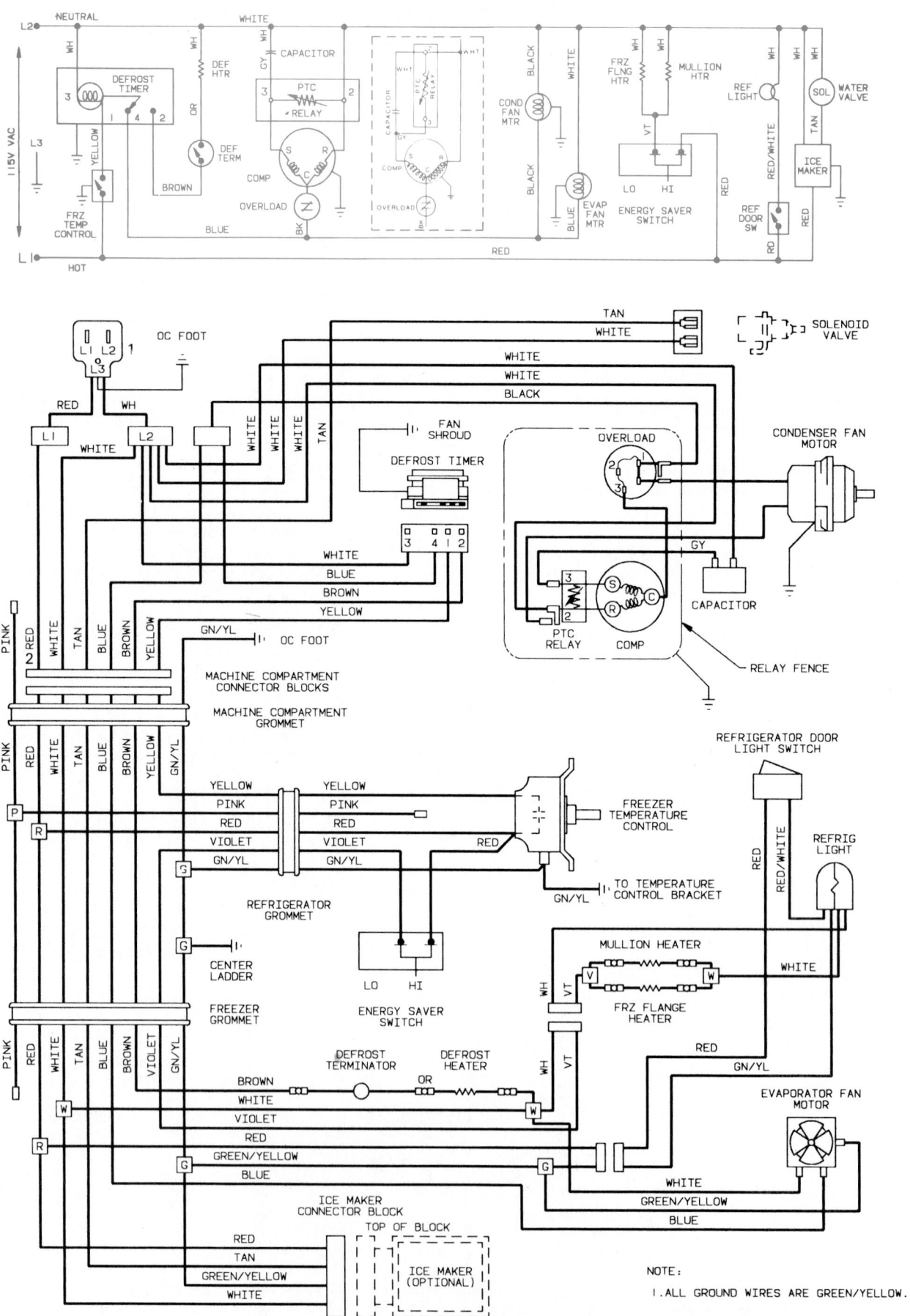

Fig. 10-13. Wiring diagram for refrigerator with frozen food compartment and automatic electric defrost. Note ladder diagram at top. Often, ladder diagrams proceed from top of page to bottom. (Amana Refrigeration, Inc.)

and the freezer compartments. These dampers regulate flow of cold air from the evaporator.

HEATERS

Several heating devices are used as driers. An electrical resistance heater at the top of the cabinet, inside the outer case, keeps the outside of the cabinet warm enough so that it will not collect condensation during damp days.

A second drier wire is placed inside the center mullion to keep its surface dry. A third is around the freezer flange (freezer door opening). A fourth is placed in the drip pan to help evaporate the condensation which flows into the drip pan after automatic defrost.

A "power saver" switch inside the cabinet provides a means of disconnecting these heaters when temperature and humidity conditions allow it. Normally, the heaters are in continuous operation.

AUTOMATIC DEFROST

The evaporator is automatically defrosted by an electric resistance heater every six hours of compressor running time. The heater is located in the fin area on the underside of the evaporator. A timer energizes the switch which turns on the defroster. A thermostat attached to the evaporator opens the defrost heater circuit to end defrosting as soon as the evaporator reaches a temperature of 50 °F (10 °C), plus or minus six degrees F (three degrees C). When 28 min. have passed after the start of the defrost cycle, the timer restores the operation of the compressor and the air circulating fan. The defrost terminator contacts close at 20 °F (-7 °C), plus or minus 8 °F (4 °C). The temperatures of the cabinet are regulated by the temperature control mounted in the rear wall of the freezer compartment. The temperature of the evaporator tubing near the end of a running cycle may vary from -13 °F (-24 °C) to -25 °F (-32 °C). The difference between the temperature of the inlet and the outlet to the evaporator will not vary more than about 3 °F (2 °C).

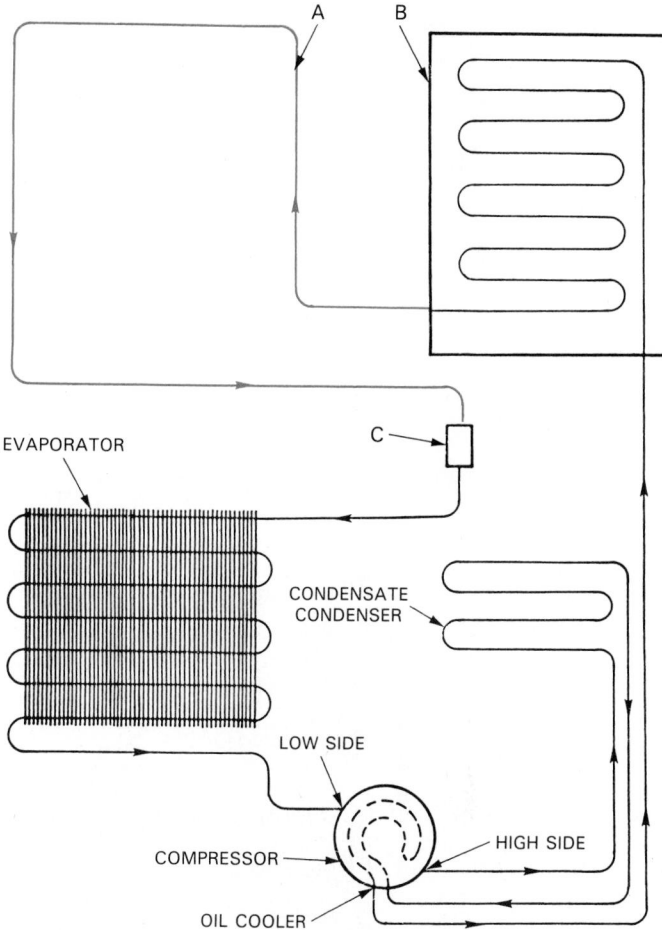

Fig. 10-14. Perimeter hot tube system. Hot tube, A, is located between condenser, B, and drier, C. (Frigidaire Company)

10-17 ELECTRICAL CIRCUITS
(Refrigerator-Freezer, Electric Automatic Defrost)

The electrical supply is through the grounded extension cord and plug shown at 1 in Fig. 10-13. The compressor and defrost timer are energized directly from the grounded extension cord.

A machine compartment connector block, shown at 2, is located in the bottom of the cabinet. It provides electrical connections to the evaporator fan, cabinet light, defrost heater, defrost terminator, mullion heater, freezer flange heater, and the power saver switch. If the cabinet is equipped with an automatic ice maker, it will be wired into the same connector block. A ladder diagram of the wiring circuit is shown at the top of Fig. 10-13.

10-18 CABINETS
(Refrigerator-Freezer, Hot Gas Automatic Defrost)

These cabinets provide the usual facilities for both fresh and frozen food storage. The thermostat is located in the fresh food compartment.

The defrost timer and the solenoid valve are housed in the cabinet base. In addition, many refrigerators pipe compressed vapor through tubing located on the inside surface of the outside shelf, particularly around the door openings. The purpose of this is to prevent condensation from forming around the front of the cabinet. It utilizes what is known as a perimeter hot tube system. The perimeter hot tube is part of the high side of the system. It provides heat needed to prevent condensation from forming around the front flanges of the cabinet. The use of the hot tube concept eliminates the need for an electrical heater. See Fig. 10-14.

The function of these coils is to keep the surface of the refrigerator cabinet dry. They are sometimes referred to as driers. The same term may be applied to electric heating units installed in the cabinet for the same purpose. *This use of the word drier should not be confused with the term drier as applied to liquid and suction lines.*

Fig. 10-15 illustrates a refrigerator with a frozen food compartment and hot gas defrost.

10-19 MECHANISMS
(Refrigerator-Freezer, Hot Gas Automatic Defrost)

The mechanism illustrated in Fig. 10-16 is typical of those used in refrigerators with a frozen food compartment and fresh food compartment using "hot gas" defrost.

EVAPORATOR

Two evaporators are used with R-12 as their refrigerant. Evaporator A is located in the freezing compartment. It

Fig. 10-15. Refrigerator with frozen food compartment and hot gas defrost. Evaporator extends along back and ends of frozen food space and lower shelf. Thermostat, butter conditioner, meat keeper, and vegetable crispers are in fresh food compartment. (Whirlpool Corporation)

extends along the back and ends of the compartment as well as along the bottom. Refrigerant (both vapor and liquid) leaving this evaporator flows into evaporator, B, located at the back of the fresh food compartment. This evaporator has a rather large accumulator. This is necessary since all refrigerant leaving the fresh food compartment evaporator should be in vapor form.

MOTOR COMPRESSOR

Vapor is drawn back to the compressor. The vapor is compressed and discharged directly into a water evaporating plate and coil assembly. This is located over the compressor and serves to evaporate the moisture drained from the evaporators during the defrost cycle. In normal operation, this compressed vapor next flows through the oil cooler line in the bottom of the compressor.

CONDENSER

From the oil cooler line, vapor flows to the vertical wire tube type condenser. Heat is given off to the surrounding air and the compressed vapor returns to a liquid state. From the bottom of the condenser, the liquid flows through the filter-drier.

CAPILLARY TUBE

From the filter-drier, the liquid then flows into the capillary tube refrigerant control soldered to the suction line. This serves as a heat exchanger. It reduces the temperature of the liquid refrigerant in the capillary tube and increases the superheat of the vapor in the suction line. From the capillary tube, the refrigerant enters the freezer evaporator at a reduced pressure. It evaporates, absorbs heat from the inside of the cabinet, and the cycle is completed.

HOT GAS DEFROST

Fig. 10-17 shows the solenoid valve open and hot gas defrost cycle in operation. In this cycle, starting with the compressor, the vapor from the evaporators is drawn into the compressor. It discharges into the water evaporating plate, helping to heat this surface. Then, hot compressed vapor flows back through the drain sump bypass line, into the freezer evaporator. From the freezer evaporator, the vapor flows into the fresh food compartment evaporator through the accumulator and back into the suction side of the compressor. Since this compressed vapor is quite warm, it quickly defrosts the evaporators, because the heat is supplied directly to the inside surfaces of the evaporators.

A defrost timer, Fig. 10-18, controls the solenoid defrost valve. The timer is located at the back lower left corner of the refrigerator cabinet. It is driven by a self-starting electric motor geared to turn the shaft slowly. The shaft completes one revolution every eight hours of compressor operation. When the timer opens the solenoid valve, the defrost cycle continues for 17 min. Then, the solenoid valve closes and the refrigerating cycle operates. The freezer compartment fan is not operating during the defrost cycle.

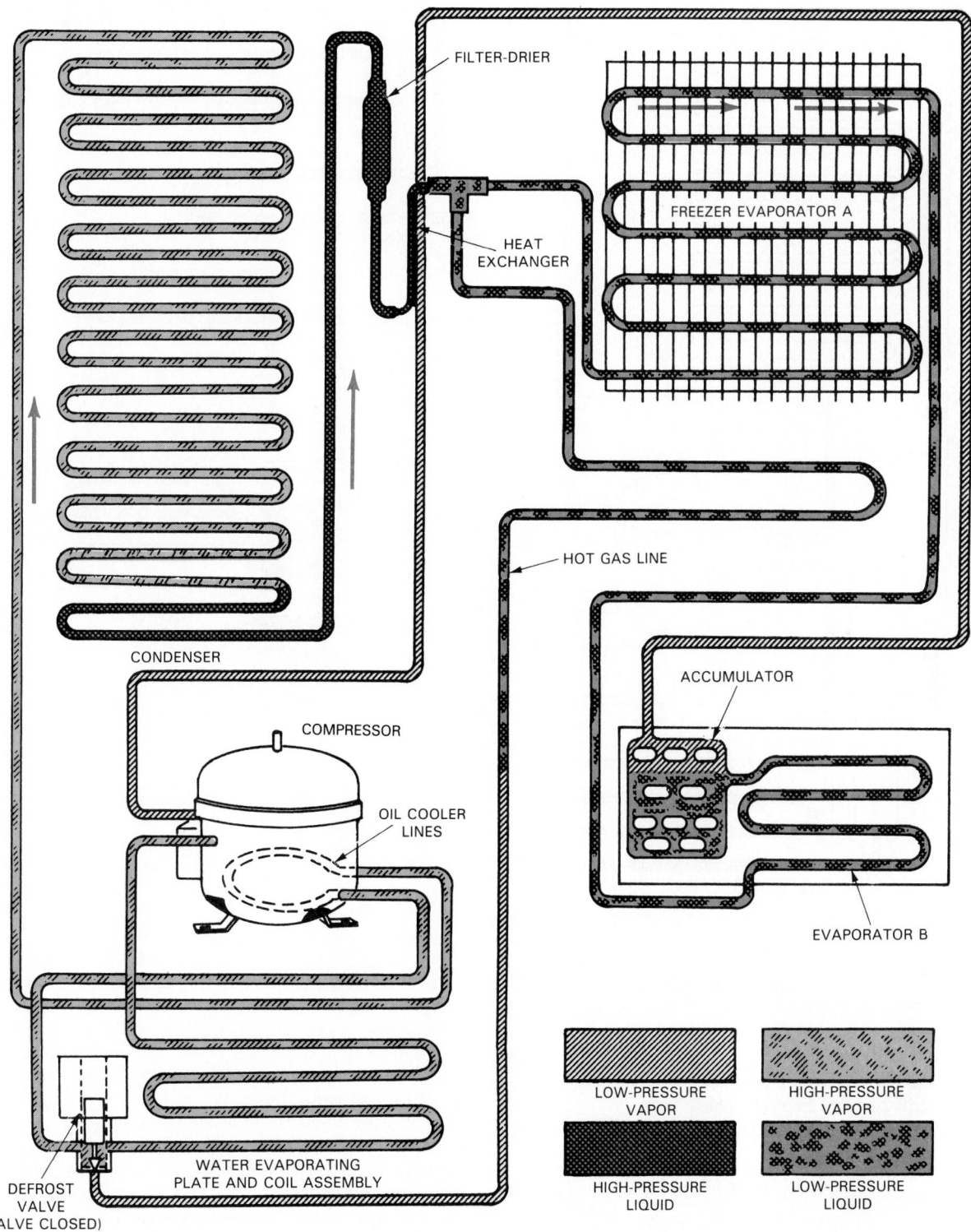

Fig. 10-16. Refrigerating cycle for hot gas defrost. Solenoid valve is closed. This closes off bypass line.

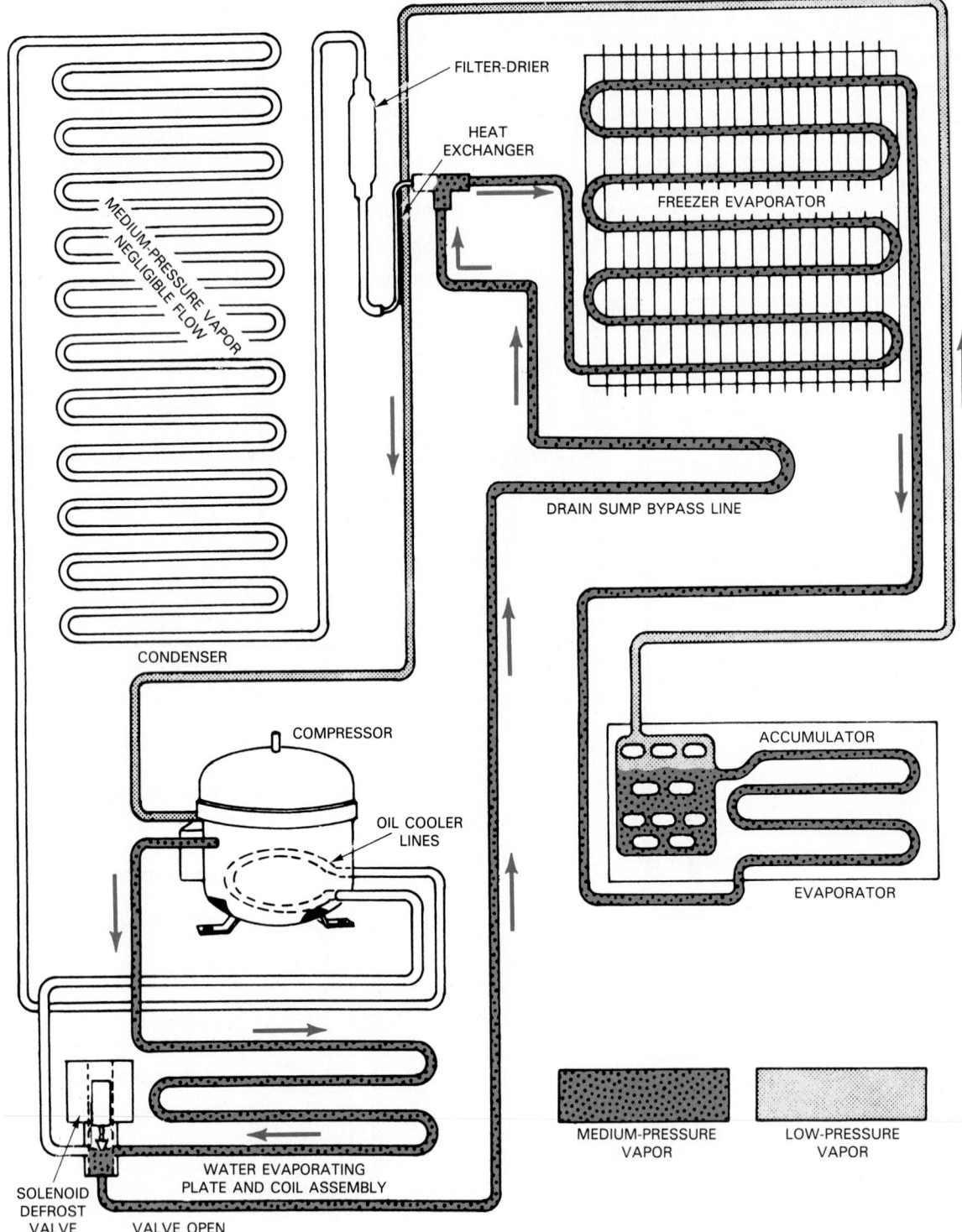

MEDIUM-PRESSURE VAPOR NEGLIGIBLE FLOW

FILTER-DRIER

HEAT EXCHANGER

FREEZER EVAPORATOR

DRAIN SUMP BYPASS LINE

CONDENSER

COMPRESSOR

OIL COOLER LINES

ACCUMULATOR

EVAPORATOR

SOLENOID DEFROST VALVE

VALVE OPEN

WATER EVAPORATING PLATE AND COIL ASSEMBLY

MEDIUM-PRESSURE VAPOR

LOW-PRESSURE VAPOR

Fig. 10-17. Defrost cycle, hot gas defrost. Solenoid valve is open and hot compressed vapor is traveling through bypass line. Heat from compressed vapor keeps drain sump free of ice. The heat melts ice from both freezer evaporator and fresh food evaporator.

10-20 ELECTRICAL CIRCUITS (Refrigerator-Freezer, Hot Gas Automatic Defrost)

Electricity is supplied through the grounded extension cord and plug shown at 1 in Fig. 10-19. The electrical supply goes to the panel mounted disconnect at 2. Various circuits are fed from this disconnect. The compressor, defrost solenoid (shown at 3), and defrost timer are ener-

gized from the four right-hand connections. If an automatic ice maker is used, the water valve, 4, is energized through the two terminals as indicated.

In addition to the hot gas defrost, there are three electrical resistance heaters in the system. The ambient compensator at 5 is attached to the insulation side of the fresh food compartment. The perimeter drier, 6, is installed in the trim of the freezer door. The drain sump heater prevents freezing of condensation from the evaporator as it moves

Fig. 10-18. Automatic defrost timer. A—Timer is connected into motor compressor circuit and provides both a clock mechanism and switching mechanism. Clock operates switch mechanism in such a way that, after eight hours of compressor operation, fan motor is turned off and defrost solenoid is energized. B—Timer motor.
(Paragon Electric Co., Inc.)

down the drain tube to the moisture evaporating pan located over the motor.

This wiring diagram provides a complete grounding system. Each metal part of the mechanism and cabinet has its own ground as specified in Para. 6-72. Fig. 10-20 is a schematic of this wiring diagram.

10-21 FROST-FREE REFRIGERATOR-FREEZER

Modern refrigerators are designed to eliminate the task of defrosting the evaporator. Frost-free refrigerators are very popular. Para. 10-14 explains how they work.

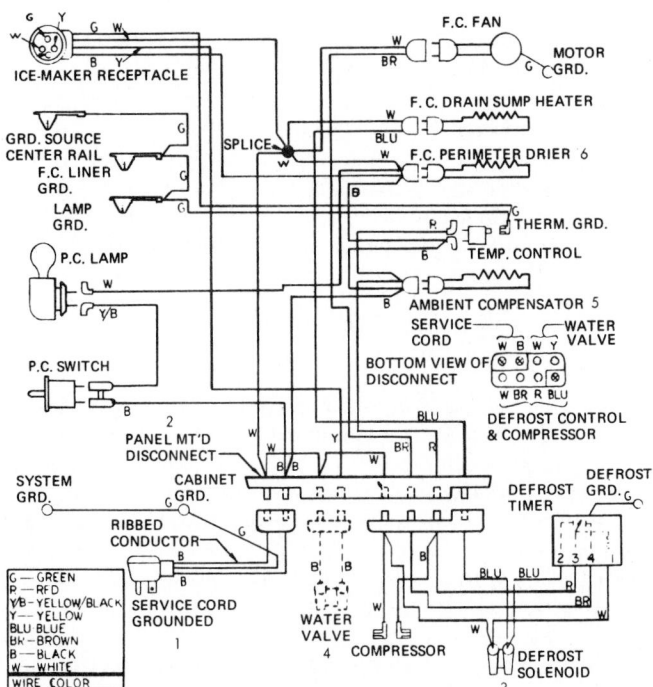

Fig. 10-19. Wiring diagram for refrigerator with frozen food compartment and hot gas automatic defrost. F.C. means freezer compartment. P.C. means produce compartment.

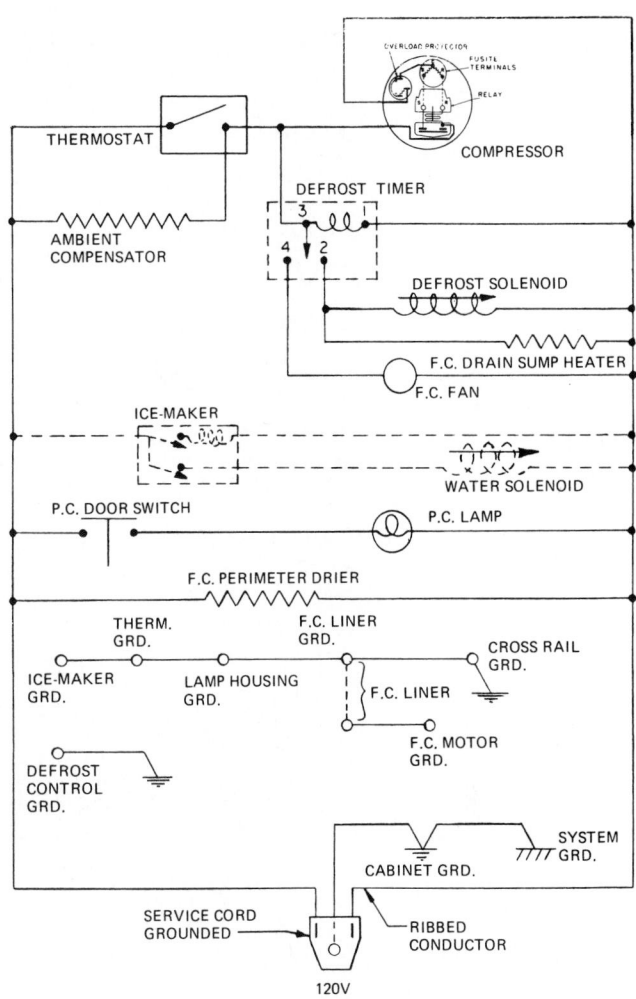

Fig. 10-20. Ladder wiring diagram, refrigerator with frozen food compartment, automatic defrost, hot gas. Note provisions for necessary grounding of the various components. F.C. means freezer compartment. P.C. means produce compartment.

10-22 CABINETS
(Frost-Free Refrigerator-Freezer)

The refrigerator in Fig. 10-21 stores frozen food at the top and fresh food at the bottom.

The evaporator is in the upper back part of the cabinet. The condenser is along the lower back part. A fan moves cold air from the evaporator in the frozen food compartment into the fresh food compartment. Another fan circulates room air through the grille at the bottom of the cabinet and over the condenser. *With this type of condenser, it is not necessary to provide any clearance space at the sides or top of the cabinet for air circulation.*

Both doors are held shut by magnets. This maintains a tight seal without the need for a mechanical latch.

The cabinet is mounted on wheels, making it easy to roll the refrigerator out for cleaning.

10-23 MECHANISMS
(Frost-Free Refrigerator-Freezer)

Fig. 10-22 is a diagram of the mechanism for this refrigerator. This illustrates the compressor, condenser, capillary tube, heat exchanger, and evaporator. Note that fans are used on both the condenser and evaporator.

319

Fig. 10-21. Refrigerator with frozen food compartment, frost-free. Frozen food compartment is cooled by very cold air from evaporator which enters through freezing compartment grille. Condenser is cooled by air which circulates in and out through bottom grille. Evaporator fan is turned off automatically when freezing compartment door is opened. (Whirlpool Corporation)

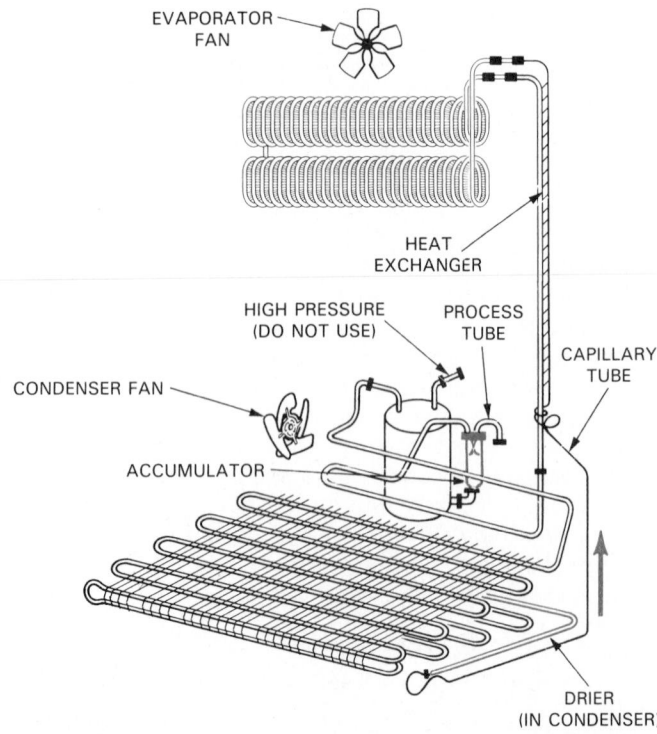

Fig. 10-22. Refrigerator with frozen food compartment, frost-free. Cycle diagram shows construction. Note accumulator and drier. (General Electric Co.)

10-24 ELECTRICAL CIRCUITS
(Frost-Free Refrigerator-Freezer)

Electricity is supplied through a grounded extension cord and plug shown at 9 in Fig. 10-23. There are two electrical resistance heating elements used in this cabinet.

The evaporator defrost is electrically heated. The freezer door contains a recess and duct heater. The heater helps eliminate moisture. In this illustration, the resistance of the various electrical components is indicated in ohms (Ω).

If an automatic ice maker is used, this electrical circuit provides for easily connecting it into the cabinet.

10-25 FROST-FREE REFRIGERATOR-FREEZER
(Side-by-Side)

The side-by-side refrigerator-freezer arrangement is very popular. The frozen food compartment stores frozen foods at a satisfactory temperature of 0 °F (– 18 °C) or below.

With the advent of electronics technology, an increasing number of features are being incorporated into the refrigerator-freezer unit. In addition to the existing ice maker hardware and freezer controls, one may find cold water, crushed ice, and liquid dispensers built into the door. See Fig. 10-24. In some, the refrigerator-freezer is being used as a kitchen entertainment center, with radio and cassette systems, digital clocks with timers, alarms, etc. Some units will also have electronic monitor consoles that control all accessories and current.

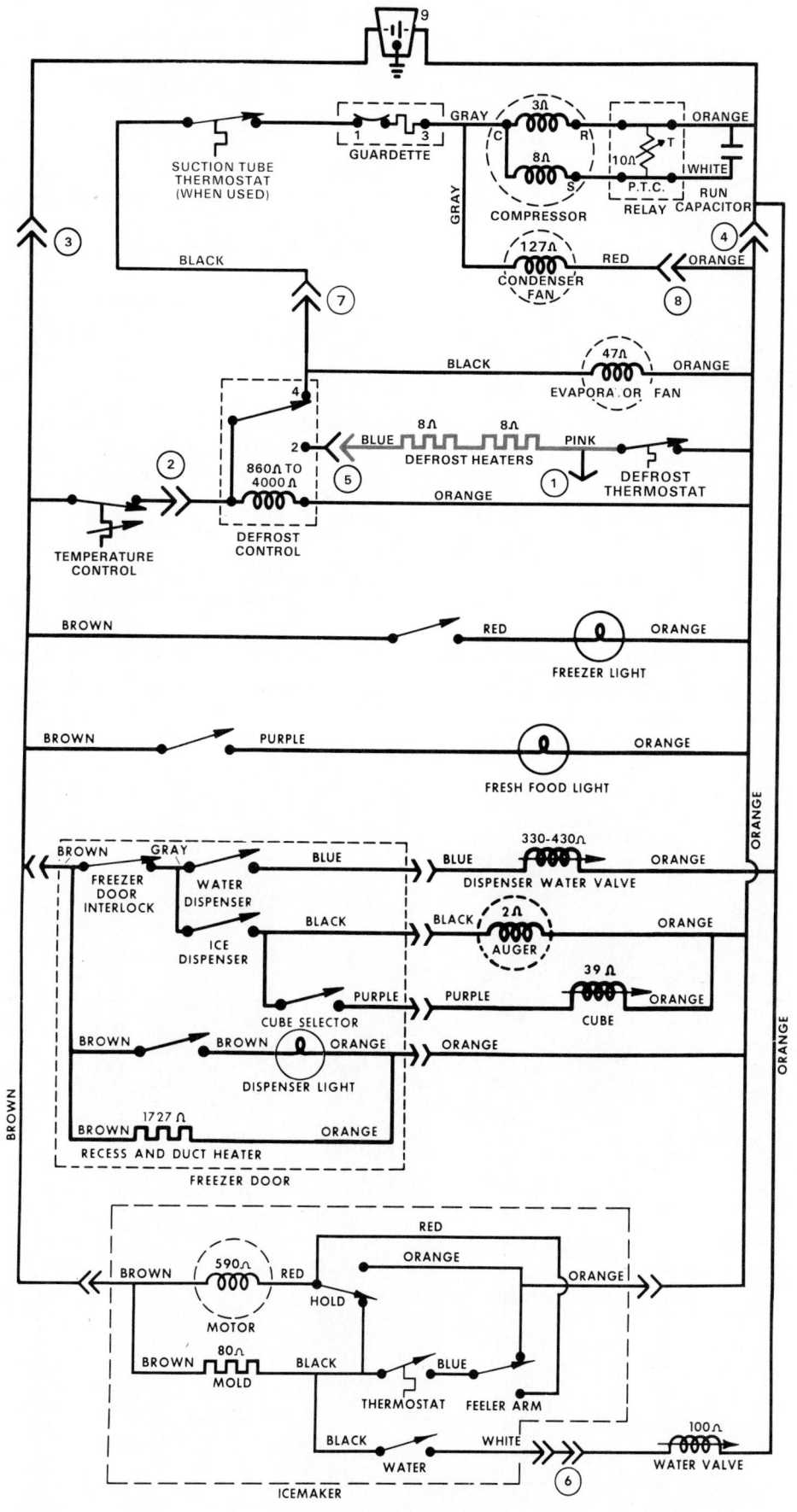

Fig. 10-23. Ladder wiring diagram for frost-free refrigerator with frozen food compartment above and fresh food compartment below.
(General Electric Co.)

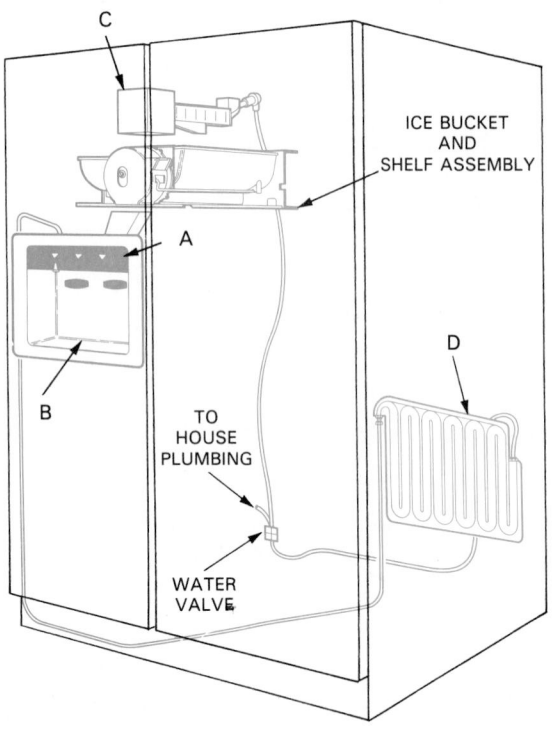

Fig. 10-24. Chilled water and ice maker system construction for side-by-side refrigerator-freezer. A—Electronic control center. B—Ice and water dispenser. C—Ice maker. D—Water reservoir.
(General Electric Co.)

10-26 CABINETS
(Frost-Free Refrigerator-Freezer, Side-by-Side, with Ice and Water Dispenser and Electronic Monitor—General Electric)

The side-by-side refrigerator/freezer is shown in Fig. 10-25. This unit provides an automatic ice and water dispenser controlled by an electronic monitor, and frozen food compartment on the left side. The fresh food compartment is located on the right side. The frozen food compartment stores foods at a satisfactory temperature of 0 °F (−18 °C) or below.

The temperature controls for both compartments are located at the top of the fresh food compartment.

The fresh food compartment defrosts on every off cycle. The freezer evaporator defrosts for about 25 min. after a six-hour accumulated compressor running time. The freezer defrost timer is located on the cabinet front near the bottom. Moisture from the evaporator surface flows down to pan resting on top of the condenser. Heat from the condenser evaporates it. The cabinet is fitted with rollers, which makes it easy to move.

10-27 MECHANISMS
(Frost-Free Refrigerator-Freezer, Side-by-Side, with Ice and Water Dispenser and Electronic Monitor—General Electric)

The evaporator, compressor, and condenser are at the back of a frost-free, side-by-side. A fan circulates air over

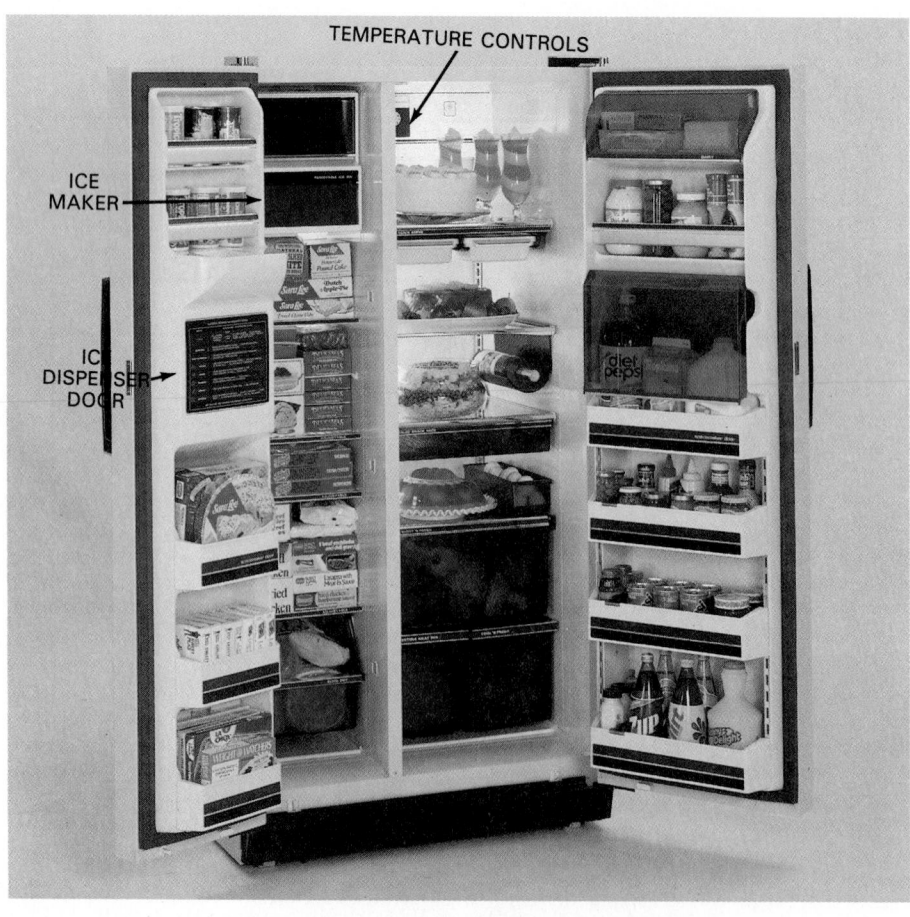

Fig. 10-25. Side-by-side frost-free refrigerator-freezer, with automatic ice maker. (General Electric Co.)

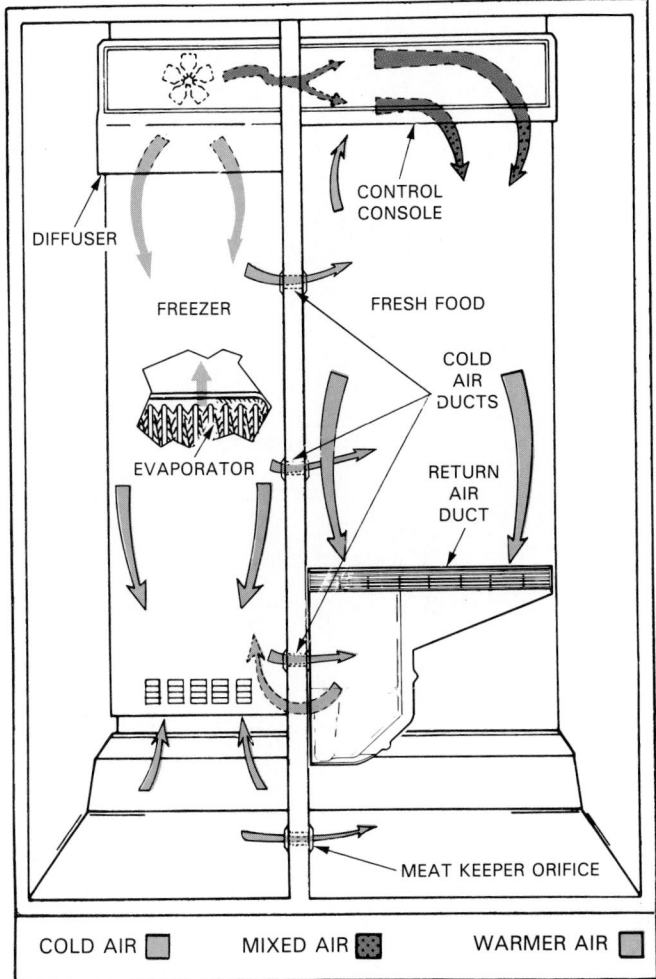

Fig. 10-26. Airflow diagram for frost-free refrigerator-freezer, side-by-side. (General Electric Co.)

the condenser. This air enters and leaves through the bottom grille.

A fan circulates very cold air from the evaporator into the freezer compartment. Damper arrangements allow some of this cold air to flow from the freezer compartment into the fresh food compartment. Warmer air in the fresh food compartment returns to the evaporator compartment. The airflow is shown in Fig. 10-26.

Refrigerant control is by capillary tube. This capillary tube is attached to the suction line, as shown in Fig. 10-27.

An electronic control console is located at the front of the freezer door, Fig. 10-28, above the ice and water dispenser. The control operates by locating the identified selection pad and touching it:

1. The ice cubes or crushed ice selection pad.
2. The dispenser light for nighttime lighting of controls.
3. The door alarm, that will beep after either the fresh food or freezer door has been opened longer than 30 sec.
4. Door open monitor, that indicates when fresh food or freezer door is open or ajar.
5. Warm temperature monitor. This monitor can indicate when the freezer temperature has been above normal for one to four hours.
6. Normal indicator, that indicates there are no failures detected by the diagnostic system.

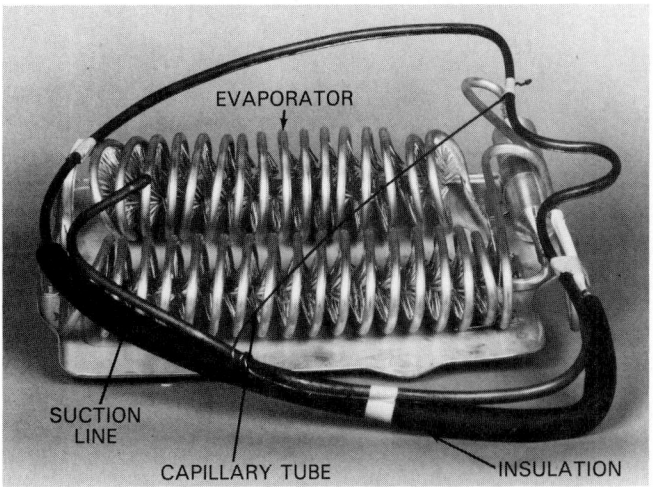

Fig. 10-27. Capillary tube and suction line connected to evaporator. (General Electric Co.)

7. Diagnostic code will be flashed on the control console when an abnormal condition exists. If more than one code function requires service, the highest priority will be displayed until corrected. Then the next highest will occur. The following codes are listed in priority:
   FF— Frozen foods should be checked for thawing.
   PF— Power has been interrupted.
   CI— Icemaker is not operating properly.
   dE— Defrost has not been detected for an extended amount of time.
   CC—Freezer temperature is warm.
8. System check/reset reviews all five diagnostic codes and will identify any abnormal condition that still exists. If there are no abnormal conditions that exist, the normal light will appear briefly.

The electronic control console is connected to the wiring in the system as shown in Fig. 10-29.

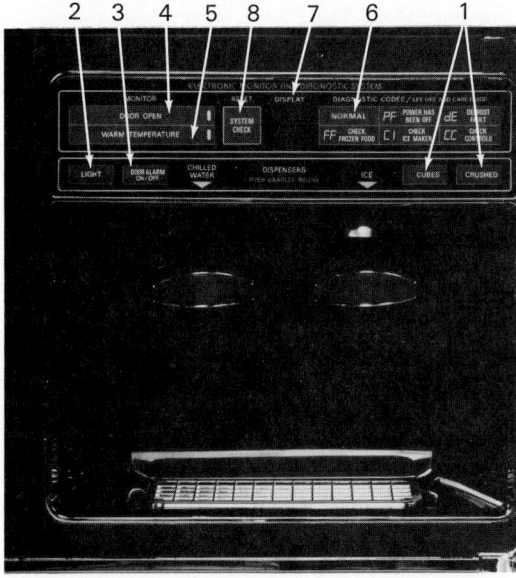

Fig. 10-28. Electronic control monitor diagnoses various operating difficulties. Monitor is used by owner and service technician to determine problems. (General Electric Co.)

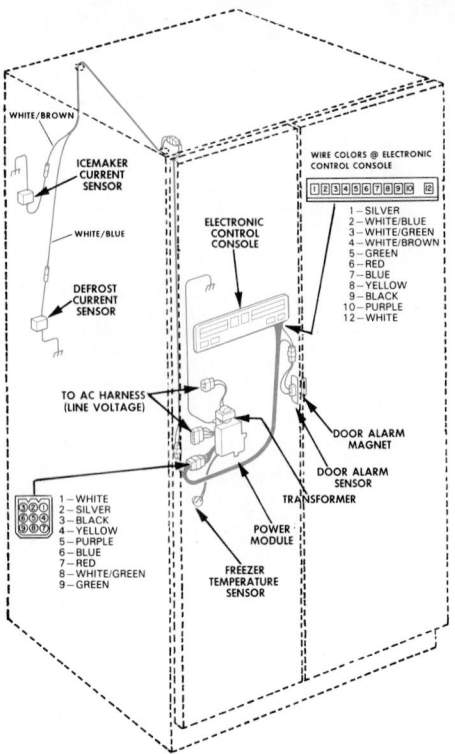

Fig. 10-29. Refrigerator with electronic control console. Note various control sensors. (General Electric Co.)

Fig. 10-31. Side-by-side refrigerator-freezer with ice and water dispenser. Frozen food temperatures are controlled by thermostat in left side, in upper left-hand compartment. Fresh food temperature thermostat is in right side. (Amana Refrigeration, Inc.)

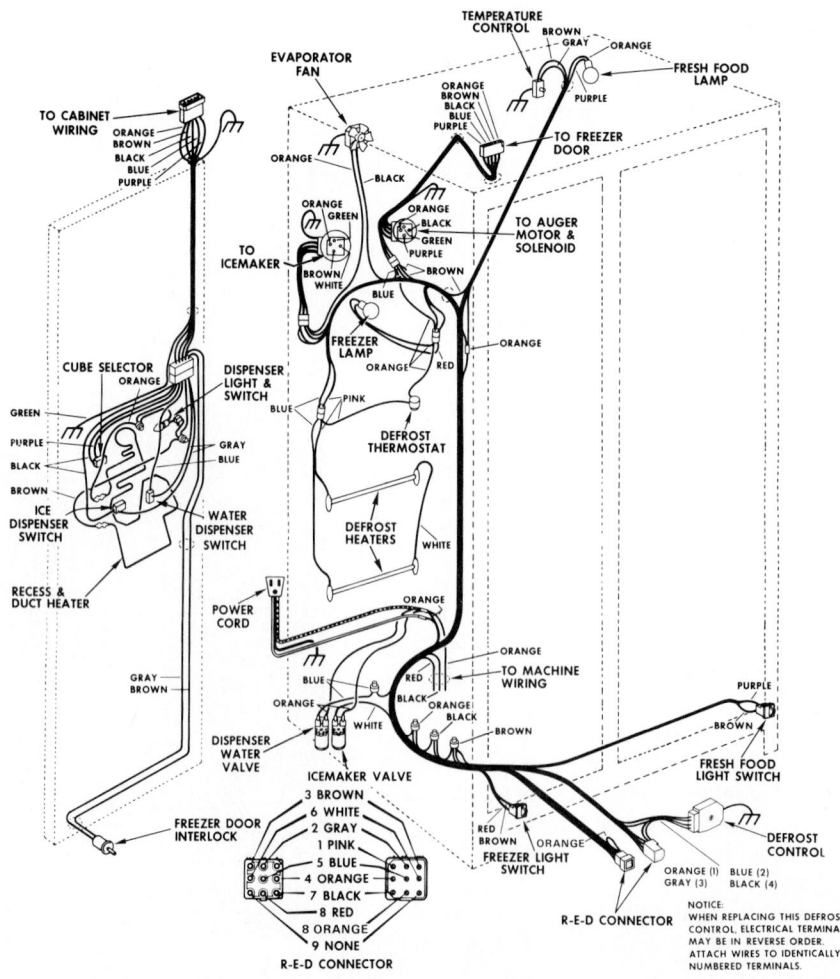

Fig. 10-30. Wiring harness diagram of side-by-side refrigerator with automatic ice maker. (General Electric Co.)

10-28 ELECTRICAL CIRCUITS
(Frost-free Refrigerator-Freezer, Side-by-Side, with Ice and Water Dispenser and Electronic Monitor—General Electric)

A grounded extension cord and plug supply electricity. Fig. 10-30 is a wire harness diagram. The individual circuits can be worked on at any time since wiring is not "foamed" in place. Note that there are heat elements for defrost heaters and a recess and duct heater. Color coding locates the wire going to each of the electrical circuits.

The automatic ice maker controls are plugged into an ice maker receptacle. This eliminates the need to disturb any of the circuits when installing or removing the ice maker.

10-29 CABINET
(Frost-free Refrigerator-Freezer, Side-by-Side with Ice and Water Dispenser—Amana)

The refrigerator in Fig. 10-31 provides a fresh food compartment on the right and a frozen food compartment on the left. Shelves in the door of the frozen food compartment make the handling of small items quite convenient.

The center left door contains an external automatic dispensing unit for ice cubes and water, Fig. 10-32.

Airflow through the cabinet is shown in Fig. 10-33. A fan draws air through the single evaporator and distributes it throughout the cabinet. Dampers control temperatures in the various parts of the cabinet.

Door switches control two lights in the refrigerator compartment and one light in the freezer compartment. A resistance wire or wires are placed inside the center mullion and around the door openings to control condensation.

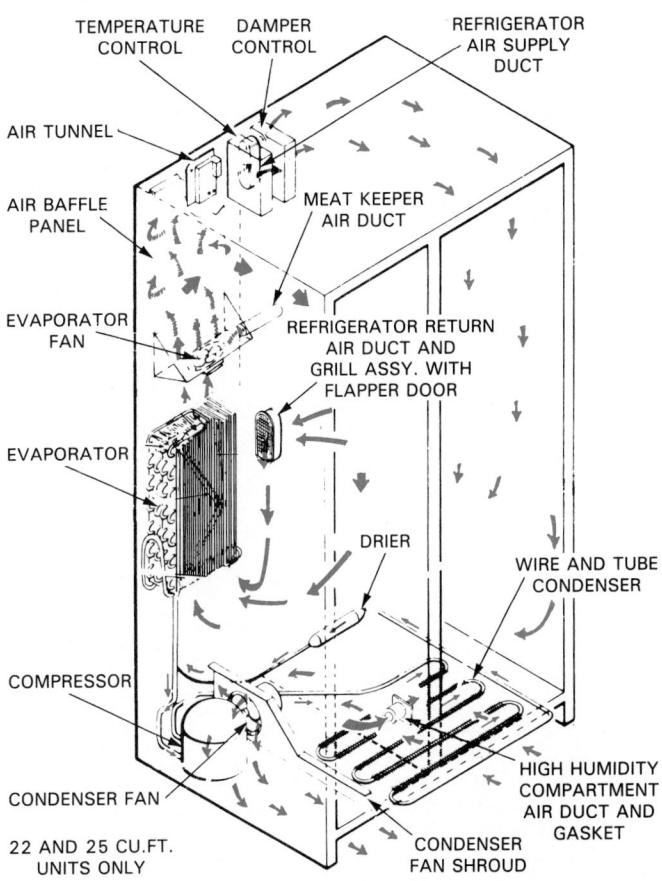

Fig. 10-33. Air movement pattern in side-by-side refrigerator-freezer. (Amana Refrigeration, Inc.)

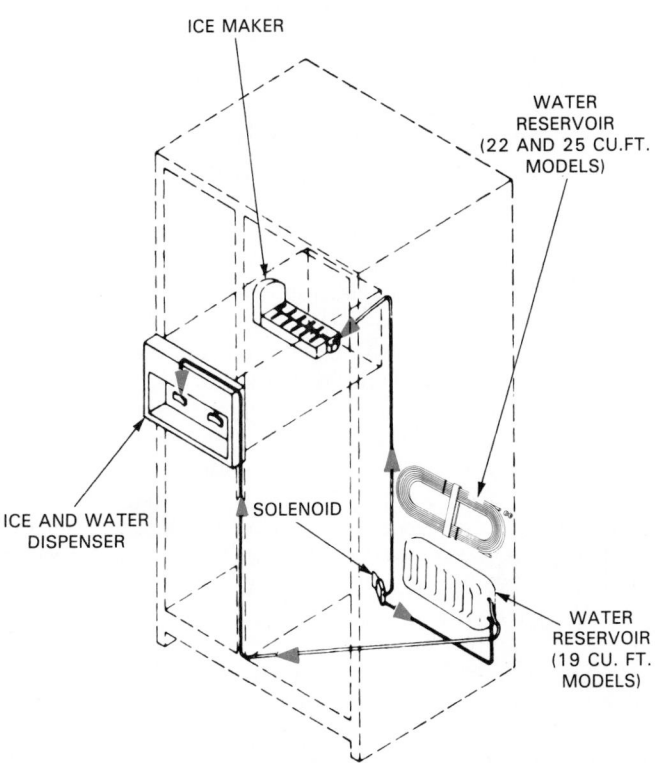

Fig. 10-32. Side-by-side refrigerator-freezer with ice and water dispenser. Note location of water reservoir in fresh food side and note water flow circuit. (Amana Refrigeration, Inc.)

Other features include magnetic door gaskets on four sides and base rollers for easy moving of the unit.

10-30 MECHANISMS
(Frost-free Refrigerator-Freezer, Side-by-Side, with Ice and Water Dispenser—Amana)

The evaporator is behind the frozen food compartment, while the compressor and condenser are in the bottom. See Fig. 10-34. Air, circulated over the condenser by a fan, enters and leaves through the bottom grille. A fan on the evaporator circulates very cold air in the freezer compartment.

Damper arrangements allow some of this cold air to flow into the fresh food compartment. The fresh food compartment acts as a return air duct from the freezing compartment back into the evaporator compartment.

Refrigerant control is by capillary tube. This tube is attached to the suction line as in Fig. 10-34, and is called the heat exchanger.

The refrigerator automatically defrosts every six hours of compressor running time. The defroster is an electric heater attached to the evaporator. It is energized by a switch which is turned on and off by a timer. A defrost terminator (thermostat) is attached to the left end plate of the evaporator and opens the heater circuit at approximately 50°F (10°C). When 28 min. have passed after the start of the defrost cycle, the timer restores the operation of the compressor and air circulating fan. The defrost terminator contacts close (reset) at about 20°F (−7°C).

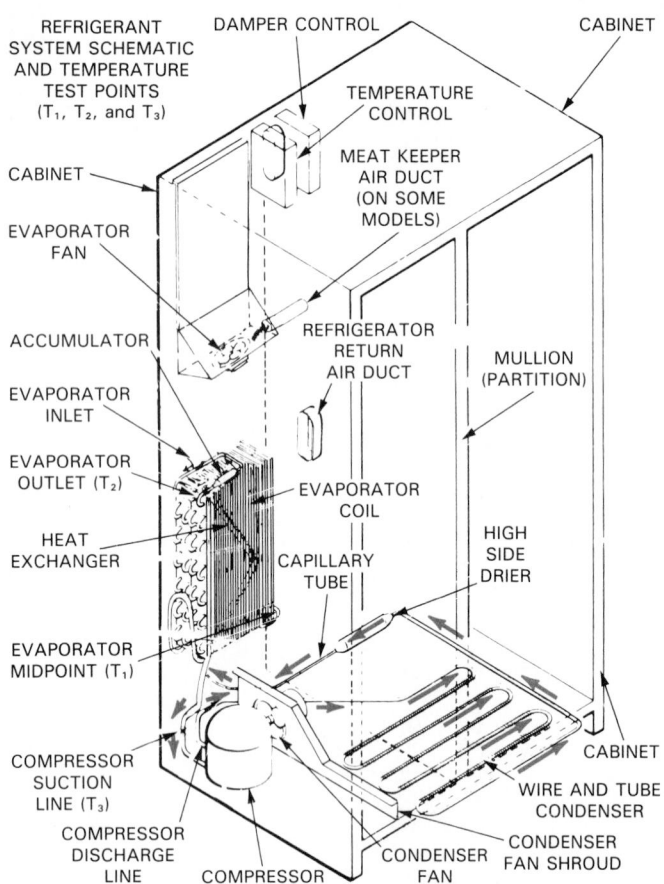

REFRIGERANT SYSTEM SCHEMATIC AND TEMPERATURE TEST POINTS (T_1, T_2, and T_3)

DAMPER CONTROL

CABINET

TEMPERATURE CONTROL

MEAT KEEPER AIR DUCT (ON SOME MODELS)

CABINET

EVAPORATOR FAN

ACCUMULATOR

REFRIGERATOR RETURN AIR DUCT

MULLION (PARTITION)

EVAPORATOR INLET

EVAPORATOR OUTLET (T_2)

EVAPORATOR COIL

HIGH SIDE DRIER

HEAT EXCHANGER

CAPILLARY TUBE

EVAPORATOR MIDPOINT (T_1)

COMPRESSOR SUCTION LINE (T_3)

CABINET

WIRE AND TUBE CONDENSER

COMPRESSOR DISCHARGE LINE

COMPRESSOR

CONDENSER FAN

CONDENSER FAN SHROUD

Fig. 10-34. Construction of side-by-side refrigerator-freezer. Note that condenser lies flat underneath bottom of refrigerator. Compressor is under freezing compartment at back. Evaporator is behind freezer compartment. The various cabinet temperatures are obtained by use of dampers which control flow of cold air into compartments. (Amana Refrigeration, Inc.)

10-31 ELECTRICAL CIRCUITS
(Frost-free Refrigerator-Freezer, Side-by-Side, with Ice and Water Dispenser—Amana)

Electricity is supplied through a grounded extension cord and plug. Note electrical circuits for this refrigerator in Fig. 10-35. Wiring is located in the foamed-in-place insulation.

There are a considerable number of heaters in the electrical circuit—defrost heaters, mullion heater, butter cavity heater, freezer flange heater, water tank heater, freezer control duct heater, and ice chute heater.

An auxiliary heater is foamed in place next to the connected heater. Should the original fail, the auxiliary may be connected into the circuit by disconnecting the existing plastic-covered male and connecting it to the extra plastic-covered male terminal. The electrical connections are located behind the evaporator covers. These heaters are in series with a switch on top of the freezer control. They may be turned off if the refrigerator is being operated in areas of extremely low humidity. This is called a power saver switch.

A color code identifies the wires going to each of the electrical circuits in the cabinet. The automatic ice maker controls are plugged into an ice maker receptacle. This eliminates the need to disturb any of the circuits in order to remove or restore the ice maker.

10-32 SOLID-STATE CONTROLLED ICE MAKER

In some of the newer models, the automatic ice cube maker controls are solid-state. They use transistors, diodes, relays, and other semiconductor components reviewed in Chapter 6. These controls are then assembled into a single printed circuit board, as shown in Fig. 10-36. Inputs (ice level, temperature, power) and outputs (motor power, switch signals) are fed into the circuit board through an edge connector.

The circuitry on the circuit board requires two types of voltages:
1. Low voltage dc signals that operate the semiconductor devices. The temperature sensor and the ice ejection cycle controls are examples of the functions directly controlled by this type of voltage signal.
2. A 120 V ac line voltage to operate the drive motor and water valve.

A thermistor is located in the front of the ice cube tray. It senses the temperature at that location. When the thermistor senses a reduction of temperature below 13°F (−11°C) it sends a signal back to the solid-state control circuit. The control system then triggers a relay which completes the circuit to drive the motor. This initiates the harvest, or ice cube ejection, cycle.

The mechanical ice ejection process is done using a cam gear mechanism. The unit in Fig. 10-37 produces ice cubes. The ice is deposited in an ice bucket or ejected out through a duct in the freezer door. General operation of the ice maker is controlled through a series of switches in the line voltage circuit to drive the motor and water valve. These switches are used to control and monitor the ice ejection process. They usually operate through a sliding pin setup which follows a cam molded into the front surface of the cam gear. The solid-state ice cube maker also has an ice level switch that stops the ice maker from ejecting new ice if the container is full.

10-33 CHEST-TYPE FREEZERS

The chest-type freezer has certain advantages. Since cold air is heavier than warm air, the very cold air in a chest-type freezer does not spill out each time the lid is opened. This stops a considerable amount of moisture from entering the cabinet. There is little air change when the cabinet is opened.

To make chest-type freezers more convenient to use, they are usually fitted with baskets that may be lifted out to provide access to frozen food packages near the bottom. Also, the lids usually have a counterbalancing mechanism which makes them easy to open. A light in the lid gives good illumination. The chest-type freezer provides the most economical type of food freezing mechanism.

Most chest-type freezers require a manual defrost. But since so little moisture enters the freezer, defrosting is usually not needed more than once or twice a year.

Defrosting may be accomplished best by unplugging the condensing unit. Then remove the stored food and place either an electric space heater or a pail or two of hot water in the cabinet. With the cabinet closed, the ice will soon drop away from the evaporator surface and will be easy to remove.

Most chest-type freezers have a drain which makes removing the moisture from the cabinet quite easy.

Domestic Refrigerators and Freezers

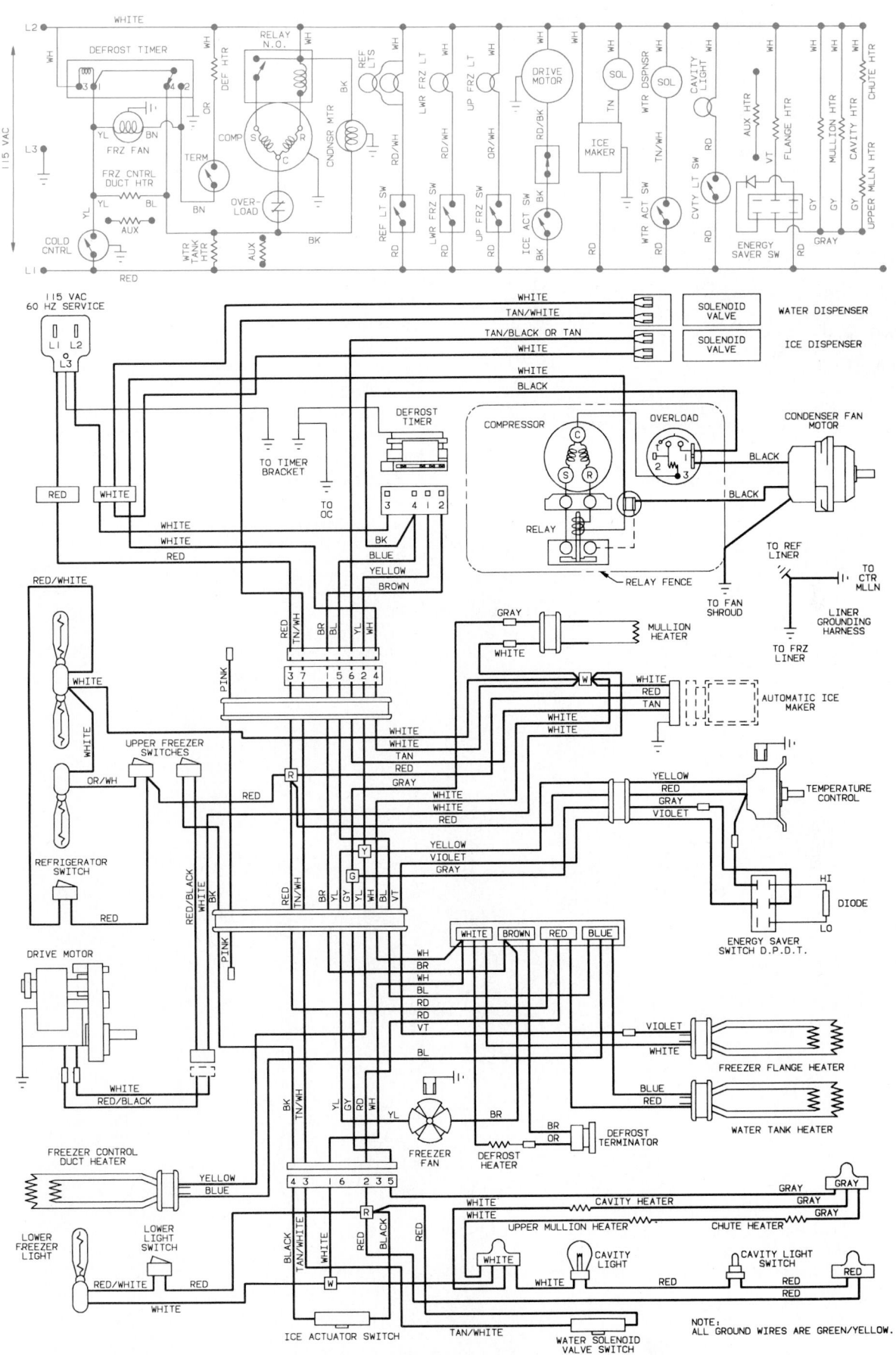

Fig. 10-35. Top portion is ladder wiring diagram for side-by-side refrigerator with automatic defrost and automatic ice maker. (Amana Refrigeration, Inc.)

Fig. 10-36. Printed circuit board for solid-state controlled ice maker. (Frigidaire Company)

Remaining moisture must be wiped out of the cabinet.

Cabinets are available in various capacities. Height and width are quite uniform. However, the length will vary with the capacity of the freezer.

10-34 CABINET, MECHANISMS, AND ELECTRICAL CIRCUITS (Chest-Type Freezer—Kelvinator)

The outer and inner shells of the chest-type freezer, Fig. 10-38, are metal. The evaporator surrounds the inner liner and is attached to it. The condenser is attached to the inside of the outer shell and completely surrounds the cabinet.

Mechanisms consist of common parts. The hermetic compressor is located at the lower right end of Fig. 10-39. The liquid refrigerant flows through the capillary tube and into the evaporator. There, evaporation of the refrigerant and cooling takes place. The compressor draws the vaporized refrigerant through the compressor and pumps it into the precooler condenser on the back wall of the freezer. Here, it releases part of its latent heat of vaporization and sensible heat of compression.

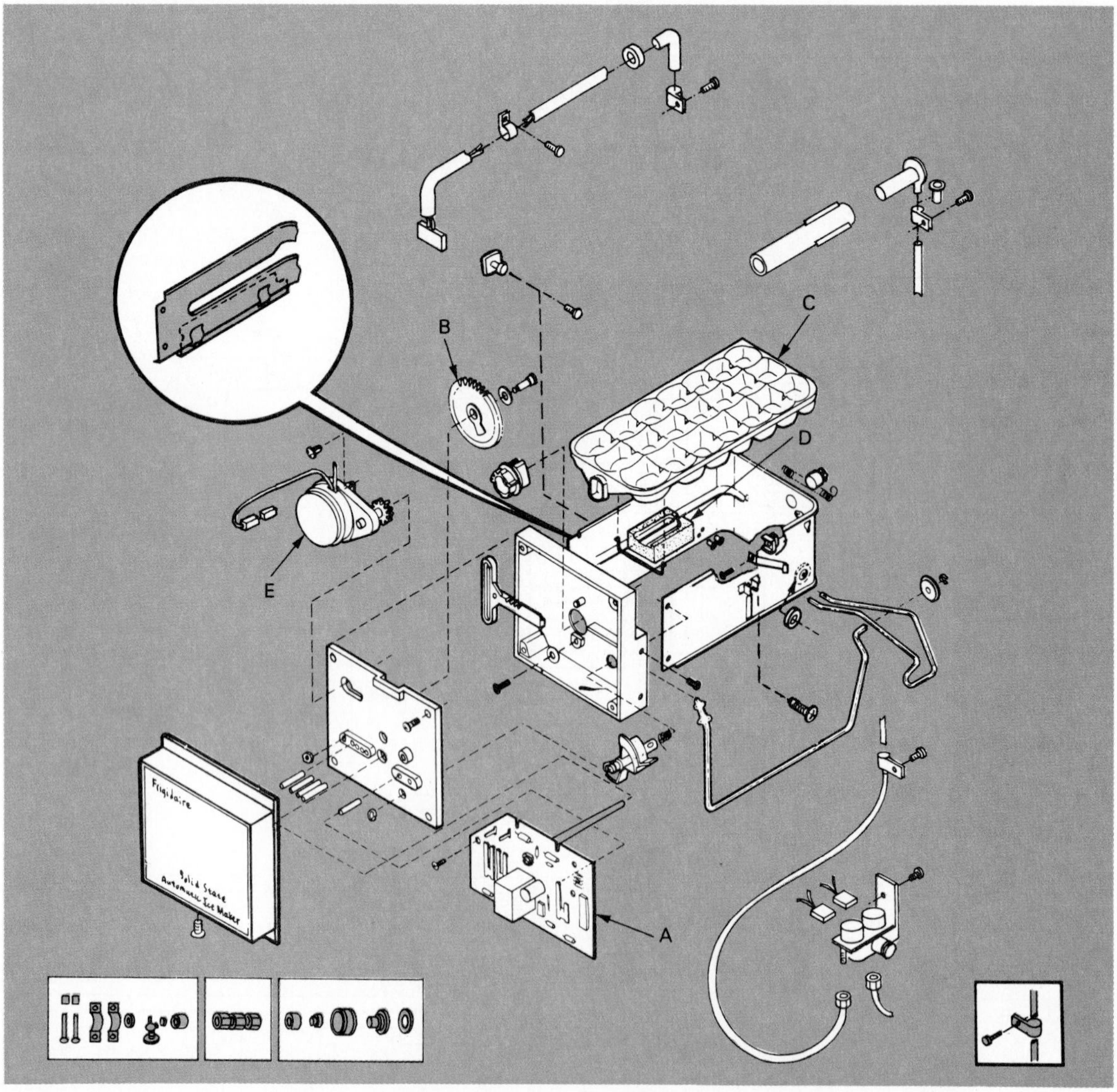

Fig. 10-37. Construction of ice dispenser. Note: A—Circuit board. B—Cam gear. C—Ice tray. D—Thermistor insulation. E—Motor. (Frigidaire Company)

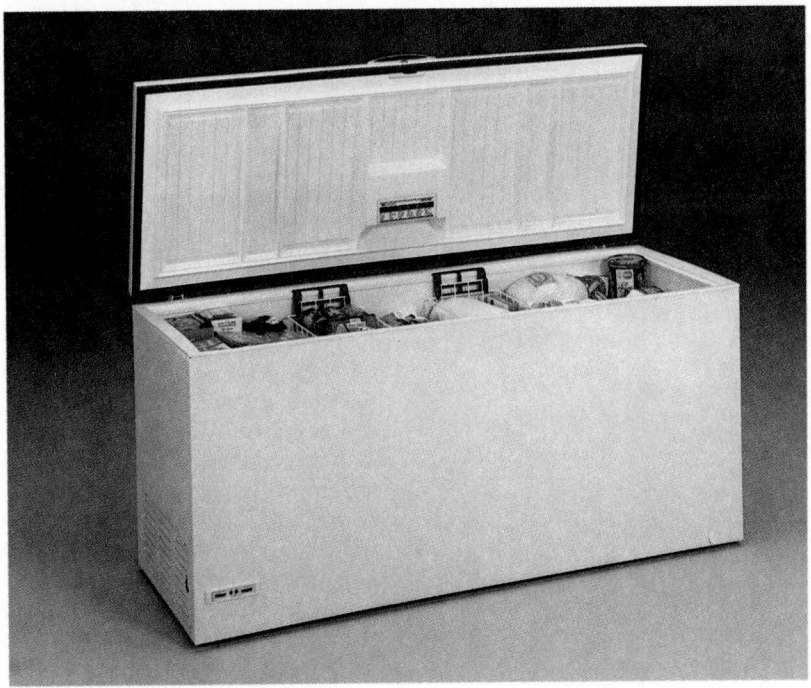

Fig. 10-38. Chest-type food freezer. The use of wire baskets makes it easy to reach items stored in bottom of freezer. (Frigidaire Company)

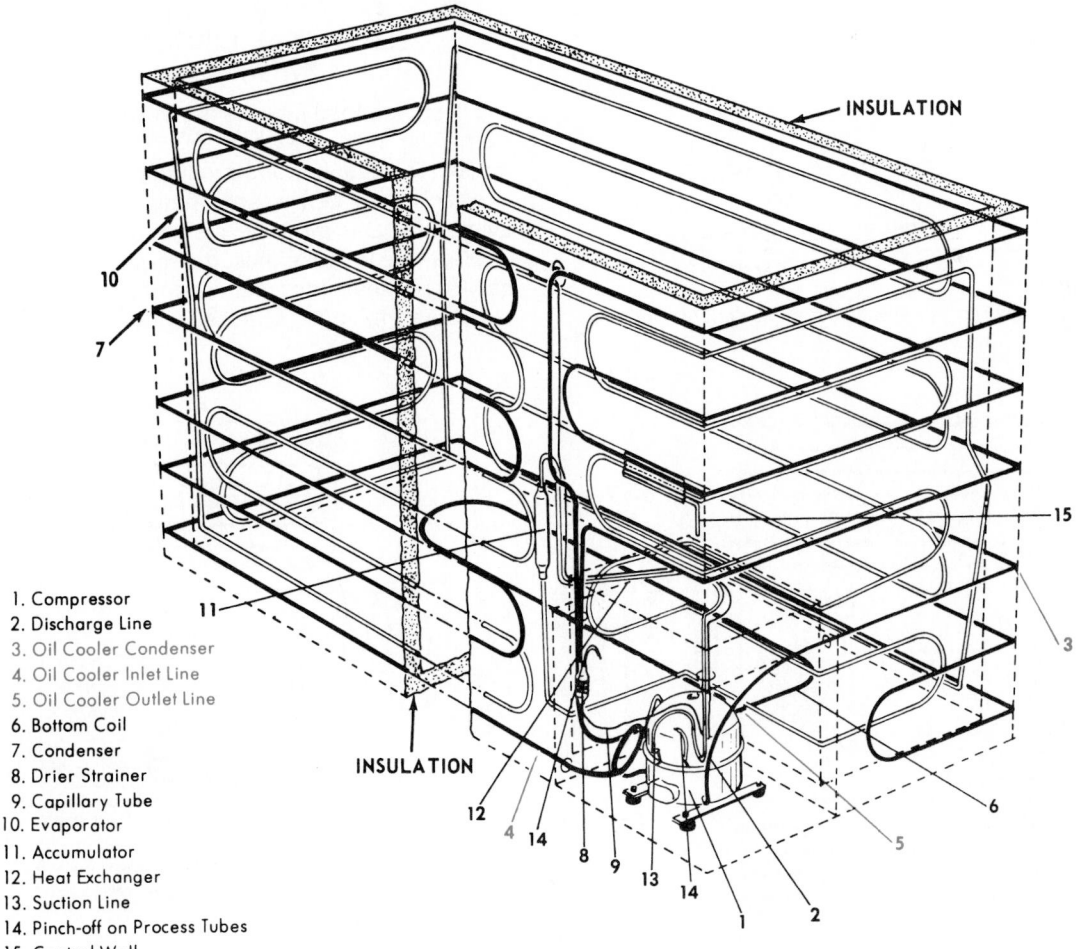

1. Compressor
2. Discharge Line
3. Oil Cooler Condenser
4. Oil Cooler Inlet Line
5. Oil Cooler Outlet Line
6. Bottom Coil
7. Condenser
8. Drier Strainer
9. Capillary Tube
10. Evaporator
11. Accumulator
12. Heat Exchanger
13. Suction Line
14. Pinch-off on Process Tubes
15. Control Well

Fig. 10-39. Chest-type freezer evaporator and condensing unit. Note special oil cooler condenser at 3 at right of cabinet, oil cooler inlet line shown at 4, and oil cooler outlet line at 5.

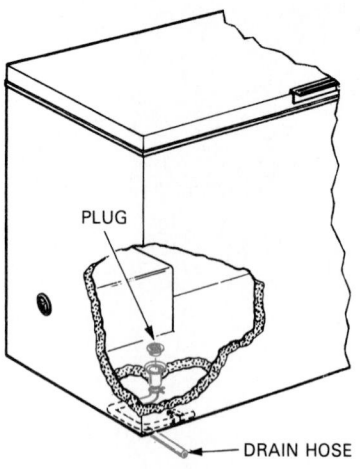

Fig. 10-40. Drain system, chest-type freezer. (Frigidaire Company)

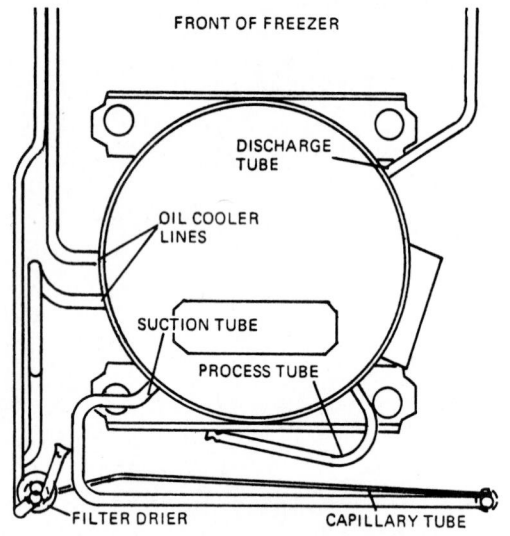

Fig. 10-41. Compressor dome showing oil cooler connections.

Since chest-type freezers are manually defrosted, condensate (water) is usually drained out through the bottom of the cabinet, as in Fig. 10-40.

From the precooler condenser, the refrigerant passes back to the machine compartment and through the oil cooling coil in the compressor dome. See Fig. 10-41. Here, additional heat is picked up from the oil. The compressed vapor then flows back to the main condenser where additional heat is released to the atmosphere. The refrigerant condenses from a high-pressure vapor to a high-pressure liquid.

Since the condenser tubes are in contact with the outer shell of the cabinet, heat from the condenser passes into the outer shell and warms it slightly. This causes a natural flow of warm air upward over the cabinet shell preventing sweating. The liquefied refrigerant collects in the bottom of the condenser tubing, flows into the filter-drier, moves into the capillary tube, on into the evaporator, and the cycle repeats. The cycle is shown in Fig. 10-42.

Electrical power is supplied through a grounding type three-prong extension plug and cord. Fig. 10-43 is a schematic wiring diagram for this freezer. The starting relay and overload protector are attached to the motor by a "push-on" type mount, Fig. 10-44. Note that all parts of the mechanism and the cabinet are grounded. The thermostat is at the end of the cabinet near the top of the compressor compartment. The dial is marked for "off," "normal," and "cold" positions.

10-35 CABINET, MECHANISMS, AND ELECTRICAL CIRCUITS (Chest-Type Freezer—Amana)

Chest-type food freezers are available in various capacities from 7 cu. ft. to 28 cu. ft. Fig. 10-45 shows a 10 cu. ft. model. The evaporator surrounds the inner metal lining and is attached to it. The schematic, Fig. 10-46, shows the evaporator and the low-side refrigerant circuit.

When the unit operates, the refrigerant flows through the capillary tube into the evaporator tubes. Since the evaporator tubing is attached to the cabinet's inside lining,

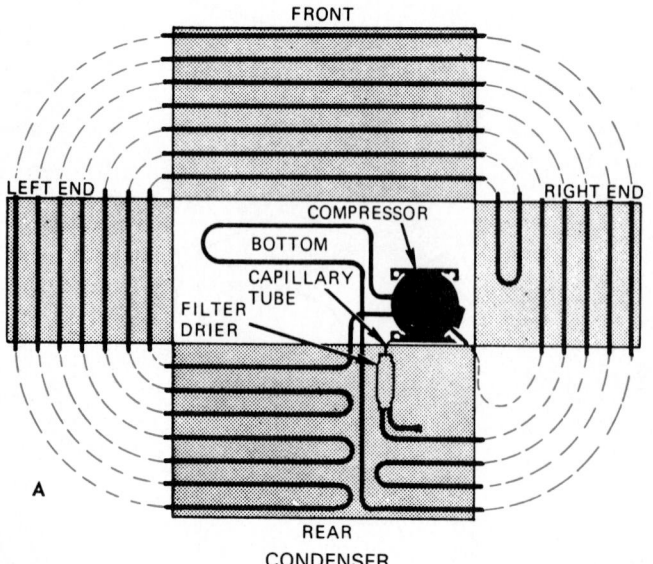

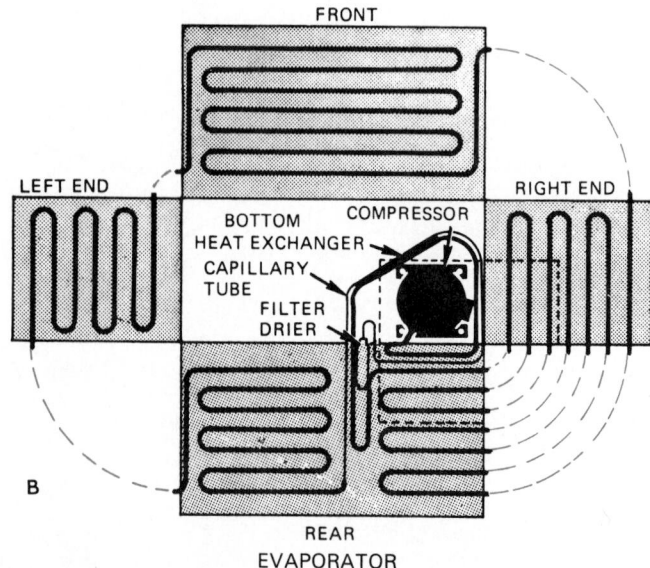

Fig. 10-42. Refrigeration cycle diagram for chest-type freezer. A—High side of cycle. Heat absorbed in evaporator is now released by condenser into surrounding atmosphere. B—Low side of cycle. Heat is absorbed by evaporator in cabinet. (Frigidaire Company)

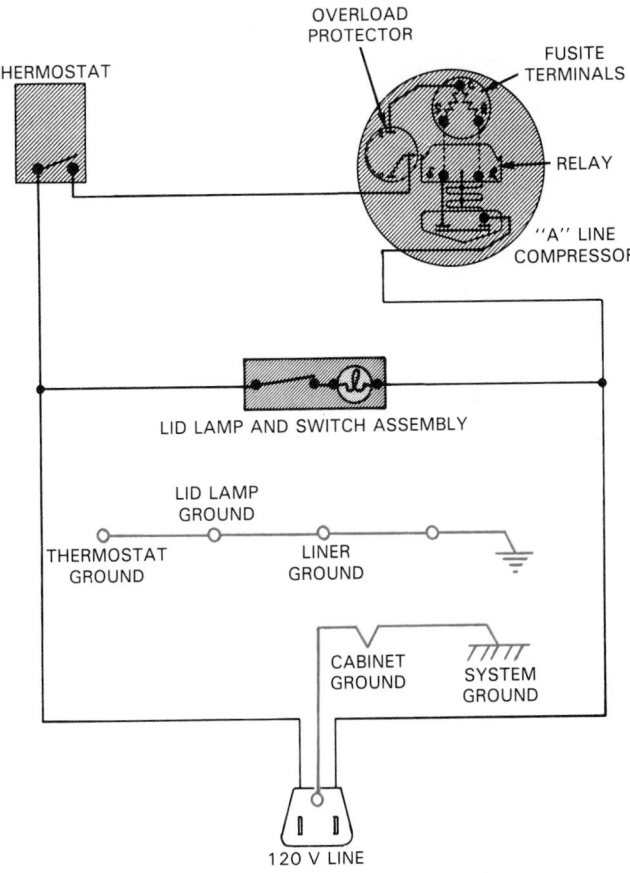

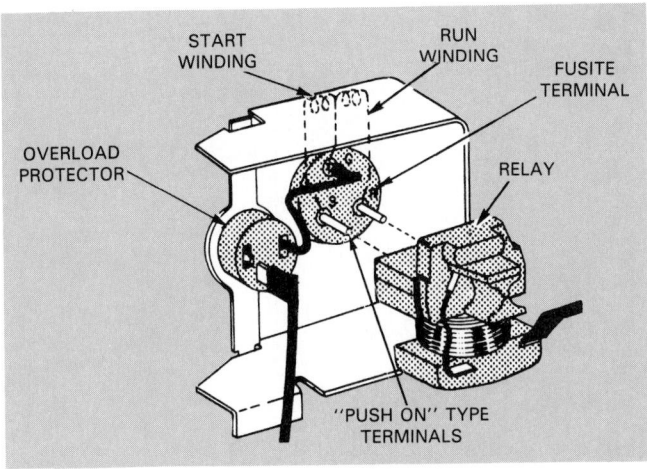

Fig. 10-44. Motor starting relay and overload protector. Motor terminals are to receive starting relay using "push-on" type terminals. Overload protector is connected in same way.

Fig. 10-43. Schematic wiring diagram for chest-type freezer.

The temperature control is wired into the motor circuit. It controls the running time of the compressor to maintain desired cabinet temperatures. The normal operating temperature range for this freezer should be between $-14\,°F$ ($-26\,°C$) and $6.5\,°F$ ($-14\,°C$). All assemblies are grounded through the green wire. This is connected to the grounding terminal of the attachment plug.

10-36 UPRIGHT FREEZERS

The upright freezer makes storage and removal of frozen food convenient. Frost-free and automatic defrost mechanisms have made these freezers very practical.

the entire inside surface of the cabinet is cooled. The refrigerant will be nearly all evaporated by the time it has passed through the evaporator tubes. Any remaining liquid flows into an accumulator placed at the end of the coil. The entrance to the accumulator is from the bottom. Thus, the accumulator holds the liquid refrigerant until it is all evaporated.

The accumulator outlet is at the top. It leads into the suction line, which connects to the inlet side of the compressor. The vapor goes to the compressor and is compressed to the high-side pressure.

The condenser is attached to the inside of the outer shell and completely surrounds the cabinet. See Fig. 10-47. High-temperature vapor from the discharge side of the compressor flows through a precooler, starting at the top of the back of the cabinet. The precooler zigzags across the back of the cabinet down to the compressor. The loop in the bottom of the compressor is immersed (completely covered) in oil. Here, the partially-cooled refrigerant picks up some heat from the oil. This helps lower the oil temperature. From here, the high-pressure vapor is carried to the top end of the cabinet. It zigzags across the ends and front of the cabinet and returns to the filter-drier, completing the high-side circuit.

Electrical power comes through a grounded type three-prong extension plug and cord. See wiring schematic in Fig. 10-48. Note warning light connected into the electrical circuit. *This warning light only indicates whether or not the electrical circuit is "hot." It does not indicate whether cabinet temperature is satisfactory.*

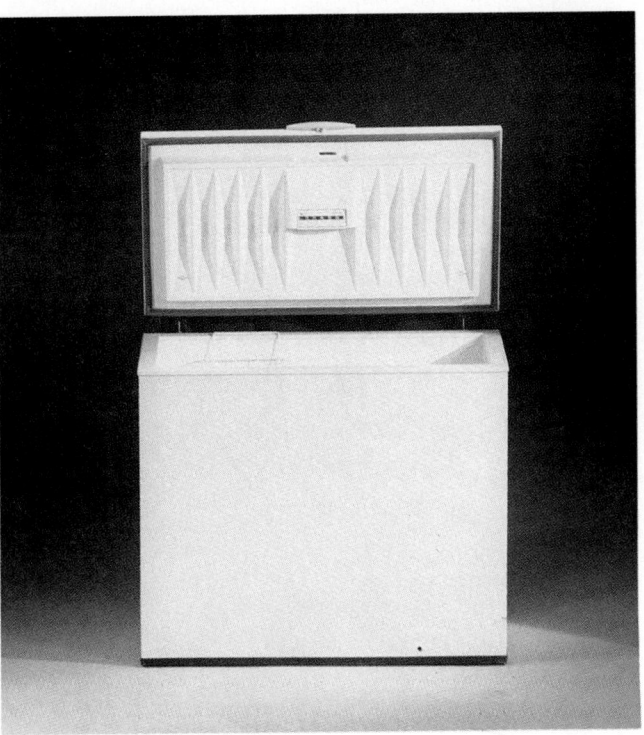

Fig. 10-45. Chest-type freezer. Power on-indicator light is at lower right-hand corner. (Fridigaire Company)

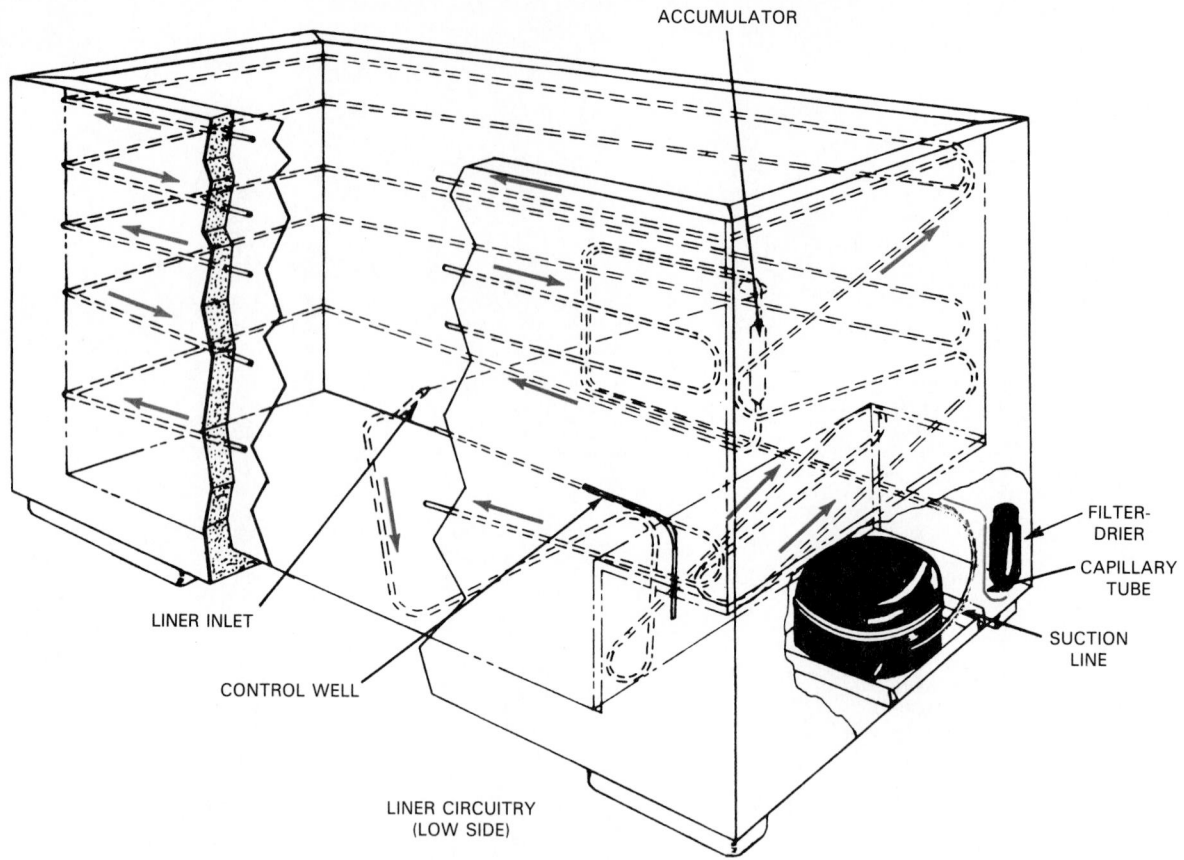

ACCUMULATOR

FILTER-DRIER

CAPILLARY TUBE

SUCTION LINE

LINER INLET

CONTROL WELL

LINER CIRCUITRY
(LOW SIDE)

Fig. 10-46. Chest-type freezer evaporator coil. Evaporator is attached to cabinet inner lining. Capillary tube refrigerant control extends from bottom filter-drier to inlet of evaporator coil.

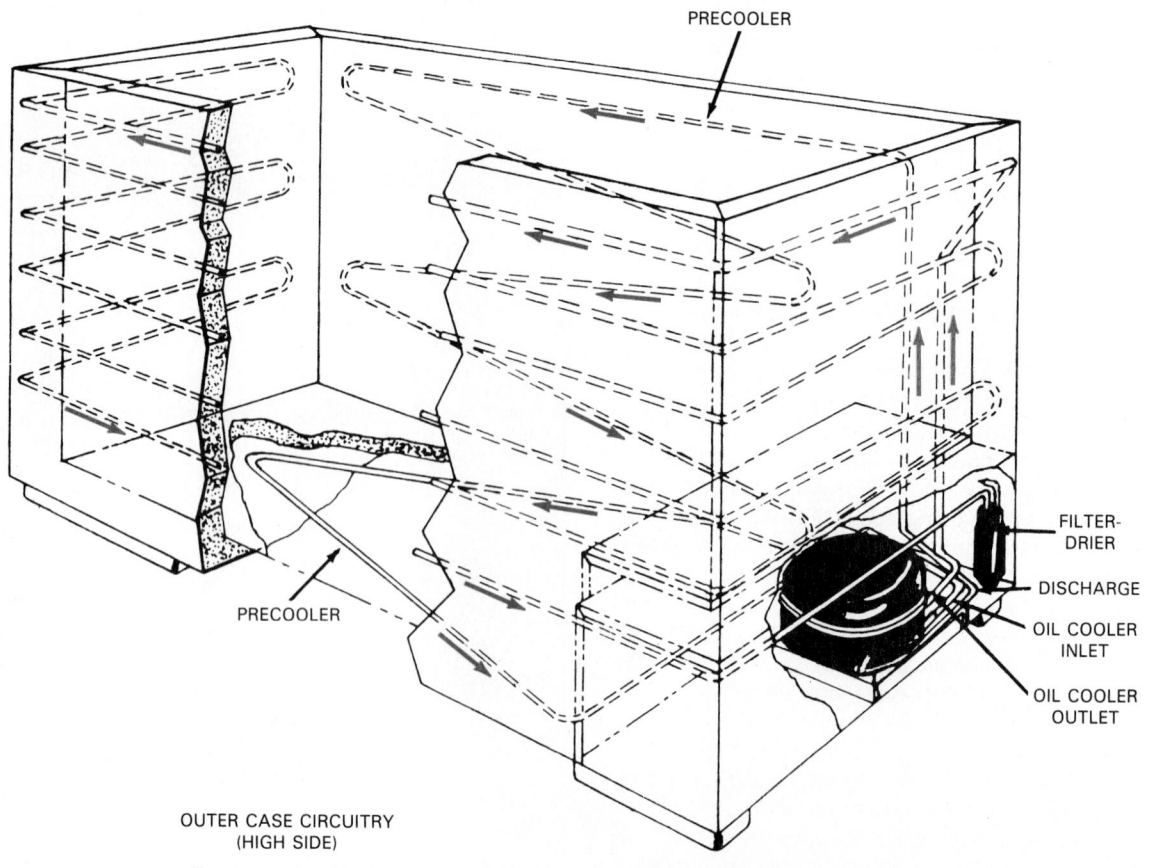

PRECOOLER

FILTER-DRIER

DISCHARGE

OIL COOLER INLET

OIL COOLER OUTLET

PRECOOLER

OUTER CASE CIRCUITRY
(HIGH SIDE)

Fig. 10-47. Chest-type freezer condenser. Coil is attached to inside surface of outer shell. Note precooler coil.

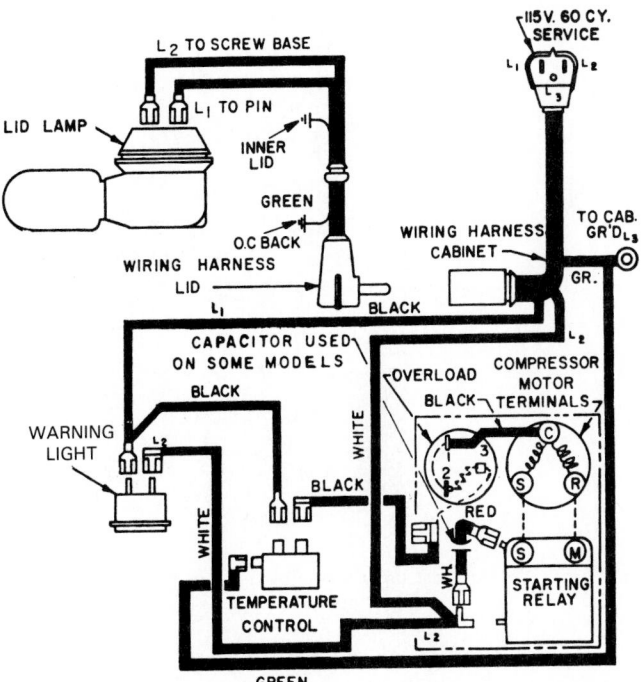

Fig. 10-48. Schematic wiring diagram, chest-type freezer. Parts are identified on illustration. (Amana Refrigeration, Inc.)

Fig. 10-49. Upright domestic freezer. (Frigidaire Company)

General construction is very similar to the upright refrigerator. However, insulation may be a little heavier, and of course, the motor control will be different.

Cabinets are available in varying capacities. However, the range is not as extensive as for chest model freezers. Capacity is lower because of height limitations.

10-37 CABINET, MECHANISMS, AND ELECTRICAL CIRCUITS (Upright Freezer—Kelvinator)

A popular upright freezer, Fig. 10-49, has outer and inner shells of enameled steel. The evaporator is located at the bottom of the cabinet. The wrap-around condenser is inside the outer shell on the sides, back, and top. Door trim is a plastic material which conducts heat poorly.

An automatic defrost cycle is activated every 12 hours of operation. The cycle diagram is shown in Fig. 10-50. Referring to the numbers on the illustration, one can follow the cycle. The evaporator, 9, is shown at the bottom of the drawing in front of the compressor. The evaporator is near the upper one-third of the cabinet. The capillary tube refrigerant control carries the liquid refrigerant from the filter-drier at 2 to the evaporator. The capillary tube, 8, is soldered to the suction line, 11. This forms a heat exchanger at 10.

Upon entering the evaporator, the liquid refrigerant evaporates and absorbs heat. A fan draws the cold air from the bottom of the cabinet through the evaporator, forcing the air to the top of the refrigerator. It is then discharged at the top of the refrigerated space. Air flows down through the refrigerated space and back up to the evaporator.

A precooler condenser, shown at 3 and 6, extends from the compressor along the back shell of the cabinet. It returns to the compressor at 5 and passes refrigerant on to the oil cooling coil in the compressor dome at 1. Here, the refrigerant picks up additional heat from the oil. The

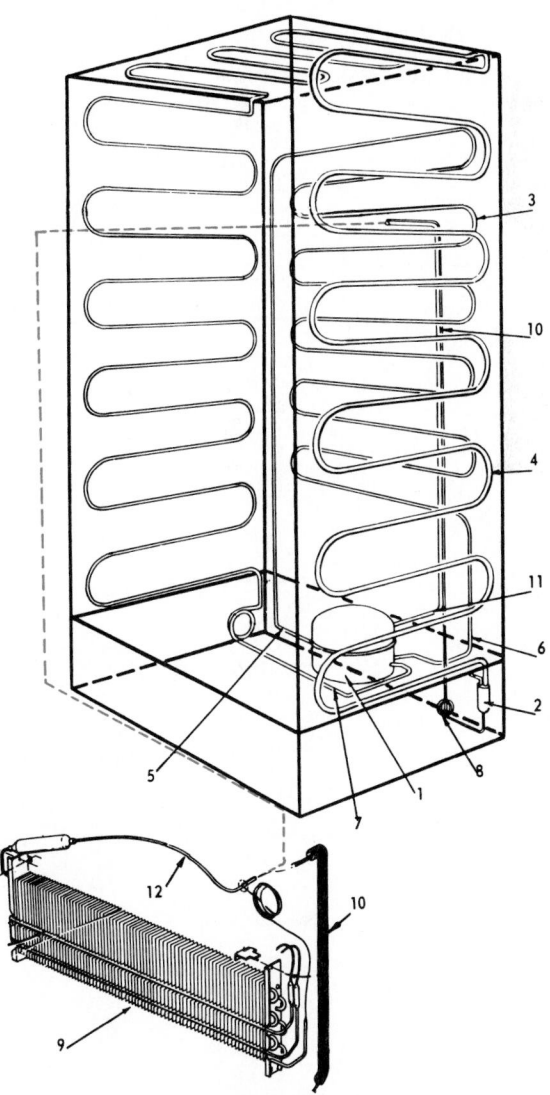

Fig. 10-50. Upright freezer cycle diagram. Note red dashed line showing connection point of evaporator tubing.

refrigerant vapor is then pumped through the tubing at 7 to the main condenser, 4. Here the remaining heat is released to the atmosphere and the refrigerant is condensed to a liquid. It flows by gravity to the filter-drier. This brings the refrigerant back to the capillary tube, where the cycle repeats.

The condenser is attached to the outer shell and warms it slightly. This causes a natural airflow upward over the cabinet shell. The heat from the condenser prevents sweating of the shell. Generally it is desirable to leave some space around the freezer. Otherwise natural convection (airflow) will be restricted.

Electrical power is supplied to the freezer through a grounding type three-prong extension plug and cord. See schematic wiring diagram, Fig. 10-51. Note that all parts of the mechanism and the cabinet are grounded (green wire extends from plug to system ground). The cabinet temperature thermostat is located at the back of the cabinet behind the second shelf from the bottom. It provides three settings: off, normal, and cold. The defrost timer is in the compressor compartment, Fig. 10-52.

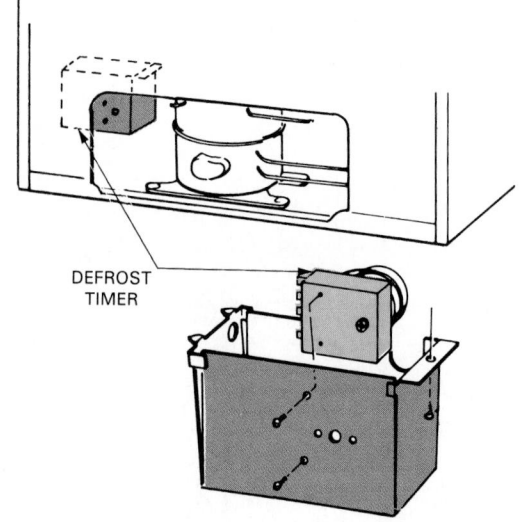

Fig. 10-52. Note defrost timer located in bottom of freezer in compressor compartment.

10-38 CABINET AND ELECTRICAL CIRCUITS (Upright Freezer—Whirlpool)

The upright freezer cabinet arrangement is similar to the one-door refrigerator. The door often has a lock. The evaporator of the manual defrost unit is usually a part of the cabinet shelves.

No-frost systems use an evaporator located behind a baffle. A fan circulates the air through the evaporator and then through the food section. The filter and drier are combined. See Fig. 10-53. Signal lights are often used to indicate when power is on and to warn the owner when cabinet temperature is above normal.

The mechanism is similar to the regular refrigerating system with a motor compressor designed for low evaporator pressures. A forced convection condenser is used with both air-in and air-out grilles located at the lower front of the cabinet.

The wiring is also similar to that on the refrigerator system. Fig. 10-54 shows the wiring of a system designed for manual defrost. Note the cabinet light, switch, and signal light, which indicates that the system has electrical power. A 23.6 VA stile heater is used. Fig. 10-55 shows a system which has a polarized circuit (ground wire, etc.). The cabinet and all other metal parts are grounded.

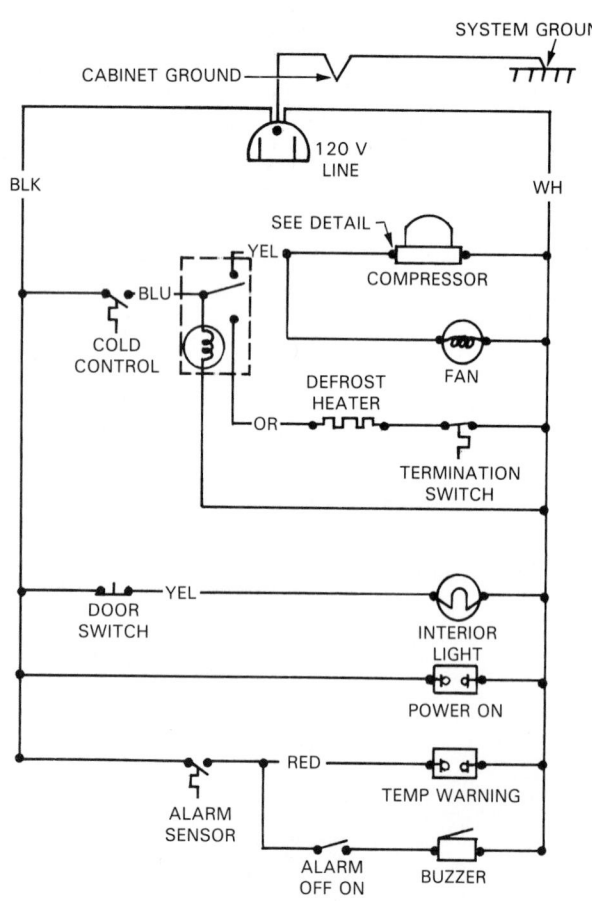

Fig. 10-51. Upright freezer ladder wiring diagram. Note that all parts are grounded through the 3-prong extension plug and cord. Compressor and fan circuits are open during defrost cycle. Note defrost heater and termination switch. (Frigidaire Company)

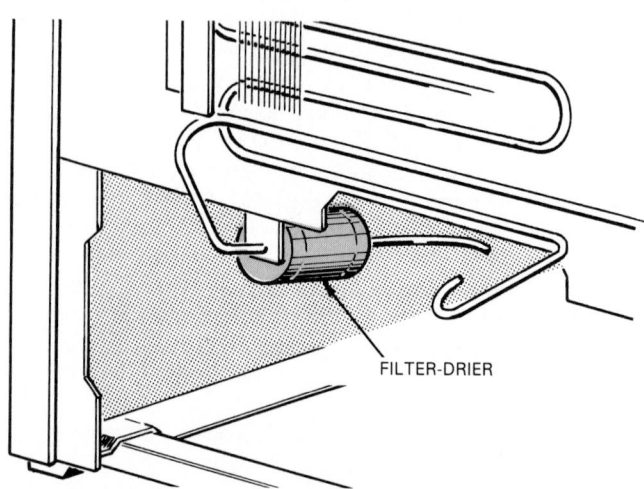

Fig. 10-53. A domestic upright freezer unit with a filter-drier. (Whirlpool Corporation)

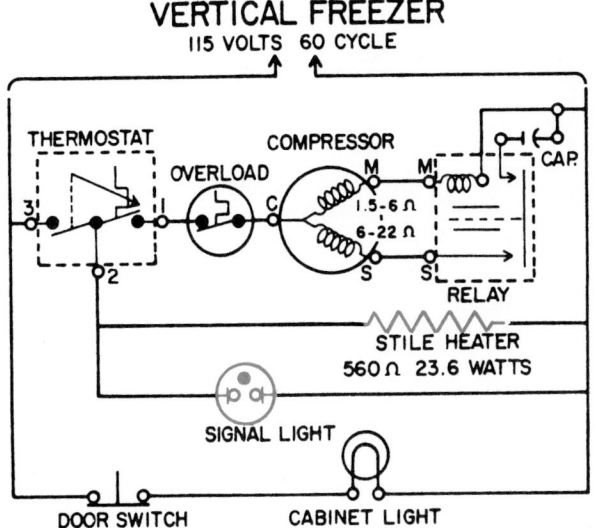

Fig. 10-54. Ladder wiring diagram for manually defrosted vertical freezer. Note stile heater and signal light which indicates power is on.

10-39 CARE OF REFRIGERATOR OR FREEZER

The owner of any domestic refrigerator or freezer should be instructed not to allow the door to remain open when the unit is in operation. The great difference in temperature between the inside of the cabinet and the room temperature will set up convection currents as soon as the door is opened. This will bring a great deal of heat into the cabinet quickly. When removing articles from, or placing them in, the cabinet, do it as quickly as possible.

The cabinet must be kept clean outside as well as inside. The condenser and motor compressor should be wiped clean at least every six months. A vacuum cleaner may be used for cleaning the lint from the condenser. The door gasket should be checked for tightness periodically.

10-40 ICE ACCUMULATION IN CABINET INSULATION

Ice accumulation is one of the main troubles with improperly or carelessly assembled units. Ice also reduces the insulating ability of the cabinet, and the unit will run more.

Such accumulation of ice in a freezer will generally be indicated by a cold spot on the outside surface of the freezer or by condensation on the outside surface. Iced up insulation will melt and drain if the freezer is shut down and allowed to warm up for a few days.

If a refrigerator or freezer has an air leak in the outer casing (shell), moisture from the atmosphere will enter and condense in the insulation. This condition will cause considerable trouble in a freezer.

If the condition is severe, the condensing unit may run continuously. Should cold (sweat) spots appear, remove breaker strips and insert lightly packed fine fiberglass insulation to fill air pockets.

Some freezers provide an opening in the inner lining which will allow any moisture in the insulation to escape into the freezer compartment and be condensed on the cold surface of the evaporator. This tends to keep the insulation dry.

10-41 BUTTER CONDITIONER

Various devices are being used to temperature-condition butter stored in the fresh food cabinet. The most common is to provide a recess in the refrigerator door which will hold a quarter lb. of butter. This recess is separated from the refrigerator compartment by a small door. Being close to

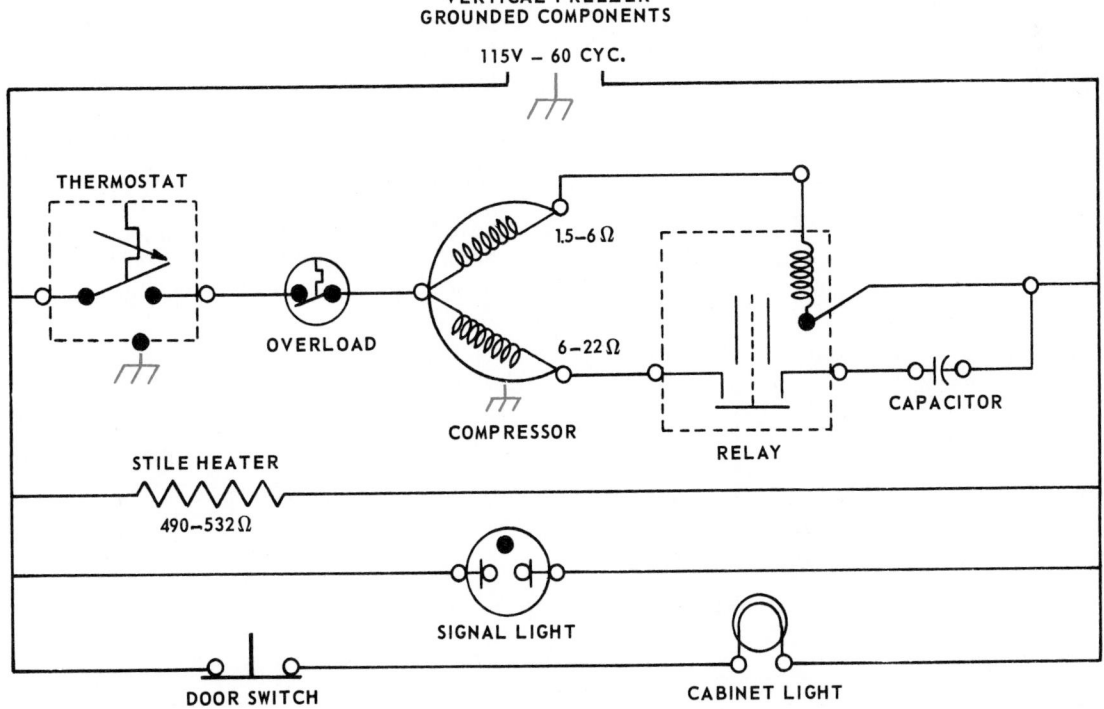

Fig. 10-55. Ladder wiring diagram for vertical freezer made according to latest electrical codes. Middle wire of power cord is grounded by a wire network to all metal parts of the cabinet, hardware, and refrigerating system.

the outer shell of the door, the recess provides little insulating effect between butter and room temperature. Some refrigerators provide an electrical resistance unit to aid in warming the butter compartment.

10-42 CABINET HARDWARE

Cabinet hinges usually use ball or nylon bearings. These require little or no lubrication. Usually they are adjustable so that the doors can be made to fit the cabinet properly. Fig. 10-56 illustrates a hinge commonly used in upright refrigerators.

Many of the older refrigerators used complicated door latch mechanisms. They were designed to draw the door firmly into place and securely latch it. Since 1958, a federal law requires that a refrigerator cabinet must be designed to be opened from the inside with no more than a 15-lb. force. Most modern refrigerators, therefore, use magnets to close and hold the door shut. No positive latching mechanism is used.

Refrigerator cabinets should be carefully leveled in order that the shelves are not tilted. Most have adjustable feet for this purpose. It is possible to adjust these feet in such

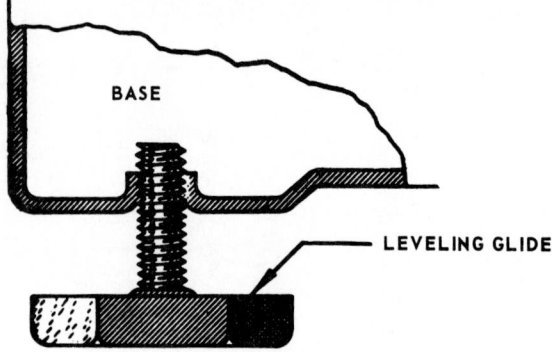

Fig. 10-57. Detail of leveling glide.

a way that the weight of the door will tend to swing it closed from any open position. Most users like this setting. Fig. 10-57 gives a detail drawing of a leveling adjustment.

Many modern refrigerator-freezers are mounted on rollers to help move the unit for cleaning and servicing.

The front rollers are usually adjustable. After the refrigerator is in place, these rollers are used to level it. See Fig. 10-58. A thumbscrew which is adjusted to touch the floor keeps the refrigerator from moving once it is in place.

Breaker strips, usually made of plastic, connect the metal outer shell (case) to the metal liner at the part of the cabinet that contacts the door. (Some refrigerators have a plastic inner liner instead of a metal liner.)

These breaker strips are molded and shaped in such a way that they snap in place. See Fig. 10-59.

A wide-bladed putty knife may be inserted to remove the outer edge of the breaker strip. (Wrap the putty knife blade with tape to prevent scratching the parts.) Sometimes pressing with the heel of the hand will separate the joint.

The liner should be room temperature to 90°F (32°C) (never warmer) to make it flexible. Use a warm dampened wiping cloth to warm it. Some breaker strips (installed before the foamed-in-place insulation is put in) are fused to the foam insulation. These strips must be broken or cut to be removed.

Chest-type freezers usually have the lid counterbalanced so that it is not necessary to hold the lid open when placing items in or removing them from the cabinet. However, the adjustment on these counterbalances is usually made so that, as the operator closes the lid, it tends to make a final closing by its own weight.

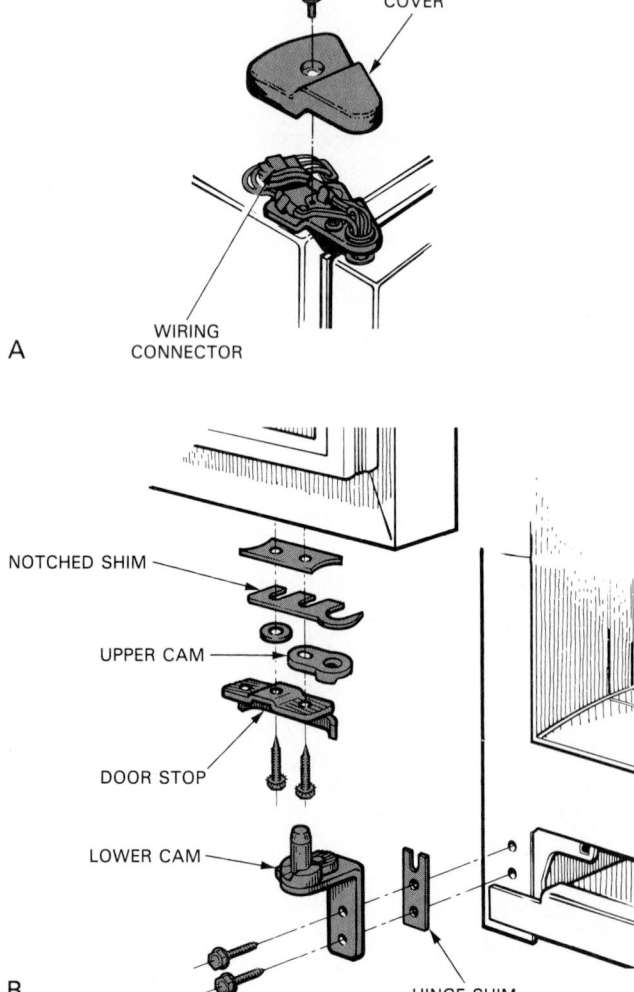

Fig. 10-56. Top, A, and bottom, B, hinges for refrigerator or upright freezer. Note wiring harness under top hinge cover in view A. (General Electric Co.)

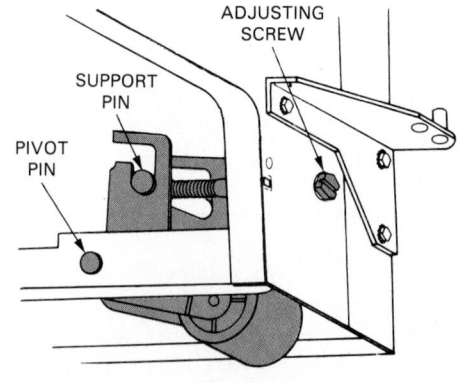

Fig. 10-58. Adjustable front roller assembly for refrigerator cabinet.

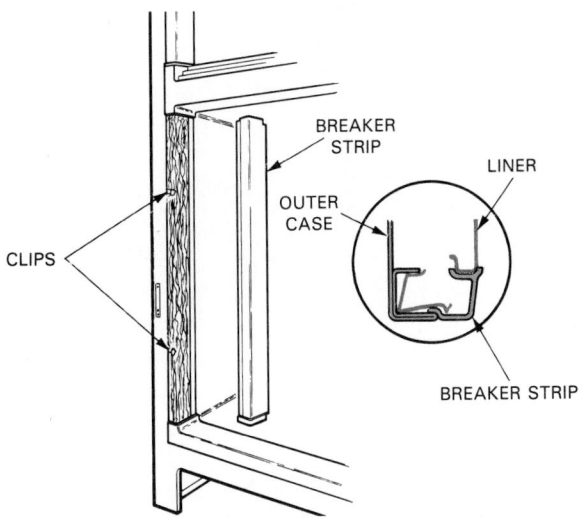

Fig. 10-59. Typical breaker strip. Note clips on outer case. (General Electric Co.)

Very little cabinet hardware repair is done on modern refrigerators. If a latch or hinge fails for any reason, it is recommended that the part be replaced with a new one.

10-43 CABINET GASKETS

Door gaskets are usually made of flexible vinyl. They usually have an air cushion design. See Fig. 10-60. Many have magnets built into the vinyl to hold the door closed. These are used in place of latch and striker plate.

Door seals may be checked using a .003 in. thick plastic feeler gauge or a thin piece of paper. To check the fit of the gasket, insert the feeler gauge or the strip of paper at several places around the door opening. With the door closed it should require a little pull on the gauge if the gasket is properly fitted. If the paper or feeler gauge falls out of its own weight, the gasket is not fitted properly.

Door gaskets may be quickly and easily checked as follows:

1. Open the refrigerator door (note carefully the pull required to open the door) about half way and allow it

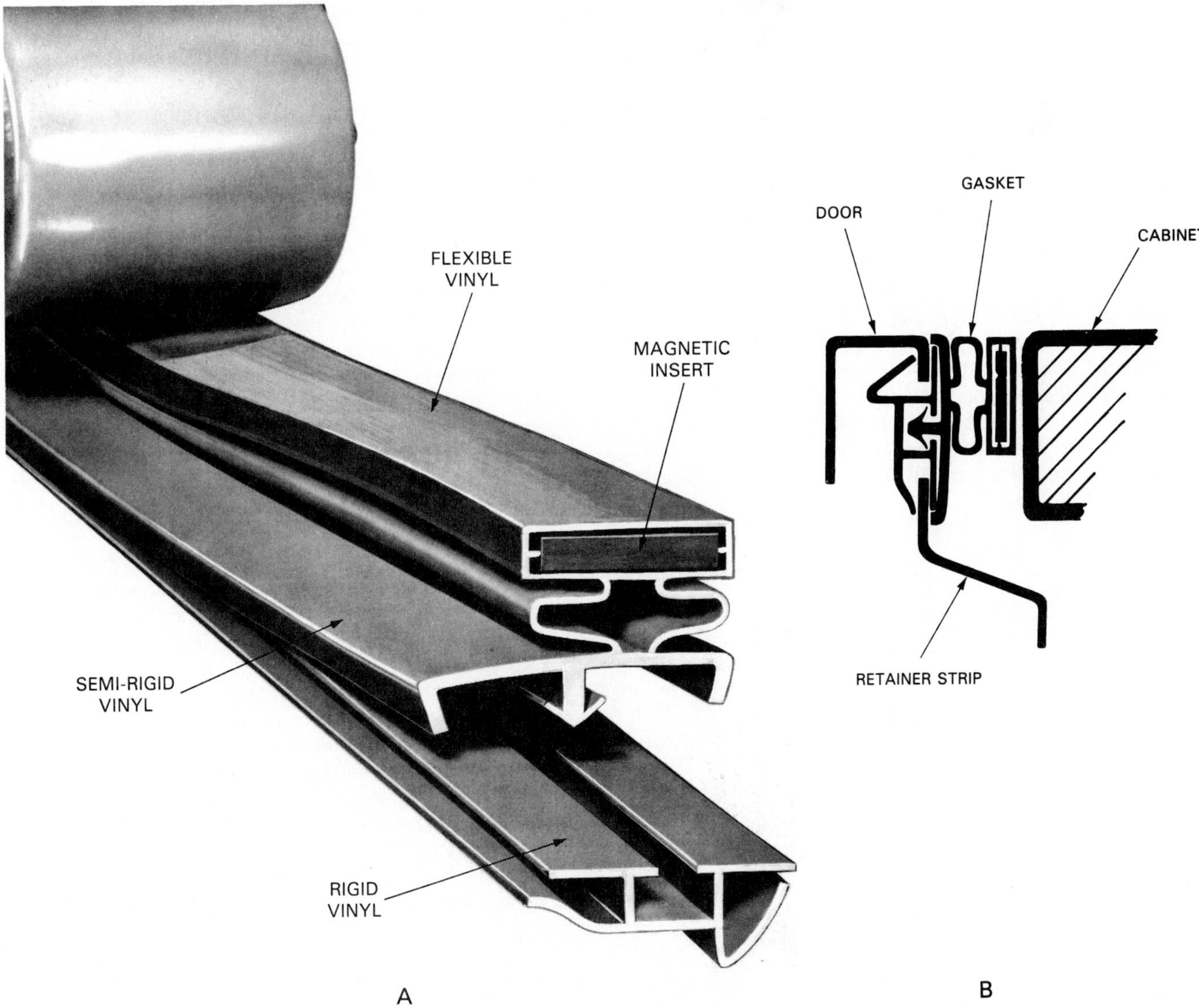

Fig. 10-60. Replacement door gaskets. A—Magnetic door gasket being rolled onto retainer strip. B—Cutaway cross section of magnetic gasket and retainer strip assembly installed in door. (Jarrow Products, Inc.)

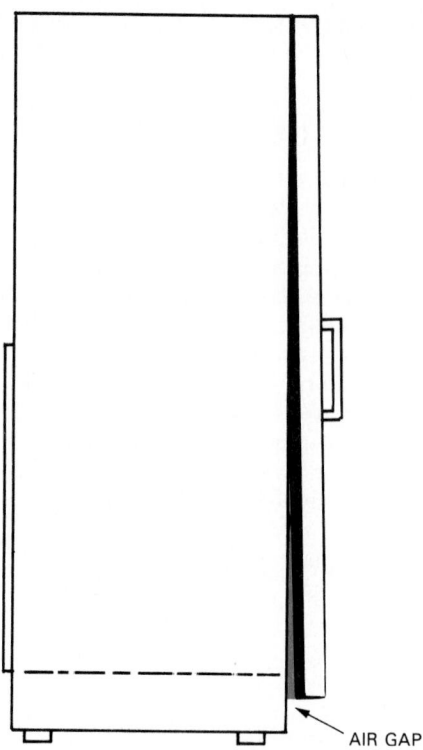

Fig. 10-61. Improperly hung or warped door.

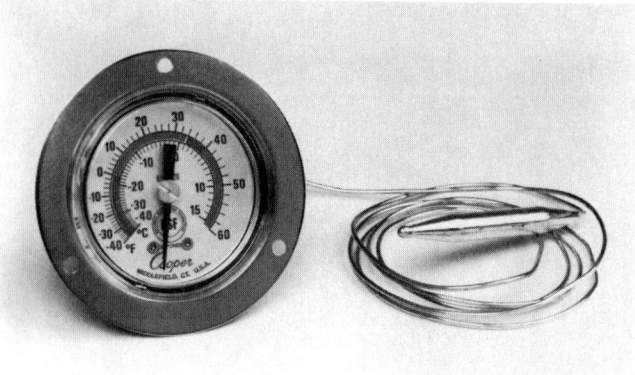

Fig. 10-63. Refrigerator-freezer cabinet thermometer. Scale reads from −40 °F to 60 °F (−40 °C to 15 °C). (Cooper Instrument Corporation)

With some refrigerators, it is possible to adjust the hinges and the door latch to correct a poorly fitted gasket.

Fig. 10-61 illustrates a door that is either warped or has improperly adjusted hinges. A warped door can usually be straightened by twisting. To check for hinge adjustment, use the test just described for proper gasket seal.

Door gaskets deteriorate with age. If the material has become hard, cracked, or broken, it should be replaced.

10-44 REPAIRING FINISHES

As indicated in Para. 10-7, cabinet finishes may be either baked-on enamel or porcelain. To repair enamel finishes, carefully sand the damaged area down to the bare metal. Be sure to remove all wax and rust.

Sand the edges of the damaged area with fine waterproof paper (6/0). The old finish should slope evenly from the surface to the center of the damaged area. This result is called ''featheredging.''

Carefully clean the sanded surface. Apply metal primer to the exposed bare metal using either a brush or a spray gun. After the primer has dried thoroughly, sand again using 6/0 waterproof paper and soapy water as a sanding lubricant. Dry thoroughly.

Spray or brush on the enamel, blending the new coat to the old finish as smoothly as possible. Allow the enamel to dry thoroughly, then sand with 6/0 paper until the edges are invisible. Fig. 10-62 shows the steps in finishing a chipped surface.

For a gloss finish, rub the entire surface, including the patched area, with rottenstone, paraffin oil, or rubbing oil. When the desired gloss has been reached, wipe off the rottenstone and oil by using a moistened cloth or chamois.

Paint spraying must be done in a fireproof, well ventilated room or booth. The air must be free from dust and the compressed air used in the spraying gun must be free of moisture and oil. Be sure to follow the paint manufacturer's recommendations when refinishing a cabinet surface.

To repair porcelain finish, a special porcelain patching material should be used. Because of the inherent nature of porcelain, its color shade will vary somewhat. A patching kit may be obtained with several colors.

The surface to be patched must be cleaned and warmed. Apply the patching material with a small, fine brush or air

to remain open for approximately 10 seconds.
2. Close the refrigerator door and allow it to remain closed for about 15 seconds.
3. Open the refrigerator door and note carefully the pull required. It should require more effort to open the door now than it did at the time of the first opening.

This is because, during the time that the door was opening, the cold air came out and warm air replaced it. On closing the door, this warmer air is cooled and contracts, with the result that the pressure inside the refrigerator cabinet will be slightly less than the room atmospheric pressure. Timing is an important element in this test, as the pressures tend to quickly balance.

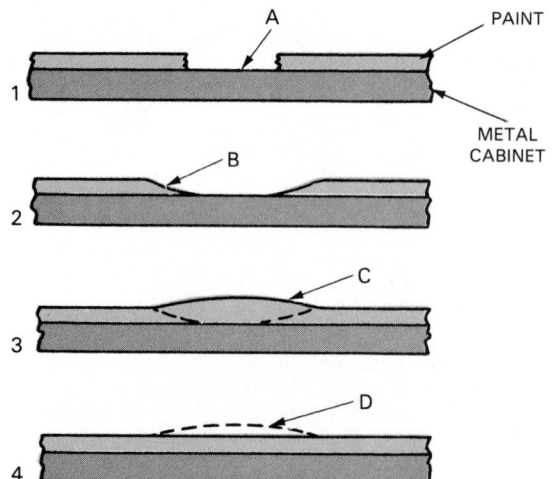

Fig. 10-62. Steps in finishing a chipped surface: A—Paint chipped from cabinet. B—Surface sanded in preparation for repairing. C—New finish applied. D—Excess finish must be sanded down to level.

brush it. When dry, the finish should be smoothed with waterproof abrasive paper and polished with a soft cloth or with rottenstone and rubbing oil.

10-45 CABINET THERMOMETERS

The recommended temperature range for a fresh food compartment is between 35 °F (2 °C) and 45 °F (7 °C). The recommended temperature range for the frozen food compartment is 0 °F (−18 °C) to −10 °F (−23 °C). Fig. 10-63 displays a thermometer which indicates temperatures within these ranges.

10-46 REVIEW OF SAFETY

The service technician's hands should not be placed near revolving fans.

Any spray painting must be done in an approved spray room or booth.

Refrigerators and freezers must be carefully handled in order not to injure those handling them or to damage the cabinets or mechanisms.

Frost and ice should be removed by heating (hot water or electric heat). *Never use a pointed or sharp metal tool to remove ice from the evaporator.* One may puncture the refrigerating unit.

Remember, if an old style refrigerator is taken out of service, the door must be removed immediately. Letting an out-of-service refrigerator stand where children may play in it, is dangerous. The children could be suffocated. Avoid this tragic mistake.

Always disconnect electric power (open the switch or pull the plug) before working on the electrical parts of the system to avoid unpleasant and perhaps fatal shocks.

Always make certain that the electrical system is grounded properly to the receptacle if an approved three-wire grounding plug is not used. Fig. 6-87 and Fig. 11-3 show how to check for proper grounding and connect a safe ground wire.

10-47 TEST YOUR KNOWLEDGE

1. What are some of the safety devices in the electrical circuit of a hermetic domestic refrigerator?
2. How is the temperature controlled in a butter compartment?
3. What controls the flow of hot gas through the evaporator for defrosting?
4. List and describe the two systems used in refrigerators for automatic defrosting.
5. What type of voltage (ac or dc) is required on a single printed circuit board of a solid-state controlled ice maker to operate the semiconductor devices?
6. By what path is the defrost water often removed during the defrost cycle in an automatic defrost refrigerator?
7. What is the most common refrigerant control system used on frozen food cabinets?
8. How often should a chest-type freezer be defrosted?
9. Why are the suction line and the capillary tube sometimes soldered together?
10. What is the best temperature range for a home chest-type freezer?
11. What kind of motor control is used on most frozen food units?
12. How does a no-frost refrigerator prevent sweating and frosting?
13. What is a drier wire?
14. List some possible locations of a condensing unit.
15. Why must an evaporator be defrosted?
16. How are cabinet door seals made airtight?
17. List two evaporator locations in the cabinet.
18. In reference to the cabinet of a frost-free side-by-side refrigerator-freezer, what happens to the defrost water?
19. What type of insulation is used in modern freezers?
20. What is meant by featheredging when repairing an enamel or lacquer finish?

Interior view of an upright refrigerator-freezer.
(Amana Refrigeration, Inc.)

Chapter 11

SERVICING AND INSTALLING SMALL HERMETIC SYSTEMS

After studying this chapter, the technician will be able to:

☐ Select proper tools and instruments needed for installation or servicing of domestic and small commercial systems.

☐ List supplies needed on a typical installation or service call.

☐ Service internal and external mechanisms using the proper tools and materials.

☐ Locate areas causing noise and make necessary adjustments to the system.

☐ Recognize trouble signals.

☐ List common external service operations.

☐ Start a stuck compressor.

☐ Demonstrate proper use of piercing valves.

☐ Check for restrictions in the system.

☐ Troubleshoot common refrigeration problems.

☐ Use the proper methods and equipment for checking electrical systems of refrigerators and freezers.

☐ Use proper safety procedures.

11-1 INSTRUMENTS, TOOLS, AND SUPPLIES

In addition to the tools and supplies listed in Chapter 2, special instruments, tools, and supplies will be needed for domestic refrigerator service work. These are listed in the following paragraphs under Instruments, Tools, and Supplies.

11-2 INSTRUMENTS

Many special testing instruments are needed for refrigeration service work. Their use will be explained as the various service operations are described. Fig. 11-1 shows a small hermetic system with a gauge manifold, vacuum pump, and charging cylinder connected to it. Some of the basic instruments required are listed below:

BASIC INSTRUMENTS

Pressure recorder.
Temperature recorder.
Off-on recorder.
Watt recorder.
Electronic sound tracer.
Electronic leak detector.
Compound gauge.
Pressure gauge.
Thermometer, −20 °F to 212 °F (−29 °C to 100 °C).
Voltmeter.
Ammeter.
Ohmmeter.
Test light 120 V, 100 W, incandescent.

11-3 TOOLS

REFRIGERANT TOOLS

High-vacuum pump.
Recovery/recycling unit.
Service cylinders for R-12, R-22, R-502, R-134a.
1 purging line 1/4 in. dia. by 15 ft. equipped with hand shut-off needle valve and check valve.
Capillary tube cleaner.
Capillary tube sizing kit.
Soldering-brazing torch, either LP fuel-air, acetylene-air, or oxyacetylene.
Hand vacuum cleaner.
Gauge manifold.
Process tube adaptors.
Bending springs.

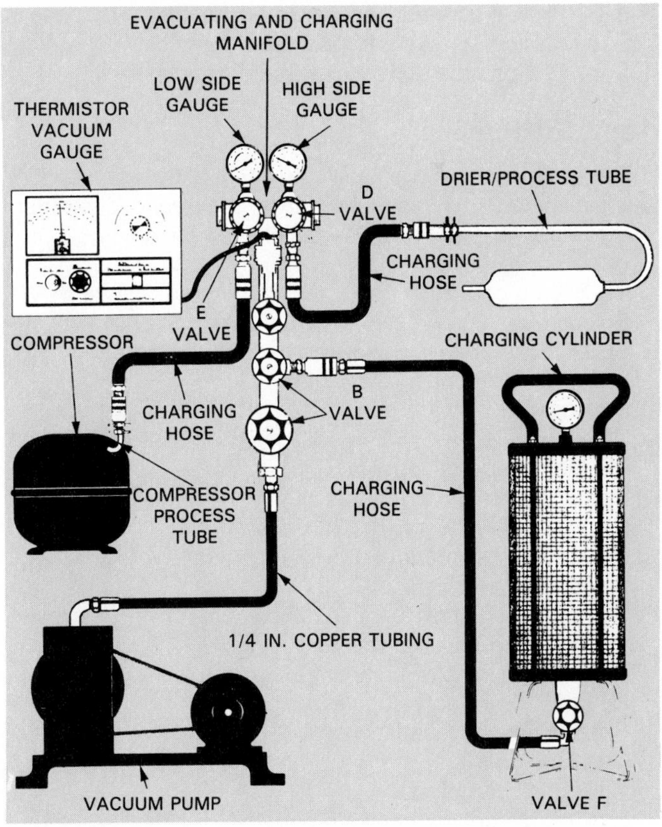

Fig. 11-1. Typical installation of service equipment manifold on hermetic system. Note refrigerant cylinder, vacuum pump, and valve arrangement. Also note adaptor connection to compressor process tube and to drier/process tube. (Amana Refrigeration, Inc.)

WRENCHES
   Torque handle 3/8 in. drive.
   Speed handle 3/8 in. drive.
   Swivel handle 3/8 in. drive.
   T-handle 3/8 in. drive for sockets.
   8 in. adjustable open-end wrench.
   Set of Allen setscrew wrenches.
   Refrigeration ratchet wrench, 3/16, 7/32, and 1/4 in.
      with square openings.
   1/2 in. 15 deg. open-end wrench.
   3/4 in. 15 deg. open-end wrench.
   7/8 in. 15 deg. open-end wrench.
   1 in. 15 deg. open-end wrench.
   1/2 in. box wrench.
   1/2 in. T-socket wrench.
   Set of sockets, 12 point, 7/16 to 1 in. with 3/8 in. drive.
PLIERS
   6 in. combination.
   Wire cutters.
   Slim nose.
HAMMERS
TUBING TOOLS
   Bending springs for 1/4, 3/8, and 1/2 in. OD tubing.
   Flaring tool 3/16 to 1/2 in. capacity.
   Tube cutter.
   Pinch-off tool.
   Swaging tool.
SCREWDRIVERS
   3 in. regular screwdriver, insulated handle.
   6 in. regular screwdriver, insulated handle.
   8 in. regular screwdriver, insulated handle.
   3 in. Phillips screwdriver, insulated handle.
   6 in. Phillips screwdriver, insulated handle.
   8 in. Phillips screwdriver, insulated handle.

11-4 SUPPLIES

Some special supplies will be needed by the service technician who makes house calls answering customer complaints. In general, these supplies will include:
   60-40 wire solder.
   95-5 wire solder.
   Solder flux.
   Silver brazing wire, 30-60 percent silver, no cadmium.
   Phosphorous-copper alloy wire.
   Brazing flux.
   Steel wool.
   Medium grade sandpaper.
   Plastic tape.
   Disposable cylinders, R-12.
   Disposable cylinders, R-22.
   Disposable cylinders, R-502.
   Coil of 1/4 in. soft copper tubing.
   Coil of 5/16 in. soft copper tubing.
   Coil of 3/8 in. soft copper tubing.
   Coil of 1/2 in. soft copper tubing.
   Copper pipe as needed.
   Capillary tubing.
   Filter-drier cartridges.
   Can refrigerant oil (spout type).
   Can refrigerant oil, 300 viscosity.
   Can refrigerant oil, 150 viscosity.
   Cleaning cloths.
   Relays.

   Capacitors.
   Motor controls.
   Refrigerant controls.
   Overload protectors.
   Light switches.
   Sealing compounds.
   Driers (flared and soldered fittings).
   Drier-filters (flared and soldered fittings).
   Sight glasses (flared and soldered fittings).
   Flared fittings (SAE)—all sizes and shapes.
   Soldered fittings—all sizes and shapes.
   Piercing valves and valve adaptors.
   Valve cores.

11-5 INSTALLING REFRIGERATORS AND FREEZERS

Proper installation is very important to the operation of a refrigerator or freezer. This includes leveling of the cabinet, correct electrical power, and good ventilation. The manufacturer ships the units carefully crated and with full written instructions on how to ship, how to uncrate, and how to install.

11-6 UNCRATING A REFRIGERATOR OR FREEZER

The carton usually has printed instructions on safe handling. These instructions should be carefully followed.

Many dealers uncrate the cabinet at the store; others do it just outside the home because most crates are too large to fit through standard doors.

Certain areas of the cabinet may be easily abused when uncrating or moving it. They are:
1. Bottom. Condensing unit may be damaged.
2. Back. Refrigeration condenser may be damaged. (Not all cabinets have condensers in the back.)
3. Door. The doors may be forced out of line or buckled.

There are two types of motor compressor dome shipping methods. One is the use of removable shipping bolts when the dome is mounted on or is suspended from springs. The other is loosening the dome shipping bolts two or three turns when the dome is mounted on synthetic flexible grommets.

Fig. 11-2 shows a hand truck with a hold-on strap. The side rails may be used as skids to aid in moving the appliance in and out of the truck and in and out of the building.

11-7 LOCATING POSITION OF REFRIGERATOR-FREEZER

It is advisable to locate the refrigerator-freezer where it will not be in direct sunlight. It should not be near an oven, heat radiator, or warm air register. Should it be necessary to locate the refrigerator-freezer near an oven or radiator, a strip of aluminum foil large enough to cover the side of the refrigerator-freezer should be hung between the refrigerator-freezer and the heat source.

The room should be large enough to provide enough air to cool the condenser. 100 sq. ft. is preferred.

11-8 ELECTRICAL SUPPLY

The electrical outlet for the refrigerator-freezer must provide the correct electrical supply. Be sure to read the electrical ratings on the refrigerator and check these against the

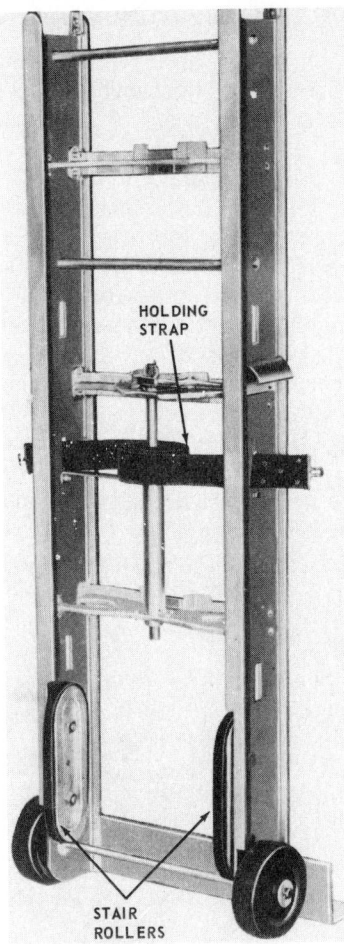

Fig. 11-2. Appliance hand truck. Strap is wound around appliance and is tightened by a wrench powering a ratchet gear.

electrical supply at the wall outlet. The modern household refrigerator-freezer may need more current than the older, simpler refrigerators and freezers.

It is best to have a separate circuit from the fuse or circuit breaker box to the refrigerator-freezer outlet. Avoid using an extension cord between the refrigerator power cord and the wall outlet. The resulting voltage may be too low.

Voltage at the refrigerator outlet can be quite easily checked with a voltmeter. The circuit capacity (wire size, etc.) is checked as follows: If at the instant of starting, the voltage at the refrigerator outlet drops more than 10 V, the wiring in the circuit is not heavy enough. A flicker in the lights at the instant the refrigerator starts is a sure sign of a poor electrical supply.

It is very important to ground the refrigerator. All removable electrical parts—fans, thermostats, timers, etc.—are already safety grounded. If the wall outlet has a three-wire socket and the unit has a three-wire prong power cord, there is grounding. Otherwise, a wire must be attached between a metal part of the cabinet and a good ground such as a water pipe. The type of male plug used on the appliance's power cord is an indication of the voltage and grounding. See Fig. 6-104 for these different plug designs.

The service technician should always check to see that there is proper grounding in the outlet box which supplies

current to any refrigerating mechanism being serviced. Refer to Fig. 11-3. As shown in view A of Fig. 11-3, take a voltmeter reading from a "live" connection in the wall receptacle to the ground connection of the receptacle. A full voltage reading will indicate that the outlet is properly grounded.

A second way to check for proper grounding is shown in view B of Fig. 11-3. Insert the leads of a 120 V incandescent light into the live terminal and the ground terminal of the receptacle. The lamp should light and should have normal brilliance.

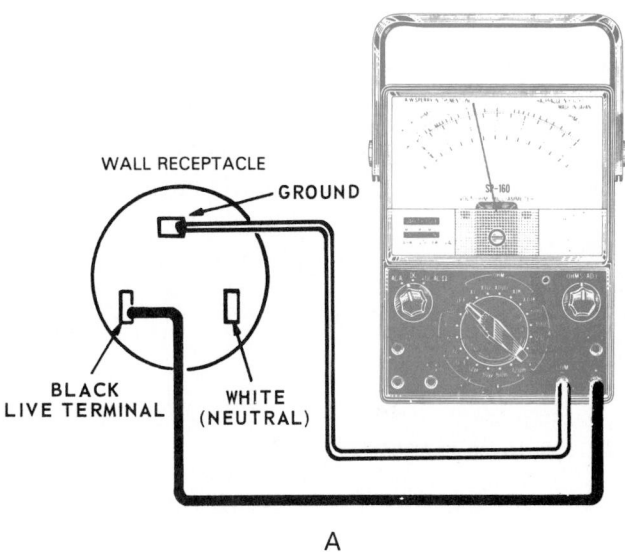

A

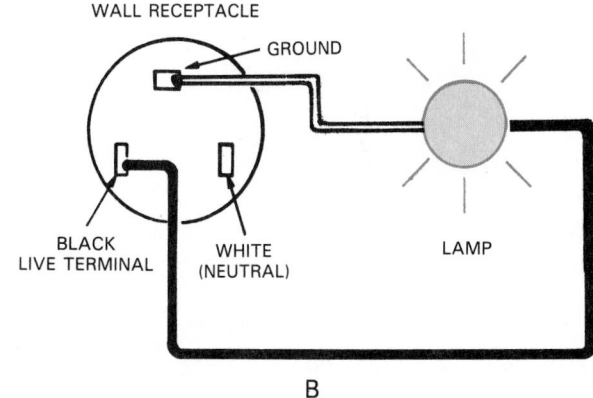

B

Fig. 11-3. How to make sure that ground terminal is actually properly connected to a ground. A—Using a voltmeter. B—Using a 120 V, 100 W incandescent lamp. (A.W. Sperry Instruments, Inc.)

11-9 INSTALLING A REFRIGERATOR OR FREEZER

Since domestic refrigerators are air-cooled, proper ventilation is very important. Yet, many kitchens are designed without providing any space for air movement around the refrigerator.

In these installations, the refrigerator must have fans to draw the cool air in at the floor level, circulate it over the condenser, exhausting the warm air back into the kitchen at, or near, floor level. It is important that nothing be placed in front of these openings to block air flow.

Condensers of many domestic refrigerators are mounted on the back. Some are protected by a shroud which provides a chimney effect to increase the rate of air flow over the condenser. With this type, air circulation must be provided at the bottom, back, and top.

Many freezers and some refrigerators use the outside shell as the condenser surface. In these cases not less than two inches of space must be provided between the refrigerator cabinet and surrounding surfaces.

The refrigerator must be carefully leveled during installation. A spirit level should be used.

First check the floor where the rear supports or legs of the refrigerator are to rest. If not level, use wood spacers. Usually, the front supports are adjustable and may be used to properly level the cabinet. Fig. 11-4 illustrates the manner of adjusting these levelers.

Test the wall outlet with a voltmeter to be certain there is electrical power. Put the temperature control in the off

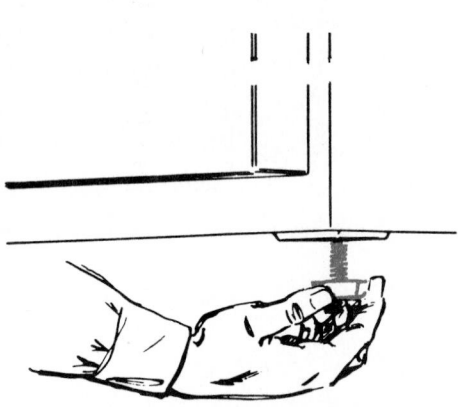

Fig. 11-4. Adjusting a leg leveler.

position. Connect electrical cord to wall outlet and test unit for running before moving refrigerator into position.

11-10 STARTING A REFRIGERATOR-FREEZER

When starting the refrigerator for the first time, it is best to set the temperature control at the middle of its range. After a few hours of operation, a thermometer in the fresh foods compartment may then be used to adjust the temperature control setting to the customer's requirements.

If the refrigerator does not start, make sure that the electrical circuit is in good condition before checking for mechanical trouble. Open and close the doors to make sure that the interior lights are functioning properly.

CAUTION: If a refrigerator-freezer which uses a capillary tube refrigerant control is stopped and then started immediately, it may fail to operate. This is not necessarily a malfunction. The motor in this type of refrigerator does not provide a large enough starting torque to overcome a high head pressure. Disconnect the refrigerator for a few minutes until the refrigerant pressure has had time to balance between the high side and the low side. Then start the system again.

11-11 INSTALLING ICE CUBE MAKER

Many domestic refrigerators have automatic ice cube makers. See Chapter 10. These units are connected to the nearest cold water line by a coil of 1/4-in. copper tubing. Refer to Fig. 11-5.

Before putting the refrigerator in its selected place, run the 1/4-in. copper line to the nearest cold water line. (Cabinet partitions or the floor may have to be drilled.) Mount a tap valve on the water line. Connect the tubing to the valve (usually a compression fitting). Connect the other end of the tubing to the refrigerator water line fitting. Allow several large loops of the tubing in back of the refrigerator to permit moving it in or out of its recess in the wall for cleaning and service.

Turn the tap valve stem in slowly and pierce the cold water pipe. Check for water leaks. Gently move the refrigerator into its recess. Be careful to avoid kinking or buckling the 1/4-in. tubing.

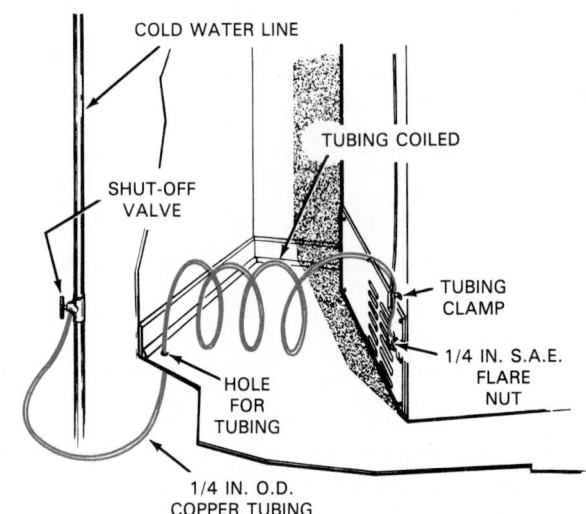

Fig. 11-5. Typical water line installation for automatic ice cube maker. Note that tubing is coiled to permit moving refrigerator without disconnecting tubing. (General Electric Co.)

11-12 LOCATING AND REMOVING NOISE

Most outside noise in the refrigerator comes from rattles. Loose baffles or ducts, tubing touching something while vibrating, uneven floor which may cause a list (leaning to one side) of the condensing unit, and fan and motor vibration—all are sources of noise. An ultrasonic leak detector, Fig. 11-6, identifies sound that is above the human hearing range. Humans hear frequencies up to 18 kHz (18,000 cycles per second). Sound above 20 kHz is termed "ultrasound." Fig. 11-7 shows a simple, homemade stethoscope.

A loose evaporator unit door, loose articles on shelves, and shelves not seated properly on supports may also cause annoying rattles. Such noise originating in the unit may indicate that it is laboring harder than it should. To determine this, test the electrical load with an ammeter or wattmeter. *One may sometimes determine if the unit is overloaded by its starting behavior. Three seconds to operate the relay is an average time.* A slower start indicates an overload.

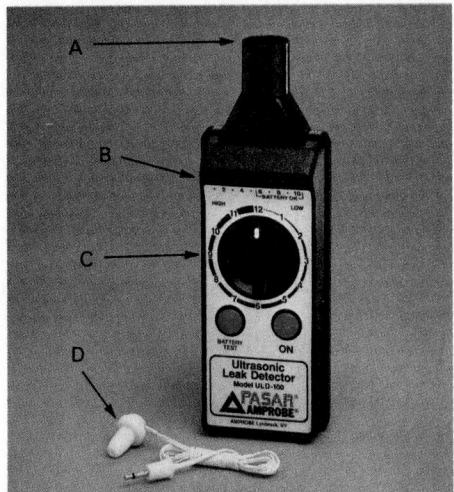

Fig. 11-6. Electronic sound detector. Note: A—Sensor horn. B—Display panel (lights up dependent upon frequency of sound). C—Sensitivity dial is set to identify sound. D—Earphone. This unit can also be used to identify refrigerant leaks.   (Amprobe Instrument)

service operations until after studying paragraphs 11-49, 11-50, and 11-51.

Another very helpful device for inspecting a refrigerator is an adjustable mirror on a long flexible extension. A mirror on a rod is shown in Fig. 11-9.

11-13 TEMPERATURE-PRESSURE CONDITIONS

Before servicing a refrigerator, a service technician should know:
1. The normal temperature in the evaporator during the operating cycle.
2. Normal pressure on the low-pressure side during the operating cycle.
3. Normal temperature of the condenser during the operating cycle.
4. Normal pressure on the high-pressure side during the operating cycle.

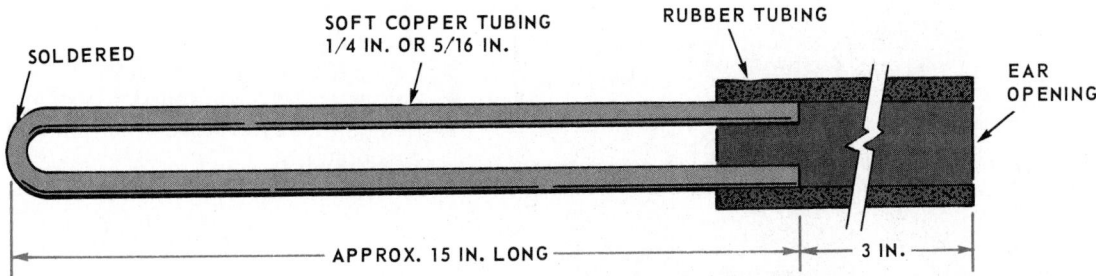

Fig. 11-7. Simple stethoscope made from section of 1/4-in. copper tubing and rubber tubing. It is excellent for locating source of many noises. Place closed end against mechanism and then put rubber tube opening to ear. (Use care around moving parts.)

Tubing, rattling against refrigerator parts, should be carefully bent away from contact. If the tubing is rigid but has a vibration or hum (harmonic vibration), this noise can be reduced by clamping rubber blocks on the tubing. See Fig. 11-8.

Loose baffles and ducts can be secured with self-tapping sheet metal screws.

If the sound tracer indicates that the noise is coming from inside the motor compressor, one should not attempt any

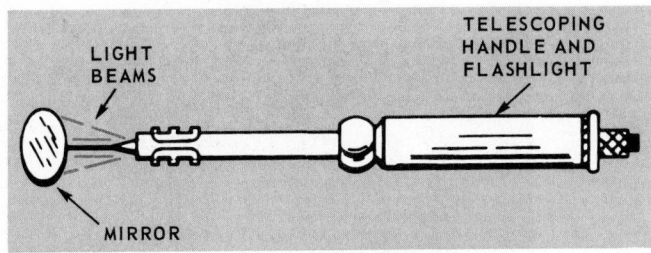

Fig. 11-9. An inspection mirror is often needed to see into hard-to-reach places. It is used to check fan alignment, motor condition, tubing location, cleanliness of condenser, condition of evaporator, etc. Avoid touching moving parts with mirror.

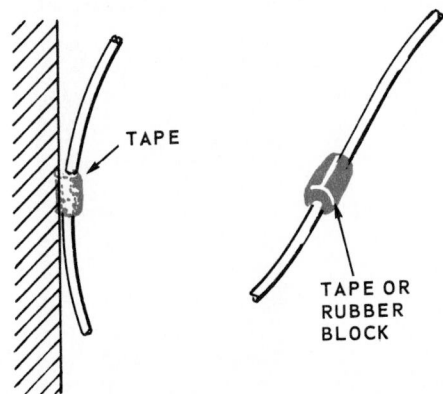

Fig. 11-8. Two ways are shown to reduce noise caused by vibrating tubing. A—Wind tape around tubing where tubing touches cabinet. B—Put tape or rubber block on tubing in center of vibrating section.

The above conditions depend upon the refrigerant being used. Temperature-pressure properties of refrigerants vary.

Fig. 11-10 lists the average temperature-pressure conditions for a domestic refrigerator-freezer evaporator and condenser. When using the table, be sure that the correct ambient temperature is used.

The most popular hermetic refrigerant is R-12. From all indications, it is being replaced by R-134a, due to the E.P.A. ruling. See Para. 9-37 and Chapter 28. In low-temperature (frozen foods) units and air conditioners, R-22 is quite popular.

To determine the operating temperatures, a temperature chart recorder having vapor-filled bulbs may be used.

TEMPERATURE-PRESSURE CONDITIONS IN THE REFRIGERATOR MECHANISM		
FRESH FOODS EVAPORATOR TEMPERATURE	R-12 AMBIENT TEMP.	
	70°F	90°F
START OF CYCLE	15	15
MIDDLE OF CYCLE	5	5
END OF CYCLE	0	0
FRESH FOODS EVAPORATOR PRESSURE, psig		
START OF CYCLE	12	12
MIDDLE OF CYCLE	8	8
END OF CYCLE	5	5
CONDENSER TEMPERATURE		
START OF CYCLE	70	90
MIDDLE OF CYCLE	100	120
END OF CYCLE	100	120
CONDENSER PRESSURE, psig		
START OF CYCLE	70	85
MIDDLE OF CYCLE	120	158
END OF CYCLE	120	158

Fig. 11-10. Average temperature and pressure conditions in domestic refrigerator using R-12 refrigerant. Unit has freezer compartment.

Fig. 11-11 illustrates such a thermometer in use.

A compound gauge and a high-pressure gauge may be used to determine the operating pressures. Fig. 11-12 shows the position of gauges for a refrigeration system.

11-14 TROUBLE SIGNALS

A fault in any part of the refrigerating mechanism will usually show up as an unsatisfactory temperature or operating condition of the refrigerator. Such conditions may include:

1. Refrigerator does not run.
2. Refrigerator runs all the time; temperatures are too cold.
3. Refrigerator runs all the time; temperatures are too warm.
4. Refrigerator runs all the time; temperatures are satisfactory.
5. Refrigerator cycles but food compartment is too warm; freezing compartment is satisfactory.
6. Refrigerator cycles but freezing compartment is too cold.
7. Motor control cuts out.
8. Refrigerator cycles satisfactorily; refrigeration is poor.
9. Refrigerator cycles; but does not freeze ice cubes.
10. Refrigerator cycles; too much ice accumulates on the evaporator.
11. Refrigerator mechanism is very noisy.

Fig. 11-11. Temperature chart recorder using vapor-filled bulbs and systems. Recorder is being used to test operating temperatures. It has two sensing bulbs. One is placed in freezer compartment and one in fresh food compartment.
(The Dickson Company)

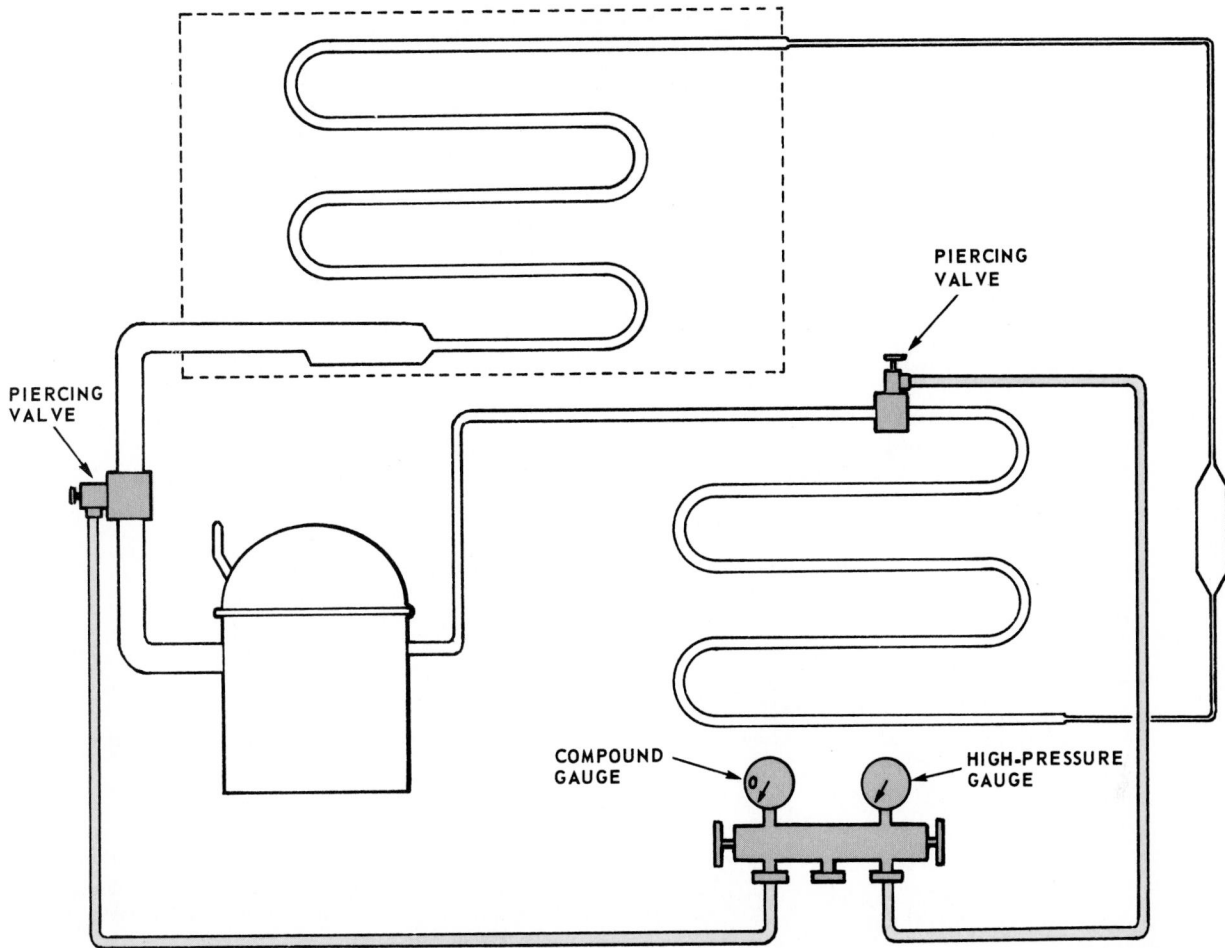

Fig. 11-12. High-pressure gauge and compound gauge are connected to hermetic system with service lines. Piercing valves are used to open service lines to hermetic system.

11-15 LOCATING TROUBLE IN THE HERMETIC REFRIGERATOR-FREEZER

Fig. 11-13 is a chart listing common troubles, their causes, symptoms, and remedies. This chart should be considered a general guide only. It does not apply to all units.

Experience and judgment are needed when one is trying to find the real cause of poor operation. *Always remember, cooling takes place only when the pressure is low enough and if there is liquid refrigerant present.*

For example, if an evaporator has the correct low pressure but is warm, there is no liquid in the evaporator. Another example, if a drier is frosting, it indicates that there is liquid and also a low pressure in the drier. (Drier is partially clogged.)

11-16 ICE ON THE EVAPORATOR

When large amounts of ice build up in the food storage space, it acts as insulation and may result in poor cooling. The cause usually is a leaky gasket seal, or the defrost system in frost-free or automatic defrost refrigerators is not operating.

Many refrigerators having a separate door on the freezer compartment also have an electric heater around the door opening to keep ice or moisture from forming there. If this heater is not working, ice build-up may keep the door from closing properly. Then too, moisture may enter and collect on the evaporator. The wiring circuit diagram for the refrigerator will indicate whether or not a "door heater" is used on the unit.

It may be that the door seal has lost its flexibility (life) or is broken. In either case, the gasket should be replaced.

A flat slip of paper inserted between the door and the cabinet should be held tightly when the door is closed. If this paper can be pulled out easily, the gasket is not tight enough. The hardware (latch and hinges) must be adjusted or the gasket replaced.

11-17 MOISTURE AND ICE IN THE CABINET INSULATION

Moisture and ice in the cabinet insulation is serious. It means there is an air leak in the outside cabinet seal or shell. The leak allows warm, moist air to enter this space. When the warm air strikes the cold inner liner, it gives up its moisture.

When this occurs in the cabinet of the fresh-food refrigerator, the insulation becomes wet and loses its heat-insulating qualities. There will be two indications pointing to this trouble:

1. The condensing unit will run more than normal.
2. The outside surface of the refrigerator will feel colder than normal wherever the insulation is wet.

TROUBLESHOOTING CHART

TROUBLE	COMMON CAUSE	REMEDY
1. Unit will not run.	Blown fuse.	Replace fuse
	Low voltage.	Check outlet with voltmeter, should check 115V plus or minus 10 percent.
		If circuit overloaded, either reduce load or have electrician install separate circuit.
		If unable to remedy any other way, install auto-transformer.
	Broken motor or temperature control.	Jumper across terminals of control. If unit runs and connections are all tight, replace control.
	Broken relay.	Check relay, replace if necessary.
	Broken overload.	Check overload, replace if necessary.
	Broken compressor.	Check compressor, replace if necessary.
	Defective service cord.	Check with test light at unit; if no circuit and current is indicated at outlet, replace or repair.
	Broken lead to compressors, timer or cold control.	Repair or replace broken leads.
	Broken timer.	Check with test light and replace if necessary.
2. Refrigerator section too warm.	Repeated door openings.	Instruct user.
	Overloading of shelves, blocking normal air circulation in cabinet.	Instruct user.
	Warm or hot foods placed in cabinet.	Instruct user to allow foods to cool to room temperature before placing in cabinet.
	Poor door seal.	Level cabinet, adjust door seal.
	Interior light stays on.	Check light switch; if faulty, replace.
	Refrigerator section airflow control.	Turn control knob to colder position. Check airflow heater.
		Check if damper is opening by removing grille. With door open, damper should open. If control inoperative, replace control.
	Cold control knob set at too warm a position, not allowing unit to operate often enough.	Turn knob to colder position.
	Freezer section grille not properly positioned.	Reposition grille.
	Freezer fan not running properly.	Replace fan, fan switch, or defective wiring.
	Defective intake valve.	Replace motor compressor.
	Air duct seal not properly sealed or positioned.	Check and reseal or put in correct position.
3. Refrigerator section too cold.	Refrigerator section airflow control knob turned to coldest position.	Turn control knob to warmer position.
	Airflow control remains open.	Remove obstruction.
	Broken airflow control.	Replace control.
	Broken airflow heater.	Replace heater.
4. Freezer section and refrigerator section too warm.	Fan motor not running.	Check and replace fan motor if necessary.
	Cold control set too warm or broken.	Check and replace if necessary.
	Finned evaporator blocked with ice.	Check defrost heater thermostat or timer. Either one of these could cause this condition.
	Shortage of refrigerant.	Check for leak, repair, evacuate and recharge system.
	Not enough air circulation around cabinet.	Relocate cabinet or provide clearances to allow sufficient circulation.
	Dirty condenser or obstructed condenser ducts.	Clean the condenser and the ducts.
	Poor door seal.	Level cabinet, adjust door seal.
	Too many door openings.	Instruct customer.
5. Freezer section too cold.	Cold control knob improperly set.	Turn knob to warmer position.
	Cold control capillary not properly clamped to evaporator.	Tighten clamp or reposition.
	Broken cold control.	Check control. Replace if necessary.

Fig. 11-13. Chart lists some common hermetic system troubles, their causes, and suggested remedies.

Fig. 11-13 continued.

TROUBLESHOOTING CHART

TROUBLE	COMMON CAUSE	REMEDY
6. Unit runs all the time.	Not enough air circulation around cabinet or air circulation is restricted.	Relocate cabinet or provide proper clearances around cabinet — remove restriction.
	Poor door seal.	Check and make necessary adjustments.
	Freezing large quantities of ice cubes, or heavy loading after shopping.	Explain to customer that heavy loading causes long running time.
	Refrigerant charge.	Undercharge or overcharge — check, evacuate and recharge with proper charge.
	Room temperature too warm.	Ventilate room as much as possible.
	Cold control.	Check control; if it allows unit to operate all the time, replace control.
	Defective light switch.	Check if light goes out. Replace switch if necessary.
	Excessive door openings.	Instruct customer.
7. Noisy operation.	Loose flooring or floor not firm.	Tighten flooring or brace floor.
	Tubing contacting cabinet or other tubing.	Move tubing.
	Cabinet not level.	Level cabinet.
	Drip tray vibrating.	Move tray — place on styrofoam pad if necessary.
	Fan hitting liner or mechanically grounding.	Move fan.
	Compressor mechanically grounded.	Replace compressor mounts.
8. Unit cycles on overload.	Broken relay.	Replace relay.
	Weak overload protector.	Replace overload protector.
	Low voltage.	Check outlet with voltmeter. Underload voltage should be 115V plus or minus 10 percent. Check for several appliances on same circuit or extremely long or under-sized extension cord being used.
	Poor compressor.	Check with test cord and also for ground before replacing.
9. Stuck motor compressor.	Broken valve.	Replace motor compressor.
	Insufficient oil.	Add oil; if unit still will not operate, replace motor compressor.
	Overheated compressor.	If compressor faulty for any reason, replace motor compressor.
10. Frost or ice on finned evaporator.	Broken timer.	Check with test light and replace if necessary.
	Defective defrost heater.	Replace heater.
	Defective thermostat.	Replace thermostat.
11. Ice in drip catcher.	Defective drip catcher heater.	Replace heater.
12. Unit runs all the time, temperature normal.	Ice builds up on the evaporator.	Check door gaskets — replace if necessary.
	Control bulb on thermostat not in contact with evaporator surface.	Place control bulb in contact with the evaporator surface.
13. Freezer runs all the time. Temperature too cold.	Faulty thermostat.	Check thermostat — test and replace if necessary.
14. Freezer runs all the time. Temperature too warm.	Ice buildup in insulation.	Remove breaker strips, stop unit, melt ice and dry insulation, seal outer shell leaks and joints and then assemble.
15. Rapid ice buildup on the evaporator.	Leaky door gasket.	Adjust door hinges. Replace door gasket if cracked, brittle or worn.
16. Door on freezer compartment freezes shut.	Faulty electric gasket heater.	Use alternate gasket heater or install new one.
	Faulty gasket seal.	Inspect and check gasket. If worn, cracked or hardened, replace it.
17. Freezer works then warms up.	Moisture in refrigerator.	Install drier in liquid line.
18. Gradual reduction in freezing capacity.	Wax buildup in capillary tube.	Use capillary tube cleaning tool or replace capillary tube.

349

In the case of a food freezer, the condensed moisture will form ice in the insulation. The symptoms will be much the same as in fresh-food refrigerators. If this condition continues, enough ice will soon be built up to cause the sides of the cabinet to buckle. The leak in the outside cabinet surface must be located and completely sealed.

Most freezers provide a small opening through the inner lining, connecting the insulated area with the inside of the freezer cabinet. Since in freezers the temperature inside of the cabinet is much lower than the insulation temperature, any moisture will tend to escape through this small opening and condense on the evaporator surface.

11-18 HERMETIC SERVICING GUIDE

To service refrigerating units successfully, one must know how they should perform when in good condition. Fig. 11-14 shows the performance data of a typical domestic unit.

Always check the system data before trying to locate the cause of the trouble. *System data is usually located on the identification plate mounted on the motor compressor.* The data for a 1/3 hp two-pole motor unit used in a combination refrigerator-freezer is shown in Fig. 11-15.

Para. 11-15 and Fig. 11-13 provide information which one should have in mind before beginning any service operation. Servicing of hermetic refrigerators may be divided into three major areas:
1. External servicing.
2. Internal servicing.
3. Overhaul of hermetic system.

11-19 EXTERNAL SERVICING OPERATIONS

Some of the more common external service operations are as follows:
1. Cabinet hardware.
2. Cleaning.
3. Noise (rattles).
4. Electrical:
   a. Power-in circuit.
   b. Thermostat.
   c. Defrost thermostat.
   d. Interior light and circuit.
   e. Fan motor and circuit.
   f. Damper controls.
   g. Motor compressor:
      Relay and overload protector.
      Capacitor.
      Motor terminals.
   h. Defroster.
   i. Defroster control and circuit.
   j. Defroster heater coil.
   k. Cabinet heaters.
   l. Evaporator fan and circuit.
   m. Condenser fan and circuit.
   n. Light circuit.
   o. Butter conditioner circuit.
5. Ice cube maker.

These external mechanisms, electrical wiring, and electrical parts can all be checked for operation quickly by inspection and by using a volt-amp-ohmmeter.

Other external service troubles can be located by checking for ice on evaporators, frost or sweat on suction lines,

	70°F AMBIENT TEMPERATURE	90°F AMBIENT TEMPERATURE	100°F AMBIENT
Cabinet Temperature	38°	40°	47°
% Operating Time	38	62	100
Cycles Per Hour	3	2	None
kWhr./24 hr.	3.8	6.0	9.9
Control Position	4	4	4
Evaporator Air Temperature	1.5°	−1°	0°
Suction Pressure (Min-Max)	2 in. Hg-13 psig	0-13 psig	13-20 psig
Watts (Complete System)	390±20	395±20	395±20

Fig. 11-14. Chart shows operating characteristics of 18 cu. ft. combination refrigerator-freezer which has 1/3 hp two-pole motor compressor. Note that kW hr. (averaged over 24 hr.) changes as ambient temperature changes.

	DATA
REFRIGERANT	R-12
CHARGE (IN OUNCES)	10 1/2
COMPRESSOR hp	1/3
COMPRESSOR SPEED rpm	3450
RUNNING AMPERES	5.6
VOLTAGE	120
PHASE	SINGLE

Fig. 11-15. Refrigerant and electrical data which is typical of information found on identification plate mounted on motor compressor.

warm or hot discharge line, ice or sweat on driers, and dirty condensers.

It is important to locate the trouble and determine the cause accurately.

11-20 DIAGNOSING EXTERNAL TROUBLES

Some hermetic units have been needlessly replaced because the service technician thought the internal mechanism was faulty when the real trouble was in the external devices.

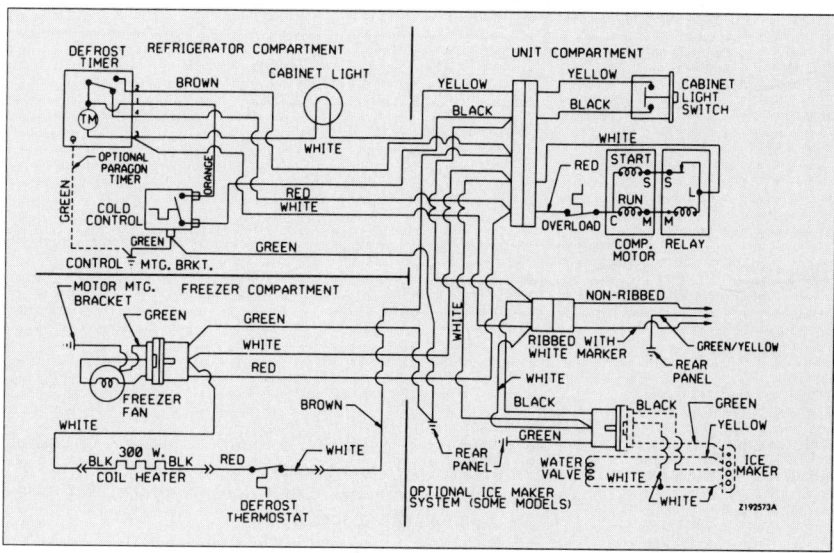

Fig. 11-17. Wiring diagram of the domestic single-door automatic defrost refrigerator with a defrost freezer fan, ice maker. Note color coding of wiring so service technician can trace the circuits.
(Frigidaire Company)

For example, if the mechanism will not start; if the unit hums but will not start; or if the unit short-cycles, the trouble could be in some part of the external electrical circuit:
1. Power-in connections.
2. Thermostat.
3. Wire terminals.
4. Relay.
5. Capacitor (if the unit has one).

Each of these devices should be checked carefully before the unit itself can be considered faulty. These parts can be checked best by removing them from the wiring system. Either of the following can then be done:
1. Parts can be checked independently.
2. A part can be temporarily replaced by a test part of the proper size to see if the unit will run. Fig. 11-16 shows a motor compressor with a relay and an overload protector mounted on it. Fig. 11-17 shows various electrical circuits typical of a domestic refrigerator.

Electrical connections must be clean and tight. If loose or dirty, they often overheat. This high temperature will discolor the connection.

The connection may be darkened by oxidation. A blue or greenish tint indicates overheating and corrosion. If the surrounding insulation is charred, overheating has occurred.

Troubles such as open circuits and grounded electrical wires can be checked easily with a test light. A test cord can be used to check four-pole motors, but two-pole motors should only be tested using a proper size relay in the circuit. These motors overheat if the starting circuit is connected for more than two or three seconds.

Chapters 6, 7, and 8 should be carefully reviewed before attempting to locate trouble in electrical units.

When checking and servicing hermetic electrical circuits, first make sure that the electrical supply at the outlet is good. Check the appliance voltage specifications.

Using a voltmeter, test the open circuit voltage. Next, plug in the appliance and check the voltage again while it is running. The open circuit voltage is likely to be slightly higher than with the motor running. However, this difference should not be more than 5 V.

A difference of 10 V or more indicates serious trouble:
1. An overload.
2. Something wrong in the motor windings.
3. Poor wiring to wall outlet.

Most refrigerators and freezers have a wiring diagram attached to the back of the refrigerator-freezer. Locate this diagram and check each circuit independently.

If the compressor fails to start:
1. Find out if electricity is reaching the motor compressor.
2. If it is, check the starting relay and circuit protectors. See Chapter 8.
3. Disconnect all wiring from the motor compressor.
4. Check the motor compressor with a manual start test cord. Such a test cord is shown in Fig. 11-18. The ground clip, labeled 4, must be fastened to the dome. *All clips or connectors should be plastic-coated or have a rubber boot over them to protect the service technician from shocks.*

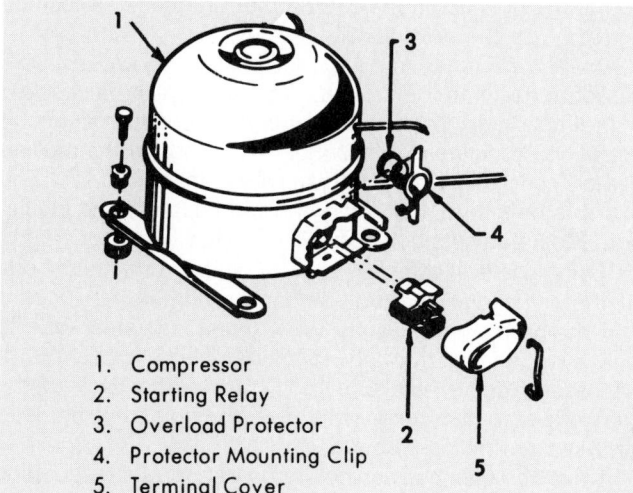

1. Compressor
2. Starting Relay
3. Overload Protector
4. Protector Mounting Clip
5. Terminal Cover

Fig. 11-16. Exploded view of starting relay and overload protector mounted on motor compressor. (WCI Freezer Division)

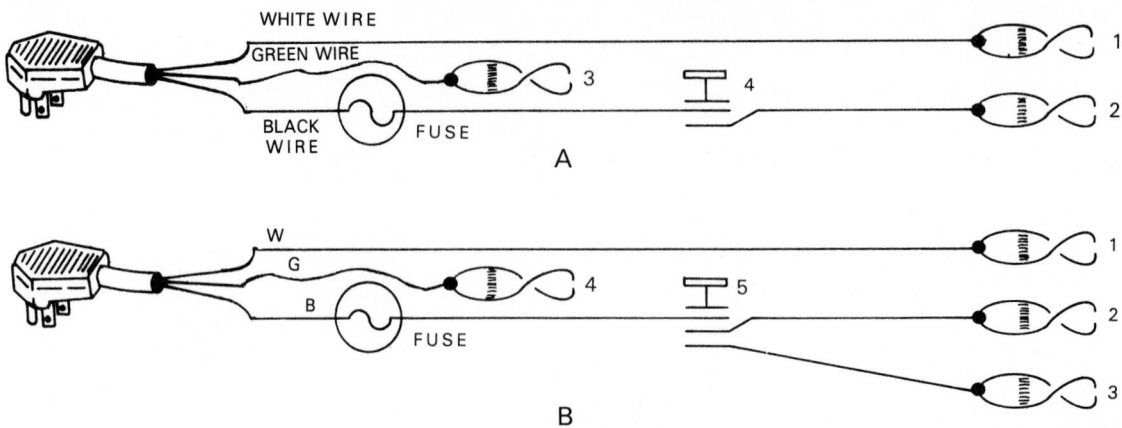

Fig. 11-18. Test cords. A—Cord for testing fan motors. Clip 1 goes to winding common terminal. Clip 2 attaches to winding power terminal. Clip 3 goes to ground. No. 4 is start and run switch. B—Test cord for hermetic motors. Clip 1 goes to hermetic compressor common terminal. Clip 2 attaches to run terminal. Clip 3 goes to start terminal. Clip 4 goes to ground. No. 5 is start and run switch. (A.W. Sperry Instruments, Inc.)

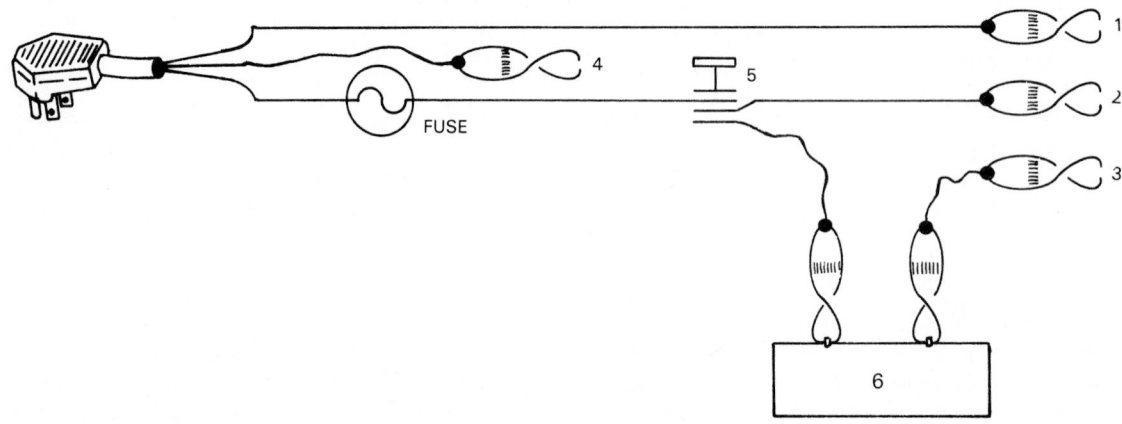

Fig. 11-19. Test cord for capacitor-start, induction-run motor. 1—Common winding terminal. 2—Running winding terminal. 3—Starting winding terminal. 4—Grounding wire and clip. 5—Start and run switch. 6—Starting capacitor. (A.W. Sperry Instruments, Inc.)

When the manual switch is pressed, the run circuit closes first, then the start circuit. After one to two seconds, lift the switch button just enough to open the start winding. If the motor operates correctly, the problem is in the external circuit. The fuse is located in the black wire of the three-prong plug.

A capacitor-start, induction-run motor is tested as shown in Fig. 11-19. The capacitor should be replaced by a new one of the same voltage and microfarad rating. An extra clip wire or lead is needed. Operate the switch as explained in the previous paragraph. *After testing, remember to short the testing capacitor by using a 20,000 ohm (20 kΩ) 2 watt resistor to avoid a shock.*

The test for a capacitor-start, capacitor-run motor is shown in Fig. 11-20. Use new capacitors of the same rating as the ones on the system being tested. Operate the switch as previously explained.

If the motor compressor works, check the electrical system up to the compressor.

If the motor does not operate when tested, further motor checks are needed, as explained in Chapter 7.

The test cords shown in Figs. 11-18, 11-19, and 11-20 can be used for checking continuity and grounds by replacing the fuse with a light bulb.

The trouble may be the evaporator motor or the condenser fan motor. These motors are usually replaced if found to be faulty.

Before removing a fan from a motor shaft, mark the position of the fan hub on the shaft to be sure the fan is located correctly on the new shaft.

An electrical failure in the mullion heater may cause a door gasket to freeze to the cabinet. The heater must be checked with a continuity light or an ohmmeter. Locate the circuit in the wiring diagram. Disconnect both ends of the mullion heater leads and test the heater for continuity.

If it is defective, look for a second heater in the insulation. Most cabinets have one. Test it also. If it is operating properly, use it. If there is no extra heating unit, install one of the same wattage (volt-ampere) rating.

If the problem is a faulty wire, use a stiff steel wire to pull new wiring through foamed-in-place insulation. If necessary, drill a hole (up to 1/2 in.) in the back of the refrigerator to help feed wires. Seal the hole after the wire or wires are pulled through.

If the complaint is cabinet temperature not responding properly to the temperature control, the temperature control mechanism may be faulty. It is advisable to use a recording thermometer. Record temperature and time over

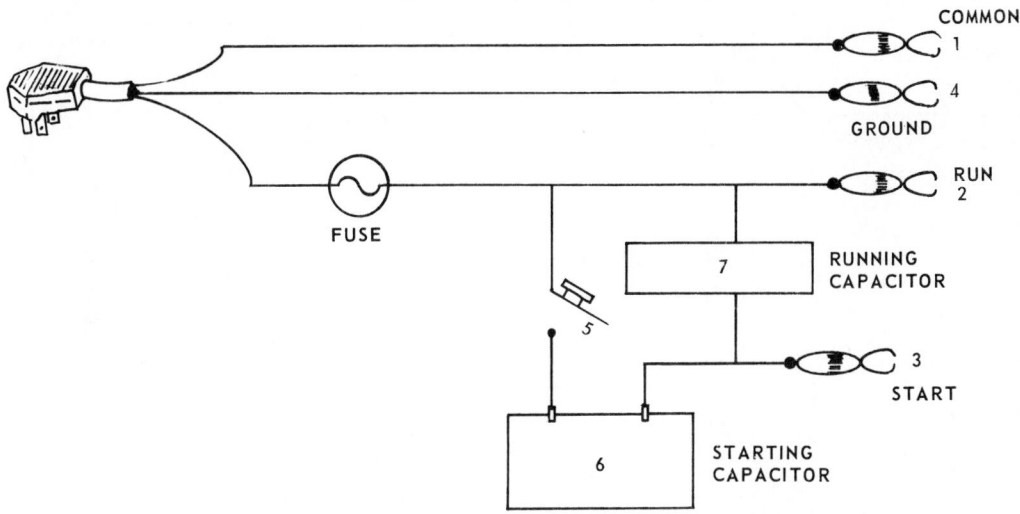

Fig. 11-20. Test cord used to check capacitor-start, capacitor-run hermetic motor. 1—Common winding terminal. 2—Running winding terminal. 3—Starting winding terminal. 4—Ground wire and clip. 5—Manual switch. 6—Starting capacitor. 7—Running capacitor. (A.W.Sperry Instruments, Inc.)

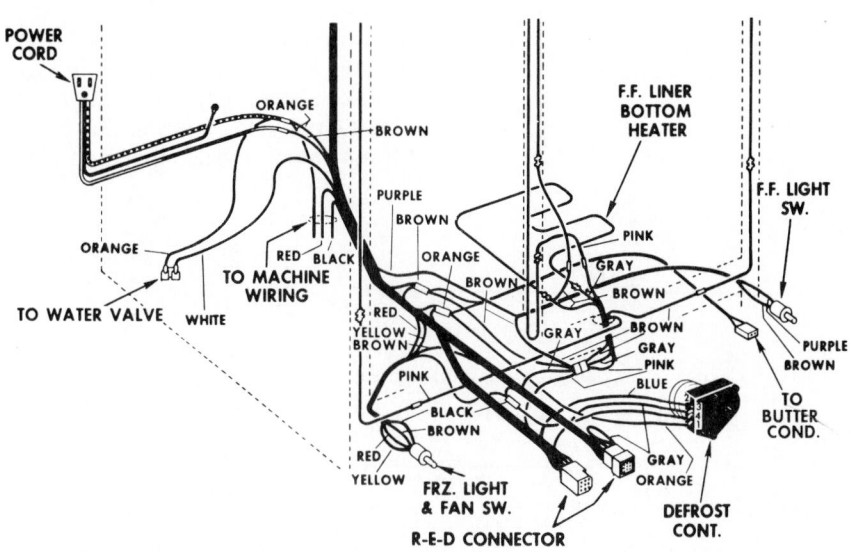

Fig. 11-21. Multi-circuit connector used for RED system simplifies testing of electrical circuits on some General Electric refrigerators.

a 24-hour period. The temperature-time chart will show if the appliance is operating properly.

11-21 ELECTRICAL TROUBLESHOOTING

The General Electric Co. has developed a rapid method (RED system) to check some of the electrical circuits of their refrigerators. Included are defrost thermostats, heaters, fans, and compressors. Simpler circuits, such as lights, being easily tested, are not included in this system.

A multiple circuit connection, mounted at the front bottom of the cabinet, is used. See Fig. 11-21. When this connector is separated, the tester is connected to it. See Fig. 11-22. A diagram of the circuits to be tested in shown in Fig. 11-23. Note number code for test connections.

Always turn the thermostat to the off position before separating the connector. The symbol for the female connectors is ⟩— and for the male connectors, →. After connecting the tester and turning the power on, the "power

on" light should be lit. If not, the power circuit needs repair.

One should refer to the General Electric data and diagrams when performing this test.

All Tecumseh motor compressor terminals are set up to read "Common-Start-Run" as one reads from left to right, going down a line at a time. See Fig. 11-24.

11-22 CLEANING THE EXTERNAL MECHANISM

The hermetic refrigerating system is basically a heat transfer mechanism. Air must circulate around and through the unit and condenser to carry away the heat. Therefore, the condenser and compressor unit must be kept as clean as possible, since dirt and lint act as heat insulation. Completely clean about every three months for economical operation and long life.

The hermetic mechanism can be cleaned where it is, by using a small vacuum cleaner or a special vacuum cleaner nozzle with a brush attachment. The vacuum cleaner

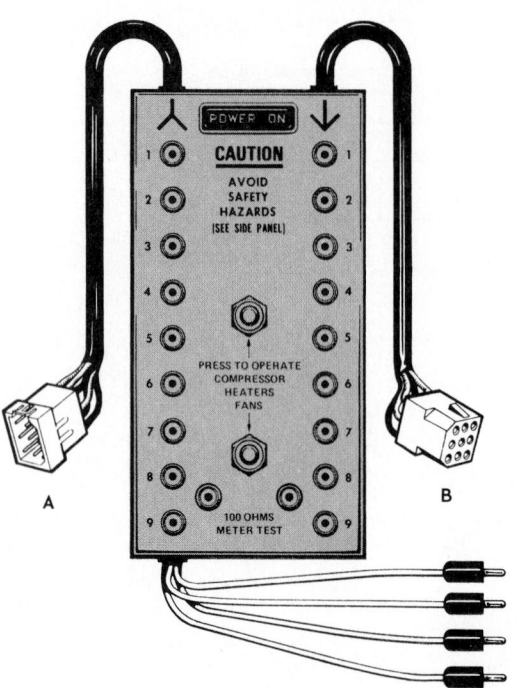

Fig. 11-22. The RED test instrument and adaptor. A—Male connector to electrical system. B—Female connector to electrical system. (General Electric Co.)

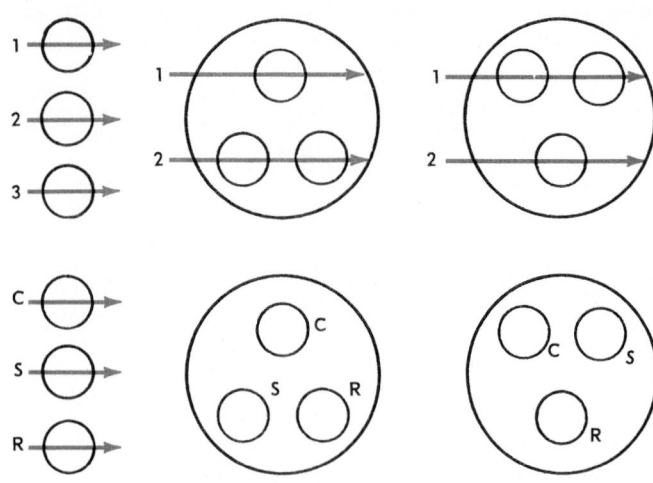

Fig. 11-24. Tecumseh motor compressor terminals are easy to identify. Referring to illustration above, read from left to right and then from top to bottom. Arrows show direction to read. Numbers show order of reading arrows.

eliminates the lint cloud that circulates in the room or settles on the floor. It is also quicker and more thorough than hand brushes or cleaning cloths. If a brush or cleaning cloth is used, place a paper or cloth on the floor under the unit.

If the unit uses a fan on the condensers, be sure to turn off the power or disconnect the unit before proceeding with

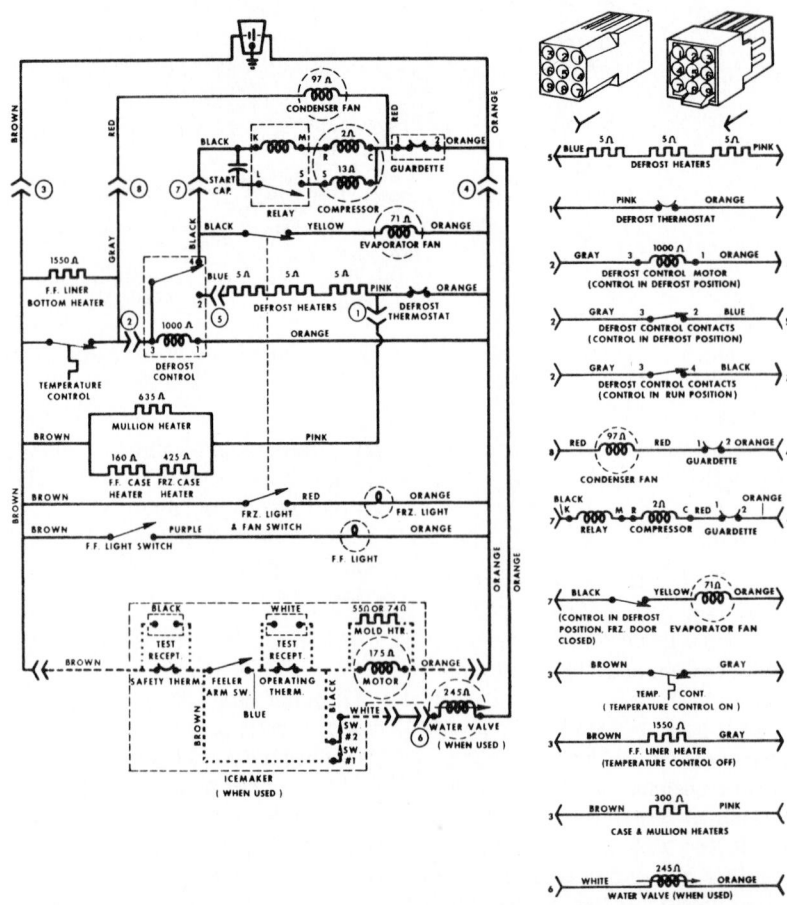

Fig. 11-23. Diagram of General Electric refrigerator circuits with number code at points where RED instrument and adaptors connect into circuits. Note: Numbers at either ends of circuits shown at right correspond to terminals in male and female plugs above.

the cleaning. In some cases it may be necessary to partially remove the unit to do a good cleaning job.

In the shop, high-pressure air, nitrogen, or carbon dioxide is often used to blow lint and other dirt from between the fins or coils or where it is difficult to reach. Wear goggles and have good ventilation. See Para. 11-59.

11-23 STARTING A STUCK COMPRESSOR

A unit which does not start when connected to the electric power may have a small piece of dirt in the compressor or the unit has not been run for a long time. To start such a compressor, three methods may be tried:

1. Disconnect the wiring to the motor compressor. Connect a test cord into the electrical circuit of the main winding. Use an extra capacitor, connecting it into the circuit as shown in Fig. 11-25. Turn on the power from one to three seconds. The extra capacitor will try to reverse the compression rotation. CAUTION: *this capacitor cannot be left in the circuit for more than a second or two. It will cause the motor to overheat.*

    If the compressor is successfully reversed, remove the reversing capacitor. Try to operate the compressor, using the normal electrical circuit. If the compressor does not start after three or four attempts at reversal as indicated above, it will usually be necessary to rebuild or

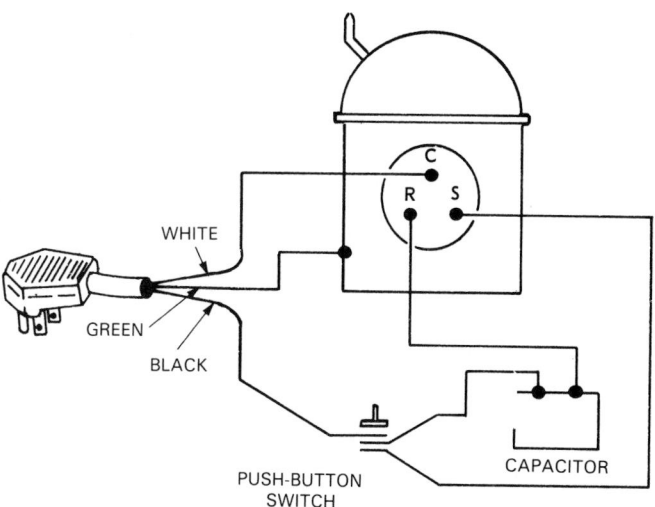

Fig. 11-25. Capacitor is being used in the running winding circuit to reverse rotation of "stuck" motor compressor. C—Common terminal. R—Running winding. S—Starting winding.

replace the motor compressor.

2. Another method is to connect the 120 V motor compressor into a 240 V power circuit using a starter cord. *One must be careful. Press the push-button switch for only a second at a time or the motor will be damaged.*

    The extra voltage may break the stuck compressor loose. If not, the motor compressor will have to be replaced.

3. An extra torque method is to connect a 240 V, 100 mfd (microfarad) start capacitor across the terminals of the run capacitor for NO MORE THAN ONE SECOND. (Count "one thousand one.") This may free the compressor.

11-24 SHORT-CYCLING

Short-cycling means that the unit starts and stops too frequently. Causes may be:
1. Thermostat not mounted securely.
2. Loose connections in the starting relay.

Avoid tapping a relay to check it. The jar may cause points to touch. This short contact may ruin the points and injure the motor. If the relay will not function correctly without being tapped, it should be replaced.

11-25 DIAGNOSING INTERNAL TROUBLES

There are many ways to find the cause of trouble inside a small hermetic system.

A lack of refrigerant is indicated when the evaporator is partially frosted, while another part is heavily frosted.

A sweating or frosted suction line means liquid refrigerant is in the suction line. This may be caused by a broken thermostat or too much refrigerant (if capillary tube is used).

Internal electrical troubles, involving the motor and connections, are very rare (about three cases out of 1000). Most internal problems come from air and moisture in the motor compressor. This causes corrosion and, eventually, a burnout.

In cases where liquid refrigerant reaches the compressor, the liquid may remove the oil as it evaporates in the crankcase and carry the oil with it into the condenser. Valves may be broken as the compressor tries to pump oil or liquid refrigerant.

A restriction on the high side (capillary tube, filter-drier, or screen) will be indicated by continuous running, no refrigeration, and a condenser cooler than normal.

Para. 11-27 through 11-33 explain how connections from the gauge manifold may be tied into a hermetic system.

11-26 INTERNAL SERVICE OPERATIONS

Internal service operations are those that do any of the following:
1. Remove any part of the hermetic system.
2. Find out if there is air in the system.
3. Discover a lack of refrigerant.
4. Check if there is a clogged filter-drier or capillary tube.

For these jobs, one must attach gauges and servicing devices. These include vacuum pumps, refrigerant cylinders, and so forth.

Before attempting any field service operations which require opening the system:
1. Thoroughly clean all connections and valve fittings.
2. Install a valve adaptor or piercing valve.
3. Install a gauge manifold.

Probably the most frequent service operations will be the following:
1. Locating and repairing refrigerant leaks.
2. Purging, charging, and recovery of refrigerant.
3. Cleaning or replacing capillary tube.
4. Replacing a compressor.
5. Replacing filter-drier, high side.
6. Installing filter-drier, low side.
7. Evacuating.
8. Adding oil.
9. Using high vacuum pump.
10. Replacing an evaporator and/or condenser.

These same service operations can also be performed in the shop. Shops have better facilities to do the job.

Before performing any service operation, the service technician should study this chapter carefully.

11-27 CONNECTING A GAUGE MANIFOLD

To check the pressure in a system, gauges must be connected to the system without allowing air, moisture, or dirt to enter. The procedure for connecting gauges to a system depends on the system design. It is different for each system. See Fig. 11-26.

A. Some systems have both a suction service valve and a discharge service valve.
B. Some have a suction service valve adaptor mounted on the compressor.
C. Some do not have any service valves but do have a process tube.
D. Some have a process tube too short or not reachable. In such systems it is necessary to attach a piercing valve to either the liquid line, the suction line, or both.

System A (two service valves) is the easiest for attaching gauges. It also permits one to check both the low-side pressure and the high-side pressure. Because this system is most common on commercial systems, its installation and

use is described in Chapter 14.

System B (valve adaptor) is described in Para. 11-30. System C (process tube) is described in Para. 11-31. System D (piercing valve) is described in Para. 11-32.

When connecting refrigerant lines or gauge manifolds to any refrigerating system, one must keep the system clean. The lines, gauges, and manifold must be free of dirt, moisture, and air. The manifold should be purged with the same refrigerant as is used in the system. The manifold and connecting lines must be purged before the system service valve is opened or before the piercing valve stem makes an opening in the tubing.

Referring to Fig. 11-27, the most popular way to purge the service lines is to loosen the line fitting on the system service valve at C, open valve B, and then open cylinder line valve E just a little. Repeat the same procedure for valves D, A, and E. The cylinder refrigerant will free or purge all the lines and the manifold of air and moisture.

Usually only one connection is made to the system and this is to the low or suction side, at valve C. The flexible line between B and C is connected to the system valve, C. However, the use of the gauge manifold allows one to check both low-side pressures (valve C is open, valve B is closed) and high-side pressures (valve D is open, valve A is closed).

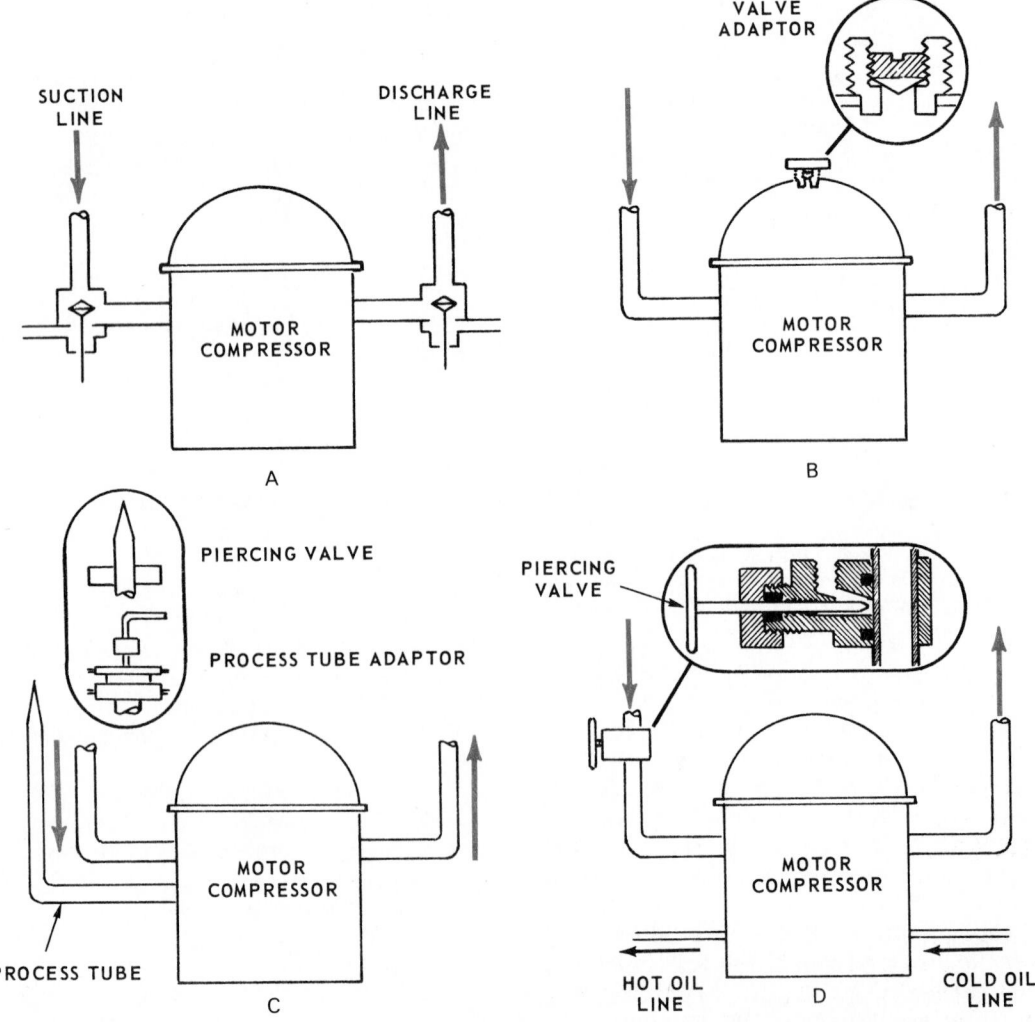

Fig. 11-26. Four different methods for connecting a gauge manifold to a hermetic system.

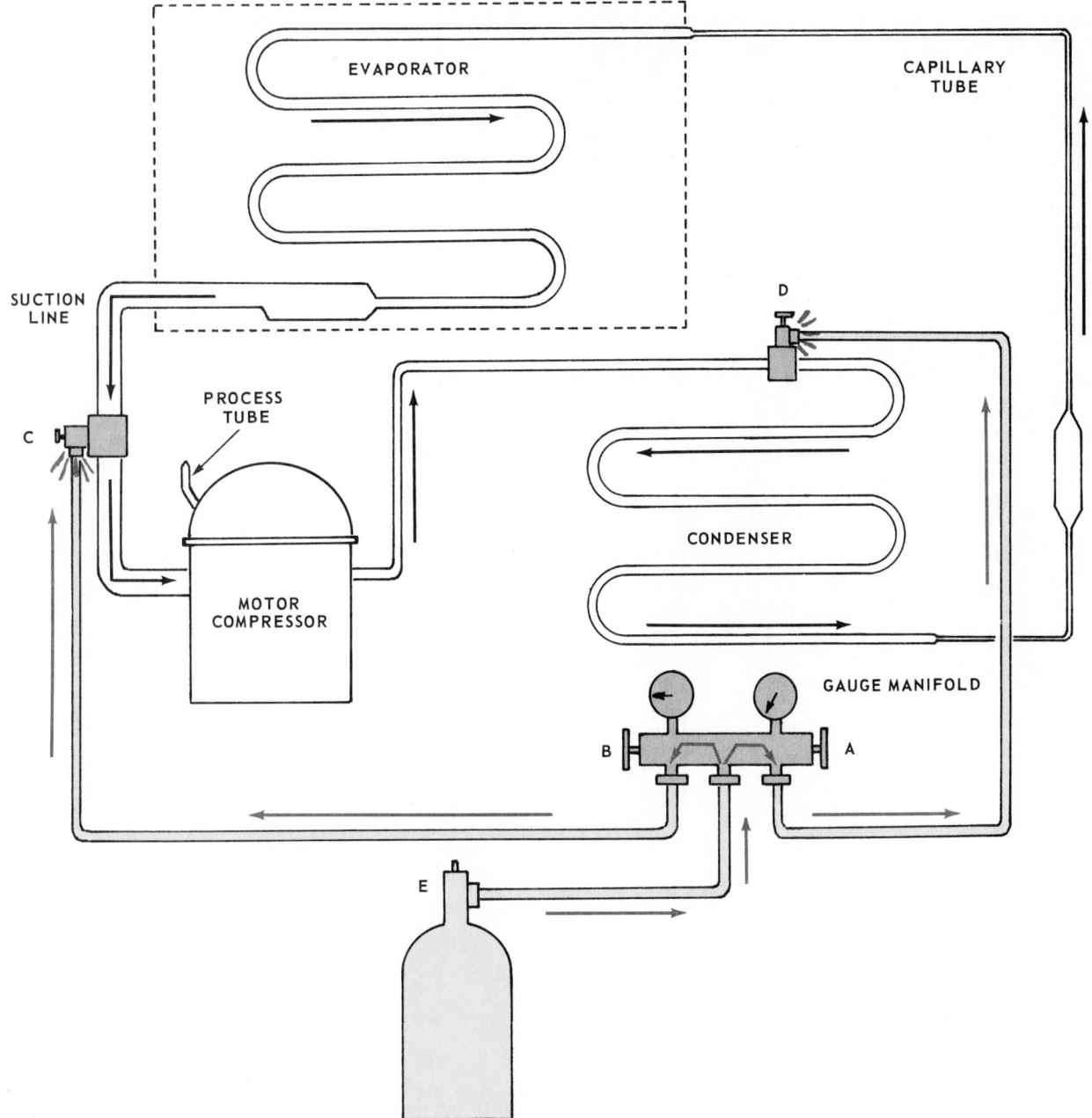

Fig. 11-27. Valves on gauge manifold are opened to purge service lines. Fittings at C and D are loosened to allow air in service lines to leak out.

The manifold also allows one to charge a system. Valves C and B are open. Cylinder valve, E, is opened slowly.

The manifold can also be used to evacuate the system. Vacuum pump line is connected to the middle connection of the gauge manifold. Valve C is open and valve B is opened.

When checking high-side pressure, use a piercing valve if the system is already charged. Braze a process tube into the condenser line if system has just been assembled and not charged.

After installing the gauge manifold (if the unit will run), operate the system through at least three operating cycles. Carefully record the suction pressures, condensing pressures, evaporator temperature, and the condenser temperature. It will be helpful if a table similar to Fig. 11-14 and/or Fig. 11-15 is recorded for future reference.

11-28 GAUGE (SERVICE) MANIFOLD TYPES AND CONSTRUCTION

It is good practice to make all gauge and service connections to a hermetic system through a gauge (service) manifold. See Fig. 11-28. There are two basic types of manifolds: the standard manifold and the block manifold. Both are available with front or side wheels. Most domestic refrigerators and freezers with hermetic systems do not provide for gauge openings. Therefore, special attaching devices must be used to make it possible to use the gauge manifold. A schematic for the manifold in Fig. 11-28, Item C, is shown in Fig. 11-29. Fig. 11-30 shows the various uses and valve positions of a gauge manifold. Fig. 11-31 shows the setup for charging a system.

Separate gauges and hand valves can be used with the

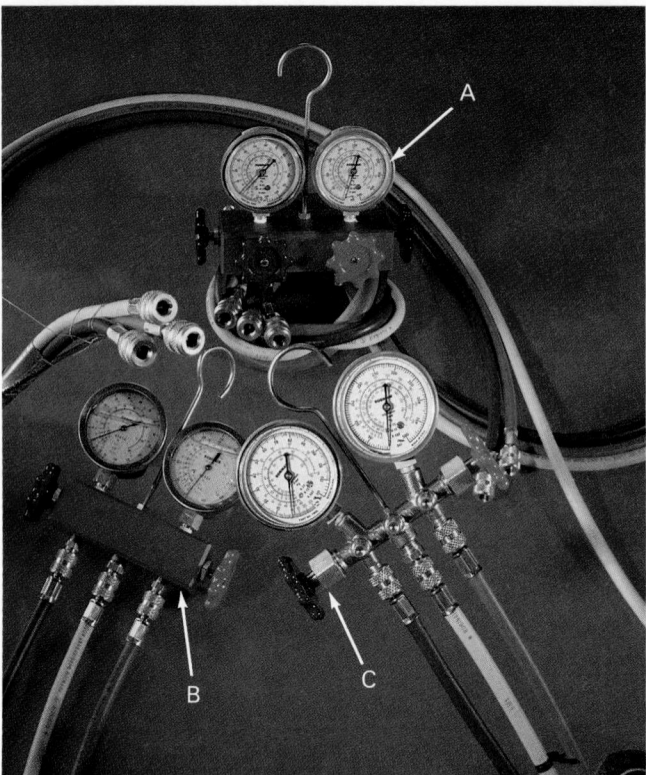

Fig. 11-28. Three types of gauge manifolds: A—Four-way manifold, front and side, used to do testing, evacuation, and recharging of a system without having to switch hoses. B—Aluminum bar manifold. C—Side wheel manifold. (Robinair Division, SPX Corporation)

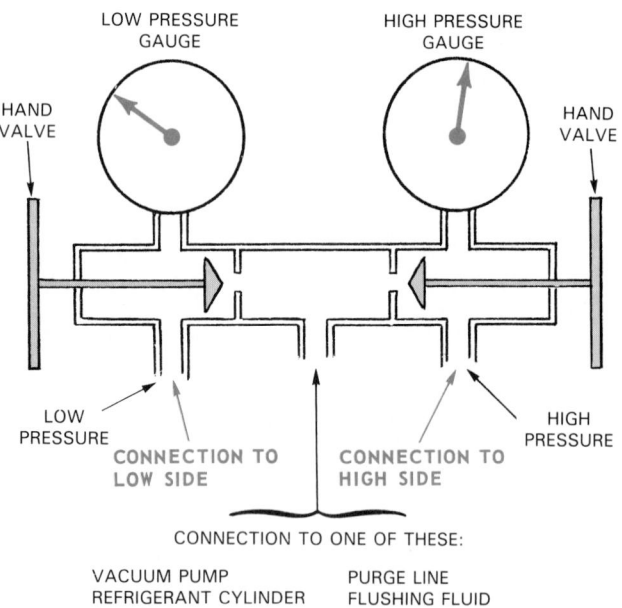

Fig. 11-29. This gauge manifold shows hand valves, gauges, and refrigerant openings. Gauges will always show a pressure reading. When low-pressure hand valve is turned all the way in, the low (evaporator) pressure can be checked. When high-pressure hand valve is turned all the way in, high (condensing) pressure can be checked. When both valves are open (turned out by twisting to the left), high-pressure vapor will flow into low side. When only low-pressure valve is open, one can charge the system or evacuate it, put oil in system or clean it.

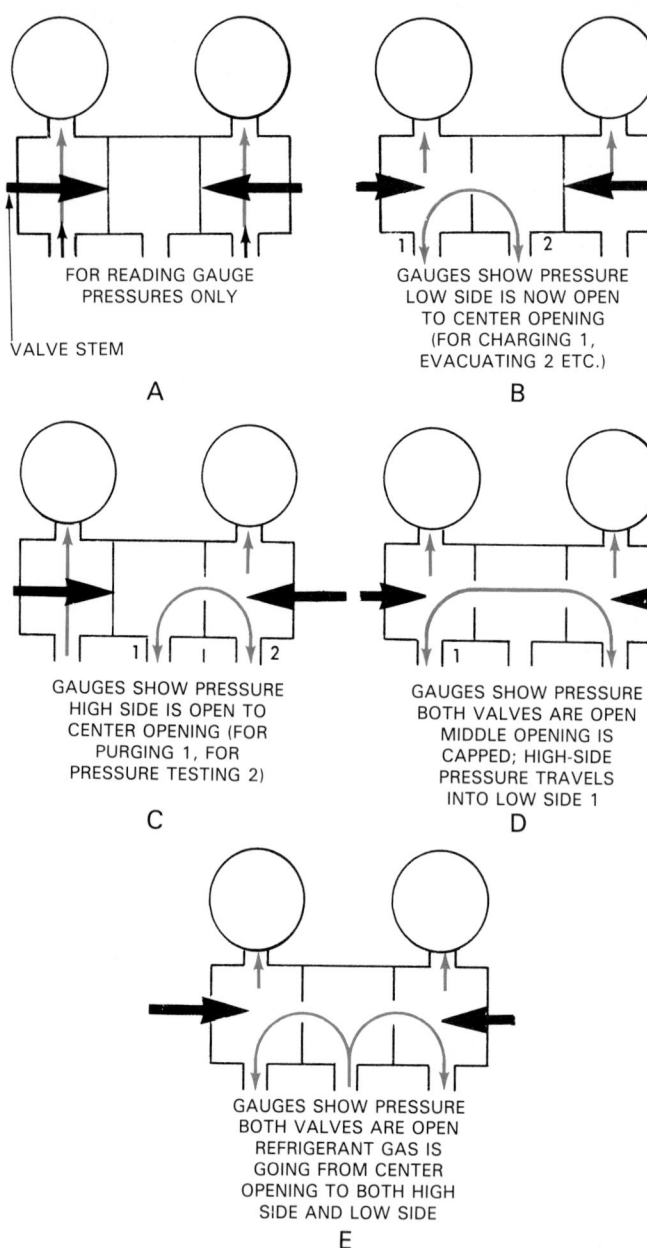

Fig. 11-30. How to use hand valves of gauge manifold for various service operations.

special service valves mounted on hermetic systems. However, to purge a system, check pressures, or to evacuate and to charge the system would mean removing these devices and using others for each operation.

A gauge manifold with two gauges, two hand valves, and three separate lengths of flexible refrigerant tubing enables the service technician to perform these operations more easily. The three flexible hoses are equipped with 1/4-in. flare fittings with synthetic rubber gaskets. Thus, connections can be made pressure-tight with finger pressure alone. Fig. 11-32 shows a design and construction of a flexible service and refrigerant line.

The service technician must learn how to use the gauge and service manifold. The hand valves on the manifold can be used for most operations. Fig. 11-33 shows a gauge manifold being used with a vacuum pump to produce a vacuum, install part, and charge a system.

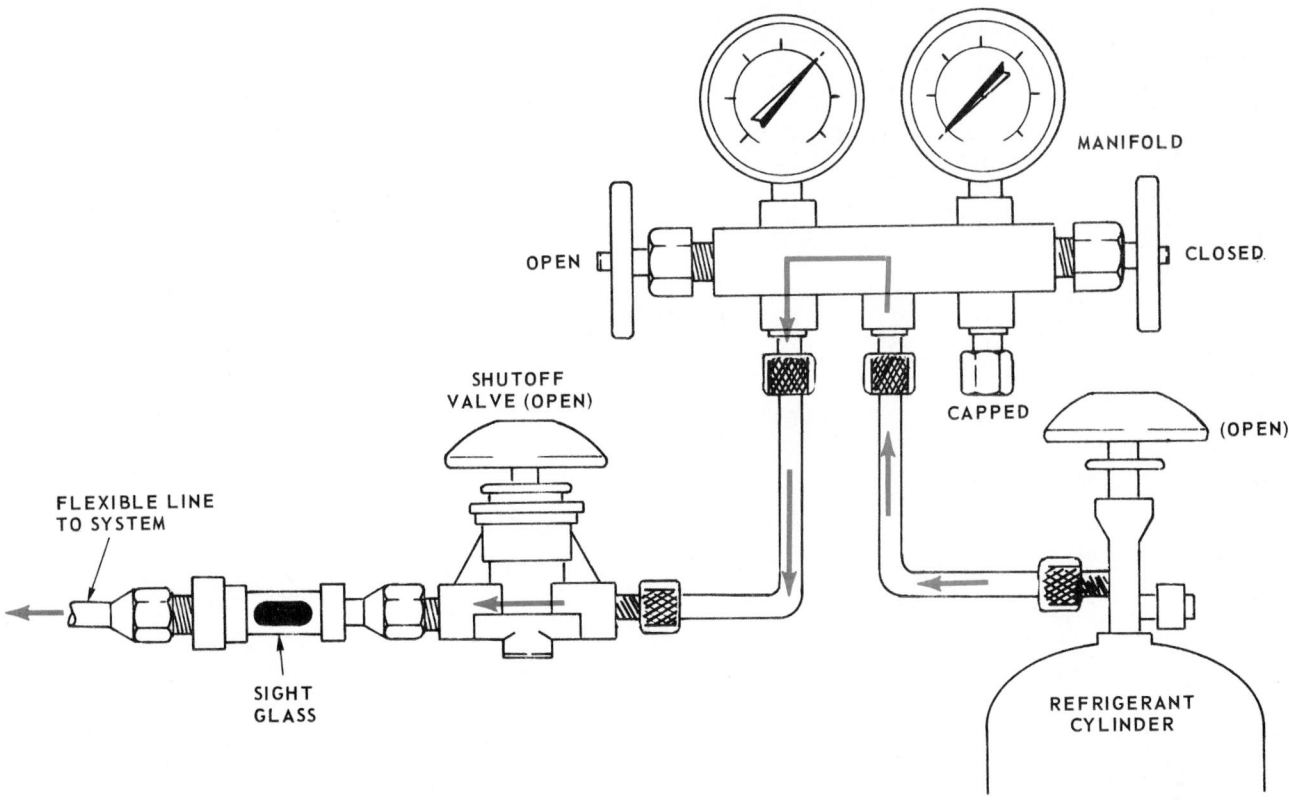

Fig. 11-31. Gauge manifold fitted with flexible refrigerant tubing. This installation is used when charging system.

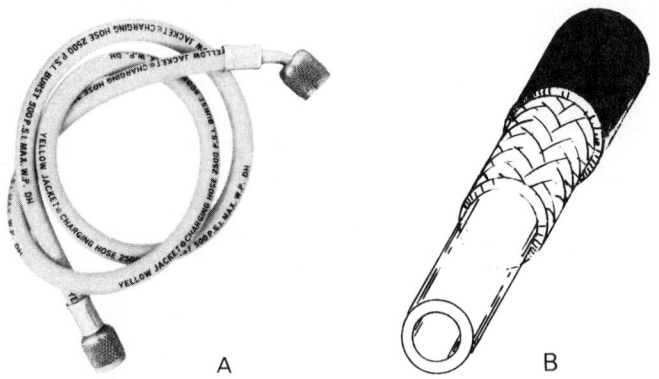

Fig. 11-32. A—Flexible charging hose with external flare connection.
B—Cutaway showing wall construction of flexible hose.
(Ritchie Engineering Company, Inc.)

11-29 HERMETIC SERVICE VALVES AND ADAPTORS

Most hermetic refrigerators do not have service valves. Some have fittings to which valves may be attached for service operations. The valves are removed when the service work has been completed.

Others have neither service valves nor provision for fitting valves to them. Where no service valves are provided, it is necessary to fit and attach valves to the mechanism. Attachments of various types are available from refrigeration supply wholesalers.

Some hermetic mechanisms have a process tube. One is shown being used in Fig. 11-34. Note the hand valves and process tube adaptor fitting. Note also that a charging

cylinder is used.

Once service valves have been mounted on a hermetic system, they can be used for many purposes:
1. To check the internal pressures.
2. To discharge the system or add refrigerant.
3. To add oil.
4. To evacuate the system.
5. To make it easier to replace driers, motor compressors, evaporators, and refrigerant controls.
6. To recharge the system.

Usually a flexible charging line is connected to the service valve adaptor and to either a hand valve or a service manifold mounted on the other end of this tubing. This makes service easier. Fig. 11-35 shows this type of service connection set up for charging a hermetic system.

Have the valve loose at F and use the cylinder vapor in G to blow out (purge) the lines. The gauge at F may be located on the compressor dome, on the suction line, or on the process tube.

11-30 SYSTEMS WITH VALVE ADAPTORS (ATTACHMENT)

Valve adaptors are one way of connecting gauges and charging cylinders to a hermetic system. Fig. 11-36 shows the part of the adaptor which is fastened to the compressor dome. The adaptor has a removable service valve as shown in Fig. 11-37. The attachment and the adaptor connect together as shown in Fig. 11-38. A service valve with two openings is shown in Fig. 11-39.

Note that for these refrigerators, the adaptor provides a means of operating the small needle valve mounted on the motor compressor. It also provides an opening for a ser-

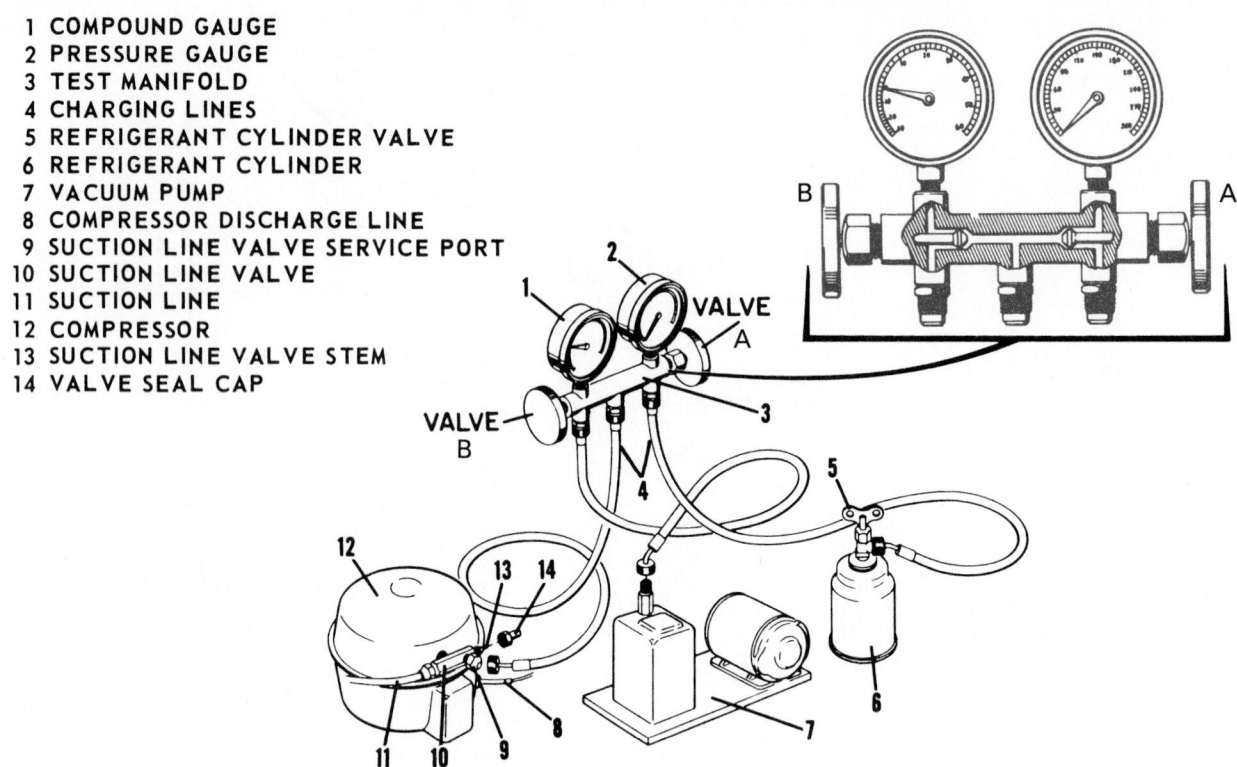

1 COMPOUND GAUGE
2 PRESSURE GAUGE
3 TEST MANIFOLD
4 CHARGING LINES
5 REFRIGERANT CYLINDER VALVE
6 REFRIGERANT CYLINDER
7 VACUUM PUMP
8 COMPRESSOR DISCHARGE LINE
9 SUCTION LINE VALVE SERVICE PORT
10 SUCTION LINE VALVE
11 SUCTION LINE
12 COMPRESSOR
13 SUCTION LINE VALVE STEM
14 VALVE SEAL CAP

Fig. 11-33. A drawing of a system being evacuated. The illustration also indicates another means of utilizing a manifold gauge. The vacuum pump is attached to (B) suction service valve line. A vacuum will be created through the center service line, which is connected to the suction service valve. Once the vacuum is obtained, the small service cylinder charges the system up to 0 or slightly above. The compressor can then be removed from system and the rest of system sealed off to avoid air entering into it.

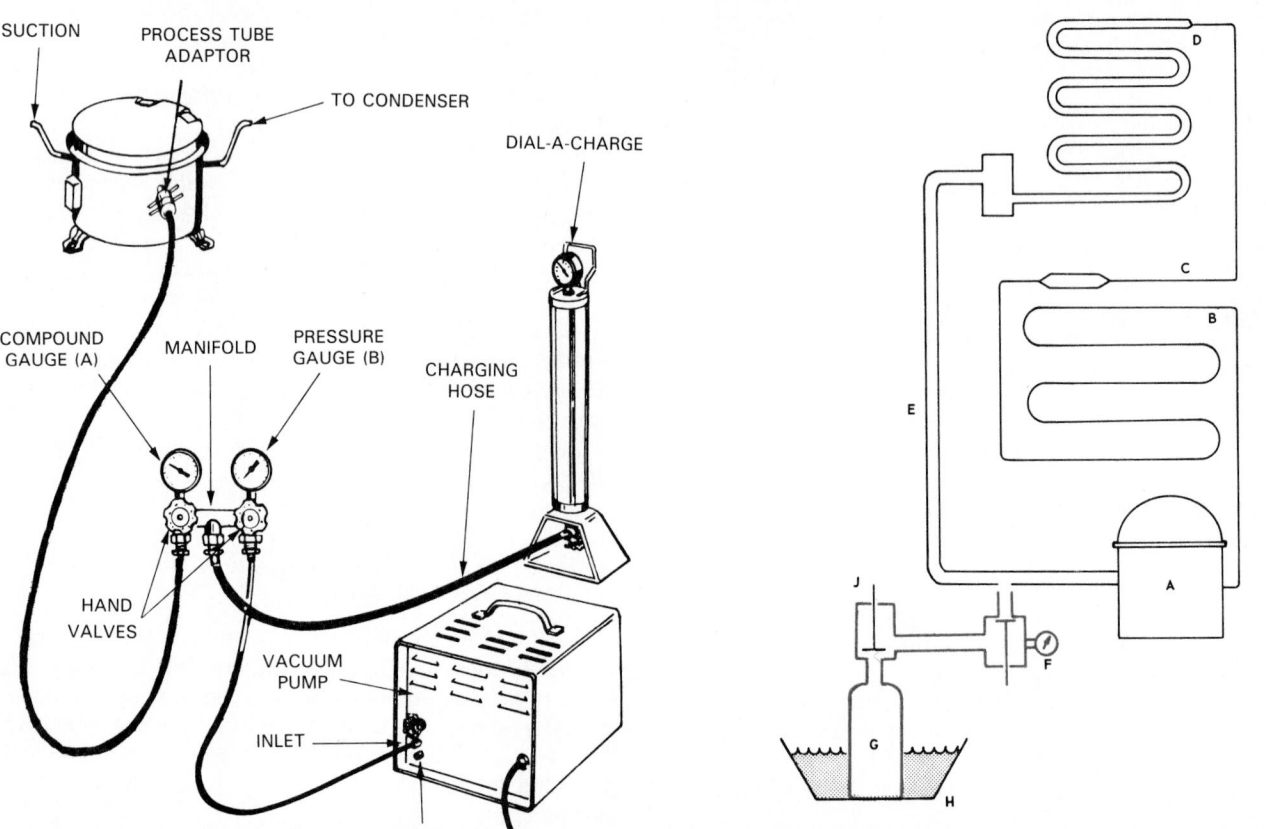

Fig. 11-34. System being charged with process tube adaptor. Note use of charging cylinder. (Frigidaire Company)

Fig. 11-35. Charging a hermetic system. A—Compressor. B—Condenser. C—Capillary tube. D—Evaporator. E—Suction line. F—Valve attachment. G—Refrigerant cylinder. H—Hot water. J—Cylinder valve.

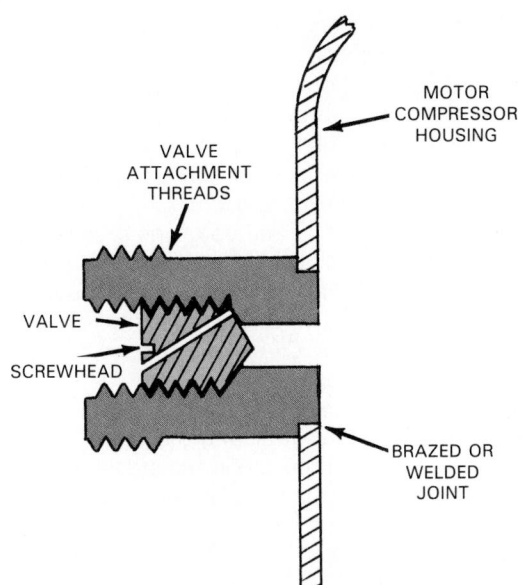

Fig. 11-36. Service valve assembly for hermetic unit. Valve attachment must be fastened in place before valve can be opened; otherwise, refrigerant will escape.

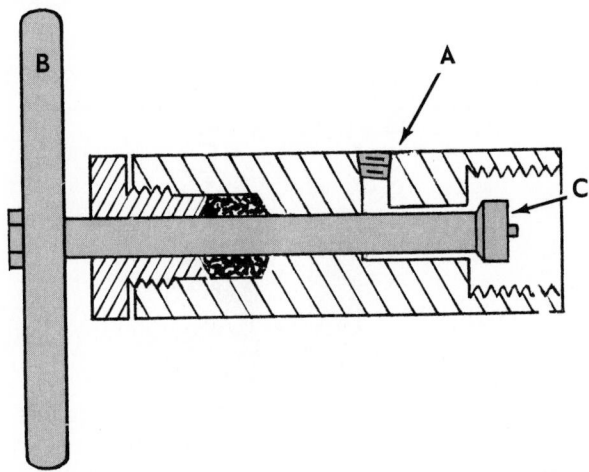

Fig. 11-37. Service valve attachment is installed on valve adaptor, which is fastened to motor compressor dome. A—Opening for gauge connection and for servicing. B—Handwheel. C—Valve adaptor needle valve drive.

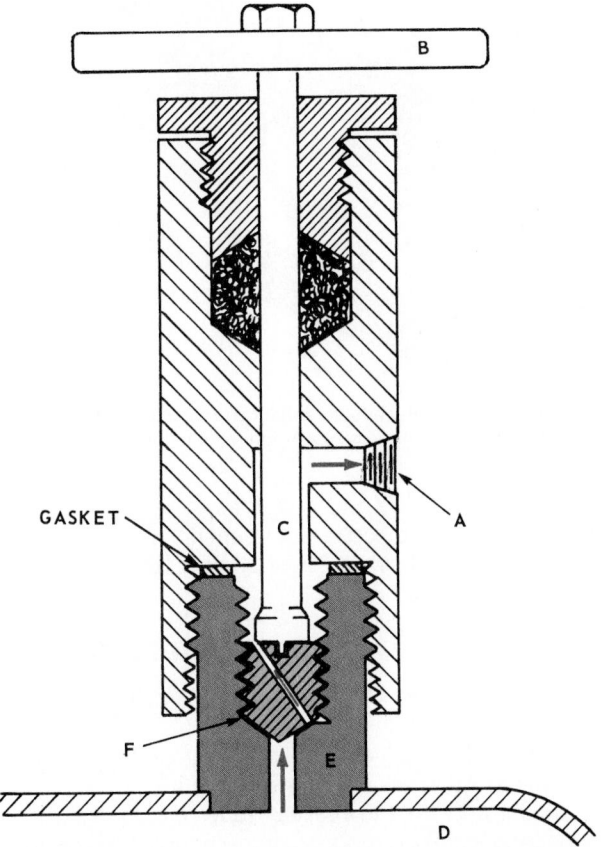

Fig. 11-38. Valve attachment as it would appear in cutaway when connected to valve adaptor. A—Opening for gauge connection and servicing. B—Handwheel. C—Valve adaptor needle valve drive. D—Motor compressor housing. E—Valve adaptor. F—Valve.

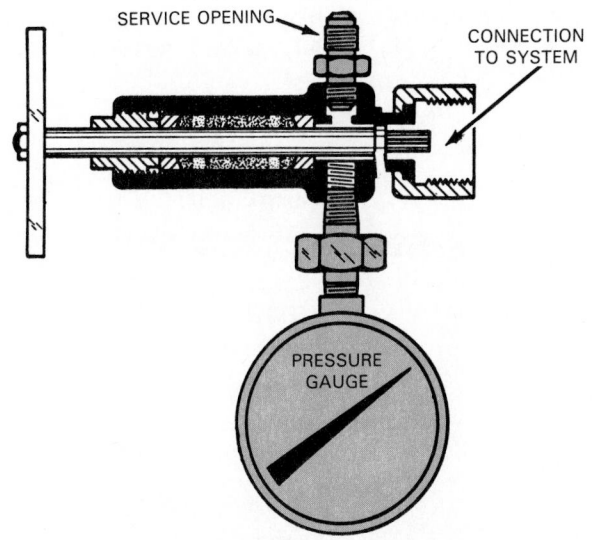

Fig. 11-39. Service valve attachment. Note that there are two openings. One may be used for pressure gauge and the other for performing service operations such as discharging, charging, and adding
(Fedders North America)

vice gauge or a gauge manifold connection. Synthetic or copper gaskets are used to seal the valve joints. An assortment of valve adaptors is shown in Fig. 11-40.

Certain procedures should be followed when using valve adaptors:
1. Clean the outside.
2. Remove the dust cap from the adaptor mounted on the motor compressor.
3. Choose the correct valve stem drive.
4. Push the service valve stem forward in the body of the valve attachment.
5. Engage the valve stem in the valve adaptor needle.
6. Thread the valve adaptor unit into the attachment body.
7. Use good gaskets.
8. Before opening the valve adaptor needle, tighten the packing unit around the valve stem.

Blow out the passages (valves, gauges, and flexible lines) using the same refrigerant as is in the system. Purge the assembly using gas from a refrigerant cylinder, leaving the flexible line fitting loose at the valve attachment. After

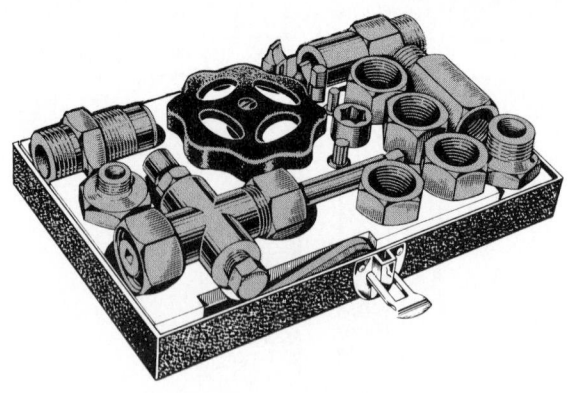

Fig. 11-40. Valve kit and adaptors that can be used on various makes of semihermetic or hermetic refrigeration units. (Mueller Brass Co.)

purging, tighten the loose connection. Always test the assembly for leaks using a 15 to 20 psig (210 to 240 kPa) refrigerant pressure.

11-31 PROCESS TUBE AND ADAPTORS

The process tube can be adapted for service of systems in two ways:

1. Install a piercing valve on the process tube. Piercing valves are discussed in Para. 11-32.
2. Cut the end off the process tube. (All the refrigerant will escape.) Then mount a process tube adaptor on the clean tube. *Caution: Always wear safety goggles when charging or discharging a system.*

The process tube left in the system at the factory is used by the manufacturer to evacuate, test, and charge the new unit. It can be used by the service technician by brazing an extension to it or by mounting a process tube adaptor on it as a means of attaching a manifold line. The adaptor kit, as shown in Fig. 11-41, makes it possible to use the

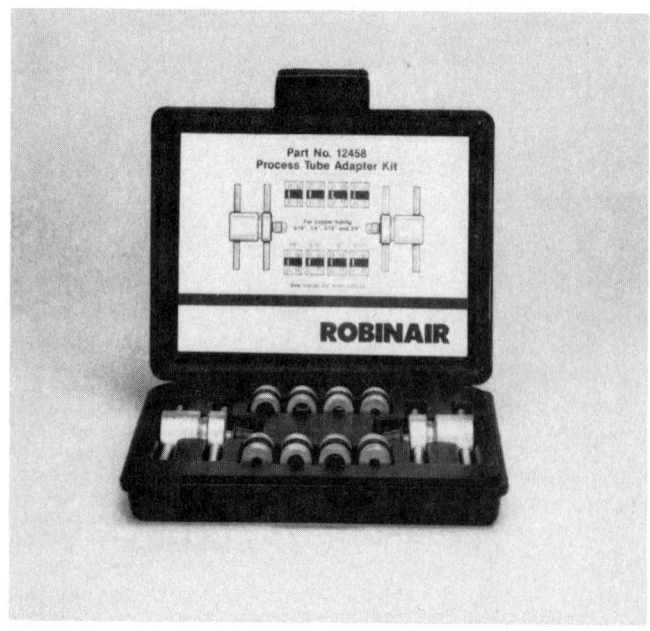

Fig. 11-41. Process tube adaptor and kit.
(Robinair Division, SPX Corporation)

process tube without soldering an extension to it or without flaring the tubing. It provides a positive seal. See Fig. 11-42.

By using adaptors of various sizes, the tool may be used on 3/16, 1/4, 5/16, or 3/8-in. copper tubing.

To attach the process tube adaptor:

1. Clean the outside of the process tube. Use clean, fine sandpaper to remove paint and dirt.
2. Using cutting pliers, cut off the tip of the process tube to release the refrigerant from the system. Use goggles! A flexible rubber tube should then be placed over the cut end of the process tube to discharge the refrigerant away from the work area. CAUTION: Always discharge a system in a well-ventilated area.
3. If the process tube is short, it may be advisable to braze an extension to it before attaching the valve adaptor.
4. Maintain some internal pressure in the system to blow out metal chips which may be formed while cutting.
5. Another method is to cut the process tube with a cutter. Wear goggles! CAUTION: Always discharge a system in a well-ventilated area.
6. Attach the adaptor to the process tube.

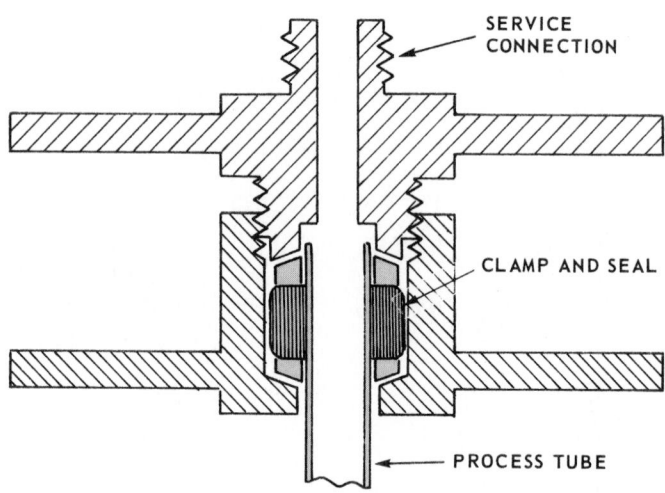

Fig. 11-42. Process tube adaptor.

After evacuating the system to dry it and to remove air, recharge it. Wait until the system operates correctly, then close off the process tube using a positive pinch-off tool. Remove the adaptor and braze the end of the process tube.

The pinch-off tool is used wherever it is necessary to seal off soft copper tubing up to 3/8 in. OD. Fig. 11-43 shows one type of pinch-off tool. It has a screw-type action shaft with a ball bearing on the end that presses against the tube. The tool is placed over the copper tubing in the same manner as a tubing cutter. The tubing should be slowly compressed by turning the pinch-off tool handle clockwise.

As the handle is turned, the ball bearing presses into the tubing and compresses it against the die on the bottom of the tool, producing a permanently pinched line. See Fig. 11-44. Take care that the pinch-off tool is not rotated too far or excessive pressure applied. It is best to leave the tool in place until the adaptor is removed and the tubing end is sealed by brazing. The pinch-off tool may be used when an emergency arises, requiring isolation of parts of the refrigeration system (such as a bad leak).

Fig. 11-43. Pinch-off tool is usable on tubing up to 3/8 in. OD. This makes good seal and also keeps tubing strong at pinch-off point.

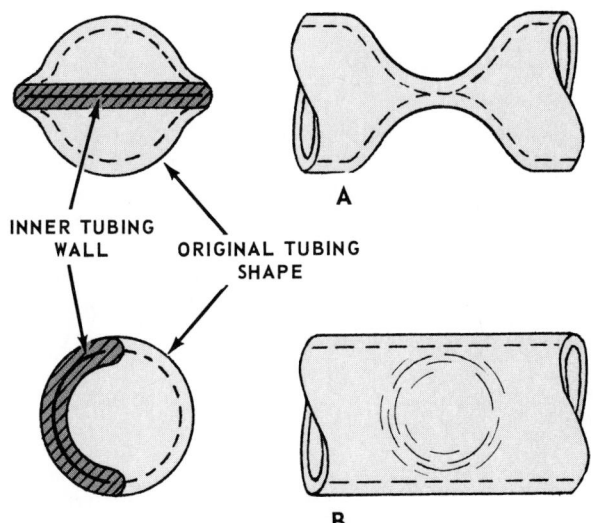

INNER TUBING WALL

ORIGINAL TUBING SHAPE

A

B

Fig. 11-44. Action of pinch-off tools. A—Pinch-off made with pliers-type tool. B—Pinch-off made by tool shown in Fig. 11-43.

11-32 PIERCING VALVES

A popular way to gain access to a hermetic system is to mount service piercing valves on the suction tubing, on the discharge tubing (tubing to condenser), on both, or on the process tube. A piercing valve is shown in Fig. 11-45. Many designs of tubing-mounted piercing valves have been developed.

There are two general types:
1. Bolted-on valves.
2. Brazed-on valves.

Fig. 11-46 shows a bolted-on valve using a bushing gasket. This valve is available in several sizes for various sizes of tubing.

Tubing should be straight and round. It should be carefully cleaned. (Do not scratch it.) Make sure there are no dents in it. Check also to see if there is enough space to operate the attachment valve and that the connecting tubing can be easily mounted on the attachment valve.

Put a little clean refrigerant oil on the tubing. Be sure the

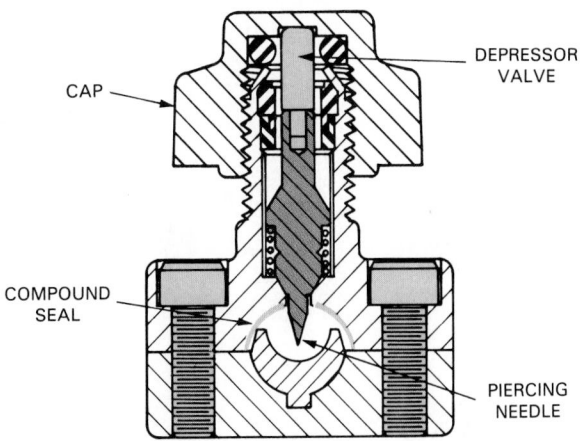

CAP

DEPRESSOR VALVE

COMPOUND SEAL

PIERCING NEEDLE

Fig. 11-45. Bolted-on piercing valve. Valve is bolted to line by two socket head screws. Note use of special compound seal. (Watsco Components, Inc.)

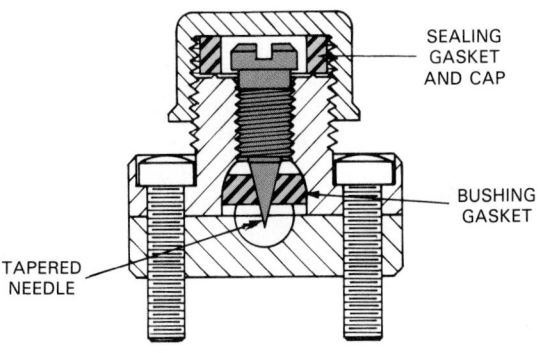

SEALING GASKET AND CAP

BUSHING GASKET

TAPERED NEEDLE

Fig. 11-46. Bolted-type tubing-mounted service valve. Gasket seals hole made by piercing needle. (Watsco Components, Inc.)

synthetic sealing washer is in place and that the needle point piercing valve stem is all the way out. Mount the valve on the tubing. Tighten the unit, clamping screws evenly. These valves are usually left on the system. The attachment valve is similar in design and construction to those shown in Figs. 11-38 and 11-39. Two types of service valve attachments are shown in Fig. 11-47.

The second type of piercing valve is brazed on. See Fig. 11-48. The braze-mounted type (saddle design) is safe to use because both the suction tubing and the condenser tubing have no liquid in them and can be heated to a brazing temperature. However, make sure there are no flammables or soft-soldered joints close to the brazing.

Be sure the tubing is straight and round at the brazing point. Clean both the saddle and tubing mating surface with clean sandpaper or clean steel wool. Remove the piercing valve stem and the gasket from the saddle. Put clean fresh brazing flux on the saddle (outer edges) or use a phosphorous-copper brazing filler rod.

NOTE: If flux is used, follow manufacturer's specifications. If phosphorous-copper brazing filler rod is used, flux is unnecessary because the phosphorous in the brazing material deoxidizes the copper surface.

Mount the saddle on the tubing. Check to see if there is room (clearance) for mounting the service valve on the tubing mounting valve. Heat both the tubing and the saddle until the filler rod material flows around the saddle.

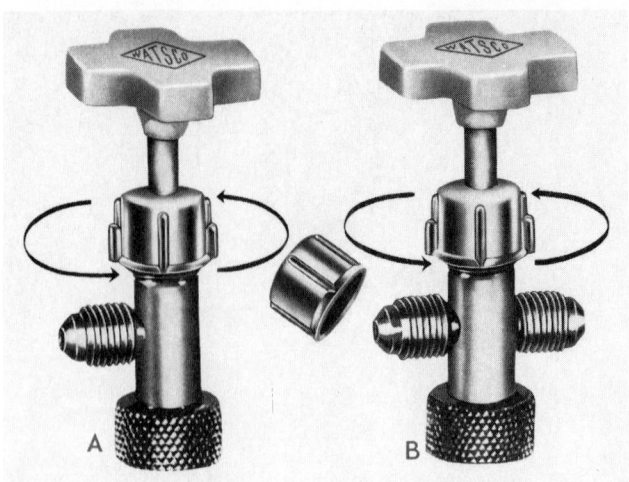

Fig. 11-47. Two types of service valve attachments used with tubing-mounted piercing valve adaptors. A—Valve with one 1/4-in. male flare service opening. B—Valve with two 1/4-in. male flare service openings. (Watsco Components, Inc.)

hard-to-see edges.

After the brazed joint has cooled, install the piercing needle and gasket. The unit is then ready for the installation of the service valve attachment as shown in Fig. 11-49. Three gasket-type piercing valves are shown in Fig. 11-50.

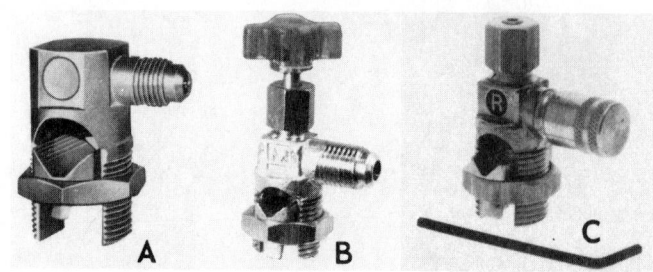

Fig. 11-50. Three types of piercing valves. A—Charge-and-tap valve. B—Hand valve type. C—Line tap type with hexagonal wrench. (Robinair Division, SPX Corporation)

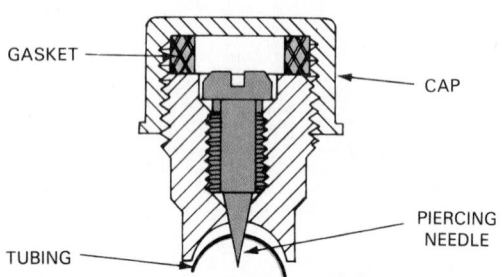

Fig. 11-48. Braze-mounted tubing piercing valve. Note use of synthetic material in sealing gaskets. Preformed brazing ring is usually used for brazing alignment and proper metal flow. (Watsco Components, Inc.)

11-33 CORE VALVES

Many systems use a Schrader core valve, Fig. 11-51, to gain access to a hermetic system. This type is similar to the valve cores used in automobile tires.

A clamp-on core valve adaptor is shown in Fig. 11-52. The flexible service tubing fitting or the service valve mounted on this fitting has a pin which depresses the core valve stem as the fitting or service device is mounted.

Some valve adaptors are threaded to the system, some are clamped to the tubing, and some are brazed to the tubing.

Some service technicians use a service valve attachment

The saddle must not move or shift during the brazing or while the brazed joint is cooling. Some service technicians hold the saddle in place with a small C-clamp during the brazing operation.

Do not overheat the tubing, as it may be weakened to the point of failure and may burst. Use goggles during the brazing operation.

Inspect the brazed joint carefully. Use a mirror to check

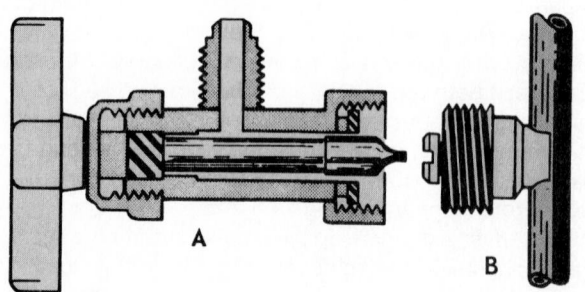

Fig. 11-49. This piercing valve, brazed onto line, may be used on hermetic refrigerator systems. Part A can be removed after servicing to discourage tampering with system. Part B remains on system. Threaded cap on piercing valve is used to protect threads and avoid tampering.

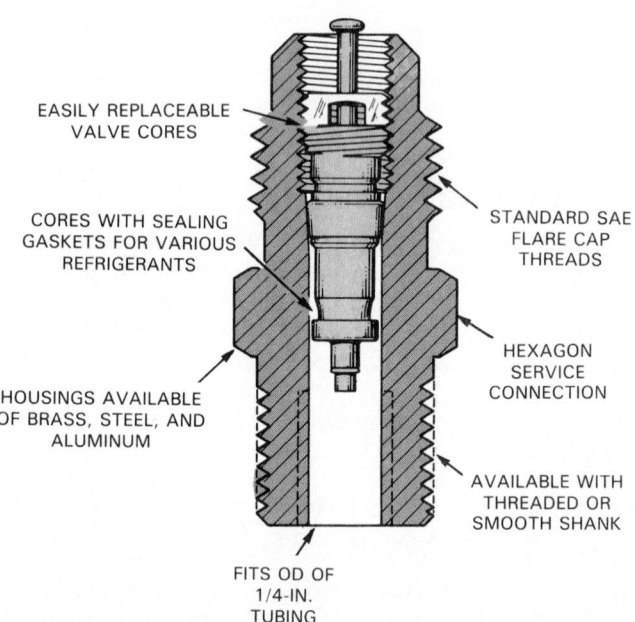

Fig. 11-51. This Schrader valve fitting may be used to connect service lines to hermetic system. When service line fitting is mounted on this fitting, a pin depresses or forces stem of valve core inward and opens system for service.

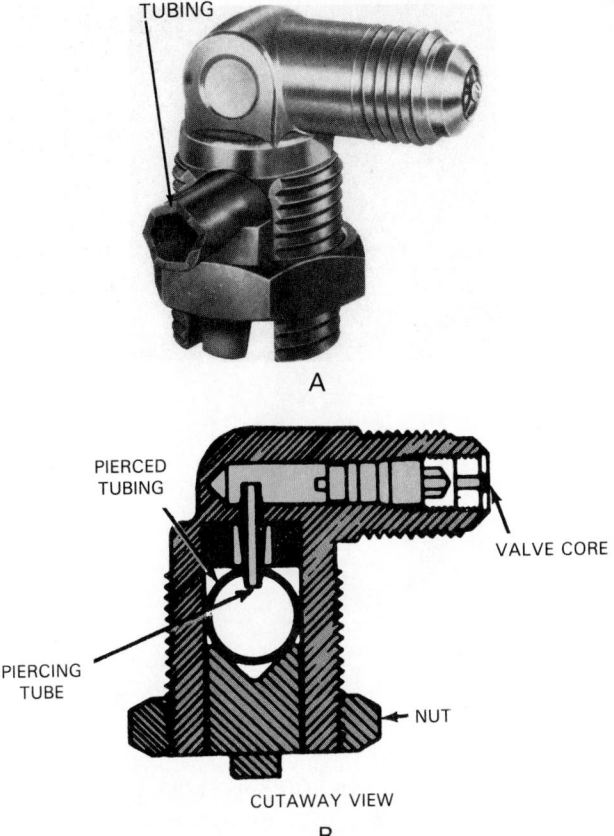

TUBING

A

PIERCED
TUBING

VALVE CORE

PIERCING
TUBE

NUT

CUTAWAY VIEW

B

Fig. 11-52. This valve-core-access valve adaptor clamps onto tubing. A—Piercing valve mounted on tubing. B—Cross section of same adaptor. Passage is opened when valve core stem is depressed by service line fitting.   (Robinair Division, SPX Corporation)

that mounts on the Schrader valve adaptor. This device has a long stem which is used to remove the valve core while evacuating the system. The main reason the core is loosened is to allow more flow of gas as when a vacuum is to be drawn on the hermetic system. Vacuum lines and fittings should be as large as possible. Fig. 11-53 shows the advantage of removing the valve core while evacuating.

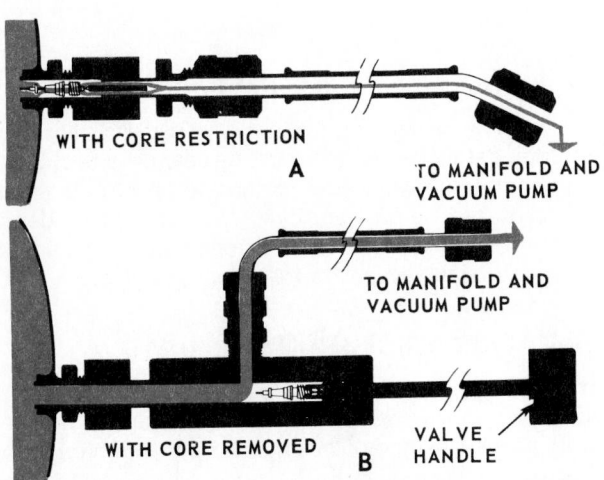

WITH CORE RESTRICTION

A

TO MANIFOLD AND
VACUUM PUMP

TO MANIFOLD AND
VACUUM PUMP

WITH CORE REMOVED   B

VALVE
HANDLE

Fig. 11-53. Evacuating passages are larger when valve core is removed from its fitting. A—Small flow with core in place. B—Large flow with core removed.

11-34 LOCATING INTERNAL TROUBLES

With gauges, thermometers, electrical instruments, and by observation, the service technician should be able to locate the cause of almost every problem in a refrigerating system.

The paragraphs that follow discuss most of the reasons why refrigeration systems will not operate correctly. Then the repair and testing of refrigeration systems will be described.

11-35 MOISTURE IN THE REFRIGERANT CIRCUIT

Moisture in the refrigerant system will cause the unit to malfunction. The moisture forms ice in the refrigerant control. This is at the point where liquid refrigerant is expanding into the evaporator. Icing closes the opening, blocking flow into the evaporator.

This condition can be recognized by several observations:
1. The system will completely defrost. Then, since the icing which caused the blockage has disappeared, the unit will work properly again. But the unit will only work until ice again forms at the refrigerant control.
2. Another symptom is decreasing pressure in the suction line. The compound gauge shows a steady decrease over several hours—even to a vacuum. Then pressure suddenly becomes normal again. This odd cycle will keep repeating.
3. If, during system shutdown, one warms the refrigerant control with a safe resistance heater (hot pad) or radiant heat bulb, the ice will melt. Should the system then begin to work properly, there is definitely moisture in the refrigerant.

Moisture in the refrigerant circuit also creates corrosion problems within the system. Refrigerants react with water molecules to form acids. The acids increase the amount of corrosion in the system.

Moisture in the refrigerant circuit may be removed by putting a drier in the liquid line. The procedure is as follows:
1. Install gauge manifold.
2. Remove refrigerant.
3. Dry and clean filter-drier connections.
4. Apply flux.
5. Heat connections.
6. Remove old drier.
7. Install new drier.
8. Braze connections.
9. Test for leaks.
10. Evacuate system.
11. Charge system.
12. Warm the refrigerant control enough to melt the ice. Drier will absorb this moisture as it circulates.

Substances are available which, when placed in the refrigerant circuit, tend to keep moisture from forming ice. A filter-drier is best. It prevents circulation of the moisture through the system and reduces the chance of oil breakdown (sludge and acid).

11-36 WAX

Manufacturers have removed as much wax from refrigeration oil as possible. Refrigeration oil is discussed in Para. 9-32. Although manufacturers have removed most of the wax from refrigeration oil, small amounts still remain.

Some oil circulates with the refrigerant. Sudden expansion at the refrigerant control, accompanied by low temperature and pressure, causes some wax to separate from the oil. The wax collects in the refrigerant control. In time, it will restrict flow or clog control completely.

Presence of a restriction can be checked with the aid of a piercing valve. Look for the following:
1. A test shows low-side pressure to be very low.
2. Liquid refrigerant shows up in the condenser.
3. The unit has not been producing any refrigeration at all.

Always clean or replace a clogged valve or capillary tube. *It is good service procedure to keep the ice or wax locked in the refrigerant control being moved.* This can be done most easily by packing the control in dry ice before removal. Another method is to very quickly open the joints after the unit is discharged. Use the same steps as in Para. 11-56, but repair and/or replace the refrigerant control. Use only the best low-wax oil when servicing frozen foods equipment.

11-37 SHORTAGE OF REFRIGERANT

Shortage of refrigerant is a common source of poor refrigeration. Because small systems only have one or two pounds of refrigerant, even the smallest leak will soon cause poor refrigeration. A leak with a loss rate of one ounce a year can be located and must be repaired.

A lack of refrigerant will be shown by:
1. A low-side pressure that is below normal.
2. Evaporator is warm or the outlet end of evaporator is warm.
3. A high-side pressure that is below normal.
4. A piercing valve mounted on the outlet of condenser, when opened, allows only gas to escape (it should be liquid).

If a shortage of refrigerant is found, there is often a leak. This leak must be found and repaired.

11-38 LOCATING REFRIGERANT LEAKS

Methods of testing for leaks vary with the refrigerant used. However, all methods have one procedure in common: apply pressure to the system with nitrogen or carbon dioxide. At the start of testing, a positive pressure (greater than atmospheric) of 5 to 30 psig (140 to 310 kPa) is necessary throughout the circuit. If no leaks are found, then test again at or above the normal condensing pressure for the refrigerant used (i.e. 90 to 135 psig [730 to 1 040 kPa] for R-12).

Check for leaks before the unit is evacuated. Moisture could enter the system through a leak during evacuation or pumpdown.

Many companies recommend using the refrigerant in the system to test for leaks. A sensitive leak detector should be employed. An electronic leak detector, shown in Fig. 11-54, is good for this purpose.

If a leak is found, it is very important to recheck the complete unit after repair has been made. This provides a check for the repair and will reveal any additional leaks.

11-39 PRESSURE-TESTING FOR LEAKS

With proper care, nitrogen or carbon dioxide may be used safely when pressure-testing for leaks. The pressure in the nitrogen cylinder is about 2000 psig (14 MPa) and in a

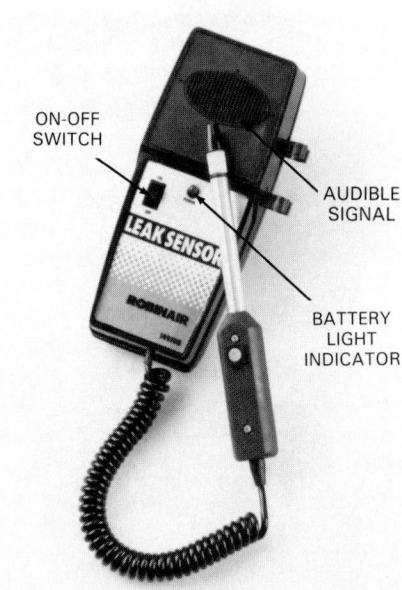

Fig. 11-54. Electronic leak detector which operates on batteries. (Robinair Division, SPX Corporation)

carbon dioxide cylinder about 800 psig (6 MPa). A pressure reducing device which has both a pressure regulator and a pressure relief valve must always be used when testing with either of these two gases. A recommended pressure regulating device is shown in Fig. 11-55.

A refrigerating system would explode if pressure were to build up on the system. Many accidents have been caused by using too much testing pressure.

Before using nitrogen or carbon dioxide to test a system, look at the name plate. In most cases, it will give recommended testing pressures. If these pressures are not known, never go over 170 psig (1 300 kPa) pressure when testing all or part of a hermetic system. See Chapter 14 for pressure testing commercial systems.

Caution: Never use oxygen or acetylene to develop pressure when checking for leaks. Oxygen will cause an explosion in the presence of oil. Acetylene will decompose and explode if it is pressurized over 15 to 30 psig (210 to 310 kPa).

11-40 LEAK DETECTING DEVICES

Leaks in a refrigerant system are usually very tiny. Therefore, detecting devices must be very sensitive. Some of the commonly used devices include bubble solutions, fluorescent dyes, refrigerant dyes, the halide torch detector, and an electronic detector. Each method has its advantage and is reviewed in the following paragraphs.

11-41 BUBBLE SOLUTIONS

The bubble method of leak detection has been very common. It uses a water-soap solution. This solution is brushed over an area suspected of leaking. Gas coming through the solution will cause bubbles.

Patented solutions are more popular to use instead of soap. These provide a stronger, longer-lasting bubble film than the soap solution. Fig. 11-56 illustrates the action of

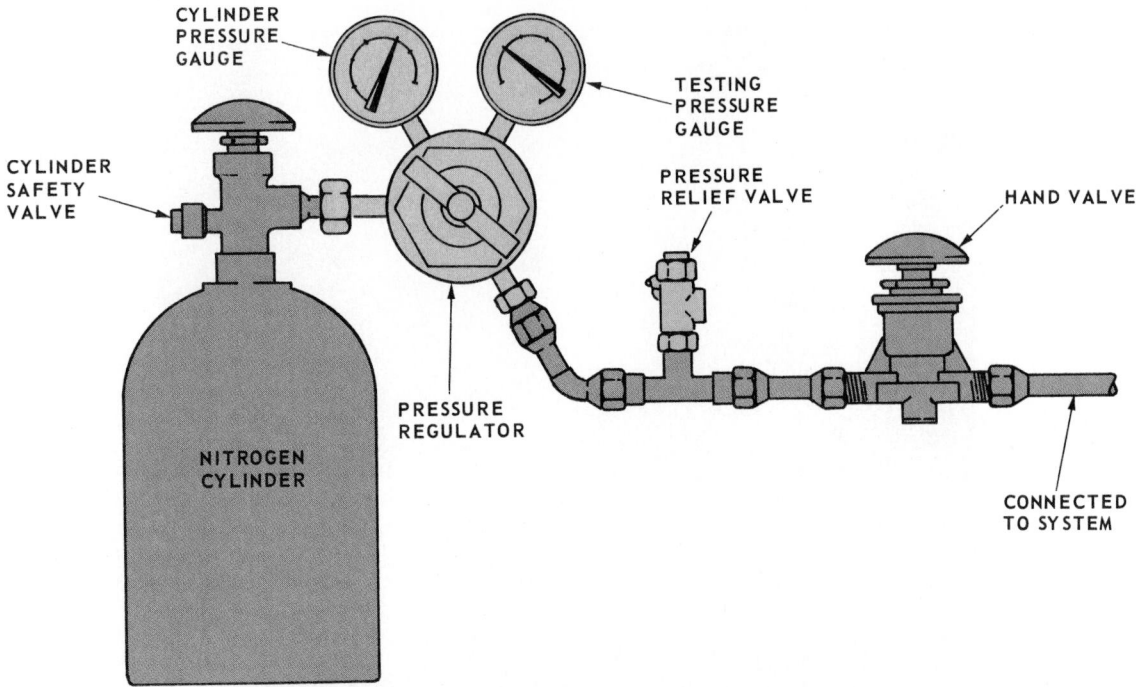

Fig. 11-55. Pressure regulator system. Note that both a regulator and pressure relief valve are used.

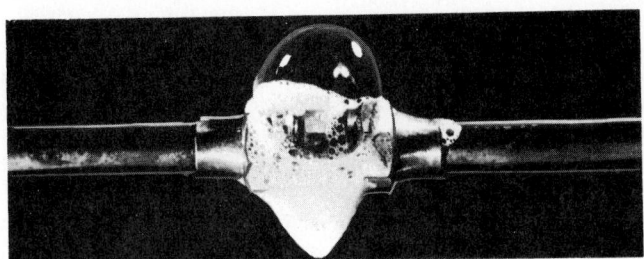

Fig. 11-56. Bubble leak test. Foam is placed on connection. Bubbles indicate refrigerant leakage.

one of these solutions in the presence of a leak. The bubble solution should be wiped off the tubing or fitting after checking for leaks.

The advantages of using the bubble method include its ease of use, low cost, and ease of application compared to instrumentation. A disadvantage is that larger leaks will blow through the solution, and no bubbles will appear.

The halide torch and electronic leak detector are difficult to use around urethane insulation. Since urethane uses refrigerant as the expander, such detection devices show a leak trace all the time. In this case, the bubble test is best.

11-42 REFRIGERANT DYE AND FLUORESCENT LEAK DETECTING

Refrigerant dye is another tool to aid in solving leakage problems. Refrigerant dye in a system will produce a bright red color at the point of leakage. Most leaks show up in a short time. However, due to oil circulation rate, 24 hr. may be necessary in some cases.

To achieve maximum leak detection, in most systems the entire refrigerant charge must be replaced with refrigerant containing the dye.

This method is dependent on the oil circulation rate and can take time to indicate leaks.

The ultraviolet fluorescent leak detection procedure is another method used. A fluorescent additive is circulated throughout the system. The refrigerant leak is found by scanning the system with an ultraviolet light, as shown in Fig. 11-57.

11-43 HALIDE TORCH LEAK DETECTOR

Alcohol, propane, acetylene, and most other torches burn with an almost colorless flame. If a strip of copper is placed in this flame, the flame will continue to be almost colorless.

However, if even the tiniest quantity of a halogen refrigerant (R-12, R-22, R-11, R-500, R-502, etc.) is brought into contact with this heated copper, the flame will immediately take on a light green color. This principle is used in halide torches to detect leaks in refrigeration systems.

A halide leak detector is shown in Fig. 11-58. The torch burner is shown at the top. One end of a rubber tube is connected into the base of the burner. The other end is free to be moved about to various parts of the refrigerator system. The rubber tube will draw air from the open end into the burner.

If the open end of the tube is brought near a leaking refrigeration connection, some of the leaking refrigerant vapor will be drawn up the rubber tube into the burner. Immediately, the color of the flame will change to green, indicating a leak.

11-44 ELECTRONIC LEAK DETECTOR

The most sensitive leak detector of all is the electronic type. See Fig. 11-59.

Three types have been used:
1. Ion source detector.
2. Thermistor type (based on change of temperature).

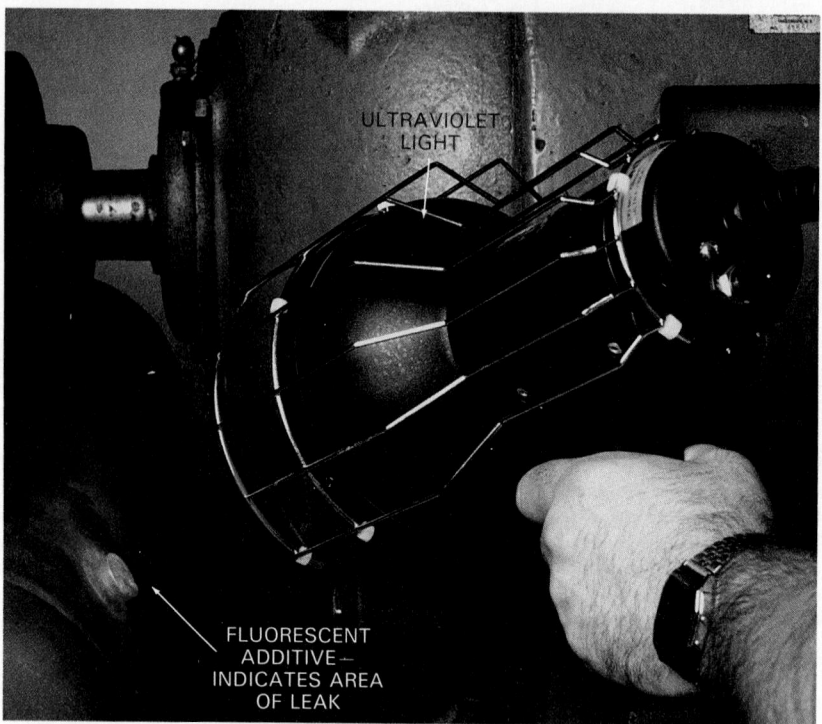

Fig. 11-57. Fluorescent leak detection system in use.
(Spectronics Corporation)

3. Dielectric type, which measures a balance in surrounding air and then responds only to halogen gas.

The principle of operation is based on the different heat conductivity of different gases. Some detectors are based on the dielectric difference of gases. The gases are run between the plates of a capacitor. The gases act as the dielectric (insulator) for each capacitor. Different frequencies of an oscillator indicate a leak.

In operation, the detector is turned on and adjusted in a normal atmosphere. The leak-detecting probe is then passed over surfaces suspected of leaking. If there is a leak, no matter how tiny, the halogenated refrigerant is drawn into the probe. The leak detector will then give out a piercing sound, or a light will flash, or both, because the new gas changes the resistance in the circuit.

This is probably the most sensitive of any of the leak-

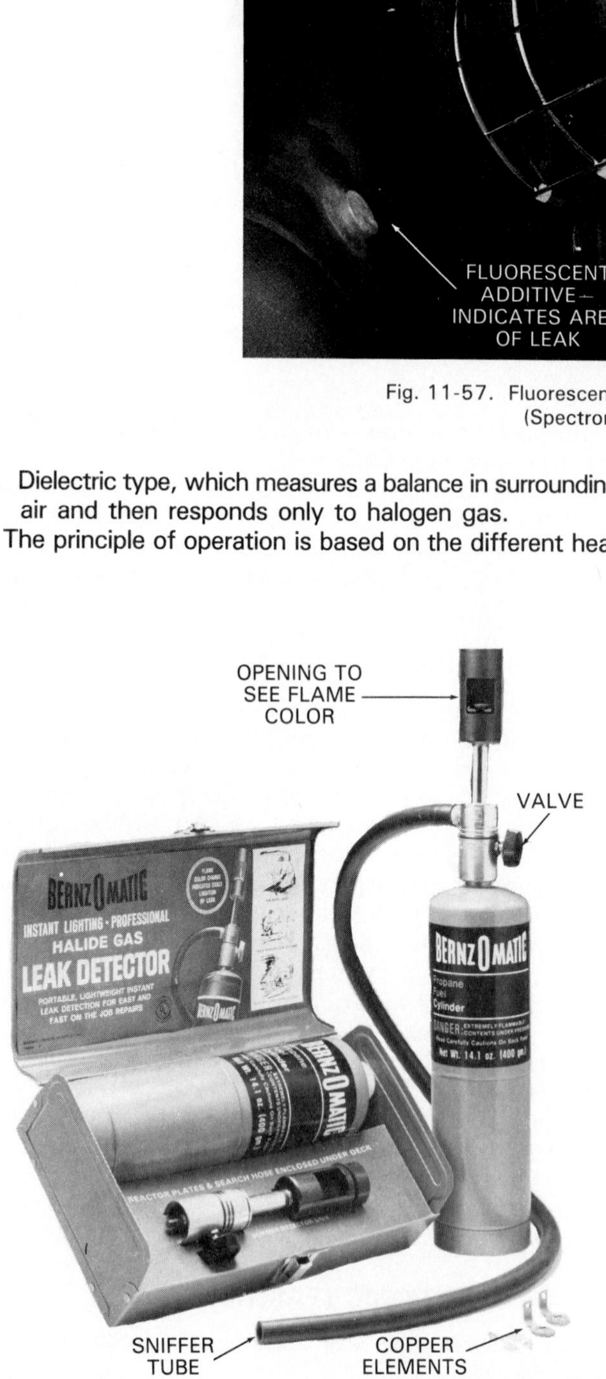

Fig. 11-58. Halide torch used to test for leaks. Green flame showing in burner opening will indicate leak at sniffer tube opening.
(Bernz-O-Matic Corp.)

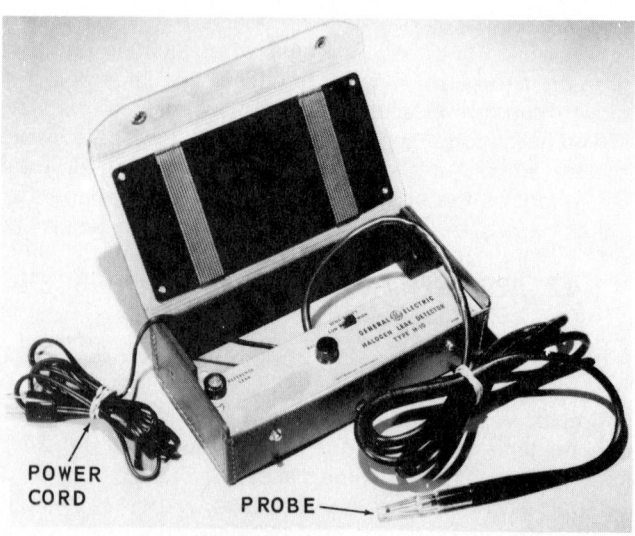

Fig. 11-59. Electronic leak detector. (General Electric Co.)

detecting devices. It uses transistorized circuitry. Flashlight batteries supply the energy. The plastic tip guard should be used only in situations that may be potentially contaminating to the sensing tip. The external construction of a popular leak detector is shown in Fig. 11-60.

When using an electronic leak detector, minimize drafts by shutting off fans or other devices which cause air movement. *Always position the sniffer below the suspected leak.* Being heavier than air, refrigerant drifts downward.

Move tip slowly (about one inch per second). One can measure this by moving tip an inch after each "one-thousand" verbal count. Tip adjusted in ambient air will only buzz; instrument will squeal when tip sniffs refrigerant—R-12, R-22, R-500, R-502, R-11, or any of the halogen family refrigerants. Remove plastic tip and clean it before each use. Avoid clogging with dirt and lint.

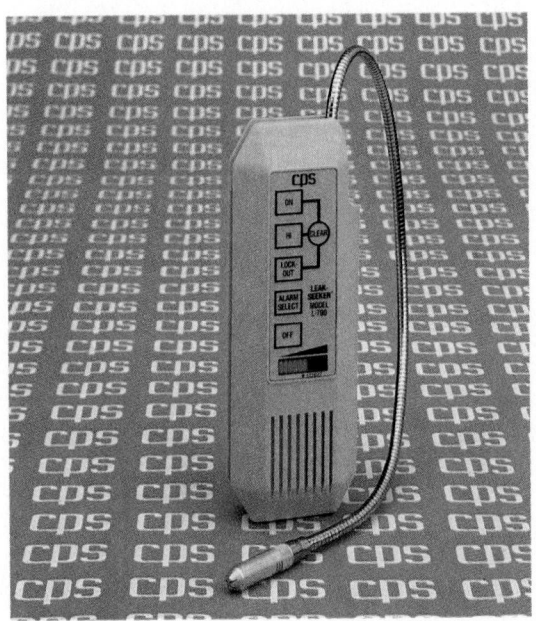

Fig. 11-60. Battery-powered electronic leak detector. (CPS Products, Inc.)

11-45 REPAIRING LEAKS

To repair a leak, remove the refrigerant from that part (the complete system, in some cases). Check the pressure to be sure it is 0 psi (no vacuum or pressure in the system). If possible, avoid soldering or brazing a system with refrigerant in it. Heat may cause a breakdown of the refrigerant.

Refrigerant systems are made of copper, steel, and/or aluminum materials. Leaks may start in any part of the system. The repair depends on the material that has failed or on the combination of materials at the leaky point.

To find out what metal is used, scrape the surface. Steel is gray-white. It is hard and magnetic. (Use a small magnet to test.) Copper is reddish in color when scraped and is nonmagnetic. Aluminum is white, soft, and nonmagnetic. Steel and copper may be brazed; aluminum may be aluminum soldered or brazed. Aluminum may be resistance welded to steel or copper or it can be repaired with epoxy cement.

Leaks most often are found at tubing connections. If they

occur at a flared connection, the tube flare is not correct, the flare nut has not been tightened securely, or threads are stripped. It is best to replace a leaking fitting by making a new flare and using new flared fittings. Leakage at a brazed or silver-soldered connection can be repaired by cleaning, coating with flux, and reheating.

Steel tubing usually has a lengthwise seam. This seam must be clean for brazing. Clean by wire-brushing lengthwise or file off enough metal to remove the seam depression.

If the fitting has been taken apart, then reflux, assemble, heat the connection, and solder or braze it in place. Avoid overheating other parts of the system. Use metal sheet, as shown in Fig. 11-61, to reflect heat away from the cabinet.

Never heat a drier. Moisture will be driven out into the system. It is better to cut tubing with pliers or tube cutter.

Alumninum evaporators can be repaired with epoxy cement. Follow instructions supplied by the manufacturer.

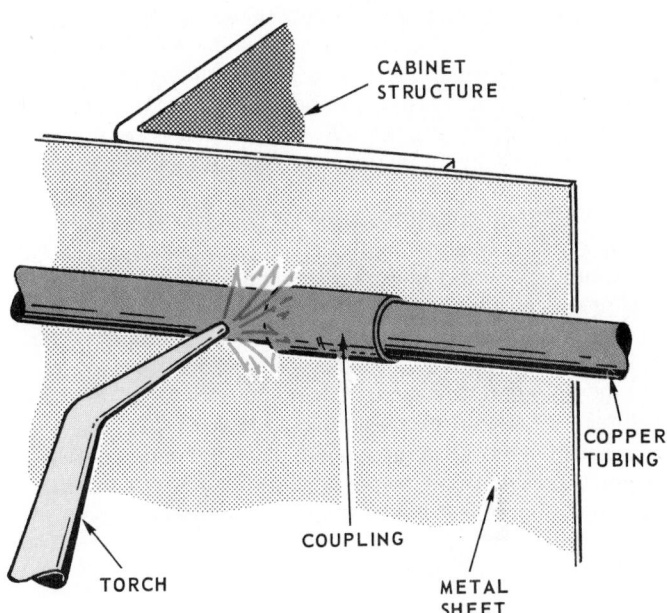

Fig. 11-61. Tubing joint prepared for brazing. Note use of metal sheet to protect other parts of cabinet and system from heat of flame.

11-46 CHARGING A HERMETIC SYSTEM

If testing indicates lack of refrigerant, this is an indication that there is, or has been, a leak in the system. Be sure the leak is repaired before adding refrigerant.

A hermetic unit needs refrigerant if there is:
1. A partially frosted evaporator.
2. A low head pressure.
3. A low low-side pressure.
4. A leak.
5. The refrigerator is running too much.

One or more of the listed conditions can indicate the need to charge the system with refrigerant. Remember that a pressure difference is needed to move the refrigerant from the refrigerant cylinder (higher pressure) into the system (lower pressure).

Methods used to add refrigerant are the following:

1. Evacuate the system.
2. Connect a refrigerant cylinder to the charging manifold. Charge only with correct refrigerant vapor.
3. The refrigerant cylinder may be heated with warm water (not over 120 °F) or with an electric heater insert. Never use an open flame for heating.
4. Install pressure gauges and valves as described and explained in Para. 11-27.

The service connection lines must be clean and free of air (which has moisture). Clean the lines by flushing refrigerant through the lines before recharging.

When charging a partially charged unit, it is best to add a small quantity of vapor refrigerant. The unit should then be allowed to cycle. A proper charge is indicated best by the frost on the evaporator. When frost starts to come down the suction line, purge out a little of the refrigerant. The system will have the correct charge.

If a system has been evacuated first, it can be charged by replacing the evacuating pump with a refrigerant cylinder or by using valves to close off the pump and then opening the connection to the cylinder.

If the system has not been evacuated, purge both lines and manifold by loosening the service line at the piercing valve and opening the left side manifold valve. When the cylinder valve is opened, the refrigerant vapor will purge air, moisture, and dirt out of the two service lines and the manifold. Tighten the service line at the piercing valve.

COMPRESSOR-RUNNING TEST 1

Start the unit and open the line service valve, gauge manifold valve, and refrigerant cylinder valve. Watch the low-pressure side gauge so that not more than 5 to 25 psig (140 to 280 kPa) is created. This pressure is controlled by adjusting the refrigerant cylinder valve. Allow the refrigerant charge to enter the system for about 3 to 5 minutes. Connection should be as shown in Fig. 11-62.

After the time lapse mentioned, close the gauge manifold valve. Allow the unit to operate and check the frost line on the evaporator. If the frost line is inadequate, repeat the

charging for short intervals, rechecking after each. The frost line must not go beyond the accumulator in the suction line.

When the proper amount of frost has been observed, close the refrigerant cylinder valve, adaptor valve, and gauge manifold valve. After closing all valves, there are some common steps listed as follows (steps one and two apply to the system which uses a separate process tube):

1. Pinch the process tube between the compressor and the adaptor valve with a pinch-off tool, as shown in Fig. 11-63.
2. Remove the adaptor valve, flatten the tube end by crimping, and braze the end of the tubing.
3. Close the adaptor valve, if one was installed on the suction line, and leave it mounted for future service operations.
4. Check for leaks using a leak detector.

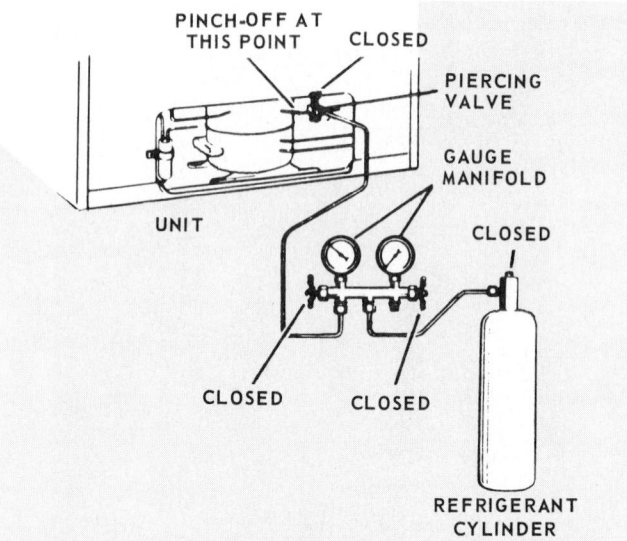

Fig. 11-63. System after piercing valve is closed along with refrigerant cylinder valve and gauge manifold valves. Note pinch-off point.

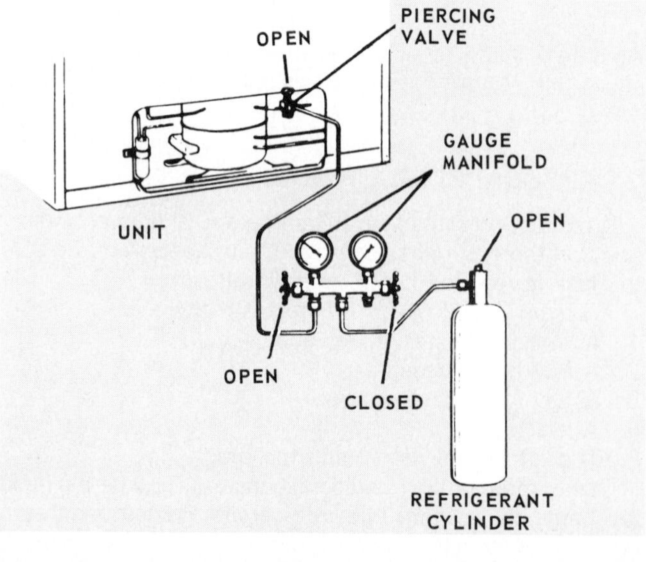

COMPRESSOR-RUNNING TEST 2

The procedures that follow rely mostly on temperature as an indicator of correct charging.

Refrigerants charged into hermetic refrigeration units must be of top quality. It is very important to transfer refrigerants in chemically clean cylinders and lines. *Always charge a unit (except large commercial units) with refrigerant vapor.*

Never charge liquid refrigerant into the low side of a domestic or small commercial unit.

Always keep the charging cylinder at room temperature or warm it only with warm water.

There is a charging device which, when mounted between the refrigerant cylinder and the low side, allows liquid refrigerant to flow from the cylinder into this charging device. The refrigerant vaporizes in this device.

When the charging device is first used, the process tube will sweat and may even frost a little. As the system becomes fully charged, this sweating and frost will go away because the suction pressure is higher. The system is now correctly charged.

A fast-reading dial thermometer provides a second check on the correct charge. The suction line temperature should

Fig. 11-62. Recharging setup after adjustment of refrigerant cylinder valve. Note open piercing valve, left-side manifold valve, and refrigerant cylinder valve.

be about 20° higher about six to ten inches away from the compressor than at the evaporator outlet.

If the temperature is lower, liquid refrigerant may enter the compressor, causing damage. If the temperature is higher, the motor compressor may overheat and "burnout."

Evacuation data

The following pertains to the charging device mounted between refrigerant cylinder and low side.

A check valve in the charging unit allows the service technician to evacuate the system easily with this unit attached to the line. The charging unit is available in three capacities: (1) Less than 1 hp; (2) 1 to 4.75 hp; and (3) 5 to 10 hp. The correct size must be used.

COMPRESSOR-RUNNING TEST 3

Systems which use a capillary tube must have an exact refrigerant charge. If the system is overcharged, the evaporator will be overcharged or flooded.

The use of accumulator spaces at the outlet of these evaporators relieves the problem somewhat, but one must be careful of the amount charged into these systems. A common method is to slowly charge these systems with refrigerant in the vapor state until the suction line starts to sweat and/or frost back. Then they are purged a little at a time until the "frost back" disappears.

CHARGE WITH EXACT AMOUNT

Another method is to completely discharge the system. Then, recharge with a cylinder containing exactly the right amount of refrigerant according to the manufacturer's recommendations.

CONCLUSIONS

The following points are important and bear repeating:
1. Always charge a system into the low side, if possible.
2. *Refrigerant should be put into the system in vapor form. Forcing liquid refrigerant into the system may damage the compressor and injure the service technician.*
3. Always remember that if a system is short of refrigerant, there is a leak. Locate and correct the leak before the system is charged.

11-47 CHARGING WITH PORTABLE CHARGING CYLINDER OR DIGITAL SCALE

A charging cylinder, view A of Fig. 11-64, with a glass tube liquid level indicator, lets a technician transfer refrigerant into a system and measure the amount on a scale. Some cylinders are electrically heated to speed up the evaporation and maintain pressure in the cylinder.

This process of electrically heating a cylinder is usually done with an electrical insert. In some cases, the compressor itself is heated, using a heat gun so the refrigerant and oil will circulate and be purged more easily. In both cases, it is extremely important that a pressure control relief valve and thermostat be used to provide the required temperature and pressure safety controls.

The system has a pressure gauge and a hand valve on the bottom for filling the charging cylinder or for charging liquid refrigerant into a system. It also has a valve at the top of the cylinder. This valve is used for charging refrigerant vapor into the system. (This is the best and safest method.) View B of Fig. 11-64 shows a digital weighing scale. Ad-

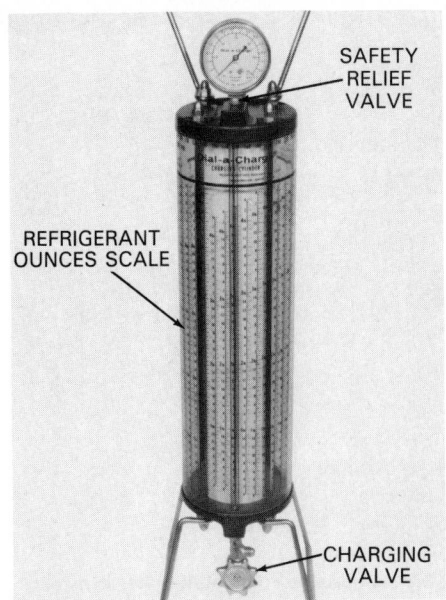

A

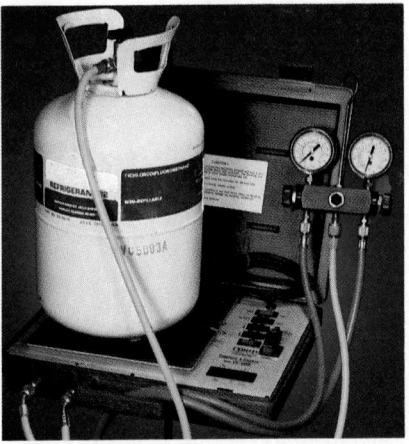

B

Fig. 11-64. Instruments for charging a system. A—Portable charging cylinder may be used to accurately charge hermetic systems. Cylinder is accurately marked in ounces of refrigerant. (Robinair Division, SPX Corporation) B—Digital scale weighs refrigerant. (CPS Products, Inc.)

ditional information concerning the use of a digital scale to charge or discharge a system can be found in Chapter 28.

The following steps are recommended for use of a portable charging cylinder after evacuation. Wear goggles and follow these steps:
1. Attach a line from the charging cylinder to the center of the gauge manifold and purge with the fitting loose at the center part of the gauge manifold. See Fig. 11-65. Tighten this connection.
2. Open the piercing valve or valve adaptor and gauge manifold valve.
3. Crack the charging cylinder valve and allow the refrigerant to enter the system, as shown in Fig. 11-66. Know what the new scale reading on the tube must be to put in the correct charge.
4. When the correct amount of refrigerant has entered the system, close the cylinder valve. Amount can be checked by reading the scale on the charging cylinder.
5. Close the piercing valve or the valve adaptor and the gauge manifold valves, as shown in Fig. 11-67.
6. Use the pinch-off tool and close off the process tube between the compressor and the valve adaptor. Leave pinch-off tool on tube until tube end has been brazed.

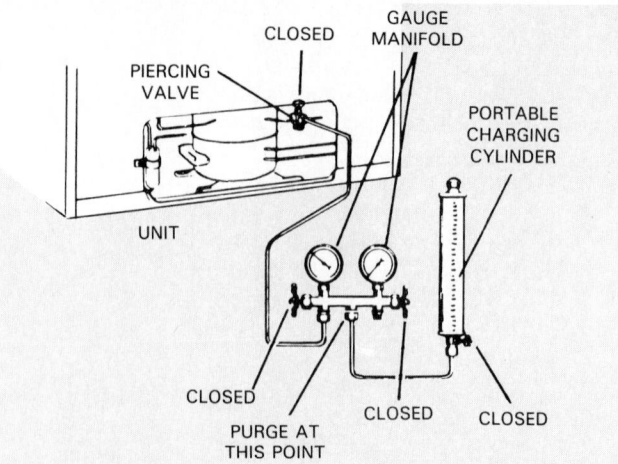

Fig. 11-65. When portable charging cylinder is used, purge charge line (after evacuating a system) by leaving fitting loose at center part of gauge manifold. If system is not evacuated, purge all lines up to piercing valve.

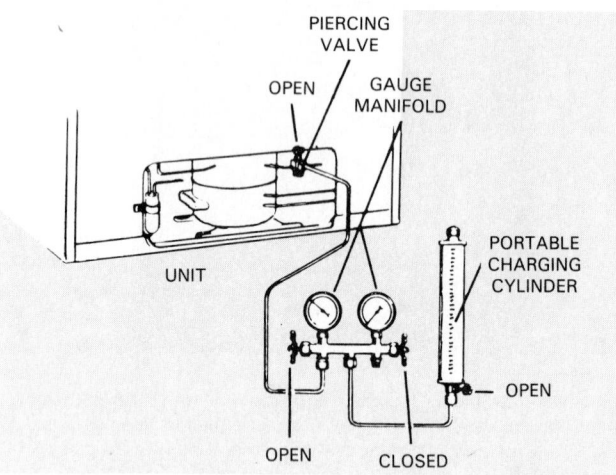

Fig. 11-66. Proper hookup for using portable charging cylinder to charge refrigerant into system.

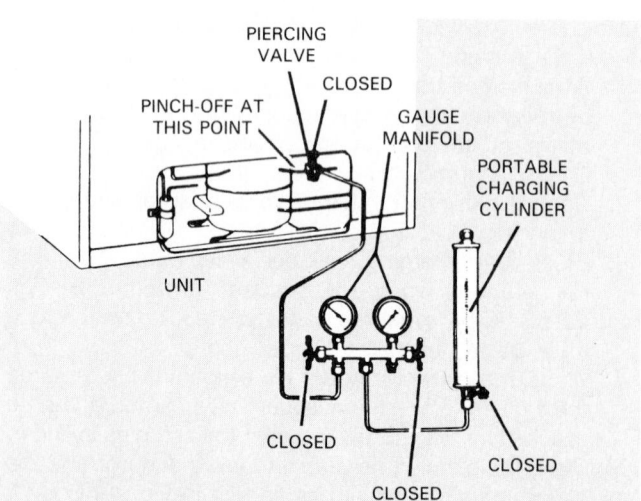

Fig. 11-67. Proper steps and valve settings for removing servicing equipment after system has been charged.

7. Remove the piercing valve or the valve adaptor.
8. If a piercing valve was used, cut off the part of tubing with hole in it. Use a pipe cutter. Wear goggles!
9. Crimp the end of the process tube.
10. Braze the end of the process tube. Wear goggles!
11. Check the system for leaks.

11-48 ADDING OIL TO THE SYSTEM

Having the correct amount of oil in a system is very important. A lack of oil will shorten the life of the mechanism, increase friction, and cause noise. An overcharge will cause the compressor to pump excessive amounts of oil, reducing its refrigerant-pumping capacity, and subject the compressor valves to severe strain.

Refrigerant oils are available in several viscosities (ease of flow at different temperatures). Be sure to follow the manufacturer's viscosity recommendations. On a service call, add oil only if there is a sign of oil leakage.

Only rarely is it necessary to add oil to a hermetic system. However, leaking refrigerant always carries some oil with it, and this lost oil should be replaced. The conventional method of adding oil to a system may be used if the hermetic unit is completely equipped with service valves. That is, oil can be siphoned in or poured in.

If the system has had a low-side leak, moisture and air may have entered the system. In this case, it is best to replace the compressor oil. Measure the amount removed and replace it with the same amount of clean, dry oil. See Chapter 14. The unit should be charged in much the same way as when adding refrigerant to the system.

Fig. 11-68 illustrates a practical transparent charging cylinder with a chart for accurately measuring the amount of refrigerant or oil charged into a system by volume. This method provides greater accuracy.

Use clean lines. Purge lines of air with clean refrigerant.

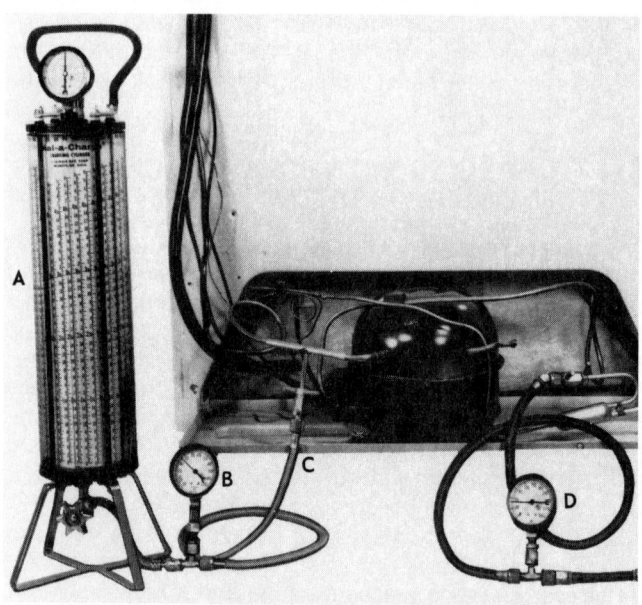

Fig. 11-68. Charging cylinder connected to low side. A—Charging cylinder. B—Compound gauge. C—Line adaptor. High-pressure gauge is connected to outlet of filter-drier by tube adaptor. D—High-pressure gauge.

There are several ways to connect a transparent charging cylinder to a system. Fig. 11-68 shows the tube connected to the low side directly, using a compound gauge and process tube adaptor. The process tube is brazed to the suction line. A method of using a vacuum pump and gauge manifold to evacuate for repairing procedures is shown in Fig. 11-69.

A pump may be used to put oil into a system. See Fig. 11-70. The charging lines must be purged to remove air, moisture, and dirt. This hand pump can build up pressures to 300 psig (2 200 kPa [2,2 MPa]). Oil can be forced into the system even when the system is under pressure.

To add oil, using the single service line technique, install as shown in Fig. 11-71. The correct amount of oil is put into the service cylinder while a small amount of refrigerant (the same as in the system) is also put into the cylinder to create a pressure. The cylinder is inverted and connected from the cylinder valve to the valve attachment by clean lines. Be sure the cylinder pressure is higher than the system pressure. Then open the cylinder valve and valve attachment. Oil will be forced into the system.

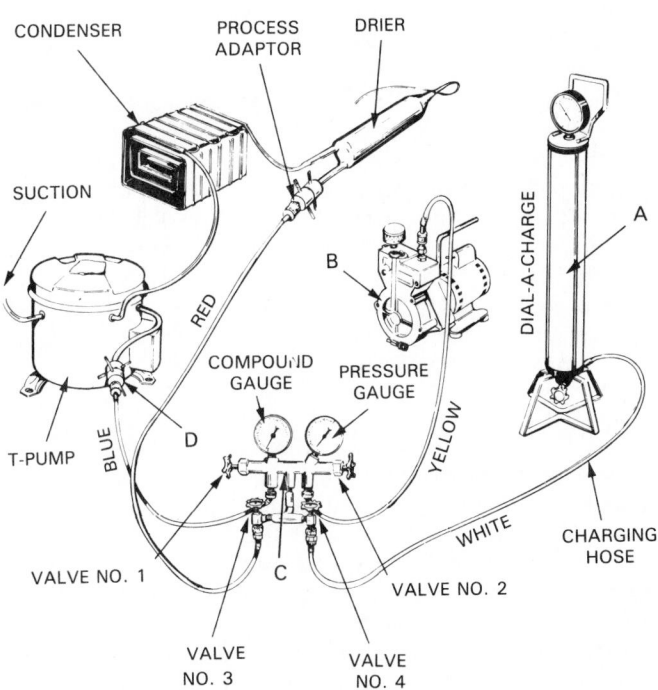

Fig. 11-69. Component parts for a general unit used for evacuating and charging a system. Note: A—Charging cylinder. B—Vacuum pump. C—Gauge manifold. D—Process tube adaptor.   (Frigidaire Company)

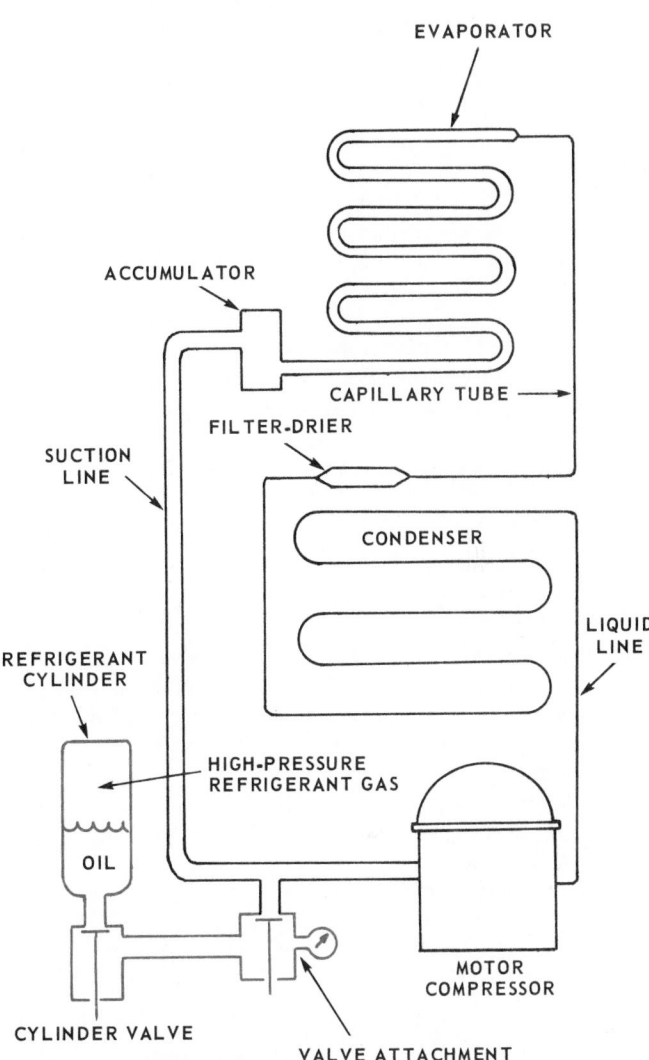

Fig. 11-71. Single line technique can be used for adding oil to small hermetic system. First, place correct amount of oil in service cylinder, along with small amount of refrigerant to create pressure. Then turn service cylinder upside down. Its pressure must be higher than system pressure so that oil will enter system.

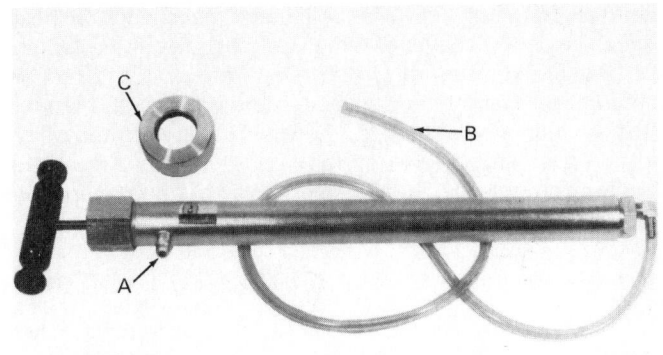

Fig. 11-70. Hand pump used to force oil into system. A—Opening to oil charging line. B—Plastic tubing for putting into refrigerant oil storage can. C—Rubber adaptor for oil can.
(Robinair Division, SPX Corporation)

11-49 REMOVING THE SYSTEM

When checking a unit, mechanical or electrical trouble is sometimes found that cannot be easily fixed. In this case, the unit should be removed from the cabinet and sent to a shop for repair.

It is important to protect the refrigerating unit while it is being moved. Cradles or crates should be used to hold mechanisms. Wooden frames and C-clamps will hold the parts down and keep them from being damaged while in transit.

The complete cabinet, if moved in a truck, should be

shrouded in a special padded blanket.

The procedure for removing the unit from the cabinet varies because some evaporators are removed from the rear of the cabinet while others are removed from the front (by way of the cabinet door).

The refrigerant lines are sometimes run just under the door frame breaker strips. These breaker strips must be removed in order to remove the lines. Do this carefully. Strips are brittle. If allowed to come up to room temperature, they will be more flexible.

These strips are usually made of a plastic material. Several other designs are shown in Chapter 10. There are several methods used to fasten the breaker strips to the cabinet shell and the liner.

The cabinets which have the evaporator removed from the rear of the cabinet are not difficult to dismantle. Skill and patience are required to remove the unit from the front of the cabinet. The hardest part is removing the breaker strips. When the mechanism is removed, be careful not to kink or buckle the refrigerant lines.

On double-door cabinets, the strips between the two compartments must be removed before the unit can be taken out.

The compressor and condenser are sometimes fastened to the rear of the cabinets by four to six mounting screws and bolts. Be careful not to damage the mechanism while removing these devices.

11-50 DIAGNOSING INTERNAL TROUBLES

Before removing any part of the hermetic system, one should be absolutely sure that the part is the one causing the problem. The parts which may cause the trouble are:
1. Motor compressor.
2. Filter-drier.
3. Refrigerant control.
   a. Capillary tube.
   b. AEV.
4. Hot gas defrosting valve.

11-51 LOCATING MOTOR COMPRESSOR FAULTS

Replacing the motor compressor is the most costly service repair. Be sure that it is a bad motor compressor before replacing it.

The fault may be either:
1. Electrical.
2. Mechanical.

Electrical faults are the most common reason for replacing a motor compressor when the unit is actually in good condition.

Other electrical problems often mislead a service technician into thinking the motor compressor is at fault. To check it, clean the outside of the compressor dome, remove the electrical connections cover, and disconnect the system wiring from the compressor: relay, capacitors, overload, and all. Check motor windings for continuity, shorts, and grounds. (Use an ohmmeter.) See Chapter 7. If unit checks out correctly, connect a starting circuit to the motor compressor. Use exactly the correct size of capacitors and overload cutout. Connect as shown in Fig. 11-72.

If system starts and operates correctly with these manual start electrical connections, the problem is in the external part of the system (wiring, thermostat, relay, or overload). If the internal electrical motor is faulty, the motor compressor must be replaced.

If the electrical system operates correctly, the compressor may not be pumping.

The best check of the motor compressor is its volt-ampere (watt) reading at normal low- and high-side pressures. If volt-ampere reading is below motor rating, the pump may be worn out. To check the compressor's pumping ability, install a piercing valve and a gauge manifold. Then pinch the suction line as shown in Fig. 11-73. Next, run the unit to determine how much of a vacuum it will pull. (It should pull between 25 and 28 in. Hg of vacuum [between 7 and 17 kPa, between 64 and 71 cm Hg].) Stop the compressor. If the vacuum starts weakening (moving toward 20 in. Hg [34 kPa, 51 cm Hg], then toward 10 in. Hg [68 kPa, 25 cm Hg]), the exhaust valves of the compressor are leaking. The motor compressor must be replaced or overhauled.

Another method is to install a piercing valve on the process tube. Then connect one end of flexible service line to valve adaptor and other end of flexible service line to compound gauge end of manifold. Purge these lines to clean them. Then pinch the suction line. Start the unit. The motor compressor should pull 16 in. Hg of vacuum (47 kPa, 41 cm Hg) in about two minutes. (Do not run any longer with no cool suction gas flowing. The motor will overheat.) Stop the compressor and observe. If low-side pressure increases, the compressor valves are leaking.

11-52 CAPILLARY TUBE SERVICE

Capillary tubes must be of the correct inside diameter (ID) and length for the capacity of the system and for the evaporator temperatures desired. Cut capillary tubing by filing a notch around tubing and then breaking it by small back-and-forth bending motions. A tube cutter will change the inside diameter too much.

Fig. 11-74(A) shows a capillary system of correct design. The system in Fig. 11-74(B) has too long a capillary tube or one with an undersized inside diameter (ID). The drawing also illustrates what happens if the system has a starved evaporator, because of a partially clogged filter-drier or capillary tube.

The amount of refrigerant in a capillary tube system is critical. Refer to Fig. 11-75. Notice the change in head pressure as the charge of refrigerant changes. If the system is undercharged as at C, the evaporator will not receive enough refrigerant and the system may run all the time. If the system is overcharged, the liquid refrigerant may flow down the suction line and may cause oil-pumping in the motor compressor. Suction line will sweat and even frost up all the way to the motor compressor.

11-53 CHECKING FOR RESTRICTED SYSTEM

To check the capillary tube, run the system for a few minutes. Stop the unit and listen where the capillary tube enters the evaporator. If there is no hissing sound, the capillary tube is clogged.

Heat the evaporator end of the capillary tube with warm water and rag. (Do not use a flame.) If the clogging is from ice, there will be a hissing sound as the ice melts. A clogged strainer or capillary tube will fill the condenser with all the

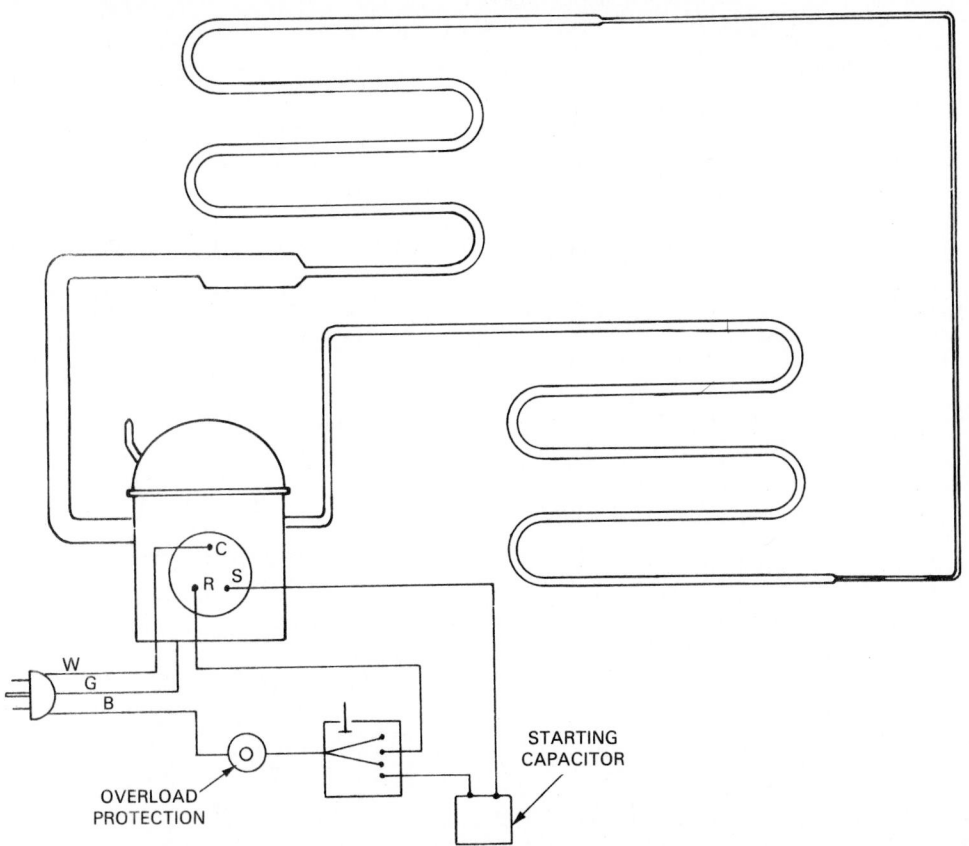

Fig. 11-72. Testing capacitor-start motor compressor after removing all electrical leads and connecting test cord.

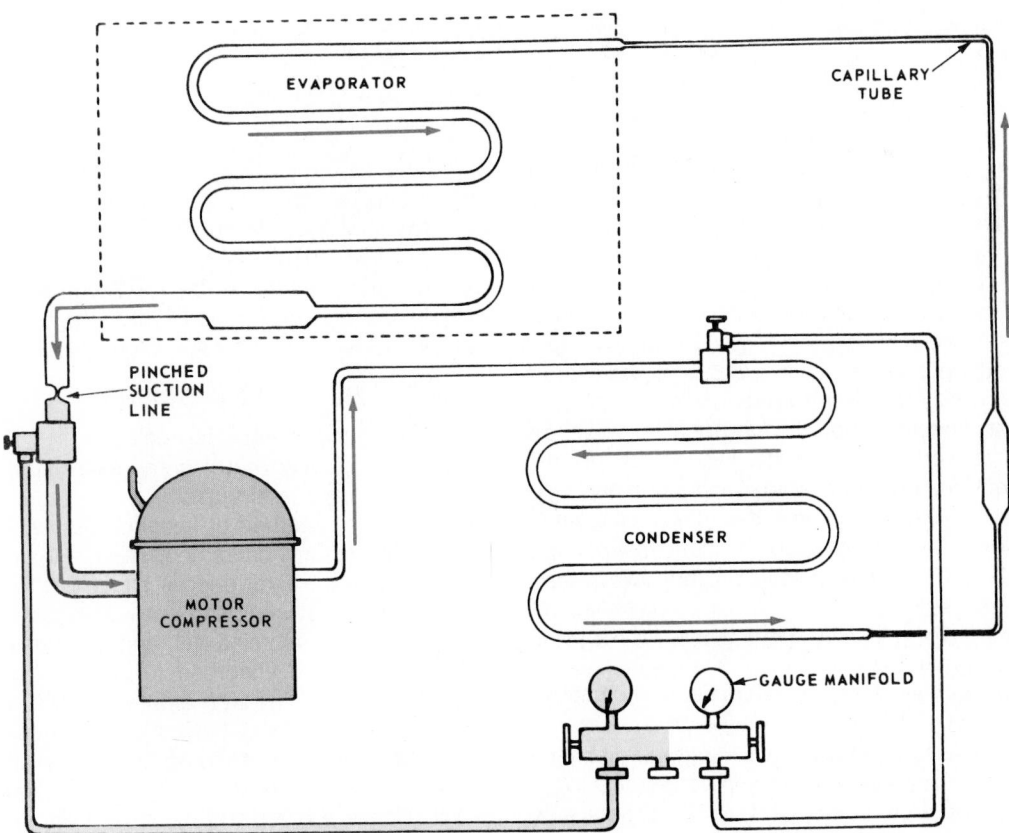

Fig. 11-73. Compressor's pumping capacity can be tested by pinching suction line. This is a very short run test or motor will overheat. Compressor should pull 16 to 26 in. of vacuum (47 to 14 kPa, 41 to 66 cm Hg; 350 to 100 mm, 350 000 to 100 000 microns) in a few seconds. Yellow area indicates vacuum in line.

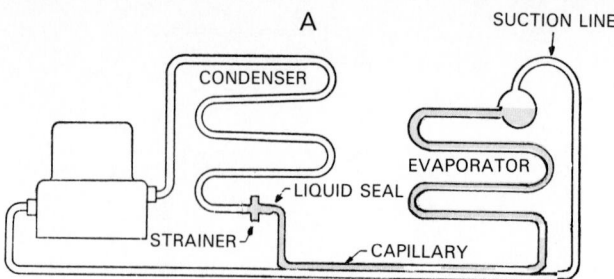

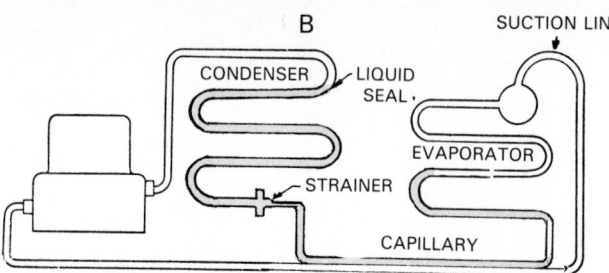

Capillary selected for capacity balance conditions. Liquid seal at capillary inlet but no excess liquid in condenser. Compressor discharge and suction pressures normal. Evaporator properly charged.

Too much capillary resistance—liquid refrigerant backs up in condenser and causes evaporator to be undercharged. Compressor discharge pressure may be abnormally high. Suction pressure below normal. Bottom of condenser subcooled.

Fig. 11-74. Effect of correct and incorrect capillary tube installation. A—Correct installation, operation normal. B—Incorrect. There is too much resistance in tube and evaporator is "starved."

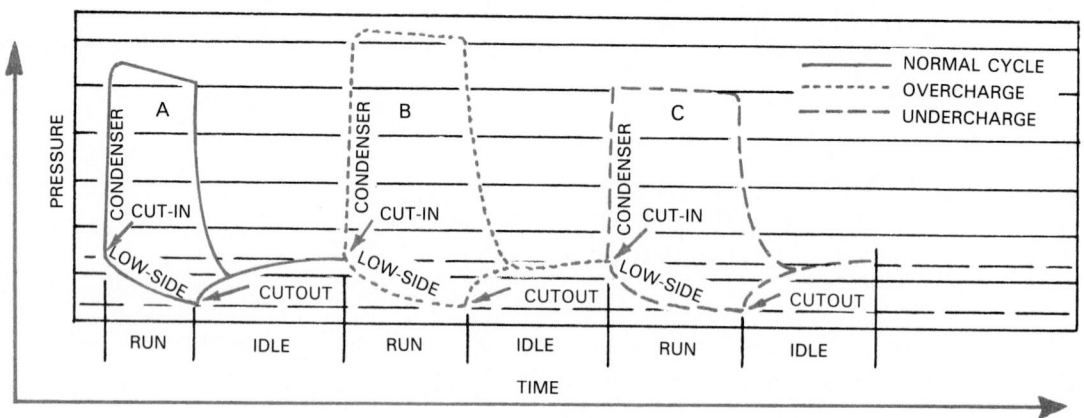

Fig. 11-75. Pressure-time cycle diagrams for three conditions in capillary tube system. A—Normal charge. B—Overcharged. C—Undercharged. Overcharge will usually cause frosted or sweating suction line.

refrigerant. The motor compressor may stop or it may overload during a start-up.

There is a quick test to determine if the system is either short of refrigerant or if the filter-drier or capillary tube is clogged. First, install a piercing valve on the suction line or on the end of the process tube of the compressor. Purge the service lines using gas from a refrigerant cylinder. Open piercing valve. If the low side is in a vacuum, the system has a restriction or is low in refrigerant.

If the defrost system is not working (ice buildup), the evaporator is not working. Check the evaporator fan. If it is working, check the defrost system (inspection and electrical test of defrost resistance wire), then check as follows:

1. Install another piercing valve at the condenser outlet. See Fig. 11-76. Open the valve just a little. Allow some refrigerant to escape.
2. If no gas escapes or if only gas escapes, the system is short of refrigerant.
3. If liquid refrigerant escapes, the filter-drier or the capillary tube is clogged.
4. To determine which component is clogged, discharge the refrigerant. Clean the connection between the drier and the capillary tube, flux it, heat, it, and separate the capillary tube from the filter-drier. System can now be checked to find out if drier or capillary tube is clogged.
5. To find out which is at fault, see Fig. 11-77. Put some vapor refrigerant in the system. Open valves C, A, and

B. If filter-drier is open, gas will come out opening D. If capillary tube is open, a small flow (because tube is small) will come out its opening at E. If either one is clogged, there will be no flow.

Clogged units must be replaced. Sometimes clogged capillaries can be opened with a high-pressure hydraulic pump. (See Para. 11-64.)

11-54 FILTER AND DRIERS

A filter-drier should be replaced:
1. When a new motor compressor is installed.
2. If the filter is clogged.

A positive method of keeping the system clean and dry inside is to install driers and filters in the refrigerant circuit. A combination filter-drier is shown in Fig. 11-78.

A solid moisture absorbent will usually do a satisfactory job. Silica gel, alumina gel, and synthetic silicates are excellent moisture absorbers. A bead silica gel gives good results. A filter-drier with a purge valve built into it is shown in Fig. 11-79. Fig. 11-80 shows the same unit installed.

Never use a liquid drying agent in a unit equipped with a solid desiccant (drying chemical). The liquid dryer chemical will release the moisture already trapped in the drier.

For the same reason, a solid drier should not be put in a system that is already using a liquid drier. To avoid this danger, all systems should be labeled to indicate which drying agent is in use. The moisture absorbent properties of

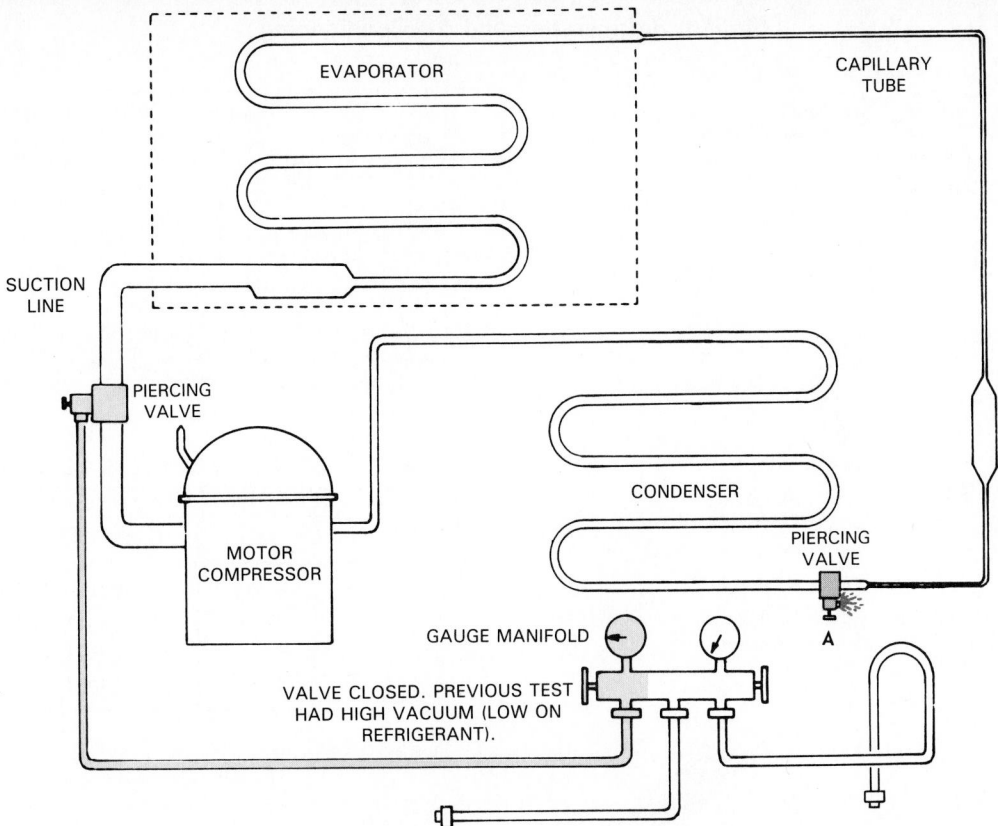

Fig. 11-76. Testing for clogged drier or capillary or for lack of refrigerant. With compressor running and valve A open, escaping gas means there is lack of refrigerant. If liquid refrigerant escapes, filter-drier or capillary tube is clogged.

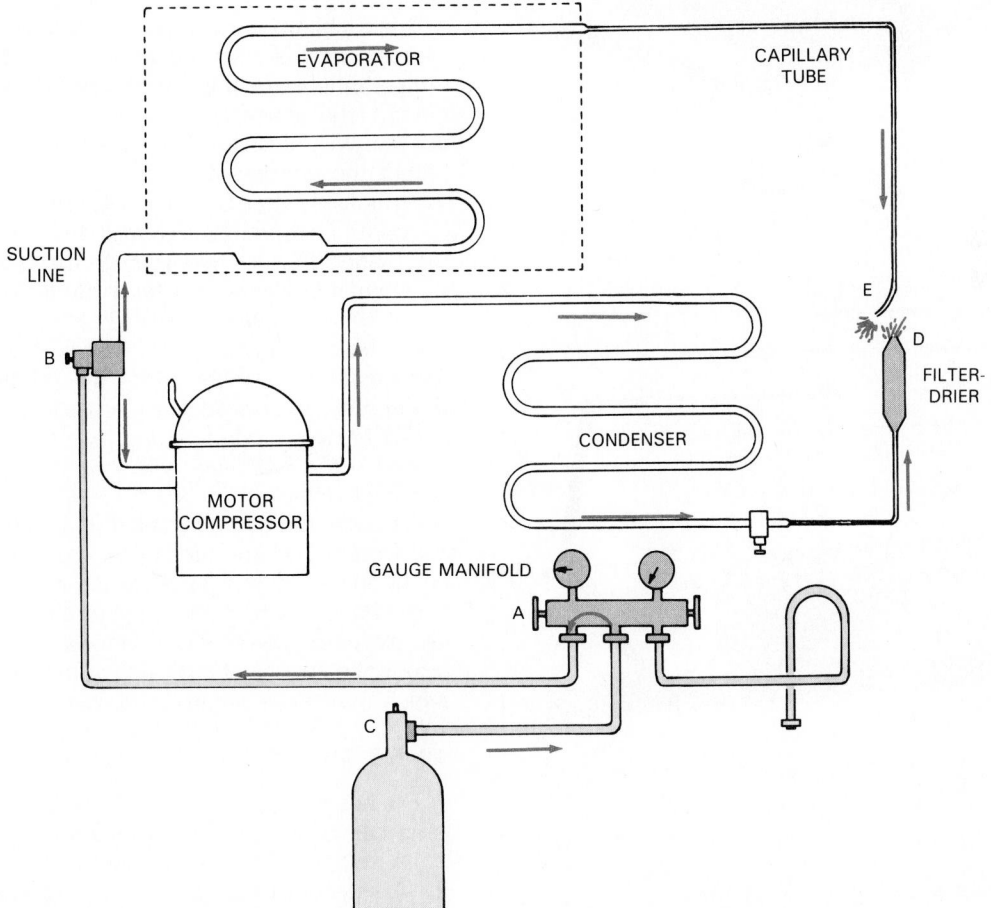

Fig. 11-77. To check if capillary tube or filter-drier is clogged, separate and check flow at D. Charge some vapor refrigerant into system by way of valves C, A, and B.

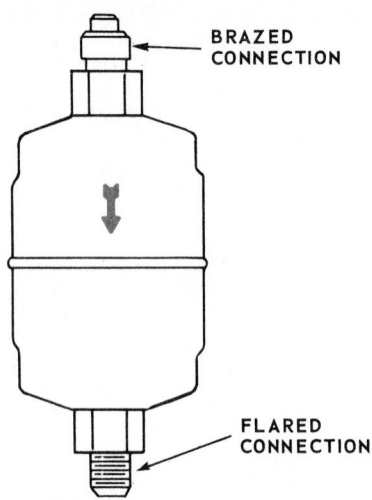

Fig. 11-78. Filter-drier which is used on small hermetic systems. Arrow shows direction of flow of refrigerant.

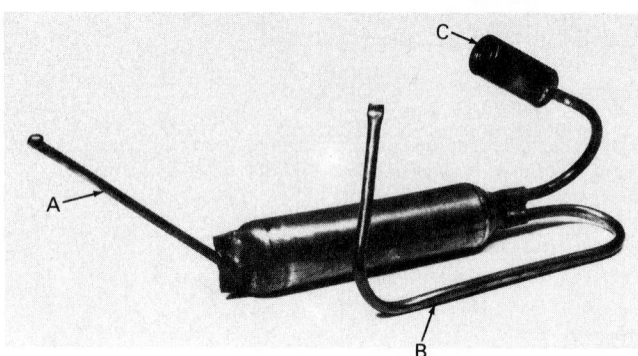

Fig. 11-79. Replacement filter-drier. A—Outlet tube. B—Inlet tube. C—Purge valve. (General Electric Co.)

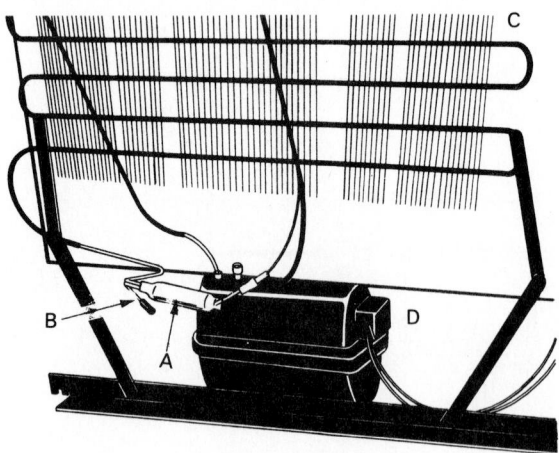

Fig. 11-80. Replacement filter-drier installed. A—Filter-drier. B—Purge valve. C—Condenser. D—Compressor.

these desiccants are shown in Fig. 11-81.

The Refrigeration Electrical Manufacturer's Association has recommended that the capacity of dehydrators be rated. Fig. 11-82 shows the recommended volume of dry-

DESICCANT (DRYING CHEMICAL)	MESH	ABSORPTION FROM LIQUID, PERCENT OF WEIGHT OF DESICCANT
SILICA GEL	8—20	16
ACTIVATED ALUMINA	8—10	12
SYNTHETIC SILICATES	8—20	16

Fig. 11-81. Moisture-absorbing ability of some dessicants in driers. Some driers have mixtures of these chemicals. These driers may be used with R-12, R-22, R-500, and R-502.

DRYER SIZES – DOMESTIC	
Cu. In.	Capacity, HP
2	1/8
3	1/6 to 1/4
6	1/4 to 1/2
9	1/2 to 3/4

Fig. 11-82. Drier capacities in cubic inches recommended for various horsepower systems.

ing agents according to horsepower of the unit. All driers are sealed by the manufacturer. Do not remove the sealing caps until just before installation.

Driers absorb water faster at lower temperatures. If at all possible, the drier should be installed just ahead of the refrigerant control. If by an accident the filter-drier becomes heated, the moisture that it has absorbed may be driven out and will recirculate with the refrigerant. A position just before the refrigerant control can have advantages:

1. It is likely that both the filter-drier and the refrigerant control become heated at the same time, reducing the chance that ice will form in the refrigerant control.
2. The filter-drier is kept far away from the heated tubing of the condenser.

A filter-drier has an arrow stamped or cast on the body to indicate the direction that the refrigerant should flow through it. Be sure that it is installed properly.

It is always recommended that a new filter-drier be installed in the liquid line whenever it is necessary to break into a system.

Filter-driers may be installed with either flared or brazed connections. Many service technicians install two filter-driers on a system after repairing it. One is placed on the high side just before the refrigerant control and one on the low side, between the evaporator and compressor. This improves the chance of removing all moisture or contaminants which may have entered the system during servicing.

11-55 HOT GAS DEFROST TROUBLES

The hot gas valve is usually operated by a solenoid. Problems can occur with a solenoid in two ways:

1. A solenoid valve stuck closed.
2. A solenoid valve stuck open, as in Fig. 11-83.
   The two solutions are as follows:
1. If the evaporator is overloaded with ice, the solenoid electric coil may have failed (open circuit) or the timer

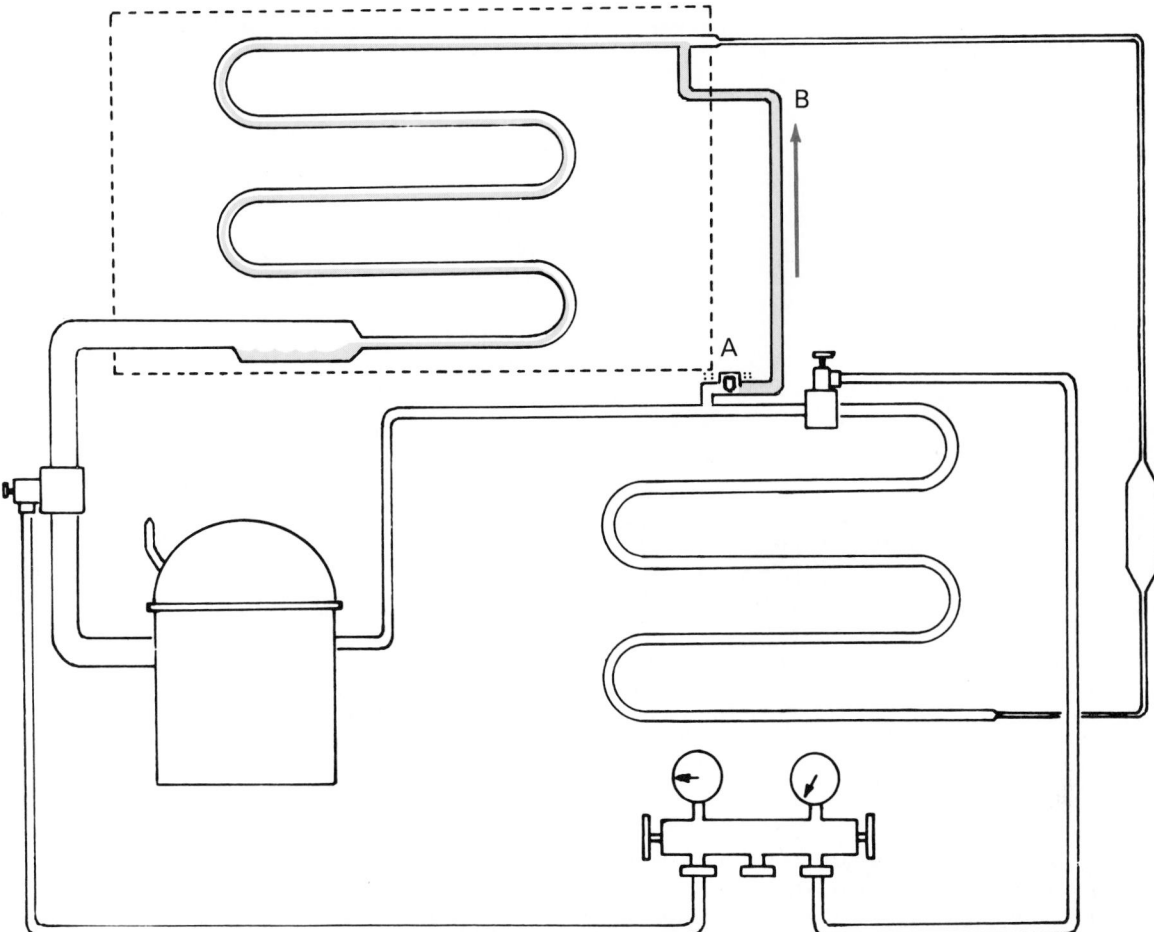

Fig. 11-83. Hot gas defrost system with solenoid valve stuck open. A—Solenoid valve. B—Hot gas line.

may not be operating correctly. Both of these problems can be checked electrically.

If the electrical system is operating correctly, the problem is probably a stuck valve stem in the solenoid. Rap the valve body while the defrost timer switch is closed. If the valve breaks loose, one can hear the surge of hot gas. The line between the solenoid and the evaporator will also become warm to the touch.

2. If the evaporator is warm and the line between the hot gas solenoid and evaporator is warm, the valve is stuck open. Refer again to Fig. 11-83. (Hot gas is shown in color.) Again, rap the valve sharply while the timer is on open circuit. If the valve closes, the low-side pressure will start to decrease immediately and the evaporator will start to cool and frost.

If the valve does not operate, it must be removed and replaced with a new one.

11-56 DISMANTLING SYSTEM

Whenever it is necessary to replace parts such as the motor compressor, condenser, capillary tube, evaporator, accumulator, filter-drier, and the like, the system should be prepared as follows:
1. Disconnect the electrical circuit.
2. Carefully clean all surfaces. It is good to clean the entire mechanism, as this will reduce the chance of dirt and other contamination entering the system.

3. Install service valve and gauge manifold.
4. Remove the refrigerant. It must be purged in accordance with the Environmental Protection Agency (E.P.A.) regulations, using a recovery/recycling unit. See Chapter 28.
5. Cut the tubing and remove the part to be replaced. Wear goggles!

11-57 REMOVING REFRIGERANT

If the unit is equipped with service valves or service valve attachment plugs, the refrigerant should be purged, using a recovery/recycling unit.

If the system is not equipped with valves, use a piercing valve or a valve adaptor. Several methods may be used. Be sure to wear goggles!
Always work in a well-ventilated area.

11-58 REMOVING MOTOR COMPRESSOR

Follow this procedure to remove the motor compressor:
1. Disconnect the electrical circuit.
2. Install a gauge manifold. Use a piercing valve if necessary.
3. Discharge the refrigerant.
4. Disconnect the lines. There are several methods:
   a. Use a pinching tool. Pinch the lines as shown in

Fig. 11-84. Wrap the tubes with cloth and bend them sharply at the pinch to separate. Fig. 11-85 shows the disconnected compressor. Wear goggles!

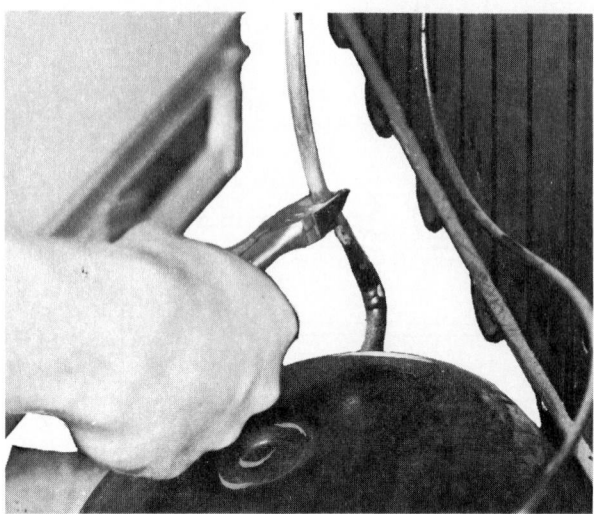

Fig. 11-84. Suction lines are separated by pinching and then breaking. Sawing is not recommended. Chips may enter system and later damage refrigerant control, compressor, or motor.

b. Clean both the suction and discharge tubing on straight sections near the compressor. Use a tube cutter. Wear goggles! *Plug the lines at once.*
c. Clean the tubing or fittings at the compressor. Put brazing flux on the connection. Heat the joint and pull the tubing out of the fittings. Wear goggles! *Plug openings immediately.*
5. Remove the motor compressor.
6. If the motor compressor has oil cooler lines, these too must be pinched and broken. The removed compressor tubing openings should be sealed (pinched and brazed).

The unit is now ready to have a replacement motor compressor installed.

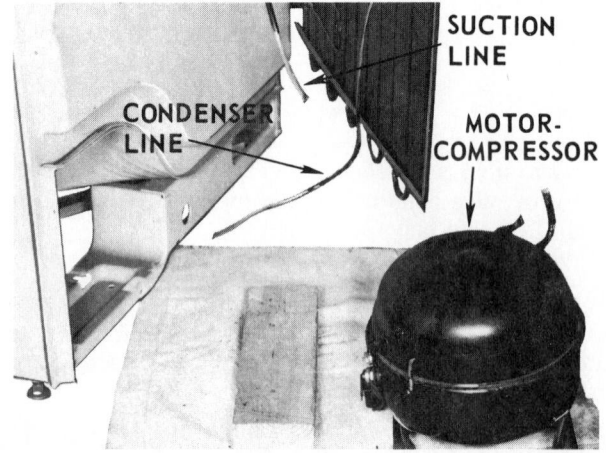

Fig. 11-85. Old motor compressor has been removed and its refrigerant lines sealed by pinching. Unit is now protected against oil spillage and internal corrosion.

11-59 CAUSE OF MOTOR COMPRESSOR BURNOUTS

A certain indication of a burnout is the strong pungent odor of the refrigerant when a piercing valve is opened just a little. Much study has gone into why motor compressors burn out. Moisture, dirt, and air in the system are possible causes. Too much current flow from inaccurate safety devices in the electrical circuit, a stiff compressor, low voltage, or a lack of refrigerant (poor cooling of motor) may also be the trouble.

High head pressure is one of the most frequent reasons for motor burnout. This pressure creates very high temperatures as the gas passes the discharge valves of the compressor. The high temperature increases chemical action, adding to or creating new corroding elements in the system. Oil breaks down and forms carbon and sludge. If the discharge line to the condenser reaches 350°F (177°C), oil breakdown is taking place.

It is very important that the condenser be large enough, that it be clean, and that air flows over it efficiently. (Fans, fan motors, ducts, air-in, and air-out must all be in good condition.)

The service technician should check the head pressure of each unit and do all the necessary things to bring this pressure down. Purge the system. Completely clean the condenser with high gas pressure (air, carbon dioxide, nitrogen) See Para. 11-39. Wear goggles! Brush the condenser. Use long bristle brushes and a vacuum. Check the air-in and air-out passages. They must be in good condition.

11-60 CLEANUP AFTER MOTOR BURNOUT

When a motor begins to burn out, it overheats. This overheating will cause the refrigerant to break down and, if moisture is present, form hydrochloric and hydrofluoric acid. Oil in this condition is said to be "acidic." The acid will cause insulation on motor windings to deteriorate and increase the motor temperature. Eventually, the motor windings will short circuit and burn out.

If a system has a motor compressor burnout, refrigerant controls should be repaired or replaced (AEV, solenoid valves, reversing valves, etc.). Flush the system with R-11 or the same refrigerant used in the system.

Do not touch the oil from a burned-out motor compressor as it will cause a severe acid burn! CAUTION: Wear goggles and rubber gloves.

If oil cooler lines must be cut, be even more careful. Do not allow oil to run on the floor. Trap it in glass containers. A burned-out motor compressor has a very unpleasant odor.

The burnout can be mild or severe. If severe, the oil will be black and acidic with a very pungent odor. If mild, oil will be clear but there will be a pungent odor and a mild acidic condition. If oil is clean and odor-free, there is no burnout. The trouble is mechanical.

After replacing a motor compressor, install two filter-driers. See Fig. 4-9. One should be in the suction line between the evaporator and the compressor. The other should be between the condenser and the liquid refrigerant line (probably at a point just before the capillary tube if a capillary tube is used).

One of the many acid test kits available may be used to determine the amount of contamination. See Chapter 14 for further information.

Flushing a system with R-11 refrigerant is still another method of cleaning a burned-out system. The R-11 can be pumped through the system after removing the motor compressor using a rotor or diaphragm pump. Instead of a pump, a cylinder of R-12 or R-22 may be used to create pressure to push the R-11 through the system.

Repeat this operation until R-11 comes out of the suction line clean. Use a pail to catch it.

R-11 is probably the best substance for cleaning and washing out the inside of the refrigerating mechanism. It has a boiling temperature of 74.8 °F (24 °C). R-11 is nonflammable and nontoxic. It leaves no noncondensable residue and does not react with the electrical insulation. R-11 can also be cleaned and reused. It is particularly recommended after a motor burnout.

One can then also purge the system with R-12 gas and even some liquid to be sure that all the R-11 is removed. Wear goggles!

11-61 INSTALLING A MOTOR COMPRESSOR

The replacement motor compressor should be an exact replacement and it must have the same capacity as the one being replaced. The motor compressor must be designed for the low-side pressure desired (low, medium, or high).

This is the recommended procedure for installing a replacement motor compressor:

1. Carefully clean the ends of the suction line, discharge line, and oil lines (about 2 in.).
2. Carefully clean the suction, discharge, and oil line connections on the motor compressor.
3. Install the piercing valve and valve adaptor and then the gauge manifold—if they have not already been installed. (See Para. 11-32.)
4. Connect the lines. Use adaptor fittings, lengths of tubing, or an expander, if necessary, in order to telescope the tubes from the motor compressor into the suction line and discharge line. See Fig. 11-86. A punch-type swaging tool may be used if an expander is not available. Telescope the tubes together, using as little flux as possible. Braze the connections.

   When brazing, keep the heat away from other brazed joints. Use cloth or special wet heat-absorbency compounds to protect the other joints. Metal sheeting will protect the cabinet, wires, and plastic parts from the flame. *CAUTION: If the motor compressor or stub lines on the replacement motor compressor are smaller in diameter than the suction line and/or discharge line of the cabinet unit, the replacement motor compressor may be too small.*
5. Cut the liquid line between the condenser and the refrigerant control. Install a filter-drier of the proper capacity. (See Fig. 11-82.)
6. Put some vapor refrigerant into the system (about 25 psig [280 kPa]) and check for leaks.
7. Connect a vacuum pump to the system through the gauge manifold. Draw as high a vacuum as possible. Now, with the vacuum pump operating, hold this high vacuum for at least an hour or again put refrigerant in and then evacuate it—*the three-step method.* (Another way to check for leaks is to close off the connection to the vacuum pump and observe. Note whether the vacuum on the system remains constant or not. If it does not, it is an indication that there is

a leak in the system.)

8. Charge a small amount of R-11 into the system and purge. This will help clean out the refrigerant lines.
9. Charge the lines with a small amount of the refrigerant used in the system. Adjust the pressure to atmospheric or very slightly above.
10. Reconnect the electrical circuit.
11. Charge the system with the correct amount of refrigerant. It is best to overcharge slightly and purge to the desired quantity. The use of the service manifold makes this operation quite easy.
12. Close all manifold valves, plug in the electrical circuit, and operate the system through several cycles.
13. Adjust the refrigerant charge as described in Para. 11-46 and Para. 11-47.
14. Remove the gauge manifold and carefully seal all openings into the system, depending upon the gauge manifold connections used.
15. It is good practice to place a recording thermometer in the refrigerator cabinet to check the operation continuously for at least 24 hours.

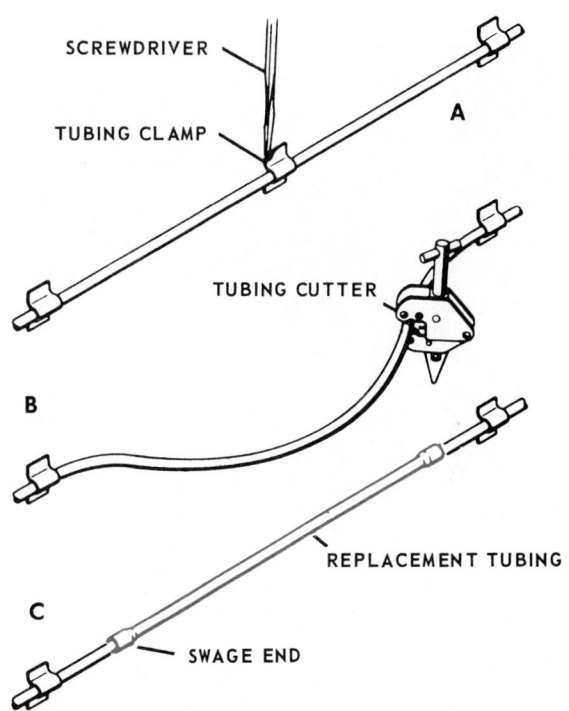

Fig. 11-86. Replacing tubing using swage ends. A—Removing damaged tubing. B—Cutting old tubing with tubing cutter. C—Installing swaged and replacement tubing. Joints are brazed.
(WCI Freezer Division)

Fig. 11-87 shows a replacement motor compressor. It has both a low-side and high-side process tube.

In Fig. 11-88, the process tube has had a tubing extension brazed to it. The right-hand process tube is brazed to the discharge tube.

There are many instances of replacement motor compressors burning out soon after they are installed. Most of these repeated burnouts are due to the system not being clean enough or because there was moisture in the system.

Fig. 11-87. Replacement compressor. A—Relay. B—Suction line process tube. C—Discharge line process tube. D—Suction line. E—Discharge line.

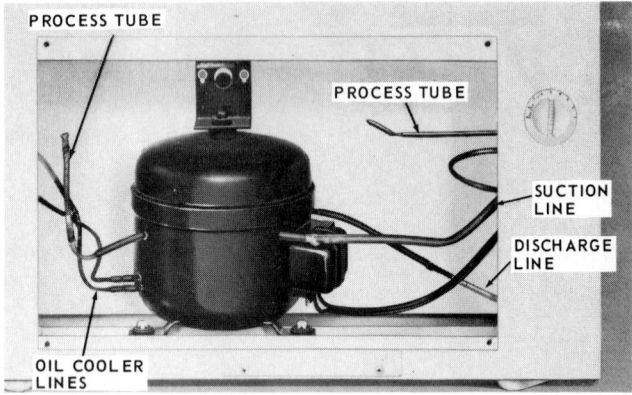

Fig. 11-88. Motor compressor replacement installed showing two process tubes. (Frigidaire Company)

Installation of both a suction line filter and a high-side filter has been found to reduce repeat burnouts.

11-62 CONDENSER TROUBLES

Condenser troubles are usually caused by leaks or the collecting of lint and dirt on the outside.

The service technician may be able to repair leaks without bringing the unit into the shop. A technician must install piercing valves, purge the refrigerant, repair the leak, test it, evacuate, charge, and check out operation. If the leak is not repairable (difficult to get at), the condenser must be replaced.

11-63 REPAIRING EVAPORATORS

Evaporators may be made of either stainless steel or aluminum. Follow the procedure recommended for repairing leaks in each type:
1. For stainless steel, either braze or weld (tungsten arc—inert gas system):
   a. Locate the leak.

b. Discharge the refrigerant.
   c. Clean the metal around the leak.
   d. Purge nitrogen (very low pressure) while brazing or welding.
   e. Polish the weld or clean the brazed joint.
   f. Test for leaks.
2. For aluminum evaporators, repair with aluminum solder, aluminum braze, aluminum weld, or a special epoxy.
   a. Locate the leak.
   b. Remove the refrigerant.
   c. Clean around the leak. The surface oxide is hard and must be removed. Repair right after cleaning as the oxide surface forms quickly. (Sand, file, and then clean with epoxy cleaner.) If leak is a large hole, use a clean small screw or metal plug to fill most of the opening.
   d. Mix the epoxy and catalyst.
   e. Apply with mixing spatula. (Be sure there is no positive pressure in the system. System must be open to atmosphere at some other opening.)
   f. Allow at least an hour for hardening.
   g. Sand the patch to a smooth finish.
   h. Test for leaks. If system still leaks at repaired joint, remove epoxy by filing and/or grinding and install a new patch.

   Aluminum foil can be used with the epoxy cement to strengthen the joint and to improve the appearance of the repair. Another method of repair is to heat the tubing and apply a paste mix (or stick) of special epoxies and resins.
3. A tube coupling used between aluminum and copper is described in Chapter 2.

Avoid brazing aluminum tubing, as it overheats too easily. Too much of the tubing will be annealed (softened), weakening the tubing walls. Aluminum tubing is usually 3003 alloy; however, 5005 and 1100 alloy are also used.

Aluminum solder, containing 92 to 100 percent zinc, is used. The melting temperatures are 700 to 800 °F (370 to 430 °C). Do not use a flux.

Aluminum may also be welded. Use an inert gas tungsten arc welding system (sometimes called TIG or GTAW).

11-64 SERVICING CAPILLARY TUBES

The service technician removes and replaces the capillary tube using the same steps as in Para. 11-56 and Para. 11-63. One step is different—the capillary tube and filter-drier are removed by loosening the brazed joints.

It is sometimes possible to repair a capillary tube by cleaning it. Proceed as follows: Disconnect the capillary tube at both ends if possible.
1. Attach the capillary tube cleaner to the outlet end of the tube.
2. Build up pressure on the tube to force the wax and dirt out. Fig. 11-89 illustrates a capillary tube cleaner attached to a capillary tube for cleaning. These cleaners are capable of building up pressure to a possible 20,000 psi (140 MPa).
3. After the capillary tube has been cleaned, continue to flush out the tube thoroughly. Use either R-11 or the refrigerant with which the system is charged.
4. Install a new filter-drier and reconnect the capillary tube into the system. If needed, new tubes must have the same inside diameter (ID) and the same length as the

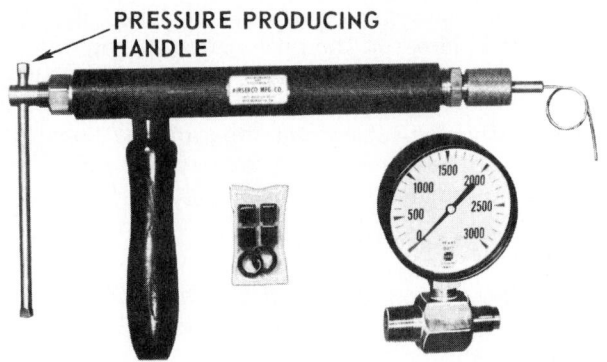

Fig. 11-89. Capillary tube cleaner used to clear obstructions from capillary tubes. It has a capillary tube adaptor fitting. Either oil or R-11 is used as pressure fluid.

one removed. Fig. 11-90 shows a tool for measuring tubing ID.

If there is an indication that wax caused the tube to become stopped up, the oil should be removed from the system. Replace it with fresh, clean, wax-free oil.

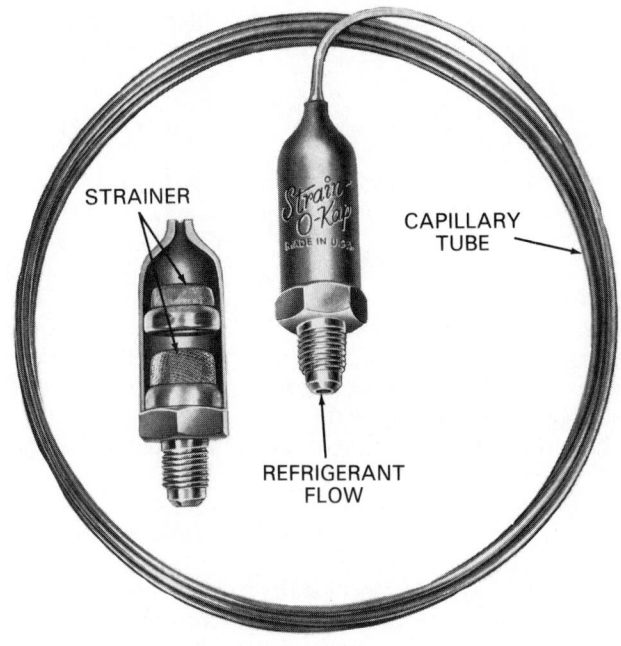

Fig. 11-91. Capillary tube, complete with strainer, meters refrigerant flow. Different strainer size is used for 1/20 to 1/4 hp and 1/3 to 1 hp units. Capillary tube can be installed by either soldering or using standard capillary tube fitting. (Watsco Components, Inc.)

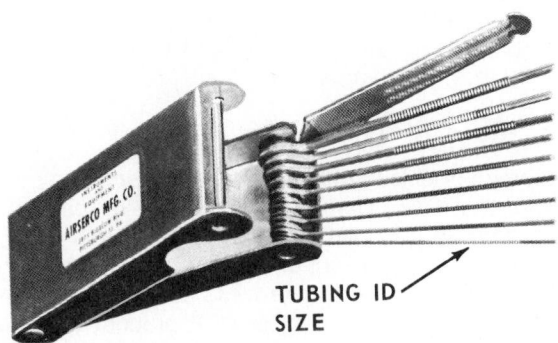

Fig. 11-90. Capillary tube sizing kit. Gauge measures ID of capillary tube. (Airserco Mfg. Co., Inc.)

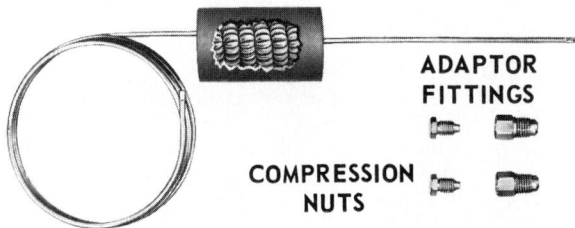

Fig. 11-92. Capillary tube refrigerant control replacement kit.

Never use a liquid drier (mainly methanol) to stop moisture from freezing at the refrigerant control. These "antifreeze" substances do not remove the moisture, they merely keep it in circulation. In many cases, they may damage the motor insulation.

Some patented replacement capillary tubes on the market may be used as capillary tube replacements when necessary. They are patented tubes arranged to be fitted either with a calibrated wire inside the tube to provide the proper refrigerant control, or the capillary tube itself is adjustable. See Fig. 11-91. Fig. 11-92 shows a replacement capillary tube with adaptor fittings.

11-65 OVERHAULING A HERMETIC SYSTEM

A complete overhaul of a hermetic system is usually done in a specialty repair shop.

The shop must be well equipped. Some of the equipment needed includes:

1. A compressor opener.
2. A welding system (a tungsten arc welder).
3. A cleaning booth.

4. A paint booth.
5. Benches and vises.
6. Storage cabinets.
7. A lathe and surface grinder.
8. A recovery/recycling unit.
9. A compressor test bench.
10. A deep vacuum system.
11. Compressed air (dry and clean).

The shop should be well ventilated with humidity kept at 35 to 45 percent range to prevent moisture getting into the internal assemblies. Para. 11-66 through Para. 11-79 explain recommended repair shop procedures.

11-66 REMOVING OIL

Oil ought to be removed if the compressor dome is cut open either by grinding or by turning on a lathe. If the sealed unit is of bolted construction, the oil may be left in until the unit is dismantled. Old oil should be drained off and carefully measured to assure replacement of the exact quantity. Replace with fresh oil of the proper viscosity. Test the old oil. If acidic, it will usually have an offensive odor and be dark in color. Acidic oil may damage motor windings. Check the windings with a megavolt meter for shorts and grounds. Handle acidic oil with extreme care. Wear rubber gloves and goggles.

11-67 OPENING THE MOTOR COMPRESSOR

Hermetic motor compressors are of two types:
1. Bolted (accessible).
2. Welded.

The bolted or accessible motor compressor overhaul is explained in Chapter 14.

The welded motor compressor is enclosed in a steel dome made of two steel stampings welded together. The process tube, suction line, discharge line, and the two oil lines (if used) are brazed to this dome.

The welded motor compressor units may be opened easily by either of the following:
1. Grinding away the welded seam.
2. Mounting the unit in a lathe and cutting the weld bead away with a cutting tool.

Fig. 11-93 shows a machine used to remove the motor compressor weld by grinding. The motor compressor is clamped to a turntable and the grinding wheel removes the weld as the compressor slowly turns. The wheel automatically adjusts itself to the contour or shape of the weld seam. The machine will remove top welds as well as side welds.

Fig. 11-93. Motor compressor opener. Special dome weld grinding machine will open motor compressor in about 20 minutes. A—Welded bead and grinding wheel. B—Machine controls. (Airserco Mfg. Co., Inc.)

The opener mechanism is housed in a steel cabinet equipped with shatterproof windows to enable the operator to watch the grinding operation. Both motors of the unit require single-phase, 120 V power and use about 11 A. The machine automatically grinds the weld.

One motor is used to operate the turntable while a second motor drives the grinding wheel. Grinding wheels may be one of the following: 1/8-, 1/4-, or 3/4- in. thickness.

Remove a minimum of material. Enough base metal must be left for the assembly to be rewelded after overhaul.

After the compressor dome weld is removed, the dome can usually be lifted off the base. Often, tubing must be unbrazed where it passes through the dome. Use carbon dioxide or nitrogen to fill the dome when heating the parts. After removing the tubing and lifting off the dome, the motor and compressor will be exposed. Clean the inside of the dome thoroughly before continuing.

11-68 CLEANING FACILITIES

Clean the outside of the unit thoroughly. Mineral spirits, obtainable from most oil companies, are used widely as a general cleaner. Carbon tetrachloride should not be used.

A special booth with a grate bottom is desirable. A booth should have drainage facilities, be equipped with a pump and spray nozzle, and be ventilated. Local fire and industrial hygiene regulations must be followed closely. Also follow OSHA (Occupational Safety and Health Administration) standards.

Several degreasing solvents are available which are quite safe. Some of their features are low toxicity (low in poisons), noncorrosive, and have a high flash point. (Flash point is the temperature at which something will burn.)

Use Class III flammables having a flash point of 140 °F (60 °C) or higher. Use self-closing cleaning pans or tanks. Tanks with a safety fuse link that closes the lid in case of fire are recommended. Suppliers can supply solvents safe to use for this type of cleaning. See Chapter 30 for the classification of cleaning solvents.

11-69 MOTOR REPAIRING

A common trouble with hermetic mechanisms is a burned (overheated) motor. The overheating may be caused by poor cooling, an overload, a short, or a ground.

Inspect by looking at it closely. One can tell the condition of the motor. A short (bare wires touching each other) will show up as a dark brown spot. A grounded spot will have the same brown spot. A completely burned-out winding will have all the insulation burned and the coils will be dark in color.

A system with a burned-out motor should be dismantled and repaired as soon as possible (two or three days). The corroding chemicals in the system will keep increasing the corrosion of the compressor and system parts.

When the windings overheat, insulation is destroyed; stator windings short and ground. If the motor starting winding or running winding is faulty, the stator must be dismantled and rewired (rewound) or replaced. The rewinding should be done by a specialist. Wire must be the same size and have the same insulation as the original.

Replacement or exchange stators are made for most hermetic motors.

11-70 REBUILDING A RECIPROCATING MOTOR COMPRESSOR

Many manufacturers provide exchange service on complete refrigeration machines. These do not include the external controls, fans, and cabinet. In many cases, if the unit cannot be repaired in the field, it is advisable to use this type service. But, if a manufacturer has gone out of business, or if it is impossible to get a replacement mechanism, overhaul procedures described in the following paragraphs may be useful.

To repair the motor compressor:
1. Check the work order. It may indicate the fault.
2. Take the compressor apart. Store the parts carefully in clean trays.
3. Clean the parts with a nontoxic, nonflammable solvent. Various spirits and/or liquids may be used. Clean the parts a second time before reassembly.

The compressor repair may include several operations:
1. If valves have been leaking, inspect, check, and if necessary, repair the valve plate and replace valve reeds.
2. If the compressor has been noisy:
   a. Hone the cylinders and replace piston.
   b. Repair piston pins and connecting rod.
   c. Repair crank journal and main bearings.

The most common compressor troubles are with valves and valve seats. The valve reed may be replaced. The valve seat should be reground using a surface grinder or a drill press. Next, lap them accurately. Replacements are available for valve reeds on most hermetic compressors.

Practically all compressor valves used on domestic refrigerators are of the diaphragm or disk type. Usually these are called flapper valves. They are made of thin spring steel and are held in place by their own spring tension and a machine screw, or by an auxiliary coil or flat spring.

Compressor bodies and valve seats are usually made of cast iron. After long use, they may be worn. It is a good policy to replace the valve reed when overhauling a compressor. The cost is small. It is also good practice to lap the valve seats.

Be sure to use fine lapping compound and clean all lapped surfaces carefully afterward. Compound left on the surfaces may ruin the compressor.

If raised valve seats are used, a plate glass surface may be used as the lapping tool, or special lapping blocks may be used. Some companies use fine polishing paper clamped to a flat plate as a lapping surface.

If the seat is in bad condition, the valve plate should be mounted in a lathe—either in a chuck or on a face plate—and the whole plate trued. Some repair shops use a surface grinder to true valve seats.

In compressors using a spring-steel-type valve, the proper surface of the valve must come into contact with the valve seat. The proper surface may be detected by the slightly turned-over edge of the opposite surface of the valve. This is caused when the valves are stamped out. The stamping process turns the edge of the valve on one side (burrs it). If this side is placed against the valve seat, it might not seal properly.

Additional information on valve repairing will be found in Chapter 14. Replacement valve plates are available and their use is recommended.

Valves must be tested before assembling the unit. Use fixtures and synthetic gaskets or a synthetic rubber vacuum cup, as shown in Fig. 11-94. Noisy valves may be caused by too much valve lift (or excessive pressure differences). The valve movement is usually measured in thousandths of an inch. Valve lift must be accurately adjusted. Measure the lift with a dial indicator before dismantling the valve. Too little lift leads to poor compressor operation and overheating.

Poor vacuum indicates a poor intake valve even if the compressor holds the vacuum in the suction line when the motor stops. The intake valve is bouncing on the valve seat.

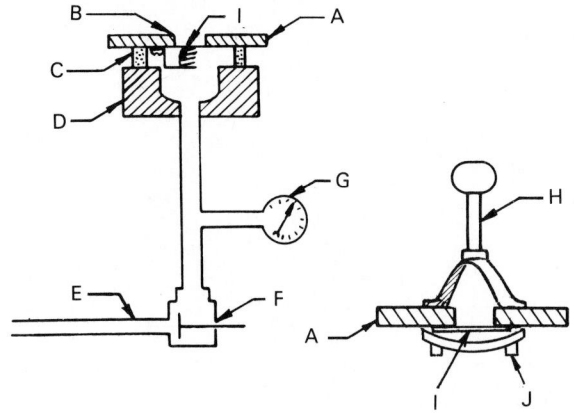

Fig. 11-94. Testing valves. A—Valve plates. B—Refrigerant oil. C—Synthetic rubber gasket. D—Fixture. E—Clean, dry compressed air. F—Air valve. G—Pressure gauge. H—Vacuum cup. I—Valve. Top left valve has coil spring. J—Valve retainer.

The valve works better in the stopped position.

Poor vacuum may also be due to worn pistons and rings or the use of a too-thick gasket (gasket works like a cylinder head gasket). Keep in mind, also, that lack of oil will cause poor pumping ability.

A check valve can be examined in much the same way as an intake valve. It is important to keep valves, pistons, connecting rods, bearings, and other moving parts as quiet as possible, so the bearing and bushing clearances must be as small as possible. Check all main bearings, connecting rod bearings, piston pin bushings, and piston clearances. Use accurate micrometers and thickness gauges for checking.

The main body or parts of the compressor are usually made of cast iron. Valves are of spring steel and the piston pins of high carbon steel. Some repair shops use drill rod as replacement piston pins. Replacements for moving parts should be the same weight as the old parts. This will keep dynamic vibrations to a minimum.

End play must be kept to a minimum to eliminate end play slap. The pump should be checked for satisfactory operation.

11-71 REBUILDING A ROTARY MOTOR COMPRESSOR

Repairing the rotating vane type of rotary compressor requires some special attention:
1. There is an adjustment in the positioning of the roller housing to locate accurately the contact spot between the concentric roller and the eccentric housing. The location of the contact point is important. In Fig. 11-95 the contact point, X, must be carefully adjusted. Four cap screws hold the housing, A, and an end plate up against the main housing. Point F is a dowel pin. The parts are assembled and the housing is tapped lightly until the rotor, B, binds when the shaft is turned by hand or by using the electric motor. Then the housing is relieved slightly by tapping opposite to X until the binding is released. Dial micrometers may be used to measure the distance.
2. Vanes at C in Fig. 11-95 must be accurately fitted to the roller, B, and housing A. These vanes must closely

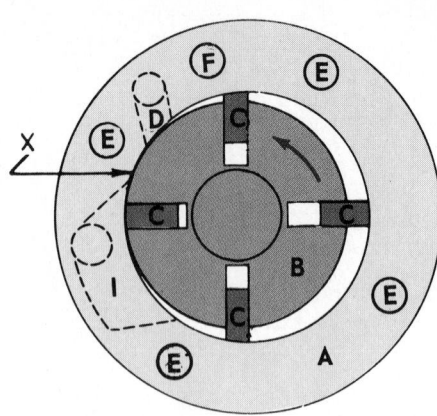

Fig. 11-95. Parts and adjustments on hermetic rotary compressor. A—Housing. B—Rotor. C—Blades. D—Exhaust port. E—Bolts. F—Dowel pin. I—Intake port. X—Contact point.

match the length of the roller, B, and the housing length, A. This is to prevent blow-by (leakage). Vanes must also fit the slots in the roller accurately. If too wide, they will bind as they pass point X. If too narrow, they will permit blow-by as they pass point X.

The single or stationary blade (divider block) type compressor has much the same fit problem as the rotating vane type. Parts must be accurately fitted together. The main bearing or bearings must be in excellent condition. The length of roller and housing must be exactly the same. The housing should be cylindrical (circular) to within .0001 (one-ten thousandth in.). There must be no evidence of scoring.

Usually the vane is spring loaded to keep it riding on the roller. If this spring is too weak, vapor will bypass the vane and cause a loud clicking noise. If it is too strong, it will place an unnecessary load on the roller and rapid wear will occur. Dowel pins are used to align housing with shaft.

Both compressor types need enough oil to keep the parts in a constant oil bath. During repair, clean all oil passages. Clean or replace the oil metering screws. Use dry compressed air.

In compressors of the stationary blade type compressor, the cylinder is mounted snugly; then the shaft is turned and the cylinder is shifted until there is equal resistance over the complete revolution. Then the cylinder bolts are tightened according to torque specifications.

11-72 ASSEMBLING COMPRESSOR

Insuring cleanliness of internal parts of the mechanism is one of the most important steps in the assembly process. The inside of the compressor and the motor windings are usually hardest to clean.

One of the best procedures is to use an air gun and a mineral spirits spray. Immersing (dunking) the parts in a mineral spirits bath is also effective. Carbon tetrachloride should not be used. After this cleaning, iron and steel parts must be oil-dipped immediately to prevent rusting.

It is essential to run the compressor after assembly and before welding the dome. It should be run for a few moments to check its pumping ability and its noise level. This can be done without danger of scoring the parts due to lack of oil.

A vacuum gauge equipped with a synthetic rubber tip can be held against the inlet opening of the unit while check-

ing its pumping ability. A similarly equipped high-pressure gauge may also be used on certain models. Running the motor provides a check of the electrical work for opens, shorts, grounds, and suitable power.

11-73 WELDING COMPRESSOR DOME

After the compressor has been checked and tested, the dome may be welded in place. Arc welding makes the best seal. Because this requires such a short time, heat does not travel to the interior of the dome, compressor, and windings.

It is a good safety practice to bleed the dome with carbon dioxide or nitrogen during welding to prevent an explosive mixture of oil fumes and air from collecting. See Fig. 11-96. Use reverse polarity coated electrodes, tungsten arc, or inert gas metal arc during arc welding to keep the motor as cool as possible. Before welding, it is important to clean the metal surfaces for at least one-half inch on each side of the weld area to prevent contamination of the weld with dirt. Dirt may form blow holes that may one day leak.

The welding should be done by an experienced operator. The welding station must be well ventilated. The arc must be shielded from the eyes of passersby.

Tack weld the dome in at least three equally spaced places before proceeding with the welding. Do all the welding with the weld in the downhand position. Turn the dome as the welding progresses.

After the unit is welded, short lengths (8 to 10 in.) of proper size tubing should be brazed to the dome. Avoid getting flux inside the dome. Cool the metal as quickly as possible after the weld is completed. Use a damp cloth or a stream of cool air. After the welding is completed, the motor should again be tested for running characteristics.

11-74 MOTOR COMPRESSOR TESTING

After the dome has been welded and the tubing brazed to it, test the assembly for leaks. Install tubing adaptors on

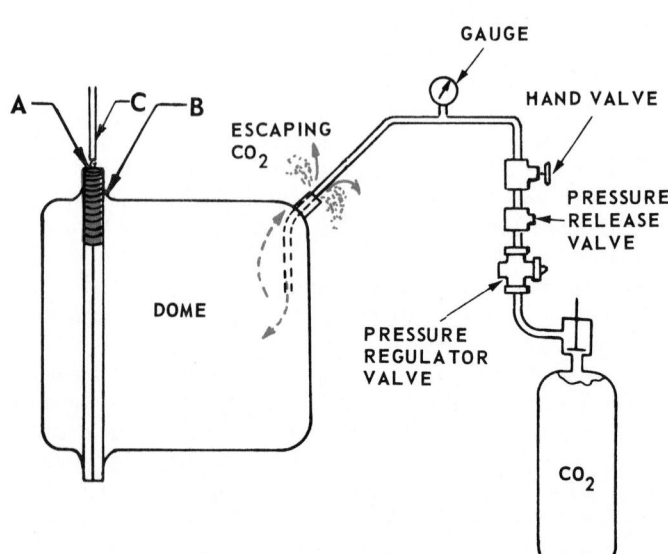

Fig. 11-96. Welding hermetic dome. A—Arc welded seam. B—Dome flange. C—Arc welding electrode. Always clean surfaces before welding. Note use of carbon dioxide or nitrogen. It acts as flushing gas to prevent collection of explosive mixtures and to reduce oxide scale inside dome.

the suction and discharge lines. Seal the oil cooler lines. Then either charge with refrigerant vapor and test for leaks or use nitrogen gas and immerse the dome in water. Never use more than 150 psi!

After the compressor has been tested for leaks, it should be filled with the correct amount of the proper viscosity oil and then tested again before it is installed. Use the correct type and proper quantity of oil recommended by the manufacturer.

The following is one way to put oil in the compressor. Connect a charging line to the process tube, valve adaptor, or suction line. Connect this tubing to the center fitting of a gauge manifold. Connect a vacuum pump to the low-side fitting of the gauge manifold. Connect one end of a tube to the high-side fitting of the manifold and put the other end into a container of fresh refrigerant oil. Open the low-side valve and close the high-side valve. Run the vacuum pump to create a vacuum in the motor compressor. Close the low-side valve and open the high-side valve. The oil will flow into the crankcase. When the unit is running, oil will start spraying through the discharge service valve once the compressor has enough oil.

Next, connect a gauge manifold to the compressor. Use a high pressure gauge calibrated from 0 to 160 psig (1 210 kPa) on the high side. Use a compound gauge calibrated to 30 in. (76 cm; 0 micron*) and 150 psig (1 140 kPa) on the low side. If the compressor tests all right, (18 in. Hg to 28 in. Hg of vacuum with a 150 psig head pressure and holds the vacuum, 46 cm Hg to 71 cm Hg with 1 140 kPa head pressure; 303 mm to 49 mm), it should be baked for 8 hr. or more at a temperature of 150°F (70°C) to 200°F (100°C) with at least a 20 in. Hg (51 cm Hg; 252 mm) vacuum on the system.

It can also be evacuated for several hours at 50 to 500 microns of pressure or lower. This evacuating will remove almost all of the air and moisture from the system.

Test benches are available which may be used to test newly repaired hermetic motor compressors. Fig. 11-97 shows an electrically-connected motor analyzer bench with flexible motor and discharge lines connected to the motor compressor. The inside construction of the analyzer bench is shown in Fig. 11-98.

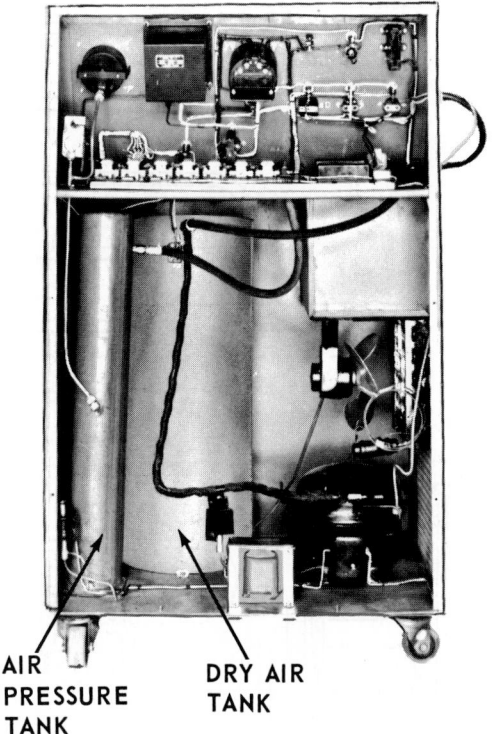

AIR PRESSURE TANK   DRY AIR TANK

Fig. 11-98. Inside construction of motor compressor analyzer. (Airserco Mfg. Co., Inc.)

After evacuating, the compressor should be charged with nitrogen or with the correct refrigerant vapor to a pressure of 15 psig (210 kPa).

The suction line, discharge line, process tube, and oil cooler lines should then be pinched and brazed if the motor compressor is to be shipped or stored.

11-75 ASSEMBLING THE HERMETIC SYSTEM

After the motor compressor has been repaired and tested, it may be installed in the system. The procedure follows:
1. Clean and test the rest of the system. Add the suction line filter-drier and liquid line filter-drier.
2. Clean the capillary tube, evaporator, suction line, oil cooler tubing, and hot gas defrost lines. This is done by forcing R-11 through the system until it comes out of the system clear. See Fig. 11-99. The filter-drier and capillary tube may have to be removed to permit enough circulation.
3. When the system is clean, install the suction line filter-drier and liquid line filter-drier. Connect suction line, discharge lines, and the oil cooler lines. Braze the joints.
4. The assembled system must be given a thorough leak test. The system can usually be pressurized to about 150 psig (1 140 kPa) with nitrogen. A bubble solution applied to joints and welds shows leaks. It may be possible to immerse parts of the system in a water tank, as suggested in Fig. 11-100.

11-76 EVACUATING A SYSTEM

A system opened for any type of repair must be completely evacuated to remove air and moisture.

Two different methods are used.
1. Deep vacuum.

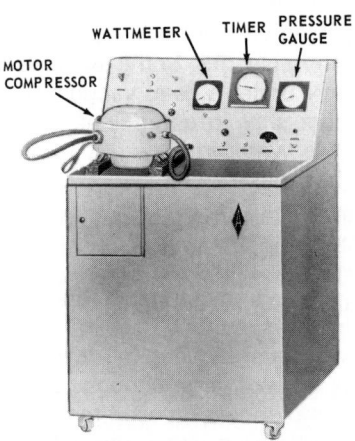

WATTMETER   TIMER   PRESSURE GAUGE

MOTOR COMPRESSOR

Fig. 11-97. Motor compressor analyzer stand which checks motor volt-amperes, pressures, and time to pump certain volume of dry air up to certain pressure (volumetric efficiency). (Airserco Mfg. Co., Inc.)

*Pressures below about 15 in. Hg are best written in microns or sometimes in mm. The mm unit here is different from cm of Hg used near atmospheric pressure. Look ahead to Fig. 11-112.

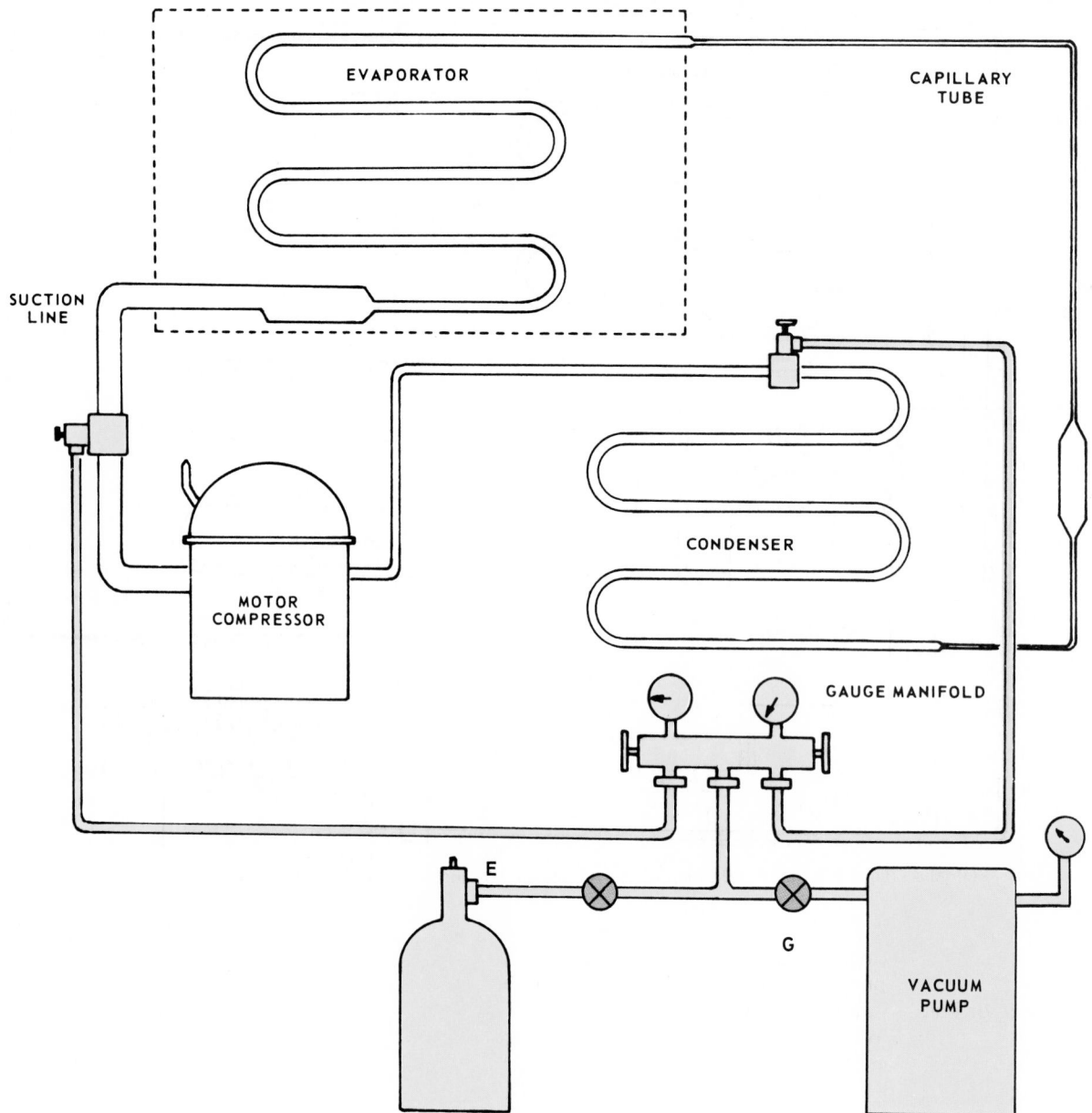

Fig. 11-102. System with gauge manifold, refrigerant cylinder, and vacuum pump ready for triple evacuation process. 1—Pressure with E. 2—Evacuate with G. Repeat three times.

Never heat nitrogen or CO_2 cylinders. The maximum cylinder temperature should be 110°F (43°C). A test setup for a triple evacuation is shown in Fig. 11-102.

The evacuating and drying of the unit is a very important part of the assembly work. The system should be as close to 100 percent clear of air, moisture, solvents, and other foreign matter as possible.

Remember that the most careful evacuating and purging will not clean a unit that was carelessly put together with dirt in the system.

See Chapter 28 for recovery and recycling procedures.

11-77 HIGH-VACUUM PUMPS

Standard reciprocating type air compressors do not create a high enough vacuum; nor are ordinary refrigeration compressors designed to produce the deep vacuum for

evacuating refrigeration systems being serviced.

There are two main types of vacuum pumps:
1. Single-stage.
2. Two-stage.

The single-stage vacuum pumps are used when the triple evacuation method is used. Two-stage vacuum pumps are used when the high-vacuum (deep vacuum) method is used.

There are two designs in high-vacuum pumps:
1. Rotary pump (oil sealed).
2. Vapor pump (diffusion type).

The rotary pump should be able to pull a 50 micron vacuum pressure. (A 50 micron vacuum is equal to .05 torr. A torr equals 1 millimeter.)

The better rotary pumps use two rotors in series (compound pump). See Fig. 11-103. Most refrigeration compressors can pull about 50 to 80 torr (50 000—80 000 microns [50-80 mm]).

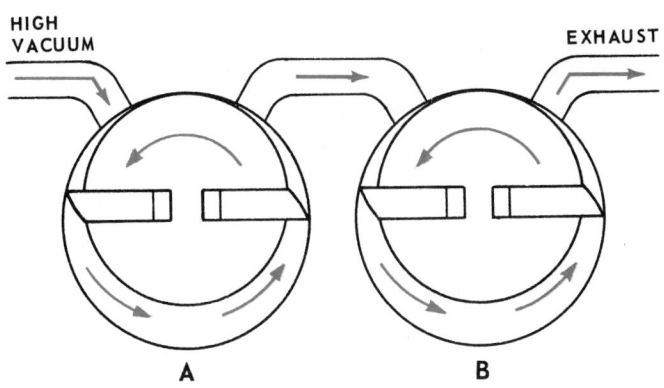

Fig. 11-103. Two-stage rotary high-vacuum pump. A—First stage. B—Second stage.

To evacuate, proceed as follows:
1. Pull vacuum with a low-vacuum pump (50 to 80 torr).
2. Switch to a high-vacuum pump.
3. Repeat pull-down to 50 microns. A rise to 300 microns in three minutes or more indicates that the refrigeration system is dry and evacuated.

Use copper tubing or special metal hose for vacuum pump connections. If standard charging or servicing hose (synthetic materials) is used, the sections tend to collapse at this high vacuum and the synthetic material tubing is volatile (will release gases).

Portable two-stage high-vacuum pumps are available which will draw down to less than 1 micron of mercury column.

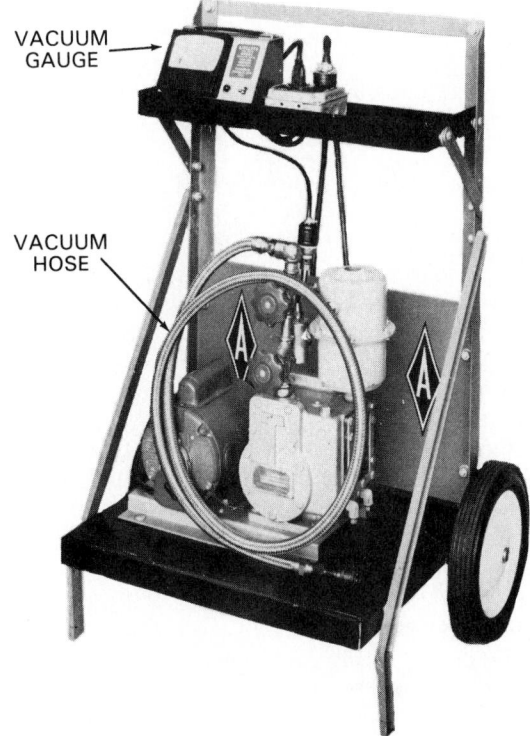

Fig. 11-104. Portable high-vacuum pump mounted on stand for transport. Note flexible metal evacuating line. This is a large capacity line and will not collapse. Two hand valves control evacuating operation. (Airserco Mfg. Co., Inc.)

A micron is 1/1000 of a millimeter (mm), 2.54 centimeters (cm) equal 1 in., 10 mm equal 1 cm. Therefore, 25 400 microns equal 1 in. One micron is close to a perfect vacuum.

A high-vacuum pump will produce a vacuum lower than 29 in. Hg (29 inches of mercury; 23 mm, 23 000 microns*). Portable high-vacuum pumps are shown in Figs. 11-104 and 11-105. Vacuums lower than 29 in. Hg are necessary for complete dehydration of (moisture removal from) the system.

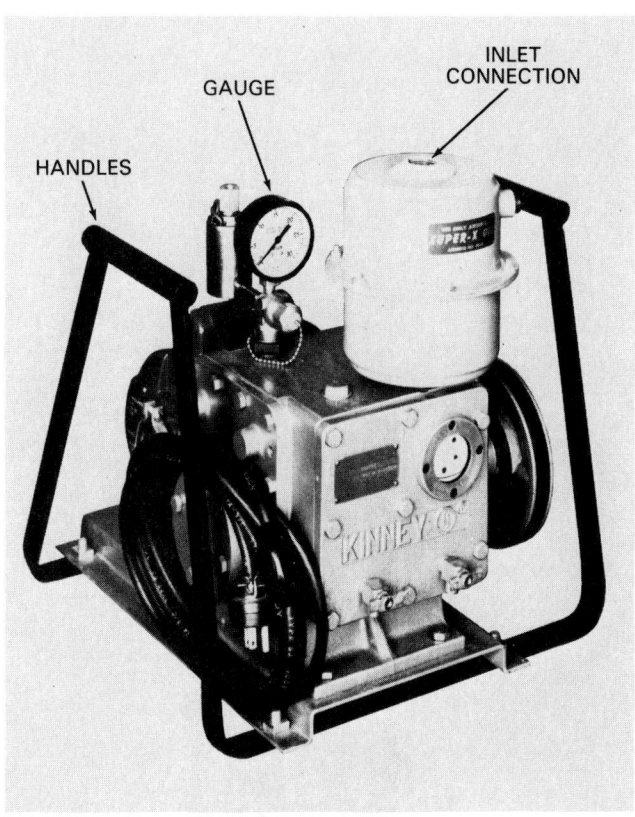

Fig. 11-105. Portable high-vacuum pump. It may be used to remove air or refrigerant from system. (Airserco Mfg. Co., Inc.)

OIL IN VACUUM PUMPS

The oil sight port permits checking the oil level, as well as the oil color. This special oil should be replaced frequently.

Oil in single-stage pumps rapidly becomes dirty when there is water and solvent vapor in the system. Water in the oil will:
1. Raise the oil level.
2. Turn the oil white and foamy. If this dirty oil is left in the pump, sludge will form in the system.

For good results, change oil before each system pump-down or test pump with valves closed and vacuum gauge connected. If pump will not pull down to high vacuum, change oil.

Compound (two-stage rotary) pumps need oil changes after about 10 pump-downs. Most of these pumps will pull a vacuum of 20 microns or better.

*Look ahead to Fig. 11-112.

HIGH-VACUUM GAUGES

To measure deep vacuum, an electronic or a solid-state thermistor vacuum gauge is used. A regular compound gauge cannot read accurately to micron levels. A high-vacuum gauge is shown in Fig. 11-106. A vacuum from 29.25 in. Hg (about 17 000 microns) to 29.9 in. Hg (540 microns) is necessary to allow the moisture inside the system to evaporate at room temperature.

The inside design of a vacuum gauge tube used on an electronic mechanism is seen in Fig. 11-107. The sensing element is a thermocouple. It is important to use the tube in an upright position to keep out foreign matter.

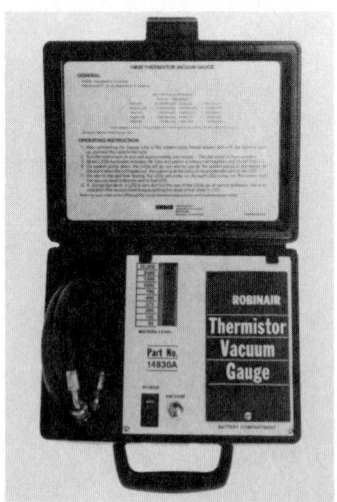

Fig. 11-106. Solid state thermistor vacuum gauge used to measure the vacuum level of a system. (Robinair Division, SPX Corporation)

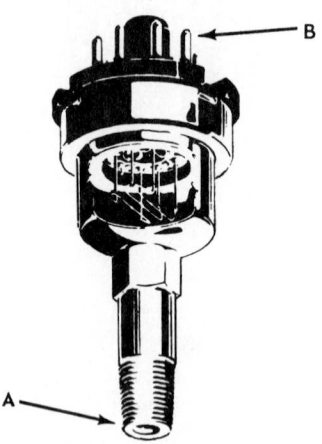

Fig. 11-107. Internal construction of high-vacuum gauge tube. A—Pipe-thread connection to system manifold. B—Electrical connections to vacuum gauge. (Airserco Mfg. Co., Inc.)

The gauge's electrical circuitry is shown in Fig. 11-108. The filament gets warmer as air pressure around it decreases. (There are fewer air molecules to remove heat.) The thermocouple gets warmer and the increase in emf is registered on the meter. The meter is calibrated in microns. The vacuum dial scale is shown in Fig. 11-109.

If the vacuum gauge reading levels off at 5 000 microns,

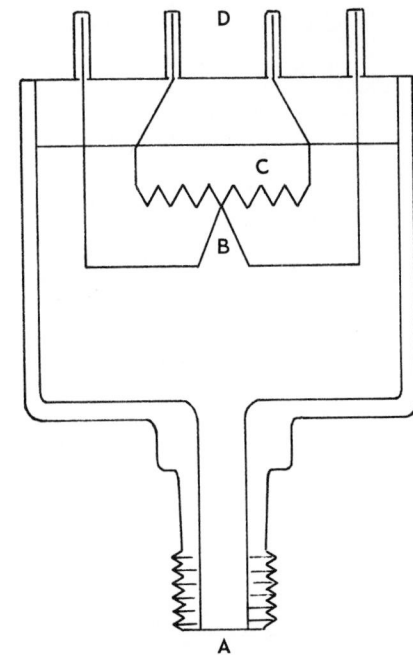

Fig. 11-108. Schematic internal design of high-vacuum gauge. A—Connection to system being checked. B—Thermocouple. C—Resistance unit. D—Electrical leads.

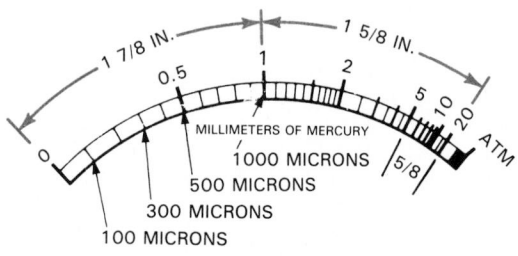

Fig. 11-109. Dial scale of high-vacuum gauge. Note that pressure between 0 and 1 000 microns has been greatly expanded for easy reading. (Airserco Mfg. Co., Inc.)

there is ice or free water in the system. Ice may be located by a cold spot, frost, or sweat on outside of system. Stop the pump; allow the ice to melt or use radiant heat.

Never allow the system pressure to enter a high-vacuum gauge.

The thermocouple vacuum gauge has two parts:
1. A meter.
2. A tube which threads into the refrigeration system. The tube should be right side up with threads down.

The tube sometimes collects vapors and/or oil. It can be cleaned by putting a solvent in the opening with an eye-dropper. Freon 23 is a good cleaner. Clean as follows:
1. Fill.
2. Rock tube gently.
3. Empty.
4. Repeat the above two or three times.
5. Clean with an alcohol rinse.

USING VACUUM PUMPS

Clean the instrument dial covers with soap, water, and facial tissues. This crystal is plastic and solvents will fog it.

If a system has a leak, the vacuum produced will not be a high vacuum. The dial needle will rise steadily when the valves are closed. If there is moisture in the system, the vacuum produced will also be less than desired. When the valve is closed, the dial needle will rise and level off at a pressure corresponding to the water vapor pressure at that temperature. Fig. 11-110 shows the two conditions of a pressure-time graph. These are made with the valve to the vacuum pump closed.

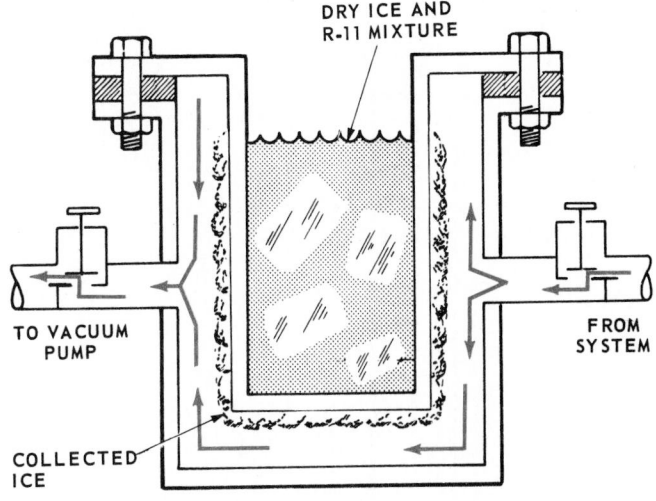

Fig. 11-111. Cold trap used to collect moisture from system being evacuated. Moisture vapor condenses on cold surface and then turns into ice.

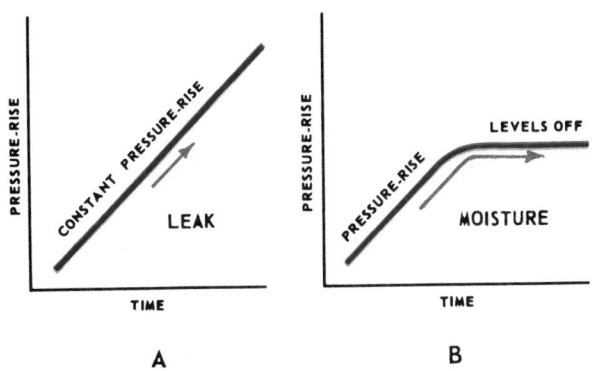

Fig. 11-110. Leak and moisture affect high-vacuum gauge readings. A—Shows effect of leak. B—Shows effect of moisture in system. (Airserco Mfg. Co., Inc.)

To make sure the dehydration (drying out) of a wet system is complete, this procedure is recommended for domestic units:
1. Connect a 250 VA (250 W) lamp (in place of a fuse) in series with a test cord.
2. Attach the cord to the compressor run and common terminals.
3. Connect to a 120 V outlet.

This procedure allows only about 30 V to go through the running windings because the lamp becomes a resistor. This warms the compressor windings, vaporizing the moisture which may be trapped in the system. *The compressor should not run during this operation.*

If a manifold is used, it must be degassed by pulling a vacuum on it for several days to remove all gases absorbed in the walls and cracks of the manifold. These residual gases may make a service technician believe there is a leak in the system. Evacuate the system from both the high side and low side.

Use as large a diameter in the vacuum line as possible. Keep the line as short as possible. Vacuum pump sizes are given in cubic feet per minute (cfm):
1. 1.5 cfm—good for domestic systems (3-5 tons).
2. 3-5 cfm—medium systems (5-100 tons).
3. 10-15 cfm—large systems (over 100 tons).

Always break the vacuum (allow air in) of a vacuum pump before storing it. Otherwise the cylinder will fill with oil and become oil locked (hard to turn).

A cold trap can be used to keep moisture out of the vacuum pump. This trap is placed in the tubing between the system and the vacuum pump. As moisture vapor contacts the cold surface, it will collect as ice. By closing the valves on each side and unbolting the trap, one can remove the ice. See Fig. 11-111.

Fig. 11-112 is a chart of the various pressures that are possible with different type pumps. Note the evaporating temperature of water at the various pressures.

During manufacture, systems are evacuated to 50 to 100 microns before oil is added. In the field, when oil is already in the unit, evacuating to 50 microns may cause some of the oil to vaporize.

Because the motor compressor depends on gas flow to cool the motor windings and compressor, avoid using it as a vacuum pump. The motor compressor may heat up and be damaged.

11-78 CHARGING REBUILT SYSTEMS

After the system has been evacuated, it is carefully charged with clean, dry oil and refrigerant.

The refrigerant charging tube, oil cylinder, connecting lines, and gauge manifold must all be as clean and as free of moisture as the system being charged.

The correct amounts of oil and refrigerant should be checked from the manufacturer's specifications. (The amount of oil was measured when the old system was dismantled.)

A combination discharge and charging station, often used to recharge rebuilt systems, is shown in Fig. 11-113.

11-79 TESTING REBUILT SYSTEMS

After the unit has been assembled, tested, evacuated, and charged with oil and refrigerant, it should be run with a thermostatic control for at least 24 hours to determine its behavior. A recording thermometer should be placed in the refrigerator during this test period to enable the technician to know what is happening during the entire test period. This very important part of repair work is often neglected. Fig. 11-114 shows a 24-hour temperature recorder. Recorders with 48-hour and 72-hour chart speeds are also available.

The sensing bulbs are put in the refrigerator and the recorder is connected to a wall plug. The one-day chart shows both the fresh food and freezer cabinet temperatures at all times. If possible, the cabinet should be placed in a

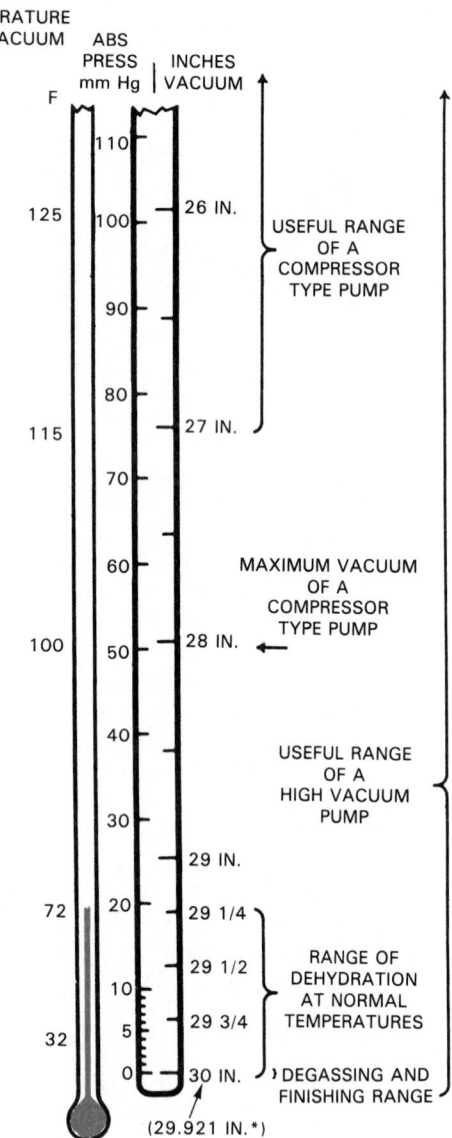

TEMPERATURE
VS. VACUUM

ABS
PRESS
mm Hg

INCHES
VACUUM

USEFUL RANGE
OF A
COMPRESSOR
TYPE PUMP

MAXIMUM VACUUM
OF A
COMPRESSOR
TYPE PUMP

USEFUL RANGE
OF A
HIGH VACUUM
PUMP

RANGE OF
DEHYDRATION
AT NORMAL
TEMPERATURES

DEGASSING AND
FINISHING RANGE

(29.921 IN.*)

*Note: 29.921 is from 76.000 cm divided by 2.540 cm per in.

Fig. 11-112. Pressure scales used during high-vacuum evacuation of system. Temperature scale shows evaporating temperature of water at these pressures. For example, water will evaporate (boil) at 72 °F (22 °C) at 20 mm Hg pressure (20 000 microns, 3 kPa, or 3 000 Pa). (Airserco Mfg. Co., Inc.)

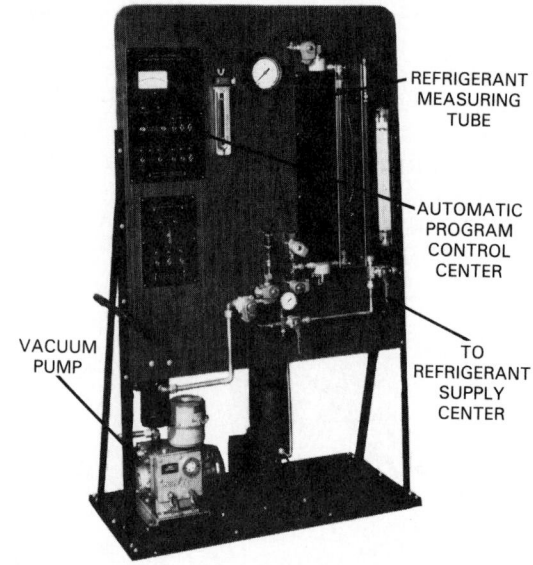

Fig. 11-113. Combination vacuum checking and charging station. (Airserco Mfg. Co., Inc.)

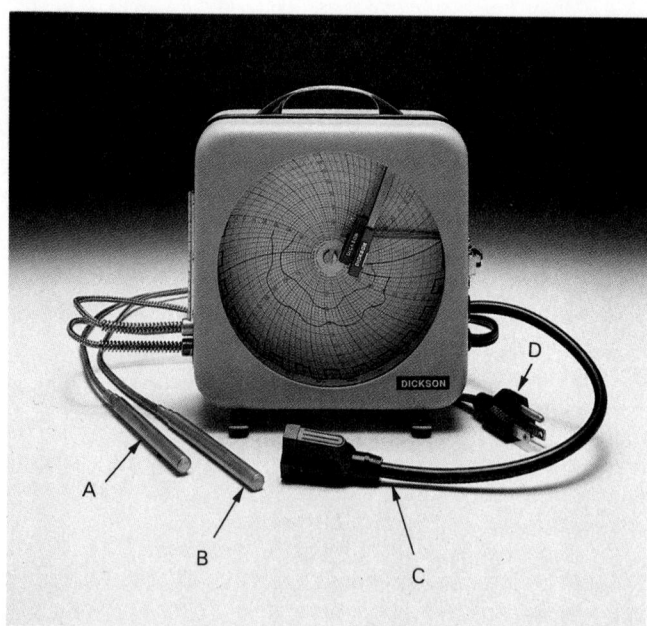

Fig. 11-114. Temperature recorder charts temperature changes of refrigerator-freezer unit over 24-hour period. A—Sensor for freezer compartment. B—Sensor for fresh food compartment. C—Series connector. D—Power connection for chart motor. (The Dickson Company)

warm room (100 °F, 38 °C) for the test. The testing room should be quiet so the unit can be checked for noise.

A 30-hour portable temperature recorder is shown in Fig. 11-115. A self-starting electric clock may be used to record the running time of the unit.

The testing should also include checking the temperature of the dome and lines. The test looks for an even frosting of the evaporator.

Use an ammeter and voltmeter or wattmeter to check the electrical section of the unit. Many service technicians use a combination hermetic compressor analyzer and electrical tester. There are numerous types available. Fig. 11-116 pictures one type of heavy-duty analyzer. It serves as a hermetic motor analyzer, three-phase tester, motor resistance tester, and potential relay tester. For additional

information on the use of electrical instrumentation, see Chapter 8 and Para. 7-50.

One way to test a system for efficiency is to cover the evaporator with insulation after installing a thermometer (thermocouple type) on one of the last coils of the evaporator. Run the system.

At the end of 30 minutes, the evaporator temperature should read as low as −25 °F (−32 °C) if the room air is a normal 70 °F to 75 °F (21 °C to 24 °C); or as low as −18 °F (−28 °C) if the room temperature is 90 °F to 100 °F (32 °C to 38 °C).

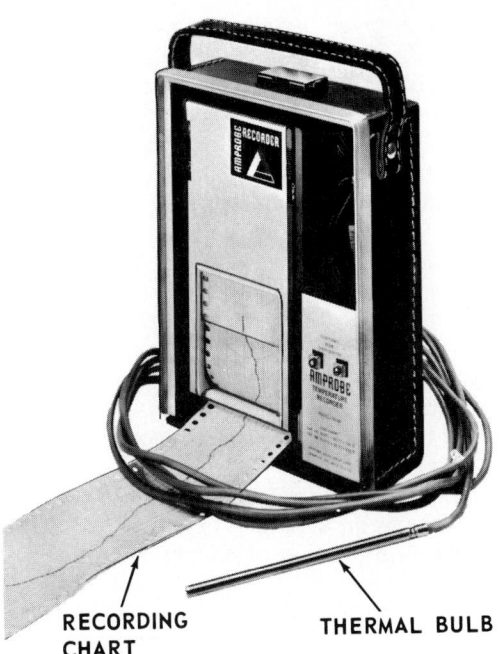

RECORDING CHART   **THERMAL BULB**

Fig. 11-115. Portable temperature recorder for up to 30 hr. recording. Thermal bulb may be placed in cabinet or be clamped to evaporator.   (Amprobe Instrument)

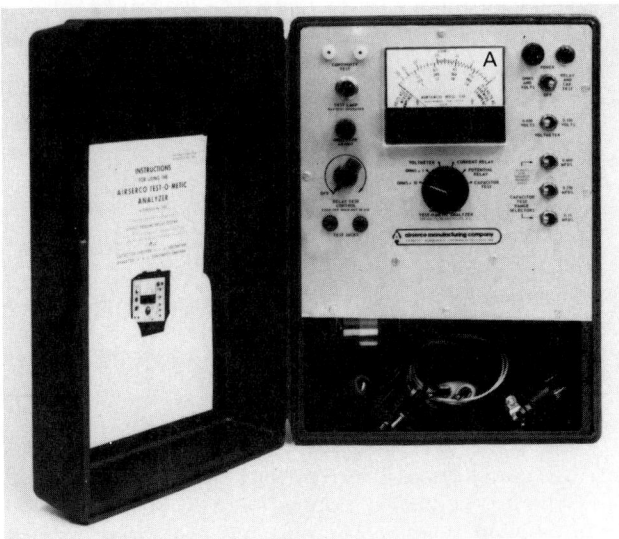

Fig. 11-116. Combination hermetic compressor analyzer and electrical tester. A—Dial has four scales: ohms, volts, microfarads, and amperes. (Airserco Mfg. Co., Inc.)

11-80 CYCLING TIME FOR REFRIGERATORS AND FREEZERS

Cycling time on home refrigerators and freezers cannot be stated very well in definite limits of time. This will vary depending on the amount of storage space being used, on the outside temperature of the box, and on the condition of the compressor.

Placing warm food in the cabinet to be frozen will also affect the cycling time. The condensing unit may run about one-third of the time. In other words, it may run five minutes and be off 10 minutes or it may run an hour and be off two hours. The important point is that any unusual change in cycling time should be investigated immediately. It may indicate that trouble is developing in the system.

11-81 SHUTTING DOWN REFRIGERATORS AND FREEZERS

When shutting down a refrigerator-freezer, special precautions should be taken to prevent rusting and to remove odor. After the electrical plug has been removed or the current shut off by a switch, allow several hours for the unit to completely defrost.

When the defrosting is complete, remove water and wash the inside of the cabinet with a solution of baking soda and water. Thoroughly dry the inside of the box. A portable heater set inside the cabinet will speed up this step. Leave the doors or lids ajar slightly to allow circulation of air during the shut down period.

CAUTION: There is a federal law in effect which states that the cabinet door must be removed from a refrigerator or freezer that is being taken out of service or is being disposed of for any reason. This law was passed because many children have been suffocated when attempting to hide in or play in an unused or carelessly discarded refrigerator or freezer. Therefore, always remove the cabinet door when taking a refrigerator or freezer out of service.

11-82 SERVICING PRECAUTIONS

Pinching lines is a practice to be used in cases of emergency only. Some service technicians follow this practice needlessly and it leads to future trouble.

One must remember that when refrigerant is added to a system, some of the oil in the system will be dissolved in the refrigerant. If the unit becomes noisy soon after the refrigerant has been added, oil should be added.

Electric motors, when installed, should have clean and tight electrical connections.

11-83 REVIEW OF SAFETY

Working safely implies three things:
1. Safety to the service technician.
2. Safe handling of tools, instruments, and equipment.
3. Proper preservation of food so that it is kept in a safe condition.

So far as the technician is concerned, there are not many great hazards in the refrigeration service work. The following items are some of the more common hazards which repair or service personnel should always recognize.

Good housekeeping is very important. Keep the work area clean. Keep oil and water off the floor.

Since the refrigerator is usually motor-driven, electrical supply to the motor and the controls presents some hazard. If not properly insulated and handled, it is possible for the operator to receive a dangerous electrical shock.

Always disconnect the electrical circuit or make sure all electrical devices are safe before starting on a job. An electrical short across a ring or wristwatch can cause a severe burn. It is best to remove rings and wristwatches when working on electrical equipment.

Many electrical shocks occur when the service technician comes in contact with an electrical current and a ground. Avoid working on any electrical circuit if standing on a damp floor or if one hand is touching a water pipe.

In some refrigerating systems, the condensing pressures run quite high. If the condenser is not losing enough radiating heat or if there is a restriction because of dust or lint, head pressure becomes high. This can become dangerous if it reaches the bursting point. The service technician should watch the gauges and shut off the system if the pressure is too high.

A hot compressor, exhaust line, or condenser can burn the operator.

When heating a charging cylinder or compressor with an electrical insert or heat gun (to speed up vaporization for a better purging result), always use a pressure control relief valve and thermostat to provide the proper temperature and pressure safety controls. (Refer to Para. 11-47.)

Freezing of skin is a hazard when handling some refrigerants. If the liquid refrigerant spills on skin, rapid evaporation may lower the skin temperature to considerably below the freezing temperature. Liquid refrigerant on the face or eyes is very critical. *Always wear goggles or a head shield when handling liquid refrigerants.*

Dropping heavy objects on feet or toes is another potential hazard. This can be avoided by using proper trucking and hoisting equipment. The operator should wear safety shoes with metal tips to protect toes.

Back injuries may be caused by attempting to lift heavy objects and by not using arm and leg muscles correctly.

Fire is always a potential hazard. Never use gasoline or any other flammable material when cleaning.

Use face shield and rubber gloves when handling oil which may be acidic. (This may be the case in hermetic systems where motor compressor has burned out.) *If burned with acidic oil, wash with water, apply ice, and see a physician right away.*

There is some danger in using common tools such as screwdrivers and wrenches. If they should slip, knuckles may be skinned. Be careful to handle tools correctly.

Vacuum pumps expose a hazard in that the rotating shaft, fans, belts, and the like, may catch on the clothing; or, if a hand is drawn into a pulley, a severe injury may result.

General safety precautions include the following:
1. Wear goggles when working on a charged system. When using a flame, be sure there are no flammables nearby. Be sure the pressure in the system is closed to atmospheric pressure before opening a system.
2. Each time an operation is planned, train yourself to think, ''Is there a possible safety hazard in the operation?'' Then take the care needed to reduce the hazard.
3. Never, under any circumstances, use carbon tetrachloride for cleaning. Its use is illegal. Remember, its effects are cumulative in the human body and its continued use may be fatal.
4. Avoid testing with excessive pressures. Test systems at 150 psig (1 100 kPa) maximum and wear goggles when testing.
5. Discharge refrigerants in a well-ventilated area.

11-84 TEST YOUR KNOWLEDGE

1. Do all hermetic systems have service valves?
2. Which refrigerants are in most common use in hermetic systems being built today?
3. What causes tubing to create noise?
4. How may a motor compressor dome be opened?

5. Describe a good way to clean the outside of the mechanism.
6. How should the hermetic mechanism be prepared for removal?
7. What is the most popular method of sealing tubing joints when servicing hermetic systems?
8. What trouble may be indicated by a warm evaporator line?
9. A frosted suction line with an excessive low-side pressure is the indication of what trouble?
10. To what part of the system is the compound gauge usually connected?
11. To what part of the system is the high-pressure gauge usually connected?
12. Identify three problems which indicate that a hermetic unit needs refrigerant.
13. Describe two tests which can be used to indicate proper receptacle grounding.
14. Describe one method of adding oil to a system.
15. What is a micron?
16. How many microns are in a millimeter?
17. How does a process tube adaptor fasten to a process tube?
18. At what temperature does water evaporate at a 28 in. Hg vacuum?
19. What is used to prevent a capillary tube from becoming blocked or plugged?
20. Why is the oil in a burned out motor compressor dangerous for a service technician to handle?
21. What happens to a system if a capillary tube of the same ID, but longer, is installed?
22. What type pump is needed to produce a high vacuum?
23. How can a stuck hermetic compressor be started?
24. Where are piercing valves installed?
25. What indicates an electrical failure in a door mullion heater, and how is the heater checked?
26. How are the service lines and manifold cleaned of air and moisture?
27. When should a service technician install a new liquid line filter-drier?
28. If the low-side pressure reads 28 in. Hg vacuum, what is wrong?
29. What is the most sensitive leak detector?
30. If the suction line is frosted, what could the trouble be?
31. How can a technician find out if the capillary tube refrigerant control is clogged?
32. What is an indication of a hot gas valve that is stuck in the closed position?
33. Is it necessary to check the service lines and gauge manifold for leaks?
34. Why is the valve core removed when evacuating a system?
35. Is the refrigerant charged into the system in the liquid form or gas form?
36. Grouped in pairs, what are the three steps of the triple evacuation method?
37. What is done to a brazed joint before it is heated and pulled apart?
38. When is a suction line filter-drier installed?
39. What kind of oil is usually found in a motor compressor burnout?
40. What kind of gauge is used to read a 25 micron pressure?

Chapter 12

COMMERCIAL SYSTEMS

After studying this chapter, the technician will be able to:
☐ Explain the differences between the mechanism of commercial refrigeration systems and domestic systems.
☐ Compare the differences between various commercial mechanisms.
☐ Describe how each mechanism (condenser, evaporator, and compressor) operates.
☐ Discuss the theory and operation of control devices.

12-1 CONSTRUCTION OF REFRIGERATING MECHANISMS

Operating fundamentals of domestic refrigeration systems also apply to commercial systems, Fig. 12-1. But many commercial systems using mechanical cycle mechanisms differ in some way from the domestic mechanism. These differences are chiefly in the following:
1. The number of evaporators connected to a single condensing unit.
2. Compressor design and size.
3. Condenser unit design and size.
4. Motor controls, both temperature and pressure.
5. Refrigerant controls, both liquid and vapor.
6. Piping.
7. Variety of evaporator designs.
8. Defrosting systems.
9. Variety of refrigerants used.

12-2 MECHANICAL CYCLE

Some large commercial compressor systems are semihermetic. These, too, can be serviced. Many single unit applications, such as bottle coolers, beverage dispensers, and ice cream cabinets use a completely hermetic system. Parts of the system are similar to the designs shown in Chapter 10. The design of the cabinet varies with its application.

In some cases, the cycles are more complicated. This is because they may include unloading and automatic defrost devices, multiple evaporators, additional controls, and more complicated piping.

12-3 COMPLETE MECHANICAL MECHANISM

Fig. 12-2 shows a single-unit mechanism. It includes:
1. High-pressure side:
   a. Compressor, usually hermetic.
   b. Condenser, usually air-cooled.
   c. Liquid receiver, when a thermostatic expansion valve or automatic expansion valve is used.

d. High-pressure safety motor control.
   e. Liquid line with drier and sight glass.
The refrigerant control is at the division point between the low side and the high side of the system. It will consist of an automatic thermostatic expansion valve or capillary tube.
2. Low-pressure side:
   a. Evaporator.
   b. Low-pressure or temperature motor control.
   c. Suction line—some with filter-driers and surge tanks.
The multiple mechanism is shown in Fig. 12-3. It includes:
1. High-pressure side:
   a. Compressor, often with an oil separator.
   b. Condenser, water- or air-cooled.
   c. Liquid receiver.
   d. High-pressure motor control.
   e. Liquid lines with a drier and a sight glass.
   f. Water valve—used with a water-cooled unit.
The refrigerant control is the division point between the high-pressure side and the low-pressure side.
2. Low-pressure side:
   a. Refrigerant controls, two or more, usually of the thermostatic expansion valve type.
   b. Evaporators, two or more. These may be any of several types: natural convection, forced convection, or submerged.
   c. Motor control, usually of the pressure type.
   d. Suction lines with drier and suction pressure regulator.
   e. Two-temperature valves for multiple temperature installation.
   f. Surge tanks for reducing rapid pressure changes.
   g. Check valves for multiple temperature installations.
There are many varieties of commercial systems. A water chiller system is shown in Fig. 12-4.

Subcooling is used on low-temperature units such as display cases, freezers, etc. The process of sub-cooling reduces the refrigerant temperature in the liquid line below the saturated temperature. The lower the temperature in the liquid line, the greater the system's heat removal capacity. This will result in a more efficient system.

Subcooling is accomplished by refrigerating the liquid line on a low-temperature system using a higher-temperature system, such as an air conditioning unit. High-temperature systems remove Btus three times more efficiently than low-temperature refrigeration systems. Two systems (the high-temperature air conditioning and the low-temperature

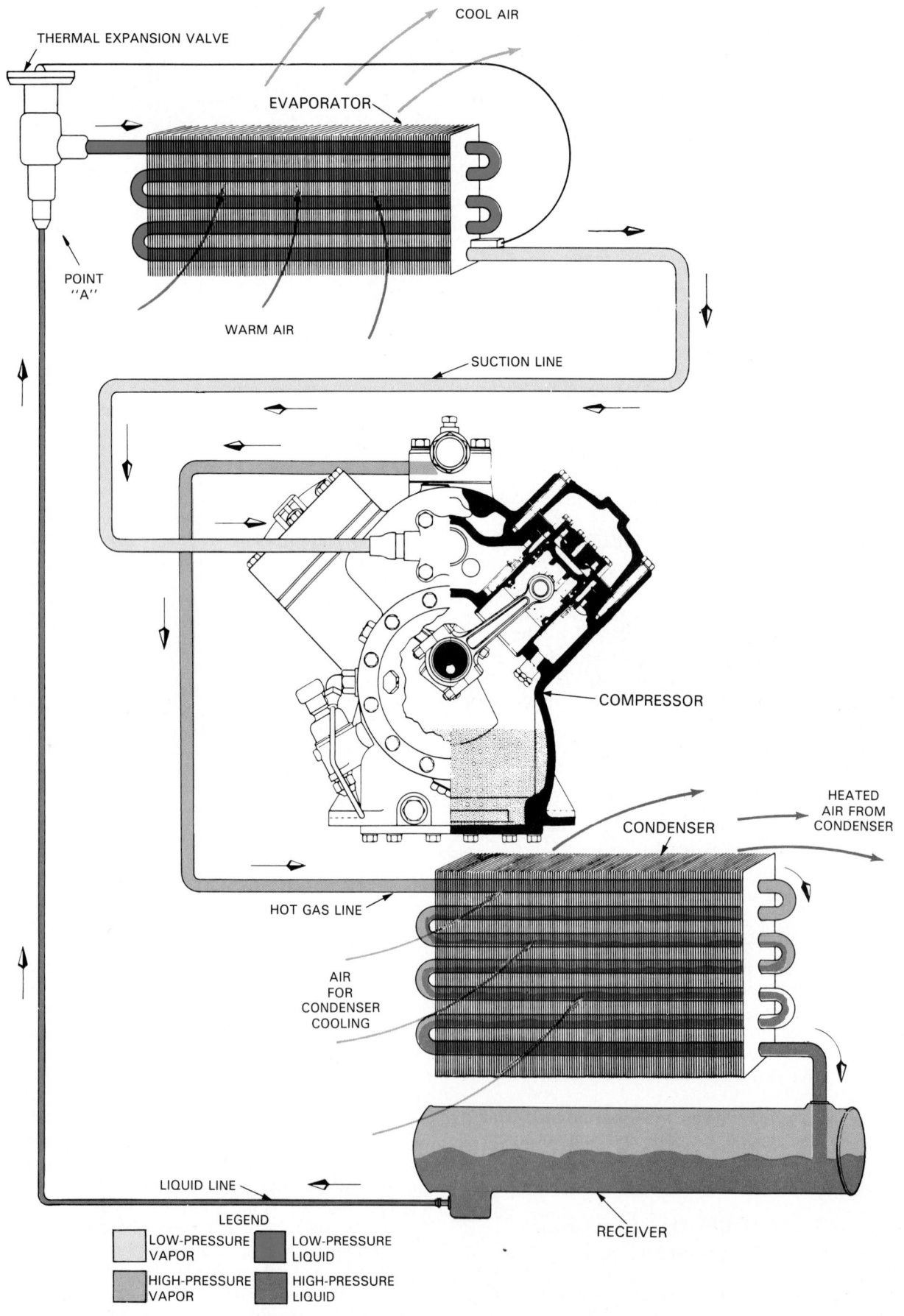

THERMAL EXPANSION VALVE

COOL AIR

EVAPORATOR

POINT "A"

WARM AIR

SUCTION LINE

COMPRESSOR

HEATED AIR FROM CONDENSER

CONDENSER

HOT GAS LINE

AIR FOR CONDENSER COOLING

LIQUID LINE

RECEIVER

LEGEND

☐	LOW-PRESSURE VAPOR	■	LOW-PRESSURE LIQUID
▨	HIGH-PRESSURE VAPOR	■	HIGH-PRESSURE LIQUID

Fig. 12-1. A serviceable commercial system with air-cooled condenser, thermostatic expansion valve and V-type compressor. Note size of compressor. (Carrier Corp., Subsidiary of United Technologies Corp.)

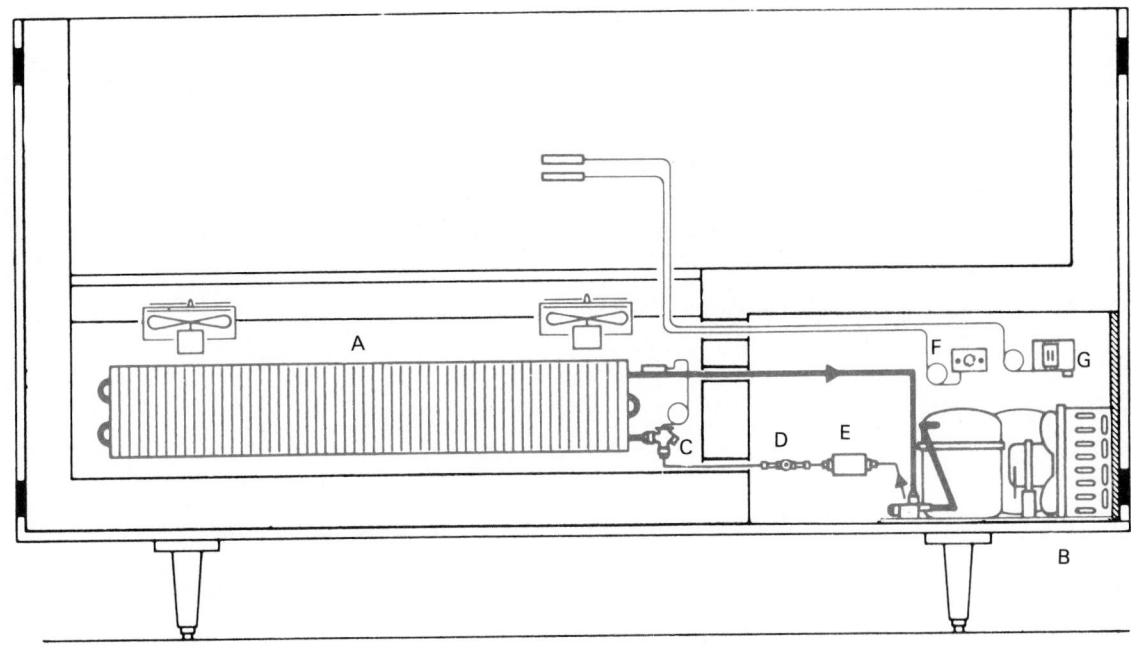

Fig. 12-2. This open top display cabinet has a self-contained refrigeration unit.
A—Evaporator. B—Condensing unit. C—Thermostatic expansion valve. D—Sight glass.
E—Drier. F—Signal thermostat. G—Cycling thermostat. (Danfoss, Inc.)

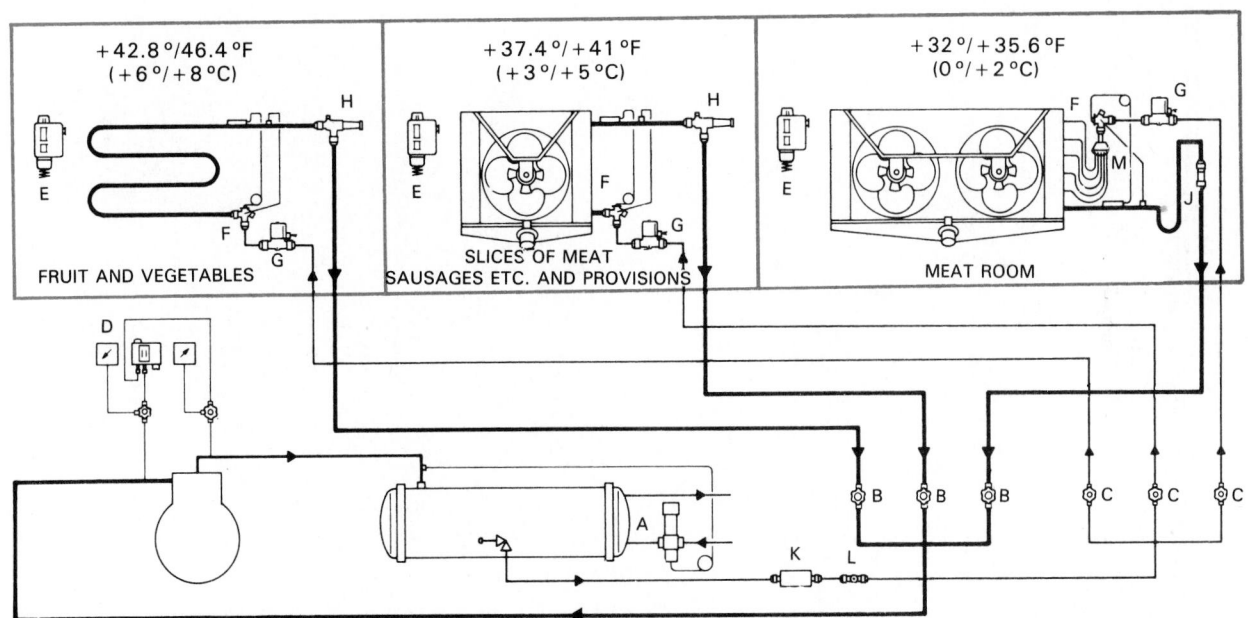

Fig. 12-3. A multiple evaporator system. A—Water valve. B—Suction line shutoff valves. C—Liquid line shutoff valve. D—Low- and high-pressure motor control. E—Thermostats. F—Thermostatic expansion valves. G—Liquid line solenoid valves. H—Two-temperature valves. J—Check valve. K—Drier. L—Sight glass. M—Distributor.

freezer cases) working together, increase the over-all efficiency of the total refrigeration process.

Subcooling systems can also be added to existing refrigeration systems in supermarkets. Fig. 12-5 shows a subcooler installation. This is done by taking the subcooler and installing it into the store's present high-temperature air conditioning system.

Condensing units are normally mounted on a steel base. In the external drive unit, the motor is mounted outside the compressor and drives the compressor either directly or with one or more belts as in Fig. 12-6.

In the hermetic unit, the motor is connected directly to the compressor. Fig. 12-7 shows a 25 hp condensing unit which has a 2 1/8-in. sweat type liquid line. It is designed for low-temperature work and has a 40,600 Btu/hr. capacity at −52 °F (−47 °C).

For large commercial installations, compressors are made with three, four, five, six, seven, or more cylinders.

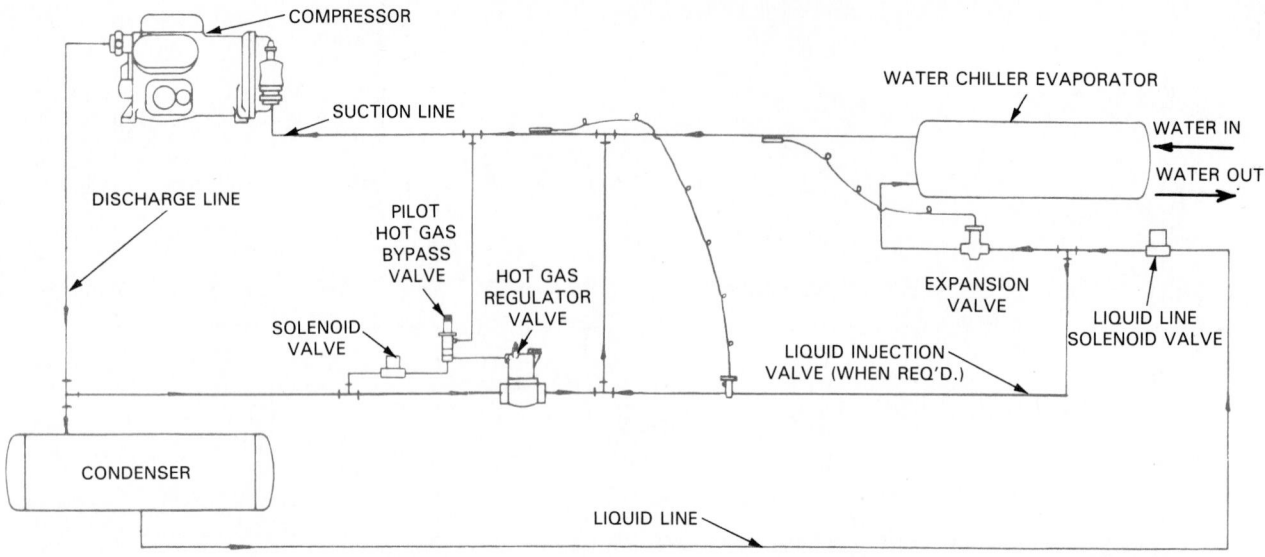

Fig. 12-4. Schematic of a water chiller. Hot gas bypass keeps low-side pressure high enough to prevent freezing of chilled water. Liquid injection keeps suction vapor cool enough to prevent overheating of compressor motor.

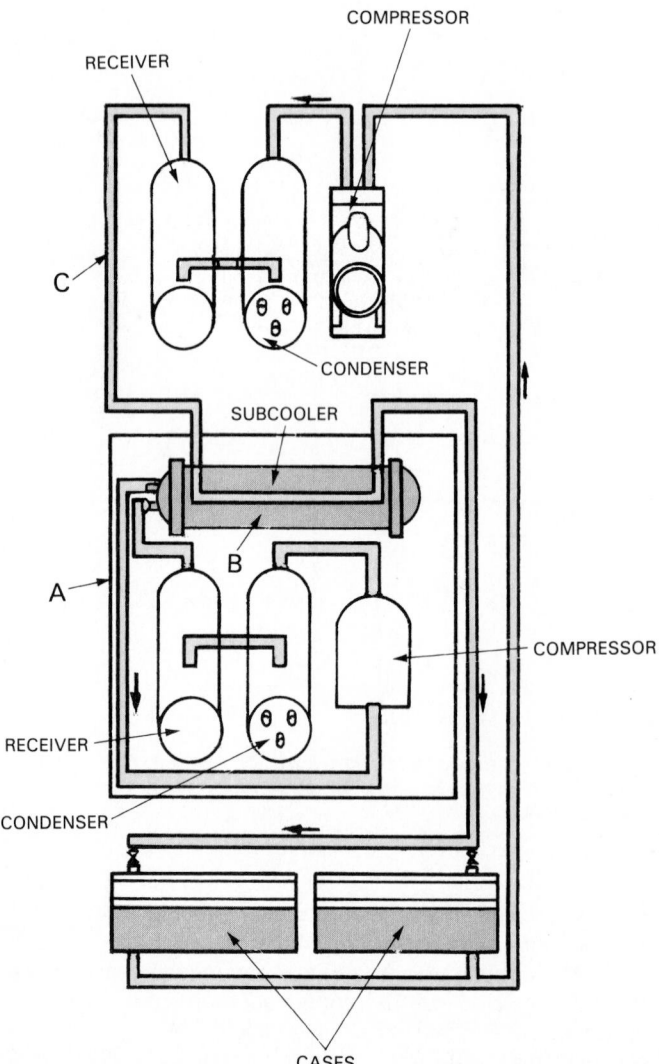

Fig. 12-5. This subcooling installation (see A) uses existing air conditioning system. Subcooler at B was installed to cool liquid lines, C, leading to the two display cases. (Standard Refrigeration Co.)

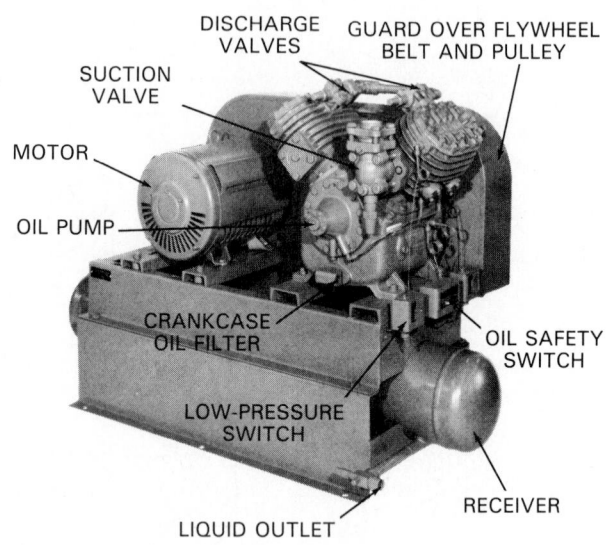

Fig. 12-6. Condensing unit with belt-driven, four-cylinder, air-cooled compressor. Unit is mounted on steel base. (Grasso, Inc.)

Compressors are also named after their cylinder arrangement. Names such as vertical single, horizontal single, 45 deg. single (inclined), vertical two cylinder, V-type two cylinder, W-type three cylinder, radial three cylinder, vertical four cylinder and V-type four cylinder, have all been used.

There are also many crankshaft arrangements. One type of serviceable hermetic motor compressor has an eccentric type crankshaft as shown in Fig. 12-8. The inside construction of a multiple-cylinder serviceable hermetic motor compressor is shown in Fig. 12-9.

To help cool the compressor and motor of a serviceable motor compressor, one company uses a fan and motor as in Fig. 12-10. This manufacturer can convert the motor compressor easily into an external drive compressor as illustrated in Fig. 12-11.

DISCHARGE LINE
MUFFLER

COMPRESSOR DISCHARGE
SHUTOFF VALVE

3/8 IN.
LIQUID
LINE TO
LIQUID
INJECTION
VALVE

LIQUID
INJECTION
VALVE

COOLER

HOT GAS BYPASS
TUBE CONNECTION
TO DISCHARGE LINE

HOT GAS
BYPASS
VALVE

SOLENOID
COIL OF
BYPASS
VALVE

HOT GAS BYPASS
ENTERS SUCTION
LINE HERE

Fig. 12-7. Large commercial chiller. Note hot gas bypass system.

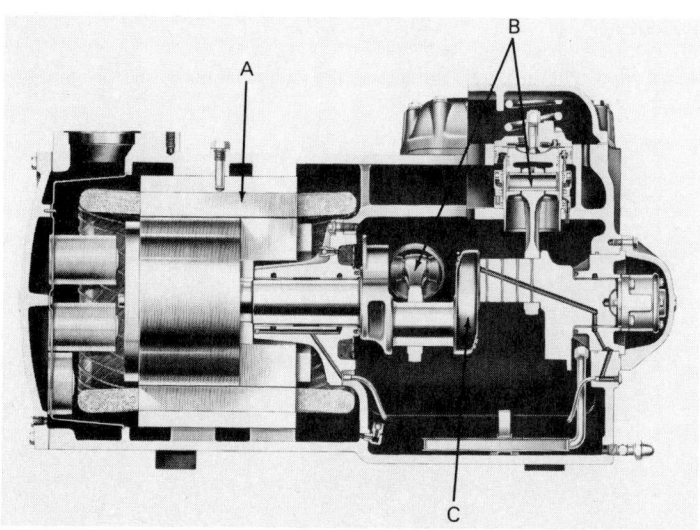

Fig. 12-8. Cutaway shows accessible four-cylinder V-type hermetic
motor compressor. A—Motor. B—Piston. C—Crankshaft.
(The Trane Co.)

A six-cylinder compressor with an external drive motor is shown in Fig. 12-12. It has a crank throw type crankshaft. A steel frame is used to hold the shell and tube condenser on the hermetic unit shown in Fig. 12-13.

12-4 COMMERCIAL HERMETIC UNITS

Several companies now produce hermetic units of 20 or more horsepower. Some of the units are the bolted

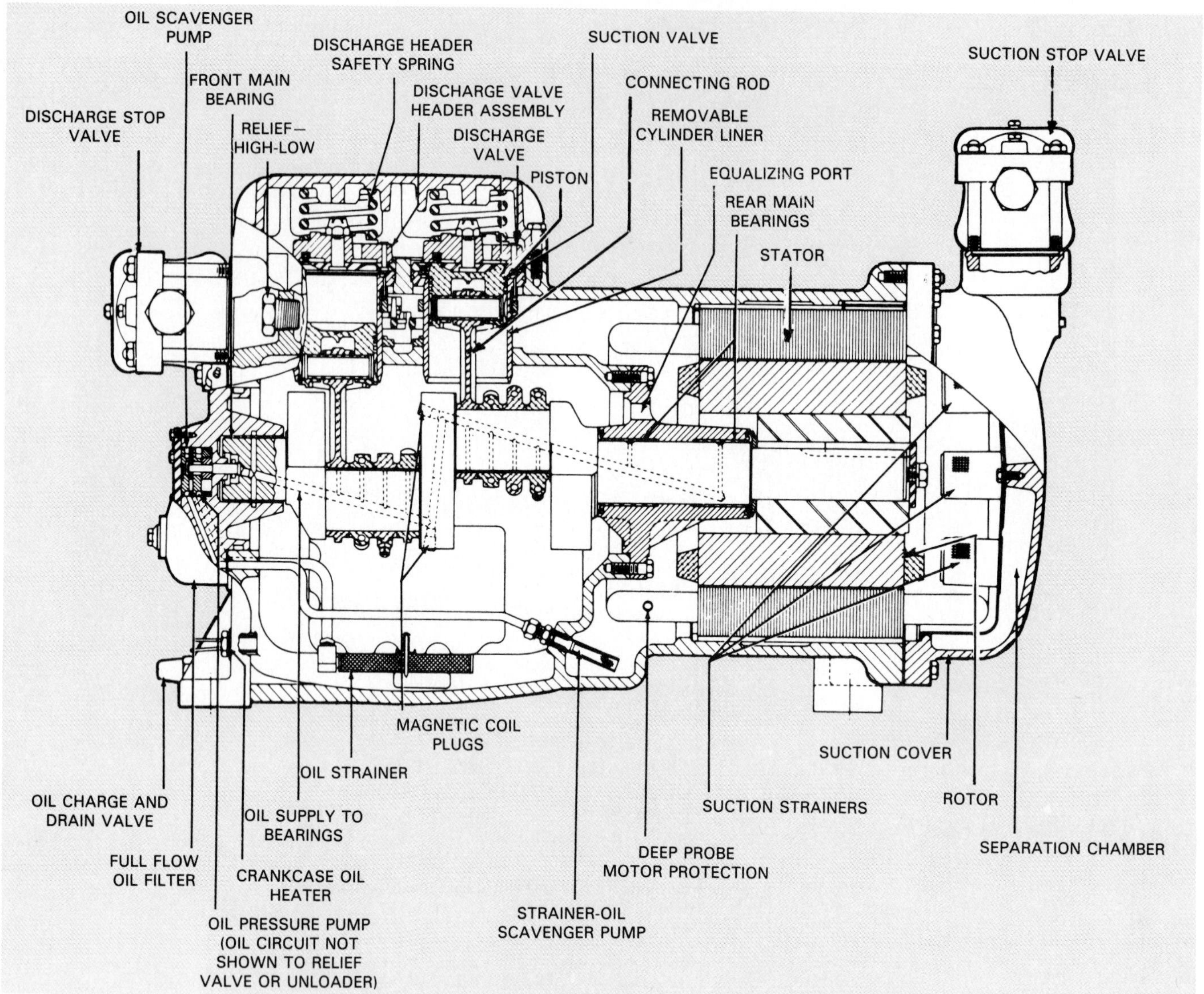

OIL SCAVENGER
PUMP

DISCHARGE HEADER
SAFETY SPRING

SUCTION VALVE

SUCTION STOP VALVE

DISCHARGE STOP
VALVE

FRONT MAIN
BEARING

CONNECTING ROD

DISCHARGE VALVE
HEADER ASSEMBLY

REMOVABLE
CYLINDER LINER

RELIEF—
HIGH-LOW

DISCHARGE
VALVE

EQUALIZING PORT

REAR MAIN
BEARINGS

PISTON

STATOR

MAGNETIC COIL
PLUGS

OIL STRAINER

SUCTION COVER

OIL CHARGE AND
DRAIN VALVE

SUCTION STRAINERS

ROTOR

OIL SUPPLY TO
BEARINGS

FULL FLOW
OIL FILTER

SEPARATION CHAMBER

CRANKCASE OIL
HEATER

DEEP PROBE
MOTOR PROTECTION

OIL PRESSURE PUMP
(OIL CIRCUIT NOT
SHOWN TO RELIEF
VALVE OR UNLOADER)

STRAINER-OIL
SCAVENGER PUMP

Fig. 12-9. Bolted type field serviceable eight-cylinder hermetic motor compressor.

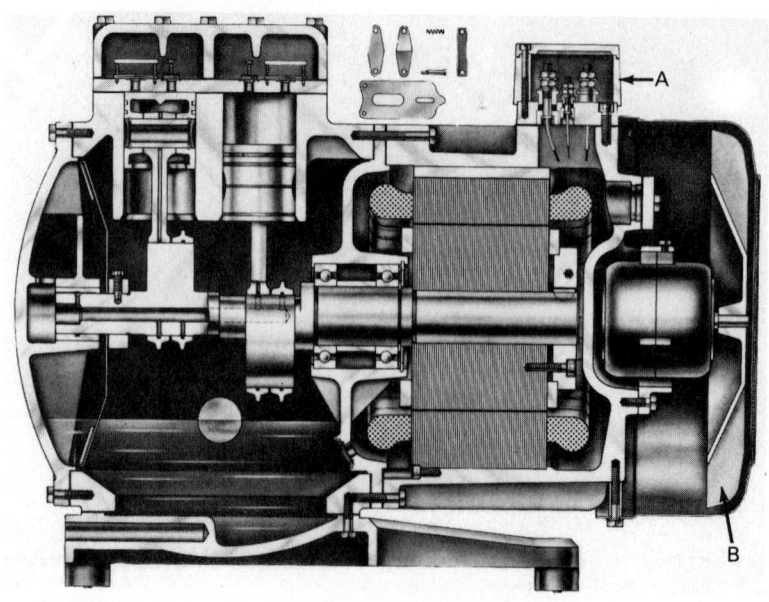

Fig. 12-10. Accessible hermetic motor compressor with four-cylinder V-type compressor. Note oil level sight glass in center of crankcase. A—Electrical terminals. B—External fan and motor force cooling air over unit.

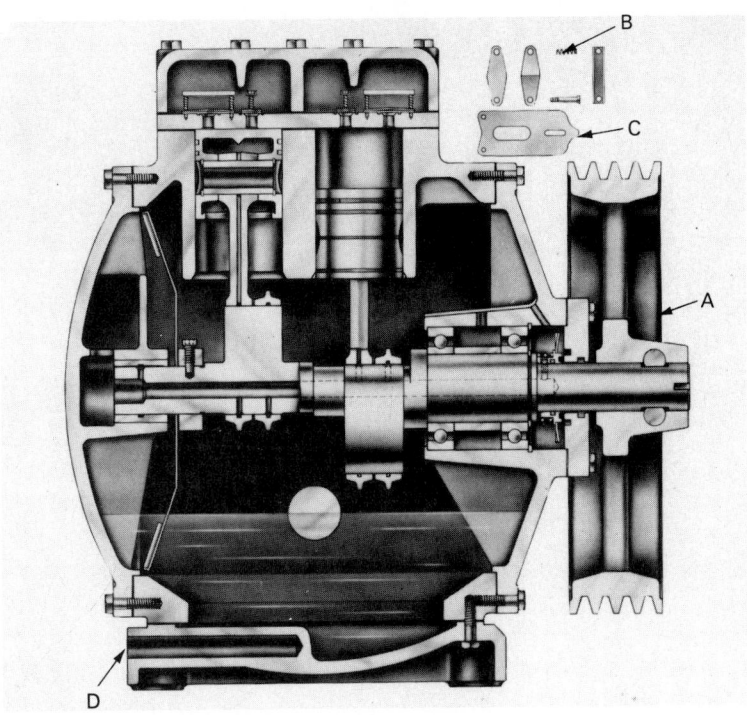

Fig. 12-11. Belt-driven external drive compressor is V-type four-cylinder model and uses an eccentric type crankshaft. A—Four-belt flywheel. B—Exhaust valve parts. C—Intake valve part. D—Opening for electric oil heater.

A

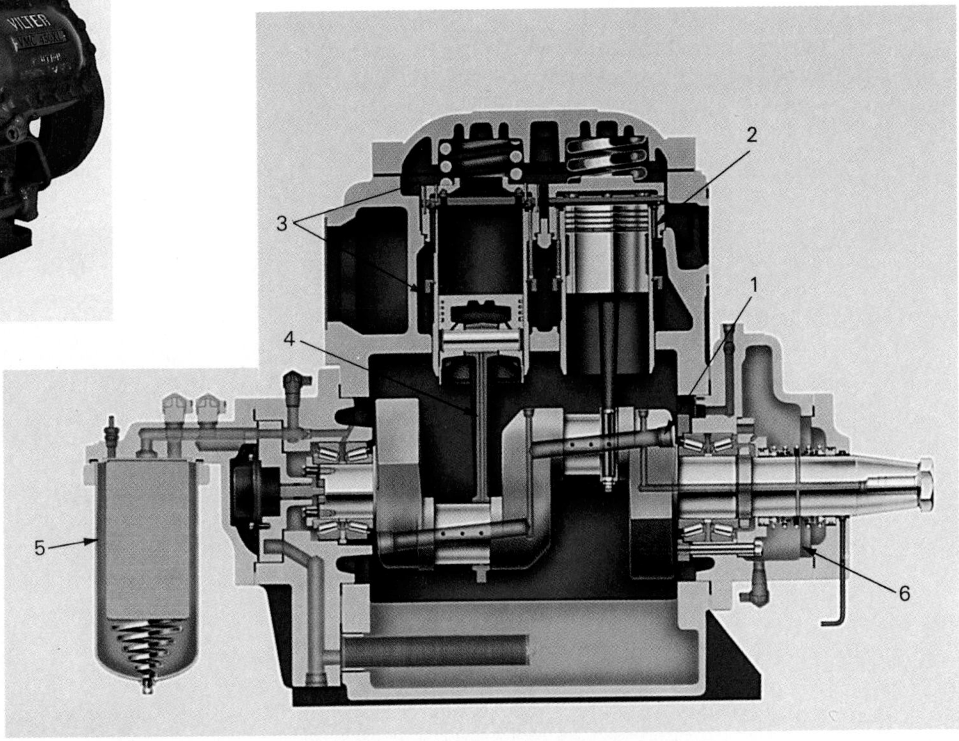

B

Fig. 12-12. Serviceable six-cylinder W-type compressor designed for external belt-driven or direct-driven engine. A—External. B—Cutaway view: 1—Oil line plug. 2—Compressor piston and rings. 3—Suction and discharge ports. 4—Piston and connecting rod. 5—Oil filter. 6—Crankshaft seal. (Vilter Mfg. Corp.)

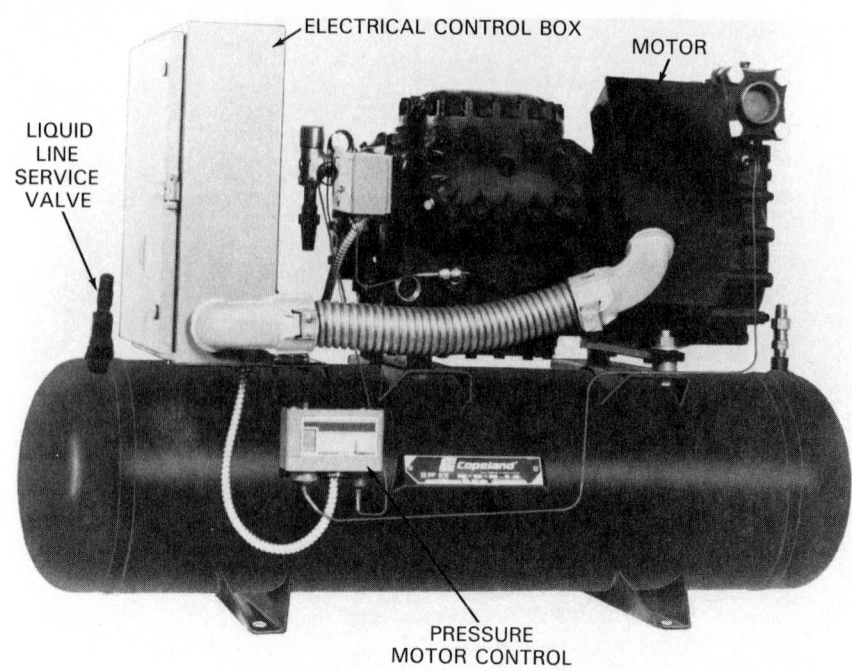

Fig. 12-13. Serviceable commercial hermetic condensing unit. Note the electrical panel. Wiring is enclosed in the flexible metal conduit because the unit is spring-mounted to allow movement. Note, also, the pressure motor control mounted on the side of the receiver. (Copeland Corp.)

assembly type. These are often called "field serviceable" or "accessible." Some units are sealed in a welded casing. Both types are equipped with service valves. They may be connected to any type of evaporator and used for many different applications.

An advantage of hermetics in the commercial field is the elimination of the crankshaft seals and belts. Because any trouble in the compressor mechanism involves both the compressor and the motor, the service technician working on these hermetic units must be very careful to keep moisture and dirt out of the system.

An outdoor hermetically sealed air-cooled condensing unit with a fan condenser, shroud, and service valve is shown in Fig. 12-14. The inside of a welded hermetic motor compressor is shown in Fig. 12-15. Smaller units have single-phase motors. Units over 1/2 hp generally have three-phase motors, Fig. 12-16.

The condensing units may be installed in many different ways. Some are mounted on the roof, others on the same floor level with the evaporator but in different rooms or

Fig. 12-14. Commercial hermetic compressor condensing unit with air-cooled condenser. (Dairy Equipment Co.)

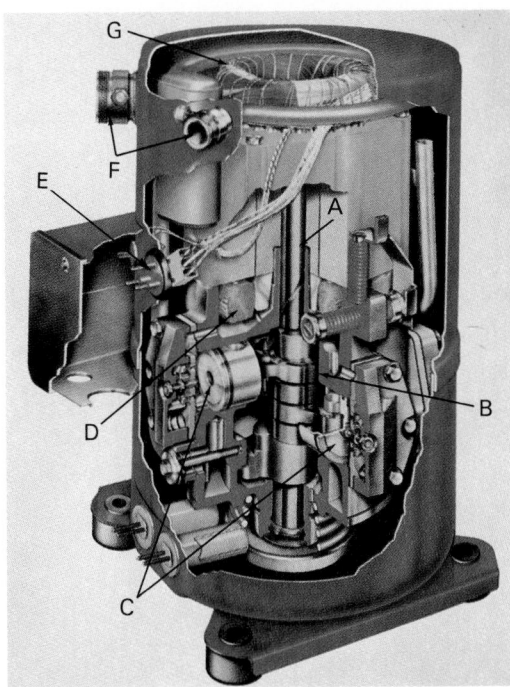

Fig. 12-15. Two-cylinder hermetic motor compressor. This type is used on air conditioning, heat pump, and commercial condensing units. A—Crankshaft. B—Connecting rod. C—Pistons. D—Motor windings. E—Electrical terminals. F—Suction and discharge openings. (Tecumseh Products Company)

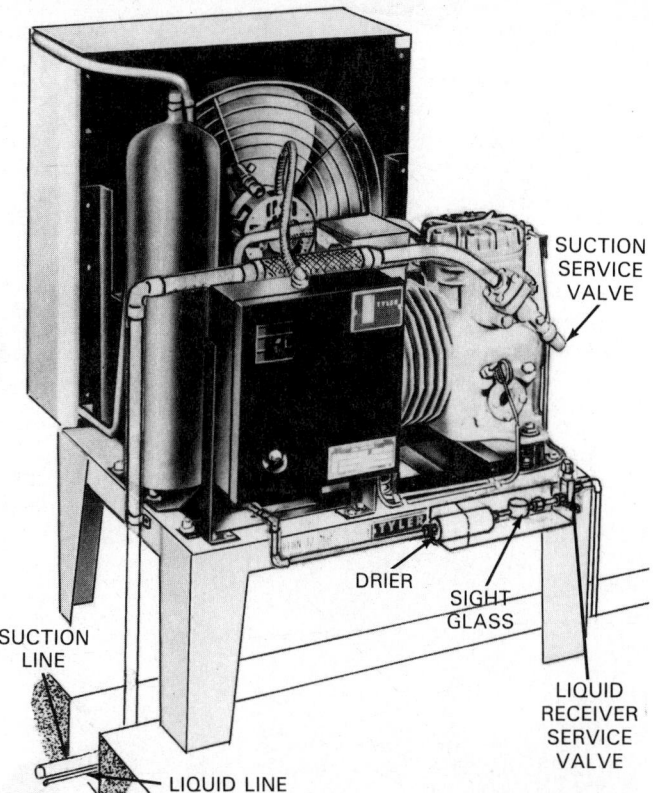

Fig. 12-17. Installed condensing unit. Note trough in floor for suction liquid lines. (Tyler Refrigeration Corp.)

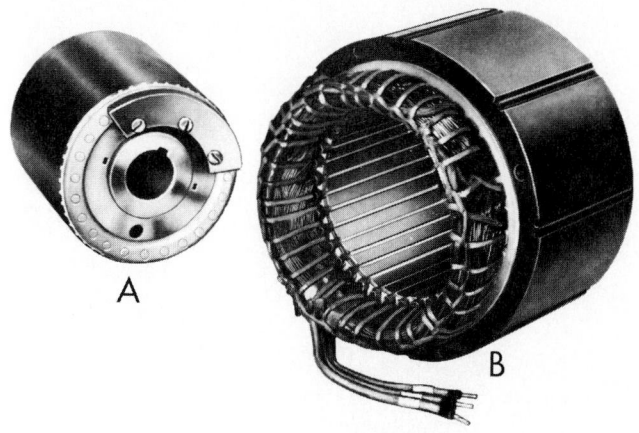

Fig. 12-16. Three-phase motor is used in larger hermetic refrigeration units. A—Rotor. B—Stator. (Emerson Electric Co.)

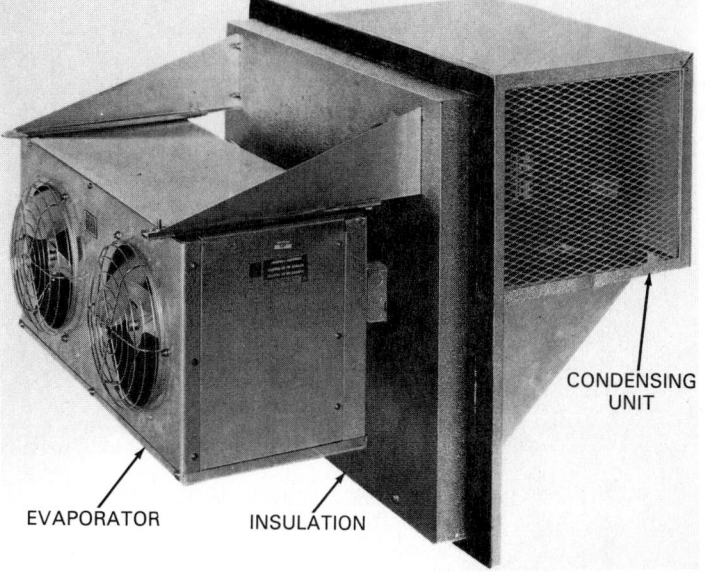

Fig. 12-18. A self-contained commercial refrigeration system. Units are available in 1/2 to 3 hp sizes. Low-temperature systems are equipped with defrosting devices. (American Panel Corp.)

outside the building. See Fig. 12-17. Note the suction service valve on the compressor, the liquid receiver service valve on the receiver, and the forced convection condenser.

A factory assembled condensing unit and evaporator combination can be installed as shown in Fig. 12-18. These straddle or plug-in units eliminate the need to put in piping between the condensing unit and the evaporator. The installation, therefore, consists of preparing the opening, mounting the unit, running the electrical lines, and opening the shutoff valves.

For large installations, two-motor compressor designs have been developed to provide greater refrigeration capacity. They are:

1. Tandem assembly motor compressors.
2. Parallel assembly motor compressors.

The tandem design connects two motor compressors together at the motor end. See Fig. 12-19. These units can be run separately for low load or together for full load. But, if one motor compressor fails, the complete system must be shut down during the service time.

Fig. 12-19. Tandem motor compressor assembly doubles refrigerating capacity of unit.   (Copeland Corporation)

Fig. 12-20. Twin parallel units with three condensing fans. System frequently used in convenience stores, specialty areas, frozen food applications, and delicatessens.   (Tyler Refrigeration Corporation)

A remote air-cooled air conditioning compressor unit is shown in Fig. 12-20.

Parallel design connects two or more units in parallel by piping, Fig. 12-21. The units also require a compressor oil piping system to make certain that all the compressors have the correct amount of oil in each crankcase while in operation.

12-5 OUTDOOR AIR-COOLED CONDENSING UNITS

To save space when air conditioning commercial buildings and homes, there is an increase in use of outdoor air-cooled

condensing units, Fig. 12-22. Air-cooled units save the cost of plumbing for water circuits and are also used where chemicals in the water make water cooling impractical. These units may be mounted on the roof, on the outside wall, or at ground level.

In such cases there are four major provisions:

1. *There must be a head pressure control if the unit is exposed to outdoor weather that may go below the operating cabinet temperature.*

2. *A method of preventing short cycling must be designed into the system.*

3. *A means must also be provided to prevent dilution of*

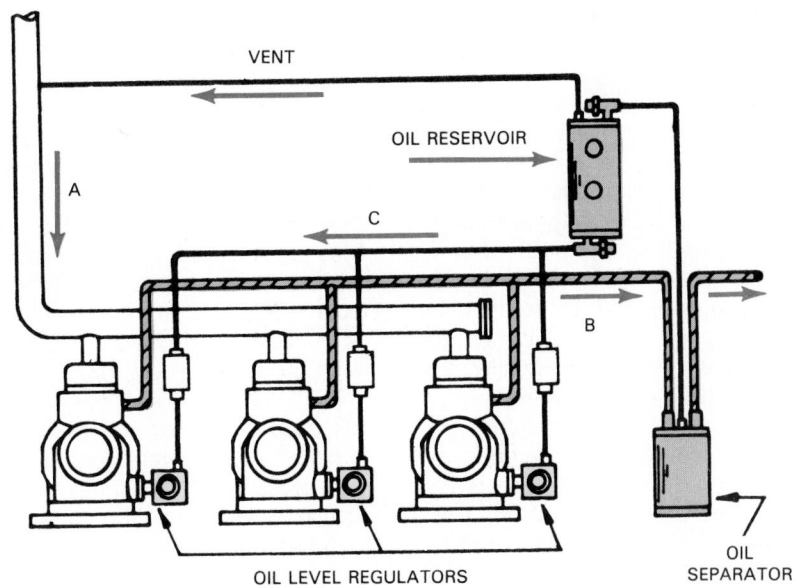

Fig. 12-21. Simple diagram of a parallel motor compressor assembly. Three motor compressors are used. Note oil reservoir and oil level regulators. Floats in these components keep oil in each compressor at the proper level. A—Suction piping. B—Discharge piping. C—Oil level piping. (AC&R Components, Inc.)

Fig. 12-22. Air-cooled condensing unit for outdoor installation (usually roof-mounted). Housing is in place. Air enters at coil in the rear and discharges through the grille in front. (BOHN Refrigeration Products, a Unit of Heatcraft, Inc.)

the compressor oil by liquid refrigerant.

4. *The completed condensing unit must be constructed and installed so it is virtually weatherproof.*

Low ambient temperatures will cause low head pressures. This pressure may drop so low it may even stop the flow of refrigerant. Four different methods will maintain pressure:

1. Partially fill the condenser with liquid refrigerant.
2. Stop or slow the condenser fans.
3. Partially or completely close the ambient air louvers.

4. Heat the condenser.

Outdoor units require about 1000 cfm (cubic feet per minute) of condenser air circulation per horsepower. They are less costly to operate than indoor air-cooled units.

A dual compressor model, Fig. 12-23, has two completely separate refrigeration systems, and they range from 6 to 35 tons and provide what is known as a partial standby system. The basic operation is that of a two-stage compressor, with each compressor activated individually from a two-stage space thermostat. It provides a two-stage cooling and a two-stage heating system, with automatic changeover.

The basic operation is as follows: System 1 turns on the first stage of the thermostat. If the load is light, the system carries the load and the compressor cycle is on and off at the call of the thermostat. An example of this would be in the cool morning, late evening, cloudy day with no sun, etc. This would cause the thermostat to signal for a compression cycle.

If System No. 1 is not adequately able to produce the function, the room thermostat, Fig. 12-24, automatically turns on the second compressor. It will then signal on and off to carry the rest of the load while Compressor No. 1 runs constantly, removing the moisture from the air. As the load begins to drop and the temperature is lowered in the facility, Compressor No. 2 will shut down, and Compressor No. 1 then cycles to carry the reduced load.

Fig. 12-25 illustrates an air-cooled packaged water chiller with a shell-and-tube type heat exchanger.

One of the main problems with outdoor units is keeping the thermostatic expansion valve operating at full capacity during cold weather. Capacity depends on the pressure difference across the valve. If condensing pressure for R-12 reduces from 102 psi, 90°F (32°C) to 56 psi, 30°F (−1°C), the valve capacity will drop. (Not enough liquid refrigerant will flow.) The refrigerated fixture temperatures may then rise too high. Also, a small pressure difference may cause short cycling of the condensing unit.

The condensing temperature and pressure may be kept at a proper operating level by a design change. The unit is made to nearly fill the condenser tubes with liquid. Just enough condensing surface is left to maintain the pressure.

Fig. 12-26 shows how a check valve and limiter valve are used to maintain a head pressure in the condenser even when the condenser operates in low outdoor temperatures. The pressure on the limitizer valve on the outlet of the condenser will not open the valve until the condensing pressure reaches the proper level. When pressure rises, the limiter valve will open and allow liquid to leave the condenser.

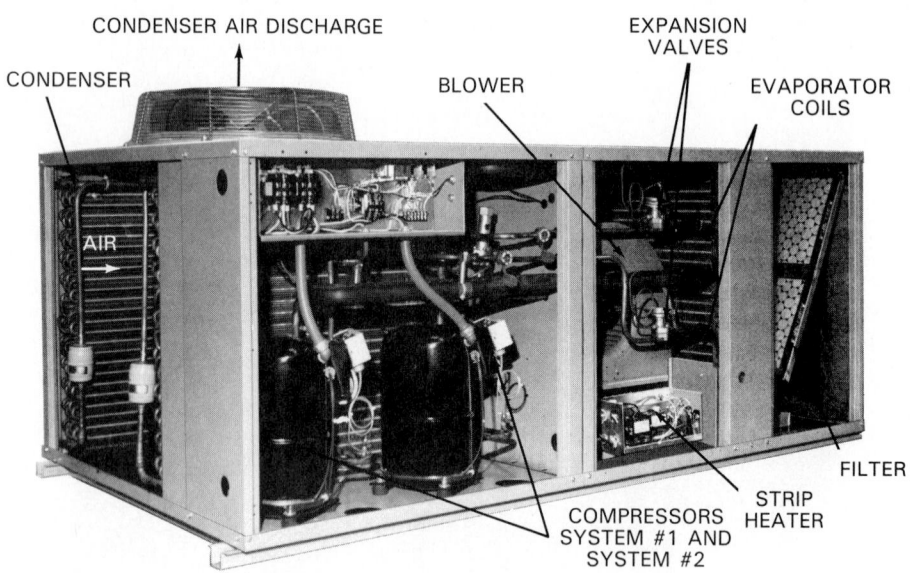

Fig. 12-23. Outdoor air-cooled condensing unit. Note twin compressors for Systems 1 and 2 and dual expansion valves. (Addison Products Company)

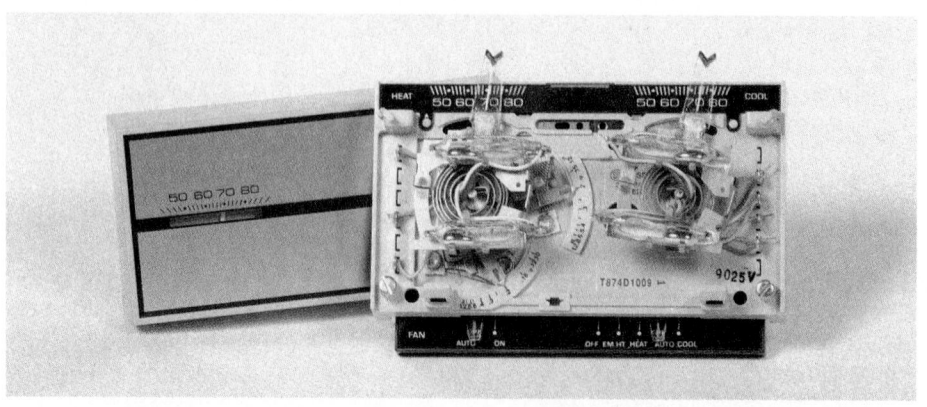

Fig. 12-24. Thermostat used with single or dual compressor air conditioner has two-stage cool, two-stage heat, automatic change-over, and dual setpoint. (Addison Products Company)

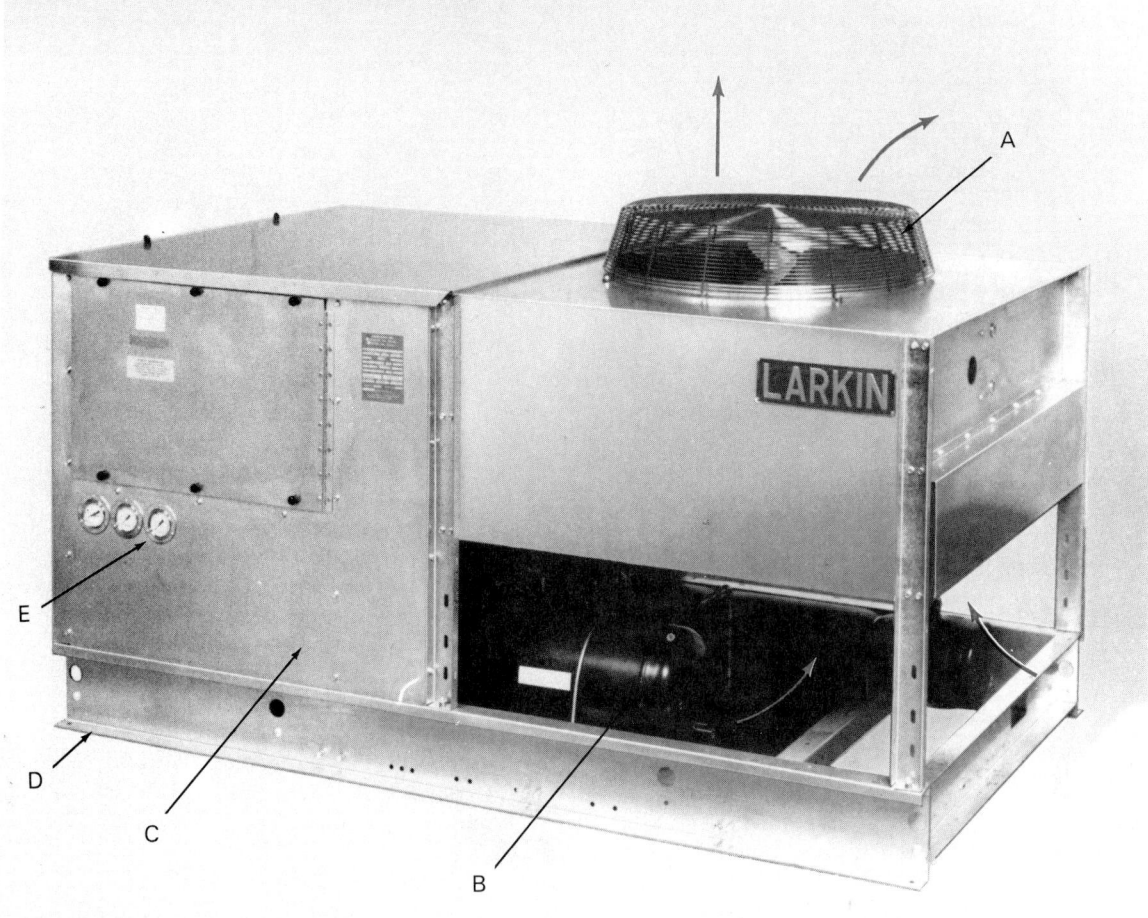

Fig. 12-25. Air-cooled packaged water chiller. A—Exhaust fan. B—Shell-and-tube heat exchanger. C—Semi-hermetic motor compressor. D—Heavy gauge steel frame. E—Suction, discharge, and oil pressure gauges. (Larkin, Inc.)

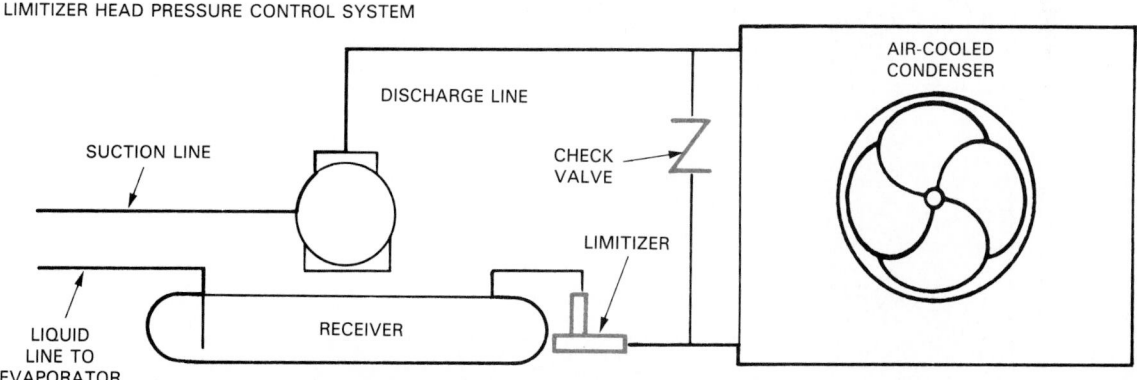

Fig. 12-26. Schematic of pressure control system on outdoor condensing system. The check valve and limiter valve ensure good condensing pressure during cold weather.

The installation specifications must be carefully checked. The receiver must hold enough liquid refrigerant to flood most of the condenser in the winter and still safely hold the refrigerant during the warm season.

Another way to keep the pressures up is to close the condenser housing airflow louvers as the pressure drops. A pressure-sensitive device does this. It is connected into the condenser tubing as shown in Fig. 12-27. This head

pressure device will move the rod out as pressures increase. The linkage will open the louvers, as Fig. 12-28 illustrates. As head pressures decrease, the rod will move back and start to close the louvers.

The condenser fans may either operate when louvers are closed or they may be shut off when the louvers near the closing point. A condensing unit with an ambient-temperature-controlled, adjustable damper is shown in Fig. 12-29.

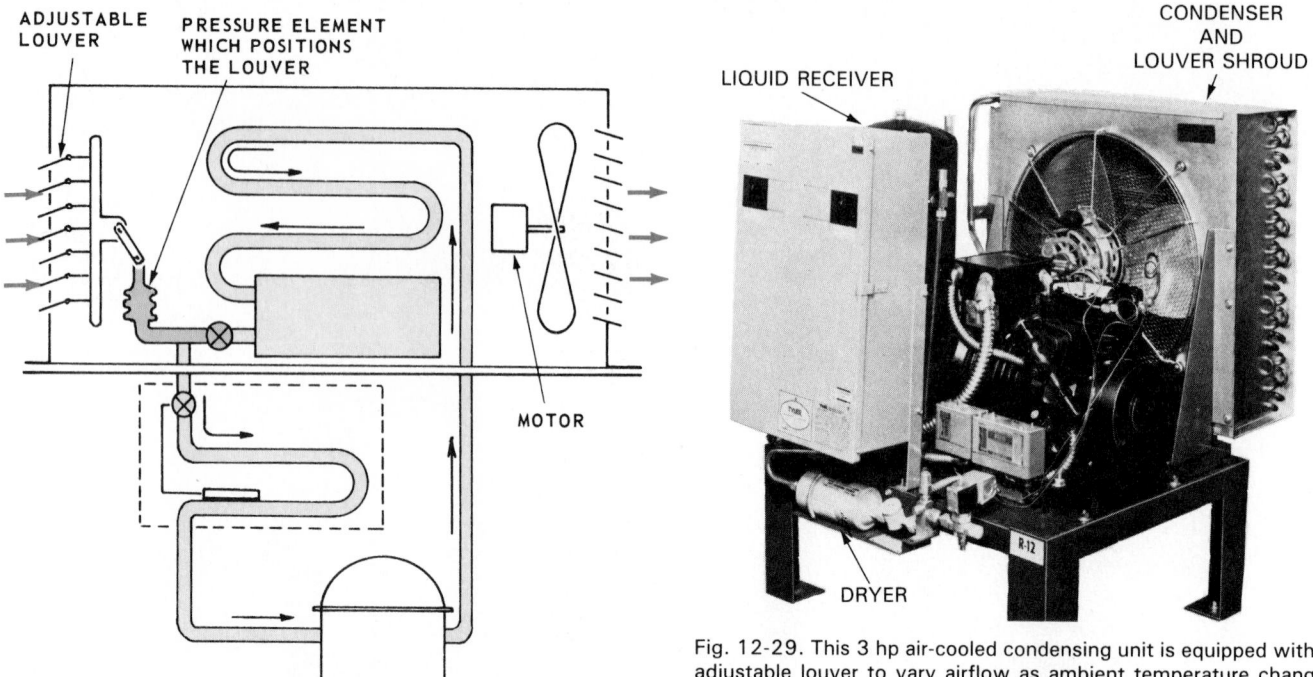

Fig. 12-27. An air-cooled condenser with pressure-operated louver. As condensing pressure decreases, louver will start to close, reducing condenser airflow.

Fig. 12-29. This 3 hp air-cooled condensing unit is equipped with an adjustable louver to vary airflow as ambient temperature changes. (Tyler Refrigeration Corp.)

speed when the head pressure drops by using electrically controlled modulated fan speeds. This system operates with

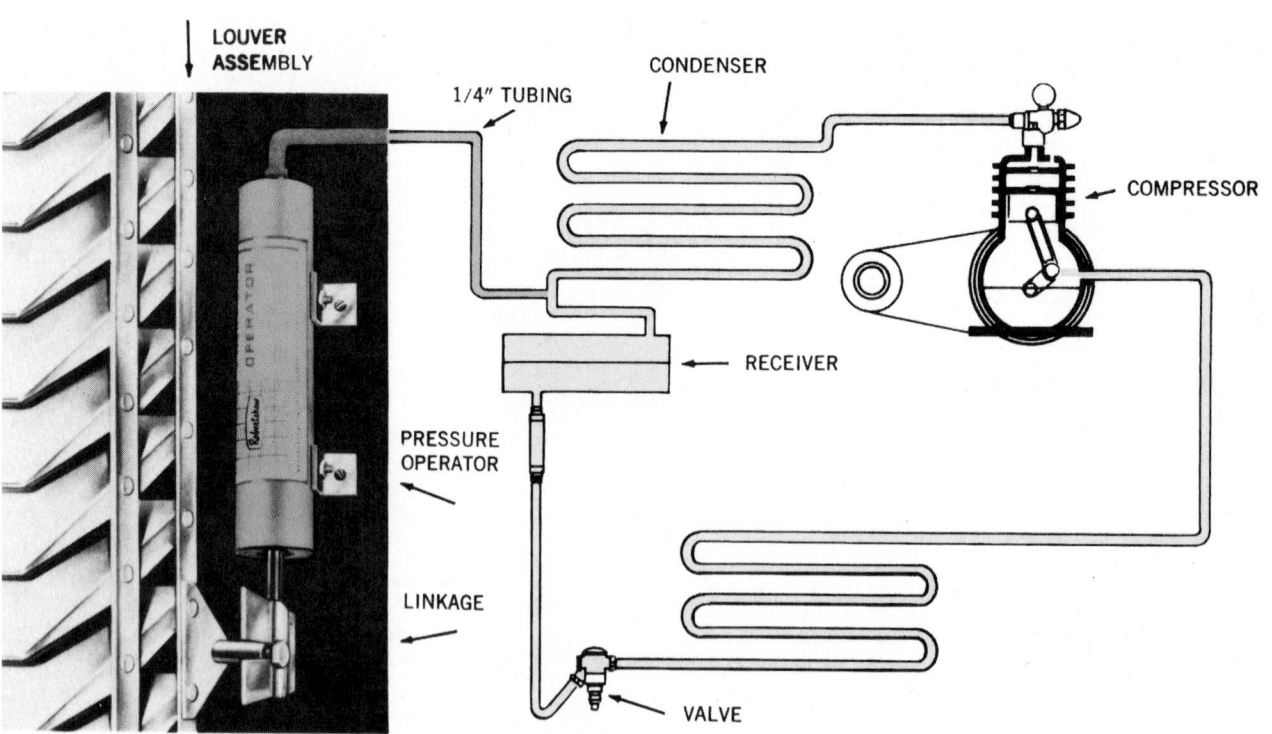

Fig. 12-28. Positioning cylinder (pressure operator) at left will react to pressure from the condenser. An increase in pressure moves cylinder and rod (connected to linkage) and causes louvers to open. (Robertshaw Controls Co.)

Other systems shut off the condenser fan when the condenser pressure falls to a minimum level, Fig. 12-30. Some shut off one or more of several fans. Others lower the fan

a thermistor type sensor on the condenser and a special fan motor. See Fig. 12-31.

To keep the receiver temperature warmer than the

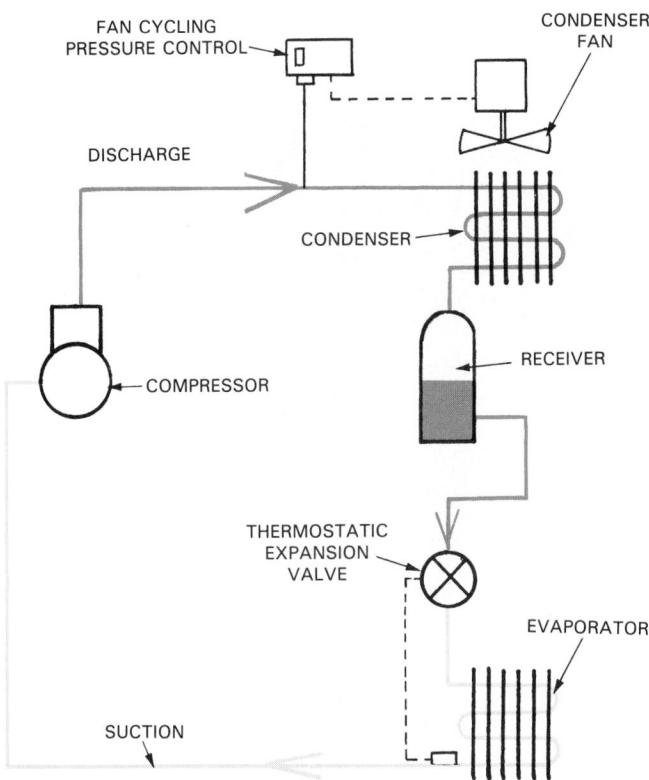

Fig. 12-30. This schematic shows refrigeration cycle with a fan-cycling pressure control which senses condenser pressure. (Ranco North America)

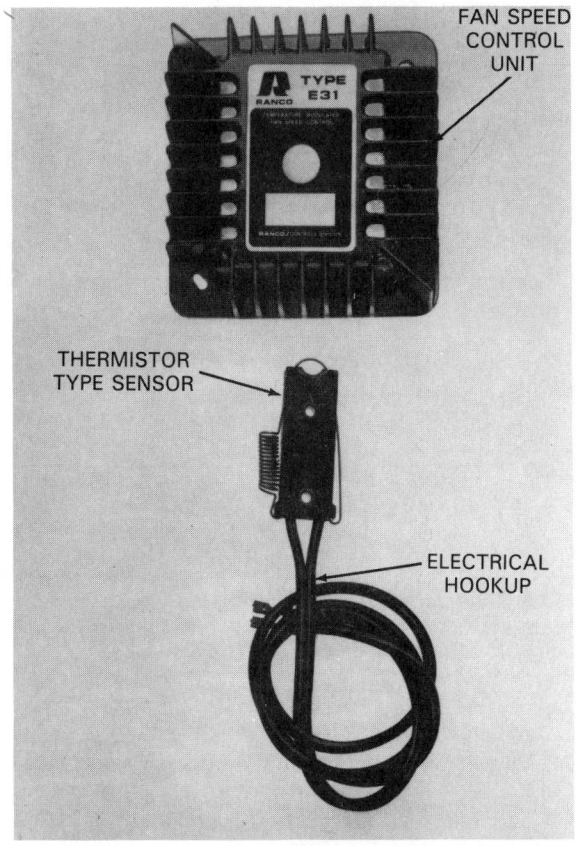

Fig. 12-31. A fan speed control unit may be equipped with a thermistor type sensor. (Ranco North America)

cabinet temperature, electric heating elements are sometimes placed in or around the receiver. If it were allowed to become too cold, the receiver would act like a condenser

Some systems use a bypass from the compressor to the receiver. This bypass feeds a certain amount of hot refrigerant vapor to the receiver to keep it warm. The bypass has a check valve mounted in it to insure only one-way refrigerant flow.

The compressor, itself, must be kept warm enough during cold weather to prevent dilution of the oil by the liquid refrigerant. Heat is provided by electric heating elements in or around the motor compressor. They are thermostatically operated to energize the heating element at about 50 °F (10 °C). This heater usually has a 100 W to 200 W capacity.

The built-in capacity to overcome low ambient temperatures may not be enough if the outdoor unit is exposed to above-normal winds. Windy conditions can prevent damper and fan operation. The cooling effect may be more than the electric heating element can overcome. The unit must be installed in a position to avoid the harmful effects of any high-velocity cold winds. It also should be weatherproofed as well as possible—particularly the electrical installation. See Fig. 12-32.

Head pressure control valves are often used. These are usually thermostat operated. A low-pressure switch may not cut in because condenser pressures are below cut-in pressures. All refrigerant may then transfer to the condenser because it will be the coldest part of the system. A check valve is often used in the condenser outlet to prevent flow of refrigerant to the cold receiver.

As noted before, some systems flood the condenser with liquid refrigerant during low ambient temperatures to maintain high enough condensing pressures. Such systems require valves. When the condenser pressure decreases, one valve closes the outlet of the condenser. The condenser then begins to fill with liquid refrigerant. This action reduces the condensing surface, and condensing pressure will rise.

Fig. 12-33 is the schematic of a system which uses a valve designed to open as the pressure in the receiver falls. This system allows hot gas to bypass into the receiver (at about 20 psi pressure difference). This raises the receiver pressure and increases the flow of liquid refrigerant to the evaporators.

The valve has two openings, B and C. As one closes, the other will open. If located in a cold place, the receiver may need an electric heating element to help the system operate efficiently. These valves must be sized to the capacity of the system. Avoid using excessive pressures when testing for leaks. The valve bellows may suffer damage. Keep the pressure at or below 200 psi.

The system usually is charged with twice as much refrigerant as the system would need without the condenser flooding feature. To permit year-around operation, the receiver must have the capacity to store all this extra refrigerant during the summer. Also, for service purposes, the receiver should be twice the normal size in order to hold all the refrigerant.

The compressor may also collect small amounts of liquid refrigerant during the unit's off cycle. A trap may be needed in the compressor discharge line and an inverted trap at the condenser outlet.

Fig. 12-32. A—Internal view of complete heating, air conditioning, and ventilating system. B—System shows the compressor and precharged refrigerant lines. (Tyler Refrigeration Corp.)

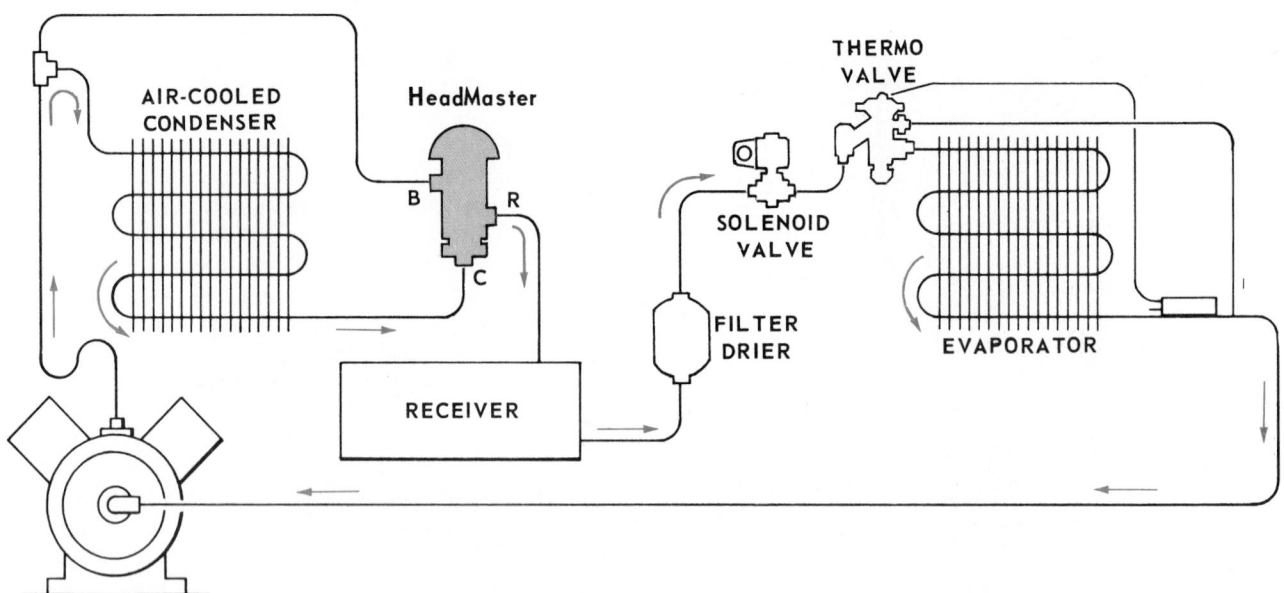

Fig. 12-33. Schematic shows condenser pressure control valve. Compressor discharge pressure is above valve setting. Opening at B is closed and flow is through condenser and openings C and R. During cold weather, opening at C is closed and openings at B and R are open. Hot gas then flows directly from the compressor into the receiver. (Alco Controls Div., Emerson Electric Co.)

12-6 THE COMPRESSOR

Commercial compressors are of two general types:
1. External drive.
2. Hermetic.

Commercial compressor designs are explained in Chapter 4. A typical small, belt-driven, external drive compressor is pictured in Fig. 12-34.

There are several types of commercial hermetic compressors. Fig. 12-35 calls attention to some parts of a bolted (serviceable) hermetic two-cylinder compressor with a force-feed lubrication system.

Some large units have either hydraulic or electric unloading devices to control the number of cylinders which

Fig. 12-34. Condensing unit with belt-driven, two-cylinder, air-cooled compressor. It uses a water-cooled condenser. (Frick Co.)

Fig. 12-35. Bolted type serviceable hermetic motor compressor. Note oil pump used for forced lubricating system. (Copeland Corporation)

are pumping. The higher the load, the more cylinders used to pump the vapor. Fig. 12-36 is a schematic of the oil circuit and describes how it is controlled to operate the compressor unloader.

Another type of hermetic compressor unit is the welded motor compressor design, nonfield serviceable. These units are built in sizes from 1/6 hp up to approximately 20 hp. Internal design varies with size and manufacturer. Some are spring mounted internally, while some use outside (external) mounting springs. The smaller units usually have one

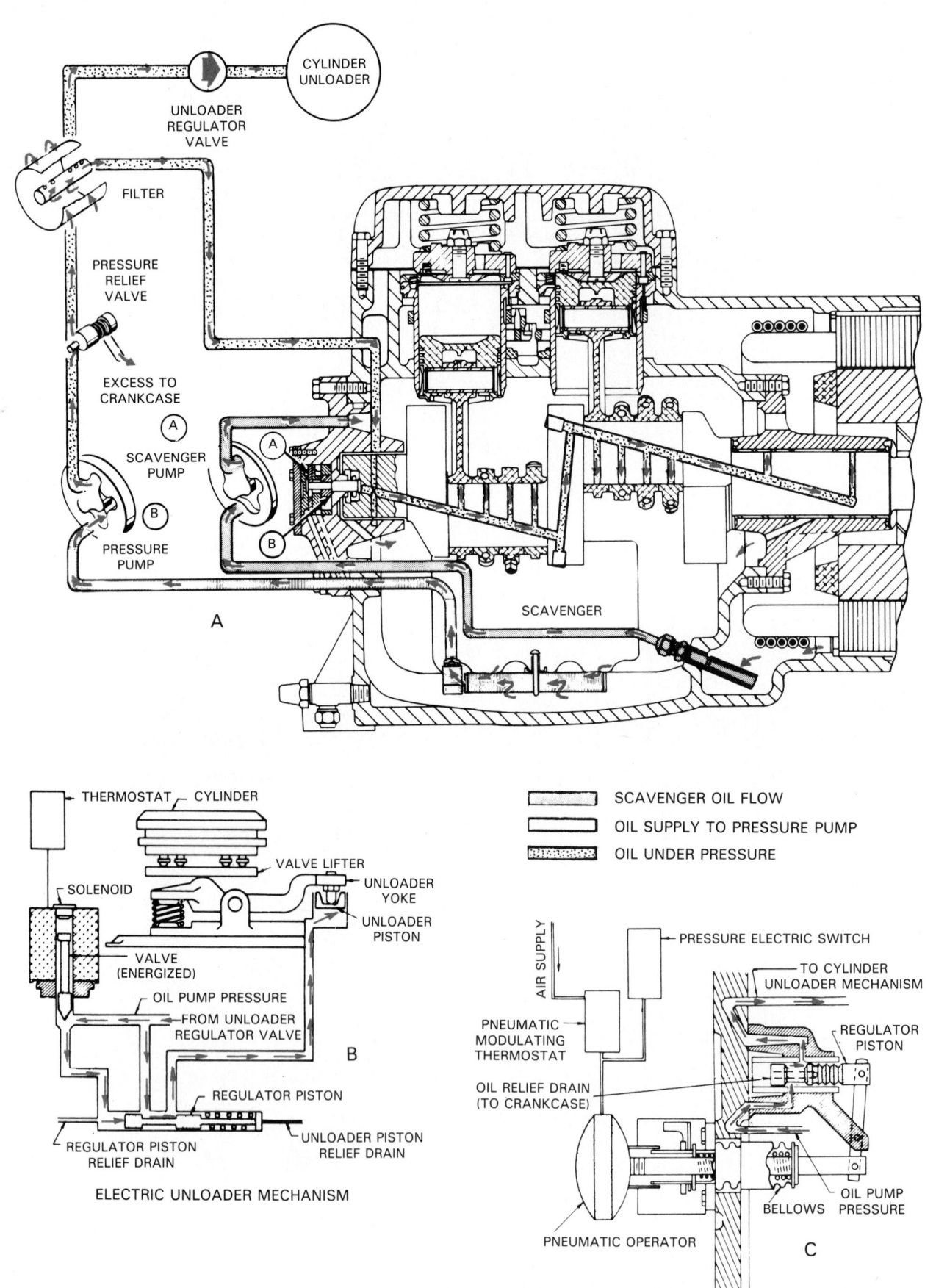

CYLINDER UNLOADER

UNLOADER REGULATOR VALVE

FILTER

PRESSURE RELIEF VALVE

EXCESS TO CRANKCASE

(A)

SCAVENGER PUMP

(B)

PRESSURE PUMP

A

SCAVENGER

THERMOSTAT — CYLINDER

SOLENOID

VALVE LIFTER

UNLOADER YOKE

UNLOADER PISTON

VALVE (ENERGIZED)

OIL PUMP PRESSURE

FROM UNLOADER REGULATOR VALVE

B

REGULATOR PISTON

REGULATOR PISTON RELIEF DRAIN

UNLOADER PISTON RELIEF DRAIN

ELECTRIC UNLOADER MECHANISM

SCAVENGER OIL FLOW

OIL SUPPLY TO PRESSURE PUMP

OIL UNDER PRESSURE

AIR SUPPLY

PRESSURE ELECTRIC SWITCH

TO CYLINDER UNLOADER MECHANISM

PNEUMATIC MODULATING THERMOSTAT

REGULATOR PISTON

OIL RELIEF DRAIN (TO CRANKCASE)

OIL PUMP PRESSURE

BELLOWS

PNEUMATIC OPERATOR

C

PNEUMATIC UNLOADER MECHANISM (ONE STAGE OF UNLOADING SHOWN)

Fig. 12-36. Oil circuit of an eight-cylinder compressor with oil take-off to operate compressor unloader. A—Oil circuit. B—Electrical control. C—Pneumatic control.

cylinder, while larger units (1/2 hp and up) have two or more cylinders. Small-unit motors may be either two- or four-pole (single-phase). Three-phase motors are generally used in larger units.

Many combinations, types, and sizes of compressors can be used to provide the pumping for a great variety of evaporator types and sizes. Each compressor has a minimum and maximum:

1. Revolutions per minute (rpm) for efficiency.
2. Compression ratio (a maximum pressure difference between low side and high side).
3. Discharge temperature.
4. Volume of gas it can pump.

Before using a certain compressor, the manufacturer's operating specifications must be known. See Chapter 15.

Many low-temperature systems use a cascade system. The first stage compressor may be a reciprocating type, but rotary units are also used. The rotary compressor pressure limit is about 45 psi across the compressor. It works very well with a compression ratio of about 4:1 and with a discharge temperature of about 200°F (93°C).

The rotary has a high volumetric efficiency. A check valve is usually placed in the discharge to prevent back-up of refrigerant during off cycle. A check valve should be placed in the oil lines for the same reason.

Compressors can have from one to twelve cylinders in many different cylinder arrangements: vertical, V, W, Y, X, or radial. See Fig. 12-37.

Internal unloaders are usually operated by oil pressure. A spring holds the intake valve open until the oil pressure builds up, causing all intake valves to operate. It is also used to reduce pumping capacity during low-load periods. Solenoid valves are mounted in the oil lines to unloaders. When the solenoid closes, the oil pressure drops in the unloader while the intake valves are kept open. See Fig. 12-38.

Low-side pressure switches operate the solenoids. A timer bypass pressure switch is used to operate the system at full capacity for about a minute each hour or two. Exter-

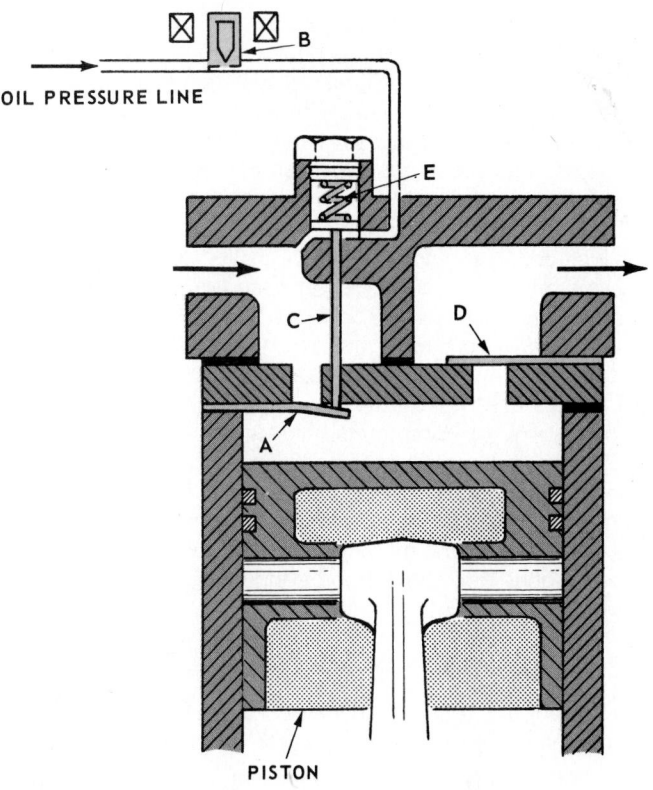

Fig. 12-38. An internal unloader for a compressor. During starting, the capacity needs to be reduced. A—Intake valve. B—Solenoid valve. C—Unloader valve. D—Exhaust valve. E—Spring.

nal unloaders use a bypass to the evaporator inlet to make sure suction vapor is cool (de-superheating).

12-7 AIR-COOLED CONDENSER

Air-cooled condensers are quite common in commercial systems. Cooling water may be too expensive or corrosive.

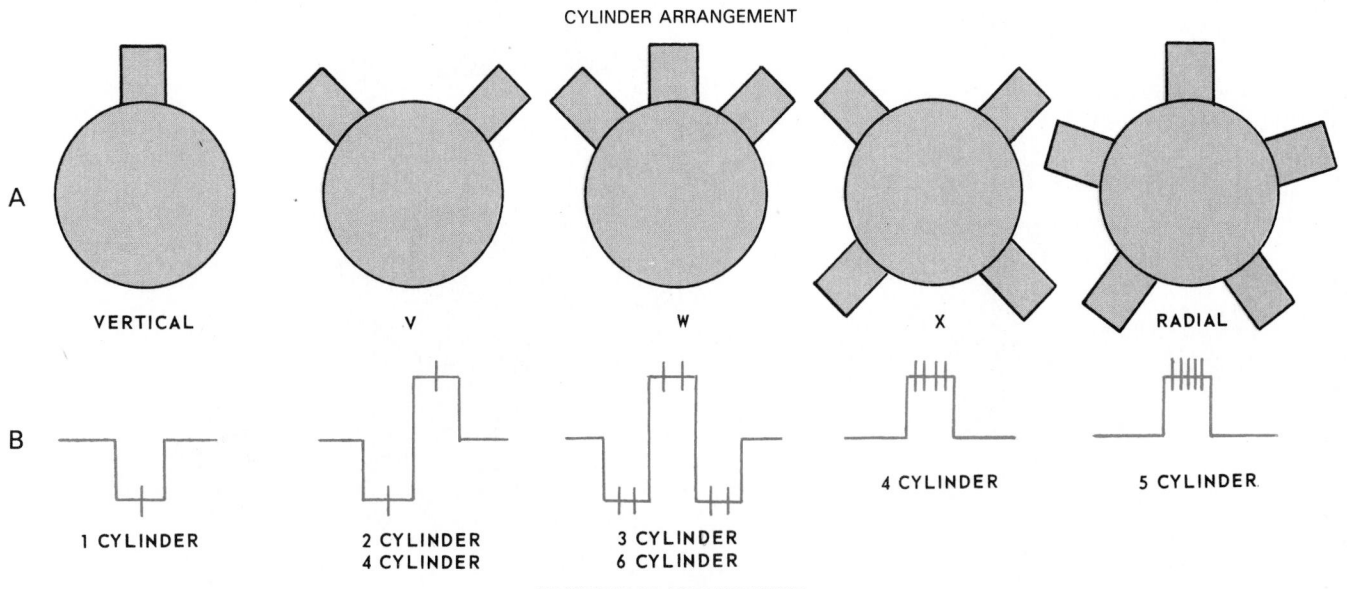

Fig. 12-37. Cylinder arrangements used for reciprocating compressors. A—Cylinder arrangement. B—Crankshaft configuration. Vertical markings on crankshaft indicate number of rods attached to each throw (offset).

Smaller units use static condensers with thermal airflow.

Larger condensers may be cooled by a big fan built onto the motor or into the compressor flywheel on external drive units. Larger hermetic units use separate motors to drive the fans. See Chapter 9 for determining condensing temperatures and pressures when an air-cooled condenser is going to be used.

The efficiency of the fan on an air-cooled condenser may be increased by placing a metal shroud around it. In Fig. 12-39, an air-cooled condenser uses two fans. Air can be drawn, induced (led into), or forced through the condensers. These condensers have fins and frequently use a double or triple row of tubes. Many fin arrangements and constructions have been used. See Fig. 12-40. Fig. 12-41 pictures another type of commercial, air-cooled condenser.

To cool the compressor head and valves, a double air-cooled condenser is sometimes used. Refrigerant leaves the compressor and passes through one condenser. Then it is led back through the motor compressor to help cool it. From there it goes into the second condenser, where it is condensed (cooled) into liquid.

12-8 OUTDOOR AIR-COOLED CONDENSERS

Outdoor air-cooled condensers may be used on systems which have the motor compressor and the liquid receiver

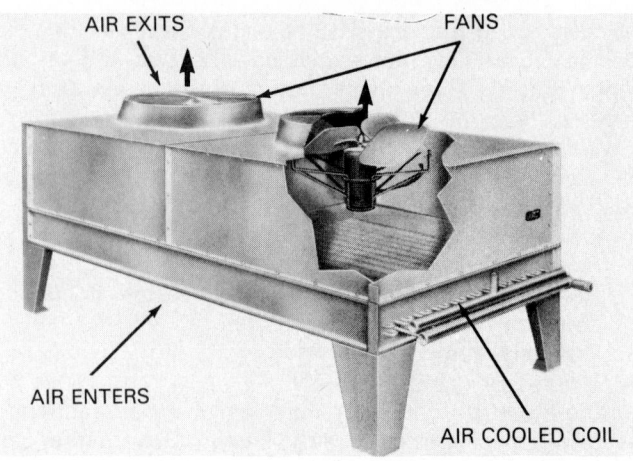

Fig. 12-39. Air-cooled condenser cutaway shows fan, motor, mount, and condenser coil arrangement. Air enters from beneath and leaves at top. (BOHN Refrigeration Products, a Unit of Heatcraft, Inc.)

inside the building. The condenser only is placed outdoors. The compressor discharge line carries the hot high-pressure vapor to the outdoor air-cooled condenser. Condensed liquid is piped back into the building.

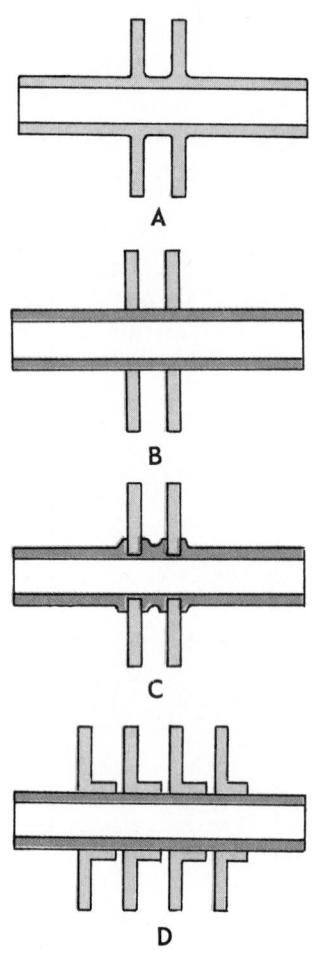

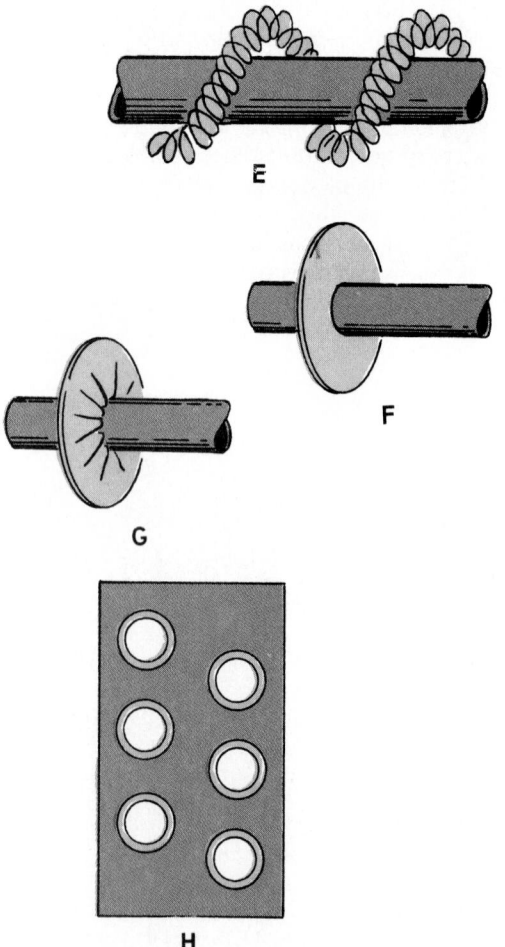

Fig. 12-40. Extended surface fin arrangements used in air-cooled condensers. A—Fin is part of tubing. B—Fin pressed on tubing. C—Fin fastened by crimped tubing. D—Fin flange pressed on tubing. E—Coiled wire used as fin. F—Circular fin. G—Crimped circular fin. H—Large multiple tubing fin.

Fig. 12-41. Commercial system air-cooled condenser. Note direction of airflow. (BOHN Refrigeration Products, a Unit of Heatcraft, Inc.)

These units use the same devices described in Para. 12-5 to protect the condenser and to maintain good head pressures when air temperature is as low as 50 °F (10 °C) or lower.

12-9 WATER-COOLED CONDENSER

Many large commercial refrigerating units use a water-cooled type condenser. This condenser is built in three styles:
1. Shell and tube.
2. Shell and coil.
3. Tube-within-a-tube.

In the first type, the refrigerant vapor goes directly from the compressor into a tank or shell while the water travels through the tank or shell in straight tubes. The second type also uses a shell but the water travels through the shell in coils of tubing.

The third type uses two pipes or tubes—one inside the other. The refrigerant passes one way through the outer pipe, while the condenser water flows in the opposite direction through the inner tube.

Water velocity should be at least 7 fps (feet per second) but no more than 10 fps. If flow is too fast, water may remove the oxide coating causing pitting. If the water velocity drops to 3 fps, scaling will occur.

Water-cooled compressors are sometimes used with water-cooled condensers. The water flow, with few exceptions, is through the condenser first, then through the cylinder head, and finally into the drain. Water flow may be regulated by an automatic water valve. See Para. 12-57.

12-10 SHELL AND TUBE CONDENSER

The shell and tube condenser is a cylinder usually made of steel with copper tubes inside. Water circulates through the tubes condensing hot vapors in the cylinder into a liquid. The bottom part of the shell serves as the liquid receiver. See Fig. 12-42.

The shell and tube condenser has some advantages. It is compact, needs no fans, and combines the condenser and receiver in one. It uses a number of straight tubes inside the receiver with a water manifold on both ends. When these manifold ends are removed, the water tubes can easily be cleaned of deposits, Fig. 12-43. This type is sometimes called a shell and pipe condenser.

12-11 SHELL AND COIL CONDENSER

The shell and coil condenser is very much like the shell and tube water-cooled condenser. It has a coil of water tubing inside the shell rather than a straight tube. It is often used in smaller commercial units. See Fig. 12-44.

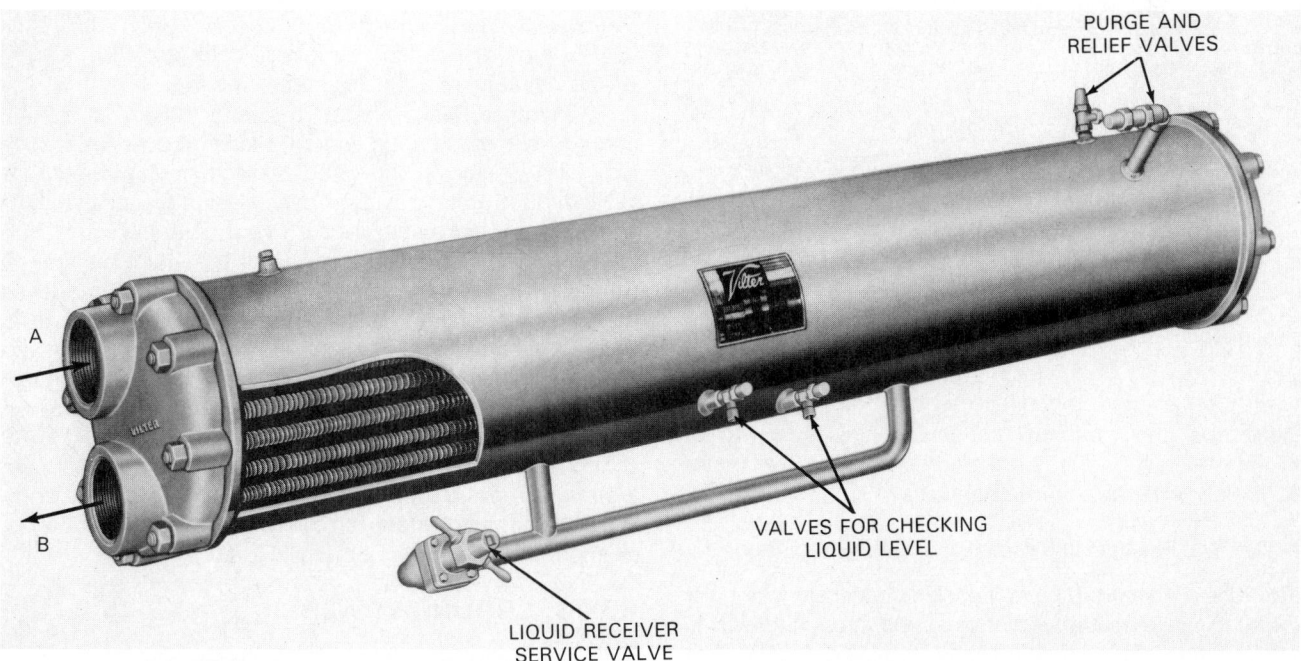

Fig. 12-42. A typical shell and tube condenser-liquid receiver. Note that water pipes run straight through and are finned for better heat transfer. Receiver ends are removable to allow cleaning access. A—Water inlet. B—Water outlet. (Vilter Mfg. Corp.)

Fig. 12-43. Cleaning a shell and tube condenser. Flexible shaft on power cleaner rotates at high speed inside a nylon casing. High-pressure water moves from the machine to the cleaning brush or tool. (Goodway Tools Corp.)

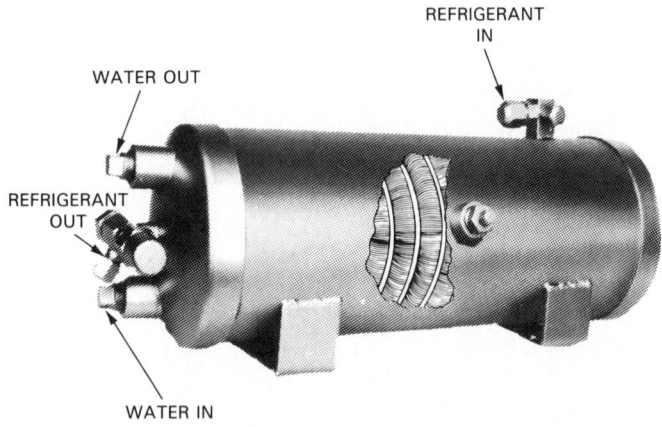

Fig. 12-44. Shell and coil water-cooled condenser. Condenser also serves as liquid receiver. (Refrigeration Research, Inc.)

While the shell and coil condenser is less costly to manufacture, it cannot be cleaned mechanically. The water tube is only cleanable with chemicals.

12-12 TUBE-WITHIN-A-TUBE CONDENSER

The tube-within-a-tube water-cooled condenser is popular because it is easy to make. Water passing through the inside tube cools the refrigerant in the outer tube, Fig. 12-45. The outside tubing is also cooled by air in the room. Double cooling improves efficiency.

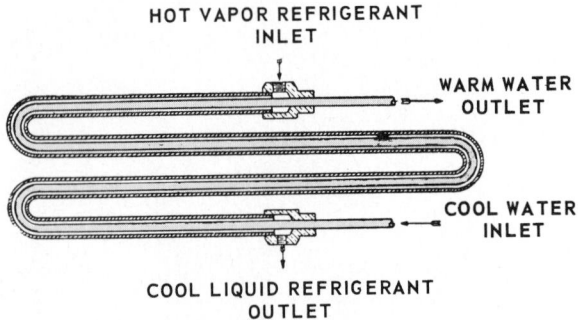

Fig. 12-45. Tube-within-a-tube condenser. Water flows through inner tube. Water flow is opposite vapor flow.

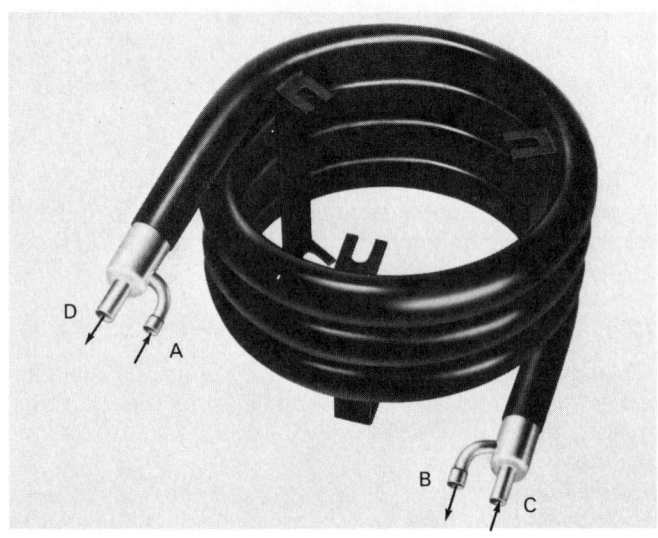

Fig. 12-46. This tube-within-a-tube condenser is shaped into a spiral. A—Refrigerant in. B—Refrigerant out. C—Water in. D—Water out. (Packless Industries)

This type of condenser may be constructed in a cylindrical, spiral or rectangular style. See Fig. 12-46.

To increase heat transfer, the inner tube, Fig. 12-47, View A, shows a 6-lead and an 8-lead grooved inner tube. Fig. 12-47, View B, shows an available option of a double-walled inner tube with a grooved design to achieve venting of refrigerant vapor in the event of a leak.

Water enters the condenser at the point where the refrigerant leaves the condenser. It leaves the condenser at the point where the hot vapor from the compressor enters the condenser. This is called counterflow design. The warmest water is adjacent to the warmest refrigerant, and the coolest refrigerant to the coolest water. This condenser can also be made with hard copper pipe. See Fig. 12-48.

The rectangular type tube-within-a-tube condenser uses a straight, hard copper pipe with manifolds on the ends. When the manifolds are removed, the water pipes can be cleaned mechanically. See Fig. 12-49.

12-13 COOLING TOWERS

In some areas, water contains chemicals making it unsuitable as a coolant. In other localities, water may be very scarce, expensive, or its use may be limited by law.

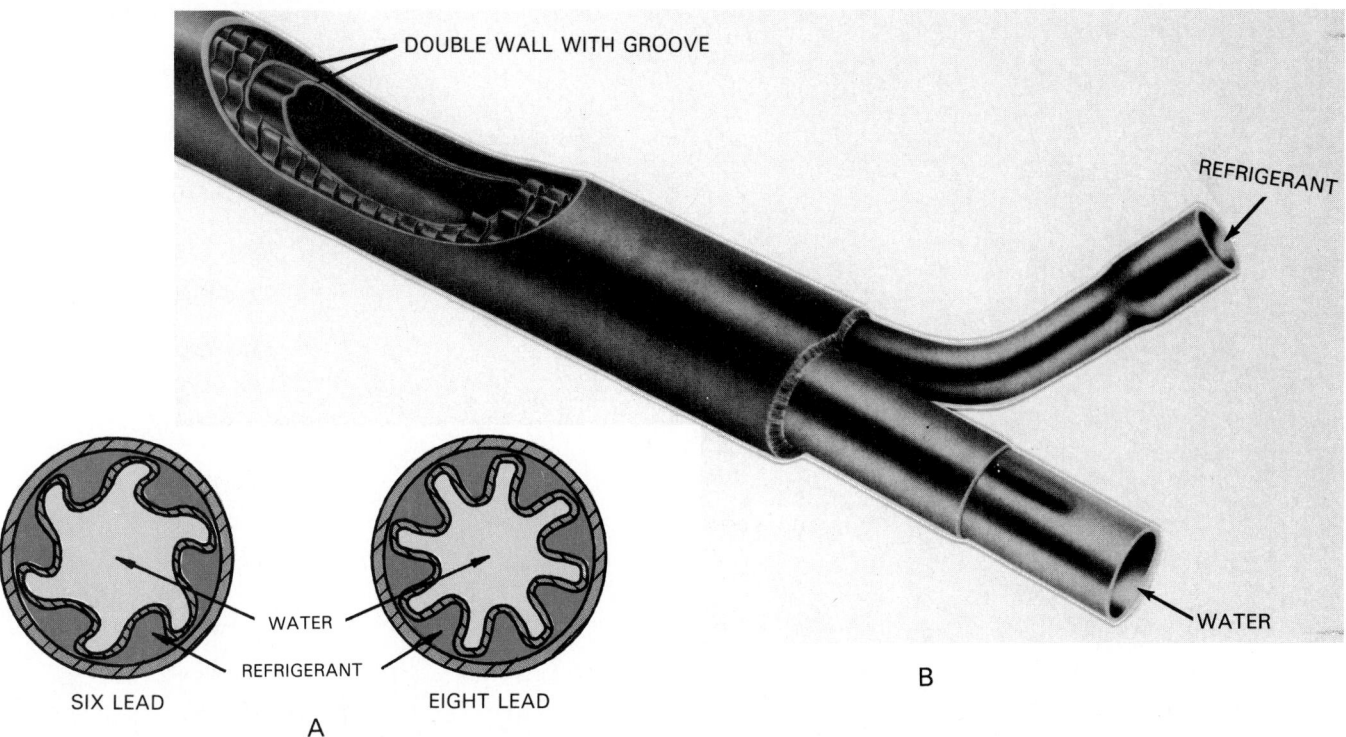

DOUBLE WALL WITH GROOVE

REFRIGERANT

WATER

WATER

REFRIGERANT

SIX LEAD

EIGHT LEAD

A

B

Fig. 12-47. Two more views of tube-within-a-tube condensers. A—End view shows six- and eight-lead (groove) designs to increase heat transfer from the refrigerant to the water. B—Cutaway of a double-walled tube-within-a-tube design. (Edwards Engineering Corp. and Packless Industries)

To permit use of a water-cooled mechanism and to save on water consumption, water-cooling towers are used. These towers serve the same purpose as the spray towers used in large industrial refrigeration systems.

A variety of cooling tower designs are shown in Fig. 12-50. Since they can be rather noisy, cooling towers should be located away from noise-sensitive areas such as offices, restaurants, and living quarters.

One system connects the water lines of the condenser to a water coil in an enclosure. A pump forces the water through the condenser and then through the coil in the tower. The tower coil is pierced with holes, and the water

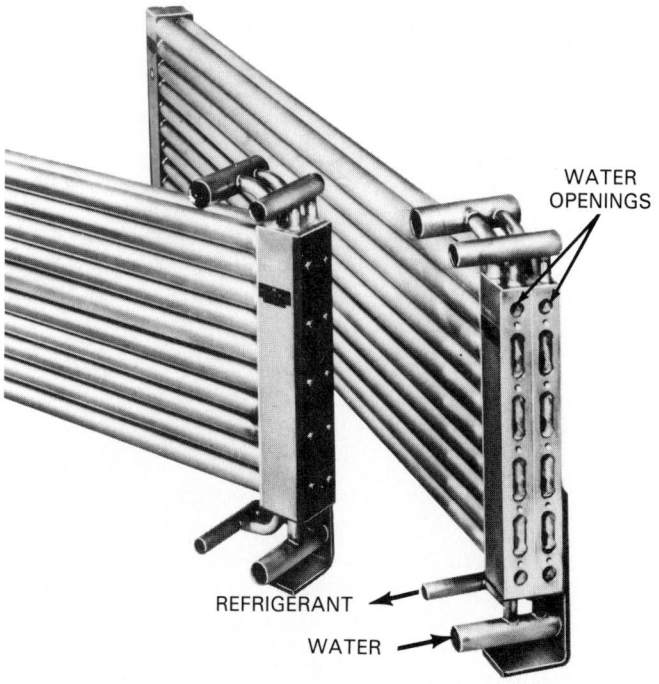

WATER OPENINGS

REFRIGERANT

WATER

Fig. 12-48. Tube-within-a-tube condenser designed to permit cleaning of water tubes (inner tube). Clean-out plate is removed in right-hand view. (Standard Refrigeration Co.)

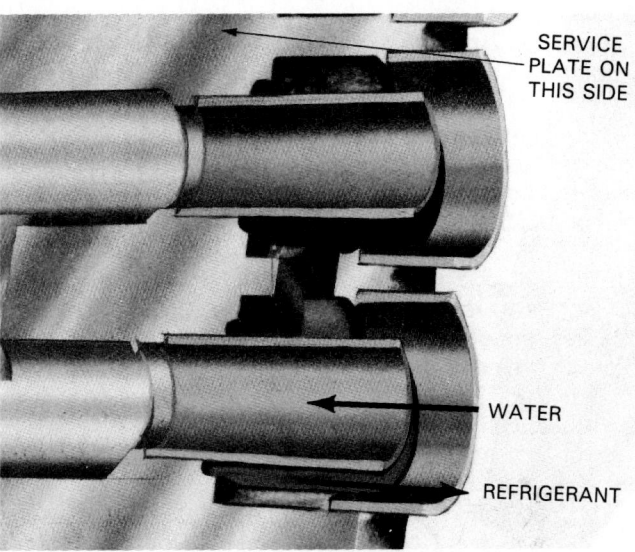

SERVICE PLATE ON THIS SIDE

WATER

REFRIGERANT

Fig. 12-49. Cutaway showing construction detail of pipe-within-a-pipe condenser. Refrigerant tubes and water tubes are brazed into header.

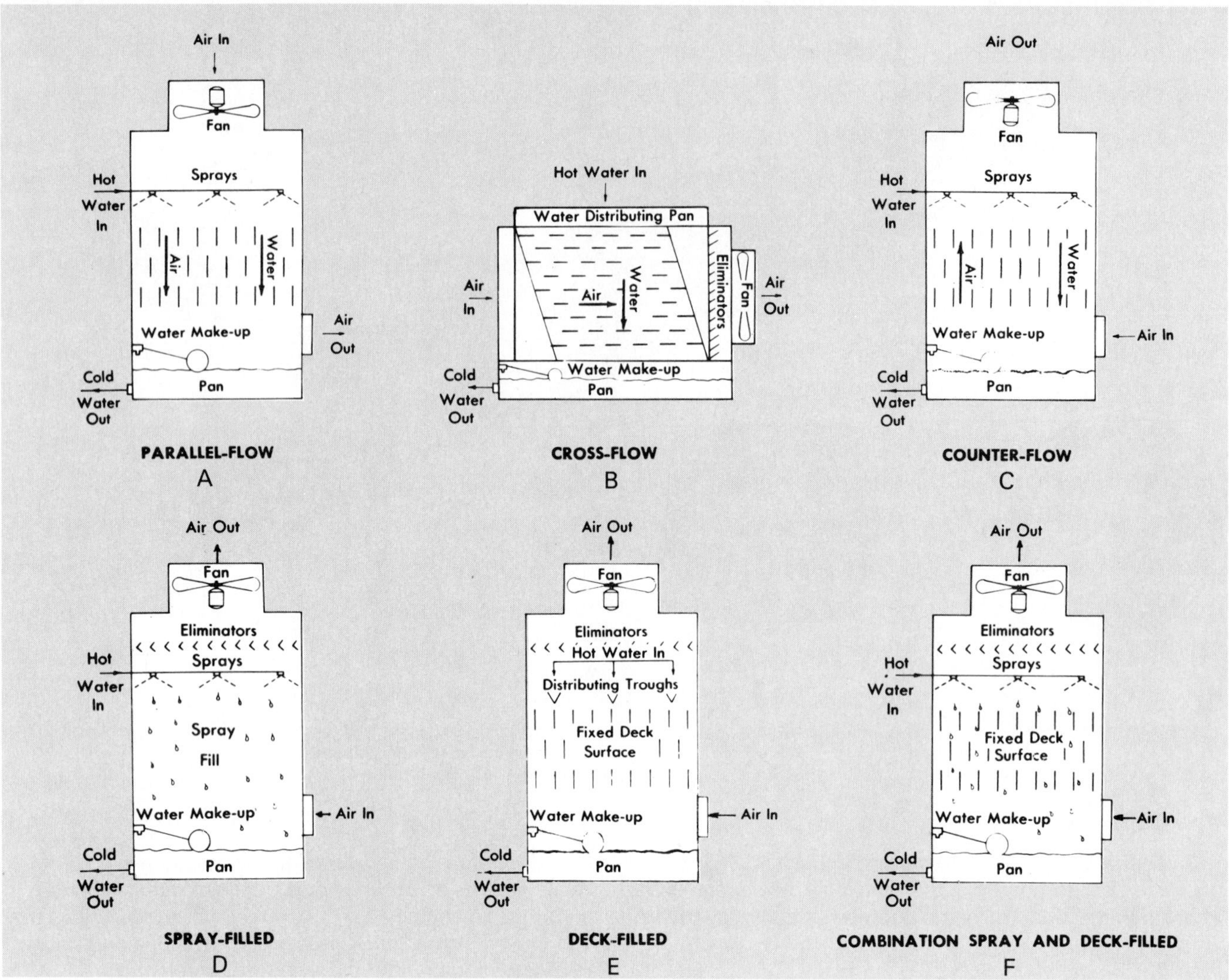

Fig. 12-50. Different types of cooling towers. A, B, and C show three different airflow patterns. D, E, and F show different methods to vaporize some of the water for cooling.

is sprayed into the enclosure.

Air, rushing through the sprayed water, evaporates some of it. Evaporation cools the remaining water to the outdoor temperature or even lower (wet bulb temperature).

In some systems, motor-driven fans control airflow through the cooling tower. Fig. 12-51 shows a cooling tower fan that is constructed primarily of fiberglass. Some recent cooling tower designs use water at a high enough pressure (15 to 60 psig) to eliminate the need for an airflow fan. The water spray at these high pressures induces the airflow to accomplish the required cooling. See Fig. 12-52.

Cooled water collects in the bottom of the enclosure and passes through a screen that removes leaves or other foreign material. Then it is recirculated through the condenser.

A float-controlled valve in the lower water pan adds more water as needed. This float operates like a refrigerant low-side float mechanism.

A drain continually bleeds some water out of the water pan to keep water hardness to a minimum. Chemicals may be added to retard rust formation, algae, fungus growth, and the like.

Recently, cooling towers have been linked to the spread

Fig. 12-51. Water flows into chamber at top of cooling tower and then by gravity through nozzles. Propeller fans provide air flow. (The Marley Cooling Tower Company)

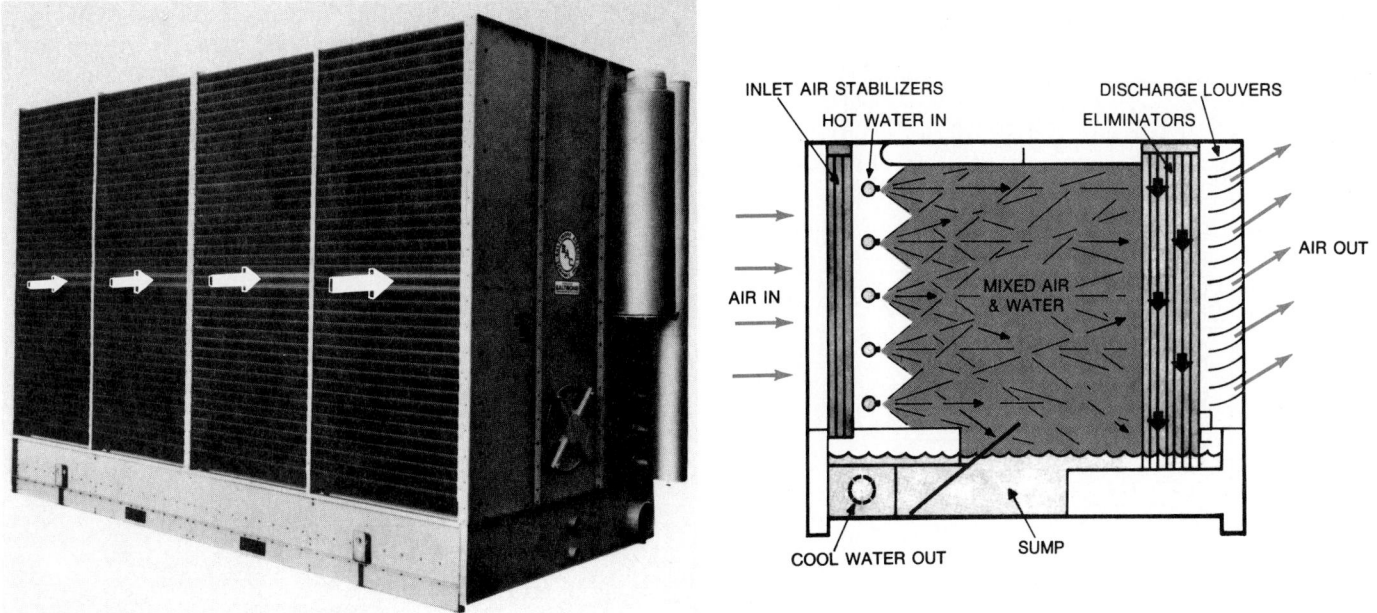

Fig. 12-52. Cooling tower using high pressure water spray. Fan is eliminated. Left. Exterior view. Right. Diagram of cooling tower. (Baltimore Aircoil Co., Inc.)

of Legionnaire's disease. Several precautionary measures are being recommended to help eliminate this problem. These include placement of the cooling towers downwind, and use of chloride compounds as disinfectants on a monthly maintenance schedule.

Cooling towers are made of corrosion resistant materials. Among these are steel (zinc dipped after assembly), copper, stainless steel, plastic, or treated wood.

The more water surface in contact with the air flowing through a cooling tower, the more efficient the cooling action. Most towers have some arrangement so that water flows over materials in thin films. This material is usually called fill. Fills are made of many materials: metal fins, wood slats, plastic, asbestos-plastic, and asbestos-cement. The shapes of the surfaces vary from Z-shaped, honeycomb, embossed, flat sheet to corrugated sheet. The cellular (honeycomb) fill is becoming very popular. The distribution system (nozzles, troughs, V-notches) must be kept clean and must distribute the water evenly to prevent scale buildup.

Ordinarily, cooling towers are in no danger of freezing while in operation. However, electric heat will keep the water temperature up during shutdowns. Immersion and convection heaters have been used for this purpose. An electric heater may also be installed in the pump circuit.

In addition to electric heat, hot water or steam can be used to prevent reservoir freeze-up. Pipes may need insulation or electric heater tape.

Overflow pipes in the cooling tower carry excess water to the drain system of the building. Some have airflow control to prevent freezing if wet or dry bulb temperature goes below 32 °F (0 °C). An air outlet thermostat operates the fan dampers.

Use only coarse screens on pump inlets and be sure all suction lines are below water level in the cooling tower. Otherwise, air may enter the suction line. The resulting drop

in pump volume can cause pump damage.

Pump outlets, on the other hand, need fine screens. The water pump should push water through the system to prevent low water pressures in the condenser tubes or pipes.

Cooling towers evaporate about two gallons of water every hour for each ton of refrigeration capacity. A gallon of water weighs about 8.3 lb. About 1000 Btu are needed to evaporate 1 lb. of water. Thus, to evaporate a gallon of water, 8.3 × 1000 or 8300 Btu are required. Two times 8300 or 16,600 Btu are required to evaporate two gallons of water.

For details of tower sizes and capacities, see manufacturers' catalogs.

12-14 EVAPORATIVE CONDENSERS

The evaporative condenser system carries the refrigerant into a condenser which is in an enclosure much like a cooling tower. In this system, as the word "evaporative" indicates, water is sprayed or drips over the condenser. This cools it. The water cycle is in the condenser cabinet only. See Fig. 12-53.

Usually, the evaporative condenser is mounted outdoors. However, it may be used indoors if air ducts are provided to the outside.

Some systems pump water to a trough above the condenser. The water then drips over coils as air is forced through them. A thermostat can be used to control the water flow. A fan blows air over the condenser whenever the condensing unit is operating.

The condenser is cooled by air alone until the condenser temperature reaches 80 °F (27 °C) or more. Water cooling is then turned on by a thermostat.

Some units subcool the refrigerant leaving the receiver by piping the receiver outlet (liquid line) into and out of the evaporative condenser. Temperature of the refrigerant can

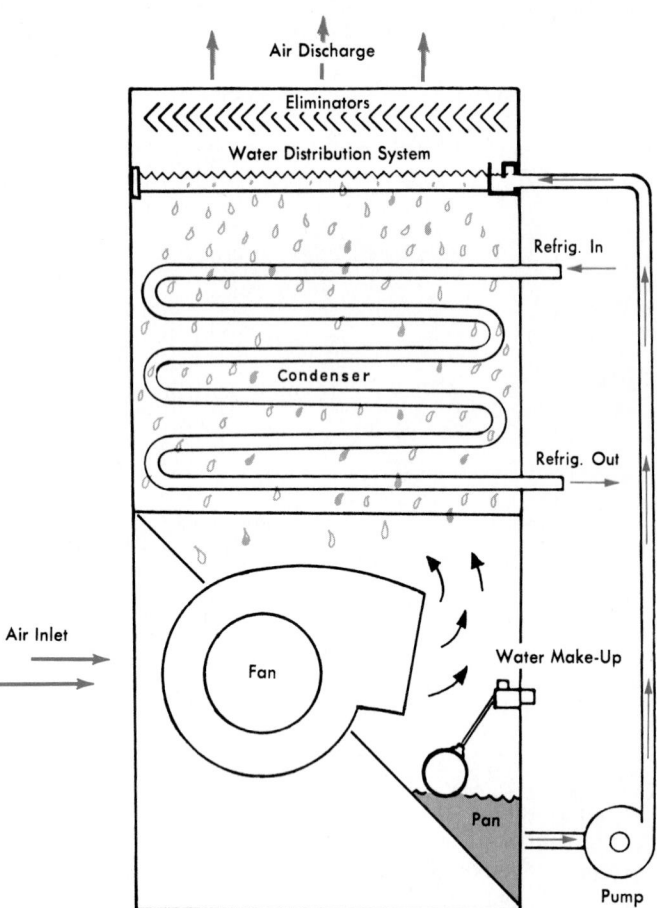

Fig. 12-53. Another evaporative condenser design. Note air and water flow. Water make-up is controlled by float valve.
(Baltimore Air Coil Co., Inc.)

be dropped 10 °F (6 °C) by subcooling. When the temperature reaches 45 °F (7 °C) or lower, the water is shut off. However, the condenser can still carry the load as an air-cooled condenser.

Some have water reservoirs inside the building. For protection from freezing weather, the reservoir must be large enough to hold all the water in the system. Fan dampers are sometimes used to decrease airflow as outside temperature drops.

12-15 LIQUID RECEIVER

The liquid receiver is a welded steel tank usually equipped with two service valves. One is a liquid receiver service valve mounted between the liquid receiver and the condenser. The other is located between the receiver and the liquid line (king valve). These two valves enable the service technician to disconnect the liquid receiver from the system separately.

Receivers should have safety devices. A thermal release plug provides minimal safety. Some receivers have both thermal and pressure releases. See Para. 12-65. A special line should be installed on relief valves to carry released refrigerant outdoors.

Receivers may be mounted either vertically or horizontally. Fig. 12-54, View A, illustrates a vertical receiver. View B illustrates a horizontal receiver with inlet valve, outlet

valve, and safety valve. The horizontal style usually hangs underneath the compressor and motor frame. Some are provided with a device—sight glass, magnet floats or valves—for determining the level of the liquid refrigerant.

If there is a water coil inside the liquid receiver, the shell is just like the receivers having no water coil. However, it is usually larger for the same size compressor unit. The receiver should be large enough to hold all the refrigerant in the system.

12-16 COMMERCIAL EVAPORATORS

Because customers demand great variety in commercial refrigeration, special evaporator designs are required for many installations. These evaporators vary from coils of tub-

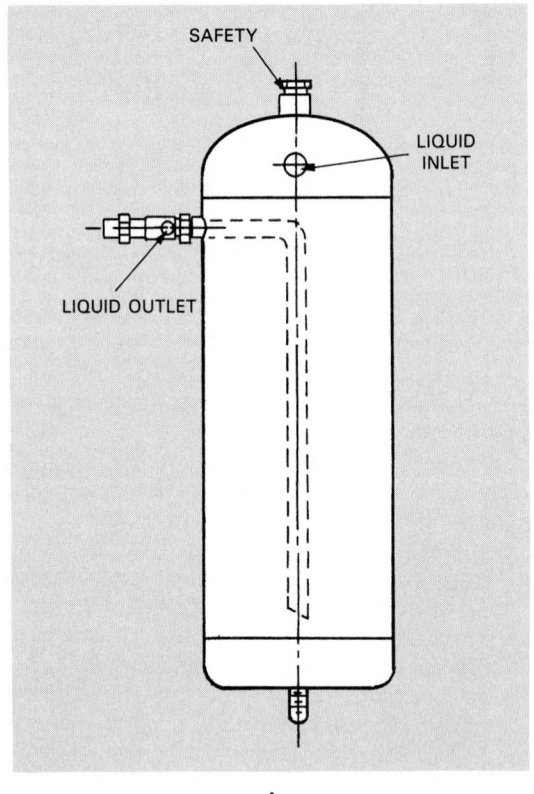

A

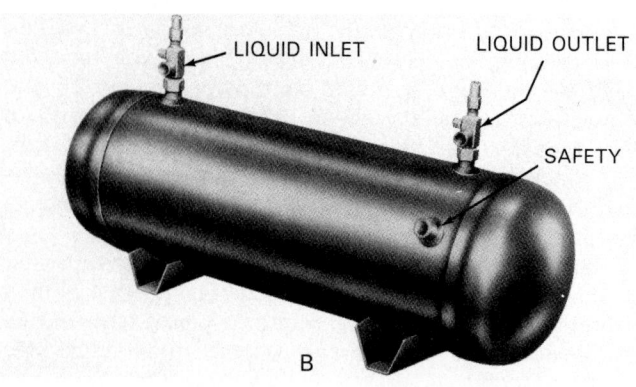

B

Fig. 12-54. Liquid receivers. A—Designed for vertical mount. B—Designed for horizontal mount. Note location of safety valves.
(Standard Refrigeration Co.)

ing immersed in a sweet water bath to forced-circulation evaporators over which air is blown by a motor-driven fan.

Evaporators may be divided into two main groups:

1. Those used for cooling air; in turn, the air cools the contents of the cabinet.
2. Those submerged in a liquid such as brine or a beverage.

Evaporators for cooling air are of two principal types:

1. Natural convection.
2. Forced convection.

In natural air convection evaporators, air circulation depends on gravitational (warm air rises—cool air descends) or thermal circulation. Natural convection, air cooling evaporators fall into three classes:

1. Frosting.
2. Defrosting.
3. Nonfrosting.

The conditions under which an evaporator must work determine its classification. The governing conditions are the desired temperature range of the cabinet and the temperature difference between the evaporator and the cabinet.

12-17 FROSTING EVAPORATORS

A frosting evaporator builds up frost continuously when in use. It usually operates at 5 °F (−15 °C) refrigerant temperature cutout and 25 °F (−4 °C) refrigerant temperature cut-in. The machine must be manually or automatically shut down from time to time to rid the system of frost.

Frost which forms on the evaporator comes from moisture in the air. This leaves the air in the cabinet dry. Dry air rapidly dries out food.

If the refrigerant temperature is below 28 °F (4 °C), heat energy of some kind is needed to defrost the evaporator. Otherwise, the evaporator must be turned off longer than the normal cycle.

Some evaporators run at extremely low temperatures to keep the fixture cool. This allows frost and ice to build up. As the frost grows thicker, it reduces the cooling efficiency of the evaporator. These evaporators are used in low-temperature and frozen food fixtures.

12-18 DEFROSTING EVAPORATORS

Many evaporators run on what is called a defrosting cycle. While the condensing unit is running, the temperature of the evaporator is low. This causes frost to accumulate on it. But after the compressor shuts off, the coil warms up above 32 °F (0 °C) and the frost melts. During the running of the compressor, the evaporator will remain at temperatures of about 20 to 22 °F (−7 to −6 °C).

This defrosting process is called air defrosting. It clears the evaporator surfaces of frost and provides efficient heat transfer. It also keeps a high relative humidity—about 90 to 95 percent. However, this sacrifices temperature differences between the evaporator refrigerant and the air in the cabinet. Greater evaporator area is needed to make up for this loss.

The defrosting evaporator sometimes presents a problem. In some installations, the top of the evaporator may defrost and moisture flows down the evaporator surface. Before it has time to escape, it may freeze around the lower parts of the evaporator. This ice accumulation on the evaporator fins may eventually block air circulation around the

evaporator and interfere with proper refrigeration. Fig. 12-55 is a sketch of a defrosting evaporator.

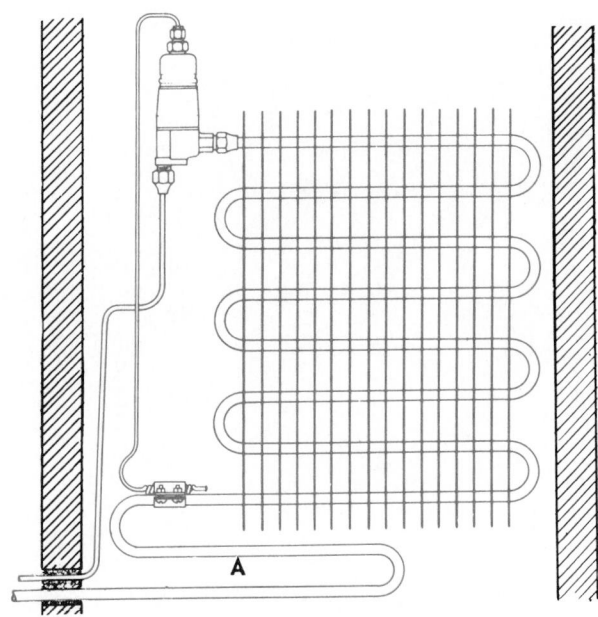

Fig. 12-55. ''Drier'' coil at A prevents sweating or frosting of suction line outside fixture. This evaporator is usually the defrosting type. Frosted surfaces transfer heat poorly.

12-19 NONFROSTING EVAPORATORS

Nonfrosting evaporators operate at temperatures which do not go much below 32 °F (0 °C). Frost, therefore, does not form on the evaporator.

Occasionally, the evaporator may build up a light coat of frost just before the compressor shuts off. This frost immediately melts on the off-cycle. These evaporators operate at a temperature of 33 °F (.5 °C) to 34 °F (1 °C). However, the temperature of the refrigerant inside the evaporator will be around 20 °F (−7 °C) to 22 °F (−6 °C). Since they do not frost over, they draw only a little moisture from the inside of the cabinet. This is a big advantage because it becomes possible to maintain an RH (relative humidity) of 75 to 85 percent in these cabinets. This helps to keep the produce fresh and stops shrinkage weight. Nonfrosting evaporators look like the one shown in Fig. 12-56.

Two types of evaporator construction are commonly used. The fin-type is shown in View A and the plate-type in View B of Fig. 12-57.

Plate-type evaporators are fabricated from two metal sheets welded together to form a series of passages through which the heat transfer media flows. A serpentine flow configuration provides high internal flow velocities and high heat transfer rates. Fig. 12-58 shows a cross section of a plate-type eutectic evaporator. See Para. 13-29.

12-20 FORCED CIRCULATION EVAPORATOR

Forced circulation evaporators are a compact arrangement of refrigerant-cooled tubes and fins. A fan, driven by an electric motor, blows air over them. Evaporator and fan are usually enclosed in a metal housing. The evaporator

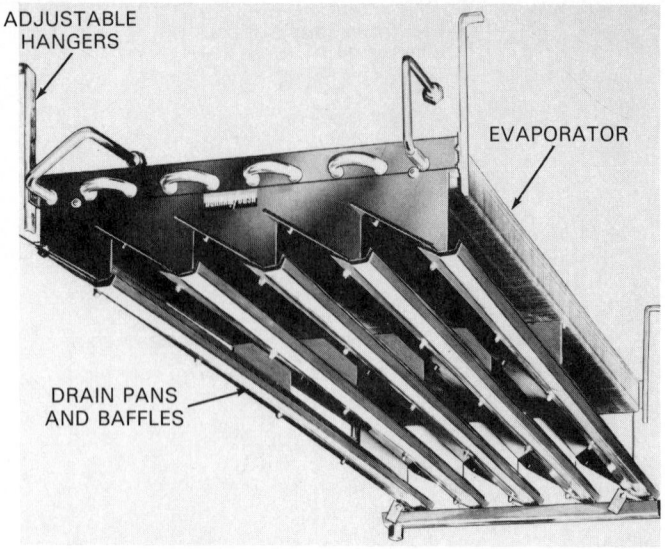

Fig. 12-56. Nonfrosting evaporator. Note baffles used to collect and remove condensate.

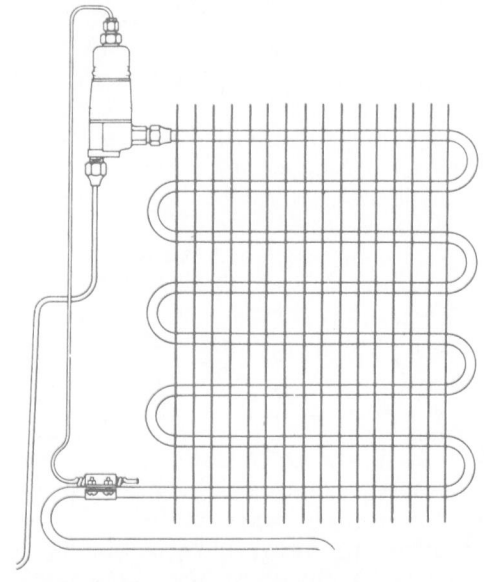

A

B

takes up little space, and needs no extra baffling, Fig. 12-59.

These evaporators, Fig. 12-60, have a tendency to cause rapid dehydration (drying) of foods unless special care is taken. If the evaporator is large, operates at a small temperature difference (10 °F to 12 °F or 6 °C to 7 °C), and the air is circulated slowly, drying will be minimized.

In installations where dehydration is not harmful and air may flow through the evaporator rapidly, small evaporators are practical. They are operated at a greater temperature difference, 20 °F to 30 °F (11 °C to 17 °C). Thermostatic expansion valve refrigerant controls are usually used with forced convection evaporators.

The fan motor may be any size from 1/50 hp up and may run continuously. It also may be controlled by the evaporator or fixture temperature. Refrigerant temperature is usually kept quite low, but rapid air circulation keeps the evaporator from frosting up. However, considerable sweating does occur and drainage must be provided in the installation.

Blower evaporators operated at low refrigerant tempera-

Fig. 12-57. Two types of evaporators are shown: A—Fin type. B—Serpentine plate coil type. (Tranter, Inc., Texas Division)

Fig. 12-58. Cross sections of a KOLD-HOLD plate-type eutectic evaporator. (Tranter, Inc., Edgefield Division)

Fig. 12-59. Typical blower evaporator designed for ceiling mounting. Air enters at back and exits at fans. Low profile of unit leaves more space in commercial refrigerator or freezer for commodities.   (BOHN Refrigeration Products, a Unit of Heatcraft, Inc.)

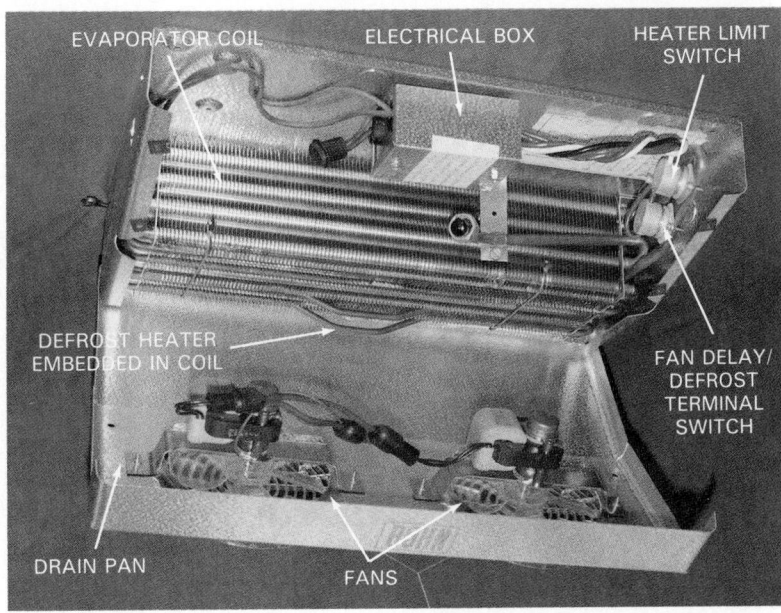

Fig. 12-60.   The inside of a typical freezer evaporator. Note defrost heater embedded in coil and the location of various controls.
(BOHN Refrigeration Products, a Unit of Heatcraft, Inc.)

tures need special defrosting care. Fin spacing is small, the evaporator surface is small, and frost or ice may interfere drastically with heat transfer.

The forced circulation evaporator will cool a refrigerator cabinet quickly. It is well suited for air conditioning as well as for refrigerator cabinets used to store bottled beverages or foods in sealed containers. Figs. 12-61 and 12-62 illustrate two of these evaporators. Figs. 12-63 and 12-64 show evaporators for large fixtures. Each has two fans for air return. A high-capacity blower evaporator used for low temperature fixtures is shown in Fig. 12-65.

A condensate pump, Fig. 12-66, may be used to provide a positive method of removing the condensate. The pump is mounted on the drain and is self-priming. It uses only about 10 W and operates continuously. This pump also is used for pumping the slightly acidic condensate produced by high-efficiency gas furnaces, Chapter 20. Therefore, there are two intakes—one for air conditioning condensate

Fig. 12-61.   Compact blower evaporator used in commercial refrigerators.
(BOHN Refrigeration Products, a Unit of Heatcraft, Inc.)

Fig. 12-62. Vertical, flat-type blower-evaporator is designed to be mounted behind either window or door frames of fixture. (Peerless of America, Inc.)

Fig. 12-63. Motor drives two propeller-type fans and cooled air exits at ends of this evaporator. (Peerless of America, Inc.)

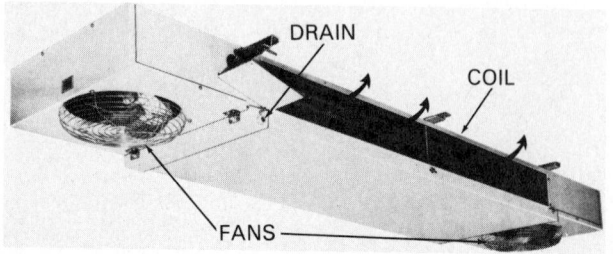

Fig. 12-64. Low-velocity blower evaporator draws air at fans and discharges it through coil from both sides. (BOHN Refrigeration Products, a Unit of Heatcraft, Inc.)

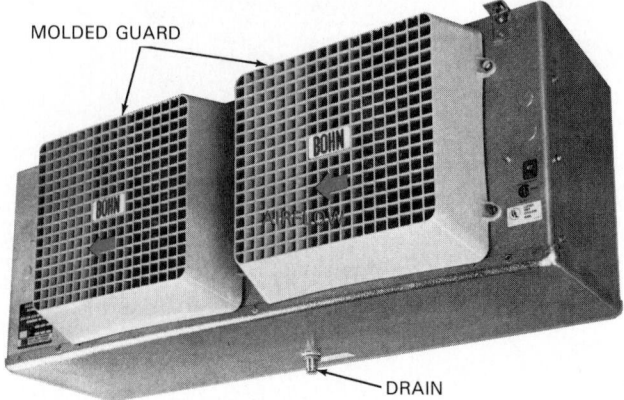

Fig. 12-65. Low-temperature blower evaporator. Air discharge at molded guards is directed for long distance to maintain constant box temperature. (BOHN Refrigeration Products, a Unit of Heatcraft, Inc.)

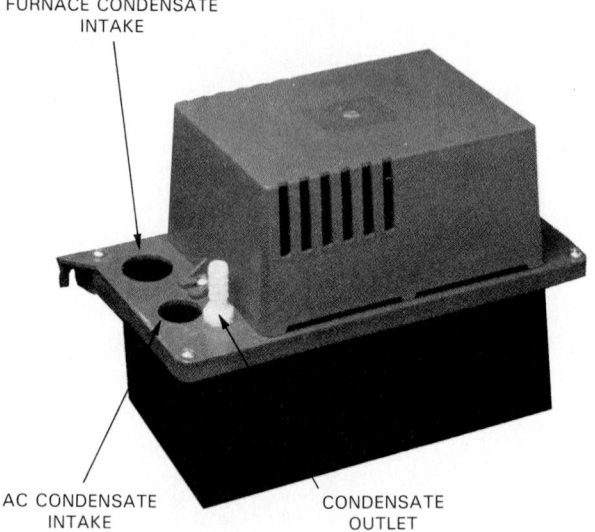

Fig. 12-66. Condensate pump is used on evaporator drains and furnaces when condensate is to be raised above drain pan level. (Beckett Corporation)

drain line, and one for the furnace.

Many evaporators use a hot gas bypass system to maintain above-freezing temperatures when the cooling load is decreased. Fig. 12-67 shows a method of providing the necessary hot gas to keep the evaporator from frosting during low head conditions.

12-21 LIQUID-COOLING EVAPORATOR

Refrigeration makes drinks more enjoyable, preserves liquid, improves manufacturing processes, and minimizes evaporation. It is used especially for water, soft drinks, alcoholic beverages, and brines.

Three types of liquid cooling evaporators are used:
1. Bottled liquids.
2. Liquids under atmospheric pressure.
3. Liquids under pressure.

When the evaporator is mounted inside the liquid being cooled, it is called an "immersed" evaporator.

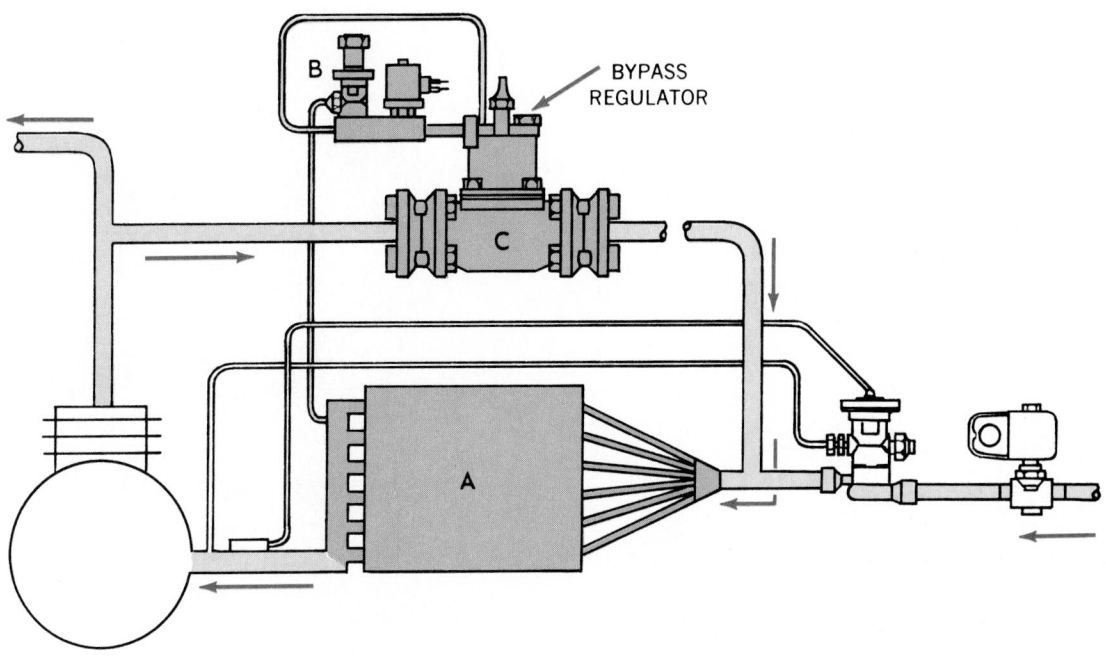

Fig. 12-67. A hot gas bypass system used to prevent a frosting evaporator during low head conditions. When the pressure in the evaporator, A, decreases to freezing conditions, the pilot valve at B opens. This opens the large valve at C, allowing enough hot gas to enter the evaporator and prevent frosting. (Alco Controls Div., Emerson Electric Co.)

12-22 IMMERSED EVAPORATOR (BRINE)

In the immersed type of liquid cooling, the evaporator may be surrounded by a brine, beverage, or water. A small, plain, tube-type evaporator, submerged in a liquid, provides good heat transfer.

Immersion makes the evaporator more efficient because liquids transfer heat to metals faster than air. An efficiency ratio of 50 or 100 to 1 is common. That is, a submerged evaporator can remove 50 to 100 Btu per hour, per degree temperature difference, per square foot of evaporator surface. Air-cooling evaporators can only remove 1 Btu under the same conditions.

These evaporators may use either a low-side float or a thermostatic expansion valve refrigerant control when used in multiple installations. Smaller, self-contained installations normally use a capillary tube refrigerant control.

The cooled liquid may be circulated and used for various purposes.

12-23 IMMERSED EVAPORATOR (SWEET WATER)

Another type of system immerses the evaporator in ordinary tap water, called a sweet water bath. The water-cooling device used in soda fountains is of this type. A coil of tubing in which the water is to be cooled and consumed by the customer may also be submerged in the same water.

This design allows the sweet water to freeze around the evaporator during what is called the nonload period. The light ice accumulation acts as a reserve of refrigeration.

These systems usually have a thermostatic expansion valve refrigerant control, Fig. 12-68. The evaporator should reach to the bottom of the bath if the temperature of the bath is to be less than 39.1 °F (4 °C). *Water, as it cools from 39.1 °F to 32 °F (4 °C to 0 °C), expands and rises. Therefore, the coldest water will be at the top.*

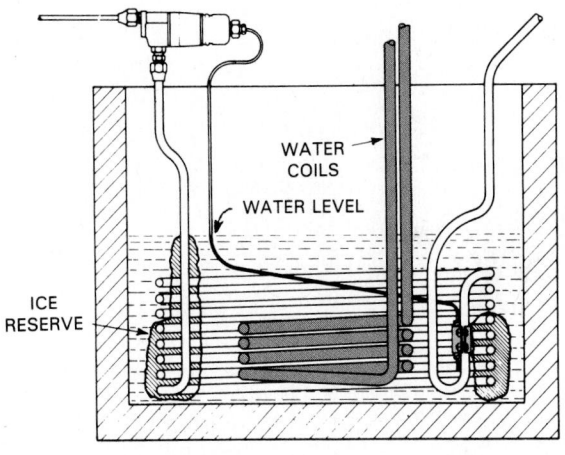

Fig. 12-68. Sweet water (tap water) bath is often used at soda fountains. Ice accumulation is controlled by location of thermostatic expansion valve thermal bulb in unit.

12-24 PRESSURE TYPE (BEVERAGE) EVAPORATOR

Included in the submerged class is the evaporator used in pressure type beverage coolers. The liquid refrigerant is carried in a tube submerged in the beverage to be cooled. The beverage itself is under pressure. Construction is similar to that of the liquid receiver type water-cooled condenser. This design is quite common for instantaneous cold water coolers. *Be careful that the temperature of the beverage is never so low that it will freeze to any extent.*

The all-metal beverage evaporator takes advantage of the high heat conductivity of aluminum. Two separate copper tubes are coiled in a helix (spiral) design. They are placed in a permanent mold and liquid aluminum is poured into the mold. After the liquid aluminum cools and becomes solid, the two tubes are completely enclosed in a hollow cylinder of aluminum, as in Fig. 12-69.

Heat transfer is excellent. Refrigerant is evaporated in one coil, and heat will flow from the liquid in the other coil.

Surrounded and encased by the casting, the coil is strong enough to sustain the pressure of water freezing in it.

Stainless steel tubing may also be used. For storing some beverages, it is the only suitable material.

Coolers having three or more separate tubes encased by aluminum are used when more than one liquid must be cooled.

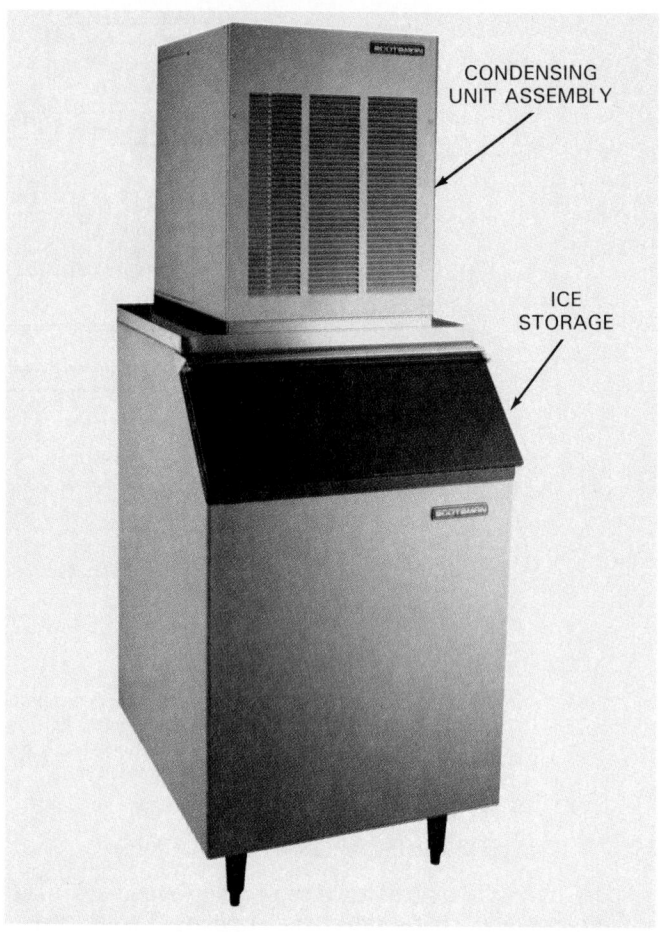

Fig. 12-70. Modular flake ice maker. Note location of evaporative condensing unit. (Scotsman Ice Systems)

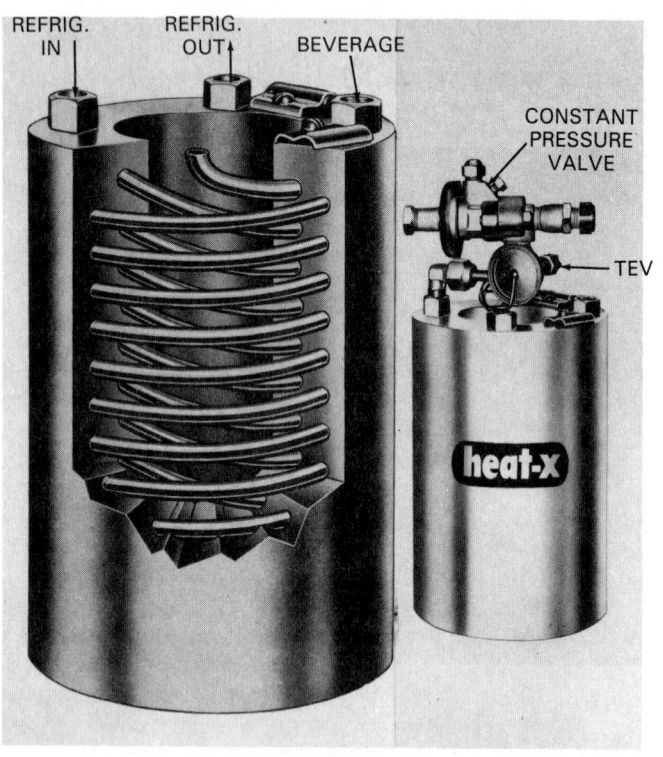

Fig. 12-69. Dry-type beverage cooling evaporator. Aluminum casting, surrounding both refrigerant and beverage coils, permits rapid heat transfer. Refrigerant control is a thermostatic expansion valve. Constant pressure valve prevents freezing of beverage. (Dunham-Bush, Inc.)

12-25 ICE CUBE MAKER EVAPORATOR

There are many different ice maker mechanisms. The simplest, used in domestic refrigeration mechanisms, freezes water in ice cube trays.

Since a large portion of artificial ice is used for beverage cooling, ice in shapes other than cubes and in smaller sizes is frequently desirable. One type produces ice flakes. A flake ice maker is shown in Fig. 12-70.

Water is made to flow over an evaporator shaped like a cylinder. The surface of the evaporator is cold (0 °F or −18 °C) so that the water freezes rapidly.

Fig. 12-71 shows the auger and evaporator of an ice flaker system. Fig. 12-72 shows how the ice is removed from the evaporator. A heavy steel auger, driven by an electric motor, cuts and scrapes the ice from the surface.

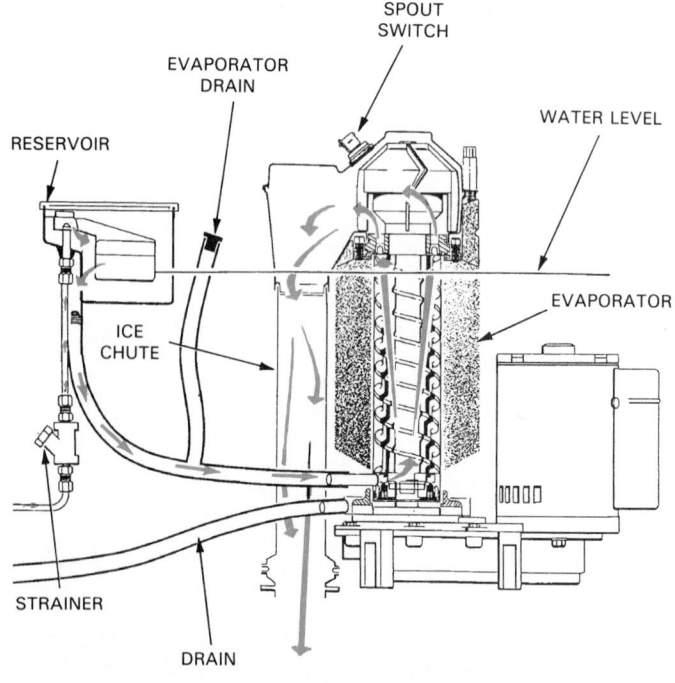

Fig. 12-71. Ice flaker system. Note water enters the reservoir, then enters the bottom of the evaporator; ice is formed, and turned into flakes as it passes through the auger. (Scotsman Ice Systems)

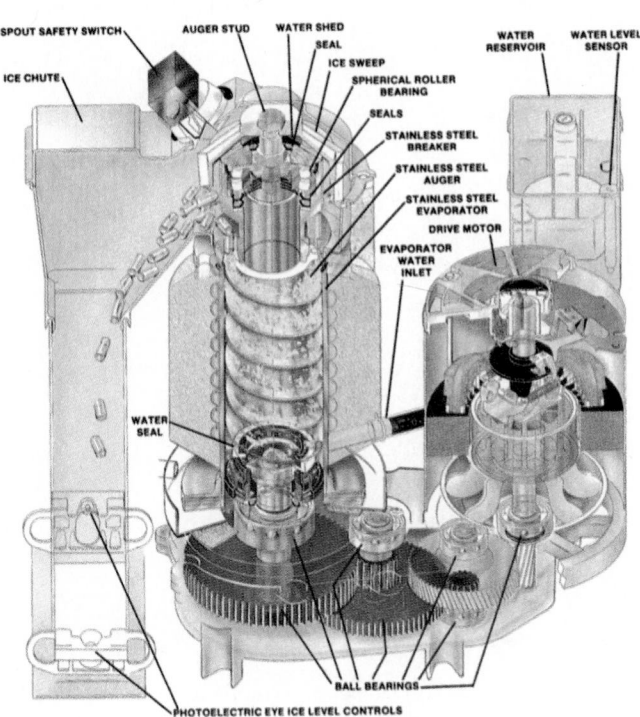

Fig. 12-72. Nugget ice machine. Note photoelectric eye used to control ice level after it is formed in stainless steel auger. (Scotsman Ice Systems)

Fig. 12-73 is a diagram of an ice flake maker using two evaporators and two harvesting augers. Ice flakes are fed into a storage bin.

Fig. 12-74 shows another cycle diagram for an ice flake

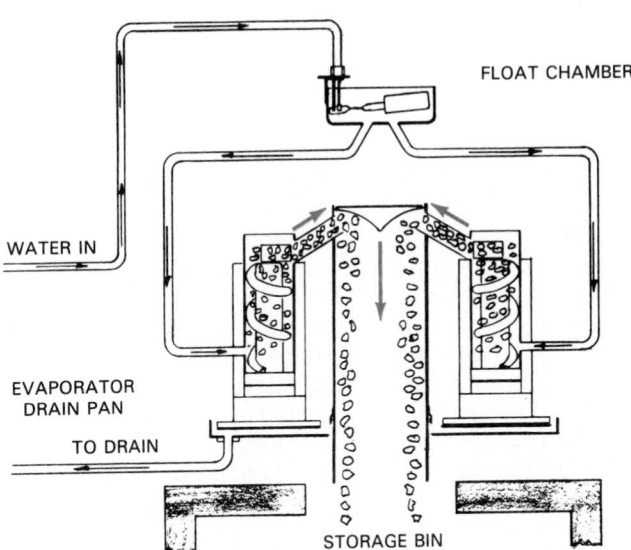

Fig. 12-73. Automatic flake ice system shows water and ice system. Water flow is float controlled. When enough ice flakes have been made, operation switch shuts off refrigerating unit and auger motors. (Ross-Temp/IMI Cornelius Americas, Inc.)

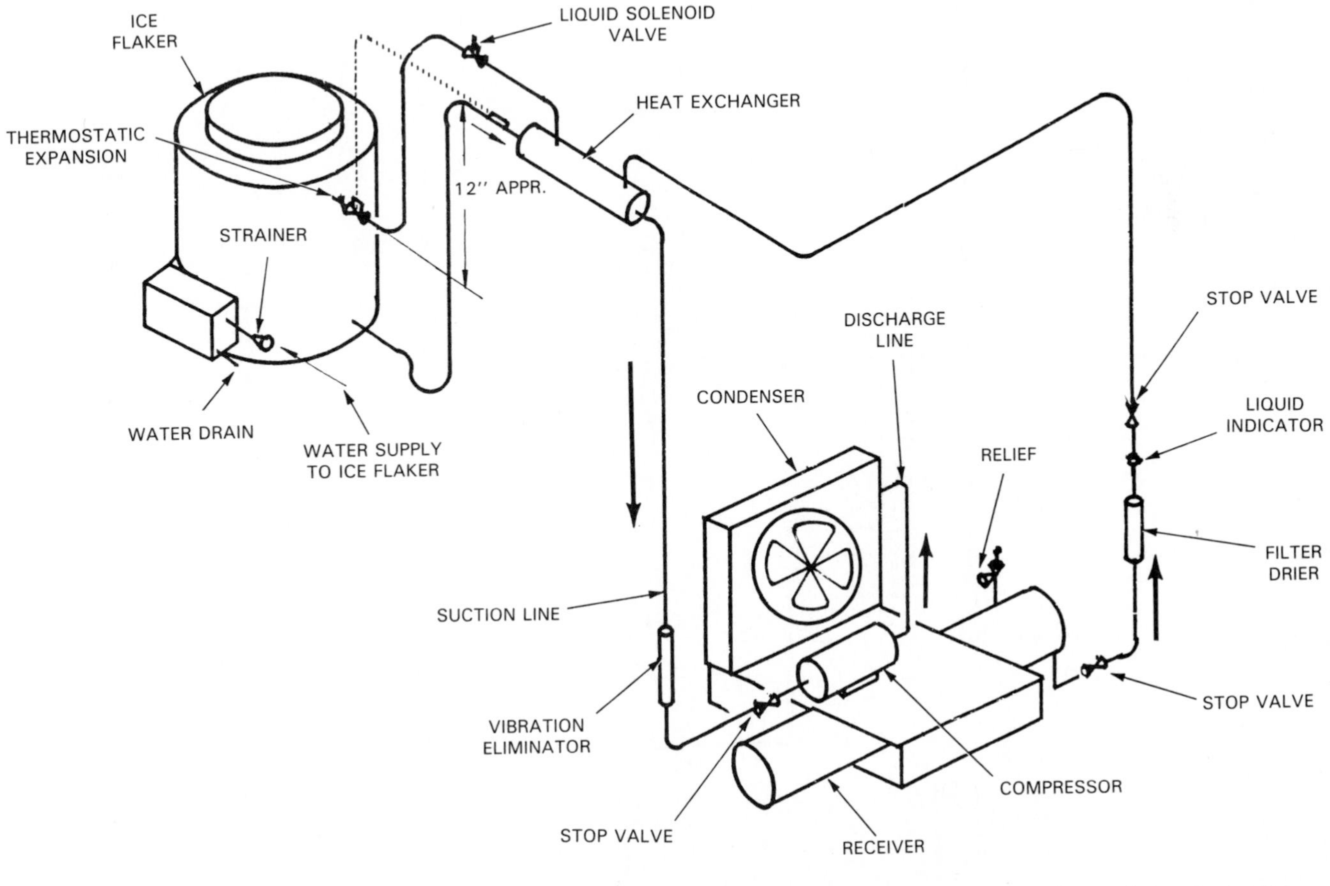

Fig. 12-74. Flake ice refrigeration system. Water circuit runs from "supply" to the water cooled condensing unit and also to the ice flaker. Return water from both condensing unit and flaker is discarded. (Howe Corp.)

maker. Fig. 12-75 shows the exterior view. All of these ice flake makers provide for float-level control of the water and a shutoff mechanism located in the storage bin. This stops ice making when the bin is full.

The square or cylindrically shaped cube with a hole in the center is formed by running water through a tube-within-a-tube. Ice forms on the inside of the inner tube. The refrigerant flows between the inner and outer tube. The cube is released from the tube during the defrost cycle.

Fig. 12-76 shows the process during the freezing part of the cycle. Fig. 12-77 shows the defrost cycle (hot gas) while Fig. 12-78 shows the water circuit with the recirculating pump and ice-cube-forming tubes.

Another style of ice cube maker is the one in Fig. 12-79. Water flows through vertical stainless steel tubes. When a hollow, square length of ice is formed, the refrigeration stops. Hot gas defrosting starts and, as the long square rods of ice slide down the tubes, they are cut into cubes. When all the tubes are empty, the refrigeration cycle starts over.

In the ice cube maker shown in Fig. 12-80, water is sprayed upward into very cold ice cube molds until the molds are filled with clear ice. When hot water is pumped around the metal cube container, the cubes fall into the storage bin. See Fig. 12-81.

Fig. 12-75. Commercial ice flake maker. Refer to Fig. 12-74 for its cycle diagram. (Howe Corp.)

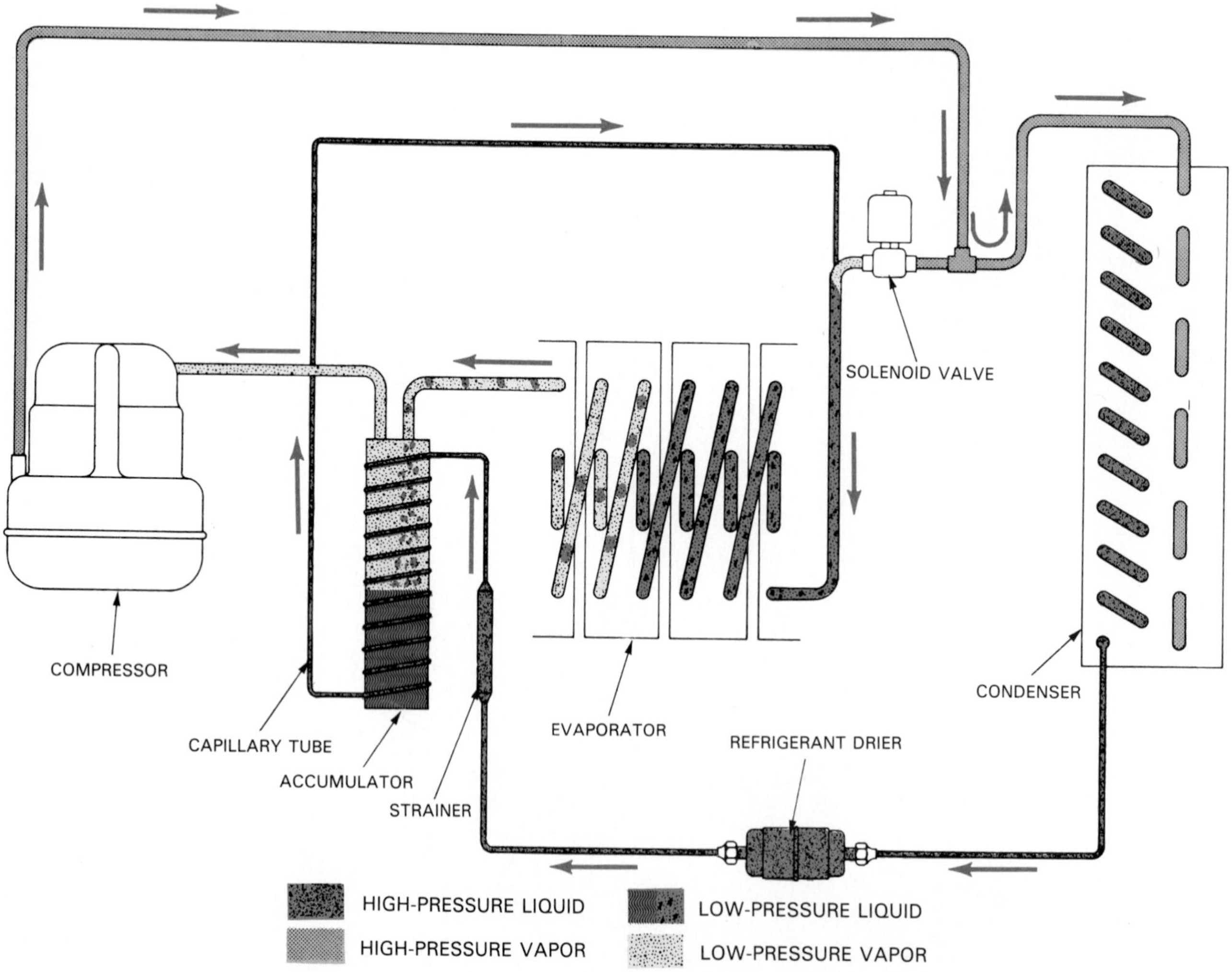

HIGH-PRESSURE LIQUID
HIGH-PRESSURE VAPOR
LOW-PRESSURE LIQUID
LOW-PRESSURE VAPOR

COMPRESSOR
CAPILLARY TUBE
ACCUMULATOR
STRAINER
EVAPORATOR
REFRIGERANT DRIER
SOLENOID VALVE
CONDENSER

Fig. 12-76. Ice cube maker refrigeration cycle during freezing process. Note capillary tube and accumulator. Cubes are released from evaporator tubes when solenoid valve opens to admit hot gas to evaporator.

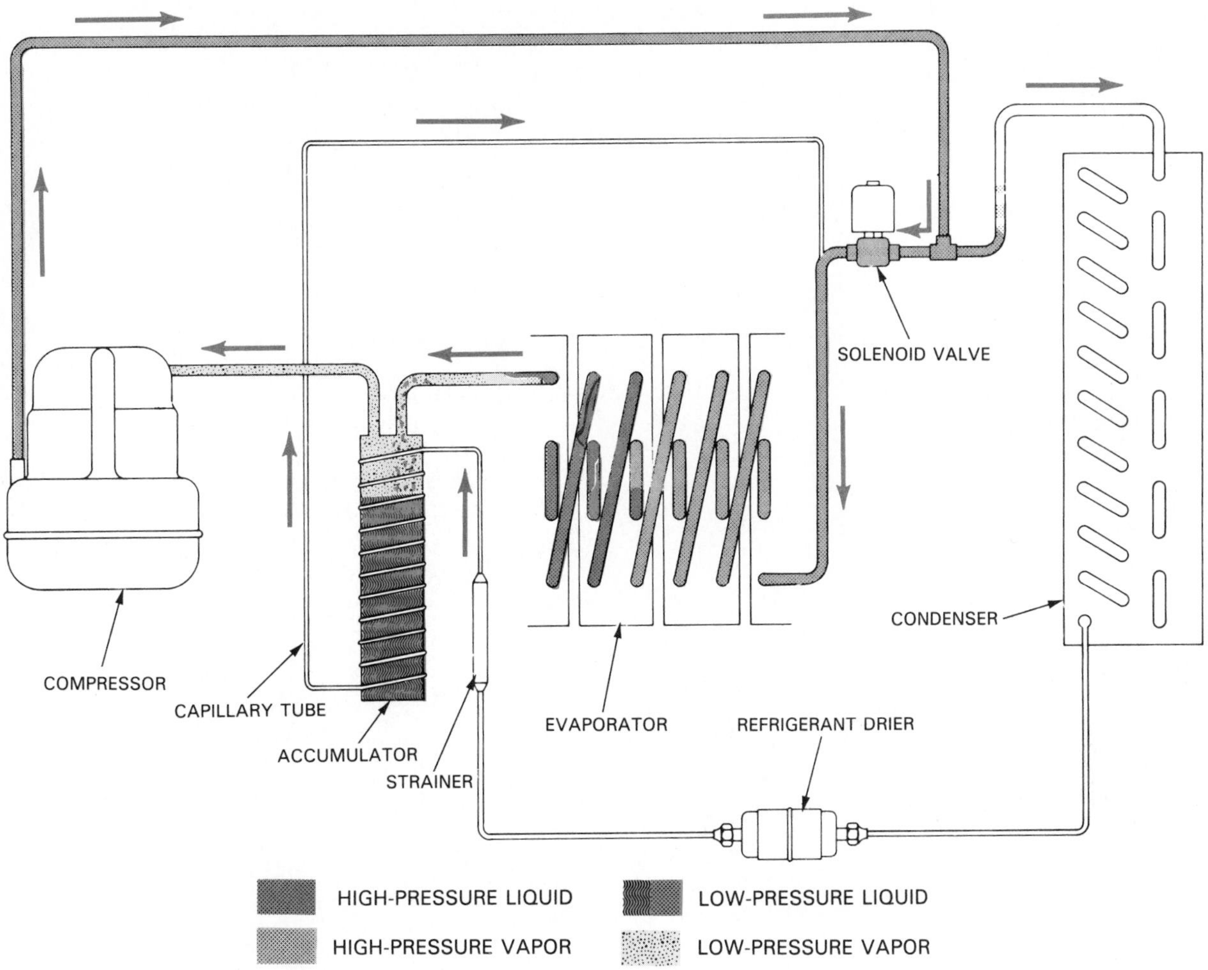

SOLENOID VALVE

COMPRESSOR

CAPILLARY TUBE

ACCUMULATOR

STRAINER

EVAPORATOR

REFRIGERANT DRIER

CONDENSER

| | HIGH-PRESSURE LIQUID | | LOW-PRESSURE LIQUID |
| | HIGH-PRESSURE VAPOR | | LOW-PRESSURE VAPOR |

Fig. 12-77. Defrost cycle of automatic ice cube maker. Solenoid valve is open. Hot gas heats and releases ice cubes which then fall into storage bin.

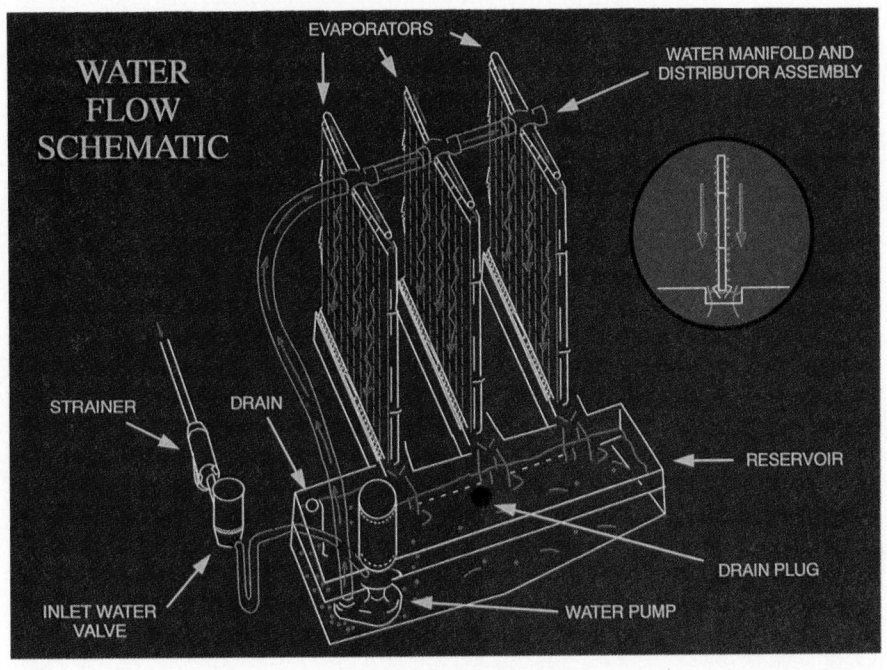

WATER FLOW SCHEMATIC

EVAPORATORS

WATER MANIFOLD AND DISTRIBUTOR ASSEMBLY

STRAINER

DRAIN

RESERVOIR

DRAIN PLUG

INLET WATER VALVE

WATER PUMP

Fig. 12-78. Water and ice circuit of automatic ice cube maker. Note inlet water strainer, water valve, evaporators, and ice cube forming tubes. (Scotsman Ice Systems)

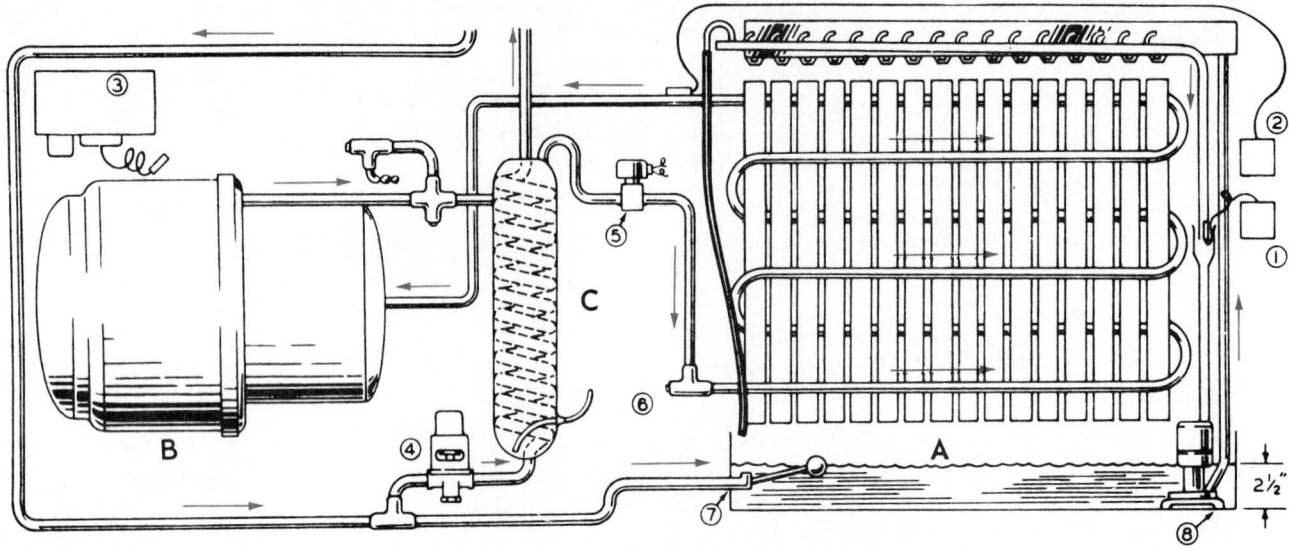

① T-1 OVERFLOW THERMOSTAT
② T-2 SUCTION LINE THERMOSTAT
③ COMBINATION BIN SAFETY THERMOSTAT & H.P. CUT-OUT
④ CONDENSER WATER REGULATING VALVE
⑤ HOT GAS SOLENOID VALVE
⑥ RESTRICTOR TUBE
⑦ MAKE-UP WATER VALVE
⑧ WATER RECIRCULATING PUMP

Fig. 12-79. Automatic ice-cube-making cycle. System uses hot-gas defrost. A—Evaporator and ice tubes. B—Compressor. C—Water-cooled condenser.

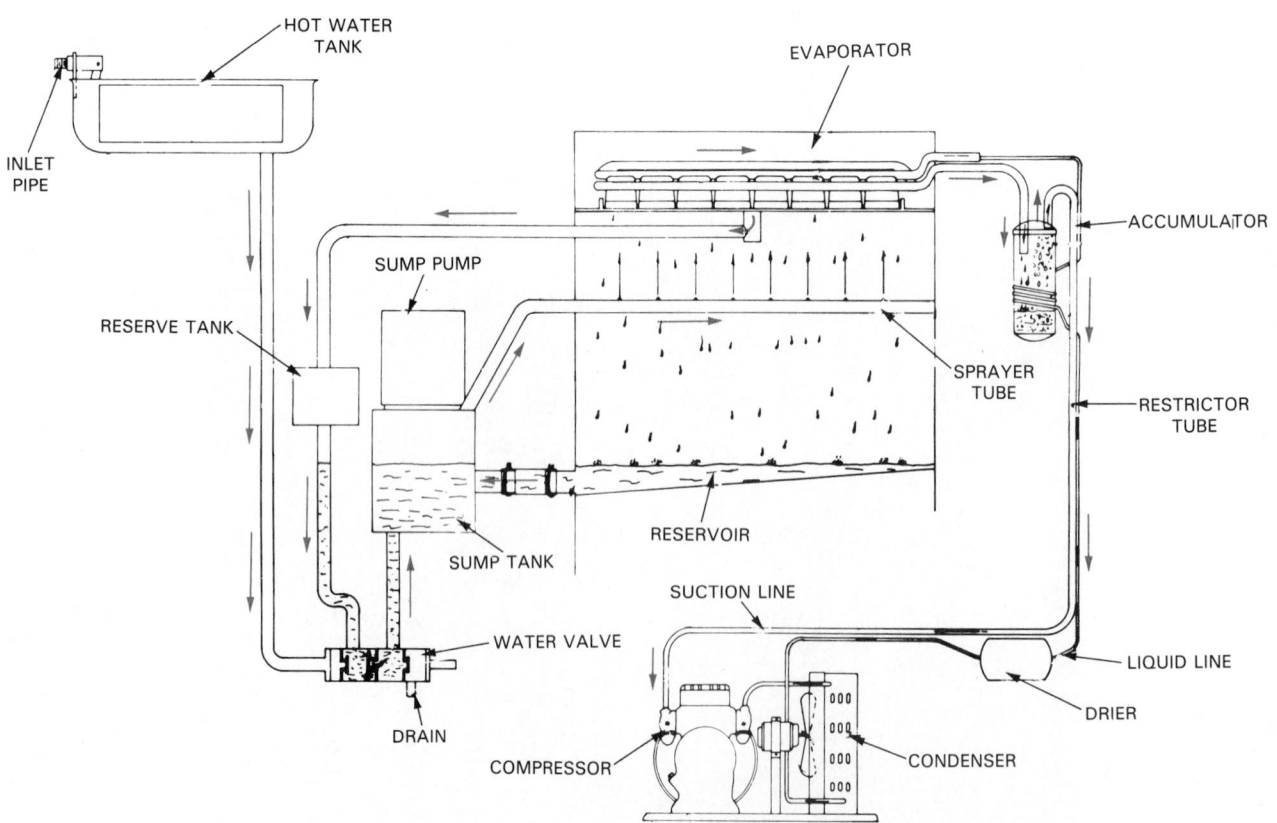

Fig. 12-80. Schematic shows freezing circuit of ice cube maker. Water freezes as it is sprayed into cold, inverted ice cube mold.

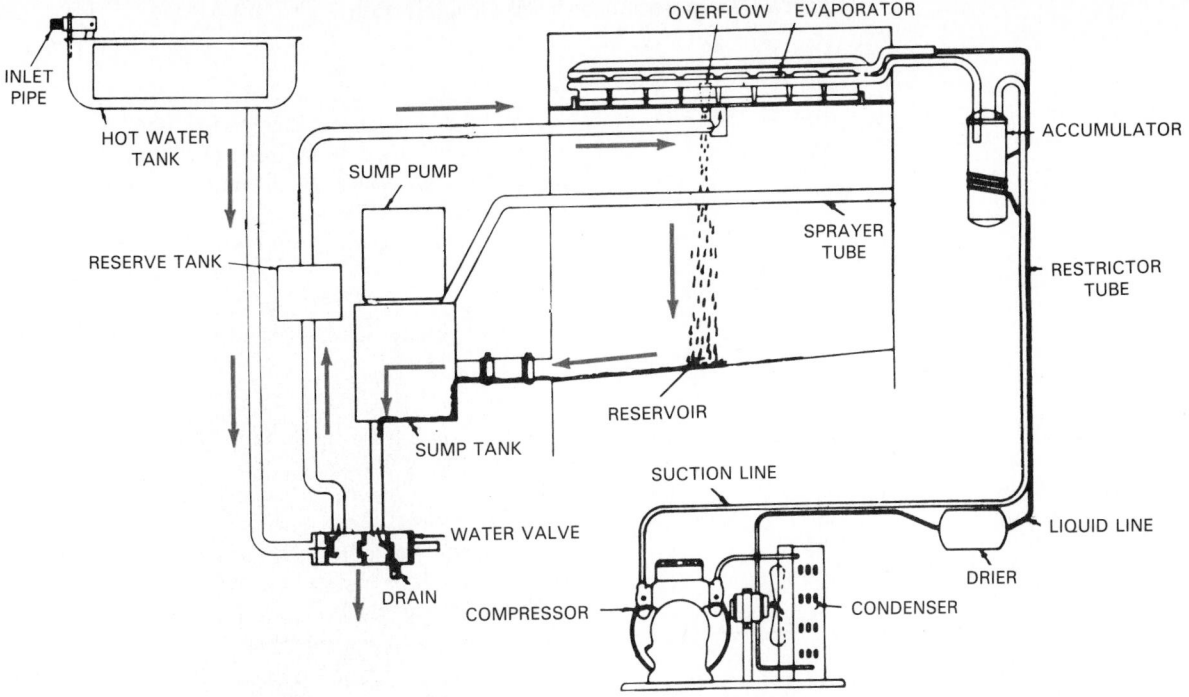

Fig. 12-81. Schematic for same unit shown in Fig. 12-80. As shown here, it is in the defrost cycle.

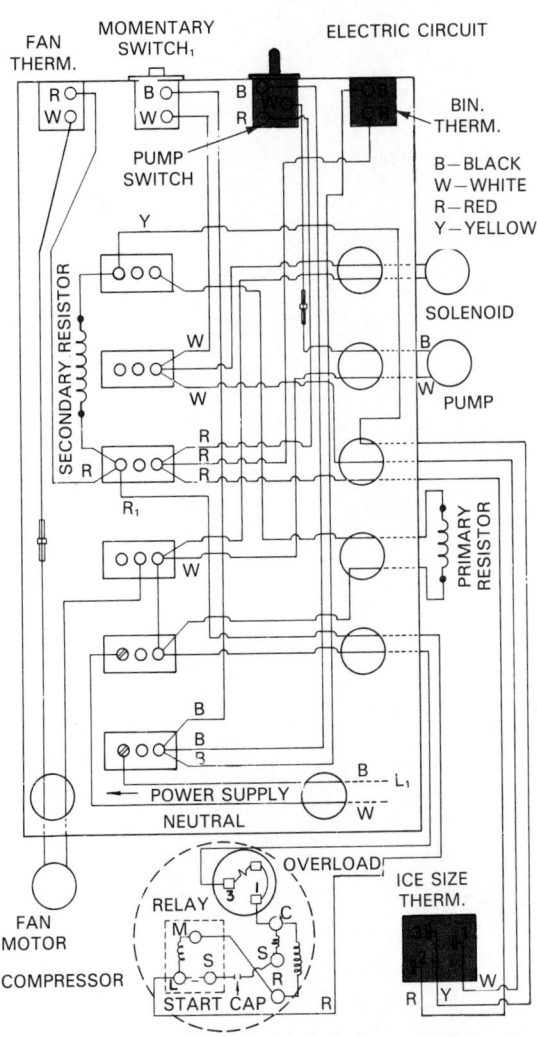

Fig. 12-82. Electrical circuits in an automatic ice maker. Note how a bin thermostat, ice-size thermostat, and a pump switch are connected into the circuitry.

Electrical circuit controls are needed with automatic ice makers. These include bin level controls, ice cube size controls, and water pump controls. The diagram in Fig. 12-82 is typical of electrical circuitry found in units just described.

The automatic unit shown in Fig. 12-83 makes solid ice cubes. Water flows over an inclined evaporator. A layer of ice is formed about 5/8 in. thick. During the defrost cycle (hot gas), this layer of ice is freed from the plate and, because of gravity, it slides over the electric grid. When the grid wires are heated, the wires cut through the plate of ice producing cubes.

Design of the electrically heated grid is shown in Fig. 12-84. Fig. 12-85 is a wiring diagram for this unit. The grid circuit has a transformer to reduce grid voltage to an

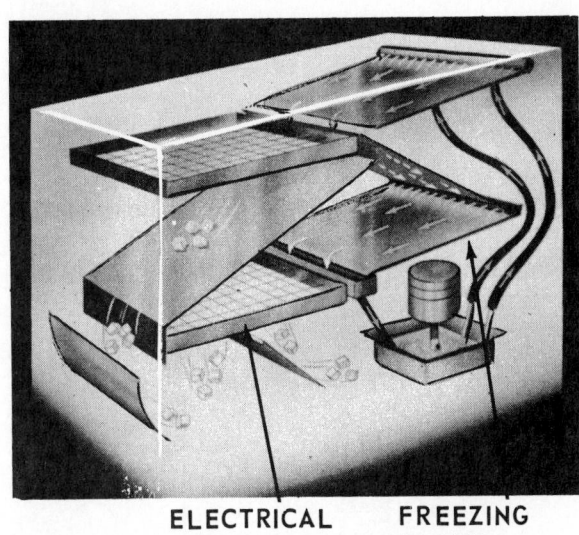

ELECTRICAL GRID   FREEZING PLATE

Fig. 12-83. Automatic ice cube maker produces slab of clear ice and then wire grid changes it into cubes.

433

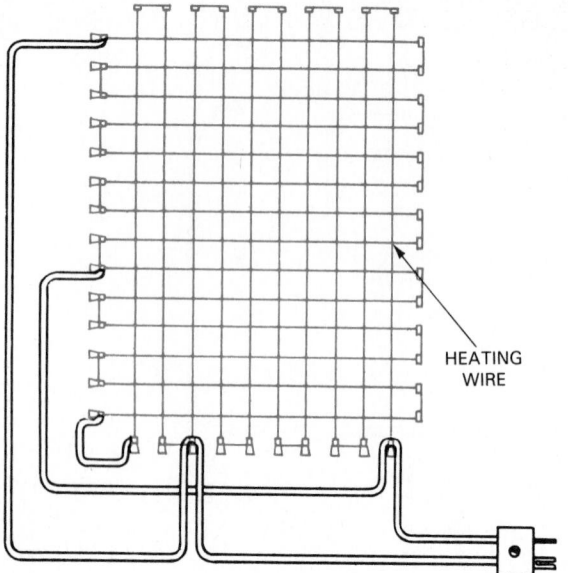

Fig. 12-84. Diagram of electrically heated grid. The hot grid melts ice slabs into cubes.

efficient level.

The most popular refrigerant for these systems is R-12. *It is very important to follow the plumbing code in piping away the drain water. Avoid any situation where a loss of water pressure in the building's water supply could cause reverse syphoning and contamination of the drinking water in the building.*

12-26 FREEZING DISPENSER EVAPORATORS

Freezing dispenser evaporators are used to keep the mix cold and then to freeze and churn the mix before it is force fed into cups for the customer. The temperatures are very critical.

The freezing evaporator is a cylinder in which a motor-driven dasher churns the mix. The refrigerant then travels to premix storage compartment. See Fig. 12-86 for wiring diagram. Note the instructions for making the electrical connections for the dasher motor.

For the actual mechanism, see Fig. 12-87. It uses a serviceable hermetic motor compressor, a water-cooled condenser, and a separate motor to drive the mixing blade in the evaporator.

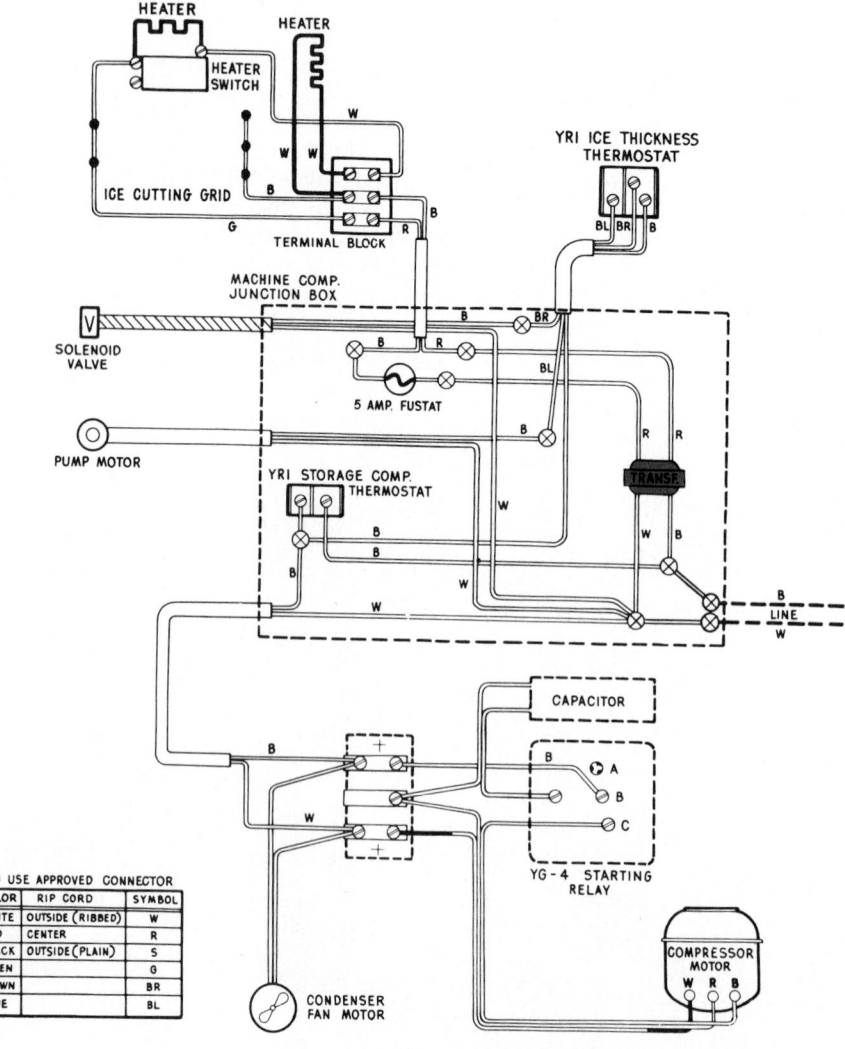

Fig. 12-85. Wiring schematic for automatic ice cube maker which uses electrically heated grid. The four grid circuits are in a parallel arrangement. Special transformer powers grid.

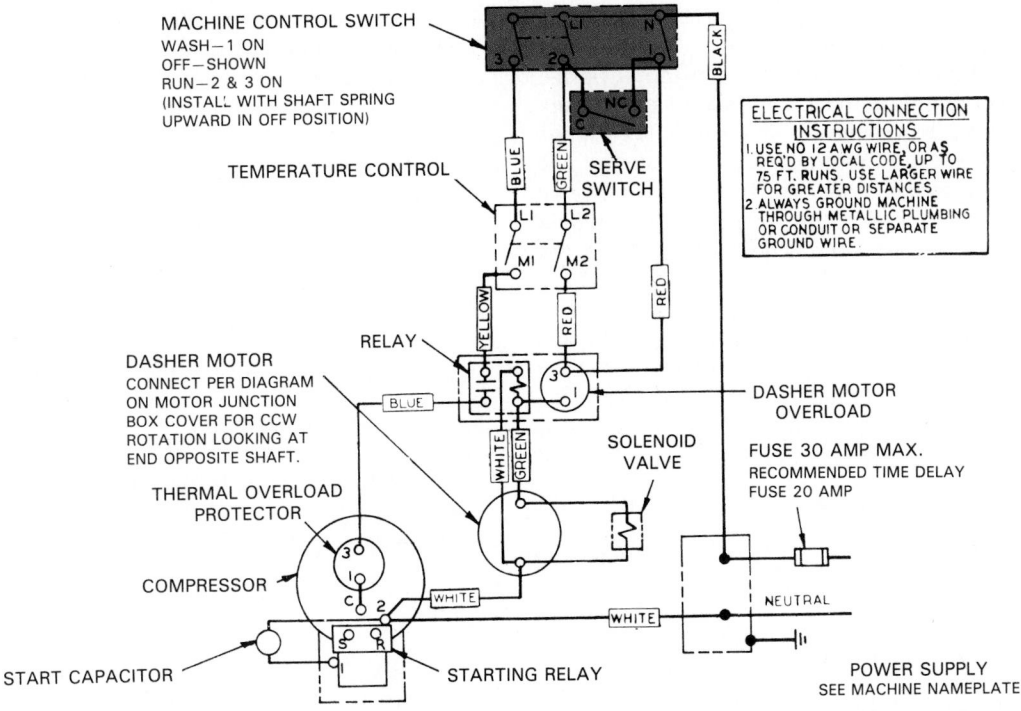

MACHINE CONTROL SWITCH
WASH—1 ON
OFF—SHOWN
RUN—2 & 3 ON
(INSTALL WITH SHAFT SPRING
UPWARD IN OFF POSITION)

TEMPERATURE CONTROL

SERVE SWITCH

ELECTRICAL CONNECTION INSTRUCTIONS
1. USE NO 12 AWG WIRE, OR AS REQ'D BY LOCAL CODE, UP TO 75 FT. RUNS. USE LARGER WIRE FOR GREATER DISTANCES.
2. ALWAYS GROUND MACHINE THROUGH METALLIC PLUMBING OR CONDUIT OR SEPARATE GROUND WIRE.

DASHER MOTOR
CONNECT PER DIAGRAM ON MOTOR JUNCTION BOX COVER FOR CCW ROTATION LOOKING AT END OPPOSITE SHAFT.

RELAY

DASHER MOTOR OVERLOAD

THERMAL OVERLOAD PROTECTOR

SOLENOID VALVE

FUSE 30 AMP MAX.
RECOMMENDED TIME DELAY FUSE 20 AMP

COMPRESSOR

START CAPACITOR

STARTING RELAY

NEUTRAL

POWER SUPPLY
SEE MACHINE NAMEPLATE

Fig. 12-86. This wiring diagram is for a soft ice cream freezing dispenser. Note control switches for serving and for washing.

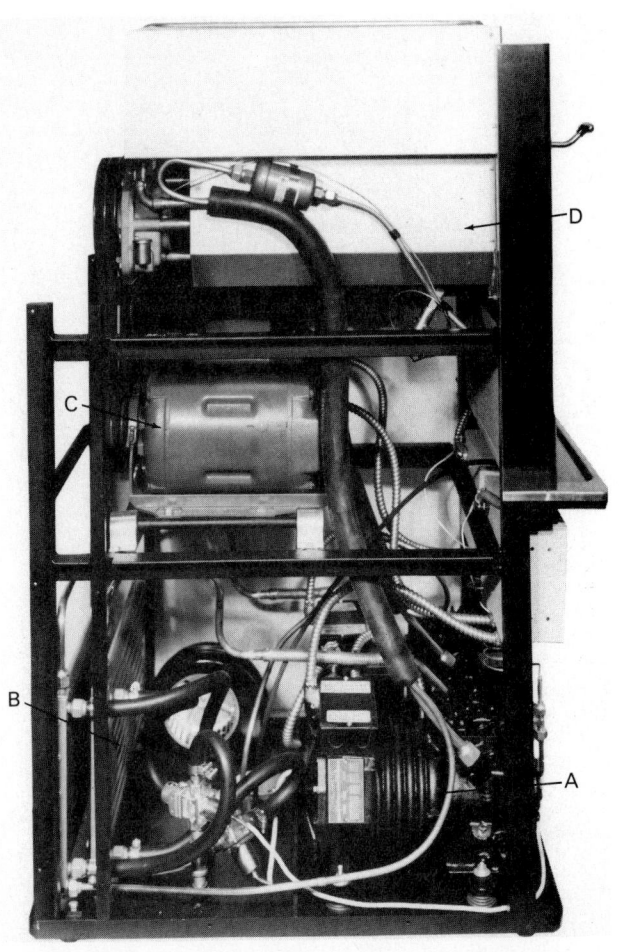

Fig. 12-87. View of freezer-dispenser with part of case removed. A—Motor compressor. B—Water-cooled condenser. C—Freezer mixing blade drive motor. D—Evaporator. (Taylor Freezer)

12-27 EVAPORATOR DEFROSTING

Many evaporators operate at temperatures below freezing. The demand for open display cases and frozen food requires these low-temperature systems. The evaporators operate at refrigerant temperatures of 0 °F (−18 °C), −10 °F (−23 °C), and even −20 °F (−29 °C). Blower evaporators are often used.

Low temperatures and small fin spacings make frequent defrosting necessary. Frost accumulation would otherwise soon clog the evaporator. Other types of evaporators also need defrosting even though not so frequently. It is desirable that this be done with very little rise in fixture temperature.

Defrosting is usually automatic. Some evaporators defrost during each "off" part of the cycle. On others, a time clock control either turns on the defrosting mechanism once a day or after a given number of hours of compressor operation.

There are six defrosting methods:
1. Hot refrigerant vapor system.
2. Nonfreezing solution system.
3. Water system.
4. Electric heater system.
5. Reverse cycle defrost system.
6. Warm air system.

These defrosting devices may either heat the evaporator internally (from the inside) or externally (from the outside) to melt the frost.

It is important to clean the evaporator, drain pans, and drain lines frequently.

12-28 EVAPORATOR MOUNTING

The evaporators are mounted in the cabinet using one of three methods:
1. Suspended from the ceiling.

2. Mounted on a pipe fastened to the vertical baffle and the wall of the cabinet.

3. Mounted on stands fastened to the horizontal baffle.

Where the evaporator is vertical, the thermostatic expansion valve may be connected to the top. This allows refrigerant and oil to flow by gravity into the suction line down to the compressor. Sometimes, the expansion valve is connected to a horizontal run of evaporator coil tubing making the refrigerant pass horizontally to come to the suction line as shown in Fig. 12-88.

If the expansion valve is fastened to the upper part of the coil permitting a gravity flow, oil binding will be negligible. However, the fins will be coldest at the top where they are in contact with the warmest air.

12-29 HOT GAS DEFROST SYSTEM

In a "hot gas" system, hot refrigerant vapor is pumped directly through the evaporator tubing. The system has a refrigerant line running directly from the compressor discharge line up to the evaporator. In some cases, this line is connected between the thermostatic expansion valve (TEV) or the capillary tube and the evaporator. The line is opened and closed by a solenoid shutoff valve.

At the predetermined time (usually 12 midnight or 1 A.M.), the time clock closes a circuit which starts the compressor, opens the solenoid valve, and stops the evaporator fan motors. Hot compressed vapor rushes through the evaporator, warms it, and then returns to the compressor along the suction line. See Fig. 12-89.

Such a system will usually defrost the evaporator in 5 to 10 minutes. To keep defrost water from freezing in the drain pan and tube, part of the hot gas defrost line or a small electric heater is installed under the drain pan and the drain pipe.

It is best to evaporate the refrigerant that condenses in the evaporator during the defrost cycle. The following steps explain how this is done:

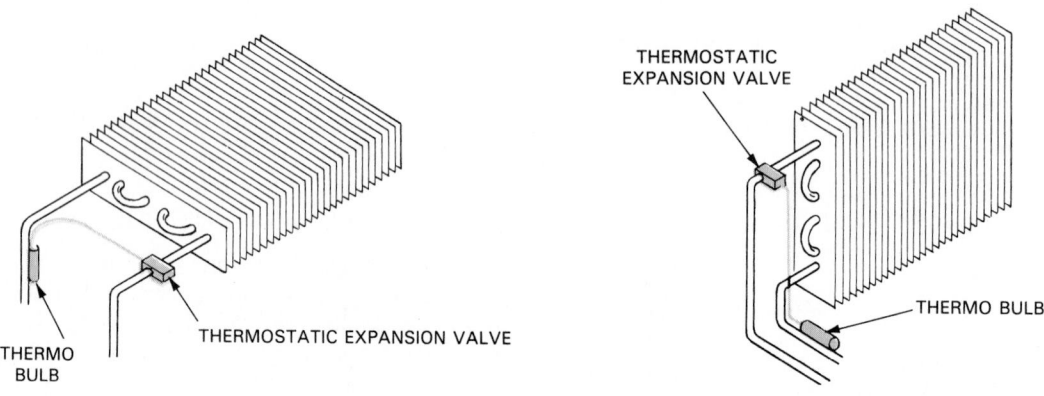

Fig. 12-88. Recommended expansion valve mounting for two different evaporator installations.

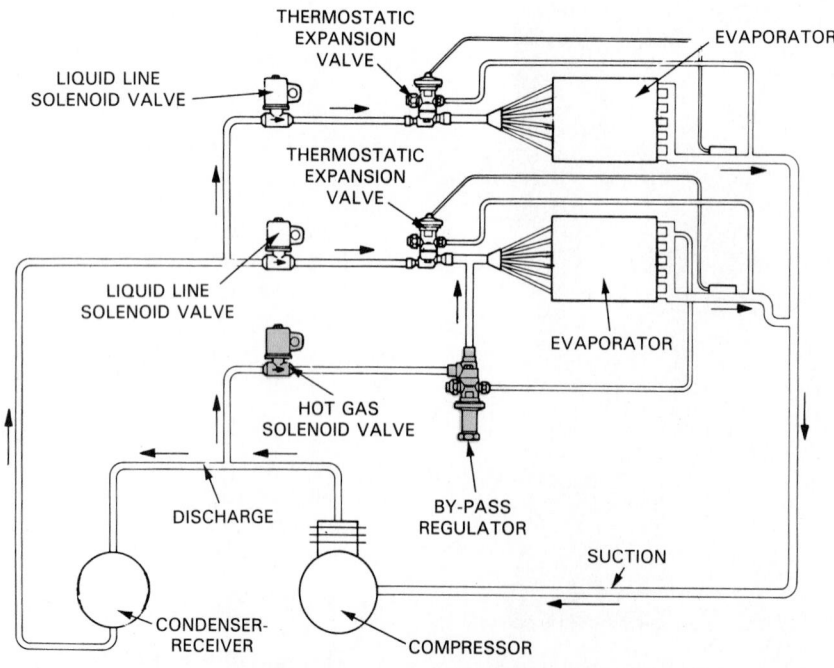

Fig. 12-89. Schematic of typical "hot gas" bypass from compressor discharge to evaporator inlet. Upper evaporator is nonfrost type. (Alco Controls Div., Emerson Electric Co.)

1. A defrost bypass puts some hot gas into the suction line to vaporize any liquid refrigerant there. Some hot gas bypass systems also use a liquid injection system. A TEV is mounted on a line between the liquid line and the suction line. Its sensing bulb is mounted on the suction line. If gas returning to the compressor becomes warm, the TEV will open and mix liquid refrigerant with the hot bypass gas. The mix is kept at 45 °F (7 °C) to 65 °F (18 °C) to make sure that the compressor will be cooled. A solenoid valve in the bypass TEV inlet shuts off the TEV when the system is operating normally or on a full load.

2. Heat is applied to vaporize the returning refrigerant. This heat is electric in some cases.

3. A special blower-evaporator may be installed in connection with the suction line. This, plus air forced over the re-evaporator, insures that only vapor can get back to the compressor. The blower works only while the unit is on defrost. *It is best to always use an accumulator mounted in the suction line to trap liquid refrigerant and vaporize it before it reaches the compressor.*

4. Sometimes the hot gas is fed backwards into the evaporator and the condensed refrigerant bypasses the TEV by means of a check valve. This forces the liquid into the receiver by way of the liquid line.

A system for forcing the hot gas backwards through the evaporator is shown in Fig. 12-90. The defrost cycle is shown in Fig. 12-91.

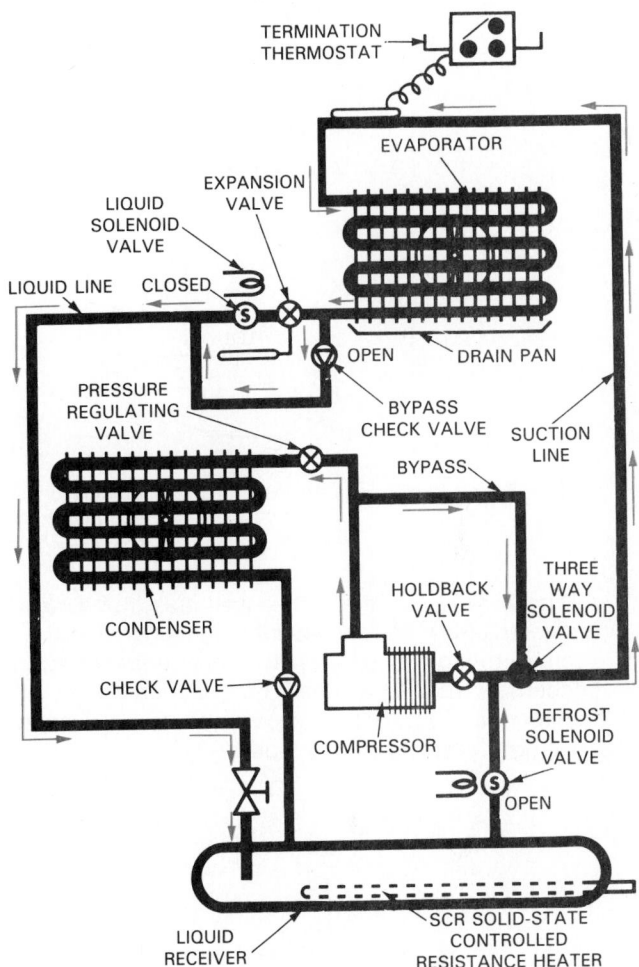

Fig. 12-91. Defrost cycle of the two-pipe hot gas defrost system. Note that hot gas from compressor and receiver is traveling back to the evaporator by way of suction line. Condensed liquid refrigerant is traveling around TEV by way of the check valve and is returning to the receiver through the liquid line. Heater in receiver provides more hot gas for defrosting.

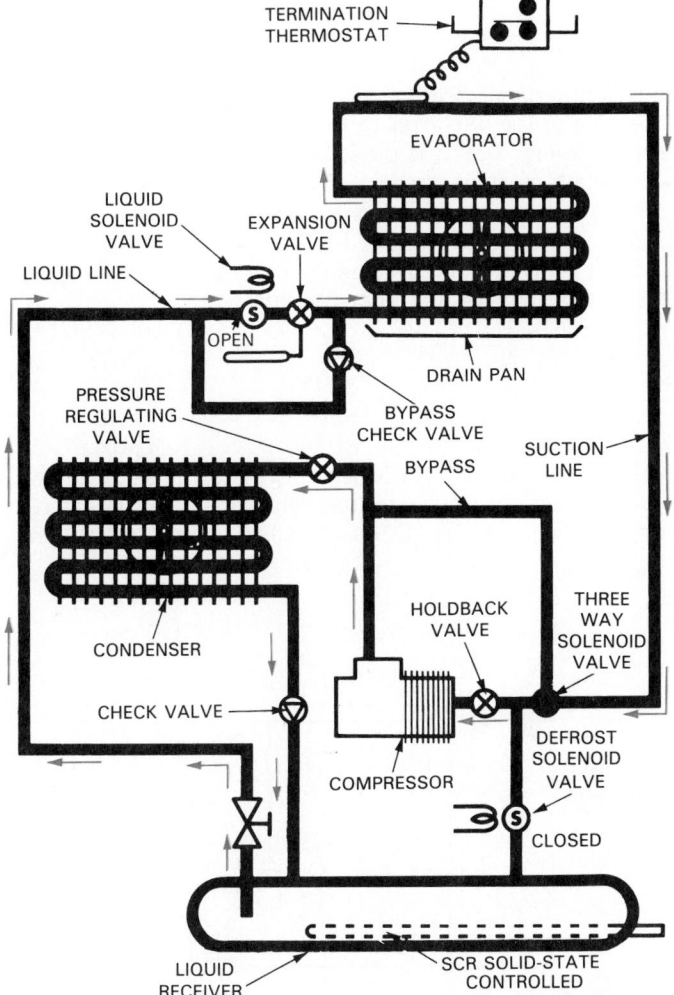

Fig. 12-90. Cooling cycle for normal two-pipe hot gas defrosting system. Note direction of refrigerant flow through compressor and to the condenser.

A timer starts the defrost action and a termination thermostat returns the system to normal refrigerating. When defrosting starts:

1. A solenoid valve opens a line from the top of the receiver to the suction line.

2. A holdback valve reduces the high-pressure gas as it goes into the compressor.

3. A three-way solenoid closes the suction line to the compressor and opens a valve to allow hot gas up the suction line to the evaporator. The hot gas warms the evaporator and then condenses.

4. The condensed refrigerant bypasses the TEV through a check valve and travels to the receiver by way of the liquid line.

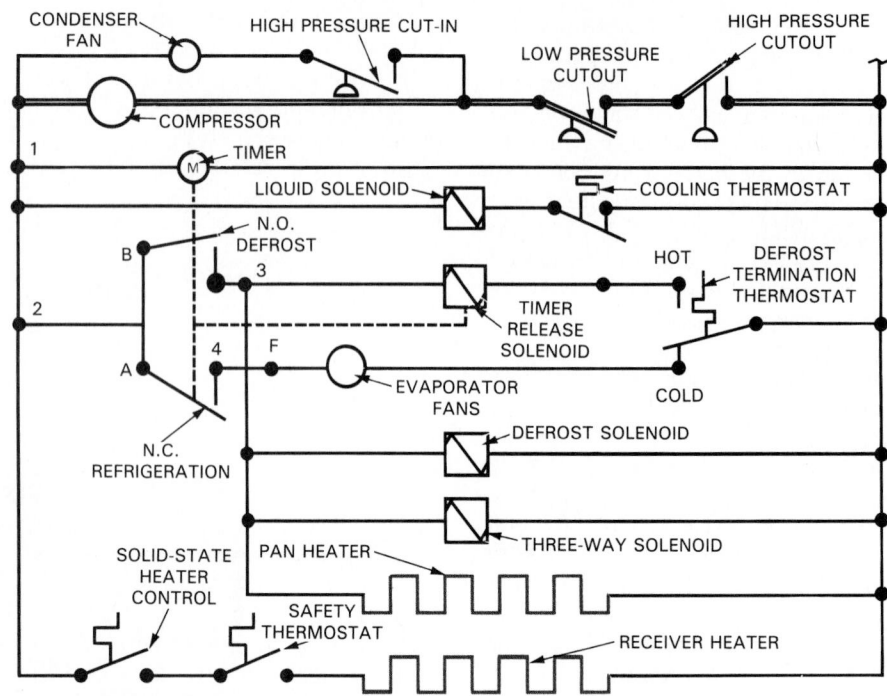

Fig. 12-92. Wiring diagram of a two-pipe hot gas defrosting system. Note the electric pan heater and electric receiver heater. Note, also, that cooling thermostat only controls liquid line solenoid.

5. A check valve in the condenser drain tube keeps refrigerant from backing up into the condenser.
6. A pressure regulating valve at the top (entrance to the condenser) maintains good hot gas pressures and temperatures. Fig. 12-92 shows the wiring diagram. One type of hot gas bypass valve has a connection to the suction line. This valve is adjusted by the low-side pressure of the vapor going to the compressor. The valve shown in Fig. 12-93 is pressure operated. It opens wider as suction pressure drops and starts to close as low-side pressure rises. The pilot-controlled valve is installed in a system as shown in Fig. 12-94.

Another type has an adjustable bellows in the sensing element to change the opening pressure:

OPENING PRESSURE	PSI	ADJUSTMENT RANGE
R-12	30	25-35
R-22	58	50-65
R-500	38	32-44

The valve sends hot condenser gas into the evaporator if the evaporator refrigerant temperature reaches freeze-up temperatures such as 26°F (−3°C). It is used where there

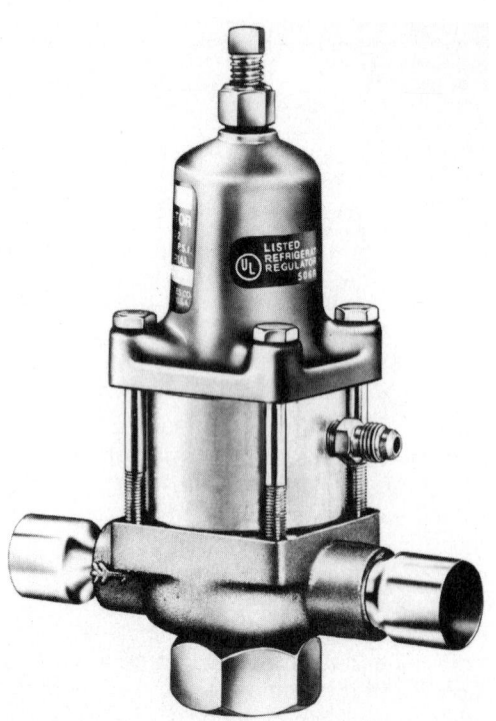

Fig. 12-93. Hot gas bypass valve. Its hot gas capacity using R-12 is 4 to 13 tons. Using R-22, it is 7.4 to 24 tons. (Refrigerating Specialties Div., Parker-Hannifin Corp.)

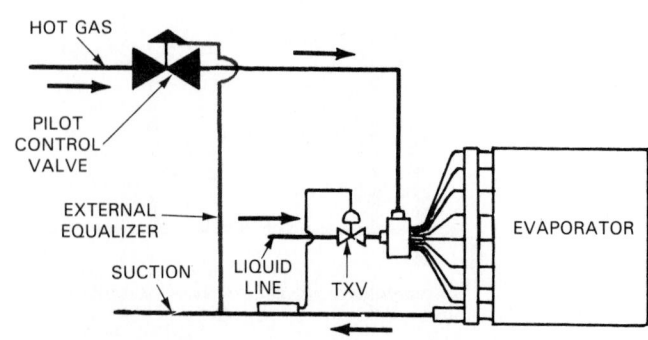

Fig. 12-94. Schematic diagram of hot gas bypass regulator. Note external equalizer line. It operates valve dependent on suction line pressure. (Refrigerating Specialties Div., Parker-Hannifin Corp.)

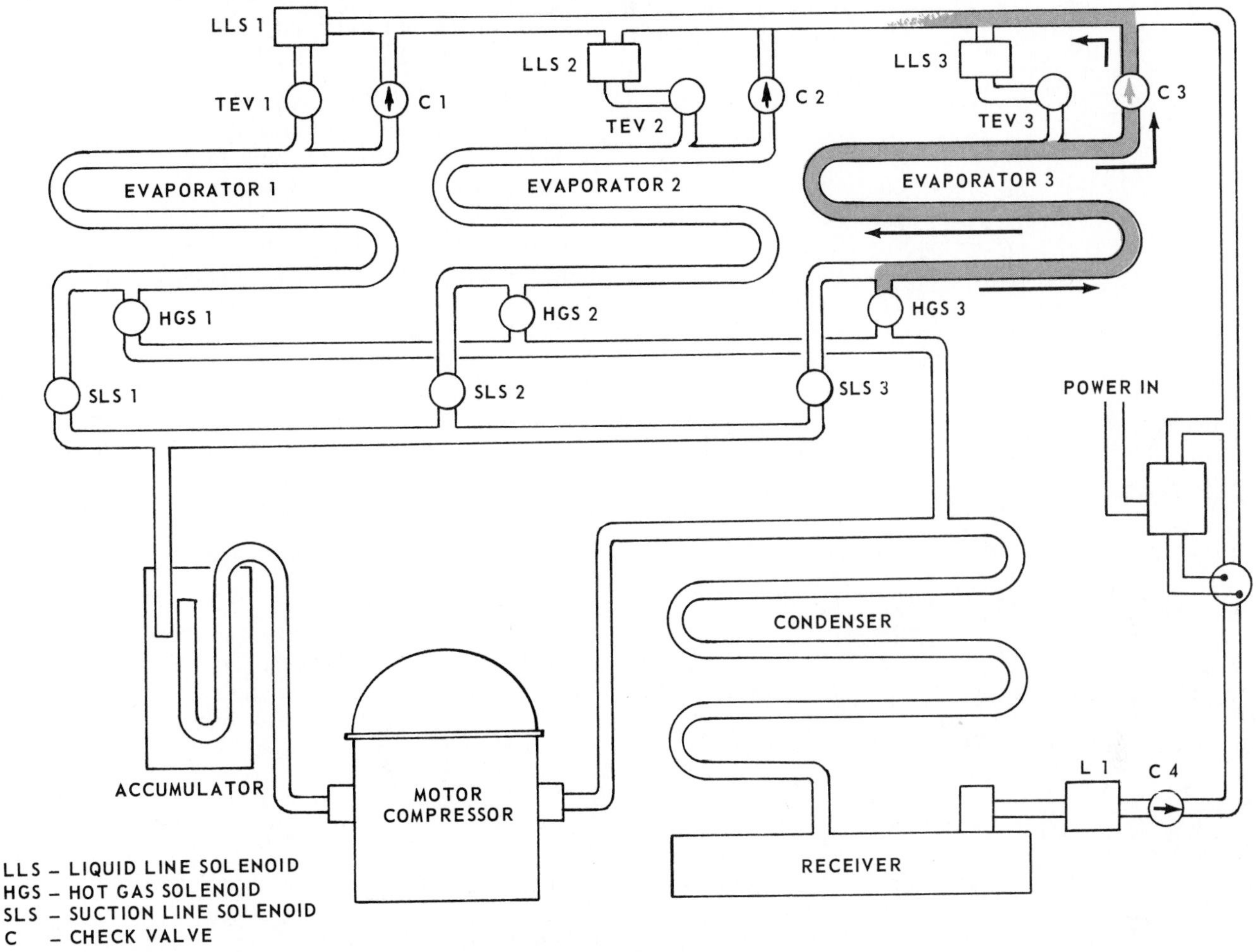

Fig. 12-95. "Hot gas" defrost system for multiple system. Timer controls each evaporator defrost at a different time. Liquid solenoid L1 closes, partially pumps down system. Hot gas solenoid on one evaporator opens suction line, solenoid valve closes and hot gas rushes into evaporator backwards. It enters liquid line through bypass check valve and feeds liquid refrigerant to other two evaporators. If Evaporator 3 is to defrost, Solenoid SLS3 closes, HGS3 opens, and LLS3 closes. Hot gas flows into Evaporator 3, condenses as it melts frost; then condensed liquid goes into liquid line by way of Check Valve C3. This liquid then moves into Evaporators 1 and 2 through TEVs 1 and 2.

LLS – LIQUID LINE SOLENOID
HGS – HOT GAS SOLENOID
SLS – SUCTION LINE SOLENOID
C – CHECK VALVE

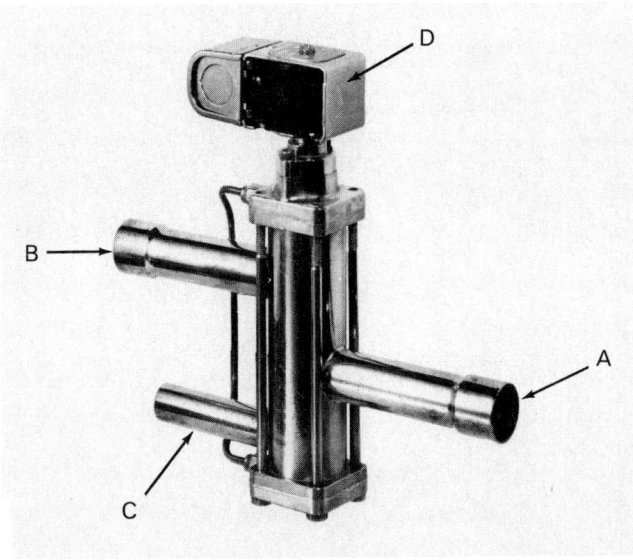

Fig. 12-96. Hot gas defrost solenoid valve. A—Evaporator connection. B—Suction line to compressor. C—Hot gas defrost connection. D—Pilot solenoid valve.
(Fluidex Division, Parker-Hannifin Corp.)

are intervals of low-heat load conditions. This system is used on medium tonnage units (5 to 30 tons).

Slugs of liquid refrigerant must be stopped from entering the compressor where they would cause damage. Re-evaporation should be almost complete before the refrigerant reaches the compressor.

In multiple systems having several evaporators, it is good practice to defrost one evaporator at a time using the other to evaporate the liquid coming from the defrosting evaporator. This method makes sure there will be no liquid refrigerant reaching the low side of the motor compressor. See Fig. 12-95.

A valve may take the place of the two suction line valves on each evaporator. See Fig. 12-96. The evaporator gas at point A travels into the valve during normal operation, and the hot gas travels out of the valve at A during defrosting. The internal construction of the valve can be seen in Fig. 12-97. When the pilot solenoid valve at A is energized, it opens and allows high-pressure gas from the discharge connection to push down on the double valve. This action closes the top valve B. It stops flow from evaporator into suction line, and opens bottom valve, C. High-pressure hot gas flows from the discharge connection up into the evaporator.

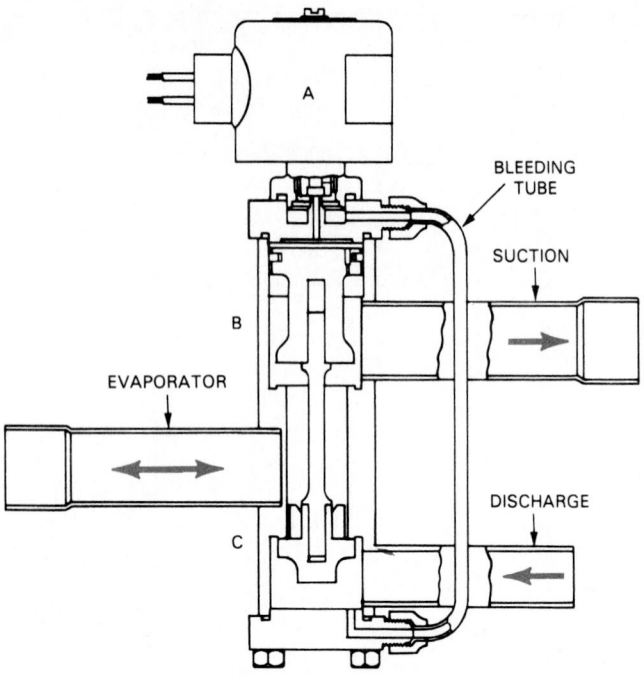

Fig. 12-97. Internal construction of a hot gas defrost solenoid valve. Note small tubing connection from discharge connection to pilot valve bleed. (Fluidex Division, Parker-Hannifin Corp.)

12-30 NONFREEZING SOLUTION DEFROST SYSTEM

The "hot fluid" defrost system has been used for years. It has a container in which a brine (a nonfreezing solution) is stored. The refrigerant vapor from the compressor is pumped through this heat storage container before it goes to the condenser.

The brine in the container may also be electrically heated. Such heating is provided during the normal running (freezing) part of the refrigerating cycle.

When the refrigerating system shuts off, the defrost timer closes a solenoid valve in a line running from the liquid line to the evaporator. This is the beginning of the defrost cycle. The evaporator fan is usually shut off. The brine solution is pumped through its own piping along the drain line, the drain pan, and the evaporator. Then it returns to its container. Fig. 12-98 shows such a defrost cycle.

12-31 WATER DEFROST SYSTEMS

Either manually or automatically, the water defrost system runs tap water over the evaporator while the system is not running. During this operation, the evaporator louvers are closed. The water is warm enough to melt the ice which drains away into the evaporator drain pan. Drainage from the water lines must be complete before the unit is turned

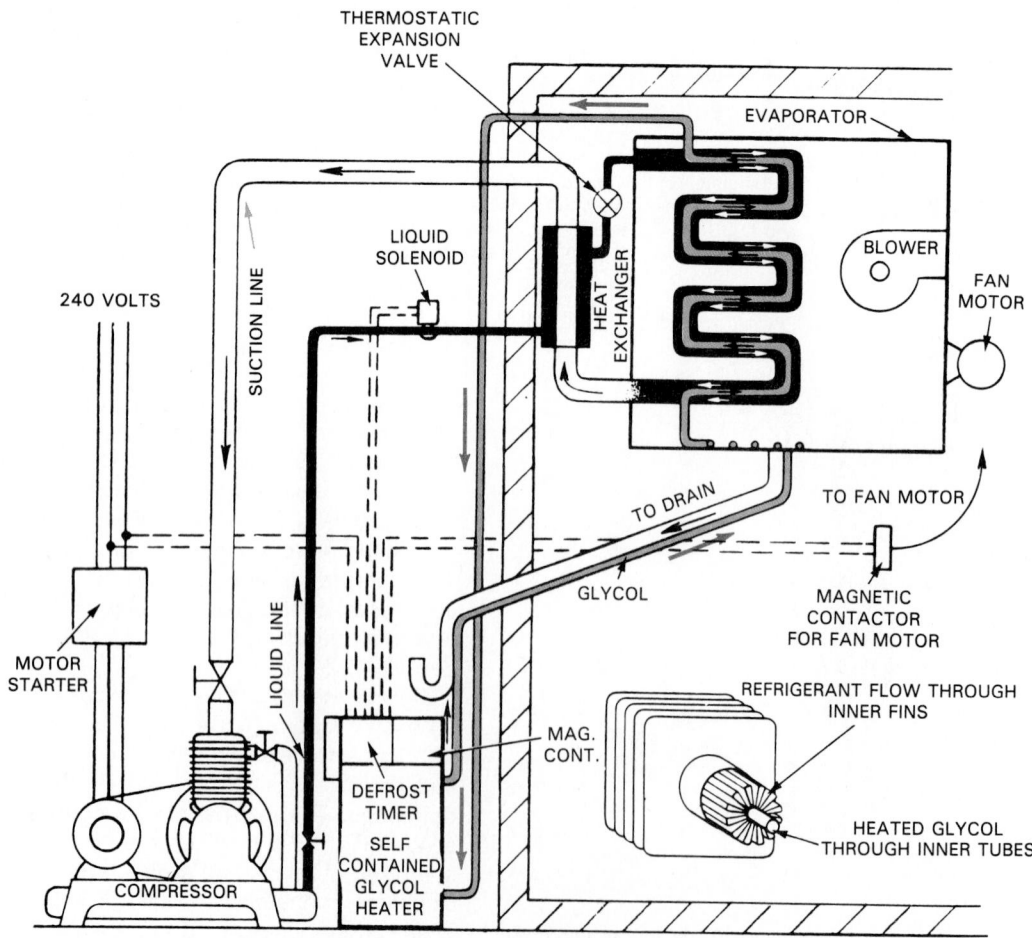

Fig. 12-98. Nonfreeze solution defrosting system. During defrost, glycol solution is pumped through inner tubing of evaporator and along the drain piping.

on or the water will freeze.

Either the water is sprayed over the evaporator or it is fed to a pan located over the evaporator. Holes in the pan feed the water evenly over the evaporator.

An electric timer provides automatic operation. Fig. 12-99 demonstrates the principle of water defrost and shows the two types of manual water defrost, as well as one automatic defrost system.

Special systems have been designed to defrost by spraying a brine over the evaporator. A pump may be employed to recirculate a lithium chloride brine. Eliminator plates are needed to prevent brine spray from passing into the refrigerated space.

12-32 ELECTRIC HEATER DEFROST SYSTEM

Electric heat is popular for defrosting low-temperature evaporators. Heating coils are installed in the evaporator, around it, or within the refrigerant passages.

One type uses resistance wire heating elements mounted underneath the evaporator, under the drain pan, and along the drain pipe. A timer stops the refrigerating unit, closes the liquid line, and pumps the the refrigerant out of the evaporator. Then the blowers and the electric heaters are turned on.

The heaters quickly melt the frost from the evaporator and the water drains away. When the evaporators are warm enough to insure that all frost is gone, a thermostat on the evaporator returns the system to normal operation.

''Pump down'' is a control system in which the thermostat operates a solenoid in the liquid line while a low-pressure switch operates the compressor. Its purpose is to prevent flow of liquid refrigerant from the evaporator to the compressor. It is especially important in cases of electric defrost.

1. When the thermostat is satisfied, it opens and the liquid line solenoid closes.
2. The compressor continues to run and continues to remove the refrigerant vapor from the evaporator and suction line.
3. When the proper low-side pressure is reached, the low-

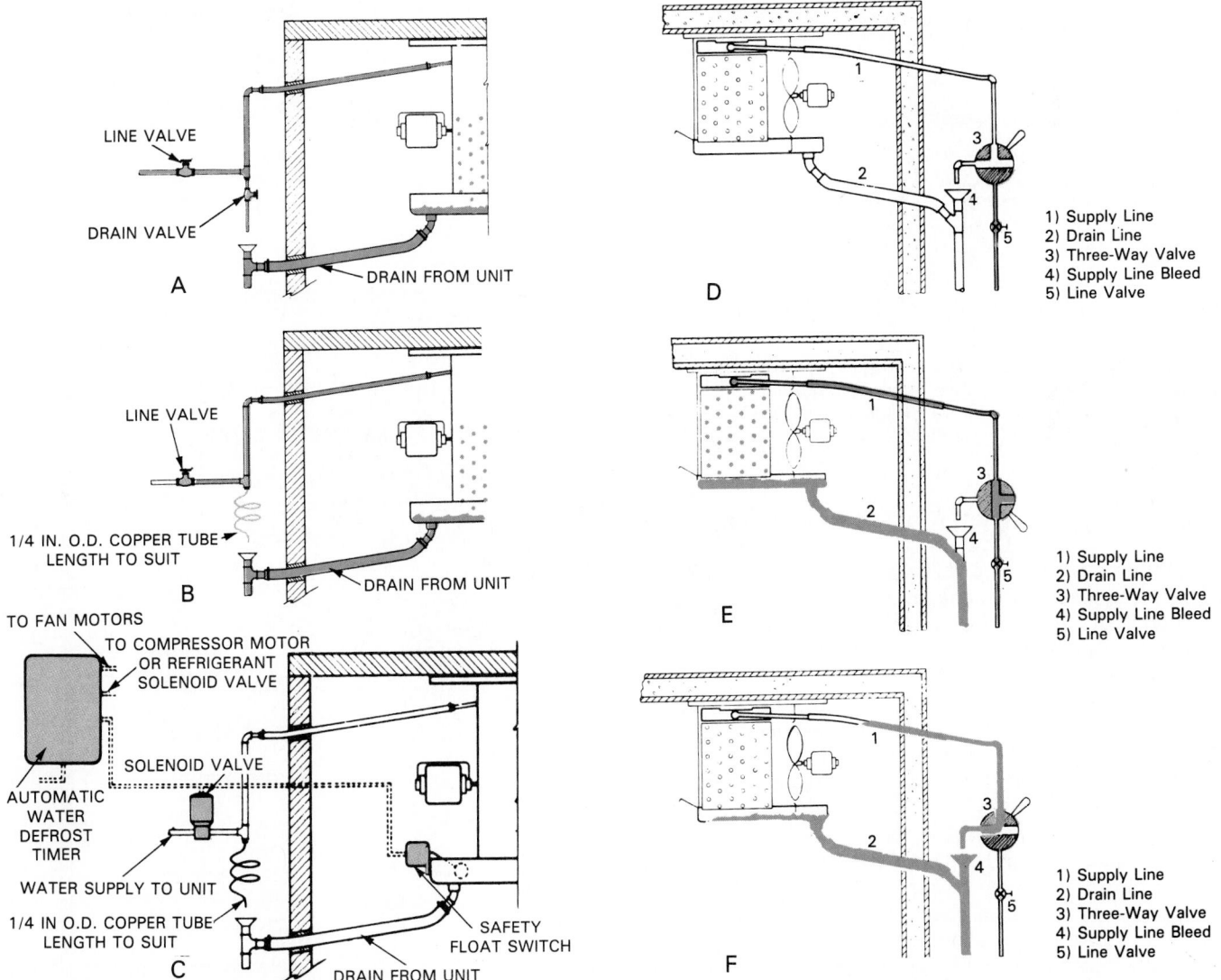

Fig. 12-99. Water spray defrost system schematics. Three methods of operation are shown. A—Manual defrost and manual drain. B—Manual defrost and automatic drain. C—Automatic defrost. There are three steps in the defrost cycle for a manual defrost and manual drain defrost system (View A): D—During refrigeration cycle. E—Defrost operation. F—Water lines and drain being cleared of water at end of defrost operation.

pressure switch opens and the compressor stops.

There should be very little refrigerant in the compressor oil. It may be necessary to use crankcase heaters to drive the refrigerant off during "off" cycles when a pump-down system is used for each cycle.

"Pump out" has a similar purpose. However, an extra relay is wired into the compressor circuit in parallel to the normal relay. It is connected to the thermostat circuit. The extra relay operates the start button on the normal starting relay. The compressor, therefore, cannot restart until the thermostat points close.

Another electric defrost system uses an immersion type electric heater to heat a separate charge of refrigerant. This warm refrigerant circulates around the evaporator in its own passageways to warm the evaporator and defrost the system. This happens while the unit is turned off.

Still another way of using the electric heater defrost system is with a double-tube evaporator. The evaporator refrigerant passes through the passageway between the tubes during normal refrigeration. Electric heating elements are inserted in the center tube. In the defrost operation, the system is stopped and the electric heating elements are turned on. See Fig. 12-100. Thereby, the evaporator tubes cause defrosting from the inside.

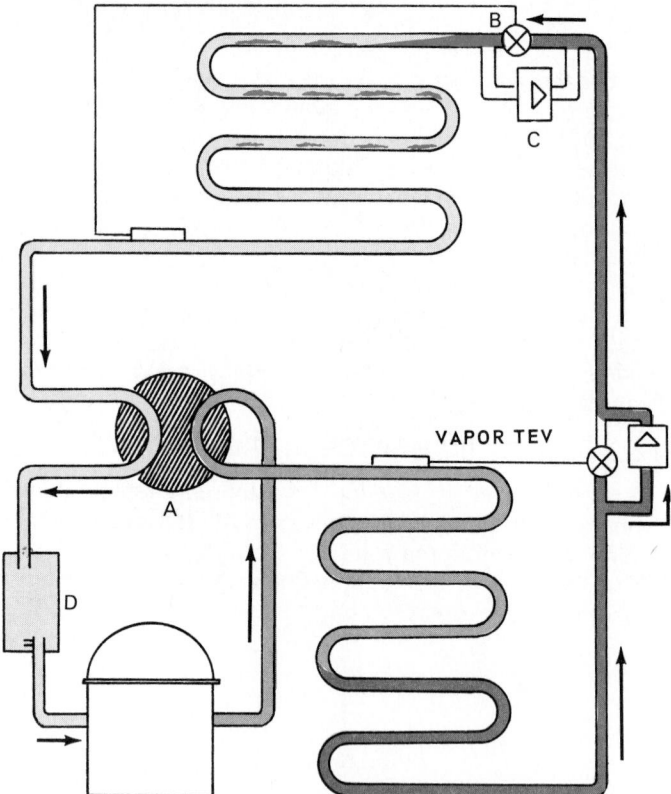

Fig. 12-101. Reverse cycle defrosting system. A—Four-way valve. Position shown is for normal refrigerating operation. B—Thermostatic expansion valve. C—Check valve. D—Accumulator.

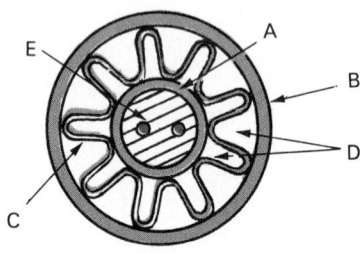

Fig. 12-100. Cross-sectional view of electric defrost system having electric heating elements installed in evaporator tubing. A—Inner tube. B—Outer tube. C—Inner fin. D—Refrigerant passages. E—Heating element.

12-33 REVERSE CYCLE DEFROST SYSTEM

Another system defrosts evaporators by reversing the flow of refrigerant. This causes the evaporator to become the condenser and the condenser an evaporator. During the time the evaporator functions as a condenser, it melts the accumulated frost.

This reversing is handled by installing a four-way valve. Chapters 3, 19, and 23 describe the reverse cycle (heat pump) in detail. See Figs. 12-101 and 12-102.

To operate on defrost, the four-way valve is turned either manually or automatically, and hot gas from the compressor travels up the suction line. It heats the evaporator when gas condenses in it and bypasses the refrigerant control by means of a check valve. It passes through the receiver. As it leaves the receiver, a check valve bypasses it through another refrigerant control into the condenser. The refrigerant evaporates in the condenser and is returned to the compressor in a vapor state.

The liquid receiver is designed to permit the reverse flow of vapor to travel over the reverse liquid in the receiver. It does not return the vapor to the condenser.

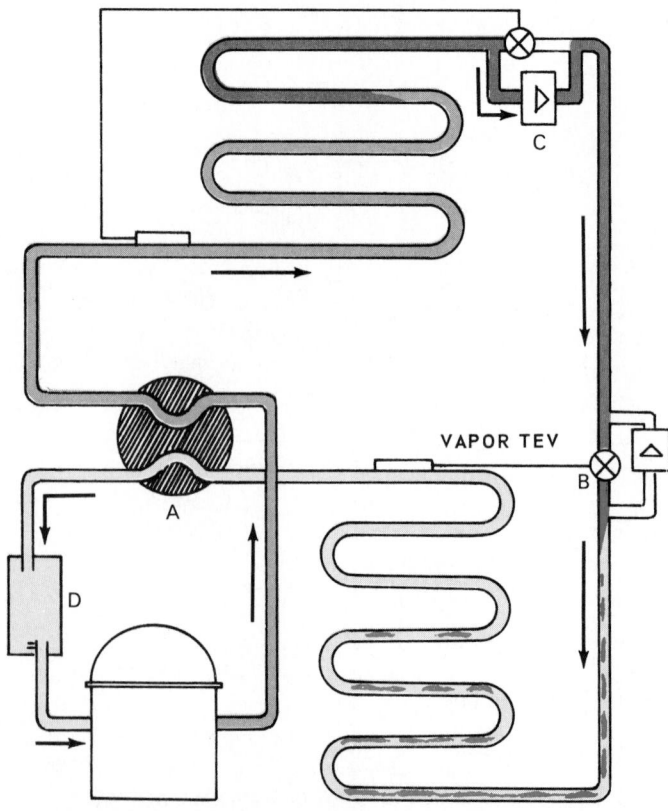

Fig. 12-102. Reverse cycle defrosting system shown in defrost position. A—Four-way valve is positioned to make evaporator serve as a condenser and the condenser as an evaporator. B—Extra TEV. C—Two check valves. D—Accumulator tank.

12-34 WARM AIR DEFROSTING

Where there is enough of it, warm air can be used to defrost low-temperature evaporators. Cabinet air at the right temperature can be used for defrosting. The cycles must be frequent enough and long enough to defrost the evaporator completely. Some installations bring in outside air for defrosting, using a controlled duct system with blowers and fan.

12-35 HEAT EXCHANGERS

A heat exchanger mounted in the suction and liquid line has three advantages:
1. It subcools the liquid refrigerant and increases operating efficiency.
2. It reduces flash gas in the liquid line.
3. It reduces liquid refrigerant in the suction line.

A heat exchanger like the one in Fig. 12-103 provides for a heat transfer from the warmer liquid in the liquid line to the cool vapor coming from the evaporator. Fig. 12-104 shows the outside appearance of a heat exchanger.

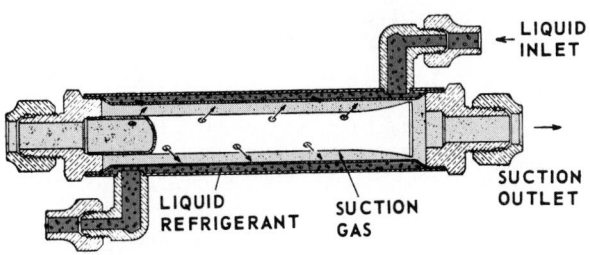

Fig. 12-103. Cross section of heat exchanger used on commercial systems. Note flared connections. (Mueller Brass Co.)

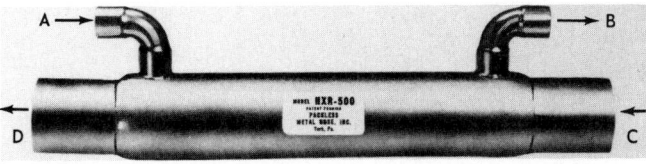

Fig. 12-104. Heat exchanger, external view. Brazed connections are used. A—Liquid in. B—Liquid out. C—Suction vapor in. D—Suction vapor out. (Packless Industries)

If the liquid is cooled 10 to 20°F (5 to 11°C) at the prevailing head pressure, it can absorb more latent heat as it changes to a vapor in the evaporator.

The reduction of flash vapor (sometimes called "flash gas") is important. Flash gas (vaporized refrigerant) comes from the sudden change of some of the liquid to a vapor as the refrigerant passes through the refrigerant control. This reduces valve capacity, increases low-side pressure drop, and reduces the amount of heat each pound of refrigerant can absorb as it evaporates. The "flash gas" cools the remainder of the liquid to the evaporating temperature.

The heat exchanger also helps prevent sweat backs or frost backs on the suction line. If there is low temperature liquid refrigerant present in the returning suction vapor, it will evaporate in the heat exchanger as it absorbs heat from the liquid line.

The subcooled liquid in the liquid line reduces the chance of flash gas forming in the liquid line—especially on warm days or if the liquid line has a long vertical run.

The pressure drop in the suction line portion of the heat exchanger should not be over 2 psi (14 kPa).

12-36 REFRIGERANT CONTROLS

For single installations involving one evaporator and one condensing unit, five types of refrigerant controls can be used: thermostatic expansion valves, automatic expansion valves, high-side floats, low-side floats, and capillary tubes. These are explained in Chapter 5.

In multiple installations which cover a greater number of uses, two types of refrigerant controls are usable. They are the low-side float and the thermostatic expansion valve. Thermostatic expansion valves are used extensively but there are also some low-side float systems.

Some of the thermostatic expansion valves have a large capacity. Usually they are constructed with a pilot valve operating a larger valve.

The thermostatic expansion valve is explained in Chapter 5. Technicians should study its design, operation, installation, care, and repair before proceeding with this chapter.

12-37 MOTOR CONTROLS

Two basic types of motor controls are used in commercial refrigeration:
1. The thermostatic type.
2. The pressure type.

These are the same ones used in domestic refrigeration. Large systems use magnetic starters operated by motor controls.

In multiple evaporator commercial work, pressure motor controls are used quite often because:
1. The low-side pressure is an indication of the temperature in the evaporators.
2. One control works well regardless of the number of evaporators connected to it.

The controls, Fig. 12-105, provide both range and differential adjustments. Explanation of various range and differential adjustments will be found in Chapter 8. Fig. 12-106 shows the internal construction of such a control.

The current draw as larger motors start is more than control contacts can handle. A motor starter is necessary for single-phase ac motors over 1 hp.

Three-phase ac motors also require a starter. The motor control operates the relay in the starter. Electrical work should be done by a licensed electrician and the work should comply with local electrical codes. Fig. 12-107 lists the recommended average pressure motor control settings for both temperature control and defrost control.

12-38 PRESSURE MOTOR CONTROL

The pressure motor control is usually mounted on the condensing unit. It is operated by low-side pressure. Some companies suggest connecting the control into the low-side suction line about 10 to 15 ft. from the compressor to reduce vibration effect on the control.

The range settings vary with the application. The cutout pressure should be set about 10°F (6°C) lower than the desired evaporator outside surface temperature. The cut-in

Fig. 12-105. Pressure motor control is typical of those used on commercial installations. (Johnson Controls, Inc.)

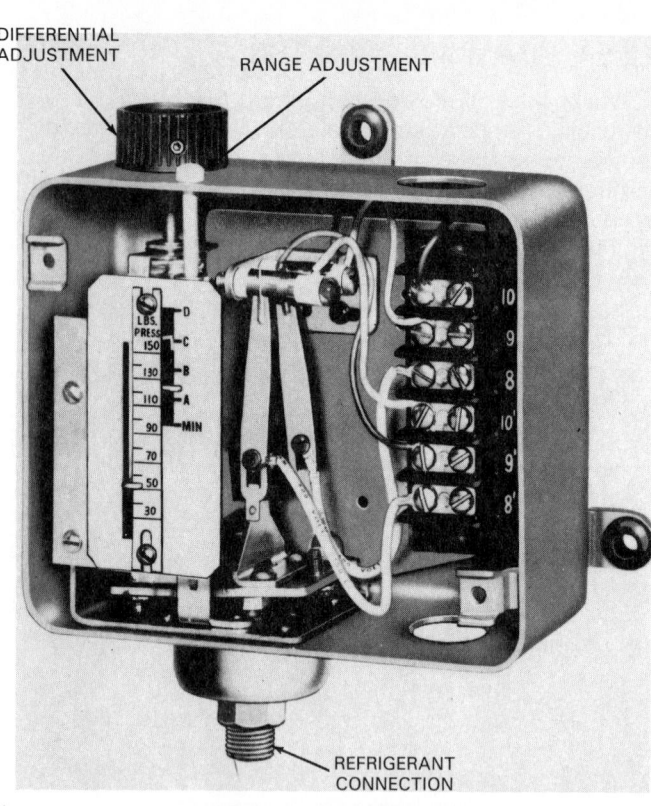

Fig. 12-106. Cover has been removed from the pressure motor control. Note adjustments, electrical connections, and pressure scale. (Johnson Controls, Inc.)

RECOMMENDED CASE TEMPERATURE AND DEFROST CONTROL SETTINGS

TYLER DEFROST CONTROLS: TC = Straight Time Clock. TG = Temperature Guard with current sensing relays resetting control. TP = Time Pressure Reset (SEE Note 1). TS = Time Solenoid reset, similar to TG and used on walk-in cooler coils. With Parallel Compressor Units, Multi-circuit timers (4,6,8 or 12 circuits) are used. Termination is by relay, thermostat, or straight time.	(X)DF (X)DFS (X)DFC2 (X)DJ DMF Frozen Food Cases	(X)DFI (X)DFIS (X)DFIC2 "DWF(I)C³ (X)DJI Ice Cream Cases	D4F D4FI Three and Four Shelf FF/IC Cases	D6F D6FI D6FL D6FH 5 & 6 Shelf FF/IC Cases	D5FG DLG Glass Door Frozen Food Cases	D5ZG DLIG Glass Door Ice Cream Cases	D5NG DSG Glass Door Refrigerators Medium Temp	DM, DMR OPEN MEAT CASES For best results, Use Thermostat & Liquid Line Solenoid System, set at 23 °F Entering Air.	DHM DHMRL Multi-Shelf Meat-Front & Rear Load	DGV(R) CGV(R) DGS(R) CGS(R) Closed Service Meat Cases	DDC (RL) DAS(S) Multi-Shelf Dairy Deli Case BLAS-RI, RL Roll-In Dairy-Deli Case	All Open Produce Cases & Reach-In Refrigerators
	(X)D7FE 12.9A	(X)DJ(I)E 8.6A	3 PHASE SUPPLY						Std. Def.	Electric			
DEFROST HEATER AMPS @208 VOLTS — 8" Case	6.9 A⁴	13.8 A⁴	12 A/Leg	18 A/Leg	3 Door 13.0 A	None	None	6.9 A Opt.	6.9 A Opt.	None	6.9 A Opt.	None
DEFROST HEATER AMPS @208 VOLTS — 12" Case	10.3 A⁴	20.6 A⁴	15 A/Leg	23 A/Leg	5 Door 19.2 A	None	None	10.3 A Opt.	10.3 A Opt.	None	10.3 A Opt.	None
PRESSURE CONTROL — CUT IN — R-12 / R-502	3-5# / 20-22#	0-1# / 13-15#	25#		4# / 20-25#	2# / 18-22#	10# / 34#	20-22# / 50-55#	20-22# / 50-55#	33# / 74#	26# / 60#	36# / 79#
PRESSURE CONTROL — CUT OUT — R-12 / R-502	5-2"Vac. / 8-12#	12-8"Vac. / 4-7#	See Note 2 / 0#		5" Vac. / 9-12#	10" Vac. / 5-7#	12-20# / 37-50#	9-15# / 31-39#	9-15# / 31-39#	13-15# / 39-41#	15-20# / 41-51#	43-52# / 63-54#
TYPICAL OPERATING RANGE OR EPR SETTING-IF USED — R-12 / R-502	4-2"Vac. / 10-12#	8-4"Vac. / 4-8#	R-12 Not Used Frozen Food 10-15# Ice Cream 6-8#		2"V-0# / 12-15#	8-4" Vac. / 5-8#	20-22# / 50-55#	14-18# / 40-47#	14-18# / 40-47#	19-25# / 49-60#	18-20# / 48-50#	22-26# / 54-61#
DEFROST CONTROL — TP, TG ELECTRIC DEFROSTS PER DAY & MINUTES FAIL-SAFE TIME	1/Day 60	1/Day 36/46³	2/F.F., 3/I.C. 36 Minutes		1/Day 60		N.A.	1 or 2/Day 36 Minutes	3/Day 36	N.A.	4/Day 36	N.A.
DEFROST CONTROL — TC TIMED OFF DEFROSTS (If Used)	N.A. = Not Applicable				N.A.		1@60	2@60	3@46	2@60	4@60	3-4@40
DEFROST CONTROL — GAS DEFROSTS PER DAY, TERM. TEMP & MINUTES FAIL-SAFE			2-3/Day, 75°F 16-20 Min.	2-3/Day, 75°F 20-26 Min.	2/Day,75°F 26 Min.		N.A.	2 to 3/Day, 70°F 16-20 Min.	3-4/70°F 16-20	N.A.	4/70°F 16-20	N.A.

NOTES: 1) TP (Pressure Reset) controls tend to erratic reset pressure response. When used, settings are: R12 Low, 45#. R12 Med-Electric Def. 52#, R12 Medium, 40#, R502 Low, 95#, R502 Med-Elect. Def. 108#. 2) These settings are for constant machine running. 3) Use Frozen Food pressure settings for (X)DWFC cases. Fail-safe is 46 minutes for both (X)DWFC & DWFIC. 4) Amps for each side on (X)DJ(I). Total for DJ or DJI is double figure shown. 5)GAS DEFROST: gas flow stops with temperature termination. Refrigeration stays off until fail-safe time expires to allow coil and pan to drain.

DEFROST CHECK LIST

1. Check to see that defrost contactor is wired to cases on that condensing unit, then check defrost time and number.
2. Check heater amps during and after defrost. Is coil clear at termination? Does limit switch open too soon?
3. Check waste outlet for proper hook-up (max. of 12' of 1" pipe, 1/4" per foot slope) and that it isn't frozen by refr. line contact.

GOOD HOUSEKEEPING—ESSENTIAL FOR GOOD REFRIGERATION!

Good housekeeping is not only necessary for sanitation, but it contributes measureably to reliability and to the quality of refrigeration. A good maintenance program is essential. Meat cases should be cleaned thoroughly once a week, other cases at least twice a year. Clean the interior with germicidal detergent. RINSE: Flush the waste outlet with hot water.

OTHER POINTS: Eliminate drafts over open cases. (Maximum allowable draft—50 FPM) DO NOT BLOCK AIR DUCTS. OBSERVE LOAD LINES!! Cases must be level to operate properly.

Fig. 12-107. Recommended motor control pressure settings. These are recommended for various case applications. Pressures are in psi. It may be necessary to change these settings somewhat for a particular installation. (Tyler Refrigeration Corp.)

*INCHES VACUUM APPLICATION	R-12		R-22		R-502	
	OUT	IN	OUT	IN	OUT	IN
FROZEN FOOD – OPEN TYPE	7*	5	4	17	9	23
FROZEN FOOD – CLOSED TYPE	1	8	11	22	17	29
ICE CUBE MAKER – FLOODED OR DRY TYPE COIL	4	17	16	37	22	47
SWEET WATER BATH – SODA FOUNTAIN	21	29	43	56	52	67
SHOW CASE – FROST CYCLE	10	25	25	50	32	60
SHOW CASE – DEFROST CYCLE	18	34	39	64	49	76
BEER, WATER, MILK COOLER	19	29	40	56	47	67
WALK-IN COOLER – DEFROST CYCLE	12	35	29	66	37	77
ICE CREAM TRUCKS, HARDENING ROOMS	2	15	12	33	17	42
VEGETABLE DISPLAY – DEFROST CYCLE	11	35	27	66	35	77
EUTECTIC BRINE TANK, ICE CREAM TRUCK	1	4	11	16	17	22
REACH-IN COOLER – DEFROST CYCLE	18	36	39	68	47	79
BEER COOLERS – BLOWER DRY TYPE	15	34	33	64	42	76
BEER COOLERS – BARE PIPE DRY TYPE – FROST CYCLE	12	27	29	53	37	64
INSTANTANEOUS BEER COOLERS	12	29	29	56	37	67
RETAIL FLORIST BOX – BLOWER COIL	26	42	51	77	61	88

Fig. 12-108. Note typical refrigeration applications and low-side pressure motor control settings for each.

pressure should be about the same as the highest allowable evaporator temperature. See Fig. 12-108.

The differential setting will vary, depending on the temperature accuracy wanted. A wide pressure difference will allow some variation in cabinet temperature and will lengthen the operating cycle interval of the condensing unit. (This means the compressor would not run as often.) A differential set to close limits will maintain a more uniform cabinet temperature but will shorten the cycling interval of the condensing unit. The unit would run more often. A common pressure difference between cut-in and cutout point is about 20 psi for R-12, 22 psi for R-22, 16 psi for R-500, and 25 psi for R-502.

12-39 THERMOSTATIC MOTOR CONTROL

The thermostatic motor control is like the pressure motor control in design except for the sensing bulb and capil' tube. See Fig. 12-109.

This type control is generally used in large single installations. However, satisfactory setups have been made in multiple installations. Hopefully, when the controlled cabinet is at the desired temperature, the others are also. These controls are also used together with a solenoid valve to control each separate cabinet in a multiple installation.

Some are made with a very close differential such as 1 °F (0.5 °C) for use in certain display cases, bulk milk coolers, frost alarms, liquid chillers, and refrigerated trucks.

Thermostatic motor controls are popular in brine cooling installations with the sensing bulb being submerged in the brine. Ice cream cabinets are a typical example. When used in single cabinet installations, the sensing bulb is usually mounted in the cabinet 4 ft. up from the floor between the cold and warm air flues and at least 2 in. from the wall. Fig. 12-110 shows a control with the cover removed.

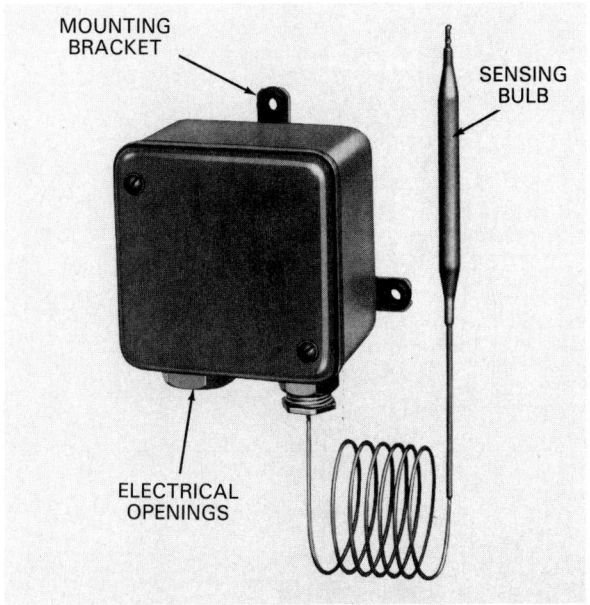

Fig. 12-109. Thermostatic motor control. This type is much like the pressure motor control except for the capillary tube and the sensing bulb attached to it. (Johnson Controls, Inc.)

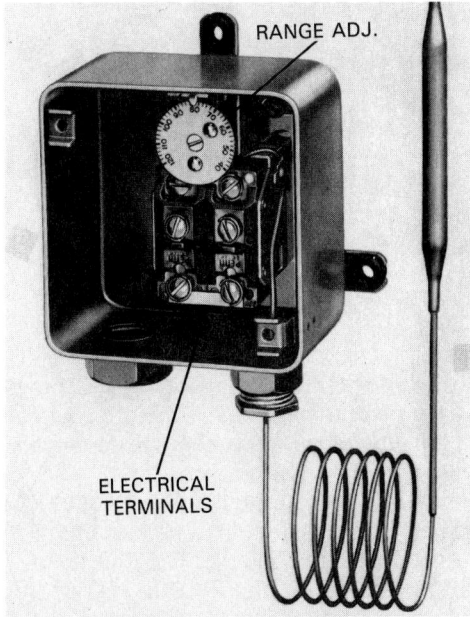

Fig. 12-110. Thermostatic motor control with cover removed. Note temperature range dial (Fahrenheit scale) and electrical terminals. (Johnson Controls, Inc.)

Some are made for wall mounting in walk-in coolers, meat storage rooms, warehouses, and florist cabinets. Some have double throw contacts (SPDT) so that the control may also operate other devices (fans and defrost systems).

12-40 SAFETY MOTOR CONTROLS

An important difference between commercial and domestic controls is the fact that many commercial electrical systems also use safety devices known as:
1. A high-pressure safety cutout.
2. An oil pressure safety cutout.

The high-pressure safety device is a bellows built into the control. It is connected to the high-pressure side of the system, Fig. 12-111. It is often connected to the cylinder head to permit easy disconnecting of the control from the system.

The bellows is attached to a plunger in such a way that, if the head pressure becomes too high from air in the system, condenser water being shut off, or other causes, the bellows will expand, push the plunger against the switch, and shut off the motor.

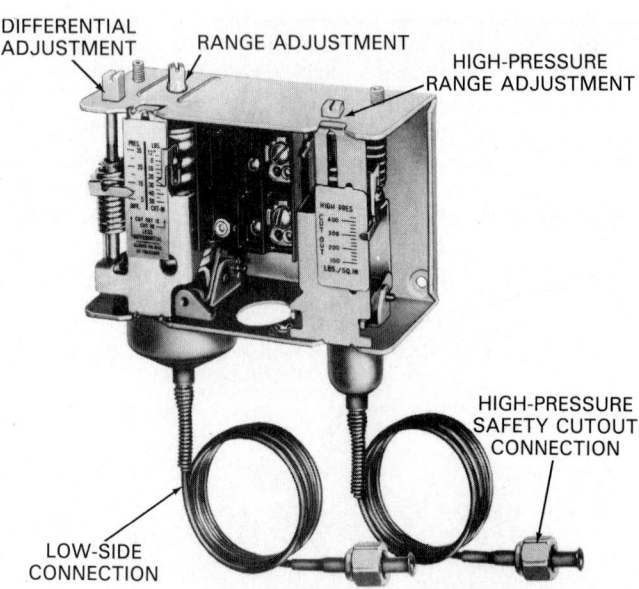

Fig. 12-111. This pressure operated motor control also has a high-pressure safety cutout. Note that it has three adjustments. (Johnson Controls, Inc.)

Action of the high-pressure safety device prevents the buildup of dangerous pressures within the system. It also prevents ruining the motor through overloading and overheating.

The control is usually set to cut out at about 20 percent above normal head pressure. In R-12 systems, the control is set at about 150 to 160 psi; R-22, 260 to 270 psi; R-502, 280 to 290 psi; and R-500, 190 to 200 psi.

The oil pressure safety cutout will shut off the electrical power if the oil pressure fails or drops below normal. It is a differential control, using two bellows. One bellows

responds to the low-side pressure and the other responds to the oil pressure. *The oil pressure must always be above the low-side pressure for oil to flow.* See Fig. 12-112.

The wiring diagram for an oil pressure safety control is shown in Fig. 12-113. The control will open the circuit if

Fig. 12-112. Commercial system pressure control with oil pressure safety cutout. It operates on difference between refrigerant and oil pressures. (Johnson Controls, Inc.)

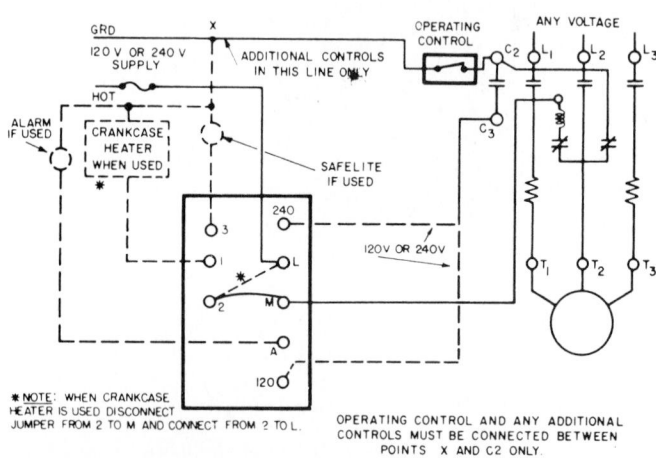

Fig. 12-113. Wiring diagram shows oil pressure safety cutout at A. Motor is three-phase. Note provisions for adding alarms, safety lights, and crankcase heater. (Johnson Controls, Inc.)

the pressure difference between the two bellows drops below the required oil pressure needed. Large commercial systems use this type. In some systems, the control points are in the compressor motor circuit. In other cases, the points will close and current is sent through a bimetal strip or a resistance heater near the bimetal strip. If this strip heats up before the pressure returns to normal, the power will be disconnected. An oil pressure safety control is shown installed on a motor compressor in Fig. 12-114.

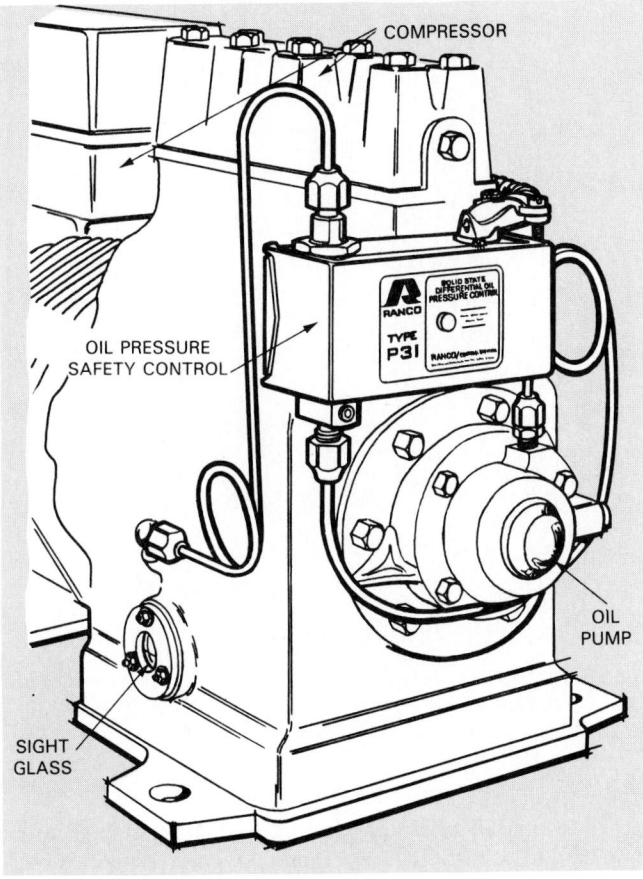

Fig. 12-114. Oil pressure safety control assembly must be installed level in a motor compressor.

Refrigerant level may be kept within safe limits by a float switch. The float may be used for signalling or it may actually control the refrigerant level.

The switch may be used to control the liquid level in:
1. Flooded surge drums.
2. Flooded shell-and-tube chillers.
3. High- and low-pressure receivers.
4. Intercoolers.
5. Transfer vessels.
6. Various kinds of accumulators including liquid recirculating types.

If the refrigerant level is too high, the float switch closes an electrical circuit. The circuit acts to allow a refrigerant flow out of the control device. If the refrigerant level is too low, the float switch will actuate (cause to work) a circuit which will allow refrigerant to flow into the system. A float control switch is illustrated in Fig. 12-115.

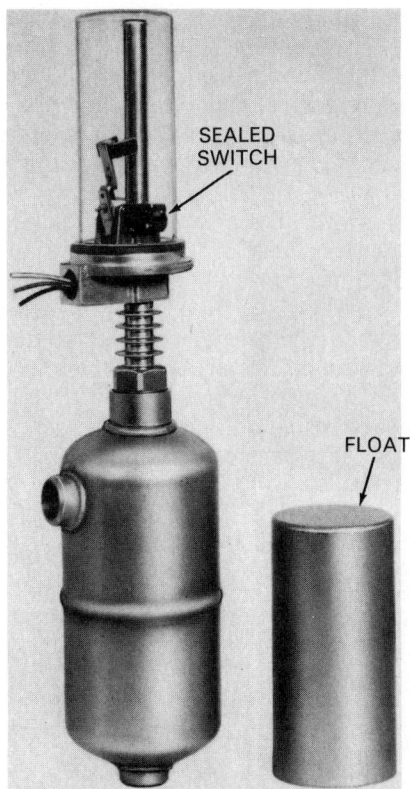

Fig. 12-115. Float-operated switch is designed to control level of liquid refrigerant in system.
(Refrigerating Specialties Div., Parker-Hannifin Corp.)

12-41 MOTOR STARTERS

The pressure or temperature control contacts—whether open or sealed—are limited in the amount of current they can safely carry. The National Electric Code and local electric codes usually set down the limitations of these controls.

However, these same commercial controls can handle larger motors (larger loads) with the help of a device called a magnetic starter (contactor).

The magnetic starter is an electromagnetic device. The magnetism is controlled by the electricity that flows through the motor control. The magnetism attracts a piece of steel (or armature). When this armature moves, it closes large

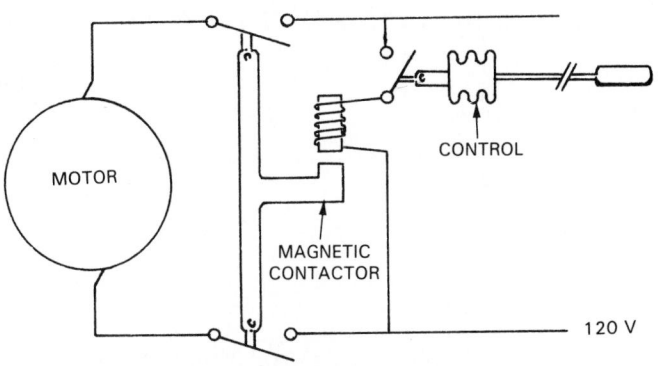

Fig. 12-116. Electrical schematic diagram for an automatic control on a magnetic starter. Circuit allows high current flow to motor without overloading control contact points.

contact points that safely carry the larger current flow needed for the large motors. Fig. 12-116 is a schematic wiring diagram of a magnetic starter.

These starters are mounted in an approved metal box with a safety access door. Some units incorporate a manual shutoff switch, fuses, and an overload thermal safety breaker switch.

The safety switch is operated by a heating element located in the black lead of the motor circuit inside the contactor box or starter. If the motor demands too much current (shorts, grounds, or overloads), this heater will bend a thermal bimetal strip in the control circuit, opening the electromagnet circuit. This action opens the main switch.

Fig. 12-117 shows a wiring diagram of a 120-240 V single-phase system using a magnetic starter. A wiring diagram for a three-phase system is shown in Figs. 12-118 and 12-119.

12-42 ICE MAKER CONTROLS

In addition to the usual refrigerant and motor controls, automatic ice cube makers or ice flake makers have controls to stop the system when the storage bin is full. The

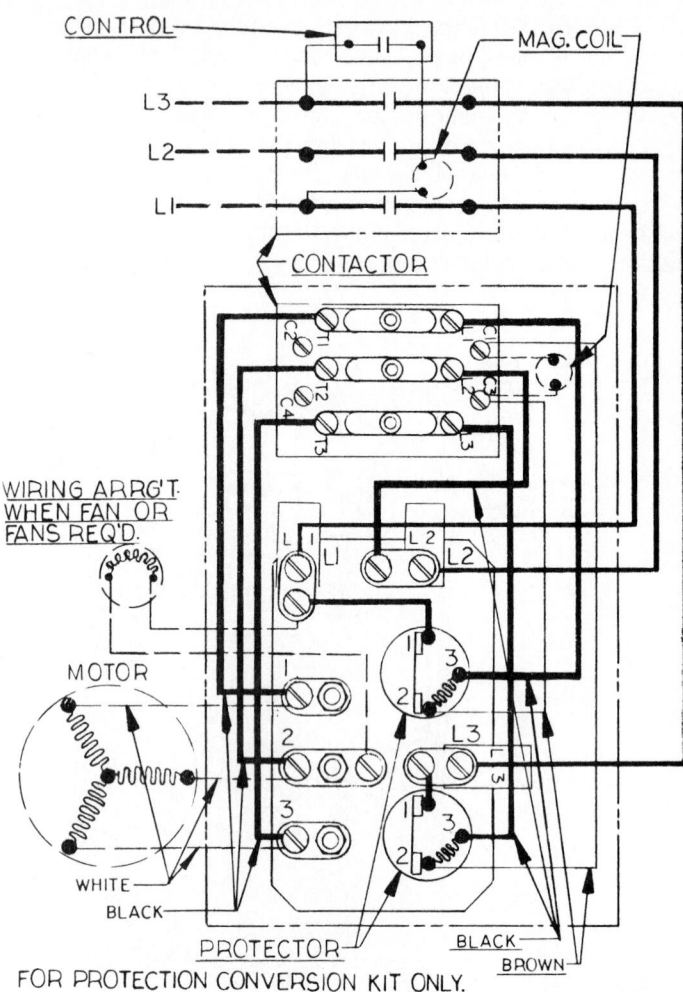

Fig. 12-118. Wiring diagram for a three-phase system using a magnetic starter. Control is wired in series with magnetic coil of contactor (starter).

control is located in the bin. It shuts off the machine until some of the ice is removed or melts. Devices used include:
1. Mechanical levers.
2. Temperature controls.

The mechanical type has a lever or a diaphragm which, when contacted (pressed) by the accumulated ice, opens a switch and stops the unit.

The temperature control shuts off the unit when the control bulb is in direct contact with the ice. Both controls are located at the top of the ice bin.

The wiring diagram, Fig. 12-120, is from a system which freezes cubes and then removes (harvests) the cubes from the freezing grid. The refrigerating system has a hot gas defrosting system, as shown in Fig. 12-121. Fig. 12-122 is an actual wiring diagram. This shows the interlocking of the controls, devices, and wiring.

12-43 VENDING MACHINE CONTROLS

Most vending machines which use refrigeration operate automatically. These machines can perform several operations:
1. Heat, cool, and select foods.
2. Accept coins to activate the dispenser system.

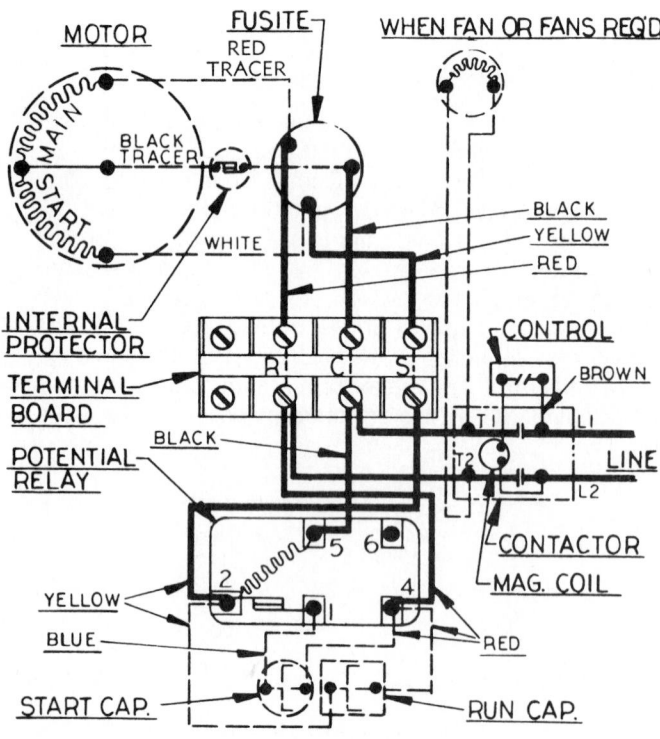

Fig. 12-117. Wiring diagram for a 120-240 V single-phase system. Note use of contactor (motor starter).

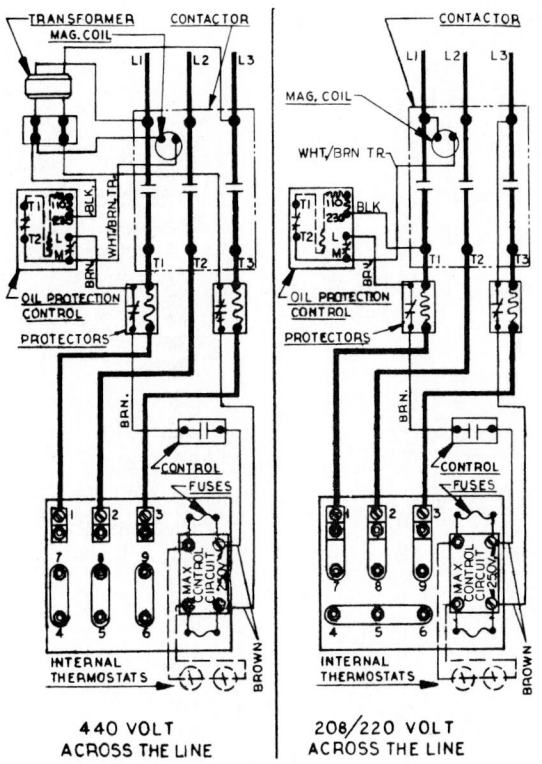

440 VOLT
ACROSS THE LINE

208/220 VOLT
ACROSS THE LINE

Fig. 12-119. These wiring diagrams show 440 V and 208-220 V circuits designed for three-phase power. Lines L₁ through L₃ each carry one leg of the three-phase voltage. (Copeland Corp.)

Some of the units automatically heat (if necessary) and move the items being dispensed (bottles, bulk fluids, packages of ice cream, and the like). Thermostats, relays, microswitches, positioning motors, and solenoids are used.

The automatic operation of this unit with its vending motor, magnets, signal lights, relays, and so on, makes an elaborate wiring system necessary. Fig. 12-123 shows eight parallel circuits being used in one dispenser. The evaporator fan operates continuously.

12-44 DEFROST TIMERS

Most automatic defrosters need an automatic device to start the defrost cycle. The unit shown in Fig. 12-124 uses a system of spring-loaded levers which are activated by trippers positioned in the 24-hour dial at the time defrosting is desired. These levers operate the switches.

Some time clocks are connected directly to electric power and will cause defrost at intervals necessary to keep the system working well. Each evaporator design has its own requirements for good operation. Some need to be defrosted during each cycle; some every few hours. Others need defrosting no more than once a day. Length of defrost is adjusted by the brass pointer on the inner dial (0-55 minutes).

Some timers are connected to electric power in parallel with the motor. The clock mechanism registers only the running time of the condensing unit. These mechanisms then start the defrost cycle after so many hours of running time.

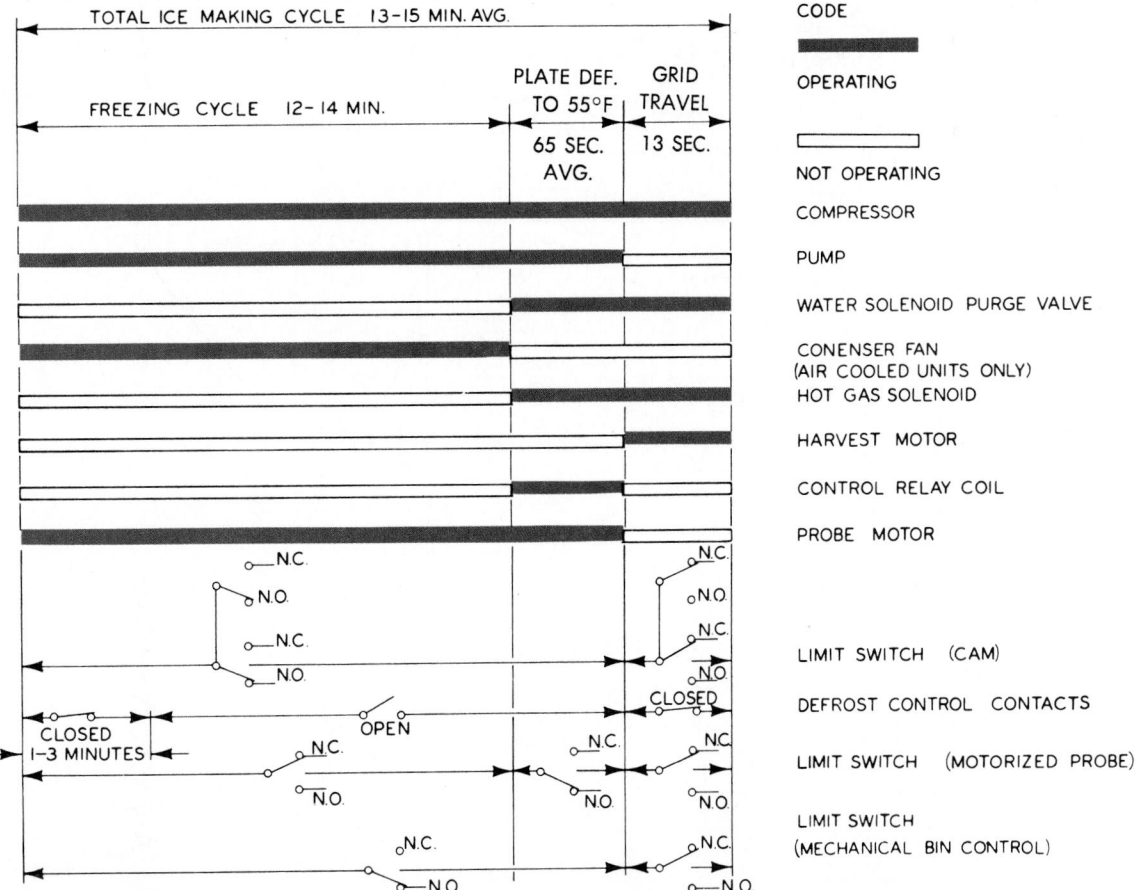

Fig. 12-120. This wiring diagram is for an ice cube maker which uses harvest motor to remove cubes from the freezing grid. (Ice-O-Matic)

449

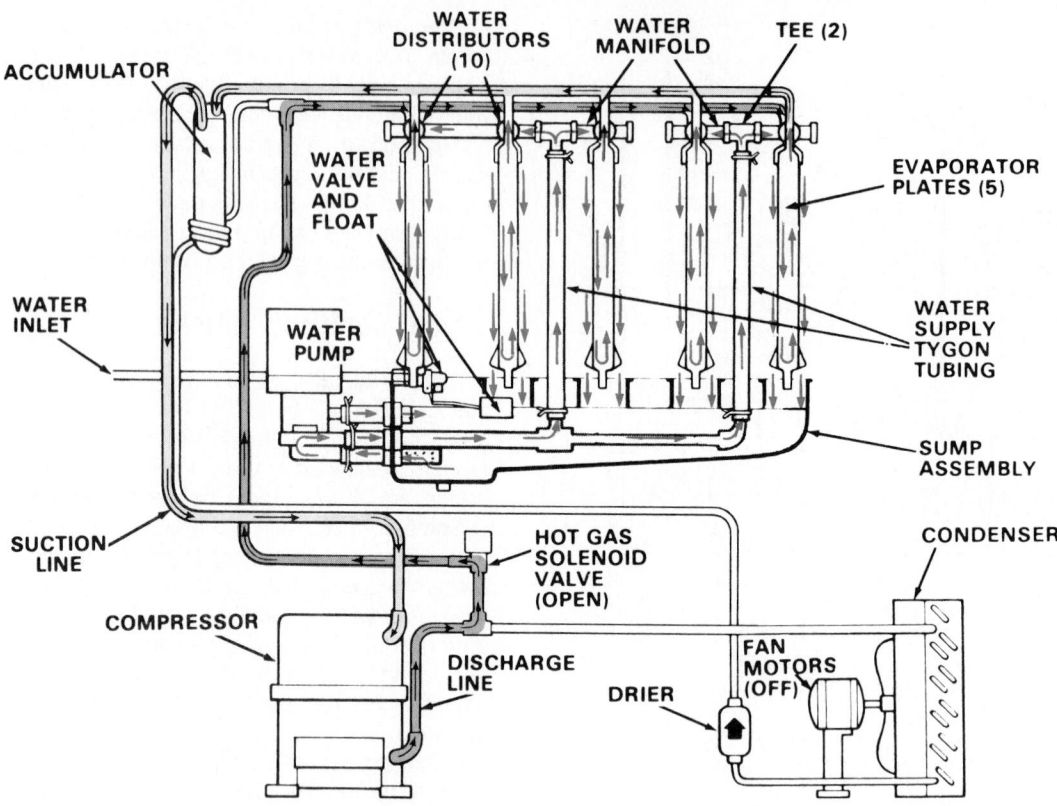

Fig.12-121. This ice cube maker refrigerating system uses a hot gas defrosting system. (Scotsman Ice Systems)

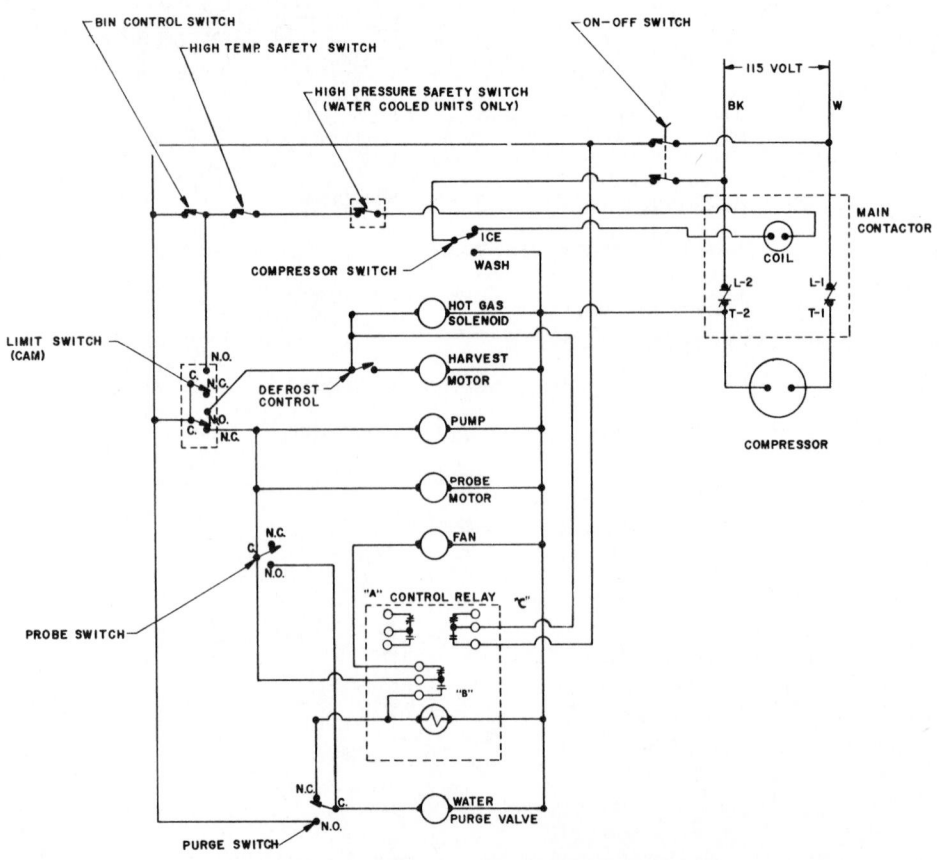

Fig. 12-122. Note number of parallel circuits in this ice cube maker diagram for hot gas defrost, probe motor, harvest motor, water purge valve, and a water pump. (Ice-O-Matic)

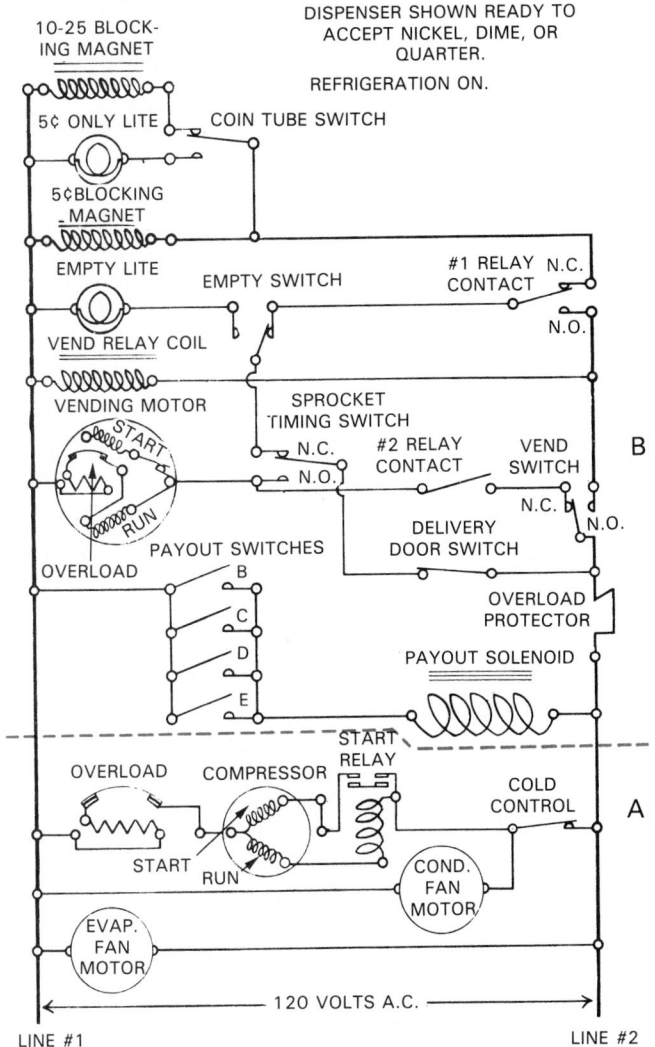

Fig. 12-123. Ladder diagram for refrigerated bottled beverage vending machine. A—Refrigerating unit wiring. B—Vending wiring.

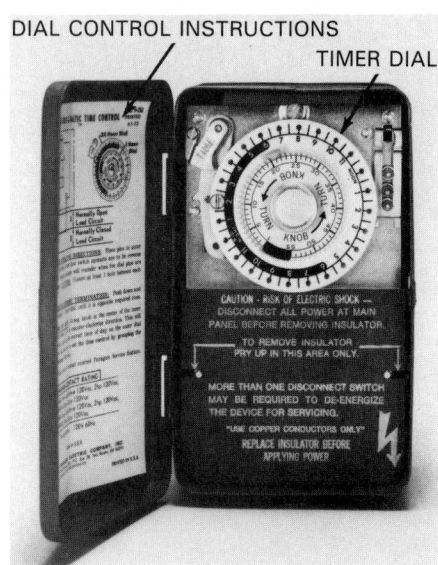

Fig. 12-124. Time switch used for controlling defrost cycles in commercial systems. Wiring diagram and instructions are located inside the cover. (Paragon Electric Co., Inc.)

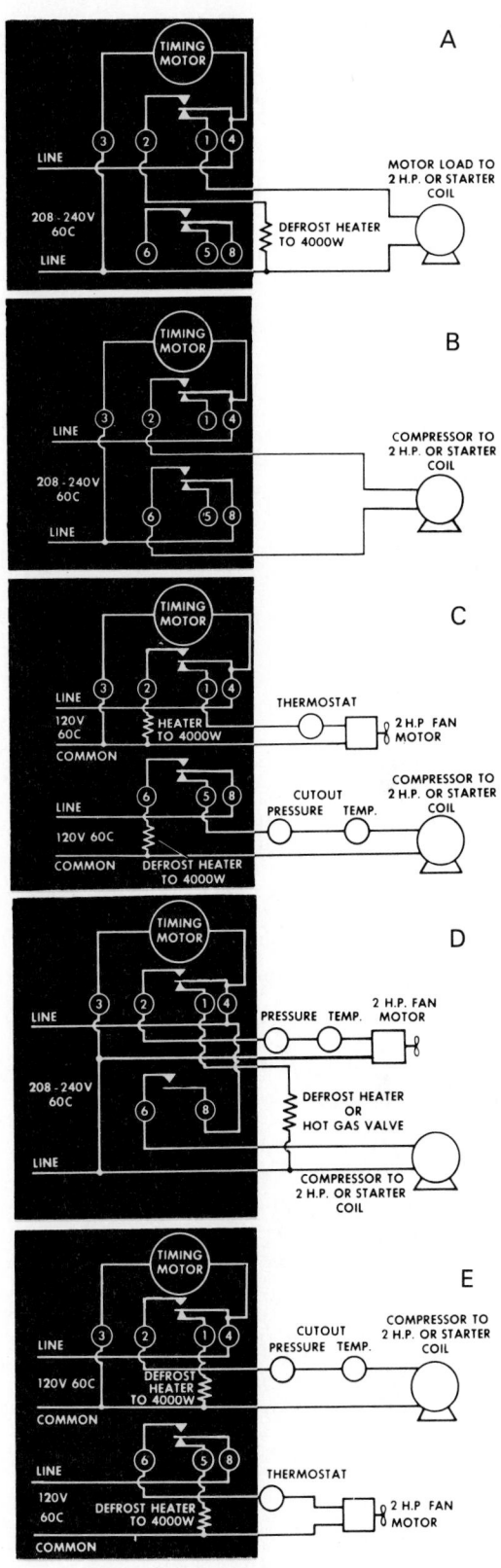

Fig. 12-125. Wiring diagrams for several types of defrost control arrangements. A—Circuit controlled by a SPDT (single-pole, double-throw) switch which activates defrost heaters as it shuts off refrigerating unit. B—Circuit controlled by a DPST (double-pole, single-throw) switch which only shuts off refrigerating unit. C—Circuit controlled by a DPDT (double-pole, double-throw) switch. It shuts off compressor and fan and turns on two defrost circuits. D—Circuit for delayed fan shutoff during defrost and for turning on one defrost circuit. E—Circuit controlled by a DPDT (double-pole, double-throw) switch for delayed fan shutoff and two defrost circuits.

The timer wiring differs with the type of defrost system. In one hot gas system, the timer energizes the solenoid bypass valve (causes it to act). It stops the fan motors, energizes auxiliary electric heater elements, and runs the compressor. It also may be used to prevent start of the normal cycle until the low-side pressure is at normal levels.

Some basic electrical circuits using timer controls are shown in Fig. 12-125.

Another type of automatic timer for defrosting is shown in Fig. 12-126. A timer starts the defrost cycle while the temperature bulb returns the unit to normal operation after the evaporator temperature is above 32°F (0°C).

The timer in Fig. 12-127 can be used with either "air defrost" or electric heat. It uses the timer motor to start the defrost action and a pressure control connected to the low-pressure side to return the system to normal operation. The electrical diagrams are shown in Fig. 12-128.

Some commercial installations use a modular multiple-circuit timer, Fig. 12-129, for defrost control. It is adjustable for handling from 1 to 12 operations during the defrost initiation. The defrost termination control is adjustable from 6 to 106 minutes in one-minute increments. Fig. 12-130 shows a time-terminated hot gas defrost system with a compressed thermostat bypass during defrost cycle. An electrical panel with four timers and six motor starters (contactors) is shown in Fig. 12-131.

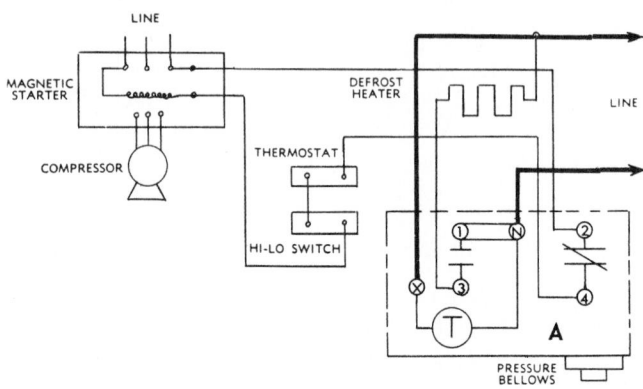

Fig. 12-128. Wiring diagram for a defrost control where timer motor starts the defrost and a low-pressure switch returns the system to normal operation. A—Timer/pressure-operated control. Timer motor at left, marked "T," is connected to line power. Pressure switch is connected to compressor circuit.
(Paragon Electric Co., Inc.)

Fig. 12-126. A timer and thermal bulb combination for controlling defrost cycle. Timer starts defrost action. Thermal sensing bulb, located on the evaporator, returns the system to normal operation after frost has melted.
(Johnson Controls, Inc.)

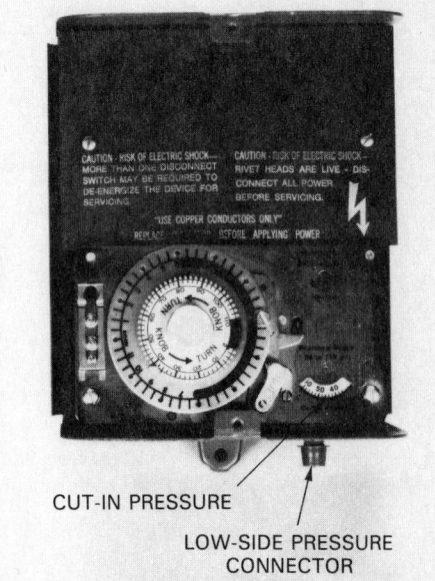

Fig. 12-127. Defrost timer with a low-side pressure-operated switch. Timer starts the defrost action and low-pressure switch returns system to normal operation. (Paragon Electric Co., Inc.)

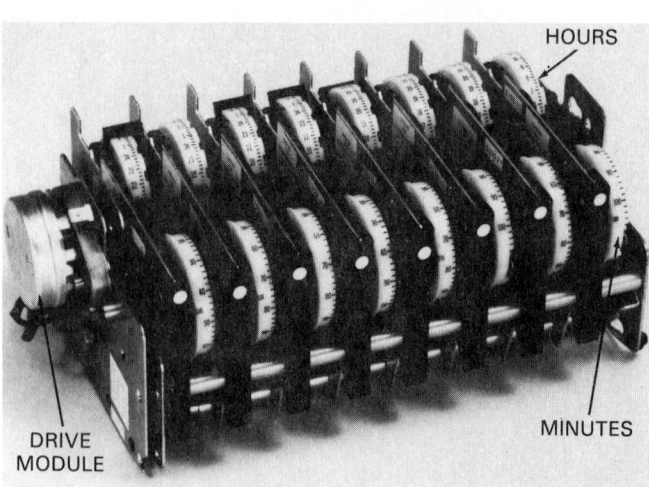

Fig. 12-129. This single-drive timer module has eight different circuits. Time setting controls are arranged in hours and minutes.
(Paragon Electric Co., Inc.)

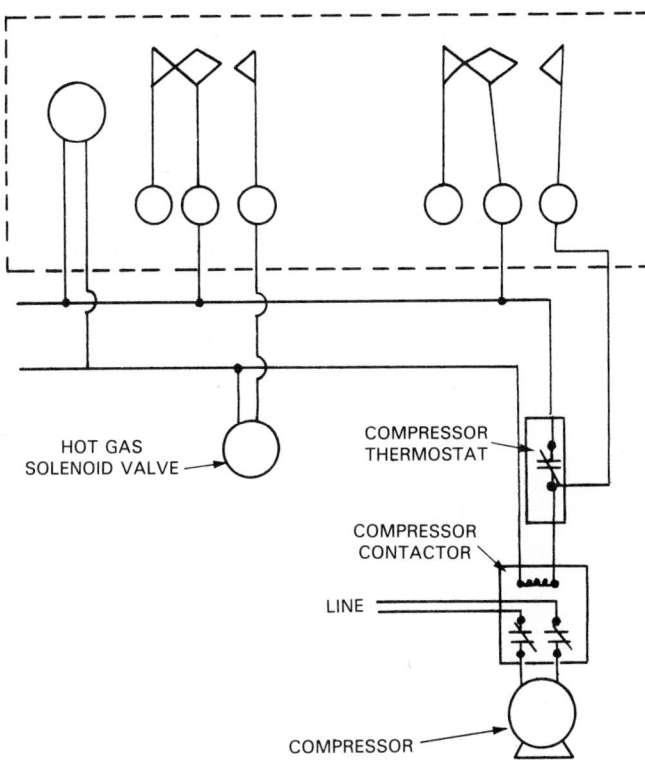

Fig. 12-131. Electric control panel with four timers. Each controls production of 10 tons of ice on each cycle.

Fig. 12-130. Typical wiring diagram for modular multiple-circuit defrost control. Note location of hot gas solenoid valve module. (Paragon Electric Co., Inc.)

Units with transistorized solid-state circuitry are also being used. A thermistor may control the defrost cycle in a display case. The thermistor measures the temperature difference of air moving through the evaporator. This control replaces a timer as defrost is controlled only by demand. Sufficient temperature differences between the air entering the evaporator and the air leaving the evaporator trigger a signal from the thermistor. If this temperature difference becomes more than 20°F (11°C) to 30°F (16°C), this

signal (electrical pulse) will start the defrost cycle. A standard thermostat returns the system to a normal cycle when the evaporator temperature measures about 40°F (4°C).

An electronic modular electric circuit defrost control used on parallel refrigeration systems in supermarkets is shown in Fig. 12-132. The unit can be used for the specific defrost control needs of the systems. Outputs may be selected as defrost, fan delay, or master hot gas defrost system. The system can handle up to sixteen different units in groups of two per module. The unit indicates if defrost terminate sensors are not functioning, or which unit missed a scheduled defrost during power outages. A proportional defrost unit attachment is shown in Fig. 12-133. This unit senses humidity and temperature changes and readjusts the defrost intervals accordingly.

Fig. 12-132. Programmable defrost control uses a three-pushbutton keypad and provides 24-hour time-of-day format. Individual programs can be inserted into control output section No. 5 through the use of output cards, two outputs per card. A—LCD display. B—Program entry/review keypad. C—Power input terminals. D—Control output section. (Paragon Electric Company, Inc.)

Fig. 12-133. A proportional defrost module. A—External view of sensor. B—Electronic sensor and defrost module. C—Mounting brackets. (Paragon Electric Company, Inc.)

12-45 VALVES, PRESSURE REGULATING

Commercial systems use many types of pressure regulating valves. Some of these valves control:
1. Evaporator pressure (two-temperature valves).
2. Crankcase pressure.
3. Discharge bypass pressure with solenoid valve control for pull down (service), to prevent freezing.
   a. Some into suction.
   b. Some into evaporator.
4. Head pressure control valve.

Some of these valves have Schrader service connections for gauge mounting.

12-46 VALVES, TWO-TEMPERATURE

In many multiple installations it is necessary to maintain different temperatures in evaporators connected in the same system. Thermostatic expansion valves may be used if the temperature differences are not over 5°F (3°C). But in some instances, such as a storage cabinet and an ice cream cabinet combination, the temperature differences are too great. A two-temperature valve is then put into the warmest evaporator suction line. This prevents pressure of the warmest evaporator from going below a safe setting.

The controlled evaporator or evaporators should not have more than 40 percent of the total load of a system. If a controlled evaporator is too large, erratic cycling will result. (See surge tanks, Para. 12-53.) If the controlled load is more than 40 percent, separate condensing units should be used.

12-47 TYPES OF TWO-TEMPERATURE VALVES

Two-temperature valves are sometimes called constant pressure valves or pressure reducer valves. They are also used to insure a constant low-side pressure. The valves have a bellows or diaphragm, a needle, and a seat. These are arranged in such a way that the bellows are operated by the pressure in the warmest evaporator.

As the compressor pumps the low side down to the desired pressure, the bellows shuts off the valve. This action stops the pressure in the warmest evaporators from going below the pressure desired. As pressure in the evaporator

builds up from vaporizing the refrigerant, the bellows again opens the valve passing vapor on to the compressor.

The pressure maintained on the surface of a quantity of liquid refrigerant determines the temperature at which the refrigerant will evaporate. The suction line valve will control the temperature of the evaporator to which the line is attached, even though the suction pressure of the compressor is considerably below the evaporator pressure.

Two general types of two-temperature valves are:
1. Pressure operated.
   a. Metering.
   b. Snap-action.
2. Temperature operated.
   a. Sensing bulb and bellows.
   b. Thermostat and solenoid.

12-48 METERING TYPE TWO-TEMPERATURE VALVE

The metering type, two-temperature valve acts more as a throttling device than as a shutoff valve. See Fig. 12-134. Fig. 12-135 shows a cross section of this valve.

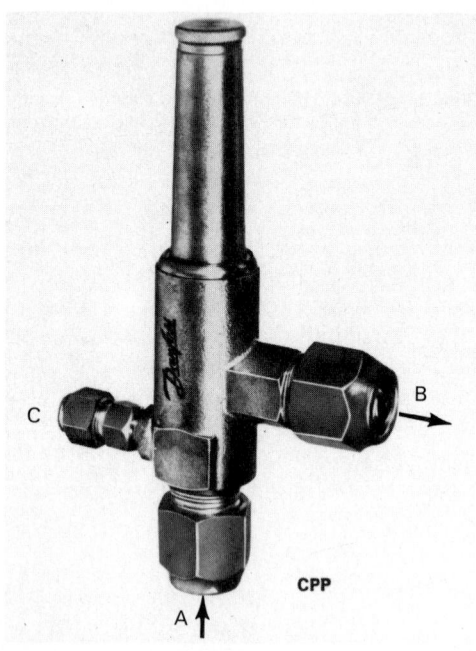

Fig. 12-134. Metering type two-temperature valve. Note the different connections. A—Refrigerant vapor in. B—Refrigerant vapor out. C—Gauge connection. (Danfoss, Inc.)

Some of these valves have a gauge opening so that the service technician can check and adjust the warmer evaporator's pressure. Having no differential, it opens and closes when the pressure varies only a fraction of a pound. (A differential means one pressure to open it and a different temperature to close it.) Because the bellows pressure area and the valve area are equal, only the adjustment spring and the warm evaporator pressure changes can operate the two-temperature valve.

The valve openings must be large enough to offer efficient vapor flow. Many of these metering controls have a

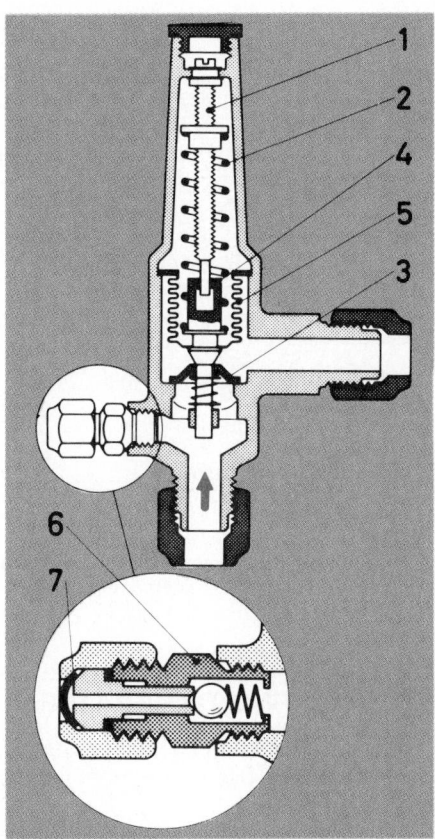

Fig. 12-135. Cross section through metering type two-temperature valve. 1—Adjusting screw. 2—Adjusting spring. 3—Valve. 4—Valve stem. 5—Bellows. 6—Service valve. 7—Cap.

small adjustment range, being especially designed to maintain temperature pressures just above crankcase pressures. See Fig. 12-136. Fig. 12-137 shows a two-evaporator system equipped with several pressure-controlled valves.

Large systems must rely on forces other than springs to control pressure for efficient operation. The large capacity

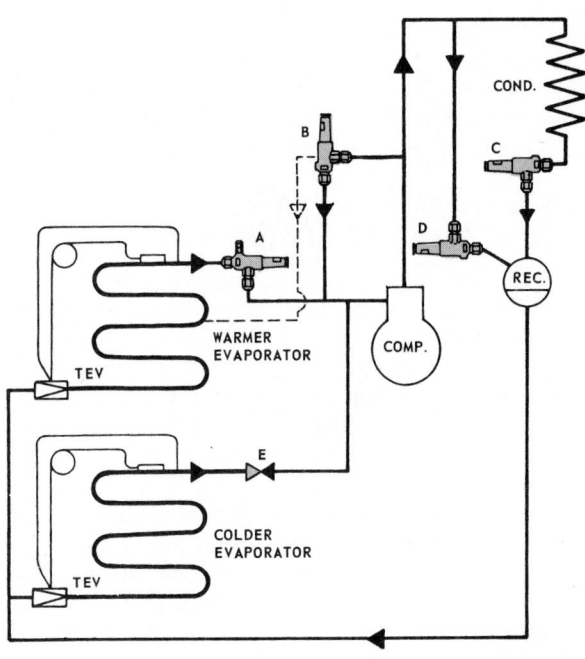

Fig. 12-137. A two-evaporator system equipped with several pressure control valves. A—Two-temperature valve (evaporator pressure regulator). B—Condenser bypass valve. C—Condenser pressure regulating valve. D—Capacity regulating valve. E—Check valve. (Danfoss, Inc.)

two-temperature valve in Fig. 12-138 has a plastic seat and connections for brazing to the suction line.

12-49 SNAP-ACTION TYPE TWO-TEMPERATURE VALVE

When a snap-action type valve closes, there is a decided rise in pressure in the warm evaporator before the valve opens again. Another important feature of this pressure-operated valve is that it has a definite cut-in pressure and temperature. It is often used where defrosting is wanted on each cycle.

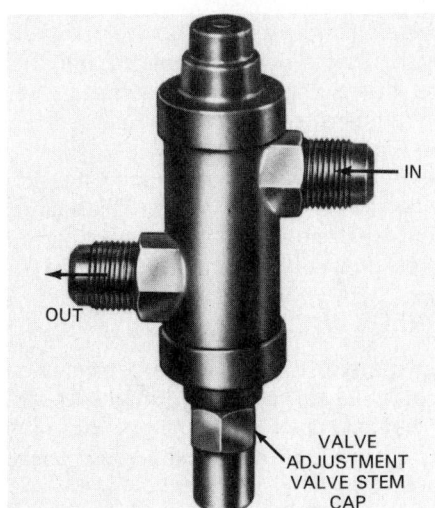

Fig. 12-136. Metering type two-temperature valve with flare fitting connections. This pressure operated throttling valve is an example of the metering type two-temperature valve.

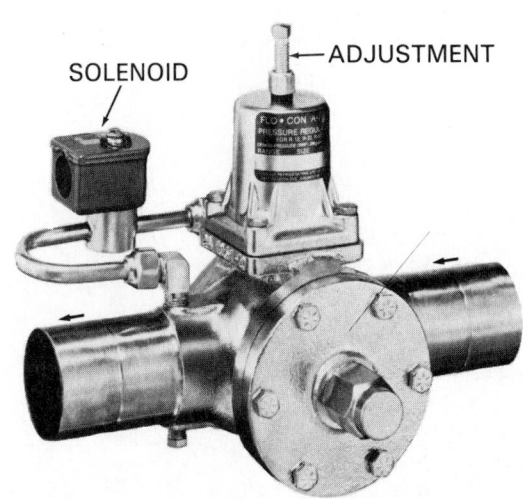

Fig. 12-138. A large capacity evaporator pressure regulator. Note use of metallic fittings for brazed connections to suction line pipe. (Refrigerating Specialties Div., Parker-Hannifin Corp.)

A snap-action two-temperature valve is normally used with multiple evaporator systems, such as a walk-in cooler and a display case, which do not operate at a wide temperature difference. The valve should be located on the suction line to the display case.

12-50 THERMOSTATIC TYPE TWO-TEMPERATURE VALVE

Another type of two-temperature valve has a temperature control and is built much like a thermostatic expansion valve. It also works much the same because it operates from the temperature of the evaporator or the air entering or leaving. It has a capillary tube and a sensing bulb much like the thermostatic expansion valve. It also has a bellows to move a rod as different pressures are created in the sensing bulb.

When the evaporator becomes cool enough, the cooling sensing bulb lowers the pressure in the bellows, and the bellows contracts. This pulls on the valve plunger and shuts off the valve. With the valve closed, pressure cannot drop any lower in the evaporator. Then, the valve controls the minimum temperatures of the evaporator.

As the evaporator warms, so does the sensing bulb. The increase in pressure is transmitted to the bellows. It expands, pushing the plunger so that the valve opens. With the valve open, the compressor is able to, once more, draw vaporized refrigerant from the evaporator. This type of valve is always located in the suction line of the warmest evaporator.

Fig. 12-139 shows a regulator which responds to the temperature of the air as the air leaves the evaporator and enters the fixture or case. It may also respond to the temperature of the air entering the evaporator.

It has a sensitive element located in this airstream. The valve body is mounted in the suction line. A liquid line connection is also made to the valve body. This provides the pressure needed to open the main valve in the TPR (temperature pressure regulator). See Fig. 12-140.

As the temperature sensing bulb warms, its increase in pressure closes a small valve and the high-side pressure in the valve decreases allowing the main piston to open the suction line more. As the sensing bulb cools, the high-pressure pilot valve is opened and the main piston is forced down, closing the suction line passage partially or completely (modulates).

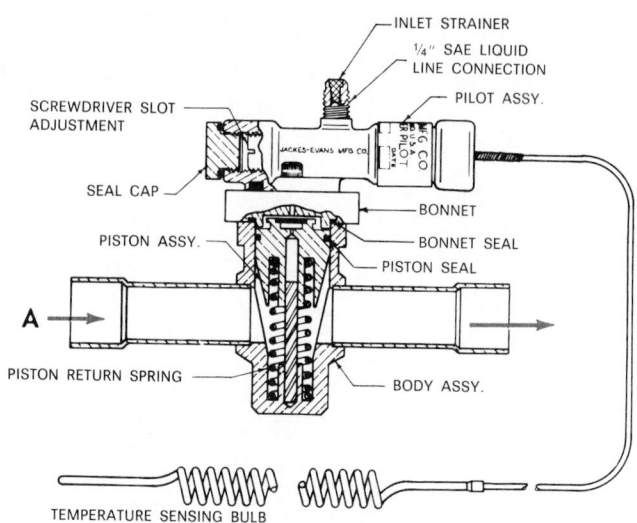

Fig. 12-140. Temperature controlled suction line evaporator pressure control. A—Suction line vapor travel.

12-51 SOLENOID TWO-TEMPERATURE VALVE

A fourth way to get various fixture temperatures in a multiple installation is to use a thermostat connected in series with a solenoid valve.

The solenoid shutoff valve is usually placed in the liquid line of the evaporator it controls. It has an electrical connection to a thermostat, as shown in Fig. 12-141.

The thermostat is operated by the fixture temperature. When the fixture reaches the correct temperature, the thermostat opens. The electric solenoid loses its magnetism and the valve closes. No more refrigerant is fed to the evaporator. The cabinet will gradually warm up until the thermostat points close, the solenoid valve opens, and refrigeration starts again.

This system of refrigeration control is based on fixture temperature. The condensing unit is controlled by a pressure type motor control. The motor will not stop until all the fixtures are cooled to their correct temperature. Some systems have the solenoid valve in the suction line to prevent removing the refrigerant from the evaporator.

This system uses a normally open solenoid and a thermostat that opens the circuit on temperature drop and closes it on temperature rise. When magnetized, the valve closes. See Fig. 12-142.

A solenoid valve is also used to stop flooding the low-pressure side during the off cycle. This solenoid is also located in the liquid line. It is electrically connected in parallel with the pressure motor control, Fig. 12-143.

12-52 CHECK VALVES

Check valves are used in refrigerating systems to prevent flow of liquid and/or vapor refrigerant in the wrong direction. They are used in two-temperature installations, in defrost systems, and to prevent vapor passage during off cycles.

In multiple installations, one condensing unit is connected to several evaporators all of which are at different temperatures. Two-temperature valves are used to obtain the desired temperatures. Check valves, Fig. 12-144, are

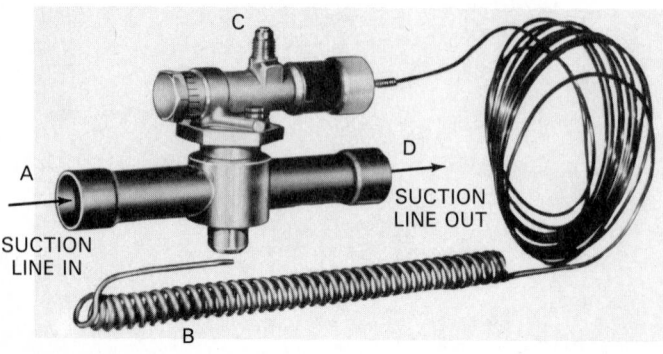

Fig. 12-139. Temperature controlled suction line metering valve. A—Suction line from evaporator. B—Temperature sensing bulb. C—High-side pressure connection. D—Suction line to compressor.

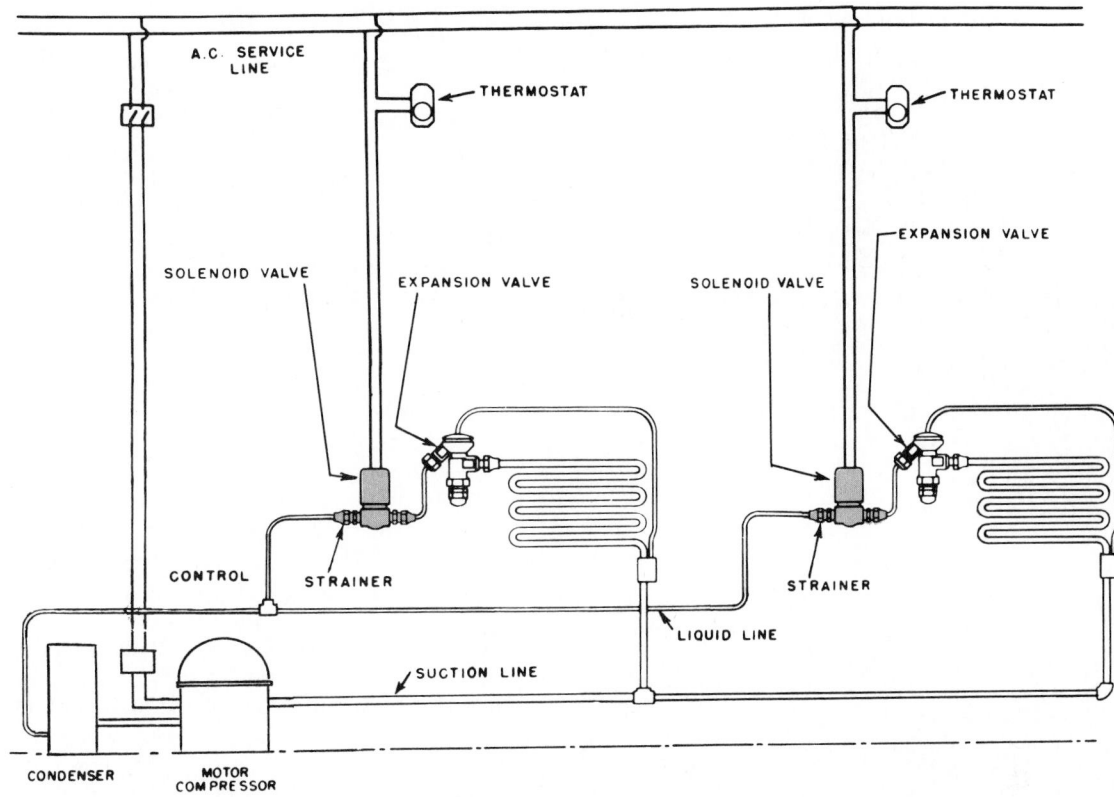

Fig. 12-141. Installation of solenoid valves. Each solenoid controls the cabinet temperature of a different cabinet. Valves and thermostats are using line voltages.

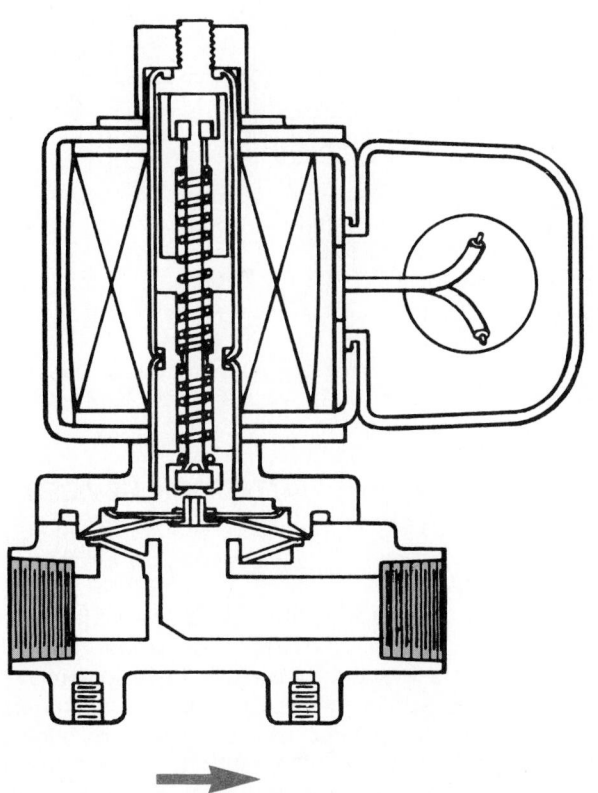

Fig. 12-142. A solenoid valve designed to open the valve when magnet is not energized and close the valve when magnet is energized. Note the threaded pipe connections at either end.
(Fluidex Div., Parker-Hannifin Corp.)

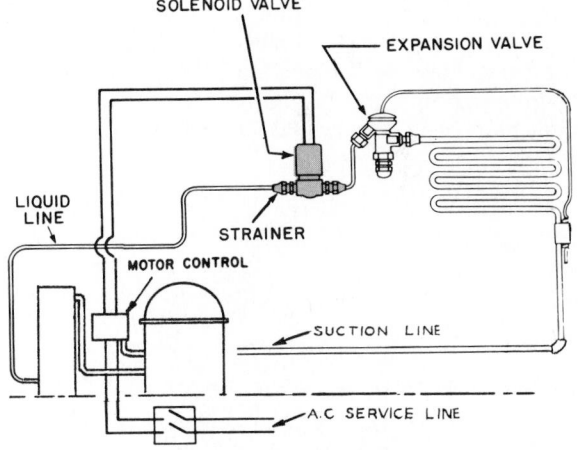

Fig. 12-143. Solenoid valve located in liquid line and controlled by pressure motor control.

sometimes put in the suction line of the coldest evaporators to prevent excess warming of the cold evaporator during the off cycle.

After the condensing unit has stopped, one of the two-temperature valves may open before the condensing unit turns on. This may flood the low side with warm refrigerant vapor. This vapor will also travel along the suction line to the coldest evaporator.

If it enters the evaporator it will start condensing, releasing its latent heat. This will make the cold evaporator defrost or, at least, warm up somewhat. The check valve, when

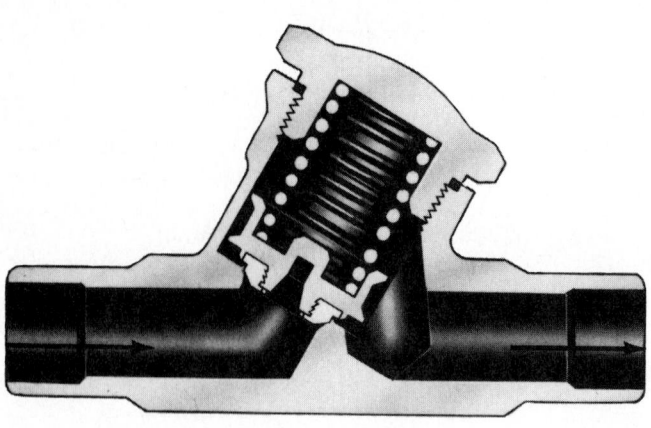

Fig. 12-144. Suction line check valve is used on colder evaporators in multiple systems. (Superior Valve Co., Division of AMCAST Industrial Corporation)

installed in the suction line of the coldest evaporator, will only allow vapor to be drawn from this evaporator.

This check valve must have a tight seat and it must open easily. If the valve is too small or if it opens with difficulty, it will act as a throttling device and cause too much pressure drop. The result will be poor refrigeration in the coldest evaporator.

Fig. 12-145 shows a large-capacity check valve. Reverse cycle systems use check valves. So do some hot-gas defrost systems.

Check valves can be a source of noise in a refrigeration system as they open and close with a metallic click or bang. There is also noise associated with the inefficient operation of a valve. If the valve does not completely close off the reverse flow, a ''hammering'' noise can be heard. In a hydronic system, this is referred to as ''water hammer.'' Valves have been designed to minimize this noise problem.

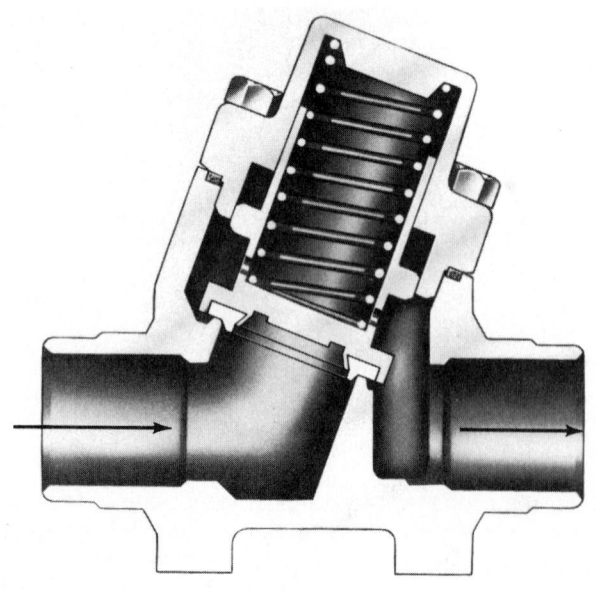

Fig. 12-145. Large-capacity check valve. Note double cylinder design which provides smoother operation. (Superior Valve Co., Division of AMCAST Industrial Corporation)

12-53 SURGE TANKS

Multiple temperature installations may short cycle a pressure-controlled condensing unit. This may be caused by the pressure fluctuations which result from the opening and closing of the two-temperature valves. Pressure is said to fluctuate when it rises and falls repeatedly and in an uncertain pattern.

The following conditions can cause a short cycle:
1. If the two-temperature valve is closed and the condensing unit cools the lowest temperature evaporator enough to open the pressure motor control, the condensing unit will stop.
2. If, just after it stops, a two-temperature valve which controls one of the warmer evaporators opens, the low-side pressure will rise rapidly. The pressure turns on the condensing unit and, thus, causes a short cycle.

To eliminate this trouble, a surge tank or a larger cylinder may be installed in the main suction line just ahead of the compressor. The surge tank shown in the Fig. 12-146 schematic is large enough to absorb a pressure buildup. Thus, if the unit is stopped and a two-temperature valve opens, the low-side pressure cannot build up quickly and short cycle the unit. The capacity of the surge tank is great enough to absorb a large volume of vapor and thereby slow down the rapid pressure changes which would make the motor control turn off and on. The line connected to the bottom of the tank leads to the compressor. It helps return the oil to the compressor.

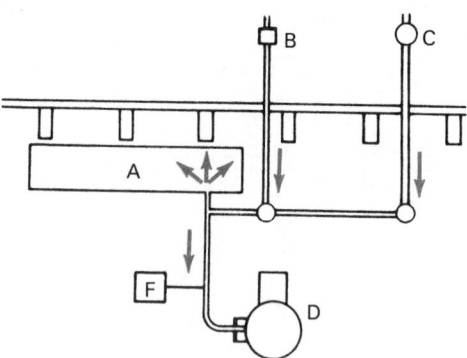

Fig. 12-146. A surge tank installation. A—Surge tank. B—Check valve. C—Two-temperature valve. D—Compressor. F—Motor control.

12-54 COMPRESSOR PROTECTION DEVICES

Many reciprocating compressors are damaged when liquid refrigerant accidentally flows into the compressor from the suction line.

The refrigerant must be in a vaporous state. This means that the vapor temperature must be higher than the temperature of the evaporating liquid in the evaporator. This increase in temperature means the vapor is superheated.

Many devices have been used to prevent or minimize entry of suction line liquid refrigerant into the compressor:
1. An accumulator in the suction line.
2. Hot-gas bypass valves to move hot gas into the suction line where the gas can evaporate any liquid.
3. Temperature-sensing devices and solenoid valves.
4. Heat exchangers to warm the suction line vapor-liquid.

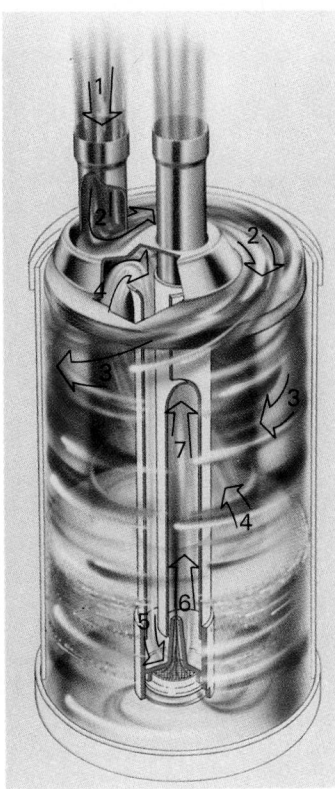

Fig. 12-147. Suction line accumulator: (1) Mixture of refrigerant vapor liquid and oil enters. (2) Swirling motion created on entering mixture. (3) Liquid strikes inside wall. (4) Refrigerant vapor and mist drawn upward, vertical motion, and then downward into tubing. (5) Flow turn 180° and upward through orifice, drawing measured amount of liquid refrigerant and oil from bottom. (6) Combination of refrigerant vapor, compressor lubricating oil, and refrigerant flow vertically, forming a mist before entering (7) the compressor suction.
(Tecumseh Products Company)

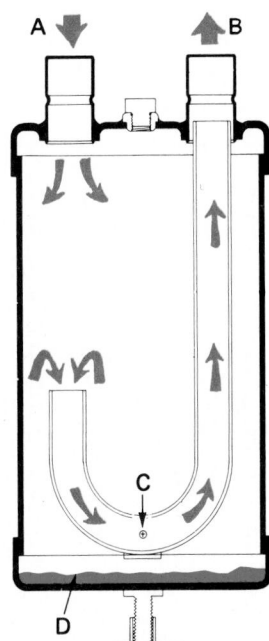

Fig. 12-148. Inside of an accumulator. A—Suction gas in. B—Suction gas out. C—Oil return aspirator hole. D—Liquid refrigerant trapped until it can evaporate. (Virginia KMP Corp.)

5. Electrical heaters to warm the suction line vapor-liquid.
6. An evaporator (blower coil) in the suction line.

When a system is using a hot-gas defrost or when a system has a sudden load change, drops and slugs of liquid refrigerant may travel in the suction line. Many systems have an accumulator in the suction line to reduce the danger of liquid refrigerant flow into the compressor, Fig. 12-147.

The accumulator will lengthen the cycling interval (gas storage). It usually has aspirating (suction) devices to return oil. Fig. 12-148 shows an internal design of an accumulator.

The hot-gas bypass device depends on the use of a temperature sensor attached to the suction line. This sensor controls a solenoid valve. This valve will open, allowing hot gas to flow into the suction line, when there is danger of liquid refrigerant flowing into the compressor.

Liquid line-suction line heat exchangers are explained elsewhere. Refer to Para. 4-7 and Para. 12-35.

Electrical heaters may be attached to the suction line and a temperature-sensing element used to turn on the current when heating of the suction line is needed.

The blower coil is usually operated by a temperature-sensing element attached to the suction line. The fan in the blower coil will be turned on when the suction line temperature indicates danger of liquid refrigerant flowing into the compressor.

A temperature-operated device or a pressure device must detect the presence of liquid quickly enough to turn on mechanisms which will stop the liquid from reaching the compressor. The best sensor is a thermistor which can react very quickly. It can be connected to an alarm circuit or operating circuit to stop the compressor before it can be damaged.

12-55 OIL CONTROL SYSTEMS

Since they must be kept lubricated, compressors run in oil. A certain amount of this oil leaves the compressor with the refrigerant vapor. It is important that this oil be prevented from moving through the system.

Various devices are designed to collect the oil and return it to the compressor. Most industrial HVAC systems use one or more of three basic oil control components: an oil level regulator, an oil reservoir, and an oil separator.

An oil level regulator controls the level of oil within the compressor. It usually uses a float-type mechanism. This allows oil flow to the compressor only when the float indicates that the oil level is low. Some compressors have an oil level regulator to control the oil in the individual compressor. See Fig. 12-149.

An oil reservoir in the system has a dual purpose. It holds the oil supply needed for the compressor. The regulator draws from this supply to replenish the oil in the compressor. Oil trapped by the separator is returned to the reservoir until it is needed again. The oil reservoir may contain two sight glasses for observation of oil level. See Fig. 12-150. The oil reservoir also contains a flare fitting for adding oil to the system.

Refrigeration systems work best when the oil is kept at a proper level in the compressor. Oil in the condenser and evaporator will reduce efficiency of the unit.

It is important to keep the oil from circulating in low-temperature installations. It thickens at these low

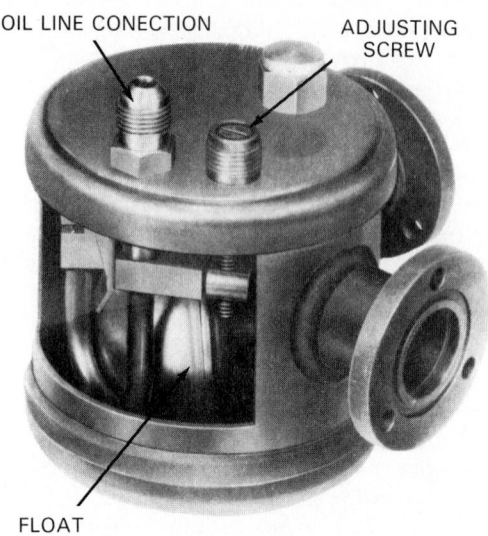

OIL LINE CONECTION
ADJUSTING SCREW
FLOAT

Fig. 12-149. The oil level regulator controls the oil level in individual compressor crankcase. Note float operated valve to hold back excess oil until the oil level in the compressor crankcase drops. (A C & R Components, Inc.)

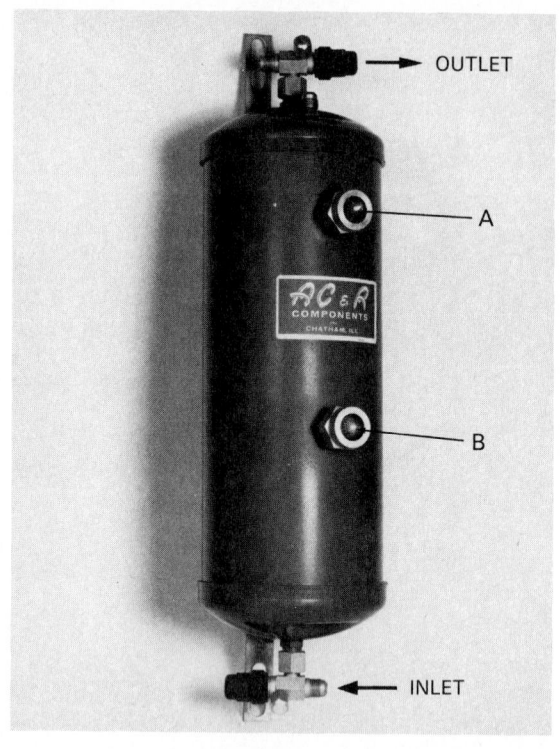

OUTLET
A
B
INLET

Fig. 12-150. Oil reservoir holds standby oil as part of the oil control system. Note sight glass ports, A and B, to observe oil level. (A C & R Components, Inc.)

temperatures and becomes difficult to move out of the evaporator.

Oil separators are designed to remove the oil from the hot compressed vapor as the vapor leaves the compressor. The oil will separate because the vapor flow slows down as it arrives in the separator. The oil will collect in the separator until a certain level is reached. Then, a float opens a needle valve. The oil then returns to the compressor crankcase.

Oil separators are also placed between the compressor and the condenser, Fig. 12-151. The separator is in-sulated to prevent it from acting as a refrigerant condenser and passing off heat to the surrounding air.

Many oil separators are serviceable (bolted construction). See Fig. 12-152. On hermetic systems, the oil-return line is usually connected to the suction line near the motor compressor.

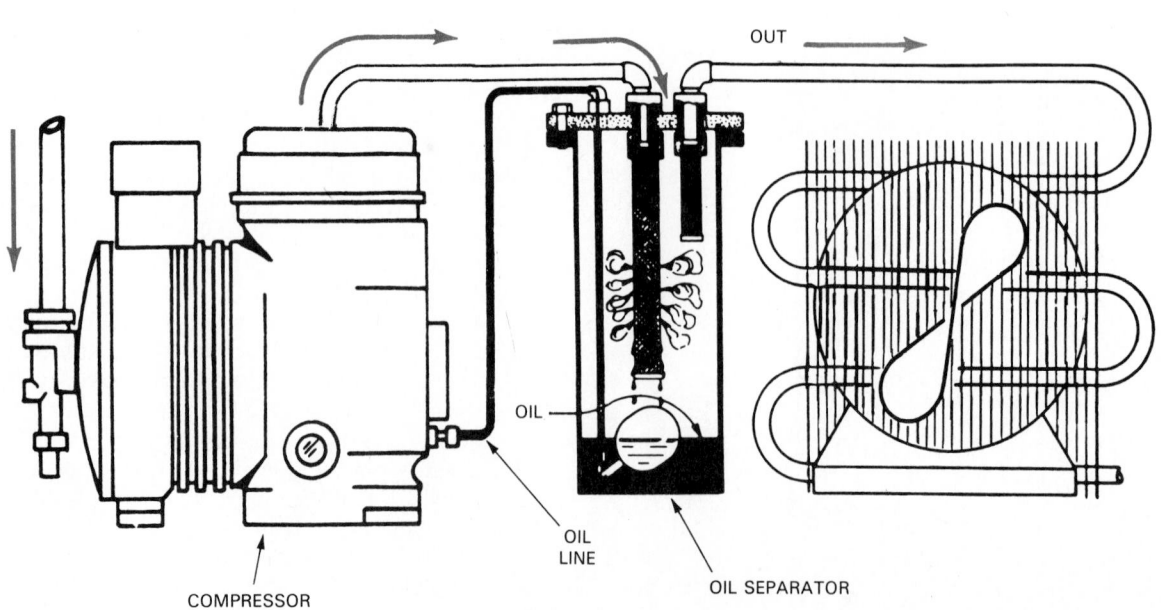

OUT
OIL
OIL LINE
COMPRESSOR
OIL SEPARATOR

Fig. 12-151. Oil separator installation. Oil is removed from high-temperature, high-pressure refrigerant and returned to compressor.

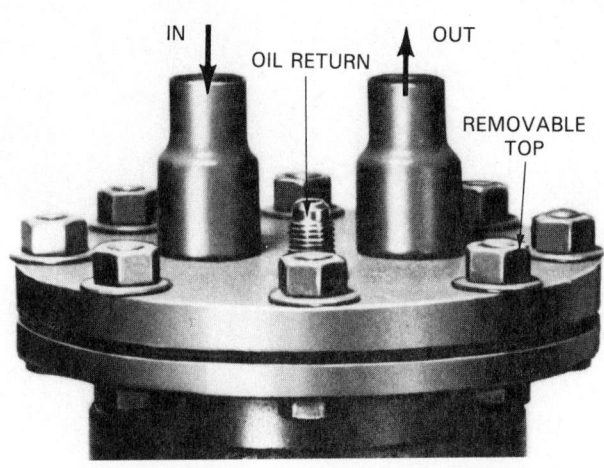

Fig. 12-152. Oil separator with bolted assembly top. This type is cleanable and serviceable. (AC & R Components, Inc.)

Fig. 12-153. Large-capacity oil separator. Refrigerant line connections are 3 1/8 in. OD. (AC & R Components, Inc.)

Liquid refrigerant may collect in the oil separator during long off-cycles or during long manual shutdown. This liquid refrigerant, as it returns by way of the oil return line to the compressor, may cause oil pumping. This action may damage the compressor. A check valve in the vapor outlet of the oil separator will reduce this danger. A filter in the oil return line will help keep the oil clean.

A solenoid valve in the oil return line is sometimes used so oil or refrigerant can return to the crankcase during the off-cycles. A thermostat controls the solenoid. The thermostat will close only when the oil separator is warm (100 to 130 °F) (38 to 54 °C).

A capillary tube, instead of a 1/4 in. OD tube, is sometimes employed to carry the oil from the oil separator to the compressor crankcase. It reduces oil (or liquid refrigerant) flow to the crankcase and also reduces the chance of oil slugging and damage to the compressor.

Large units, up to 150 tons capacity, can use the oil separator shown in Fig. 12-153. It has a 16-in. diameter shell. Note oil return located at the bottom.

12-56 COMPRESSOR LOW-SIDE PRESSURE CONTROL VALVES

Starting a compressor places a heavy load on the motor. It has to overcome inertia of the moving parts (objects at rest, tend to stay at rest) as well as high crankcase pressure. (In fact, crankcase pressure may be at its highest just then.) It is important, then, to use higher horsepower. Even so, the motors are usually taxed to the limit at the moment of starting, especially against normal or above-normal head pressures.

A compressor low-side pressure control is used on some installations. This keeps low-side pressures in the crankcase at a reasonable level, even though the rest of the low-side pressure may be high. It may be called a reverse metering two-temperature valve because it never permits the crankcase pressure to exceed a certain safe value.

Crankcase pressure-regulating valves are needed where the compressor runs too long before the low side drops to a pressure which does not overload the compressor. If the suction line pressure is higher than the safe pressure, the valve shuts the suction line off from the compressor.

The valve body is usually made of brass, the diaphragm or bellows of phosphor bronze, and the needle and seat of wear-resisting steel alloy. See Fig. 12-154. In Fig. 12-155 the valve is shown in a typical installation, in the suction line between the evaporator and the compressor. On some applications, it may be necessary to locate other system components, accumulator, after the valve.

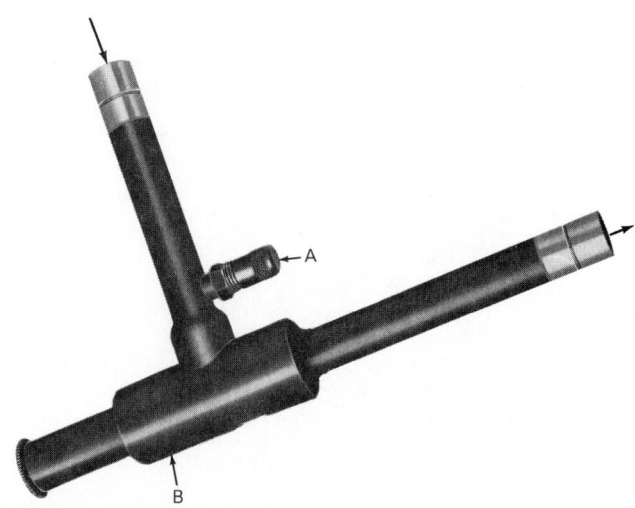

Fig. 12-154. Crankcase regulating valve with adjustable pressure. Arrow indicates flow of refrigerant. A—Access valve. B—Valve adjustment. (Alco Controls Div., Emerson Electric Co.)

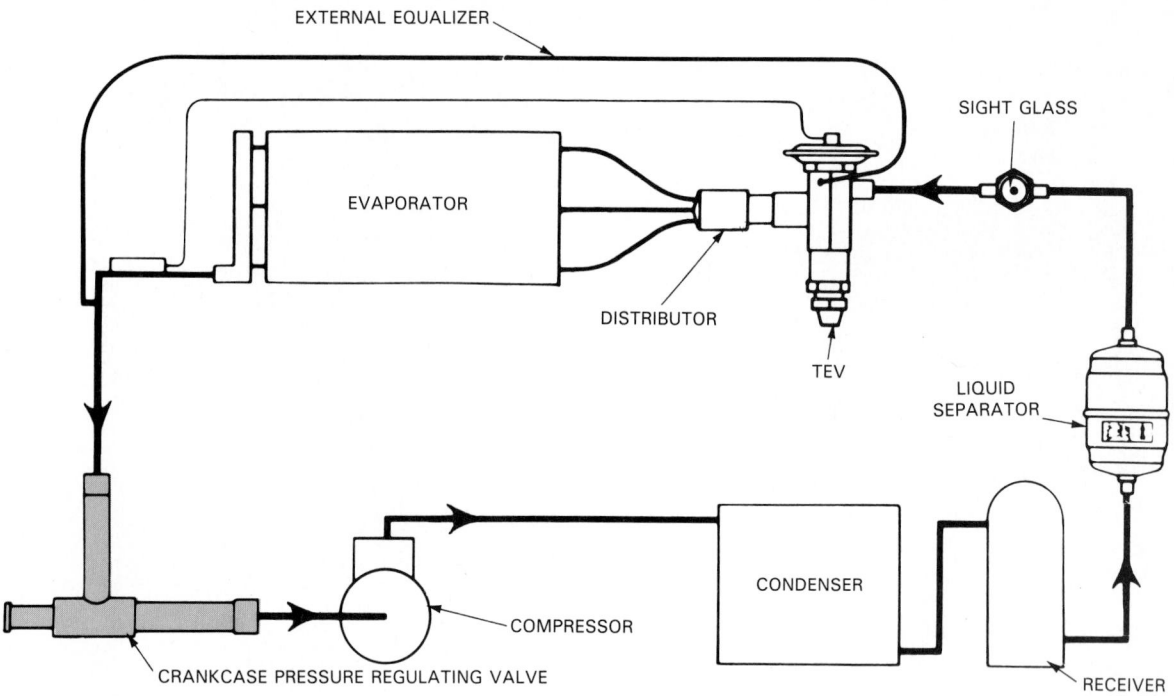

Fig. 12-155. Note location of the crankcase pressure regulating valve in the suction line between the evaporator and compressor. (Sporlan Valve Co.)

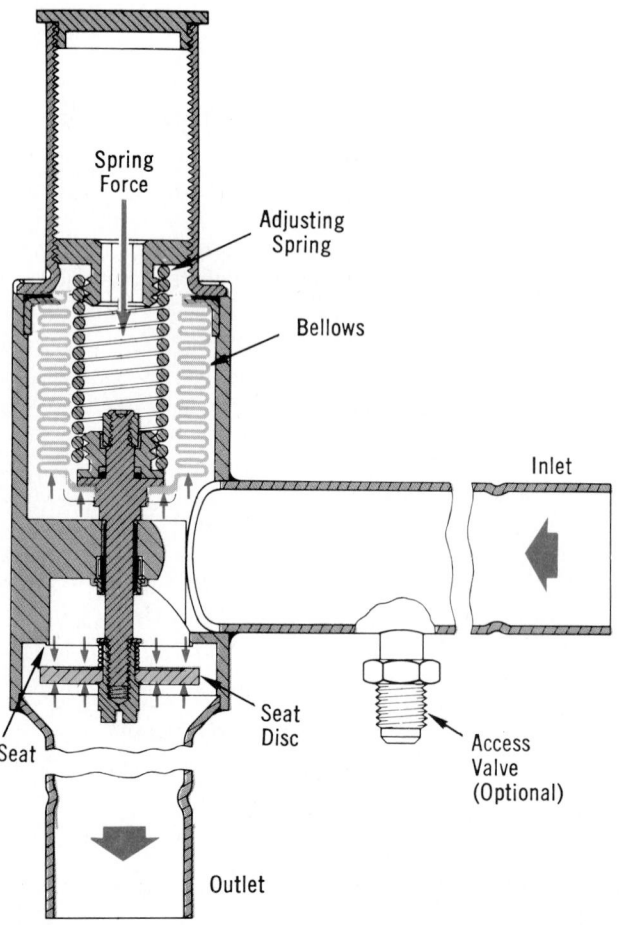

Fig. 12-156. Cutaway of crankcase pressure-regulating valve. Note inlet pressure on bellows and on seat disk. Also note outlet pressure on seat disk. Arrows indicate which direction vapor is flowing. (Sporlan Valve Co.)

The pressure regulating valve prevents overloading of the compressor motor by eliminating the crankcase pressure during and after the defrost cycle or after the normal shut-down period.

The crankcase pressure-regulating valves are sensitive to the outlet pressure, and the compressor, crankcase, or suction pressure. They close on the rise of the outlet pressure. Fig. 12-156 illustrates the operation of this valve. Note that the inlet pressure is exerted on the underside of the bellows and the top side of the seat disk.

A refrigerating system suction line with a pilot-operated suction regulator valve is shown in Fig. 12-157.

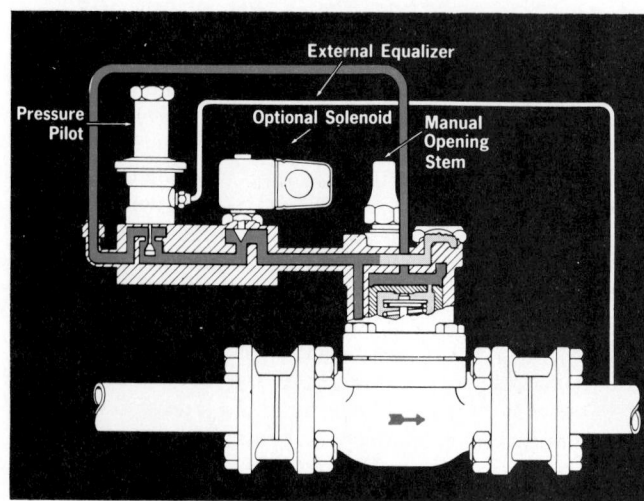

Fig. 12-157. Pilot-operated compressor low-side pressure control. Pilot valve releases pressure above main piston when compressor pressure reaches safe level. This opens main valve allowing evaporator vapor to move to compressor. (Alco Controls Div., Emerson Electric Co.)

12-57 WATER VALVES

Some commercial units of 1/2 hp and up use water-cooled condensers where good, inexpensive water is available. Less power is needed to drive the condensing unit than for the same size air-cooled installation. This is due to better heat transfer as well as the lower condenser temperatures and pressures made possible in a water-cooled condenser. The saving in electrical power makes up, somewhat, for the cost of the large amounts of water used for cooling.

The water valve turns the water on and off as needed. But it also varies the amount of water as required. Three types of water valves are used:

1. Electric.
2. Pressure.
3. Thermostatic.

It is good practice to install a strainer in the water inlet to the valve. See Fig. 12-158.

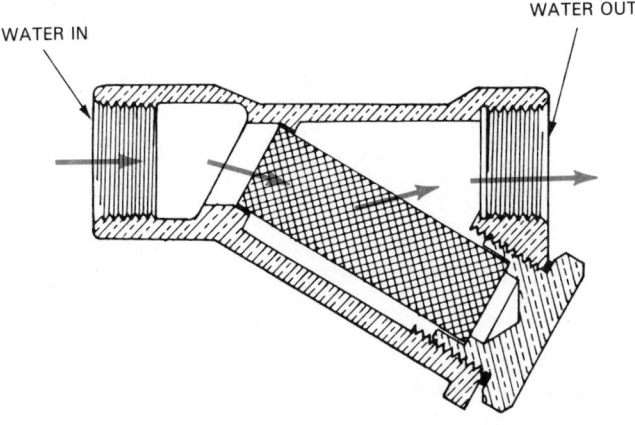

Fig. 12-158. Water line strainer. Screen is removable for cleaning. (Superior Valve Company, Division of AMCAST Industrial Corporation)

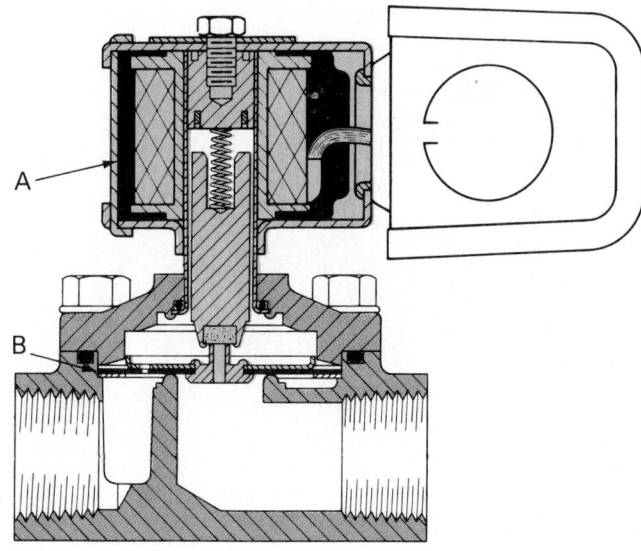

Fig. 12-159. Solenoid-operated water valve with diaphragm. A—Solenoid. B—Diaphragm. (Sporlan Valve Company)

12-58 ELECTRIC WATER VALVE

Electrically operated water valves are of two principal types: solenoid activated and motor operated.

A water valve is located between the water supply and the condensing unit. Usually, it is mounted on the condensing unit base. The moment the motor starts, this valve opens. When the motor circuit is opened the solenoid is de-energized and the valve closes. See Fig. 12-159.

Electric water valves consume a small amount (6 to 10 W) of current while in operation. Fig. 12-160 illustrates two typical electric water valve circuits. One uses a low-voltage solenoid valve, and the other a 120 V solenoid valve. Most valves require 120 V.

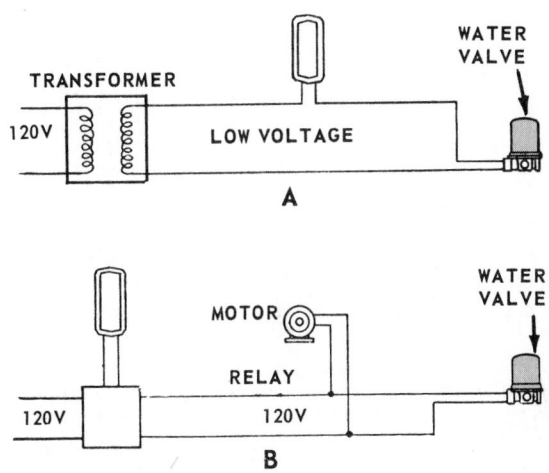

Fig. 12-160. Solenoid-operated water valve wiring diagrams. A—Solenoid uses low voltage. B—Solenoid uses 120 V.

A solenoid water valve of larger capacity is shown in Fig. 12-161. The body of the valve is brass and is made with either threaded or soldered connections. The plunger is made of noncorrosive steel. The valve seats are usually made of brass or bronze and the valve face is of a special rubber composition.

Water flow is constant in this type of control. The valve stem is loosely connected to the plunger to permit a shock action to open the valve. Gravity and water pressure close the valve when the power is shut off. Large-volume water flow may also be controlled by motor-operated valves, Fig. 12-162. An advantage in the use of the electrically controlled water valve is that it may be removed or replaced without disturbing the refrigeration system.

The inside of the motor-actuated water valve is shown in Fig. 12-163. Pipe joints are unions to enable easy removal of the valve. The screen may be serviced by removing the cap on the bottom of the valve. These valves have capacities varying from 1/2 to 4 in. pipe size.

12-59 PRESSURE-OPERATED WATER VALVE

The pressure-operated water valve is the most popular. It is a bellows attached to the high-pressure side of the system, preferably to the cylinder head. This bellows operates the water valve, as shown in Fig. 12-164.

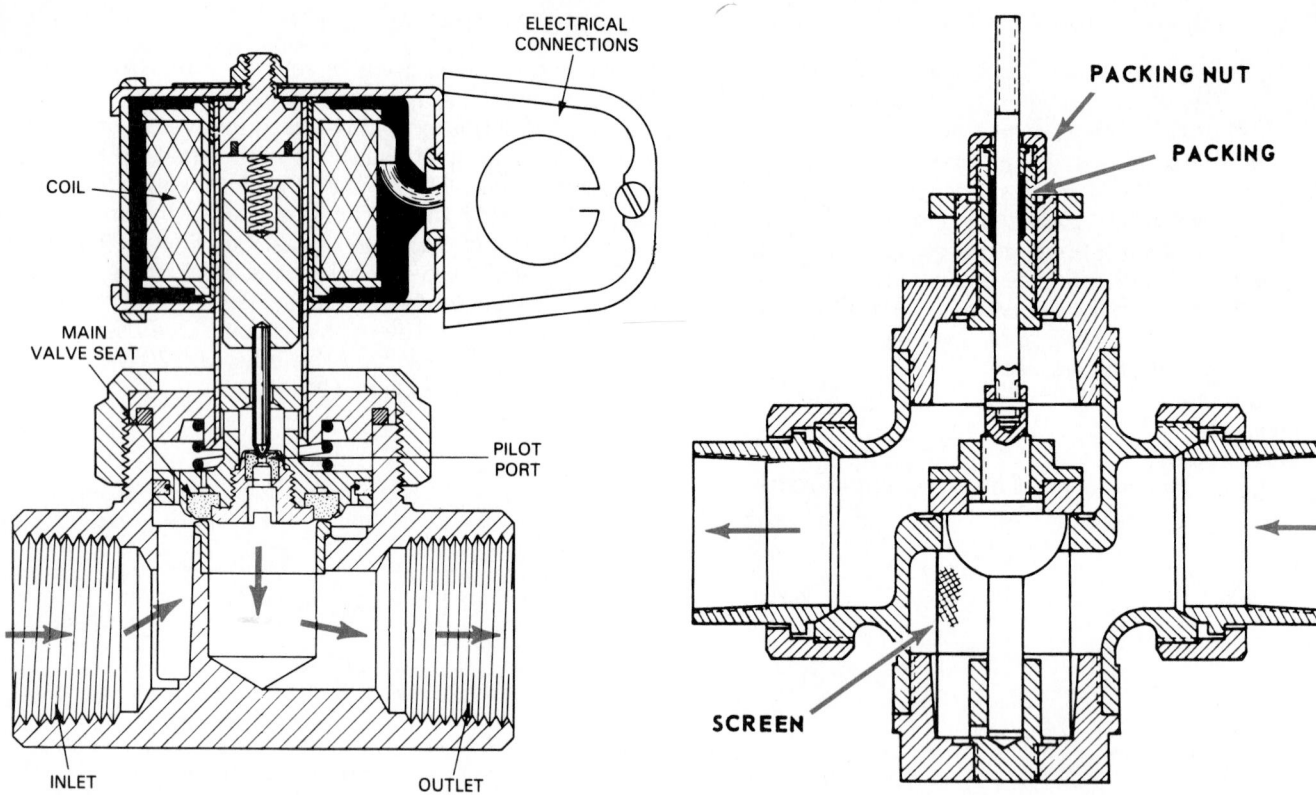

Fig. 12-161. Pilot-operated electric water valve used on large installations. Note that solenoid valve, when open, only decreases water pressure above large piston. (Sporlan Valve Co.)

Fig. 12-163. Water valve body of motor-actuated valve. Note direction of flow, screen, valve stem packing, and packing nut.

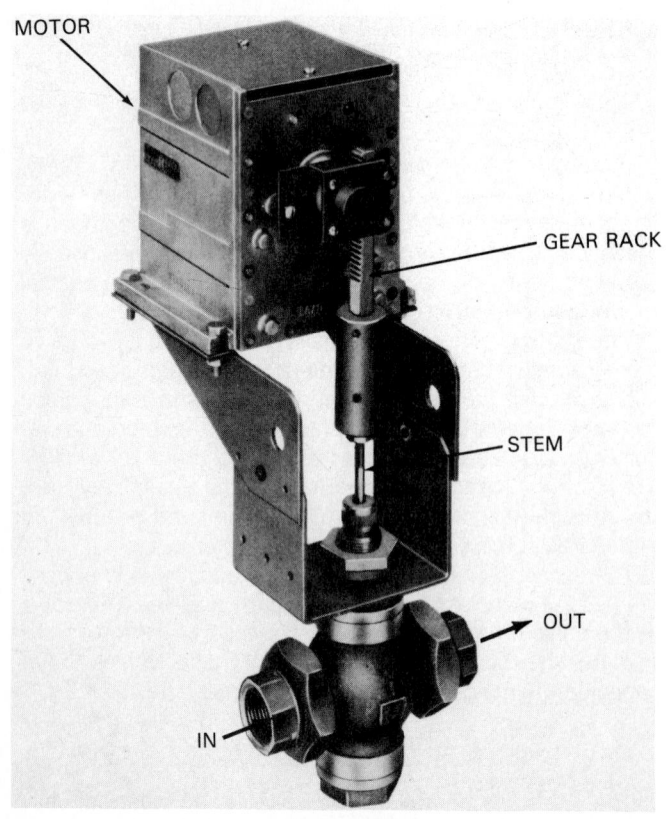

Fig. 12-162. Motorized water valve. Motor raises or lowers valve stem. Note gear rack connected to stem. (Johnson Controls, Inc.)

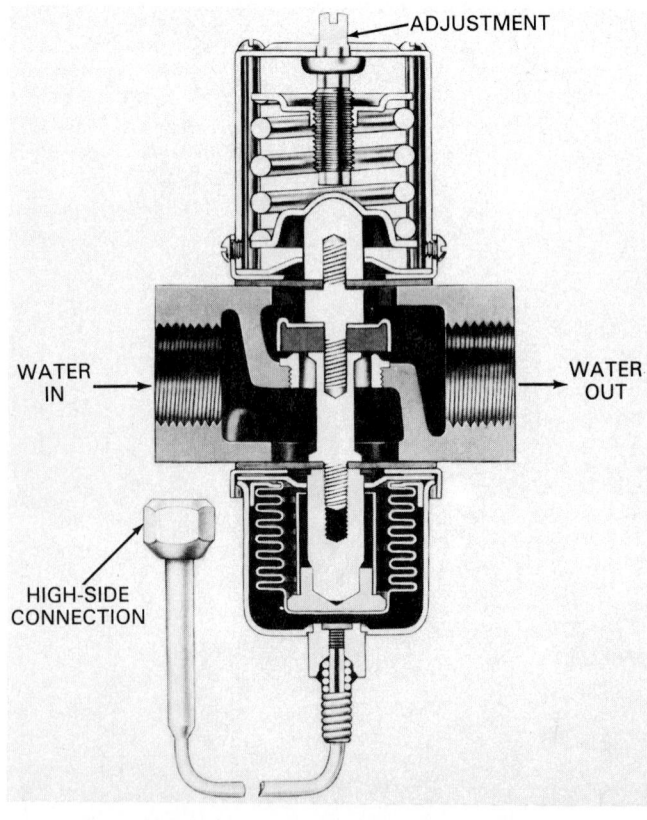

Fig. 12-164. Pressure-controlled water valve is connected to high-pressure side at compressor head. Rate of water flow is adjusted by spring pressure (top) on valve. Water flows through valve from right to left. (Johnson Controls, Inc.)

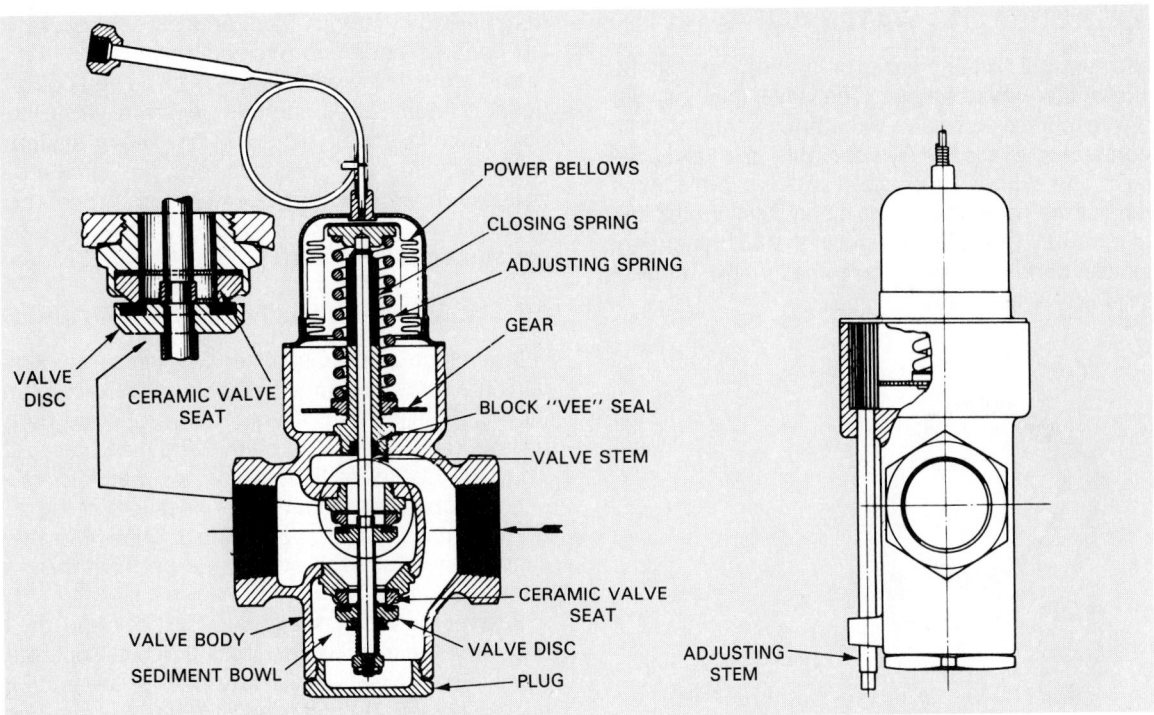

Fig. 12-165. Large capacity pressure-operated water valve. Double valve and seat arrangement balances force from water pressure. One valve is opened and the other is closed by water pressure.

As condenser pressure rises, the bellows in the water valve contracts. The valve is opened by any of various mechanisms, depending on the specific water valve. Water flows into the condenser to cool the compressed vapor. The valve opens the water circuit only when the water is needed—as the pressure rises. It will keep increasing the water flow just as long as there is a tendency for an increase of pressure in the high side.

These valves may be adjusted by adjusting a heavy spring which presses against the bellows. The valves are set to open at definite head pressures. The pressure depends on the temperature of the water and the refrigerant used. See Chapter 14.

Some pressure-controlled water valve designs require opening the system to remove the valve. Others may be removed without disturbing the refrigeration system.

Water flow can be modulated with this valve. (Modulate means to vary the amount.) As the condensing pressures and temperatures increase, the valve opens farther. When the pressures and temperatures drop, the water flow decreases.

The valve faces are a hard rubber composition, Bakelite, or fiber. The seat is usually made of copper or brass. The valves are equipped with either a packing gland or with a bellows at the point where the water valve stem goes into the water valve body. The packing must be adjusted occasionally to keep it from leaking. Some valves have a bellows in place of a packing.

These valves usually do not depend on the pipe for support, but have a mounting arrangement or flange. The water-in and the water-out connections are clearly labeled. Valves are usually threaded for standard pipe connections. Most are constructed so that water pressure tends to keep the valve closed. Fig. 12-165 shows a large-capacity valve

used on 1-in. lines. This valve has a gear mechanism for adjusting the pressures. The pressure-operated double water valve in Fig. 12-166 controls flow in two separate circuits.

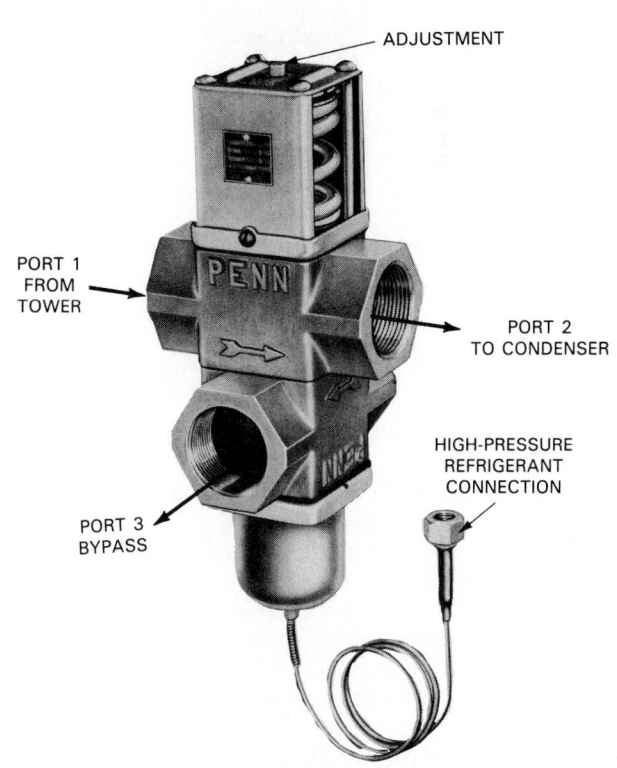

Fig. 12-166. This double water valve is pressure operated. Note direction of water flow on each valve body.
(Johnson Controls, Inc.)

12-60 THERMOSTATIC WATER VALVE

The thermostatic water valve is controlled by the temperature of the exhaust water. The valve, itself, is identical to the pressure water valve except for a thermostatic element connected to the bellows operating the valve. See Fig. 12-167. The element is charged with a volatile liquid. The power bulb is mounted in the condenser water line. Pressure created by the volatile liquid in the bulb opens the valve when the condenser water becomes warm. It closes the valve as the water cools.

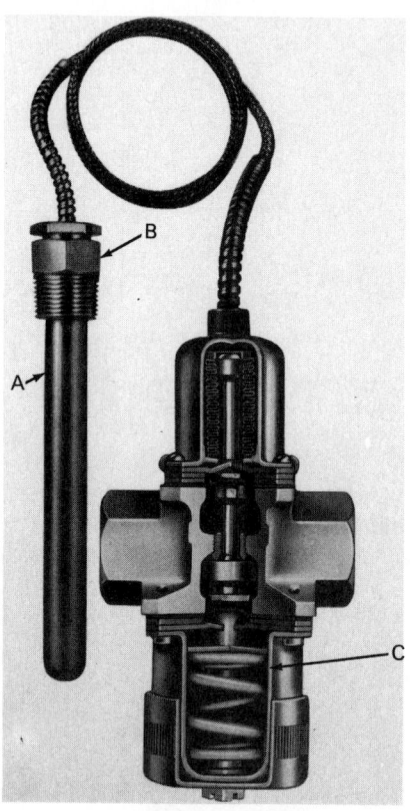

Fig. 12-167. Thermostatic water valve has sensing bulb shown at A. It is fitted inside water outlet piping, using fitting at B. Temperature adjustment is at C.

12-61 MANUAL VALVES

Manual servicing valves used on commercial refrigerating systems help the service technician:
1. To determine the operating pressures.
2. To charge or discharge a system.
3. To remove any part of the system without disturbing the other parts.

These hand valves and service valves must resist corrosion and withstand frequent opening and closing without leaking. They should be built of the best materials. Valve stems and packing must be handled with care.

12-62 CONDENSING UNIT SERVICE VALVES

Many of the condensing units are equipped with servicing valves of both the two-way and one-way type. See Chapter 2. Some of these valves are quite large. The valve stems

may be 3/8-in. across flats, and larger because liquid lines are as large as 3/8-in. OD.

Some of the larger systems may be equipped with additional service valves such as separate valves for installation purpose and/or for servicing. Many systems have a valve between the condenser and the liquid receiver. The gauge connections may be 1/4- or 1/8-in. pipe. Some systems use Schrader valves to connect gauges and to perform service operations. See Chapter 11.

12-63 MANUAL INSTALLATION VALVES

In addition to the usual service valves, multiple installations are usually equipped with what are called hand shutoff valves, Fig. 12-168. These valves operate by hand and, by law, must be located so that they may be easily turned. They may be classified as riser or manifold valves.

In multiple installations, it is best to run the suction line from the compressor to a manifold. Then have the individual suction lines for each evaporator go from this manifold to the evaporators.

Between each of these suction lines and the manifold, and mounted into the manifold, is a hand-operated shutoff valve. This valve permits any one of the suction lines to be closed without interfering with the operation of the others. A similar manifold device is also provided for the liquid line. These valve groupings are usually mounted in a steel box or cabinet or on a special valve board near the condensing unit.

12-64 RISER VALVES

A riser valve is another type of shutoff valve. It is hand-operated with three openings to which refrigerant lines may be connected. Two of these openings are in line with each other on opposite sides of the valve. The third is a little closer to the valve wheel and is at right angles to the other two openings.

By turning the hand valve in, the opening at right angles to the other two is closed. This construction permits mounting of the valve in either a liquid or suction line. The tech-

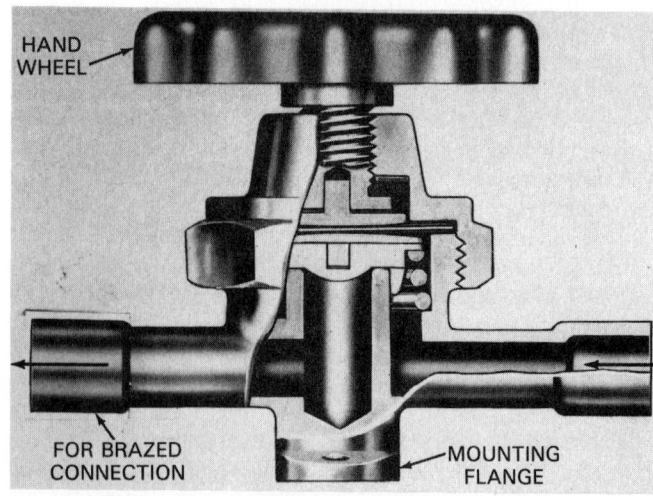

Fig. 12-168. Manual shutoff valve used on multiple installations. Valve uses diaphragm in place of packing. Piping openings are in line. (Henry Valve Co.)

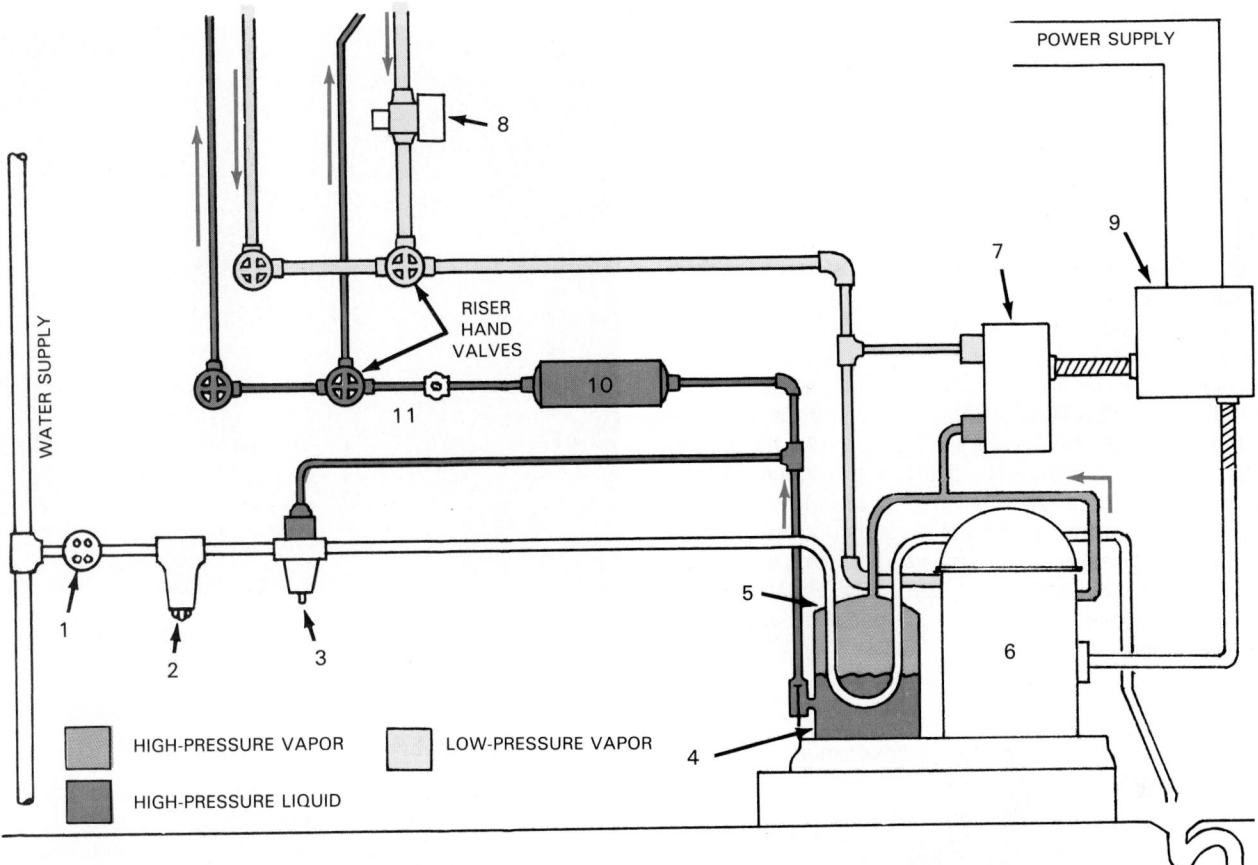

Fig. 12-169. Typical multiple installation showing location of important parts. 1—Water shutoff valve. 2—Strainer. 3—Water valve. 4—Liquid receiver. 5—Condenser. 6—Compressor. 7—Motor control. 8—Two-temperature valve. 9—Magnetic starter. 10—Drier. 11—Sight glass. Note this installation uses four riser valves.

nician can then connect another evaporator to it. This may be shut off from the remainder of the system by turning the valve in all the way. Fig. 12-169 shows a multiple installation using two liquid line riser valves and two suction line riser valves.

Service valves are usually made of drop forged brass to reduce seepage through the valve. The valve stem may be either brass or steel. Packing around the valve stem may be asbestos, lead and graphite, or the valve may be a packless type. This is the type that uses a bellows or a diaphragm as a sealing device rather than packing. Some valves have self-seating features. This means the valve is easily seated again by tapping the valve stem into the seat. The valve seat is made of a soft lead alloy or Monel metal.

12-65 RELIEF VALVES

A refrigerating system, regardless of size, is a sealed system. It is a pressure container. The pressures vary, but, during shutdowns, fires, extreme temperature conditions, or faulty electrical controls, high pressures could cause some part of the system to explode.

To prevent extreme, dangerous pressures, relief valves are mounted on the units, usually on the liquid receiver. The National Refrigeration Code and most local codes make this a requirement if the unit is a certain tonnage or more, or if the amount of refrigerant exceeds specified minimums or if the internal volume is large enough. Hand valves must not be placed between the system and the relief valve.

The relief devices are of three principal types:
1. Fusible plug.
2. Rupture disk.
3. Spring-loaded valve.

The fusible plug is shown in Fig. 12-170. It is threaded

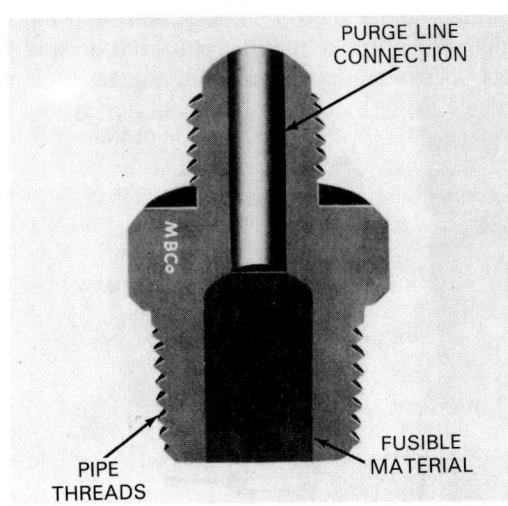

Fig. 12-170. Fusible plug for liquid receivers. Note flared fitting at outlet for connecting purge line that carries refrigerant outdoors. (Mueller Brass Co.)

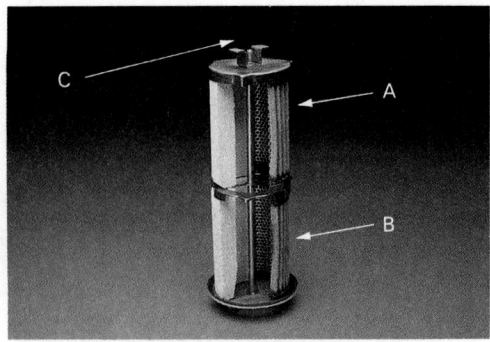

Fig. 12-184. This replacement cartridge is made in two sections (A and B). Note the nut (C) at end of shaft that is removed when replacement cartridges are inserted.
(Alco Controls Division, Emerson Electric Company)

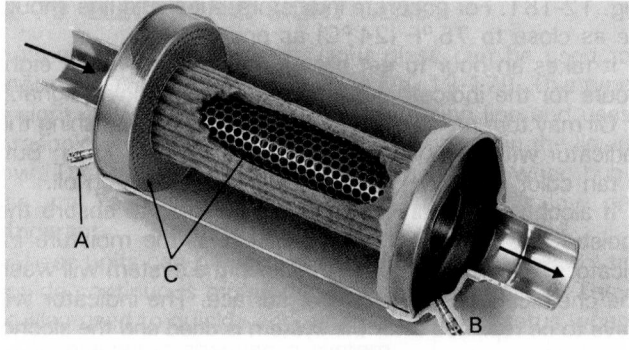

Fig. 12-185. Suction line filter-drier. Note A and B, Schrader valves, at both ends. The filter-drier has been cut open to show the strainer (screens), C, in the center end.
(Alco Controls Division, Emerson Electric Company)

Driers will handle the first three tasks. A moisture indicator is required for the fourth.

Driers should be left in the system permanently since oil loses its moisture slowly. Also, insulation in hermetic compressors and in small crevices may release moisture over a long period of time. A drier is like a sponge; however, it can become saturated and leave the refrigerant still wet if the drier is too small. A moisture indicator is the only sure means of recognizing a wet condition.

Remember that R-22 driers must be three to five times as large as those needed for an equal quantity of R-12. The greater the ability of a refrigerant to hold water, the larger the drier required. R-500 driers need to be as large as R-22 driers and R-502 driers need to be as large as R-12 driers.

Most driers have shaped cores of two or three drier (desiccant) materials. The cores are designed for efficient flow and efficient drying. They are shaped, then fused together at high temperatures into a porous ceramic structure which must be:
1. Noncorrosive.
2. Nonsoluble.
3. Nonreactive with oil.
4. Capable of absorbing moisture.
5. Capable of filtering out particles down to 10 micron in size (0.0004 in.).
6. Neutralizer for hydrochloric and hydrofluoric acid.

12-73 FILTER-DRIERS (SUCTION LINE)

Filter-driers are often mounted in the suction line to prevent foreign particles of over 5 microns in size as well as acids, sludge, and moisture, from entering the compressor. Fig. 12-185 shows the inside of a filter designed for suction line use. Strainers (screens) are usually made of Monel metal.

Only two things should be allowed inside a refrigeration system: clean, dry refrigerant and good, dry oil. A system which is clean, dry, and acid-free will run almost indefinitely without corrosion, freeze-ups, oil breakdown, or hermetic motor burnouts. In such a system, there is nothing to filter and plugging is impossible. A clean, dry, acid-free system remains factory bright and trouble-free in operation.

A normal system is completely clean. A dirty system is faulty and is just as much a mechanical failure as a faulty valve plate or connecting rod. A large-capacity suction line filter-drier is shown in Fig. 12-186.

A suction line filter-drier should be replaced if pressure drop is excessive. Replace for R-12 and R-500 refrigerants if it exceeds 2 psi for low-temperature units, or up to 8 psi for high-temperature units.

	Low Temp. psi	Medium Temp. psi	High Temp. psi
R-12 and R-500	2	6	8
R-22 and R-502	3	9	14

Replace for R-22 and R-502 refrigerants if pressure drop exceeds 3 psi for low-temperature units, or up to 14 psi for high-temperature units.

The density of the gas increases more with a pressure increase than it does with a temperature rise. The best way to know that a system is dry is to use and depend on a moisture indicator.

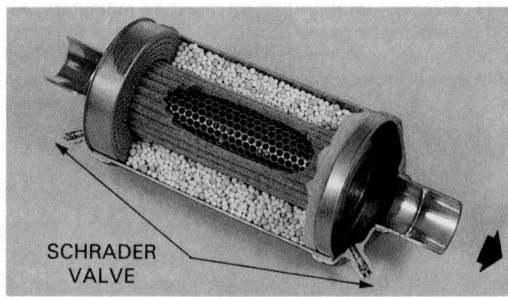

SCHRADER VALVE

Fig. 12-186. Large capacity suction filter-drier. Note Schrader valves used to check the pressure drop through the filter-drier.
(Alco Controls Division, Emerson Electric Company)

12-74 ENGINE-DRIVEN SYSTEMS

Natural gas, gasoline, and propane engines may be used to drive refrigerating compressors. The advantages are a variable compressor speed to produce flexible capacity and a comparatively low operating cost. Such units are available in four to 75 ton capacities. Engine-compressor units of one to five tons capacity are available for use on truck units and for air conditioning.

Pressure controls are usually used. The pressure control is connected to the engine's throttle. It is placed in the low-side suction line. The linkage is such that as the suction pressure increases, the engine's throttle is opened to increase the compressor's speed to increase the rate of refrigeration. As the temperature in the evaporator drops, the engine will slow down. This should result in a balance between the engine's speed and low-side pressure. It will give the desired temperature in the refrigerated space.

12-75 REVIEW OF SAFETY

Commercial systems vary considerably in size.

Small, self-contained units must be handled with all the care and safety described in the reviews of safety in earlier chapters.

As the units become larger, safety precautions become increasingly important both because the investment in the machines is greater and repairs are more costly. The larger machines are also more dangerous. The energy output of the larger moving parts and the larger refrigerant containers is potentially dangerous.

Closing the compressor discharge valve on a 10-ton capacity unit while it is operating would almost instantly ruin the compressor or rupture a gasket. Carelessly opening a receiver valve may cause the loss of hundreds of pounds of refrigerant while possibly injuring the service technician. Trapping liquid refrigerant in any part of the system with no gas space may cause sufficient hydraulic pressure to burst the container.

A technician must be positive that the inside pressures are atmospheric and that there is no liquid present before opening any part of the system. Goggles should ALWAYS be worn when working on any unit.

Open the electrical circuits and lock the switches before working on a refrigerating system if no power is needed.

Local and national refrigeration and electrical codes must be followed when servicing all systems. Follow OSHA standards.

It is not safe to work on any part of the system unless the pressure and temperature are known and the condition of the refrigerant (liquid or vapor) inside that part and the fundamentals of working on that system. See Chapters 11 and 14.

Pressure and temperature relief devices are installed on the units for equipment protection, user's protection, and service technician's protection. They should be frequently checked for accuracy and kept in good operating condition.

Never use cylinder oxygen to test any device for leaks. Use either refrigerant, carbon dioxide, or nitrogen. CAUTION: See Para. 11-39.

12-76 TEST YOUR KNOWLEDGE

1. What are the advantages of a nonfrosting evaporator?
2. Why must high-pressure motor cutouts be used with water-cooled condensing units?
3. What part of the compressor usually contains the intake valves?
4. Why is it advisable to connect the high-pressure motor cutout into the cylinder head of the compressor?
5. Name the three types of water valves.
6. Name the various two-temperature valves.
7. What are the advantages of water-cooled condensers? Of air-cooled condensers?
8. What are the advantages of forced circulation evaporators?
9. Why is a pressure-type motor control usually used in multiple evaporator installations?
10. In multiple installations which use two-temperature valves, where should the check valves be placed?
11. What percentage of a refrigeration load may be placed on an evaporator controlled by a two-temperature valve?
12. What is the primary principle of the operation of an evaporative condenser?
13. Are liquid receivers equipped with safety devices? Why or why not?
14. What is inside the inner tube of a tube-within-a-tube condenser?
15. How is cast aluminum used in a pressure-type evaporator?
16. What is the counterflow principle in tube-within-a-tube condensers?
17. Why is a pressure limiter valve used on some outdoor air-cooled condensers?
18. What principle is used to produce clear ice in an automatic ice cube maker?
19. How is water used to defrost a system?
20. In a hot gas defrost system, why must the drain pan and the drain pipe be heated during defrosting cycles?
21. What is the purpose of the check valve in the reverse cycle defrost system?
22. Why is a motor starter necessary?
23. Why is an oil separator insulated?
24. Why do some "hot gas" defrost systems reheat the refrigerant before it returns to the compressor?
25. Which type water valve will not vary the water flow as the refrigeration load changes?
26. Why is a float valve used with a cooling tower?
27. Name two types of electric heater defrost systems.
28. What is the purpose of a surge tank?
29. What is a sweet water bath?
30. Where are vibration dampers installed and why are they needed?
31. How is an auger used to produce flake ice?
32. What shuts off an ice cube maker when the bin is full?
33. How is an electric grid used to produce ice cubes?
34. What does the evaporator temporarily become during the defrosting action of a reverse cycle system?
35. What is the purpose of an accumulator in a hot gas defrost system?
36. Why are some systems pumped down before the electric defrost starts?
37. When should a suction line filter-drier be replaced? How is this measured?
38. How does an oil pressure safety cutout switch operate?
39. What is a tandem compressor assembly?
40. How are adjustable louvers and fans used on outdoor air-cooled condensers?

This total environmental control center is designed to provide heating, air conditioning, and relative humidity control for large commercial buildings. (Tyler Refrigeration Corp.)

A combination hot water, cold water, ice cube, and beverage/food storage unit. System has two separate compressors—one for the water and one for the refrigerated compartment. (Ebco Manufacturing Company)

Chapter 13

COMMERCIAL SYSTEMS APPLICATIONS

After studying this chapter, the technician will be able to:
☐ Discuss the various types of commercial refrigeration systems and their applications.
☐ Explain the differences between commercial and industrial refrigeration applications.
☐ List various types of commercial and industrial applications.

The industrial field is sometimes confused with commercial refrigeration. Industrial refrigeration uses refrigerating machines which need an attendant, usually a licensed refrigeration engineer, on the job constantly. Industrial plants usually have manually-operated refrigeration machines. They are commonly used by large storage houses, packing houses, industrial plants, ice cream manufacturing, frozen food processing plants, and for ice making.

13-1 COMMERCIAL CABINET CONSTRUCTION

Commercial cabinets are designed and constructed to suit the service required of them. Surfaces are either metal or plastic and finishes are formulated for easy cleaning. Structural members are steel, capable of supporting the evaporator and condensing unit. Insulation is usually polystyrene or urethane in slabs or foamed in place.

The capacity of the evaporator and condensing unit is such that adequate refrigeration is possible under the most severe service conditions. Heat leakage, in some cases, may cool the cabinet surfaces enough to cause some moisture to condense on them. To avoid this condition, some cabinets have a resistance heating strip around these surfaces to warm them.

Many commercial cabinets are designed to be used with a remote condensing unit. These units may be connected to several cabinets of different temperatures. Most condensing units are air cooled but some are water cooled.

13-2 GROCERY CABINET (REACH-IN CABINET)

Grocery cabinets have been used for many years to keep perishable products at a satisfactory temperature. They range in size from 20 cu. ft. to 100 cu. ft. inside volume (net capacity). They have from one to three doors with magnetic gaskets. Door widths vary from 30 to 85 in. Height of the cabinet is from 5 1/2 to 6 1/2 ft. Fig. 13-1 shows a typical reach-in unit.

Fig. 13-1. Two-section hinged door reach-in cabinet. It has self-contained condensing unit mounted at top. (Hobart Corp.)

The inside of a cabinet is shown in Fig. 13-2. A blower evaporator is mounted on the top and cooled air is distributed through a vertical duct. Fig. 13-3 is a cross section through the same refrigerator.

The space holding the evaporator is usually called the bunker. Blower type evaporators are very popular for grocery cabinets. Hardly any other type is now used. Fig. 13-4 shows a unit with the condensing mechanism located in the base while a blower evaporator is installed inside the cabinet.

The evaporator is usually located in the upper center of the cabinet. The insulation is most commonly foamed in place. Exteriors are aluminum, stainless steel, or vinyl.

The temperatures are about the same as in domestic cabinets with a minimum of 35 °F (2 °C) and a maximum

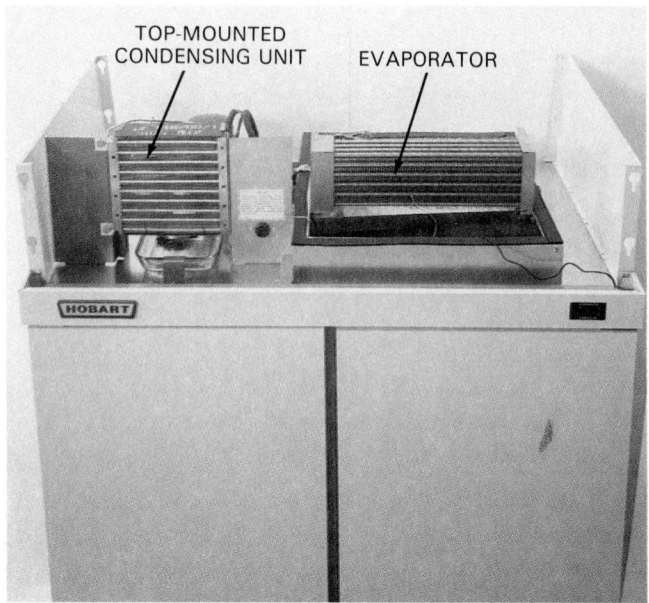

Fig. 13-2. Reach-in refrigerator cabinet with top-mounted condensing unit and evaporator. Cover is removed to show these mechanisms. (Hobart Corporation)

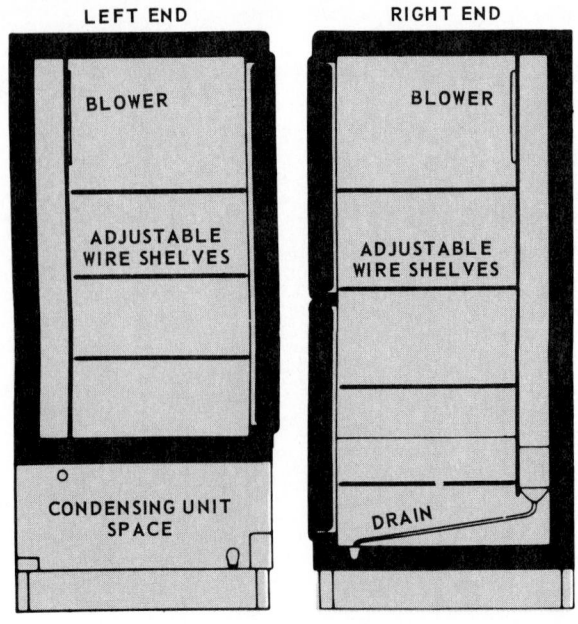

Fig. 13-4. Cross section of a reach-in cabinet. Note location of blower.

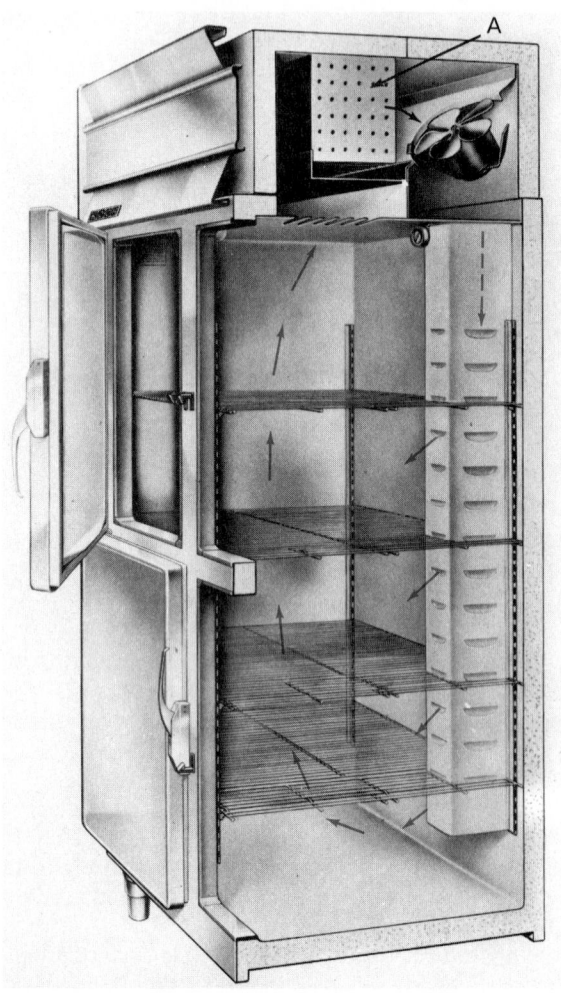

Fig. 13-3. Cross section of reach-in refrigerator. Note wall construction and airflow. A—Top-mounted evaporator.

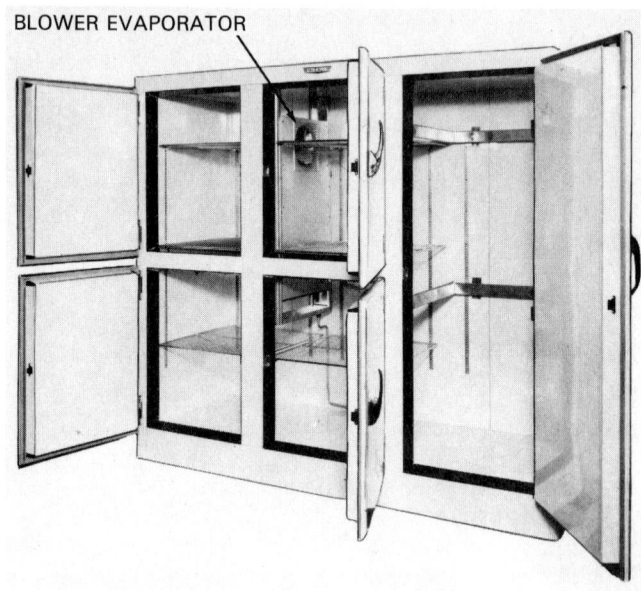

Fig. 13-5. This reach-in cabinet is designed to store and refrigerate a variety of foods.

of 45 °F (7 °C). A relative humidity of about 80 percent is necessary for salads, desserts, and fresh foods.

A reach-in cabinet for small grocery stores and markets is shown in Fig. 13-5. The blower type evaporator is mounted in a vertical position at right angles to the half doors. The evaporator takes little space. The blower circulates the cold air to provide an even temperature through the cabinet. This cabinet uses a remote condensing unit. The full-height door permits storing of beef quarters.

13-3 WALK-IN CABINET

Establishments such as restaurants and supermarkets, where perishable products are stored, use walk-in cabinets.

These cabinets have large doors and windows and are sometimes classified as butcher boxes. Sizes of these cabinets vary but two heights are usually considered standard: 7 ft. 6 in. and 9 ft. 10 in. outside dimensions. See Fig. 13-6. These boxes are the "knockdown" type. This means that they may be taken apart for easier moving. See Fig. 13-7.

Typical walk-in cooler sizes are:

Length	Width	Height
7 ft.	5 ft.	9 ft. 10 in.
8 ft.	6 ft.	9 ft. 10 in.
8 ft.	8 ft.	9 ft. 10 in.
9 ft.	7 ft.	9 ft. 10 in.
12 ft.	10 ft.	9 ft. 10 in.
6 ft.	5 ft.	7 ft. 6 in.
6 ft.	6 ft.	7 ft. 6 in.
7 ft.	6 ft.	7 ft. 6 in.

Many cabinets are made with metal linings and exteriors. Galvanized steel or aluminum is the usual metal. Vinyl, porcelain, and stainless steel are also used extensively.

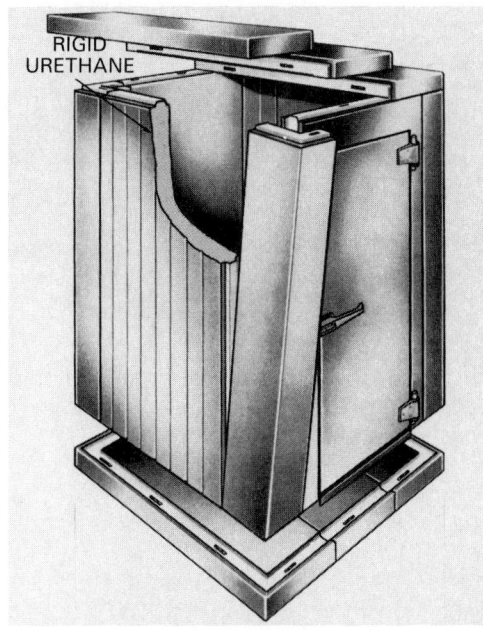

Fig. 13-7. Many walk-in refrigerators are prefabricated. Sections are locked together on site.   (Bally Engineered Structures, Inc.)

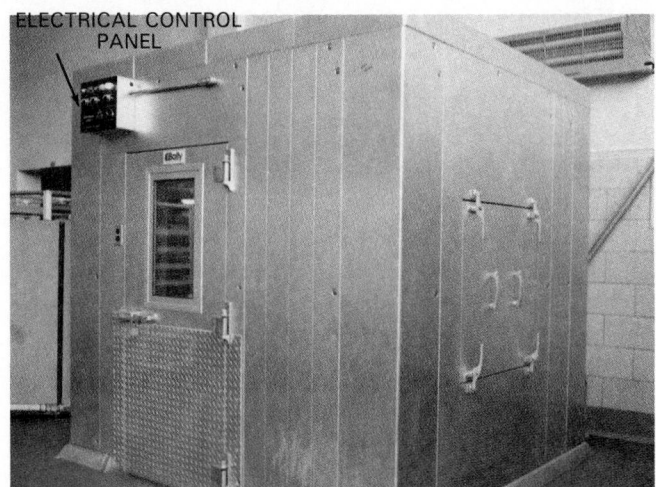

Fig. 13-6. An institutional walk-in cooler with remote outside system. Cooler has a 37 °F holding temperature with an R-502 air-cooled thermal balanced semiautomatic system. Note location of electrical control panel.   (Bally Engineered Structures, Inc.)

Cabinet doors are usually of the same construction as the box and are gasketed to make the box airtight. See Fig. 13-8. Door latches must be accessible from the inside for safety, Fig. 13-9. The doors may also be provided with heating wires along the edge to eliminate sweating and freezing.

Some walk-in cabinets are dual temperature. These cabinets have both a regular temperature compartment and a frozen foods compartment.

Cabinets may have additional reach-in doors, usually with two, three, or four panes of glass. Instead of insulation, these doors have two or three dead air spaces arranged in such a way that they are airtight. Plate glass is usually used. Special chemicals, such as calcium chloride, keep the spaces between the panes free from moisture.

Newer walk-in coolers use rigid polyurethane foamed-in-place insulation. Foamed between the inner and outer walls, such insulation produces a very strong wall and ends the

Fig. 13-8. Prefabricated walk-in refrigerator door. Heater wires are built in around door openings to eliminate condensation and freezing. A—Thermometer. B—Locking mechanism. C—Condensate control. (Bally Engineered Structures, Inc.)

need for metal framing. Insulation is usually 4-in. thick.

Walk-in cabinets usually have a lighting system. Some have a wall-mounted evaporator. It is separated from the main part of the cabinet interior by a vertical baffle. Forced convection evaporators are popular.

The temperature in this type of cabinet depends on its use. For meat or fresh produce storage, a temperature between 35 °F (2 °C) and 40 °F (4 °C) is needed. Relative

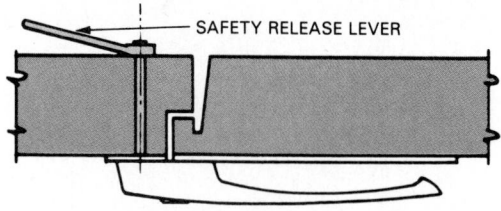

Fig. 13-9. Inside safety release lever is connected to door latch. Inside lever can be used for emergency door opening.

humidity should be about 80 percent. Air movement is necessary. Ultraviolet lamps may also be used to help keep down bacteria and mold growth.

Because overexposure to ultraviolet rays is dangerous, persons working near these lamps must be protected from the rays. Otherwise, the lamps must be turned off when anyone is in the cabinet.

Some type of drain is recommended. Fig. 13-10 shows a common method of installing the drain in a prefabricated walk-in which sits on top of a permanent floor.

For milk storage, beverage cooling, and other service in which the dehydration of foods is not important, colder temperature may be used as desired and less attention may be paid to relative humidity. Blower evaporators are commonly used in these installations. Walk-in cabinets are also used for storing frozen foods (walk-in freezers).

13-4 FLORIST CABINET

Florist cabinets vary in size and construction. They may be either self-contained or walk-in and differ from the grocery cabinet in three principal ways:

1. The cabinet temperature may be kept higher than in the other types of boxes. Temperatures between 38°F (3°C) and 40°F (4°C) are common.
2. Insulation, because the lesser temperature difference, generally is only 1 to 2 in. (2.5 to 5 cm) thick.

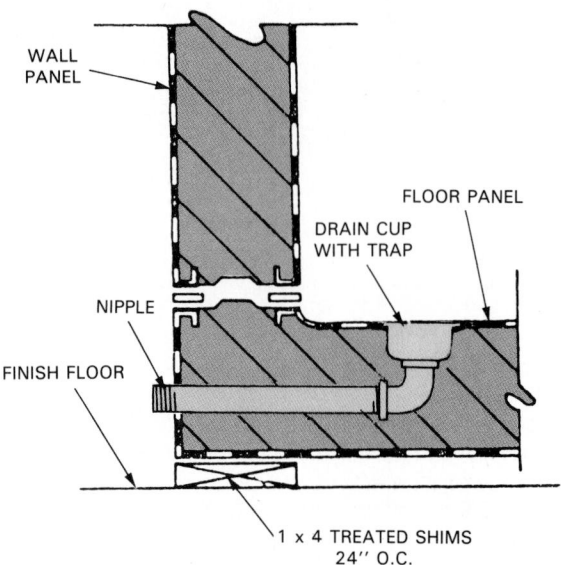

Fig. 13-10. Walk-in cooler drain connection. Note that connection is part of prefabricated bottom section.

Fig. 13-11. Florist's refrigerated display cabinet with overhead evaporator. (Buchbinder, Chicago, IL)

3. The cabinet is usually made with large window surfaces, permitting the display of cut flowers, Fig. 13-11.

Humidity is important in the florist cabinet. It is normally kept at 90 to 95 percent although a minimum of 80 percent is acceptable. This is necessary in order to retard evaporation from the surface of the leaves and blooms.

The evaporators have large cooling surfaces to keep the humidity as high as possible. Natural convection evaporators are used in most cases. Also, the motor controls allow little variation in the cabinet temperature. Many florist cabinets have odor-removing devices to prevent contamination of the flowers. An activated carbon filter, Fig 13-12, containing potassium permanganate, may be used to reduce mold growth, neutralize ethylene, and extract odors given off by flowers. In order to assure effectiveness; there should be good airflow across the filter. It should have its own circulating fan, or it should be so placed that the cooler fan pulls air across it.

13-5 DISPLAY CASES

To display produce to best advantage, stores frequently use a refrigerated display case. This case is equipped with glass fronts so the purchaser can see the articles. At the same time, the food is safely refrigerated.

Temperature in the case is determined by its usage. Fig. 13-13 shows the recommended temperature for some common applications. Display case lighting is usually installed outside the glass case so heat generated by the lights

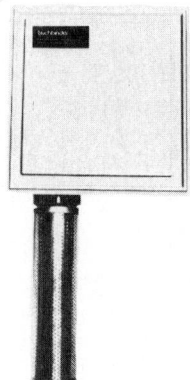

Fig. 13-12. Ethylene purifier automatically filters out odors and ethylene gas that may exist where flowers are stored. (Buchbinder, Chicago, IL)

sist of coils of tinned tubing so placed that each shelf is individually cooled.

13-7 DOUBLE-DUTY CASE

Some cases have additional storage space beneath the display section of the counter. This also is refrigerated. The evaporators are usually connected in series. Such cases usually serve as a temporary container for food or produce which is transferred to a walk-in storage cabinet overnight. Therefore, temperatures may be kept at 40°F (4°C) to 45°F (7°C) in both compartments.

Evaporators used in these installations must, necessarily, be narrow. They are made with fins as small as 1 1/4 in. wide. Some of the shelf evaporators are the plain tubing type.

Many of these display cases are now using blower

Temperatures in Display Refrigerators		
	Temperature, °F	
Type Fixture	Minimum[a]	Maximum[b]
Meat, unwrapped		
Display area	35	37
Storage compartment	34	36
Meat, wrapped		
Display area	28	30
Storage compartment	28	30
Produce, display area	35	40
Produce, storage compartment	35	40
Dairy	35	38
Frozen food	[b]	−5
Ice cream, dough, juice	[b]	−12

[a] These temperatures are air temperatures, with thermometer in the refrigerated airstream and not in contact with the product.
[b] Minimum temperatures for frozen foods and ice cream are not critical; maximum temperature is important for proper preservation of product quality.

Fig. 13-13. These temperatures are recommended for display of certain foods. (Reprinted from ASHRAE Guide and Data Book)

will not increase the refrigerating load.

Display cases vary in design, length, and height. Three types are:
1. Glass enclosed display case only.
2. Glass enclosed display case and enclosed storage cabinet.
3. Open display case.
   a. Fresh produce.
   b. Frozen foods.
   c. Fresh meats.
   d. Dairy products.

The display case is sometimes classified by the location of the evaporator: overhead, end, base.

13-6 SINGLE-DUTY CASE

One popular display case uses an overhead evaporator. In this case, the main evaporators are mounted in the upper portion of the display space under the shelf which forms the top. See Fig. 13-14. This provides good refrigerating temperatures all the way through the display space.

Some cases have shelf evaporators called auxiliary evaporators. These are located under the shelves and con-

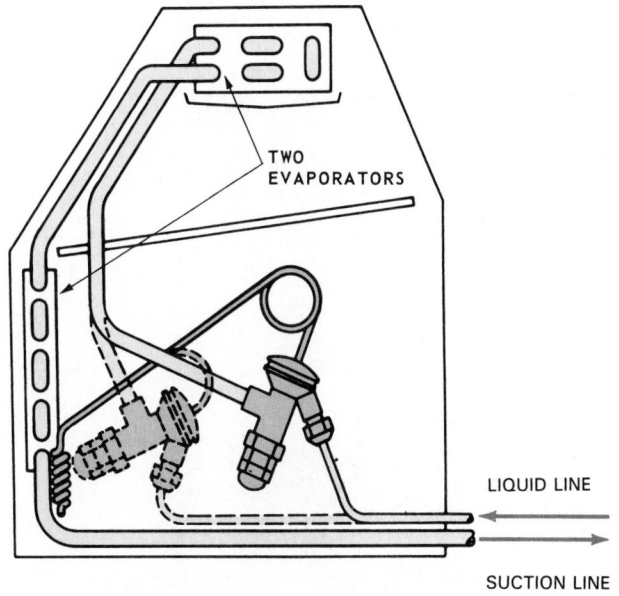

Fig. 13-14. Cross section of glass-enclosed display case. Two evaporators are connected in series; one is overhead and one is at back of case. Note use of thermostatic expansion valve.

evaporators for cooling. They take little space and, because of the circulating air, provide even refrigeration temperatures throughout the display case.

13-8 OPEN DISPLAY CASE

For easier customer self-service, open display cases are commonly used in markets. These cases may have storage space in the base of the unit. Storage space at the top is open. The walls, or the upper part of the walls, may be enclosed in three to four layers of glass.

The higher temperature case, such as is used for fresh meats and dairy products, does not present any special evaporator problems. Blower evaporators are used, and ducts carry the cold air through grilles at the rear of the case at the same level as the refrigerated foods. See Fig. 13-15. The warm air returns down the front of the case. An open display case for meat and deli products is shown in Fig. 13-16. Note location of food display in relation to the evaporator.

Many supermarkets have open display cases for produce. These cases are kept at about 40°F (4°C) and at a high humidity. See Fig. 13-17. If dry air circulates over the contents, it will remove some of the moisture, spoiling the appearance and decreasing the weight of the produce.

A cross section of an air curtain open display case is shown in Fig. 13-18. An installed, operating, unit is shown in Fig. 13-19.

An airflow meter (see Chapter 18) or a chemical smoke may be used to check the airflow patterns in these open cases. The airflow curtain should not touch the shelves or the products. A cross section through a dairy/deli open

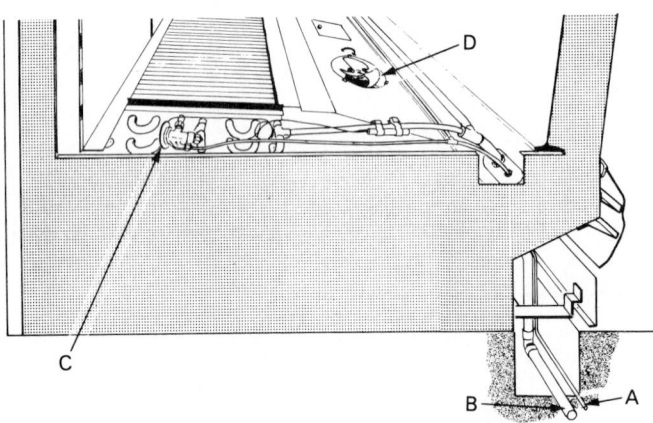

Fig. 13-15. Display case installation. Note trough in floor. It provides space to run refrigeration piping and electrical conduit. A—Liquid line. B—Suction line. C—Evaporator. D—Fan.

Fig. 13-17. Open display case with canopy. Canopy mirror (at back) helps display produce. (Tyler Refrigeration Corp.)

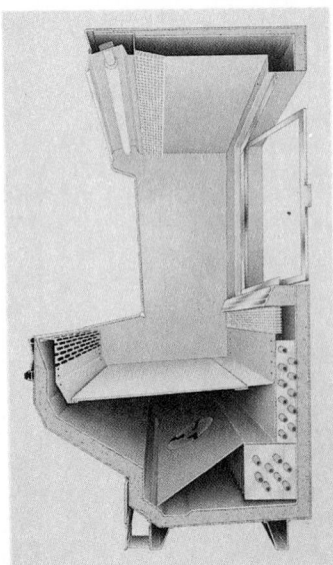

Fig. 13-16. Cross section of open meat display case. Note evaporator located to right of meat tray. (Tyler Refrigeration Corp.)

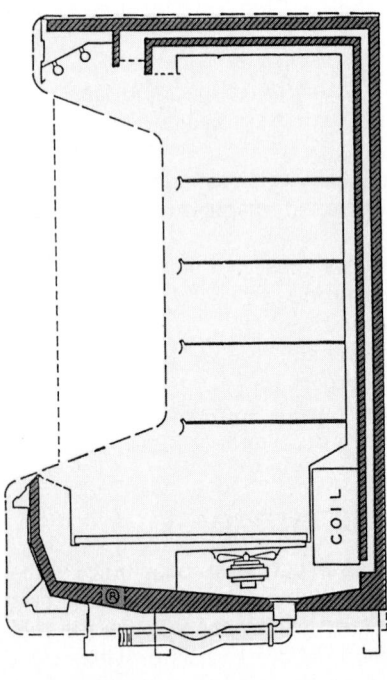

Fig. 13-18. This case is designed for display of dairy products and delicatessen items. (Kysor/Warren Division of Kysor Industrial Corp.)

Fig. 13-19. Air curtain display case. Units like this are designed for deli and dairy foods. (Kysor/Warren Div. of Kysor Industrial Corp.)

display case is shown in Fig. 13-20. Some of the electrical circuits of an open display case are shown in Fig. 13-21.

13-9 OPEN FROZEN FOOD DISPLAY CASE

Storing and displaying frozen foods in either open or closed cabinets presents some problems. Fig. 13-22 shows such a display case. Temperatures near 0 °F (−18 °C) must be maintained. The evaporators, therefore, must operate at −10 °F to −15 °F (−23 °C to −26 °C). Heater wires are installed along those parts of display cases that could otherwise collect condensation from the air.

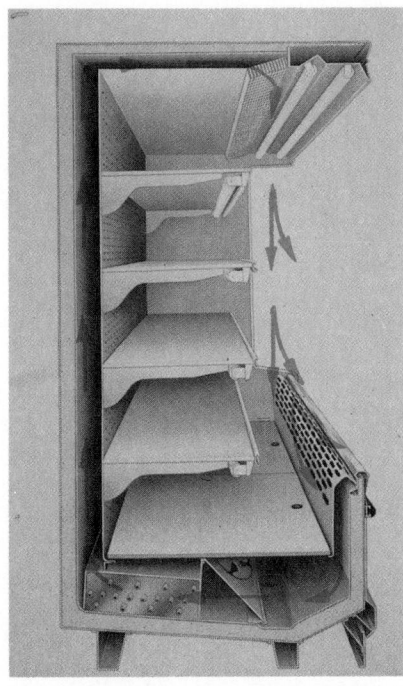

Fig. 13-20. Cross-sectional view is of open display case using blower evaporator. Note airflow pattern which carries cooled air over displayed foods. (Tyler Refrigeration Corp.)

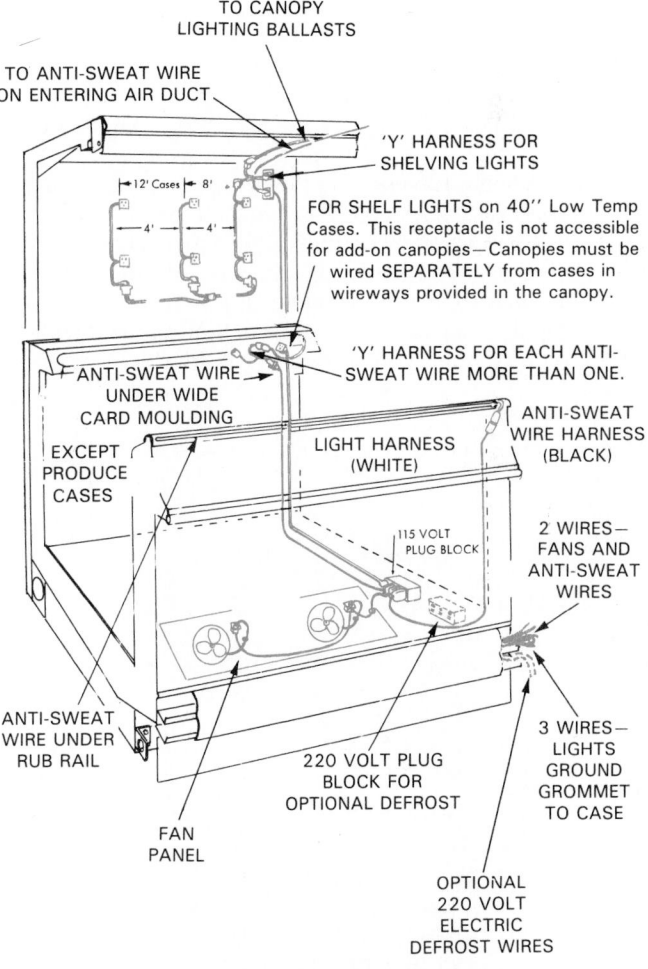

Fig. 13-21. Electrical wiring system of a display case is designed to handle lighting, heating elements, and fans. This system is typical of most open display cases.

Frozen food storage and display cases are constructed in both chest and upright cabinet styles. Fig. 13-23 shows

481

air curtain of the case, causing higher operation costs and defrosting problems. Several cases are usually connected end-to-end in supermarkets. Total electrical load must be carefully checked to provide enough service.

Need for low temperature presents a difficult evaporator defrosting problem. The evaporator must be defrosted at least once a day. This must be done as quickly as possible to prevent the case from warming up too much. The defrosting must be done automatically. A time clock is usually used. It operates a hot gas defrosting system or an electric heater defroster device. See Chapter 12 for details concerning these systems.

Some cabinets use two or three air curtains. The principle of operation is shown in Fig. 13-24. The construction of such a cabinet is shown in Fig. 13-25.

13-10 FROZEN FOOD STORAGE CABINET

The frozen food storage cabinet may be either a chest or upright type insulated with four to six in. (10 to 15 cm) of polyurethane. The doors or access openings are also heavily insulated while double gaskets are usually provided for better sealing. See Fig. 13-26. Electrical resistance strip heaters usually surround door frames and other parts which sweat.

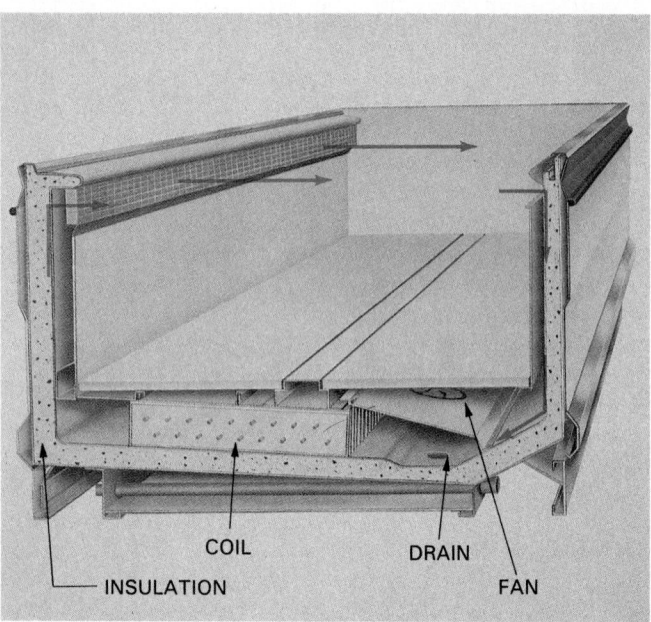

Fig. 13-22. Construction details of open frozen foods case. Notice airflow curtain across top of case as shown by arrows. (Tyler Refrigeration Corp.)

an upright case.

Since temperatures are very low in these cases, openings are gasketed or sealed. For the same reason, insulation is thick and carefully hermetically sealed. Chest types are more popular because the top openings prevent spillage of cold air when the case is being used.

Open cases must be protected from drafts produced by grilles, unit heaters, and fans. Drafts will interfere with the

Fig. 13-23. An upright frozen foods display case. Note, especially, the fans and airflow pattern. (Tyler Refrigeration Corp.)

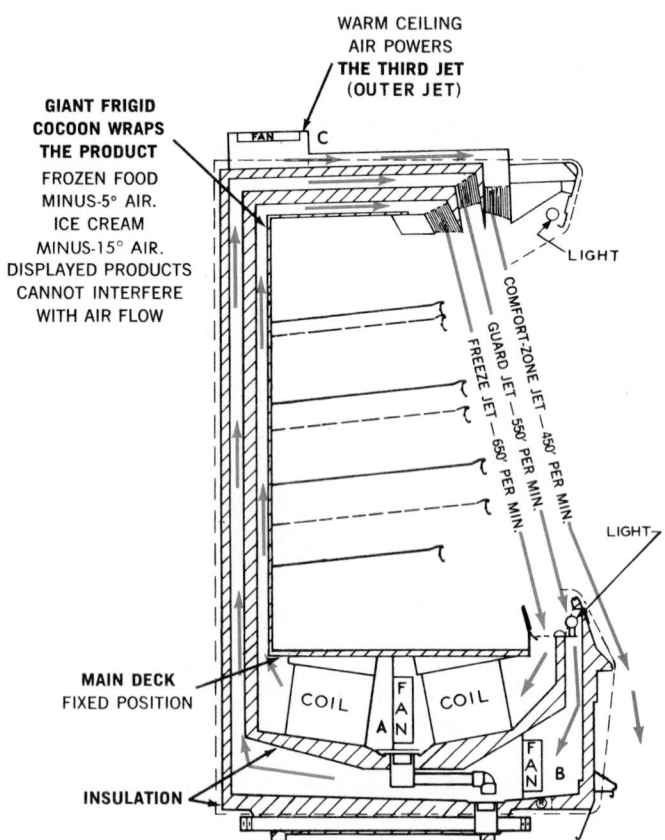

Fig. 13-24. A frozen foods case with three curtains of air. Note the three separate fan systems. Inner curtain is powered by fan at A, middle curtain by fan at B, outer curtain by fan at C. (Kysor/Warren Division of Kysor Industrial Corp.)

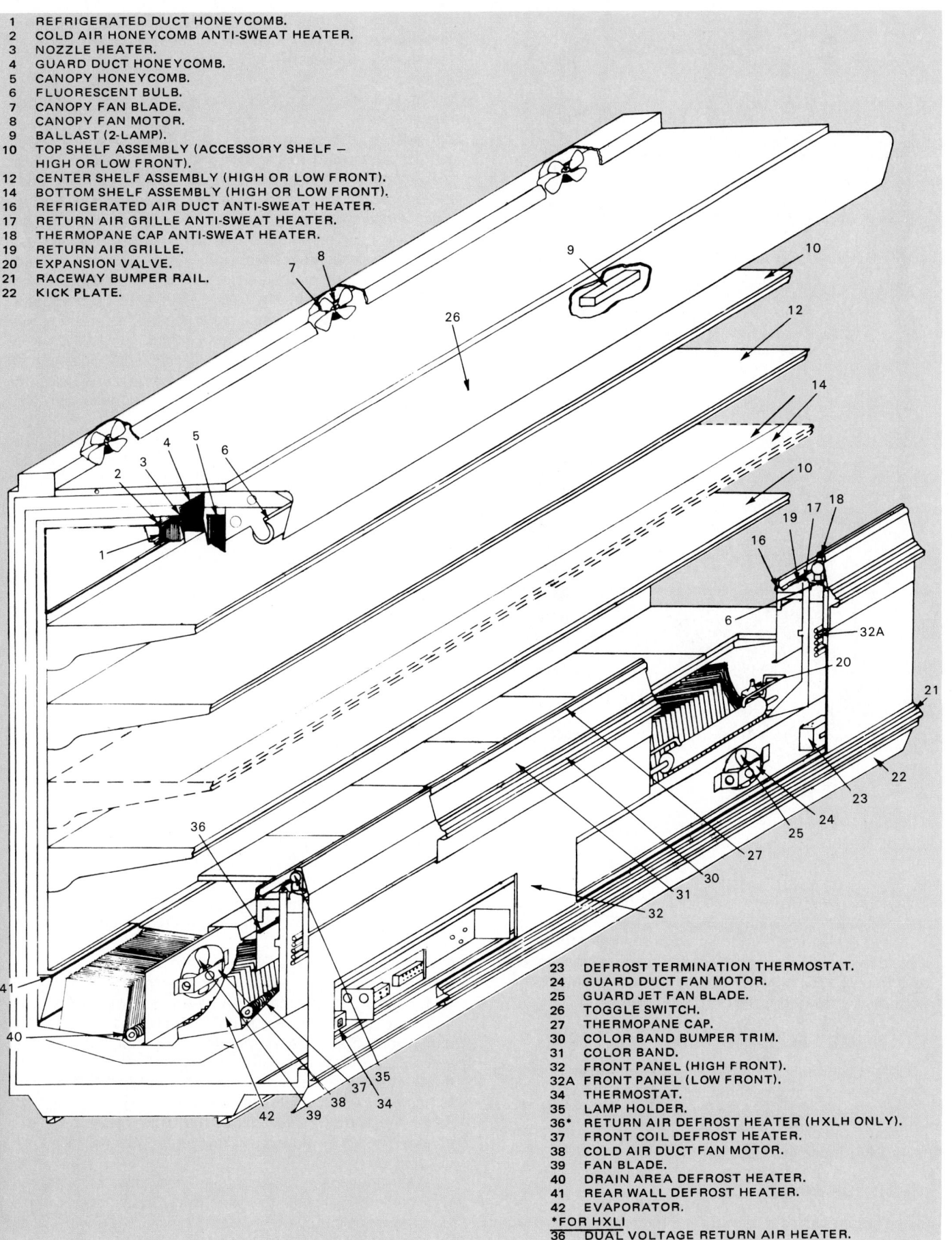

1   REFRIGERATED DUCT HONEYCOMB.
2   COLD AIR HONEYCOMB ANTI-SWEAT HEATER.
3   NOZZLE HEATER.
4   GUARD DUCT HONEYCOMB.
5   CANOPY HONEYCOMB.
6   FLUORESCENT BULB.
7   CANOPY FAN BLADE.
8   CANOPY FAN MOTOR.
9   BALLAST (2-LAMP).
10 TOP SHELF ASSEMBLY (ACCESSORY SHELF —
    HIGH OR LOW FRONT).
12 CENTER SHELF ASSEMBLY (HIGH OR LOW FRONT).
14 BOTTOM SHELF ASSEMBLY (HIGH OR LOW FRONT).
16 REFRIGERATED AIR DUCT ANTI-SWEAT HEATER.
17 RETURN AIR GRILLE ANTI-SWEAT HEATER.
18 THERMOPANE CAP ANTI-SWEAT HEATER.
19 RETURN AIR GRILLE.
20 EXPANSION VALVE.
21 RACEWAY BUMPER RAIL.
22 KICK PLATE.

23   DEFROST TERMINATION THERMOSTAT.
24   GUARD DUCT FAN MOTOR.
25   GUARD JET FAN BLADE.
26   TOGGLE SWITCH.
27   THERMOPANE CAP.
30   COLOR BAND BUMPER TRIM.
31   COLOR BAND.
32   FRONT PANEL (HIGH FRONT).
32A FRONT PANEL (LOW FRONT).
34   THERMOSTAT.
35   LAMP HOLDER.
36* RETURN AIR DEFROST HEATER (HXLH ONLY).
37   FRONT COIL DEFROST HEATER.
38   COLD AIR DUCT FAN MOTOR.
39   FAN BLADE.
40   DRAIN AREA DEFROST HEATER.
41   REAR WALL DEFROST HEATER.
42   EVAPORATOR.
*FOR HXLI
36   DUAL VOLTAGE RETURN AIR HEATER.

Fig. 13-25. Construction details of open frozen foods case. Three air curtains protect food.

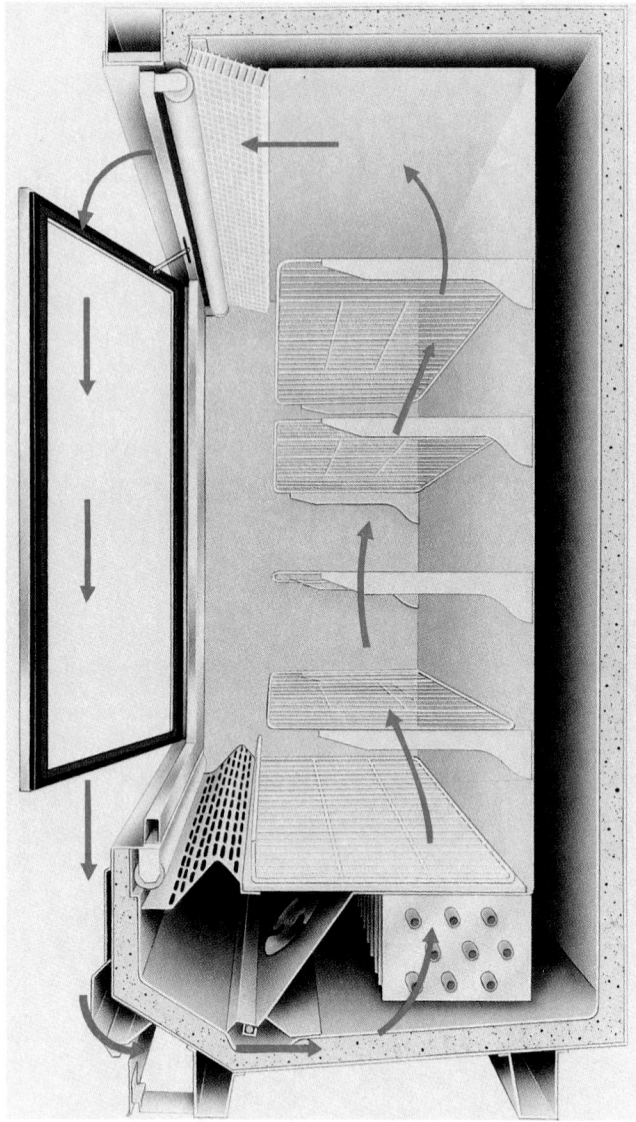

Fig. 13-26. A frozen food closed-door display case. Note the flow of the air. (Tyler Refrigeration Corporation)

Stainless steel is often used for the inner liner. The outer liner is usually aluminum or stainless steel.

These cases are for storage purposes only; the food is moved to display cases as needed. Storage cabinets operate at 0 °F (−18 °C) or lower. The refrigerating mechanism is normally installed on top of the cabinet.

13-11 FAST FREEZING CASE

Cases used for freezing foods rapidly are similar to storage cases except that temperatures are maintained at about −20 °F (−29 °C), and food is placed as close to the freezing plates as possible. Some cases use refrigerated shelves to provide more heat transfer surface.

13-12 ICE CREAM CABINET

Ice cream cabinets have a steel framework with a sheet metal exterior. Modern cabinets use polyurethane insulation about 3-in. (8 cm) thick.

The size of the sleeve or tank holder is standard. Therefore, construction of the various makes of bulk ice cream cabinets is similar. The size ranges from one to 12 sleeves.

The bulk ice cream cabinet should be kept at approximately 0 °F (−18 °C). If the temperature is too cold, it is difficult to scoop out the ice cream. Also, too-low temperatures tend to crystallize the ice cream.

Dry type evaporators are used with either a capillary tube, a thermostatic expansion valve, or an automatic expansion valve refrigerant control. Some of the evaporators are tinned tubing wrapped around and soldered directly to the sleeves. Others are sheet metal with refrigerant passages formed in them. The sleeve covers (tops which must be raised to get to the ice cream) are also standardized by manufacturers. Since sleeve openings are at the top, there is no spilling of cold air from the opened cabinet.

Some ice cream cabinets are self-contained. This means that the refrigerating machine or condensing unit is built into one end of the cabinet. Other cabinets are made with the condensing unit separate (remote type).

In addition to the chest type ice cream cabinet, the upright and open display types are used. Packaged ice cream should be kept at about −10 °F (−23 °C) in order to retain its firmness.

13-13 SODA FOUNTAIN

The soda fountain provides a compact unit for storing and dispensing ice cream, cold water, beverages, and syrup. It also makes and stores ice. It is usually attractive and its design makes serving easy.

A built-in ice cream cabinet usually occupies one part of the fountain. Another part contains a water-cooling mechanism.

The evaporator outlet tubing from the ice cream cabinet and the drinking water cooler is passed around the syrups, keeping them relatively cool. Beverages are cooled to the same temperature as the water.

Soda fountains are often difficult for the technician to service. Being compact, they leave little space in which to work.

Syrups should be kept at about 45 °F (7 °C). Water temperature should range between 32 °F and 50 °F (0 to 10 °C). Ice cream, as mentioned, should be kept between 0 °F and 10 °F (−18 °C and −12 °C).

A diagram, Fig. 13-27, shows the complete cycle of a soda fountain refrigerated by one condensing unit. The soda fountain has an ice cream compartment, syrup rail, bottle compartment, and beverage coolers. It uses a thermostatic expansion valve. The ice cream evaporator has a check valve in the suction line. The syrup and bottom compartment evaporator temperature is controlled by the two-temperature valve, while the beverage cooler has a solenoid liquid line shutoff valve. Note the sight glass, heat exchanger, and drier mounted in refrigerant lines near the condensing unit.

Many drive-ins, soda fountains, and other public areas use drink dispensers. A dispenser for ice and various beverages is pictured in Fig. 13-28. Figs. 13-29 and 13-30 show the fluid circuits and system schematic diagrams for a beverage dispenser.

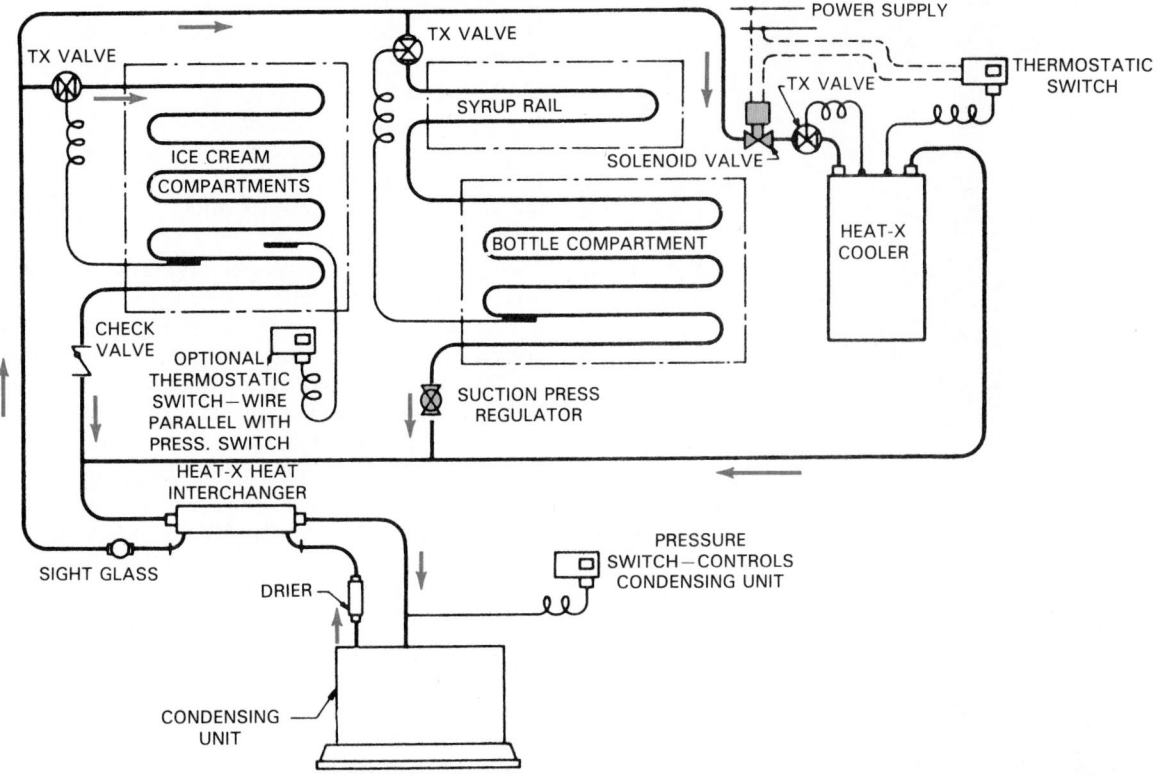

Fig. 13-27. Complete soda fountain cycle diagram. Note solenoid valve, suction pressure regulator, and how these valves control temperatures in various parts of installations. (Dunham-Bush, Inc.)

Fig. 13-28. Ice and beverage dispenser. (Scotsman Ice Systems)

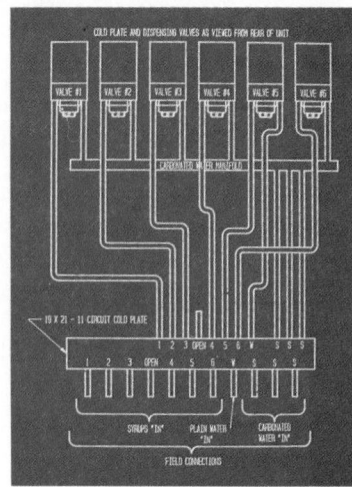

Fig. 13-29. Fluid circuits of drink dispenser. Carbonated water and syrups are mixed in electric heads. Note post-mix valve which combines chilled syrup and carbonated water; also, separate ice dispensing system. (Scotsman Ice Systems)

13-14 DISPENSING FREEZERS

Special applications of refrigerating systems are required in soft ice cream making machines. Temperature requirements will vary depending on product:

1. Frozen carbonated beverages, 25 °F (−4 °C).

2. Slushes, 28 °F (−2 °C).
3. Fruit or water ices, 10 to 20 °F (−12 °C to −7 °C).
4. Soft serve, 21 °F (−6 °C).
5. Milkshakes, 27 °F (−3 °C).
6. Sherbets, 20 °F (−7 °C).

This unit uses a large refrigerating machine to fast freeze or cool the mix before it is fed to the freezing cylinder. The same or a separate motor drives the stirring mechanism (dasher).

A 1/2 hp (373 W) refrigerating unit can fast freeze one gallon of custard in about six minutes.

A combination unit is illustrated in Fig. 13-31. Fig. 13-32 shows the construction of a freezing dispenser cabinet.

Some units are continuous in operation because the mix in the freezer often has to be kept within 1 or 2 °F (0.5 to 1 °C) of its correct temperature. Thermostatic expansion valves are usually used as the refrigerant control.

The machine is usually adjusted to deliver the custard or sherbets at 20 °F (−7 °C). A total of 3 hp (2 238 W) can operate the refrigerating unit and drive the dasher. A unit with a capacity of 12.5 gal (48 liters) per hour usually has a 2 hp (1 492 W) dasher motor and a 3 hp (2 238 W) refrigerating unit. The larger units are water cooled.

The quality of the mix is very important. Many problems thought to be in the refrigeration have turned out to be poor mixes. The frozen carbonated beverage is about one part syrup and four parts carbonated water. It is served at 22 to 26 °F (−6 to −3 °C). Health codes require keeping the

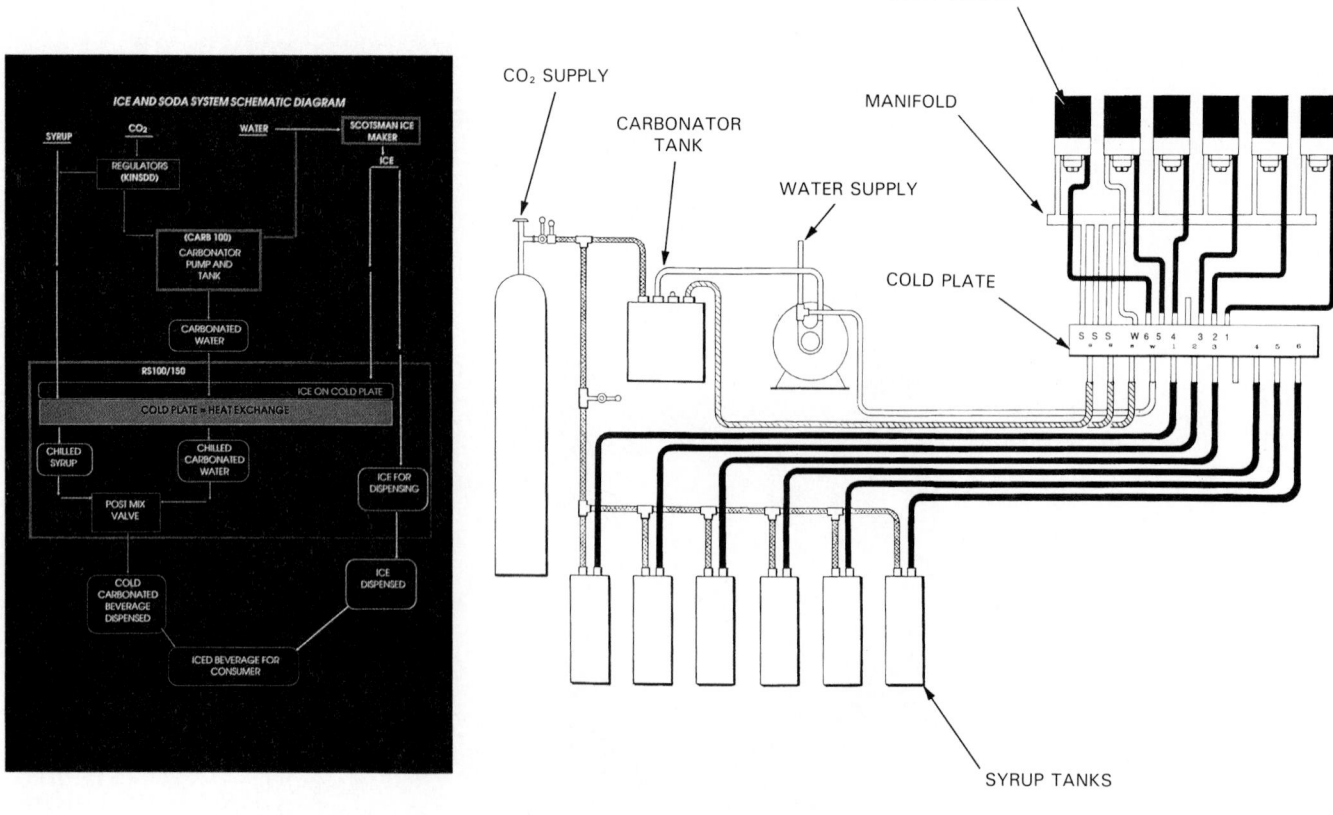

Fig. 13-30. Beverage dispenser diagrams: A—Fluid circuit diagram. Beverage syrup, plain water, and carbonated water inlets pass through the cold plate, then to the dispensing valves, where carbonated manifold is located. Ice dispensing system is separate from soda system. B—Soda system schematic. (Scotsman Ice Systems)

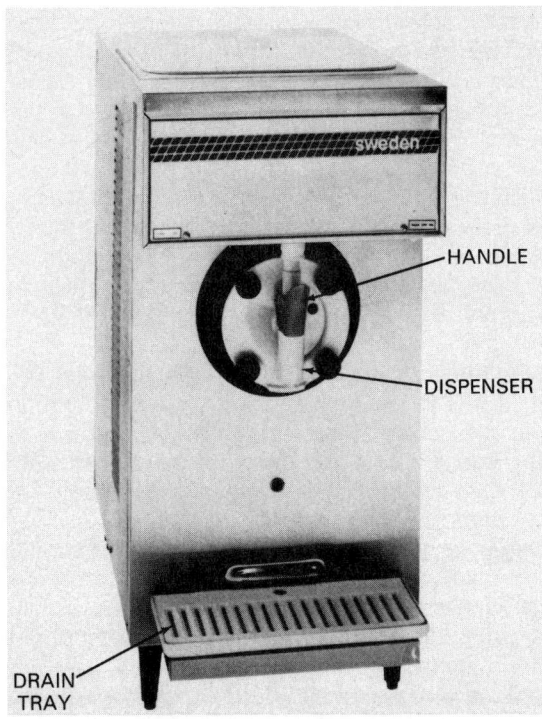

Fig. 13-31. Air-cooled soft serve dispensing unit. Note names of its various parts. (Sweden-Alco Dispensing Systems, A Div. of Alco Foodservice Equipment Co.)

mix containers as well as the freezing cylinder sanitary.

A mix storage or supply cabinet is usually found with three units to store the mix until it is used.

A commercial application of the dispensing freezer is the "shake maker." This machine is filled with the desired shake mix. The temperature of the "shake maker" is controlled with the use of a thermostat. The temperature control is based on the consistency of the mix. As the mix is frozen, the torque required to drive the agitator increases. An idler on the belt drive side is connected to a microswitch. When the mix is frozen to the desired consistency, the belt tightness will move the idler and open the microswitch which shuts off the machine. See diagram in Fig. 13-33.

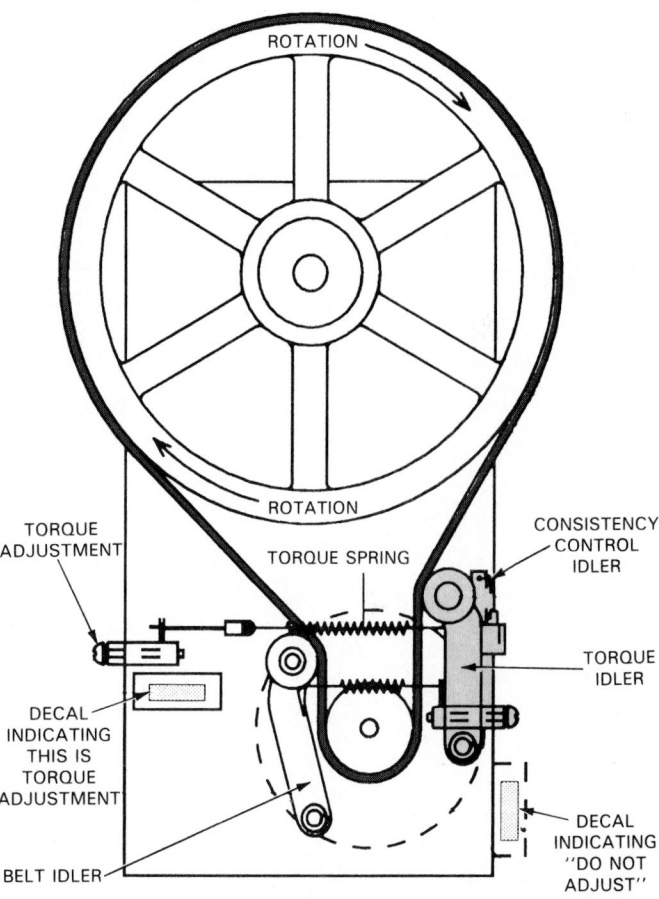

Fig. 13-33. Shake maker consistency control. As mix hardens, tight side of belt moves consistency control idler and finally opens motor circuit. (Sani-Serv)

Fig. 13-32. A dispensing freezer. A—Dispensing opening. B—Heavy insulation around freezing cylinder. C—Rear product seal. D—Dasher. E—Scraper blade. (Sweden-Alco Dispensing Systems, A Div. of Alco Foodservice Equipment Co.)

13-15 WATER COOLER

The water cooler cabinet usually has a sheet metal housing attached to a steel framework. Inside this sheet metal housing there is usually a condensing unit located near the floor. Above is the water-cooling mechanism. The latter is the only part insulated with foamed plastic. The insulation usually is specially formed and is between 1- and 2-in. (4 to 5 cm) thick.

Some water coolers also have a heater providing hot water. Some also have a refrigerated compartment for storing milk, soft drinks, and food.

This cabinet is made so that one or more sides may be easily removed for access to the interior. The basin of the water cooler is generally made of porcelain-coated cast iron, porcelain-coated steel, or stainless steel.

Heat exchangers are frequently used on water coolers. Some make use of the low temperature of the waste water to precool the fresh water line to the evaporator. See Fig. 13-34, items 6 and 8. The temperature of the cooled water is usually maintained at 50°F (10°C). The thermostat controls the temperature of the cooled water.

A water cooler using a plumbing supply and drain connection must be installed according to the National Plumbing Code and local codes. The plumbing should be concealed. A hand shutoff valve should be installed in the fresh water line and a drain pipe, at least 1 1/4 in. in diameter, should be provided. The bubbler opening must be above the drain to eliminate the chance for accidental syphoning of the drain water back into the fresh water system.

The tap water model uses a variety of evaporator designs. Fig. 13-35 shows a combination evaporator water-cooling tank. Fig. 13-36 is the wiring diagram for a water cooler having a water heating service.

Water temperatures are set by a person checking a cooler. If it is in the heat-treating room of a tool and die factory, the cooled water should be 50 to 55°F (10 to 13°C). For offices, the temperature should be 50°F (10°C). Para. 15-26 lists temperatures. *Use clean materials at flow parts.*

In large business establishments, office buildings, or factories, multiple water coolers, instead of individual ones, are popular. These coolers have one large condensing unit supplying refrigeration to many bubblers (water fountains). Bubbler construction may vary.

Some water coolers use a self-contained water supply. See Fig. 13-37. Large glass containers of water, delivered on a regular basis, are used for the supply. The cooling system is similar to that used in other water cooler models.

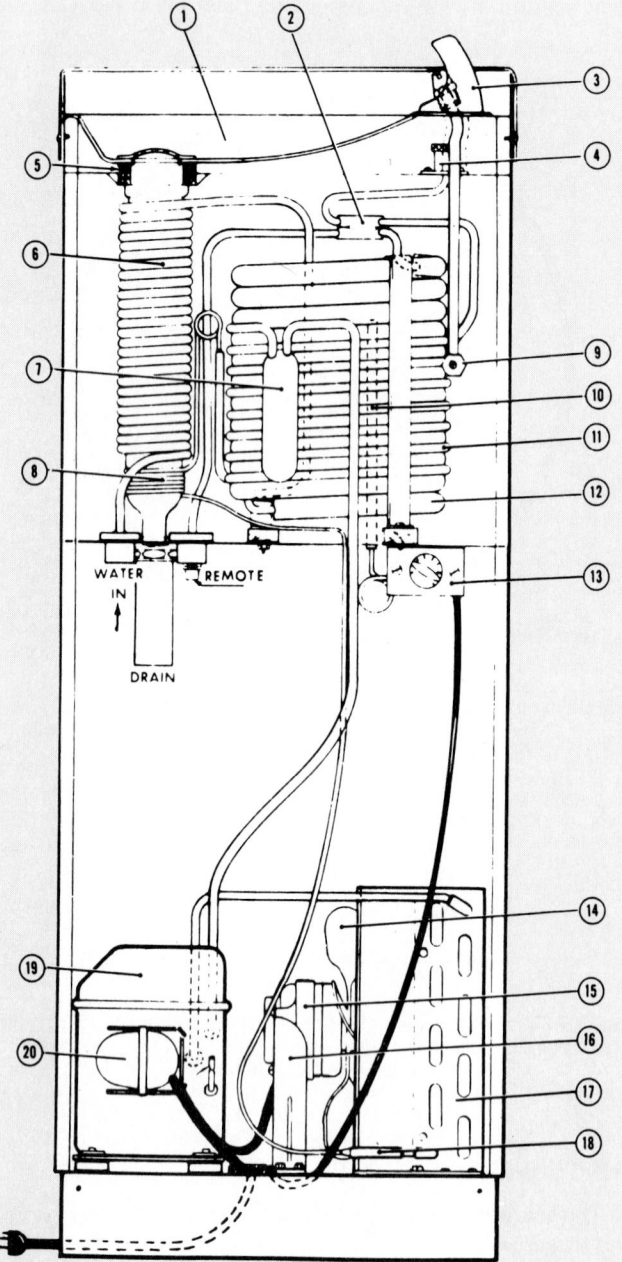

Fig. 13-34. Self-contained water cooler. 1—Top. 2—Cold water distributor. 3—Bubbler guard. 4—Glass filler connection. 5—Drain gasket. 6—Precooler assembly. 7—Accumulator. 8—Capillary tube. 9—Water valve. 10—Thermostat bulb well. 11—Evaporator. 12—Water-cooling coil. 13—Thermostat. 14—Fan blade. 15—Fan motor. 16—Fan bracket. 17—Condenser. 18—Liquid refrigerant strainer. 19—Compressor. 20—Relay, overload protector. (Temprite Div. of Elkay Mfg. Co.)

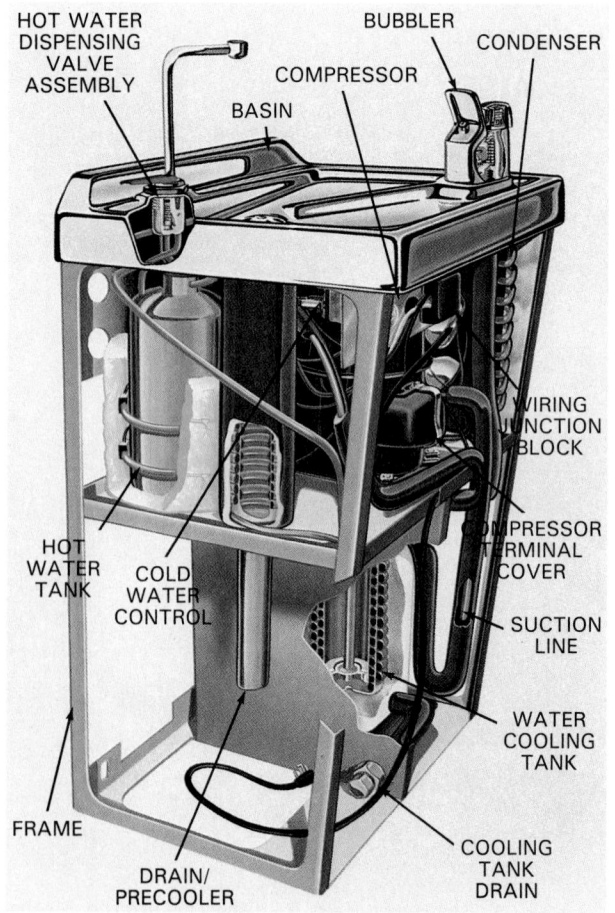

Fig. 13-35. Another view of self-contained water cooler. Note combination hot water and cooler. (Ebco Manufacturing Company)

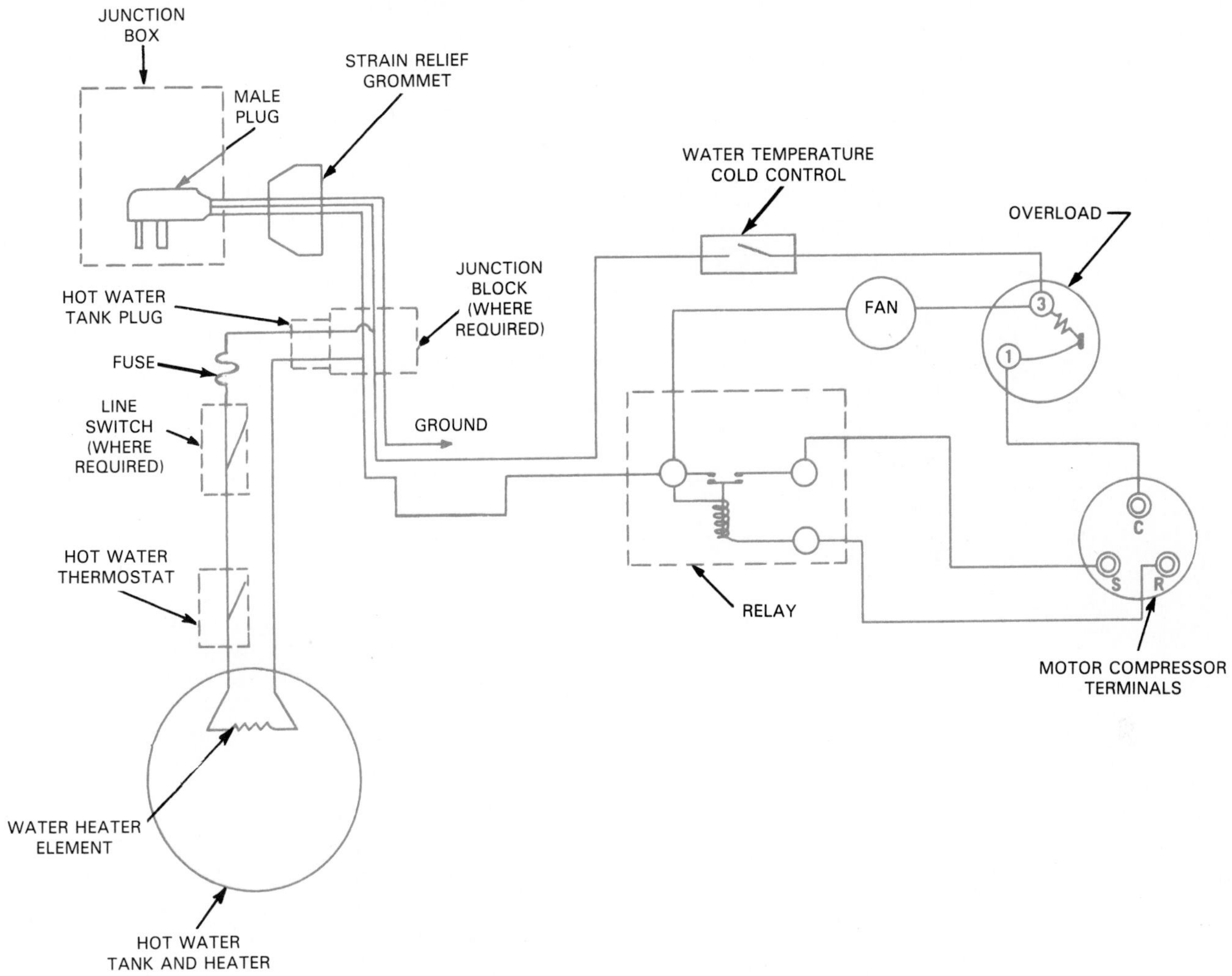

JUNCTION
BOX

MALE
PLUG

STRAIN RELIEF
GROMMET

WATER TEMPERATURE
COLD CONTROL

OVERLOAD

FAN

HOT WATER
TANK PLUG

JUNCTION
BLOCK
(WHERE
REQUIRED)

FUSE

LINE
SWITCH
(WHERE
REQUIRED)

GROUND

HOT WATER
THERMOSTAT

RELAY

MOTOR COMPRESSOR
TERMINALS

C

S   R

WATER HEATER
ELEMENT

HOT WATER
TANK AND HEATER

Fig. 13-36. Wiring diagram for water cooler which also provides hot water. Fan at upper right is a condenser fan. (Elkay Mfg. Co.)

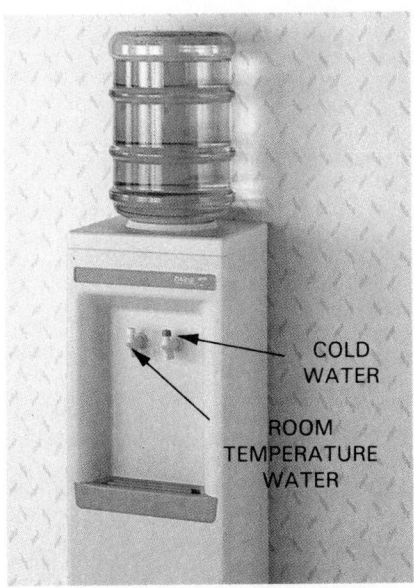

Fig. 13-37. Water cooler with self-contained water supply. One outlet is for chilled water. Second outlet is for room temperature water. On some units, the second outlet provides heated water.   (Ebco Mfg. Co.)

COLD
WATER

ROOM
TEMPERATURE
WATER

13-16 MODULAR REFRIGERATION SYSTEMS

Many party/specialty stores use flexible refrigeration systems, Fig. 13-38. The components with the glass door storage units can be used in numerous combinations with the refrigeration units. The refrigeration unit shown cools up to four storage units. The system has forced air circula-

Fig. 13-38. Modular refrigeration system with cooler units between cabinets.   (Stevens Lee Company, Silver King Division)

tion, automatic defrost, adjustable temperature control, and automatic condensate evaporator. Fig. 13-39 illustrates an under-counter refrigeration unit used in delicatessens and small restaurants.

Fig. 13-39. Under-counter refrigeration unit. (Stevens Lee Company, Silver King Division)

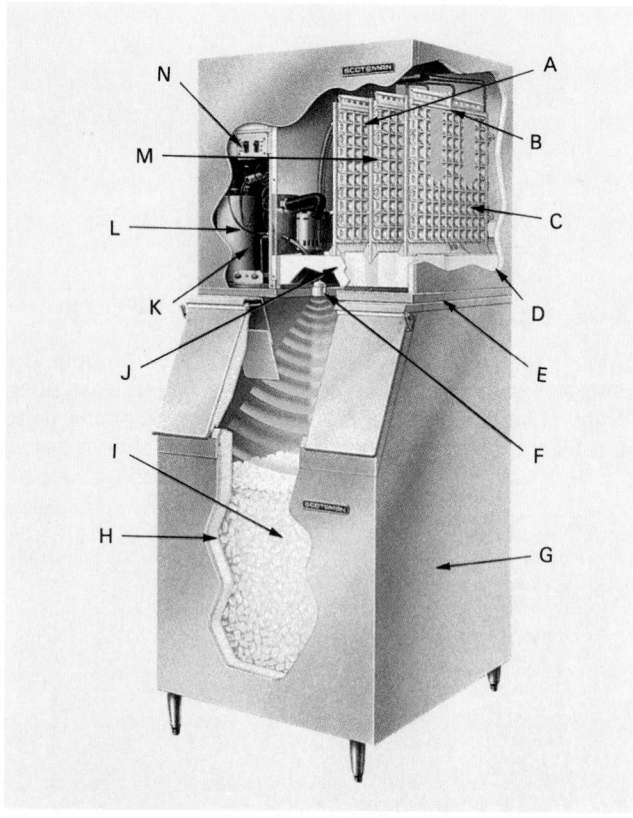

Fig. 13-40. Automatic ice maker. A—Modular evaporator plate. B—Water distribution tubes. C—Evaporator plate. D—Insulated cabinet. E—Base. F—Sonar bin control. G—Nonmagnetic stainless steel. H—1 1/2 in. foam insulation. I—Contour ice cubes. J—Frontal access for service. K—Suction line accumulator. L—Compressor—R-22 refrigerant. M—Water control. N—Adjustable bin control. (Scotsman Ice Systems)

13-17 AUTOMATIC ICE MAKER

Automatic ice makers are now widely used commercially. Several different types are available. They automatically control water feed, freeze the water, empty the ice into storage facilities, and shut down when the storage space is full.

The ice cubes or chips formed are clear and sanitary, since only flowing water is used. Cloudy ice cubes are caused by entrapped air.

Floats and solenoids control water flow. Switches operate the storing action when ice is made. Electrical heating elements, hot water, hot gas defrosting, or mechanical devices remove the ice from the freezing surfaces.

Urethane foam, polystyrene, or fiberglass provides cabinet insulation. Freezing surface and bin storage basin are made of stainless steel.

Fig. 13-41. Flake or chipped ice unit. It has a 1/2 hp (373 W) condensing unit which is air cooled. It will store 255 lb. (115 kg) of ice and produces 442 lb. (199 kg) of ice in 24 hours. (Scotsman Ice Systems)

Some are self-contained while others use remote condensing units. Both supply and drain plumbing are needed.

Capacity can vary among units from a few pounds up to many tons per day. Capacities, of course, decrease as water temperature and/or ambient air temperature increase. Ice cube maker water circuits and ice-freezing parts should be cleaned at least once a year.

A typical automatic ice cube maker is shown in Fig. 13-40. Another unit, Fig. 13-41, automatically makes chipped or flaked ice. Each of these units produces approximately 900 lb. (408 kg) of ice a day. Fig. 13-42 shows a combination ice cube and cold water dispensing unit used in restaurants.

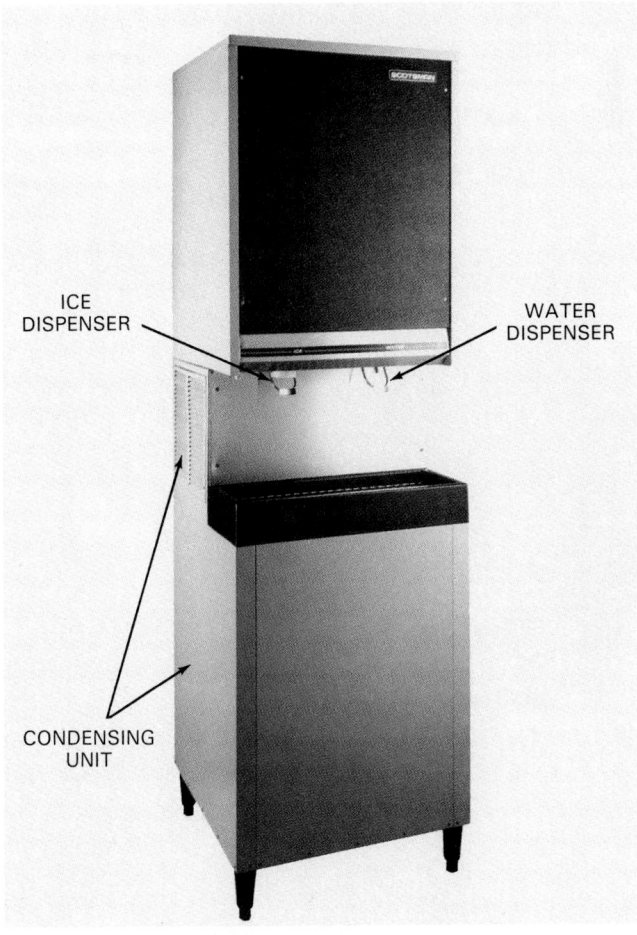

Fig. 13-42. Automatic ice cube maker and cold water dispenser. Technician will need to be familiar with the condensing unit as well as the ice and water dispensers. (Scotsman Ice Systems)

Fig. 13-43. Can and bottle beverage refrigerated coin-operated vending machine. Cabinet doors are open to show storage compartment, controls, and refrigeration mechanisms.

13-18 VENDING MACHINES

Refrigerated vending machines for foods or beverages are becoming increasingly popular. Cold drinks with ice, soft drinks, ice cream, cold food, cold milk, and frozen desserts are all automatically dispensed. See Fig. 13-43.

Refrigerating systems employed in these units are typical. Most use capillary refrigerant controls, hermetic motor compressors, and defrosting devices.

The electrical system generally is part of the overall system that electrically transfers the material and operates the coin and currency devices.

Vending machines have many components necessary for dispensing:
1. Coin and currency device.
   a. Acceptors.
   b. Rejectors.
   c. Changers.
   d. Steppers and accumulators.
2. Carbon dioxide systems.
3. Cup dispensers.
4. Heating systems.
5. Refrigerating systems that are usually air-cooled using fan evaporators. Some have defrost systems.
6. Transfer systems.

13-19 MILK COOLER

By law, milk must be cooled to 50 °F (10 °C) within an hour after being taken from the cow. Furthermore, it must be stored at 40 °F (4 °C) or lower.

Bacterial growth in milk is dramatically affected by temperature. During a 24-hour period, bacteria count will increase to 2400 at 32 °F (0 °C), to 2500 at 39 °F (4 °C), to 3100 at 46 °F (8 °C), to 11,600 at 50 °F (10 °C), to 180,000 at 60 °F (16 °C), and to 1,400,000,000 at 86 °F (30 °C). Bulk type coolers will handle this specific cooling requirement.

The stainless steel bulk cooler shown in Fig. 13-44 has an evaporator in the base. Coolers of this type have a 450- to 6000-gal. (1 703 to 22 712 L) capacity. R-22 is the refrigerant used in this type of system.

Usually air cooled, the condensing unit is mounted outside the milk room. *It should not be put in the vacuum pump room.* Air flowing over the condenser can be drawn from the milk room, however. In winter, where permitted, this warm air can be ducted to heat the milk room.

Coolers should always be properly grounded. Use a separate safety ground wire the same size as the power wires or larger. These units also typically have a crankcase heater.

13-20 REFRIGERANT-TO-WATER HEAT RECOVERY SYSTEM, HOT WATER

A heat recovery system produces and stores hot water by transferring the heat from the condenser to cold water.

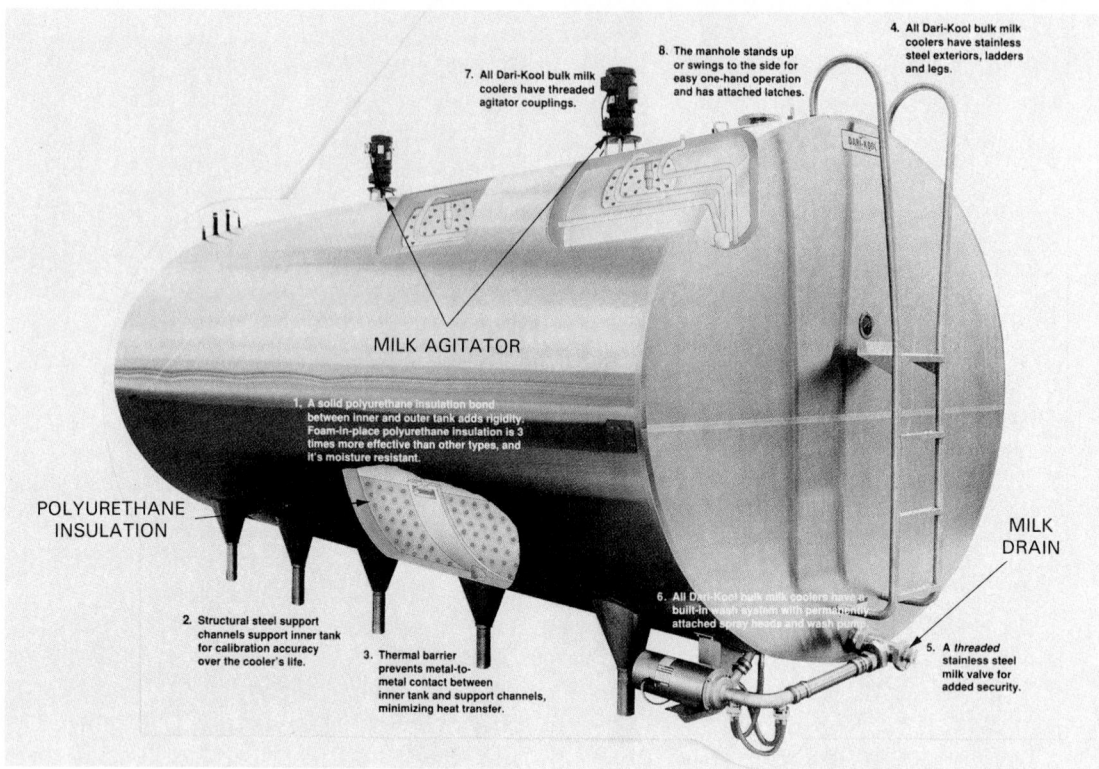

7. All Dari-Kool bulk milk coolers have threaded agitator couplings.

8. The manhole stands up or swings to the side for easy one-hand operation and has attached latches.

4. All Dari-Kool bulk milk coolers have stainless steel exteriors, ladders and legs.

MILK AGITATOR

1. A solid polyurethane insulation bond between inner and outer tank adds rigidity. Foam-in-place polyurethane insulation is 3 times more effective than other types, and it's moisture resistant.

POLYURETHANE INSULATION

MILK DRAIN

2. Structural steel support channels support inner tank for calibration accuracy over the cooler's life.

3. Thermal barrier prevents metal-to-metal contact between inner tank and support channels, minimizing heat transfer.

6. All Dari-Kool bulk milk coolers have a built-in wash system with permanently attached spray heads and wash pump.

5. A threaded stainless steel milk valve for added security.

Fig. 13-44. Bulk milk cooler. Direct expansion evaporator is located in the base of the unit. (Dairy Equipment Co.)

See Fig. 13-45. This type of unit is adaptable to most refrigeration systems. The tanks are constructed of vertical plates that allow the refrigerant to flow through the plates

and provide a rapid transmission of heat to the water. This system is often used as a supplement to existing hot water systems in supermarkets, restaurants, and other commercial applications.

13-21 MILK DISPENSERS

Many food service places dispense milk from bulk containers. Cans or plastic bags holding three to five gallons (11-19 liters) are installed in dispensers. These units meet all health and sanitation codes. Milk is kept at about 36 °F (2 °C) by a hermetic refrigerating system. Reserve milk containers are kept in a walk-in cooler or in a milk storage refrigerator.

13-22 LABORATORY REFRIGERATED INCUBATORS

An interesting application of refrigeration helps maintain correct temperatures in incubators. Fig. 13-46 shows an incubator which can maintain constant temperatures using a refrigerating system for cooling and electric heating elements for heating. Such a unit can maintain any constant temperature between 36 °F (2 °C) and 158 °F (70 °C)..

13-23 BAKERIES

Many raw products used by bakeries must be kept at refrigerated temperatures to preserve or improve quality. Bakeries must store:
1. Perishable raw materials.
2. Perishables during interrupted processing.
3. Finished products.

Fig. 13-45. A two-unit refrigeration system, shown at left, is connected to a water heat recovery system at right. (Dairy Equipment Co.)

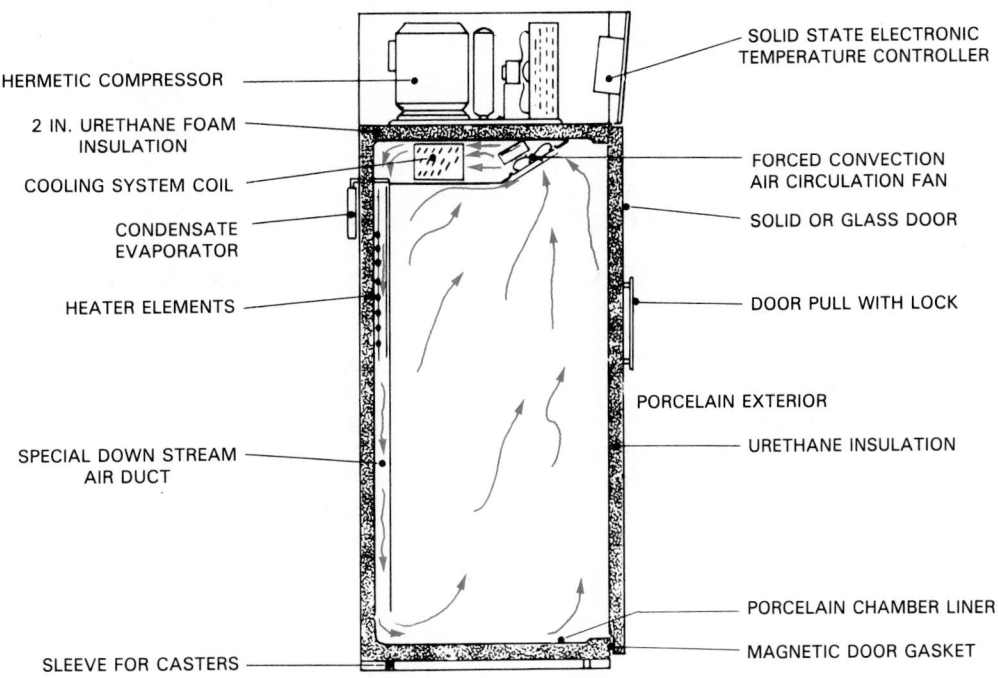

HERMETIC COMPRESSOR

2 IN. URETHANE FOAM INSULATION

COOLING SYSTEM COIL

CONDENSATE EVAPORATOR

HEATER ELEMENTS

SPECIAL DOWN STREAM AIR DUCT

SLEEVE FOR CASTERS

SOLID STATE ELECTRONIC TEMPERATURE CONTROLLER

FORCED CONVECTION AIR CIRCULATION FAN

SOLID OR GLASS DOOR

DOOR PULL WITH LOCK

PORCELAIN EXTERIOR

URETHANE INSULATION

PORCELAIN CHAMBER LINER

MAGNETIC DOOR GASKET

Fig. 13-46. Incubator has hermetic refrigerating system and electric heat. Temperatures are accurate within 1 °F or 0.5 °C. (Rheem Mfg. Co., Scientific Products Div.)

Frozen ingredients must, of course, be stored. Even the water and flour used for bread making must be cooled during certain periods of the year. Health codes require these practices. Creams and custards can be kept longer at a cool temperature. Fig. 13-47 shows a reach-in cabinet designed for baking use.

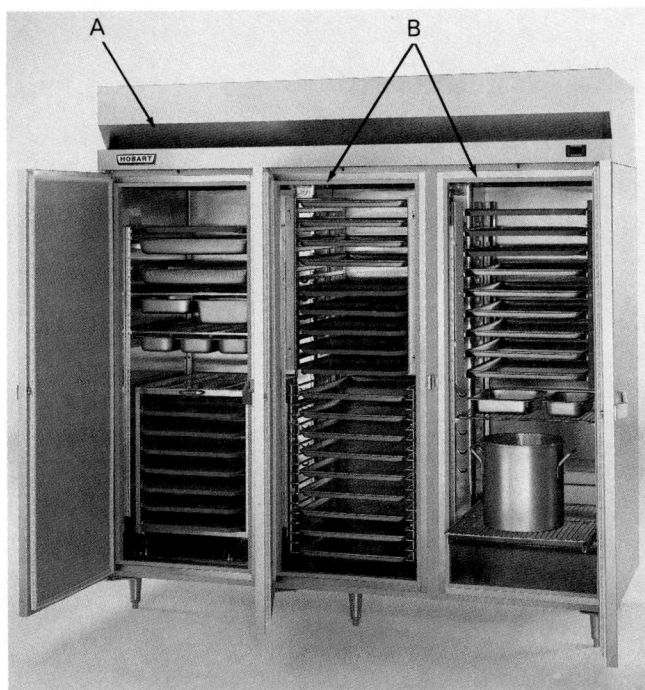

Fig. 13-47. This reach-in cabinet is designed for bakery use. A—Condensing unit and blower evaporator. B—Cold air duct. (Hobart Corp.)

Both normal and low-temperature refrigeration is used. Normal refrigeration is suitable for butter, eggs, coconut, cream, fat, meat, margarine, nuts, yeast, and dough retardation. Low-temperature systems are needed to freeze baked goods which are sold frozen. For example, if bread is fast frozen to −1 °F (−18 °C), it will remain fresh for almost a month.

Air conditioning using refrigerating equipment is also used in bakeries since controlled temperature and humidity is important in many baking processes.

13-24 FUR STORAGE

Furs are stored at a low temperature to prevent moth damage. Moths in the egg, larvae, or adult stage will thus be destroyed. In common practice, furs are first cooled to approximately 15 or 20 °F (−9 to −7 °C) then warmed to above 50 °F (10 °C) for 24 to 48 hours. After this cycle—found to be the most effective—the furs are stored at 35 to 40 °F (2 to 4 °C).

The fur storage cabinet is usually constructed like a cold storage room. Walls are concrete, thoroughly moisture sealed with asphalt, corkboard insulation, or its equivalent. An inside door is built like a refrigerator door; the outer, like a vault door.

Blower evaporators are used to force air circulation around and into the furs. The humidity must be kept at approximately 55 percent to prevent drying of the skins. Vaults should be fireproof.

13-25 INDUSTRIAL APPLICATIONS

Refrigeration has a variety of applications in manufacturing processes. Smaller units are automatic. Three common uses are:

1. Cooling of water which, in turn, cools electrodes on

resistance welders.
2. Cooling of quenching liquids used for cooling metals in heat treating.
3. Cooling compressed air.

Compressed air must be dry since moisture may rust and corrode air tools or spoil paint spraying. Air, therefore, is cooled to keep its dew point (water condensing temperature) above the coldest spot in the air system. This prevents moisture in the lines.

Air is cooled with a refrigerating system to bring it below its critical pressure dew point. Then it is reheated. Pressure dew points are about 50 °F (10 °C) above atmospheric dew points at 100 psi air pressure. In general practice, the compressed air is cooled to about 35 to 50 °F (2 to 10 °C). A 3000 cfm (1.4159 m³/s) air compressor will need about a 20 ton (55.2 kW) capacity refrigerating machine.

If a refrigerating system is near flammable or explosive materials, the system must be explosion proof. Sealed lights and sealed contact points are required.

Refrigerators and freezers of all types are used in research. They are used to maintain constant temperature, constant humidity, and low-temperature control. Low-temperature units capable of −140 °F (−96 °C) are available. These systems usually use 5-in.

(13 cm) insulation and a cascade of systems (two systems connected in a series).

13-26 INDUSTRIAL FREEZING OF FOODS

Industrial freezing of food is carried on in two principal types of establishments:
1. Processing plants.
2. Locker plants.

Processors of frozen foods have freezing centers in many large food-producing areas. For example, fish is packed and frozen along the coast and then shipped to all parts of the country.

A locker plant is a smaller unit designed to prepare, freeze, and store a great variety of products. Refrigerating equipment in processing and locker plants vary considerably, but the plan for freezing the food is similar.

Fig. 13-48 shows the flow of produce through a typical plant. The food is weighed and checked for purity and suitability for freezing. Then it moves to the processing room. Here meats are cut, fowl are cleaned and dressed, vegetables blanched, and the various items packaged. Next, the packed foods are sent to the freezing section where they are completely frozen and readied for storage.

Fig. 13-49 shows a typical locker plant floor plan. An

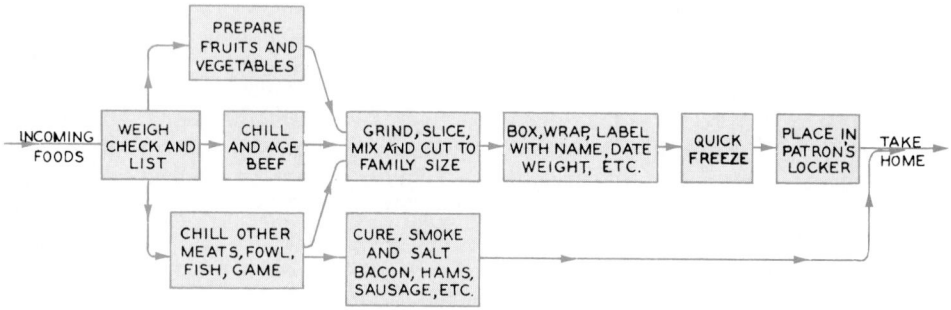

Fig. 13-48. Flow chart of food moving through preparation process in a freezing plant.

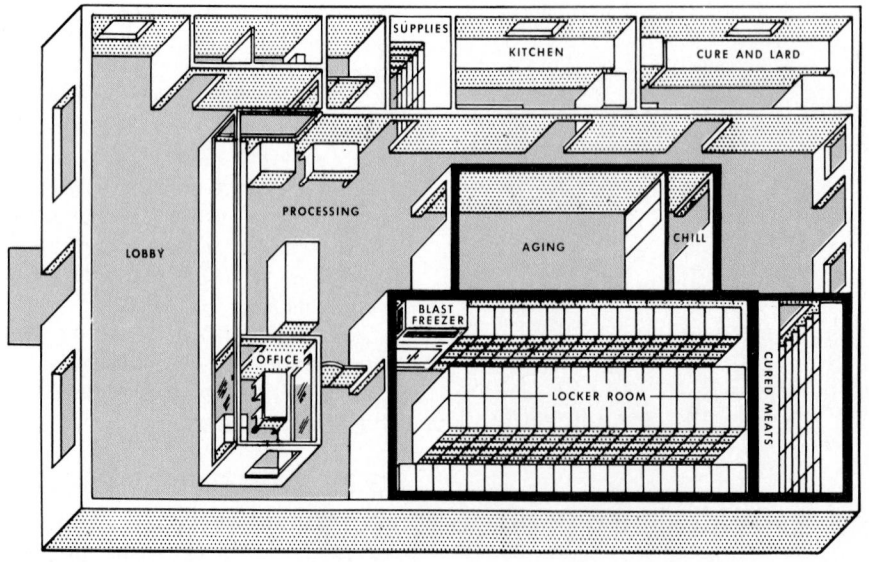

Fig. 13-49. Typical locker plant layout. Special blast freezer fast freezes food before it is stored in locker room.

ultraviolet ray lamp is sometimes placed in the aging room to kill bacteria.

High humidity is very important in rooms where food is cured and stored. Meat especially tastes better and does not lose weight if relative humidity is kept as close to 100 percent as possible. Temperature should be near 39 °F (4 °C), as this gives the best humidity results with a non-frosting evaporator.

Processing plants attempt to freeze food as fast as possible. They also attempt to expose as much of the food as possible to the lowest possible temperature. A fast-freezing system usually uses a track to hold the produce as it moves through an ultralow-temperature chamber for fast freezing.

Some fast food freezing systems use liquid nitrogen or carbon dioxide to turn perishable fresh food into long-lasting frozen food. This is commonly referred to as cryogenic food freezing. Temperatures of − 320 °F (− 196 °C) are obtained and freezing is instantaneous. This method of quick freezing causes little or no damage to the food. See Para. 17-11.

13-27 FREEZE DRYING

Freeze drying consists of the following four steps:
1. Food is frozen.
2. Frozen food is put into a vacuum chamber, Fig. 13-50.
3. Frozen food is heated to change its frozen water (ice) into vapor. It does not go through the liquid state, since below 4.7 mm Hg. water will change directly from ice to vapor. Freeze drying is usually done between 50 and 500 microns of pressure.
4. Vapor is condensed in the form of frost on a cold evaporator at about − 40 °F (− 40 °C) in the chamber. This step saves pumping about 100,000 cu. ft. (2 832 m³) of water vapor to form 1 lb. (.5 kg) of vapor at 50 microns.

After the product has dried, the evaporator is exposed to heat at atmospheric pressure. The frost melts and is removed. Dry nitrogen is used to raise the chamber pressure from 50 microns to atmospheric pressure.

13-28 INDUSTRIAL STORAGE OF FROZEN FOODS

Storage requirements of most frozen foods are about the same. A temperature of 0 °F to − 20 °F (− 18 °C to − 29 °C) is desirable and a variation of 2 to 3 °F (1 to 2 °C) is normally allowed. A high temperature differential causes harmful ''breathing'' and volume changes. Humidity in storage rooms should be as high as possible. Storage areas may be cooled by blower evaporators, direct contact plates, or brine coils.

In smaller plants, the customer enters the refrigerated area and pulls out a lockable drawer to get frozen food. In other locker plants where special construction makes it possible, frozen foods are delivered from the storage area to the customer without the need to enter the cold area.

13-29 TRUCK REFRIGERATION

Truck refrigeration requires special trailer bodies and refrigerating units. Such bodies are 9 ft. to 18 ft. (3 m to 6 m) long, and use one refrigerating system of 1 1/2 to 2 hp (1 120 to 1 490 W).

Bodies should be light and well insulated. Construction must be sturdy so that constant vibration and rough handling will not destroy the insulating value of the walls. Fig. 13-51 illustrates a refrigeration system used on a trailer system. The main components are the compressor, the air-cooled condenser, the expansion valve, and the direct expansion evaporator. These systems commonly use R-22, R-500, R-502, and R-12 as a refrigerant. Fig. 13-52 illustrates the cooling and heat-defrost cycle of a diesel-powered unit.

Various insulating materials have been used. Most trailers have all-metal bodies with insulation thicknesses depending

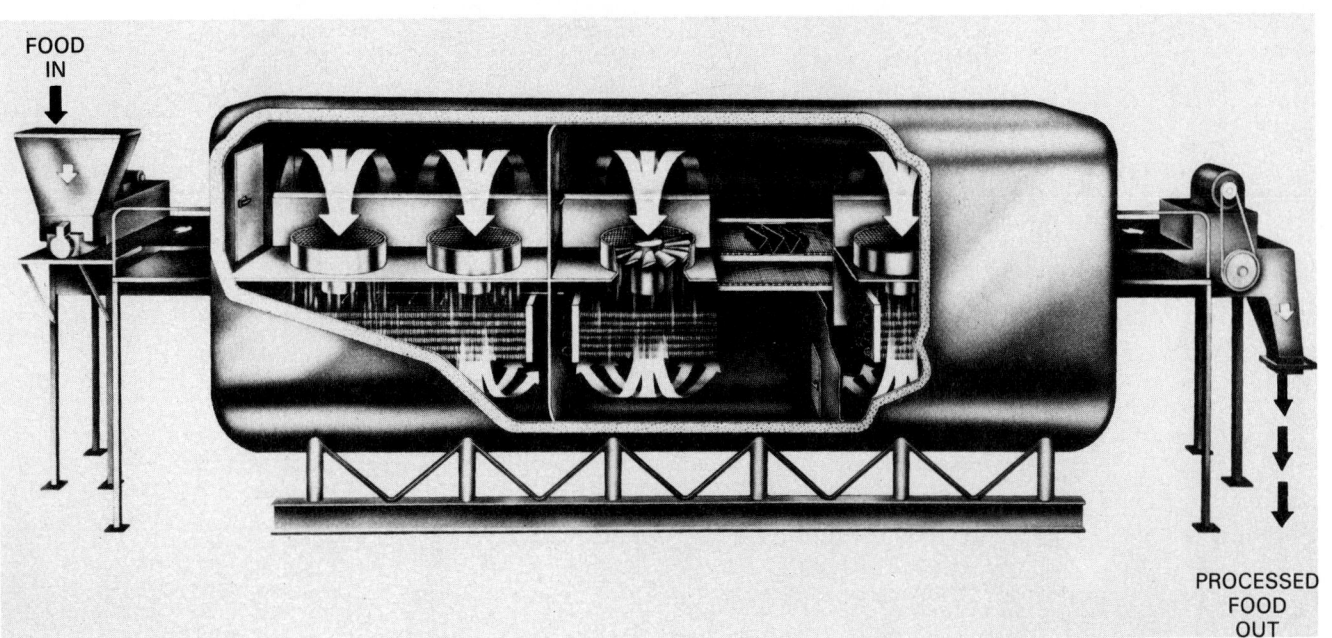

Fig. 13-50. Automatic track food freezer. Air in the chamber is about − 30 °F (− 34 °C). Food moves from left to right on the conveyor belt.

on the application. Foamed-in-place insulation is most often used. Fig. 13-53 shows such insulation being installed in the side of a trailer body.

The most interesting thing about trailer refrigeration is the great variety of applications. Each application must be

Fig. 13-51. Refrigerated trailer. Refrigeration system may be either electric, gasoline, or diesel powered.
(Carrier Transicold Division, Carrier Corp.)

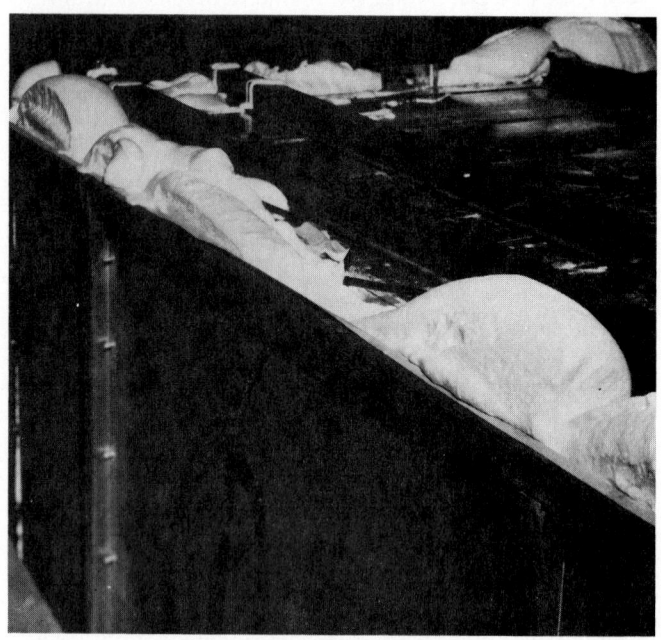

Fig. 13-53. Truck body is being insulated with plastic foam insulation. Excess will be trimmed away.

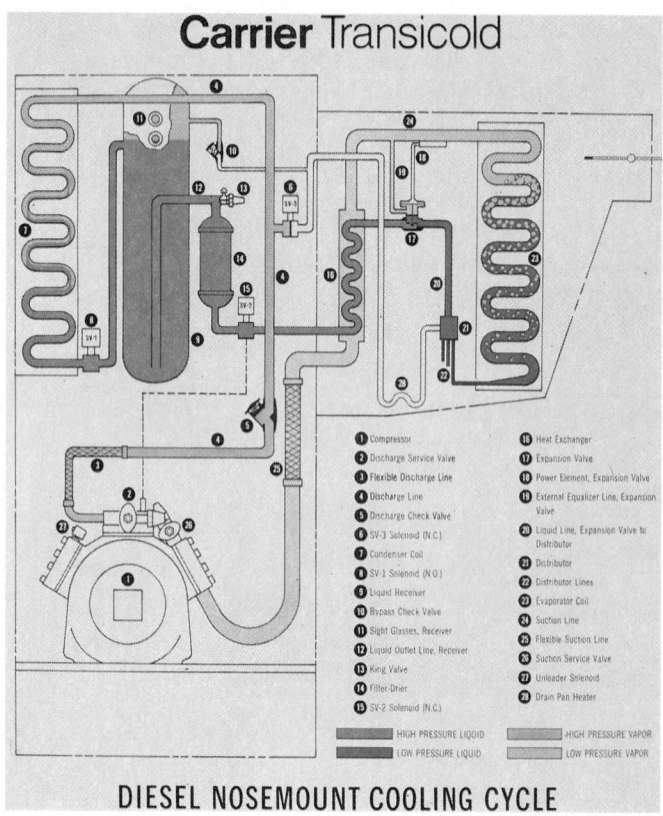

DIESEL NOSEMOUNT COOLING CYCLE

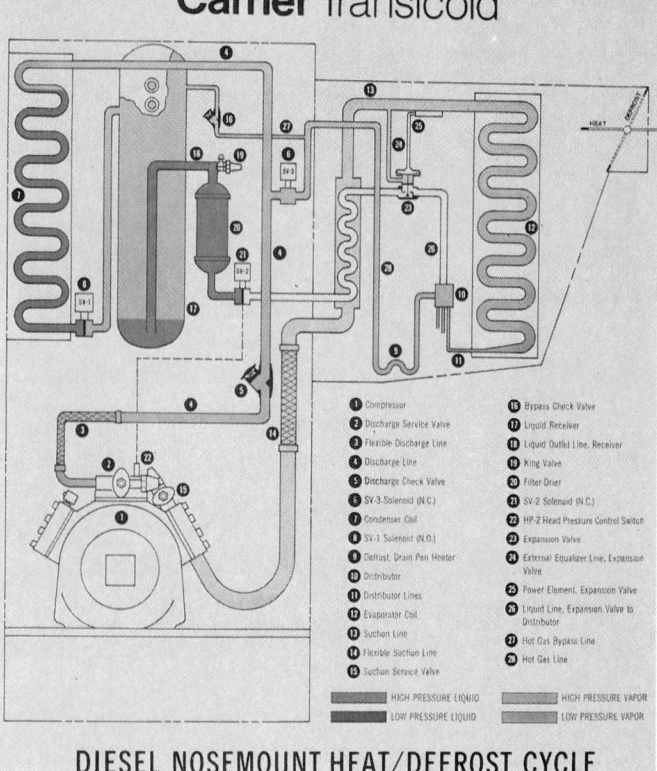

DIESEL NOSEMOUNT HEAT/DEFROST CYCLE

Fig. 13-52. Diagrams of cooling system using a diesel-powered gas engine. A—Cooling cycle. B—Heat/defrost cycle.
(Carrier Transicold Div., Carrier Corp.)

studied before a temperature may be recommended. A truck using dry ice for refrigeration must be insulated for −109 °F (−78 °C). An ice cream truck must be insulated for −15 °F (−26 °C), while fresh foods require insulation for only 32 to 35 °F (0 to 2 °C) temperatures. Due to aspiration of these materials, fresh produce, flowers, and fruits need accurate control of temperature, humidity, and ventilation. (Aspiration means the tendency of the materials to lose their fluids to the surrounding air.)

There are four main refrigeration systems:
1. Ice.
2. Dry ice.
3. Mechanical:
   A—Blower system.
   B—Hold-over eutectic plate.
4. Expendable refrigerant (see Chapter 17).

Ice and dry ice are seldom used today—ice, because of its weight and load requirements; and dry ice, because of the difficulty to control and maintain specific temperatures.

Mechanical refrigeration is similar, in most cases, to typical refrigerating units. The major difference is in the nature of the compressor drive. Two common drives are:
1. Engine-driven electric generator and motor.
2. Separate gasoline or diesel engines.

Electric generator and motor types use standard voltages, cycles, and phases which permit plugging into a wall outlet in the garage if the trailer must be kept cold while off the road. Units driven by gasoline engines are automatically controlled to start and stop as the system requires.

Some trailers use a dual refrigeration system. Fig. 13-54 shows a refrigerated truck with a refrigerating unit mounted over the cab. This is a standby unit used only when the truck is parked for longer periods. The engine-driven compressor provides refrigeration while on the road.

Fig. 13-55 shows the inside of the standby motor-driven compressor compartment. In addition to this condensing unit, a compressor mounted above (and driven by) the engine may be used. See Fig. 13-56. Most of these systems use a hot gas defrost. Some larger commercial units utilize a diesel fuel for the unit mounted. In some systems a generator provides power for the evaporator fans.

Typical of a finished installation mounted over the truck cab is the one shown in Fig. 13-57.

Eutectic plates operate without the use of mechanical assistance during normal operating. A small electrical freon condensing unit is normally connected at night to freeze the eutectic solution. This stores enough energy for positive cooling throughout the day. The plates basically act like a battery with stored energy. The eutectic plates are constructed of small diameter tubing for circulating freon. Fins are attached to the tubing to conduct heat from the freon into the eutectic solution. The assembly is inserted between two steel pans. The condenser is vacuum filled with eutectic solution that will freeze at a specific temperature. Various sizes are available.

Another form of truck refrigeration is the use of liquid nitrogen, which does not need a condensing unit. It provides excellent low-temperature refrigeration. At atmospheric pressure, it boils at −320 °F or at 140 R (−196 °C or 77 K). Temperatures from absolute 0 up to −168 °F (0 absolute to −111 °C) are considered to be in the cryogenic range. See Chapter 17.

13-30 RAILWAY CAR REFRIGERATION

Refrigerated railway cars, long used to transport perishable goods, often have mechanical refrigeration. Car construction and insulation is similar to that of the truck trailer.

Two general types of cooling are used on trains: freight

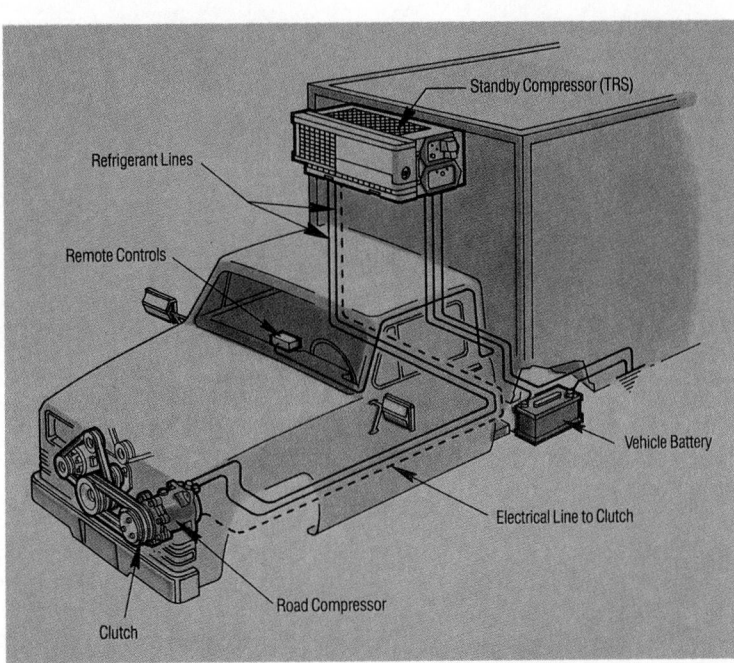

Fig. 13-54. Truck refrigeration cooling system with standard built-in compressor on engine and standby compressor above cab. (Carrier Transicold Div., Carrier Corp.)

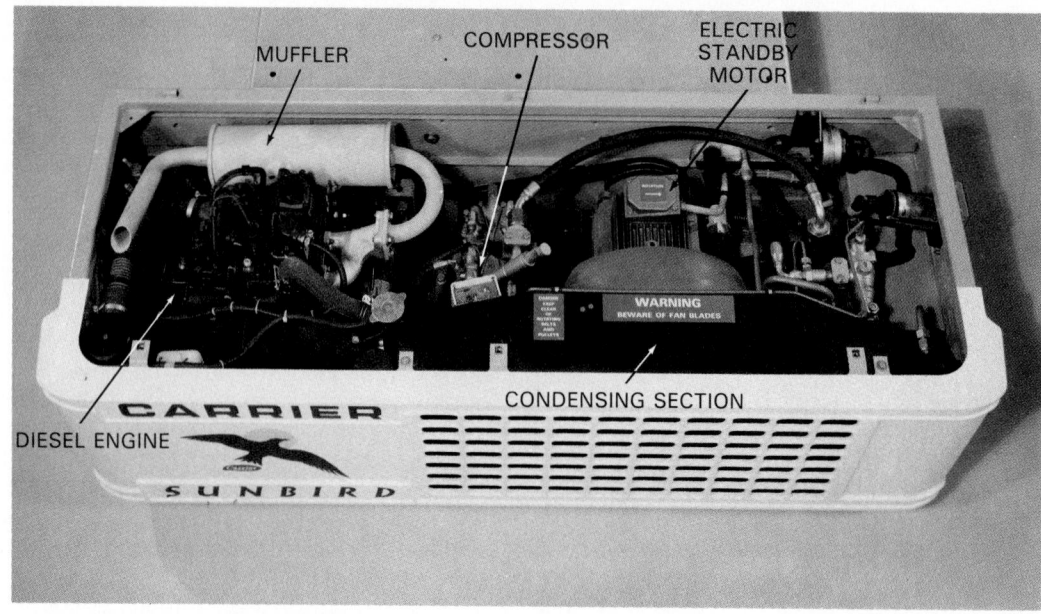

Fig. 13-55. Truck refrigeration unit. Note electric standby motor. (Carrier Transicold Div., Carrier Corp.)

Fig. 13-56. This refrigeration compressor is driven by the truck's engine. (Carrier Transicold Division, Carrier Corp.)

Fig. 13-57. Completed over-cab installation of a condensing unit of refrigerated truck. (Carrier Transicold Div., Carrier Corp.)

refrigeration and air conditioning in passenger cars.

Like truck refrigeration, freight refrigeration involves four methods:
1. Ice.
2. Dry ice.
3. Mechanical (used exclusively on passenger cars).
4. Chemical (expendable refrigerant).

Compressors in mechanical installations are usually driven from the car axle while rail cars are in motion and by an electric motor while cars are stopped or being held in the rail yard.

Some train refrigeration systems use the absorption system. Others have used the steam jet system.

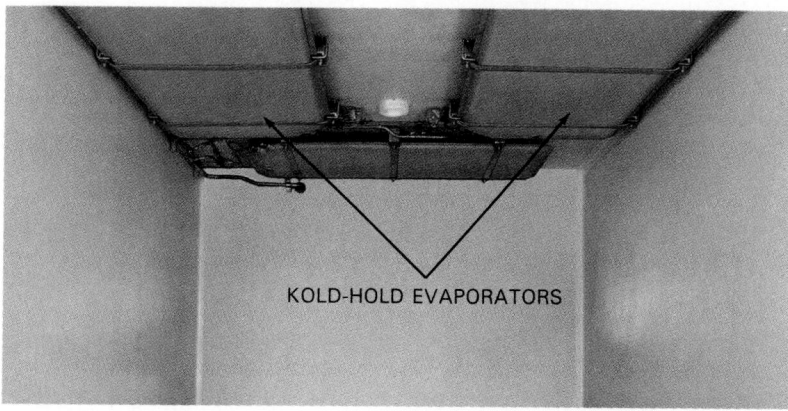

Fig. 13-58. Two KOLD-HOLD plate evaporators installed in a refrigerated truck body. (Tranter, Inc.—courtesy of Hackney Brothers Body, Wilson, NC)

13-31 MARINE REFRIGERATION

Basically, marine refrigeration equipment is the same as land-type refrigeration. However, refrigerants are restricted to those which are nontoxic and nonflammable. All cabinets are carefully sealed to exclude the possibility of moisture entering the insulation. Refrigerant lines must be installed to permit some vibration and movement without danger of line or joint failure.

Saltwater-cooled units use piping, valves, and condenser made of materials designed to minimize the corrosive effects of salt water.

While some automatic refrigerating equipment is used aboard ship, many of the machines are semiautomatic and are interconnected so they can be used either together or separately. Thus, capacity can be adjusted to varying loads. R-12 and R-22 are popular as refrigerants in units located aboard ship.

13-32 SNOW MAKING

Many ski slopes add to the natural snow on their runs with artificial snow. Artificial snow is a water spray into which compressed air is added so that the water is broken up into fine mist and freezes very rapidly. See Fig. 13-59.

The low temperature required for snow making is not created by refrigeration mechanism but comes from the surrounding atmosphere. The temperature must be 30°F (−1°C) or lower, although the maximum temperature which may be used depends on the relative humidity.

If relative humidity is low, snow may be made at temperatures as high as 35°F (2°C). This is possible because, with low humidity, some of the water evaporates and, in doing so, absorbs heat. This reduces the temperature of the water droplets below the freezing temperature and ice crystals (snow) are formed.

In general, the ground must be frozen before snow

Fig. 13-59. Snow-making equipment being used on a ski slope. The system breaks up H_2O into a fine mist which freezes rapidly. (Larchmont Engineering)

making is practical. Usually, manufactured snow only supplements natural snow.

13-33 CRYOGENIC APPLICATIONS

Cryogenics is a branch of physics. It relates to the mechanics for producing very cold temperatures. It covers refrigeration in the range of $-250\,°F$ to $-460\,°F$ or $-157\,°C$ to $-273\,°C$ (0 K).

Such cold temperatures are used to liquefy and separate gases such as oxygen, nitrogen, and argon. They are also used to liquefy hydrogen and helium.

Three methods are used to create these low temperatures:
1. Expansion process with heat exchangers (Joule-Thompson process).
2. Expansion process with heat exchangers and with gases performing work.
3. Multiple cascade system.

Cryogenics may be used to condense and solidify obnoxious and harmful gases from exhaust products of industrial processes. One example is removing sulphur and ammonia gases from the exhaust gases of burning coke.

Cryogenic techniques are used in the propulsion of liquid fuel rockets and for preserving tissue, blood, and whole organs.

Special care must be taken when operating and servicing cryogenic equipment. Liquids are ultracold and will severely injure anyone coming in contact with them. Certain parts of the system are also ultracold and could cause injury.

Caution: The service technician should always wear goggles and gloves when servicing any part of cryogenic equipment. Temperatures being very low, liquid refrigerant will instantly freeze any part of the body it is allowed to touch.

In the presence of such low temperatures, special materials must be used in construction. These include nickel (stainless) steels, copper alloys, and aluminum metals. Electrical devices must be of special design and construction.

Lubrication, too, is a special problem because of the extreme cold. Either there is no lubricant at all or a dry type is used (molybdenum sulphide, for example).

13-34 REVIEW OF SAFETY

The best way to protect oneself and the public is to check the unit before working on it to be sure the mechanism and cabinet are installed according to local and national fire, electrical, plumbing, building, and safety codes.

Move cabinets on rollers. Avoid pinching hands or allowing the cabinets to fall on the feet. Wear safety shoes.

Floor load limits should be checked before heavy commercial equipment is installed. Wiring and plumbing must be adequate, securely mounted, and of correct materials. Remember, one must call attention to code violations which might result in serious damage to either the physical structure or to personnel.

Board of Health regulations apply to all equipment which is used around food and beverages. These regulations must be carefully followed. Moving of large commercial cabinets calls for extreme caution. Use handling equipment appropriate for the job and always work carefully.

13-35 TEST YOUR KNOWLEDGE

1. What cabinet temperatures are recommended for the following commercial installations: water-cooling, a walk-in cooler, a florist's cabinet, and an ice cream cabinet?
2. Why is it necessary that the doors and windows in commercial cabinets be airtight?
3. What is the purpose of dead air spaces formed between the three- and four-pane windows in these cabinets?
4. How is moisture kept from between the glass in multiple pane windows?
5. Why are display case lights sometimes located outside the case?
6. A water cooler using a plumbing supply and drain connection must be installed according to the restrictions of what codes?
7. Why are many commercial refrigeration installations of the multiple evaporator type?
8. To what temperature should water be cooled:
   a. If the bubbler were located in the heat treating room of a tool and die factory?
   b. In an office building?
9. What can be used to help prevent bacteria and mold growth?
10. Why must the humidity be kept high in produce storage cabinets?
11. How is refrigerated air kept from spilling from open display cases?
12. What is a double-duty display case?
13. Where are evaporators located in open display cases?
14. How are frozen foods open display case evaporators defrosted?
15. What are the two basic types of ice makers?
16. What should the milkshake temperature be in a freezing dispenser?
17. Why must milk be cooled immediately after it has been taken from the cow?
18. Are evaporators sometimes mounted outside the refrigerator cabinet?
19. What is the purpose of an activated carbon air filter?
20. How are certain parts of a frozen foods case kept from sweating?
21. What causes cloudy ice cubes?
22. Why are some soft ice cream making machines designed for continuous operation?
23. How can electric resistance heat be used in making ice cubes?
24. What two power sources drive truck refrigeration compressors?
25. What is a dual trailer refrigeration system? Describe its operation.
26. How many different airstreams are common in a frozen foods case?
27. Some water coolers serve two other purposes. What are they?
28. How often should an ice maker be cleaned?
29. What two parts of a freezing dispenser for soft ice cream are refrigerated?
30. Is food heated while being freeze dried?

Chapter 14

SERVICING AND INSTALLING COMMERCIAL SYSTEMS

After studying this chapter, the technician will be able to:
☐ List and describe the types of commercial installations.
☐ Explain the difference between noncode and code installations.
☐ Demonstrate troubleshooting techniques.
☐ Explain the proper procedures used in replacing or repairing defective commercial components.
☐ Demonstrate procedures for evacuating and recharging a refrigerating system.
☐ Write service estimates.
☐ Remove, test, repair, or replace various compressors.

Installing and servicing commercial units is a very important part of the refrigeration industry. If equipment fails, companies may suffer severe losses.

14-1 TYPES OF COMMERCIAL INSTALLATIONS

Commercial refrigeration installations vary considerably. Following are the classifications of installations commonly used:
1. Self-contained unit:
   a. Hermetic.
   b. Conventional.
2. Remote condensing unit:
   a. Single cabinet.
   b. Multiple cabinet.

An installation may be as small as a 1/20 hp self-contained unit or as large as a 140 hp unit. Larger units must be assembled on the premises.

Multiple units must be installed to handle the refrigeration load efficiently while eliminating hazardous conditions that might lead to accidents. Two major concerns should be durability and neatness. Many cities have laws and codes covering certain refrigeration systems. Furthermore, most refrigeration manufacturing companies have rules for installing their equipment.

In cities and rural communities not restricted by code, installations are usually made at minimal cost. One must, therefore, consider two types of installations:
1. Noncode.
2. Code.

Where there is no local code, it is good practice to follow the code of the closest city.

14-2 NONCODE INSTALLATIONS

Definite procedures should be followed when assembling a system. This safeguards against mistakes and reduces errors through carelessness. Refrigeration service departments claim that carelessness causes more than 90 percent of the servicing difficulties.

Assume the units are of a size that will efficiently handle the heat load. The installation must then take full advantage of the design. Tubing, safety valves, and protective devices should be placed into the system where they will produce operating efficiency, permanency, and safety.

All refrigeration installation and service work should be performed with correct tools. Fig. 14-1 shows some that are commonly used. Chapters 2 and 11 describe in more detail such tools and supplies.

14-3 INSTALLING CONDENSING UNITS

First the technician must determine where to place the condensing unit with respect to the cabinets and cases. This location should be as close to the cabinets as possible. A central location is best.

Installation should progress in the following order:
1. Put cabinets in place.
2. Locate place for condensing unit and install it.
3. Install evaporators.
4. Install valves and controls.
5. Install tubing.
6. Check for leaks.
7. Evacuate system per E.P.A. regulations.
8. Charge system.
9. Start system.
10. Check operation of unit and get 24-hour temperature and pressure records of unit in operation.

Cabinets are bulky and sometimes difficult to handle. A dolly, as shown in Fig. 14-2, is handy for cabinet moving.

It is recommended that the condensing unit be put in the basement or in a room next to the one in which the cabinet is located. *Avoid putting the condensing unit where it will be exposed to the sun or to low or freezing temperatures.* Locating condensing units in the same room with the counters and cabinets is not recommended because of the heat and noise they produce.

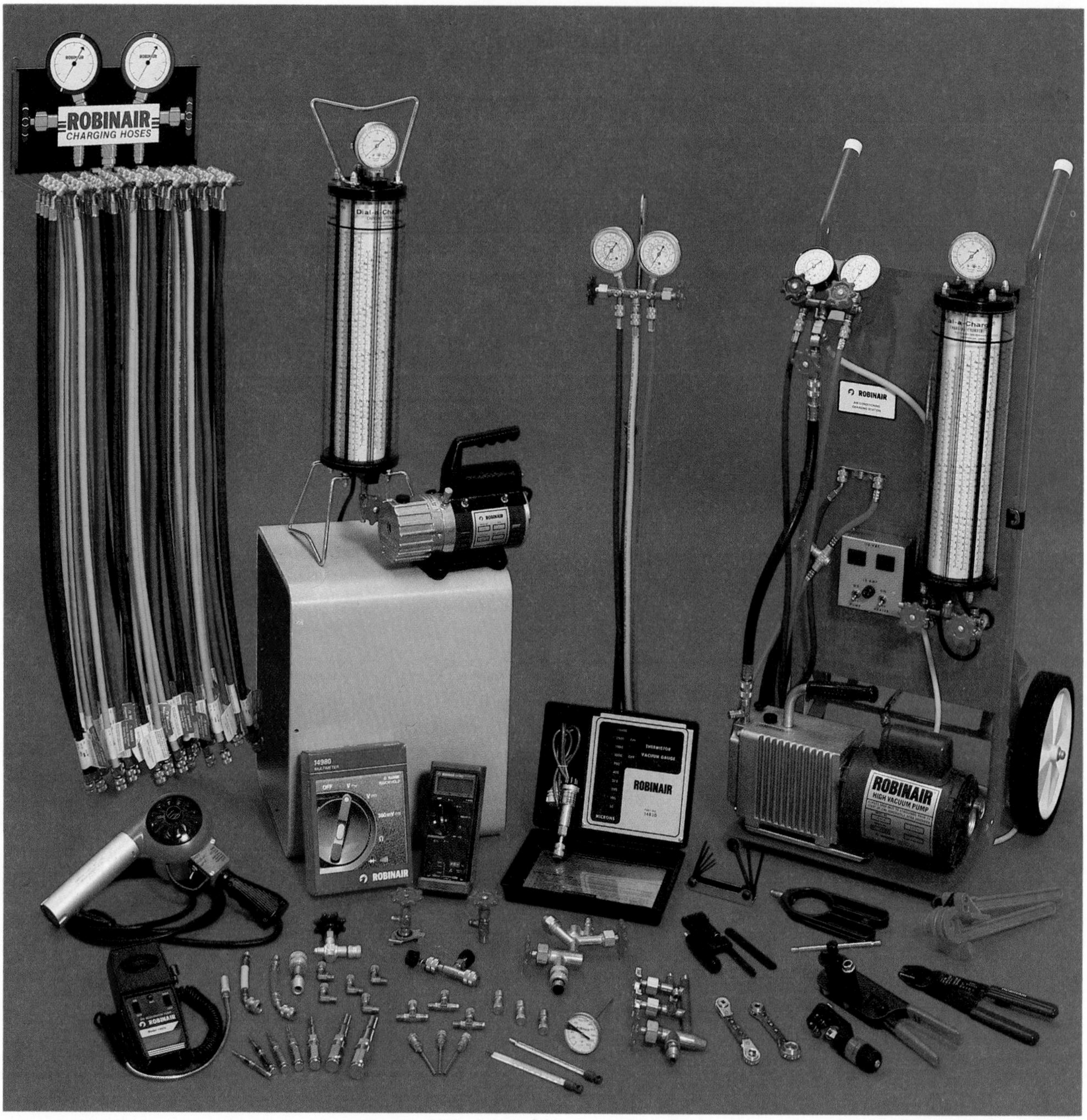

Fig. 14-1. The installation and servicing of commercial refrigeration units is not greatly different from working on domestic units discussed in earlier chapters. The tools and instruments listed in Chapters 2 and 11 can also be used for commercial systems. Pictured here are some of the commonly used tools the technician should have on hand at any installation or service call. From actually handling the tools and from studying Chapters 2 and 11, many of the tools will be recognizable. (Robinair Division, SPX Corporation)

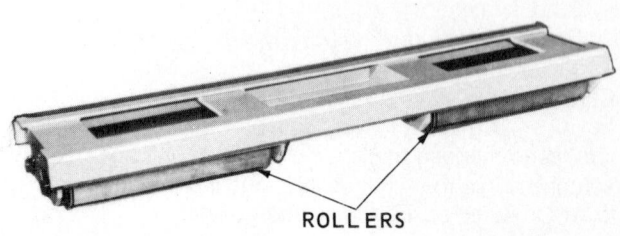

Fig. 14-2. One or more of these dollies under cabinet enables a service technician to easily move and locate heavy, bulky cabinets. Dolly shown will carry four tons. (Airserco Mfg. Co. Inc.)

There will be some vibration produced by the running condensing unit. Fig. 14-3 illustrates a vibration absorbing type compressor mounting. Protect the condenser by putting a heavy wire cage around and over it. Install a valve and accessory board on the wall just above the condensing unit to support valves, drier motor controls, two-temperature valves, electrical boxes, and service instruction cards.

In some cases, the condenser is installed away from the compressor, too. In these instances, traps are sometimes used in the compressor discharge line to keep the oil in the condenser (slant discharge lines down also) and away from the compressor discharge valve. See Fig. 15-80.

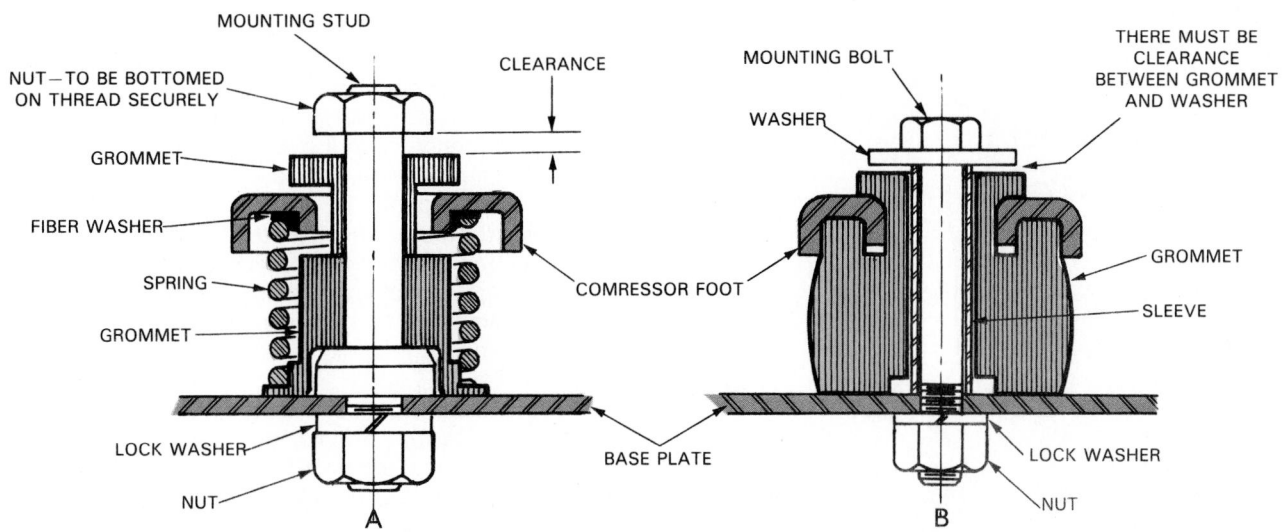

Fig. 14-3. Hermetic motor compressor mountings are designed to absorb vibration. A—Synthetic rubber grommet and spring. B—Synthetic rubber grommet only. (Tecumseh Products Co.)

14-4 INSTALLING EVAPORATORS

Evaporators for commercial refrigeration installations should be carefully mounted, leveled, and firmly fastened. Type of mounting depends on evaporator type. Fig. 14-4 shows a display case evaporator.

Blower evaporators are usually equipped with hanging brackets, as in Fig. 14-5. Evaporators are usually fastened to the ceiling in florist cabinets and walk-in coolers. A plumb line is used to locate the holder positions. A cardboard template is useful for locating the mounting brackets.

The hanger is attached to the ceiling of the cabinet. Hydraulic or pneumatic adjustable-height platforms lift and hold the evaporators in place until they can be fastened. Regardless of the devices used to mount the evaporator, the unit should be carefully rechecked after mounting to make sure it is level.

All natural convection evaporators should be properly baffled. Fig. 14-6 shows one type baffle and the hangers that fasten it to the evaporator.

Display counter evaporators are usually supported by stands. These stands or brackets should be provided with leveling adjustments in all directions.

14-5 INSTALLING TUBING

Tubing in noncode installations is usually run along the walls and ceiling with supports at intervals frequent enough to keep tubing straight and firmly fastened. Special clamps are available as tubing fasteners but a galvanized 1/2-in.

Fig. 14-5. Small blower evaporator. A—Drain connector. B—Drain pan. C—Hanger. Evaporator must be mounted level for efficient operation and good drainage. (Peerless of America, Inc.)

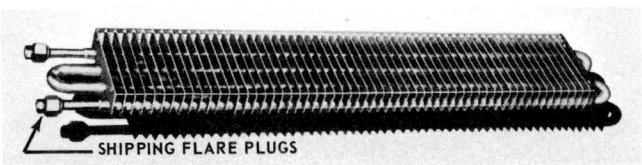

Fig. 14-4. Shelf evaporator. Aluminum fins and copper tubing are mechanically bonded. Fins are offset. (Peerless of America, Inc.)

Fig. 14-6. Evaporator baffle and drain pan. It is compact, permits good circulation and condensate drainage. A—Hangers. B—Individual baffle drains. C—Collector drain.

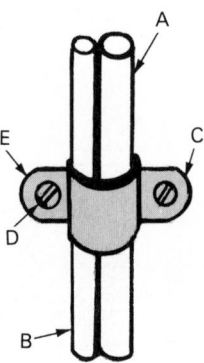

Fig. 14-7. A method of fastening tubing to the wall. A—Suction line. B—Liquid line. C—Clamp. D—Wood screw. E—Plastic tape.

conduit clamp, Fig. 14-7, is sufficient for most situations. The tubing should be insulated or protected from these clamps. Short wrappings of plastic tape will prevent chafing and galvanic action (an erosion or eating away of material caused by two different metals touching in moist air).

Where tubing runs through a floor or wall, it should be protected by short runs of conduit or flexible metal tubing (Greenfield). The ends should be sealed with a sealing compound to avoid chafing and other troubles. See Fig. 14-8. In all cases, tubing should be run horizontally and vertically with neat bends.

The liquid line presents no difficulties regarding slope and position. Suction lines must drain toward the compressor. Low spots in the suction lines will accumulate return oil which may eventually form a liquid slug in the tubing. Such a slug carried to the compressor will produce disturbance in the crankcase and may cause temporary oil pumping.

If tubing must slant upward from the evaporators to the condensing unit, the run should be made with a steady downward slant of the tubing to a certain point. Then a short U-bend (two street ells) should be made. The U-bend will act as an oil trap, ensuring a positive return of oil to the compressor.

Never run tubing near sources of heat, such as hot water lines, steam lines, or furnaces. Heat will reduce efficiency.

Copper tubing usually comes in 50-ft. coils. It is dehydrated and sealed at the ends by the manufacturer.

In the average small commercial installation, 1/4-in. tubing is used for the liquid line and 1/2-in. for the suction line. See Chapter 15 for correct line sizes.

During installation, tubing should be kept clean. Never put it aside with ends open if it is not to be used for a period of five minutes or more. A practical method of installing tubing is to uncoil about 10 ft. of it at a time, unrolling the coil along the floor. Then run the tubing up through floor openings from underneath, gradually working it into place.

Quarter-inch tubing should not be difficult to install. But 1/2-in. diameter must be carefully handled to prevent buckling when it is bent. A tube bender should be used. See Chapter 2.

In a noncode installation where individual suction lines and liquid lines run to main lines, T-fittings may be used and the valves for shutting off the individual evaporators may be located near the evaporator.

Valves, driers, or other heavy objects should not be supported by tubing. These items should be mounted on the wall or some other support. These connections may be of the SAE flare type or of streamline soldered (sweat) fittings.

Newly installed tubing should be sealed immediately after flare or streamline connections are made to keep it clean. The tubing should be attached permanently to the supports along which it runs.

Always try to run tubing so that supports will protect the tubing from accidents. Many service technicians use a sponge rubber covering over the tubing which serves both as a protection and as insulation, Fig. 14-9. This covering

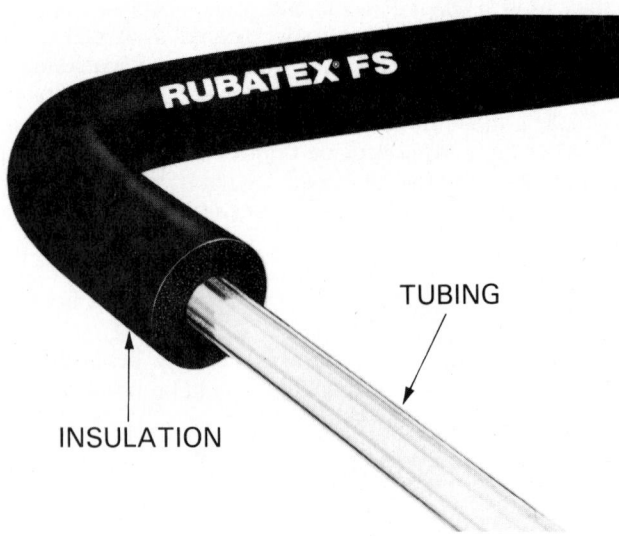

Fig. 14-9. Rubber insulation mounted on suction line. Insulation is usually installed before tubing or pipe connections are made. (Rubatex Corp.)

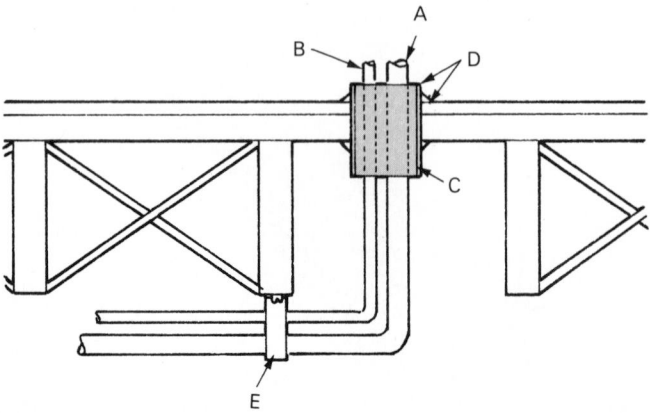

Fig. 14-8. Use a sleeve to protect tubing where it goes through a wall or floor. A—Suction line. B—Liquid line. C—Sleeve. D-Sealing compound. E—Clamps.

must be placed on the tubing before assembly unless the insulation is split.

At tubing connections or where valves have been installed, it is common practice to cover them with insulating tape, as in Fig. 14-10. Be careful to place the tubing where it will not be damaged by handling of articles in the room. Do not put loops or unsupported bends in the tubing except at the condensing unit.

Fig. 14-10. Foamed plastic insulating tape is wrapped around valve to prevent sweating or frosting. (Rubatex Corp.)

Install a drier and sight glass in the liquid line at the condensing unit. A vibration eliminator, as shown in Fig. 14-11, should be included in both lines. One horizontal loop of soft copper tube may be made in the suction and liquid lines if a suitable vibration eliminator is not available. Carefully study the manufacturer's installation and service procedures to make sure that every assembly and adjustment is correct.

14-6 ELECTRICAL CONNECTIONS

Refrigeration units must have enough electrical power of the correct type to operate motor, controls, and solenoids. Chapters 6, 7, and 8 explain electrical fundamentals and electrical power variables.

Before installing a condensing unit, be sure the electrical voltage, cycle, and phase of the compressor motor are the same as the electrical power source. Smaller units—1/8 to 1/4 hp—usually use 120 V single-phase ac. Medium size units—1/4 to 1 hp—may use 208 V or 240 V single-phase or three-phase ac. Larger units may use 240 V or 440 V three-phase ac.

Fans—both evaporator and condenser—must be checked to be sure the identification plate data matches power available. This is true as well for solenoid valves and controls.

Wire size is important. *Wire should have a capacity 50 percent over the load it will carry.*

Condensing units have wiring diagrams either fastened

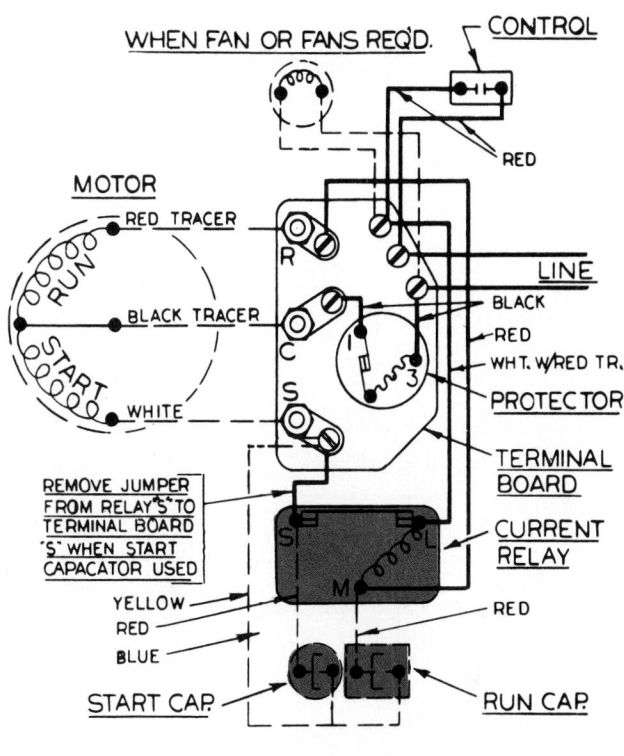

2 TERMINAL PROTECTOR-SINGLE POLE LINE CONTROL CURRENT RELAY

WHEN FAN OR FANS REQ'D.

Fig. 14-12. Wiring diagram of single-phase unit using current relay, run capacitor, and start capacitor. (Copeland Corp.)

to them or supplied in the shipping crate. Fig. 14-12 shows a typical wiring diagram for a single-phase circuit. Fig. 14-13 shows a three-phase refrigeration system wiring diagram. Correct connections are extremely important. Do not turn on the electrical power until the circuits are correct and all connectors are clean and tight.

Overloaded electrical circuits are dangerous. They may cause unit burnouts or electrical wiring fires.

A complete condensing unit installation is shown in Fig. 14-14. Use of an ohmmeter is strongly recommended to check all circuits for continuity before turning on the power.

14-7 GAUGE MANIFOLD

A big help to service technicians is the service gauge and testing manifold. With this piece of equipment, they are able to check low- and high-side pressures, charge and discharge a system, check pressures, add oil, bypass the compressor, unload gauge lines of high-pressure liquid and vapor, as well as perform many other operations without replacing regular gauges. See Chapter 11 for gauge manifold information.

A typical manifold, as shown in Fig. 14-15, has two gauge openings, three line connections, and two shutoff valves that separate the outside connections from the center line connection. Fig. 14-16 shows a compound gauge having a pressure scale and three different refrigerant temperature scales (for R-12, R-22, and R-502).

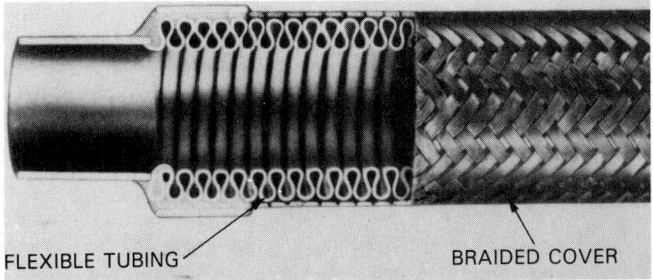

FLEXIBLE TUBING   BRAIDED COVER

Fig. 14-11. Flexible tubing is used for vibration damping. One of these is usually mounted in suction line near condensing unit. (Anamet, Inc., Anaconda Metal Hose)

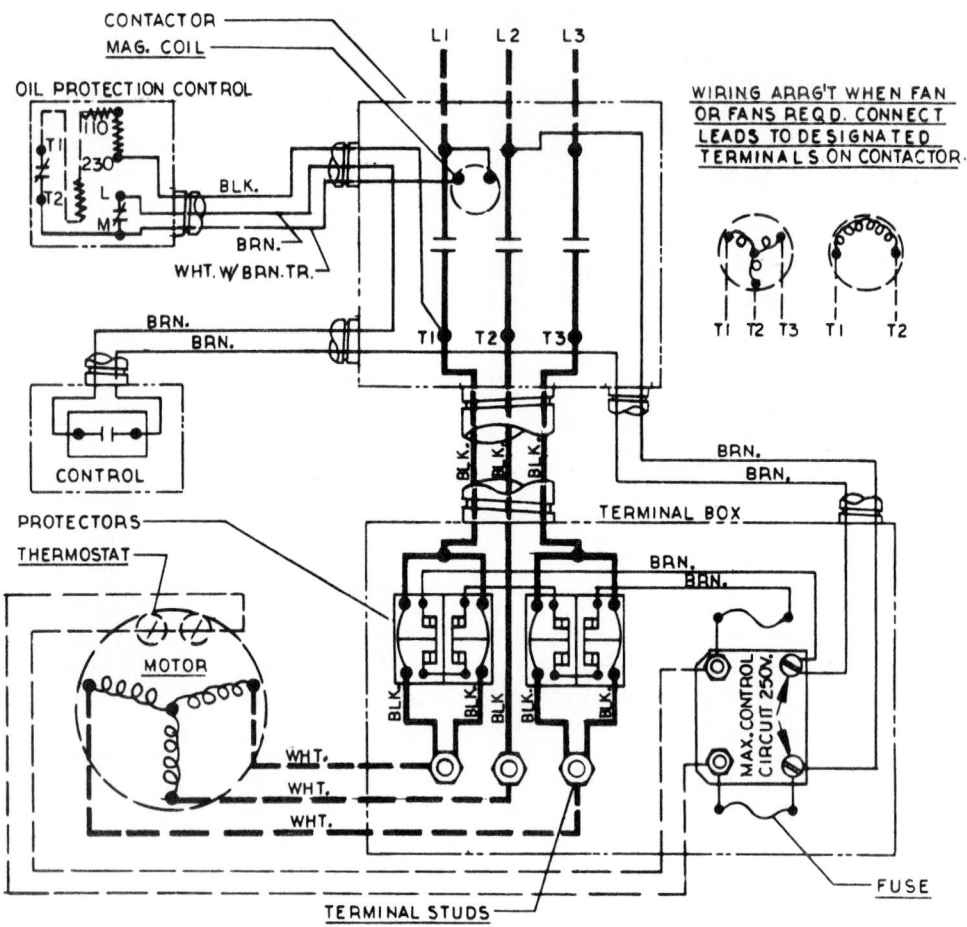

Fig. 14-13. Wiring diagram of three-phase motor refrigeration unit. Note contactor, control, oil protection control, and motor circuit protectors.

The manifold has 1/4-in. square drive valve stems or is equipped with handwheels. The three line attachment fittings are usually 1/4-in. MF (male flare).

Fig. 14-14. Commercial unit with dual compressors. A—Vibration eliminators. B—Sight glass. C—Filter-drier. (Anamet, Inc., Anaconda Metal Hose)

In Fig. 14-15, a 1/4-in. copper tubing or a flexible line connects the manifold to the SSV (suction service valve) shown at D and the DSV (discharge service valve) at C. Because most service valves on the compressor have 1/8-in. FP (female pipe) gauge openings, two 1/8-in. MP (male pipe) by 1/4-in. MF (male flare) half unions are installed in the service valves. Be sure the compressor service valve stems are turned all the way out and that the outside of the valve is cleaned, before removing the pipe line plugs and installing the fittings.

Lines from the manifold are attached to these fittings. The line attached to the SSV at D should be left one to two turns loose while the line to the DSV should be tightened. Then open both the manifold valves at A and B 1/4 to 1/2 turn and cap the middle opening, E.

Now turn the DSV C stem in 1/8 to 1/4 turn for just a moment (crack the valve). A surge of high-pressure refrigerant will then rush through the lines and the manifold and purge to the atmosphere at the loose connection at D, the SSV. This connection may then be tightened. Purging is necessary to remove air and moisture from the manifold and lines.

Carefully test for leaks while the manifold and its lines are under high pressure. Correct any leak immediately.

Various service and testing may be performed after the testing manifold has been installed.

1. Observe operating pressures by:
   Closing valve A by turning all the way in.
   Closing valve B by turning all the way in.

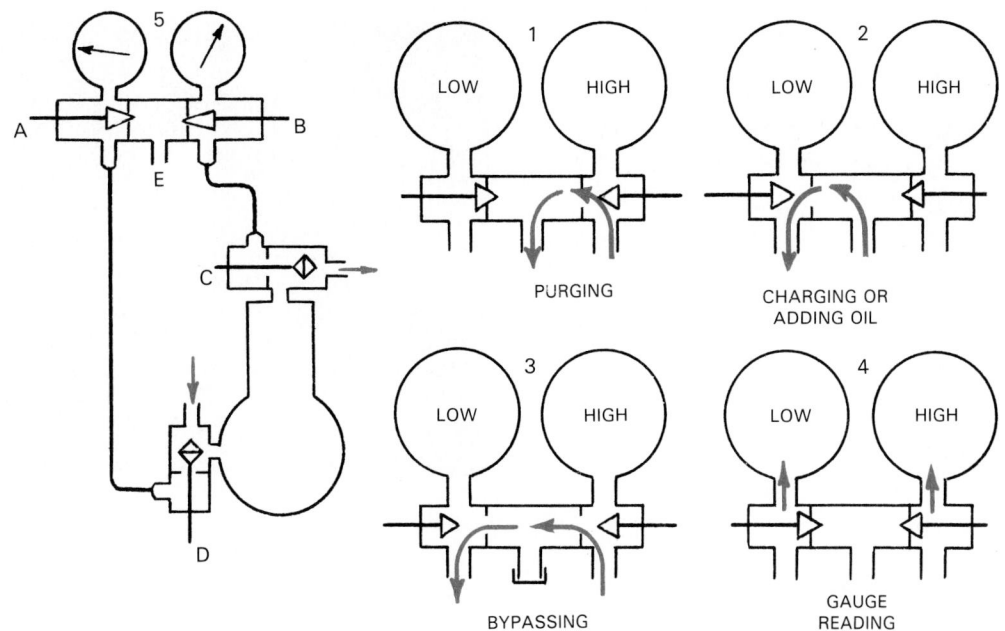

Fig. 14-15. Schematic of gauge manifold installation on external drive compressor with service valves. A—Manifold suction valve. B—Manifold discharge valve. C—Compressor discharge service valve. D—Compressor suction service valve. E—Service opening. 1—Purging. 2—Charging or adding oil. 3—Bypassing. 4—Gauge reading. 5—Both manifold valves are turned all the way in. System is pumping vapor and both low- and high-side pressures are being read.

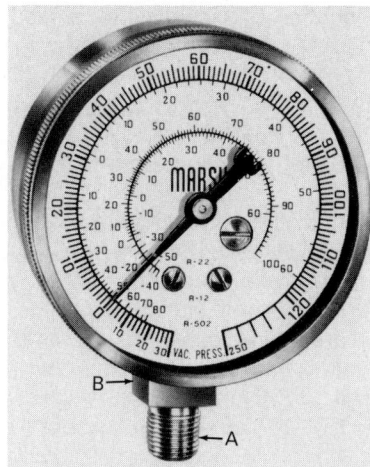

Fig. 14-16. A compound gauge 30 in.—0—120 psi with a retarded scale to 250 psi. Notice three refrigerant temperature scales. These are for R-12, R-22, and R-502. A—Connection is 1/8 in. male pipe thread. B—Wrench flats. (Marsh Instrument Co.)

Cracking open back seat of valve C.
Cracking open back seat of valve D.
2. Charge refrigerant into system by:
Connecting refrigerant cylinder to E (vapor only).
Opening valve A.
Closing valve B.
Closing front seat of valve D slowly.
3. Purge condenser by:
Closing valve A.
Opening valve B.
Cracking open valve C.
4. Charge liquid refrigerant into high side by:
Connecting refrigerant drum to E.

Closing valve A.
Opening valve B.
Mid-positioning valve C.
5. Build up pressure in low side for control setting or to test for leaks by:
Sealing E with seal cap.
Opening valve A.
Opening valve B.
Back seating then cracking open valve C.
Mid-positioning valve D.
6. Charge oil into compressor by:
Connecting oil supply to E.
Opening valve A.
Closing valve B.
Turning valve D all the way in.

After completing service operations, the manifold is removed from the system. This must be done without losing refrigerant or admitting air. Turn the DSV at C all the way out. Then open both manifold valves A and B 1/4 to 1/2 turn. This arrangement will move all the high-pressure refrigerant from the line and the high-pressure gauge and put it into the low side. Now turn the SSV stem at D all the way out and turn both manifold valve stems all the way in. Remove the lines from the service valve. Use soft synthetic fittings for finger tight connections. See Fig. 14-17. Remove the fittings from the service valves, install the service valve gauge opening plugs, and tighten them. Immediately plug the lines and all other openings on the manifold to keep out dirt, moisture, and air. See Fig. 14-18.

14-8 SERVICE VALVES

In systems with service valves, the valves must be leakproof where the valve stem goes into the valve. The packing is made of lead, graphite, and other materials. Packings differ with different valve designs. When replacing them,

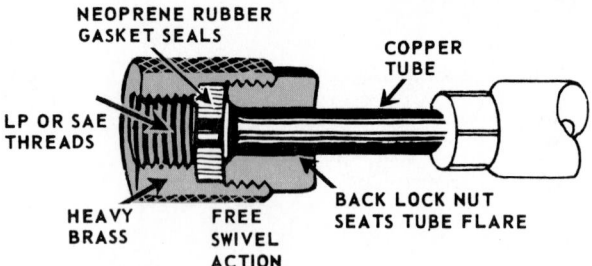

Fig. 14-17. Speed coupling used on charging and purging lines. Synthetic rubber gasket produces leakproof joint when connection is finger tightened.

Fig. 14-18. Gauge manifold set, equipped with three flare plug hose holders. These allow the flexible lines to be attached when not in use. This method is used to keep service lines clean. (Robinair Division, SPX Corporation)

one must be sure the proper packing is used.

A replacement service valve is shown in Fig. 14-19. It is a four-bolt mounting style service valve. Surfaces are slightly recessed for ease of bolt tightening.

Practically all service valves have a drop-forged brass body and a steel stem. Stems have a tendency to rust and score the valve gland or packing. Always clean and oil a valve stem before turning it. See Fig. 14-20.

A good way to lessen this corrosion, especially in damp locations, is to fill the valve body with clean and dry refrigerant oil before replacing the plug, each time the service valve is used. This oil should be the specified refrigerant oil for that machine.

Service valves on commercial installations must be kept in good condition. The technician can do three things to assure good service and valve life:

1. Fit the wrench to the valve stem.
2. Maintain the packing so that the service valve will not leak.
3. Oil the threads of the gauge connections each time gauges are used.

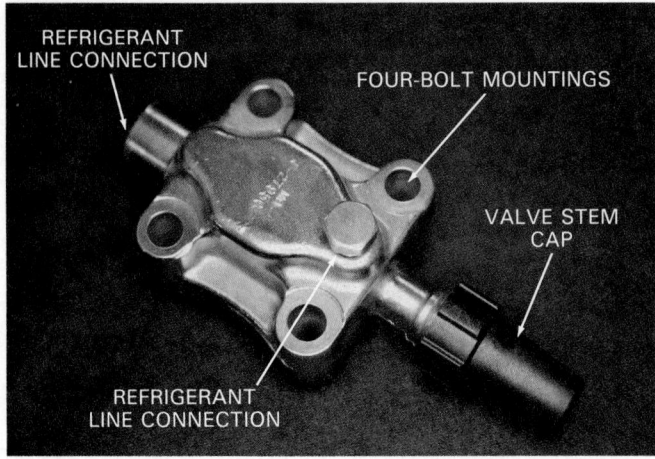

Fig. 14-19. Compressor service valve, four-bolt style. Note refrigerant line connection. (Mueller Brass Company)

Occasionally, after a period of use, these service valves must be replaced. After flexible line fittings have been mounted in the gauge opening of the valve a number of times, the pipe threads in the valve gauge openings may become worn and leak. If the fittings inserted in these gauge openings are given a thin coat of solder, this trouble can often be eliminated.

When cracking the valve, always use a fixed wrench (not a ratchet wrench). This is done so the valve may be quickly closed again if necessary.

Occasionally a service valve will be found in such bad condition as to be useless. In such cases remove the refrigerant or isolate it in another part of the system and replace the valve.

The third valve of the external drive system is the liquid receiver service (LRS) valve. See Fig. 14-21. Some of the LRS valves are three-way to enable the service technician to charge liquid refrigerant into the system.

When using a system service valve, remove the valve cap—if the valve has one—and loosen the service valve packing nut one turn. Next, clean the valve stem before turning it. Turn the valve stem back in about 1/16 of a turn.

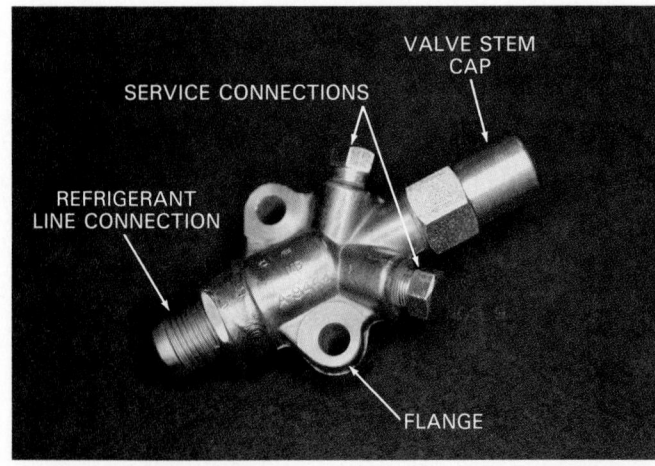

Fig. 14-20. Refrigeration unit service valve with two service openings, two-bolt flange, and valve stem cap. (Mueller Brass Company)

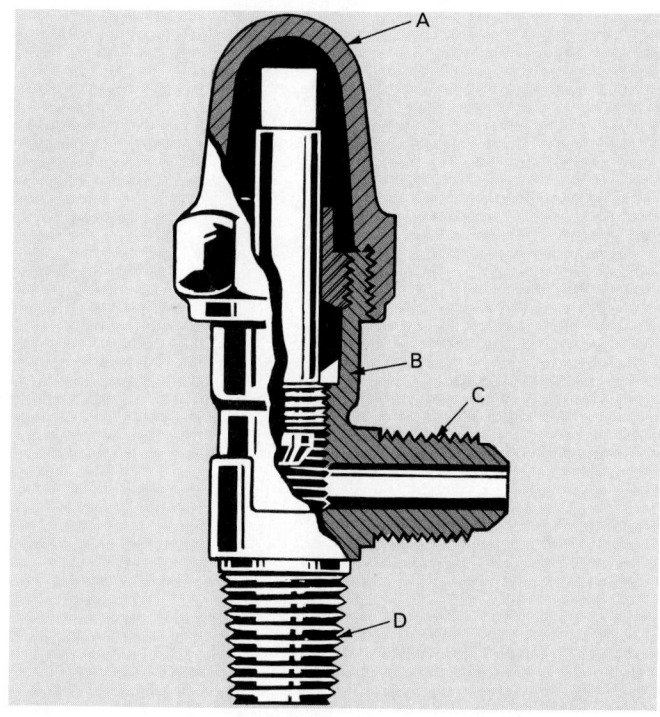

Fig. 14-21. A liquid receiver service valve. A—Cap. B—Packing. C—Connection for liquid line (SAE flare). D—Connection to liquid receiver (pipe threads).
(Superior Valve Company, Division of AMCAST Industrial Corporation)

Tighten the packing nut and replace the valve stem cap.

The purpose of turning the valve stem back in slightly is to prevent the valve from "freezing" against its seat. Such a condition sometimes leads to broken valve stems.

When installing the gauge opening plug, tighten the plug firmly. Never tighten a cold gauge plug into a hot service valve. This may result in freezing of the plug to its seat. When using a service valve wrench on these valves, apply the turning force gradually. Adjustable end or fixed open end wrenches are not recommended for service valve stems. *Only the special socket wrenches—sometimes called keys—are to be used.*

If the gauge plug is "frozen" in the service valve, it can be loosened by first heating the outside of the service valve body with flame from a torch. *Be careful not to overheat!* This heating will cause the valve body to expand and will weaken the body thread grip on the plug. The wrench can then be used to loosen the valve stem.

Access valves are often used on commercial refrigerating systems. They may be mounted at the evaporator outlet or the liquid line inlets just ahead (downstream) of the refrigerant control. They may also be mounted on both sides of the automatic valves in the system (solenoid, bypass valves, hot gas defrosting valves, and driers). The old policy of having a minimum of connections has given way to the need for more convenient service.

14-9 TESTING FOR LEAKS

After a system is assembled, it must be checked for leaks. Before trying to locate leaks, build up a pressure in all parts of the system. Two methods may be used:
1. Using an inert gas.
2. Using refrigerant under pressure.

In case a low-pressure refrigerant is used, or if the local code specifies a pressure test above the refrigerant's vapor pressure, some other gas may be used for testing. Carbon dioxide, nitrogen, or argon are satisfactory. However, the pressure may be dangerous. See Para. 11-39.

Caution: Never use oxygen, air, or any flammable gas for this purpose. An explosion may occur.

This testing should include the liquid line, suction line, and all other parts installed by the installer. The only exception is the condensing unit itself. It has been pressure tested at the factory. Install a high-pressure gauge only. (A compound gauge may be ruined by the pressure.) After building up a medium (30 to 100 psi or 311 to 794 kPa) pressure, close the cylinder valve.

If the pressure gauge shows no drop in pressure after an hour or more, raise the test pressure to 170 psi (1 277 kPa) and test again. Do not exceed the pressures prescribed by the code, as a too high pressure may rupture some part of the system. If pressure shows no decrease during a 1- to 24-hour period, the system is safe to operate.

Purge the test gas from the system, evacuate by the deep vacuum, two- or three-step vacuum method, and charge the system. The unit should be ready to operate.

Testing for leaks using the system's own refrigerant is the most common noncode practice. It is convenient; there is no need for an inert gas cylinder and the leak testers are a standard part of each service technician's tool kit.

To do this, proceed as follows.

Install a pressure gauge in the system. The liquid line valve should be opened just enough to build up a 15 to 30 psi (207 to 311 kPa) pressure throughout the system.

Test for leaks using one or more of the following:
1. Soapsuds.
2. Halide torch.
3. Electronic leak detector.
4. Liquid tracer.

Chapter 11 describes these tests. A combination halide torch leak detector, soldering and brazing unit is shown in Fig. 14-22. This apparatus is both an oxygen-acetylene and an air-acetylene unit which can be used for welding, brazing, soldering, or leak testing. The acetylene cylinder has a halide leak testing device connected to it.

If no leaks are detected at low pressure, increase system to full pressure of refrigerant (vapor only) and test again.

If a leak is found, purge the system to atmospheric pressure, open the system at the leak point, inspect all parts, replace any defective parts, clean, and assemble. If a soldered or brazed joint is leaking, flux, heat, and take the joint completely apart. Clean and assemble, then repeat the leak detecting procedure. If no leaks are found, this part of the unit is ready to operate.

When blowing out lines and/or pressure testing with nitrogen or carbon dioxide, use an accurate pressure regulator and a relief valve designed to open at 180 psi (1 346 kPa). One should not exceed 170 psi (1 277 kPa) for CO_2 or nitrogen, while testing. See Chapter 11 for safe use of high-pressure gases.

Some motor compressor domes are designed to operate under low-side pressure. Tecumseh Products Company warns "Do not exceed 170 psi! Dome may burst!"

14-10 EVACUATING SYSTEM

After it is known that the system is leakproof, remove all air and moisture from the system. Air is pumped out of

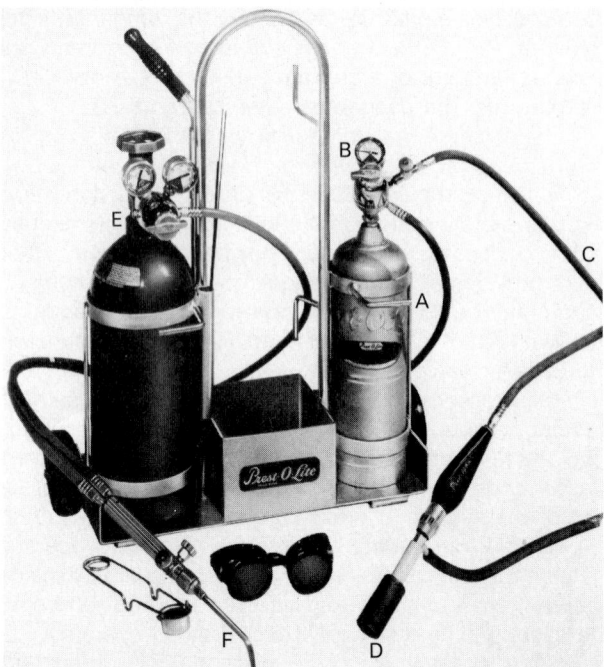

Fig. 14-22. Combination halide leak detector and soldering-brazing unit. A—Acetylene cylinder. B—Regulator. C—Hose. D—Halide leak detector used for checking leaks of most refrigerants. Flame will turn blue-green if there is a leak. E—Oxygen cylinder. F—Oxyacetylene welding and brazing torch.

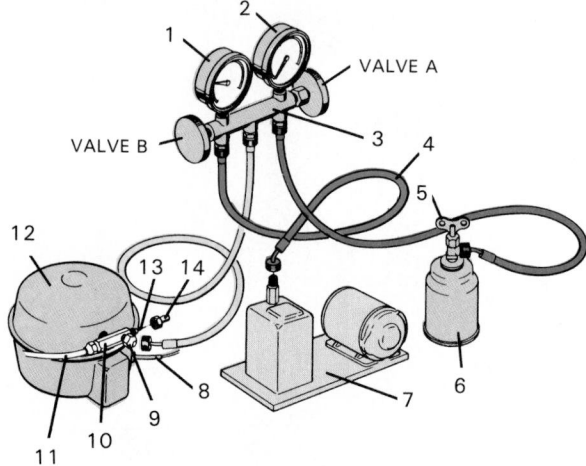

1. COMPOUND GAUGE
2. PRESSURE GAUGE
3. TEST MANIFOLD
4. CHARGING LINES
5. REFRIGERANT CYLINDER VALVE
6. REFRIGERANT CYLINDER
7. VACUUM PUMP
8. COMPRESSOR DISCHARGE LINES
9. SUCTION LINE VALVE SERVICE PORT
10. SUCTION LINE VALVE
11. SUCTION LINE
12. FREEZER COMPRESSOR
13. SUCTION LINE VALVE STEM
14. VALVE SEAL CAP

Fig. 14-23. Gauge manifold, vacuum pump, and refrigerant cylinder connected to small commercial hermetic motor compressor.

the lines and the evaporator with a vacuum pump.

Avoid pumping refrigerant vapor into the room where the condensing unit is located. Refrigerant vapors may be harmful to people in the room and will interfere with leak detecting. Follow E.P.A. regulations.

Connect a gauge manifold. Open both the discharge service valve and the suction service valve and pump a vacuum on the complete system. Air being removed will be discharged through the vacuum pump.

Fig. 14-23 shows a gauge manifold and a vacuum pump connected to a small commercial hermetic system. See Chapter 11 for evacuating methods.

A 3 cu. ft./min. vacuum pump is large enough for systems up to 10 hp. *Pressure drop is very important! The service lines must be as large and as short as possible.*

It takes eight times longer with a 1/4-in. line as it does with a 1/2-in. line, and it takes twice as long through a 6-ft. line as through a 3-ft. line.

Use heat lamps, electric heaters, and blowers to provide heat. Avoid using a torch flame because it may cause local high temperatures which may decompose oil, insulation, and refrigerant.

When the pump is shut off (valve closed), pressure will rise a little at once (pressure drop equalizing). Take a reading one minute after closing valve and again, 30 minutes later. If there is no pressure rise, it indicates that the system is sealed and is also free of moisture.

14-11 CHECKING SYSTEM BEFORE STARTING

If the motor control has not already been adjusted for installation, this should be done before the system is put in operation. The settings of the motor controls will vary with the demands of the cabinets and with various kinds

of refrigerant used. See Chapter 8.

Be sure the water is turned on—if it is a water-cooled condensing unit—and check fuses in the electric circuit for proper size.

It is good practice to install recording thermometers, voltmeters, and ammeters on a unit for the first 24 to 48 hours of its operation. Records will make adjustments easier.

14-12 CHARGING COMMERCIAL SYSTEMS

When charging a system, refer to the manufacturer's directions if they are available. The manufacturer has designed and tested the products under various operating conditions and has developed specific charging procedures. In general, there are two basic methods used to charge a system:
1. Low-side method.
2. High-side method.

In the low-side method, charging small quantities of refrigerant into commercial systems is similar to charging domestic machines. It is usually done by charging into the low side (vapor method).

To charge a commercial external drive system equipped with service valves, the storage cylinder should be attached to the gauge manifold, Fig. 14-24. Evacuating and charging apparatus combinations are popular with service technicians. Fig. 14-25 illustrates a combination vacuum pump and charging unit equipped with charging cylinder, gauge manifold, vacuum pump, and vacuum gauge.

Charging lines must be clean and purged to rid them of air and moisture. Connections must be tested for leaks prior to the actual charging operation. *Remember to wear goggles*

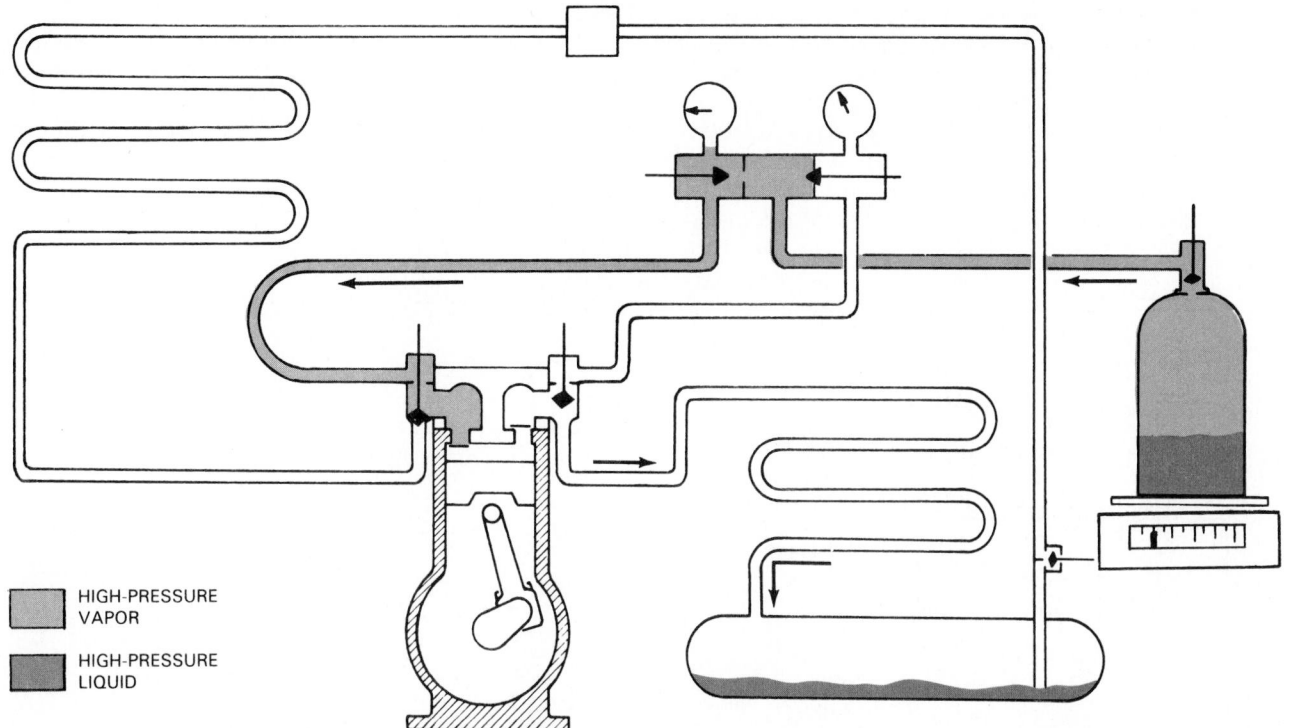

HIGH-PRESSURE
VAPOR

HIGH-PRESSURE
LIQUID

Fig. 14-24. Method of charging small external drive system with refrigerant vapor. Refrigerant cylinder is connected to manifold center opening. After purging charging lines, the SSV is turned almost all the way in. Unit is started and cylinder valve is opened just enough to keep low-side pressure within normal operation safe limits. Scale indicates amount of refrigerant being put into system.

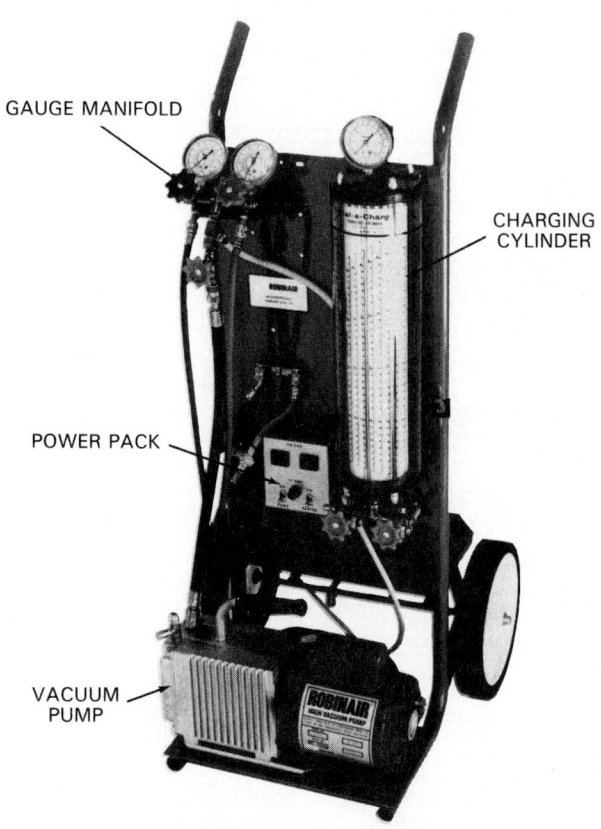

GAUGE MANIFOLD

CHARGING
CYLINDER

POWER PACK

VACUUM
PUMP

Fig. 14-25. Combination vacuum pump and charging unit.
(Robinar Division, SPX Corporation)

when transferring refrigerants.

In the low-side method, the principle of operation is to use the service cylinder as a temporary evaporator in the system. As the compressor runs, it will remove refrigerant vapor from the cylinder as well as from the evaporator.

Charging may be speeded up by partly closing the suction service valve to reduce flow from the regular evaporators and speed the evaporation from the service cylinder. Hot water may be applied to the service cylinder to help speed the evaporation. Never use a torch to warm a cylinder. The low-side pressure should be kept at normal levels. Too high a pressure may overwork the compressor. Pressures which are too low may cause oil pumping.

The low-side method insures clean refrigerant due to the distilling action during evaporation of the refrigerant. *A service technician must be present at all times during the charging. A service cylinder must not be left connected into a system.*

It is very important that liquid refrigerant not be allowed to reach the compressor. The liquid is not compressible—and the compressor valves, and even the bearings and rods, may be ruined if the compressor should pump liquid.

Although it is not usually recommended, some service technicians do put liquid refrigerant into the high-pressure side of the system. The compressor should not run while this charging is being done. Larger systems are equipped with a liquid charging valve on the receiver. This is a dangerous practice because dynamic hydraulic pressures are possible, which may rupture the lines, causing considerable damage. However, this method can be used to put the initial charge into a system, if done very carefully. If one inverts a cylinder and it has a higher pressure than the system, liquid refrigerant will be forced into the system.

One reason this practice is discouraged is that if the compressor exhaust valve is leaking, liquid may enter the cylinder and damage the compressor when started.

If the unit is water-cooled, the pressure in the liquid receiver, with the water flowing, will be sufficiently below that of the pressure in the cylinder to permit opening of the two valves after the charging line has been purged. The pressure difference will force refrigerant from the cylinder into the system.

If the unit is air-cooled, pressure in the refrigerant drum must be increased. This may be done by using the compressor to pump vaporized refrigerant into the cylinder, increasing its pressure. In detail, this method is as follows:

1. Connect the refrigerant cylinder to the gauge manifold with a flexible charging line, as shown in Fig. 14-26. Never use a disposable container here. It may explode!

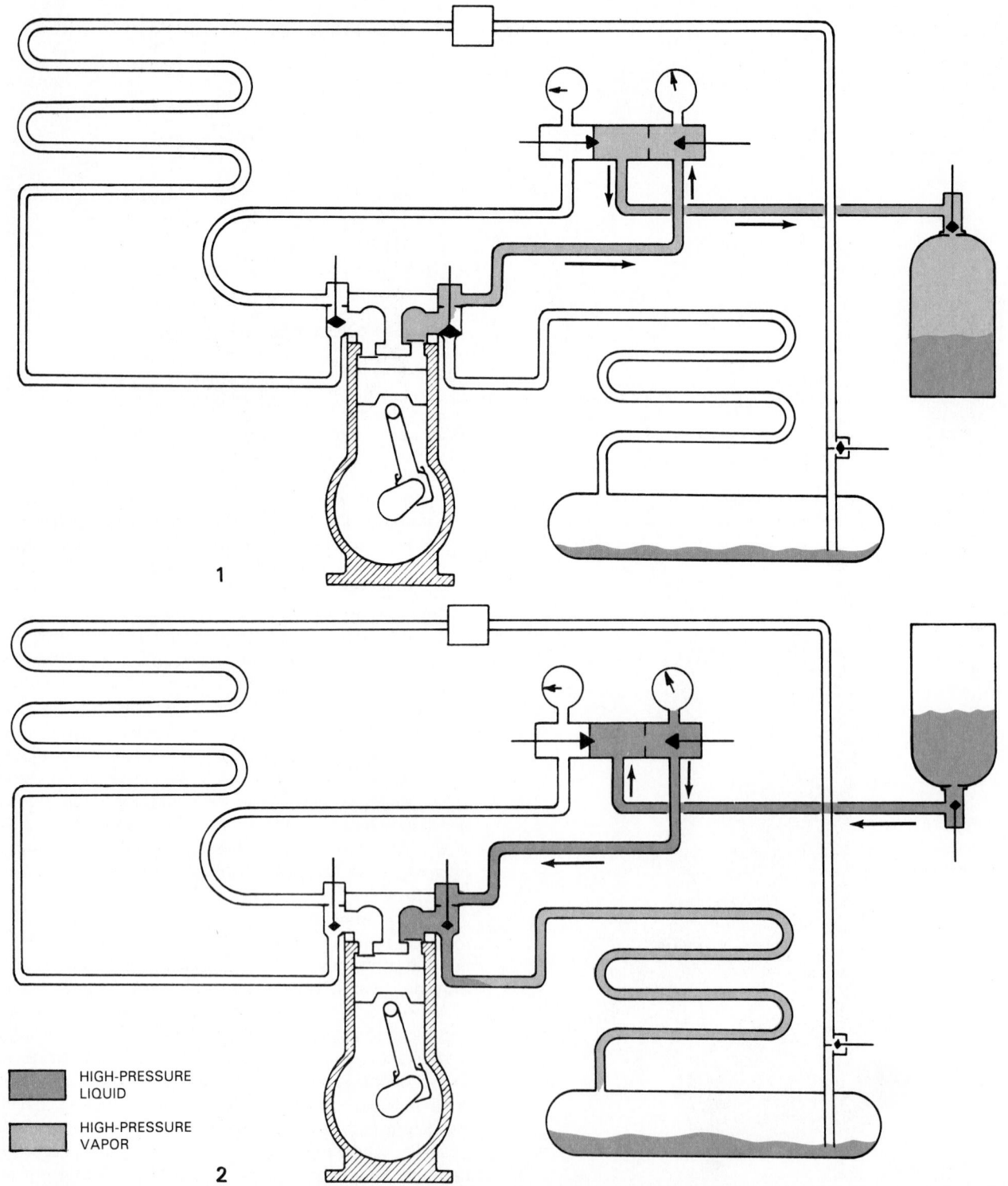

HIGH-PRESSURE
LIQUID

HIGH-PRESSURE
VAPOR

Fig. 14-26. Charging small external drive system through high-pressure side. In part 1, above, compressor is used to build up slightly higher than condensing pressure in cylinder. Then cylinder is inverted and DSV turned out a turn or two as in part 2, below. Small pressure difference will force cylinder refrigerant liquid into condenser receiver.

2. Run the compressor for a few revolutions with the discharge service valve turned all the way in until a pressure of 20 to 30 psi (242 to 311 kPa) above the condenser pressure is built up in the cylinder.
3. Stop the compressor.
4. Invert the refrigerant cylinder. (Be careful not to injure the line.)
5. Turn the discharge service valve part way out.

High pressure on the surface of the refrigerant in the cylinder will force liquid into the system. While the liquid is flowing into the high side, a gurgling sound may be heard. If this sound stops abruptly, it means that the cylinder has been emptied. Use this method only if all the refrigerant has been removed from the system. Fig. 14-27 is a table of approximate refrigerant amounts which may be added to the system.

General information:
1. The method of checking refrigerant charge is to charge refrigerant vapor into the low side of the system until the evaporator has its normal amount. The liquid line is at room temperature, the bubbles cease at the sight glass, and there is no hissing sound at the refrigerant control valve. If one or two extra pounds are put in the unit, this will become a reserve and will be stored in the liquid receiver. The smaller the unit, the less the amount of refrigerant needed as a reserve.
2. Another method of charging liquid refrigerant into a system is to use the liquid receiver service valve—if this valve is a three-way valve. See Fig. 14-28. Connect the cylinder to the service opening of the valve, purge the line, turn the cylinder upside down, open the cylinder valve, and then turn the receiver service valve (king

valve) in several turns. The liquid refrigerant will flow into the receiver and into the liquid line. See Fig. 14-29.
3. A hermetic or serviceable system, not equipped with service valve, is charged as described in Chapter 11. Always remember that a running hermetic motor requires refrigerant vapor for cooling the windings or it will overheat. For this reason, it is best to charge these systems with vapor refrigerant into the low side.

14-13 STARTING A SYSTEM

One should follow a planned procedure when starting a new system or starting a system which has been shut down for a period of time. Avoid overloading electrical circuits, compressor, and motor.

Make a check first on the electrical characteristics of the power-in circuit.
1. Be sure the phase is correct.
2. Be sure the voltage is correct.
3. Be sure the power leads are large enough. The electric utility company will assist.
4. Connect a voltmeter and an ammeter into the circuit. Recording types are preferred.
5. Install the gauge manifold to check pressures.
6. If the condensing circuit is water-cooled, be sure the water circuit is turned on.

By using hand valves, load the compressor with a normal back pressure during start-up. Remember, it is just as hard on the compressor to have too low a suction pressure as too high (oil pumping).

Once the unit is started, check the electrical meters, the pressure gauges, and the water flow as soon as possible. Shut down the unit at the first sign of trouble. If possible, record the electrical characteristics, the pressures, and the temperature during the first 24 hours to one week. These charts will serve as a good future reference.

To start the system with the full load of all evaporators on may overload the compressor because the TEVs will open. Therefore, all evaporator liquid line hand shutoff valves should be closed and the condensing unit started. Immediately open just one of the liquid line manifold valves slowly. The low-side pressure should be only a little over the cut-in pressure. It is an important precaution in refrigeration never to overload the compressor even for a short time.

After one evaporator has been opened a few minutes, the evaporator has a chance to cool down somewhat. This tends to make the expansion valve choke off the refrigerant flow. It also gives the compressor a chance to gradually reduce its load. The other evaporators may then be brought into service in the same way—one at a time.

Another way is to almost close the SSV all the way in and then gradually turn it out, keeping the compressor low side at normal pressure.

After starting the unit, the service technician should check high- and low-side pressures, amount of water flow in case it is a water-cooled system, and operation of each individual expansion valve. The technician should also determine if the TEV adjustment is correct for each evaporator. That is, frost or sweating on the suction line will indicate whether or not the expansion valve is opened too far or not far enough. Another important procedure or routine to be followed at this time is to determine whether or not the system has enough refrigerant. This can be checked as described in Para. 14-30.

	R—12 Flooded	Dry	R—22 Flooded	Dry	R—502 Flooded	Dry
1/2 hp Unit	3	1 1/2	3	1 1/2	3	1 1/2
1 hp Unit	6	3	6	3	6	3
1 1/2 hp Unit	9	4 1/2	9	4 1/2	9	4 1/2
2 hp Unit	12	6	12	6	12	6

Fig. 14-27. Approximate pounds of refrigerant which may be safely added to system that is low on refrigerant.

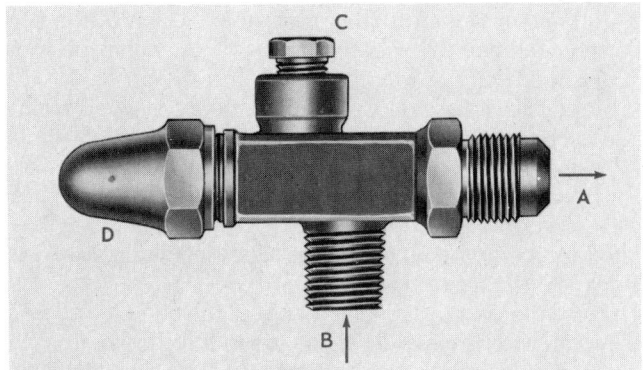

Fig. 14-28. Liquid receiver service valve with service connection. A—Connection to liquid line (flare). B—Connection to receiver (pipe threads). C—Service connection (1/8-in. pipe). D—Valve stem cap.

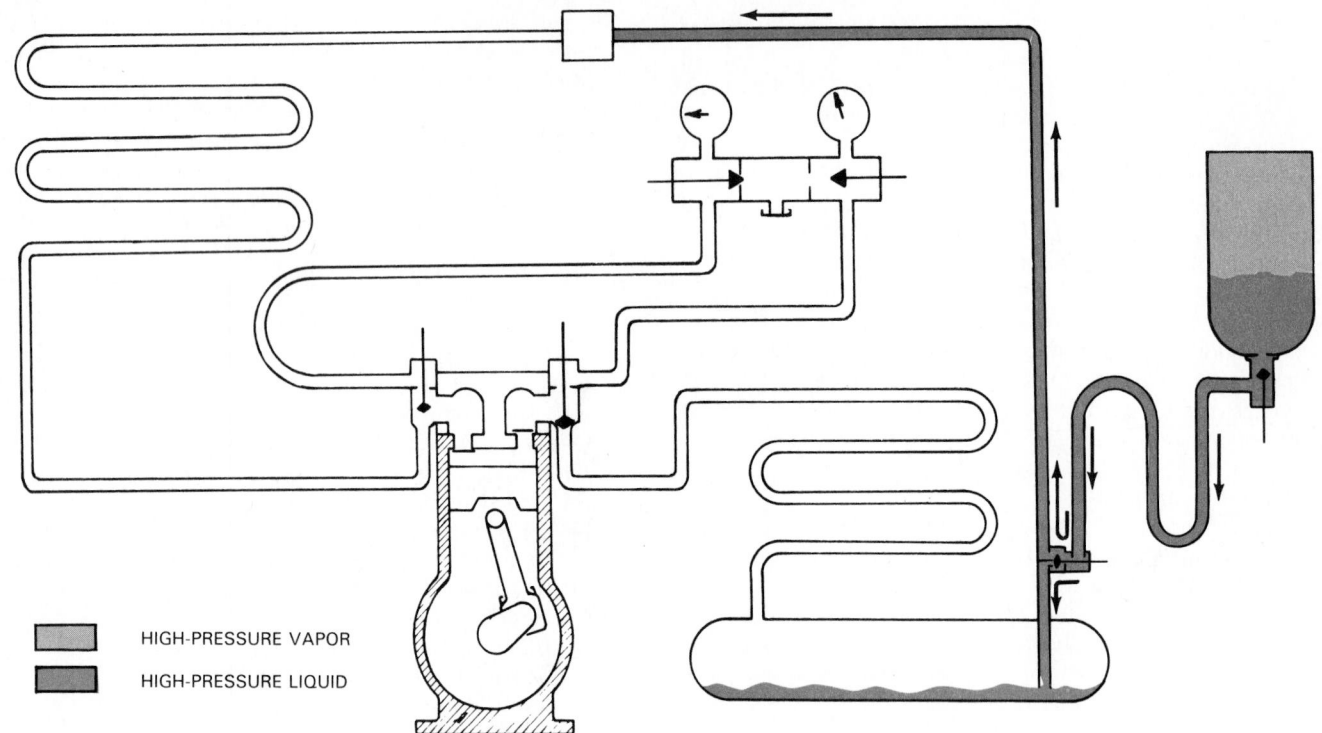

HIGH-PRESSURE VAPOR

HIGH-PRESSURE LIQUID

Fig. 14-29. Charging system through two-way liquid receiver service valve.

Test for leaks after the unit has operated for 24 hours. Maintain records for use during future maintenance or service operations.

14-14 CODE INSTALLATION

Most localities have definite codes or rules and regulations covering the installation of refrigerating equipment. Domestic systems and some other small capacity self-contained systems are usually not included because these units use only small quantities of refrigerant.

In commercial installations (where the units are assembled on the premises, or where the horsepower or refrigerant needs exceed certain set limits), there are requirements to insure uniform performance and a safe installation. Installation codes also protect the purchaser from careless installations.

The following are some of the high points of these codes:
1. Only licensed refrigeration contractors may install commercial equipment.
2. A permit must be obtained for each installation.
3. Each installation must be inspected by civic authorities.
4. Lines must be labeled as to refrigerant used.
5. Certain safety devices must be installed in the system.
6. The condensing unit must be installed in a safe place.
7. Electrical and plumbing work must conform to code and must be done by licensed electricians and plumbers.
8. Systems must be tested under certain pressures on both high side and low side, and must be free from leaks.

The national code and local codes are based on experiences with many installations. Codes provide safety for installer, owner, user, and public. They should be carefully followed.

14-15 INSTALLING A CONDENSING UNIT

Most codes require the condensing unit to be placed where it cannot be damaged. It should have a protective cage around it (small unit) or it should be placed in a separate room. The area should be well ventilated to permit escape of refrigerants should the unit develop a leak. Windows provide this for smaller units, while a forced exhaust is needed for larger units. Screen the openings to prevent insects and other objects from entering. Larger units must also be protected from fire damage by fire-resistant, self-closing doors. The condensing unit must be electrically grounded. All electrical and plumbing work should be done by a licensed contractor.

If the unit has motors and fans to cool the air-cooled condenser, the shipping blocks used to protect the motor and fan must be removed. Spin the fan by hand to be sure it runs freely.

To prevent violent rupturing or an explosion of the condensing unit due to excessive pressure, the code specifies:
1. High-pressure cut-outs to stop the motor.
2. Pressure relief valves or rupture disks to dissipate discharge slowly. These safety openings are piped outside by way of copper pipe connected by brazed joints. Spring-loaded safety valves and/or fuse plugs are used (where the unit may become overheated due to fire).

All refrigerant lines should be permanently labeled with signs identifying refrigerant.

Install the condensing unit where all the parts are accessible for maintenance and service. Keep them away from the sun and away from other heat sources such as steam pipes, hot air grilles, and ovens.

Water lines at the unit are either soft copper or flexible plastic pipe. Allow enough piping to permit some movement of the condensing unit. See Para. 14-84 about adding extra oil for long suction and liquid line runs.

Be careful! If a system has too much oil, it will pump liquid oil and be damaged.

14-16 INSTALLING AN EVAPORATOR

The code recommends limits of refrigerant for evaporators that would expose people to the refrigerant in case of leaks. On some installations, large tonnage evaporators installed in air ducts must be cooled with a brine rather than refrigerant.

The evaporator should be mounted firmly in the cabinet and protected to avoid damage to the system.

The evaporator should be leveled when installed. A spirit level may be used.

If the evaporator has motors and fans, remove the shipping blocks which protect them. Hand spin the fans to be sure they run freely. The evaporator should be electrically grounded, if it has a motor and fan. A commercial evaporator using a motor-driven fan is shown in Fig. 14-30.

14-17 INSTALLING REFRIGERANT PIPING

Code specifications require strong piping for refrigerant lines. These should be type K, the strongest, or type L. Piping should be protected by adequate guards. Some codes recommend that hard copper pipe, where it is exposed,

should have at least .065 in. wall thickness. Joints in the refrigerant piping must be placed so they can be easily inspected. The joints must be made with strong fittings and the brazing material used must be of excellent quality.

To avoid pinching or crimping the piping, the code recommends that piping always be supported by the building structure. Pipe should not be run across joists or studs where unsupported sections can be damaged. It is recommended that the piping be at least 7 1/2 ft. above floor level when run across a room. Fig. 14-31 shows a piping system for a roof-mounted refrigerating system.

The suction line should be mounted with a slight drop in horizontal runs toward the compressor. This provides for proper oil return.

It is important for noise control and for piping that experiences rapid temperature changes (defrosting hot gas lines) to install flexible sections in the pipe. See Fig. 14-32.

14-18 FITTINGS

Commercial system capacity has increased steadily during the past few years. This is especially true with comfort cooling installations in air-conditioning units and in supermarkets.

Sizes of the liquid and suction lines in these installations may be as large as 2 and 6 in. OD. Brazed joints with sweat fittings are used. These fittings, described in Chapter 2, are

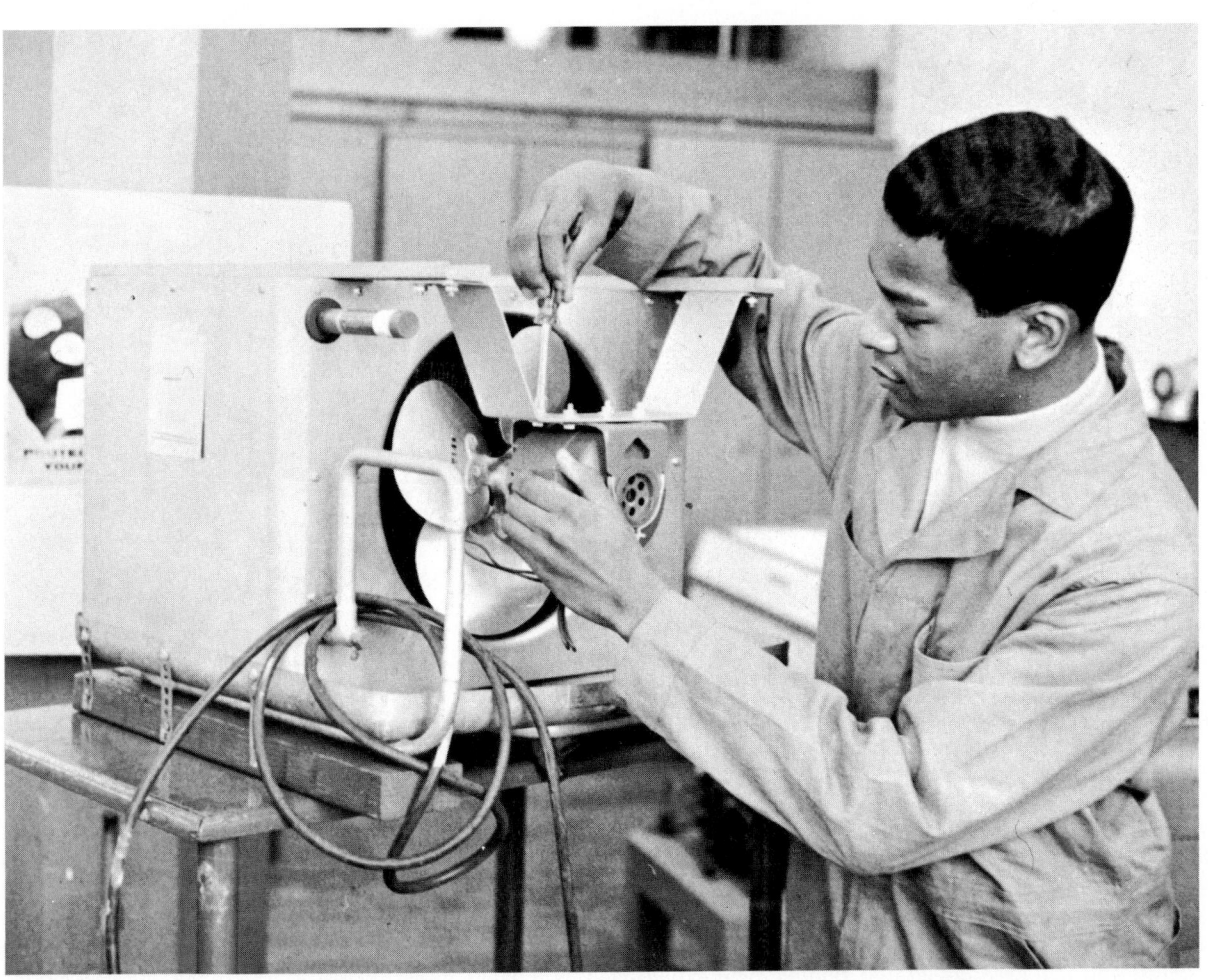

Fig. 14-30. Mounting motor fan on commercial evaporator. (Detroit Public Schools)

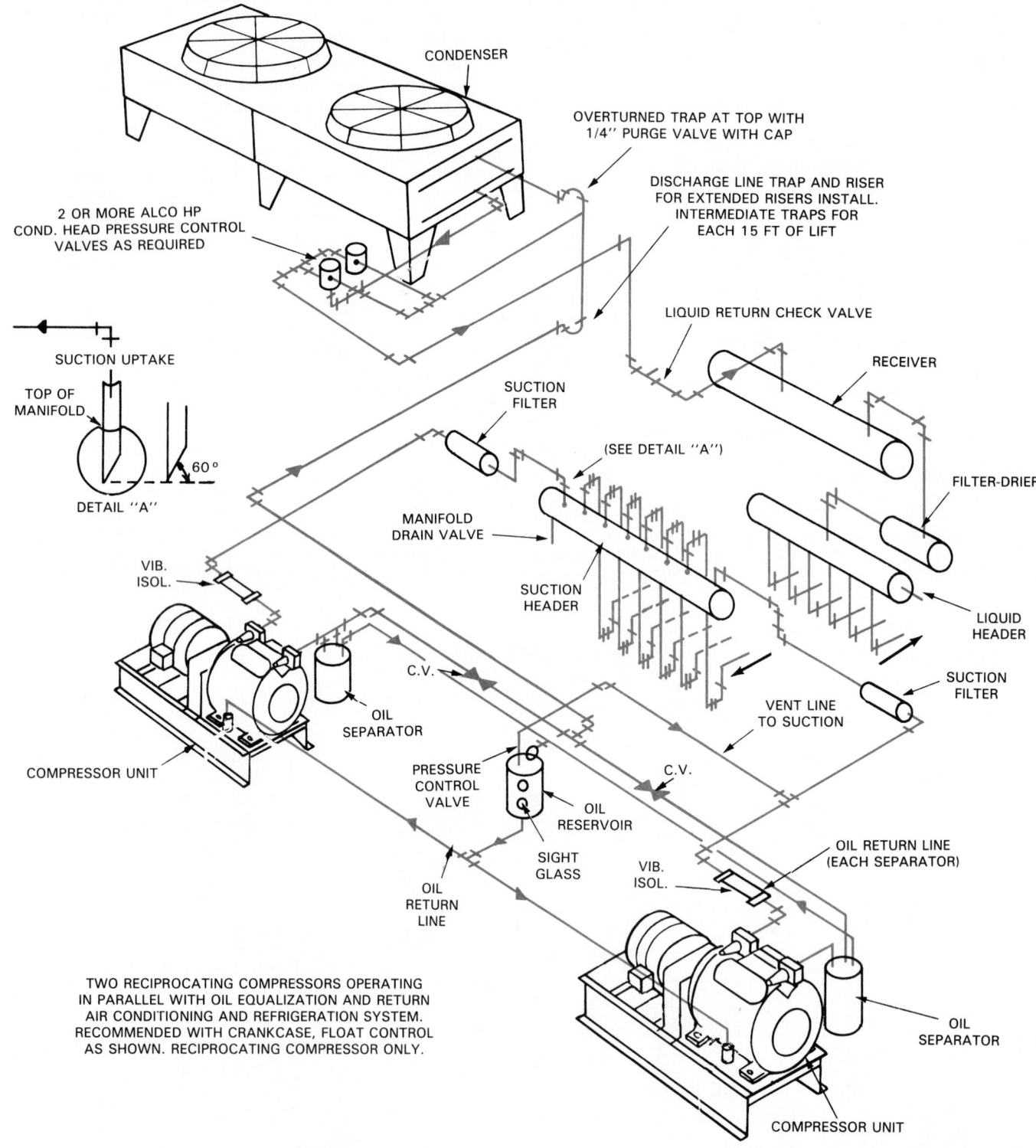

CONDENSER

OVERTURNED TRAP AT TOP WITH
1/4" PURGE VALVE WITH CAP

DISCHARGE LINE TRAP AND RISER
FOR EXTENDED RISERS INSTALL.
INTERMEDIATE TRAPS FOR
EACH 15 FT OF LIFT

2 OR MORE ALCO HP
COND. HEAD PRESSURE CONTROL
VALVES AS REQUIRED

LIQUID RETURN CHECK VALVE

RECEIVER

SUCTION UPTAKE

SUCTION
FILTER

(SEE DETAIL "A")

FILTER-DRIER

TOP OF
MANIFOLD

60°

DETAIL "A"

MANIFOLD
DRAIN VALVE

SUCTION
HEADER

LIQUID
HEADER

VIB.
ISOL.

SUCTION
FILTER

C.V.

VENT LINE
TO SUCTION

OIL
SEPARATOR

C.V.

COMPRESSOR UNIT

PRESSURE
CONTROL
VALVE

OIL
RESERVOIR

SIGHT
GLASS

VIB.
ISOL.

OIL RETURN LINE
(EACH SEPARATOR)

OIL
RETURN
LINE

TWO RECIPROCATING COMPRESSORS OPERATING
IN PARALLEL WITH OIL EQUALIZATION AND RETURN
AIR CONDITIONING AND REFRIGERATION SYSTEM.
RECOMMENDED WITH CRANKCASE, FLOAT CONTROL
AS SHOWN. RECIPROCATING COMPRESSOR ONLY.

OIL
SEPARATOR

COMPRESSOR UNIT

Fig. 14-31. Schematic piping diagram for commercial refrigeration system. It uses a roof-mounted, air-cooled condenser, two motor compressors, and suction and liquid header, each connected to six refrigerant lines. (Dunham-Bush, Inc.)

made of drop-forged or extended copper, with a recess large enough to receive the hard copper pipe.

It is important that the brazing of hard copper pipe joints be expertly done otherwise considerable trouble may result (bad joints and leaks). Hard drawn copper pipe which is seamless usually has a greater wall thickness than annealed copper tubing. It comes in 10 or 20 ft. lengths rather than in rolls. Ends are either capped or plugged. When making an installation of this kind, use those fluxes and solders recommended by the manufacturer. Joining surfaces should be clean and ends of the tubing should be square. This will prevent flux or joining metal from running into the tubing.

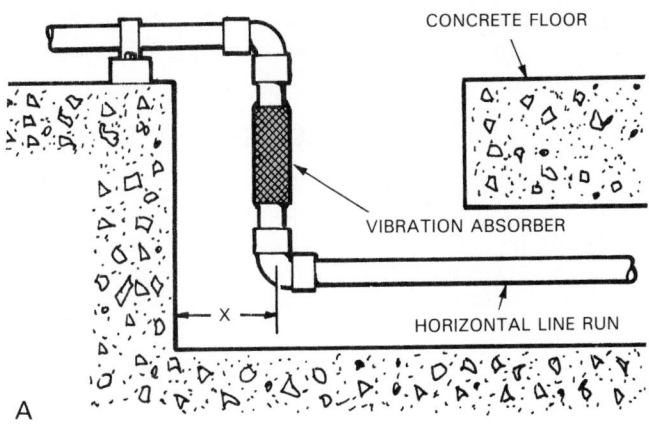

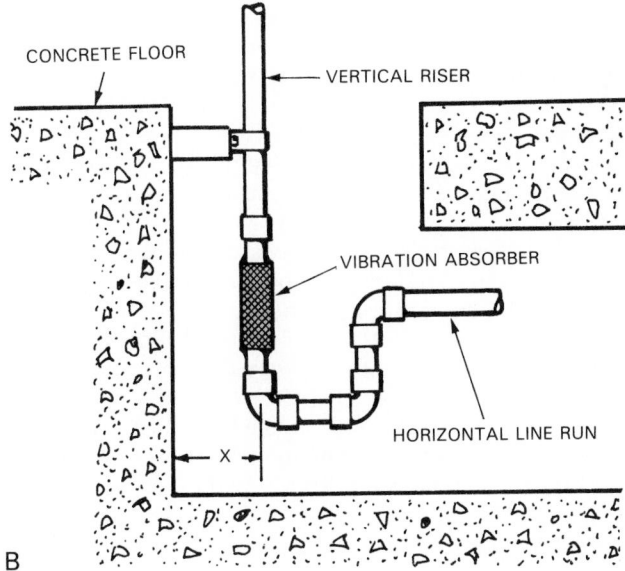

Fig. 14-32. Two instances showing how flexible vibration absorbers are used. Allow a space of 1 1/4 in. at X for each 100 ft. per 100 °F (55 °C) temperature change. A—Horizontal piping. B—Vertical piping.

14-19 SPECIAL TUBING

Copper tubing normally comes with no special finish provided for the inside or the outside surface. Such tubing will corrode if it must be run through liquid, food, beverages, or through air saturated with acid fumes or corrosive elements. Where sanitation is of primary importance, tubing with a tinned surface may be used.

Where the tubing is used to convey beverages, such as beer, soft drinks, and carbonated beverages between kegs and dispensers, many local codes require the use of special tubing. Stainless steel is usually specified.

14-20 MULTIPLE EVAPORATOR PIPING

There are two common methods of installing piping in a multiple installation.

In one method there is a common liquid line and common suction line. The various evaporator liquid and suction lines tap into it at the most convenient points.

Another method is to use a clustering system. Here, various lines are brought to a common point and connected through a hand valve to a manifold. A large suction and liquid line is run from the manifold to the compressor.

This method is not always practical. For example, where one evaporator is at some distance from the box, it would require a duplication of long runs. The manifold of a code installation is located on a wall near the condensing unit.

Keep in mind always that every fitting used, and every bend put into the refrigerant and suction lines, cuts down the efficiency of the installation. Limit their number as much as possible.

Remember to remove inner parts of valves, driers, filters, and sight glasses while tubing is being brazed to them, or wrap the part with a wet cloth or some heat absorber. Do not allow moisture to enter the valve.

14-21 WELDING EQUIPMENT

Gas welding equipment needed for refrigeration installation consists of an oxygen cylinder, acetylene cylinder, regulators and gauges, hose, and a torch, as shown in Fig. 14-22. This equipment may be used for soldering, brazing, and welding the various parts of refrigeration systems.

Before doing any welding, the local code on welding should be thoroughly studied and understood. *Never operate a welding outfit near flammables.*

CAUTION: Never use oxygen, acetylene, or any other welding fuels for the purpose of developing pressure in refrigeration tubing, piping, or equipment.

Carbon dioxide, nitrogen, and argon are safe if a pressure regulator and a pressure relief valve are used. They may be used for developing pressures in refrigeration lines.

Never use excessive pressures with any gas. A severe explosion may result.

14-22 BRAZING EQUIPMENT

The refrigeration service technician uses brazing for many jobs. Air-acetylene torches furnish a clean flame at a temperature of 2500 °F (1 370 °C). With compressed air, the torch flame temperature is about 2500 to 2800 °F (1 370 to 1 540 °C). Acetylene is supplied in cylinders of 40 cu. ft. capacity or 10 cu. ft. capacity.

Detailed instructions on the construction and use of acetylene-air brazing equipment are in Chapter 2. A kit of this equipment is shown in Fig. 14-33.

It is important to follow these safety precautions:
1. Acetylene pressure should never be over 15 psi (207 kPa). Higher pressure may cause an explosion because

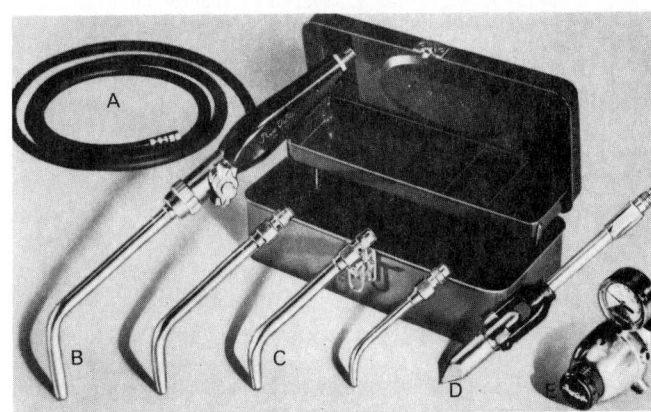

Fig. 14-33. Air-acetylene soldering and brazing kit. A—Hose. B—Large tip and handle. C—Other tips. D—Soldering copper. E—Regulator.

acetylene is not stable at higher pressures.

2. Always use the cylinder in a vertical position. It has a porous filler wetted with acetone in which the acetylene is dissolved. If the cylinder is lying down while in use, some acetone may flow out. Acetone causes a dirty flame and may grease up the regulator and valves.

Fuel air torches that use propane or other high-temperature fuel are also used. They are light and work in any position. See Fig. 14-34. Propane fuel is usable down to −10°F (−23°C). For safety:

1. Keep the flame away from any combustible substance such as oil, wood, paper, paint, cleansing fluids, methyl chloride, barrels, or cylinders which may have contained flammable material at one time. Use sheet metal or board to protect surfaces that might be discolored or scorched when using the torch, such as assembling piping along a wall.

2. Always light the torch with a flint lighter. Matches or a cigarette lighter may bring your hand too close to the flame.

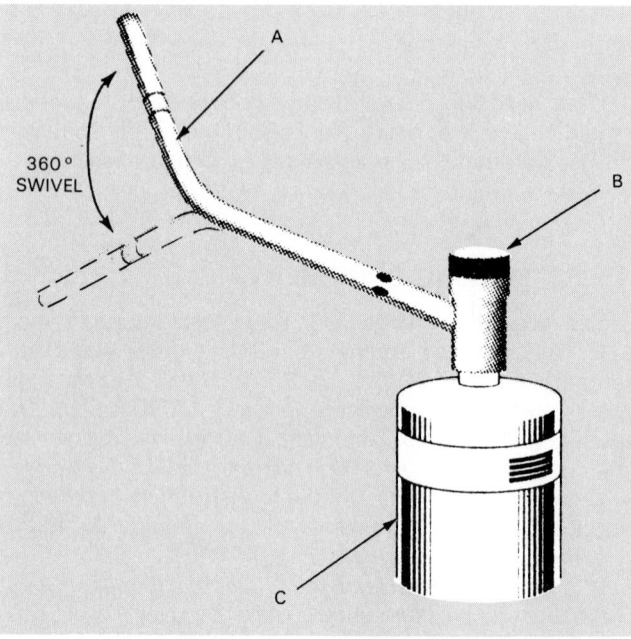

Fig. 14-34. Soldering and brazing torch. A—Tip that swivels 360° for soldering in difficult areas. B—Pressure regulator and adjustment. C—Disposable fuel cylinder. (Goss Inc.)

14-23 REFRIGERANT LINE VALVES

All code line valves are to be hand operated. Codes require that these valves be so constructed that anyone may shut them off without the need of special tools. The valve usually has a handwheel mounted permanently on it. These valves are provided with brackets that may be attached firmly to a panel. Two styles are available:

1. Three-way valves.
2. Two-way valves.

One type of three-way valve shuts off just one of the three connections to the valve. The other two, remaining uncontrolled, permit the passage of refrigerant to the rest of the system.

The other type valve is two-way and stops the flow of refrigerant when turned in clockwise. Fig. 14-35 shows a two-way valve for brazed joints.

14-24 TESTING CODE INSTALLATIONS

Code authorities require that permits be obtained before an installation can be made or a major service operation performed on commercial units. Specifications of the proposed job must be presented. Permits are not issued unless the specifications presented meet code requirements.

On completion of the work, an inspector is called. Approval must be given by that person before the unit may be run. Some codes require that refrigeration installation personnel be licensed.

The inspector, upon reaching the premises, checks the installation to see if all the work has been done according to specifications and code. Then the system is tested, primarily for leaks and safety. This requires building up the normal pressures in the high and low sides of the system. Nitrogen is usually used for this. These pressures vary with the kind of refrigerant in the system. Fig. 14-36 gives the recommended minimum test pressures to be used for each refrigerant.

Low-side and high-side test pressure are usually specified by the code. One must avoid using higher pressures because the system may rupture.

The same refrigerant that will be used in operation is first charged into the system in a vapor form to create a low pressure (20 to 50 psi or 240 to 450 kPa). The system is then tested for leaks. Large leaks are easily detected at these low pressures. If there is no leak, nitrogen is then used to build up the pressures to the code pressure.

The technician or inspector must be very careful! First, the refrigerant cylinder should be disconnected to prevent nitrogen backing up into it. Second, there must be a hand shutoff valve, a pressure regulator, a pressure gauge, and

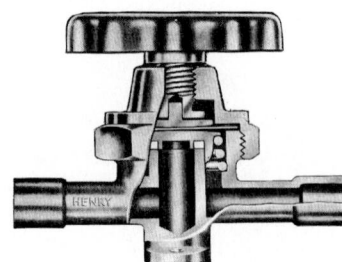

Fig. 14-35. Two-way hand valve designed for brazed connections. (Henry Valve Co.)

MINIMUM DESIGN — TESTING PRESSURE			
	LOW SIDE	HIGH SIDE	
REFRIGERANT		EVAP. OR WATER COOLED	AIR COOLED
R-12	85	127	169
R-22	144	211	278
R-500	102	153	203
R-502	162	232	302
R-717	139	215	293

Fig. 14-36. Recommended minimum design testing pressures are based on Safety Code for Mechanical Refrigeration.

a pressure relief valve in the nitrogen charging line. The relief valve should be adjusted to open 1 or 2 psi above the test pressure. A safe method for using nitrogen to pressurize a system is shown in Fig. 11-55.

After pressures are built up in the piping, it is good practice to rap each brazed and mechanical joint with a rubber mallet to make sure the joint will be leakproof under working conditions. (Paint or flux may otherwise temporarily stop a leak.) If no leaks are found, leave the nitrogen pressures in the system for 24 hours.

With nitrogen, the soap bubble test is used to check for leaks. If some R-12 or R-22 is mixed with the nitrogen, the halide torch or electronic leak detector can be used (mix about 1/4 lb./ton of refrigeration).

If a brazed joint leaks, take it apart to repair it. Put flux on the joint before heating it to keep the brazing material clean. Take the joint completely apart and then assemble and braze again.

If no leaks are indicated at the pressures established, the inspector sometimes checks further by producing a vacuum on all of the system. If this vacuum is maintained over a certain period of time, the installation is approved.

After approval of the system, the installation technician should make a record of the running behavior of the unit for at least 24 hours. Fig. 14-37 shows a recording thermometer. Any variations in temperature reveals need for adjustments.

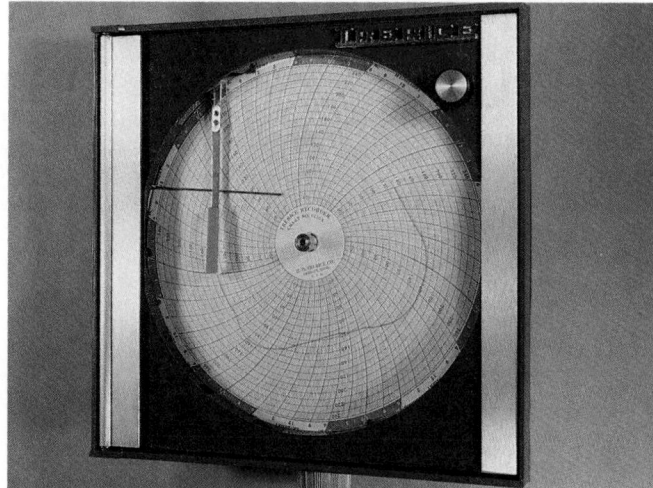

Fig. 14-37. A 24-hour, 7-day recording thermometer. Chart should be dated and kept for future reference. (H.O. Trerice Co.)

14-25 SERVICING COMMERCIAL UNITS

Modern commercial refrigerating units are available in a great variety of forms. Chapters 12 and 13 describe the design, construction, and operation of various mechanisms.

Small units, such as self-contained beverage coolers of hermetic design, are serviced in the same way as the domestic systems. Details are given in Chapter 11.

Some commercial installations use an external drive system with motors and belts. Many use hermetic condensing units.

The servicing of the larger commercial systems is controlled in most communities by the local refrigeration code.

Any major repairs or changes to a commercial system can only be done by licensed contractors. When completed, their work must be checked by the local community refrigeration inspector.

Any plumbing and electrical service work should be subcontracted to licensed plumbers and electrical contractors.

The servicing of commercial installations is much like working on domestic units. However, the use of multiple evaporators on a single compressor is common. Unloading and defrosting systems add to service complications.

The troubles encountered come under various headings. Examples include no refrigeration, continuous running, high cost of operation, poor refrigerating temperatures, and frosted suction lines.

Fig. 14-38 illustrates a service technician measuring the pressure at a suction line service valve. He also is checking the evaporator boiling temperatures from the suction line through a digital thermometer mounted to the line.

14-26 SERVICE EQUIPMENT

Two major items of concern are:
1. Obtaining and using high quality tools.
2. Keeping complete records of each job.

Most companies provide a panel truck or pickup truck equipped with major items such as:
1. Vacuum pump, recovery/recycling unit.
2. Tubing and piping.
3. Combination soldering, brazing, and welding outfit.
4. Supply of replacement parts and materials.
   a. Controls.
   b. Fittings.
   c. Oil.
   d. Refrigerant.
5. Leak detectors—especially electronic tester.
6. Electrical testing instruments.

The service technician is usually expected to furnish his or her own hand tool kit. Chapter 2 describes many of the tools needed. These should be of good to excellent quality. Three good habits will speed up the work:
1. Keep tools clean. This action will result in better and faster work while stretching tool life.
2. Keep tools together on the job—in a tool kit or in the truck. They should be organized so that the service technician can pick up the desired tool in the correct way without looking.
3. Use good lighting. Keep an extension cord and light that can be safely mounted.

14-27 GENERAL SERVICE INSTRUCTIONS

Servicing, troubleshooting, and diagnosing a refrigerating system involve common sense plus a thorough knowledge of refrigeration fundamentals. To operate correctly, a system must have the following capabilities:
1. Cooling (low side).
   a. Enough liquid refrigerant must be in the evaporator.
   b. Pressure in the evaporator must be low enough so that the liquid will boil at the correct temperature.
   c. Heat from the items being cooled must transfer to the liquid refrigerant in the evaporator.
2. Condensing (high side).
   a. Vapor must be pumped into condenser at the correct pressure and temperature.

b. Heat must be removed from condenser (clean condenser, airflow, or water flow).

c. There must be enough vapor space (heat transfer surface) in the condenser.

3. Refrigerant flow in liquid line. Line must be large enough with minimum restrictions (pinched pipe, partially clogged screens, filters, or drier). Only liquid refrigerant should be in the liquid line.

4. Vapor and oil flow in the suction line. Only a small pressure drop is allowable. The screens and drier must not be restricted in any way.

The diagnosing starts with the owner's report. Then the service technician should check the low-side and high-side pressures and the evaporator temperature. Check the sight glass for bubbles. Feel the suction line. It should be cool. Feel the liquid line. It should be the temperature of the surrounding air (ambient).

If refrigeration equipment has been exposed to flooding (a flooded basement or a flood), it must be carefully reconditioned before one attempts to start it.

Clean and dry all of the outside of the equipment. Use a detergent and bacteria cleanser. Replace all open motors or have them completely reworked.

Replace all external electrical parts. If one attempts to clean and reuse them, one must use an electric insulation leak inhibitor.

Replace capacitors, relays, overload devices, and limit switches. Clean compressor terminals and spray with electrical insulation leak inhibitor. Check the electrical system completely with an ohmmeter. Check especially for grounds.

14-28 SERVICING CONDENSING UNIT

Condensing units come under several divisions:
1. Open (external drive) type compressor.
2. Serviceable hermetic motor compressor (field serviceable compressors).
3. Welded hermetic motor compressor (nonfield serviceable compressors).

Compressor types may be:
1. Reciprocating.
2. Rotary.
3. Centrifugal.
4. Screw.

The condensers used may be either:
1. Air-cooled.
2. Water-cooled.

The variety of mechanisms and applications is a great challenge to the service technician. Fortunately, in spite of the great variety, there are certain basic problems all these condensing units have in common:
1. Compressor efficiency.
2. Condenser efficiency (air-cooled or water-cooled).
3. Refrigerant charge.
4. Refrigerant cleanliness.
5. Electric circuit problems.

The compressor may be tested for efficiency as described in Chapter 11.

An air-cooled condenser gives the same symptoms when there is a lack of refrigerant as those explained in Chapter 11. Water-cooled types present a different problem. The water flow should be so adjusted that the temperature rise is no more than 15 deg. as the water goes through the condenser. The water passages must be clean.

If the unit is belt driven, belts should be checked for alignment and tautness.

A decidedly metallic pounding sound occurring regularly in the compressor should be looked into carefully. Check for low oil level or worn parts.

The amount of refrigerant in the system should be checked carefully as explained in Para. 14-30. The motor control should be inspected to determine whether it trips freely and whether the points—if any—are clean. Dirty or pitted contact points should be replaced.

14-29 SERVICING EXTERNAL DRIVE SYSTEMS

As all open (external drive) type compressors are similar, the general instructions which follow will apply to nearly all of them. There are many external drive compressor systems in use. The external drive compressor is either belt driven or a direct drive.

14-30 CHECKING REFRIGERANT CHARGE

The correct refrigerant charge is very important. Several methods may be used to determine whether or not a refrigerator has enough refrigerant.

In undercharged systems the motor operates continuously, the motor compressor is overloaded, and there is poor refrigeration. A lack of refrigerant will show up in an increase in liquid line and drier temperature. A heated drier will release some of its moisture and cause a wet system.

Overcharge will cause excessive head pressure in TEV systems. Liquid refrigerant will be forced into the compressor in capillary tube systems.

A dry or expansion valve system is more difficult to check for refrigerant amounts. The appearance of the valve body may be the first sign of low refrigerant. Under normal conditions the body of the valve frosts over evenly, as far back as the liquid line nut. But when the system has too little refrigerant, the expansion valve body next to the liquid line will not frost. This frost method cannot be used for above-freezing evaporator operation conditions.

A common method of determining the amount of refrigerant is to check how much refrigerant is actually in the liquid receiver and condenser. One way to find this out is to determine the high-side head pressure. If the unit is water-cooled, head pressure should correspond to refrigerant temperatures about 10 °F (-12 °C) higher than the temperature of the water leaving the condenser. The temperature of the water, in this case, should be checked as it leaves the condenser, not at the end of a long drain pipe. If head pressure indicated on the gauge is below normal as much as 10 psi (173 kPa), lack of refrigerant is indicated.

A popular way to check for sufficient refrigerant charge is through the use of a sight glass in the liquid line. The sight glass can be installed into the line (pipe fitting type) or clamped on the line (electronic type). See Chapter 12 for details of both types. A sight glass allows one to check for the presence of bubbles in the liquid line. Bubbles indicate insufficient refrigerant.

Fig. 14-39 shows a sight glass which also indicates moisture. Fig. 14-40 pictures a see-through sight glass for larger liquid lines.

At low head pressure, bubbles may appear regardless of

Fig. 14-38. The service technician checking the system pressures and temperatures, utilizing the proper tools necessary to service the system. (John Fluke Mfg. Co., Inc.)

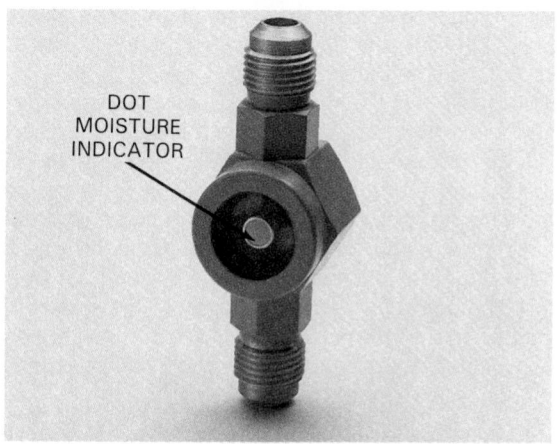

Fig. 14-39. Sight glass which also indicates if refrigerant in system is wet or dry. Green indicates a dry system. Yellow indicates wet system. Bubbles indicate lack of refrigerant. (Sporlan Valve Company)

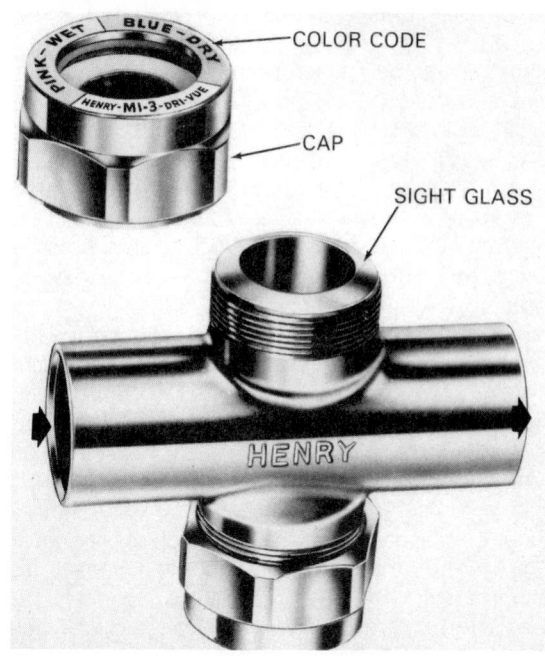

Fig. 14-40. Double port or "see-through" sight glass for larger liquid lines. Unit also indicates dryness of refrigerant. Device has brazed connections. Note dryness code on the seal cap (pink for wet, blue for dry). (Henry Valve Co.)

the amount of refrigerant in the system. If no bubbles appear in the sight gauge, the machine probably has enough refrigerant.

However, if a restriction is in the line ahead of the sight gauge, bubbles may appear even though there is sufficient refrigerant in the system. If possible, the sight glass should be mounted between the drier and the liquid line (upstream from the drier).

Some machines are equipped with refrigerant liquid level indicators such as petcocks mounted in the side of the liquid receiver at definite heights. If the petcock is opened and liquid refrigerant comes out, the level of the refrigerant is at least up to this height. Two petcocks are usually provided. When opened, vapor should come from the top petcock and liquid refrigerant from the lower one.

In the liquid receiver type of water-cooled unit, where the water coils are located within the receiver, the amount of refrigerant in the system may be checked by determining the temperature difference on two different points of the receiver shell. The part of the receiver filled with hot vapor and the part filled with cold liquid refrigerant will be indicated by a temperature difference. This may be easily checked by feeling the receiver with the hand.

One more method for finding the quantity of liquid refrigerant in the liquid receiver involves turning off the cooling water to the condenser and allowing the compressor to operate. If the liquid line warms up quickly, it indicates there is insufficient refrigerant. Another indication is change in head pressure after the water is shut off. It rises quickly. If the compressor is stopped and head pressure drops rapidly, it indicates presence of too little liquid refrigerant in the liquid receiver.

Still another method of checking for the quantity of liquid refrigerant involves shutting the machine down and purging the liquid receiver. Boiling of the refrigerant in the liquid receiver when the pressure is reduced will cause that part of the receiver filled with liquid to get cold, sweat, and perhaps frost over. This method should be used only as a last resort because it wastes refrigerant. It may also freeze the water in a water-cooled unit.

Lack of refrigerant is likely due to a leak in the equipment. A careful check should be made of all joints and parts that could possibly leak. Do this before the unit is recharged and put into service. See Chapter 11.

14-31 REMOVING SYSTEM PART

When part of a system needs service, empty the refrigerator cabinet or put contents to one side and cover them. Spread papers or a tarpaulin around and under the mechanism.

Be careful of all surfaces. Porcelain is brittle. Chipping or cracking may necessitate replacing a complete panel. Do not soil enamel finishes with oil or grease.

Tools and materials should be in a safe place to prevent injury from tripping. Always arrange for good lighting.

When removing any part of any system, there are general steps to follow:

1. Remove all refrigerant from part to be opened.

2. Balance pressures in parts just evacuated to 0 psi (101.3 kPa).
3. Isolate parts to be opened from the rest of the system.
4. Clean and dry joints to be opened.
5. All refrigerant openings should be immediately plugged as soon as they are opened.

Fig. 14-41 shows an elementary unit with the location of the three main service valves marked.

Refrigerant is removed by installing a gauge manifold in the system, by proper adjustment of the service valves, and by operation of the compressor.

Removal of any part of the refrigerant is usually a matter of drawing a low pressure (less than atmospheric pressure) on the part to be dismantled in order to evaporate the refrigerant from it. Then pressure is equalized to 0 psi or 101.3 kPa. (This is called balancing with atmospheric pressure.)

The low pressure removes the refrigerant, while the equalizing or balancing prevents a rush of air into the mechanism when the system is opened. This last step is very important.

To begin removal of refrigerant, close the inlet service valve to the part to be removed. Run the compressor until the gauge shows a 0 psi (103.5 kPa) or a slight vacuum. Stop the compressor and then, after opening the inlet service valve until the gauge reads zero, close the inlet service valve to that part. Close the outlet service valve to the part. Clean and dry the joints. Remove the part. *Always plug all refrigerant openings immediately after removing the part to keep out dirt and moisture.*

For example, suppose one wished to remove the compressor evaporator or TEV. The refrigerant, then, is stored in the liquid receiver. The liquid receiver service valve is closed. The compressor is run until no liquid refrigerant is in the liquid line, evaporator, or suction lines, Fig. 14-42.

When servicing a refrigerating mechanism, the internal part of the machine must be kept as chemically clean as possible. Moisture causes acids, sludge, and freezes in low temperature passages. Dirt (solids) clog screens and cause wearing of control valves, compressor valves, and seats.

14-32 REMOVING REFRIGERANT FROM A SYSTEM

If there is no place to store the refrigerant in the system, the refrigerant must be:

1. Stored in an outside cylinder.
2. Purged, removed from a system by recovery/recycling equipment.

Refer to Fig. 14-43.

To remove refrigerant from a system:

1. Attach a line from a storage cylinder (if one is to be used) to the middle opening of the manifold.
2. Purge the line leading from the cylinder by sealing the line at the cylinder but leaving it loose at the manifold.
3. Crack the cylinder valve. Escaping gas will force the air out of the line.
4. Seal the line at the manifold. Close the discharge line of the compressor by turning the discharge service valve all the way in. Test for leaks and start the compressor. The pressure should not exceed the normal condensing pressure for the particular refrigerant. Excessive head pressures may be avoided by cooling the refrigerant cylinder with ice or water, or by running the compressor intermittently. (This means intervals of running interrupted by short stop periods.)
5. Allow the compressor to run with all but the discharge service valve open.

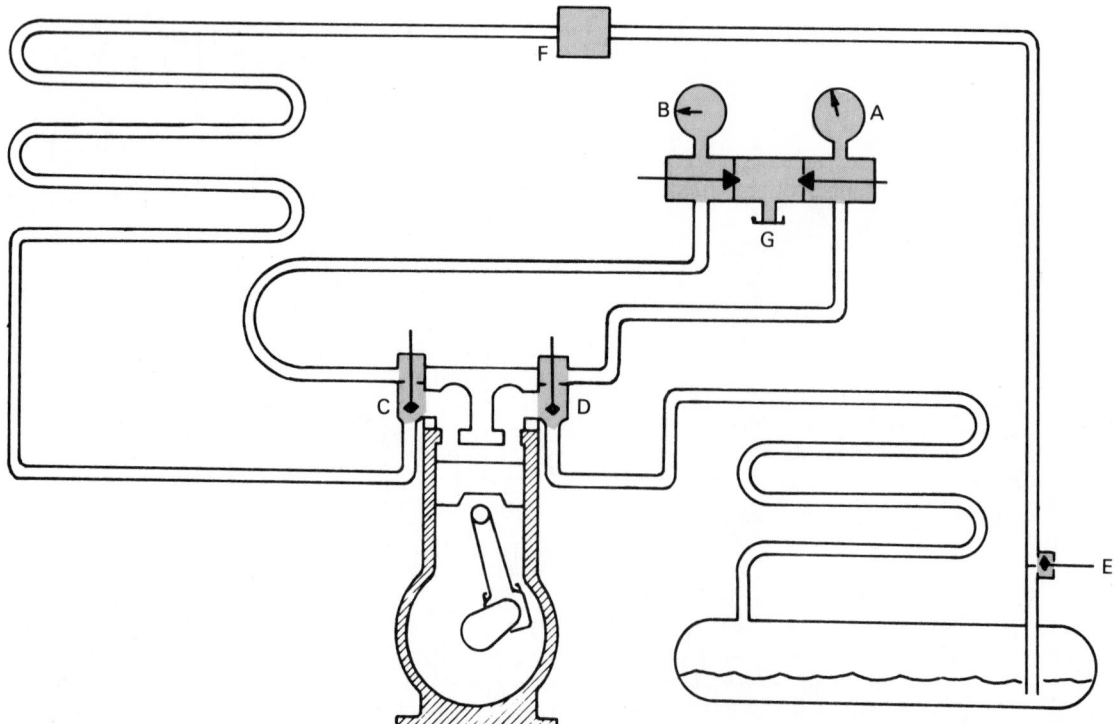

Fig. 14-41. Elementary system shows location of gauges and service valves. A—High-pressure gauge. B—Compound gauge. C—Suction service valve. D—Discharge service valve. E—Liquid receiver service valve. F—Expansion valve. G—Gauge manifold.

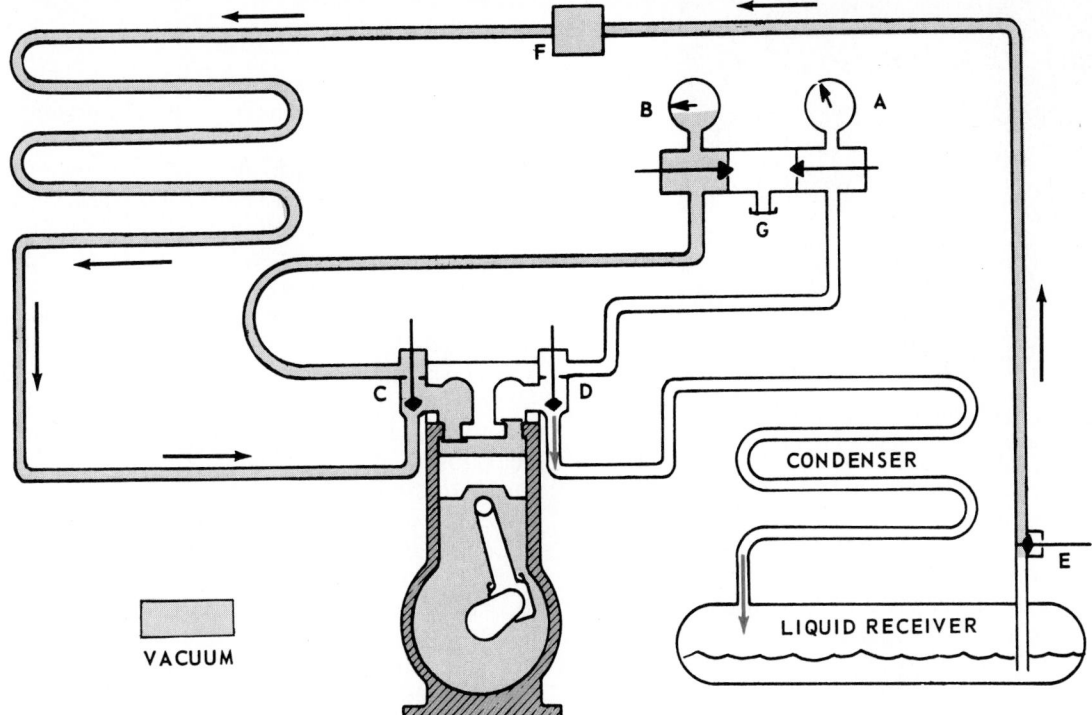

Fig. 14-42. Compressor is evacuating liquid line, evaporator, suction line, and compressor crankcase. Refrigerant is being stored in condenser and liquid receiver. Valves A and B are closed, valves C and D are in mid-position, and valve E is closed.

6. Shut the compressor off after a constant low pressure has been maintained for several minutes. Never allow the system to pump oil, as the hydraulic pressures may cause serious damage to the compressor and lines.

7. The operation may be speeded up by cautiously applying heat to the liquid receiver and to the evaporator. Use a heat lamp or warm water. *Never use a torch, as it may melt the fuse plugs and brazed joints.* Never allow any part or spot to become too warm to touch with the hand.

8. After the refrigerant is all pumped from the system and placed in a storage cylinder, stop the compressor.

9. Open the low-side manifold valve until 0 psi (103.5 kPa) is indicated on the compound gauge. This action returns enough vapor refrigerant into the system to balance the pressure in the entire system. Any part of the system may now be removed. Wear goggles! Always leave the gauge manifold connected to the system until the system is opened. One must know the pressure in the system as it is first taken apart. As mentioned before, clean and dry all the connections to be opened.

10. Immediately upon removal of any parts, the refrigerant openings should be carefully plugged.

The unnecessary release of chlorofluorocarbons (CFCs) into the atmosphere is a concern of all service technicians. Local laws and regulations must be taken into consideration before the release of any refrigerant of any type into the atmosphere. The exhausting of refrigerant to the air by the service technician is dependent upon the type of refrigerant and the laws pertaining to it.

When purging refrigerant into a clean refrigerant cylinder, it is good practice to attach a purging line made of 1/4-in. copper tubing to the manifold center opening. This purging line should have a hand needle valve and a check valve mounted in it at the manifold end. The hand needle valve should be located between the check valve and the manifold. During purging, the manifold high-pressure valve is opened.

The purpose of the hand valve is to control the amount of gas purged. The check valve prevents backing up of air or moisture into the unit after it has been completely purged.

Because all refrigerants being purged have an oil content, purging should be done into an oil trap. *Always purge in a well ventilated space.*

This method cannot be used for ammonia (R-717), because of the odor. See Para. 9-16. Most communities forbid purging refrigerant into a sewer system. Check local codes.

If the refrigerant is to be put back into the same machine or, if facilities are available for distilling it, it may be stored temporarily in a clean refrigerant cylinder. Remember that the refrigerant will always have an oil content. Some large companies save all refrigerant, redistill it, and process it for further use. This is good practice from the standpoint of economy and ecology.

Federal laws govern chemical substance disposal. The United States Resources Conservation and Recovery Act (RCRA) of 1976, its amendments of 1984, and the E.P.A. regulations, 1990, strictly classify refrigerant disposal. Large quantities of refrigerants must be stored in steel drums and moved to a registered waste disposal site. See Chapter 28 for further details concerning refrigerants containing CFCs and HCFCs.

14-33 CHECKING EXTERNAL DRIVE COMPRESSOR

In most cases of refrigeration failure, it is advisable to check the compressor first to determine if it is operating satisfactorily. Good service technicians will also look for other indications of troubles as they check the compressor.

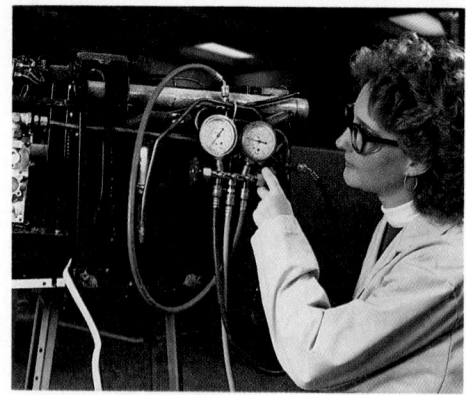

Fig. 14-43. Service technician removing refrigerant from a residential air conditioning system. Note the proper use of safety glasses. (Southeast Oakland Vocational Education Center)

cannot produce a vacuum of greater than 20 in. Hg. (34 kPa) against the normal head pressure, it should be overhauled.

A worn piston or cylinder is indicated by a clicking noise which is somewhat duller than the noisy valve indication mentioned before. Worn connecting rods and main bearings are quite noisy when the compressor is running with a low-suction pressure.

The compressor must pump a specified quantity of gas at a certain pressure difference to do the work necessary. This is difficult to check, so the methods just described are used as secondary checks. Some repair shops use a shop-mounted tank into which the compressor pumps air while being tested. If the time required by a compressor to pump to 150 psi (1 139 kPa) is recorded for each size compressor, a relative volumetric efficiency check is possible.

Compressor testing methods:

1. Vacuum producing ability.
2. Ability to hold high pressure.
3. Ability to hold both vacuum and head pressure.

If the compressor exhaust valve leaks, there are two ways to check it:

1. The high head pressure will leak back through the exhaust valve and produce a pressure above atmospheric on the compound gauge. Therefore, if the compound gauge creeps up above 0 psi (101.3 kPa) when the compressor is idle, it is a sign that the exhaust valve needs repair. If there is a leak on the low side (seal) of the compressor, the pressure will only rise to approximately 0 psi.

2. Turn the discharge service valve stem all the way in. If the discharge valve leaks, the pressure—as indicated on the high-pressure gauge—will decrease. This is

However, they should be certain that the compressor is in satisfactory operating condition.

The amount and the condition of oil in the compressor is important. Some compressors have an oil level sight glass or port. Others have a plug located at the proper oil level.

Two of the most common causes of compressor trouble are faulty valves and seals. Noisy valves may be detected by a sharp clicking noise in the compressor as it operates.

Leaky valves may be detected as follows:

1. Install the gauge manifold and test for leaks. Turn the suction service valve stem all the way in to close the suction line. Turn the power on and off for a few seconds at a time until the danger of pumping oil is stopped. Then allow the compressor to run, Fig. 14-44.

2. Record the best vacuum obtainable against the normal head pressure for the refrigerant being used. Also record the time. In most cases, if the compressor

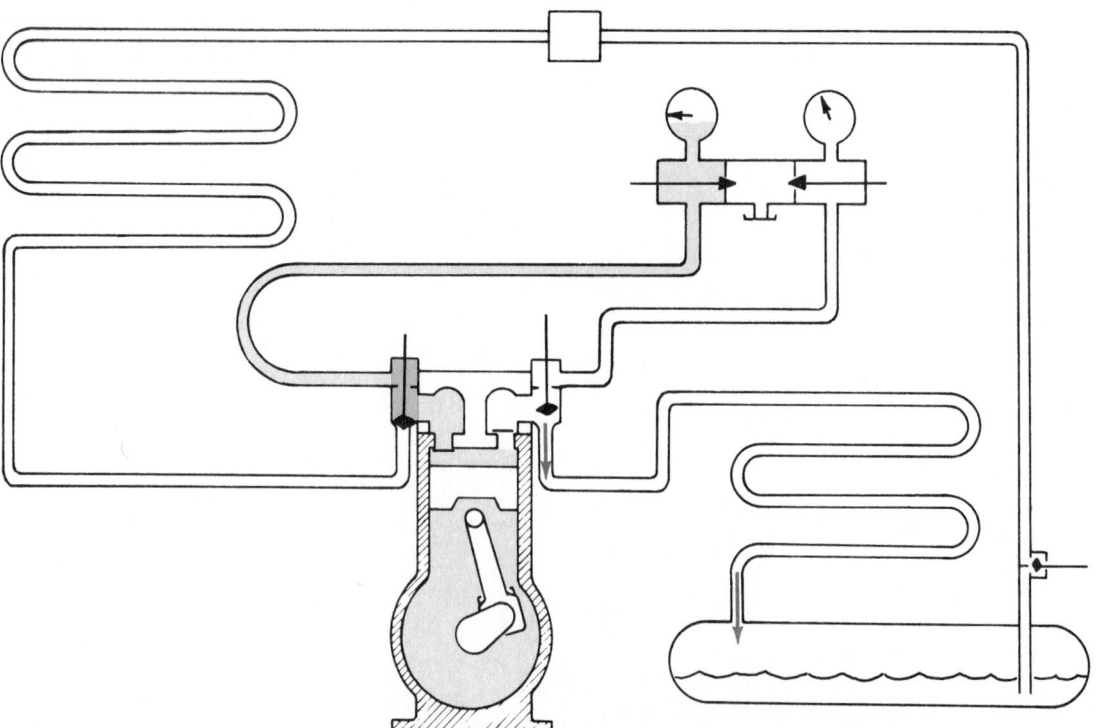

Fig. 14-44. Efficiency test for an external drive compressor. Close suction service valve and run compressor. Pumping efficiency will be indicated by maximum vacuum obtainable and time it takes to develop this vacuum.

especially true if the compressor is turned by hand, because the pressure will drop as the piston goes down. Thus, if the gauge pressure fluctuates considerably, it indicates a leaky exhaust valve. If the pressure merely increases and does not drop back to any extent, the exhaust valve is not leaking.

An intake valve leak is indicated by the inability of the compressor to produce a high vacuum. However, the vacuum produced is maintained after the compressor is shut off, provided the exhaust valve is holding.

This may also be due to too thick a gasket or worn piston and rings. Lack of oil will also result in poor pumping ability.

To determine whether the crankshaft seal is leaking, close the suction service valve and pump as high a vacuum as possible on the crankcase of the compressor. Then turn the discharge service valve all the way in, as shown in Fig. 14-45. Keep the compressor running. If there is a low-side leak, the head pressure (indicated on gauge at A in Fig. 14-45) will gradually increase with the running of the compressor. This indicates that gas or air is being drawn in on the low side of the compressor.

A seal leak is usually noticeable by traces of oil at the seal or on the floor underneath. Leak detectors may also be used to check a seal leak. (Crankcase pressure must be over 0 psi.)

14-34 REMOVING SERVICE VALVE

Occasionally a service valve stem will break or the threads will strip. The valve must be replaced.

If it is the suction service valve, one must remove all the refrigerant from the evaporator unit and then balance the pressure—unless the evaporator is furnished with shut off hand valves.

To remove a discharge service valve or a liquid receiver service valve, the refrigerant must be removed from the entire system. Do not pinch the lines to replace valves, as weakened tubing will shortly cause trouble. Service technicians have successfully replaced these valves by super-cooling the refrigerant in the system with dry ice. When dry ice is packed around the refrigerant containing parts of the system and when the gauges show atmospheric pressure, the system can be opened. CAUTION: Wear goggles!

14-35 REMOVING COMPRESSOR

When a compressor must be removed, the procedure is this:

1. Install the gauge manifold, as shown in Fig. 14-46.
2. Carefully test for leaks. Note connection at E is for fastening the vacuum pump to the manifold. Opening has a Schrader valve which closes the opening unless a depressor is connected to the opening and is used for purging or charging.
3. Turn the suction service valve at D all the way in, closing off the suction line.
4. Start the compressor but let it run for only a moment in order to prevent oil pumping. (Oil in the crankcase may bubble vigorously as the refrigerant boils out.) Pumping of oil is indicated by a pounding noise in the compressor and should be avoided.
5. After starting and stopping the unit two or three times, it may finally be run continuously. Keep the unit running for a few minutes after a constant vacuum is reached on the suction gauge.
6. Stop the compressor. Open the two manifold valves at A and B (Fig 14-46) to allow the high-pressure

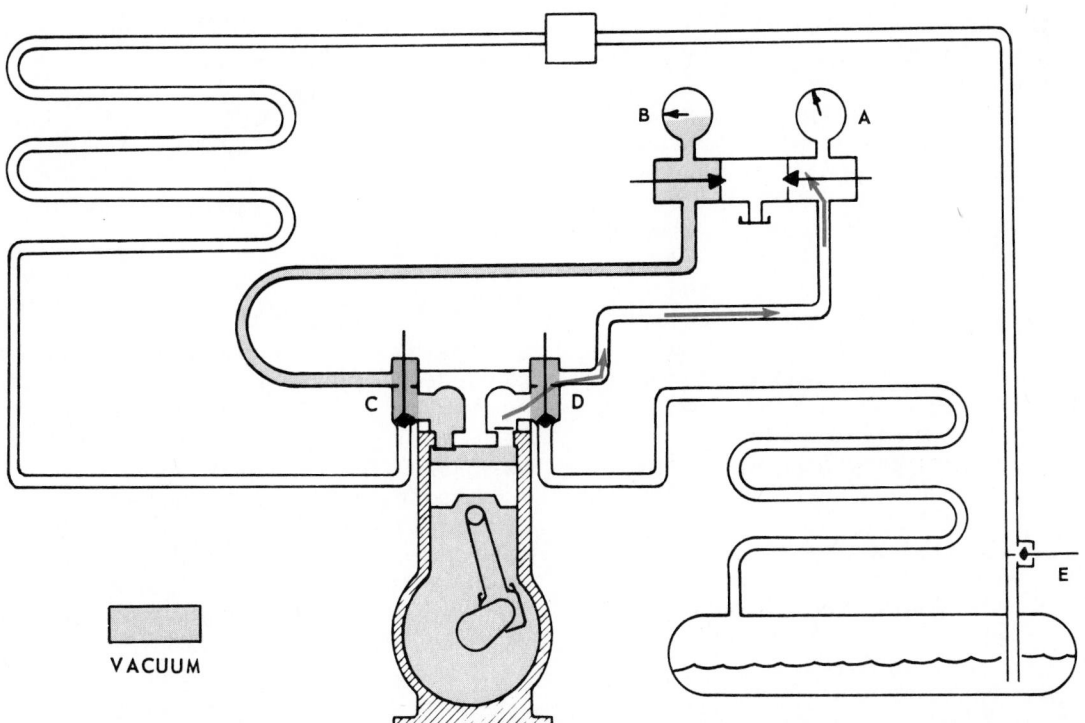

VACUUM

Fig. 14-45. Testing compressor for low-side leaks. This is done by determining best vacuum it can create with suction service valve at C turned all the way in, then turning discharge service valve at D all the way in. If air or vapor are entering compressor, high pressure will increase.

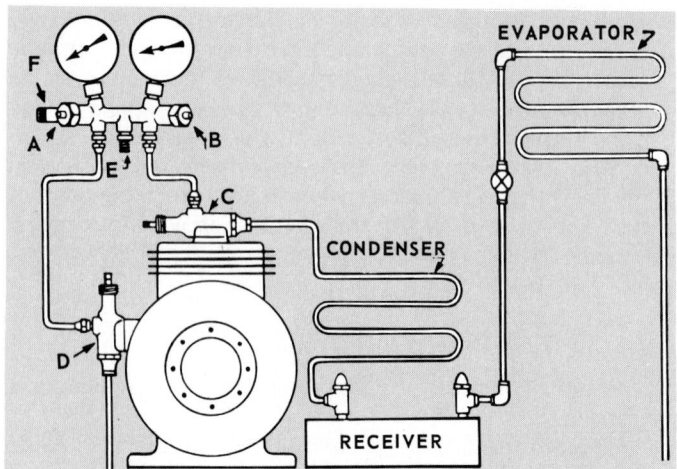

Fig. 14-46. Refrigeration system with gauge manifold installed. A—Manifold low-side valve. B—Manifold high-side valve. C—Compressor discharge service valve (DSV). D—Compressor suction service valve (SSV). E—Vacuum pump connection. F—Manifold purging and charging connection. (Mueller Brass Co.)

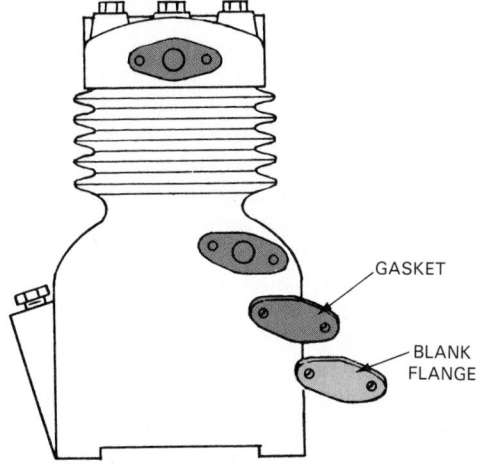

Fig. 14-47. Blank flange used to close compressor valve opening while compressor is being moved.

vapor to build up the crankcase pressure to 0 psi (101.3 kPa). Then turn the discharge service valve stem at C all the way in.

7. Some service technicians crack the suction service valve at D until the compound gauge reads 0 to 1 psi (equalizing the pressures). Shut off the electric power and lock the switch in the open position. Close manifold valves.

8. Joints should be cleaned with a grease solvent and dried before opening. Unbolt the suction service and discharge service valves from the compressor. *Do not remove the suction and discharge lines from the compressor service valves.* Immediately plug all openings through which refrigerant flows using dry rubber, "cork" stoppers, or tape.

9. Disconnect bolts that hold compressor to base and remove belt. Compressor is ready now for removal.

10. The oil should be drained immediately and compressor refrigerant openings plugged. Fig. 14-47 shows a blank flange. These are best for plugging openings. Do not reuse old oil if it is discolored.

To keep the crankshaft seal from being abused, never rest the compressor weight on the flywheel. Always place the compressor on a block so that the flywheel hangs free.

If possible, remove the flywheel before removing the compressor. Any undue strain on the flywheel may injure the crankshaft and/or the crankshaft seal.

The flywheel can be removed with a universal flywheel puller. Supplying a little heat to the flywheel hub will help while the wheel puller is drawn up snugly. Fig. 14-48 shows a universal type flywheel puller.

When the compressor from a larger unit must be removed for overhaul, the handling of the compressor—because of its weight—presents a problem. When lifting a compressor, the service technician should avoid strain from assuming an awkward position. Use care not to slip on oil or loose tools. Carts and small hydraulic hoists are available for moving heavy compressor. Compressors are usually reconditioned by companies specializing in this work.

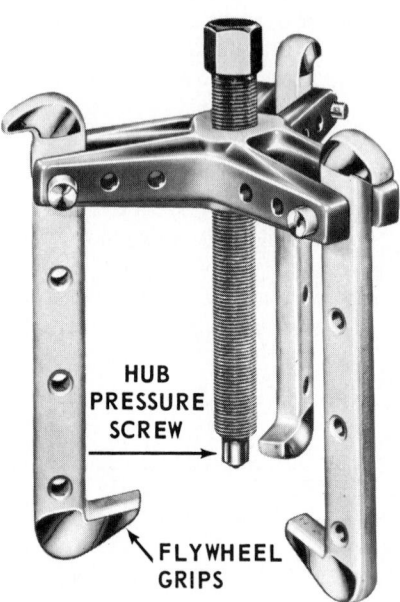

Fig. 14-48. Universal flywheel puller.

14-36 OVERHAULING THE COMPRESSOR

If only the valve plate needs repairs or reconditioning, the service technician can do this without removing the compressor. Para. 14-39 explains how to remove a compressor head and the valve plate; how to recondition a valve seat, assemble a valve plate, and test it.

Frequently, crankshaft seals can also be replaced and/or repaired on the job without removing the compressor. First, the technician removes the flywheel and the old seal. Para. 14-38 explains how to install a crankshaft seal kit.

In cases where a compressor has been removed and is to be reconditioned, this procedure should be followed:

1. Identify (tag) the compressor.
2. Clean the outside of the compressor.
3. Prepare work order form.
4. Take compressor to shop.
5. Take compressor apart.

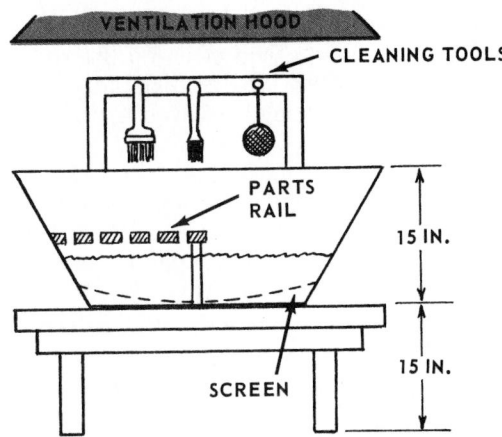

Fig. 14-49. Ventilated cleaning station.

6. Clean all parts. Place parts in tagged trays. Fig. 14-49 shows a cleaning station.
7. Order replacement parts and tag parts needing repairs. (All precision parts must be precision inspected with dial indicators and micrometers. These parts must fit to tolerances of .001 to .003 in.)
8. Recondition parts.
9. Assemble. (Gasket surfaces must be flat, clean, and free from burrs.)
10. Test compressor.

Many means have been used to clean parts of a refrigerator. Each method has its advantages. The cleaner should be a good moisture absorber and remove oil and grease quickly. It should be nontoxic, nonflammable, and should evaporate quickly.

R-11 is a good cleaning solvent, being nontoxic and nonflammable. It is an excellent cleaning solvent for use in flushing systems which have been contaminated by a hermetic motor burnout. It leaves no noncondensible residue and has no reaction with insulation. R-11 can be recleaned and reused. It has a high boiling point of 74.8 °F (23.7 °C). *Note that R-123 is being considered as a replacement for R-11 per E.P.A. regulations.*

Also available commercially is Virginia No. 10, a degreasing solvent. It has low toxicity, is noncorrosive, and has a high flash point of 165 °F (74 °C).

Hand wire brushes and power wire brushes are excellent tools for removing scale and crusted dirt, but grease and oil should be removed first. Be sure to wear goggles.

Fig. 14-50 shows a cleaning station that uses acid solution.

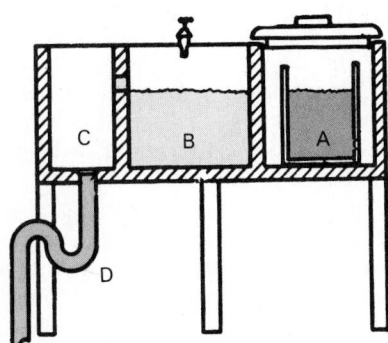

Fig. 14-50. Acid type cleaning station. A—Weak acid. B—Rinsing basin. C—Drain (overflow from B). D—Trap.

14-37 REPLACING BEARINGS

Compressor bearings may be of either plain, ball, or roller type.

Only rarely must any work be done on the connecting rod or crankshaft bearings of compressors. These are usually plain bearings or bushings. Should work be necessary, it is a replacement and reaming process similar to automobile repair. Old bearing sleeves, usually of brass or bronze, are removed by first splitting them with a cape chisel. The new sleeve is pressed into place and line reamed to fit.

A bushing pressed into a blind hole can be removed by using hydraulic pressure. Fill the cavity with grease and then insert a shaft the same size as the ID of the bushing. Cover the shaft with a cloth to protect against flying grease, then hit the shaft a sharp blow. The hydraulic pressure created will usually push out the bushing.

Always measure a crankshaft at the journals (where it turns in the bearings) for size, taper, out-of-roundness, and/or bent crankshaft. Fig. 14-51 shows a dial indicator used to determine the variation in a crankshaft. A variation of more than .001 in. makes reconditioning the journal or replacement of the crankshaft necessary.

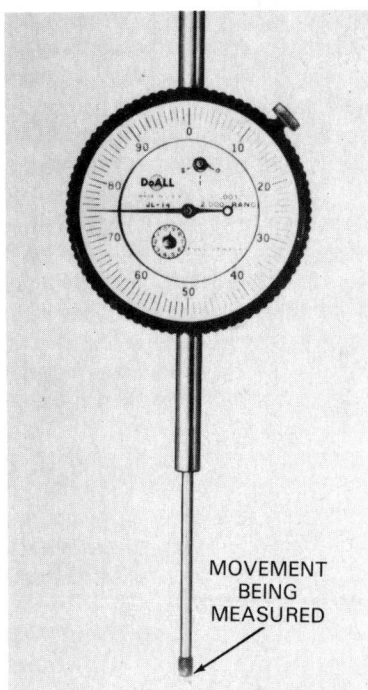

MOVEMENT BEING MEASURED

Fig. 14-51. Dial indicator used to check clearance, roundness, and accuracy of surfaces. (DoALL Co.)

A method of checking shafts for trueness is illustrated in Fig. 14-52. Close clearances are usually held in fitting compressor bearings.

One of the most common sources of noise in a compressor is the piston pin. This pin must be replaced in practically all overhauls. Tolerances of .0005 in. are not too small. The fit must be very snug.

Hand operated, adjustable reamers may be used to ream the piston and connecting rod in line. If the pin itself is badly worn, a new one must be used. Some automobile piston

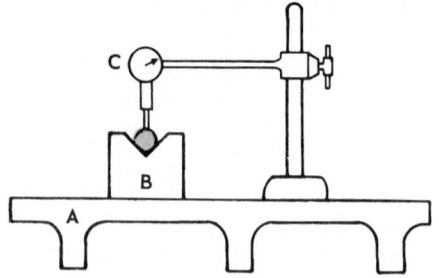

Fig. 14-52. Checking shafts, rollers, and cylinders for trueness. A—Surface plate. B—Blocks. C—Dial indicator.

pins are usable in some of the larger refrigerating compressors. Badly scored eccentric connecting rod bearings must be replaced.

14-38 REPAIRING CRANKSHAFT SEAL

Crankshaft seal design and construction is described in Chapter 4. Repair or replacement required depends on the seal design.

In Sylphon seals, a squeaky noise may be caused by running the compressor with a dry seal surface. A leaky seal may be caused by a scored seal surface.

A noisy seal will soon become a leaky one if not attended to. The trouble may be fixed by the usual process of replacing or lapping the seal surface. It may sometimes be repaired by tapping the seal box lightly with a hammer. Another remedy is to wrap the bellows with oil-soaked wool yarn or string.

A leaking seal may be detected by the usual leak test. Air in the system may be the result of a leaky seal, if the system operates below atmospheric pressure on the low side. This may be detected by a high head pressure in those systems having a below atmospheric low-side pressure. A seal leak will usually cause a lack of refrigerant in systems using an above-atmospheric low-side pressure. The usual symptoms of this trouble are oil under the compressor, high power bill, constant running, and poor refrigeration.

The contact surfaces of the Sylphon ring and the crankshaft shoulder must be perfectly square and polished. To lap in a Sylphon seal properly, the work should be done with a special tool made of case-hardened steel with a ground surface, as shown in Fig. 14-53. Use oil-saturated lapping compound of fine texture. The crankshaft surface, and the surface of the Sylphon that comes in contact, should not be scratched and should have a burnished appearance to give satisfactory service.

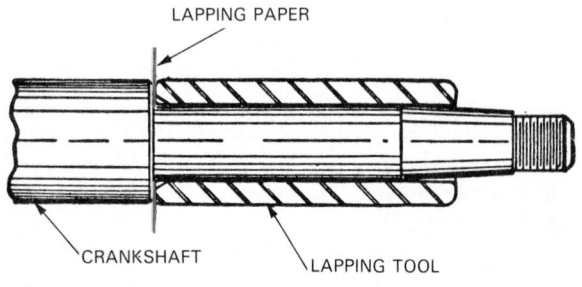

Fig. 14-53. Lapping crankshaft seal shoulder.

A second crankshaft seal shoulder is repaired by putting the crankshaft in a lathe and polishing the shoulder face with a high speed grinder. The amount ground away must be small because the case hardening is only .015 in. to .030 in. deep. Once this case hardening is ground through, the seal will not wear long.

After the shaft is ground, it must be lapped in the lathe. Use phosphor bronze or cast-iron lapping blocks and a fine lapping compound with a light, even pressure. The seal ring or the seat may be lapped also. Fig. 14-54 shows a kit used for lapping. Seal ring is on pad between pans. The use of a lapping block is shown in Fig. 14-55.

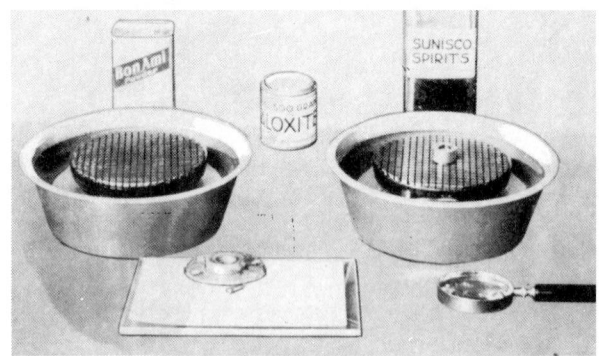

Fig. 14-54. Equipment and supplies needed to lap seals and valve plates. Note magnifying glass used for inspection and for checking cleanliness needed for accurate lapping. Lapping block is shown in pan. (Fedders North America)

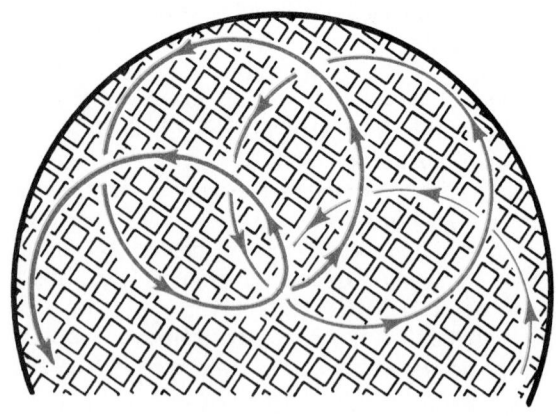

Fig. 14-55. Correct lapping procedure is a circular traveling motion of seal faces or valve plates over the lapping block above.

When assembling a new or rebuilt seal, a special tool may be used so that the seal mechanism will be properly aligned with the crankshaft and seal housing. Fig. 14-56 illustrates a seal alignment tool.

Many service technicians use replacement seal assemblies. These assemblies do not require grinding and polishing. The kit replaces both the crankshaft shoulder and the seal ring.

Fig. 14-57 shows one type of replacement seal. The rotating seal ring is fastened to the crankshaft by a synthetic circular gasket (neoprene or flexible, oil-immune plastic). Another type of seal is shown in Fig. 14-58.

To install a replacement seal:

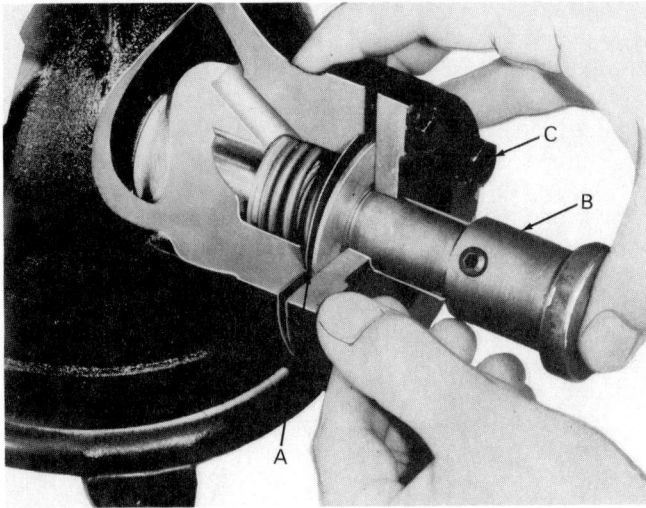

Fig. 14-56. Tool used to align shaft seals. A—Seal. B—Alignment tool. C—Seal ring.

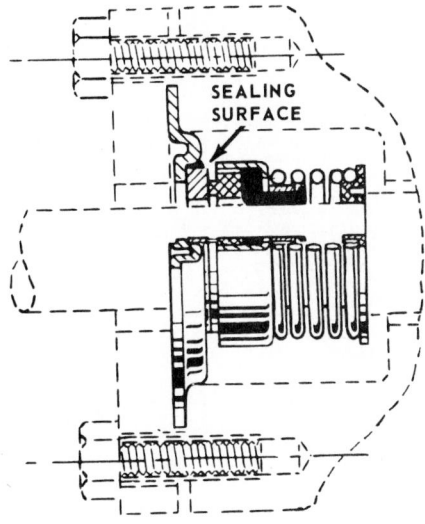

Fig. 14-57. Replacement seal for conventional compressors. Seals of this type are also used on many new compressors.

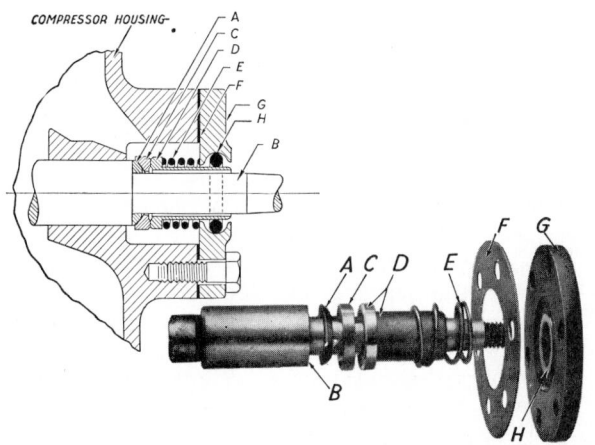

Fig. 14-58. Replacement seal. Such units are available for most external drive compressors and pumps. Letters on photo correspond to letters on cross-section above photo. (Chicago Valve Plate and Seal Co.)

1. Remove original seal.
2. Clean the shaft with lintless, clean cloth.
3. Put clean refrigerant oil on seal surfaces.
4. Install the seal parts in correct order (varies with type of seal).

14-39 REPAIRING COMPRESSOR VALVE PLATE

Carefully clean the valve plate. Resurface the valve seat with great precision. Minor repairs can be made using a lapping block and fine grinding compound. But extensive wear (erosion and pits) is removed best by grinding the complete valve plate surface on a surface grinder. Grinding should continue until the seat is in good condition. Then finish it by lapping the surface.

The complete valve plate is usually replaced if a new or reconditioned one is available. Fig. 14-59 shows a replacement type valve plate. These units are available for most compressors.

If possible, replace the valve. Chapter 11 describes both the precautions to follow and the testing of these valve plates. Valves are also described in this chapter. There is no practical way to repair the disks or reeds, so they must be replaced if leaking. The drawing in Fig. 14-60 shows two views of a disk valve.

Exhaust valve carbon deposits (coking) indicate that the compressor had a discharge temperature too high for the oil used. The system should be checked for excessive head pressures after the repaired compressor is installed. The trouble may be a low suction pressure, or a dirty or undersize condenser.

14-40 ASSEMBLING COMPRESSOR

Always use new gaskets when assembling a compressor. Old gaskets lose their ability to compress. Lead or special composition gaskets are often used. Thickness of the lead

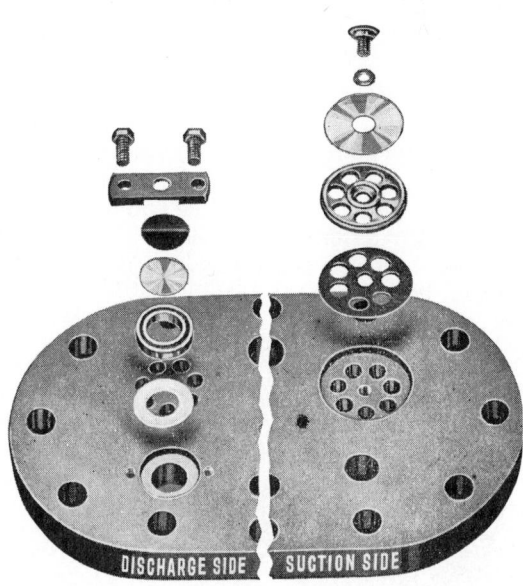

Fig. 14-59. Replacement valve plate. Special valve plates are available for most conventional refrigeration compressors. Valves and valve seats are removable and can be easily repaired. Left. Discharge or exhaust valve assembly. Right. Plate turned over to reveal intake (suction) valve assembly. Note multiple intake openings and single exhaust port. (Chicago Valve Plate and Seal Co.)

529

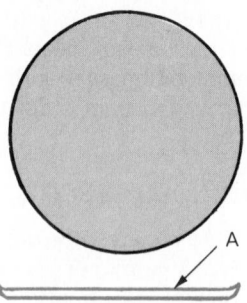

Fig. 14-60. Disk valve used frequently as an exhaust valve in compressor. Surface A should be away from the valve seat; otherwise, small burr may cause valve leak.

is usually between .010 and .020 in. Composition gaskets, when used, must be of the same thickness as gaskets removed and must be thoroughly dry (dehydrated).

If a gasket has to be made, remember the gasket must be exactly the same thickness as the original. If too thick, it will reduce the compressor efficiency. If too thin, it may cause an annoying knock.

If a compressor has been "frozen" due to a high head pressure or moisture in the refrigerant, it should be carefully cleaned and the piston and cylinder burnished. This precaution should remove all foreign substances.

When a compressor is overhauled, only new refrigerant oil should be put in the crankcase. The compressor must be thoroughly dehydrated (baked) for eight to 24 hours at 200°F (93°C) while subjected to a high vacuum. Only after this has been done can it be used. Manufacturer's specifications should be followed carefully.

There should be no noticeable end play in the crankshaft. Check manufacturer's specifications for allowable crankshaft end play. See Fig. 14-61. Some have spring-loaded end thrust bearings. Main bearings must be in line.

The crankshaft seal must be clean when assembled. A drop of refrigerant oil should be put on the two sealing surfaces. Parts must be carefully aligned on the shaft. They also must be free to move to allow the sealing surfaces to press together.

The cylinder head, end bearing housing, seal plate, and crankcase must be fastened to the cylinder evenly. Draw up or tighten each cap screw a little and alternate across the center of the assembly until all are tightened evenly. Use a torque wrench for final tightening. Careless tightening may warp or break parts.

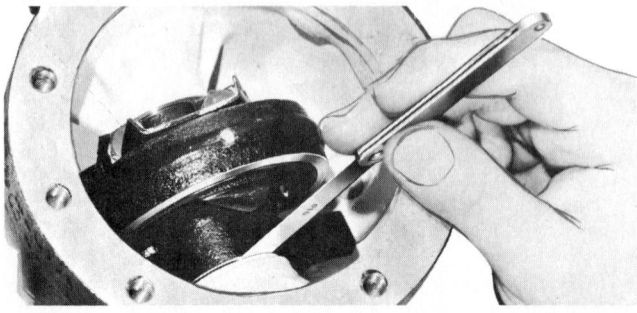

Fig. 14-61. Checking end play of crankshaft. Thickness gauge is .010 in.

All brass and copper parts of a compressor can be cleaned. A weak acid solution is used. Sometimes called a muriatic acid, this is 78 percent water and 22 percent commercial hydrochloric acid, with 1.19 specific gravity or 1/4 oz. inhibitor powder per gallon.

To make the mixture, first put the water in an acid-proof container, then add the inhibitor. Stir until mixed, then slowly add the acid. (The solution will become warm as the acid is added.) A buffing wheel can also be used to polish the parts.

Wear goggles and rubber gloves for acid work and buffing. Use tongs to handle parts being cleaned in the acid bath. Gases are formed as the solution reacts with the deposits. Good ventilation is required to prevent breathing problems and respiratory damage.

It takes from 12 to 24 hours for the solution to thoroughly clean the parts.

14-41 TESTING REPAIRED COMPRESSORS

After a compressor has been repaired, it should be tested, dehydrated, sealed, and painted. Test for leaks and pumping efficiency. This may be done more easily on a shop stand, as shown in Fig. 14-62.

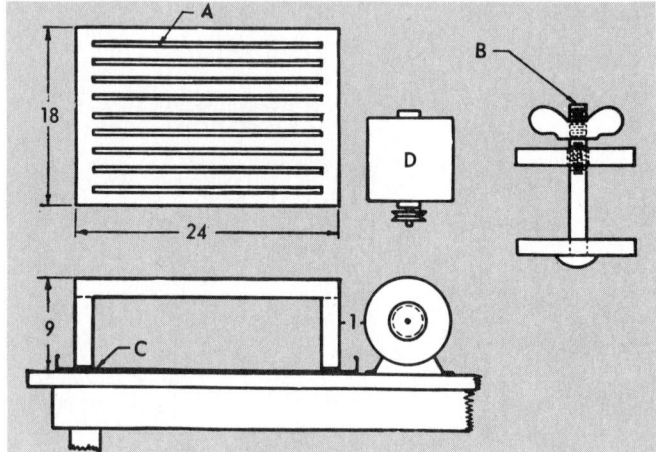

Fig. 14-62. Compressor testing stand. A—Universal clamping plate. B—Clamp down bolts, wing nut, and bars. C—Oil pan. D—Motor.

To test for leaks on the low-pressure side of a compressor, such as at gaskets, suction service valve, or crankshaft seal, one of these methods may be used.

1. Close the suction service valve and draw as high a vacuum on the compressor as possible. Then turn the discharge service valve all the way in. Keep the compressor running. If the head pressure rises gradually, air is being drawn into the low side of the system.

2. A better way is to balance pressures in the crankcase and turn the discharge service valve all the way in. Remove the discharge service valve gauge plug and connect a 15 in. length of copper line to this opening or assembly, as shown in view A of Fig. 14-63. Then immerse the end of the copper line into a glass bottle partly filled with oil.

A gauge manifold can also be used during this test. If, when the compressor is running, the tube continuously causes bubbles in the oil, air is being admitted to

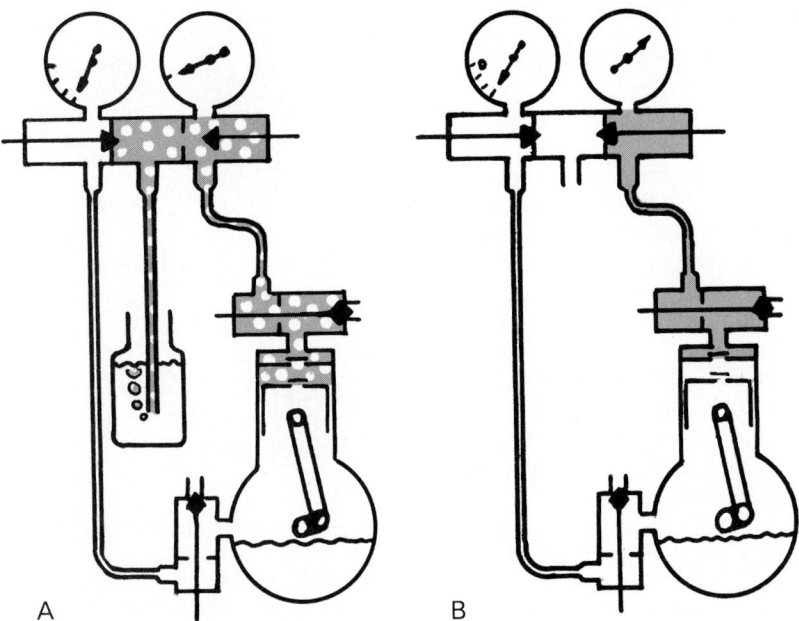

Fig. 14-63. Testing for leaks. A—Testing compressor for low-pressure side leaks. Bubbles indicate air is leaking into compressor at seal, gaskets, and/or fittings. B—Testing compressor exhaust valve for leaks. If high-pressure gauge reading drops while compressor is idle, vapor is leaking back into compressor low side.

the low side of the compressor. If there are no leaks, the bubbling will stop immediately after the compressor is started.

To locate the leak, put refrigeration oil around one joint at a time. If air is leaking in at that point, air bubbles will cease while oil is being drawn in instead of air.

Exhaust valve leaks may be located one of three ways:

1. By turning the discharge service valve all the way in after mounting the high-pressure gauge or a gauge manifold as shown in view B of Fig. 14-63.
2. When the compressor is turned over (revolved) a few turns and the head pressure rises rapidly.
3. If the exhaust valve leaks, the pressure will decrease when the compressor is stopped. Any decrease in pressure will indicate a leaking exhaust valve.

Compressor efficiency (intake valve, piston ring fit, and valve action) is checked best by running the compressor at a constant low-side pressure, while checking the time required to pump a head pressure in a certain size of cylinder.

Another efficiency test is to observe the amount of vacuum a compressor will produce against a standard head pressure. If service valves are not available, a universal flange can be bolted to the compressor in place of the service valve. This is shown in Fig. 14-64.

Prior to storage (short or long term), a gauge should be installed on the crankcase. Once the crankcase is leak checked and proven tight, a 25-lb. to 30-lb. freon charge should be pumped into the crankcase. This prevents moisture contamination of interior parts during storage.

14-42 INSTALLING EXTERNAL DRIVE COMPRESSOR

Before installing a compressor, the condensing unit base should be cleaned. Be certain that all the hold-down bolts

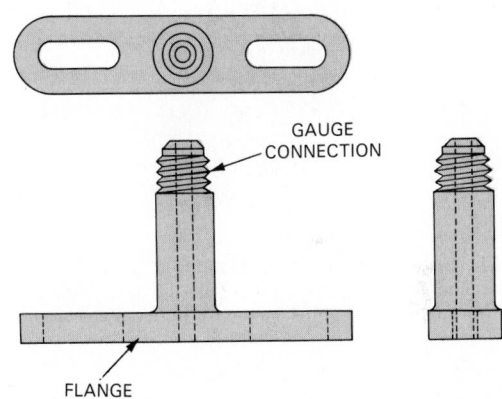

Fig. 14-64. Flange attachment takes place of service valve on compressor, while compressor is being tested.

are used and are tightened evenly.

If hold-down bolts are difficult to hold in place, masking tape may be used to hold them temporarily. String may be used to pull the bolts into position. A universal socket wrench is handy for holding bolts not readily accessible.

Compressor and motor must be carefully aligned. The compressor shaft and motor shaft must be parallel. Flywheel and pulley must be in line.

Belt tension is important. The belt should be tight enough to allow only 1 in. of belt deflection with approximately 25 lb. force on it. Multiple belts should all deflect the same amount. New gaskets should always be used when mounting service valves on a compressor.

Mount the compressor on a flat level base, test with a thickness gauge and use shim stock if necessary. An uneven base will put stress on the crankcase and may cause damage or misaligned parts.

14-43 SERVICING HERMETIC COMPRESSORS

Two types of hermetic compressors used in commercial refrigeration are:
1. Welded motor compressors.
2. Bolted motor compressors.

The welded motor compressors may or may not have service valves. Bolted units usually have them.

The bolted motor compressor units are tested, removed, overhauled, retested, and installed in a manner similar to the conventional compressor. Since the motor is built into the housing, it is usually tested and reconditioned at the same time as the compressor.

Welded motor compressors very seldom have service valves in the smaller units. Such valves must be mounted in the system. One method of doing this is to use a line tapping valve, as shown in Fig. 14-65. Larger units usually do have two-way compressor service valves.

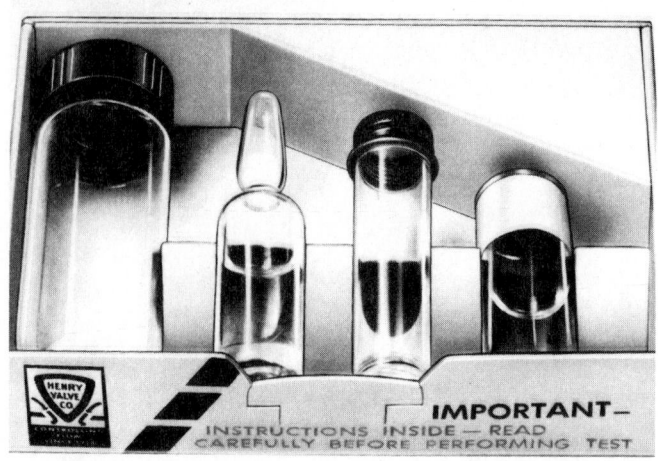

Fig. 14-66. Kit tests oil for acidity. Oil sample is put in container D. Oil and acid test solutions in containers B and C are poured into container A. After shaking, container A is put on bench for a minute or so to allow liquids to separate. If liquid at bottom is clear, oil is acid. If liquid at bottom is pink, oil is OK. (Henry Valve Co.)

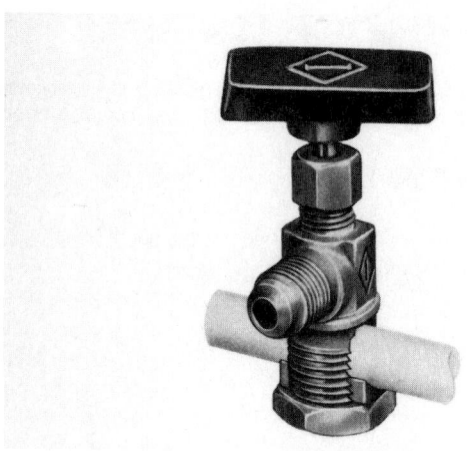

Fig. 14-65. Special valve can be mounted on suction line to pierce refrigerant line for pressure testing, charging, and/or discharging purposes. (Imperial Eastman, Imperial Division)

Chapter 11 describes the testing, removing, reconditioning, and installing of welded hermetics.

The condition of the oil in hermetic compressors is very important. It should be tested to determine its acid level. The presence of acid in the oil indicates the oil is breaking down and reacting with the moisture and refrigerant it has contacted. This is usually caused by breaks in the motor winding insulation causing localized hot spots. This check can be done by removing an oil sample and using an oil test kit. See Fig. 14-66. If the oil is acidic, the compressor should be thoroughly inspected.

An oil sample is easily drained from the serviceable hermetics. Welded hermetics usually require removing the motor compressor and draining some oil out of the suction line opening.

When replacing the motor compressors, always use an exact replacement.

14-44 SERVICING CONDENSERS

Condenser service procedures depend on the type of condenser. See Para. 14-47 and Para. 14-49.

In all cases, heat transfer surfaces must be clean. This is true for both the surfaces in contact with the refrigerant and the surfaces in contact with the cooling medium (air or water).

There must be enough refrigerant vapor space (area) to remove the heat from the vapor. For example, if a condenser has liquid refrigerant in it (is overcharged), cooling will be reduced. If the amount of air or water flowing is not enough, the condenser cannot do its job. If the air temperature or the water temperature is above normal, the condenser capacity is decreased; the condenser temperature will rise and the head pressure will increase.

If there is air in the system, it will collect in the condenser. (It cannot condense and will be held back by the liquid trap in the receiver or at the lower end of the condenser.) Each pound of air pressure will increase the head pressure by 1 lb. This increase in pressure will reduce the pumping efficiency of the compressor and will increase the condensing temperatures.

Excessive head pressures are very hard on a system. High exhaust temperatures will cause formation of sludge, carbon—even acid—and will injure reed valves.

High head pressures may be caused by:
1. Excessive low-side pressure.
2. Poor cooling by air or water.
3. Air in the system.
4. Overcharge.

Check the low-side pressure with a low-side pressure gauge. If the trouble is poor cooling of the condenser, make the following checks and adjustments:
1. Air cooled: Condenser must be clean—very clean. Use high pressure jet of air.
   Wear goggles!
   Use mechanical scrubbing or high-pressure water (jet) with detergent.
   Some air-cooled condensers may be cleaned by the use of a vacuum cleaner.
2. Water cooled: Is there enough water flow? Check outlet temperature and inlet temperature. Rise should not be over 15 °F (8 °C). Is there scale in water tube? Inspect

inside of water tube. Clean with cleaning solution.
For air in the system:
1. Purge the condenser while compressor is stopped.
2. Take precautions against freezing. This is necessary with water coolers and water chillers.

For an overcharged condition, purge the system while compressor is stopped.

Remember that the purpose of the condenser is to remove heat. The condenser will fail to do its job if the heat transfer surfaces are inefficient or if the heat removing medium (air or water) is not in the correct volume or temperature.

14-45 SERVICING AIR-COOLED CONDENSERS

The correct pressure in an air-cooled condenser may be determined by first adding 30°F or 35°F (−1° to −1.6°C) to the air temperature to get the refrigerant temperature on the inside of the condenser. Then, using this corrected temperature, refer to the refrigerant charts in Chapter 9 for the correct head pressure.

If pressure is above normal and there is enough air movement (both air-in passages and air-out passages must be free), there is a good chance there is air in the system. There is also a chance that the unit is overcharged. If the pressure is below normal, it is possible that the unit is undercharged.

Most commercial air-cooled condensers are of the forced convection type. They have one or more fans for moving air through the condenser. Some larger fans are belt driven and others are direct driven. Fans, motors, and belts need regular maintenance and service. Belt service operations are described in Chapters 7 and 22 as well as Para. 14-42.

Multiple fan condensers sometimes have sequenced fans. This means that, when more condensing is needed, all fans operate. As the condensing load decreases, first one fan is shut off, then the second, and so on. Sequence controls should be checked if the fans fail to operate.

Other systems use variable speed fans. Some outdoor air-cooled condensers have thermostat controlled louvers. They are powered to partly close or completely close as the outdoor temperature decreases. If the head pressure is too low, the refrigerant control capacity is decreased.

14-46 REMOVING AIR-COOLED CONDENSER

If a condenser is leaking or needs replacing for some other reason, it usually must be removed from the system. Before removal, the liquid refrigerant must be removed from the condenser and pressure must be adjusted to atmospheric pressure.

Usually, the procedure is this (be sure to wear goggles): Close the valve between the condenser and the liquid receiver. Purge the condenser by removing the gauge plug from the discharge service valve. Then open the valve until the condenser pressure is down to atmospheric.

Be careful. There may be oil in the condenser. It is best to connect a purge line to the gauge opening of the valve and run the purge line outdoors and into a container to trap the oil.

If there is no shutoff valve between the condenser and the liquid receiver, refrigerant can be saved by pumping it into a cylinder, as shown in Fig. 14-67. The refrigerant cylinder usually has to be cooled during this operation or its pressure will quickly rise to dangerous levels. Run the compressor intermittently and put the cylinder in a tub or bucket of running cold water or ice water.

Some service technicians put a spare condenser between the compressor and the service cylinder. This speeds the condensing operation.

The liquid receiver may need to be heated to vaporize the liquid in it. Use warm water. Never use an open flame.

If the refrigerant is not to be saved, purge the refrigerant outdoors. Be sure to use an oil trap in the purging line.

To remove a condenser, first clean the condenser as well as possible. Brushes, vacuum cleaner, air or nontoxic refrigerant jets, carbon dioxide, and/or nitrogen jets may be used. Wear goggles! *CAUTION: See Para. 11-39.*

Most air-cooled condensers are housed in a protective shroud which also serves as an air duct. On some of the larger units these sheet metal parts are heavy. Handle with care. Gloves and safety shoes are recommended. Be sure to save sheet metal screws and/or assembly bolts in a container.

Fans, fan brackets, belts, and motors may need to be removed on some units. These parts should be labeled and stored for reuse. Be sure the fan blades are not nicked or bent. This may put them out of balance and decrease their efficiency.

If electrical connections are removed, label them. Use masking tape and a marking pencil.

Always clean the connections before disconnecting the condenser from the unit. Immediately plug the refrigerant openings to keep the internal refrigerant passages clean and to avoid oil spills as the condenser is moved.

Avoid damage to condenser fins. Wood or cardboard protectors taped over the corners of the fins will provide protection. Because fins are sharp, always use gloves when lifting or carrying a condenser.

14-47 REPAIRING AIR-COOLED CONDENSER

A leaking condenser can be repaired. First clean the condenser and flush the inside of the refrigerant tubes. If a brazed joint is leaking, clean the outside of the joint, put flux on it, heat it, and take the joint apart.

Clean the brazed surfaces, flux the male part of the joint, assemble, support the joint, and braze. Remove the flux.

If a tube is cracked, remove the damaged part and replace with a new tube section. Braze the new part in place.

Fins may be straightened using a fin comb, Fig. 14-68.

To test a condenser, plug one end of the condenser and connect a refrigerant cylinder to the other end. Build up a refrigerant vapor pressure in the condenser and test for leaks. Use one of the following methods:
1. Bubble test.
2. Halide torch.
3. Electronic leak detector.
4. Immerse condenser in water.

Inspect fittings (and flares, if used). These connections must be in good condition.

14-48 INSTALLING AN AIR-COOLED CONDENSER

Always protect the condenser fins and condenser tubing, including return bends. Mount the condenser securely in its frame. Install it as level as possible. Connect the condenser to the compressor and liquid receiver, if used.

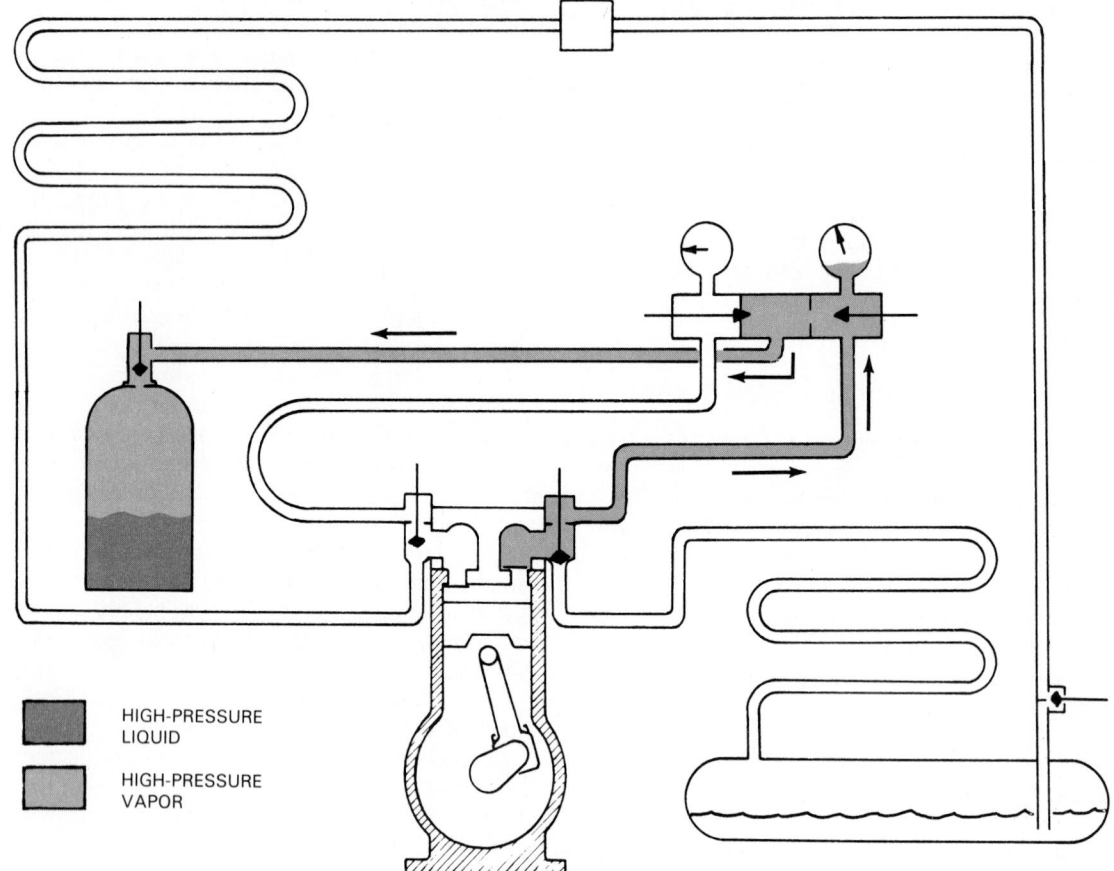

HIGH-PRESSURE LIQUID

HIGH-PRESSURE VAPOR

Fig. 14-67. One way to discharge a system. Refrigerant is being pumped into refrigerant cylinder.

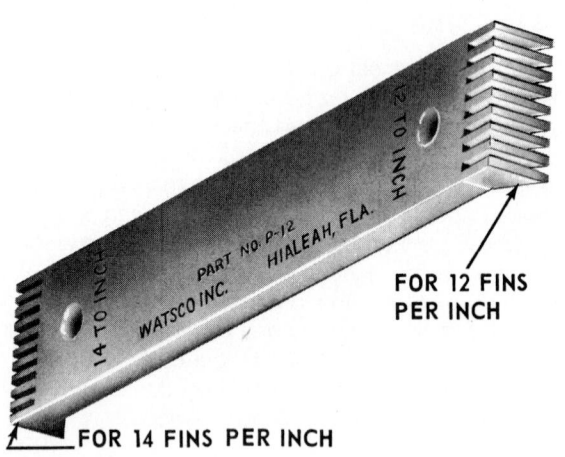

PART NO. P-12 WATSCO INC. HIALEAH, FLA.

12 TO INCH

14 TO INCH

FOR 12 FINS PER INCH

FOR 14 FINS PER INCH

Fig. 14-68. Plastic comb used to straighten condenser and/or evaporator fins. (Watsco Components, Inc.)

Flare connectors should be carefully aligned. The fittings should be in line so they are not under tension or forced in any way. If the fittings are out of line or are under strain, threads on the fittings or the flare may be damaged. Brazed connections must also be carefully aligned before brazing.

The surface to be brazed should be cleaned with clean steel wool, wire brush, or dry sandpaper just before assembling the joint. Flux should be put on the outside of the male part of the fitting only. The joint should be supported during the brazing operation. See Chapter 2 for instructions on brazing. Use asbestos or metal sheets to protect other parts of the unit from the brazing flame. Be careful when working with asbestos.

Use the refrigerant in the system or a service cylinder to purge the air from the condenser. (Purge outdoors or out of the room if possible.) Build up a pressure of about 15 psi (207 kPa) in the condenser and test for leaks. If a leak is found, do not attempt to patch the leak. Instead, take the joint completely apart and do it over. If no leak is found, increase the pressure to approximately 100 to 170 psi (794 to 1 277 kPa) and test for leaks again.

If no leaks are found, install fans, belts, and motors as needed to operate the system. All these parts should be cleaned prior to assembling.

Run the system. *Keep clear of the fan, belts, and pulleys as they may cause serious injuries.* Check refrigerant charge and operation of the unit.

Test for leaks again. (Shut the unit off to stop airflow.) If the unit is OK, install the shroud or casing. These parts should be securely fastened or rattles and inefficient airflow may result.

14-49 SERVICING WATER-COOLED CONDENSERS

Water removes heat from metal surfaces about 15 times more rapidly than air. Therefore, water-cooled condensers are much smaller than air-cooled condensers. Since the water is usually colder than air, the condenser temperature and pressure can be lower.

Usually the condenser is designed to permit a water temperature rise of 10 °F (6 °C) as it goes through the condenser. If the water-in temperature is high, the condenser temperature will be high and vice versa. Take care to avoid freezing the water circuit of water-cooled units.

If the unit is water-cooled, add 10 °F to 15 °F (6 °C to 8 °C) to the temperature of the water as it is leaving the condenser to determine what the refrigerant temperature should be. The correct head pressure may then be taken from a refrigerant chart. The unit must be running for these conditions to hold true.

If the head pressure exceeds this value by more than 5 lb., stop the compressor. Purge the system through the discharge service valve gauge opening for 10 to 15 seconds. Then run the condensing unit again.

To determine whether the trouble is excess refrigerant or air in the system, stop the unit and purge as before for 15 to 20 seconds. If the pressure drops somewhat, the trouble was air in the system.

If the pressure does not drop, continue to purge the unit until that part of the condenser and liquid receiver which is full of liquid refrigerant becomes cold. *Do not allow the temperature to go lower than 32 °F (0 °C) or the water tubes may freeze and burst.*

Some condensing units have small valves that can be used to check liquid levels. A liquid level sight glass is installed in some large units. It is connected to the top and bottom of the receiver. When the pressures are equalized, the liquid level in the sight glass will be the same as that in the receiver. This will quickly reveal the level of the refrigerant in the condenser and liquid receiver. One may easily judge whether this is the correct amount.

Always leak test the water leaving the condenser; there may be a leak between the water tubing and the refrigerant in the system.

An excess of oil in a system is indicated by erratic refrigeration and a constant slugging or pumping of oil in the compressor, especially just as it starts. (Erratic means that sometimes it works and sometimes it does not work.)

A common water-cooled condenser problem is formation of deposits from the water on the tubing walls. The minerals normally found in solution in water (carbonate, sulphate, lime, iron, etc.) are drawn to the walls of the condenser tubing. This occurs in an electrical process set up due to the opposite charges of the minerals and the tubing. These deposits then act as an insulating layer. See Fig. 14-69. If this layer cannot be removed, the condenser must be replaced.

Cleaning condensers has two different operations:
1. Preventing scale.
2. Cleaning the system.

A sulphuric acid-chromate solution is good scale prevention in open systems such as cooling towers and evaporative condensers. However, chromates are a poison and no significant amounts should be allowed in waterways. (Wear goggles, rubber gloves, and a rubber bib apron!)

Sulphuric acid in weak solution reacts with steel, while sulphuric acid in strong solution reacts with copper. It is best not to use it at all for scale prevention.

Careless mixes of sulphuric acid and water can ruin a copper piping system and a steel structure very fast. Some

Fig. 14-69. Scale deposits. A—Tubing with treated water. B—Tubing using hard, untreated water. (Scale Free Systems, Inc.)

technicians do use acids to descale, but it must be done as fast as possible, then cleaned as fast as possible. Wear goggles and rubber gloves.

For cleaning condensers, use only prepared chemicals from a good company. Carefully balanced cleaners such as inhibited muriatic acid or sulphuric acid do a good job without damage to the equipment, if properly used.

One can recognize a corroded condenser by checking the liquid line temperatures of the refrigerant. If the amount of refrigerant is correct and if the other troubles just mentioned are not found, a corroded condenser will produce a hot liquid line in a water-cooled condenser. Temperature and pressure in the condenser will be considerably above that to be expected. If all other possible causes of excessive head pressure are eliminated, a badly corroded or dirty condenser is probably the cause.

Soft scale deposits can be removed from some water-cooled condensers by using a power-driven wire brush, as demonstrated in Fig. 14-70. This can usually be done without removing the condenser from the system. However, the water circuit must be closed and the unit shut down. Always use new gaskets and tighten assembly cap screws evenly.

When the condenser water tubes have a hard lime or iron deposit, they may be cleaned with the acid solution described in Para. 14-40. The procedure is as sketched in Fig. 14-71. CAUTION: Always wear goggles, rubber gloves, and a rubber bib apron when using acid solution.

Fig. 14-72 shows cleaning being done with a forced cir-

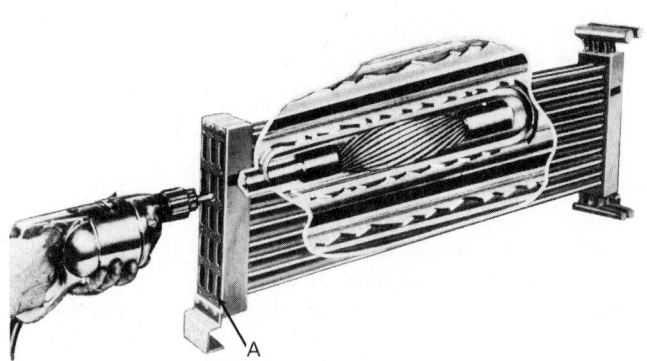

Fig. 14-70. Tube-within-a-tube condenser being cleaned with power-driven wire brush. A—End plates removed. (Standard Refrigeration Co.)

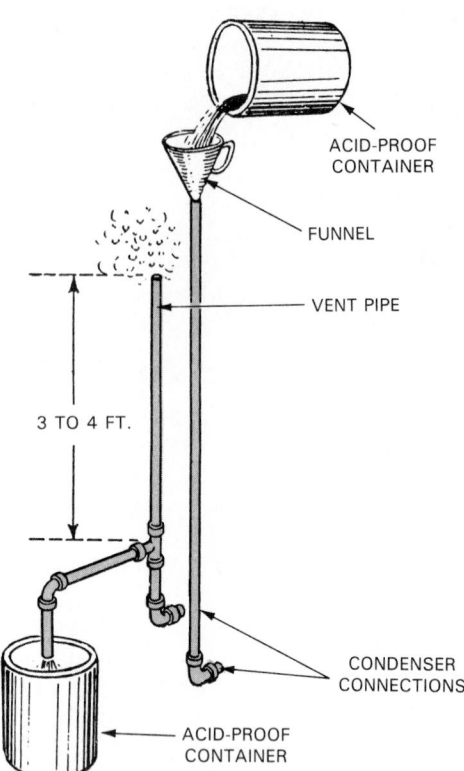

Fig. 14-71. Cleaning water-cooled condenser water tubes. Using dilute hydrochloric acid solution, connect vent pipe to upper condenser connector as shown.

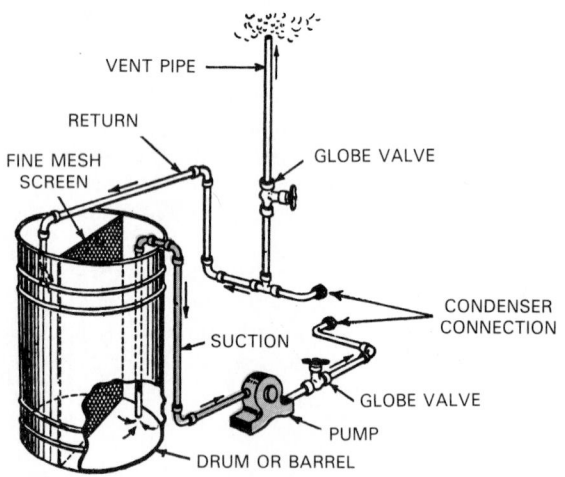

Fig. 14-72. Condenser water tube cleaning using a forced circulation system for acid solution. Note screen in drum to keep scale from entering pump. Drum must be acid proof. Use ceramic, crock, glass, or glass-lined tank.

culation system. The pump must be designed for an acid solution. Fig. 14-73 pictures a pump which can be put into the solution. Note the use of plastic tubing instead of pipe.

Warm solutions will react faster than cold. But do not heat above a warm temperature. Avoid spilling solution on the skin, clothing, or floor.

Crocks and barrels used must be made of acid-resisting materials. Wood or ceramic materials are acceptable. *Do not use galvanized or aluminum containers.*

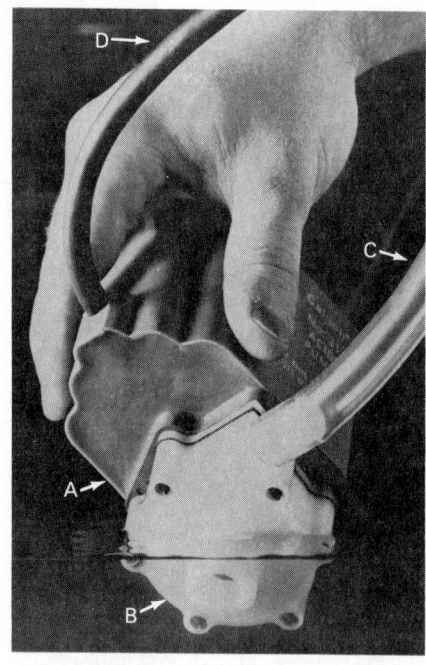

Fig. 14-73. Pump designed to pump cleaning acids through the water tubes of water tube condensers. A—Motor. B—Pump. C—Acid fluid outlet. D—Power cord. (Calgon Corp.)

There are "scale free" systems now available which effectively eliminate the electrical process which causes deposits. Fig. 14-74 illustrates a scale free system which exposes an electrolyte, in the form of a positive ground rod, to the water in the system being treated. This, in turn, picks up the electrical energy from the water, grounds it to the outside of the boiler or condenser, and keeps the inside tubing free from scale. This electrical energy applied to the outside of the container also causes existing scale or deposits to go back into solution, where they are then carried through the system and discarded.

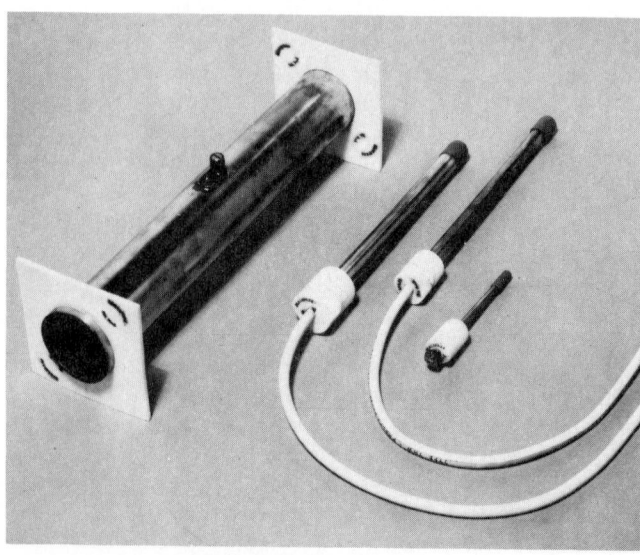

Fig. 14-74. Four different size units, used in equipment that has water as a heat exchanger. These units will remove and prevent the formation of scale. (Scale Free Systems, Inc.)

14-50 REMOVING WATER-COOLED CONDENSER

A leaking condenser should be removed and repaired or replaced. Removing a water-cooled condenser is similar to removing an air-cooled condenser except for the disconnection of the water line. First, close off the water circuits; then disconnect the water lines and drain all the water from the condenser. (It may freeze.) Then remove the refrigerant from the condenser and isolate the condenser. Do not allow freezing temperatures.

If a pressure-operated water valve is used, leave it installed in the system, if possible, because of its refrigerant tubing connection.

Clean the outside of the condenser, wipe or mop the water, and clean the condenser connections. Dry the connections thoroughly. If the connections are mechanical, use wrenches of proper size. *Wear goggles.* Plug the refrigerant openings at once using good plugs. Either synthetic rubber expanding plugs or flared plugs are recommended.

If the condenser is large and heavy, use a lifting device or have two or more individuals handle it.

14-51 REPAIRING CONDENSERS AND RECEIVERS

When either a condenser or a receiver does not work properly, it is usually cheaper to replace the unit than to repair it. Welding or brazing is sometimes used to repair leaks, but it requires careful work. If possible, weld or braze all joints to insure a lasting, leakproof joint.

A welded shell type condenser can be cut open, a new water coil installed, and the shell rewelded. But this is done only in an extreme emergency. The reason is this: pressure vessels should be made under well controlled conditions by a pressure vessel certified welder. Then they must be tested to at least twice the operating pressure.

Liquid receivers in most commercial systems serve as refrigerant storage cylinders. They usually have a welded steel shell.

The shell and coil condenser—liquid receiver that has a water coil built into it—sometimes develops leaks. Because of the corrosive action of the water and refrigerant under certain conditions, the copper tubing used to carry the water is eventually eaten through. The leaking tube lets refrigerant from the system into the cooling water. A leak of this kind may be found by checking for refrigerant at the water drain. Leaks sometimes occur at the joints where the water-cooling coil is attached to the liquid receiver. Such damage may be due to abuse or to corrosive action. In such a case, the condenser should be replaced with a new one.

If new parts cannot be found, a fairly satisfactory repair may be made as follows:
1. The water tubing—if eaten through within the liquid receiver—must be removed.
2. The liquid receiver should be mounted on a lathe and the end of the receiver cut open. This permits removing the old water coil and putting in a new one. The replacement unit must have the same length of tubing as the one removed, or the capacity of the condenser will be changed. The new coil is usually made up by winding it on a drum mounted on a lathe. The tubing is then put in the liquid receiver and the joints are brazed. The end of the liquid receiver that was cut off may now be replaced and welded after the interior has been cleaned thoroughly.

A bolted condenser can be dismantled quite easily to replace the water tubing.

All liquid receivers above a certain size must be equipped with safety release valves. Receiver repair should be attempted only with permission of local inspectors.

14-52 INSTALLING WATER-COOLED CONDENSER

Installing a water-cooled condenser is similar to the installation of an air-cooled unit. The condenser mounting, joint brazing, and leak testing should be done with the same care. On a water-cooled unit, the leak test should include the exhaust water.

Connect water lines according to the local plumbing code. Test the water circuit for leaks also. All parts should be cleaned before assembly. This includes the assembly devices. Be sure to test for leaks after the unit has run for a few hours and the system is in normal operation.

14-53 SERVICING COOLING TOWERS

Evaporative condensers and cooling towers also collect deposits from the cooling water. These deposits must be removed periodically or they will act as insulation. Deposits may be reduced by using water softening chemicals. Such chemicals can be bought from wholesale supply companies.

The treatment of water is necessary. Chemicals in the water are measured by a pH factor. The scale of pH is from 1 to 14, with 1 through 7 indicating an acid solution and 8 through 14 indicating a basic condition. Chemicals may be added to the water to create a 7 or 8 condition in the water.

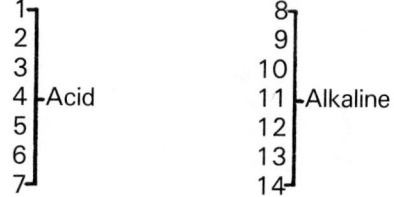

pH
Intensity factor (parts per million in quantity)

```
1┐                    8┐
2│                    9│
3│                   10│
4├Acid               11├Alkaline
5│                   12│
6│                   13│
7┘                   14┘
```

When testing water for pH factor (acidic or basic), the temperature of the water is important. The warmer the water, the more active the reaction to probes or color testers. It is best to test the water between 70 °F and 80 °F (21 °C to 27 °C). Water near boiling is about 15 percent more active. Very cold water (near freezing) will give readings about 5 percent below the true value.

Other chemicals may be used to lessen algae, mold, and slime growths. If deposits have formed, they can be removed by scraping or by using a weak acid solution. This is followed by a soda solution rinse and wash. A water-cooled condenser must be protected by electric heater or automatic drain controls if it is exposed to freezing temperatures.

If a water-cooled condensing unit is to be shut down, and perhaps be exposed to below freezing temperatures, the condenser coils must be completely emptied of water. This may be done by blowing out the coils with air, nitrogen, or carbon dioxide. *Do not exceed a 50 to 60 psi (449 to 518 kPa) pressure or the system may be damaged.*

The water drain valves should be left open to allow drainage of residual water in the piping. Be sure the drain plug of the circulating pump is removed and left loose.

Cooling towers need regular maintenance. Once a year repair any corrosion spots. It is good practice to do the following about once a month:

1. Inspect fan and motor bearings; oil sleeve type; grease (with water inhibitor) ball bearing types.
2. Inspect belt tightness and alignment; adjust, if necessary.
3. Clean strainer.
4. Clean and flush pump.
5. Inspect water level; adjust float, if necessary.
6. Inspect spray nozzles and clean if necessary.
7. Inspect water level bleed; it must be working.
8. Inspect air inlet screens and clean, if necessary.
9. Inspect water for algae, leaves, or other dust particles.

During cool weather, cooling towers and evaporative condensers create a fog exhaust. The high humidity air leaving the unit condenses enough moisture (dew point condition in the air) to create this fog. The fog condition can be reduced by cutting down cooling tower operation in cold weather. It can also be reduced by increasing airflow and reducing water flow.

Cooling tower drain lines should be as carefully planned and installed as the refrigerant and water lines. They must be large enough to be easily cleaned and must have clean-out connections. The joints must be leakproof. Piping must have a down slope all the way (1/4 in. per foot for horizontal runs). If a rise is unavoidable, a pump must be used.

All lines exposed to a freezing temperature should be insulated and heated. Heating tape can be used.

Most outdoor air-cooled condensers, cooling towers, and evaporative condensers use motor-driven fans. The motor bearings, belts, and fan bearings are exposed to wide temperature changes, moisture, and dirt. Then plain bearings require a special lubricant that will not wash out under moisture conditions. Silicone oils and greases are very satisfactory. The same lubricants are also good for ice makers, water pumps, and hydronic heating pumps.

Good city water systems deliver water with about 120 parts per million (ppm) of dissolved solids. This water can be cycled through a cooling tower only about six times before scale starts to form. For this reason, most water must be treated. Moreover, there must be a bleed-off device to keep the evaporating water cycle and system clean.

Water which evaporates should be in vapor form. If any small droplets of water are also exhausted from the cooling tower, these contain solids and add to air pollution.

One of the new ways to treat water is the electrostatic method. The device exposes untreated water to an electrostatic force. The force loosens the bond in the scale forming chemical. The loosened bond prevents scales from forming and prevents the impurities from combining. It even loosens scales already formed on metal surfaces.

The basic theory behind water treatment is to reduce the ion content in the water to a condition where the salts and foreign matter are made electrically neutral. These materials then remain in solution.

The electrostatic treatment must be adjusted to the raw water condition. Electrical instrumentation and chemical analysis is used to determine the adjustment.

14-54 SERVICING ICE MAKERS

Ice makers have refrigerating systems similar to other cooling applications. However, they have special water

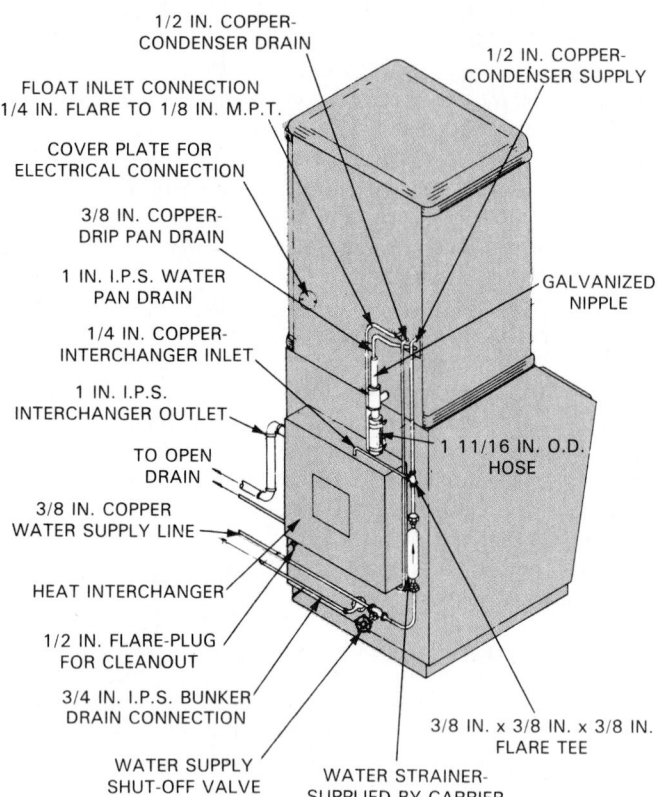

Fig. 14-75. External plumbing installation for automatic ice maker.

circuits, defrosting devices, and ice cube or flakes moving devices. Fig. 14-75 shows an ice maker piping installation. Many of these units have a water pump for recirculating the water used for making ice. Fig. 14-76 shows a water pump and motor installed. Just in front and below the motor, is the float control for water makeup. Motors, pumps, and water level float controls should be checked frequently. Fig. 14-77 shows one type of installation.

Those parts of the ice maker which are in contact with water should be cleaned about once each month. Special commercial cleaners for ice makers are available.

14-55 SERVICING WATER VALVES

Water-cooled condensers often require attention because of incorrect water flow. This trouble is sometimes due to the water valve or to the screens in the water circuit. See Chapter 12 for details on how these valves are built.

The main job of the water valve is to provide water when the unit is running and to stop the water flow when the unit is idle. Some trouble caused by water valves are:

1. Too little water flow.
2. Too much water flow.
3. Water flow does not stop when the unit is idle.

A water control valve will only operate correctly if the installation is correctly made. Water must also be clean.

14-56 RESTRICTED WATER FLOW

If water flow is too little, the trouble might be:

1. Leaking valve.
2. Clogged or partially clogged screen.
3. Chattering valve.

Fig. 14-76. Water circulating pump and water float control for ice maker. (Ross-Temp/IMI Cornelius Americas, Inc.)

4. Valve adjustment turned out too far.
5. Sediment-bound valve.
6. Leaking bellows.

In addition to these, the water-cooling system may present other problems. Some water-cooled systems use a length of rubber or plastic hose between sections of the water pipe.

This hose is run along the wall and the condensing unit water lines to eliminate the transmission of the condensing unit vibration into the plumbing system of the building and the breakage of tubes from this vibration.

The cold water inlet hose connection presents no difficulties. This hose will ordinarily give many years of service. However, the water outlet hose may decompose and clog.

Occasionally, someone may partially or completely shut off the water supply by closing the hand-operated valve installed in the system. A service technician should always put signs near the shutoff valves warning of the effect on the refrigerating unit if these valves are closed.

There are two common complaints which indicate troubles with water circulation. They are a lack of cooling in the condensing unit and too great a consumption of water.

If the water circulation is stopped, the refrigerating system will start to short cycle. The high-pressure switch causes this since these systems are always provided with a high-pressure safety motor control for this purpose. As the head pressure of the machine builds up (due to a lack of cooling), a pressure is soon reached which will cause the control to open a switch. This stops the motor. Once the motor is stopped, the head pressure drops rapidly. This permits the high-pressure control to turn the motor on again.

Short cycling will continue unless the trouble is remedied. Such a condition is a severe strain on the motor. Furthermore, it does not provide satisfactory refrigeration.

14-57 TOO MUCH WATER FLOW

Too much water flow will give satisfactory refrigeration, but more water will be used than needed, increasing the cost of operation. Three things may cause too great a water flow:
1. Water pressure too high.
2. Leaking water valve.
3. Water valve out of adjustment and holding a valve too far open.

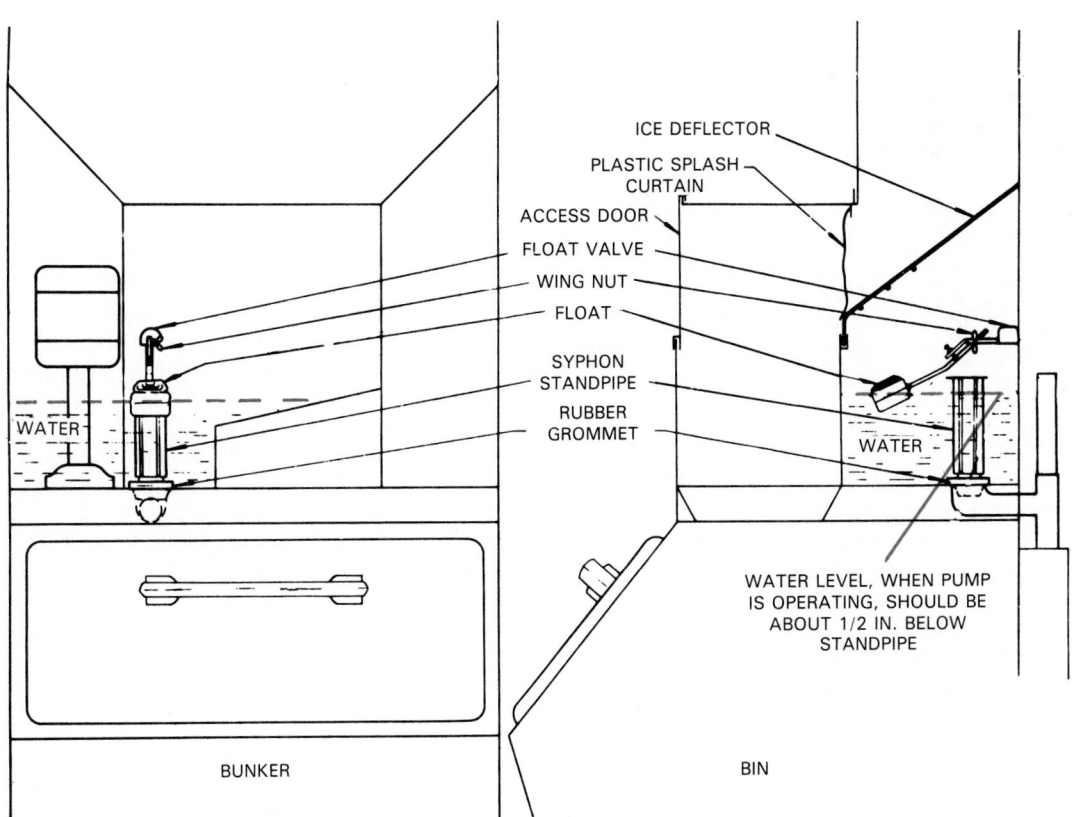

Fig. 14-77. Water pump and water level control for typical automatic ice maker.

Water pressure that is too high is seldom found, unless the water supply pressure is uncontrolled. If found in one machine, it will perhaps be true for all the machines in that locality.

To determine whether the condition is due to a leaking valve or to a valve open too far, find out if anyone has been working on the machine. If the machine has not been tampered with and if trouble has just started, the trouble is usually a leaking water valve. A further indication of this is a continuous flow of water on the off cycle. Quite often a leaking water valve may be caused by foreign matter lodging between the valve and its seat. One can usually dislodge this material by flushing the valve. The flushing may be done by using a screwdriver to pry the valve open several times.

14-58 TRACING WATER CIRCUIT TROUBLES

In water circuit troubles, one must determine whether the water valve is at fault, or if it is some other part of the water circulating system. To locate the exact source of the trouble, disconnect the joints where the hose is fastened to the rest of the system.

Disconnecting the water outlet pipe from the machine unit will indicate if the water is flowing as far as this point. If so, the connection should be resealed and the other end of the pipe disconnected from the wall pipe. If the water does not flow up to this point, the trouble is probably in the drain pipe of the system.

Determine whether the water is coming as far as the water valve by disconnecting the pipe used as the inlet connection. If the water flows through the pipe, the sources of trouble are the water valve and the condenser proper. To check these sources, reseal the inlet pipe to the system and disconnect the water valve from the condenser.

If the water flows through the water valve, but does not go through the condenser, the trouble is a major one, requiring replacement or cleaning of the condenser water tubes. However, if the water does not flow through the water valve, the water valve must be disconnected from the system and repaired or replaced.

Water hammer is a very noisy condition that is easily recognized. A single, distinct thump or rap is heard in the pipes just as a valve is closed. Generally, this condition can be corrected by placing a short, vertical pipe in the water line just ahead of the valve. It provides a column of air which absorbs the shock of flowing water suddenly being required to stop.

14-59 REMOVING WATER VALVE

To remove an electrical solenoid water valve from the system, open the hand switch that controls the circuit to the motor. Remove the water valve wires from the motor circuit. If these wires are soldered and taped, it will be necessary to unsolder or cut them. Most connections are of the slip-on type.

Before disconnecting the water valve from the water system, shut off the water supply at the hand valve. If the service technician does not have a replacement valve on hand, the water system may be temporarily connected without a water valve; the water flow can be regulated with the hand shutoff valve.

Some pressure-operated water valves are difficult to remove from the system because the valve is connected to the high-pressure side of the condensing unit. The pressure tube for these valves is usually connected into the cylinder head of the compressor. However, some manufacturers connect this tube into the liquid line. Sometimes this tube has a hand shutoff valve. If so, removal of the valve is simple.

If the tube is connected to the cylinder head of the compressor, the following procedure is suggested:

1. Install gauge manifold.
2. Turn suction service valve all the way in.
3. Run compressor until pressure in crankcase reaches 0 psi. *Note: Be very careful of oil pumping which will sometimes occur before 0 psi can be reached.*
4. Heat the water valve line and the water valve bellows carefully with a heat lamp for three or four minutes until both are quite warm to the touch. This operation will move the liquid refrigerant (which has condensed in this tube and valve) back into the condensing unit. Then, only a small quantity of high-pressure vapor will be left in this tube.
5. Turn discharge service valve all the way in.
6. Bypass the high pressure in the manifold and water valve refrigerant line into the low side by opening both manifold valves.
7. Clean joints to be opened.
8. Disconnect pressure tube from water valve. Wear goggles!
9. Plug the refrigerant pressure tubing openings immediately.
10. Gently heat water valve again. Often a quantity of liquid refrigerant becomes oil bound within the bellows chamber and releases with explosive force a few minutes after the valve is opened to the atmosphere. Heating will drive it out.
    Note: Be very careful not to point the refrigerant openings toward anyone because of the danger of being hit by the refrigerant. It is best to wrap the refrigerant openings with several layers of heavy toweling to absorb any refrigerant being thrown from the mechanism.
11. Shut off water supply.
12. Disconnect water valve from water line and replace it with a good one or connect water lines directly. Now the water valve is ready to be dismantled and repaired.

Some water valves are designed to permit removal of the valve body without disturbing the refrigerant connections. To disconnect one of these valves, shut off the water and disconnect valve body from water lines and bellows body.

Thermostatic water valves or motorized water valves are easily removed; only electrical connections need to be broken and the water circuit closed.

14-60 REPAIRING WATER VALVES

It is better to replace than repair a worn water valve. Sometimes a water valve only needs cleaning. Wire brushes and a muriatic acid solution work best. Use the same precautions as when cleaning the water tubes of a condenser.

If no replacement valve is available, the valve and valve seat must be repaired. The valve seat, brass in most cases, may be lapped in a manner similar to lapping a compressor valve seat.

The valve is usually made of fiber, rubber, or Bakelite material. It should be replaced. In cases of emergency, however, this valve may be trued up by using a fine grade of sandpaper backed up by a level surface.

Occasionally, where the valve stem passes into the valve body proper, a packing gland is used to seal the joint. This packing is usually composed of graphite, lead, and other materials. If the packing nut has been turned all the way down and this joint still leaks, the packing should be replaced.

Electric water valves may have a faulty electrical coil (either shorted or with an open circuit). Replace it with a coil of the same electrical properties (voltage and wattage). Thermostatic water valves may lose the element charge. If so, replace.

14-61 INSTALLING AND ADJUSTING WATER VALVES

After cleaning and repairing a pressure-operated water valve, it should be tested and adjusted before being placed in service. If the maximum temperature of the water supply is 75 °F (24 °C), the valve should be adjusted to open at the following pressures: 87 psig (704 kPa) for R-12; 144 psig (1 097 kPa) for R-22; 152 psig (1 152 kPa) for R-717; 112 psig (876 kPa) for R-500; and 173 psig (1 297 kPa) for R-502.

If the water-in temperature is above or below 75 °F, the valve should be adjusted to its correct opening pressure by referring to Fig. 14-78.

A valve should also be tested for leaks at the same time it is being adjusted. To do this, connect an air pressure line to the inlet water opening of the valve. To adjust and test the pressure operating bellows, which controls the water flow, connect another air line and a pressure gauge to this fitting. No air should flow through the water valve until the correct control bellows pressure has been reached. Adjustment may be made to obtain this condition.

Solenoid water valves are nonadjustable. Thermostatic water valves may be tested by using an adjustable temperature well to change the bulb temperature and air pressure used to check the valve operation.

After installing the water valve, check it for leaks (both water and refrigerant). Outlet water temperature, water flow, and condensing pressure should also be checked.

14-62 SERVICING DIRECT EXPANSION EVAPORATORS

''Dry'' evaporators are coils using either an automatic expansion valve (AEV) or a thermostatic expansion valve (TEV) refrigerant control. See Chapter 5.

These evaporators must be clean both inside and out for good heat transfer. They must contain just enough liquid refrigerant at the proper vapor pressure to provide the required cooling. Air or water being cooled must flow in and out of the evaporators efficiently. The evaporator must be leakproof and of the proper size.

A service technician should check these conditions. Ideally, pressure at the inlet of the evaporator (just after the AEV or TEV) should be measured as well as pressure at the outlet.

However, most service technicians check only the low-side pressure at the compressor suction service valve and think that the evaporator pressure is close to this pressure. The pressure drop due to friction in the tubing and bends can be checked by reading the low-side pressure when the unit is running and then reading it again just as the compressor stops. The rise in pressure is the pressure drop. Normally this pressure drop will be 2 to 3 psi.

Evaporator temperature should be checked too. Thermometers can be mounted on tubing using spring-loaded clip-on thermometer holders. Superheat setting of the thermostatic expansion valve can be checked using thermometers. The best setting is when the bulb temperature varies the least while the unit is running. Location of the liquid in the evaporator can also be determined this way.

Frost accumulation acts as an insulation and also tends to reduce the airflow. If found near the TEV, it usually means too great a superheat adjustment along with low-suction pressures. Spotty frost usually means uneven airflow over the evaporator or that some defrosting elements are not working.

Airflow through the evaporator can be checked with an anemometer. This is also called an air velocity meter. See Chapter 18. If the air outlet or inlet is too small, the evaporator will be starved. Air-in and air-out temperatures can also be checked. The air temperature will usually drop about 15 °F (−9 °C) as it passes through the evaporator.

When checking for leaks, the fan and unit should be shut off. The low-side pressure should be at least 15 psig (207 kPa) when testing for leaks.

14-63 REMOVING THE EVAPORATOR UNIT (DRY SYSTEM)

If a dry evaporator needs to be removed, first install a gauge manifold and test for leaks. Start the compressor and close the LRSV (liquid receiver service valve). Run the compressor until atmospheric pressure or a constant vacuum has been produced. Continue running the compressor until the evaporator and liquid line are warm. At this point, all the liquid refrigerant is removed. To speed up this operation, heat the evaporator carefully with a heat lamp or hot water. Never allow it to get more than warm to the touch.

	WATER TEMP. °F										
	50	55	60	65	70	75	80	85	90	95	100
REFRIGERANT	HEAD PRESSURE (PSIG)										
R-12	56	62	68	74	80	87	93	101	108	117	125
R-22	95	104	113	123	133	144	155	158	180	194	208
R-717	98	108	119	130	140.5	152	164	177	191	205	220
R-500	71	78	86	94	103	112	121	131	142	153	165
R-502	116	126	137	148	160	173	186	200	214	230	246

Fig. 14-78. Table of head pressures for systems with various refrigerants at various inlet water temperatures.

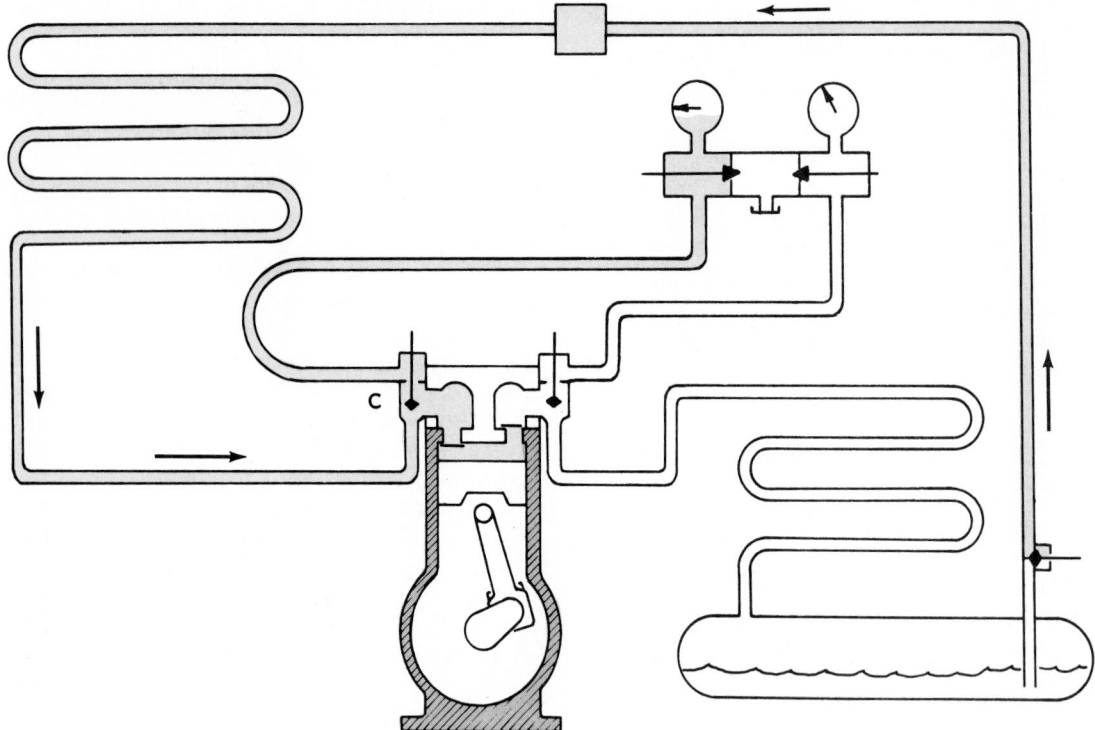

Fig. 14-79. Liquid line, evaporator, and compressor crankcase being pumped down. Refrigerant is being pumped into condenser and receiver.

A balanced (atmospheric) pressure in the evaporator may be obtained by either warming the unit or by bypassing high pressure back through the gauge manifold. Fig. 14-79 shows an evaporator being pumped down. Turn the suction line service valve, C, all the way in, closing the suction line after balancing.

Check the suction line; if it has a suction pressure regulator in it, or a solenoid valve, be sure it is open.

If a "hot gas" injection unit or a liquid injection unit is connected to the suction line, be sure the solenoid control valves for these units are shut off.

If the system has hand shutoff valves for each evaporator of a multiple installation, use these valves instead of the compressor service valves. Close the liquid line valve first and pump refrigerant out of the evaporator. Be sure there is 0 psi or slightly more in the evaporator. Close the suction line hand valve and the evaporator is ready for removal.

Shut off the electric power to the fan and liquid line solenoid valve (if there is one). Remove the casing or shroud of the evaporator carefully.

If electric defrost elements are mounted in or on the evaporator, disconnect them.

Clean and dry the suction line where it is connected to the evaporator. Also clean the inlet connection. Then unfasten the suction line and liquid line from the evaporator. Plug the openings with appropriate fittings. Wear goggles!

14-64 REPAIRING DIRECT EXPANSION EVAPORATOR

Repairs are usually limited to:
1. Repairing leaks.
2. Repairing or replacing fittings.
3. Straightening fins.

4. Replacing defrosting elements.
5. Repairing or replacing hangers.
6. Repairing or replacing fins and/or motors.

Where leaks occur, completely dismantle that part and clean the surfaces. If it is a brazing repair, follow the procedure explained in Chapter 2.

Always anneal an old tube before flaring it. Fins can be straightened using a fin comb or wide-jaw pliers.

Electrical defrosting elements should be checked for continuity. Terminals and insulation should be inspected also. Rusty or bent hangers and abused hanger assembly bolts should be replaced.

Check the fan and motor for vibration: check tightness of the fan on the motor shaft, motor end play, motor bearing wear, and condition of lubricant. Small faulty motors should be replaced. Larger motors can be rebuilt. (See Chapter 7.)

All parts should be cleaned before assembly. The evaporator is usually assembled on the job. If leaks have been repaired, the evaporator should be leak tested before it is installed.

14-65 INSTALLING DIRECT EXPANSION EVAPORATOR

In the event the evaporator has been removed, several important things must be remembered during the assembly to insure proper operation. After bolting the evaporator back into the refrigerator and leveling it, remove the plugs on the refrigerant openings and attach the liquid line and suction line to the unit. Be careful during these operations that no moisture enters the lines. It is good practice to dry the surfaces of lines and evaporator before removing seals.

The thermostatic expansion valve (TEV) should be installed using a new gasket—if it is of the bolted type. Connect a vacuum pump and evacuate the evaporator, suction line, and compressor. Evacuate the evaporator a second time. Test for leaks with 5 to 25 psig (138 kPa to 276 kPa) pressure; test for leaks again at ambient refrigerant pressure (high-side pressure) before operating the evaporator.

The evaporator may be dehydrated more completely by heating it to a fairly high temperature of 175 to 200 °F (79 to 93 °C) as it is evacuated. This drives out any moisture that may be present. Heat lamps may be used for this purpose.

After the evaporator has been installed and tested for leaks, install the electrical connections for the defrost units. Install the fan and motor. The electrical connections should be tight and moisture-proof. Operate the defrost unit and the fan.

Assemble the casing or shroud. Start the unit and check for normal operation.

14-66 INSTALLING EXPANSION VALVES

Mount the expansion valve and evacuate the liquid line, evaporator, and suction line. Carefully test for leaks by first purging, then building up a refrigerant vapor pressure. Install the fan and motor, if used. Open liquid receiver valve or liquid line hand valve. Start the compressor and observe its operation.

In multiple systems, all dry evaporators using expansion valves should be installed with individual shutoff valves for both the liquid and suction lines to each evaporator.

Attempts to adjust thermostatic expansion valves in an effort to maintain too great a difference in temperatures in various cabinets in the system gives rise to erratic operation. This is true, particularly in the evaporators which are closed off the most. To fix this—in addition to the thermostatic expansion valve—install one or more two-temperature valves in the proper places in the suction line.

Use gauges and thermometers to check for superheat setting. When the unit is operating correctly, connect the defrost wires and, if used, install casing and shroud.

In multiple commercial installations, in which finned evaporators are used, the expansion valve is sometimes attached to the evaporator with an SAE flared connection. The flare nut in such an installation must be shellacked or sealed from moisture. This is done after the installation has been made and before the unit starts to operate. Otherwise, ice may form between the nut and the tubing. In a short time the tube will collapse or break. This condition can also occur at the point where the suction line fastens to the evaporator. Fig. 14-80 shows the pinching operation of ice formation between the flare nut and the tubing.

Other methods have been devised to stop moisture from collecting behind flare nuts. One has a rubber seal at the end of the flare nut. Another method consists of drilling holes through the flare nut. The idea is that moisture will drain out and, if ice does form, it will release its pressure through the holes in the nut rather than against the tubing.

Short shank flare nuts should be used in places where frosting occurs. Fig. 14-81 shows a short flare nut with openings across the threads for moisture escape. As mentioned before, the best type of connection for use in the interior of the cabinets is a brazed flanged connection.

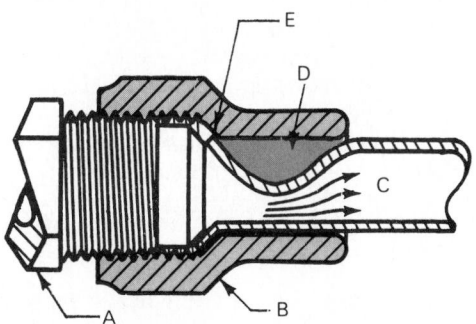

Fig. 14-80. Illustration shows what happens when moisture freezes between nut and tubing. A—Fitting. B—Flare nut and tubing. C—Restricted refrigerant opening. D—Ice formation. E—Flare being pulled out of place.

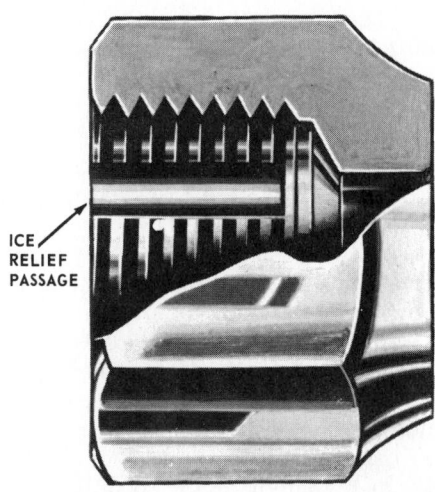

Fig. 14-81. Special flare nut used to prevent ice accumulation between flare nut and tubing. (Superior Valve Company, Division of AMCAST Industrial Corporation)

14-67 ADJUSTING THERMOSTATIC EXPANSION VALVES

The sensing element of the thermostatic expansion valve should be clamped to the suction line at the point where it attaches to the evaporator. It is possible to make bench adjustments of this valve.

In service, the thermostatic expansion valve may be adjusted as follows:
1. With the refrigerating condensing unit operating, note the temperature of the evaporator.
2. If the evaporator is too warm, adjust the control to allow more refrigerant into the evaporator.
3. If the suction line frosts up, adjust the control to reduce the refrigerant flow. Most service technicians adjust the thermostatic expansion valves according to the MSS (minimum stable signal) point. See Chapter 5. The amount of superheat is properly taken care of by this adjustment.

14-68 DRY EVAPORATOR REFRIGERANT CONTROLS

The operation of refrigerant controls in commercial systems is described in Chapter 5.

All types of refrigerant controls may be found in commercial systems. If the system has a self-contained hermetic unit, it may be serviced as described in Chapter 11.

The multiple evaporator installations, however, and the larger commercial units, do have several other features that a service technician should know.

14-69 SERVICING THERMOSTATIC EXPANSION VALVES

Dry systems which use the thermostatic expansion valve refrigerant control may be of the single evaporator or multiple evaporator type. The design and operation of this valve is explained in Chapter 5.

The service technician should check the complete system, install the gauge manifold and check for pressure. Check the amount of refrigerant through the sight glass. Check liquid line solenoid valves, two-temperature valves, "hot gas" or liquid injection systems, driers (both suction line and liquid line), temperature or pressure motor controls and electrical supply.

The service technician can check the evaporator by appearance, sound, and temperature of the expansion valve. If the evaporator is frosting back so the suction line is frosted, it may be due to the following troubles:

1. Needle may be leaking.
2. Control may be adjusted for too little superheat.
3. Valve may have the incorrect thermal bulb charge.
4. Valve orifice may be too large.
5. Power element may be attached loosely to suction line.
6. Power element may be located in too warm a position.
7. Dirt in system may be holding valve open.
8. An external equalizer valve may be needed.
9. Screen may be clogged.

It is difficult to determine which of these troubles is responsible for the problem. The best method is to systematically check everything it could be and rule them out one by one. The location of the power element may be easily checked and its attachment to the suction line noted. Recommended procedure is to place the thermostatic bulb on top of the suction line rather than beside or below it. In this position the liquid in the bulb will make good thermal contact with the suction line. Fig. 14-82 shows two recommended thermal bulb locations.

It is rare to find a thermostatic control in which someone has tampered with the adjustment. Usually customers will not try to set them. Therefore, do not attempt to readjust the control at first. Check for other troubles instead. A leaking needle or valve cannot be repaired. It should be replaced.

A starved evaporator (one giving poor cabinet temperatures while frosting unevenly or not sweating properly) may be due to the following:

1. Clogged screen in expansion valve which may give no refrigeration.
2. Loss of refrigerant from power element in expansion valve which will give erratic (undependable) refrigeration.
3. Moisture in the system which may sometimes give good refrigeration and then none. The moisture will freeze in the orifice to expansion valve and close it, then defrost.
4. Wax in valve. This wax is from the oil and its presence means that the oil used was for a different temperature range or was improperly prepared for refrigeration service.
5. Needle stuck shut. (This is a rare occurrence.)

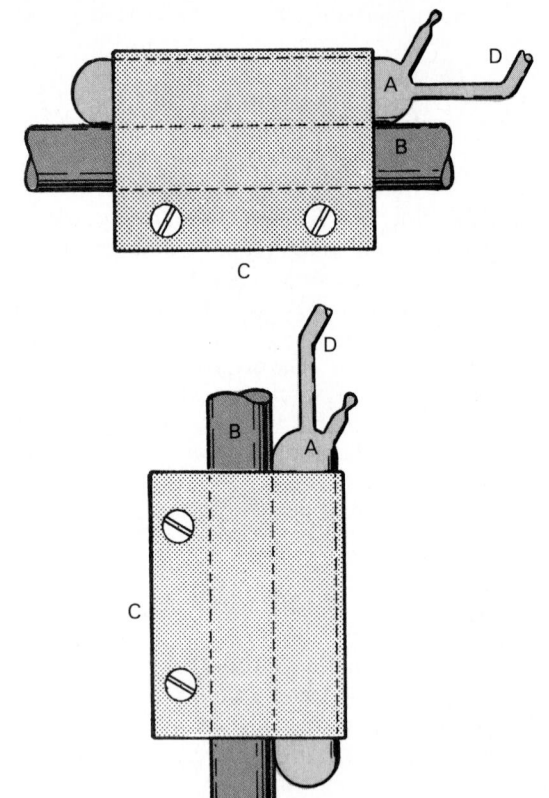

Fig. 14-82. Recommended locations for thermostatic bulbs on suction lines to obtain best operation. Bulb should be on top of horizontal suction line. It should have closed end on bottom when mounted on vertical suction line. A—Thermal bulb. B—Suction line. C—Clamp. D—Thermal bulb capillary tube.

6. Under-capacity valve orifice.

When an expansion valve' gives trouble, it is usually recommended that a new valve be installed.

Too much refrigerant flow will be indicated by a sweating or frosted suction line beyond the thermal bulb position. This condition may be caused by:

1. Thermal bulb loose from suction line.
2. Thermal bulb in warm airflow.
3. TEV orifice is too large.
4. TEV needle stuck open.
5. Undersize evaporator.
6. Thermal bulb has wrong charge.
7. Pressure drop in evaporator is too great.

If the bulb is loose, is mounted wrong, or is in a warm air stream, remove it; clean it and the tubing; remount it firmly and insulate it if necessary.

An oversize TEV should be replaced with one of the correct size. When a needle is stuck open, the best remedy is to replace the valve.

Only rarely does one find an undersize evaporator (replace it), a TEV with the wrong charge (replace), or too great a pressure drop (replace evaporator).

14-70 REMOVING EXPANSION VALVE

The service technician should remove a faulty expansion valve and replace it with one in good condition. The troublesome valve may then be checked in a shop equipped for this purpose and the trouble accurately determined.

Remove the valve using the procedure in Para. 14-63.

14-71 REPAIRING CLOGGED TEV SCREEN

A clogged screen may be easily detected by poor refrigeration, sweating, or frosting near the TEV only and no refrigerant sound.

Removing an expansion valve which has a clogged screen means removing both the liquid and vaporized refrigerant from the lines to be opened. The liquid line may be carefully heated with a heat lamp, driving the liquid refrigerant back to nearest shutoff valve. This valve then is closed. The evaporator is already evacuated (indicated by a warm evaporator). If low-side pressure is at atmospheric pressure or higher, the suction line valve may be closed. The screen may be removed after cleaning and drying the connections.

Clean the screen or replace with a new one. This is important. Fine-mesh copper or stainless steel screens may be cleaned fairly successfully with air pressure and a safe solvent. But the best way is by heating it. This must be very carefully performed or the screen will be burned. *Never allow an expansion valve to go into service without a screen being placed in the liquid line entrance.*

Install the screen, assemble the TEV, install it in the system, evacuate the air, test for leaks, and return the evaporator to normal operation by opening the suction and liquid line valves.

14-72 TESTING AND ADJUSTING THE THERMOSTATIC EXPANSION VALVE

Thermostatic expansion valves sometimes get out of adjustment; also, certain installations may require a periodic readjustment of this valve. A simple method of testing these valves in the shop will be explained.

A regular service kit having the necessary equipment and setup required is shown in Fig. 14-83. The equipment required is as follows:

1. Service cylinder of R-12 or R-22. In the shop a supply of clean dry air at 75 to 100 psig (621 to 794 kPa) pressure can be used in place of the service cylinder. The service cylinder or air is needed only to supply pressure.

2. High-pressure and low-pressure gauges. The low-pressure gauge should be accurate and should be in good condition so that the pointer does not have too much lost motion. A high-pressure gauge is recommended in order to show the pressure on the inlet of the valve.

3. A small quantity of finely crushed ice is necessary. One of the most convenient ways of handling this is to keep it in a thermos bottle with a large neck. If such a container is completely filled with crushed ice, it will easily last for 24 hours. Whatever the container is, fill it completely with crushed ice. Do not attempt to make this test with the container full of water and a little crushed

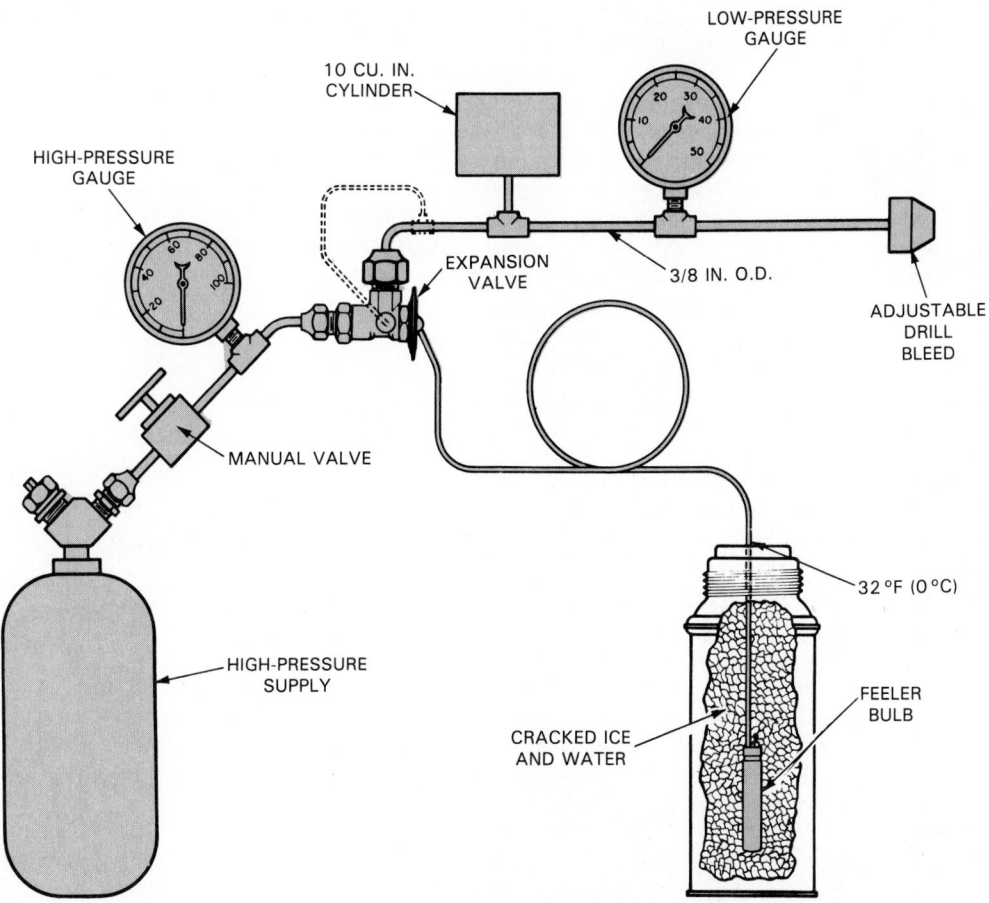

Fig. 14-83. Recommended setup for testing and adjusting thermostatic expansion valve. Note that liquid refrigerant is supplied from storage cylinder. Control bulb is in thermos bottle filled with cracked ice at temperature of 32 °F (0 °C).

ice floating around on the top (32 °F [0 °C] ice floats at the top, while 39 °F [3.8 °C] water sinks to the bottom). Proceed as follows to adjust the valve:

1. Connect the valve as shown in Fig. 14-83. The adaptor on the expansion valve outlet provides a small amount of leakage through the No. 80 drill orifice opening. A 10-cu. in. tank is used to reduce pressure fluctuations.
2. Insert the bulb in the crushed ice and allow it to cool.
3. Open the valve on the service cylinder. Be sure the cylinder is warm enough to build up a pressure of at least 70 psig (587 kPa) on the high-pressure gauge connected in the line to the valve inlet.
4. The expansion valve can now be adjusted. The pressure on the outlet should equal the pressure for the refrigerant at 22 °F (5.5 °C). The water and ice mixture is 32 °F (0 °C) and the sensitive bulb is also 32 °F. If the superheat is to be 10 °F (−12.2 °C), then the temperature of the refrigerant will be 22 °F and the pressure must be adjusted to match this 22 °F temperature. The pressure on the outlet gauge should be different for various refrigerants as follows:

R-12,	22 psig (255 kPa)
R-22,	45 psig (414 kPa)
R-500	29 psig (304 kPa)
R-502	55 psig (483 kPa)

When making the adjustment, be sure there is a small amount of leakage through the low-pressure orifice.

5. Tap the body of the valve lightly in order to determine if the valve is smooth in operation. The needle of the low-pressure gauge should not jump more than one pound.

To test the needle for leaks, close the orifice to stop the leakage and determine if the expansion valve closes off tightly. If the valve is in good condition and not leaking, the pressure will increase a few pounds and then either stop or build up very slowly. With a defective, leaking valve, the pressure will build up rapidly until it equals the inlet pressure.

To test the power element, remove the power element bulb from the crushed ice. Warm it up by hand or by putting it in water at about room temperature. The pressure should increase rapidly if the power element is operating.

Another method of checking and adjusting a thermostatic expansion valve (TEV) makes use of a device with a clamp for the TEV bulb, an accurate thermometer, and a small cavity for refrigerant. With this instrument mounted on the TEV bulb and the cavity charged with liquid R-12, the small screw adjusts boiling pressure. Therefore, the temperature of the R-12 and the bulb temperature can be quickly and accurately lowered and raised to check the control adjustment and to readjust it if necessary. Fig. 14-84 shows the instrument in use.

Note: With vapor charged expansion valves (charged with a small quantity of refrigerant), the amount of charge in the power element is limited and the pressure will not build up above the specified pressure. This pressure is always marked on the valve body and must be considered when testing vapor-charged valves.

The body bellows is tested with high pressure showing on both gauges as outlined in the preceding paragraph. A leak (escape of vapor) is detected by using a leak detector. *When making this test, it is important that the body of*

Fig. 14-84. Thermostatic expansion valve being tested. Note refrigerant cylinder, thermometer, low-pressure gauge, high-pressure gauge, bulb temperature control device at A, and expansion valve being tested.

the valve has a fairly high pressure. Gauge connections and other fittings must be tight to eliminate leakage at other points. Leakage can also be detected by use of soap suds.

14-73 REPAIRING EXPANSION VALVES

Three things likely to go wrong with an expansion valve are:

1. Needle and seat become worn.
2. Screen gets dirty; bellows leak. (See Para. 14-71.)
3. Thermal element loses charge. See Para. 14-72. If charge is lost, one must replace thermal element or the complete valve. Occasionally, one finds such trouble as a broken spring, but this is rare and the troubles are easy to locate.

As there are many models and types of TEV's on the market, only general instructions are given here. One type is shown in Fig. 14-85. A worn needle and seat is usually due to a lack of refrigerant in the system. Refrigerant does not have a chance to condense before it passes through the expansion valve orifice and this dry, hot gas rapidly cuts or erodes the needle and seat.

The best remedy for a worn needle and seat is replacement. If replacements are not available, the needle must be reground or restored in a lathe or drill press until the

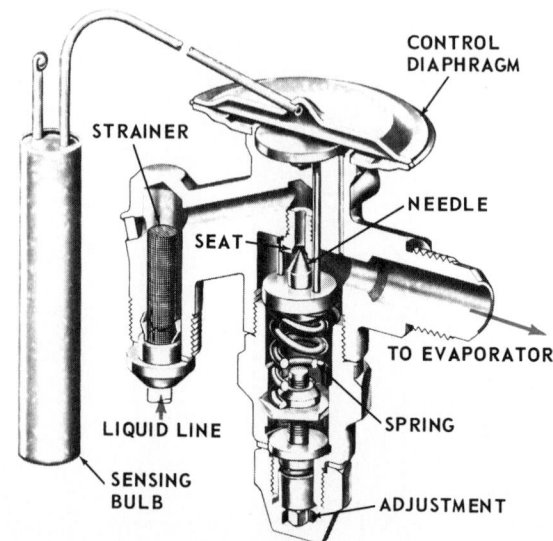

Fig. 14-85. A thermostatic expansion valve with a removable screen, needle, and valve seat. (Alco Controls Div., Emerson Electric Co.)

shoulder that has been worn into the needle surface disappears. It is important that the same taper be kept on the needle point.

The seat, which is nearly always made of softer metal than the needle, may usually be filed with a dead smooth file until the older surface against which the needle was seated has disappeared. The seat must be filed square at right angles to the center line of the needle. A file guide is best for this purpose. One may tap the new needle into the new seat during assembly of the valve and get good results.

A leaking bellows may be due either to breaking down of the soldered joint, or to a fracture in the bellows or diaphragm itself. A bellows is a difficult thing to repair. A special fixture is needed to hold all the parts in the proper position. When the technician encounters a bellows in this condition, it is best to replace the expansion valve.

14-74 INSTALLING TWO-TEMPERATURE VALVES

The pressure-operated two-temperature valves should be installed in the suction line of the warmer evaporators. They may be connected at any place on the suction line because their operation is usually not affected by the distance from the evaporator. Frequently valves of this kind are found in the soda fountain. Some are located near the condensing unit. Solenoid valves for two-temperature operations are mounted in the liquid line of the warmer evaporators.

This type of installation may be improved by mounting a check valve in the suction line of the coldest evaporator. This prevents the backing up of the higher pressure gases into it. Also, a surge tank should be mounted near the condensing unit and connected between the compressor and the main suction line. This will cut down on fluctuations (rapid up and down changes) of the low-side pressure.

In most code installations, the two-temperature valve and check valve must be mounted near the condensing unit. The surge tank must be mounted on the condensing unit base.

14-75 SERVICING TWO-TEMPERATURE VALVES

As explained in Chapter 12, the two-temperature valves are automatic. They maintain a higher refrigerant pressure in one or more evaporators of a multiple evaporator system. This pressure is higher than pressures in the remainder of the system.

Four types of two-temperature valves are:
1. Metering type.
2. Snap-action type.
3. Thermostatic type.
4. Solenoid valve, which is thermostat controlled.

The four troubles commonly encountered with pressure valves are:
1. Leaky needle.
2. Valve stuck shut.
3. Valve out of adjustment.
4. Frost accumulation on bellows.

If the valve is leaking, the warmer evaporator will be too cold and there will be danger of freezing. This cold temperature may also be due to the two-temperature valve being adjusted too close and at too low a pressure.

To determine which of the two troubles is present, check to see if the valve has been adjusted recently. If the valve has not been tampered with, the trouble is very likely a leaking needle. If the valve has been tampered with, one must readjust it by using a thermometer to get the correct evaporator temperature. The adjusting nut should not be turned more than a half turn at a time. A 15-minute interval should be allowed between each adjustment to permit the evaporator to completely respond to the new pressure.

For accuracy in making adjustments on two-temperature valves, a low-pressure gauge should be installed in the evaporator side of the valve. Sometimes such a gauge opening is available, but in many cases it is not. Service work on such valves will be made easier if the service technician, when installing two-temperature valves, will install a shutoff valve with a gauge opening in the suction line. This will permit the use of a gauge to check the evaporator low-side pressure.

It is rare to have a valve stuck shut. It may be easily recognized by a lack of cooling in the warmer evaporator and an adequate refrigerant supply to the two-temperature valve (but not through it). This refrigerant supply may be checked by cracking the flare nuts on the high-pressure side of the two-temperature valve. Some of these two-temperature valves are provided with screens. A clogged screen will be indicated by a warmer cooling condition of the evaporator with symptoms similar to the stuck needle condition.

Frost will accumulate on the bellows only when the valve is located in or near a freezing compartment. The valve should be removed from the freezing compartment and the bellows covered with light grease.

The same troubles are encountered with the snap-action, two-temperature valve as with the metering type. However, most snap-action valves have a gauge connection in the nature of a one-way service valve mounted on the two-temperature valve body. This gauge connection makes adjustment of the valve simple and therefore it can be determined if the valve is leaking or out of adjustment.

Thermostatic two-temperature valves offer the same troubles, causes, and remedies as the other two. In addition, they have troubles resulting from the thermostatic element. These troubles are the same as mentioned for thermostatic expansion valves. They are:
1. Loss of charge from thermostatic element.
2. Frost on bellows.
3. Poor power element contact with evaporator.
4. Pinched capillary tube.
5. Wrong adjustment.

These troubles are checked in much the same way as one would check the thermostatic element in thermostatic expansion valves.

An electric solenoid valve in the liquid line of the warmer evaporator and electrically connected to a thermostat in the cabinet may have the following troubles:
1. Needle stuck open.
2. Needle stuck shut.
3. Thermostat troubles:
   a. Points stuck together.
   b. Open circuit.
   c. Out of adjustment.
4. Poor wiring.
5. Open solenoid winding.
6. Burned-out solenoid winding.

Felt filters may be used. These are about 1/8 in. thick and are made of special material. Wool batt is also used as a filter in some driers. One company has a specially processed coarse cotton yarn wound in a diamond pattern over a metal frame. One of the latest filters makes use of powdered metal pressure castings.

Driers and filters to protect the motor compressor from burnout may be installed in the suction line. They should always be used after a hermetic motor burnout. Dirt in a system may cause serious damage. The small solid particles act as an abrasive. They will wear the needles and seats of refrigerant controls, valves, and valve seats of compressors, bearings, and pistons of compressors. These particles may also wear through the motor insulation and cause a burnout.

Fig. 14-88 shows a filter for a large system. It has a bolted assembly to permit replacement of the filter element. The service connection allows a service technician to check the pressure drop. If this gauge reads more than 2 psig (117 kPa) higher than the compressor compound gauge with the unit running, the filter element should be replaced.

Assume the refrigerant is 2 percent oil.

Each circulation of 10 lb. of refrigerant × .02 = .2 lb. oil. If the unit has 1 qt. of oil or about 1.5 lb.:

$$\text{It takes } \frac{1.5}{\frac{2}{10}} = \frac{15}{2} = 7.5$$

circulations of the refrigerant to clean most of the oil. This means 2 min. × 7.5 = 15 minutes to clean most of the oil.

A suction filter-drier is a must if oil is added to a system. The oil additives in the two oils may react and make a sludge. All of the common refrigerants can be successfully dried with a drier in either the liquid or suction line.

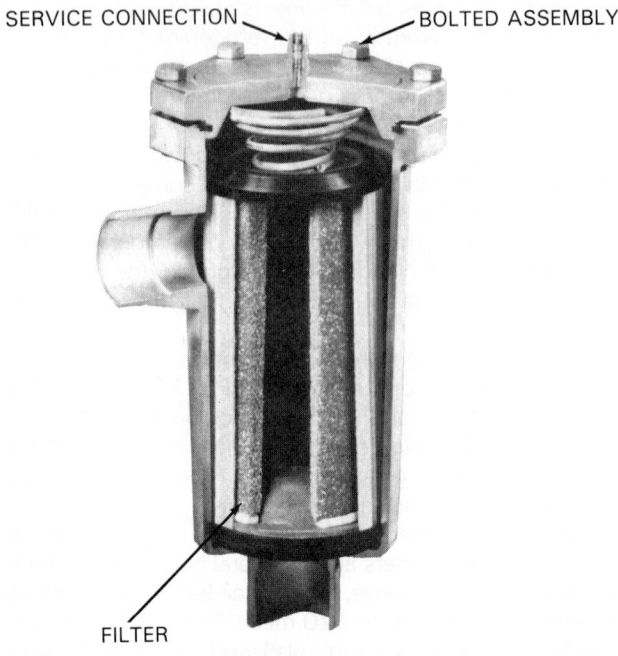

SERVICE CONNECTION — BOLTED ASSEMBLY

FILTER

Fig. 14-88. Suction line filter with bolted flange construction permits easy replacement of filter element. Service connection is equipped with Shrader valve. (Superior Valve Company, Division of AMCAST Industrial Corporation)

14-82 SERVICING BURNOUTS

A motor burnout is possible with any system using a welded motor compressor or a bolted assembly motor compressor. Moisture, dirt in the system, plus too-high temperature (warmest spot is usually motor windings) may cause an acid condition to develop. In time, the condition could lead to a motor winding short (burnout). See Chapter 11.

To prevent this, the system must be kept free of moisture and dirt. To detect acid formation, the system should be checked regularly.
1. Use a sight glass with moisture indicator.
2. Take an oil sample often. Test it with an oil test kit. The oil sample must be kept sealed until tested. See Para. 14-43, Fig. 14-66.

If a burnout occurs, the motor compressor must be replaced. If it is a small hermetic, discard the refrigerant. *Be careful: the oil coming out with the refrigerant may be acidic.* In large systems equipped with shutoff valves, it may be possible to save the refrigerant. Purge the motor compressor. Use goggles, rubber gloves and ventilate the space. The oil may be acidic and cause serious burns—do not get on skin. The fumes may also be irritating and toxic.

The system can be reconditioned by flushing the complete system with R-11. Use regulated CO_2 pressure to circulate the refrigerant or use a separate pump.

After cleaning, install the new motor compressor; test for leaks; evacuate the system to 50 to 500 microns. Install a drier in the suction line; test for leaks and evacuate again. Charge the system.

Connect the electrical wires. Never solder leads to the compressor terminal. The glass may crack or the terminal may come loose, causing a leak.

Start the unit and check operation. Make frequent oil acid tests. Replace the drier if the oil sample is discolored or shows an acid trace.

14-83 TRANSFERRING REFRIGERANTS

It is good economy to purchase refrigerants in 50-lb. and 100-lb. quantities. This practice, however, requires that refrigerant be transferred from the large cylinders to smaller service cylinders used by service technicians. Large cylinders are equipped with two types of valves:

1. A regular valve which releases vapor only when the cylinder is upright and the valve open. This cylinder must be turned upside down to remove liquid refrigerant.

2. A special valve with two valve stems and handles. The handle marked "gas" releases vapor. The other handle marked "liquid" allows liquid to escape from the cylinder when opened. See Chapter 9.

If the cylinder has the first type valve, liquid refrigerant is removed from it following certain steps.

Turn the cylinder upside down in a special stand and connect it to a small cylinder with a horizontal tubing or flexible charging line at least four feet long. Purge this line and then open both valves after placing the small cylinder on scale.

When transferring refrigerant from a storage cylinder to a service cylinder, never fill a service cylinder completely full of liquid refrigerant. This is particularly dangerous in cold weather since liquid refrigerant expands and contracts greatly with a rise in temperature. If a service cylinder is filled completely full with cold liquid refrigerant and then brought into a warm room, the liquid tends to expand and the resulting increase in hydrostatic pressure may burst the service cylinder.

When the small cylinder is charged with the correct amount, as indicated on the scale, close the valve on the large cylinder. Carefully warm the line with a heating tape or a heat lamp. If neither is available, simply rub the line to warm it. This will force the liquid out of the line. Next, close the valve on the small cylinder. Never allow the line to become more than warm to the hand. Wear goggles.

If the cylinder has a double valve, the same procedure is used except that the cylinder is kept upright.

A charging board provides a more accurate means of transferring refrigerant. This system uses a glass cylinder or measuring tube to measure the amount of refrigerant. The scale reads in pounds and usually has scales for R-12, R-22, R-500, and R-502. Fig. 14-89 shows a complete

Fig. 14-90. Disposable refrigerant container. Containers range from 1 to 30 lb. capacity. (Du Pont Company)

14-84 ADDING OIL TO SYSTEM

When a compressor runs too warm or is noisy and it is found that the trouble is a lack of refrigerant oil, there are several ways to add oil. Remember, the oil and all the oil transfer equipment must be clean and dry. The oil must be the proper viscosity for the compressor, the refrigerant, and the low-side temperature.

1. The most rapid method is to attach tubing equipped with a hand valve to the middle opening of the gauge manifold. After purging the tubing, immerse it in a clean, dry glass jar nearly filled with refrigerant oil. Then run the compressor. After drawing a vacuum on the low side by turning the suction service valve all the way in, crack the manifold suction valve. Oil will then be drawn into the crankcase. It is important that some of the oil in the glass container be left in the container so the filling tube is always immersed in the oil. Otherwise, air will be drawn into the system. The reason a glass container is used, is to enable the service technician to observe how much oil has been added to the unit. It is a safe policy never to add more than 1/4 pt. of oil at a time to smaller units.

2. Another method of adding oil to the system is to evacuate the crankcase. Then equalize the pressures, remove the oil plug of the crankcase housing, and add the oil. Replace the oil plug and then evacuate the compressor.

3. Oil can also be forced into the system by putting the oil in a service cylinder first (draw in by using an evacuated cylinder). Then build up a pressure in the cylinder with refrigerant vapor through the gauge manifold. Invert the cylinder and, by using the low-side manifold valve, the oil can be forced into the compressor.

4. Special pumps are available to hand pump oil into a compressor, even against a high low-side pressure. Because some compressor oil circulates around the system with

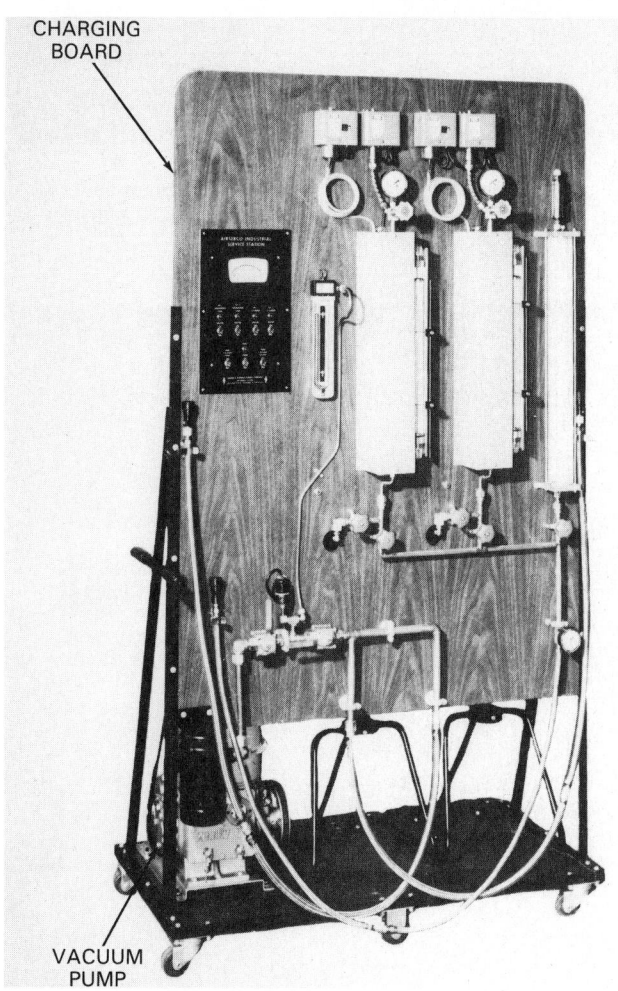

CHARGING BOARD

VACUUM PUMP

Fig. 14-89. Complete evacuating and charging system. The 8-cfm pump is connected to both the high side and low side of system. Large evacuating lines and vacuum breaker are used. (Airserco Mfg. Co., Inc.)

evacuating and charging service station. In many cases—especially with small commercial systems—it is better to use disposable containers of refrigerant rather than transferring refrigerant from one cylinder to another, Fig. 14-90.

the refrigerant, one must add oil to the compressor if the refrigerant lines are over 30 ft. long (suction and liquid). Add about 3 fl. oz. of oil for each 10 ft. of tubing installed.

14-85 SERVICING ELECTRICAL CIRCUITS

More and more electrical devices are being used on refrigerating systems. With the addition of electrical defrost systems, solenoid valves, multi-units with electrical interlocks, crankcase heaters, internal motor winding protectors, and various other accessories, a service technician must be knowledgeable about electrical devices and electrical circuits. Chapters 6, 7, and 8 explain the fundamentals of electricity, electric motors, and electric controls.

It is important that the service technician have a wiring diagram of the system being serviced. Certain items should always be checked:

1. Are the wires large enough?
2. Is there electrical power up to the machine?
3. Is the voltage correct? (It must not be too low.) See Fig. 14-91.

Fig. 14-92. Measuring current flow to motor. Current flow should not be more than motor rating.   (Amprobe Instrument)

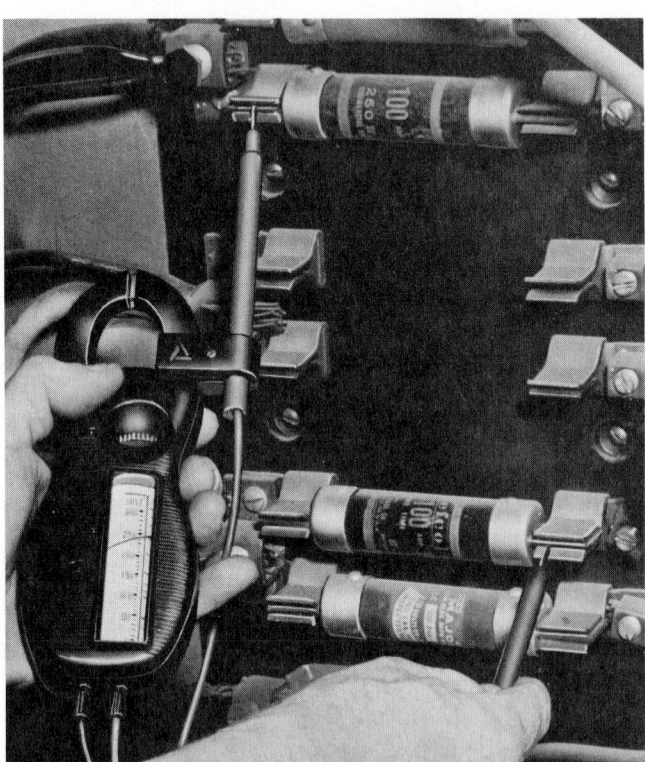

Fig. 14-91. Checking voltage at main load center.
(Amprobe Instrument)

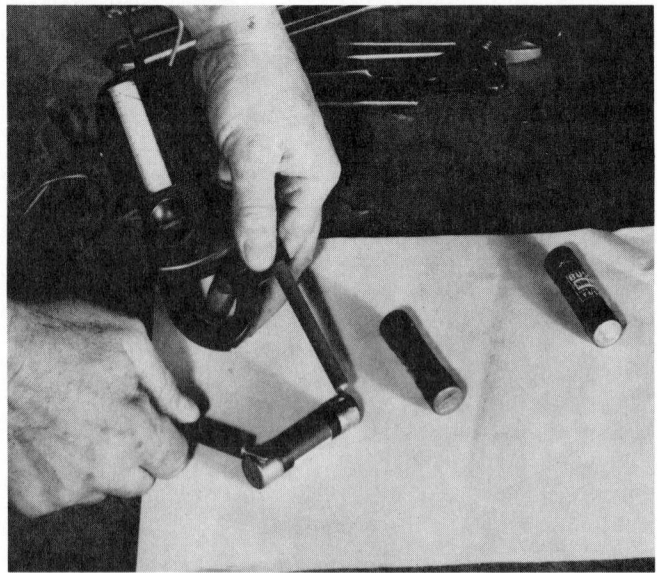

Fig. 14-93. Using ohmmeter to check continuity of fuses.
(Amprobe Instrument)

4. Is the current draw correct? See Fig. 14-92.

With the power on, the current draw and the voltage at the unit can be checked. If the unit will not run, turn off the power and check the circuits for continuity. Use an ohmmeter. See Fig. 14-93.

Locate the break in the circuit continuity by measuring sections of the circuit with the ohmmeter.

1. Controls.
2. Motor. (If circuit is open and motor is warm, wait at

least an hour to allow thermal cut-out points to close.)

If the motor hums but will not start, check the starting capacitor with a capacitor tester. Do not start a unit without having overload cut-out in the circuit.

14-86 SERVICING EXTERNAL DRIVE MOTORS

Motors used on external drive commercial systems usually vary in size from 1/12 hp for fans to 15 hp for compressors. Air conditioning systems require motors of 1/3 hp to 25 hp. These motors are connected to either 120-240 V single-phase or 240 V three-phase lines.

In addition to the condensing unit motors, commercial systems use motors for fans, water pumps, mixers in ice cream machines, and the like.

Many localities require that a licensed electrical contractor remove, repair, and install these motors. However, the refrigeration service technician must be able to diagnose motor troubles to locate the fault. But, motor work should be subcontracted.

Motor troubles can be traced to:
1. Mechanical troubles.
2. Electrical troubles.

Mechanical troubles are faults in the bearings and pulleys, out of alignment and excessive end play. Electrical troubles may be further classified as:
1. Internal troubles.
2. External troubles.

The sound of the motor can indicate trouble. Under normal conditions a motor will make a steady low hum. But, in case of worn bearings, rubbing armatures, dry bearings, or lack of voltage, erratic beats will be heard in the humming and the rotor may chatter. If the condition of the motor is in doubt, it should be thoroughly checked as described in Chapter 7.

With the motor running, make sure the rotor position is between the two extremes of the rotor end play. If the rotor operates against one extreme of end play, it means that it is trying to assume its magnetic center. Because it cannot, it is running inefficiently. The end play should never exceed 1/16 in. This may be adjusted by using end play washers which are obtainable at electrical supply houses.

Adequate lubrication of the motor bearings is necessary. Normally motors should be oiled twice each year. Use electric motor oil (about SAE 30). Too much oil is just as bad as not enough. Most motors are equipped with overflow openings which eliminate most of the effects of too much oil. Many lubricants are now available in aerosol cans. These spray lubricants cause a problem around refrigerating systems, because the pressurizer in the container is usually R-12. If one tests for leaks after using one of these aerosol containers, the halide torch or electronic leak detector will falsely indicate a leak in the system.

Bearing temperatures should be checked after operation. Use a thermometer.

Thoroughly clean the motor occasionally. Dust, dirt, and grease accumulations should be removed from inside and outside the motor. Clean commutators and brushes, if used. They must make good contact.

The brush throw-out mechanism should move freely. Brush releasing mechanisms of small motors can be checked by mounting a V-belt on the motor pulley and putting a load on the motor by pulling on the belt.

A torque stand, or dynamometer, is the best means of determining the real capacity of a motor.

A noisy motor may be caused by a loose pulley, loose fan on the pulley, or loose flywheel. These items should be checked when a noise complaint is received.

Fan motors are usually of the shaded-pole type. The most common trouble is worn bearings. Location of these motors usually results in a lack of attention to oiling. Many of these motors are designed to not need lubrication, but practice has proven that many do need lubrication periodically.

The pump motors and mixer motors are serviced much like the methods described.

Always be sure the motor is wired correctly and that the voltage at the motor is enough. Always test a motor for grounds and always ground the frame of the motor.

14-87 REMOVING ELECTRIC MOTORS

To remove an electric motor, first check for the presence of a fuse or circuit breaker in the system and disconnect the circuit. Then, disconnect the power line and remove the wires from the motor terminals.

Label the terminals for easier assembly later. Next, loosen hold-down bolts which attach the motor to the base.

If it is a belt-driven unit, such as a compressor drive or large fan drive, remove the belt from the flywheel first, then from the pulley. The motor can then be lifted out. Do not allow the fan to hit the condenser or catch on the belt. Use a puller to remove the pulley from the motor shaft after loosening the lockscrew.

Fan motors are sometimes difficult to remove. It is best to loosen and remove the fan. Generally, the fan hub is locked in place on the motor shaft by means of an Allen setscrew. Fig. 14-94 shows an Allen setscrew wrench set designed to work in hard-to-reach places.

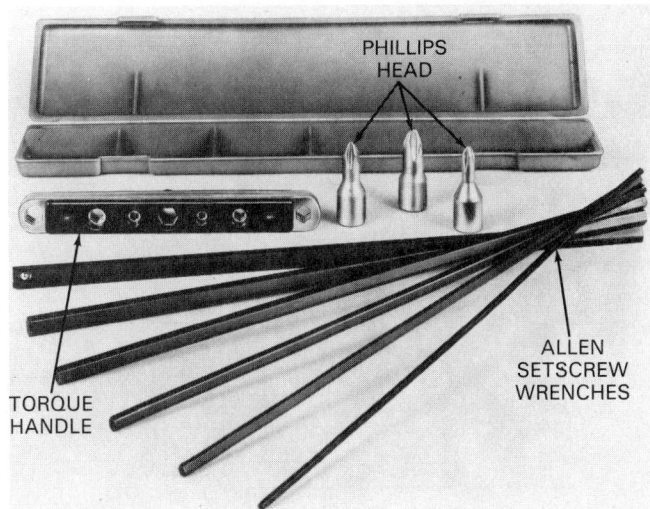

Fig. 14-94. Set of Allen setscrew wrenches and Phillips screwdrivers with turning handle. These tools can reach screws in recess up to 9 in. in depth. (Watsco Components, Inc.)

14-88 REPAIRING AND TESTING ELECTRIC MOTORS

Conventional motors are described in Chapter 7.

The motor should be cleaned outside, completely dismantled, and cleaned inside. Replace the internal switches on capacitor start motors. The bearing walls should be cleaned of oil and the bearings carefully inspected. If they are scored or have more than .001 in. clearance, they should be replaced (ream to size in line). If the contact points are scored or worn, install new points.

When assembling the motor, do not force the parts. Be sure the end bells fit into place. Keep turning the rotor by hand during the assembly to detect any binding before damage is done.

Test a motor before installing:
1. Run under no load.
2. Run under load.

Run the motor, check it for end play (1/16 in. max.), check bearing temperatures, and listen for noises other than the normal hum.

Run the motor under load. Fig. 14-95 shows a torque arm and pulley for motor testing. Fig. 14-96 shows a motor being tested on it. The electrical load of the motor is checked by connecting meters as shown in Fig. 14-97.

When checking three-phase motors, measure the voltage across each compressor terminal when running. The voltage should not vary over 3 percent. If it does, check again at the power-in panel. If satisfactory there, the problem is in the system wiring (loose or dirty connection or wrong wire size). If difference is more than 3 percent at the power-in panel, call the electric utility and have them correct the fault.

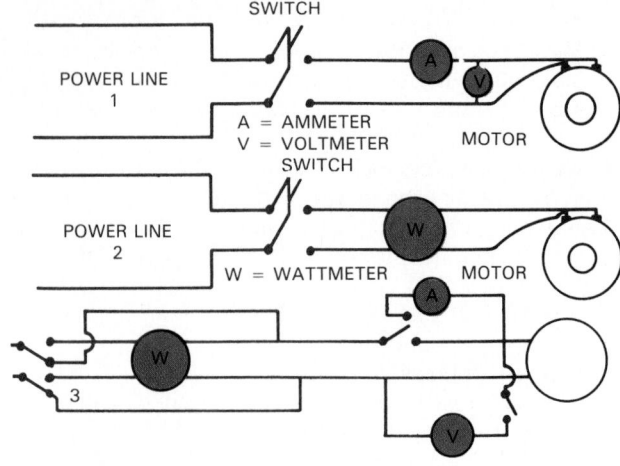

Fig. 14-97. Electric motor testing. 1—Using voltmeter and ammeter. 2—Using wattmeter. 3—Using either voltmeter or combination of voltmeter and ammeter.

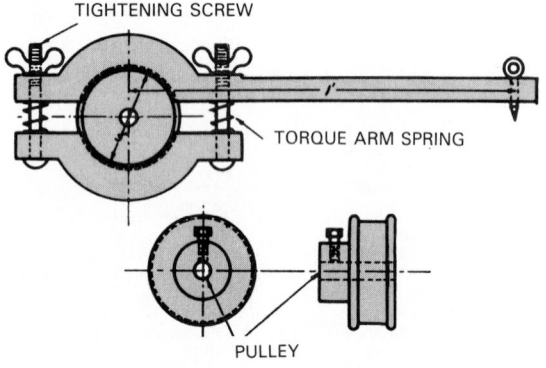

Fig. 14-95. Torque arm and pulley used for testing conventional motors under load.

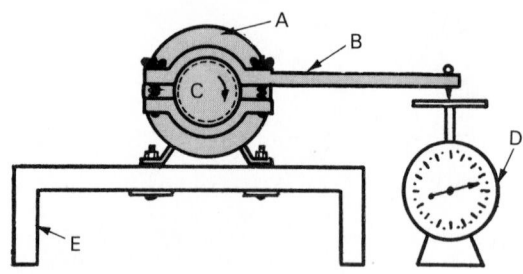

Fig. 14-96. Motor being tested under load. A—Motor. B—Torque arm. C—Motor pulley. D—Torque scale. E—Motor stand.

14-89 INSTALLING EXTERNAL DRIVE CONVENTIONAL TYPE ELECTRIC MOTOR

External drive type compressor motors may drive a compressor directly or by means of a belt. In either case, the motor must be carefully aligned when installed. *Always lock out the power before starting work.*

Four hold-down bolts are usually used. After installing the bolts loosely, install the belts. Use a lever to move the motor until the belts are tight and the motor is in line. While holding the motor in this position, tighten at least two of the hold-down bolts. Then tighten the others. Motors are usually fastened to their mounting base with four nut, bolt, and washer combinations. Motors of 1/2 to several hp are heavy and it is hard to put on the four bolts at one time.

Usually, either the motor base or the base float it is bolted to, have slots for the bolts. Often it is very difficult to reach under the mounting base to hold the bolt in place or to put a wrench on it.

Some service technicians use caulking compound to hold the bolts in place until the motor is installed and the washers and nuts are started on the bolts. Make clean, tight electrical connections.

In the direct drive units, set the motor on its part of the stand, assemble, and install the coupling. Check the alignment carefully because the motor shaft center must be the same height as the compressor shaft center. The two shafts must be in alignment when looking down on them. A dial gauge should be used to check the alignment. Install and tighten the bolts. Water pump motors require the same care.

Fan motor installations sometimes require installing the fan on the shaft before the motor is installed.

If possible, always turn the motor by hand to determine if the assembly will rotate freely. Do this before unlocking the power and turning it on.

14-90 SERVICING HERMETIC MOTORS

Servicing hermetic motors is basically the same as servicing conventional motors. Chapter 7 describes the principles, design, construction, operation, and servicing of such motors.

As with all energy devices, first check to determine if the power has all the correct characteristics. Voltmeters, ammeters, demand meters, wattmeters, and power factor meters can be used for this purpose. Check the electrical properties of the motor and then check the power to make sure the motor is receiving:
1. Correct voltage.
2. Correct amount of current. See Fig. 14-98.

Voltage must be within 10 percent of the motor rating. (A 120 V motor should have 108 V circuit voltage minimum.) This voltage should be read with the motor run-

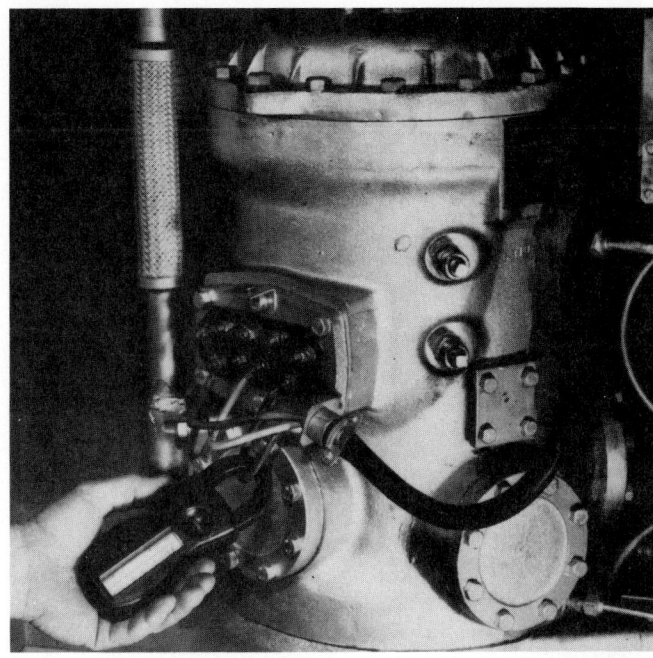

Fig. 14-98. Checking current flow to serviceable hermetic motor using ammeter. (Amprobe Instrument)

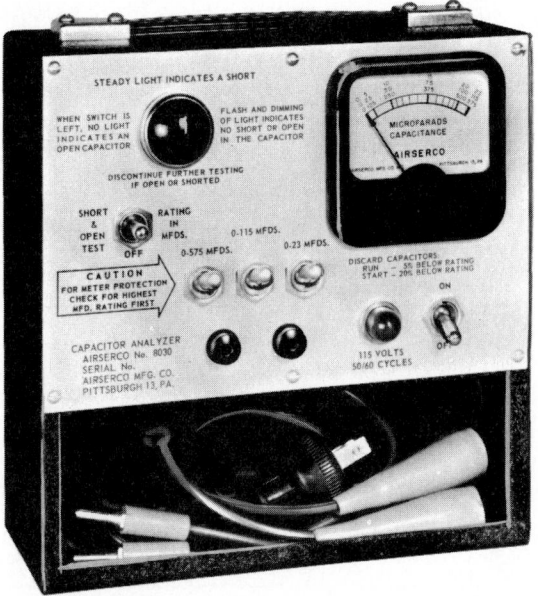

Fig. 14-100. Capacitor tester in range of 0 to 575 microfarads. Also tests for open and short circuits. (Airserco Mfg.Co., Inc.)

ning. Voltage higher than required is not as critical, unless it is 20 percent or more over the rating.

The correct amount of current is vital to good operation. Again, a good instrument is needed. An ammeter of the terminal type (shunt unit) should never be connected across the line (in parallel). It must always be put in series (interrupt one wire only) with the electrical device being checked.

Carefully check external wiring and electrical controls for correct operation before assuming the motor is at fault. An ohmmeter is recommended to check continuity in these circuits. Check each circuit separately. Disconnect if there is a chance of parallel circuits. Be sure the power is locked off.

Capacitors should be checked with a capacitor tester, as shown in Figs. 14-99, 14-100, and 14-101. Study Paragraphs 6-46, 7-39, and 7-40. Do not test capacitors by shorting after charging, as this method is not accurate enough.

The motor should be checked for:
1. Open circuits. (Motor should be cool or internal overload in open position will give false readings.)
2. Shorted windings. (Ohmmeter readings should be compared to manufacturer's specifications.)

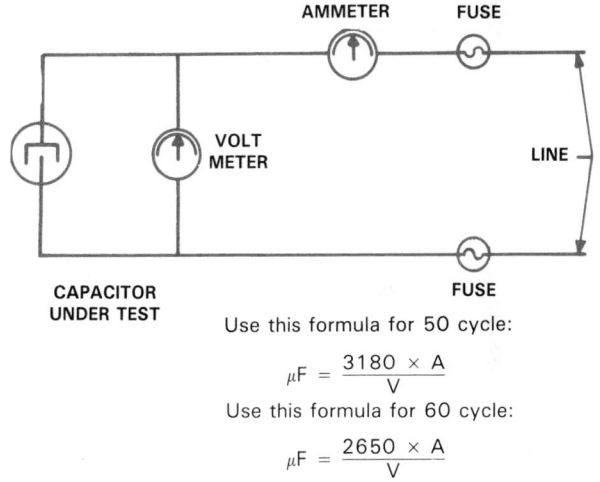

Use this formula for 50 cycle:

$$\mu F = \frac{3180 \times A}{V}$$

Use this formula for 60 cycle:

$$\mu F = \frac{2650 \times A}{V}$$

Note: μF is mathematical symbol used for microfarads.

Fig. 14-101. Wiring diagram and formula used for determining capacity (capacitance) of capacitors. (Copeland Corporation)

3. Grounded windings. First check with ohmmeter, then with a 500 V (megavolt) circuit tester. Insulation breaks can only be checked accurately with this high voltage tester. *Handle carefully to avoid shocks.*

Most companies report that a high percentage of motor compressors returned labeled ''faulty motor'' actually have good motors. This false diagnosis indicates the need for careful checking.

The chart on page 556 is a list of typical motor and circuit troubles, their causes, and their remedies.

14-91 REMOVING HERMETIC MOTOR COMPRESSORS

If it is certain that the motor in a hermetic motor compressor unit is faulty, it must be removed. Removal procedure follows:

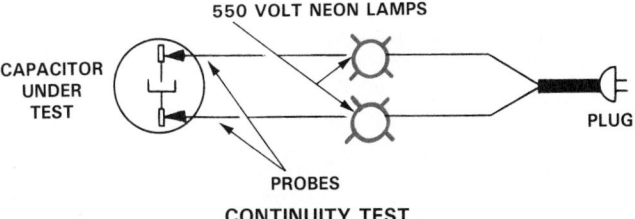

Fig. 14-99. Testing continuity of capacitor with neon lamps. Lamps will light to full intensity if capacitor has continuity. Be sure circuit is of same voltage as capacitor. Higher voltage may cause capacitor to explode. (Copeland Corp.)

HERMETIC COMPRESSOR SERVICE CHART

PROBLEMS AND CAUSES	REMEDIES
Compressor will not start — no hum.	
1. Open line circuit.	1. Check wiring, fuses, receptacle.
2. Protector open.	2. Wait for reset — check current.
3. Control contacts open.	3. Check control, check pressures.
4. Open circuit in stator.	4. Replace stator or compressor.
Compressor will not start — hums intermittently (cycling on protector).	
1. Improperly wired.	1. Check wiring against diagram.
2. Low line voltage.	2. Check main line voltage, determine location of voltage drop.
3. Open starting capacitor.	3. Replace starting capacitor.
4. Relay contacts not closing.	4. Check by operating manually. Replace relay if defective.
5. Open circuit in starting winding.	5. Check stator leads. If leads are all right, replace compressor.
6. Stator winding grounded (normally will blow fuse).	6. Check stator leads. If leads are all right, replace compressor.
7. High discharge pressure.	7. Eliminate cause of excessive pressure. Make sure discharge shut-off and receiver valves are open.
8. Tight compressor.	8. Check oil level — correct binding condition, if possible. If not, replace compressor.
9. Weak starting capacitor or one weak capacitor of a set.	9. Replace.
Compressor starts, motor will not get off starting winding.	
1. Low line voltage.	1. Bring up voltage.
2. Improperly wired.	2. Check wiring against diagram.
3. Defective relay.	3. Check operation — replace relay if defective.
4. Running capacitor shorted.	4. Check by disconnecting running capacitor.
5. Starting and running windings shorted.	5. Check resistances. Replace compressor if defective.
6. Starting capacitor weak or one of a set open.	6. Check capacitance — replace if defective.
7. High discharge pressure.	7. Check discharge shutoff valves. Check pressure.
8. Tight compressor.	8. Check oil level. Check binding. Replace compressor if necessary.
Compressor starts and runs but cycles on protector.	
1. Low line voltage.	1. Bring up voltage.
2. Additional current passing through protector.	2. Check for added fan motors and pumps connected to wrong side of protector.
3. Suction pressure too high.	3. Check compressor for proper application.
4. Discharge pressure too high.	4. Check ventilation, restrictions and overcharge.
5. Protector weak.	5. Check current — replace protector if defective.
6. Running capacitor defective.	6. Check capacitance — replace if defective.
7. Stator partially shorted or grounded.	7. Check resistances; check for ground — replace if defective.
8. Inadequate motor cooling.	8. Correct cooling system.
9. Compressor tight.	9. Check oil level. Check for binding condition.
10. Unbalanced line (three-phase).	10. Check voltage of each phase. If not equal, correct condition of unbalance.
11. Discharge valve leaking or broken.	11. Replace valve plate.
Starting capacitors burnout.	
1. Short cycling.	1. Reduce number of starts to 20 or less per hour.
2. Prolonged operation on starting winding.	2. Reduce starting load (install crankcase pressure limit valve), increase voltage if low — replace relay if defective.
3. Relay contacts sticking.	3. Clean contacts or replace relay.
4. Improper relay or incorrect relay setting.	4. Replace relay.
5. Improper capacitor.	5. Check parts list for proper capacitor rating — mfd. and voltage.
6. Capacitor voltage rating too low.	6. Install capacitors with recommended voltage rating.
7. Capacitor terminals shorted by water.	7. Install capacitors so terminals will not be wet.
Running capacitors burnout.	
1. Excessive line voltage.	1. Reduce line voltage to not over 10 percent above rating of motor.
2. High line voltage and light load.	2. Reduce voltage if over 10 percent excessive.
3. Capacitor voltage rating too low.	3. Install capacitors with recommended voltage rating.
4. Capacitor terminals shorted by water.	4. Install capacitors so terminals will not be wet.
Relays burnout.	
1. Low line voltage.	1. Increase voltage to not less than 10 percent under compressor motor rating.
2. Excessive line voltage.	2. Reduce voltage to maximum of 10 percent above motor rating.
3. Incorrect running capacitor.	3. Replace running capacitor with correct mfd. capacitance.
4. Short cycling.	4. Reduce number of starts per hour.
5. Relay vibrating.	5. Mount relay rigidly.
6. Incorrect relay.	6. Use relay recommended for specific motor compressor.

1. Remove the refrigerant as described in Chapter 11 and as described in this chapter for serviceable hermetics.
2. Open the main circuit switch and lock the switch in open position. Tag the switch to inform others why the switch is locked open.
3. Disconnect the wires from the motor compressor (label them or use color code).
4. Clean outside of motor compressor.
5. Disconnect lines (wear goggles!). The type of disconnect depends on assembly. Unbolt service valves, open brazed joints by heating or cutting the lines. Use tube cutter only to avoid getting chips into the system.
6. Remove motor compressor. Do not tilt or oil may be spilled. Avoid lifting over 50 lb. If it is a large, heavy unit, use a lifting machine (tripod or fork lift).
7. Plug refrigerant openings.

14-92 REPAIRING HERMETIC MOTORS

Removing a hermetic motor, repairing the bearings, or replacing the windings is a specialty job. Most service technicians replace the complete assembly by getting a replacement unit, either new or rebuilt.

The actual overhaul of a hermetic motor compressor is described in Chapter 11.

14-93 INSTALLING HERMETIC MOTOR COMPRESSORS

A replacement hermetic motor compressor is usually furnished with a starting relay, capacitors, overload protectors, and other accessories. Use an exact replacement. These motor compressors are either designed for low, medium, or high low-side pressures. Use the same type as the one removed. The refrigerant openings are closed with service valves. Or, the unit is provided with short tubing ends, brazed in place with the ends of the tubing crimped and brazed.

Carefully mount the motor compressor in place. Use all the safety precautions (lifting, safety shoes, protecting floors, and equipment). Install the mounting bolts. The springs and grommets must be in the correct position and the hold-down bolt or nut must be positioned correctly.

Install the electrical devices (overload and starting relay) and the electrical wires. All connections must be clean and tight. All wires, including insulation and wire terminals, must be in good condition.

Avoid connecting aluminum wires to copper wires or copper terminals. Rapid corrosion takes place. Special adaptors must be used or the joint will corrode.

Install suction and condenser lines. Units using service valves are installed the same way as described for a conventional compressor.

If brazed connections are used, identify which tubing stubs are the suction connections, the discharge connections, the process tube, and the oil cooler connections. Cut the tubing stubs with a tube cutter. Select connector fittings or swage the tubing. Flux the outside of the joints brazed to the compressor dome. Connect the system lines to the compressor lines. Clean the inside of the suction line and condenser line. Braze the joints. *The area must be well ventilated during brazing.* Install a liquid line drier and a suction line drier. Clean the joints with warm water to remove flux.

Install a gauge manifold using charging stub or a valve mounted on the suction line and liquid line.

Evacuate the system, put some vaporized refrigerant in (enough to build a 15 psi pressure), test for leaks (repair any, if found), evacuate to a 50 to 500 micron range for several hours, and then charge the system.

Turn on the power, run motor, check the temperatures, pressures, and electrical power. If any operation is not normal, be sure to diagnose and remedy it before leaving the job. Remove the service connections and braze these joints. Clean up the area.

14-94 SERVICING MOTOR CONTROLS

Four main types of motor controls are used in commercial refrigeration: low-side pressure motor control, thermostatic motor control, high-side safety motor control, and an oil pressure safety control. Troubles encountered with these include:
1. Corroded points.
2. Broken mercury bulb.
3. Switch out of adjustment.
4. Corroded or broken operating springs.
5. Unit not level.
6. Leaking bellows.

Motor controls normally work year in and year out without giving trouble. However, in cases such as unit overloading, the resultant short cycling will rapidly deteriorate the contact points in the control, or will so overload the mercury contact that it will crack and be destroyed. Corroded points should be replaced. They can be temporarily repaired by cleaning with fine sandpaper or a clean fine mill file. Never use emery cloth. A broken mercury bulb must be replaced.

An out-of-adjustment switch is often the result of tampering. A pressure control may be easily checked by installing the gauge manifold. Using the compressor as a vacuum and pressure pump, check the cut-in and cut-out points. Three different methods may be used to build up pressure in the crankcase after the compressor is run to the cut-out point of the control:
1. The suction line may be cracked open again.
2. A bypass may be run from the discharge service valve of the compressor to the suction service valve. (Use the gauge manifold.)
3. A refrigerant service cylinder may be used containing the same kind of refrigerant attached to the gauge manifold. Many service technicians carry a hand vacuum and pressure pump in their tool kit. This tool allows rapid check of pressure controls, Fig. 14-102.

A thermostatic motor control is more difficult to reset. An approved method is to use an ice bath and thermometer

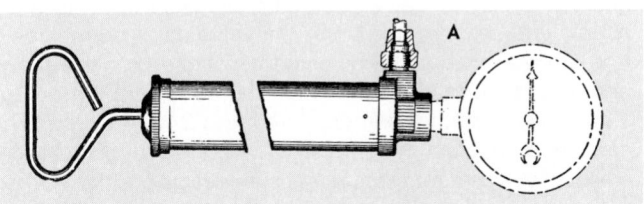

Fig. 14-102. Hand vacuum and pressure pump used for adjusting motor controls. A—Connection to control being tested and adjusted.

or a temperature bath, as shown in Fig. 14-103. The control may then be set to cut in or out at the temperature desired.

The high-pressure motor control presents few difficulties because it is adjusted to work only under extreme conditions. The most common troubles are leaks in the bellows or at the joints of the controls. Occasionally, however, this control may be short cycling the refrigerating mechanism. This short cycling is due to excessive head pressure.

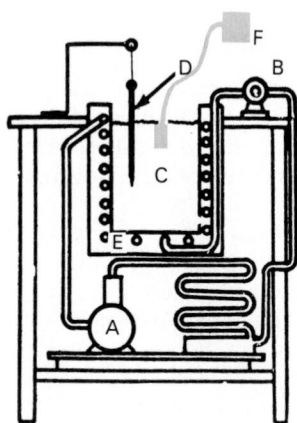

Fig. 14-103. Temperature bath system. Cooling system used to control liquid bath temperature. May be used for checking thermostat cutout and cut-in temperatures and for checking thermostatic expansion valves. A—Compressor unit. B—Refrigerant control (AEV). C—Temperature bath. D—Thermometer. E—Evaporator. F—Thermostat being tested.

To check a high-pressure motor control element, connect it to the gauge manifold and turn the discharge service valve all the way in. Running the compressor will produce a head pressure sufficient to cut out the control. The high-pressure gauge will record the pressure at which the cutting out takes place.

Many large compressors use oil pressure lubrication. If oil pressure fails or falls below a certain pressure above the low-side pressure, there is danger of damaging the compressor. An oil pressure safety control protects these systems. It will cut out if the oil pressure drops to a dangerous level.

14-95 SERVICING SOLENOID VALVES

Solenoid valves develop both electrical problems and refrigerant troubles. If the electrical connections are dirty or loose, the coil may not create enough magnetism to raise the valve. Usually the valve should be mounted with the coil on top and with the valve level, or they may stick or chatter. Solenoid valves sometimes will develop a leaky needle and seat. In this case, the valve must be replaced. It is important that the solenoid be of the proper voltage and amperage rating; otherwise, the coil may burn out. A 120 V valve cannot be connected to a 240 V circuit. The service technician must also be careful to place most solenoid valves right side up and in a vertical position. Otherwise, they will not operate.

To remove a valve from the system, one must follow the same pump-down procedure as for a TEV.

14-96 SERVICING LIQUID LINE

The liquid line contains several important items needing inspection by the service technician:
1. Size of the liquid line.
2. Hand shutoff valves.
3. Sight glass.
4. Moisture indicator.
5. Screen filter.
6. Drier or dehydrator.
7. Vibration absorber.
8. Connections.
9. Solenoid valves.
10. Joints.
11. Pinched or buckled pipe or tubing.

When diagnosing troubles of a system, the service technician's first problem is to determine if each part of the system is the right capacity. It is important that the liquid line be as large as the condensing unit liquid receiver valve connections. Check to be sure that reducer fittings have not been used. See Chapter 15 for recommended liquid line sizes. Many large units use various sizes of liquid lines as the line feeds to various evaporators.

All service technicians must remember to have at least one end of the liquid line valves open when servicing the unit. Otherwise, a temperature rise may create a very high hydrostatic pressure. This pressure may burst the line. Be especially careful if a solenoid valve, clogged screen, or clogged dehydrator is in the line.

Check the full length of the line for line condition. The line must be protected from abrasion and abuse as objects are moved. The line should be well supported along its full length.

When admitting liquid refrigerant into a liquid line by opening the liquid receiver service valve, always open the valve slowly. A sudden rush of liquid may injure the screen and may pack the desiccant in the drier so firmly that it will soon clog.

Test all joints for leaks. Check the temperature of the liquid line; it should be close to room temperature along its full length.

A partially clogged screen or drier is indicated by a lower-than-normal temperature at the outlet. Sweating and even frosting may be seen. There will also be an excessive pressure drop. This may cause bubbles in the sight glass.

14-97 SERVICING THE SUCTION LINE

Servicing the suction line is much like servicing the liquid line. Parts to be inspected are:
1. Size of the suction line.
2. Hand shutoff valves.
3. Vibration absorber.
4. Check valves.
5. Two-temperature valves.
6. Constant-pressure valves.
7. Filter-drier.
8. Muffler.
9. Accumulator.
10. Joints.
11. Pinched or buckled suction line.

Suction line size is important. If it is too small, it will cause too much pressure drop and high gas velocities will cause noise. The line should be the size of the suction service valve connection or the suction line connection on the hermetic dome.

On multiple installations, the suction line is smaller for each evaporator than the main suction line. For example, the most remote evaporator may have a 1/2 in. OD size, the next size 5/8 in. OD, the next 1 in. OD, the next 1 1/2 in. OD. The line at the compressor may be 2 in. OD. See Chapter 15.

The pressure drop should be checked by installing a gauge at the most remote evaporator and one at the compressor. The pressure drop should be approximately 2 psi (14 kPa). If more, line sizes should be increased.

The pressure change across two-temperature valves and constant pressure valves should be checked. Their pressure drops are separate from the line pressure drops and are not included in the psi pressure drop.

Also determine the pressure drop across the suction line filter-drier. A large pressure drop here indicates a partially clogged filter-drier. In this case, the drier should be replaced.

Test for leaks with 15 psi pressure or more in the lines.

14-98 SERVICE NOTES

Pinching lines is a practice to be used in cases of emergency only. Some service technicians follow this practice needlessly and it only leads to future trouble. Almost all systems are provided with enough valves to service them. However, pinching lines used to service hermetic units, prior to brazing the stub, is common practice.

Remember when adding refrigerant to a system that some of the oil will be dissolved in the refrigerant. *If a unit becomes noisy soon after the refrigerant has been added, some refrigerant oil should be added.*

Driers, if heated while in the system (because of a refrigerant shortage, for example), will release water to the system. Install a new drier if the moisture indicator signals moisture.

Crankshaft seals may leak if the compressor has been idle for a long time. Turn the compressor over by hand a few times to allow oil to seep between the rubbing metal surfaces. Also put an ounce of special refrigerant detergent oils into the crankcase to help eliminate this problem.

14-99 SUMMARY OF REFRIGERATOR MECHANISM SERVICING

If this chapter has been studied carefully, one should recognize the following important things:
1. Liquid refrigerant must be removed from that part of the mechanism to be overhauled.
2. It is necessary to equalize pressures in the unit before dismantling. This avoids a rush of air into the unit or of refrigerant out of the unit upon opening the lines.
3. All refrigerant openings should be plugged immediately after dismantling.
4. Put in new gaskets wherever gaskets are used.
5. When reassembling, remove all the air and moisture from the lines and from whatever part has been open to the air. This may be done by purging, by evacuating and purging, or by deep evacuating.
6. Keep the inside of the system clean.
7. To remove any part of the system, one must close the nearest valve between that part and the liquid receiver. Evacuate the unit, by means of the compressor, into

the condenser and liquid receiver.

It is usually advisable to remove a large overhaul to the shop. In the meantime, replace the unit with a temporary one so the owners may have the use of the system.

Where possible, a service technician should review the wiring diagram and the specific service manual before working on a refrigerating system. The complex nature of some of the systems requires that a service manual be used.

14-100 PERIODIC INSPECTIONS

The capital investment in a commercial refrigeration installation amounts to hundreds of dollars. Because of the way most mechanisms are built, one problem in the system will cause others. It is important, therefore, that all commercial machines be completely checked over periodically. The service technician should use a systematic method of doing this. Then no detail is overlooked. All inspections should cover such things as:
1. Electrical connections.
2. Motor and safety devices.
3. Compressor noises.
4. Amount of refrigerant.
5. Dryness of refrigerant.
6. Oil level.
7. Water flow.
8. Gas leaks.
9. Coil conditions.
10. Supports for tubing.
11. Coil supports.
12. Cleanliness.
 For a conventional condensing unit, one should check:
1. Belt condition.
2. Belt alignment.
3. Belt tightness.
 For a hermetic condensing unit, one should check:
1. Overload cutout.
2. Relay.
3. Capacitors.
 It establishes good will and also builds up a good contact file to prepare a check sheet, one copy of which should be given to the owner. This check sheet is a time-tried system that prevents overlooking important items.

14-101 LOCATING TROUBLES

Methods of testing to locate sources of trouble are based on the operating principles of the mechanism. By checking the pressures, temperatures, running time, and current or voltage, the service technician is soon able to pinpoint which part of the system is causing the trouble.

The service technician must have a thorough knowledge of the fundamentals of refrigeration and of the cycles before he or she can become reliable and competent at trouble tracing and repair. *Troubles in a refrigerator mechanism must be located before dismantling.* This keeps the cost of servicing at a minimum and assures proper operation of the unit after repair and assembly.

Methods of locating troubles vary with the type of system—whether it has direct expansion or a capillary tube. The call for service should indicate what the trouble is. The owner will probably say that it costs too much to operate, that it is not freezing but is running continuously, or that it is freezing but running continuously. From these complaints, the service technician may usually get an idea what the trouble is. Always verify these statements by checking over the refrigerator before attempting any troubleshooting or service work.

In trouble tracing, first classify the type of service call and then determine what caused the trouble described in the service call. The following troubleshooting pointers have been prepared to help the service technician. Naturally, it is impossible to give every detail, but once the technician learns the method of tracing trouble, there should be no difficulty.

One should check all of the following things in a refrigerating mechanism before deciding what is the trouble:
1. Low-side pressure.
2. High-side pressure.
3. Temperature of evaporator.
4. Temperatures of liquid and suction lines.
5. Amount and dryness of refrigerant.
6. Running time of mechanism.
7. Probability of leaks.
8. Noise.

Several basic fundamentals make locating trouble easier. When there is poor refrigeration, or no refrigeration, either or both of two things can be wrong:
1. There is little or no refrigerant.
2. The pump is not moving the refrigerant. Pressures are not correct.

If there is no refrigerant, there will be no liquid refrigerant in the evaporator. This means that the refrigerant has leaked out or is being held in a certain part of the system by clogged needles, clogged screens, and pinched lines. Clogging causes a high vacuum reading on the low side.

If there is a lack of refrigerant throughout the system, there will be a hissing sound at the refrigerant control indicating the refrigerant passages are not closed. The sight glass will show bubbles.

A hissing sound at the refrigerant control always indicates a lack of refrigerant because the dry gas going through the restriction will cause the gas noise.

If the pump is not functioning, the low-side pressure will be above normal, while the condenser and discharge line from the compressor will be below normal temperature.

To determine what is responsible for a poor condensing condition, install the gauge manifold and determine the head pressure. Compare this pressure with what the pressure should be for the refrigerant being used.

14-102 LITTLE OR NO REFRIGERATION AND UNIT RUNS CONTINUOUSLY

DIRECT EXPANSION SYSTEM (TEV)

If the unit has lost all refrigerant, there is, naturally, no refrigeration. To test for this, install gauges and determine evaporating or low-side pressure. If this pressure is normal, the unit probably has little or no refrigerant. If the compound gauge indicates a high vacuum—20 in. Hg. (34 kPa) or more—it means:
1. The expansion valve is adjusted so that it draws this vacuum.
2. The expansion valve is frozen closed.
3. The system has a clogged screen.

Clogging can be caused by moisture freezing at the refrigerant control and stopping the flow of refrigerant. The results are the same as a needle stuck closed, or a clogged

screen. There is one difference. After the system warms above 32 °F (0 °C) at the valve, frozen moisture will melt and normal refrigeration will return.

The only sure cure for the moisture condition is to remove the moisture by installing a new drier. If it is suspected that moisture in the valve has caused the clogging, heat the valve using a heat lamp. This moisture problem occurs with all refrigerants that do not chemically combine with the water. These refrigerants include, for example, R-12, R-22, R-500, and R-502.

A high vacuum may also be caused by a clogged or restricted suction line filter. In either case, it is not allowing refrigerant to flow through. There will be a high pressure in the evaporator which gives continuous running with little or no refrigeration.

If the compound pressure gauge shows a high pressure on the low side—that is, a pressure which will not allow the refrigerant to evaporate at a low temperature—the trouble may be an expansion valve stuck open or out of adjustment. If this is the trouble, there may also be a frosted or sweating suction line. It simply shows that the refrigerant is going into the low side too fast and the liquid will flood both the evaporator and the suction line.

High pressure may also be due to an inefficient compressor. If the expansion valve is stuck open, it may be due to dirt on the needle. To remedy this, one may flush the valve. Do this by alternately opening and closing the liquid receiver service valve or liquid line hand shutoff valve. Surges of liquid will rush past the expansion valve needle, cleaning it.

CAPILLARY TUBE

If the capillary tube is restricted or completely clogged, if there is moisture frozen in the tube, or if the screen is clogged, no refrigerant can pass into the evaporator. This stoppage will cause a high vacuum reading, the evaporator will be warm, and the condenser will be cool with a normal or low head pressure. If refrigerant is low, the capillary tube will have a hissing sound just as the compressor shuts off. In addition, low- and high-side pressure will be below normal.

An inefficient pump will be indicated by an above-normal low-side pressure and a normal or below-normal head pressure.

14-103 NO REFRIGERATION: UNIT DOES NOT RUN

Where there is no refrigeration and the unit does not run, the trouble is probably somewhere in the electrical circuit. A test light or a voltmeter will tell if power is being supplied to the motor.

If there is power, the motor may be burned out or the circuit open. An ineffective temperature control—one having a leaking power element, for instance—may prevent the motor from starting. It may be checked out as described in Chapter 8.

Further checking of the circuit may show a manual switch or overload in the off position. Or an internal overload switch may be open.

If the motor compressor is hot, this may well be the trouble. It should be checked with a test light or an ohmmeter. Basically, the problem is in the motor or the electrical power circuit.

14-104 MOTOR RUNNING CONTINUOUSLY WITH NORMAL OR TOO MUCH REFRIGERATION

If the motor compressor runs continuously, and the unit produces normal or too much refrigeration, the probable cause of the trouble is a faulty motor control. It will not cut out at the correct temperature. Remember, there is a relationship between the control cut-out point and the evaporating pressure on the low side. If the control is adjusted to cut out at a temperature or pressure lower than the evaporating pressure, the control cannot stop the electric motor.

DIRECT EXPANSION SYSTEM (TEV)

An undercharged system may provide just enough refrigerant to fill the evaporator partially, but not enough to operate the temperature control cut-out point and shut down the unit. In this case one may get normal refrigeration, but the unit never shuts down. This trouble may be indicated for certain by the way frost or sweat collects on the outlet tubing of the evaporator and seeing if it reaches as far as the TEV sensitive bulb. Another indication of refrigerant shortage is a warm liquid receiver and liquid line. The sight glass should show bubbles.

The trouble may also be caused by an overcharge of refrigerant or air in the condenser. Excessive head pressures lower the efficiency of the compressor so much that continuous operation results.

A leaking expansion valve will sometimes give normal refrigeration but will not allow the pressure to drop to where the motor control will cut out the motor. An improperly adjusted expansion valve may cause the same trouble. This will create a frosted or sweating suction line and an above-normal, fluctuating low-side pressure. (Fluctuating means rapid changing from higher to lower and back.)

An inefficient compressor is another possible cause and can be checked by use of gauges. Expansion valve troubles and the compressor troubles may be checked as mentioned earlier in this chapter.

CAPILLARY TUBE

If there is too much refrigerant in the system, the excess will collect on the low side and may enter the suction line. This liquid may prevent the compressor from producing a low enough pressure to operate the thermostat. Therefore, the unit will run continuously and will produce either a normal refrigeration effect or, more likely, excessive refrigeration.

A slight lack of refrigerant will cause a partially refrigerated evaporator and the refrigerated part may not be close enough to the thermostat to cause it to shut off the motor.

14-105 SHORT CYCLING

Short cycling means that the system runs and then stops every few minutes. It is first necessary to locate the control which is turning the system on and off:
1. Temperature control.
2. Overload controls.
3. High-pressure safety control.
4. Oil pressure safety control.

Short cycling may come about from a rapid pressure rise on the low side of the system where an expansion valve is leaking. This leak will also cause a frosting or sweating of the suction line.

Most units are equipped with an overload device in the electrical system. If the motor becomes too hot or uses too much current, these safety devices will stop the motor and then restart it after it cools. A temperature control which is out of adjustment—that is, one with a small differential—will also cause a short cycle.

If the refrigerator has a pressure motor control, short cycling may be caused by either a leak in the refrigerant control or poorly seated compressor valves. In either case, the pressure on the low side will rise rapidly during the off part of the cycle causing the motor to start.

Machines equipped with a high-side pressure safety control will sometimes short cycle if the condensing pressure becomes too high because of a high condensing temperature.

14-106 NOISY UNIT

Noise can come from three principal sources: compressor, electric motor, and the mounting of the complete condensing unit. A compressor is noisy when it pumps oil or when the valves, the piston pin, the connecting rod, and piston have become worn. Sometimes when the compressor gets very warm, it will develop knocks. These are usually hard to remedy. Lack of oil in the compressor may cause it to be noisy.

The metal shaft seal used on most of the open type compressors occasionally becomes noisy and gives out a shrill squeal. This is usually caused by a lack of oil at the seal. If not remedied, it will soon score the seal and cause it to leak.

A conventional electric motor may have noises such as a fan roar, bearing squeak, or motor rumble. Occasionally, if the motor is loaded too heavily, the starting winding does not cut out and will cause continuous noisy operation. If allowed to run this way, the motor will burn out. End play in an electric motor is necessary, but too much will cause a dull knock.

In a conventional unit, belt noise may result from a dry belt or pulleys that are out of line. It may be stopped by using a dressing recommended for belts and by realigning the pulleys. *Do not use oil.*

Sometimes the whole machine unit will vibrate excessively, producing a rumbling sound as the unit runs, shaking the cabinet disagreeably. This is probably due to poor mounting or spring suspension. Another cause may be too little movement in the suction and liquid lines. Or, some obstruction may have been put in the compartment which destroys the action of the shock and the noise-absorbing mounting of the condensing unit.

Excessive head pressure will make a unit vibrate more than normal. A badly worn needle or seat will sometimes make a chattering noise while the unit is in operation.

14-107 REFRIGERATION SERVICE CONTRACTING

It is good business to offer contracts for maintenance and service. Many large companies have developed such contracts, and even larger independents are now offering their customers this type of servicing.

The usual contracting plan offers a definite monthly or weekly rate, for which the service company agrees to keep the refrigerating mechanism in good condition. This charge may or may not cover parts.

The success of such a plan depends on large volume in order to offset cost of maintaining extremely bad installations. Contracts may be on a time and materials basis. Two features of a service contract which appeal to the purchaser are the 24-hours available service clause and an absolute guarantee of work done.

If one has a service contract, a procedure sheet or record sheet should be used to prove service and to insure thorough inspection.

This check sheet should clearly indicate the date and the name of the service technician making the call. In addition, this sheet should include:

1. Test for leaks.
2. Check refrigerant charge:
   a. Head pressure.
   b. Low-side pressure.
3. Check oil charge.
4. Check water valve.
5. Check water drain.
6. Check and lubricate motor.
7. Check belt condition and tension.
8. Clean evaporator.
9. Clean condenser.
10. Straighten fins.
11. Voltage reading.
12. Wattage reading.
13. Check circulating fans.

Service records are absolutely essential if one wishes to establish a permanent business. These records should contain details of the ownership, machine, type of work done, and materials used. This record enables "check backs" if the system does not operate correctly. Furthermore, it establishes sales prospects as systems become older.

14-108 SERVICE ESTIMATES

Many organizations operating refrigerating equipment ask for bids when repair, replacement, or service is required. A service organization bidding on such work should have a person who specializes in estimating work of this kind. This specialist should be thoroughly acquainted with cost of materials, service problems, labor costs, and must be able to judge the time necessary to do the repair. Records kept of service and maintenance work are used as a guide in making estimates.

Needless to say, estimates must include overhead expenses, such as rent, equipment obsolescence, office and shop services, and advertising.

A pleasing personality combined with rapid and accurate estimating ability is essential.

14-109 REFRIGERANT RECOVERING AND RECYCLING

The law as set forth by the Federal Government has produced a decline in production of fully halogenated chlorofluorocarbon refrigerants, as explained in Para. 9-37. This has created the need for refrigerant recovery and reclamation procedures. CFC's R-12, R-22, R-500, and R-502 cannot be reused from old or damaged refrigeration systems. The vapor must be cleaned.

Whereas, in the past, the refrigerants had been vented to the atmosphere, the new governmental emission standards concerning chlorofluorocarbons prohibit this. It now is recovered and recycled by the use of recovery management systems. See Fig. 14-104. The refrigerant is processed through two filters which are accessible for replacement at the top of the unit. Oil trap removes contaminated oil from the system's refrigerant and can be drained. Sight glass indicates condition of refrigerant after all filtering and oil separation has been completed. When the refrigeration system is completely evacuated, the reclaimer will shut down automatically on low suction pressure. The unit does not have internal storage capacity and must be connected to refillable storage cylinders.

Recycling as performed by most of the machines on the market today reduces the contaminants through oil separation and filtration. Many of these units are designed to pump down the system and return the recycled refrigerants to the same system.

The standard procedures for the use of refrigerant recovery, recycling, and reclamation equipment as used by the technician are explained in Chapter 28.

14-110 REVIEW OF SAFETY

A refrigeration service engineer must always be alert to safety. Refrigerating systems have hazards arising from pressure, electricity, power devices, heat, flames, heavy objects, and climbing.

The safety program must include:

1. Safety for the mechanism.
2. Safety for the items being refrigerated.
3. Safety for the operator, the installation, the service technician, and people who may be near the mechanism.

All parts in a refrigerating system must be absolutely clean before they are installed in the system.

Always know what is inside a pressure vessel and know the pressures. Always wear goggles when working on a pressure vessel (refrigerating unit) and when there is danger of flying particles.

Never breathe fumes of any kind. Do not neglect the use of the gas mask when working in a refrigerant-laden atmosphere. This protective measure also applies to cleansing bath fumes, soldering fumes, brazing fumes, and welding fumes. It is true that one's body can and will get rid of certain amounts of strange chemicals and fumes, but some chemicals and fumes accumulate in the body and there may be no ill effects felt for years. Good ventilation is of vital importance.

Avoid exposure to electrical shocks. Keep open electrical terminals covered. Do not work on electrical circuits in damp or wet surroundings.

Avoid spilling liquid refrigerant on any fixture, finished surface, or floor. It may ruin the finish. Avoid contact with the liquid refrigerant, especially on the body and eyes. A freeze burn will result.

Put guards on powered moving objects such as flywheels, belts, pulleys, and fans. Use the leg muscles, not the back, when lifting.

Before using a flame for leak testing, soldering, brazing, or welding, have a fire extinguisher handy. Remove all combustibles from the area and provide good ventilation. A dry chemical fire extinguisher is one of the best for all types of fires. When using a flame to perform tests or make repairs, protect surrounding objects and surfaces with sheet metal or some other flame-resistant shield.

Always test used compressor oil for acid content before allowing any of it to touch the skin. A severe acid burn may result.

Never use air, oxygen, or any fuel gases for developing pressure in a system. If gas, other than refrigerant, is desired, use carbon dioxide, nitrogen, helium, or argon at controlled pressures, with a pressure relief valve, Para. 11-39.

Always wear goggles when handling refrigerants, when opening a refrigerating mechanism, or at any time when there is danger from flying liquids. Wear rubber gloves when handling substances which may have an acid content. Never use carbon tetrachloride for any cleaning operation. This caution is given because at one time it was common practice to use carbon tetrachloride for cleaning refrigerant parts. Such use is now illegal.

Use care when handling capacitors. A charged capacitor can deliver a severe shock.

The Occupational Safety and Health Act (OSHA) is now in effect with regulations. It is also known as the William Steiger Act of 1970. Most of the regulations became mandatory in March, 1973.

Many of the regulations are based on standards developed by the American National Standards Institute (ANSI); National Fire Protection Association (NFPA); Walsh Healy Act (noise); The Service Contract Act; American Society for Testing and Materials (ASTM); and the American Conference of Governmental and Industrial Hygienics (ACGIH).

The Act covers almost every present safety standard plus about everything that can be sensed. Among areas covered are:

1. Walking and working surfaces.
2. Means of exit (egress).
3. Powered platforms.
4. Occupational health and environment control.
5. Hazardous materials.
6. Personal protective equipment.
7. General environment controls.
8. Medical and first aid.
9. Fire protection.
10. Compressed gases.
11. Material handling.
12. Machinery and machine guarding.

Any employee can ask for an inspection. The inspector can impose fines. The employee cannot be fired for requesting an inspection. The employer is responsible. If safeguards are present but the employee does not use them, the employer will still be fined. If this happens, however, the employer has excellent grounds for releasing the employee who is not using the safety precautions.

Some of the OSHA priorities during inspection are:

1. Asbestos.
2. Carbon monoxide.
3. Cotton dust.
4. Lead.
5. Silica.

Asbestos and carbon monoxide are of special importance to refrigeration and air conditioning.

Some other OSHA concerns are:
1. Beryllium.
2. Heat.

3. Mercury.
4. Ultraviolet radiation.
5. Fibrous glass.
6. Trichloroethylene.
7. Chromic acid.
8. Paratheon.

Heat, ultraviolet radiation, fibrous glass, and trichloroethylene are of special interest in refrigeration and air conditioning.

The Threshold Limit Values (TLV) are important. Usually this is the upper safety limit for the hazard over an eight-hour exposure period. Most have been established by ANSI (American National Standards Institute).

Noise is important. The OSHA lists 90 dB on dBA noise scale as maximum for eight hours of exposure, 92 dB for six hours, 95 dB for four hours.

For seated mental performance for eight hours, the room temperature maximum should be 79 °F (26 °C) for men and 76 °F (24 °C) for women. For a four-hour exposure, it can be as high as 87 °F (31 °C). This regulation will have an impact on the ventilation and air conditioning industry.

Above all else, report any injury—no matter how slight—in writing. Send the report to your employer and keep a copy.

14-111 TEST YOUR KNOWLEDGE

1. Why do code installations require safety release valves on condensing units?
2. Why should hand valves be provided in the refrigerant lines to each individual evaporator in a multiple evaporator system?
3. Why is it necessary to run refrigeration tubing parallel with beams rather than across them?
4. Why is it necessary to mount evaporators absolutely level?
5. Where are two-temperature valves usually located?
6. Where may soft tubing be used in a code installation?
7. Explain the method for removing air, through evacuating, from a system.
8. Why is it necessary to put a drier on a new system?
9. What would be the purpose of a drier placed in a suction line?
10. Why are hard drawn copper tubing and strong fittings used?
11. List some of the ways that a lack of refrigerant in a TEV installation is indicated.
12. What may be wrong if a thermal expansion valve evaporator suddenly starts to frost or sweat excessively at the expansion valve, but the evaporator near the suction line connection is dry?
13. How are flare nuts protected so moisture cannot get under them and freeze?
14. What steps should be taken when removing any part from a system?
15. How can the needle in a TEV be tested for leaks?
16. Are commercial systems normally charged through the high-pressure side or low-pressure side?
17. What safety precautions should a service technician follow when charging a system through the low side?
18. If tubing must slant upwards from the evaporator to the condensing unit, how should the run be made?

19. What is the purpose of mesh screen or strainers in a drier?
20. Why must a system be very carefully checked for leaks if a lack of refrigerant is discovered in the system?
21. How is a leaking exhaust valve detected in an external drive compressor?
22. What special precautions must be followed when installing a suction line?
23. What will happen if the lowest temperature evaporator check valve leaks?
24. What cleaning fluid can be used to clean joints before they are opened?
25. What may cause a water-cooled condensing unit to short cycle?
26. When removing a part from a commercial system, where should the refrigerant be stored?
27. List the procedure for lapping a valve plate.
28. Why are suction service valves and discharge service valves called three-way valves?
29. What are four common thermostatic two-temperature valve troubles?
30. What two things can most sight glasses indicate to a service technician?
31. What should be done to the main switch before one works on the electrical parts of a system?
32. Why is a high-pressure relief valve or rupture disk used on condensing units?
33. What is put on brazed condenser joints before they are taken apart?
34. What may happen if moisture collects on the outside of the bellows of a two-temperature refrigerant control?
35. What happens if a suction line is undersized?
36. What is done to a water-cooled condenser before the refrigerant is purged from it?
37. How should the contaminated hermetic compressor oil be handled?
38. What are the symptoms of a leaking compressor crankshaft seal using an above-atmospheric low-side pressure?
39. When should one wear goggles when working on a refrigerating mechanism?
40. What should one do if a leaky refrigerant joint is located?
41. What are two uses for an ohmmeter?
42. Why do some compressors have a sight glass?
43. How may one put oil in a compressor?
44. What may one do if refrigerant lines are noisy?
45. What treatment should be given to water-cooled condenser water?
46. What is the chemical that is sometimes used to clean water lines?
47. What causes a hermetic motor compressor burnout?
48. Why should a suction line have a slight down slope toward the compressor?
49. What is a refrigerant charging board? How does it indicate the weight of the refrigerant?
50. How often should the water circuit of an ice maker be cleaned?
51. What happens if the TEV orifice is too large?
52. How may a fuse be tested for continuity before installing it?
53. What is done with the refrigerant when a hot gas bypass valve is replaced?
54. Can any 1/2 hp hermetic motor compressor replace a worn out 1/2 hp unit? Why or why not?
55. When a system is purged, what else leaves the system besides refrigerant?
56. What happens when a drier is heated?
57. Can liquid refrigerant be charged into the low side of a system? Why or why not?
58. What happens to the liquid line when a drier is almost clogged?
59. Can aluminum electrical leads be connected to copper leads? Why?
60. What are some of the troubles that may occur with motor controls?

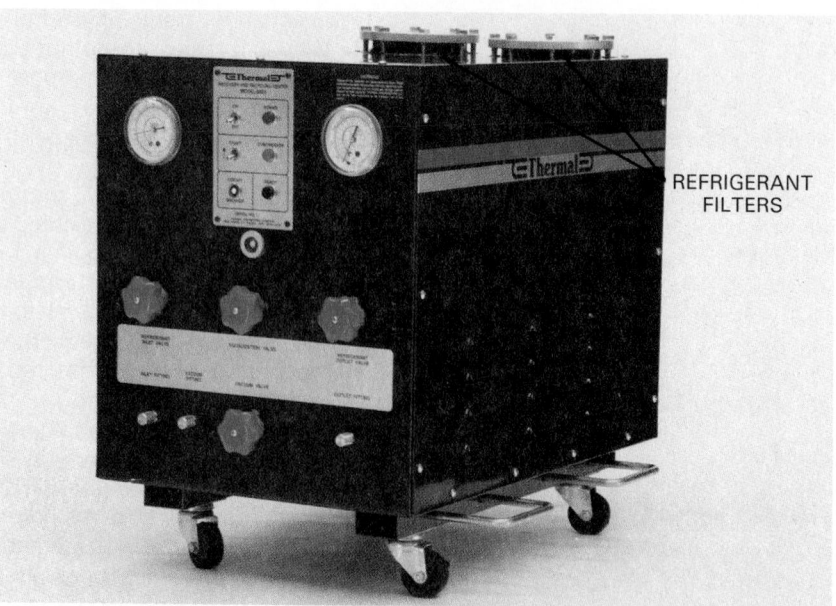

Fig. 14-104. Refrigerant recovery system designed to be used with systems containing R-12, R-22, R-500, and R-502, in either liquid or vapor form. (Thermal Engineering Company)

Chapter 15

COMMERCIAL SYSTEMS HEAT LOADS AND PIPING

After studying this chapter, the technician will be able to:

☐ Discuss system balance and explain four important factors in balancing commercial systems.

☐ Explain and calculate heat loads.

☐ List the individual loads that make up the total heat load.

☐ Demonstrate proper use of tables in computing heat loads.

☐ Correctly size system components using manufacturers' tables.

☐ Discuss and calculate energy efficiency ratings (EER).

To have good refrigeration system performance, four main items must be matched (balanced or made equal to each other):

1. Heat load. Determine the total amount of heat that must be removed for each 24 hours. Amount of heat is measured in Btus. A Btu is the amount of heat needed to raise the temperature of 1 lb. of water one degree Fahrenheit.

2. Condensing unit. Determine what size condensing unit is needed to handle the heat load. To do this, determine whether the unit is to run 16, 18, or 20 hours of each 24 hours.

3. Evaporator. Determine evaporator capacity required to handle the heat load. The evaporator can remove heat only while the condensing unit is running. Therefore, its capacity must be based on the same hours of operation as the condensing unit.

4. Total system. Consider such factors as water supply, temperature control devices, refrigeration line sizes, air circulation, and humidity control. Correctly install the components according to code.

When determining the heat load, two main factors must be considered:

1. Heat leakage into the cabinet. Heat leakage is affected by: amount of exposed surface, thickness and kind of insulation, and temperature difference between inside and outside of cabinet.

2. Usage or service heat load of the cabinet. This load is determined by: temperature of articles put into the refrigerator, their specific heat, generated heat, and latent heat, as the requirements demand. Another consideration is the nature of the service required. This includes air changes (determined by the number of times per day that doors of the refrigerator are opened) and heat generated inside by fans, lights, and other electrical devices.

Total heat load is the sum of:

1. Wall heat transmission load.
2. Air change load.
3. Product load.
4. Miscellaneous loads.

The selection of a condensing unit is usually made from manufacturers' tables of condensing unit capacities. Evaporators are selected from manufacturers' specifications for capacities to balance the capacity of the condensing unit. Also affecting the selection of the evaporator are the capacity of the refrigerant control, type of temperature control, arrangement for air circulation, and specific duty.

The installation of all commercial refrigeration equipment involves a technical understanding of the variables. This is a determining factor in the operation of the system.

15-1 HEAT LOAD

The total heat load consists of the amount of heat to be removed from a cabinet during a certain period. It is dependent on two main factors:

1. Heat leakage load.
2. Heat usage or service load.

The heat leakage load or heat transfer load is the total amount of heat that leaks through the walls, windows, ceiling, and floor of the cabinet per unit of time (usually 24 hours).

The heat usage or service load is the sum of the following heat loads per unit of time (usually 24 hours):

1. Cooling the contents to cabinet temperature.
2. Cooling of air changes.
3. Removing respiration heat from fresh or "live" vegetables and from meat.
4. Removing heat released by electric lights, electric motors, and the like.
5. Removing heat given off by people entering and/or working in the cabinet.

In this chapter, heat load calculations are in the U.S. Conventional System (pounds, Btu, temperatures °F, foot, etc.). Conversion factors used to make these calculations in the SI System (kilograms, kilocalories, temperatures °C, centimeters, etc.) are explained in Chapter 30.

15-2 HEAT LEAKAGE VARIABLES

Research organizations, manufacturers, and refrigeration associations have determined the amount of heat leakage through walls and other heat loads. Charts and tables based on these calculations are used by engineers and technicians.

Five factors (variables) which affect heat leakage are:

1. *Time.* The longer the period of time, the more heat will leak through a certain wall. The standard time unit is the 24-hour period in refrigeration situations, while the one-hour period is used in air conditioning situations.

2. *Temperature difference.* The difference in temperature is an important factor in the heat leakage into a container. The greater the temperature difference, the more heat will leak or transfer through the wall. One might compare this idea to pressure: the more pressure, the more water will flow through an opening. The room temperature usually chosen is the average summer temperature. In the United States, it varies between 90 and 105°F (32°C and 40°C). See Fig. 15-1. This value can be reduced to 75°F or 80°F (25°C or 27°C) if the room is air conditioned.

3. *Thickness of insulation.* The thicker the insulation, the less heat will flow through it. Twice as much heat will leak through a wall with 1-in. insulation than through a wall having 2-in. insulation.

4. *Kind of insulation.* The kind of insulation or the material used is important. Expanded polystyrene (foam), for instance, will insulate approximately six times better than wood. Some insulations, however, are more costly than others.

5. *External area of cabinet.* The more area through which heat may leak, the greater the heat flow. To compare this with water flow, the size of a pipe determines how much water will flow through it. The bigger the pipe, the more water will flow.

The common unit used for determining the heat flow is the total square foot area. This area is always measured on the outside of the cabinet.

15-3 HEAT LEAKAGE—K FACTOR

To bring together the values just discussed, standards have been developed which are used by refrigerating companies. In preparing these standards, the variables have been reduced to unit values. The unit values, in turn, are used to indicate heat leakage of the wall.

The unit or basic values are the thermal conductance (symbol is K) obtained for an area of insulation one square foot in size, one inch thick, with a temperature difference of 1°F over a period of time of either one hour or 24 hours. Values obtained represent the amount of heat flow through the insulation under these conditions.

Unit values vary with the kind of insulation. This material has no air film or liquid film on either side. The symbol is K.

If the insulation is less than or more than 1 in. thick, the heat leakage will be different. In this case, the symbol is K_T. Example:

$$K_T = \frac{K_1}{\text{thickness}}$$

By definition:

K_T = conductance, total
K_1 = conductance for 1 in. thickness.

If thickness is 2 in.:

$$K_T = \frac{K_1}{2} \qquad K_T = \frac{1}{2}K_1$$

SUMMER DESIGN TEMPERATURES					
STATE	DESIGN DRY BULB °F	°C	STATE	DESIGN DRY BULB °F	°C
ALABAMA	95	29	MASSACHUSETTS	90	32
ALASKA	74	23	MICHIGAN	88	31
ARIZONA	105	41	MINNESOTA	90	32
ARKANSAS	98	37	MISSISSIPPI	97	36
CALIFORNIA:			MISSOURI	98	37
LOWER	86	30	MONTANA	88	31
MIDDLE	94	34	NEBRASKA	97	36
UPPER	83	28	NEVADA	95	35
COLORADO	92	33	NEW HAMPSHIRE	90	32
CONNECTICUT	88	31	NEW JERSEY	92	33
DELAWARE	93	34	NEW MEXICO	95	35
DIST. OF COL.	94	34	NEW YORK	90	32
FLORIDA:			NORTH CAROLINA	95	35
UPPER	96	36	NORTH DAKOTA	93	34
LOWER	93	34	OHIO	90	32
GEORGIA	95	35	OKLAHOMA	102	39
HAWAII	87	31	OREGON	90	32
IDAHO	94	34	PENNSYLVANIA	92	33
ILLINOIS:			RHODE ISLAND	87	31
UPPER	95	35	SOUTH CAROLINA	95	35
LOWER	97	36	SOUTH DAKOTA	95	35
INDIANA	95	35	TENNESSEE	96	36
IOWA	95	35	TEXAS	101	38
KANSAS:			UTAH	95	35
UPPER	97	36	VERMONT	87	31
LOWER	100	38	VIRGINIA	95	35
KENTUCKY	95	35	WASHINGTON	90	32
LOUISIANA	98	37	WEST VIRGINIA	94	34
MAINE	88	31	WISCONSIN	90	32
MARYLAND	94	34	WYOMING	90	32

Fig. 15-1. Table of summer design temperatures. These may be used as ambient temperatures when calculating heat leakage loads.

If thickness is $\frac{1}{2}$ in.:

$$K_T = \frac{K_1}{\frac{1}{2}} \qquad K_T = K_1 \div \frac{1}{2}$$

$$K_T = K_1 \times 2 \qquad K_T = 2K_1$$

A special formula is needed to find heat leakage or thermal conductance through a composite wall, Fig. 15-2. The formula for computing the K_T or total heat conductance factor follows:

Resistance to heat flow is known by the symbol R. If the same heat is flowing through two substances, the total resistance is equal to the resistance of each substance.

$R_T = R_1 + R_2$
R_T = resistance total
R_1 = resistance of substance 1
R_2 = resistance of substance 2

R is the opposite, or reciprocal, of K, or $R = \frac{1}{K}$. In the formula, it would be:

$$\frac{1}{K_T} = \frac{1}{K_1} + \frac{1}{K_2}$$

K_T = conductance, total
K_1 = conductance, substance 1
K_2 = conductance, substance 2

Example:

K_1 = .6 and K_2 = .2

$$\frac{1}{K_T} = \frac{1}{.6} = \frac{1}{.2} \qquad \frac{1}{K_T} = \frac{1}{\frac{6}{10}} + \frac{1}{\frac{2}{10}}$$

$$\frac{1}{K_T} = \frac{10}{6} + \frac{10}{2} \qquad \frac{1}{K_T} = \frac{5}{3} + \frac{5}{1}$$

$$\frac{1}{K_T} = \frac{5}{3} + \frac{15}{3} \qquad \frac{1}{K_T} = \frac{20}{3}$$

$$K_T = \frac{3}{20} \qquad\qquad K_T = .15$$

$$R = \frac{1}{K_T} \qquad\qquad R = \frac{1}{.15}$$

$$R = \frac{1 \times 100}{.15 \times 100} = \frac{100}{15} = 6.7$$

If the wall is made up of three different materials, Fig. 15-3, overall heat leakage (K) is found as follows:

$$K_T = \cfrac{1}{\cfrac{\text{thickness of material 1}}{\substack{\text{conductivity factor} \\ \text{material 1}}} + \cfrac{1}{\cfrac{\text{thickness of material 2}}{\substack{\text{conductivity factor} \\ \text{material 2}}} + \cfrac{1}{\cfrac{\text{thickness of material 3}}{\substack{\text{conductivity factor} \\ \text{material 3}}}}}$$

where Th_1 = thickness of material 1
Th_2 = thickness of material 2
Th_3 = thickness of material 3
and K_1 = conductivity factor for material 1
K_2 = conductivity factor for material 2
K_3 = conductivity factor for material 3
then the formula becomes:

$$K_T = \cfrac{1}{\cfrac{Th_1}{K_1} + \cfrac{Th_2}{K_2} + \cfrac{Th_3}{K_3}}$$

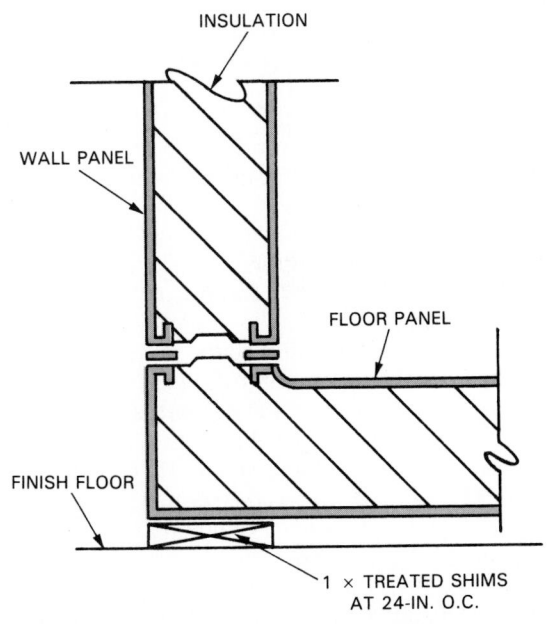

Fig. 15-2. Cross section of a walk-in cooler wall. Note system for sealing joint. (Tyler Refrigeration Corp.)

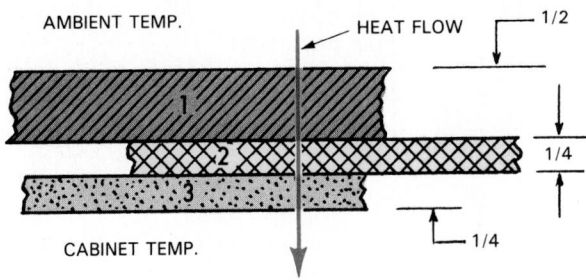

Fig. 15-3. A composite insulated panel or wall. Insulating materials, labeled 1, 2, and 3, are of three different types and thicknesses.

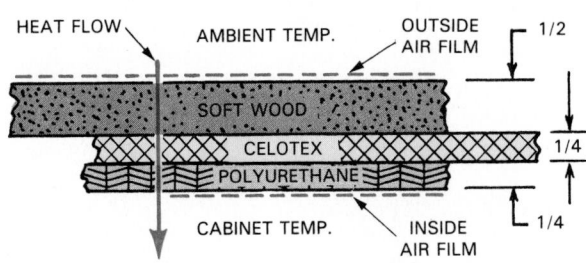

Fig. 15-4. Cross section shows a composite insulating panel made up of various materials at specified thicknesses with an air film on both sides.

To solve for the conductivity for the panel shown in Fig. 15-4, proceed as follows:

$$K_T = \cfrac{1}{\cfrac{\text{thickness A}}{\text{K for wood}} + \cfrac{\text{thickness B}}{\substack{\text{K for} \\ \text{Celotex}}} + \cfrac{\text{thickness C}}{\substack{\text{K for} \\ \text{polyurethane}}}}$$

From Chapter 30, K values are:

Material	K value
Wood	= .80 for 1-in. thickness
Celotex	= .31 for 1-in. thickness
Polyurethane	= .160 for 1-in. thickness

However, in the example, the wood is 1/2 in. thick, the Celotex is 1/4 in. thick, and the polyurethane is 1/4 in. thick.

Substituting these values in the above formula:

$$K_T = \cfrac{1}{\cfrac{.5}{.80} + \cfrac{.25}{.31} + \cfrac{.25}{.160}} =$$

$$= \cfrac{1}{.625 + .807 + 1.563} = \frac{1}{3.0}$$

K_T = .333, which is the conductivity for the panel in Btu/hr/ °F/sq. ft.

An air film that clings to the outer and inner surfaces of the cabinet adds to the insulating value of the walls of the cabinet. This added resistance is calculated in the following formula in which the outside air film (F_O) is considered to have a heat transfer value of 6.00 and the inside-wall air film (F_I) has a value of 1.65:

K_T = unit of conductivity for materials of a composite nature

U = unit of conductivity for materials of a composite nature plus the effect of the air clinging to both the outside (F_O) and the inside (F_I) walls.

If the insulating value of the air clinging to the walls is considered, the formula becomes:

$$U = \frac{1}{\frac{1}{F_O} + \frac{Th_1}{K_1} + \frac{Th_2}{K_2} + \frac{Th_3}{K_3} + \frac{1}{F_I}}$$

If the value of F_O = 6.0 and the value of F_I = 1.65, then the problem in Fig. 15-4 may be solved as follows:

$$U = \frac{1}{\frac{1}{6.0} + \frac{.5}{.80} + \frac{.25}{.31} + \frac{.25}{.160} + \frac{1}{1.65}} =$$

$$\frac{1}{.166 + .625 + .807 + 1.563 + .606} =$$

$$\frac{1}{3.767} = .27 \text{ Btu/hr/}^\circ\text{F/sq. ft.}$$

The value of K, as computed in the previous problem, is .33. The value of U, as computed in this problem, is .27. This shows the additional insulating effect of the air films.

This type of computation is complicated and slow. For this reason, standard tables for computing heat leakage are generally used.

15-4 AIR CHANGE HEAT LOAD

Air that enters a refrigerated space must be cooled. Air has weight and it also contains moisture. When air enters the refrigerated space, heat must be removed from it.

By Charles's Law, air which enters and is cooled reduces in pressure. If the cabinet is not airtight, air will continue to leak in. Also, each time a service door or a walk-in door is opened, the cold air inside, being heavier, will spill out the bottom of the opening allowing the warmer room air to move into the cabinet. The actions of material moving in or out of the cabinet, and a person going into or leaving a cabinet, result in warm air moving into the space. This movement is sometimes called infiltration of air.

Fig. 15-5 shows accepted air change volume values for refrigerated cabinets of various internal volumes. Fig. 15-6 shows the total heat (sensible + latent) to be removed from this air, depending on various outside conditions and refrigerator temperatures.

15-5 PRODUCT HEAT LOAD

Any substance which is warmer than the refrigerator it is placed in will lose heat until it cools to the refrigerator temperature.

Three kinds of heat removal may be involved:
1. Specific heat.
2. Latent heat.
3. Respiration heat.

The total product heat load would be the sum of these three heat loads. An example of all three heat loads would be moist head lettuce at 55°F (13°C) being put into a 35°F (2°C) refrigerator. The lettuce must be cooled—specific heat—to 35°F (2°C). Some of the moisture on the lettuce will evaporate and collect on the evaporators (latent

heat). The lettuce, being a live vegetable, would absorb carbon dioxide and release oxygen. This change would release heat energy (respiration heat). Meats go through a slow bacteriological change and, during this action, heat is released (respiration heat).

1. An example of specific heat: If bottled beverages at 50°F (10°C) are placed in a 35°F (2°C). This action is a specific heat problem. If the bottles or bottle cartons were moist, there would be a moisture evaporating problem (latent heat).
2. An example of latent heat: If meat at 50°F (10°C) is placed in a refrigerator, and cooled (frozen) to 0°F (−18°C), the meat cools to about 27°F (−3°C), freezes, and then it cools to 0°F (−18°C). The latent heat of freezing of the meat is considerable. For fresh lean beef, it is 100 Btu per pound. Fig. 15-7 shows the specific heats and latent heats of various refrigerated products. In addition, it recommends storage temperatures and relative humidity.
3. An example of respiration heat: If lettuce is stored at 40°F (4°C), each pound will release 7.99 Btu/24 hr. or 15,980 Btu/ton. Fig. 15-7 shows the respiration heat of some of the more common vegetables and fruits.

15-6 MISCELLANEOUS HEAT LOAD

All sources of heat not covered by heat leakage, product cooling, and respiration load are usually listed as miscellaneous heat loads. Some of the more common miscellaneous heat loads are: lights, electric motors, people, defrosting heat sources, and sun (solar heat).

1. Lights located in the refrigerated space will release heat. For example, a 100-watt lamp will give off 342 Btu in one hour or

    342 × 24 = 8208 Btu/24 hr.

    If the workday is eight hours (only time light is on), the heat load would be

    342 × 8 = 2736 Btu/24 hr.

2. On the average, electric motors release 2545 Btu/hp/hour. The amount of heat released also depends on motor efficiency; the larger the motor, the more efficient it is. Fig. 15-8 shows the heat given off by motors and devices they drive. Because forced convection evaporators usually have motors and fans, it should be noted that the total heat release of such a motor is about 4250 Btu/hp/hr. for sizes from 1/8 to 1/2 hp. For example, a 1/8-hp motor-fan would release:

    4250 Btu/hp/hr.

    4250 Btu/hr. × 1/8 × 1 =

    4250 ÷ 8 = 531 Btu/hr.

    531 × 24 = 12,744 Btu/24 hr. if motor runs continuously.

3. People inside a refrigerated space release heat at varying rates depending on what they are wearing (insulation), the temperature of the cabinet, and on how hard they are working. Fig. 15-9 shows a variation of 720 Btu/hr./person at 50°F (10°C) to 1400 Btu/hr./person at −10°F (−23°C). For example, if one person worked in a 30°F (−1°C) refrigerator for eight hours, the heat load would be:

    950 Btu/hr. × 8 hr. = 7600 Btu

4. Many refrigerating units have defrosting heat sources, especially if the fixture temperature is 32°F (0°C) or lower. Whether the defrost heat source is electric, hot

VOLUME CU. FT.	AIR CHANGES PER 24 HR.	VOLUME CU. FT.	AIR CHANGES PER 24 HR.
200	6.5	6,000	44.0
300	5.5	8,000	34.5
400	4.9	10,000	29.5
500	3.9	15,000	26.0
600	3.5	20,000	23.0
800	3.0	25,000	20.0
1,000	2.7	30,000	17.5
1,500	2.3	40,000	14.0
2,000	2.0	50,000	12.0
3,000	1.6	75,000	9.5
4,000	1.4	100,000	8.2
5,000			7.2

NOTE: For heavy usage multiply the above values by 2. For long storage multiply the above values by 0.6.

Fig. 15-5. Average air changes per 24 hours for storage rooms. Values take into account door openings and air filtration.　(ASHRAE Guide and Data Book)

Heat Removed in Cooling Air to Storage Room Conditions, Conventional Units (Btu/ft³)

Storage Room Temp, F @ 80% rh	Temperature of Outside Air, F							
	85		90		95		100	
	Relative Humidity, %							
	50	60	50	60	50	60	50	60
65	0.45	0.64	0.68	0.9	0.93	1.20	1.21	1.51
60	0.66	0.85	0.89	1.12	1.14	1.41	1.42	1.71
55	0.85	1.04	1.08	1.31	1.33	1.60	1.61	1.91
50	1.03	1.22	1.26	1.49	1.51	1.78	1.79	2.09
45	1.19	1.39	1.43	1.66	1.68	1.94	1.95	2.25
40	1.35	1.55	1.59	1.81	1.83	2.10	2.11	2.41
35	1.50	1.70	1.74	1.96	1.99	2.25	2.26	2.56
30	1.64	1.84	1.88	2.10	2.13	2.39	2.40	2.70

Storage Room Temp, F @ 80% rh	Temperature of Outside Air, F							
	40		50		90		100	
	Relative Humidity, %							
	70	80	70	80	50	60	50	60
25	0.39	0.43	0.69	0.75	2.02	2.24	2.54	2.84
20	0.52	0.56	0.82	0.89	2.15	2.38	2.68	2.97
15	0.65	0.69	0.95	1.01	2.28	2.50	2.80	3.10
10	0.77	0.82	1.08	1.14	2.40	2.63	2.93	3.22
5	0.89	0.94	1.20	1.26	2.52	2.75	3.05	3.34
0	1.01	1.05	1.31	1.38	2.64	2.86	3.16	3.46
−5	1.13	1.17	1.43	1.49	2.76	2.98	3.28	3.58
−10	1.24	1.29	1.55	1.61	2.88	3.10	3.40	3.70
−15	1.36	1.41	1.67	1.73	2.99	3.22	3.52	3.81
−20	1.48	1.52	1.78	1.85	3.11	3.34	3.64	3.93
−25	1.60	1.64	1.90	1.97	3.23	3.45	3.75	4.05
−30	1.72	1.76	2.03	2.09	3.35	3.58	3.88	4.17

Fig. 15-6. Chart gives total heat removed to cool storage room air under varying conditions of humidity and temperature. Values of heat removed are in Btus per cubic foot.　(ASHRAE Guide and Data Book)

HEAT LOAD OF VARIOUS REFRIGERATED PRODUCTS

Product	Quick Freeze Temp.	Storage Temp. Long	Storage Temp. Short	Humidity % R.H.	Specific Heat Above Freezing	Specific Heat Below Freezing	Latent Heat	Freezing Point	Respiration BTU/lb. Per Day
Apples	—15	30-32	38-42	85-88	0.92	0.39	91.5	28.4	0.75
Asparagus	—30	32	40	85-90	0.95	0.44	134.0	29.8	....
Bacon, Fresh		0-5	36-40	80	0.55	0.31	30.0	25.0	....
Bananas		56-72	56-72	85-95	0.81	....	108.0	30.2	4.18
Beans, Green		32-34	40-45	85-90	0.92	0.47	128	29.7	3.3
Beans, Dried		36-40	50-60	70	0.30	0.237	18	....	....
Beef, Fresh, Fat	—15	30-32	38-42	84	0.60	0.35	79	....	....
Beef, Fresh Lean	—15	30-32	38-42	85	0.77	0.40	100	....	....
Beets, Topped		32-35	45-50	95-98	0.90	....	....	26.9	2.0
Blackberries	—15	31-32	42-45	80-85	0.89	0.46	125	28.9	....
Broccoli		32-35	40-45	90-95	0.93	....	....	29.2	....
Butter	+15	....	40-45	....	0.64	0.34	15	15.0	....
Cabbage	—30	32	45	90-95	0.93	0.47	130	31.2	....
Carrots, Topped	—30	32	40-45	95-98	0.87	0.45	120	29.6	1.73
Cauliflower		32	40-45	85-90	0.90	....	....	30.1	....
Celery	—30	31-32	45-50	90-95	0.95	0.48	135	29.7	2.27
Cheese	+15	32-38	39-45	....	0.70	....	....	....	....
Cherries		31-32	40	80-85	0.85	....	118	28.0	6.6
Chocolate Coatings		45-50	....	....	0.3	....	....	....	....
Corn, Green		31-32	45	85-90	0.86	....	....	29.0	4.1
Cranberries		36-40	40-45	85-90	0.91	....	....	27.3	....
Cream		34	40-45	....	0.88	0.37	84	....	....
Cucumbers		45-50	45-50	80-85	0.93	....	....	30.5	....
Dates, Cured		28	55-60	50-60	0.83	0.44	104	....	....
Eggs, Fresh	—10	30-31	38-45	....	0.76	0.40	98	31.0	....
Eggplants		45-50	46-50	85-90	0.88	....	....	30.4	....
Flowers		35-40	....	85-90	....	....	....	....	....
Fish, Fresh, Iced	—15	25	25-30	....	0.82	0.41	105	30.0	....
Fish, Dried		30-40	....	60-70	0.56	0.34	65	....	....
Furs		32-34	40-42	40-60	....	....	....	....	....
Furs, To Shock		15	15	....	....	....	....	....	....
Grapefruit		32	32	85-90	0.92	....	111	28.4	0.5
Grapes		30-32	35-40	80-85	0.92	....	111	27.0	0.5
Ham, Fresh		28	36-40	80	0.68	0.38	87	....	....
Honey		31-32	45-50	....	0.35	0.26	26	....	....
Ice Cream	—20	....	0-10	....	.5-.8	0.45	96	....	....
Lard		32-34	40-45	80	0.52	0.31	90	....	....
Lemons		55-58	....	80-85	0.91	0.39	190	28.1	0.4
Lettuce		32	45	90-95	0.90	....	....	31.2	8.0
Liver, Fresh		32-34	36-38	83	0.72	0.42	94	....	....
Lobster, Boiled		25	36-40	....	0.81	0.42	105	....	....
Maple Syrup		31-32	45	....	0.24	0.215	7.0	....	....
Meat, Brined		31-32	40-45	....	0.75	0.36	75.0	....	....
Melons		34-40	40-45	75-85	0.92	0.35	115	28.5	1.0
Milk		34-36	40-45	....	0.92	0.46	124	31.0	....
Mushrooms		32-35	55-60	80-85	0.90	....	....	30.2	....
Mutton		32-34	34-42	82	0.81	0.39	96	29.0	....
Nut Meats		32-50	35-40	65-75	0.30	0.24	14	20.0	....
Oleomargarine		34-36	....	....	0.65	0.34	35	15.0	....
Onions		32	50-60	70-75	0.91	0.46	120	30.1	1.0
Oranges		32-34	50	85-90	0.89	0.40	91.0	27.9	0.7
Oysters		....	32-35	....	0.85	0.45	120.0	....	....
Parsnips	—30	32-34	34-40	90-95	0.82	0.45	120.0	28.9	....
Peaches, Fresh		31-32	50	85-90	0.92	0.42	110	29.4	1.0
Pears, Fresh		29-31	40	85-90	0.90	0.43	106	28.0	6.6
Peas, Green		32	40-45	85-90	0.80	0.42	108	30.0	....
Peas, Dried		35-40	50-60	....	0.28	0.22	14	....	....
Peppers		32	40-45	85-90	0.90	....	....	30.1	2.35
Pineapples, Ripe		40-45	50	85-90	0.90	....	127	29.9	....
Plums		31-32	40-45	80-85	0.83	....	115	28.0	....
Pork, Fresh		30	36-40	85	0.60	0.38	66	28.0	....
Potatoes, White	—30	36-50	45-60	85-90	0.77	0.44	105	28.9	0.85
Poultry, Dressed	—10	28-30	29-32	....	0.80	0.41	99	27	....
Pumpkins		50-55	55-60	70-75	0.90	....	....	30.2	....
Quinces		31-32	40-45	80-85	0.90	....	....	28.1	....
Raspberries		31-32	40-45	80-85	0.89	0.46	125	30.0	3.3
Sardines, Canned		....	35-40	....	0.76	0.410	101	....	....
Sausage, Fresh		31-36	36-40	80	0.89	....	....	....	....
Sauerkraut		33-36	36-38	85	0.91	0.47	128	....	....
Squash		50-55	55-60	70-75	0.90	....	....	29.3	....
Spinach		32	45-50	85	0.92	....	....	30.8	....
Strawberries	—15	31-32	42-45	80-85	0.92	0.48	129	30.0	3.3
Tomatoes, ripe		40-50	55-70	85-90	0.95	....	135	30.4	0.5
Turnips		32	40-45	95-98	0.90	....	....	30.5	1.0
Veal	—15	28-30	36-40	....	0.71	0.39	91	29	....

Fig. 15-7. Temperature, specific heat, and latent heat data for some common foods. These can be used in determining heat loads. (Dunham-Bush, Inc.)

Heat Equivalent of Electric Motors			
	Connected Load in Refrigerated Space[a]	Motor Losses Outside Refrigerated Space[b]	Connected Load Outside Refrigerated Space[c]
Motor hp	Btu/hp·h	Btu/hp·h	Btu/hp·h
1/8 to 1/3	4600 ±	2550	2100 ±
1/2 to 3	3800 ±	2550	1300 ±
5 to 20	3300 ±	2550	800 ±

[a]For use when both useful output and motor losses are dissipated within refrigerated space; motors driving fans for forced circulation unit coolers.

[b]For use when motor losses are dissipated outside refrigerated space and useful motor work is expended within refrigerated space; pump on a circulating brine or chilled water system; fan motor outside refrigerated space driving fan circulating air within refrigerated space.

[c]For use when motor heat losses are dissipated within refrigerated space and useful work expended outside of refrigerated space; motor in refrigerated space driving pump or fan located outside of space.

Fig. 15-8. Heat released by operating electric motors. Note different operating conditions and how they affect heat released.
(ASHRAE Guide and Data Book)

Heat Equivalent of Occupancy	
Refrigerated Space Temperature, °F	Heat Equivalent/ Person, Btu/h
50	720
40	840
30	950
20	1050
10	1200
0	1300
−10	1400

Note: Heat equivalent may be estimated by $q_p = 1295 - 11.5t(°F)$

Fig. 15-9. Heat released by a person in the cooled space.
(ASHRAE Guide and Data Book)

gas, or water, the defrosting operation adds heat to the interior of the refrigerator. The amount of heat is difficult to determine because most of the defrosting heat is removed in the defrost drain water. Add approximately 10 percent of the defrosting heat input as part of the heat load.

5. If part of the refrigerator is exposed to the sun, the heat from this source must be considered. Add the following to the room or ambient temperature: for a dark surface, add about 10 °F (6 °C). If it is a medium colored surface, add 5 °F (3 °C). If a light surface, add 3 °F (2 °C) to the ambient temperature.

15-7 CABINET AREAS

The area of a cabinet is measured from the outside. There are six surfaces: four walls, the ceiling, and the floor. Usually the floor and ceiling have the same area; opposite walls are the same area, also. To determine the total outside area:

1. Multiply width by length, then multiply by two. These areas are the areas of the floor and ceiling of the cabinet.
2. Multiply width by height, then multiply by two. These areas are the areas of ends of the cabinet.
3. Multiply length by height, then multiply by two. These areas are the areas of sides of cabinet. Add these three

values to determine the total external area of the cabinet.

By formula:

L = Length   W = Width   H = Height

W × L × 2 = area of ceiling and floor

W × H × 2 = area of ends

L × H × 2 = area of sides

Total external area = sum of the three areas

Most companies compute total area based on the outside of the cabinet. This is done because the exterior is easier to measure and the results are on the safe side.

After computing the external area of the cabinet, subtract the window area to obtain the area of the insulated surface. Window areas are calculated from the measurements of the outside edges of the window frame and must be considered separately. Total external area minus the window area equals the insulated area.

Use the accompanying table to find the amount of heat that will leak through the insulation per square foot of area per 24 hours for that particular type of wall construction for the temperature difference. Consider a wall made of steel paneling on both sides with a 4 in. slab of cork insulation (or its equivalent) in between. The tables in Fig. 15-10 will reveal that at a temperature difference of 60 °F (95 °F − 35 °F) 108 Btu will leak through every square foot during a 24-hour period. Expanded polystyrene insulation would either be 33 percent less or about 2 1/2-in. thick for the same heat loss.

Some synthetic material insulation values are:

	K
Expanded rubber, rigid	0.22
Glass fiber, organic bonded	0.25
Expanded polystyrene (extruded), plain	0.25
Expanded polystyrene (extruded), R-12 expanded, 1-in. thick or greater	0.19
Expanded polystyrene, molded beads	0.28
Expanded polyurethane, R-11 expanded, 1-in. thick or greater	0.16
Silica aerogel, loose fill	0.17

The glass leakage table will give values for the heat leakage through one square foot of glass. If the cabinet has double glass, 660 Btu will leak through at a temperature difference of 60 °F. Adding the two heat leakages will give the total heat leakage into the cabinet.

Example: A walk-in cooler is 10 ft. × 9 ft. × 8 ft. high with two double-pane glass windows 1 1/2 ft. × 2 ft. The box is kept at 35 °F (2 °C) in a room with a summer design temperature of 95 °F (35 °C). The wall construction consists of 4-in. cork (or its equivalent) with metal on each side (or it could be 2 1/2 in. of expanded polystyrene). The windows are of double-pane construction. The temperature difference is 95 minus 35 = 60 °F.

Solution:

Walls: 10 × 9 × 2 = 180 sq. ft. (ceiling and floor)
          9 × 8 × 2 = 144 sq. ft. (ends)
          10 × 8 × 2 = 160 sq. ft. (sides)
          484 sq. ft. of total area

Windows: 1 1/2 × 2 × 2 = 6 sq. ft. of window
          484 − 6 = 478 sq. ft. of insulated wall

From table, Fig. 15-10:

1 sq. ft. of the wall allows 108 Btu per 24 hr.

108 × 478 sq. ft. = 51,624 Btu per 24 hr. through the walls

Heat Gain Factors (Walls, Floor and Ceiling)
Btu per (sq ft) (24 hr)

Insulation	Temp difference (ambient temp minus storage temp), F deg																		
Cork or equivalent in.	1	40	45	50	55	60	65	70	75	80	85	90	95	100	105	110	115	120
3	2.4	96	108	120	132	144	156	168	180	192	204	216	228	240	252	264	276	288
4	1.8	72	81	90	99	108	117	126	135	144	153	162	171	180	189	198	207	216
5	1.44	58	65	72	79	87	94	101	108	115	122	130	137	144	151	159	166	173
6	1.2	48	54	60	66	72	78	84	90	96	102	108	114	120	126	132	138	144
7	1.03	41	46	52	57	62	67	72	77	82	88	93	98	103	108	113	118	124
8	0.90	36	41	45	50	54	59	63	68	72	77	81	86	90	95	99	104	108
9	0.80	32	36	40	44	48	52	56	60	64	68	72	76	80	84	88	92	96
10	0.72	29	32	36	40	43	47	50	54	58	61	65	68	72	76	79	83	86
11	0.66	26	30	33	36	40	43	46	50	53	56	60	63	66	69	73	76	79
12	0.60	24	27	30	33	36	39	42	45	48	51	54	57	60	63	66	69	72
13	0.55	22	25	28	30	33	36	39	41	44	47	50	52	55	58	61	63	66
14	0.51	20	23	28	28	31	33	36	38	41	43	46	49	51	54	56	59	61
Single glass	27.0	1080	1220	1350	1490	1620	1760	1890	2030	2160	2290	2440	2560	2700	2840	2970	3100	3240
Double glass	11.0	440	500	550	610	660	715	770	825	880	936	990	1050	1100	1160	1210	1270	1320
Triple glass	7.0	280	320	350	390	420	454	490	525	560	595	630	665	700	740	770	810	840

Note: Where wood studs are used multiply the above values by 1.1

Fig. 15-10. Heat gain factors for walls, floor, and ceiling. (ASHRAE Guide and Data Book)

From table, Fig. 15-10:
1 sq. ft. of window allows transfer of 660 Btu per 24 hr.
660 × 6 sq. ft. = 3960 Btu per 24 hr. through the windows
Or a total heat leakage of:
51,624 + 3960 = 55,584 Btu per 24 hr.

15-8 CABINET VOLUME

Cabinet volume is the volume based on the inside dimensions of the cabinet. The volume is used to help find out the air changes and product load.

In the sample cabinet, which is 10 ft. long × 10 ft.

CABINET AREAS, VOLUMES, AND THICKNESS OF INSULATION

CABINET		INTERNAL VOLUME 8 FT. HIGH CAPACITY (CU. FT.)								INTERNAL VOLUME 10 FT. HIGH CAPACITY (CU. FT.)						
Lg. & Wd.	Outside Sq. Ft.	WALL THICKNESS							Outside Sq. Ft.	WALL THICKNESS						
		4″	4½″	5″	6″	7″	8″	10″		4″	4½″	5″	6″	7″	8″	10″
5x 5	210	137	131	124	112	101	90	71	250	174	167	159	144	131	117	93
5x 6	236	169	161	154	140	127	114	91	280	215	206	197	180	164	148	120
5x 7	262	201	194	184	168	153	138	111	310	256	248	236	216	198	179	146
5x 8	288	233	224	214	196	179	162	131	340	296	286	274	252	232	210	172
6x 6	264	209	201	193	175	160	145	119	312	266	256	247	225	207	188	156
6x 7	292	248	238	228	210	193	176	146	344	316	304	292	270	249	228	192
6x 8	320	286	277	267	245	226	207	173	376	364	353	342	315	292	269	228
6x 9	348	325	315	305	280	259	238	200	408	414	402	390	360	335	309	263
6x10	376	364	353	343	315	292	269	227	440	463	451	439	405	378	350	299
6x12	432	444	432	419	385	358	331	281	504	555	546	536	495	463	430	370
7x 7	322	294	283	272	254	234	214	180	378	374	361	348	326	302	278	237
7x 8	352	341	329	317	294	273	252	214	412	434	420	406	378	353	328	281
7x 9	382	386	374	362	334	312	290	248	446	492	477	463	430	403	377	326
7x10	412	433	420	407	374	346	318	282	480	551	536	521	481	448	413	371
7x12	472	527	512	497	454	414	374	348	548	670	653	635	583	535	486	458
8x 8	384	394	382	369	343	320	296	253	448	501	487	473	441	413	385	333
8x 9	416	448	434	420	392	367	341	294	484	570	553	538	504	474	443	386
8x10	448	503	587	471	441	413	385	335	520	641	748	603	567	534	500	441
8x12	512	610	591	573	539	506	473	417	592	776	755	734	692	653	615	548
8x14	576	718	697	675	637	594	561	499	664	914	889	864	818	768	730	656
9x 9	450	510	489	469	448	420	392	341	522	649	623	600	576	543	510	449
9x10	484	570	554	537	504	473	443	386	560	725	706	686	647	612	575	508
9x12	552	694	674	654	616	581	545	476	636	883	859	836	792	752	708	626
9x14	620	814	793	771	728	687	647	566	712	1035	1011	987	935	888	840	745
10x10	520	638	620	602	567	534	500	440	600	870	790	770	729	680	650	579
10x12	592	776	755	734	693	655	617	547	680	988	962	939	890	847	802	720
10x14	664	912	889	866	818	775	733	654	760	1158	1132	1110	1050	1005	954	860
12x12	672	946	919	893	848	804	760	680	768	1203	1172	1144	1090	1038	988	895
12x14	752	1110	1086	1052	1001	951	900	809	856	1411	1382	1348	1289	1230	1170	1060
14x14	840	1304	1269	1235	1180	1126	1072	968	952	1660	1619	1568	1518	1458	1394	1272

Fig. 15-11. Table of cabinet external areas and internal volumes (capacity).

wide × 8 ft. high, the walls are 4 in. thick. Therefore, the internal or inside width is 10 ft. minus 4 in. minus 4 in. (there is a wall at each end).

10 ft. − (4 in. + 4 in.) = inside width
or 10 ft. − 8 in. = inside width
or 9 ft. − 4 in. = inside width
or 9 1/3 ft. = inside width

The same method is used to calculate other internal dimensions. The internal dimensions are:

Width 9 1/3 ft., length 9 1/3 ft., height 7 1/3 ft.

The inside volume =

9 1/3 × 9 1/3 × 7 1/3 =

$$\frac{(9 \times 3) + 1}{3} \times \frac{(9 \times 3) + 1}{3} \times \frac{(7 \times 3) + 1}{3} =$$

$$\frac{27 + 1}{3} \times \frac{27 + 1}{3} \times \frac{21 + 1}{3} =$$

$$\frac{28}{3} \times \frac{28}{3} \times \frac{22}{3} = \frac{17,248}{27}$$

$$\frac{17,248}{27} = 638.8 \text{ cu. ft.}$$

Usable inside volume is the total inside volume minus shelves, racks, and evaporator space. For practical purposes and to be on the safe side, the total internal volume is used when figuring heat loads. Fig. 15-11 uses a net (internal) volume.

15-9 TOTAL HEAT LOAD

Information given in the previous paragraphs can be illustrated by the following problem:

If a metal sheathed walk-in cabinet 10 ft. long × 10 ft. wide × 8 ft. high with 4 in. thick walls is in an 85 °F (29 °C) 50 percent relative humidity (RH) room, and it cools 2000 lb. of fresh, lean beef from 60 °F to 35 °F each day, and the evaporator has two 1/10-hp motors and the cabinet has two 40-watt lamps (8 hr.), and one person works in the cabinet 8 hr. each day, what is the total heat load?

1. Heat leakage load = 55,584 Btu/day (refer to the end of Para. 15-7).
2. Air change load = volume × air changes/24 hr. × Btu/cu. ft. (heat to be removed, cooling air from 85 °F [29 °C] 50 percent RH to 35 °F [2 °C] 60 percent RH)
   Air change load = 638 (see Para. 15-8) × 24 (see Fig. 15-5) × 2.09 (see Fig. 15-6) = 32,000 Btu per day.
3. Product load = weight × spec. heat (see Fig. 15-7) × temp. diff.
   Product load = 2000 lb. × .77 spec. heat × 25 °F temp. diff. (14 °C) = 38,500 Btu per day.
4. Miscellaneous load =
   a. The two motors' loads (continuous operation) = No. of motors × Btu/hp/hr. (see Fig. 15-8) × hp × hr.
   2 × 4250 × 1/10 × 24 = 20,400 Btu
   b. The two lamps' loads = No. of lamps × watts × hr. of operation × 3.42 Btu/watt
   2 × 40 × 8 × 3.42 = 2189 Btu
5. Occupancy load = No. of persons × hours of work × heat equivalent per hour (see Fig. 15-9)

$$1 \times 8 \times \frac{950 + 840}{2}$$

$$1 \times 8 \times \frac{1790}{2}$$

$$1 \times 8 \times 895 = 7160$$

Therefore: Total heat load = 55,584 + 32,000 + + 38,500 + 20,400 + 2189 + 7160
= 155,833 Btu/24 hr.

Per 16 hr. = $\frac{155,833}{16 \text{ hr.}}$ = 9,739 (3/4-ton load) Btu per hr. based on 16 hours of system operation.

The 16 hours of running time will provide a system with 50 percent reserve capacity. If 30 percent reserve capacity is desired, select the equipment which will operate 18 hours per day to handle the load. If the fixture is in an air-conditioned room, and the ambient temperature is 75 °F (24 °C) the year around, one may select equipment to run 20 hours per day.

15-10 DETERMINING HEAT LEAKAGE USING TABLES (SHORT METHOD)

A method used by some manufacturers to determine heat leakage into a cabinet is shown in Fig. 15-10. The table in Fig. 15-11 gives the external area and volume of a cabinet, based on actual experiments and investigations.

To use the tables, proceed as follows (using a walk-in refrigerator box as a sample problem): Visit the establishment and obtain all the data possible about the cabinet and the service. Determine exterior dimensions of the box, dimensions of windows, dimensions of wall, thickness of insulation, kind of insulation, number of panes in windows, how much business user does, temperatures user desires in cabinet, the average summer temperature for the locality, and the highest possible water temperature (if a water-cooled installation is to be made).

Specification sheets are available for tabulating data needed for the selection of proper equipment. A sample sheet is shown in Fig. 15-12.

Using cabinet size 9 ft. × 10 ft. × 8 ft. without windows, Fig. 15-11 shows that the area is 484 sq. ft. In Fig. 15-10, note that the Btu leakage per sq. ft./24 hr. (4-in. thickness, 60-deg. temperature difference) is 108 Btu.

Heat leakage = total sq. ft. × leakage per sq. ft.
Heat leakage = 484 × 108 = 52,272 Btu/24 hr.

15-11 DETERMINING USAGE LOAD USING TABLES (SHORT METHOD)

The total heat load of the refrigerator cabinet depends upon the heat leaking through the walls and windows. It is also affected by the heat to be removed from articles in the cabinet, the air change, and other sources of heat. This heat is called heat usage, or service load. Heat usage is caused by changes of air in the cabinet, by produce to be cooled, by lights and motors which may be used inside the box, and by the occupancy of the box.

Refrigeration equipment manufacturers have developed a standard that gives a fairly accurate estimate of the usage heat load. With this method, the cabinet is classified according to the type of service to be performed: florist's cabinets, grocery boxes, normal market coolers, fresh meat cabinets, and restaurant short order cabinets. From experience, these companies have found that cabinets used for the same general line of business hold rather closely to the same usage heat load.

This load depends on four basic factors:

1. Temperature difference between exterior and interior of cabinet.
2. Volume of cabinet (internal).
3. Type of service.
4. Time.

It is possible to determine the usage heat load of an installation. One must know the amount of food put into the refrigerator, how many times the door is opened, and how long the employees are inside the cabinet. This is a difficult process. If not carefully done, errors are bound to appear in the results.

Data in the tables are based on 1 cu. ft. content at various temperature differences. To determine the usage heat load, proceed as follows:

1. Use temperature difference of the same value used for heat leakage into the cabinet.
2. Figure the volume of the cabinet from inside dimensions.

REFRIGERATION SALES ENGINEERS DATA SHEET

Name _____ Type of Business _____ Date _____

Address _____ City _____ Zone _____ State _____

Person Contacted _____ Title _____ Phone _____

Fixture No. 1-Make _____ Fixture No. 2-Make _____ Fixture No. 3-Make _____

Use _____ Model _____ Use _____ Model _____ Use _____ Model _____

Temperature _____ _____ _____

Humidity _____ _____ _____

Width _____ _____ _____

Length _____ _____ _____

Height _____ _____ _____

Construction _____ _____ _____

Insulation:
　Kind _____ _____ _____

　Thickness _____ _____ _____

Glass:
　Area _____ _____ _____

　No. of Panes _____ _____ _____

Produce _____ _____ _____

Lights _____ _____ _____

Motors _____ _____ _____

Sun Load _____ _____ _____

No. of People in Refrigerator ____ _____ _____

Unusual Temperatures _____ _____ _____

Unusual Service _____ _____ _____

Remarks _____ _____ _____

Use Reverse Side for Sketch of Installation

Salesman _____

Fig. 15-12. This data sheet is typical of those used by sales engineers in recording information prior to a refrigeration installation.

3. Determine the type of service for which the cabinet is being used, such as average, heavy, or long storage. The service load is also dependent on cabinet size. The smaller the cabinet, the more heat load is caused by service. A case for meat storage, for instance, may be in a small neighborhood store or in a supermarket.

4. Time, 24 hours.
   a. After the total volume of the box has been found (from Fig. 15-11), the load for each cubic foot is determined by using the table in Fig. 15-13.
   b. If the 9 × 10 × 8-ft. cabinet appears to have average service with a temperature difference of 60 °F

USAGE HEAT GAIN, BTU PER 24 HR. FOR ONE CU. FT. INTERIOR CAPACITY

Volume Cu. Ft.	Service*	Temperature Difference °F (Ambient Temp. Minus Storage Room Temp.)										
		1	40	50	55	60	65	70	75	80	90	100
20	Average	4.68	187.	234.	258.	281.	305.	328.	351.	374.	421.	468.
	Heavy	5.51	220.	276.	303.	331.	358.	386.	413.	441.	496.	551.
30	Average	3.30	132.	165.	182.	198.	215.	231.	248.	264.	297.	330.
	Heavy	4.56	182.	228.	251.	274.	297.	319.	342.	365.	410.	456.
50	Average	2.28	91.	114.	126.	137.	148.	160.	171.	182.	205.	228.
	Heavy	3.55	142.	177.	196.	213.	231.	249.	267.	284.	320.	355.
75	Average	1.85	74.	93.	102.	111.	120.	130.	139.	148.	167.	185.
	Heavy	2.88	115.	144.	158.	173.	188.	202.	216.	230.	259.	288.
100	Average	1.61	64.	81.	84.	97.	105.	113.	121.	129.	145.	161.
	Heavy	2.52	101.	126.	139.	151.	164.	176.	189.	202.	227.	252.
200	Average	1.38	55.	69.	76.	83.	90.	97.	103.	110.	124.	138.
	Heavy	2.22	90.	111.	122.	133.	144.	155.	166.	178.	200.	222.
300	Average	1.30	52.0	65.	71.5	78.	84.5	91.	97.5	104.	117.	130.
	Heavy	2.08	83.2	104.	114.	125.	135.	146.	156.	166.	187.	208.
400	Average	1.24	49.6	62.	68.2	74.4	80.6	86.8	93.	99.2	112.	124.
	Heavy	1.96	78.4	98.	108.	118.	128.	137.	147.	157.	176.	196.
500	Average	1.21	48.4	60.5	66.6	72.6	78.7	84.7	90.7	96.8	109.	121.
	Heavy	1.87	74.8	93.5	103.	112.	122.	131.	140.	150.	168.	187.
600	Average	1.17	46.8	58.5	64.	70.	76.	82.	88.	94.	105.	117.
	Heavy	1.85	74.0	92.5	102.	111.	120.	130.	139.	148.	167.	185.
800	Average	1.11	44.4	55.5	61.1	66.6	72.2	77.7	83.3	88.8	100.	111.
	Heavy	1.76	70.4	88.0	96.8	106.	115.	123.	132.	141.	158.	176.
1,000	Average	1.10	44.0	55.0	60.5	66.	71.5	77.	82.5	88.	99.	110.
	Heavy	1.67	66.8	83.5	91.9	100.	108.	117.	125.	134.	150.	167.
1,200	Average	.995	39.8	49.8	54.7	59.7	64.7	69.7	74.7	79.6	89.6	99.5
	Heavy	1.58	63.2	79.0	86.9	94.8	103.	111.	119.	126.	142.	158.
1,500	Average	.920	36.8	46.0	50.6	55.2	59.8	64.4	69.	73.6	82.8	92.
	Heavy	1.50	60.0	75.0	82.5	90.0	97.5	105.	113.	120.	135.	150.
2,000	Average	.835	33.4	41.8	45.9	50.1	54.3	58.5	62.7	66.8	75.2	83.5
	Long storage	.775	31.0	38.8	42.6	46.5	50.4	54.3	58.1	62.	69.8	77.5
3,000	Average	.750	30.0	37.5	41.3	45.0	48.8	52.5	56.2	60.0	67.5	75.0
	Long storage	.576	23.0	28.8	31.7	34.6	37.3	40.3	43.2	46.1	51.8	57.6
5,000	Long storage	.403	16.1	20.2	22.2	24.2	26.2	28.2	30.2	32.2	36.3	40.3
7,500	Long storage	.305	12.2	15.3	16.8	18.3	19.8	21.4	22.9	24.4	27.5	30.5
10,000	Long storage	.240	9.6	12.0	13.2	14.4	15.6	16.8	18.0	19.2	21.6	24.0
20,000	Long storage	.187	7.48	9.35	10.3	11.2	12.2	13.1	14.0	15.0	16.8	18.7
50,000	Long storage	.178	7.12	8.90	9.79	10.7	11.6	12.5	13.4	14.2	16.0	17.8
75,000	Long storage	.176	7.04	8.80	9.68	10.6	11.5	12.3	13.2	14.1	15.8	17.6
100,000	Long storage	.173	6.92	8.65	9.52	10.4	11.2	12.1	13.0	13.8	15.6	17.3

*For average and heavy service, product load is based on product entering at 10 deg. above the refrigerator temperature; for long storage the entering temperature is approximately equal to the refrigerator temperature. Where the product load is unusual, do not use this table.

Fig. 15-13. Usage heat gain table. It uses average storage, heavy storage, and long storage variables.

TEMPERATURE DIFFERENCE IN DEGREES FAHRENHEIT	USE OF REFRIGERATOR			
	FLORIST	GROCERY OR NORMAL MARKET	MARKET WITH HEAVIER SERVICE OR FRESHLY KILLED MEATS	RESTAURANT SHORT ORDER
40°	40.0	65.0	95.0	120.0
50°	50.0	80.0	120.0	150.0
60°	60.0	95.0	145.0	180.0
70°	70.0	114.0	167.0	210.0
80°	80.0	130.0	190.0	240.0
90°	90.0	146.0	214.0	270.0

Fig. 15-14. This chart is based on usage heat gain in Btu per 24 hours for one cu. ft. interior capacity. (ASHRAE Guide and Data Book)

(33 °C), and it has a volume of 504 cu. ft. the amount of heat to be removed from each cubic foot will be 78 Btu per 24 hours. Multiply this value by the total volume in cubic feet, and a fairly accurate estimate of the service or usage load may be obtained. The table in Fig. 15-13 gives the heat usage over a period of 24 hours, since this time is the established standard.

Heat usage = usage Btu per cu. ft. × volume in cu.ft.
Heat usage = 78 × 504 = 39,312 Btu/24 hr.

Another table based on type of usage of the cabinet is shown in Fig. 15-14. Its values can be substituted for the values in Fig. 15-13 when exact use of cabinet is known.

15-12 TOTAL HEAT LOAD USING TABLES

The total heat load is the sum of the heat leakage load and the heat usage load.

Total heat load = heat leakage + heat usage
From the previous example:

Total heat load = heat leakage + heat usage
= 52,272 (Para. 15-10)
+ 39,312 (Para. 15-11)
Total heat load = 91,584 Btu/24 hr.

The addition of the heat leakage and usage will give the total heat load upon the cabinet for a certain set period of time. This value may be listed either as Btu per 24 hours, or Btu per 1 hour.

Btu/1 hr. = Btu/24 hr. ÷ 24
= 91,584 ÷ 24
= 3816 Btu/hr.

However, in refrigeration applications, the unit to be installed should be big enough to remove this heat in less than 24 hours. This gives extra capacity for heavy loads, wear in the unit, and allows time for defrost operations.

For fixtures above 32 °F (0 °C), it is generally understood that the system should operate 16 hours out of 24 (two-thirds of the time). For fixtures below 32 °F (0 °C), unit should operate 18 hours out of 24 (three-fourths of the time).

For example, if the system were to operate 16 hr./day, the service technician would divide the total Btu heat load for a 24-hr. period by 16.

For 16-hr. running:

$$\frac{91,584}{16} = 5724 \text{ Btu/hr. for each hour of running}$$

For 18-hr. running:

$$\frac{91,584}{18} = 5088 \text{ Btu/hr. for each hour of running}$$

15-13 EVAPORATOR AND CONDENSING UNIT CAPACITIES

After calculating heat load, it is necessary to determine what size evaporator will furnish adequate refrigeration.

Some important things to remember are:
1. The evaporator removes heat from the cabinet only when the condensing unit is running.
2. The refrigerating unit usually runs from 16 to 20 hours out of each 24 hours. This means that the unit must have a refrigerating capacity in 16 hours of operation equal to the total heat load in 24 hours. If the refrigerator is installed in an air-conditioned room, the room temperature can be lowered or the running time of the unit can be increased (for example, from 16 hours to 18 or 20 hours).

The evaporator capacity is determined by three conditions:
1. Cabinet temperature.
2. Refrigerant temperature.
3. Space allowed for the evaporator.

Condensing unit capacity depends on three conditions:
1. Low-side pressure.
2. Condensing cooling medium (air or water).
3. Size of compressor and condenser.

It is more important to balance the capacity of the evaporator to the capacity of the condensing unit than to balance either to the heat load of the cabinet. When balancing the capacity of the condensing unit and the evaporator, calculations for each must be based on the same low-side pressure.

The same low-side pressure is used because:
1. The capacity of the evaporator increases as the evaporator temperature decreases.
2. The capacity of the condensing unit decreases as the low-side pressure decreases. See Fig. 15-15.

From Fig. 15-16 it can be seen that this particular evaporator matches the condensing unit at a low-side pressure of 31 psi. The evaporator/condenser combination will remove 12,500 Btu/hr. when the evaporator refrigerant temperature is 32 °F (0 °C) and the pressure is 31 psi. The temperature difference for a 42 °F (6 °C) cabinet in this case is 10 °F (6 °C).

The manufacturers of evaporators and condensing units list the capacities of their products in Btu for either one hour, 16 hours, or 18 hours of operation. Fig. 15-17 lists typical condensing unit capacities at condensing temperatures of 110 °F (43 °C) and 120 °F (49 °C). Fig. 15-17 also gives four different evaporating temperatures of −30 °F, −15 °F, 20 °F, and 40 °F (−34 °C, −26 °C, −7 °C, and 4 °C).

Note that only at a 40 °F (4 °C) evaporator refrigerant temperature does the hp equal the tonnage capacity. Referring to Fig. 15-17, for example, a 3-hp condensing unit, operating at 40 °F evaporator refrigerant temperature and 110 °F (43 °C) condensing temperature, has the capacity of 36,300 Btu/hr. The capacity of a 1-ton machine is 12,000 Btu/hr. Therefore, 36,300 ÷ 12,000 = 3-ton (plus) capacity.

15-14 CONDENSING UNIT CAPACITIES

Before choosing a condensing unit, first decide if it is to be water-cooled or air-cooled. Next, determine whether it is to be a hermetic unit or an external drive unit. Then find out what electric power is available (120, 208, or 240 volt, single-phase; 220 or 440 volt, three-phase).

Fig. 15-18 is a table of data on a typical external drive condensing unit. The air-cooled unit is used with a 1-hp motor and a 2 cylinder compressor with a 2-in. bore and a 2-in. stroke.

If the evaporator was selected on the basis of a 15 °F (8 °C) temperature difference and a 16-hour running time, this means that if the cabinet is to operate at 38 °F (3 °C) the refrigerant temperature will be 38 minus 15 °F or 23 °F (−5 °C). If it is an R-12 refrigerant unit, the condensing unit capacity must be matched to the evaporator capacity at a low-side pressure of 23.2 psig (261.5 kPa).

The table, Fig. 15-17, lists the average hourly capacity of various horsepower capacity condensing units. This clearly shows the increase in capacity of a condensing unit as the low-side pressure (and temperature) increases, providing the head pressure remains fairly constant. It also shows the effect of condensing temperature (and pressure) on the capacity of a condensing unit.

The capacity of hermetic condensing units can also be found in tables provided by manufacturers.

15-15 EVAPORATOR INSTALLATIONS

Practically all refrigerator cabinets are built to be used with mechanical refrigeration. These cabinets are specially designed to improve the cooling efficiency of the cabinet with evaporators. The refrigeration service engineer should know the theory of air circulation in the cabinet, and what is supposed to happen during the operation of the unit. The following deals with the study of efficient baffling and correct air circulation in cabinets.

Baffles are surfaces, or air ducts, which increase the efficiency of the airflow through the evaporator and throughout the cabinet. They direct the airflow so that it is carried all around the interior of the box, leaving no dead or warm air spots.

A typical evaporator and baffle arrangement is shown in Fig. 15-19. The colder air coming from the evaporator is made to flow down the center of the cabinet, while the

CONDENSING UNIT		EVAPORATOR	
LOW-SIDE TEMP.	BTU/HR.	TEMP. DIFF.	BTU/HR.
40	6650	1	400
35	6100	10	4000
30	5600	12	4800
25	5100	15	6000
20	4650		
15	4200	300 sq. ft. surface	
10	3800	natural convection	
5	3400	evaporator.	
0	3000		
− 5	2600		
−10	2250		
−15	1900		
−20	1550		
−25	1250		
−30	950		

90 °F ambient air.
Add 6% for 10 °F drop in air temperature, subtract 6% for each 10 °F rise in ambient temperature.
Liquid line 1/4 in.
Suction line 5/8 in.
Approximately one hp

Fig. 15-15. Condensing unit and evaporator capacity variations with temperature and pressure. Evaporator temperature difference is the difference between refrigerant temperature and cabinet air temperature.

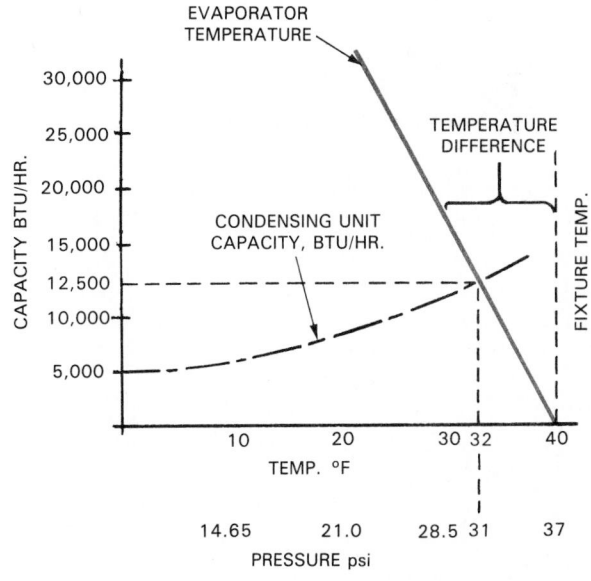

Fig. 15-16. Graph shows relative effect on evaporator and condensing unit capacity of different temperatures inside the evaporator.

warmer air is directed up the walls back to the evaporator. The design must be scientifically proportioned to insure that air circulation is unrestricted, with no objects blocking airflow.

HP	AVERAGE COMPRESSOR CAPACITIES BTU/HR							
	EVAPORATING TEMPERATURES °F							
	−30°		−15°		+20°		+40°	
	CONDENSING TEMPERATURE °F							
	110°	120°	110°	120°	110°	120°	110°	120°
2	5,200	4,500	9,100	8,200	18,000	16,800	22,800	21,200
3	9,000	8,300	14,300	13,200	22,100	20,800	36,300	34,200
5	14,200	12,500	24,800	22,400	41,700	39,300	62,400	58,500
7 1/2	25,000	20,000	31,000	28,000	53,000	48,000	87,000	81,700
10	31,000	26,000	43,600	44,800	81,000	75,000	120,000	112,000
15	42,600	37,500	74,400	67,200	111,000	102,000	171,600	160,000
20	56,000	44,700	82,000	71,000	154,000	142,000	235,000	218,000
25	70,000	56,000	96,000	85,000	188,000	174,000	283,000	263,000
30	80,000	67,000	116,500	102,500	225,000	210,000	349,000	324,000
40	94,000	75,000	155,000	135,000	325,000	306,000	439,000	406,000
50	122,000	100,000	188,500	159,500	375,000	350,000	585,000	550,000
60	168,000	134,000	240,000	220,000	450,000	420,000	710,000	670,000
70	196,000	156,000	272,000	239,000	571,000	534,000	800,000	742,000
75	210,000	167,000	291,000	256,000	582,000	542,000	855,000	795,000
80	224,000	178,000	310,000	273,000	622,000	578,000	900,000	842,000
90	252,000	201,000	349,000	307,000	750,000	700,000	1,027,000	955,000
100	280,000	223,000	388,000	341,000	777,000	723,000	1,170,000	1,100,000

NOTE: The above figures are only approximate and based on catalog ratings of leading compressor manufacturers. For precise figures, refer to catalog of your compressor manufacturer.

Fig. 15-17. Condensing unit capacities in Btu per hour at various evaporator temperatures and at two different condensing temperatures.

| CONDENSING UNIT CAPACITIES | | |
COMPRESSOR RPM	REFRIGERANT (EVAPORATOR) TEMPERATURE	BTU/HR
475	45	11,000
	40	10,200
	35	9,370
	30	8,530
	25	7,740
540	25	8,700
	20	7,960
	15	7,200
	10	6,430
	5	5,740
	0	5,100
665	0	5,780
	− 5	5,700
	−10	4,430
	−15	3,820
	−20	3,280
	−25	2,800

Fig. 15-18. Btu/hr. capacity table for a 1 hp condensing unit which is air cooled and designed to operate at 90 °F ambient temperature at various low-side temperatures.

Any horizontal baffle or drain pan should be insulated because the top surface may be in contact with cold air while the under part of the baffle may be in contact with relatively warm air. If it were not insulated, this temperature difference might cause condensation. Eddy currents of air (small circular flows of air) would also disturb airflow in the cabinet.

Multiple-baffled evaporators of the natural convection type are often used. The air flows around the cabinet because of the relative weights of the cold air and the warm air (density). Warm air is lighter per cubic foot of volume and, therefore, rises in the box. This natural circulation must not be blocked or the box temperature will not be constant. Baffling the evaporators promotes this natural circulation of the air and speeds it up. Baffles may also serve as drain pans.

15-16 EVAPORATOR LOCATIONS

Many cabinets do not have room for overhead evaporators. A walk-in cooler with an exterior height of approximately 7 1/2 feet is too low. This height requires the use of a blower evaporator. Two common types of mountings are:

1. Evaporator may be placed in an upper corner of the cabinet as far away as possible from the entrance door. See Fig. 15-20.

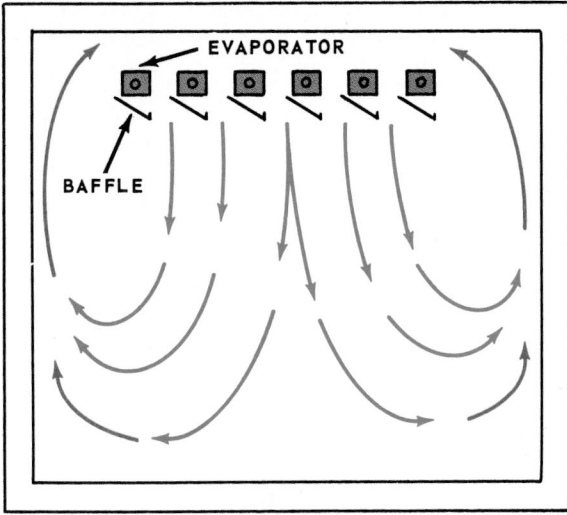

Fig. 15-19. Airflow pattern for a typical refrigerant cabinet. Baffles direct cold air downward while warm air rises and circulates through the evaporator.

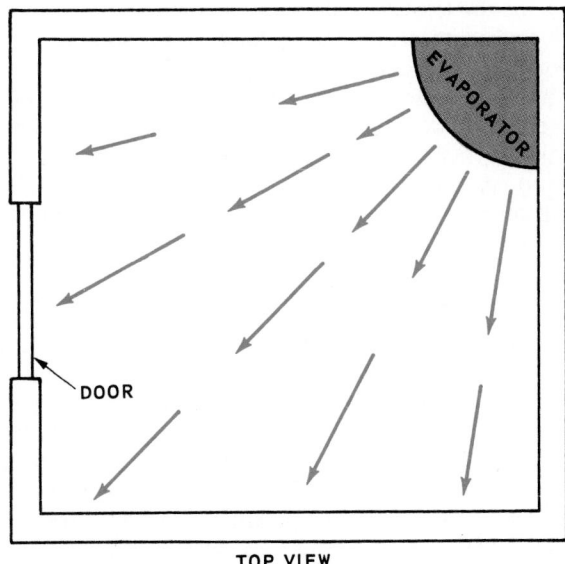

Fig. 15-20. Blower evaporator mounted in upper corner of walk-in cooler. Note air distribution relative to door.

2. Wall evaporators may be mounted against the wall opposite the windows or the reach-in doors of the cabinet, Fig. 15-21.

The evaporators are shrouded and have a motor-driven fan to circulate air. They also have built-in drain pans. Also see Figs. 15-22, 15-23, and 15-24.

15-17 EVAPORATORS FOR DISPLAY CASES

Evaporator designs for closed- and open-type display cases have to overcome a difficult air circulation problem.

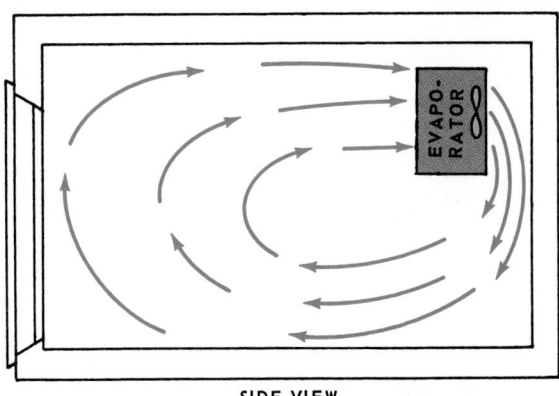

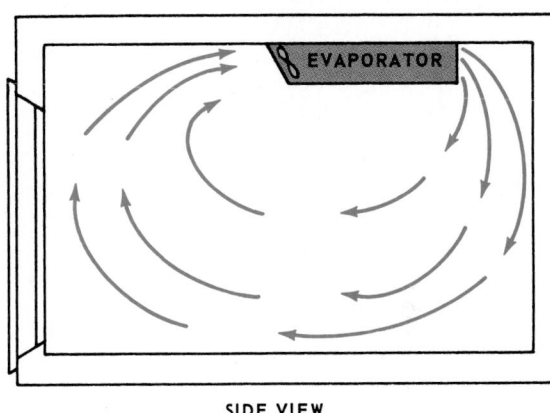

Fig. 15-22. Where there is too little clearance for overhead installation, blower evaporators may be installed on walls of walk-in refrigerators. Installation should always be opposite reach-in doors or windows.

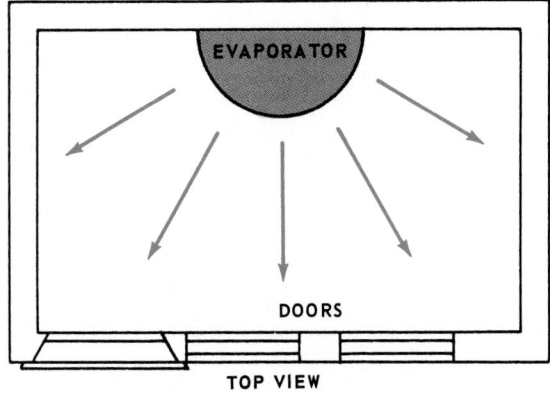

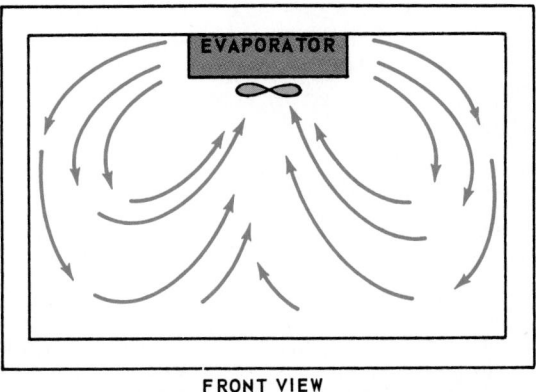

Fig. 15-21. Note airflow patterns with blower evaporator mounted on center back wall of walk-in cooler.

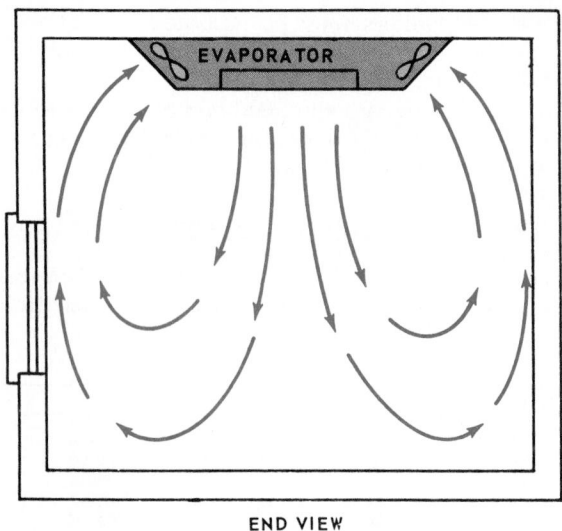

Fig. 15-23. Two motor-driven fans reduce distance air must circulate in cabinet.

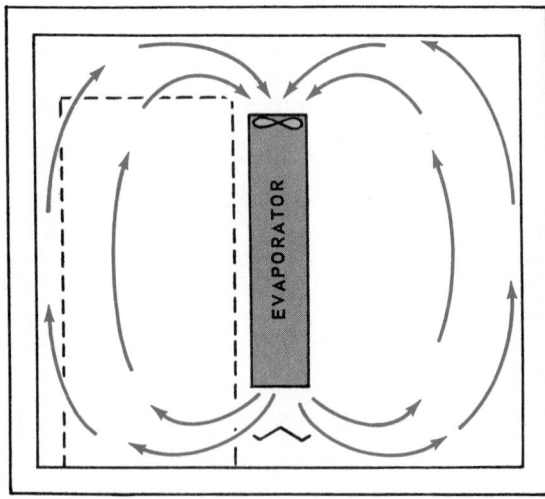

FRONT VIEW

Fig. 15-24. Air circulation pattern is shown for a reach-in cabinet with blower evaporator mounted behind mullion or door frame.

The cases are narrow and there is less room for the evaporator. Many of the cases are open. Fig. 15-25 shows a single evaporator installation for a double-duty case.

A display case using a blower evaporator is shown in Fig. 15-26. Note the location of the motor-driven fan.

A design for an open display case is shown in Fig. 15-27. Notice how ducting is used to control the flow of the refrigerated air.

15-18 EVAPORATOR TYPES

Many kinds of evaporators have been used in mechanical refrigeration. However, two basic types are in use:
1. The air-cooling evaporator is used to cool the air within the cabinet directly.
2. The liquid-cooling evaporator is used to cool a liquid, which may be consumed or used to cool other substances.
The air-cooling evaporator is classified as the dry type.

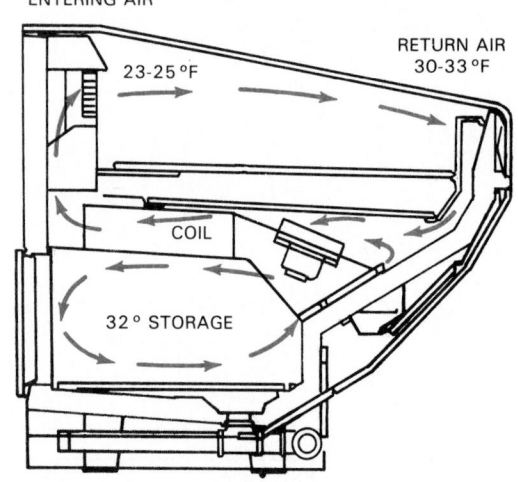

Fig. 15-25. A cross section shows the airflow pattern of a double-duty case. Note the temperature differences. (Tyler Refrigeration Corp.)

It may be the frosting, defrosting, or nonfrosting type. Some air-cooled units are forced-circulation evaporators.

Liquid-cooling evaporators are of two types: the submerged evaporator and the tube-within-a-tube evaporator.

In commercial refrigeration, certain kinds of evaporators are more popular than others. Among these are the dry nonfrosting air-cooling evaporator, forced circulation dry evaporator, and submerged flooded or dry evaporator. See Chapter 12.

15-19 AIR-COOLING EVAPORATOR THEORY

The theory of the air-cooling evaporator involves heat transfer from the air circulating over the evaporator to the refrigerant as follows: As the warmer air comes in contact with the evaporator, air molecules strike the fins and release some of their energy to the fin (transfer some heat to it). This heat, in turn, travels through the fins, then through the tubing and comes in contact with the liquid refrigerant on the inside, transferring the heat energy to it.

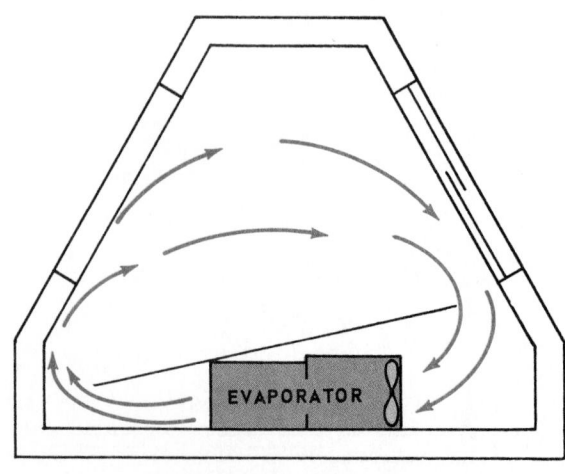

Fig. 15-26. This diagram of a closed display case shows a blower evaporator located below display shelf.

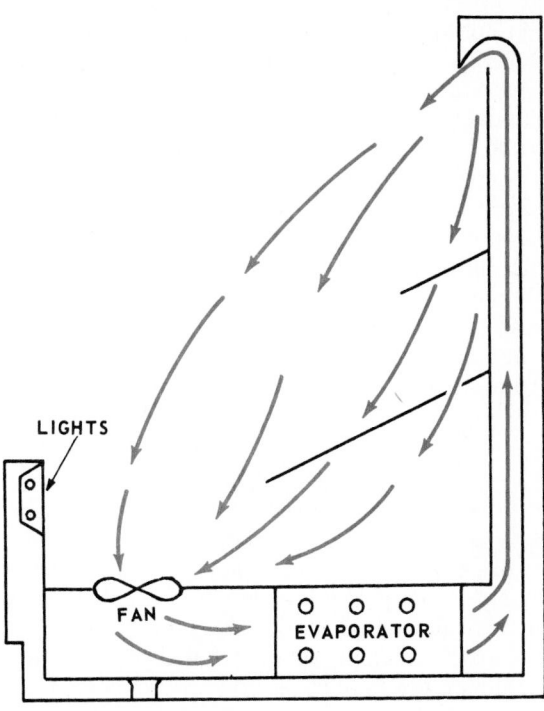

Fig. 15-27. Open display case with a blower evaporator located in base of cabinet.

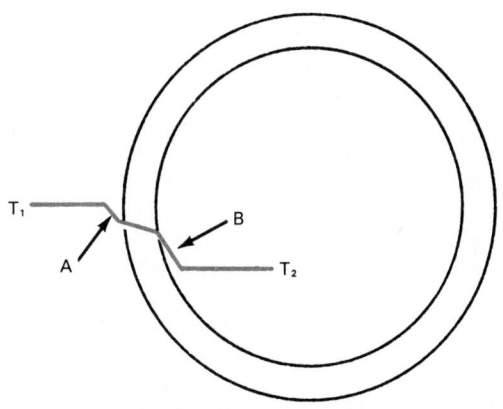

Fig. 15-28. Heat transfer from air, A, surrounding an evaporator to the evaporator, then through the metal to refrigerant, B, inside. T_1 is cabinet air temperature; T_2 is refrigerant temperature.

Because air density is low, heat transfer is slow. After reaching the metal, the heat travels efficiently and rapidly. Upon reaching the interior surface of the tubing, it again has difficulty in reaching the refrigerant in the system. Gas bubbles clinging to the internal surface, along with an oil film, reduce the heat flow.

15-20 EVAPORATOR CAPACITIES

One of the laws of thermodynamics (heat in action) is that heat always flows from an object at a higher temperature to an object at a lower temperature. As in the case of heat leakage, the amount of heat transfer depends on five variables:
1. Area.
2. Temperature difference.
3. Heat conductivity of the material.

4. Thickness of material.
5. Time.

The kinds of materials used in evaporators are of utmost importance because they must be good heat conductors.

Heat may be transmitted through various materials. For air evaporators, Fig. 15-28, the heat must pass through air film, A, on the metal surface; then it travels through the metal tubing and the oil or liquid refrigerant film on the inside of the evaporator tubing to refrigerant, B.

If the air is moved rapidly, heat flow to the metal is greater because more air contacts the metal per unit of time and the air film is thinner. If the oil or refrigerant film is moved faster, the film will be thinner due to greater movement. Generally, the denser the fluid, the greater the heat flow; the faster the fluid motion, the greater the heat flow.

The U factors (heat transferability) for several types of evaporators follows:
1. Natural convection evaporators—about 1 Btu/sq. ft./°F/hr.
2. Blower evaporators—about 3 Btu/sq. ft./°F/hr.
3. Liquid-cooling evaporators—about 15 Btu/sq. ft./°F/hr.

These values are fairly accurate if the temperature difference is taken between the average air or liquid temperature and the refrigerant temperature. See points T_1 and T_2 in Fig. 15-28.

For natural convection evaporators, the temperature difference usually selected is 10°F (6°C). For example, a 45°F (7°C) cabinet would have a refrigerant temperature in the evaporator of 35°F (2°C).

The smaller the temperature difference, the higher the relative humidity can be kept. For example, 10°F to 12°F (6°C to 7°C) temperature difference will keep a 75 to 90 percent RH, while a 20°F (11°C) to 30°F (17°C) temperature difference will keep a 50 to 70 percent RH.

If the evaporator refrigerant temperature goes below 28°F (−2°C), frost will form on the evaporator. The cycle off time must be long enough to permit defrosting, with the air at 35°F (2°C). If refrigerant temperature goes below 28°F, a defrost system must be used.

15-21 EVAPORATOR AREA

When calculating or determining the capacity of evaporators, it is best to rely on the manufacturer's specifications. They obtain their heat capacity values from actual experiments. Such things as poor circulation, frosted fin condition, air turbulence around the evaporator, and even the amount of moisture in the air will affect the capacity of the evaporator.

To calculate the external surface area of an evaporator, consider the following:
1. Both surfaces of the fin.
2. Outside surface of the tubing (disregard the area where it comes in contact with the fins).
3. External surface of the tubing bends.

For example, find the area of a 10-ft. long evaporator with 6- × 8-in. fins .025-in. thick, having 1/2-in. fin spacings and using two 5/8-in. tubes 4 in. apart. See Fig. 15-29 and proceed as follows:
1. Area of one fin:
   Each fin is 8 in. × 6 in.
   8 in. × 6 in. = 48 sq. in.
   There are two sides to each fin:
   48 in. × 2 in. = 96 sq. in.

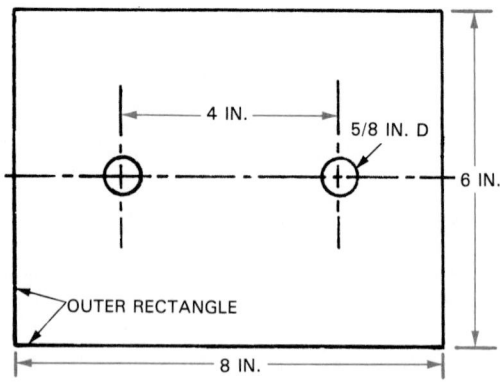

Fig. 15-29. Evaporator fin specification for sample problem. Fin is .025 in. thick.

2. However, there are two 5/8-in. diameter holes in each fin. Therefore:

Fin area =
   area of each side of fin minus area replaced by tubing
Area of a circle = πR^2
(R = radius and π = 3.1416)
$2R$ = diameter = D

$$\pi \left(\frac{D}{2}\right)^2 = \pi \frac{D^2}{2^2} = \pi \frac{D^2}{4}$$

$$\text{Area} = \frac{\pi \times D^2}{4} = \frac{\pi \times \left(\frac{5}{8}\right)^2}{4}$$

$$= \frac{\pi \times \frac{25}{64}}{4} = \frac{\pi \times 25}{4 \times 64} = \frac{3.1416 \times 25}{4 \times 64}$$

$$= \frac{78.54}{256} = .307 \text{ sq. in.}$$

However each hole takes out two fin surfaces this size
.307 × 2 = .614 sq. in.
There are two holes in each fin
.614 sq. in. × 2 = 1.228 sq. in.

3. Outer rectangle fin edge area, Fig. 15-29:
28 in. × 0.025 = 0.70 sq. in.
The 0.70 sq. in. adds to the fin area. Find the area for one fin as in the following process:
Heat removing area for one fin =
96 sq. in. − 1.228 + 0.70 = 95.472 sq. in.

4. The total number of fins:
10 ft. × 12 in./ft. = 120 in. long
2 fins per in. = 120 × 2 = 240 fins
240 fins + 1 extra fin at end = 241 fins

5. Total fin area = 95.472 × 241 = 23,008.7 sq. in.

6. Area of two tubes 5/8 in. D; 10 ft. long:
10 ft. × 12 in./ft. = 120 in.
There are two tubes: 120 in. × 2 = 240 in.
The circumference of the tube is
Cir. = π × diameter
   = π × 5/8 in.
   = 3.1416 × 5/8 in.
   = 1.9635 in.
The area = length × circumference
   = 240 in. × 1.9635 in.
   = 471.24 sq. in.

7. The actual tube area is decreased by thickness of fins:

Fin contact area = π5/8 × .025 × 241 =
1.9635 × .025 × 241 = 11.83 sq. in.
There are two holes: 2 × 11.83 = 23.66
Actual tube area 471.24 − 23.66 = 447.58 sq. in.

8. Tube bend area = length of bend × circumference length × number of bends (1)
(2-in. radius)
Length = π × D = 3.1416 × 4 in. = 12.5664
but it is only 1/2 of a circle so 12.5664 ÷ 2 = 6.2832
Circumference of tube =
5/8 in. × 3.1416 = .625 × 3.1416 = 1.96
Tube bend area = 6.2832 × 1.96 = 12.3 sq. in.

9. Total area: 23,008.7 = total fin area
          447.6 = actual tube area
     +   12.3 = tube bend area
          23,468.6 sq. in.
or in sq. ft. 23,468.6 ÷ 144 = 162.98 sq. ft.

The evaporator is unable to remove heat from the cabinet when the compressor is not running. Therefore, the heat removing capacity of the evaporator is calculated on the same running time as that of the compressor.

Heat transfer problem: What is the capacity of a natural convection evaporator having an external area of 15 sq. ft. with a refrigerant temperature of 22 °F (−16 °C), if the average cabinet temperature is 42 °F (6 °C)?

Solution:

First, it is known that 1 sq. ft. will handle 1 Btu per °F per hr. The refrigerant temperature is 22 °F (−6 °C); with the air temperature passing over the evaporator at 42 °F (6 °C), the temperature difference is 20 °F (11 °C). If 1 °F temperature difference will handle 1 Btu, 20 °F (11 °C) temperature difference will handle 20 Btu. Multiply this value by the number of sq. ft. to find the total capacity of the evaporator per hr. In this case:

Btu = 15 × 20 = 300 Btu per hr.

It is claimed that the maximum effective distance that the fin should extend from a natural convection evaporator is 3 in.

Fig. 15-30 shows an evaporator constructed of rippled

Fig. 15-30. Section through tubes and fins of an evaporator. A—Tubing (primary surface). B—Fins (secondary surface).

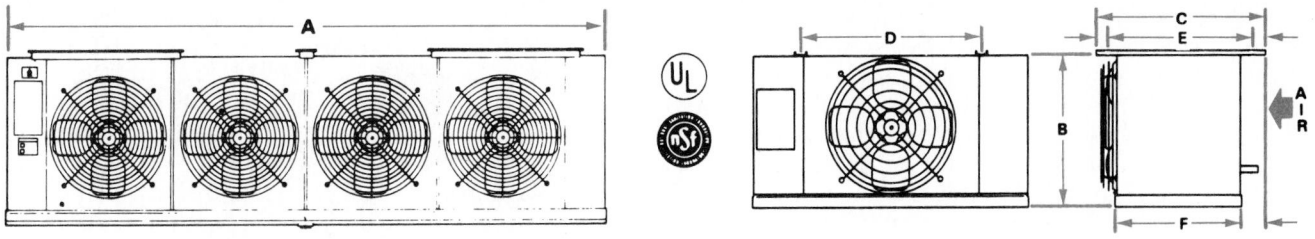

PERFORMANCE DATA

MODEL NO.	CAPACITY BTUH 10°TD	12°TD	15°TD	CFM	FAN(S) DIA.	AIR THROW FT.	HP	RPM	BTUH HEAT	TOTAL AMPS. 115-1-60	230-1-60	APPROX. R-12 CHARGE
HDC- 44-1	4,400	5,280	6,600	700	12 (1)	20	1/25	1,550	342	1.8	0.88	2.9
HDC- 56-1	5,600	6,720	8,400	700	12 (1)	20	1/25	1,550	342	1.8	0.88	3.9
HDC- 68-1	6,800	8,160	10,200	1,400	14 (1)	20	1/4	1,075	1,060	3.8	1.9	3.6
HDC- 81-1	8,100	9,720	12,150	1,400	14 (1)	20	1/4	1,075	1,060	3.8	1.9	4.2
HDC- 94-1	9,400	11,280	14,100	1,700	18 (1)	25	1/4	1,075	1,060	3.8	1.9	4.6
HDC- 113-1	11,300	13,560	16,950	1,700	18 (1)	25	1/4	1,075	1,060	3.8	1.9	4.9
HDC- 150-1	15,000	18,000	22,500	2,600	18 (1)	25	1/4	1,075	1,060	3.8	1.9	6.6
HDC- 200-1	20,000	24,000	30,000	3,300	20 (1)	30	1/4	1,075	1,060	3.8	1.9	8.5
HDC- 280-1	28,000	33,600	42,000	3,600	20 (1)	30	1/4	1,075	1,060	3.8	1.9	9.6
HDC- 400-2	40,000	48,000	60,000	5,200	18 (2)	45	1/4	1,075	2,120	—	3.8	11.0
HDC- 560-2	56,000	67,200	84,000	6,600	20 (2)	50	1/4	1,075	2,120	—	3.8	13.0
HDC- 640-2	64,000	76,800	96,000	7,800	18 (3)	55	1/4	1,075	3,080	—	5.7	15.0
HDC- 800-2	80,000	96,000	120,000	10,000	20 (3)	60	1/4	1,075	3,080	—	5.7	19.0
HDC-1120-2	112,000	134,000	168,000	13,300	20 (4)	65	1/4	1,075	4,140	—	7.6	24.0

DIMENSIONAL DATA

UNIT MODEL NO.	CONNECTIONS COIL INLET	COIL SUCT.	EXT. EQL.	DRAIN	DIMENSIONS A	B	C	D	E	F	HEAT EXCHANGER PART NUMBER	APPROX. SHIP. WT.
HDC- 44-1	1/2 FN	5/8 ODS	1/4 MF	3/4 FPT	26-5/8	17-5/16	18	17	16	13-3/8	72.35211	58
HDC- 56-1	1/2 FN	5/8 ODS	1/4 MF	3/4 FPT	26-5/8	17-5/16	18	17	16	13-3/8	72.35211	56
HDC- 68-1	1/2 FN	5/8 ODS	1/4 MF	3/4 FPT	30-5/8	17-5/16	20	21	18	15-3/8	72.35211	80
HDC- 81-1	1/2 FN	5/8 ODS	1/4 MF	3/4 FPT	30-5/8	17-5/16	20	21	18	15-3/8	72.35212	90
HDC- 94-1	1/2 FN	7/8 ODS	1/4 MF	3/4 FPT	34-5/8	25-5/16	20	25	18	15-3/8	72.35212	100
HDC- 113-1	1/2 FN	7/8 ODS	1/4 MF	3/4 FPT	34-5/8	25-5/16	20	25	18	15-3/8	72.35213	104
HDC- 150-1	1/2 FN	7/8 ODS	1/4 MF	3/4 FPT	34-5/8	25-5/16	23	25	21	18-3/8	72.35213	106
HDC- 200-1	1/2 FN	7/8 ODS	1/4 MF	3/4 FPT	42-5/8	25-5/16	23	33	21	18-3/8	72.35214	130
HDC- 280-1	1/2 FN	1-1/8 ODS	1/4 MF	3/4 FPT	42-5/8	28-5/16	23	33	21	18-3/8	72.35214	136
HDC- 400-2	7/8 ODS	1-1/8 ODS	1/4 MF	1 FPT	64-5/8	29-1/2	31	48	—	21-3/16	72.35214	390
HDC- 560-2	1-1/8 ODS	1-1/8 ODS	1/4 MF	1 FPT	72-5/8	29-1/2	31	48	—	21-3/16	72.35215	412
HDC- 640-2	1-1/8 ODS	1-5/8 ODS	1/4 MF	1 FPT	84-5/8	29-1/2	31	48	—	21-3/16	72.35215	524
HDC- 800-2	1-1/8 ODS	2-1/8 ODS	1/4 MF	1 FPT	91-1/8	32-1/2	31	51-3/4	—	21-3/16	72.35216	620
HDC-1120-2	1-5/8 ODS	2-1/8 ODS	1/4 MF	1 FPT	116-5/8	32-1/2	31	*77-1/8	—	21-3/16	72.35217	700

Fig. 15-31. Manufacturers develop performance and dimensional data for evaporators. The tables shown are for a 35°F (2°C) evaporator.

fin surfaces and with a flange contact between the prime surfaces (tubing) and the fin.

A typical evaporator capacity table for nonfrosting evaporators is shown in Fig. 15-31.

15-22 EVAPORATOR DESIGN

Many types of evaporators are available. Included among the types are some which use the following combinations of material: copper tubing and aluminum fins, copper tubing and copper fins, and steel tubing and aluminum fins (for ammonia R-717).

Usually, the evaporator fins are securely bonded to the tubing. However, construction varies. Some manufacturers make the fin to fit the tubing with a drive fit; others use some mechanical devices to attach the fins firmly to the tubing. Some expand the tubing with a mandrel or by hydraulic pressure to force it against the fin. See Fig. 15-32. A method of bonding tubing to off-center fins is shown in

Fig. 15-33. Another method of mounting fins on tubing while spacing them is shown in Fig. 15-34.

Fin spacing varies between 1/2 and 1 1/2 in. for natural convection evaporators, and 1/16 to 1/4 in. for forced convection. This spacing is a means of varying the capacity

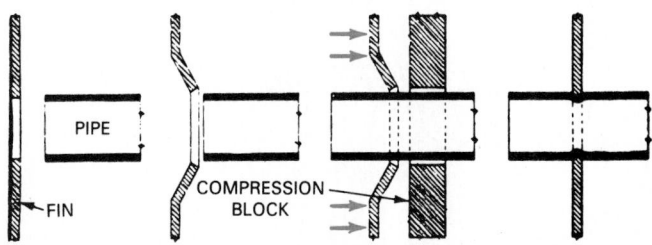

Fig. 15-32. Evaporator fins can be mechanically bonded to tubing. Fin is dished to expand tube hole before tubing is inserted. Mandrel flattens fin, firmly pressing fin into tubing surface.

of the evaporator. It is also used to compensate for the depth of the evaporator. The deeper the evaporator, the greater the fin spacing (to minimize air restriction). Evaporators which have 6- to 8-in. depth usually have 1-in. spacing, whereas those with 18- to 20-in. depth have 1 1/2 in. spacing. Fin spacing of 1 in. or less is said to decrease air turbulence. The tubing used usually is 5/8 in. OD although 3/4 in. OD tubing is used in large evaporators.

Some companies use one continuous piece of tubing for the complete evaporator; others make the bends separately and then braze straight lengths to them. Some manufacturers use devices inside the tubing to swirl the refrigerant for improved heat transfer to the boiling refrigerant.

Fittings which connect tubing to the evaporator are usually 1/2 in. OD tubing brazed to the 5/8-in. tubing, then flared with an external nut mounted on it. Some manufacturers use a 1/2-in. male flare fitting brazed to the end fin. Fig. 15-35 shows two types.

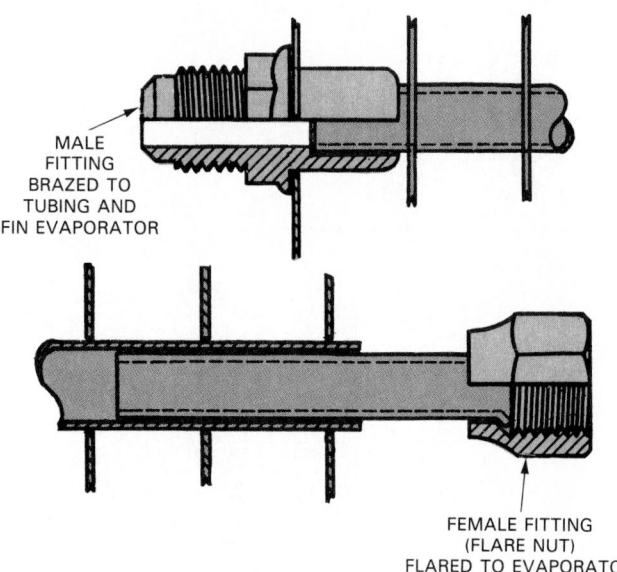

Fig. 15-35. Construction of coupling devices (fittings) used in attaching refrigeration lines to evaporator.

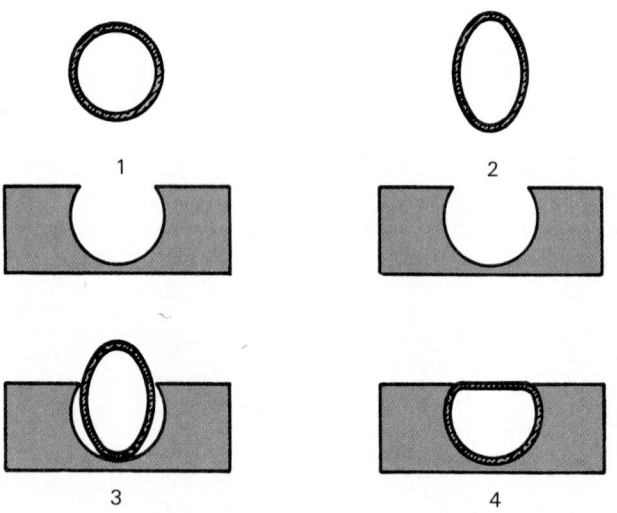

Fig. 15-33. Note method of mechanically bonding tubing to off-center fins. 1—Original tubing and fin. 2—Tubing is formed into elliptical shape. 3—Tubing is inserted into fin opening. 4—Fixture holds fins while tube is pressed into fin-opening shape.
(Peerless of America, Inc.)

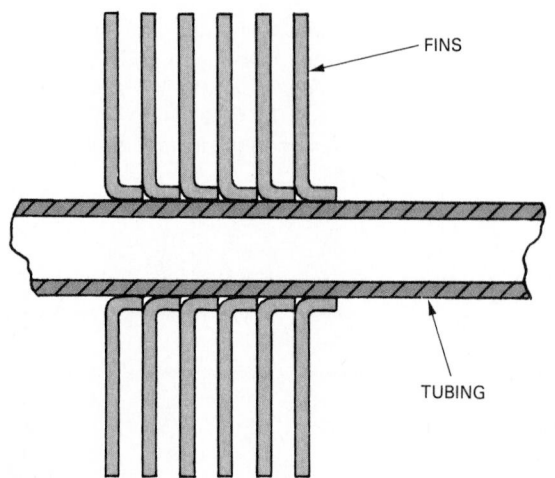

Fig. 15-34. How flanged fins can be mounted on a section of tubing. Flanges automatically space fins.

15-23 FORCED-CIRCULATION AIR-COOLING EVAPORATOR CAPACITIES

A forced-circulation evaporator is one having an electric fan mounted adjacent to it to increase the airflow. Velocities of 44 feet per minute (fpm) to 2000 fpm are permissible, with 1000 fpm being the average. Sometimes, variable speed fan motors are used. Draining facilities for condensation removal must be built into the unit.

Because of the large amount of air striking the evaporator per unit of time, the capacity of the evaporator in Btu per square foot per °F per hour is increased remarkably. The values vary with the air speed. The table in Fig. 15-36 gives these values.

Evaporators are finned, and the spacing of the fins varies considerably. See Para. 15-22. The fins must be kept straight and equally spaced or airflow will be reduced. See Chapter 14 for service instructions.

15-24 LIQUID EVAPORATOR CAPACITIES

Capacities of liquid evaporators, regardless of type, are calculated by using a base factor of about 10-120 Btu per square foot per °F per hour. The U factor (heat transferability) varies, however, with the fluid velocity, evaporator construction, and total temperature difference. Fig. 15-37 shows the average values for a certain evaporator.

Some evaporators are used to cool a brine solution. Since this solution never freezes, there are no frost accumulation problems. Many submerged evaporators use a sweet water bath to provide an ice holdover around the evaporator. This maintains good capacity during peak loads.

The capacity of evaporators varies depending upon whether or not they are of the frosting type. Ice formation around the evaporator reduces its capacity because the heat must go through this extra material. Furthermore, it makes the heat travel through one extra contact surface. This, too, reduces efficiency. Liquid cooling evaporators are used for beverage cooling, for water chillers used on some air conditioners, and for cooling brine solutions.

Commercial Systems Heat Loads and Piping

FORCED CIRCULATION EVAPORATOR CAPACITIES

SURFACE IN SQ. FT.			PARAL-LEL PATHS	COOLING CAPACITIES BTU PER HOUR		MOTOR		FAN	
TUBE	FIN	TOTAL		15°F TD*	25°F TD*	HP	SPEED RPM	DIAMETER	CFM
2.94	32.39	35	3	2200	4500	1/80	1800	12"	620
5.88	63.0	68	5	4100	7200	1/80	1800	12"	465
7.94	70.06	78	3	5200	9000	1/10	1140	15 1/2"	1200
13.2	116.8	130	5	8300	12300	1/10	1140	15 1/2"	1000
14.7	83.3	98	3	6500	10500	1/8	1140	17"	2020
17.2	145.8	163	5	10000	14000	1/8	1140	17"	1715

* TEMP. DIFFERENCE

Fig. 15-36. Table lists forced-circulation evaporator capacities.

15-25 SPECIAL EVAPORATOR CAPACITIES

Many different types of evaporators are in use. Among them are brine spray units, intermediate refrigerant evaporators, and cold plates. They may be constructed from cast metal or iron pipe. Whatever the type, they have the same problems as the ones previously discussed.

The metal used as the refrigerant-carrying device does not have much effect upon the heat transfer capacity of the evaporator. Its conductivity is relatively much greater than the contact between the evaporator and the air. The capacity of the cast metal and the iron pipe evaporator may be calculated exactly the same as the method described in Para. 15-21.

In brine spray systems, the brine is forced through a pipe extending into the cooling chamber. The pipe is perforated with a number of fine holes. The brine sprays out of these holes and mixes with the air flowing over the baffle, removing the heat.

The capacity of these systems is calculated on the temperature difference between the air in the box and the brine. It is safe to assume a high efficiency of heat transfer for brine spray installations.

The Baudelot Cooler (milk cooler) runs water, or the liquid to be cooled, over refrigerant cooled pipes or plates. The liquid being cooled is in the open and can be easily controlled. Icing is not critical and, therefore, the liquid can be cooled close to its freezing temperature. See Para. 15-56.

The cooling capacity of cold plates may also be determined by the area, temperature difference, and the materials. The heat transfer for air to metals is about 1 Btu/°F/sq. ft./hr. Cold plates usually contain a eutectic solution which freezes at a certain temperature. The plate remains at this constant temperature until all the eutectic solution has melted. The refrigerating system may or may not be operating at the time the eutectic solution is absorbing heat (solution is melting).

15-26 WATER-COOLING LOADS

Finding the refrigeration load of a water-cooling installation is a combination of a specific heat and a heat leakage problem:

1. Water is cooled to temperatures which vary upward from 35°F (2°C). The amount of heat removed from the water to cool it to a certain temperature is a specific heat problem.

2. Water, being maintained at these low temperatures, results in a heat leakage from the room into the water. This part involves the heat leakage portion of the installation.

The two major items to be solved in a water-cooling installation are:

1. How much water is to be consumed at the temperature difference desired. Fig. 15-38 gives the values of these two variables.

2. Water temperature for drinking purposes should be regulated according to the type of work the consumers are performing. The heavier the work, or the warmer

EVAPORATOR HEAT TRANSFER DATA

WATER VELOCITY FT./MIN.	TOTAL TEMPERATURE DIFFERENCE				
	6	8	10	12	15
	BTU TRANSFER/SQ. FT./HR./°F				
150	67	76	83	90	97
200	83	95	103	110	118
250	97	109	115	122	129
300	103	115	123	130	138

Fig. 15-37. Heat transfer in Btu per sq. ft. per hr. per °F for typical flooded liquid evaporator using 5/8 in. OD tubes. Total temperature difference is between refrigerant and water temperatures.

Usage	Final Temp. Required °F.	Total Amount of Water Used and Wasted
1. Office Building—Employees	50	1/8 gallon per hour per person
2. Office Building—Transients	50	1/2 gallon per hour for each 250 persons per day
3. Light Manufacturing	50 to 55	1/8 gallon per hour per person
4. Heavy Manufacturing	50 to 55	1/4 gallon per hour per person
5. Restaurant	45 to 50	1/10 gallon per hour per person
6. Cafeteria	45 to 50	1/12 gallon per hour per person
7. Hotels	50	1/2 gallon per day per room (14 hr. day)
8. Theaters	50	1 gallon per hour per 75 seats
9. Stores	50	1 gallon per hour per 100 customers per hour
10. Schools	50 to 55	1/4 gallon per hour per student
11. Hospitals	45 to 50	1/12 gallon per day per bed

Note—Total amount of water used and wasted varies with type of installation and kind of service. This table will serve as a basis for determining cooler capacity required.

Fig. 15-38. Recommended values for providing cooled drinking water for various public places and places of work. (Temprite Div. of Elkay Mfg. Co.)

the room temperature, the warmer the drinking water must be.

The amount of water consumed varies quite a bit in different applications. By using the table in Fig. 15-38, the exact heat load is easily determined.

Example: Item No. 4 points out that for heavy manufacturing, drinking water should be kept within 50 to 55 °F (10 to 13 °C). Also, 1/4 gallon of water per hour per person will be consumed.

A production foundry may be classed as heavy manufacturing. If the foundry employs 50 men for a period of eight hours, the water load per day would be 50 × 8 × 1/4, or 100 gallons of water to be cooled each eight hours.

If the incoming water is at a temperature of 75 °F (24 °C) in the pipes, it must be cooled 20 °F (11 °C) to reach 55 °F (13 °C). Since there are 8.34 lb. of water in 1 gallon, the specific heat load can be computed as follows:

Btu = specific heat × weight × temperature difference
Therefore:

Btu = 1 × 100 × 8.34 × 20
    = 16,680 Btu per 100 gallon water

The heat leakage for this particular problem is determined by the external area of the insulated parts of the system. Insulation one to three in. thick is common for water-cooling insulations, with ice water insulation being standard at 1 1/2 in. Heat leakage for water-cooling installations is calculated identically with that of heat leakage for cabinets. The example just described deals only with a unit installation.

Many water-cooling installations involve the circulation of refrigerated water to various fountains. To maintain satisfactory temperatures, the heat leakage load in a case of this kind is calculated on the basis of gallons of water per hour circulated through the system. The table in Fig. 15-39 illustrates another method of computing this load.

15-27 ICE CREAM FREEZING AND STORAGE LOAD

Ice cream is manufactured from milk, solids, fat, sugar, gelatin, and water. After mixing, the product is cooled to about 27 °F (−3 °C), then frozen. Next, it is cooled rapidly to approximately −20 °F (−30 °C). Brick ice cream is maintained between 0 to 5 °F (−18 to −15 °C). Bulk ice cream is held at 5 to 12 °F (−15 to −11 °C).

The heat values of various ice creams vary. See Fig. 15-7. Specific heat of the mix before freezing is 0.80; latent

heat, at 27 °F (−3 °C), is about 96 Btu/lb. Specific heat of frozen ice cream is 0.45 Btu/lb. Weight of the original mix is about nine lb. per gallon. On freezing, it expands and comes to a density of five lb. per gallon if simply flavored, and to six lb. per gallon if it contains fruits or nuts.

Next, determine the size of the refrigerating unit needed to keep ice cream in its cabinet during dispensing. Normally, an ice cream cabinet is designed to hold brick ice cream and flavored ice creams next to the evaporator. However, many cabinets need separate evaporators for the two types. Fig. 15-40 shows an ice cream display cabinet. Metal-finished cabinets with 2 to 3 in. of urethane are used.

The ice cream is delivered at the correct temperature. The heat load, therefore, is composed only of leakage and air entering when the covers are removed. The table in Fig. 15-11 may be used to calculate heat leakage. Use the outside area and add 20 percent to take care of the cover openings.

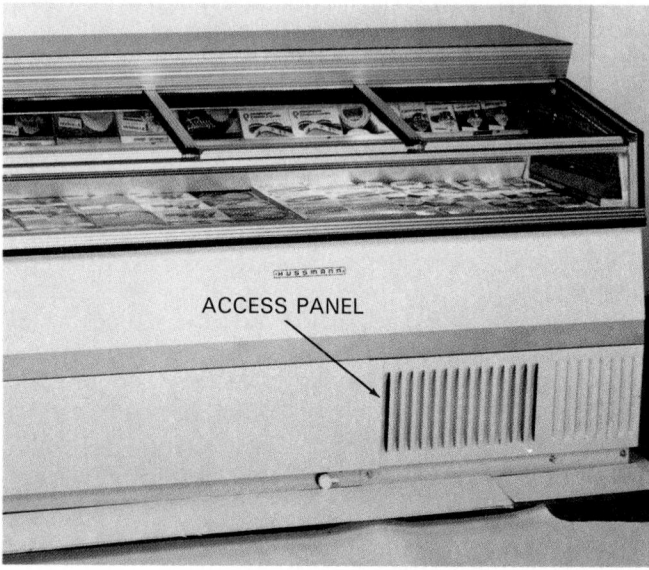

Fig. 15-40. Ice cream display cabinets such as this one are used to merchandise packaged ice cream. Note access panel to condensing unit assembly. (Hussmann Corporation)

15-28 THERMODYNAMICS OF THE REFRIGERATION CYCLE

Thermodynamics is the science which deals with the relationships between heat and mechanical action. The refrigeration compression cycle is based on thermodynamics.

Every technician should know how to determine what size compressor is needed to produce a certain amount of refrigeration. He or she should also know how large a motor is needed to drive this compressor. To understand how these values are determined, the heat behavior of the refrigerant must be understood.

The refrigerant cycle is simple. The refrigerant is let into the evaporator in the liquid state and near room temperature. Some of the refrigerant vaporizes under the low pressure in the evaporator and cools the remainder to the desired refrigerating temperature. Then, as the remainder of the refrigerant evaporates, it removes heat from the

Gallons per Hour To Be Circulated per 100 Feet of Pipe to Hold Temperature Rise Within 5° F. Add This Amount to Usage								
PIPE SIZE	**TEMPERATURE DIFFERENCE BETWEEN ROOM AND CIRCULATING WATER**							
	20°	**25°**	**30°**	**35°**	**40°**	**45°**	**50°**	**55°**
1/4"	4.8	6.3	7.3	3.2	9.7	10.9	12.3	13.6
3/8"	5.5	6.8	8.2	9.6	11.0	12.3	13.7	15.0
1/2"	6.3	8.0	9.5	11.2	12.7	14.3	16.0	17.6
3/4"	6.7	8.4	10.1	11.8	13.5	15.2	16.9	18.6
1"	7.3	9.1	10.9	12.8	14.6	16.5	18.5	20.6
1 1/4"	8.6	10.4	12.5	14.6	16.6	18.7	20.4	22.1
1 1/2"	9.0	11.2	13.5	15.7	18.0	20.2	22.5	24.7

Fig. 15-39. This table shows heat gain through insulated cold-water pipes. Add the amounts to the usage in Fig. 15-38. (Temprite Div. of Elkay Mfg. Co.)

evaporator and, therefore, from the cabinet. The total amount of heat absorbed is the total LATENT HEAT of vaporization. The amount of heat absorbed from the cabinet and evaporator is the EFFECTIVE LATENT HEAT.

The refrigerant vapor formed during evaporation passes down the suction line. As this happens, the vapor decreases a little in pressure (usually 2 psi [14 kPa]) and increases in temperature about 10 °F (6 °C). Vapor warming up or increasing in temperature after it has vaporized is called ''superheating of the vapor.'' The degree of superheat is the temperature difference between the temperature of the vapor at the compressor and its evaporating temperature.

The compressor then takes the slightly superheated vapor and converts (compresses) it to a high-temperature, high-pressure vapor. This condensing temperature sometimes becomes as high as 250 °F (121 °C), depending upon the refrigerant and the conditions.

The superheated vapor passes to the condenser and, if its temperature is higher than ambient temperature (water or air), it transfers enough of its heat to the air or water to cool to its vapor pressure-temperature. If the vapor pressure-temperature is above that of the water or air ambient temperature, the refrigerant vapor starts losing some latent heat of evaporation. The quantity of heat it loses determines the amount of the vapor that will condense into a liquid. After it has become liquid, the refrigerant cools down close to the water or air temperature.

The refrigerant then goes to the refrigerant control, where the pressure is reduced. The refrigerant is cooled as it vaporizes (called ''flash gas'') and the rest vaporizes to remove heat from the cabinet. Thus the refrigerant cycle is repeated.

Fig. 15-41 shows this cycle taking place. It also indicates the temperatures in various parts of the refrigerating system.

15-29 PRESSURE-HEAT DIAGRAM

The following discussion of refrigerant behavior is based on one pound of refrigerant regardless of its state (liquid or vapor). The discussion deals only with the pure refrigerant, neglecting the effect of lubricating oils and other influences.

Fig. 15-42 charts the behavior of one pound of refrigerant in a refrigerating machine. The horizontal scale shows the amount of heat present in one pound of refrigerant at all times and under all conditions. The vertical scale shows the pressure imposed upon it.

The graph shown in Fig. 15-42 is commonly called a pressure-heat chart. It is also called a pressure-enthalpy chart. The heat (Btu) in the pound of refrigerant is usually measured from saturated liquid refrigerant at −40 °F (−40 °C). The pressures will be different for each kind of refrigerant.

Note in the pressure-heat chart that as the refrigerant vaporizes at the lower constant pressure, it passes horizontally from B to D. This line indicates the vaporization of the refrigerant from a liquid into a vapor in the evaporator. The distance D to E represents the heating of this vapor into a superheated condition as it passes down the suction line. Note that only a few Btus of heat have been added and that the pressure has decreased a little.

Point E is the condition of the vapor when it moves into the compressor and is compressed. It is then compressed to F. Note how the pressure increases rapidly and how a

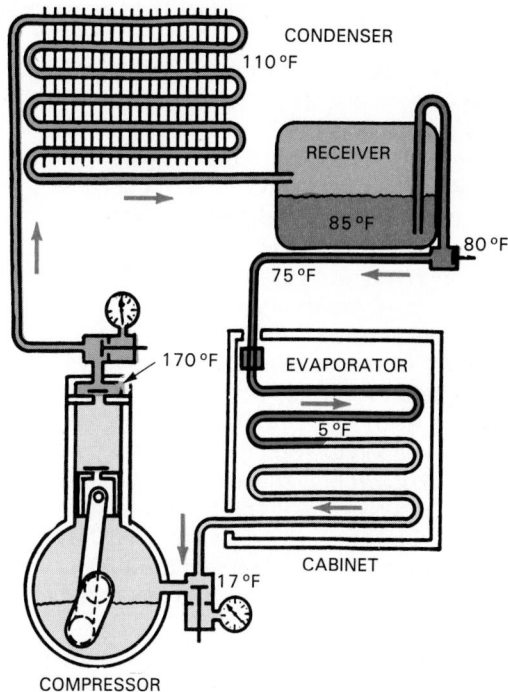

Fig. 15-41. A refrigerating system schematic shows the approximate temperatures of refrigerant in various parts of the system.

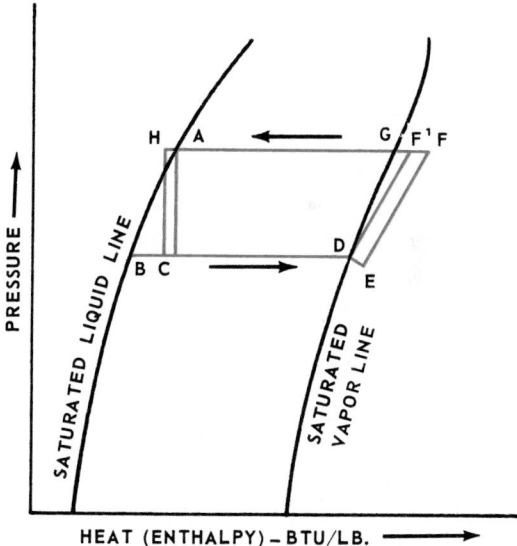

Fig. 15-42. Pressure-heat diagram for a refrigerant. Saturated liquid curve represents heat in liquid refrigerant at various pressures before it will start vaporizing. Saturated vapor curve represents division between superheated gas and point where gas starts condensing into liquid.

few Btus of heat are added to the vapor as the vapor is compressed from E to F. The vapor leaving the compressor is considerably superheated. See Para. 15-34.

Point F represents the condition of the vapor as it leaves the exhaust valve of the compressor. The distance between F and G is the cooling down of this superheated vapor to the point where it starts to condense.

At G the vapor has no superheat and is 100 percent saturated vapor. The line G to A represents the condensation of the refrigerant in the condenser from a vapor into a liquid.

Point A represents the amount of heat in the liquid and the pressure imposed on the liquid as it forms in the condenser. From A to H is the loss of heat from the liquid as it passes along the liquid line to the refrigerant control. This action (sometimes called "subcooling") is because the liquid refrigerant cools to room temperature.

Line H to C represents the throttling of the liquid upon passing through the refrigerant control orifice. The cycle is now ready to be repeated for the pound of refrigerant.

Note that the distance C to D does not represent the total LATENT HEAT of the liquid at the low-side pressure condition. This means that R-12, which has a latent heat of 70 Btu per pound at 5°F (−15°C), will not remove all of this heat from the evaporator. Some of it (approximately 19 Btu) is used to cool down the remaining liquid refrigerant to the 5°F (−15°C) temperature. The 51 Btu remaining is called the EFFECTIVE LATENT HEAT or the EFFECTIVE REFRIGERATING CAPACITY. This means that 19/51 of the pound (37 percent) of the R-12 flashes into gas at the refrigerant control.

The pressure-heat chart in Fig. 15-42 graphically shows physical property changes in a pound of a refrigerant as it passes around the refrigerating cycle. A thorough knowledge of this chart is helpful to the engineer and service technician.

15-30 PRESSURE-HEAT AREAS

The pressure-heat chart in Fig. 15-43 is divided into three main areas. To the left of the saturated liquid line, all of the pound of refrigerant is liquid. Between the saturated liquid line and the saturated vapor line, the refrigerant is a mixture of liquid and vapor, as shown in the boxes in the drawing.

Close to the saturated liquid line in Fig. 15-43 the refrigerant is almost all liquid; close to the saturated vapor line, the pound of refrigerant is almost all vapor. Note that the vapor in the area to the right of the saturated gas line is in a superheated condition.

15-31 CONSTANT VALUE LINES OF PRESSURE-HEAT CHART

Many facts can be read from the chart in Fig. 15-44. Along any vertical line such as A, the heat in one pound of refrigerant is constant (the same). Along any horizontal line such as B, the refrigerant has constant pressure. The line C, along which the temperature reading is constant, is almost vertical in the liquid area, horizontal in the liquid and vapor area, and slopes down and to the right in the superheated vapor area.

Refrigerant quality means how much of the one pound of refrigerant is liquid and how much is vapor. Ten percent (10%) quality means that the pound of refrigerant is 10 percent vapor and 90 percent liquid. The line showing the same or constant quality is at D.

15-32 EFFECT OF PRESSURE ON LATENT HEAT

The value of the latent heat of the refrigerant when vaporizing, or when condensing in a refrigerating machine, differs at different pressures. At the lower pressure (vaporizing), Fig. 15-45, the TOTAL LATENT HEAT to be added

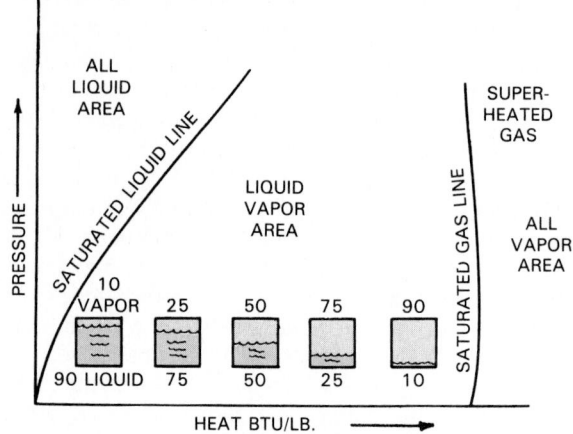

Fig. 15-43. Pressure-heat diagram. Note that as heat is subtracted, refrigerant becomes a liquid. As heat is added, refrigerant becomes a vapor.

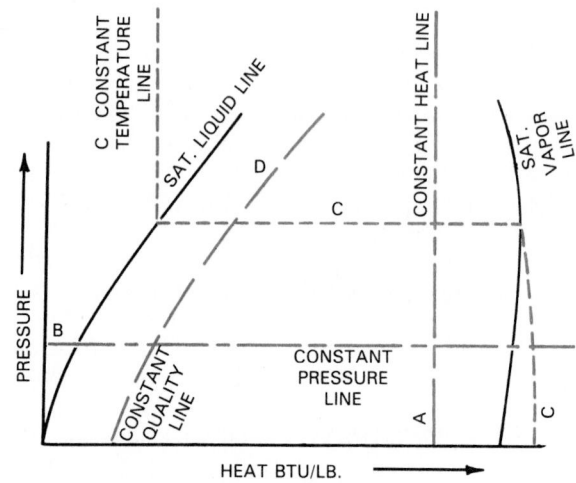

Fig. 15-44. Pressure-heat diagram. A—Constant heat condition with pressure change. B—Constant pressure and condition of refrigerant with changing heat content. C—Constant temperature with changing pressure and heat.

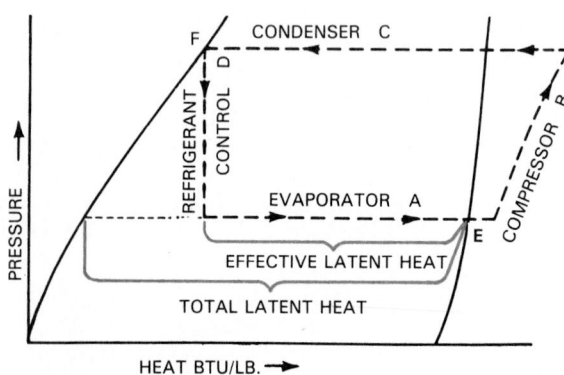

Fig. 15-45. Pressure-heat diagram. A—Liquid is boiling in evaporator and pressure is constant as heat is being added to refrigerant. B—Compressor raises pressure to condensing pressure; heat of compression is added and temperature rises. C—Condenser cools hot vapor to saturated vapor line, then condenses vapor into a liquid. Pressure is constant and heat is removed. D—Refrigerant control reduces pressure quickly to evaporating or low-side pressure. Temperature of remaining liquid drops as some liquid refrigerant "flashes" into vapor.

to (be absorbed by) the liquid refrigerant to vaporize it is more than that needed to be subtracted or removed from it to condense it into a liquid at the higher pressure. This is because the liquid, when formed, is at a higher temperature than the liquid at the vaporizing pressure. This difference is represented by the specific heat of the refrigerant multiplied by the temperature difference for the two conditions.

The EFFECTIVE LATENT HEAT of the refrigerant, when vaporizing, is the total latent heat at the low pressure minus the difference in the heat content of the liquid at the high (condensing) pressure, minus the heat content of the liquid at the low (evaporating) pressure. This is because the high-pressure liquid, when throttled in the refrigerant control, must be cooled down to a low-pressure-temperature liquid before it can vaporize and remove heat from the surrounding substances. Part of the liquid vaporizes so it can cool the remaining liquid to the lower temperature. The vapor formed during this operation is called "flash gas."

The effective latent heat is the total heat at E in Fig. 15-45 minus the heat of the liquid at F. The effective latent heat is an average value because low-side pressure and high-side pressure vary somewhat during the operation of the system. Other conditions, such as oil in the system, and system efficiences affect the ideal cycle.

For R-12 the total latent heat of vaporization at 5 °F (−15 °C) is 79 Btu per pound. However, its actual heat-absorbing ability or effective latent heat is only about 51 Btu per pound. These values are based on the standard evaporating temperature of 5 °F (−15 °C) and standard condensing temperature of 86 °F (30 °C). This reduction in heat absorption is due to the fact that some heat is absorbed by the flash gas when refrigerant pressure is reduced at the refrigerant control.

15-33 SATURATED VAPOR

A saturated vapor is a refrigerant vapor under conditions which permit some of the vapor to condense when a little heat is removed from it. Another definition: *Saturated vapor is a substance in vapor form in the presence of some of its own liquid.* For example, a saturated vapor is the vapor in a refrigerant cylinder half full of liquid refrigerant.

When the refrigerant vaporizes in the evaporator, it is saturated vapor at first. However, as this vapor passes down the suction line to the compressor, it usually becomes warmer by 5 °F to 15 °F (3 °C to 8 °C). This additional heat and increase in temperature is called superheating the low-side vapor. That is, it is raising the vapor above its saturated condition for this pressure, and the vapor will now obey Charles' or Boyle's Laws. See Chapter 30. In Fig. 15-43, any vapor in the space marked "liquid vapor area" is saturated vapor.

15-34 SUPERHEATED VAPOR

A superheated vapor is a vapor under conditions where the volume and/or pressure of the vapor decreases, with no condensation, when some heat is removed from it. For example, the vapor that the compressor handles is always superheated—unless a condition arises where the last drop of liquid refrigerant evaporates just as the refrigerant travels past the exhaust valve of the compressor.

After the low-pressure superheated vapor enters the compressor, it is compressed. The energy put into it by the compressor tremendously increases the temperature and pressure on the vapor. The amount of superheat increases.

Superheating of the vapor lowers the efficiency of a machine. The less the superheating, the more efficient the machine will be. The heat added to the vapor is the mechanical energy of compression being converted into heat energy. By knowing how much heat has been added, the technician may calculate the size of the motor necessary to drive the compressor.

There are three places in the refrigeration cycle where superheating usually takes place: in the suction line, a low-temperature superheat; in the compressor; and in the top part of condenser, a high-temperature superheat. These three conditions are shown in Fig. 15-46.

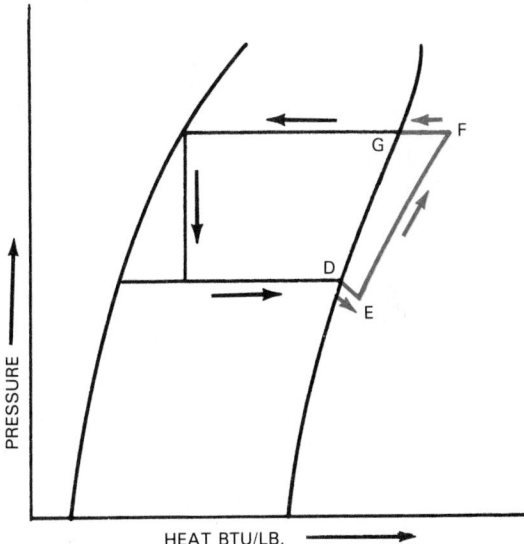

Fig. 15-46. Superheated vapor sections of refrigeration cycle. D to E—Superheat added to one pound of refrigerant vapor as it travels through suction line from evaporator to intake valve of compressor. E to F (condition at exhaust valve)—Superheat added as vapor is compressed. F to G—Superheat removed in top portion of condenser.

15-35 SPECIFIC HEAT

The specific heat of a substance is the amount of heat necessary to raise the temperature of one pound of that substance 1 °F.

Substances may exist in three different states (solid, liquid, and vapor). Every substance has three different values for its specific heat, depending on whether it is a solid, a liquid, or a vapor.

There are two kinds of specific heat of vapor:
1. Specific heat when under a constant pressure.
2. Specific heat when confined to a constant volume.

A vapor under constant pressure has a greater specific heat value than that of the same vapor under constant volume. This is because the vapor heated with a constant pressure upon it will expand and do external work (such as increasing the size of a balloon). This external work naturally requires an additional quantity of heat.

When a compressor compresses one pound of the refrigerant vapor, it does not add heat to it under a constant pressure or constant temperature condition. This state in the compressor is called adiabatic compression. *Adiabatic compression means that no heat has been removed from or added to the vapor as it was compressed.* Actually, a refrigeration compressor operates almost adiabatically because the compression takes place so rapidly.

The specific heat of liquid refrigerants varies considerably, depending on pressure imposed upon them. As shown in Fig. 15-47, the pressure to which the liquid refrigerant is subjected in the condenser, after it has condensed at D, must be determined before calculating how much heat must be removed from one pound of the liquid at this temperature to further cool it to room temperature C.

Example: One pound of R-12 condensing at 125 °F (52 °C), and then cooled to 75 °F (24 °C) in the line, must lose:

Heat of liquid at 125 °F = 37.28 (Fig. 9-5)
Heat of liquid at 75 °F = 25.20
37.28 − 25.20 = 12.08 Btu/lb. must be removed

After the refrigerant passes through the throttling valve, it is subjected to a lower evaporating pressure. The specific heat of liquid under the evaporating pressure, shown at B, must be determined to find out how much heat must be removed to cool it down to vaporizing temperature. If refrigerant goes through the refrigerant control while at its condensing temperature, it will have A to D liquid specific heat/lb.

15-36 CASCADE SYSTEM

To produce extremely low temperatures efficiently, two refrigerating systems may be used instead of one. The two systems are connected in series (cascade system). That is, the evaporator of the higher pressure-temperature cycle (first or low stage) removes the heat from the condenser of the lower pressure-temperature cycle (second or high stage). Fig. 15-48 shows the principle of this type system on a pressure-heat diagram.

Many cascade systems use a different refrigerant for this low-temperature system than for the high-temperature system. Actually, the evaporator of the high-temperature system must remove all the condensing heat of the low-temperature system. A¹ to B¹ should equal E to D.

15-37 TWO-STAGE COMPRESSOR

Some refrigerating systems, especially ultralow temperature systems, use two compressors connected in series to pump the very low pressure suction line vapor up to the condensing pressure and temperature condition. The first stage, usually a large cylinder, pumps the vapor up to a midpoint on the compression curve, then the compressed vapor is cooled but stays vaporized. The second cylinder then compresses the cooled intermediate vapor to the final pressure-temperature condition. Fig. 15-49 shows an approximate pressure-heat cycle.

Two-stage compressors are used when the compression ratio is more than 10:1. That is, if the low-side pressure is 0 psi and the head pressure is 210 psi, the ratio is:

$$\frac{\text{Head pressure abs}}{\text{Low-side pressure abs}} = \frac{210 + 15}{0 + 15} = \frac{225}{15} = 15:1$$

In this case, a two-stage compressor would be used.

Stage 1: $\frac{45 + 15}{0 + 15} = \frac{60}{15} = 4:1$

Stage 2: $\frac{210 + 15}{45 + 15} = \frac{225}{60} = 3.75:1$

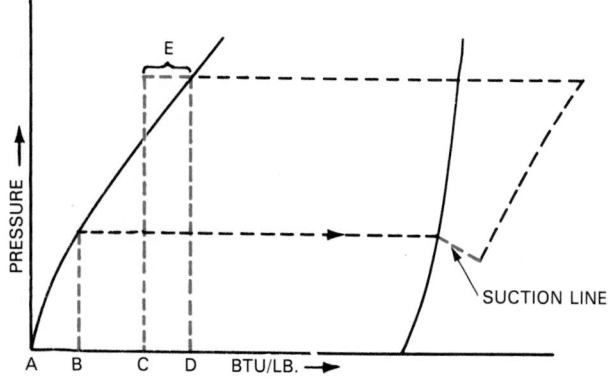

Fig. 15-47. Pressure-heat diagram. By using heat exchanger or installing liquid and suction lines together, heat increase in suction line vapor comes from heat decrease in liquid line. E indicates cooling of liquid in receiver, liquid line, and by heat exchangers. Note gain in effective latent heat from C to D.

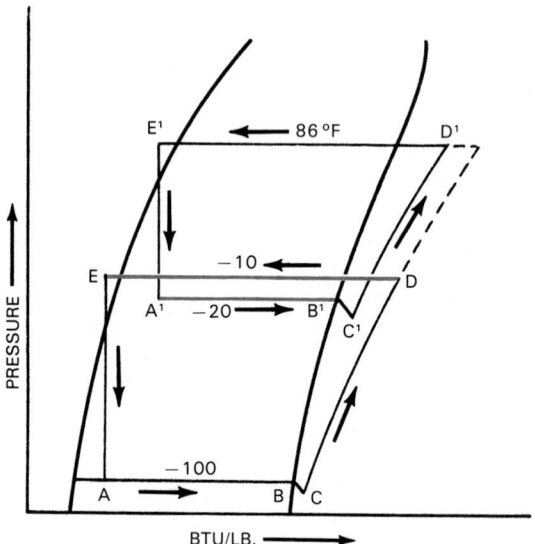

Fig. 15-48. Pressure-heat diagram for cascade system used to obtain ultralow temperatures. Evaporator A¹ to B¹ removes heat from condenser D to E.

15-38 BYPASS CYCLE

The "hot gas" bypass may be used for any of these purposes:
1. Defrost evaporators.
2. Prevent suction pressure from going too low (when cooling load decreases).
3. Keep liquid refrigerant from entering compressor.

Many refrigerating systems use an automatic bypass system. Two automatic bypass types are:
1. "Hot gas" bypass.
2. Liquid bypass.

The "hot gas" bypass cycle used to defrost an evaporator is shown in Fig. 15-50. The "hot gas" is traveling through the evaporator from A to B to C.

The cycle for "hot gas" bypass for low-pressure control is shown in Fig. 15-51. The bypass line (controlled by a solenoid valve and a pressure valve) is piped from the "hot gas" part of the condenser into the suction line near the

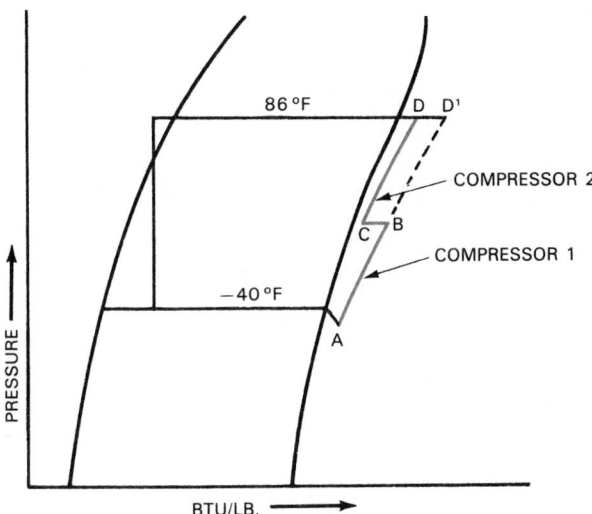

Fig. 15-49. Cycle of a two-stage compressor system. Compressor 1 (first stage) compresses vapor from A to B. Vapor is cooled in heat exchanger (air to water) from B to C. Compressor 2 (second stage) then compresses vapor to condensing pressure from C to D. This action reduces amount of heat of compression at final stage D to D¹ and also reduces superheat temperature at compressor 2 exhaust valve.

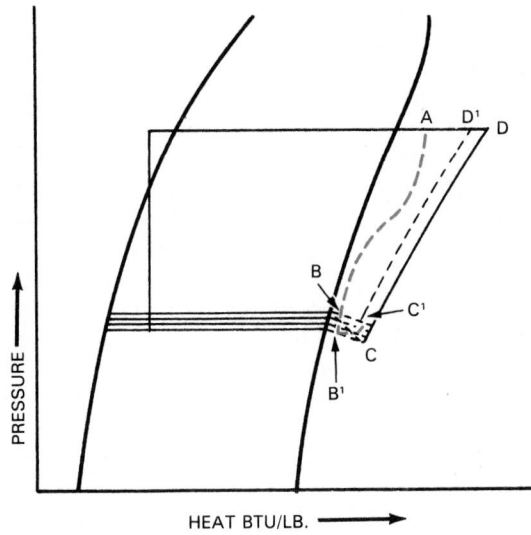

Fig. 15-51. "Hot gas" bypass system allows low side to maintain normal low-side pressure. If pressure tends to drop to point B¹, bypass circuit A to B opens and brings low-side pressure up to normal at C¹. Without bypass, low-side pressure would tend to operate at C.

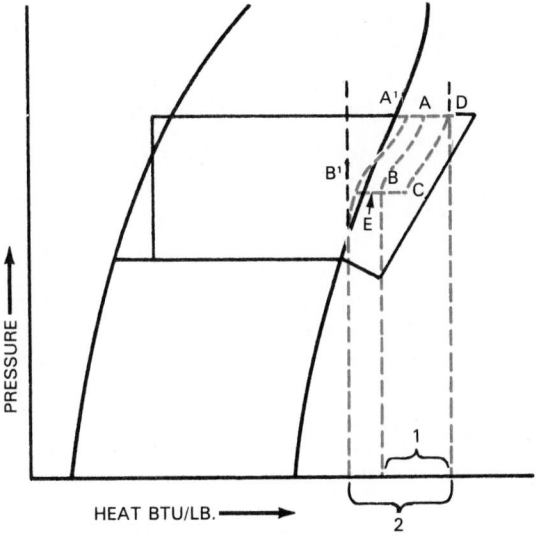

Fig. 15-50. "Hot gas" defrosting cycle. Heat lost from D to B as shown at 1 is heat used to defrost the evaporator. If defrost hot gas is cooled too much as at B¹, it will become partly liquid. This could cause liquid to enter the compressor, so an accumulator is put at E to insure that only vapor can reach compressor. Defrost cycle is A to B to C to D. Maximum heat for defrosting is shown at 2, unless bypass gas is allowed to condense and is then vaporized (in another evaporator of a multiple system or in a special defrost evaporator).

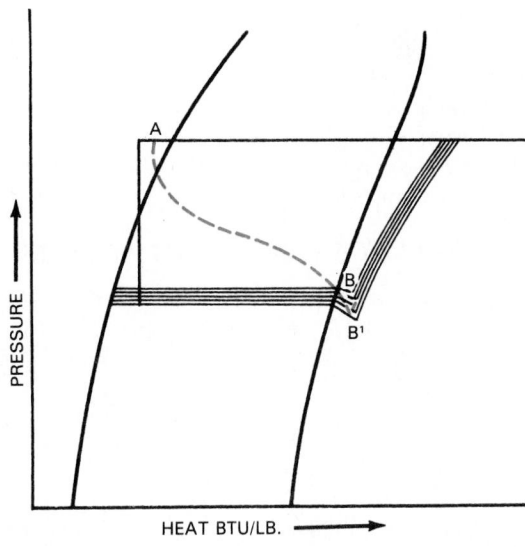

Fig. 15-52. Liquid refrigerant bypass A to B is used to maintain normal low-side pressures. The five parallel evaporator pressure lines and compression lines show changes in the cycle as the low-side pressure changes from cut-in to cutout pressure.

compressor. The bypass circuit is controlled by a pressure control connected to the suction line. The bypass action will return the compression line to approximately C¹ to D¹. The four horizontal evaporator lines represent how the low-pressure side changes from cut-in pressure to cutout pressure.

A liquid bypass cycle is shown in Fig. 15-52. If the low-side pressure drops to B¹ because the evaporator is not picking up enough heat, the suction pressure control will open a solenoid valve. This permits liquid refrigerant to bypass

the refrigerant control and evaporator, and feed directly into the suction line at B. The liquid evaporates quickly and maintains a definite low-side pressure.

15-39 PRACTICAL PRESSURE-HEAT CYCLE

The refrigeration cycles described in previous paragraphs are based on 5°F (−15°C) evaporating temperature and 86°F (30°C) condensing temperature.

Practical cycles for various other refrigerating systems are somewhat different. For example:

1. With an air-cooled condensing unit used for long term frozen food storage, the evaporator refrigerant temperature will be −10°F (−23°C).

2. With summer design ambient temperatures of 95 °F (35 °C), the condensing temperature will be 95 °F (35 °C) + 30 °F (17 °C), or 125 °F (52 °C).

In Fig. 15-53, note the difference in compressor performance and refrigerant temperatures between suction line refrigerant temperatures entering the compressor at 0 °F (−18 °C) and 60 °F (16 °C). NOTE: Surface of suction line must be above dew point temperature of air or suction line will sweat, collect moisture, and/or collect frost.

An air conditioning (comfort cooling) cycle is shown in Fig. 15-54. A water-cooled unit with 70 °F (21 °C) and, therefore, 80-90 °F (27-32 °C) refrigerant condensing temperature is similar to the standard cycle.

15-40 SERVICING: REFRIGERATION TROUBLESHOOTING

To locate trouble, the service technician must be able to accurately determine what is going on inside a refrigerating system. The system is sealed. So the technician uses gauges to check the pressure and thermometers to measure evaporator, line, and condenser temperatures. He or she also uses the system sight glass to check the amount of refrigerant and its dryness.

Much of the investigation has to be by logic. The technician needs to know what is supposed to be going on inside the system. Then she or he must be able to visualize the behavior of the refrigerant and what each part of the system is supposed to do. The pressure-heat diagram provides considerable aid in this area.

The following paragraphs show the effect of some of the more common troubles on the pressure-heat cycle.

15-41 EFFECT OF LACK OF REFRIGERANT

If the system is undercharged, each pound of refrigerant does not completely liquefy before it passes through the refrigerant control, as shown at A in Fig. 15-55. The result is threefold:

1. Effective latent heat is reduced by the amount indicated by the "loss." Therefore, refrigeration is poor.

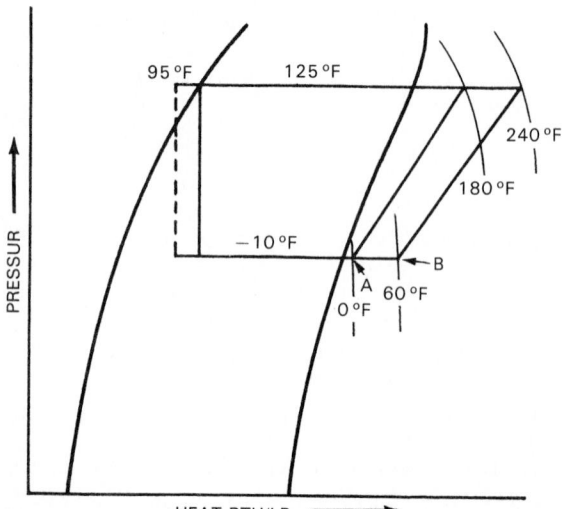

Fig. 15-53. A typical cycle of an air-cooled system for refrigerating frozen foods. Cabinet is kept at 0 °F (Refrig. −10 °F). Air temperature is 95 °F (35 °C). Note effect of the temperature change of suction vapor entering the compressor at Point A (0 °F) and Point B (60 °F).

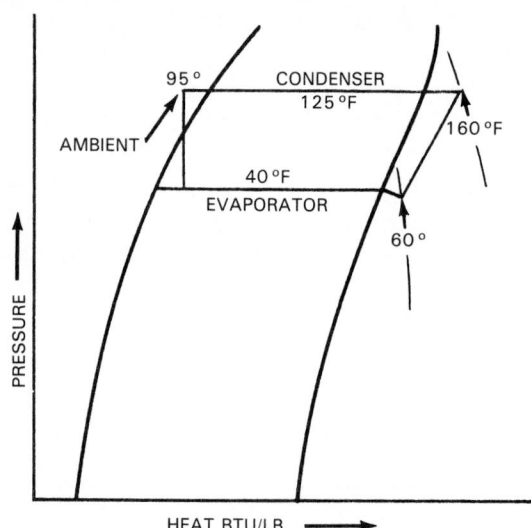

Fig. 15-54. Typical air conditioning comfort cooling cycle with an evaporator temperature of 40 °F and an ambient temperature of 95 °F.

2. Some vapor now passes through the refrigerant control, reducing the refrigerant control capacity.
3. As this vapor passes between the needle and seat at a high velocity, it increases the wear on the refrigerant control needle and seat.

15-42 EFFECT OF AIR IN SYSTEM

Air in the refrigerating system increases the total head pressure. Total head pressure will equal the refrigerant condensing pressure plus the pressure of the air in the condenser. The refrigerant will have to condense at the higher temperature and pressure.

Because total head pressure is higher, the compressor has to pump the vapor to a higher temperature and pressure. The extra work performed by the compressor is illustrated in Fig. 15-56. The heat added to do this is a "loss." The cylinder head (especially the exhaust valve) and the top tube of the condenser will be at above-normal condensing temperatures. This may also harm the oil.

15-43 EFFECT OF HEAT EXCHANGER

After the suction line vapor leaves the evaporator, it travels down the suction line and into the compressor. During this part of the cycle, the vapor usually warms up somewhat, as shown at A in Fig. 15-57.

Since the low-pressure vapor picks up heat in most cycles, system efficiency can be improved if excess heat is removed during some part of the cycle. The heat exchange is done by putting the suction line in contact with the liquid refrigerant line just before the liquid goes into the refrigerant control.

Fig. 15-57 illustrates the removal of heat at B. The result is a gain in effective latent heat and a reduction in "flash gas," which will serve to increase the life of the refrigerant control.

15-44 EXCESSIVE CONDENSING PRESSURE

If the condenser is undersize or dirty (internally or externally), the head pressure and condensing temperature will rise. Fig. 15-58 shows a cycle diagram in which condensing pressure is above normal. The higher temperature will cause the compressor to pump to this higher pressure and temperature, and the added heat of compression is shown as a "loss" at A. If the liquid does not subcool to room temperature, there is an added loss in effective latent heat and an increase in flash gas. See B in Fig. 15-58.

15-45 EXCESSIVE SUCTION LINE PRESSURE DROP

If the pressure of the vapor going into the compressor decreases, the compressor will pump less weight of vapor per stroke and, therefore, per minute. The less vapor pumped, the lower the capacity of the system.

Fig. 15-59 shows the effect of excessive suction line pressure drop on the cycle. Note that as the volume of the vapor increases there is more volume per pound of vapor and the vapor picks up heat, increasing its temperature. Therefore, the exhaust valve temperature increases and the condenser must remove more heat from each pound of vapor. When the exhaust valve temperature becomes too high, there is danger that the oil will deteriorate. The effect of a partially clogged filter-drier in the suction line is shown in Fig. 15-60.

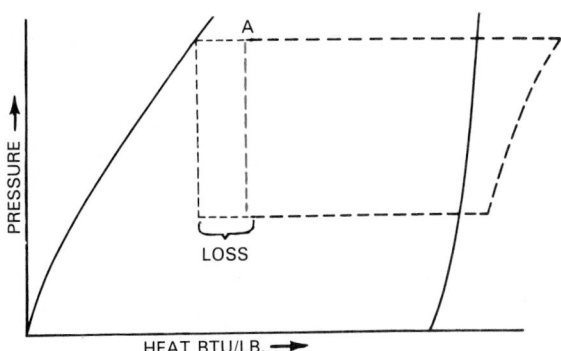

Fig. 15-55. Pressure-heat diagram shows effect of insufficient refrigerant in system. Note loss of effective latent heat. This loss means unit will have to run longer to remove same amount of heat.

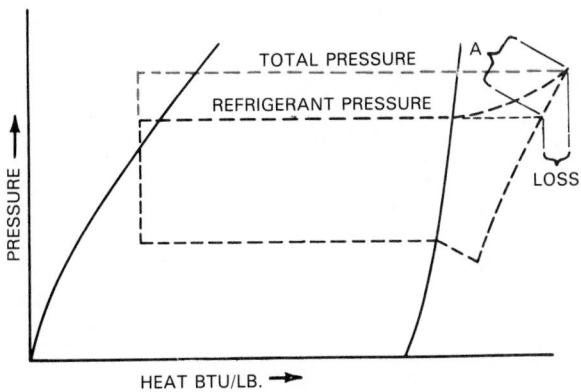

Fig. 15-56. Pressure-heat diagram shows effect of air in system. Point A indicates increase in cylinder head and exhaust valve temperature. The "loss" is heat energy put into the vapor by compressor. The waste is in the extra electrical energy used by the electric motor.

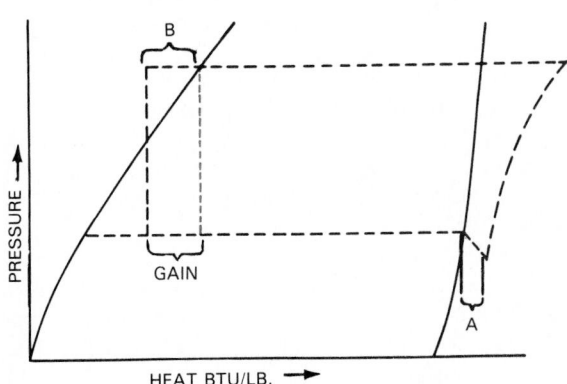

Fig. 15-57. Pressure-heat diagram shows effect of use of heat exchanger. A—Slight amount of decrease in intake pressure and an increase in temperature. B—Amount of effective heat gain. In addition, there is a reduction of flash gas which improves operation of refrigerant control. Heat exchanger also minimizes chance of liquid refrigerant in suction line reaching compressor.

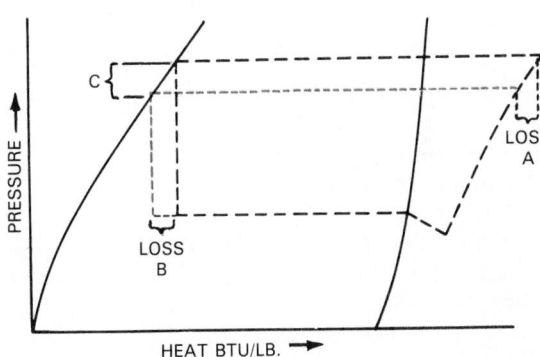

Fig. 15-58. Pressure-heat diagram shows effect of a dirty or undersize condenser or an above-average room temperature. A—Loss due to unnecessary added heat of compression. B—Loss in effective latent heat of liquid. C—Loss due to work done to compress vapor at higher pressure.

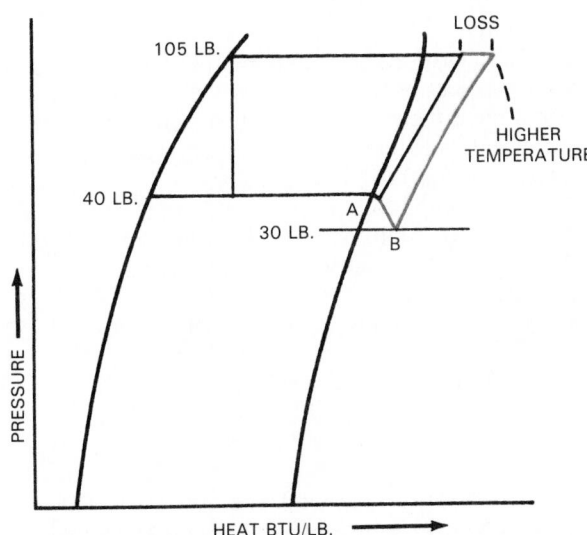

Fig. 15-59. Typical refrigeration cycle is shown when there is an excessive pressure drop and temperature rise in suction line A to B. This causes excessive temperature at the exhaust valve and cylinder head.

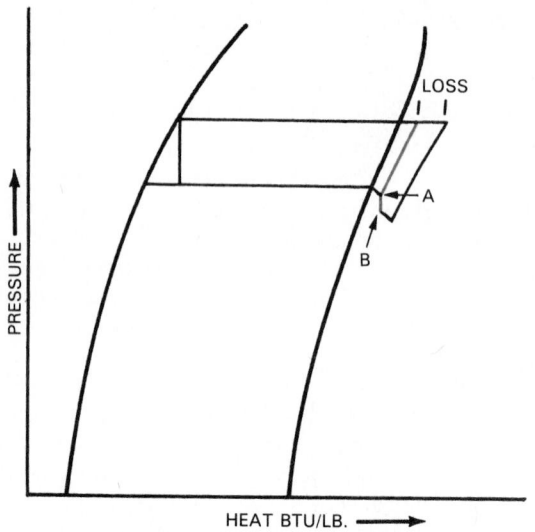

Fig. 15-60. Effect of partially clogged suction filter and/or drier. Drastic pressure drop occurs from A to B.

15-46 SYSTEM CAPACITY

There are several ways to find out the capacity of a system:
1. Measure the refrigerant flow:
   a. Amount—weight.
   b. Flow meters.
2. Measure the heat gain of the condenser cooling water.
3. Measure the heat gain of the air flowing over the condenser.

Next, record the condensing temperature and pressure, and the evaporating temperature and pressure of the refrigerant in the cycle. Use the refrigerant's pressure-heat diagram to find the Btu pickup in the evaporator.
1. If the pounds/hour of refrigerant flow is known:
   a. Draw the system cycle on the pressure-heat diagram.
   b. Multiply the Btu/pound heat pickup by the pounds circulated to find the Btu removal capacity of the system.
   NOTE: Take the temperature of the refrigerant in the liquid line just before it reaches the refrigerant control. Use a good thermometer in a thermometer well on the line or a thermocouple thermometer fastened to the line.
2. Determine the heat gain in water:
   lb./hr. × temp. change × spec. heat = Btu lb./hr. = gal./hr. × 8.34 lb./gal. (1 gal. water = 8.34 lb.)
   Temp. change = temp. out minus temp. in
   specific heat = 1
   Example: If a system uses 300 gal. per hour of 70 °F (21 °C) water that warms to 80 °F (27 °C) at the condenser outlet, how much heat is being removed?
   (300 gal. × 8.34) × (80 °F-70 °F) × 1 = Btu
   2502.00 × 10 × 1 = Btu
   25,020 = Btu
   A one-ton machine would be about 12,000 Btu/hr. Then, 25,020 divided by 12,000 equals approximately two. So this will be approximately a two-ton machine.
3. Measure air temperatures into and out of the condenser. Measure the average air velocity in ft./hr. in or out (smoothest airflow). Take as many as 16 readings at different places across the airflow; then average them. (See Chapter 18.)

Measure the condenser's length and width in inches.
Air temp. rise = air temp. out − air temp. in
Condenser area = in. in width × in. in length =

$$\frac{\text{Area in sq. in.}}{144} = \text{area sq. ft.}$$

Volume of air = area in sq. ft. × average velocity (ft./hr.)
About 14 cu. ft. of air = 1 lb.
Spec. heat of air = .24 Btu/lb. °F
Btu = weight of air/hr. × spec. heat × temp. change

$$= \frac{\text{volume of air/hr.}}{14 \text{ cu. ft./lb.}} \times .24 \text{ Btu/lb. }°F \times (\text{temp. out}$$
minus temp. in)

Example: An air-cooled condenser has an air velocity of 300 ft./min., an air-in temperature of 80 °F and an air-out temperature of 90 °F. The condenser measurements are 25 in. × 40 in. Calculate the Btu/hr. released by the condenser.

Btu = weight of air/hr. × spec. heat × temp. change
Weight of air:
1. Condenser area = 25 × 40 = 1000 sq. in.

$$\frac{1000}{144} = 6.94 \text{ sq. ft.}$$

2. Volume = 6.94 sq. ft. × (300 ft./in. × 60 min./hr.)
   = 6.94 × 18,000
   Volume = 124,920 cu. ft./hr.
   Weight = 124,920 ÷ 14 cu. ft./lb.
   = 8922.86 lb. of air/hr.
   Btu = 8922.86 × .24 × (90 − 80)
   = 8922.86 × .24 × 10
   = 8922.86 × 2.4
   = 21,414.86 Btu (about 1 3/4-ton system)

15-47 COMPRESSOR CAPACITIES

The compressor is the heart of the refrigerating machine. It uses mechanical energy, produced by an electric motor, to pump the refrigerant through the cycle. The refrigerant picks up heat at one place and releases it at another place.

The most efficient construction possible is to have a compressor just large enough to handle the necessary amount of refrigeration. If the compressor is too large, energy is lost in excess friction, starting, etc. If the compressor is too small, it will not produce enough refrigeration.

Basically, the compressor must remove the vapor fast enough from the evaporator to enable the refrigerant to vaporize at the correct low pressure. To do this, it must remove the refrigerant vapor as fast as heat goes into the evaporator and vaporizes the refrigerant.

The method of determining compressor size may be simply stated: an evaporator is usually designed to remove the 24-hour heat load in a 16- or 18-hour running period. The amount of time depends on the factors described in previous paragraphs.

Assume that the effective heat removing ability of the refrigerant is 60 Btu per pound. This means that as each pound of refrigerant vaporizes in the evaporator, it picks up 60 Btu of heat from the evaporator. To remove this much heat, the compressor must handle all the vapor formed. Refrigerant tables give these values called ''specific volumes.'' The specific volume values mean that at certain pressures one pound of refrigerant, as it is vaporizing, will form a certain number of cubic feet of vapor.

Assume, for example, that 1 lb. of R-12 vaporizing at 9.17 psig (167 kPa) and 0 °F forms 1.637 cu. ft. of vapor in 10 minutes. The compressor, then, must remove 1.637 cu. ft. of vapor from the evaporator in the same period.

The size of the compressor needed to do this depends on the volume of vapor pumped per revolution of the compressor. This is determined by the bore, stroke, number of cylinders, speed of the compressor (rpm), and its volumetric efficiency.

As the crankshaft of the compressor completes one revolution, the piston reaches bottom dead center of its travel.

The low-pressure vapor fills up the space between the top of the piston and the head of the cylinder. The piston compresses this vapor on the upstroke and pushes it through the exhaust valve into the high-pressure side of the system. The volume handled in each case is the volume displaced by the piston as it moves from top dead center to bottom dead center. This volume may be calculated by the following formula:

$$V = \frac{\pi D^2}{4} \times S \times N \times R$$

V = Volume in cubic inches
D = Diameter of cylinder in inches
S = Length of stroke in inches
N = Number of cylinders
R = Rpm (revolutions per minute)

This formula for volume simply calculates the area of the piston head, $\frac{\pi D^2}{4}$, and multiplies it by the length of the stroke. The answer is the displacement volume in cubic inches. Next, this figure is multiplied by the number of cylinders, then by the revolutions per minute of the compressor to give the total volume in cubic inches pumped per minute.

Example: How much vapor will a two-cylinder compressor pump if it has a 2 in. bore, a 2 in. stroke, and operates at 400 rpm?

Using the formula:

$$V = \frac{\pi D^2}{4} \times 2 \times 2 \times R$$

$$= \frac{3.1416 \times 4}{4} \times 4 \times 400 = 3.1416 \times 4 \times 400$$
$$= 3.1416 \times 1600 = 5026.56 \text{ cu. in./min.}$$
1728 cu. in. = 1 cu. ft.
The volume in cu. ft. will be

$$V = \frac{5026.56}{1728} = 2.9 \text{ (plus) cu. ft./min.}$$

Compressor problem:
Calculate the bore and stroke of a two-cylinder compressor operating at 1750 rpm, which will compress in 10 minutes the refrigerant vapor formed by vaporizing 10 lb. of R-22 at 5 °F. Note from the table that vaporized R-22 at 5 °F occupies 1.2434 cu. ft./lb.
Solution:
V = 12.434 cu. ft. = 12.434 × 1728 cu. in. ft./lb.
= 21,485.95 cu. in.

Volume pumped per min. $= \frac{21,485.95}{10} =$
2148.6 cu. in./min.
Using the formula:

$$V = \frac{\pi D^2}{4} \times S \times N \times R = 2148.6 \text{ cu. in.}$$

Compressors are usually designed with a bore equal to the stroke (S = D). So the formula becomes:

$$V = \frac{\pi D^3}{4} \times N \times R =$$

$$D^3 = \frac{V \times 4}{\pi \times N \times R}$$

$$D^3 = \frac{2148.6 \times 4}{\pi \times 2 \times 1750} = \frac{2148.6 \times 2}{\pi \times 1750}$$

$$D^3 = \frac{2148.6}{\pi \times 875}$$
D³ = .8 cu. in.
D = .93 in. (approx.)

Therefore, this compressor has a bore of .93 in. and a stroke of .93 in. However, this value is the theoretical amount of vapor pumped by the compressor. The actual amount will be less and will depend on the volumetric efficiency of the compressor. See Para. 15-48.

15-48 VOLUMETRIC EFFICIENCY

Volumetric efficiency is the ratio between the volume actually pumped per revolution, divided by the volume calculated from the bore and stroke.

If the refrigerating unit is maintaining a 150 psig (1 139 kPa) head pressure and a 0 psig (101 kPa) low-side pressure, the following things happen:

1. When the piston is on its upward stroke, it compresses this 0 psi vapor until the pressure of the vapor in the cylinder reaches 150 psi. When this pressure is reached, the vapor should start passing through the exhaust valve into the condenser. However, in addition to reaching high-side pressure, it must overcome exhaust valve spring force or weight. This means a slight additional increase in pressure is required.

2. After the piston reaches top dead center, there is still a little vapor remaining at a high pressure between the piston head and the exhaust valve. The space it occupies is necessary for clearance between the piston and cylinder head. Otherwise, the piston would pound against the cylinder head at top dead center.

   The residue of vapor is under a high pressure (150 psi or more). When the piston goes down to receive a new charge of vapor, this high-pressure vapor expands and partially fills the cylinder chamber. The residue vapor decreases the amount of vapor that may move into the chamber from the low-pressure side of the system. The necessary space, or ''clearance volume,'' varies between 4 and 9 percent of the piston displacement.

3. At speeds of 300 to 3400 rpm or more, the piston is traveling so fast that the inertia or weight of the vapor prevents it from filling the cylinder chamber completely, so there are losses. The pressure in the cylinder may be 2 in. Hg. (95 kPa), instead of 0 psig. The resistance to vapor flow through the valve openings is called ''wire drawing.'' The pressure in the cylinder never gets as

high as the pressure in the suction line during the suction stroke. The higher the speed of the compressor, the less vapor will be pumped per stroke.

4. Just as the exhaust valve offers a restriction to the vapor flow, so does the intake valve with its force and the weight of the valve parts.

5. The compressor runs at a warm temperature. Some of this heat warms the vapor as it enters the cylinder, causing an expansion of the vapor that also keeps a complete load of vapor from entering the cylinder.

6. Other losses, such as the leaking of the vapor past the piston and rings into the crankcase, etc., also explain why compressors cannot pump the amount of vapor calculated by the bore and stroke formula.

For small compressors used in domestic refrigeration (bore and stroke of about 1 1/2 in.), the volumetric efficiency varies between 40 and 75 percent, with 60 being an average value. The larger commercial compressors, depending on their size and speed, have volumetric efficiencies between 50 and 80 percent, with 70 percent being an average value. A value of 60 to 65 percent volumetric efficiency should be used if the unit is air-cooled.

Example: If the compressor described in Para. 15-47 has a volumetric efficiency of 60 percent, the size of the compressor would be increased as follows:

D^3 = .8 (as calculated in Para. 15-47)

D^3 = corrected = $\dfrac{.8}{.60}$ = 1.33 cu. in.

D = cube root of 1.33
= 1.1 × 1.1 × 1.1
= 1.33 in.

D corrected = 1.1 in.

Therefore, a bore and stroke of 1.1 in. × 1.1 in. would be required. Note that the correction for volumetric efficiency was made on the displacement volume of the cylinder, not on the calculated bore and stroke.

The volumetric efficiency of a compressor depends on the difference between its low-side pressure and high-side pressure. For instance, the compressor in an R-12 system used for domestic purposes will be more efficient than if it were converted into an ice cream system. This is because the decrease in low-side pressure from 10 psig (173 kPa) down to 10 in. Hg. (68 kPa) with the same head pressure reduces the actual pumping capacity of the compressor. The low-pressure vapor expands when the cylinder is filled at low-side pressure. Therefore, only a small amount, by weight, is pumped. All of the various items affecting efficiency—like increasing head pressure, increasing speed, using thicker gaskets, and overheating the compressor—will reduce the pumping efficiency of the compressor.

15-49 COEFFICIENT OF PERFORMANCE

Coefficient of performance (cop) is the ratio of output divided by input. In refrigeration work: the output is the amount of heat absorbed by the system; the input is the amount of energy required to produce this output.

The cooling effect in Btu values in a refrigeration cycle compared to the Btu equivalent of the energy put into the system is called the coefficient of performance.

For example, if one pound of refrigerant has an effective latent heat of 50 Btu, and the compressor pumping energy

is equivalent to 10 Btu per pound, the coefficient of performance is 50 to 10 or 5:1.

The heat input by the compressor is less than the electrical energy put into the motor. This is because the motor is not 100 percent efficient, and there are also compressor friction losses. Usually, the overall coefficient of performance will be approximately 60 percent of the theoretical. The actual coefficient, then, is near 3:1 rather than 5:1.

This means, however, that three times the amount of heat would be obtained from a heat pump by using the compressor than by using electricity to produce the heat. This explains the advantage of a heat pump. It also explains why hot gas defrost is used in some large systems. The cost of the extra piping and valves is soon recovered in the savings in cost of defrosting.

15-50 MOTOR SIZES

The size of an electric motor necessary to drive a compressor in a refrigerating machine may be calculated in two different ways:

1. Mean effective pressure method. The horsepower (hp) of the motor may be calculated by determining the hp put into the compressor. This hp is based on the speed of the compressor and the mean effective pressure (mep) of the vapor in the compressor.

2. Heat input method. Motor size may be determined by using the amount of heat added to the vapor in the compressor as being the energy taken out of the motor.

15-51 MOTOR SIZE—MEAN EFFECTIVE PRESSURE METHOD

The mean effective pressure (mep) of the vapor is the median (average) pressure bearing down on the piston head. It is the pressure to be overcome by the electric motor when driving the compressor.

The mep of the vapor is determined by a formula which uses:

1. Low-side pressure.
2. High-side pressure.
3. Ratio of specific heat at constant pressure to specific

heat at constant volume $\dfrac{C_P}{C_V}$ for the refrigerant used.

The formula for mep and solution will be found in Chapter 30, Technical Characteristics.

15-52 MOTOR SIZE—HEAT INPUT METHOD

From the pressure heat charts, one can determine the amount of heat (in Btus) added to a vapor when compressed by the compressor. Referring to the R-12 pressure-heat chart in Chapter 9, note that approximately 10 Btu are added to the one pound of gas if the low-side pressure is 10.81 psig (178 kPa) and the high-side pressure is 93.2 psig (747 kPa).

It is known that 2545.7 Btu per hour is equal to 1 hp and is also equal to 746 watts.

Example: Calculate the hp required to drive the compressor just discussed. Add 10.0 Btu per pound. Suppose 10 lb. of refrigerant are compressed in two minutes.

Solution:

The Btu rate per hour = $\dfrac{100}{2}$ × 60 = 3000 Btu per hour

The hp required (Btu method) $= \frac{3000}{2546} = 1.18$ or, in round numbers, 1.2 hp

hp = 1.2 hp

This is the mechanical equivalent of the heat energy put into the vapor. If the compressor friction were 0, this would be the size of the motor necessary to drive the compressor. However, about 50 percent must be added to this value to allow for compressor friction and motor losses. Therefore, this system would require about:

1.2 × .50 = .6
1.2 + .6 = 1.8 or 2 hp

15-53 MOTOR EFFICIENCY

Theoretically, an electric motor should produce 1 hp of mechanical energy for every 746 watts of electrical energy put into it. That is, a 1-hp electric motor on a 120 V circuit should only consume 6.8 A. Such efficiency, however, is not possible because of bearing friction, magnetic eddies, magnetic air gaps, and the power factor of the motor.

The efficiency of the motor is the mechanical energy delivered at the motor shaft divided by the power input to the motor.

As the size of the motor increases, the efficiency increases. For small domestic motors of approximately 1/6 hp, the efficiency of the motor is only 40 to 60 percent. This is because friction losses and air gap losses are constant, even though the size of the motor increases. Large motors have an efficiency of 90 to 95 percent.

Example: The motor example used in Para. 15-52.

Solution: With 1.8 hp needed, the electrical input to the motor would need to be more. If 1.8 hp is 75 percent of motor input, the motor input would be:

Total input × .75 = 1.8 hp

Total input $= \frac{1.8}{.75} = 2.4$ hp

2.4 × 746 watts/hp = 1790 watts input

15-54 CONDENSER CAPACITIES

The calculation of the heat transfer capacity of a condenser is similar in many ways to the problem of figuring the capacity of an evaporator. The condenser must remove the heat from the vapor fast enough so that just as much vapor condenses in the condenser as is being pumped into it by the compressor in a given unit of time. When this condition is reached, the head pressure will have built up until the temperature rises to the point where the heat removed will equal the heat put into the condenser.

The problem of figuring the capacity of the condenser varies according to the type of condenser being used. Condensers may be divided under the following headings:

1. Air-cooled:
   a. Plain tubing.
   b. Finned tubing:
      Natural convection.
      Forced convection.
2. Water-cooled:
   a. Tube and shell type.
   b. Pipe and shell type.
   c. Tube-within-a-tube type.

The methods of calculating condenser capacities are explained in Para. 15-55 and Para. 15-56.

15-55 AIR-COOLED CONDENSER CAPACITIES

The capacity of an air-cooled condenser may be calculated by:
1. Using the total external area of the condenser to compute its heat dissipating ability.
2. Basing computations upon the frontal area of the condenser.

Using the total external area of the condenser for dissipating heat depends upon the following variables:
1. External area.
2. Temperature difference.
3. Time.
4. Air velocity.

Using these values, the capacity of an air-cooled condenser varies between 1 and 4 Btu per sq. ft. per °F per hour. The effect of air velocity is to increase the condenser's capacity as the air speed is increased. The fans drive air through the condenser at speeds between 400 to 1000 ft. per minute. When an air speed of 400 ft. per minute is used, a 2.5 Btu per square foot per °F per hour value will be found satisfactory. This value will increase up to approximately 4 Btu with a 1000 ft. per minute air velocity. The calculation of the area of the condenser is the same as that for a finned evaporator. (See Para. 15-21.)

Example: If a condenser has 60 sq. ft. of surface with a heat removal of 2.5 Btu/sq. ft./hr./°F, what refrigerant temperature is necessary to dissipate 5000 Btu per hour if the room temperature is 75 °F?

Formula:
Area × Btu/sq. ft./hr./°F × temp. diff. = Btu/hr.
60 × 2.5 × temp. diff. = 5000
150 × temp. diff. = 5000

temp. diff. $= \frac{5000}{150}$

temp. diff. = 33.3 °F.

Assuming an ambient temperature of 75 °F, the refrigerant temperature = 33.3 + 75 = 108.3 °F.

The same problem using a 75 sq. ft. condenser:
75 × 2.5 × temp. diff. = 5000

temp. diff. $= \frac{5000}{187.5}$

temp. diff. = 26.7 °F.

Therefore:
The refrigerant temperature = 26.7 + 75 = 101.7 °F.

The heat to be removed by the condenser for each pound of vapor is the heat content of the vapor as it leaves the compressor, minus the heat of the liquid at the condensing pressure.

If a condenser is under-capacity, the compressor head pressure will rise in proportion in order to dissipate the required amount of heat. Therefore, condensers of various sizes can be used with the same compressor. If too small a condenser is used, it will result in a decrease in compressor efficiency, an increase in motor load and a decrease in the life of the unit. The examples given illustrate this principle.

When the capacity of the condenser is based upon frontal area, it is claimed that the air being blown through the condenser is removing heat only from the surface which it strikes directly. Also, the turbulent flow against the rear surfaces reportedly makes the heat removal from these surfaces negligible.

A single row tube condenser has a total area 20 times its frontal area. The capacity per sq. ft. of frontal area naturally is greater than the value stated. It is between 6 to 10 Btu per sq. ft. per °F per hour, depending on the air speed. For air cooling, the dry bulb temperature of the room should be used in the calculation.

15-56 WATER-COOLED CONDENSER CAPACITIES

The capacity of a water-cooled condenser is high because of the good thermal contact between the cooling medium and the refrigerant. Different types of water-cooled condensers are in common use. See Fig. 15-61. Capacity will vary with the type used.

The heat transfer varies directly with the amount of water passed through the condenser. If the water flow is fast, more heat will be removed; if the flow is slow, heat removal will be less. At 50 ft. per minute, water will remove about 185 Btu/sq. ft./hr./°F. At 200 ft. per minute, the water will remove about 330 Btu/sq. ft./hr./°F. The heat-removing capacity of these units varies between 30 and 50 Btu per sq. ft. per °F per hour in the smaller machines. For machines of one-ton capacity or more, this value may be increased up to 90 Btu per sq. ft. per °F per hour or 330 + 90 = 420 Btu/sq. ft./°F/hr.

In addition to heat removal by water, the air-cooling surface of the condenser must also be calculated to reach the correct capacity. Include such things as external area of the shell, or the external area of the refrigerant tubing in a tube-within-a-tube type.

To determine the temperature difference between the cooling medium (water) and the refrigerant, use the following factors: For refrigerant temperature, use the saturation temperature of the refrigerant at the existing head pressure. For water temperature, take the average between the water-in and water-out temperature. Example: A shell and tube type water-cooled condenser is required to remove 5000 Btu per hour. How much tubing 3/8 in. OD must be put into the receiver to remove this heat if the water supply is 70 °F and the outlet water is 80 °F? How many gallons of water per hour must be circulated?

Consider the refrigerant temperature at 100 °F. Assume the heat removing capacity of the condenser to be 40 Btu per sq. ft. per °F per hour.

Formula:

Tube area = condenser capacity = area (sq. ft.) × temp. diff. deg. F × Btu rate × time

Area = circumference × length

Area = π D × length

Area in square feet = $\dfrac{\pi D \times \text{length (in.)}}{144}$

Cooling towers are a very efficient type of water-cooled condenser. One pound of evaporating water removes about 1050 Btu and cools the remaining water. Ideally, the evaporating water will cool it to the wet bulb temperature. See Chapter 18. But, practically, it cools the remaining water to some temperature above the wet bulb temperature. This can be measured with a thermometer.

About 3 percent of the water is evaporated and must be made up by using a float valve water feed as a control.

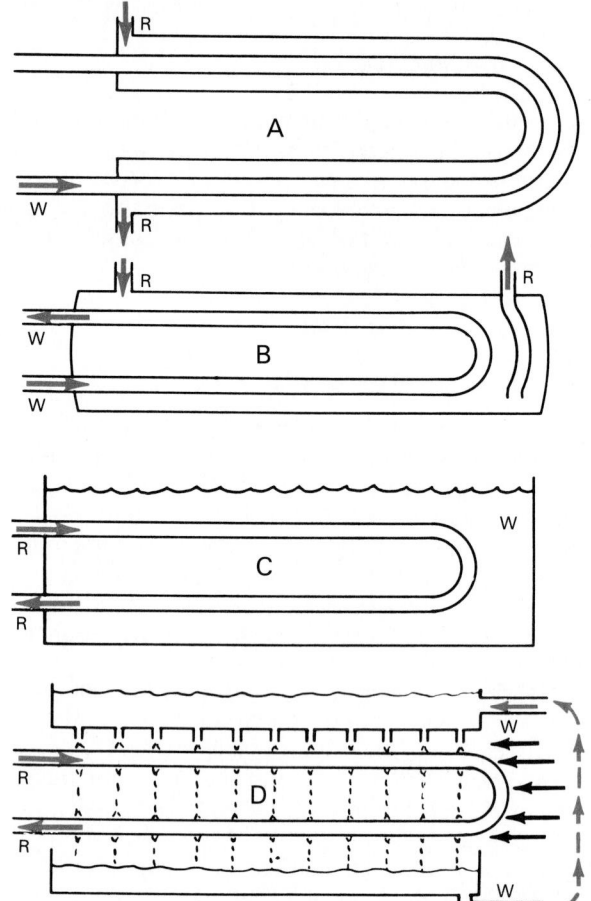

Fig. 15-61. Water-cooled heat exchangers. A—Tube-within-a-tube. B—Shell and tube. C—Tank. D—Baudelot. R—Refrigerant. W—Water.

The float valve also makes up for run-off water which is also about 3 percent (used to keep water chemicals to a minimum). The water pump used to circulate this water should be about 1 percent of the condensing unit horsepower, or about 2 percent if the water pipes are long (high total head).

15-57 LIQUID RECEIVER SIZES

Liquid receivers for a commercial system should be 15 percent larger than all the liquid volume in the system. This

RECOMMENDED LIQUID RECEIVER VOLUMES					
		WEIGHT — LB.			
HP	VOLUME CU. IN.	REFRIGERANT			
		R-12	R-22	R-500	R-502
1/2	150	6.8	6.2	5.9	6.3
3/4	225	10.3	9.3	8.9	9.4
1	300	13.7	12.4	11.9	12.9
1 1/2	450	20.5	18.6	17.9	19.3
2	600	27.4	24.8	23.8	25.8
3	750	35	32	31.8	33.0
5	900	41	37	35.5	38.5
7 1/2	1500	70	64	61.6	66.0

Fig. 15-62. Minimum net recommended liquid receiver volume for four common refrigerants: R-12, R-22, R-500, and R-502. (ASHRAE Guide and Data Book)

practice is recommended for service operations and as a safety measure should the refrigerant circuit become restricted. (The restriction could be a clogged filter or a clogged screen).

Fig. 15-62 shows recommended minimum sizes of receivers based on horsepower capacity of the system. The

receivers may have to be larger than this, depending on the refrigerant, piping lengths, and other factors.

Liquid receivers are a service addition to a system. The systems would operate efficiently without them, but practical problems of refrigerant reserve and a convenient service storage makes them a common part of most systems.

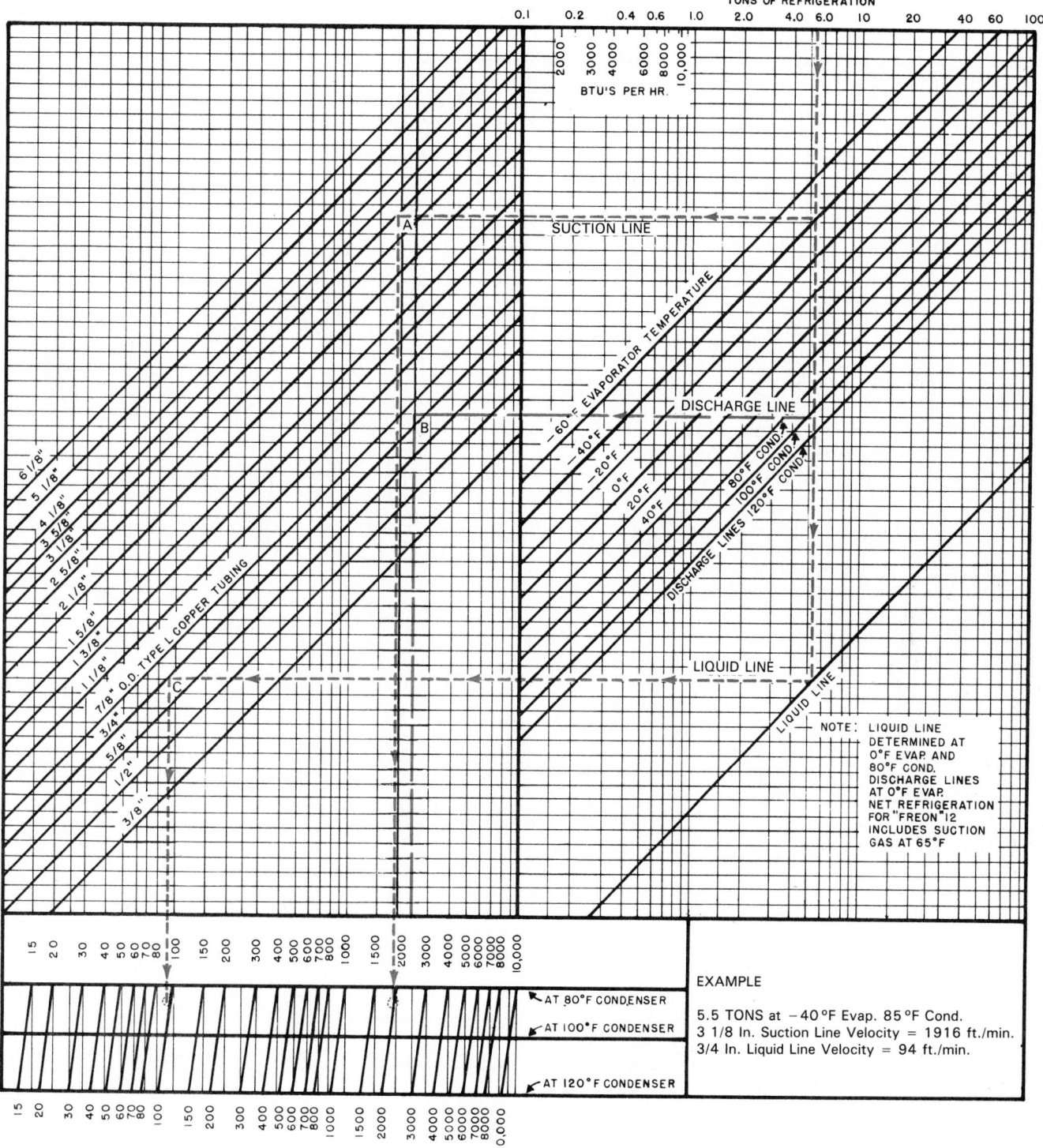

Fig. 15-63. Chart shows liquid and suction line sizes for R-12 systems from 0.1-ton to 100-ton capacity.
(Du Pont Company)

15-58 REFRIGERANT LINES AND PIPING

The liquid lines, suction lines, compressor discharge lines, and "hot gas" lines on refrigerating machines must be large enough to handle the amount of the liquid or vapor required. To calculate the capacities of these lines, first determine the maximum velocity allowed in the line. Then, knowing the amount of liquid or vapor to be handled, the internal cross section of the line may be calculated.

Generally speaking, for R-12, R-22, R-500, and R-502, liquid velocities are about 100 fpm; suction lines have about 1500 fpm, discharge lines about 3000 fpm, and hot gas lines about 3000 fpm.

Fig. 15-63 is a graph of refrigerant line capacities for

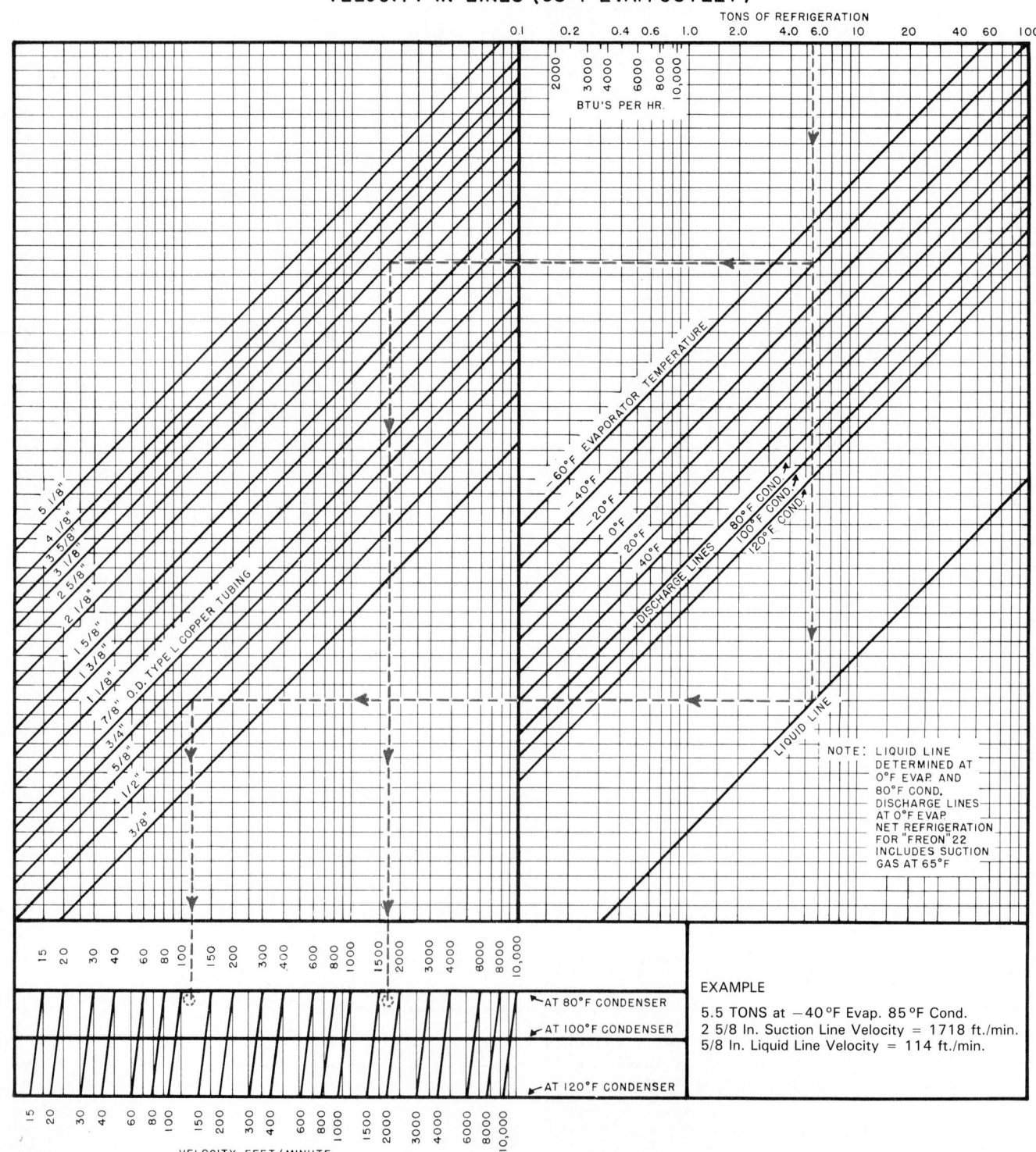

"FREON" 22 REFRIGERANT
VELOCITY IN LINES (65°F EVAP. OUTLET)

EXAMPLE

5.5 TONS at −40°F Evap. 85°F Cond.
2 5/8 In. Suction Line Velocity = 1718 ft./min.
5/8 In. Liquid Line Velocity = 114 ft./min.

NOTE: LIQUID LINE DETERMINED AT 0°F EVAP. AND 80°F COND. DISCHARGE LINES AT 0°F EVAP. NET REFRIGERATION FOR "FREON" 22 INCLUDES SUCTION GAS AT 65°F

Fig. 15-64. Chart used to find liquid and suction line sizes for systems using R-22.
(Du Pont Company)

R-12. For a 6-ton unit, suction line diameter, with a −40 °F (−40 °C) evaporating temperature and a velocity of 2000 fpm, is 3/8 in.; liquid line diameter is 3/4 in.; discharge line diameter is 7/8 in.

Fig. 15-64 is for R-22. Fig. 15-65 is for R-500.

15-59 REFRIGERANT (LIQUID) LINE CAPACITIES

Velocities in the liquid line of a refrigerating unit vary with the density of the liquid and with its viscosity. These

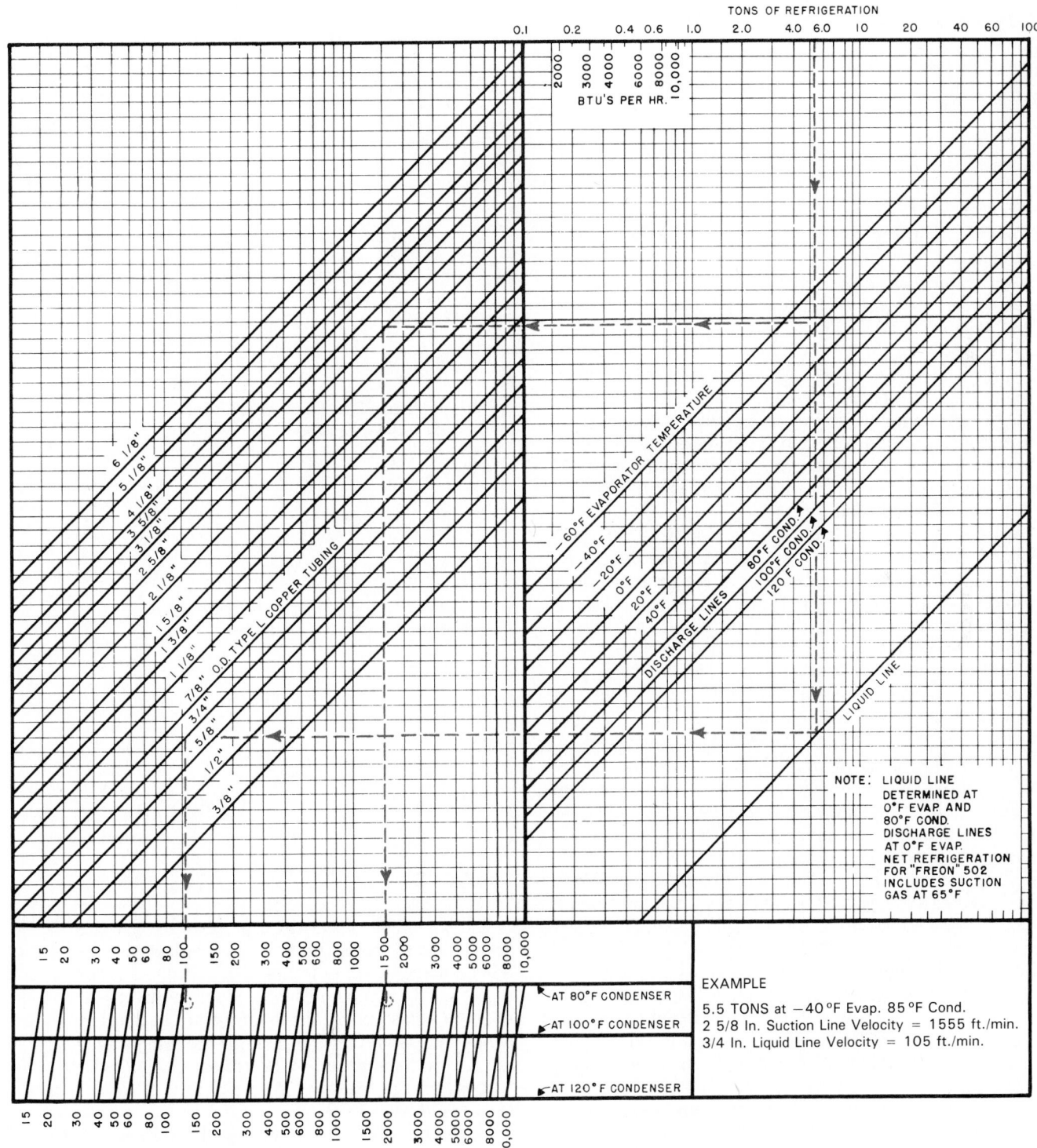

Fig. 15-65. Chart for finding liquid and suction line sizes for systems using R-502.
(Du Pont Company)

velocities may vary between 50 and 200 ft. per minute (fpm), depending on the refrigerant used. (R-12 should have velocities no greater than 100 fpm.)

Example: If 75 cu. in. of liquid were used a minute, the internal cross-sectional area of the liquid line to keep line velocity at 100 ft. per minute or below would be:

$$\text{Cross-sectional area} = \frac{\text{volume in cu. in./min.}}{\text{velocity in in./min.}}$$

$$= \frac{75}{100 \times 12}$$

Cross-sectional area = .063 sq. in.

$$\text{Area} = \frac{\pi D^2}{4} \quad D^2 = \frac{\text{area} \times 4}{\pi} \quad D = \sqrt{\frac{\text{area} \times 4}{\pi}}$$

$$\text{The inside diameter} = \sqrt{\frac{.063 \times 4}{\pi}} = .28 \text{ in.}$$

Use 3/8-in. OD tubing (ID = .307).

It is important that refrigerant-carrying lines have sufficient capacity. The cost of increasing the tubing size is so small in comparison to the total cost of the machine that there is no real necessity for calculating tubing size to close limits. At least a 10 to 20 percent oversize liquid line is recommended.

If the liquid line is too small, or has too many restrictions, pressure drop may reduce refrigerant flow at the capacity of the refrigerant control below the capacity of the evaporator. Extremes of this condition are revealed by sweating or frosting liquid lines when excessive pressure drops occur. These drops are caused by partially clogged driers and strainers or pinched lines.

The pressure drop in a liquid line carrying R-12 is shown in Fig. 15-66. Note that if 1/4-in. OD liquid line were used for a 12,000 Btu/hr. or one-ton load, the pressure drop would be 0.42 psi per ft. (9.5 kPa/m). A 100-ft. equivalent length liquid line would then have a total pressure drop of 42 psi (290 kPa). A 10 m line has a drop of 95 kPa. (Equivalent length is the actual length of the piping, plus the pressure drop in the bends and fittings, as expressed in feet.) Looking ahead, Fig. 15-73 lists values that should be added to the tube length for each fitting or valve.

If normal head pressure were 125 psig (966 kPa), the pressure in the liquid line near the thermostatic expansion valve would be 125 minus 42 or 83 psig (676 kPa). At 83 psi, the boiling temperature is 79°F (26°C) and sweating might occur in a humid 90°F (32°C) room.

In large systems, bear in mind how much refrigerant is stored in the liquid line. This amount also affects pressure based on the weight of the liquid (static head), as shown in Fig. 15-67.

Bends and fittings increase the resistance to the fluid flow. It is claimed that there is as much friction to flow in a 90-deg. elbow as there is in five ft. of straight tubing of the same size. The friction bends, fittings, and normal fluid flow through the tubing must be calculated when figuring fluid velocities.

In multiple installations that use series-connected evaporators, Fig. 15-68, the liquid line usually varies in diameter as the number of evaporators it feeds changes. Note in the diagram that two 1/2-in. OD liquid lines do not feed from a 1-in. OD line. Instead, the cross-sectional areas

Load Btu/hr.	\frac{1}{4}"	\frac{3}{8}"	\frac{1}{2}"	\frac{5}{8}"	\frac{3}{4}"	\frac{7}{8}"	1\frac{1}{8}"
3,000	.035						
6,000	.120	.011					
9,000	.250	.021					
12,000	.420	.036					
18,000		.075	.010				
24,000		.127	.016				
36,000		.260	.033	.012			
48,000		.450	.054	.020	.010		
60,000			.080	.030	.014	.009	
84,000			.150	.054	.025	.015	
120,000			.280	.100	.049	.028	.009
240,000				.350	.160	.095	.029
360,000					.340	.200	.058
480,000						.340	.100

Fig. 15-66. Pressure drop in an R-12 liquid line in psi per foot of tube. Table is based on size of tube and load in Btu/hr. Total pressure drop will be indicated pressure drop per foot multiplied by length in feet.

\frac{1}{4}"	\frac{3}{8}"	\frac{1}{2}"	\frac{5}{8}"	\frac{3}{4}"	\frac{7}{8}"	1\frac{1}{8}"
.015	.043	.086	.134	.202	.269	.458

Fig. 15-67. Refrigerant charge in pounds per foot of liquid line. (Dunham-Bush, Inc.)

are added: Line 1 area + line 2 area = line 3 area. The wall thickness is ignored.

$$\frac{\pi D_1^2}{4} + \frac{\pi D_2^2}{4} = \frac{\pi D_3^2}{4}$$

$$D_1^2 + D_2^2 = D_3^2$$

$$\sqrt{D_1^2 + D_2^2} = D_3$$

$$\sqrt{\left(\frac{1}{2}\right)^2 + \left(\frac{1}{2}\right)^2} = D_3$$

$$\sqrt{\frac{1}{4} + \frac{1}{4}} = D_3$$

$$\sqrt{\frac{2}{4}} = D_3$$

$$\sqrt{\frac{1}{2}} = D_3$$

$$\sqrt{.5} = D_3$$

$$.70 = D_3$$

The liquid line presents no problem except size, unless it has a considerable static head (vertical run). In case of a large static head, pressure at the refrigerant control end of the liquid line must be high enough to maintain pressures above the flash point of the refrigerant at the temperature at the refrigerant control. Due to the weight of the liquid refrigerant, the pressure in a vertical liquid line will drop. The amount of the pressure drop per foot of vertical rise is shown in Fig. 15-69.

Example: If a 50-ft. rise in a 3/4-in. R-12 liquid line is needed, the pressure drop will be .56 × 50 = 28 lb. If pressure at the condensing unit is 100 lb., pressure at the 50-ft. level will be 100 minus 28 = 72 lb. If the

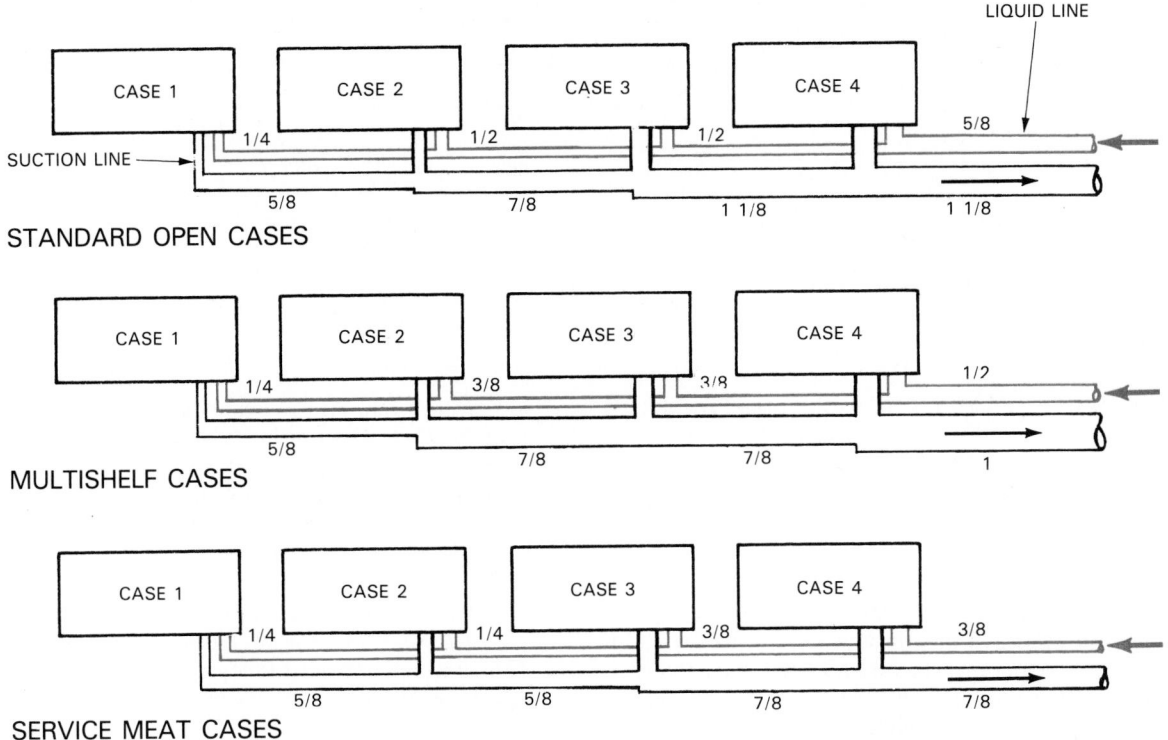

Fig. 15-68. Liquid line sizes. The size of liquid line must increase as the number of evaporators increases. Line sizes shown are for normal-temperature refrigeration installations. Note change in diameter of liquid lines.

liquid line temperature goes over 87 °F, some flash gas may form in the liquid line. (See R-12 tables in Chapter 9.) It is best to arrange the liquid line piping in multiple evaporator installations to make any flash gas formed go to the nearest evaporator. See Fig. 15-70. This design prevents one evaporator from receiving all the flash gas.

15-60 REFRIGERANT (SUCTION) LINE CAPACITIES

Suction line velocities usually are between 1500 and 2000 feet per minute (fpm), as shown in Fig. 15-71. As the refrigerant evaporates in the evaporator, the vapor will go down the suction line if its pressure is somewhat higher than the vapor pressure at the suction side of the compressor. There must be a pressure difference before a fluid or vapor will flow. A compound gauge mounted on the compressor will show that the pressure is lower at the compressor than it is in the evaporator while the unit is running. Just as the compressor stops, pressure at the suction side will rise to evaporator pressure. The rise at the

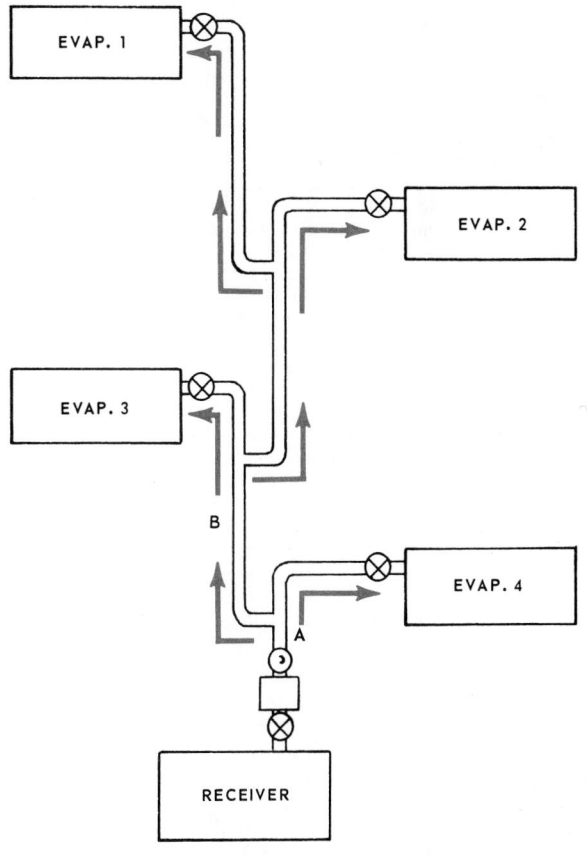

Fig. 15-70. Multiple evaporator liquid line installation. This design distributes flash gas to the nearest evaporator rather than to the most remote evaporator. Example: Flash gas at A will go to Evaporator No. 4. Flash gas at B will go to Evaporator No. 3, etc.

PRESSURE DROP IN VERTICAL LINES	
REFRIGERANT	PRESSURE DROP PER FOOT OF VERTICAL COLUMN
R-12	.56
R-22	.50
R-500	.50
R-502	.50

Fig. 15-69. Pressure drop in vertical liquid refrigerant line. This table is for refrigerant at 100 °F (38 °C). At lower temperatures, pressure drop will be slightly more.

	ALLOWABLE VELOCITIES IN FEET PER MINUTE		
REFRIGERANT	LIQUID LINE VEL. (LIQUID)	SUCTION LINE (VAPOR)	CONDENSER VEL. (VAPOR)
Ammonia	100–250	4000–5000	5000–6000
R-12	80–100	1500–1800	1800–2250
R-22	100–125	1500–2000	1800–2200
R-500	100–125	1500–2000	1800–2200
R-502	100–125	1500–2000	1800–2200
Water	100–250	30–50 (Liquid)	. . . .

Fig. 15-71. Table of allowable refrigerant velocities in feet per minute. Note that there is very little difference in allowable velocities of the different refrigerants.

gauge indicates the pressure difference in the low side of the system.

To understand why there is a pressure drop in the suction line, consider that:

1. The oil must be returned to the compressor by means of the flowing vapor in the suction line.
2. When two or more evaporators are connected to one compressor, resistance to flow in the different lines may cause the pressure on the surface of the refrigerant in the different evaporators to vary as much as 2 to 3 psi.

Fig. 15-72 shows suction and liquid line systems for multiple refrigerating systems with four cases. The friction in a 1/2-in. OD suction line is approximately 0.25 in. Hg per 10 ft. of length (0.2777 kPa per meter) with a gas velocity of 1000 ft. per minute. The varying pressure drops pose the problem of balancing evaporator capacities.

Total pressure drop becomes greater:

1. As the length of the suction line increases.
2. As the cross-sectional area of the tubing decreases.
3. As the number of bends and fittings in the suction line increases.

Fig. 15-73 shows the values to be added to the actual length of the pipe in order to get the equivalent length.

The correct size of suction lines is important in refrigerating units. An excessive pressure drop in a suction line is similar to operating the compressor at a lower pressure, thereby reducing its capacity.

Example: In an R-12 system, a pressure drop of 2 psi (14 kPa) at −15 °F (2.45 psig, 118 kPa) is the same as trying to work at −20 °F (0.7 psig, 106 kPa), and the condensing unit capacity is lowered about 11 percent.
(2.45 + 14.696) − (0.653 + 14.696) =
17.146 − 15.349 = 1.797
1.797 ÷ 17.146 = 0.105 or 11 percent

The oil return should be taken care of by sloping the suction line consistently downward from the evaporator to the compressor to permit the oil to drain naturally into the compressor. If a low spot is built into the suction line, the oil will accumulate there. Fig. 15-74 shows how this oil buildup will decrease the cross-sectional area of the tubing, causing an orifice action that decreases the efficiency of the gas flow. Furthermore, when this low spot eventually

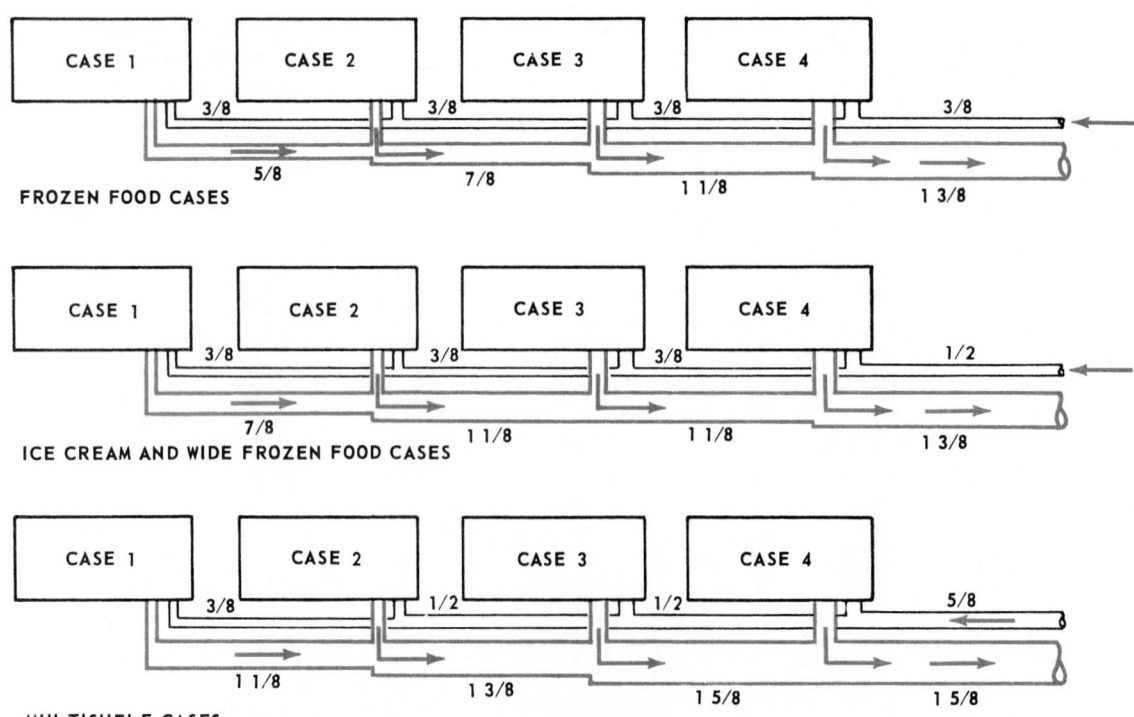

Fig. 15-72. Three typical supermarket piping diagrams for low-temperature fixtures. Note how size of suction lines increases as vapor accumulates from each case!

FITTING	TUBE AND PIPE SIZE, OD													
	¼	⅜	½	⅝	¾	⅞	1⅛	1⅜	1⅝	2⅛	2⅝	3⅛	3⅝	4⅛
	FEET TO BE ADDED FOR EACH FITTING													
Valve	1.5	1.5	2	2	2.5	3.0	4.0	5.0	6.0	7.5	9.0	11.0	13.0	15.0
Elbow (90°)	.75	.75	1	1	1.5	1.5	2	2.5	3	4.0	5.0	5.5	6.5	7.5
Tee	1.5	1.5	2	2	2.5	3.0	4.0	5.0	6.0	7.5	9.0	11.0	13.0	15.0

Fig. 15-73. Table for amount of resistance of valves, elbows, and tees over straight lengths of pipe. Values of each fitting shown should be added to length of pipe in order to obtain equivalent length. Note that valve has as much resistance as 2 ft. of 1/2 in. pipe.
(Dunham-Bush, Inc.)

becomes filled with oil, the pressure difference builds up and oil is slugged into the compressor. Slugging of the oil into the crankcase of the compressor accelerates oil pumping momentarily and may damage the compressor.

Fig. 15-75 shows the capacity of various sizes of suction lines using R-12 refrigerant. Note that a 1-in. nominal

Fig. 15-74. How vapor flow may become restricted as oil collects in a low spot in a suction line. A—Suction line. B—Refrigerant vapor. C—Oil.

or 1 1/8-in. OD suction line can carry from 1/2 to 3 tons of capacity, depending on the pressure drop. However, the best choice would probably be between .02 and .03 psi pressure drop per foot, and this pipe should be used for 2- to 2 1/2-ton units. If the suction temperatures are lower or higher than 20 °F, the pressure drops must be corrected. See Fig. 15-76. Denser vapor should have lower velocities, and vice versa.

Fig. 15-77 shows a method of finding suction line, discharge line, and liquid line sizes.

The capacity of the installation is usually known in Btu/hour or in tons of refrigeration. Correct suction line size can be estimated by first getting an approximate size from Fig. 15-75. Next, calculate the equivalent length of pipe from Fig. 15-73. Then correct for temperature using the factors given in Fig. 15-76.

Example: To determine suction line size for a 5-ton system at 0 °F, allow for a total 2 psi pressure drop. The suction

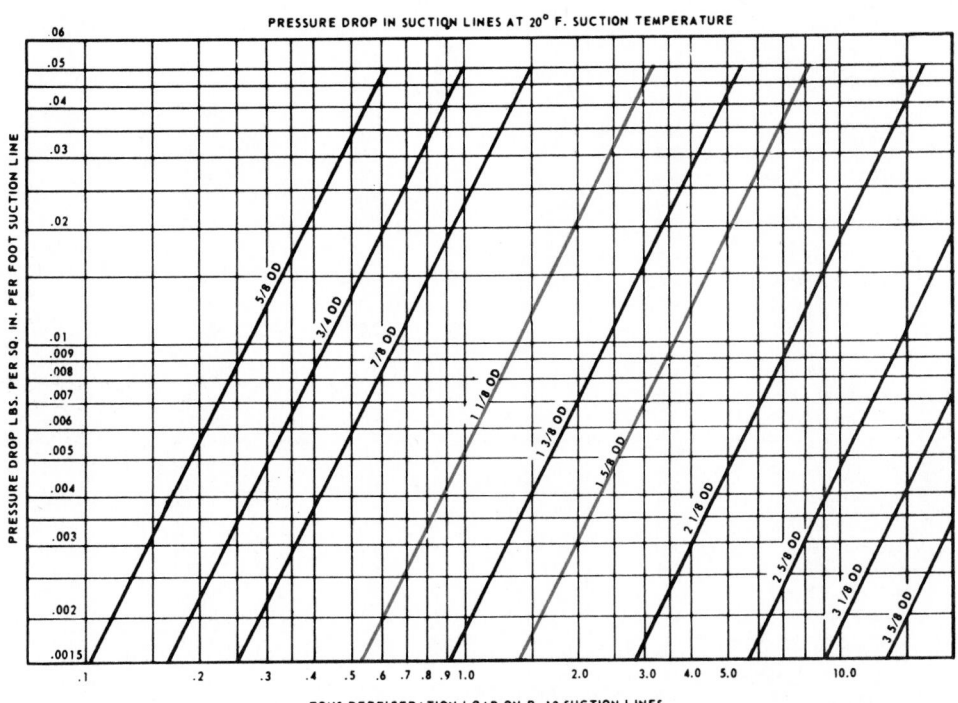

Fig. 15-75. Graph of suction line capacities for R-12 refrigerant. Choose suction line size by using a pressure drop of .02 to .03 psi per foot. Chart is based on 20 °F. If lower temperatures are used, larger suction lines are needed, and vice versa.

CORRECTION FACTORS FOR OTHER SUCTION TEMPERATURES										
Suction Temp	—40	—30	—20	—10	0	10	20	30	40	50
Correction Factor	2.70	2.28	1.90	1.62	1.38	1.18	1.00	0.88	0.75	0.64

Fig. 15-76. Table of correction values for pressure drops in a suction line. If suction temperatures exceed 20 °F, pressure drop is decreased. If temperatures are 0 °F, pressure drop increases by 1.38. Equivalent length is to be multiplied by correction factor.

line has 30 ft. of straight run, six 90 deg. elbows, one tee, and one valve, at 0 °F. Assume that 1 1/8-in. OD suction line is to be used:

Suction line	30 ft.
6 elbows × 2	12 ft.
1 tee × 4	4 ft.
1 valve × 4	4 ft.
The equivalent length	50 ft.

If a total of 2 psi pressure drop is desired, then 2 ÷ 50 = .04 psi per foot.

However, this line operates at 0 °F instead of 20 °F. Therefore, .04 ÷ 1.38 = .029 psi per foot of length.

Now, referring to Fig. 15-75, a 5-ton load with a .029 psi per foot pressure drop needs a 1 5/8-in. OD pipe for the suction line.

It is important to always use piping as large as the fittings on the evaporator and on the compressor. If a compressor suction line connection is 1-in. OD, it is advisable to use this size piping. If the liquid receiver liquid line connection is 1/2-in. OD, use this size.

The oil return to the compressor by way of the suction line is a critical problem in refrigeration systems.
1. Returning oil is a mixture of oil and refrigerant.
2. The mixture thickens as the temperature drops.
3. The mixture travels mainly along the inside wall of the tubing or piping.
4. The mixture travels by gravity and by the action (velocity) of the refrigerant vapor.
Observe these cautions:
1. Keep the mixture as fluid as possible.
2. Slope the horizontal suction line downward toward the compressor.
3. It is important to leave enough refrigerant vapor velocity to push the mixture along the pipe.

Many experiments have shown that the velocity in a horizontal line should be at least 500 to 750 fpm. If the refrigerant vapor must flow up a suction line, the velocity in the vertical tubing or pipe must be at least 1000 to 1500 fpm. (The vapor velocity must overcome both gravity and viscosity.)

	RECOMMENDED REFRIGERANT LINE SIZES					
	COMPRESSOR CAPACITY BTU/HR.	LENGTH OF RUN				
		15 FT.	25 FT.	35 FT.	50 FT.*	100 FT.*
		TUBE DIA.	TUBE DIA.	TUBE DIA.	TUBE DIA.	TUBE DIA.
SUCTION LINE	18,500—20,000	5/8	5/8	5/8		
	20,000—22,000	5/8	5/8	5/8		
	22,000—24,000	5/8	5/8	5/8	3/4	3/4
	24,000—34,000	5/8	5/8	3/4	3/4	3/4
	38,000—40,000	3/4	3/4	3/4	7/8	7/8
	40,000—44,000	3/4	7/8	7/8	7/8	7/8
	44,000—51,000	7/8	7/8	7/8	7/8	7/8
	53,000—66,000	7/8	7/8	7/8	1 1/8	1 1/8
LIQUID LINE	18,500—20,000	5/16	5/16	5/16		
	20,000—22,000	5/16	5/16	5/16		
	22,000—24,000	5/16	3/8	3/8	3/8	3/8
	24,000—34,000	5/16	3/8	3/8	3/8	3/8
	38,000—40,000	5/16	3/8	3/8	3/8	3/8
	40,000—44,000	3/8	3/8	3/8	3/8	3/8
	44,000—51,000	3/8	3/8	3/8	3/8	3/8
	53,000—66,000	1/2	1/2	1/2	1/2	1/2
DISCHARGE LINE	18,500—20,000	5/16	3/8	3/8		
	20,000—22,000	3/8	3/8	3/8		
	22,000—24,000	3/8	3/8	3/8	1/2	1/2
	24,000—34,000	3/8	3/8	1/2	1/2	1/2
	38,000—40,000	3/8	1/2	1/2	1/2	1/2
	40,000—44,000	3/8	1/2	1/2	1/2	1/2
	44,000—51,000	3/8	1/2	1/2	1/2	5/8
	53,000—66,000	1/2	1/2	5/8	5/8	3/4

(1) These recommendations are based on the use of standard refrigeration tubing with .028 or .032 wall thickness.
(2) Line sizes listed are outside tube dimensions.
(3) These suggestions do not include consideration for additional pressure drop due to elbows, valves or reduced joint sizes.
  * = Add 3 fluid ounces for each 10 ft. of pipe over 35 ft.

Fig. 15-77. Suction line, discharge line, and liquid line sizes. Selection is based on system capacity, length of pipe, and use of R-12.

Be sure to install small U traps at the base of the vertical up-flow suction lines. (Small traps prevent a large slug of oil returning to the compressor during start-up of system.)

The viscosity of refrigerant oil determines how easily it flows. See Chapter 30.

Oil containing dissolved refrigerant has a lower viscosity (flows easier). As the oil travels in the suction line, it becomes warmer (suction superheat) but it also loses some of its dissolved refrigerant. Tests show that the viscosity of the oil actually increases as it travels through the suction line toward the compressor. This means that the longer the suction line, the more careful the technician must be to provide proper suction line velocities and oil traps.

In low-temperature systems, the refrigerant dissolved in the oil is the one main factor that keeps the viscosity low enough to allow the return of oil.

Example: 150 USU oil at −20°F has 100,000 USU viscosity. With R-12 dissolved in it, however, 150 USU oil has a viscosity of about 50 USU and it flows with relative ease. As temperatures in the suction line rise, viscosity of the oil first increases (as refrigerant content decreases), then it finally starts to decrease (as the oil becomes warmer).

Refrigerant vapor velocity will vary in the suction line as the heat load changes. At maximum heat load, the amount of vapor produced will be maximum; vapor velocities will be high and the oil return will be good. But, as the vapor volume decreases and the compressor unloads (one· or more compressors stop if it is a compound system or one or more cylinders of a modulated compressor stop pumping), vapor velocity will drop and oil return will be more difficult.

One solution to suction line problems under varying heat loads is to use a double suction line, one with an oil trap and one without. See Fig. 15-78. When the system is at full capacity, suction lines A and B will carry the refrigerant vapor at about 1500 ft./min. As load decreases and compressor pumping is reduced, the vapor velocity will slow and the oil trap, C, will fill with oil, closing line B. Now the vapor velocity in suction line A will be high enough to carry the oil back to the compressor. When the system returns to full capacity, the oil at C will be moved back to the compressor by way of line B.

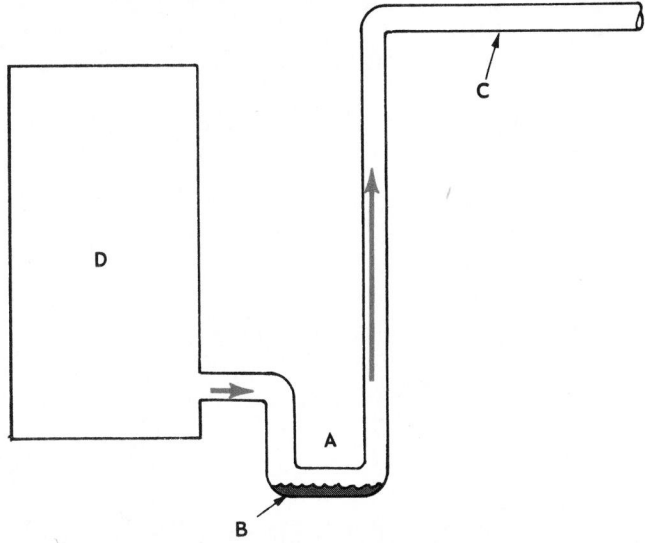

Fig. 15-79. On vertical rises of a suction line, an oil trap must be installed at the low point of the rise so oil cannot enter compressor. A—Oil trap. Note that it is below the compressor suction line outlet. B—Collected oil. C—Horizontal pipe must slope toward the compressor about 1/4 in. for every 10 ft. of pipe. D—Evaporator.

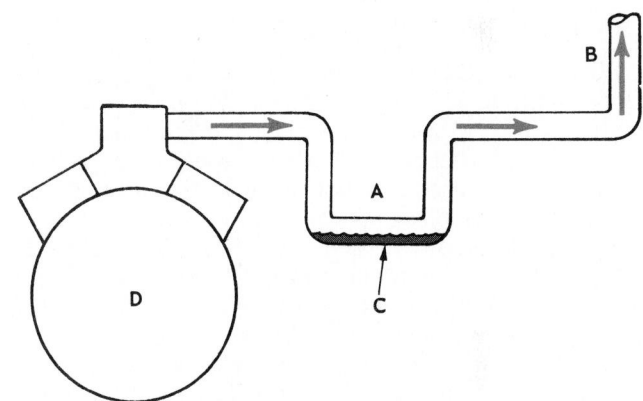

Fig. 15-80. An oil trap installed in the discharge line of a compressor having a remote condenser keeps oil from draining back to exhaust valves of the compressor. A—Oil trap. B—Discharge line. C—Oil collects here during off cycle. D—Compressor.

Remember, all horizontal suction lines must slope toward the compressor at about 1/4 in. in every 10 ft. If the suction line rises (leaves the compressor vertically for a distance), a trap must be installed at the bottom of the rise. See Fig. 15-79.

Discharge lines rising from compressors to remote condensers must also have an oil trap if vertical rise is 8 ft. or more. The trap keeps oil from returning by gravity flow to fill the space above the exhaust valves of the compressor during the off cycle. This condition could damage the valves when the compressor first starts up. See Fig. 15-80. A horizontal discharge line must slope toward the condenser about 1/4 in. in every 10 ft. length of pipe.

Systems using compound compressors have an oil balance to maintain. Otherwise, one compressor may collect too much and, then, will pump oil. Another compressor, too low on oil, may be damaged. Fig. 15-81 shows oil piping which provides equal distribution of oil.

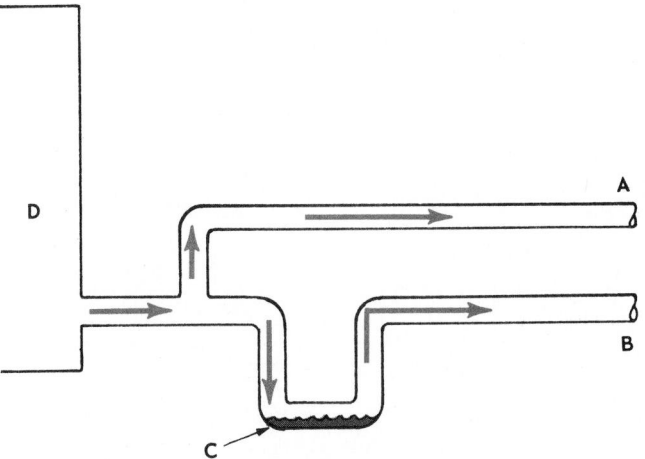

Fig. 15-78. Double suction line. A—Suction line direct to compressor. B—Suction line with oil trap. C—Oil trap. D—Evaporator.

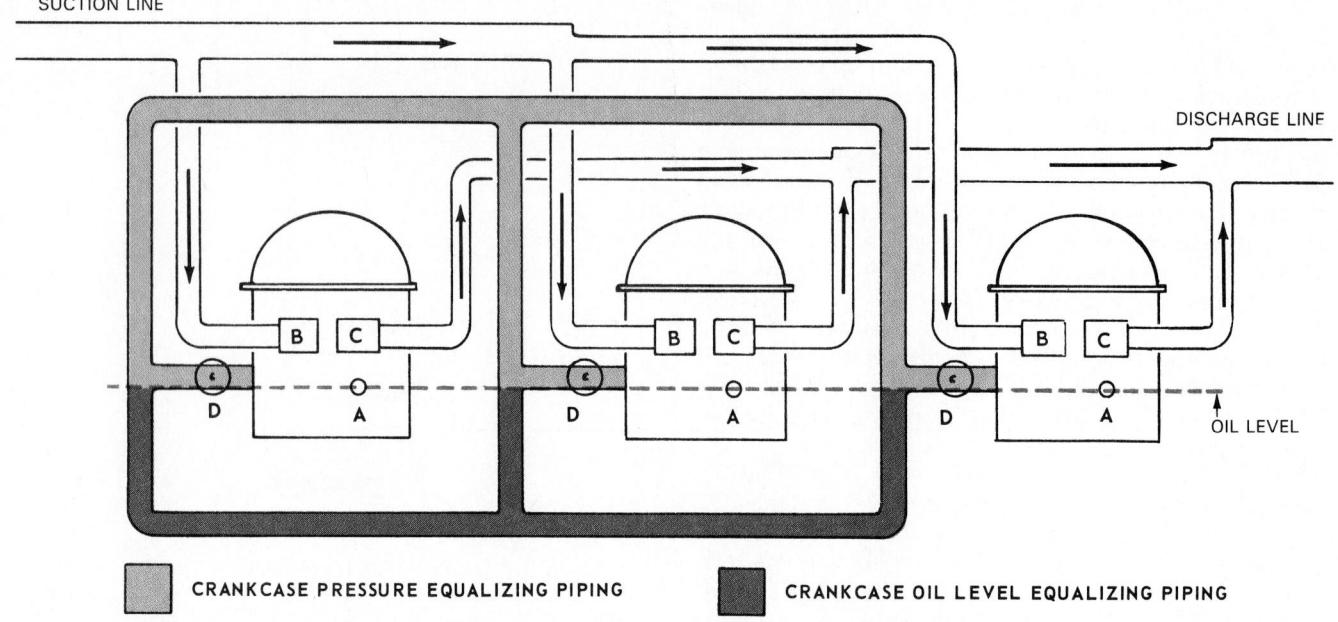

Fig. 15-81. Diagram shows a piping system used to keep an equal amount of oil in each motor compressor crankcase. A—Oil level sight glass (one on each compressor). B—Suction service valve. C—Discharge service valve. D—Shutoff valve (closed only when removing motor compressor).

15-61 DISCHARGE LINE PIPING

When compressor discharge vapor is piped to a remote condenser, it is possible that the condenser may become warmer than the compressor during the off-part of the cycle. When this happens, refrigerant vapor may move back from the condenser and condense in the head of the reciprocating compressor.

If the compressor discharge valves leak, liquid refrigerant will collect in the cylinders. This could result in the compressor pumping liquid refrigerant on start-up. This condition would reduce lubrication of the pistons and valves or even break them. If the compressor valves do not leak, the collection of liquid refrigerant in the cylinder head may cause damage when the compressor starts up because of dynamic hydraulic pressure on the compressor head and piping. A check valve installed in the discharge line near the condenser will eliminate this potential danger.

15-62 REFRIGERANT CONTROL CAPACITY

Two popular refrigerant controls are the thermal expansion valve (TEV) and the capillary tube. Both control refrigerant to the evaporator from the liquid line.

The size of the TEV's orifice (opening) is controlled by a needle and must be carefully calculated. See Fig. 15-82. TEV orifice size depends on the shape of the opening, viscosity of liquid passing through it, and upon the pressure difference as the fluid passes through the orifice.

These TEV orifice sizes, however, become fairly standard for domestic and commercial machinery. Orifice openings of 0.93 in. and .156 in. have become most popular. If large orifices are needed, multiple installations of expansion valves are used.

If the orifice is undersize (too small), not enough refrigerant can pass through the valve and the evaporator will be starved. Also, the evaporator pressure will drop too rapidly.

If the orifice size is too large, the valve will feed too much refrigerant too fast, causing a "sweat back" or "frost back"

down the suction line. The result will be a "searching" or "hunting" action, causing alternate flooding and starving of the evaporator.

The pressure difference is important. As the difference increases, TEV capacity increases. Therefore, if head pressure is high, the valve may feed refrigerant too fast and cause a sweat back or frost back. If the pressure is too low, the valve will feed too little refrigerant and the evaporator will starve.

Causes of low pressure may be:
1. Head pressure is low.
2. Liquid line is too long or has too many bends or fittings.
3. Liquid line is too small.

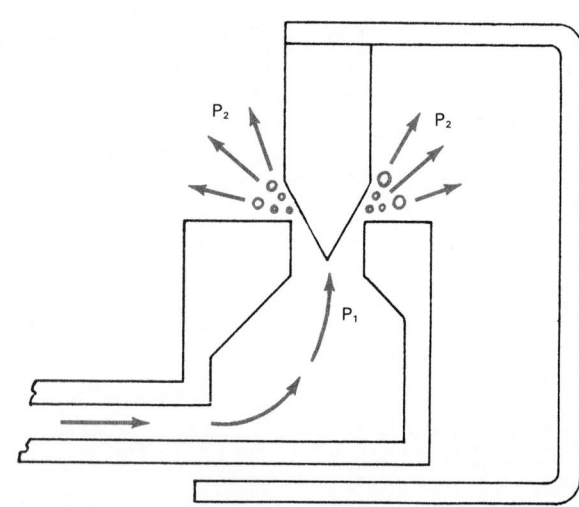

Fig. 15-82. Action of refrigerant as it passes through orifice of an automatic or thermostatic expansion valve. Liquid refrigerant at P_1 (high-side pressure) is forced through orifice and almost at once reaches P_2 (low-side pressure). As pressure changes, some liquid (about 30 percent) instantly changes into vapor (flashes). This vaporizing action cools remaining liquid to evaporator refrigerant temperatures.

COMPRESSOR CAPACITY BTU/HR.	CONDENSER TYPE	NORMAL EVAPORATING TEMPERATURE			
		−10 TO +5	+5 TO +20	+20 TO +35	+35 TO +50
R-12 LOW TEMPERATURE					
200–300	STATIC (FAN)	16 FT. − .026 IN.	10 FT. − .026 IN.		
300–400	STATIC (FAN)	12 FT. − .026 IN.	12 FT. − .031 IN.		
400–700	STATIC	12 FT. − .031 IN.	12 FT. − .036 IN.		
	FAN	10 FT. − .031 IN.	10 FT. − .036 IN.		
700–1,100	STATIC	12 FT. − .036 IN.			
	FAN	10 FT. − .036 IN.			
1,100–1,300	STATIC	10 FT. − .036 IN.			
	FAN	8 FT. − .036 IN.			
1,300–1,700	STATIC	12 FT. − .042 IN.			
	FAN	10 FT. − .042 IN.			
1,700–2,000	STATIC	12 FT. − .049 IN.			
	FAN	10 FT. − .042 IN.			
2,000–3,000	FAN	10 FT. − .054 IN.	15 FT. − .059 IN.		
3,000–4,000	FAN	10 FT. − .059 IN.	12 FT. − .064 IN.		
4,000–4,500	FAN	12 FT. − .064 IN.	12·FT. − .070 IN.		
4,500–5,000	FAN	10 FT. − .070 IN.	12 FT. − .080 IN.		
5,000–7,000	FAN	10 FT. − .059 IN. (2 PCS.)	12 FT. − .064 IN. (2 PCS.)		
7,000–9,000	FAN	10 FT. − .064 IN. (2 PCS.)	10 FT. − .070 IN. (2 PCS.)		
9,000–12,000	FAN	10 FT. − .070 IN. (2 PCS.)	12 FT. − .080 IN. (2 PCS.)		
12,000–15,000	FAN	10 FT. − .070 IN. (3 PCS.)	12 FT. − .080 IN. (3 PCS.)		
R-22 LOW TEMPERATURE					
1,000–2,000	FAN	10 FT. − .036 IN.	12 FT. − .042 IN.		
2,000–3,000	FAN	12 FT. − .042 IN.	15 FT. − .049 IN.		
3,000–4,000	FAN	10 FT. − .054 IN.	15 FT. − .059 IN.		
4,000–5,000	FAN	10 FT. − .064 IN.	15 FT. − .070 IN.		
R-12 MEDIUM AND HIGH TEMPERATURE					
1,400–1,600	FAN		12 FT. − .036 IN.	8 FT. − .036 IN.	8 FT. − .042 IN.
1,600–1,800	FAN		10 FT. − .036 IN.	12 FT. − .042 IN.	
1,800–2,500	FAN		12 FT. − .042 IN.	12 FT. − .049 IN.	8 FT. − .049 IN.
2,500–3,500	FAN		10 FT. − .042 IN.	10 FT. − .049 IN.	
3,500–4,000	FAN		12 FT. − .049 IN.	10 FT. − .054 IN.	
4,000–5,000	FAN		10 FT. − .054 IN.	10 FT. − .059 IN.	
5,000–6,000	FAN		12 FT. − .059 IN.	12 FT. − .064 IN.	
6,000–7,000	FAN		10 FT. − .059 IN.	10 FT. − .064 IN.	12 FT. − .070 IN.
7,000–10,000	FAN		12 FT. − .070 IN.	12 FT. − .080 IN.	
			12 FT. − .054 IN. (2 PCS.)	10 FT. − .059 IN. (2 PCS.)	
10,000–13,000	FAN		12 FT. − .059 IN. (2 PCS.)	10 FT. − .064 IN. (2 PCS.)	
13,000–16,000	FAN		12 FT. − .070 IN. (2 PCS.)	10 FT. − .080 IN. (2 PCS.)	
16,000–25,000	FAN		12 FT. − .080 IN. (2 PCS.)	10 FT. − .085 IN. (2 PCS.)	
25,000–40,000	FAN		10 FT. − .070 IN. (4 PCS.)	12 FT. − .080 IN. (4 PCS.)	
40,000–60,000	FAN		10 FT. − .070 IN. (5 PCS.)	12 FT. − .080 IN. (5 PCS.)	

Table title: **RECOMMENDED CAPILLARY TUBE LENGTH AND DIAMETER**

Fig. 15-83. Capillary tube sizing. Length and diameter of the capillary tube is based on the kind of refrigerant, system capacity, type of condenser, and evaporator temperature. These capacities are based on the assumption that not less than 3 ft. of the capillary tube is attached to the suction line (heat exchanger). There is no subcooling of the liquid below ambient temperature. (Tecumseh Products Co.)

Capillary tube capacity is determined by the pressure difference, length of the tube, and inside diameter of the tube. Fig. 15-83 lists suggested capillary tube sizes for two different temperature applications. It covers low-temperature and medium-to-high temperature evaporators, based on capacity of the system in Btu/hr.

15-63 ENERGY EFFICIENCY RATIO

By law, refrigeration and air conditioning units sold for household use must be rated for efficiency by the U.S. Department of Commerce. This rating must be indicated on the machine. See Fig. 15-84.

The EER or energy efficiency ratio is the rated cooling capacity of a unit in Btu per hour divided by the electrical power in watts.

An energy efficiency ratio is used to evaluate an HVAC unit much the same way that an MPG (miles per gallon) rating is used to evaluate automobiles. The higher the value, the more efficient the machine.

Window air conditioners have EERs varying from 5.4 to 9.9. An EER approaching a value of 10.0 indicates a very efficient machine.

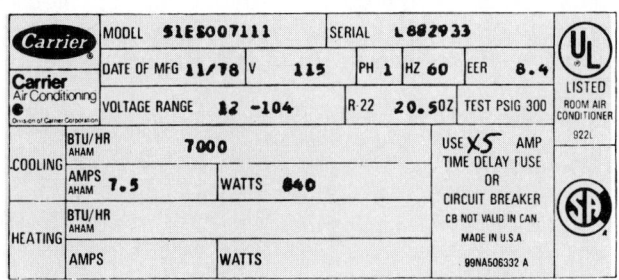

Fig. 15-84. Label, indicating cooling capacity and efficiency, must be carried by all air conditioning and refrigeration units. (Carrier Corp., Subsidiary of United Technologies Corp.)

For the label shown in Fig. 15-84, the EER was calculated as follows:

EER = 7000 Btu/hr. ÷ 840 watts = 8.33

The Coefficient of Performance (COP) of a machine can be found by multiplying the EER by a factor of 0.293:

COP = EER × 0.293.

The COP for the machine indicated is:

COP = 8.33 × 0.293 = 2.44.

Example: Find the COP and EER for a refrigerator which has a cooling capacity of 10,000 Btu/hr. and requires 800 watts of electrical energy:

Solution:

EER = 10,000 ÷ 800 = 12.5

COP = EER × 0.293 = (12.5) × (0.293) = 3.66

Example: What is the EER of a 2-ton air conditioner which required 1.5 kW of electrical power?

Solution: A ton of refrigeration is 12,000 Btu/hr. The cooling rate is then 2 × 12,000 = 24,000 Btu/hr. The EER is then: EER = 24,000 ÷ 1500 = 16.

CONVERTING EER OF HEAT PUMP USED IN HEATING CYCLE TO COP OF SAME HEAT PUMP USED IN COOLING CYCLE

Sometimes one wants to convert the EER of a heating machine to the COP of a cooling machine. Suppose the heat output of the heat pump is 50,000 Btu/hr. and the electrical input is 4 kW.

EER = 50,000 ÷ 4000 = 12.5

COP = (EER × 0.293) − 1

COP = (12.5 × 0.293) − 1 = 3.66 − 1 = 2.66

The formulas are more useful with all numbers in Btu/hr.:

Q_{HOT} = Heat given off at the high temperature

Q_{COLD} = Heat removed from the cool space

$Q_H − Q_C$ = W: Work done by compressor plus heat given off by compressor to refrigerant

$$EER_Q = Q_H/W = \frac{50,000}{4000} \times 0.293 = 3.66$$

$$COP = Q_C/W = \frac{36,340}{4000} \times 0.293 = 2.66$$

The number Q_C was obtained as follows:

$Q_C = Q_H − W$ = 50,000 − 4000 (3.415 Btu/hr./ watt)

= 50,000 − 13,660 = 36,340 Btu/hr.

Note that W = 13,660 Btu/hr. COP = EER_Q − 1.

The rating EER (energy efficiency ratio) is being modified. It is now called the "seasonal energy efficiency ratio." The SEER or seasonal energy efficiency ratio is a more complicated formula which rates the machine over a wide range of operating conditions. The result is that this rating is more representative of performance throughout the cooling season. SEER ratings are now required in the industry, and manufacturers are converting to the new ratings.

15-64 REVIEW OF SAFETY

Because excessive temperatures and pressures are dangerous, refrigeration service technicians cannot guess at piping, condenser, evaporator, and motor sizes to be used on a refrigeration system. Use manufacturer's specification sheets and recommendations in all cases.

Carefully compute the sizes of each item according to the methods described in this chapter. Improper sizing of the unit, or any part of the unit, may create a damaging or dangerous condition.

Always carefully review any calculations related to pressures, velocities, and capacities. An error might cause too high a pressure or a system failure.

15-65 TEST YOUR KNOWLEDGE

1. What is a Btu?
2. Name two miscellaneous heat loads.
3. What is total heat load?
4. What is latent heat?
5. What is specific heat?
6. Why is it important when calculating the heat load of a cabinet to know how the cabinet will be used?
7. Why is the heat leakage into a cabinet based upon the external area?
8. What is the purpose of a baffle?
9. What is usage heat load?
10. Name five types of synthetic material insulations.
11. What is the purpose of the ice holdover in the sweet water bath?
12. Which principle (or law) of thermodynamics is used to explain evaporator operation?
13. What is the specific heat of frozen ice cream?
14. At what temperature should brick ice cream be maintained?
15. What is meant by the superheating of the vapor in the suction line?
16. What is the "effective latent heat" of a refrigerant? Why is it different from latent heat?
17. What superheats vapor passing through the compressor?
18. Why may one use the Btu added to the vapor as it goes through the compressor to calculate the size of the motor to drive the unit?
19. What is meant by volumetric efficiency, and what variables influence it?
20. What happens to the compressor capacity as the low-side pressure increases?
21. What is meant by motor efficiency, and what are the usual efficiencies?
22. Explain how heat travels from one carrying medium to the next, starting with the refrigerated cabinet and continuing to the suction line vapor, describing the ease of transfer from one medium to the other.
23. What is the product heat load?
24. How many heat leakage surfaces does a cabinet have?
25. How does evaporator capacity vary as refrigerant temperature decreases?
26. If evaporator capacity is "rated" at a temperature difference of 10 °F (5.5 °C), what two temperatures are involved in establishing this rating?
27. What is a nonfrost evaporator?
28. What is the coefficient of performance?
29. Why are two suction lines used in some single evaporator systems?
30. What is the best way to determine liquid line and suction line sizes?

Chapter 16

ABSORPTION SYSTEMS PRINCIPLES AND APPLICATIONS

After studying this chapter, the technician will be able to:

☐ Explain the difference between absorption and compression refrigeration systems.

☐ Explain and describe the operation of each type of absorption system.

☐ Demonstrate procedures for servicing absorption systems.

The absorption system is different from the compression system. It uses heat energy instead of mechanical energy to make a change in the conditions necessary to complete a refrigeration cycle. The absorption system may use natural gas, LP gas, kerosene, steam, or an electric heating element as a source of heat.

The system has few moving parts. Smaller units have moving parts only in the heat source valves and in controls which are used. Some larger units also use circulating pumps and fans.

16-1 THE ABSORPTION SYSTEM

The condenser, receiver, and evaporator (cooling coil) are quite similar to those used in the compression system. The compressor has been replaced by a heater and generator. Systems shown have been simplified by leaving out various controls. These will be covered later. Fig. 16-1 illustrates a basic absorption system. It is the liquid/absorbent type. It uses a water-cooled condenser.

Fig. 16-2 illustrates the fundamentals of a basic absorption system. This diagram shows the solid absorbent type.

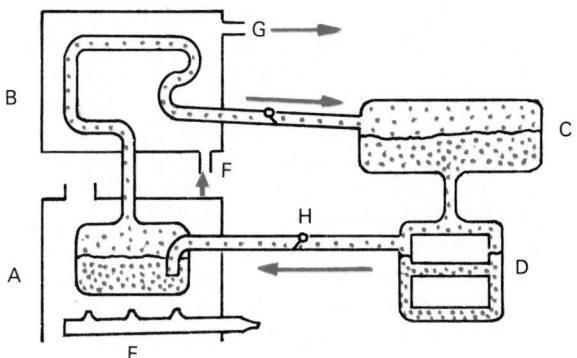

Fig. 16-1. Simple liquid absorbent refrigerating unit. A—Generator. B—Condenser. C—Receiver. D—Evaporator. E—Burner. F—Water in. G—Water out. H—Check valve.

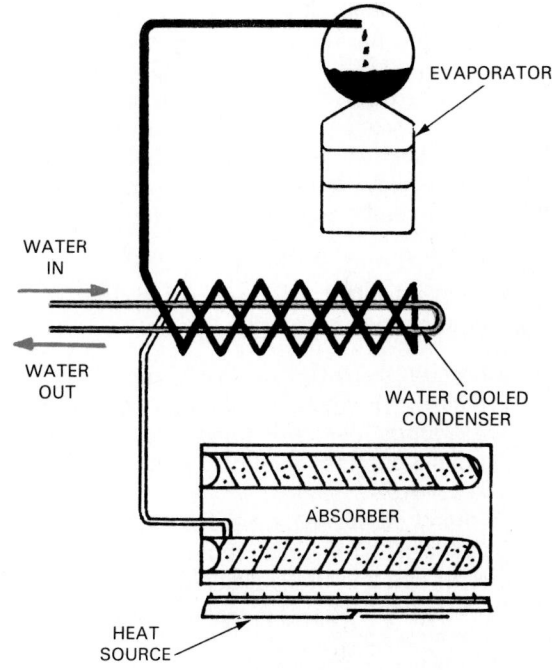

Fig. 16-2. Elementary solid absorbent refrigeration unit. Note water-cooled condenser.

16-2 TYPES OF ABSORPTION SYSTEMS

Absorption systems are based on several combinations of substances which have an unusual property. One substance will absorb the other without any chemical action taking place. It will absorb the other substance when cool and release it when heated. *If the substance is a solid, the process is sometimes called adsorbing; if the substance is a liquid, the process is called absorbing.*

There are two types of absorption refrigerators. One uses a solid adsorbent material; the other uses a liquid absorbent.

Absorption systems are further classified as:

1. Intermittent systems.
2. Continuous systems.

Absorption systems have had several applications:

1. Domestic.
2. Recreational vehicle.
3. Hotel rooms.
4. Industrial.
5. Air conditioning.

Absorption systems are also identified by heat source:
1. Kerosene.
2. Natural gas.
3. Steam.
4. Electrical heat.
5. Solar energy.

Some absorption units used in family trailers and mobile homes may be heated either electrically or by LP fuel.

16-3 PRINCIPLE OF THE SOLID ADSORPTION SYSTEM

Solid adsorption systems operate on principles discovered by Michael Faraday in 1824. Through experiments, he succeeded in liquefying ammonia, which scientists had believed to be a "fixed" gas. That is to say, it was considered impossible to change it to either a solid or a liquid.

He exposed the ammonia vapor to silver chloride, a powder. When the silver chloride had taken all the vapor it could adsorb, he applied heat and got a liquid. But when the heat was removed, he discovered that the liquid soon began to "boil," vaporize, and draw heat from its surroundings. The present-day adsorption system uses this same phenomenon. Faraday's experiment is described in Chapter 3 and Fig. 3-21.

16-4 PRINCIPLE OF THE CONTINUOUS ABSORPTION SYSTEM

The absorption system uses ammonia, water, and hydrogen. When it provides refrigeration constantly, it is called a continuous absorption system. A continuous refrigerating cycle operates automatically through the use of automatic controls.

Although many companies have variations of the basic system, the principle of operation remains the same. When the burner is lighted and its heat applied to the generator at 1 in Fig. 16-3, ammonia vapor is released from the solution. This hot vapor passes upward through the percolator tube at 2. As the hot ammonia vapor rises through this tube, it carries the solution to the upper level of the separator at 3.

Most of the liquid solution settles in the bottom of the separator and flows into the absorber. The hot ammonia vapor, being light, rises to the top of the tube, marked 4, into the condenser. The hot ammonia vapor then condenses into a liquid. The ammonia is now in a pure state and it flows by gravity into the evaporator.

Because a liquid will always seek its own level, the liquid ammonia flows through the liquid ammonia tube and spills into the evaporator. There it forms in large shallow pools on a series of horizontal baffle plates.

The hydrogen gas that is being fed to the evaporator in large quantities permits the liquid ammonia to evaporate at a low pressure and at a low temperature (Dalton's principle). During this process of evaporation, the ammonia absorbs heat from the food compartment of the refrigerator and causes the water in the ice cube containers to freeze. The more hydrogen and less ammonia, the lower the temperature. The evaporated ammonia mixes with the hydrogen gas.

Meanwhile, a weak solution of ammonia and water is flowing by gravity from the separator, at 3, down to the top of the absorber. Here it meets the mixture of hydrogen gas and ammonia vapor coming from the evaporator. The

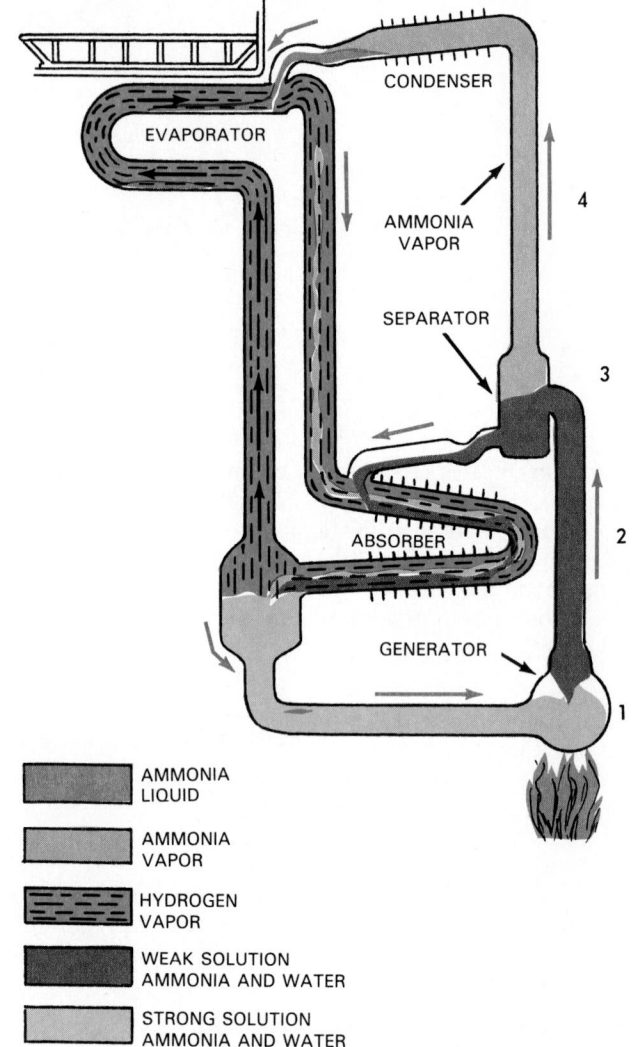

AMMONIA LIQUID

AMMONIA VAPOR

HYDROGEN VAPOR

WEAK SOLUTION AMMONIA AND WATER

STRONG SOLUTION AMMONIA AND WATER

Fig. 16-3. Air-cooled continuous refrigeration cycle. Note water circuit, ammonia flow, and hydrogen circuit.

weak and fairly cool solution absorbs the ammonia vapor. The hydrogen gas is left free since hydrogen will not mix with water. Because the hydrogen is also very light, it now rises to the top of the absorber and returns to the evaporator.

Being air cooled, the absorber has fins. The cooling of the weak solution helps it to reabsorb the ammonia gas out of the mixture of ammonia vapor and hydrogen gas. As the weak water solution reabsorbs the ammonia vapor, considerable heat is liberated. The air-cooled fins remove this heat to permit refrigeration to continue.

The ammonia liquid and water mixture flows back to the generator, where it again starts its cycle.

The apparatus is a welded assembly. There are no moving parts to wear out or get out of adjustment. Since the total pressure throughout the cycle is about 400 psi (2 864 kPa) at a room (ambient) temperature of 100 °F (38 °C), construction must be rugged to insure a long life.

To produce a temperature of 0 °F (−18 °C) in the evaporator, the ammonia must boil at 15.7 psi (211.8 kPa). This means the hydrogen must make up the remainder of the pressure (384.3 psi or 2 652.2 kPa), if the total

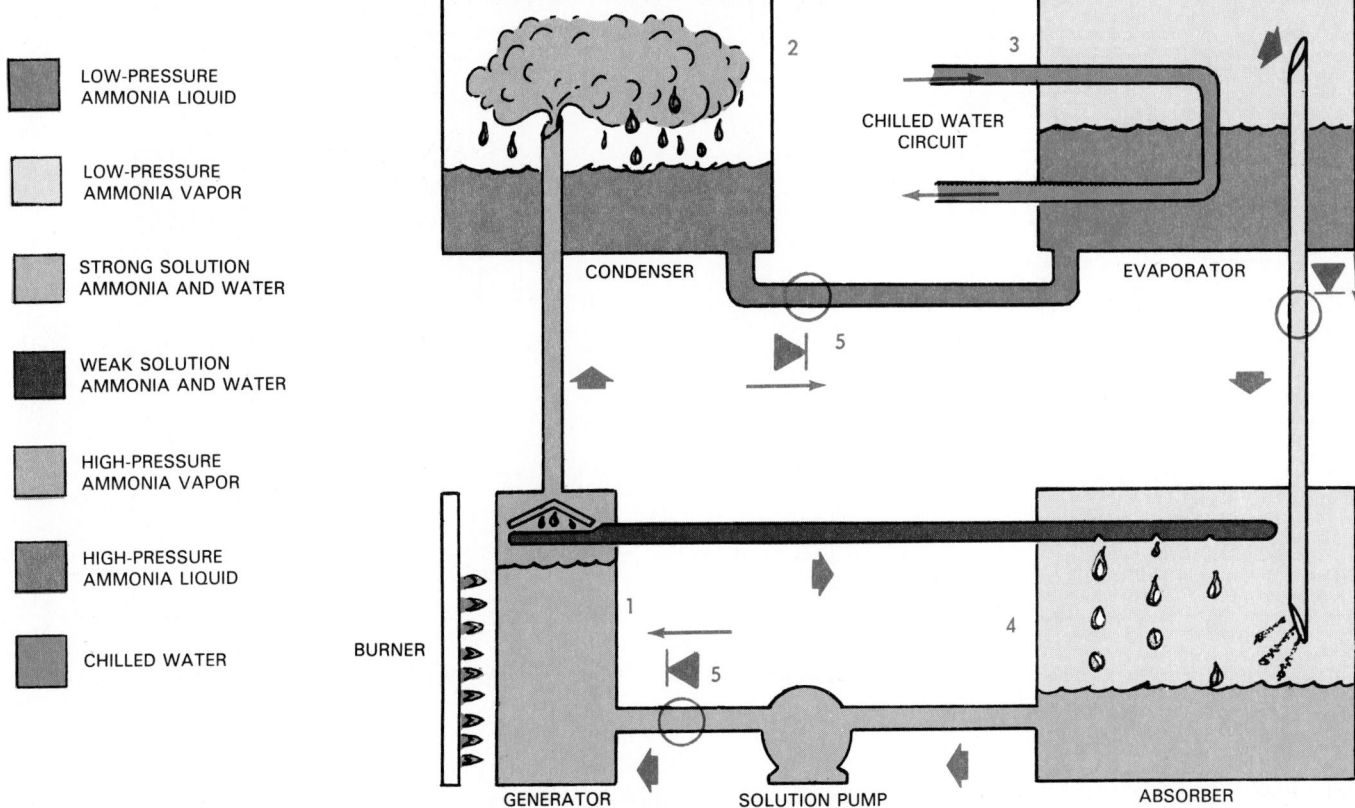

Fig. 16-4. Continuous absorption system uses pump to maintain pressure difference between low-pressure side and high-pressure side of system. Same pump transfers strong-in-ammonia, weak-in-water solution. 1—Generator. 2—Condenser. 3—Evaporator. 4—Absorber. 5—Check valves.   (The Dometic Corporation)

pressure is 400 psi. This refrigerator is considered to be unique among domestic refrigerators.

16-5 CONTINUOUS ABSORPTION SYSTEM WITH PUMP

The continuous absorption refrigerating system with a pump, Fig. 16-4, uses ammonia as the refrigerant and an aqueous ammonia solution as the absorbent. Any means of heating can be used, but natural gas, steam, or LP gas are the most popular.

The system operates under two pressures. The high-side pressure is from 200 to 300 psi (1 484 to 2 174 kPa) and the low-side, 40 to 60 psi (380 to 518 kPa). The high and low sides are separated by check valves, liquid traps, a pump, or other controlling devices.

The operational system can be divided into four sections including generator, condenser, evaporator, and absorber.

The generator at 1 in Fig. 16-4 is heated by a vertical burner. Heat causes the liquid to boil and the ammonia in the liquid turns into vapor. The vapor will rise up through the tube to the air-cooled condenser. In the condenser, heat from the vapor is removed by cooler air passing across it. The vapor will condense into a liquid, which then acts as the refrigerant.

The liquid refrigerant now passes, at a high pressure, to the evaporator at 3. In the evaporator, water carrying heat from the cooled area passes through tubes. The heat from the water tubes is transferred to the refrigerant liquid. The water in the tubes returns to the area that needs to be

cooled. Being at a low temperature, the water can absorb heat from the area that is to be cooled.

The heat that the refrigerant has absorbed from the chilled water circuit causes it to boil and turn into a vapor. This vapor refrigerant is drawn back to the solution-cooled absorber at 4, where the heat is sent to the outside air.

The liquid refrigerant is then pumped back—preheated— by the solution pump to the generator, where the procedure is repeated.

16-6 PRINCIPLE OF INTERMITTENT ABSORPTION SYSTEM

A convenient refrigerator cycle for localities having neither gas nor electricity is the Superfex and Trukold cycle. The Superfex cycle is about the same as the Faraday principle but it has some different features.

In Fig. 16-5, ammonia is mixed with water in a sealed tank or generator shown at A. Underneath, a kerosene burner, M, heats it. Heat from the burner drives the ammonia, in vapor form, out of the mixture. This vapor is forced up the pipe marked D and through a condenser, E. This is immersed in water from a tank at B on top of the refrigerator.

Cooling effect of the water causes the ammonia vapor to change back into a liquid (condenses) at the high generating pressure. This liquid ammonia drops through a pipe into the liquid receiver, C. From here it passes to the evaporator, K, which is surrounded by a brine at H. The liquid receiver is insulated at F to prevent this container from

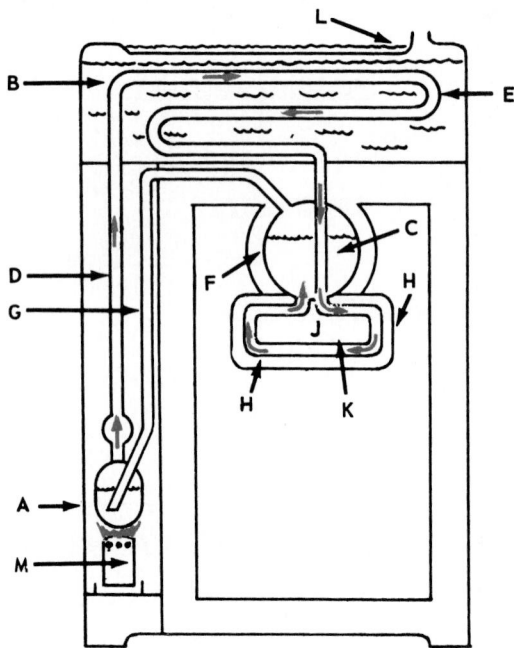

Fig. 16-5. Generation, or heating interval, in typical intermittent type absorption refrigerator.

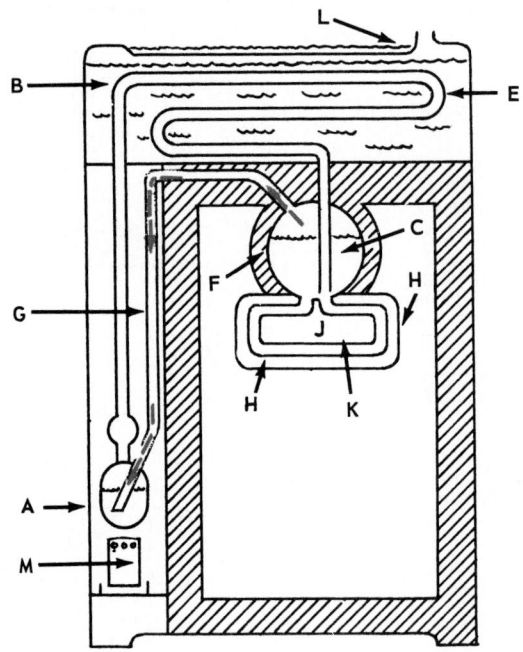

Fig. 16-6. Refrigerating part of cycle in intermittent type absorption refrigerator.

overcooling the food compartment by acting as the evaporator.

The process continues for a short time until the kerosene is used up and the burner goes out.

As the absorber cools to room temperature, the ammonia will evaporate at a low temperature in the evaporator. This is because, as the generator cools, it tends to reabsorb the ammonia vapor. This reduces the pressure and permits the liquid ammonia in the evaporator to boil at low temperature. This evaporation causes the cooling effect or refrigeration required to preserve the contents of the food compartment. See Fig. 16-6.

In other words, heat from the oil burner drives the ammonia from the generator, A, into the evaporator, K, in a short time. The ammonia in the evaporator vaporizes and passes back to the generator slowly over a period of 24 to 36 hours. The vaporization of the ammonia in the evaporator produces a refrigerating effect.

For additional efficiency in unusually hot climates, or for handling extra large loads, a depression, L, in the top of the condenser tank may be filled with water. The water will evaporate rapidly and aid the cooling of the condenser.

Absorption mechanisms are provided with a fuse plug. The fuse plug will release the charge from the mechanism if the temperature of the unit becomes excessive (175°F to 200°F or 79°C to 93°C), as might be experienced in a fire. This prevents the mechanism from exploding.

16-7 COOLING PROCESS

Absorption refrigeration is widely used in recreational vehicles and residential applications where the use of electricity is important. The absorption system is unique in refrigeration since it involves no moving parts and is virtually noiseless. Absorption refrigerators are used extensively in

motels. A typical unit is shown in Fig. 16-7. Continuous absorption types of refrigerators have four main sections: the generator (boiler), shown as A; condenser, D; evaporator, E; and absorber, J. The four sections are connected by steel tubes. The entire system is welded together. The necessary heat for generation is obtained by using either a gas burner or electric heating element at the generator (boiler), A.

The system is charged with ammonia, water, and hydrogen. The amount of this combined solution is at a pressure which will allow the ammonia to condense at room temperature.

When the unit begins operation, some of the ammonia is in a rich solution with water in the vessel. Heat is applied to the *boiler*, A, raising the solution temperature to 350°F (177°C). Some of the ammonia rises through the vapor pump. The liquid falls back through the boiler and liquid heat exchanger. It then flows back to the vessel, as a weak solution, through the *absorber*, J.

When leaving the vapor pump, B, the ammonia vapor is about 300°F (149°C). The ammonia gas is mixed with steam and the water is condensed out of the solution in the rectifier, C.

The pure ammonia vapor then flows through the *condenser*, D, at room temperature. After the ammonia condenses, the liquid falls in the precooler, I. As the liquid ammonia reaches the *evaporator*, E, the solution begins to evaporate into the hydrogen. This cools the freezer section to between −24°F and 0°F (−31°C and −18°C).

The hydrogen ammonia mixture then drops back down through the return pipe and into the reservoir to begin another cycle.

It is important to understand that the entire cycle is carried out entirely by gravity flow of the refrigerant. *It is important that the unit remain in a level, upright position. It is important that the heat generated in the absorber be*

Absorption Systems, Principles and Applications

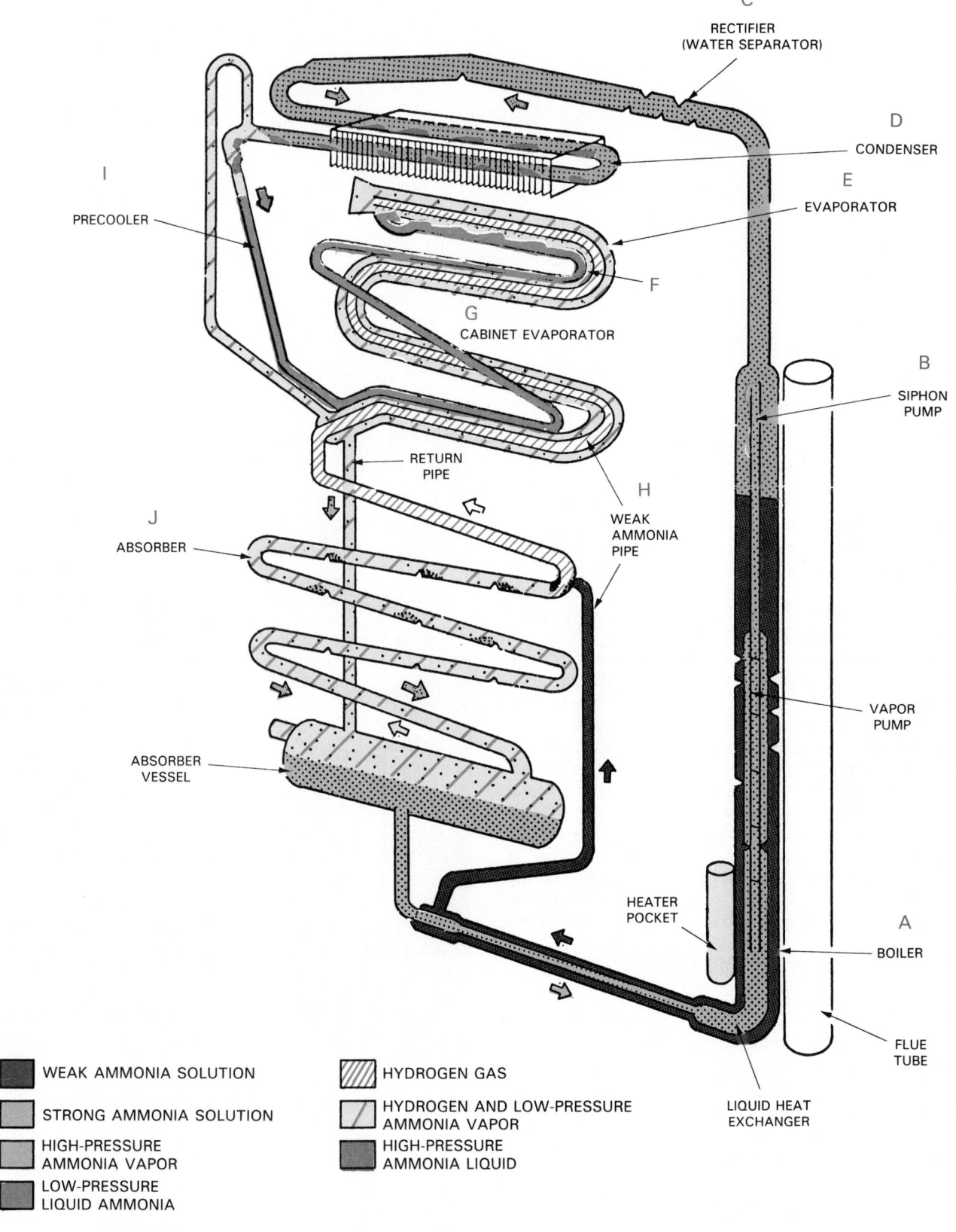

Fig. 16-7. This absorption system is used in recreational vehicles. Leveling of system is not as critical as for older machines. Internal syphon pump helps reduce liquid blockage in cooling coil. System can operate on any heat source, such as 120 volt ac, 12 volt dc, propane, butane, or kerosene. (The Dometic Corporation)

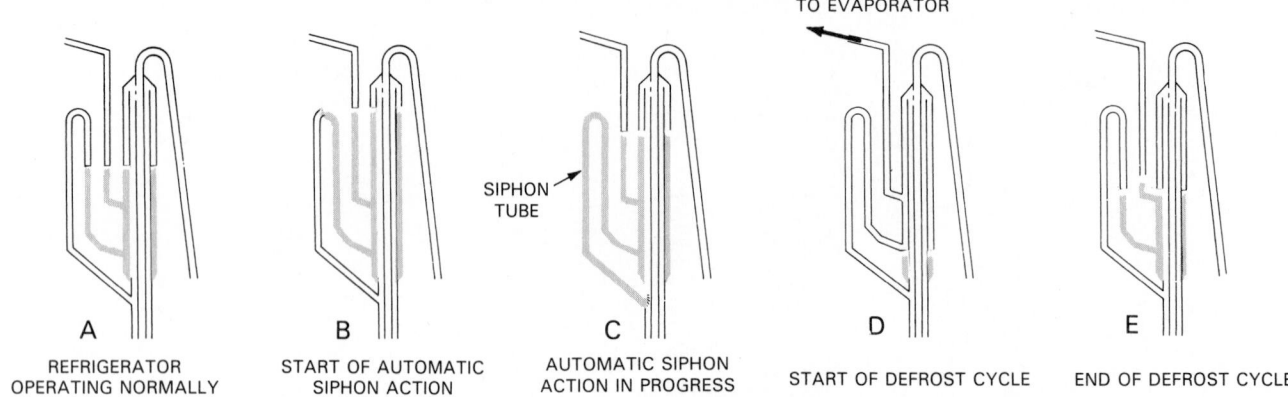

Fig. 16-8. One type of automatic defrost system for domestic continuous cycle absorption system. Liquid gradually builds up in right side of siphon tube and every 15 to 24 hours spills over and allows hot vapor to move into higher temperature evaporator. Defrosting stops when enough liquid collects in outer tube to close tube to evaporator. (Electrolux AB)

removed and that the heat removed by the condenser be carried away by the surrounding atmosphere.

16-8 AUTOMATIC DEFROSTING

Automatic defrosting on domestic continuous absorption units can be done by the hot gas method.

Hot gas is brought from the generator directly to the fresh food evaporator rather than to the freezer evaporator. This hot gas melts the ice on the evaporator fins. Defrost water runs into a drip tray. The operation is controlled by a siphon tube in the boiler (generator). See Fig. 16-8.

The bypass pipe outlet from the siphon chamber during normal operation is closed by the strong ammonia-water liquid solution pictured in view A. During a normal cycle, the solution collects slowly in the chamber until it reaches the siphon tube outlet—about every 15 to 25 hours. See view B. The siphon action then empties the siphon system at view C of its liquid and allows the hot gas from the boiler to go directly from the generator to the evaporator, in view D. This circulation will continue for about 30 minutes until the solution fills the siphon system again and covers the outlet to the bypass pipe in view E. The circulation of the hot gas will repeat when the liquid level in the siphon chamber rises high enough to repeat the siphon action.

16-9 CONTINUOUS ABSORPTION SYSTEMS CONSTRUCTION

There are several types of continuous absorption systems. Fig. 16-9 illustrates a typical two-door domestic refrigerator with a frozen food compartment. This unit uses LP gas or electricity as its source of heat. The internal system with ice formation in the freezer is shown in Fig. 16-10.

Some units are heated with fossil fuels. Some are heated with electricity. An electrically heated generator is shown in Fig. 16-11.

Either fuel gas or electricity can be used to heat the unit shown in Fig. 16-12. Electricity may be either 12 V (battery) or 120 V (house current). A switch, as shown in Fig. 16-13, is used to change over from 12 V to 120 V.

The combination gas/electrical systems use two thermostats, one when gas is used, the other when electricity is used. Wiring diagrams for the 120 V and the combination 12 V and 120 V systems are shown in Fig. 16-14.

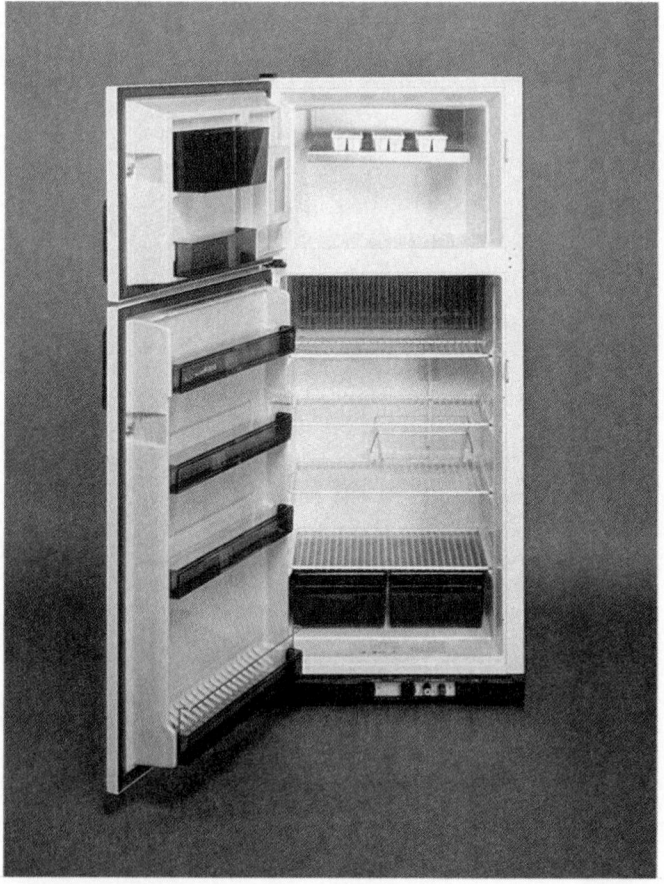

Fig. 16-9. A domestic absorption refrigeration system, working on LP gas or electricity. Note series of controls at base: electricity or gas selector dial, intermittent spark ignition for reignition of gas flame in emergencies, and thermostat. (SIBIR)

16-10 INSTALLING AN ABSORPTION REFRIGERATOR

Installation of the absorption refrigerator depends on whether it is to be placed in a home, a recreational vehicle, or a mobile vehicle.

The absorption refrigerator, like all refrigerator

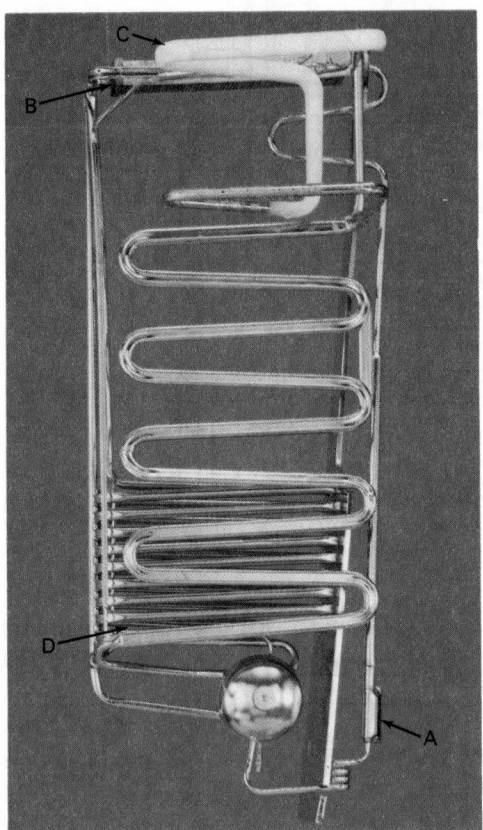

Fig. 16-10. Domestic absorption system. It contains four main sections. A—Generator (boiler). B—Condenser. C—Evaporator. D—Absorber. Note ice formation at freezer, evaporator. (SIBIR)

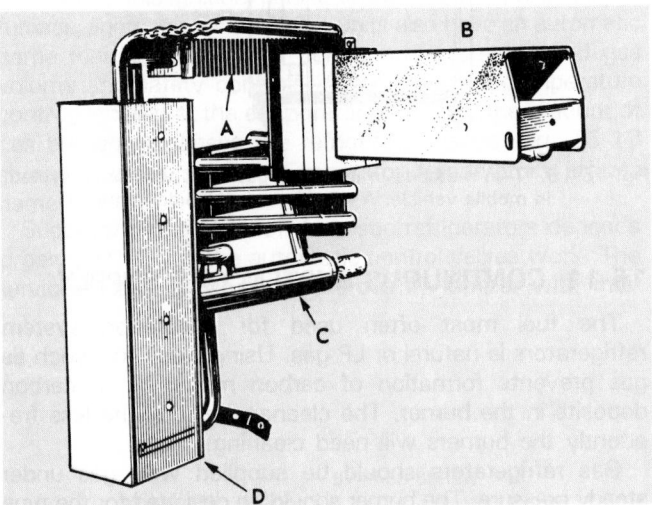

Fig. 16-11. Continuous absorption refrigerating mechanism using electricity as heat source. A—Condenser. B—Evaporator. C—Absorber. D—Electrically heated generator.

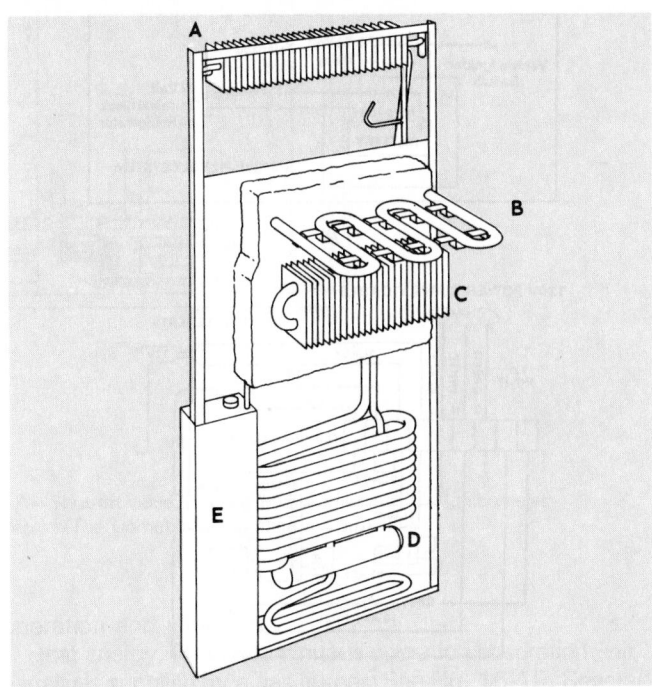

Fig. 16-12. Continuous operation absorption unit can be heated by fuel gas or electricity. A—Condenser. B—Low-temperature evaporator. C—High-temperaure evaporator. D—Absorber. E—Generator.

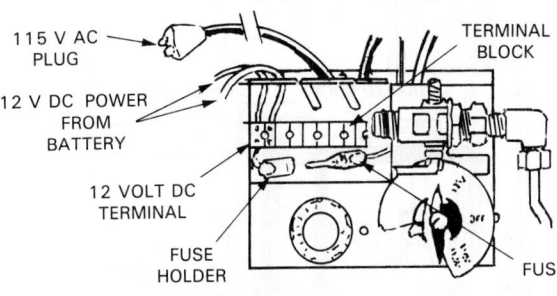

Fig. 16-13. Electrical switch and fuse used to shut off electrical power or change from 120 V to 12 V. If switch is on 12 V when system is connected to wall outlet, fuse at A will open circuit (blow) to protect heater from damage. (Dometic Corp.)

mechanisms, moves heat from inside of a cabinet to outside of a cabinet. This warmed air must be removed from near the cabinet to allow cooler air to continue to receive heat from the condensers, as shown in Fig. 16-15.

Kerosene, natural gas, or LP gas, when burned, forms carbon dioxide gas (harmless) and steam vapor (harmless). *If burner is not burning all the fuel, carbon monoxide may*

be formed and is dangerous! To allow good airflow past the burner, these products of combustion must be moved away from the cabinet. Because warm gases rise, the absorption cabinets must have proper airflow space beneath, in back of, and over the top of the cabinet.

Inside the mechanism, liquids flow by gravity. The unit must be properly leveled or movement of liquids and gases will be uncertain. Some units have a level indicating device. This allows for easier positioning during installation and for quick level checks during maintenance.

The condenser, absorber, and flues of the unit must be kept clean to allow proper airflow and flue gas flow. The condenser, burner, and absorber should be cleaned at least twice a year, and more often, if necessary.

Absorption refrigerators which use a fuel gas and are installed in trailers, mobile homes, or motor homes, must be level when parked. They must also remain as level as possible when being moved. Avoid having the wind blow directly against the vents on the outside when the vehicle is parked. The refrigerator must be securely fastened to the floor. An

The pull-down time for this refrigerator is from two to six hours, depending on the ambient temperature. The small propane cylinder will provide approximately 70 hours of continuous operation.

Another portable absorption refrigerator is shown in Fig. 16-25. The gas cylinder, electrical connections, and con-

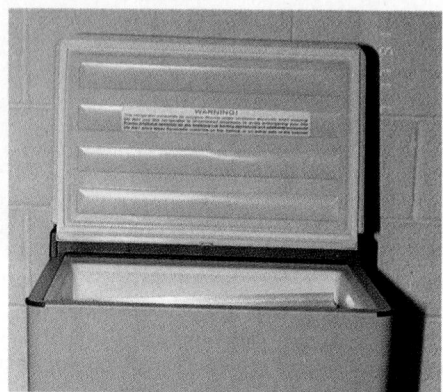

Fig. 16-25. Portable absorption refrigerator. Heating source used in this refrigerator may be either LP gas, 120 V ac or 12 V dc current. (Dometic Corp.)

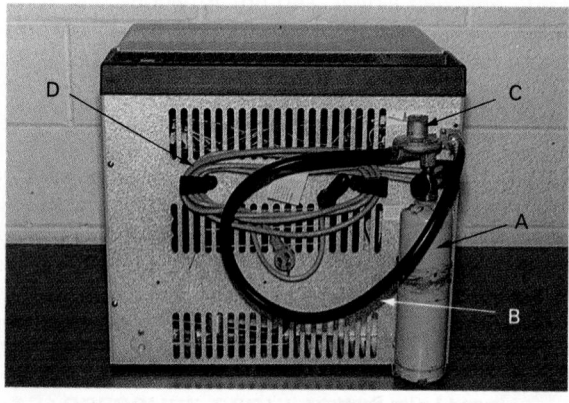

Fig. 16-26. Portable absorption refrigerator. A—Propane fuel cylinder. B—Flexible gas hose. C—Pressure regulator. D—Electrical power cord. Refrigerator can be operated with either gas flame or electric heating unit of 12 V or 120 V. (Dometic Corp.)

trols for this refrigerator are shown in Fig. 16-26.

When operating from a propane cylinder:
1. Attach gas connection to gas inlet fitting.
2. Open valve on gas bottle.
3. Press the red button on the safety valve and hold it for 10 to 15 seconds. (This clears air from the gas line.)
4. Light match. Press red button again and apply flame to burner. Keep depressed for 20 seconds after lit.

The safety valve will automatically shut off gas supply, should the flame go out for any reason. A pressure regulator will maintain an 11-in. (279-mm) water column pressure on the burner. When using the gas burner, be sure no combustible material or vapors are near the refrigerator.

When operating with electricity, separate leads are supplied for the 120 V ac and the 12 V dc connection.

16-15 ABSORPTION REFRIGERATORS FOR MOBILE HOMES

Mobile homes and travel trailers often use an absorption type refrigerator. The units are usually designed with both an electric heating element and a gas burner to heat the generator of the continuous unit. Gas heat is used when electricity is not available. The refrigerator must be mounted level. If not, gravity-controlled flow of fluids will not operate well.

The installations must be carefully designed with air ventilation to cool the condenser and to provide air for the flame and outside exhaust for combustion gases.

An absorption system requires a long cool down time (8 to 10 hours) when being started up. Manufacturers usually recommend starting the unit the night before it is to be loaded with food.

16-16 RESIDENTIAL ABSORPTION AIR CONDITIONERS

The number of absorption type air conditioning systems used in residential and commercial buildings has increased. Their basic operation is similar to that described in Para. 16-5. One of the major changes for such use is the addition of a pump to transfer the weak solution from the absorber to the high side of the cycle. Either a hydraulic diaphragm pulse pump or an electric motor driven magnet pump is sealed in the system.

A typical chilled water residential air conditioning system is shown in Fig. 16-27. The generator has a gas burner which heats a mixture of ammonia and water. The boiling point of ammonia is lower than that of water. Therefore, it becomes a vapor and flows through the line marked 1, through the rectifier, and then to the condenser as a high-temperature, high-pressure gas.

As outside air passes over the condenser, it removes heat from the ammonia. The ammonia condenses to a liquid and passes through the line marked 2, to the precooler. The precooler acts as a heat exchanger and reduces the temperature of the liquid ammonia before it reaches the evaporator. It also heats the cold ammonia vapor leaving the evaporator through line 3.

When liquid ammonia leaves the precooler, pressure drops as it passes through a restrictor into the evaporator. Here it picks up heat from the chilled water circuit and boils to ammonia vapor. Ammonia vapor in line 3 passes through the precooler to the absorber heat exchanger. In the generator, most of the ammonia boils out, leaving a weak solution. This solution leaves the generator at a high pressure. It passes through restrictors which meter flow and separate high- and low-pressure sides of system.

Restrictors are in line 4 leading to the absorber heat exchanger. Here solution temperature is lowered more by heat transfer. At the absorber end, heat is removed by outside air and absorption of ammonia is completed.

The solution travels from the absorber, through the condenser, to the solution pump through line 5. The pump forces it at a high pressure to the rectifier through line 6. It picks up heat from the surrounding hot ammonia vapor along the way. As it continues on to the absorber heat exchanger through line 7, more heat is picked up. This preheated solution returns through line 8 to the generator. The cycle begins all over again. The same action occurs

AIR COOLED CYCLE OF OPERATION
SOLUTION-PUMP TYPE

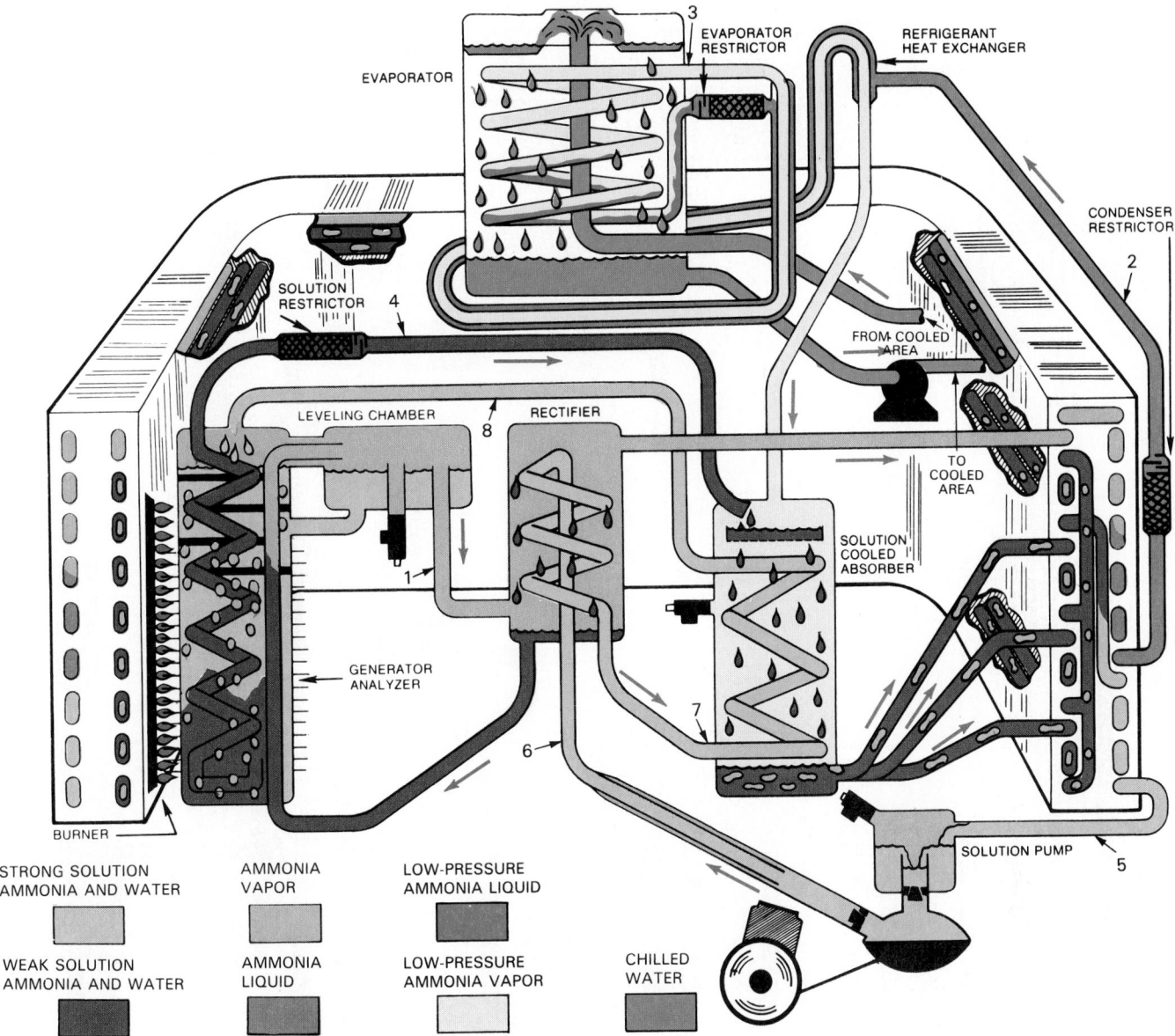

EVAPORATOR
EVAPORATOR RESTRICTOR
REFRIGERANT HEAT EXCHANGER
3

SOLUTION RESTRICTOR
4
CONDENSER RESTRICTOR
2

FROM COOLED AREA

LEVELING CHAMBER
8
RECTIFIER
1

TO COOLED AREA

GENERATOR ANALYZER
SOLUTION COOLED ABSORBER

7

6

BURNER

SOLUTION PUMP
5

STRONG SOLUTION
AMMONIA AND WATER

AMMONIA
VAPOR

LOW-PRESSURE
AMMONIA LIQUID

WEAK SOLUTION
AMMONIA AND WATER

AMMONIA
LIQUID

LOW-PRESSURE
AMMONIA VAPOR

CHILLED
WATER

Fig. 16-27. Absorption refrigeration cycle. This refrigeration cycle uses pump to circulate ammonia liquid through cycle.
(The Dometic Corporation)

in larger units except that two burners may be used.

Another system cycle of slightly different design is shown in Fig. 16-28. The weak solution (strong in ammonia content) is returned to the generator by the solution pump. The burner heats it and drives off the vapor. The vapor, a mixture of ammonia and water, flows through the generator assembly and comes into direct contact with the weak solution coming into the generator. During the process, the vapor is partially purified. Passing through the rectifier assembly, the vapor is further purified as it comes into contact with the cooler, weak solution flowing in the opposite direction through a coil.

The rectifier contains rings and, as the ammonia vapor passes through, it comes into direct contact with the rings. The rings, having a large surface contact area, help remove any water vapor left in the ammonia vapor.

The purified ammonia vapor then flows into the condenser tube cooled by air moving across the condenser. As the hot ammonia vapor in the condenser gives up heat to the flow of air, it is liquefied.

Liquid refrigerant leaves the condenser and passes through the first restrictor, where pressure and temperature drop somewhat. It then flows through the outer tube of the liquid suction heat exchanger, giving up heat to cooler ammonia vapor flowing through the inner tube in the opposite direction.

From the liquid suction heat exchanger, the liquid ammonia enters the second restrictor, where pressure and temperature are further reduced. The low-pressure, low-temperature liquid refrigerant then flows through the chiller coil (evaporator). Here a glycol and water solution cascades down across the evaporator, giving up heat and vaporizing

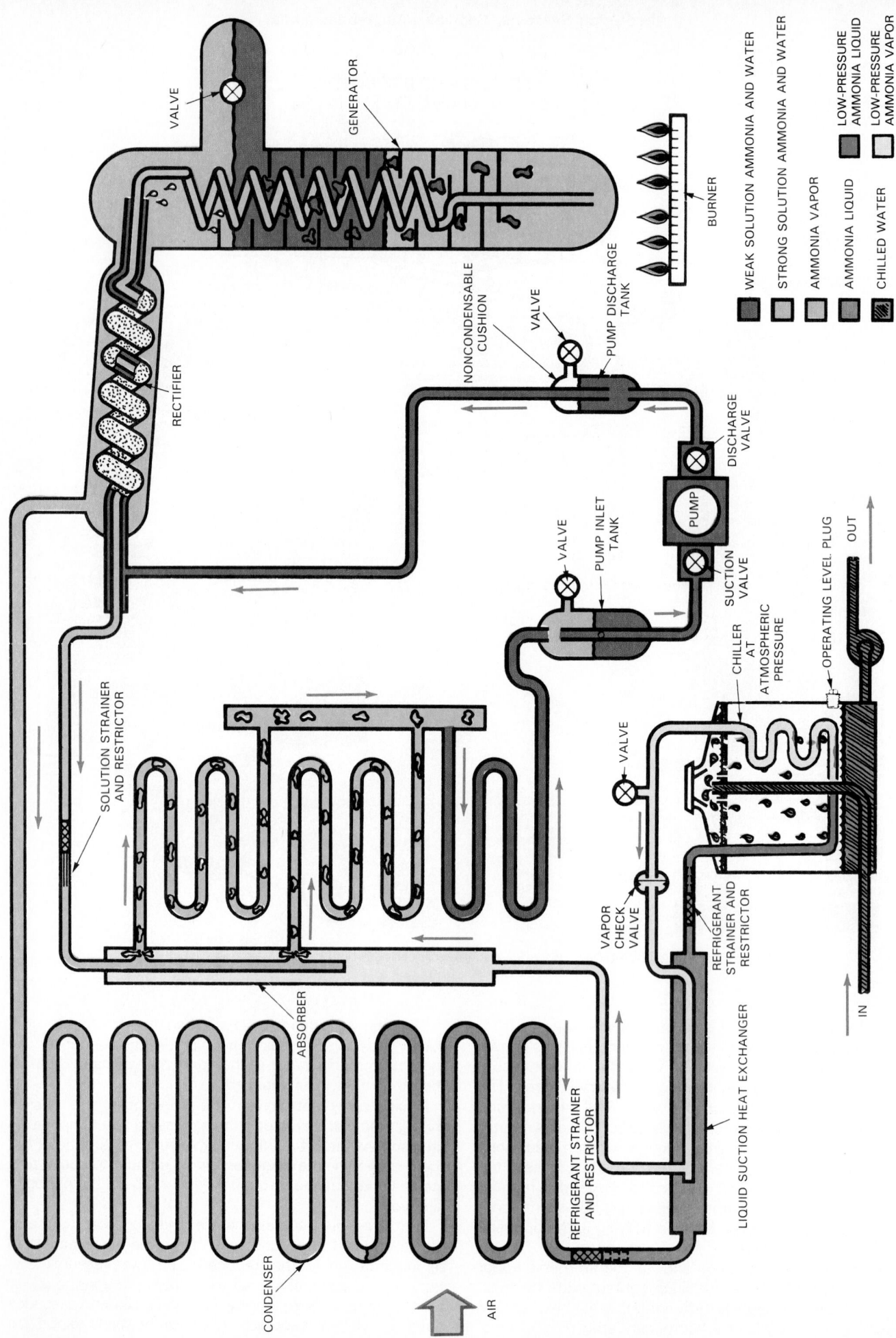

VALVE

GENERATOR

RECTIFIER

VALVE

NONCONDENSABLE CUSHION

PUMP DISCHARGE TANK

DISCHARGE VALVE

PUMP

VALVE

PUMP INLET TANK

SUCTION VALVE

CHILLER AT ATMOSPHERIC PRESSURE

OPERATING LEVEL PLUG

OUT

IN

VALVE

VAPOR CHECK VALVE

REFRIGERANT STRAINER AND RESTRICTOR

LIQUID SUCTION HEAT EXCHANGER

SOLUTION STRAINER AND RESTRICTOR

ABSORBER

REFRIGERANT STRAINER AND RESTRICTOR

CONDENSER

AIR

BURNER

WEAK SOLUTION AMMONIA AND WATER

STRONG SOLUTION AMMONIA AND WATER

AMMONIA VAPOR

AMMONIA LIQUID

CHILLED WATER

LOW-PRESSURE AMMONIA LIQUID

LOW-PRESSURE AMMONIA VAPOR

Fig. 16-28. Absorption cycle used for small air conditioning systems. Notice service valves on system.

liquid refrigerant.

The refrigerant vapor leaves the chiller and passes through the inner tube of the liquid suction heat exchanger. In passing, it picks up heat from the liquid refrigerant flowing through the outer tube in the opposite direction. Finally, the vapor enters the absorber header. This completes the refrigeration circuit.

Hot, strong solution, left behind as the ammonia vapor was driven out of the weak solution in the generator, passes up through a coil in the generator. The strong solution leaves the generator and enters the inner coil of the rectifier. Here the strong solution comes into thermal contact with the weak solution, flowing in the opposite direction through the outer coil. As the strong solution leaves the rectifier, it passes through the strong-solution restrictor. Here the restrictor reduces the solution from high-side to low-side pressure.

As the strong solution leaves the restrictor, it enters the absorber header. This is a tube-within-a-tube. The inner tube contains the strong solution; the outer tube contains ammonia vapor which is being returned from the chiller (evaporator). The strong-solution tube contained within the absorber header has two small holes that will allow strong solution to flow out and come into direct contact with the ammonia vapor surrounding it. At this point the strong solution and ammonia vapor begin to form a weak solution.

Strong solution and ammonia vapor then leave the absorber header by way of the two tubes and begin to flow into the absorber. Throughout the absorber, the strong solution completely absorbs the vapor, forming a weak solution.

As the weak solution leaves the absorber, it is picked up by the solution pump, which will move the solution back to the high-pressure side of the unit. As the weak solution leaves the solution pump, it passes through the rectifier, picking up heat from strong solution vapor flowing in the opposite direction.

By preheating the weak solution as it flows through the rectifier, the need for heat input is reduced and the overall efficiency of the cycle is increased. The weak solution leaves the rectifier and drips back into the generator analyzer assembly to start another cycle.

16-17 RESIDENTIAL ABSORPTION AIR CONDITIONER CONSTRUCTION

In this self-contained unit, Fig. 16-29, the insulated evaporator cools a glycol and water solution. The solution then circulates through a heat exchange coil in the furnace bonnet. It may also be used in a separate air circulation system within the building. The absorption system is shown in Fig. 16-30. Controls are shown in Fig. 16-31.

The absorption system shown in Fig. 16-30 can also be utilized as part of a multiple unit load system. Typical applications are in large office buildings, strip malls, and efficiency apartments. An internal temperature-sensing device monitors the return water temperature. When temperature is higher than normal, indicating increased need for cooling, another unit is ignited to bring the chilled water temperature down to proper level. The chilled water is directed by means of valves to the proper fan coil, thereby accomplishing the proper zone control.

The low pressure and temperature of the chilled water system allows the use of PVC. Gas units are single-phase

Fig. 16-29. Exterior view of gas-fueled absorption system used for residential air conditioning system. (The Dometic Corporation)

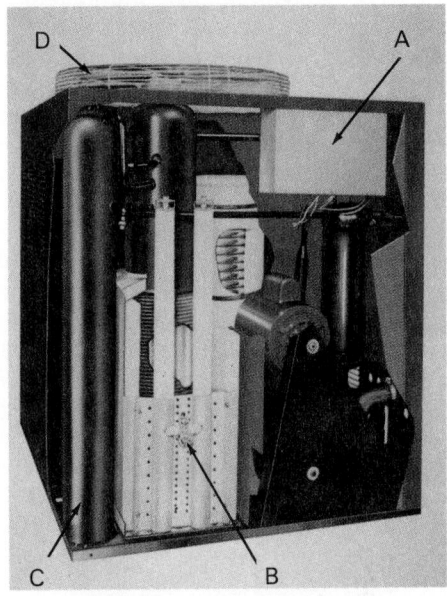

Fig. 16-30. Absorption system air conditioner with housing removed. A—Controls. B—Burner. C—Solution pump. D—Exhaust fan. (The Dometic Corporation)

Fig. 16-31. Controls of absorption air conditioner are easy to reach for service when housing is removed. A—Gas burner. B—Gas controls. C—Electrical controls. (The Dometic Corporation)

power, 115 or 230 volt. Individual units range from 3 to 25 tons.

These systems are made of steel and aluminum. Use of copper or copper alloys is very dangerous. An explosion may result.

16-18 RESIDENTIAL ABSORPTION AIR CONDITIONER SERVICE

Most residential absorption systems are serviceable. They are equipped with service valves. However, one should be trained by the manufacturer before attempting to service these systems.

Ammonia is toxic and flammable when mixed at certain ratios with air. Wear a face shield or safety goggles.

Ammonia reacts with some metals. Use only steel or aluminum tubing, gauges, fittings, and manifolds.

Fig. 16-32 shows a system equipped with four service valves. Valves A and D are on the low-pressure side. Valves C and E are on the high-pressure side of the system. Valve C is not shown. Valves D and E are gauge mounts for checking pressures on the system. Refer back to Fig. 16-28. It also shows the location of the four service valves.

Anhydrous ammonia is available in 25-lb. cylinders. The cylinder has both a vapor and a liquid valve.

The system is charged with a solution of ammonia, distilled water (pH 6.0+), and a corrosion inhibitor. A solution cylinder is used to charge the system with the solution. This cylinder usually has about a 45-lb. capacity. The solution charge is about 35 lb. of distilled water, about seven packages of inhibitor, and 15 lb. of anhydrous ammonia.

The solution cylinder is filled as follows:

1. The solution of distilled water and inhibitor (yellow in

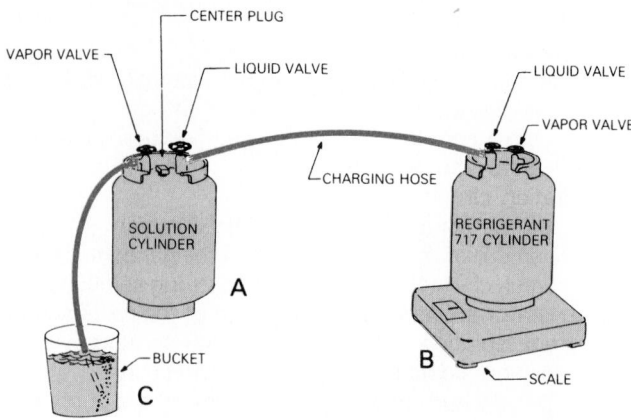

Fig. 16-33. Cylinders set up for recharging an absorption system. A—Solution cylinder has liquid valve, vapor valve, and center plug. It is being charged with liquid ammonia from cylinder at right (1 lb. ammonia for each 2 lb. distilled water). B—Ammonia cylinder. C—Pail holds water and a purge line connected to gas valve of cylinder A. Purging decreases pressure in A to allow flow from B. (The Dometic Corporation)

color) is put in the solution cylinder first through the fill plug. If there is a white powder in solution, discard it and make a new solution.

2. Fig. 16-33 shows a solution cylinder being charged with anhydrous ammonia. The pail holds some water and is connected to the vapor valve of the solution cylinder. Vapor is purged from the solution cylinder if necessary to lower the pressure so that anhydrous ammonia will flow into the solution cylinder. The water in the pail will absorb the small amount of ammonia purged. Note that the charging line is connected to the liquid valve of the anhydrous ammonia cylinder.

The system must have the correct pressures, the correct amount of anhydrous ammonia, and the correct amount of distilled water and inhibitor.

To check the pressures, install a 100 percent steel constructed gauge manifold, steel lines, and steel fittings. Do not use copper or brass. Fig. 16-34 shows a gauge manifold installed. Note the steel manifold. It is constructed of standard steel fittings and steel valves. The manifold operates exactly like the manifolds described in Chapters 11 and 14.

The four service valves on the system are used for a number of service operations.

Valve A:
1. Checks absorber pressure (low-side pressure).
2. Purges ammonia vapor.
3. Adds ammonia liquid or vapor.
4. Adds solution.
5. Reduces system pressure to atmospheric pressure.

Valve C:
1. Checks high-side pressure.
2. Checks solution level.
3. Removes excess solution.
4. Adds solution after repairs.
5. Reduces system pressure to atmospheric.

Valve D:
1. Purges air.
2. Adds air.
3. Adds solution.
4. Removes solution.

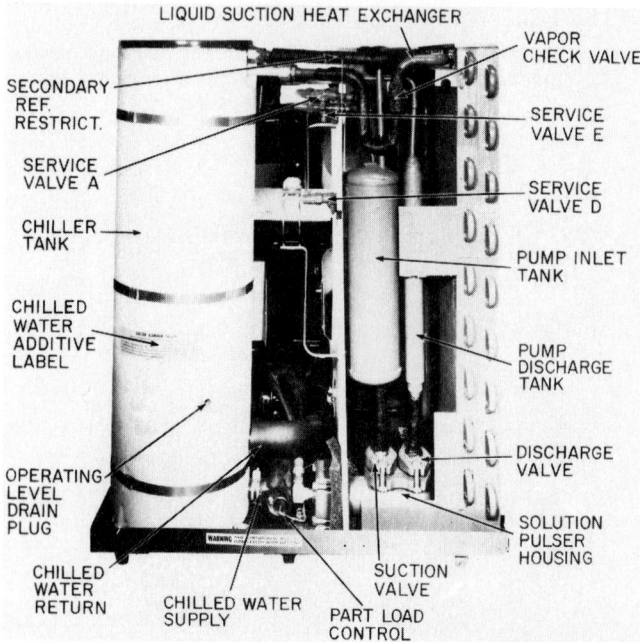

Fig. 16-32. Absorption system for air conditioning. Service valve A is mounted on solution-cooled absorber. Service valve D is mounted on pump inlet tank. Service valve E is mounted on pump discharge tank. Service valve C is not shown.

Valve E:
1. Removes large amounts of solution.
2. Determines if discharge chamber has proper amount of noncondensables.

Fig. 16-34. All-steel gauge manifold is connected to valve C (high-pressure side) and valve A (low-pressure side). Valve E is for removal and checking of solution.

16-19 COMMERCIAL ABSORPTION SYSTEM

Absorption systems are used successfully for air conditioning comfort cooling installations. But such systems may also be used for heating.

Some units use the ammonia-water-hydrogen continuous cycle. Others use water as the refrigerant, and various chemicals as the absorber.

In a system using water as the refrigerant and lithium bromide as the absorber, Fig. 16-35, steam heat applied to the generator percolates water vapor (red dots) and weak solution up to the separator. The liquid lithium bromide, shown in black, then flows by gravity through the heat exchanger to the absorber where it absorbs the evaporated water. The strong solution (black dots) settles to the bottom of the absorber and returns to the generator after passing through the heat exchanger. The pressure difference is maintained by the pressure head of the lithium bromide liquid.

The water vapor (red dots) in the separator rises to the

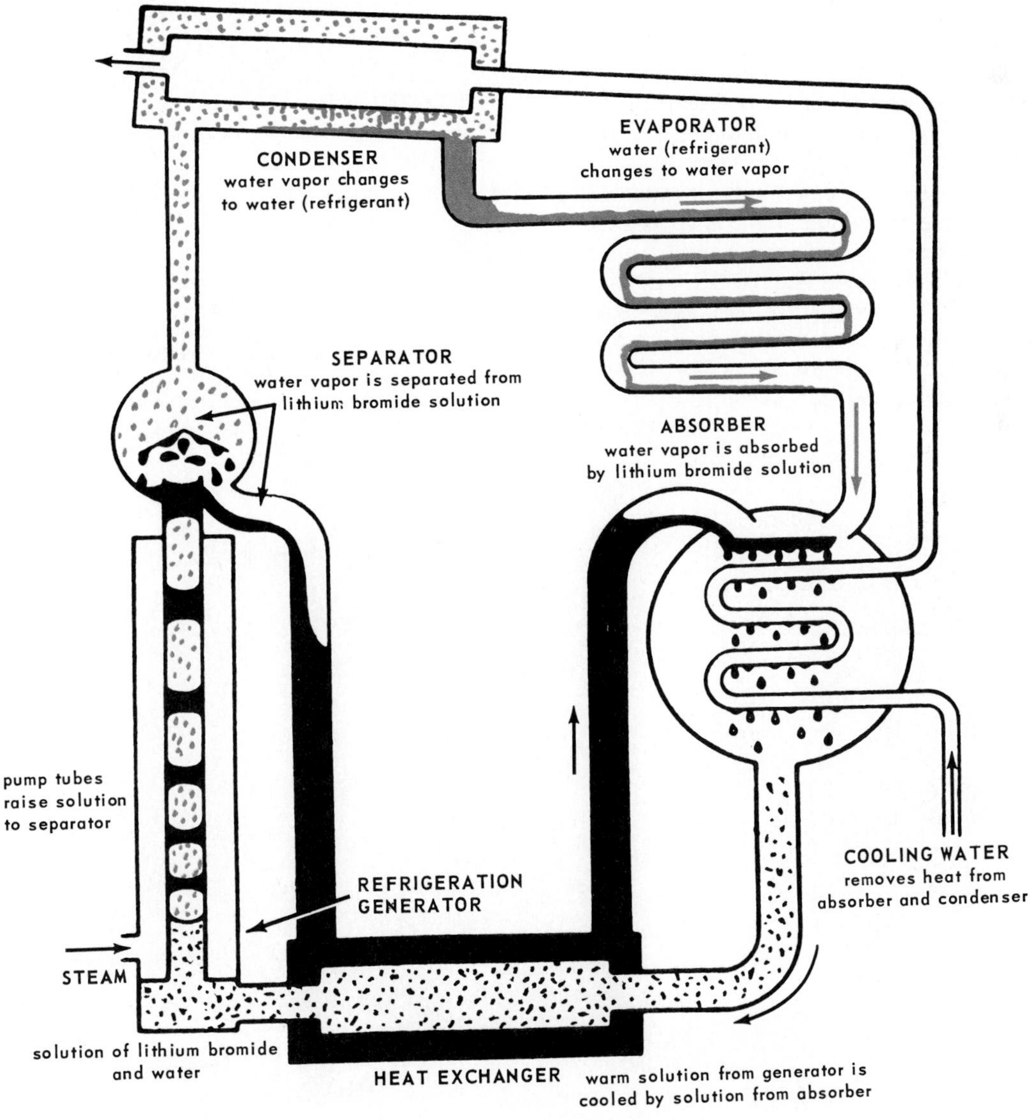

CONDENSER
water vapor changes
to water (refrigerant)

EVAPORATOR
water (refrigerant)
changes to water vapor

SEPARATOR
water vapor is separated from
lithium bromide solution

ABSORBER
water vapor is absorbed
by lithium bromide solution

pump tubes
raise solution
to separator

**REFRIGERATION
GENERATOR**

COOLING WATER
removes heat from
absorber and condenser

STEAM

solution of lithium bromide
and water

HEAT EXCHANGER warm solution from generator is
cooled by solution from absorber

Fig. 16-35. Absorption refrigeration cycle which uses water as refrigerant and lithium bromide as absorbent.

condenser where it is condensed and becomes water. The water flows by gravity through an orifice into the evaporator. The water evaporates at a low temperature due to a near-perfect vacuum in the system. The water vapor formed is absorbed by the lithium bromide (black). Note that the absorber and condenser are both cooled by water coils. The condenser water is then taken to a cooling tower where it is cooled and used over again. The condensing pressure is about 50 to 60 mm Hg (about 1 psia or 6.9 kPa), and the evaporating pressure is 8 to 10 mm Hg (about 0.17 psia or 1.2 kPa). Lithium chromate is often used as a corrosion inhibitor. A typical cooling tower is shown in Fig. 16-36. More detail is shown in Chapter 12.

16-20 ABSORPTION UNIT FOR AIR CONDITIONING AND HEATING

The application of absorption refrigerating systems in comfort cooling air conditioning and heating is increasing. Absorption units have advantages in solar energy systems. Solar energy, as a source of heat, can cool buildings when used in absorption systems. In other installations, where steam heat is used in winter, the steam becomes a heat source for absorption cooling in the summer.

These systems are also used to produce chilled water. The chilled water, in turn, may be used as quenching baths, drinking water, and as a special coolant to bring down the working temperature of welding tips. An absorption system for chilling water and heating is shown in Fig. 16-37. The cooling cycle is shown in Fig. 16-38, view A.

The refrigerant which is dispersed in the evaporator extracts the heat from the chilled water and is vaporized. The chilled water then passes through the system.

The heating cycle is shown in Fig. 16-38, view B. During the heating cycle, the evaporator functions as a condenser. Hot water in the evaporator tubes absorbs the heat given off during condensation of the refrigerant. The heated water is circulated throughout the system.

Absorption system heat can be supplied by exhaust gas from another industrial or residential system. This results in large efficiency gains.

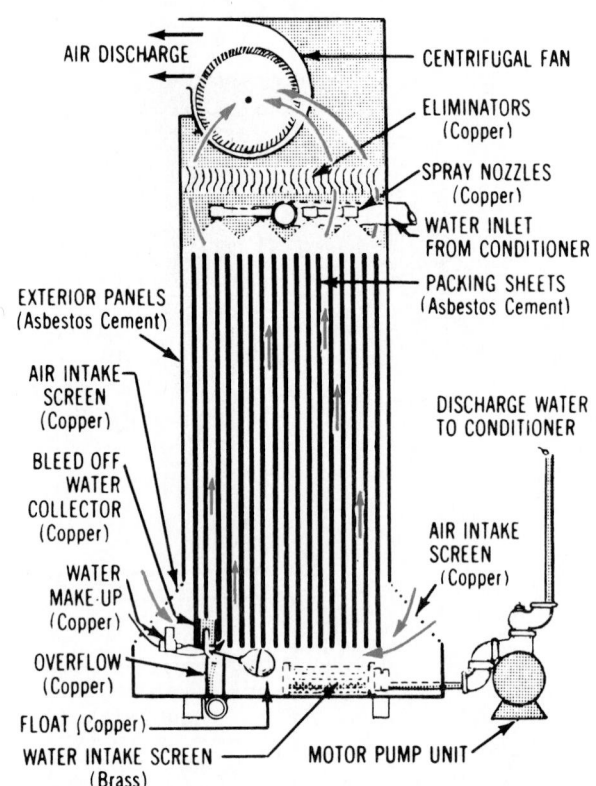

Fig. 16-36. Cooling tower used to cool condenser and absorber cooling water. Tower evaporates about 15 percent of condenser water and, in so doing, cools rest of water down to wet bulb temperature of air. It consists of water sprays, asbestos sheets, overflow tubes, make-up water float valve, and water pump. Eliminator plates keep water from being drawn into fan. Air enters at bottom and leaves at top.

Fig. 16-37. Large capacity, double effect absorption chiller and heater. System produces cooling and heating.

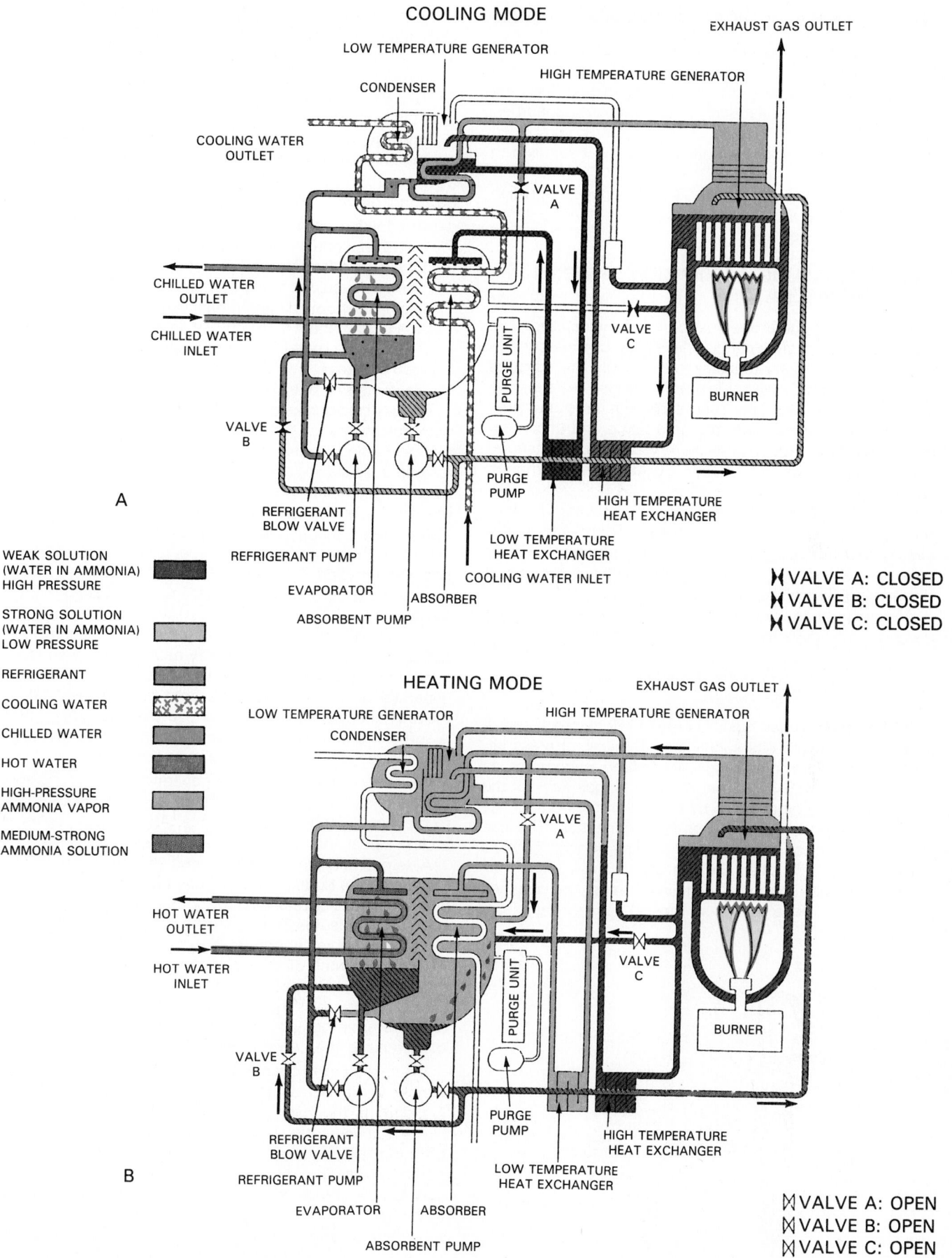

COOLING MODE

EXHAUST GAS OUTLET

LOW TEMPERATURE GENERATOR

HIGH TEMPERATURE GENERATOR

CONDENSER

COOLING WATER OUTLET

VALVE A

CHILLED WATER OUTLET

CHILLED WATER INLET

VALVE C

PURGE UNIT

BURNER

VALVE B

A

REFRIGERANT BLOW VALVE

REFRIGERANT PUMP

PURGE PUMP

HIGH TEMPERATURE HEAT EXCHANGER

EVAPORATOR

ABSORBER

LOW TEMPERATURE HEAT EXCHANGER

ABSORBENT PUMP

COOLING WATER INLET

WEAK SOLUTION (WATER IN AMMONIA) HIGH PRESSURE

STRONG SOLUTION (WATER IN AMMONIA) LOW PRESSURE

REFRIGERANT

COOLING WATER

CHILLED WATER

HOT WATER

HIGH-PRESSURE AMMONIA VAPOR

MEDIUM-STRONG AMMONIA SOLUTION

VALVE A: CLOSED
VALVE B: CLOSED
VALVE C: CLOSED

HEATING MODE

EXHAUST GAS OUTLET

LOW TEMPERATURE GENERATOR

HIGH TEMPERATURE GENERATOR

CONDENSER

VALVE A

HOT WATER OUTLET

HOT WATER INLET

VALVE C

PURGE UNIT

BURNER

VALVE B

PURGE PUMP

HIGH TEMPERATURE HEAT EXCHANGER

REFRIGERANT BLOW VALVE

REFRIGERANT PUMP

LOW TEMPERATURE HEAT EXCHANGER

EVAPORATOR

ABSORBER

ABSORBENT PUMP

B

VALVE A: OPEN
VALVE B: OPEN
VALVE C: OPEN

Fig. 16-38. Schematic diagrams of an absorption system. A—Cooling cycle of chiller/heater. B—Heating cycle of chiller/heater.

16-21 EFFICIENCY OF ABSORPTION SYSTEMS

In absorption cooling systems, performance is evaluated in two ways:

1. The energy efficiency. This is the cooling effect produced divided by the heat energy supplied to the absorber.
2. The effectiveness. This is the cooling effect produced divided by the work equivalent to the heat supplied to the absorber.

The second evaluation helps in comparing absorption systems with vapor compression systems where the input energy is work, not heat.

16-22 INSTALLING ABSORPTION SYSTEM REFRIGERATORS

Absorption refrigerators must be carefully installed. The absorption unit must be carefully leveled to operate correctly. It will be necessary to install a gas supply line from the house gas piping to the refrigerator. The gas line must be tested for leaks. Use only soap suds! Gas pressure must be carefully adjusted. The minimum flame and maximum flame adjustments must be made. A water column manometer is usually used. See Chapter 20 for more information on fuel gas servicing.

Some of the refrigerators have electric circuits. The electrical service must be carefully checked. Some city codes specify that the fuse plug opening in the refrigerant system must be vented to the outside to prevent any chance of discharging the refrigerant into the house.

Always provide enough air inlet and exhaust for proper combustion, condenser, and absorber cooling.

16-23 SERVICING ABSORPTION REFRIGERATORS

When servicing gas-fired absorption refrigerators, check to be sure that the gas supply is at the correct pressure. Check the gas pressure using a water-filled manometer, as shown in Fig. 16-39. (The amount of gas fed to the refrigerator may be checked by the size of the flame.) The safety valve body has a manometer connection. The flue must be kept clean to allow good transfer of heat. Brushes should be used to clean the flue.

Fins on the ammonia condenser must be cleaned periodically to make sure there is good heat removal from these surfaces.

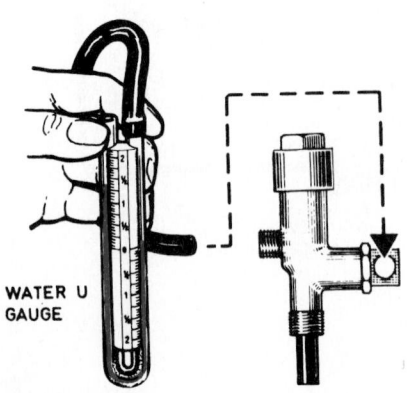

WATER U GAUGE

Fig. 16-39. Water-filled manometer measures gas pressure to burner. (General Electric Co.)

If a service call indicates that the refrigerator is too cold, it should be checked as follows:

1. First, check the temperature control dial. It may be set too cold.
2. The temperature of the evaporator unit may be lower than that indicated by the temperature control dial setting. A time-temperature graph of the evaporator temperature should be taken.

Perhaps the most common service call will be "little or no refrigeration." Following are some possible causes:

1. Overloaded cabinet.
2. Improper condensing temperatures.
3. Little or no heating of the generating unit.
4. The gas supply has been turned off or restricted. If the line has become clogged, resulting in a small consumption of gas, there is, of course, little or no refrigeration. This trouble may be traced by checking the pressures at the burner.
5. Restricted or dirty gas flue.

The gas-fired and kerosene-fired refrigerators are equipped with flues to direct the hot gases around and away from the generating units. Occasionally, these flues may be restricted by placing the refrigerator too close to the wall, by placing objects over the opening, or by having some obstruction fall into it. These flues must be kept clean to insure proper functioning of the refrigerator.

If either the condenser or absorber is dirty or lint-covered, poor refrigeration will result. This is due to poor airflow around these components. Wipe, brush, or vacuum away these accumulations.

It is normal, after a period of operation, for the generator flue to become coated with sooty deposits. When this occurs, rapid transfer of heat from the gas flame to the generator is not possible. This soot deposit must be removed periodically (usually once every one to two months is sufficient) to insure proper refrigeration. Frequent cleaning also reduces gas consumption of the unit.

When scraping the flue of a generator or removing soot from any surface of the generator, care should be taken to prevent damage to the surface. Always put papers or a cloth under the refrigerator when cleaning flues.

A temporarily unused absorption unit may not freeze. Make sure the burner is lit. Air may have filled the gas line and it may take several tries to light the burner. If the burner is lit, the problem is likely due to blockage within the unit. Manufacturers recommend tilting to remove the blockage. After 10 minutes of operation, tilt the refrigerator to the right for about 30 seconds. Then tilt it to the left for 30 seconds. Do this three to four times, then put the refrigerator in the upright, level position. If it still does not cool, replace the cooling unit.

If the system is overheated, the pipe going to the condenser will overheat and the percolation pump will stop working. If the paint on the pipe to the condenser is blistered, overheating has taken place. To remedy, shut off the heat, allow the system to cool, turn the unit upside down several times to put the fluids in their proper place, and then restart with a lower heat input to the boiler (generator).

If there is a leak, a yellow deposit will collect at the leak. If the leak occurs at the evaporators, an ammonia odor will be noticeable. A burning sulphur candle will produce a white smoke at an ammonia leak.

16-24 SERVICING LITHIUM BROMIDE SYSTEMS

Evacuation of the system is necessary after the system is opened:
1. To be able to reach 40 °F (4 °C).
2. To remove noncondensables.

Evacuation is needed if the pressure in the system is 1 in. Hg. (25,400 microns) or more. The pressure is determined by a manometer which is connected by means of a service valve.

Lithium bromide is a nontoxic, nonflammable, nonexplosive, and chemically stable substance which is used as a liquid. It can be handled in open containers but becomes corrosive when exposed to air. It may irritate skin, eyes, and mucous membranes. Octyl alcohol is sometimes added to reduce surface tension of lithium bromide because it acts as a wetting agent.

Important note: "Strong" (concentrated) solution means strong in ability to absorb. "Weak" (dilute) means weak in ability to absorb.

Sixty-five percent lithium bromide by weight will start to crystallize at 110 °F (43 °C). Solution must not be allowed to reach high concentrations or low temperatures which allow crystallization.

The typical charge is a:
1. Lithium bromide solution   120 gal.
2. Inhibitor charge   1 pt.
3. Refrigerant (water)   35 gal.
4. Octyl alcohol (two ethyl hexanol)   1 gal.

The solution becomes thicker as the amount of lithium bromide increases. This will cause a greater temperature difference between refrigerator temperature and chilled water temperature. If solution concentration gets too high, the refrigerant will turn solid and must be dissolved. If the absorber becomes too cold (below 85 °F or 29 °C), solidification can also occur.

The system is charged with R-13 vapor to test for leaks (not soluble in water) and an electronic leak detector is used. Then the system is evacuated completely. Helium is also used, but it requires the use of a special leak detector or soap bubbles.

16-25 REVIEW OF SAFETY

The refrigerant most commonly used in the small absorption refrigerating units is ammonia. Its odor is pungent (sharp or irritating) and tends to restrict breathing. It is toxic and injurious to the skin and eyes. Avoid puncturing the system or creating too high a pressure in the system or a leak may result.

Caution: Never cut or drill into an absorption refrigerating mechanism. The high-pressure ammonia solutions are dangerous and may cause blindness if the fluid gets into the eyes.

Many of the absorption units are heated with LP-Gas or natural gas. The gas piping system must be leakproof. Always use soapsuds to check for leaks. Never use an open flame such as a match. An explosion may occur. The burner flues should be cleaned periodically or a poor flame may result.

The flame safety device should be checked. To do this, smother the flame and check to determine if the safety valve closes.

The condenser duct system and the condenser should be cleaned at least each six months or excessive condenser pressures may result.

Some absorption systems use electrical current as well as fuel gas. The usual precautions should be used in handling these circuits.

It is a good idea to ground these refrigerators to eliminate any danger of receiving a shock should a circuit become grounded to the frame of a cabinet or part of the mechanism.

16-26 TEST YOUR KNOWLEDGE

1. Who discovered the absorption principle?
2. What localities would probably need kerosene-fired intermittent absorption refrigerators?
3. Name three substances which have been used in absorption refrigerators to absorb refrigerant gas.
4. In absorption refrigerators, does the liquefication of the refrigerant depend upon compression?
5. What is the primary difference between a conventional compression air conditioning system and an absorption system?
6. What purpose does hydrogen serve in the continuous absorption system?
7. Why is the storage cylinder or receiver in the intermittent absorption refrigerator insulated?
8. Why do absorption systems have a fuse plug?
9. How can burning more gas in the continuous absorption cycle produce more cold?
10. What is the combined solution used inside a continuous absorption cycle mechanism?
11. How is the cabinet temperature adjusted in the continuous absorption cycle refrigerator?
12. Why must the absorption unit be level?
13. Is kerosene sometimes used as the fuel for continuous systems?
14. Why is a pressure regulator not required in the base of the cabinet for LP systems?
15. How is the temperature regulated in a piped gas continuous system?
16. Why doesn't the flame go out when the cabinet is cold enough in a continuous LP gas system?
17. What are two basic causes for too little refrigeration in a continuous system?
18. How are some continuous cycle systems automatically defrosted?
19. What is the purpose of lithium bromide in an absorption system?
20. Do any absorption systems operate under a vacuum?
21. What does the solution pump do in an ammonia water five-ton air conditioner?
22. What causes the liquid refrigerant to flow in a domestic absorption system?
23. Does the generator percolate a weak-in-ammonia and strong-in-water solution?
24. What method of defrosting can be used in absorption systems?
25. Name three sources of heat than are used in absorption systems.

Cryogenic food freezing in batch-type production. (Compressed Air Magazine)

Chapter 17

SPECIAL REFRIGERATION SYSTEMS AND APPLICATIONS

After studying this chapter, the technician will be able to:
☐ Discuss the operation and application of expendable refrigerant systems.
☐ Define and discuss principles and operation of: systems of liquefying gas; thermoelectric refrigeration; vortex tube cooling and heating; jet cooling; multistage cooling systems; heat pipe; immersion freezing; cryogenic refrigeration, and the Sterling cycle.

17-1 EXPENDABLE REFRIGERANT SYSTEMS

Use of liquid nitrogen or liquid carbon dioxide for cooling transportation vehicles (truck bodies) is rapidly increasing. Expendable refrigerant systems are also used in the cooling of railroad cars and shipping containers used to transport perishable items. Basic system uses a liquid nontoxic low-temperature refrigerant. It is the same as any vapor system but has no condensing unit. The low-cost liquid can be used as a refrigerant, then is released to the atmosphere. This is called chemical or open-cycle refrigeration.

Latent heat of vaporization of the refrigerant determines its heat-absorbing ability. The three most common refrigerants and their latent heat of vaporization are shown in Fig. 17-1. All are fairly high in latent heat of vaporization. If properly processed, they create little air pollution. Refrigerant is supplied in large cylinders.

Two basic cooling mechanisms are in common use. One is cold plate cooling, the other is spray cooling.

17-2 EXPENDABLE REFRIGERANT EVAPORATOR SYSTEM

Liquid refrigerant is kept in large metal insulated cylinders. These are really large thermos bottles. Sometimes they are located in the front of the cargo vehicle, as shown in Fig. 17-2. Each unit has a temperature control providing a temperature range of $-20\,°F$ ($-29\,°C$) to $60\,°F$ ($16\,°C$).

The temperature control is connected to a temperature sensor much the same as in a standard thermostatic motor control. As temperature rises, the switch operating the control valve is opened and liquid refrigerant flows into the evaporators. The evaporators may be blower coils, plates, or eutectic plates. (See Chapter 12.) As liquid refrigerant passes through the plates, it vaporizes. Vapor is forced through the plates by the pressure difference. When the desired temperature is reached, the refrigerant valve closes. The used vapor leaves the evaporator at approximately the same temperature as the air in the cargo space. With this method, no refrigerant mixes with the air in the interior of the vehicle.

17-3 EXPENDABLE REFRIGERANT SPRAY SYSTEM

Transport vehicles are effectively cooled by spraying liquid nitrogen or carbon dioxide directly into the refrigerated space. The nitrogen turns into vapor inside the cargo area.

Fig. 17-2. Expendable refrigerant system. Two nitrogen cylinders located inside truck body are connected by a manifold to regulators and to temperature control solenoid valves. Vaporizing liquid nitrogen flows into vaporizers or cold plates to refrigerate truck box.

REFRIGERANT	BOILING TEMPERATURE		LATENT HEAT OF VAPORIZATION Btu/lb.
	°F	°C	AT 32 °F (0 °C) AND 0 psi
AMMONIA	−28.1	−33	590.4
CARBON DIOXIDE	−109.3	−78.5	275
NITROGEN	−320.5	−196	173

Fig. 17-1. Heat-absorbing ability of expendable refrigerants.

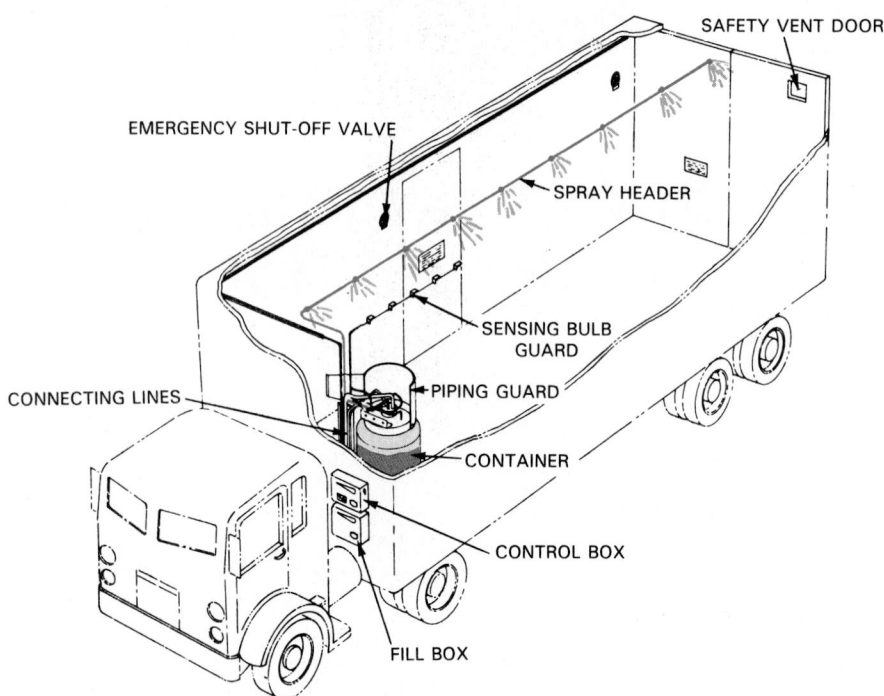

Fig. 17-3. An expendable refrigerant system for a refrigerated truck. Liquid nitrogen is in the insulated container. The container is installed vertically inside the truck body.

Fig. 17-3 shows a complete spray system with the cylinder inside the truck body. For details, see Fig. 17-4.

The liquid spray method has many of the same parts as the cold plate method—liquid containers, control box, and fill box, for example. It also requires additional devices not necessary in the plate method, such as spray headers, emergency switches, and safety vents. Another type of spray cooling system is shown in Fig. 17-5. Details of the horizontal cylinder system are shown in Fig. 17-6.

This is how the system operates:
1. Liquid nitrogen is pumped into the storage cylinder by way of the fill box. (Fig. 17-7 shows the details of the container piping connections.)
2. When the containers are filled and the cargo space is loaded, the temperature is selected at the main control. A temperature sensing device anticipates temperature changes in the cargo space. See Fig. 17-8.
3. When the cargo temperature rises above the setting, the temperature controller opens the liquid line solenoid valve. This allows liquid refrigerant to enter into the spray header where it becomes a vapor and maintains the desired temperature.

Some units have two or more containers—a primary container and a secondary container—which are filled in series. As the first or primary container is filled, liquid refrigerant will overflow into the second container and so forth. Fig. 17-9 is a wiring diagram for a typical installation.

The spray header system is a perforated pipe usually mounted along the roof at the center of the cargo compartment. The nitrogen tanks are equipped with safety valves which will release nitrogen to the outside of the vehicle if the pressure in the containers rises above 22 psi (255 kPa).

All units have a safety vent, Fig. 17-10, which allows gas to exhaust to the atmosphere when inside cargo space pressure increases above atmospheric. This safety vent is a spring loaded device and will close when excess pressure has been released. In addition, each door has a safety switch connected to it which will automatically shut down the unit when a door is opened and before anyone can enter. A system which maintains four compartments in the truck body at different temperatures is shown in Fig. 17-11.

Some truck bodies have electric heaters. These heaters are used when the truck is exposed to temperatures below the temperature wanted inside the truck. See Fig. 17-12.

Always read the warning signs on refrigerated vehicles before entering them. Caution: The temperature of liquid nitrogen as it comes from the spray nozzles—depending on the thermostat setting—is much below 0 °F (−18 °C). Were it to hit any part of a human body, the flesh would be frozen instantly. Be sure that no living animal or human is in the refrigerated space when the doors are closed.

The spray cooling system which uses nitrogen or carbon dioxide has other advantages beyond ease in providing necessary refrigerating temperatures. Since these gases take the place of oxygen in the storage space, fruits, vegetables, meats, and fish—either in transit or in storage—are preserved by the inert atmosphere.

The inside of the expendable refrigerant systems must be kept free of dirt and moisture. To pressure test the piping, use only nitrogen or helium. CAUTION: See Para. 11-39.

17-4 OPEN CYCLE AMMONIA

There are two different methods of using ammonia in open cycle refrigeration:
1. Burn the exhaust ammonia vapor in the atmosphere.
2. Burn the ammonia in an internal combustion engine.

Ammonia has a very high latent heat of vaporization. This makes it a good heat-absorbing substance. Large quantities of commercial grade ammonia are used in agriculture. This

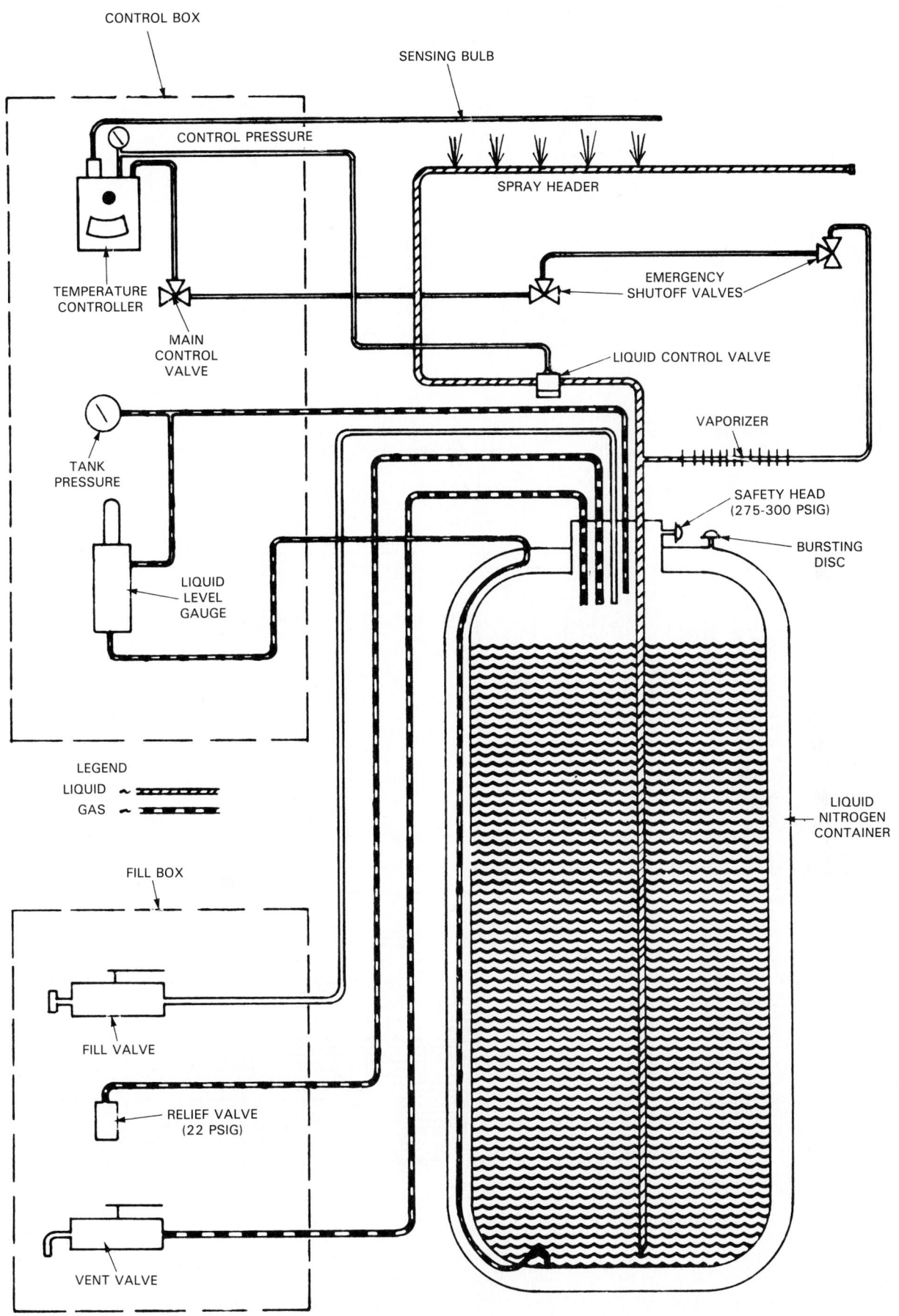

CONTROL BOX

SENSING BULB

CONTROL PRESSURE

SPRAY HEADER

TEMPERATURE CONTROLLER

EMERGENCY SHUTOFF VALVES

MAIN CONTROL VALVE

LIQUID CONTROL VALVE

VAPORIZER

TANK PRESSURE

SAFETY HEAD (275-300 PSIG)

BURSTING DISC

LIQUID LEVEL GAUGE

LEGEND

LIQUID ~

GAS ~

LIQUID NITROGEN CONTAINER

FILL BOX

FILL VALVE

RELIEF VALVE (22 PSIG)

VENT VALVE

Fig. 17-4. A vertical container expendable refrigerant system. Note controls, safety devices, and filling piping. (Union Carbide Corp., Linde Div.)

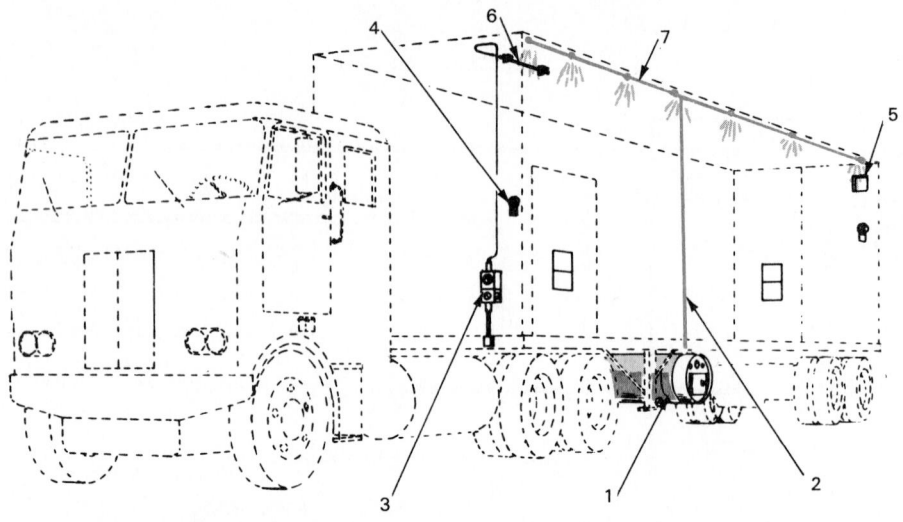

Fig. 17-5. An expendable refrigerant system using nitrogen. 1—Horizontal container mounted under truck body. 2—Liquid line. 3—Controls. 4—Shutoff valve. 5—Safety vent door. 6—Sensing bulb. 7—Spray header.

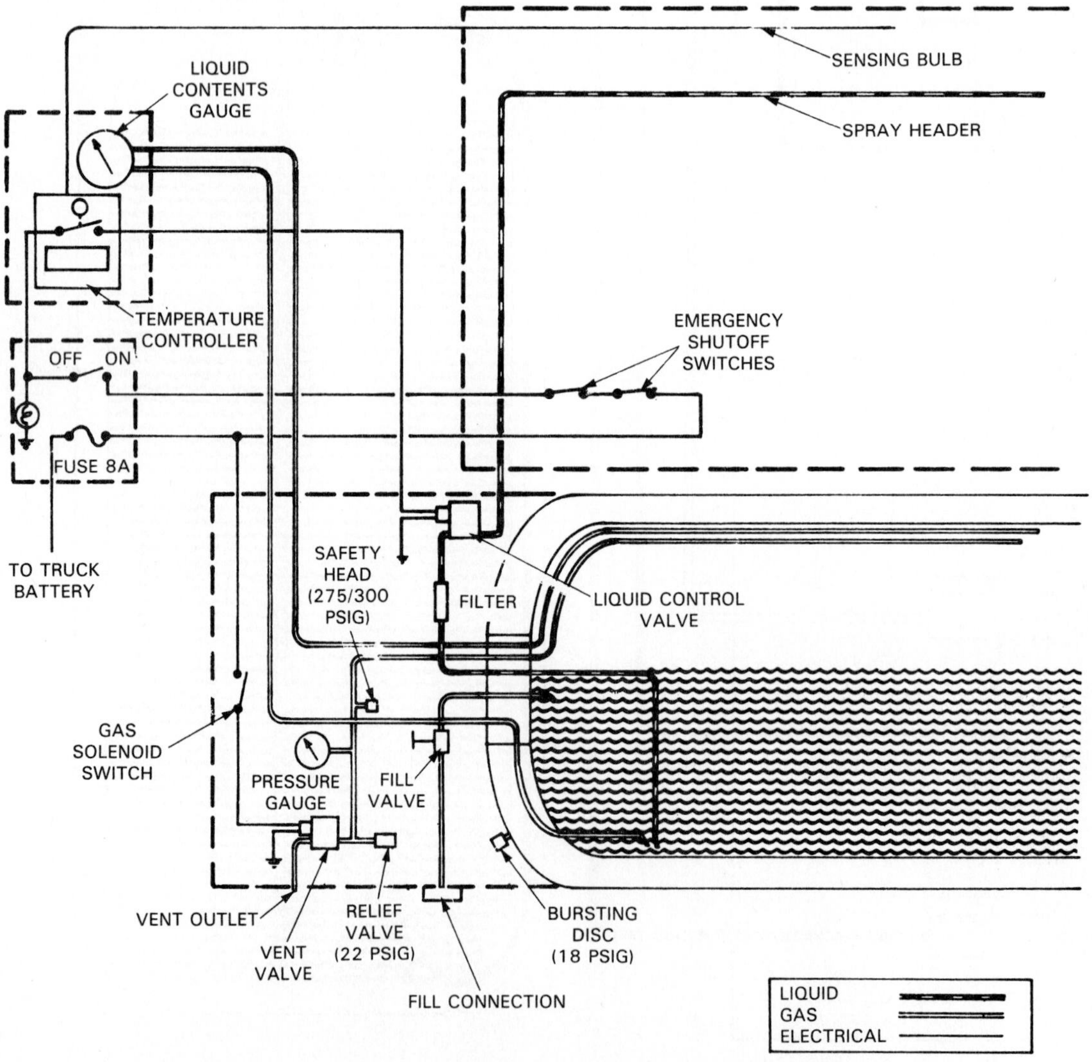

Fig. 17-6. Expendable refrigerant system showing piping, safety devices, and controls.

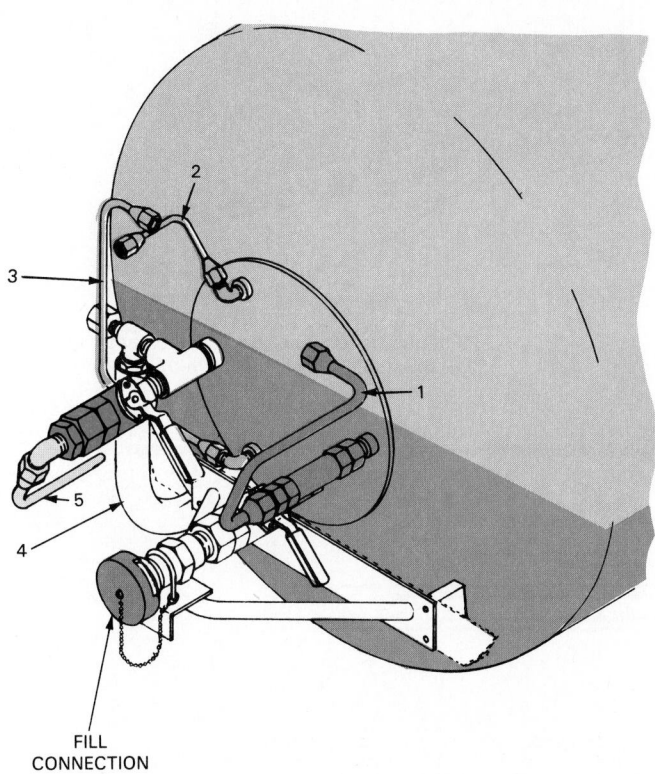

Fig. 17-7. Piping connections to container of expendable refrigerants. 1—Liquid line connection. 2—Liquid level gauge connection, vapor phase. 3—Liquid level gauge connection, liquid phase. 4—Polyethylene tubing. 5—Line assembly. (Union Carbide Corp., Linde Div.)

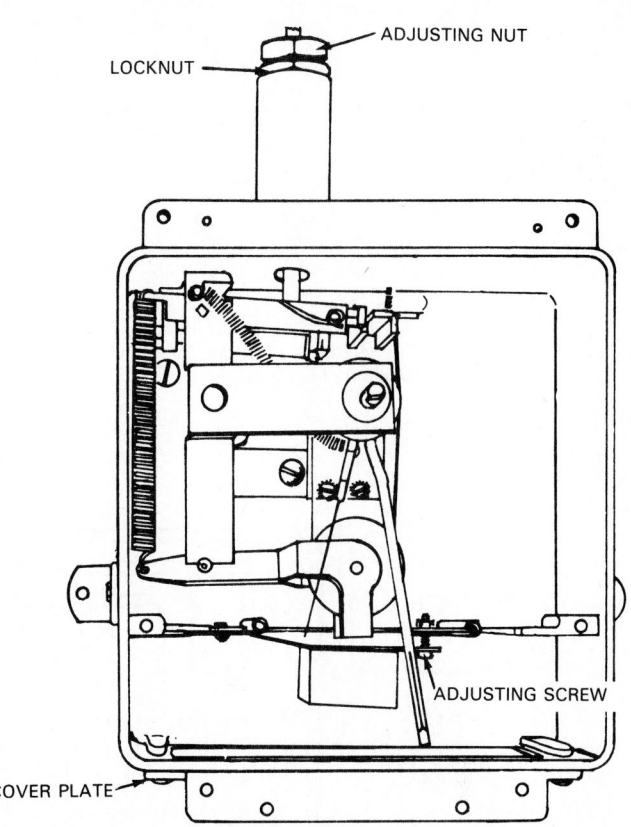

Fig. 17-8. Temperature controller used with expendable refrigerant systems. Unit has both temperature (range) and differential adjustments.

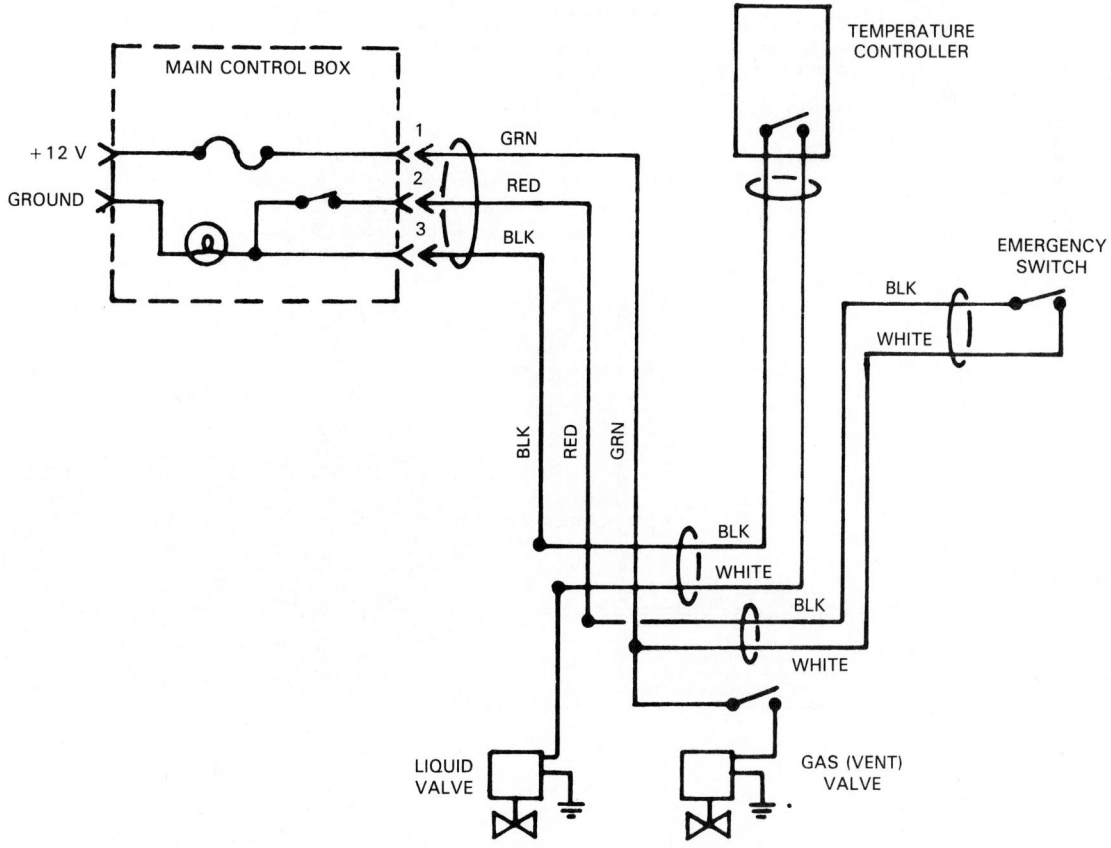

Fig. 17-9. Diagram of electrical circuit for automatic expendable refrigerant system.

A
B

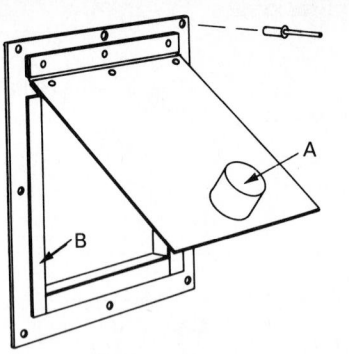

Fig. 17-10. Safety vent. Small door will open automatically if refrigerant pressure in truck body increases. A—Weight. B—Magnetic gasket normally holds door shut.

HEATER TAPES
TRUCK FLOOR
HEATER TAPES

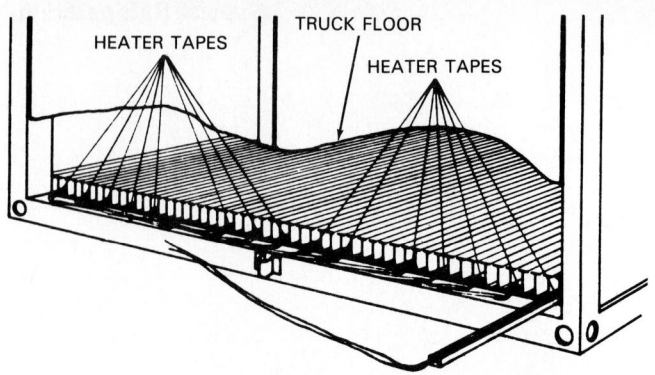

Fig. 17-12. Electric heating system for truck bodies. (Union Carbide Corp., Linde Div.)

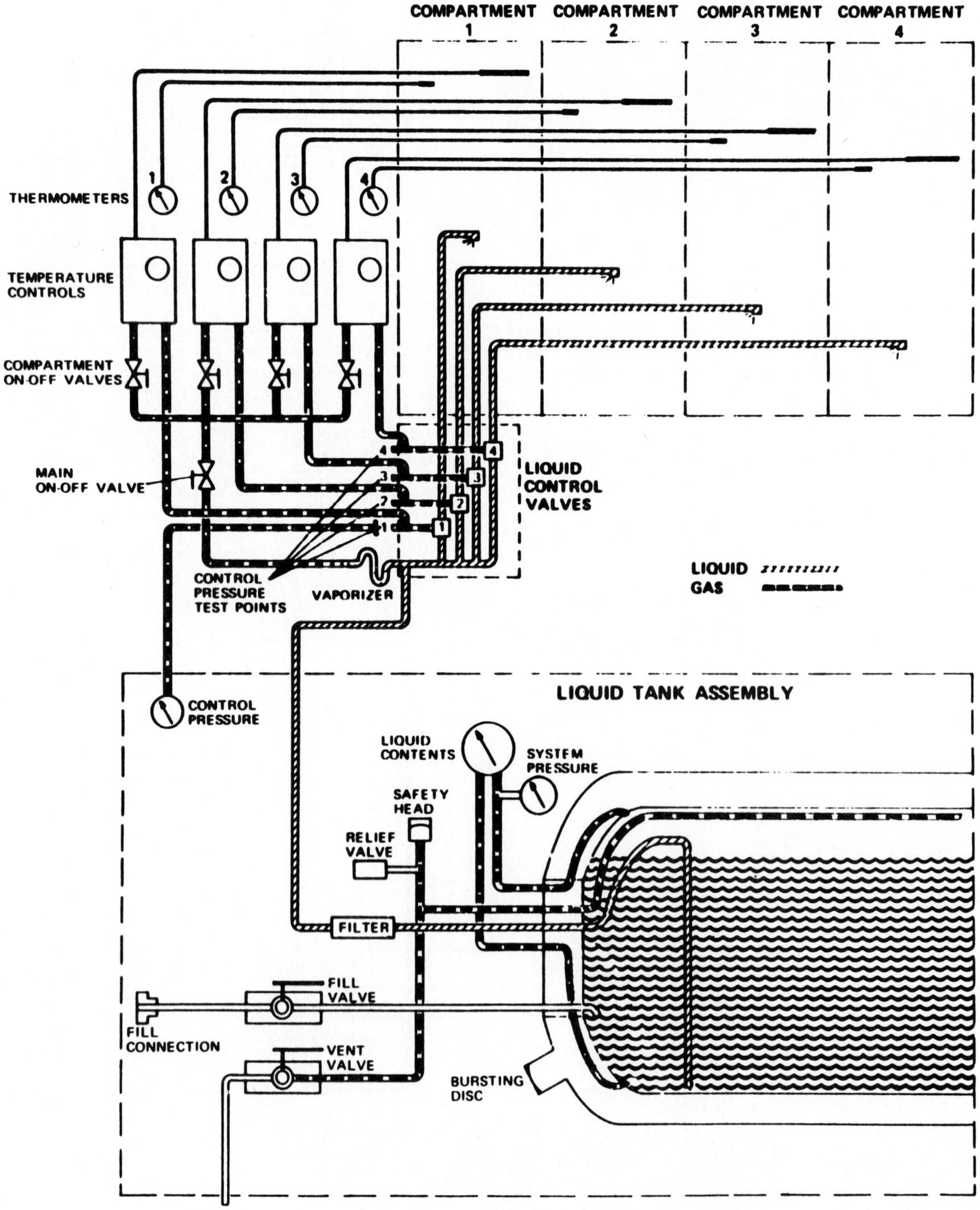

COMPARTMENT 1
COMPARTMENT 2
COMPARTMENT 3
COMPARTMENT 4
THERMOMETERS
TEMPERATURE CONTROLS
COMPARTMENT ON-OFF VALVES
MAIN ON-OFF VALVE
LIQUID CONTROL VALVES
CONTROL PRESSURE TEST POINTS
VAPORIZER
LIQUID
GAS
LIQUID TANK ASSEMBLY
CONTROL PRESSURE
LIQUID CONTENTS
SYSTEM PRESSURE
SAFETY HEAD
RELIEF VALVE
FILTER
FILL VALVE
FILL CONNECTION
VENT VALVE
BURSTING DISC

Fig. 17-11. Expendable system with four separately controlled sprayers and controls. Four compartments in truck body can be kept at four different temperatures.

makes possible a rather inexpensive product.

Ammonia vapors are very irritating and offensive. They cannot be tolerated by humans or animals. In an open cycle refrigeration system using ammonia, the exhaust must be processed in some way so that the irritating quality is removed.

The simplest way to do this is to burn the vapor in an open flame. The chemical formula is NH_3; this means that ammonia is made up of nitrogen and hydrogen. When ammonia burns in an open flame, the hydrogen combines with oxygen from the air and forms water. The odorless nitrogen is released to the atmosphere.

Another system of processing expendable ammonia vapor is to burn the vapor in an internal combustion engine. In this way the vapor produces some fuel and, as in the open flame, the products of combustion will be released as nitrogen and water vapor.

17-5 COOLING SYSTEMS FOR LIQUEFYING GASES

Refrigerants are usually chemically produced in gaseous form. They must then be liquefied for storage and transportation. Natural gas, used as a heating energy fuel for many homes and businesses, may also need to be put into a liquid form to assist in transporting and storage.

Fig. 17-13 shows a diagram of the typical process for liquefying R-12. The gas at 200°F (93°C) and 400 psi (2 864 kPa) is expanded into an evacuated chamber, where some of it evaporates. The remaining liquid, at 0°F (−18°C), is drawn off into a storage tank. A heat exchanger uses the cold vapor produced to cool the incoming gas before expansion. These simple refrigeration systems are typical of methods used to liquefy gases such as nitrogen and oxygen as well.

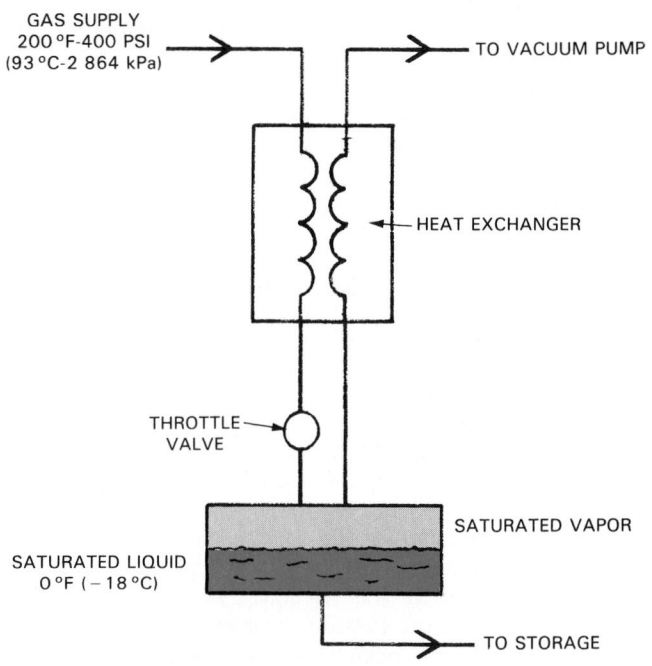

Fig. 17-13. Gas liquefication process using the saturated vapor evacuation method. This method is typical of systems used to liquefy common gases and refrigerants in chemical plants.

Natural gas, which is mainly methane, occurs naturally in gaseous form underground. In some instances, the natural gas must be stored and then transported long distances. A good example is the storage of natural gas on ships and the subsequent distribution throughout the world. In these cases, the natural gas is liquefied. This is done by a "cascade" cooling process shown in Fig. 17-14. The

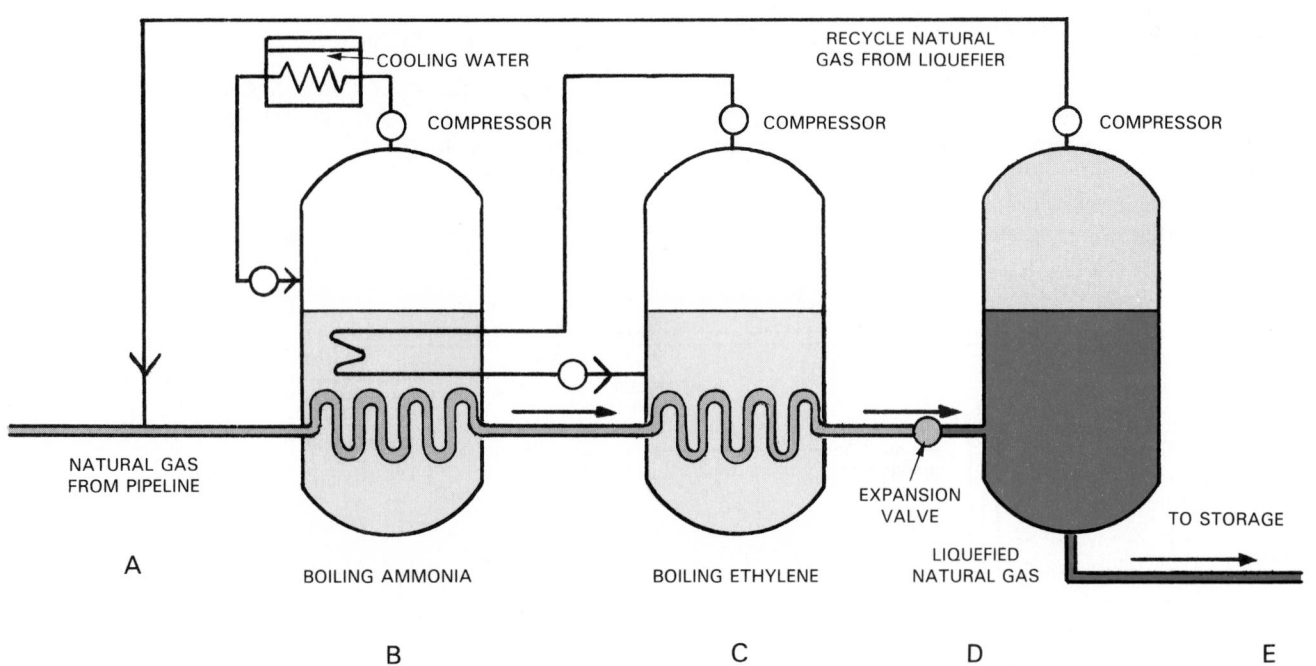

Fig. 17-14. A cascade process used for liquefication of natural gas. A—Gas at high pressure. B and C—Gas is cooled in turn by boiling ammonia and boiling ethylene. D—Natural gas is then reduced to a low pressure (through an expansion valve) and becomes a liquid. E—Liquefied natural gas then goes to storage.

natural gas from the pipeline, A, is cooled, using boiling ammonia, B, and boiling ethylene, C. This is followed by the evaporative cooling of the gas itself, D. This last step is similar to the process used to liquefy refrigerants. The liquid methane is then put into storage or piped directly into holding tanks. Subcoolers, similar to those used for liquid refrigerants, are used to maintain the liquid form of the methane during shipping.

17-6 THERMOELECTRIC REFRIGERATION

The thermoelectric process is a means of removing heat from one area and putting it in another area, using electrical energy rather than refrigerant as a "carrier." It has been used mainly in portable refrigerators and luxury-type stationary domestic refrigerators, water coolers, and for cooling the scientific apparatus used in space explorations and in aircraft.

Another application of the thermoelectric principle is in computer systems or electronic component cooling. In general, these components are cooled using fans and finned components. In larger systems, chilled water or refrigerant circulation may be necessary. However, as computer systems continue to become smaller, the use of a thermoelectric cooling system is becoming more common. As with any cooling system, the thermoelectric system and its specifications must be carefully reviewed in advance to make sure the cooling capacity it will deliver equals or exceeds the capacity that is required.

Thermoelectric cooling requires none of the conventional equipment necessary in a vapor system. There is no compressor, evaporator, condenser, or refrigerant. In fact, there are no moving parts; the unit is silent, compact, and requires little service.

Fig. 17-15 shows an electrical circuit diagram for the power supply of a typical thermoelectric cooler.

The input of 120 V ac current is stepped down in a transformer to 20 V ac. This current passes through a rectifier and is changed to 20 V dc. The direct current is then passed through the thermoelectric module so that the junction inside the refrigerator becomes cold and the junction outside the refrigerator is warmed. Principles covering operation of the thermoelectric refrigerator are explained in Chapter 6.

Thermoelectric cooling units, when used in refrigeration, are called modules. A module consists of several cold and hot junctions in series. The diagram of such a module is shown in Fig. 17-16. *The letters P and N, used in thermoelectric application, do not refer to current polarity positive (+) and negative (−).* P and N, in thermoelectric units, refer to the properties of the semiconductor materials. The materials are also designated as positive or negative, depending on how the semiconductor electrons behave under the influence of current flow. Construction of a

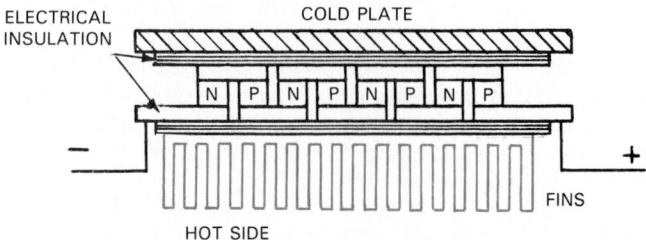

Fig. 17-16. Assembled thermoelectric module. Note use of fins (hot side) to speed up removal of heat which has been absorbed from surface of the cold plate. (Koolatron Industries)

module attached to a refrigerator is shown schematically in Fig. 17-17. The direction of the direct current flow into the module determines whether the junction will be warmed or cooled.

The reversing switch is usually made a part of the electrical circuit. This makes it possible to cause the cabinet to either cool the food or warm it.

The use of thermoelectric refrigerators is increasing, even though they have a low coefficient of performance (COP). The unit has versatility. It can operate on 12 V dc power or from a 110 V ac power adaptor. This has made the unit popular for picnics and camping use. One of these refrigerators is illustrated in Fig. 17-18.

17-7 VORTEX TUBE

The vortex tube is an interesting device capable of providing both cooling and heating at the same time. See Chapter 19. Its source of energy is compressed air. It converts the compressed airflow into two streams of air, one hot and one cold.

The amount of hot or cold air released from the outlets

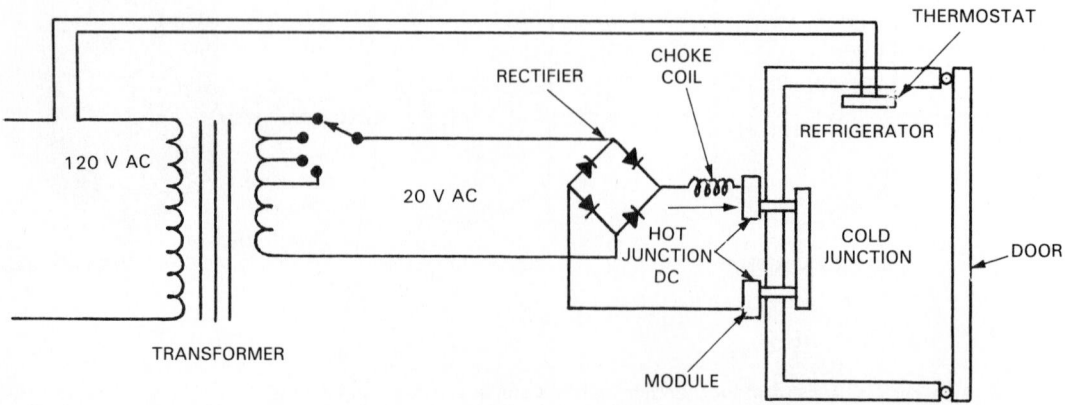

Fig. 17-15. Electrical circuit for thermoelectric refrigerator.

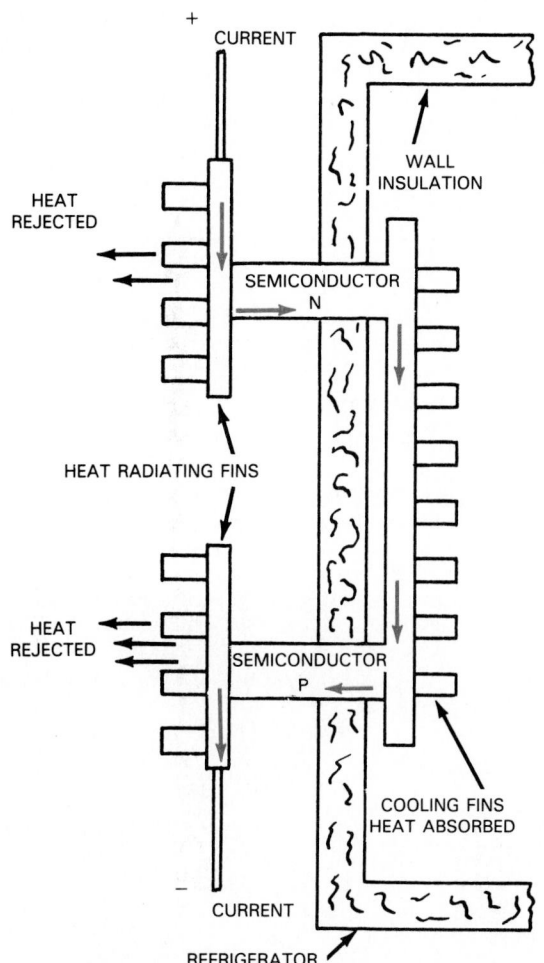

Fig. 17-17. Diagram of simplified thermoelectric system used in cooling small areas. Note flow of current to produce cooling within box.

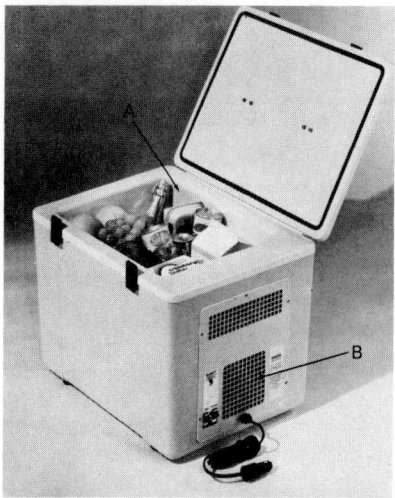

Fig. 17-18. Small portable refrigerator operates from normal 120 V household current or from 12 V battery. A—Cooling surface. B—Airflow openings for cooling hot junctions. (Koolatron Industries)

can be varied. For example, a unit with a 100 psi (794 kPa) air supply at 70 °F (21 °C) can be adjusted to cool half the air to − 29 °F (− 34 °C) while heating the other half to 91 °F (33 °C).

Fig. 17-19 is a schematic drawing of a vortex tube. Compressed air enters at A. It then goes into a number of small nozzles where it will lose some of its original high pressure. As it expands, it moves at near sonic speed (velocity). This is shown in Fig. 17-20.

Nozzles are arranged so that the air is injected tangentially (around the inner surface) to the circumference of the generation chamber. This makes the air swirl or spin like a cyclone. The control valve at the end of the hot air tube controls the flow of both the heated air and the cooled air.

The position of the control valve determines how much air will leave the hot end and how much air will be forced out at the cold end.

There are two streams of spinning air going through the generation chamber. The outside layer moves toward the hot outlet and the inside layer moves toward the cold outlet. In this process, the air traveling through the center becomes very cold and leaves the vortex tube at the cold opening, providing the required cooling.

Vortex tube cooling has many different applications. Since is must depend upon supply of compressed air at fairly high pressure, it is often used where the air exhaust—either hot or cold—for the vortex tube can be used for other purposes also. It is particularly desirable in locations where both ventilation and cooling are needed.

A common application is the cooling of workers' clothing. A diffuser air vest, worn under a protective cover, is frequently used to distribute the tempered air over the upper body, where it is needed most. The vest has tiny holes that allow cool or warm air from the vortec to escape over the upper body, Fig. 17-21. A worker wearing the diffuser air vest while spraying in a confined area is shown in Fig. 17-22.

If the entire body is to be cooled or heated, a total system is used that distributes the air over the upper and lower parts of the body. A total air respiratory system, consisting of a helmet and a garment, is shown in Fig. 17-23.

Many industries use the vortex tube for cooling tool bits and in similar applications where the use of liquid coolants

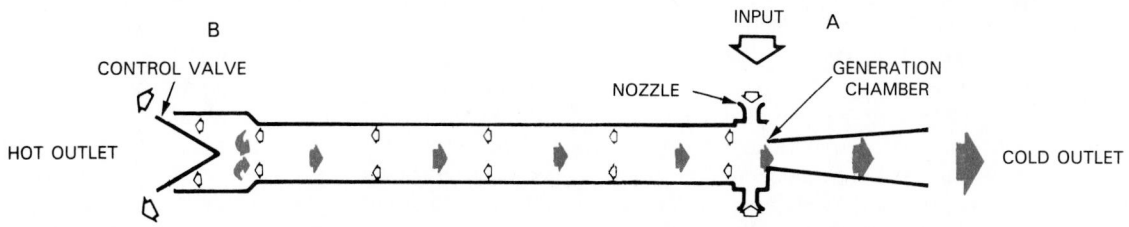

Fig. 17-19. Schematic drawing of vortex tube. A—Compressed air inlet. B—Control valve. (Vortec Corp.)

25 CFM OF COMPRESSED AIR AT 100 PSI (794 kPa) AND 100 °F (38 °C) IS
FED TO NOZZLES, ACCELERATES TO SONIC SPEED

AIR ENTERS CHAMBER TANGENTIALLY, CREATES
CYCLONE SPINNING AT 500,000 RPM

75% OF AIR SPIRALS INWARD, EXPANDS AND COOLS TO 40 °F (4 °C)

REMAINING 25% OF AIR CHURNS
IN TUBE, HEATS UP TO 270 °F (132 °C)

VALVE REGULATES
HOT AIR FLOW

COLD AIR
EXHAUSTS

AIR TRIES TO SPEED UP TO 5,000,000 RPM WHILE SPIRALING
INWARD, IS RETARDED BY AIR COLUMN IN TUBE, FORCIBLY
TURNS COLUMN WITH 1/2 HORSEPOWER

HOT AIR
EXHAUSTS

Fig. 17-20. Diagrammatic view of vortex tube. Note flow of air at 100 psi (794 kPa) and 100 °F (38 °C) into nozzles. Air is exhausted at temperature of 40 °F (4 °C) at cold end. Cold air is shown by the broken line and arrows.   (Vortec Corp.)

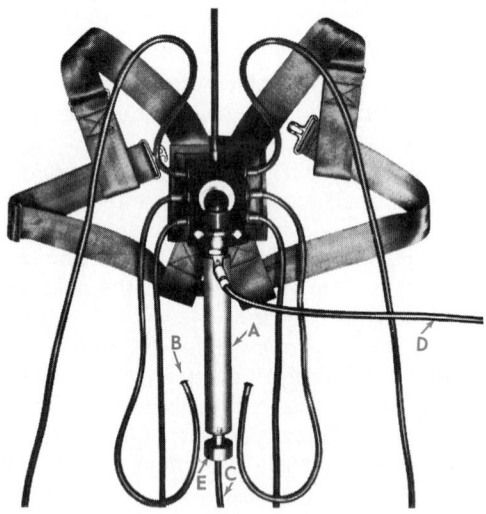

Fig. 17-21. Vortex tube assembly ready to be fitted to miner's clothing. A—Vortex tube device. B—Cooled air outlet (seven). C—Exhaust warm air. D—Compressed air lines. E—Temperature adjustment valve.

Fig. 17-22. A paint sprayer using a diffuser air vest in a hot confined area. Note location of A—compressed air line; B—vortec tube; and then C—into back of vest.   (Vortec Corporation)

would be undesirable. It is particularly useful in installations where the stream of cold air may be used for both cooling and removing chips or in installations where the exhaust air will provide good air for the operator to breathe. Typical vortex tubes are illustrated in Fig. 17-24.

17-8 JET COOLING SYSTEMS

A jet pump, which consists of a centrifugal pump and ejector, can replace the compressor in some air conditioning and refrigeration systems. Two applications involve *steam jets* and *refrigerant jets.* The refrigerant jet uses R-11 or R-12 as the working fluid.

The principle of operation for a steam jet system is based on the fact that water under high vacuum boils at a relatively low temperature. This causes evaporation to occur and reduces the temperature. Chapter 19 shows a table of water boiling temperatures under various vacuum pressures. Since water is the refrigerant in this type of system, only

temperatures down to about 40 °F (4 °C) are possible. This type of cooling system is found in beverage cooling distilleries, chemical plants, and other installations where a steam power plant is needed for other uses. The exhaust steam, from a high-pressure steam operating machine, may be just right for use in a steam jet application.

In a refrigerant jet, refrigerant vapor at high pressure and temperature flows through the ejector nozzle and carries refrigerant gas to the condenser. Fig. 17-25 shows a systems diagram of an air conditioning system using a refrigerant jet pump. These units have low efficiency and require a large condenser to remove heat. This type of cooling system is found in commercial installations where waste heat is available.

Both the steam and refrigerant jet cooling systems show how energy conservation can be incorporated into current high waste applications.

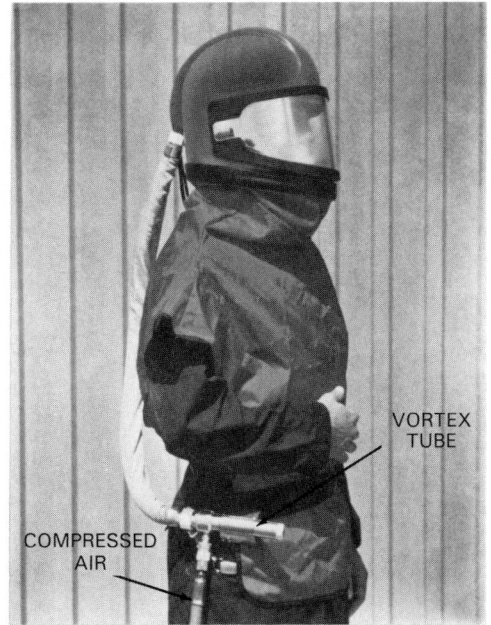

Fig. 17-23. Garment and hood fitted with vortex tube cooling and heating device. (Vortec Corp.)

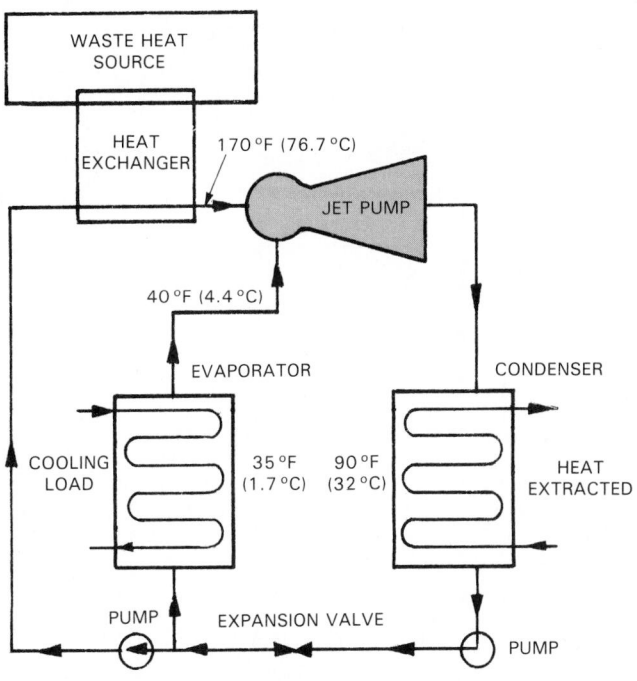

Fig. 17-25. Refrigerant jet pump air conditioning system.

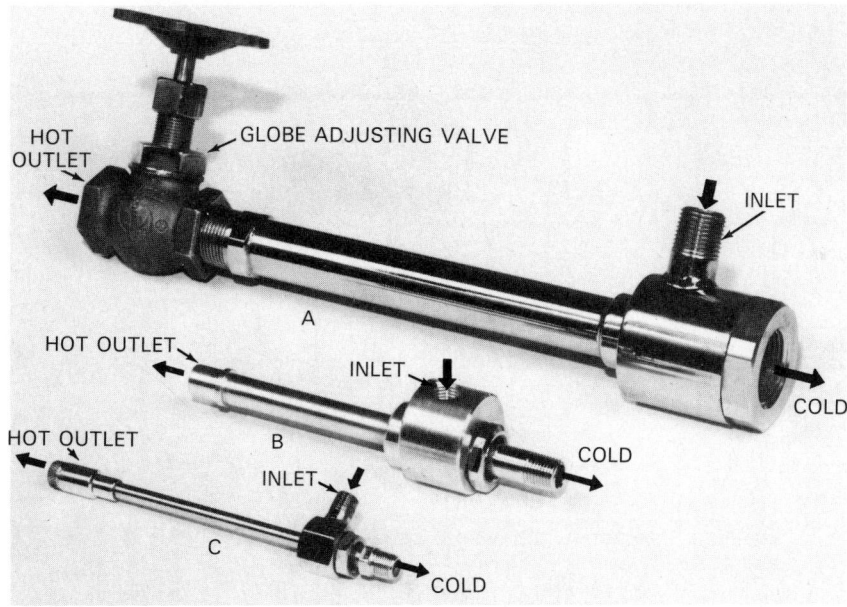

Fig. 17-24. Three vortex tubes. A—Large capacity vortex tube, 13 3/4 in. long, with 3/8 in. diameter globe adjusting valve. B—6 27/32 in. vortex with smaller capacity. C—6 1/2 in. vortex tube. (Vortec Corp.)

17-9 MULTISTAGE SYSTEMS

Multistage systems are used where ultralow temperatures are desired but cannot be obtained economically through the use of a single-stage system. This is because the compression ratios would be too high to get the necessary evaporating and condensing vapor temperatures.

The name, multistage, applies to any refrigeration system which contains more than one stage of compression. There are two general types: cascade and compound.

In the cascade system, as shown in Fig. 17-26, two separate refrigerant systems are interconnected so that the evaporator from one unit is used to cool the condenser of the other unit. This arrangement allows one of the units to operate at a lower temperature and pressure than would be possible with the same size single-stage system.

Cascade refrigeration systems may be used to produce temperatures below −250°F (−157°C). See Cryogenics in Para. 1-56. This system is actually two independent units. It allows, if desired, the use of two different refrigerants.

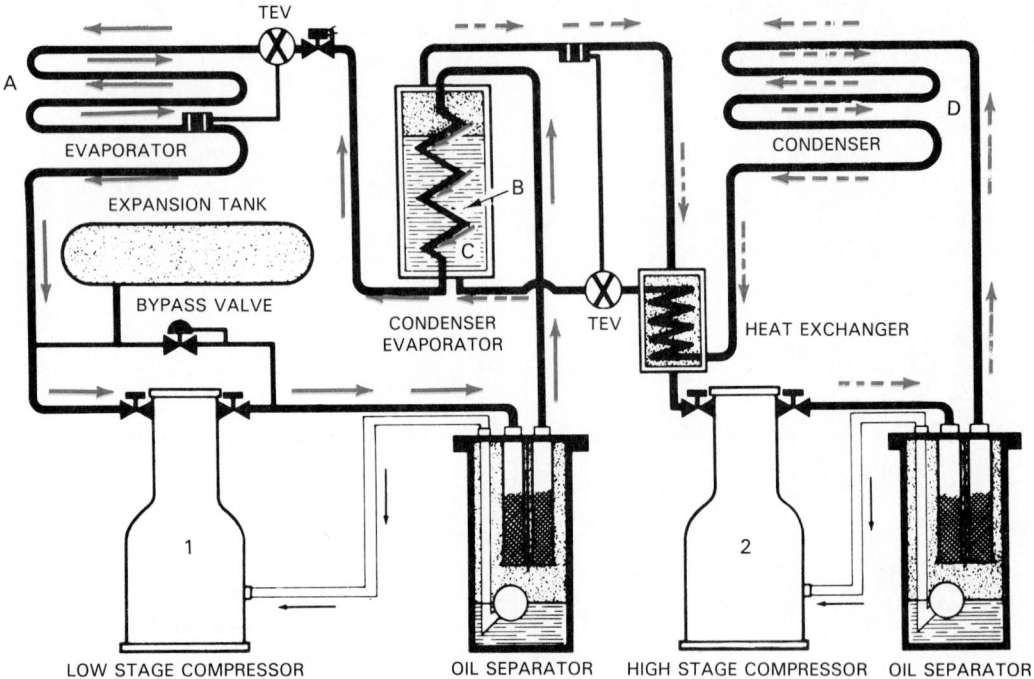

Fig. 17-26. Cascade refrigerating system. Condenser B of system No. 1 is being cooled by evaporator C of system No. 2. This arrangement enables ultralow temperatures in evaporator A of system No. 1. D—Condenser of system No. 2. TEV—Refrigerant controls. Note use of oil separators to minimize circulation of oil.

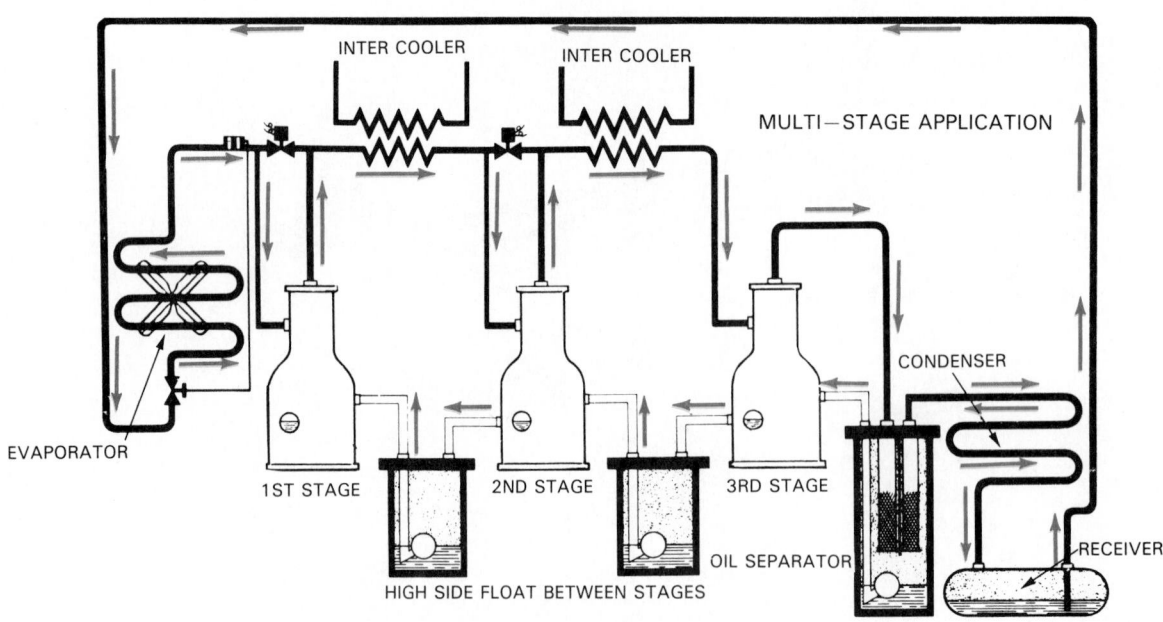

Fig. 17-27. Multistage refrigerating system using three compressors (stages). Compressor No. 1 pumps vapor into intercooler and then into intake of compressor No. 2. This operation is repeated between second and third stages. In third stage, refrigerant vapor is further cooled and travels to evaporator for specific cooling use.

Compound systems obtain low temperatures by using several compressors connected in series to the same refrigeration system. Fig. 17-27 is a diagram of such a unit. It can increase the performance and efficiency of low-temperature refrigeration systems. Vapor in the evaporator of a low-temperature system has a high specific volume at a low temperature. In a single-stage system, this would re-quire a longer-than-normal compressor piston stroke operating at high speeds. Because of the temperature it would also reduce the volumetric efficiency so that it would cost too much. In the compound system, the first stage compressor is larger than the secondary stage compressor. In each stage, the compressor gets smaller. This is because each higher stage handles denser vapor.

Compound refrigeration systems using two-stage compression equipment can be used to produce temperatures from −20 °F (−29 °C) to about −80 °F (−62 °C). If three-stage equipment is used (three compressors in series), temperatures down to −135 °F (−93 °C) can be maintained.

17-10 HEAT PIPE

R.S. Gaugler developed the basic principle of the heat pipe in 1942. More recently, it has been used in aerospace work as well as in industrial and domestic applications. Its purpose is to transfer heat from one location to another.

The heat pipe is an evacuated, hermetically sealed chamber containing volatile working fluid. Its inside walls are lined with a porous substance called a wick. Fig. 17-28 is a cross-section of such a pipe.

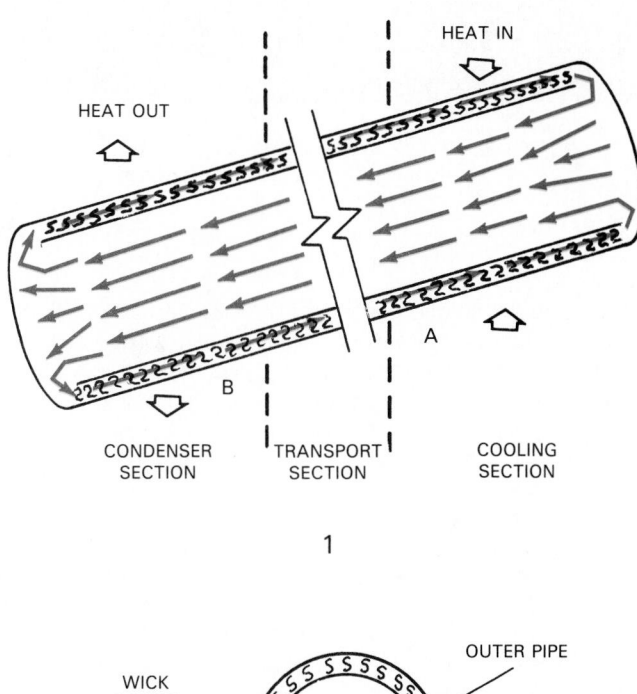

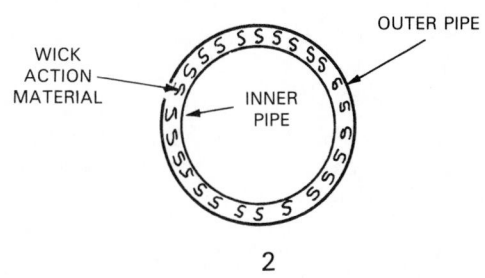

Fig. 17-28. 1—Basic design of heat tube. Fluid evaporates at A as heat travels into tube. Vapor moves length of inner tube to condenser B where fluid releases enough heat to become liquid. Liquid then travels through wick (by capillary action) back to evaporator. Transport section does not gain or lose heat. 2—Cross-section shows end view of tube.

Heat is applied to one end (heat in). The liquid vaporizes and vapor flows to the opposite end. When the heat is removed, the vapor condenses into a liquid again. The liquid returns to the evaporator or hot end of the pipe through the wick. This completes the cycle.

Return of fluid as a liquid to the evaporator section is normally accomplished by a gravity return. The pumping ac-

tion of the wick is the major breakthrough. This occurs as a result of the liquid vaporizing in the evaporator, thereby leaving voids in the wick's porous structure. The attractive force between the wick material and liquid, combined with the surface tension of the liquid, fills these voids with liquid next to them, thereby allowing liquid to continuously flow through the wick from the condenser and to the evaporator section.

The heat pipe has been used in aerospace projects where it cools instruments or devices. Another successful application has been found in cooking. A heat pipe (pin) may be inserted into a piece of meat and, when placed in a conventional oven, cooks the meat from the inside.

Still another application of the heat pipe is for heat recovery from the flue of a furnace or boiler. With the heat pipe, the high stack temperature in a flue may provide additional heat for basements, garages, and the like. Fig. 17-29, view A, shows how a heat pipe can be used to recover heat from exhaust air during the heating season. In view B, it cools incoming air in the cooling season.

The Alaskan pipeline provides another good application for heat pipes. The pipeline is used to transport oil above ground across Alaska, Canada, and the United States. The pipeline itself is supported on pilings which have been driven into the ground. The designers faced the problem of freezing and thawing which shifted the pilings, subsequently causing cracks in the pipeline. Their solution was to incorporate liquid ammonia heat pipes into the pilings. In cold weather, the ammonia vapor at the top of the heat pipe

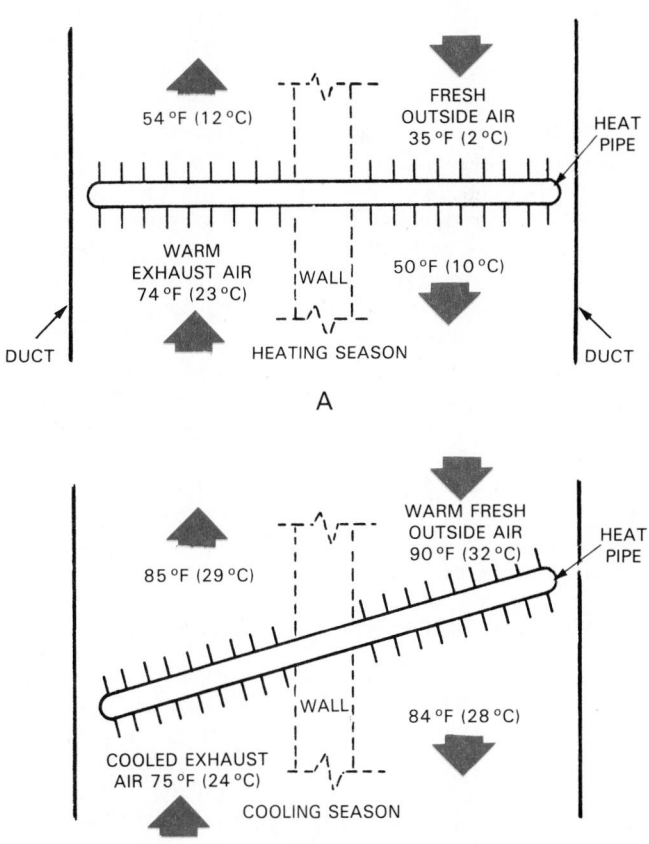

Fig. 17-29. Heat pipe application. Pipe is used to transfer heat between building exhaust air and fresh incoming air.

is cooled, then condenses and flows down below ground level. The liquid ammonia, below ground level, evaporates and removes heat from the surrounding ground. This insures that the area surrounding the piling will remain in a deep freeze even throughout the warmer months.

17-11 IMMERSION (FAST FREEZE)

Immersion freezing consists of dipping articles to be frozen into liquid refrigerant.

For fast freezing of food, the usual refrigerant is liquid R-12, specially prepared for this purpose. For some very low-temperature applications, liquid carbon dioxide or liquid nitrogen may be used. See Fig. 17-30.

The temperature is so low that refrigerant should never be allowed to touch the worker. It would result in immediate freezing of the skin.

In this freezing system, some liquid refrigerant boils and is vaporized in absorbing heat from the food. Some larger installations recover the vaporized refrigerant. R-12 refrigerant is recovered by using a refrigerated, finned coil placed over the liquid refrigerant. This will condense the vapor so it may be collected and returned to the refrigerant storage tank. With liquid carbon dioxide or nitrogen, however, the refrigerant is lost and becomes expendable.

17-12 CRYOGENIC REFRIGERATION

Cryogenic is the term which represents temperatures so cold that they are beyond the experience of most people. Cryogenic temperatures are between $-459.7\,°F$ (absolute zero) $(-273\,°C)$ and $-250\,°F$ $(-157\,°C)$. Many substances are completely different in character at those temperatures. For example, a steel block dropped on the floor at $-350\,°F$ $(-212\,°C)$ will shatter like glass.

The science of cryogenics has produced many useful instant freezing and refrigeration techniques. The scientific community uses nitrogen at $-350\,°F$ $(-212\,°C)$ to instantly freeze tissue or other fast decaying items for future study. Liquid oxygen, in addition to being a convenient way to store millions of tons of rocket fuel, is also used as a sub-zero chiller for steel, to make it less prone to warpage.

Liquid nitrogen, poured out into more common ambient temperatures, immediately becomes a vapor with a temperature of $-320\,°F$ $(-195\,°C)$. This process is used in some cargo trucks requiring deep freeze conditions. In addition to being the agent for immersion freezing (Para. 17-11), liquid nitrogen is also used to "freeze dry" foods. Freeze drying is a cryogenic process where foods are instantly frozen and then exposed to high levels of vacuum to remove all ice crystals. The resultant food, if kept completely dry, no longer requires any refrigeration.

17-13 REFRIGERATED CONTAINERS

Refrigerated containers are being used aboard ships, trucks, railroad cars, and airplanes. Perishable commodities are often gathered in the fields or orchards and stored immediately in refrigerated containers. Some containers may be as large as 8 x 8 x 20 ft. For long distance shipment, the containers may have a refrigeration mechanism such as an evaporator and a condensing unit.

Condensing units may be driven either with an electric motor or with an internal combustion engine. For short distances, a cold plate with a eutectic solution is used. (See

Fig. 17-30. Liquid nitrogen immersion freezing unit used to freeze cooked shrimp. (Compressed Air Magazine)

Chapter 30.)

The eutectic solution is frozen by the evaporator which is part of the cold plate. It is not necessary to operate the condensing unit during the short trip, as the eutectic solution in the cold plate provides considerable cooling for the short time the shipment is in transit.

Containers used on shipboard may be refrigerated by being connected to the ship's central refrigerating system. If the refrigerating mechanism uses an electric motor and condensing unit, it may be plugged into an outlet and driven by power aboard the ship. In such cases, the condensing unit is usually water-cooled. When containers are stored on the dock or in a warehouse, the condensing unit may be plugged into the local power supply.

17-14 STERLING CYCLE

A refrigerating cycle originally developed in 1816 by Robert Sterling is now being used in some refrigeration installations which operate at $-110\,°F$ $(-79\,°C)$ down to $-300\,°F$ $(-184\,°C)$. This cycle, when used as a three-stage system, can produce temperatures down to $-450\,°F$ $(-268\,°C)$. These units are very compact. Some of the gases used are helium and hydrogen.

The ideal system will pick up heat only at the lowest temperature and discard it only at the highest temperature. There can be no heat gain or heat loss between these two temperatures.

The Sterling cycle is almost as good because it conserves the energy and uses it in another part of the cycle. The Sterling cycle was adapted for refrigeration by John Herschel in 1834. The first practical machine was built in 1845.

The system uses one cylinder and two pistons with a stationary regenerator in between the pistons. See Fig. 17-31. In view A, piston No. 2 is stationary (standing still) and up to the regenerator, while piston No. 1 is at the beginning

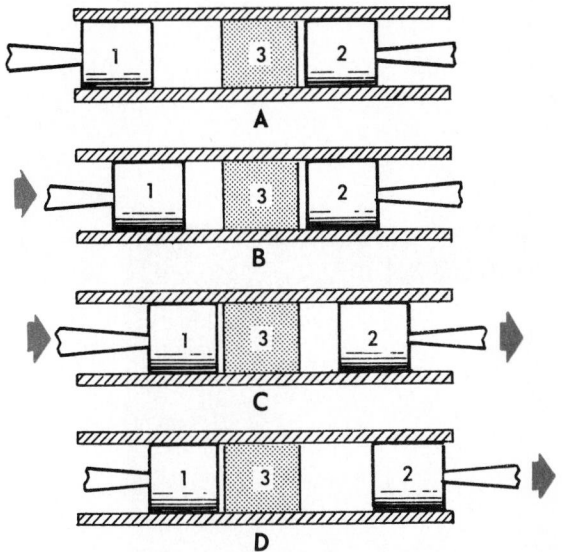

Fig. 17-31. Basic actions of Sterling Cycle system. 1—Left piston. 2—Right piston. 3—Regenerator. A—Start position. B—Compressing gas and cooling. C—Moving gas through regenerator. D—Expanding gas (it absorbs heat). Pistons now return to positions in A.

of compression. Then in view B, piston No. 1 compresses the gas and this gas is cooled (no temperature rise).

In view C, piston No. 1 now completes its stroke. At the same time, piston No. 2 moves to the right. The volume of trapped gas thus remains the same. The regenerator collects heat during this operation. In view D, piston No. 2 moves to the right and the gas cools by expansion. This gas is heated by warming the right side of the cylinder. Finally both pistons move together and return to their position in view A.

During this time, all the gas passes through the regenerator. Heat is added to the gas from the regenerator during this action.

In practice, the left end of the cylinder is water-cooled and the right end of the cylinder is the cooling unit. The regenerator must allow flow of the gas and it must have a good heat-absorbing ability.

17-15 REVIEW OF SAFETY

When working with electrical equipment, it is always recommended that the equipment being tested be fully grounded through either the use of a polarity plug or a ground wire. See Chapter 8.

Local and national refrigeration and electrical codes should be followed when servicing and installing all units.

Before checking a thermoelectric system, get a wiring diagram of the system to be certain that the polarity is not reversed.

When checking expendable refrigerant systems, always make certain that the safety doors are opened and the truck body vented before entering the conditioned space. Before entering a unit that is being cooled, make certain that all refrigerant flow control valves have been closed.

As has been repeated throughout the text, a service technician should always wear goggles when checking a unit which uses a refrigerant.

It is not safe to work on any part of an expendable refrigerant system unless one knows the required pressures and the nature of the safety valves and controls used in the system.

As indicated in Para. 17-9 or 17-12, temperatures in the cryogenic range are below $-250\,°F$ ($-157\,°C$). These temperatures are dangerous. The rate at which heat will be removed from the surface of the body at these temperatures is so great that the flesh may be severely frozen before one feels the cold.

Remember, most expendable refrigeration systems use carbon dioxide or nitrogen. Humans and animals cannot live in atmospheres of either of these substances.

In handling any type of refrigerants, the operator should wear gloves, face shield or goggles, and make sure that no liquid refrigerant is ever allowed to touch the skin. It is particularly dangerous to handle such refrigerants as liquid nitrogen, liquid air, and liquid carbon dioxide.

17-16 TEST YOUR KNOWLEDGE

1. What are the advantages of a thermoelectric system as compared to a compression system?
2. Does a thermoelectric system use alternating current throughout the system?
3. How can a thermoelectric module used for cooling be converted into a heating unit?
4. What is the main refrigerant used in an expendable refrigerant cycle?
5. Name two basic types of expendable refrigerant systems available today.
6. What is the purpose of the safety vent in an expendable refrigerant system cabinet?
7. Name one application of a cascade system.
8. In most portable cargo containers, what is the primary cooler used for short distances?
9. What is the main advantage of the containerized system used for transporting goods?
10. What are the least number of compressors a multistage system will use?
11. How are the refrigerating systems in a cascade system placed in reference to each other?
12. What three refrigerants are used in a fast-freeze system?
13. How is the fluid in a heat pipe returned to the evaporator or heat source?
14. What determines the amount of hot air released from a vortex tube system?
15. What is used as a coolant in a vortex tube system?
16. What system is used to change natural gas to liquid?
17. What supplies the energy to a steam jet cooling system?
18. Can a Sterling cycle refrigeration system produce very low temperatures?
19. What does the regenerator in a Sterling cycle do?
20. Why does an expendable refrigerant spray system require an electrical system?

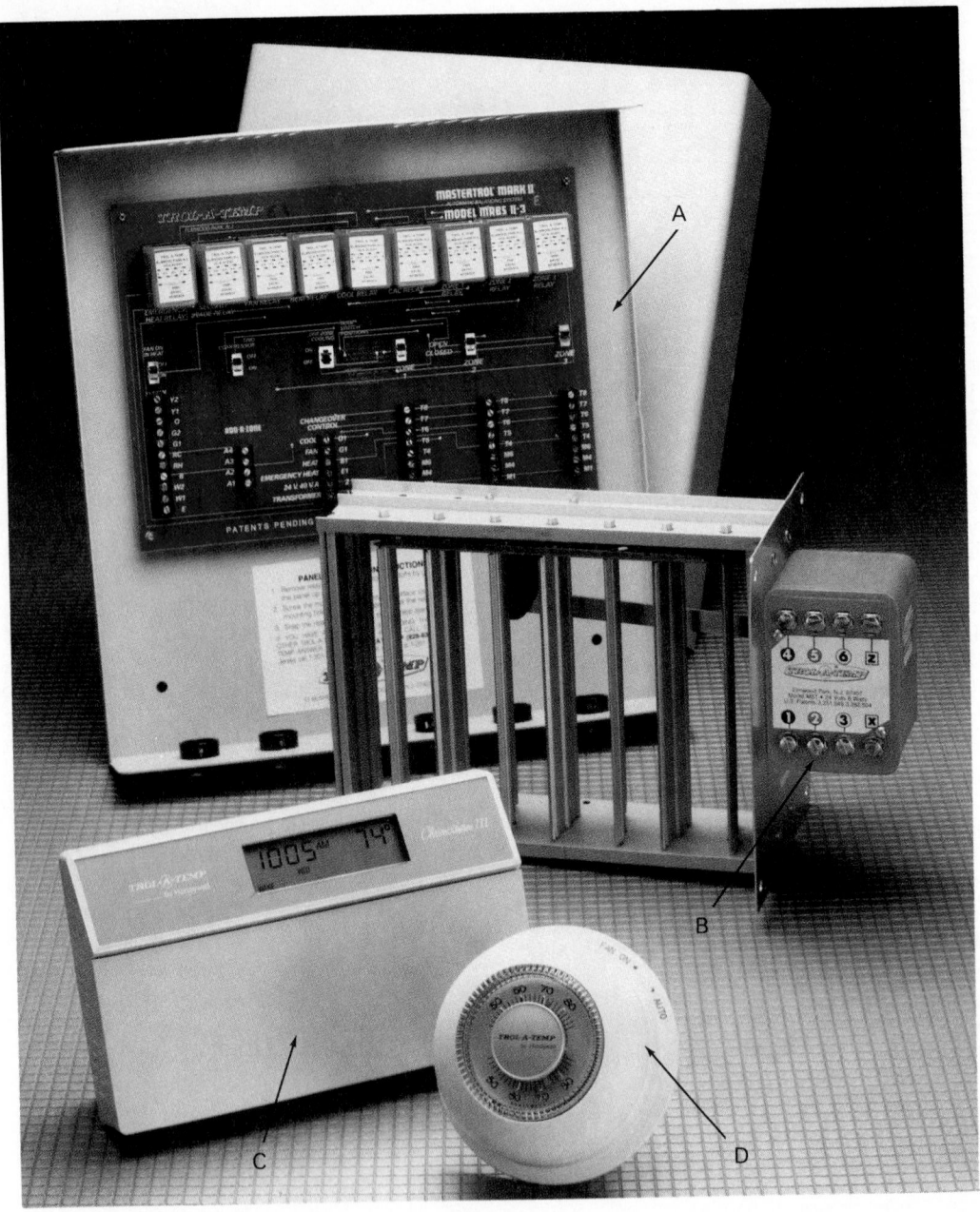

Zoned heating and cooling controls: A—Multi-zone control panel for controlling single-stage or multi-stage heating and cooling system or heat pump. B—Round thermostat provides two options—fan continuously on or automatic. Basically a manual temperature control. C—Programmable thermostat that can be temporarily changed from the program. D—Variable volume blade damper, operated by 24-volt damper motor. (Trol-A-Temp, Division of Trolex Corporation)

Chapter 18

FUNDAMENTALS OF AIR CONDITIONING

After studying this chapter, the technician will be able to:
☐ Define the principles of air conditioning and other terms important to air conditioning systems.
☐ Discuss the physical principles of air movement and humidity.
☐ List the important factors involved in the operation of an air conditioning system.
☐ List and explain the factors of air conditioning which affect comfort, health, and the methods of conditioning air for these purposes.
☐ Demonstrate the proper ability to use various instruments such as a psychrometer, dry bulb thermometer, hygrometer, pitot tube, recorders, manometers, and barometers.
☐ Read and interpret psychrometric charts and scales.

18-1 DEFINITION OF AIR CONDITIONING

The American Society of Heating, Refrigerating and Air Conditioning Engineers (ASHRAE) defines air conditioning as: "The process of treating air so as to control simultaneously its temperature, humidity, cleanliness, and distribution to meet the requirements of the conditioned space."

As indicated in the definition, the important actions involved in the operation of an air conditioning system are:
1. Temperature control.
2. Humidity control.
3. Air filtering, cleaning, and purification.
4. Air movement and circulation.

Complete air conditioning provides automatic control of these conditions for both summer and winter.

Temperature control for winter heating conditions requires automatic control of the heating source as a means of maintaining desired room temperatures.

Temperature control for summer cooling conditions requires automatic control of the refrigerating system to maintain the desired room temperatures.

Humidity control for winter conditions usually requires automatic control of the addition of moisture to the heating system (by use of a humidifier).

Humidity control for summer conditions requires the automatic control of dehumidifiers. Usually, this is done at the time the air to be cooled is passed over the cold evaporator surfaces.

In general, air filtering is the same for both summer and winter air conditioning. Air filtering equipment usually consists of very fine porous substances. Air is drawn through them to remove contaminating particles. Filters using activated carbon and electrostatic precipitators may be added to the usual filtering mechanisms to improve air cleaning. The air pollutants, and methods used to remove them from the air, will be covered in later paragraphs.

Many industries air condition their plants for two reasons: for the comfort provided and for more complete control of manufacturing processes and material. Better control of manufacturing temperatures and relative humidity improves the quality of the finished product.

18-2 AIR—ATMOSPHERE

Air is an invisible, odorless, and tasteless mixture of gases which surround the earth. Air surrounding the earth is called the atmosphere. It extends above the earth about 400 miles and is divided into several layers. The layer closest to the earth is called the lower atmosphere. It extends from sea level up to about 30,000 ft. The next layer, called the troposphere, is from 30,000 to 50,000 ft. The layer extending from 50,000 ft. up to 200 miles is called the stratosphere. The layer from 200 miles upward is called the ionosphere.

Atmospheric air is a mixture of oxygen, nitrogen, carbon dioxide, hydrogen, sulphur dioxide, water vapor (moisture), and a very small percentage of rare gases. Fig. 18-1 gives the percentages of these gases, both by volume and by weight. Each of these gases behaves as though it occupied the space alone (Dalton's Law).
1. Oxygen. The atmosphere is approximately 23 percent oxygen by weight. Oxygen readily combines with many substances. When fuels such as wood, coal, or oil are burned, the oxygen of the atmosphere combines with the carbon and hydrogen in the fuel to form carbon dioxide and water. The oxygen in the atmosphere is replenished by growing plants. The roots absorb moisture from the soil; the leaves absorb carbon dioxide from the air. Part of the moisture (H_2O) absorbed by the

Name	Chemical Symbol	DRY AIR	
		Amount by Weight %	Amount by Volume %
Nitrogen	N_2	75.47	78.03
Oxygen	O_2	23.19	20.99
Carbon Dioxide	CO_2	.04	.03
Hydrogen	H_2	.00	.01
Water	H_2O	.00	.00
Dust		.00	.00
Rare Gases		1.30	.94

Fig. 18-1. Gases and substances that make up air in the atmosphere.

plant combines with the carbon in carbon dioxide. Cellulose, the plant structure (stem, leaves, flowers, etc.), is produced in this way. Oxygen from the water intake of the plant is released by the leaves. Note in Fig. 18-1 that, because oxygen (O_2) is a heavier gas, it has a higher percentage by weight than by volume.

2. Nitrogen. About three-fourths of the earth's atmosphere by weight consists of nitrogen, a gaseous element that does not readily combine with other substances. If combined with other elements, nitrogen usually is unstable and tends to separate from the other elements. Compounds of nitrogen make up most explosives. Nitrogen is combined commercially with hydrogen to form ammonia. Ammonia produced in this way is the basis of most fertilizers. Ammonia is also an important refrigerant (NH_3[R-717]). Liquid nitrogen obtained by cooling of air is also a special purpose expendable refrigerant.

3. Carbon dioxide. The atmosphere contains approximately 0.03 to 0.04 percent carbon dioxide. Carbon dioxide is a combination of carbon and oxygen. Absorbed by growing plants, it becomes one of the "building blocks" in the development of plant cells.

4. Hydrogen. Hydrogen (H_2), a very light gas, does not show in weight percentage. However, it is shown as volume in Fig. 18-1. Hydrogen makes up a very small part of the atmosphere. It is present in most fuels. When burned, it combines with oxygen to form water (H_2O) in steam and vapor form.

5. Sulphur dioxide. The most common gaseous contaminant, sulphur dioxide is formed by combustion of fuels which contain sulphur. Many large power plants now have facilities for removing sulphur from these fuel sources, and also for removing sulphur dioxide from the stack gases.

6. Water vapor (moisture). The amount of water vapor in the atmosphere varies with the temperature. It is not indicated in percentage, but rather by the term "relative humidity."

7. Rare gases. Rare gases make up from 0.9 to 1.3 percent of the atmosphere by weight. These gases include neon, argon, helium, krypton, and xenon. Some are used in light bulbs and tubes and in certain industrial processes.

In addition to these substances, air contains a variety of contaminants. These are so variable that they cannot be given any definite value in a table. However, contaminants in the air are of great importance in air conditioning.

18-3 PHYSICAL PROPERTIES OF AIR

Air has weight, density, temperature, specific heat, and heat conductivity. In motion, it has momentum and inertia. It holds substances in suspension and in solution.

Air pressure at the surface of the earth is due to the weight of the air above the earth. Air pressure decreases as altitude increases, due to the reduction of the weight of the air above. Air presses against the earth at sea level with a force of 14.7 psi (101 kPa).

Because air has weight, energy is required to move it. Once in motion, air has energy of its own (kinetic energy). The weight of moving air turns windmills. The mills convert the kinetic energy to mechanical energy.

The kinetic energy of air in motion is equal to half the mass of the air multiplied by the square of the velocity (speed). Velocity is measured in feet per second or meters per second. According to Bernoulli's Equation, increasing the velocity decreases the pressure. In a tornado, the velocity is very high, reducing the pressure. The low pressure in a tornado causes much of the damage to buildings. Outside pressure is lowered rapidly. Pressures inside the building push outward in an "explosion" of air.

Tiny particles of dust may be picked up and held in suspension in moving air for long periods of time. It is possible to measure the amount of particles so suspended.

The density of air varies with the atmospheric pressure and humidity. One pound of air at standard conditions (14.7 psi, 69.8 °F) occupies 13.341 cu. ft. One kilogram occupies 0.83285 m^3. Air has a density of 0.07496 lb./cu. ft. (1.2007 kg/m^3). The density of gases is indicated in pounds per cubic foot or in kilograms per cubic meter.

Air temperatures may be measured with either the Fahrenheit scale or the Celsius scale. Under ordinary conditions, the familiar glass-stemmed thermometers are satisfactory. Expanding metals (solids) such as bimetal strips or rods are also used for ordinary circumstances. When making measurement of very low temperatures, thermocouple thermometers or resistance temperature detectors are used. Thermocouple thermometers may be used for measuring high air temperatures. Thermistor thermometers and pyrometers are also popular.

The specific heat of air is the amount of heat required to raise the temperature of one pound of air one degree Fahrenheit or one kilogram of air one degree Celsius. The specific heat of air at sea level is 0.24 Btu per pound.

Air is a poor conductor of heat. For this reason, air spaces are often used for insulating purposes.

For computation purposes, certain pressure, temperature, and density values are required. The requirements are defined under Standard Air in Chapter 30.

18-4 HUMIDITY

Humidity is a term used to describe the presence of moisture or water vapor in the air. The amount of moisture that the air will hold depends upon the temperature of the air. Warm air will hold more moisture than cold air.

The amount of humidity in the air affects the rate of evaporation of perspiration from the body. Dry air causes rapid evaporation, which makes the surface feel cool. Moist (humid) air prevents rapid evaporation of perspiration, making it feel warmer than the temperature indicated by a thermometer. Remember that this moisture (humidity) is in vapor form, and it is invisible.

How much moisture the air will hold is shown in Fig. 18-2 and explained in Para. 18-5.

18-5 RELATIVE HUMIDITY

Relative humidity (rh) is a term used to express the amount of moisture in a given sample of air in comparison with the amount of moisture the air would hold if totally saturated at the temperature of the sample. Relative humidity is stated in a percentage, such as 30 percent, 75 percent, 85 percent, etc.

Referring to the water vapor saturation graph in Fig. 18-2, Point B contains 111 grains of moisture per pound of dry air at 85 °F. The saturated condition at C for the same temperature is 183 grains of moisture per pound of air.

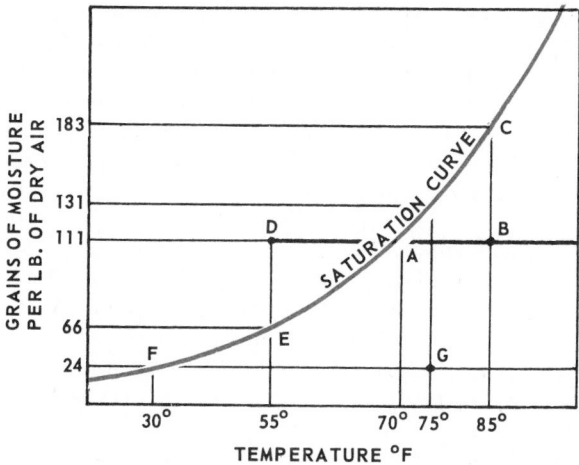

Fig. 18-2. Typical water vapor saturation curve for air. As temperature increases, amount of moisture that air will hold also increases.

Therefore, the relative humidity at Point B is:

$$\frac{111}{183} \times 100 = rh$$

$$.6 \times 100 = rh$$

$$60 \text{ percent} = rh$$

Line A to B in Fig. 18-2 represents what happens when saturated air is warmed. Point D represents what happens when saturated air is cooled. The distance D to E represents the moisture condensed out of the air, since saturated air at the same temperature will hold only 66 grains of moisture. The amount condensed is 111 minus 66 = 45 grains.

A typical outdoor condition in winter is represented at Point F in Fig. 18-2. Air is taken indoors at 30°F (−1°C) and 100 percent relative humidity. It holds 24 grains of moisture. If this air is heated to 75°F (24°C) and no moisture is added, its new condition will be as shown at G. The saturated condition at Point G would be 131 grains. Since the original air had only 24 grains of moisture, the relative humidity is 24 ÷ 131 × 100 = 18.3 percent.

18-6 DRY BULB THERMOMETER

Human comfort and health depend a great deal on the air temperature. *In air conditioning, the air temperature indicated usually is the dry bulb temperature (db) taken with the sensitive element of the thermometer in a dry condition.* It is the temperature measured by thermometers in the home.

18-7 WET BULB THERMOMETER

If a moist wick is placed over a thermometer bulb, the evaporation of moisture from the wick will lower the thermometer reading (temperature). This temperature is known as the "wet bulb" temperature. If the air surrounding a wet bulb thermometer is dry, evaporation from the moist wick will be more rapid than if the air is quite moist. Fig. 18-3 compares dry bulb temperature and wet bulb temperature taken at the same place and at the same time.

When the air is saturated with moisture, no water will evaporate from the cloth wick, and the temperature on the wet bulb thermometer will be the same as the reading on

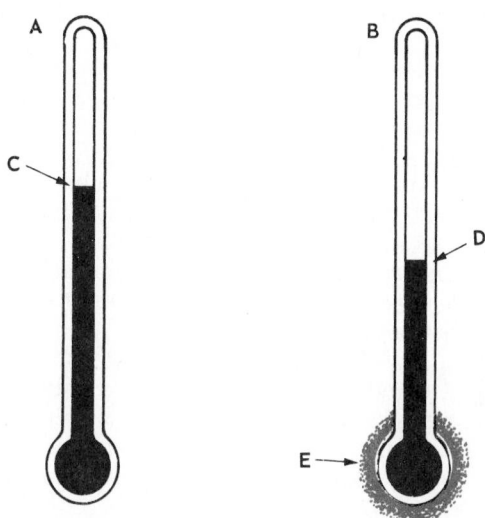

Fig. 18-3. Dry bulb and wet bulb thermometers. A—Dry bulb thermometer. B—Wet bulb thermometer. C—Dry bulb temperature. D—Wet bulb temperature. E—Wick surrounding wet bulb. Note that temperature shown on wet bulb thermometer is considerably lower than dry bulb thermometer.

a dry bulb thermometer near it.

However, if the air is not saturated, water will evaporate from the wick. In doing so, it will lower the wick temperature. Then heat will flow from the mercury to the wet wick and the reading will be lower.

The accuracy of the wet bulb reading depends on how fast the air passes over the bulb. Speeds up to 5000 ft./min. or 60 mi./hr. are best but dangerous if the thermometer is moved at this speed. Also, the wet bulb should be protected from heat radiation surfaces (radiator, sun, electric heater, etc.). Errors as high as 15 percent may be made if the air movement is too slow, or if too much radiant heat is present.

A hygrometer is an instrument used to measure the amount of moisture in the air. The hygrometer uses both a dry bulb thermometer and a wet bulb thermometer. By using a psychrometric chart, the relative humidity can be found. See Para. 18-15 and Para. 18-16.

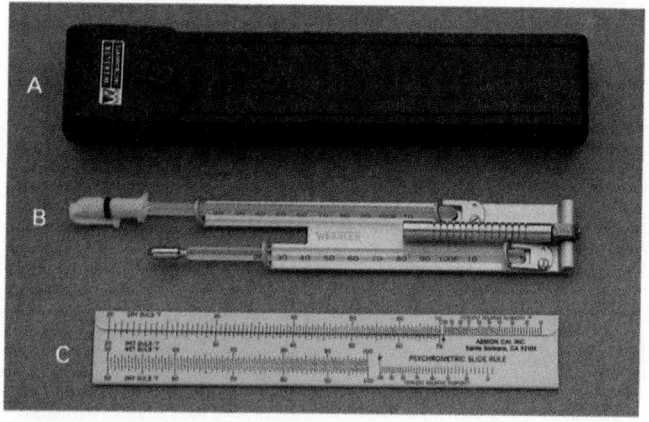

Fig. 18-4. Sling psychrometer. A—Wet wick mounted on thermometer. B—Dry bulb thermometer. C—Slide rule. (Abbeon Cal, Inc.)

18-8 PSYCHROMETER

To ensure that the recorded wet bulb temperature is accurate, airflow over the wet bulb should be quite rapid. A device designed to whirl a pair of thermometers, dry bulb and wet bulb, is called a sling psychrometer. See Fig. 18-4. This instrument consists of two thermometers, a wet bulb and a dry bulb. To operate, saturate the wick on wet bulb and whirl. When temperature stops dropping, read the two thermometers and place the wet bulb temperature over the dry bulb temperature scale on a slide rule. Arrow will indicate the relative humidity.

There are certain places in which it is difficult to spin the psychrometer (narrow passages, etc.). To obtain accurate results in these places, an aspirating psychrometer is used. With this instrument, Fig. 18-5, the air sample is blown over the wet and dry bulb thermometer with suction created by an air pump.

A battery-operated aspirating psychrometer is shown in Fig. 18-6. It has illuminated thermometer scales and a fan which draws air over the thermometer sensitive bulbs.

18-9 DEW POINT

Dew point is the temperature below which water vapor in the air will start to condense. It is also the 100 percent humidity point. The relative humidity of a sample of air may be determined by its dew point. Different methods may be used to find the dew point.

Dew point temperature can be determined with fair accuracy by placing a volatile fluid in a bright metal container, then stirring the fluid with an air aspirator. A thermometer placed in the fluid will indicate the temperature of both the fluid and the bright metal container. While stirring, carefully note the temperature at which a mist or fog appears on the outside of the metal container. This indicates the dew point temperature. Flammable or toxic volatile fluids must not be used for this experiment. R-11 refrigerant is safe to use.

A commercial instrument for determining dew point is illustrated in Fig. 18-7. This unit can measure dew point temperatures from room temperatures to as low as $-80\,°F$ ($-62\,°C$). The principle of operation is to pump a sample of air into the observation chamber of the instrument. The pressure is above atmospheric.

The pressure ratio gauge on the right in Fig. 18-7 adjusts for this pressure, then the valve is manipulated to exhaust the air. The lighted observation window will indicate a fog when the sample is cooled to its dew point. This window is lighted and a "sunbeam" effect is noted if any fog exists. The pressure ratio determines the dew point temperature.

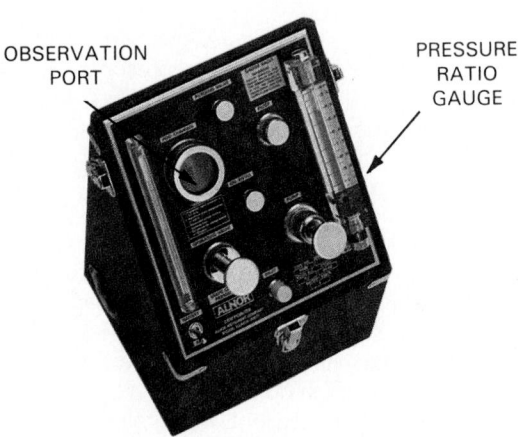

Fig. 18-7. Instrument for determining dew point temperature. Note observation port. (Alnor Instrument Co.)

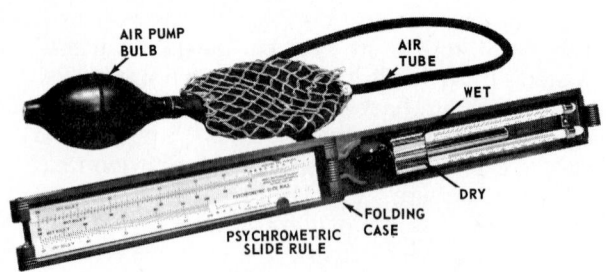

Fig. 18-5. Aspirating psychrometer. Air samples are drawn over thermometer bulbs by air pump. Note handy calculating slide rule in case.

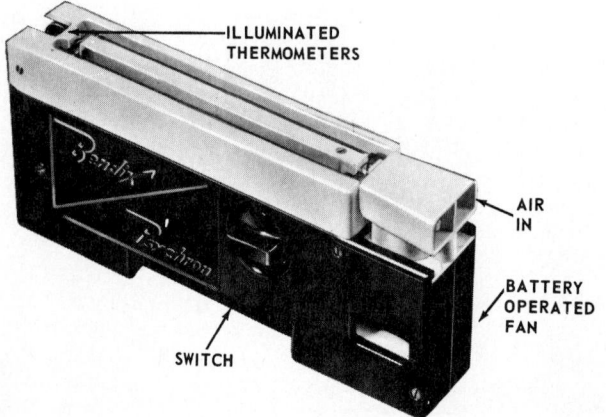

Fig. 18-6. Aspirating psychrometer. Shown is battery-powered unit with illuminated thermometer scales. Motorized fan draws air over wet bulb and dry bulb thermometers.

A window during the winter heating season offers a good example of dew point. Fig. 18-8 shows the surface temperature which will cause condensation (dew point) for various humidity conditions. The two room temperatures used are $70\,°F$ and $80\,°F$ ($21\,°C$ and $27\,°C$).

18-10 INDICATORS OF LOW HUMIDITY

Low atmospheric humidity will be indicated by an increase in the amount of noticeable electrostatic energy. As one moves about and touches grounded metal objects, a spark jumps from the hand or fingers to the object. Also, human hair tends to become unmanageable. Furniture joints shrink and become loose. Woodwork, such as doors and floors, crack open. The surface of the skin becomes dry, and membranes in the nose tend to become dry. To feel more comfortable, it is usually necessary to raise the ambient temperature (db) above normal.

RELATIVE HUMIDITY OF AIR (PERCENT)	DRY BULB TEMPERATURE OF SURFACE WHEN CONDENSATION STARTS	
	70 °F ROOM AIR TEMP.	80 °F ROOM AIR TEMP.
100	70	80
90	67	77
80	64	73
70	60	69
60	56	65
50	51	60
40	45	54
30	37	46
20	28	35

Fig. 18-8. Table gives temperature to which surface must be cooled to have condensation start. Table is based on ambient temperature of air at either 70 °F or 80 °F (21 °C or 27 °C).

18-11 HUMIDITY MEASUREMENT

Humidity measurement using a dry bulb and wet bulb thermometer is explained in Para. 18-6 and Para. 18-7. The use of the dry bulb and wet bulb thermometer means one must use psychrometric charts.

Instruments have been developed which give a direct reading of relative humidity. See Fig. 18-9. The operation of these instruments depends upon the property of some substances to absorb moisture and then change their shape or size, depending upon the relative humidity of the atmosphere. Human hair, wood, fiber element, and other substances may be used.

It is also possible to measure relative humidity electronically. This is done by using a substance in which the electrical conductivity changes with the moisture content. Such an instrument is shown in Fig. 18-10. In operation, the sensing element is placed in the space in which the relative humidity is to be measured.

Fig. 18-11 illustrates an easy-to-use electronic relative humidity measuring instrument. To use the meter, press the on-off button, A, and the display panel appears. Select the desired temperature reading °C or °F (B). Then select the

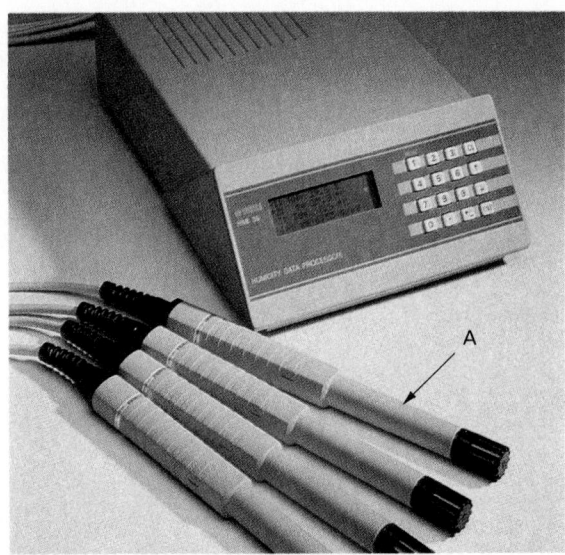

Fig. 18-10. Relative humidity data processor contains a microprocessor. Unit measures relative humidity and related quantities and temperatures in specially air conditioned areas such as computer rooms, laboratories, etc. Unit can have up to four probes (A) to measure the relative humidity and/or temperature in ambient air. From these measurements, the humidity data processor calculates the dew point, mixing ratio, and absolute humidity. (Vaisala, Inc.)

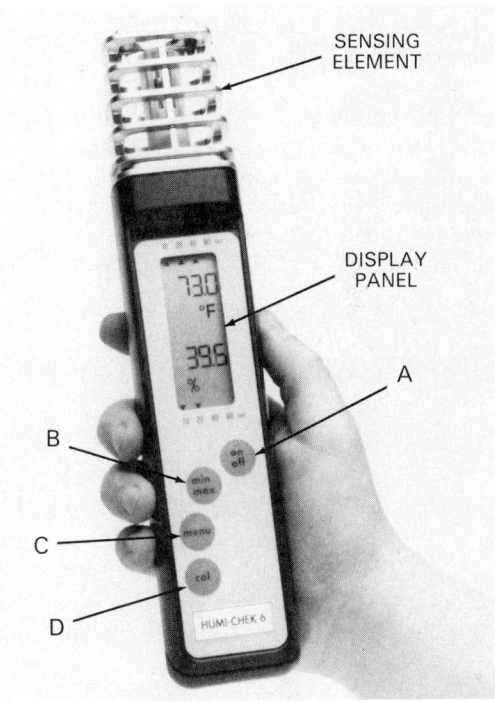

Fig. 18-11. Electronic hygrometer with buttons. A—On-off; B—Temperature; C—Function (relative humidity, dew point); D—Accuracy and recalibration. (Rosemount Analytical, Inc.)

Fig. 18-9. Wall-type hygrometer and temperature indicator is calibrated in percent of relative humidity. A—Air movement openings. (Abbeon Cal, Inc.)

desired relative humidity or dew point, C. The reading will then appear. Button D is used to measure the accuracy of the readings and to recalibrate it when necessary.

As in measuring pressure and temperature, it is sometimes helpful to have an extended reading in a controlled space. A recorder that indicates the moisture and temperature is shown in Fig. 18-12. A seven-day recording type

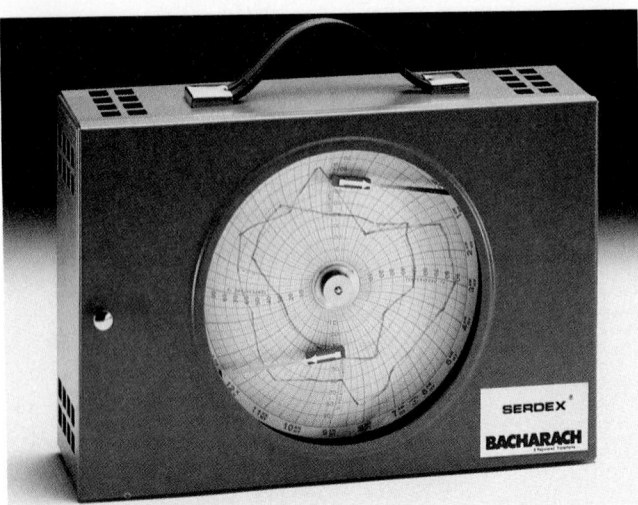

Fig. 18-12. A seven-day humidity/temperature recorder. The unit has temperature tolerances from $-35\,°F$ to $+130\,°F$, $-37\,°C$ to $+54.5\,°C$. Note marker pens in red and blue; one indicates the humidity and the other the temperature. (Bacharach, Inc.)

Fig. 18-14. Seven-day recorder for both temperature and relative humidity. These records are important as way to check efficiency of air conditioning systems. (Bristol Babcock, Inc.)

dry-bulb and wet-bulb thermometer is shown in Fig. 18-13. When using this instrument, it is necessary to refer to a psychrometric chart to find the relative humidity.

Fig. 18-14 pictures a seven-day instrument that records both temperature and relative humidity. Fig. 18-15 shows how temperature and relative humidity are charted by a 24-hr. recorder.

A different type of temperature and relative humidity recorder is shown in Fig. 18-16. The charts, printed on stiff paper, move down as the recording proceeds. A common humidity-sensitive element is usually made of multiple strands of human hair.

The unit in Fig. 18-17 is a portable temperature and

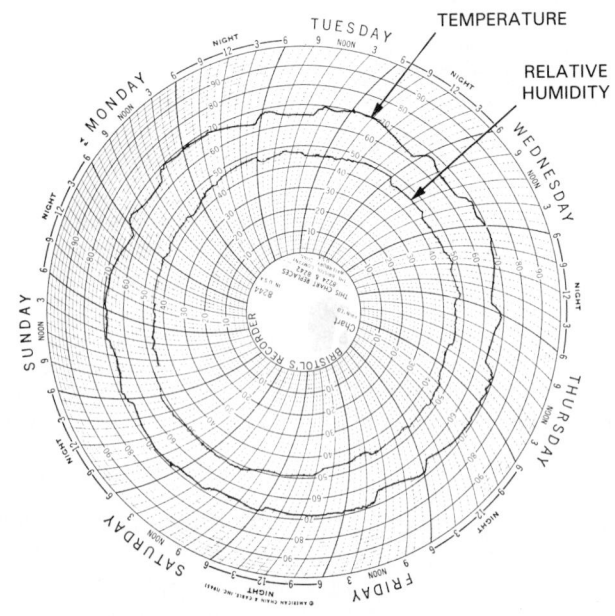

Fig. 18-15. Chart from 24-hr. temperature and relative humidity recorder. Note how relative humidity changes during 24-hr. period. (Bristol Babcock, Inc.)

Fig. 18-13. Self-contained seven-day or 24-hour recorder. Measures and records temperature variations from 0 °F to 130 °F, and humidity variations from 0% to 100% RH. Unit operates with animal membrane element for relative humidity and has a bimetal coil for temperature. Note different marker pens on graph for the two records. (Bacharach, Inc.)

relative humidity recorder. It can be set to record both dry bulb temperature and relative humidity for one day or for one week.

18-12 HYGROSCOPIC SUBSTANCES— DESICCANTS

Substances that have the ability to absorb moisture from the air are called desiccants. Some common desiccants are:

Fig. 18-16. Temperature-relative humidity recorder. Various time clock records are available, such as 10-hr. or 30-hr. charts. A—Roll-type chart. B—Air openings.

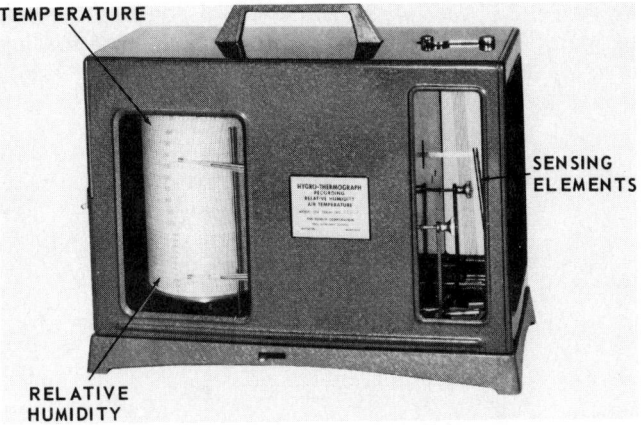

Fig. 18-17. Portable temperature and relative humidity recorder.

activated alumina, silica gel, calcium sulfate, and zeolites. Many desiccants can be reactivated (dried out) by heating.

Many instruments are packaged in containers with a package of desiccant. The desiccant tends to absorb the moisture in the container and keeps the instrument dry to reduce corrosion.

18-13 HUMIDITY CONTROLS

Health studies indicate that humidity control is an important factor in air conditioning.

Humidity controls are used to keep the relative humidity of air conditioned rooms at a satisfactory level. These controls determine the hygrometric state of the air.

Humidity controls operate during the winter heating season to add moisture to the air to keep the humidity approximately constant.

Humidity controls operate in the summer to remove moisture from the air. For the removal of moisture, the humidity control usually operates an air bypass to vary the airflow over the evaporators. These controls usually operate electrically to regulate solenoid valves or dampers. The control element may be a synthetic (made by humans) fiber

or human hair, which are sensitive to the amount of moisture in the air. Fig. 18-18 shows construction principles of a humidity control device.

In computer rooms, and other installations which require close humidity control, thermo-humidigraphs (temperature and humidity recorders) are fitted with alarms which will alert attendants in the event the temperature or humidity fails to remain at the proper level. Fig. 18-19 shows a thermo-humidigraph fitted with alarms.

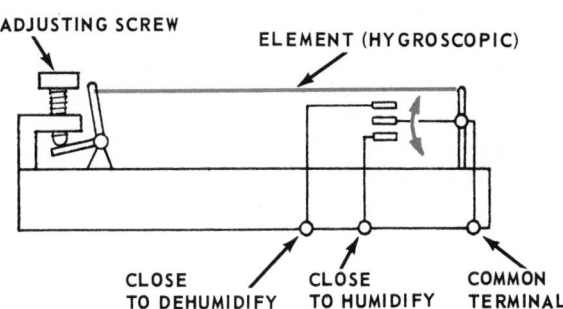

Fig. 18-18. Schematic diagram of relative humidity control, showing operating mechanism.

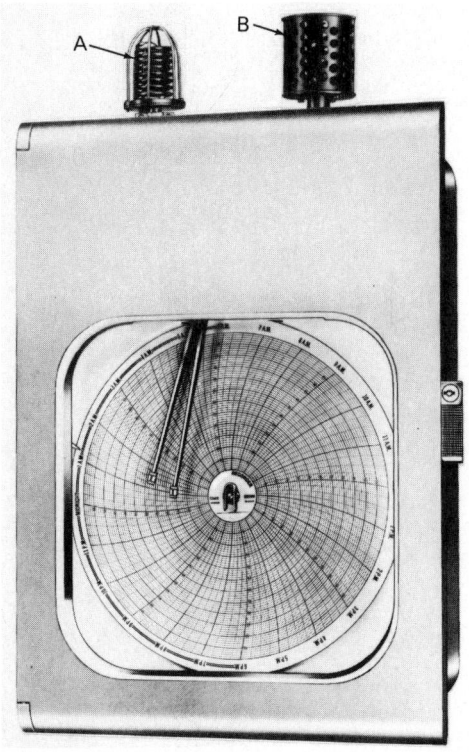

Fig. 18-19. Temperature-relative humidity recorder can be fitted with contact points. Points may be connected to electric alarm to provide signal if temperature or relative humidity is not kept within required limits. A—Temperature sensor. B—Relative humidity sensor. (Bristol Babcock, Inc.)

18-14 PSYCHROMETRIC PROPERTIES OF AIR

Psychrometry is the science and practice of dealing with air mixtures and their control. The science deals mainly with

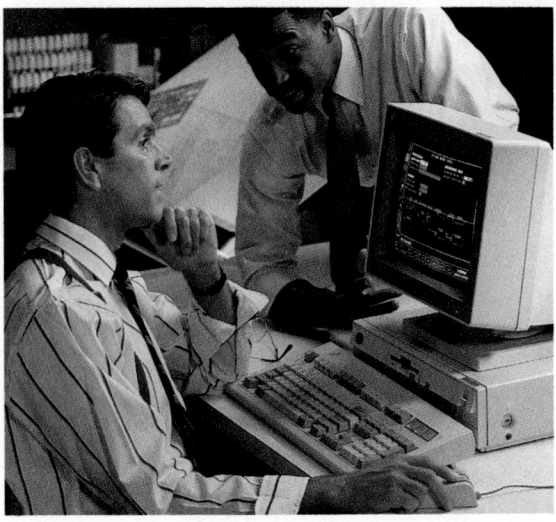

Fig. 18-20. Computerized building automation system controls energy, comfort, fire, and security. (Johnson Controls, Inc.)

dry air and water vapor mixtures. Fig. 18-20 shows a computer terminal used to determine and control the condition of the air in a large building complex.

Psychrometry deals with the specific heat of dry air and its volume. It also deals with the heat of water, heat of vaporization or condensation, and the specific heat of steam in reference to moisture mixed with dry air.

Tables and graphs have been developed to show the pressure, temperature, heat content (enthalpy), volume of air, and its steam content. The tables and charts are based on one pound of dry air, plus the water vapor to produce the air conditions being studied.

A standard pressure of 29.92 in. Hg (76 cm Hg) is used as the standard atmospheric pressure.

18-15 PSYCHROMETRIC CHARTS

The psychrometric chart is a graph of the properties (temperature, relative humidity, etc.) of air. It is used to determine how these properties vary as the amount of moisture (water vapor) in the air changes. A basic psychrometric chart is shown in Fig. 18-21. The horizontal scale (abscissa) is the dry bulb temperature, while the vertical scale (ordinate) represents water vapor pressure. Fig. 18-22, views A through D, show lines on the psychrometric chart which represent constant conditions.

Fig. 18-22, view A, shows a line of constant dry bulb temperature. This is always a vertical line and is usually in units of degrees Fahrenheit.

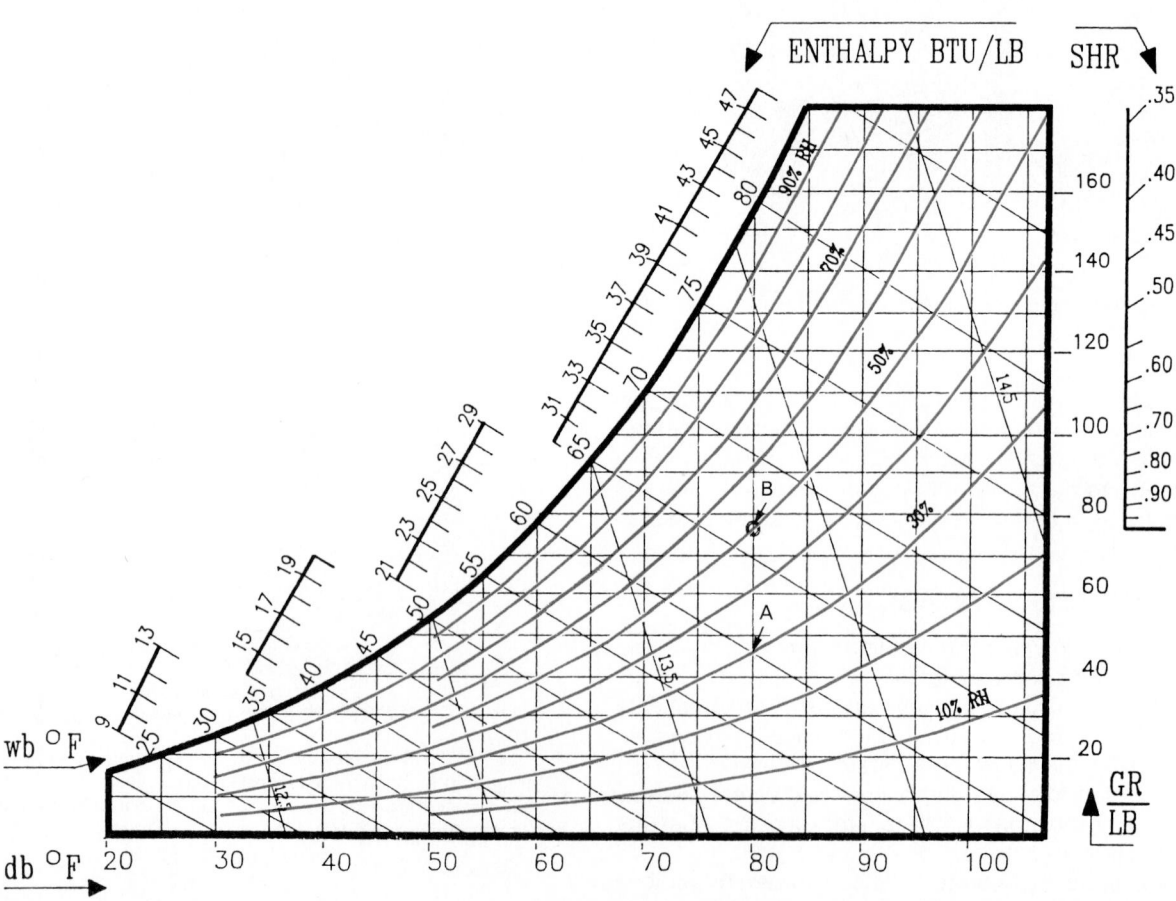

Fig. 18-21. Psychrometric chart. Red lines indicate relative humidity in percent. Dry bulb (db) temperature is shown at bottom. Wet bulb (wb) temperature is on uppermost curve. Right side gives grains/lb. (GR/LB). Point A indicates relative humidity of 30 percent. Note reading of 13.5 cu. ft. near center. This is volume of 1 lb. of air at dry bulb temperature and grains/lb. given at crossing points. See Fig. 18-34. Also note ''SHR'' scale, each line of which angles away from central comfort zone point B. SHR means Sensible Heat Ratio.
(Reprinted from AIR CONDITIONING CONTRACTORS OF AMERICA'S [ACCA] Basic Installation Manual by permission of ACCA)

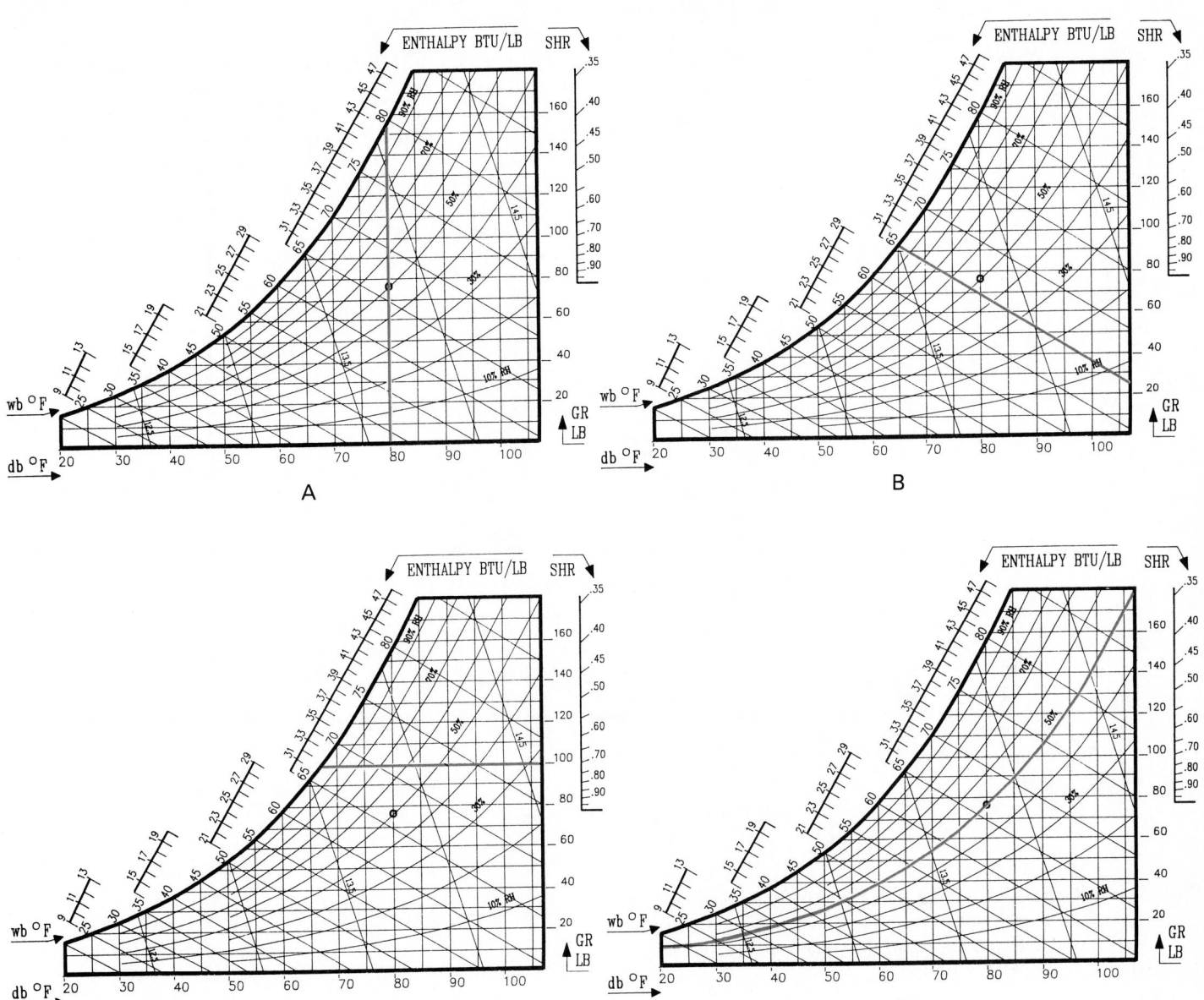

Fig. 18-22. Constant psychrometric conditions. A—80 °F constant dry bulb temperature. B—65 °F constant wet bulb temperature. C—100 grains of water/lb. of dry air constant water pressure. D—50 percent constant relative humidity.

Fig. 18-22, view B, shows a line of constant wet bulb temperature. This is also the line of constant enthalpy. Wet bulb temperature is usually shown in degrees Fahrenheit and enthalpy is shown in Btus per pound.

Fig. 18-22, view C, shows a line of constant water vapor pressure. The units for water vapor pressure are either pounds of water per pound of dry air or grains of water per pound of dry air.

Fig. 18-22, view D, shows a line of constant relative humidity. Relative humidity is always shown in percent. The 100 percent relative humidity line is also known as the dew point or saturation temperature line.

Each point on the psychrometric chart represents air at a specific set of conditions. The following examples refer to the points A through D on Fig. 18-23.

Example: Dry bulb temperature is 75 °F: If wet bulb temperature is 60 °F, what is the relative humidity?

Follow the vertical line corresponding to the 75 °F dry bulb temperature. Then follow the 60 °F wet bulb temperature line. These lines cross each other at Point A. This point is just above the 40 percent relative humidity line. Therefore, the correct answer would be about 41 percent relative humidity.

Example: Determine the dew point for a sample of air in which the dry bulb temperature is 80 °F and the relative humidity is 60 percent.

Find the point at which the 80 °F dry bulb line crosses the 60 percent relative humidity line. This point is labeled B. If the air represented by this point were cooled without a change in moisture content (represented on the psychrometric chart as a horizontal line), the dew point line would be intersected at about 66 °F. In Fig. 18-23, this is labeled point C.

Therefore, 66 °F is the dew point for a sample of air

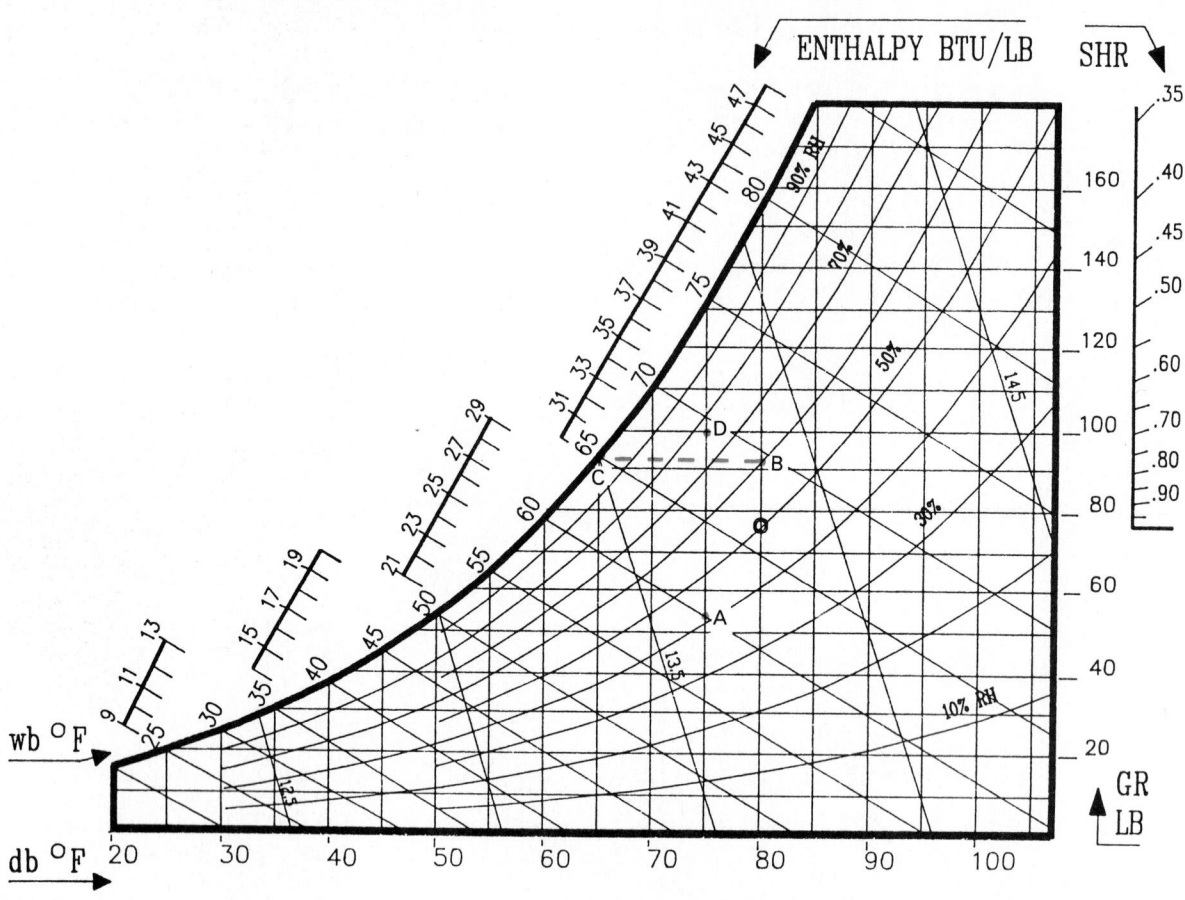

Fig. 18-23. Psychrometric chart showing specific conditions.

in which the temperature is 80 °F and the relative humidity is 60 percent. This could represent a typical summer evening. Dew would appear on surfaces as the 66 °F temperature was reached.

Example: Find the relative humidity when the dry bulb temperature is 75 °F and the humidity (or water vapor pressure) is 100 grains per pound of dry air.

First, find the vertical line representing a constant dry bulb temperature of 75 °F. Travel along that line until it crosses the horizontal line representing 100 grains of moisture per pound of dry air. The intersection point is labeled Point D. This point falls between the 70 percent and 80 percent relative humidity lines. The answer would be a relative humidity of about 77 percent.

This chart should be studied carefully. It provides a simple way for determining the various conditions of air. *In using the psychrometric chart, remember that the warmer the air, the more moisture it will hold. Also, as pressure is reduced, air absorbs more moisture.*

18-16 USING THE PSYCHROMETRIC CHART

Many of the air conditioning problems in this text will involve the use of a psychrometric chart. A chart can be used to show what is happening during a specific heating, ventilating, and air conditioning (HVAC) process.

Psychrometric charts give a considerable range of temperature and humidity conditions. As explained earlier,

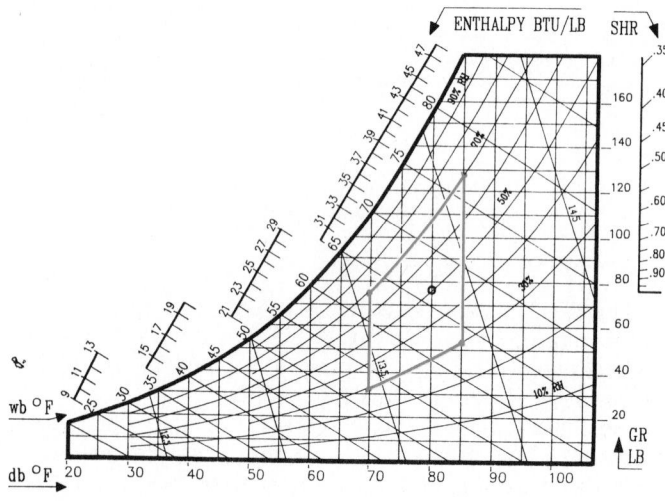

Fig. 18-24. Psychrometric chart showing comfort zone.

the human body will be comfortable under a variety of combinations of temperature and humidity. This is shown in Fig. 18-24. Most people are comfortable in an atmosphere with the relative humidity between 30 and 70 percent and the temperature between 70° and 85 °F. These points are represented by the area outlined in Fig. 18-24.

The reason for the existence of the HVAC industry is that

nature does not always provide those ideal conditions reviewed above. The HVAC system must modify existing conditions, using the heating, cooling, humidification, and dehumidification processes to provide a desired condition.

These processes can be modeled on the psychrometric chart. See Fig. 18-25. For example, the current air condition is: dry bulb temperature of 40 °F and relative humidity of 30 percent. This is Point A in the illustration. The desired condition is 75 °F and 50 percent relative humidity. This is represented by Point B. The HVAC system will then have to provide the processes represented by the colored lines connecting points A and B. This is a good way to visually show what the capability of a system must be.

The psychrometric chart can then be used to plot the actions of the evaporators, heaters, chillers, and so on, in a HVAC system. Further study of psychrometrics can result in equations representing all the processes used in the conditioning of air. This provides the scientists and engineers in the HVAC industry the basics for design and evaluation of new systems.

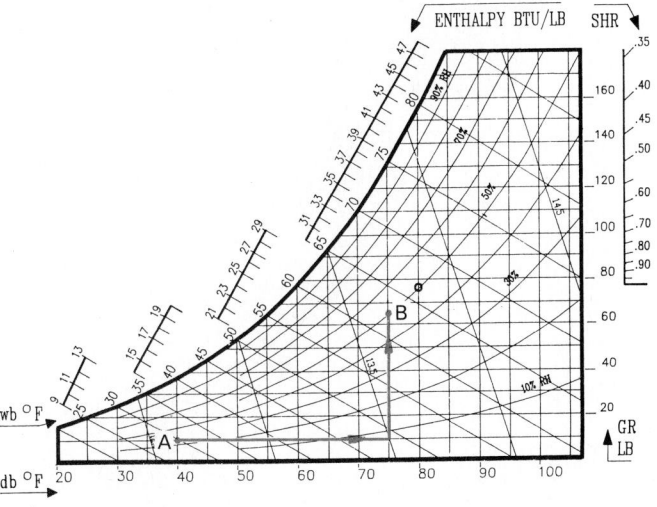

Fig. 18-25. HVAC process modeled on psychrometric chart.

18-17 VAPOR BARRIERS

Water vapor flows easily through all porous substances. Water in vapor form will remain a vapor as long as its temperature is above the dew point. However, as soon as its temperature drops to the dew point, the water vapor will condense into water droplets.

In modern housing, water vapor is kept from passing through walls and toward surfaces where it might condense by using moisture-proof materials. Aluminum foil, plastic sheeting, etc., are used. These materials form a vapor barrier which keeps water vapor from passing from warm surfaces to cold surfaces. Vapor barriers should always be installed on the warm side of a heated space.

The lack of proper vapor barriers is often indicated by the peeling of the paint on a house near kitchen and bathroom areas. The moisture from these areas travels through the wall structure and, on contacting the cold undersurface of the paint, forms droplets of water. This condition, with freezing and thawing, may cause the paint to peel.

18-18 AIR MOVEMENT

Air movement is an important condition affecting comfort. If cool, dry air is circulated past a warm body, heat flow from the body will speed up and evaporation will increase. This tends to cool the body.

During cold days in winter, early spring, or late fall, a person exposed to outside atmospheric conditions often feels much colder than the thermometer indicates. This chilling effect is due to wind velocity and relative humidity. The term "wind chill" applies to this uncomfortable feeling. See Para. 18-30.

Air movement in a conditioned space is also very important. Air movement is very necessary to supply fresh air to a controlled space. If the air moves too fast, persons feel uncomfortable (a draft). If the air movement is too slow, the air becomes stale (contaminated) and lacks oxygen.

18-19 AIR VELOCITY MEASUREMENT

Outside air velocity (wind) is measured in miles per hour (mph) or in knots.

Air velocity (distance traveled per unit of time) is usually expressed in feet per minute (fpm). If the air velocity is multiplied by the cross-sectional area of a duct, it is possible to calculate the volume of air flowing through the duct in cubic feet per minute.

Different methods may be used to measure air velocity:
1. Anemometer (rotating).
2. Velocimeter (swinging vane).
3. Velocity pressure (pitot tube).
4. Anemometer (hot wire).

The rotating anemometer, the direct-reading velocimeter, and the pitot tube are not accurate at very low air velocities.

18-20 ANEMOMETER—ROTATING AND HOT WIRE

If a small propeller is placed in an airstream, it will revolve as the air flows past the blades. If the propeller is connected to a dial calibrated in feet, it will indicate the feet of flow. See Fig. 18-26. Devices of this type are called "anemometers."

Anemometers generally have a start lever and a return-to-zero lever. To use the instrument, carefully place it in the airstream at right angles to the airflow. Allow it to reach a constant speed (takes about one minute). Then trip the registering mechanism and, at the same time, start a stopwatch. Record the reading and the time. From this data, compute the velocity of the air in feet per minute. Divide the number of feet by the elapsed (passed) time. For example, if the reading is 236 for 1/2 min., the velocity will be 472 ft/min.

It is advisable to take several readings, then compute the average to insure greater accuracy.

Another type of anemometer is shown in Fig. 18-27. The dial will indicate airflow from HVAC grilles in cubic feet per minute (cfm). The operator now can calculate the number of Btus going into the space through each grille by multiplying by the appropriate temperature factor. The device takes account of the grille area which is entered as data into the instrument.

The operation of the hot wire anemometer depends upon the cooling effect of air flowing over an electrically heated wire. A hot wire instrument is shown in Fig. 18-28.

Fig. 18-26. Anemometer used for measuring airflow. Large dial shows airflow up to 100 ft./sec.; lower left dial reads in 100-ft. graduations (spaces); lower right dial reads in 1000 ft. graduations. (Taylor Scientific, Consumer Instruments, Sybron Corp.)

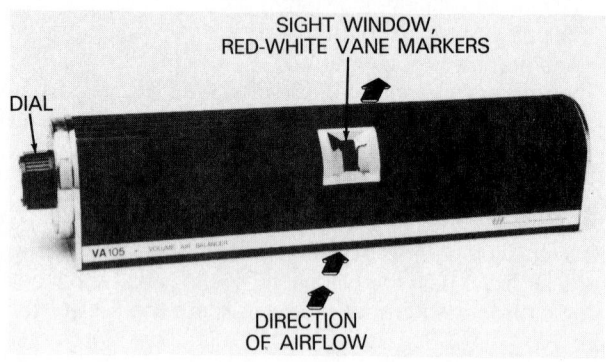

Fig. 18-27. Anemometer reads air velocity in cfm. (TIF Instruments, Inc.)

Fig. 18-28. Direct-reading, air-velocity-indicating instrument of hot-wire type. (Anemostat Products Div.)

18-21 VELOCIMETERS (SWINGING VANE)

Some velocity measuring instruments read in feet per minute (fpm). See Fig. 18-29. Service technicians often use these instruments to avoid the arithmetic which is necessary when using the pitot tube instrument.

In using a swinging vane velocimeter, incoming air pushes on a small vane that tilts at different angles as the air velocity or speed increases. The instrument is put directly in the airstream with the left side facing the airflow. The velocimeter shown in Fig. 18-29 has two velocity scales, 0-200 and 0-800 feet per minute.

For velocity readings where it is difficult to place the instrument in the airstream, special jets are used to adapt the instrument to these conditions.

Fig. 18-29. Air flow velometer provides fast measurements. Instrument is held in the air stream. (Alnor Instrument Company)

The direct-reading instrument can be used to measure duct air velocities, as shown in Fig. 18-30. Note that a special jet is attached to the air inlet of the instrument with a flexible tube.

The velocimeter is also used to measure air velocities in main ducts and branch ducts. See Fig. 18-31. An instrument of this type is necessary to balance air distribution systems.

The instrument is calibrated for use at a temperature of 68°F. Corrections may be made if the duct temperature is not at 68°F. Formula for correction:

$$fpm = \frac{460 + T}{460 + 68} \times instrument\ reading$$

(T = temperature Fahrenheit of air in duct.)

18-22 VELOCITY-PRESSURE (PITOT TUBE)

The velocity-pressure method of measuring air velocity makes use of an instrument called a pitot tube, Fig. 18-32. Air contacting the nose of the pitot tube creates a total

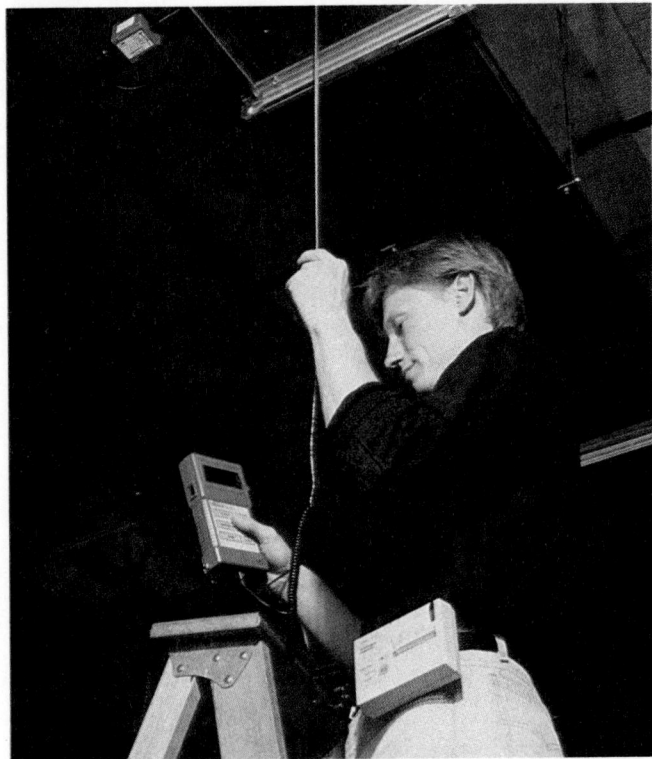

Fig. 18-30. Technician measuring air velocity at duct. (TSI Incorporated)

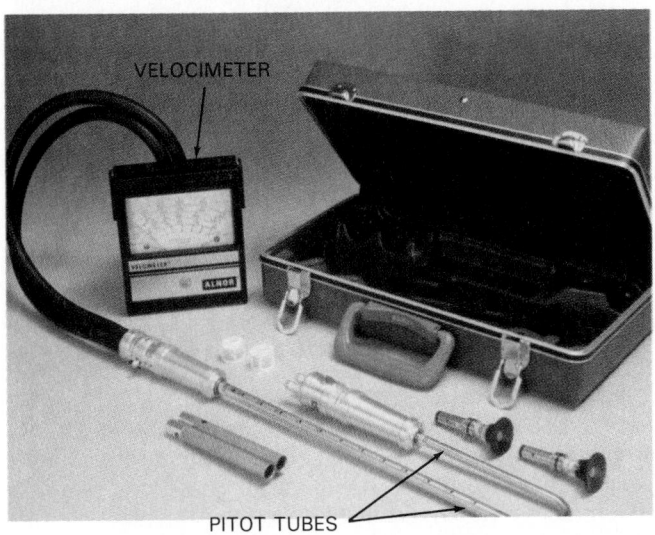

Fig. 18-31. Direct-reading airflow meter which determines air velocity inside a duct. (Alnor Instrument Co.)

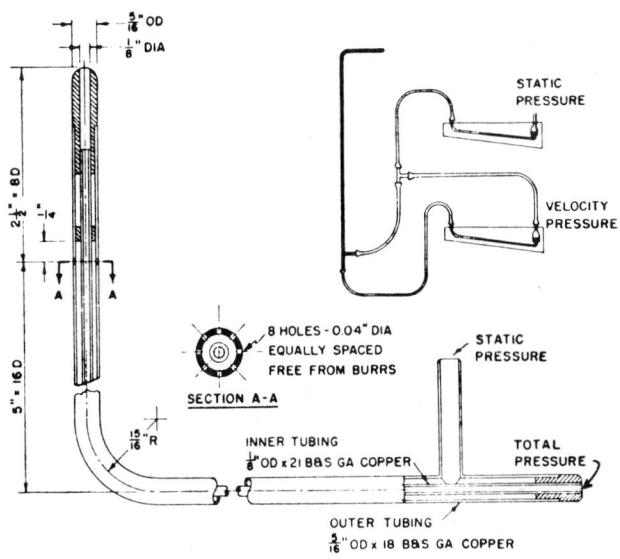

Fig. 18-32. Pitot tube connected to two inclined manometers. The unit measures air velocity. (ASHRAE Guide and Data Book)

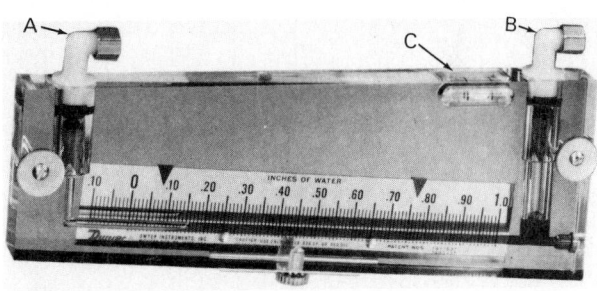

Fig. 18-33. Inclined gauge for use with pitot tube. A—Total pressure connection. B—Static pressure connection. This gauge may also be used for measuring filter pressure drops. C—Built-in spirit level. Liquid level must be adjusted to a zero reading to level the unit. (Dwyer Instruments, Inc.)

Example: If the velocity-pressure is 1 in. of water, what is the velocity?

Velocity = 4050 $\sqrt{1 \text{ in.}}$
Velocity = 4050 × 1
Velocity = 4050 ft./min.

If the velocity-pressure is .25 in. of water, what is the velocity?

Velocity = 4050 $\sqrt{.25}$
Velocity = 4050 × .5
Velocity = 2025 ft./min.

The constant 4050 is for approximately 80°F at a 500 ft. altitude. Other values are shown in Fig. 18-34. The constant changes are based on density of the air. The manometer must be mounted level to obtain accurate readings.

To obtain correct velocity readings in a duct, take several readings in various parts of the duct and average the readings. The recommended method to use is shown in Fig. 18-35. The rectangular duct is divided into equal areas, then the pitot tube is put in the center of each small area. Finally, the 16 resulting velocities are averaged to obtain the overall average velocity.

pressure. The outer tube, with the holes on the side, measures the static pressure. When these two pressures are connected to the end of a manometer, the difference in the pressures is the velocity-pressure. An inclined manometer, Fig. 18-33, is used with the pitot tube.

This pressure difference is measured in inches of water. Formula:

Velocity = 4050 × square root of velocity-pressure in inches of water.

Volume Cu.ft./lb.*	Velocity Constant
11.5	3720
12.1	3818
13.2	3980
**13.4	4010
14.1	4118
15.1	4260
16.2	4410
17.1	4530

*Values for any conditions may be read from the psychrometric chart.
**Standard.

Fig. 18-34. Velocity correction factor changes with change in air density (effect of temperature and altitude).

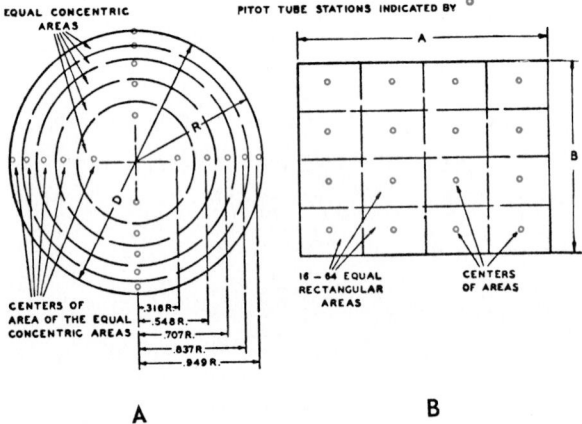

Fig. 18-35. Locations for velocity readings in duct. Average of readings will produce average duct velocity. A—Round duct (20 readings). B—Rectangular duct (16 readings). (ASHRAE Guide and Data Book)

The round duct is more difficult to measure, because it must be divided into equal circular areas. See Fig. 18-35 again. The location of each of the points is as recommended by the American Society of Heating, Refrigerating and Air Conditioning Engineers (ASHRAE).

Rotating, turbulent airflow can affect the ability of the pitot tube to measure the true pressures. The pitot tube should be used only where the duct is very long. It is best if the length of the duct downstream of the measuring location is a minimum of 10 duct diameters. If precise measurements are required, air straightening vanes should be located upstream from the pitot tube.

18-23 DRAFTS AND DRAFT MEASUREMENT

If the heating and cooling arrangement in a structure is set up in such a manner that air flows through the occupied space at a rate of over 15 to 20 feet per minute, occupants of the space will feel uncomfortable. They feel a draft.

It is difficult to develop accurate instruments to measure drafts. The usual method is to use a smoke generator and a stopwatch to time the flow of the smoke through the space being analyzed.

18-24 VENTILATION

Ventilation is a term applied to changing the air in a workplace or living place. In any space occupied by people, the breathing of the air reduces the oxygen content. Activities in the space may add some pollutants to the environment. The most economical way to maintain the health and comfort conditions of the space is by replacing the air. This is done by bringing in outside air to ventilate the space.

Sometimes it is desirable to quickly replace all the air in the confined space. This is done by opening windows and doors to flush the space completely with 100 percent outside air.

In most heating systems, provisions are made to continually replace a small percent of the air in the conditioned space. This is done by slowly exhausting some of the air and bringing in makeup air. In homes, the gradual change may not be noticeable, since there is always a small amount of air entering and leaving the home. This movement of air is by way of the cracks around windows and doors and through doorways each time doors are opened.

A considerable amount of air also filters in and out of a building because some building materials are porous (many tiny holes). The amount of infiltration depends chiefly upon the wind velocity and the temperature difference inside and outside the building. Air tends to enter the building on the upwind side and leave the building on the downwind side.

Since warm air is lighter than cold air, it tends to rise in a room. If the building has more than one story, the warm air tends to rise from the lower floors to the upper floors. The rising air will create a slight pressure that causes some warm air to escape through the upper surfaces of the building. This lost air is replaced by cold air entering the lower levels.

With summer air conditioning, the opposite situation occurs. Cold air tends to flow downward and may leave the building at the lower levels. The cold air, then, is replaced by warmer air entering at the upper levels.

Whenever air is exhausted from a space, such as a room in a home or factory building, air must be brought into the room to replace the air exhausted. If the outdoor temperature is either too high or too low for comfort, the air brought into the area to replace exhausted air must be conditioned:
1. Either heated or cooled.
2. Cleaned.

Replaced air is conditioned to provide a comfortable room environment. This conditioned air brought into the room is called "makeup air."

A structure which maintains an inside air pressure slightly above atmospheric pressure is said to have a positive pressure. A structure which maintains an air pressure slightly below atmospheric pressure is said to have a negative pressure.

Fuel-burning furnaces, stoves, and fireplaces operating in the winter tend to cause a negative pressure in a building. This means that there will be a considerable amount of air leakage through the walls and through cracks. Positive pressure can be maintained only if a fan or blower of some kind is used to bring in fresh air.

With heated tall structures, there may be a tendency for upper rooms to be slightly above atmospheric pressure, and lower rooms to be slightly below atmospheric pressure, because warm air rises.

18-25 CLIMATE, OUTDOOR—INDOOR

Climate usually is defined as the weather conditions of a region. These conditions include temperature, humidity, sunshine, pressure, and air movement.

It is evident that outdoor climate cannot be affected much by any type of air conditioning (heating, cooling, humidifying). In an enclosed space, however, the above factors may be controlled and an "indoor" climate provided to meet any desired conditions.

Indoors, it is possible to completely control the factors which determine comfort in an enclosed space. There is a definite relationship between comfort and the conditions of temperature, humidity, and air movement. Fig. 18-36 illustrates the constant comfort condition with varying temperatures and humidities. Many homes and workplaces are completely air conditioned.

Increasing the air movement tends to give a cooling effect on the human body. If the heating system provides too much air movement (over 15 to 20 fpm), it may be necessary to increase the temperature somewhat in order to maintain a comfortable indoor climate.

18-26 WEATHER

Weather usually is defined as the conditions in the atmosphere such as temperature, wind velocity and direction, clouds, moisture, and atmospheric pressure. Weather affects the need for and requirements of air conditioning.

18-27 AIR TEMPERATURE

Air temperatures in the United States vary from a low of about −55°F (−48°C) to a high of around 120°F (49°C). The normal, desirable temperature is said to be 72°F (22°C).

Normally, the temperature of the human body is 98.6°F (37°C). Skin temperature is lower, about 91°F (33°C). In temperate zones, the average atmospheric temperatures in winter are below the body temperature, so clothing is required to help conserve the body heat. Also, heat needs to be added to the occupied space so that the occupants may be comfortable.

The human body can lose heat easily after the air temperature falls below 98.6°F (37°C). The body can lose heat at air temperatures above 98.6°F (37°C) by evaporation of perspiration and respiration from the body.

Heating the air in some instances and cooling the air in other instances is necessary in order to maintain temperatures that are comfortable. The specific heat of dry air is .24 Btu per lb. Either for heating or cooling, then, energy is required to bring about the desired temperatures.

18-28 TEMPERATURE CONTROLS

When heating an air conditioned space, the amount of heat supplied to the conditioned space regulates the temperature. The temperature-regulating devices control the flow of heat-carrying media (substances). Usually, the heat-carrying medium is either warm air, warm water, steam, or water vapor. Typical heat sources are gas flame, oil flame, electric resistance, or coal fire.

When cooling an air conditioned space, the amount of cooling will depend upon the temperature of the cooling surface (evaporator), the rate of flow of air over the cooling surface, and the initial temperature of the air in the conditioned space. All of these variables may be controlled.

18-29 SUN HEAT LOAD FUNDAMENTALS

Radiant heat (light) from the sun furnishes a tremendous amount of heat energy. If a glassed-in surface is exposed to this light energy, the energy will enter a space and become heat. Since glass is a rather poor conductor of heat, the heat that comes in as a ray of light is trapped in the room as heat energy.

The surfaces of buildings exposed to sunlight are also heated by the sun's rays. This heat source must be taken into account when designing both the heating and the cooling requirements of air conditioning systems. Since many building materials are poor conductors of heat, there is a lag or delay between the time the radiant heat energy strikes the building surface and the time the heat enters the air conditioned space.

Color has a considerable effect on the amount of heat absorbed from the sun's rays. Black and red absorb much

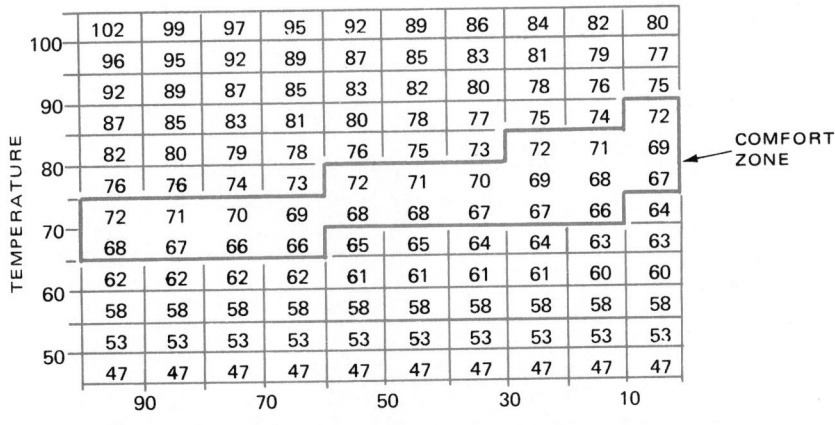

Fig. 18-36. Equivalent temperatures (similar to effective temperature). Note comfort zone. Area inside red lines indicates usual temperature and relative humidity range in which most people are comfortable. Note that with high relative humidity, one is comfortable at lower temperature than temperature desired for low relative humidity conditions.

more heat than white and yellow. Likewise, surfaces that radiate heat are much more efficient if painted in dark colors than if painted in light colors. Light-reflecting surfaces, such as polished metal, chrome, and nickel plate, do not absorb heat easily; neither do they efficiently radiate heat from their surfaces.

18-30 WIND CHILL INDEX (CHILL FACTOR)

During the winter months, the wind chill index, or "chill factor," combines temperature and wind speed, Fig. 18-37. The chill factor is calculated and released by the United States Weather Bureau through the usual weather forecasting channels.

The chill factor is based on both temperature and wind speed in mph. For example, at a temperature of 0 °F and a wind speed of 10 mph, the chill index temperature is −22 °F. Human flesh exposed to the atmosphere freezes at about −25 °F, making the chill factor an important consideration.

Fig. 18-37 shows the chill factor in degrees corresponding to the wind speed. Note that, at a temperature of 0 °F, with a wind speed of 40 mph, the chill factor mounts to −54 °F. This temperature will quickly freeze exposed flesh.

18-31 BEAUFORT SCALE

The Beaufort Scale is frequently used by the Weather Bureau in indicating wind velocity. Fig. 18-38 gives wind velocity values and effects according to the Beaufort Scale.

Increasing the wind velocity increases the heat loss of a heated structure. The calculated heat load for a structure should include provisions for the maximum wind velocity expected for the area.

18-32 COMFORT CONDITIONS

Comfortable conditions result from a desirable combination of temperature, humidity, air movement, and air cleanliness. However, one may have comfort under varying values of these factors. For instance, high relative humidity which tends to be uncomfortable may be counteracted by a relatively low temperature and rapid air movement. In many homes in wintertime, a low relative humidity is compensated for by an increase in room temperature and slight air movement. Fig. 18-39 illustrates what is commonly accepted as the comfort zone for the various conditions.

A more technical graph showing the comfort zones for both winter and summer is illustrated in Fig. 18-40. This comfort zone represents a considerable area. However, experiments indicate that any point in this area gives approximately equal comfort under equal conditions of clothing and work. These areas are sometimes defined as effective temperature (ET). *Effective temperature is the combined effect of dry bulb temperature, wet bulb temperature, and air movement, which provides an equal sensation of warmth or cold.*

The human body is able to accustom itself only to a certain amount of change in a given length of time.

WIND CHILL INDEX

		AMBIENT TEMPERATURE													
	°F	40	35	30	25	20	15	10	5	0	−5	−10	−15	−20	−25	−30
	°C	4	2	−1	−4	−7	−9	−12	−15	−18	−21	−23	−26	−29	−32	−34
WIND VELOCITY IN MPH		EQUIVALENT TEMPERATURE IN STILL AIR														
CALM (0)	°F.	40	35	30	25	20	15	10	5	0	−5	−10	−15	−20	−25	−30
	°C.	4	2	−1	−4	−7	−9	−12	−15	−18	−21	−23	−26	−29	−32	−34
5	°F.	37	33	27	21	16	12	7	1	−6	−11	−15	−20	−26	−31	−35
	°C.	3	1	−3	−6	−9	−18	−14	−17	−21	−24	−26	−29	−32	−35	−37
10	°F.	28	21	16	9	2	−2	−9	−15	−22	−27	−31	−38	−45	−52	−58
	°C.	−2	−6	−9	−13	−17	−19	−23	−26	−30	−33	−35	−39	−43	−47	−50
15	°F.	22	16	11	1	−6	−11	−18	−25	−33	−40	−45	−51	−60	−65	−70
	°C.	−6	−9	−11	−17	−21	−24	−28	−32	−36	−40	−43	−46	−51	−54	−57
20	°F.	18	12	3	−4	−9	−17	−24	−32	−40	−46	−52	−60	−68	−76	−81
	°C.	−8	−11	−16	−20	−23	−27	−31	−36	−40	−43	−47	−51	−56	−60	−63
25	°F.	16	7	0	−7	−15	−22	−29	−37	−45	−52	−58	−67	−75	−83	−89
	°C.	−9	−14	−18	−22	−26	−30	−34	−38	−43	−47	−50	−55	−59	−64	−67
30	°F.	13	5	−2	−11	−18	−26	−33	−41	−49	−56	−63	−70	−78	−87	−94
	°C.	−16	−15	−19	−24	−22	−32	−36	−41	−45	−49	−53	−57	−61	−66	−70
35	°F.	11	3	−4	−13	−20	−27	−35	−43	−52	−60	−67	−72	−83	−90	−98
	°C.	−11	−16	−20	−25	−29	−33	−37	−42	−47	−51	−55	−58	−64	−68	−72
40	°F.	10	1	−6	−15	−22	−29	−36	−45	−54	−62	−69	−76	−87	−94	−101
	°C.	−12	−17	−21	−26	−30	−34	−38	−43	−48	−52	−56	−60	−66	−70	−74

Fig. 18-37. Wind chill index reveals that increasing wind velocity greatly increases chill effect. With thermometer reading of 0 °F (−18 °C) and wind velocity of 20 mph, effect on human body is same as it would be if person were in temperature of −40 °F (−40 °C). (Michigan Farmer)

BEAUFORT NUMBER	WIND VELOCITY		OBSERVED WIND EFFECTS	TERMS USED IN USWB FORECASTS
	MPH	KNOTS		
0	LESS THAN 1	LESS THAN 1	CALM; SMOKE RISES VERTICALLY	
1	1–3	1–3	DIRECTION OF WIND SHOWN BY SMOKE DRIFT; BUT NOT BY WIND VANES	LIGHT
2	4–7	4–6	WIND FELT ON FACE; LEAVES RUSTLE; ORDINARY VANE MOVED BY WIND	
3	8–12	7–10	LEAVES AND SMALL TWIGS IN CONSTANT MOTION; WIND EXTENDS LIGHT FLAG	GENTLE
4	13–18	11–16	RAISES DUST, LOOSE PAPER; SMALL BRANCHES ARE MOVED	MODERATE
5	19–24	17–21	SMALL TREES IN LEAF BEGIN TO SWAY; CRESTED WAVELETS FORM ON INLAND WATERS	FRESH
6	25–31	22–27	LARGE BRANCHES IN MOTION; WHISTLING HEARD IN TELEPHONE WIRES; UMBRELLAS USED WITH DIFFICULTY	STRONG
7	32–38	28–33	WHOLE TREES IN MOTION; INCONVENIENCE FELT WALKING AGAINST WIND	
8	39–46	34–40	BREAKS TWIGS OFF TREES; GENERALLY IMPEDES PROGRESS	GALE
9	47–54	41–47	SLIGHT STRUCTURAL DAMAGE OCCURS; LEAVES, BRANCHES BLOWN FROM TREES	
10	55–63	48–55	SELDOM EXPERIENCED INLAND; TREES UPROOTED; CONSIDERABLE STRUCTURAL DAMAGE OCCURS	WHOLE GALE
11	64–72	56–63	VERY RARELY EXPERIENCED; ACCOMPANIED BY WIDESPREAD DAMAGE	
12 OR HIGHER	73 OR HIGHER	64 OR HIGHER	VERY RARELY EXPERIENCED; ACCOMPANIED BY WIDESPREAD DAMAGE	HURRICANE – TYPHOON

Fig. 18-38. Beaufort Scale of wind velocity.

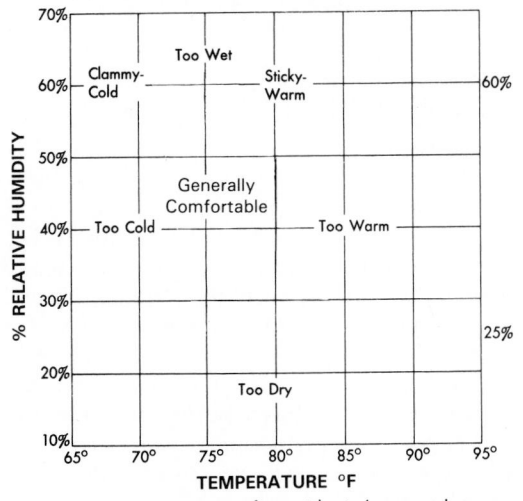

INDOOR CLIMATE COMFORT CHART

Air motion continuous at five to eight air changes per hour.

Fig. 18-39. Indoor comfort chart. Most people will feel comfortable at temperature and relative humidity indicated in center. (Lennox International, Inc.)

Therefore, it becomes necessary to regulate air conditioning equipment so that it will produce only a certain output. While this will be less comfortable, it will not subject the person to too great a shock on entering this conditioned space, or when going out into the normal outdoor atmosphere.

In the summertime, air conditioned buildings are usually maintained at a temperature which is not more than 10 degrees below the outside temperature. Some people are quite sensitive to thermal shock when entering or leaving an air conditioned space. The danger of this thermal shock is lessened if the difference between inside and outside temperatures is reduced, or if a person will put on a sweater or coat when entering a cooled air conditioned space, and remove it when returning to the warm outdoors.

Fig. 18-40 indicates that, in the summer, most people are comfortable between 72 °F db (dry-bulb temperature) and 90 percent rh (relative humidity) up to 87 °F db and 23 percent rh. During the winter, most people feel equally comfortable between 66 °F db and 70 percent rh to 80 °F db and 20 percent rh.

Experiments show that the average person is most comfortable under normal temperature and humidity conditions

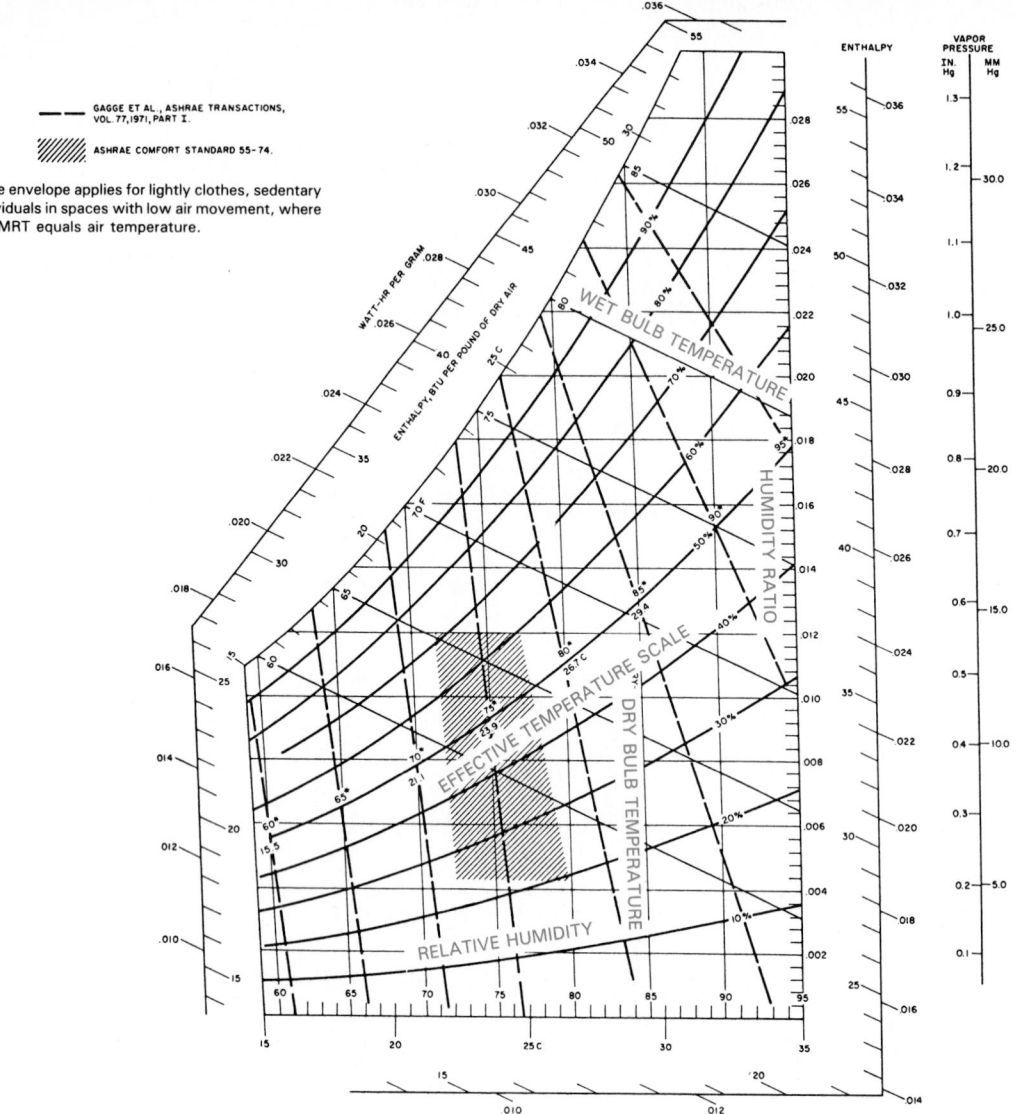

Fig. 18-40. Graph of comfort zone. Note dry-bulb and wet-bulb temperature lines, and relative humidity line. (ASHRAE Guide and Data Book)

if the skin surface temperature is approximately 91 °F (33 °C). This skin temperature is usually maintained in cold weather by wearing clothing. In hot weather, the temperature is maintained by the evaporation of moisture (sweat) from the skin surface and by radiation from the skin surface.

The skin temperature may drop below this figure in hot, humid weather because of rapid evaporation of moisture from the skin surface. Even though this skin temperature may be considerably below what is considered to be a comfortable temperature, the person is not uncomfortable because the heat generated in the body is being released by moisture evaporation from the skin surface.

Human ills due to thermal environment are sometimes called "thermal disorders." In cold climates, it is possible for the body temperature to drop a few degrees below normal. This is chiefly due to lower metabolism.

High temperatures may cause human illness, particularly if the high temperature is accompanied by high humidity. Heat stress is being investigated by the Occupational Safety and Health Administration. The measure of heat stress is done using a 6-in. copper sphere, painted black, with a dry bulb thermometer inserted until the sensitive bulb is at the center of the sphere.

At 79 °F (26 °C) WBGT (wet bulb globe temperature),

a person should only work one half of the time for the first five days. A person should take salt tablets moderately if the WBGT is 79 °F (26 °C) or higher. Many studies have proven the health benefit obtained from a properly air conditioned living space.

18-33 COMFORT-HEALTH INDEX (CHI)

The American Society of Heating, Refrigeration and Air Conditioning Engineers recognizes a Comfort-Health Index. Fig. 18-41 indicates the temperature, the sensation, and the effect on the physiology and health of the body. From this chart, it may be seen that, under long exposure to very hot conditions, the human body attempts to adjust to the conditions by increasing sweating and the flow of blood. These physiological conditions may result in an increased danger of heat strokes and cardiovascular difficulty.

The chart shows that, at comfortable temperatures, there is no sensation of either warmth or cold. Also, there are no apparent physiological effects, and the body is in the condition of normal health. Moving down in temperature to very cold conditions, the body is uncomfortable. Also, physiologically, the body attempts to correct this condition by shivering. From the standpoint of health, this may cause an increase in mortality, particularly in older people.

COMFORT-HEALTH INDEX

NEW T_eff SCALE °C °F	TEMPERATURE LEVEL	COMFORT RANGE	PHYSIOLOGICAL RESPONSE	HEALTH EFFECT
	LIMITED TOLERANCE	LIMITED TOLERANCE	BODY HEATING FAILURE OF REGULATION	CIRCULATORY COLLAPSE
40	VERY HOT	VERY UNCOMFORTABLE		
100	HOT		INCREASING STRESS CAUSED BY SWEATING AND BLOOD FLOW	INCREASING DANGER OF HEAT STROKES CARDIOVASCULAR EMBARRASSMENT
35				
90	WARM SLIGHTLY WARM	UNCOMFORTABLE	NORMAL REGULATION BY SWEATING AND VASCULAR CHANGE	
30				
80	NEUTRAL	COMFORTABLE	REGULATION BY VASCULAR CHANGE	NORMAL HEALTH
25	SLIGHTLY COOL			
70	COOL	SLIGHTLY UNCOMFORTABLE	INCREASING DRY HEAT LOSS URGE FOR MORE CLOTHING OR EXERCISE (BEHAVIORAL REG.)	
20				
60 15	COLD			INCREASING COMPLAINT FROM DRY MUCOSA AND SKIN (WATER VAPOR PRESSURE < 10 mm Hg)
50 10	VERY COLD	UNCOMFORTABLE	VASOCONSTRICTION IN HANDS AND FEET SHIVERING	MUSCULAR PAIN IMPAIRMENT OF PERIPHERAL CIRCULATION

Fig. 18-41. Comfort-Health Index indicates sensory, physiological, and health responses by people to prolonged exposures. (ASHRAE Handbook of Fundamentals)

18-34 DEGREE DAYS

Degree days is a term used to help indicate the heating or cooling needed for any given day. Calculations are based on a temperature of 65 °F (18 °C). The degree day is computed by taking the mean (average) of the highest temperature and the lowest temperature for a day and subtracting it from 65 °F (18 °C).

Example: The lowest recorded temperature for a certain day was 28 °F (−2 °C).

The highest recorded temperature for the same day was 36 °F (2 °C).

The mean temperature for the day was

$$\frac{28 + 36}{2} = 32 °F \qquad 65 - 32 = 33 \text{ degree days (F)}$$

If the temperature conditions went on for two days, the result would be 66 degree days. If Monday has 30 degree days and Tuesday has 20, the result is 50 degree days for the two days.

In Celsius degrees:

$$\frac{-2 + 2}{2} = 0 °C \text{ or } 18 °C - 0 °C = 18 \text{ degree days (C)}$$

Degree days may be added by weeks, months, or for a season to give a comparison of the heating needs for different years.

18-35 AIR CONTAMINANTS

Air contains substances other than those described in Para. 18-2. Called contaminants, these substances are not normal in the atmosphere. Most of them are in some way a detriment to comfort, health, and desirable industrial environment.

There are three general classes of contaminants:
1. Solid.
2. Liquid.
3. Gases and vapors.

Solid particles are kept in suspension in the air by air

currents. They may be classified into four general groups: dust; fumes; smoke; and pollen, bacteria, and molds.

Dust is a result of wind, of a sudden earth disturbance, or of mechanical work on a solid. The origin of dust can be animal, vegetable, or mineral. Dust particles are usually over 600 microns in size (about 0.024 in. in dia.). See Chapter 30 for the definition of micron. See Fig. 18-42.

Fumes are solids formed by condensation and solidification of materials that are ordinarily solids but have been put into a gaseous state (usually an industrial or chemical process). These particles are about 1 micron in size.

Smoke is due to poor combustion. Solid particles are carried into the atmosphere by gaseous products of combustion. Particles vary in size from .1 to 13 microns.

Pollen, bacteria, and molds are living substances.

In addition, there are liquid impurities in the air. Two of the most common are mists and fogs.

Mists are small liquid particles, mechanically ejected into the air by splashing, mixing, atomizing, etc.

Fogs are small liquid particles formed by condensation. Fogs mean that the atmosphere has reached the saturation state for that particular chemical. A fog may consist of minute particles of water, sometimes contaminated with sulphur dioxide, fumes, smoke, and dust particles. It may form a very obnoxious atmosphere to be breathed by either humans or animals.

There are also gases and vapor contaminants which may act like true gases. There is little difference between these two impurities. Vapors are gases that have condensing temperatures and pressures close to normal conditions.

Not all contaminants are objectionable or harmful. Perfumes and deodorizers have been used for years to make air more pleasant to smell or to conceal objectionable odors.

Pollen grains come from vegetation growth such as weeds, grasses, and trees. Their presence in the air is usually responsible for hay fever, rose fever, and other respiratory conditions. Removal of pollen from the air has been an important contribution of air conditioning. These particles vary in size from 10 to 50 microns.

Bacteria are microorganisms that are responsible for the transfer of many diseases. Many manufacturing processes require the removal of these bacteria. Hospital rooms and some refrigerators use bacteria-removing devices.

Mold is a growth of minute fungi which forms on vegetable and animal matter. Many typical air conditioning applications may provide a favorable environment for the growth and development of many molds (particularly if some moisture is present). There have been instances where the spores from these molds have caused illness to the occupants of the air conditioned space.

Mold-killing sprays are available which may be used to spray air conditioning evaporators to make them free of any possible mold development.

18-36 COMMON AIR CONTAMINANTS— MICRON SIZES, FILTERS

In filtering out contaminants, the size of the particle determines to a great extent the nature of the filter required. Fig. 18-42 illustrates some common contaminants in relation to their micron size. Refer to Chapter 30 for information concerning the size of a micron.

Many different types of filters may be used for removing solid particles from the air. See Chapter 22.

18-37 POLLUTANTS

The Federal Clean Air Act of 1963 gives the United States Department of Health, Education and Welfare power to establish and enforce standards for clean air. The pollutants, in addition to those previously named, include particulates, carbon monoxide, photochemical oxidants, and nitrogen oxides. The photochemical oxidants result from the effect of sunlight on hydrocarbons and nitrogen oxides, which causes them to react to produce smog. The word smog combines the terms "smoke" and "fog." Terpene, a hydrocarbon released from growing trees, is sometimes considered a pollutant. Methane comes from the decomposition of vegetable matter.

Particulates include fogs, mists, molds, pollen, dust, fly ash, asbestos, and larger bacteria. In addition to the vapors and particulates, the environment often includes bacteria, viruses, and fungi. Remember that outside air under the most favorable conditions always includes some of the above substances. In some cases, the construction and operation of some air conditioning equipment may increase rather than decrease these pollutants. For instance, humidifiers kept at a lukewarm temperature may provide an excellent breeding place for bacteria. Also, surfaces of dehumidifiers may become breeding places for bacteria and fungi. This means that these surfaces must be sterilized from time to time to maintain a safe operating condition.

Sulphur dioxide, which frequently results from the burning of coal, gas, and oil, is a common gaseous pollutant. Hydrogen sulphide results from some industrial processes, particularly papermaking. In addition, chlorine, odors from paints, insecticides, and other volatile solvents release polluting vapors into the atmosphere. Many vapor-caused illnesses may be difficult to identify. However, it has been established that when patients are removed from these environments their illnesses disappear.

Carbon monoxide is the result of incomplete combustion of fuel. A common source of carbon monoxide is automobile engine exhaust. Fuel-burning furnaces also produce carbon monoxide, and it is present in the combustion chamber, heat exchanger, flue, and stack. It is very dangerous. It is an odorless, tasteless, and colorless gas.

The effect of carbon monoxide on people is that it produces headaches, nausea, and vomiting. However, if a person is exposed to a large amount of carbon monoxide, it is possible that none of these symptoms will have time to develop. Instead, the person suddenly becomes unconscious, leading to death. Carbon monoxide replaces the oxygen in the red corpuscles.

Nitrogen oxide is formed at high temperatures. The nitrogen and oxygen atoms combine to form the compound nitrogen oxide. This combination can take place in automobile engine combustion. This substance, nitrogen oxide, is one of the substances common in photochemical oxidants which produce smog. Nitrogen oxide is unstable, meaning that it changes back to nitrogen and oxygen easily at lower temperatures.

Organic vapors are a major source of air pollution. Fig. 18-43 illustrates an instrument that is used to measure the intensity of odor-causing molecules in trace concentrations. The portable odor monitor uses a highly sensitive metal oxide thermal conductivity sensor. Numerous readings are initially made to set the quality standards or acceptable odor levels. Once this is

MICRON SIZES OF CONTAMINANTS

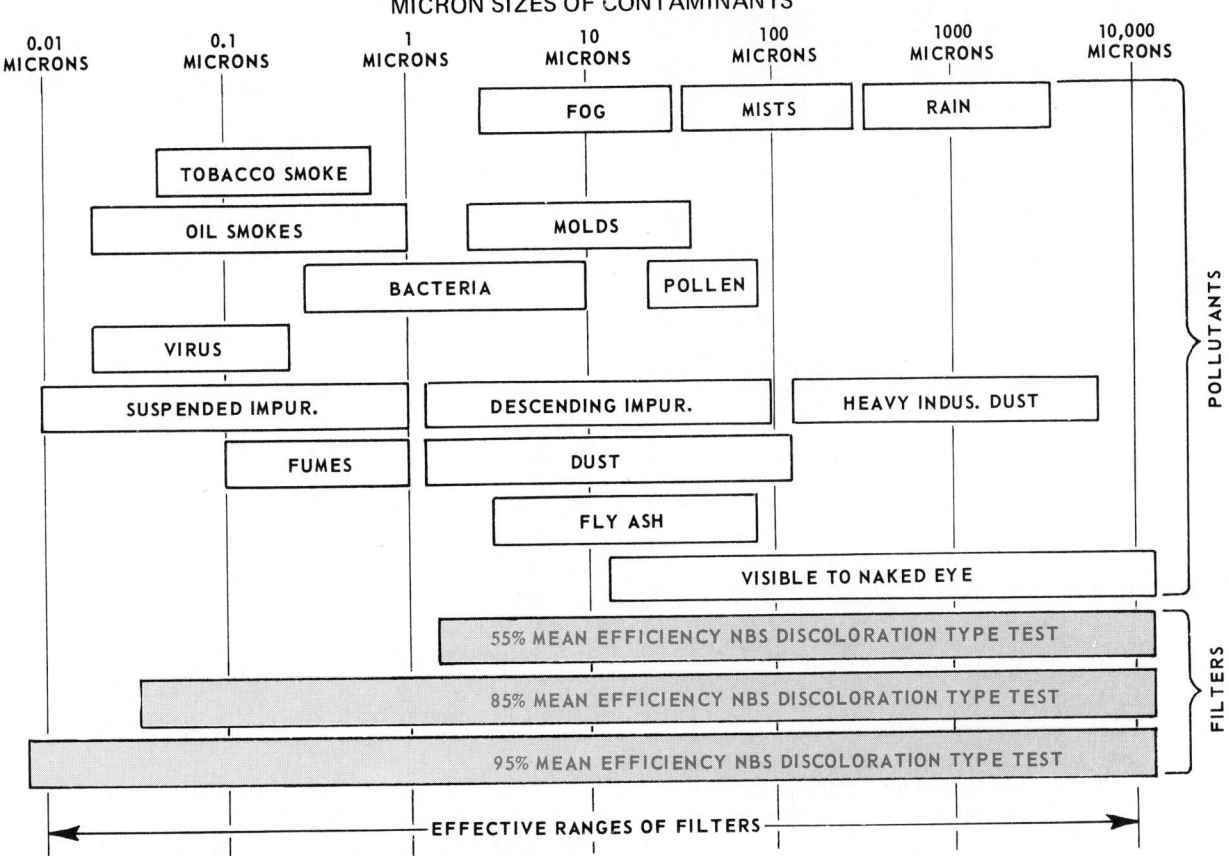

Fig. 18-42. Micron sizes of common air contaminants. (Arcoaire/Snyder General Corp.)

established, the monitor will then identify the odor level at that time.

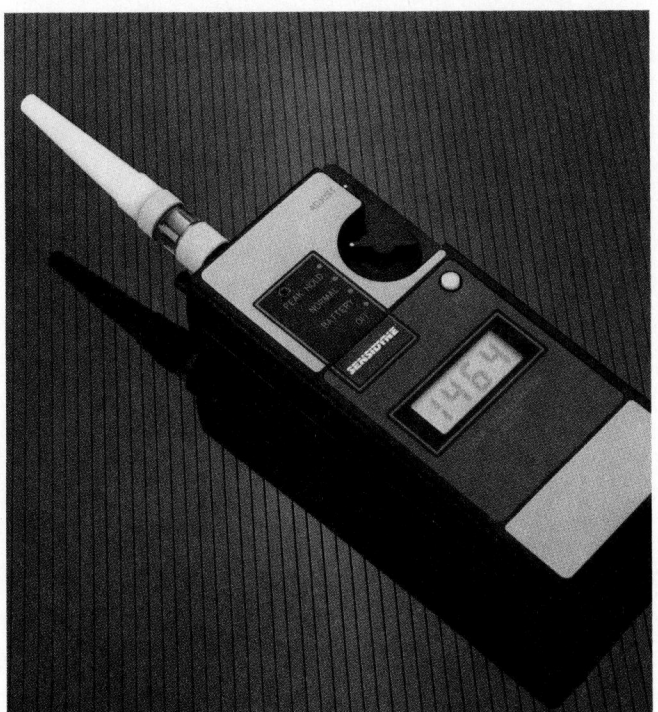

Fig. 18-43. Portable odor monitor. Note digital display, which indicates odor concentration level. (Sensidyne, Inc.)

18-38 OZONE

Ozone is a form of oxygen produced in nature photochemically. The chemical formula for oxygen is O_2. Ozone is O_3.

In the upper atmosphere, ozone is made by ultraviolet light reacting with oxygen. Or, it may be produced by an electric discharge in air (lightning). Ozone is a disinfectant. It is sometimes used to purify water, maintain a sterile atmosphere, and to remove odors in such places as cold storage rooms and hospitals.

No universal agreement exists concerning the benefits or possible hazards in the use of ozone in a conditioned space. It is generally considered that a very small amount of ozone in the air is beneficial. However, in larger concentrations, it may be detrimental to health. Ozone may be one of the reasons for irritation caused by smog. There is no evidence that ozone is accumulatively harmful.

It is generally considered that a concentration of ozone of 0.1 parts per million (ppm) is the maximum permissible for eight-hour exposure. For continuous occupancy, ozone should not exceed 0.01 ppm. The effect of ozone increases with temperature. The effect doubles for each 15 °F (8 °C) increase in temperature. It is said that the use of some electronic air cleaners may slightly increase the ozone content of the conditioned space.

Reliable instruments have been developed for measuring the concentration of ozone in a conditioned space. This measurement is made in parts per million (ppm) or in parts per hundred million (pphm). A typical ozone monitor is illustrated in Fig. 18-44.

Fig. 18-44. Ozone monitor indicates ozone content of area in 0 to 9.99 parts per million (ppm) or 0 to 999 parts per hundred million (pphm). (Mast Development Co.)

18-39 MEASURING FILTER EFFICIENCIES

Two common methods of measuring filter performance are described in ASHRAE Standard 52-76:

1. Atmospheric Dust Spot Efficiency.
2. Synthetic Dust Weight Arrestance.

Atmospheric dust spot efficiency is the measurement of the ability of the device to remove atmospheric dust from the test air. It is the measurement of equal flow rates on both sides of the filter being tested.

Special filter paper targets are located on both sides of the tested filter. The efficiency is calculated based upon the quantity of air drawn through the target filter paper, the amount of light transmission through the target filters, and the difference in light transmission for the two target filters. The amount of air and light transmission through the target filter papers will decrease during the test due to a buildup of dust.

Synthetic dust weight arrestance is a measure of a filter's ability to remove synthetic dust from test air. It is calculated based upon the weight of the synthetic dust which passes through the filter being tested, compared to the weight of the amount fed into the filter.

Another filter efficiency test is called the DOP Smoke Penetration Method. The name of the test comes from the name of the testing particles DiOctyl Phthalate. This test is used mainly with high efficiency filters. Particles of 0.3 microns are sprayed into the inlet duct of the filter being tested. Small sample white filters collect some of this dust from the airstream before the filter being tested (inlet). Other sample filters collect dust from the air leaving the filter being tested. The difference in the sampling filters by color or weight determines the filter efficiency. For example, the outlet sample filter may collect only 1 percent of the particles as the sample filter in the inlet. This filter would then have 99 percent efficiency.

18-40 POLLEN COUNT

During certain times of the year, some plants create a concentration of pollen in the atmosphere which may be irritating to many people. The plants that most commonly cause the trouble are ragweed, timothy, and, to a lesser extent, some flowering plants such as goldenrod and roses.

Pollen count may be determined by exposing a surface coated with an adhesive to the atmosphere for a period of 24 hours. The number of pollen grains in a square centimeter are then counted, and this becomes the pollen count for the past 24 hours.

18-41 THERMOMETERS—AIR CONDITIONING

Common thermometers regularly used by refrigeration and air conditioning service technicians are illustrated and explained in Para. 2-53. The air conditioning technician frequently needs special thermometers to accurately determine the operating temperatures. A quick-reading instrument is shown in Fig. 18-45. It has a scale from 0 to 600 °F (− 18 to 316 °C) and operates on the thermocouple principle. Other temperature scales are available. To use the thermometer, place the end of the adjustable probe against the surface where the temperature is to be measured. The temperature will be shown on the scale.

Two types of electric thermometers are shown in Fig. 18-46. These units may be either battery or 120 V ac powered. The probe reacts quickly and accurately. The scale is calibrated in both Fahrenheit and Celsius degrees.

A recording-type thermometer helps locate malfunctions by making 24-hour or 7-day temperature records. Fig. 18-47 illustrates a recording-type thermometer.

The wet globe thermometer shown in Fig. 18-48 has

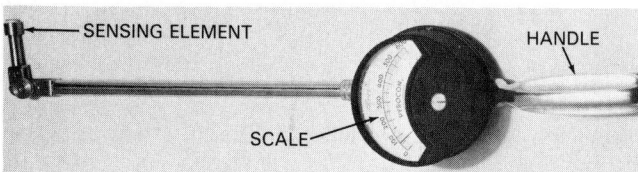

Fig. 18-45. Temperature-measuring instrument used to find air and surface temperatures. (Alnor Instrument Co.)

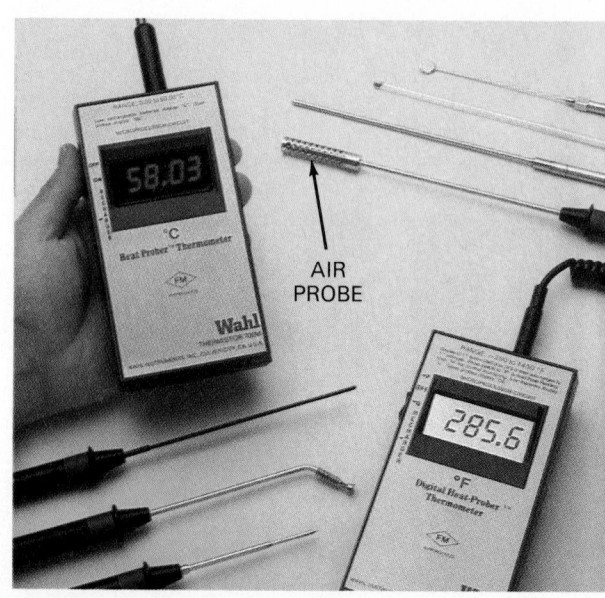

Fig. 18-46. Digital sensor probes for measuring temperature. Left. Thermistor-type thermometer. Right. Platinum-RTD (Resistance Temperature Detector) type. Note the air probe with ventilated shield around tip. (Wahl Instrument, Inc.)

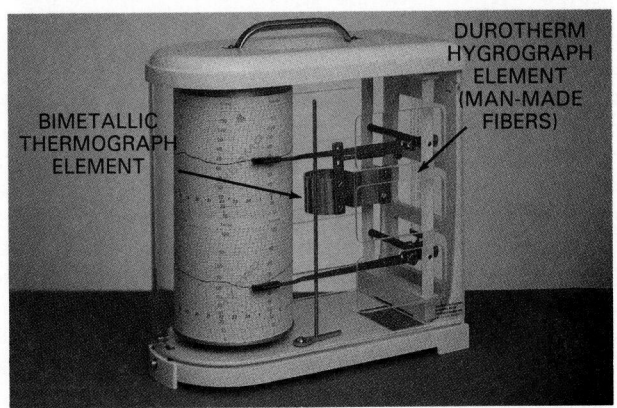

Fig. 18-47. Thermo-hygrograph showing location of the temperature and humidity elements. Arm at top records the temperature; arm at bottom records the humidity. (Abbeon Cal, Inc.)

Fig. 18-48. Wet globe thermometer. Black-cloth-covered sphere will reach temperature which is balance of wet bulb cooling ability, outgoing radiation, and incoming radiation. A—Cloth-covered sphere. B—Dial thermometer. (BOTSBALL, Howard Mfg. & Consulting, Inc.)

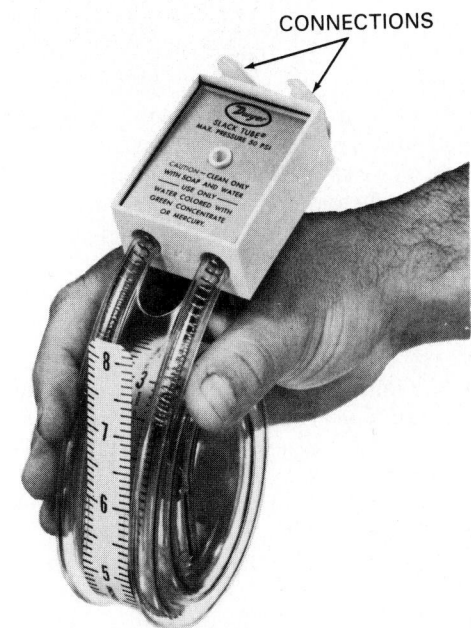

Fig. 18-49. Manometer often used for measuring air pressure in ducts. Flexible tube permits easy storage. (Dwyer Instruments, Inc.)

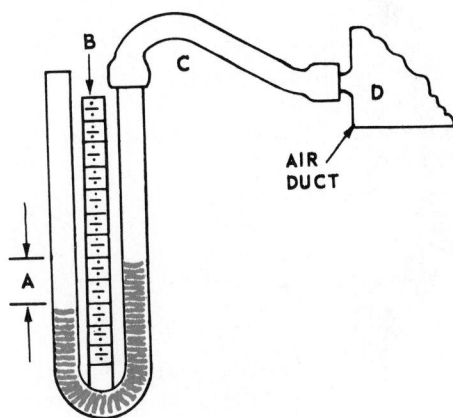

Fig. 18-50. Simple manometer in operation. A—Pressure is indicated by difference in liquid level in two sides of manometer. Usually, pressure is measured in inches. B—Scale in inches. C—Rubber connecting tube. D—Pressure being measured.

been developed to measure overall comfort conditions in hot workplaces. The instrument consists of a 2 3/8-inch hollow copper sphere that is painted black and covered with a double layer of black cloth. A five-inch aluminum tube connects to the copper sphere. This tube is filled with water and is capped at the other end. It keeps the globe wet. The stem of a dial thermometer passes through the centerline of the water reservoir tube and into the globe. When placed in a hot area, the globe is warmed by the surrounding air and by heat radiating from hot surfaces. It is cooled by evaporation from the globe surface, which depends on air velocity and relative humidity.

The wet globe reaches an equilibrium temperature after a few minutes, at which time the heating and cooling effects are in balance. The dial thermometer reading will indicate the wet globe temperature and provide a direct physical measurement of human stress factors in the thermal environment. The wet globe thermometer provides an excellent index of human responses to heat.

18-42 MANOMETERS

The principle of operation of the manometer is explained in Para. 1-17. A manometer used in air conditioning work is shown in Fig. 18-49. Made with flexible tubing, it can be rolled or folded into a small space for carrying.

Fig. 18-50 illustrates a method of connecting a manometer to an air duct to determine its pressure. To measure duct pressures, one usually needs a water manometer. Scale B usually is movable, making it easier to adjust for the neutral point. Sudden pressure changes must be avoided or the liquid may be forced out of the manometer.

Some manometers measure the pressure difference between two different places in a duct. An example of this is a manometer used to measure the pressure drop across a filter in an airflow system.

Manometer scales are based on the following data:
14.7 psi = 29.92 in. Hg = 34 ft. water
1 in. Hg = .491 psi
1 psi = 2.035 in. Hg
1 psi = 2.31 ft. water

1 ft. water = .432 psi
1 in. water = .036 psi

A dial-type manometer is shown in Fig. 18-51. These manometers may have different scales indicating pressure from 0 to 5 in. of water or 0 to 5 psi.

The inside of a manometer is shown in Fig. 18-52. Fig. 18-53 shows a dial-type manometer in use. In Fig. 18-53, two probes allow a comparison of readings.

18-43 BAROMETERS

Barometers are used to measure atmospheric pressure. The simple mercury barometer is illustrated and explained in Para. 1-17. Barometers used in air conditioning usually are of a different type. The barometric pressure is measured by the deflection of a bellows or diaphragm. A recording-type barometer commonly used in air conditioning work is shown in Fig. 18-54. This instrument has a selectable rotation period of 1 day, 7 days, or 31 days.

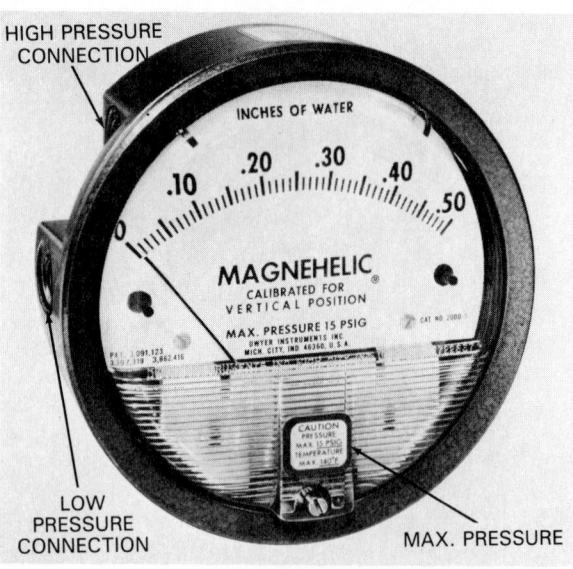

Fig. 18-51. Dial-type manometer. Note that calibration is in inches of water. Maximum pressure is 15 psig. (Dwyer Instruments, Inc.)

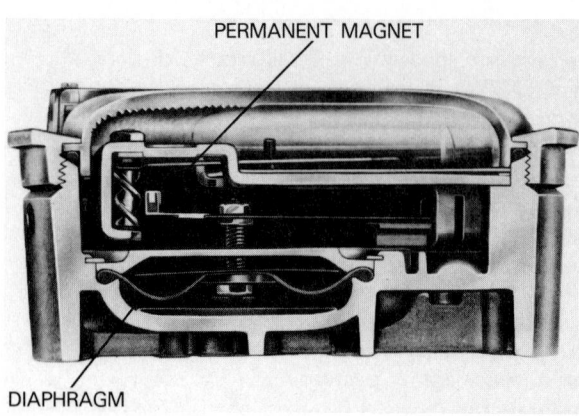

Fig. 18-52. Cross section of dial-type manometer. Moving diaphragm moves permanent magnet. Permanent magnet rotates helix (spiral) attached to needle. Diaphragm separates two pressure chambers. (Dwyer Instruments, Inc.)

18-44 HEAT INSULATION

In order to economically maintain desired air conditioned temperatures in climates of either extreme hot or cold, it is desirable to use materials which do not transfer heat readily. One method is to reduce heat conductivity. Usually, spaces in the structure can be filled with insulating material, which helps to stop or prevent the flow of heat through the structure. Heat flow is measured in Btus per square foot of area per degree F per hour temperature difference between the inside and outside. This is called the U factor. Problems in heat leakage are worked out in Chapter 26. Modern buildings are usually insulated with one of the following: mineral wool, expanded mica, balsam wool, urethane, and sometimes cork in either sheet or granular form. Usually, it is desirable to use insulating materials that are either nonflammable or at least burn very slowly.

18-45 HEAT SINK

A warm body is always giving off heat rays. There are two common results:

1. If the heat rays strike another body or surface of the

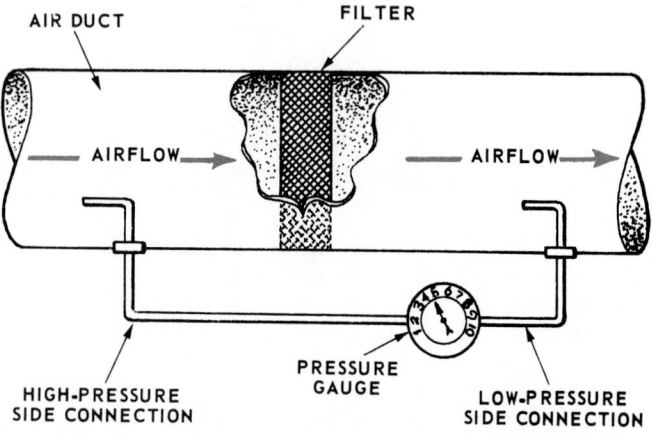

Fig. 18-53. One of many uses of dial-type manometer. The greater the pressure difference, the more resistance (clogging) at filter.

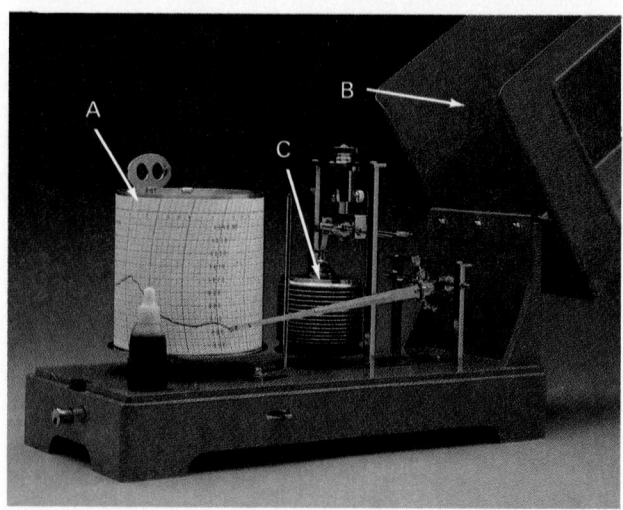

Fig. 18-54. A 24-hr. recording barometer (barograph) with cover lifted. A—Recording chart. B—Cover. C—Sensitive element. (Qualimetrics, Inc.)

same temperature, the rays bounce back or reflect back. Since there is no increase or decrease in the amount of heat, the body being struck by the radiation remains at the same temperature.

Imagine that the body giving off the heat rays is totally surrounded by a surface at the same temperature in the radiating body. Then all the surfaces, including the central body, receive back as much radiation as they give off. All the surfaces and the central object remain at the original temperature.

2. If the radiant heat strikes a surface colder than the radiating body, the heat rays do not all bounce back. Some of the radiant heat is absorbed into the colder surface, and this surface becomes what is called a "heat sink."

On a cold day, heat in a heated room will flow from the room through cold window surfaces. If one sits close to a window under such conditions, one will feel cold due to the loss of heat into the heat sink. This will also be true if the walls of the room are not well insulated. If the walls are cold, they become a heat sink.

18-46 STRATIFICATION

If there is no air movement within a room, the air may tend to stratify. That is, the cold air will sink to the floor and the warmer air rise to the ceiling. By providing a certain amount of air movement in the room, the air will be stirred up so that a more uniform temperature will exist throughout the room. Air movement is accomplished by means of fans located in air conditioners or in air ducts.

If the thermostat is located quite high in a room with no air movement, the temperature difference (because of stratification) will be more noticeable than if the thermostat is located nearer to sitting level in the room.

This also applies when using summer air conditioning to cool a room.

18-47 HEAT EXCHANGE

Methods of heat transfer are covered in some detail in Chapter 1. However, some additional information on heat exchange may be necessary in the study of air conditioning. Four types of heat exchange are possible:
1. Radiation.
2. Convection.
3. Evaporation.
4. Conduction.

The four types of heat exchange are described as follows:
1. *When radiant heat exchange takes place, heat is removed or it travels from one body that is surrounded by an environment at a higher temperature to one at a lower temperature.* Radiant heat exchange means that the body is radiating heat. If the heat being radiated strikes a body or substance at a lower temperature, this heat is lost to the lower temperature substance. If, on the other hand, a body is surrounded by surfaces at a higher temperature, its temperature will tend to increase. Therefore, it may be necessary to mechanically cool this body to maintain its temperature. This is because the radiant heat it receives is greater than the amount it gives out.
2. A body may either gain or lose heat by convection. A person sitting at a desk loses heat by convection. The

warm body tends to warm the surrounding air, and the heat rises and floats away. This is one of the values of clothing; it slows up the convective flow of heat from the human body.

3. Evaporative heat exchange takes place from the human body. Moisture is fed to the skin from the sweat glands, and evaporation of this moisture tends to lower the skin temperature. Respiration also exhausts moisture from the body. The evaporative moisture constitutes a considerable heat exchange from the human body. Evaporative heat exchange can be considered a form of convection, since the evaporated moisture is carried away along with its heat content.

4. Heat exchange by conduction occurs within:
   a. Metal tubing and metal parts of an air conditioning system.
   b. Metal fins and ductwork of an air conditioner or furnace.

The principles of heat exchange are used to provide a more comfortable living environment. Rather than supplying a small heat source at a high temperature, the tendency now is to heat or warm walls, floors, and ceilings to a medium temperature. These large warmed surfaces do not absorb the heat of the body, but bounce back the heat, which gives a person the feeling of a comfortable temperature. This also applies to comfort cooling or summer air conditioning.

18-48 NOISE

The air conditioning system must be designed for handling both the heat load and the fresh air required. It must do these functions in a manner that will not be annoying to the occupants. Two causes of annoyance are objectionable noise and drafts.

Noise is unwanted sound. Complaints of unpleasant noise connected with air conditioning are often directed to equipment vendors and service technicians.

The noise problem can be divided into three types:
1. Noise source.
2. Noise carriers.
3. Noise amplifiers or reflectors.

The noise source is an audible vibration. This vibration may start in the heating unit, cooling unit, fan mechanism, air turbulence, duct panels, duct hangers, or grilles.

Sound or noise is produced by movement of an object. This movement may be caused by vibration of the object or by the movement of air against the object, as in air conditioning ducts.

A vibrating duct panel will create alternate waves of low-pressure and high-pressure air and produce a sound similar to the hum of a mosquito.

Another common complaint is noise caused by high speed air traveling through the ducts, causing air turbulence. Often, this is the result of an undersize unit or duct, and the blower has been speeded up in an attempt to make up for the unit's small size.

Noise or vibration carriers are rigid structures that carry vibrations to places where they may be annoying. Floors, ceilings, ducts, doors, and pipes may carry these vibrations.

Noise amplifiers or reflectors are usually hard, smooth surfaces in a conditioned space. Walls, ceilings, floors, and furnishings may pick up a small vibration and reflect it at such a frequency and in such a direction that all or certain parts of the space may be made uncomfortable. Problems involving acoustics and maintenance of a low decibel noise level are constantly being studied. Improvements are being made.

Soft fabrics such as drapes and curtains and fabric covered furniture are noise absorbers. Felt-lined or soft-insulation-lined ducts also absorb noise.

Some communities have codes regulating how noisy a mechanism may be. For example, one city is requiring that the decibel level of a window unit or outdoor condensing unit must not exceed 50 decibels at a 10-foot distance. More sound-deadening devices may be needed to meet this standard.

The air velocity (feet/minute) permissible is somewhat dependent on the type of building being air conditioned—hospital, church, hall, residential, etc. Where noise is a factor, the velocity should be kept to a minimum. If the velocity cannot be decreased, noise may be reduced by using acoustical discharge chambers or by lining or wrapping the ducts with sound-absorbing material such as felt or other soft material.

18-49 NOISE MEASUREMENT

Sound waves are rapid changes of air pressure. The amplitude or strength of the sound pressure waves can be measured. Sound strength and sound pressure level (SPL) are rated in decibels (dB). In engineering practice, two words are used in connection with the definition of sound:

1. Sound strength, in decibels, is the total amount of sound coming from a unit.
2. Sound pressure is the strength, in decibels, of sound after it travels a specified distance from a source.

For instance, a truck engine gives off a measurable sound strength at the engine. Sound, after it has traveled 50 feet to a storefront, is measured in sound pressure.

The number of vibrations per second of sound waves is also measurable. The usual measuring instrument is an amplifying microphone. The international unit for sound frequencies is the Hertz (Hz), which means cycles per second (cps).

Sound travels in sound waves through the air. Sound travels from its source in all directions, and its strength diminishes with the distance from the source. The first measurement of the strength of sound is usually taken about three feet from its source. The amplitude of the sound waves will be reduced by the cube of the distance that the receiving or recording instrument is away from the sound source.

Sound pressure is shown in Fig. 18-55. Increasing the sound frequency tends to increase the apparent loudness,

Source	Sound Pressure Pa	Sound Pressure Level dB re 20μPa	Subjective Reaction
Military jet takeoff at 100 ft	200	140	Extreme
Artillery fire at 10 ft	63.2	130	danger
Passenger's ramp at jet airliner (peak)[a]	20	120	Threshold of pain
Loud rock band[a]	6.3	110	Threshold of discomfort
Platform of subway station (steel wheels)	2	100	
Unmuffled large diesel engine at 130 ft	0.6	90	Very loud
Computer printout room[a]	0.2	80	
Freight train at 100 ft	0.06	70	
Conversational speech at 3 ft	0.02	60	
Window air conditioner[a]	0.006	50	Moderate
Quiet residential area[a]	0.002	40	
Whispered conversation at 6 ft	0.0006	30	
Buzzing insect at 3 ft	0.0002	20	
Threshold of good hearing	0.00006	10	Faint
Threshold of excellent youthful hearing	0.00002	0	Threshold of hearing

Fig. 18-55. Sound pressure measured in pascals (Pa). (ASHRAE Guide and Data Book)

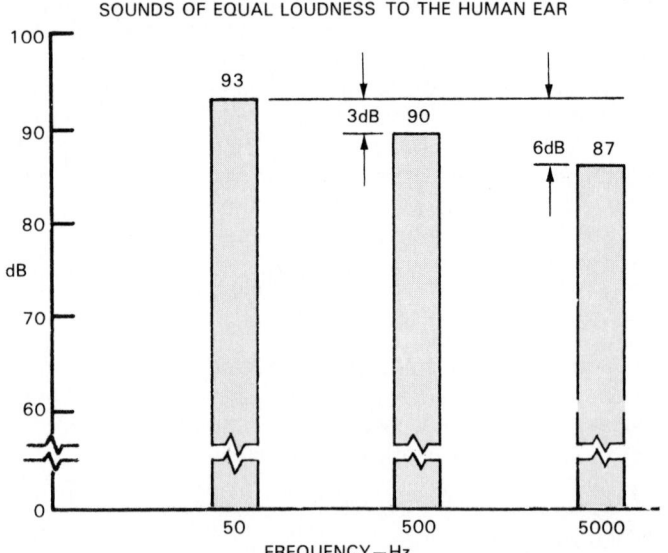

SOUNDS OF EQUAL LOUDNESS TO THE HUMAN EAR

Fig. 18-56. Effect of frequency on ear's ability to hear sounds. These three sounds appear of same strength to human ear, yet decibel rating for each of the frequencies is different.

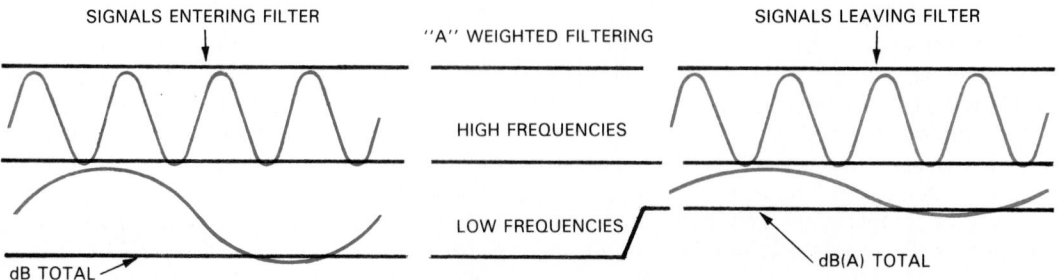

Fig. 18-57. Effect of using "A" filter in microphone circuit. Left. Unfiltered sound waves. Right. Sound waves after passing through "A" filter. Note that only low frequency waves have been reduced in amplitude. (Vickers, Inc.)

because the human ear does not respond equally to all frequencies. This effect is illustrated in Fig. 18-56. For most people, sounds in the frequency range of 1000 to 4000 Hz are the easiest to hear.

In measuring the loudness of sound, the meters generally read in dB(A) or dBA. The dBA scale loudness means that a standardized "A" filter has been placed in the microphone circuit to reduce the intensity of the low frequencies. Fig. 18-57 illustrates the effect of using an "A" filter in the microphone circuit.

A comparison of loudness measurements using the dB and the dBA scales shows the effect of this filter in Part A in Fig. 18-58. Note that equal loudness requires a higher dB rating in the lower frequencies than the dBA rating in the lower frequencies. Part B shows one city's code standards.

Laws regulating permissible sound or noise levels usually are written around the "A" scale. The Walsh-Healy Act of 1969 establishes limits on exposure time under which workers may work at various sound levels. A table of these limits is shown in Fig. 18-59.

An instrument for measuring sound level (dBA) is illustrated in Fig. 18-60. Note that Item A, the measuring instrument, can be used with Item B, a memory storage unit, and Item C, a printer that provides a hard copy (a printout) of the noise data. Based on OSHA (Occupational Safety and Health Act) standards, the instrument measures continuous, intermittent, and impulse noises in the range from 80 to 130 dBA. It may be used as a personal noise dosimeter by being worn in the pocket or on the belt. The microphone in turn is attached to the operator's shirt, close to the ear for hearing. When used as a sound level meter, the instrument is hand-held, and the microphone is directed at the incoming sound waves. The data storage, B, stores 60-second integrated sound level averages and maximum sound pressure levels within each 60-second period for a maximum of 32 hours. Item C, printer unit, is a battery-operated thermal printer. It provides an immediate printout of the data processed and stored in Item B.

The meter shown in Fig. 18-61 makes sound level measurements over a range of 40 to 140 dB. It provides

LOUDNESS SCALES

A

FREQUENCY	dB	dB(A)
5000 Hz	87	87
500	90	87
50	93	87

RECOMMENDED MAXIMUM SOUND

B

FREQUENCIES (Hz)	DECIBELS	
	NIGHT	DAY
63	66	76
125	59	69
250	52	62
500	46	56
1000	42	52
2000	40	50
4000	38	48
8000	37	47

Fig. 18-58. A—Effect of using "A" filter on sound loudness for the three frequencies shown in Fig. 18-56. B—City code maximum noise level (Milwaukee, Wisconsin)

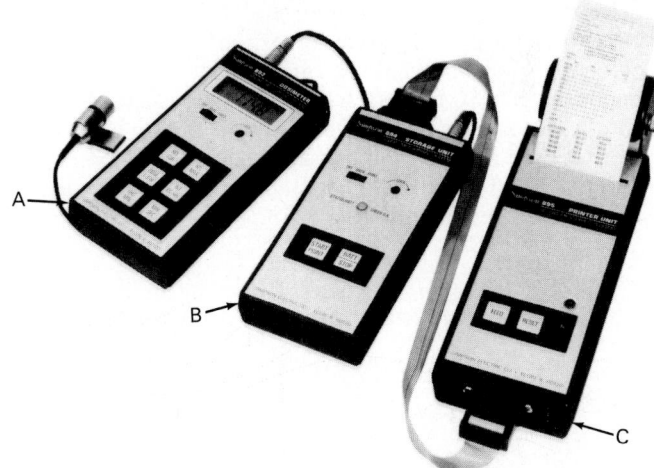

Fig. 18-60. Instruments used for measuring sound level. A—Noise dosimeter. Unit measures sound level in dBA. B—Memory storage unit. C—Printer that gives printout of noise data. (Simpson Electric Co.)

TOLERANCE TO SOUND

NOISE LEVEL dBA	MAXIMUM EXPOSURE HRS.
90	UNLIMITED
90 TO 92	6
92 TO 95	4
95 TO 97	3
97 TO 100	2
100 TO 102	1.5
102 TO 105	1
105 TO 110	0.5
110 TO 115	0.25
ABOVE 115	NONE

Fig. 18-59. Limits of human tolerance to sound. Note that there is no limit if dBA is 90 or below. Also, unprotected human ear should not be exposed to sound over 115 dBA. (Walsh-Healy Act of 1969)

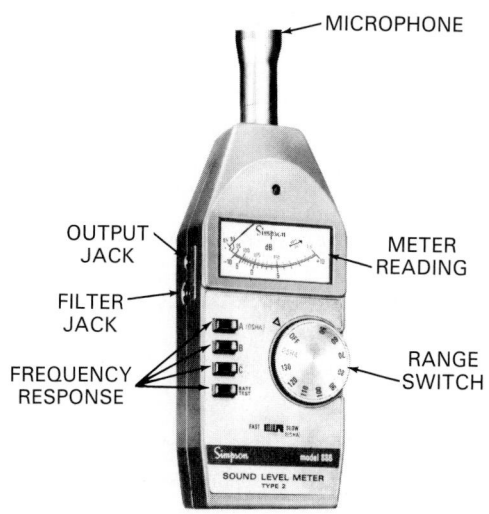

Fig. 18-61. Sound level meter measures sound pressure level over range of 40 to 140 dB. (Simpson Electric Co.)

an output jack which permits the use of a meter with a recorder. The unit has a jack for use with an "A" filter.

Usually, a noise is made up of a number of different vibrations per second (not a pure tone). It is possible, with the use of filters, to separate noise into octave bands. The sound level in each octave band is recorded in decibels. By adding the decibels for all the bands, the sones value is obtained:

Octave: A series of eight tones extending from a given tone to a tone on the eighth degree from the given tone.

Sone: A calculated sound loudness rating.

The sone rating is particularly useful in comparing machine noise levels.

Noise sources include fans, compressors, high velocity air, the ac hum of a motor and high-velocity refrigerant flows (especially at sharp pipe turns). Fig. 18-62 illustrates some typical ratings and their corresponding sound pressure ratings.

Sound pressures in the 0 to 90 dB range generally are not objectionable. Sound levels in the 100 to 160 range are very objectionable, particularly in the upper end of this range.

To protect the hearing of people in noisy situations, earmuff-type hearing protectors are used. These protectors will attenuate (reduce the level of) the noise by about 20 percent for low frequencies, 40 percent for medium frequencies, and 30 percent for high frequencies. The federal safety law known as the Occupational Safety and Health Act (OSHA) requires the use of hearing protectors in certain places. Always wear them when working on or near noisy machinery. See Fig. 18-63.

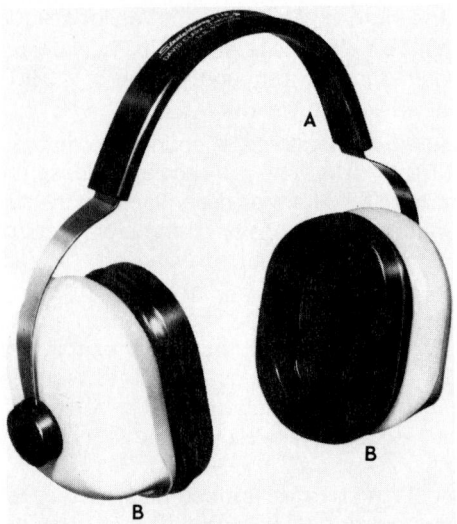

Fig. 18-63. Hearing protector should be used when working on or near source of painful noise. A—Headband. B—Ear covers. (David Clark Co., Inc.)

18-50 ECOLOGY—ENVIRONMENT

"Ecology" has come to mean more than the initial definition, which states: "Ecology is the branch of biology which treats the relationships between organisms and their environment." In this text, ecology means air temperature, relative humidity, pollution, air movement, oxygen content, and percent of noxious vapors. Webster defines environ-

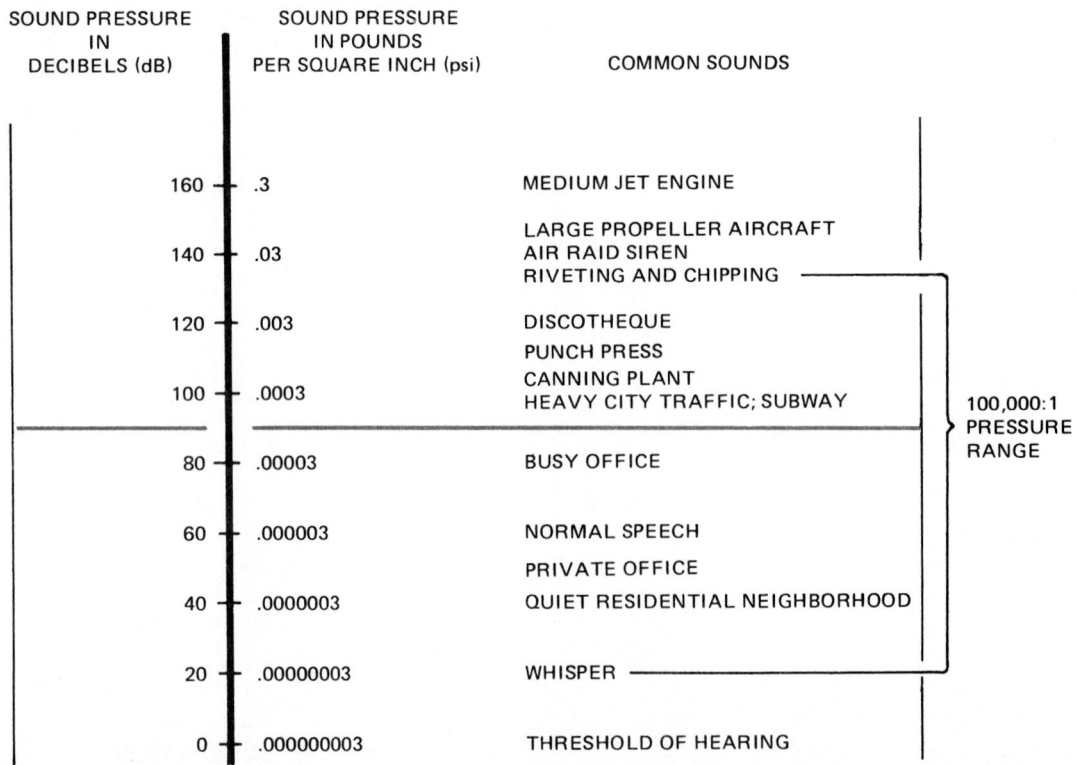

Fig. 18-62. Decibel (dB) rating of some common sounds. Line in color indicates usual upper limit to which human ear may be continuously exposed. (Vickers, Inc.)

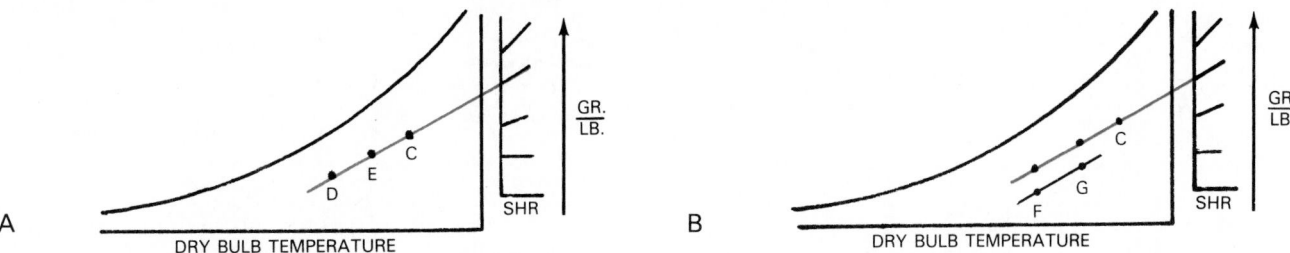

Fig. 18-64. Drawing developed by technician to indicate sensible heat ratio. Points D, E, F, and G are predetermined. D is a point determined by the dry bulb temperature (T_{D1}) and wet bulb temperature. E is determined by the dry bulb temperature (T_{D2}) and wet bulb temperature. Illustration B with points F and G is another example of determining a sensible heat ratio.

ment as "the aggregate of surrounding things, conditions, or influences."

18-51 SENSIBLE HEAT RATIO

The sensible heat ratio (SHR) is a scale given on the right side of a psychrometric chart. The SHR is the ratio of the sensible heat to the total added heat when moving from one point to another on the psychrometric chart. Therefore, the SHR is the quantity of sensible heat divided by the total heat. The formula is as follows:

$$SHR = \frac{(T_{D2} - T_{D1})}{(T_{D2} - T_{D1}) + 0.05236\left(RH_2 e^{\frac{T_{D2}}{28.116}} - RH_1 e^{\frac{T_{D1}}{28.116}}\right)}$$

See Para. 28-70.

The scale is used as shown in Fig. 18-64. Connect the two points, D and E, by a line and extend the line to the scale on the right. Then read the SHR. For points F and G, draw a line parallel to the line FG, going through point C. Extend the line to the SHR scale and obtain a reading.

18-52 REVIEW OF SAFETY

Air to breathe should be as clean as possible and have the correct oxygen content. It is absolutely essential that none of the fumes (products of combustion) become mixed with the air being sent to the rooms. Air conditioners must provide enough fresh air to the rooms being conditioned to keep the oxygen content of the air within allowable limits.

Kitchen ventilating fans tend to produce a slightly lowered pressure in a house. Leakage into the house must make up for this exhausted air. Under certain conditions of prevailing winds and building construction, it is possible that products of combustion may be drawn back into the house if the air pressure is too low. It is wise to provide some air leakage into the house. Air leakage can be designed into the heating, ventilating, and air conditioning (HVAC) system.

Work or experiments performed in connection with this chapter require the use of instruments. Conditions of safety must be met. If the instruments are connected to an electrical or compressed air supply line, carefully check the installations so there is no danger to the operator handling these supplies.

Instruments used in connection with work or experiments in this chapter are very delicate. They must be handled carefully, never dropped, and many must be kept in an upright position. Some instruments, such as hygrometers and psychrometers, are made by using some glass tubing. Use care in handling these instruments. Do not break the

tubing and perhaps cause the operator to be cut by broken glass.

Many of these instruments are very expensive. If connected incorrectly or not handled carefully, the instrument may not read accurately or it may be severely damaged. The person handling the instrument must use great care.

In this chapter, the operation and use of instruments are explained in some detail. This information and knowledge will be used in performing experiments and work in Chapters 20, 21, 22, and 23.

18-53 TEST YOUR KNOWLEDGE

1. What four important actions are involved in a complete air conditioning system?
2. Is air conditioning used only for human comfort?
3. What gases make up the living portion of the atmosphere?
4. In what form does water exist in the air?
5. What is fog?
6. What is humidity?
7. What is relative humidity?
8. What do the dry bulb temperature and the wet bulb temperature indicate about any particular air sample?
9. Define psychrometry.
10. How does air movement affect one's comfort?
11. List two ways to remove moisture from the air.
12. What happens to the moisture absorption properties of air as the temperature decreases?
13. What should be done prior to whirling a sling psychrometer?
14. What values are constant along a horizontal line of the psychrometric chart?
15. Does air contain moisture below 32 °F (0 °C)?
16. When air is heated, what happens to relative humidity?
17. Under what conditions are most people comfortable in the summer?
18. How large are dust particles?
19. What are fumes?
20. What is the percentage of oxygen in the atmosphere?
21. A pitot tube involves what three pressures?
22. Why is a vapor barrier needed in building walls?
23. What is the moisture condition in air when the dew point is reached?
24. Does vegetation provide oxygen in the atmosphere?
25. What is a hygroscopic material?
26. What is a common sensor material in a hygrometer?
27. The chemical formula for oxygen is O_2. What is the chemical formula for ozone?
28. At what velocity does a draft become uncomfortable?
29. How many dew points can one sample of air have?
30. What is an aspirating psychrometer?

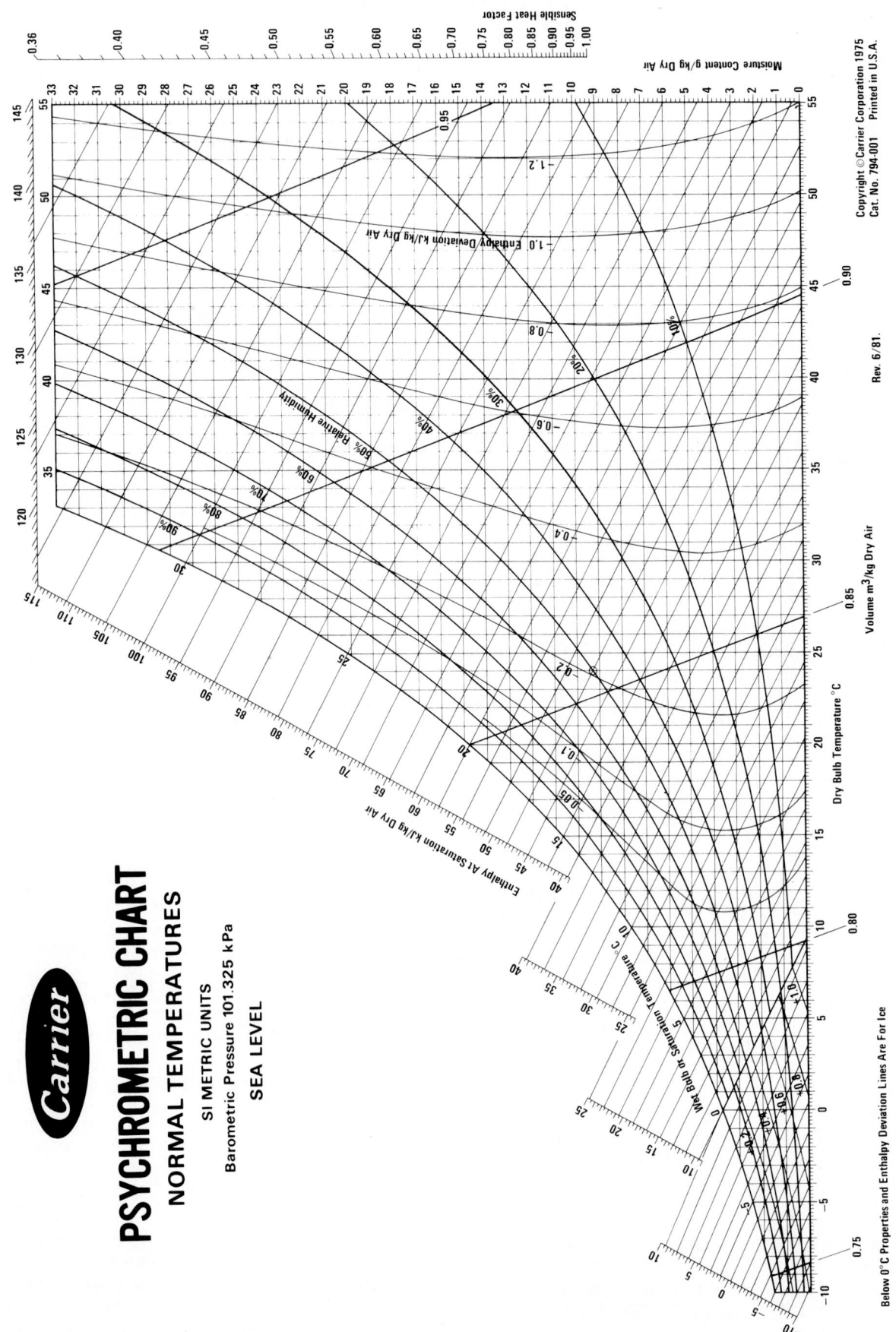

PSYCHROMETRIC CHART

NORMAL TEMPERATURES

SI METRIC UNITS
Barometric Pressure 101.325 kPa

SEA LEVEL

Carrier

Sensible Heat Factor

Moisture Content g/kg Dry Air

Enthalpy Deviation kJ/kg Dry Air

Relative Humidity

Volume m³/kg Dry Air

Dry Bulb Temperature °C

Enthalpy At Saturation kJ/kg Dry Air

Wet Bulb or Saturation Temperature °C

Below 0°C Properties and Enthalpy Deviation Lines Are For Ice

Copyright © Carrier Corporation 1975
Cat. No. 794-001   Printed in U.S.A.

Rev. 6/81.

Metric psychrometric chart. (Carrier Corp., Subsidiary of United Technologies Corp.)

Chapter 19

BASIC AIR CONDITIONING SYSTEMS

After studying this chapter, the technician will be able to:
☐ Discuss the design of different air conditioning systems.
☐ Name and describe various systems, explain their differences by type of fuel or energy used, by transfer medium, and by type of heating/cooling components used.
☐ Explain in specific detail how each system operates.
☐ List various types of systems and explain their applications.

Detailed application, installation, maintenance, and servicing data of each system will be explained in later chapters. These chapters discuss heating, ventilating, and air conditioning (HVAC) systems.

19-1 GRAVITY WARM AIR FURNACE— FUEL GAS ATMOSPHERIC BURNER

Fuel gas, controlled by a pressure regulator, is fed to the burner in this heating system, Fig. 19-1. Fuel is under constant low pressure. An atmospheric-type burner is used. Fuel may be natural gas, propane, LP gas, or artificial gas.

The room thermostat controls the operation of the burner through a solenoid-controlled gas valve. A pilot light, which burns continuously, ignites the gas in the burner whenever the solenoid valve opens the gas line. Some systems use electronic ignition. A thermocouple is connected in series with a safety gas control solenoid. It will shut off the gas if the pilot light goes out and the thermocouple cools.

Products of burning, carbon dioxide and water vapor, flow through the stack into the chimney. An air break helps keep a constant pressure in the combustion (burning) chamber.

The heat made in the combustion chamber (heat exchanger) is conducted (carried) through the combustion chamber wall and is carried or radiated into the air surrounding the combustion chamber.

Air around the combustion chamber heats up and naturally rises. It flows through the warm air ducts into the rooms through the warm air registers. As air cools in the rooms, it becomes heavier and flows down through the cold air duct and back into the bottom of the furnace.

A high-limit control (safety stat) is located in the bonnet of the furnace. (The bonnet is the sheet-metal chamber where heat collects before being distributed, as suggested in Fig. 19-2.) The high-limit control will automatically shut off the gas if the bonnet temperature goes higher than the high-limit control temperature setting.

A room thermostat checks the room temperature, and responds as needed. The thermostat operation keeps the temperature of the room within about 2 °F (1.1 °C) of the desired temperature.

19-2 FORCED WARM AIR—FUEL GAS POWER BURNER

In this heating system, Fig. 19-2, fuel gas is fed to the burner under constant low pressure controlled by a pressure regulator. The fuel gas is burned in a power-type burner.

The room thermostat controls the operation of the burner through a solenoid-controlled gas valve. The thermostat also turns on a combustion air blower which forces air into the combustion chamber. A pilot light, or an electric spark, ignites the gas in the burner at the instant the solenoid valve opens the gas line and the power burner starts.

The flame heats a thermocouple which, in turn, controls a safety shutoff valve. A thermocouple solenoid, located above the pilot light, will shut off the gas control if the pilot light goes out.

Products of combustion flow through the stack into the chimney. An air break helps keep constant pressure in the combustion chamber.

Heat generated in the combustion chamber (heat exchanger) is conducted through the combustion chamber wall and is radiated into the air surrounding the combustion chamber. As the air around the combustion chamber heats, it rises and warms the bonnet fan control.

As soon as the bonnet temperature is high enough, the fan in the cold air duct return starts and moves air through the heating system. This air is drawn from the cold air register in the floor above, through the air filter, and then through the furnace. Warm air is distributed through the ducts and through the warm air registers or diffusers into the space to be heated.

A high-limit control will automatically shut off the burner if the bonnet temperature exceeds the high-limit control temperature setting. The high-limit control is sometimes a part of the bonnet fan control.

19-3 FUEL GAS ATMOSPHERIC BURNER— HYDRONIC SYSTEMS

This heating system burns fuel gas under low and constant pressure. The system is shown in Fig. 19-3. An automatic pressure regulator maintains constant gas pressure on the atmospheric type burner. The room thermostat controls the operation of the water pump. The pump circulates the warm water through the room radiators and returns it to the boiler.

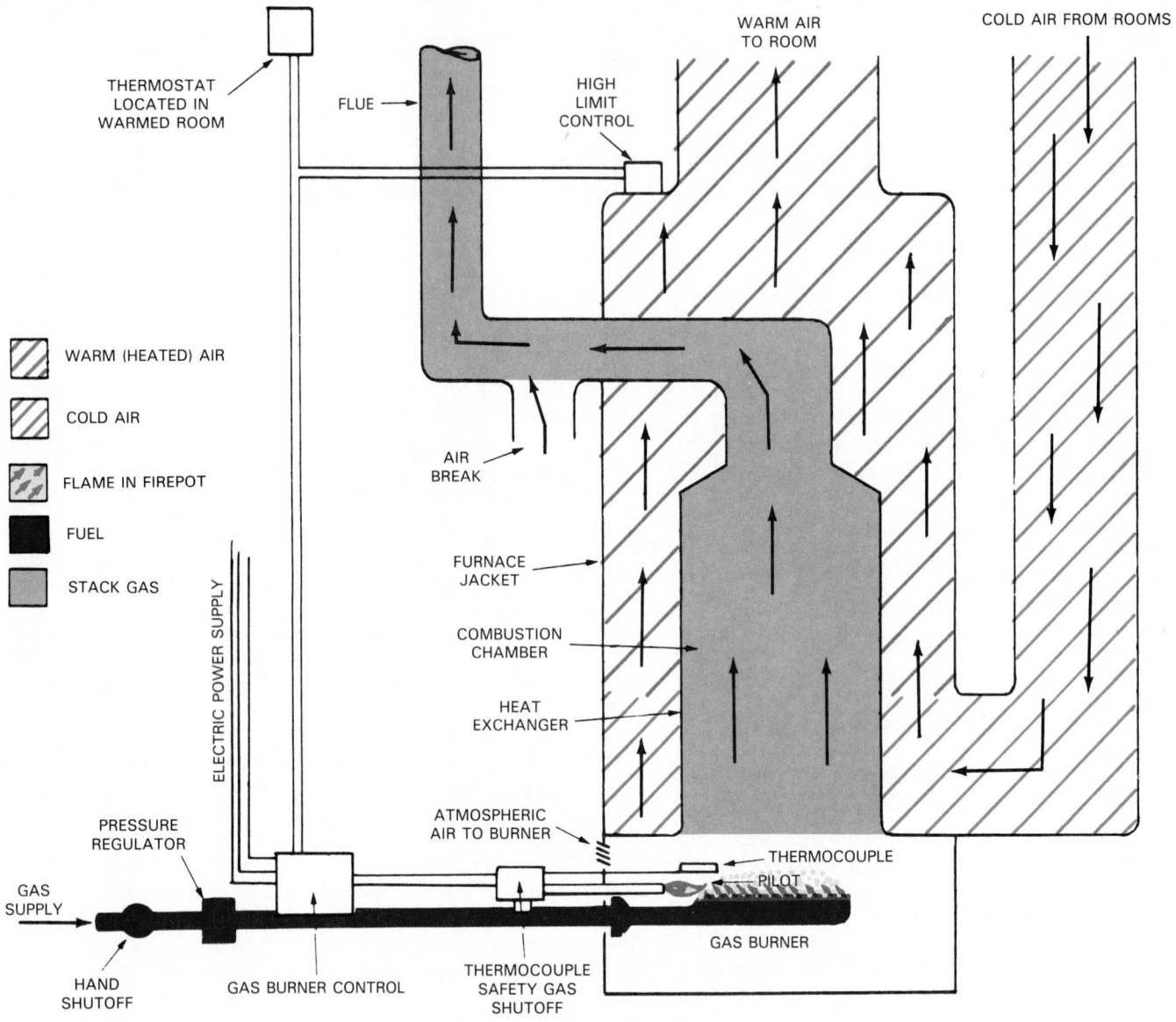

Fig. 19-1. Gravity warm air furnace. Air heated in furnace rises. Colder air from rooms sinks to take its place. This natural convection circulates air through rooms.

Water temperature in the boiler is controlled by a temperature- and pressure-sensing element in the top of the boiler. This sensing element is connected into the electrical system in such a way that the burner turns on when the temperature drops below the required level. It also turns off when the temperature reaches this level. A pilot light ignites the burner. A thermocouple, located at the pilot light, is connected into the electrical controls. The control circuit used with the thermocouple automatically shuts off the gas if the pilot light goes out.

The heat generated (made) in the combustion chamber is carried through the boiler wall and into the water. The products of burning (smoke) flow through the stack into the chimney. An air break helps maintain a constant pressure in the combustion chamber.

A high-limit control (safety stat) is attached to the warm water outlet of the boiler. The high-limit control automatically shuts off the gas if the water temperature or pressure get too high.

The system also has a pressure relief valve which prevents the buildup of dangerous pressures in the boiler. An expansion tank permits the water to expand and contract in volume as it heats and cools. The expansion tank should be located at the highest place in the heating system. It is possible to use a pressure regulator in place of an expansion tank.

19-4 OIL BURNER—FORCED WARM AIR

Where fuel oil is the heat source, a gun-type oil burner throws a flame into a firepot lined with refractory (fire-resistant) material. Fig. 19-4 shows this type of furnace.

Fuel oil is stored in a tank, either inside or outside the building. Fuel is pumped into the burner nozzle under a pressure of about 100 psig (790 kPa).

A room thermostat controls operation of the burner. When heat is required, the thermostat operates a relay. It closes the electrical circuit to the burner motor. When the burner starts, a high-voltage transformer is connected into the electrical circuit. Sparks jump the gap of two electrodes located at the edge of the atomized (broken into small drops) fuel spray of the burner nozzle. This spark ignites the fuel, causing a continuous flame in the firepot as long as the burner is operating.

A stack thermostat senses the temperature of the gases leaving the furnace. Its control is such that if, for any reason, the atomized fuel is not ignited after a few seconds of pump operation, the pump will stop. Normally, a manual reset will have to be operated before it will cycle again.

A temperature-sensing device in the furnace bonnet will start the blower or fan in the cold-air duct as soon as the bonnet temperature reaches its desired setting. A temperature-controlled limit switch is placed in the bonnet. It will open the circuit and stop the burner if the bonnet temperature goes too high.

19-5 OIL BURNER—HYDRONICS

In oil-fired hydronic heating systems, fuel oil is burned in a gun-type burner. See Fig. 19-5. The firepot is lined with refractory material. Fuel oil is stored in a tank which may be located outside the building. Fuel oil is drawn through a filter and pumped into the burner nozzle under a pressure of about 100 psig (790 kPa).

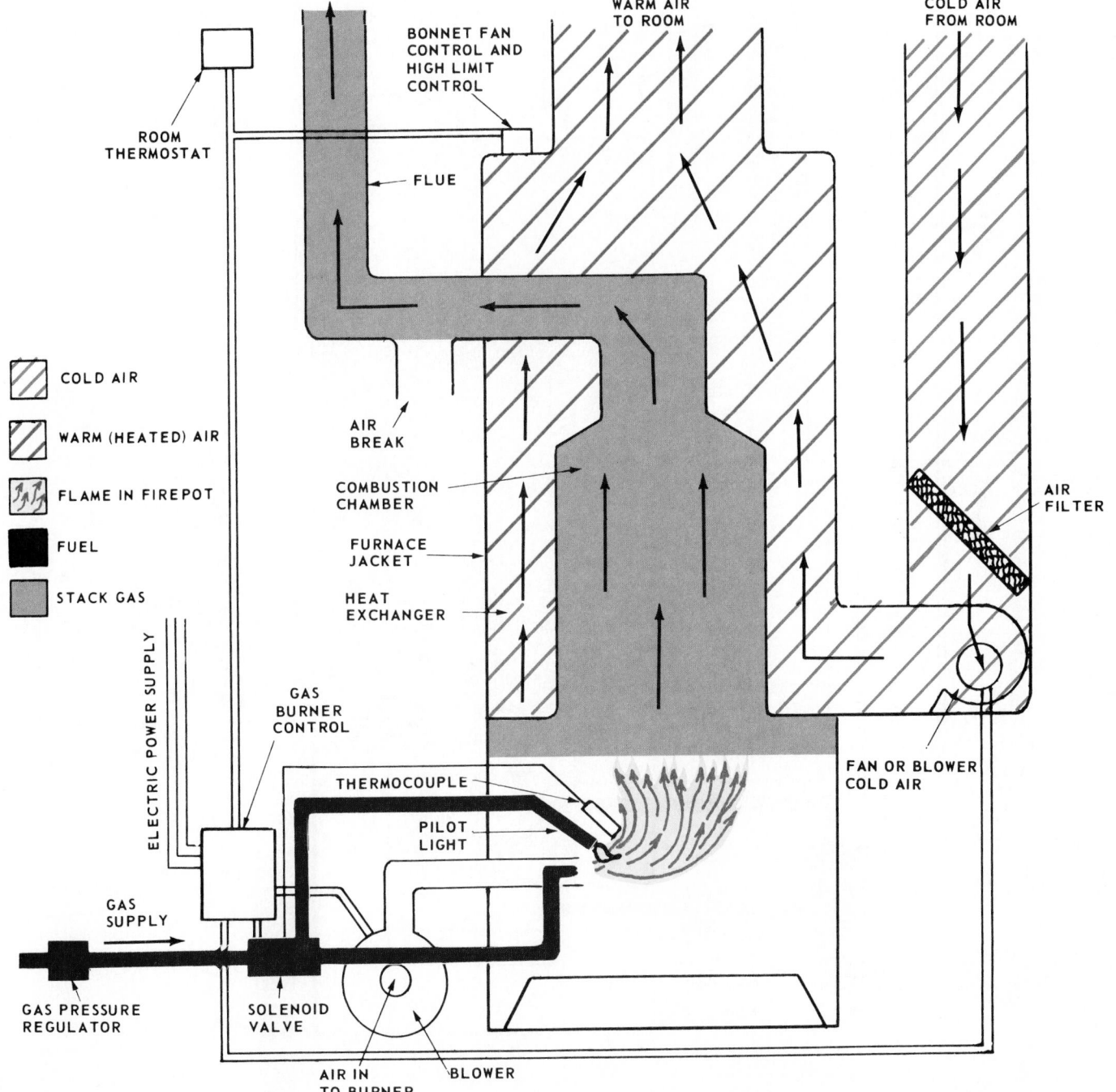

Fig. 19-2. Forced air circulation heating system using gas fuel power burner.

A room thermostat controls the water pump which circulates the warm water through the room radiators and returns it to the boiler. The temperature of the boiler water is controlled by a temperature- and pressure-sensing element in the boiler top. This sensing element is connected into the electrical system. It turns the burner on when the temperature drops below the required level, and shuts it off when the temperature reaches the desired level.

As the burner starts, a high-voltage transformer is con-

nected into the electrical circuit. Sparks jump across the spark gap just at the edge of the atomized fuel spray of the burner nozzle. This spark ignites the atomized fuel, causing a flame in the firepot as long as the burner is operating.

A stack "stat" senses the temperature of the gases leaving the furnace. If the atomized fuel is not lighted after a few seconds of pump operation, the pump will stop. Probably a manual reset will have to be operated before it will cycle again.

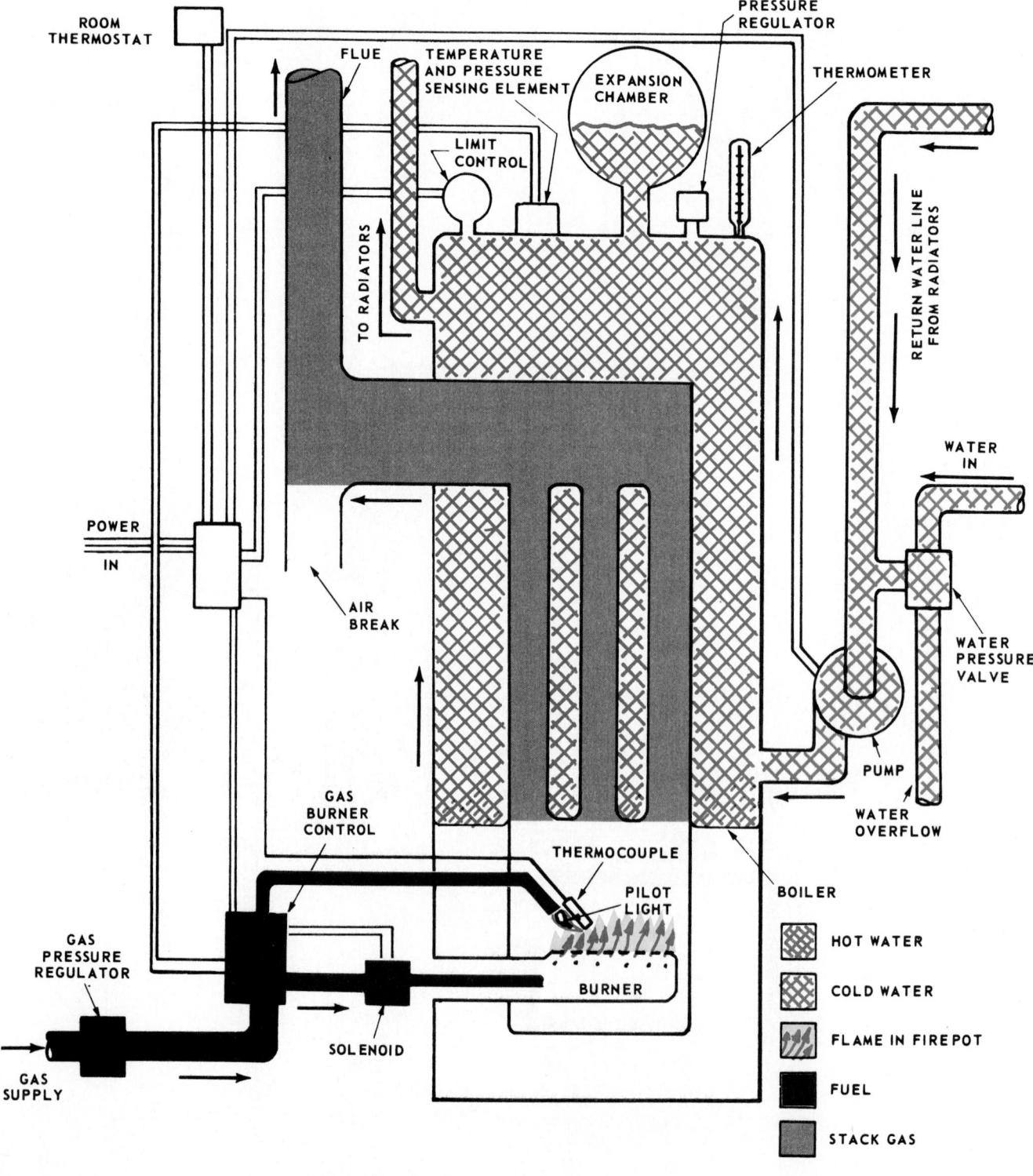

Fig. 19-3. Hydronic heating system using atmospheric fuel gas burner as source of heat.

Heat from the combustion chamber is conducted through the boiler wall into the water. The gases from burning fuel flow through the stack into the chimney. An automatic draft regulator helps maintain a constant pressure in the combustion chamber (firepot).

A high-limit control (safety stat) is attached to the boiler chamber (or sometimes to the warm water outlet of the boiler). The high-limit control automatically shuts off the fuel if the water temperature or pressure get too high. A pressure relief valve is also mounted on the boiler to keep pressures down to a safe level.

An expansion tank is used to take care of expanding (warm) or contracting (cool) water. Air in the tank acts as a cushion.

19-6 FORCED WARM AIR—ELECTRIC RESISTANCE HEAT

Resistance heating units sometimes replace the oil- or gas-fired flame. Through a system of relays, the room thermostat energizes (turns on) the heating unit when heat is needed. When the desired room temperature is reached, the

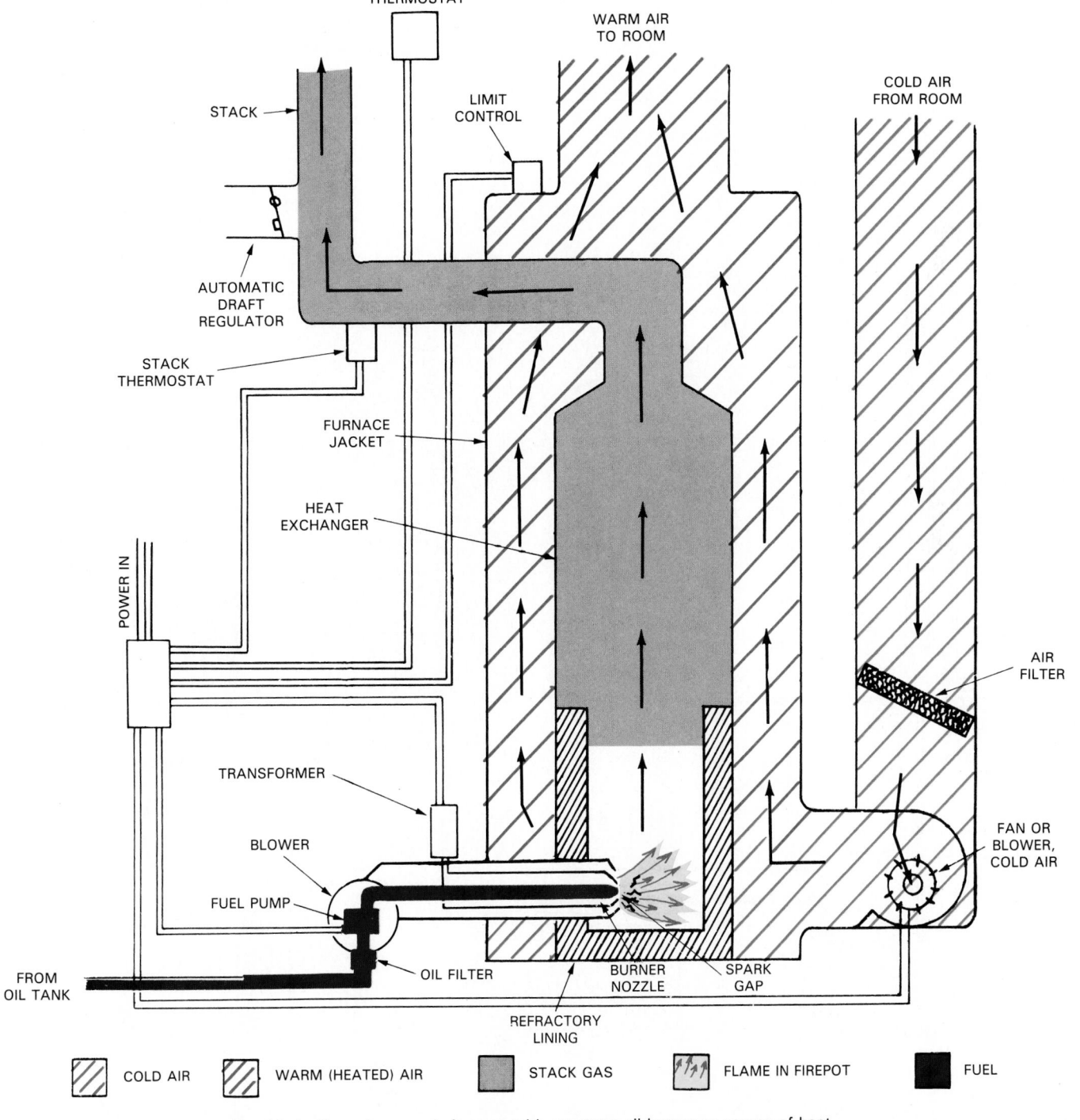

Fig. 19-4. Forced warm air furnace with gun-type oil burner as source of heat.

same thermostat turns off power to the resistance unit. Fig. 19-6 is a sketch of such a furnace.

Warm air is distributed by a blower. It forces the air through the resistance unit where it picks up heat. The blower then distributes the heated air to the registers.

Some controls for a furnace that uses electricity to heat air combine the functions of several controls. The control for the furnace shown in Fig. 19-6 is a combination bonnet, high-limit, and blower control. The control turns on the blower as soon as the bonnet temperature reaches a predetermined setting. The blower will continue to run as long as the bonnet temperature remains above a minimum temperature setting. In the event of poor air circulation through the bonnet, causing the bonnet temperature to ex-

ceed a predetermined setting, the high-limit control will shut off the resistance unit. This protects the furnace and ducts from overheating.

A filter is placed between the cold-air duct and blower. A humidifying device is usually placed in the warm air duct leading out of the furnace. The humidifying device operates whenever the blower is operating.

Individual circuit breakers should be put in the power lines of electrical resistance heating units. No other appliance or lights should be connected into these circuits. This applies to all heating and cooling system circuits.

There is an advantage of heating forced air with an electric resistance element. This furnace does not require a stack or chimney.

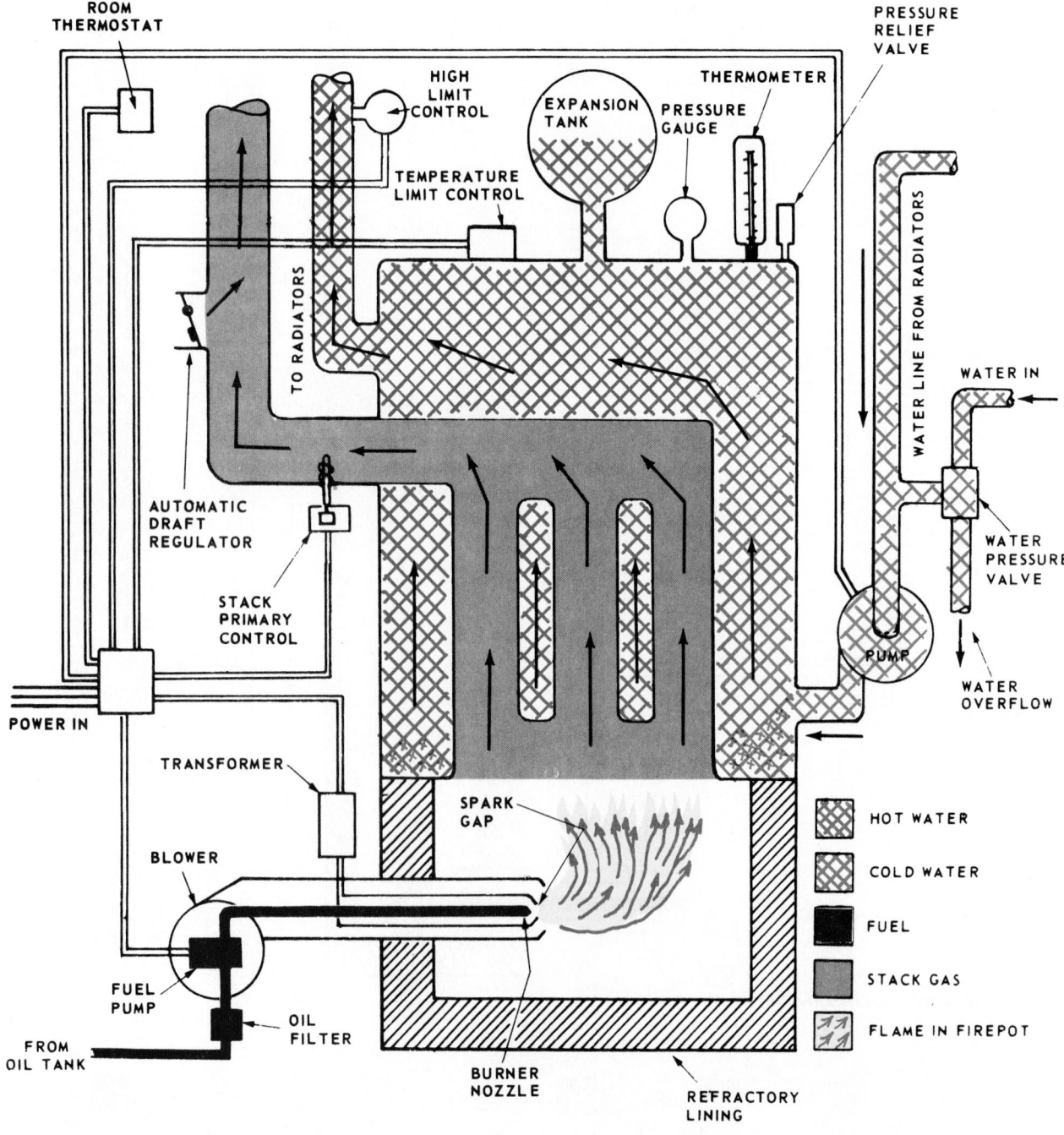

Fig. 19-5. Hydronic heating system with gun-type oil burner as source of heat.

19-7 HYDRONICS—ELECTRIC RESISTANCE HEAT

In hydronic systems, the electrical resistance heating units are inside the boiler. The boiler is very small. See Fig. 19-7.

A high-limit and safety control is attached to the boiler chamber or warm-water outlet of the boiler. It automatically turns on one or all three stages of the resistance heating units when the temperature of the water in the top of the boiler drops below a minimum setting. It also turns off the electrical heating units when the temperature of the water in the top of the boiler reaches the upper setting. The same control becomes a safety device, shutting off heating elements if no water is circulating through the radiators.

In some cases there are pressure controls as well. They will shut off the heating unit if the water pressure in the boiler exceeds a preset limit. A pressure gauge is used on the furnace shown in Fig. 19-7. With electrical contacts connected to a gauge, it can turn off the heater if required.

The room thermostat controls the operation of the pump or pumps which force the warm water through the room radiators. With this system, it is possible to use more than one pump. In this case, a separate thermostat controls the temperature at the space served by each pump.

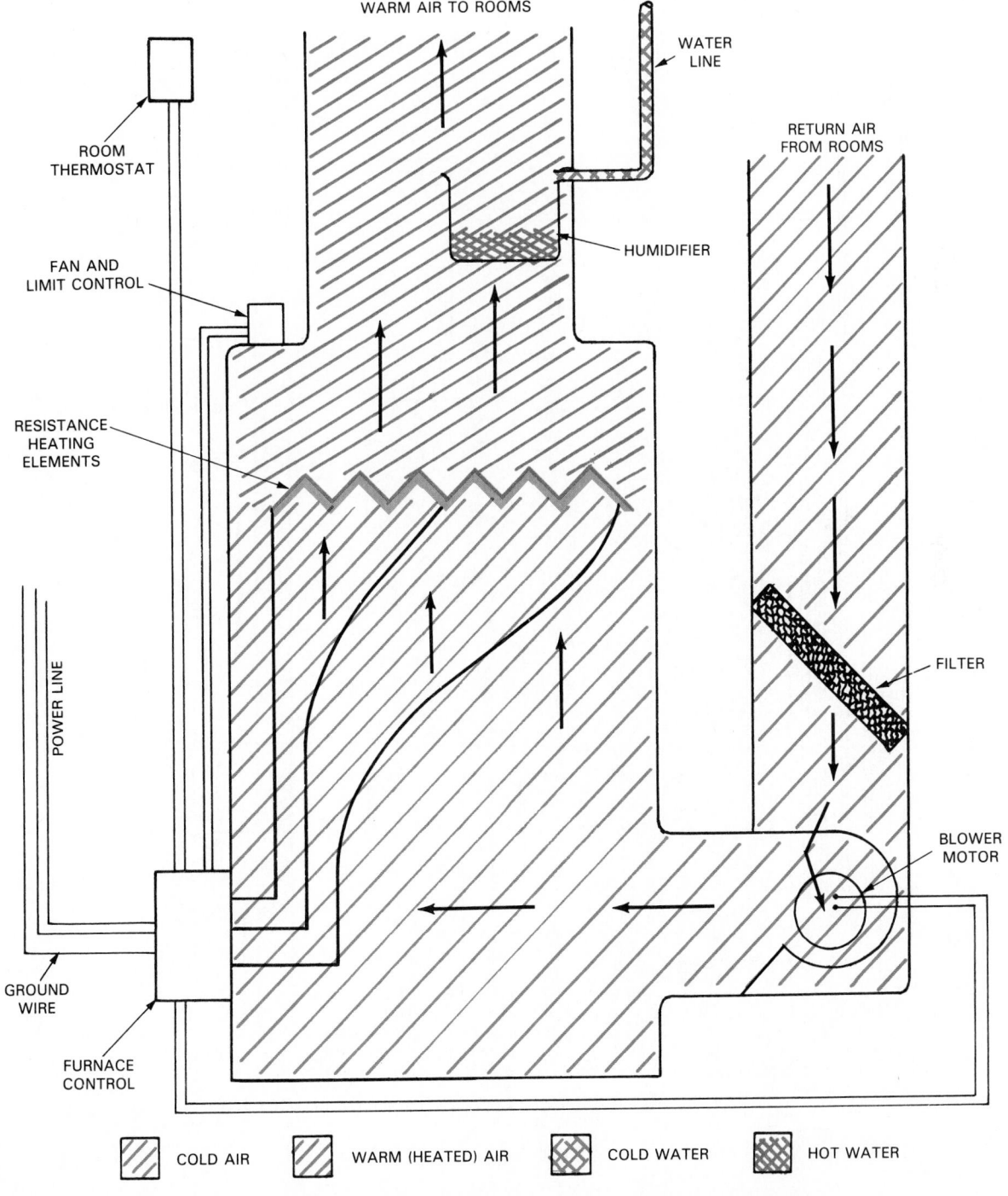

Fig. 19-6. Forced warm air heating system with electric heating elements as source of heat.

Individual circuit breakers should be installed in the power lines of electrical resistance heating units. No other appliance should be connected into these circuits. As with the system described in Para. 19-6, no stack or chimney is required.

19-8 ROOM HEATING UNITS— ELECTRICAL RESISTANCE

Where this heating system is installed, electrical resistance units are located in each room. Electrical power is brought to the units from an electrical heating power panel. The power supply is usually 240 V. One advantage

of this type of heating is that the temperature of each room is regulated by its own thermostat. There are four different ways that the control may be accomplished. These different ways are illustrated in Fig. 19-8 at A, B, C, and D.

At A, electrical power is taken directly from the power panel to the baseboard heating unit. The baseboard heating unit has an individual thermostat attached to it. Electrical power is supplied to the thermostat and, if heating is required, the thermostat connects the power supply to the resistance heating unit.

At B, a room thermostat controls the power supply at the power panel. When heat is required, a relay in the power

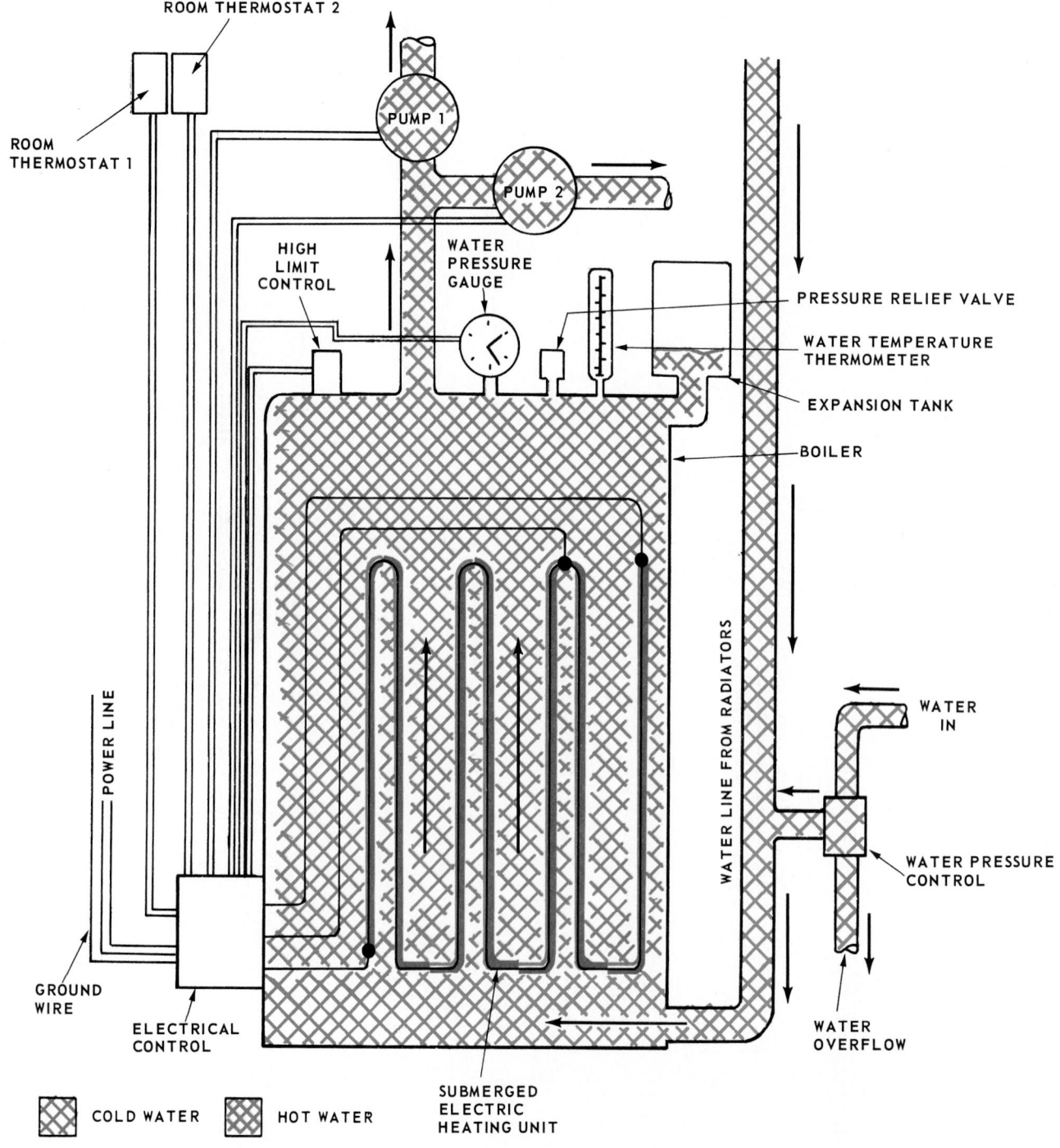

Fig. 19-7. Hydronic heating system using electric heating elements as source of heat.

panel connects the baseboard resistance unit to the electrical power supply.

At C, a room thermostat is mounted on the wall. The power supply from the power panel is connected through the thermostat to the baseboard heating unit.

In the case of A and C, all of the current used by the heater flows through the thermostat points. In the case of B, the room thermostat is handling only a small amount of low-voltage current. The relay, connected to the thermostat and located in the power panel, switches the current to the room resistance heater.

The electric heating units are always grounded. The green wire usually indicates the cabinet ground.

At D, the unit is a resistance heater and a fan. The thermostat controlling this unit is usually mounted on top of the heater. It controls the current to both the heating unit and the fan. All of the fan and heater current goes through the thermostat points. The fan either runs any time that the heating unit is on or it can be run separately.

Circuit breakers should be installed in the power line to each electrical resistance heating unit. No other appliances should be connected into these circuits.

19-9 AIR CONDITIONER, COOLING—WINDOW OR THROUGH-THE-WALL

Window or through-the-wall air conditioners consist of three basic parts:
1. A hermetic compressor.
2. Condenser.
3. Evaporator using a capillary tube refrigerant control.

In the schematic diagram, Fig. 19-9, dark red indicates high-pressure liquid refrigerant; dark blue, low-pressure liquid refrigerant; light blue, low-pressure vapor; and light red, high-pressure vapor.

Liquid refrigerant collects in the lower coils of the condenser and flows through the capillary tube refrigerant control into the evaporator. When the unit is in operation, the evaporator is under low pressure. The liquid refrigerant rapidly boils and picks up heat from the evaporator surface. A motor-driven fan draws air from inside the room, and pulls it through a filter. The fan forces the air over the evaporator. Here it is cooled and goes back into the room. Arrows in Fig. 19-9 show the airflow pattern.

Low-pressure vapor is drawn from the evaporator through

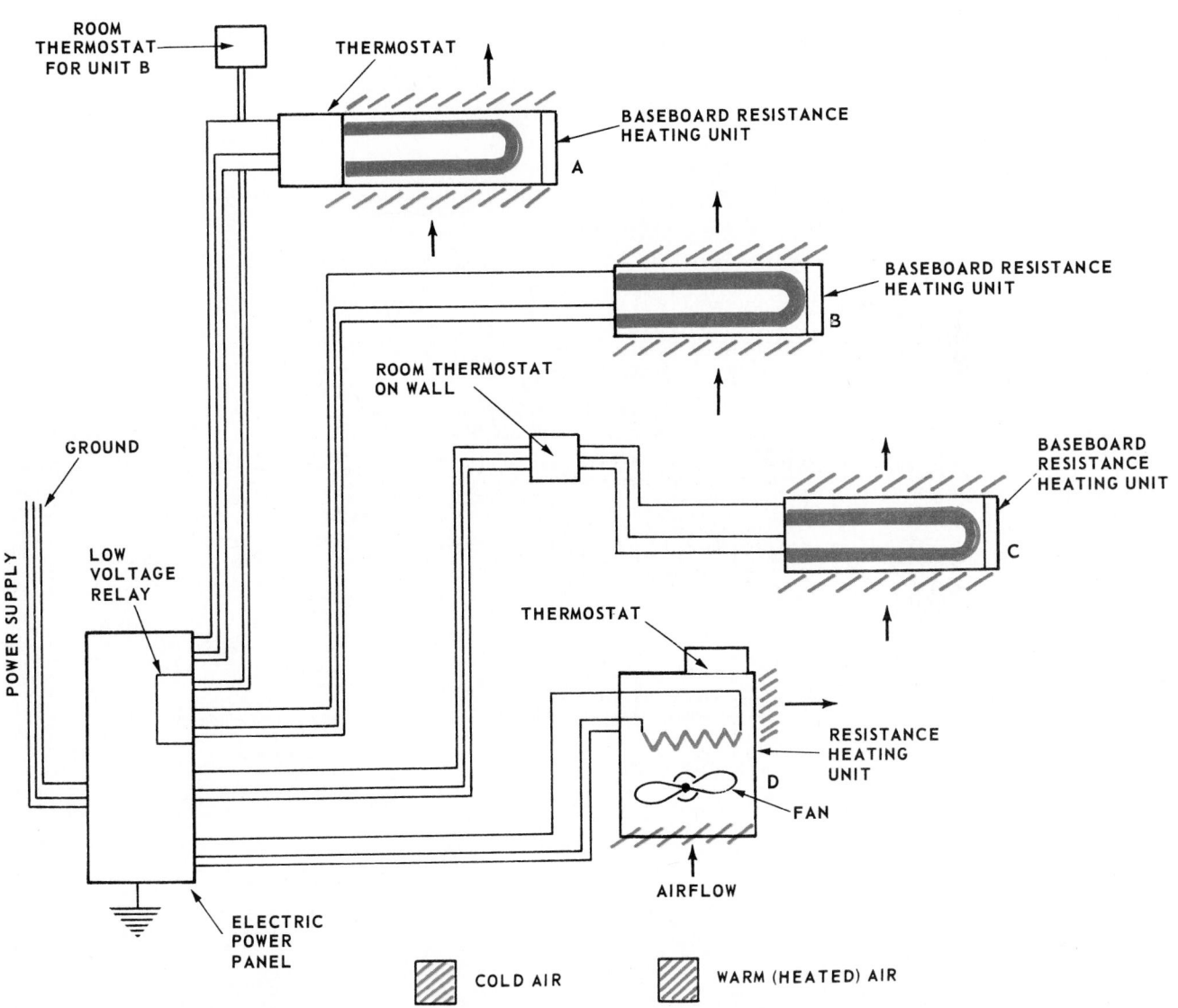

Fig. 19-8. Four different types of electric heating systems. These units provide separate temperature control for each room.

the suction line back to the compressor. Compressed to the high-side pressure, it is forced into the condenser to be cooled and condensed to a liquid. The cycle then repeats.

An adjustable thermostat, mounted on the control panel, provides the necessary control. The thermostat has an on-and-off switch.

The compressor and condenser are in the part of the unit which is outside of the building. The compressor and condenser are mounted in such a way that the fan in the compressor-condenser compartment draws outdoor air in, circulates it over the condenser, and discharges it outside.

Air flowing through the evaporator is cooled and, to some extent, dehumidified. Moisture which collects on the evaporator drains to a drip pan under the evaporator. In some machines it flows into a pan in the compressor compartment. Here, in evaporating, it helps cool the compressor and condenser.

19-10 AIR CONDITIONER, COOLING—WINDOW OR THROUGH-THE-WALL WITH ELECTRIC HEAT

Generally, a window or through-the-wall air conditioner consists of a hermetic motor, compressor, condenser, evaporator, and capillary tube refrigerant control. In the air

conditioner shown in Fig. 19-10, electric resistance heating units are included for cold weather use. During cold weather:

1. The refrigerating mechanism is turned off.
2. The electric resistance heating units are turned on.
3. The room air fan is turned on. The same fan circulates warm air in cold weather and cooled air in warm weather.

These air conditioners are usually connected to 240 V circuits. A control provides a choice of temperatures.

19-11 CENTRAL AIR CONDITIONER, COMPLETE SYSTEM—GAS HEATING, COMPRESSION SYSTEM COOLING, WITH HUMIDITY CONTROL

Fuel gas, burned in an atmospheric burner, is used for heating in this air conditioning system. A compression system using an A-frame evaporator in the furnace plenum chamber provides cooling. Figs. 19-11A and 19-11B show the system in heating and cooling operations.

The condensing unit is located outside the building. A single combination heating and cooling thermostat is often used. A humidistat controls the relative humidity in the conditioned space.

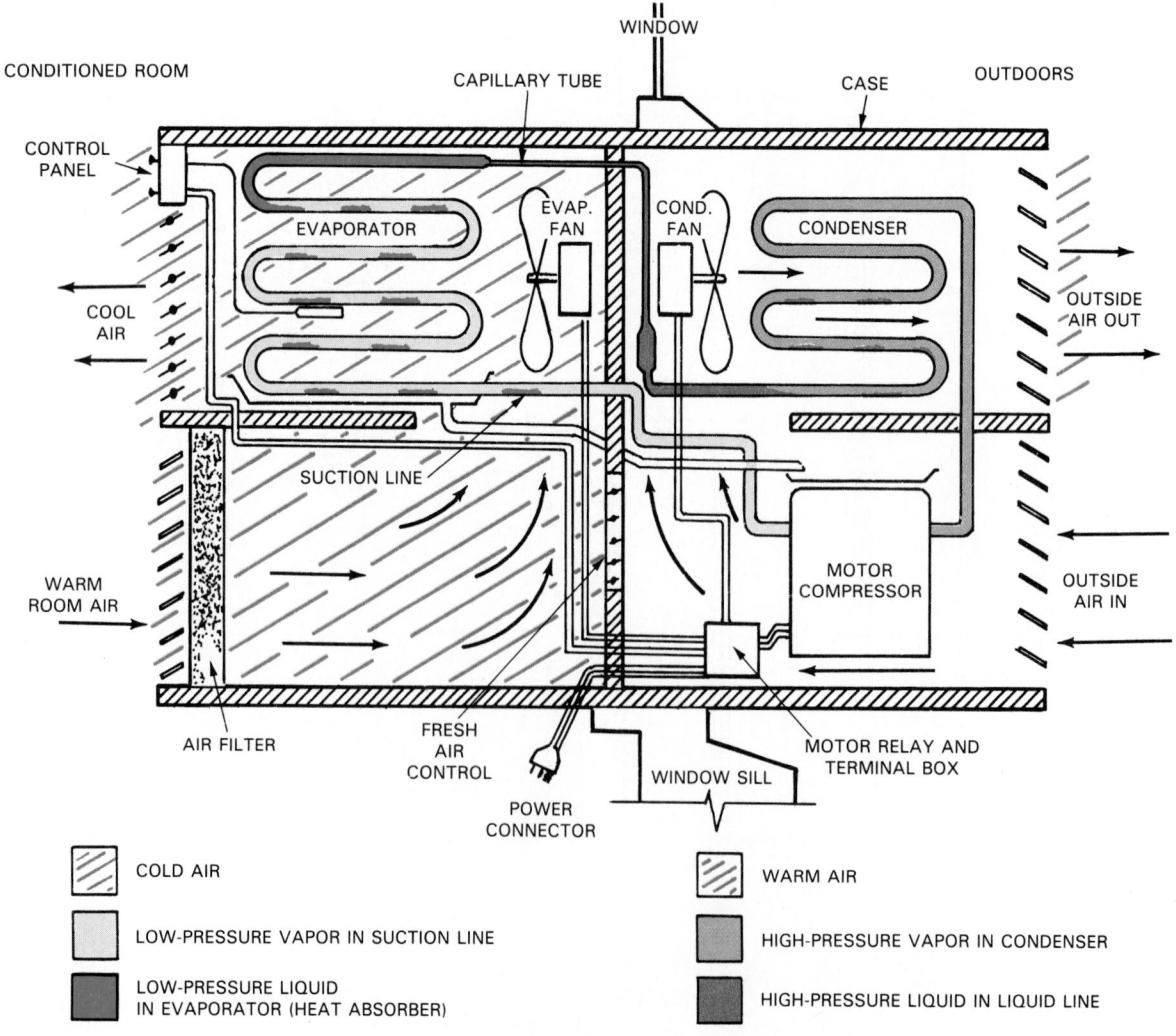

Fig. 19-9. Window-type air conditioning comfort cooling system.

In winter, a humidifier in the plenum chamber adds moisture to the heated spaces. Details of the humidifier are shown in Fig. 19-11C.

Summer humidity is controlled by condensation of moisture on the evaporator. A drain removes this moisture.

Warm air from the furnace is forced into the rooms by a blower. This is located beneath the filter in the cold air return. A control in the top of the furnace turns on the blower when the desired bonnet temperature is reached. This control will also turn off the furnace if the temperature in the bonnet goes higher than a predetermined setting. This is a safety device to keep the furnace from overheating. Electrical power to the furnace is turned on and off by a control panel located on the outside wall of the furnace.

An arrangement is sometimes provided to bring in outside fresh air as needed. This may be controlled either thermostatically or manually.

The pilot light is controlled by a thermocouple connected in series with a solenoid valve in the gas supply line. If the pilot light goes out, the gas to the burner will be shut off.

In summer, a centrally-located thermostat may call for cooling. The same airflow used for heating is used for cool-

ing (the distribution is the same). However, instead of the forced air passing through a heated chamber, it passes across the cooled evaporator. This lowers the temperature of the air and, at the same time, it removes some moisture to reduce the humidity. Air cleaning is done by the filter in the incoming air duct before air reaches the blower.

19-12 ABSORPTION CYCLE

Most large absorption air conditioning systems use water as the refrigerant and an LiBr (lithium bromide) water solution as the absorber. Fig. 19-12 is a schematic diagram of such a system.

Steam or hot water heats the water and lithium bromide solution. The water turns to water vapor which is then condensed by a water-cooled condenser. The water then flows into the evaporator where it evaporates and is absorbed by the lithium bromide at the absorber.

Three pumps maintain the pressure difference. Pump No. 1 moves the solution which is strong in LiBr (sometimes this solution is described as weak in water) back to the absorber and removes more strong LiBr solution from the

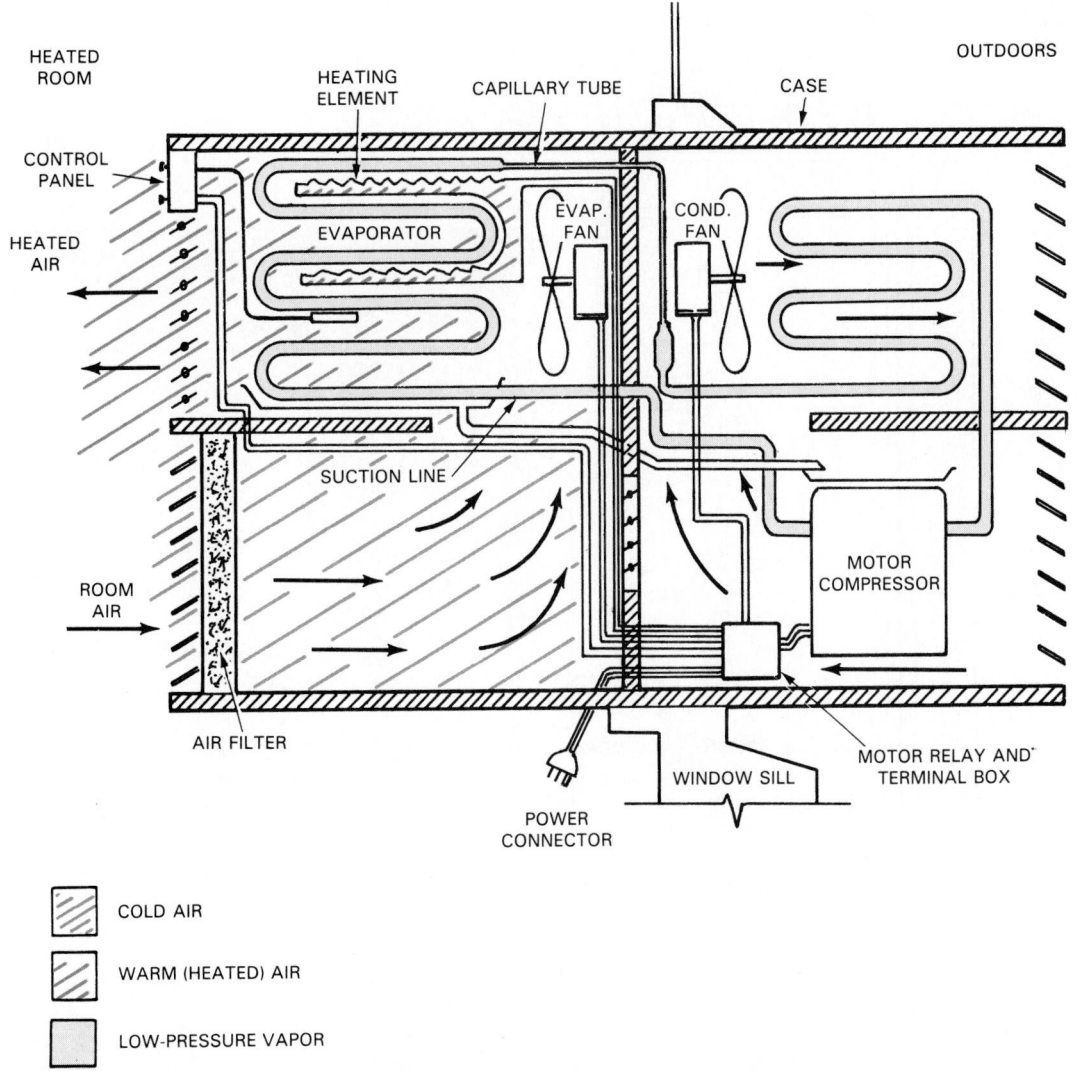

COLD AIR

WARM (HEATED) AIR

LOW-PRESSURE VAPOR

Fig. 19-10. Window or through-the-wall air conditioner with electric heating elements. These provide heat during cold weather. Unit is shown heating air in winter.

concentrator. Pump No. 2 recycles the water not evaporated in the evaporator back to the spray heads in the evaporator. Pump No. 3 moves the solution which is weak in LiBr up to the concentrator.

The temperature changes in the system are marked on the drawing. Cooling water leaves the evaporator at 44 °F (7 °C) and travels through the cooling coils located in the rooms to be air conditioned (comfort cooled). The water then returns to the evaporator at 54 °F (12 °C). The lithium bromide solution always stays in liquid form. The condenser cooling water also cools the absorber.

19-13 EVAPORATIVE CONDENSER

Many air conditioning systems use water-cooled condensers. An evaporative condenser, as shown in Fig. 19-13, may be used to cool the condenser vapor.

In this system, a conventional motor compressor, condenser, liquid receiver, drier, thermostatic expansion valve, and evaporator are used. The hot compressed refrigerant vapor is piped to the evaporative condenser. This part of the system is usually located on the roof or outside the building, as shown.

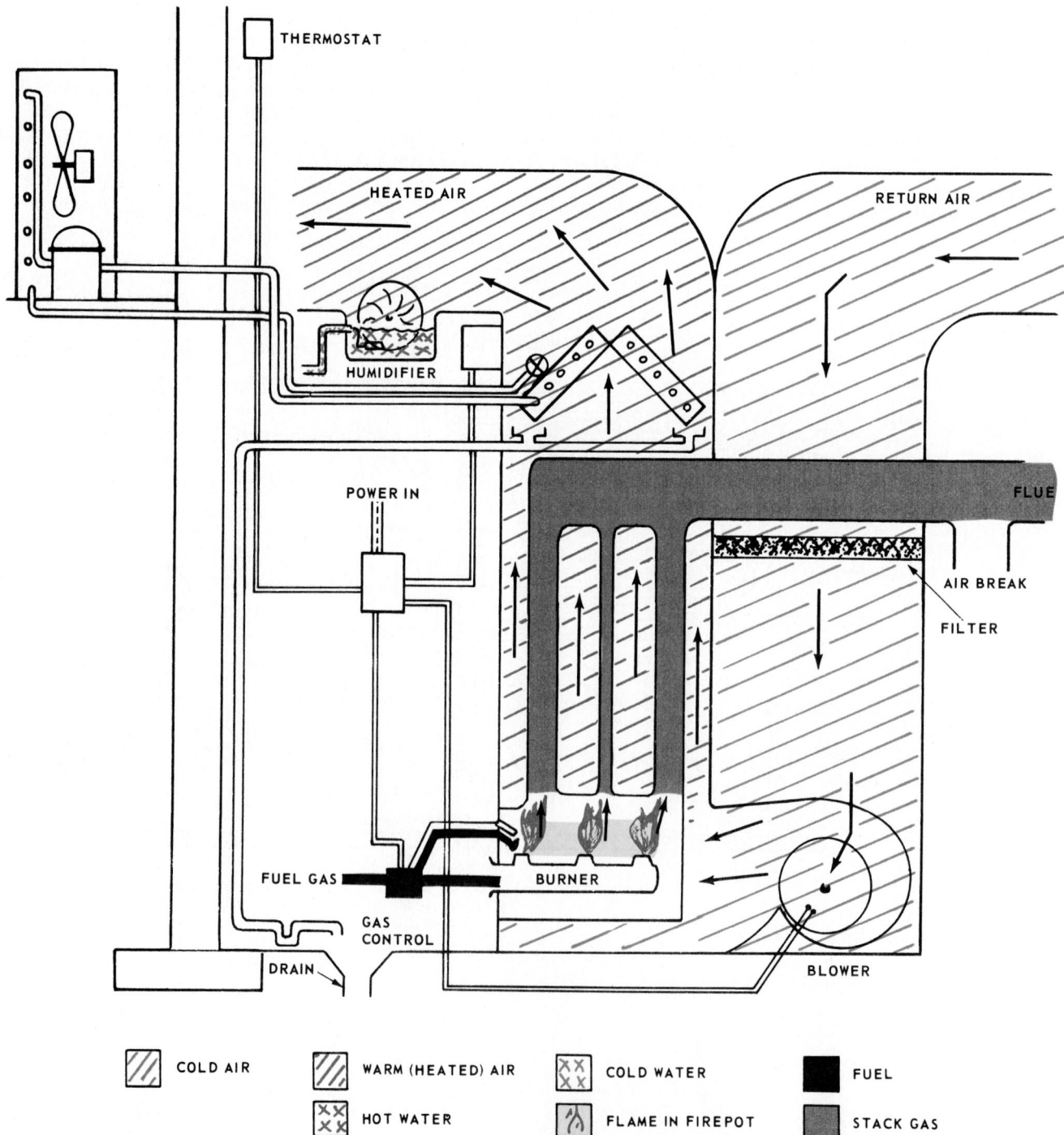

Fig. 19-11A. Complete air conditioning unit provides both heating and cooling. Winter operation is shown. Winter heating is supplied by gas burner. Humidity is supplied by humidifier in plenum chamber. Same blower and filter are used for both summer and winter operation.

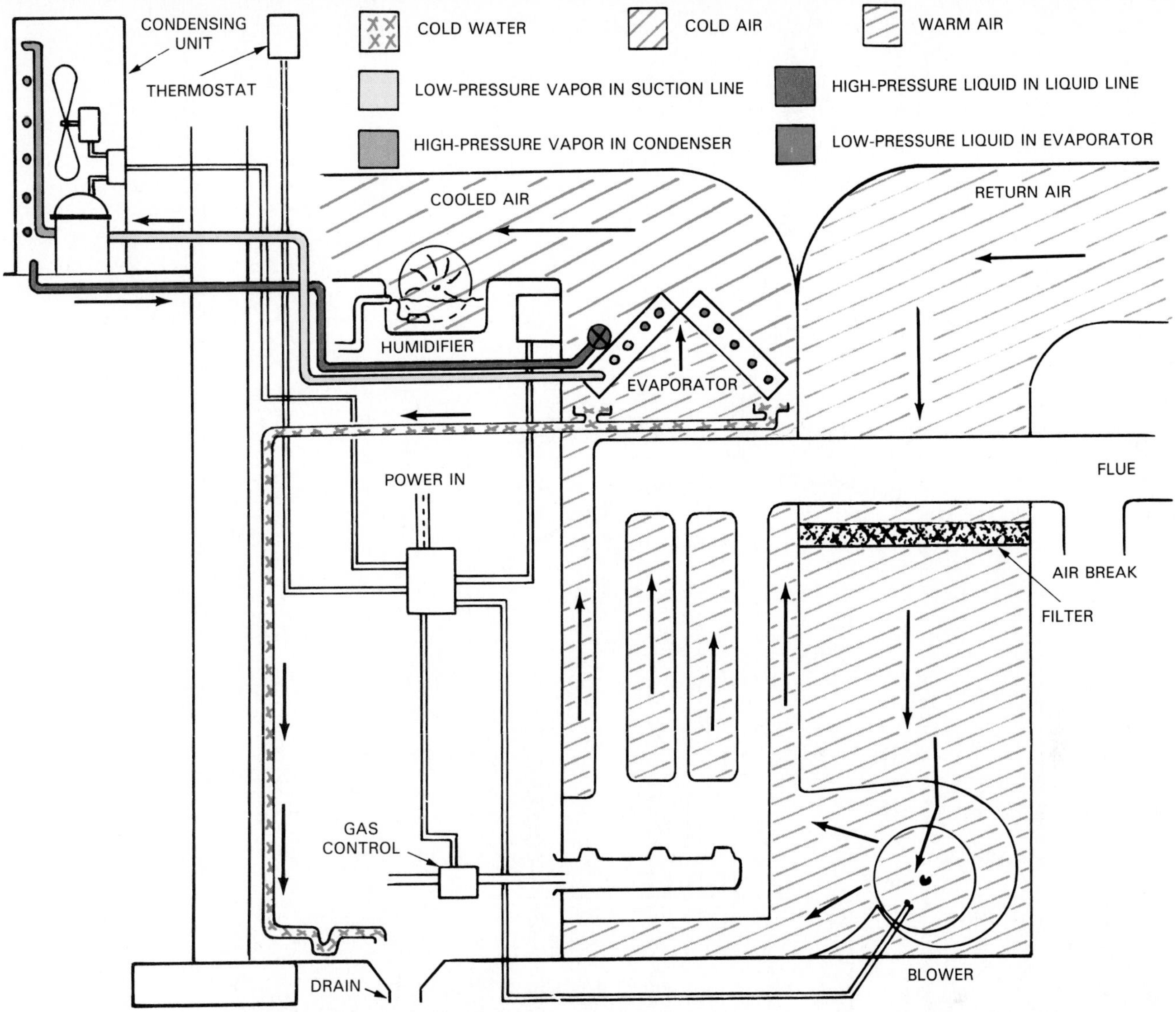

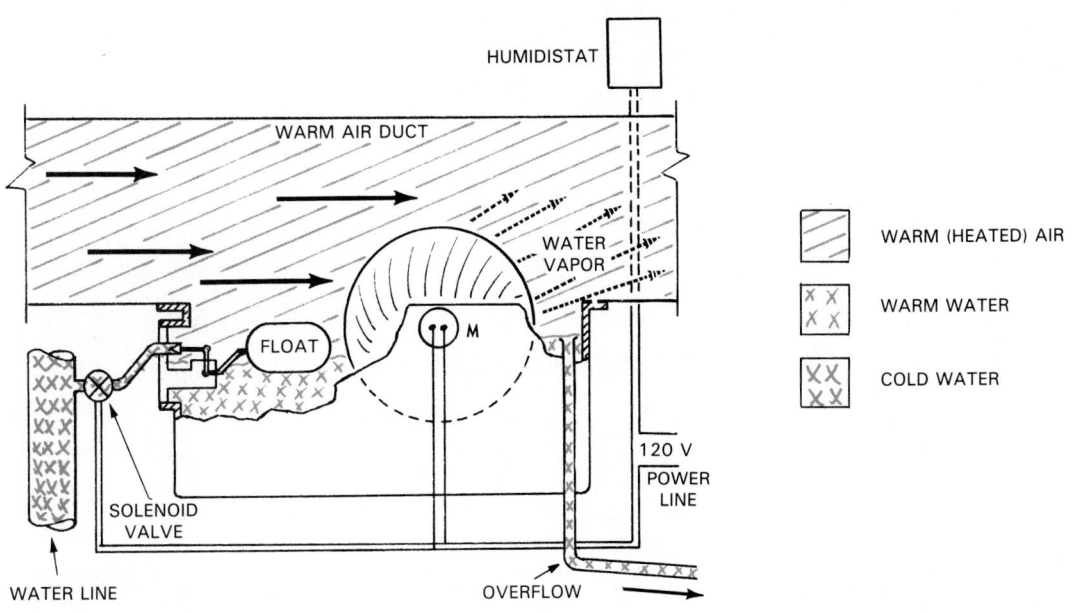

CONDENSING UNIT

THERMOSTAT

COLD WATER

COLD AIR

WARM AIR

LOW-PRESSURE VAPOR IN SUCTION LINE

HIGH-PRESSURE LIQUID IN LIQUID LINE

HIGH-PRESSURE VAPOR IN CONDENSER

LOW-PRESSURE LIQUID IN EVAPORATOR

COOLED AIR

RETURN AIR

HUMIDIFIER

EVAPORATOR

FLUE

POWER IN

AIR BREAK

FILTER

GAS CONTROL

BLOWER

DRAIN

Fig. 19-11B. Complete air conditioning system during summer operation. An A-frame evaporator in plenum cools air forced through it by blower. Outside condensing unit disposes of heat absorbed in evaporator. Summer humidity is removed by condensing of moisture on evaporator surface. Drain tube carries away condensed moisture.

HUMIDISTAT

WARM AIR DUCT

WATER VAPOR

FLOAT

M

WARM (HEATED) AIR

WARM WATER

COLD WATER

120 V POWER LINE

SOLENOID VALVE

WATER LINE

OVERFLOW

Fig. 19-11C. Warm-air-duct humidifier. Water level is controlled with float. Humidistat operates motor and solenoid water valve. Broken arrows indicate air with moisture.

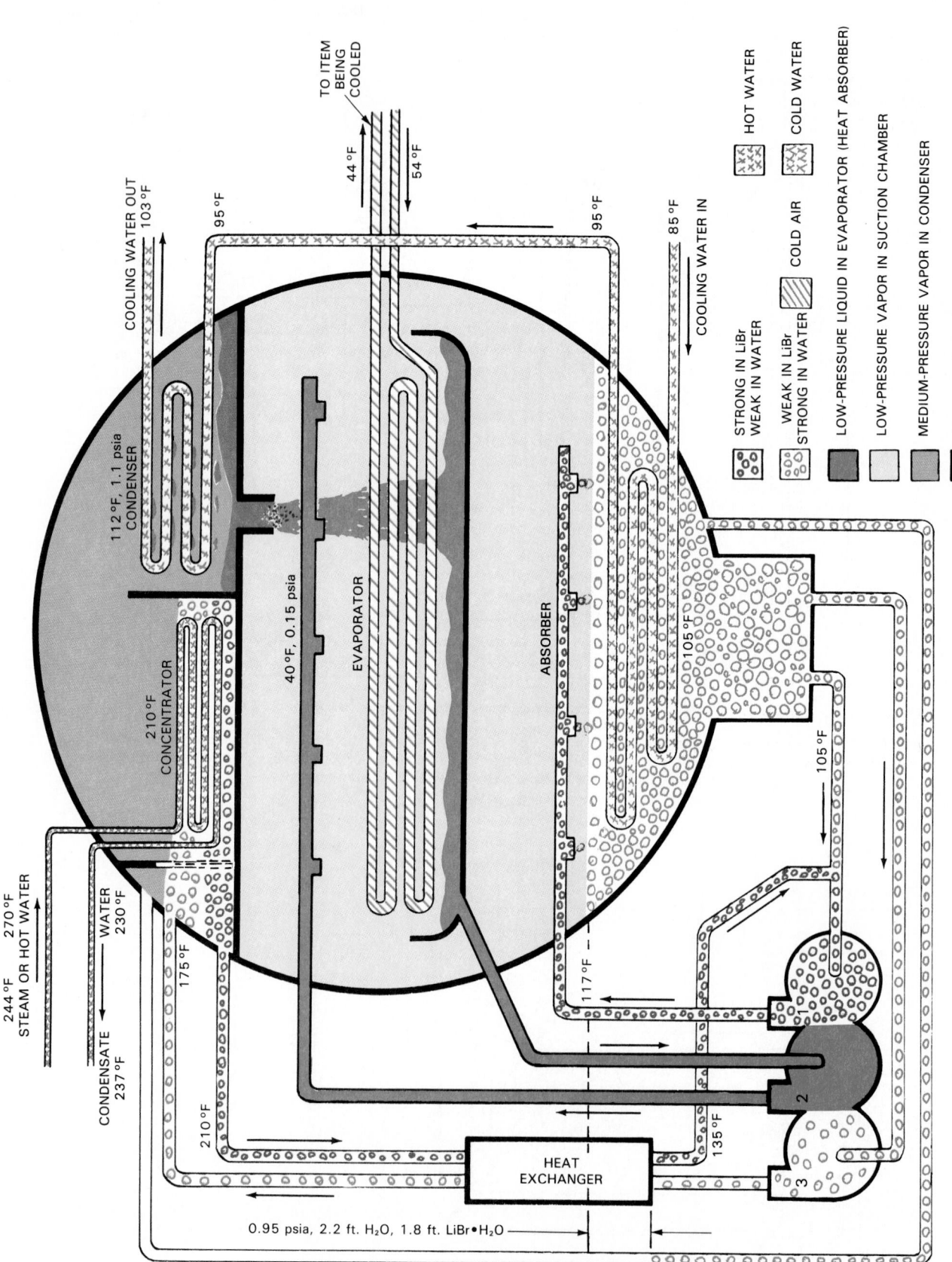

Fig. 19-12. Absorption-type air conditioning system. It uses water as refrigerant and lithium bromide solution as absorber. (The Trane Co.)

In this mechanism, the water supply is piped to a holding tank. A float mechanism maintains a constant level of water in the tank. A water pump circulates and sprays water over the refrigeration condenser.

A fan draws in air through the side of the evaporative condenser housing and forces it upward through the top. The water droplets are cooled by evaporation and then flow over the condenser. Some water is used up by the evaporative process. This is automatically replaced using a holding tank and a float mechanism. A pressure motor control is used on the refrigeration motor compressor in this instance.

19-14 COOLING TOWER

Many refrigeration and air conditioning systems have water-cooled condensers. These are very efficient and do not take very much space. Often water-cooled condensers have tap water circulated through them. This water is then discharged into the sewer. Such an arrangement uses large amounts of water and may be expensive. Moreover, many places do not allow the use of tap water for cooling air conditioner condensers.

In such cases, cooling towers can be employed to cool the water. The cooled water is recirculated through the

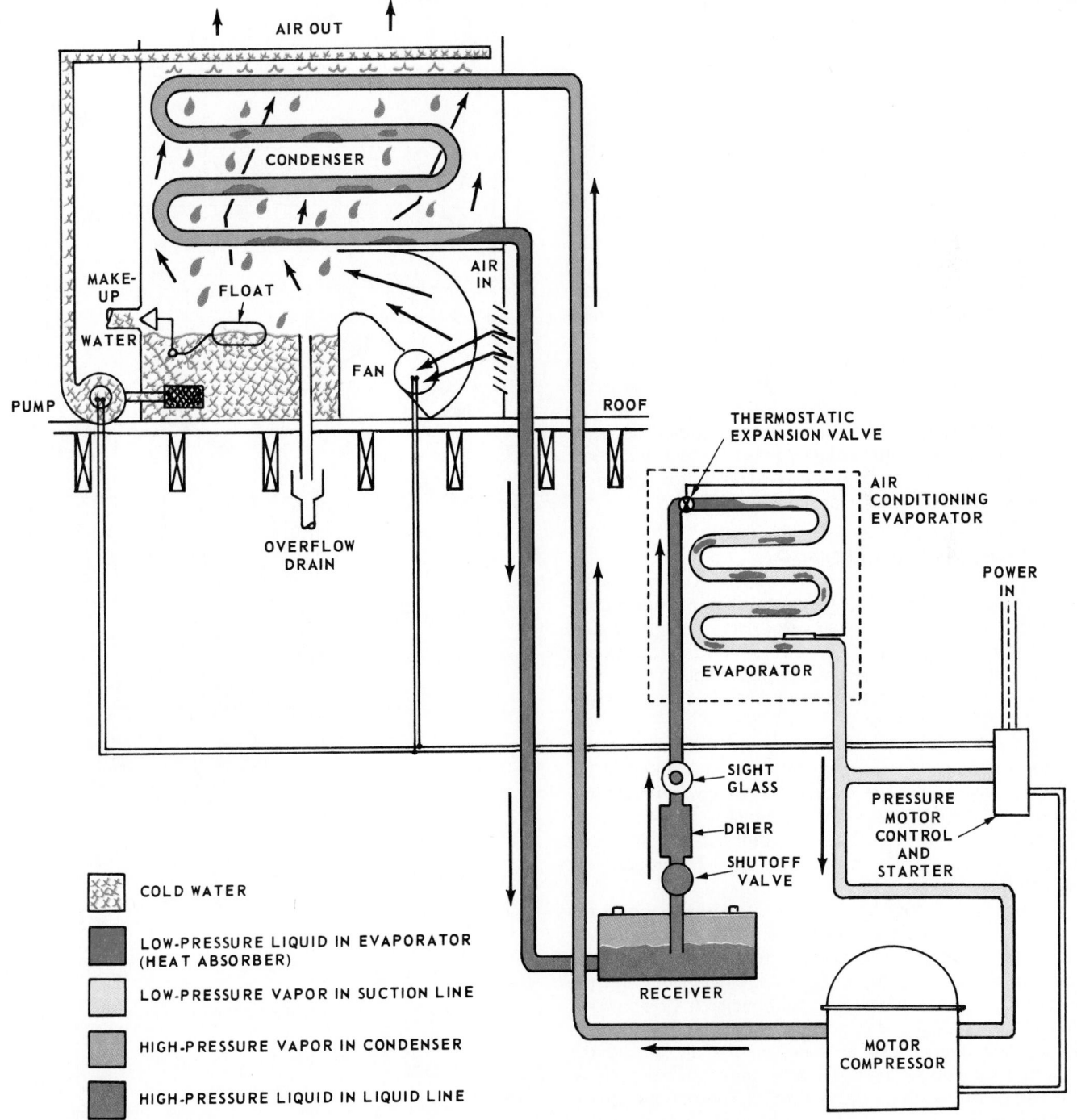

Fig. 19-13. Air cooling unit which uses an evaporative condenser. Note that condenser is located outside of conditioned space.

condenser and sometimes through the outer shell of the compressor. Some makeup water will be required to replace the water lost by evaporation. A schematic of a water cooling tower is shown in Fig. 19-14A.

The cooling tower is a housing or shed into which air is drawn. It has a water spray arrangement and baffles. The sprayed water is exposed to the stream of air and becomes cool. A float mechanism connected to the water spray maintains a constant water level in the water reserve tank. The pump circulates the cooled water through the water-cooled refrigerant condenser. Water is sometimes sprayed over the baffles.

Cooling towers are available in a great range of sizes. Small ones may cool the water-cooled condensers for home air conditioners. Very large ones are required for cooling the condensers used in large steam power plants. Fig. 19-14B shows a complete system.

19-15 ROOM HUMIDIFIER

Room humidifiers may be used to maintain relative humidity. These humidifiers are housed in a cabinet which is located in the room or space in which humidity is to be increased. The cabinet is supplied with air-in and air-out louvers. A fan circulates air through the cabinet. A water pan or trough is used. A rotating screen or filter (wetted surfaces) dips into a water pan or trough and then exposes the wetted surfaces to the air stream. A typical humidifer is shown in Fig. 19-15.

An electric heating element is sometimes used to warm the water for greater evaporation. The controls consist of an on-and-off switch, a relative humidity control, and an indicator light to signal the need for more water. The water in the trough may be renewed by pouring more water into it. It may be connected to the building water supply by means of an automatic float mechanism.

19-16 ROOM DEHUMIDIFIER

Typically, a dehumidifier consists of a hermetic compressor, condenser, and evaporator using a capillary tube refrigerant control. See Fig. 19-16. In the schematic diagram, dark red indicates high-pressure liquid; dark blue, low-pressure liquid; light blue, low-pressure vapor; and light red, high-pressure vapor.

Liquid refrigerant collects in the lower coils of the condenser and flows through the filter into the capillary tube. Then it moves into the evaporator which is under low pressure. In the evaporator, the liquid refrigerant boils rapidly and picks up heat from the evaporator surface. A motor-driven fan forces large amounts of air through the evaporator.

Because of the low temperature of the evaporator, the moisture carried in the air condenses on the evaporator surfaces. The moisture drips to the bottom of the evaporator and into the condensate trough. Air flowing through the evaporator is both cooled and dehumidified. Cooled air is then forced through the condenser, where it cools the condenser and again picks up heat, so the air leaving the dehumidifier is about the same temperature as it was when it entered. Although air discharged by the dehumidifier is nearly the same temperature as intake air, it has a lower relative humidity.

Low-pressure vapor is drawn from the evaporator through the suction line to the compressor. It is again compressed to high-side pressure and is forced into the condenser. Here it is cooled and becomes a liquid. The cycle is repeated.

In addition to an on-and-off switch, dehumidifiers usually have two other controls. One is for humidity. It permits the dehumidifier to operate until the desired relative humidity is reached; then the control shuts the machine off. The other is a frost control element placed in the suction line between the evaporator and the compressor. It stops the motor compressor at a high enough temperature so the evaporator will not freeze over and stop the flow of air through it.

In the drawing, arrows in black show the direction of airflow through the dehumidifier. A fan is commonly used to move the air.

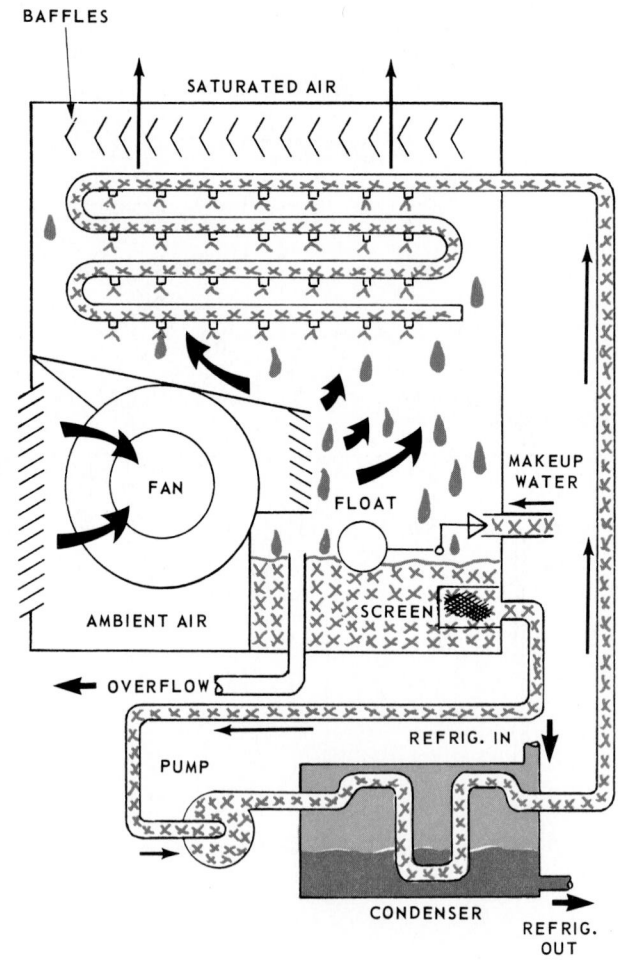

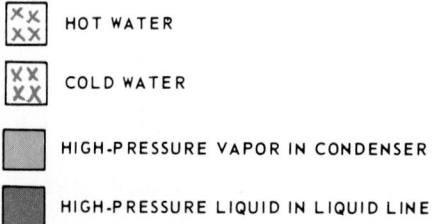

Fig. 19-14A. Typical cooling tower application. Recirculated water is cooled several degrees in tower. Then water is circulated through refrigerant condenser to cool it.

19-17 HEAT PUMP—AIR-TO-AIR

Used in homes and in some industries, the "heat pump" is a heat-moving mechanism. Heat is absorbed in an evaporator in one location and released through a condenser in another location. The system can reverse its operation so that the evaporator becomes the condenser and the condenser becomes the evaporator. Heat flow is reversed.

Thus, using a special reversing valve, the mechanism either heats or cools the conditioned space. *The flow through the compressor is always in the same direction.*

Fig. 19-17A shows the flow through the valve causing the conditioned space to be heated. Fig. 19-17B shows the valve in position to cool the conditioned space.

Heat pumps use compression-type refrigerating mechanisms, similar to a food-refrigerating or air conditioning

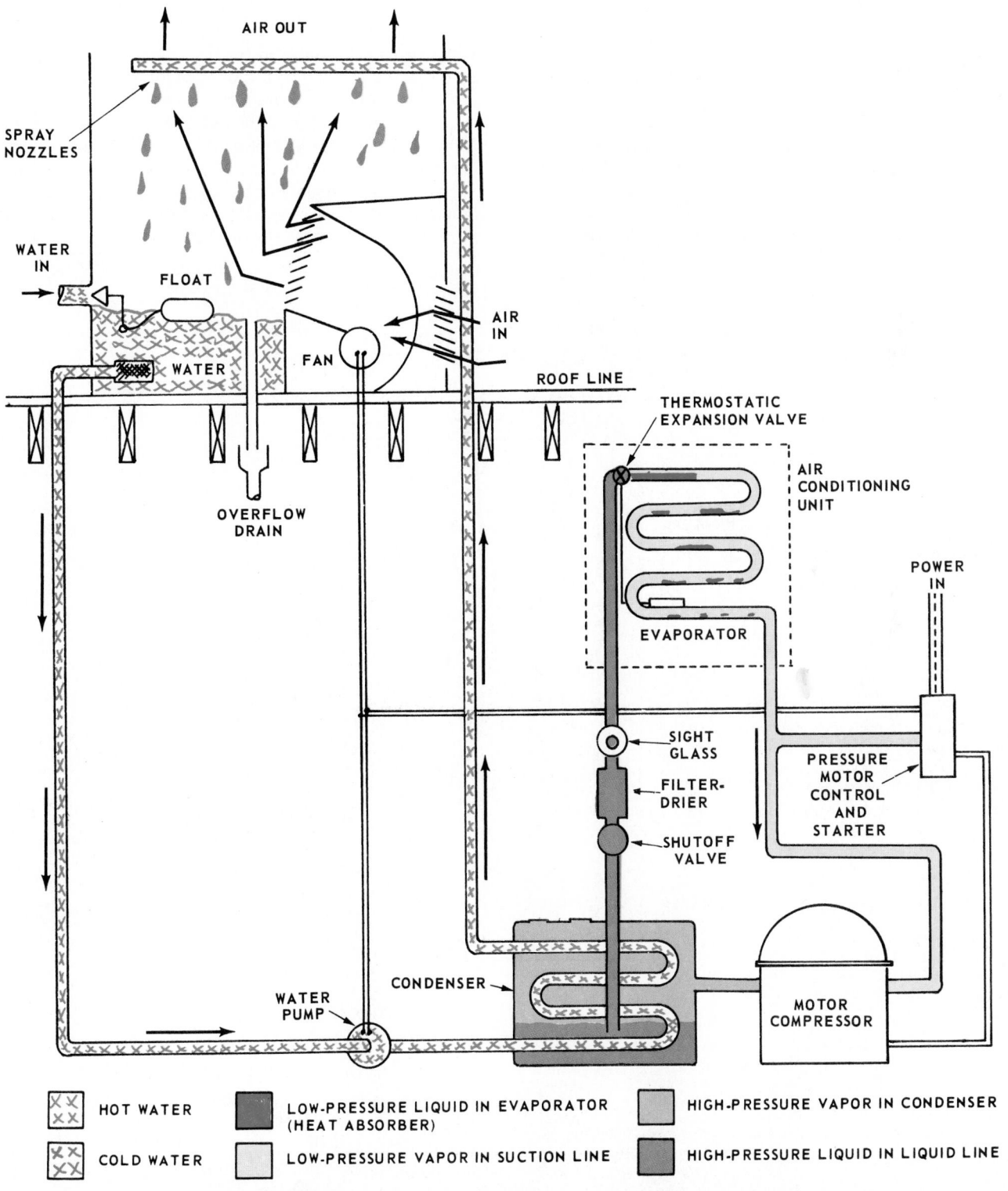

XX / XX	HOT WATER	■	LOW-PRESSURE LIQUID IN EVAPORATOR (HEAT ABSORBER)	▨	HIGH-PRESSURE VAPOR IN CONDENSER
XX / XX	COLD WATER	▨	LOW-PRESSURE VAPOR IN SUCTION LINE	▨	HIGH-PRESSURE LIQUID IN LIQUID LINE

Fig. 19-14B. Air conditioning comfort cooling installation using cooling tower.

mechanism. Heat pumps have two heat transfer surfaces—one located inside the conditioned space and the other out-of-doors.

On the heating cycle, Fig. 19-17A, the outdoor coil becomes an evaporator while the indoor coil becomes the condenser. In operation, liquid refrigerant enters the outdoor coil, picks up heat from out-of-doors and is vaporized. The vapor is drawn into the compressor, is compressed to a high temperature and is pumped into the indoor coil. Since its temperature is higher than the indoor temperature, heat will be released into the room.

Compressed refrigerant vapors will condense upon giving up their heat of vaporization and will return to a liquid state. The liquid then flows back through the capillary tube into the evaporator, and the cycle is repeated. Since the outdoor coil is colder than the outdoor surrounding air, ice may form on it if the outdoor temperature is rather low. Therefore, outdoor coils are fitted with de-ice controls. These de-ice controls operate in either of two ways when ice forms:

1. They automatically turn on electric heating units.

2. They turn off the compressor, allowing the evaporator surface to warm up and melt the ice.

On the cooling cycle, Fig. 19-17B, the reversing valve causes the coil in the conditioned space to become an evaporator. Refrigerant flows through the capillary tube into the evaporator. The liquid refrigerant boils, absorbing heat. Vapor from the boiling refrigerant is drawn into the compressor where it is compressed. The heated vapor is pumped into the outdoor coil, which has become a condenser.

Since the air surrounding the outdoor coil is cooler than the compressed vapor in the coil, the compressed refrigerant vapor gives up its heat to the outside air. It condenses and flows to the bottom of the condenser as liquid refrigerant. From here it flows through the capillary tube into the bottom of the evaporator. From this point the cycle is repeated. Motor-driven fans on both coils aid heat flow from coil surfaces.

Heat pump installations are ideal for locations where the heat load in winter is almost the same as the cooling load in summer. *Air-to-air installations are most satisfactory*

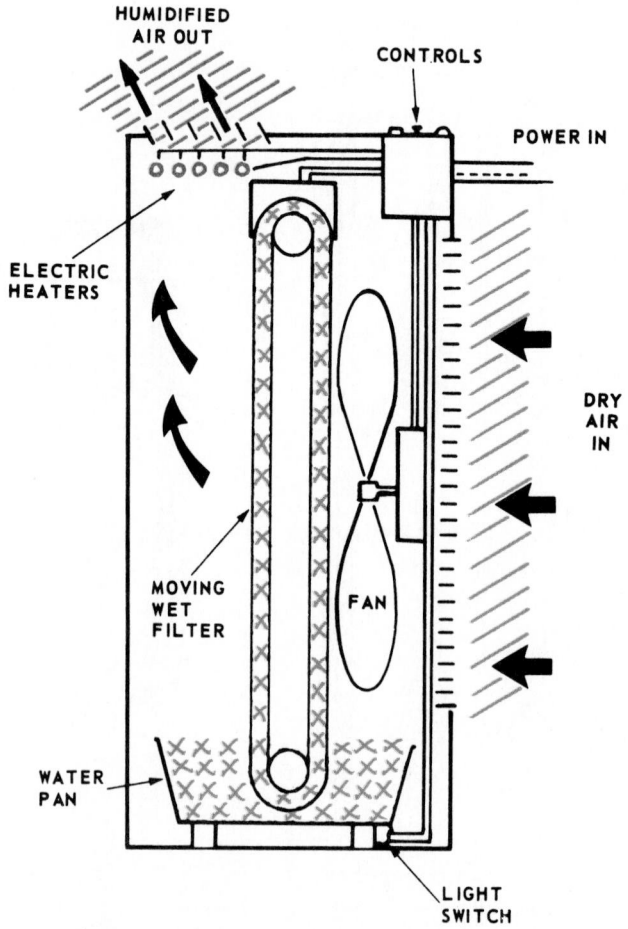

Fig. 19-15. Room humidifier. Porous belt slowly moves through water in water pan. Fan forces air through moist porous belt and relative humidity increases. Some units have electric heaters to reheat humidified air. Humidistat controls operation of unit. Signal light goes on when water pan is empty.

when the ambient air temperature in the winter remains above—or only occasionally below—the freezing temperature.

19-18 HEAT PUMP—WITH ELECTRIC HEATERS

Air-to-air heat pump installations operate efficiently when the outside air temperature is above freezing. However, when the outside temperature drops down to or below freezing, efficiency drops off rapidly.

To make up for this loss of efficiency, the indoor section is often fitted with electric resistance heating units. When the thermostat calls for more heat than the heat pump is able to deliver, the electric heating elements will turn on.

Heat pump operation for the heating and cooling cycle is the same as explained in Para. 19-17. Fig. 19-18 is a heat pump cycle diagram for a unit with an auxiliary (additional) electric heating system. Note the resistance heating units in the indoor section.

19-19 HEAT PUMP—GROUND OR WATER COIL

As explained in Para. 19-18, the efficiency of the air-to-air heat pump depends greatly on outdoor temperature. To improve this efficiency, some installations use a coil buried in the ground beneath the frost level, rather than a coil in the atmosphere. If the coil is buried at some depth and a long enough coil is used, the efficiency of the heat pump may be very good. A schematic diagram of the cycle is shown in Fig. 19-19A.

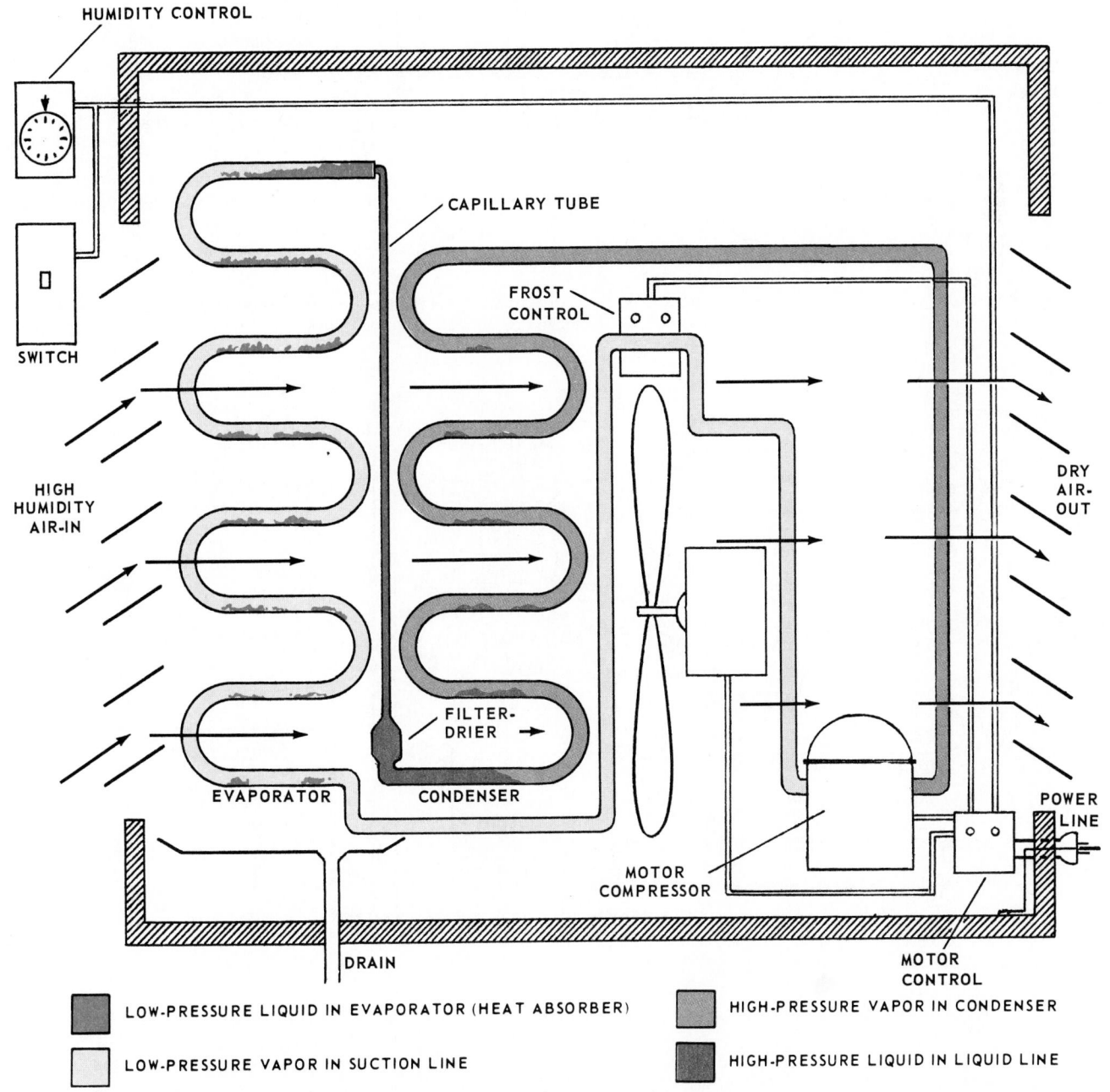

Fig. 19-16. Room dehumidifier. Room air is cooled as it flows through evaporator. Water vapor is condensed on evaporator surface and drains away. Air is reheated as it flows through and cools condenser.

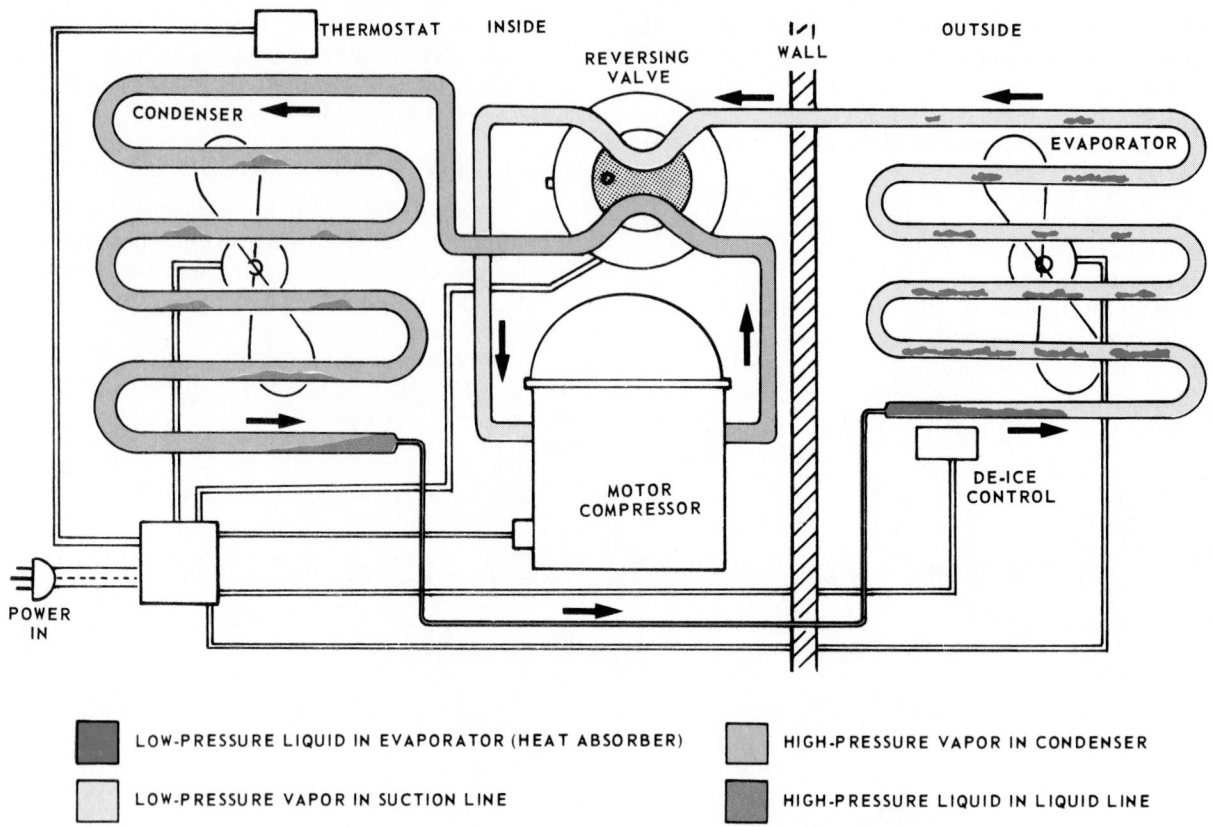

Fig. 19-17A. Air-to-air heat pump illustrating heating cycle. Reversing valve is set so that coil on outside acts as an evaporator. Heat absorbed in evaporator is released by condenser inside house.

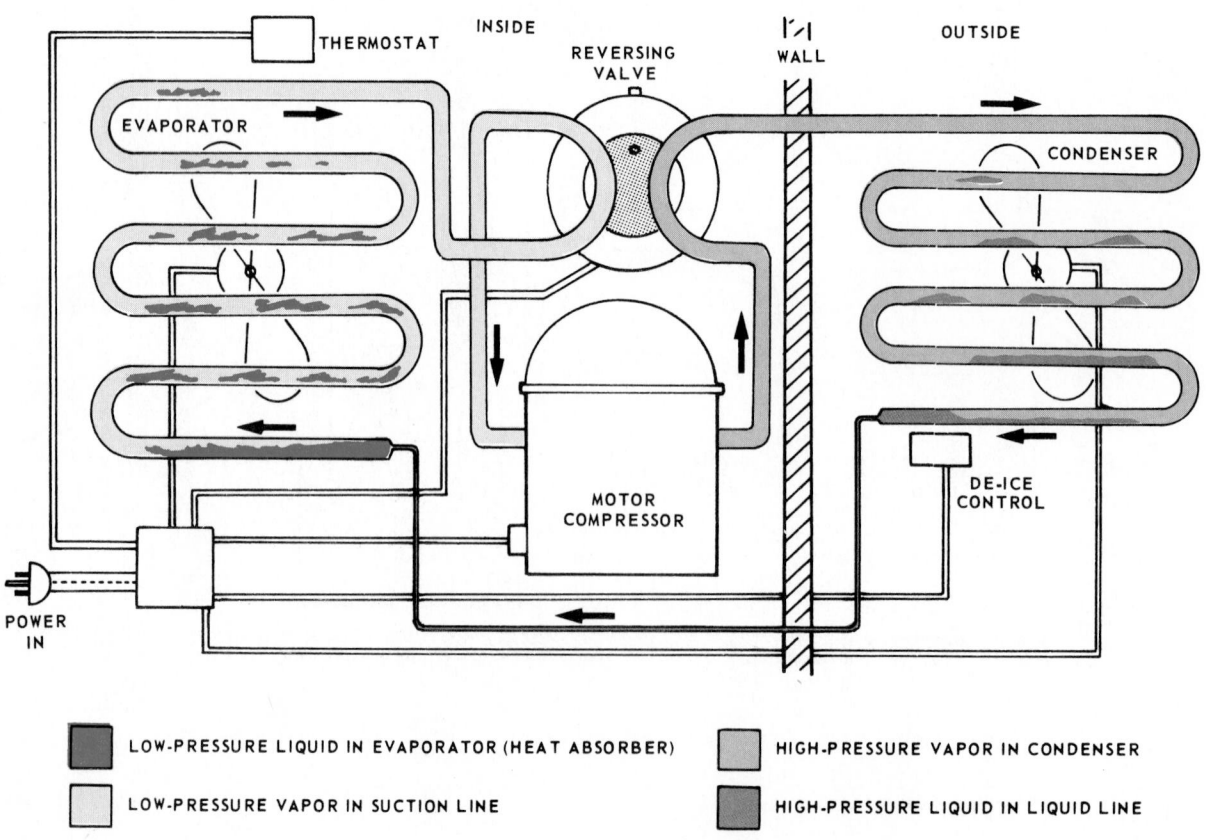

Fig. 19-17B. Air-to-air heat pump during cooling cycle. Valve is set so that coil on inside acts as an evaporator. Heat absorbed in evaporator is released by condenser outside house.

On the heating cycle, liquid refrigerant flows through a refrigerant control and into the ground coil. Since the refrigerant in the ground coil is under low pressure, it boils, absorbing heat from the ground surrounding the coil.

The vaporized refrigerant is then drawn into the compressor. It is compressed and discharged into the condenser, which, in this case, is the heating coil for the heating system. The condenser changes the vaporized refrigerant to a liquid. The refrigerant gives up its heat to the room air. The liquid refrigerant returns to the refrigerant control to repeat the cycle.

The same mechanism may be used to cool the building in summer. The cycle is reversed to move heat from the building to the outdoors. In this case, the inside coil serves as the evaporator and the ground coil becomes the condenser. The ground absorbs the heat from the vaporized refrigerant. Fig. 19-19B illustrates the heat pump with the valves set for cooling the conditioned space.

The four-way valve is electrically controlled by the thermostat. If heat is called for, the valve will allow fluid flow as indicated in Fig. 19-19A. If cooling is needed, the flow will be as shown in Fig. 19-19B. In each case, the compressor refrigerant flow is in the same direction. The suction side and discharge side of the compressor are always

the same. The cycle change is accomplished entirely by operation of the four-way valve.

Heat pump installations are even more efficient if the ground coil is placed in a spring or in a flowing well with water at about 50 °F (10 °C). Some heat pump installations have been successful using a coil placed in the bottom of a lake.

19-20 AUTOMOBILE AIR CONDITIONING

Automobile air conditioners use a refrigerating mechanism to cool the air inside the vehicle. The equipment consists of a refrigerating compressor driven by the engine, a condenser located in front of the radiator, a liquid line to the refrigerant control, an evaporator, a fan, and a duct system to circulate the air inside the vehicle. Temperature control is based chiefly on the temperature of the air flowing through the evaporator. Fig. 19-20 illustrates a typical automobile air conditioner. To air condition a moving vehicle presents some problems not found in the usual refrigeration or air conditioning installation. Since the compressor is driven by the engine, its speed will change as the engine speed changes. The cooling capacity of the system is usually sufficient to take care of the cooling load under the most

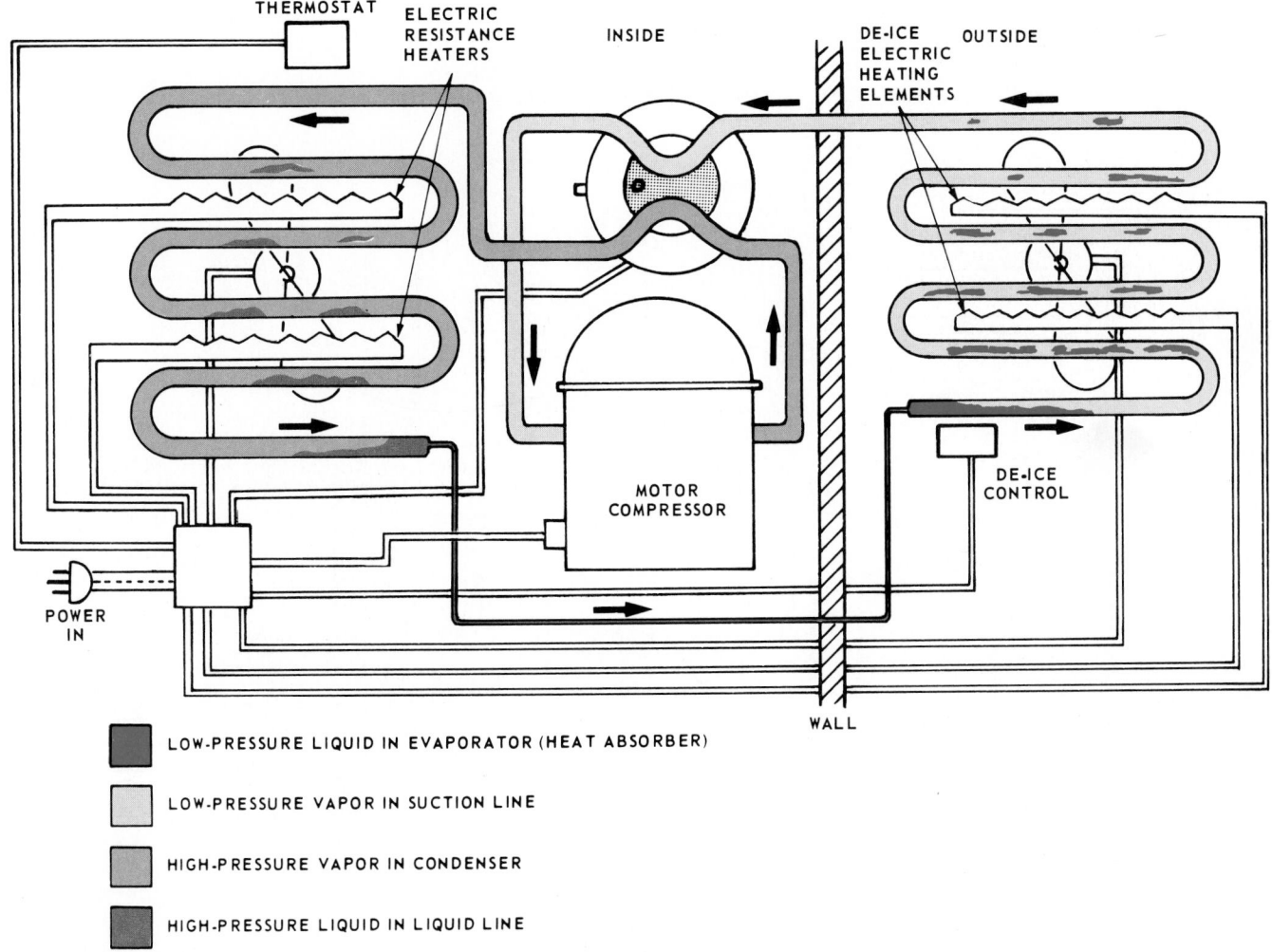

LOW-PRESSURE LIQUID IN EVAPORATOR (HEAT ABSORBER)

LOW-PRESSURE VAPOR IN SUCTION LINE

HIGH-PRESSURE VAPOR IN CONDENSER

HIGH-PRESSURE LIQUID IN LIQUID LINE

Fig. 19-18. Air-to-air heat pump with electric resistance heating elements. Heating cycle is on in this diagram. Electric heaters in refrigerant condenser area provide additional heat if needed.

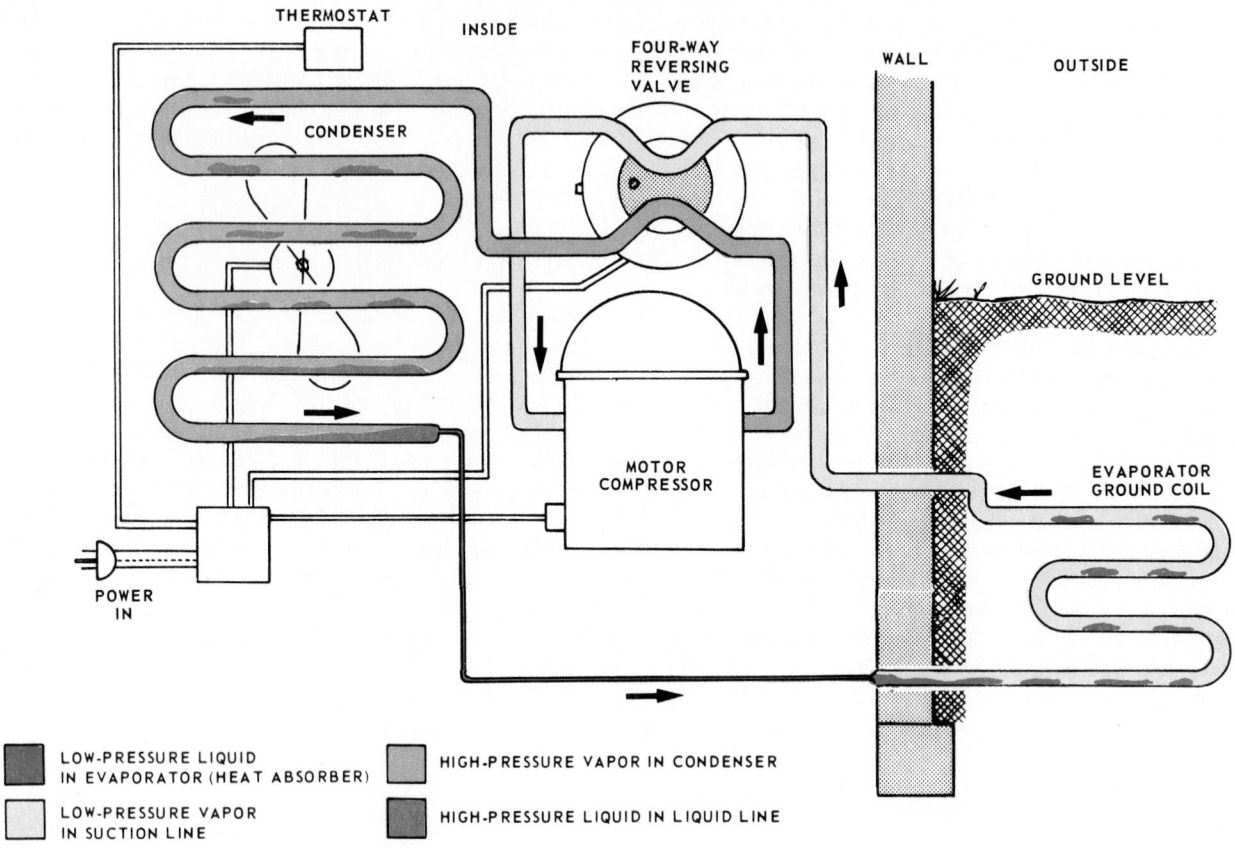

Fig. 19-19A. Heat pump using ground coil (or coil in a well or lake). Heating cycle is shown.

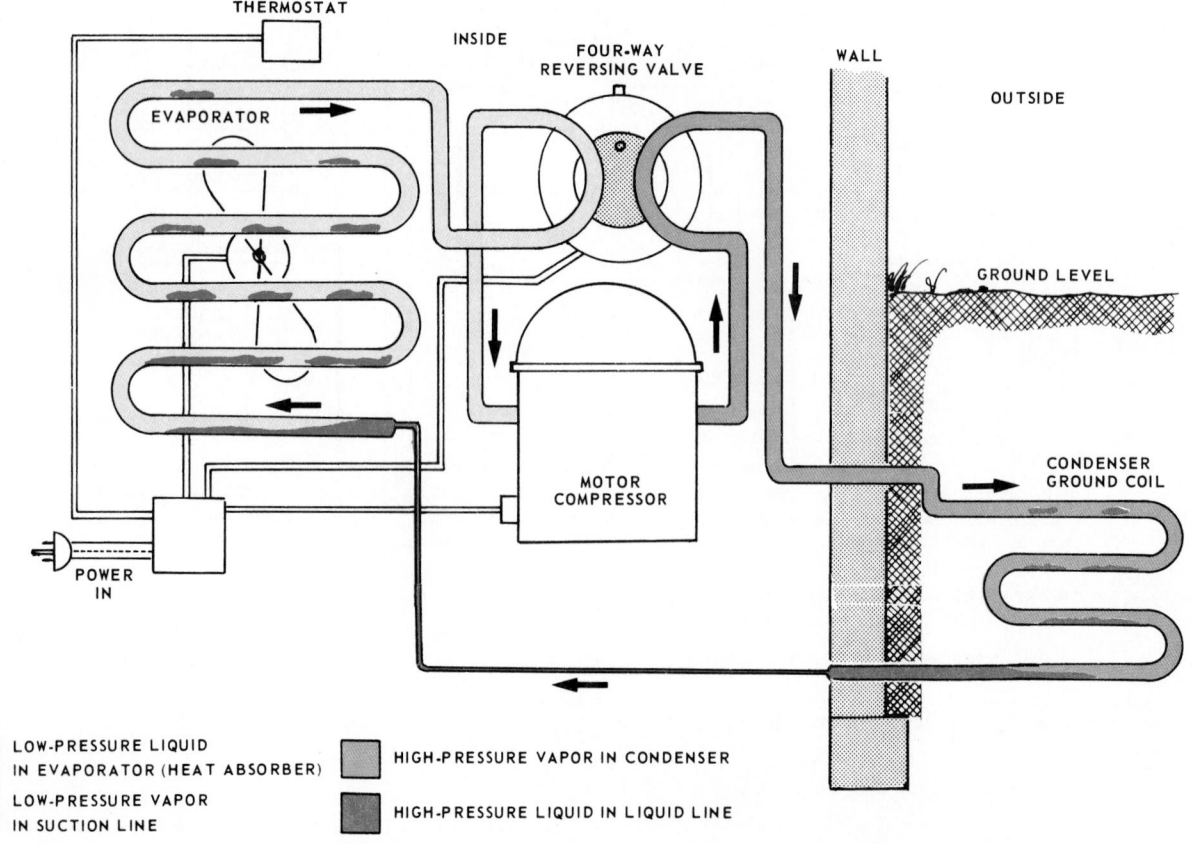

Fig. 19-19B. Heat pump using ground coil (or coil in well or lake). Cooling cycle is shown.

unfavorable conditions of temperature and speed. As a result, under normal driving conditions, the system has much more capacity than is needed.

This problem is solved by providing a magnetic clutch in the hub of the compressor drive pulley. This clutch is controlled by a thermostat. When the passenger compartment temperature has been brought down to the desired level, the thermostat releases the magnetic clutch on the compressor drive pulley, and the compressor stops turning. When cooling is again needed, the magnetic clutch engages the compressor.

The compressor draws power from the engine when the magnetic clutch is engaged. A cutout switch is used to disengage the compressor when more power is needed for acceleration.

There is still another problem with air conditioning an automobile. The evaporator condenses moisture from the air. If the evaporator temperature is maintained at or below freezing temperature, this moisture will freeze and adhere to the evaporator. Soon the evaporator will be completely frozen over. No air can circulate through it. This problem is solved by placing a suction throttling valve in the suction line. This valve maintains the pressure in the evaporator

slightly above the pressure at which the temperature of the boiling refrigerant in the evaporator will cause the moisture to freeze to the evaporator surface.

The preceding description of automobile air conditioning has been kept brief. See Chapter 27 for full details concerning automobile air conditioning.

19-21 STEAM JET COOLING

Steam jet cooling uses water as the refrigerant. Pressure on the surface of water is reduced to lower its boiling temperature. This is shown in the table in Fig. 19-21A. At a pressure of 0.2 psia (10 mm Hg), the boiling temperature of water is 53 °F (12 °C).

A steam jet is shown in Fig. 19-21B. An ejector sucks or draws water vapor from the surface of the water in the evaporator, causing the pressure in the evaporator to drop. The ejector reduces the pressure in the evaporator to a point at which the water will vaporize at the desired temperature. While vaporizing, it absorbs heat and cools the rest of the water in the evaporator.

Steam pressure at the ejector nozzle should be about 150 psia (1 030 kPa). Sometimes the steam is condensed at

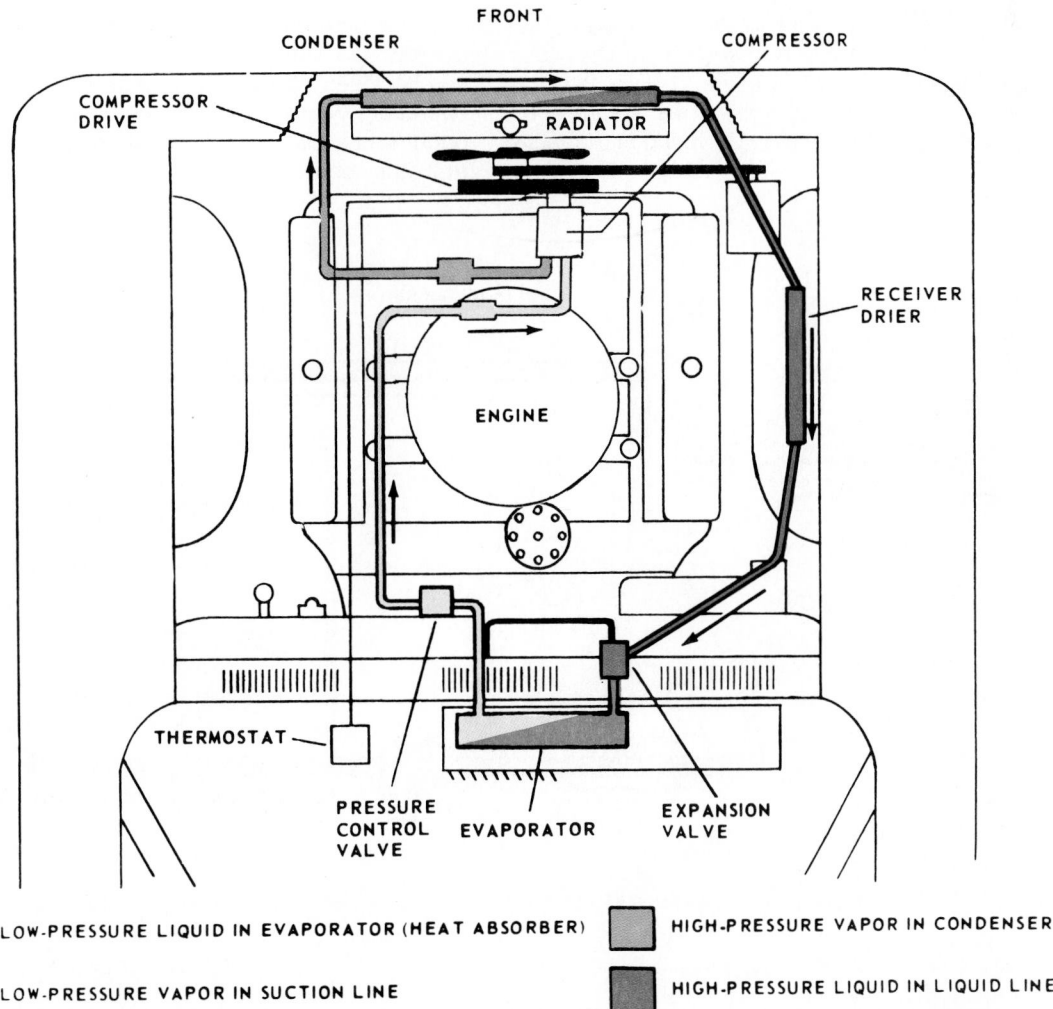

Fig. 19-20. Automobile air conditioner. Dark red indicates high-pressure liquid refrigerant; dark blue, low-pressure liquid refrigerant; light blue, low-pressure vapor; light red, high-pressure vapor. Compressor is driven by engine. Condenser coil is located ahead of automobile radiator. Evaporator is located in duct system in passenger compartment.

WATER BOILING TEMPERATURES

PSIA	BOILING TEMPERATURE		PSIA	BOILING TEMPERATURE	
	°F	°C		°F	°C
.1	35	2	5.	162	72
.2	53	12	6.	170	77
.3	64	18	7.	177	81
.4	73	23	8.	183	84
.5	80	27	9.	188	87
.6	85	29	10.	193	89
.7	90	32	11.	198	92
.8	94	34	12.	202	94
.9	98	37	13.	206	97
1.	102	39	14.	209	98
2.	126	52	14.7	212	100
3.	141	61	15.	213	101
4.	153	67	20.	228	109

Fig. 19-21A. Table shows boiling temperature of water at various pressures. Note that pressures are in pounds per square inch absolute (psia). Atmospheric pressure is 14.7 psia (101.3 kPa).

another location for the following reasons:
1. To recover some heat from the steam.
2. To recover the water in the steam.
3. To reduce the pressure so that the pressure will not "back up" into the steam jet cooling chamber. The pressure in the condenser, not shown in the illustration, will be about 3 psia (21 kPa, 160 mm Hg). The pressure, 3 psia, corresponds to the steam condensing temperature, 141 °F (61 °C), found in the table in Fig. 19-21A.

Evaporation of some of the water in the evaporator reduces the temperature of the remaining water. Pumps circulate this cold water at 40° to 70°F (4° to 21°C), to the area to be cooled.

Because of the need for a large supply of steam under a fairly high pressure and for a large supply of water for cooling the condenser, these systems usually have a large capacity—100 tons and over.

Steam jet systems are often used in air conditioning and for cooling water used in certain chemical plants for gas absorption. The cooling temperatures provided by the steam jet mechanism are usually between 40° and 70°F (4° and 21°C). Temperatures below 40°F (4°C) are impractical due to the danger of freezing.

Steam jet cooling is also used to remove water from diluted solutions that may contain juices. Orange juice can be concentrated in this way. By replacing the water with orange juice and not providing makeup water, steam jet cooling cools and concentrates the juice. The process does not boil juice at temperatures near 212°F (100°C). This keeps vitamins at full strength.

19-22 VORTEX TUBE COOLING

An interesting device called a vortex tube uses compressed air to produce low temperature. Pressurized air is directed smoothly along the inner surface of a tube or cylinder. One end of the tube is completely open. The other end is closed off except for a small-diameter tube. During operation, warm air leaves through the unrestricted (wide open) end while cold air leaves through the small tube at the other end.

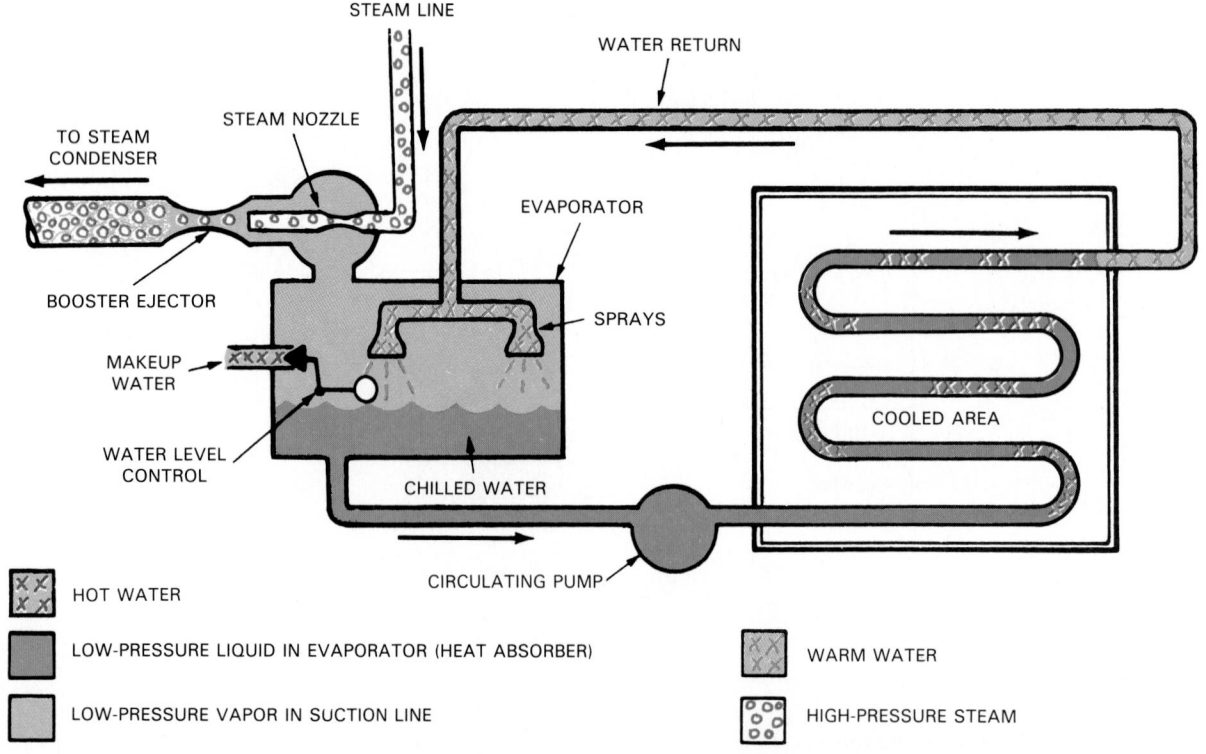

Fig. 19-21B. Steam jet refrigeration. Steam escaping through nozzle in ejector causes low-pressure condition over surface of water in evaporator. This low pressure on surface of water causes it to evaporate rapidly, absorbing heat and reducing temperature of water in evaporator. Chilled water is circulated where needed for cooling. If items in cooled area are above 57 °F (14 °C), some liquid water will turn into water vapor inside cooled area coil. If sprayheads are restricters, no boiling occurs in cooled area.

Fig.19-22 shows the operation of the vortex tube, at D, C, and E. There are three openings. One is an inlet while the other two, as explained before, are outlets. The inlet opening at C is a jet nozzle connected to the compressed air source B. This jet injects the air into the tube at an angle.

Due to the design of the jet and the high pressure of the air, the air swirls rapidly in a corkscrew pattern inside the large tube. Both openings, D and E, are at or near atmospheric pressure. The temperature of the air, as it leaves through tube E, will be greatly reduced (see green). Temperatures below zero are obtained with this device.

The principle of operation of the vortex tube is as follows. High-velocity air stays in the outer circles of tube D in Fig. 19-22. Low-velocity air stays near the center of the vortex. The low-velocity air contains air molecules which are moving slower, on the average, than molecules in the high-velocity region. Thus, the center of the vortex is cooler. (Temperature is just a measure of the average speed of molecules. Lower speed means lower temperature.)

There is a reason that the low-velocity air stays near the center of the vortex. Circles near the center are very small. Air cannot achieve much speed due to the center area becoming more and more like a point. At the exact center, all velocities are zero. Another way to say this is that near the center, each circle of the airstream consists of opposite flows on opposite sides of the circle. These opposite flows cancel each other.

The vortex system is particularly useful where both fresh air and cooling are desired at the same time. A large quantity of compressed air must be used.

A typical application of the vortex tube principle is in the cooling of protective suits used by industrial workers who must do their jobs in bad atmospheres or in very hot places.

Suits cooled by vortex tubes are used when a worker is required to wear heavy protective clothing, such as when sandblasting, cleaning a brick building, etc. In most cases, devices of this kind are needed only when the worker is performing a particular job.

Vortex tube devices are designed to operate on a continuous basis. No thermostatic control of any type is used. However, there is a manual control that allows the vortex tube operator to adjust outlet air temperature as needed.

19-23 EVAPORATIVE COOLING

Evaporative cooling is often used in the summertime in bright sunshine when relative humidity is low. For example, evaporative cooling can maintain a safe temperature for plants growing in greenhouses. See Fig. 19-23A.

One end of the greenhouse has a lattice of fibers such as excelsior (fine curled wood shavings). Water pipes with small holes are along the top of the lattice. The water from the small holes flows downward, wetting the lattice material. A fan at the other end of the greenhouse draws air through the lattice and discharges it outdoors.

Due to the evaporation of the water on the lattice surfaces, the air entering the greenhouse will be cooler than the outside air. Since plants do best under conditions of high relative humidity, this type of cooling is ideal. It provides a high relative humidity as well as a lower temperature inside the greenhouse.

Foundries sometimes use this system where a few degrees of cooling are desirable. In such installations, the evaporation takes place in a structure on the roof of the building. The cooled air is brought down into the work area.

It is also possible to provide a shallow pool of water, 2 or 3 in. deep, on a flat roof for cooling purposes. Where

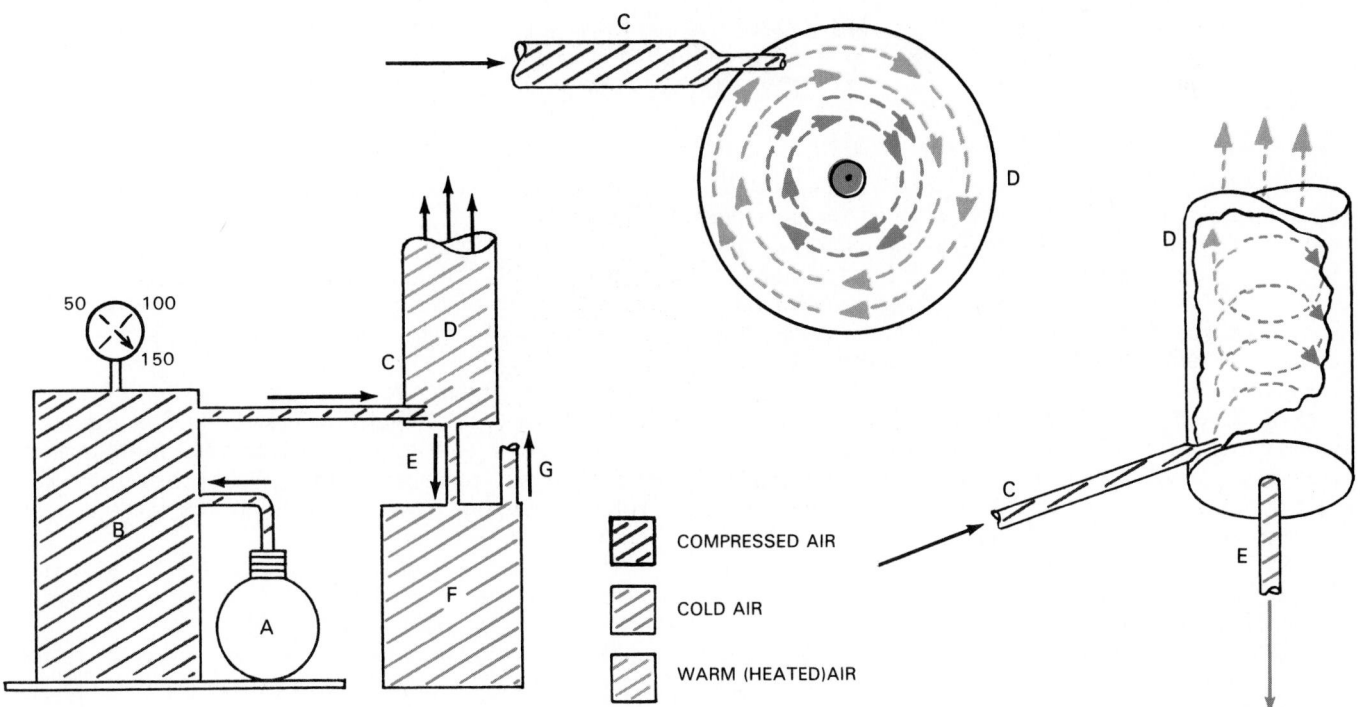

Fig. 19-22. Simplified illustration of vortex tube refrigeration. A—Air compressor. B—Compressed air storage tank. C—Compressed air nozzle. Cold air is produced at E, and flows into space F, which is to be cooled. Extra air from space F is pushed out through passageway G. Warm air from outer circles of vortex region is expelled through large tube D.

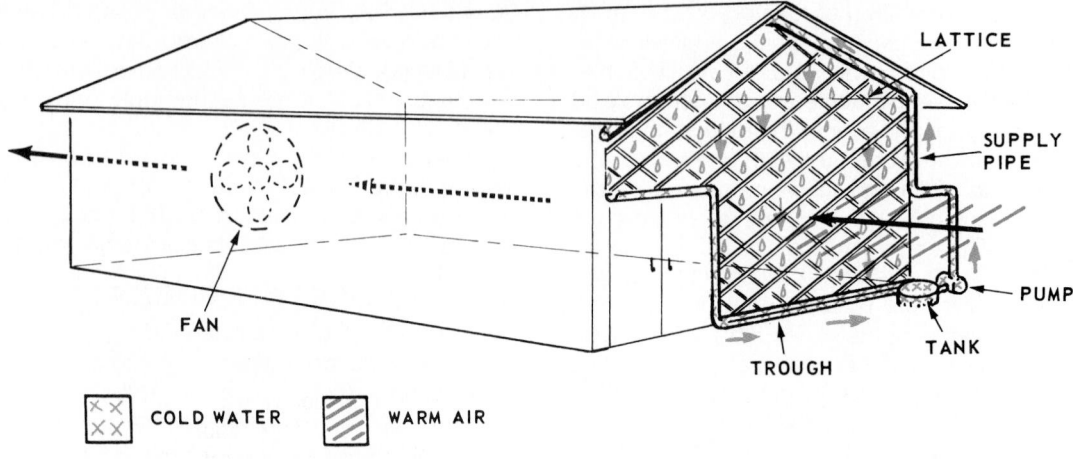

COLD WATER   WARM AIR

Fig. 19-23. Top. Evaporative cooler used as a swamp cooler for greenhouse. Bottom. Lowering the roof surface temperature through the use of a fine mist of water on the roof surface. (Sprinkool Systems, Inc.)

there is bright sunshine for much of the day, the heat load on the building may be high. If the relative humidity is low most of the time, the ponded roof will help greatly in maintaining a comfortable temperature inside the building.

Wet Roof/Roof Spraying are at times referred to as swamp cooling. This type of cooling is primarily used where ambient air temperatures are high and relative humidity is low.

Roof spraying is based on the relatively simple fact that when water is sprayed on a hot roof, it will change its state and, in so doing, will absorb a relatively large amount of heat from the roof's surface. The evaporation, and therefore cooling, will continue as long as the roof's surface temperature is greater than the wet-bulb temperature of the ambient air. The surface temperature varies, depending on the latitude, month, etc.

There are numerous methods of wet roof cooling. Some systems are designed so that the roof is a trough or pond, and the entire surface remains wet and the water is replaced as it changes phases.

Another method is to spray the roof with fine droplets of water, Fig. 19-23B. This brings the surface temperature of the roof close to the wet-bulb temperature of the am-

bient air above it. This will reduce the heat gain through the roof.

Various controls in the system monitor the roof surface temperature and control the amount of water that is sprayed on the roof so that it can be evaporated from the roof according to its temperature at that time. This coordination of the amount of water used with the roof temperature and ambient air temperature will allow the water to evaporate without any type of overflow or stagnation of the water.

The major intent of a roof cooling system is to maintain a roof surface temperature that will be close to the wet-bulb temperature. This will, therefore, minimize the heat flow through the roof.

The higher the temperature, the greater the amount of water vapor that can exist with air in the atmosphere at any relative humidity.

The evaporization will continue even when the relative humidity is 100%, provided that the heat source (roof) that is in contact with water is higher than the wet bulb of the air/vapor combination adjacent to the surface.

The evaporative roof cooling can be used as a single system or as part of a mechanical air conditioning system or energy management program.

19-24 RADIANT HEATING

Radiant heating provides a very comfortable living environment with little or no equipment in sight. The most common type consists of electric heating wires embedded in the plaster of ceilings or walls, in the floor of a room, or in some combination of the three. Fig. 19-24 illustrates a typical radiant heating installation.

With this installation, the surface is slightly warmed. A thermostat, mounted on the wall, controls the current flow through the heating wires, thus controlling the temperature.

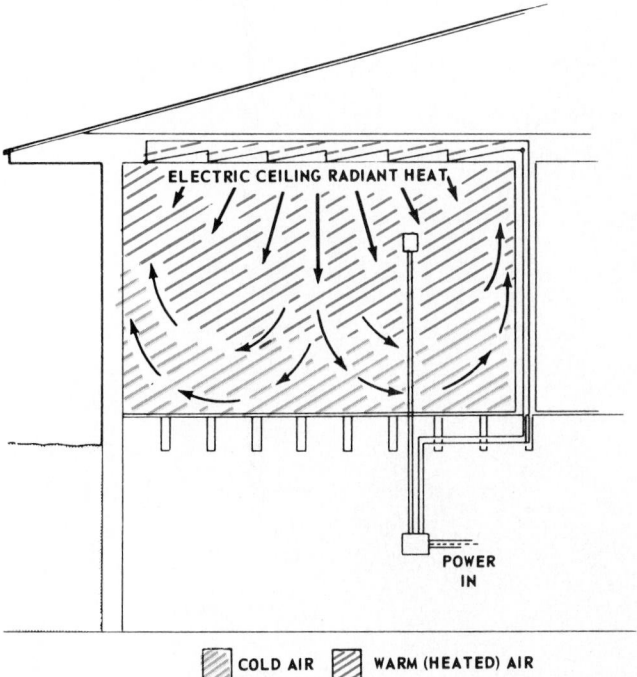

Fig. 19-24. Radiant heat is supplied by electrical resistance heating unit embedded in plaster on ceiling. Heavy insulation is needed with radiant heat installation.

The amount of heat radiated by this type of heating system can heat a room.

This type of heating works the opposite of a heat sink. (See Para. 18-45.) In the heat sink, heat radiated from the human body is lost to surrounding surfaces. With radiant heating, heat radiated from the surrounding surfaces is absorbed by the human body. The body is comfortable in ambient temperatures somewhat lower than 66 °F (19 °C).

In order to maintain these building surface temperatures, it is necessary to heavily insulate the walls, floors, or ceilings which use radiant heat. Often, radiant heat installations are supplemented by other heat sources to provide a complete heating and air conditioning system that has enough heating capacity.

19-25 REVIEW OF SAFETY

It is very important to be experienced (have some on-the-job training as an assistant to a service technician) before installing, maintaining, or servicing air conditioning systems.

It is vital to know the regional and local legal codes and regulations in force before doing any type of work on air conditioning systems.

Whenever one works with equipment which has electrical circuits, fuels, vapor, or liquids under pressure, safety is the first thing to be considered.

One should always use instruments to check the equipment. Only by knowing what the voltages are, or what the pressures are, can one be reasonably certain of one's safety.

Avoid using any ignition source where there is any chance of a fuel being present. A friction spark, an electric spark (opening or closing switches), or a flame can ignite a mixture of fuel vapors and air which could be fatal and also damaging to property.

Chapters 20, 21, 22, 23, 25, and 26 all have specific safety precautions relating to air conditioning systems.

19-26 TEST YOUR KNOWLEDGE

1. Where is steam jet refrigeration most used?
2. What is the most common application of the heat pump?
3. In automobile air conditioning, how is the compressor usually driven?
4. What is the basic principle of operation of a room air conditioner installed in a window opening?
5. In automobile air conditioning, why is it necessary to use a magnetic clutch on the compressor drive pulley?
6. What changes from a liquid to a vapor in a humidifier?
7. Will a dehumidifier, operating in a room, change the temperature within the room?
8. In a heat pump installation, is the direction of flow of refrigerant vapor reversed through the compressor when the cycle is reversed from heating to cooling?
9. What is a common application of the vortex tube cooling system?
10. Why are some furnaces known as gravity-type?
11. Can an air filter be used on a forced warm air heating plant?
12. What is a hydronic heating system?
13. What fluid action takes place to provide cooling in a cooling tower?
14. Why is an automatic draft control needed on an oil burner installation?
15. Name three common ways of heating the water in a hydronic heating system.
16. In a window or through-the-wall air conditioner, what becomes of the heat absorbed by the evaporator?
17. In a central air conditioning system, where is the condensing unit usually located?
18. In an air-to-air heat pump installation, where is the heat for warming the room obtained?
19. Why is it necessary to provide supplementary electric heating on some heat pump installations?
20. Is steam jet refrigeration recommended for use where temperatures below freezing are desired?
21. How is a constant water level maintained in a cooling tower?
22. Where are the heating elements usually placed in radiant heating installations?
23. What is the purpose of an air break?
24. What device senses whether a pilot light is on or not?
25. What is the purpose of a suction throttling valve in an automobile air conditioner?

A service technician installing a high efficienty furnace.　(Lennox International, Inc.)

Chapter 20

AIR CONDITIONING SYSTEMS HEATING AND HUMIDIFYING

After studying this chapter, the technician will be able to:
☐ Identify and explain differences in various types of heating systems.
☐ List the steps in performing a furnace inspection.
☐ Explain the service procedures for various types of heating systems.
☐ Describe the installation procedures for various types of heating systems.

20-1 TYPES OF SYSTEMS

Air conditioning systems that are able to perform all air conditioning functions are not in common use. Most systems do only part of the job. The two most difficult jobs to be performed are satisfactory air cleaning and relative humidity control.

Most so-called air conditioning systems are only partial air conditioning systems. That is, the system is designed for heating, humidifying, cleaning, and distribution; or, it is designed for cooling, dehumidifying, cleaning, and distributing. The operation of these systems usually is automatic.

20-2 TYPES OF HEATING AND HUMIDIFYING SYSTEMS

There are many types of heating equipment in use. Regardless of type, the heating source must be economical and safe. Systems which require a minimum of attention by the user are most desirable.

Sources of heat may be classified by fuels: oil, gas, and electric. Electric heat includes resistance, light (radiant), or heat pump. Wood and coal-fueled furnaces are becoming obsolete and are not covered in great depth in this text. See Para. 20-40.

The three most commonly used heat sources as applied to furnaces are shown in Fig. 20-1.

Humidifying devices provide a means of turning water into water vapor and mixing this vapor with air in the occupied space. Fig. 20-2 shows six methods:
1. Exposing a large surface of water to air being humidified.
2. Spraying atomized water into air being humidified.
3. Water tray on radiator.
4. Water tray in top of warm air furnace.
5. Wetted revolving screen in a warm air duct.
6. Space humidifier, not part of a heating system. Water is vaporized by electric heat.

Heat energy may be distributed by:
1. Circulated air through ducts.
2. Warm water thermal circulation.
3. Hydronic systems; warm water circulated by means of a pump.
4. Steam lines and radiation.
5. Electric heat radiation from electric grids or infrared lights.

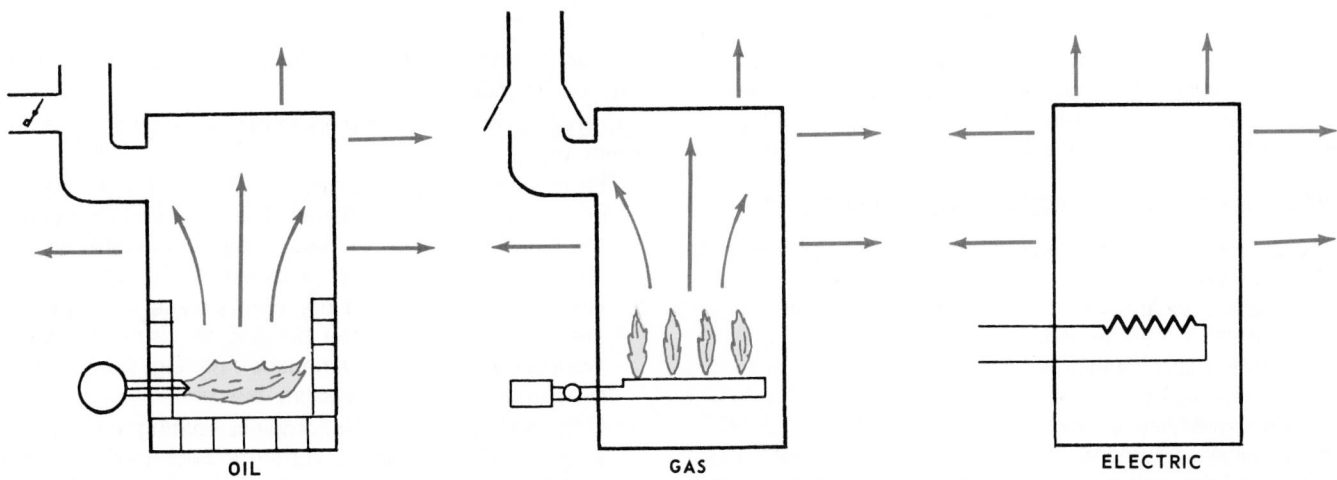

Fig. 20-1. Three popular types of heat sources used in furnaces.

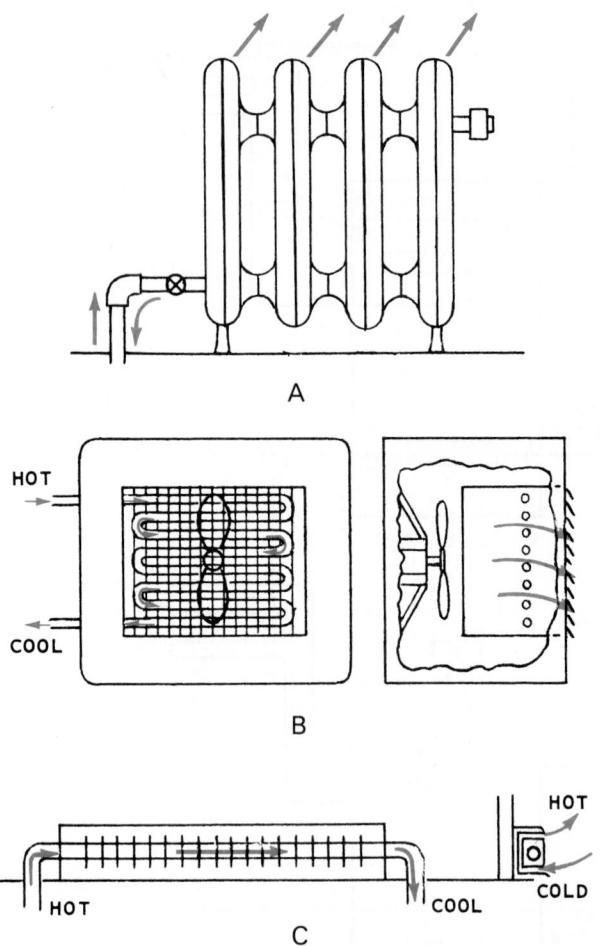

Fig. 20-4. Methods of transferring heat to an occupied space, using steam or warm water as heating medium. A—One-pipe radiator using steam or vapor heat. B—Forced air heating coil. C—Baseboard convector using warm water.

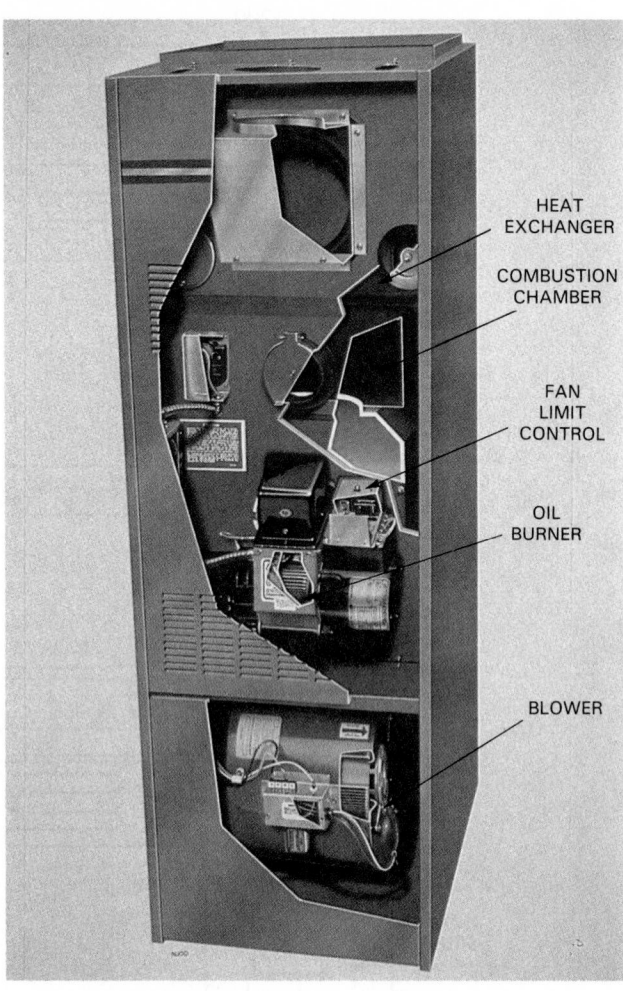

Fig. 20-6. Sectional view of warm air furnace using oil burner. Blower is multispeed and distributes warm air evenly.
(Tempstar Heating and Cooling Products, Inter-City Products Corp.)

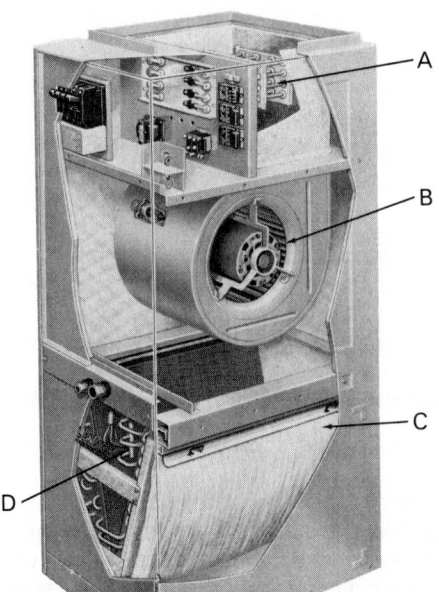

Fig. 20-5. Electrical heating and cooling unit. Electricity in heating chamber provides winter heating. Evaporator coil provides summer cooling. A—Heating element. B—Blower and motor. C—Filter. D—Evaporator coil. (Lennox International, Inc.)

following circumstances:
1. If combustion is delayed or ceases.
2. If the furnace overheats.
3. If air, steam, or water stop circulating.
4. When the heated space becomes warm enough.

The efficiency of a furnace is reduced if, during the "off" part of the cycle, the air which usually flows through the firepot to support combustion is permitted to travel into the furnace, through the combustion chamber, through the heat exchanger, through the flue, and up the chimney. This air will tend to cool these parts to room temperature; then the furnace must reheat them during the next "on" part of the cycle. Forced draft or induced draft combustion air devices reduce this loss. Examples are gun-type oil burners and forced convection (power) gas burners.

20-5 FURNACE DESIGN AND CONSTRUCTION

Furnace design is based on the fuel used and on the heat transfer medium (air, water, or steam).

Furnace construction includes a combustion chamber (except electric furnaces), a fuel feeding and burning device, a flue, and a heat exchanger. The combustion chamber

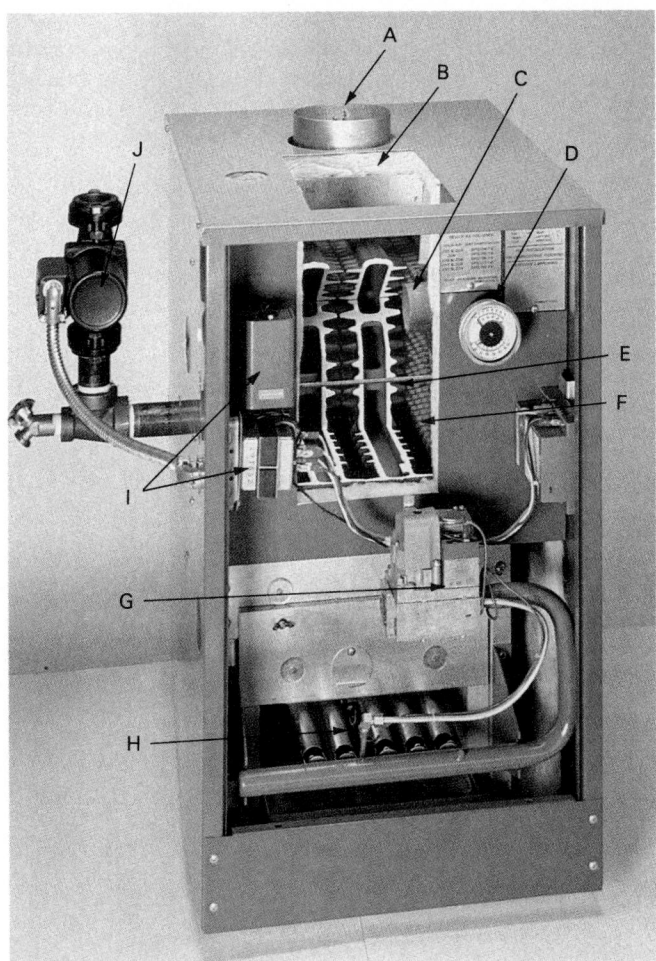

Fig. 20-7. Gas-fired hot water boiler as part of hydronic heating system. A—Flue collar. B—Insulated jacket. C—Cast iron nipple joins sections. D—Pressure-temperature gauge. E—Cast iron sections. F—Pinned heating surface. G—Gas valve. H—Stainless steel burners. I—Controls. J—Self-lubricating circulator. (Burnham Corporation)

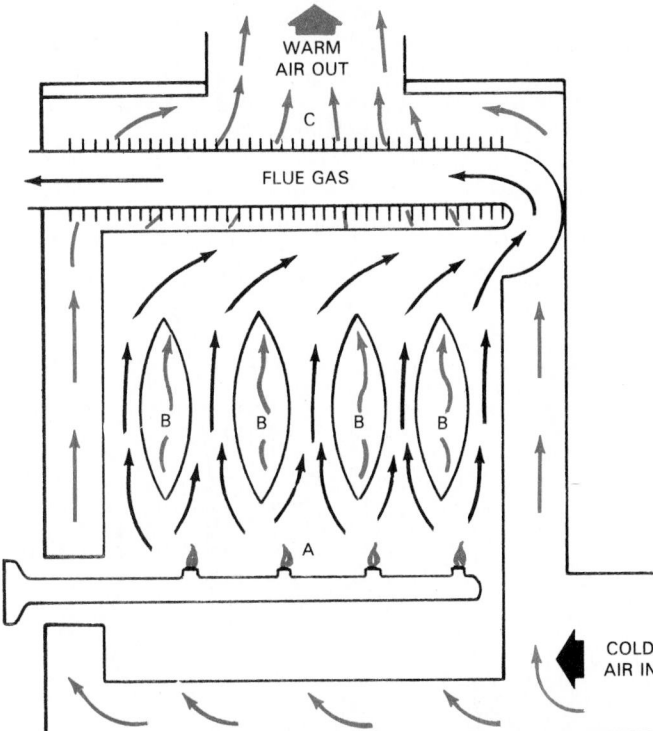

Fig. 20-8. Warm air furnace design. A—Combustion chamber. B—Clamshell-shaped heat exchanger. C—Flue heat economizer.

must be leakproof. It must provide efficient heat transfer to the air or water. It must be able to change temperatures from room temperature to almost 2000 °F (1 093 °C) with expansion or contraction stresses which may cause cracks.

The heat transfer surface must be large enough to remove enough heat from the combustion gases and from radiation to reduce the stack or flue temperature to an efficient level. The heat transfer surface usually consists of the combustion chamber surface and the heat exchanger. See Fig. 20-8. Some systems design the flue with the heat exchange surface. Some even have part of the flue installed in the cold air or cold water return to improve efficiency.

The design of the combustion chamber, heat exchanger flue, and chimney is determined by the type of furnace and fuel used.

Heat exchangers have been made of cast iron. Today, most are made of 12 to 14 gauge steel with a ceramic coating on those surfaces contacted by the combustion gases. Fig. 20-9 shows some basic designs.

Some heat exchangers are made of stainless steel. In all cases, the heat exchanger must be designed to eliminate noise as it expands and contracts. The unit must be corrosion resistant and rust resistant. Moisture from pilot light

condensation and from a comfort cooling evaporator creates rust problems. A poorly placed or maladjusted humidifer may also expose the heat exchanger to moisture. If the burner is incorrectly placed, a hot spot may form on the heat exchanger and shorten its life. Most manufacturers ripple the metal and place the welds where expansion and contraction will not cause the heat exchanger to crack and leak (very dangerous because of fumes).

Poor airflow will cause a heat exchanger to overheat. A dirty filter, loose fan belts, and/or undersize ducts will reduce airflow. Corrosive vapors in the circulated air also will increase the corrosion of furnace heat exchangers.

Furnaces are going through a great change due to technological advances and pollution reduction requirements. Most modern warm air furnaces have two-speed motors driving the blower that circulates air through the rooms: high speed for summer comfort cooling; slow speed for winter heating. Some motors are three- and four-speed units to enable the service technician or owner to select the speed wanted for the air volume (cfm) desired.

Some furnaces no longer require standard size chimneys, but use 2-in. diameter PVC (polyvinyl chloride) pipe instead. The flue gases are cool enough that the pipe is not melted. The furnace is very efficient, since little heat is lost through the flue.

The flow of combustion gases in a flue and chimney has an effect on the efficiency of a heating system. Fuel losses of 4 to 15 percent are possible, plus air pollution. Fuel may be lost up the flue without being burned.

20-6 GAS FURNACES

Gas is being used at an ever-increasing rate as a heating fuel. Natural gas is piped into the building from an underground distribution network. Usually, it is fed to the furnace at 4- to 6-in. (100 to 150 mm) water column pressure. A pressure regulator reduces this pressure to approximately 2 in. (50 mm). Note that about a 27-in. (0.7 m) water column pressure equals 1 psi.

A solenoid valve generally is used to control the main gas flow to the furnace burners. This solenoid valve may be operated by a typical household voltage. Some operate through a transformer at a reduced voltage (24 V). Some solenoid valves are operated with current generated by a

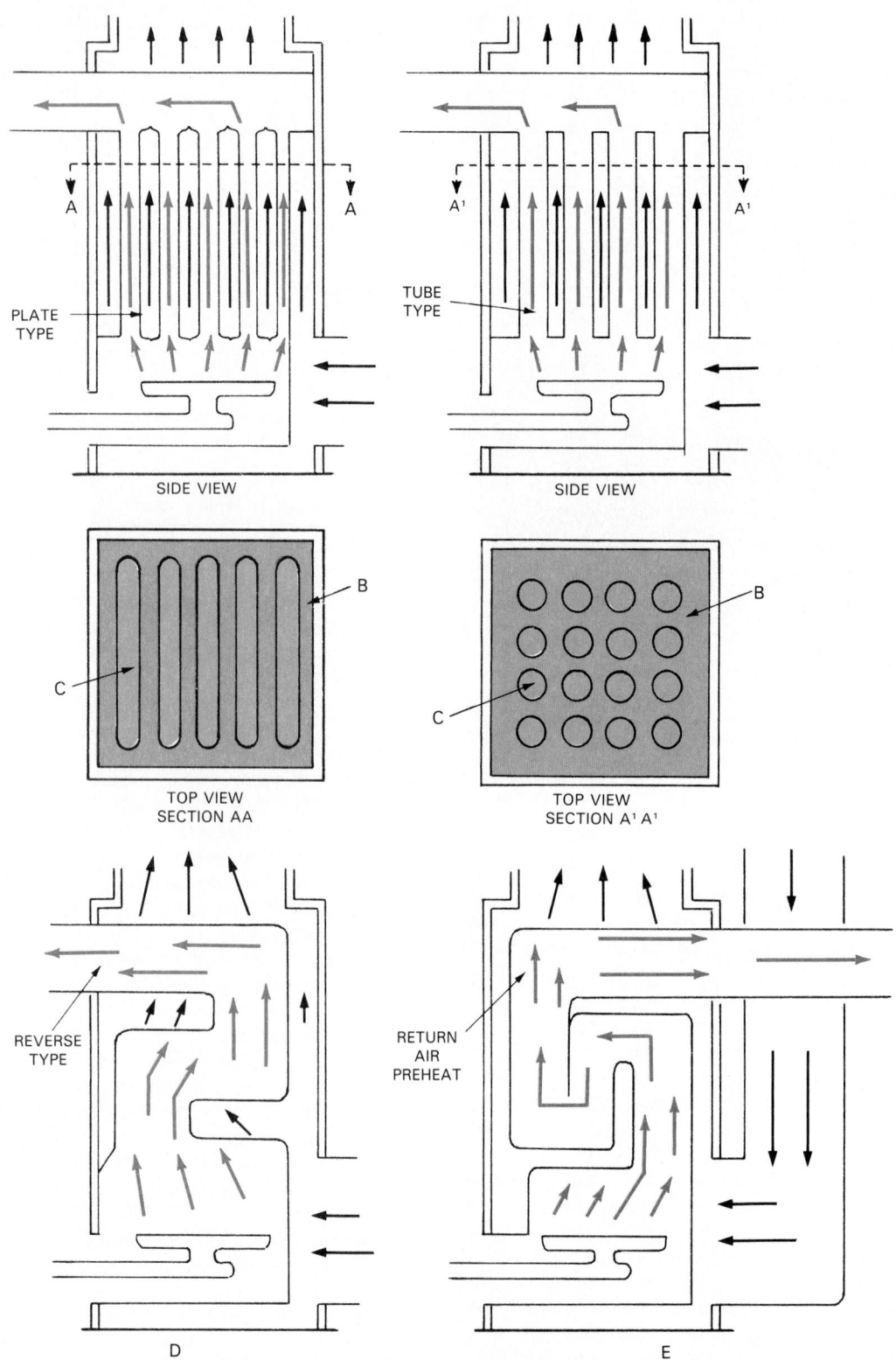

Fig. 20-9. Four types of heat exchangers for warm air furnaces. AA and A¹A¹—Section lines. B—Air. C—Flue gas. D—Flue gas direction in upper pipe is opposite to that in pipe immediately below it. Construction is less expensive than plate or tube type. E—Complex series of bends makes more efficient use of heat in flue gas than in view D.

thermocouple located near the pilot light. These do not require any connection to the house current. See Chapter 25.

The burner may be made of steel pipe or a casting. It may be either of the multiple jet or orifice type, or it may be the one-opening type with a deflector plate. An elementary gas burner is shown in Fig. 20-10. These units usually operate on the Bunsen burner principle.

A complete gas-fired boiler is shown in Fig. 20-11. Note the four burners and limit control.

20-7 HIGH EFFICIENCY GAS FURNACES

Many new furnaces are being introduced to the market that have very high efficiency ratings. Four new designs are:
1. Condensing, gas forced-air.
2. Ethylene glycol and water solution, gas forced-air.
3. Heat pipe, gas forced-air.
4. Pulse, gas forced-air.

These new furnaces have ratings between 85 and 95 percent efficiency. This compares with conventional ratings of between 60 and 65 percent efficiency.

There are some principles that are common to many of the new furnace designs. The most basic is recycling. Recycling has always been associated with energy conservation, whether one is referring to the recycling of aluminum cans, the recycling of waste exhaust heat in an automobile turbocharger, or the recycling of furnace combustion gases. (Note: most diesel engines and some gasoline engines use a turbocharger.) In many of the new furnace designs, efficiency is gained by passing the flue gas through one or more additional heat exchangers to use as much latent (from condensing the flue water vapor) heat as possible.

Fig. 20-12 illustrates a maximum efficiency furnace. Fresh air from the sealed combustion system enters the burner module, where ignition takes place. The hot gases

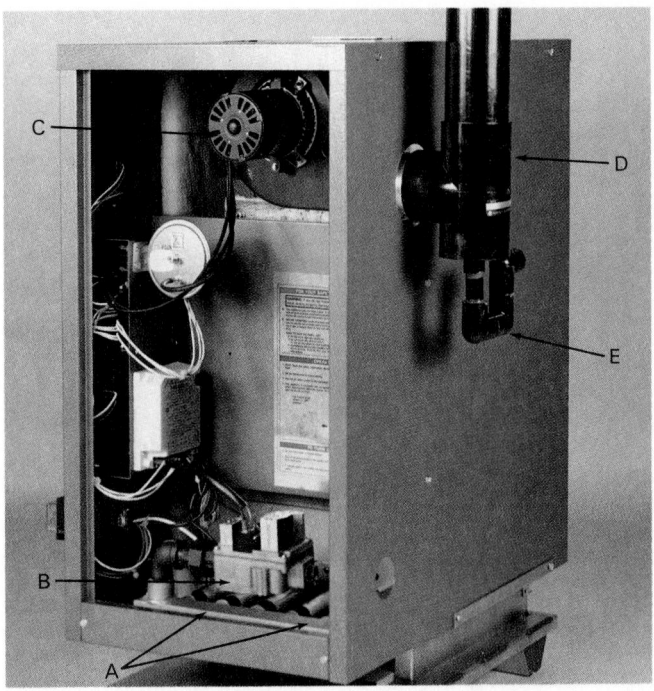

Fig. 20-11. Four-burner gas-fired boiler. A—Burner. B—Gas valve. C—Fan. D—Vent pipe. E—Condensate drain trap. (The Peerless Heater Company)

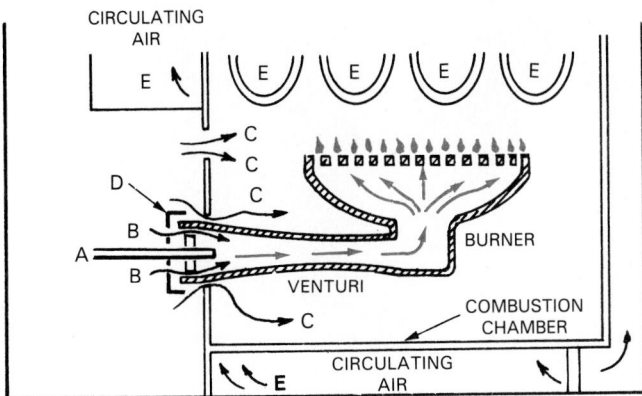

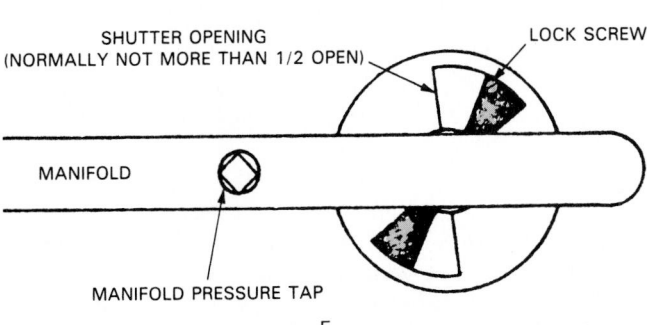

Fig. 20-10. Schematic of gas burner installed in warm air furnace. A—Fuel gas. B—Primary air. C—Secondary air. D—Primary air adjustment. E—Circulating air for heating. F—How to adjust primary air. Shutter is on end of venturi tube.

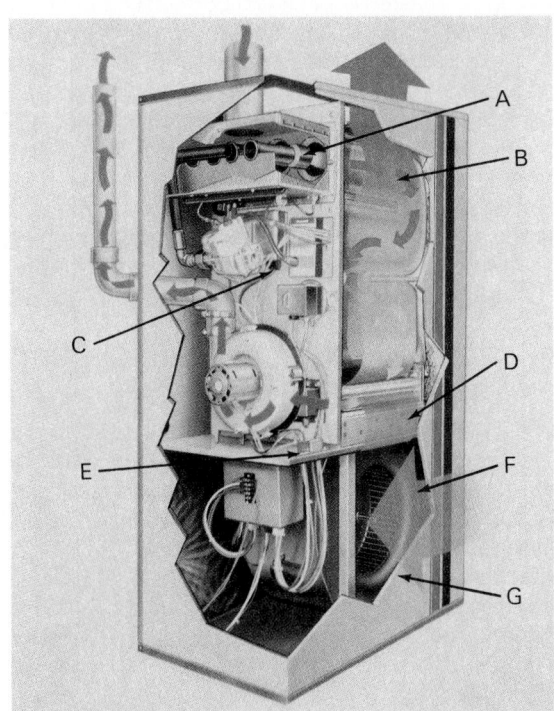

Fig. 20-12. High efficiency gas furnace. A—Sealed combustion—fresh air carried from outdoors to furnace heat exchanger. B—Stainless steel primary heat exchanger. C—Safety controls. D—Condensing coil. E—Plug-in controls. F—Fan. G.—Fiberglass insulation to retain heat inside furnace.   (Yukon Energy Corporation)

are drawn through heat exchanger and condensing coil. Cooled exhaust gases are vented outside. The circulating air blower forces air from the return air ducts to the coil and up alongside the heat exchanger into the warm air ducts.

A second method commonly used to increase furnace efficiency is the use of a pulse combustion process in place of the continuous combustion that is found in conventional furnaces. The repeated ignition of the gas and air mixture (using a spark plug or residual heat as the igniter) results in a more efficient, hotter combustion. See Para. 20-11. This combustion process usually operates at about 60 cycles per second. These furnaces also use the resulting pressure pulses to assist in the movement of the combustion gases, resulting in additional efficiency gains.

Efficiency gains are possible with careful use of flue heat. The following paragraphs detail the operating characteristics of four common high efficiency furnaces.

20-8 CONDENSING FURNACE

The basic concept of a condensing gas forced-air high technology furnace is to extract the latent heat that is lost in a conventional gas forced-air furnace. The latent heat of the water vapor in the flue gas is used.

The condensing furnace produces a cool exhaust gas and does not require a chimney, as is necessary in conventional gas furnaces. For combustion, outside air enters through a polyvinyl chloride (PVC) pipe. The pipe may be run through either a side wall or the roof.

The condensing gas furnace, Fig. 20-13, uses a gas flame, A, to heat the basic heat exchanger.

The combustion gas blower, C, draws the hot combustion gases through the primary heat exchanger, F. Instead of allowing hot gases produced by combustion to be vented out the chimney, the hot gases are drawn down by the blower into a secondary heat exchanger, E, where the latent heat is extracted.

When air is forced across the primary and secondary heat exchanger by the circulating room air blower, D, the air picks up the heat that would normally have been lost in a conventional gas furnace. The hot air, created by the secondary heat exchanger, is then circulated through the furnace duct system. The exhaust gases, low in temperature, are vented through a PVC pipe attached to the exhaust vent, B.

20-9 ETHYLENE GLYCOL WATER SOLUTION FORCED-AIR FURNACE

The ethylene glycol water solution forced-air furnace is different from other forced-air furnaces. The difference is that the glycol unit operates with a glycol water solution that is heated and circulated through a system. The system contains four significant parts:
1. Heat transfer module.
2. Ethylene glycol water circulating pump.
3. Primary coil.
4. Recuperative coil.

The furnace, Fig. 20-14, runs as follows: If thermostat calls for heat, electricity goes to heat transfer module, 1. A transfer module, Fig. 20-15, is a ceramic igniter, a burner, and safety controls. A module has a 50 percent glycol and 50 percent distilled water solution that passes through the circular tube and fin structure within the heat transfer

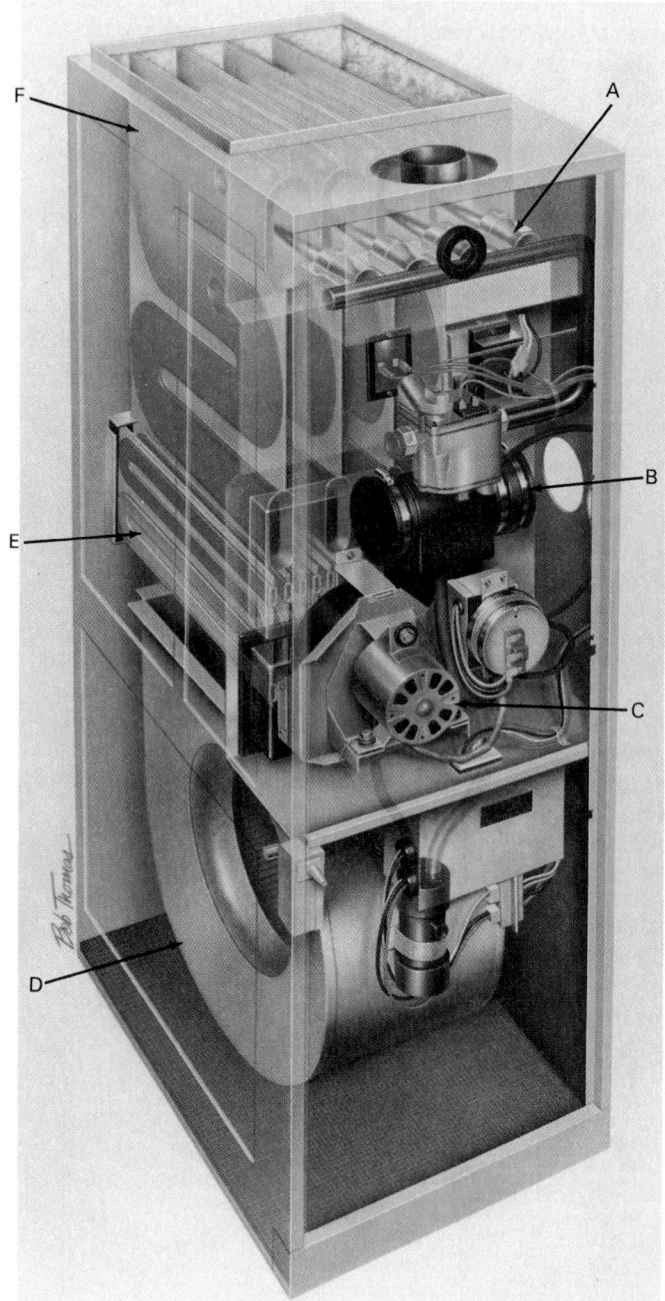

Fig. 20-13. Sectional view of condensing furnace. Combustion products are 85 to 90 percent condensed. A—Burners. B—Exhaust vent. C—Combustion gas blower. D—Circulating room air blower. E—Secondary heat exchanger. F—Primary heat exchanger. (BDP Company)

module. This provides more efficient heating. When operating, electricity passes through the ceramic igniter, heating the tip to 2500°F (1 400°C). The combustion blower mixes gas and air, blowing it into the stainless steel burner inside the heat transfer module. As gas and air pass over the hot igniter, ignition occurs and combustion begins.

Intense heat is rapidly absorbed by hundreds of steel fins surrounding the burner. The solution of glycol and distilled water is circulated through the tubing in the steel fins. The glycol and distilled water solution absorbs heat from the fins

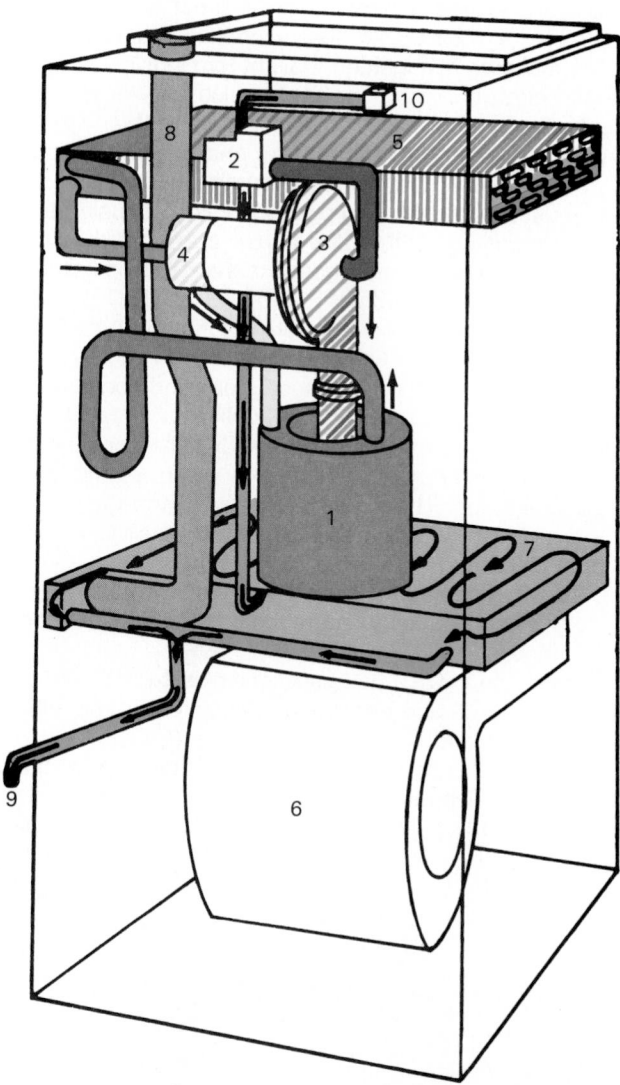

Fig. 20-14. Schematic drawing of ethylene glycol water solution forced-air furnace. (Amana Refrigeration, Inc.)

and leaves the heat transfer module at about 180 °F (80 °C). A solution pump, 4, circulates the hot solution at atmospheric pressure through a copper tubing and aluminum fin heating coil, 5. As this coil is heated by the solution, a blower, 6, moves cool air through the coil. The heated air is circulated through the furnace duct system. The cooled glycol and distilled water solution is pumped from the coil back to the heat transfer module and the cycle is repeated, until the set temperature is reached.

The exhaust gases from the heat transfer module are circulated through a recuperative coil, 7. The furnace blower moves cool return air through the warm coil, cooling the gases inside the recuperative coil. This cooling causes the water vapor in the exhaust gas to condense, releasing its latent heat.

The exhaust gas, 8, leaves the furnace through a plastic vent pipe at 105 to 115 °F (41 to 46 °C). The condensate water is drained from the furnace, 9. The water is mildly acidic and must be drained. At the end of the heating cycle a water valve, 10, opens, and water automatically flushes away any remaining condensate.

20-10 HEAT PIPE GAS FORCED-AIR FURNACE

The heat pipe gas furnace, Fig. 20-16, consists of sealed pipes which are nearly vertical. The furnace has a power burner that mixes air and gas. The mixture is forced down through a multiport burner head into the combustion

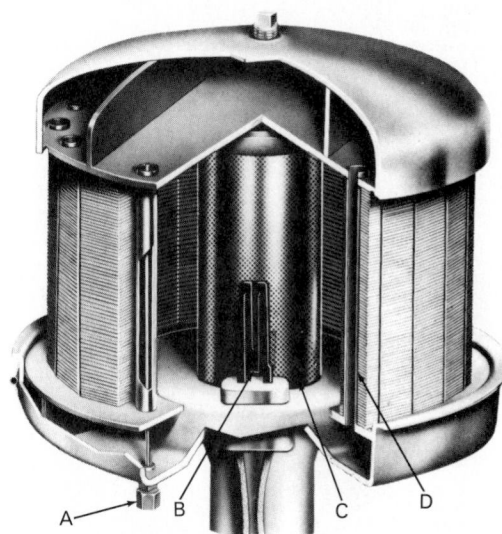

Fig. 20-15. Heat transfer module (HTM) that is central part of ethylene glycol water solution furnace. A—Limit control. B—Ceramic igniter. C—Burner. D—Solution tubes. (Amana Refrigeration, Inc.)

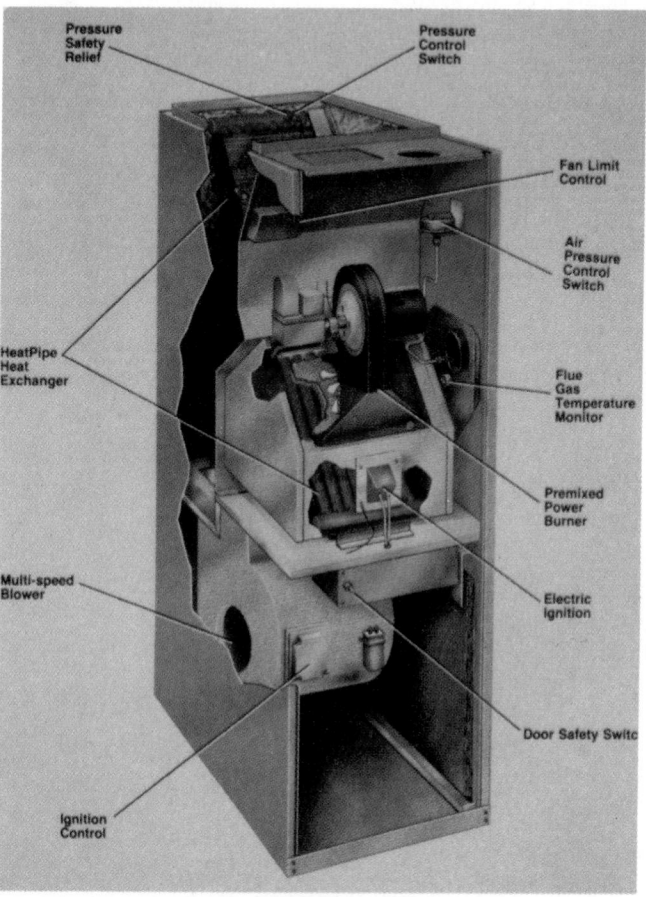

Fig. 20-16. Sectional view of heat pipe furnace. Notice location of various components.
(Central Environmental Systems, York International Corp.)

chamber, where an electric igniter lights it. The lower end of the heat-pipe heat exchanger is within the combustion chamber. The heat from the burning gas-air mixture heats the liquid inside the heat pipes. The burner forces exhaust gases out. Therefore, the heat pipe gas furnace does not require a natural draft effect as in a conventional chimney. The liquid in the heat pipes boils and the vapor rises within the heat pipes. The furnace blower circulates return air over the upper end of the heat pipe heat exchanger. The air absorbs heat from the heat pipes, cooling the vapor, and the vapor releases its latent heat into the circulating air. The warmed air is circulated to the house through the duct system. The condensate flows by gravity to the low end of the heat pipes and the cycle repeats itself.

20-11 PULSE COMBUSTION FURNACE

The pulse combustion process was first discovered in the 1900s, but is recent in the heating industry. The concept is used in forced air condensing furnaces, Fig. 20-17. The concept is different than that in the conventional atmospheric burner furnace. The pulse furnace does not have an open flame, pilot burner, main burner, or conventional flue or chimney.

In a furnace using the pulse combustion process, combustion air that is 100 percent from the outdoors is brought into the unit through a 2-in. PVC pipe. The combustion is

started in an enclosed chamber from a direct spark ignition device. See Fig. 20-18. A sensor checks if ignition has begun and allows five ignition trials before closing the gas valve and control circuit. The sensor also checks for loss of combustion and will shut down the system. Many other safety sensors are part of the intake and exhaust outlets.

After ignition, the heat from combustion passes through the combustion chamber, tail pipe, exhaust decoupler, and heat exchanger coil. All of the above are located in the system's airstream. In the process, exhaust temperatures drop from 1200°F to 350°F (650°C to 180°C). As exhaust gases are forced through the fin-and-tube heat exchanger, water vapor is condensed out to recover the latent heat of combustion. Exhaust gases and condensate at temperatures from 100°F to 120°F (40°C to 50°C) are vented into a plastic "T" connection. The condensate exits from one side of the "T" into a 1/2-in. plastic condensate drain pipe. The exhaust exits from the top of the "T" into a 2-in. PVC pipe that is vented outside. A conventional chimney is not required.

Fig. 20-19 illustrates the basic pulse combustion process:

1. The gas and air supply enters the combustion chamber and mixes.
2. To start the cycle, the spark igniter is turned on and ignites the gas and air mixture. This creates the initial combustion and is referred to as one "pulse."
3. The positive pressure from the resulting combustion closes the flapper valves and forces exhaust gases through a tail pipe. These combustion products are

Fig. 20-17. Pulse gas furnace upflow model. A—Tail pipe. B—Combustion chamber. C—Condenser coil. D—Flue vent. E—Outdoor air intake.   (Lennox International, Inc.)

Fig. 20-18. Pulse combustion process heat exchanger assembly. A—Gas intake. B—Air intake. C—Spark igniter. (Lennox International, Inc.)

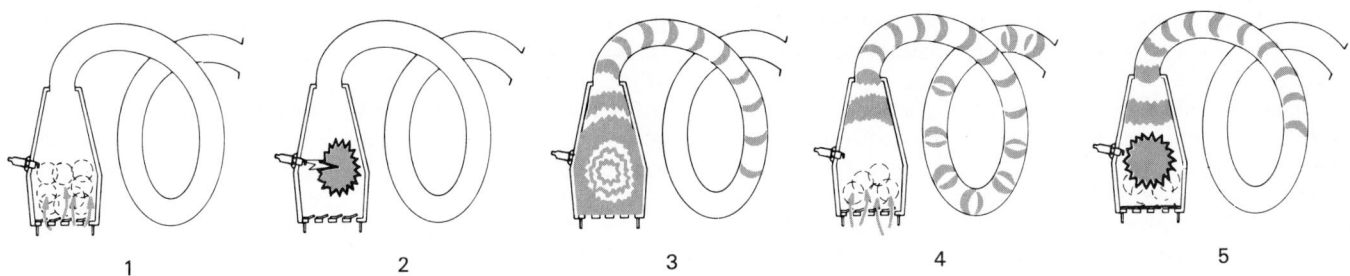

1   2   3   4   5

Fig. 20-19. Pulse combustion cycle.   (Lennox International, Inc.)

vented outdoors through a 2-in. PVC pipe installed vertically or horizontally.

4. The venting of the exhaust gases creates a negative pressure in the chamber. This opens the flapper valves, drawing in more gas and air for the next ignition.

5. At the same time, some of the pressure pulse is deflected back from the tail pipe. This causes the new gas and air mixture to ignite. This is referred to as another "pulse." After the first few seconds, the spark igniter and air blower are turned off, because the combustion process is self-sustaining. No spark is needed.

Steps 4 and 5 are repeated 60 to 70 times per second. This forms consecutive "pulses" of 1/4 to 1/2 Btu each.

Before and after each heating cycle, a small blower purges (flushes out with air) the combustion chamber. This provides fresh air for the next mixture of air and fuel gas.

The units are designed for use with natural gas or LP-Gas. Para. 20-14 discusses mixing ratios for natural gas and LP-Gas.

20-12 VENTING OF FURNACES AND CHIMNEY OR EXHAUST GASES

The flow of combustion gases in a flue and chimney has an effect on the efficiency of a heating system. Fuel losses of 4 to 15 percent are possible. Combustion gases leaving the chimney can also contribute to air pollution.

The flow of combustion gases affects the amount of air entering the furnace for combustion purposes. This flow is affected by the difference in pressure between the combustion air entering a furnace and combustion air leaving the flue or chimney. Both the pressure in the building and atmospheric pressure affect the flow of combustion gases. The temperature of the combustion gases also has an effect (too cold—slow flow, too hot—fast flow).

In order to remove the exhaust gases (the products of combustion) from a conventional furnace, it must be vented. This vent system consists of a flue pipe which joins a chimney that is vented to the outside atmosphere. The flue pipe reaches a temperature of between 500 to 600 °F (260 to 320 °C). Therefore, 30 to 40 percent efficiency of the furnace is exhausted out the chimney.

In the new high technology furnaces, the conventional chimney is not needed. The flue gases leave the furnace at 115 to 118 °F (46 to 48 °C) as a result of the high efficiency heat exchanging processes. Exhaust gases are vented to the outside through a plastic (polyvinyl chloride or PVC) pipe. Outdoor air is brought into the combustion chamber through a PVC pipe. See Fig. 20-20. Instruments needed to determine the efficiency of a furnace are covered in Chapter 22.

A draft control device is used to keep the flue pressure constant. With this device, it is possible for a heating system to maintain proper combustion air, even with changes in atmospheric pressure. The pressure may change in the building if the flue temperature changes or if the wind changes the flue pressure dynamically. The draft control improves the flow of flue gases by varying the amount of dilution air which goes into the flue as the flue pressure tries to change. A design of a draft control is shown in Fig. 20-21. Fig. 20-22 shows a draft control for an oil-burning furnace.

Combustion gas vent pipes (flues) should always be the same size as the furnace vent opening. The horizontal length should not exceed 20 ft. (6 m), and it is best to have it as short as possible. Any horizontal run should slant up (pitch) toward the chimney 1/2 in. for each foot of pipe (42 mm for each meter), to improve flue gas flow. Minimize the use of elbows, since they add resistance to flow.

The chimney or vent pipe should extend at least 2 ft. (0.6 m) above the highest part of the roof. This helps

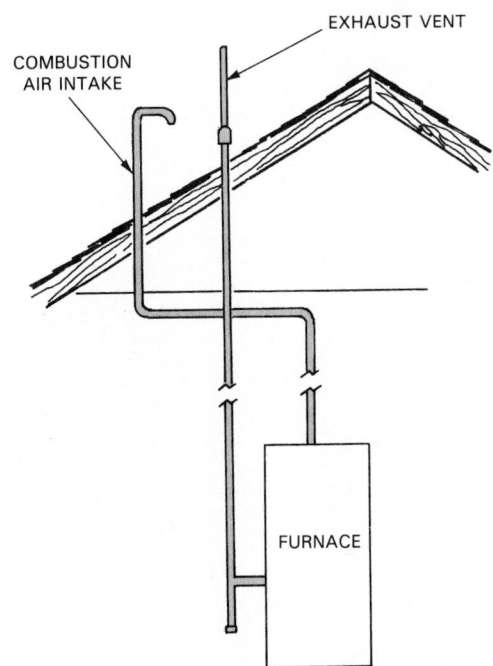

COMBUSTION AIR INTAKE

EXHAUST VENT

FURNACE

Fig. 20-20. Location of vent and intake for high technology furnace. Combustion air intake and exhaust vent may be installed vertically or horizontally through a wall.

prevent back pressure in the chimney under certain wind conditions. The taller the pipe, the better the draft (upward flow).

A pressure differential occurs in a chimney due to the large change in temperature between the combustion gases and the outside air. A typical chimney with a 450 °F (230 °C) temperature average, and 40 °F (4 °C) outside temperature will produce 0.14 in. of water draft pressure if the chimney is 30 ft. (9 m) high (measured from floor of furnace room).

The draft control must be located so that the flue gases will not flow against it. See Fig. 20-23. All controls must be located between the furnace and the draft control.

Gas furnaces use an air break to control flue gas flow and to prevent back pressure from a gust of wind from reaching the furnace flame or pilot light. Fig. 20-24 shows an air break installed on a gas furnace. Some installations

use a draft diverter on the chimney. Fig. 20-25 shows a diverter for a round stack. A gas furnace control used on some furnace installations is shown in Fig. 20-26.

For higher efficiency, vent dampers for gas furnaces are often used. They are designed to restrict or block the opening during the time the burner is not operating. This is done

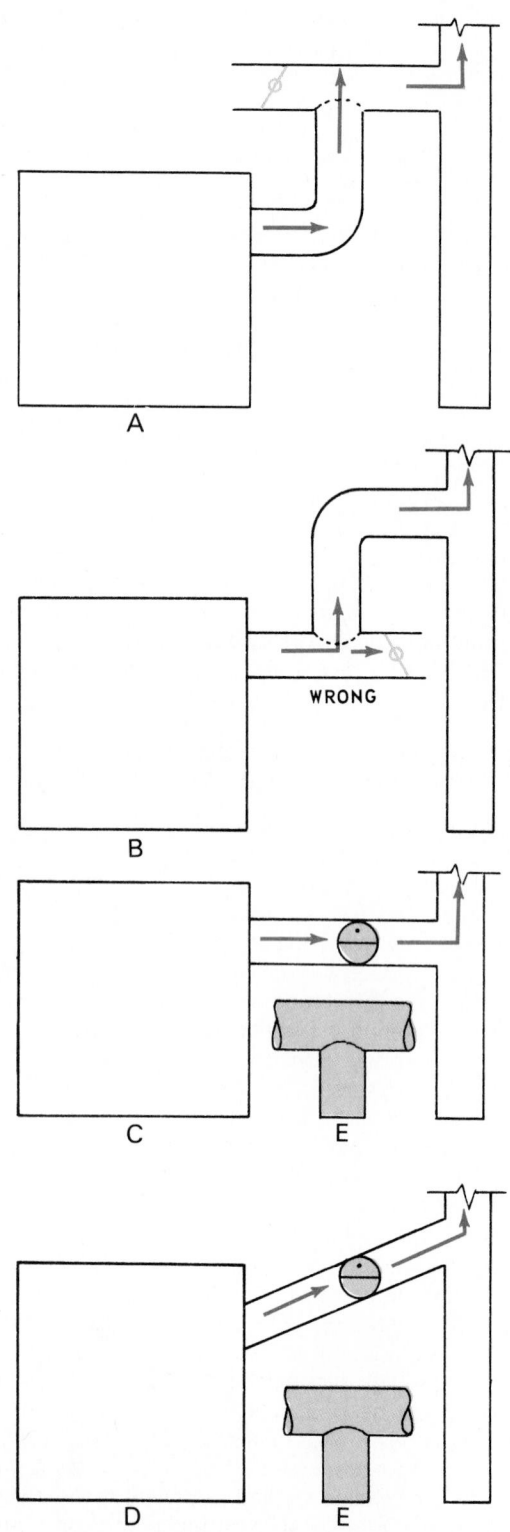

Fig. 20-23. Locating draft control in flue. A—Good location. B—Wrong location (hot gases flow against draft control). C and D—Designs for oil furnaces. E—Top views of C and D.

Fig. 20-21. Schematic of draft regulator as used on coal, coke, and/or oil furnaces. Valve will open if draft tends to increase and will close as flue stack draft decreases. This action keeps a constant draft in the furnace.

Fig. 20-22. Draft control for oil- or coal-fired furnaces and boilers. A—Adjustable weight for low, medium, or high draft. B—Gate. (The Field Controls Co.)

by installing the damper between the draft hood and vent pipe leading to the chimney, Fig. 20-27. The damper can be installed in a vertical or horizontal position.

The mechanical vent damper, Fig. 20-28, operates on a bimetal expansion concept. The four bimetal quadrants will remain closed when the furnace or blower is off and there is no heat being generated. When the thermostat calls for heat and the burner is ignited, the hot flue gases cause the bimetal quadrant to flex to an open position. This allows the exhaust gases to go up the vent damper and out the chimney. When the burner shuts off, the bimetal quadrants return to the closed position.

Fig. 20-29 illustrates an electric draft control vent damper that prevents warm air from going up the chimney when the furnace or blower is off. When the thermostat calls for

Fig. 20-26. Gas furnace draft control mounted on furnace flue. A—Amount of flue draft is adjusted by washer weights on chain. (The Field Controls Co.)

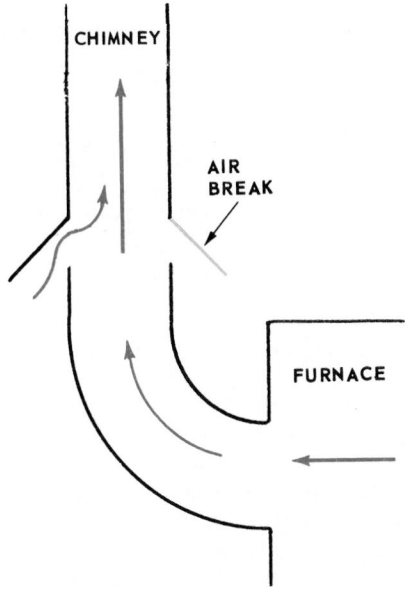

Fig. 20-24. Air break system used on gas furnace flues (stacks) to maintain constant draft in furnace.

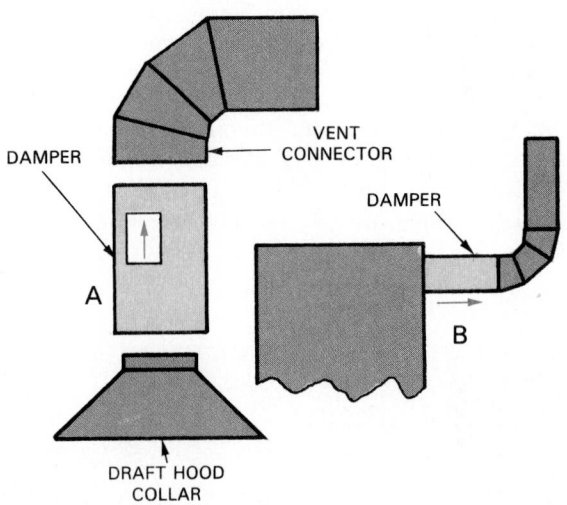

Fig. 20-27. Vent damper installation. A—Vertically between draft hood collar and vent collar. B—Horizontally between furnace and vent connection. (American Metal Products Co., Div. of MASCO Corp.)

Fig. 20-25. Draft diverter designed for use on gas furnace stack. (Excelsior Mfg. & Supply Corp.)

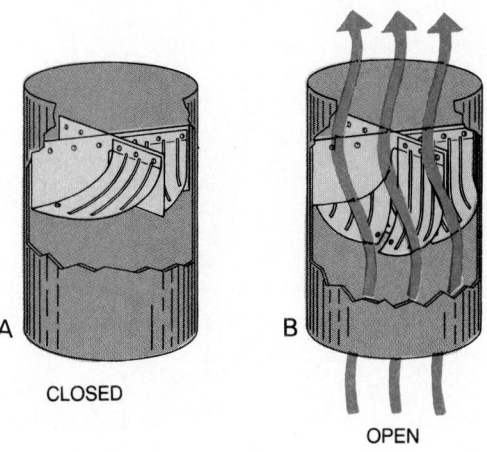

Fig. 20-28. Bimetal quadrant draft control. A—Closed position. B—Open position. (American Metal Products Co., Div. of MASCO Corp.)

heat, the damper blade moves to the open position. When the heat demand has been satisfied, the damper blade returns to the closed position.

Always make certain that there is no leak from the furnace flue or chimney into the building. Carbon monoxide is deadly, and carbon dioxide reduces oxygen in the house. Check the flue draft control to be sure the air is entering the flue. One safe way to do this is by hanging thin strips of light aluminum foil near the flue opening. If the foil bends into the flue, the draft control is good. If the foil bends away from the flue, the gases are entering the house. Air out the building at once. Determine the cause of the backflow (partially blocked chimney, etc.). Repair at once.

If the flue gas flow is too slow, a flue or smoke-pipe booster fan can be installed near the chimney. These fans run only when the furnace is operating.

20-13 FUEL GASES

There are three types of fuel gases:
1. Natural.
2. Manufactured.
3. Liquefied petroleum (LP-Gas).

Natural gas is obtained from gas deposits in the ground. Manufactured gas is made by distilling or cracking coal or oil, and by other processes. Natural gas consists of about 84 percent CH_4 (methane) and 16 percent C_2H_6 with a heating value of 1000 to 1100 Btu/cu. ft. To burn 1 cu. ft. of natural gas, 8 cu. ft. of air is required. However, some excess air is needed, so about 11 cu. ft. of air is used for each cu. ft. of natural gas (30 percent excess air).

Fig. 20-29. Electric draft control with damper in open position. Note blade position indicator on shaft.
(Johnson Controls, Inc.)

As the gas burns, it yields about 1 cu. ft. carbon dioxide, 1 cu. ft. nitrogen, 2 cu. ft. of water vapor and about 25 to 50 percent excess air. If there is insufficient primary air (refer to Para. 20-14), the burning will produce a yellow flame (incomplete combustion). If there is too much primary air, the flame will be noisy, and it will jump around above the burner. If the stack has CO (carbon monoxide), more primary air or secondary air is needed (secondary air is usually fixed).

Manufactured (coal) gas varies in content. It contains about 50 percent H_2 (hydrogen), 8 percent CH_4 (methane) and other gases. It has a heating value of approximately 500 to 600 Btu/cu. ft.

Liquefied petroleum gas (LP-Gas) usually is propane with a little butane added. It can be liquefied, stored, and transported in cylinders or tanks. LP-Gas vaporizes easily and is changed into its gas form before it is burned. It has a heating value varying between 2500 and 3200 Btu/cu. ft.

Propane boils at $-40\,°F$ $(-40\,°C)$ at atmospheric pressure. Butane boils at $32\,°F$ $(0\,°C)$ at atmospheric pressure. It is fed to the burner at about 11-in. water column pressure.

Propane and butane are dangerous if carelessly used. Both are heavier than air and will collect in the firebox and in the basement. All these gaseous fuels have an odor added. If an odor is detected, the gas is present. If gas odor is found, again, shut off the main fuel valve and thoroughly vent the area.

20-14 GAS BURNERS

Gas burners usually have a simple design. Gas is fed through an orifice and mixed with a certain amount of air (primary air). This mixture passes to the burner head, where combustion takes place and where the gas mixes with the secondary air. As much as 35 percent excess air is fed to the burner to insure thorough combustion.

There are several types of gas burners:
1. Atmospheric injection.
2. Luminous flame.
3. Power burner.

These burners are used in:
1. Unit heaters.
2. Furnaces.

Furnace-type gas burners may be used with warm air, hot water, or steam. The warm air furnace may be one of the following three types:
1. Airflow up through heat exchanger.
2. Airflow down through heat exchanger.
3. Airflow across the heat exchanger.

There must be a burner orifice correction made when the altitude is over 2000 ft. (610 m). As altitude increases, the size of the burner orifice must be smaller. There is less air and, as a result, less oxygen. Therefore, less fuel must be fed through the orifice at one time. The capacity of the furnace will decrease as the altitude increases.

The burner system consists of a manual shutoff valve, pressure regulator, automatic shutoff valve, control valve, manifold, burner spuds (nozzles) and adaptors, orifices, primary air inlet, burner head, and pilot valve.

Some gas burners feed the fuel gas and primary air mixture through a manifold:
1. To a series of holes.
2. To a series of narrow slots.

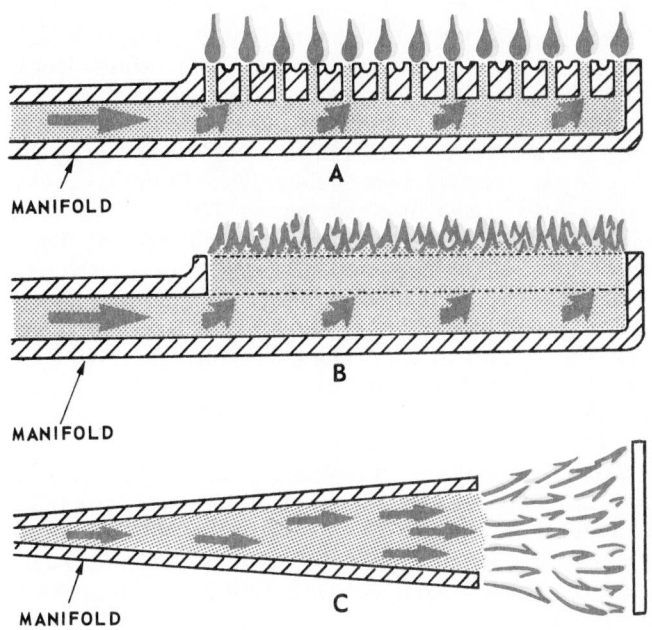

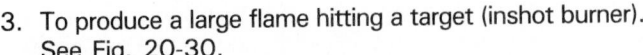

Fig. 20-30. Three gas burners of atmospheric type. A—Burner has holes. B—Burner has slots. C—Burner flame hits target plate.

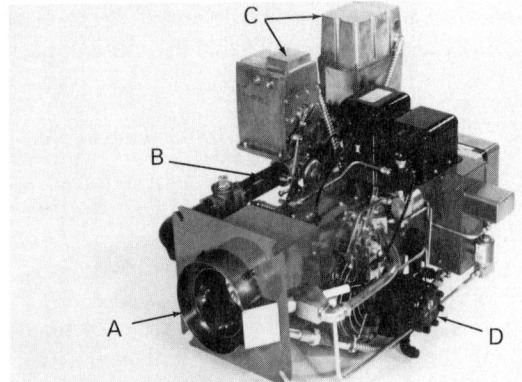

Fig. 20-32. A replacement type gas/oil burner for commercial and industrial applications. A—Burner tube. B—Gas line. C—Controls. D—Oil pump. (Midco International, Inc.)

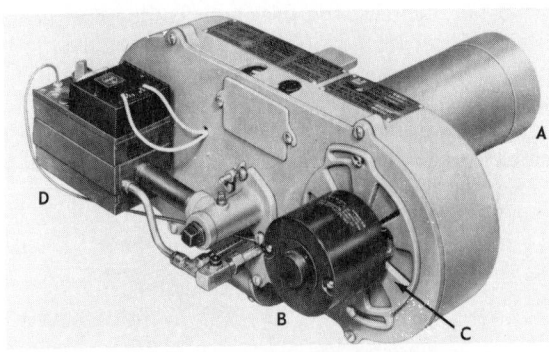

Fig. 20-33. Forced draft gas burner for furnaces. Unit is replacement for oil burner. A—Burner tube. B—Blower motor. C—Adjustable air inlets. D—Controls. (White-Rodgers Div., Emerson Electric Co.)

3. To produce a large flame hitting a target (inshot burner). See Fig. 20-30.

The power burner uses a blower to force both primary air and secondary air into the burner. See Fig. 20-31. The burner tube usually has angular vanes to spin or twirl the flame for more efficient burning.

Fig. 20-32 shows a burner which can be used for gas or oil in large commercial and industrial applications. A power conversion gas burner is shown in Fig. 20-33. A different design power gas burner is shown in Fig. 20-34.

Many commercial buildings now have combination gas furnaces and cooling systems installed on the roof. Many use fans to force or draw air through the combustion chamber. Keep these fans clean. No stack is needed. Fans reduce the chance of downdrafts. Filters are used to clean the combustion air, and they require service. Clean the heat exchange surfaces each year.

To check gas pressure to the manifold (main burner must be on), install a water manometer by attaching it to the

manifold gauge opening located on the outlet of the pressure regulator (1/8-in. pipe).

It is possible to use LP-Gas as a standby fuel source in place of natural gas if the LP-Gas is mixed with air before it is burned in the heating device. The mixing of the proper amount of air with 1 cu. ft. of LP-Gas (in vapor form) allows use of the same primary air opening, the same secondary air opening, and the same burner holes or slots. In other words, the furnace needs no adjustment if fed the air and

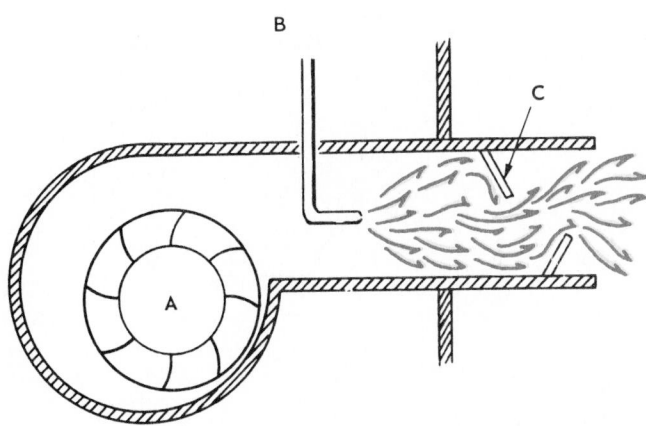

Fig. 20-31. Power burner. A—Blower air inlet. B—Fuel gas inlet. C—Deflector plates.

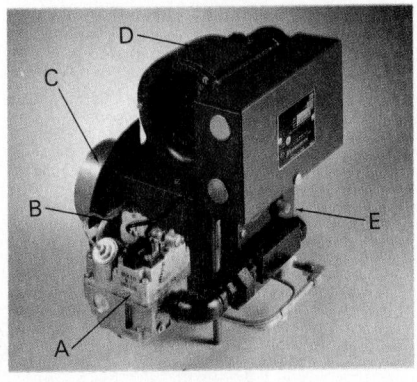

Fig. 20-34. Gas conversion replacement for oil type burner. A—Gas control. B—Gas pressure regulator. C—Nozzle. D—Blower. E—Electronic Control. (Midco International, Inc.)

LP-Gas mixture. The final air and LP-Gas mixture should have about 1000 Btu/cu. ft. to be usable in a natural gas furnace.

20-15 PILOT LIGHTS AND ELECTRIC IGNITION

A gas furnace is designed to heat a building to a comfortable temperature regardless of the outdoor temperature. Usually, the gas furnace must be shut off at certain times to prevent overheating. The controls required are explained in Chapter 25.

When the main gas flame is shut off, some device must be used to ignite the fuel gas and air mixture when heat is needed. Two devices are used:
1. Pilot light.
2. Electric ignition.
   a. Spark.
   b. Glow coil.

Pilot lights keep burning. Then, when the thermostat turns on the heating system, the pilot light ignites the gas. See Fig. 20-35. The pilot light obtains its fuel gas from a tube attached to the combination pressure regulator and gas control valve.

The pilot light is also equipped with a safety device, either a thermal fluid bulb or a thermocouple. The thermal bulb allows the main valve to open only if the pressure is high enough (heated by the pilot flame). This safety device generally is connected to the main gas regulator.

The thermocouple unit operates when the thermocouple generates electricity. When a thermocouple is heated by the pilot flame, about 25 millivolts are generated to operate a pilot solenoid valve (one volt equals 1000 millivolts, or 1000 mV). Only when the thermocouple is heated will the solenoid valve stay open and allow the main gas valve to open.

One system has a 24 V ac through the flame, which rectifies the current to dc. It uses the same principle as a vacuum tube rectifier, where electrons will flow from a point to a flat surface, but not from a flat surface to a point. If the flame goes out, no dc will flow.

Some thermocouple assemblies consist of several thermocouples electrically connected in series. These units generate from 125 to 750 millivolts of electricity, which is powerful enough to operate the main solenoid valve. Therefore, the furnace is independent of the electrical circuit in the building.

In large systems, the thermocouple operates only a small pilot solenoid valve in the main gas valve. The thermocouple current holds the valve open. If the pilot is out or if the pilot flame is low, the valve will close.

A pilot flame should be blue. If it is yellow, the primary air inlet to the pilot is probably dirty and should be cleaned. The pilot should be cleaned at least once a year.

To test, shut off the pilot flame while the burner is on. The main burner should shut off in two minutes. The pilot light flame should touch the middle of the thermocouple 3/8 to 1/2 in. (10 to 12 mm) from the thermocouple tip. The tip should be very hot and dull red. Keep pilots at the edge of the main burner flame, Fig. 20-36. Location varies with various manufacturers.

An electronic spark igniter (about 19,000 V) may be used to keep the pilot light burning. The ignition spark starts either in less than a second or after five minutes (to allow gas to empty from furnace) if the flame goes out. It stops if the pilot light burns (usually in 20 seconds). If flame does not go on, ignition shuts off in 15 seconds.

Electric ignition depends on an electric spark to light the gas. The system includes electrodes, transformer, flame sensor, safety switch, and purge timer. It is used where pilot lights would be difficult to service or where air drafts may blow out a pilot flame.

Electric ignition systems can operate in temperatures of −40 °F (−40 °C). The gas valve and pressure regulator

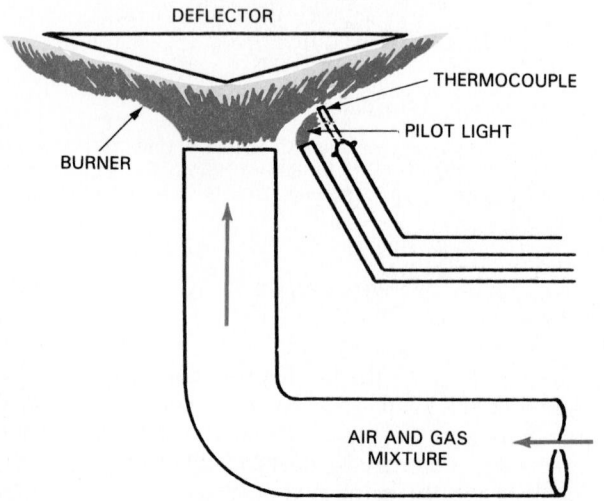

Fig. 20-35. Schematic drawing of safety thermocouple and pilot light for gas furnace. Thermocouple generates small electric current which actuates a control. Control will shut off gas supply to furnace if pilot light is extinguished.

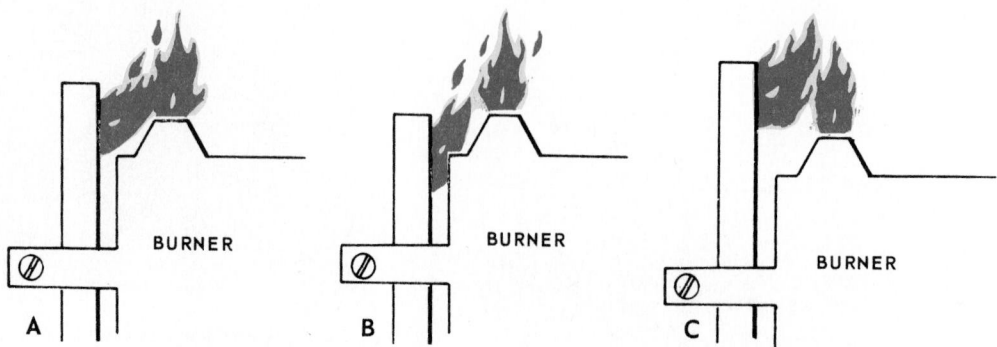

Fig. 20-36. Pilot light flame positions. A—Correct. B—Too low. C—Too high.

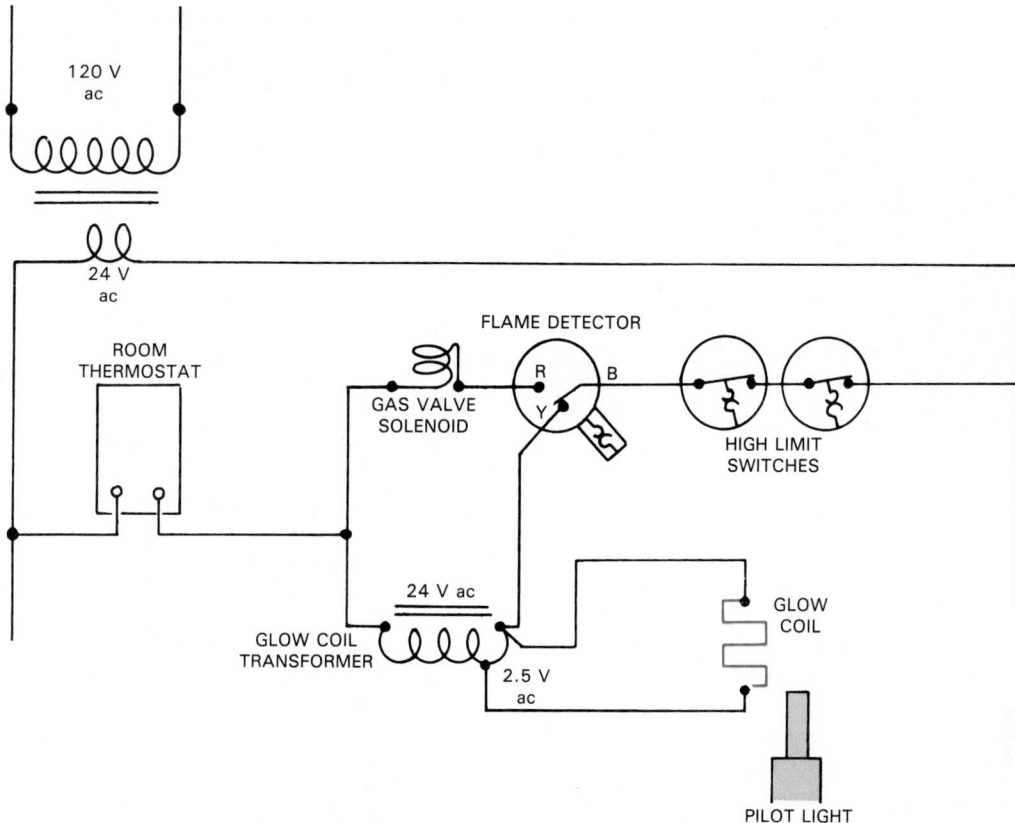

Fig. 20-37. Glow coil circuit. When points close, glow coil heats to a red temperature and ignites pilot. Flame detector heats up and moves switch from Y to R. With switch position at R, switch opens main gas valve and burner starts.

must also be designed for low-temperature operation. From 4000 to 12,000 V are used with a 0.040 in. (1 mm) to 0.190 in. (5 mm) gap. The electrodes are extended so that the spark gap is within 1/4 in. (6 mm) of the gas stream just outside the burner. One electrode is insulated, the other is grounded. These electrodes can operate satisfactorily at high temperatures. The ceramic insulators are designed for about 1200 °F (650 °C). If the main flame goes out, the flame sensor cools enough in four seconds to operate the system shutoff controls. This unit is of rod-and-tube design and is about 4 1/2 in. (11 cm) long.

Still another method is used to ignite a gas burner in an automatic system. It makes use of a glow coil, Fig. 20-37. When the thermostat points close, a transformer feeds 2.5 V ac to the glow coil. As soon as the glow coil turns red, it ignites the pilot. The flame detector then opens the glow coil switch circuit and turns on the main gas valve. The burner fuel gas is ignited.

When working on this system, always open the main electrical switch. The 24 V control voltage is not very dangerous, but one could contact the 120 V supply.

20-16 THERMOCOUPLE CIRCUITS

A thermocouple may be used to prevent the main gas valve from opening or to close it if the pilot light goes out. The principle of operation of the thermocouple is described in Para. 6-65.

The hot junction of the thermocouple is located in the pilot flame. A small electrical current is generated in the thermocouple, Fig. 20-38. The current from the thermocouple

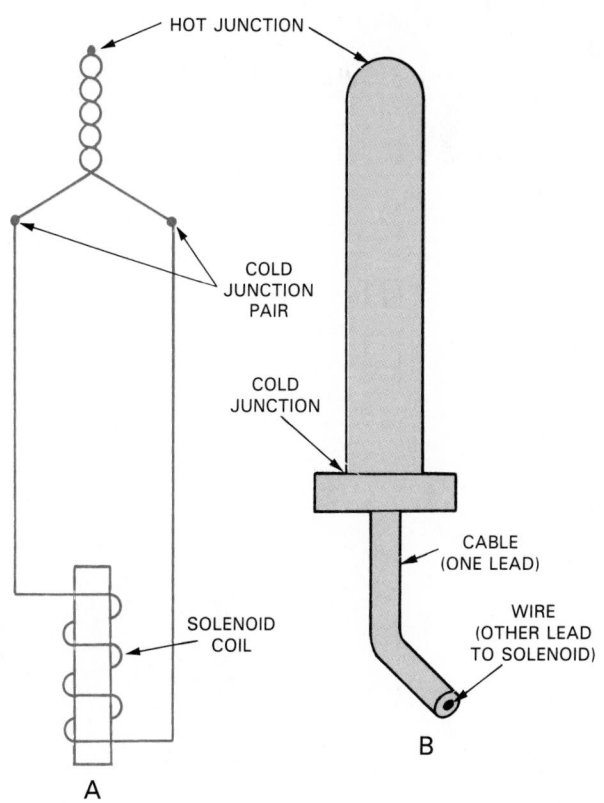

Fig. 20-38. Thermocouple. A—Schematic of electrical circuit for thermocouple pilot light safety control. B—Construction of thermocouple tip.

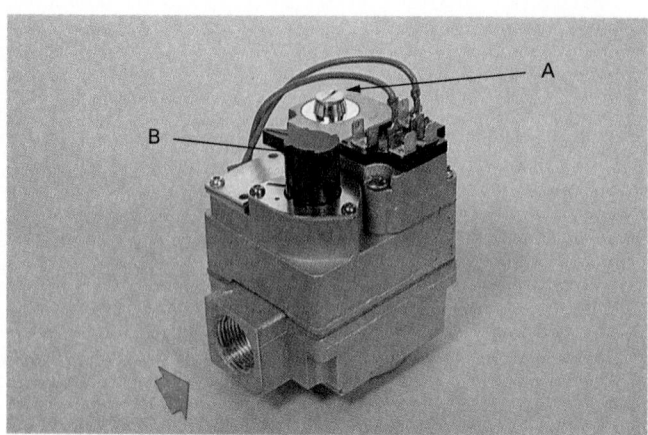

Fig. 20-39. Gas control with several functions. It has shut-off valve, pressure regulator, gas flow control, and pilot light. A—Terminals. B—Pilot solenoid valve.
(White-Rodgers Division, Emerson Electric Company)

opens the pilot solenoid valve in the gas supply line when the thermostat contacts are closed. The pilot solenoid valve operates the main gas valve.

On open circuit, the thermocouple generates 30 mV (recall that one volt equals 1000 millivolts). A millivolt reading of two millivolts or more above or below this reading indicates a faulty thermocouple. The pilot solenoid requires at least seven millivolts to operate. If the hot junction heating flame goes out, the magnetic coil will lose its power in one minute if it is operating correctly (although some codes allow as much as three minutes for system shutdown in domestic and small commercial units).

Aluminum tubing is preferred for pilot light gas connections (copper has a flaking action inside the tubing).

Aluminum tubing must be protected against electrolysis (usually by a plastic exterior coating).

If the pilot flame is too high (there is an adjusting screw), the thermocouple may be overheated and ruined. If the main burner flame contacts the thermocouple, it will be destroyed.

20-17 GAS CONTROLS

The air-fuel gas ratio to the burner must be carefully maintained. Since the fuel gas must be supplied at a constant pressure, a pressure regulator generally is mounted in the inlet to the gas manifold. The regulator operates much like an expansion valve, opening and closing a valve in response to the outlet pressure. This pressure is measured in inches of water column. Most pressure regulators are adjustable. However, they should be adjusted only when a water column manometer is connected to the outlet.

The pressure regulator has a tapped hole in the bonnet for two reasons:

1. To insure that there is atmospheric pressure on the outside of the regulator diaphragm for good pressure control.
2. To permit tubing to be installed from this bonnet opening to the outdoors or to the pilot light. This is a safety device in case the regulator diaphragm leaks. Escaping gas will then go outdoors or be burned at the pilot light. The best practice is to pipe it outdoors with a weatherproof outlet.

Most pressure regulators are built into a complete gas control. See Fig. 20-39. The internal design is shown in Fig. 20-40. The main valve is opened by using a solenoid-controlled orifice. The solenoid is energized when the thermostat calls for heat. The cycling valve opens and gas pressure travels to the bottom, (underneath the large diaphragm in Fig. 20-40), opening the main valve. The gas

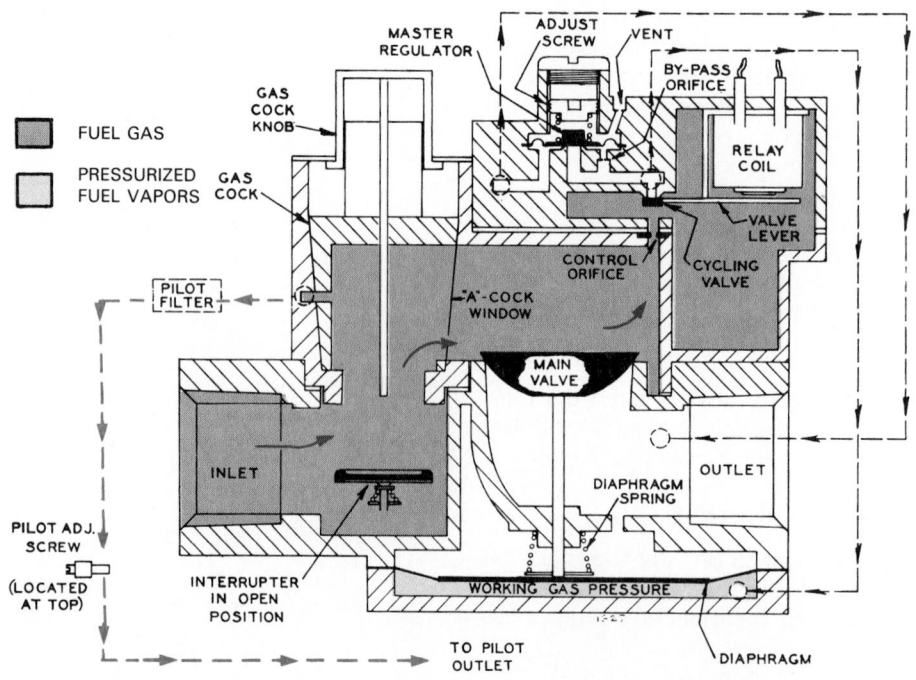

Fig. 20-40. Sectional view drawing of gas control in closed (no flow) condition. Note main valve in center, which is in closed position.

from the cycling valve is shown in Fig. 20-41. After the thermostat opens and the relay coil is de-energized, the cycling valve closes, Fig. 20-42, and gas pressure under the diaphragm decreases. This causes the main valve to close. Refer to Fig. 20-40 again.

20-18 GAS BURNER INSTALLATIONS

Most cities have code requirements covering the installation of heating equipment. It is important to know this code and carefully follow it. Gas is dangerous. See Para. 20-13.

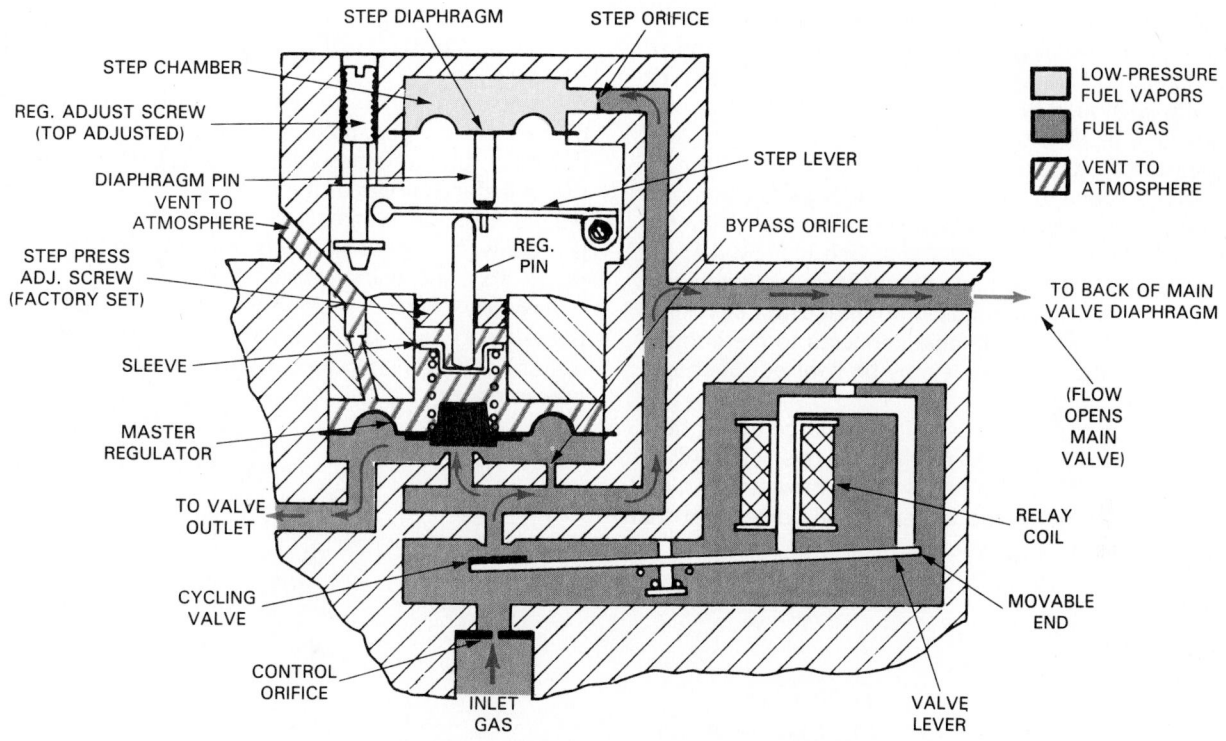

Fig. 20-41. Bypass control system for opening gas control main valve. Relay coil is energized. (White-Rodgers Div., Emerson Electric Co.)

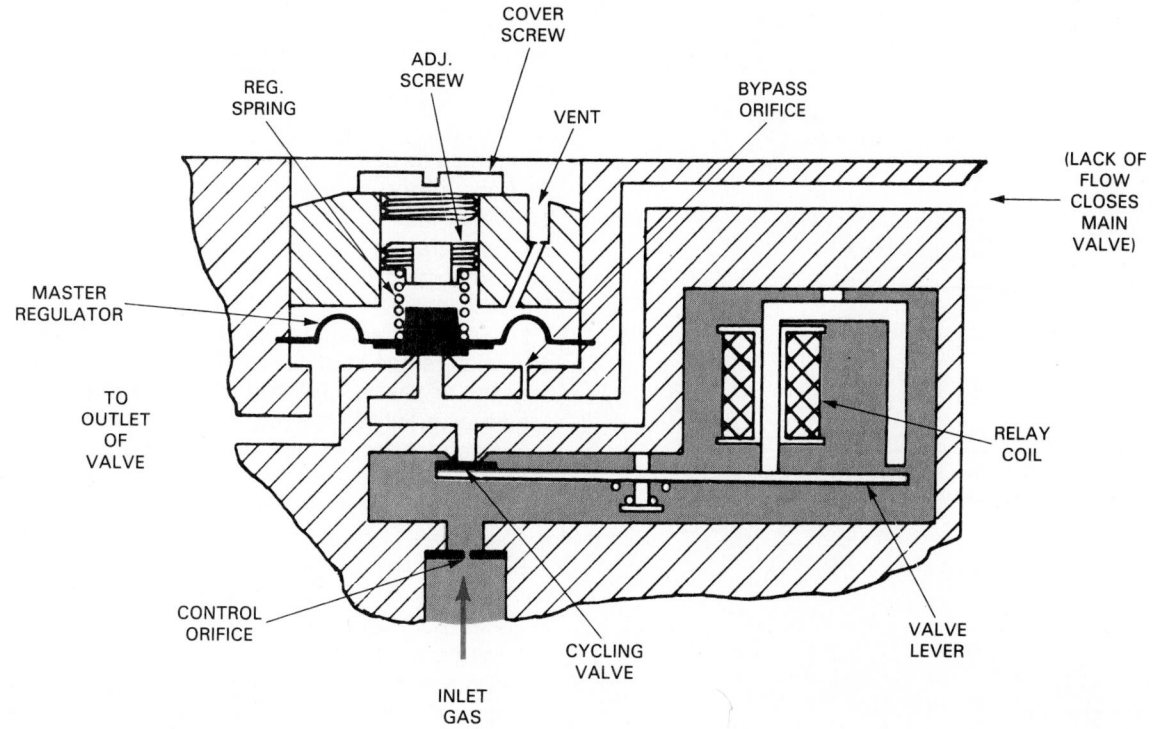

Fig. 20-42. Bypass control system used to close main valve. Relay coil is not energized. (White-Rodgers Div., Emerson Electric Co.)

Furnace capacities are certified by the American Gas Association. This information is given on furnace identification plates.

Furnace capacities decrease at about four percent for each 1000 ft. of elevation above sea level due to the decrease in atmospheric pressure (air becomes less dense). This rating change is effective above 2000 ft. elevation.

When installing gas line pressure regulator valves and control valves, cut the pipe accurately and apply the thread sealing compound correctly. See Fig. 20-43. Tighten pipe joints with a moderate torque. The regulator and valve housings are usually made of a zinc die casting, and excessive force will crack them.

It is good practice to install a pipe drip leg in the gas pipe to the furnace to keep dirt and moisture from entering the pressure regulator and gas controls. See Fig. 20-44.

When threading the pipe into the pressure regulators and controls, place the wrench holding the regulator or control near the opening where the pipe connects to the control. Otherwise, the die cast bodies may be bent out of shape, and may even crack.

When starting a gas furnace unit, first purge the air from the installed gas piping up to the furnace gas valve. Allow the purge air (it will have some gas mixed in with it) to clear the room. Then turn on the furnace hand valve, light the pilot light, and start the system.

Only use soap suds when checking for gas line leaks. Use of a match or any open flame may cause an explosion. Another safe practice when looking for gas leaks is to turn off the electric power and use a spark-proof flashlight to check the soap suds areas.

Gas burners burn with a blue flame when the primary and secondary air adjustments are correct. A yellowish flame indicates a lack of primary air, and perhaps a lack of secondary air. A collection of soot usually indicates a lack of secondary air.

Flame propagation (speed of burning) is important. The primary air and gas mixture must flow slightly faster than the flame burns. Otherwise, flashback will occur and the gas will start burning at the primary inlet (spud). Flashback occurs if too much primary air is used or if fuel gas pressure is too low. If primary airflow is too fast, the flame will blow away from the burner (called "lifting").

Annually check the primary air inlet, fuel gas pressure, flame color, pilot lights, and burners. Fig. 20-45 shows a water column manometer which is used to measure gas pressures.

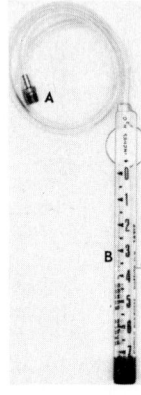

Fig. 20-45. Manometer used to measure gas pressures up to 7 1/2 in. water column. Large scale enables easy reading. A—Connection to gas line. B—Scale.

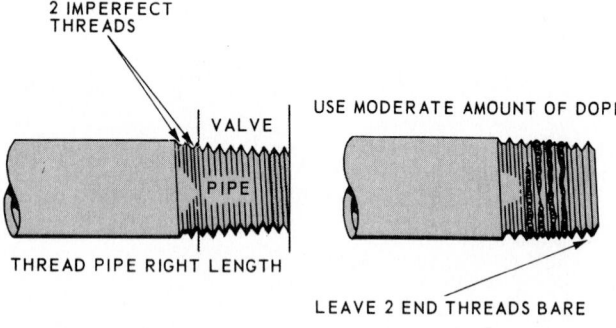

Fig. 20-43. Diagram shows proper way to put pipe compound on threads. (Honeywell, Inc.)

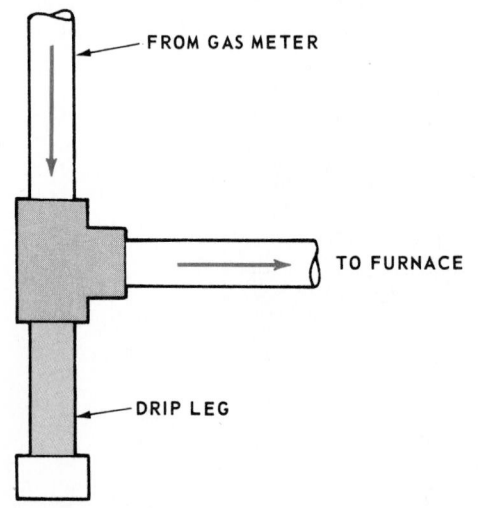

Fig. 20-44. Drip leg installed in gas pipe to furnace. Drip leg will trap dirt and moisture.

Incomplete combustion may be the result of:
1. Poor mixing of fuel gas and air.
2. Partial lack of air.
3. Temperature too low to produce ignition, or to keep combustion going.
4. Back draft.

Flue gas temperatures must be maintained above the condensation temperature (dew point) of the flue gas. To check the flow of gases in the flue and in the stack, use a smoke candle or a lighted match. See Fig. 20-46.

For natural gas, the recommended stack temperature is 300°F to 900°F (150°C to 480°C), depending on the length of the chimney. Values for LP-Gas are approximately the same.

Excessive temperatures in any part of the heating system will quickly erode the unit. It is best if no part of the system exceeds 830 to 1230°F (440 to 660°C).

In all cases, there must be a sufficient air supply to the furnace for combustion. Approximately one square inch of opening is needed for each 1000 Btu per hour capacity of the furnace. For example, a 28,000 Btu per hour furnace will require an opening 7 in. by 4 in.

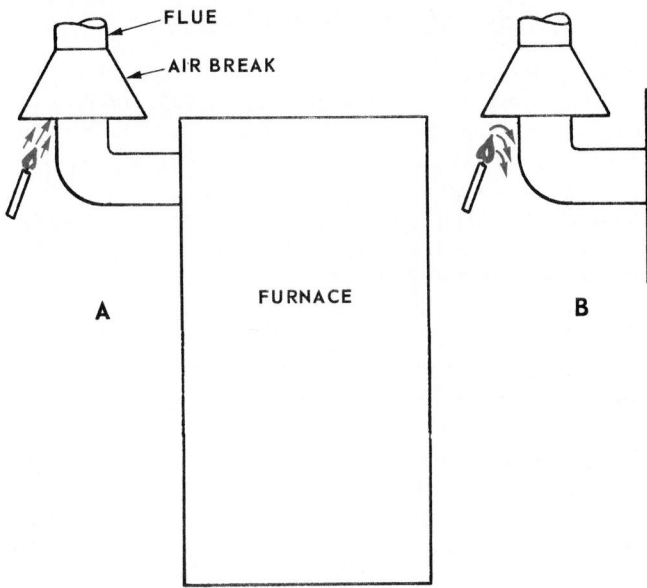

Fig. 20-46. Checking draft in flue and chimney when furnace is on. A—Smoke from smoke candle should be drawn into air break and up chimney. B—Smoke being blown away from flue indicates back draft.

20-19 SERVICING: STARTING A GAS FURNACE

The service technician should take the following steps in starting a gas furnace:

1. Adjust thermostat to its lowest setting.
2. Determine type of system (forced air, water, or steam).
3. Check installation for completeness, including flue, controls, wiring, and water level. (Water level of boiler is controlled by owner.)
4. Read lighting instructions.
5. Close all manual supply valves, both pilot and main supply.
6. Inspect combustion chamber for holes, cracks, and water leaks. Use a light, or make a salt spray test (see Para. 20-21). Do not start unit if any of these conditions are present.
7. Locate pilot. Use a spark-proof flashlight.
8. Smell and listen for gas inside combustion chamber. If gas is present, do not proceed until this condition is corrected.
9. If pilot gas line is connected before main gas shutoff, light the pilot.
10. If pilot gas line is after main shutoff, close pilot valve, open main shutoff, and listen for gas entering the furnace. No gas should enter. If leakage is noticed, turn main valve off, determine cause of leakage, and make repair. If no leakage is noticed, open pilot valve and light pilot.
11. Check pilot flame for size, color, stability, and position to insure quick, safe lighting of main burner. If pilot flame is yellow, clean pilot light.
12. Check main burner, including position of any baffles.
13. Determine type of safety system. If manual, reset engage mechanism.
14. Turn on main shutoff valve to check thermostat. Burner should not come on. Then turn main valve off.
15. Turn thermostat above room temperature.
16. Turn on main valve.
17. Check the off time; if pilot light flame is shut off, main burner should shut off in 2 min. or less. Para. 20-15 introduces the concept of off time.
18. Inspect all controls and operate them.
19. Check gas pressure with a manometer.
20. Adjust thermostat to temperature customer desires.

20-20 SERVICING: GAS FURNACE

To locate the source of a problem in a gas furnace, use the following steps:

I. Gas valve doesn't operate.
   A. Pilot light not burning like it should.
      1. Low gas pressure.
      2. Excessive draft.
      3. Partially clogged pilot orifice or filter.
   B. Pilot lights, but goes out when reset button is released.
      1. Loose or dirty thermocouple.
      2. Defective thermocouple.
      3. Defective power unit.
      4. Loose or dirty electrical connection.
   C. Pilot lights, but will not burn normally.
      1. Yellow flame, because pilot primary air is partially clogged.
      2. Blue wavering flame, caused by too much draft.
      3. Flame too small, because orifice is partially clogged or pressure is too low.
      4. Flame blows off pilot light orifice, because pressure is too high.
      5. Flame too sharp in shape, because pilot light orifice is too small.
II. Gas valve operates, but poor heating.
   A. Delayed ignition.
      1. Weak pilot.
      2. Pilot not mounted correctly.
      3. Pilot flame not located near burner.
   B. Flashback to spud.
      1. Too much primary air.
      2. Pressure too low in manifold.
      3. Burner inserts damaged.
   C. Not enough heat.
      1. Thermostat set too low or out of calibration.
      2. Some heat source placed near thermostat.
      3. Limit control cutout switch set too low.
      4. Furnace dirty; soot, lint, or dirty filter.
      5. Ducts dirty.
      6. Low manifold pressure.
      7. Pilot goes out while unit operates.
   D. Too much heat.
      1. Thermostat set too high or out of calibration.
      2. Burner keeps operating when thermostat opens.
   E. Soot and/or fumes.
      1. Puffs back because of delayed ignition or overfiring.
      2. Not enough primary air.
      3. Firepot or heat exchanger leaks, resulting from not enough duct air or too much heat.
   F. Cycles too often.
      1. Thermostat anticipator loose or burned out.
      2. Limit control turns burner off too soon.
   G. Does not cycle often enough.
      1. Faulty thermostat.
      2. Dirty contacts.
      3. Loose or dirty electrical connections.

20-21 SERVICING: FURNACE INSPECTION (WARM AIR)

The condition of the furnace should be checked at least once each year. Usually, this checkup is made before the heating season starts:

1. Clean and light the pilot.
2. Test pilot safety.
3. Check and oil fan motor.
4. Check blower and fan belt; oil blower bearings.
5. Test fan control operation.
6. Test limit control operation.
7. Test gas valve operation.
8. Check burner adjustment.
9. Check thermostat operation.
10. Check filters (usually replace).
11. Check heat exchanger (usually clean it).
12. Check combustion chamber for cracks:
    a. Spray a salt water solution (6 tsp. salt per pint of water) into the combustion chamber with the flame lighted.
    b. Direct a small propane torch flame into warmed air inside furnace jacket. If combustion chamber is cracked, color of torch flame will turn from blue to yellow.
    c. With warm air systems, combustion chamber must be airtight. Any leakage through combustion chamber may allow products of combustion to enter occupied space through warm air registers.
13. Operate system and make a stack gas analysis.
14. Make adjustments if necessary.

20-22 WARM AIR HEATING SYSTEM

The forced warm air system or mechanical warm air system uses a motor-driven blower to increase the flow of heated air to the needed areas. This system usually includes an air filter. Fig. 20-47 shows a forced warm air furnace with a gas fuel heating unit.

With this heating arrangement, a system of ducts delivers warm air to the various spaces to be heated. The ducts are carefully sized to provide the correct amount of heat to each room. See Chapter 22 for duct sizing. Another system of ducts returns the cool air to the furnace for reheating. To provide the correct relative humidity conditions, a humidifier usually is installed in the airflow, either in or near the furnace.

All ducts collect dust and dirt during use. They should be vacuum cleaned about each five years, using special equipment designed for this purpose.

20-23 HYDRONIC HEATING SYSTEM

Hot water systems for carrying heat to occupied spaces have been in use for many years. In some systems, the hot water circulates by thermal convection. The circulating water is under atmospheric pressure, and an expansion tank allows changes in volume.

Most hot water systems use a circulating pump. The pump increases water flow and carries more heat per unit of time to the room heat transfer units. Having a pump in the system permits the use of a smaller boiler. Fig. 20-48 shows a cross section of a gas-fired domestic water boiler. This particular unit can be vented in two different ways: 1) a natural venting process, and 2) a power-vented package. The natural venting may be vented into fireclay tile-lined masonry chimneys. The power vent is designed to vent directly through the wall.

A commercial hydronic heating oil, gas, or combination gas/oil boiler is shown in Fig. 20-49. Switches on the burner control permit the choice of fuel being burned, either gas or light oil.

Hydronic under-floor heating systems are designed in

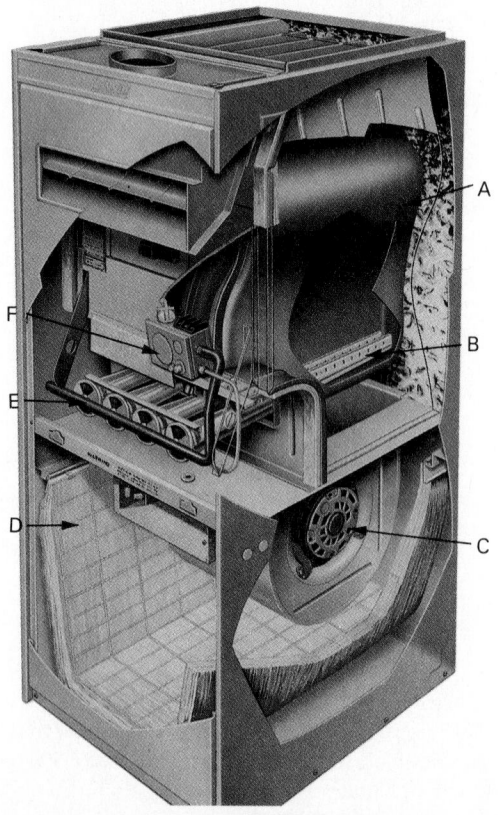

Fig. 20-47. Four-burner gas-fired warm air furnace. A—Heat exchanger. B—Burners. C—Blower. D—Air filter. E—Primary air adjustment. F—Controls. (Lennox International, Inc.)

Fig. 20-48. Cross section view of a gas-fired hot water boiler. A—Non-metallic venting. B—Insulated jacket. C—Draft inducing fan. D—Labeled wiring. E—Pressure-temperature gauge. F—Cast-iron sections. G—Gas valve. H—Stainless steel burners. I—Controls. J—Self-lubricating circulator. (Burnham Corporation)

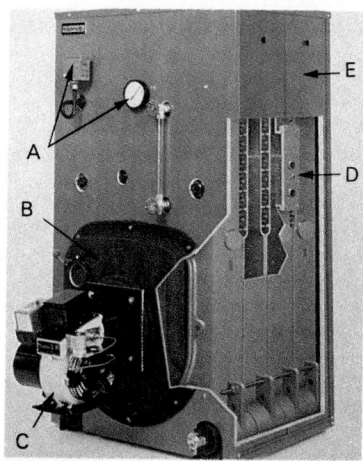

Fig. 20-49. Commercial steam boiler. A—Controls for adjustment and maintenance. B—Burner mounting plate with flame observation port. C—Burner. D—Tankless heater. E—Rear flue outlet. (Burnham Corporation)

numerous ways. Two basic concepts are: 1) Concrete slab floor, in which tubing is placed, enforced in mesh, and then slab is poured. 2) Utilization of wood with aluminum plates; in residential setting with suspended wooden floors, tubing is run within the flooring joists. See Fig. 20-50.

Fig. 20-51 shows a circulating pump of the centrifugal type. A shaft seal is located where the pump shaft leaves the casing.

A hermetic type of hydronic pump is shown in Fig. 20-52. It will pump 10 gpm at a 6-ft. water head. The motor is a permanent split capacitor type. Its rotor turns in the water, and the shaft seal has been eliminated. Fig. 20-53 shows the inner structure of the hermetic pump.

To remove the pump, turn the shutoff valves all the way in. A piping arrangement when one pump is used is shown in Fig. 20-54. A system with three zones is illustrated in Fig. 20-55. It has a pressure relief valve, which is required in all pressurized heating systems.

Fig. 20-50. Hydronic heating system being applied on concrete slab floor in large commercial installation. (Wirsbo Company)

Fig. 20-52. Exterior view of hydronic pump. A—Motor. B—Capacitor. C—Inlet shutoff valve. D—Outlet shutoff valve. Cutaway view is shown in Fig. 20-53.

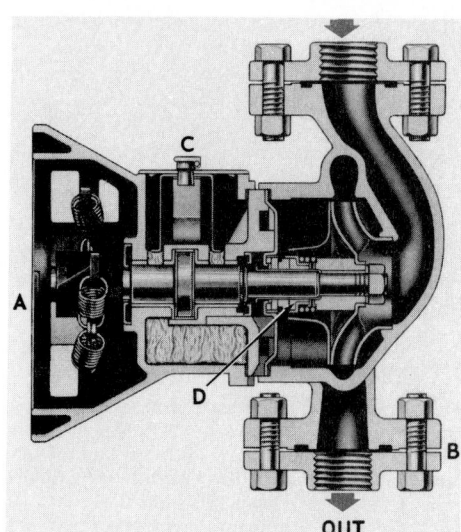

Fig. 20-51. Cross section of centrifugal hydronic system pump. A—Motor mount. B—Flange for threaded pipe connection. C—Oil cup. D—Seal. (Bell and Gossett, ITT)

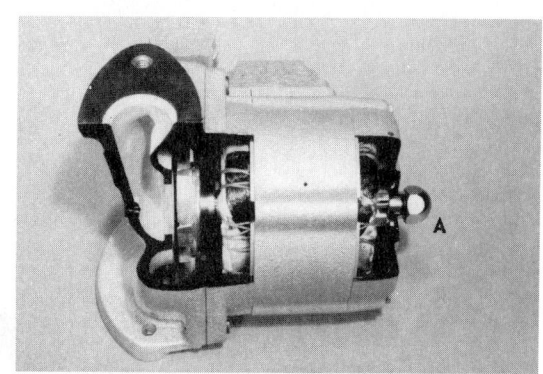

Fig. 20-53. Internal view of hermetic hydronic pump. A—Turn acorn nut all the way out to check for proper rotation. Then turn nut all the way back in.

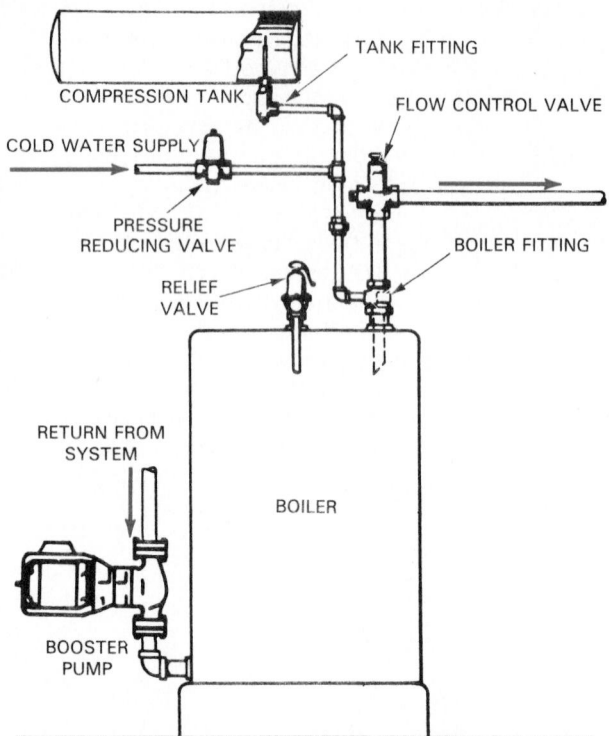

Fig. 20-54. Diagram of hydronic system shows piping and location of pump, compression tank, and relief valve. (Bell and Gossett, ITT)

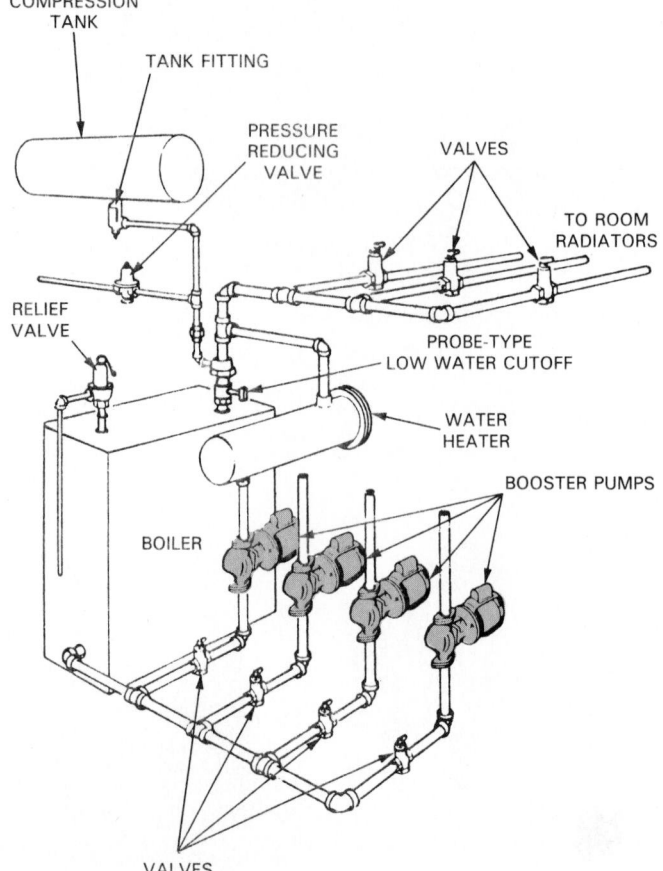

Fig. 20-55. Three-zone hydronic system with four booster pumps. Fourth pump circulates water through heater to provide hot water. (Bell and Gosset, ITT)

Water heating systems may use one of several different temperature control devices:

1. Single control, which starts and stops the pump.
2. Zone control, using two or more controls. Each control operates one pump.
3. Individual radiator controls for individual room control.

Fig. 20-56 illustrates a hydraulic type radiator control. Temperature control is adjustable. Note that the control operates the flow valve, and the sensitive bulb is located in the cold air inlet to the radiator. It is a modulating control (variable volume flow) that can be used for either steam systems or hot water systems (hydronic).

A hot water system using a flow switch control system is shown in Fig. 20-57. The main circuit is shown in solid

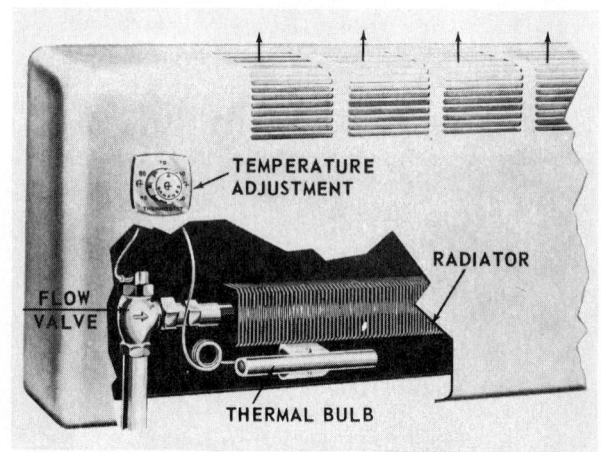

A

B

Fig. 20-56. Individual thermostatic radiator valve commonly used for one-pipe or two-pipe hot water heating systems in residential, commercial, or institutional buildings. A—Thermal control valve mounted on convector heating system. This control provides individual room temperature control with either circulating hot water (hydronic) or steam as heat source. B—Thermostatic radiator valve. 1) Drop in ambient temperature causes the valve to open. 2) This increases supply of hot water or low-pressure steam entering the radiator. Rise in ambient temperature causes valve to close, decreasing supply of hot water or low-pressure steam entering radiator. (Danfoss Inc.)

lines. If the water ceases to flow, the flow switch will open the electrical circuit and shut off the burner. Some systems use a recirculating system within the boiler, as shown in the broken line piping diagram. This recirculation of water maintains a more constant water temperature within the boiler. If water flow stops in this water circuit, the flow switch shown will shut down the system.

Fig. 20-58 shows a hot water system used to heat the incoming outside air of a ventilating system. If the water flow stops, a flow switch will shut the damper, turn on the alarm, and sometimes shut off the burner, depending on the number of other radiators.

A system which mixes outside air within return air is shown in Fig. 20-59. The recirculated air and fresh air intake are controlled together to ensure proper air conditions in the occupied space.

In many buildings other than homes, it is desirable to maintain different temperatures in different rooms or parts of the building. This is also desirable in some homes. For instance, the bedrooms may be kept at a different temperature than the living room; the laundry at a different temperature than the kitchen. Rooms are kept at different temperatures by thermostatically controlling the heating or cooling media for these areas.

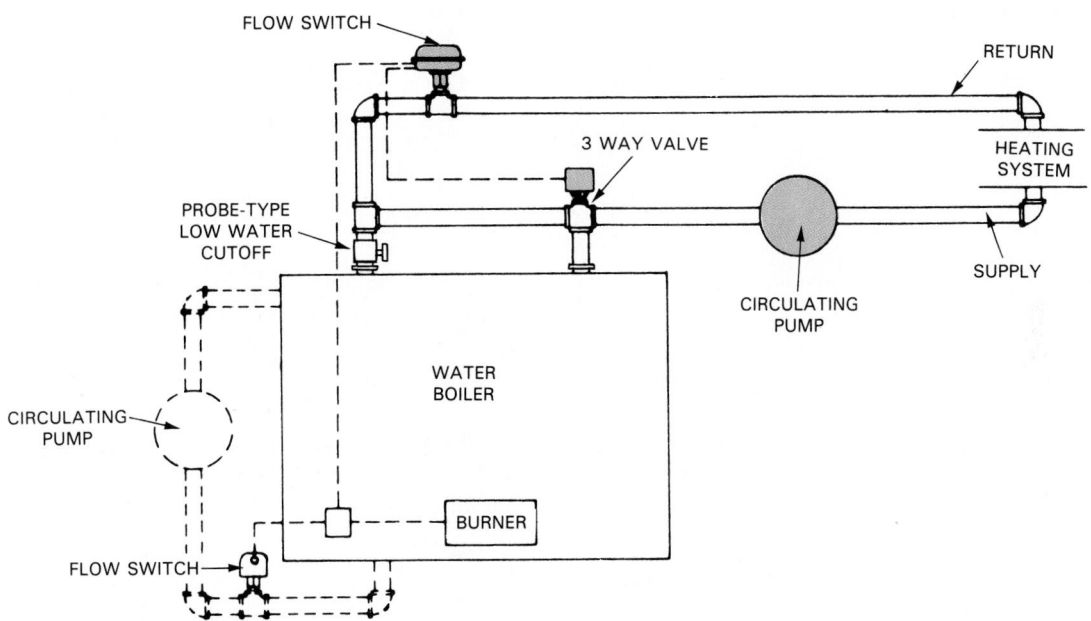

Fig. 20-57. Schematic diagram of hot water system using circulating pump, three-way valve, and flow switch. (ITT McDonnell & Miller)

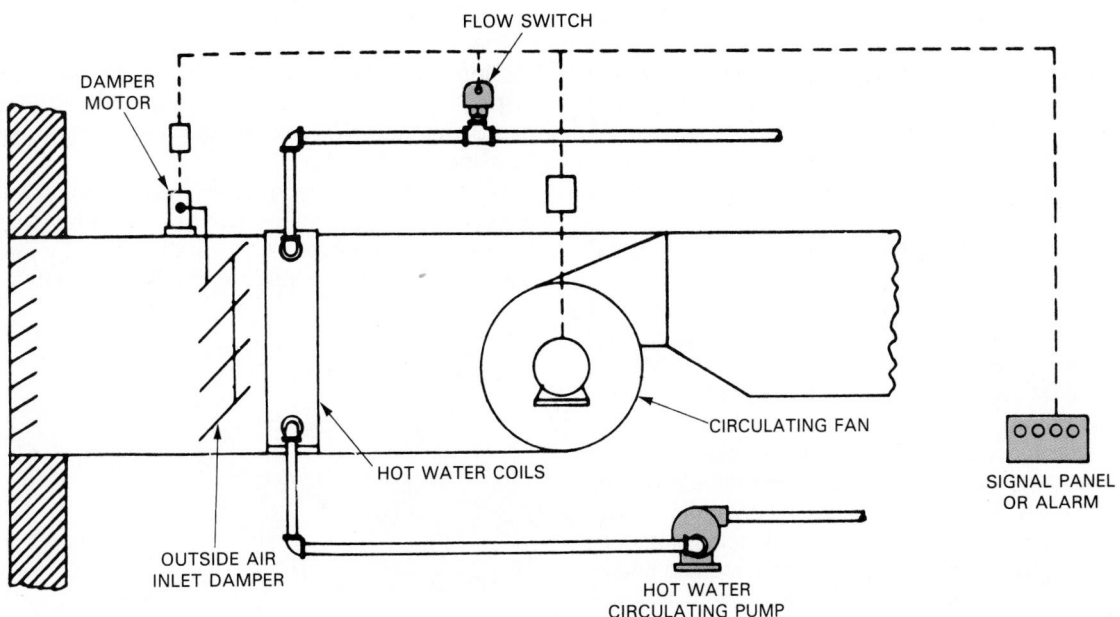

Fig. 20-58. Water heating system used to heat incoming air of ventilating system. Note motorized damper control, flow switch, and signal panel. (ITT McDonnell & Miller)

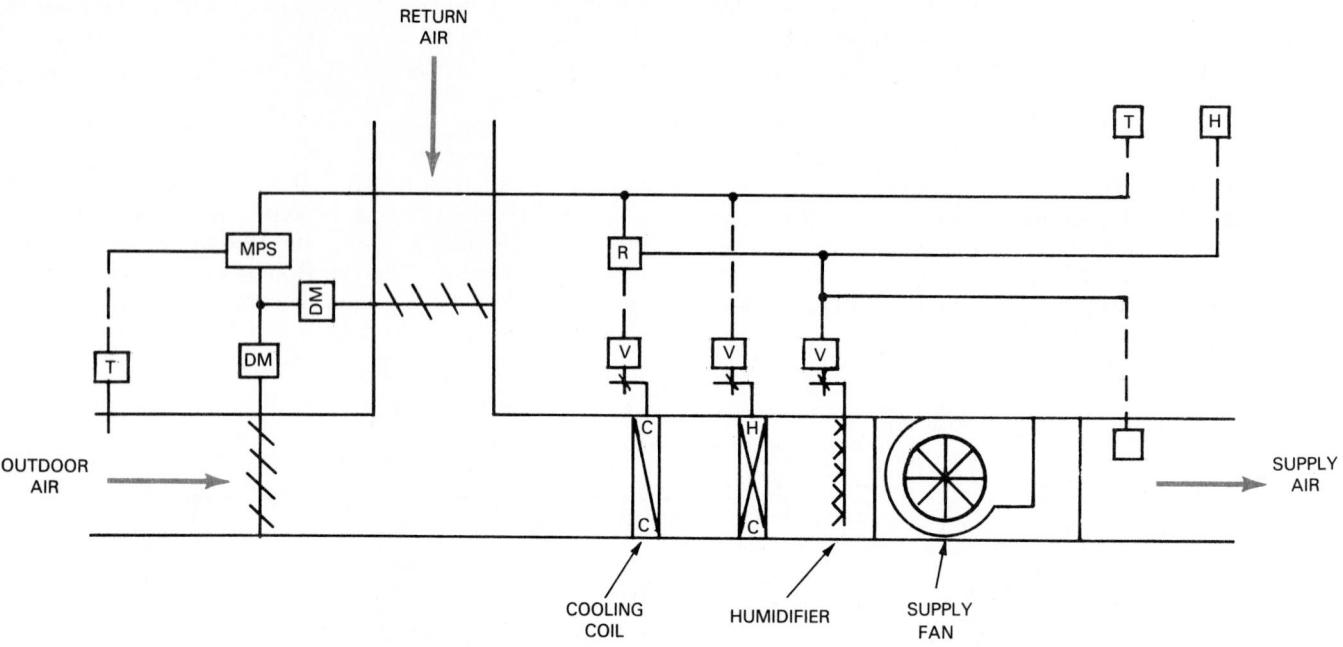

Fig. 20-59. Diagram of controlled air system. Outdoor air and return air are proportionately mixed to obtain correct atmosphere. (ASHRAE Guide and Data Book)

Hydronic systems are controlled in four basic ways:
1. The heat is turned on at the same time that the pump is turned on.
2. The heat is on continuously and the pump is cycled to provide heat.
3. The pump is operated continuously and the zone valves are cycled.
4. The zone valves are cycled and turn on the pump or heater.

In forced circulation closed systems, the high-temperature water is above atmospheric pressure and smaller pipes can be used. The heat load is based on 20 degrees differential between water-in and water-out temperature. The top of the tubing should be even. It is best to use eccentric reducer fittings. Fig. 20-60 suggests a reason for using eccentric reducer fittings.

Two-pipe systems are the most common:
1. Direct return (less piping). Each circuit is a different design.
2. Reverse return. Easier to balance and each pipe circuit is the same length. See Fig. 20-61.

When installing hot water heating systems, one must allow for pipe expansion. Expansion for steel is 3/4 in./100 ft./100°F; for copper, 1 1/16 in./100 ft./100°F. Where riser pipes connect to a horizontal run, allow for a flexible joint. Also install an expansion joint at the boiler.

When a hydronic system delivers hot water to two or more heating units (radiators), the water flow must be balanced to insure that each radiator receives its design quantity of water per unit of time. This balancing is usually done by installing gate valves in the piping. Opening and/or closing these valves obtains balance. Flow meters are used to accurately measure quantity of flow.

Water used in hydronic systems usually has substances added which lower the freezing temperature and raise the boiling temperature. These substances may also keep the water from forming deposits in the pipe. Tap water has impurities that may cause:
1. Scale.
2. Corrosion.
3. Embrittlement.

Scale is formed from salts in the water. The salts settle on metal surfaces as the water passes through temperature changes. These salts should be removed before the water enters the heater; or, chemicals should be added to form a sludge with these salts. The salts can then be purged from the system.

The salts are usually calcium carbonate, calcium sulphate, calcium chloride; magnesium carbonate, magnesium sulphate, magnesium chloride; sodium carbonate, sodium hydroxide, and silica oxide. Iron and manganese may also form deposits in the boiler.

Corrosion takes place when the water is acidic or gases are dissolved in the water. Corrosion is reduced by neutralizing the acid condition with an alkali and by removing the gases by de-aeration (release of gases dissolved in a liquid).

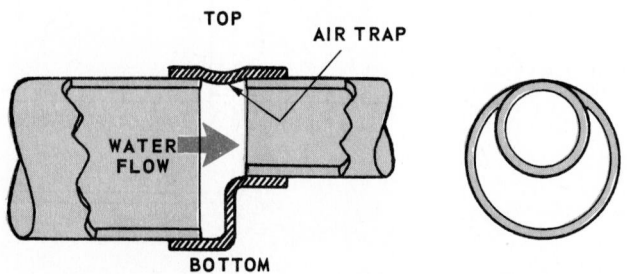

Fig. 20-60. Eccentric fitting in water circulation systems used to reduce danger of air pockets when pipe size is reduced.

732

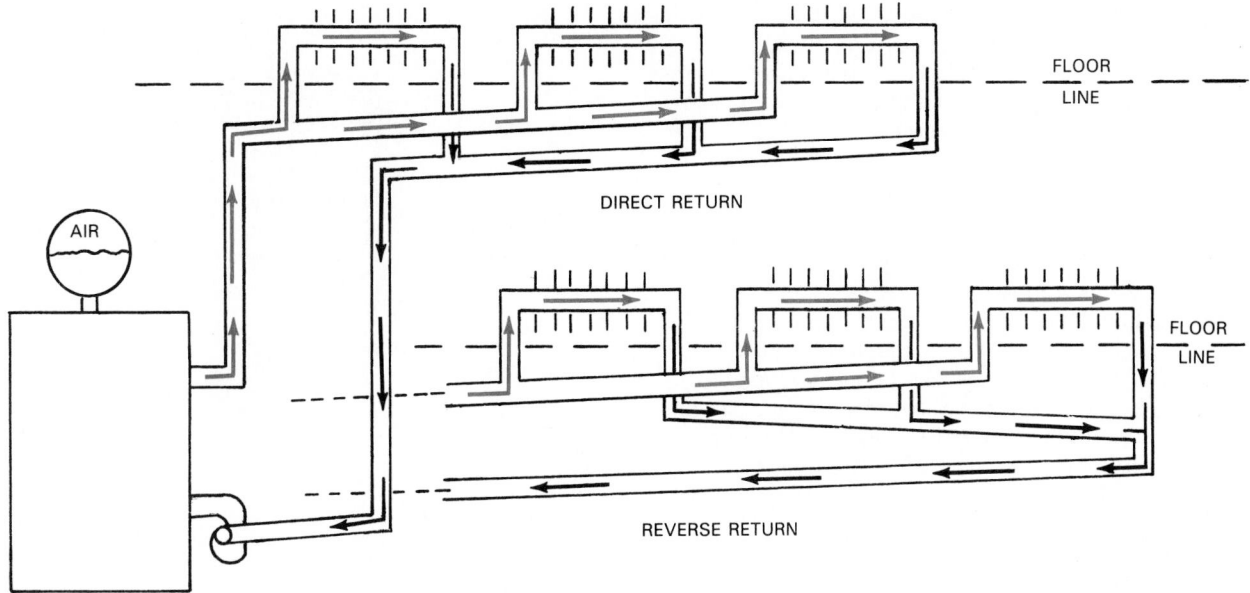

Fig. 20-61. Horizontal pipes in hydronic steam system should always slope toward boiler. Amount of slope is exaggerated in drawing. Total length of pipe run for each radiator (both hot water and return water) should be of equal length. Note use of equal total lengths in reversed return system.

Chemical scavengers and corrosion inhibitors are also used.

Water can be contaminated in the following ways:

1. The usual dissolved gases are hydrogen sulfide, carbon dioxide, and oxygen.
2. Organic matter and oil cause foaming of the water.

Some companies specialize in the treatment of boiler water. Substances can be added to water to reduce the effects of impurities. These are available at contracting supply houses.

Embrittlement causes metal failure along drum seams, under rivets, and at tube ends. Water flashing to steam through any small leaks in these highly stressed areas allows any sodium hydroxide in water to concentrate. See Fig. 20-62. Embrittlement can be slowed by maintaining low alkalinity (low hydroxide), avoiding leaks at stressed metal, and using special inhibiting agents.

Safe levels of impurities in boiler water for various chemicals are shown in Fig. 20-63. The pH level should be about 10. See Para. 14-53.

20-24 INSTALLING HYDRONIC SYSTEMS

The boilers for hydronic systems must be mounted level. All local code requirements must be checked and must be followed.

A new boiler should be boiled out before it is put in service. The water to be used should be analyzed and a preventative maintenance solution should be added. A card with a record of past maintenance needed along with a list of future treatment procedures should be included.

After a system is installed, and before putting it into service, always flush the system. The fluxes, pipe joint compounds, and cutting oils sometimes form gases in a system. Dirt, sand, steel thread chips, sawing chips, filing chips, and solder bits may erode the system and clog screens. Dirt and chips also ruin valves and pump seals.

Fill the system with water. Add about one pound of tri-sodium phosphate for each 50 gallons of water. Circulate

for about four hours, then drain. Clean the screens and fill the system with water. It is ready to operate.

Unless the pump seals leak and the vents are used quite often, avoid treating the water with chemicals. The chemicals may injure seals and valves. Avoid using

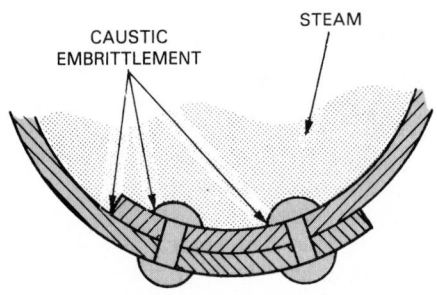

Fig. 20-62. Cross section view of steam container (pipe or boiler) shows where caustic embrittlement can take place. Activity is greatest where metal is stressed and where caustic material (sodium hydroxide) can collect.

MAXIMUM ALLOWABLE IMPURITIES IN BOILER WATER

CHEMICAL NAME	CHEMICAL SYMBOL	PPM
SODIUM SULPHITE	Na_2SO_3	1.0
SODIUM CHLORIDE	$NaCl$	10.0
SODIUM PHOSPHATE	Na_3PO_4	25.0
SODIUM SULPHATE	Na_2SO_4	25.0
SILICA OXIDE	SiO_2	0.20
TOTAL DISSOLVED SOLIDS		50.0

Fig. 20-63. Maximum allowable amount of certain impurities in good quality boiler water listed in parts per million (ppm).

phosphates and polyphosphates, and do not use over 300 ppm of chromates and over 500 ppm of nitrites.

Organic growth can be controlled by using sodium pentachlorophenate. It is best to consult a water treatment expert before attempting boiler water treatment.

20-25 SERVICING HYDRONIC SYSTEMS

Air in a hydronic system is a common cause of trouble. Air will cause noise in the system, reduce the system water supply, and interfere with water circulation. The air acts as a brake on the circulation of water. Series systems can be purged of air by putting an outlet hose in a bucket. When bubbles stop appearing, the air has been removed.

Other systems must have a manual or automatic air vent at the high points in the system. See Fig. 20-64. A standpipe (drain) should be provided for each air vent.

Thermostats are another common source of heating problems for these reasons:
1. Vibration.
2. Poor contacts.
3. Broken wires.
4. Improper temperature settings.
   Other common problems include:
1. Air in hydronic systems using natural gas. Pump motor failure and water leaks around the pump packing due to air in the system.
2. Water pump motor and pump bearings (noise) made of carbon or teflon (water lubricated).
3. Cavitation.
4. Turbulence.
5. Poor supports (noise).
   Hydronic heating problems include:
1. Uneven heating. (Room-by-room heat loss calculation needed. Balancing valves needed.)
2. Velocity.
3. Vibration of parts of system.
4. Pipe expansion noises (do not clamp tightly; pipes should not touch edges of openings through floors). Expansion joints are of considerable help.
   Some controls needed are:
1. Air control devices (very much needed).
2. Air vents (piping pitched up to the vent). Automatic air vents are a problem source.
3. Balancing valves (to control flow rates).
4. Check valves that prevent reverse flow during low heat load periods.
5. Gate valves.
   Venting a system requires more water to be added to a system. This water contains air in solution, plus some corrosive chemicals. It is best to trap this new air in the compression tank. Avoid venting a compression tank if at all possible. The air compressed in this tank permits the water volume in the system to expand and contract, and it permits higher water temperatures.

When the system is working properly, one should be able to hold a hand on the compression tank. Higher temperatures indicate problems, and the relief valve may open. Because the compression tank water is cooler, it absorbs air before the water flows into the boiler. When in the boiler, the water is heated and the air is released. Some air may go into the pipes and end up in the vents. The tank may gradually lose its air, then the relief valve will open. If the relief valve spills water on each heating cycle, the compres-

Fig. 20-64. Automatic air vent for water heating systems (also on chilled water systems) is installed at high point of convector, baseboard units, or radiator. A—Vent. B—Connection to system. (Maid-O'-Mist Div.)

sion tank lacks air. Most air is released from the water when the water velocity is lowest, and where the temperature is highest (in the boiler).

Another problem is when the heat is off. The water will cool and the pressure will drop. The reducing valve will allow more fresh water (containing more air) into the system. If this continues, the inside of the system will become corroded from the chemicals that were in the makeup water.

One safety caution deserves mention. Cut the exhaust end of the pipe from the relief valve at an angle to prevent someone plugging it or capping it.

20-26 PREPARING BOILER SYSTEM FOR HEATING SEASON

Use the following checking procedure on a boiler system before the heating season begins:
1. Clean burner (gas or oil).
2. Clean nozzle of oil burner. Use a cloth and solvent. Do not use wire brushes (bristles may scratch orifice).
3. Clean and adjust electrodes. Inspect insulation. Replace if cracked.
4. Clean flame detector lens. Operate by closing fuel valve; detector controls should lock out (lock in the off position).
5. Clean pilot light if a gas burner is used.
6. Tighten all connections.
7. Oil motor.
8. Check motor temperatures. If warmer than normal, cleaning or new bearings may be necessary.
9. Inspect tubes for soot or fly ash. Clean.
10. Inspect breeching (top of boiler flue). Clean.
11. Cycle controls. Shut off water feed to check for low water cutoff.
12. Operate all valves and cocks to check operating condition.
13. Operate safety and relief valves.

20-27 STEAM HEATING SYSTEMS

Steam heating is a means of distributing heat to occupied areas. Steam is generated in a boiler. The steam (vapor),

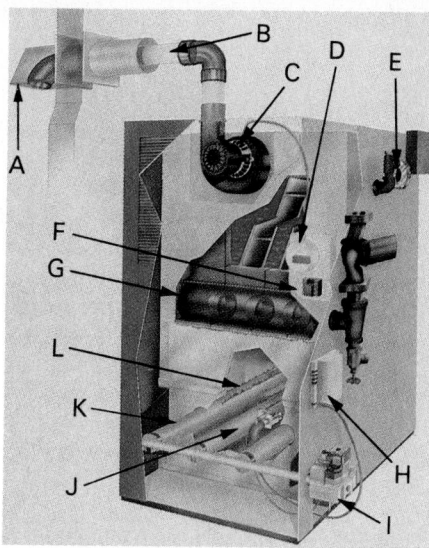

Fig. 20-65. Gas-fired direct vent hot water boiler. A—Outdoor cap. B—Vent piping. C—Exhaust fan. D—Positive vent safety shut-off. E—Pressure relief valve. F—Relay operated safety circuit. G—Heat exchanger. H—Primary ignition module. I—Safety gas valve. J—Electronic pilot ignition sensor. K—Stainless steel burners. J—Flame.
(Utica Boilers, Inc.)

Fig. 20-66. Water heater being mounted in steam boiler. (Weil-McLain, a Division of The Marley Company)

being lighter, travels to the upper parts of the piping circuit. This steam is at 212 °F (100 °C) or higher, except for vacuum systems (rare). As it releases its heat to the occupied area, the steam condenses. The condensed water, being heavier, returns to the boiler.

The steam releases about 1000 Btu for each pound that condenses. The heat exchange devices located in the room are called "radiators."

Two basic systems are in use. The single-pipe system uses the same pipe to carry steam to the radiators and return the condensate. The two-pipe system uses one pipe to carry the steam to the radiator and another pipe to return the condensate.

For domestic purposes, these systems operate at low pressures or at a partial vacuum. The units are tested at 50 psi (450 kPa) for safety purposes. Commercial and industrial systems use progressively higher pressures that approach 1000 psi (7 MPa).

A steam boiler must have pressure safety valves, water level gauge, pressure gauge, and temperature gauge. Fig. 20-65 shows a steam boiler.

A water heating unit (tankless heater) in a boiler is shown in Fig. 20-66. The unit heats only as much water as is needed: water is heated as it flows, instead of being stored.

20-28 STEAM HEATING INSTALLATION

Steam heating systems are excellent heating systems. When room radiators are used, the system does not produce humidity or humidify or clean the air. Air circulation is by thermal movement (hot air rises). Separate humidity systems and cleaning systems must be used.

Many installations have the radiator installed in a forced convection duct. This duct system may also have a filtering system and humidifying system.

All steam heating systems must be installed according to code regulations. As with hot water boilers, the steam boiler must be mounted level. The pipes must be mounted with a slope down to the boiler. The piping must be designed to provide for expansion of the pipe. Air vents

must be located at the high points of the system. Each radiator should have an air vent and a steam trap.

The steam trap keeps the steam in the radiators and only allows condensate to return to the boiler. There are three types of steam traps (not discussed further):
1. Mechanical.
2. Thermostatic.
3. Impulse.

After installation, the system must be flushed to remove all dirt. The system must be leak-tested as per code regulations before it may be operated. After system startup, perform the safety inspection detailed in Para. 20-30.

20-29 SERVICING STEAM HEATING SYSTEM

Servicing a steam heating system should be done with great care. Escaping steam or condensate can cause severe body burns. A boiler will explode if the steam pressure is permitted to exceed the boiler safe pressure.

Check the water level gauge and the pressure gauge. Shut down the system at once if the water level does not show in the level gauge, or if the pressure gauge is above normal.

Servicing of the gas burner has been explained in previous paragraphs. Para. 20-39 suggests components to check in oil burner servicing.

If one radiator of the system is cool while the others are hot, this radiator is not receiving steam. This problem may be caused by:
1. Thermal valve to radiator is closed (thermostat for valve or valve not working).
2. Radiator may be air bound (air vent not working).
3. Radiator may be filled with condensate (steam trap not working).

Lightly rapping the control with a rubber mallet may loosen the valve momentarily and allow the system to work. If this happens:
1. Shut down system.
2. Reduce pressures to atmospheric by purging. (BE CAREFUL! Stand to one side to prevent burns from steam or condensate water.)
3. Replace thermal valve, air vent, or steam trap.
4. Start up system.

If the system is low on water, and the boiler is hot, cool the boiler before opening or repairing the boiler feed valve. Cold water may crack a hot boiler section.

Some boilers are made with tubing heat exchangers. These boilers may use either a water tube or a finned tube design.

Short cycling times cause rapid temperature changes in the boiler. After several thousand cycles, the boiler walls or tubes may crack.

20-30 SERVICING: STEAM HEATING SAFETY INSPECTION

A steam heating system should be checked once each month during the heating season.

Check the water level. The water level gauge must show that the boiler water level is at one-third to one-half full. If the water level is higher, drain the system to the correct lower level. If the water level is low, but still shows in the level gauge, add water until the level is correct. Close the fill valve completely. If no water shows in the water level gauge, shut the system off at once. Add water only after the boiler has cooled.

The water level sight glass may show water if its openings are clogged. Trust the sight glass reading only if the system is clean, and if the sight level glass and its connections have been cleaned recently. Water level petcocks (when used) are a good way to check the sight level glass tube reading. Be sure the petcocks do not spill steam or hot water on anyone. Wear goggles!

Operate the safety (relief) valve to be sure it is not stuck closed. Again, be careful not to allow the escaping steam to hit anyone. Wear goggles! If no steam or water comes out, shut off the system at once and have the relief valve serviced or replaced.

Check the low level shutoff by opening the drain valve. The shutoff should operate right away by cutting off all fuel and electrical power. If the shutoff does not operate, shut down the unit and service the low water level shutoff.

20-31 FUEL OILS

Fuel oils vary considerably. Generally, they contain about 85 percent carbon (C), 12 percent hydrogen (H), and various other elements in the remaining 3 percent. During combustion, carbon and hydrogen combine with the oxygen (O) in the air to produce carbon dioxide (CO_2) and water (H_2O). Fuel oil grades which are established by the U.S. Department of Commerce conform to ASTM (American Society of Testing Materials) specifications. A very low sulphur (S) content is very important since the sulphur turns into corrosive gases and liquids.

Fuel oil grades 1 and 2 are used in domestic and small commercial furnaces. Grade 1 is used in pot-type oil burners; Grade 2 is the most popular domestic fuel oil (about 140,000 Btu/gal.). It has a flash point of 100 °F (38 °C), a Sayboldt viscosity of 37.9 (compared to Grade 4 with a viscosity of 125). Grades 4, 5, and 6 are used in industrial applications. The heavier oils 3, 4, 5, and 6 provide slightly more heat per gallon.

Products of combustion should be carbon dioxide (CO_2) and water in vapor form. Actually, there is also carbon monoxide (CO) (very toxic—dangerous), plus sulphur dioxide (SO_2) and sulphur dioxide vapors. About 106 lb. of air are required for each gallon of Grade 2 fuel oil consumed. Multiplying 106 × 14 (cu. ft. per lb.) equals about 1500 cu. ft., which is the quantity of air that must be fed to a furnace for each gallon of fuel oil consumed. This means that 1500 cu. ft. of air must enter the building for each gallon burned (about each two hours of oil burner running time for the average home).

It takes about 104 cu. ft. of air (7.43 lb.) to burn 7 lb. of oil (1 gal.). The 7 lb. of oil in one gallon are made up of 6 lb. of carbon and 1 lb. of hydrogen. The 106 lb. of air equal about 1500 cu. ft. of air. Of the 106 lb. of air, about 84 lb. are nitrogen. Nitrogen does not produce burning, but acts as a gas to lower the temperature and to waste some heat as it is warmed and goes up the chimney.

Of the 106 lb. of air, about 22 lb. are oxygen (O) that combine with the oil to form about 20 lb. of CO_2 and 9 lb. of water (steam).

When perfect combustion takes place, about 15 percent of the flue gas volume is CO_2. This level is not reached with oil burners. Because of the heavy carbon molecules in the oil (soot- and smoke-formers), excess air is used to burn these carbons more completely. So, generally, enough excess air is fed to the firepot to reduce the CO_2 to about 10 percent. If 100 percent excess air is used, the CO_2 content reduces to 7.5 percent. CO_2 is measured first because it is easier to measure.

In the flame, the hydrogen always burns first. It burns completely, then the carbon starts to burn. This action causes pulsations (pressure waves) in the flame. Also, the carbon first turns into CO, then into CO_2. This, too, may cause pulsation.

Combustion gases vary. For good combustion, excess air must be used. Therefore, considerable nitrogen (from the air), some oxygen, carbon dioxide, steam, and impurities go up the stack. About 2000 cu. ft. of air is used per gallon of oil (400 cu. ft. of oxygen). These gases may be moved up the stack either by natural convection (common in domestic and small commercial units), by forcing with a fan or blower (forced draft), or by drawing the gases up the chimney (induced draft).

It is important to keep flue gases warm. Otherwise, condensation will take place in the stack and flue, causing severe corrosion. One corrosion agent will be sulphurous acid (H_2SO_3), which corrodes steel rapidly and discolors brick and stone. Most good fuel oil additives will: reduce sulphur dioxide about 50 percent; keep the flue and chimney cleaner (65 percent cleaner); cause less opaque (visible) smoke; reduce soot blowing of tubes. If there is incomplete combustion, the flame is white rather than the normal orange color (more complete combustion).

An oil furnace in good condition should not release visible smoke from the flue, chimney, or stack. However, there may be soot deposits and fly ash, which should be removed annually. The soot may be removed by using air pressure, mechanical cleaning, vacuum cleaning, or chemicals.

Oil sludge which clogs filters and nozzles may be caused by bacteria which multiply if water is present in the oil. An additive can be used to kill the bacteria.

The combustion chamber must be kept in good condition. Deposits on the refractory (asbestos cement) lining must be kept to a minimum.

Fuel oil additives reduce deposits in the combustion chamber, heat exchangers, and flue.

Sulphur trioxide is more odorous than sulphur dioxide and is minimized by using additives and a higher temperature.

Number 2 fuel oil viscosity changes from between 50 and 100 at 0 °F (−18 °C) to 35 and 45 at 70 °F (21 °C). This means that gun-type oil burners may have pumping and combustion problems when the oil is cold.

Efforts to reduce air pollution have increased the use of

Fig. 20-67. Base-mounted gun-type oil burner. A—Combustion head nozzle with air-injection holes. B—Electrical control, pulse ignition transformer, recycle primary control, and cad cell flame detector. C—Oil pump. (The Carlin Company)

Number 2 fuel oil distillate in commercial buildings and in industry. This grade burns more completely and cleaner than Number 3, 4, 5, and 6 fuel oils. Some fuel oils are called distillate, because they are products of a distillation or cracking process at the oil refinery. The Number 1 and Number 2 oils are called distillates because they are oils which were vaporized, then condensed in the refining system.

20-32 OIL FURNACES

The most common type of oil burner is the gun type. Rotary-type and pot-type oil burners are rarely used.

The gun-type burner forces oil under pressure through an orifice having a controlled size. The oil is broken into finely divided particles (atomized), mixed with air, and forced into the combustion chamber by a blower. Fig. 20-67 shows a gun-type oil burner. Unit has air-injection combustion head with radial air-injection holes in nozzle.

Remember that oil will not burn while it is in the liquid form. It must be vaporized and turned into a gas. To vaporize oil, heat must be added to the oil (latent heat of vaporization). The oil will turn into a gas quicker and easier if it is finely divided (sprayed). This spraying process is called atomizing. The minimum atomization pressure is 75 psi (620 kPa). Gun-type oil burners accomplish atomizing by forcing oil into a twisting, spiraling, and turbulent air stream. A small amount of heat (electric spark) will turn a few of the finely divided particles into gas and the burning will start.

Some large industrial furnaces use combination oil and gas burners. Switches on the burner control permit the choice of fuel.

Pulsation in an oil furnace is usually caused by positive pressure in the combustion chamber (not enough draft). Draft should be 0.02 in. (0.5 mm) water pressure. Too much positive pressure may be caused by:
1. Too much air (air shutter open too much).
2. Chimney too small, partially blocked, or not high enough. The chimney must be 2 ft. (61 cm) higher than the highest point of the building.
3. A faulty nozzle (distorted flame pattern).

Oil on the floor of the furnace room is dangerous. It may be caused by an air leak in the oil suction line (air causes drip at nozzle). It may be caused by loose compression fittings or unions, or by a pump seal leak. Compression fittings should never be used on oil systems.

Check for air in the system by connecting a pressure gauge. If the gauge needle fluctuates, it is a sign of air. For small leaks: put oil outlet tube in a bottle of oil; bubbles will indicate an air leak in the suction line. The mechanical resonance of 2 1/2- and 3-in. gauges is close to gearset frequency.

If there is soot in the boiler flue passages:
1. Clean.
2. Clean blower blades with a brush; clean blower tube.

An oil furnace blowback is usually caused by delayed ignition. The most common reasons for blowback are:
1. Electrodes improperly spaced.
2. Distorted pattern away from electrodes.

If the oil nozzle is in poor condition, replace it with an exact replacement unit (same orifice, spray angle, and solid or hollow cone as originally used).

Large oil burner furnaces may have a metal combustion chamber. Smaller furnaces use refractory cement liners for the combustion chamber. This cement consists of 80 percent dry asbestos, 20 percent Portland cement, and enough water to make the mixture workable. Avoid putting this cement on the edge of or inside of the air cone (cement will change the air pattern and cause inefficient burning). It is best to fill the tube with a rag while applying the cement.

Many refractory liners are preformed, then installed. Fig. 20-68 shows a burner installed in a boiler.

20-33 GUN-TYPE OIL BURNER

Gun-type oil burners are available in two types:
1. High-pressure type.
2. Low-pressure type.

In the high-pressure type, oil is fed under 100 to 300 psi (790 to 2 200 kPa), to a nozzle. Air is forced into the furnace through a tube that surrounds this nozzle. Usually, the air is twisted in one direction; the oil spray is given a twist in the opposite direction. Fig. 20-69 shows a gun-type oil burner. The nozzle should be carefully centered in the housing. The ignition transformer provides a high voltage spark between two electrodes located near the front of the nozzle. Fig. 20-70 shows the nozzle and electrode assembly.

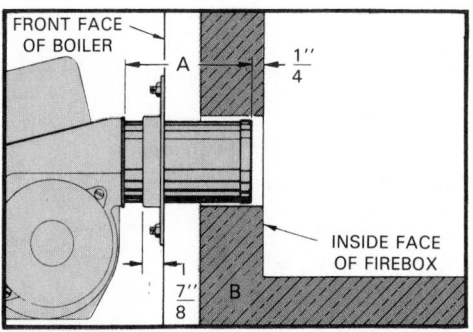

Fig. 20-68. Typical gun-type oil burner installation in boiler. A—Length of air tube. B—Refractory insulation. (R.W. Beckett Corp.)

Fig. 20-69. Flange-mounted gun-type oil burner. A—Transformer. B—Oil pump. C—Air adjustment. D—Motor. (R.W. Beckett Corp.)

The low-pressure type burner uses oil at 1 to 4 psi (110 to 130 kPa). Oil is mixed with air before it reaches the nozzle.

The main parts of a gun-type burner are: motor, oil pump, fans, nozzle, choke, air tube, and ignition system. Fig. 20-71 shows the various parts of a gun-type oil burner.

The choke is a tapered down or smaller opening at the end of the air tube (large tube that is part of the pair marked No. 1 in Fig. 20-71). The choke is located just past the oil nozzle. The choke has swirl strips (vanes) to increase the twist and turbulence of the air to obtain better mixing of the oil spray and air for more efficient burning. The flame shape can be changed by moving the choke closer to or away from the nozzle.

The air tube has a disk mounted inside it. This disk disturbs the airflow and creates air turbulence for better mixing. This disk is usually called the static pressure disk.

The oil moving through the nozzle is also given a twisting movement as it travels through very small holes drilled at an angle to the nozzle.

The oil burner motor is usually a split-phase 1/6 hp unit that provides power for both the fan and the fuel pump. The motor is electrically connected to the master oil burner stack control and uses 120 V 60 cycle electricity. Fig. 20-72 shows an adjustable air band that adjusts the firing range of the burner.

Motor speeds may be 1750 rpm (60 Hz) fan speed and pump speed, or 3450 rpm (60 Hz) fan speed and pump speed, depending on whether the motor is a 4-pole or 2-pole type. On a 50 Hz input, a 2-pole motor will run at 2850 rpm.

The fan is usually the radial flow type with adjustable air inlet openings. The openings are adjusted until the flame burns a yellow color.

Some excess air is needed to insure enough oxygen for the oil (the flame action is fast!) and also to allow for airflow decrease between furnace inspection and cleaning (fan blades pick up lint and airflow will decrease). Excess air will slow down evaporation of oil droplets.

Adjusting an oil burner to obtain a proper flame can be misleading. A dirty nozzle, impingement (striking the choke vanes), or oil leakage during the "off" part of the cycle will

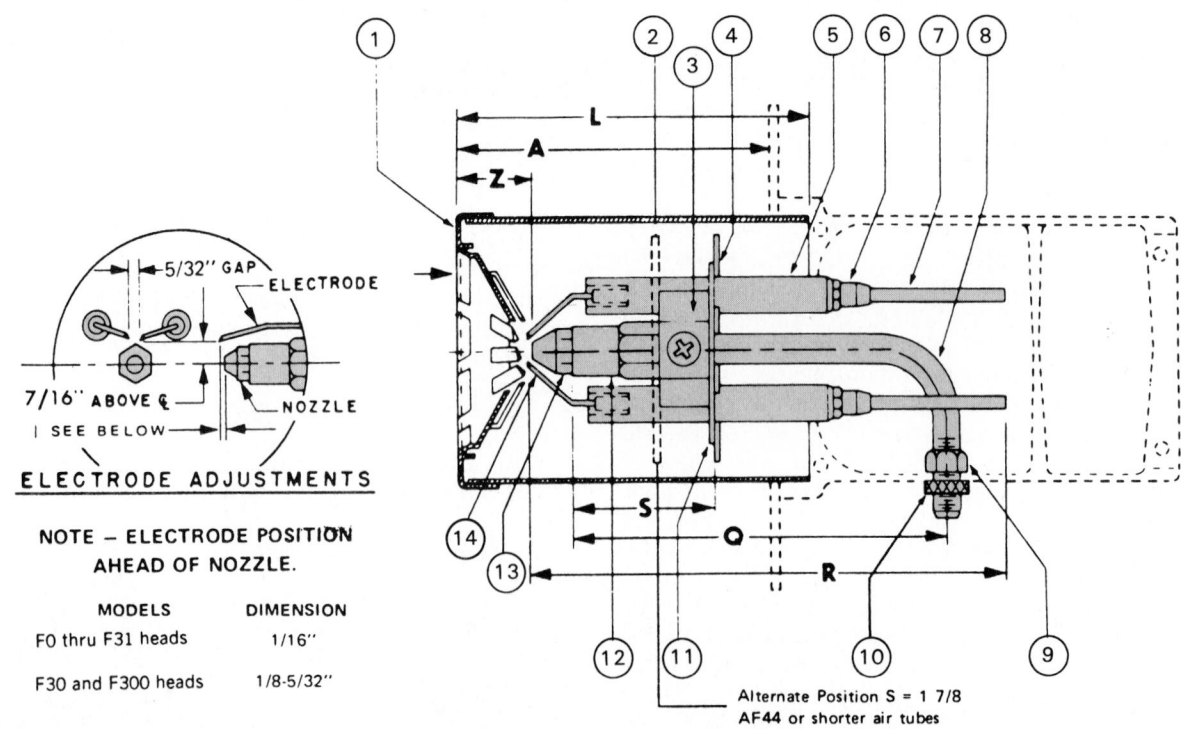

FIg. 20-70. Design of air tube, oil nozzle, and electrode assembly of gun-type oil burner: 1—Burner head. 2—Air tube. 3—Electrode clamp. 4—Centering spider. 5—Porcelain. 6—Electrode rod extension adaptor, as required. 7—Electrode rod extension, as required. 8—Nozzle line and vent plug. 9—Bulkhead fitting kit. 10—Locknut bulkhead fitting. 11—Static plate, static plate holding screws. 12—Nozzle adaptor - single. 13—Nozzle. 14—Electrode rod and tip.   (R.W. Beckett Corporation)

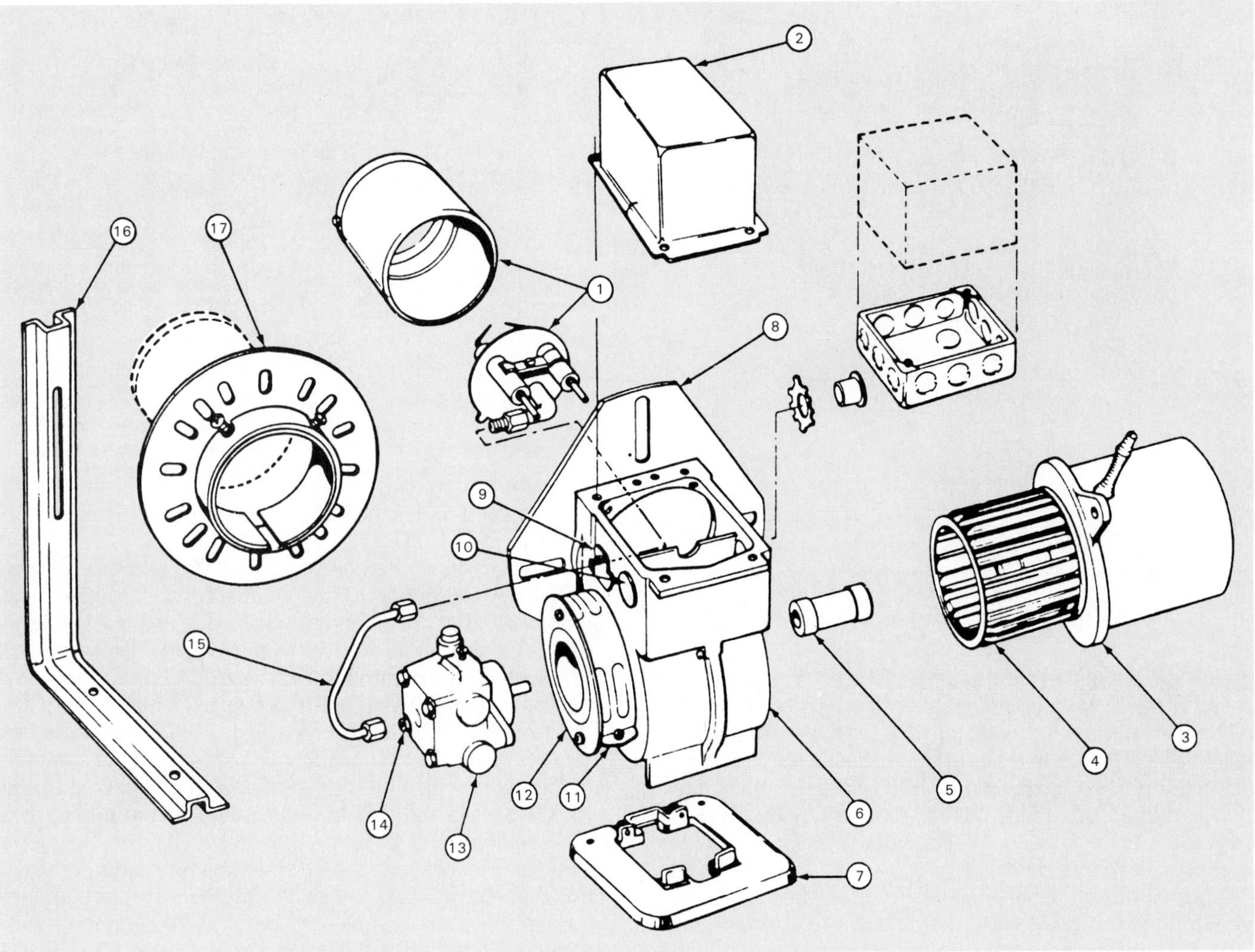

Fig. 20-71. Exploded view of gun-type oil burner: 1—Air tube combination. 2—Ignition transformer. 3—Drive motor. 4—Blower wheel. 5—Flexible coupling. 6—Burner housing assembly, with inlet bell. 7—Pedestal support. 8—Unit flange, or square plate. 9—Nozzle line escutcheon plate. 10—Hole plug-wiring box. 11—Bulk air band. 12—End air shutter. 13—Fuel unit. 14—Pump outlet fitting. 15—Connector tube assembly. 16—Extended pedestal. 17—Adjustable mounting flange.   (R.W. Beckett Corporation)

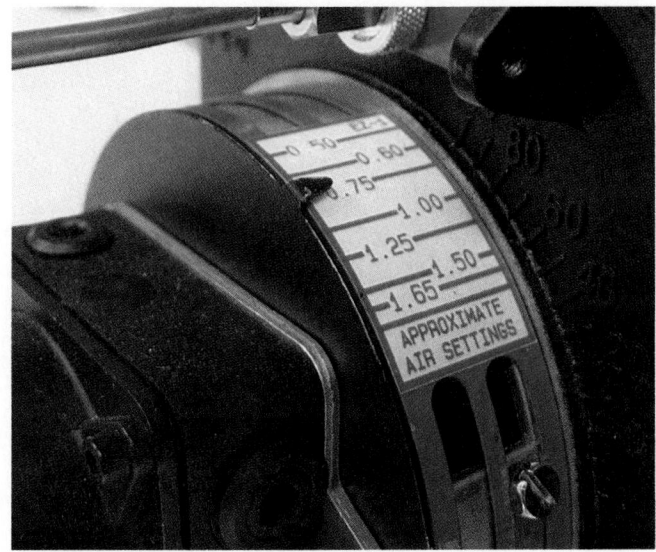

Fig. 20-72. Adjustable air band. The desired firing rate is set on air scale with adjustable arrow. This provides instant air setting for firing rate of burner.   (The Carlin Company)

tend to make a flame look like it needs more air when it does not.

The only good way to check an oil burner is with instruments (O analyzer, CO₂ analyzer, smoke test, etc.).

Safety devices are installed to avoid spraying unburned oil into a furnace (DANGEROUS). Safety devices also prevent continuous operation of the oil pump in case of ignition failure or if the oil flame goes out. The stack control is one method. See Chapter 25 for oil burner controls and wiring circuits.

20-34 GUN-TYPE OIL BURNER PUMPS

Several types of fuel oil pumps are used in gun oil burners, including the gear type and the rotary type. The rotary oil pump is more common, but the gear type is more easily serviced.

ROTARY-TYPE FUEL OIL PUMP

Rotary pumps come in either single-stage, Fig. 20-73, or two-stage models. The internal construction of a single-stage rotary fuel oil pump is shown in Fig. 20-74.

Fig. 20-73. Single-stage gun-type oil burner fuel pump. A—Oil inlet connections. B—Pressure regulator. C—Oil outlet connection. D—Air bleed and gauge connection. Note arrow on top for shaft rotation direction. (Suntec Industries, Inc.)

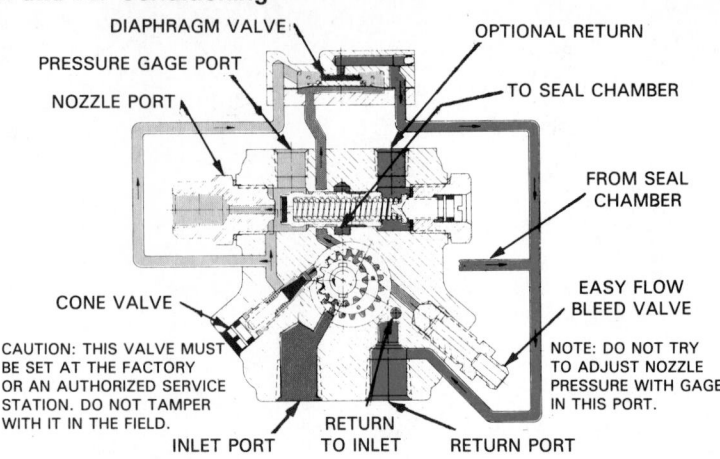

Fig. 20-75. Diagram of oil flow through single-stage oil pump. The four colors represent the different pressure levels: Green—Suction or inlet pressure. Red—Gear pressure. Yellow—Nozzle pressure. Blue—Return pressure. (Suntec Industries, Inc.)

The oil supply system carries fuel oil from the tank through a filter in the line, through the inlet screen in the oil burner, to the pump, and into the pressure regulator and relief valve. The pump rotates counterclockwise and oil flow is from top to bottom in Fig. 20-73. Note the shaft direction arrow on top in Fig. 20-73. As shown in Fig. 20-75, the oil leaves the lower central axis of the pressure regulator and passes to the gun nozzle.

Another design for a single-stage rotary fuel oil pump is shown in Fig. 20-76. Note the use of a filter screen. The inlet is at the bottom left and the return/bypass is at the bottom right. The outlet is at the top left.

Many systems are equipped with the two-stage fuel oil pump when the two-pipe system is used and part of the

oil is returned to the fuel tank. The principle of operation is shown in Fig. 20-77.

The selection of a single- or two-stage pump is a function of the limitations of the oil itself. Most single-stage pumps are sold for inlet vacuums of 7 in. Hg (78 kPa) single-pipe and 10 in. Hg (68 kPa) two-pipe. When fuel oil is subjected to vacuums in excess of 10 in. Hg, it starts to come apart. At 15 in. Hg (51 kPa), it is a mixture of foam and clean oil. Two-stage pumps, Fig. 20-78, are ported so that, at 15 in. Hg, the second or pressure stage's inlet is submerged in clean oil. The intake from the tank is at the top. Oil passes through the first stage of the pump, into the regulator, and back to the tank. The second stage removes only oil from the strainer chamber and pumps it into the nozzle. Excess oil is returned to the strainer chamber, in some pumps, and back to the tank in others.

Details of the relief valve are shown in Fig. 20-79. Oil pressure creates a force against the piston. When this force equals the compression spring force, the piston moves down and permits oil to flow back into the pump inlet.

Fig. 20-74. Single-stage rotary fuel pump for gun-type oil burners. A—Shaft. B—Shaft seal. C—Pump rotor. D—Pump housing. (Suntec Industries, Inc.)

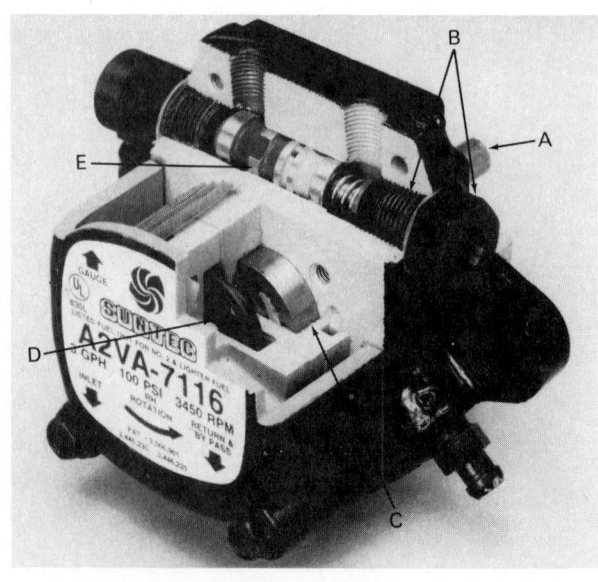

Fig. 20-76. Cutaway of single-stage rotary fuel oil pump for oil burners. A—Shaft. B—Pressure adjusting screw. C—Single-stage gear set. D—Shaft seal. E-Piston. (Suntec Industries, Inc.)

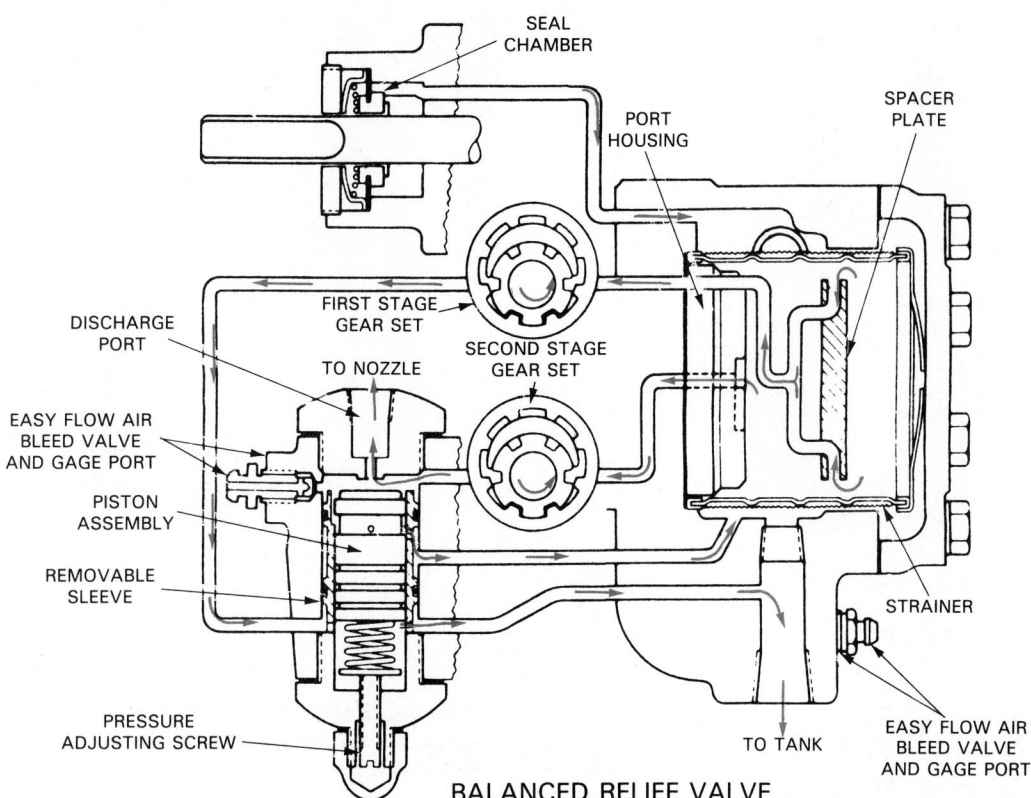

Fig. 20-77. Two-stage fuel oil pump for oil burners, used with two-pipe system from storage tank.

GEAR-TYPE FUEL OIL PUMP

The gear-type oil pump is available in both single-stage and two-stage models. Fig. 20-80 shows the external appearance of a single-stage gear pump. Single-stage pump operation is shown in Fig. 20-81. In A, the fuel oil enters the single-stage unit and fills the front chamber. The rotating blades filter the oil as it passes from the front chamber to the suction sides of the gears. In B, the oil then goes from the lower suction side to the upper pressure side of the gears and flows into the valve. At a predetermined pressure, the valve piston moves and the oil flows out the nozzle port. In C, the surplus oil returns to the front chamber through the surplus return passage. Oil lubricating the internal shaft and seal returns to the front chamber through the seal drain.

The nozzles generally are the 80 deg. type giving a hollow cone spray pattern and with a capacity of 0.75 to 1.75 gph (gallons per hour), or the 60 deg. hollow cone type of 1.75 to 12.00 gph (1 U.S. gallon = 3.79 liters). To maintain oil temperature and viscosity during off cycle in ambient situations, a nozzle line heater, Fig. 20-82, is used on some oil burners. The oil is heated to 120°F to 130°F at the oil nozzle. This improves the ignition and overall combustion.

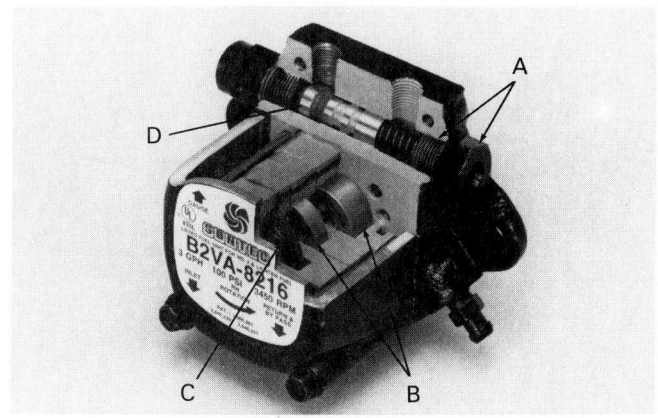

Fig. 20-78. Two-stage fuel oil pump used with two-pipe system from storage tank. A—Shaft. B—Double-stage gear set. C—Shaft seal. D—Piston. (Suntec Industries, Inc.)

Fig. 20-79. Cutaway view of relief valve for fuel oil pump. A—Outlet to nozzle. B—Pressure regulating screw. C—Pump shaft. D—Mounting flange. E—Oil pressure release line. (Suntec Industries, Inc.)

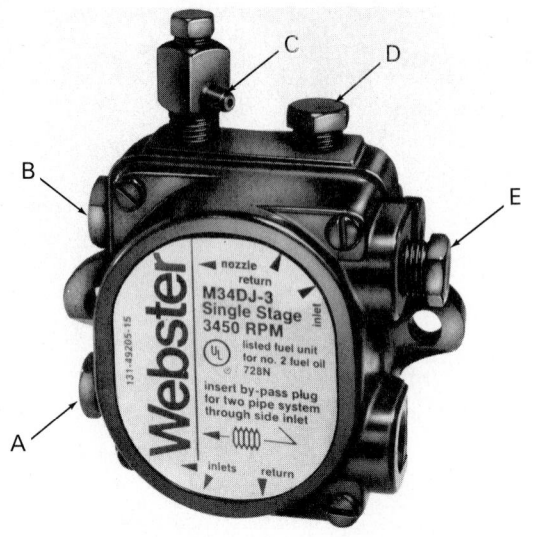

Fig. 20-80. Single-stage gear pump for gun-type oil burners: A—Inlet port. B—Nozzle port. C—Vent plug. D—Alternate return port. E—Alternate inlet port. (Webster Heating and Specialty Products, Inc.)

Fig. 20-82. Nozzle line heater attached to combustion head. (The Carlin Company)

The gun-type oil burner is an efficient heating unit. However, it must be properly maintained to give peak performance. An experienced service technician should check, clean, and adjust the system each year. Some of the important items to check are shown in Fig. 20-83.

20-35 ELECTRICAL IGNITION

Gun-type oil burners generally use electrical ignition. Ignition controls are described in Chapter 25. The system includes a transformer and two electrodes. The transformer is mounted on the oil burner. It transforms 120 V ac to about 10,000 V ac. The ignition system must raise the oil temperature to 700° (370 °C) for burning to take place. The electrodes, made of stainless steel, are mounted in ceramic insulators. No part of the electrodes should be less than 1/4 in. (6 mm) away from metal parts.

The electrode ends are positioned in front and above the nozzle. As the atomized oil swirls out of the nozzle and mixes with the turbulent air, a spark jumps between the electrode ends and ignites the mixture. The ignition may be continuous while the oil burner is in operation, or it may operate only until the fuel ignites.

The electrode gap should be between 1/8 in. (3 mm) and 3/16 in. (5 mm). The electrode ends should be approximately 1/2 to 5/8 in. (13 to 16 mm) above the nozzle and 5/16 to 1/2 in. (8 to 13 mm) in front of the nozzle. For over 45 deg. nozzles, this last dimension should be approximately 1/2 in. (13 mm). For 30 deg. nozzles, the electrode ends should be 5/16 in. (8 mm) in front of the nozzle. Refer to the manufacturer's service manual for exact setting specifications. The porcelain insulators must be kept clean or the high voltage will short. The spark should jump a 1 in. (25 mm) gap with the blower off.

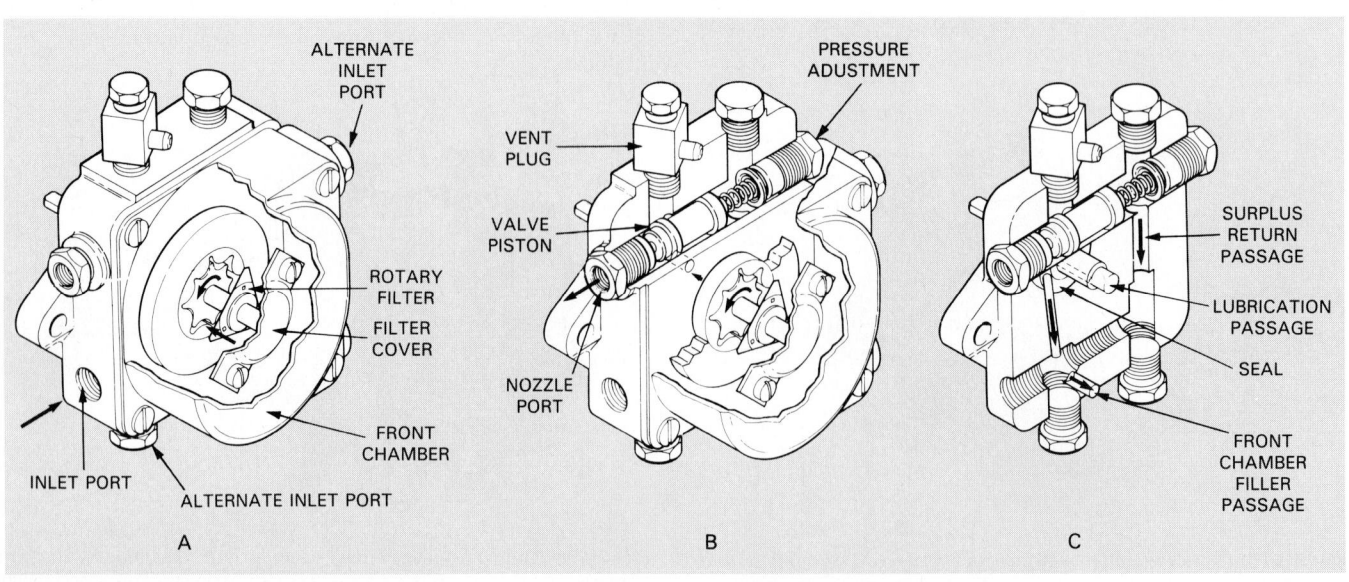

Fig. 20-81. Single-stage gear pump unit operation. (Webster Heating and Specialty Products, Inc.)

Fig. 20-83. Six main items to be checked at start of each heating season. 1—Shutoff valve and line filter. Replace filter cartridge. 2—Check and clean nozzle assembly. Follow manufacturer's recommendations. 3—Clean strainer using clean fuel oil or kerosene. 4—Check connections for tightness. 5—Insert pressure gauge into pressure port. Start burner and adjust pressure setting to manufacturer's specifications, usually about 100 psi (790 kPa). 6—Pressure gauge reading for correct pressure setting. (Suntec Industries, Inc.)

A flame mirror, Fig. 20-84, can be used to observe ignition action and spray action, to check if operation is normal.

Weak ignition, wrong position of the electrodes, or poor insulation may cause delayed ignition and a puffback. Fig. 20-85 shows the interior of an ignition transformer case. A transformer and line voltage testing instrument are shown in Fig. 20-86. A puffback is the ignition of a large amount of vaporized oil in the firepot. Sometimes, it will blow soot into the furnace room and into the living quarters, making a major cleaning job necessary. In no case should the electrode ends be touching the oil spray. If they do, the electrodes will become carbon coated.

Moisture in oil will retard combustion and may even cause a "flameout." A flameout is when combustion stops, but fuel continues to feed into the combustion chamber. If moisture is causing poor combustion, continuous ignition is usually recommended.

20-36 OIL TANK INSTALLATION

Oil burners must be installed with great care. A complete installation includes a 200- to 1000-gallon tank, hand shutoff valve, filter and trap combination, and copper tubing oil line. Fig. 20-87 shows an installation for a one-pipe

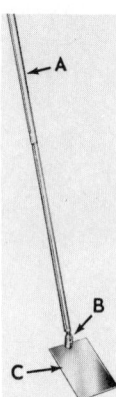

Fig. 20-84. Inspection mirror used to check nozzle condition, electrode condition, and flame. A—Telescoping handle. B—Hinge. C—Metal mirror.

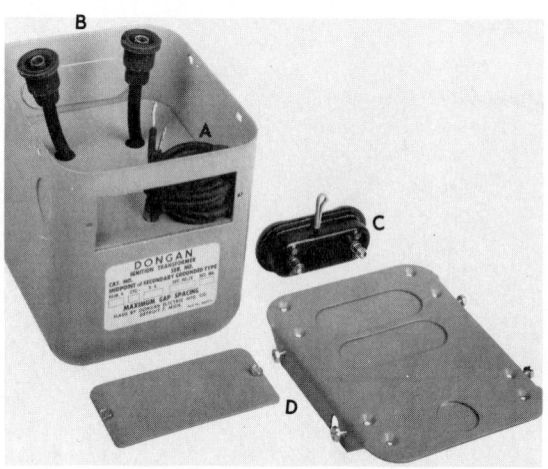

Fig. 20-85. Ignition transformer used on gun-type oil burners. A—Primary leads. B—Secondary leads. C—Secondary bushing. D—Cover plates.

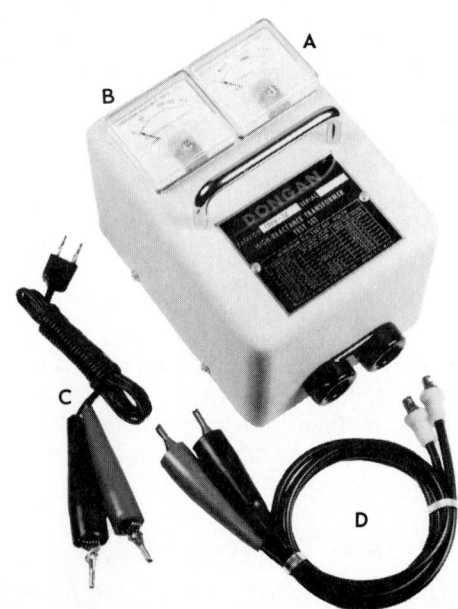

Fig. 20-86. Ignition transformer testing instrument. A—High voltage meter. B—Line voltage meter. C—Line testing leads. D—Transformer high voltage testing leads. (Dongan Electric Mfg. Co.)

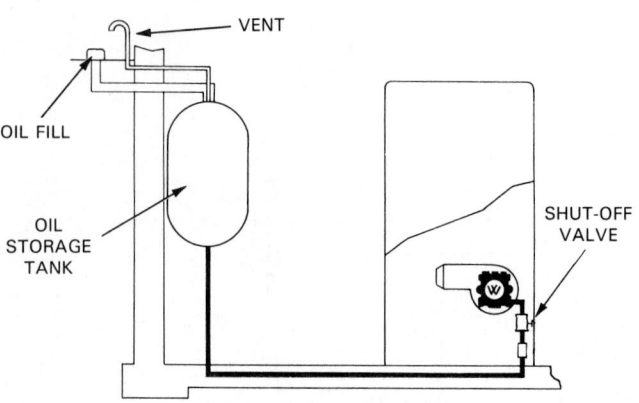

Fig. 20-87. Typical gun-type oil burner installation. Note fill pipe, vent pipe, and oil line installation. (Webster Heating and Specialty Products, Inc.)

system with the storage tank located in the room with the furnace. Remember that the storage tank should be at least 7 ft. (2 m) away from the furnace. In this installation, oil feeds by gravity to the oil burner. The storage tank should be elevated less than 25 ft. (7 m) above the burner to keep the feed line pressure below 10 psi (170 kPa), or a pressure reducing valve must be used.

In some installations, the fuel tank is placed outside the building, sometimes underground. Some local codes allow only 3-5 psi (120-140 kPa) on the pump inlet. The two most common installations are:

1. With the tank above the oil burner, as in a residence with a basement, Fig. 20-88.
2. With the tank below the level of the oil burner, as in a home without a basement, Fig. 20-89.

These installations should have the tank located within a reasonable distance of the oil burner. On runs of 50 to 100 ft. (15 to 30 m), 3/8-in. (10 mm) tubing should be used. For runs of 200 to 300 ft. (60 to 90 m), 1/2-in. (13 m) tubing should be used. The manufacturer's specifications should be checked if the oil must be raised above the tank.

An oil tank should be installed with a slight slope away from the oil line connection to provide a low spot in the

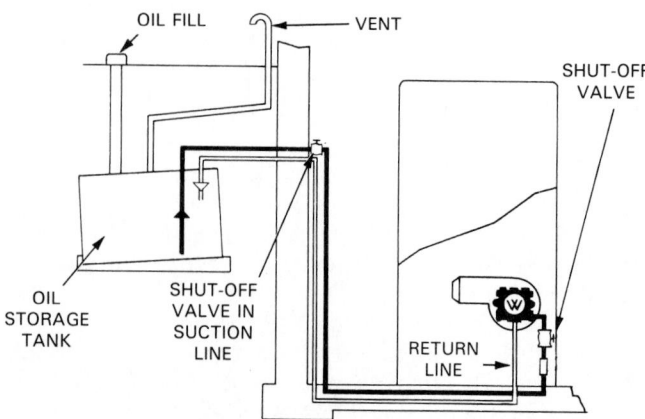

Fig. 20-88. Gun-type oil burner installation with storage tank installed underground, but above oil burner. Note two oil lines. (Webster Heating and Specialty Products, Inc.)

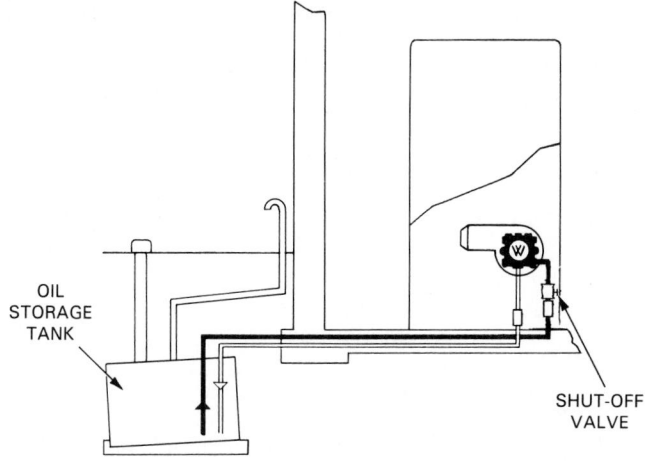

Fig. 20-89. Gun-type oil burner installation having storage tank underground and below level of oil burner. (Webster Heating and Specialty Products, Inc.)

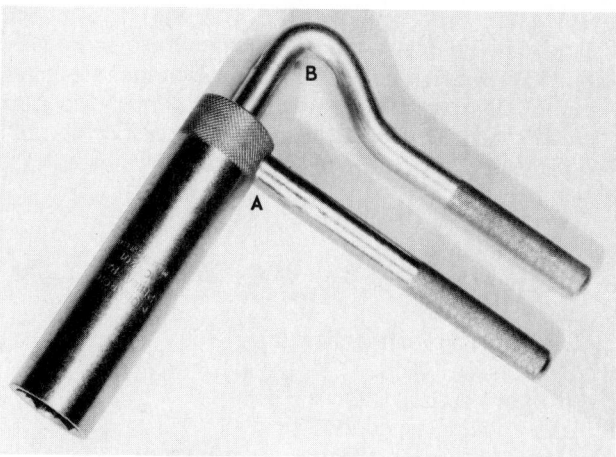

Fig. 20-90. Nozzle removing and installing tool. A—Nozzle tube socket wrench and handle. B—Nozzle socket wrench and handle. (Monarch Mfg. Works, Inc.)

tank for dirt and water to collect. The vent pipe is very important:

1. It provides atmospheric pressure inside the tank and permits volatiles to escape.
2. It must be designed with a 180 deg. bend (to keep out dirt and rain).
3. The opening should be above the highest possible snowfall or other blockage.

The oil fill cap should always be in place except when filling the tank. The cap helps keep the fuel oil clean and reduces the chances of fire or explosion.

Always use a pipe thread compound on the pipe threads of the fittings. This compound should be of the oil-resistant, nonhardening type. During storage and while installing tubing, keep the tubing ends sealed with tape to keep out dirt and moisture. Remove the tape at the time the tubing is connected. The 3/8- or 1/2-in. (10 or 13 mm) OD copper tubing is attached to the fittings with standard SAE 45 flares. Flaring techniques are described in Chapter 2. Tight, leakproof connections are essential. Never use Teflon® tape.

The oil lines in the tanks should be mounted so the tubing opening is 3 to 4 in. (7.5 to 10 cm) above the bottom of the tank. If the system has a return oil line, this line does not need to go near the bottom of the tank for light oils (Number 1 or Number 2). It should go within 4 to 5 in. (10 to 13 cm) of the tank bottom for the heavy fuel oil to keep the oil more fluid. Never use compression fittings with fuel oil systems.

20-37 OIL BURNER INSTALLATION

The oil burner installation must be made in accordance with local codes and the manufacturer's instructions. The burner must be the correct height above the bottom of the combustion chamber. Burners are mounted either on adjustable legs or on a flange bolted to the furnace. The burner air tube and nozzle must be inserted into the combustion chamber the exact distance the manufacturer recommends. The opening into the furnace must be carefully sealed (some have flange adaptors) to prevent air leaks.

The nozzle must be the correct size, and it must be in good condition. The size of the hole in the nozzle and the

amount of oil pressure determines the rate at which fuel oil is burned and, therefore, the rate at which heat is produced. The size of the nozzle selected must match the heating requirements of the heated space. If the nozzle is too small, the burner may not heat the space adequately. If the nozzle is too large, there will be a tendency for the burner to come on and off quite frequently.

Nozzles are usually supplied with a fine filter at their fuel oil inlet. The filter is designed to eliminate the possibility of dirt entering and plugging the nozzle.

Be careful not to twist the tube or move the nozzle tube out of line. A tool for safely removing the nozzle is shown in Fig. 20-90.

20-38 STARTING AN OIL BURNER INSTALLATION

Before starting the oil burner, air should be removed from the lines and the pump. A vent plug (air bleeder fitting) is mounted in the pump housing. Usually, this vent plug seals the port used for pressure gauge installations.

If enough oil and air collects in the combustion chamber and is ignited, anything may happen—anything from a puff of flame to an explosion that may wreck the building and maim or kill. Inspect the firepot. If oil is present, shut off oil valves and vent the combustion chamber. Remove the oil (by means of a suction pump, rags, etc.) until all danger of oil fumes is gone. Use explosion-proof flashlight.

Air in an oil line will form bubbles, which could result in:

1. Oil not being pumped.
2. Blowbacks.
3. Flame failures.

The line must be completely purged of air. A two-pipe system reduces the chance of air remaining in the system. However, air can still be trapped in high spots in the line. A leak in the oil line will almost always cause air-in-line troubles.

Always check the fuel oil nozzle to be sure it is the correct size, and that it is in the center of the gun air duct. The electrodes must be kept clean and in correct relation to the nozzle. Fig. 20-91 shows a typical oil burner nozzle. These nozzles come in various capacities, all based on gallons per hour (gph) at 100 psi (790 kPa) (from 0.40 to 28 gph). Some nozzles are large enough to feed 100 gph. Remember that a 1 gph nozzle delivers 140,000 Btu/hr. If the overall efficiency is 60 percent, the useful heat would be 84,000 Btu/hr. Poor oil delivery may be the result of the main filter, the pump filter, or the nozzle filter being partially clogged. Check all three filtering devices when servicing the unit.

Flame failure may be caused by one or more of the following:

1. Oil tank out of oil.
2. Oil tank not vented.
3. Clogged filter in oil line.
4. Ice in fuel line.
5. Loose oil line connection (air in line).
6. Dirt in supply line.
7. Water in supply line.
8. Loose wiring or connections.
9. Motor not running (check reset button).
10. Defective pump.
11. Pump losing prime.

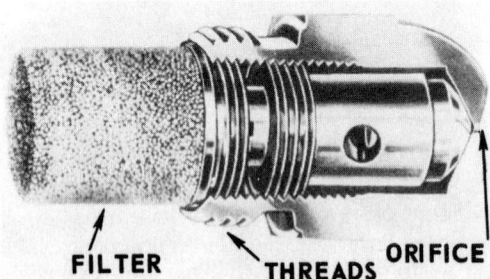

Fig. 20-91. Stainless steel nozzle used with gun-type oil burners. Note fine filter at entrance to nozzle. (Monarch Mfg. Works, Inc.)

Fig. 20-92. Gun-type oil burner flame. A—Refractory (asbestos cement lining) insulation. (The Carlin Company)

Fig. 20-93. Method of reversing direction of rotation of oil burner motor. This is accomplished by reconnecting two wires as shown by broken lines.

12. Changing pressure or low pressure at pump (slipping coupling).
13. Clogged nozzle.
14. Damaged nozzle.
15. No spark at electrodes:
    a. Loose wiring.
    b. Bad transformer.
    c. Low voltage.
    d. Crack in electrode porcelain.
    e. Electrodes carboned.
    f. Electrode spacing too far or too close.
    g. High voltage wiring loose.
16. Improperly installed bypass plug.

Proper flame appearance is luminous (mainly yellow). If there is insufficient air, the flame turns dull orange or red, and there may be smoky tips to the flame. Fig. 20-92 illustrates a properly adjusted flame.

The draft in the firepot is measured by the air pressure drop (in the firepot). It should be about 0.02 to 0.05 in. (0.5 to 1.3 mm) of water. (Use an inclined water tube manometer.) See Fig. 1-16. This check will also help determine if the automatic draft is working properly.

Some oil burner motors are reversible. Fig. 20-93 shows the method of reversing one type of oil burner motor. Controls for oil burners are described in Chapter 25. Testing instruments are described in Chapter 22.

Inspect the electrode wires. If they are cracked or brittle, replace the wires. Inspect the electrode tubular ceramic insulators. If the ceramic tubes are cracked, replace them.

Most soot deposits in a combustion chamber collect when the unit first starts. The more often the unit starts (cycles frequently), the greater the soot deposit. A correctly sized oil burner which operates less frequently will leave less soot in the furnace and stack.

If an oil line is dirty or clogged, blow it out using nitrogen gas (always use a pressure regulator and relief valve) or R-12 or R-22 refrigerant gas. Never use compressed air or oxygen, because a violent explosion may result.

20-39 SERVICING FUEL OIL BURNER

Oil burner problems, symptoms, and possible causes are given in the following troubleshooting outline.

I. Burner motor does not start, starts and locks out (CAD cell shuts off the control switch), or cycles.
  A. Does not start.
    1. Relay does not close (will not close or contacts dirty).
    2. Safety lockout stays open.
    3. Bad relay coil.
    4. Low voltage.
    5. Open high limit control.
    6. Broken wires or loose connections.
    7. Relay transformer open.
    8. Thermostat open (dirt on contacts, loose or dirty connections).
    9. Stack switch open.
    10. Heat sensing contacts out of place or open.
    11. Motor overload open (burned out, or has dirty contacts).
  B. Starts, but locks out.
    1. No fuel oil out of nozzle.
      a. Clogged.
      b. Pressure too low.

c. Pump not working.
d. Loose motor coupling.
e. Air leaks in fuel line.
f. Fuel oil line hand valve closed.
g. Strains or screens clogged (filter, pump screen, or nozzle strainer).
h. The pressure regulator in the pump body is stuck open.
i. Vent on fuel oil tank closed.
j. Empty fuel oil tank.

2. Fuel oil coming out of nozzle but no ignition.
a. Electrodes not positioned correctly.
b. Insulators cracked.
c. Ignition wires worn, loose, or with dirty connections.
d. Transformer not operating.
e. Primary wires worn, loose, or with dirty connections.
f. Low line voltage.

3. Fuel oil to nozzle, ignition OK, but no flame.
a. Clogged nozzle.
b. Clogged nozzle strainer.
c. Nozzle loose.
d. Pressure too low.
e. Fuel oil too heavy (wrong oil or too cold).
f. Excessive air or too much draft.
g. Electrodes in wrong position.

4. Flame only burns a few seconds.
a. Flame sensor not in correct position.
b. Stack switch not operating correctly.
c. Excessive air or air too cold.
d. Flame is too lean.

C. Cycles, but not on lockout.
1. Thermostat differential too close.
2. Anticipator set too close.
3. Limit switch set too low.
4. Overfired (reaches high limit temperature too quickly).

II. Burner does not operate correctly.
A. Smoke, soot, odors, and/or pulsating sound.
1. Wrong oil pressure.
2. Flame touches combustion chamber.
3. Not enough draft.
a. Dirty chimney.
b. The draft control is out of adjustment or it is stuck open.
c. Dirty flue.
d. Either the combustion chamber or the heat exchanger leaks.
4. Poor mixing of air and oil.
a. Nozzle is worn, loose, dirty, or drips.
b. Oil pressure too low or high.
c. Poor air velocity and turbulence.
d. Not enough air (shutter closed too much, fan binding, or tight bearings).

B. Puffs back.
1. Water in oil.
2. Delayed ignition.
a. Electrodes not positioned correctly or loose.
b. Insulators carbonized.
c. Nozzle worn, loose, dirty, or drips.
d. Voltage drop when burner starts.
e. Oil pressure too low or too high.

f. Transformer leads loose or dirty.
g. Transformer not operating correctly.
h. Excessive air or high draft.

C. Noise.
1. Loose fan.
2. Loose shutter.
3. Worn pump.
4. Dirty strainer.
5. Air in oil line.
6. Transformer hum.
7. Draft control vibrates.
8. Motor coupling worn.
9. Motor and pump not lined up correctly.
10. Relay contacts not seating tightly.
11. Oil suction line restricted.
12. Motor mounting loose.
13. Tight motor bearings.
14. Tank hum.

D. Fuel oil consumption is too high.
1. Nozzle is worn, loose.
2. Combustion chamber is dirty.
3. Too much combustion air (heat escapes up flue due to high flow of flue gases).
4. Poor mixing of air and oil.
5. Not enough draft over fire.
6. Air leaks into combustion chamber.
7. Oil pressure too high or too low.
8. Overfired, as noted by a high stack temperature.

There are two ways to check combustion efficiency: carbon dioxide (CO_2) analysis of flue gas; temperature of flue gases.

1. Use a carbon dioxide analyzer to check a sample of flue gas. It should be 10 to 12 percent CO_2 without visible smoke. If the reading is too low, it means too much air:
With   6 percent CO_2, 155 percent excess air is used.
With   8 percent CO_2,   85 percent excess air is used.
With 10 percent CO_2,   50 percent excess air is used.
With 12 percent CO_2,   26 percent excess air is used.

2. If the temperature of the flue gas is too high, much heat is being wasted. If the temperature of the flue gas is too low, water vapor will condense in the flue or chimney and the small amount of sulphur will form sulphurous acid (H_2SO_3), which is very corrosive.

With a 10-12 percent CO_2 stack analysis, combustion efficiency is as follows:

TEMP. °F	TEMP. °C	PERCENTAGE OF EFFICIENCY
1000	538	65 to 69
800	427	70 to 73
600	316	76 to 79
400	204	82 to 84

These are flue gas temperatures minus furnace room air temperature.

20-40 SOLID FUEL HEATING

In some areas of the country, an abundant supply of solid fuel exists. This is usually in the form of wood or coal. In an effort to conserve energy, some homeowners are looking to solid fuel to provide some or all of their heating energy needs. Commercial systems are available which replace the natural gas or oil heating source with a wood- or coal-burning source. This can be found in hydronic systems as well as forced-air systems.

While the use of wood or coal may be more economical in some situations, it usually requires a considerable amount of attention by the homeowner. It takes time to load the fuel, stoke the fire, clean the ashes, and maintain the flue and chimney. Solid fuels burn dirtier than conventional fuels, which means there is more soot in the flue and chimney.

A solid fuel heating unit is commonly used as part of a more conventional furnace system. An example of this is a system which utilizes a heat exchanger in the fireplace. This heat exchanger is connected in series with the conventional ductwork in the heating system within the home. When a fire is in the fireplace, the heat exchanger draws the latent heat (from the water vapor in the fireplace flue gas) and distributes the warm air through the forced air ductwork.

20-41 ELECTRIC HEAT

The use of electricity to heat homes, stores, commercial buildings, and factories is steadily growing in popularity.

Some advantages of electric heat are:
1. Low first cost.
2. Electric heating devices need no oxygen and, therefore, need no air supply.
3. Highest temperatures needed are below the ignition temperature of most materials. Therefore, the system is considered safer than other heating systems.
4. Because of the absence of combustion and combustion gases, there is less danger of toxic conditions arising. No chimney is needed.
5. Equipment normally requires less space.
6. Individual room temperature control is easily obtained.
7. Electric heat is very clean.

The electric heating process has some disadvantages:
1. The cost per unit of heat is higher than for some other fuels.
2. Humidity control problems may occur.
3. Added electrical circuits are required.

20-42 PRINCIPLES OF ELECTRIC HEATING

Electricity is a form of energy. Since energies can be changed to other forms of energy, electrical energy can be changed to heat energy.

Heating with electricity can be done either directly or indirectly:
1. Direct heat.
   a. Resistance heating, accomplished by passing a fluid over an electrically heated element. (Air is the fluid, although some units use water.)
   b. Radiant heating, accomplished by heating an element to a temperature high enough to give off heat.
   c. Thermoelectric heating. (See Chapter 17.)
2. Indirect heat, by using a heat pump. (See Chapter 23.)

20-43 APPLICATIONS OF ELECTRIC HEATING

Electric heating has a wide range of applications. It may be used for heating operations in industrial processes, such as fast drying of paints and melting low temperature metals or metal alloys. It has been used for domestic and commercial cooking and baking. It is often used for providing hot water.

For residential use, electric heating has a growing use as:
1. The only source of heat.
2. Supplementary heat, even though some other system provides much of the heat in the residence, or where the heat load is low. Heat pump systems may use electric resistance heating in climates where the heating energy required during cold weather is more than the heat pump can supply.
3. Resistance heating is used to provide heat in parts of a building that are not well heated or unsafely heated by the standard system. Resistance heating also may be used to heat additions to buildings where the present system does not have enough capacity to heat the addition, or where the extension of the present system would be too costly.

20-44 PRINCIPLES OF ELECTRIC RESISTANCE HEATING

In electric resistance heating, metals are generally used as heating elements. The metals are designed to permit a certain current to flow at either 120 V or 240 V to provide the heat required.

Some units are designed to operate at incandescent (glowing) temperatures; some are mounted in protected cabinets. Types of systems used are baseboard units or wires installed in floors, walls, and/or ceilings. The systems which do not glow are described as nonincandescent temperature units.

The higher the temperature of the heating element, the smaller the space it needs to occupy. Some units are designed with high temperature heating elements to release both air (convection) heating energy and radiant energy.

Electrical energy is changed into heat energy in the following ratios:

$$1 \text{ watt} = 3.415 \text{ Btu/hr.}$$
$$100 \text{ watts} = 341.5 \text{ Btu/hr.}$$
$$1000 \text{ watts} = 3415 \text{ Btu/hr.}$$

The voltage multiplied by the amperage flow in a circuit equals the watts.

At 20 amperes maximum input, a 120 V circuit will provide:

$$2400 \text{ watts } (20 \times 120 = 2400) \text{ or}$$
$$8196 \text{ Btu/hr. } (2400 \times 3.415 = 8196)$$

Example: A home with a need for 50,000 Btu/hr. for heating, therefore, needs the following amount of electricity (given first in kilowatts, second in kilowatt-hours, third in amperes):

$$\text{Btu/hr.} = (\text{Btu/hr.})/\text{watt} \times \text{watts}$$

In better form:

$$\text{Watts} = (\text{Btu/hr.})/[(\text{Btu/hr.})/\text{watt}]$$
$$\text{Watts} = 50,000/3.415 = 14,640$$

Then:

$$14,640 \text{ watts} = 14.64 \text{ kilowatts}$$

To find the electricity usage in kWhr., multiply by 1 hr. as follows:

$$14.64 \text{ kilowatts for an hour} = 14.64 \text{ kWhr.}$$

To find the electricity usage in amperes, do the following (assuming a 240 V supply):

$$\text{Amperes} = \frac{\text{watts}}{\text{volts}} = \frac{14,640}{240}$$
$$= 61 \text{ A}$$

For most computations, 1 watt equals 3.4 Btu/hr.

20-45 BUILDING DESIGN FOR ELECTRIC RESISTANCE HEATING

In some cases, converting a coal, oil, or gas-fired heating system in a building into an electric resistance heating system is possible and practical if the building is modified to reduce heat transfer and air infiltration. The walls and ceiling should be insulated as thoroughly as possible. Basement walls or the slab of buildings without a basement must be insulated. Windows should be double glazed. Wood or plastic window and door frames are preferred, rather than metal. Walls should have 4-in. insulation; ceilings, 7-in. insulation. The floor slab should be insulated 4 in. thick and 42 in. deep. The basement wall should have 2- to 4-in. insulation. Follow the latest specifications available from manufacturers.

Where electric resistance heating is desired, the building should be designed to reduce heat losses to cut the operating cost to a minimum.

When possible, advantage should be taken of solar heat as it contacts the east, south, and west exposures.

Humidity control may require dehumidification rather than humidification. The building is likely to be very tight, which lets the relative humidity build up from cooking, laundry, bathing, pets, and other sources. Compared to sources of heat that require burning, there is no draft with electric resistance heating to help pull excess water vapor out and replace it with dryer outside air.

20-46 ELECTRIC RESISTANCE HEATING ELEMENTS

There are three types of electric resistance heating wires:
1. Open wire.
2. Open ribbon.
3. Tubular cased wire.

Fig. 20-94 shows some heating wire designs. Looped wires are used in some electric resistance heating elements to transfer heat better than straight wires. A looped wire also helps the manufacturer get a controlled resistance.

The open wire type usually consists of nickel chromium resistance wire mounted on ceramic or mica insulation. The open wires must be carefully protected to keep them from being contacted by metal objects and/or by humans or animals to avoid the danger of burns or electrical shock.

Ribbon type resistance heating wires are made of the same material as open wires, and they are mounted in the same general way. The ribbon system, too, must be carefully covered by a grid to prevent burns or shock. The ribbon design provides more surface exposure for air contact.

The tubular heating element is similar to the heating elements used in electric stoves. Usually, nickel chromium resistance wire is surrounded by a magnesium oxide powder. The wire and powder are enclosed in a heat- and corrosion-resistant steel tube.

The tube-encased wire design protects against electrical shock. However, the element may reach high temperatures. To increase the heating surface, and to reduce the danger of high-temperature wiring, tubular covered elements are sometimes placed in fin-type aluminum castings.

There are types of electric resistance heating elements that are similar in shape to a tube-encased element. These heating elements are usually made as follows:

1. A helical wound coil of resistance wire is supported by

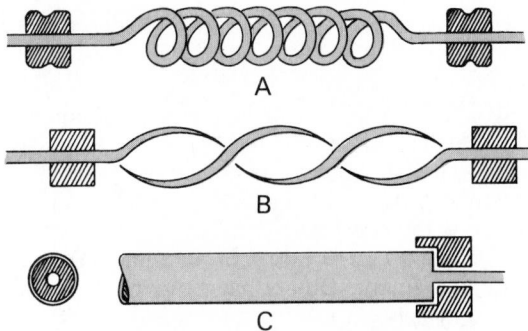

Fig. 20-94. Three types of electric resistance heating elements. A—Open wire. B—Open ribbon. C—Tube-encased wire.

ceramic insulators. These open coils are made of nickel chromium (nichrome) wire and are wound on insulation (ceramic and mica). Ceramic disks shaped like doughnuts with the helical coil passing through it are common.

2. Metal foil is expanded to form meshlike resistance heating strips. These operate on lower temperatures (black heat range about 1450°F [790°C]) than the resistance wire type.

3. A small diameter coil of resistance wire is inserted into a metal tube and powder forms the insulation. The metal tube is pressed, and the powder (magnesium oxide) becomes rigid. During manufacture, these tubes are formed into many shapes. Heated to 1550°F (840°C), these tubes become dull red.

20-47 BASEBOARD ELECTRIC RESISTANCE HEATING

Baseboard heating is a popular form of natural convection heating. It has the electrical resistance heating unit mounted in a casing that is designed to efficiently move air over the heating element by natural convection. See Fig. 20-95. Warmed air is lighter (air expands when heated) and rises. The colder air, being heavier, settles to the lower opening and enters the unit to replace the rising heated air. The units used are shaped much like a regular baseboard. They are mounted on the wall close to the floor, usually under windows. Fig. 20-96 shows a section of a baseboard heater. The construction of a baseboard heater is shown in Fig. 20-97.

In most cases, a baseboard heating unit should be mounted on the wall. If built into the wall, dust in the heated air coming from the unit may streak the wall and make frequent cleaning needed. It is important to keep the air passages clear to prevent poor airflow. The temperature of the heating element may become too high if the air passages are blocked.

Each baseboard heating unit may be thermostatically operated, which permits individual room temperature control. The units are easy to install, Fig. 20-98, and take up a minimum of space. Since there are no moving parts, they are noiseless. Installation of a safety (limit) switch in each unit is strongly recommended. This switch will open the electrical circuit if any part of the heater reaches a temperature above normal. See Chapter 25 for information about controls.

The variation of room temperature will be greater using baseboard electric heat than with ceiling cable heat. The baseboard heat temperature variation between the floor and ceiling can be 4 to 15 °F (2 to 8 °C) with about two heating periods per hour. Cycles of five to six per hour are recommended. The ceiling cable temperature variation between the floor and ceiling can be 4 to 5 °F (2 to 3 °C), with about one heating period per hour (as recommended). With ceiling cable heat, the floor above the cable is warm.

Baseboard units are available in lengths from approximately 36 to 100 in. (0.9 to 2.5 m). Some units have only one heating element. Others have two or more elements connected in parallel.

On most installations, it is good practice to keep the current load to 20 A or less per circuit. The use of 240 V circuits, where practical, is desirable. Smaller wires can be used in 240 V circuits than in 120 V circuits.

20-48 ELECTRIC FURNACES

Electricity may be used to provide heat for either warm air or hydronic central heating systems. These furnaces have capacities of 34,000 Btu/hr. (10 kW) up to 120,000 Btu/hr. (35 kW).

A forced warm air furnace that uses tubular heating elements is shown in Fig. 20-99. This furnace can be installed for upflow, downflow, or horizontal airflow. The heating elements are used in stages of 5 kW or 10 kW. Their use is sequenced (turned on one at a time). The electrical circuits are shown in Fig. 20-100. Note that the power in is 240 V and controls are 24 V.

A hydronic boiler is shown in Fig. 20-101. This boiler has a capacity of 34,130 Btu/hr. (10 kW) to 81,900 Btu/hr. (24 kW). It is a very efficient unit measuring about 10 1/2 in. wide by 21 3/4 in. high and 23 3/4 in. long (about 25 by 50 by 60 cm).

Electric heat furnaces designed for warm air usually have several heating elements. These elements vary from 5 to 10 kW each (17,000 to 34,000 Btu). These elements are usually sequenced by a relay panel which energizes them about 30 seconds apart, one after the other. The thermostat is usually the low-voltage type with two heat stages. These heat stages are put into operation from 0.5 to 1.5 °F (0.3 to 1 °C) apart.

Humidifiers are rarely needed with electric heat.

The total furnace output should be about 20 percent over the design heat load. The extra 20 percent can be wired

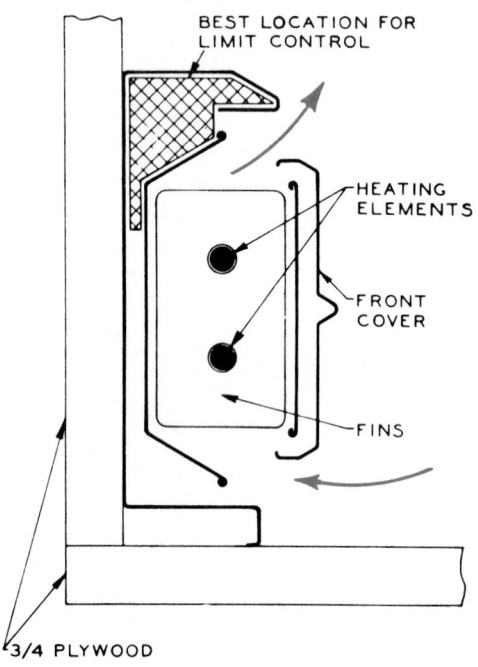

Fig. 20-95. Common construction of baseboard natural convection electric resistance heating unit.
(White-Rodgers Div., Emerson Electric Co.)

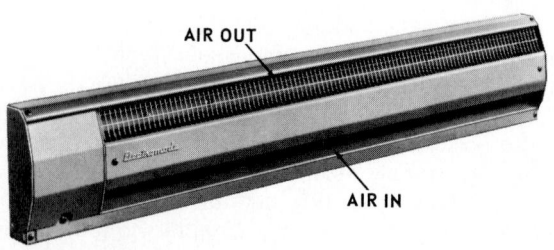

Fig. 20-96. Section of baseboard electric resistance heating unit. Covered left end may be used to house junction box and thermostat.

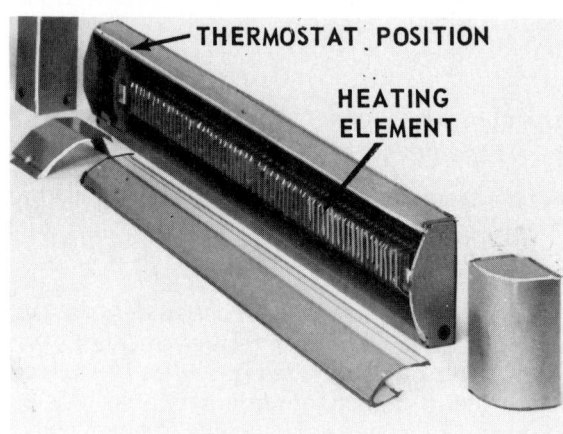

Fig. 20-97. Baseboard electric resistance heating unit with front cover removed. Note thermostat location, corner block, and end piece.

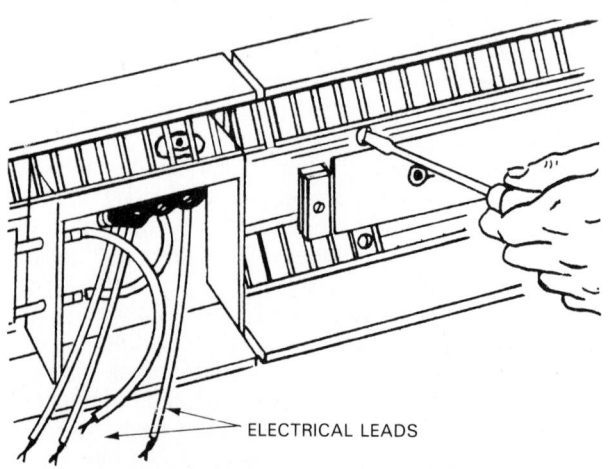

Fig. 20-98. Baseboard electric resistance heating unit with decorative panel removed, showing mounting screw.

Fig. 20-99. Forced warm air furnace having tubular electric resistance heating elements. 1—Warm air outlets. 2—Insulated jacket. 3—Fan and limit controls. 4—Electric resistance heating elements. 5—Operating controls. 6—Fan outlet. 7—Heating chamber. 8—Fan and motor. 9—Filter. 10—Furnace base.

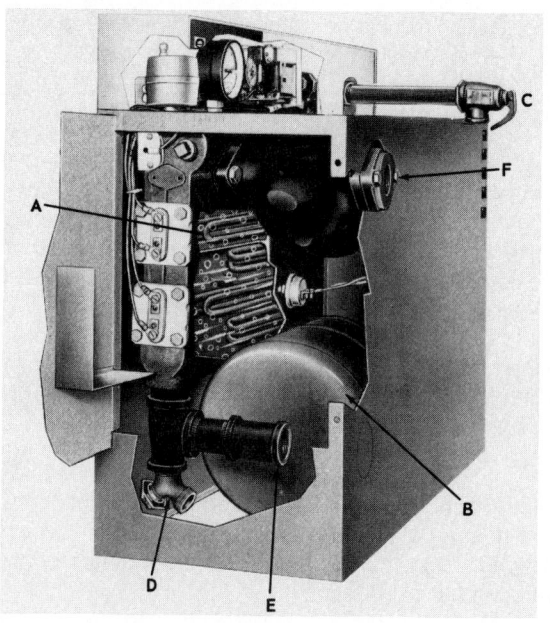

Fig. 20-101. Electrically heated hydronic boiler. A—Heating element. B—Expansion tank. C—Relief valve. D—Drain valve. E—Water return. F—Water outlet.

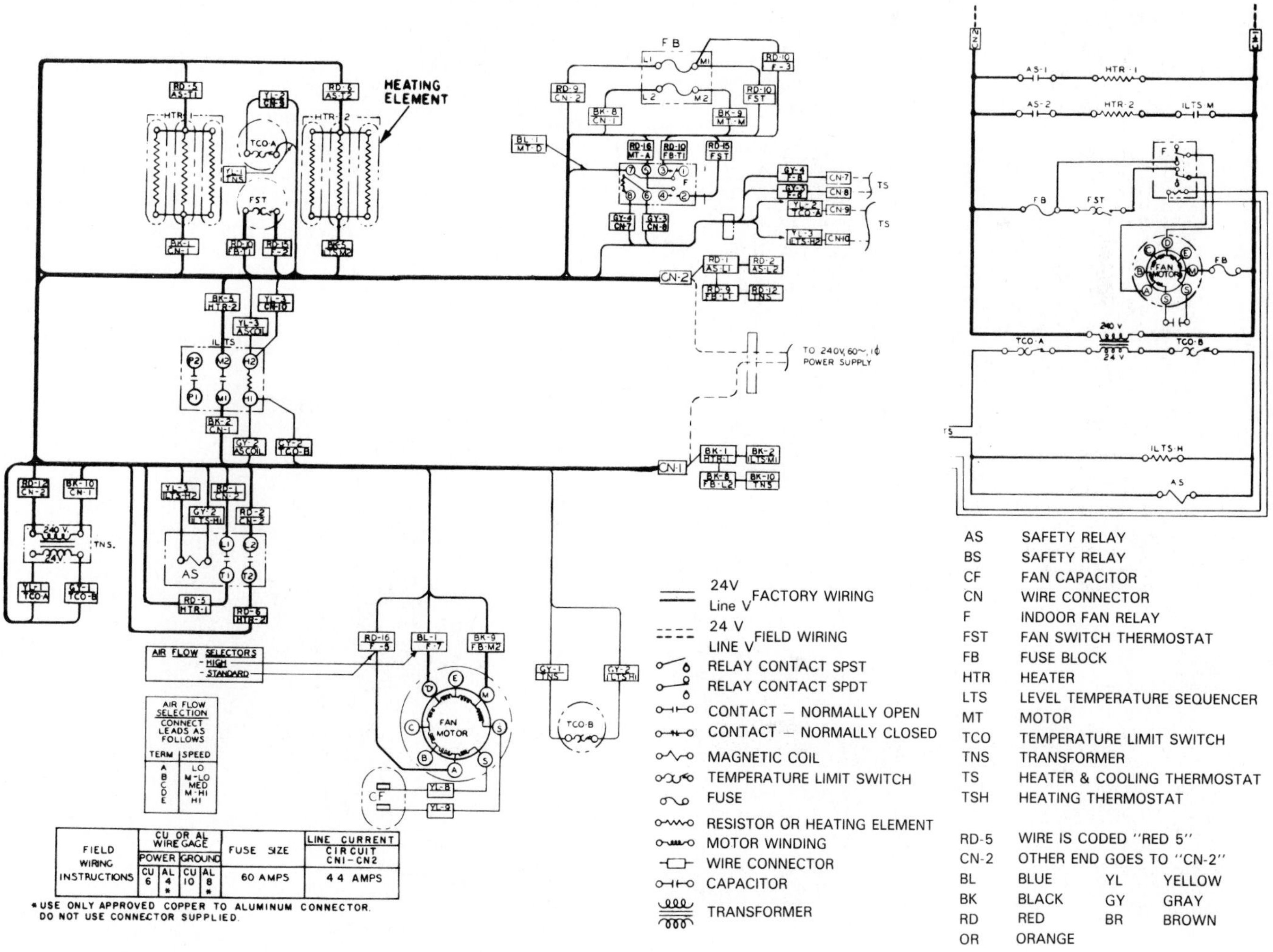

AS	SAFETY RELAY	
BS	SAFETY RELAY	
CF	FAN CAPACITOR	
CN	WIRE CONNECTOR	
F	INDOOR FAN RELAY	
FST	FAN SWITCH THERMOSTAT	
FB	FUSE BLOCK	
HTR	HEATER	
LTS	LEVEL TEMPERATURE SEQUENCER	
MT	MOTOR	
TCO	TEMPERATURE LIMIT SWITCH	
TNS	TRANSFORMER	
TS	HEATER & COOLING THERMOSTAT	
TSH	HEATING THERMOSTAT	
RD-5	WIRE IS CODED "RED 5"	
CN-2	OTHER END GOES TO "CN-2"	
BL	BLUE	YL YELLOW
BK	BLACK	GY GRAY
RD	RED	BR BROWN
OR	ORANGE	

AIR FLOW SELECTORS
- HIGH
- STANDARD

AIR FLOW SELECTION CONNECT LEADS AS FOLLOWS	
TERM	SPEED
A	LO
B	M-LO
C	MED
D	M-HI
E	HI

FIELD WIRING INSTRUCTIONS	CU OR AL WIRE GAGE		FUSE SIZE	LINE CURRENT CIRCUIT CN1–CN2
	POWER	GROUND		
	CU AL	CU AL	60 AMPS	4 4 AMPS
	6 4	10 8		

* USE ONLY APPROVED COPPER TO ALUMINUM CONNECTOR. DO NOT USE CONNECTOR SUPPLIED.

24V Line V	FACTORY WIRING
24 V Line V	FIELD WIRING
	RELAY CONTACT SPST
	RELAY CONTACT SPDT
	CONTACT — NORMALLY OPEN
	CONTACT — NORMALLY CLOSED
	MAGNETIC COIL
	TEMPERATURE LIMIT SWITCH
	FUSE
	RESISTOR OR HEATING ELEMENT
	MOTOR WINDING
	WIRE CONNECTOR
	CAPACITOR
	TRANSFORMER

Fig. 20-100. Wiring diagram of electric furnace, with schematic diagram. (General Electric Co.)

and controlled to turn on only during those very rare times when the heat load exceeds the design load.

20-49 DUCT HEATERS

Spaces can be heated by installing electric heating elements in existing duct systems or in ducts of comfort cooling installations. It is recommended that the air be very clean to avoid odors due to particles burning on the electric resistance heating elements. Air should also be clean to avoid corrosion of the heating element wires.

Fig. 20-102 shows a unit, complete with controls, designed for a duct installation. A duct heating unit being installed in a duct system is shown in Fig. 20-103.

20-50 SUPPLEMENTARY ELECTRIC HEATING

Electricity is often used to heat building additions which have a heating plant without enough capacity to carry the extra load. Also, it is used where extending the present system would be difficult.

Radiant heat panels built into the wall, baseboard heaters, and resistance heating wire imbedded in the ceiling or wall plaster may be used for this purpose. Added bedrooms, family rooms, and utility rooms may be heated this way.

Fig. 20-104 shows a unit which may be used either as a primary heating source or as a supplementary heating source. Air enters at the top, and a fan forces the air down over the electrically heated aluminum element. Air leaves the unit through the lower part of the grille. Figs. 20-105

Fig. 20-104. Forced convection electric resistance heating unit.

and 20-106 show another design, while Fig. 20-107 shows a schematic wiring diagram for the unit.

20-51 RADIANT HEAT

Radiant heat usually is the impact of the infrared band of light energy waves against an object. These rays are at a frequency of 900 MHz to 2500 MHz (M = million) and wavelength of 4.0 microns or less. The object absorbs the rays and becomes warmer. When a furnace door is opened, even if one stands back from the door, the heat impact which is felt is usually an infrared ray impact. This principle can be used for comfort heating. The energy source may

Fig. 20-102. Electric resistance heating unit designed for duct installation. A—Heating elements. B—Control circuit and housing. (Brasch Mfg. Co., Inc.)

Fig. 20-103. Electric resistance heating unit being installed in roof duct. (Tutco)

Fig. 20-105. Wall-mounted electric resistance heater which may be used for either primary heating or supplementary heating. It has a thermostat, off-and-on switch, and fan switch.

Fig. 20-106. Parts of electric resistance space heater. A—Frame. B—Front cover. C—Heating element and fan. D—Back case.

be any fuel, although gas-fired and electrically heated elements are the chief sources.

Radiant energy sources may be either:

1. A lamp source or high-temperature electric element or ceramic source.
2. A low-temperature electric cable or hot water source.

If radiant heat rays are focused on an individual having several square feet of surface to absorb the rays, this person can be kept quite comfortable even though the ambient temperature is below the comfort range. A large warehouse with a few small areas where employees work is a typical example of where radiant heat is often used to good advantage.

Radiant heat decreases as the square of the distance. That is, an object twice as far away from the radiant heat source will receive only one fourth as much heat.

20-52 RADIANT CEILING HEATING

Radiant ceiling installations are most popular for homes. In these installations, electric heating cables are enclosed in the ceiling plaster. The ceiling cables are about 500 to 5000 watts (1700 to 17,000 Btu/hr.), and they have a 1/8- to 1/4-in. diameter.

Ceiling cables usually release 2.75 watts per foot and are spaced on 1 1/2-in. centers. Their temperature is about 150 °F (70 °C). This heats the ceiling surface to about 120 °F (50 °C) on 1 1/2-in. centers, but wider spacing will reduce the ceiling surface to about 100 °F (40 °C). Drywall uses lower watts/foot ratings (about 2.2 watts/foot). About 60 percent of the heat released is by radiation.

Floor cables are heavier and are rated at 2.75 watts/foot, with a 1 1/2- to 3-in. thickness of concrete. About 45 percent is radiant heat.

20-53 RADIANT LAMPS AND GLASS PANEL HEATERS

Some electrically heated fixtures use quartz lamps that are either Vycor or metal sheathed. Some use open resistance wire or ribbon. See Fig. 20-108. Quartz lamps usually consume 800 to 2500 watts (2700 to 8500 Btu/hr.) of power. The lamps and wires reach temperatures of approximately 1200 °F (650 °C).

Radiant heat lamps may have either 90 deg. and 60 deg. or 45 deg. reflection. They radiate about 50 watts per square foot from at least two directions. Install one watt for each 1 °F (0.5 °C) desired above lowest expected temperature (minimum 12 W per sq. ft.). Add five percent for each foot distance if the source is farther than 10 feet (3 m) from a person.

When these lamps are used outdoors, use wind shields and about two watts/degree above coldest temperature expected.

Quartz tubes usually use a nickel chromium coil. They are about 60 percent efficient. The air-filled open tube operates at 1500 to 1700 °F (820 to 930 °C), and it gives off infrared rays which are in the visible part of the spectrum. A quartz lamp filled with inert gas operates at 4000 °F (2 200 °C). It uses a tungsten coil and is about 85 percent efficient. It gives off infrared rays and visible rays in the yellow part of the spectrum.

Glass panel heaters are also available. They give off about 60 percent radiant, 40 percent convection heat. The glass is borosilicate. Electrical conductors are installed on the back of the glass and covered with a reflector surface. The element operates at about 500 °F (260 °C). The glass surface

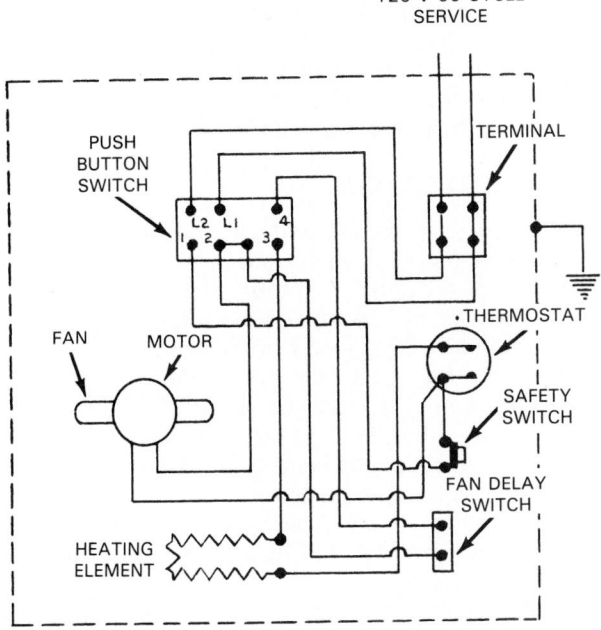

Fig. 20-107. Wiring diagram of wall-mounted heater.

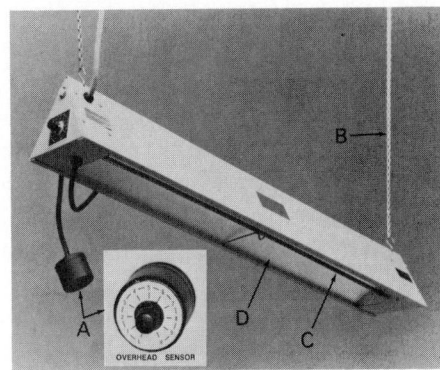

Fig. 20-108. Radiant heat lamp. Unit uses element like that used in electric range. Element is called a metal sheath heat element. A—Overhead sensor. B—Hanger. C—Metal sheath heat element. D—Reflector. (Kalglo Electronics Co., Inc.)

Fig. 20-109. Service technician connecting line to terminals of duct-mounted electric resistance heating unit.   (Tutco)

is about 350°F (180°C). Glass panel heaters are used on the wall under windows.

A useful application for radiant heat in the temperate zones is for snow melting. Generally, 100 to 200 watts (340 to 680 Btu/hr.) are needed for each square foot of surface to be serviced.

Gas may also be used for radiant heating. Gas-fired radiant heat usually is provided by heating ceramic elements to incandescence and using a reflector to focus this heat. In gas-fired units, about 50 percent of the heat energy is converted into radiant heat. These units operate at about 700°F to 1600°F (370°C to 870°C).

20-54 HEATING COIL INSTALLATION

Electric heating elements must be installed in accordance with both the electrical codes and the manufacturer's recommendations.

The electric heating installation must be checked to be sure it is safe with reference to fires, safe with reference to humans and animals, and it must be efficient.

Baseboard heaters must have unhampered air circulation. The unit heaters must not be installed dangerously close to flammables. Furnace units must be shielded electrically and heat protected. All metal parts of the units must be grounded.

The electrical service must use wire sizes according to the voltage and amperage of the circuit. The circuit must be properly fused and provided with adequate limit and safety controls. All controls must be for the correct voltage and current. Fig. 20-109 shows electrical connections being made to an open wire, supplementary heating coil that has been installed in a duct.

20-55 SERVICING HEATING COIL

Electric resistance heating requires little service. Air passages over the heating units must be kept clean. Grilles, ducts, heating coils, and fins should be cleaned at least once each year. Brushing and vacuuming are recommended. It is extremely dangerous to use flammable fluids for cleaning electric resistance heating units.

Check the terminals for tightness and cleanliness. A

voltmeter or ohmmeter helps check if connections are loose or corroded. The ohmmeter is preferred, because it may be used for checking with the power off.

The thermostat, limit control, and relay are possible service problems. If the circuit does not function, the following typical electrical circuit diagnosis is recommended:
1. Is there power to the fuse box or circuit breaker box?
2. Is the fuse in good condition and are the connections electrically good?
3. Is the thermostat operating? (Check opening and closing temperatures.)
4. Is the limit switch operating? (Check opening and closing temperatures.)
5. Are the relay coils in good condition and operating?
6. Are the relay contact points clean and operating?
7. Does the electrical heating coil circuit have continuity?

Some troubles and possible causes are as follows:
A. Blower turns on and off. If heater strips heat and cool as blower runs and stops, the thermostat is short cycling. Perhaps too high rating of the anticipator.
1. Limit control may be opening and closing.
2. Incorrect low voltage.
3. Motor overload may be opening and closing.
B. Blower runs, but there is not enough heat.
1. Dirty filters.
2. Voltage too low.
3. Only some heater elements are energized.
a. Open fuse or circuit breaker.
b. Element burned out.
4. Sequence switch not operating.
5. Second stage of two-stage thermostat not operating (second-stage cycle must add to heating capacity).
C. No power or low voltage.
1. Sequence switch defective (open).
2. Thermostat open.
3. Low line voltage.
4. Low transformer output voltage.
5. Motor not operating.
a. Defective capacitor.
b. Defective internal overload.
D. Proper voltage at motor.
1. Defective motor (open circuit).
2. Defective overload (open circuit).
3. Defective motor capacitor (shorted or open).

20-56 UNIT HEATERS

Many stores, commercial buildings, and factories use unit heaters to heat certain rooms or spaces. These heaters can be gas-fired, oil-fired, or use hot water coils or steam coils. They are suspended from the ceiling and use a motor-fan to force the heated air in a controlled direction. Many units have adjustable louvers to help direct airflow. The different sizes handle from 300 cfm to about 6000 cfm. Their capacities range from 20,000 Btu/hr. to 360,000 Btu/hr. Gas-fired and oil-fired unit heaters use a flue to carry the products of combustion outdoors.

Special unit heaters are available for vertical downward warm airflow and also for high velocity airflow (about 2500 fpm) against large door openings to keep out cold outside air. Usually, they are mounted about three feet above and four feet away from the door opening. They are often used on car and truck doors or shipping and receiving doors. The

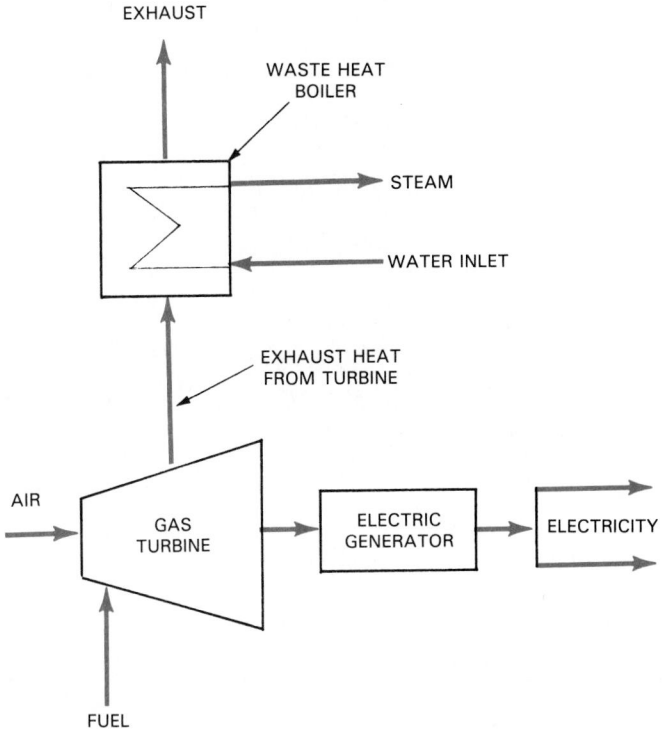

Fig. 20-110. Simple cogeneration system. Waste heat from gas turbine is used to produce steam, which may be used for heating or for absorption cooling systems.

A dry atmosphere causes dry skin, breathing dryness, and loss of moisture from hygroscopic materials such as wood natural fibers (wood furniture and woodwork) and most foods. Dry wood cracks and furniture joints get loose. A dry condition also creates static electricity conditions. Moisture must be added to the air. Moisture sources are from plumbing devices, cooking, and perspiration. In addition, moisture in the air can be increased and controlled by using humidifiers.

Humidifiers add water vapor (low temperature steam) to the air. If the return to a warm air furnace is about 60 °F

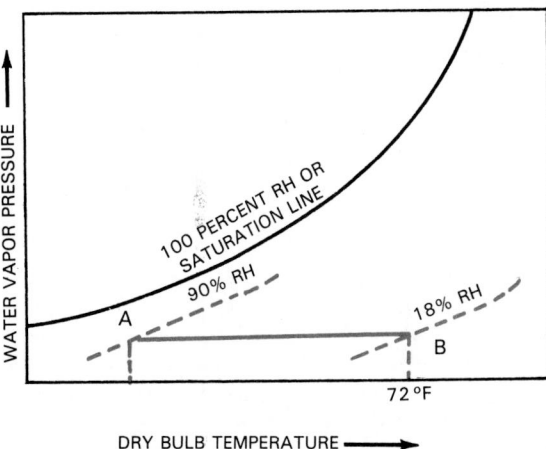

Fig. 20-111. Graph showing decrease in relative humidity. A—Starting point. Air sample is heated from 30 °F (−1 °C) at 90 percent RH, to 72 °F (22 °C) at 18 percent RH. B—Result for process carried out without adding any water vapor to air sample.

unit is operated by either a door switch or a thermostat connected in parallel. Doors less than 8 ft. high and 10 ft. wide usually are not protected by these devices. Some users operate the fan only during the summer months to help keep out dust and insects.

20-57 COGENERATION

Large industrial plants, commercial buildings, and shopping malls have installed energy systems that generate electricity and supply heat. These are called cogeneration systems. This type of system has been installed where there is a need for heat year-round.

Most installations use gas turbines, Fig. 20-110, to generate electricity. The waste exhaust heat is used to produce steam for industrial processes or building heat. The exhaust gases leave the turbine at temperatures high enough (about 700 °F [370 °C]) to run a boiler at 75 percent of its usual efficiency.

In some installations, the waste heat is used to drive absorption cooling systems. These are called on-site generating systems, combined-cycle plants, or total energy systems.

20-58 HUMIDIFIERS

When air is heated, it can absorb more water vapor. Human comfort requires a relative humidity (RH) of about 35 percent. When outside air at 30 °F (−1 °C), 90 percent RH, is heated to 72 °F (22 °C), its RH drops to about 18 percent. See Fig. 20-111 for information from a psychrometric chart.

(16 °C) and 25 percent RH, and the furnace heats the air to 140 °F (60 °C), a humidifier may be used to add moisture to the warmed air. This heated air is then mixed with the air in the room. In Fig. 20-112, A to B indicates the air being heated. From B to C, this warm air is passing over the humidifier (total heat is constant). Between C and A, the heated and the humidified air are mixed. D indicates the final condition of the air as it is delivered to the conditioned space. Remember, it requires about 1000 Btu to vaporize each pound of water (7000 grains).

Most humidifiers in warm air systems are part of the furnace or the ductwork. Fig. 20-2 shows some basic designs.

The proper humidity level reduces static electricity and the drying and cracking of furniture and wood trim. Fig. 20-113 shows the operating principle of a humidifier. Evaporation takes place as the heated air passes through the evaporator pad. The pad is wetted by water metered through a solenoid valve and then distributed over the pad. Water that has not evaporated drains to the bottom of the humidifier and is drained out. Humidified air is then returned to the heating system and enters the living area. This type of unit has a small fan motor and solenoid water valve. Fig. 20-114 shows the exterior of this type of humidifier.

Some humidifiers use a vibrating object to atomize water (break up into tiny droplets). The vibrating object can be a piezoelectric crystal operating at a high frequency of about

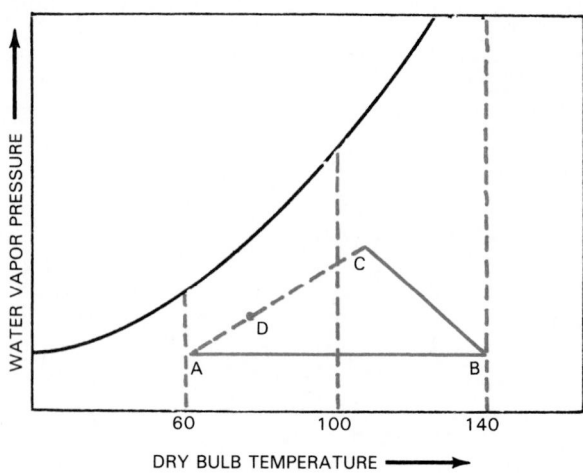

Fig. 20-112. Psychrometric chart indicates warm air recirculating heating cycle. A—Cold air return. A to B—Heating in furnace. B to C—Humidifying air. C to A—Mixing of air with room air. D—Final conditions after mixing.

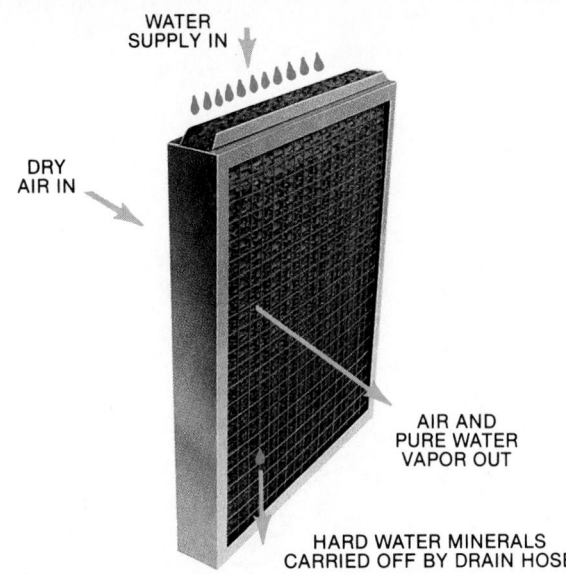

Fig. 20-113. Water flow through a basic filter humidifier. The flow of air is governed by the furnace blower and the operation of a small fan motor. (General Filters, Inc.)

100 kHz (100,000 cycles per second), Fig. 20-115. "Piezo" is pronounced "pea-ay-zo." "Piezoelectric" means vibration due to electric waves in a flat crystal plate. The unit produces moist air which feels cool. The air feels cool because partially evaporated droplets are vaporized on contact with a warmer object like the human body. If the air cools the body too much, the air must be heated in the building heating system before being distributed.

Humidifiers easily can be added to warm air heating systems. Hydronic heating systems, steam heating systems, and most electric heating systems, however, require a separate cabinet-type humidifier if the needed relative humidity is to be maintained.

Humidifiers of various types are used:
1. Plate type humidifier (low capacity).
2. Rotating drum type (for restricted spaces).
3. Rotating disk type.
4. Fixed filter type.
5. Fan type.
6. Plenum/warm air duct humidifier (slings the water).
7. Plenum/duct electric type.
8. Ultrasonic (piezoelectric) type.

Humidistats are used to control the level of relative humidity. Too much humidity may cause swelling of hygroscopic materials (wood products) and may cause condensation on cold surfaces such as windows, window frames, and doors. Water vapor may also condense inside the outside walls.

Excess humidity should be avoided because mold and rot can occur at 70 percent RH. Wet objects take a long time to dry. Wood window frames rot from water dripping off of cold glass.

If outside air at 70 percent RH filters into a building and is heated to 72 °F (22 °C), this heated air will have the relative humidity shown in Fig. 20-116. If outside air at 90 percent RH and 30 °F (−1 °C) is heated to 72 °F, it will have a relative humidity of 19 percent. If it starts at 100 percent and 30 °F, it will be 20.7 percent at 72 °F. All water services in the home, plus perspiration and respiration of the humans and animals in the home, will increase the relative humidity. Even if these moisture resources dou-

ble the relative humidity, it would still be too low in all cases except for 30 °F (−1 °C) outside air. To use the chart in Fig. 20-116 starting at outside relative humidities other than 70 percent, use the following example with 100 percent outside RH (converting the first line of the chart):

$$\frac{100\%}{70\%} \times \text{INSIDE RH} = 1.43 \times 2 = 2.86 \text{ percent.}$$

Humidifiers are used to add the needed moisture (water vapor) to the room air. A ranch home 25 ft. x 60 ft., with 8 ft. ceilings, has a volume of 12,000 cu. ft. and requires about four (tight building) to 16 (loose building) gallons per day, depending on the number of changes of air in the building:
1. A tight building has 0.5 air changes/hour.
2. An average building has 1.0 air changes/hour.
3. An average loose building has 1.5 air changes/hour.
4. A loose building has 2.0 air changes/hour.

Regular water (city mains or well water) contains different foreign matter. If used, this foreign matter must be treated or removed. Fig. 20-117 shows a humidifier with a water level indicator, which can be used to drain the water out of the humidifier.

The quality of the water varies with its source.
1. Soft water:
   a. Natural, untreated water with low mineral content has about 5 grains of hardness per gallon and no chlorides. The natural source is rain water.

Fig. 20-114. Humidifier with evaporator pad. Unit has a solenoid valve to evenly distribute water over pad and has a motor and fan for air movement. (General Filters, Inc.)

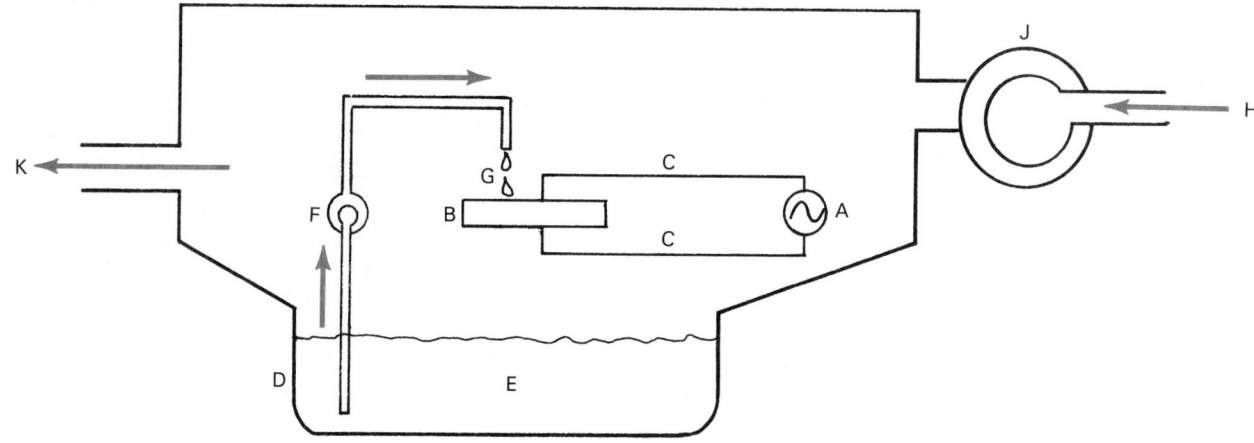

Fig. 20-115. Crystal vibrates to break water into mist. Crystal is called a piezoelectric crystal. A—Electric wave generator. B—Crystal plate (size is exaggerated for clarity). C—Connecting wires. D—Water pan. E—Liquid water. F—Pump. G—Dripping water. H—Air inlet. J—Blower. K—Air outlet.

b. Softened water is treated by the ion exchange process (water softener) to remove hardness and minerals. Then the unwated mineral ions (charged atoms) are replaced with water-soluble sodium salts.
2. Demineralized water has been treated to remove minerals.
3. Medium hard water is untreated water with 5 to 15 grain/gallon of mineral content (well water).
4. Very hard water (well water) is untreated water with over 15 grain/gallon of mineral content. Hard water is tested with a substance called "green soap." Green soap will not form suds in water that has 10 grain/gallon or more of minerals.

The perfect water for humidifiers would be distilled water, which does not put any foreign vapors into the air. In addition, distilled water does not leave any deposits in the humidifier or in the duct system.

20-59 SERVICING AND INSTALLING HUMIDIFIERS

Humidifiers should be carefully installed. Mount the water reservoir so it is level in all directions (use a spirit level). The humidifier should have the capacity to add enough moisture to the building at design conditions. See Chapter 26 for the gal./day needed.

The water feed line should be tapped into a cold water line in the same manner as an ice cube maker line. See Chapters 11 and 14. The overflow drain tube (bleeder tube) should slant down all the way to the outlet. The outlet should be at least one inch above the open drain.

If the unit is power driven by a fan or rotary unit (or both), the electric system can be low-voltage controlled by a humidistat and relay, or it can be controlled by a 120 V humidistat. Electrical installation should be done according to electrical codes in effect. A grounded receptacle is required.

After installation, the unit should be turned on (electrical and water) and checked for proper operation. Every month during operation, look for bacterial growth and scale buildup. Water treatment compounds are available at hardware stores and HVAC supply dealers.

20-60 ELECTRIC HUMIDIFIERS

When a house or business facility is electrically heated, there is a tendency to have too much relative humidity. The building is very tight and humidity builds up. Water vapor sources include:
1. Cooking.
2. Wash basins.
3. Lavatories.

RELATIVE HUMIDITY (PERCENT)			
OUTSIDE TEMPERATURE	OUTSIDE RH	INSIDE RH	SAFE INSIDE RH
−10	70	2	20
0	70	5	25
10	70	7	30
20	70	11	35
30	70	15	40

Fig. 20-116. Chart shows relative humidity change in air as it is brought into building from outside; also recommended inside relative humidity based on outside temperature. Building inside temperature is 72 °F (22 °C). Example: 0 °F (−18 °C) outside air at 70 percent relative humidity when brought into building and heated to 72 °F will have relative humidity of 5 percent. It should have relative humidity of 25 percent.

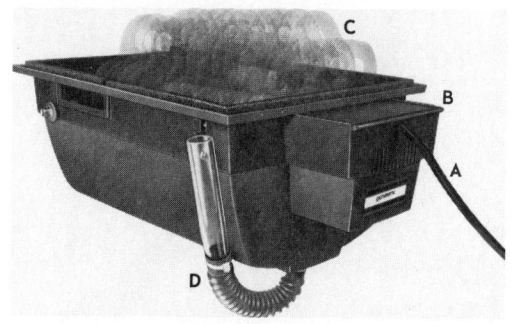

Fig. 20-117. Humidifier designed to fasten to bottom of duct. A—Motor to drive moisture pickup wire mesh disks. B—Flange for mounting unit to duct. C—Wire mesh moisture disks. D—Water level tube and water drain.

4. Respiration.
5. Perspiration.
6. Laundry.
7. Showers and/or bathing.
8. House plants.
9. Pets.

A dehumidifier may be needed if the relative humidity is too high. See Chapter 21. However, most heated buildings need humidifying. A water-vapor-producing device is usually needed during the winter season in the temperate zones. An electric humidifier may be used for this purpose.

Buildings which have steam heated, hydronic heated, or warm air heated equipment are examples of installations that can use an electric humidifier. Warm air heating units sometimes are equipped with a humidifier that uses the warm air as a source of heat. However, if the humidifier water is separately heated, the amount of humidification can be more accurately controlled.

The electric humidifier has the advantage of ease of installation, flexibility of location, and accuracy of relative humidity control. Fig. 20-118 shows an electric humidifier designed for use in a duct or plenum chamber. This unit has 1400 watt electric heating coil which is controlled by an adjustable humidistat. Water level is controlled by a pan-type float which operates a switch in the solenoid water valve circuit. Fig. 20-119 shows the construction of the electric humidifier.

20-61 CABINET-TYPE HUMIDIFIERS

The use of humidifiers is required in many situations that are separate from and independent of the heating system. They are popular when a hydronic or steam heating system is used. Many are used with warm air heating systems.

Independent humidifying units are housed in an attractive cabinet and usually consist of a large wheel or drum which has a rim of porous material. A humidistat controls a fan motor, which drives the large wheel that is wetted

Fig. 20-118. Electric humidifier designed for installation in warm air duct. A—Plate for mounting on duct. B—Electrical line and control cord. C—Drain cap. D—Water inlet. (Auto Flo Co., Div. of Masco Corp.)

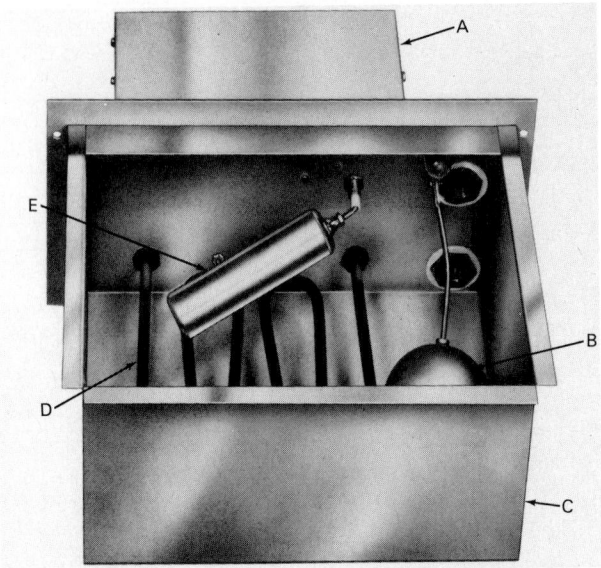

Fig. 20-119. Construction of electric humidifier. A—Electric controls. B—Water float. C—Water tank. D—Heater. E—Switch float. (Auto Flo Co., Div. of Masco Corp.)

as it dips into a water reservoir. As the air flows through the wet, porous rim, most of the moisture evaporates and adds moisture (water vapor) to the air. See Fig. 20-120. Some of the units have an electric heater to warm the outlet air. The water reservoir is manually filled with water. A float-operated switch will turn on a signal light when the water level is low. The unit is easily maintained. The water tank and the porous wheel rim can be removed and cleaned.

20-62 MOBILE HOME AIR CONDITIONING

Most mobile home heating is with LP-Gas, although some mobile home heating equipment may use oil. These furnaces usually heat the mobile home through an air distribution system. Ducts for this type of system are shown in Fig. 20-121. An evaporator may be placed in the plenum chamber. A condensing unit is located outside the mobile home to provide cooling through the heating duct installation.

Fig. 20-122 shows a combination heating and cooling system commonly used in mobile homes, apartments, or other small areas. Systems are self-contained and are installed through window openings or through the wall.

20-63 REVIEW OF SAFETY

Safety must be strongly emphasized when installing and operating heating systems. Oil and fuel gas furnaces are fires in a confined area.
1. Fuel must be stored safely.
2. Fuel must be fed in a safe manner to firepot of furnace.
3. Means must be taken to shut down the unit:
   a. If fuel flow ceases.
   b. If any part of system overheats.
   c. If products of combustion are blocked from leaving the building. There is a monitor, called a "spill switch," which fits into the flue. If the upper flue is blocked, the unit heats up and shuts off the furnace.

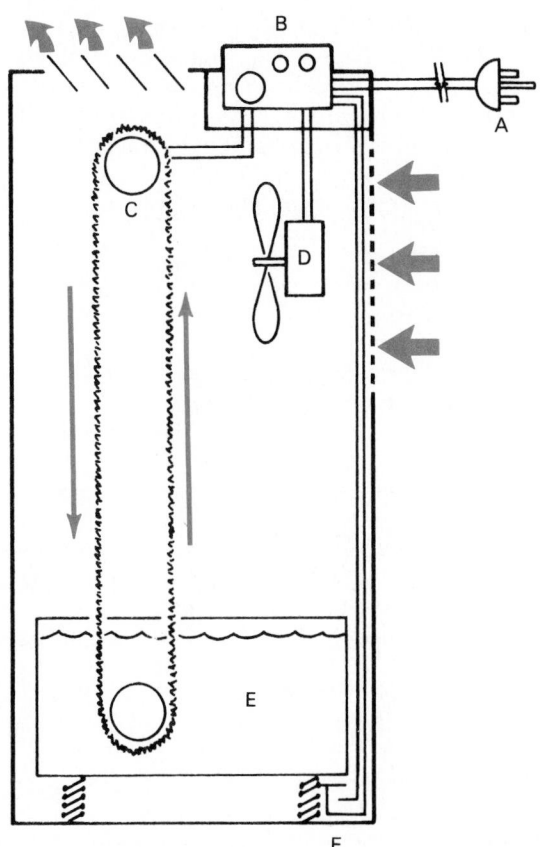

Fig. 20-120. Cabinet-type humidifer in which a porous filter revolves and picks up moisture. Air blown through porous material vaporizes some of the water, raising relative humidity in room. A—Power cord. B—Humidistat, switch, indicator lights. C—Moisture belt and filter. D—Fan and motor. E—Water tank. F—Light switch and light: when light is lit, this indicates that water tank is empty.

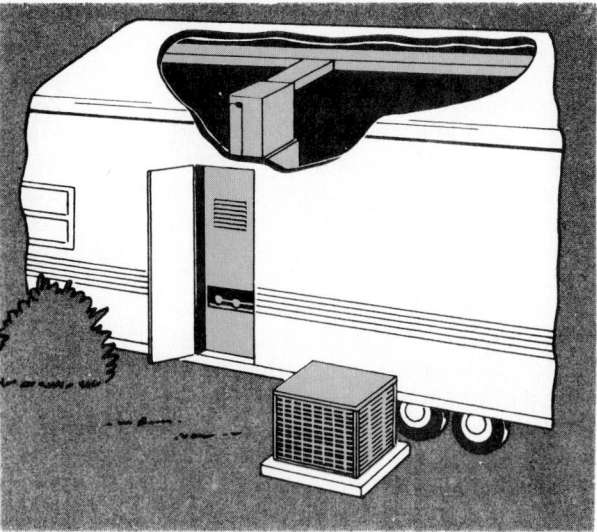

Fig. 20-121. Mobile home year-round air conditioner. Condensing unit for summer cooling is located on ground outside home.

Avoid tampering with safety controls; they must be in excellent condition and properly adjusted. Tampering may cause delayed ignition, and an explosion may damage equipment and injure persons. Avoid turning on the electric ignition or lighting a pilot light until the firepot has been examined for gas or oil.

Remember, if fuel is in the presence of air and an ignition source exists, fire will result.

Know exactly what is being done. Follow manufacturers' specifications carefully. Always follow the building and safety codes in effect in the locality.

Electric heating is considered to be a safe heating method. The elimination of flames, pilot flames, electric spark ignition, and sparks reduces many sources of danger. There is, however, possible danger from electrical shock and the chance of burns or of combustion from the heating elements.

All electrical devices must be installed according to local, state, and national electrical codes. The equipment should have Underwriters Laboratory approval (UL). The installation should be made by a licensed electrician.

All nonelectrical metal parts of the unit must be safely grounded. Combustibles must not come in contact with heating elements. The heating elements must be mounted to eliminate the chance of persons or animals contacting the heating grids or elements.

Special gas heaters for tents and cabins should be installed with a stack and a safety pilot light which will shut off the fuel gas flow in case the pilot flame is extinguished.

Testing for gas leaks should be done with soapsuds. If a leak is suspected, do not use open flames. Turn off the electric power. Use an explosion-proof flashlight only.

Be careful when inspecting for a source of leaking fuel oil. Turn off the electrical power. Use an explosion-proof flashlight only.

The propellant in a hair spray aerosol can and the chemical methylene chloride found in varnish removers have an intensely corrosive effect on metal surfaces. This action is accelerated with an increase in temperature, so it may shorten the life of the heat exchanger in warm air furnaces.

An improperly vented furnace, regardless of the kind of fuel used, presents two dangers. If the combustion is incomplete, a considerable amount of carbon monoxide may be generated. This is a very poisonous gas, and inhaling it may result in illness and death. In low concentrations, the

Fig. 20-122. Through-the-wall combination heating and cooling system. (Suburban Manufacturing Company)

symptoms may be headaches, fatigue, dizziness, or loss of muscular control.

If the combustion is complete, a considerable amount of carbon dioxide (CO_2) may be formed. If CO_2 is not vented, it replaces the oxygen in the air. Animals require oxygen in the air they breathe, so if carbon dioxide has replaced the oxygen, the body will suffer from the lack of oxygen.

All heating furnaces should be equipped with safety devices. If, for any reason, the fuel fed into the combustion chamber is not ignited, the safety device will immediately shut down the furnace. This is to avoid an explosion in the event the furnace and flues were to become filled with combustible gas and ignited. All furnaces must be equipped with a temperature limit control which will shut off the burner if the bonnet temperature, water temperature, or steam pressure exceeds a predetermined safety setting.

Before lighting the pilot light in a gas furnace, shut off the gas supply and allow the furnace door to remain open for not less than five minutes. If there is any combustible gas in the furnace, it will be vented by this procedure, making it safe to light the pilot.

Any gas odor or any sign of oil in the combustion chamber is a strong signal to be super-cautious.

A hot water or steam system is a pressure vessel. The safety devices must work or the plant may explode. Always operate the safety devices as a part of system inspection. Low water failures are a common cause of trouble; always inspect the low-water controls and safety drains. A low-water control can be tested by deliberately removing some boiler water (with the heat source off, of course). The switch should shut off the fuel and electrical power. Be careful when removing water which might be under pressure (removal is called a blowdown). Let the unit cool for 1 hour before opening valves.

Never use compression fittings on an oil-burning system.

Many materials are both flammable, explosive, and/or toxic (harmful) to living things. But all these materials are safe until a threshold limit value (TLV) is reached. For example: The TLV of a gas may be 200 ppm while the lowest explosive limit (LEL) is 6.7 percent (67,000 ppm).

The National Fire Protection Agency explains explosive limits as the minimum amount of the gas in air or oxygen that can be ignited by an ignition source and keep on burning (self-propagating). Remember that slow burning is a fire, while very rapid burning is an explosion. Air out a furnace chamber before lighting the pilot light.

Many devices have been used to detect the presence of these gases, including birds, animals, plants, and instruments.

Many electrical instruments are now available for not only quickly detecting the presence of these gases, but also measuring the amount of these gases in the space being tested. These instruments have greatly increased the safety

conditions. However, there are many situations in service work where these instruments are not available or the conditions change quickly. *One of the best health safeguards and fire preventions is to completely and thoroughly ventilate the space in which one is working.*

20-64 TEST YOUR KNOWLEDGE

1. Name three types of heat sources.
2. Identify two major differences between a pulse-combustion-type furnace and a conventional atmospheric furnace.
3. What is the heating value of domestic heating oil?
4. Name three types of fuel gases.
5. Why must stack temperatures be kept over 300 °F (150°C)?
6. Do gas furnaces have the same capacity at all altitudes?
7. How much excess air is needed for combustion?
8. Why is humidifying needed during the heating season?
9. What is a hydronic heating system?
10. How much ''heat-per-hour'' (Btu/hr.) is obtained from one watt?
11. Name three different gases found in a flue.
12. What is the advantage of a 240 V electric resistance heating circuit over a 120 V circuit?
13. What grade fuel oil is used in most domestic gun-type oil burners?
14. What is forced draft and induced draft?
15. Why is a filter needed?
16. What is the purpose of the draft control device?
17. What is the main cause of stack corrosion?
18. In addition to being clean, what are three advantages of electric heat?
19. How may soot be removed from a furnace flue or a boiler stack?
20. How does a gun-type oil burner obtain combustion air?
21. What is the color of the combustion of a well-adjusted gun-type oil burner?
22. How many oil filters or screens does a gun-type oil burner system use?
23. What safety device is part of a pilot light?
24. Where is a thermocouple used in a fuel gas system?
25. What is the color of a correctly adjusted and properly burning fuel gas flame?
26. What happens if there is air in a hydronic system?
27. What is the most common service problem with hydronic system pumps?
28. Where should the pilot light flame be when it is measured from the thermocouple?
29. What is meant by zone control heating systems?
30. What common furnace flue or boiler flue gas is most dangerous to humans and animals?

Chapter 21

AIR CONDITIONING SYSTEMS COOLING AND DEHUMIDIFYING

After studying this chapter, the technician will be able to:
☐ Explain the principles of air conditioning.
☐ Classify types of comfort cooling systems and recognize variations between them.
☐ Properly install and service window air conditioners.
☐ Properly install and service console air conditioners.
☐ Define and calculate energy efficiency ratings.

Refrigeration is the basis of the comfort cooling part of air conditioning. All comfort cooling systems use one of the standard refrigerating cycles, standard refrigerants, standard types of compressors, condensers, piping, refrigerant controls, motor controls, and evaporators. The refrigerant evaporating temperature of most comfort cooling systems is about 40 to 50°F (4 to 10°C). This chapter describes most designs which use mechanical refrigeration for cooling and dehumidifying.

21-1 PRINCIPLES OF ATMOSPHERE COOLING

As explained in Chapter 18, one's comfort as well as the success of certain industrial operations depends on temperature and relative humidity. For example, computers will operate only in special temperature and relative humidity conditions. Usually, a ''clean room'' must be provided. Manufacturers of medicines and biological products require special conditions. Any industrial operation dealing with hygroscopic (water-absorbing) materials and processes must be carried out in controlled conditions of relative humidity.

The first mechanical atmosphere cooling and humidity control used cooled water, both to reduce the temperature of the air and dehumidify it. Air was passed over water-cooled coils. When relative humidity was not a concern, air was passed through cooled water sprays.

Fig. 21-1 shows the operation of typical cooling units. Return air is mixed with some fresh air. Then this air mixture is filtered and cooled. Moisture is removed before it is redistributed into the building.

Cool air leaving the evaporator is at 100 percent relative humidity. This saturated air, as it mixes with air in the conditioned space, warms up somewhat. Thus, relative humidity is brought down to a comfortable level.

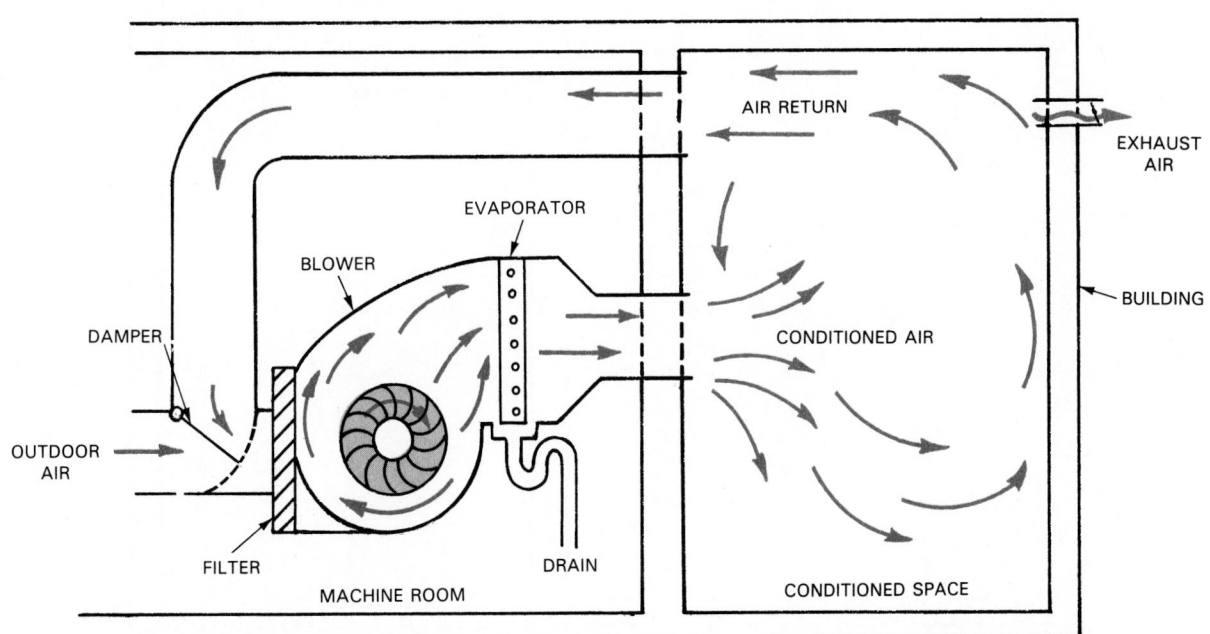

Fig. 21-1. Basic operation of conditioner which cools air and removes moisture from outdoor air and recirculates air. Centrifugal blower, shown in cutaway view, is used.

There is a more positive way to control relative humidity. It involves bypassing some of the return air into the air conditioner outlet to warm up the cooled air before it leaves the duct system. This method is shown in Fig. 21-2.

21-2 COOLING

In a cooling cycle, the dry bulb (db) temperature of the air is lowered. When this happens, as in Fig. 21-3, A to B, the relative humidity increases. Some moisture should be removed to make this air comfortable. Moisture can be removed by either of two methods:
1. Dehydrate the air with chemicals.
2. Cool the air down to the saturation curve at C and then remove moisture by condensing it on a cool surface. See line C to D, Fig. 21-3. The distance from C to D is the drop in vapor pressure or grains of moisture removed.

Reheating along a horizontal line, D to E, will decrease the humidity. However, what most often happens is that the air leaving at D is mixed with the room air which is at some intermediate condition between 81 °F (27 °C), and 100 °F (38 °C). The mixture meets on the line between point D and point A.

If a third of the air, by weight, is passed through the evaporator, the mixed air temperatures will be a third of the way from A to D—that is to F. The system in Fig. 21-2 mixes the air inside the air conditioner and also brings fresh air into the air conditioner.

21-3 COMFORT COOLING SYSTEMS

Several types of comfort cooling systems are in common use. They can be classified by arrangement of the mechanism:
1. Self-contained or unit coolers.

2. Remote (controlled from a distance).
The self-contained system includes:
1. Window units.
2. Through-the-wall units.
3. Cabinet units.
Remote units are of two types:
1. The condensing unit is remote. The evaporator is installed in the room to be conditioned or in the main duct.
2. The central air conditioning plant. The condensing unit and the evaporator are installed away from the place being conditioned. A cooled brine or water is circulated to heat exchangers in the various spaces to be conditioned.

21-4 COOLING EQUIPMENT FOR AIR CONDITIONING PURPOSES

Cooling equipment usually consists of evaporators. These are kept at refrigerant temperatures of 40 to 50 °F (4 to 10 °C) and the air to be conditioned is moved through them.

There are several ways to cool the air:
1. By mechanical refrigeration.
2. By absorption refrigeration. (Absorption system air conditioning is described in Chapter 16.)
3. With ice.

Ice may be the cooling medium, particularly if cooling is only needed for a few days of the year. Cold water from streams or wells may also be used. The water should be 50 °F (10 °C) or cooler to produce satisfactory dehumidification (removal of moisture from the air).

Most air conditioning installations use automatic mechanical or absorption refrigeration for cooling. The mechanical system consists principally of:
1. Condensing unit.
2. Evaporator.
3. Motor-driven fans.

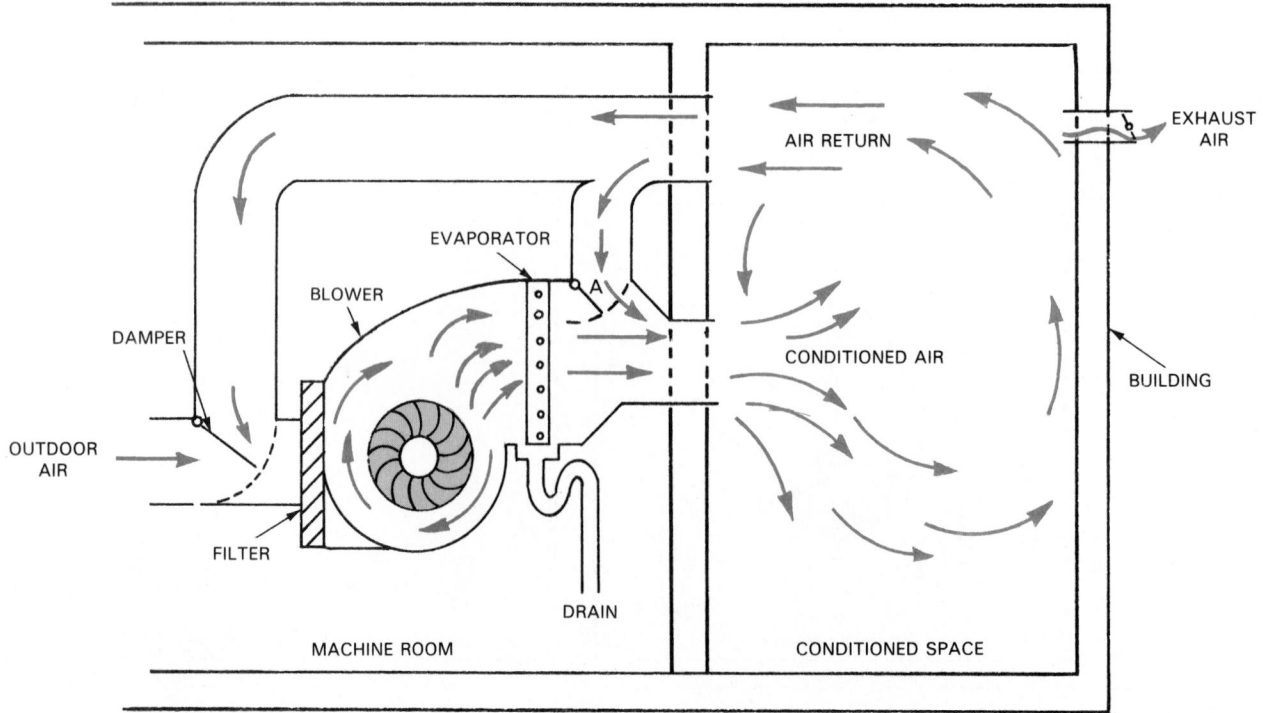

Fig. 21-2. Comfort cooling installation with return air bypass, A, for relative humidity control.

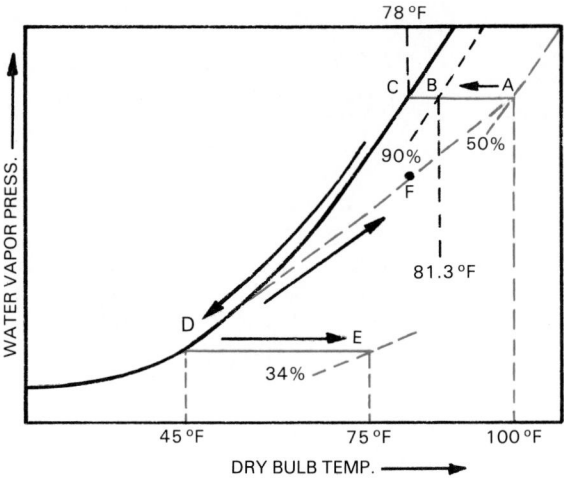

Fig. 21-3. Cooling cycle on psychrometric chart. A—Condition of outside air. B—Partly cooled air. C—Air cooled to saturation. D—Air cooled to remove some moisture. E—Dehydrated air reheated. (A condensing coil can reheat the air. Refer to Fig. 21-41.) F—Result of mixing treated and untreated air.

4. Filters.
5. Ducts and airflow controls.
6. Motor controls
7. Temperature and humidity controls.
8. Piping.
9. Refrigerant.

21-5 UNIT COMFORT COOLERS

Unit or self-contained air conditioners are of two types:
1. Those providing comfort cooling only for summer.
2. Those for both summer cooling and winter heating. (See Chapter 23.).

Both types have a complete refrigeration plant. This includes condensing unit, refrigerant valves, and evaporators.

Filtering equipment is also a part of the system. Individual room thermostats provide control. These units may also be classified as window or in-the-wall units, or console units.

The window units and in-the-wall units are air-cooled, easily mounted, and operate from 120 V or 240 V single-phase circuits. Their capacities vary from 4000 Btu/hr. to 40,000 Btu/hr.

Console units may be either air-cooled or water-cooled. They are installed in the room to be conditioned or in an adjacent room with short ducts to deliver air and provide for return air.

21-6 WINDOW UNITS

The window or wall-mounted comfort cooler is very popular. The window unit mounts on a windowsill, and installation is relatively easy. The condenser is located in the section of the cabinet that is outside the building. Outside air is forced over the condenser by a fan. Inside the room, another fan draws air in through a filter and forces it over the evaporator.

The two airflow fans may be driven by the same motor or each may have its own motor. Fig. 21-4 shows the airflow in a window air conditioner. Fig. 21-5 illustrates a modern window air conditioner. The internal construction of a unit is shown in Fig. 21-6.

Elements of a window comfort cooler are shown in Fig. 21-7. A more detailed schematic is provided in Fig. 21-8.

Window units are available in several types. One type cools and filters the air and has a fresh air intake. Another type has these same devices but, in addition, has an electrical resistance heating unit to furnish heat. A third type uses a reverse cycle system (heat pump) to permit the use of the refrigerating units both for comfort cooling and heating. See Chapter 23.

Window units may be obtained to fit double-hung windows or casement windows. Units can be installed in special wall openings.

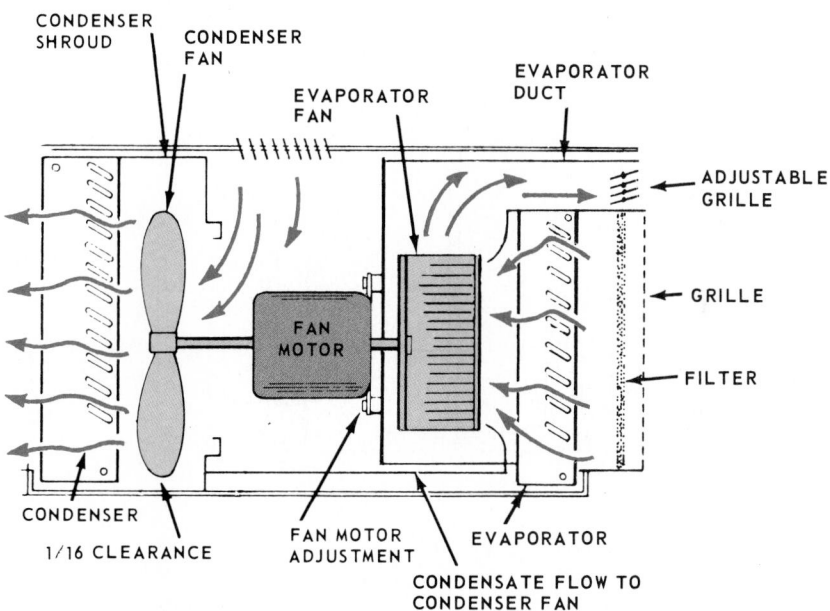

Fig. 21-4. Drawing of air circuits in window air conditioner. Axial-type fan is used at condenser. Evaporator air circuit uses radial flow fan. Same motor drives both fans.

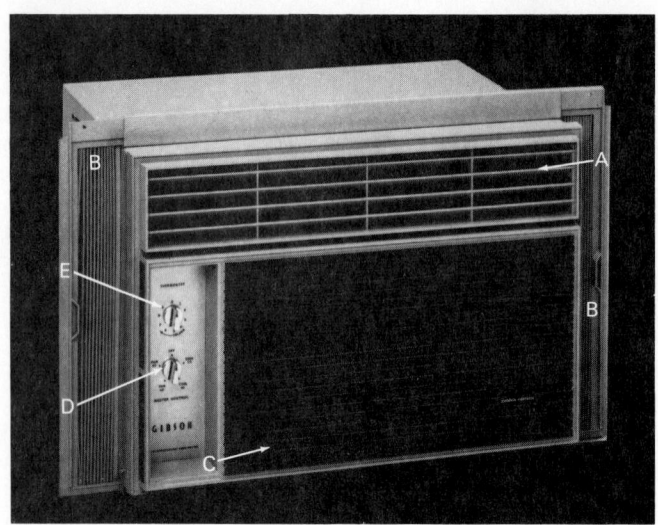

Fig. 21-5. Window air conditioner. A—Cooled air outlet. B—Expandable partitions. C—Air inlet. D—Circulating or cooling fan control. E—Temperature control. (Frigidaire Company)

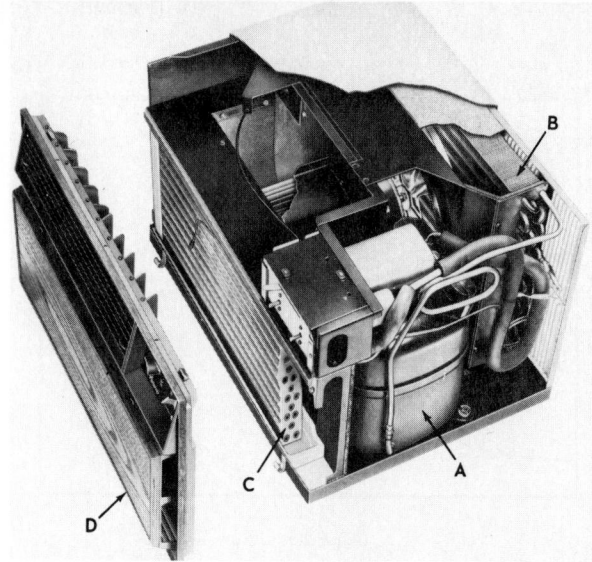

Fig. 21-6. Internal construction of window-mounted air conditioner. A—Hermetic motor compressor. B—Condenser. C—Evaporator. D—Removable front panel. Return grille, filter, and outlet grille are built into front panel.

The condensate from the evaporator is often drained to the base of the motor compressor and the condenser where it helps to cool these parts. A capillary tube or a bypass-type AEV refrigerant control is usually used.

Some units change the cooled airflow from side to side as the unit runs. This is done by rotating angle deflector plates. The plates are mounted on a shaft which is turned by an air-operated rotor in the exhaust air, Fig. 21-9.

The systems are controlled by thermostats. The sensitive bulb is usually mounted at the inlet of the evaporator. A differential of about 5°F (3°C) is normal.

If the part of the bulb farthest from the evaporator is insulated, the bulb will respond better to the evaporator temperatures. It will cool sooner and stop the unit before it overcools. It will also stop the unit if the evaporator ices up and will prevent it from starting again until the ice melts.

Caution: If a unit is shut off and then immediately turned on again, it may stall and damage the motor compressor. Let high pressures dissipate.

21-7 INSTALLING WINDOW UNITS

Window units must be installed with the outside tilted down for condensate drain. They must be securely fastened in place to prevent the unit from falling out of the window. All edges should be sealed to minimize air infiltration, and the window must be secured in the proper position.

When units are installed in windows, metal plates, rubber gaskets, and sealing compounds are used to seal the unit into the opening. The unit and the parts needed to install it are shown in Fig. 21-10. Fig. 21-11 shows a windowsill with a leveling bracket and security bracket

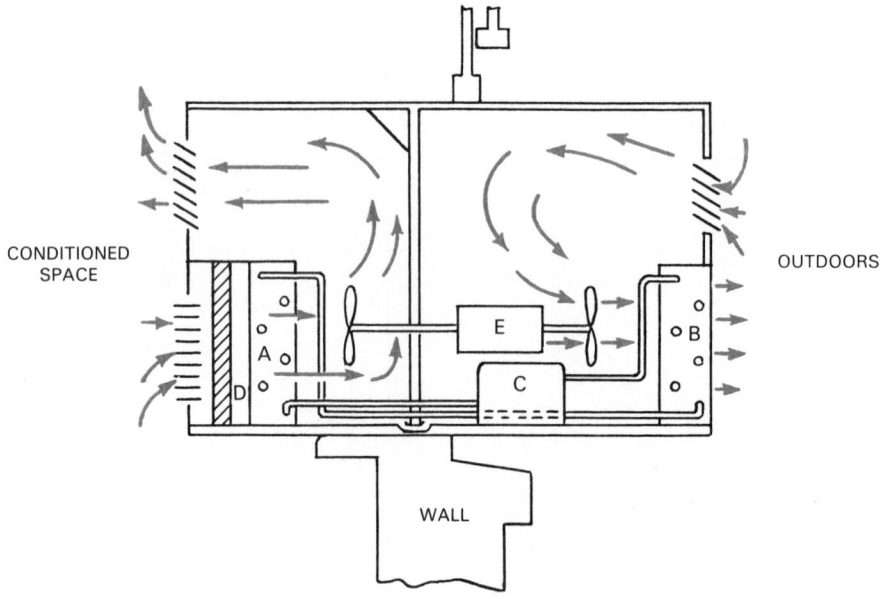

Fig. 21-7. Drawing of window comfort cooler installation. A—Evaporator coil. B—Condenser. C—Motor compressor. D—Filter. E—Fan motor.

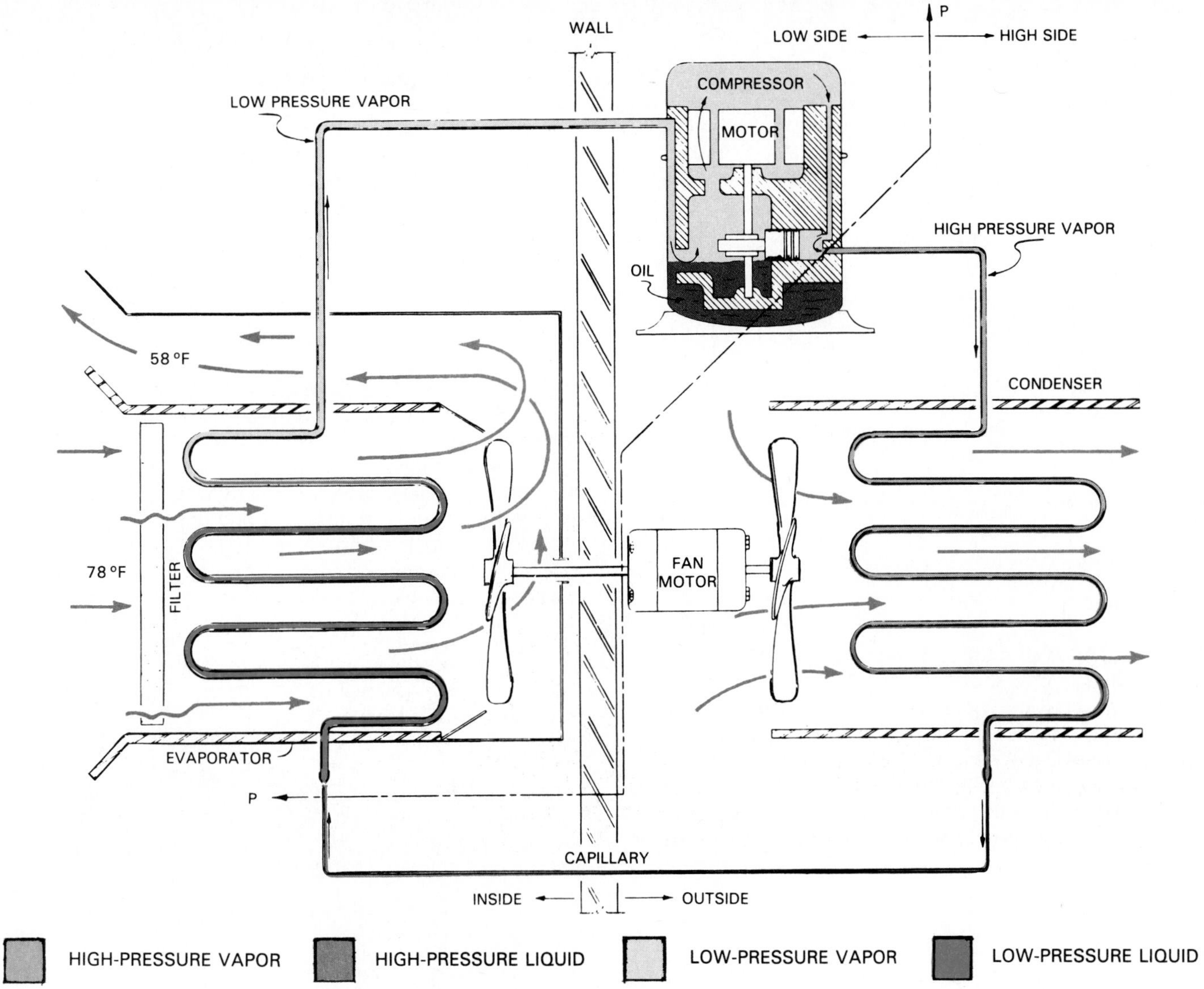

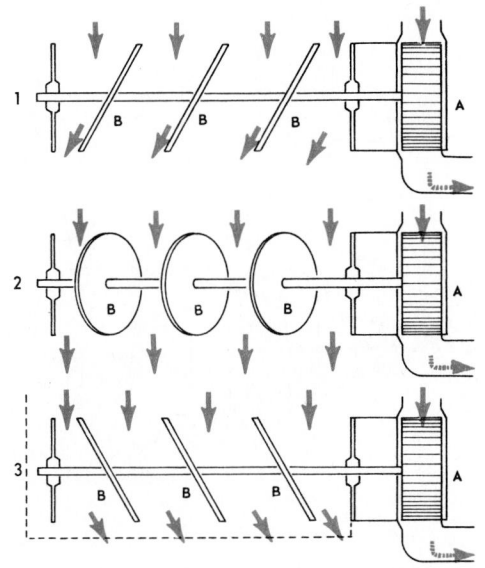

Fig. 21-8. Drawing of window or in-the-wall comfort cooling unit. Note conditioned airflow and separate condenser airflow.

WALL

LOW PRESSURE VAPOR

LOW SIDE ← P → HIGH SIDE

COMPRESSOR
MOTOR
OIL

HIGH PRESSURE VAPOR

CONDENSER

58 °F

78 °F

FILTER

FAN MOTOR

EVAPORATOR

P

CAPILLARY

INSIDE ← → OUTSIDE

HIGH-PRESSURE VAPOR   HIGH-PRESSURE LIQUID   LOW-PRESSURE VAPOR   LOW-PRESSURE LIQUID

Fig. 21-9. Oscillating grille deflector. Grille is behind conditioned air outlet for an air conditioner which continually sweeps air from side to side as long as fan runs. Air turbine, A, revolves. As it does, it turns deflector plate at B. 1—Air is deflected to left. 2—Air flows directly ahead. 3—Air flows to right.

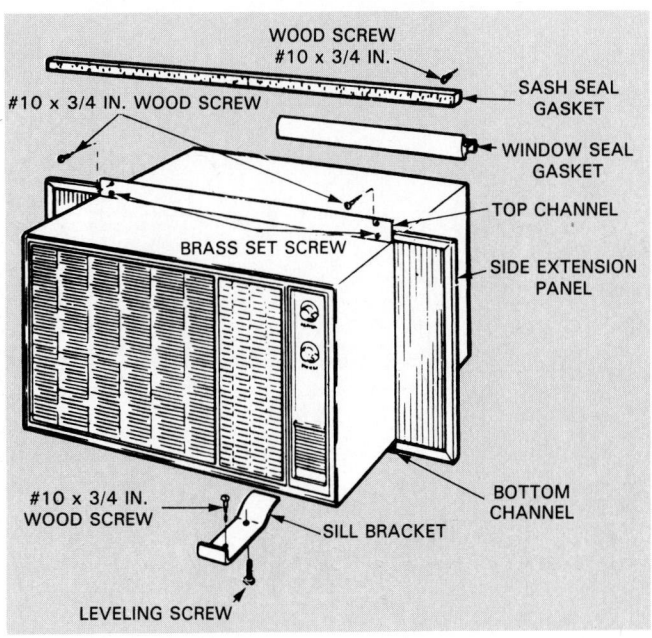

Fig. 21-10. Window unit showing necessary parts to mount unit safely in window and seal openings.

WOOD SCREW #10 x 3/4 IN.

#10 x 3/4 IN. WOOD SCREW

SASH SEAL GASKET

WINDOW SEAL GASKET

TOP CHANNEL

BRASS SET SCREW

SIDE EXTENSION PANEL

#10 x 3/4 IN. WOOD SCREW

SILL BRACKET

BOTTOM CHANNEL

LEVELING SCREW

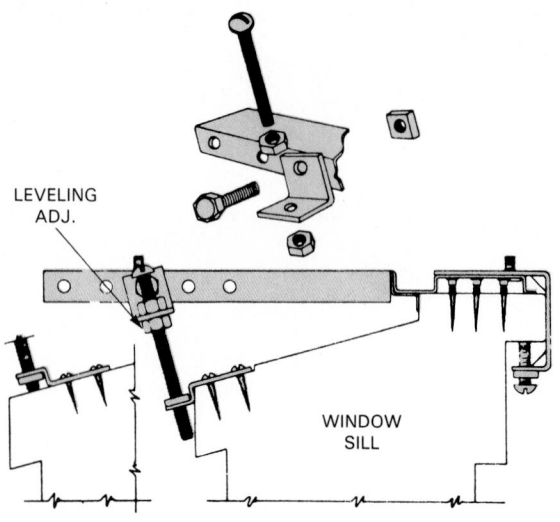

Fig. 21-11. Bracing is used to hold window air conditioner on sill.

filler boards are put in place, as shown in Fig. 21-13. Manufacturers each have slightly different methods for mounting window units.

Where the lower sash is raised to make room for the air conditioner, an air gap will exist between the two sashes. This opening may be sealed with a sponge rubber or styrofoam strip, as shown in Fig. 21-14. Notches help make the strip fit.

The housing must be securely fastened in place before the unit is put in place. The filler boards (between the unit housing and side of the window) are usually sealed with sponge rubber strips or styrofoam and held in place with sheet metal screws and with spring clips. Fig. 21-15 shows one method of installing them. A typical window unit installation in a casement window is shown in Fig. 21-16.

The inside mechanism is heavy. It should be moved using a dolly or special carrier. Avoid moving or lifting the unit by using the tubing or coils as hand grips. Carry the unit by holding onto the bottom pan.

Avoid forcing the unit into the casing. Check to be sure the refrigerant lines and the wiring are free and clear as the

mounted. Another method of bracing and leveling a comfort cooling unit is shown in Fig. 21-12.

The unit housing should be adjusted to tilt downward about 1/4 in. on the outside. This is enough to provide condensate drainage. A sponge rubber or plastic strip is usually placed between the housing and the windowsill to help make a leakproof joint. After the sill brackets are installed and the unit housing installed, the rubber seal strips and the

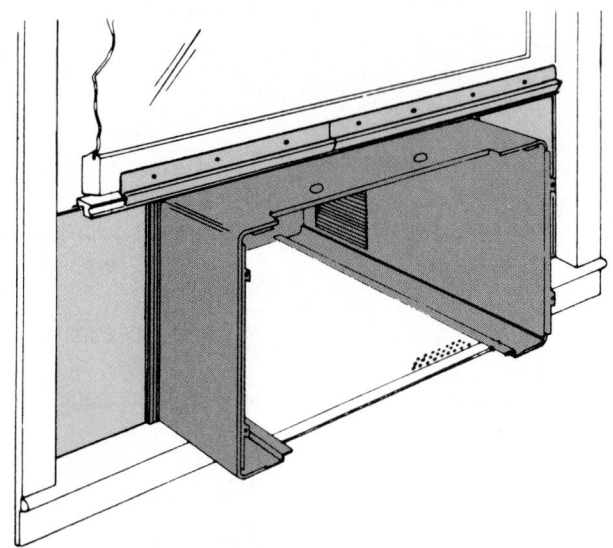

Fig. 21-13. Window air conditioning casing installed, showing rubber seal strips and filler boards.

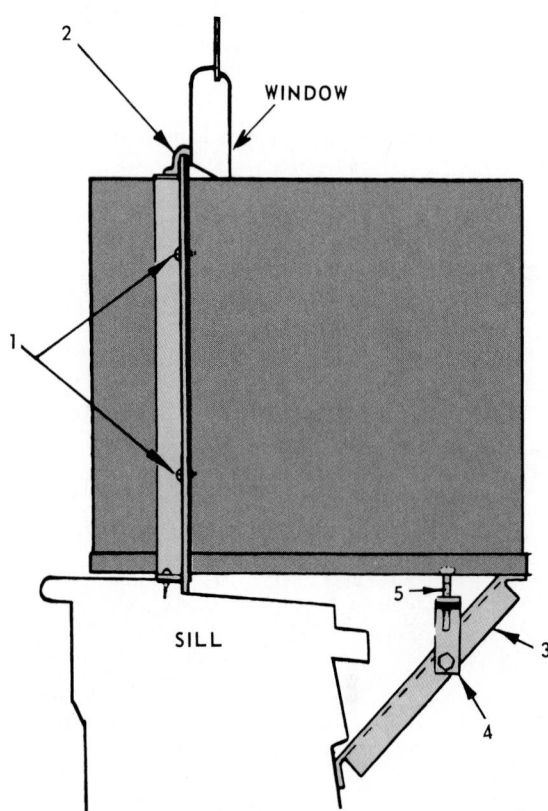

Fig. 21-12. Window unit with indoor flush mounting. 1—Sheet metal screws to hold side closure panels. 2—Gasket. 3—Bracket. 4—Clamp pivot. 5—Clamp adjustment.

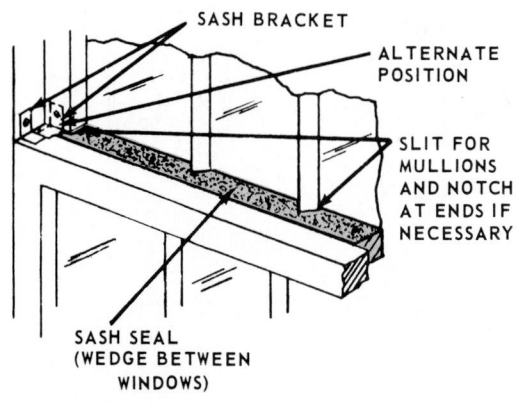

Fig. 21-14. Sponge rubber seal placed between upper edge of lower sash and upper sash of double-hung window. Sash bracket keeps lower sash locked.

unit moves into the casing. The front grille, filter, and control knobs are easily installed.

As a final step, check all joints for tightness. Caulk seams which show light leaks or which one suspects may not be airtight.

When making the electrical hookup, use a separate circuit. A polarized plug (one with a ground wire) is required.

Thermostats are used with most window units. They are adjustable to cut out between 56 °F (13 °C) and 60 °F (16 °C) and cut in between 77 °F (25 °C) and 80 °F (27 °C). Their differentials vary between 3 °F (2 °C) to 8 °F (4 °C). If a thermostat fails, the unit will not start. To test the operation of a thermostat, cover the air-out and air-in with a cloth. The air will now recirculate into the unit and the temperature will quickly drop to the cutout temperature. Use a thermometer.

Units which mount through the wall are popular in new apartment units. There is no interference with windows and comfort cooling can be provided as desired. Fig. 21-17 shows a typical installation.

21-8 SERVICING WINDOW UNITS

Servicing window units is similar to the servicing of hermetic refrigerating units. Chapters 11 and 14 describe most of the servicing operations.

Some of the external service operations are as follows:
1. Semiannual cleaning or replacement of the filter (usually done by the owner). Fig. 21-18 shows a filter design.
2. Annual cleaning of the evaporator, condenser, fan blades, fan motor, motor compressor, and casing. The unit is removed from its casing for these operations, as shown in Fig. 21-19.
3. Inspect fan motor or motors and lubricate them unless they have sealed-for-life bearings. Always wipe away

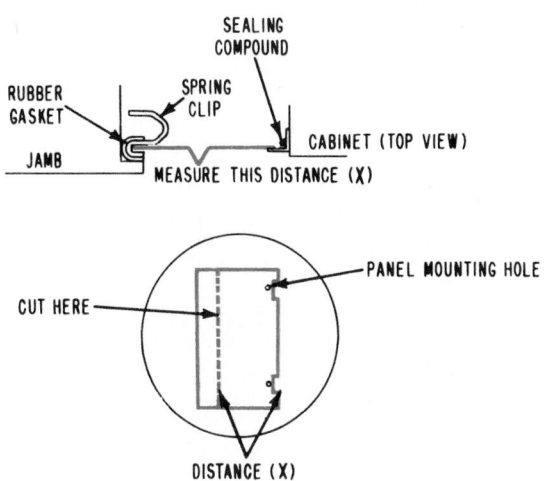

Fig. 21-15. Filler panel between unit housing and window casing.

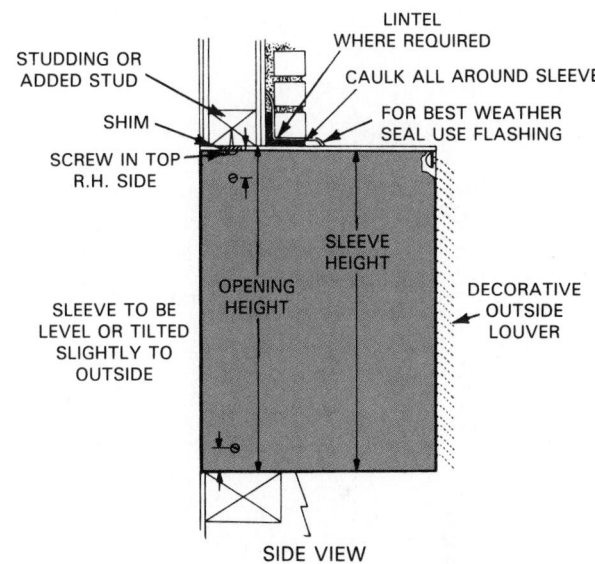

Fig. 21-17. Typical through-the-wall unit air conditioning installation.

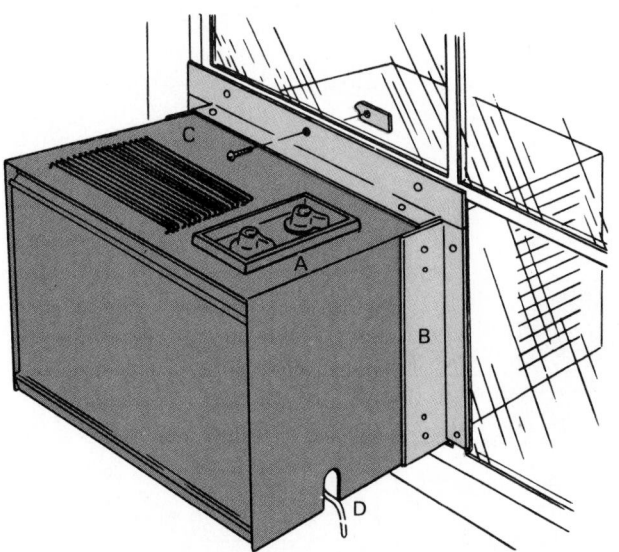

Fig. 21-16. Method of mounting window unit in casement window. A—Controls. B—Angle plate, usually enameled steel, fastens to both casing and window frame. C—Machine screw holds angle plate to window frame. D—Electrical cord. (Frigidaire Company)

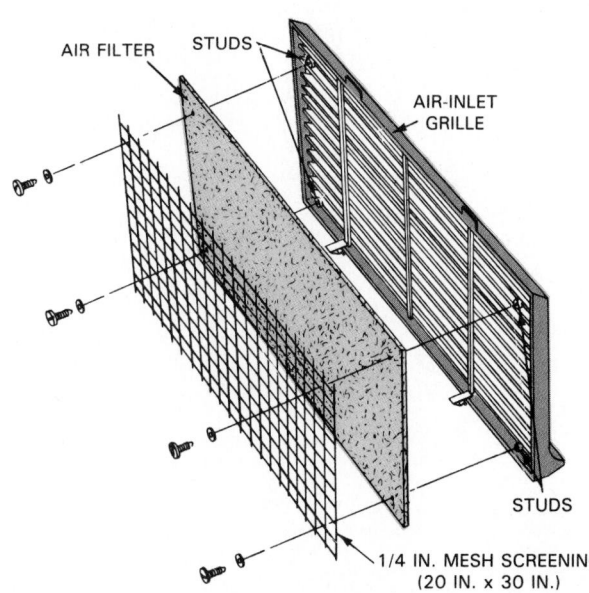

Fig. 21-18. Typical filter installation for window air conditioner.

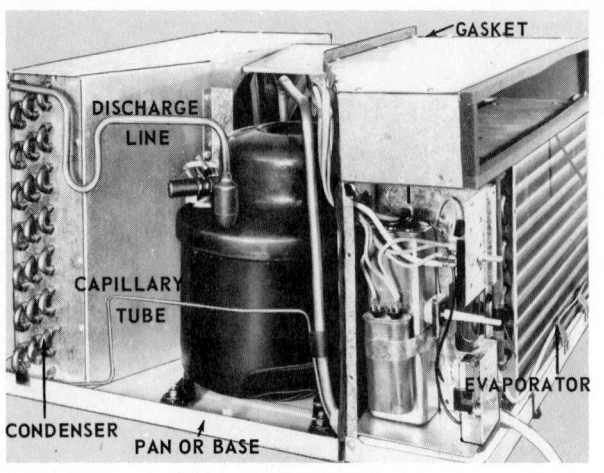

Fig. 21-19. Window air conditioner unit removed from its cabinet and ready for cleaning. (Frigidaire Co.)

excess oil. Oil mist on the fan blades collects lint and reduces air movement efficiency.

Place a tarpaulin or newspapers on the floor and remove or tie back curtains or drapes before cleaning the unit. Use a commercial model vacuum cleaner with a brush-equipped nozzle to clean the inside of the cabinet.

Finned evaporators and condensers are difficult to clean. The fin spacing prevents the vacuum brush from reaching the lint and dirt. In such cases, plastic blades along with a powerful vacuum will remove most of the dirt. Dirt must be removed if the unit is to continue working well. *Never use metal blades for cleaning; they may cause leaks.*

If the unit can be taken outside (and this is always more desirable), a powerful water and detergent spray does a good job. Coils must be cleaned thoroughly. Fins, if bent, should be straightened.

When servicing fan motors, make certain that fans are tight on the shaft. They should be carefully positioned in

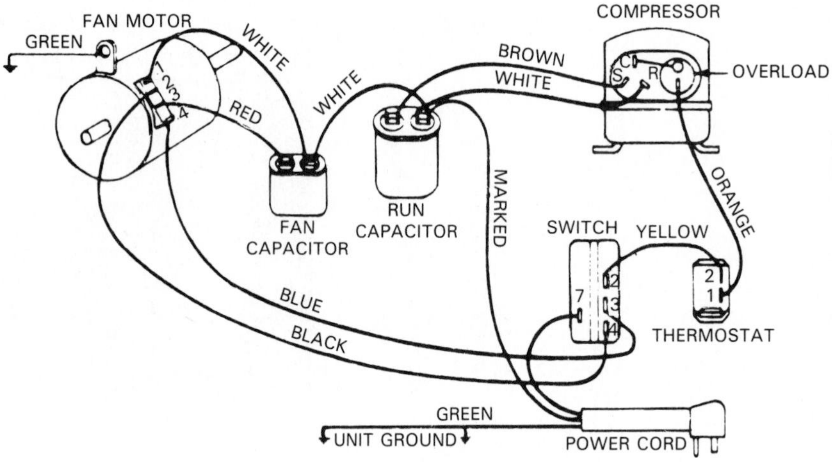

SWITCH	CONTACTS		
POSITION	2	3	4
Off	O	O	O
Normal Fan	O	C	O
Super Fan	O	O	C
Normal Cool	C	C	O
Super Cool	C	O	C
C—CLOSED			
O—OPEN			

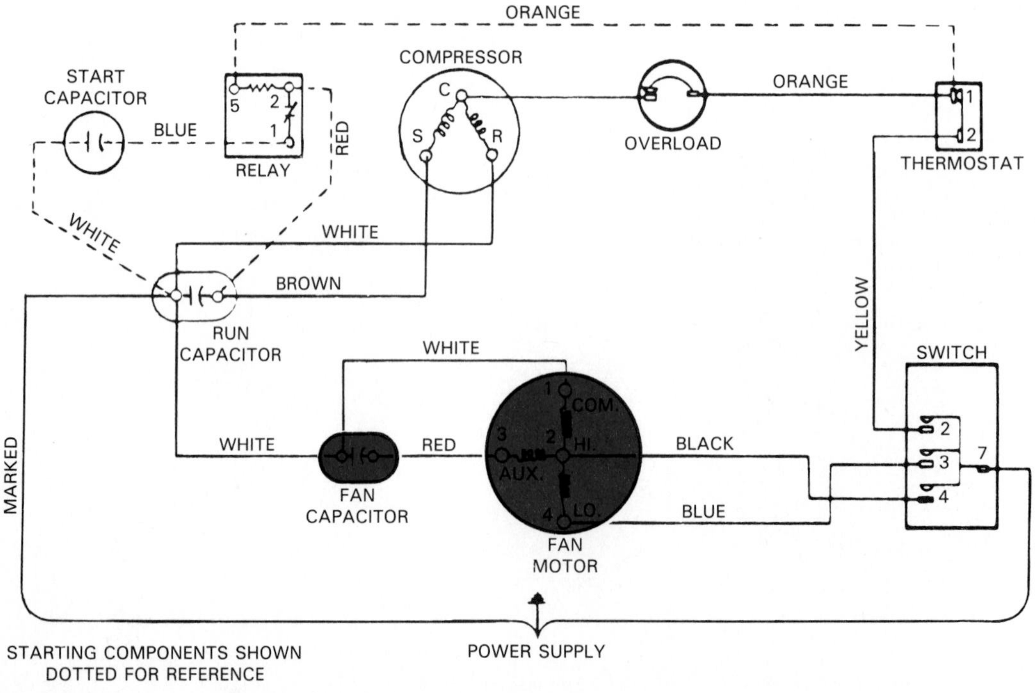

Fig. 21-20. Wiring diagram of 120 V window air conditioner with 7000 Btu/hr. capacity. Note two-speed fan motor and fan capacitor. (Fedders North America)

the shroud for efficient air movement. Avoid bending the fan blades or twisting them. An off-balance fan will soon wear out the motor bearings and will be noisy because of vibration. Replace an abused fan.

Inspect the drain. It must be clean. Remove lint from the drain hole and tube using a soft wire. Check all bolts, nuts, and screws for tightness.

Before replacing the unit in the cabinet, run it to check for noise. Find its source and stop the noise.

Always put a cloth over the outlet of the air conditioner when it is first started after cleaning. Loosened dirt that the vacuum cleaner failed to pick up will be blown out the adjustable grille.

The wiring of a window unit is very similar to other refrigerating units. See Fig. 21-20. External electrical servicing procedures are usually the same as for domestic and commercial units except that:

1. Fan motors usually have two or three speeds.
2. Some systems have three capacitors: starting capacitor, running capacitor, and fan motor capacitor.

Shown in Fig. 21-21 is a unit with a starting capacitor and a running capacitor. An arrangement with three capacitors is shown in Fig. 21-22. The position of the fan control switch, thermostat, capacitors, and wiring are seen in Fig. 21-23.

The wiring diagram in Fig. 21-24 is for a window unit with a variable speed fan motor and a thermostat. Fig. 21-25 shows a wiring diagram for a three-speed fan motor system of 21,000 Btu/hr. (6.15 kW) capacity using a 240 V circuit.

Testing of outside electrical parts is described in Chapters 6, 7, and 8. Fan and compressor motor testing is described

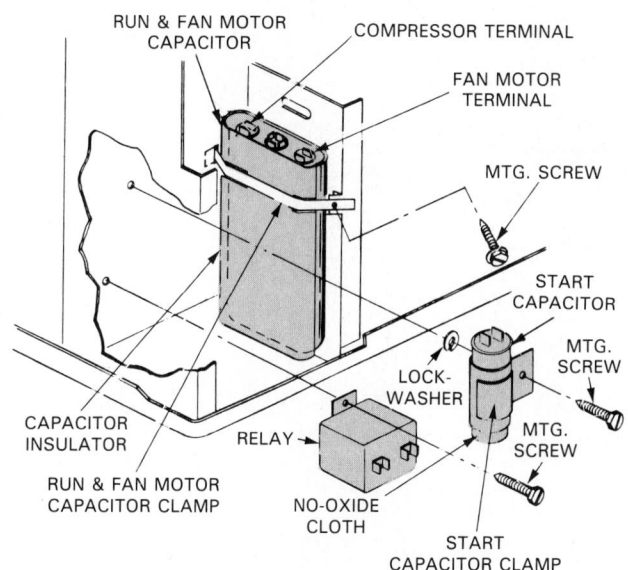

Fig. 21-22. Arrangement of start, run, and fan-motor capacitors. Note that fan-motor-run capacitor and compressor-motor-run capacitor are in same container.

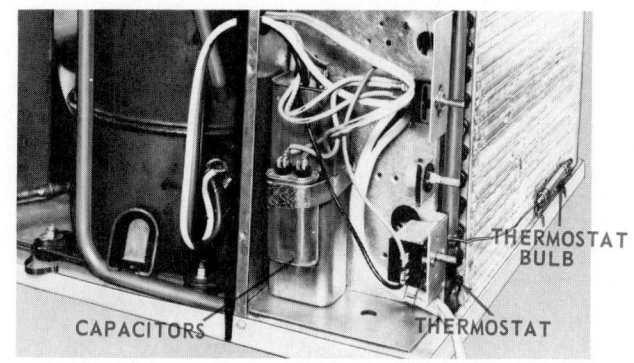

Fig. 21-23. Window unit showing location of capacitors and thermostat.

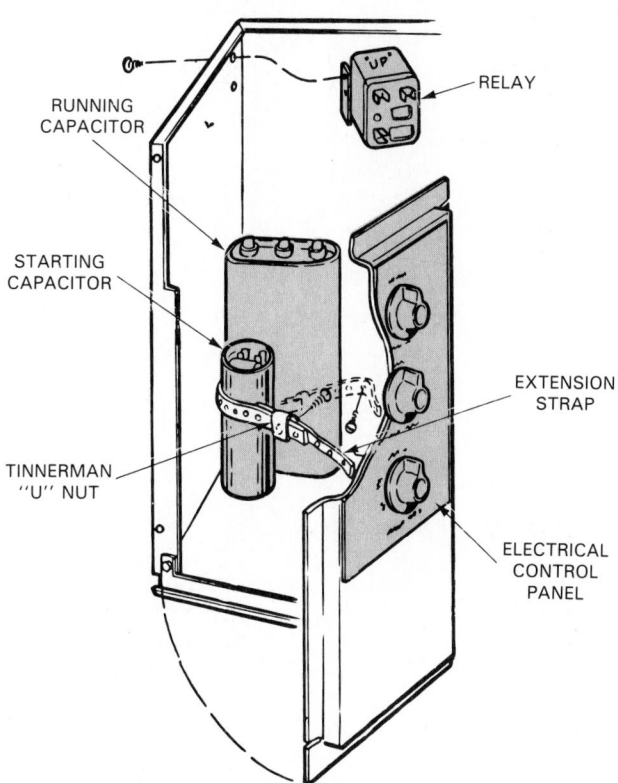

Fig. 21-21. Window unit showing location of parts such as capacitors, control panel, and relay.

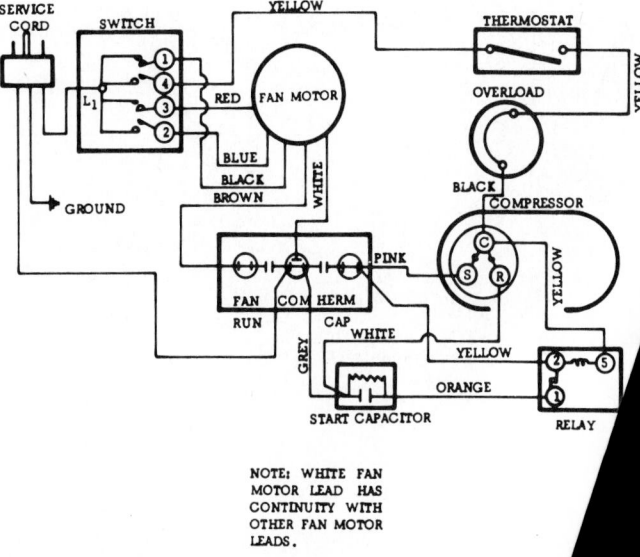

NOTE: WHITE FAN MOTOR LEAD HAS CONTINUITY WITH OTHER FAN MOTOR LEADS.

Fig. 21-24. Wiring diagram of window unit which uses R-22. about 11.7 A at power factor of 90 percent.

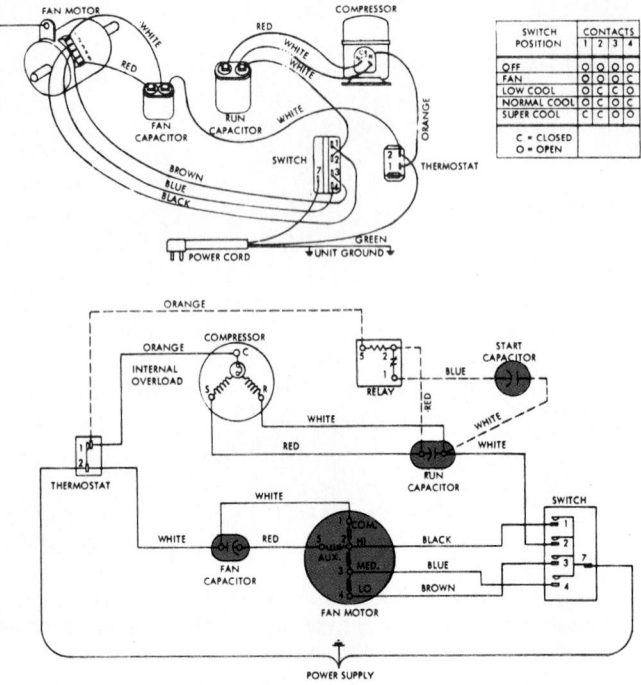

SWITCH POSITION	CONTACTS			
	1	2	3	4
OFF	O	O	O	O
FAN	O	O	O	C
LOW COOL	O	C	C	O
NORMAL COOL	O	C	O	C
SUPER COOL	C	C	O	O

C = CLOSED
O = OPEN

Fig. 21-25. Wiring diagram of three-speed fan system.
(Fedders North America)

in Chapter 7. All but the motor compressor can be repaired on-the-job. Before doing internal service work on the unit, be sure the malfunction is not in the external circuit. Test for power in. Check the thermostat, the relay, the capacitors, and the overload protectors (both electrical and temperature).

Troubles inside the unit may include:
1. Lack of refrigerant.
2. Stuck compressor.
3. Inefficient compressor.
4. Clogged refrigerant circuit.
5. Short circuit, open circuit, or grounded motor windings.

The condition of the motor can be checked with a continuity light or with an ohmmeter. To check for lack of refrigerant or clogged refrigerant lines, install a gauge manifold as shown in Fig. 21-26.

Determining (diagnosing) trouble is explained in Chapters 11 and 14. The unit should be moved to the shop if the motor compressor needs repairs. The motor compressor can be replaced on the owner's premises. If the unit lacks refrigerant, locate the leak and repair it before recharging.

Many window air conditioners use PSC (permanent split capacitor) compressor motors. These motors do not use a relay for starting. If supplied voltage is low (10 percent or more), they will start with great difficulty. Many service technicians install a starting capacitor and a relay to over-

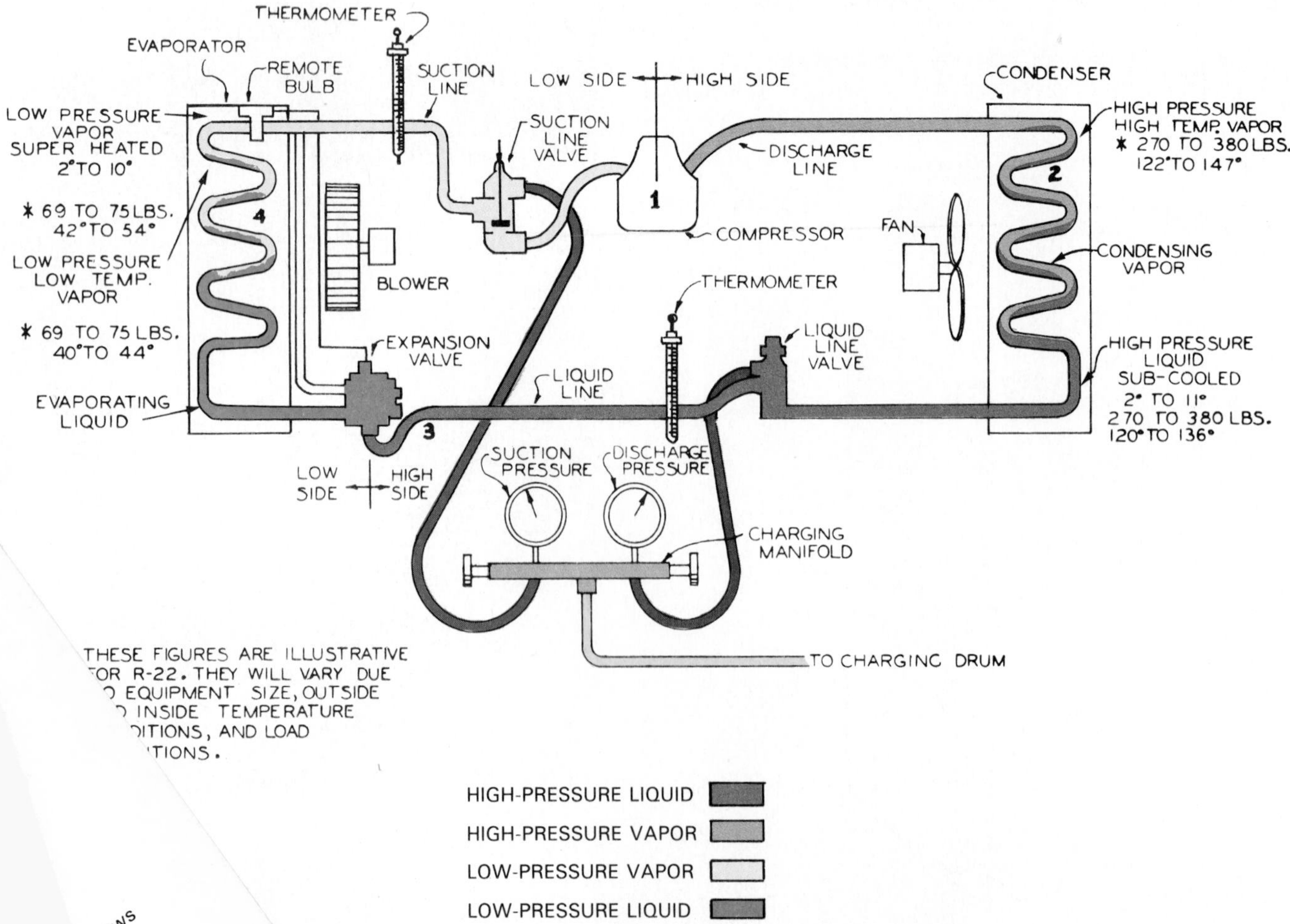

THESE FIGURES ARE ILLUSTRATIVE FOR R-22. THEY WILL VARY DUE TO EQUIPMENT SIZE, OUTSIDE AND INSIDE TEMPERATURE CONDITIONS, AND LOAD CONDITIONS.

HIGH-PRESSURE LIQUID	
HIGH-PRESSURE VAPOR	
LOW-PRESSURE VAPOR	
LOW-PRESSURE LIQUID	

21-26. Air conditioning unit cycle diagram showing gauge manifold installed.
(Goodman Manufacturing Corporation)

RECOMMENDED RATINGS FOR ADD-ON STARTING CAPACITORS	
RUNNING CAPACITOR MICROFARADS	SPECIAL STARTING CAPACITOR MICROFARADS (Add-on) *
20	18
25	18 or 25
30	25
35	25
40	25 or 45
45	45
50	45

*Most manufacturers specify Microfarad rating of special start capacitor for each of their units. Follow manufacturer's specifications.

Fig. 21-27. Table of special starting capacitor sizes to be used on PSC (permanent split capacitor) motor compressor when starting difficulties are found. PSC motors are introduced in Para. 7-20.

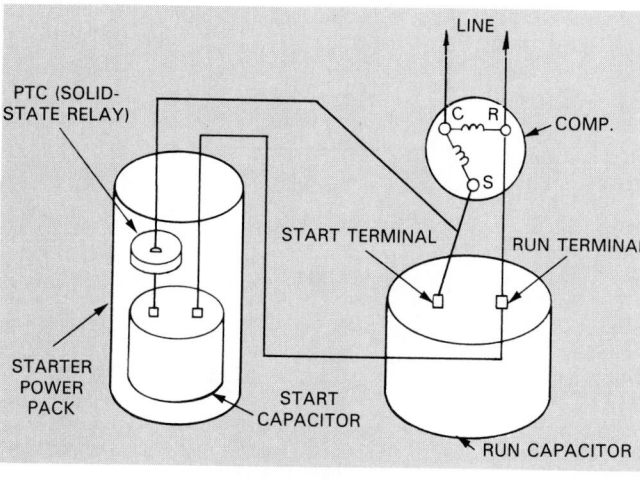

A

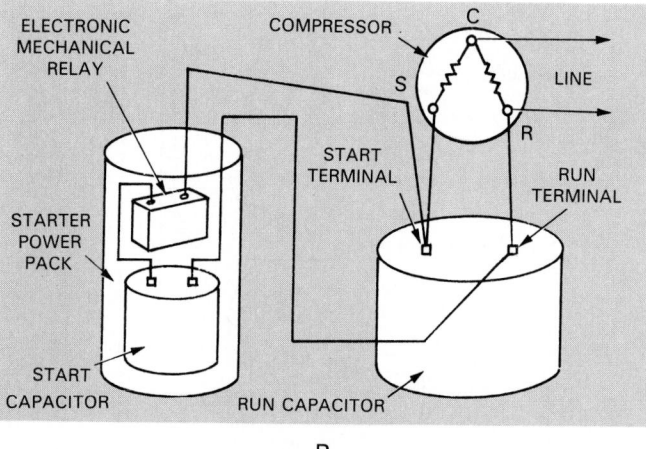

B

Fig. 21-28. Starting kits for PSC motors. A—This starting kit uses a PTC (Positive Temperature Coefficient) solid-state relay. As current enters the PTC, the resistance increases by 1000 times and literally chokes off current to the starting winding. This sharp increase in resistance occurs within 1/2 second. B—This starting kit is a combination start capacitor and electronic-mechanical relay with normally open contacts. When the internal coil is energized, a solid-state timer deactivates the coil in about 1/2 second. This type of relay does not depend upon ''pick-up'' or ''drop-out'' voltages as the common potential relay operates. (Sealed Unit Parts Co., Inc.)

come this problem. The capacitor and the relay must be exactly the right size for the motor.

The best way to determine the correct size is to follow the manufacturer's recommendation. If this is not available, the table in Fig. 21-27 will help in selecting the correct electrical starting system.

When the motor compressor reaches a satisfactory speed, the starting capacitor needs a potential-type relay to open the starting capacitor circuit. Special starting kits, Fig. 21-28, are available for PSC motor compressors.

Window units are often removed during the winter season. But, if this is inconvenient, the unit can be winterized by blocking the air-in and the air-out grilles with cardboard or flexible plastic sheeting. A storm sash can be custom built to fit around the air conditioner or one may use plywood held in place with caulking or rubber grommets.

Replacement capillary tubes for window units must be very accurately picked for size. Fig. 21-29 gives correct sizes when R-22 refrigerant is used.

If the window unit drips water into the room, it is not correctly installed. Check the slope of the unit from inside to outside with a spirit level. It must slope to the outside (condenser edge) about 1/4 in. Condensate water will then run to the depression of the unit base under the condenser fan and condenser. Make sure the drain hole is open, then level the unit along its other dimension. Finally, recheck the unit installation for airtight sealing in the window opening.

21-9 CONSOLE AIR CONDITIONERS

In console air conditioners, whole systems are mounted in a cabinet. They vary in capacity from 2 hp to 10 hp. Such units are often used in small commercial establishments such as restaurants, stores, and banks.

Console models may have either water-cooled or air-cooled condensing units. Air-cooled models, needed in some localities because of water restrictions, must have air ducts to the outdoors for condenser cooling.

COMPRESSOR CAPACITY Btu/Hr.	CAPILLARY SIZE		COIL CIRCUITS	
	SHORT	LONG	3/8 IN. TUBE	1/2 IN. TUBE
4500	36 IN. x .042	80 IN. x .049	1	
5000	25 IN. x .042	64 IN. x .049	1	
5500	20 IN. x .042	52 IN. x .049	1	
6000	40 IN. x .049	75 IN. x .054	1	
6500	35 IN. x .049	65 IN. x .054	1	
7000	28 IN. x .049	52 IN. x .054	1	
8000	36 IN. x .054	65 IN. x .059		1
9000	28 IN. x .054	48 IN. x .059	2	1
10,000	36 IN. x .059	64 IN. x .064	2	1
11,000	28 IN. x .059	50 IN. x .064	2	1
12,000	40 IN. x .064	68 IN. x .070	2	1
13,000	32 IN. x .064	56 IN. x .070	2	1
14,000	44 IN. x .070	70 IN. x .075	2	1
15,000	36 IN. x .070	56 IN. x .075	3	2
16,000	30 IN. x .070	48 IN. x .075	3	2
17,000	38 IN. x .075	65 IN. x .080	3	2
18,000	35 IN. x .075	55 IN. x .080	3	2
19,000	28 IN. x .075	48 IN. x .080	3	2
20,000	40 IN. x .080	58 IN. x .085	3	2

Fig. 21-29. Capillary tube sizes for window air conditioners which use R-22 refrigerant. Coil circuits are the size of tubing in evaporator. Each unit has one capillary tube. (Tecumseh Products Co.)

Fig. 21-30 shows a water-cooled console unit. Return air enters the lower grille. Cooled air is discharged at the upper grilles. Ducts can be connected to portions or all of the upper section. These are needed when partitions interfere with cooled air distribution. The condensing unit is mounted in the bottom of the console. Air blowers are in the middle, while the evaporator is in the top of the cabinet.

Console units also have adjustable fresh air intakes and evaporator bypass controls as shown in Fig. 21-31. All must have drains to remove the condensate flowing from the evaporator.

Most of the console models have a complete refrigerating system, filtering system, and evaporator. Fig. 21-32 shows a console unit. Fig. 21-33 shows a large self-contained comfort cooler with an air-cooled condenser.

Water-cooled units require plumbing connections to both fresh water and a drain. The drain also receives the moisture condensed out of the air by the evaporator in the summer. One company has developed three openings in one rubber hose for this purpose. This permits moving the unit quickly from room to room. Such units usually do not provide winter conditioning facilities.

21-10 INSTALLING CONSOLE AIR CONDITIONERS

Console units are factory-assembled. After moving the unit into place and leveling it, plumbing and electrical connections are required. Such connections must conform to local codes.

Usually the motor compressor is hermetic and the refrigerant control is generally a thermostatic expansion valve.

The unit should be thoroughly checked. The air temperature both in and out, the electrical load, and the operating pressures should be checked and recorded for future reference.

21-11 SERVICING CONSOLE AIR CONDITIONERS

Panels must be removed to work on internal parts of the unit. Periodic maintenance duties include replacing the filter or cleaning it, cleaning the evaporator and fins, cleaning the fan motor and oiling it (unless it has sealed bearings), and

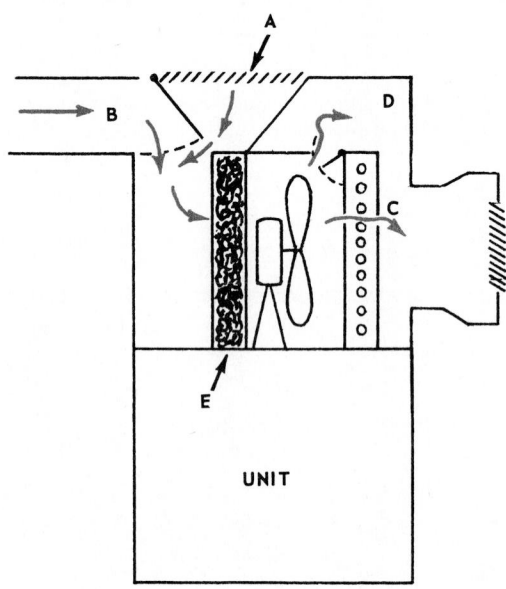

Fig. 21-31. Air circulation in console comfort cooling air conditioner. A—Recirculated air. B—Fresh (outdoor) air. C—Cooled air. D—Recirculated untreated air. E—Filter.

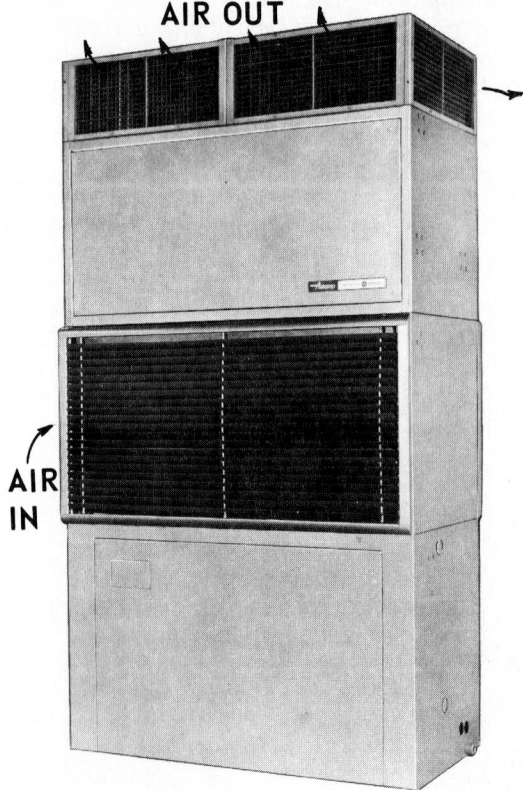

Fig. 21-30. Self-contained console-type comfort cooling air conditioner has water-cooled condenser. Upper grille is cooled air outlet.

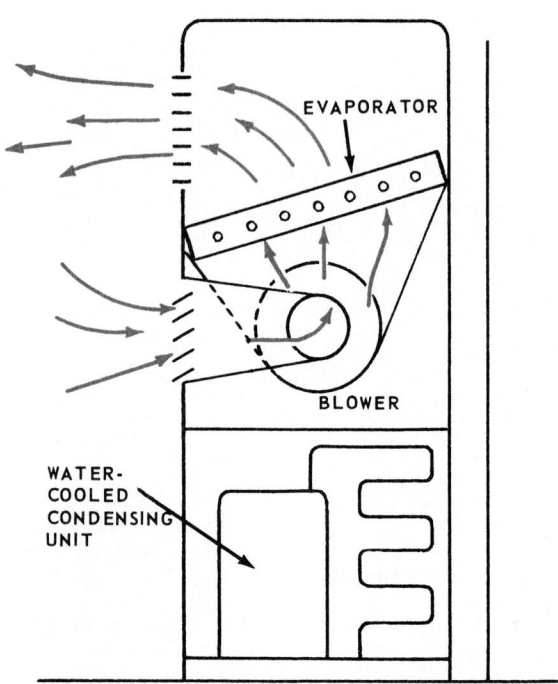

Fig. 21-32. Drawing of console-type air conditioner.

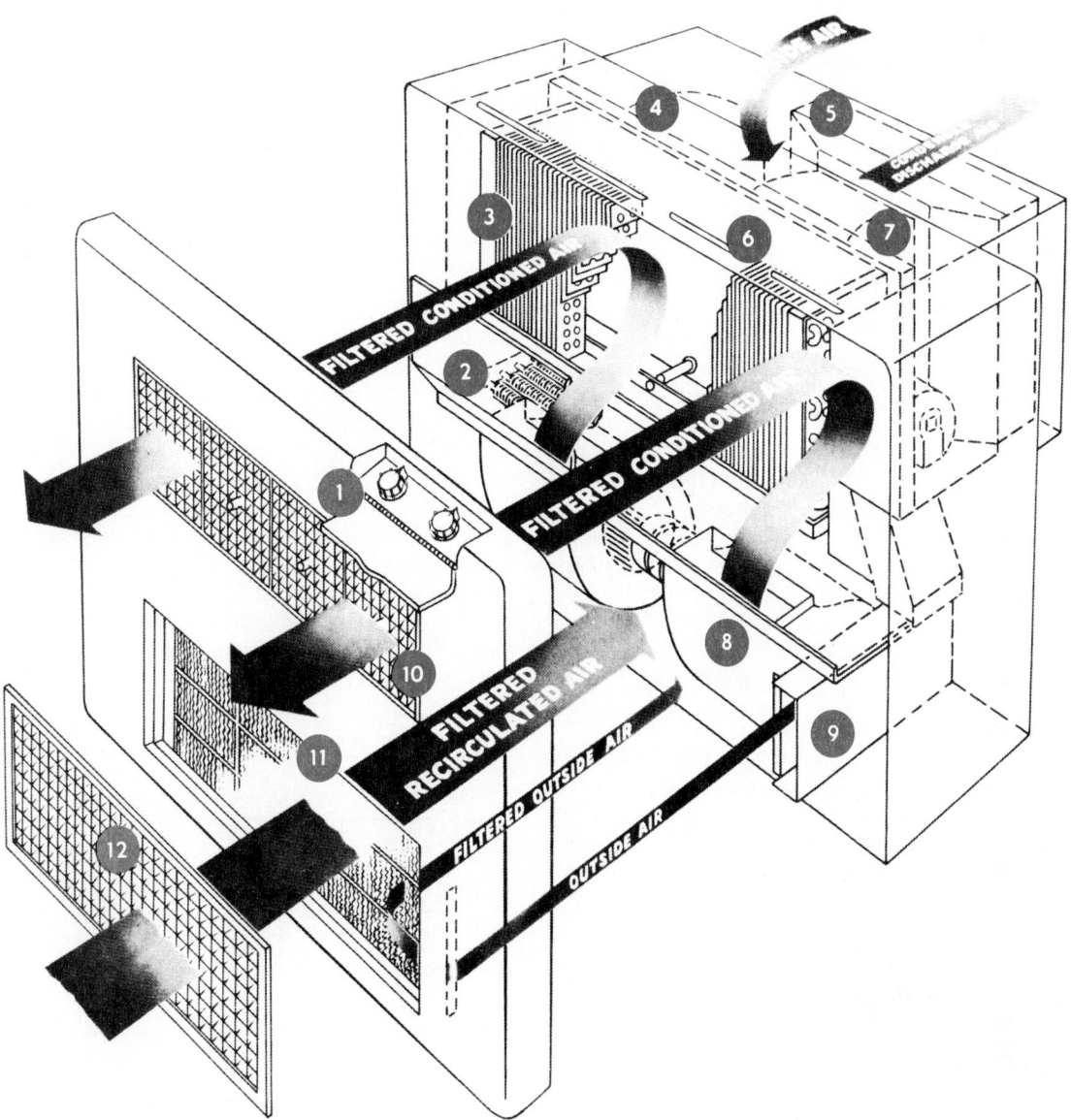

Fig. 21-33. Console unit with air-cooled condenser. 1—Controls. 2—Heating coil. 3—Evaporator. 4—Compressor. 5—Air-cooled condenser. 6—Filtered-conditioned air diverters. 7—Condenser fan. 8—Evaporator fan. 9—Ventilating air duct. 10—Discharge grilles. 11—Air filter. 12—Return air grille.

cleaning the drain pan and drain tube. The inner lining of the cabinet sometimes gathers lint. This should be removed by vacuuming.

Servicing of the refrigerating unit and the condenser water circuit is explained in Chapter 14. *It is important to check the refrigerant charge, the operation of the thermostatic expansion valve, and the water flow.*

A regular maintenance schedule is necessary if the owner is to receive long and satisfactory service from the air-conditioning system. Different parts of the system must be checked more frequently than others. A record should be kept of all checks including both the data and the date. Check the following weekly:

1. V-belts.
2. Fan speeds.
3. Pump speeds.
4. All standby units.
5. Water leaks.
6. Controls (pressure, temperature, and airflow).
7. Lubrication.
8. Canvas connectors on ducts.
9. Cooling tower.
10. Water treatment.
11. Bleed off.

Monthly checks should be made on these:

1. Refrigerating system (charge, purge, test for leaks; check strainers and dryers).
2. Filters
3. Humidifier.
4. Safety valves.
5. Cooling tower pump.
6. Duct dampers, registers, and diffusers.
7. Piping (insulation, vibration, and wear).

Every six months:

1. Clean fans and casings.
2. Clean duct registers and diffusers.

Every year:
1. Efficiency check of compressors.
2. Efficiency check of pumps.
3. Damper operation.
4. Clean water circuits.
5. Operate all hand valves.

Every two years:
1. Inspect condenser wet surfaces.

21-12 REMOTE COMFORT SYSTEMS

In remote air conditioning systems, the refrigerating equipment is located away from the space to be conditioned. These units vary in capacity from two tons to thousands of tons.

Some units which condition the air do not use long refrigerant piping runs, Fig. 21-34. Conditioned air is distributed by ducts to the space or spaces.

A water-cooled system which has service valves, a tube-within-a-tube condenser, and a water valve is shown in Fig. 21-35. A chilled water system used in combination with a duct distribution system is shown in Fig. 21-36.

Coolant lines are run to air conditioning units in each room to be air conditioned. Each room unit has a thermostat. A solenoid valve usually controls the flow to each unit.

This valve is connected to the room temperature control. This controls both a fan and the solenoid coolant valve in the room unit. Chilled water starts to flow to the room unit when the fan starts in that unit. It is shut off when the fan stops. Air filters are also sometimes installed in these units.

21-13 RECREATIONAL VEHICLE AIR CONDITIONING

Air conditioners similar to the home window units are being used more and more on recreational vehicles and larger pleasure boats.

Self-contained units are usually mounted on the roof. See Fig. 21-37. The three main sections of a roof-mounted unit are shown in Fig. 21-38. Removable covers make cleaning easy.

Since a 120 V electrical supply is needed, most can be used only when the vehicles are parked or the boats moored at the dock. But, where the vehicle or boat has an ac generator, the air conditioning system can be used while on the road. Boat units have some corrosion problems. For this reason, marine air conditioners use a great deal of copper and plastic components.

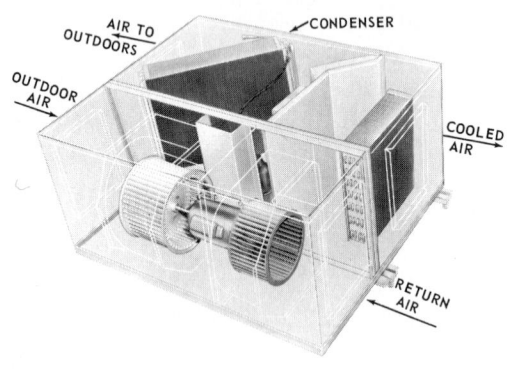

Fig. 21-34. Air-cooled refrigerating unit for duct distribution of comfort-cooled air. Central heating ducts and blower are used on right side in addition to right side refrigerating blower. (Fedders North America)

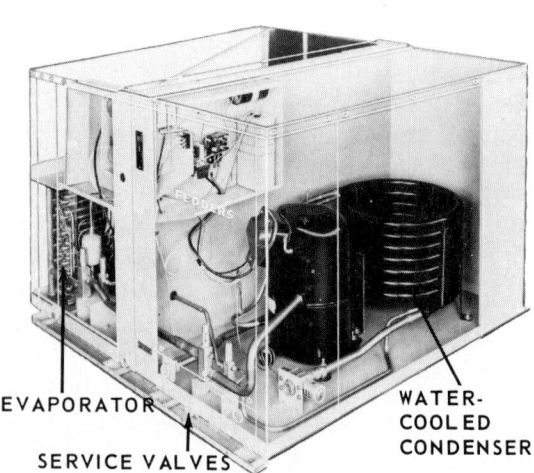

Fig. 21-35. Water-cooled condensing unit designed for central comfort cooling.

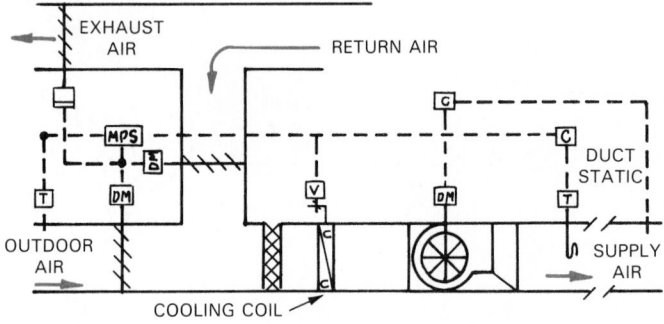

Fig. 21-36. Control of space temperature by chilled water coil. (ASHRAE Guide and Data Book)

Fig. 21-37. Recreational vehicle with twin rooftop air conditioners. A—Air conditioner. B—Storage compartment. (The Dometic Corporation, Inc.)

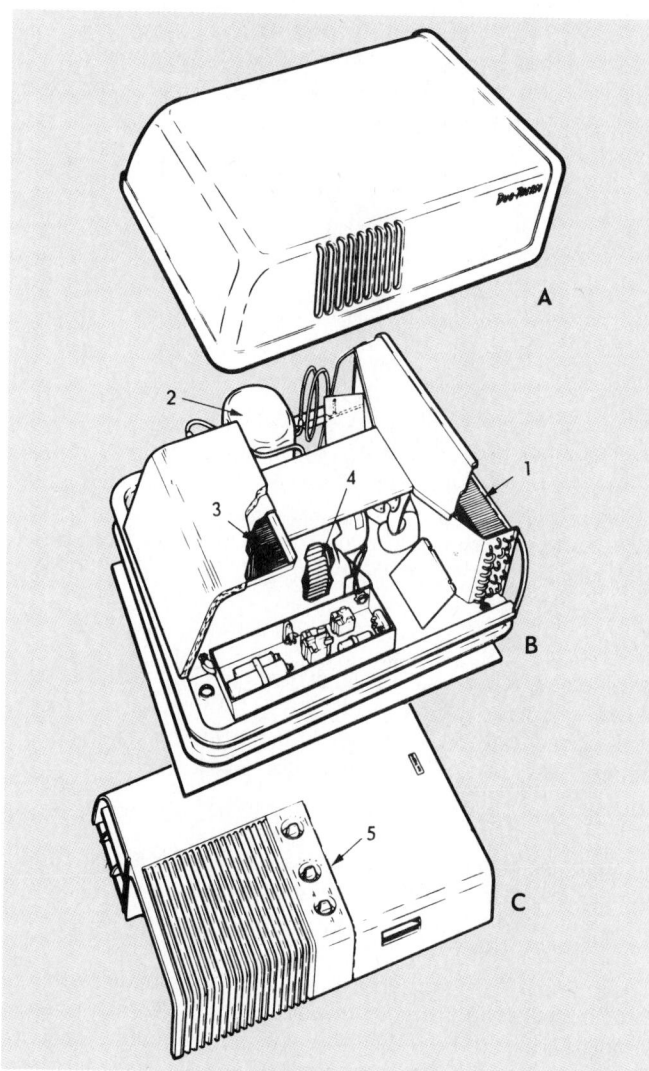

Fig. 21-38. Mobile air conditioning system has compact design. A—Outside cover. B—Cooling system. C—Interior unit. 1—Condenser. 2—Motor compressor. 3—Evaporator. 4—Fan. 5—Controls. (Duo-Therm, by The Dometic Corporation, Inc.)

21-14 EVAPORATIVE COOLING

In dry climates, evaporative comfort cooling is very desirable and practical. If air at 105 °F (41 °C) and 20 percent relative humidity is moved rapidly over water at the same temperature, some of the water will evaporate and the remaining water can cool to as low as 55 °F (13 °C). See the psychrometric chart in Chapter 18. The water will cool down to a range of about 65 °F (18 °C) to 70 °F (21 °C). If other air is forced around the water container and then distributed into the space to be cooled, it will cool down the air leaving the unit to around 75 °F (24 °C) or 80 °F (27 °C) at about 40 percent relative humidity. The conditioned room would be quite comfortable. Fig. 21-39 illustrates such a system.

These units use two fans, one for the outside air and one for the conditioned air. Small units have been built to mount in the windows of automobiles.

21-15 STORING COLD FOR COMFORT COOLING

There are many situations where comfort cooling is only needed part of the day. Stores, funeral homes, churches, tourist homes, and theaters need cooling for only an eight-hour period.

In such cases, one can use a smaller capacity system. Cold is stored during 16 hours of operation. This allows the smaller unit to have twice as much air cooling capacity during the eight hours it is needed.

These storage systems may be tanks of water, cooled to 32 °F (0 °C) with some ice forming. They may also be tanks using a eutectic solution—see Chapter 30—designed to freeze or partly freeze at a low temperature.

The tanks or plates store up the effect of the refrigeration during the time cooling is not needed. When the extra cooling is needed, the tanks or plates can release the stored cooling, adding it to the normal cooling of operation.

21-16 DEHUMIDIFYING EQUIPMENT

A mechanism which dehumidifies (removes moisture) is known as a dehumidifier. Such equipment depends upon

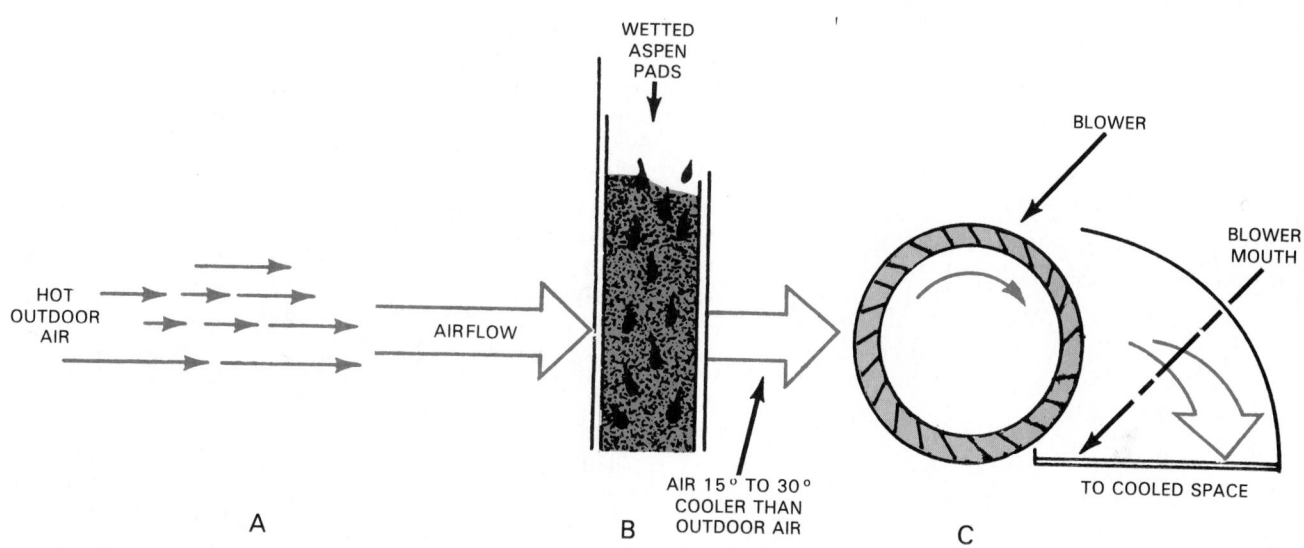

Fig. 21-39. Evaporative cooling process. A—Outside air. B—Filtering through water-saturated evaporative medium and cooled by evaporation. C—Circulated by blower.

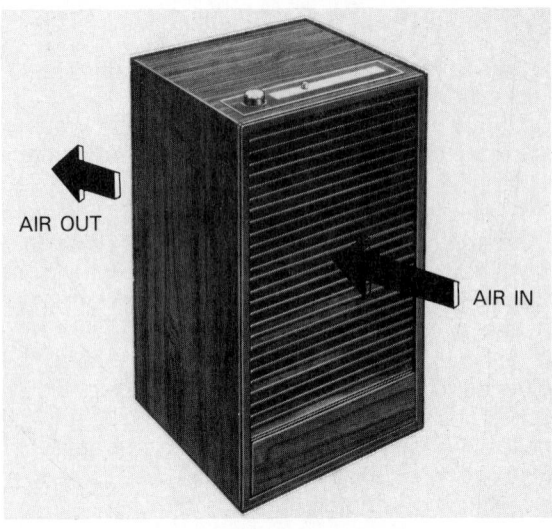

Fig. 21-40. Outside view of dehumidifier. Air enters in front of cabinet and is forced out at back. Note arrows.

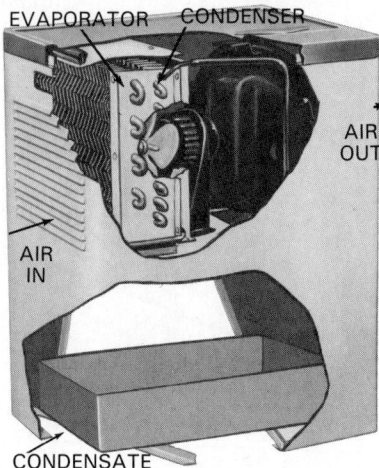

Fig. 21-41. Section view of dehumidifier. First pass of finned coil near fan is cooling and dehumidifying coil, while next pass is condenser. See Fig. 21-3. Pail, pan, or direct drain can be used to remove condensate.

a cold coil over which air is blown. Moisture is condensed out by coming in contact with the cold surface.

When coil surfaces are at a temperature below the dew point of the air, moisture will condense out of the air. The coil surface temperature must be kept above freezing. Frost or ice formation would block airflow.

A dehumidifier, as shown in Fig. 21-40, is usually a small hermetic refrigerating system that has both a condenser and evaporator in a cabinet. Air is drawn over the evaporator. As the air touches the cold surface of the evaporator, it cools below its dew point. Water condenses out of the air and collects on the evaporator. The cooled air is then moved over the condenser, Fig. 21-41, to reheat the air to a reasonable relative humidity.

The device is used to "dry" the air. It is useful in basements and other damp places.

These units usually have a humidistat. This is a device that senses moisture in air. A container is used to collect the condensate or a drain tube carries it off. Fig. 21-42 shows a humidistat and other controls.

In some installations, certain chemicals are used which can absorb moisture from the air. Where used, the chemicals are usually cycled in such a way that first the moisture from the air is absorbed into the chemical. Then the chemicals are heated. Moisture driven from the chemical is exhausted out-of-doors, and the chemicals are ready to once more absorb moisture.

21-17 AIR CONDITIONER COEFFICIENT OF PERFORMANCE (COP)

The air conditioner coefficient of performance (COP) is measured by the amount of cooling (heat absorbed) in proportion to the amount of electrical energy used. (See Chapters 15 and 30.)

The formula is: $COP = \dfrac{Q_C}{\text{watts}} \times 0.293$

Q_C is the heat absorbed, in Btu/hr. The watt rating is the electrical power used.

In an attempt to conserve energy, the United States Department of Commerce, National Bureau of Standards, has recommended that an energy guide label be attached

to all air conditioners. This label gives the information needed to calculate the COP.

Suppose that the following data is given for an air conditioner: 8000 Btu/hr. capacity; 860 watts.

$$COP = \frac{8000}{860} \times 0.293 = 2.726$$

Suppose that the following data is given for an air conditioner: 14,000 Btu/hr. capacity; 4.1 kW.

$$COP = \frac{14,000}{4100} \times 0.293 = 1$$

The number 1 means that the unit is absorbing as much heat in the evaporator as the equivalent electrical energy supplied (refer to Para. 20-44). The unit "dumps" 28,000 Btu per hour at the condensing coil (28,000 is the heat from cooling plus the heat due to compression). Although the COP is not near the more common values from 3 to 6, the unit provides enough cooling from the electrical energy supplied to be practical. A COP near zero would not be practical.

The efficiency of an air conditioning unit goes up as the COP value increases. Generally, an air conditioner with a high COP rating will be less expensive to operate than a lower rated air conditioner that produces the same amount of cooling action.

Therefore, when planning to purchase an air conditioner, the COP of the unit should be an important consideration. It is best if the COP is from 4 to 8.

The COP and EER are being combined in the SEER (Seasonal Energy Efficiency Ratio). See Para. 15-63 and Chapter 30.

21-18 REVIEW OF SAFETY

All of the safety practices described in Chapters 11 and 14 also apply to comfort cooling units.

Window units should be handled with care. Heavy units should be moved and lifted with hand trucks and lifts. Many window units are installed in upper floors. With double-hung windows, the upper sash is lowered to the air conditioner and helps keep the air conditioner in place.

Sometimes the upper sash is opened accidentally. Then, since most of the weight of the air conditioner is outside

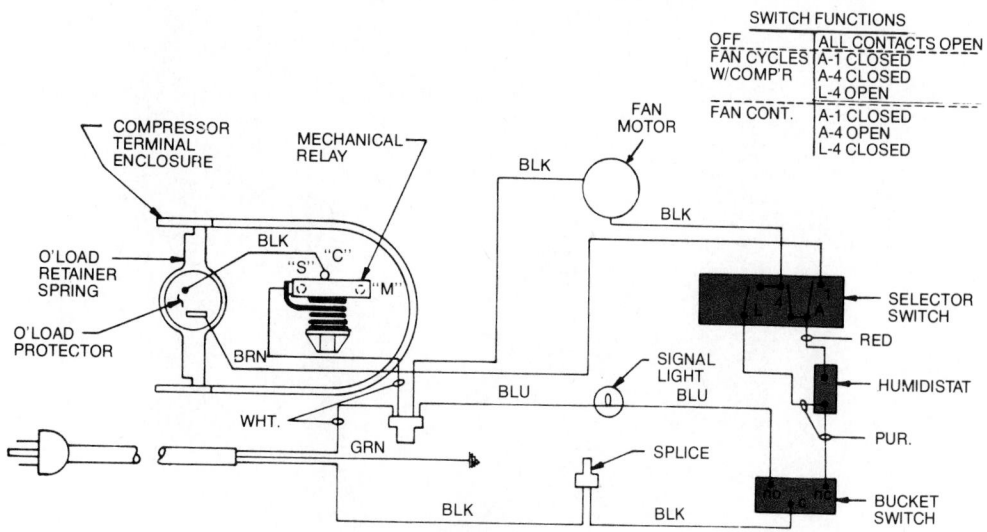

SWITCH FUNCTIONS	
OFF	ALL CONTACTS OPEN
FAN CYCLES W/COMP'R	A-1 CLOSED A-4 CLOSED L-4 OPEN
FAN CONT.	A-1 CLOSED A-4 OPEN L-4 CLOSED

Fig. 21-42. Wiring diagram for dehumidifier. Note selector switch, humidistat, and bucket switch.

the building, the unit may fall to the ground. It should be securely attached to the windowsill and braced! When removing mechanisms, be careful not to drop them. They may be slippery. Safety shoes are recommended. Carefully follow installation instructions supplied by the manufacturer.

Remote systems should be sturdily mounted. Suction lines and liquid lines should be protected from abuse.

Always review instructions on larger units before performing any service work or installation work on them.

Window air conditioners are usually available for use on either 120 V or 240 V circuits. It is advisable—particularly with the higher Btu ratings—to use a 240 V circuit. This is because the voltage drop between the power panel and the air conditioner will be less when using a 240 V motor. A separate air conditioner circuit should be provided. Make certain all systems are properly grounded.

Whenever it becomes necessary to add a refrigerant to a system, always be sure that the refrigerant added is the same as the refrigerant already in the unit.

Always wear goggles when testing for leaks and adding refrigerant.

Much of the sheet metal used in air conditioning has very sharp edges. The service technician must be careful not to cut hands or fingers when making repairs.

21-19 TEST YOUR KNOWLEDGE

1. What refrigerant controls are usually used on air conditioning evaporators?
2. Why are solenoid refrigerant control valves used on some air conditioning evaporators?
3. Why is air leaving an evaporator considered to be damp?
4. Can air be dehumidified by any other method than by cooling?
5. How is condensate handled in a unit comfort cooler?
6. What air is used to cool the condenser of an air-cooled comfort cooler?
7. What is the relative humidity of the air just as it leaves the evaporator?
8. How many fans do the most common window-type comfort cooling units have?
9. How many coils does air pass through in a dehumidifier?
10. What component of a dehumidifier is often used to reheat the air after moisture is removed?
11. Do window comfort coolers have different types of electrical extension plugs? Why?
12. How are the installation joints sealed?
13. Why is a window comfort cooler slanted toward the outside of the home?
14. What is the normal cutout setting of a window unit thermostat?
15. Why is the dew point reading important when using a dehumidifier?
16. Can some of the return air be used to reduce the humidity of the cooled air? If so, how?
17. Is it possible to have a hydronic comfort cooling system?
18. Describe the two ways window air conditioners may be installed. What protects the unit in winter?
19. How is the cooled air moved from side to side as the window unit operates?
20. What may happen if a window unit is shut off and then immediately turned on again?
21. If the window comfort cooling unit will not start, what should one check first?
22. Why should one avoid bending or twisting fan blades?
23. How many capacitors does a window comfort cooler have?
24. How can a console air conditioner be installed for condenser air cooling?
25. What is happening if a window unit drips water inside the room?
26. What electrical power is most used with mobile air conditioning systems?
27. What chamber is usually located in the upper part of a console air conditioner?
28. Why is a drain needed on an air-cooled comfort cooling system?
29. Can one use a duct system with a chilled water system? Explain.
30. How does a bypass air system operate?

OUTDOOR
CONDENSING
UNIT

WALL-
MOUNTED
UNITS

A

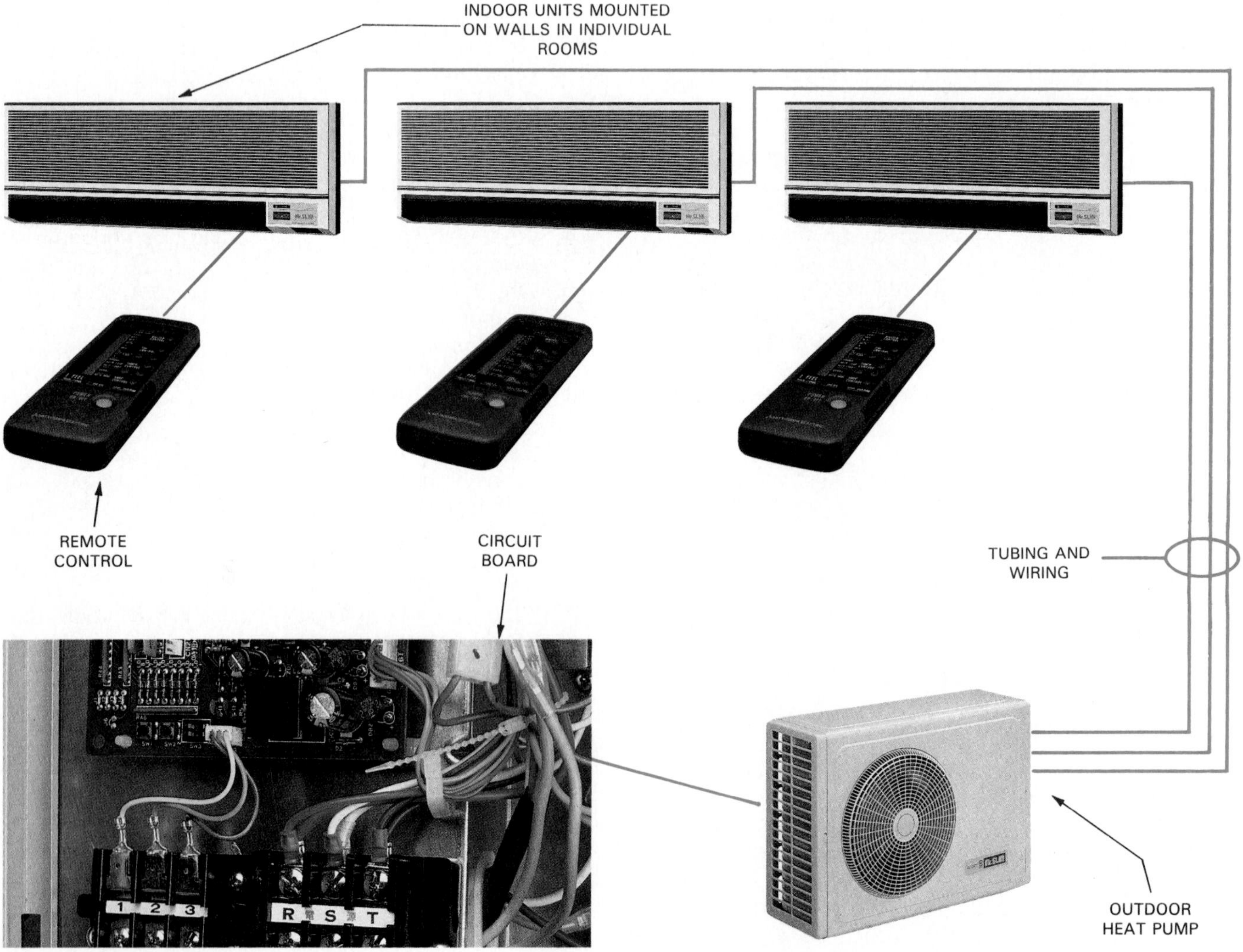

INDOOR UNITS MOUNTED
ON WALLS IN INDIVIDUAL
ROOMS

REMOTE
CONTROL

CIRCUIT
BOARD

TUBING AND
WIRING

1 2 3

R S T

OUTDOOR
HEAT PUMP

B

A multizone ductless split air conditioning system which uses a single outdoor condensing unit. A—Three wall-mounted units are controlled by remote controls. Each unit can be maintained at independent temperatures. (Burnham Corporation) B—Remote multizone heat pump system.   (Mitsubishi Electronics America, Inc.)

Chapter 22

AIR CONDITIONING SYSTEMS—DISTRIBUTING, CLEANING, AND INSTRUMENTS

After studying this chapter, the technician will be able to:
☐ Give basic ventilation requirements of conditioned space.
☐ List and describe the types of air duct systems.
☐ Relate air distribution to duct sizing and design.
☐ Size and construct ducts for a balanced system.
☐ Compute total pressure drop in a ducted system.
☐ Discuss types and classifications of fans.
☐ Explain principles and methods for cleaning air.
☐ Service different types of air cleaning systems.
☐ Demonstrate proper use of various instruments in checking: airflow, draft control, liquid flow, and electric current.

When a radiator system or a room convector system— such as a steam heating plant or a hot water plant—is used, air distribution is simple. The heat exchange units are located along the outside walls. During the heating season the heated air rising from the radiator along the wall mixes with the cold air adjacent to the cold wall. Then natural air currents (convection) distribute the air mixture throughout the room. Many heating systems use motor-driven fans to help circulate the air.

During the cooling season, chilled water may be circulated in these same convectors while a blower moves air over the coils. Many installations are combinations of a hot water (hydronic) system and a warm air duct system.

22-1 CONDITIONED AIR

An air conditioning system, properly designed and installed, will provide proper amounts of conditioned air the proper temperature and relative humidity. The distributed air must:
1. Be clean.
2. Provide the proper amount of ventilation.
3. Carry enough heat to warm conditioned spaces or absorb enough heat to cool the conditioned spaces.

22-2 WEIGHT OF AIR

Air has definite weight. Although it is invisible, its gases have a definite mass and it takes energy to move it. Fig. 22-1 gives the weight of air under various temperature and relative humidity conditions. One lb. (0.454 kg) of dry air at 70 °F (21 °C) at standard atmospheric pressure will occupy a space of 13.35 cu. ft. (0.378 m³). See Fig. 18-21. If there is 50 percent relative humidity, 13.51 cu. ft. (0.383 m³) of air and moisture mixture weighs 1 lb. (0.454 kg). Because air is a gas, it responds closely to Boyle's and Charles's Laws. Therefore, as the temperature rises, it takes more cubic feet to weigh one pound. As the pressure drops, it takes more cubic feet to weigh one pound. As relative humidity increases, it takes more cubic feet to weigh one pound. Each water molecule weighs less than each nitrogen or oxygen molecule. A gas made up only of water molecules would weigh 0.63 times as much as an equal volume of dry air.

22-3 HEAT IN AIR

Because air is a physical substance, it can carry heat. It will remove heat from or take it to a space.

Air Temp. °F	Cu.ft./lb. Dry Air	Lb./cu.ft. Dry Air	Cu.ft./lb. 50% Saturated	Lb./cu.ft. 50% Saturated	Cu.ft./lb. 100% Saturated	Lb./cu.ft. 100% Saturated
0	11.58	.08635	11.585	.08632	11.59	.08628
50	12.84	.07786	12.915	.0774	12.99	.0769
70	13.347	.07492	13.51	.074	13.68	.0731
100	14.10	.0709	14.585	.06856	15.07	.06635
120	14.60	.0685	15.55	.0643	16.50	.0606
150	15.3	.0652	17.7	.0565		
200	16.7	.0600				

Fig. 22-1. Weight of air at various temperatures and relative humidities. Water vapor weighs less than an equal volume of dry air.

The psychrometric properties already studied in Chapter 18 show how heat content can change as the temperature and relative humidity change. The specific heat of dry air is 0.24 Btu/lb. °F. Converted to metric, this is equal to 1.004 j/g °C (1.004 kJ/kg °C). The additional heat due to the moisture in the air varies considerably, depending on the amount of saturation. For example, from 0 °F (−18 °C) to 100 °F (38 °C), 24 Btu are added to 1 lb. of dry air. There may be 0.04293 lb. of moisture added to this 1 lb. of dry air to saturate it, but the heat in this moisture is 47.4 Btu (latent heat and sensible heat, from steam tables). The total heat in 1.04293 lb. is 71.4 Btu. However, as far as distributing the air is concerned, only sensible heat needs to be taken into account. This is because vaporizing or condensing of water should not take place in the ducts, or in the room being conditioned.

To find some of the above values in metric requires more than the usual accuracy in rounding numbers. The temperature difference is from −17.777 °C to 37.777 °C, which amounts to 55.555 °C. The calculation is done as follows:

$$1.00416 \times 55.555 \times 0.4536 \text{ kg} = 25.305 \text{ kJ}$$

The number 1.00416 was obtained as follows:

0.24 Btu/lb. °F × 1.8 °F/°C × 2.32444 (kJ/kg) /(Btu/lb.) = 1.004158 kJ/kg °C

There are 7000 grains to a pound; thus, because it requires about 1000 Btu/lb. (2 324 kJ/kg or 2.324 MJ/kg) to change water to water vapor, each grain of water changed to vapor requires a latent heat of 1000/7000 or 0.143 Btu per grain. One grain equals about 0.065 grams. If condensation takes place in a cooled air duct or on the outside surface, some heat will be released. The temperature of the air delivered to the conditioned space may be changed. It may cause a failure in the operation of the equipment. Only sensible heat changes should take place outside of the heating or cooling system.

22-4 BASIC VENTILATION REQUIREMENTS

As noted before, air is a mixture of gases. Normally air contains about 21 percent oxygen. A human system requires that a certain oxygen content be contained in the air:
1. To maintain life.
2. To be comfortable.

If a room is tightly sealed, any human in that room would slowly consume the oxygen and increase the amounts of carbon dioxide, water vapor, and various impurities. This could cause drowsiness or even death.

One must remember that space for human living must have air with a good oxygen content and that this air must be kept at a reasonable temperature. *It is of utmost importance that fresh air be admitted to provide the oxygen.*

In the past, this fresh air entered the space by infiltration (leakage) from the outside at door and window openings and through cracks in the structure. However, modern construction is reducing this air leakage. The air conditioning apparatus, then, must furnish fresh air. Modern units have a controlled fresh-air intake. This fresh air is conditioned and mixed with the recirculated air before it reaches the room.

Some conditioned air leaves a building through doors, windows, and other construction joints. Some also leaves by exfiltration. (This means leaking out or being blown out by mechanical means.) Any kind of exhaust fan removes conditioned air. Some of this air is replaced by infiltration on those sides of the building exposed to wind pressure.

It is best to bring in replacement fresh air through a makeup air system. When this is done:
1. The makeup air can be cleaned.
2. The makeup air can be cooled or heated.
3. A positive pressure can be maintained in the building to keep out airborne dirt, dust, and pollen. (A negative pressure reduces the efficiency of exhaust fans and of a fuel-fired furnace.)
4. A definite amount of fresh air is brought into the building for health purposes (oxygen content).

Certain areas of a building should have a slightly less positive pressure (5 to 10 percent) than the rest of the building to reduce the spread of odors. Such areas would include the kitchen, lavatories, and where certain industrial operations produce fumes.

The presence of positive or negative pressures can be measured using a manometer. See Para. 22-40.

The amount of fresh air required depends on the use of the space and the amount of fresh air admitted by infiltration. One basic rule is to provide at least 15 cfm of fresh air per person to provide enough oxygen and to remove carbon dioxide. If six people occupy a space of 10,000 cu. ft., they would need 90 cfm of fresh air (6 × 15 cfm = 90 cfm). It would take 10,000/90 = 111 minutes to completely replace the air in the space. This is 1.85 hours. This seems rather slow, but remember that the purpose is not to replace the air quickly. It would be costly to use heat as fast as needed to replace the air immediately. Some engineers say that a 10,000 cu. ft. space for six people is not efficient use of space. A better choice of space is 2160 cu. ft. Then it would take 2160/90 = 24 min. (0.4 hr.) to replace the air. Note that very little heat is used for such a small room.

One must remember that the air can be handled either to produce positive pressure (higher than atmospheric pressure) in a building or negative pressure (below atmospheric pressure). A positive pressure will eliminate infiltration of air from outside or from other spaces. It is done by using special air intakes to the blowers. A positive pressure assures that all air entering a building can be filtered and cleaned before reaching the occupied space. Negative pressure increases the infiltration at windows and doors. This air is untreated and may be dirty.

Residential homes which use fuel-burning furnaces need air for combustion. Combustion air, leaving by way of the chimney, might give the interior of the house a slightly negative pressure. View B of Fig. 22-2 shows a diagram of negative pressure conditions in a home.

If the amount of impurities in the air—such as odor, smoke, and bacteria—is great enough to require air cleaning, the remedy may be either ventilation, using fresh air, or improved air cleaning.

Ventilation is usually based on air changes per hour for the conditioned space. If the space is 1000 cu. ft., for example, three changes per hour would mean 3000 cu. ft. per hour or 50 cu. ft./min. Three changes every hour is the minimum for a residence during the heating season. As high as 12 changes an hour (in the above case, 200 cu. ft./min.) are recommended for cooling. Fig. 22-3 shows typical air changes for both the heating season and the cooling season.

It is good practice to keep the air blowers running all the time to provide good ventilation to all parts of the building.

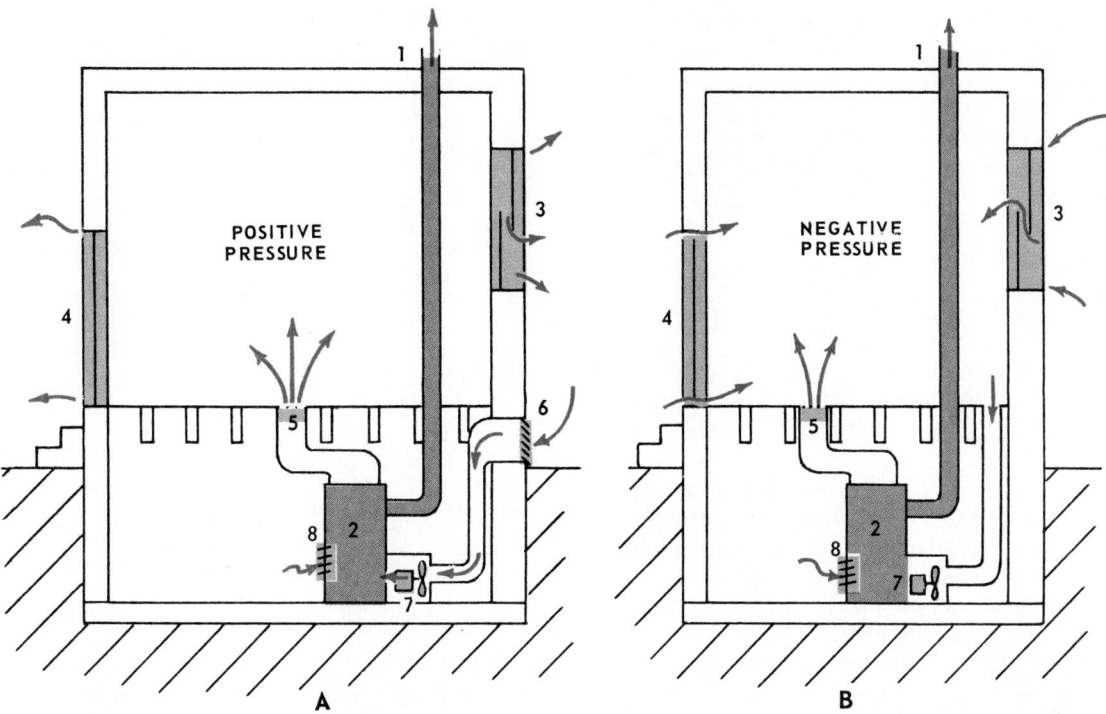

Fig. 22-2. Simplified diagram of airflow into and out of building during heating season. A—Positive air pressure. B—Negative air pressure. 1—Chimney. 2—Furnace. 3—Window. 4—Door. 5—Warm air grille. 6—Fresh air intake. 7—Fresh air fan. 8—Furnace draft control. Red arrows indicate airflow—both within building and into and out of building.

Use	Air Changes/Hour	
	Heating	Cooling
Homes	3–6	6–9
Offices Stores	5–8	6–12
Public Assembly	5–10	6–12

Fig. 22-3. Recommended air changes for various types of occupancies.

Variable speed blowers are sometimes used. They provide more air movement when the heating or cooling system is running; less movement when the systems are off.

An adequate air supply is the best way to control comfort. Body comfort is controlled by evaporation, convection, radiation, and respiration. One must, therefore, control the temperature of the walls, floors, or ceilings to make sure they are not too warm or cold (radiation). One must also supply enough air to promote good respiration, evaporation, and convection. If the specific conditions are not known, it is best to design for 2 cfm/sq. ft. and/or 12 changes/hr. It is also very important to remember that people occupying a closed space give off considerable heat. A sleeping person gives off about 200 Btu/hr., while a person doing heavy work gives off up to 2400 Btu/hr. One Btu = 252 calories.

Another way to determine ventilation requirements is to design for 4 cfm to 6 cfm of fresh air per person and for about 25 cfm to 40 cfm of recirculated air per person. This means the system should handle a total of 29 cfm to 46 cfm per person. One cfm = 0.0283 cu. m/min.

22-5 NOISE

Air distribution systems must be designed to circulate a certain required amount of clean air. Yet, the movement of air must not annoy occupants or cause them discomfort. Two common problems that must be overcome are objectionable noise and drafts. Noise is produced by movement or vibration of an object.

Three factors affect noise. These are:
1. Noise source.
2. Noise carriers.
3. Noise amplifiers or reflectors.

The noise source is a vibration that is loud enough to be heard. This vibration may originate in the heating system, cooling system, or fan mechanism. A vibrating duct panel will create alternate waves of low-pressure and high-pressure air and produce sound.

Duct noise can be very disturbing. Noises may be:
1. A high pitch sound usually caused by an air velocity that is too high or by air hitting sharp metal edges.
2. A low pitch rumble usually caused by fan and motor sounds traveling along the duct system.
3. A popping sound when the unit starts or stops. This is caused by expansion (becoming larger) or contraction (shrinking) of the duct as it warms up or cools.

To locate the source of the high-pitched sound, remove the grille or diffuser. If the noise stops, it is caused by sharp edges in the grille. If it continues, the air velocity is too high (use an anemometer to measure) or there is a sharp edge in the duct system. Locate the problem and correct it.

In measuring the loudness of sound, the unit used is the decibel. Decibel meters register noise level. See Chapter 18. *Decibels (dB) refer to the frequency of the pressure fluctuations in the air and the amplitude or size of these vibrations. Airborne sound is usually expressed in cycles per second (cps).*

Both mechanisms and the motion of matter cause noise in systems. Sources include:

1. Fans.
2. High-velocity air traveling through ducts and causing turbulence.
3. The ac hum of a motor.
4. Sucking and throbbing noises produced by compressors.
5. High-velocity refrigerant flow, especially at sharp bends in piping.

Noise caused by high-speed air is often the result of an undersize unit or duct. Such noise problems develop where the blower has been speeded up to compensate (make up for) the duct's or the unit's lack of capacity.

Noise or vibration carriers are rigid structures that carry vibrations to places where they may be annoying. Floors, ceilings, ducts, doors, and pipes may carry vibrations.

Hard, smooth surfaces in the space being conditioned often reflect or amplify sound (make it louder). Walls, ceilings, floors, and furnishings may pick up a small vibration and reflect it at such a frequency and in such a direction that all or certain parts of the space may be made

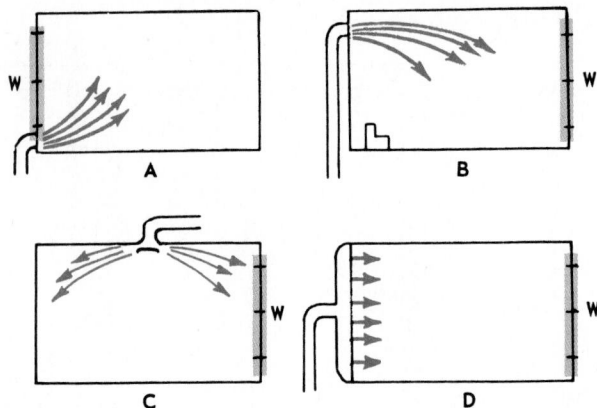

Fig. 22-4. Location of heating grilles minimizes drafts in living areas of room. A—People are exposed to drafts. B—High velocity air is above living level. C—Center location permits lower grille velocity. Higher velocity is above living level. D—Ideal large grille opening. W—Windows.

uncomfortable. Problems involving acoustics (absorption and reflection of sound) and maintenance of a low-decibel noise level are constantly being studied and improved.

Soft fabrics such as drapes and curtains and fabric-covered furniture are noise absorbers. Felt-lined or soft-insulation-lined or covered ducts also absorb noise.

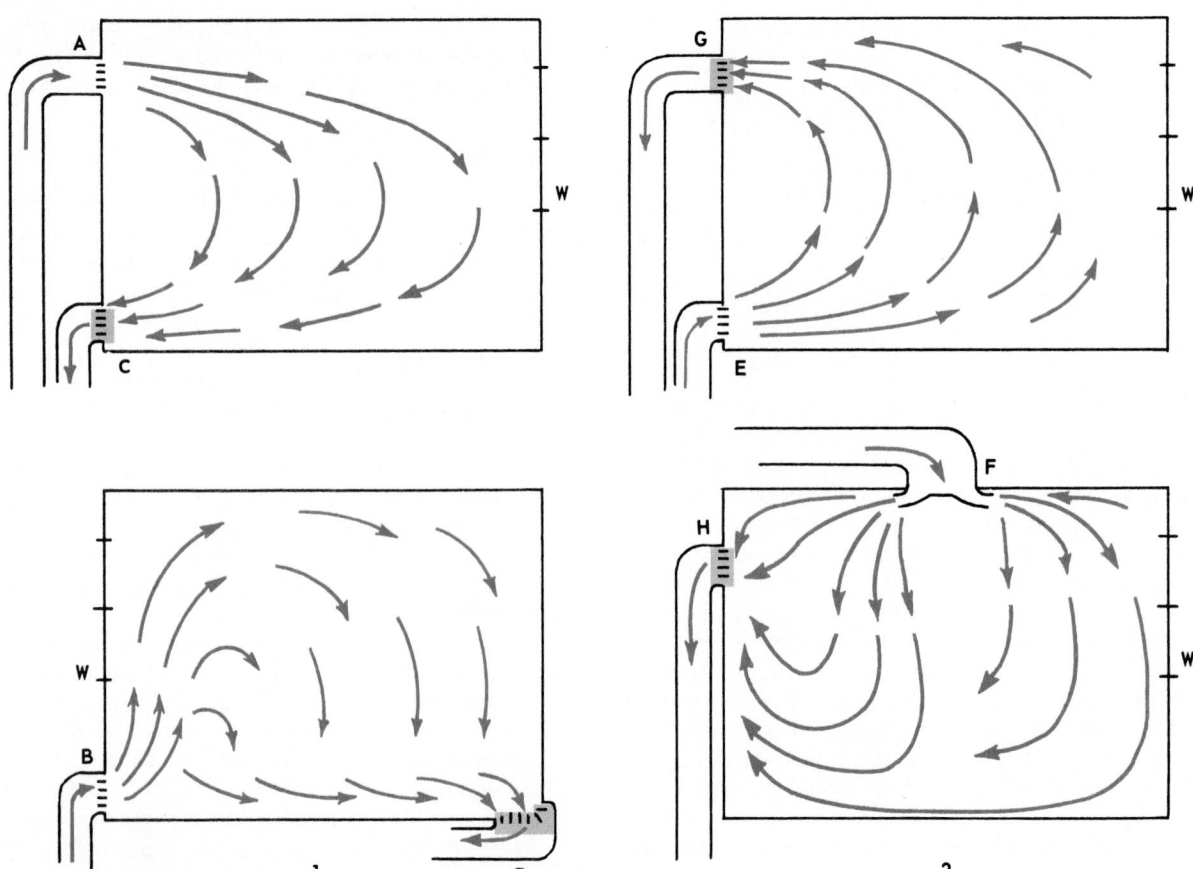

Fig. 22-5. Location of return air grilles in residential installation. 1—For heating: A and B warm air in, C and D cold air return. 2—For cooling: E and F cold air in, G and H warm air return. W—Windows. Some backflow into grille in bottom right view is useful for air mixing (see Fig. 22-8). Note that these short rooms can have both air outlet and air inlet on same side of room.

22-6 DRAFTS

It is a relatively simple matter to provide ducts and fans large enough to give a room the correct amount of air for conditioning. However, the air must enter the room, and circulate to all parts of it, without interfering with flow of air to the air return. Neither should there be objectionable drafts or noise.

When air moves past people faster than 25 fpm (about 1/3 mph), most people feel an annoying draft. This means that, if the air flows faster than 1/3 mph through the length of a 25-foot living room, an uncomfortable feeling results.

To have a grille outlet designed to throw the air into the room a distance of 8 to 13 ft., a velocity of 500 fpm (about 5.5 mph) is needed. Therefore, to keep that part of the space occupied by humans at a 25 fpm velocity, the location of grilles or outlets must be carefully selected.

To move air across a long space at a reasonable velocity, the location of the air returns is important. *The air returns for a long room should be located on the opposite side of the space from the air entrance.*

Air returns should be located high on the wall if warm air return is desired (cooling season) and low on the wall (or in the floor) if cold air return is desired (heating season). Fig. 22-4 shows some typical airflow patterns. The return air openings often used are shown in Fig. 22-5.

22-7 STRATIFICATION OF AIR

Warm air tends to rise. Cold air tends to settle. In a room where the air is not deliberately moved, the air will assume levels according to its temperature, as shown in Fig. 22-6. This is called stratification. *Air in an occupied space must be kept moving in order to eliminate stratification.*

It is important to place thermostats and humidistats at the proper level because of this stratification. Stratification tends to make smoke haze hover in layers. The layers make

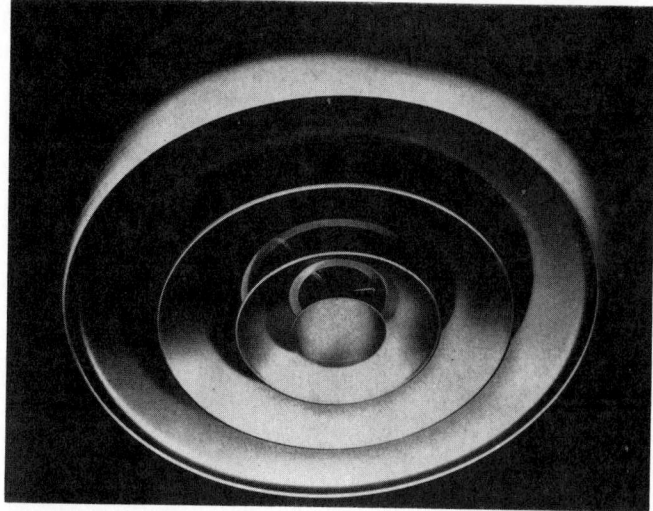

Fig. 22-7. Ceiling diffusion grille distributes air in all directions in occupied space.   (Anemostat Products Div.)

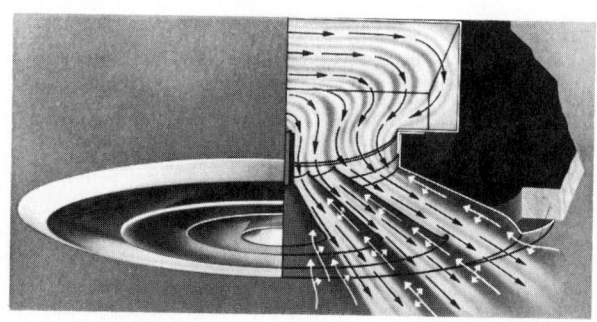

Fig. 22-8. Airflow and air mix of ceiling grille. Black arrows indicate airflow from duct. White arrows indicate room air moving into grille to mix with duct air. (Anemostat Products Div.)

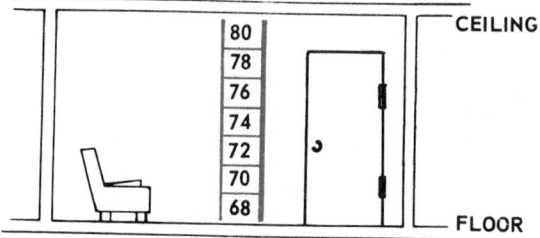

Fig. 22-6. Various temperature levels (stratification) found in room with little or no air circulation.

it difficult to get rid of smoke.

Unfortunately, some grilles are poorly located. Then air moves only in certain parts of the room and becomes stagnant (not moving) in other parts. Then, too, furnishings obstruct air movement in the room. For this reason, and to enable higher grille velocities, some grilles are located 6 ft. high in the room or in the ceilings. In these locations, the grilles must be attractive in appearance or concealed entirely. In Fig. 22-7, a diffusion grille promotes mixing of some of the room air with the entering air in the grille. The mixing principle is shown in Fig. 22-8.

22-8 AIR DUCTS

To deliver air to the conditioned space, air carriers are needed. These carriers are called ducts. They are made of sheet metal or some structural material that will not burn (non-combustible).

Ducts work on the principle of air pressure difference. If a pressure difference exists, air will move from the higher pressure area to the lower pressure places. The greater this pressure difference, the faster the air will flow.

Ducts are made of many materials. Pressure in the ducts is small, so materials with a great deal of strength are unnecessary. Originally, hot air ducts were thin, tinned sheet steel. Later, galvanized sheet steel, aluminum sheet and, finally, insulated ducts made from materials such as fiberboard, were developed. Passageways formed by studs or joists are sometimes used for return air where a fire hazard does not exist.

There are three common classifications of ducts:
1. Conditioned-air ducts.
2. Recirculating-air ducts.
3. Fresh-air ducts.

Ducts commonly used for carrying air are round, square, or rectangular. See Fig. 22-9. Round ducts are more effi-

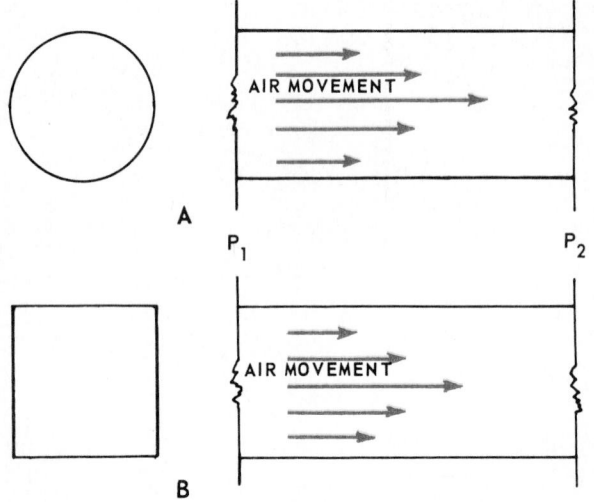

Fig. 22-9. Shape of ducts varies. A—Round. B—Rectangular. Arrows indicate that the closer air is to inside surfaces of duct the slower the flow. P_1 and P_2 indicate pressure along a duct. To create movement, P_1 must be greater than P_2.

cient based on volume of air handled per perimeter distance (distance around). That is, less material is needed for the same capacity as a square or rectangular duct. Resistance to airflow is also less. Since a round duct has less outside surface area than a rectangular duct of equivalent capacity, it will have a much lower heat loss or gain.

The square or rectangular duct conforms better to building construction. It fits into walls and ceilings better than round ducts. It is easier to install rectangular ducts between joists and studs.

Tables have been developed to compare carrying capacities of rectangular and round ducts. See Figs. 22-10 and 22-11. There are several round duct equivalent sizes to

choose from. The one selected depends on the one side dimension desired. For example, ducts 11 in. high may be wanted to improve appearance, or to fit in between joists (a 14 in. distance). Fig. 22-11 shows the equivalents.

The quantity of sheet metal used for a rectangular duct may be a deciding factor. The amounts of material for round and rectangular ducts can be compared.

An 18 in. round duct has a perimeter (distance around) of:
diameter $\times \pi$ = circumference (perimeter).
$18 \times 3.1416 = 56.55$ in.

A rectangular duct of equal capacity is 17×16 in. It has a perimeter twice the width (W) plus twice the depth (D), or $2W + 2D$ = perimeter.

$17 + 17 + 16 + 16$ = perimeter
$34 + 32 = 66$ in.
66 in. $- 56.55 = 9.45$ in. more of duct material
per unit of duct length.

In other words, for every foot of length, the rectangular duct requires 12 in. $\times$ 9.45 in. or 113.4 sq. in. more of material than the round duct.

22-9 TYPES OF DUCT SYSTEMS

Air ducts deliver air to a room or rooms and then return the air from this room or the rooms to the heating (furnace) or cooling (evaporator) system. Fig. 22-12 shows a typical residential air duct system.

There are several types of supply duct systems:
1. Individual round pipe system.
2. Extended plenum system.
3. Reducing trunk system.
Some of the installations, like that shown in Fig. 22-12, are combinations of two of the systems or all of the systems.

Return air systems are usually of two types:
1. Single return system.
2. Multiple return system.

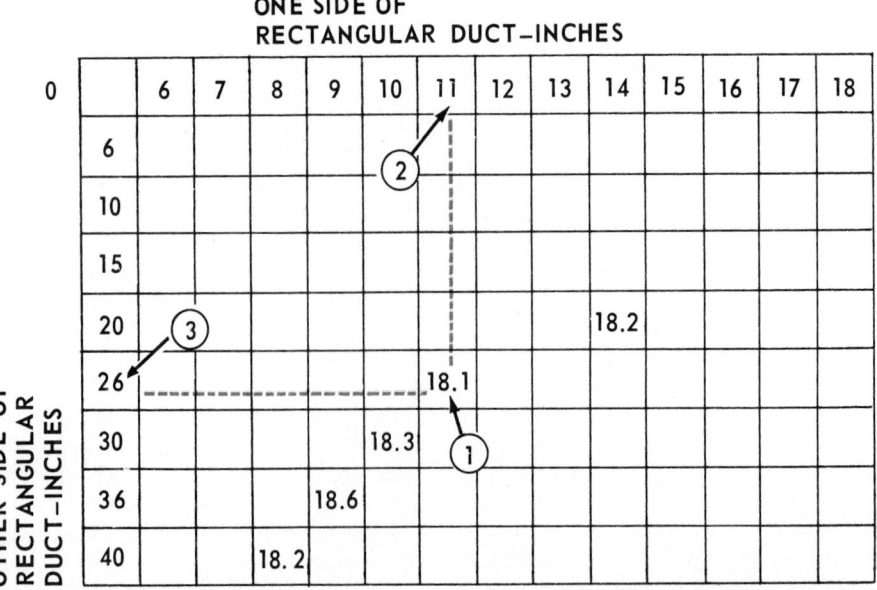

Fig. 22-10. Chart for converting round ducts to square or rectangular duct measurements. Numbers represent inches. 1—Locate round duct size (diameter in inches). 2—Go up vertically and read number at top in corresponding square. This value is for one side of rectangular duct. 3—Moving horizontally from round duct size in middle, read number at left side. This value is measurement of other side of rectangular duct.

Circular Equivalents of Rectangular Duct for Equal Friction and Capacity[a]

Lgth Adj[b]	Length of One Side of Rectangular Duct (a), in.																
	4.0	4.5	5.0	5.5	6.0	6.5	7.0	7.5	8.0	9.0	10.0	11.0	12.0	13.0	14.0	15.0	16.0
3.0	3.8	4.0	4.2	4.4	4.6	4.7	4.9	5.1	5.2	5.5	5.7	6.0	6.2	6.4	6.6	6.8	7.0
3.5	4.1	4.2	4.6	4.8	5.0	5.2	5.3	5.5	5.7	6.0	6.3	6.5	6.8	7.0	7.2	7.5	7.7
4.0	4.4	4.6	4.9	5.1	5.3	5.5	5.7	5.9	6.1	6.4	6.7	7.0	7.3	7.6	7.8	8.0	8.3
4.5	4.6	4.9	5.2	5.4	5.7	5.9	6.1	6.3	6.5	6.9	7.2	7.5	7.8	8.1	8.4	8.6	8.8
5.0	4.9	5.2	5.5	5.7	6.0	6.2	6.4	6.7	6.9	7.3	7.6	8.0	8.3	8.6	8.9	9.1	9.4
5.5	5.1	5.4	5.7	6.0	6.3	6.5	6.8	7.0	7.2	7.6	8.0	8.4	8.7	9.0	9.3	9.6	9.9

Lgth Adj[b]	Length of One Side of Rectangular Duct (a), in.																				Lgth Adj[b]
	6	7	8	9	10	11	12	13	14	15	16	17	18	19	20	22	24	26	28	30	
6	6.6																				6
7	7.1	7.7																			7
8	7.6	8.2	8.7																		8
9	8.0	8.7	9.3	9.8																	9
10	8.4	9.1	9.8	10.4	10.9																10
11	8.8	9.5	10.2	10.9	11.5	12.0															11
12	9.1	9.9	10.7	11.3	12.0	12.6	13.1														12
13	9.5	10.3	11.1	11.8	12.4	13.1	13.7	14.2													13
14	9.8	10.8	11.4	12.2	12.9	13.5	14.2	14.7	15.3												14
15	10.1	11.0	11.8	12.6	13.3	14.0	14.6	15.3	15.8	16.4											15
16	10.4	11.3	12.2	13.0	13.7	14.4	15.1	15.7	16.4	16.9	17.5										16
17	10.7	11.6	12.5	13.4	14.1	14.9	15.6	16.2	16.8	17.4	18.0	18.6									17
18	11.0	11.9	12.9	13.7	14.5	15.3	16.0	16.7	17.3	17.9	18.5	19.1	19.7								18
19	11.2	12.2	13.2	14.1	14.9	15.7	16.4	17.1	17.8	18.4	19.0	19.6	20.2	20.8							19
20	11.5	12.6	13.5	14.4	15.2	16.0	16.8	17.5	18.2	18.9	19.5	20.1	20.7	21.3	21.9						20
22	12.0	13.0	14.1	15.0	15.9	16.8	17.6	18.3	19.1	19.8	20.4	21.1	21.7	22.3	22.9	24.0					22
24	12.4	13.5	14.6	15.6	16.5	17.4	18.3	19.1	19.9	20.6	21.3	22.0	22.7	23.3	23.9	25.1	26.2				24
26	12.8	14.0	15.1	16.2	17.1	18.1	19.0	19.8	20.6	21.4	22.1	22.9	23.5	24.2	24.9	26.1	27.3	28.4			26
28	13.2	14.5	15.6	16.7	17.7	18.7	19.6	20.5	21.3	22.1	22.9	23.7	24.4	25.1	25.8	27.1	28.3	29.5	30.6		28
30	13.6	14.9	16.1	17.2	18.3	19.3	20.2	21.1	22.0	22.9	23.7	24.4	25.2	25.9	26.6	28.0	29.3	30.5	31.7	32.8	30
32	14.0	15.3	16.5	17.7	18.8	19.8	20.8	21.8	22.7	23.5	24.4	25.2	26.0	26.7	27.5	28.9	30.2	31.5	32.7	33.9	32
34	14.4	15.7	17.0	18.2	19.3	20.4	21.4	22.4	23.3	24.2	25.1	25.9	26.7	27.5	28.3	29.7	31.0	32.4	33.7	34.9	34
36	14.7	16.1	17.4	18.6	19.8	20.9	21.9	22.9	23.9	24.8	25.7	26.6	27.4	28.2	29.0	30.5	32.0	33.3	34.6	35.9	36
38	15.0	16.5	17.8	19.0	20.2	21.4	22.4	23.5	24.5	25.4	26.4	27.2	28.1	28.9	29.8	31.3	32.8	34.2	35.6	36.8	38
40	15.3	16.8	18.2	19.5	20.7	21.8	22.9	24.0	25.0	26.0	27.0	27.9	28.8	29.6	30.5	32.1	33.6	35.1	36.4	37.8	40
42	15.6	17.1	18.5	19.9	21.1	22.3	23.4	24.5	25.6	26.6	27.6	28.5	29.4	30.3	31.2	32.8	34.4	35.9	37.3	38.7	42
44	15.9	17.5	18.9	20.3	31.5	22.7	23.9	25.0	26.1	27.1	28.1	29.1	30.0	30.9	31.8	33.5	35.1	36.7	38.1	39.5	44
46	16.2	17.8	19.3	20.6	21.9	23.2	24.4	25.5	26.6	27.7	28.7	29.7	30.6	31.6	32.5	34.2	35.8	37.4	38.9	40.4	46
48	16.5	18.1	19.6	21.0	22.3	23.6	24.8	26.0	27.1	28.2	29.2	30.2	31.2	32.2	33.1	34.9	36.6	38.2	39.7	41.2	48
50	16.8	18.4	19.9	21.4	22.7	24.0	25.2	26.4	27.6	28.7	29.8	30.8	31.8	32.8	33.7	35.5	37.2	38.9	40.5	42.0	50
52	17.1	18.7	20.2	21.7	23.1	24.4	25.7	26.9	28.0	29.2	30.3	31.3	32.3	33.3	34.3	36.2	37.9	39.6	41.2	42.8	52
54	17.3	19.0	20.6	22.0	23.5	24.8	26.1	27.3	28.5	29.7	30.8	31.8	32.9	33.9	34.9	36.8	38.6	40.3	41.9	43.5	54
56	17.6	19.3	20.9	22.4	23.8	25.2	26.5	27.7	28.9	30.1	31.2	32.3	33.4	34.4	35.4	37.4	39.2	41.0	42.7	44.3	56
58	17.8	19.5	21.2	22.7	24.2	25.5	26.9	28.2	29.4	30.6	31.7	32.8	33.9	35.0	36.0	38.0	39.8	41.6	43.3	45.0	58
60	18.1	19.8	21.5	23.0	24.5	25.9	27.3	28.6	29.8	31.0	32.2	33.3	34.4	35.5	36.5	38.5	40.4	42.3	44.0	45.7	60
62		20.1	21.7	23.3	24.8	26.3	27.6	28.9	30.2	31.5	32.6	33.8	34.9	36.0	37.1	39.1	41.0	42.9	44.7	46.4	62
64		20.3	22.0	23.6	25.1	26.6	28.0	29.3	30.6	31.9	33.1	34.3	35.4	36.5	37.6	39.6	41.6	43.5	45.3	47.1	64
66		20.6	22.3	23.9	25.5	26.9	28.4	29.7	31.0	32.3	33.5	34.7	35.9	37.0	38.1	40.2	42.2	44.1	46.0	47.7	66
68		20.8	22.6	24.2	25.8	27.3	28.7	30.1	31.4	32.7	33.9	35.2	36.3	37.5	38.6	40.7	42.8	44.7	46.6	48.4	68
70		21.1	22.8	24.5	26.1	27.6	29.1	30.4	31.8	33.1	34.4	35.6	36.8	37.9	39.1	41.2	43.3	45.3	47.2	49.0	70
72			23.1	24.8	26.4	27.9	29.4	30.8	32.2	33.5	34.8	36.0	37.2	38.4	39.5	41.7	43.8	45.8	47.8	49.6	72
74			23.3	25.1	26.7	28.2	29.7	31.2	32.5	33.9	35.2	36.4	37.7	38.8	40.0	42.2	44.4	46.4	48.4	50.3	74
76			23.6	25.3	27.0	28.5	30.0	31.5	32.9	34.3	35.6	36.8	38.1	39.3	40.5	42.7	44.9	47.0	48.9	50.9	76
78			23.8	25.6	27.3	28.8	30.4	31.8	33.3	34.6	36.0	37.2	38.5	39.7	40.9	43.2	45.4	47.5	49.5	51.4	78
80			24.1	25.8	27.5	29.1	30.7	32.2	33.6	35.0	36.3	37.6	38.9	40.2	41.4	43.7	45.9	48.0	50.1	52.0	80
82				26.1	27.8	29.4	31.0	32.5	34.0	35.4	36.7	38.0	39.3	40.6	41.8	44.1	46.4	48.5	50.6	52.6	82
84				26.4	28.1	29.7	31.3	32.8	34.3	35.7	37.1	38.4	39.7	41.0	42.2	44.6	46.9	49.0	51.1	53.2	84
86				26.6	28.3	30.0	31.6	33.1	34.6	36.1	37.4	38.8	40.1	41.4	42.6	45.0	47.3	49.6	51.7	53.7	86
88				26.9	28.6	30.3	31.9	33.4	34.9	36.4	37.8	39.2	40.5	41.8	43.1	45.5	47.8	50.0	52.2	54.3	88
90				27.1	28.9	30.6	32.2	33.8	35.3	36.7	38.2	39.5	40.9	42.2	43.5	45.9	48.3	50.5	52.7	54.8	90
92					29.1	30.8	32.5	34.1	35.6	37.1	38.5	39.9	41.3	42.6	43.9	46.4	48.7	51.0	53.2	55.3	92
96					29.6	31.4	33.0	34.7	36.2	37.7	39.2	40.6	42.0	43.3	44.7	47.2	49.6	52.0	54.2	56.4	96

Fig. 22-11. Chart determines sizes of rectangular ducts needed to equal carrying capacity of round ducts. To use, find diameter of round pipe in chart. Then find one side of rectangular duct by reading up. Find other side by reading left to first row of numbers representing other side of rectangular pipe. (ASHRAE Guide and Data Book)

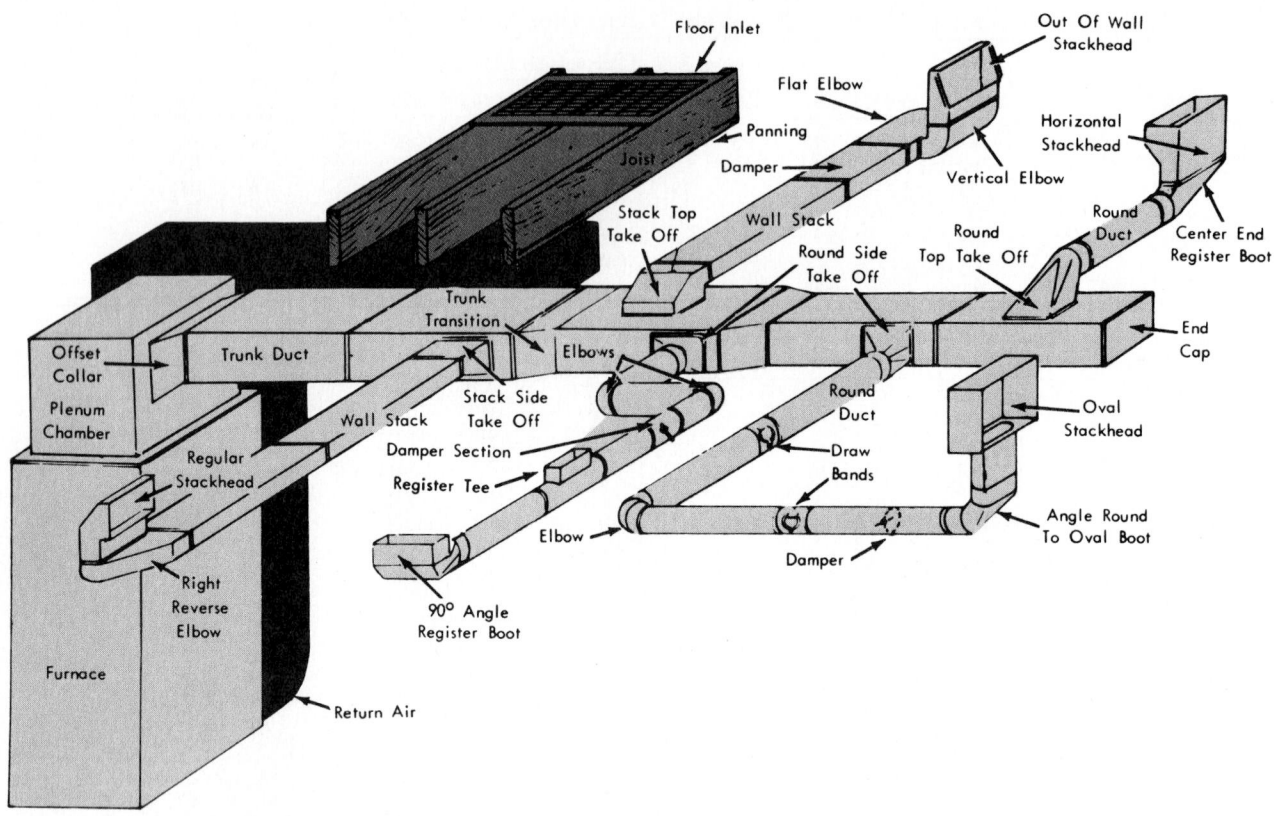

Fig. 22-12. Residential layout showing most sheet metal parts used in conditioned air duct system.
(Reprinted from AIR CONDITIONING CONTRACTORS of AMERICA'S [ACCA] Basic Installation Manual, by permission of ACCA)

The return systems can also be combinations of the two systems. Fig. 22-13 shows the basic designs. Some types of duct connections are shown in Fig. 22-14.

Duct systems may be installed in basements, in crawl

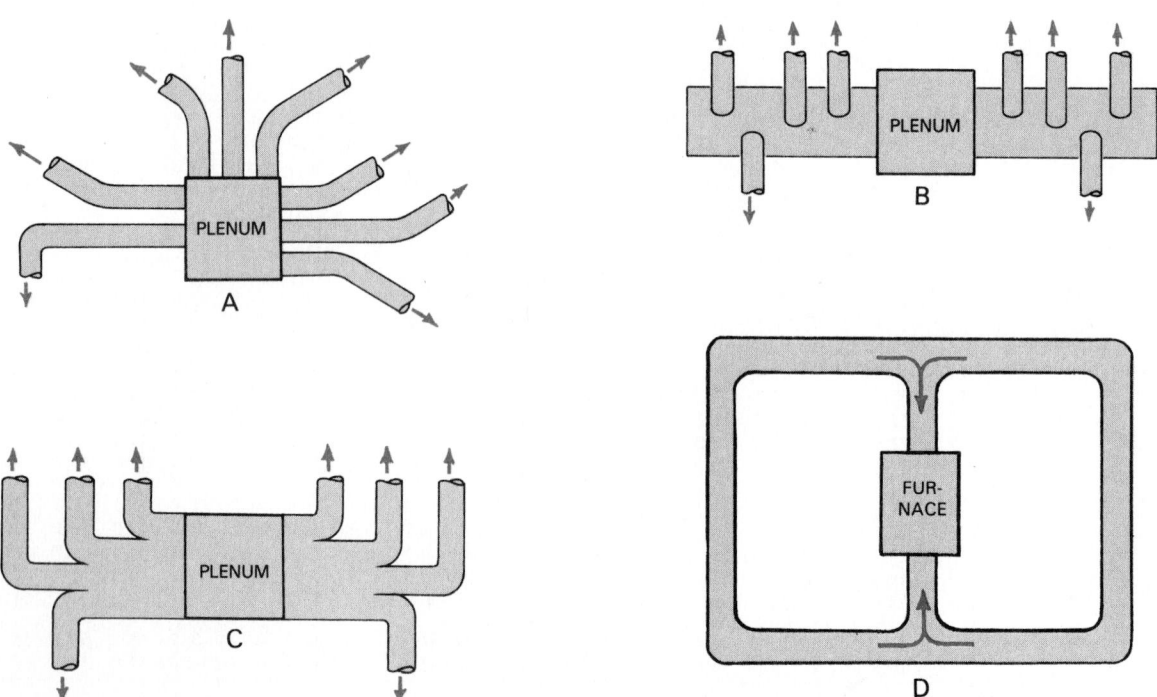

Fig. 22-13. A, B, C, are drawn as top views of three types of duct systems often used. A—Individual round pipe system. B—Extended plenum system. C—Reducing trunk system. D—Multiple return system is shown in view from top.

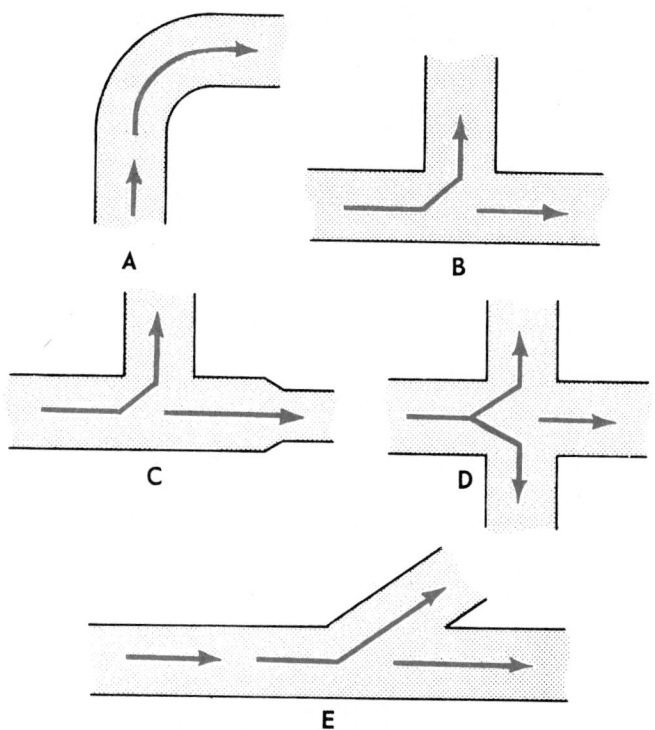

Fig. 22-14. Typical branch duct designs. A—Elbow. B—T-fitting. C—Reducing T. D—Cross. E—Lateral.

spaces, in attics, and in concrete floors (slabs) of homes without basements. Fig. 22-15 shows an overhead duct system.

In basements, the conditioned-air main duct is run across and just under the floor joists. The branch ducts, round or rectangular, are then run between the joists to the grille or diffuser openings. The return ducts usually use the joists and the floorboards for three sides of the branch ducts (panned joist space) and then run the main return duct alongside the conditioned air duct.

When the duct system is located in a crawl space or attic, the ducts are usually insulated. Otherwise heat loss would be too great.

When installed in the concrete slab, ducts are usually made of metal, plastic, or ceramic. The branch ducts usually connect to a perimeter duct. Grilles or diffusers are connected to the perimeter duct at intervals along the floor. See Fig. 22-16. A downflow type furnace is used as shown in Fig. 22-17.

Some buildings are designed with unit ventilators that automatically draw air from the conditioned space and, at the same time, bring in makeup air from the outside.

False ceilings are often used to conceal piping, wiring, heat exchangers, and ducts. The false ceiling may have holes to allow movement of conditioned air into the room below, or diffusers can be used. Space between the false ceiling and the real ceiling may be used as a fresh air plenum

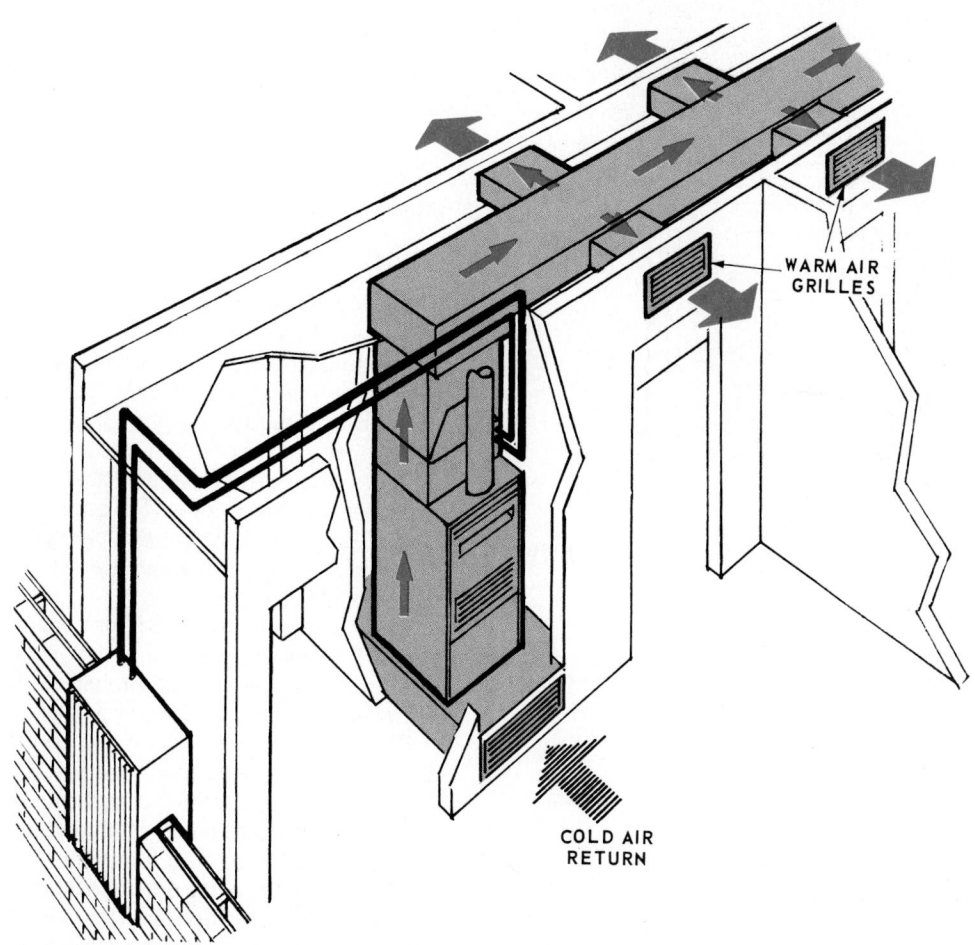

Fig. 22-15. Overhead extended plenum duct system.   (Lennox International, Inc.)

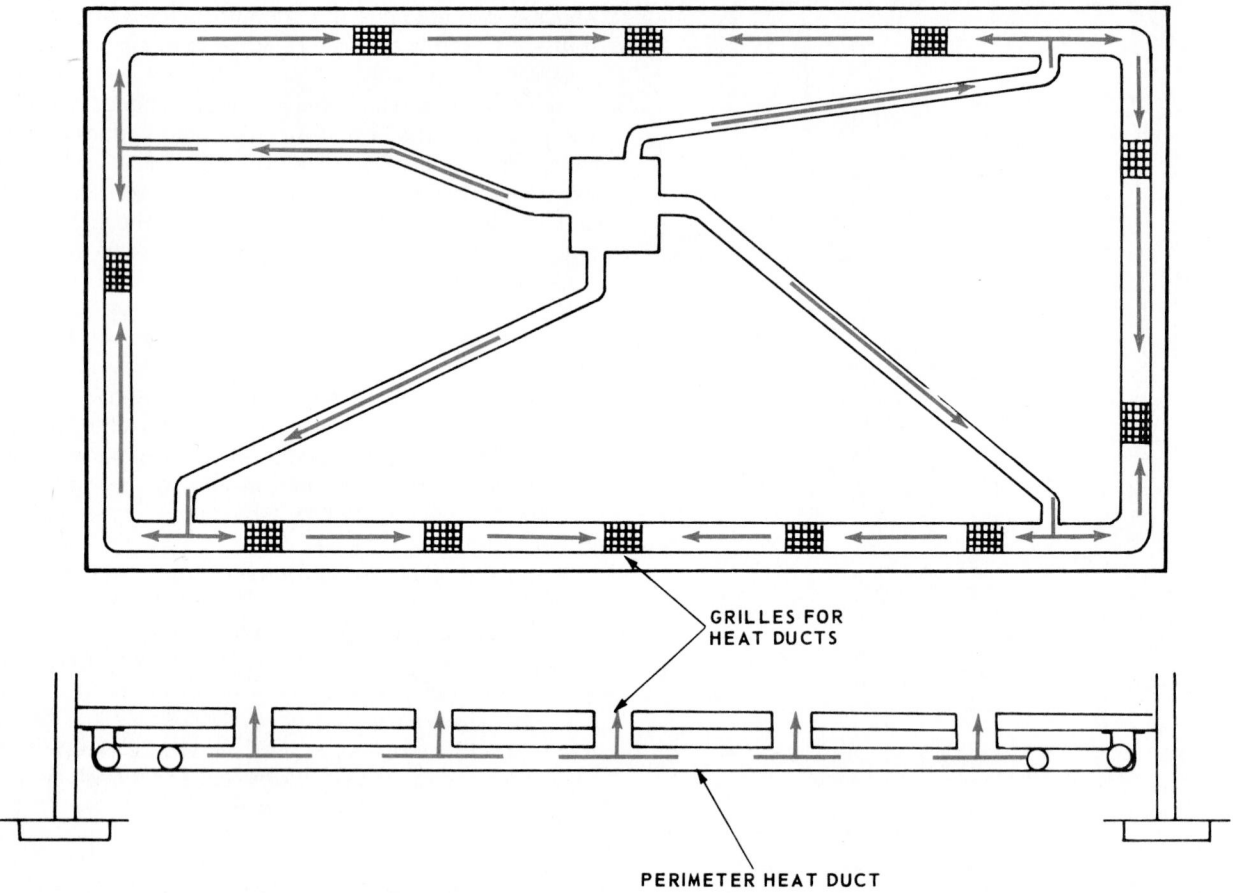

Fig. 22-16. Top view of perimeter loop system for conditioned air ducts. Ducts are cast into concrete floor slab of building.

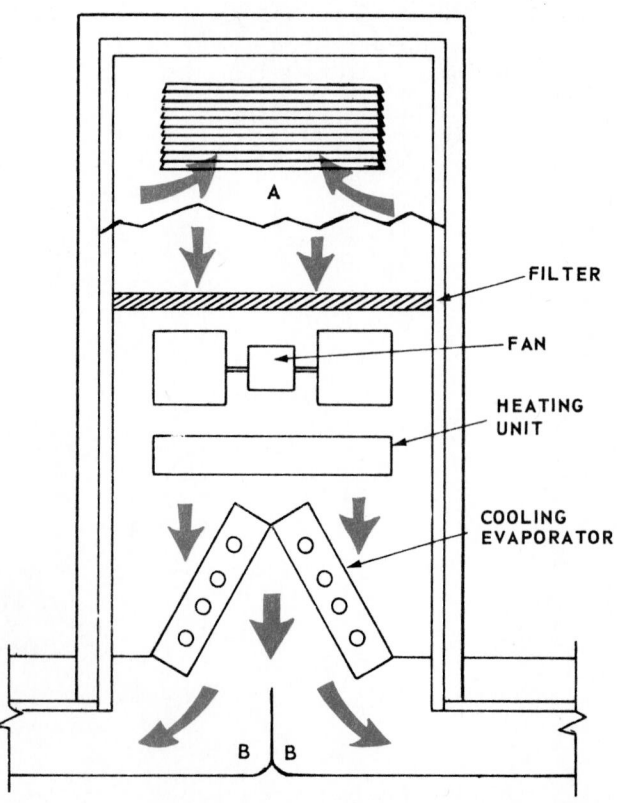

Fig. 22-17. Downflow furnace. A—Air return. B—Air outlet.

chamber. Some systems use heated panels to provide either radiant heat or radiant cooling. See Fig. 22-18.

22-10 DUCT CONSTRUCTION

Ducts may be made of metal, wood, ceramic, or plastic materials. Metal ducts are used for warm air distribution and for exhaust air ducts. The metal is usually sheet steel coated with zinc (galvanized steel). Some ducts are made of aluminum to reduce weight. Sheet lead is used when the duct must carry corrosive gases.

Ducts made of aluminum, glass fiber, or plastic are not approved in some codes for duct installations where fire spread is possible. Many flexible ducts made of wire and fabric are not permitted where fire could spread through them to other areas.

Sheet metal brakes and formers are used for making ducts. Elbows and other connections such as branches are designed using geometric principles. Sheet metal work is a specialty trade and is done by skilled sheet metal workers.

Sheet metal ducts expand and contract as they heat and cool. Fabric joints are often used to absorb this movement. To prevent most of the fan and furnace noise from traveling along the duct metal, fabric joints should also be used where ducts fasten to a furnace or air conditioner. But, in fact, most duct joints are made of sheet metal.

The length of the duct can be altered in many ways. Fig. 22-19 illustrates a combination air electrical cutting tool being used by a technician to alter the length of a circular duct.

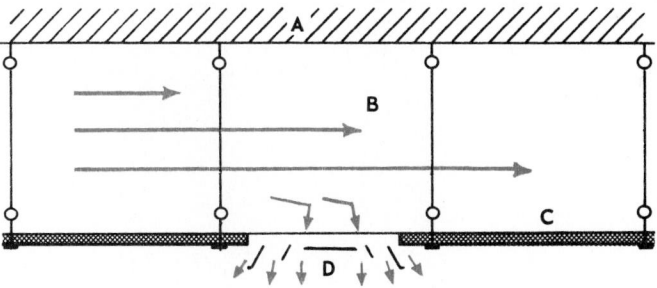

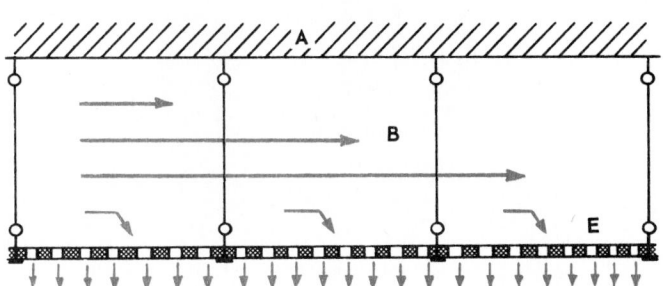

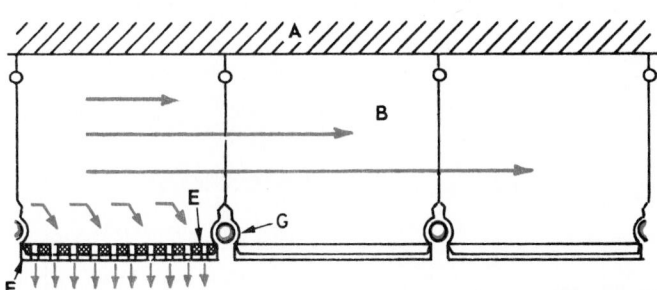

Fig. 22-18. False ceiling can be used to form plenum chamber for either cool or warm air. A—Ceiling. B—Plenum chamber. C—False ceiling. D—Diffuser. E—False ceiling with fresh air holes. F—False ceiling with metal panels heated or cooled by water at G.

This type of cutting tool is used for cutting duct work that is up to 3/16ths of an inch in thickness.

Several types of sheet metal joints have been developed. See Figs. 22-20 and 22-21. The joint should be airtight and strong. Many joints are riveted for added strength and tightness. Fig. 22-22 shows a tool used to rivet from one side only.

To use this tool:

1. Drill a hole.
2. Place the rivet blank in the hole.
3. Attach the tool.
4. Work the handle which withdraws the expander.
5. When the rivet head is completely formed, the expander stem will break off flush with the inside surface of the rivet. The cost of the tool is quite low considering the great saving in time when making riveted connections.

Many of these joints are also sealed with special duct tape to make them leakproof. Sealants are put in the duct seam for the same purpose.

Fig. 22-19. Service technician using electrical air combination cutting tool to alter length of circular duct.
(Hypertherm, Incorporated)

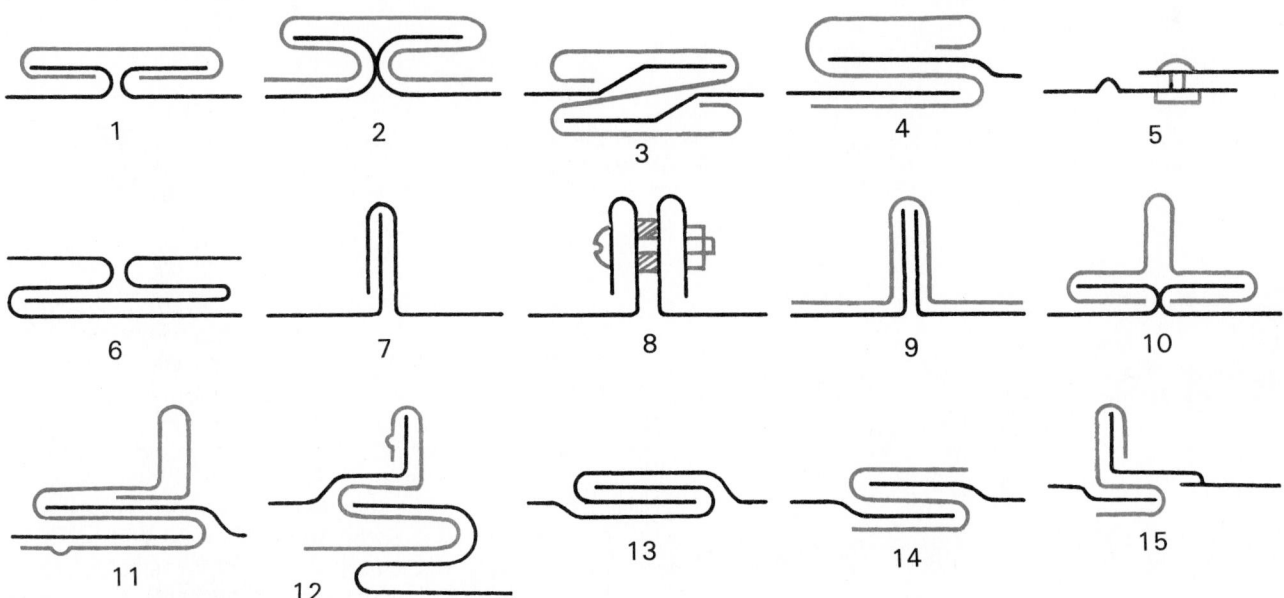

Fig. 22-20. Some sheet metal duct joint designs. 1—Drive cleats. 2—Double ''S'' slip. 3—Reinforced ''S'' cleat. 4—Button punch snap lock. 5—Button punch, rivet, or screw. 6—Inside slip joint. 7—Standing seam. 8—Flanged joint with gasket. 9—T-connector. 10—Reinforced ''on center'' drive cleat. 11—Standing ''S'' cleat. 12—Angle slip with upward offset. 13—Pittsburgh lock. 14—''S'' slip. 15—Pocket slip.

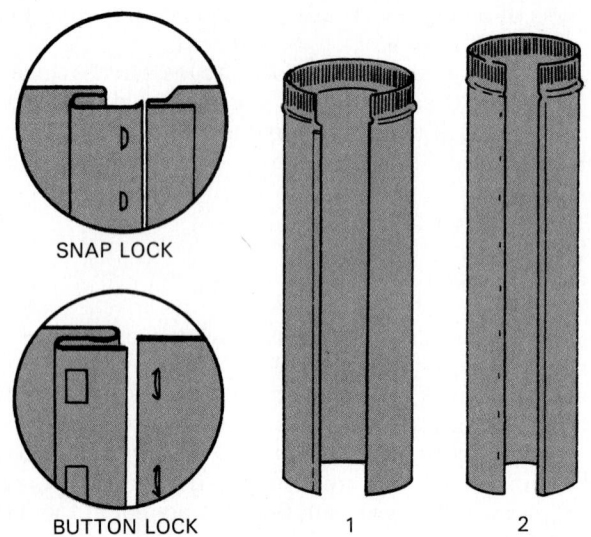

Fig. 22-21. Two seamlock methods. 1—Snap lock. 2—Button lock. (Reprinted from AIR CONDITIONING CONTRACTORS of AMERICA'S [ACCA] Basic Installation Manual, by permission of ACCA)

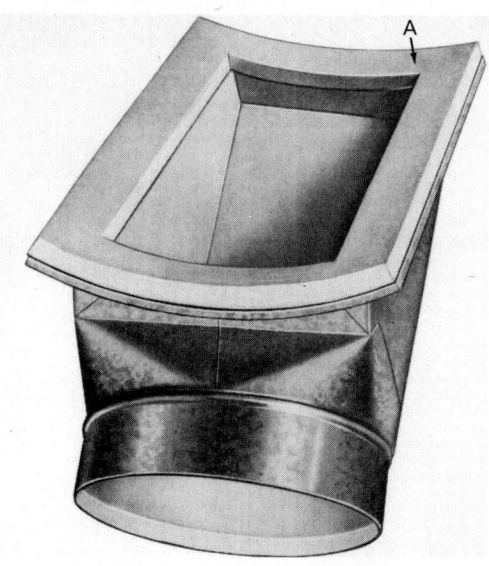

Fig. 22-24. Round duct branch elbow made of sheet metal with inside insulation. A—Insulation. (Americal Seal-Tite)

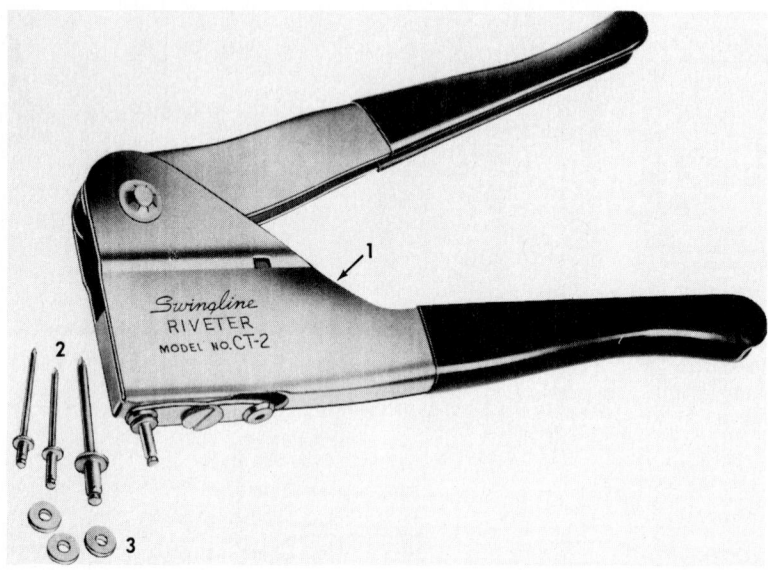

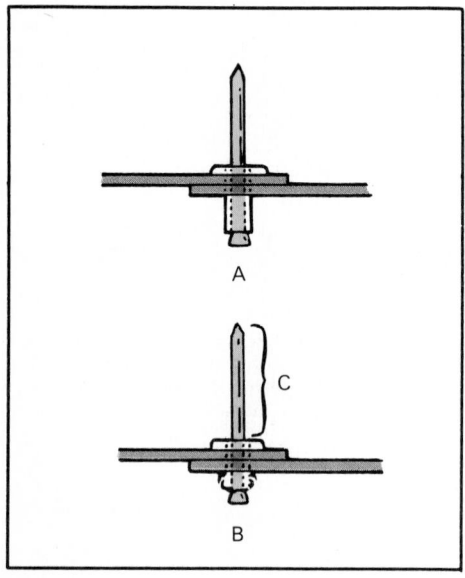

Fig. 22-22. Useful tool for riveting sheet metal and duct work in places where it would be difficult to install a solid rivet, such as areas with access only to the outside. Rivet is often called a "blind" rivet. 1—Riveting pliers. 2—Rivets. 3—Washers. Inserts illustrate riveting procedure. A—Hole drilled and rivet inserted. B—Riveting completed. C—This part is cut off or breaks off. (Desa International)

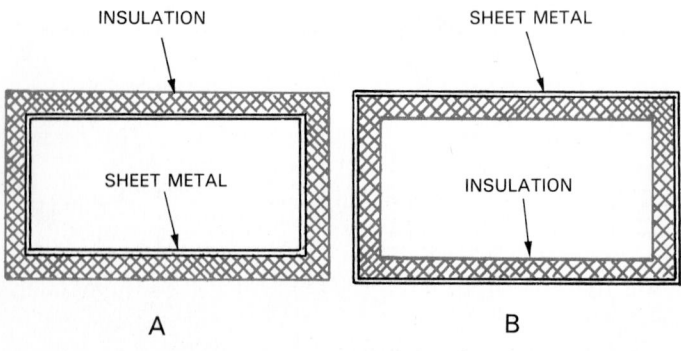

Fig. 22-23. Insulated sheet metal duct. A—Insulation outside the duct. B—Insulation inside the duct.

Duct joints in forced warm air system should be as leakproof as possible. Leaking joints cut down on air volume delivered as outlets at the end of long runs. Connect all duct sections with good sheet metal joints.

Leaks may also make cold air ducts inefficient. Joints and seams should also be sealed the same as warm air ducting.

If duct systems are noisy, the air velocity may be too high. There may be turbulence, and metal edges may produce wind noise. Some ducts are lined with an acoustic material to produce more silent duct system. Duct sections equipped with silencers are also available. Such units act as mufflers.

Many ducts are insulated—either on the inside or outside—to reduce noise as well as heat transfer. See Fig.

A

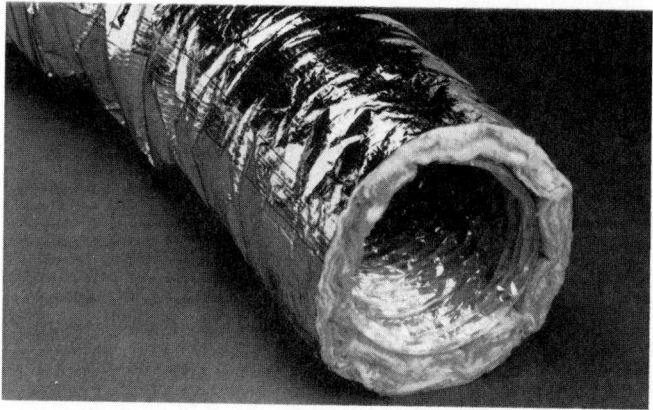

B

Fig. 22-25. Resin bonded fiberglass air duct system (A) used in residential and commercial cooling, heating, or dual-temperature air-handling systems. (Knauf Fiber Glass GmbH) B—Flexible duct insulated with fiberglass insulation. (Owens-Corning Fiberglas Corp.)

Fig. 22-26. Flexible fiberglass duct being installed into rectangular duct air-handling system. (Owens-Corning Fiberglas Corp.)

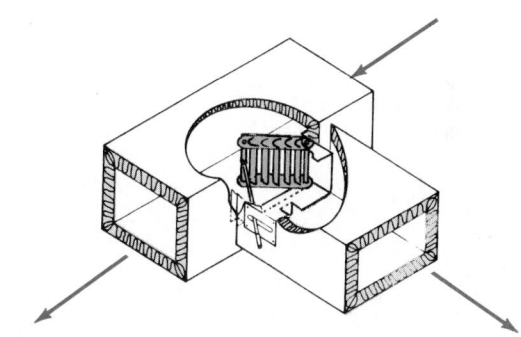

Fig. 22-27. Fiberglass duct connection with adjustable vane air diverter.

22-23. The insulation is fastened to the duct with adhesives. In some cases metal clips hold the insulation in place. Fig. 22-24 shows a sheet metal connector insulated on the inside.

Round ducts made of plastic and spring steel wire are often used. They need no elbows and are easily installed around other piping, beams, and joints. This type of ducting generally relies on self-inflation, which should be considered when determining system requirements. The plastic used will not burn.

Some ducts are made of nonmetals such as fiberglass, urethane, plastic, or sheetrock. See Figs. 22-25 and 22-26. Fig. 22-25, Part B, shows a flexible duct for use as an air duct or for supply and return air lines in residential and commercial heating. A flexible bend allows it to be used for gradual bends needed when connecting air ducts to diffusers or when routing air ducts through spaces with many obstructions. Fig. 22-26 shows a flexible duct being installed into a rectangular air-handling system. A duct with a branch and a vane air diverter is shown in Fig. 22-27.

Duct material must not erode (release) any loose material into the airstream. Ducts made of asbestos or fiberglass or ducts lined with these materials should be plastic coated. *Asbestos fibers and fiberglass particles released into the air are harmful to health.* Some ducts are made with a sheet

metal inner wall and a sheet metal outer wall with insulation and sound absorbent material in between.

Any reduction in the size of a duct should be tapered. The taper should be approximately 1 in. for every 4 to 7 in. in length. When changing from small to large, use a change rate of 1 in. in every 7 in. of length.

Rectangular ducts must be carefully chosen. If the width is much more than the height, the amount of metal needed for the cross-sectional area becomes excessive. The ratio of the wide side to the narrow side is called the aspect ratio. For example, if a duct is 18 in. wide and 6 in. deep, the aspect ratio is $\frac{18}{6}$ or 3 to 1. Costs of making the duct also increase.

To show how much more metal is used as the aspect ratio increases: Assume a duct is 10 by 10 in. (aspect ratio 1:1). It has a cross-sectional area of 100 sq. in. while the distance around (perimeter) is 10 + 10 + 10 + 10 = 40 inches. Another duct of 100 sq. in. is made 20 in. wide and 5 in. deep. It has a perimeter of 20 + 5 + 20 + 5 = 50 in. or a 25 percent increase in metal for each unit of length. Sheet metal contractors classify ducts by their aspect ratios and estimate costs by class. Fig. 22-28 shows some common aspect ratios.

One should not try to put a duct in a stud or joist space when its size is equal to the inside dimensions of the stud

DUCT CLASS (ASPECT RATIO)	WIDE SIDE INCHES	PERIMETER
1	6-18	24-72
2	12-24	36-72
3	26-40	70-106
4	24-88	60-220
5	48-90	116-216
6	90-144	210-336

Fig. 22-28. Duct aspect ratio. This ratio is determined by dividing width of duct by its height.

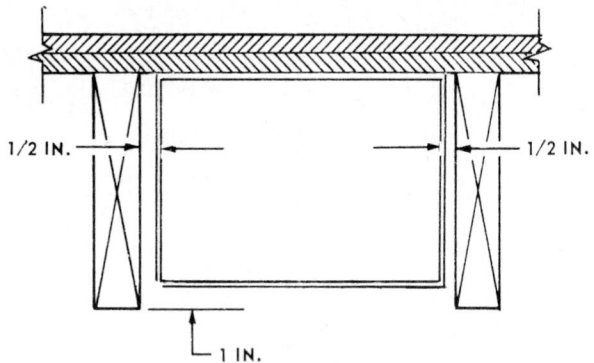

Fig. 22-29. Duct installation between floor joists. Note clearance space allowed between duct and joists.

ROUND DUCT DIAMETER	COMMERCIAL SHEET STEEL GALVANIZED GAUGE	RESIDENTIAL SHEET STEEL GALVANIZED GAUGE
Up to 12	26	30
13 to 18	24	28
19 to 28	22	
27 to 36	20	
35 to 52	18	

Fig. 22-30. Recommended gauge thicknesses for round metal ducts.

RECTANGULAR DUCTS WIDE SIDE INCHES	COMMERCIAL		RESIDENTIAL
	SHEET STEEL GALVANIZED	ALUMINUM	SHEET METAL GALVANIZED
Up to 12	26	.020	28
13—23	24	.025	26
24—30	24	.025	24
31—42	22	.032	
43—54	22	.032	
55—60	20	.040	
61—84	20	.040	
85—96	18	.050	
Over 96	18	.050	

Fig. 22-31. Prescribed gauge thicknesses for rectangular metal ducts.

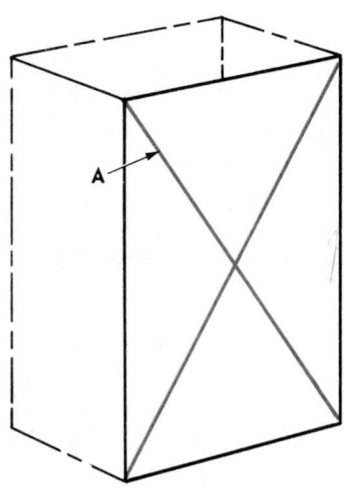

Fig. 22-32. Break lines (ridges) are formed on large sheet metal panels or large ducts to make them more rigid. A—Typical break line.

22-11 DUCT SIZES

Before determining the size duct needed to carry air to a room, one must know the volume of air that is to be delivered. This volume depends on the amount of heat the air must deliver to the room during the heating season and the heat removed in the cooling season.

Normally these calculations are not difficult. However, conditions such as split heating systems (those where the same duct is delivering both heated and cooled air) complicate the problem. (A split system can use radiators or convectors, or use hydronic baseboards. It can have a separate mechanical ventilation system.) *The amount of air delivered to a room must always equal or exceed the minimum fresh air ventilation requirements.*

To reduce duct size and to save space in a building, smaller size ducts are now being used. Such ducts operate with about twice the normal air velocity. This increase in pressure and velocity requires more powerful fans. In turn, there is more noise.

The size of the ducts must be based on the heating needs in the colder climates and on the cooling needs in warmer parts of the country. Where both heating and cooling are required, the same duct system serves both. To compensate for the higher air volumes needed for air conditioning, the capacity of the system is increased by increasing the fan speed about 20 percent.

or joist space. There must be room for fittings and allowance for an out-of-alignment condition. It is better to make the ducts 1/2 in. smaller both ways. The recommended clearance varies with the type of system. Ducts from solar collectors (or heat exchangers) have more clearance. Fig. 22-29 illustrates a satisfactory duct installation between floor joists.

Fig. 22-30 gives the recommended gauge thicknesses for various sizes of round ducts. Fig. 22-31 lists recommended gauge thicknesses for various size rectangular ducts. Ducts going through floors should be protected on the corners by angle iron. Large sections of sheet metal or ducts must be cross broken. This reduces panel vibration. See Fig. 22-32.

The use of controlled fresh air in a building is increasing. It has been found that natural infiltration is unreliable and too variable. It is very necessary that this fresh air be filtered. The outdoor intake end of the fresh air duct should be fitted with a strong screen and bar combination to keep out debris, birds, animals, and insects, and prevent illegal entry.

22-12 AIR VOLUMES FOR HEATING

If the furnace is the only source of heat for a room, only three things need be known to be able to calculate the air volume:
1. Heat load.
2. Room temperature.
3. Duct temperature.

The heat load can be determined by methods described in Chapter 26. The room temperature is decided by the designer. Normally, the temperature is 72 °F (22 °C) dry bulb temperature (dbt).

The duct air temperature is more difficult to decide. If a low duct temperature is used, large air volumes will be necessary to carry enough heat. If high duct temperatures are used, the furnace will have to operate with higher chimney (stack) temperatures, and the ducts may have to be insulated.

Engineers recommend that the grille temperatures be at least 125 °F (52 °C) and that the duct air temperature be near 140 °F (60 °C). The lowest temperature at which these results are possible depends on the lengths of the ducts. Knowing that the specific heat of air is 0.24 Btu/lb. °F, the weight of air needed is easily found by using the specific heat equation:

Room heat load = 0.24 × wt. of air × the temperature difference.

For example:

What is the weight of air needed for one hour if a room has a heat load of 20,000 Btu/hr.?

With a room temperature of 72 °F (22 °C) and a duct temperature of 140 °F (60 °C), the temperature difference is 68 °F (38 °C).

$$20,000 = 0.24 \times \text{wt. of air} \times 68$$

$$\frac{20,000}{0.24 \times 68} = \text{wt. of air}$$

$$\frac{20,000}{16.32} = \text{wt. of air}$$

1225.5 lb. = wt. of air per hour
Divide by 60 to obtain lb./min.
20.425 lb. = wt. of air per min.

The air volume must be determined before proceeding. To find the volume, first find the volume of one pound of air at the room and duct temperature.

If chart does not read as high as 140 °F (60 °C), this volume is calculated using Charles's Law (Chapter 1), knowing volume at 72 °F (22 °C) dbt and 50 percent relative humidity (13.55 cu. ft.). Obtain this value from the psychrometric chart, Fig. 18-21. One lb. of air has volume of 15.28 cu. ft. One kilogram of air has volume of 0.9539 m³.

$$\frac{V_o}{V_m} = \frac{Fa_o}{Fa_m}$$

$$\frac{13.55}{\text{Vol.}} = \frac{460 + 72}{460 + 140}$$

$$\frac{13.55}{\text{Vol}_m} = \frac{532}{600}$$

$$13.55 \times 600 = \text{Vol.} \times 532$$

$$\frac{13.55 \times 600}{532} = \text{Vol.}$$

$$\frac{8130}{532} = \text{Vol.}$$

15.28 = Vol. (15.294 if convert from Para. 22-2)
The volume of air/min. is 20.425 lb./min. × 15.28 cu. ft.
Vol. = 312.094 cu. ft./min. (8.8 m³/min.). The reason the calculation is done this way is as follows: 15.28 cu. ft. is the volume of 1 lb. of air at 140 °F (60 °C) (one may say it is 15.28 cu. ft./lb.). Any physics or chemistry textbook says that multiplying a number given in lb./min. by a number given in cu. ft./lb. produces a number given in cu. ft./min. The "lb." above the line has cancelled the "lb." below the line. Therefore: 20.425 lb./min. × 15.28 cu. ft./lb. = 312.094 cu. ft./min.

Now, to determine the duct size, two separate items must be considered. If the space is limited, the area of the duct is already fixed. For example, if the duct is to run between studs in a partition, the space available is no more than 14 × 2 1/3 in. (This fits between 2 × 4 studding on 16-in. centers.) Such a duct has an area of:

14 × 3 1/4 = 45.5 sq. in. or 293.6 cm²
Convert this to square feet, dividing by 144, thus:

$$\frac{45.5}{144} = 0.316 \text{ sq. ft. or } 0.029 \text{ m}^2$$

Using the volume of 312.09 cu. ft./min.:

The velocity in ft./min. multiplied by 0.316 sq. ft., must provide a volume of 312.09 cu. ft./min.

Since velocity is not known, it must be found by dividing volume by area, thus:

$$\text{Velocity} = \frac{312.09}{0.316} = 988 \text{ ft./min. or } 301 \text{ m/min.}$$

Since this velocity would produce air turbulence noise, two ducts measuring 14 × 3 1/4 in. must be used.

The resulting velocity will now be 494 ft./min. (151 m/min.). This should be satisfactory.

22-13 AIR VOLUME FOR COOLING

To get the air volume needed, it is better to use large ducts, less air pressure (velocity) and, therefore, less power. Duct cost is a one-time cost, but a larger fan and motor mean continuous higher power costs. The higher velocity needed with smaller ducts also creates more noise.

A short method for determining air volume is to use 1 cfm for each square foot of floor space, excluding basement. If a home has 1500 sq. ft. (139 m²), based on outside dimensions, the fan capacity should be 1500 cu. ft. (43 m³) per minute.

To determine the cooling load, consider that the typical uninsulated home may need 12,000 Btu of cooling per hour (1 ton) for each 400 sq. ft. (37 m²) of floor space. The

1500 sq. ft. (139 m²) should then use $\frac{1500}{400} \times 12,000$

Btu = 45,000 Btu/hr. (4 ton).

Number of air changes are important; there should be 6 to 10 per hour. The volume of the 1500 sq. ft. room with an 8-ft. ceiling is 1500 × 8 = 12,000 cu. ft. (340 cu. m). The 1500 cu. ft./min. fan will move 1500 × 60 min./hr. = 90,000 cu. ft./hr.

$$\frac{90,000 \text{ cu. ft./hr.}}{12,000 \text{ cu. ft.}} = 7.5 \text{ air changes per hour.}$$

Return air ducts must be as efficient as the delivery air ducts. The return air is warmer, is at a lower pressure, and, therefore, occupies more volume. Return duct work should be about 20 percent larger in cross-section area than the delivery duct because it is induced (low pressure) flow.

If the delivery duct to a room has a cross section of 30 sq. in. (194 cm²), the return air duct should be 30 + (30 × 0.20) = 30 + 6 = 36 sq. in. (232 cm²) or 30 × 120 percent = 30 × 1.2 = 36 sq. in.

The air volume for cooling is calculated in much the same way as for heating. Use the same specific heat equation. (See Chapter 1 and Para. 22-12.)

Btu = sp. ht. of air × wt. of air × (temp. diff.).

Knowing heat load, specific heat of air, and temperature difference, the weight of air needed can be determined. If the weight is known, the air volume can be determined and the duct sizes selected.

For example, suppose a computer room has only 400 sq. ft. (37 m²). Recall that a room with 400 sq. ft. has a heat load of one ton of cooling (12,000 Btu/hr.). With a specific heat value for air equal to 0.240 Btu/lb. °F, the air temperature in the duct and the air temperature in the room must be determined. If the designer wants 75 °F (24 °C) in the room and the comfort cooling unit is designed for 65 °F (18 °C) duct air temperature:

$$12,000 \text{ Btu/hr.} = 0.24 \times \text{wt. of air} \times$$
$$\text{(diff. between 75 °F and 65 °F)}$$

$$12,000 \text{ Btu/hr.} = 0.24 \times \text{wt. of air} \times 10$$

$$\frac{12,000}{0.24 \times 10} = \text{wt. of air}$$

$$\frac{12,000}{2.4} = \text{wt. of air}$$

$$5000 \text{ lb./hr.} = \text{wt. of air}$$

$$\frac{5000}{60} = 83.33 \text{ lb./min. (37.8 kg/min.)}$$

At 60 °F (16 °C) db and 60 percent relative humidity (from the psychrometric chart), 13.4 cu. ft. (0.38 m³) of air weighs 1 lb.

To change 83.3 lb./min. to cfm:
Cu. ft. per min. = 83.3 × 13.4 = 1120 cu. ft./min.
or (32 m³/min.)

To determine the duct size for this volume of air moving into the room at 500 ft./min., refer to Figs. 22-33 and 22-34. Locate the point where the 1120 cu. ft./min. line crosses the 500 ft./min. velocity line. This point will show the several alternative sizes:

1. A 20 in. (51 cm) round duct (at 0.018 in. of friction loss). Note: 20.3 must be rounded down to the next duct size that is available.
2. A 22 × 16 in. (56 × 41 cm) rectangular duct (using conversion table in Fig. 22-11).
3. A 28 × 13 in. (71 × 33 cm) duct.

22-14 AIR CIRCULATION

In warm air heating, three basic systems are used to circulate the air:
1. Gravity.
2. Intermittent forced air.
3. Continuous forced air.

The gravity system is no longer popular. Too much energy is lost before the air gets to the room being heated.

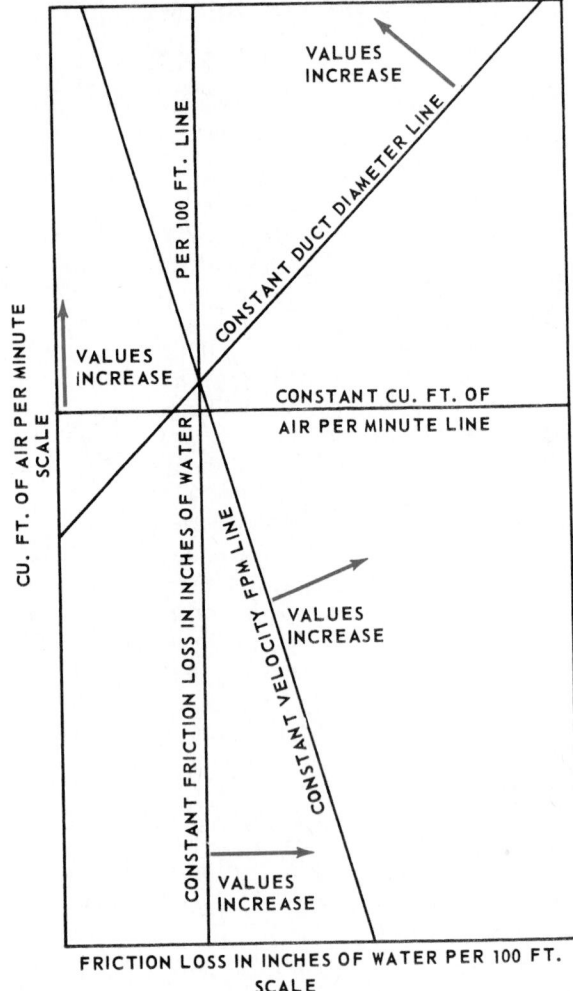

Fig. 22-33. Diagram shows how to read value lines in Fig. 22-34. Air volume per minute is read off scale on left-hand side. By locating air volume per minute lines, one can find duct diameter, velocity, and friction by following along proper lines to scales on edges.

Most installations use the intermittent forced air system. A thermostat in the furnace plenum chamber is used to control the fan.

Becoming more popular is the continuous blower system. It provides a more constant temperature in rooms.

Systems designed to provide cooling as well as heating need additional capacity to move air. This is because a cubic foot of cooled air will not change the room temperature as much as the same amount of warmed air.

The reason for this is the temperature difference. The warmed air comes out of the duct many degrees warmer than the room air it is replacing; the cooled air is not all that much cooler than the room air it is replacing. Therefore, greater quantities of it need to be moved into the room to get the desired effect. Also, the cooled air has a high relative humidity. People have difficulty getting cool when the relative humidity is high. More airflow is needed to cool than to heat.

The high volume of cooled air needed is somewhat offset by the lower cooling load. In average conditions, a 30 to 50 percent airflow increase is required. Either of two methods will increase the airflow when needed:

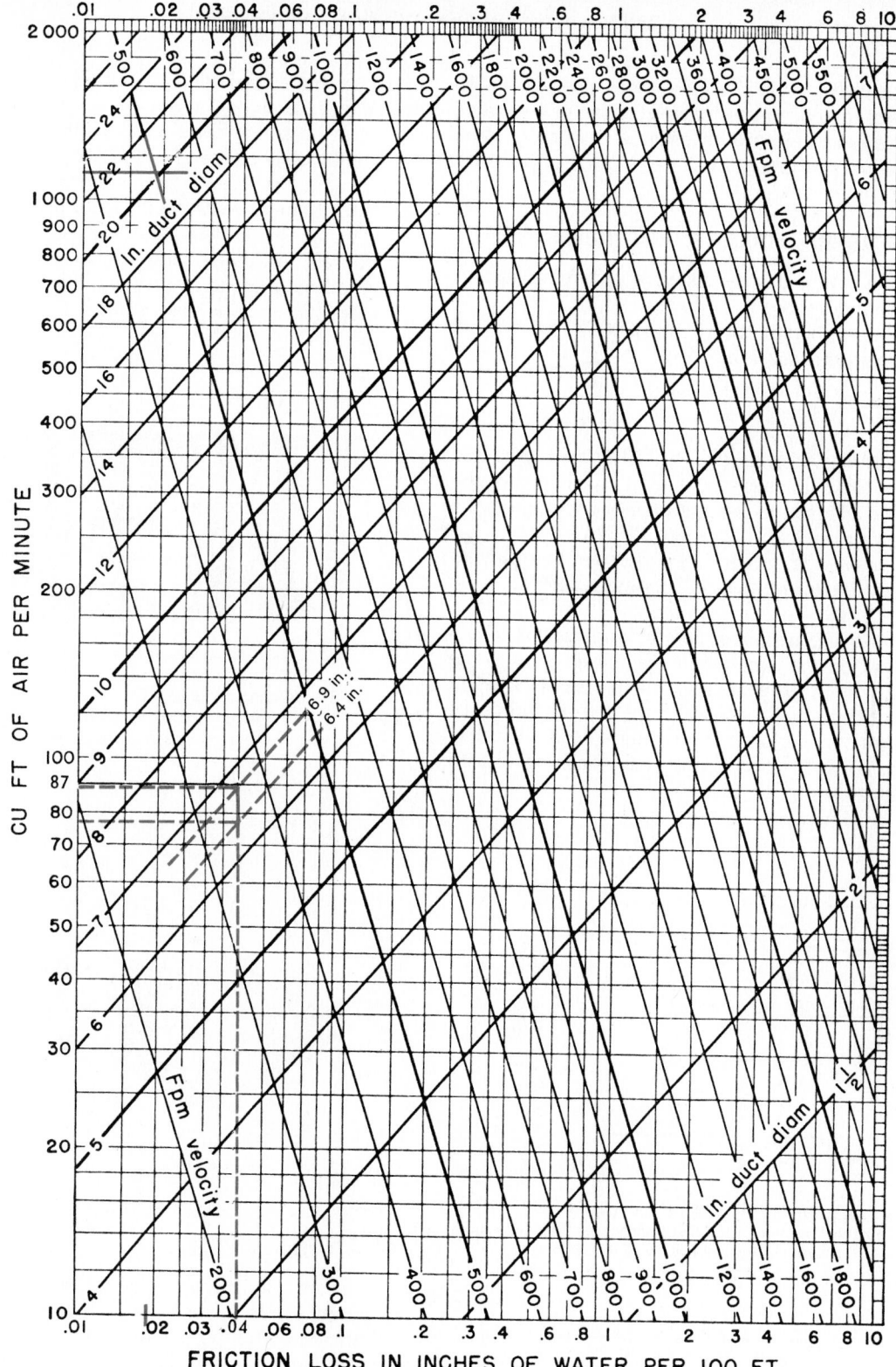

Fig. 22-34. Friction chart for low volumes of 10 to 2000 cfm airflow in ducts. (ASHRAE Guide and Data Book)

1. Use a two-speed blower motor (for directly-driven blowers).
2. Install a two-speed pulley on the motor if a belt-driven unit is used.

22-15 ROOM AIR MOVEMENT

Air entering a conditioned space through ducts must circulate throughout the room without causing annoying drafts. This depends on the number and size of the air inlet grilles as well as on the velocity (speed) of the air moving through them.

Air coming into the room pushes against and mixes with air already in the room. *Combined air that moves at 150 ft. per minute is called primary air. The distance the air from the grille travels before it slows down to 50 ft. per minute (terminal velocity) is called the throw. The outlet velocity is the speed of the duct air as it leaves the grille.*

The overall size of the grille is not important. The total

area of the air openings in the grille determines the grille capacity. The spread of the air that leaves the grille is very important. The return air grille or grilles should be located where the slowest movement of air takes place in the room (stratified air).

22-16 DIFFUSERS, GRILLES, AND REGISTERS

Room openings to ducts have several different devices:
1. To control the airflow.
2. To keep large objects out of the duct.
These opening devices are called:
1. Diffusers.
2. Grilles.
3. Registers.

Diffusers deliver widespread fan-shaped flows of air into a room. Some diffusers cause the duct air to mix with some room air in the diffuser. Grilles are usually used as covers for the return air duct and have no adjustments.

Registers are used to deliver concentrated airstreams into a room. Many have one-way or two-way adjustable air stream deflectors.

Grilles control the distance, height, and spread of air-throw, as well as the amount of air. Grilles offer some resistance to airflow. Grille cross-section pieces block about 30 percent of the air. For this reason—and also because they must use slow air movement to reduce noise—the duct cross section is usually enlarged at the grilles. Fig. 22-35 shows a typical warm air grille.

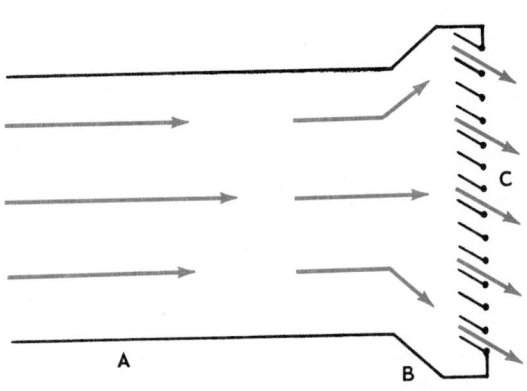

Fig. 22-35. Typical warm air grille. Many are supplied with controls for adjusting size of opening, and to redirect airflow. A—Duct. B—Enlarged duct. C—Grille with adjustable vanes.

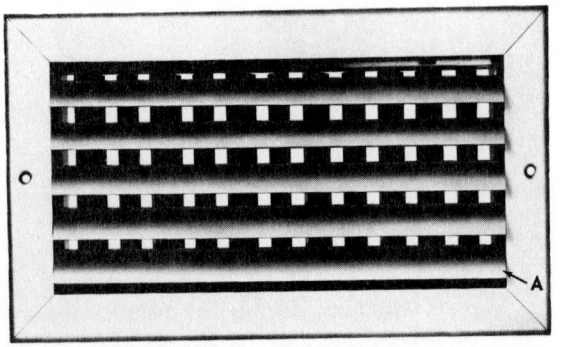

Fig. 22-36. Rectangular grille. A—Adjustable airflow direction vanes. (Hart & Cooley, Div. of Clevepak)

Grilles have many different designs. Some are fixed and can direct the air only in one direction. Others are adjustable and can be set to send air in different directions.

A rectangular grille with adjustable direction airflow vanes is shown in Fig. 22-36. Some grille designs have air vanes to direct the airflow three different directions at one time. See Fig. 22-37. Others are made with adjustable dampers, horizontal flow vanes, and vertical flow vanes. A ceiling-mounted rectangular air diffuser, Fig. 22-38, directs air in four directions along the ceiling.

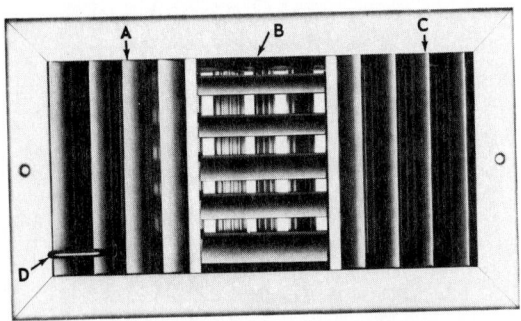

Fig. 22-37. Rectangular grille with adjustable airflow vanes. A—Vanes sweep flow in horizontal direction. B—Vertical flow. C—Horizontal flow. D—Adjustment.

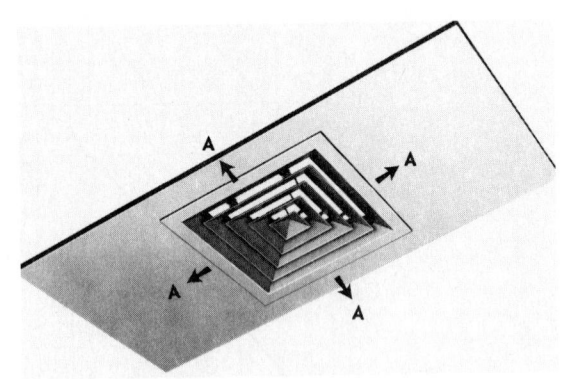

Fig. 22-38. Ceiling-mounted air diffuser. A—Air flows four ways.

22-17 DAMPERS

Without some way of controlling airflow in forced air systems, some spaces would receive too much air while others would not get enough. One method of getting even air distribution is through the use of duct dampers.

These dampers balance airflows or they can shut off or open certain ducts for zone control. Such a damper is shown in Fig. 22-39. Some are located in the diffuser or grille, some are in the duct itself. They are of three types:
1. Butterfly.
2. Multiple blade.
3. Split damper.

All three types are shown in Fig. 22-40. There are two types of multiple blade dampers: the parallel blade type, as in Fig. 22-41, and the opposed blade type, Fig. 22-42. A detail drawing of the parallel blade design is shown in Fig. 22-43. When installing a damper, always draw a line on

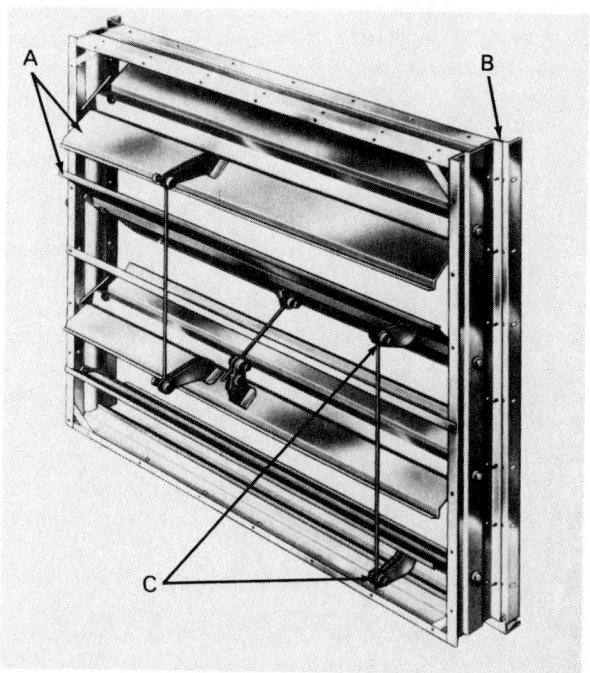

Fig. 22-39. Four-vane adjustable damper controls airflow in duct. These dampers may be operated either manually or automatically. A—Damper vanes. B—Frame attached to duct. C—Linkage. (Honeywell, Inc.)

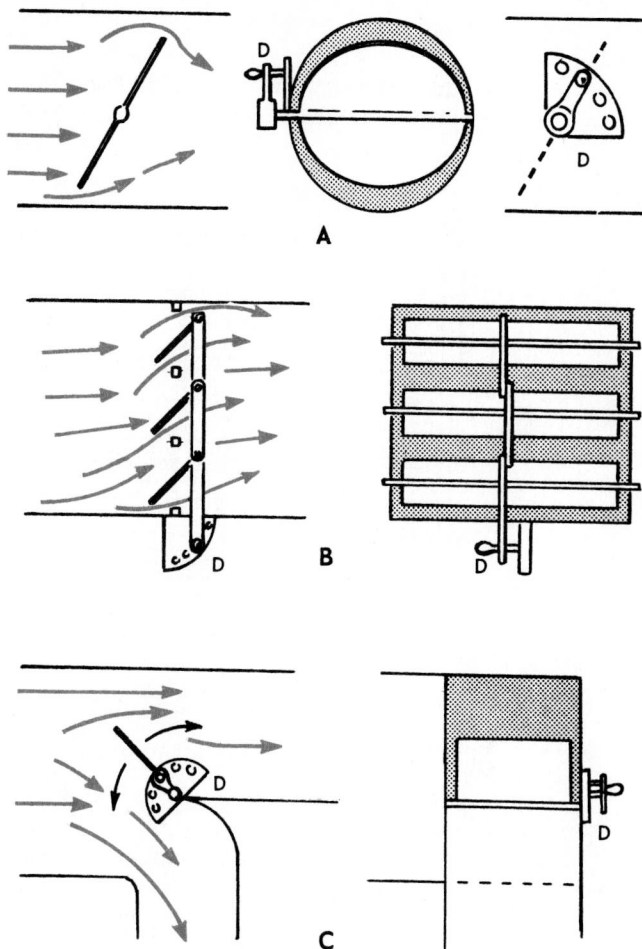

Fig. 22-40. Three types of duct airflow controls. These dampers adjust airflow volumes to help service technician balance system. A—Butterfly damper. B—Multiple vane damper. C—Split damper. D—Adjustment handle.

the end of the damper shaft that extends out of the duct to show the position of the damper. See Fig. 22-44. Labels may help show the position of the damper blades.

For accurate air control, these dampers should be tight-fitting with minimum leakage. Many are automatically controlled for either zone heating or cooling. Automatic controls are also used to mix two airflows for either fresh air and recirculated air mixes, for humidity control, or for

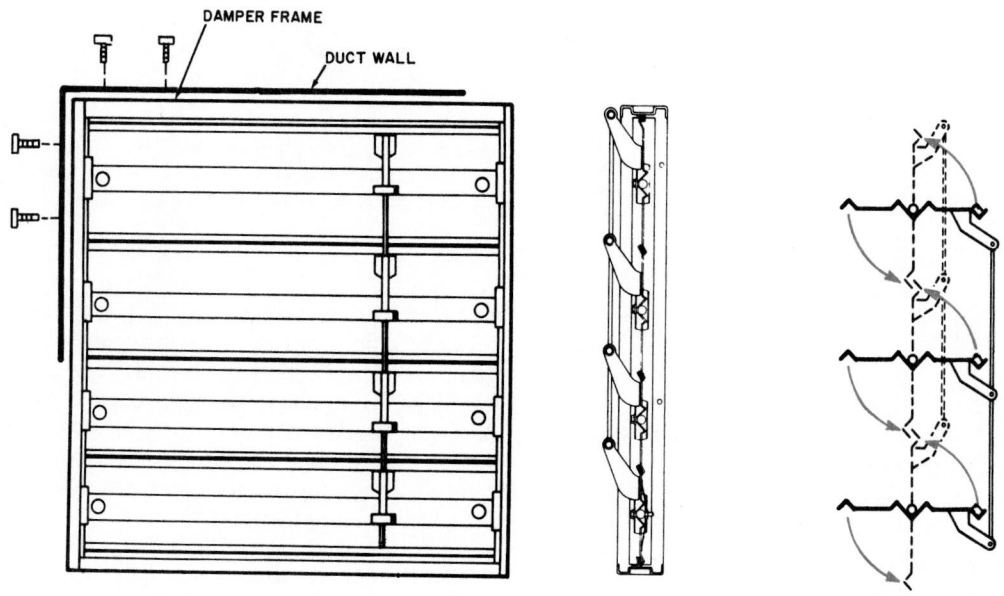

PARALLEL BLADE ACTION

Fig. 22-41. Construction detail of parallel blade damper. (Honeywell, Inc.)

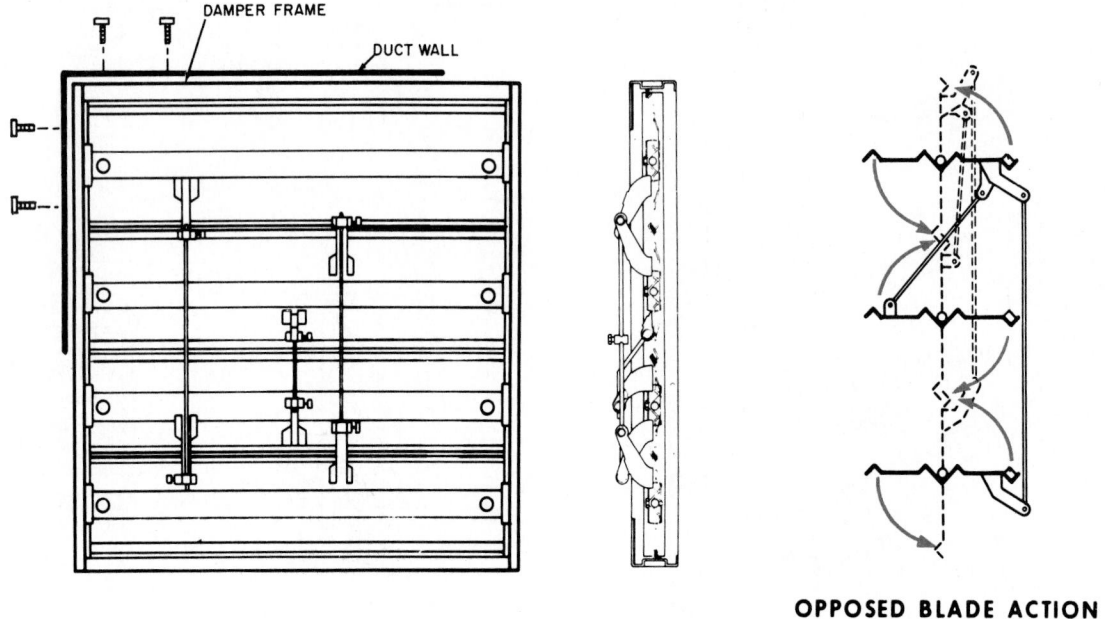

OPPOSED BLADE ACTION

Fig. 22-42. Opposed blade damper with blades open. (Honeywell, Inc.)

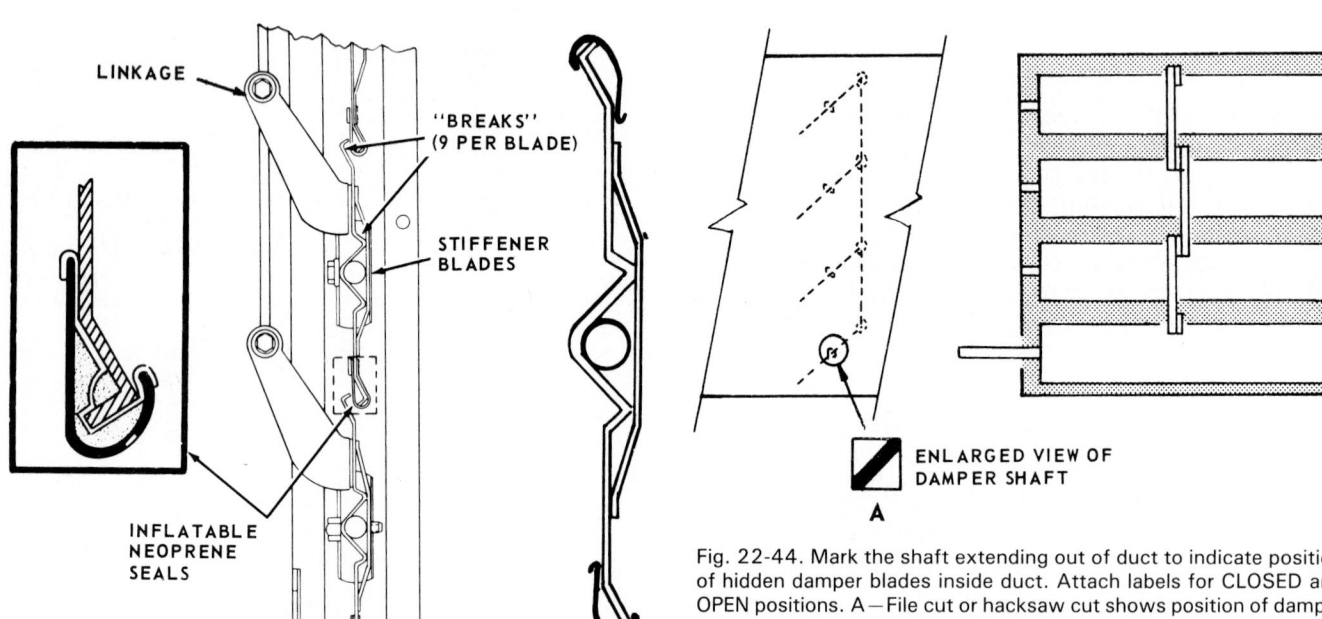

Fig. 22-43. Damper blades are of parallel design. (Honeywell, Inc.)

Fig. 22-44. Mark the shaft extending out of duct to indicate position of hidden damper blades inside duct. Attach labels for CLOSED and OPEN positions. A—File cut or hacksaw cut shows position of damper blades.

temperature control. Fig. 22-45 is a diagram of typical damper installations. These multiple dampers are interlocked by controls to provide different mixes and to maintain correct total airflow.

22-18 FIRE DAMPERS

Automatic fire dampers should be installed in all vertical ducts in commercial and industrial buildings. Ducts, especially vertical ones, will carry fumes and flames from fires. These dampers should be inspected and tested at least once a year to be sure that they are in good operating condition.

Ducts going into or through a fire wall must have fire dampers. These openings are rated as follows:
1. Class A—opening will hold back a fire indefinitely.
2. Class B—fire dampers may be used when a two- to four-hour hold is required. This class is approved for most general installations.
3. Class C—fire dampers are used where a one-hour hold is required.

The following dampers are fail-safe units:
1. Spring-loaded to close.
2. Weight-loaded to close.

Vertical shafts serving two or more floors must be enclosed in a fire partition and fire dampers must be used. Fig.

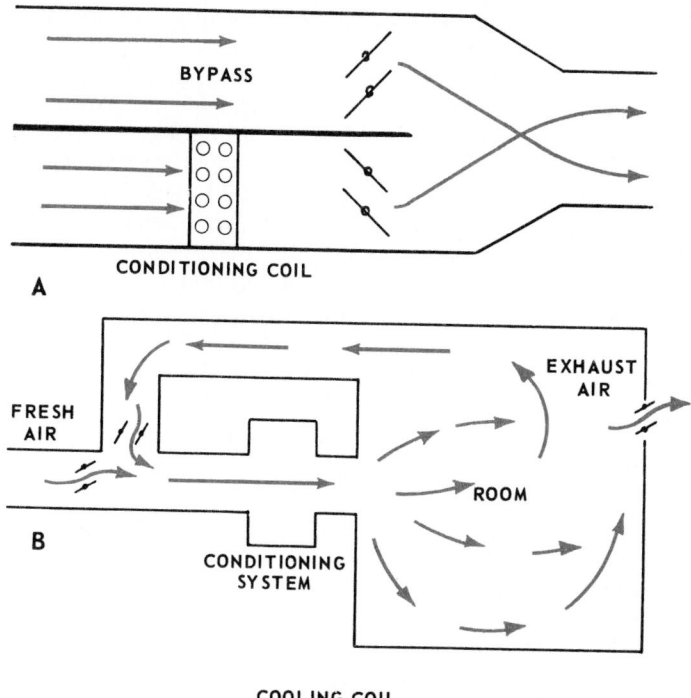

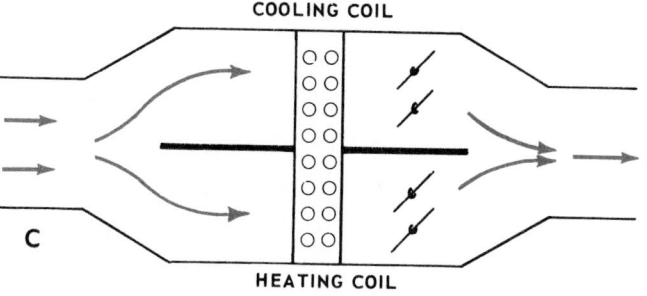

Fig. 22-45. Dampers control airflow. A—Bypass to control temperature and relative humidity. B—Dampers to control fresh air, exhaust air, and bypass. C—Controlling air over either cooling or heating coil.

Fig. 22-46. Fire damper in closed position. A—Blades. B—Duct frame. (Air Balance Inc.)

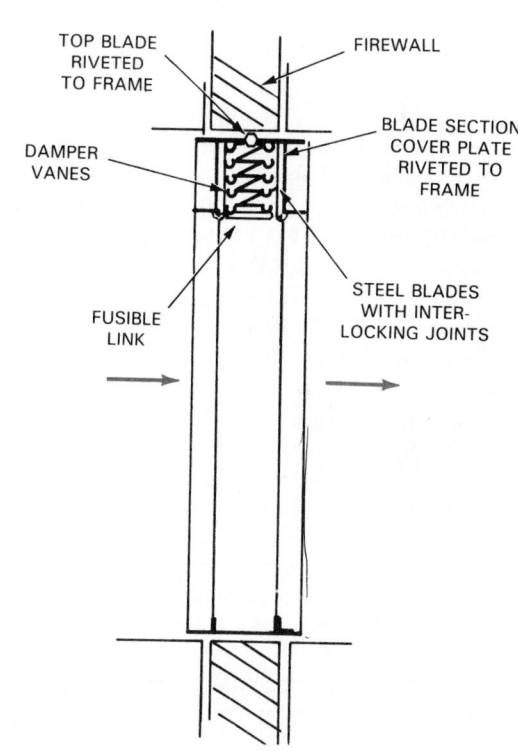

Fig. 22-47. Fire damper in open position. If link is heated, it will melt and vanes will fall, closing duct.   (Air Balance Inc.)

22-46 shows a fire damper. Ducts of less than 20 sq. in. (129 sq. cm) area do not require a fire damper.

Fire dampers are usually held open by a fusible link. Heat will melt the link and damper will close either by gravity, weights, or springs. See Fig. 22-47. Some fire dampers have electronic sensors which operate a closing mechanism. In these dampers, the damper blade latches operate either pneumatic (compressed air) or electric power devices which close the damper.

Smoke dampers use a photoelectric device to detect smoke. An electronic device will trip a holding device and the damper will close.

Before making a duct installation, local smoke and fire damper regulations must be checked.

22-19 DUCT CALCULATIONS

In a heating or cooling system where a duct serves more than one room, the duct must be designed so that each room being served receives the correct amount of air. If the distribution is not balanced, one room will be too warm while another will be too cold.

Two methods used to calculate the proper size plenum chambers, main ducts, branch ducts, and grilles are:

1. Unit pressure drop system.
2. Total pressure drop system.

22-20 UNIT PRESSURE DROP SYSTEM

Air forced through a duct follows the path of least resistance. Many air conditioning duct systems have several openings (grilles) for the air to escape from the duct. A duct with low resistance will allow most of the air to flow through it, while other ducts with higher resistance will not carry the correct amount of air.

in the past, many duct installations were made that fed too much air to some rooms and did not heat or cool other rooms sufficiently.

The unit pressure drop calculating system uses the same pressure drop for each length of duct throughout the system.

For example, suppose that the total heat load during the heating season is 80,000 Btu/hr. and that there are six rooms with heat loads as follows:

1. Living Room   = 25,000 Btu/hr.
2. Dining Room   = 15,000 Btu/hr.
3. Kitchen   = 5000 Btu/hr.
4. Bathroom   = 8000 Btu/hr.
5. Bedroom No. 1 = 15,000 Btu/hr.
6. Bedroom No. 2 = 12,000 Btu/hr.

Referring back to the problem in Para. 22-12, one can work out the air volume needed to heat these six rooms. Recall that the specific heat of air is 0.24 and that the volume of one pound of air is 15.28 cu. ft. Therefore:

cu. ft./Btu for a 68 °F (38 °C) change =

$$\frac{15.28}{0.24 \times 68} = 0.936 \text{ cu. ft./Btu}$$

$$= 0.936 \text{ ([cu. ft./hr.]/[Btu/hr.])}$$

Divided by 60 (to get the answer into minutes) = 0.0156 ([cu. ft./min.]/[Btu/hr.])

Knowing the amount of heat that must be carried to each room per minute, one can work out the air volumes required per minute for each room:

1. Living Room   25,000 × 0.0156 = 390 cu.ft./min.
2. Dining Room   15,000 × 0.0156 = 234 cu.ft./min.
3. Kitchen   5000 × 0.0156 = 78 cu.ft./min.
4. Bathroom   8000 × 0.0156 = 124.8 cu.ft./min.
5. Bedroom No. 1 15,000 × 0.0156 = 234 cu.ft./min.
6. Bedroom No. 2 12,000 × 0.0156 = 187.2 cu.ft./min.

The total air volume is 1248 cu. ft./min.

To determine duct sizes that will handle the air volumes specified in this problem, data must be obtained about airflow. Figs. 22-34 and 22-48 are friction air charts for straight ducts. Values were obtained by research. These charts have four variables:

1. Friction loss in inches of water on the horizontal scale (equal value lines are vertical).
2. Cubic feet of air/min. on the vertical scale (equal value lines are horizontal).
3. Velocity on scale lines that slant down to right.
4. Round duct diameter, which is on scale lines that slant down to left.

To continue with the problem, the main duct must handle 1248 cu. ft./min. To keep the velocity to a low noise level, a friction loss of 0.04 in. water column per 100 ft. (1 mm water column per 30 m) should be used.

On the chart, these two values meet and show that the velocity will be 700 ft./min. (213 m/min.), and the main round duct will be 18 in. (46 cm) in diameter.

Using the same friction loss for the branch ducts, one gets the following round duct sizes:

1. Living Room   = 530 ft./min. and 12 in. dia. (30.5 cm)
2. Dining Room   = 480 ft./min. and 10 in. dia. (25.4 cm)
3. Kitchen   = 350 ft./min. and 7 in dia. (16.3 cm)
4. Bathroom   = 400 ft./min. and 8 in. dia. (20.3 cm)
5. Bedroom No. 1 = 480 ft./min. and 10 in. dia. (25.4 cm)
6. Bedroom No. 2 = 440 ft./min. and 9 in. dia. (23.0 cm)

(In converting these diameters to metric, remember that 1 in. = 2.54 cm.)

These velocities are reasonably low and the system would work. However, the total pressure drop system is a more accurate method. See Para. 22-21.

Round duct diameters may be changed to rectangular duct sizes using the table in Fig. 22-11. *When changing round duct sizes to rectangular duct sizes, remember that partition ducts cannot exceed 3 1/4 in. deep and 14 in. wide.* All ducts in the basement should have the same depth for better appearance. This also makes concealment of the ducts easier where the basement is used as recreation space or living quarters. In most installations, basement ducts should not exceed 8 in. in depth.

22-21 TOTAL PRESSURE DROP SYSTEM

A more accurate method of calculating proper sizes of ducts is based on having the same total pressure drop from the fan to each outlet. Fig. 22-49 shows the duct system used with the rooms as calculated in Para. 22-20. To keep the various ducts identified, it is good practice to letter each different size of duct.

In Fig. 22-49, suppose that the following air volumes must be carried:

Duct	Air Volumes in cu. ft./min.
A	1404
B	930
C	790
D	526
E	439
F	220
G	220
H	474
I	211

To be sure that the correct air volume leaves each outlet, each duct preceding an outlet must have a correct, equal amount of total air pressure drop.

The method followed is to determine the longest and most complicated duct. This combination includes ducts A, B, C, D, E, and F.

Assume a total pressure drop of 0.04 in. (0.1 cm) of water. This total means that the pressure drop to each room outlet must be 0.04 in. For example, the opening to the bathroom is the shortest overall distance. It must have the same total pressure drop as the longest run through F.

An important part of this duct design is that the bends and elbows must be considered when determining pressure drop. *The pressure drop of one elbow is equal to 10 diameters of the duct.* Assuming there is one large bend above the furnace and that the grilles are located at the 7-ft. level in the room, the total length of duct A, B, C, D, E, and F is approximately:

Elbow (19 × 10)	15 ft.	(4.6 m)
A	12 ft.	(3.8 m)
Elbow (16 × 10)	13 ft.	(4.0 m)
B	2 ft.	(0.6 m)
C	8 ft.	(2.4 m)
D	2 ft.	(0.6 m)
E	4 ft.	(1.2 m)
Elbow (9 1/2 × 10)	8 ft.	(2.4 m)
F	8 ft.	(2.4 m)
Elbow	8 ft.	(2.4 m)

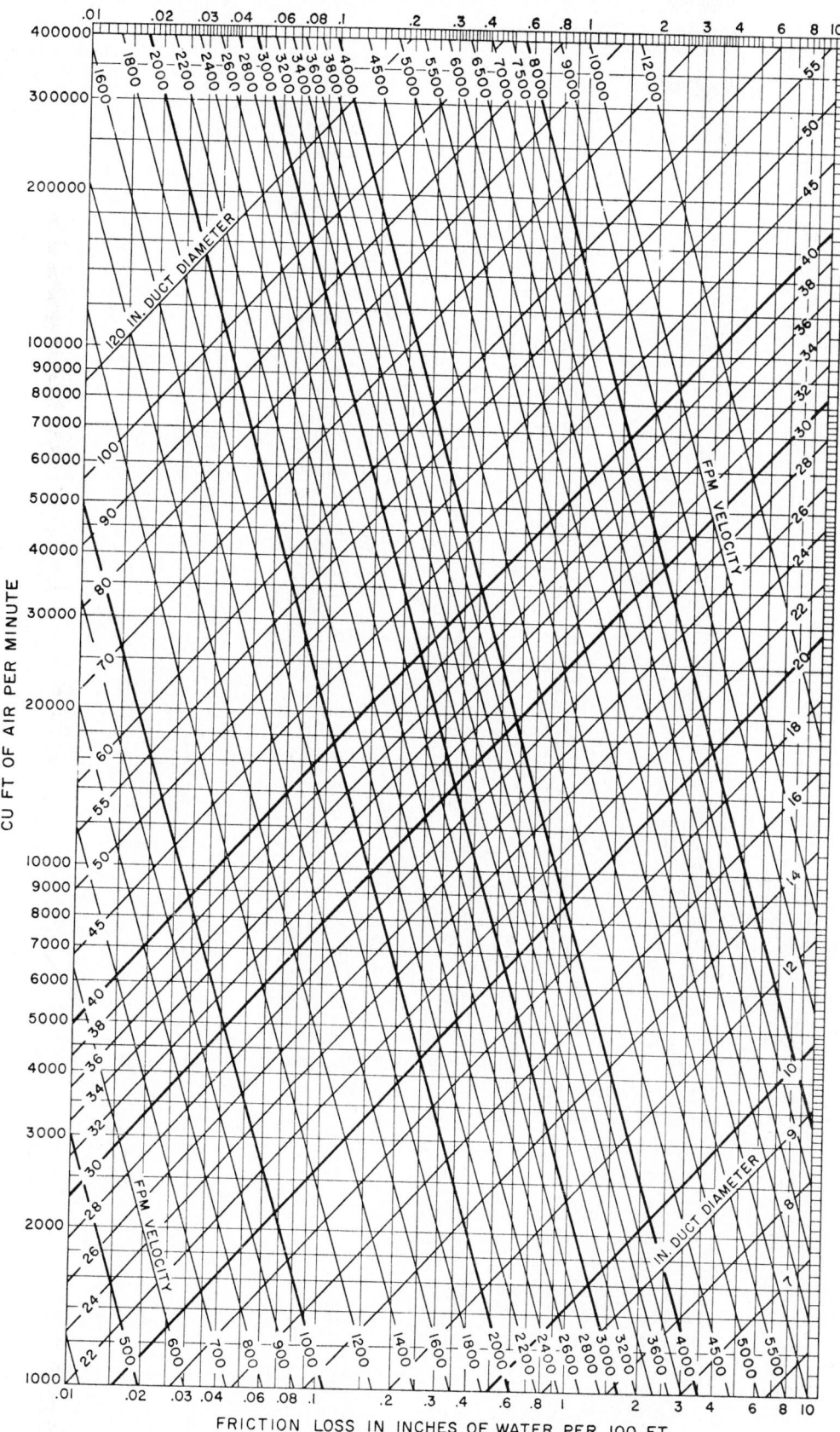

(Based on Standard Air of 0.075 lb per cu ft density flowing through average, clean, round, galvanized metal ducts having approximately 40 joints per 100 ft.)

Fig. 22-48. Friction chart for high-volume airflow in ducts. (ASHRAE Guide and Data Book)

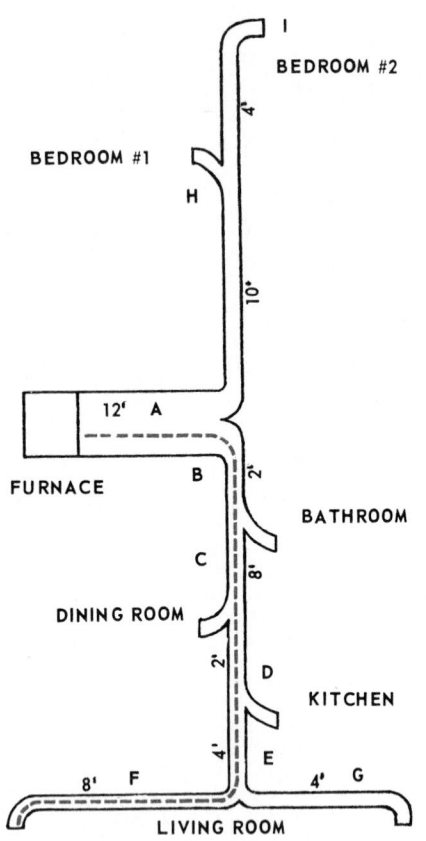

Fig. 22-49. Typical duct installation. Longest air path is shown in red.

The preceding column of numbers is added to find the effect of all of the elbows and straight sections. The total length (equivalent) is 95 ft. Recall that the total pressure drop was 0.04 in. The 0.04 in. drop occurred over 95 ft. This 0.04 must be converted to the pressure drop for 100 ft. If a 95 ft. section gives 0.04, then a 100 ft. section gives more than 0.04. Proceed as follows:

$$\frac{100}{95} \times 0.04 = \frac{20}{19} \times 0.04 = \frac{0.8}{19} = 0.042 \text{ in. (0.107 cm) water}$$

The drop for 100 ft. is 0.042 in. (0.107 cm) of water. If there had been a greater difference in the lengths than that between 95 and 100, the pressure for 100 ft. would have been much greater.

However, more important than the factor related to a 100 ft. length is the pressure drop provided by each section of the longest duct:

Pressure drop in each part equals total pressure drop multiplied by the ratio of length of part to longest equivalent length.

Pressure drop for each part	= 0.04 × $\dfrac{\text{Length of part}}{\text{Total equivalent length}}$
Elbow (19 × 10)	= 0.04 × $\dfrac{15}{95}$ = .0063
A	= 0.04 × $\dfrac{12}{95}$ = .0050
Elbow (16 in. × 10)	= 0.04 × $\dfrac{13}{95}$ = .0055
B	= 0.04 × $\dfrac{2}{95}$ = .0008
C	= 0.04 × $\dfrac{8}{95}$ = .0034
D	= 0.04 × $\dfrac{2}{95}$ = .0008
E	= 0.04 × $\dfrac{4}{95}$ = .0016
Elbow (9 1/2 × 10)	= 0.04 × $\dfrac{8}{95}$ = .0034
F	= 0.04 × $\dfrac{8}{95}$ = .0034
Elbow	= 0.04 × $\dfrac{8}{95}$ = .0034
Vertical rise	= 0.04 × $\dfrac{7}{95}$ = .0029
Elbow	= 0.04 × $\dfrac{8}{95}$ = .0034
Total	= 0.0399 =

0.04 in. (0.1 cm) of water column pressure drop.

Knowing the pressure drop in each part of the longest duct, now determine the pressure loss up to each branch duct. Then, from this value and the length of the branch duct, determine the pressure loss per 100 ft. (30.5 m) for the branch duct.

For example, consider the kitchen duct:

The pressure loss up to the kitchen branch duct is the

Vertical rise	7 ft. (2.1 m)
Elbow	8 ft. (2.4 m)
Total	95 ft. (28.9 m)

It is worthwhile to look at how some of the numbers in the preceding column were obtained. The elbow that is 16 in. in diameter is important for seeing how calculations are done. Section B has a flow of 930 cu. ft./min., which is about 2/3 of 1404. (1404 is the total flow. The total flow passes through section A, the duct that is 19 in. in diameter. The other 1/3 goes to section H, which carries 474 cu. ft./min. Note that 930 + 474 = 1404.) Since the flow through section B is 2/3 of the total, the cross-sectional area of B should be about 2/3 of that for A. The diameter for B is found as follows:

$$B = \sqrt{2/3 \times 19 \times 19} = 15.51 \text{ in.}$$

This value of the diameter for duct B is then rounded to 16 to account for different velocities of air in each duct. ($B = \sqrt{[730 \text{ fpm}/670 \text{ fpm}] \times 2/3 \times 19 \times 19} = 16.2$). By using similar steps for 220 cu. ft./min., the method gives 9 1/2 as a duct diameter ($\sqrt{[730/460]} \times 220/1404 \times 19 \times 19 = 9.5$). The numbers are also obtained from the graph in Fig. 22-34 with 930 cu. ft./min., 220 cu. ft./min., and 0.04 in. of water pressure drop. The 0.04 is used, since each duct acts like it extends back to the furnace, as if it had its own isolated duct wall inside the main duct. (Such ducts would all be parallel inside the main duct, with the total of all the "isolated" ducts adding up to exactly the cross-sectional area of the main duct.) One can imagine each duct extending 100 ft. back, so that it will provide 0.04 in. of loss per 100 ft.

sum of all pressure losses along the way. Thus, 0.0063 + 0.0050 + 0.0055 + 0.0008 + 0.0034 + 0.0008 = 0.0218, which is the total pressure drop.

If the total pressure drop to the outlet at the kitchen must equal 0.04 in. (0.1 cm) water, then 0.0400 − 0.0218 = 0.0182 as the pressure drop in the kitchen branch. Assuming 87 cu. ft./min. (2.5 cu. m/min.) volume (526 − 439 = 87) and 0.04 in. (0.1 cm) pressure drop, the kitchen branch has length equal to the following:

Elbow (6.9 in. × 10)	= 6 ft.	(1.8 m)
Riser	= 7 ft.	(2.1 m)
Elbow (6.9 in. × 10)	= 6 Ft.	(1.8 m)
Total	19 ft.	(5.7 m)

The number 0.04 is used because the kitchen branch acts like an isolated duct extending back to the furnace. One allows a kitchen-to-furnace length anywhere from 52 to 100 ft. The error in making the allowance is only about 11 percent (17 ft. replaces 19 ft.). The graph has been used only to find 6.9 in. for the elbow diameter. The 7 ft. rise is given.

If the pressure drop in 19 ft. is 0.0182, the pressure drop

$$\text{per 100 ft.} = 0.0182 \times \frac{100}{19} = \frac{1.82}{19} = 0.096 \text{ in.}$$

water/100 ft. From the graph, using a volume of 87.73 cu. ft./min. and the resistance of 0.096, the following data is obtained: Size = 5.8 in. (14.7 cm) dia. Velocity = 530 ft./min. (161.5 m/min.).

Compare this method with the unit pressure drop values. Notice how these values differ. One calculates comfort cooling air in about the same way. If the cold air duct is exposed to warm, moist air, condensation on the outside surface of the duct may cause corrosion. Moisture may also drip on structural parts and cause damage. In such cases, ducts should be insulated.

22-22 RETURN AIR DUCTS

Return air ducts are important. Flow of air through these ducts is almost always the result of the "pulling" action of a fan or blower. If the return airflow is not matched with the airflow into a room, the flow of air in cubic feet per minute will not be properly balanced.

If there is more return air than air going into a room, there may be a negative pressure in that room. Thus, more air-in will be used by this room. In turn, other rooms may starve for air. During the heating season, rooms starved for air will be too cold.

Duct return grilles should be placed in the stratified (stagnant) air zone of a room. During the heating season, this area is along the floor and, during the cooling season, this place is near the ceiling. Ideally there should be two places for return air grilles. In all cases, the place is the maximum distance from the air-in grille.

22-23 ELBOWS

Air has inertia. That is to say, air has weight and it obeys Newton's laws of motion. Once set in motion, air tends to continue on in the direction it is moving. In addition, air is compressible and, because of these laws, air in motion has the following characteristics.

It takes energy to make airflow change its direction. The air wants to flow in a straight line. On turns, therefore, it

crowds against the outside, as shown in Fig. 22-50. View A shows a typical elbow. It has a short bend radius. Pressure drop as the air goes through the elbow is about 10 times greater than in an equal length of duct. View C shows a better airflow design but cost of the duct and the room needed to install it make it impractical for many installations. View B is a turbulent air duct design, while view D is the same type of duct with a much better airflow design. The duct elbow at E has vanes placed at the bend. These vanes, located at point F in view E, help reduce the pressure drop.

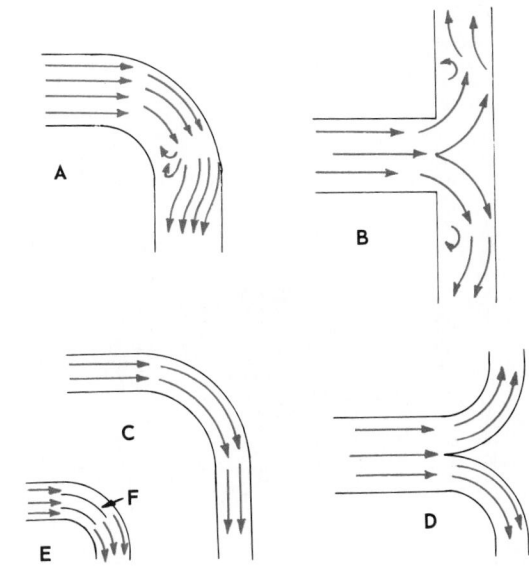

Fig. 22-50. Airflow in some duct bends and elbows. A—Turbulence in air. B—Turbulence in air. C—Smooth airflow. D—Smooth airflow. E—Smooth airflow using vanes at F.

22-24 BALANCING THE SYSTEM

Balancing means sizing the ducts and adjusting the dampers to insure that each room receives the correct amount of air. Conditioned air must be fed in the right amounts to each different room of a multiple room system (home or office). Also the correct amount of air must be returned. If the system is not balanced, rooms will maintain different temperatures; some ducts will be noisy; some will have incorrect relative humidity; and some will have stale air.

To total air balance (TAB) a system, one must measure the air velocity leaving each grille and determine how much "free" area the grille or diffuser has. The "free" area is the actual size of the air openings.

To balance a system do the following:
1. Inspect the complete system; locate all ducts, openings, and dampers.
2. Open all dampers in the ducts and at the grilles.
3. Check the velocities at each outlet.
4. Measure the "free" grille area.
5. Calculate the volume at each outlet.

   Velocity × area = volume

$$\text{fpm} \times \frac{\text{area in sq. in.}}{144} = \frac{\text{cu. ft./min.}}{\text{(volume per minute)}}$$

6. Total the cu. ft./min.
7. Determine the floor areas of each room. Add to determine total area.
8. Find out the proportion each room should have.

$$\frac{\text{Area of room}}{\text{Total floor area}} \times \text{total cfm} = \text{cfm for room}$$

9. Adjust duct dampers and grille dampers to obtain these values.
10. Recheck all outlet grilles.

In some cases, it may be necessary to overcome excess duct resistance by installing an air duct booster. These are fans used to increase airflow when a duct is too small, too long, or has too many elbows. A booster fan is shown in Fig. 22-51.

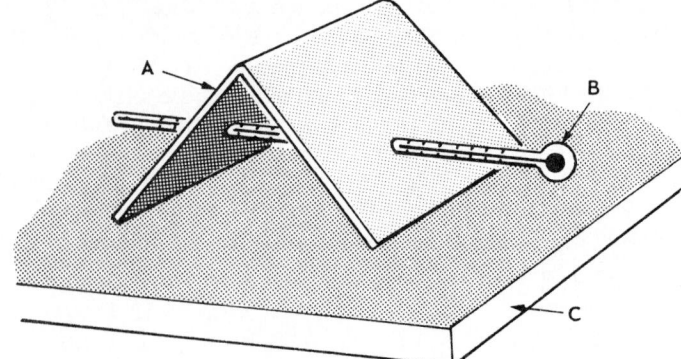

Fig. 22-52. How to use glass stem thermometer to measure room temperature. One is placed in each room. A—Folded heavy paper. B—Thermometer. C—Table top.

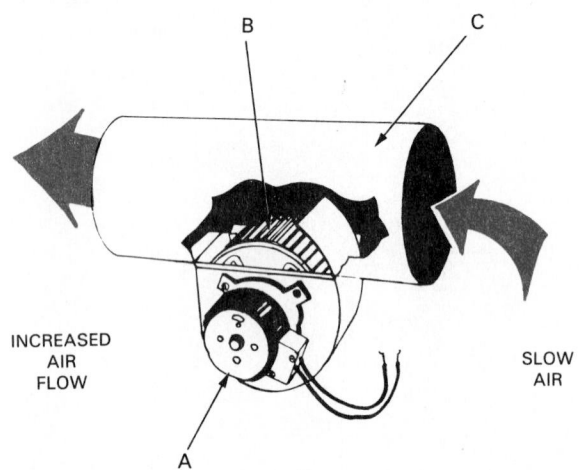

INCREASED AIR FLOW

SLOW AIR

Fig. 22-51. Duct fan increases airflow in air ducts. A—Motor. B—Fan. C—Duct. (Tjernlund Products, Inc.)

An effective but simpler technique can be used to balance airflow to the different rooms:

Mount accurate thermometers in folded cardboard as shown in Fig. 22-52. Place one of these thermometers in each room. Locate them on a table away from sunlight, lamps, or any extra heat source. Adjust dampers until each room has the temperature desired. It is best to allow several hours for the system to adjust to any damper change.

Remember, if one room is too warm, one or more other rooms should be too cool. Close damper to the warm room a little and open dampers a little to the room or rooms that are too cool.

22-25 SPECIAL DUCT PROBLEMS AND DUCT MAINTENANCE

Sometimes ducts need to be painted on the outside for good appearance if the ducts will not be concealed. If the ducts are made of galvanized (zinc-coated) steel, paint will not stick very well. A treatment with vinegar or other weak acid will etch the surface so that paint will stick. If any ducts are outdoors, the weather will etch the surface in one year. The duct may then be painted. Some contractors store ducts outside until they are needed, to make use of natural weathering.

It is best if duct systems have no "dead ends" where dust and stagnant air can collect. If a duct must have a dead space, a duct door for cleaning may be needed. Any section of duct, straight or not, will collect some dust. Reversing the direction of airflow can loosen some of the dust, which can be caught in a filter.

Aluminum-to-sheet-steel joints encourage corrosion. This is due to electro-chemical action between the two different metals. Joints between different metals should be avoided. However, if such a joint must be made, slip a thin strip of zinc or magnesium into the gap between the two different metals (this takes some skill to wedge it into the small gap, which is often only 0.010 in. on each side). The zinc or magnesium will corrode before any damage is done to the aluminum or steel duct.

City and state codes require smoke detectors in residential or commercial structures. Units must be in the proper places to be effective. High-velocity air from ducts can keep them from working. When a long horizontal run of duct ending in an outlet grille is not concealed, the smoke detector should be hung on the underside of the duct, just behind the outlet grille. With other duct designs, such as residential wall, floor, or ceiling grilles, with the duct concealed, smoke detectors should be put as far away from the grilles as possible. Detectors should never be put in corners where there is almost no airflow.

22-26 FANS

Air movement is usually produced by some type of fan. Usually fans are located in the inlet of the air conditioner. Air can be moved by either creating an above-atmosphere pressure (positive pressure) or a below-atmosphere pressure (negative pressure). All fans produce both conditions; the air inlet to a fan is below atmospheric pressure, while the exhaust of the fan is above atmospheric pressure. See Fig. 22-53. *The air feed into a fan is called induced draft and the air exhaust from a fan is called forced draft.*

Fans are constructed of metal and of plastic. There are several types of fans, but the two most popular types are:
1. Axial flow (propeller).
2. Radial flow (squirrel cage).

The construction of the fan shows which type it is. If air flows along the direction the axle is pointing, it is called axial flow. If the flow is at right angles to the axle (radius),

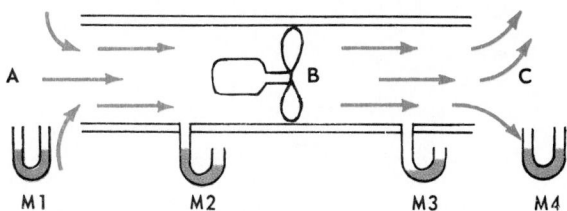

Fig. 22-53. Pressure conditions in simple duct and fan installation. A—Intake. B—Fan and motor. C—Exhaust. M1—Atmospheric pressure. M2—Negative pressure. M3—Positive pressure. M4—Atmospheric pressure.

it is called radial flow. This is shown in Fig. 22-54.

A three-blade axial flow fan is shown in Fig. 22-55. The rotation is clockwise. The axial flow fan is usually direct-driven by mounting the fan blades on the motor shaft. These

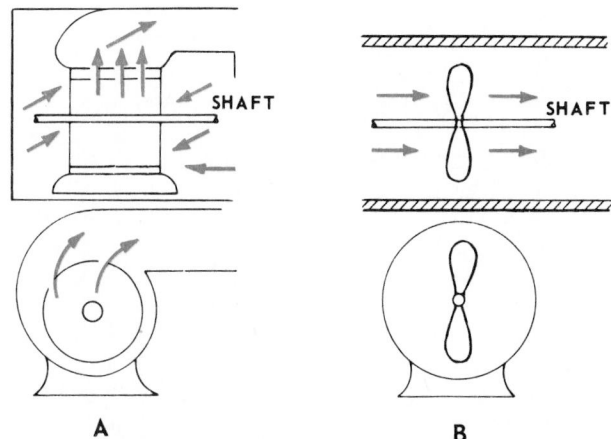

Fig. 22-54. Principal types of fans. A—Radial flow. B—Axial flow.

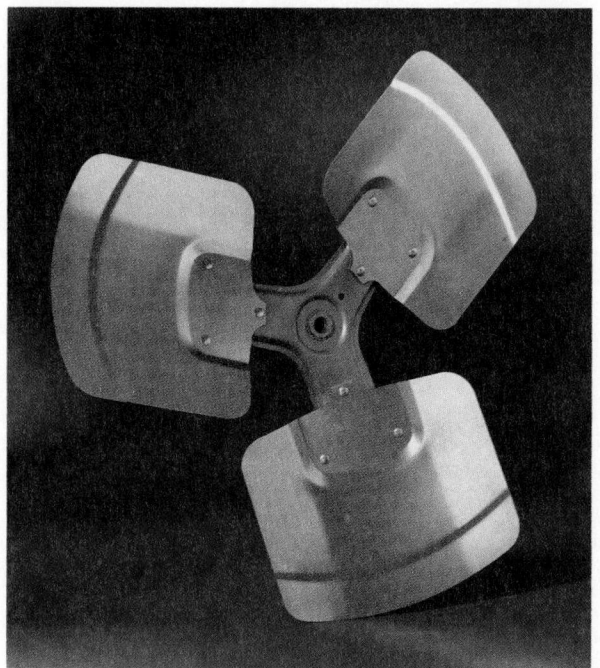

Fig. 22-55. Blade for axial flow fan. (Revcor, Inc.)

fan blades should be handled carefully. If they are bent or twisted, the fan should be replaced.

Part of a radial flow fan is shown in Fig. 22-56. A belt-driven radial flow fan assembly is shown in Fig. 22-57. The radial flow fan is most often used on large installations. It is either directly driven or belt-driven. A belt-driven unit is shown in Fig. 22-58.

Fig. 22-56. Rotor for radial flow fan. (Torrington Research)

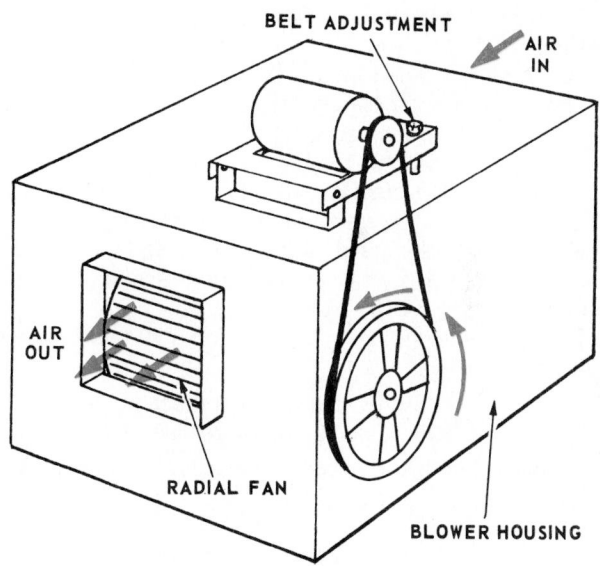

Fig. 22-57. Typical blower housing using belt-driven radial flow fan. Note direction of rotation of fan pulley and belt adjustment.

These units are made in the four classes of air systems which are listed by the Air Movement and Control Association, Inc. (AMCA):

Low pressure, Class I	Up to 3 3/4 in. (9.5 cm) water column total pressure.
Medium pressure, Class II	Up to 6 3/4 in. (17 cm) water column total pressure.
High pressure, Class III	Up to 12 1/4 in. (31 cm) water column total pressure.
High pressure, Class IV	Over 12 1/4 in. (31 cm) water column total pressure.

Fig. 22-58. Radial flow belt-driven fan shown in complete assembly including housing and filters. A—Motor. B—Fan belt. C—Fan.

Centrifugal flow or radial flow fans are made in different designs:
1. Backward-inclined blades (small units).
2. Forward-inclined blades (large units). See Fig. 22-59.
   Static pressure increases as the square of the ratio of change of the cfm.
   1000 to 2000 cfm = two times.
   2×2 (2^2) = four times more static pressure.
This means that the static pressure is the square of the ratio change of rpm. How to measure the pressure of a fan is shown in Fig. 22-60.

To determine the fan capacity for a furnace, use the following formula:

$$cfm = \frac{Btu/hr.\ output\ of\ furnace}{1.0505 \times air\ temperature\ rise\ in\ {}^\circ F.}$$

For cooling, use 400 cfm/ton of capacity. The total pressure

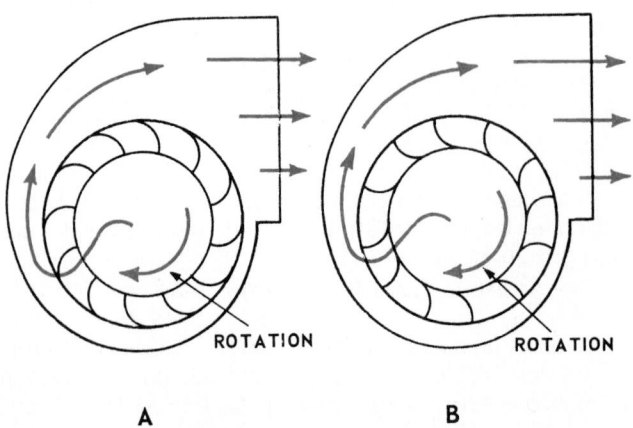

Fig. 22-59. Radial flow fan designs. A—Backward-inclined blades. B—Forward-inclined blades.

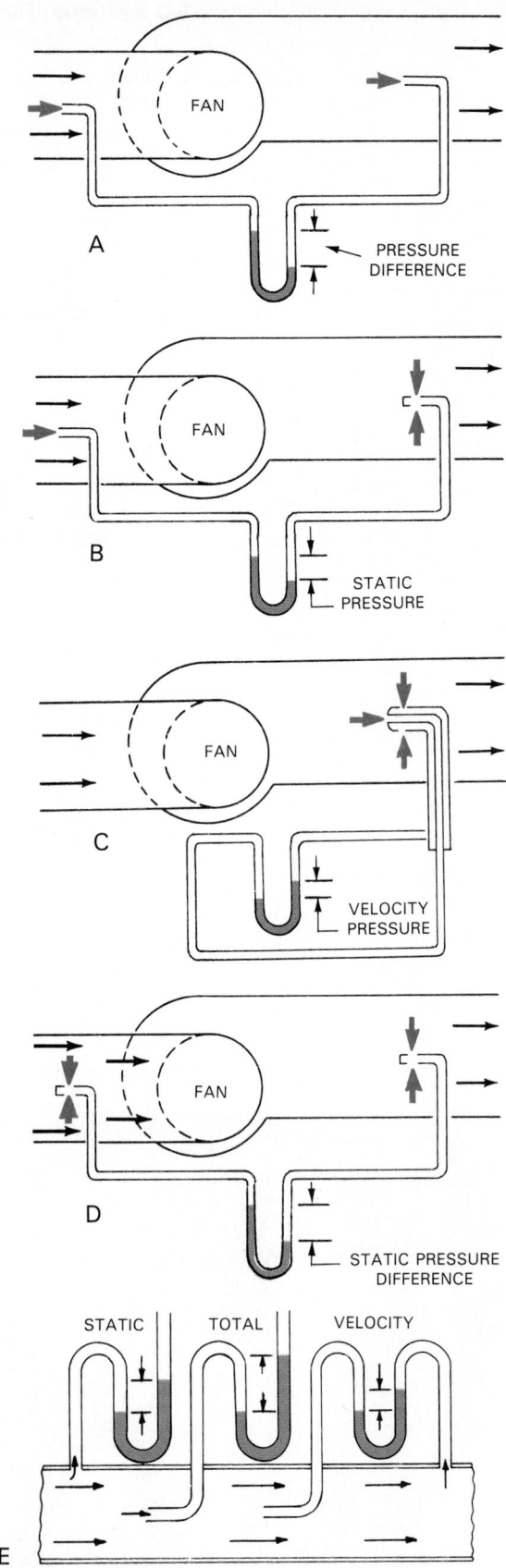

Fig. 22-60. Ways to measure performance of fan. A—Pressure difference across fan. Inlet velocity pressure equals outlet velocity pressure (due to conservation of mass). B—Static pressure. C—Velocity pressure. Pressures are related as follows: C = B − A (A, B, C are the pressures from views A, B, C). Note also that B = C + A, which means that B acts like a total pressure. Refer to Fig. 18-32. 500 ft./min. gives 0.015 in. H_2O. D—Static pressure difference. E—Some technicians measure pressure relative to room air pressure. Note that V = T − S.

drop in the ducts should be about 0.2 in. of water column. (This equals pressure rise across the furnace.) A furnace fan should provide 0.4 to 0.5 in. of water column for furnaces with a cooling unit installed. (This leaves 0.2 in. for all but the cooling unit).

If possible, belt tension should be on the lower belt section to provide more efficient belt drive, as shown in Fig. 22-61. Belt tension is about right when the belt can be pushed out of line a distance equal to its width. It is often desirable to change fan speeds to obtain more or less airflow in cubic feet per second. One way to vary speed is to use adjustable (variable pitch) pulleys. The one shown in Fig. 22-62 is for a two-belt drive.

Any fan which moves air collects lint and dirt, reducing the efficiency of the fan. Dirt should be removed every six months. Remove the fan and scrape, rub, or vacuum the dirt off the blades. Blades will collect dirt even more quickly. If fan bearings are oiled too much, this extra oil coats the fan blades. Each bearing requires but one or two drops of oil each year. See Chapter 7 for more fan motor information.

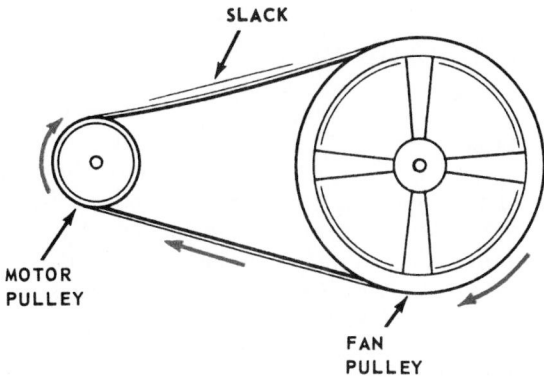

Fig. 22-61. Motor pulley and fan pulley. Note tension on lower part of belt and slack at upper. This is how properly tensioned belt should look in operation. One should be able to push slack area down about 1/2 in. Some technicians allow slack area to go down by an amount equal to width of belt.

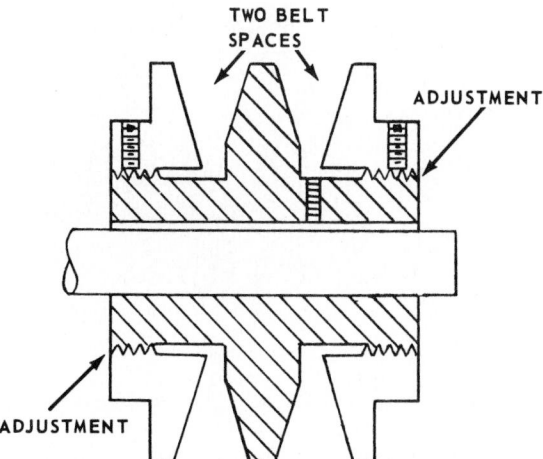

Fig. 22-62. Variable pitch pulley. Note location of setscrews. Pulley is used on dual belt system and can provide fan variation of about 100 rpm through adjustment.

22-27 ATTIC VENTILATION

It is important to properly ventilate the attic space. The air within the attic affects the conditions of the structure. To prevent the attic from having a mildew odor or becoming extremely hot, some type of ventilation is required.

During a hot summer day, it is possible for the temperature of an unventilated attic to reach 150°F (66°C). This would make it difficult to maintain comfortable temperatures in the structure.

A common method of ventilating an attic is to use louvers or vents. This allows the fresh air to enter and the internal air to escape, while preventing water, insects, and other objects from entering.

Many buildings use exhaust fans to remove extremely hot air that collects in attic spaces above insulated ceilings during the summer. Some attic fans may also be used to bring cooler evening air into a building.

These fans have large capacities ranging from 1000 cfm to 4000 cfm. it is usually desirable to have a fan large enough to make a complete change of air in the building every 8 to 10 minutes.

Example: A building measures 30 ft. × 60 ft. with an 8-ft. ceiling.

The volume = Length × width × height
= 60 × 30 × 8
= 1800 × 8 = 14,400 cu. ft.

An exhaust fan of 1440 cfm capacity will change air every 10 minutes.

$$\frac{\text{cu. ft. of space}}{\text{cu. ft./min.}} = \frac{14{,}400}{1440} = 10 \text{ minutes/change.}$$

22-28 BASEMENT FAN

Since basements tend to be cool and damp in the summer, mold and odors are a problem. An exhaust fan will reduce the dampness and mold growth. This fan should remove basement air from the floor level and exhaust it outdoors.

One method of installing such a fan is to remove a basement window pane and install an exhaust fan with an inlet duct leading down to floor level.

22-29 AIR CLEANING

Air pollution is a growing problem as urban areas increase in population and industries expand. Cleaning the air has become an important part of air conditioning.

Air contaminants, as all foreign matter is called, include solids, liquids, gases, and vapors. Efficient air conditioning systems will remove 75 to 95 percent of these contaminants.

Solid particles, kept in suspension in the air by air currents, fall into three general groups:
1. Dust. Usually, this results from wind, a sudden earth disturbance, or from mechanical work on some solid. Dust can have its origin in animal, vegetable, or mineral matter.

    Dust particles are usually over 600 microns in size (about 0.024 in. in diameter). Coal dust particles are usually 1 to 100 microns, while atmospheric dust is 0.001 to 30 microns in size.
2. Fumes. These are formed from materials that are ordinarily solids but have been put into a gaseous state,

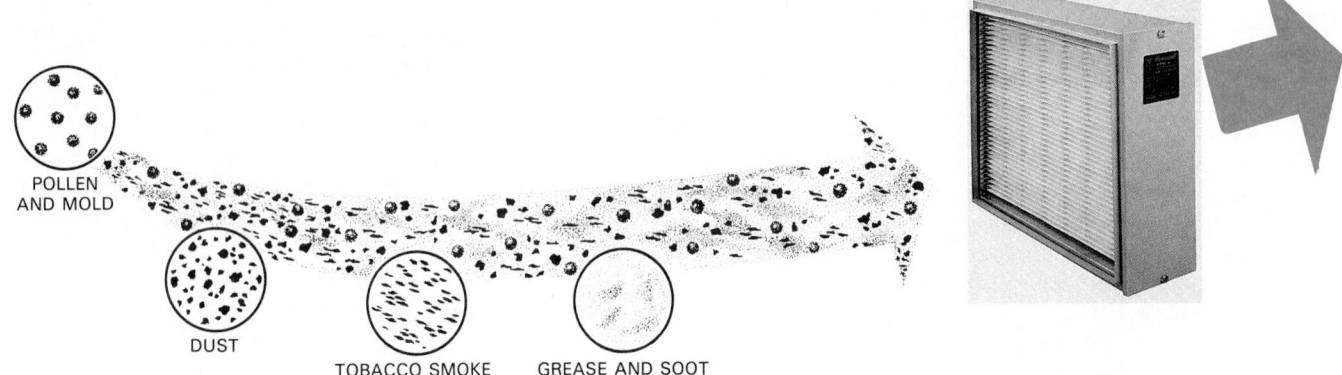

Fig. 22-63. Air cleaner that removes pollen and mold, dust, grease, soot, tobacco smoke, and other foreign air pollutants and irritants. (General Filters, Inc.)

POLLEN AND MOLD

DUST

TOBACCO SMOKE

GREASE AND SOOT

usually by an industrial or chemical process. Such particles are about 1 micron in size.

3. Smoke. Caused by incomplete combustion, smoke consists of solid particles carried into the atmosphere by the gaseous products of combustion. These particles vary in size from 0.01 to 13 microns. Oil and tobacco smoke particles range from 0.01 to 1 microns in size.

Gases are on the increase as air contaminants. They include carbon monoxide, sulphur oxides, nitrogen oxides, and hydrocarbons. These gases form smog when combined with water vapor.

There are also liquid impurities in the air. Two of the most common are:

1. Mists. These small liquid particles are mechanically ejected into the air by splashing, mixing, or atomizing.
2. Fogs. These small liquid particles are formed by condensation. Fogs indicate that the atmosphere has reached the saturation state for that chemical.

A third classification of air impurities act as true gases:

1. Vapors.
2. Gases.

There is little difference between these two impurities. Vapors are gases that have condensing temperatures and pressures close to normal conditions.

Not all impurities are objectionable or harmful. Perfumes and deodorizers have been used for years to either make air more pleasant to smell or to cover up bad odors.

Special applications for air cleaning may be provided for: pollen (10 to 1000 microns), bacteria, and mold. The best air conditioning practice is to clean the air of these materials.

Pollen grains come from vegetation growth such as weeds, grasses, and trees. Their presence in the air is usually responsible for hay fever, rose fever, and other respiratory conditions. These particles vary in size from 10 to 15 microns. Spores vary in size from 10 to 30 microns.

Bacteria are microorganisms responsible for the transfer of many diseases. Many manufacturing processes require the removal of these bacteria. Hospital rooms and some refrigerators use bacteria-removing devices.

Air may be cleaned in many ways, depending on the foreign matter contaminating it.

1. To remove solids such as dust, soot, and smoke, one may use:
   a. Centrifugal force (for large particles).
   b. Washing the air (for particles that are wettable).

c. Screens (to block the larger particles).
d. Adhesives. The air impinges on (strikes against) a air stick to the adhesive. Fig. 22-63 shows a filter air cleaner which has a humidity condition indicator that indicates when the filter should be changed.
e. Electrostatic (electrically charging the particles and adhering these particles to an opposite charge surface). Most of these cleaners have a screen to trap large particles, an electronic unit to remove particles as small as 0.001 micron, and a mat to trap the electron-treated particles. They are usually equipped with a pressure drop indicator and controls. Fig. 22-64 illustrates an electronic air cleaner with a water wash system. This uses hot water to wash the unit through a spray manifold. Water is released into the manifold and forced through a series of spray jets. The dirt is flushed off and drains to the bottom of the unit.

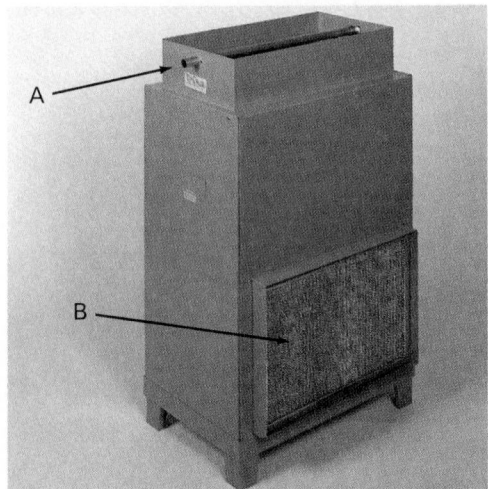

A

B

Fig. 22-64. Water wash electronic air cleaner. A—Water wash manifold. B—An after filter.
(Electro-air/White-Rodgers Division, Emerson Electric Company)

2. To remove liquids:
   a. Liquid absorbents (chemicals to absorb or react with the liquid).

b. Deflector plates.

c. Settlement chambers.

3. To remove gases and vapors (these are molecular size impurities):

a. Condensation (cool the contaminant gas to its dew point and remove as a liquid).

b. Chemical reaction (to react with the gas).

c. Dilution.

It is possible to remove almost 100 percent of the contaminants in the air, but to do so is expensive. Removal of 90 to 95 percent is much more common and practical.

Filter efficiency is measured by:

1. Total weight of dirt it collects.

2. Size of the smallest particle it will remove.

3. Checking for discoloration on the exhaust side of the filter being tested.

Many rooms gradually collect a brown-yellow color on walls, windows, and light-colored drapes. This deposit does not come from the furnace or ducts. It is carried in the vapors from smoking and open cooking such as for meats. These gases and small particles collect on the cooler surfaces of a room, especially when air movement is slow.

22-30 ADHESIVE FILTERS

Adhesive filters are made of various fibers—glass, cotton, synthetic material, and aluminum. There are two classes of these filters.

1. Class 1: fire resistance when clean.

2. Class 2: nonfire resistant.

Most homes use Class 2 filters.

Fibers of adhesive filters are coated with adhesive liquid or oil. Air is forced to change direction and lose speed as it passes through the filter. This results in trapping of the particles of lint and dust as they contact the adhesive surfaces. The filter material is also packed tighter at the outlet side of the filter to improve its dirt-holding capacity.

These filters will remove as much as 90 percent of the dirt if they do not become "loaded," or if air velocity is not too high. The more common filters are of the throwaway or disposable type. See Fig. 22-65. *These filters should be renewed twice each year or more frequently if the dust conditions are high.* The frames are usually made of rustproof steel or cardboard with wire reinforcement. Fig. 22-66 shows how a fiber filter is installed in a furnace.

Another method used to determine if a filter needs replacement is to use a water manometer. The two manometer openings are connected to measure the airflow on the two opposite sides of the filter. The filter should be replaced if the pressure drop exceeds 0.5 in. of water across the filter.

The system is usually designed to allow the filter pressure drop (resistance) to be about a fourth of the total pressure drop (pressure rise across the fan).

For example, if the total pressure rise across the fan is 4.0 in. (10 cm) of water column, the allowed pressure drop across the filter is 1.0 in. of water column. These pressures are measured with a water manometer. See Chapter 18.

Filters are tested in laboratories (hot DOP) and they can also be tested on the job (cold DOP). See Para. 18-39. One test is used to find out how many 0.3 micron size particles the filter can remove. The test is made by measuring the interference with diffused (scattered) light.

To increase filtering surface or area in filters, many

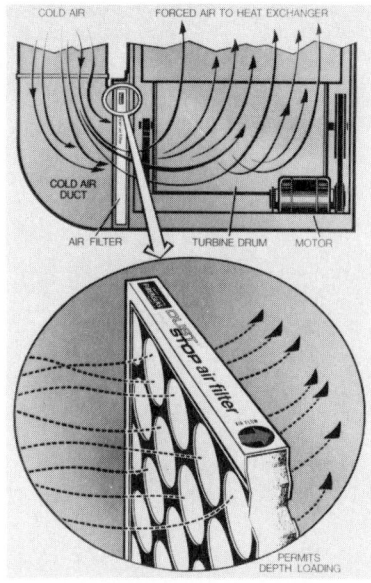

Fig. 22-65. Throwaway paper frame glass fiber filter.

Fig. 22-66. Throwaway-type filter being installed in a furnace. A—Air return. B—Filter. C—Furnace.

different designs are used. A popular method to increase the area is to use pockets to trap the air, as in Fig. 22-67.

Most electronic filter systems also have cleanable filters. Remove the filter. Be sure the power is off first! Clean with a mild detergent-warm water mixture, rinse, and replace. Turn on the power. These filters last indefinitely.

22-31 THROWAWAY FILTERS—SERVICING

Filters should be replaced when they lose their efficiency or when they are so clogged that they produce too much pressure drop across the filter.

Visual inspection is one way to decide that filters need replacement. If they have turned black, if the frame is bent or warped, or if the filtering medium is punctured, replace the filter. If the housing shows signs of corrosion, clean it by sand blasting and repaint.

Checking the pressure drop across the filter, as described in Para. 22-30, is another way to decide whether the filter

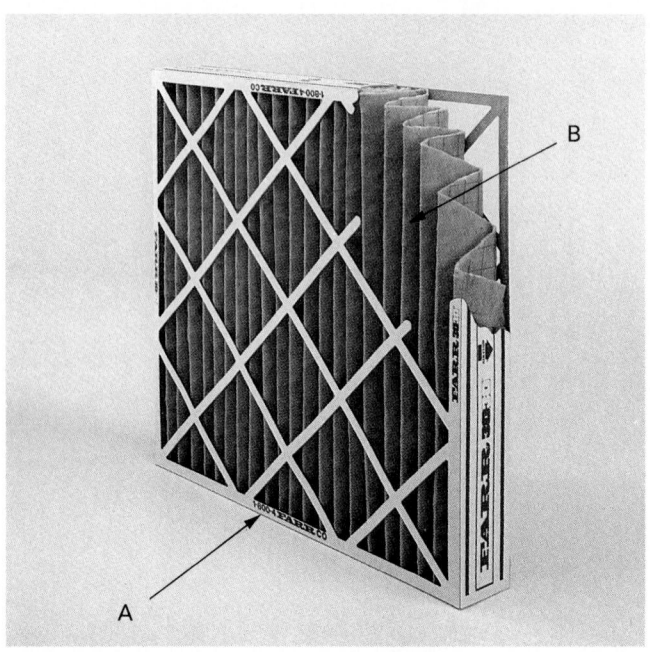

Fig. 22-67. Pocket-type disposable air filter. Air pockets increase filtering surface. A—Frame. B—Folded filtering material. (Farr Co.)

ELECTRICAL CHARGE	MATERIAL
+ END	ASBESTOS
	GLASS
	HUMAN HAIR
	NYLON
	WOOD
	LEAD
	SILK
	ALUMINUM
	PAPER
	COTTON
	STEEL
	SEALING WAX
	HARD RUBBER
	ACETATE RAYON
	NICKEL COPPER
	BRASS SILVER
	SYNTHETIC RUBBER
	ORLON
	SARAN
	POLYETHYLENE
	TEFLON
− END	SILICONE RUBBER

Fig. 22-68. Ability of various materials to generate electricity. Materials high on list, when brought into contact with materials low on list, provide efficient generation of static electricity. Contact requires friction.

should be replaced. *When the pressure drop across the filter is more than 25 percent of the pressure drop across the fan, the filter should be changed.*

Always replace filters with the arrows (printed on the frame) pointing in the direction of airflow. The side towards the blower has more adhesive and must be on the air-out side of the filter. If this is not done, the filter will quickly load with dirt and clog.

When replacing filters, make these two checks:
1. Inspect the filter for tears or holes. Place a strong light on one side of the filter and look through the filter from the other side.
2. Use a manometer to check the pressure drop.

22-32 ELECTROSTATIC (ELECTRONIC) THEORY OF CLEANING

Static electricity has an important effect on dirt and dust clinging to walls, drapes, and ceilings of rooms. Static electricity is created by two surfaces coming into rubbing contact and then separating.

Rubbing a cloth on a nonconductor will build up surplus electrons on one of the nonconductors (negative charge). The other material will lack electrons (positive charge). This excess of electrons is called static electricity. When this static electrical charge jumps an air gap, the spark can cause a fire or explosion.

Static electricity also attracts dirt and dust to vertical and overhead surfaces.

Ionizing the air will neutralize this static electrical charge. (Ionizing means to break it down into positive and negative particles or charges.) When this is done, dirt and dust will settle to the floor and the danger of a static electricity spark is over.

Instruments are used to measure static electricity. They can "read" both a lack of electrons (positive charged) and a surplus of electrons (negative charged).

Materials vary in their ability to generate static electricity. Fig. 22-68 lists a variety of materials and their abilities to generate static electricity. The farther apart the substances are on the list, the greater their ability to generate static electricity. The material at the top assumes a positive (+) charge, while the material at the bottom assumes a negative (−) charge.

There are three ways to ionize air:
1. Power.
2. Nonpower.
3. Nuclear.

The power unit is the electronic filter. A nonpower unit uses metal to remove the static electrical charge. The nuclear power unit gives off a double positive or alpha particle.

Basically, the electrostatic filter puts a static electrical charge on all particles that pass through it. These charged particles are then attracted to collector plates with an opposite electrical charge.

Usually the air is first passed through a throwaway filter to remove most of the larger particles of dirt. Then it is fed through the electrostatic filter. In this filter, the air first passes through a highly ionizing field. A wire with a high positive voltage is suspended between ground wires.

Ions passing through the air space put a positive electrical charge on any particle that attempts to pass through the ionizing field. This particle is then drawn to the grounded plates (negative potential relative to the wires).

Note: On domestic and commercial units, the wires are positive and the plates are grounded. On large industrial units, the wires are negative and the plates are grounded.

Potentials of as high as 12,000 V are used. Fig. 22-69 shows a diagrammatic view of an electrostatic filter. A typical wiring diagram is shown in Fig. 22-70.

Because of the high voltages used, the electrostatic filter may be dangerous. The units should be designed to shut

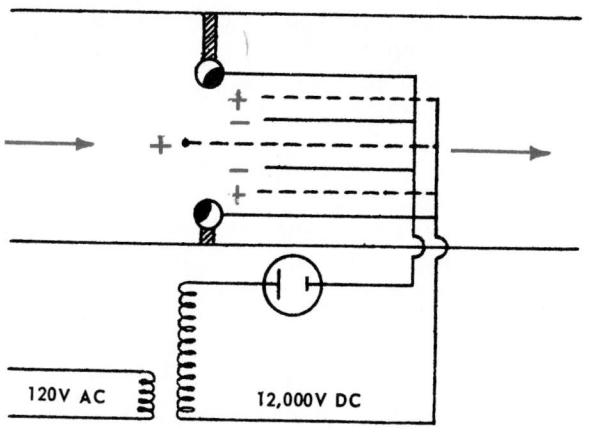

Fig. 22-69. Electrical circuit and airflow in simple electrostatic-type air filter for forced-air furnace.

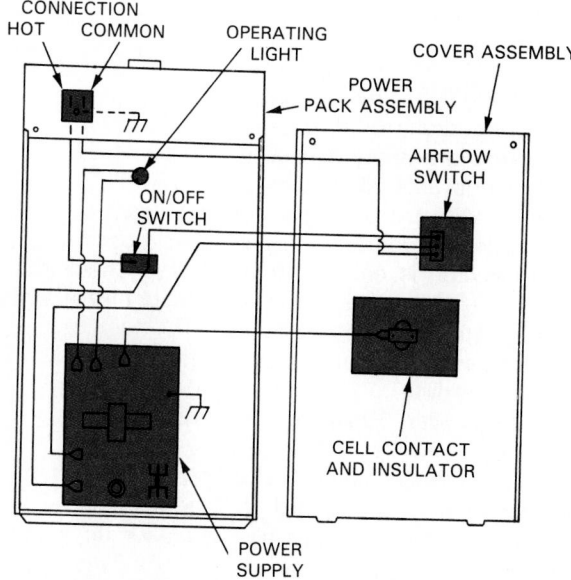

Fig. 22-70. Wiring diagram of electronic air filter. (White-Rodgers Div., Emerson Electric Co.)

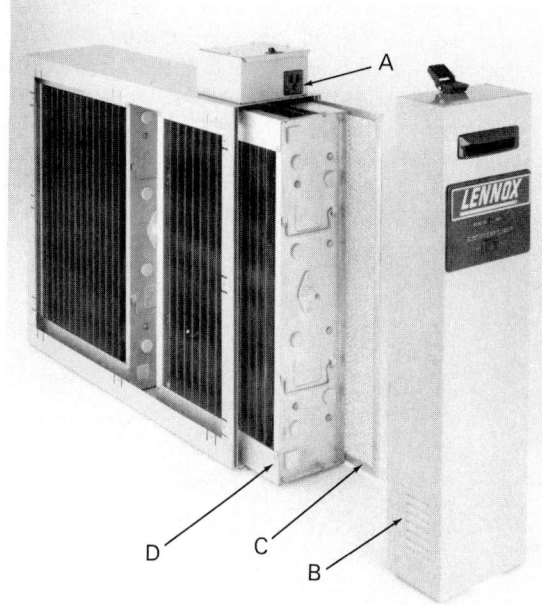

Fig. 22-71. Electronic filter designed for installation in air duct. A—Switch. B—Access panel. C—Protective screen. D—Electronic cell unit. Note airflow direction on label. (Lennox International, Inc.)

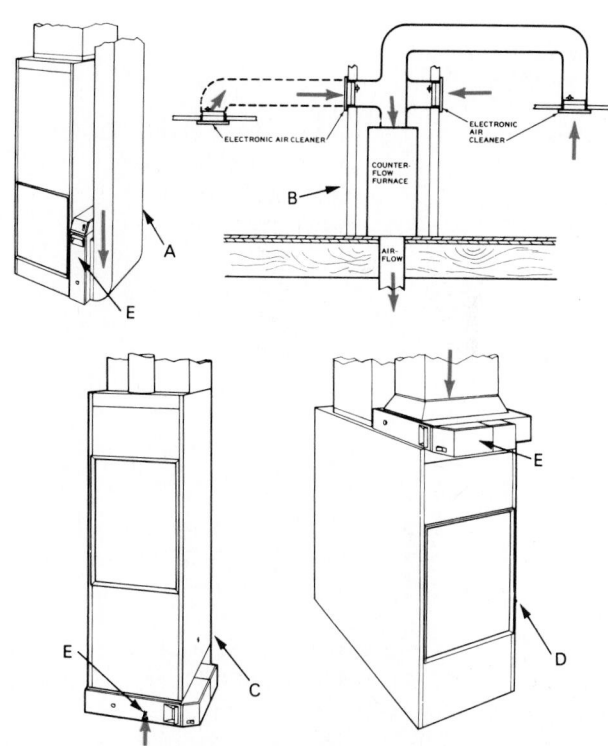

Fig. 22-72. Various installation positions for electronic filter. A—Cold air return from the side. B—Independent room units. C—Upflow. D—Downflow. E—Electronic filter. (Honeywell, Inc.)

off automatically when the service doors are opened to gain access to the units. Fig. 22-71 shows an electronic cleaner designed for duct installation.

Electronic filters may be installed in duct systems and are also used as separate cabinet units. The filters can be placed in the return airflow ducts in several ways. Fig. 22-72 shows different positions of a filter at a furnace or in room air ducts.

The electronic cells and protective screens must be cleaned every two to three months. Material collected is black in color. Continue cleaning with hot water, until cleaning water is clear. Use recommended detergents and water-detergent solutions.

Electrostatic air cleaners utilized in ceilings and in portable units operate on the basis that airborne particles that are being circulated by heating and air conditioning systems are removed by electronic air cleaners. This includes smoke, dust, mold, spores, pollen, and bacteria. See Fig. 22-73. 1)—Internal fan draws the polluted air into the electrostatic air cleaner. 2)—Prefilter screen filters and collects large par-

ticles. 3)—Ionizing section has fine wires which give the incoming airborne particles a positive charge. These particles are attached to the collecting area, where they are held by negatively charged plates of the collecting cell (4), just like magnets attract and collect iron filings. 5)—The clean, purified air is recirculated throughout the room.

A ventilation system may be used as another means of

controlling indoor air from excessive humidity (due to bathrooms, laundry rooms, shower rooms, etc., which causes condensation on windows and walls); odors (from cooking, smoking, or other household activities, which create unpleasant-smelling air). A central ventilation with heat recovery system is shown in Fig. 22-74. The heat recovery ventilation system is designed to bring air in from the outside, pre-heating it in the winter or cooling it in sum-

mertime. It is designed to blend outdoor air into the home by the use of a fan.

Electronic filters have four main parts:
1. Frame.
2. Power supply.
3. Prefilter and airflow distributor.
4. Electronic cell.

The unit must be installed level and plumb for proper drainage and more efficient airflow. The hot water line to the washer should have a strainer.

Airflow should be evenly distributed across the face of the air cleaner for maximum efficiency. If the unit is near an airflow elbow, use movable air vanes or baffles.

Excessive lint interferes with electronic air cleaner (EAC) operation. A fine-mesh screen or filter should be installed ahead (upstream) of the filter.

22-33 SERVICING ELECTRONIC FILTERS

Electronic air cleaners (EAC) need service when:
1. Unit does not arc.
2. Meter (if used) reads low.
3. Trouble lights remain on.
4. Strong ozone odor is detected.
5. Rooms are dusty and dirty.
6. The unit arcs all the time.

Check with the owner on the last cleaning of the filter. Be sure the power doors or panels are closed. Be sure the power switch is on. Check the fuses. Check the meter readings. (Refer to the service manual or owner's manual.) These meters will show if:
1. Conditions are normal.
2. Filter is dirty.
3. Filter is wet. (Operate "dry" switch.)
4. There is an electrical failure.

Some meters will show if electrical trouble is in the power source circuit or in the high-voltage circuit.

If the trouble is in the high-voltage circuit, inspect and electrically test the "power pack" capacitors and collecting cell ionizing wires. Fig. 22-75 shows an electronic filter with the collector cell and power factor being checked. A kilovoltmeter probe is being used to check the high voltage circuit to determine the location of the faulty cell.

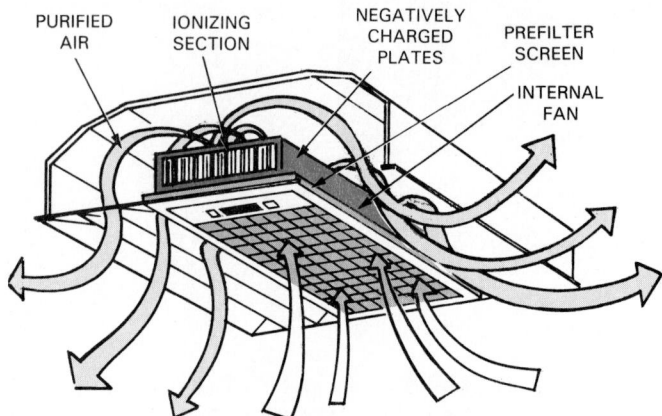

Fig. 22-73. Electrostatic design ceiling filter. (Tectronic Products Company, Inc.)

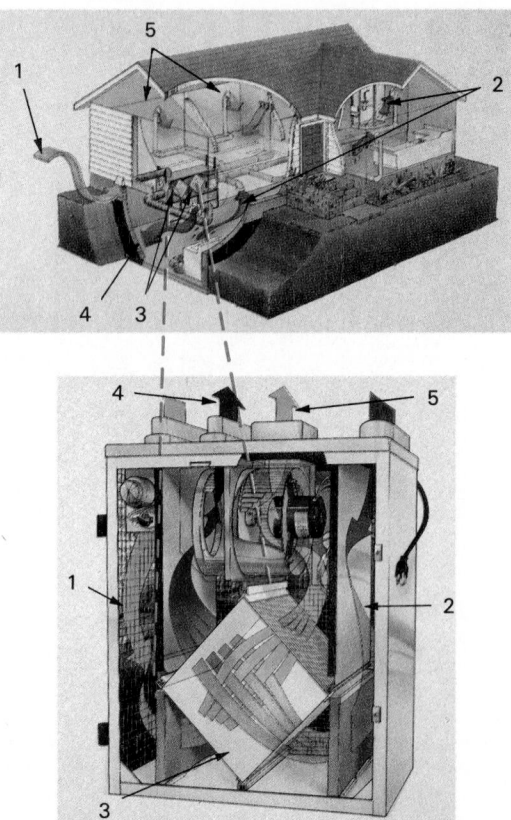

Fig. 22-74. Recovery ventilators in a normal installation. Note: 1—Outdoor air enters through vent and through filter and heat exchanger core. 2—Fan draws exhaust air from bathrooms, laundry, etc., and through alternate passages of the core. 3—Core allows two air streams to pass close to each other, and heat is transferred. Energy transfer warms or cools incoming air. 4—Exhaust air continues through ducting to the outside. 5—Fresh air is distributed throughout the home. (Conservation Energy Systems)

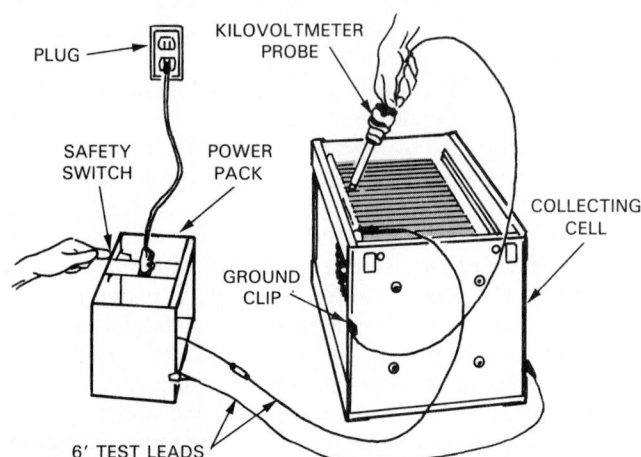

Fig. 22-75. Electrostatic filter being checked. Note test leads from power pack to collecting cell. Make sure that power pack is functioning prior to testing collecting cell. (White-Rodgers Division, Emerson Electric Company)

In the power pack, inspect the low side first with either a test light or voltmeter. Test each part starting with the wall outlet or power source.

Then check the transformer, rectifiers (ac to dc), and the capacitor resistors (the resistors should discharge the capacitors in about 10 seconds). The capacitor can be checked by replacement. Fig. 22-76 shows tools and instruments used during servicing.

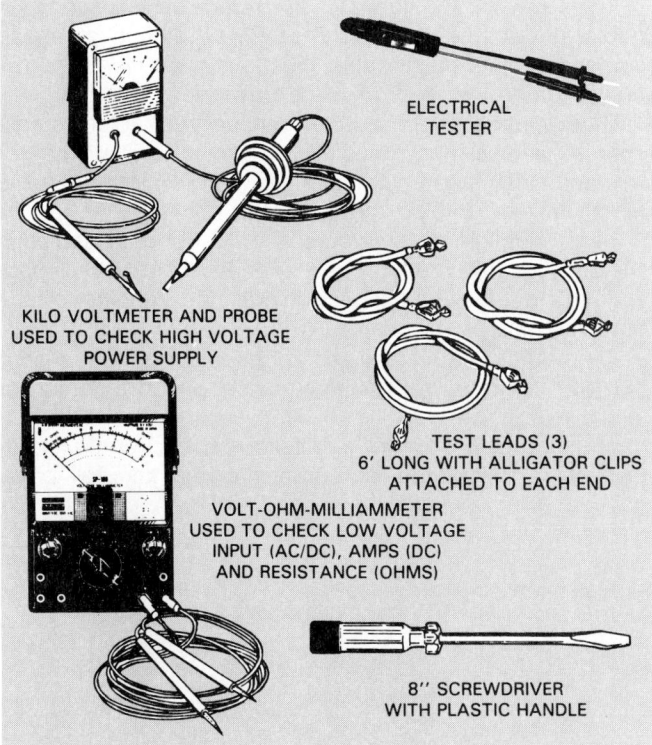

KILO VOLTMETER AND PROBE USED TO CHECK HIGH VOLTAGE POWER SUPPLY

ELECTRICAL TESTER

TEST LEADS (3) 6' LONG WITH ALLIGATOR CLIPS ATTACHED TO EACH END

VOLT-OHM-MILLIAMMETER USED TO CHECK LOW VOLTAGE INPUT (AC/DC), AMPS (DC) AND RESISTANCE (OHMS)

8'' SCREWDRIVER WITH PLASTIC HANDLE

Fig. 22-76. Instruments, leads, and tools used to test electrical circuits of electronic air filter. (A.W. Sperry Instruments, Inc.)

In the collector section, inspect the bent plates, plates out of position, dirt bridging the gap between ionizing wires and the plates, broken insulators, and broken wires. Plates must be straight. Remove and replace broken insulators and ionizing wires.

One should inspect the building and complete air-handling system. New carpeting, for example, may temporarily cause an overload on the filter. Leaking duct systems and untreated concrete floors are all unusual high-load conditions. Dusty construction work in the vicinity may also overload the unit.

Electronic air cleaners should be cleaned on a planned schedule. Frequent washing of a unit is not harmful, but a neglected unit will not clean air effectively.

The filter washing procedure for units without built-in wash systems is as follows:
1. Turn the electronic air cleaner, furnace, and blower power off.
2. Remove lint screens and ionizing collecting cells. Do not run the system without replacing the screens and cells.
3. Clean, using hot soapy water, and rinse thoroughly.
4. Replace lint screens and ionizing collecting cells after

they have dried thoroughly. If proper drying does not occur, it may cause electrical arcing when system is turned on.

A properly operating unit will be indicated by black water when the cell is cleaned. A properly operating unit allows only fine white dust to leave the ducts. If a cheesecloth placed over a grille becomes discolored, the EAC is not working properly.

22-34 DIRT ON WALLS AND DRAPES

Dirt collecting on walls, ceilings, curtains, and drapes of a conditioned space is always a problem. In most cases, this dirt does not come from the ducts but is already in the room. Room air movement, also called convection air current, is responsible for carrying it to the room surfaces.

The collection of dirt around warm-air grilles is called thermal precipitation. Warm air coming out of the grille picks up dirt from the room air. As this air hits cooler surfaces, the dirt settles (precipitates) on the surfaces. This precipitation takes place on windows also. The cooler the surface, the more dirt it collects; therefore, insulation and storm sashes reduce the amount of dirt settling out of the air.

"Clean" rooms are now in use for surgery, for research, and for manufacturing, repairing, and servicing critical items such as instruments. Computer disk drives benefit from clean rooms.

A clean room uses four ways to maintain the "clean" requirements:
1. Extremely high efficiency filters.
2. Laminar airflow.
3. Anticontamination devices.
4. Higher pressure in the clean room than outside the room. This helps prevent dust and vapors from entering.

Ideally, the future of residential living will be air cleaning similar to the "clean room" standards. Home computers require clean conditions.

Filters are ineffective against gases. Activated charcoal will adsorb gases and so will suitable liquids. Water removes gases such as SO_2 and NO_2 effectively. Water saturated with calcium hydroxide removes CO_2. At least 99.6 percent of the CO_2 is removed.

22-35 WATER SPRAYS

Large air conditioners use water sprays to remove wettable solid contaminants, liquid contaminants, and water soluble gas contaminants from the air. Some of these gases are sulphur dioxide, nitrogen oxides, and carbon monoxide. Water does not remove soot.

Usually the water is sprayed in a pattern which produces 100 percent duct cross-section coverage. A drain pan catches the water, while eliminator plates in the duct collect any water droplets which travel down the duct. The water drain pan is usually equipped with a float-controlled makeup water connection.

A continual overflow drainoff is used to remove dust and dirt as it collects on the surface of the water. Water in the drain pan is recirculated by a centrifugal pump. A screen is located at the pump inlet to prevent dirt particles from clogging the spray nozzles. These air washers are popular during the heating season.

Care must be taken to avoid freezing temperatures in the spray chamber. A preheat coil is usually used to keep the temperatures above freezing. The water spray, in addition to cleaning the air, also serves as a humidifier.

Comfort cooling systems which condense moisture out of the air make use of the wet evaporator surfaces as a filtering device. A rinse device is used to remove the collected dirt.

22-36 ODOR

Vapors and odors frequently form a large part of atmospheric air contaminants. Most odors are gases, and filters, even electrostatic ones, will not remove them.

Some odors can be removed by cooling the gases to their condensation or freezing temperature. Some are removed by oxidation (and by ultraviolet ray treatment).

Others can be removed by chemically combining them with other chemicals, by diluting them with air, or by absorbing them into a liquid or adsorbing them into a solid (holding on the surfaces of cavities).

Both vapors and odors can be removed with activated charcoal. Activated alumina with potassium permanganate has also been used. There are other chemicals which have the ability to absorb and/or destroy odors.

22-37 CARBON FILTERS

A filter made of activated carbon will remove solid particles as well as odor-causing gases and bacteria. This type of filter is being used in air conditioners and in refrigerators. Fig. 22-77 shows an activated carbon filter assembly. The carbon in activated charcoal form is made from various substances, including such materials as carbon from refining petroleum, coconut shells, and other carbon-producing substances. This charcoal will adsorb (hold on the surface) as much as 50 pecent of its weight in foreign gases.

It is possible to rejuvenate used carbon filters. However, this is usually done by the manufacturers, who remove the carbon and process it for reuse.

22-38 ULTRAVIOLET LIGHT

Ultraviolet light of 14 000 microwatt/cm² will kill most bacteria in a fraction of a second. Such ultraviolet lamps are available. *One should avoid looking at or being exposed to these rays. The effects are harmful when one is exposed to them for any length of time.*

The lamps are installed in the return air duct. Access doors must be equipped with safety-off switches in case someone opens these doors. The rays must cover the full cross-section of the duct to be effective.

Ultraviolet lights are used to reduce the bacteria, mold spores, viruses, and other microorganisms. The types of microorganisms and quantity killed is dependent upon the length of exposure to the ultraviolet rays.

Ultraviolet light duct fixtures, Fig. 22-78, are installed in the duct with an interlocking access door. The access door is required for scheduled cleaning and changing of the ultraviolet lamps.

A view port, with plastic and/or glass lenses that absorb germicidal ultraviolet rays, may be installed in the duct for visual checking.

The fixtures are available in various sizes. For maximum protection, install the lamp tubes across the full cross

Fig. 22-77. Activated carbon filter and air purifier. A—Frame.B— Activated carbon elements.

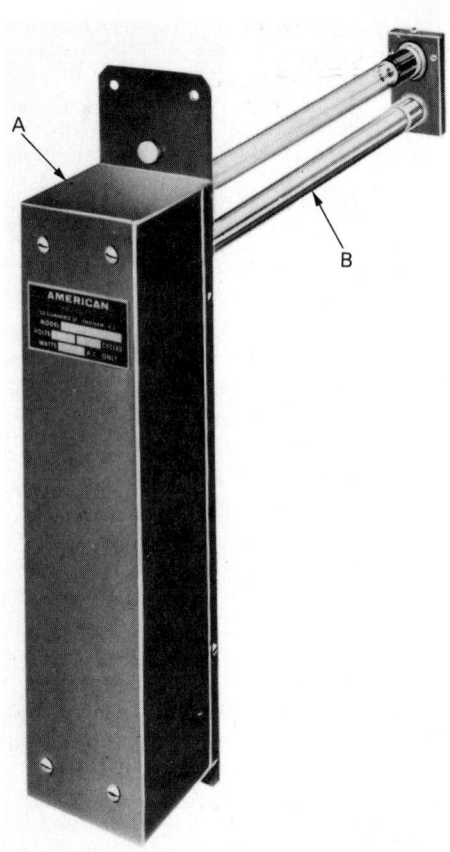

Fig. 22-78. Ultraviolet ray unit designed for duct installation. A—Electrical power supply. B—Two tubular ultraviolet ray lamps. (American Ultraviolet Co.)

section of the duct, Fig. 22-79.

Units normally are equipped with non-ozone-producing lamps. Applications that require both odor control and germicidal protection are installed using ozone-producing lamps. The ozone level should not exceed 0.05 ppm in a 24 hr. period.

Lamps should be installed as close as possible to the end of the duct where the air is entering the room. Fixtures may also be installed in the return ducts of recirculating systems.

Lamps should remove 90 percent of the bacteria in domestic and commercial systems and 99 percent in hospital operating and recovery rooms. Systems that remove more than 99 percent of the bacteria are costly and do not allow any human exposure.

The useful life of a lamp is approximately 7500 hours. The lamps should be cleaned at least once a month.

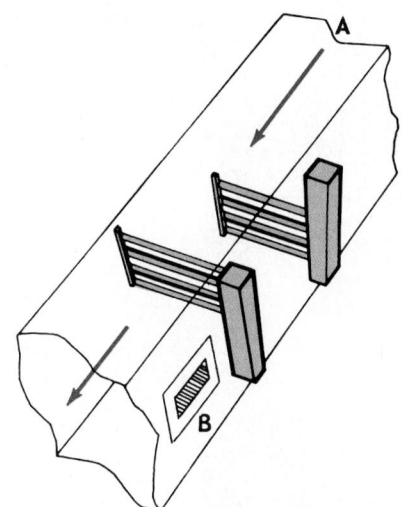

STERILE CONDITIONING UNITS INSTALLED
ACROSS AIRFLOW OF TYPICAL
AIR CONDITIONING SYSTEM.

Fig. 22-79. Two banks of ultraviolet lamps installed in cold air return duct. Rays are directed the length of duct. A—Airflow. B—Protected inspection opening.

22-39 AIR CURTAINS

Air curtains are often used at garage doors, delivery docks, doorways to areas used very little, and similar boundaries. Air curtains may be used on doors opened frequently for either winter use in cold climates or for summer use in hot climates.

A powerful blower is connected to a source of warm or cold air. As a door is opened, this air is directed through openings which provide a narrow stream of warm or cooled air across the entire door area. This air flow stops any natural flow of air from the building to the outside air or from the outside air into the building.

22-40 INSTRUMENTS

Instruments are important to refrigeration, heating and air conditioning installation, and service personnel. It is almost impossible to check on the electrical characteristics of a unit without using ammeters, voltmeters, test lights, and ohmmeters. Instruments are explained in Chapters 6, 7, and 8.

What takes place inside a refrigerating system is almost impossible to check without pressure gauges and thermometers. These instruments are discussed in Chapter 2, 11, and 14.

Instruments should be used well within their scale range. They will maintain accuracy longer if use never exceeds more than a half to two-thirds of their scale range.

In heating systems, the combustion process and the flue gas properties can only be accurately determined by the use of instruments. A set of these instruments is shown and identified in Fig. 22-80.

The flow of liquids should be checked as carefully as possible. The velocity and amount of water flowing in hot water systems, chilled water systems, condensers, and cooling towers should be accurately measured.

The amount and velocity of refrigerant liquids is important to know during maintenance and repair work. These flow meters can be installed in the piping temporarily or permanently.

Because air is invisible, instruments are important to measure its flow and pressure conditions. Thermometers and pressure gauges have already been explained. Some different instruments especially useful for airflow study are:
1. Manometer.
2. Barometer.
3. Pitot tube.
4. Anemometer.
5. Smoke as a velocity indicator.
6. Hot wire anemometer.

A velocity meter is shown in Fig. 22-81. This instrument is used with a pitot tube. The inclined scale is for low pressures and velocities. The vertical scale is used for high pressures and high velocities.

Many of these air-measuring instruments are explained in Chapter 18. Chapter 18 suggests ways to position air-measuring instruments over grilles.

22-41 RECORDING INSTRUMENTS

Service technicians can easily take "spot" checks of volts, amperes, and watts in a circuit or parts of a circuit. However, most causes of electrical faults may occur when the service technician is not there with the needed instruments. It is possible to use recording instruments. These instruments may provide a 24-hour record or even a seven-day record of electrical values as well as temperature, pressure, and relative humidity. Chapter 18 illustrates these latter instruments.

There are two basic charts in use:
1. Circular chart.
2. Strip chart (used in roll form).

There are two types of electrical recording instruments:
1. Galvanometric.
2. Potentiometer.

The galvanometric type uses a meter movement (usually d'Arsonval-type) and is most used by service technicians. The potentiometer type uses a potentiometer with a variable (what is being measured) against a standard value.

The galvanometric type has a meter needle or pointer which makes an impression on a moving piece of paper (circular or strip).

A trace on paper is produced in several ways:

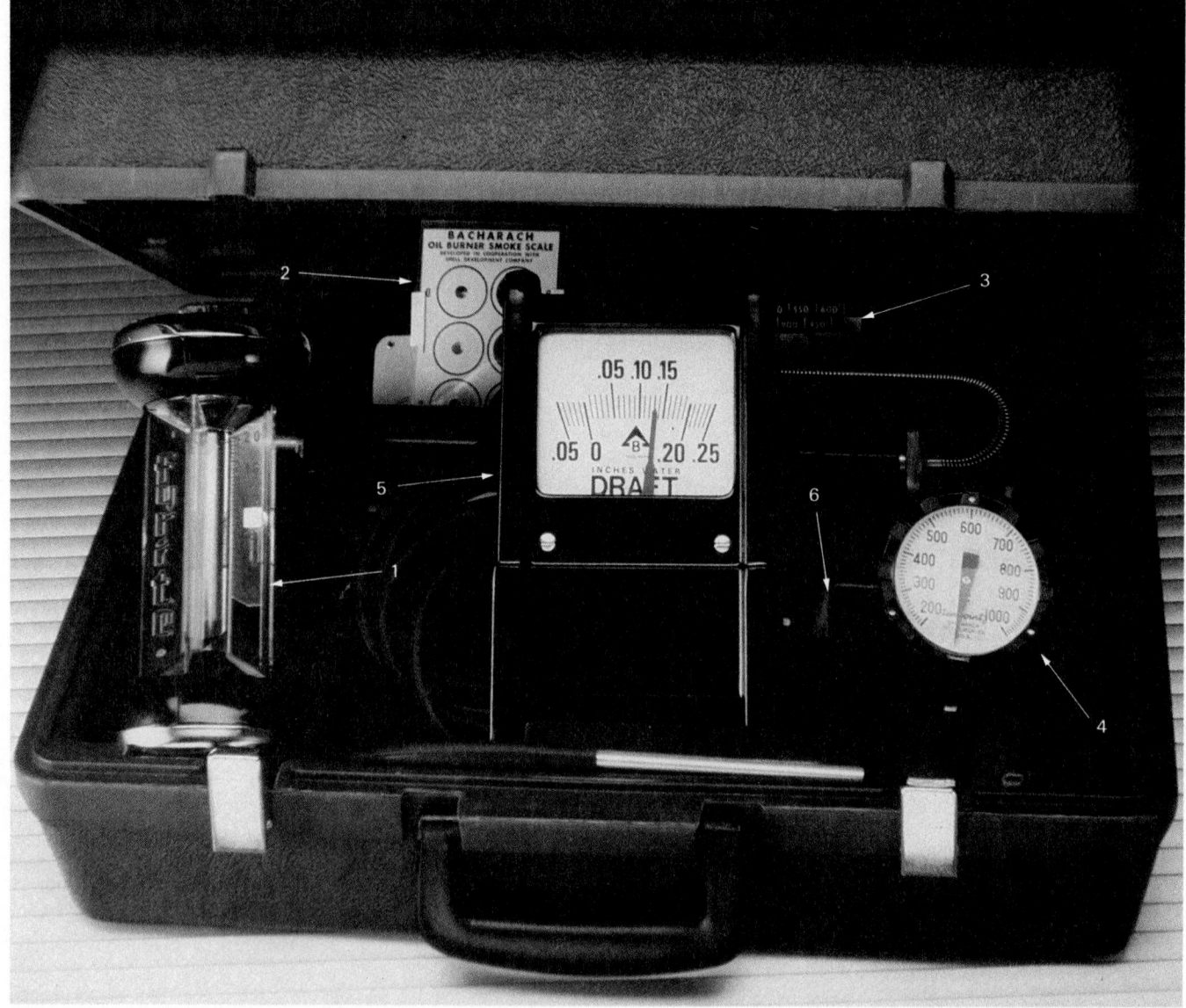

Fig. 22-80. Service technician's oil burner combustion testing kit. It is used for testing oil burner efficiency during installation or service. The test kit contains: 1—CO_2 indicator used to examine flue gas contents. 2—A flue test checker paper strip and test pump. 3—Slide rule used to determine combustion efficiency and stack loss after the CO_2 and stack test. 4—A dial thermometer. 5 and 6—The 6-hr. draft gauges. (Bacharach, Inc.)

1. Ink trace.
2. Pressure trace (paper sensitive to pointer pressure).
3. Electrical trace (paper conducts electric current and paper color changes).
4. Thermal trace (the pointer is hot and the paper is sensitive to heat).

The paper is moved by a spring-loaded clock mechanism or a geared timer motor.

22-42 DRAFT CONTROL INSTRUMENTS

Efficient and safe combustion in a furnace needs accurate draft control (control of flue gas movement). Flue gas flow depends on two things:

1. The density of flue gas versus the density of the air.
2. The pressure difference between the inside of the building and the outside of the building.

Flue gas conditions vary according to the type of fuel used (oil or gas). High-efficiency furnaces require a small flue gas flow, since heat must not be lost up the flue.

A draft gauge is generally used to determine the efficiency of flue gas (combustion gas) flow. Fig. 22-82 shows a draft gauge being used. The stack temperature is also an indicator of draft efficiency. Stack temperatures vary from 300 °F to 900 °F (150 °C to 480 °C). Fig. 22-83 shows a stack thermometer being used to find the temperature of the flue gas.

22-43 COMBUSTION EFFICIENCY INSTRUMENTS

Combustion efficiency can be determined by measuring the amount of carbon dioxide (CO_2) in the flue gas. A

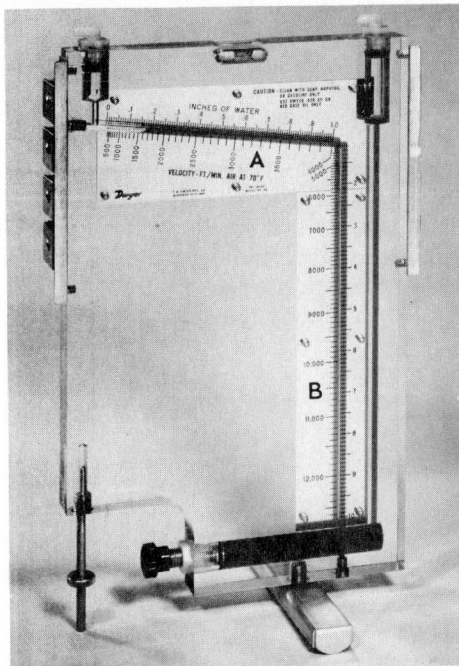

Fig. 22-81. Manometer used with pitot tube. A—Scale for velocities. B—Scale for pressures. (Dwyer Instruments, Inc.)

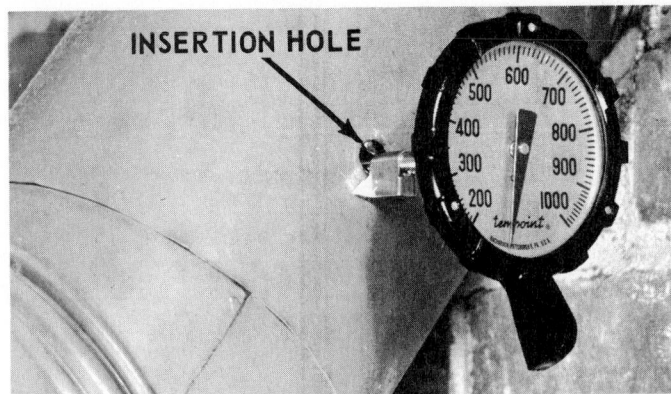

Fig. 22-83. Stack thermometer used to determine flue gas temperature. Note small hole made for putting instrument in flue.

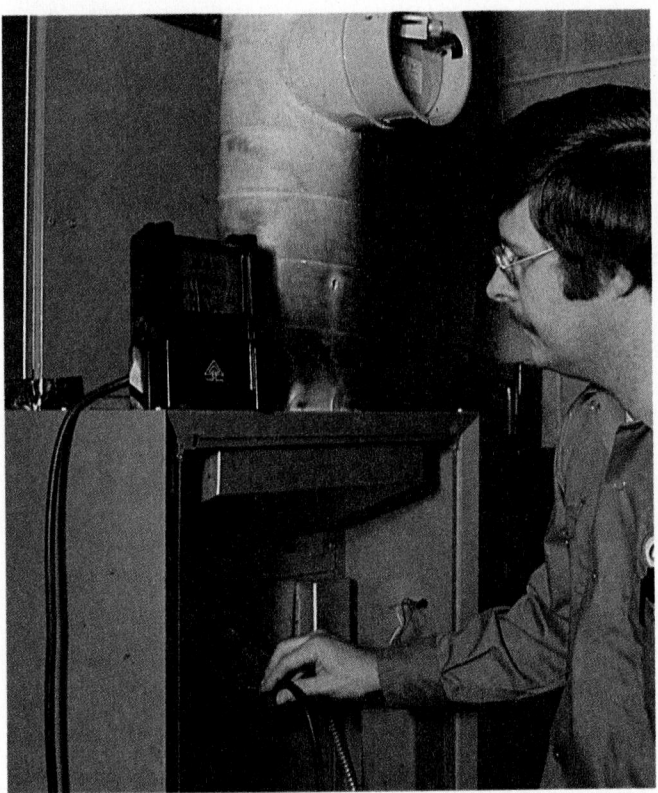

Fig. 22-82. Draft indicator for residential and commercial heating systems. It is used to identify draft problems. (Bacharach, Inc.)

Fig. 22-84. Service technician measuring carbon monoxide level of residential plants. (Bacharach, Inc.)

sample of the combustion gas is exposed to a chemical which absorbs carbon dioxide only. Fig. 22-84 shows a portable digital analyzer. The probe is inserted into the air being sampled and digitally displays readings of the carbon monoxide levels.

If 10 cc (cubic centimeters) of the flue gas reduces to 9 cc of gas after exposure to the chemical, the flue gas contained 1 cc or 10 percent of carbon dioxide.

If the gas cools during testing, Charles's Law affects the answer. When the gas cools, its volume reduces, and the technician will get a reading that is too high.

The flue gas temperature before and after must be known and a correction must be made for accurate results. Tables are provided by equipment manufacturers.

Systems will vary in CO_2 content. Some are operating correctly with as low as 8 percent CO_2; some as high as 12 percent CO_2. The manufacturer's service manual will give the correct amount for the system being tested. *Remember, a clean flame is essential together with the correct CO_2*

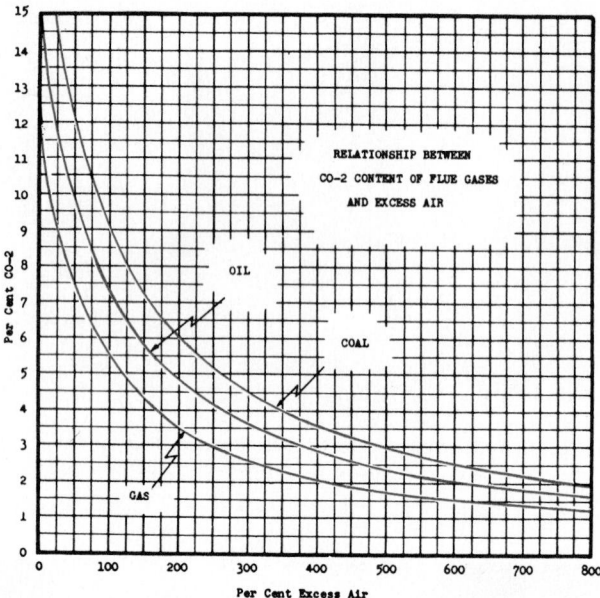

Fig. 22-85. Graph shows change in stack gas carbon dioxide content for various fuels as amount of excess air changes from 0 to 800 percent.

cian can next find the stack temperature. Again, the combustion efficiency of the furnace is determined through the use of a chart, as shown in Fig. 22-86.

Some analyzing instruments use electronic sensors, circuits, and indicators. They are used to test for air pollution, boiler efficiency, and stack gas analysis. These instruments use thermal conductivity and a combustion chamber with catalysts to measure the gas sample. See Fig. 22-87, which shows an orsat apparatus. An orsat

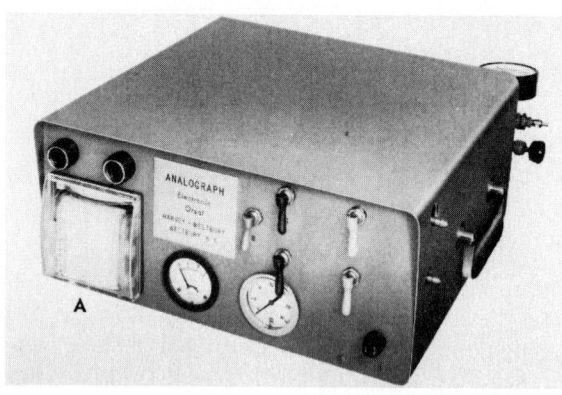

Fig. 22-87. Electronic orsat apparatus measures air pollution and flue gases. Silica gel and molecular sieves partition and adsorb (hold on a surface) different gases. Test takes about 3 min. A—Recorder. (Harvey-R.J. Instrument Corp.)

reading. Fig. 22-85 shows the CO_2 content of the flue gases and excess air for three common types of fuel. When the amount of CO_2 has been determined, the service techni-

COMBUSTION EFFICIENCY CHARTS

To determine combustion efficiency follow the proper CO-2 column to the intersection with the stack temperature column. For example 11% CO-2 & a 500° stack temperature produce an efficiency of 79% for coal & 80.5% for oil.

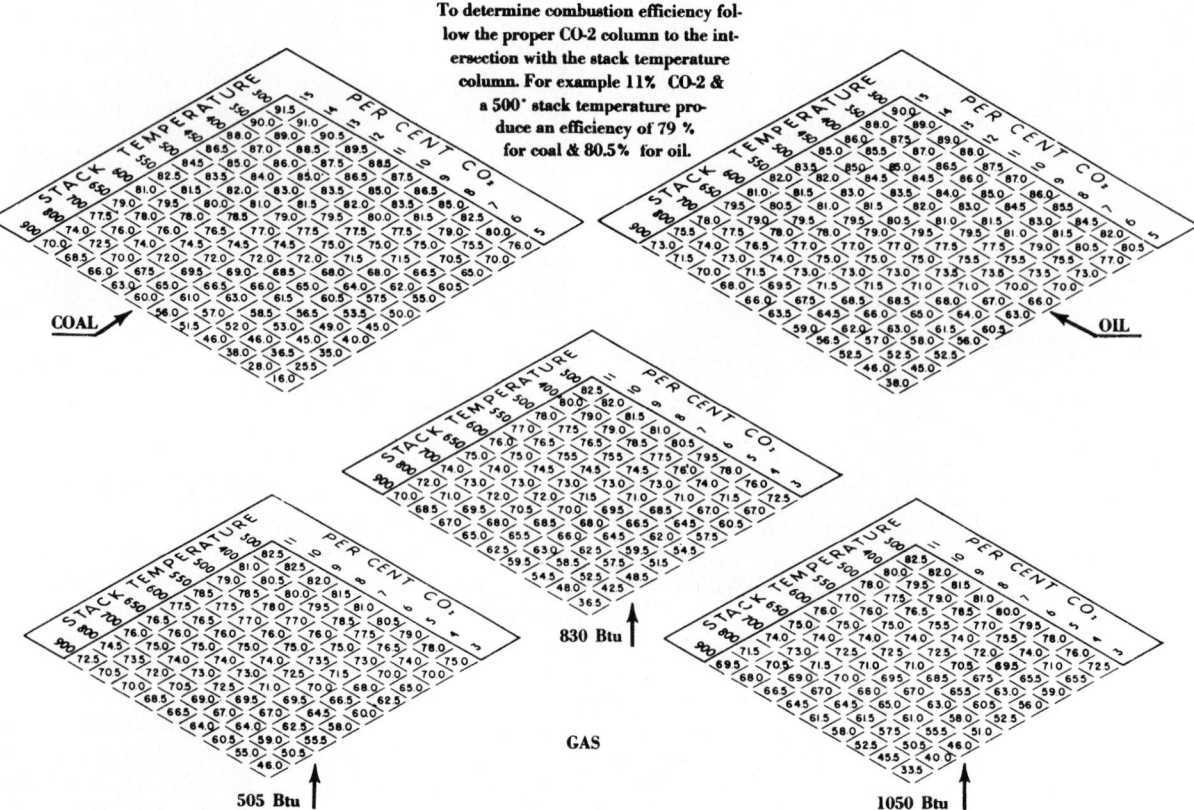

Fig. 22-86. Charts show combustion efficiencies at various stack temperatures and carbon dioxide percentages. Chart gives efficiencies for coal, oil, and three different Btu gas qualities. Follow directions given with chart.

apparatus consists of pipettes (glass tubes with narrow ends) and a burette (a graduated glass tube that shows liquid or gas quantity). Some instruments use air to carry the sample; some use helium, argon, or nitrogen. These instruments can accurately analyze for hydrogen, oxygen, carbon monoxide, methane, ethane, and carbon dioxide.

Separate silica gel and molecular sieves are used to partition and adsorb certain gases. A pump is used to move the gases and a built-in recorder records the results. In about three minutes, this unit will give accurate flue gas analysis for oxygen, carbon monoxide, and carbon dioxide.

22-44 SMOKE TEST

A smoke test is an excellent way to check combustion efficiency such as air-fuel ratio, primary air, secondary air, and draft. Several tests are used.

The first test is an empirical test (one which depends on the averages from experience and observation) rather than a scientific rule. A white filter paper is inserted in the flue. The flue gas sample is drawn through the filter, and the smoke deposit on the filter paper is compared with a sample chart as a standard.

Another method inserts a tube in the flue and an aspirator bulb pulls flue gas samples through a filter mounted in a fixture (about 30 aspirator bulb squeezes are made). The smoke deposit on the filter is then compared to a master comparison scale. This scale (the Ringelmann Scale) rates smoke samples by numbers from 1 to 4.

Photoelectric cells may also be used to check smoke density in large systems or in laboratories. A smoke monitoring system is shown in Fig. 22-88. The reflector can be mounted in chimneys from 1.5 ft. diameter to 10 ft. diameter. The photoelectric signal is amplified and the Ringelmann Scale reading is shown on the upper left instrument.

The readings can also be recorded. An alarm signal is given when the smoke density reaches a preset maximum. This unit can also shut down the system or start up flue blowers or afterburners. Fig. 22-89 is the wiring diagram of a typical pilot-ignited oil or gas burner using a photo-

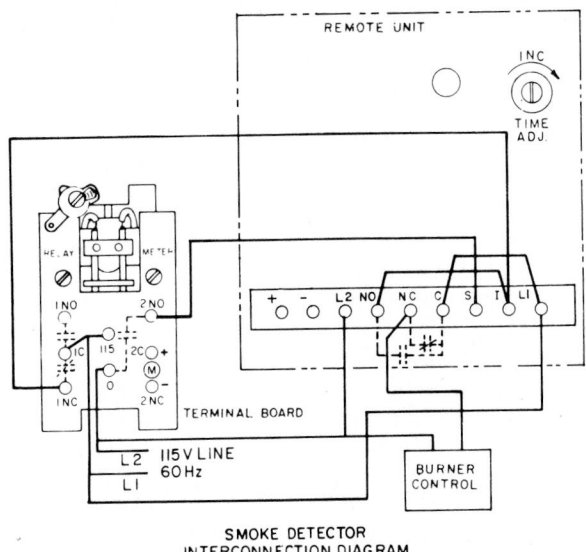

Fig. 22-89. Wiring diagram for smoke detector and burner control. (Allen Bradley Co., Presence Sensing Products, Fireye Division)

electric safety control. In no case should there be smoke appearing at the chimney exit.

22-45 AIR VOLUMES

Because air is invisible, instruments are usually used to measure its flow. Chapter 18 explains most of these instruments.

A direct-reading instrument for cubic feet per minute (cfm) has been developed. First, the grille area is measured in square inches. This area is measured to the outside edge of the grille openings and does not subtract the grille mesh or cross bars. The method of measuring airflow in cubic feet per minute is shown in Fig. 22-90. The area in square

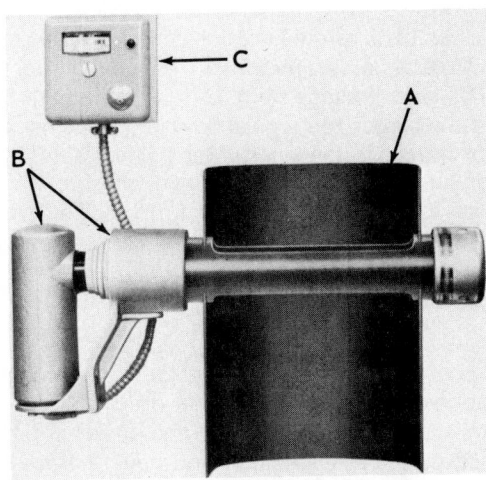

Fig. 22-88. Photoelectric system measures smoke density in chimney. A—Chimney. B—Photoelectric cell and beam tube. C—Indicating instrument calibrated in Ringelmann Scale. (Allen Bradley Co., Presence Sensing Products, Fireye Division)

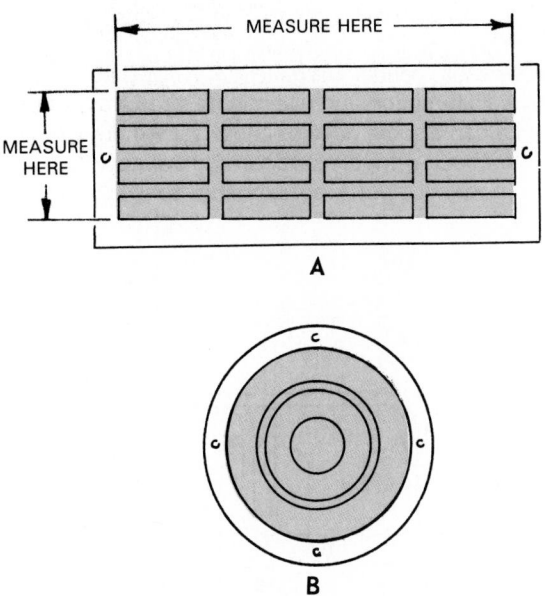

Fig. 22-90. Area to be measured when figuring volume of air leaving grille. Notice that area does not include rim of grille, but does include strips of metal inside. A—Rectangular grille. B—Round grille.

inches is entered as data into the instrument. The device takes this factor into account as it computes the reading from the quantity of feet per minute.

22-46 VISIBLE AIRFLOW INDICATORS

Drafts of 15 to 25 ft./min. are allowable and are comfortable for most installations. If air movement is less than this, air stagnation results. If it is more, persons exposed to the draft are uncomfortable.

To determine the amount of the draft and the direction, the most successful method has been to use smoke (visible vapor). Smoke generators release small puffs of smoke into the space being tested. The distance the puffs move, in 30 to 60 seconds, is observed. Several readings must be taken and averaged to obtain a degree of accuracy.

One type of smoke generator is shown in Fig. 22-91. Each of the two bottles contains a liquid. The aspirator forces the vapors from the two bottles to mix at the nozzle. The mixing of the two gases forms a white smoke that has a density not much greater than the density of air. The liquids used are diluted hydrochloric acid and aqua ammonia. The smoke formed is ammonium chloride. Caution: Both hydrochloric acid and aqua ammonia are dangerous to use. Wear goggles, rubber gloves, and a respirator when handling this material.

A mixture of titanium tetrachloride and air moisture forms a dense, lasting white smoke and can be used for checking air movement and air leaks. The mixture is somewhat toxic. Wear goggles, gloves, and a respirator.

A zinc stearate powder, when mixed with air, may also be used to form a white, small cloud and can be used for checking air movement and air leaks. The mixture is not very toxic.

Other devices can be used to add smoke to the air:
1. Smoke sticks.
2. Smoke guns.
3. Smoke candles.

Sticks and candles are ignited and placed in the intake of an air distribution system with filter removed. Distribution of air and air balance can then be observed in the system.

A three-minute candle will generate 40,000 cu. ft. of visible smoke. Half-minute and five-minute sizes are also made. The smoke can also be used to check for window and door leaks or to check refrigerator door and window seals. The smoke is nontoxic, but long exposure to it is not recommended. It is likewise harmless to clothing or the contents of a building normally exposed to air. Avoid using too much of the smoke.

The smoke is produced by a zinc chloride mist with a trace of carbon in it. While they usually produce white smoke, candles producing yellow or orange smoke are also available.

22-47 WATER ANALYSIS INSTRUMENTS

Whenever a test of water is involved, its condition is expressed in "pH." This factor indicates the activity of the hydrogen ion whenever there is moisture present. The term "pH" followed by a number is used to indicate whether water tends to be acidic or alkaline. Distilled water at 77 °F (25 °C), neither acidic nor alkaline, has a pH of 7. It is said to be neutral. Numbers above 7 and up to 14 express increasing alkalinity. Numbers decreasing from 7 (6 to 0) indicate increasing acidity.

Since electrical properties of water change as pH changes, acidity or alkalinity of water can be measured with electronic sensors, circuits, and meters. An increase in pH causes a decrease in millivolts. A water analysis instrument has an amplifier and two electrodes that are immersed in the water.

22-48 FLOW METERS

Flow meters have many different designs and are made for many purposes. The gas meter and the water meter are the most common. Both are used to determine consumption.

All heating and cooling systems use moving air, moving liquids, or both (condenser water, hydronic water, and chilled water). It is important that flows be known if one is to install the equipment correctly and service it accurately.

Airflow is measured with pitot tubes and electronic direct reading meters. See Chapter 18.

The flow of liquids in pipes can also be measured. A variety of liquid flow meters are available. These meters are usually accurate to 1 percent.

One is known as the turbine flow meter. It has a turbine (wheel) turned by a moving liquid. A magnetic pickup gives an electrical pulse each time the turbine wheel makes one turn. These electrical pulses are connected to an electronic circuit with a timing device and a motor. This meter, Fig. 22-92, can be calibrated in cu. ft./min., gal./min., or cc/sec.

Fig. 22-93 shows a meter for measuring water flow. This type measures flowage from 1/4 gal./min. up to 5 gal./min. Fig. 22-94 shows the typical flow in gallons per minute for each of the scale markings. Many flow meters are based on a square root rule. For example, the upper half of the graph in Fig. 22-94 follows this formula: FLOW = $\sqrt{(S/2)}$ where S is the scale marking read on the meter.

22-49 ELECTRICAL METER

An electrical meter such as a VOM (volt-ohm-milliammeter) or a digital multimeter is a useful instrument for diagnosing (finding the cause of) problems with electrical motors, electrical controls, and power supply circuits. Some meters test capacitors and have a scale that gives the capacitance value of a capacitor being tested. See Fig. 22-95. Some meters have an "autoranging" feature which automatically sets the proper range for the best reading on the meter.

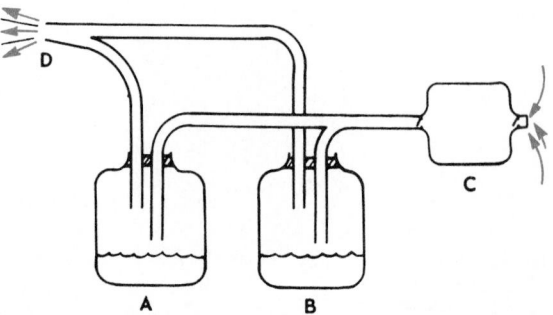

Fig. 22-91. Smoke generator used in determining airflow. A—Hydrochloric acid container. B—Aqueous ammonia. C—Rubber aspirator bulb. Smoke is released at nozzle D.

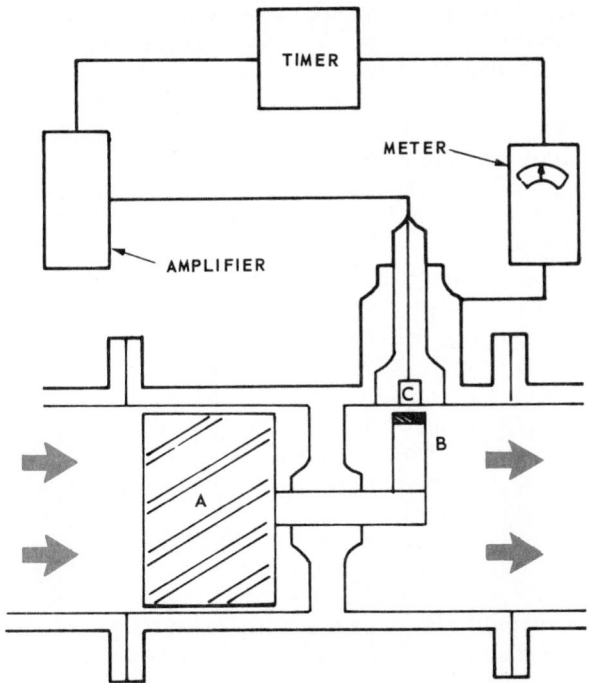

Fig. 22-92. Turbine flow meter. As liquid turns turbine wheel at A, magnet, B, induces an electrical pulse in pickup at C. This pulse is then amplified and timed. Pulse is converted in digital processor. Converted pulse operates meter which registers flow.

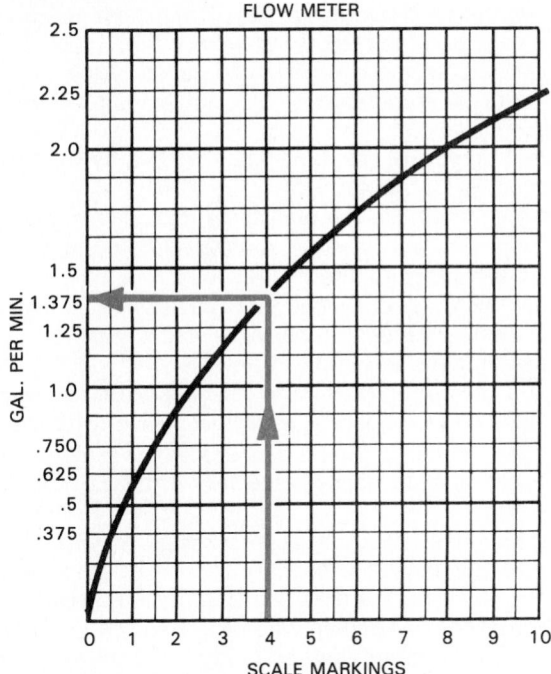

Fig. 22-94. Typical flow meter graph. Amounts of flow in gallons per minute are shown for each scale marking on meter. For example, when orifice moves to the 4 on marking scale, flow is 1.375 gal./min. Upper part of curve is nearly a square root curve. Formula is as follows: $FLOW = \sqrt{(S/2)}$ where S is the scale marking. Use formula for scale readings between 7 and 13.

Fig. 22-93. Flow meter indicates water flow and measures volume of water flow. (The Henry G. Dietz Co., Inc.)

22-50 REVIEW OF SAFETY

It is especially important to remember that the conditioned air must contain enough oxygen to support life. The carbon dioxide content must be kept to a minimum.

Always be careful when working with or handling metal duct material. Use gloves with metal inserts when handling the material. Use stepladders with nonskid bases.

Be sure that the pressure in a duct is low before opening a duct door. If the door bursts open, it may injure someone.

Fans, motors, and belts are potential safety hazards. When these units are operating, protective shields or guards should be provided for protection. When adjusting, be sure the main power switch is off and locked in the off position before handling.

Be careful that objects do not fall into a revolving fan. The object (nut, bolt, tool) may become a dangerous projectile. Always spin a fan by hand to check if it is free to move before turning on the power.

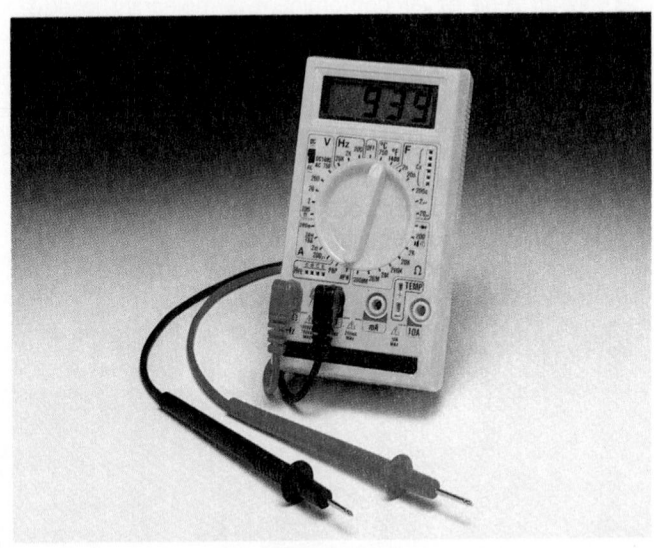

Fig. 22-95. Digital multimeter with capacitance measurement feature saves time when troubleshooting electrical motors and electrical controls. (The Dickson Company)

Electrostatic air filters require high voltage to charge the dust particles. Be sure that the current is turned off and the resistors have dissipated the capacitor charge before servicing electrostatic air filters.

All air conditioning equipment is provided with safety controls which cut out the burner if the bonnet (plenum chamber) temperatures get too high. If filters become clogged, the airflow may be reduced, which may cause the temperature limit control to cut off the heat source.

Fan belts and fan motors sometimes fail. A failure in the fan drive will result in an overheated furnace. Check to make sure that the temperature limit controls are in satisfactory condition.

Avoid exposure to ultraviolet lights. Eyes may be damaged by ultraviolet light.

Always determine the pressure in that part of a system on which work is to be done. Always measure the temperature of the parts of the system which will be repaired, adjusted, or touched. Do not guess pressure or temperature.

Use instruments to check electrical circuits. Never assume that the power is off.

Use a master instrument to check any test instrument that has been dropped. If necessary, have it repaired.

Carbon monoxide (CO) is dangerous! It is always present in the combustion gases, especially if combustion is not complete.

A substance called palladium chloride may be used to measure the presence of CO. To use it, place a small amount, the size of a dime, on 2 in. by 2 in. plastic tabs. The substance will darken when exposed to CO. Amounts of CO can be determined by color:

30 to 70 ppm will cause slight darkening.

80 to 120 ppm cause a grey color.

Over 130 ppm result in a black color.

22-51 TEST YOUR KNOWLEDGE

1. Is it possible to have a hydronic system and duct system combination?
2. What is meant by negative pressure?
3. What type of air contaminant is molecular in size?
4. What is the volume of 1 lb. of 72 °F (22 °C) air at atmospheric pressure if it has 50 percent relative humidity? (For greater accuracy, compare Para. 22-2 with Para. 22-12.)
5. Does air have specific heat?
6. Should air blowers run all the time?
7. To what does decibel refer?
8. When a comfort cooling air duct has condensation on its outer surface, does the duct air become warmer?
9. How much fresh air should each person receive?
10. What four methods cool a person?
11. What is the voltage of the ionizing wires of an electronic filter?
12. How efficient are 1-in. throwaway filters?
13. Why is an air velocity of 25 ft. per minute important?
14. Does air have inertia?
15. List two ways smaller ducts may be used to distribute heated air and maintain good heating season temperatures.
16. What is meant by the unit pressure drop system?
17. What is meant by 0.5 in. (1.3 cm) water column?
18. What is meant by a total pressure drop system?
19. What amount of oxygen is normally contained in air?
20. What is induced draft and forced draft?
21. What is the first step in servicing electrostatic filters?
22. Why are variable pitch pulleys used?
23. What is the purpose of a water spray in an air conditioning system?
24. What is the purpose of the fire damper?
25. List three uses for air ducts.
26. What is the purpose of a grille?
27. Why must one be careful when using ultraviolet light?
28. What is the principal impurity removed by an activated carbon filter?
29. How is electricity used to remove dust from the air?
30. How are most electronic filters cleaned?
31. Where does CO_2 come from in flue gas?
32. What is the CO_2 content of oil furnace flue gas at 100 percent excess air?
33. How does a smoke tester operate?
34. What is a cause if the temperature limit control shuts off in an electronic filter system?
35. What is the purpose of ultraviolet lights?
36. How can a galvanized steel duct be prepared for painting?
37. When using a CO_2 indicator to measure combustion efficiency by a reduction in volume, will the contraction of the flue gas sample by cooling give a higher or a lower reading than the true reading?
38. What is an elbow?
39. How can a draft be detected?
40. What is the purpose of a damper?

Translucent air duct, approximately .040'' thick and weighing 1/2 lb. per lineal foot. (Solar Components Corporation)

Chapter 23

HEAT PUMPS AND COMPLETE AIR CONDITIONING SYSTEMS

After studying this chapter, the technician will be able to:
☐ Describe the basic operation of a heat pump and list its principal parts.
☐ Explain the operation of the heat pump reversing valve.
☐ Identify on a heat pump diagram the changes in refrigerant flow when the reversing valve is changed from heating mode to cooling mode or vice versa.
☐ Discuss types of outdoor coils and their applications.
☐ Perform routine maintenance and service on heat pumps.
☐ Explain total energy systems.

23-1 HEAT PUMP

Heat pump theory rests on the principle that heat will move from a higher temperature to a lower temperature. Thus, if a heat transfer coil can be kept at a lower temperature than its surroundings, it will pick up heat.

If the evaporator of a refrigerating system is mounted outdoors and is operated at a refrigerant temperature of 0 °F (−18 °C), it will remove heat from the air even when the outside temperature is 10° to 15 °F (−12 °to −9 °C). If, after it has evaporated, the refrigerant is compressed to a temperature of 120° to 140 °F (49° to 60 °C), the hot refrigerant will release heat to the surrounding space (inside of a building).

Then, if, by using a system of valves, the evaporator is changed into the condenser and the condenser becomes the evaporator, heat can be removed from the living zone during hot weather and discharged outdoors. Fig. 23-1 shows the pressure-heat diagram for a heat pump.

From the above explanation, one can see that evaporators and condensers are heat transfer devices. They can be used for cooling (picking up heat) or heating (releasing heat).

The principle of using a refrigeration unit as a heating mechanism too was first proposed by Lord Kelvin over 100 years ago. Such a device was not built for over 30 years.

The heat pump is sometimes called a reverse-cycle mechanism. However, the cycle is not actually reversed, only the evaporator and condenser are interchanged. Therefore, the name "reverse-cycle" is not technically correct.

The heat pump may be used for many purposes. Heating water and heat recovery from industrial processes are but two. Even the defrosting of evaporators, using the hot gas defrosting method, is a form of heat pump. Both the compression system and the absorption system can be adapted as heat pumps.

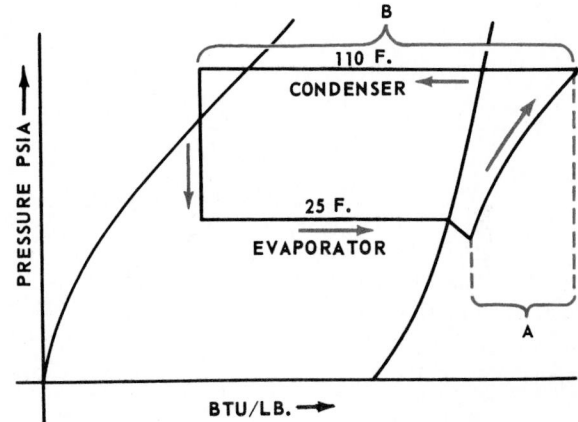

Fig. 23-1. Typical heat pump cycle used for heating. Refrigerant evaporates outside at 25 °F (−4 °C) in 35 °F (2 °C) air. Same refrigerant condenses at 110 °F (43 °C) in condenser in air duct. A—Heat of compression. B—Heat released to house.

Fig. 23-2 illustrates a typical ground source heat pump installation operating on a heating cycle. Fig. 23-3 illustrates the same heat pump operating on a cooling cycle. The basic principles are described in Chapter 19.

Large systems of 100- to 1000-ton capacity are in use as are window units of 1/2 to 2 tons. Self-contained systems of 2 to 25 tons are common.

23-2 HEAT PUMP OPERATION

Operation of the heat pump is like any other compression cycle. The principal parts of the systems are:
1. Compressor.
2. Condenser.
3. Liquid line.
4. Two refrigerant controls.
5. Evaporator.
6. Suction line.
7. Motor control.
8. Reversing valve.
9. Two check valves.

Notice that two check valves, a reversing valve, and two refrigerant controls are needed in the simpler of the heat pumps to change from summer cooling to winter heating. Fig. 23-4 shows a heat pump operating as a comfort cooler. The liquid refrigerant bypasses the TEV at the lower left.

GROUND SOURCE HEAT PUMP—WINTER CYCLE

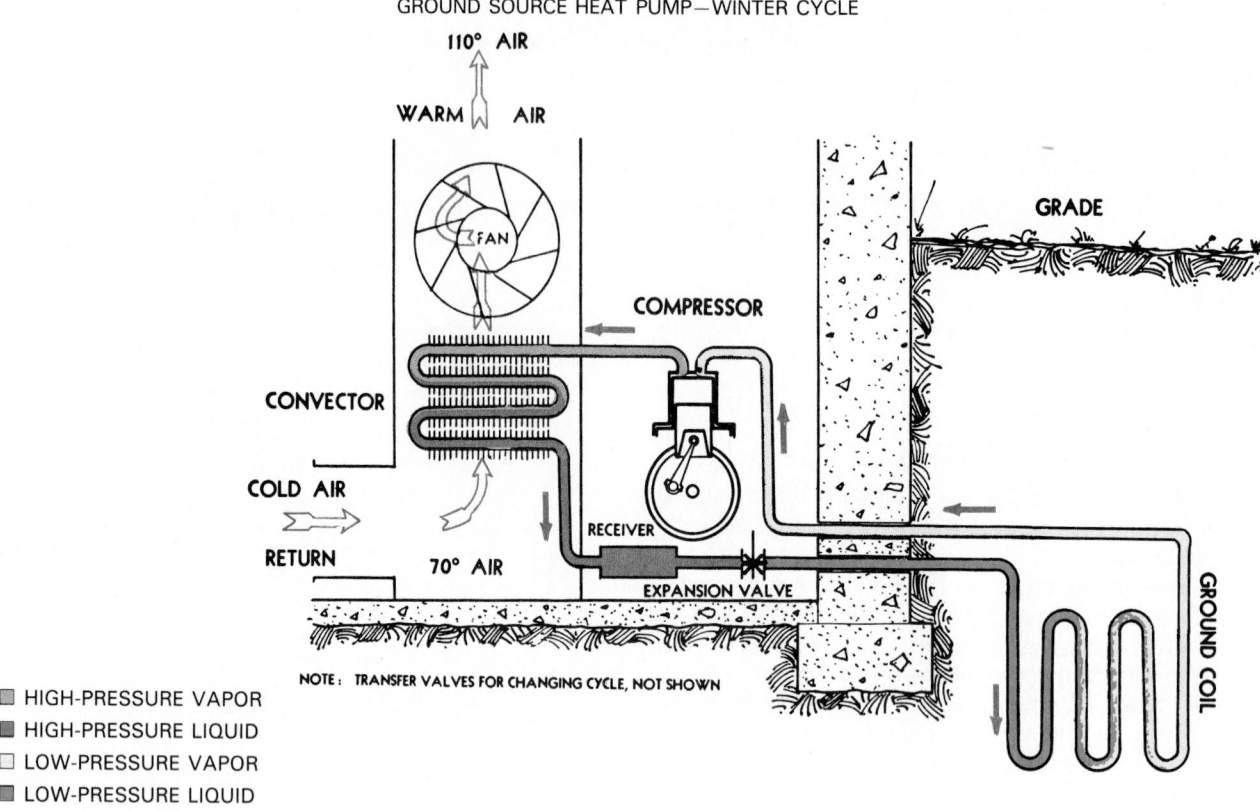

HIGH-PRESSURE VAPOR
HIGH-PRESSURE LIQUID
LOW-PRESSURE VAPOR
LOW-PRESSURE LIQUID

Fig. 23-2. Diagram illustrates principle of heat pump heating cycle using ground coil.

GROUND SOURCE HEAT PUMP—SUMMER CYCLE

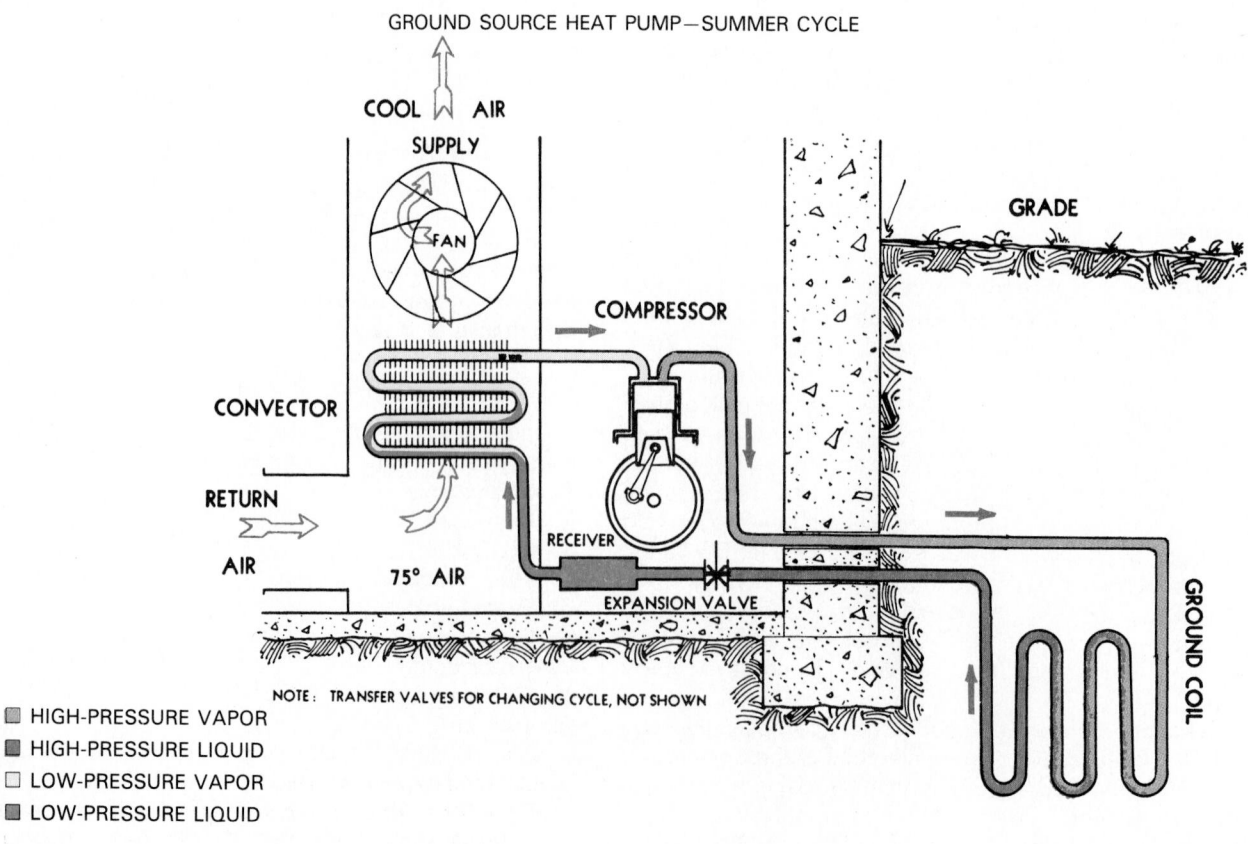

HIGH-PRESSURE VAPOR
HIGH-PRESSURE LIQUID
LOW-PRESSURE VAPOR
LOW-PRESSURE LIQUID

Fig. 23-3. Ground coil heat pump operating on cooling cycle.

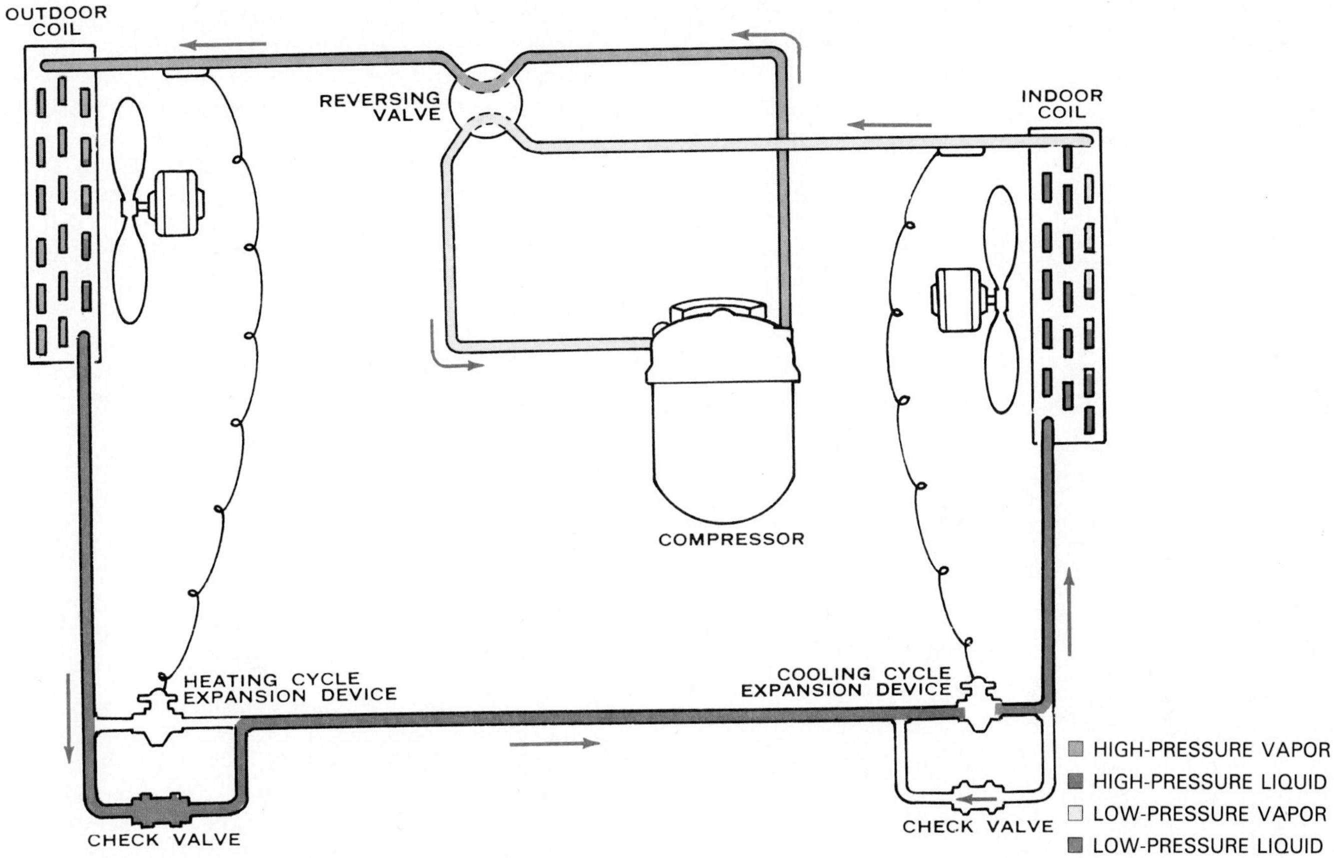

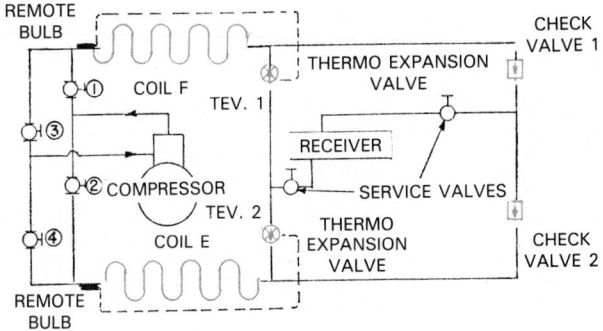

Fig. 23-4. Heat pump. Both heat transfer coils are blower coils. TEV refrigerant controls are used. Note four-way reversing valve. System is operating as comfort-cooling unit with outdoor coil as condenser and indoor coil as evaporator.

(It goes through the check valve bypass.) The TEV at the lower right is the refrigerant control in use.

By turning the reversing valve 90 deg., refrigerant flow is reversed in all lines except the two lines leading into the compressor. These valves are often called four-way valves because of the number of openings.

23-3 HEAT PUMP CYCLES

The heat pump operates in two different cycles:
1. Heating cycle.
2. Cooling cycle.

The same mechanism is used for both cycles, but the travel of refrigerant is reversed to change from cooling to heating. Fig. 23-5 shows a basic heat pump system. Either hand valves or thermostatically controlled valves are used to reverse the cycle.

During the heating cycle, heat is removed from the ambient (surrounding) air and released inside the house. Remember that heating is not usually needed until the outdoor temperature is less than 65 °F (18 °C). Heat from appliances and people living in the house usually makes up this small difference.

For example, if the following conditions prevail,

OUTSIDE (AMBIENT) CONDITIONS		INSIDE CONDITIONS	
Temp.	Humidity	Temp.	Humidity
50 °F	80%	72 °F	50%
10 °C		22 °C	

Fig. 23-5. Heat pump has hand valves to permit manual changing of system from heating to cooling. Coil F is the outside coil and coil E is the inside heat transfer surface. During cooling season, valve 1 is open; valve 2 is closed; valve 3 is closed; valve 4 is open. Check valve 1 is open; check valve 2 is closed. TEV 1 is not working; TEV 2 is working. During heating season, valve 1 is closed; valve 2 is open; valve 3 is open; valve 4 is closed. Check valve 1 is closed; check valve 2 is open; TEV 1 is working; TEV 2 is not working.

(Alco Controls Div., Emerson Electric Co.)

the outdoor coil will act as an evaporator and pick up heat from outdoors. This heat is released in the house, as shown in Fig. 23-6.

The heating cycle of the heat pump becomes less efficient as outdoor temperature drops below the freezing level. This action, along with an increase in the heating load as the outside temperature lowers, creates problems in those colder climates where temperatures cool down to 20 °F

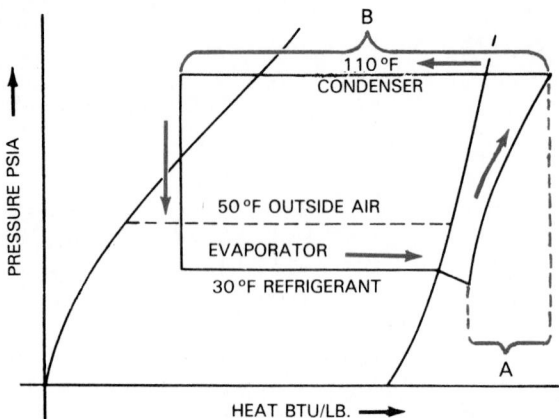

Fig. 23-6. Pressure-heat chart **for** heat pump serving as heating system with outside temperature at 50 °F (10 °C). Refrigerant is evaporating at 30 °F (−1 °C). Refrigerant is condensing at 110 °F (43 °C). Btu ratio of B to A is Energy Efficiency Ratio in Heat Units (EER$_Q$). EER$_Q$ is B/A, which is about 4.

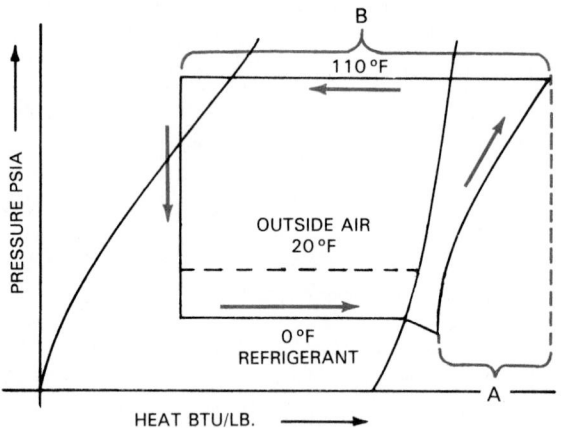

Fig. 23-7. Pressure-heat chart of heat pump heating cycle in operation when outside (ambient) temperature is 20 °F (−7 °C). This requires a 0 °F (−18 °C) refrigerant temperature. Note increase in heat of compression compared to Fig. 23-6. EER$_Q$, which is B/A, is less than that for Fig. 23-6.

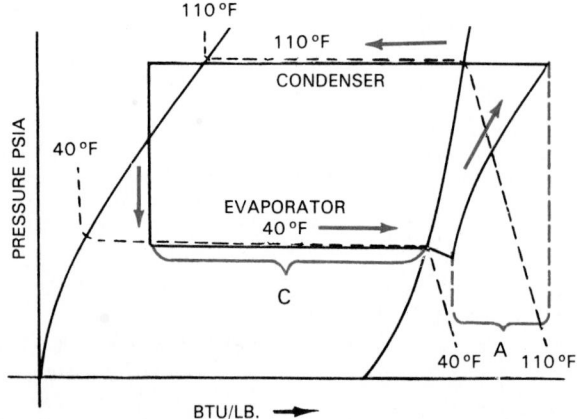

Fig. 23-8. Pressure-heat chart shows cooling cycle for heat pump. A—Heat energy of compression. C—Heat energy removed from air (cooling and dehumidifying). Note that A is about 1/3 of C, meaning that 3 times more heat is moved than is used to pump. This is a Coefficient of Performance (COP) of 3:1.

(−7 °C) or lower. Fig. 23-7 shows such a condition with the outside temperature at 20 °F requiring a refrigerant temperature of 0 °F (−18 °C).

Note that A has increased with very little increase in B, which means that the energy efficiency ratio in heat units (EER$_Q$) is less. Also, with the refrigerant boiling at 0 °F (−18 °C), the evaporator will frost rapidly. Some way of frequent defrosting would have to be found.

If the ambient conditions are within 10 °F (6 °C) and 10 percent relative humidity (RH) of the inside conditions, very little treatment is needed due to the heat lag and time lag in controlling the variables (the sun's heat on a south wall in the morning may correct a problem of low temperature by afternoon; likewise, temperature and relative humidity in a very large room may hold to desired levels overnight). But if the ambient temperatures were to increase over this amount, for example:

OUTSIDE (AMBIENT) CONDITIONS		INSIDE CONDITIONS	
Temp.	Humidity	Temp.	Humidity
85 °F	75%	72 °F	50%
29 °C		22 °C	

a cooling cycle would be required. Heat and moisture would be removed from the inside of the house and released outdoors as shown in Fig. 23-8.

On the psychrometric chart, the apparatus would affect the air conditions as shown in Fig. 23-9. Point A is the condition of the ambient (surrounding) air and represents air at 85 °F (29 °C) and 75 percent RH. The line from A to B shows how the air is cooled to 100 percent relative humidity. Line B to C shows how the air is further cooled and how moisture is removed from the air.

If 100 percent fresh air is being conditioned, point C represents the air as it leaves the evaporator. Then, as the air mixes with the air in the house, or as it mixes with air being brought into the duct system (recirculated air), point D is reached.

23-4 HEAT PUMP EFFICIENCY

Heat pump units have been designed for single rooms, for complete houses, and for industrial uses. The unit takes

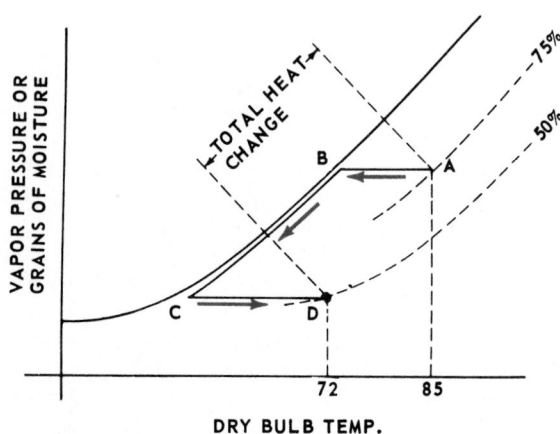

Fig. 23-9. Psychrometric chart shows effect on air in house when using cooling cycle of heat pump.

the place of separate comfort cooling and heating apparatus.

Cooling load is about 60 percent of heating load for most of the United States. Above the 37 deg. parallel, the heating load is greater and the cooling lower. Below this line, the cooling load is greater while the heating is lower.

The source of energy is usually electricity. To furnish 50,000 Btu per hour directly from electrical resistance heating would require 50,000 Btu/hr. This can be converted to kilowatts (kW) as follows:

Since 1 kW = 3412 Btu/hr.,

kW = 50,000 ÷ 3412 = 14.6, thus,

50,000 Btu/hr. = 14.6 kW

Electricity is sold by the kilowatt hour (kWh). First find the cost for 1 hr. as follows—multiply the kW by 1 hr:

14.6 kW × 1 hr. = 14.6 kWh

The cost per kilowatt hour varies from 3.04¢ to 19.23¢. At five cents a kilowatt hour, the heat load would cost

14.6 × 0.05 = 0.73 for 1 hr.

For one day, the cost would be

0.73 × 24 = $17.52/day.

For one 30-day month, the cost would be

$17.52 × 30 = $525.60/mo.

This cost appears high. However, the actual cost is much less than this, as the heat load for a house averages much less than this value. The average outdoor temperature is much higher than the design temperature.

The 50,000 Btu/hr. is based on 70 °F (21 °C) indoor and 0 °F (−18 °C) outdoor. If the average outdoor temperature is 35 °F or 2 °C (quite common), the heat load would be 25,000 Btu/hr. (This is from $50,000 \times [1 - \frac{35}{70}]$.) The cost would, therefore, be cut in half, or $262.80/mo. The heat pump reduces this cost considerably.

Electricity drives the heat pump compressor. The refrigeration cycle—if the proper temperatures are used—permits the condenser to release three to four times as much heat as it takes in electrical energy to drive the compressor. The ratio of the heat delivered at the condenser to the electrical energy used to drive the compressor is important. However, it is easier to discuss power than energy. The following formula is used (see Para. 15-63):

$$EER_Q = \frac{Btu/hr.}{watts} \times 0.293$$

The factor EER_Q is called the energy efficiency ratio in heat units (Q is the thermodynamic symbol for heat). The number 0.293 is from 1/3.412 Btu/hr. per watt. This EER_Q means that one kilowatt of electrical power driving the compressor can, by using the heat pump, release not 3412 Btu/hr., but 3412 × 3, or 10,236 Btu/hr. The steps just done are described in the following formula:

Heat released in Btu/hr. = kilowatts × 3412 × EER_Q

In the preceding cost example, with an outdoor temperature of 0 °F (−18 °C), the EER_Q will be equal to one. (The EER_Q cannot be less than one.) The heat pump would then cost as much to run as electrical heating. If the temperature outside is 35 °F (2 °C), the EER_Q would be approximately 1.5. Therefore, the operating cost of the heat pump would be $175.20 per month—a savings of $87.60 per month compared to electric heat. The cost for the preceding example is:

Cost = $262.80/mo. ÷ EER_Q

Suppose a 25,000 Btu/hr. heat pump costs $3000 to install, not including the ductwork. This is about $1000 more than an equivalent gas furnace. Then at a savings of $87.60/mo., the savings relative to a gas furnace is estimated as 1/3 of $87.60, or $30. The payback period is:

$$\frac{\$1000}{\$30} = 33 \text{ months} = 2.7 \text{ yr.} = 3 \text{ yr.}$$

A payback period of 3 yr. is a little long, but the purchase should still be practical. The cost per kilowatt hour ranges from 3¢ to 19¢. From 5¢ to 19¢, the savings will be multiplied by about 4, but the payback period stays the same. (It does not help if the electricity to run the compressor costs as much as 20¢ per kWh.)

Note: a heat pump is designed for cooling. Its heating ability is left to whatever quantity is the result. With an EER_Q = 1.5, the COP = 0.5 (see Para. 15-63). With Q_{COLD} = 16,600 Btu/hr., $Q_{HOT} = 16,600 \times \frac{1.5}{0.5} =$ 49,800 Btu/hr. This suggests that the 50,000 Btu/hr. heating unit may be needed after all (instead of 25,000 Btu/hr.).

This EER_Q can be further increased by using a heat source that is warmer than the outside air. For example, well water, lake water, and the ground itself may be used to provide the heat for the heat pump.

If a well furnishing water at 60 °F (16 °C) can be found, the EER_Q can be raised to as much as 4 or 5. This factor would lower the cost of heating a great deal. For EER_Q = 5, the cost is $52.56, a savings of approximately $210 per month compared to the sole use of electric resistance heating.

23-5 HEAT PUMP SYSTEMS

Use of heat pumps for residential heating and cooling is increasing. Hermetic units have been developed in many designs and in a wide range of sizes.

The hermetic system is ideal for heat pump installations because it is so simple to install. Some are equipped with service valves, suction and discharge mufflers, and other special features which provide quiet, reliable operation and long life.

The coil mounted inside the house is usually a standard finned type with a blower. The outside coil comes in a variety of designs. The choice depends on the medium (substance) in which the coil is placed—air, water, or earth.

The "outside" coil is not always located outside the structure. It may be located inside but ducted to the outside if it is an air coil. See Para. 23-10.

23-6 HEAT PUMP COMPRESSORS

Heat pump compressors have a standard construction. Units up to 5 hp use compressors of constant capacity. Larger systems sometimes use a modulating type. The trend is to use hermetic systems.

Because the pumping load varies greatly during the day and with the change of seasons, variable capacity compressors are being used in larger systems. These compressors change capacity by operating valves which unload one or more compressor cylinders into clearance pockets. See Chapters 4 and 12.

Heat pump compressors must be designed to operate during the unusual conditions of reversing the cycle. Motors should be protected with internal temperature thermostats. The compressor must be designed to handle some liquid slugging without injury. Crankcase heaters are often used to protect the compressor from liquid refrigerant buildup during low temperature operating periods. Suction line accumulators are often used to protect the motor compressor.

23-7 HEAT PUMP MOTORS

Heat pumps of 1/3- to 1-ton capacity use standard single-phase motors. They usually have a starting capacitor. Single-phase motors should be operated at 240 V, if possible, so that smaller wires can be used. Three-phase motors are preferred in units over one-ton capacity, mainly for electrical economy.

Motors in hermetic systems are usually well insulated. They can tolerate some voltage change and have temperature sensor protectors built into the motor windings.

23-8 HEAT PUMP MOTOR CONTROLS

A double set of automatic controls is usually required for the heat pump. One thermostat must be designed for heating during cold weather conditions. Another is needed for warm-weather cooling. In most cases, these two controls are mounted in one casing or housing. Humidistats are not used on all models. However, they are needed for complete automatic control.

Lines are normally fitted with brazed or welded connections. Brazing is the most popular. Flexible connections are usually installed in the lines at the compressor to eliminate noise and to absorb some vibration.

Usually there is a receiver to hold the extra refrigerant needed when the system is on a cooling cycle. An accumulator is often used on the main suction line to minimize the chance of liquid refrigerant reaching the compressor.

23-9 HEAT PUMP REFRIGERANT CONTROLS

Either thermostatic expansion valve or capillary tube refrigerant controls may be used with heat pumps. Fig. 23-10 illustrates a heat pump system using a capillary tube.

A device called a restrictor provides the same function as a capillary tube.

Two thermostatic expansion valves or capillary tubes may be used on a heat pump system. The thermostatic expansion valve is usually the pressure-limiting type. Only one valve or tube will operate as a control valve at any time.

Heat pump coils may be placed in air, in the ground, or in a well or lake. If an air coil is used, the valve bulb on the outlet of the coil being used as the evaporator should be cross charged to provide more efficient operation at low air temperatures. The changeover valves are much simpler in units that use the capillary tube control, because the flow of refrigerant can be reversed through the tube for pressure-reducing purposes. In this type of installation, a strainer must be installed at both ends of the tube. Both the thermostatic expansion valve (TEV) and the capillary tube control are described in Chapter 5.

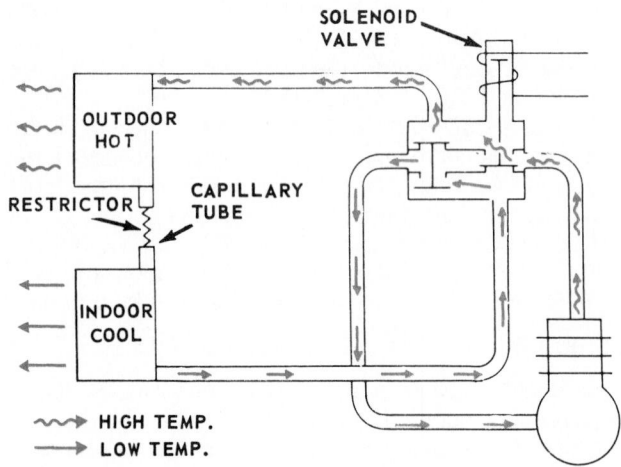

Fig. 23-11. Heat pump with capillary tube refrigerant control operating on cooling cycle (cool indoor coil). Bent arrows mean high temperature. Straight arrow lines mean low temperature. Solenoid valve assembly is four-way type. High pressure sets left pair of check valves in position. (Ranco North America)

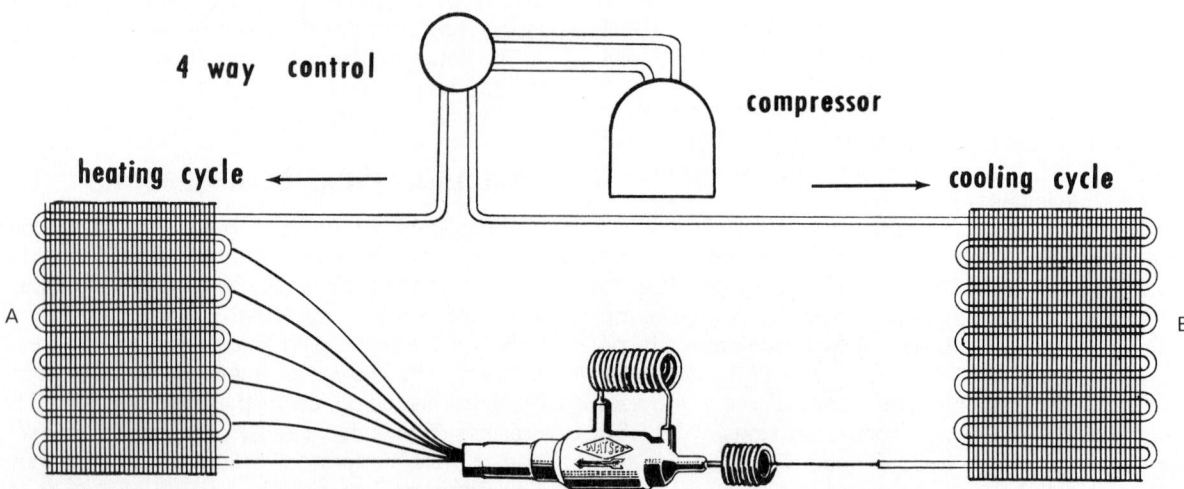

Fig. 23-10. Heat pump system using capillary tube refrigerant controls. A—Indoor coil. B—Outdoor coil. Arrows show refrigerant flow to the left when system is used for heating and to the right when system is used for cooling. (Watsco Components, Inc.)

Fig. 23-11 is a diagram of a heat pump using a capillary tube and a solenoid reversing valve. As shown, it is operating on the cooling cycle. The wiring diagram in Fig. 23-12 shows the control contacts in position for this cooling cycle. It also shows the interlocking of the four-way solenoid-operated reversing valve and the compressor.

Fig. 23-13 is a simple diagram of a heat pump using a capillary tube and a solenoid-operated reversing valve. It is shown operating on a heating cycle. The wiring diagram in Fig. 23-14 shows the control contacts for this heating cycle. Note the change in the upper contact position of the reversing lever.

The thermostat in Fig. 23-15 operates the four-way valve and also controls temperature. Note the three terminals. The wiring of heat pumps depends mainly on whether the motor is connected directly to the line or whether some kind of starter (magnetic or transformer) is used.

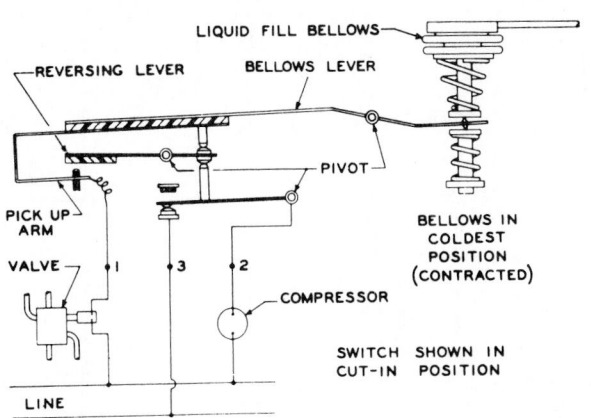

Fig. 23-14. Electrical connections during heating cycle of heat pump.

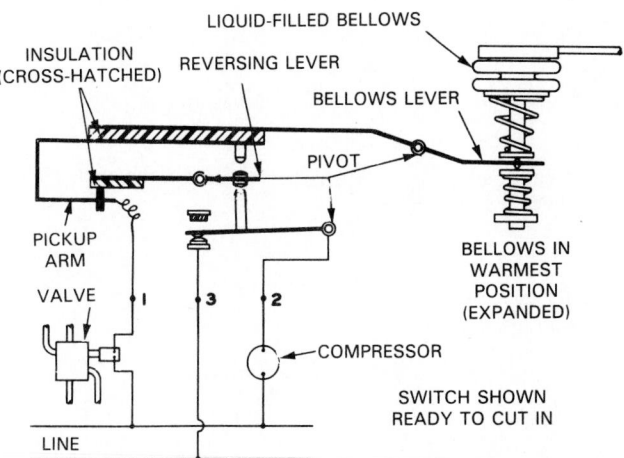

Fig. 23-12. Wiring diagram of heat pump shows electrical connections for cooling cycle. When switch connected to point 3 is open (in its upper position), whole system is off. When reversing lever is pushed upward, its left end tends to bring pickup arm down. The tension on pickup arm makes it easier for bellows to pivot bellows lever. This will energize solenoid early: solenoid will change valve to heating cycle.

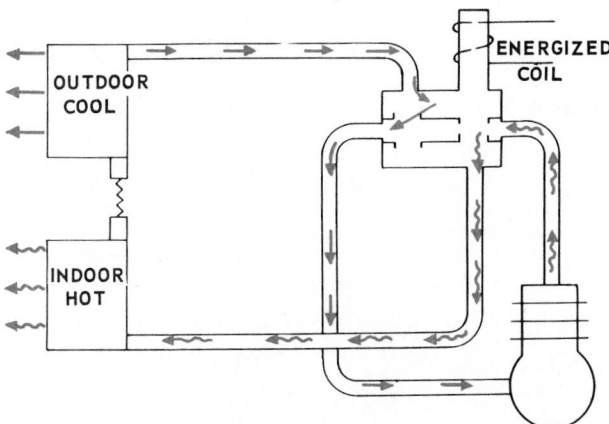

Fig. 23-13. Heat pump operating on heating cycle. Crooked arrows indicate high tempeaure. Straight arrows indicate low temperature. Note use of four-way valve. High pressure holds left pair of check valves in raised position. Heat transfer in valve body increases superheat.

23-10 HEAT PUMP REVERSING VALVES

Many different types of special reversing valves are used in heat pumps. If refrigerant flow is manually reversed, at least six one-way valves are needed. The special reversing valves may be operated automatically, manually, or electrically (through solenoids).

Some units use three-way valves either manually or electrically operated. These valves have one opening to the compressor, one to the condenser, and one to the evaporator. Two of these valves are needed to run the unit.

Other units use a four-way valve to reverse the flow of refrigerant. The valve is operated by the movement of one valve stem. This stem closes and opens several ports in one valve body. Its operation may be manual or electrical. The complete system is easily reversed with one of these valves. They are popular in small tonnage conditioners such as window units and other air-to-air systems.

One kind of four-way valve operated by pressure is shown in Fig. 23-16. A solenoid valve controls the pressure at the top portion of the reversing valve. When the solenoid pilot valve is de-energized, the compressor high-side pressure pushes on the top of the four-way valve. The reversing valve slider is lowered, producing the heat cycle. Coil A, the indoor coil, is the condenser, while coil B, the outdoor coil, is the evaporator.

When the solenoid pilot valve circuit is closed, either manually or by a thermostat, the solenoid valve is lifted and the slide inside the four-way valve comes up. Refrigerant flow to the coils is reversed.

The heat pump has a thermostatic expansion valve and a check valve on each coil. When the system is reversed, the refrigerant flow bypasses the thermostatic expansion valve of the coil serving as the condenser.

During the cooling cycle, the upper thermostatic expansion valve is used. The refrigerant cannot travel through the companion check valve.

Fig. 23-17 illustrates this same system operating as a cooling unit. Coil B becomes the condenser and is releasing heat to the outdoors. Coil A becomes the evaporator.

Another style of reversing control uses a lever inside of a solenoid pilot valve, as shown in views B and C of Fig. 23-18. The pilot valve controls a four-way slide. The solenoid coil is energized on the heating cycle. The pilot slide

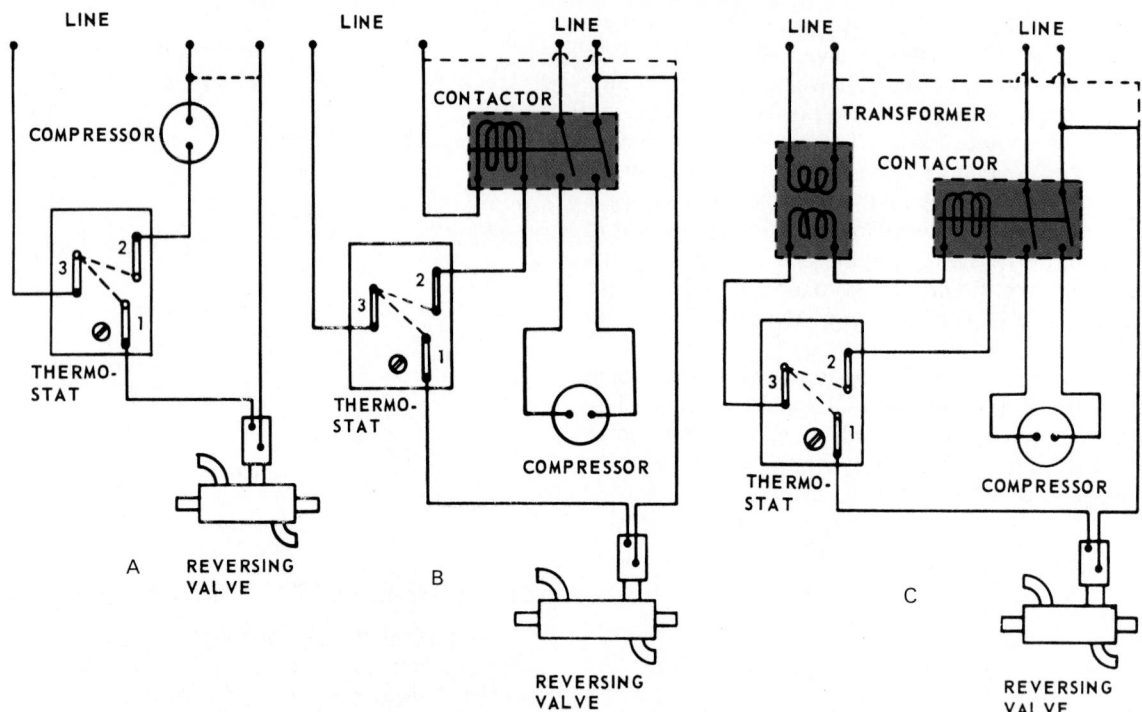

Fig. 23-15. Three wiring diagrams for heat pumps. A—Thermostat controls motor circuit. B—Circuit uses line voltage coil magnetic contactor. C—Circuit uses low-voltage coil magnetic contactor.

HEATING CYCLE

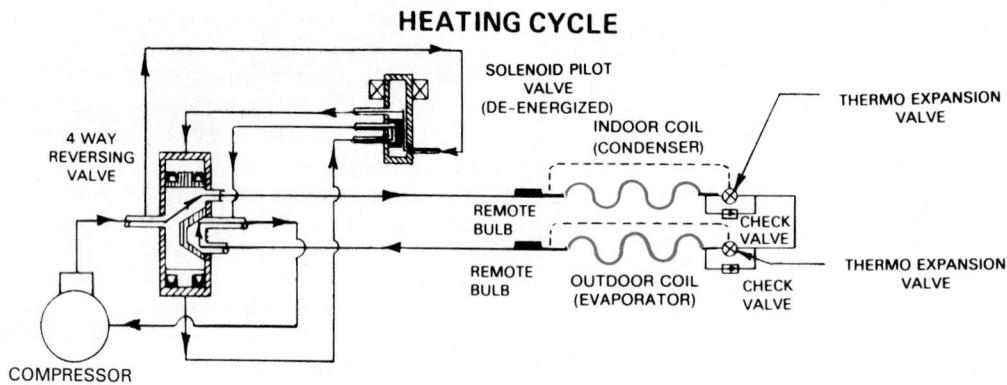

Fig. 23-16. Heat pump diagram. System uses four-way reversing valve. Solenoid pilot valve is de-energized during heating cycle. (Alco Controls Div., Emerson Electric Co.)

COOLING CYCLE

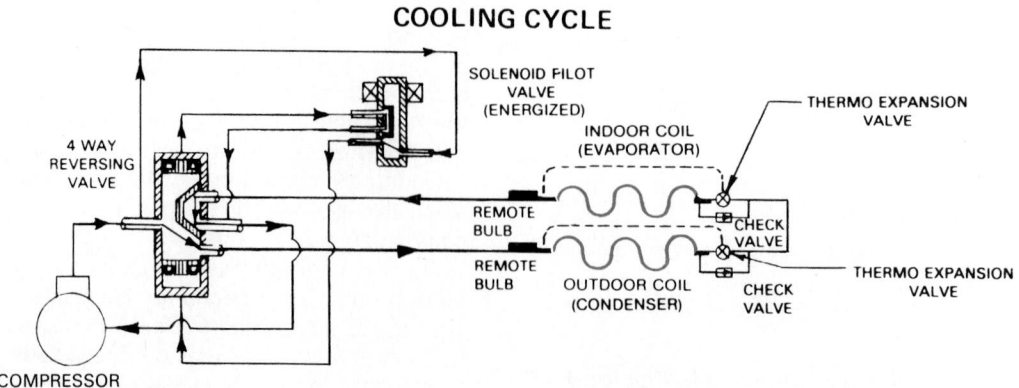

Fig. 23-17. Heat pump diagram shows four-way reversing valve in use operating on cooling cycle. Note how functions of indoor and outdoor coils reverse from heating cycle shown in Fig. 23-16. (Alco Controls Div., Emerson Electric Co.)

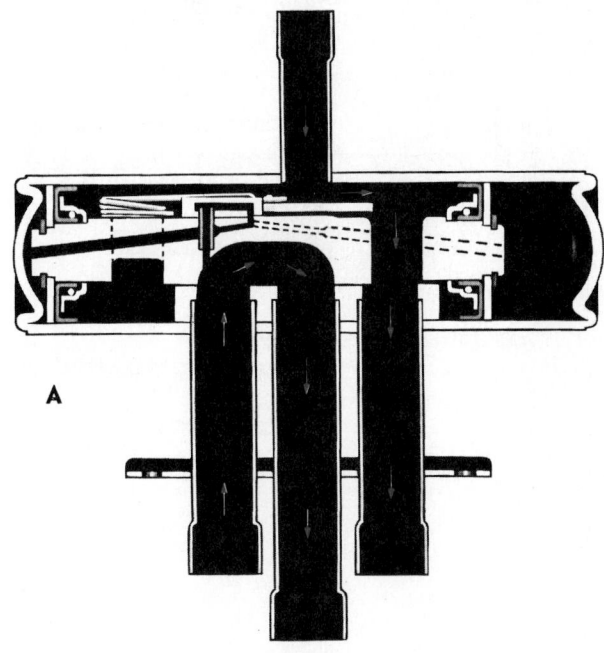

A

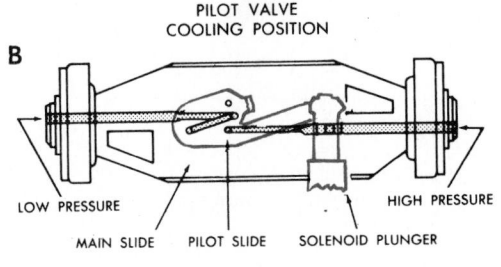

PILOT VALVE
COOLING POSITION

B

LOW PRESSURE   HIGH PRESSURE

MAIN SLIDE   PILOT SLIDE   SOLENOID PLUNGER

PILOT VALVE
HEATING POSITION

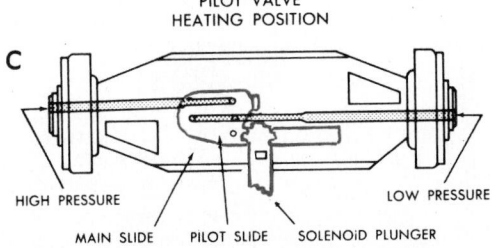

C

HIGH PRESSURE   LOW PRESSURE

MAIN SLIDE   PILOT SLIDE   SOLENOID PLUNGER

Fig. 23-18. Cutaway of pilot valve mechanism used to operate four-way reversing valve. Note position of solenoid plunger in heating position when solenoid is energized. A—Side view of four-way valve. B—Pilot mechanism and piston in cooling position. C—Pilot mechanism and piston in heating position. Note how high pressure changes place with low pressure. Holes in pilot slide lead to compressor inlet and outlet.

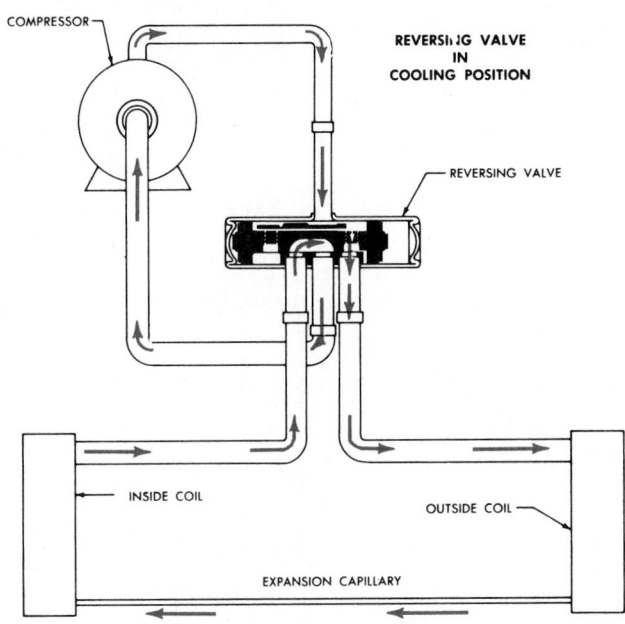

Fig. 23-19. Heat pump cycle uses reversing valve. Note flow of refrigerant when valve is set for cooling cycle.

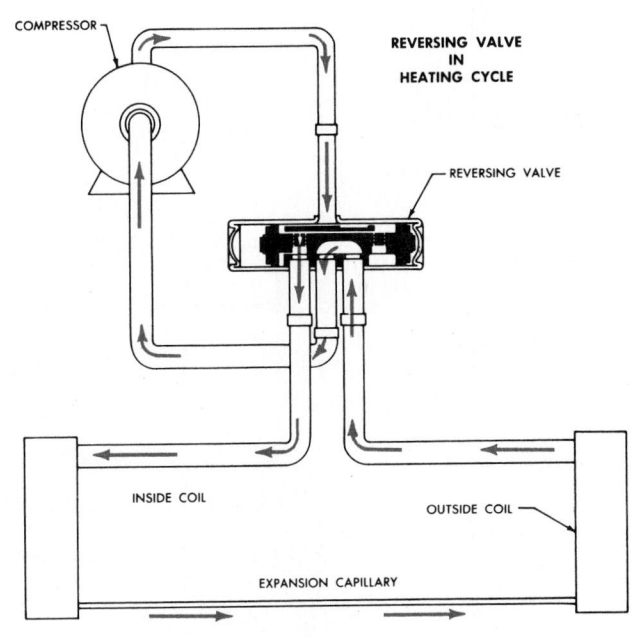

Fig. 23-20. Heat pump circuit with reversing valve set for heating cycle.

pivots and changes the flow of pressure. Pressure moves the main slide seen in view A of Fig. 23-18. In the cooling cycle, the solenoid is not energized. Fig. 23-19 shows the reversing valve installed in a heat pump system. The system is on a cooling cycle. When the valve is in position for a heating cycle, the system is set up as in Fig. 23-20.

Figs. 23-21 and 23-22 show a main sliding valve with bleed holes. With the bleed hole system, one less pipe is needed to the pilot valve.

Some systems use a flow check piston as an expansion valve instead of a capillary tube. A sketch of a flow check piston assembly is shown in Fig. 23-23. Liquid flow from the condenser is metered into the distributor of the

evaporator through a hole in the center of the piston. With flow in this direction, the flow check piston provides an even distribution of refrigerant into the evaporator. If the flow is reversed, the piston moves back, providing a free flow of refrigerant without throttling.

This device is useful in heat pumps where two flow check pistons are used in the flow line between the evaporator and condenser. In heat pumps used for heating in winter, Fig. 23-24, and cooling in the summer, the flow is reversed. One flow check piston then acts as the expansion device while operating in the heat pump mode. The other operates as the expansion valve when in the cooling mode, when

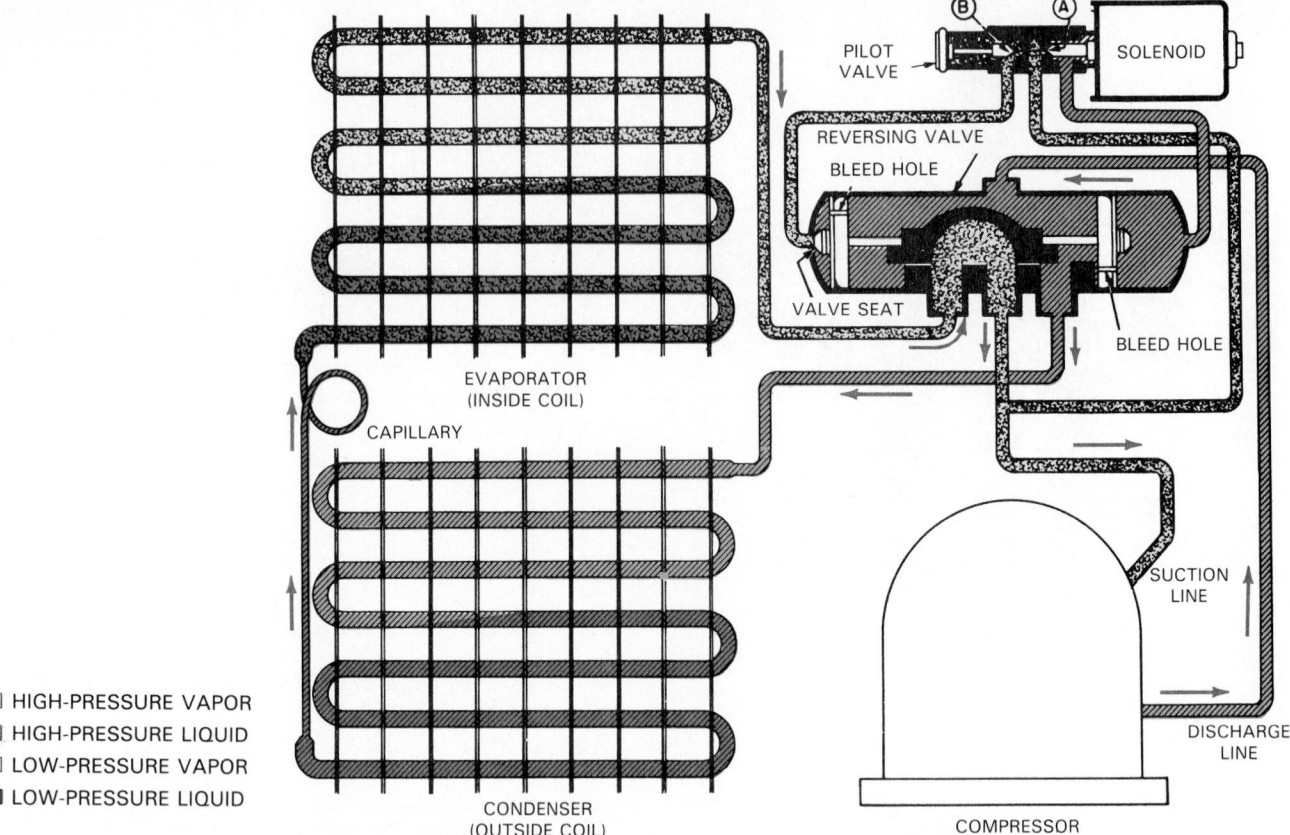

HIGH-PRESSURE VAPOR
HIGH-PRESSURE LIQUID
LOW-PRESSURE VAPOR
LOW-PRESSURE LIQUID

Fig. 23-21. Heat pump uses capillary tube refrigerant control and solenoid-valve-controlled four-way valve. Indoor coil is serving as evaporator. Note bleed holes in main slide. With bleed hole system, only three tubes to pilot slide are needed.

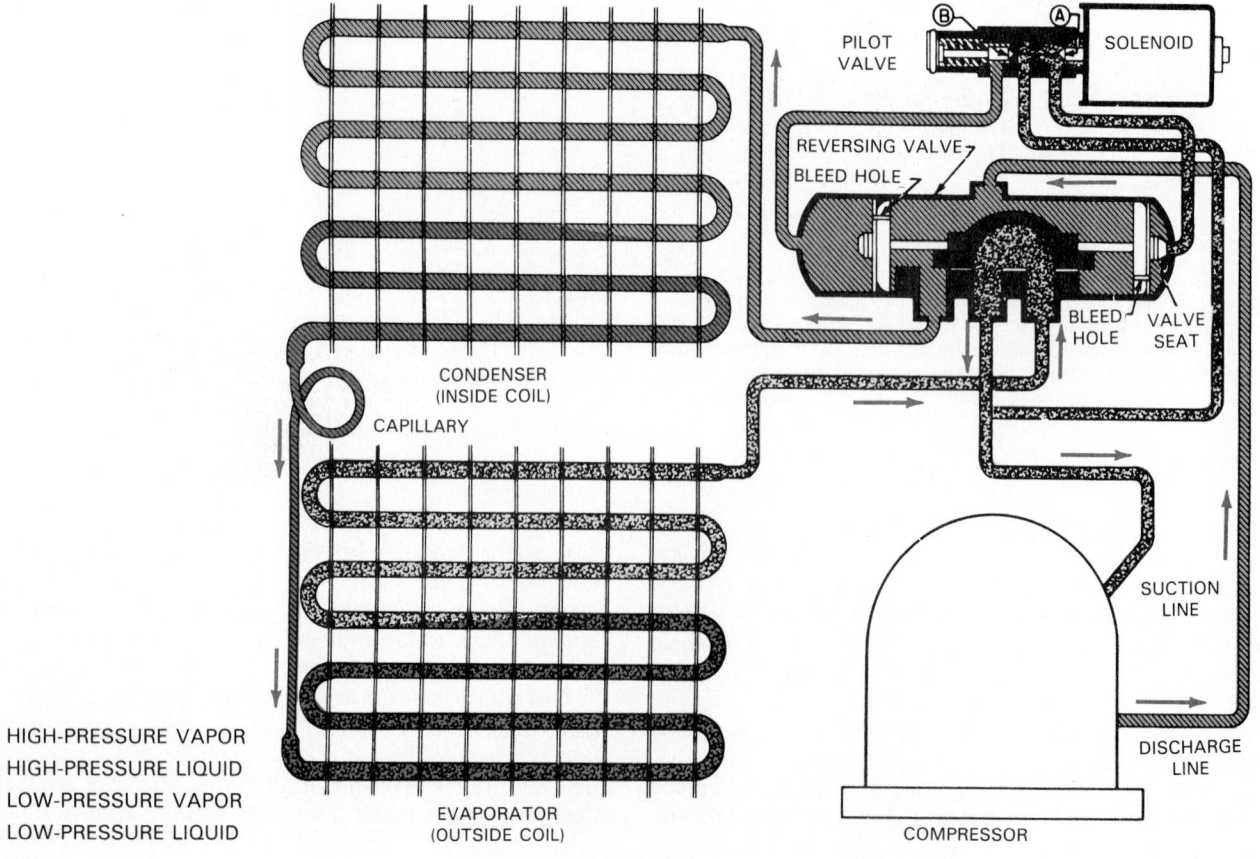

HIGH-PRESSURE VAPOR
HIGH-PRESSURE LIQUID
LOW-PRESSURE VAPOR
LOW-PRESSURE LIQUID

Fig. 23-22. Capillary-tube-type heat pump using indoor coil as condenser.

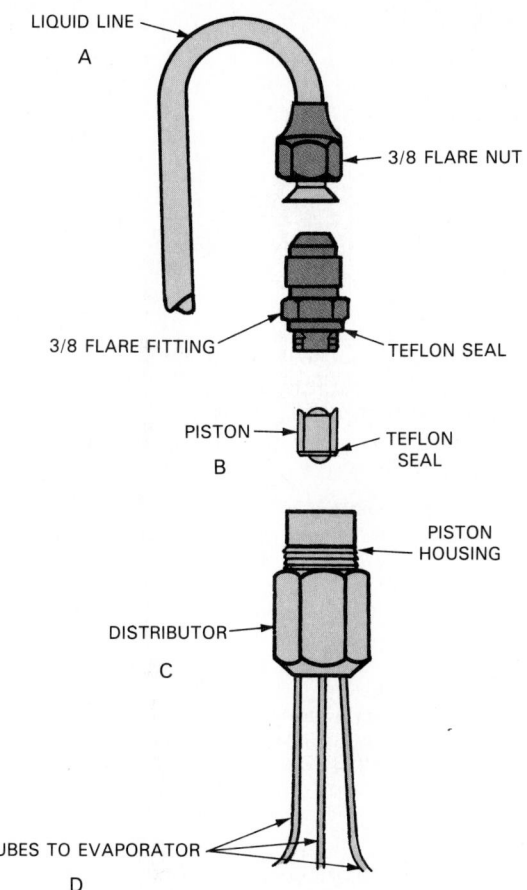

Fig. 23-23. Flow check piston. Refrigerant flows from liquid line A through piston B, through distributor C, and through tubes leading to evaporator D. (Rheem Mfg. Co., Air Conditioning Div.)

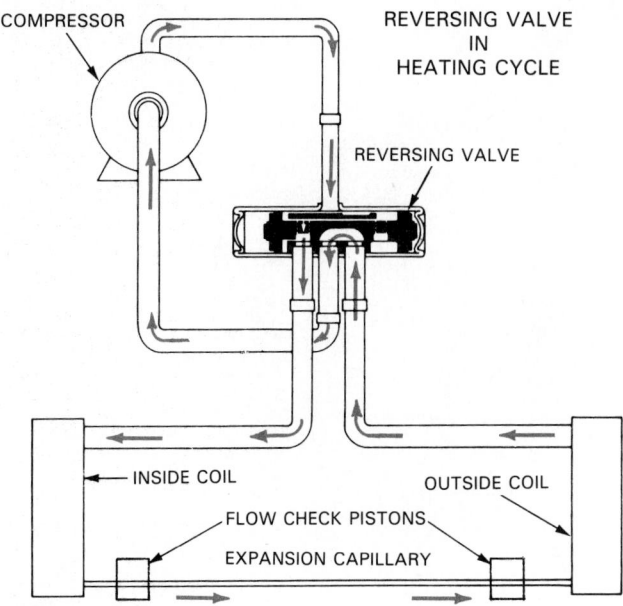

Fig. 23-24. Heat pump circuit with flow check pistons set for heating cycle.

flow is reversed.

Fig. 23-25 is a diagrammatic sketch of a self-contained heat pump system. Two separate circulating fans are used. In each illustration, the white coil is the warm one (condenser) and the black coil is the evaporator.

These units are typically installed either in the basement or on the first floor of a home, usually in a utility room. Both coils and the motor compressor are installed in a casing.

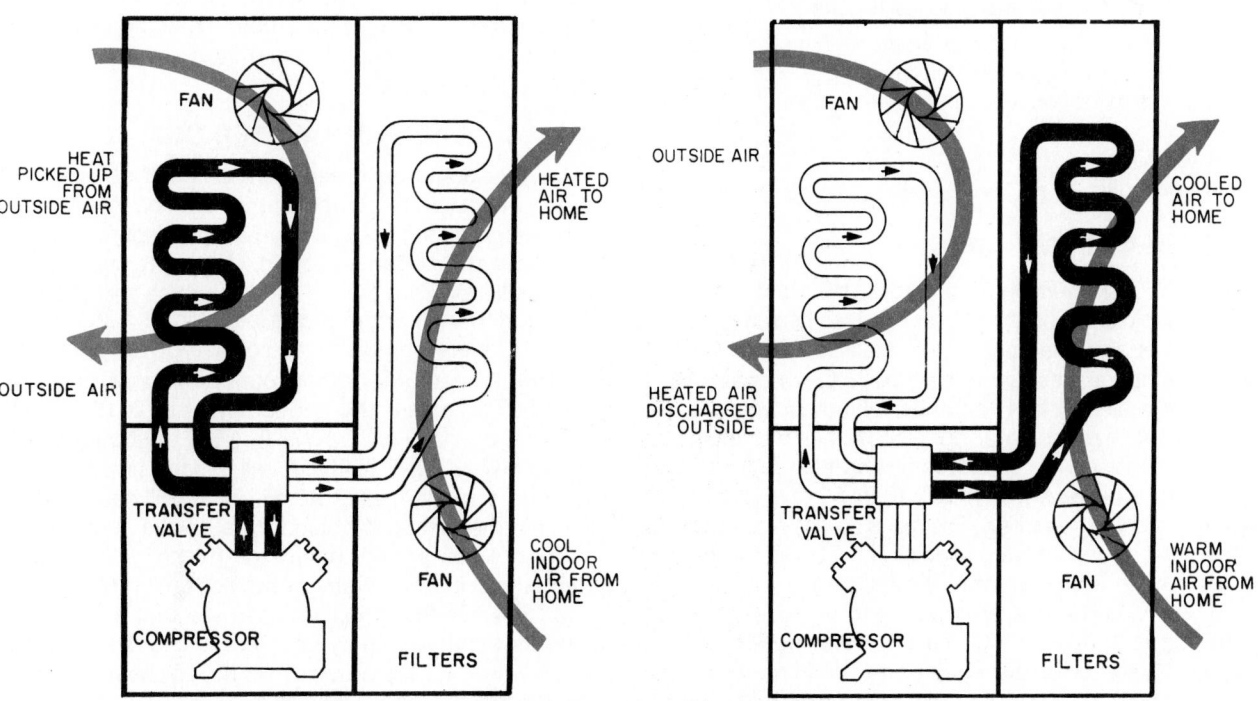

Fig. 23-25. Air and refrigerant circuits in heat pump during heating and cooling operations.

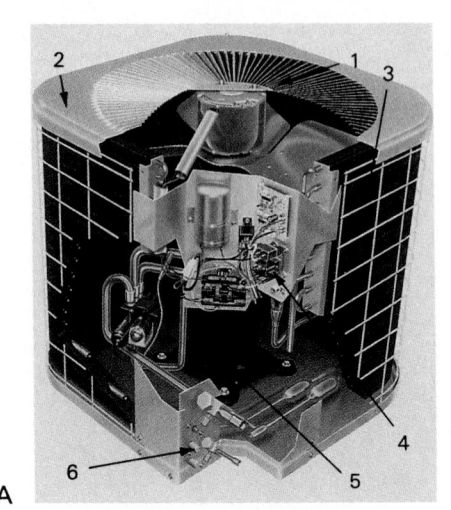

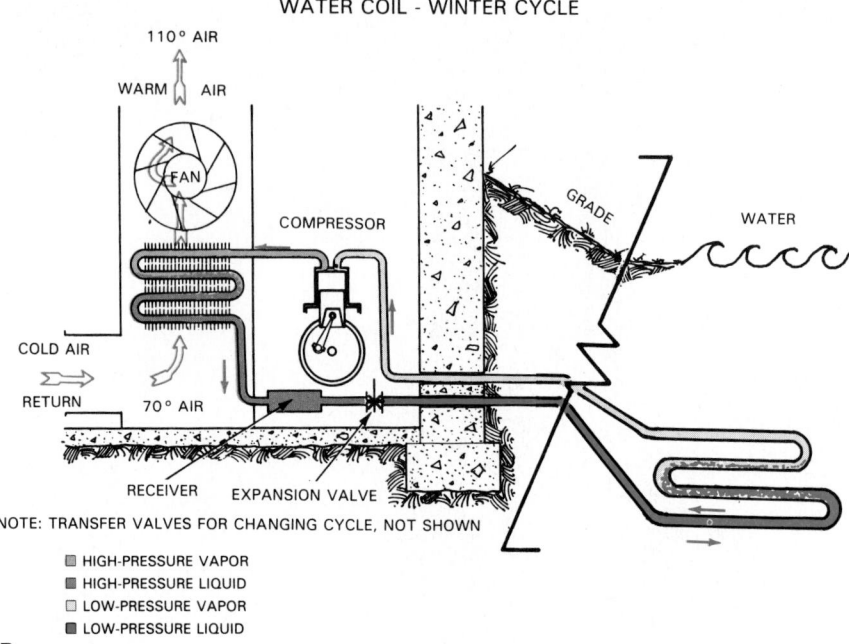

WATER COIL - WINTER CYCLE

NOTE: TRANSFER VALVES FOR CHANGING CYCLE, NOT SHOWN

▨ HIGH-PRESSURE VAPOR
▨ HIGH-PRESSURE LIQUID
☐ LOW-PRESSURE VAPOR
▨ LOW-PRESSURE LIQUID

Fig. 23-26. Two types of heat pumps, air and water. A—Five-ton heat pump air coil. 1—The exhaust air discharges upwards. 2—Cabinet. 3—Condenser coil. 4—Solid-state defrost system to prevent ice build-up. 5—Scroll compressor. 6—High-pressure switch shuts unit off if pressure gets abnormally high. (Tempstar Heating and Cooling Products, Inter-City Products Corporation) B—Diagram illustrates principle of heat pump using water coil.

The indoor air and the outdoor air are ducted to the self-contained unit.

23-11 HEAT PUMP CONDENSER AND EVAPORATOR COILS

The coil mounted inside the structure is normally a standard finned coil with a blower.

The outside coil is available in several designs. The type used is determined by the substance (water, stone, etc.) into which the coil releases its heat or from which it picks up its heat. Coils, classified by the type of external heat source used to obtain the heat of evaporation, are:
1. Air coil.
2. Water coil.
3. Ground coil.

23-12 HEAT PUMP AIR COIL

Of all the heat pump outdoor coils, the least expensive is one that picks up heat from (or releases heat to) the outside air. It is also the easiest to install. In climates where outdoor temperatures range from 20 °F (−7 °C) to 110 °F (43 °C), the air coil has a number of advantages.

A standard heat transfer coil, the air coil has tubing for a primary surface with extended fins bonded to the tubing. Along with a blower, it is mounted in a housing that protects if from the weather. The housing may be mounted on the outside wall of the building, Fig. 23-26, view A, on the roof, or in a shelter near the building.

A typical wiring and piping diagram for a split system heat pump unit is shown in Fig. 23-27, view A. The outdoor unit has a vertical air discharge with the compressor in a separate compartment. The indoor component is a vertical upflow-type unit. The flow is up from the basement. Fig.

23-27, view B, shows six types of heat pump unit installations and the airflow for each.

During the heating cycle, the outside coil tends to frost and ice up in certain weather conditions. To prevent this, most heat pumps have a special thermostat mounted on the coil. If frost or ice starts to form, this control will shut off the system and electric heating coils will start a defrost action. In one method, the cycle will reverse long enough to defrost the coil. Some systems use a timer. It operates the system on defrost for a time. During defrost, the drain also must be heated to make sure drain water can flow away from the unit.

23-13 HEAT PUMP WATER COIL

A heat pump becomes more efficient as the condenser and evaporator temperature approach each other. Therefore, during the heating cycle, if a source warmer than the ambient air can be found, the pump will work better.

Installations have been tried in which the outside coil is lowered to the bottom of the lake, Fig. 23-26, view B. The coil so located has a more consistent temperature.

The temperature of maximum density of water is 39 °F (4 °C). This means that the temperature of the water at the bottom of a lake will be 39 °F (4 °C) or above. If the water cools below 39 °F (4 °C), it will expand and rise to the surface. A temperature of 32 °F (0 °C) will cause it to freeze. This is the reason that ice forms on the surface of a lake.

The lake would have to freeze solid before the temperature of the water at the bottom could be lower than 39 °F (4 °C). From this, one can see that a lake—even one covered with ice—may be a reservoir of heat. So too with well water; it may provide an efficient heat pickup unit for heating and a good heat dissipator during the cooling cycle. The cost of the well is a disadvantage, but this will prob-

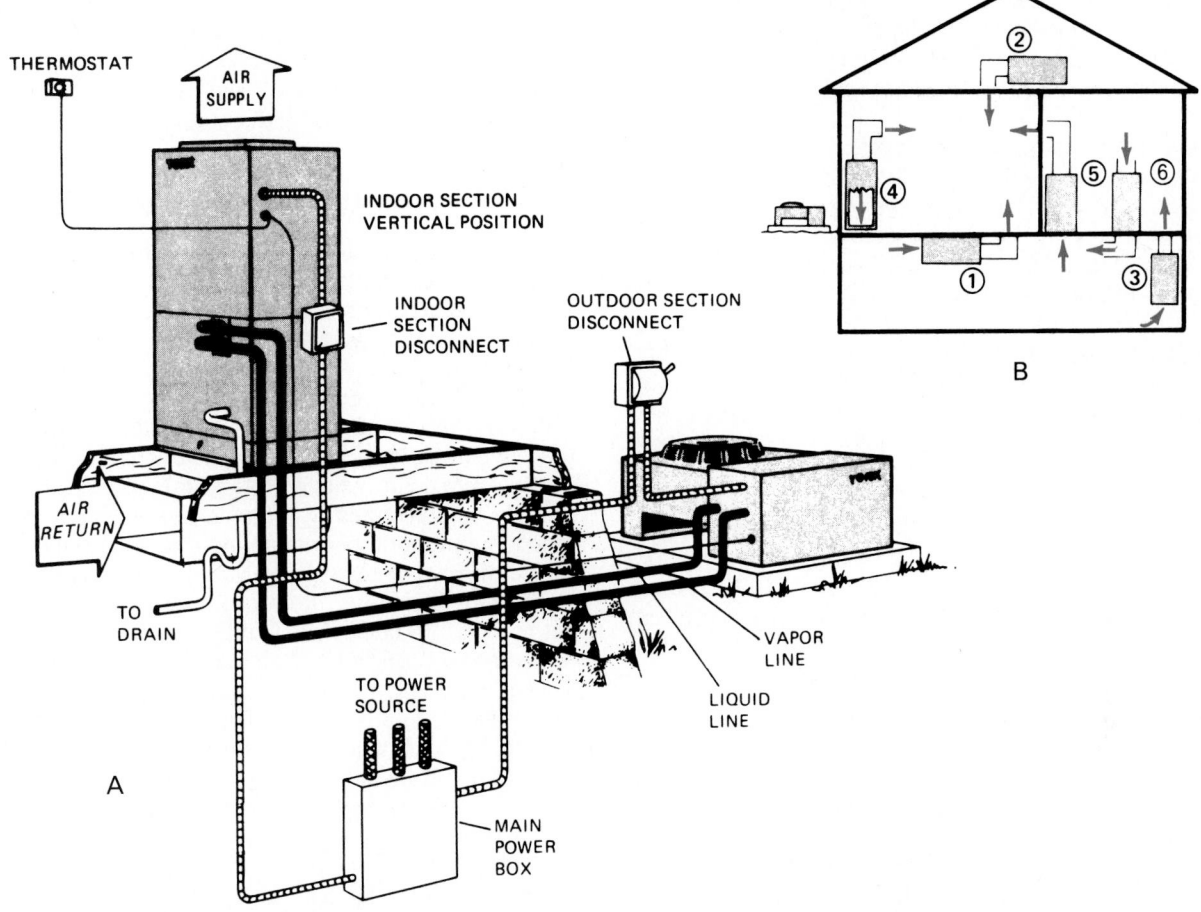

Fig. 23-27. Split system heat pump diagrams. A—Wiring and piping drawing. B—Drawing of six types of split systems. 1—Horizontal suspended. 2—Horizontal; attic or crawl space. 3—Upflow wall-mounted. 4—Upflow side return. 5—Upflow; basement or crawl space return. 6—Counterflow. (Central Environmental Systems, York International Corp.)

ably soon be offset by the lower cost of operation.

In one popular method, well water is drawn from the well, used, and quickly returned. A tube-within-a-tube or a shell-and-tube heat exchanger is used.

With well water at 60°F (16°C), a condensing temperature of 80°F (27°C) can be used during a cooling cycle. An evaporator temperature of 40°F (4°C) can be used during the heat cycle. Often, flowing wells or springs are used to supply water for heat pumps.

When a lake or ground water coil installation is used, one of the major concerns has been the chemical nature and mineral content of the water. A coil made of cupronickel (copper and nickel alloy) is corrosion resistant and desirable for water installations.

23-14 HEAT PUMP GROUND COIL

Regardless of latitude or air temperature changes, the ground temperature at a depth of four to six feet changes very little. Most ground temperatures average between 40°F (4°C) and 60°F (16°C).

A coil with sufficient heat transfer surface, buried four to six feet deep, can be used as an outdoor coil for both heating and cooling cycles. Fig. 23-2 illustrates a ground coil system as it operates during the winter. Ground coils are installed horizontally to the grade although drawings show a simpler view. Also, in commercial units, the air

return is a split-air system with a fresh air makeup duct.

Fig. 23-3 illustrates the same basic system used as a comfort-cooling mechanism. In practice, valves would be used to send the condensed refrigerant through the liquid receiver. From there the refrigerant would go through the expansion valve from left to right before the reduced-pressure refrigerant enters the evaporator in the duct. For example, the receiver would be between the two check valves in Fig. 23-4.

Installations have been made using a combination of air coils and either a lake coil, well coil, or ground coil. The air coil is used alone when the outdoor temperature permits efficient operation. But, when the outside air becomes too cold or too warm, the auxiliary coil may be connected into the heat pump air coil system.

23-15 HEAT PUMP DEFROSTING

The defrosting cycles on many heat pumps begin with a 5 min. period of supplementary electric heat. Also in common use today are defrost controls based upon temperature differentials, pressure differentials, and combined time and temperature differentials. Another recent method uses a unit that depends upon the electrical loading of the heat pump's condenser fan motor due to ice blockage. The unit electrically measures the mechanical load of an induction motor. The cutout signal is obtained from a remote thermistor clipped to the liquid line or from a pressure or temperature control. This method is used in order to maintain a comfortable indoor temperature (by not defrosting too long) during the defrosting process, which is the reversal of refrigerant flow.

The reversing of the heat pump cycle, for defrosting when the compressor is running, causes a strain on the compressor and motor. A heat pump on low ambient (outdoor) temperatures cools the compressor motor with liquid refrigerant. (The suction pressure is high enough that liquid remains.) If the defrosting system fails to return the liquid rapidly from the outdoor coil at the start of the defrost cycle, it fails to provide adequate and rapid defrosting and motor cooling. Hot and cold spots can form on the motor windings and laminated steel cores, leading to high thermal stresses. Also, when the unit shifts to defrost, a large surge of liquid is returned to the compressor, often causing liquid floodback or slugging.

To compensate for the liquid line floodback, and to make the refrigerant return more even, suction line accumulators are used by many manufacturers. See Fig. 23-28. The accumulators act as a receiver for the excess liquid and return it to the compressor in a vapor form just above its condensing temperature at the low suction pressure.

23-16 SUPPLEMENTAL RESISTANCE HEATERS

It is standard practice to size heat pumps to handle the cooling load. This size may or may not be enough to handle heating needs.

In southern climates, a unit big enough for cooling can probably supply enough heat with little supplementary aid. However, in very cold weather, additional heat must be sup-

plied from another source. The supplementary electric resistance heating unit is one such source. It requires little space. Blowers distribute the heated air evenly throughout a room. Such units may be installed in the blower-coil section of the heat pump or in the supply air duct of an indoor system.

23-17 HEAT PUMP INSTALLATION

A heat pump is nearly the same as a refrigerating system. Most of the installation instructions in Chapters 11 and 14 apply to heat pumps as well. The units must be carefully leveled. Electrical service must be of the correct voltage and phase. Wiring must be large enough to carry the load without a critical voltage drop.

In cold climates, the heat pump should be installed above the predictable total snow height for that area or approximately 8 to 18 in. above ground. It is also desirable in extremely cold climates to install a defrost timer instead of standard outdoor thermostats. Always advise the owner to shovel the snow away from the condensing unit.

The temperature at the room heating register is usually set at 105° (41 °C). This is much lower than the register temperature (125 °F [52 °C]) in oil or gas heating systems. Heat pumps will, therefore, take longer to heat up a cold room than a furnace system would. This causes many first-call customer complaints about new heat pump installations.

Window and wall units require the same casing mounting techniques described in Chapter 21. It is important to closely follow the manufacturer's instructions.

Fig. 23-29 pictures a heat pump unit which can be installed in an exterior wall of an office or apartment. This type extracts "natural heat" from the outside in the winter for heating the room or apartment and cools it in the summer by absorbing heat from the room and discharging it to the outside atmosphere.

Through-the-wall heat pumps are easy to maintain. The heat pump chassis may be removed for servicing. The individual heat pump unit is well adapted to apartment units. The owner of a multiple unit does not have a central air conditioning system. Each tenant can set the thermostat to give the heating or cooling desired without affecting others living in the building.

23-18 SERVICING HEAT PUMP

Because the heat pump system is a refrigerating unit, the service techniques of troubleshooting and repair are much the same as the service described in Chapters 11 and 14.

Routine maintenance requires that voltage, current, and refrigerant pressures be checked. The heat pump should be cleaned. Blow out the coils and fins, clean the duct passages, and oil the fan bearings and fan motor bearings. Clean the fan blades. The unit should be run on both cycles to check the operation of the reversing valve.

Service calls often start with a customer complaint. Some common complaints are:
1. Unit will not operate on either cooling or heating.
2. Unit runs too much.
3. System is noisy.

The technician must find the cause of the trouble and repair it. For example, a lack of refrigerant indicates a leak. The leak must be located and repaired. A four-way valve that will not operate must be replaced.

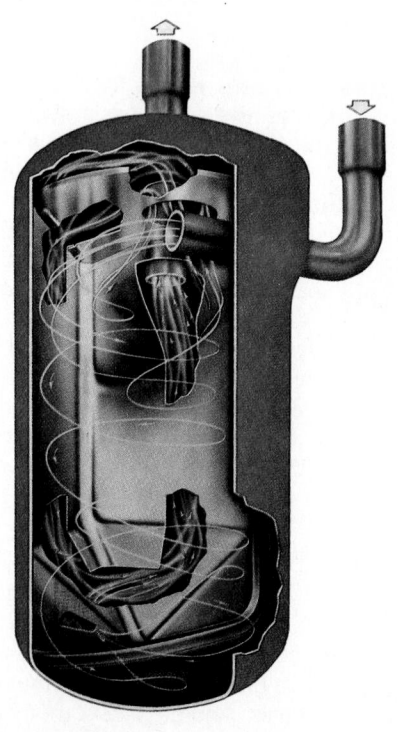

Fig. 23-28. Suction line accumulator. Note that inlet is near top of accumulator and suction line vapor is made to swirl as it leaves accumulator at top. This prevents liquid refrigerant from entering compressor. (Tecumseh Products Co.)

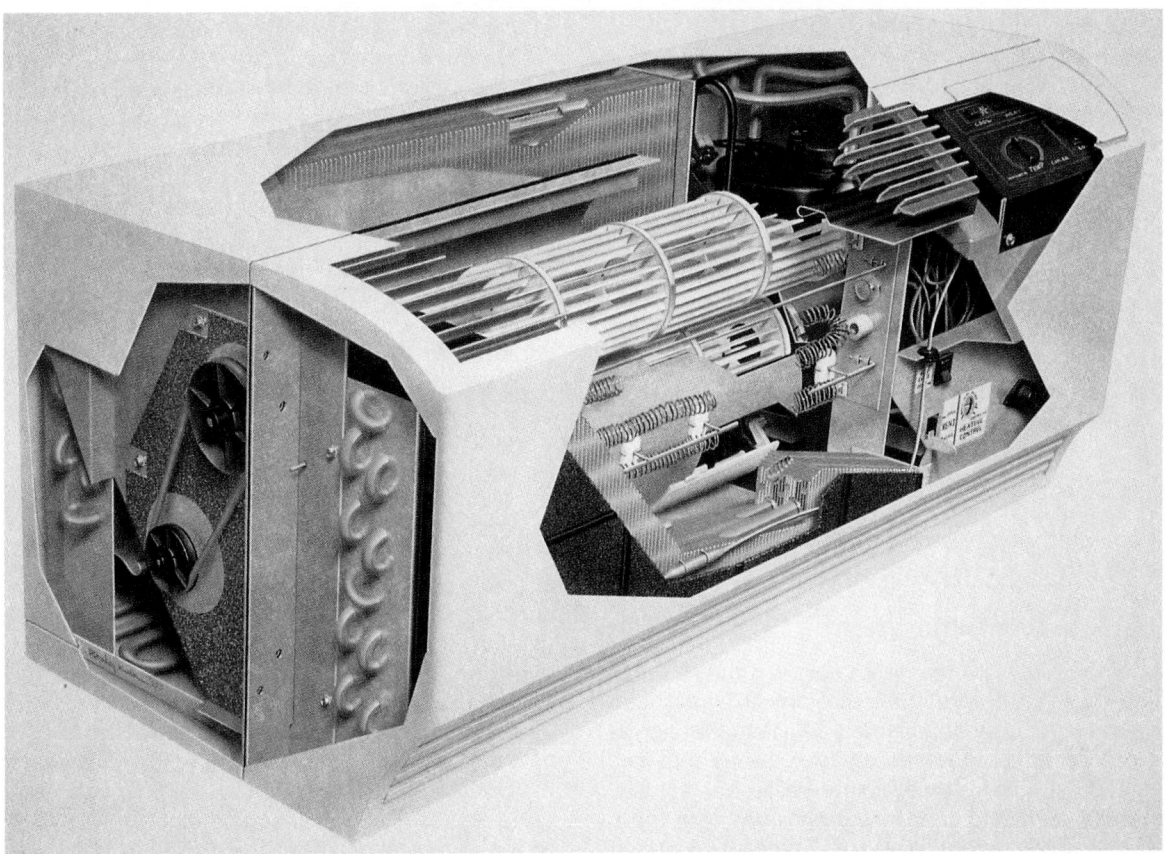

Fig. 23-29. Package terminal air conditioning and heat pump unit that can be installed through wall. Supplementary electrical resistance heating elements maintain desired amount of heating during unusually cold weather. System has rotary compressor, blower wheels.   (ZoneAire, Inter-City Products Corporation)

Fig. 23-30. Two-speed heat pump operates at low speed for normal loads and at high speed for heavy loads.   (Lennox International, Inc.)

Certain checks should be routinely made:
1. Check defrost control.
2. Check and lubricate condenser fan motors.
3. Check refrigerant charge.
4. Check all electrical connections for tightness.

23-19 TWO-SPEED HEAT PUMPS

Two-speed heat pumps, Fig. 23-30, are produced by some manufacturers to get a higher seasonal energy efficiency ratio (SEER). For normal heating or cooling loads, the compressor operates at low speed. Under exceptional conditions when the outdoor temperature is very high, the compressor will operate at high speed for maximum cooling. Likewise, when the outdoor temperature is very low, the compressor will operate at high speed for maximum heating.

One of the major differences between a single-speed motor and a two-speed motor is its adaptability. At high speeds, it operates as a two-pole motor; at low speeds, it is a four-pole motor.

23-20 HI/RE/LI SYSTEM

Another type of heat pump cycle is used in the Hi/Re/Li system. The system differs from the conventional heat pump system in many ways. Fig. 23-31 illustrates some of the main operational differences. They are as follows:
1. The location of the refrigerant control is not on the evaporator but on the condensing unit, with the thermal bulb mounted on the liquid line. Therefore, liquid is controlled as it leaves the condenser rather than when it enters the evaporator.
2. The accumulator/heat exchanger in the suction line allows more subcooling of the liquid refrigerant going to the evaporator. The accumulator/heat exchanger separates liquid-vapor and allows dry vapor and oil to enter the compressor because the liquid is boiled off by the heat exchanger.
3. A heat exchanger on the suction line (a pair of tubes in contact) subcools the condensed liquid refrigerant before it enters the expansion valve. Through subcooling, the amount of "flash gas" and pressure drop is lowered.
4. The evaporator is completely flooded and utilized for cooling, since is it not required to superheat "suction gas" as it is in the normal system to avoid slugging.
5. The condenser system can be reduced because it is not necessary to store refrigerant during periods of low load.

The Hi/Re/Li system uses a reversing valve to change the direction of flow of the refrigerant to fit the needs of either heating or cooling. In addition, it uses a check valve to direct the flow of refrigerant in the proper direction dependent upon whether a heating or cooling cycle is desired. Fig. 23-32 is a diagram of this system in its cooling cycle.

When charging the system, it is best to first remove all of the refrigerant. Then charge it with the correct amount as indicated in the manufacturer's manual. Install service valves on the suction lines and discharge lines fairly close to the compressor to get the correct operating pressures.

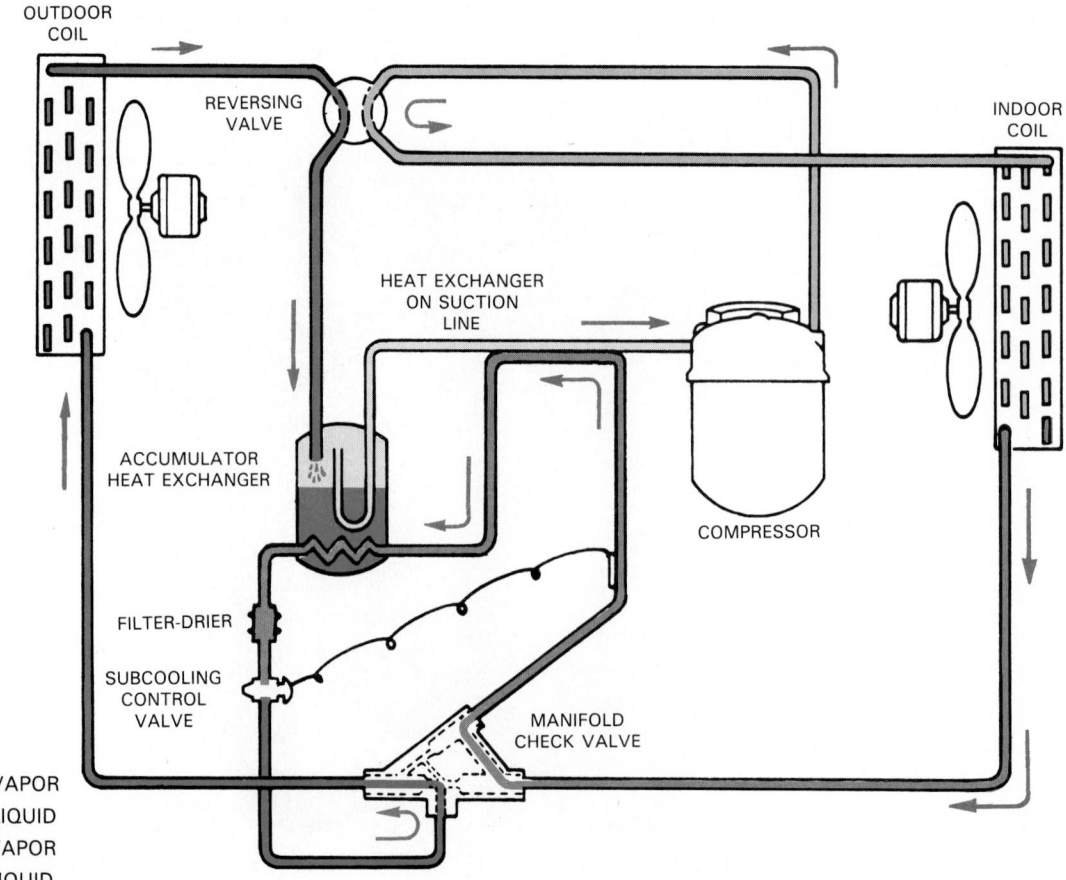

Fig. 23-31. Hi/Re/Li heat pump system operating as heating system.

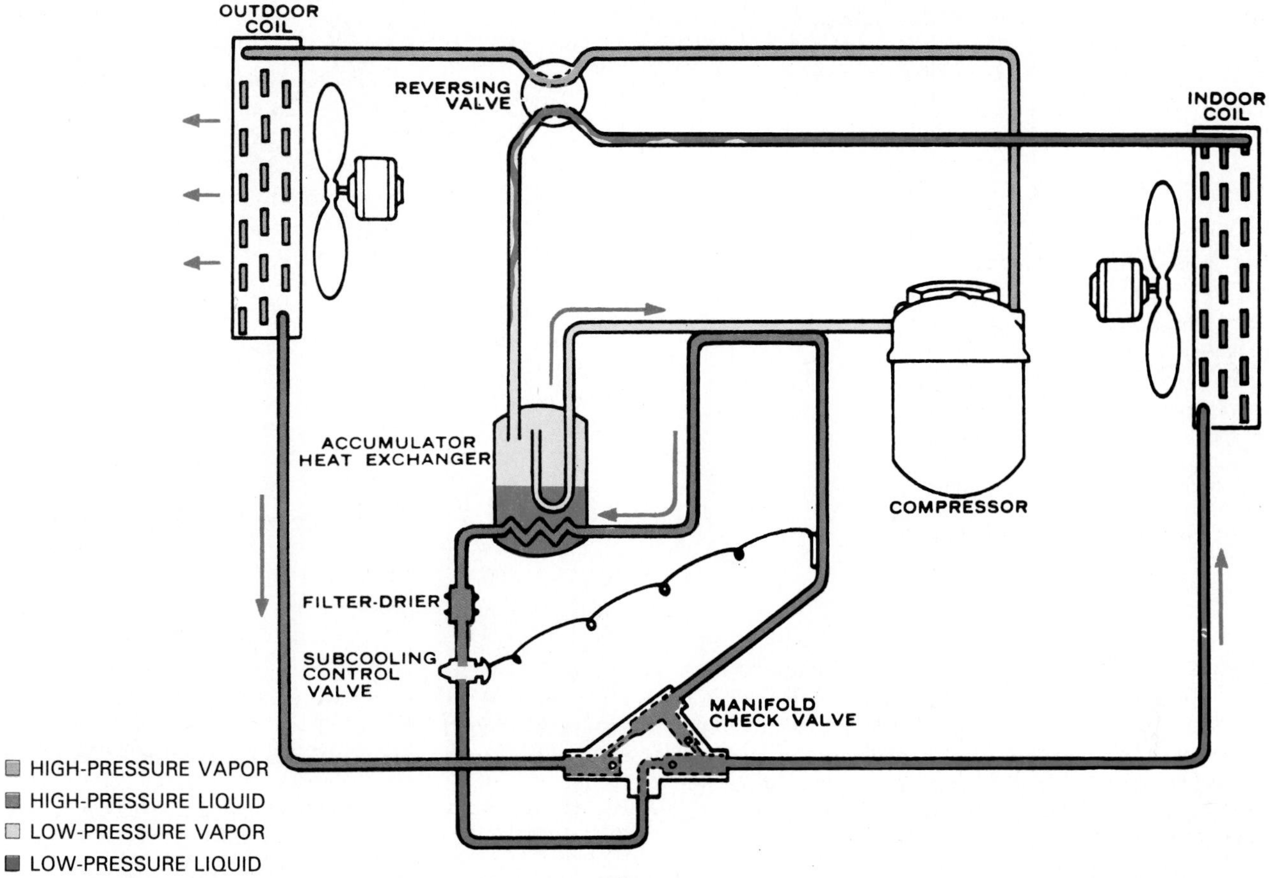

OUTDOOR COIL

REVERSING VALVE

INDOOR COIL

ACCUMULATOR HEAT EXCHANGER

COMPRESSOR

FILTER-DRIER

SUBCOOLING CONTROL VALVE

MANIFOLD CHECK VALVE

☐ HIGH-PRESSURE VAPOR
■ HIGH-PRESSURE LIQUID
☐ LOW-PRESSURE VAPOR
■ LOW-PRESSURE LIQUID

Fig. 23-32. Hi/Re/Li heat pump system operating as cooling system. Note how manifold check valve changes or reverses refrigerant flow.

23-21 HEAT PUMPS AND SOLAR-HEATED LIQUID OR SOLAR-HEATED AIR

Solar heat systems (Chapter 24) and heat pumps are being linked for residential use. Several designs are manufactured.

One system works with air solar collectors. These collect heat which can be:
1. Routed directly into the residential area.
2. Stored in a bin of rocks or in a thermal storage tank.

Air in some units gives up heat to a liquid storage area. When there is no direct solar heat, the storage area will supply heat for the living area. See Fig. 23-33.

When the storage temperature drops below 100°F (38°C), the heat pump will operate and produce the required temperature.

An air-to-water heat exchanger located at the output of the air collectors can provide the domestic hot water heat, Fig. 23-34. In the summer, the heat pump will operate to provide the desired cooling with a separate outside coil.

23-22 COMBINING LIQUID SOLAR COLLECTORS WITH HEAT PUMPS

Some solar heat and heat pump combinations being manufactured use a fluid to transfer the heat collected. The fluid is often water with ethylene glycol added.

The heat pump will operate when the energy efficiency ratio in heat units (EER_Q) is 2.0 or better and the ambient

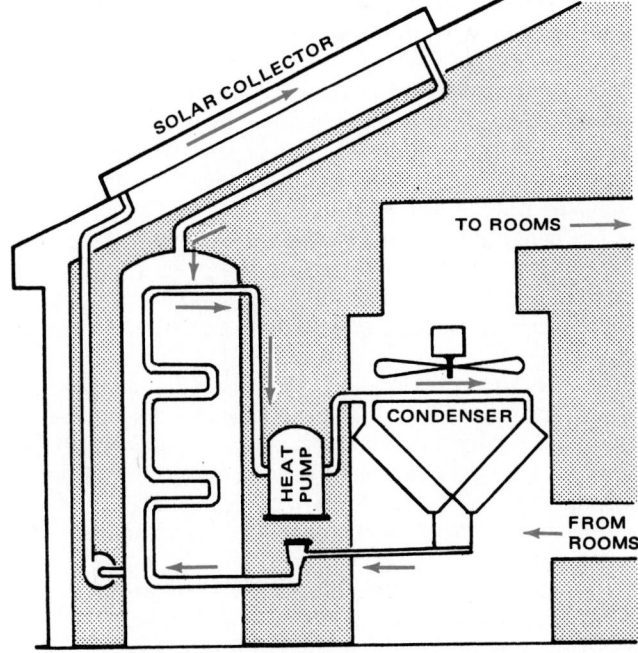

SOLAR COLLECTOR

TO ROOMS →

CONDENSER

HEAT PUMP

FROM ROOMS

Fig. 23-33. Solar collector is used with heat pump. Solar collector system pump circulates air from solar collector on roof to space surrounding heat exchanger tubes in storage tank. Heat pump evaporator is in storage tank. Condenser is in plenum chamber of furnace. Heat pump evaporator picks up collected heat in storage tank. Heat pump condenser gives up heat in furnace plenum chamber.

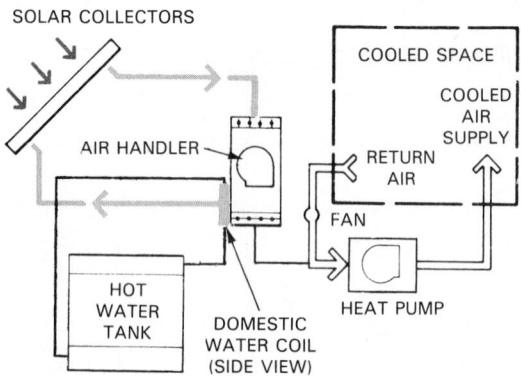

Fig. 23-34. This air solar collector and heat pump system can provide domestic hot water supply. Lower air handler air vanes to heat pump are closed. Hot air goes left to domestic water coil. In summer, cooling from heat pump is separate from solar circulation system. In winter, when heat pump heats room on right, heat pump gets some heat from air which has gone through solar collector.

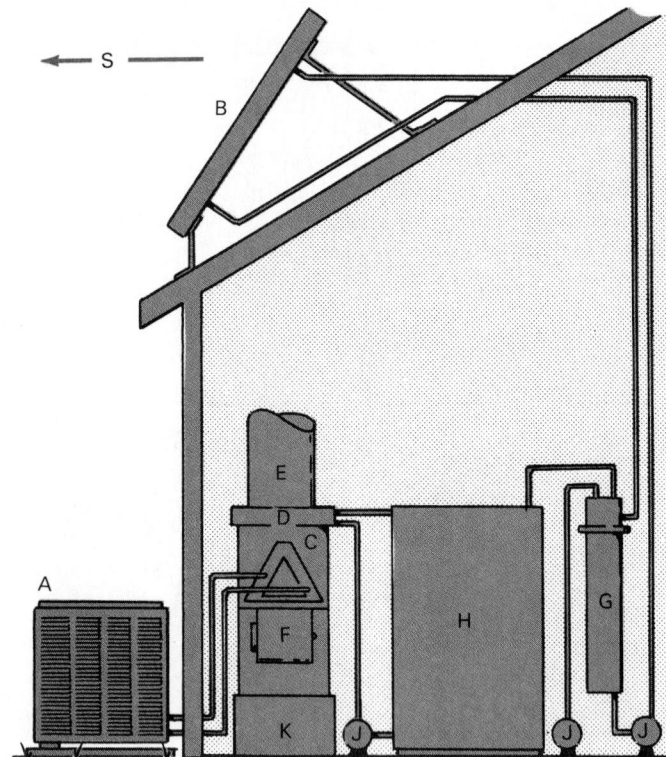

Fig. 23-35. Liquid solar collector/heat pump combination provides both summer cooling and winter heating. A—Outdoor section. B—Solar heat collector panel. C—Indoor heat exchanger. D—Hydronic coil. E—Air supply duct. F—Blower. G—Liquid-to-liquid heat exchanger. H—Heat storage tank. J—Pumps. K—Return air duct.

(outdoor) temperature is 20 °F (−7 °C) or higher. Some heat pumps use supplemental electric resistance heaters when the outdoor temperature drops below 25 °F (−4 °C). This increases operating costs. A liquid solar collector/heat pump combination, Fig. 23-35, is designed to provide heating and cooling at minimum operating costs. When the EER_Q is lower than 2.0 and the ambient temperature is less than 20 °F (−7 °C), the solar panels will provide the needed heat. The EER_Q is 2.0 at 20 °F (−7 °C). When the ambient temperature is 35 °F (2 °C), the EER_Q = 2.5, and the cost of heating is 40 percent of the cost of electric resistance heating.

The heat pump operates when the temperatures are at or above the balance point. This is the point at which the heating capacity of the heat pump is equal to the heat loss of the home.

The solar panels are sized to meet only the supplemental heating needs of the heat pump. The heat from the solar panels is stored in a water storage tank. The stored heat is used when needed. An electric immersion heater provides supplemental heat if needed.

The heat pump will draw heat from the solar panels until the outdoor temperature drops to the point at which supplementary heat is required. Then the supplemental heat will be provided by the solar energy that has been collected and stored.

The system is a standard heat pump unit installed in a residential home. The solar panel collector is installed outside, and a hydronic coil is placed in the supply air duct. The hydronic coil receives its heat from the thermal energy storage tank which holds the heated water.

The solar supplemented heat pump system combines five basic systems:

1. A standard heat pump system that will provide cooling when required.
2. A standard heat pump system that will produce heat and operate whenever the EER_Q is above 1.0. (The EER_Q is never less than 1.0. For the unit shown in Fig. 23-35, one assumes that the EER_Q = 1.0 for any temperature at 0 °F [−18 °C] or lower.)
3. A solar circuit. This unit consists of solar panels containing a solution of ethylene glycol and water that

passes through the shell side of a tube-and-shell heat exchanger. The heat that has been collected by the solar panels is not stored in the heat exchanger.
4. A storage circuit. A thermal energy storage tank uses a storage pump to circulate water through the tube side of the heat exchanger. Here the water is heated and returned to the storage tank.
5. A hydronic circuit. A hydronic pump circulates the heated water from the storage tank to the hydronic coil in the supply duct.

The EER_Q of a heat pump can be large, although not as large as the theoretical value. The theoretical limit is given in the following formula:

$$EER_Q = \frac{T_H}{T_H - T_C}$$

T_H is the temperature of the condenser and T_C is the evaporator temperature. All temperatures are in degrees Kelvin. As an example, use a hot-side temperature of 72 °F (295.2 K) and a cold-side temperature of 20 °F (266.3 K):

$$EER_Q = \frac{295.2 \text{ K}}{295.2 \text{ K} - 266.3 \text{ K}} = 10.2$$

With an EER_Q of 10, the cost of heating is 1/10 of that for electric resistance heating.

23-23 HEAT PUMP WATER HEATERS

Heat pumps are replacing water heaters in both domestic and industrial installations. The pump removes heat from the interior of a building and delivers it to a hot water tank while using about half the energy of an electric water heater.

The condenser is placed around the outside of the water tank and the evaporator is on top of the tank, Fig. 23-36. The heat removed from the building must, of course, be replaced by the furnace during the heating season.

Some heat pump water heaters on the market include a heat recovery unit for conditioning ventilation air to a house. In the winter, the cooled room air leaving the evaporator is used to preheat the makeup air required for ventilation.

In summer operation, when the outside air is warmer than room air, the outside air supplies heat to the evaporator. The air cooled by the evaporator may be used to help cool the house.

In summer operation, inside air must leave the house at the same rate as it is brought in from the outside. The air inside the house can be exhausted directly to the outside. A vent with a natural outward flow can be used to save more indoor air than a fan would save. A fan might create too much negative pressure.

Sometimes dehumidification of basement air is required. The evaporator can be used to condense water out of the basement air.

23-24 COMPLETE AIR CONDITIONING SYSTEMS

The complete air conditioning system controls the temperature, relative humidity, air movement, and air cleanliness. Details of the construction and location of the components of the complete air conditioning system vary with the design of the system. Some of these are described in the following paragraphs.

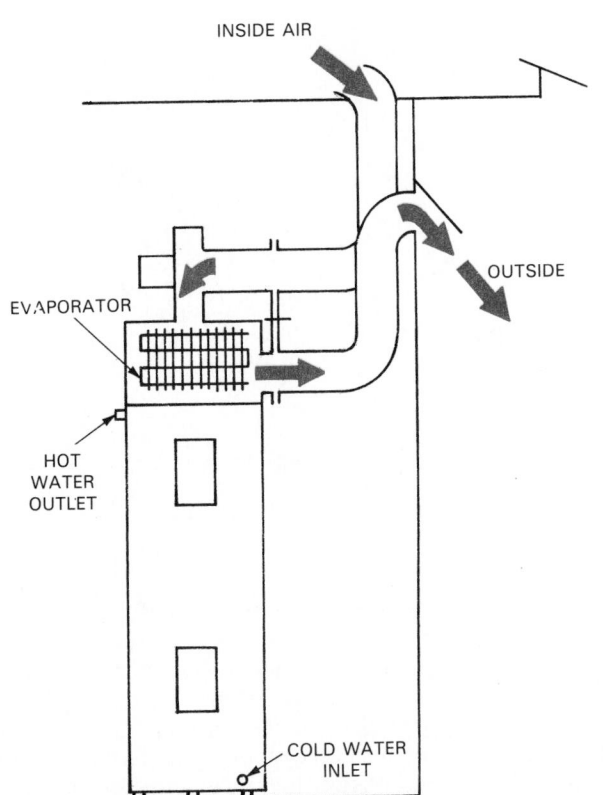

Fig. 23-36. Heat pump water heater has evaporator on top of water tank. Airflow shown is for winter operation. Heat is provided by heat pump condenser inside water tank. (DEC International, Inc.)

23-25 THROUGH-THE-WALL SYSTEMS

The complete through-the-wall system (heating, cooling, filtering, and moving air) is mounted in the wall. Also called a unitary system, it is popular in moderate climates. Some models are mounted flush to the outside of the building while some extend outside a few inches. They must be mounted firmly to the studding of the building and leveled in all directions. Joints must be weatherstripped and leakproof. Use caulking or gaskets.

Through-the-wall units which also have electric heat usually require 240 V power. Gas-fired or oil-fired systems use 120 V power. Sometimes a separate 240 V power circuit is used for the refrigerating system.

When cooling the indoor area, all of these units trap the comfort-cooling condensate and evaporate it to help cool the air-cooled condenser. The plenum chambers of some are designed to allow direct airflow into the room. Others use a separate duct for air distribution.

Through-the-wall units must be installed away from combustible material. They provide fresh air intake, combustion air intake, and a chimney flue. They are factory assembled.

23-26 OUTSIDE COMPLETE SYSTEMS

One type of all-season air conditioning for homes has all the systems—heating, cooling fan, and filtering—factory-assembled into one unit. The unit is installed outdoors. Then it is connected to the duct system of the house or building. A conditioned air duct connection and a return duct connection are all that is needed. These systems are an adaptation of rooftop units made for commercial and industrial use. See Fig. 23-37.

The advantages are:
1. The heating system, cooling system, and fan system are factory-assembled and tested.
2. The installation consists of mounting the unit, connecting the electric power, the gas line, and the two duct connections. Fig. 23-38 shows two views of this system.

23-27 ROOFTOP UNITS

Heat pumps are best used for both heating and cooling. Less equipment is needed for a combined unit. Rooftop systems for heating and cooling were first developed for flatroof commercial structures such as supermarkets and office buildings.

There are three types:
1. Heating only.
2. Cooling only.
3. Both heating and cooling.

Being factory-assembled and tested, the systems are economical. They are also easily installed once hoisted to the site.

There are some disadvantages. The roof must be strong enough to carry the extra weight. Openings may develop leaks. Units must be designed for ease of service even in very cold or bad weather. They are also exposed to winds which may affect their operation. The service technician must carry tools and supplies to the roof. Equipment must be adequate to service the unit in all kinds of weather.

Fig. 23-39 shows a typical rooftop unit for both heating and cooling. Another assembly is shown in Fig. 23-40. Many commercial units come completely equipped with

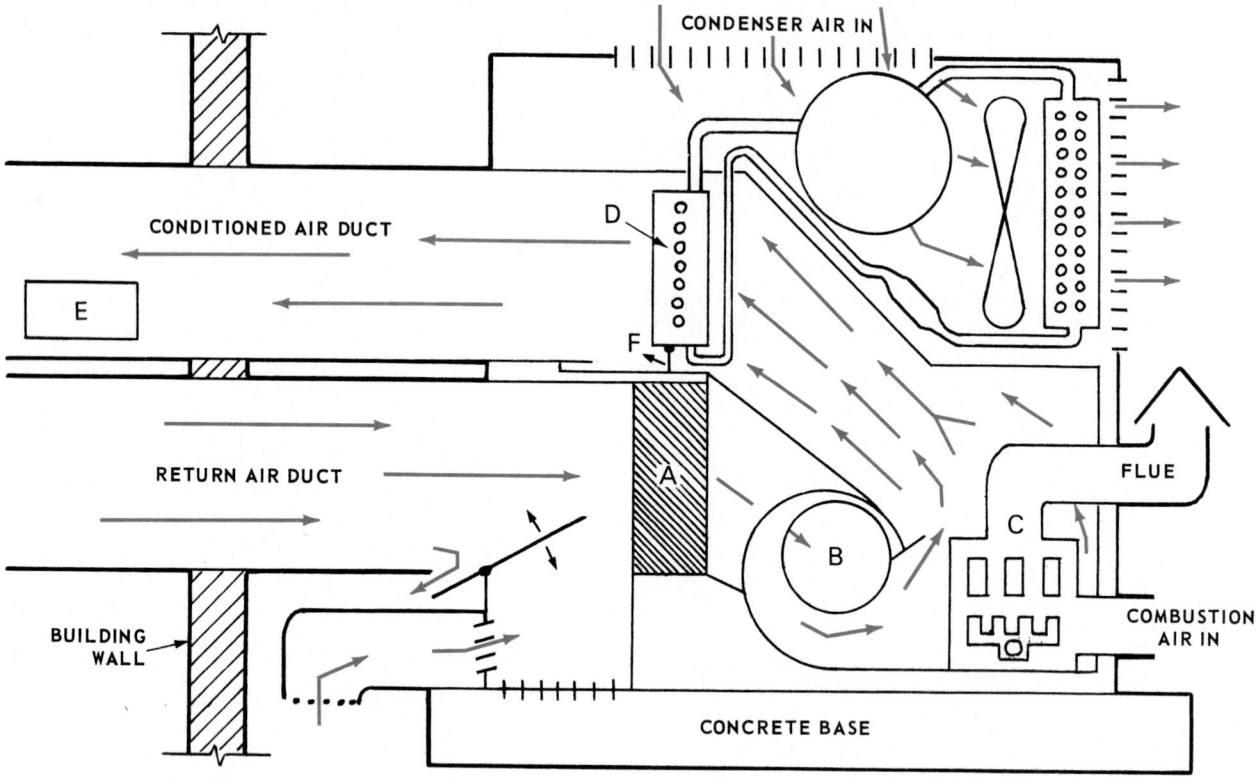

Fig. 23-37. All-season air conditioning system installed outside house or office. A—Filter. B—Fan. C—Furnace. D—Cooling system evaporator. E—Humidifier. F—Air bypass for relative humidity control during cooling.

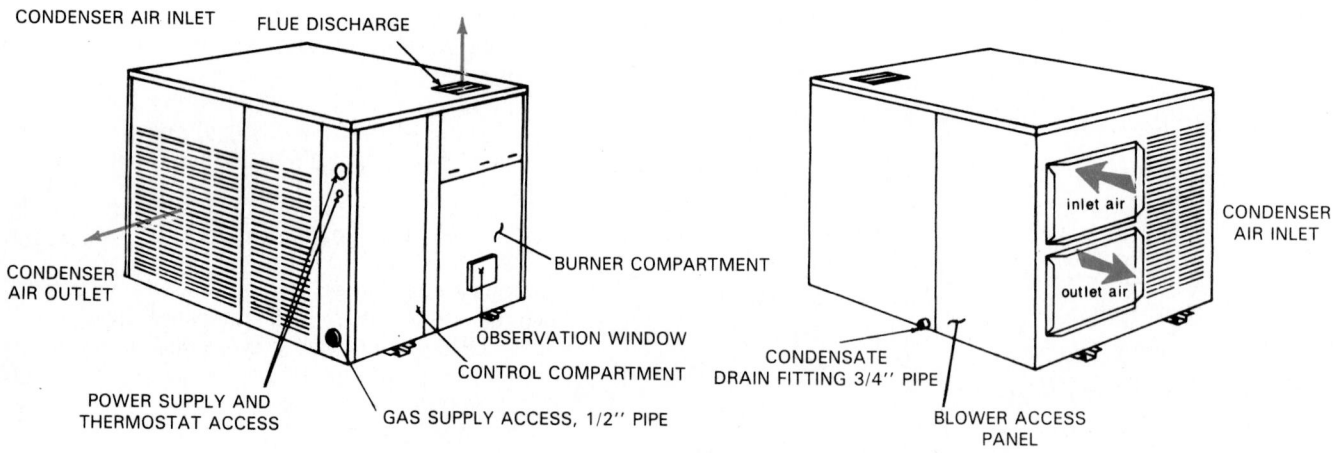

Fig. 23-38. Complete air conditioning system designed for connection to building air conditioning system.
(Fedders North America)

factory-installed heating units. Fig. 23-41 shows a tubular heat exchanger made of aluminized steel. Gas furnaces usually use an electric ignition system because gas pilot flames tend to blow out during high winds or gusts. These units are completely weatherproofed since they are exposed to all kinds of weather. Access doors for inspection are designed to operate safely under high wind conditions. Fig. 23-42 is a pictorial drawing of the refrigerating mechanism of a rooftop unit. A complete heating and cooling system is shown in Fig. 23-43.

Rooftop units are sometimes oil fired. Fuel is stored in

an underground tank. A lift pump draws oil to a second smaller tank. This tank may be at the same level as or slightly lower than the oil burner. Under no circumstances should the tank be more than 10 ft. below the burner.

Electric heat may also be used in rooftop units. In Fig. 23-44, the electric heating elements are partially removed from the housing. These heating elements are available in about 20, 40, and 60 kW capacities. Heating elements are carefully protected by ceramic insulators. The circuit has fuses, fusible links, and thermal cutouts.

Air is controlled by dampers. Sensing elements in the

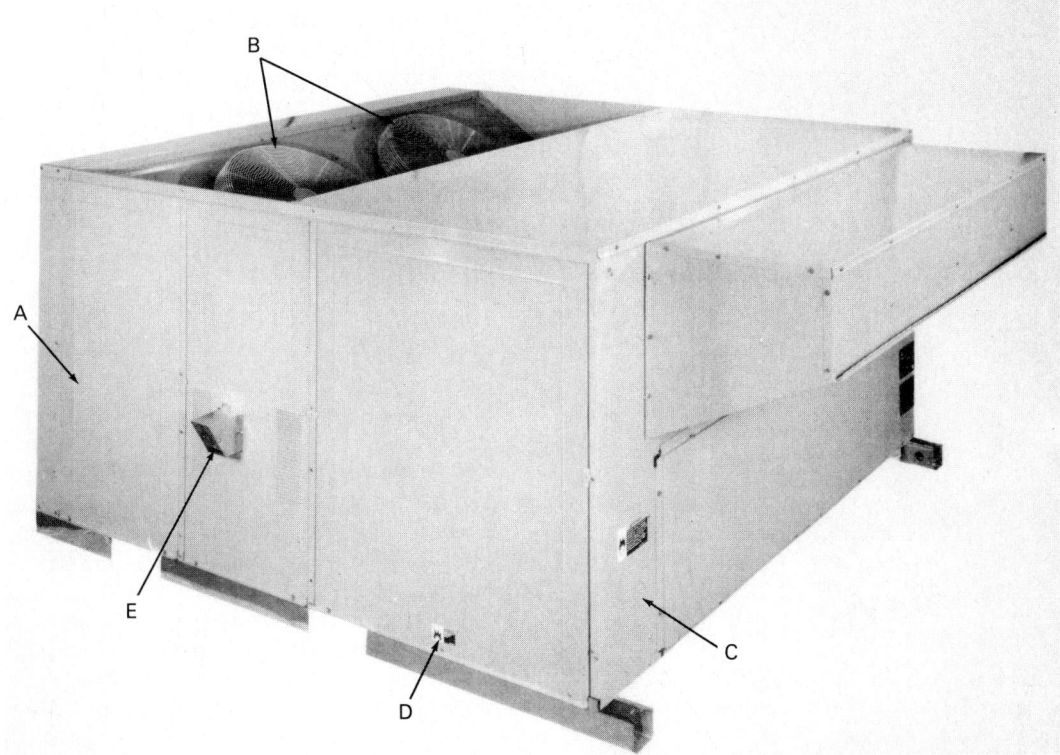

Fig. 23-39. Rooftop unit provides both heating and cooling. This unit has gas-fired furnace. A—Condenser. B—Fan air outlets. Fans move outdoor air through condenser. C—Electrical connection knockouts. D—Gas inlet. E—Furnace flue. (BDP Company)

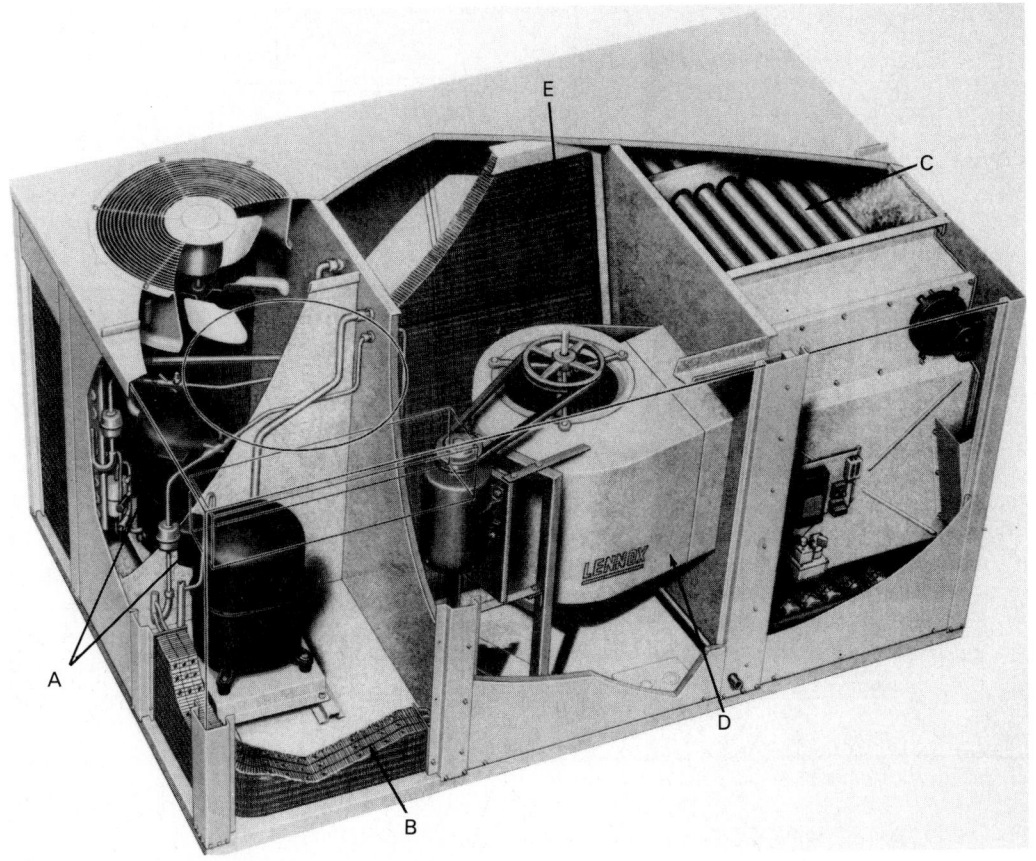

Fig. 23-40. Cutaway of rooftop unit. A—Twin compressors. B—Condenser. C—Heat exchanger for gas burner. (See Fig. 23-41.) D—Blower. E—Filter. (Lennox International, Inc.)

Fig. 23-41. Gas heat section of single-package cooling and gas heating system. (Lennox International, Inc.)

mixed-air duct control a motor which operates:
1. The return damper.
2. The conditioned air damper.
3. The exhaust air damper.
4. The fresh air damper.

Fig. 23-45 shows the air control system. The air is filtered as in other air circulation systems. See Chapter 22.

23-28 ROOFTOP UNITS—INSTALLATION

Rooftop units must be installed to provide a leakproof roof connection. Fig. 23-46 shows how the rooftop system is connected to a duct distribution system. The installation is usually made as shown in Fig. 23-47.

One method used to make the roof leakproof is shown in Fig. 23-48. Notice the use of acoustical material for deadening sound. The resilient (flexible) gasket installed between the rooftop base rail and the roof edge is very important. There may be leaks if this gasket is not tight. A duct installation connected to a rooftop unit is shown in Fig. 23-49.

Rooftop units are heavy. Riggers are usually required for installing, removing, and replacing them.

The electrical service must comply with local codes. The control system is usually the responsibility of the refrigera-

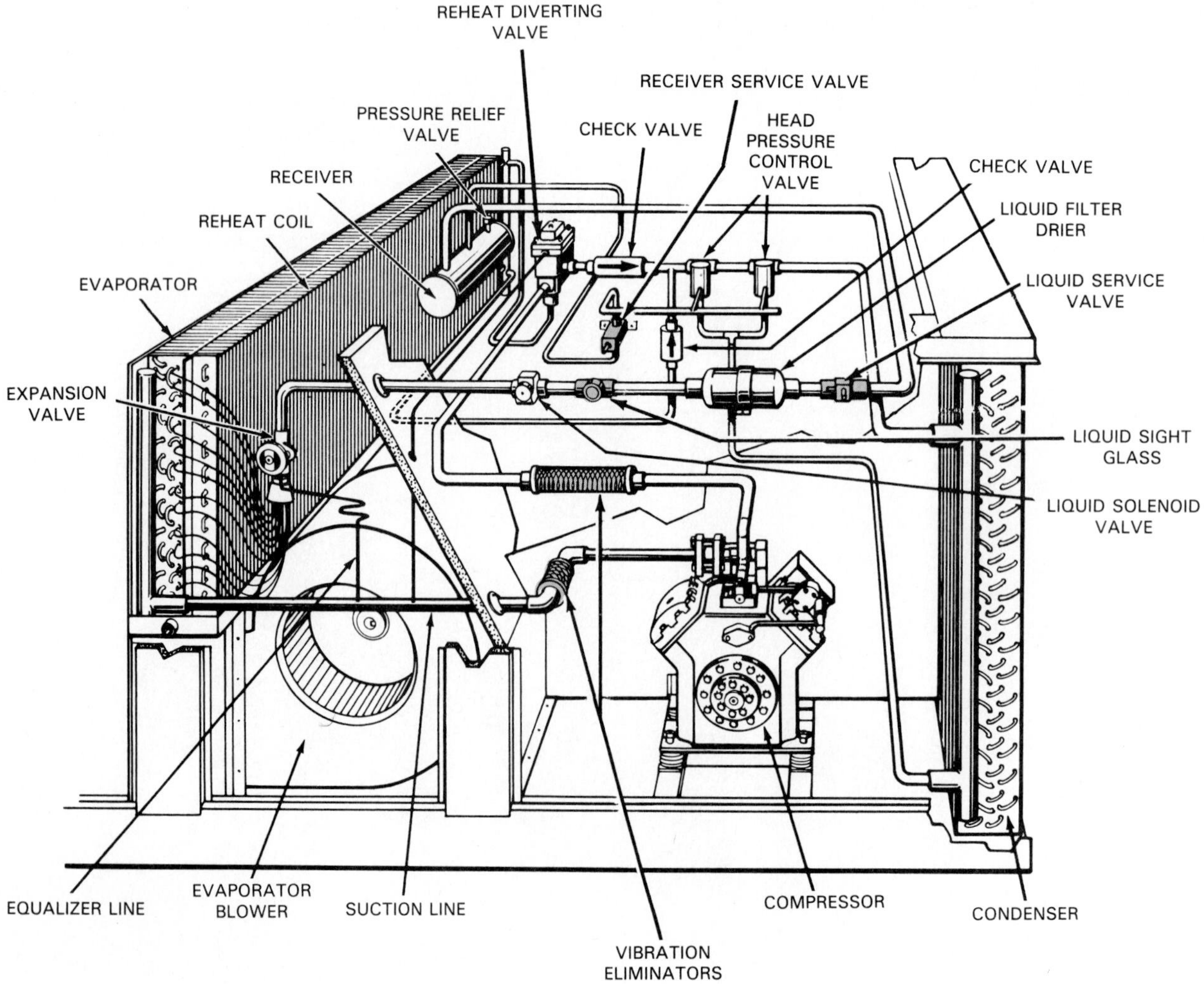

Fig. 23-42. Refrigerating circuit of rooftop system. Note service valves, sight glass, and vibration eliminators.

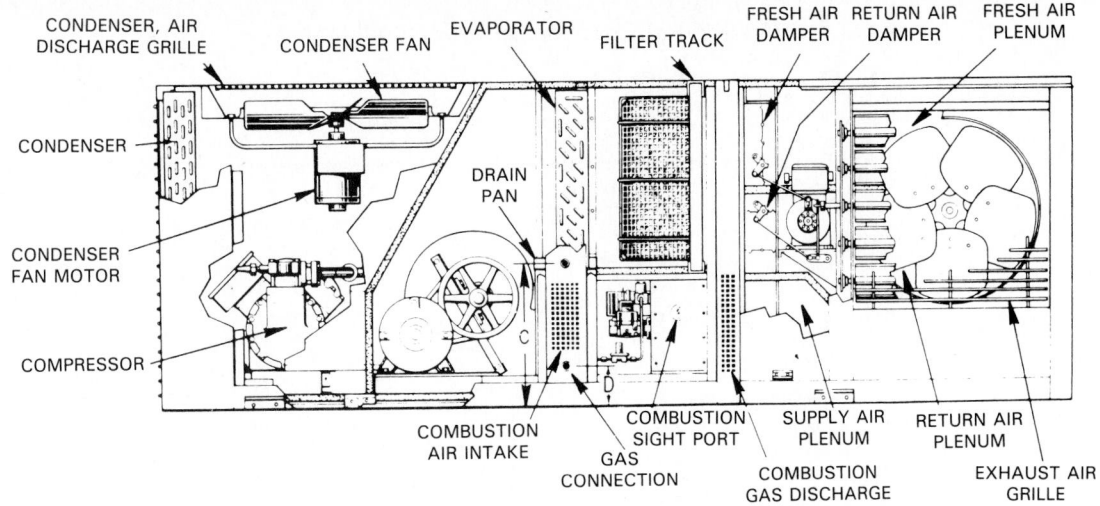

Fig. 23-43. Combination heating and cooling rooftop system.

Fig. 23-44. Rooftop system is equipped with electric heat. A—Control circuit and housing. B—Electric resistance heating elements.

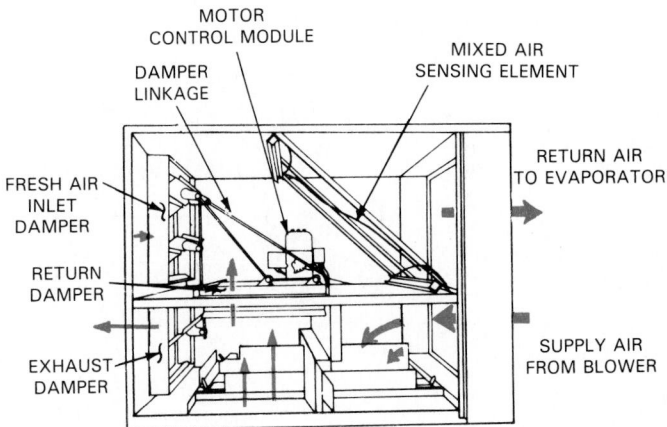

Fig. 23-45. Air control system for rooftop unit. Dampers are automatically controlled so that they always provide desired air mixture.

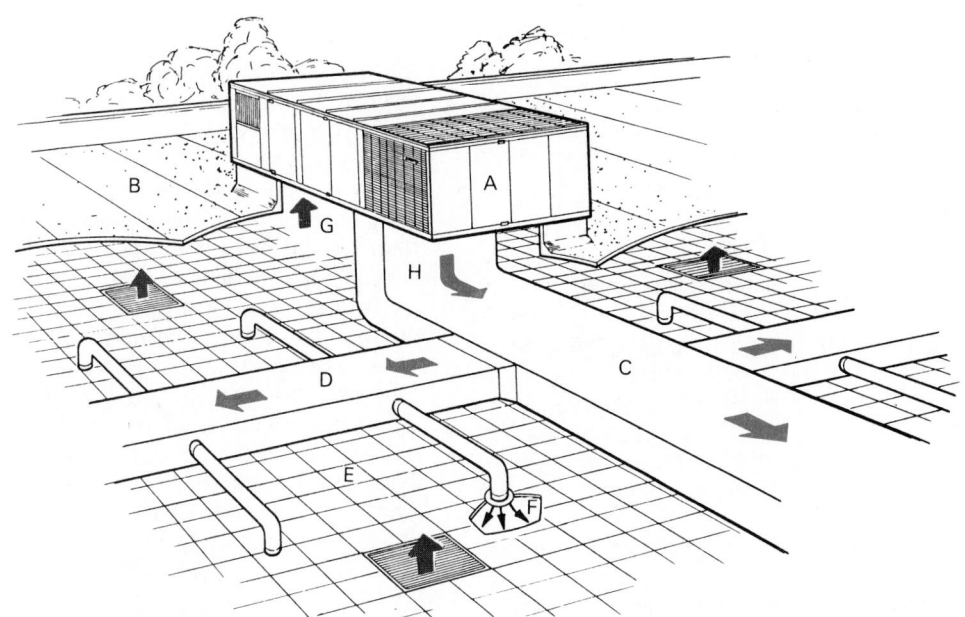

Fig. 23-46. Rooftop unit connected to duct distribution system. A—Rooftop unit. B—Roof line. C—Main duct. D—Branch duct. E—False ceiling. F—Diffuser. G—Cool air returning in winter. H—Heated air flowing out in winter. (Lennox International, Inc.)

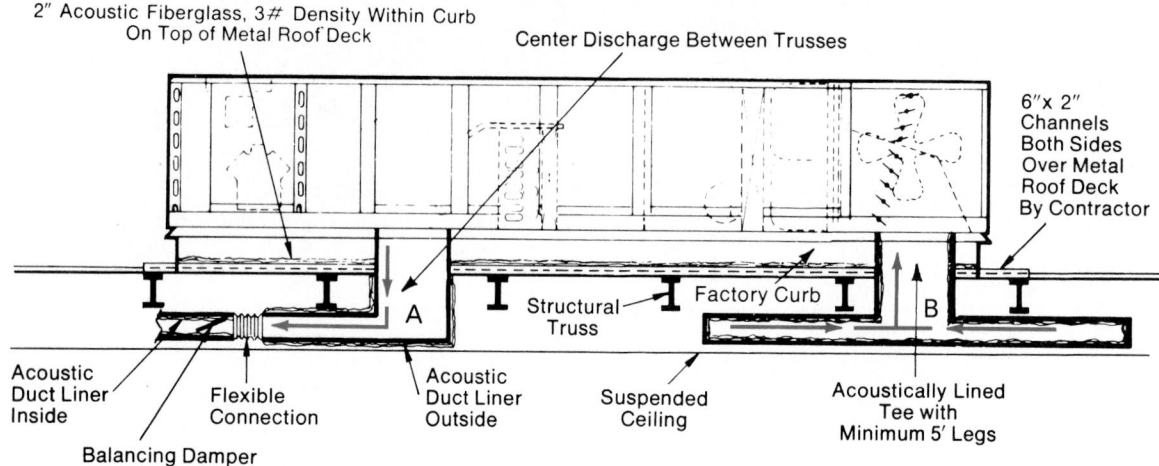

Fig. 23-47. Installation details of rooftop complete air conditioning system. A—Conditioned-air duct. B—Return-air duct.

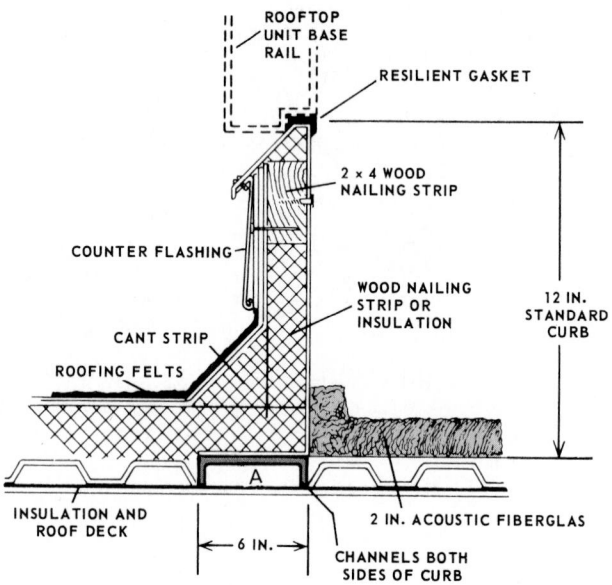

Fig. 23-48. Method used to make roof leakproof when rooftop unit is installed. A—Roof reinforcement.

tion installation technician. A system using electronic controls is shown in Fig. 23-50.

When the rooftop system has a gas furnace, the gas piping must be installed according to code. Pipes must be supported and protected. Fig. 23-51 shows a typical installation for a 1 1/4-in. gas line. Note that two hand shutoff valves are used. One is located outside the rooftop system casing to permit emergency shutdown of the furnace in case of accident or fire.

Some systems use piped-in hot water or steam to provide the heating required. The piping of these systems must be carefully installed to avoid air traps and freezing. Hot water systems usually use a nonfreezing solution of water and ethylene glycol. Fig. 23-52 shows the piping for a steam system.

23-29 SERVICING ROOFTOP UNITS

Certain chapters have been devoted to servicing different systems. For more information on:
1. Refrigerating systems, refer to Chapter 14.
2. Heating systems, refer to Chapter 20.
3. Blowers and filters, refer to Chapter 22.

However, there are some very important service operations unique to rooftop systems. *It is very important that the service technician use all possible safety precautions when climbing to the roof.*

If a portable ladder is used, it must be securely and firmly based on the ground. It must be inclined (leaned against the building) at an angle that will keep it from falling away from the building. It must extend two or three rungs above the roof edge.

The service technician must use both hands on the ladder when climbing or coming down the ladder. The portable ladder should not be used in a high or gusty wind.

Be especially careful if it is raining or snowing. Lift tool boxes, refrigerant cylinders, and other objects with a rope. The rope should be guided from below also.

When the roof is wet or snow-covered, be extremely cautious while working on or near electrical circuits. Hinged panels must be secured in the open position or the wind may swing them violently and injure someone. Loose panels must be held down securely to prevent a wind blowing them off the roof. If the system has a gas or oil burner, avoid breathing fumes coming out of the flue.

When working on a rooftop unit in cold and/or windy weather, use gloves and hand warmers. Some prefer portable heaters. Set up a windbreak to provide wind shelter. If the wind is very high, use a safety belt attached to a solid part of the unit.

When working on a rooftop unit in hot, sunny weather, shield the tools from the sun. Metal may become so hot it will burn the hands. Use gloves when handling the heated panels. Wear a hard hat and goggles. Wear rubber gloves when working on electrical circuits.

23-30 CENTRAL AIR CONDITIONING SYSTEMS

Central air conditioning systems are of two types:
1. Unitary.

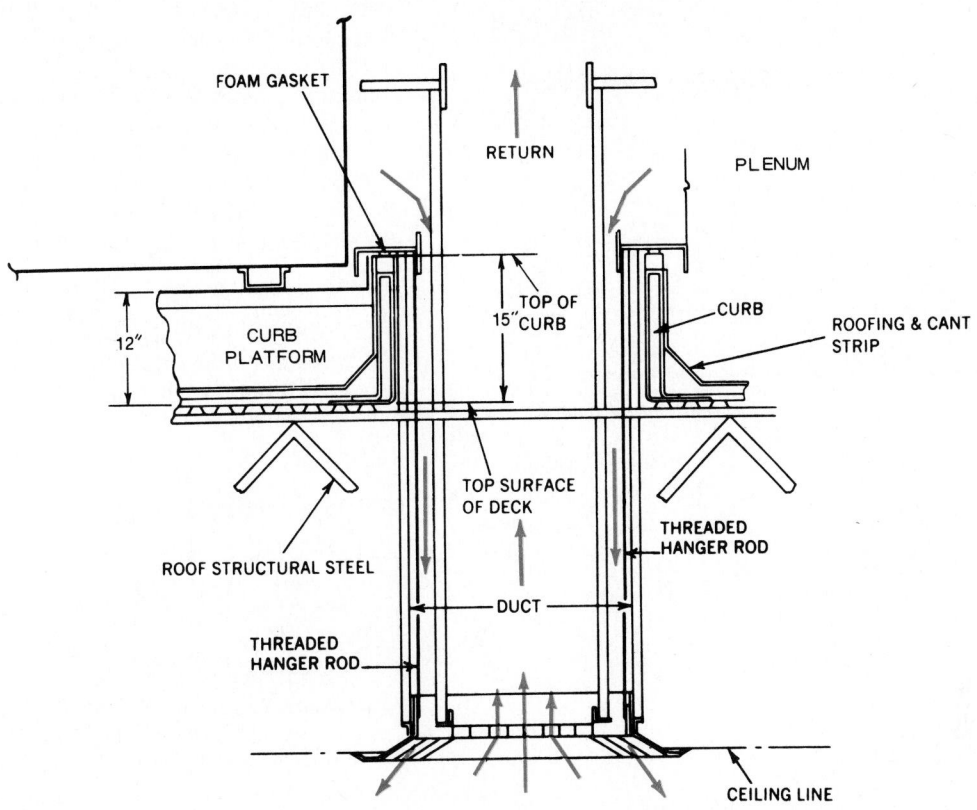

Fig. 23-49. Method of connecting ducts to rooftop unit.

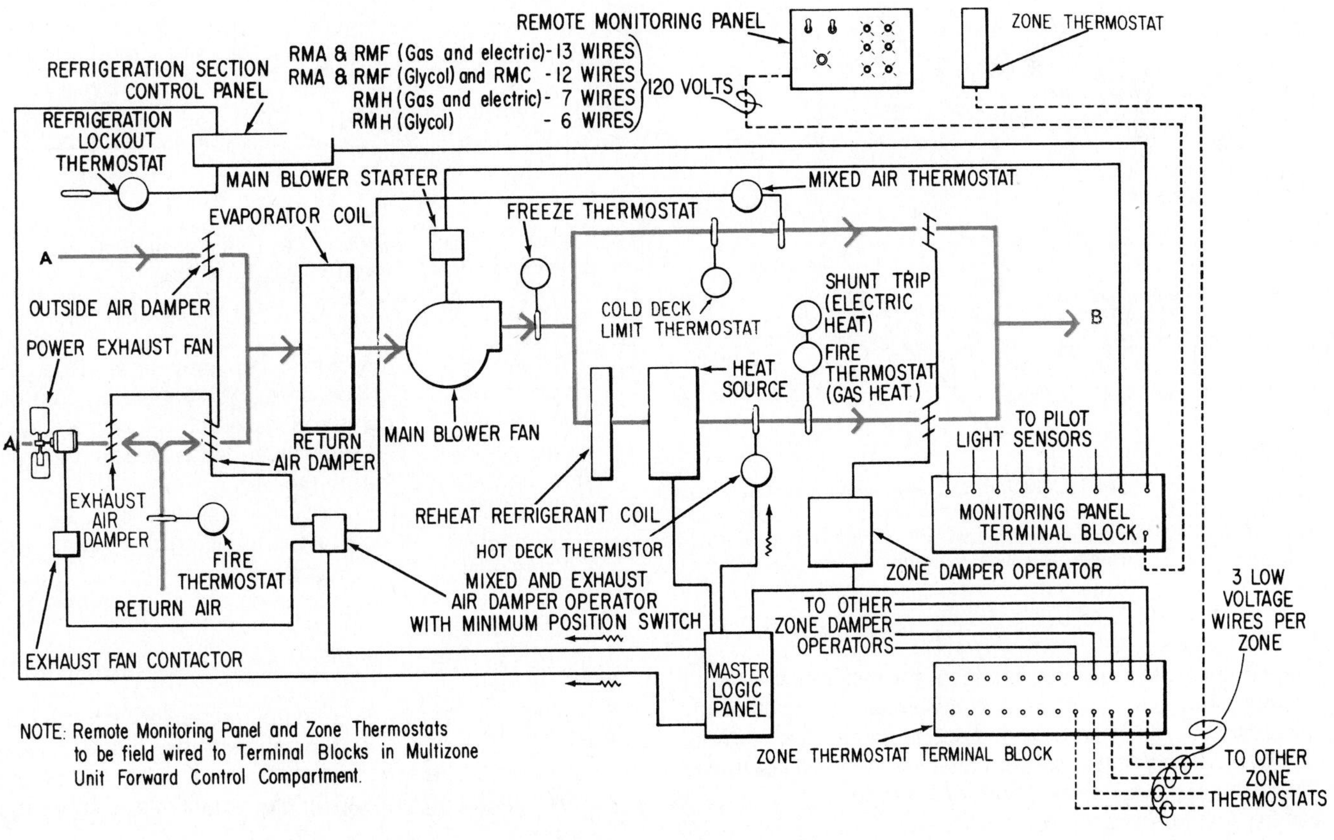

FOAM GASKET

RETURN

PLENUM

12"

CURB PLATFORM

TOP OF 15" CURB

CURB

ROOFING & CANT STRIP

TOP SURFACE OF DECK

ROOF STRUCTURAL STEEL

DUCT

THREADED HANGER ROD

THREADED HANGER ROD

CEILING LINE

REMOTE MONITORING PANEL

ZONE THERMOSTAT

REFRIGERATION SECTION CONTROL PANEL

RMA & RMF (Gas and electric)-13 WIRES
RMA & RMF (Glycol) and RMC -12 WIRES
RMH (Gas and electric)- 7 WIRES
RMH (Glycol)   - 6 WIRES

120 VOLTS

REFRIGERATION LOCKOUT THERMOSTAT

MAIN BLOWER STARTER

MIXED AIR THERMOSTAT

EVAPORATOR COIL

FREEZE THERMOSTAT

A

OUTSIDE AIR DAMPER

SHUNT TRIP (ELECTRIC HEAT)

COLD DECK LIMIT THERMOSTAT

POWER EXHAUST FAN

FIRE THERMOSTAT (GAS HEAT)

B

A

HEAT SOURCE

RETURN AIR DAMPER

MAIN BLOWER FAN

TO PILOT LIGHT SENSORS

EXHAUST AIR DAMPER

REHEAT REFRIGERANT COIL

MONITORING PANEL TERMINAL BLOCK

FIRE THERMOSTAT

HOT DECK THERMISTOR

ZONE DAMPER OPERATOR

RETURN AIR

MIXED AND EXHAUST AIR DAMPER OPERATOR WITH MINIMUM POSITION SWITCH

3 LOW VOLTAGE WIRES PER ZONE

EXHAUST FAN CONTACTOR

TO OTHER ZONE DAMPER OPERATORS

NOTE: Remote Monitoring Panel and Zone Thermostats to be field wired to Terminal Blocks in Multizone Unit Forward Control Compartment.

MASTER LOGIC PANEL

ZONE THERMOSTAT TERMINAL BLOCK

TO OTHER ZONE THERMOSTATS

Fig. 23-50. Electronic control system for rooftop unit. A—Return air. B—Conditioned air. Red arrows indicate direction of airflow.

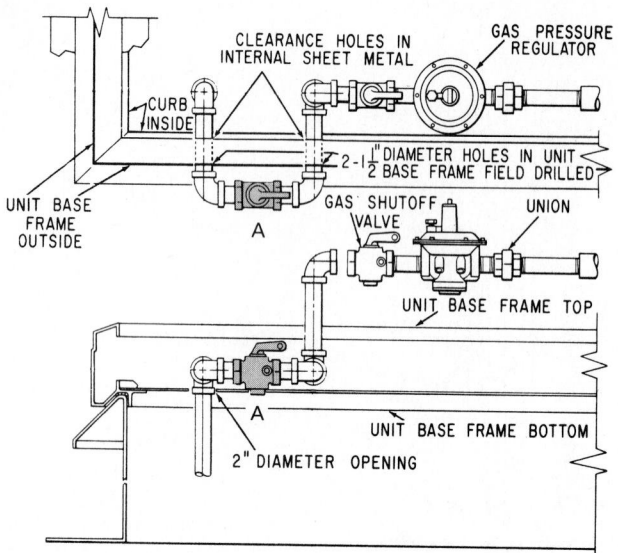

Fig. 23-51. Fuel gas line installation used on rooftop system. A—Hand shutoff valve is mounted outside casing for easy access.

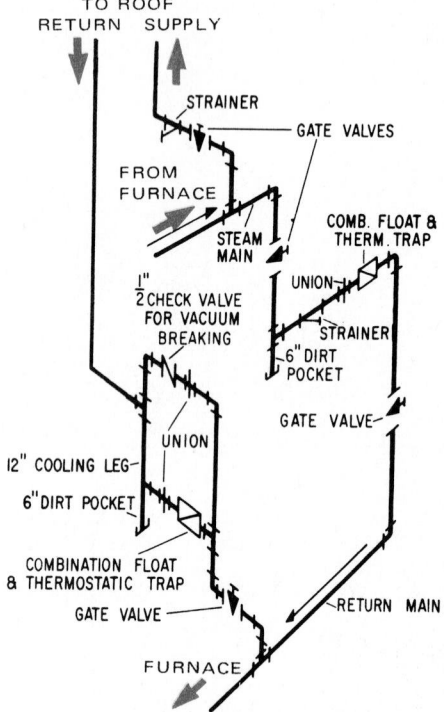

Fig. 23-52. Steam pipe installation for steam heating system used on rooftop complete air conditioning system. More than one steam path is provided. This helps prevent restricted flow if one section is partially closed. Balance is improved.

2. Field-erected.

Central unitary systems are ideal for residential air conditioning. They are a complete, manufactured package ready for assembly. All internal wiring and piping has been done. The condensing unit is located away from the evaporator. There are three evaporator designs in use:
1. An A-type evaporator.
2. Slant-type evaporator.
3. Flat-type evaporator for horizontal airflow.

Three methods are used to install central air conditioning:
1. One can buy the condensing unit, evaporator, controls, and tubing, and assemble the conditioner on the customer's premises. Most of these systems are installed in forced-warm-air heating systems. Fig. 23-53 is a drawing of such an installation.

These units vary in capacity from 1 1/2 hp (approximately 12,000 Btu/hr. or 12 650 kJ/hr) to 7 1/2 hp (60,000 Btu/hr. or 63 260 kJ/hr). Assume the coefficient of performance (COP) is 3.14. $COP = \dfrac{T_C}{T_H - T_C}$

An oil-fired furnace, complete with a comfort-cooling A-type evaporator, electronic air filter, humidifier, and blower system, is shown in Fig. 23-54.

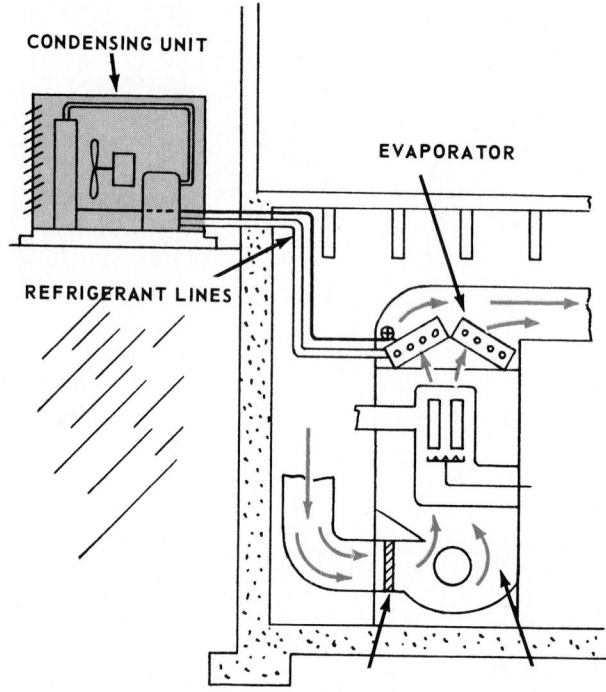

Fig. 23-53. Drawing shows residential central comfort-cooling system installed in forced-air furnace. Condensing unit is mounted outdoors.

2. The unit can be ordered completely assembled and charged, including condensing unit, evaporator, controls, and tubing. The complete assembly is shipped as a single package. The tubing needed is usually carefully wrapped around the evaporator.

Work required at the point of installation includes:
a. Uncrating and carefully unwinding the tubing from the evaporator.
b. Installing the condensing unit and evaporator in their proper places.
c. Making the electrical and control connections needed.

This type of system requires some very careful handling, since the condensing unit is charged with refrigerant. The service technician must be very careful while uncrating the tubing to avoid kinks which might later crack and leak.

Fig. 23-55 shows an A-type evaporator and Fig. 23-56 a slant-type evaporator. A condensing unit is

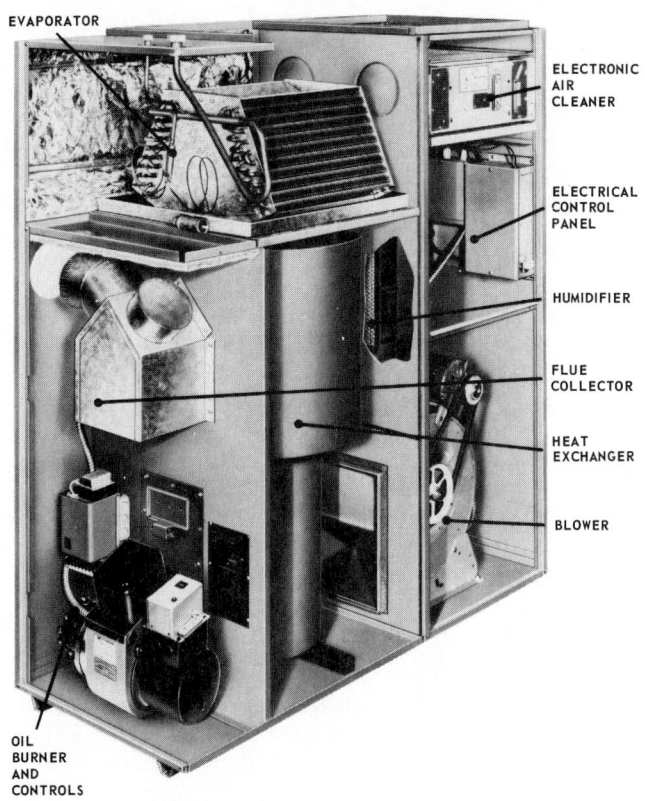

Fig. 23-54. All-season air conditioner has oil burner, evaporator, electronic air cleaner, and humidifier.   (Williamson Corporation)

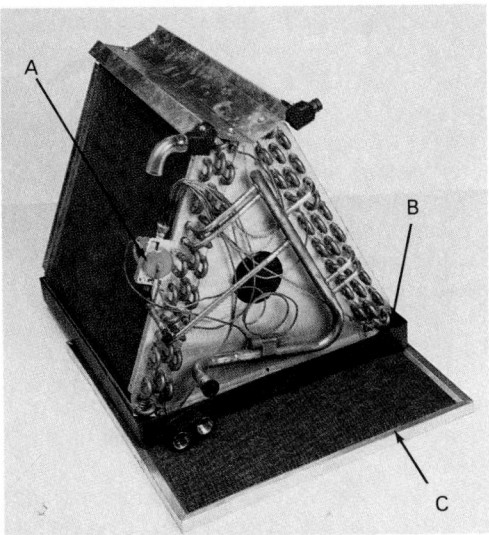

Fig. 23-55. An A-shaped evaporator designed for installation in warm air furnace plenum chamber. A—Thermostatic expansion valve. B—Pan for collection of condensate from coil. C—Air filter that can be removed for cleaning.   (BDP Company)

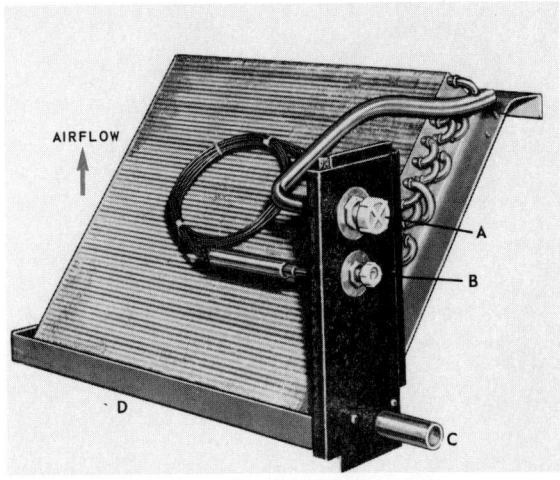

Fig. 23-56. Slant-type evaporator is designed for installation in plenum chamber of warm-air furnace. A—Suction line connection. B—Liquid line connection. C—Drain connection. D—Drain pan.

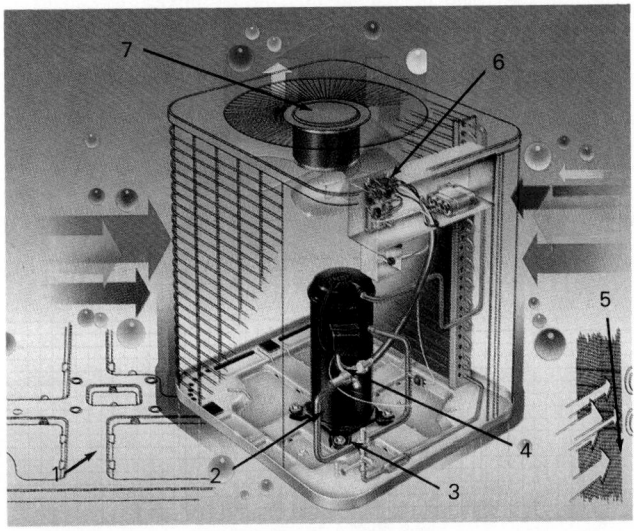

Fig. 23-57. Central air conditioning unit with horizontal inlet air and vertical air discharge. 1—Base pan for unit. 2—Scroll compressor. 3—Liquid line valve. 4—Suction service valve. 5—Air flow through condensing coil. 6—Controls. 7—Condenser fan motor. (BDP Company)

shown in Fig. 23-57. Fig. 23-58 shows a completely assembled system.

3. The technician can get a completely charged evaporator, condensing unit, and lines. However, the condensing unit, evaporator, and connecting tubing are shipped as separate items. See Fig. 23-59. Line set C uses flare

fittings at the evaporator coil A, and compression or sweat fittings at the condensing unit B. Other manufacturers connect the parts with "quick-couplers." This system is easy to assemble since the quick-couplers may be connected without losing refrigerant or getting air into the system. The complete system is shown in Fig. 23-60.

Some systems use the liquid line as the capillary tube. This larger bore (ID) tubing reduces the chance of clogging from dirt or moisture. *It is very important not to shorten or lengthen this liquid line capillary tube combination when installing this unit.*

Many systems use evaporators with aluminum fins mechanically bonded to copper tubing. Some condensers and/or evaporators use aluminum spines fastened to aluminum tubing to reduce corrosion. Plastic grilles are often

849

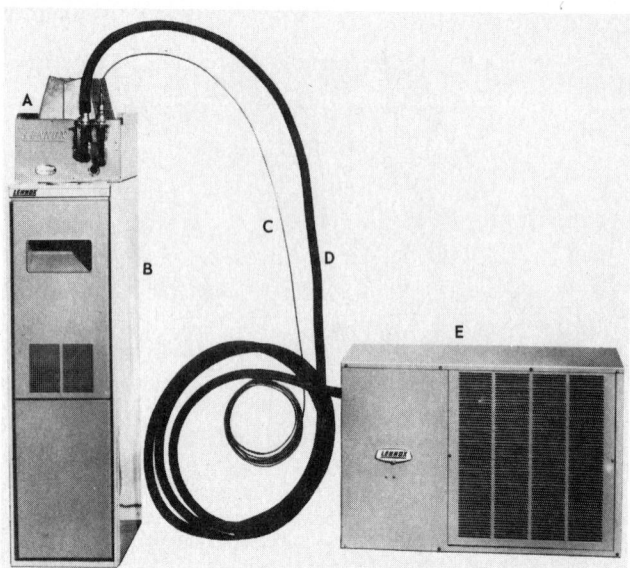

Fig. 23-58. Central comfort-cooling system is for a home. A—Evaporator. B—Plenum chamber of warm-air furnace. C—Liquid line. D—Insulated suction line. E—Outdoor air-cooled condensing unit. (Lennox International, Inc.)

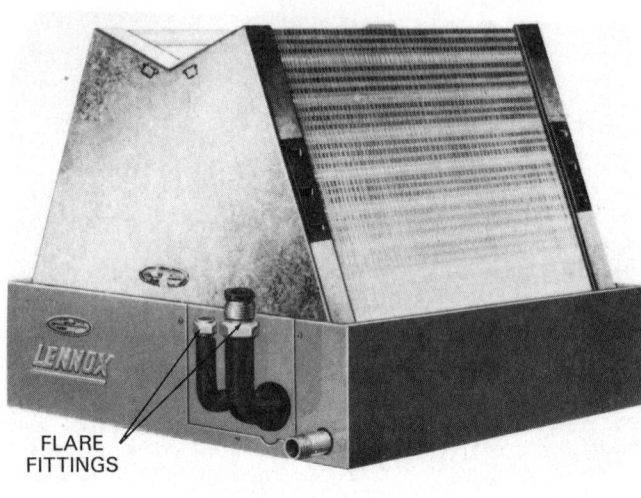

FLARE FITTINGS

A

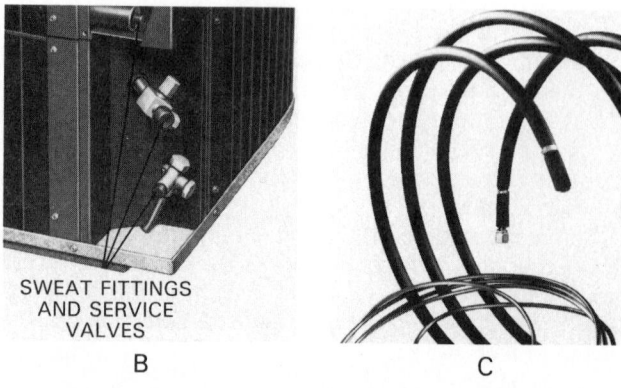

SWEAT FITTINGS AND SERVICE VALVES

B   C

Fig. 23-59. Domestic central comfort-cooling system uses flare fittings and sweat fittings. A—Evaporator coil has flare fittings. B—Sweat fittings are used on condensing unit. C—Line set. (Lennox International, Inc.)

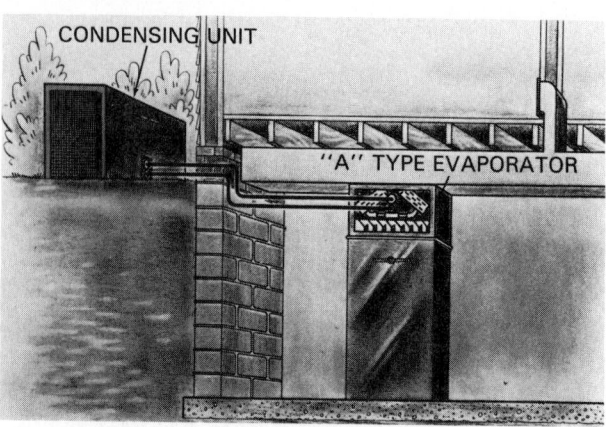

CONDENSING UNIT

"A" TYPE EVAPORATOR

Fig. 23-60. Typical installation of complete air conditioning system can have units at different levels. Evaporator is below condenser for best flow in liquid line. However, oil may not go back to motor compressor easily; some installers put U-bend in suction line to assist oil return. U-bend allows oil to collect, to be pushed through as a single "plug." (Aeroquip Corp.)

used on the condensing unit to avoid corrosion problems.

Fig. 23-61, Part A, shows the internal construction of a heat pump condensing unit which operates with a variable speed computer controller that controls the compressor and outdoor fan speed. Unit also has an electronic demand defrost which prevents ice from building up on the outdoor coil.

Electronic thermostat, Fig. 23-61, Part B, monitors the indoor and outdoor units *concurrent* to the correct operating speed. The LCD on the thermostat face shows the approximate operating speed, indoor temperature, outdoor temperature, and whether the unit is in a heating or cooling mode.

Field-erected air conditioning systems are systems in which all the components—motor, compressor, receiver, evaporator, piping, and controls—are assembled and erected at the spot where the system is to be used. Chapter 14 covers most of the installation instructions.

There are a variety of central field-erected systems. Some are large systems which heat and cool a number of buildings or various parts of a large building. Others are field-erected systems which service one domestic building or a small commercial building.

These systems may:

1. Cool and/or heat air which is then distributed by a dual system.
2. Cool and/or heat water which is then pumped to heat exchangers in the spaces to be conditioned. The cabinets in the spaces have fins, filters, and controls in them.

Automatic controls make it possible for a system to change from heating to cooling. The use of outside air is also controlled if the outside air is a degree or so above the heating thermostat setting or a degree or so below the cooling thermostat setting. The system then becomes an air-distribution and air-cleaning system only. This permits the greatest economy of operation.

Other systems use more fresh air as the outside temperature approaches the temperature desired inside the system. Solid state controls, operated by thermistors, make this possible.

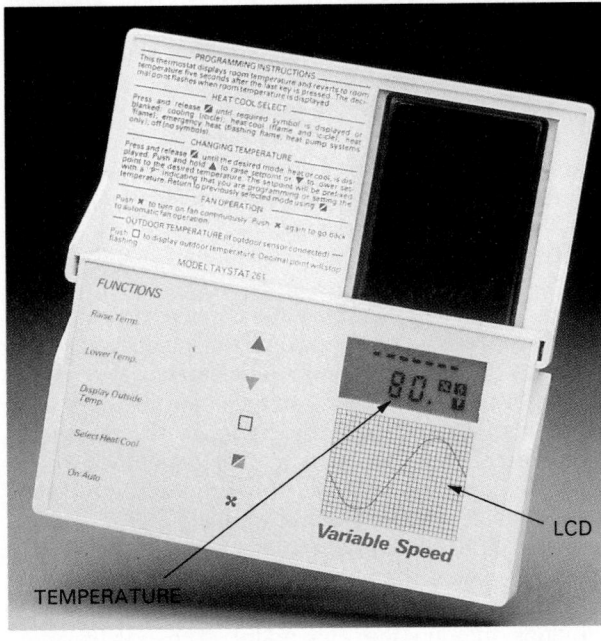

FAN

AIR DISCHARGE
GRILL

VARIABLE
SPEED
COMPUTER
CONTROLLER

ELECTRONIC
DEMAND
DEFROST

VARIABLE
SPEED
COMPRESSOR

A

AIRFLOW
LOUVERS
TO COILS

CONDENSER
COIL

TEMPERATURE

Variable Speed

LCD

B

Fig. 23-61. Heat pump condensing unit with variable speed compressor and computer controller (A). Electronic wall thermostat is shown at B. Note the location of temperature indicator and LCD, which indicates operating speed, indoor temperature, outdoor temperature, and heating or cooling mode identification. The system can be altered by adjusting the thermostat, raising the temperature, lowering the temperature, taking outside temperature readings on the thermostat, or selection of heating, cooling, continuous fan movement, or automatic fan movement. (The Trane Company, Unitary Products Group)

23-31 QUICK-CONNECT COUPLINGS

Self-sealing couplings enable manufacturers to produce precharged refrigeration and air conditioning units—along with the necessary tubing—in separate packages. These separate units may be assembled at the installation site and are ready to operate without evacuating, charging, or cleaning.

The self-sealing coupling fittings are brazed directly to the tubing; flared joints are not needed. There are two types of quick-connect fittings:

1. Those which can be connected and disconnected many times with very little loss of refrigerant. This type is very seldom used.
2. Those which can only be quick-connected once. This type uses diaphragms. When the fittings are attached, the diaphragms are punctured to allow refrigerant to pass. These couplings cannot be disconnected unless the refrigerant is first removed from the system.

In the first type, when the couplings are separated, independent springs force valves in both halves of the coupling to close. This prevents the escape of refrigerant.

To assemble either type of quick-connect fitting, align the couplings and tighten the coupling nut. This draws the coupling together and pierces the sealing diaphragm of the coupling internally so the refrigerant can flow. Fig. 23-62

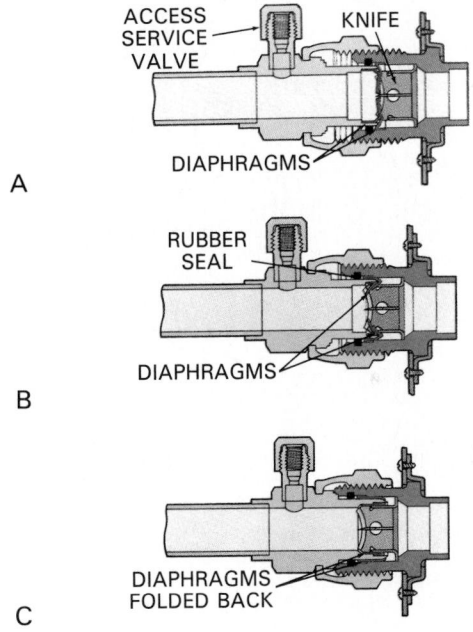

ACCESS
SERVICE
VALVE

KNIFE

A

DIAPHRAGMS

RUBBER
SEAL

DIAPHRAGMS

B

DIAPHRAGMS
FOLDED BACK

C

Fig. 23-62. Quick-connect and disconnect coupling with access service valve. Three views are shown: A—Knife edge aligns and begins piercing diaphragm. B—Partially assembled. C—Connected and refrigerant passage open. (Aeroquip Corporation)

shows three views of a quick-connect coupling. Fig. 23-63 is an exterior view of an assembled quick-connect as it would appear on an installation.

Quick-connect fittings are used mostly on precharged residential air conditioning systems and precharged transportation units. Units are usually charged at the factory with the condensing unit, refrigerant lines, and evaporator being charged separately. Fig. 23-64 shows a liquid line that has an access (service) port.

It is recommended that the gasket which joins the quick-connect-disconnect fittings be covered with clean, dry refrigerant oil just before assembly. Avoid excessive wrench pressure because a distorted fitting may leak. Fig. 23-65 shows two wrenches being used to tighten the fittings.

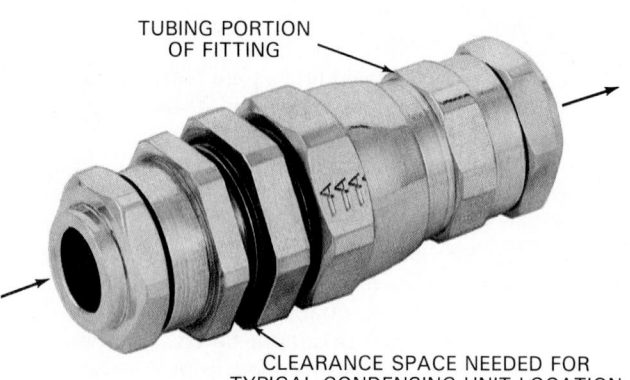

TUBING PORTION
OF FITTING

CLEARANCE SPACE NEEDED FOR
TYPICAL CONDENSING UNIT LOCATION

Fig. 23-63, Assembled quick-connect-and-disconnect fitting.
(Aeroquip Corp.)

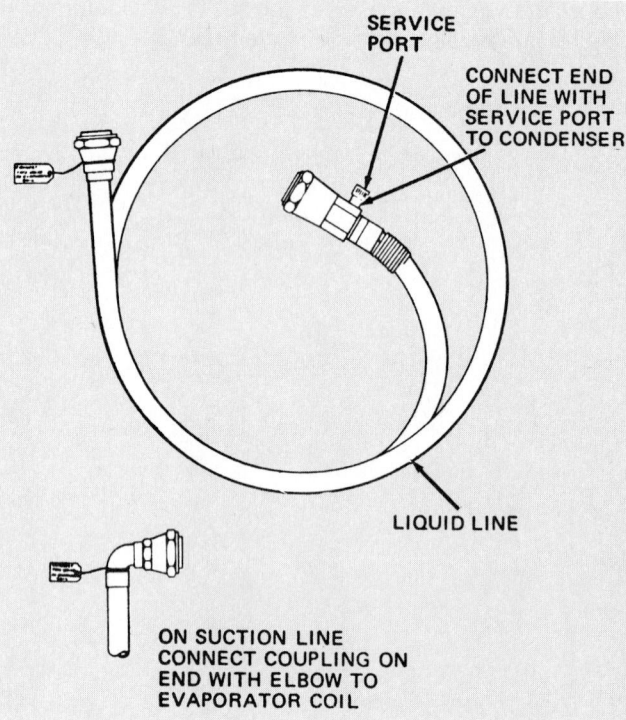

SERVICE
PORT

CONNECT END
OF LINE WITH
SERVICE PORT
TO CONDENSER

LIQUID LINE

ON SUCTION LINE
CONNECT COUPLING ON
END WITH ELBOW TO
EVAPORATOR COIL

Fig. 23-64. Quick-coupler liquid line. Service port is used for making
service manifold high-side pressure gauge connection.
(The Coleman Co., Inc.)

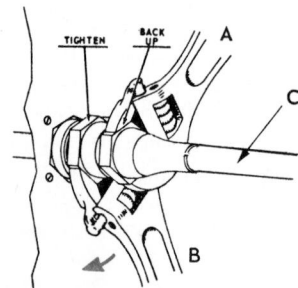

TIGHTEN   BACK UP   A

C

B

Fig. 23-65. Correct way to tighten a quick-coupling. Wrench A is held
firmly while wrench B is turned. Note that sleeve closest to wall mounting turns, but sleeve on tubing C does not turn. Use of two wrenches
prevents twisting of tubing, C.

Most quick-connects-and-disconnects will reseal themselves
several times.

23-32 INSTALLING RESIDENTIAL CENTRAL AIR CONDITIONING SYSTEMS

*Cooling systems should be installed only in furnaces less
than 15 years old.* Older furnaces will need to be replaced.

Units are assembled on-site in four steps:
1. Install condensing unit.
2. Install evaporator.
3. Install suction and liquid lines.
4. Install electrical wiring.

The condensing unit uses outdoor air to cool the condenser. Some homes have water-cooled condensers.

Many installation methods have been used. Some units
are mounted inside the building with ducts bringing outdoor
air to the condenser and discharging warm air outside. Some
units have been mounted on an outside wall. A more
popular practice is to mount the unit on a concrete slab or
prefabricated slab 12 to 24 in. (31 to 61 cm) from the
building. A concrete slab at least 4 in. (10 cm) thick and
reinforced with steel mesh is recommended. Fig. 23-66
shows various installation methods. It is desirable for the
outlet air from the condensing unit to move in the same
direction as the prevailing summer winds.

Location of the condensing unit is very important. The
suction line, liquid line, and power lines should be as short
as possible. The condensing unit should be carefully located:
1. Away from neighbors (noise).
2. Away from bedrooms (noise).
3. At least 24 in. away from wall (air circulation).
4. Away from inside corners (air circulation).
5. Not under eaves (airflow).
6. Beyond overhang (air circulation).
7. Away from patio (noise).

Fig. 23-67 shows alternative condensing unit locations.

The condensing unit should be mounted level. Fig. 23-68
shows a typical installation with clearance spaces needed.

The evaporator is mounted level and firm in the bonnet
or plenum chamber of the furnace. The design of the
evaporator and its condensate drain depend on the type of
furnace (upflow, downflow, or horizontal flow). Removable
panels are needed for periodic cleaning and servicing as required. Fig. 23-69 shows a slant evaporator installation with
blank-off plates shaped like a box.

The plenum chamber is blanked off to make sure all the
air goes through the evaporator. The condensate drain
should be piped to an open drain with an air break at the
drain. If the drain pipe is plastic, keep it away from the
warmer parts of the furnace. Some technicians install a 4-in.
U-trap in the drain line to stop airflow through the line. The
drain pan is built into the evaporator as is the drain connection. Check the local building code for proper installation of all units. In some installations, it may be necessary
to install a drain pump to remove the condensate to the
outdoors. Fig. 23-70 shows an A-type evaporator being installed in a furnace plenum chamber.

Suction and liquid line connections may be:
1. Flared connections.
2. Brazed connections.
3. Quick-connect-and-disconnect couplings.

The condensing unit in Fig. 23-71 has flared or brazed

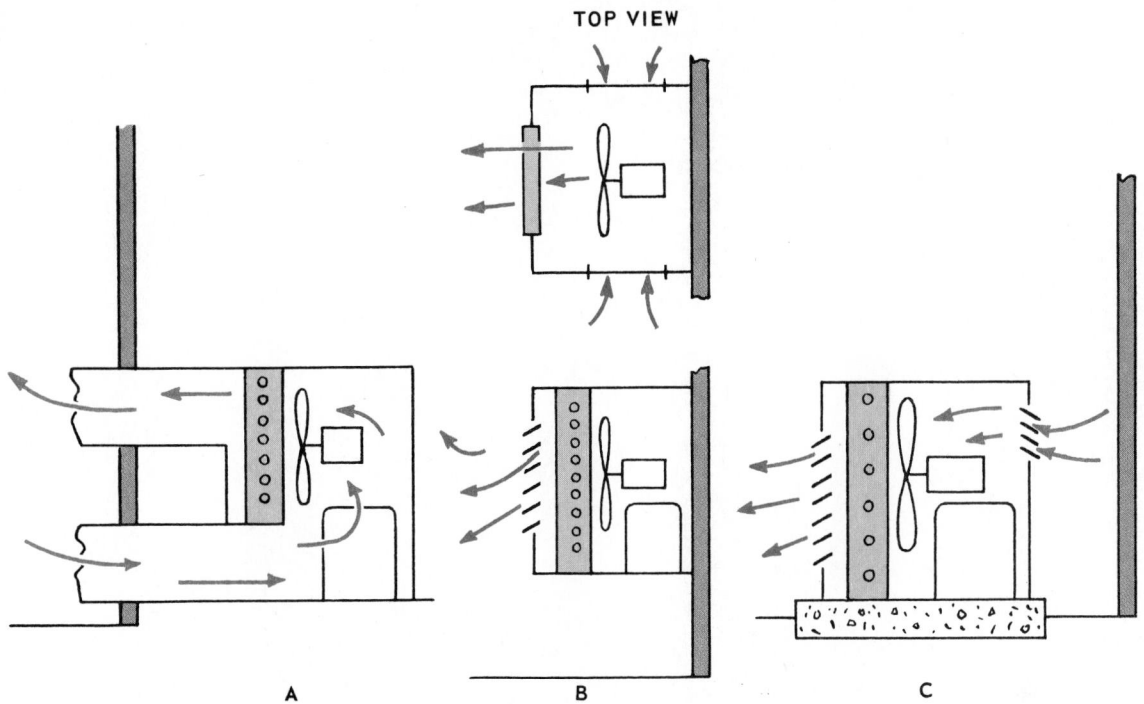

Fig. 23-66. Three types of air-cooled condensing units for residential air conditioning installations. A—Condensing unit inside building. B—Hung on outer wall (usually through window). C—On concrete slab outside building.

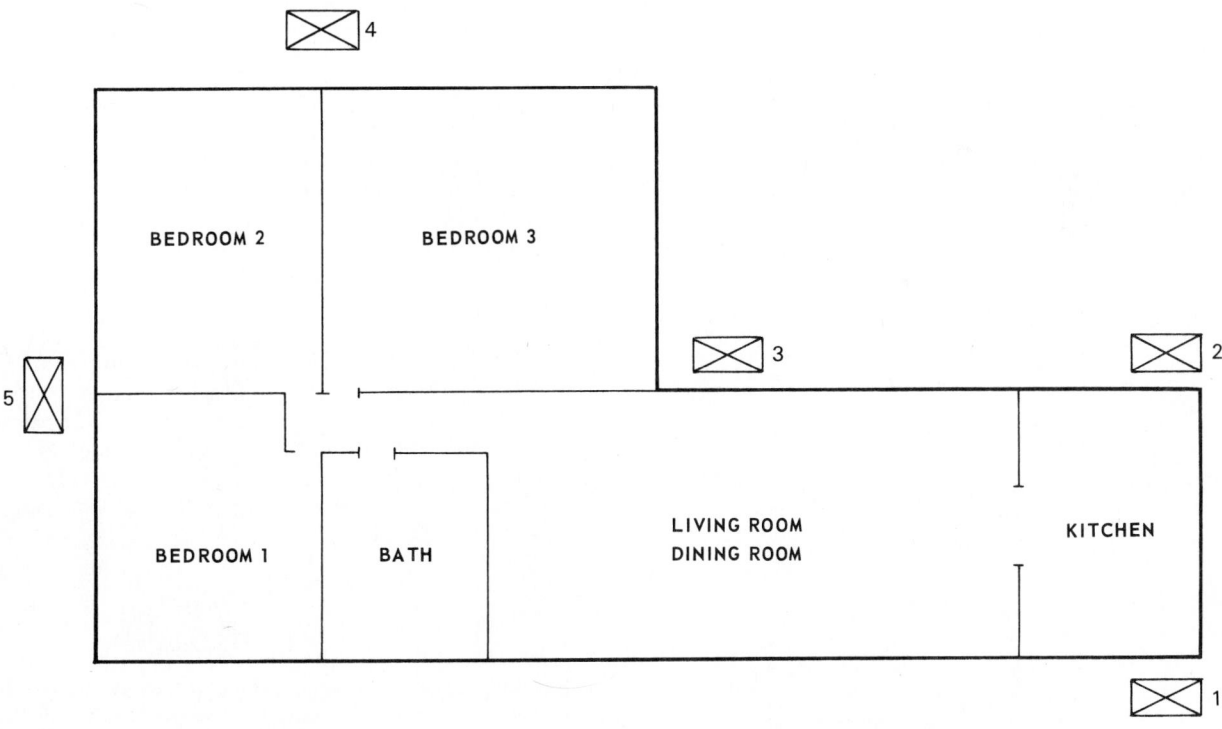

Fig. 23-67. Choose location of outdoor air-cooled condensing unit for best airflow and least noise. 1 and 2 are in good positions; 3 is not recommended, as it is in air pocket and near bedroom; 4 and 5 are also too near bedrooms.

tubing connections equipped with service valves. The refrigerant control is a thermostatic expansion valve. This unit would be installed as described in Chapter 14.

Many condensing units are installed above the evaporator.

Therefore, a U-bend should be put in the suction line (to assist oil return). The suction line should slope downward slightly toward the condensing unit.

A filter-drier and a sight glass should also be put in the

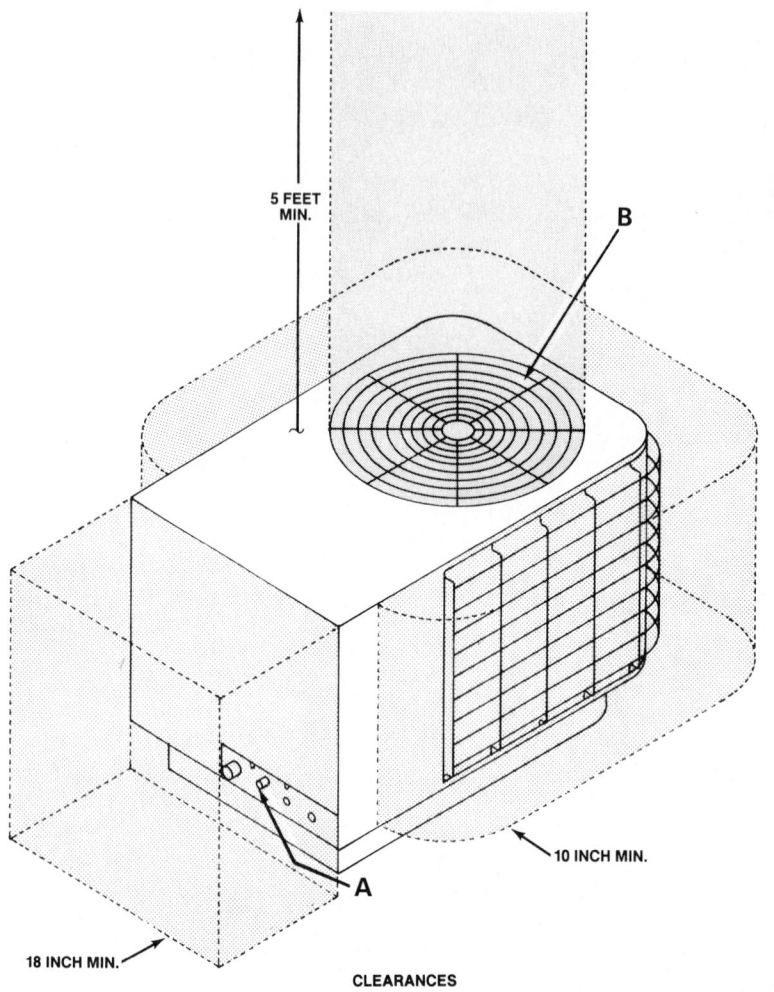

5 FEET MIN.

B

A

10 INCH MIN.

18 INCH MIN.

CLEARANCES

Fig. 23-68. Side view of typical heat pump installation shows minimum clearances to maintain around units. A—Refrigerant line connections. B—Fan outlet needs 5 ft. clearance. (The Coleman Co., Inc.)

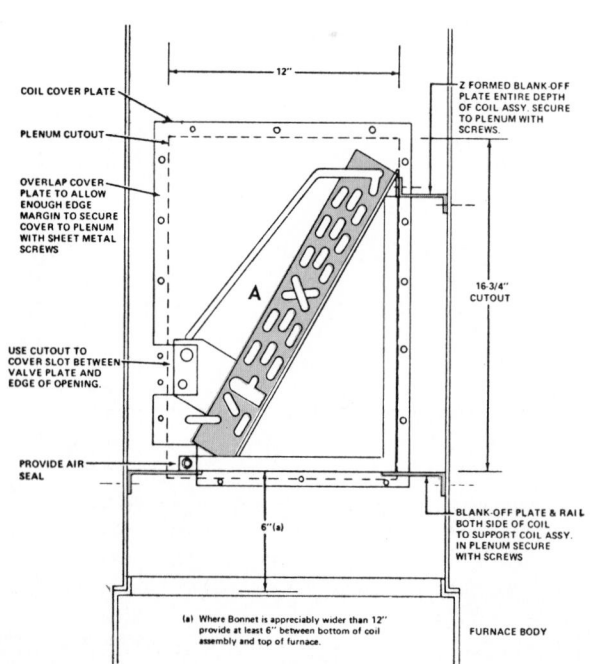

COIL COVER PLATE

PLENUM CUTOUT

OVERLAP COVER PLATE TO ALLOW ENOUGH EDGE MARGIN TO SECURE COVER TO PLENUM WITH SHEET METAL SCREWS

USE CUTOUT TO COVER SLOT BETWEEN VALVE PLATE AND EDGE OF OPENING.

PROVIDE AIR SEAL

12"

Z FORMED BLANK-OFF PLATE ENTIRE DEPTH OF COIL ASSY. SECURE TO PLENUM WITH SCREWS.

16-3/4" CUTOUT

A

6"(a)

BLANK-OFF PLATE & RAIL BOTH SIDE OF COIL TO SUPPORT COIL ASSY. IN PLENUM SECURE WITH SCREWS.

(a) Where Bonnet is appreciably wider than 12" provide at least 6" between bottom of coil assembly and top of furnace.

FURNACE BODY

Fig. 23-69. Slant evaporator installed in furnace plenum. Note blank-off plate to make all air go through evaporator. A—Evaporator.

REFRIGERANT TUBING

FURNACE PLENUM

Fig. 23-70. An A-type evaporator being installed in plenum chamber of upflow warm-air furnace. (Fedders North America)

liquid line. Many service technicians also place a filter-drier in the suction line to prevent motor compressor burnouts.

That part of the suction line installed inside the building should be insulated. Use 1/2-in. line for hot, humid conditions and 1/4- to 3/8-in. for normal conditions. Without insulation, moisture from the air will condense on the suction line and drip. Some installers insulate all of the suction

Fig. 23-71. Condensing unit designed for residential central comfort-cooling system. A—High-side and low-side service valves. B—Motor compressor. C—Filter-drier. D—Condenser. E—Fan. F—Controls. (Williamson Corporation)

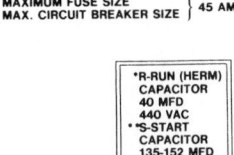

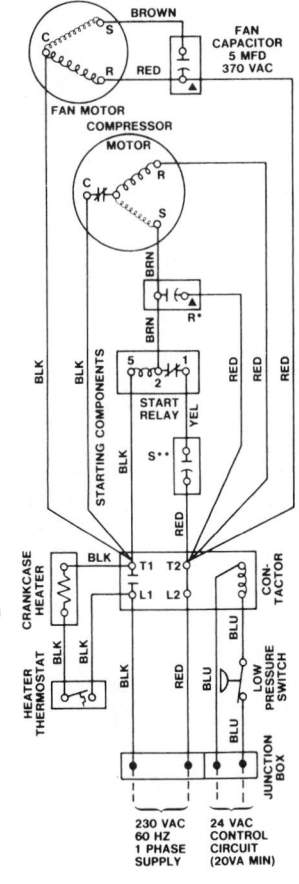

MAXIMUM FUSE SIZE
MAX. CIRCUIT BREAKER SIZE } 45 AMPS

*R-RUN (HERM)
CAPACITOR
40 MFD
440 VAC
* *S-START
CAPACITOR
135-152 MFD
330 VAC

▲ IDENTIFIED TERMINAL

NOTE:
WHEN REPLACING CAPACITORS, BE SURE IDENTIFIED
TERMINAL IS CONNECTED TO LINE SIDE. METALLIC FILM
CAPACITORS MAY NOT HAVE AN IDENTIFIED TERMINAL.

DUAL CAPACITOR MAY BE USED AS ALTERNATE.

FACTORY WIRING SHOWN SOLID, FIELD WIRING SHOWN
DOTTED. FOR ELECTRICAL COMPONENT RATINGS SEE
UNIT NAMEPLATE.

DANGER — SHOCK HAZARD
TURN OFF ELECTRICAL POWER BEFORE SERVICING TO
PREVENT POSSIBLE DAMAGE TO THE EQUIPMENT AND
POSSIBLE PERSONAL INJURY.

NOTE:
WHERE POWER SUPPLY HAS (1) 240 VOLT CONDUCTOR
AND (1) NEUTRAL CONDUCTOR, CONNECT L2 OF
CONTACTOR TO NEUTRAL.

Fig. 23-72. Wiring diagram for central comfort-cooling system condensing unit.   (The Coleman Co., Inc.)

line to avoid absorbing heat which will just have to be rejected at the condenser. Some efficiency is lost because of the extra heat, and the motor compressor is not cooled as well. Lines should be supported and free of kinks. Openings in the furnace duct and the wall should be sealed with weatherproof, nonhardening sealing compound and tape.

When the quick-connect system is used, the lines are first installed. The correct length of prefabricated line should be used. When the quick-connects are made, the unit is ready to operate.

In all cases, the system should be thoroughly tested for leaks while the pressures inside the system are near ambient temperature-pressure conditions. Use soap bubbles, a halide leak detector, or an electronic leak detector.

The electrical installation must follow the wiring diagram furnished with the unit. Fig. 23-72 shows a wiring diagram for a single-phase 240 V ac system.

The electrical circuit must follow the National Electrical Code and local codes. Consult the local electrical utility concerning the primary service capacity. With 240 V circuits, smaller wires can be used.

23-33 SERVICING: INSPECTING RESIDENTIAL CENTRAL AIR CONDITIONING SYSTEMS

To maintain a heat pump central heating/cooling/humidifying/dehumidifying/ventilating unit and keep it economical, a few procedures are recommended. The complete system should be inspected and serviced each season before the system is put in use.

Services which the owner can perform include:
1. Energize crankcase heater 24 hours before starting system.
2. Clean condenser and fans.
3. Check dampers in ducts.
4. Replace filters.
5. Lubricate motor and fan bearings.

6. Check fan belt for cracks and glaze (replace if necessary).
7. Check and clean drain pans and drain pipe.
   Services done by the service technician include:
8. Clean thermostat points.
9. Check system pressures.
10. Check voltage and full load amperage.
11. Check refrigerant charge.
12. Check suction line sweating or frosting.

23-34 SERVICING RESIDENTIAL CENTRAL AIR CONDITIONING SYSTEMS

The economy and dependability of a system depends on proper servicing. Servicing procedures and troubleshooting diagnosis for residential central systems are similar to those described in Chapters 11 and 14.

Check for leaks, proper refrigerant charge, malfunctioning refrigerant controls and motor controls, and moisture in the system. Some systems use service valves. Fig. 23-73 shows one type of valve. Fig. 23-74 is a cross section showing the internal construction. An access type of service valve with a quick-connect tubing connection is shown in Fig. 23-75.

Often the system uses the same blower, motor, filter, and duct system used during the heating season. It is essential that the blower be cleaned once a year; that the motor be lubricated (a few drops of 30 SAE oil) once or twice

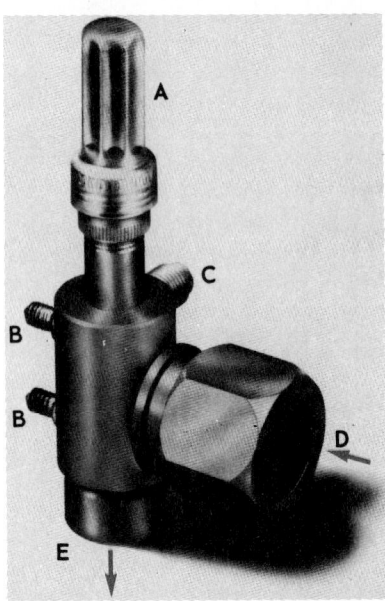

Fig. 23-73. Suction line service valve for residential central air conditioner. A—Valve stem cover. B—Mounting studs. C—Service connection. D—Suction line connection. E—Connection to compressor. (Chatleff Controls, Inc.)

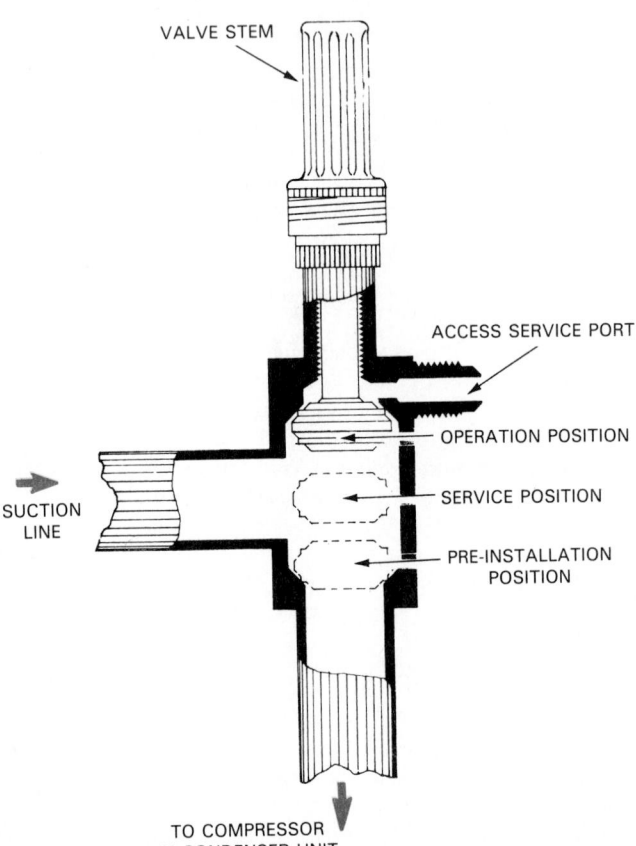

Fig. 23-74. Internal construction of central system air conditioner suction service valve.

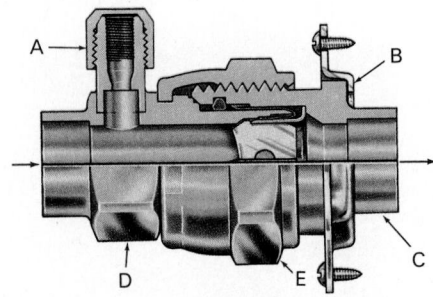

Fig. 23-75. Access-type service valve and connection. A—Access service valve. B—Mounting flange. C—Male coupling. D—Female coupling. E—Quick-connect fitting for precharged tubing. (Aeroquip Corporation)

of cooling season).

The motor-blower speed should be increased for summer comfort cooling. Belts on belt-driven blowers should be inspected and replaced if glazed or cracked.

The condensing unit should be cleaned once a year. The condenser, especially, should be blown clean of lint and fins straightened. A carbon dioxide blower and/or vacuum cleaner may be used for cleaning the unit. The A-style evaporator should be cleaned of any lint and its fins straightened, if bent.

Check condensate drainage. Any condensate which escapes the drain pan may drip on the furnace heat exchanger and corrode (rust) it. The best way to clean these coils is to remove the coil, plug all connections at once, and then either steam clean or use hot water and detergent to clean the fins and the tubes. Steam is the best cleaner, although high-pressure warm water and detergent will do a fair job.

Do not adjust the thermostat too low for summer cooling. The evaporator may freeze the condensate. This will stop airflow through the evaporator. The evaporator may continue to collect a lot of ice.

A split air-conditioning system with the compressor and condenser located outside the home is shown in Fig. 23-76. The compressor condensing unit is attached to the evaporator. The evaporator is installed in the plenum chamber of the warm air furnace.

There may be several reasons for service calls:
1. No heat or insufficient heating. See Chapter 20.
2. No cooling or insufficient cooling. Refer to Chapters 11, 14, and 21.
3. Relative humidity too high. See Chapter 20.
4. Air in house is stuffy (stale). See Chapter 22.
5. Excessive noise. See Chapters 20 and 21:
   a. Indoors.
   b. Outdoors.
6. High cost of operation. Refer to this chapter and Chapters 20 and 21.
7. Unit will not start. See Chapters 20 and 21.

23-35 AIR CIRCULATION SYSTEMS AND RELATIVE HUMIDITY CONTROL

Good, complete air conditioning systems must provide heat, remove heat, clean, and circulate the air. Most systems accomplish each of these. However, many systems do not completely control the relative humidity. To

a year; and that the filter be replaced or cleaned at least twice a year (at beginning of heating season and beginning

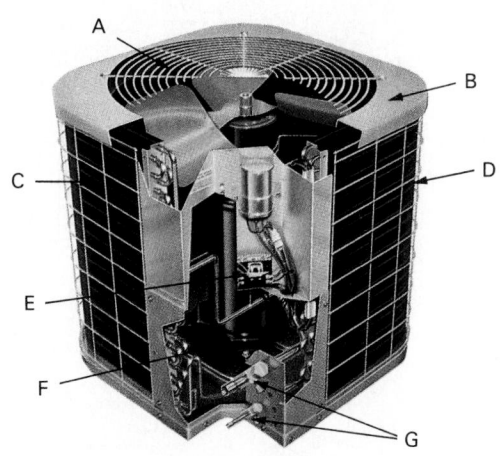

Fig. 23-76. A four-ton central air conditioner using a scroll compressor. Condenser coil has copper tubes. Note junction box where all electrical components and switches are located. A—Top discharge. B—Cabinet. C—Condenser coil. D—Inlet grille. E—Junction box. F—Scroll compressor. G—Sweat/valve fittings.
(TempStar Heating and Cooling Products, Inter-City Products Corporation)

control the relative humidity, one must have in the system two devices ready to be used at any time, in either winter or summer:

1. A device to add water vapor to the air. See Chapter 20.
2. A device to remove water vapor from the air. See Chapter 21.

Another method is to always have a supply of cool air with normal relative humidity (50 percent) and a supply of warm air with normal relative humidity (50 percent). By mixing these air volumes, one can produce the temperature and relative humidity conditions needed. On the psychrometric chart, the process moves up or down along a relative humidity line.

Duct systems are used when the heating and cooling system components are remote (far away from the space to be conditioned). The cost of extending refrigerant lines or having separate relative humidity treatment units may be more than the cost of extending ducts.

Some systems confine the air distribution to the space being conditioned. A cabinet with fans, filters, grilles, and registers is located in the room.

23-36 TWO-DUCT SYSTEMS

The two-duct system uses two supply ducts, one with cool dehumidified air and the other with warm humid air. These separate airstreams are mixed just before they reach the space to be conditioned. Through dampers that control and balance air, each different space in the building can be conditioned as needed. A single-duct air return is used. See Fig. 23-77. The mixture of the two airflows takes place in the duct section at B.

The air control is excellent in these systems. However, the ducts are large in cross-section and take up valuable space. Some space savings may be obtained by using high-velocity ducts, but a noise problem then develops.

23-37 FOUR-PIPE SYSTEMS

Four-pipe complete air conditioning systems have a hot water supply pipe, a chilled water supply pipe, and two

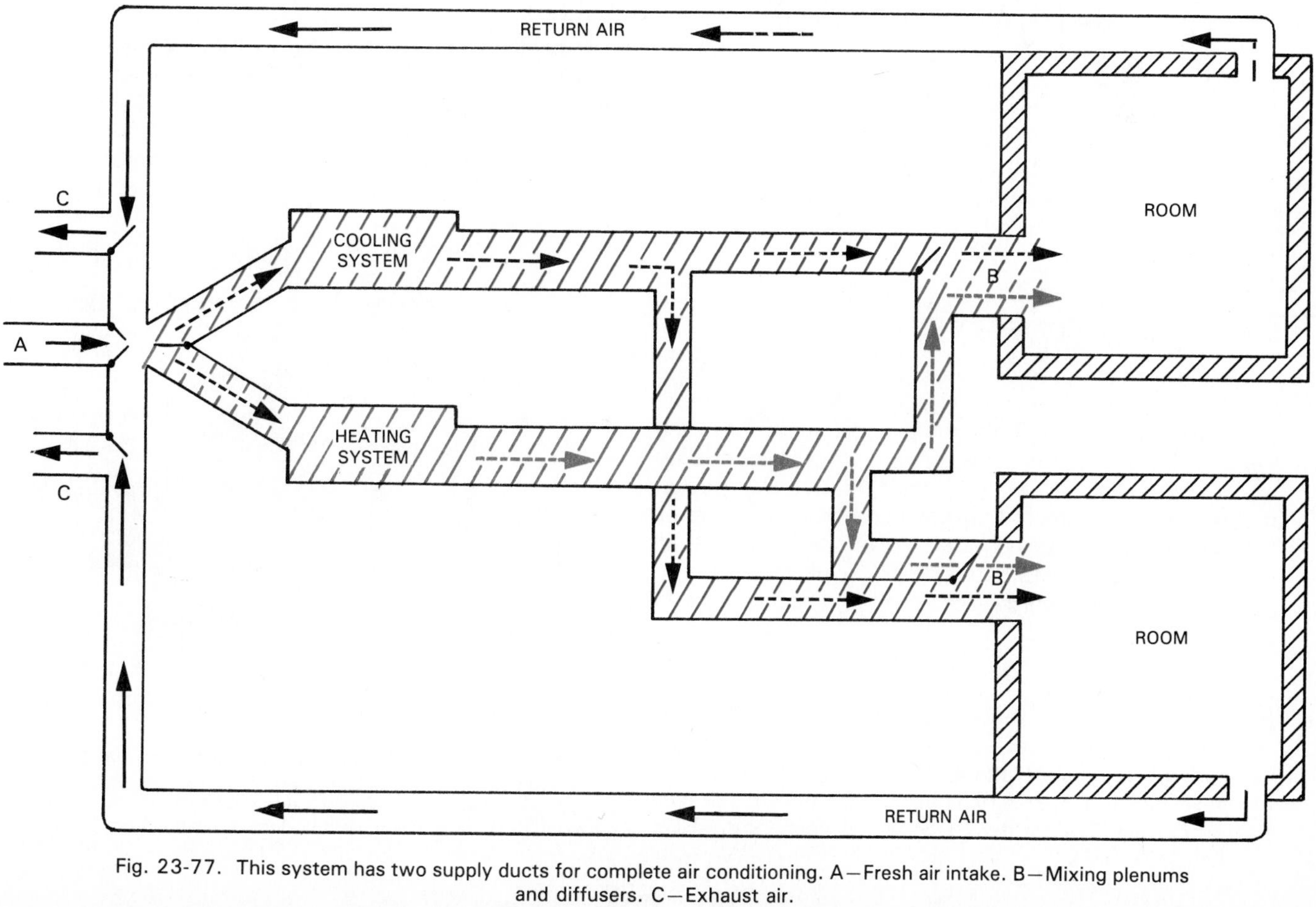

Fig. 23-77. This system has two supply ducts for complete air conditioning. A—Fresh air intake. B—Mixing plenums and diffusers. C—Exhaust air.

return pipes. Four-pipe systems are of two types. One has separate heating and cooling coils in the space to be conditioned. The other uses the same coil for both warm water and the chilled water.

A system using separate heating and cooling coils is shown in Fig. 23-78. The heating circuit is completely separate from the cooling circuit. In this case, the heating fluid (heat carrier) may be either water or steam. One two-way valve is used for each coil.

Some systems use the same space heat transfer coil for both heating and cooling, as shown in Fig. 23-79. Two separate three-way control valves are needed for each heat transfer coil in the conditioned space.

Hot water and chilled water pipes are insulated. The pumps operate only when the thermostats in the conditioned spaces call for heating or cooling.

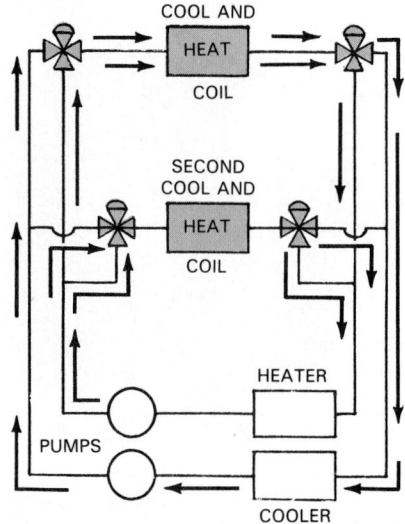

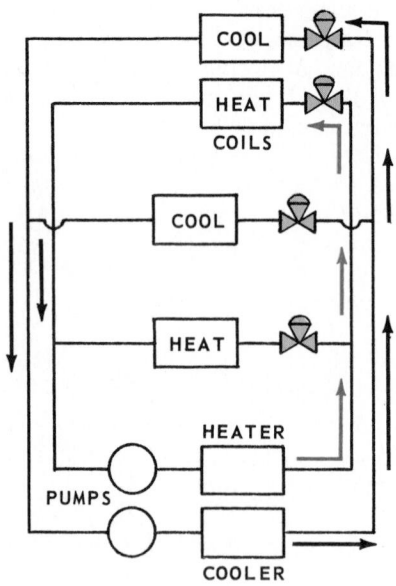

Fig. 23-78. Complete four-pipe system uses separate heating and cooling coils. Only one thermostatically-controlled valve is used for each heat transfer coil. Heating circuit and cooling circuit are separate.

23-38 LARGE COMFORT SYSTEMS

Central station comfort cooling, part of a complete system, is available in many styles. One unit uses shell-and-tube construction in both the water-cooled condenser and water-chiller evaporator. This unit has a variable capacity system which unloads (lets refrigerant into clearance pocket or sends it through more than one condenser) cylinders as the load decreases. These hydraulic controls also unload the compressor to minimize starting load (usually allows a leak past suction valve, which keeps the liquid backed up into the suction line).

Another air-conditioning system circulates chilled water to the various cooling coils in the multiple installation. Refrigeration systems of this type are often referred to as "chillers." Chilled water systems are used for many industrial processing installations. The evaporator of this type of cooling system has a heat exchanger which removes heat from a liquid rather than from air. This evaporator-heat

Fig. 23-79. Complete four-pipe system uses water or water-and-glycol mixture to move heat. Same coil is used for heating and cooling. Note four valves (two at each heat transfer unit). These valves, controlled by room thermostats, may be off-on valves or modulating-type valves.

exchanger is sometimes referred to as a "cooler." A chiller system containing a cooler is shown in Fig. 23-80. The unit is cooled by a microprocessor control and operated by a screw-type compressor. A compressor of 10- to 15-ton capacity is shown in Fig. 23-81.

A water chiller using a screw-type compressor with a microprocessor is shown in Fig. 23-82. The internal construction of a motor compressor of a water chiller with a six-cylinder unit, two cylinders in each bank, is shown in Fig. 23-83.

Many large comfort cooling installations use centrifugal-type compressors. Large centrifugal units are frequently designed with capacities of 100 to 2000 tons. Their basic design is shown in Fig. 23-84. These systems use low-pressure refrigerants, and the evaporator operates at below-

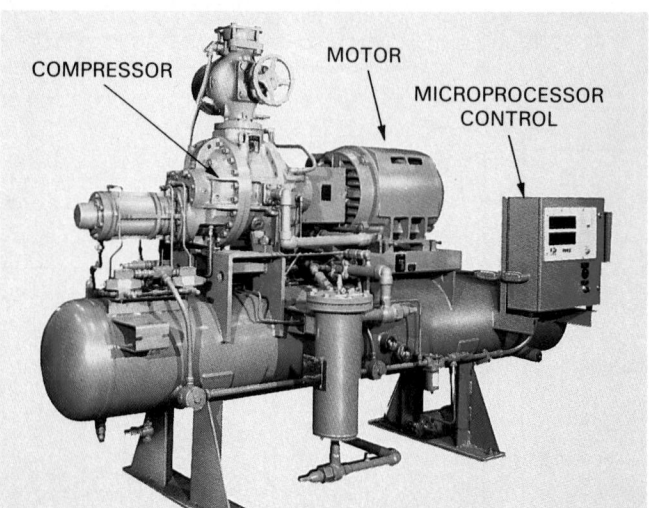

Fig. 23-80. A water chiller unit with screw-type compressor typical of those used for industrial processes. Some units are part of a complete industrial process heating-and-cooling system. Other units are part of complete building heating-and-cooling system. (Frick Co.)

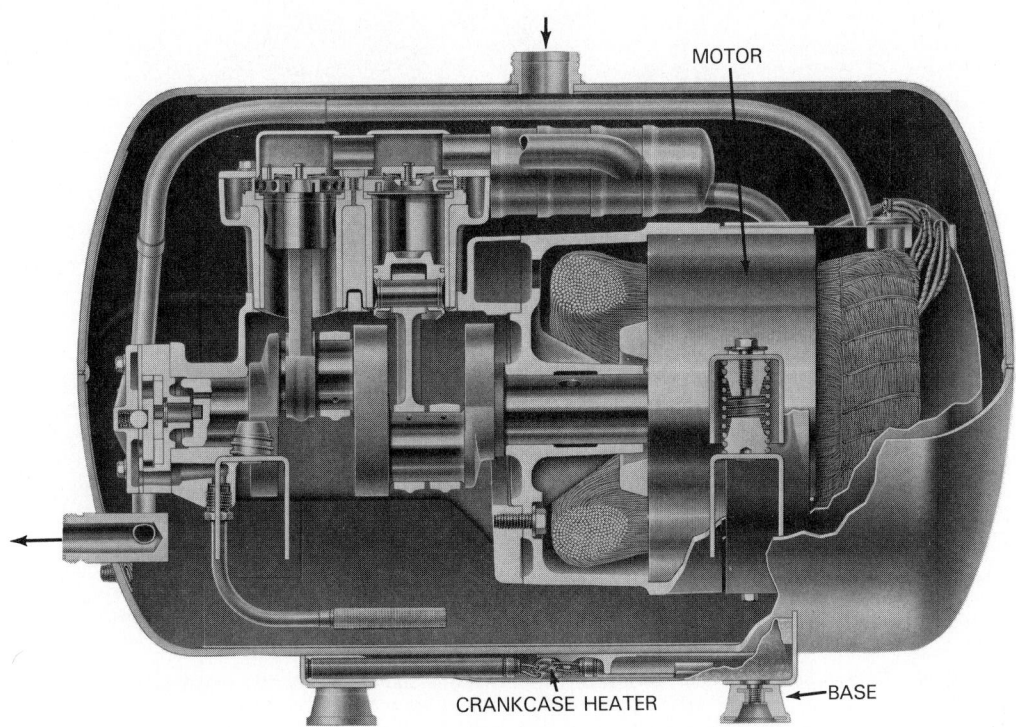

Fig. 23-81. Cross section of "Hermeticom" hermetic compressor. It is used in systems ranging from 10-ton to 15-ton capacity. Note crankcase heater connections at bottom.

Fig. 23-82. Microprocessor-controlled chiller package using a screw-type compressor. (Carrier Transicold Division, Carrier Corporation)

atmospheric pressures. Both condenser and evaporator are the shell-and-tube type. Water lines vary in size from 6 ips to 14 ips (internal pipe size). The compressor is a two-stage centrifugal unit, driven by a hermetically sealed motor. The capacity is controlled by inlet vanes to the two-stage centrifugal compressor. The vanes may be either electronically or pneumatically controlled and hydraulically operated. During starting, vanes are closed to reduce the starting load. Fig. 23-85 shows the compressor construction. This compressor has a forced lubrication system. A separate motor drives the oil pump. The compressor motor is a three-phase unit of 208, 240, 440, 480, 550, 2300, or 4160 volts. Note the bolted construction for service purposes.

These systems have an automatic purging device for removing noncondensible gases. Design and construction details of the condenser and the evaporator are shown in Fig. 23-86. Because of their capacity, these units must have thorough, accurate control.

Persons responsible for the operation of these units should receive thorough training in their correct operation. A drawing of the complete wiring and piping system of one of these units is shown in Fig. 23-87. These units are also used to cool process liquids. The evaporator, compressor suction line, and chilled liquid lines are always insulated.

Another large comfort system concept that has been used is thermal energy storage. The thermal energy storage concept utilizes large amounts of electricity at off-peak hours. These are low electrical demand points which normally occur at night, when utilities have excess generating capacity and charge a lower rate. A cooling system stores the on-peak cooling requirements during the off-peak hours. This shifts all refrigeration energy requirements to off-peak hours and reduces the size of the refrigeration system.

An ice-based system which builds and stores ice and utilizes low-cost off-peak electrical energy is shown in Fig. 23-88. The stored ice is built on Temp-Plate® evaporator plates located above the water ice storage. When ice reaches approximately 5/16 of an inch in thickness, it is released for storage in the tank. The ice is then also used as a water chiller system.

During the ice-building/chilling mode, Fig. 23-89, refrigerant is compressed, condensed, and then passed through a high-side float to a liquid receiver. The refrigerant then passes through a plate evaporator and back to the compressor. Water passes over the evaporator, causing ice to form on the outside surfaces of the evaporator. During

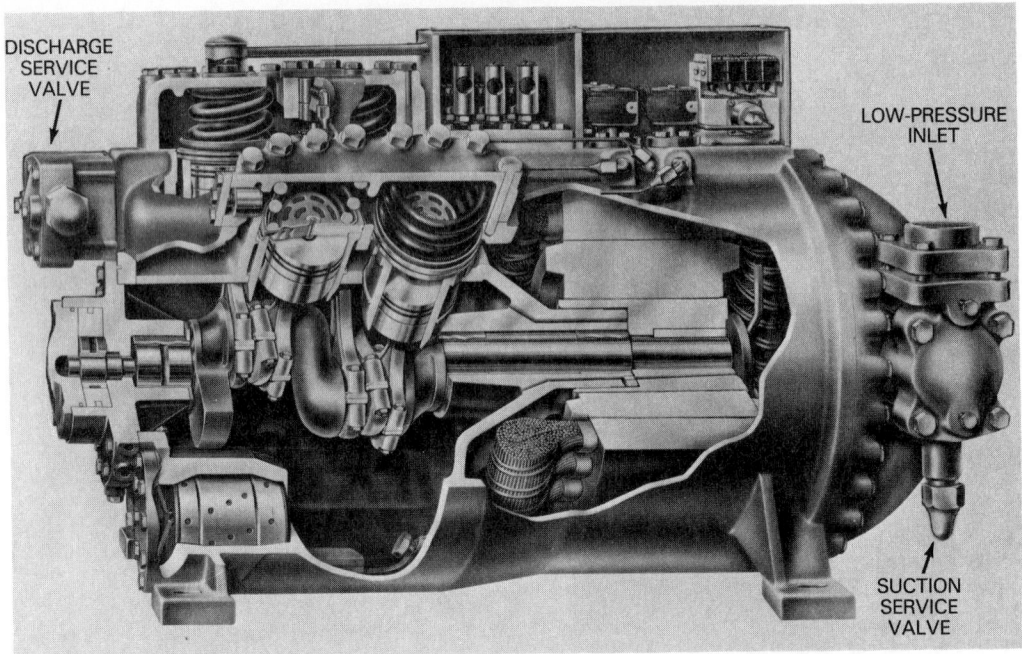

DISCHARGE SERVICE VALVE

LOW-PRESSURE INLET

SUCTION SERVICE VALVE

Fig. 23-83. Cutaway of six-cylinder serviceable hermetic motor compressor. Note arrangement of three piston rods on each crank throw.

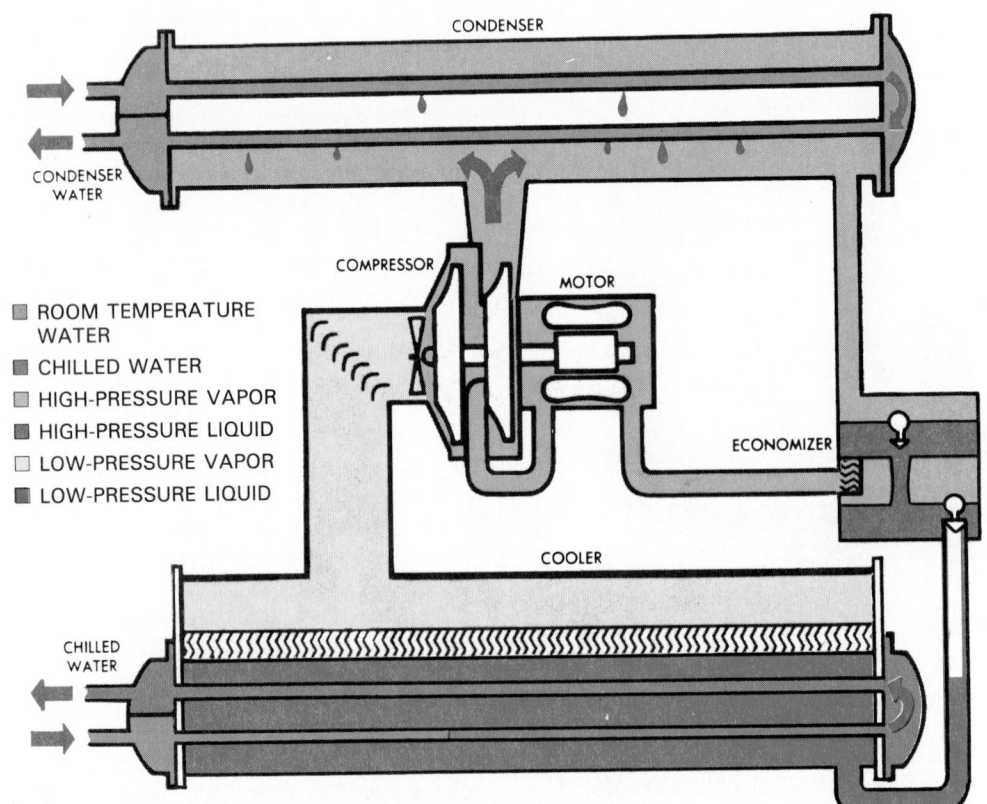

CONDENSER

CONDENSER WATER

COMPRESSOR

MOTOR

■ ROOM TEMPERATURE WATER
■ CHILLED WATER
■ HIGH-PRESSURE VAPOR
■ HIGH-PRESSURE LIQUID
□ LOW-PRESSURE VAPOR
■ LOW-PRESSURE LIQUID

ECONOMIZER

COOLER

CHILLED WATER

Fig. 23-84. Centrifugal compressor for hermetic chilled water system uses rotors instead of pistons. Chilled water is under great pressure to move it through chemical reaction tank spraybars.
(Carrier Corp., Subsidiary of United Technologies Corp.)

the ice harvesting mode, Fig. 23-90, the refrigeration system uses a hot-gas defrost cycle. Hot refrigerant gas travels through the evaporator for a few seconds. This causes the ice built up on the evaporator to fall into the storage tank. The chilled water in the storage tank is then circulated for air conditioning or refrigeration use.

CAPACITY
CONTROL
MOTOR
GAS
INLET
TWO STAGE
COMPRESSOR
OIL PUMP

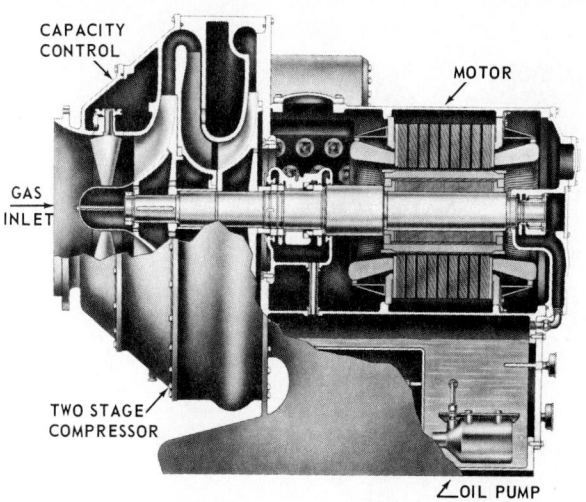

Fig. 23-85. Two-stage centrifugal compressor. Oil pump is driven by separate power source.
(Carrier Corp., Subsidiary of United Technologies Corp.)

A
B
CONDENSER
C
D
EVAPORATOR

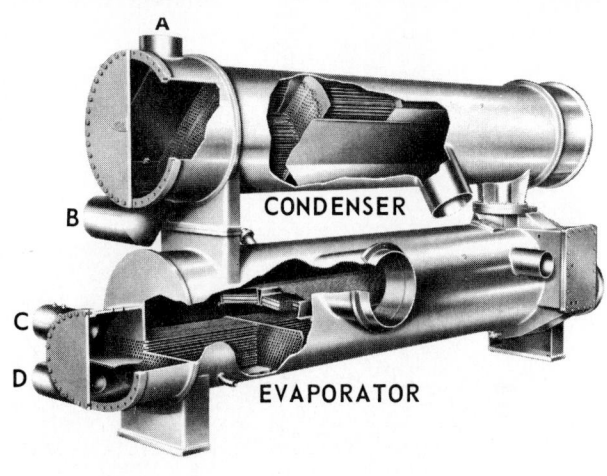

Fig. 23-86. Condenser and evaporator of large chilled water system which uses centrifugal compressor. A—Condenser water in. B—Condenser water out. C—Chilled water out. D—Chilled water return.
(Carrier Corp., Subsidiary of United Technologies Corp.)

NOTES:
1. Enclosed in a separate and grounded metallic conduit.
2. Separate 115-volt source for controls, unless transformer is furnished with compressor motor controller.
3. Condenser water temp control recommended when machine must operate with tower water below 55 F
ENCLOSED ITEMS
ONLY WHEN SPECIFIED
COOLING
TOWER
FAN
STARTER
CONDENSER
WATER
PUMP
STARTER
CHILLED
WATER
PUMP
STARTER
PILOT
RELAY
2
WIRES
I WIRE
2
WIRES
4
WIRES
I WIRE
6 WIRES
COMPRESSOR
MOTOR
CONTROLLER
COOLING TOWER
SEE
NOTE 3
OIL
PUMP
STARTER
FUSED
DISCONNECT
SEE NOTE 2
3 OR 6 WIRES
(TO COMPRESSOR
MOTOR)
3 WIRES
5
WIRES
VENT
I WIRE
2 WIRES
3 WIRES
(TO OIL PUMP MOTOR)
6 WIRES
2 WIRES
I WIRE
2 WIRES
3 WIRES
115 V TO OIL HEATER
AND THERMOSTAT
VENT
CONDENSER
2 WIRES
REFRIGERANT LOW-TEMP
CUTOUT
6 WIRES
2 WIRES
EACH
VENT
SEE
NOTE
I
SEE
NOTE I
OIL HEATER AND
THERMOSTAT
LEGEND
WATER PIPING
POWER CABLE
CONTROL WIRING
CONDENSER
WATER PUMP
COOLER
CHILLED
WATER
TO LOAD
FROM
LOAD
CHILLED
WATER
PUMP
FOLLOW UP
POTENTIOMETER
AND VANE
SWITCH
LOW OIL PRESSURE CUTOUT
OIL COOLER WATER SOLENOID VALVES
GUIDE VANE CONTROL SOLENOID VALVES
IN
COMPRESSOR
BASE
COMPRESSOR MOTOR HIGH-TEMP CUTOUT
Wiring and piping shown are general points-of-connection guides only and are not intended for or to include all details for a specific installation.
All wiring must comply with applicable local and national codes.
All piping must follow standard refrigerant piping techniques.
CHILLED WATER FLOW SWITCH
CHILLED WATER LOW-TEMP
CUTOUT AND RECYCLE CONTROL
CHILLED WATER CONTROL BULB
DRAIN

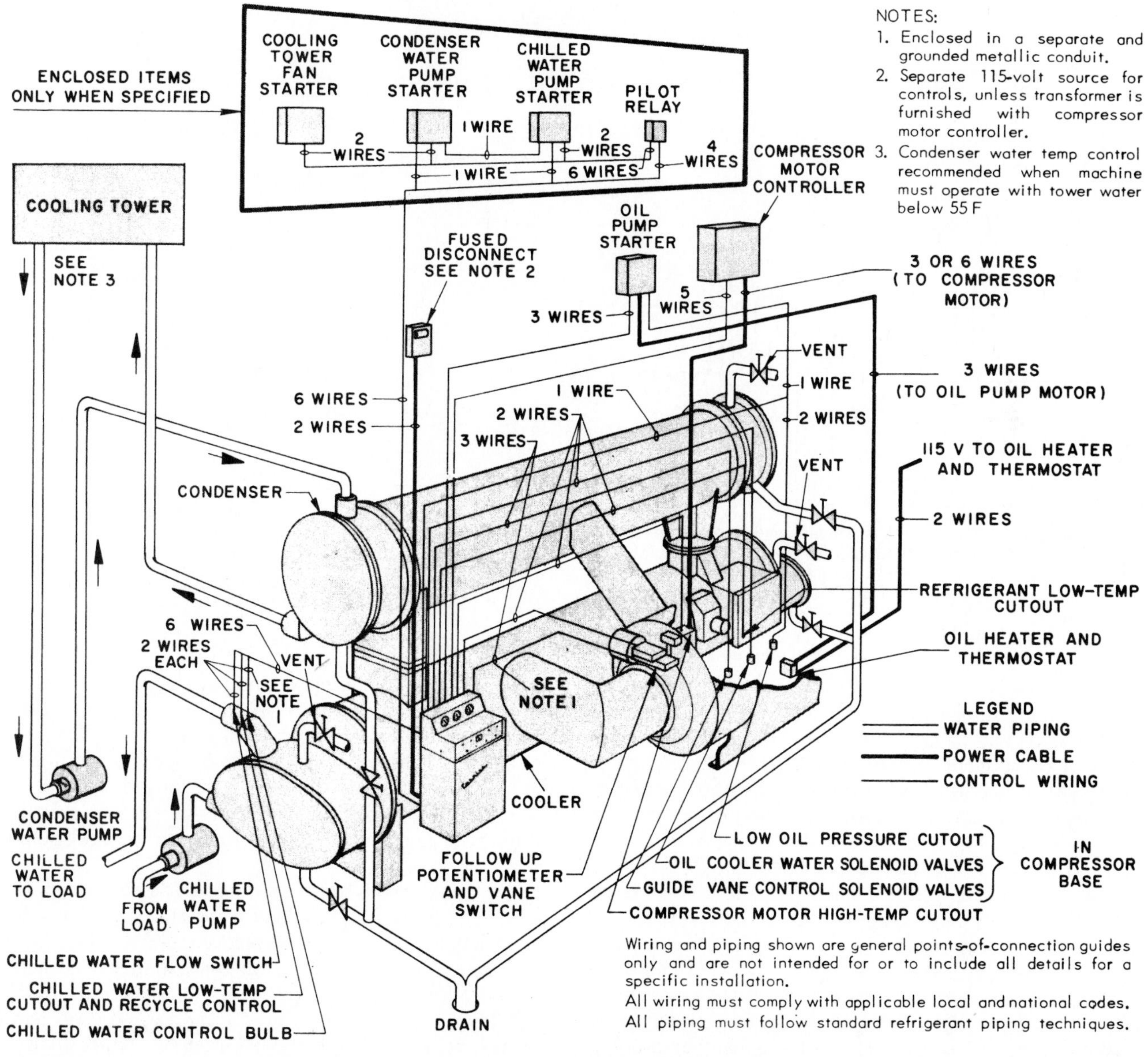

Fig. 23-87. Drawing shows wiring and piping of large-capacity water chiller. (Carrier Corp., Subsidiary of United Technologies Corp.)

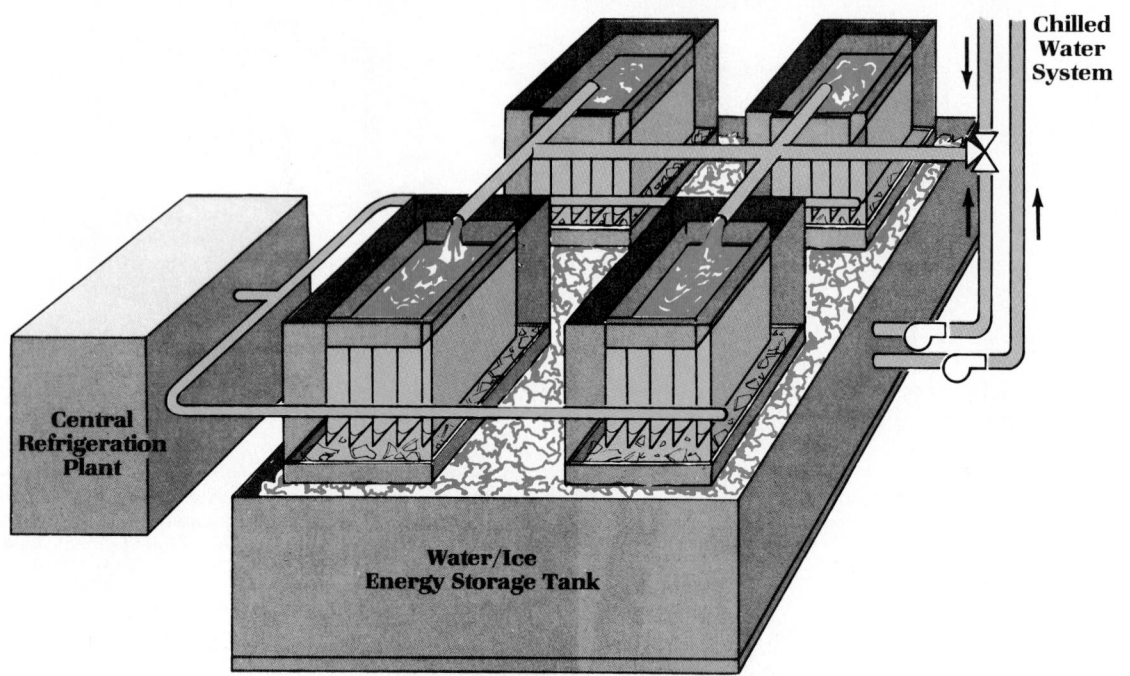

Fig. 23-88. An ice harvester/chiller module with a central refrigeration plant. (Paul Mueller Company)

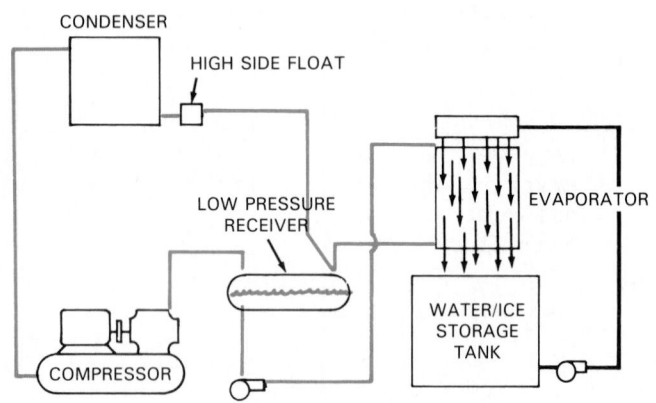

Fig. 23-89. An ice-building/chilling mode. Note flow of refrigerant to produce ice. (Paul Mueller Company)

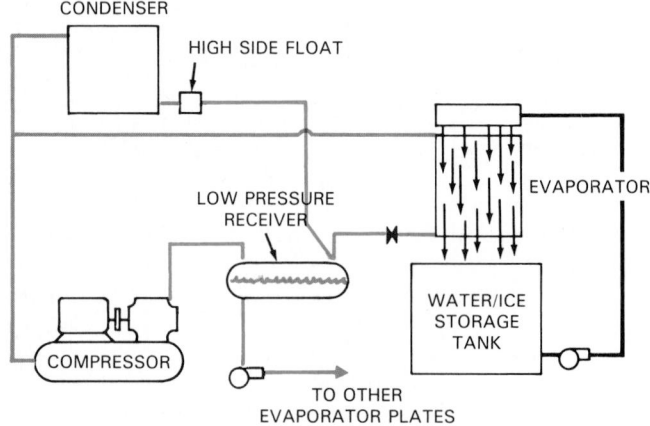

Fig. 23-90. Ice harvesting mode. Note flow of hot refrigerant vapor to evaporator plates, releasing sheets of ice from evaporator. (Paul Mueller Company)

23-39 TOTAL ENERGY

In recent years, many large buildings have been constructed using "total energy" or "single energy" systems. All the energy-use devices are designed to use up all the energy of combustion before discarding the byproducts.

The system generates whatever electricity is needed in the building. The controls used in this type of system are reviewed in detail in Chapter 25.

The advantage of a total energy system is twofold. First, a lower fuel rate can be obtained because of the large volume of fuel supplied. Secondly, more complete use is made of the energy released by burning this fuel.

In a total energy system, electricity is generated with reciprocating gas engines, gas turbines, or steam turbines. The engines drive electric generators in the building. Hot water from the engine's cooling jacket is used as a prime source of heat. The exhaust gas is another major heat

source. A gas engine or turbine uses about 33 percent of its fuel energy to generate electricity. Another 30 percent ends up in the water-cooling jacket and about 30 percent is released in the exhaust gases. About 7 percent is used to heat the lubricating oil or is lost by radiation.

Heat in the jacket hot water, in the exhaust gases, and in the hot lubricating oil can be converted to useful purposes such as:

1. Air conditioning (comfort cooling). Absorption systems often use the exhaust heat. At present, because exhaust gases cannot be cooled below 250 °F (121 °C), only about 50 or 60 percent of the exhaust gas heat can be used.

2. Exhaust heat from turbines may also be used to raise the temperature of water used by hydronic heating systems during the heating season.

3. The heat from the water jackets can be used to furnish hot water.

In a total energy system, the heat produced by lighting, solar radiation, people, and outside air are all taken into account. Each is a heat source in a total energy system. The warm exhaust air, during the heating season, is used to warm the cold replacement fresh air entering the building. Cooled exhaust air in the summer cools the incoming warm replacement fresh air.

Many aspects of total energy systems have been used. For example, heat in one system from the return air ducts of the heating system is released through channels in lighting fixtures. The cool air passes across the lighted fixtures, where it is heated. The needed heat for that room is allowed to enter through a duct. The surplus heat is passed back to the primary duct system for use where needed in the building.

An example of a total energy system is one used in a multi-apartment complex. A separate hot and chilled water plant is used for each independent apartment complex. A single electric power generator is used for the entire apartment complex.

The hot and chilled water plant provides the heating, air conditioning, and domestic hot water. A four-pipe heating and cooling system is used in each complex. The domestic water is circulated to each use area and then back for reheating. The cooling may use high-efficiency gas-fired air cooled absorption chillers.

Three types of systems are in use:
1. Fuel is burned to create steam; the steam drives a turbine to power an electric generator.
2. Fuel is burned to drive an electric generator:
   a. Gas engine.
   b. Turbine engine.
3. Nuclear fuel is used to create steam.

In most cases, the exhaust gases are used to heat one or more of the following:
1. Hot water for heating or for consumption, or both.
2. Absorption cooling systems.
3. To heat or cool the fresh air intake.

In most cases, exhaust air (either cooled or heated) is used to cool or heat incoming fresh air. Likewise, waste cool or warm water is most often used to cool or heat incoming fresh water. Heat from lighting is sometimes used to heat water or air.

23-40 DISTRICT HEATING AND COOLING SYSTEMS

Systems of central utilities that provide heating and cooling have recently been developed for full or partial sections of cities and towns. They are common in Scandinavian countries, Russia, and Eastern Europe. They are becoming more economical and popular in the United States. The largest system is in the city of St. Paul, Minnesota.

In these systems, waste heat from an electric power plant or heat from an incinerator or large boiler facility is used to heat water that is then piped to buildings.

In the summer, the waste heat is used in absorption cooling system chillers. The cold water produced is then used for air conditioning.

23-41 REVIEW OF SAFETY

Working on a complete air conditioning system involves most of the dangers which exist in refrigerating systems and heating systems.

Always wear goggles when working on these systems. Be sure the electrical service is off (and locked off) before working on the electrical circuits or on electrically powered parts such as the controls and fans.

Always install pressure gauges when checking the system, charging it, or adding oil.

When working on rooftop units, use only a safe ladder and use it safely. Never carry equipment, tools, or supplies up or down a ladder. Lift and lower these items using a rope with a safety tie-down handled by someone on the roof and another person on the ground. Use a guide rope to keep items from swinging or turning while items hang. Avoid live electrical circuits when the roof is wet or snow-covered.

Avoid having loose panels or loose light materials on the roof during a high wind or when it is gusty. Fasten the hinged panels of a rooftop unit to prevent uncontrolled swinging of these panels during a wind.

23-42 TEST YOUR KNOWLEDGE

1. How can heat be removed from 10 °F air?
2. What are the two operating cycles of a heat pump?
3. Under what conditions is a heat pump most efficient?
4. What is the energy efficiency ratio in heat units? Is it different from the coefficient of performance?
5. Is it less expensive to heat with electrical resistors or with a heat pump? Why?
6. How many Btu/hr. are generated by one kW through direct electrical resistance?
7. What are the types of heat pump outdoor coils?
8. List one disadvantage of each type of outside heat pump coil.
9. What type of heat pump coil is used indoors?
10. What are the three ways to install the condensing unit of a residential central system?
11. For what purpose is the high speed used in a two-speed heat pump?
12. What does the TEV control in the Hi/Re/Li system?
13. What does the check valve do in the Hi/Re/Li system?
14. Describe the five systems which are combined in a solar-supplemented heat pump system.
15. Why is the capillary tube especially advantageous to use in heat pumps?
16. In a heat pump, how many thermostatic expansion valves are used?
17. How many refrigerant connections does a four-way reversing valve have?
18. Explain the operation of a four-way valve used to reverse the refrigerant flow in a heat pump.
19. Describe the two ways a solenoid is used to operate a reversing valve.
20. How many check valves are used in a heat pump system equipped with thermostatic expansion valves?
21. What is done to remove frost or ice from an air outdoor coil?
22. How does well water release heat to or pick up heat from the heat pump?

23. What is used to protect the motor compressor during low temperature conditions?
24. What two electrical devices does a heat pump thermostat control?
25. Where is an A-shaped evaporator used?
26. What parts of a warm-air furnace may be used with a central comfort-cooling system?
27. What is carried in the ducts of a two-duct system?
28. What is carried in each pipe of a four-pipe system?
29. What is the reason for having a supply of cool air with normal relative humidity (50 percent) and a supply of warm air with normal relative humidity (50 percent)?
30. How many wrenches should be used when tightening a quick-coupler connection?
31. Why do rooftop gas burners usually use electric ignition?
32. What is district heating and cooling?
33. Why are absorption systems used in a total energy system?

Chapter 24

SOLAR ENERGY

After studying this chapter, the technician will be able to:
☐ Explain the nature of solar energy and the physics of energy conversion.
☐ Discuss ways of collecting and converting solar energy for space and water heating and for producing electricity.

24-1 ENERGY FROM THE SUN

The sun radiates energy to space as a result of nuclear fusion processes at its surface. The solar energy (sun) input to the earth and its atmosphere heats the earth. Solar energy is also one of the causes for weather changes.

The solar heating of the atmosphere is the cause of winds. The sun is therefore one indirect source of wind energy. This energy can be converted to mechanical or electrical energy by wind machines. Ocean waves are caused by solar energy. The hydroelectric energy generated by water turbines at Niagara Falls, Grand Coulee Dam, and Hoover Dam is directly related to the sun. Evaporated water is condensed in the atmosphere, and the resulting rain fills reservoirs. The reservoir water is then used to drive the water turbines in hydroelectric power plants.

Wood burned in stoves and fireplaces gives back, in the form of heat, solar energy absorbed by the tree as it grew. The photochemical processes used to change solar energy to fuels are the same processes by which fossil fuels (coal, oil, and natural gas, for example) are formed.

Most energy sources are the indirect result of solar energy conversion processes. In this chapter we will describe methods to convert solar energy directly to useful heat and electrical energy. This energy can then be used as the power source for cooling systems.

Solar energy is best illustrated with one basic fact. When a dark surface of about one square meter is exposed to bright sunlight, it will absorb about one horsepower (746 watts) of energy.

24-2 THE NATURE OF SOLAR ENERGY

Visible light seen on a sunny day is only a small fraction of the radiant energy present. Our eyes are sensitive to about 25 percent of the energy in solar radiation.

Solar energy or solar radiation is electromagnetic energy. The energy in a beam of radiation can be described by thinking of the beam as a stream of particles called photons. The photons in a beam of solar radiation have a wide range of energies. Some photons have low energy (infrared) and

others have high energy (ultraviolet). Any beam of solar radiation has a mixture of many different photon energies.

Solar radiation has both particle (photon) and wave characteristics. In describing the conversion of solar energy to electricity, it is helpful to refer to the energies of photons. In describing the conversion of solar energy to heat, it is common to refer to it as "energy waves." When describing the principles of solar energy conversion, whichever characteristics are easiest to understand (photons or energy waves) will be used.

An electromagnetic wave, as shown in Fig. 24-1, has both wavelength and amplitude. The wavelength is the distance between two peaks. The amplitude is the height of the wave shown. Think of the radiation wave as an ocean wave approaching a beach. The energy of a wave for any given amplitude depends upon the wavelength. The shorter the wavelength, the greater the energy of the wave of radiation.

The energy of a single light wave can be expressed by the formula:

$$E = \frac{hc}{L}$$

E = Energy.
h = Planck's constant, which is 6.626×10^{-34} watt seconds squared (W·s²).
c = Speed of light, which is 3×10^8 meters per second (m/s).
L = Wavelength in meters.
The energy (answer) is given in watt seconds (W·s) or joules (J) per wave.

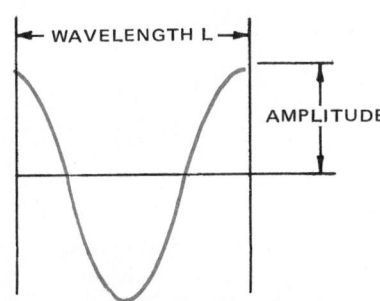

Fig. 24-1. Electromagnetic radiation is described as a wave with wavelength, L, and amplitude as shown.

In solar energy work, the wavelength of a light wave is usually expressed in microns. A micron is a millionth of a meter. Light radiation with wavelengths greater than 0.73 micron is called infrared radiation. Ultraviolet radiation is radiation with wavelengths less than 0.4 micron. Light radiation with wavelengths between 0.4 and 0.73 microns is termed visible light.

The solar energy flux (flow) at different wavelengths as received by the earth is shown in Fig. 24-2. The area under the curve represents the total energy flow. Outside the atmosphere this is approximately 1.35 kW/m². This is known as the solar constant.

In Fig. 24-2, a typical solar energy flow at the surface of the earth is shown as the black line. The atmosphere absorbs and reflects much of the solar energy due to the presence of water vapor and other gases, particularly carbon dioxide and oxygen.

At the earth's surface, about 45 percent of the incident radiation is visible light. Going from outside the atmosphere to the earth's surface, most of the ultraviolet radiation is absorbed by ozone. The infrared radiation is partly absorbed by both water vapor and carbon dioxide.

On a clear day, the energy flow at the earth's surface will be about 0.9 kW/m². Therefore, 0.45 kW/m² of the solar energy does not reach the earth's surface (1.35 − 0.9 = 0.45 kW/m²).

The radiation given off by a hot object changes with its temperature. At a low temperature, an object like a stove appears black. This indicates that most of the radiation given off is infrared radiation, outside the visible range of wavelengths. As the object is heated, it gives off more and more radiation in the visible range. It may first appear as a red glow. With further heating, it will change from red to yellow or white. When the object gets hotter, the wavelength at which the most radiation is given off becomes shorter and shorter.

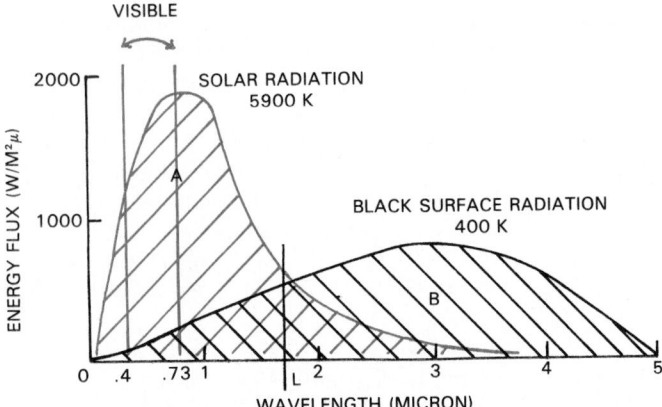

Fig. 24-3. Illustration of the comparison of the energy of radiant rays from the sun with the heat energy of a heated object. An ideal collector would absorb the solar energy with wavelength less than L but not give off radiation for wavelength greater than L.

Fig. 24-3 shows how the radiation given off by an object would change as its temperature increases from 400 K to 5900 K. At 5900 K, much of the radiation is in the visible wavelength range from 0.4 to 0.73 microns. At 400 K, only a small part of the radiation is visible. The sun's surface is at about 5900 K, so that most of the solar radiation is at wavelengths less than 3 microns. An object at 400 K gives off much less radiation. However, as is shown, most of this radiation is at wavelengths greater than the visible wavelengths (in the infrared).

The peak of the solar radiation curve occurs at the wavelengths of yellow light. Thus the sun appears to be yellow. This shift in the distribution of radiation with temperature is critical to the design of solar collectors, as noted in Para. 24-7.

24-3 PASSIVE AND ACTIVE SOLAR ENERGY SYSTEMS

There are two types of solar energy systems, passive and active. Passive solar energy systems depend on the solar radiation striking directly on the surface or area to be heated. The best example of passive solar heating is the conventional greenhouse. In a greenhouse, the energy flows through the glass into the area where the plants are growing. Passive solar energy systems usually require no auxiliary pumps or blowers to distribute the heat collected.

A house designed with a large area of windows on the south wall allows solar energy to be used during winter months. This can also be considered a passive solar heating system.

In active solar energy systems, the solar energy is absorbed into a collector. The energy is then transferred from the collector, stored, and distributed by an auxiliary circulation system.

In many of the new residential solar applications, a combination of both passive and active systems is used. This is referred to as a hybrid solar energy system.

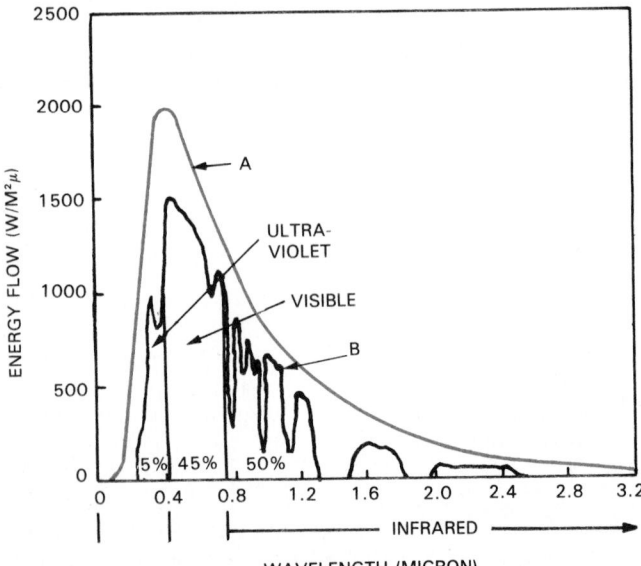

Fig. 24-2. Solar radiation at different wavelengths of light. Curve A illustrates the solar radiation outside the atmosphere, which is approximately a 5900 K source temperature. Curve B illustrates an approximate radiation flow at the earth surface, indicating the atmospheric absorption of light. Also indicated are the approximate percentages of energy in the ultraviolet, visible, and infrared radiation regions.

24-4 REQUIRED COMPONENTS OF A SOLAR ENERGY HEATING SYSTEM

A solar energy heating system must have parts which will provide the following functions:

1. Collect solar radiation.
2. Circulate heat from the collector and move it to the space being heated.
3. Store heat for later circulation when solar radiation is insufficient to heat the space.

Passive solar systems have parts which collect the radiation. However, they depend on natural radiation and convection to circulate the heat through the space being heated.

Active solar heating systems have mechanisms which can store the collected heat. Then, during nights or on cloudy days, a circulating system draws heat from storage and moves it to the space.

Existing houses usually require an active solar heating system. But, in the design of new homes, passive systems can be more easily incorporated. They require less energy input because pumps and blowers are not needed to circulate heat.

24-5 SOLAR COLLECTORS

The rays from the solar energy flux are converted to heat upon striking a dark surface. This applies to visible rays, infrared rays, and ultraviolet rays.

Heat from this solar radiation can be trapped. The simplest trap is an insulated black surface. A surface looks black because it absorbs visible light. Black surfaces will absorb infrared and ultraviolet radiation as well. Coatings of lampblack or fine carbon powder produce surfaces which are very close to black. In the sun they will get much hotter than white or shiny surfaces, because they absorb more radiation.

If an object is insulated on the back and placed in the sun, it will absorb radiant energy and get hot. The temperature will increase until the radiation emitted by the object is equal to the radiation received from the sun. If the object is black, the emitted radiation will be distributed over the spectrum, Fig. 24-3. In this figure, the energy from the sun striking the surface is shown in red. The energy emitted by the surface is shown in black. When the surface reaches its maximum temperature, the area under the red curve will equal the area under the black curve.

The maximum surface temperature a black object can reach in full sunlight is about 253 °F (123 °C) or 396 K. One way to increase the temperature involves trapping the radiation emitted by the surface with a glass cover over the black surface. Low-iron tempered glass is used, which has a higher transmission of heat than float glass. A sketch of a solar collector using this trapping principle is shown in Fig. 24-4. The glass allows radiation with wavelength less than about 2 microns to pass through and be absorbed by the black surface. The black surface at temperatures below 396 K (123 °C) emits most of its energy at wavelengths greater than 2 microns. This radiation is not transmitted out by the glass and is trapped inside the collector. Temperatures of 600 K (327 °C) are obtainable in collectors of this type. Fig. 24-5 is a drawing of a typical flat plate solar collector showing construction details.

24-6 SELECTIVE ABSORBER SURFACES

Special absorber surfaces are sometimes used to increase the temperature of a collector. Such a design is often called a selective absorber surface. Acting much like the combined glass plate and absorber surface, this design absorbs most

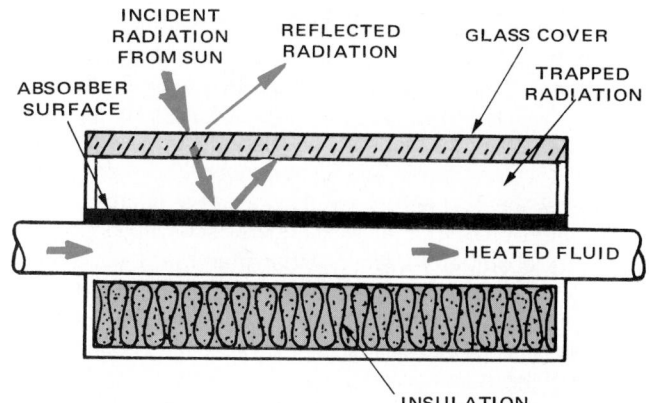

Fig. 24-4. Solar collector elements. Trapping of radiation by glass cover and insulation on the back of the collector is illustrated. Solar radiation passes through the glass, but the radiation from the absorber is trapped.

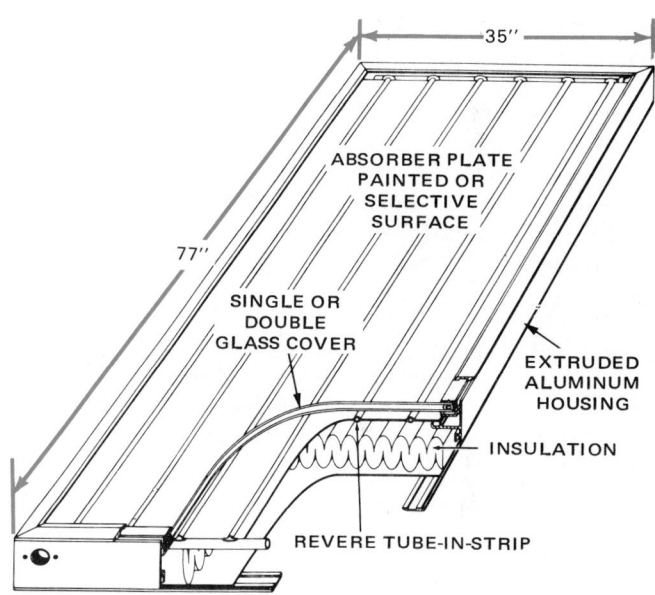

Fig. 24-5. Drawing shows dimensions and design of typical solar collector.

of the radiation from the sun which has wavelengths less than some "cutoff wavelength" (around 2 microns). It does not, however, absorb longer wavelength radiation.

It can be shown that a surface which absorbs poorly at a given wavelength also emits (gives off) poorly at that wavelength. The surface then emits infrared radiation very poorly and the temperature increases.

Many special paints and surfaces are being developed with this characteristic. They selectively absorb most of the sunlight but not infrared rays. Their performance is specified (rated) by an absorption to emission ratio. This ratio is the absorptivity to sunlight or radiation with wavelength shorter than the cutoff wavelength, divided by the absorptivity for longer wavelength radiation. (Absorptivity means ability to absorb.) Commercial surface materials are produced which have an absorption to emission ratio of about 20 to 1.

The way these surfaces work is similar to the absorption of sound by a surface with small holes in it. Sound

867

waves with wavelengths smaller than the holes are absorbed by passing into the holes. Sound waves with wavelengths greater than the hole size are reflected.

Selective absorber surfaces do not absorb as much energy as black surfaces at wavelengths shorter than the cutoff wavelength and may appear gray. This is more than offset by the fact that they do not allow energy at long wavelengths to escape. Most newer solar collectors use selective absorber surfaces rather than black surfaces.

The combination of a glass cover and a selective absorber surface gives the best solar collector performance. Multiple glass covers with special antireflecting coatings also increase the radiation trapping efficiency.

The closer the radiation is to being perpendicular (at right angles) to the glass surface of the absorber, the more it is transmitted through the glass. The radiation not transmitted is reflected and lost for heating purposes.

Collectors with a single glass cover and a selective absorber surface have been shown to be as efficient in trapping radiation as collectors with two glass covers and a black absorber surface.

24-7 GOOD ABSORBER SURFACES FOR SOLAR ENERGY COLLECTORS

The solar collector absorber surface should be black or very dark gray. It should be slightly roughened—somewhat rougher than an eggshell condition. A rough gray surface is a better absorber than a black shiny surface. A smooth and shiny surface would reflect the radiant energy away, not absorb it.

The collector surface is usually best applied as an electrodeposition process. (In electrodeposition, metallic particles are applied to another metal surface through use of an electric current.) This process bonds the surface material to the metal conduction plate. Paints are generally not satisfactory as they will peel and crack at the high temperature of the collector.

24-8 SOLAR COLLECTOR COVERS

A transparent cover is used on most solar collectors. This cover has three basic functions:
1. To protect the absorber surface from the weather.
2. To transmit sunlight to the absorber surface.
3. To prevent the escape of heat collected by the absorber surface.

Glass is the most widely used cover material. It provides good light transmission and remains clear indefinitely. Typical glass covers are 1/8 to 1/4 in. thick. Special glass materials transmit more sunlight and allow less collector heat to escape. Iron-free glass is generally used for this purpose.

Glass poses problems in sealing the cover to the other parts of the collector. Glass and metal do not expand at the same rate with temperature change. A flexible sealing material or rubber gasket is necessary at the glass-to-metal joint to avoid breaking of the glass as the materials expand and contract.

Plastic cover materials usually soften at high temperatures. Plastic also becomes brittle and opaque from absorption of ultraviolet light.

Some covers are formed in a bubble shape rather than a flat shape. This provides more structural support and

collects more radiation when the sun is near the horizon. Some special collectors are formed as a tube and are sealed with a vacuum inside. These collectors are very efficient and do not lose much of the heat collected. Fig. 24-6 shows a typical vacuum tube collector. They are used primarily in high-temperature applications, as they are much more efficient at higher temperatures.

Periodic washing of the glass will improve the light transmission. Most installations, however, require washing only once or twice a year.

24-9 SOLAR ENERGY STORAGE SYSTEMS

Large bodies of water are a natural solar energy storage system. The surface of a lake may be frozen over in the winter, but in the spring, as the sun shines on it, the ice melts and the water warms. This heat then is given to the atmosphere in the fall or winter as the water on the surface again freezes.

This system is often used for storing heated water in an insulated container for later use. One very common example is the heating of water in a tank located on the roof of a house. The tank is connected to the water system in the house. Many homes, particularly in the southern part of the United States, use this system entirely for domestic hot water.

Rocks may also be used for heat storage. If the sun shines on a big stone, the surface of it will be warmed and may stay somewhat warm throughout the night.

The specific heat of water is about 1 Btu/lb. °F and that of the rock about 0.25 Btu/lb. °F (See Para. 1-31). The specific heat of water in SI metric is 4.187 kilojoules per kilogram degree kelvin. The equation is written kJ/kg·K. This means that 1 lb. of water will store as much heat as 4 lb. of rocks if they are both heated to the same temperature. The same heat stored in crushed rock requires about four times the volume of a water tank.

Another type of storage medium used is a series of polypropylene rods in an enclosed honeycomb structure. This allows for efficient movement of air.

Heat can be stored in a chemical. Sodium sulphate crystals are sometimes used. They give up heat when they form crystals from a water solution.

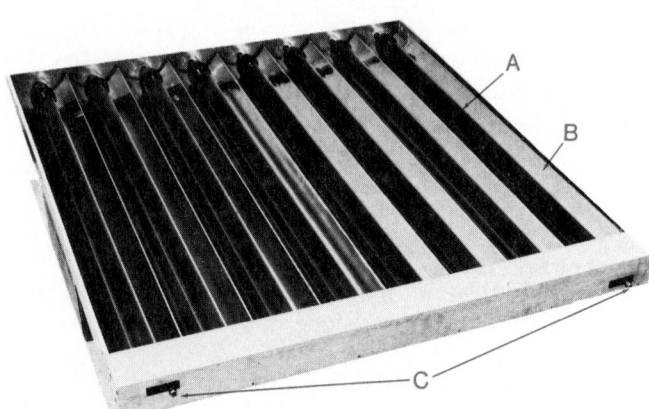

Fig. 24-6. Vacuum tube solar collectors are used for high-temperature application. A—Vacuum tubes. B—Reflectors. C—Hydronic connectors. (General Electric Co.)

24-10 HEAT ENERGY STORAGE IN A CLOSED WATER SYSTEM

To use solar energy more efficiently, closed system water heating is employed. The system gathers, circulates, and stores absorbed energy in water. The amount of solar energy absorbed depends upon the area and color of the absorbing surface exposed to the sun. In a complex system, collector lenses are used to heat the water to a very high temperature.

As a practical application, a heat-absorbing solar energy system usually consists of:
1. A large solar panel facing the sun.
2. Water circulated through the panel to storage.
3. A large insulated holding tank.

Heat from the water may be used by either of two methods:
1. The water may give up its heat directly to the space.
2. The water becomes a source of heat for a heat pump. The pump will then be used to transfer the heat to the space as desired.

Ethylene glycol or propylene glycol is added as an antifreeze to protect the system from freezing in colder climates. In mild climates a drain-down system is used which drains the collector on cold nights.

Heated water may be stored in a large insulated tank. Typical solar heating systems have storage tank capacities from 1.3 to 2.5 gallons for every square foot of collector surface area.

24-11 HEAT ENERGY STORAGE IN A WARM AIR SYSTEM

Solar energy may also be absorbed at the collector and transferred to warm air as heat. Air is circulated through the solar panel, and the heat is stored in crushed stone. The crushed stone is held in place by an insulated container. The air heated by the sun is circulated through the container. At night, or at other times when heat is needed, air in the house is circulated through the heated, crushed stone. The stone then gives up its heat to the house.

The crushed stones should be from 1 to 2 in. in diameter. Too much energy is required to circulate the air around smaller stones. On the other hand, stones that are too large will not be heated completely. Only part of the stone will be effective in storing heat.

Fig. 24-7 illustrates a rock storage container. The size of this container is about 0.3 to 1.2 ft.³ in volume for a square foot of collector area. The height of the container is usually limited to less than 6 ft. to allow the air to be circulated from the bottom to the top with a small blower.

There must be an empty space at the top and bottom of the container to provide uniform airflow through the rock storage.

It is important that the rock storage be heated by flowing the air from the top to the bottom of the container. When heat is taken out of the rocks, the flow of air should be from the bottom to the top and then out. The reason for this is that the temperature of the rocks becomes hotter at the top when being heated. The rocks at the bottom may still be relatively cool. When heat is to be removed, the air should be as warm as possible. This is achieved by the reverse flow of air, as shown in Fig. 24-7.

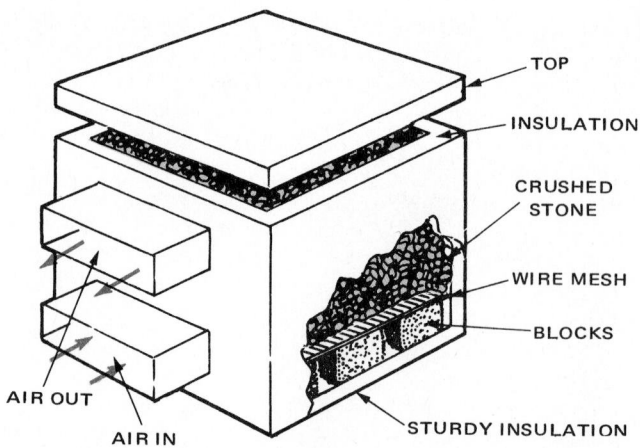

Fig. 24-7. Typical crushed rock heat storage container. Room air, blown in bottom, comes out top heated.

24-12 ANGLE OF THE COLLECTOR

It would be ideal if a mechanism could be arranged to always present a collector surface exactly perpendicular to the rays of the sun. A device to provide this aiming—except for extreme conditions—is too expensive. The position of the collector surface, therefore, usually remains fixed. The angle of the sun's rays to the horizon varies during the year. It is, therefore, necessary to establish an angle for the collector which will give the best average collecting effect. The sun angle in midwinter is 66.55° minus the site latitude angle. The angle in midsummer = 113.45° — latitude.

If the solar energy is used chiefly for home heating, the angle needs to be set near a position to absorb the greatest amount of energy during the heating season. This setting is about 45 to 55 degrees from the horizontal for use across the mid-section of the United States.

If the collector is to be used during all seasons to provide hot water, the collector angle should be 35 to 45 degrees to the horizontal. For year-round heating, the angle from the horizontal should be the site latitude angle or up to 10 degrees less. The collector should face south, but an angle from 30 degrees east or west of south only reduces the energy collected by about 10 percent.

Naturally, it is important that there be no obstructions, such as buildings or trees, which might create a shadow across the collector surface. Simple shadow-indicating devices are available to aid in determining the effect of obstructions at a specific site. There are also meters available which will read out the intensity level of the solar energy at a particular location. See Fig. 24-8.

24-13 HEAT INSULATION OF THE COLLECTOR SURFACE

A collector surface will both absorb solar energy and release heat energy unless it is properly insulated. The solar energy coming into the collector is changed to heat. The surface will then radiate heat the same as any other warmed surface. This means the collector surface needs to be insulated to cut down as much as possible the radiation of heat to the atmosphere. This can best be done by having one or more layers of glass over the front surface. This will hold in the energy as it is converted from solar radiation

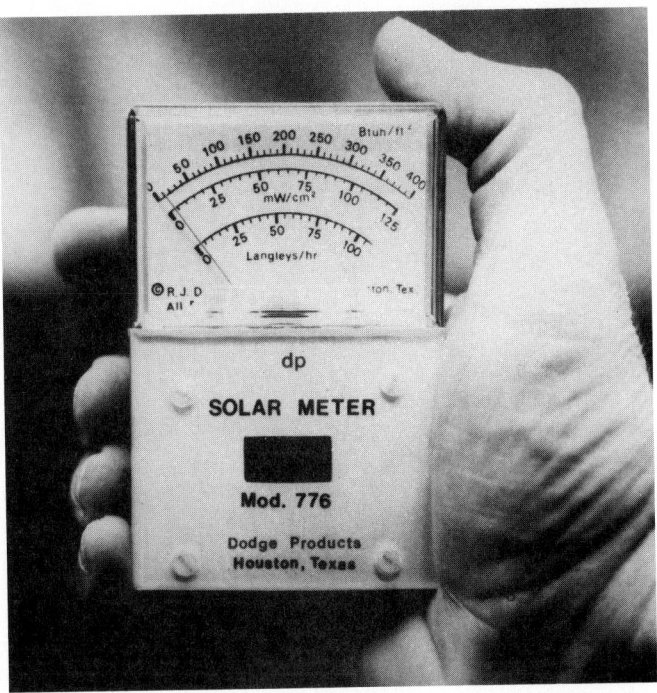

Fig. 24-8. Hand-held solar intensity meter is used to estimate the total heat rate of solar radiation into heat collector. (Dodge Engineering)

to sensible heat. The back surface must also be insulated to hold the heat from escaping through the back. About 4 in. of conventional insulating material will usually provide sufficient insulation on the back surface.

24-14 SOME TYPICAL SOLAR SPACE HEATING INSTALLATIONS

There are three types of effective solar heating installations in common use:

1. Natural convection closed water system. Heat is absorbed in the collector. Since warm water tends to rise, the storage tank must be located above the collector. The storage tank should be quite heavily insulated to eliminate heat loss. The radiator through which the warm water circulates to heat the desired space must be located with its inlet about even with the warm water supply. As the heat from the warm water is radiated into the room, the water will be cooled. It will then flow down and be carried back to the bottom of the heat storage tank. There is no temperature control with this system. With bright sunshine the temperature will continue to rise. The heated space could become uncomfortable.

2. Forced convection in a closed water system. In this system a pump circulates the water. The warm water storage tank may be located where it is convenient without regard for location of the collector or the radiator. Water is heated in the collector and is circulated to the storage tank and radiator. The circulation through the radiator may be controlled with a thermostat in order to maintain a certain room temperature. Fig. 24-9 illustrates the components of a forced convection solar heated closed water system. The differential thermostat turns off the pump when the temperature in the solar

collector is lower than the temperature in the storage tank. It is usually set to turn on the circulating pump when the temperature of the collector is from 8 °F to 10 °F (4 °C to 6 °C) warmer than the water in the storage tank. This prevents the pump from turning on and off frequently and assures that cold water from the collector is not mixed in the tank.

3. Warm air heating system. A warm air heating system using forced air circulation is illustrated in Fig. 24-10. The solar energy is converted to heat in the solar collector. It must be insulated so that only a small portion of the heat is lost. The heated air is circulated through the rock bed. This bed consists of crushed stones approximately 1 to 2 in. in size. Air from the collector is circulated through the rock bed. Heat is stored in the rocks. When heat is needed in the room, air is circulated through the warmed rock bed, which gives up its heat to the room air. Rocks for this purpose are very carefully washed and cleaned, so that no dust or sediment circulates through the room. A thermostat controls the airflow. Supplementary heat, as needed, may be provided as indicated in Para. 24-17.

24-15 SOLAR DOMESTIC WATER HEATING

Perhaps the most-used solar heating system is one which supplies heated hot water for domestic supply. The heated water for dishwashing, showering, and washing can be supplied easily using a solar heating system to preheat the water. If the water supply is from a well, the supply temperature is fairly constant throughout the year, ranging from 40 to 50 °F (4.5 to 10 °C). If the supply is through a large municipal water supply system, the water is either drawn from a large reservoir or storage tanks. In some locations the temperature of the water may range from 35 °F (2 °C) in the winter to 80 °F (27 °C) in the summer. The heat required to supply domestic hot water at 120 to 140 °F (49 to 60 °C) can vary considerably throughout the year, dependent upon geographic location.

Solar domestic water heating systems are more cost effective than space heating. There is a demand for hot water in a home the year-round, whereas space heating is only used in the winter season.

Two types of solar water heating are in common use:
1. One-tank system.
2. Two-tank system.

The one-tank system consists of a single hot water storage tank. The water, preheated in the solar collector, is circulated through a heat exchanger in the bottom of the hot water heater tank. An auxiliary heater coil in the top of the tank provides the final heating to bring the water to the desired temperature. This system requires replacement of the present hot water tank in retrofit (see Chapter 32) situations.

The two-tank system is shown in Fig. 24-11. It uses a separate storage tank in which the water is preheated by the solar heated liquid through a heat exchanger. Water is circulated to the conventional hot water heater from this storage tank. This system with a separate storage tank is easily added to the present hot water system in a house.

The two-tank system is usually from 30 to 50 percent more efficient than a single-tank system. The cost is usually higher than a single-tank system if a new separate water heater must be purchased.

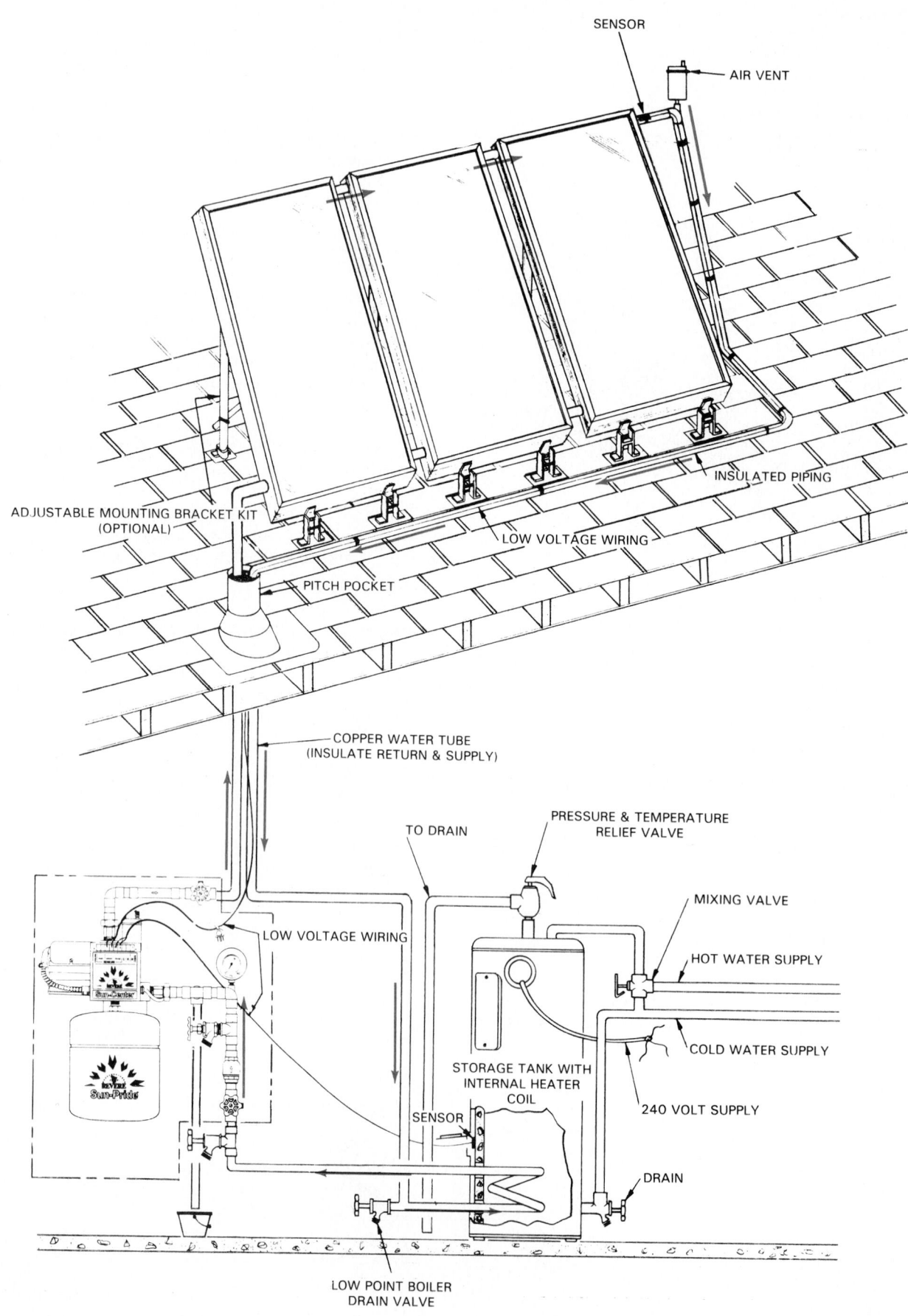

Fig. 24-9. Components in forced convection solar heated closed water system. One tank stores heat and supplies hot water.

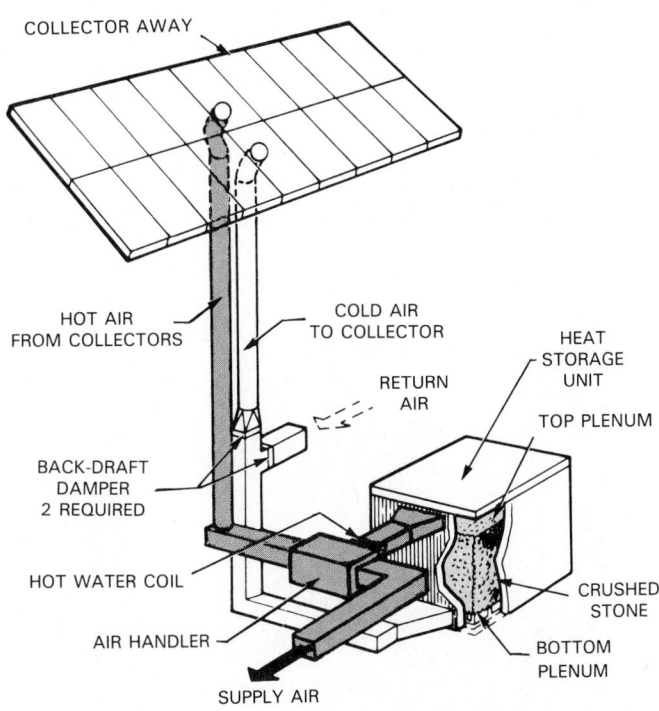

Fig. 24-10. Warm air solar heating system with forced air circulation. Heat is stored in crushed stone.

Domestic hot water systems are designed to economically supply from 40 to 75 percent of the required hot water heat. Auxiliary heating supplies the final heat to reach to 120 to 140°F (49 to 60°C), the temperature required.

Domestic hot water use is approximately 25 gallons a day for each person in a family. Solar collector sizes range from 1/2 to 2 sq. ft. per gallon of hot water required per day, depending on location. In a two-tank system, the low temperature storage tank is usually designed so that it is only about half the size of the main storage tank.

24-16 SOLAR HEATING OF WATER FOR SWIMMING POOLS

Heating of water for swimming pools can be used to extend the swimming season in many areas. The solar heating system for a swimming pool can be very simple and inexpensive. Swimming pool solar heaters can be just a set of black or transparent plastic bags. The reason for the simplicity is that:

1. No storage is required.
2. The water pressure that the collector must withstand can be much lower than in space or domestic water heating systems.
3. No freezing protection is necessary.

These swimming pool heaters require only a low pressure pump, plastic solar collectors, and piping.

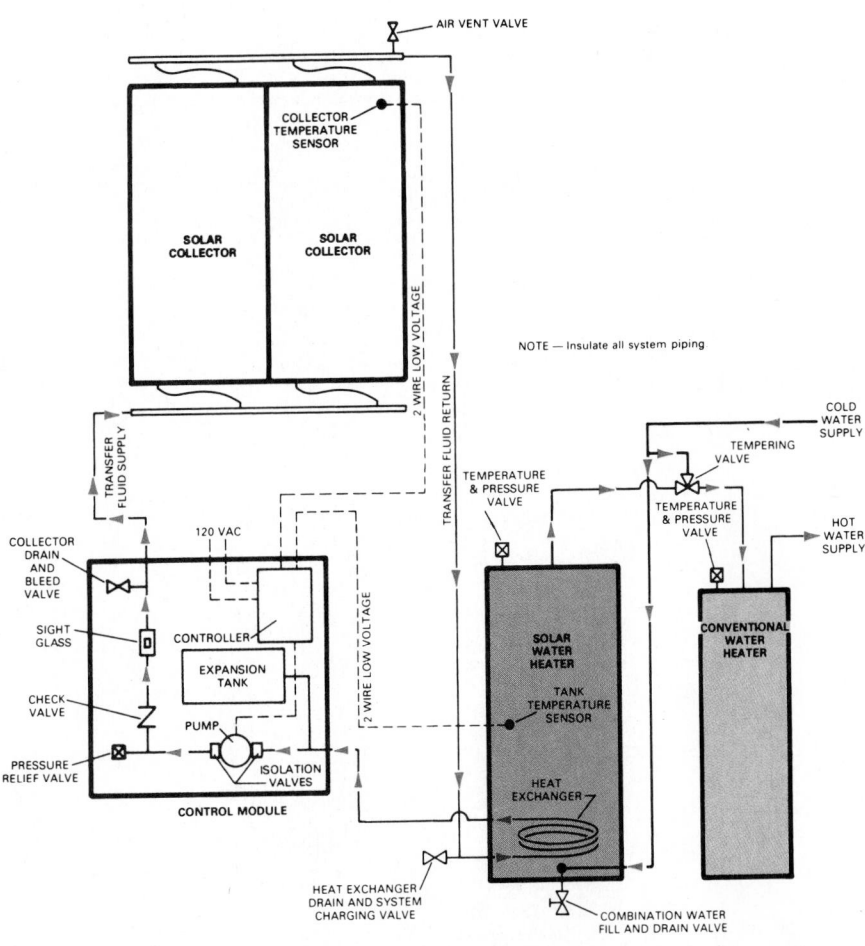

Fig. 24-11. Two-tank domestic hot water heating system. Conventional water heating tank supplies auxiliary heat.

The size of a solar collector for a swimming pool is usually approximately half the surface area of the swimming pool to provide good spring and fall heating. The efficiency of swimming pool solar heaters can be 70 to 80 percent because of the low temperature (80 °F or 27 °C) water that the collectors are required to supply.

24-17 SUPPLEMENTARY HEAT

There may be times when the sun is not supplying enough heat energy to the collector to maintain a comfortable room condition, even with storage of heat. In such cases, supplementary heat must be supplied. This may be through the use of electric resistance heating, the application of a gas or oil burner, or a heat pump.

In some domestic solar hot water heating systems, the water is preheated by solar energy, then heated to the temperature required by a supplementary heater.

24-18 SUPPLEMENTARY ELECTRIC HEATING

Electric resistance heating may be used to supplement solar heat. The nature of this installation will be governed by the nature of the heating system to which it is applied. With a warm air heating system, a direct electric resistance radiator may be installed in the air duct. A thermostat may be used to control the amount of electric heat needed.

In a hot water space heating system, an electrical heating element may be installed in an auxiliary tank to provide supplementary heating when the heat stored in the main tank is exhausted. This is indicated in Fig. 24-11.

In some heating situations, individual room resistance heating radiators may be installed.

24-19 SUPPLEMENTARY OIL AND GAS HEATING

In the event that oil or gas is used for supplementary heat, a separate furnace is installed, as shown in Fig. 24-12. The burner will be controlled by a thermostat, which will turn on the burner when the amount of heat from the solar source is below the amount of heat required. If the system uses warm water radiators, the burners may be used to heat the water in an auxiliary tank.

24-20 HEAT PUMPS

The use of a heat pump is probably the most efficient method of supplying additional heat in a solar heat system. The heat pump may also be used in connection with an air conditioning system to take away heat. The heat pump uses a refrigerant fluid the same as an air conditioner. (See Chapter 23.) It can be arranged to either add heat to a room or absorb heat from a room. Fig. 24-13 illustrates a heat pump installation.

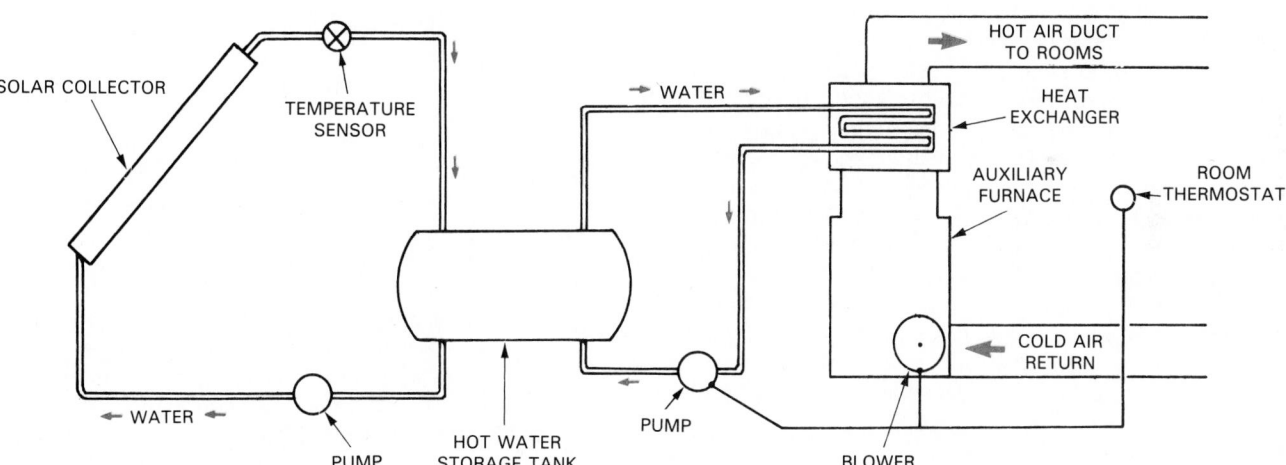

Fig. 24-12. A schematic of a liquid solar collector system used in forced air heating application. The heat exchanger is shown located in the hot air duct of a conventional furnace which supplies auxiliary heat.

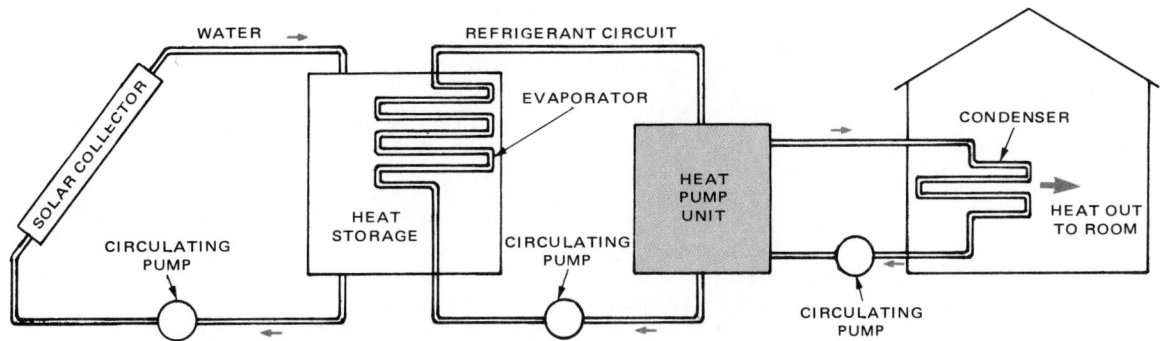

Fig. 24-13. Solar heating with supplementary heat supplied by a heat pump. Heating cycle is shown.

The use of a heat pump in connection with a solar energy collector is the most efficient means of maintaining a desired temperature with the fluctuating heat supplied by the solar collector. Theoretically, a heat pump can be efficient on a 4-to-1 basis; that is, the amount of heat delivered can be four times the heating value of the electrical current required to drive the mechanism. In application, this theoretical value is never reached. A ratio of 3-to-1 is considered to be very good; that is, three times as much heat is obtained from the heat pump as the heating value of the electricity consumed if converted directly to heat. In many practical applications, this ratio may not be over 1.5-to-1.

The heat pump is a natural device to be used in connection with solar heat. It is possible to use solar energy to warm water in a forced convection system. The solar energy may not bring the temperature of the water up to a level where it could maintain a comfortable room temperature. Through the use of a heat pump, it is possible to extract heat from the heat storage system and increase its temperature to a comfortable level. In this case, electrical energy is being used to supplement the solar energy, but two to three times as much heat is obtained as would be obtained by direct electric resistance heating.

24-21 SOLAR ENERGY COOLING SYSTEMS

Solar energy may also be used for cooling. This is usually done by using absorption system refrigeration. See Chapter 16. Absorption systems require a heat source. The heat is used to drive the refrigerant out of a solution. The sun can supply the heat required to operate absorption cycles.

Usually the solar energy must be concentrated by lenses or mirrors to increase the solar radiation temperature to whatever is required by the absorption system.

At present, solar energy is chiefly used for heating. The required mechanisms for cooling are quite expensive.

Experimental systems using solar energy as the heat source have been built. The system elements are shown in Fig. 24-14.

Experiments are also being made with solar dynamic cooling systems. In these systems, the solar energy is converted to electrical or mechanical energy. This energy is then used to drive a compressor. Concentrator collector systems, using lenses or mirrors, are required to provide the high temperatures needed for efficient operation.

Solar cooling systems have the highest input when the cooling requirements are greatest. Energy storage requirements are minimized and the energy is available when needed most. In hot, sunny regions of the world, they are becoming competitive with conventional cooling systems.

24-22 CONVERTING SOLAR ENERGY TO ELECTRICITY

Solar energy can be converted directly from radiant energy to electricity. The most common device used for this conversion is the solar cell (also called a photovoltaic device). Solar cells, although expensive, are a preferred electrical energy source at remote locations and where the power requirements are relatively small. The power to run remote weather stations and communications stations is often supplied by solar cells rather than fueled engines. They are maintenance-free and do not require fuel to supply power. Other applications include situations where a small amount of power is required. These include bridge and pipeline electrochemical corrosion protection systems.

24-23 SOLAR CELL CONSTRUCTION

The components of a typical solar cell are shown in Fig. 24-15. The principal elements required are two semiconductor materials called an N-type semiconductor and a P-type semiconductor. An N-type semiconductor carries current by means of electrons or Negative charges. A P-type semiconductor carries charges by Positive charges, sometimes referred to as "holes." "Holes" refers to the particle remaining after an electron has been removed from an atom.

Photons in the light striking a solar cell give up their energy to electrons at the junction between the P-type and N-type materials. The electrons with excess energy are then stored in the N material, giving it a net negative charge and leaving a net positive charge on the P-type material. Electrodes are attached to each material and conduct the charge to an external load circuit.

P-type and N-type materials in most solar cells are mainly silicon. The difference between the two materials is that the silicon of the N-type material contains a very small quantity of a material which, compared to silicon, has an excess of electrons. Nitrogen, phosphorus, or arsenic are materials having an excess of electrons.

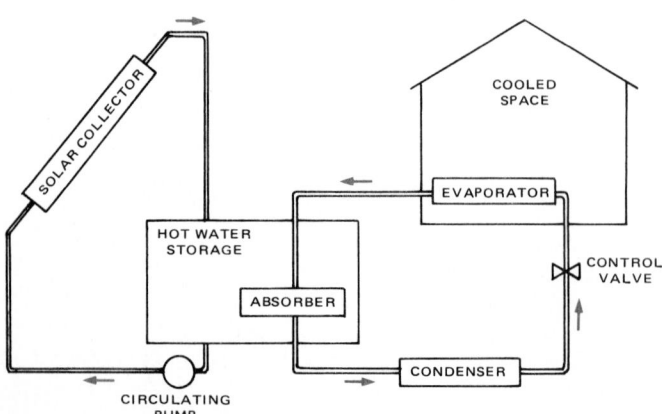

Fig. 24-14. Solar energy air conditioning system utilizing an absorption refrigeration unit. Solar energy supplies heat for the absorption unit. A minimum external power supply is necessary in this cooling unit. Cooling cycle is shown.

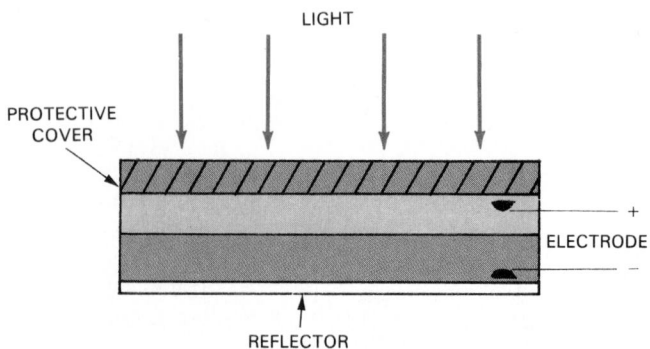

Fig. 24-15. Solar cell schematic indicating the physical components. The sketch is not to scale. The cells are very thin compared to their length and width.

The silicon of the P-type material has a small quantity of material which, compared to silicon, has a deficit of electrons. Aluminum, boron, gallium, or indium are such materials.

The electrodes on the back are also reflectors, so that any light that goes through the cell is passed back through again so that more light can be absorbed.

The P-type material must also be transparent to allow the light to reach the junction between the P- and N-type materials.

Silicon power cells are presently made with a thin slice of silicon. A small amount of Boron dopant added to the molten silicon during the crystallization process makes this material a P-type semiconductor. A small amount of phosphorous is diffused into the front surface of the wafer, creating a shallow N-type layer in the silicon. This creates the N-P junction which makes the silicon wafer a solar cell. Metallic contacts are then deposited on the front and back of the cell to allow for connections and to improve the flow of current out of the cell. An antireflective coating is also applied to improve the absorption of light by the cell. Finished solar cells are then connected together in series and parallel to provide the desired voltage and current output. The assembly of cells is then encapsulated with weatherproof plastics to a sheet of tempered glass and framed with anodized aluminum. A weatherproof junction box is then added to allow for external connections to the module. The resulting photovoltaic module will withstand even the most severe environments with a design life in excess of twenty years.

24-24 PHOTOVOLTAIC SOLAR CELL APPLICATIONS

Electrically, solar cells operate similarly to batteries. A single cell produces a voltage of approximately one volt. The voltage depends on the material of the cell, ranging from 1/2 volt to 1 1/2 volts. For a specific solar cell material, the voltage will remain constant under normal operating conditions. It does not change very much with light intensity. However, under operating conditions, the current produced increases as the sunlight intensity increases. Therefore, the power output, which is the product of the voltage times the current, increases with intensity.

In situations where the sunlight is focused to increase the intensity, the effects of increasing the temperature of the cell can be harmful to the power output and efficiency. Under these conditions, cooling of the solar cells is beneficial. High intensity solar cells, like those shown in Fig. 24-16, are designed to operate at solar intensities up to 50 times normal sunlight.

Solar cells are available commercially for various applications in a wide variety of configurations and sizes. A 12-volt module is shown in Fig. 24-17. This particular cell is specifically designed to power a water pump for circulating fluid through hot water solar collectors. They are ideal power sources for those situations where the demand for power increases as sunlight intensity increases. As a power source for use in warm climate cooling applications, they have the ideal characteristic of producing increased power as sunlight intensity increases.

Photovoltaic solar cells are used as the primary energy source for most space satellites. They are a key factor in the development of communication satellites used for

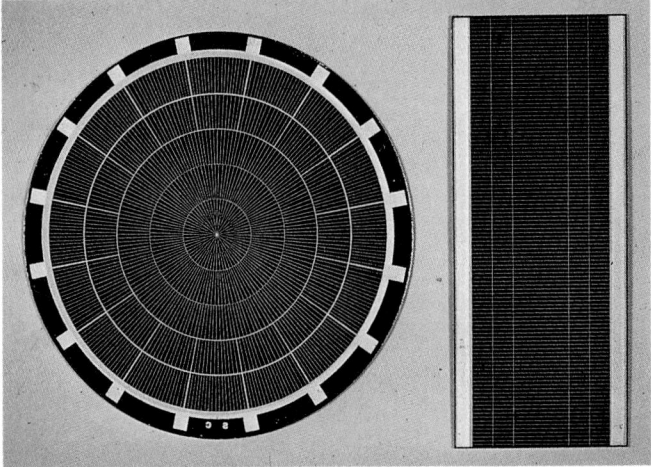

Fig. 24-16. Photograph of typical solar cells which convert solar energy into electrical energy. (Solarex Corporation)

Fig. 24-17. Photovoltaic module for powering a 12-volt circulating pump. Unit has film of polycrystalline silicon as the absorber. (Solarex Corporation)

telephone, radio, and television. Another application is powering hand calculators that operate from interior room lighting as well as in sunlight. They are also widely used to provide electricity for waterpumping, lighting, and telecommunications in remote areas.

Solar cells are connected in series similar to batteries, in order to obtain higher voltages. The total voltage produced is then the sum of the voltages produced by each cell. The electrical circuit may be as simple as shown in Fig. 24-18. This indicates a cell connected to a dc motor. The motor would then deliver power depending on the sunlight intensity. (At night the motor would not run at all.)

Solar cells usually need to be connected to an electrical storage device like a storage battery. A motor can draw power from the solar cell directly in bright sunlight and from the storage battery when there is little or no sunlight. The

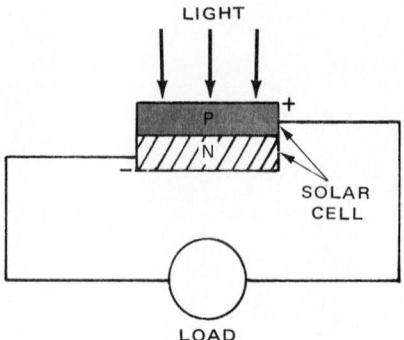

Fig. 24-18. Simple solar cell circuit. Power produced will depend on sunlight intensity.

control system would be designed to store electrical energy in the battery in the brightest sunlight. This energy would then be drawn out of the battery when light intensity is low, as on cloudy days or at night.

An electrical circuit to provide controlled power would include a storage battery. It would be connected as shown in Fig. 24-19.

In electrical energy storage systems that utilize batteries like that shown, it should be remembered that the battery cannot be charged at the same time it is delivering power. Thus, the total power to the load will not be greater than the solar cell or battery alone can provide.

A solar cell produces dc power like a battery. If ac power is required to operate an electrical appliance or ac motor, an inverter must be added to the circuit. Modern inverters use solid state semiconductors to produce ac power from dc power. They have an efficiency of about 90 percent. Inverters are expensive, however, and dc motors are often used directly to avoid their cost. See Para. 6-59.

24-25 SOLAR CELL PERFORMANCE

Solar cell performance depends heavily upon the intensity of sunlight and the temperature of the solar cell. Fig. 24-20 shows a typical performance curve for a working cell. The electrical power delivered by the cell increases with the

sunlight intensity. If intensity continues to increase, the output power eventually levels off as shown. The leveling off is the result of excessive heat buildup in the cell. This heat can be carried away by circulating a cooling fluid in back of the cell or cells.

Current output can be increased by using a focusing collector to increase the light intensity on the cell. Presently, most practical applications are designed without a focusing system. Experimental focusing systems with intensity up to 40 times full sunlight have operated successfully when the cells are cooled. But most research is directed at producing cells which do not require focusing. Since a focusing system must follow or "track" the sun across the sky, it is expensive.

While the performance curve indicates that power output increases with solar intensity, the accompanying increase in power becomes smaller and smaller the greater the intensity. Often, manufacturers of solar cells will indicate solar cell efficiency in their specifications. Fig. 24-21 shows how this efficiency varies with solar intensity.

The efficiency is defined as the output power divided by the sunlight intensity. This indicates that the efficiency can be very high at very low light intensity but that it decreases rapidly as the light increases.

Manufacturers' specifications of efficiency should show the solar intensity at which this efficiency is achieved.

Typical solar cells at low light intensity can have an efficiency of 20 percent. At full sunlight, the efficiency may be only 15 percent.

A solar energy system having a concentrating collector with cooled cells can be used to provide both electrical energy and heating. These systems are now in the experimental stage.

24-26 REVIEW OF SAFETY

There may be some hazards connected with the handling of solar energy equipment. Some plumbing connections may be required. If the pipes are not well made, and connectors are not tight, some leakage may result. Also, some plastic materials may not withstand the temperatures and pressures to which they are subjected over a long period of time.

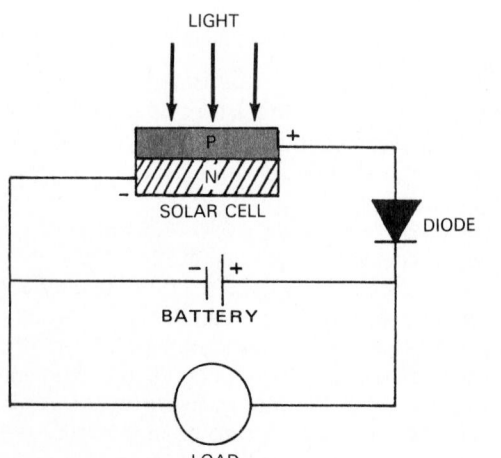

Fig. 24-19. Solar cell circuit provides for electrical energy storage in a battery. But it still provides a constant flow of energy to a load.

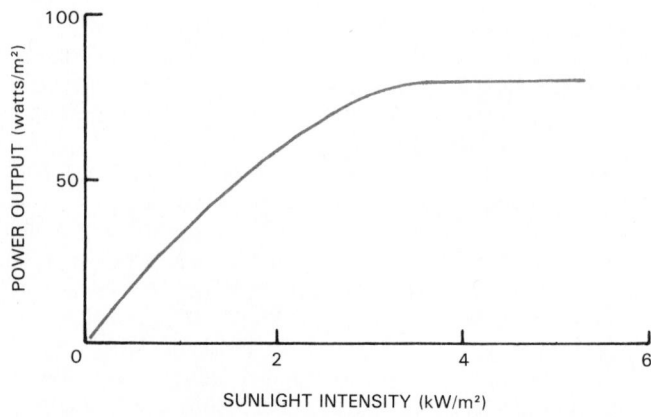

Fig. 24-20. Typical solar cell performance curve. This illustrates how the power output increases as the sunlight intensity increases. At high light intensity, a limit is reached where increased light intensity does not increase output due to heat effects.

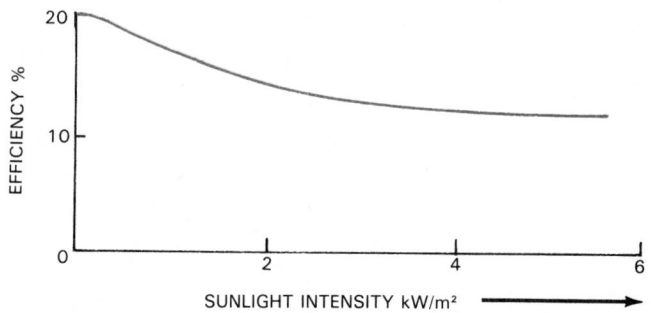

Fig. 24-21. Efficiency of a typical solar cell varies with light intensity.

Some attention needs to be given to the use of dissimilar (unlike) metals in a system. Electrolysis may corrode the two materials at the point where they are in contact.

Some insulating materials used with solar systems are flammable. Use care when soldering, brazing, or welding near these materials.

Caution must be used in domestic solar water heating systems that use ethylene glycol or other additives to make sure that there are no leaks into the water in the hot water tank. A pressure regulation system which maintains the hot water at higher pressure than the fluid in the solar collector circulation helps provide some protection.

When installing solar collectors, protect the eyes from bright sunlight reflecting off surfaces. Some vacuum tube solar collectors can also implode (burst inward) if broken. Wear eye protection when installing collectors.

24-27 TEST YOUR KNOWLEDGE

1. Is solar energy a new source of energy?
2. Are all solar rays visible?
3. What sun rays are absorbed by the atmosphere?
4. Wavelengths of the sun's rays are measured in microns. What is a micron?
5. What percentage of the sun rays are visible?
6. What are the names of the sun rays that are not visible?
7. Do the invisible rays produce any heat?
8. What is the approximate value in watts of the sun rays on a clear day?
9. What is a passive solar energy system?
10. Can the sun's rays be converted directly into electricity?
11. What is a solar collector?
12. Is the wavelength of solar energy the same as the wavelength of heat rays?
13. Why is it necessary to provide heat insulation to solar collectors?
14. What color absorbs solar energy best?
15. How may solar energy be trapped?
16. What is a selective absorber surface?
17. What is the best angle for the sun's rays to strike the absorber?
18. What should be the nature of the surface of a solar collector?
19. Why is glass usually used to cover a solar collector?
20. Why are some solar collectors sealed in a vacuum tube?
21. What are common ways of storing solar energy?
22. Is solar energy stored as solar energy or as heat?
23. What is the advantage of using 1- to 2-in. diameter crushed stone as a heat storage system?
24. How may water be used as a heat storage system?
25. What are the advantages of using water as a heat storage system?
26. What solution may be used in cold climates to eliminate freezing in the liquid-filled collector?
27. If water is used for a heat storage system, how much water volume should be supplied for each square foot of collector surface?
28. At what angle to the horizontal should the collector surface be set, in the midsection of the United States, where the collected energy is to be used chiefly for home heating?
29. Why is it necessary to insulate the backs of most solar collectors?
30. When should a differential thermostat be used?
31. What is the most common heat storage system used with a warm air heating system?
32. Can solar energy be used to heat domestic water?
33. Does the most efficient domestic warm water system use one or two tanks?
34. What are some of the advantages of using solar energy for heating swimming pools?
35. What is meant by supplementary heat?
36. What is the most efficient method of providing supplementary heat in a solar system?
37. What is a solar dynamic cooling system?
38. What is a heat pump?
39. Can solar energy be used for cooling?
40. Why is solar energy not being employed more often to create electricity?
41. Name the main substance used in making a solar cell.
42. What is the voltage provided in a solar cell?
43. How should solar cells be connected to produce a higher voltage?
44. Should solar cells be kept cold or heated?
45. Is the solar cell efficiency greatest with high or low sunlight intensity?

Facility management system provides environmental control, energy management, lighting control, fire management, security functions, and overall facility monitoring. (Johnson Controls, Inc.)

Chapter 25

AIR CONDITIONING AND HEATING CONTROL SYSTEMS

After studying this chapter, the technician will be able to:
□ Define the terms: controls, controller, thermostat, heat anticipator and humidistat.
□ List the various applications of controls in an air conditioning system and tell what conditions they regulate.
□ Name the different types of control systems.
□ Describe the operation of various devices in the control systems for heating and comfort cooling.
□ Explain the application of a computer to air conditioning control.
□ Read control circuit diagrams.
□ Troubleshoot, service, and repair control systems.

Modern homes, offices, and workplaces are kept comfortable and healthy with automatically controlled temperature, humidity, air movement, air cleaning, and air sterilization. Two important developments have improved heating and cooling systems:
1. Automatic controls now operate systems, Fig. 25-1.
2. Electronic circuits control and operate the automatic systems. See Fig. 25-2.

25-1 CONTROLS

Controls commonly mean devices that operate or regulate electrical and mechanical systems. Controls are used on heating systems, cooling systems, humidifying systems, dehumidifying systems, combustion, and flue systems. Usually, each control device is designed to respond to a certain condition. Examples of this include those that regulate:
1. Temperature.
2. Pressure.
3. Liquid or gas flow.
4. Liquid level.
5. Timed operations.
These controls have made it possible to develop safe, automatic systems.

25-2 CONTROLLER — CONTROL SYSTEMS

Controllers are groups of controls and circuits that accurately and automatically operate a device. Components of a controller may include primary controls, operating controls, and limit controls. All of these components will be explored in more detail as parts of typical controllers are discussed later in this chapter.

A control system includes the controller, the operating device or devices, and the conditioned area. (See Para. 8-3.)

Several types of control systems are in use. They include:
1. Electric.
2. Pneumatic.
3. Electronic.
4. Fluidic.
5. Combinations.

Fig. 25-1. Computer systems are now used for automatic HVAC control. (Johnson Controls Inc., Systems and Services Div.)

Fig. 25-2. Direct digital modules for building control. Package unit controller controls both start and stop and staging outputs. Variable air volume (VAV) controller controls terminal units, electric and hot water reheat as well as fan-powered systems. (Robertshaw Controls Co.)

These control mechanisms automatically turn systems on and off, modulate (adjust) certain operations, and signal conditions. Such mechanisms are interlocked with safety devices to limit unwanted conditions. The basic principles of many of these devices have been described in Chapters 6 and 8.

Control systems of interest in the heating and cooling industry are those which are used to operate:
1. Heating systems.
2. Air conditioning systems.
3. Total energy management systems.

There are a number of operating controls necessary in a typical air conditioning system. These controls include:
1. Temperature controls.
2. Humidifying controls.
3. Airflow controls.
4. Filter controls.
5. Defrost controls.
6. Limit or safety controls.

25-3 THERMOSTATS

Probably the most common controller in the heating and cooling system is the thermostat. *A thermostat contains an operating control which starts and stops a system when correct temperature conditions are reached.*

Some thermostats may modulate (increase or decrease the effect) instead of starting and stopping the system. An introduction to the various types of thermostats follows as does a review of some of the basic components of these devices, and an in-depth review of specific thermostats.

25-4 THERMOSTAT TYPES

There are many types of thermostats that control heating and cooling systems. Two basic types are:
1. Heating thermostat.
2. Cooling thermostat.

These may be combined into one unit which is called a heating-cooling thermostat. Some combination thermostats change over manually; others do it automatically.

Time-operated thermostats are coupled with a clock mechanism. They will change the off-on settings for night and reset it again for day. Some units will also change the on-off settings for different days of the week (off during Saturdays and Sundays, for example).

Automatic furnaces and other heating devices are controlled by safety devices which turn the system on and off or stop an operation if abnormal conditions arise. These devices are mainly electrical, activated (tripped) by temperature, pressure, or time. Relays are often used either to control full voltages by low voltage signals or to interlock signal devices for safe operation. See Fig. 25-3.

All systems use room thermostats. Warm air systems use a bonnet safety thermostat to shut off the system if the plenum chamber overheats. If automatic humidity is desired, humidistats are used.

Oil burners sometimes use a stack thermostat which will shut off the burner if the stack temperature does not rise within a few seconds after the oil burner is turned on.

Pressurestats shut off steam systems if pressure becomes dangerously high. Each type of heating system has its own special automatic devices.

There are three types of heating thermostats:
1. Control of electrical circuits.
2. Control of air circuits (pneumatic).
3. Control of combination electric and pneumatic circuits.

Some of the newer thermostats use solid-state electronic devices. These include triacs, transistors, and amplifiers which control the functions of systems such as:
1. Power circuits.
2. Airflow.
3. Water flow.
4. Steam flow.
5. Damper operation.

This solid-state control will provide the best operation for highest efficiency of the system. These same systems will indicate, by instruments, lights, and recorders, the conditions of all the variables in the system.

Components of a solid-state thermostat usually include a thermistor temperature sensor, a potentiometer temperature adjustment, and an SCR (silicon controlled rectifier).

There are many types and models of thermostats designed for heating and cooling systems:
1. Voltage:
    a. Low voltage types — 24 V.
    b. Line voltage types — 120 V, 120/240 V or 240 V (also used for 208 V).
2. Points:
    a. SPST (single-pole single-throw) two wire.
    b. SPDT (single-pole double-throw) three wire.
3. Selector switches:
    a. None.
    b. Winter — summer.
    c. Heat — fan.
    d. Heat on — heat off.
4. Thermometer:
    a. Yes.
    b. No.
5. Solid-state thermostats.

Thermostats are supplied with or without heat anticipators. Their use varies. Some applications are:
1. Air — outside, inside, combination.

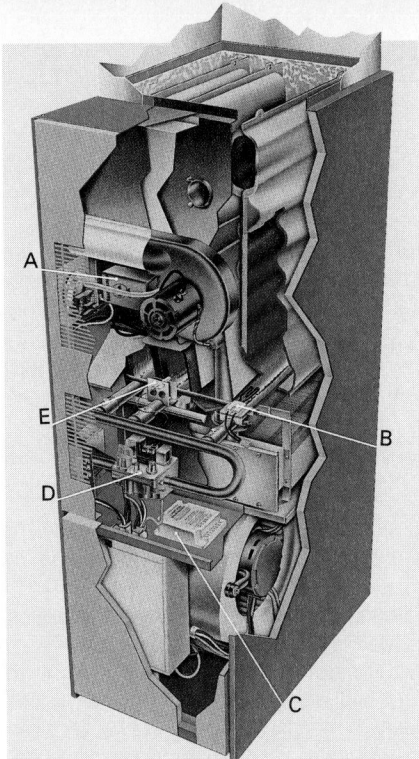

Fig. 25-3. Forced hot-air furnace. A—Fan and limit control. B—Silicon carbide igniter. C—Microprocessor control. D—Gas valve. E—Flame sensor. (White-Rodgers Division, Emerson Electric Company)

2. Cooling coil.
3. Heating coil.
4. Fan (exhaust or fresh air).
5. Fan — coil.

25-5 THERMOSTATS — OPERATION

The primary device in a thermostat is the component which reacts to temperature change. There are several different types:

1. Bimetal strip.
2. Rod and tube.
3. Bellows or diaphragm.
4. Electrical resistance.
5. Hydraulic.

Fig. 25-4 shows the elements for these five types. Another type of element for heating (boilers and hot water) systems is shown in Fig. 25-4, view 6.

Thermostats are further classified as:

1. Line voltage thermostats.
2. Low voltage thermostats.

Usually, line voltage thermostats are used directly in the electrical operating circuit as shown in Fig. 25-5. They are designed to be mounted on the baseboard or on the wall four to five ft. above the floor. Most have a manual "off" switch.

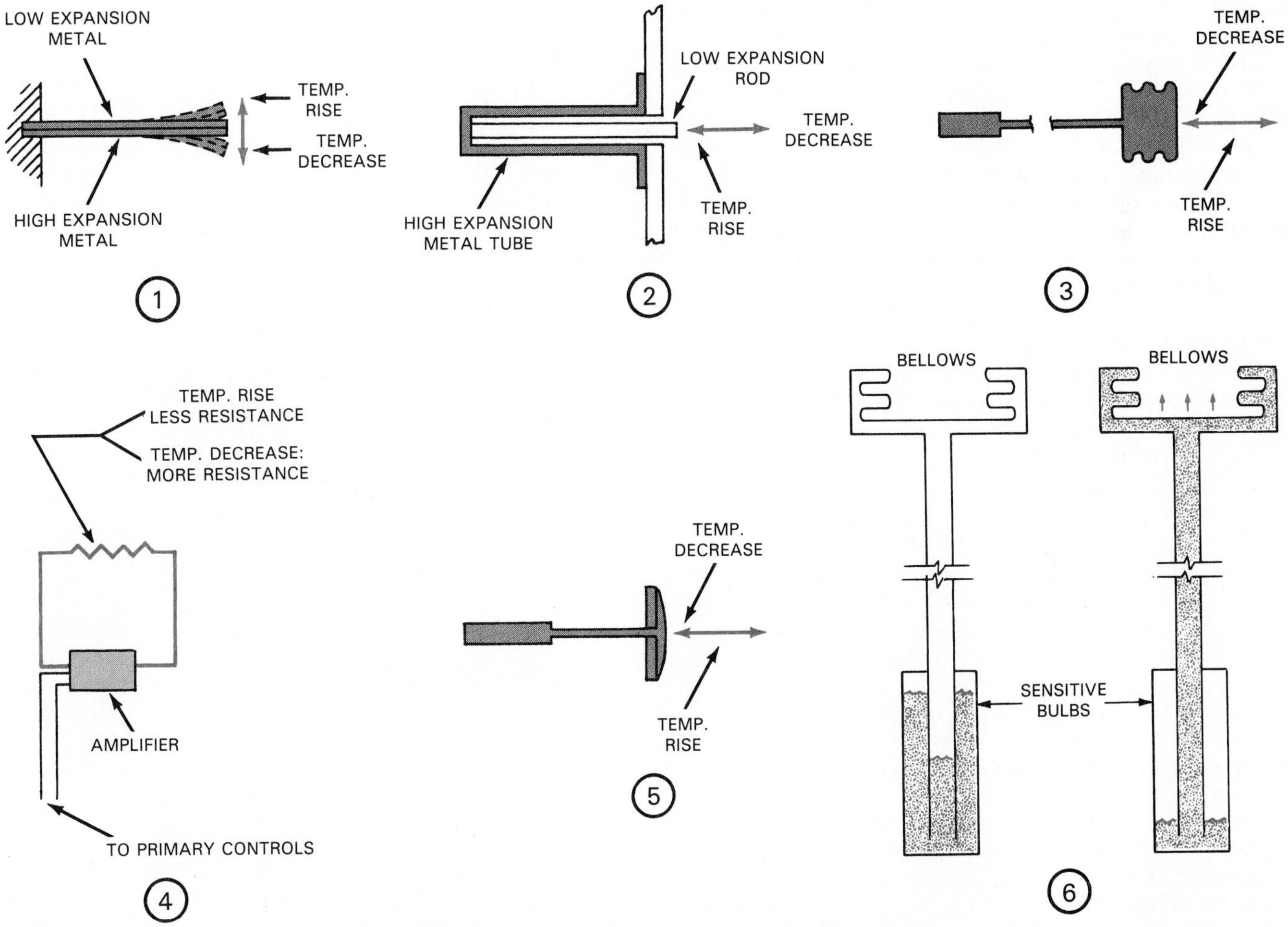

Fig. 25-4. Five basic types of thermostat operating elements: 1—Operates on principle that two different metals will expand at different rates. 2—Operates on principle that metal will expand when heated. 3—Operates on principle that gas will expand with heat. 4—Operates on principle of change of resistance in conductor or semiconductor with change in temperature. 5—Hydraulically operated diaphragm (100 percent liquid). 6—Temperature-sensing element for heating units such as boilers. Sensitive bulb is placed where temperature is to be controlled. As bulb warms, it forces liquid into bellows area and moves bellows.

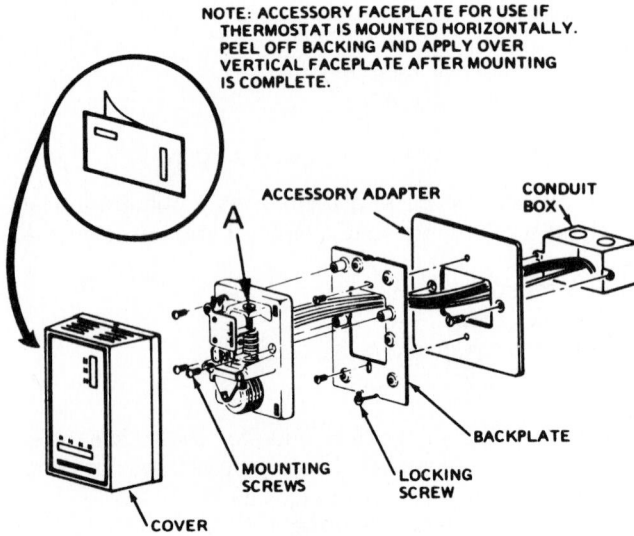

NOTE: ACCESSORY FACEPLATE FOR USE IF THERMOSTAT IS MOUNTED HORIZONTALLY. PEEL OFF BACKING AND APPLY OVER VERTICAL FACEPLATE AFTER MOUNTING IS COMPLETE.

Fig. 25-5. Installation details for line voltage, heavy-duty thermostat. It controls electric heating circuit. A—Temperature adjustment. (Honeywell Inc.)

Low voltage thermostats are found in a low voltage circuit which is connected to a solenoid coil or relay switch. *These thermostats will be damaged if connected into a 120 V line circuit.* They draw electrical power from a step-down transformer.

If thermostats have a large range of operation, they should have ambient temperature correction. For example, a thermostat designed to operate from 50°F to 250°F (10°C to 121°C), will not be accurate throughout this range unless it has a built-in correction. The correction is needed because all parts of the thermostat expand as they become warmer. The correction can be made by using a temperature-compensating bimetal strip which balances the expansion by moving slightly in the opposite direction.

Some three-wire thermostats have two sets of contacts. One set is for "pull-in" and the other is for "hold-in." The design has more stable contact connections.

The bimetal strip is usually wound in a spiral to get more length. This is needed to give more contact movement with each degree of temperature change.

These controls are designed to reduce contact points bounce (causes arcing) and contact "walking" (sliding over each other). The latter action also causes arcing.

Mercury contacts are used with the bimetal strip. Sealed tubes prevent dust, lint, and oxygen from wearing the contacts. Inside the sealed glass tube is a ball or globule of liquid mercury and two or more electrical probes which are fused into the glass. Fig. 25-6 shows an SPST mercury switch; Fig. 25-7 shows an SPDT switch.

The mercury tube thermostat wiring diagram shown in Fig. 25-8 is for a zone valve. When connections No. 5 and No. 4 are closed, the motorized zone valve opens. When the bimetal of the thermostat tilts the mercury tube and connections No. 5 and No. 6 are closed, the motorized valve closes. A bimetal-operated thermostat with four mercury switches is shown in Figs. 25-9 and 25-10. Two are for stage No. 1 and stage No. 2 cooling. Two are stage No. 1 heating and stage No. 2 heating. The wiring connections for this control are shown in Fig. 25-11, while the schematic wiring diagram is as shown in Fig. 25-12.

Some heavy-duty thermostats are designed to carry high current. They can carry as much as 16 A at 120 V and will not be damaged by as much as 96 A locked-rotor amperes for a short time.

In the thermometer-type thermostat, two leads are inserted in a mercury-filled glass column where they contact the inner tubing in which the mercury column moves. When the mercury column rises as the temperature increases, it will close the connection between the two contacts. This closed circuit then activates a warning system.

25-6 THERMOSTATS — LINE — ELECTRIC

Line thermostats are either 120 or 240 V. They are designed to control as much as 22 A. However, many electric heating experts recommend low voltage thermostats for most electric heating systems.

Wiring to line thermostats must be installed according to the local code. Thermostats may have a fan switch mounted on the side. The fan has an off-on automatic setting. "Off" position means blower is off. "On" position means the fan is on continuous operation. "Auto" (automatic) means the fan will be controlled by the fan thermostat in the furnace plenum chamber.

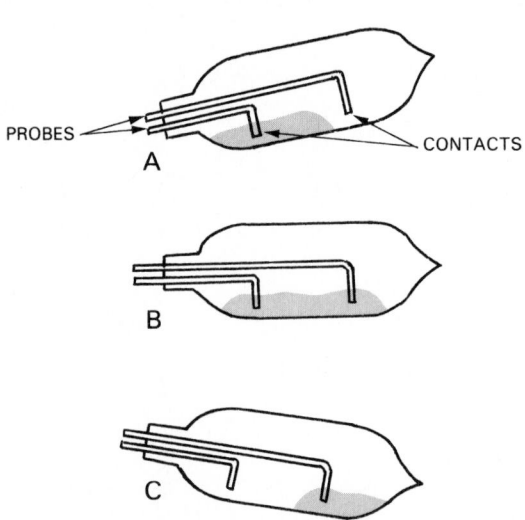

Fig. 25-6. How mercury switch works. Mercury completes circuit between contacts and must be level for proper operation. A—Off for heating. B—On for heating or cooling. C—Off for cooling system.

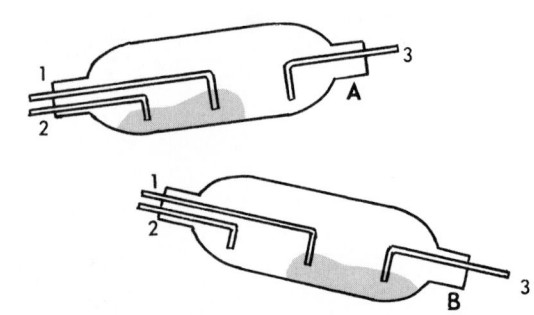

Fig. 25-7. A single-pole, double-throw (SPDT) mercury switch. Mercury tube is usually tilted or moved by bimetal coil. A—Leads 1 and 2 are connected by mercury to complete circuit. B—Leads 1 and 3 are connected by mercury globule.

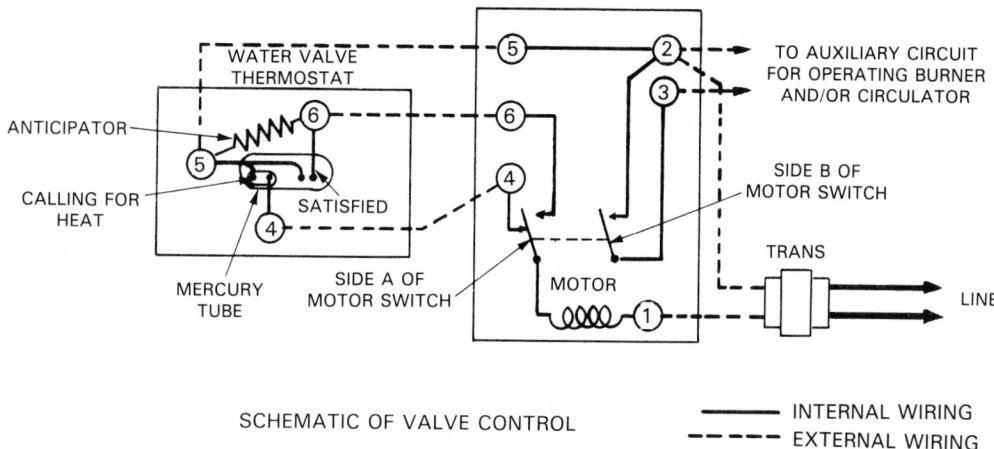

SCHEMATIC OF VALVE CONTROL

——————— INTERNAL WIRING
- - - - - - - EXTERNAL WIRING

Fig. 25-8. Single-pole, double-throw thermostat mercury switch. This thermostat opens and closes motorized zone valve in hydronic or steam heating system. It can also control zone valve in chilled-water cooling system.

The thermostat operates the electrical load directly or through a line voltage operating control. The control is designed for 31 to 300 V and 24 A maximum for noninductive circuits (resistances only — no coils such as solenoids or transformers or motor windings are in the circuit).

Some are designed for 31 to 300 V with six different amperage capacities for inductive circuits (solenoids, transformers, and motors). These capacities are: 3, 6, 8, 10, 14, and 16 A at 120 V, full load.

The capacities are: 18, 36, 48, 60, 84, and 96 A at 120 V — locked rotor (starting load) on motor.

At 240 V, these ampere ratings are reduced by half. Note the following thermostat classifications and see Fig. 25-13:

A. Single line break (single-pole) thermostat:
   1. Heating only.
   2. Cooling only.
B. Heating and cooling.
C. Double line break disconnect thermostat.
D. Double line break cycling thermostat.
E. Two-stage thermostat.
F. Two-circuit thermostat.

25-7 THERMOSTATS — LOW VOLTAGE

Low voltage thermostats carry only a small amount of electrical power. This is an advantage because line voltage thermostats have problems with electrical power heating the thermostat and opening the contacts too soon. This extra heating temperature rise is called thermostat droop.

Example: Thermostat is set for 72°F (22°C) and room is at 68°F (20°C). As the room heats up, the thermostat becomes warmer than room temperature because of electric resistance heat of electrical parts in the thermostat. The thermostat then reaches 72°F (22°C) and cuts off too soon (about the time the room is 69 or 70°F).

Low voltage thermostats should use No. 18 American Wire Gauge (AWG) for distances up to 50 ft. No. 16 AWG wire is specified for distances over 50 ft.

Many thermostats have three or four wires. Each is a different color. Red, white, green, yellow, and black are popular. *Be sure to wire as indicated in the wiring diagrams provided by the manufacturer. This is required.*

Fig. 25-9. Staging thermostat controls two stages of heating and two stages of cooling. A—Thermostat temperature adjustment marked "C" for cooling and "H" for heating. B—Dial thermometer. C—Heat-cool automatic switch. D—Fan switch.
(White-Rodgers Div., Emerson Electric Co.)

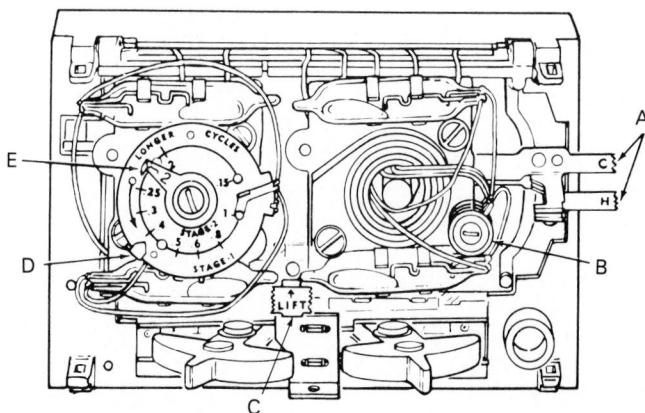

Fig. 25-10. Internal construction of staging thermostat. Two heating control mercury tubes are on the left. Two mercury tubes controlling cooling are on the right. A—Levers for adjusting temperature. B—Cooling anticipators. C—Bronze tab must be lifted to raise cover. D—First stage heat anticipator. E—Second stage heat anticipator.
(White-Rodgers Division, Emerson Electric Company)

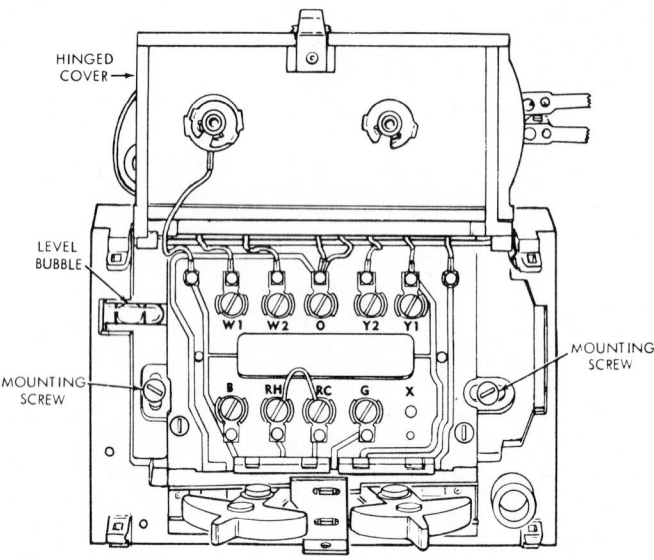

Fig. 25-11. Wiring terminals of a staging thermostat. Note identification on terminals. (White-Rodgers Div., Emerson Electric Co.)

These controls are rated as follows:

0 to 15 V — Maximum of 5 A
15 to 30 V — Maximum of 3.2 A

They control the applied energy to a low voltage control of a Class 2 (power) circuit.

The thermostat mounting usually requires other parts. The subbase is the part to which the thermostat is attached. It contains the lead terminals. Sometimes it holds other parts such as switches and indicator lamps. Beneath the subbase is the wall plate. This is a part attached to an outlet box or wall.

A millivolt thermostat may be used with a self-generating electrical source. For example, a multiple thermocouple located near a gas pilot light will generate the needed current to operate the thermostat.

There are two groups and 12 classifications of millivolt thermostats. Group 1 thermostats are those with fixed or adjustable anticipators. Group 2 are thermostats without anticipators.

The 12 classifications are:

1. Heating only (one-stage).
2. Cooling only (one-stage).
3. Heating and cooling, manual changeover (one-stage).
4. Heating and cooling, auto changeover (one-stage).
5. Heating only (two-stage).
6. Cooling only (two-stage).
7. Heating (one-stage); cooling (two-stage); manual changeover.
8. Heating (one-stage); cooling (two-stage); auto changeover.
9. Heating (two-stage); cooling (one-stage); manual changeover.
10. Heating (two-stage); cooling (one-stage); auto changeover.
11. Heating (two-stage); cooling (two-stage); manual changeover.
12. Heating (two-stage); cooling (two-stage); auto changeover.

25-8 HEAT ANTICIPATORS

An interesting problem in room heating thermostats is that the room temperature always tends to rise above the thermostat setting after the thermostat points have opened and the burner has stopped. This action is called "overshoot" of the thermostat. This action is the result of the heat in the furnace.

Even though some heat is generated by the electric circuit in the thermostat, a small heating coil (resistor) is also placed in the thermostat. These small resistance coils are usually called heat anticipators. During the heating part of the cycle, the thermostat is always about 1 °F warmer than the room. If the thermostat is set for 74 °F (23 °C), the thermostat will open when the room temperature is actually 73 °F. Then, after the thermostat has opened, the room temperature will still rise to 74 °F, because of the heat in the furnace.

Fig. 25-14 shows an electric heating thermostat equipped with an anticipator. The heat anticipator is connected in series with the operating contact points.

Both heating and cooling anticipators for thermostats are of two types:

1. Fixed anticipators.
2. Adjustable anticipators.

They are most often used on low voltage thermostats and on heating-cooling combination thermostats. The size of the anticipator (current capacity) is related to the current (amperes) that flows through the thermostat contacts. This current varies between .15 to 1.0 amperes, depending on the controls connected to the thermostat.

Fixed anticipators' current flow should equal the control ampere flow. The adjustable anticipator should be regulated to the same ampere flow as the control circuit. The current flow of the control circuit is shown on the label of the control.

If the anticipator has a higher rating than the control circuit, it will warm the thermostat more slowly since there will be less anticipator heat. The reverse is also true. Anticipators are used in 24 V systems and also in 750 millivolt systems.

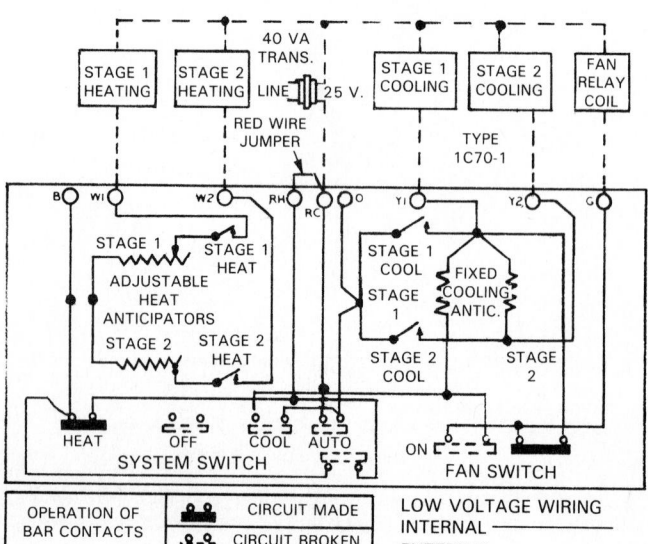

Fig. 25-12. This schematic is for the staging thermostat of a heating and cooling system.

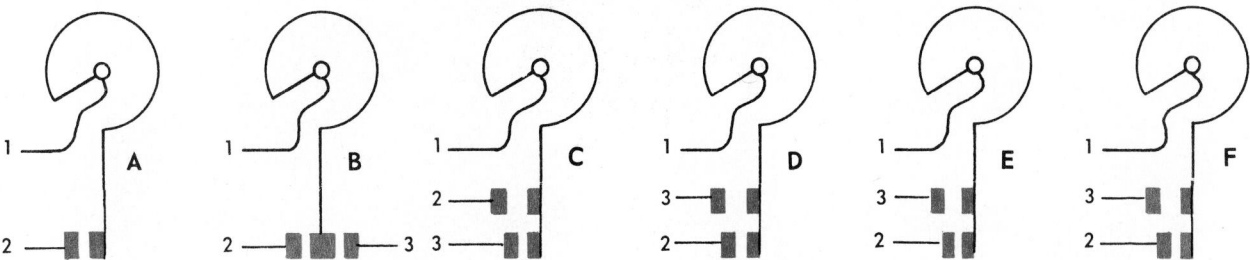

Fig. 25-13. Schematic for thermostat classification. A—Single line break thermostat. B—Heating-cooling thermostat. C—Double line break disconnect. D—Double line break cycling. E—Two-stage thermostat. F—Two-circuit thermostat. 1—Common lead. 2—First controlled lead. 3—Second controlled lead.

Cooling thermostats may have a heat anticipator connected in parallel with the thermostat points. The anticipator becomes warm during the off cycle and will close the thermostat points about 1 °F (.56 °C) warmer than the room temperature. If control is set to cut in at 78 °F (26 °C), the control will actually start the system at 77 °F (25 °C). This will allow the system to start cooling the room before it becomes too warm and humid.

Different anticipator heaters are used depending on the system. Fig. 25-15 shows the outside of a combination thermostat with a heat anticipator. The inside of this thermostat is shown in Fig. 25-16. The more residual heat in the heating system, the larger the capacity of the anticipator. Fig. 25-17 shows a heating thermostat wiring schematic diagram.

At the end of the off part of the heating cycle, when the thermostat points close, it takes time for the furnace to heat and to move this heat to the room. The room temperature may fall about 1 °F during this start-up lag. *This action is called "system lag."* The cut-in temperature is usually set about 1 °F above the lowest temperature desired.

25-9 HEATING THERMOSTATS

This paragraph introduces the basic operation of a heating thermostat. Para. 25-14 reviews the more recent electronic versions of these thermostats. The heating thermostat in a typical residential system usually uses open contact points to control either a 24 or 120 V primary circuit.

Fig. 25-18 shows the inside of a typical room thermostat designed to mount on a wall. A coiled bimetal strip reacts to temperature change. It has a range of 35 or 50 °F low setting to 90 or 95 °F high setting.

These thermostats usually operate on a .5 °F to 2 °F differential or smaller. This allows close temperature control. Fig. 25-19 shows more details of a thermostat operated by a bimetal strip.

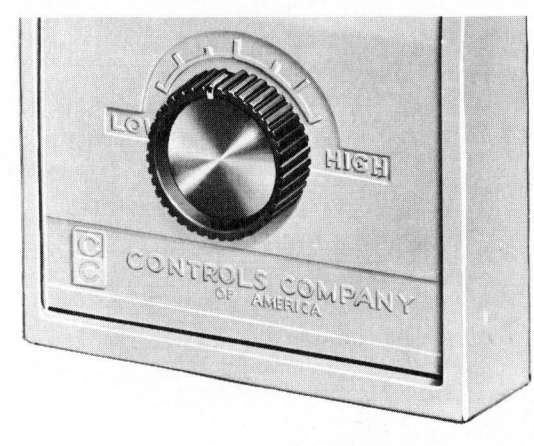

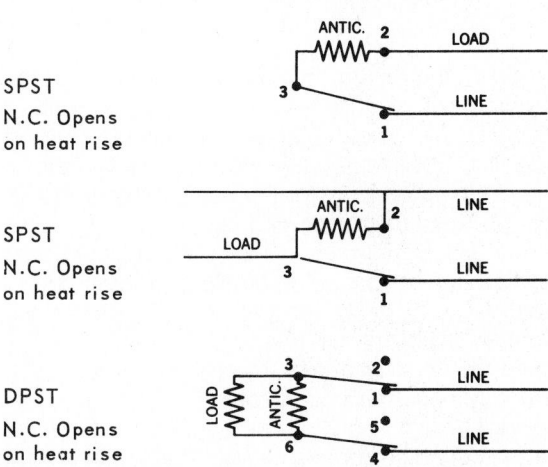

Fig. 25-14. Thermostat used in electrical heating system. Top. External view. Bottom. Wiring diagrams for thermostat. Note that it is equipped with an anticipator. The SPST indicates that control has single-pole single-throw switch.

Fig. 25-15. Combination heating and cooling thermostat which has a heat anticipator. A—Switch setting for heating. B—Off position for heating-cooling switch. C—Switch position during cooling season. D—Control switch for fan operation. Top scale is for setting temperature. Scale adjusts for temperatures between 50 to 90 °F. Bottom scale indicates actual room temperature (72 °F).

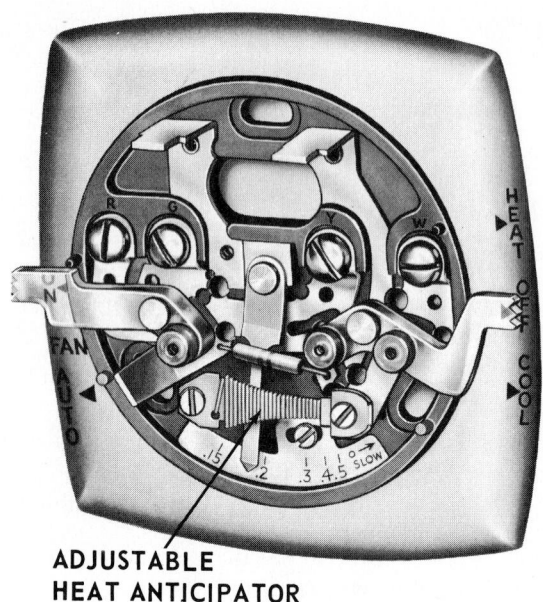

ADJUSTABLE
HEAT ANTICIPATOR

Fig. 25-16. Inside view of thermostat. Note variable heat anticipator. (Johnson Controls, Inc.)

BIMETAL
ELEMENT

Fig. 25-18. Inside view of another thermostat shows coiled bimetal element. (Johnson Controls, Inc)

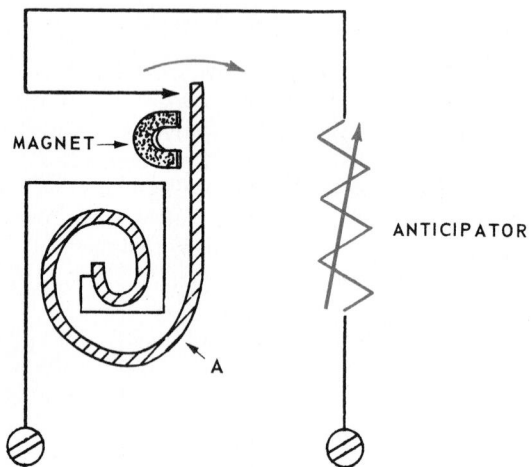

Fig. 25-17. Schematic wiring diagram is of a heating thermostat that is equipped with an anticipator. A—Bimetal strip.

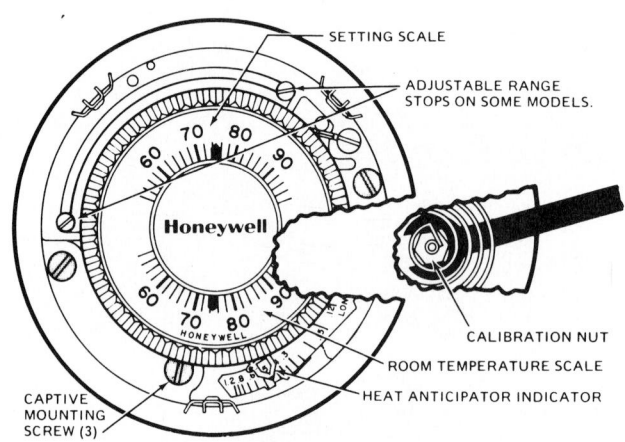

Fig. 25-19. More detail of a heating thermostat which uses a bimetal strip. (Honeywell, Inc.)

Generally, the sensitive element is a bimetal strip. A small magnet is sometimes used to produce snap action in the points. A range adjustment is usually a direct force on the bimetal, while the differential usually consists of moving the small magnet either closer to or farther away from the bimetal strip.

Some thermostats are hydraulically operated. Fig. 25-20 shows a diaphragm mechanism. The diaphragm moves as a liquid expands and contracts. The dial adjusts the setting and the small magnet produces snap action in the thermostat.

Electronic measurement of an electric resistance element may be used for temperature control. It is possible to get differentials of .01 deg. using this method.

Thermostats are rated by the voltage they carry and also by the controls to which they are electrically connected. These controls open the circuit when the temperature rises and complete the circuit when the temperature falls.

One 24 V thermostat has three wires connected to it. The wires make a contact on temperature drop and close a holding circuit. They will open the circuit on a rise of temperature. This thermostat must be used with a relay. The relay opens and closes the 120 V circuit for the oil burner, gas burner, or electric heater.

Another 24 V thermostat has separate controls. One opens the contacts on rise in temperature and one closes contacts on a drop in temperature. It is used on gas burners, oil burners, or electric heaters.

A third type is a two-wire 24 V thermostat. It controls the 24 V furnace devices directly.

A wiring circuit is shown in Fig. 25-21 for a control which uses a mercury bulb contact device. The circuit is designed for forced warm-air heating.

If lint should lodge between the points, an open circuit will result. To clean contact points, use a piece of clean paper. If a mercury tube is used instead of contact points,

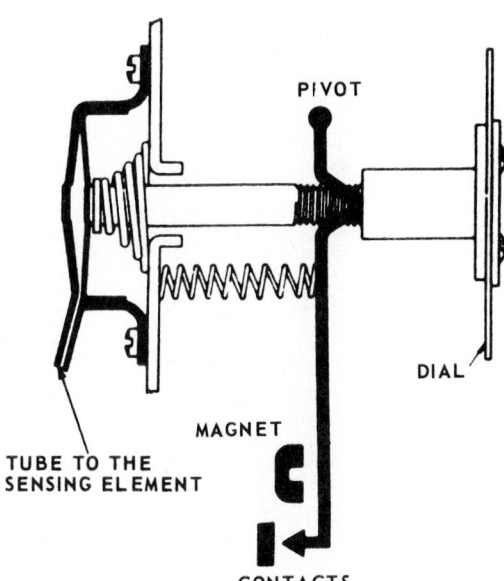

Fig. 25-20. Diagram shows side view of hydraulically operated thermostat. (White-Rodgers Div., Emerson Electric Co.)

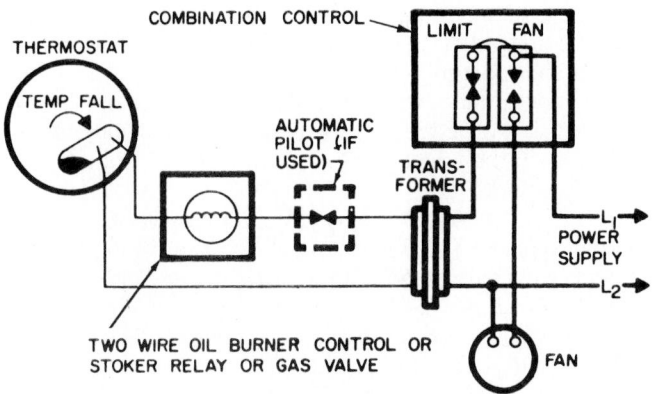

Fig. 25-21. Wiring diagram of a circuit for a heating thermostat. Note that automatic pilot and fuel control are in the low-voltage (24 V) circuit (to left of transformer).

the thermostat must be carefully leveled. Use a plumb line or spirit level.

Another style of thermostat has been developed for air conditioning systems. These units have easily operated settings. They are designed with all the various types of electrical bases to enable the thermostat to control heating, cooling, or both.

Some heating-cooling systems use a clock-operated thermostat. It changes the temperature range during the night and returns it to normal during the day. Fig. 25-22 shows a clock-operated thermostat powered on 24 to 30 V. When programming the thermostat, the display shows the programmed time/temperature settings selected. When programming is completed, the liquid crystal display (LCD) shows time and current room temperature. Fig. 25-23 shows wiring diagrams for three different installations using two- and three-wire clock thermostats.

Frequently a unit will short cycle because the thermostat is exposed to vibration (on a shaky wall, or stairs). *Mount the thermostat to a firm structure.*

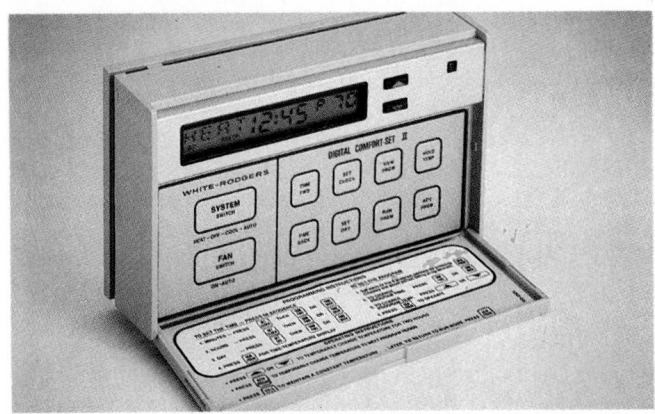

Fig. 25-22. Electronic digital thermostat. This type may be used for either heating or cooling, or for a combination system. Unit has digital display for time and temperature, and fan operation. (White-Rodgers Division, Emerson Electric Company)

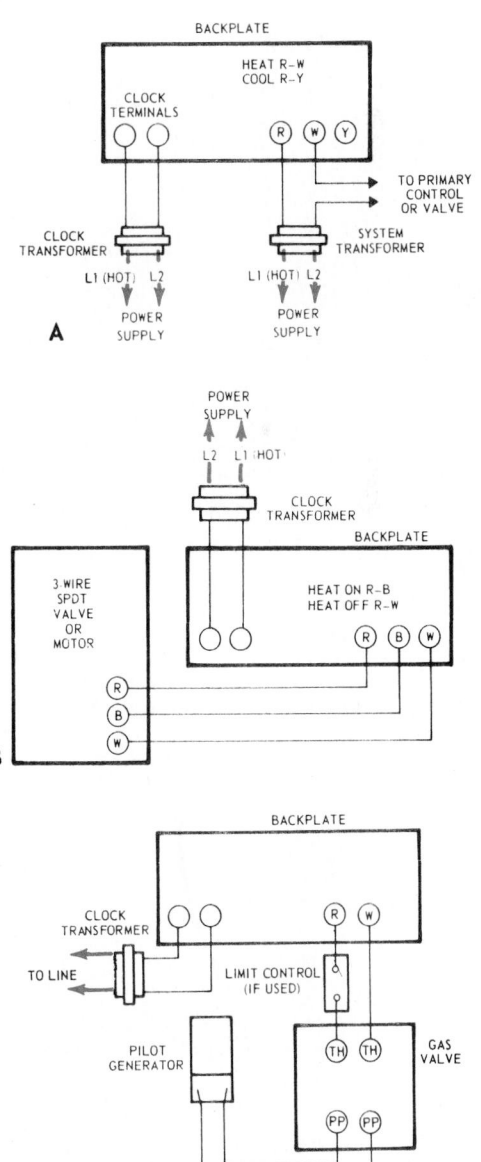

Fig. 25-23. Three different clock thermostat wiring diagrams. A—Two-wire low-voltage system. B—Three-wire system. C—Multivolt pilot generator system. (Honeywell, Inc.)

The thermostats just described are used primarily in forced air systems. The operating features of a thermostat used in an electrical heating system are very similar. In general, a line-type thermostat is more commonly used when trying to control electric heat. It is usually rated at 5000 W and 120 V ac. See Fig. 25-24. The inside of such a control, operated by a heat-sensitive bulb, is shown in Fig. 25-25.

Units are equipped with safety controls known as temperature limit switches. Large-capacity units use a relay along with low voltage thermostats to control high-wattage loads. Some prefer to use low voltage thermostats with a relay on all electric heating systems. It reduces "heat lag." The sequence control is sometimes used also. It is made up of a block of relays. Each relay controls a separate heating circuit. The relays are activated one at a time. This sequence timing prevents placing the full electrical load on the main building circuit at one time.

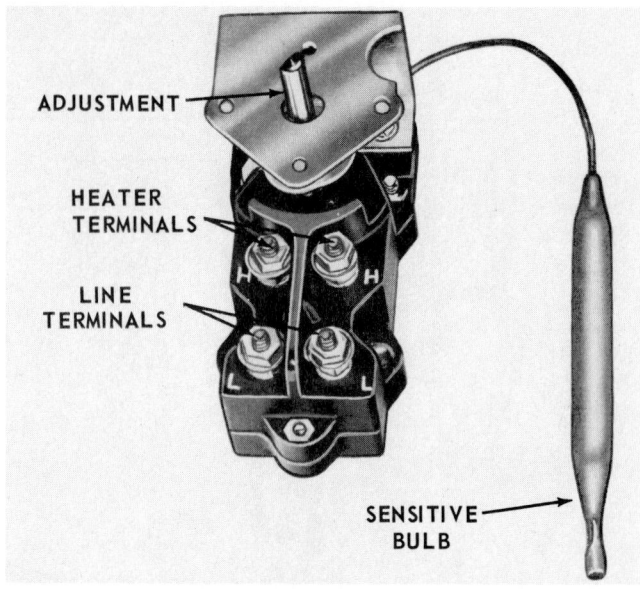

Fig. 25-26. Hydraulic-type thermostat used on electric resistance heating. It controls both electrical leads to heating unit.

Some codes require that both leads to the heating element be opened by thermostatic control. Fig. 25-26 shows such a thermostat.

Thermostats mounted on baseboard heaters are usually encased in attractive covers and have a calibrated dial mounted on the adjustment knob. Fig. 25-27 shows such an installation.

Some line voltage thermostats produce a wide swing of temperatures causing room temperatures to vary too much. Others may have a temperature droop and the temperature of the room goes too low before heating effect is felt.

For example, if the thermostat is set for 75 °F with a 1 °F differential but room cools to 70 °F and then warms to 80 °F, the 10 °F difference is too much swing.

If the thermostat is set for 75 °F with a 1 °F differential and room only heats to 65 °F when the thermostat shuts off, the 10 °F below normal is a droop. This condition is usually caused by too much heat release in the thermostat from wiring or too much heat from the anticipator.

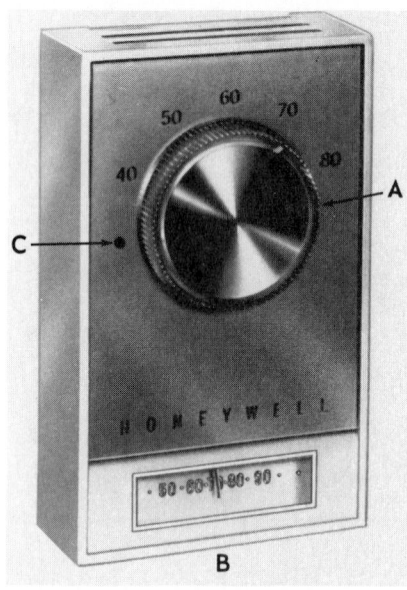

Fig. 25-24. Two-switch wall thermostat used for controlling electric heating circuit. It is line voltage control. One switch is operated by thermostat and one opens manually when setting is turned to "off." A—Temperature setting. B—Thermometer. C—Off-position.

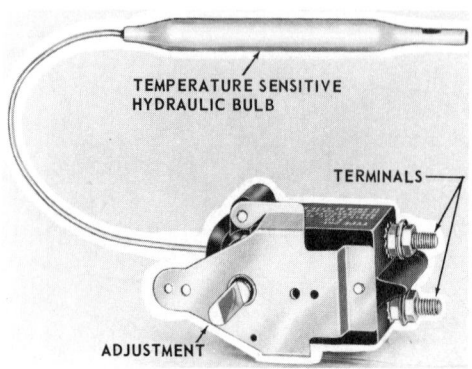

Fig. 25-25. Hydraulic type thermostat used on electric resistance heating units. Rated at 240 to 277 V, it can control up to 5000 W. Its 1 1/2-degree differential provides uniform temperature control. (White-Rodgers Div., Emerson Electric Co.)

Fig. 25-27. A thermostat designed to be mounted on electric resistance baseboard heater casings. (White-Rodgers Division, Emerson Electric Company)

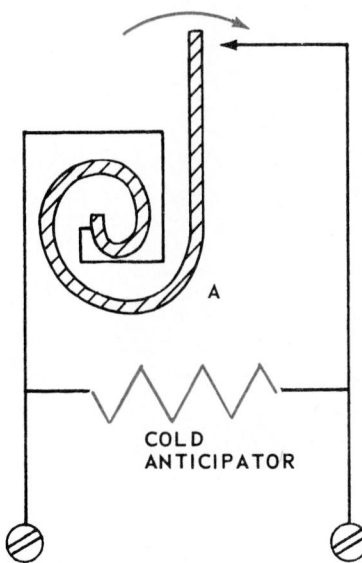

Fig. 25-28. Schematic wiring diagram of cooling thermostat equipped with cold anticipator. A—Bimetal strip.

When adjusting a thermostat, be careful to avoid heating the thermostat with the hands or by breathing on the element. If the thermostat is warmed above the room temperature, wrong settings of the thermostat will result.

25-10 COOLING THERMOSTATS

Comfort cooling thermostats are similar in design to heating thermostats except that the contacts open as the room cools and close as the room warms up.

The popular bimetal thermostats usually have a 1 °F (.6 °C) differential. Both the 24 V and 120 V controls are available. Some units have cold anticipators. These small electric resistance elements are in parallel with the bimetal strip. They tend to close the bimetal-controlled points just a little before the unit reaches the room cut-in temperature.

Fig. 25-28 shows a simplified wiring diagram of a cooling thermostat. The cold anticipator warms the bimetal strip during the off cycle but is bypassed on the on cycle. This heat during the off cycle tends to turn the system on just a little ahead of normal.

25-11 COMBINATION THERMOSTATS

Some thermostats have control mechanisms for both heating and cooling. These thermostats are used with heat pumps or with other installations having both a heating and a cooling system. Fig. 25-29 is a simplified wiring diagram for a combination thermostat using a bimetal strip, heat anticipator, and cold anticipator.

The appearance of a combination heating-cooling thermostat is shown in Fig. 25-30. This control is for 24 to 30 V electrical circuits. Inside view is shown in Fig. 25-31.

A thermostat using a bellows operating mechanism is shown in Fig. 25-32. See its construction in Fig. 25-33.

Controls are made that go beyond control of cooling and heating. Some regulate humidity, indicate the condition of the filter, and control an odor system. These controls usually operate on a 24 V circuit.

Other combination heating and cooling thermostats use individual system and fan control switches. Fig. 25-34

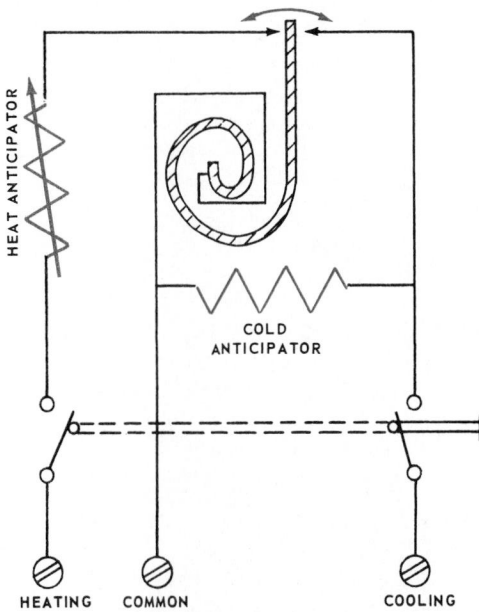

Fig. 25-29. Schematic of bimetal heating-cooling thermostat. Unit is equipped with both a heat and a cold anticipator.

Fig. 25-30. A 24 V thermostat that may be used both for heating and cooling. Fan control switch is at A and a heating or cooling switch is located at B. Range adjustment is on upper temperature scale. Bottom scale indicates room temperature.   (Honeywell, Inc.)

Fig. 25-31. Construction of combination heating-cooling thermostat, shown in Fig. 25-30. Filter warning light is shown at A. Bulb lights when filter becomes dirty.

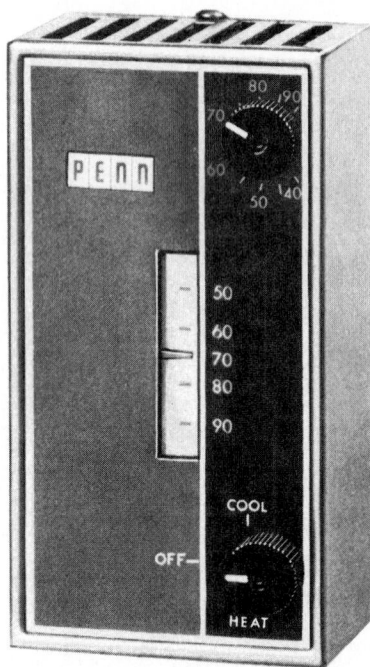

Fig. 25-32. Combination thermostat used for controlling both heating and cooling units. (Johnson Controls, Inc.)

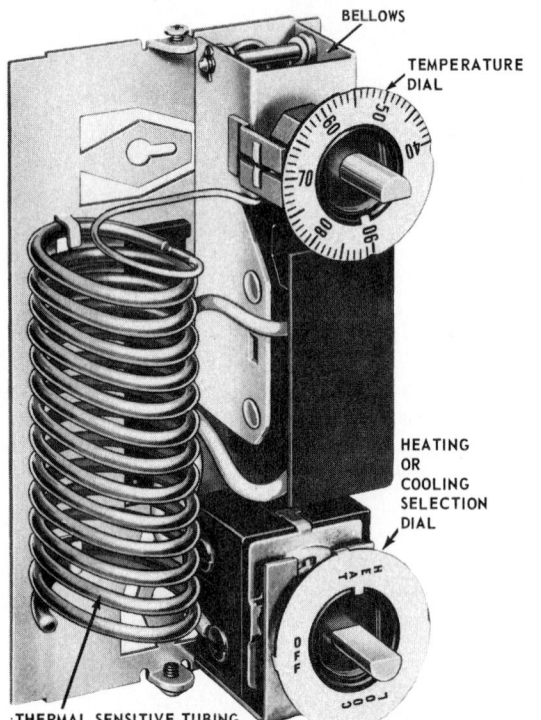

Fig. 25-33. Looking at the inside of a cooling-heating thermostat. (Johnson Controls, Inc.)

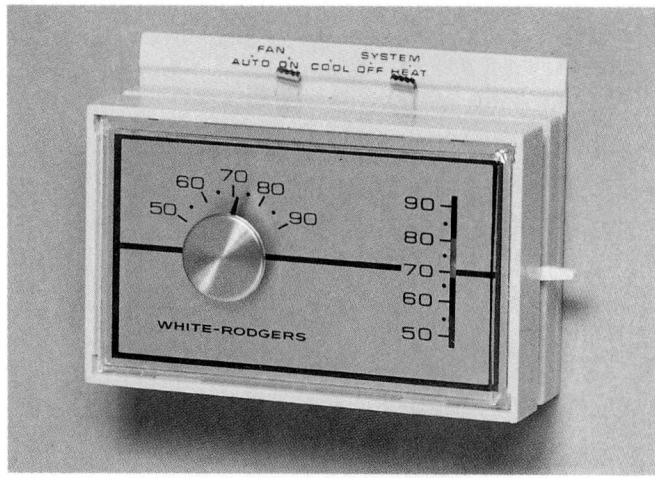

Fig. 25-34. Combination heating-cooling thermostat. Note separate controls for heating, cooling, and fan. (White-Rodgers Div., Emerson Electric Co.)

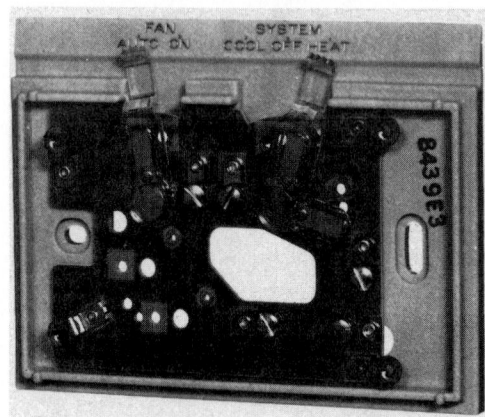

Fig. 25-35. Inside view of combination heating-cooling thermostat shown in Fig. 25-34. (White-Rodgers Div., Emerson Electric Co.)

with a modulating gas control, it keeps the heating temperature range within 1 °F. Cooling is held within a 3 °F range. The thermostat is shown in Fig. 25-37. An electronic system is mounted on the thermostat base.

25-12 TIMER THERMOSTATS

Many air conditioning installations can be automatically clock controlled. For example, cooling systems installed in offices need not run at normal temperatures on Saturdays and Sundays or during other nonworking hours. Many users of air conditioners want the units to start functioning at a certain time before the premises are occupied.

Heating systems, too, can provide reduced temperatures during those hours or days the building is not occupied.

Fig. 25-38 illustrates an automatic timer that operates on a seven-day schedule. The "trippers" on the edge of the time disk can be set to turn on the air conditioner and shut if off again at any set time each day.

25-13 MULTISTAGE THERMOSTATS

Multistage thermostats are usually designed for low voltage. They will operate two or three separate circuits in

shows this type thermostat with a calibrated temperature-setting dial and a thermometer. The thermostat base, showing the mounting holes, terminals, and switches, is shown in Fig. 25-35.

An electronic circuit using a combination thermostat is shown in Fig. 25-36. Five leads go to the thermostat. Used

5 WIRE SYSTEM

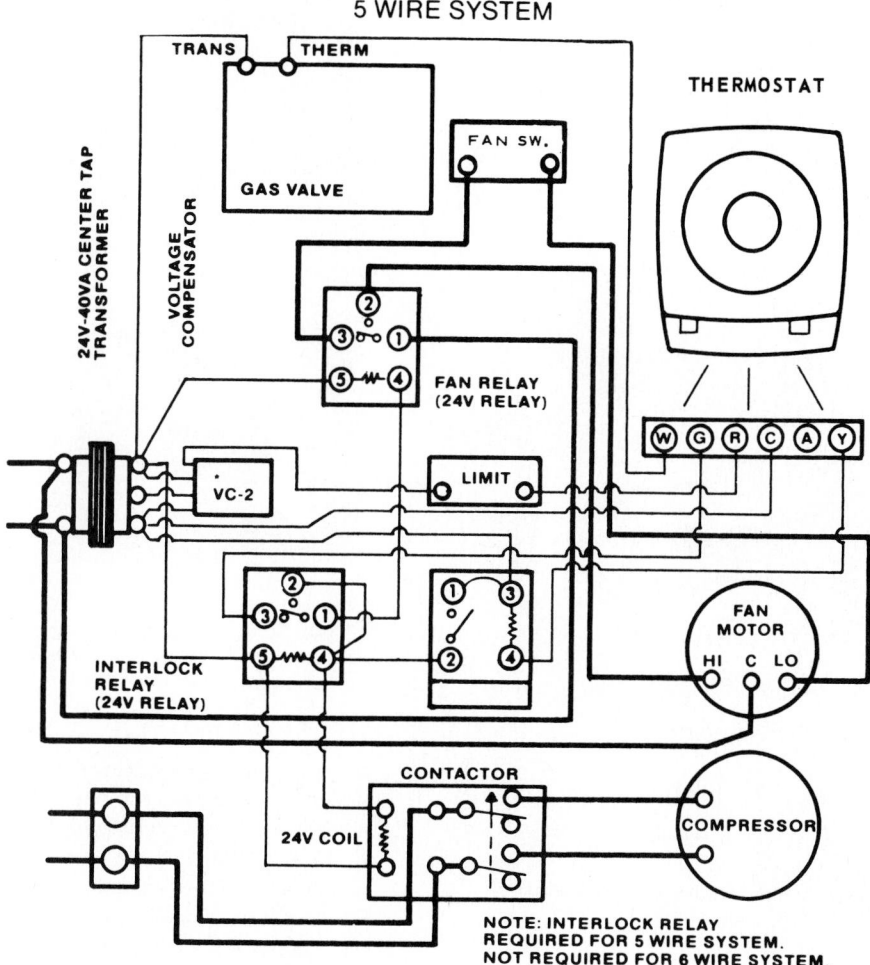

Fig. 25-36. Electrical wiring diagram for an electronic thermostat. (Robertshaw Controls Co.)

sequence and are made for either heating (close on temperature fall) or for cooling (close on temperature rise).

Fig. 25-39 is a schematic of a multistage thermostat designed to control two stages of heating and one stage of cooling. Three mercury tube switches (SPST) are used. Notice the four electrical leads to the thermostat.

The thermostat operating element is a bellows. Fig. 25-40 shows a two-stage heating thermostat.

25-14 ELECTRONIC THERMOSTATS

Advances in electronic circuitry and controls have revolutionized the thermostat and its capabilities. Electronic thermostats use the same basic operating principles as earlier models. The various sensing, switching, timing, and staging functions, however, are all accomplished with electronic components introduced in Chapter 6. Multiple-feature

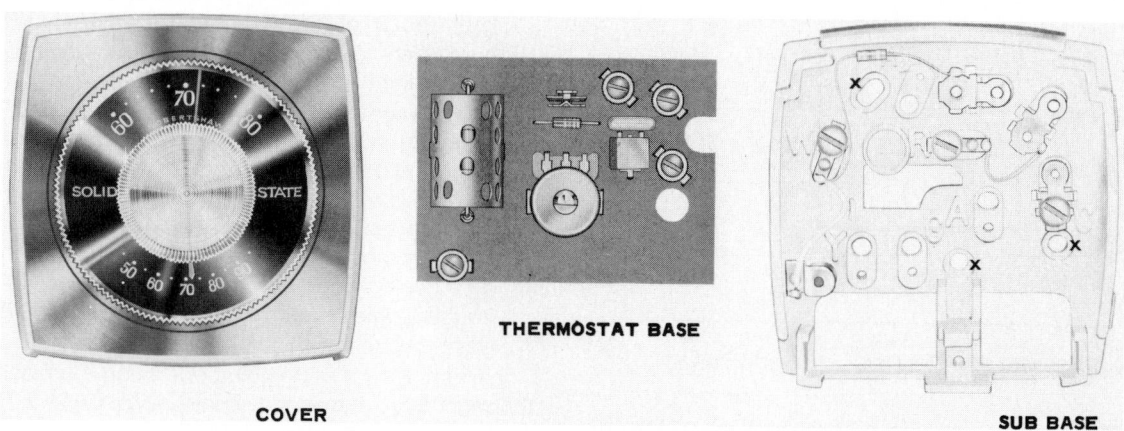

Fig. 25-37. Electronic thermostat used on combination heating and cooling system.

models using transistors, relays, amplifiers, electronic integrated circuit chips, and microprocessors are now in use.

The appearance of an electronic thermostat is different from the standard mechanical thermostat. See Fig. 25-41.

A typical unit can perform the following functions:

1. *Fully programmable* temperature settings. In general, various temperatures can be programmed to achieve

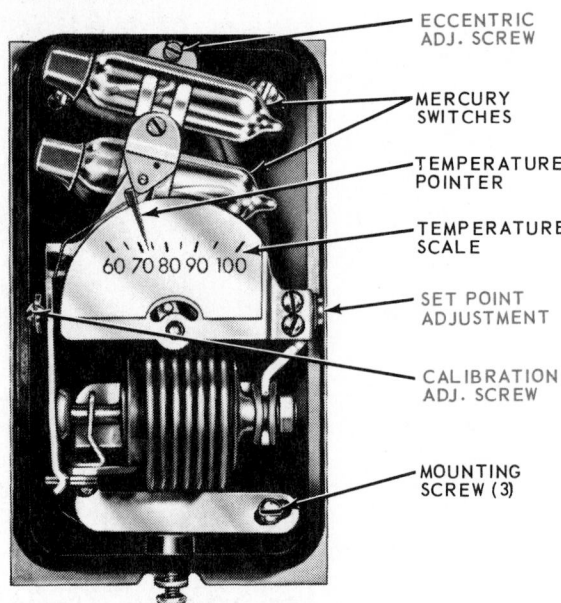

Fig. 25-40. Two-stage thermostat operated by sealed bellows element and using mercury switches. The eccentric adjusting screw adjusts the differential which is usually 1 °F.

Fig. 25-38. Timer used to control operation of heating or cooling system on seven-day schedule. It can be used for different circuit control every three hours up to a total of 28 actions a week. A—Trippers mounted on disk. B—Day indicator. (Paragon Electric Co., Inc.)

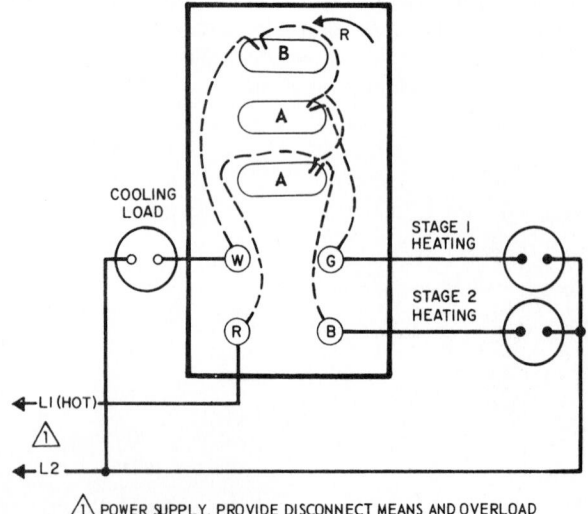

⚠ POWER SUPPLY. PROVIDE DISCONNECT MEANS AND OVERLOAD PROTECTION AS REQUIRED.

Fig. 25-39. Schematic for a multistage thermostat. It controls two heating stages and one cooling stage. A—Heating contacts. B—Cooling contacts. (Honeywell, Inc.)

maximum comfort throughout the day. These settings can also be varied for each day of the week.

2. *Multistage and combination control* combining the features in Para. 25-11 and 25-13. A single control device can control various stages of heating and cooling. For heat pump applications, programmable thermostat designed specifically for heat pump use should be used.

3. *Override control.* This feature results in a preset temperature condition. An example would be a vacation setting. A single temperature might be desired for those periods of time when no one is home.

4. *Digital display.* The room temperature, desired temperature, time, and date are available for display at the push of a button.

5. *Battery backup.* Most of the electronic thermostats are in the low-voltage category (vs. line voltage—see Para. 25-6 and 25-7).

Most models have a battery backup. This prevents loss of the entire program in case of a power failure. Additional features that can be found on some models include:

1. Maintenance requesting. The unit can automatically keep track of the number of days that the filters, humidifiers, etc., are in use and then automatically signal with a light or tone when maintenance is required.

2. Outside temperature anticipator. Some electronic thermostats will use outside temperature, combined with warm-up or cool-down time of the heating and cooling system, and adjust the operating cycles accordingly. Outside temperature can be digitally displayed.

3. Security protection. The thermostat is made to respond to a certain code before it allows changes to be made to the programmed temperature settings. This helps prevent tampering by unauthorized personnel.

Instructions for setting and operating a typical electronic thermostat are shown in the fold-down cover in Fig. 25-41.

The electronic thermostats, due to their precision components, offer a precise temperature control. Usually

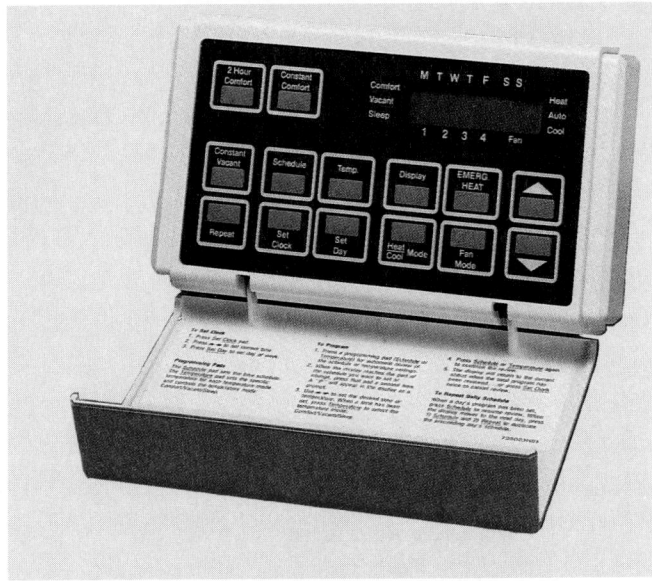

Fig. 25-41. A typical electronic thermostat that indicates present temperature, time, and heating or cooling mode.
(Harper-Wyman Company)

temperature can be controlled within 1 degree of the set temperature.

The greatest advantage of an electronic thermostat is that it offers significant energy savings without reducing overall comfort. This can be accomplished through simple preprogrammed temperature settings. This requires very little effort from the owner once the unit has been programmed.

25-15 HYDRONIC THERMOSTATS

Hot water or hydronic heating systems also require thermostats. While these thermostats provide functions similar to the thermostats discussed, they are constructed differently. Hot water thermostats are more easily adapted to individual room (or radiator) use, Fig. 25-42, and result in considerable energy savings when compared to a single control unit system.

25-16 PORTABLE THERMOSTATS

Portable thermostats are available which may control a furnace or air conditioner. The thermostat electronically activates a responder on the furnace or air conditioner up to a distance of about 120 ft.

The thermostat may be moved from room to room and responds to the temperature of that room. Since it is an electronic battery-powered unit, it has no moving parts. It has an accurate adjustable differential from 0 to about 5 °F (0 to about 3 °C) with a range of 45 to 90 °F (7 to 32 °C).

Once every minute, the thermostat senses the temperature. If heating (or cooling) is necessary, the thermostat sends a radio signal to the responder.

25-17 THERMOSTAT GUARD

One of the biggest problems with thermostats is their ease of adjustment. Whenever there is more than one person in the building, there is a chance that two or more per-

sons will change the thermostat setting. These actions may produce uncomfortable results, may be irritating, and may even be dangerous.

One solution is to use a thermostat that can only be adjusted with a special key. This key is controlled by one person. Another method is to cover the thermostat with a clear plastic lockable cover.

25-18 SERVICING THERMOSTAT AND THERMOSTAT LOCATION

The location of a thermostat is important. It should be placed in an average-temperature location such as an inner wall about 5 ft. from the floor where it is out of the way of furniture and beyond the reach of small children.

The location of a thermostat should be changed if it is affected by:
1. Drafts, or dead air spots behind doors and in corners.
2. Hot or cold air from ducts.
3. Radiant heat from the sun or an appliance.
4. Concealed pipes and chimneys.
5. Unheated areas behind a thermostat, such as an outside wall.

If the heating or cooling system in a building is not working, the thermostat may be at fault. In general, a thermostat has warning signals and comes with a diagnostic chart to follow when troubleshooting, Fig. 25-43.

Lacking this, a good first step is to check all electrical connections and to make sure that all switches are operating. Continuity should then be verified between the thermostat and the furnace or cooling system.

If no problems are found, and the heating or cooling system, itself, has no apparent problems, the temperature-sensing device within the thermostat is probably at fault. Checking this device externally can be difficult. This condition can best be verified if another thermostat temporarily set in place corrects the problem. In most cases, the easiest and most economical means of correcting this is to replace the thermostat.

Electronic thermostats are somewhat easier to service in that they may provide some simple self diagnostics (i.e., battery condition, no programmed settings, etc.). In addition, electronic components are usually built onto circuit boards that can be removed and replaced. See Chapter 6.

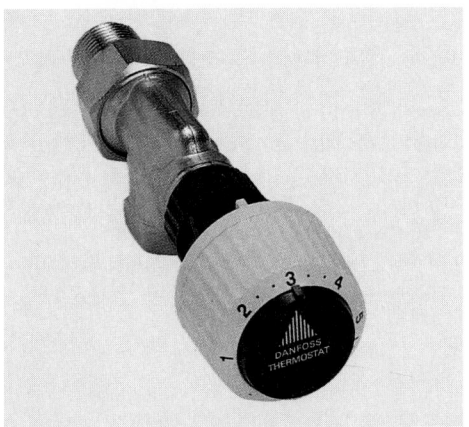

Fig. 25-42. Self-contained thermostatic valve is used for temperature control of individual rooms in a hot water or hydronic heating system.
(Danfoss, Inc.)

XIV. TROUBLE-SHOOTING GUIDE AND VOLTAGE REFERENCES

Problem Observed	Possible Cause	Corrective Action
1. No display	No ac voltage to sub-base.	Measure power at TB-1 pins 1 and 2. Should be 22 VAC to 30 VAC.
	If auxiliary transformer is used.	Check for power output. Check P-3 hook-up for 22 VAC to 30 VAC.
	No power to unit.	Check power hook-up. Check for defective wiring.
	Blown fuse (F-8).	Replace fuse with 1.0 A.
	HVAC transformer defective.	Replace transformer with correct type.
	J-1 not connected to sub-base.	Connect J-1.
	Damaged pins on J-1 connector.	Straighten pins if possible, if broken replace controller.
	Subbase defect.	Replace subbase.
	Controller defect.	Replace controller.
2. No temperature display or wrong temperature displayed	Unit not calibrated.	Calibrate per Section XIII.
	Temperature Sensor Open.	Replace temperature sensor board.
	Temperature Sensor Cable damaged or not connected.	Check temperature sensor cable for damage or check for correct connection. If damagaed, replace controller.
	Temp cal was used.	Press RESET for 10 seconds.
3. Display on, but none of the function keys work.	Controller defect.	Remove and replace controller.

Fig. 25-43. Troubleshooting guides are normally provided by manufacturers of the thermostat. This excerpt is typical. (Dupont Energy Management Co., Inc.)

25-19 CONTROLLERS — HEATING AND AIR CONDITIONING

A controller is the group of controls and circuits used to accurately and automatically operate a device. The various components in a controller are divided into three categories. See Fig. 25-44. These include:
1. Primary controls.
2. Operating controls.
3. Limit controls.

In the heating and cooling industry, these components are combined to form control systems used to operate:
1. Heating systems.
2. Air conditioning systems.
3. Total energy management systems.

The following paragraphs will explain, in detail, some of the basic controller components. Then they will review the more common systems used today.

25-20 RELAYS

The basic construction and operation of a relay is covered in Chapters 6 and 8. The relay is commonly used as a primary control in a control system. The advantage of using relays is that the current required for operating a system does not have to flow through the controls in that system. Thus, the main circuit can be made as short and as direct as possible. Relays are sometimes called contactors.

The solenoid which creates magnetism to close the points in the heating circuit or circuits may be energized (turned on) by either 24, 120, or 240 volts. Voltage of the thermostat and the relay solenoid coil must be the same. Fig. 25-45 shows a schematic wiring diagram of a relay which controls five separate electric heating circuits and one fan motor circuit.

When the solenoid is energized, it moves (activates) a lever which operates all of the snap switches. The snap switches are calibrated (set) to make and break (close and open) in sequence. The snap switch controlling the fan motor makes (closes) first and breaks (opens) last.

A type of relay commonly referred to in control systems is the lockout relay. The lockout relay shuts down a circuit and keeps it from restarting when any of the safety control devices (motor overloads, high- or low-pressure switches, etc.) have opened. Repairing the problem and closing the safety device will not allow the circuit to restart. The power control circuit to the lockout relay must be interrupted. Fig. 25-46 shows a schematic diagram of a lockout relay circuit.

894

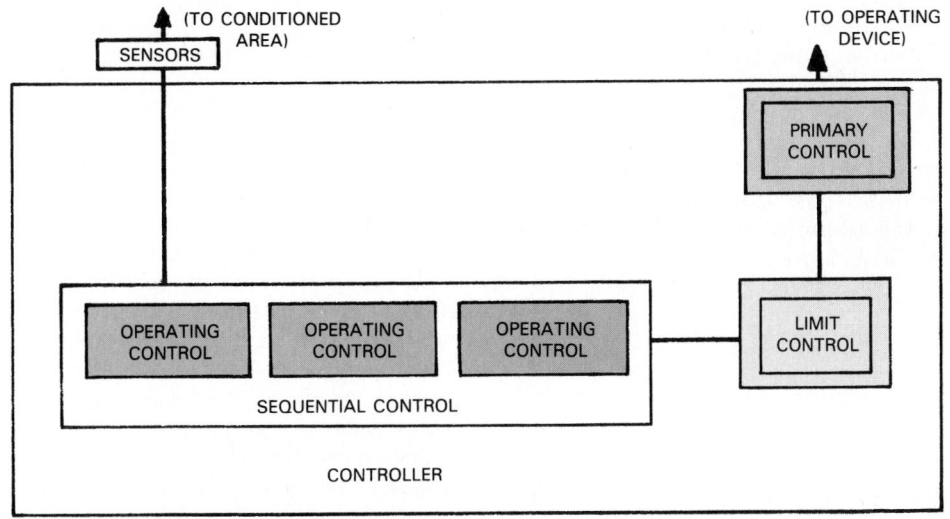

Fig. 25-44. Controller block diagram. The three major control components—operating, limit, and primary—are indicated. A series of operating controls is shown as a sequential control.

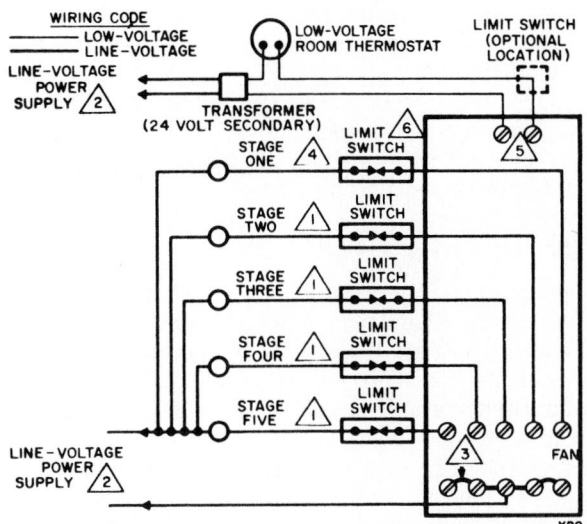

Fig. 25-45. Wiring diagram for relay circuit controlling five electric resistance heaters. In upper left, the "2" indicates the power supply to a step-down transformer. In lower left, "2" indicates line voltage power supply. Solenoid coil connections are shown at "5."

25-21 PRIMARY CONTROLS

Primary controls are devices in a control system which safely turn on and operate the system on command from the operating controls. These primary controls differ from one another depending on the type of heating and/or cooling system used.

The kind of controls used depends on the type of heat energy used: oil, gas, or electricity. Another factor is the kind of heat distribution system used: steam, water, or air. These controls will be explained.

Fig. 25-47 is the wiring diagram for a gun-type oil burner system. Note thermostat, oil burner control, and limit control. An oil burner circuit used with a circulation pump (hydronic system) is shown in Fig. 25-48. The heavy wires are high-voltage (120 V) lines. The light wires are low-voltage 24 V lines.

Fig. 25-49 shows a wiring diagram for a gun-type oil burner installation that is somewhat different. The circulator motor is turned on by the room thermostat when it calls for heat.

Gas furnaces use many varieties of electrical devices.

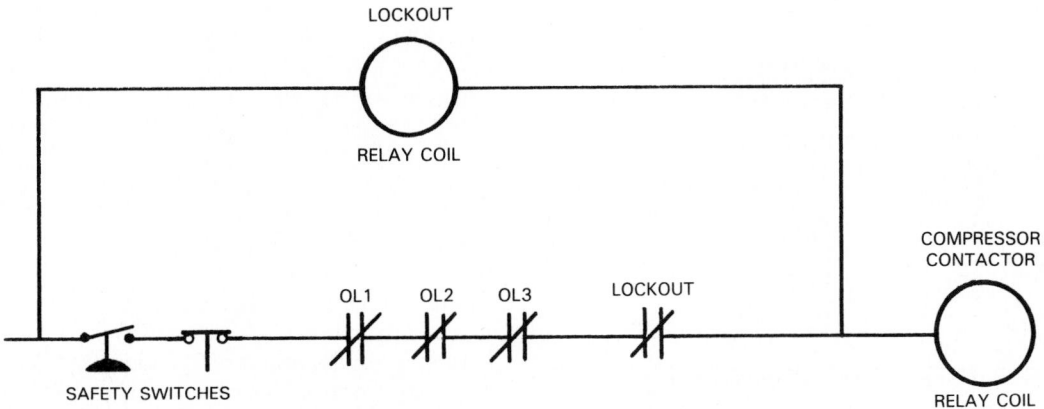

Fig. 25-46. Schematic diagram is for a lockout relay circuit. It closes whenever a safety switch opens.

Some operate on 120 V, some on 24 V, and some on current generated by a thermocouple (25 to 700 millivolts).

The wiring diagram in Fig. 25-50 is designed for a solenoid-operated gas line valve. It is installed on a hydronic heating system.

Wiring diagram as shown in Fig 25-51 is used on a system with a diaphragm gas valve. Circulator motor operates all the time the burner is on. Fig. 25-52 shows another circulator motor wiring arrangement.

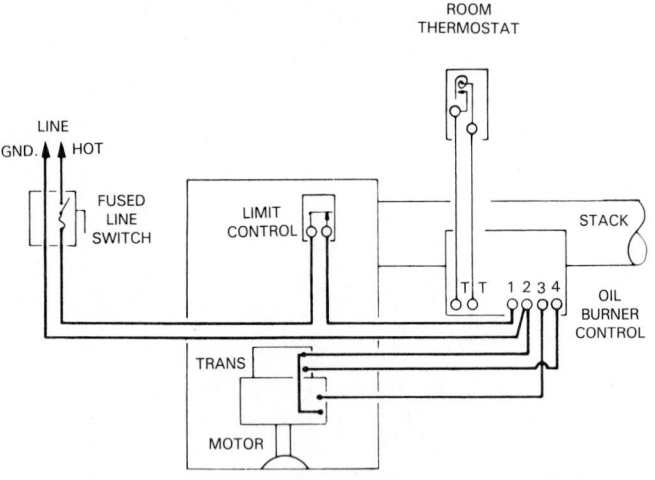

Fig. 25-47. This wiring diagram represents a control system for a gun-type oil burner. (White-Rodgers Div., Emerson Electric Co.)

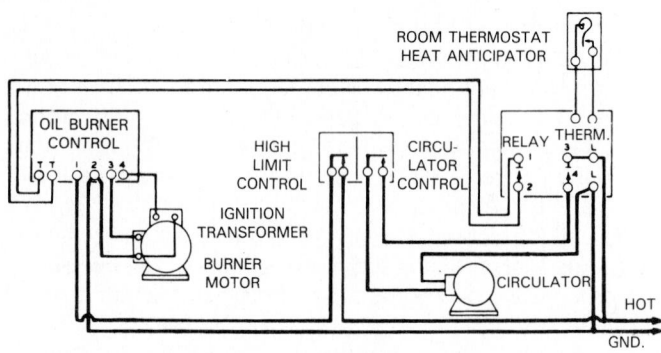

Fig. 25-48. Wiring diagram for gun-type oil burner, hydronic installation. Separate thermostat controls water circulating pump.

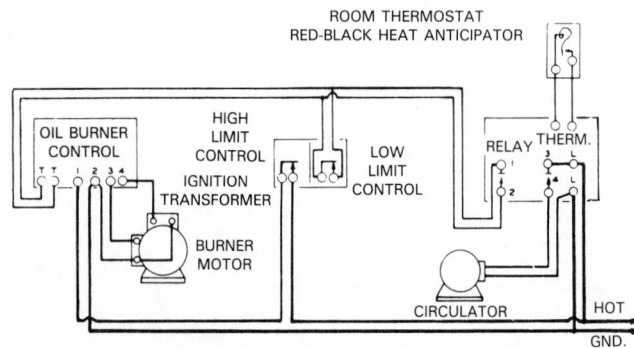

Fig. 25-49. Wiring diagram for gun-type oil burner used in a hydronic installation having a motor-driven circulator pump. Room temperature controls hydronic pump operation.

25-22 OPERATING CONTROLS — SEQUENTIAL

An operating control is the device used to control the many start-ups and stops within a heating or cooling cycle. A sequential control is a series of operating controls used in a preset order when starting or stopping a heating or cooling cycle.

Sequential controls are used to vary heat input during the heating season or to vary cooling capacity during the cooling season. Large electrical resistance heating units usually have several heating coils. Such high kilowatt loads should never be placed on electrical service lines all at once; it might cause a voltage drop and flickering lights, radio interference, or TV flicker. Sequenced controls, when heat is called for, will close the circuit to one heating element or grid at a time until all of the heating circuits are functioning. This is called "sequencing."

The electrical controls of an electric boiler are shown in Fig. 25-53. Devices used to sequence multiple heating elements include:

1. Thermal element switches. The switches stay open until one element heats up; then the switch for the next element closes. This action continues until all the circuits are operating.
2. Relays. As each one closes, it activates the relay for the next heating element.
3. Timers. As they rotate through a series of contacts, circuits are closed one at a time.

25-23 LIMIT CONTROLS

A limit control is generally a safety feature used to maintain the maximum and minimum conditions that are acceptable within a system. Usually, these controls, upon detecting an unsafe condition, will either shut the system down or revert to some safe condition. They are also frequently used to activate a warning device.

For example, heating devices must have built-in safety features to limit the highest temperature the device may reach. If, for any reason, the electric heating baseboard chamber, the hydronic system, or furnace plenum chamber reaches a dangerous temperature, pressures created in the limit switch element will open the switch to protect the equipment and the building. Fig. 25-54 shows a wiring diagram of an electric heating element, thermostat, and limit-control circuit.

One type of limit control mounts the temperature-sensitive element the full length of the baseboard chamber. Fig. 25-55 shows such a safety switch. The limit control is usually set to open the circuit at about 20 to 40°F (11 to 22°C) above the highest thermostat setting.

In warm air furnaces, the limit control is generally a coiled bimetal strip. It is connected in series with the fuel controls. Located in the plenum chamber, it is usually adjusted to open the circuit if the furnace bonnet or plenum chamber reaches about 190°F (88°C). It cuts in again when the system cools to about 160°F (71°C).

It will have a fixed differential in most cases. The cutout setting is usually about 40°F above the cut-in temperature of the fan control. The controls have either a 25°F (14°C) or a 35°F (19°C) nonadjustable differential.

Hot water (hydronic) limit controls are usually mounted in the boiler. The thermal element is in the boiler water.

Some sensing devices are bimetal and some have a hydraulic sensing element. These controls usually have a fixed differential of about 5 °F (3 °C) and a range of 100 °F to 225 °F (38 °C to 107 °C). In most instances they are adjusted to about 150 °F (66 °C).

Water-temperature limit controls are connected in series with the thermostat. They open the circuit when the temperature rises to the cutout setting.

Steam systems usually use a pressure limit control. The control is connected by a pipe fitting to the top of the boiler.

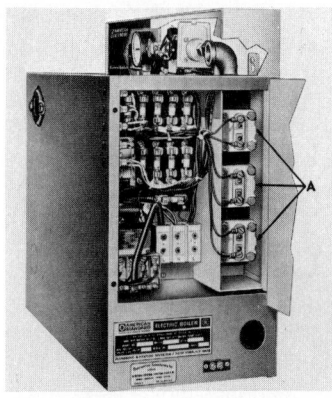

Fig. 25-53. Panel is opened to show the electrical controls of an electrically heated boiler. A—Heating element.

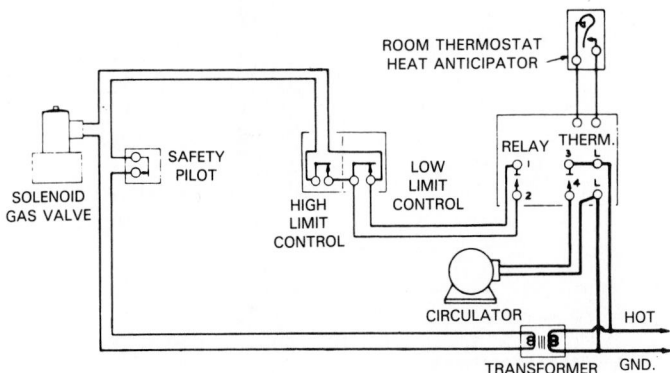

Fig. 25-50. Solenoid-operated gas line valve. This is a low-voltage type. Note step-down transformer at lower right-hand corner.

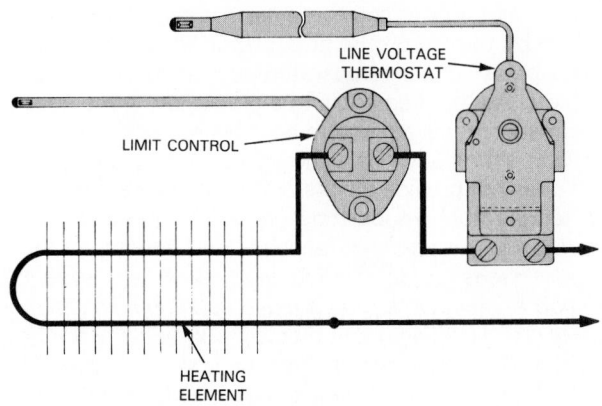

Fig. 25-54. Diagram of electric resistance heating circuit. Note that thermostat and limit control are connected in series with power line.

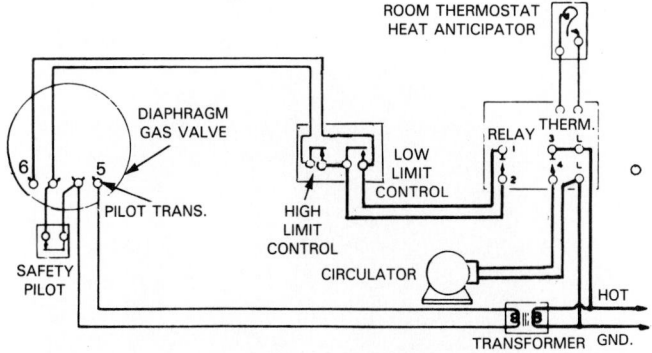

Fig. 25-51. Wiring diagram for diaphragm-type control on gas fired furnace. Note that room thermostat controls water-circulating pump. System uses low-limit control. Terminals 5 and 6 are the inputs to the gas valve operating mechanism.

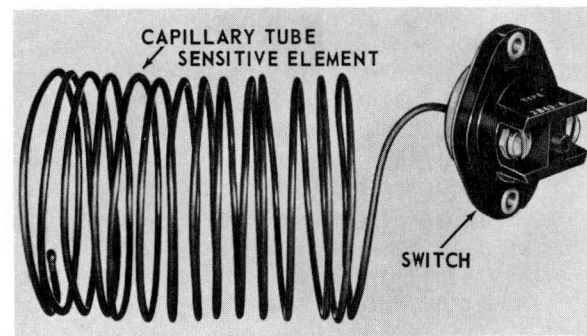

Fig. 25-55. Safety limit switch may be connected in series in electric heater circuit. If temperature exceeds safe limit, switch will open circuit.

When the steam pressure in low pressure systems rises to between 0 psi and 10 psi, the operating circuit is opened and the boiler cannot operate until the pressure drops. Cast iron steam boilers usually operate between 2 to 5 psi. One psi = 6.89 kPa.

25-24 FEEDBACK CONTROLS

Feedback is the information a controller uses as an indicator that a particular operation has taken place.

Information concerning the conditioned area is always

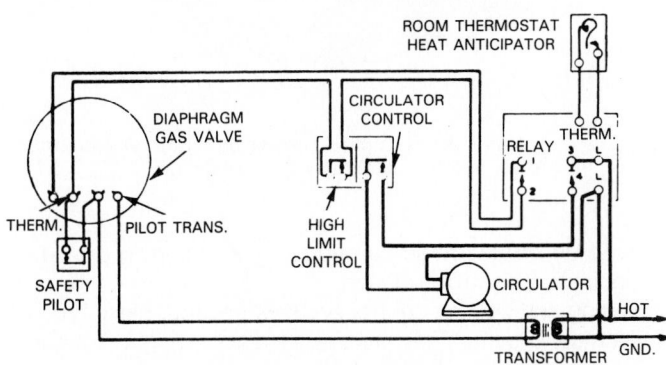

Fig. 25-52. Wiring diagram of diaphragm-type gas valve heating control. Heavy lines represent line voltage; light lines represent low voltage from step-down transformer.

being monitored by the controller's sensors. This information is then compared to the desired conditions that are set in the controller. It is this information about the conditioned space that is fed back to the controlled device. An example is the signal sent to a damper from a thermostat indicating that an adjustment must be made to correct the actual temperature.

Feedback information usually results in one of three actions within the control system. These can be categorized by the type of feedback involved:

1. Positive—the feedback confirms that the control system is maintaining the proper condition and the system will continue to operate without change.
2. Negative—the feedback indicates to the control system that its operation must be modified to maintain a programmed condition.
3. Abort—a type of negative feedback. The control system will revert to a known safe condition or shut down.

In most air conditioning systems, the conditioned space is heated by turning on the heat source to warm the area. When a preset upper temperature is reached, the heat is turned off. As the area cools to a preset low temperature, the heat source will again turn on. This on-off type of control action alternately heats the area and allows it to cool, which does not provide a constant temperature. For a constant temperature, modulating controls are needed.

Modulating controls use feedback and constantly make smooth changes of a continuous type in the system's operation to maintain a programmed condition. Modulating controls provide a heat input that is proportional to the heat loss. This is a proportional type of control action. This modulating control will maintain a constant temperature in the conditioned area. This type of control may be applied to either heating or cooling. To modulate means to vary as follows: 0, 1, 2, 3, 4, 5, etc. Modulation is contrasted with a "0, 1, 0, 1" or on-off cycle. Solid-state electronics make possible this accurate modulating of temperature.

The modulator is usually controlled by a thermostat, a remote bulb and bellows or diaphragm, pressurestats, or humidistats. These controllers, in most cases, operate a variable potentiometer. The variable resistance of the potentiometer is operated by the controller.

The modulator motor usually combines a reversible capacitor motor with a balancing relay, a feedback potentiometer, and a gear train. See the schematic wiring diagram of the motor control in Fig. 25-56. A three-wire thermostat controls the current fed to the balancing relay.

Motors are also designed especially to move only to two separate positions. These two-position motors use cams to open and close switches when the driven mechanism is operated. For such use, motors are usually equipped with electric brakes. They are operated by the motor position switches and lock the rotor in position. Fig. 25-57 shows a solid-state controlled-positioning motor.

25-25 CONTROL CIRCUITS

A typical control circuit will use many of the components of electrical and electronic circuits covered in Chapters 6 and 8. These components are combined to form primary, sequential, limit, and feedback parts of the total circuit. A block schematic of a typical home heating and air conditioning control system is shown in Fig. 25-58. Each block

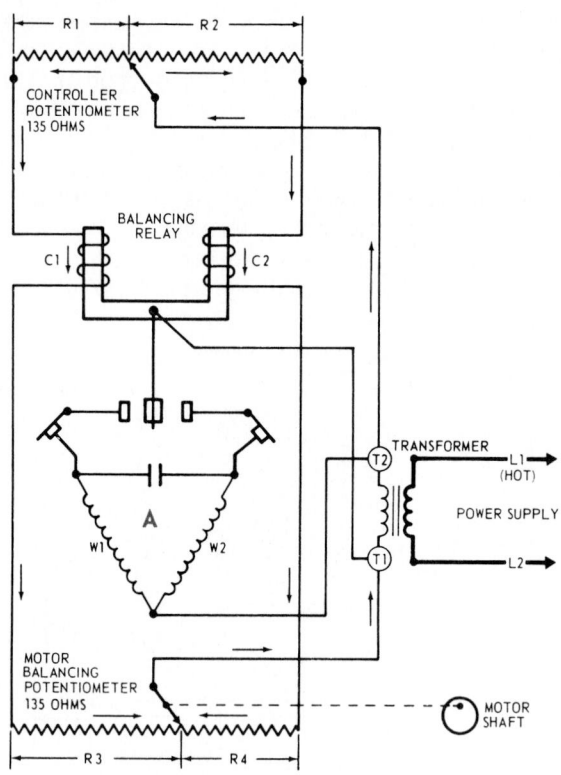

Fig. 25-56. Wiring diagram for modulating electric motor used for controlling valves and dampers. A—Motor windings. (Honeywell, Inc.)

Fig. 25-57. A solid-state controlled-positioning motor. Note motor shaft at left. (Johnson Controls, Inc.)

is made up of electronic components arranged to provide the indicated control function.

Heating and cooling control systems use four basic types of electrical circuits. They are classified according to voltage requirements:

1. The 120 V circuit. (In a few cases, 240 V as well as multiphased circuits are used for large commercial and industrial systems.)

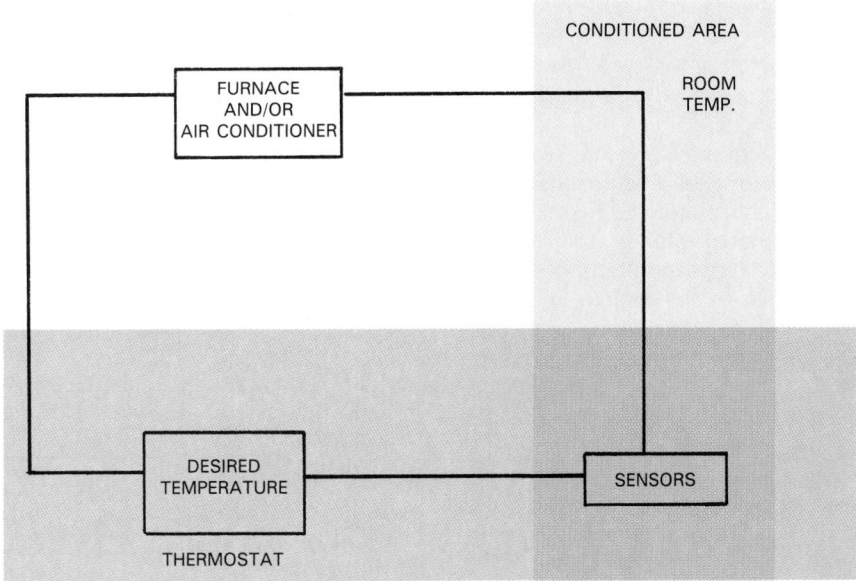

Fig. 25-58. Control system block diagram for a typical home HVAC system. Each block represents specific electronic circuitry.

2. A 24 V circuit. (These circuits often operate relays which, in turn, control the main electrical circuits.)
3. Thermocouple circuits (a few millivolts to about 700 millivolts) used in some pilot light safety devices and in some gas heating systems.
4. Electronic circuits which use approximately 5V to operate components such as: thermistors, diodes, and transistors. These circuits are designed with overload safety devices. These may be fuses and/or circuit breakers.

Smaller motors usually have built-in overload protection. Larger motors use external overload devices usually located in the magnetic starter. *In all cases, the wiring, components, and protection systems found in control circuits must conform to local electrical codes.* Paragraphs 25-26 through 25-38 will present some typical control systems and review their basic circuitry.

25-26 GAS FURNACE CONTROLS

Gas furnaces have primary controls to insure safe starting and safe operation of gas burners. Controls used to operate a gas fuel heating system include:
1. Room thermostat or thermostat with outdoor air-sensing adjustment.
2. Pilot light temperature sensor, such as a thermocouple.
3. Limit controls.

Fig. 25-59 shows three wiring diagrams for a gas furnace. Note location of the limit controls in the line.

These controls allow gas to flow only if the pilot light is burning. A thermal element is located near the pilot light. It must either produce thermocouple electrical energy or develop a sensitive bulb pressure to open valves that allow main gas flow. See Fig. 25-60. This thermal element will also shut off the system if the pilot light stops working during the operating cycle of the unit.

If a blower or water pump motor is used in the distributing system, it is turned on by a relay. Such a relay may have instant action or there may be a short delay.

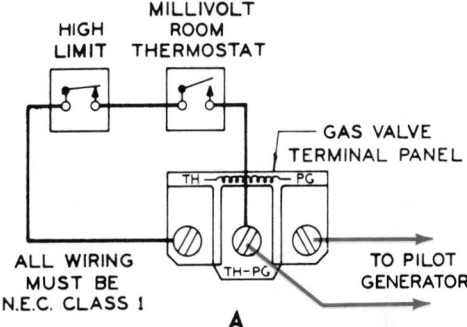

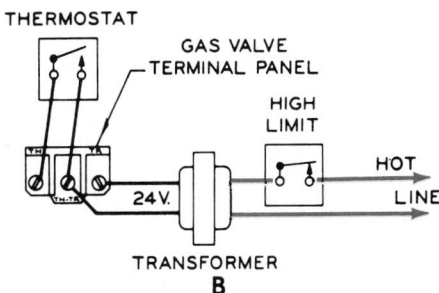

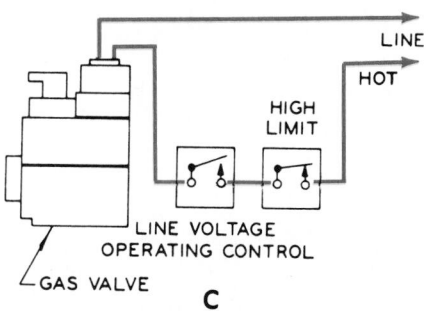

Fig. 25-59. Electrical circuits used with gas controls. A—Millivolt circuit. B—A 24 V circuit. C—A 120 V circuit. A high-limit control is used on each. (White-Rodgers Div., Emerson Electric Co.)

Primary controls operate on a series of electric interlocks. All conditions must be safe before the interlocks are in the correct position to allow the system to operate. Mechanical or thermal sensors are usually considered adequate for the smaller capacity domestic burners.

Commercial and industrial systems use more elaborate controls due to the larger flow of fuel. Flame sensors are generally used. These shut off the system by reacting very rapidly (in parts of a second) should a flame fail. The sensors are electronic. They use a flame rod and a photocell. The photocell is either sensitive to the radiant energy or ultraviolet rays of the flame. Some installations use a lead sulphide cell which responds to the infrared rays from the gas flame.

Electric ignition for gas systems is used when the gas furnace is located outside the building or is in a hard-to-reach position. In this system, an electrical spark ignites the gas at the main burner. It also automatically shuts off the gas supply to the main burner if the main burner does not ignite. Fig. 25-61 shows an electrical ignition system for a

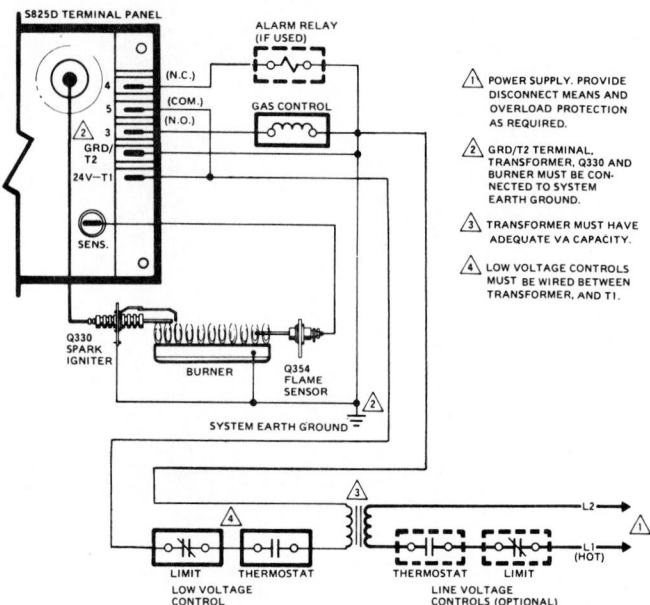

Fig. 25-62. Wiring circuit for gas furnace electric ignition system. "N.C." means normally closed. "N.O." means normally open. (Honeywell, Inc.)

gas burner. The wiring diagram shown in Fig. 25-62 is for such a system.

The main gas line to a gas furnace has four valves:
1. Hand shutoff valve.
2. Pressure regulator.
3. Safety valve operated by the pilot flame, see Fig. 25-63.
4. Automatic gas valve operated by thermostat.

Originally these valves were separate units. About 1957, the pilot safety valve and the automatic gas valve were made into one unit or body.

About 1959, the pressure regulator, the pilot safety valve, and the automatic gas valve were made into one unit or body, mainly for ease of installation, fewer joints to leak, space-saving purposes, and economy. Fig. 25-64

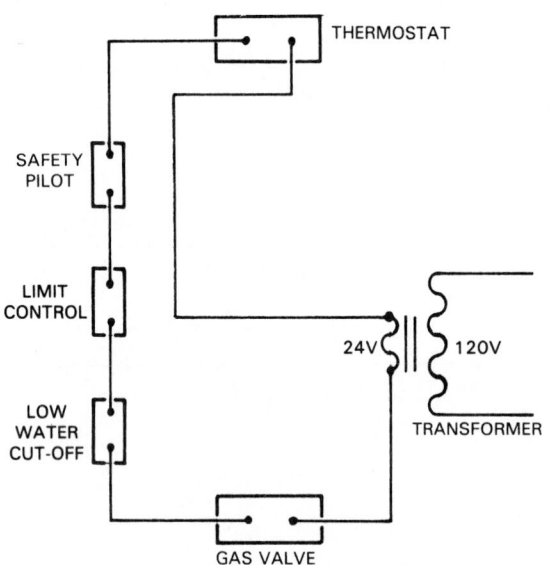

Fig. 25-60. Schematic is for low-voltage electrical circuit used on gas furnace without blower.

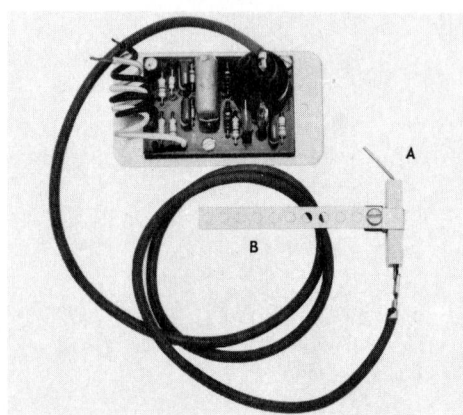

Fig. 25-61. Solid-state electric ignition system for gas furnace burner. A—Spark gap electrodes. B—Mounting bracket.

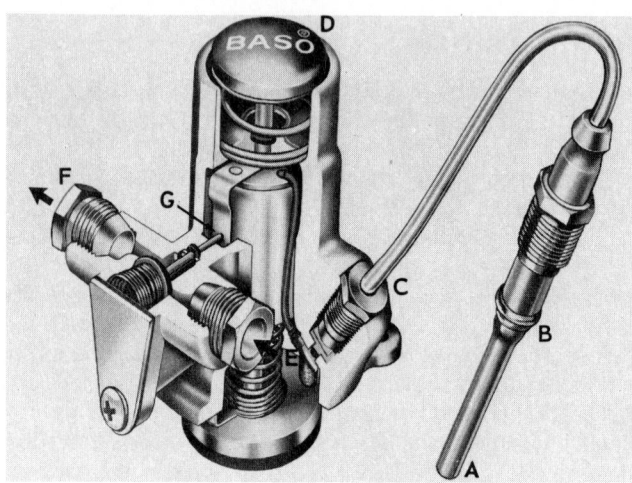

Fig. 25-63. Automatic safety pilot thermocouple construction. A—Thermocouple hot junction, mounted in pilot flame. B—Cold junction and mounting. C—Thermocouple connection to valve body. D—Reset button for pilot gas. E—Pilot flame gas in. F—Pilot flame gas out. G—Pilot flame gas valve operating lever.

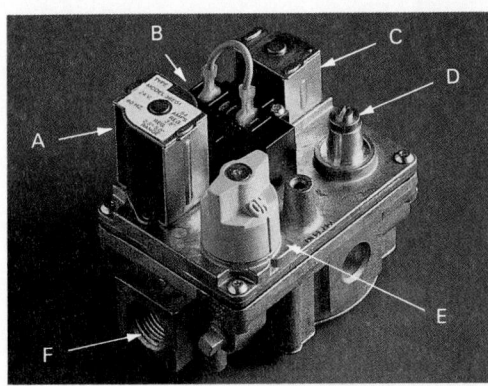

Fig. 25-64. Combination gas valve and automatic pilot. A—Redundant/pilot solenoid. B—Terminal panel combinations. C—Main solenoid. D—Regulator adjusting screw. E—Manual gas knob. F—Gas inlet. (White-Rodgers Division, Emerson Electric Company)

shows a combination valve. This combination gas control is often called a "CGC." Fig. 25-65 is a cross section of the combination gas control.

The automatic gas valve portion is operated by either a solenoid, an electrical resistance-heated bimetal blade, or by a sensing bulb, capillary tube, and bellow combination (hydraulic thermal element).

The advent of electronics technology combined with continuously regulated safety improvement has resulted in many recent advances in the gas furnace control system. Electronic circuits, sometimes used to control heating systems, can vary (modulate) the size of the gas flame. The modulating system uses a solid-state thermostat. It also uses a thermistor and several transistors.

The size of the gas flame depends on a temperature difference between the thermostat setting and room temperature. The flame is larger with a greater temperature difference and gets smaller as room temperature approaches the thermostat setting. The unit starts up the flame at about 20 to 50 percent of capacity; then it adjusts the flame to temperature differences. The system's three main control parts are the thermostat, an amplifier, and a modulating gas valve.

Fig. 25-66 illustrates a wiring circuit used with a makeup air system. The 120 V ac power is reduced to 24 V ac, using a transformer. A rectifier in the amplifier changes this current to dc. Three control devices are used:
1. The remote temperature selector.
2. The discharge air sensor.
3. The duct stat (for safety).

These controls signal the amplifier which, in turn, operates the solenoid and modulating regulator on the main gas burner line.

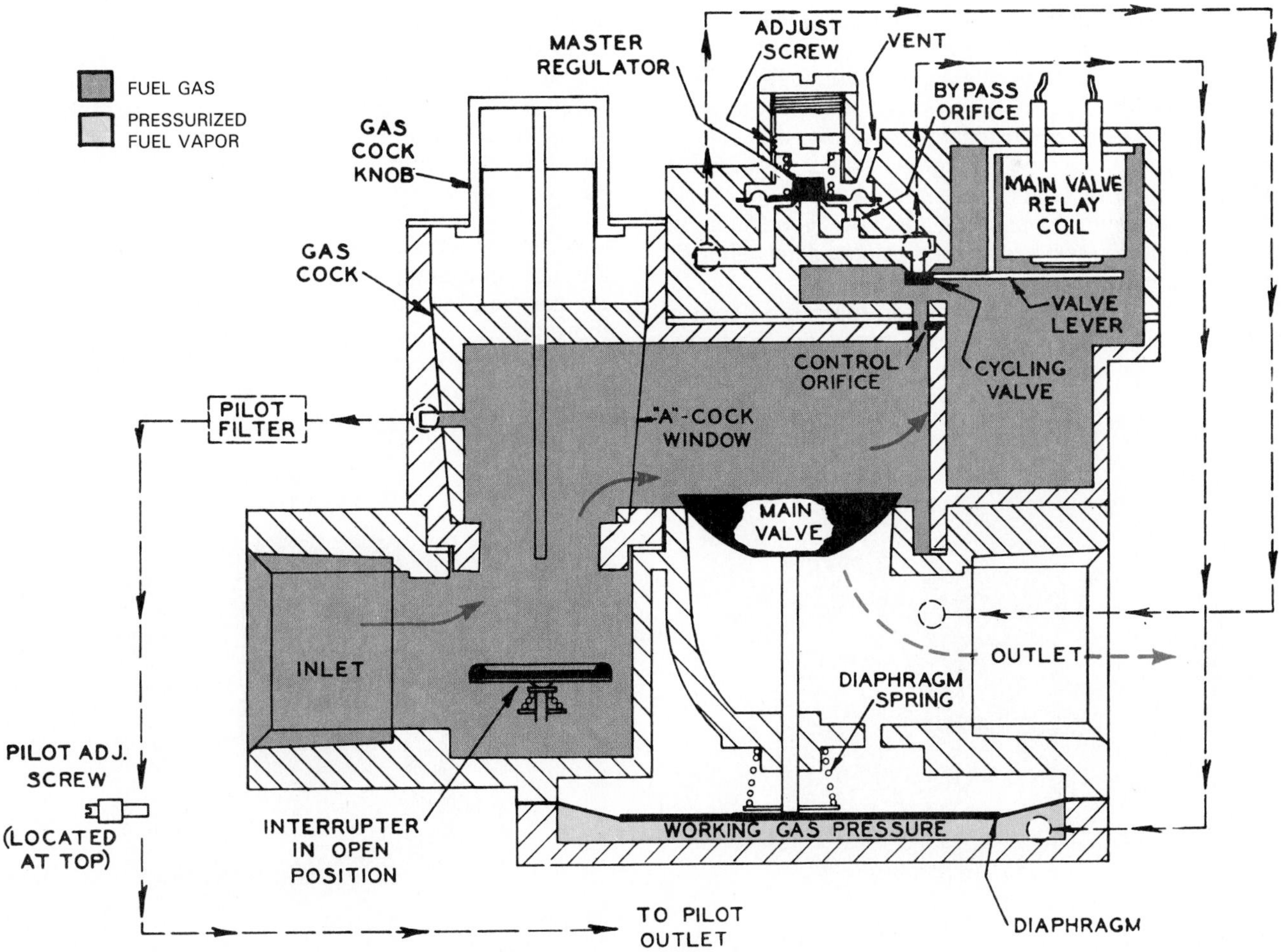

FUEL GAS

PRESSURIZED FUEL VAPOR

MASTER REGULATOR

ADJUST SCREW

VENT

BYPASS ORIFICE

GAS COCK KNOB

MAIN VALVE RELAY COIL

GAS COCK

VALVE LEVER

CONTROL ORIFICE

CYCLING VALVE

PILOT FILTER

"A"-COCK WINDOW

MAIN VALVE

OUTLET

INLET

DIAPHRAGM SPRING

PILOT ADJ. SCREW

(LOCATED AT TOP)

INTERRUPTER IN OPEN POSITION

WORKING GAS PRESSURE

DIAPHRAGM

TO PILOT OUTLET

Fig. 25-65. Combination gas control in cross section. It has hand shutoff valve, pilot light control, bypass-operated main valve, and pressure regulator. (White-Rodgers Div., Emerson Electric Co.)

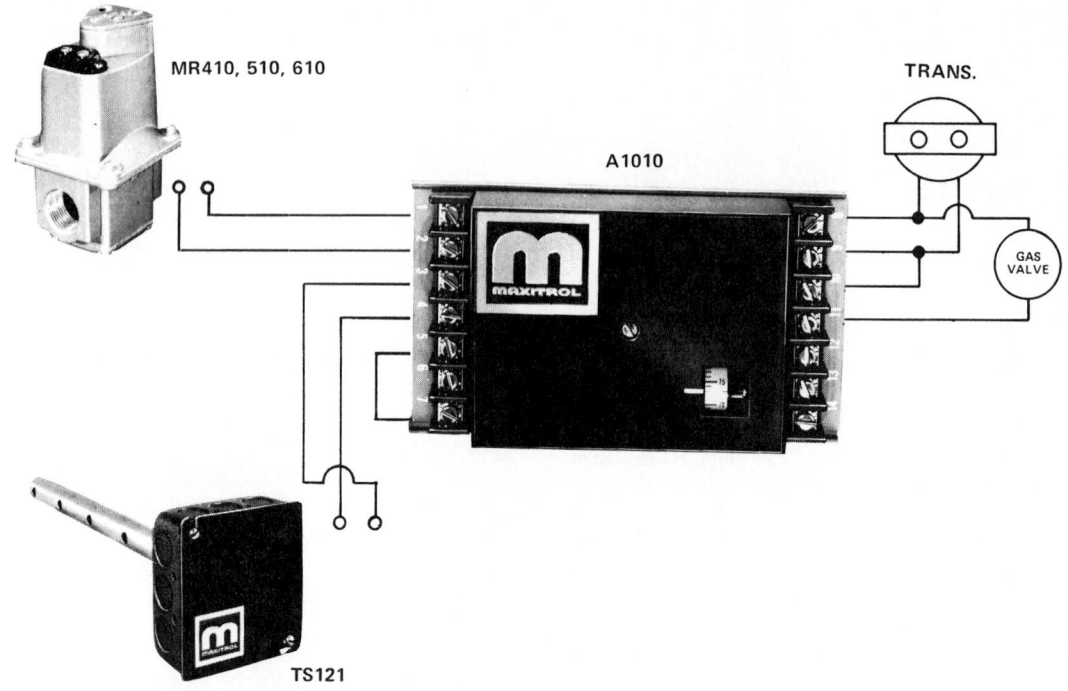

Fig. 25-66. Wiring diagram shows indirect fired makeup air application. (Maxitrol Co.)

Fig. 25-67 shows the amplifier unit. It holds the rectifiers and solid-state components for the control circuits. It also contains an adjustable potentiometer for calibration.

The remote temperature selector is shown in Fig. 25-68. This unit contains an adjustable potentiometer. It sets the temperature level of the discharge air being sensed by a thermistor located in the discharge air sensor, Fig. 25-69.

The modulator/regulator valve is shown in Fig. 25-70. This is the valve which varies the gas flow. In addition, an automatic solenoid valve is needed to completely shut off the fuel supply.

Direct current to the modulator controls the amount of gas flow. The less the current flow, the higher the flame.

The duct stat is connected in series with the solenoid valve and is used as a safety device in case too high a duct temperature is reached.

Space heating is accomplished by combining the remote temperature selector and discharge air sensor into a single wall-mounted Selectrastat, shown in Fig. 25-71.

25-27 OIL FURNACE CONTROLS

In the gun-type oil burner, a primary control usually starts the burner motor. If the flame goes out or if the flue temperature becomes too high, the control will also stop the unit.

Some primary units are mounted in the flue. Upon a signal from the thermostat for heat (closing of points), the primary control will start the gun-type oil burner motor and turn on the ignition.

Fig. 25-67. Solid-state electronic amplifier may be placed at any convenient location. A—Electrical terminals. B—Cover plate. C—Remote temperature selector. (Maxitrol Company)

Fig. 25-68. Remote temperature selector is not temperature sensitive and may be placed in any convenient location. (Maxitrol Company)

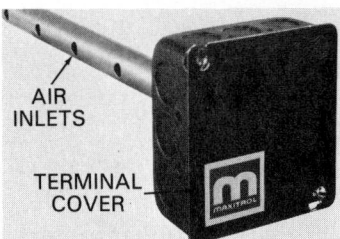

Fig. 25-69. Discharge air sensor. It sends signals to the amplifier of any temperature change from set point. (Maxitrol Co.)

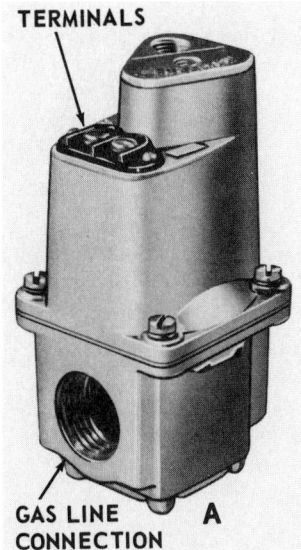

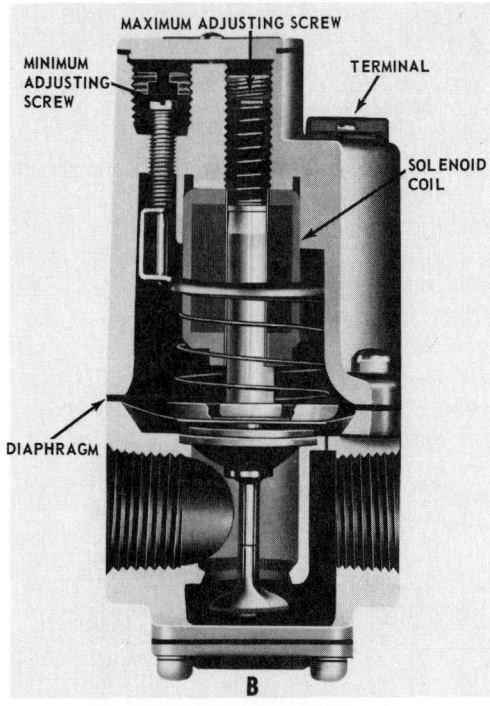

Fig. 25-70. Modulator/regulator valve performs both regulation and modulation to vary burner flame size (or burning rate). A—Outside view. Note gas line connection and electrical terminals. B—Cutaway of same modulator/regulator valve. Note adjusting screws used to vary amount of pressure required to operate valve.
(Maxitrol Co.)

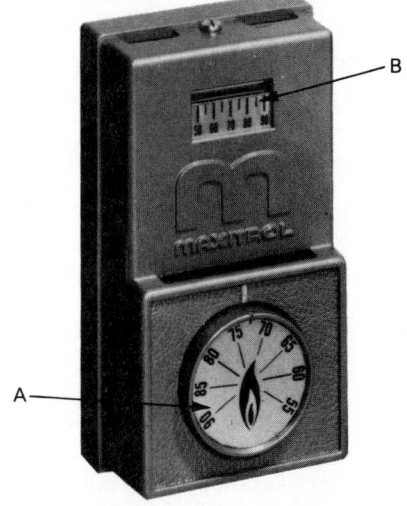

Fig. 25-71. Combination modulating makeup air thermostat and selector. A—Temperature selector. B—Temperature. (Maxitrol Co.)

This control is shown in Fig. 25-72. It has a temperature-sensing element which will shut down the unit unless a fast temperature rise in the flue takes place in a few seconds (indicating that the oil is burning).

This same sensor will constantly check for flame temperatures and will shut off the system if the flame goes out. Or, it will shut off the system if the thermostat, or any one of the limit controls, opens the circuit.

Fig. 25-73 shows an electrical circuit for a gun-type oil burner with continuous ignition. Fig. 25-74 is an intermittent ignition system.

Many oil burners also use a flame-out safety control. The type shown in Fig. 25-75 is one used with large oil and gas burners. It utilizes an ultraviolet flame detector or flame

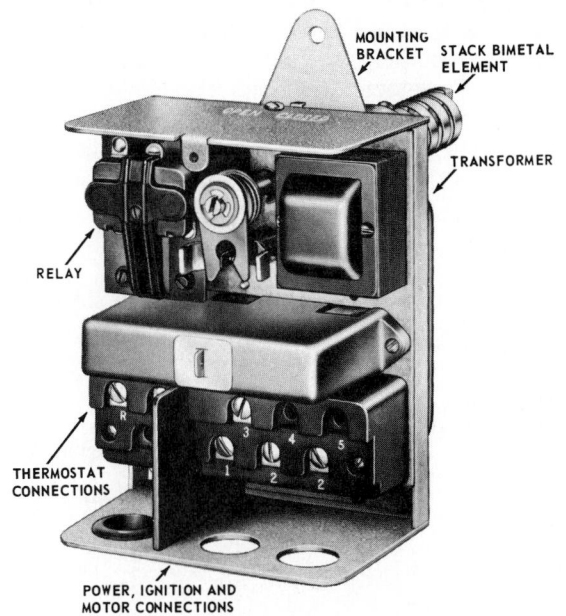

Fig. 25-72. Oil burner primary control. This control will cycle burner, operate electric ignition system, shut off unit if ignition fails, and scavenge unit after each cycle.

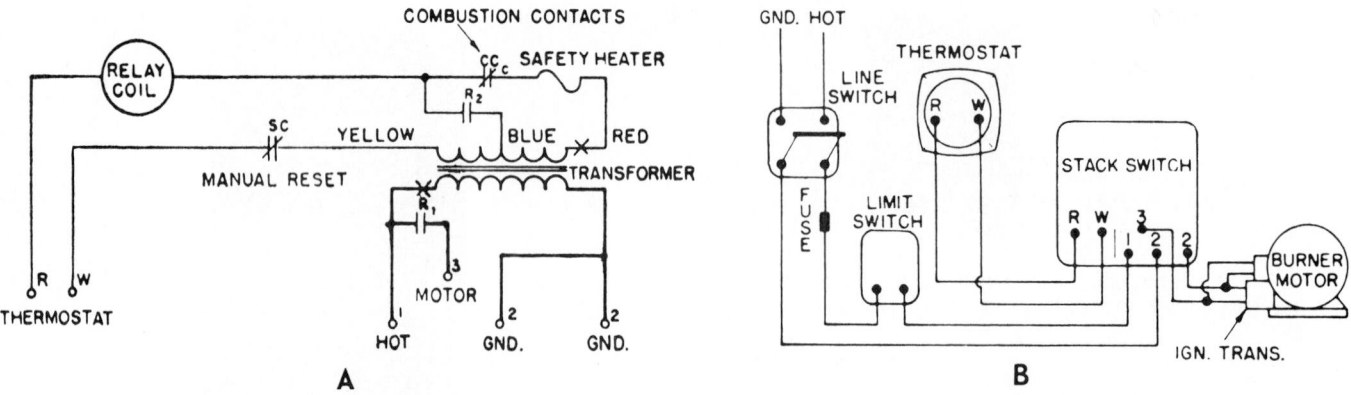

Fig. 25-73. Wiring diagrams of continuous ignition gun-type oil burner stack control. A—Internal wiring diagram. B—External wiring diagram.

rod/photocell detector. If combustion is delayed or the flame goes out, the control safely shuts down the burner and energizes an alarm terminal. The scanner is mounted in the wall of the firebox and is aimed at the oil or gas burner. The circuit for this safety device is shown in Fig. 25-76.

New models of oil burners use solid-state controls. Ignition (electrical) may be either continuous or intermittent. Fig. 25-77 shows a gun-type oil burner primary control. One model operates on low-voltage input while another model operates on line voltage. Both controls use a low-voltage photosensitive flame detector. When the thermostat calls for heat, the gun oil burner motor turns on, the oil pump starts, and the ignition is turned on. If the fuel does not ignite in about 40 seconds, the flame detector stops the burner motor and closes the solenoid valve, if one is used. Fig. 25-78 shows the circuits.

A solid-state primary control, as shown in Fig. 25-79, can be used as a replacement for older model primary controls. *The technician must be absolutely certain that the power is off before installing the unit.*

The control is wired as shown in Fig. 25-80. The cadmium cell must be very carefully mounted. The cell must "see" the flame as well as be heated over 140 °F (60 °C). The correct mounting is shown in Fig. 25-81. Another photocell flame detector mounting is shown in Fig. 25-82. The flame detector must be lined up with an opening in the static pressure disk to be able to "see" the flame.

The transformers have a primary winding of 120 V, 240

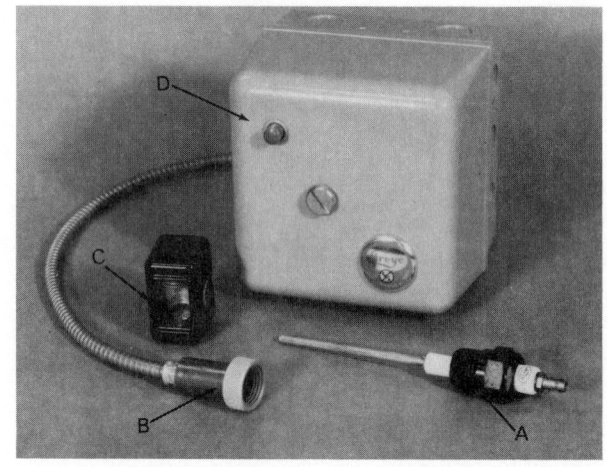

Fig. 25-75. Flame-out safety control used with large oil or gas furnace. It is operated with ultraviolet rays. A—Flame rod. B—Scanner. C—Photocell. D—Control.
(Allen Bradley Co., Presence Sensing Products, Fireye Division)

V, or 208 V. The secondary winding usually provides 10,000 V. Some systems use 12,000 V.

It is recommended that, when 10,000 V transformers need replacement, a 12,000 be used. This is advisable especially in cold air or cold oil situations or when line voltage drops are known or suspected.

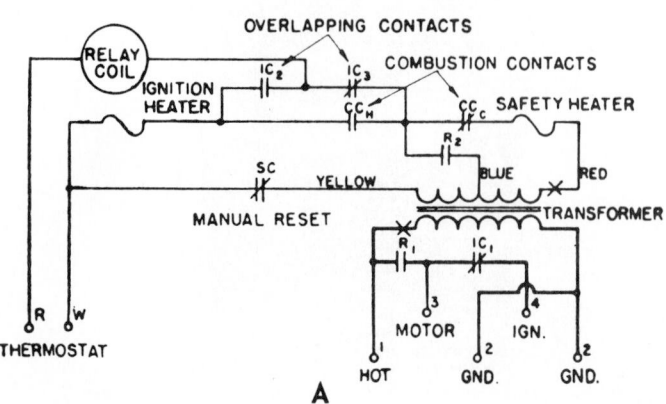

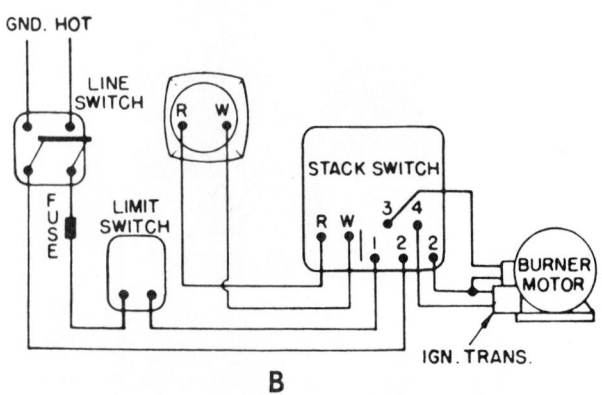

Fig. 25-74. Internal and external wiring diagrams of an intermittent ignition system used on a gun-type oil burner stack control. A—Schematic wiring diagram for intermittent ignition stack control. B—External wiring diagram for same control.

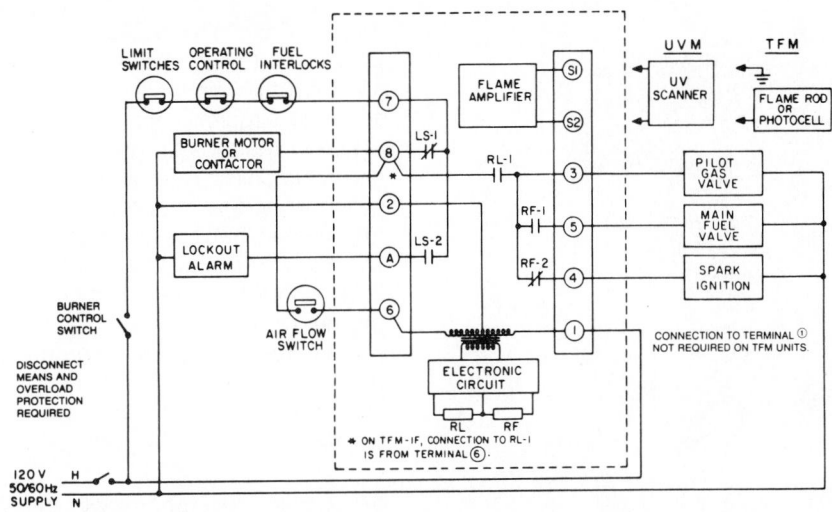

Fig. 25-76. Wiring diagram of a typical pilot-ignited oil or gas burner using ultraviolet-sensitive safety device. (Allen Bradley Co., Presence Sensing Products, Fireye Division)

Fig. 25-77. Low-voltage solid-state gun-type oil burner primary control. It has a continuous ignition system and a photosensitive unit for flame detection. If ignition fails, the flame detector will shut off the system in two seconds. A—Photocell. B—Leads to furnace.

25-28 ELECTRIC HEAT CONTROLS

Baseboard units usually have individual thermostats. Primary controls are most often relays and limit controls. Units using blowers or fans have fan controls. Such controls may operate from a separate thermostat or may be connected in parallel with the heating element.

Central systems normally use sequence relays as primary controls. Blowers operate the same as on baseboard units. In addition, a safety control is usually provided which will shut off the heating elements if the blower fails to operate or if air fails to circulate.

25-29 INFRARED HEAT CONTROLS

Electric infrared heat lamps may require as much as 32 kilowatts (kW). The heating unit may have as many as 16, 2000 W lamps (32,000 W). The electric circuit, in such

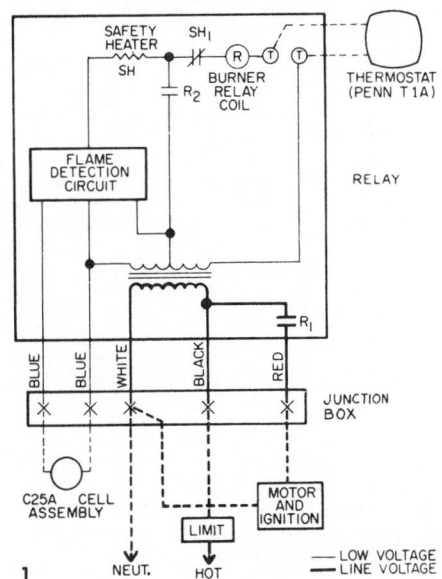

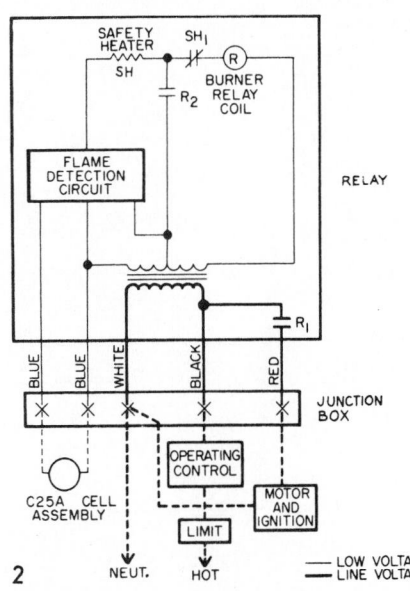

Fig. 25-78. Wiring circuits for solid-state control equipped with photosensitive flame detector. 1—Low-voltage system. 2—Line voltage system.

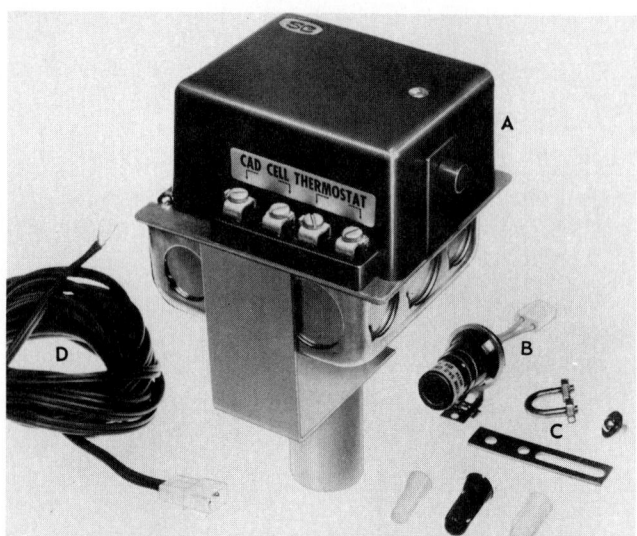

Fig. 25-79. Solid-state primary control for gun-type oil burner. A—Primary control. B—Cadmium cell flame detector. C—Cadmium cell mounting bracket and fittings. D—Connecting lead.

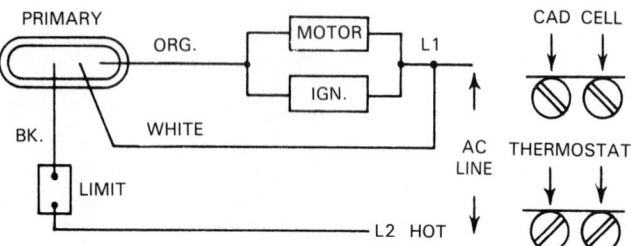

Fig. 25-80. Schematic wiring diagram for solid-state primary control for gun-type burner. Note that current is supplied to motor and igniter at same time.

cases, is controlled by a large-capacity relay (contactor). At 240 V ac, the line capacity will need to carry 175 A. The 175 A main circuit is usually divided into four or more separate circuits, each with a relay (contactor).

Either thermostat or solid-state sensors energize the operating coil of the contactors. The contactors are usually sequenced so that only one closes at a time. (Sequencing means that the contactors are set to close one after the other — not all at once. This is usually done by a time-delay on each contactor switch which operates the coil of the next contactor.)

Solid-state temperature controls modulate the current flow with thermistors and triacs. They reduce part of each sine wave of the ac flow to maintain a constant temperature. See Chapter 6.

25-30 HYDRONIC SYSTEM CONTROLS

The primary control feature (in addition to the temperature control) in a water or hydronic heating system is the water level. Water level controls are especially important on steam heating systems.

The control is generally a switch turned on and off by a float. If the water level drops near the danger point, the float drops far enough to open the electrical circuit to the operating controls. This stops the boiler.

The switch generally operates in two stages. Lowering of the float will trip a switch to turn on the feed water pump or feed water solenoid valve. If the float drops still more, the system is shut off.

Some water level controls are of the probe type. This type immerses two electrodes in the water. As long as water covers both electrodes, a small current flowing in the water between the two will energize a holding relay and allow the system to run. If the water level falls below the upper probe,

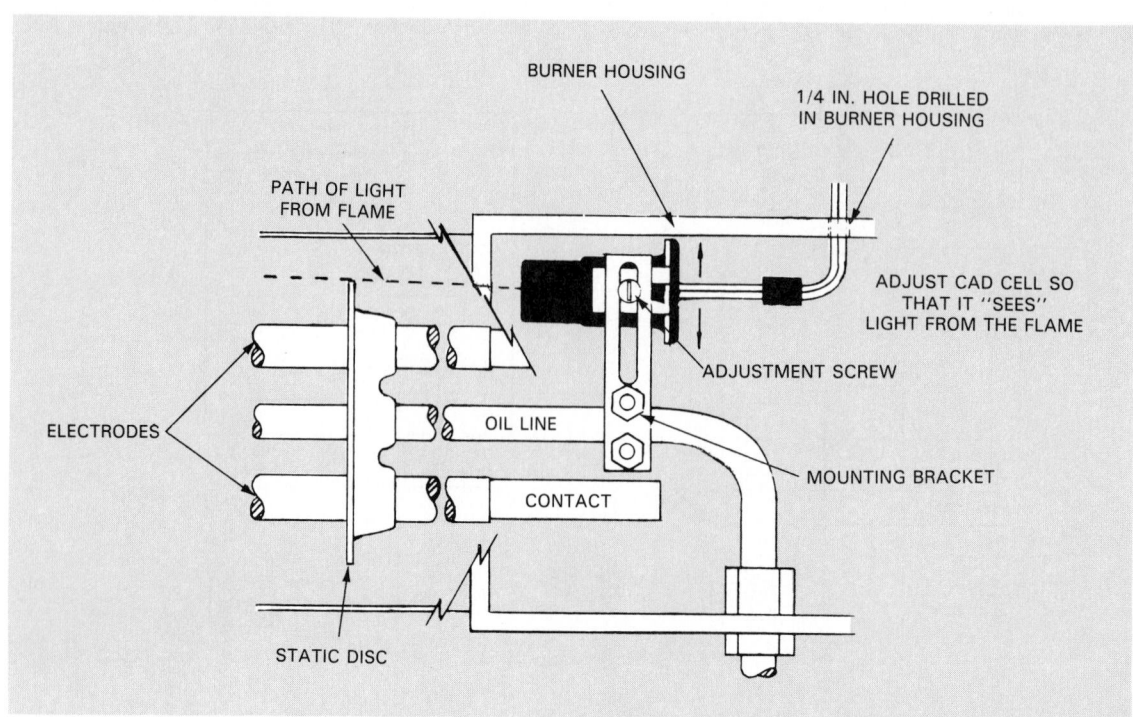

Fig. 25-81. Method of mounting photoelectric flame sensor in gun-type oil burner.

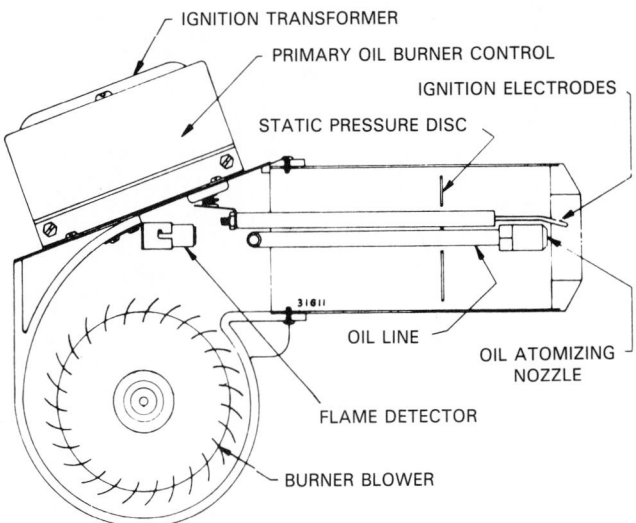

Fig. 25-82. Flame sensor detector mounted in gun-type oil burner. (White-Rodgers Div., Emerson Electric Co.)

the current flow will cease and the operating controls will shut down.

The operating principle of these two water level controls is shown in Fig. 25-83. Avoid air in the system.

25-31 COMFORT COOLING CONTROLS

Controls for comfort cooling are of the same basic types as those in heating. There are operating controls, primary controls, and limit controls.

The operating controls are thermostats, pressurestats, and humidistats. Primary controls include motor starters and starting relays. Limit controls are overload circuit breakers, thermal overloads, internal motor overloads, refrigerant pressure limit controls, and oil pressure limit controls. Most of these controls are described in Chapters 8 and 12.

The two most popular refrigerant controls are the thermostatic expansion valve (TEV) and the capillary tube. These controls are described in Chapters 5 and 12.

The schematic diagram in Fig. 25-84 is for a circuit used in a comfort cooling unit. This system uses a low-voltage, two-wire thermostat. The thermostat controls a relay which will close the motor circuit. If the motor cannot be connected directly to the line, and if pressure safety devices are to be put in the system, the wiring will be somewhat like that shown in Fig. 25-85. The high-pressure safety cutout is wired in series with the starter coil. It will open the circuit if pressures become too high.

Some systems cycle on command from a low-side pressure control. The thermostat operates a solenoid valve mounted in the liquid or suction line. When the thermostat temperature is satisfied, the solenoid valve will close. When the low-side pressure drops enough, the motor circuit will be opened by means of the pressure control connected to a magnetic starter, as shown in Fig. 25-86. The unit will then stop. A high-side switch is also provided in this control. This control will stop the compressor in the event the high-side pressure exceeds a preset limit.

The controls found in comfort cooling systems are:
1. Thermostat (24 V or 120 V) or thermistor sensor.
   a. One-stage.
   b. Two-stage.
2. Evaporator icing control (freeze-up control); two-wire 24 V or 120 V to control dampers, valves, or compressors when evaporator approaches 32 °F.
3. Multiple compressor sequence starting controls.
4. Multicylinder compressor unloading sequencing controls.

Generally, large air conditioning evaporators use the thermostatic expansion valve for refrigerant control. Some installations use several such valves on one large evaporator to get maximum efficiency. Self-contained systems — especially those hermetically built — may use the capillary tube refrigerant control. These controls are described in Chapter 5.

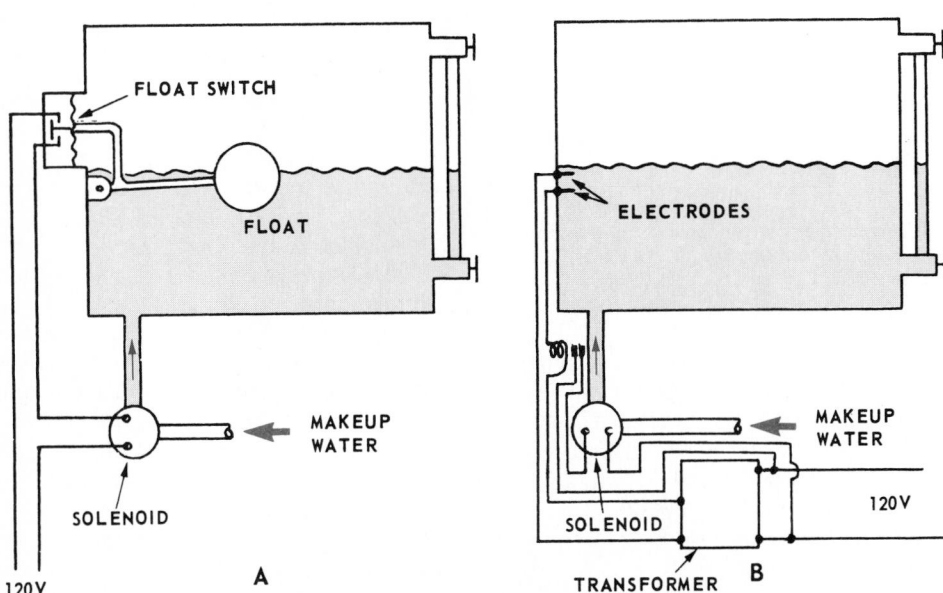

Fig. 25-83. Two types of water level controls. A—Float control. Float valve will open makeup water valve when water level drops. Some valves are connected to electric switches which will shut off the unit if water level gets too low. B—Probe-type water level control. Solenoid opens when current stops traveling across electrodes.

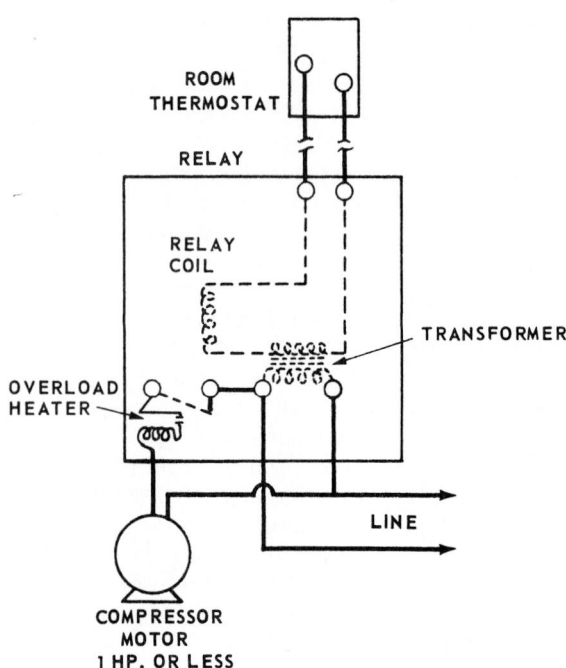

Fig. 25-84. Wiring diagram of thermostat relay combination for small comfort cooling units.

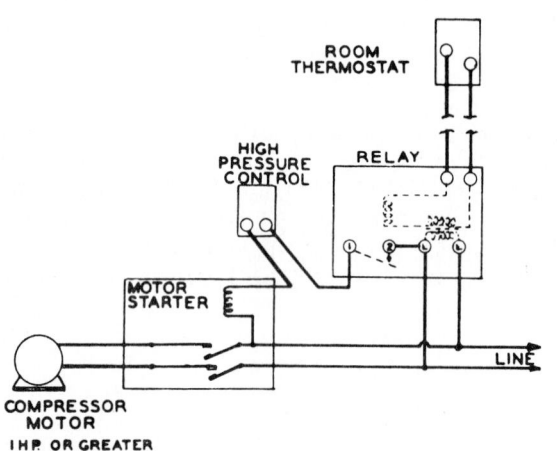

DIAGRAM USING A LARGE REFRIGERATION UNIT.

IF THE INSTALLATION HAS A BLOWER, IT IS USUALLY WIRED TO RUN CONTINUOUSLY DURING THE COOLING SEASON.

Fig. 25-85. Wiring diagram is for a comfort cooling unit which uses high-pressure safety cutout and motor starter. (White-Rodgers Div., Emerson Electric Co.)

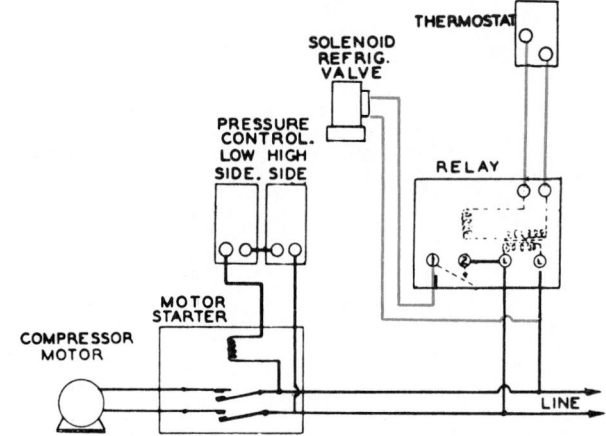

Fig. 25-86. Comfort cooling system wiring diagram. Note that thermostat operates solenoid valve. System cycles as low-side pressures vary.

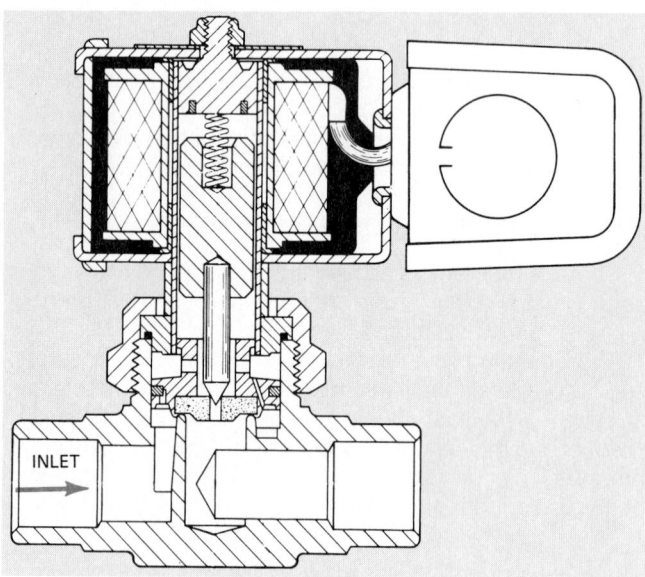

Fig. 25-87. Cross section of solenoid refrigerant control valve used in liquid lines on air conditioning systems. (Sporlan Valve Co.)

In addition to the refrigerant control used on automatic refrigeration systems, a solenoid refrigerant valve is sometimes placed in the liquid line. This automatically stops the flow of refrigerant to the evaporator:
1. The instant the condensing unit stops.
2. When the low-side pressure control opens.
   This is done in order that the evaporator:
1. Will not become flooded with refrigerant while the condensing unit is idle.
2. Will be pumped down.
   Fig. 25-87 illustrates a solenoid refrigerant control valve. This valve uses 5 W at 120 V ac. It has a .100 in. diameter

orifice. It has a refrigerating capacity, at a 5-lb. pressure drop across the orifice, of 1.3 ton for R-12, 2.1 ton for R-22, and 1.62 ton for R-500.

In addition to the operating controls (thermostats), comfort cooling systems have primary controls and limit controls.

Larger refrigerating units are usually equipped with pressure controls. The low- and high-pressure controls are usually designed to lock the circuit open if any unusual pressures occur. The operator must then manually turn the system on. This allows a careful check for faults.

The internal construction of a combination low- and high-pressure control is shown in Fig. 25-88.

Comfort cooling systems use several types of limit controls:
1. Motor limit controls.
2. Pressure limit controls.
3. Temperature limit controls.
4. Fluid flow limit controls.

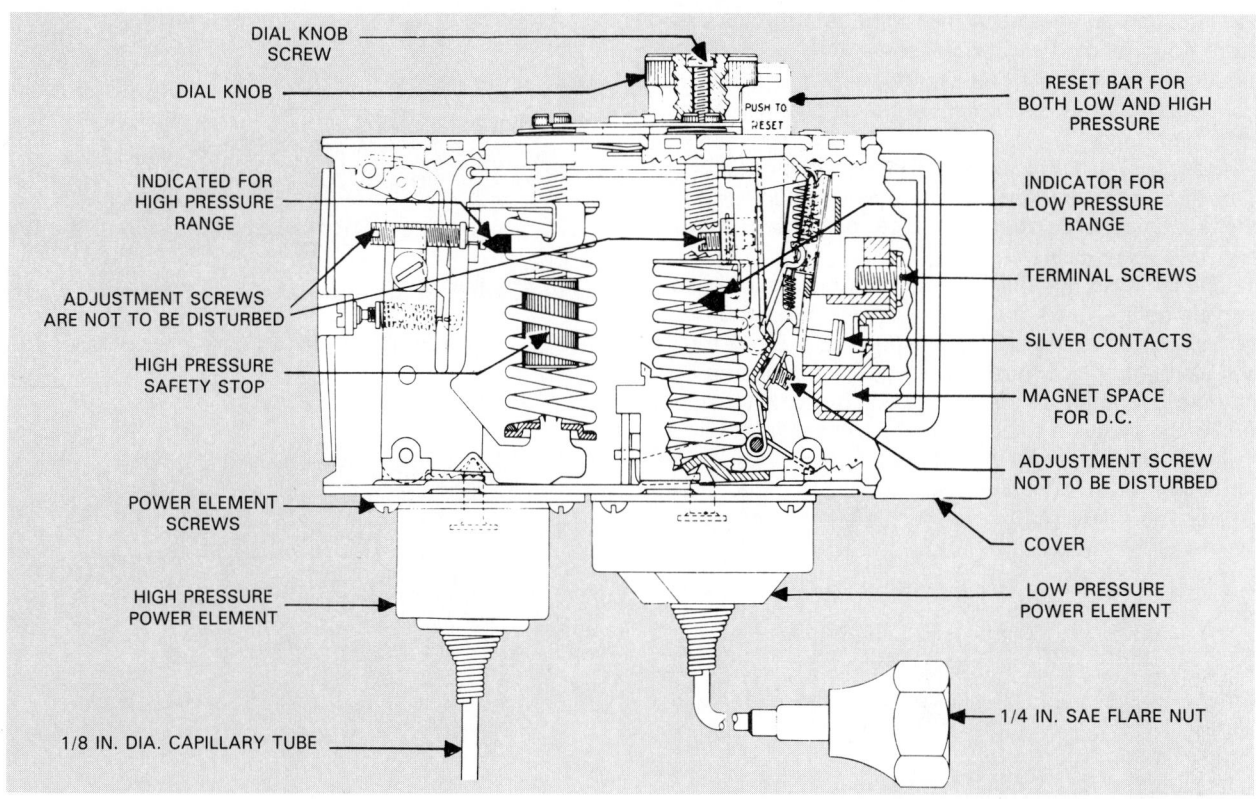

Fig. 25-88. Inner mechanism of air conditioning condensing unit pressure motor control. (Ranco North America)

Motor limit controls are described in Chapter 8. Such controls will stop the unit if current draw becomes too high or if motor temperature rises to dangerous levels.

Another type of temperature limit control (other than the motor thermistor or bimetal protector) is an anti-icing control on the evaporator. Should ice accumulate there, this control will open and stop the system.

Fluid flow controls stop the system in case chilled water

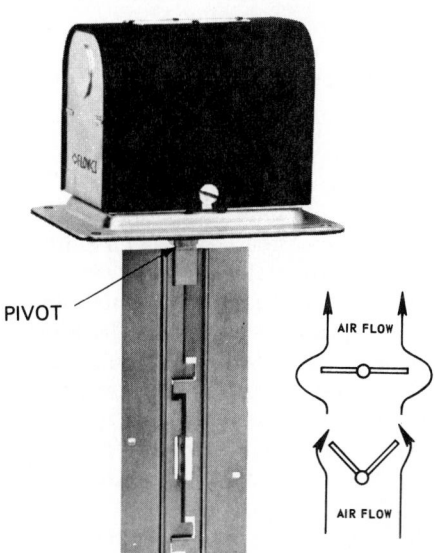

Fig. 25-89. Airflow signal switch prevents damage to system from too-high air velocities. If velocity gets too high, it will move the paddle, tripping the electrical switch. (ITT McDonnell & Miller)

flow ceases or if conditioned airflow stops or slows to inefficient amounts. Fig. 25-89 shows an airflow signal switch or shutoff switch or both. If airflow is greater than standard, a paddle mounted on a pivoted arm moves back through a small arc, tripping an electrical switch. Fig. 25-90 shows a similar switch for liquid flow such as chilled water or condenser water.

25-32 HUMIDITY CONTROLS

Humidity control systems are almost always electric on smaller units. Pneumatic systems are usually used for large systems.

The electrical system uses a humidistat control and an electrical power source. Either a solenoid or a motor operates a valve or damper.

The pneumatic system uses a humidistat control, piping, and a vacuum or pressure source. The controlled vacuum or pressure then acts upon a diaphragm-operated valve or a diaphragm-operated damper.

Humidity is most often made higher by adding water vapor to the air. It is most often lowered by cooling air below its dew point temperature. This condenses moisture out of the air.

25-33 HUMIDISTATS

A humidistat is the control device in a humidity control system. It responds to humidity changes and, in doing so, opens or closes a control system.

The sensing element of the humidistat is called a hygroscopic element. The most commonly used hygroscopic elements are: human hair, wood, nylon ribbon, membranes, and electronic solid-state sensors. Wood, human hair, membranes, and nylon ribbon stretch as the moisture content of the air increases.

In electronic solid-state sensors, the sensor changes its resistance as the moisture content of the air changes. Some sensors used are:
1. Hygroscopic salt (for example, lithium chloride).
2. Carbon particles imbedded in a hygroscopic material. In these substances, the resistance decreases as the humidity increases.

The change in size or shape or electrical resistance of the sensing element is used to operate a switch or a pneumatic system. Fig. 25-91 shows a humidistat which uses a multiple hair element. The control should be kept dust free and the cover must permit free air circulation over the element.

Fig. 25-92 shows a typical electronic humidistat which varies its output amperage based on an internal resistance that changes with humidity. See Para. 18-7 and Para. 18-11 for more information on humidity measurement.

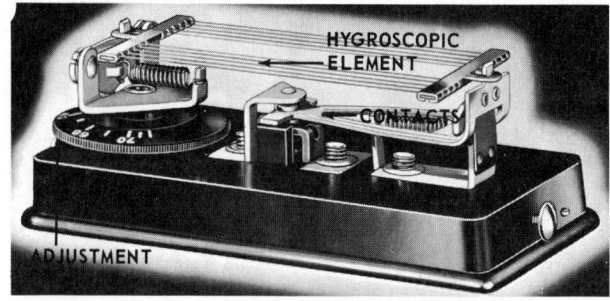

Fig. 25-91. A humidity control. Operating mechanism is shown.

Fig. 25-92. Electronic humidistat. A—Element varies resistance with changes in humidity. (Action Instruments, Inc.)

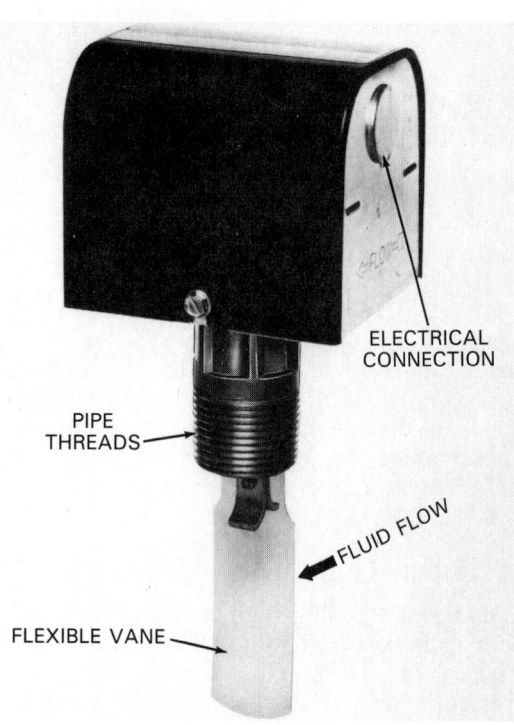

Fig. 25-90. Liquid flow switch. If chilled water or condenser water flow is not sufficient, this unit will close signal circuit, shut off the unit, or both. (ITT McDonnell & Miller)

25-34 ELECTRONIC CLEANER CONTROLS

Electronic cleaners have an electrical circuit which converts 120 V ac to about 9000 V dc. This circuit also has safety devices such as door interlocks and service devices such as automatic washing and diagnostic circuits.

Fig. 25-93 shows an electronic cleaner electrical circuit. Note the two step-up transformers, the door interlock, the full wave rectifiers, and the neon light circuit. The actual wiring diagram is shown in Fig. 25-94.

The electronic cleaner is connected into the electrical circuit of a heating-cooling system which uses a single speed fan. There is 120 V service to the cleaner and 240 V service to the fan motor. This is shown in Fig. 25-95.

High voltages are used in these electronic cleaners. They should be serviced only by one who has had special training on the model being serviced. The automatic water-cleaning system has a washing cycle with detergent, a rinsing cycle, and a drying cycle. See Fig. 25-96.

Service to electronic cleaners includes checking rectifiers and high dc voltage. Fig. 25-97 shows how the rectifiers are removed and how the high voltage is checked. Note the heavy insulators on the meter leads to protect against this very dangerous voltage.

25-35 FAN CONTROLS

Forced air heating systems have a thermostat on the plenum chamber to start and stop the fan motor (blower). The fan starts only when the plenum chamber reaches a certain temperature (about 140 °F or 60 °C). The blower continues to run when the heating unit shuts off until the plenum chamber temperature drops to the fan thermostat cutout temperature — about 80 to 90 °F (27 to 32 °C). Fig. 25-98 shows a fan control. This control has a temperature range of 70 to 150 °F and a differential of 15 to 55 °F. These controls are built to carry as much as 14 amperes at 120 V.

In some cases, the fan control and the limit control are in the same casing. Some fan controls have a manual switch to operate the fan continuously during the hot summer months. Recent developments in this area include combination motors and fan controls that operate the motors at two or more speeds or at modulating speeds. Air moving at 20 to 25 fpm (6 to 8 m/min.) or faster will produce a noticeable draft.

25-36 AIRFLOW CONTROLS

Airflow controls constantly regulate air volume and temperature. If outside air is too cold, thermostats will cause power-operated devices to close the outside air damper. Thermostats will also react if the mix of recirculated air and outside air is out of balance. They will open one damper and close the other just enough to produce the correct mix. Damper motors are usually used, although a heated vapor element may also be used, as may air pressure or vacuum (pneumatic).

If air temperature is too high or too low (in the airflow to the room, in the recirculated air, in the fresh air, or in the exhaust air), the thermostat will react to properly adjust the dampers.

Pneumatic motors, controlled by an outside thermostat, or electric modulating motors controlled by the outside temperature are used to vary the supply of air as outside temperature changes.

One motor for operating dampers uses power to open the damper. It works against spring pressure and holds the dampers open until the thermostat points open. Spring pressure then closes the damper.

The motor operates the damper through a gear reduction train. Fig. 25-99 shows such a damper-control motor and shaft.

Modern control devices that regulate the volume of air in a distribution system are referred to as Variable Air Volume (VAV) controllers. These devices use electronic components and are usually part of a larger computer-controlled HVAC system.

25-37 PNEUMATIC SYSTEMS

Pneumatic (air) systems are often used to control air conditioning. Thermostats control a pressurized air line. The air in this line can, in turn, operate pneumatic motors (a piston and cylinder or a diaphragm). Motors, in turn, operate dampers, valves, and switches.

The system's two main parts include sensing devices and pneumatic controllers.

The operating pressures used are either 12 or 24 psi. Some systems operate at a vacuum.

The 12 psi system actually uses 3 to 15 psi operating pressure (regulator reduces 20 psi supply pressure). The 24 psi system actually uses 3 to 27 psi operating pressure.

The supply air pressure for these systems is as follows:
12 psi system = 18 psi to 20 psi (227 kPa-241 kPa)
24 psi system = 30 psi to 35 psi (310 kPa-344 kPa)
$$([1 \text{ psi plus } 15] \times 6.9 = 110 \text{ kPa})$$

Air pressure controls are often used in large commercial and industrial systems. These control systems should be thoroughly checked each month.

25-38 DISTRIBUTION CONTROLS

All of the heating and air conditioning control systems reviewed so far in this chapter would not benefit the user

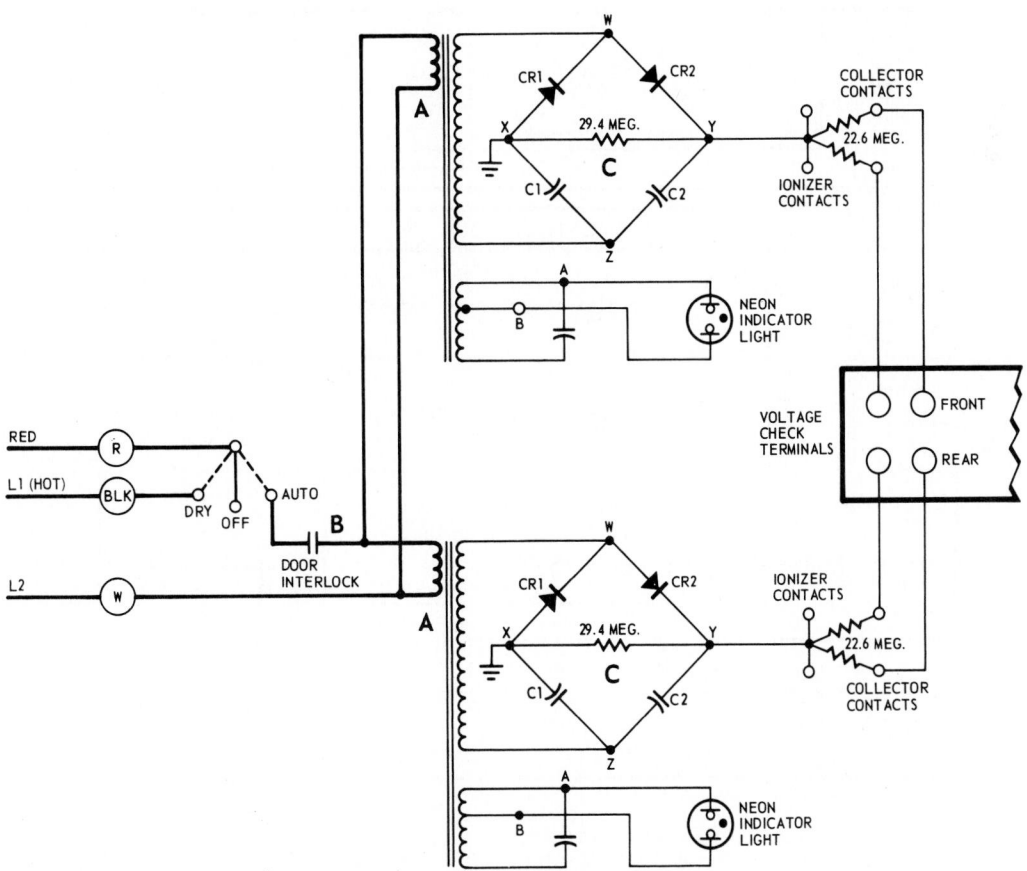

Fig. 25-93. Wiring circuit for an electronic air cleaner. A—Transformers. B—Door interlock (safety switch). C—Rectifiers. (Honeywell, Inc.)

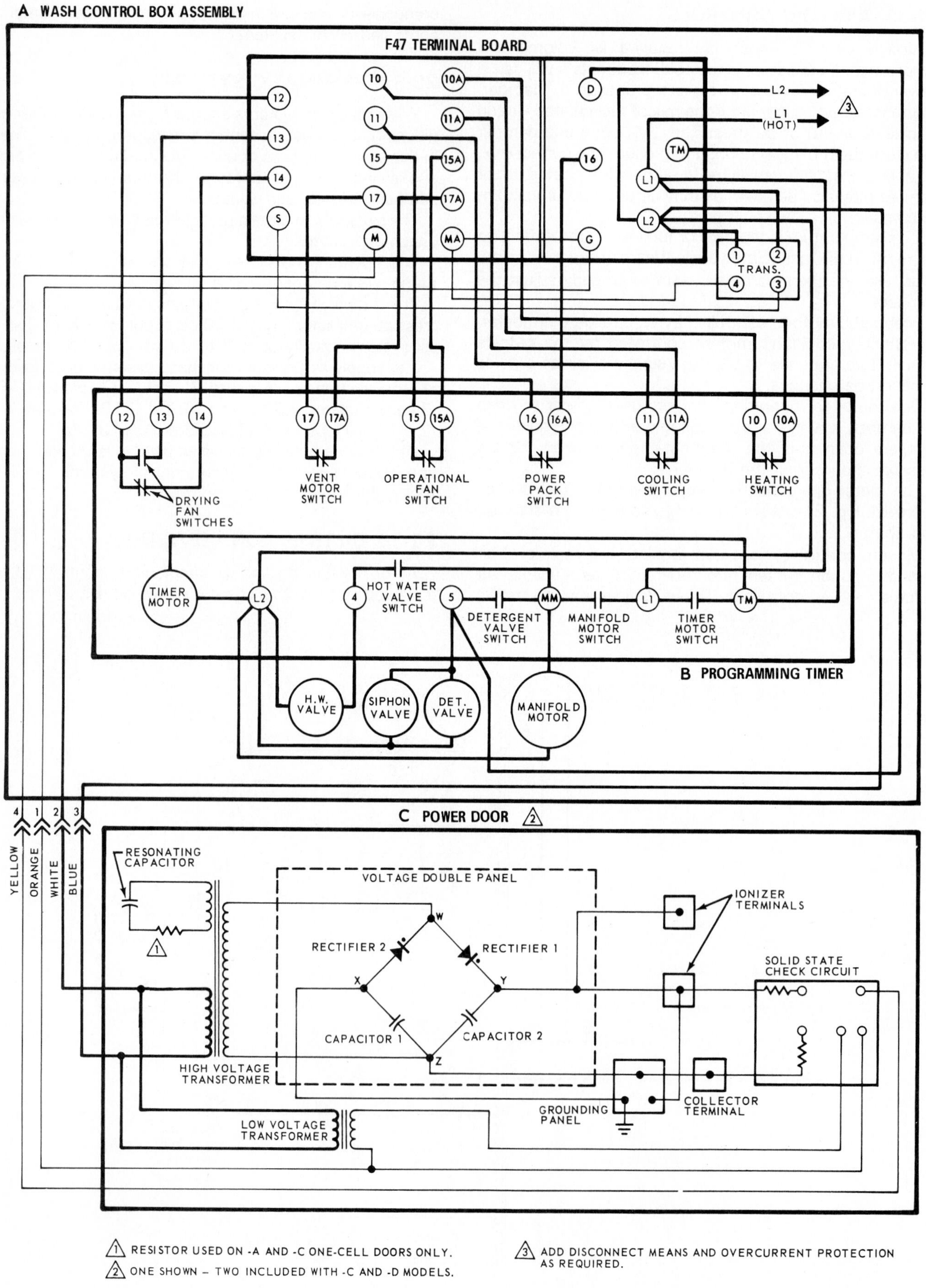

A WASH CONTROL BOX ASSEMBLY

F47 TERMINAL BOARD

12 13 14 S M

10 11 15 17 MA

10A 11A 15A 17A

D TM L1 L2

16 G

L2 →
L1 (HOT) →

⚠3

TRANS.
1 2
4 3

12 13 14

17 17A

15 15A

16 16A

11 11A

10 10A

DRYING FAN SWITCHES

VENT MOTOR SWITCH

OPERATIONAL FAN SWITCH

POWER PACK SWITCH

COOLING SWITCH

HEATING SWITCH

TIMER MOTOR

L2

4

HOT WATER VALVE SWITCH

5

DETERGENT VALVE SWITCH

MM

MANIFOLD MOTOR SWITCH

L1

TIMER MOTOR SWITCH

TM

B PROGRAMMING TIMER

H.W. VALVE

SIPHON VALVE

DET. VALVE

MANIFOLD MOTOR

4 1 2 3

YELLOW ORANGE WHITE BLUE

C POWER DOOR ⚠2

RESONATING CAPACITOR

⚠1

VOLTAGE DOUBLE PANEL

IONIZER TERMINALS

RECTIFIER 2

W

RECTIFIER 1

X Y

CAPACITOR 1

CAPACITOR 2

Z

HIGH VOLTAGE TRANSFORMER

SOLID STATE CHECK CIRCUIT

GROUNDING PANEL

COLLECTOR TERMINAL

LOW VOLTAGE TRANSFORMER

⚠1 RESISTOR USED ON -A AND -C ONE-CELL DOORS ONLY.

⚠2 ONE SHOWN — TWO INCLUDED WITH -C AND -D MODELS.

⚠3 ADD DISCONNECT MEANS AND OVERCURRENT PROTECTION AS REQUIRED.

Fig. 25-94. Complete wiring diagram of an electronic air cleaner. The letters A, B, and C indicate the three main circuits. Note that the power door uses direct current in part of the circuit.

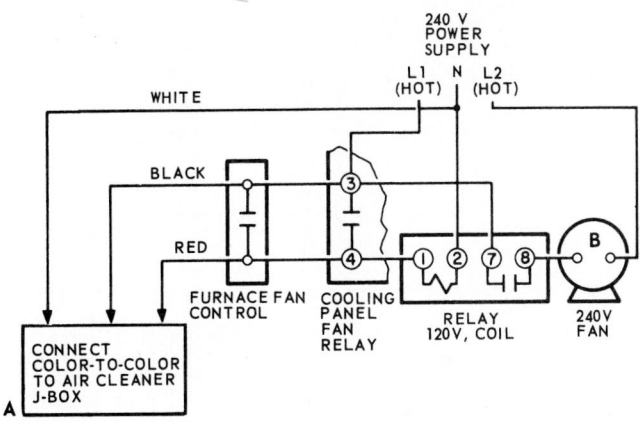

Fig. 25-95. A 120 V service electronic cleaner is connected to a 240 V system. A 240 V fan motor is used. A—Electronic cleaner. B—Fan motor.

unless there is a method for controlling the distribution of the air, water, or whatever medium is used.

Distribution controls are those systems which help to evenly and efficiently transfer the heating or cooling medium to the area where it is needed.

These controls insure that steam, water, or air is properly circulating in the system. In steam systems, the zone control valves are either electrically, pneumatically, or hydraulically operated upon a signal from a thermostat.

In hot water (hydronic) systems, pumps and valves must work in the proper sequence upon command from the operating controls.

In warm air systems, the movement of the air in the complete system or in part of the duct system may be controlled by using separate thermostats. These thermostats may operate the blower motor, duct controls, or the furnace primary controls.

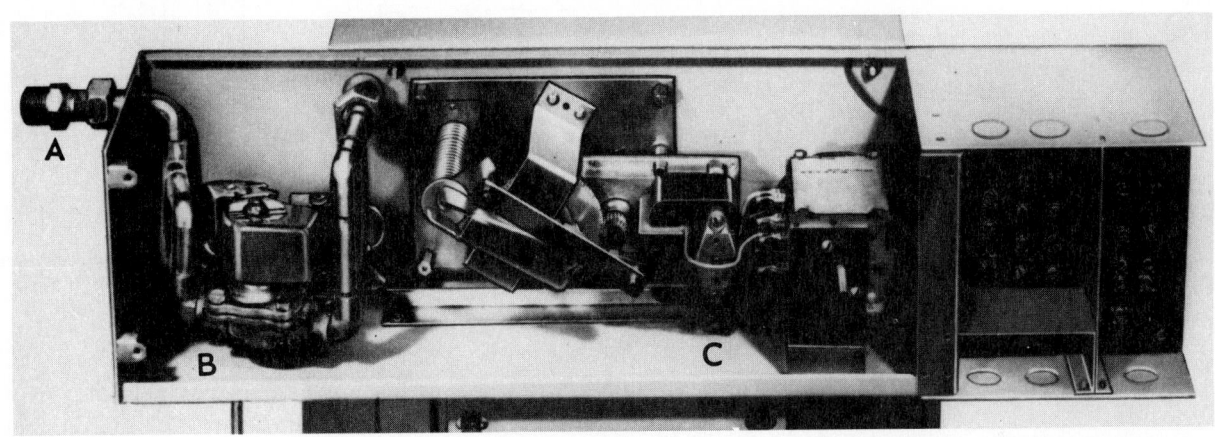

Fig. 25-96. Electronic cleaner washing controls. A—Warm water in. B—Solenoid. C—Timer.

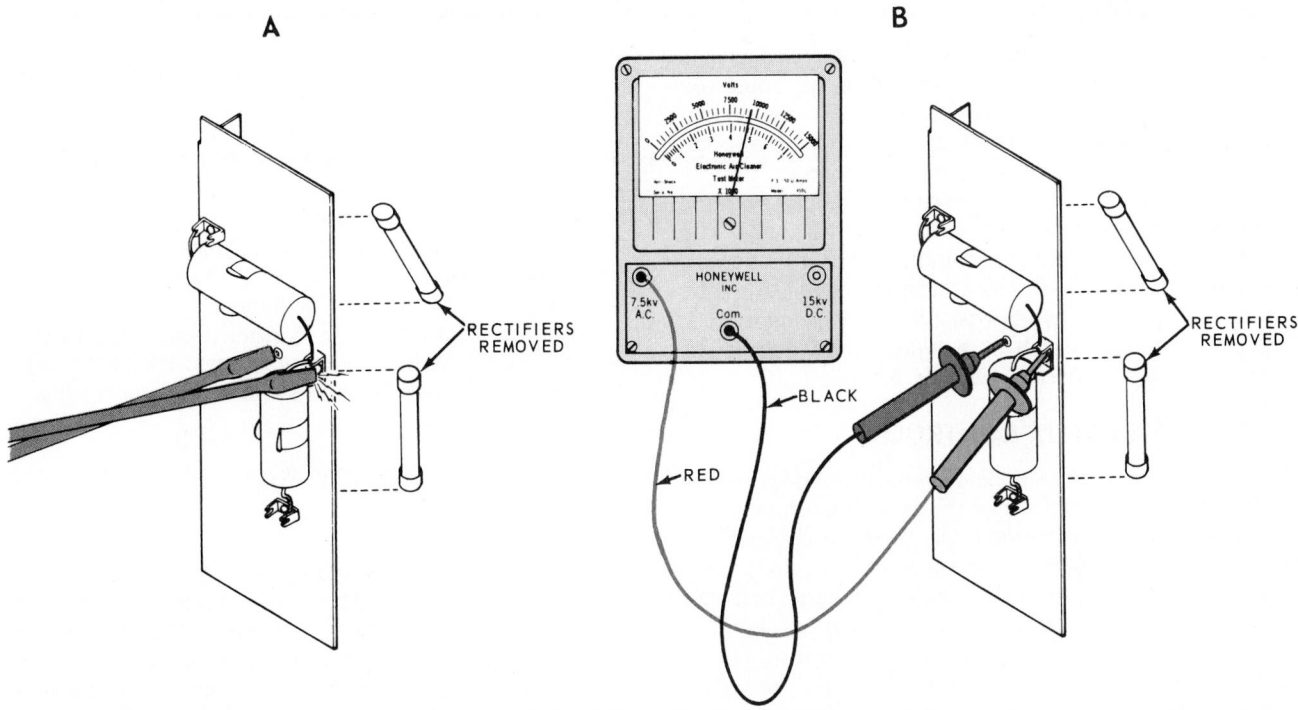

Fig. 25-97. Servicing an electronic air cleaner. A—Removing the rectifiers. B—Testing the high voltage.   (Honeywell, Inc.)

Fig. 25-98. Fan control thermostat for forced warm air furnace. A—Body of thermostat is mounted on plenum chamber. B—Thermal element is located inside plenum chamber. (White-Rodgers Division, Emerson Electric Company)

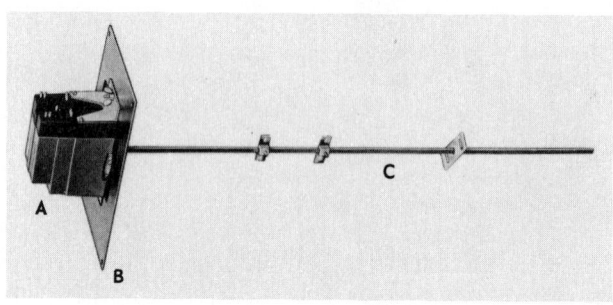

Fig. 25-99. An electric motor damper control. A—Motor. B—Mounting plate. C—Damper shaft. (White-Rodgers Div., Emerson Electric Co.)

25-39 DUCT CONTROLS

Safe distribution of conditioned air to the occupied spaces is important in warm air systems. This air must be distributed in the proper amounts to condition the space. It must not be too cold or too hot. It must have enough fresh air (for oxygen). The air must be flowing.

Many warm air units are zone controlled. These systems have dampers to regulate the flow of air to the zones. The dampers open and close upon command of a thermostat. Some dampers are powered by motorized valves, some by pneumatics, some by hydraulics, others by solenoids.

Fig. 25-100 is a sketch of a typical duct damper. It shows four ways to control damper action.

25-40 SPLIT-SYSTEM CONTROLS

More and more air conditioning systems are designed to provide either hot or cold water to the heat exchange unit in the space to be conditioned. Or, they are designed to provide either hot or cold air to the space to be conditioned.

The three-pipe system, for example, will carry both hot and cold water to a heat exchanger. The third pipe is a return line for either. Fig. 25-101 is a diagram of a thermostat connected to the two valves of a split system.

Many water heating or cooling systems bypass the heat exchange coil when no heating or cooling is needed. This ac-

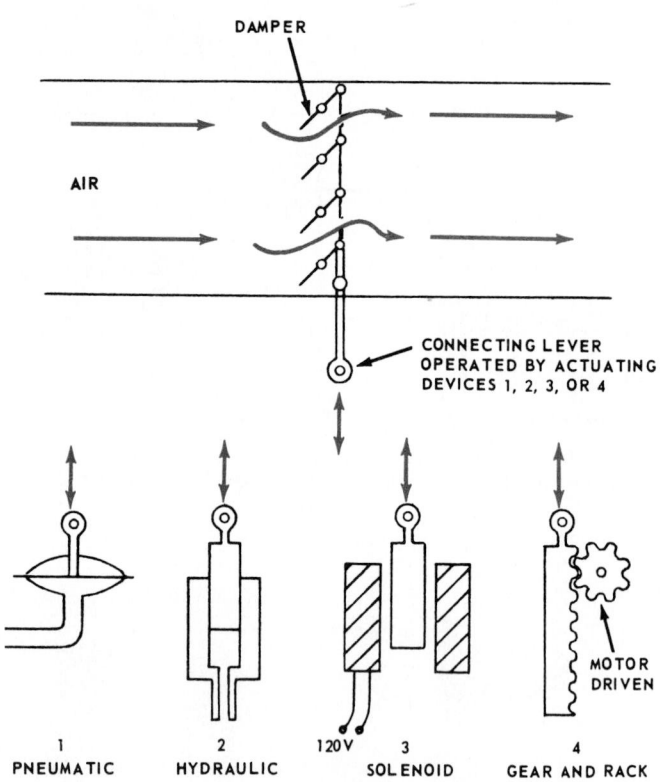

Fig. 25-100. Duct damper for controlling airflow and power devices used to operate it. 1—Pneumatic motor. 2—Hydraulic motor. 3—Solenoid. 4—Motor-driven gear and rack.

tion is obtained through a special solenoid valve. Fig. 25-102 shows such a valve.

25-41 ZONE CONTROLS

Many heating and cooling systems have zone controls. They maintain each zone at the desired temperature. For heating, the zone controls may activate an electrical heating circuit, or control the flow of conditioned air, the flow of hot water, or the flow of steam.

Zone cooling is arranged by using damper controls in ducts or automatic valves in the chilled-water circuit. These zone controls include:
1. A thermostat or sensors.
2. Proportional thermostats or sensors.
3. Damper motors.
4. A proportional damper positioner.

The new electronic zone control devices automatically sense temperature in a number of zones. Then they regulate dampers within a distribution system. A device of this type is shown in Fig. 25-103.

25-42 TOTAL ENERGY MANAGEMENT SYSTEMS

The various control systems and components reviewed so far in this chapter relate to the specific operation of a heating, cooling, or HVAC system within a building or residence. Most buildings in the United States were built when energy was relatively cheap and no one understood the urgency to conserve. Energy sources, types, and cost have changed dramatically, and attempts are now being

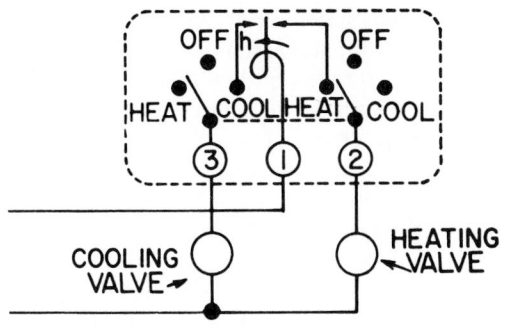

Fig. 25-101. Combination thermostat is connected to two valves of a split system. One is for cooling; the other is for heating. (Johnson Controls, Inc.)

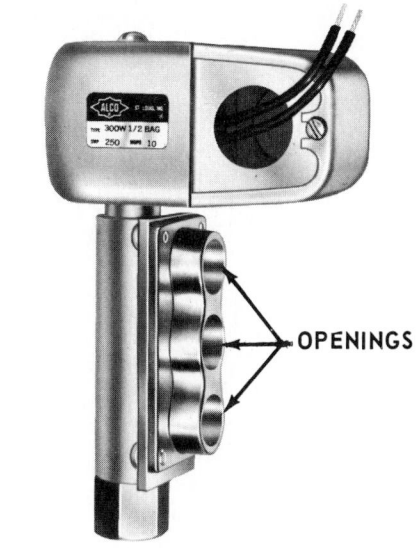

Fig. 25-102. Three-way valve. It is designed to bypass cold or hot water as necessary.

made to modify older buildings and design new buildings so that their operation requires less energy. Fig.25-104, automation system, illustrates modular controllers for commercial and industrial control applications. The system has numerous small stand-alone controllers, Fig. 25-105.

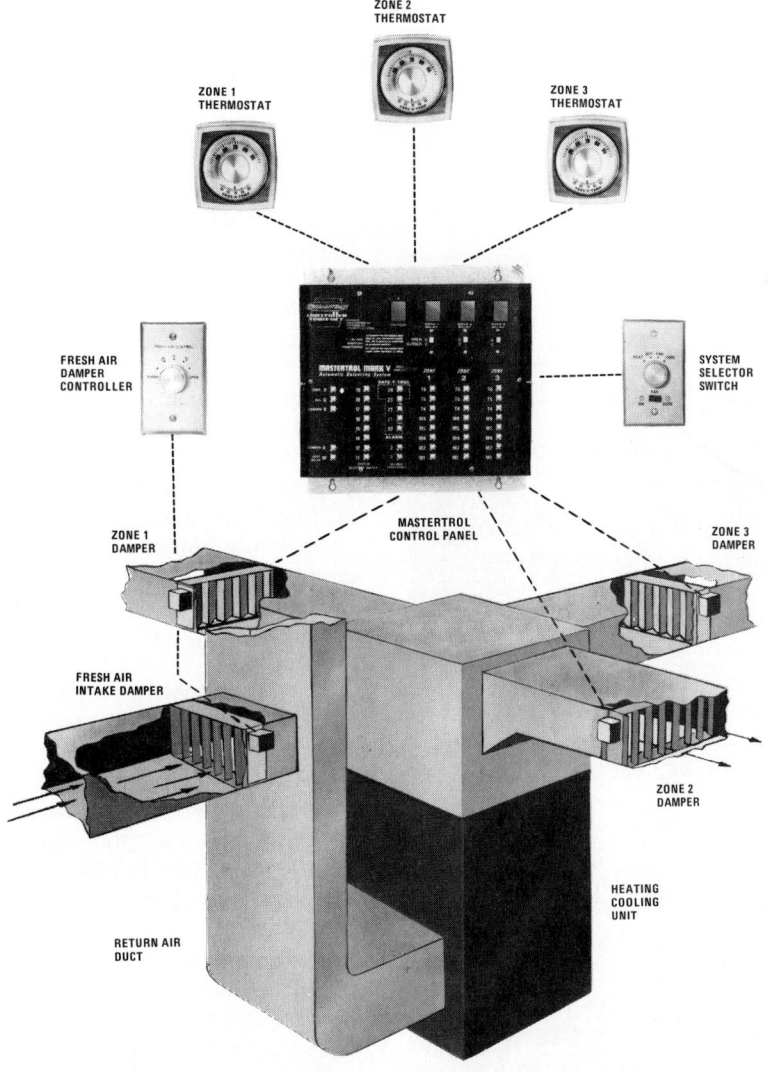

Fig. 25-103. Master control panel such as this is used to control heating, cooling, and air distribution in three different zones. (Trol-A-Temp, Division of Trolex Corporation)

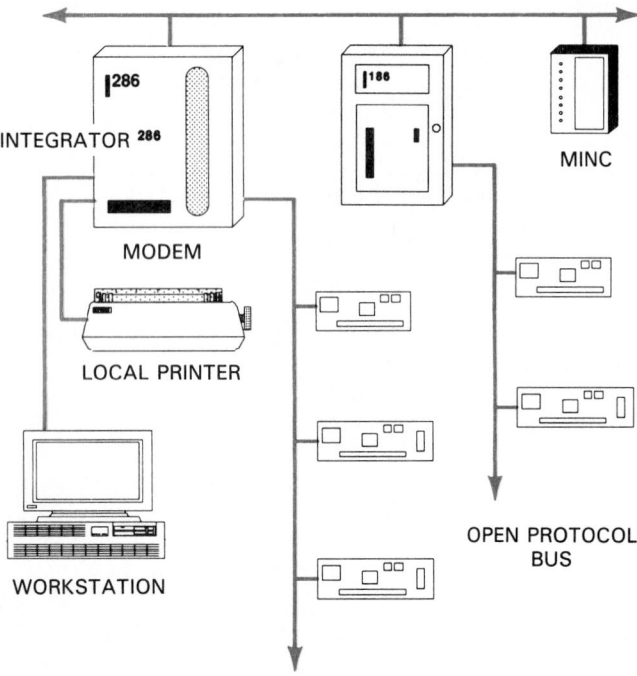

Fig. 25-104. Management control system. Major controller is the Integrator, #286. This controller can be connected with up to 128 internal input/output points. Unit adjacent to it, #186, is a stand-alone control and can be located throughout the facility and network back to #286. MINC (modular integrated network controller) is a smaller stand-alone controller connected back to the #286.
(Teletrol Systems, Inc.)

Fig. 25-105. A master control system used for commercial and industrial applications. (Teletrol Systems, Inc.)

Total Energy Management (TEM) is a new energy conservation concept. It follows the premise that each building should be viewed in terms of its total energy consumption, rather than the energy requirements of each separate system: heating, cooling, lighting, etc. Ideally, each system within the building would be optimized (changed to get maximum efficiency). However, each building represents many energy consumption systems interacting with each other. Compromises can be made in the total energy used. This has been referred to as "smart" building.

The single greatest advantage of a Total Energy Management system is that it provides the owner or operator of a building the flexibility to conserve energy and cut costs. At the same time it meets the needs of the building's users.

The following paragraphs will present background on: energy consumption, various types and functions of typical Total Energy Management systems, system installation and usage, and a section on servicing.

25-43 ENERGY CONSUMPTION

Energy use in a building is determined primarily by the climatic conditions of the region, the working environment, and the type of equipment required by those in the building.

The level of energy consumption is determined by three main systems in a typical building:
1. *Energized systems* required for heating, cooling, lighting, machine operation, etc.
2. *Nonenergized systems* such as floors, walls, roof, windows, etc.
3. *Human systems* including personnel who work in the building and any other people who may use the building in a typical day.

Each of these systems can be modified to achieve significant savings of energy. Energized systems are the easiest to define in terms of direct energy usage. Therefore, the tendency is to concentrate conservation efforts in this area. This may result in total energy consumption greater than that which could be achieved if all systems are involved in the conservation effort. The areas of heating and cooling usually consume the largest single "block" of energy.

Heating and cooling should not be thought of as adding or removing heat from room air to achieve a given temperature. Rather, heating and cooling are the processes of providing for a heat loss or a heat gain in a building. Heat loss and gain occur at the same time in a building. One is usually greater than the other, depending on the outside temperature.

The factors which influence heat loss or gain include:
1. *Infiltration,* which involves passage of outside air into the building through doors, cracks around windows, etc.
2. *Transmission,* which is the heat loss or gain through such building components as ceilings, walls, floors, windows, and other building components.
3. *Ventilation,* or forced airflow, by design, between the inside and outside of the building.
4. *Lighting,* in direct proportion to the wattage involved.
5. *Solar,* as a result of the intensity and direction of the sun's rays.
6. *Equipment* (which generates heat during operation).
7. *Occupants,* and their activities.

Heat loss and gain factors have been only briefly presented here to indicate that heating and cooling alone are complex problems in a Total Energy Management system. Complexity increases dramatically with the number of systems to be integrated. Chapter 26 covers these items in greater depth. There, actual heat-load calculations are discussed.

25-44 ENERGY MANAGEMENT SYSTEM TYPES AND FUNCTIONS

There are three general types of energy management systems.
1. Localized controllers.
2. Remote controllers.
3. Centralized computerized control.
More on these systems is to be found in Para. 25-45 through 25-47.

Typical energy management functions can be classified in about the same way as the control system components in Para. 25-19. These included primary, sequential, and limit controls.
1. Primary functions would include:
   a. Timed operation.
   b. Demand control.
   c. Temperature-compensated duty cycling.
   d. Air temperature cutoff.
   e. Duty cycling.
2. Sequential functions are associated with increasing efficiency in a Total Energy Management system. They include:
   a. Economizer.
   b. Boiler temperature.
   c. Air distribution.
3. Limit functions are operations associated with monitoring. Examples include:
   a. Boiler profile operation.
   b. Maintenance time scheduling.
   c. Safety alarms.
   d. Chiller profile operation.
   e. Efficiency scheduling.
   Some miscellaneous additional functions within a Total Energy Management system may include:
1. Security.
2. Fire detection.

25-45 DIRECT DIGITAL CONTROL

Direct Digital Control (DDC) is the use of a digital computer (see Para. 6-69) to perform the required automatic control operations in a Total Energy Management system. Each of the three types of Total Energy Management system controllers (localized, remote, and centralized computer) use direct digital control. The size of the computer varies from a small microprocessor to a minicomputer or mainframe computer, depending on the number and complexity of the operations required.

25-46 LOCALIZED CONTROLLERS

Localized controllers are used in control systems which provide independent and relatively low-cost control for specific systems or equipment. Each local controller is independent. It controls its specific system and has no interaction with any other controlling device.

Typical localized controllers include:
1. Time clocks.

2. Alarms.
3. Local optimization devices (for example, those that control dampers based on zone temperatures).
   Localized controllers are used for relatively simple HVAC system control where installation time allowed is minimal, where individual system efficiencies are more important than total system management, and where minimum cost is desirable. See Fig. 25-106.

25-47 REMOTE CONTROLLERS

A remote controller operates differently from a localized controller in that more than one energy-consuming device can be controlled. It can also be located some distance from the devices it controls. These controllers, Fig. 25-107, are used to minimize overall energy demand.

The newer remote controllers are programmable through the use of microprocessors. See Para. 6-67 and 6-69. Due to the number of units typically controlled and the lower cost of microprocessor devices, this is the most common type of Total Energy Management system in use today. An illustration of this type of controller is shown in Fig. 25-108.

Functions normally found in a remote controller include:
1. Remote start/stop. The controller turns systems and devices on and off at certain times.

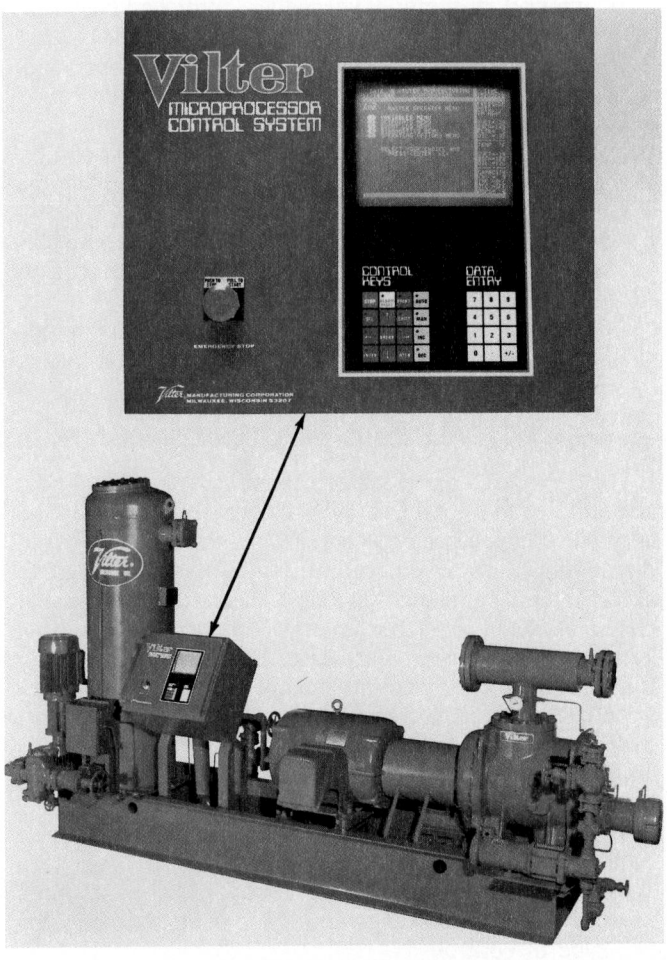

Fig. 25-106. This localized controller is used to operate a screw compressor. (Vilter Mfg. Corp.)

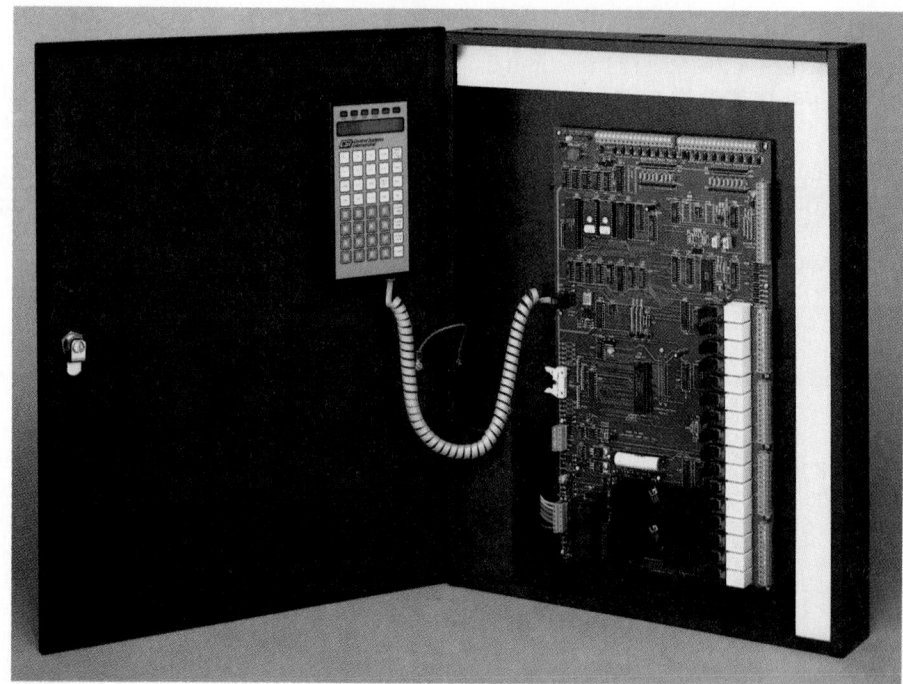

Fig. 25-107. A remote energy controller. Many energy-consuming devices can be controlled at the same time. (Control Systems International)

2. Optimized start/stop. Devices are controlled based on some preprogrammed schedule to minimize energy use.
3. Status monitoring. It indicates if the system is operating.
4. Alarms. They indicate if a system is operating incorrectly.
5. Demand control. The overall demand for electrical power is monitored and modified for minimum energy consumption.
6. Duty cycling. Systems are turned on and off in cycles to maintain minimum energy consumption with little effect on the building's users.

A remote controller is used when multiple functions and systems (approximately 50 maximum) are to be controlled.

25-48 CENTRALIZED COMPUTERIZED CONTROL

The centralized computerized control system is the most elaborate of the Total Energy Management systems. One or more centralized computers make control decisions based on operating data, programmed information, and data already stored in the memory bank. This system is the most costly of the three Total Energy Management systems: localized, remote, and centralized. However, it does offer the widest range of control functions. In large complex structures, this centralized system will result in the best overall control of total energy consumption. Fig. 25-109 shows a central control site for this type of Total Energy Management system.

The centralized computerized controller is used in most newly constructed building complexes. It also is being installed in many older structures which use large amounts of energy. School buildings and office complexes can use this type of control.

Functions of a centralized computerized controller include all tasks basic to the remote and localized controllers as well as additional capabilities. Some of the systems available today will monitor utility usage and verify billing as well as schedule and notify personnel of all maintenance functions.

This centralized computerized Total Energy Management system is required when energy reduction and control of a complex system is needed. The use of this system may require on-site availability of personnel knowledgeable in the maintenance and programming of the system. Fig. 25-110 shows a programming station for a computerized system.

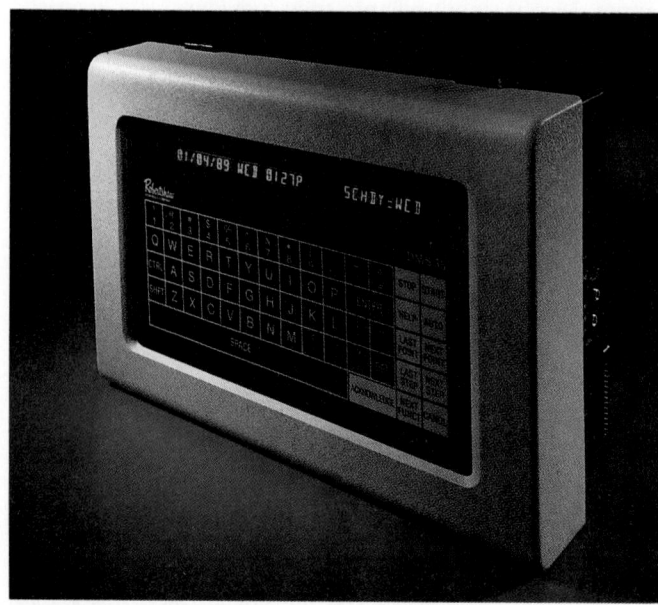

Fig. 25-108. Close-up of a stand-alone energy management and temperature control system having direct digital control capabilities. (Robertshaw Controls Co.)

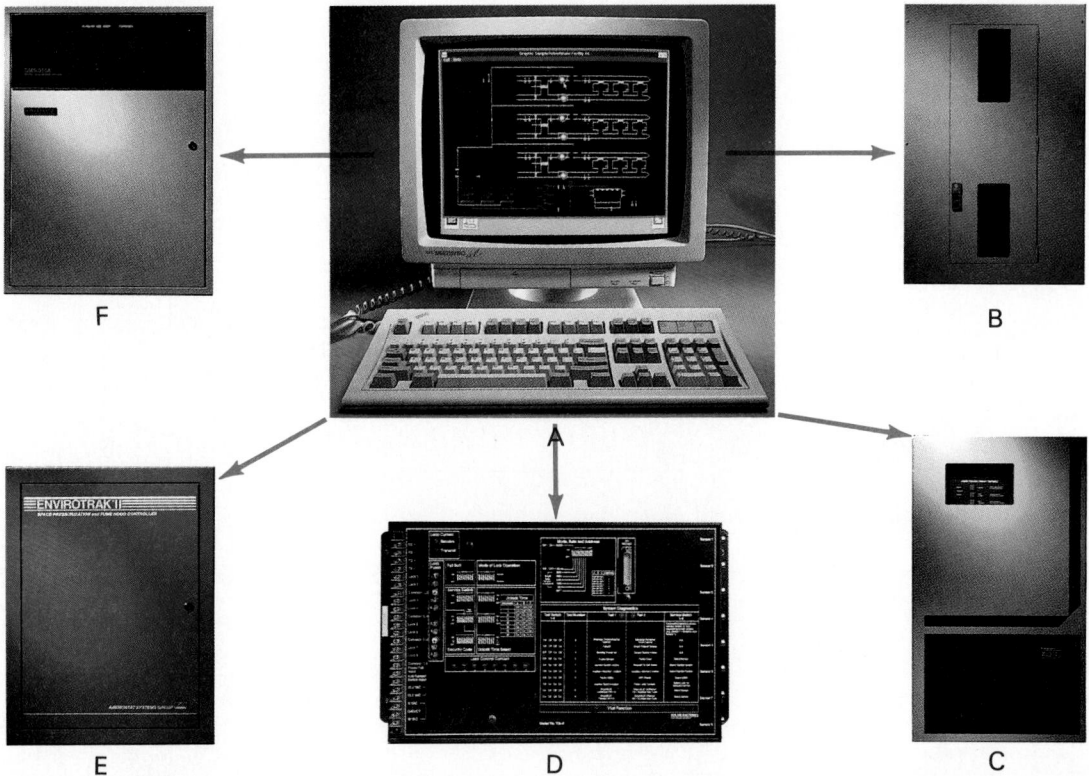

Fig. 25-109. Facility Integrator™ for controlling environment. A—Color monitor with keyboard, IBM PS/2 computer. B—General Electric Co. lighting panel. C—Cerberus Pyrotronics' fire and smoke detector safety panel. D—Schlage Electronics' door access panel. E—Anemostat's Envirotrak panel (laboratory exhaust fume hood control). F—Robertshaw's DMS 350A Digital Management System: Controls heating, ventilating, and air conditioning in facility.   (Robertshaw Controls Co.)

The centralized computerized control system is classified into two general categories based on system construction. They are:

1. *Packaged,* in which the complete system is furnished by a manufacturer who also usually provides some level of service and follow-up after installation.
2. *Hybrid,* in which a system is composed of components from several manufacturers. These systems may be designed by personnel within the company or through an outside firm. The same persons may then assume total system responsibility.

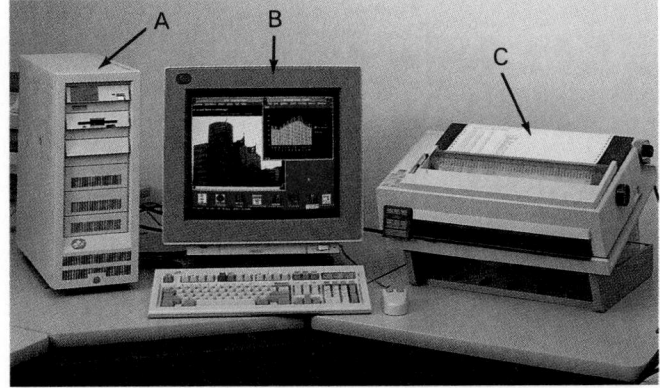

Fig. 25-110. A typical program station for a facilities management program. A—Facility integrator (FI) for control of building or buildings. B—Computer and status display screen. C—Printer. (Robertshaw Controls Co.)

25-49 SYSTEM DETERMINATION AND USAGE

The first step in the installation of a Total Energy Management system is to determine the necessity. This is done through the use of an *Energy Audit.* The audit is needed to develop an *Energy Utilization Index* (EUI).

An energy audit is the process of accurately determining the current energy consumption for a given area. Initially the person doing the audit walks through the structure in question and notes all areas where waste and inefficiency are obvious. Then an accurate recording of all energy consumed during the audit (usually a month or two) follows. An example of the audit form used in this process is shown in Fig. 25-111. The energy utilization index is determined by dividing the energy consumption in BTUs by the square footage of conditioned space in the building. This calculation is shown in the lower right corner of the energy audit form in Fig. 25-111.

Additional information is then gathered as part of the total energy audit. This includes:

1. Previous year's energy use data.
2. Weather data (for accurate comparisons).
3. Building data such as plans or accurate floor layouts.
4. Equipment operation logs.
5. Other information on any aspect of energy consumption within the structure.

The sum of all this information is then processed (computer programs are available) to determine the overall status of energy management for the structure. Then, based on the cost, time, and efficiency, a Total Energy Management system is chosen.

Building _____

Month	Heating Deg. Days	Cooling Deg. Days	Electricity						Purchased		
			kWh	kWh/ Deg. Days	kW Demand		Cost		M (lbs.)	M (lbs.)/ Deg. Days	lbs/hr
					Actual	Billed	Total	Per Ut.			Actual
1	2	3	4	5	6	7	8	9	10	11	12
Jan.											
Feb.											
March											
1st Quarter											
April											
May											
June											
2nd Quarter											
July											
Aug.											
Sept.											
3rd Quarter											
Oct.											
Nov.											
Dec.											
4th Quarter											
Total Per Year											

Building Data

Gross Conditioned Area (ft)^2 _____

Gen. Notes: _____

_____

_____

Annual Energy Consumption In Btu

Quantity

1. _____ kWh
   Electricity

2. _____ (M) lbs
   Purchased Steam

3. _____ MCF
   Natural Gas

4. _____ Gallons
   Oil

5. _____
   Other Fuel

© National Electrical Contractors Association

Fig. 25-111. This is a standard energy management form used in conducting energy audits. The energy utilization index (EUI) is calculated in the lower right-hand corner. (Form copyrighted by National Electrical Contractors Assoc., additional copies available)

_____ Year

Steam			Fuel							Total Energy Cost
			Oil			Check Gas ☐ Coal ☐ Other ☐			Fuel/ Deg. Days	
Demand	Cost		Quant. (Gal.)	Cost		Quant.	Cost			
Billed	Total	Per Unit		Total	Per Unit		Total	Per Unit		
13	14	15	16	17	18	19	20	21	22	23

Conversion Fac.   Btu/Yr

X   3413   =   _____

X   1,000,000.   =   _____

X   1,030,000.   =   _____

X   #2- 138,700   =   _____
    #6- 149,700

X   _____   =   _____

6. Total   _____

Energy Utilization Index

$$EUI = \frac{\text{Total Energy Consumption Btu/yr}}{\text{Gross Conditioned Area (ft)}^2}$$

= _____ Btu/ft^2/Yr

Fig. 25-111. Continued.

Once the system has been selected and installed, some computer programming will be required. This can vary from setting temperatures to implementing a complex computer program. These programs can be changed as needed. When the system is operating, it is important to verify that the Total Energy Management system continues to perform as designed.

25-50 SERVICING: CONTROL SYSTEM DIAGNOSTICS AND REPAIR

The diagnosis and repair of a control system problem is the key to efficient system operation. Control systems of a simpler type, such as basic gas or oil furnace control, are more difficult to diagnose and repair. These systems usually involve more mechanical and individual electrical components. The more advanced Total Energy Management systems involve electronic circuit boards and offer computer diagnostic advantages.

Manufacturers of all levels of controls usually offer a troubleshooting guide. Following is an example of the troubleshooting instructions for an HVAC control system failure:

System fails to start:

1. Check the fuses or circuit breakers and the power-in with a suitable test light or voltmeter.
2. Check the main switch and thermostat switch to make sure they are in the closed position.
3. Check the thermostat setting. It must be on a setting which will close thermostat switch.
4. Check the contactor. It must be pulled in. If not, the contactor circuit is open:
   a. Contactor open but not buzzing. Contactor coil not powered. Check control circuit using voltmeter and ohmmeter.
   b. Contactor open and buzzing. Contactor coil is operating, but armature not pulled in. Either

TROUBLESHOOTING GUIDE		
PROBLEM	**POSSIBLE FAULT**	**PROCEDURE**
1. No LED Display on Display Panel.	STEP 1. Loss of 12 VAC input supply.	Take a voltage reading across terminals 12 V and 12 C on the terminal strip of the interface P.C. board. There should be a reading of 12 VAC. If not present, check for a tripped circuit breaker. Verify primary input (120 VAC) at the electrical outlet and 12 VAC output at the secondary of the transformer. If 12 VAC is verified, proceed to STEP 2.
	STEP 2. Blown 12 VAC fuse on interface board.	Check for improper wiring or shorts to ground and replace fuse if necessary. If the fuse has been replaced and there is still no display or if fuse was not blown, proceed to STEP 3.
	STEP 3. ON/OFF power switch is in the OFF position.	Slide switch to the UP position. If switch is in UP (ON) position, proceed to STEP 4.
	STEP 4. Loose 20-conductor ribbon cable connection.	Check the 20-conductor ribbon cable connection on the processor and interface P.C. boards to see that it is properly inserted. If it is, proceed to STEP 5.
	STEP 5. Faulty interface P.C. board.	Take a voltage reading across capacitor TB1 located on the lower left-hand corner of the interface P.C. board. There should be a reading of 12 VAC. If there is no voltage, replace interface P.C. board (See Service Section). If there is voltage, proceed to STEP 6.
	STEP 6. Faulty microprocessor P.C. board.	Take a voltage reading across the 12 VAC quick disconnects located on the upper left hand corner of the microprocessor P.C. board. There should be a reading of 12 VAC. If there is no voltage, replace the 20-conductor ribbon cable. If there is voltage, replace the microprocessor P.C. board. (See Service Section.)
2. LED Display frozen. Clock will not advance and program modes cannot be accessed.	STEP 1. Defective microprocessor P.C. board.	Replace microprocessor P.C. board. (See Service Section.)

Fig. 25-112. This excerpt is from a typical energy management system troubleshooting guide. Note that complete replacement of a circuit board is sometimes recommended.

the armature is stuck or control coil voltage is too low.

    c. Contactor closed and motor hums. Motor is powered but will not start. Check capacitors, if used, for low voltage; check for excessive pressures in system.

    d. Contactor closed and motor does not hum. Motor is not powered. Check for open circuit by continuity test.

Fig. 25-112 shows part of a troubleshooting guide for a Total Energy Management system. Note, as was mentioned in Chapter 6, that repair recommendations may sometimes include the replacement of a given circuit board within the control device.

25-51 REVIEW OF SAFETY

Electronic cleaner controls are of a high voltage and should be serviced only by a technician who has had special training in the model being serviced.

Safety controls for a heating or a cooling system should never be removed and bypassed to keep a system operating. Dangerous conditions or damage to the system may result.

The handling of heating or cooling controls does not present any great shock hazard to the service technician. The voltage and amperage levels are quite low. However, if the instrument readings are not read or interpreted properly by the technician, the installation may be faulty. If this should occur, the equipment could be seriously damaged and persons within the structure may be placed in danger of serious injury.

Before working on any part of a system, always determine the pressure in that part of the system. Carelessness in this regard could result in injury if the system is opened. To avoid serious burns, always measure the temperature of the parts of the system which are to be repaired, adjusted, or touched. Never guess at pressure or temperature.

Use instruments to check electrical circuits. Never assume that the power is off.

Remember that even a few milliamperes can injure or kill a person. Always use rubber gloves and rubber boots or shoes when working on electrical devices or circuits. Tools, such as wrenches, pliers, and screwdrivers should have insulated handles.

25-52 TEST YOUR KNOWLEDGE

1. What refrigerant controls are generally used on air conditioning evaporators?
2. What is a pressurestat?
3. What is a bimetal thermostat?
4. Why is a battery back-up required in an electrical thermostat?
5. What is the approximate voltage of a low-voltage thermostat?
6. What type thermostat must be carefully leveled?
7. Do millivolt systems have thermostats?
8. What is a pneumatic control?
9. Why is a timer a part of some thermostats?
10. Why is a flame detector used in an oil burner system?
11. What is a limit control?
12. How does a rod and tube thermostat operate?
13. At what voltages do line thermostats operate?
14. What is the usual bonnet or plenum chamber limit control cutout temperature setting?
15. Describe how a heat anticipator operates in a cooling thermostat.
16. Describe a heat anticipator.
17. Is there such a control as a combination heating and cooling thermostat?
18. If a flow switch has two sets of contact points, describe the use of each set of points.
19. Name at least two things to consider when locating a thermostat.
20. Name the three types of controls that make up a controller system.
21. Modern control devices regulating the volume of air in a distribution system are known as what?
22. What is Total Energy Management?
23. How is a heat anticipator wired?
24. Is a mercury tube switch with two leads called an SPST switch?
25. How does a hydraulic thermostat operate?
26. What does a 24 V gun-type oil burner thermostat operate when the points close?
27. How are thermometers used as thermostats?
28. What is "system lag" in a heating system?
29. How many controls must be mounted in the gas line of a gas-fired furnace?
30. What is an energy audit?

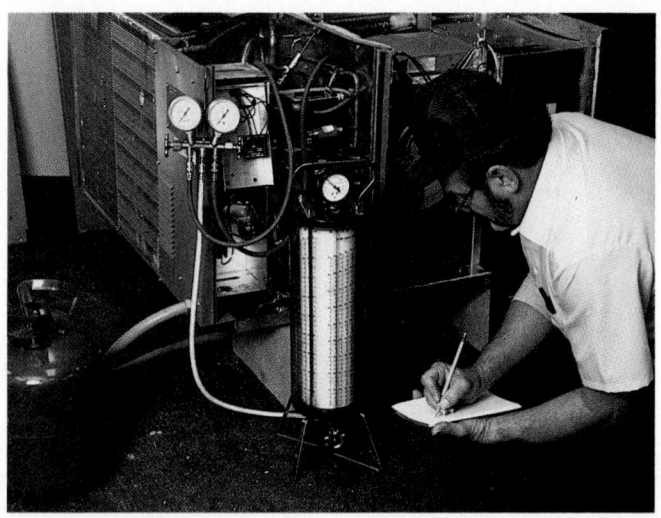

Service technician checking operating pressures, which technician will then compare with operational specifications.
(York International Corp., Central Environmental Systems)

Refrigeration systems use many types of control valves and other necessary equipment. A—Replaceable catch-all filter-drier. B—Ammonia solenoid valve. C—Sealed model suction filter. D—Replaceable suction filter. E—Sealed model suction line filter-drier. F—Sealed model catch-all filter-drier. G—Evaporator pressure regulating valve. H—Discharge bypass valve. I—Ammonia thermostatic expansion valve with level control. J—Three-way heat reclaim valve. K—Solenoid valve. L—Normally open solenoid valve. M—Ammonia thermostatic expansion valve. N—Evaporator pressure regulating valve. O—Balanced port TEV. P—Thermostatic expansion valve. Q—Nonadjustable solenoid valve. R—Pressure differential valve. S—Solenoid valve with manual lift stem. T—Acid test kit. U—Thermostatic expansion valve. V—Refrigerant distributors. W—Moisture and liquid indicators.　　(Sporlan Valve Co.)

Chapter 26

AIR CONDITIONING SYSTEMS HEAT LOADS

After studying this chapter, the technician will be able to:
☐ Define heat load and identify its sources for both heating and cooling of space.
☐ Determine heat loads through the use of U or R factors, square footage, and design temperature charts.

Heating systems require an installation that will give enough heat to keep the occupied space at a comfortable temperature and relative humidity. This requirement should hold true even at record low temperatures for the given locality. In much of the United States, typical conditions are: a 70°F (21°C) inside temperature when it is 0°F (−18°C) outside with a 15 mph wind blowing.

26-1 HEAT LOADS

An air conditioning system must put enough heat into a space to make up for the heat losses during heating. It must remove as much heat as the space accumulates during cooling.

Whenever a temperature difference exists, heat energy will flow from the higher temperature level to the lower temperature level. In the case of heating, it is necessary to retard this heat flow as much as possible, since the amount of heat lost must be replaced. In the case of cooling, the system must remove the amount of heat gained.

The most common method of computation is to determine the maximum heat load (lost or gained) for a period of one hour. Charts help make the computation.

Heat load calculation manuals are available through numerous sources. The Air Conditioning Contractors of American (ACCA) have the *Load Calculation for Residential Winter and Summer Air Conditioning*, known as *Manual J*, and the *Load Calculation for Commercial Summer and Winter Air Conditioning*, known as *Manual N*.

26-2 TYPES OF HEAT LOADS

There are several major heat loads:
1. Heat that is transmitted through walls, ceilings, and floors (conduction):
   a. From inside to outdoors (heating).
   b. From outdoors to inside (cooling).
2. Heat necessary to control moisture content in the air:
   a. Adding moisture (humidifying requires additional heat [added as latent heat]).
   b. Removing moisture (dehumidifying requires removal of heat).
3. Conditioning the air that enters the building by leakage and for ventilation.

4. The sun also produces heat in buildings, directly through the windows, and by heating the surfaces it strikes (a cooling load).
5. Energy devices—such as light fixtures, electric motors, electric stoves, or gas stoves—all produce heat. People, too, release a considerable amount of heat.
In all cases, the heat load can be described as either sensible heat load (temperature change) or latent heat load (moisture), evaporating or condensing.

26-3 HEAT LOADS FOR HEATING

Heat loads for heating consist of all means by which heat will be lost from a building or to the warming of cooler substances brought into the building. This heat transfer generally is called heat loss. See Fig. 26-1.

The two major heat losses are:
1. Conduction through walls, ceilings, and floors.
2. Air leaking out of the building (exfiltration), and that which leaks into the building (infiltration).

Combustion air going out the flue from gas or oil furnaces, or from fireplaces, is also a heat loss. Normally, all other heat losses are ignored since they are relatively too small to affect the size of the unit to be installed.

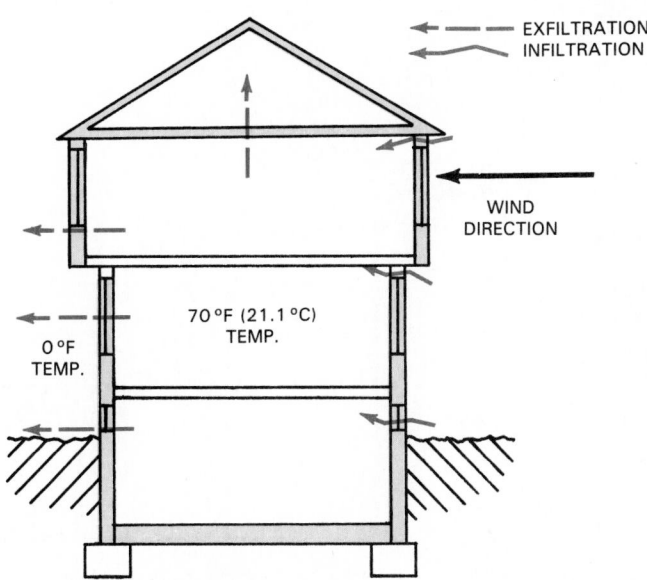

Fig. 26-1. Large heat losses occur during heating season due to cold air filtering into building and warm air filtering out. Heat is also lost through walls, floors, windows, and ceilings.

925

Humidification requires heat, utilizing about 1000 Btu per lb. to vaporize water.

Heat gains from lights, motors, appliances, and people are often taken into account in commercial and industrial structures, but usually not in domestic heating calculations.

26-4 HEAT LOADS FOR COOLING

There are definite sources of heat gain in warm weather:
1. Heat leakage into the building.
2. Air leakage into the building.
3. Ventilation air.
4. Sun load.
5. Heat from appliances and lights.
6. Heat gain from occupants.

Heat gain is the term applied to heat added to a space that is being cooled. This heat must be removed to keep the temperature and relative humidity at the values desired.

Heat gain in warm weather is produced by heat conduction through the walls, ceilings, floors, windows, and doors of the enclosure. Also, heat enters the room by way of infiltrated air. People or animals in the room also give off heat.

Miscellaneous sources of heat are electrical devices (lights and motors), gas-burning devices, and steam tables. Another source of heat that may be considerable in some cases is the sun or sun effect. See Fig. 26-2.

26-5 HEAT LEAKAGE

Heat leakage is the heat that is conducted through the walls, ceilings, and floors of a building. To compute the total heat leakage, first determine the area of each type of surface through which the heat is leaking. Next find the U value for each type of surface. See Para. 26-7 for typical U values. The U value represents the amount of heat that will pass from one side of a wall to the other. The values of heat leakage are for each square foot, each degree F temperature difference, and each hour. The total heat leakage is found by multiplying the heat leakage areas by their respective U values. The U value may also be referred

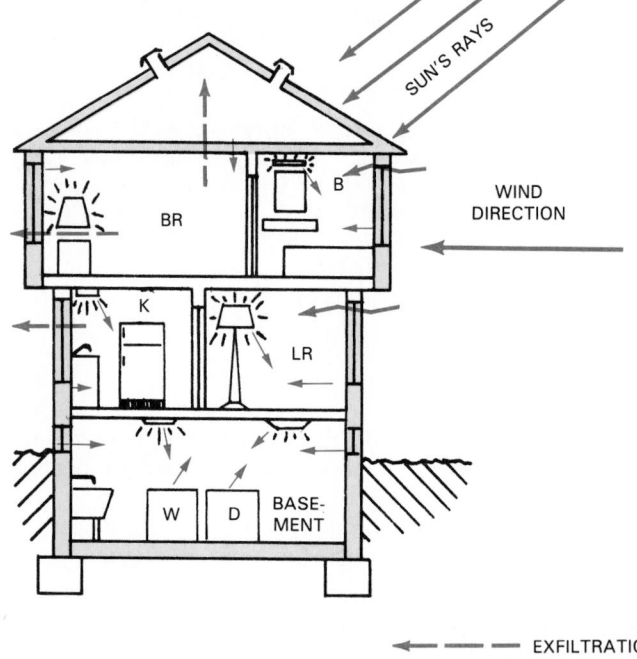

Fig. 26-2. Heat is gained in building during cooling season. Note heat leakage, air infiltration, sun load, lights, appliances, and moisture sources. B—Bathroom. BR—Bedroom. K—Kitchen. LR—Living room. W—Washer. D—Drier.

to as the heat transfer coefficient.

Another way to figure heat leakage is to first determine the thermal resistance of the structure, then use this value to compute the amount of leakage. *Thermal resistance is known as the R value. It is the reciprocal of conductance (C) or the overall heat transfer (U).*

All building materials and structures have been carefully measured and tested in laboratories. From this information, one can obtain the amount of heat that will transfer through almost any enclosure surface being used today. Fig. 26-3

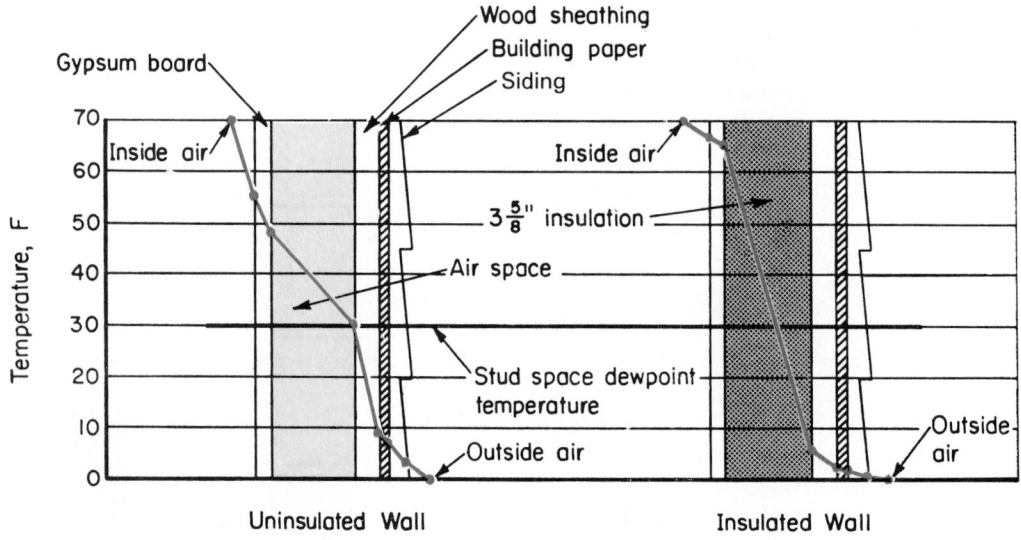

Fig. 26-3. Temperature change through uninsulated wall and insulated wall. Uninsulated wall allows temperature of inside wall surface to drop as low as 55 °F (13 °C). Insulated wall maintains temperature of inside wall surface at 67 °F (19 °C).

shows how the temperature changes during the heating season for a typical wood siding residence wall, both noninsulated and insulated. This heat transfer is called conductivity.

Since there are three general conductivity conditions, the following terms are used:

The letter K represents the Btu that will be transmitted through one square foot of the wall (or surface) in one hour if there is a temperature difference of one degree F, if the material is one inch thick.

1. The units of K are Btu·in./sq. ft./°F/hr.
2. The letter C is used to indicate heat transfer through a wall made of different substances.

$$\frac{1}{C} = \frac{X_1}{K_1} + \frac{X_2}{K_2} + \frac{X_3}{K_3}$$

X is the thickness of a material in inches.

$$C = \frac{1}{\dfrac{X_1}{K_1} + \dfrac{X_2}{K_2} + \dfrac{X_3}{K_3}}$$

3. The letter U is used to represent heat leakage from the air on one side of the wall to the air on the other side of the wall. The meaning and values for U are explained in Para. 26-7.

26-6 HEAT TRANSFER RATE

Heat transfer rate is the amount of heat that is conducted through a structure for a given unit of time. It is usually represented by a Q and is expressed in Btu/hr. The total heat transfer rate is found by taking the total heat leakage number (heat transfer coefficient) and multiplying it by the temperature difference.

Heat Transfer Rate =
Total Heat Leakage × Temperature Difference
$$Q = (U \times Area) \times (T \text{ outside} - T \text{ inside})$$

26-7 U FACTOR FOR COMPUTING HEAT LEAKAGE

In computing heat transfer, the letter U has almost the same value as the letter C. The value of U, however, represents the additional insulating effect of the air film that always exists on each side of the surface. See Fig. 26-4.

$$U = \frac{1}{\dfrac{1}{F_i} + \dfrac{X_1}{K_1} + \dfrac{X_2}{K_2} + \dfrac{X_3}{K_3} + \dfrac{1}{F_o}}$$

where F_i is the heat transfer through the inside dead air film and F_o is the heat transfer through the outside air film.

The U value is a common term used to indicate the amount of heat transferred through a structure (wall):
U = Btu/sq. ft./°F/hour.

This U value is based on a 15 mph wind on the outside and a 15 fpm (1/6 mph) draft on the inside wall surface.

The U value for almost every type of construction can be obtained from reference data books published by the American Society of Heating, Refrigerating and Air Conditioning Engineers (ASHRAE). Fig. 26-5 is a condensed table for some of the more common constructions. Also refer to Para. 26-8 for U values.

Given the U factor, the design temperature conditions of 70°F (21°C) indoors and 0°F (−18°C) outdoors, and the area, calculate the heat load as follows:

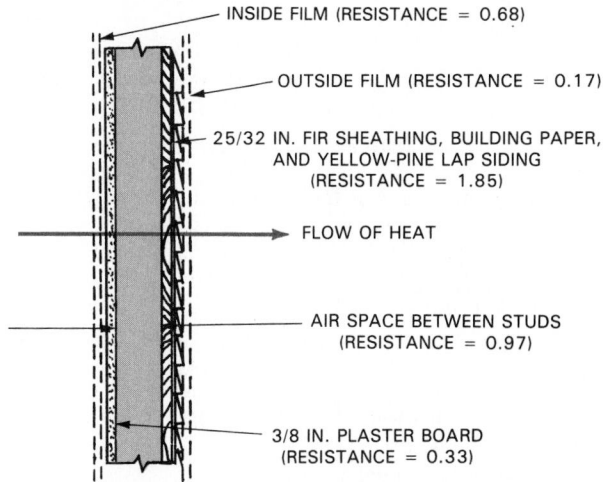

Fig. 26-4. Drawing of outside wall section shows inside air film and outside air film. C values are reciprocal of the R value of material, or C = 1 ÷ R. (Edison Electric Institute)

Heat load = area × temp. diff. × U factor
Total heat transfer (Q) = U × total surface × temp. diff.
Example: A structure has 400 sq. ft. of surface. The temperature difference is 70°F. The structure has a brick veneer wall, no insulation, and has a U factor of 0.25.
Solution: This U value means that 0.25 Btu will transfer through each square foot of wall for each one degree F temperature difference in one hour.

Total heat transfer (Q) = 400 × (70°F − 0°F) × 0.25
= 400 × 70 × 0.25
= 28,000 × 0.25
(Q) = 7000 Btu/hr.

Example: If the total surface area is 1200 square feet, heat transfer can be determined as follows:
Q = 1200 × (70°F − 0°F) × 0.25
= 1200 × 70 × 0.25
= 300 × 70
Q = 21,000 Btu/hr.

The outdoor or ambient temperature is different for each locality. Fig. 26-6 shows design conditions used to calculate heat loads for heating and cooling in various regions.

In the metric system, the same method of computing heat leakage is used but with metric units.
Example: Celsius (Centigrade) temperatures are used. Square meters are used for area and U values are based on calories/square meter/C/hour or watts. However, the trend is to calculate the heat load in watts instead of calories/hour. Fig. 26-7 shows typical heat leakages listed in watts per square foot. To obtain the watt load per degree F, divide the "35 column" by 35. To convert watts to Btu/hr., multiply watts by 3.414368.

Some common metric system heat transmission units are: joules/second; kilocalories/hour; watts. These values are usable on square meter or square centimeter areas.

Heat transmission using watts per square meter is probably the most popular method. The watt unit is both a US conventional unit and a metric unit. It may be easily applied to heating and cooling unit capacities.

The watt per degree Celsius unit is written W/°C. For a square meter, it is written W/m²/°C. Conversions between

HEAT LOADS

CONSTANTS FOR HEAT TRANSMISSION (U AND R VALUES)

Expressed in Btu per hour per square foot per degree temperature difference, based on 15 mph wind velocity.

MASONRY CONSTRUCTION

GENERAL WALL CLASSIFICATION	MASONRY THICKNESS 8" U	R
BRICK — PLAIN		
1/2" WALLBOARD	.29	3.51
1" POLYSTYRENE, 1/2" WALLBOARD	.13	7.54

FRAME CONSTRUCTION

	U	R
WOOD SIDING ON 1" WOOD SHEATHING, STUDS, GYPSUM BOARD 1/2"	.23	4.40
WOOD SIDING, SHEATHING, STUDS, GYPSUM BOARD 1/2", 1" POLYSTYRENE	.07	14.43
WOOD SIDING, SHEATHING, STUDS, 1/2" FLEXIBLE INSULATION, GYPSUM BOARD 1/2"	.15	6.7
WOOD SIDING, SHEATHING, STUDS, ROCK WOOL FILL, LATH AND PLASTER	.072	13.9
NOTE: FRAME WALLS WITH SHINGLE EXTERIOR FINISH SAME AS WALLS WITH WOOD SIDING		
STUCCO, WOOD SIDING, STUDS, GYPSUM BOARD	.30	3.32
STUCCO ON 25/32" RIGID INSULATION, STUDS, 1/2" RIGID INSULATION AND PLASTER	.20	5.0
STUCCO ON 1/2" RIGID INSULATION, STUDS, ROCK WOOL FILL, LATH AND PLASTER	.074	13.5

CONCRETE FLOORS AND CEILINGS

	U	R
4" THICK CONCRETE, NO FINISH	.65	1.54
4" CONCRETE, SUSPENDED PLASTER CEILING	.37	2.7
4" CONCRETE, METAL LATH AND PLASTER CEILING, HARDWOOD FLOOR ON PINE SUB-FLOORING	.23	4.35
4" CONCRETE, HARDWOOD AND PINE FLOOR, NO CEILING	.31	3.23

PITCHED ROOFS

	U	R
ASPHALT SHINGLES ON WOOD SHEATHING	.44	2.30
ASPHALT SHINGLES, 1" FLEXIBLE INSULATION	.136	7.93
ASPHALT SHINGLES, POLYSTYRENE 1"	.132	7.56

WINDOWS AND SKYLIGHTS

	U	R
SINGLE GLASS	1.04	.96
SINGLE GLASS AND STORM WINDOW	.56	1.79
DOUBLE GLASS, INTERMEDIATE AIR SPACE 1/2"	.58	1.73
HOLLOW GLASS TILE WALL. 6" x 6" x 4" BLOCKS	.60	1.67

BRICK VENEER ON FRAME CONSTRUCTION

	U	R
BRICK VENEER, 1" WOOD SIDING, STUDS, GYPSUM BOARD 1/2"	.27	3.71
RIGID INSULATION, STUDS, GYPSUM BOARD 1/2"	.25	4.00
BRICK VENEER, 1" WOOD SIDING, STUDS, 1" POLYSTYRENE INSULATION, GYPSUM BOARD 1/2"	.07	14.43
BRICK VENEER, 1" WOOD SIDING, STUDS, ROCK WOOL FILL, LATH AND PLASTER	.074	13.5

INTERIOR WALLS

	U	R
NOTE: IN GENERAL FOR COOLING COMPUTATIONS, BASE THE CALCULATIONS FOR HEAT GAIN FROM ADJOINING NON-CONDITIONED ROOMS ON A DIFFERENTIAL EQUAL TO 1/2 THE DIFFERENTIAL TO OUTSIDE.		
GYPSUM BOARD 1/2" ON BOTH SIDES	.31	3.23
GYPSUM BOARD 1/2" BOTH SIDES, FOAM INSULATION	.08	13.26

FLAT ROOFS WITH BUILT-UP ROOFING

DECK MATERIAL	NO CEILING U	R	METAL LATH AND PLASTER CEILING U	R
PRECAST CEMENT TILE	.81	1.23	.43	2.33
PRECAST CEMENT TILE, 1" INSULATION	.24	4.7	.19	5.26
4" THICK CONCRETE	.72	1.39	.40	2.50
4" THICK CONCRETE, 1" INSULATION	.23	4.35	.18	5.55
2" WOOD	.32	3.13	.24	4.17
FLAT METAL ROOFS	.95	1.05	.46	2.18
FLAT METAL ROOFS, 1" INSULATION	.25	4.00	.19	5.26

FRAME FLOORS AND CEILINGS

	U	R
HARDWOOD AND PINE FLOORING ON JOISTS, METAL LATH AND PLASTER CEILING	.23	4.37
ROUGH PINE FLOOR, WOOD LATH AND PLASTER CEILING	.28	3.57
NO FLOOR, LATH AND PLASTER CEILING	.62	1.61
NO FLOOR, METAL LATH AND PLASTER CEILING, 3 5/8" ROCK WOOL FILL	.079	12.65
NO FLOOR, LATH AND PLASTER CEILING, 1" FLEXIBLE INSULATION	.17	5.9

Fig. 26-5. U and R values are given for walls, ceilings, floors, and partitions for various types of construction and for various thicknesses. (ASHRAE)

TABLE 1 CLIMATIC CONDITIONS FOR THE UNITED STATES

Col. 1	Col. 2		Col. 3		Col. 4	Winter,[b] °F Col. 5		Summer,[c] °F Col. 6			Col. 7	Col. 8			Prevailing Wind Col. 9			Temp. °F Col. 10	
State and Station[a]	Lat.		Long.		Elev.	Design Dry-Bulb		Design Dry-Bulb and Mean Coincident Wet-Bulb			Mean Daily	Design Wet-Bulb			Winter		Summer	Median of Annual Extr.	
	° '		° '		Feet	99%	97.5%	1%	2.5%	5%	Range	1%	2.5%	5%	Knots[d]			Max.	Min.
ALABAMA																			
Montgomery AP	32	23	86	22	169	22	25	96/76	95/76	93/76	21	79	79	78	NW	7	W	98.9	18.2
ALASKA																			
Juneau AP	58	22	134	35	12	− 4	1	74/60	70/58	67/57	15	61	59	58	N	7	W		
ARIZONA																			
Phoenix AP (S)	33	26	112	01	1112	31	34	109/71	107/71	105/71	27	76	75	75	E	4	W	112.8	26.7
ARKANSAS																			
Little Rock AP (S)	34	44	92	14	257	15	20	99/76	96/77	94/77	22	80	79	78	N	9	SSW	99.0	11.2
CALIFORNIA																			
Sacramento AP	38	31	121	30	17	30	32	101/70	98/70	94/69	36	72	71	70	NNW	6	SW	105.1	27.6
COLORADO																			
Denver AP	39	45	104	52	5283	− 5	1	93/59	91/59	89/59	28	64	63	62	S	8	SE	96.8	−10.4
CONNECTICUT																			
Hartford, Brainard Field	41	44	72	39	19	3	7	91/74	88/73	85/72	22	77	75	74	N	5	SSW	95.7	−4.4
DELAWARE																			
Dover AFB	39	08	75	28	28	11	15	92/75	90/75	87/74	18	79	77	76	W	9	SW	97.0	7.0
DISTRICT OF COLUMBIA																			
Washington, National AP	38	51	77	02	14	14	17	93/75	91/74	89/74	18	78	77	76	WNW	11	S	97.6	7.4
FLORIDA																			
Tallahassee AP (S)	30	23	84	22	55	27	30	94/77	92/76	90/76	19	79	78	78	NW	6	NW	97.6	20.9
GEORGIA																			
Atlanta AP (S)	33	39	84	26	1010	17	22	94/74	92/74	90/73	19	77	76	75	NW	11	NW	95.7	11.9
HAWAII																			
Honolulu AP	21	20	157	55	13	62	63	87/73	86/73	85/72	12	76	75	74	ENE	12	ENE		
IDAHO																			
Boise AP (S)	43	34	116	13	2838	3	10	96/65	94/64	91/64	31	68	66	65	SE	6	NW	103.2	.6
ILLINOIS																			
Springfield AP	39	50	89	40	588	− 3	2	94/75	92/74	89/74	21	79	77	76	NW	10	SW	98.1	−7.2
INDIANA																			
Indianapolis AP	39	44	86	17	792	− 2	2	92/74	90/74	87/73	22	78	76	75	WNW	10	SW	96	−7
IOWA																			
Des Moines AP	41	32	93	39	938	−10	− 5	94/75	91/74	88/73	23	78	77	75	NW	11	S	98.2	−14.2
KANSAS																			
Topeka AP	39	04	95	38	877	0	4	99/75	96/75	93/74	24	79	78	76	NNW	10	S	101.8	−6.4
KENTUCKY																			
Lexington AP (S)	38	02	84	36	966	3	8	93/73	91/73	88/72	22	77	76	75	WNW	9	SW	95.3	−.5
LOUISIANA																			
Baton Rouge AP	30	32	91	09	64	25	29	95/77	93/77	92/77	19	80	80	79	ENE	8	W	98.0	21.4
MAINE																			
Augusta AP	44	19	69	48	353	− 7	− 3	88/73	85/70	82/68	22	74	72	70	NNE	10	WNW		
MARYLAND																			
Baltimore AP	39	11	76	40	148	10	13	94/75	91/75	89/74	21	78	77	76	W	9	WSW		
MASSACHUSETTS																			
Boston AP (S)	42	22	71	02	15	6	9	91/73	88/71	85/70	16	75	74	72	WNW	16	SW	95.7	−1.2
MICHIGAN																			
Lansing AP	42	47	84	36	873	− 3	1	90/73	87/72	84/70	24	75	74	72	SW	12	W	94.6	−11.0
MINNESOTA																			
Minneapolis/St. Paul AP	44	53	93	13	834	−16	−12	92/75	89/73	86/71	22	77	75	73	NW	8	S	96.5	−22.0
MISSISSIPPI																			
Jackson AP	32	19	90	05	310	21	25	97/76	95/76	93/76	21	79	78	78	NNW	6	NW	99.8	16.0
MISSOURI																			
Jefferson City	38	34	92	11	640	2	7	98/75	95/74	92/74	23	78	77					101.2	−6.1
MONTANA																			
Helena AP	46	36	112	00	3828	−21	−16	91/60	88/60	85/59	32	64	62	61	N	12	WNW	95.6	−23.7
NEBRASKA																			
Lincoln Co (S)	40	51	96	45	1180	− 5	− 2	99/75	95/74	92/74	24	78	77	76	N	8	S	102.0	−12.4
NEVADA																			
Carson City	39	10	119	46	4675	4	9	94/60	91/59	89/58	42	63	61	60	SSW	3	WNW	99.2	−5.0
NEW HAMPSHIRE																			
Concord AP	43	12	71	30	342	− 8	− 3	90/72	87/70	84/69	26	74	73	71	NW	7	SW	94.8	−16.0
NEW JERSEY																			
Trenton Co	40	13	74	46	56	11	14	91/75	88/74	85/73	19	78	76	75	W	9	SW	96.2	4.2
NEW MEXICO																			
Sante Fe Co	35	37	106	05	6307	6	10	90/61	88/61	86/61	28	63	62	61				90.1	−1.2
NEW YORK																			
Albany AP (S)	42	45	73	48	275	− 6	− 1	91/73	88/72	85/70	23	75	74	72	WNW	8	S		
NORTH CAROLINA																			
Raleigh/Durham AP (S)	35	52	78	47	434	16	20	94/75	92/75	90/75	20	78	77	76	N	7	SW	97.7	12.2
NORTH DAKOTA																			
Bismarck AP (S)	46	46	100	45	1647	−23	−19	95/68	91/68	88/67	27	73	71	70	WNW	7	S	100.3	−31.5
OHIO																			
Columbus AP (S)	40	00	82	53	812	0	5	92/73	90/73	87/72	24	77	75	74	W	8	SSW	96.0	−3.4
OKLAHOMA																			
Oklahoma City AP (S)	35	24	97	36	1285	9	13	100/74	97/74	95/73	23	78	77	76	N	14	SSW		
OREGON																			
Salem AP	44	55	123	01	196	18	23	92/68	88/66	84/65	31	69	68	66	N	6	N	98.9	15.9

Fig. 26-6. Design temperature conditions used to calculate heat loads for heating and cooling in various regions of the United States. (Adapted by permission from the ASHRAE Handbook 1989 Fundamentals)

TABLE 1   CLIMATIC CONDITIONS FOR THE UNITED STATES

State and Station[a]	Lat. ° '	Long. ° '	Elev. Feet	Winter,[b] °F Design Dry-Bulb 99%	Winter,[b] °F Design Dry-Bulb 97.5%	Summer,[c] °F Design Dry-Bulb and Mean Coincident Wet-Bulb 1%	Summer,[c] °F 2.5%	Summer,[c] °F 5%	Mean Daily Range	Design Wet-Bulb 1%	Design Wet-Bulb 2.5%	Design Wet-Bulb 5%	Prevailing Wind Winter Knots[d]	Prevailing Wind Summer Knots[d]	Temp. °F Median of Annual Extr. Max.	Temp. °F Min.
	Col. 2	Col. 3	Col. 4	Col. 5	Col. 5	Col. 6	Col. 6	Col. 6	Col. 7	Col. 8	Col. 8	Col. 8	Col. 9	Col. 9	Col. 10	Col. 10
PENNSYLVANIA																
Harrisburg AP	40 12	76 46	308	7	11	94/75	91/74	88/73	21	77	76	75	NW 11	WSW	96.5	3.7
RHODE ISLAND																
Providence AP	41 44	71 26	51	5	9	89/73	86/72	83/70	19	75	74	73	WNW 11	SW	94.6	− .5
SOUTH CAROLINA																
Columbia AP	33 57	81 07	213	20	24	97/76	95/75	93/75	22	79	78	77	W 6	SW	100.6	16.2
SOUTH DAKOTA																
Pierre AP	44 23	100 17	1742	−15	−10	99/71	95/71	92/69	29	75	74	72	NW 11	SSE	105.7	−20.6
TENNESSEE																
Nashville AP (S)	36 07	86 41	590	9	14	97/75	94/74	91/74	21	78	77	76	NW 8	WSW		
TEXAS																
Austin AP	30 18	97 42	597	24	28	100/74	98/74	97/74	22	78	77	77	N 11	S	101.6	19.7
UTAH																
Salt Lake City AP (S)	40 46	111 58	4220	3	8	97/62	95/62	92/61	32	66	65	64	SSE 6	N	99.4	− .1
VERMONT																
Burlington AP (S)	44 28	73 09	332	−12	− 7	88/72	85/70	82/69	23	74	72	71	E 7	SSW	92.4	−16.9
VIRGINIA																
Richmond AP	37 30	77 20	164	14	17	95/76	92/76	90/75	21	79	78	77	N 6	SW	97.9	9.6
WASHINGTON																
Olympia AP	46 58	122 54	215	16	22	87/66	83/65	79/64	32	67	66	64	NE 4	NE		
WEST VIRGINIA																
Charleston AP	38 22	81 36	939	7	11	92/74	90/73	87/72	20	76	75	74	SW 8	SW	97.2	2.9
WISCONSIN																
Madison AP (S)	43 08	89 20	858	−11	− 7	91/74	88/73	85/71	22	77	75	73	NW 8	SW	93.6	−16.8
WYOMING																
Cheyenne	41 09	104 49	6126	− 9	− 1	89/58	86/58	84/57	30	63	62	60	N 11	WNW	92.5	−15.9

[a]AP following the station name designates airport temperature observations. Co designates office locations within an urban area that are affected by the surrounding area. Undesignated stations are semirural and may be compared to airport data.

[b]Winter design data are based on the 3-month period, December through February.
[c]Summer design data are based on the 4-month period, June through September.
[d]Mean wind speeds occurring coincidentally with the 99.5% dry-bulb winter design temperature.

Fig. 26-6. Continued.

HEAT LOSS—WATTS/SQ. FT.

CONSTRUCTION	DTD	35	45	55	65	75	85	95	105
EXTERIOR WALLS									
Masonry (8″ concrete block)		3.60	4.63	5.66	6.68	7.71	8.74	9.77	10.80
Masonry (8″ concrete block; ½″ gypsum board) with R-7 (2¼″) Fiberglas		1.05	1.35	1.65	1.95	2.25	2.55	2.85	3.15
Brick Veneer (½″ gypsum sheathing, ½″ gypsum board)									
with R-11 (3½″) Fiberglas		.75	.97	1.18	1.39	1.61	1.82	2.04	2.25
with R-13 (3⅝″) Fiberglas		.65	.83	1.02	1.20	1.39	1.57	1.76	1.94
Frame (Woodrock siding ½″ gypsum sheathing, ½″ gypsum board)									
with R-11 (3½″) Fiberglas		.85	1.09	1.33	1.58	1.82	2.06	2.30	2.54
with R-13 (3⅝″) Fiberglas		.75	.97	1.18	1.39	1.61	1.82	2.04	2.25
Frame (wood clapboard or shingles, plywood or wood fiber sheathing, ½″ gypsum board)									
with R-11 (3½″) Fiberglas		.70	.90	1.10	1.30	1.50	1.70	1.90	2.10
with R-13 (3⅝″) Fiberglas		.60	.77	.94	1.11	1.28	1.45	1.62	1.79
WINDOWS									
Single Light		12.65	16.19	19.73	23.28	27.07	30.61	34.41	37.95
Double Glazed		6.80	8.70	10.61	12.51	14.55	16.46	18.50	20.40
Single with Storms		5.80	7.42	9.05	10.67	12.41	14.04	15.78	17.40
DOORS									
Door only		5.10	6.53	7.96	9.38	10.91	12.34	13.87	15.30
with Storm Door		3.40	4.35	5.30	6.26	7.28	8.23	9.24	10.20
FLOORS									
Concrete Slab No insulation		8.30	10.67	13.04	15.41	17.78	20.15	22.52	24.89
with R-7 (1″) Zer-O-Cel Urethane Perimeter (watts/lin. ft. exposed edge)		2.40	3.08	3.77	4.45	5.14	5.82	6.51	7.19
Wood (⅝″ plywood) over vented space									
with R-13 (3⅝″) Fiberglas		.65	.83	1.02	1.20	1.39	1.57	1.76	1.94
with R-19 (6″) Fiberglas		.50	.64	.78	.93	1.07	1.21	1.35	1.49
over unheated basement									
with R-7 (2¼″) Fiberglas		.95	1.22	1.49	1.76	2.03	2.30	2.57	2.84
with R-11 (3½″) Fiberglas		.70	.90	1.10	1.30	1.50	1.70	1.90	2.10
CEILINGS									
with R-19 (6″) Fiberglas		.49	.63	.77	.91	1.05	1.19	1.33	1.47
with R-22 (6½″) Fiberglas		.45	.58	.70	.83	.96	1.09	1.22	1.34
FIREPLACES (Watts)									
tight damper		205	264	322	381	440	498	557	615
average damper		510	657	803	949	1095	1241	1387	1533
BASEMENT WALL, above grade* (U = .10, 70°F basement)		1.02	1.31	1.60	1.89	2.18	2.47	2.76	3.05

NOTE: The below grade portion of a heated basement must be calculated separately since heat loss relates to ground or ground water temperature rather than air temperature. The U value of .10 represents typical concrete construction with insulation and a furred finish wall. Add above-grade figure to below-grade figure for total basement heat loss.

*BELOW GRADE
HEAT LOSS, WATTS/SQ. FT. (U = .10, 70°F basement)

Ground Water[†] Temperature, °F	Floor	Wall
40	.879	1.758
50	.586	1.172
60	.293	.586

[†]Ground water temperature is available from the local weather bureau.

Fig. 26-7. Table of heat leakage values lists watts per square foot. The DTD (Design Temperature Difference) is listed from 35 °F (19 °C) to 105 °F (58 °C) in vertical column. Note that from first column to last column, every number, including DTD, is multiplied by 3.00.

the units are possible:

1. For either U or C values:

   To change US conventional units to metric units, multiply Btu/hr./sq. ft./°F by 5.674 to obtain W/m²/°C. To change metric units to US conventional units, multiply W/m²/°C by 0.1762 to obtain Btu/hr./sq. ft./°F.

2. For R values:

   To change US conventional units to metric units, multiply hr. sq. ft. °F/Btu by 0.1762 to obtain m² °C/W. To change metric units to US conventional units, multiply m² °C/W by 5.674 to obtain hr. sq. ft. °F/Btu.

26-8 R FACTOR FOR HEAT LEAKAGE

R represents the thermal resistance of a given material. See Figs. 26-4 and 26-5 for R values of some common construction materials. Heat transfer and heat leakage calculations may use R. It is also called "ru," which means resistance unit.

If U is the heat transfer, then R is the reciprocal of heat

transfer. The symbol R is the reciprocal of C or $R = \frac{1}{C}$ and,

in case of overall heat transfer, $R = \frac{1}{U}$. For a composite wall (a typical building), the total R equals the sum of the individual reciprocals of the C values.

$$R_T = \frac{1}{C_1} + \frac{1}{C_2} + \frac{1}{C_3} + \frac{1}{C_4} + \frac{1}{C_5} \text{ or}$$

$$R_T = R_1 + R_2 + R_3 + R_4 + R_5$$

This method is becoming popular because any composite wall can be easily calculated. By knowing the individual R values, they can be totaled. Then, U will equal the reciprocal of the total R. The R values can be found by taking reciprocals of the heat conductance (C values from tables), or by using tables showing the R values. See Fig. 26-8. Example: R values for a typical brick veneer wall are as follows:

	R
Outside air film	0.17
Face brick veneer	0.39
Wood siding and building paper	0.86
Air space	0.97

CONSTRUCTION	(R) RESISTANCE VALUE
1. SURFACE (STILL AIR)	.68
2. AIR SPACE	.97
3. GYPSUM WALLBOARD 3/8 IN.	.32
4. OUTSIDE SURFACE (15 mph WIND)	.17
5. FACE BRICK	.39
6. CONCRETE BLOCK 4 IN.	1.11
7. URETHANE INSULATION	9.1
8. SIDING (WOOD) 1/2 IN. x 8 IN.	.85
9. BUILDING PAPER	.06
10. WOOD SHEATHING	.98
11. WOOD FLOOR 1 IN.	.98
12. LINOLEUM OR TILE	.05
13. ASPHALT SHINGLES OR PLYWOOD	.95

Fig. 26-8. Table of typical thermal resistance (R) values for various parts of a building. (ASHRAE Guide and Data Book)

1/2-in. plaster (0.09) on gypsum lath (0.32)	0.41
Inside air film	0.68
Total R =	3.48

$$U = \frac{1}{R} = \frac{1}{3.48} = 0.287$$

Since conservation of energy becomes more and more important, it is recommended that homes and apartments have thermal insulation. See Fig. 26-9.

THERMAL INSULATION VALUES		
AREA	U	R
CEILING WITH UNHEATED SPACE ABOVE	.08	12
EXTERIOR WALL	.07	14
WALL WITH UNHEATED SPACE ON ONE SIDE	.10	10
FLOOR OVER UNHEATED SPACE	.07	14

Fig. 26-9. Some recommended thermal insulation values are listed for homes and apartments.

A comparison of US conventional and metric system values follows:

	US CONVENTIONAL	METRIC
Specific heat at constant pressure	Btu/lb./°F	kj/kgK
Internal film coefficient	Btu/hr. sq.ft. °F	W/m²K
Total heat flow	Btu/hr.	watts or kcal/hr.
R—Total resistance to heat flow	hr. sq. ft. °F/Btu	m²K/W
U—Overall heat transfer coefficient	Btu/hr. sq. ft. °F	W/m²K
Velocity	ft./min.	m/s
Density of fluid in tube	lb./ft.³	kg/m³ (Pa/v²)

The v in the chart means velocity.

26-9 WALL HEAT LEAKAGE AREAS

In addition to finding the several U values or R values for the building structure, the area of the walls with each different U or R value will need to be calculated:

Wall heat leakage = U × wall area × temp. diff.

Areas to be measured are the outside dimensions of the building. These measurements will result in slightly higher heat leakage loads than if inside dimensions are used. U values based on outside dimensions are on the safe side.

To estimate the heat load, measure the complete building: walls, windows, ceilings, and floors.

To measure the walls, take the outside length and width of the house and the inside ceiling height. To determine the total surface area, first measure the distance around the house. This will be the length plus the width, plus the length, plus the width. L + W + L + W = perimeter.

When these values are added and multiplied by the wall height, the total wall area is obtained. For example, a house 24 ft. × 32 ft. (outside), as shown in Fig. 26-10, with an 8 ft. ceiling, has a total area of:

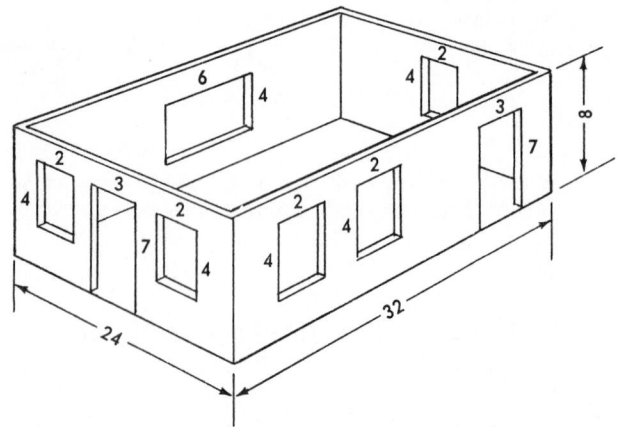

Fig. 26-10. Drawing of one-story home shows wall, window, and door areas. Building is 32 ft. (9.8 m) long by 24 ft. (7.3 m) wide. Room height is 8 ft. (2.44 m). Five windows are 4 ft. (1.22 m) high by 2 ft. (0.61 m) wide. Two doors are 7 ft. (2.13 m) high by 3 ft. (0.914 m) wide. Large window is 4 ft. (1.22 m) high by 6 ft. (1.83 m) wide.

Perimeter = L + W + L + W
= 32 + 24 + 32 + 24
= 112 ft.
Area = perimeter × height
Area = 112 × 8
Area = 896 sq. ft.

This is the total wall area; window and door areas must be subtracted. In the metric system, the total wall area is found by using the same formula:

Area = perimeter × height
Area = (9.8 m + 7.3 m + 9.8 m + 7.3 m) × 2.44 m
Area = 34.2 m × 2.44 m
Area = 83.4 m²

26-10 WINDOWS

The area of each window, to be used in a heat leakage calculation, is determined by measuring the opening in the wall. This would be the distance to the brick edges as shown in Fig. 26-11.

Window construction varies considerably. Windows may

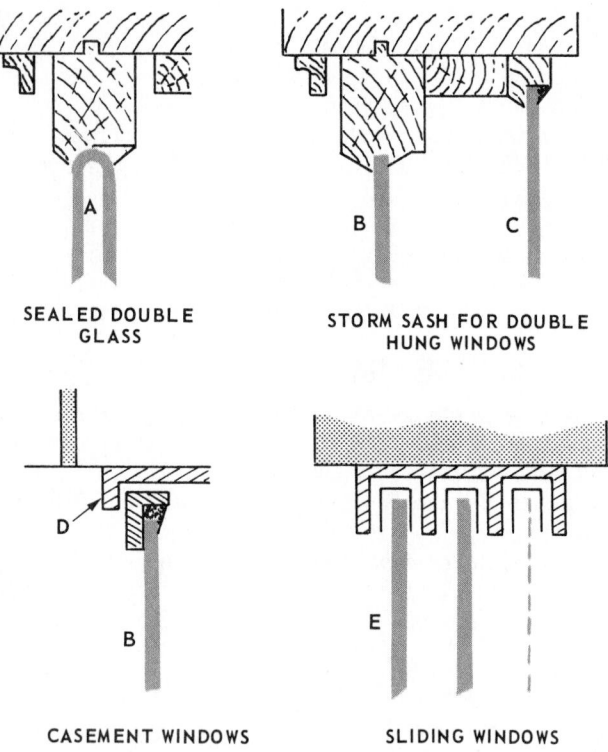

Fig. 26-12. Window construction. A—Dry air. B—Primary glass. C—Storm glass. D—Metal sash. E—Sliding window and storm with screen.

be single pane, double pane using a storm window, or permanent double pane. Recent energy efficient windows may even have three permanent layers of glass. See Fig. 26-12.

The permanent window (single or double pane) is called the primary window. An additional framed pane of glass, set into place to provide added insulation, is called a "storm" window or sash.

The most efficient window construction is the permanent double or triple pane. Two or three panes of glass with sealed air spaces between panes provide excellent insulation. This air space is dehydrated, evacuated, and then usually filled with nitrogen or some other dry gas to prevent sweating (condensation). The registered trade name for this type of window is Thermopane.

Windows are installed in a variety of ways. Some possibilities include fixed ("picture windows"), single or double hung (where either one or both sashes move up and down), or sliding horizontal or casement (hinged on one side and open out with a crank or similar mechanism). The frame around windows may be made of wood, metal or, more recently, a vinyl clad aluminum, to minimize frosting.

The R and U values of window pane assemblies are shown in Fig. 26-13. Remember that R stands for total resistance to heat flow and U means overall heat leakage.

Another problem with windows and walls is that water vapor in warm air will condense on cool or cold surfaces. Condensation occurs due to a combination of both temperature and relative humidity (dew point). For example, if air at 70 °F (21 °C) and 40 percent relative humidity contacts a window at 45 °F (7 °C), condensation will take place. This process is shown on the psychrometric charts in Chapter

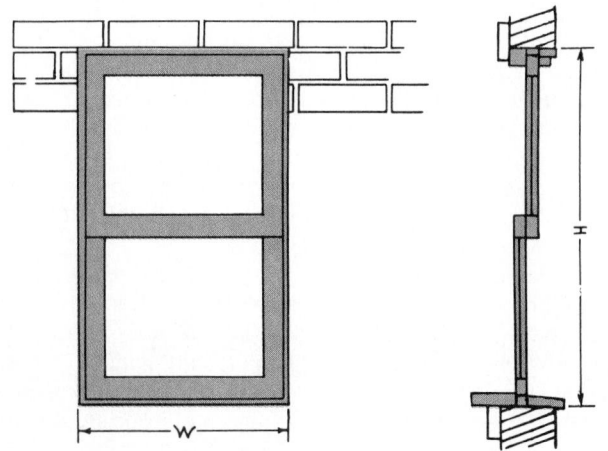

Fig. 26-11. Typical double-hung window showing width and height of window opening.

R AND U VALUES

TYPE WINDOW	U	R
Single Glazed	1.04	.96
Double Pane Insulating Glass	.49	2.0
Triple Glazed	.32	3.1

Fig. 26-13. Thermal resistance (R) and overall heat leakage (U) of various window pane assemblies are specified. (Andersen Corp.)

18, and in the table shown in Fig. 18-8.

To prevent condensation, reduce the relative humidity or raise the temperature of the glass surface. Lowering the relative humidity in the home may not be practical or comfortable. Another solution would be to add a vapor barrier between the inside surface of the window and the room air. This will also serve as added insulation. Many companies make products to provide a vapor barrier on a temporary basis throughout the cold season.

26-11 DOORS

The area used in calculating heat leakage for a door is the height times the width of the door opening.

Doors are constructed in a variety of designs. They may be made of solid wood or of wood veneers over foam cores. Recent energy efficient doors are constructed of a metal shell which is filled with insulation. Some doors have windows as part of their design. Large sliding glass doors are called patio doors. They are responsible for 20 percent of solar heating and heat leakage in some homes.

Proper door-to-wall sealing is very important to minimize heat leakage. The use of a rubber weatherstripping is common practice. On the newer metal doors, a magnetic type weatherstripping, similar to that used in refrigeration, can also be used.

26-12 WINDOW AND DOOR HEAT LEAKAGE

When computing the wall heat leakage area (see Para. 26-9), add the area of the doors in the outside walls to the area of the windows. Then, subtract this amount from the total wall area.

Example: In Fig. 26-10, there are five windows measuring 2 ft. × 4 ft., two doors measuring 3 ft. × 7 ft., and one window measuring 4 ft. × 6 ft.

Total area = 2 × 4 × 5 = 8 × 5 = 40 sq. ft.
3 × 7 × 2 = 21 × 2 = 42 sq. ft.
4 × 6 × 1 = 24 = 24 sq. ft.
Total opening area = 106 sq. ft.

The total wall area is 896 sq. ft. (See Para. 26-9.)

Subtracting 106 sq. ft., the net wall area is 896 sq. ft. − 106 sq. ft. = 790 sq. ft.

The two values, 106 sq. ft. of window area and 790 sq. ft. of wall area, can be used later to find the heat load on the building.

Using the metric system, the window and door areas are found as follows:

Total area = W × H × No.
= 0.61 m × 1.22 m × 5 = 0.744 × 5 = 3.72
= 0.914 m × 2.13 m × 2 = 1.947 × 2 = 3.89
= 1.22 m × 1.83 m = 2.23 = 2.23
Total window and door area = 9.84 m²

26-13 CEILINGS

Ceilings generally are made by fastening drywall to the joists. Variations in construction will not change the heat transfer to any great extent. Fig. 26-14 shows several typical ceiling constructions. If the joists do not have a floor over them, or if there is no insulation between the joists, heat leakage will be considerable.

See Fig. 26-5 for U values for ceilings.

In Fig. 26-10, the ceiling area is calculated as follows:

Ceiling area = W × L
= 24 ft. × 32 ft.
= 768 sq. ft.

In the metric system, the ceiling area is measured in square meters as follows:

Ceiling area = W × L
= 7.3 m × 9.8 m
= 71.5 m²

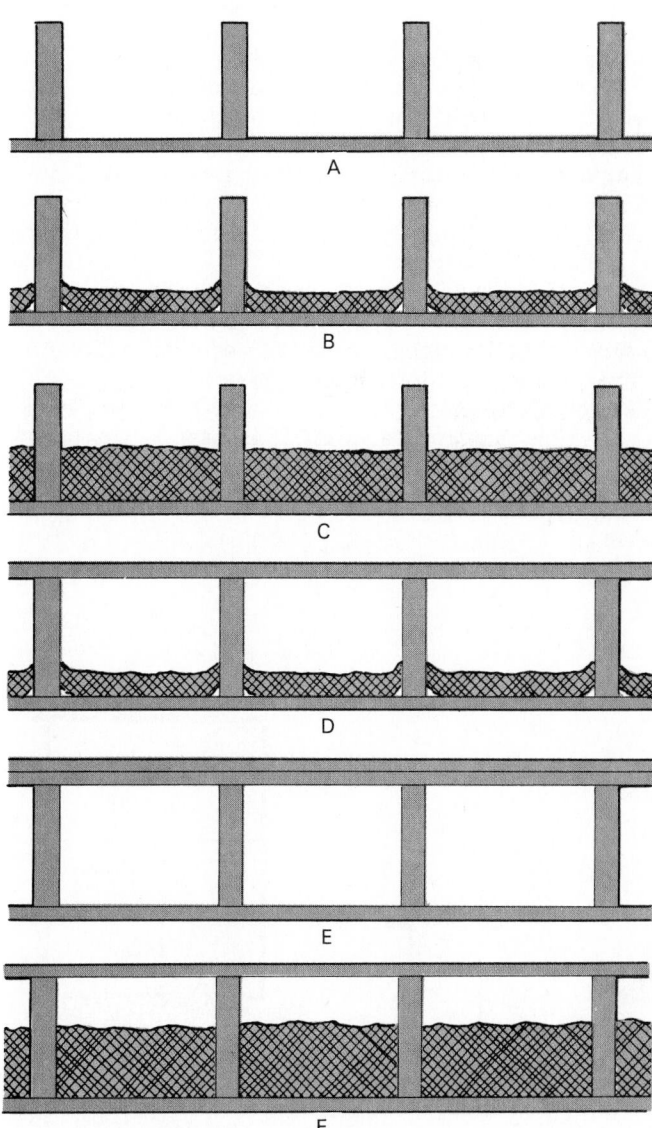

Fig. 26-14. Ceiling construction. A—No floor or insulation. B—No floor, 2-in. insulation. C—No floor, 4-in. insulation. D—Floor, 2-in. insulation. E—Double floor, no insulation. F—Floor, 6-in. insulation.

26-14 DESIGN TEMPERATURES

Design temperatures shown in Fig. 26-6 are the result of considerable testing and the accumulation of much data. A study of the table reveals a country-wide range of design temperatures. Contact the local weather bureau or local chapter of the American Society of Heating, Refrigerating and Air Conditioning Engineers (ASHRAE) for data.

Always choose design ambient temperatures on the low side. Heating plants that are overworked cause excessive stack and chimney temperatures and may cause fires. Oversized units will be less efficient and waste energy.

The design temperature is never as low as the lowest temperature recorded for the area, since these extreme lows are usually of short duration. The residual heat in the building enables the furnace based on the design temperature to handle the load in most cases.

ASHRAE has a new method of listing outside design temperatures (ODT). In this method, the ODT varies with latitude and elevation. (Refer to Fig. 26-6.) New listings have been selected for most urban areas. When working with design temperatures, use the latest values (available from the local ASHRAE chapter). This may result in an 11 percent equipment capacity savings.

ASHRAE charts give three different values for each locality, Fig. 26-6. The lowest temperature is for small (domestic and office) uninsulated buildings. The 99 percent temperature means that the outdoor temperature is at or above this temperature 99 percent of the time. It can be used for well constructed and well insulated buildings having a standard number of windows.

The 97 1/2 percent means that the outside temperature is above the value listed 97 1/2 percent of the time. This value is used for large buildings with considerable thermal capacity and small total window area.

Example: Detroit once used −10 °F as ODT for all buildings. Now it is recommended that 0 °F be used for small, uninsulated buildings; +4 °F (99 percent factor) for well constructed, insulated buildings with a standard window area; +8 °F (97 1/2 percent factor) for large buildings with a standard number of windows.

If IDT (Inside Design Temperature) is 72 °F, the 0 °F means a 72 °F temperature difference (TD). The 99 percent (4+) means a 68 °F TD. The 97 1/2 percent means a 64 °F TD.

$$\frac{72 - 64}{72} \times 100 = \frac{8}{72} \times 100 = 11 \text{ percent.}$$

This means an 11 percent savings in equipment size.

26-15 BUILDING CONSTRUCTION

The construction of a building or residence will be a factor when designing or servicing an HVAC system. A building's design may affect:
1. The type of heating or air conditioning system.
2. The air or water distribution network.
3. The HVAC installation.

An HVAC technician should have a basic understanding of some of the more common construction practices. A typical residence can be classified as either a single-floor type (ranch style), a two-floor type (colonial style), or a combination of both (split-level style). Each style, in turn, may be built over a basement, a crawl space, or a cement slab. Each type of structure requires careful consideration when installing or servicing an HVAC system. In addition, the materials used in construction (brick, wood, aluminum, insulation, etc.) all have significant effects on the heating and cooling of the structure.

The HVAC technician should be able to look at a blueprint or floor plan and find information useful in making heating or cooling calculations. This would include dimensions, with locations of ductwork and electrical wires. A typical floor plan with some dimensional information is shown in Fig. 26-15. Note the air conditioner at the center.

The following paragraphs deal with some specific construction differences and their effects on heat load calculations. Recent designs and new materials have reduced heat losses and gains.

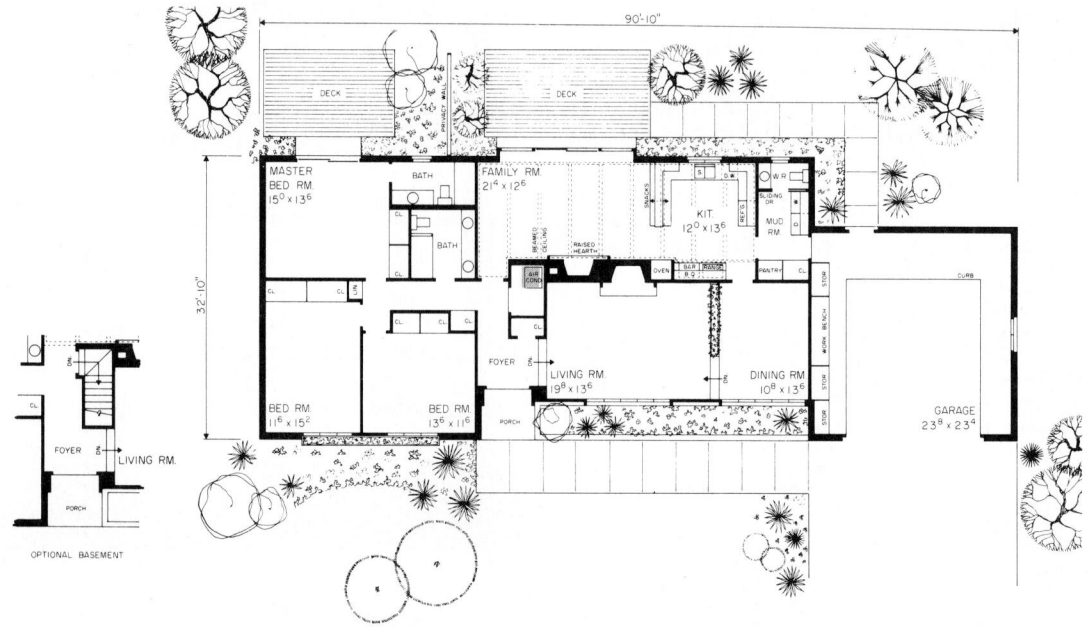

Fig. 26-15. Typical home floor plan shows dimensional information. Note air conditioner at center. (©Home Planners, Inc.)

26-16 ROOF DESIGN AND CONSTRUCTION

The type of roof construction, ventilation, and covering all have an effect on the heat loss through the roof of a building. The most common roof construction in residences today is the pitched roof that uses a prefabricated truss system, Fig. 26-16. The slope of a pitched roof is defined as the vertical rise or height of the roof compared to the horizontal run, Fig. 26-16. Slope is given as ''____ in 12.'' For example, a typical ranch style home will have a roof slope of 4 in 12. If you were in the attic and walked a distance of twelve feet in the direction of the roof pitch, the roof height would have increased by four feet. The pitch of a roof is defined as the ratio of the vertical rise to the span (twice the run). It is given as a fraction. For example, if the total roof rise is 4 ft. and the total span is 24 ft., the pitch is 1/6 (4/24 = 1/6). The pitch of a roof is deter-mined by the climate and the designer's interior space needs. The more snow that falls, the greater the pitch should be. This minimizes the snow weight at the center. A split-level is more likely to have a higher pitch to allow more headroom on the second story.

Recently, roof pitch has become an aspect of energy conservation as well. A building's designer must take into account such things as the space required for adequate insulation when using a cathedral ceiling, or what roof pitch would allow an efficient solar panel system.

Proper ventilation is also important in the roof enclosed area. Attics are used to vent household air through various ceiling and exhaust fans in the home. Most attic areas are vented so that, in the summer, cool input air can come from the overhangs and be exhausted through roof vents or gable vents. See Fig. 26-17.

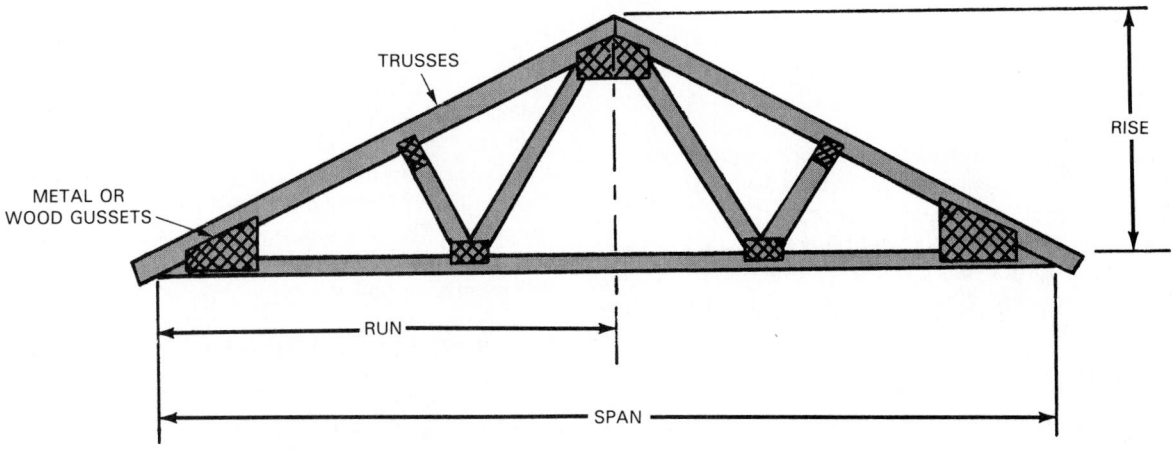

Fig. 26-16. Prefabricated roof truss. Unit is called a Fink truss.

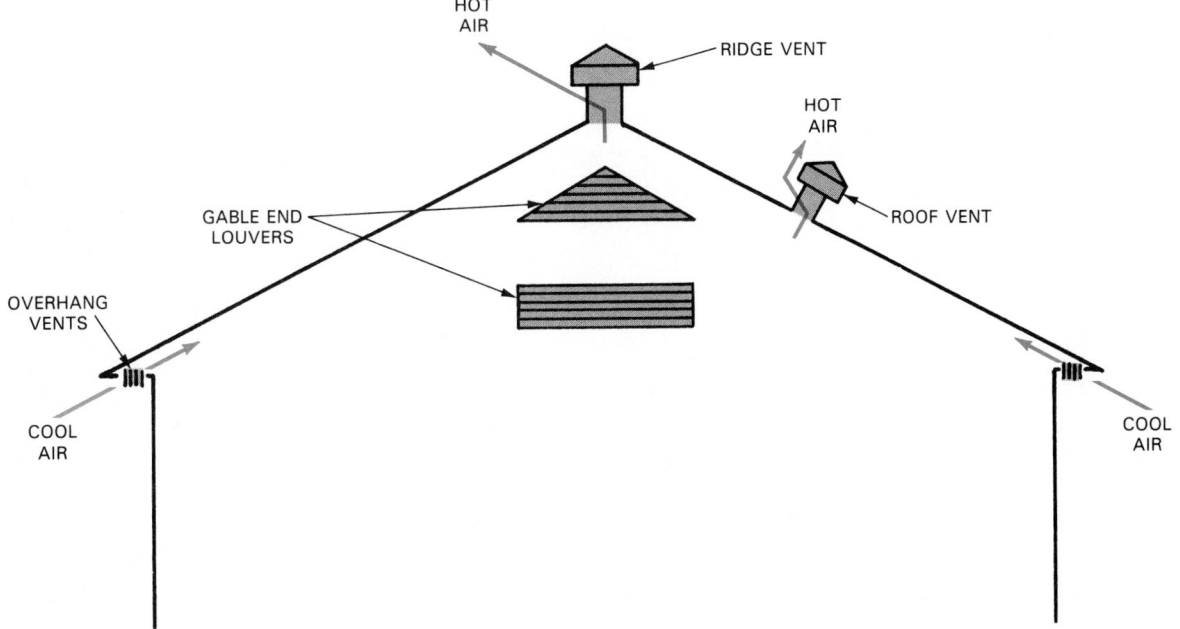

Fig. 26-17. Roof and attic ventilation. If attic can be kept cool in summer, less cooling will be need-ed inside building. A second use of ventilation is to help keep snow from melting on top of roof, running down, and freezing at overhang. Ice dam on overhang can force water under shingles.

Roof coverings are chosen based on the surrounding climate. Dark asphalt shingles help to absorb the sun's heating effect during winter months. The same feature, however, would work against cooling in the summer.

Roof coverings for pitched roofs are installed over sheathing. Insulation can be used beneath the sheathing.

Roof coverings for flat roofs can consist of many layers, as in built-up roofing. Insulation can be a part of the roof covering.

The R-values for pitched roofs are higher than those for flat roofs. Refer to Fig. 26-5. The design must take this into account, along with the color of the roof covering.

26-17 WALL CONSTRUCTION

Building wall construction has changed in the past few years to reduce heat leakage and moisture passage through the wall structure. Fig. 26-18 shows typical brick veneer outside wall construction. During the heating season, the inside vapor barrier is necessary; in summer, the outside vapor barrier is required. If the dew point temperature in winter (based on the indoor relative humidity) is reached between the two vapor barriers in Fig. 26-18, the location of the dew point must not be at the inside vapor barrier, or condensation will take place on the room side of the inside barrier. If the dew point temperature in summer (based on the outdoor relative humidity) is reached between the two barriers, the dew point must not be at the outside vapor barrier, or condensation will occur on the exterior side of the outside barrier. For all weather conditions during the year, it is best if the dew point location is halfway between the two barriers. This means to avoid the following:

1. Trying to get the indoor relative humidity above 21 percent when the outdoor temperature is −20°F (−29°C) or lower.
2. Trying to get the indoor temperature below 67°F (19°C) when the outdoor conditions are higher than 98°F (37°C) and higher than 97 percent relative humidity. In older homes, an approximate 1-in. air space

separates the brick veneer from the sheathing. Vapor barriers were not commonly used.

Where vapor barriers are used, they should be as tightly sealed as possible, even to the extent of tarring the breaks in the seal. The barriers may be made of tarred paper, aluminum foil, or plastic film. Aluminum foil has a reflection value as well as being a vapor-tight seal. Ideally, a hermetically sealed (airtight) insulation should be used in areas that have dew point temperatures that occur during either the heating or cooling season.

Another type of wall construction, more common in warm areas, is the stucco wall. Stucco is a thin coating of cement. It is spread over concrete blocks or a metal mesh which is nailed to a wood stud wall. This type of finish can be applied by hand-troweling or machine spray.

A variety of all wood wall constructions are also found. In most newer construction, all wood walls are made of some type of hardwood siding, fit together over the sheathing. The remainder of the wall is similar to a brick construction. Some of the more decorative constructions include siding made from cedar and redwood.

Aluminum siding is nearly as common as the brick exterior. Aluminum siding is available in a variety of colors, textures, and styles, and is usually installed over the sheathing material. Newer aluminum siding applications have a layer of insulation bonded to the back side of the aluminum. The aluminum is installed in an overlapping pattern with holes provided on the bottom edges of the siding for moisture release.

26-18 BASEMENT HEAT LOSS

Heat losses or gains for basements vary widely. The heat loss for a basement built about five feet into the ground is shown in Fig. 26-19. The deeper the basement goes in-

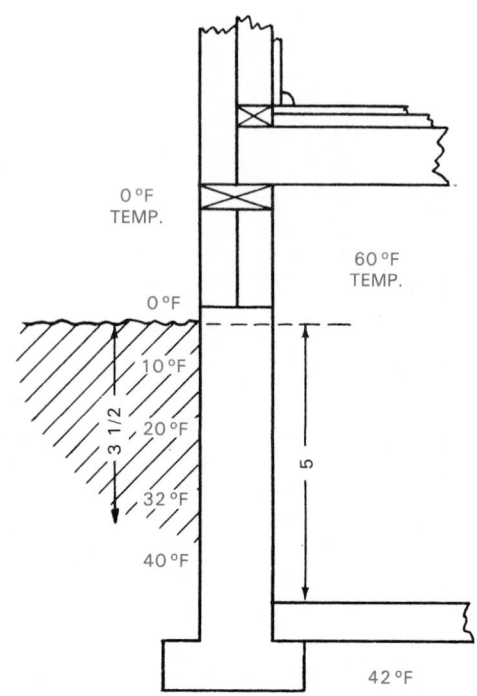

Fig. 26-18. Brick veneer wall construction. A—Brick. B—Outside vapor barrier. C—Sheathing. D—Stud and insulation. E—Inside vapor barrier. F—Drywall. G—Approximate dew point location.

Fig. 26-19. Temperature conditions and construction of building with basement.

to the ground, the less the heat loss. It is usually assumed that a basement is at 60 °F (16 °C). This means a floor loss and a basement wall loss. Leakage through the basement floor is usually not calculated. The usual heat leakage load is calculated through the first floor of the building (the basement ceiling).

Buildings built on a concrete slab have heat losses of a different nature than those built with a basement. See Fig. 26-20. With the building on a slab, the heat loss above ground is calculated in the typical manner. One method for

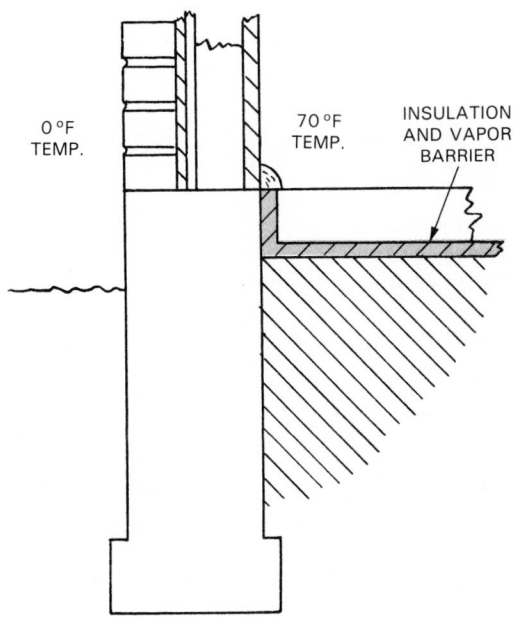

Fig. 26-20. Temperature conditions and construction of building built on concrete slab.

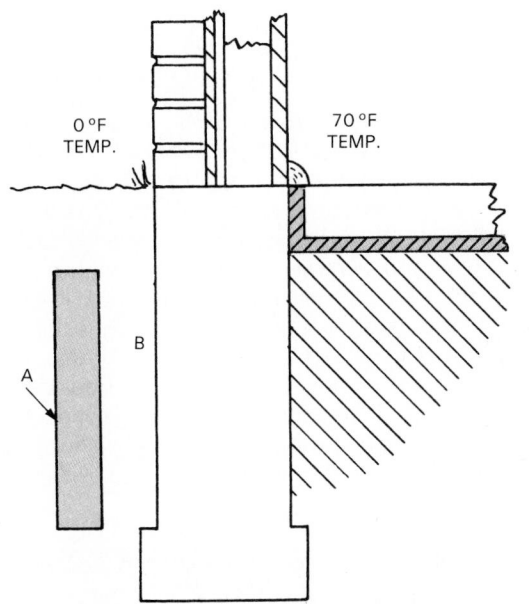

Fig. 26-21. A method is shown for preventing ice and frost from forming around perimeter of building built on slab. A—Rigid urethane insulation. B—Install as close to building as convenient.

minimizing the ice and frost which forms around the outside of a slab building is to use a rigid urethane slab at least 2 in. thick installed 2 to 4 ft. in the ground. Fig. 26-21 shows a typical installation. Heat losses for the slab design are usually calculated by determining the perimeter of the building and multiplying the total length by 18 Btu/hr. for each foot of length (0 °F [−18 °C] design temperature).

Another popular type of foundation is that which leaves just enough space between the floor and the ground to allow access. This is referred to as a crawl space. The earth floor of a crawl space should have a vapor barrier on it (plastic sheeting, roofing paper, etc.). In addition, the floor can be insulated from underneath to provide maximum thermal protection. A crawl space must have sufficient venting to minimize moisture problems in the summer and also minimize the amount of cold air entering in the winter. Vents with dampers are installed to serve this dual purpose.

Recent construction practices call for vapor barriers (plastic sheeting) between the basement walls and the surrounding ground. Refer to the recommendations of the American Society of Heating, Refrigeration and Air Conditioning Engineers for additional data for other conditions.

26-19 UNHEATED SPACES

In many buildings, certain areas such as closets, hallways, and attics are unheated. These spaces receive their heat from heat leakage through partitions, ceilings, and floors. Generally, the unheated areas are assumed to be at a temperature halfway between the indoor design temperature and the ambient design temperature.

26-20 INFILTRATION

Since buildings are not airtight, air will leak in if there is any difference between inside air pressure and outside air pressure. Air also leaks out of the building under the same conditions.

The air pressure difference is usually caused by wind. Parts of the building that the wind is pressing against are those areas through which the air leaks in. The remaining areas are those areas through which the air leaks out, Fig. 26-22. Each arrow in Fig. 26-22 shows a possible direction air leakage can take.

Infiltration has one very important role in air conditioning. It provides the fresh air necessary for living comfort and health. It is important to insure that sufficient fresh air is entering a building during both the heating season and cooling season for health purposes.

During the heating season, any cold air that filters in must be heated, and air that leaks out represents lost heat. During the cooling season, warm air that filters in must be cooled, and cooled air that filters out is like a heat gain (some cooling effect is lost). If the building can be sealed, this infiltration and exfiltration can be minimized. However, care must be taken to provide enough fresh air for ventilation and combustion purposes.

Another way to prevent unwanted infiltration is to maintain a positive air pressure within the building. The pressurized air will filter out at the cracks and openings in the building. With this practice, a special fresh air intake is needed, so incoming air can be conditioned before it is admitted to the rooms in the building.

Infiltration calculations can be based on the total volume

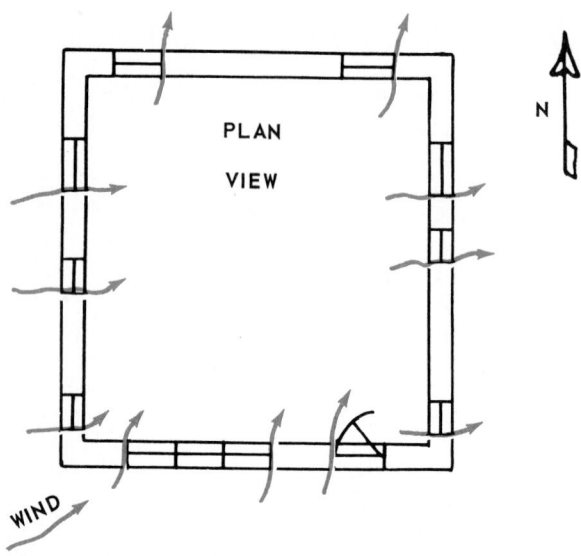

Fig. 26-22. Diagram illustrates how wind direction affects air leakage into and out of house.

SURFACE	AREA	R	U	TEMP. DIFF.	HEAT LEAKAGE Btu/hr.
WALL, GROSS	996	0.73	1.38		
WINDOW	116	0.89	1.13	70	9176
WALL, NET	880	4.0	0.25	70	15400
CEILING	768	1.61	0.62	35	16666
FLOOR	768	2.94	0.34	25	6528
				TOTAL	47770

Fig. 26-24. Typical heat load calculation for 24 x 32-ft. home having 8-ft. ceiling height.

of the building, or by measuring the length and size of all the cracks in the building. Fig. 26-23 lists the air changes in buildings. If a building has a volume of 10,000 cu. ft., it will have at least 10,000 cu. ft. of fresh air infiltration per hour. If six people occupy this space, there is 10,000 ÷ 6 or 1667 cu. ft. per hour for each person, or 1667 ÷ 60 = 27.8 cu. ft./min. (0.79 m³/min.), which is a good ventilating value. If this building is constructed with vapor barriers, and all doors and windows are fitted with weatherstripping, the air change will be reduced considerably. It may even be reduced to the point of unsafe ventilation (too little oxygen in the air).

TYPES OF SPACE	NO. OF AIR CHANGES/HR.
1 Side Exposed	1
2 Sides Exposed	1 1/2
3 Sides Exposed	2
4 Sides Exposed	2
Entrances	2-3

Fig. 26-23. Approximate number of air changes desirable per hour for various room exposures.

26-21 TOTAL HEAT LOAD FOR HEATING

It is best to set up total heat load calculations in table form. Fig. 26-24 shows a typical heat load calculation for a 24 by 32-ft. (7.3 by 9.8-m) house. Refer to Fig. 26-5. Note that the temperature difference for the ceiling is only 35°F (19°C). Also consider that the roof serves as added insulation and keeps the winter attic temperature higher than the outdoor temperature. The attic temperature can be accurately calculated by making the heat leaking into

the attic in winter equal the heat leaking out.

Ceiling area × (70°F − [attic temp.]) × U_c = Roof area × (attic temp. − 0°F) × U_r.

Most homes have an 8-ft. ceiling, although many homes are now being built with 7 1/2-ft. (2.3 m) ceilings. The lower ceiling helps avoid trapping hot air near the ceiling. The hot ceiling air can be used for heating.

Each heat leakage value is obtained by means of the following formula:

Heat leakage = area × U × temp. diff.

It is sometimes desirable to calculate the total heat loss for one degree, then multiply this value by the design temperature difference to obtain the total heat loss.

A very fast method used to estimate total heat loads is shown in Fig. 26-25. Note that the heat load is based on room volume. The table also includes cooling. A method for estimating duct sizes is shown in Fig. 26-26. This table is used as a companion to the heat load table. For a given room volume, a recommended supply and return duct size is given. Outlet and return grille areas are also shown. More information on the calculation of proper air distribution systems is found in Chapter 22.

Standard worksheets are available for calculating total heat load. See Fig. 26-27.

26-22 TOTAL HEAT GAIN FOR COOLING

Heat gain calculations for a building to determine the total cooling load are similar to the calculations for heat loss.

The temperature difference is based on the locality being considered. The indoor temperature usually is designed to be 75°F (24°C) at 50 percent relative humidity (RH). Therefore, if the summer design temperature is 100°F (38°C), the temperature difference is 25°F (14°C). This temperature difference is for load calculations only. In practice, a 10 to 15°F (6 to 8°C) difference is recommended.

Miscellaneous heat sources must be considered. Sun load, electrical load, and occupants are large enough sources of heat to be included in the heat load calculations. The following paragraphs describe some of the specifics of these heat sources.

26-23 WINDOW HEAT LOAD FOR COOLING

Heat flow through ordinary window glass is approximately three times as great as heat flow through ordinary residential roofs and ceilings. The U factors (see Para. 26-7, Fig. 26-5, and Fig. 26-13) are: ordinary glass, 1.04;

RESIDENTIAL FORCED AIR SYSTEM DESIGN GUIDE (For Estimating Purposes Only)

ROOM VO. CU. FT.	4 A/C CFM	WINTER HEAT LOSSES HEATING BTU @ SUPPLY °F.						SUMMER HEAT GAINS COOLING BTU @ SUPPLY °F.		
		120°	130°	140°	150°	160°	170°	60°	55°	50°
200	14	750	910	1060	1210	1360	1510	300	375	450
300	20	1080	1300	1510	1730	1945	2160	430	540	650
400	27	1460	1750	2040	2330	2625	2915	580	730	875
500	34	1830	2200	2570	2940	3300	3670	735	920	1100
600	40	2160	2590	3025	3455	3890	4320	865	1080	1290
700	47	2540	3025	3550	4060	4565	5080	1015	1270	1530
800	55	2970	3565	4155	4750	5345	5940	1190	1485	1780
900	60	3240	3880	4535	5185	5830	6480	1295	1620	1940
1000	65	3500	4210	4915	5615	6320	7000	1400	1755	2100
1100	75	4050	4860	5670	6480	7290	8100	1620	2000	2430
1200	80	4320	5200	6040	6910	7780	8640	1730	2160	2600
1300	87	4700	5635	6570	7520	8455	9400	1880	2350	2820
1400	95	5130	6155	7180	8200	9235	10260	2050	2560	3080
1500	100	5400	6480	7560	8640	9720	10800	2160	2700	3240
1600	107	5775	6930	8090	9245	10400	11550	2310	2890	3460
1700	113	6100	7320	8540	9760	10950	12200	2440	3050	3660
1800	120	6480	7775	9070	10360	11665	12960	2590	3240	3880
1900	125	6750	8100	9450	10800	12150	13500	2700	3370	4050
2000	135	7300	8750	10200	11660	13120	14600	2920	3645	4370
3000	200	10800	12960	15120	17280	19440	21600	4320	5400	6480
4000	265	14300	17170	20035	22890	25750	28600	5720	7150	8580
5000	335	18100	21700	25325	28940	32560	36200	7240	9040	10850
6000	400	21600	25920	30240	34560	38880	43200	8640	10800	12960
7000	465	25100	30130	35150	40170	45200	50200	10040	12550	15060
8000	535	26900	34670	40450	46220	52000	53800	10760	14440	17330
10000	670	36180	43410	50650	57890	65125	72360	14470	18090	21700
12000	800	43200	51840	60480	69120	77760	86400	17280	21600	25920
14000	935	50500	60590	70680	80780	90880	101000	20200	25240	30290
16000	1065	57500	69000	80510	92010	103520	115000	23000	28750	34500
18000	1200	64800	77760	90720	103680	116640	129600	25920	32400	38780
20000	1335	72100	86500	100920	115340	129760	144200	28840	36040	43250
25000	1670	90200	108215	126250	144290	162320	180360	36070	45090	54100

(Note: first column also labeled vertically "MAX. AIR ON ONE OUTLET")

BASED ON 70 °F RETURN AIR — BASED ON 80 °F RETURN AIR

Fig. 26-25. Approximate heat load chart for winter heating and summer cooling is based on volume of conditioned space. (Detroit Edison Co.)

residential roofs, 0.31. Therefore, air conditioning of areas containing a large amount of ordinary glass can become a problem. To reduce the heat conductivity through glass, storm sash is used. To reduce the solar heat through glass, it is advisable to use special types of glass which have high heat-reflecting qualities.

Special heat-absorbing glass can reduce the solar heat load by as much as 30 percent. Another method is to use glass tinted a bluish gray to reduce the solar glare and cooling load.

Roof extensions over a window will reduce the area ex-

posed to the sun. Double glazed windows also exposed to sun rays reduce solar heat absorption approximately 15 percent.

Awnings to shade glass windows exposed to the sun are also recommended. They can reduce the heat load 55 percent.

26-24 SUN HEAT LOAD

The heat energy that comes from the sun adds considerable heat load to the total load during the summer. The

RESIDENTIAL FORCED AIR SYSTEM DESIGN GUIDE

ROOM VOLUME CU. FT.	AIR DISTRIBUTION SIZES: SUPPLY DUCT (IN.) R'ND. (DIA.)	SUPPLY DUCT (IN.) EQUIV.	OUTLET (IN.) FLOOR	OUTLET (IN.) WALL	OUTLET (IN.) CEILING (DIA.)	RETURN (IN.) GRILLE	RETURN DUCT (IN.) R'ND. (DIA.)	RETURN DUCT (IN.) EQUIV.
200	4	4 1/2 x 3			4	6 x 10	6	8 x 4
300	4	4 1/2 x 3			4	6 x 10	6	
400	4	4 1/2 x 3			4	6 x 10	6	
500	4	4 1/2 x 3			4	6 x 10	6	
600	5	10 x 2 1/4	2 1/4 x 10		4	6 x 10	6	
700	5	8 x 3 1/4	2 1/4 x 10	4 x 10	6	6 x 10	6	
800	5	5 x 4	2 1/4 x 10		6	6 x 10	6	
900	6	14 x 2 1/4	2 1/4 x 10		6	6 x 10	6	
1000	6	10 x 3 1/4	2 1/4 x 10		6	6 x 10	7	8 x 6
1100	6	8 x 4	2 1/4 x 12		6	6 x 10	7	
1200	6	6 x 5	2 1/4 x 12	10 x 6	6	6 x 10	7	
1300	6	6 x 5	2 1/4 x 12		6	6 x 10	7	
1400	7	14 x 3 1/4	2 1/4 x 14		6	6 x 14	8	8 x 7
1500	7	11 x 4	2 1/4 x 14	12 x 6	8	6 x 14	8	
1600	7	8 x 5	4 x 10	14 x 6	8	6 x 14	8	
1700	7	7 x 6	4 x 10	14 x 6	8	6 x 14	8	
1800	7		4 x 12	14 x 6	8	6 x 14	8	
1900	7		4 x 12	14 x 6	8	6 x 14	8	
2000	7		4 x 12	14 x 6	8	6 x 14	8	
3000	7 1/2	13 x 4				8 x 14	10	8 x 11
4000	9	8 x 8				6 x 24	11	8 x 13
5000	10	8 x 11				6 x 30	12	8 x 16
6000	11	8 x 13				8 x 30	13	8 x 18
7000	11 1/2	8 x 14				8 x 30	14	8 x 22
8000	12	8 x 16				18 x 18	15	8 x 24
10000	13	8 x 18				18 x 18	16	8 x 28
12000	14	8 x 22				18 x 24	18	8 x 36
14000	14 1/2	8 x 24				24 x 24	20	8 x 46
16000	15	8 x 26				24 x 24	20	
18000	16	8 x 30				24 x 30	20	
20000	17	8 x 34				24 x 30	22	8 x 60
25000	18	8 x 39				24 x 30	22	

Fig. 26-26. Duct sizes based on estimated values found in Fig. 26-25. (Detroit Edison Co.)

sun's rays in the northern hemisphere shine on the east wall, south wall, west wall, and on those roof sections that are exposed to its rays. Therefore, when computing total heat load, the heat from the sun must be considered on the east wall in the morning, on the south wall all day long, and on the west wall in the afternoon, as shown in Fig. 26-28.

The sun releases different amounts of heat to surfaces, depending upon the part of the world in which the building is located. The approximate maximum heat pickup or heat gain from the sun is 330 Btu per hr. per sq. ft. (97 watts/sq. ft. [1 040 W/m^2]). This condition exists for a black surface at right angles to the sun's rays near the equator (tropic). Any other color or any surface at an angle to the sun's rays will receive less than this amount of heat.

At the 42nd parallel (a line going through New York City, Cleveland, and Salt Lake City), the maximum heat from the sun's rays is about 315 Btu per hr. per sq. ft. (92 watts/sq. ft. [993 W/m^2]). Much of the heat from the sun is reflected back into the atmosphere. The heat gain through windows (that must be removed with air conditioning cooling) is listed in Fig. 26-29 for windows facing different directions.

If the windows are not protected with awnings, use a temperature of 15 °F (8 °C) higher than outside ambient temperature to get correct results. The effect of the sun shining on walls may also be taken care of by adding 15 °F (8 °C) to the ambient temperature.

The approximate values obtained by using the 15 °F (8 °C) temperature correction generally are usable.

Customer's Name _____ Address _____

City _____ State _____ Zip _____ Telephone Number _____

WINTER: Inside Design Temp _____ °F—Outside Design Temp _____ °F = Heating Temp Difference _____ °F

SUMMER: Outside Design Temp _____ °F—Inside Design Temp _____ °F = Cooling Temp Difference _____ °F

HEATING		COMMON DATA SECTION		COOLING	
BTUH LOSS	HEATING FACTOR	SUBJECT	SQ. FT.	COOLING FACTOR	BTUH GAIN
	FROM TABLE E	GROSS WALL		FROM TABLE E	
		DOORS & WINDOWS (Table A or B)			
		NET WALL			
		CEILING			
		FLOORS			

Infiltration Btu/hr	=	Heating Table D	×10×	1.1/60 ×	Volume (Cu. Ft.)	Volume (Cu. Ft)	×	1.1/60	× △T ×	Cooling Table D	=	Infiltration Btu/hr
	=		×0.18333	×			×	0.01833	×	×	=	

		SUB-TOTAL BTUH LOSS (per 10°F)			
x		ADJUSTMENT FACTOR (Table C)			
		TOTAL BTUH LOSS			
		PEOPLE_____ × 300 BTUH GAIN (Assume 2 persons per bedroom)			
		APPLIANCES BTUH			1200
		SUB-TOTAL BTUH GAIN (room sensible only)			
x		DUCT LOSS/GAIN FACTOR (Table F)			x
		SUB-TOTAL BTUH (Sensible Gain)			
		MOISTURE REMOVAL (sub total × 1.3)			× 1.3
		TOTAL BTUH LOSS/GAIN			

TABLE A—HEATING—DOORS & WOOD FRAME WINDOWS (PER 10°F)

For sliding glass doors - use factors for the same type window construction.

Window & Door Types	Frames			x Area	= Btuh Loss
	Wood	TIM	Metal		
Single Pane Clear	9.90	10.45	11.55		
With Storm	4.75	5.25	6.50		
Double Pane Clear	5.51	6.09	7.25		
With Storm	3.41	3.85	4.90		
Triple Pane Clear	3.80	4.39	5.46		
Jalousie Single	—	—	11.0		
Single w/storm	—	—	5.0		
Skylights Single	11.07	11.69	12.92		
Double	6.65	7.35	8.75		
Door Wood Only	4.60	—	—		
Wood w/storm	3.20	—	—		
Urethane Core (R-5)	—	—	1.90		
Urethane Core (R-5) w/storm	—	—	1.70		
			TOTALS		

TABLE B — COOLING — DOORS & WINDOWS

Factors assume windows have inside shading by draperies or venetian blinds and sliding glass doors are treated as windows.

	SINGLE GLASS			DOUBLE GLASS			TRIPLE GLASS			X Area	= BTUH GAIN
	TEMP. DIFF.			TEMP. DIFF.			TEMP. DIFF.				
Direction	15°	20°	25°	15°	20°	25°	15°	20°	25°		
N	18	22	26	14	16	18	11	12	13		
NE & NW	37	41	45	31	33	35	26	27	28		
E & W	52	56	60	44	46	48	38	39	40		
SE & SW	45	49	53	39	41	43	33	34	35		
S	28	32	36	23	25	27	19	20	21		
Skylights	164	168	172	141	143	145	132	136	140		
Wood ①	8.6	10.9	13.2	8.6	10.9	13.2	8.6	10.9	13.2		
Metal ②	3.5	4.5	5.4	3.5	4.5	5.4	3.5	4.5	5.4		

① For wood doors and polystyrene core metal doors
② For urethane core metal doors

TOTALS

TABLE D — INFILTRATION MULTIPLIERS
Winter Air Changes Per Hour

Floor Area	900 or less	900-1500	1500-2100	over 2100
Best	0.4	0.4	0.3	0.3
Average	1.2	1.0	0.8	0.7
Poor	2.2	1.6	1.2	1.0

For each fireplace add:		Best	Average	Poor
		0.1	0.2	0.6

Summer Air Changes Per Hour

Floor Area	900 or less	900-1500	1500-2100	over 2100
Best	0.2	0.2	0.2	0.2
Average	0.5	0.5	0.4	0.4
Poor	0.8	0.7	0.6	0.5

TABLE C — ADJUSTMENT FACTORS — (HEATING)

°F. Temperature Diff.	30	40	50	60	70	80	90
Adjustment Factor	3	4	5	6	7	8	9

Fig. 26-27. Use this worksheet to calculate heat losses and gains. (Continued, next page.) (The Trane Company)

20. What is the usual design wind velocity?
21. What is conductance?
22. Are the outside or the inside dimensions used when a person is determining wall areas?
23. Which has the greatest heat flow resistance, inside air film or outside air film?
24. Describe perimeter heat loss for grade level slabs.
25. What is a vapor barrier?

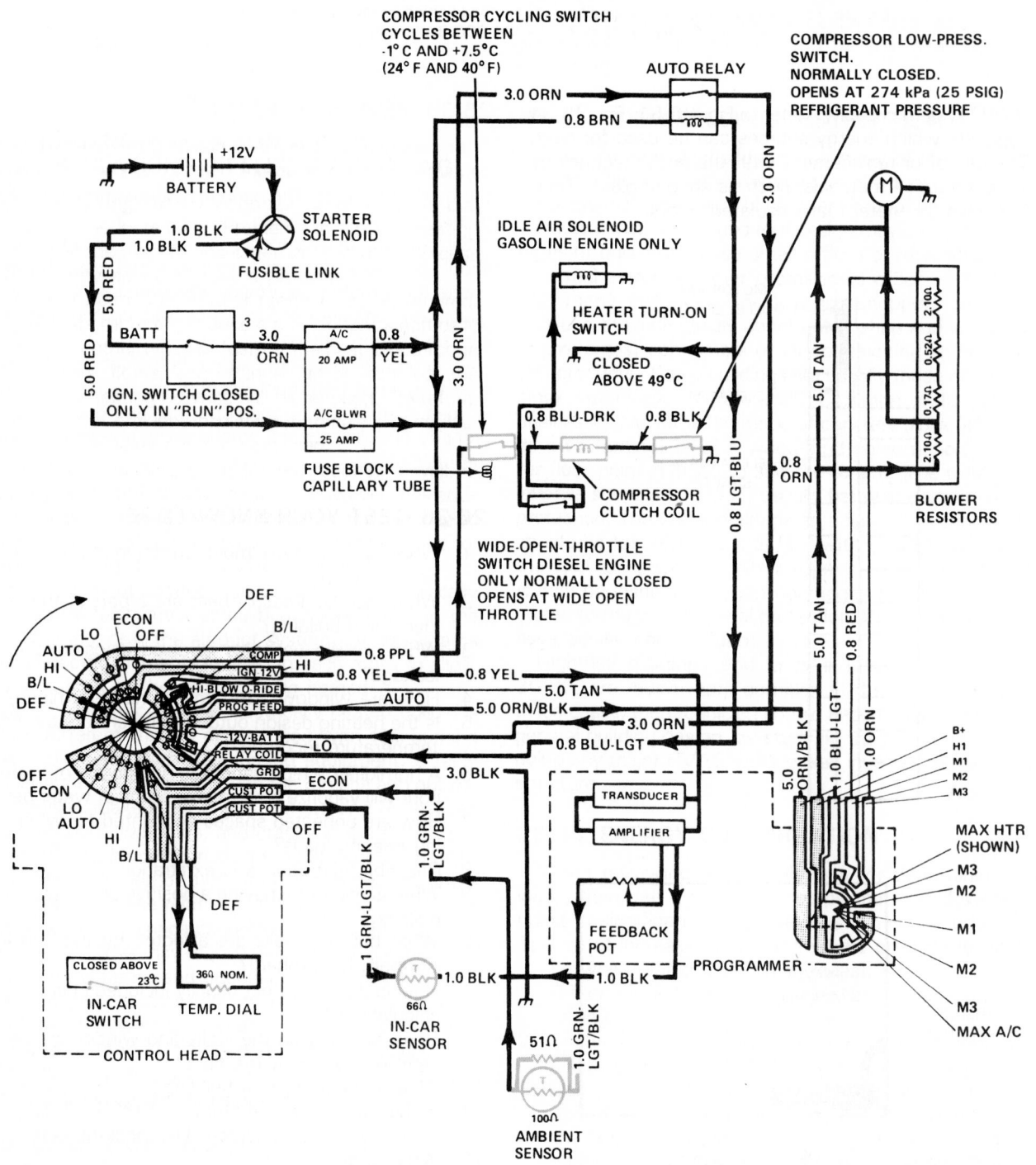

Typical automobile automatic temperature control system. Note how in-car switch, temperature dial, in-car sensor, and ambient temperature sensor are connected in circuit. Also note compressor cycling switch with capillary tube (orange) and compressor low-pressure switch (orange). These switches control compressor clutch coil (orange). (Buick Motor Division, General Motors Corporation)

Chapter 27

AUTOMOBILE AIR CONDITIONING

After studying this chapter, the technician will be able to:
☐ Explain how automotive air conditioning varies in design and application from stationary systems.
☐ List and describe the types of automotive compressors.
☐ Service and repair automotive air conditioning units.

Many of the principles discussed in earlier chapters apply also to automotive air conditioning. There are many unique applications and slight variations in design that the technician should understand.

The automobile air conditioner uses a refrigerating system driven by the car's engine to furnish cooling action. Warm engine coolant usually provides heat to the passenger compartment when needed.

To fully understand this chapter, study Chapters 1, 2, 4, 5, 6, 8, 9, 14, 18, and 19 first. This will provide a solid foundation in fundamentals and insure a better grasp of the principles of automotive air conditioning systems.

Conditioning the air in the interior of a car involves heating, cooling, and dehumidification. The heat to warm the passengers is usually provided by circulating hot coolant (water and antifreeze mixture) through a heater core (small radiator-like device under dash). The engine water pump forces the warm coolant through heater hoses and inside of the heater core.

When cooling is needed, the air conditioning system is brought into operation. It uses an evaporator inside a plenum chamber (enclosure or housing under dash) to cool the air circulated through the passenger compartment. The car's engine, using a belt, drives the compressor to pump R-12 through the system.

As will be learned, the car's heating system and air con-

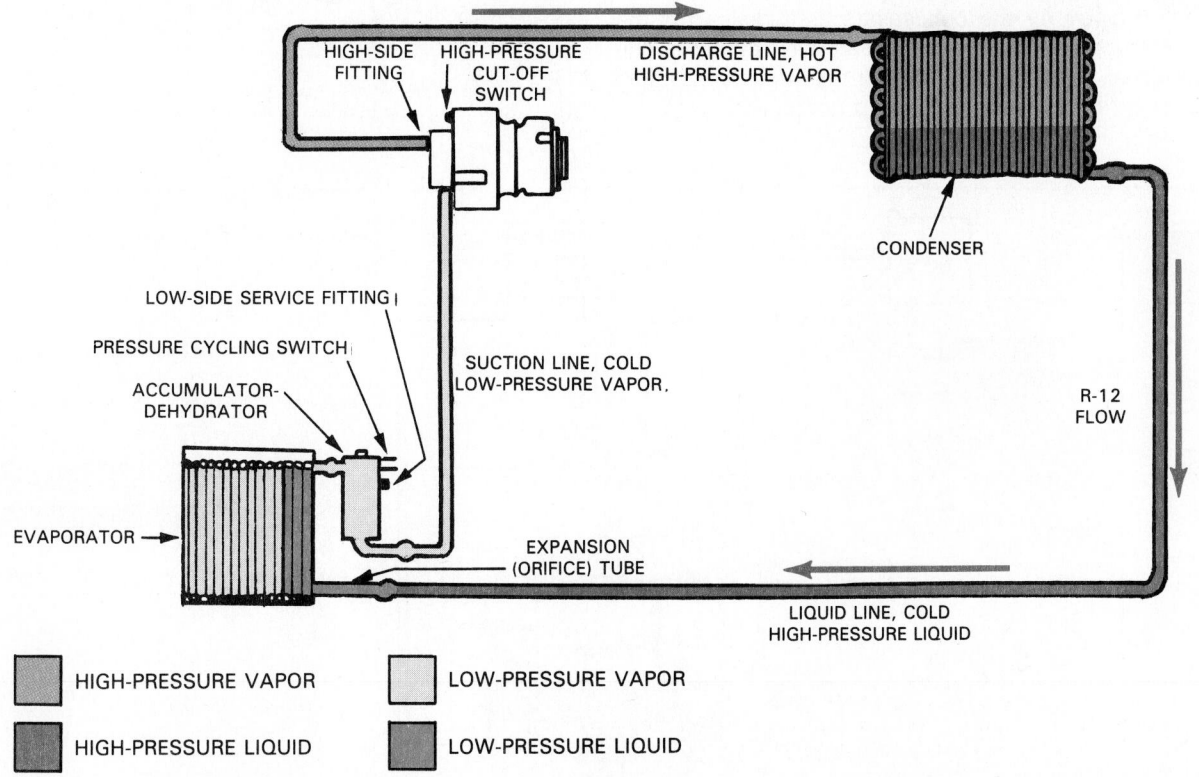

Fig. 27-1. Basic refrigeration cycle of a typical air conditioning system. Note refrigerant flow.   (Buick Motor Div., General Motors Corp.)

ditioning (refrigeration) system work together to provide comfort for the passengers. Both manual and automatic type systems are used on late model cars.

27-1 PURPOSE OF AUTOMOTIVE AIR CONDITIONING

An automobile passenger compartment is relatively small. Yet, when a car travels at high speed on a hot summer day, it will require considerable refrigerating capacity to keep the interior at a comfortable temperature. Likewise, the same car, traveling on a cold winter day, will require considerable heating capacity.

The mechanism and controls of a factory-installed air conditioning system are arranged to ease the task of selecting and controlling temperature. Fig. 27-1 shows the makeup of a typical automotive refrigerating system used for air conditioning. Fig. 27-2 shows a basic automatic temperature central air conditioning system. The system maintains the preset temperature by the in-car sensor. Ambient sensor and solar sensor detect the outside air and sunlight. The interior temperature is detected by the interior sensor. These signals are sent to the computer where they are converted to signals to operate the double vacuum valve (DVV). This, in turn, sets the air conditioning and heater settings. The computer signals the DVV and provides the proper vacuum, which is the difference between the preset and the detected temperature. The vacuum actuates the water valve that controls the warm water through the heater core.

The vacuum also actuates the power server that changes the air damper position. This maintains the interior at the desired temperature.

Operating the air conditioner in the summer reduces the humidity of the air inside the car. In addition, moisture (condensate) formed on the evaporator surfaces collects much of the dust and pollen in the air. These entrapped particles are carried away by the condensate as it drains off the evaporator underneath the car. In this way, the air conditioner cleans, dehumidifies, and controls the temperature of the air in the passenger compartment.

27-2 AUTOMOTIVE AIR CONDITIONER OPERATION

Another illustration of an air conditioning system for automobiles is shown in Fig. 27-3. The compressor is mounted on the engine and is driven by a belt. The condenser is mounted ahead and to one side of the car radiator. This allows cool air to flow over the condenser.

In operation, liquid refrigerant flows from the condenser to the liquid receiver which dries and filters the refrigerant. The liquid refrigerant then travels through a refrigerant control device and into the evaporator. When in the evaporator, the R-12 is vaporized and absorbs heat. The vaporized refrigerant finally flows back through the suction line to the compressor.

Meanwhile, a blower (fan) forces inside or outside air through the evaporator. The resulting cool air is circulated to the interior of the car by means of ducts and grilles.

Low-pressure refrigerant vapor enters the compressor through the low side. The vapor is drawn into the cylinder

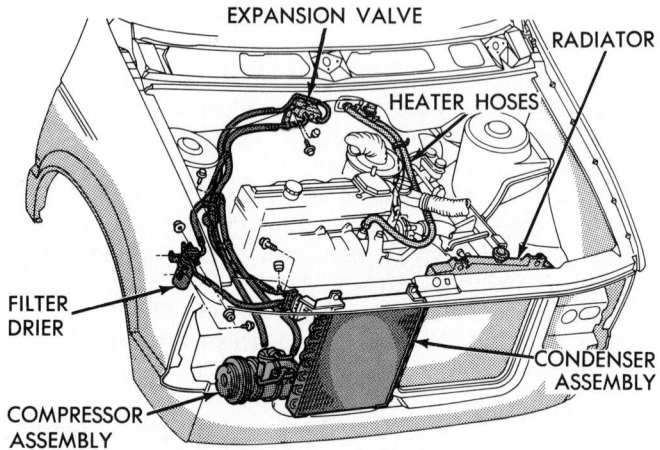

Fig. 27-3. Note location of automotive air conditioning system components. (Chrysler Corporation)

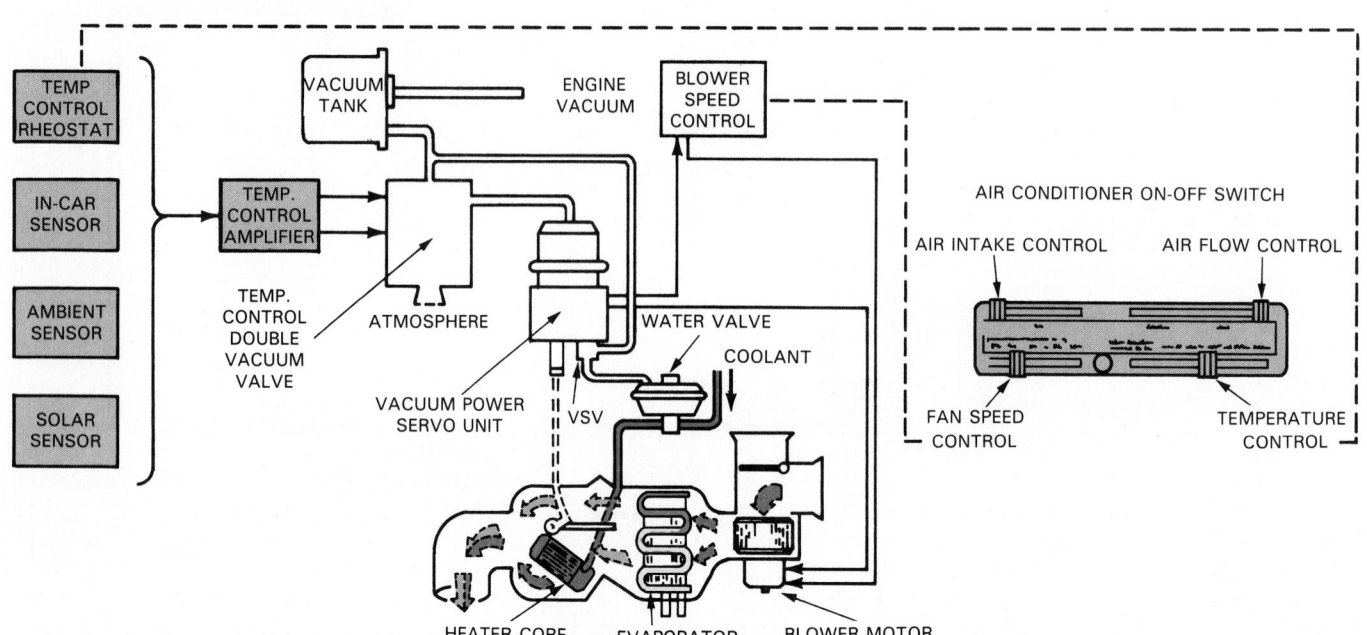

Fig. 27-2. Automatic temperature control air conditioning system. Interior car temperature is maintained by the temperature control, in-car sensor, ambient sensor, and solar sensor. (Toyota Motor Corporation)

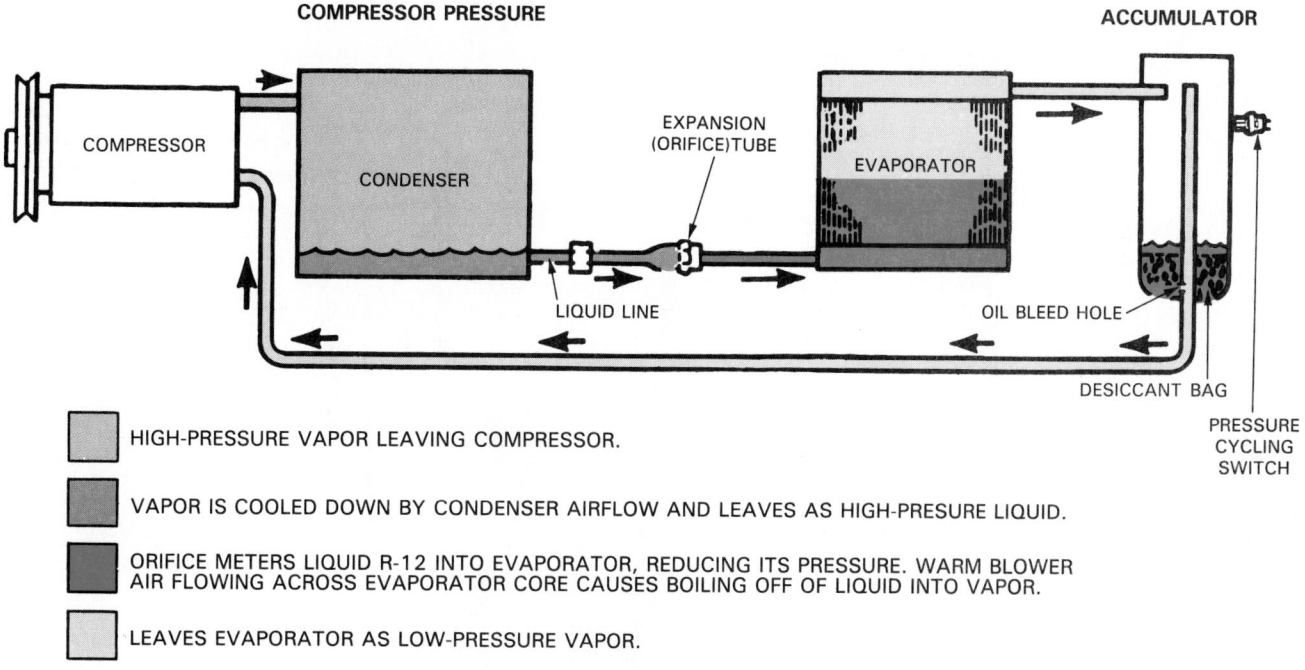

COMPRESSOR PRESSURE

ACCUMULATOR

COMPRESSOR

CONDENSER

EXPANSION (ORIFICE) TUBE

EVAPORATOR

LIQUID LINE

OIL BLEED HOLE

DESICCANT BAG

PRESSURE CYCLING SWITCH

HIGH-PRESSURE VAPOR LEAVING COMPRESSOR.

VAPOR IS COOLED DOWN BY CONDENSER AIRFLOW AND LEAVES AS HIGH-PRESURE LIQUID.

ORIFICE METERS LIQUID R-12 INTO EVAPORATOR, REDUCING ITS PRESSURE. WARM BLOWER AIR FLOWING ACROSS EVAPORATOR CORE CAUSES BOILING OFF OF LIQUID INTO VAPOR.

LEAVES EVAPORATOR AS LOW-PRESSURE VAPOR.

Fig. 27-4. Automobile air conditioning system diagram shows flow of refrigerant in each part of cycle. (Buick Motor Div., General Motors Corporation)

and is compressed by the piston; then it is discharged through the high side. The heat of compression and the latent heat of vaporization absorbed by the refrigerant are given up to the air flowing over the fins of the condenser. The refrigerant is again liquefied, and moves to the liquid receiver or accumulator. Fig. 27-4 shows the condition of the refrigerant in each part of the refrigeration cycle.

If the air conditioning system ran continuously, the temperature in the car would drop to an uncomfortable level and the evaporator would frost over. To prevent this condition, most systems use a magnetic clutch mechanism that allows compressor "free wheeling." The clutch is operated by a thermostat or pressure-cycling switch, which opens the electrical circuit to disengage the compressor. This allows the compressor pulley to rotate while the compressor crankshaft remains stationary.

In cold weather, when the air conditioning system is not turned on, the refrigerating system does not operate because the clutch electromagnet on the compressor is not energized. Instead, hot water is circulated through the heater core. The blower and most of the duct system components are utilized for heating as well. See Fig. 27-5. A thermostat controls coolant flow through the engine's cooling system. By opening only when coolant is up to operating temperature, it allows quick engine warm up.

27-3 OPERATING CONDITIONS

The automobile air conditioner must provide comfort and control conditions in the car during cold, mild, damp, and hot weather. It must provide heating, defogging, and deicing. It must also remove dust, smoke, and odor.

Because the compressor is belt-driven from the engine, compressor speed will vary with engine speed. The system must have enough capacity to provide sufficient cooling at idling speed on the hottest day, in the sun, and under side-

wind conditions. This setup will provide considerable excess capacity for normal driving speeds, particularly under cool weather conditions.

Varying weather conditions can cause problems, both with the control of the temperature and the refrigerant flow (both liquid and vapor) within the system. If the compressor is operating, and little or no refrigeration is needed, the low-side pressure may drop too low.

Decreasing low-side pressure lowers evaporator temperature. The evaporator surface temperature should not be allowed to drop below 33 °F (0.5 °C). If the evaporator should operate at a temperature of 32 °F (0 °C) or lower for any length of time, its surface may become covered with ice. This will stop air circulation through the evaporator and cool air cannot enter the passenger compartment.

Also, operating the system with low-side pressure too low may cause oil pumping. This condition may damage the compressor valves and, if continued, may burn out the compressor.

Various cycling systems and mechanical systems have been devised to overcome these problems. Remember that fresh air ducts must be closed during high-heat loads in the summer to obtain maximum cooling.

A typical automobile air conditioning system will cool an automobile from 110 °F (43 °C) down to 85 °F (29 °C) in about 10 minutes. The inside of the car may reach 150 °F (66 °C) when parked in the sun with the windows closed. The greatest heat load or heat gain is the sun load and heat conducted through the car windows.

Automobile air conditioning systems use anywhere from no fresh air (all air recirculated) to 100 percent fresh air. The fan or blower motor uses approximately 200 watts or 15 A and delivers from 250 to 275 cfm.

The operation of an air conditioning system may reduce fuel economy by as much as 10 percent. This is primarily due to the power needed to turn the compressor shaft.

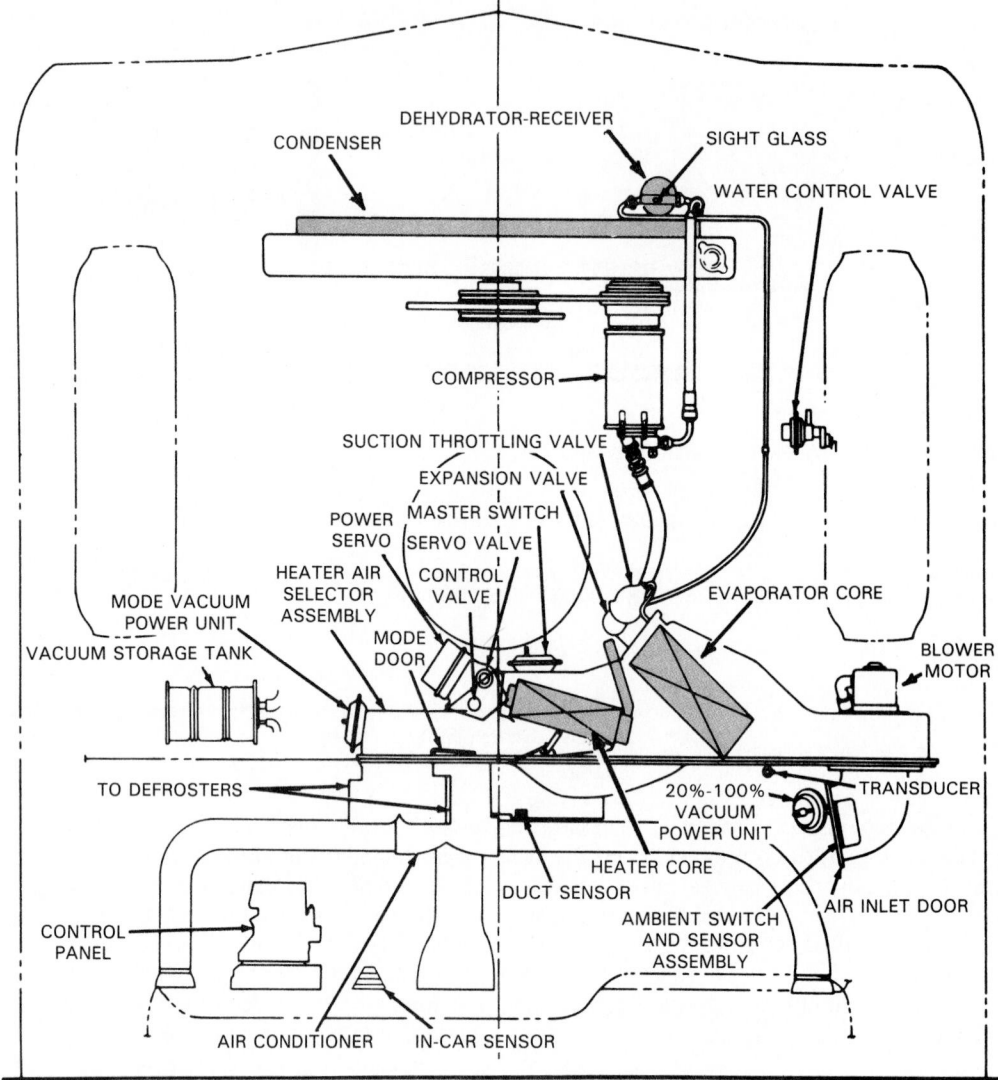

Fig. 27-5. Top view of year-round, all season air conditioning system showing main components. (Cadillac Motor Car Div., General Motors Corp.)

27-4 COOLING CAPACITY

Automobile air conditioning systems range in size or cooling capacity. Their output range is similar to the one- to four-ton residential or commercial units. A cooling capacity of 12,000 Btu/hr is minimum. This is equivalent to a one-ton machine. See Para. 1-38. Capacities up to 48,000 Btu/hr are available for special applications—station wagons or vans, for example.

The capacity of the air conditioning system should match car size. Undercapacity may result in inadequate cooling in hot weather. Overcapacity is uneconomical and causes frequent cycling. The systems are usually designed to keep the inside of the automobile 15° to 20°F (9° to 12°C) below the outside (ambient) temperature when the car is traveling about 30 mph (48 km/hr.).

Fig. 27-6 shows how the horsepower required to operate the air conditioning system varies as car speed changes. As the automobile speeds up, the capacity of the compressor will increase. As it slows down, capacity will decrease. This variation in output is somewhat parallel to the changing heat load, except when the car is parked or is in slow moving traffic. One solution to this problem is the variable displacement compressor, which is covered in Para. 27-9. At these critical times, compressor capacity may be below normal. A partial solution is to idle the engine at a higher speed; another is to travel in traffic in a lower gear to obtain higher engine speeds.

Larger air conditioning systems can consume as much as 8 hp (6 kW) from the engine at high speeds. Capacity at this speed will be approximately 48,000 Btu/hr for a four-ton equivalent system. This means that 2 hp (1.5 kW) is used for each ton of refrigeration. Compare this to the use of 1 hp (0.75 kW) for each ton of refrigeration in a motor-driven, constant-speed compressor comparably built and with the evaporator and condenser more ideally located.

27-5 MAGNETIC CLUTCH

Automotive air conditioning compressors have a mechanism that permits the engine to run with the compressor disengaged. A clutch is used to engage or disengage the compressor belt drive pulley and the compressor shaft. The clutch is operated by forcing a clutch disk against the

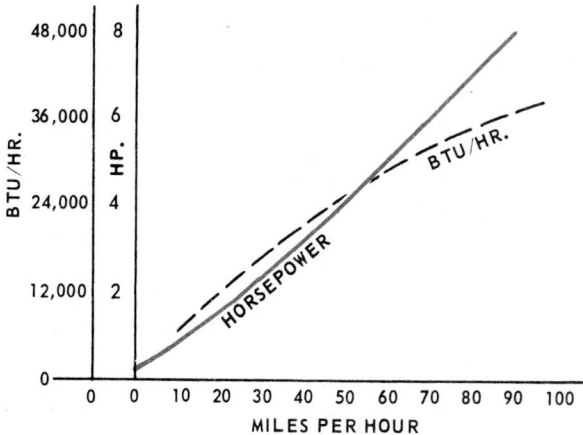

Fig. 27-6. Curves show relationship between car speed, heat load, and horsepower required to drive automobile air conditioning system.

pulley through the use of electromagnetism. It is controlled by a thermostatic switch or a high- or low-pressure cycling switch. This device is called a magnetic clutch. Its principle of operation is shown in Fig. 27-7. Its general construction, including the magnetic field circuit, is in Fig. 27-8.

Two basic designs of magnetic clutches have been used:

1. A revolving magnetic coil that turns when the compressor revolves. It has two carbon brushes in contact with two copper rings mounted on the coil.
2. A stationary magnetic coil mounted on the compressor body. It has two electrical leads, one from the thermostatic or low-pressure switch and one to ground. Fig. 27-9 shows the design of one type of stationary coil electromagnetic clutch.

When the temperature of the return air to the evaporator is brought down to a predetermined setting, a thermostat or low-pressure cycling switch disengages the compressor clutch. This causes the pulley to "free wheel" on its shaft, and the compressor stops.

There are several makes of electromagnetic clutches on the market. The electric current needed to magnetize the various clutches is approximately the same. They draw 2

to 3 A at normal voltage.

Some systems use a thermostat, sensor, or low-side pressure valve to open the compressor clutch circuit when the ambient (outside) temperature is 32 °F (0 °C) or lower. This prevents the evaporator from frosting and freezing.

27-6 COMPRESSOR PROTECTION AND CONTROL SWITCHES

Several devices are used to protect the compressor from extremely high- and low-pressure conditions. A high-pressure cutoff switch protects the compressor from operating at extreme high pressures. The high-side pressure switch is usually mounted on the compressor housing. When the high-side pressure exceeds a predetermined point, the switch opens and cuts off current to the compressor clutch. When the pressure drops, the switch closes and compressor operation resumes.

The low-pressure cutoff switch is normally mounted on the accumulator. It prevents compressor damage by disengaging operation when the refrigerant charge in the system is lost due to a leak. When the low-side pressure falls below a predetermined point, the low-pressure condition is sensed by the low-pressure cutoff switch. The switch then opens, thereby turning off the compressor clutch to prevent internal compressor damage.

Some air conditioning systems use a wide open throttle switch to disengage the compressor during periods of rapid acceleration. Its purpose is to reduce engine load or power loss when needed, as when passing other cars, climbing

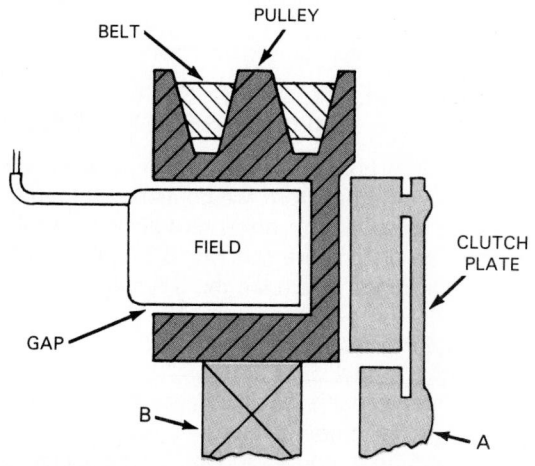

Fig. 27-7. Schematic of an electromagnetic clutch. Note stationary magnetic field. A—Clutch plate. B—Pulley rides on bearing.

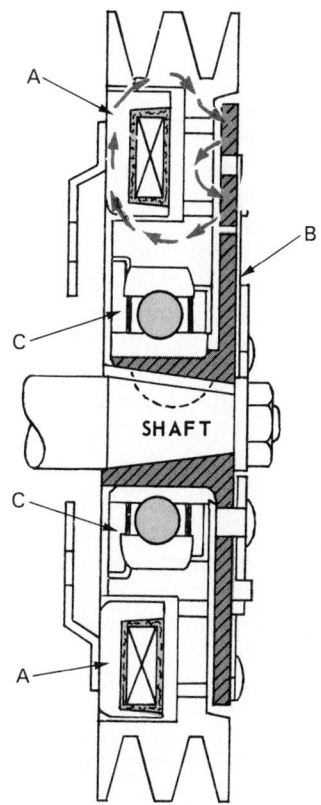

Fig. 27-8. Compressor pulley with stationary electromagnet, showing magnetic field. A—Electromagnetic coil. B—Clutch disk. C—Bearings.

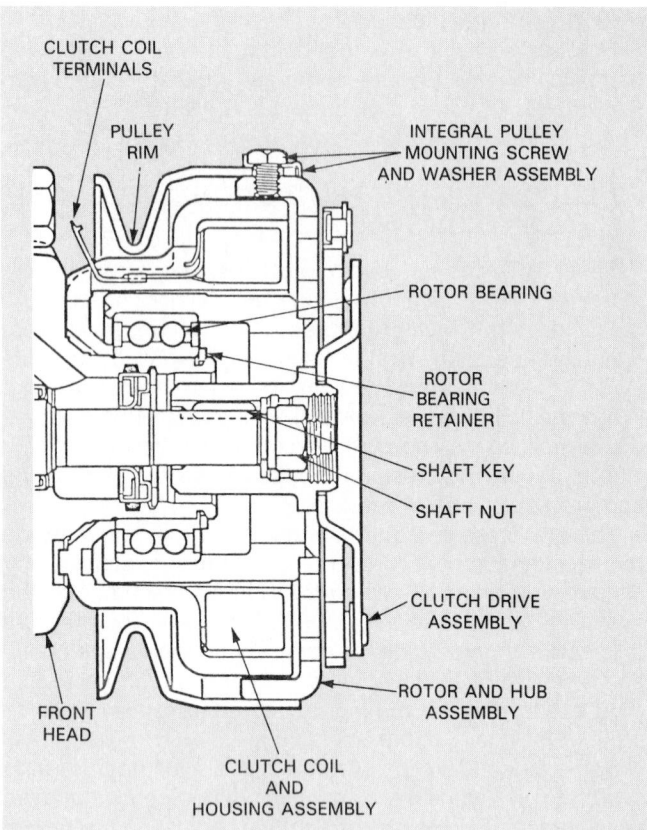

Fig. 27-9. Cutaway view of automotive air conditioning compressor magnetic clutch. When stationary coil is energized, armature revolves with pulley and turns compressor shaft. (Pontiac Motor Div., General Motors Corp.)

a steep grade, or entering a highway. When the driver presses the gas pedal to the floor (wide open throttle), the switch opens the circuit to the compressor clutch. This shuts the system down temporarily so there is more engine power for acceleration.

The wide open throttle switch normally mounts on the carburetor or throttle body (fuel injected engine). It contacts the throttle linkage and is only open with the throttle fully depressed. The compressor works normally (switch closed) under typical driving conditions. The wide open throttle switch is usually wired in series with the magnetic clutch on the compressor.

27-7 COMPRESSOR TYPES

The operation of the automobile air conditioning compressor is similar to compressors used in nonautomotive applications. Its function is to compress low-pressure, low-temperature refrigerant vapor into a high-pressure, high-temperature gas.

There are three basic types of compressors in general use in automotive air conditioning. Each of these uses a reciprocating (back-and-forth motion) piston arrangement:
1. Two-cylinder reciprocating compressors.
2. Swash plate compressors.
3. Scotch yoke compressors.

Most automotive compressors are semihermetic. This type can be internally serviced.

27-8 TWO-CYLINDER COMPRESSOR

Conventional reciprocating compressors used in automotive air conditioning usually contain two pistons in a parallel V-type arrangement. These compressors use a special 500 viscosity refrigerant oil. The amount varies from three to seven ounces, depending on the model. A lack of oil may cause bearing, seal, and valve problems.

The two-cylinder compressor is usually constructed of die cast aluminum. See Fig. 27-10. The pistons are attached to a connecting rod which is driven by the crankshaft. The crankshaft is connected to the compressor clutch assembly, which is driven by an engine belt.

Most conventional compressors contain high- and low-side service valves. These valves are for attaching gauges and/or servicing the system. See Para. 4-9 and Para. 4-11.

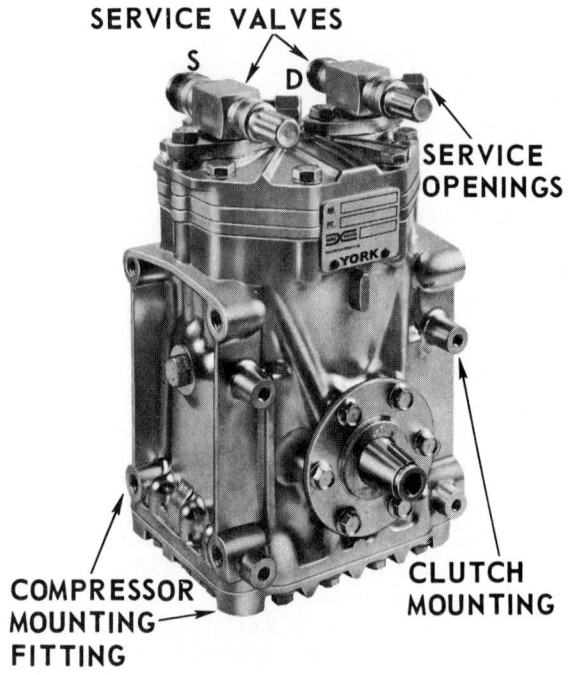

Fig. 27-10. Two-cylinder, reciprocating compressor. Compressor body has variety of mounting holes to permit vertical, horizontal, or inclined mounting. S—Suction service valve. D—Discharge service valve. Service valves are flange mounted, using two cap screws.

27-9 SWASH PLATE COMPRESSOR— VARIABLE DISPLACEMENT COMPRESSOR

The swash plate compressor, also known as wobble plate compressor, is different from the conventional type. In the swash plate compressor, piston motion is parallel to the crankshaft. The pistons are connected to an angled swash plate using ball joints. There are three types of swash plate compressors:
1. Five-cylinder model.
2. Six-cylinder model.
3. Five-cylinder variable displacement model.

The five-cylinder model is pictured in Fig. 27-11. Its internal construction is shown in Fig. 27-12. Each cylinder has a set of reed valves for intake and exhaust. The cylinders utilize connecting passages to form one common

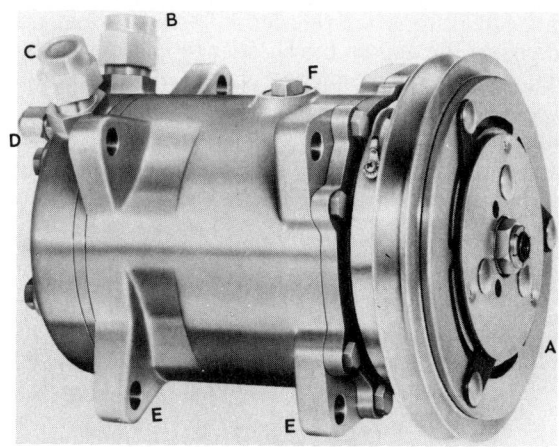

Fig. 27-11. Five-cylinder, swash plate compressor. A—Clutch. B—Suction line connection. C—Discharge line connection. D—Discharge access service port. E—Mounting flanges. F—Oil filler plug. (Abacus International)

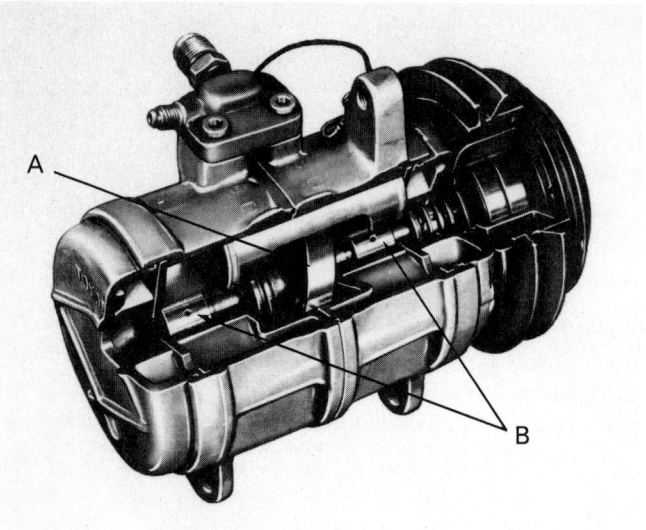

Fig. 27-13. Six-cylinder, swash plate compressor. Note location of swash plate (A) and pistons (B). (Chrysler Corporation)

high-side port and one low-side port. These ports are located at the rear of the compressor. Also located at the rear of the compressor is a high-side pressure relief valve.

A six-cylinder swash plate compressor is shown in Fig. 27-13. It consists of three sets of opposing cylinders and three double-acting horizontal pistons. An axial plate, pressed to the shaft, drives the pistons. Fig. 27-14 shows the parts of the compressor.

Some automobile air conditioning systems use a variable position swash plate compressor. The variable displacement swash plate is different from other swash plate compressors. It uses a plate connected to a hinge pin that allows the swash plate to vary its angle, Fig. 27-15. The angle of the swash plate is controlled by a bellows valve that senses suction pressure. During high load conditions, the swash plate angle is large and the compressor displacement is maximum. During low load conditions, the swash plate angle is smaller and compressor displacement is reduced.

Fig. 27-16 shows the parts of a five-cylinder variable displacement swash plate compressor. It has a low-pressure

Fig. 27-12. Note internal construction of typical five-cylinder, swash plate compressor. A—Piston. B—Swash plate. C—Pulley bearing. (Abacus International)

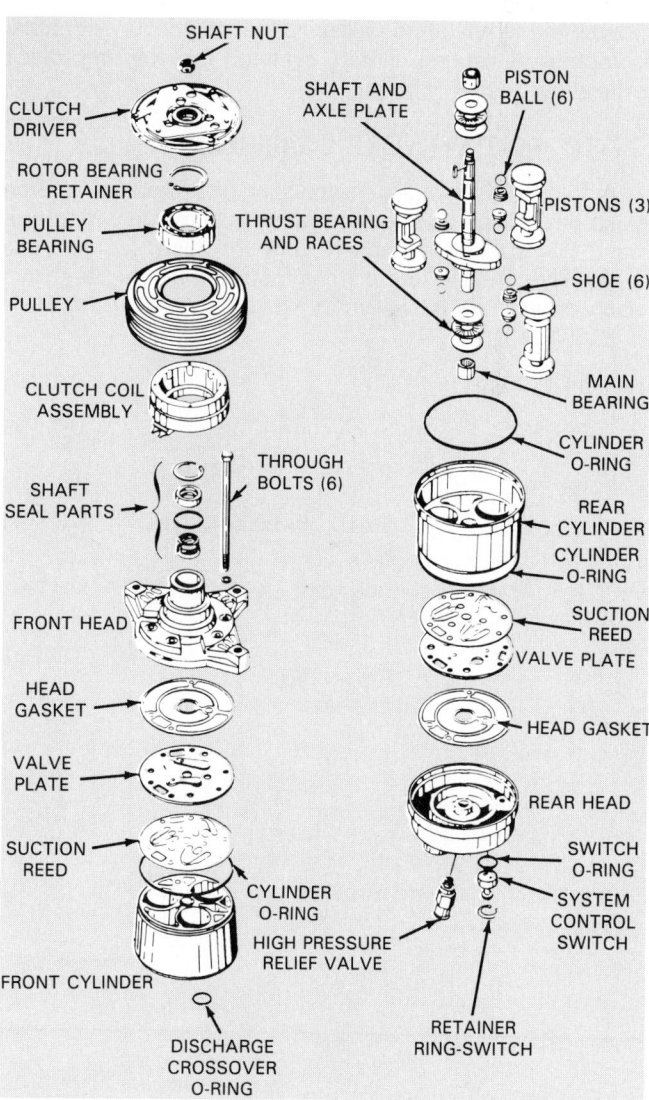

Fig. 27-14. Internal parts of swash plate compressor. Unit has six cylinders. (Cadillac Motor Car Div., General Motors Corp.)

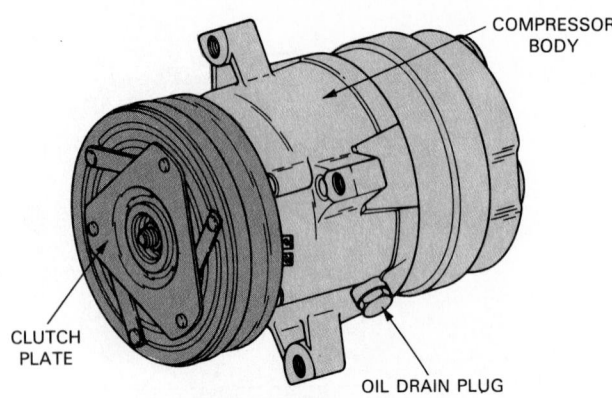

COMPRESSOR
BODY

CLUTCH
PLATE

OIL DRAIN PLUG

Fig. 27-15. Five-cylinder, variable displacement swash plate compressor. Note location of oil drain plug and clutch. (Cadillac Motor Car Div., General Motors Corporation)

Opposed pistons are pressed into a yoke that rides on a slide block located on the shaft eccentric. Rotation of the shaft provides reciprocating motion with no connecting rods, Fig. 27-18.

Refrigerant flows into the crankcase from the rear. It is drained through the reeds attached to the piston top during the suction stroke. Refrigerant is then discharged

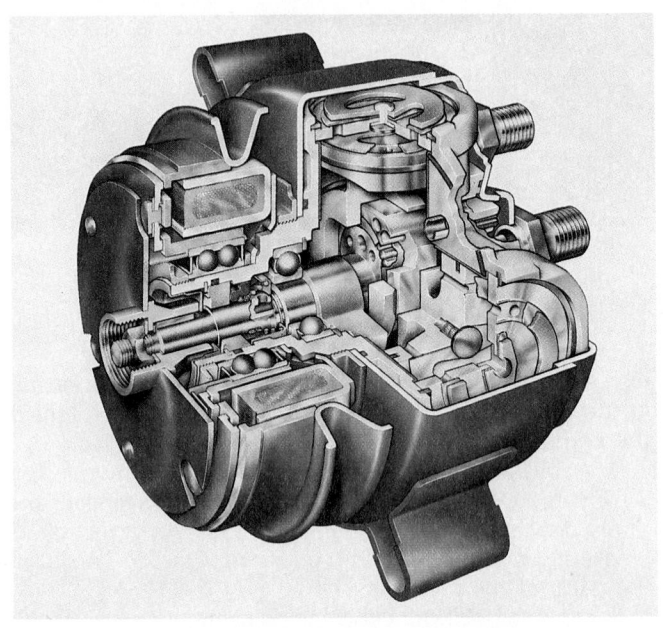

Fig. 27-17. Four-cylinder, scotch-yoke radial compressor can be mounted in any position, through 360° around crankshaft. (Tecumseh Products Company)

safety cutoff switch at the rear of the compressor. This switch is designed to disengage the clutch during low refrigerant conditions. The five-cylinder variable displacement compressor is designed to compensate for engine power and temperature variations that can occur with other compressors during clutch cycling, thereby improving efficiency.

27-10 SCOTCH YOKE COMPRESSOR

A four-cylinder radial compressor of a modified scotch yoke design changes rotary motion into reciprocating motion. The basic mechanism of a modified scotch yoke assembly contains four pistons mounted 90 degrees from each other. This is shown in Fig. 27-17.

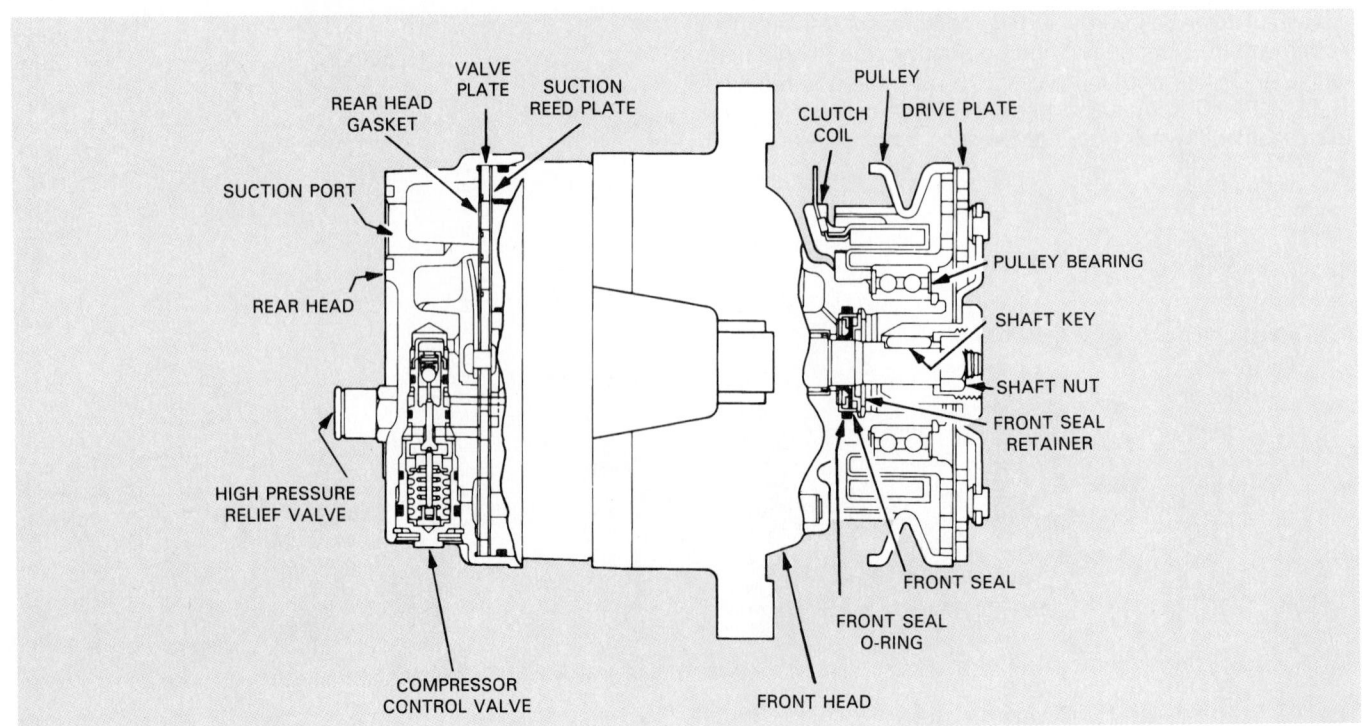

VALVE
PLATE

REAR HEAD
GASKET

SUCTION
REED PLATE

PULLEY

CLUTCH
COIL

DRIVE PLATE

SUCTION PORT

PULLEY BEARING

SHAFT KEY

REAR HEAD

SHAFT NUT

FRONT SEAL
RETAINER

HIGH PRESSURE
RELIEF VALVE

FRONT SEAL

FRONT SEAL
O-RING

COMPRESSOR
CONTROL VALVE

FRONT HEAD

Fig. 27-16. Internal parts of five-cylinder, variable displacement swash plate compressor.   (Cadillac Motor Car Div., General Motors Corp.)

through the valve plate and flows out the connector block at the rear.

The radial compressor is shorter in length and larger in diameter than most compressors.

Fig. 27-19 illustrates a typical radial compressor's internal parts.

27-11 COMPRESSOR SEAL

All automotive air conditioning compressors have a crankshaft seal. In some cases, sealing is done by means of a carbon ring with a smooth surface rubbing against a flat, smooth, cast-iron surface. This ring is bolted to and sealed on the body of the compressor. Some compressors use a rotating carbon ring rubbing against a stationary carbon ring. A synthetic rubber O-ring seals the joint between the carbon ring and the shaft. Carbon seals are often used for their excellent lubrication properties. Teflon seal surfaces are used on some late model compressors.

Fig. 27-20 shows a typical crankshaft seal. This seal must be leakproof between −60 °F and 250 °F (−51 °C to 121 °C), and from one micron vacuum to several hundred pounds of pressure.

Another type of crankshaft seal is shown in Fig. 27-21. Seals of this design may be used as a replacement for worn seals on automotive compressors.

Many manufacturers recommend operating the air conditioning system for a few minutes at least once a week throughout the year. If the compressor is out of operation for a long period of time, the oil may drain from the compressor seal surfaces. If these sealing surfaces become dry, refrigerant vapor may leak out through the seal. The compressor seal is the most common point of R-12 leakage in automotive applications.

In case of refrigerant loss, it is possible to purchase a sealed container with the correct amount of refrigerant for any make or model of car. This container may also provide the correct amount of refrigerant oil for the system. To make use of these containers, it is necessary to first drain all the refrigerant and oil from the compressor. Then add the refrigerant and oil from the container.

27-12 BELTS

Compressor rotation is accomplished by the use of a "V" or ribbed belt driven from the automobile engine. The belt is mounted on the pulley of the electromagnetic clutch and is driven by a pulley on the engine crankshaft. An adjustable idler pulley may be used to maintain correct belt tension.

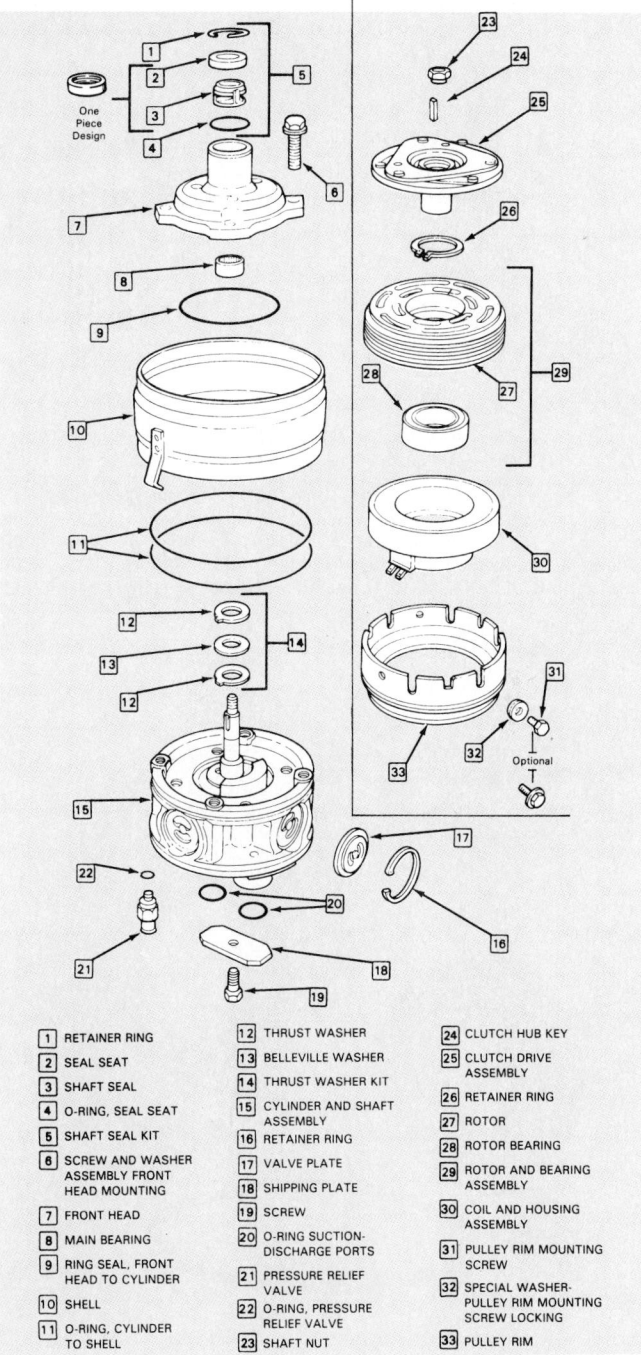

1 RETAINER RING	12 THRUST WASHER	24 CLUTCH HUB KEY
2 SEAL SEAT	13 BELLEVILLE WASHER	25 CLUTCH DRIVE ASSEMBLY
3 SHAFT SEAL	14 THRUST WASHER KIT	26 RETAINER RING
4 O-RING, SEAL SEAT	15 CYLINDER AND SHAFT ASSEMBLY	27 ROTOR
5 SHAFT SEAL KIT	16 RETAINER RING	28 ROTOR BEARING
6 SCREW AND WASHER ASSEMBLY FRONT HEAD MOUNTING	17 VALVE PLATE	29 ROTOR AND BEARING ASSEMBLY
7 FRONT HEAD	18 SHIPPING PLATE	30 COIL AND HOUSING ASSEMBLY
8 MAIN BEARING	19 SCREW	31 PULLEY RIM MOUNTING SCREW
9 RING SEAL, FRONT HEAD TO CYLINDER	20 O-RING SUCTION-DISCHARGE PORTS	32 SPECIAL WASHER-PULLEY RIM MOUNTING SCREW LOCKING
10 SHELL	21 PRESSURE RELIEF VALVE	33 PULLEY RIM
11 O-RING, CYLINDER TO SHELL	22 O-RING, PRESSURE RELIEF VALVE	
	23 SHAFT NUT	

Fig. 27-19. Internal parts are of a four-cylinder radial compressor. (Cadillac Motor Car Div., General Motors Corp.)

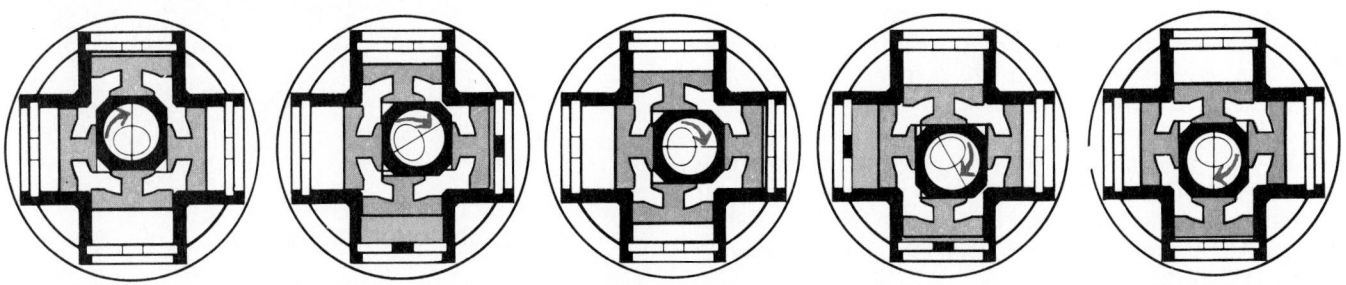

Fig. 27-18. Four-cylinder, scotch-yoke mechanism. Note rotation of shaft upon slide block to produce compression of refrigerant. (Tecumseh Products Company)

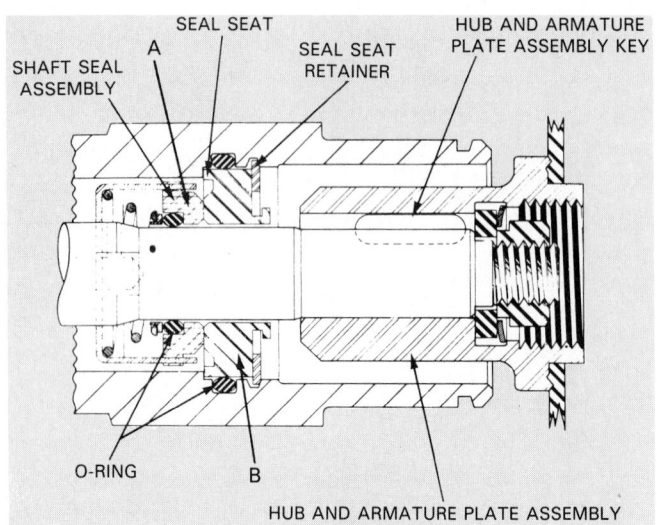

Fig. 27-20. Compressor crankshaft seal. A—Rotating carbon ring. B—Cast iron seat. Rubbing surfaces should be smooth, flat, and accurate to 0.000001 in. Note use of O-rings to seal two sealing rings in place.

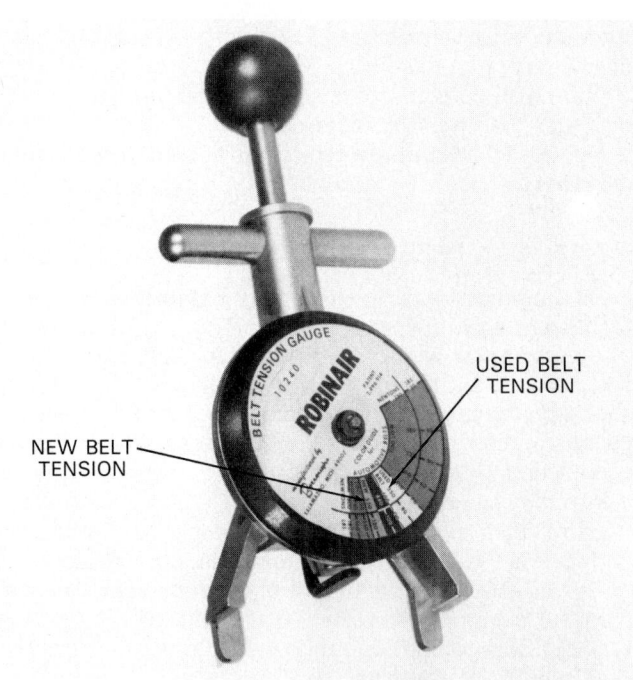

Fig. 27-22. Belt tension gauge. Note increments for new or used belts. (Robinair Division, SPX Corporation)

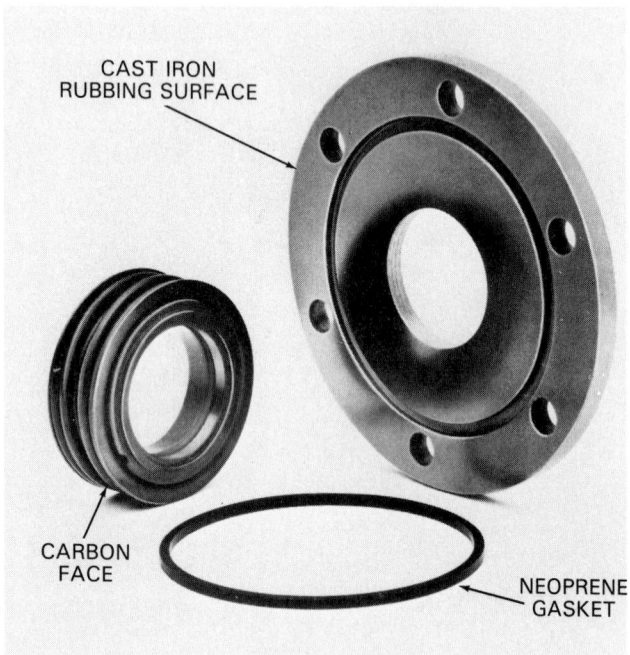

Fig. 27-21. Replacement crankshaft seal for automotive compressors.

The compressor belt must be designed and constructed for automotive type service. It must run true on the pulleys, and the pulleys must be in line. The belt must have the correct tension for efficient operation and long life.

For a new belt, the approximate tension should be from 90 to 120 lb. Use a belt tension gauge to measure tension, as shown in Fig. 27-22. Always recheck a new belt for tension a day or two after installation. A used belt (when overhauling a system that has been in use) should have a 70- to 90-lb. tension.

One way to roughly check belt tension is to apply firm hand pressure in the middle of the longest belt span with

the engine stopped. If the belt is correctly adjusted, it should depress about 1/2 in. (13 mm) out of line. Also, a correctly tensioned belt will twist 1/4 to 1/2 in. (6 to 13 mm) using a firm grip and firm twisting with fingers.

The belt may also power the water pump and the fan, as shown in Fig. 27-23. Some fans have a clutch that con-

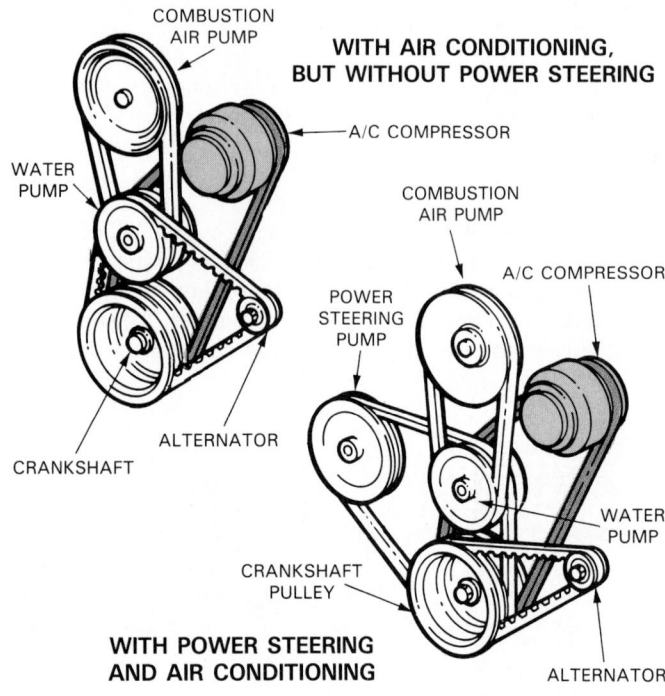

Fig. 27-23. Belts used in two typical air conditioning installations, one with power steering. Note compressor location. (Pontiac Motor Div., General Motors Corporation)

958

nects the fan to the engine only after the engine compartment has reached a correct temperature. This helps speed engine warm-up.

27-13 CONDENSERS

The automotive condenser is mounted in front of the engine radiator. It is usually constructed of copper or aluminum with multiple rows of finned tubes, Fig. 27-24. Heat moves from the condenser into the airflow set up by the moving vehicle and by an electric or engine-driven fan. This removal of heat causes the vaporized refrigerant to change to a liquid as it passes through the condenser.

The discharge line from the compressor to the condenser may have a vibration absorber mounted on it, Fig. 27-25, or the line may be flexible.

Most automotive air conditioning systems also use a muffler, located between the discharge side of the compressor and the condenser. The muffler acts to reduce the pumping noise of the compressor. Look at Fig. 27-26.

27-14 ACCUMULATOR AND RECEIVER

Many automotive systems use a low-side accumulator. See Fig. 27-27. The accumulator is connected to the evaporator outlet. After the refrigerant passes through the evaporator, it separates liquid from vapor. The accumulator retains the liquid and releases mostly vapor to the compressor. This insures that liquid refrigerant will not enter and damage the compressor.

The accumulator is usually constructed of aluminum and includes a desiccant (moisture absorbing material) to remove any contamination from the system. Accumulators may also have a service or Schrader valve for gauge manifold hook-up. Most accumulators are sealed and must be replaced if defective.

Some automotive air conditioning systems use a receiver or receiver-drier located between the condenser and

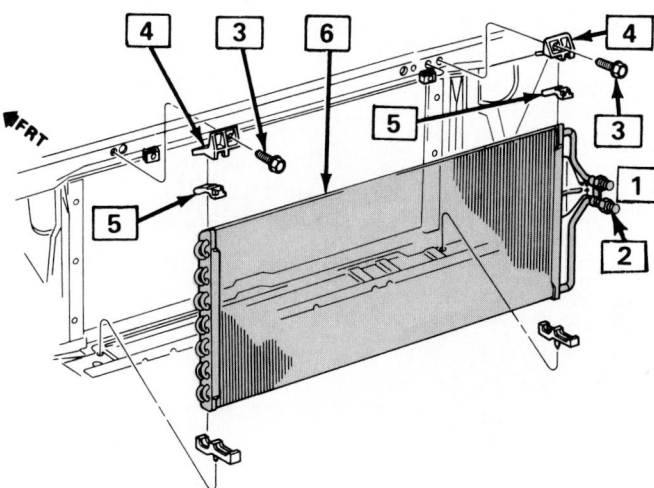

Fig. 27-24. Typical condenser mounting system: 1—High-side pressure fitting to condenser. 2—High-side pressure fitting from condenser. 3—Screws. 4—Brackets. 5—Insulators. 6—Condenser. (Cadillac Motor Car Division, General Motors Corporation)

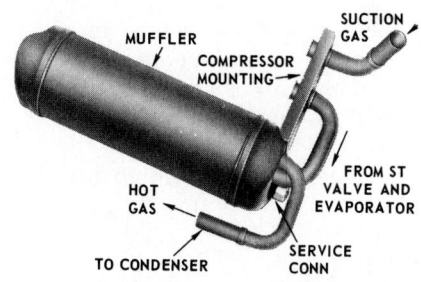

Fig. 27-26. Muffler used on automotive air conditioning system. Note compressor mounting, service connection, and how muffler is installed in system.

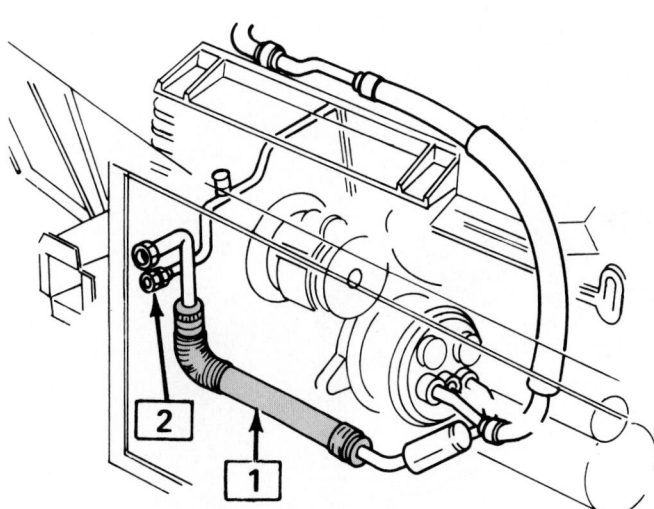

Fig. 27-25. An air conditioning system with a vibration damper mounted between compressor and condenser (1). The high-side pressure fitting for condenser is indicated at (2). (General Motors Corporation)

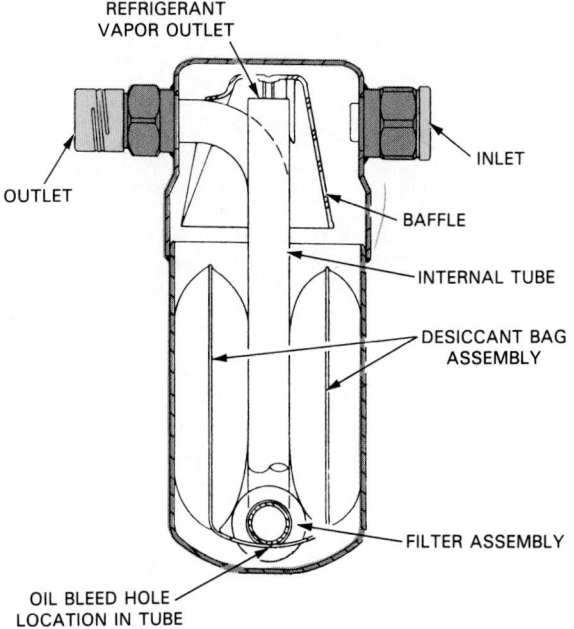

Fig. 27-27. Low-side accumulator. Note inlet and outlet. (Cadillac Motor Car Div., General Motors Corporation)

evaporator. The receiver also allows for some changes in refrigerant charge and liquid volume caused by expansion and contraction of refrigerant as temperatures change. Its main purpose is to serve as a storage container for the liquid refrigerant that enters from the condenser. A receiver-drier is shown in Fig. 27-28.

27-15 REFRIGERANT LINES

Special flexible refrigerant lines are used in automobile air conditioning applications. Their purpose is:
1. To carry liquid refrigerant from the receiver-drier to the evaporator expansion valve (liquid line).
2. To carry vapor refrigerant from the evaporator to the compressor (suction line).
3. To carry hot compressed vapor from the compressor to the condenser.
4. To carry liquid refrigerant from the condenser to the liquid receiver-drier (on some units).

Flexible refrigerant lines are also called hoses. They are commonly covered with a braid to protect them against injury. These hoses are designed and constructed to be flexible and vibration proof.

Refrigerant lines are made of steel or copper, or the lines may be flexible. Double-flare fittings are sometimes used where units must be disconnected for servicing. Flexible lines also use other types of fittings. Refrigerant lines should be carefully routed to prevent them from rubbing against any part of the car. Wear and corrosion would quickly cause leaks at the points of contact.

Refrigerant lines are fastened to the system parts in various ways:
1. Flared fitting.
2. O-ring fitting.
3. Hose clamp fitting.

These fittings are shown in Fig. 27-29.

The flexible lines or hoses vary in size from 3/8 to 5/8 in. depending on the capacity of the unit and on the state of the refrigerant. Vapor-carrying lines are larger. The size of the lines must match the fittings supplied by the manufacturer, so that system capacity will not be reduced.

Refrigerant lines should have large bends to prevent restriction of flow. They should also be supported, grommeted, and clamped to prevent wear by chafing and to prevent them from touching hot engine parts. Leave enough slack in the hose at the compressor end of the lines to allow for movement of the engine on its motor (vibration absorber) mounts.

The lines come equipped with caps and plugs to keep the inside of the lines clean and dry. Remove these plugs

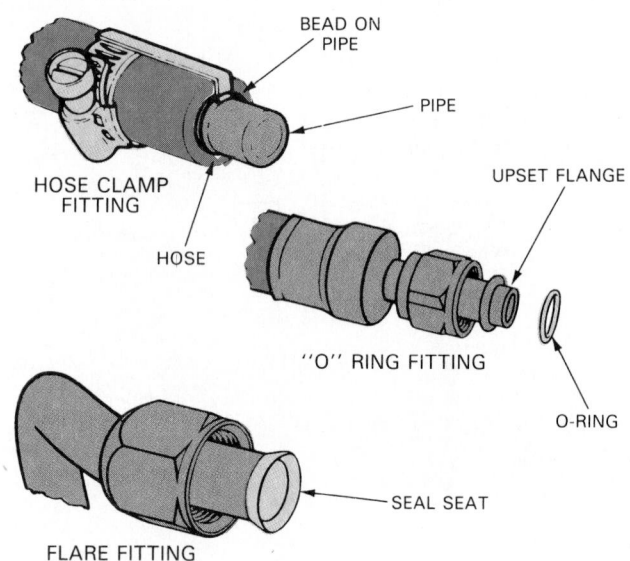

Fig. 27-29. Typical refrigerant line connection fittings are hose clamp, O-ring, and flare. (Harrison Radiator Div., General Motors Corp.)

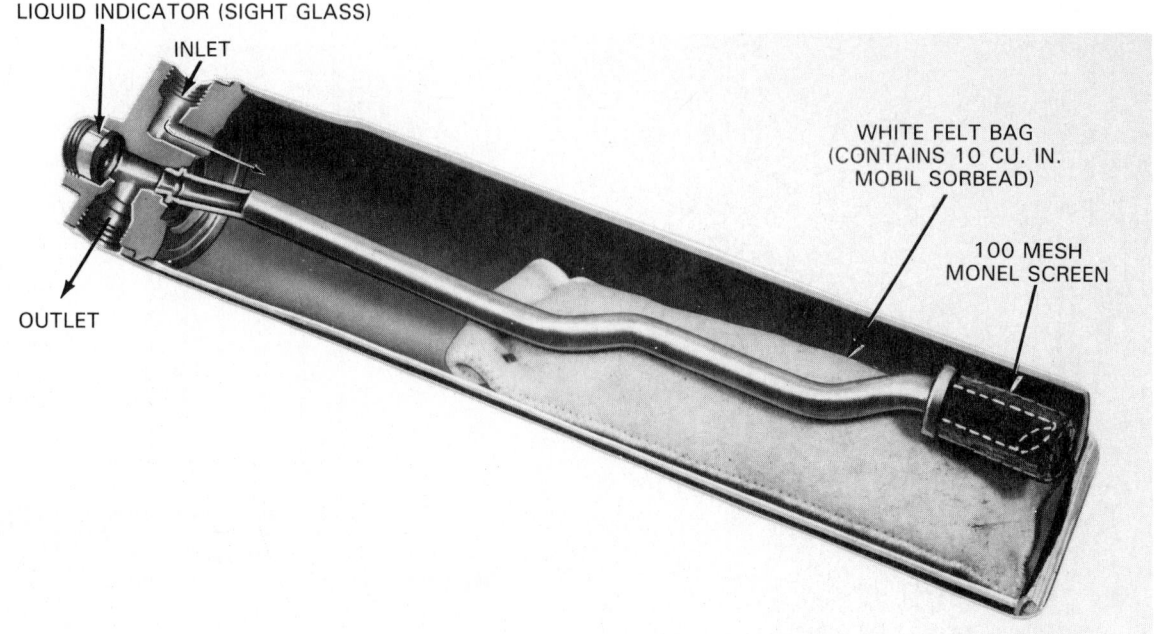

Fig. 27-28. Cross section of combination liquid-receiver, filter, drier, and sight glass. Drier material is in felt bag.

and caps just before installing the lines. Place clean refrigerant oil on all assembly threads and fittings just before assembly. This will help provide a leakproof connection.

27-16 EVAPORATOR AND HEATER CORE

The evaporator is often mounted next to the heater core assembly. Both are enclosed in a plenum chamber to allow blower or fan action.

Evaporators are usually constructed of aluminum and are used to both cool and dehumidify the passenger compartment air. An evaporator, blower, and duct installation mounted under the hood and dashboard is in Fig. 27-30.

In the evaporator, low-pressure liquid changes to low-pressure vapor. This change of state absorbs heat and cools the passenger compartment. Many evaporator assemblies include an accumulator for storage of liquid refrigerant.

The heater core is usually made of copper, similar to a small engine radiator. It is mounted within the blower housing. The heater core is attached to the engine through an inlet and an outlet hose (heater hoses). An ethylene-glycol and water mixture is circulated from the engine to the heater core to warm the passenger compartment. The same blower fan is often used for the evaporator and heater core.

27-17 METERING DEVICES

Automotive air conditioning metering devices are similar in operation to other types of refrigeration and air conditioning metering devices. They must allow high-pressure, high-temperature liquid refrigerant to pass through an orifice and then rapidly expand in volume. This expansion results

in a lowering of refrigerant pressure, and the refrigerant begins to boil, thus absorbing heat.

Three types of common metering devices used in automotive air conditioning are:

1. Thermostatic expansion valve (TEV).
2. H-type expansion valve.
3. Orifice (expansion tube).

Thermostatic expansion valves are used to throttle the amount of refrigerant entering the evaporator. TEV's operate on the temperature of the evaporator outlet and the low-side pressure. Chapter 5 explains the principle, installation, and service of this type of valve.

Many automotive TEV's have a pressure equalizer connection. There are two basic types of TEV pressure equalizers—the externally equalized and the internally equalized. The only difference between the two is that the external line is connected to an evaporator pressure control valve or to the evaporator outlet line. This provides a means of sensing evaporator outlet pressure. The internal equalizer senses evaporator pressure internally through the equalizer pressure passage. See Fig. 27-31.

The sensing bulb should be mounted at the outlet of the evaporator. This sensing bulb must be tightly clamped to the outlet and must be well insulated from engine heat.

H-type expansion valve operation is similar to that of the thermostatic expansion valve. Many H-type expansion valves contain a low-pressure cut-off switch in the base of the unit, Fig. 27-32. Care must be taken not to scratch gasket surfaces when servicing the H-valve. Fig. 27-33 illustrates an H-type valve and a cycling clutch switch.

The expansion tube orifice is located at the inlet of the

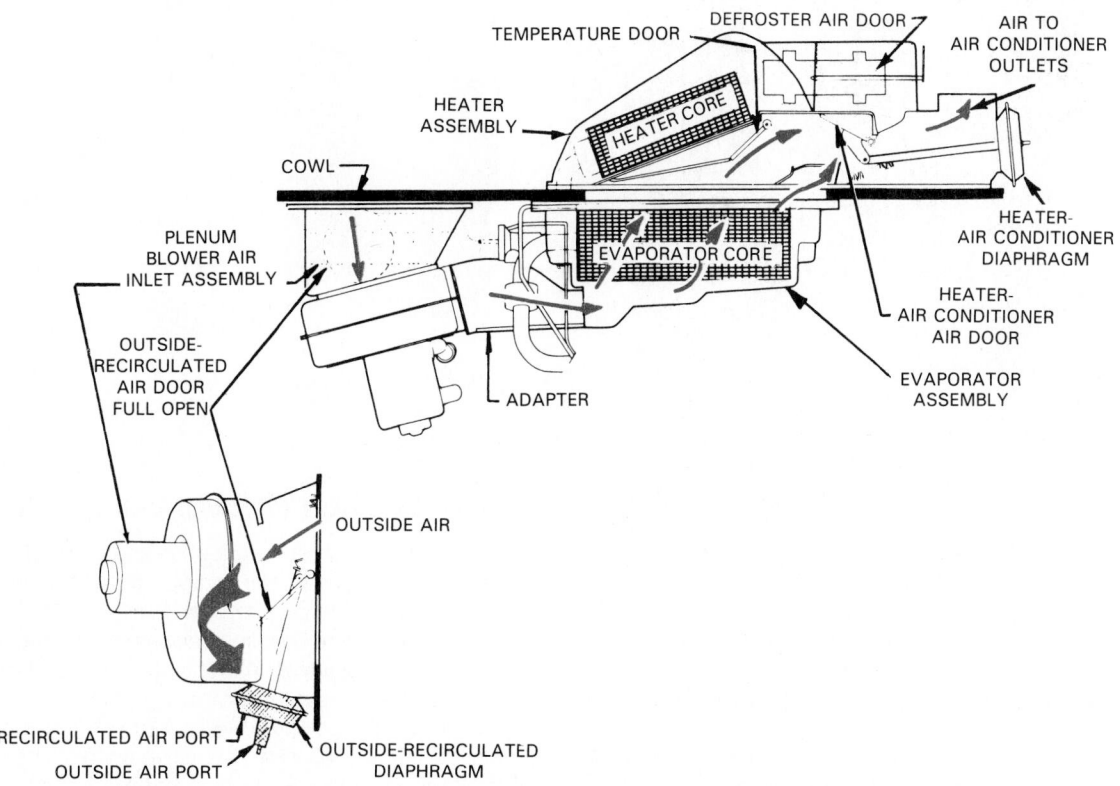

Fig. 27-30. Typical installation showing arrangement of blower, evaporator, heater coil, ducts, and controls. Heater core is under dashboard. Evaporator is under hood.

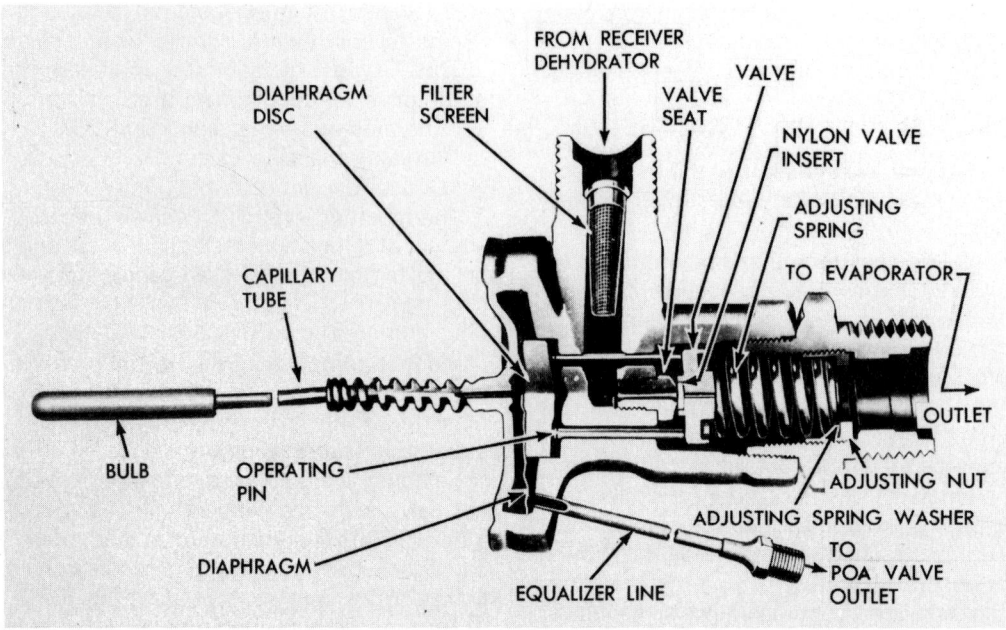

Fig. 27-31. Nonadjustable thermostatic expansion valve. Note inlet connections and oulet connections.

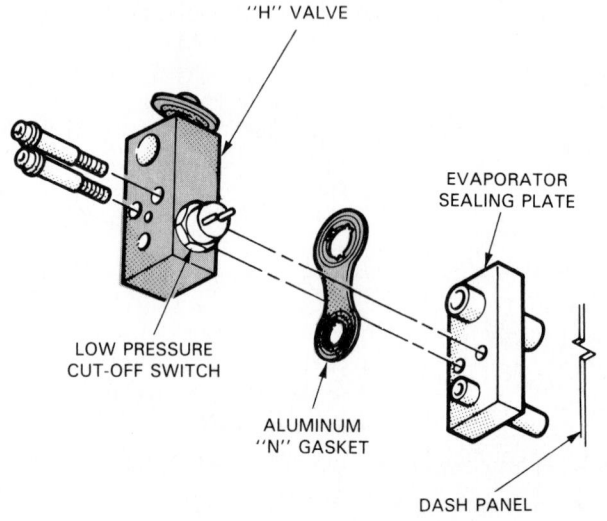

Fig. 27-32. Internal construction of an H-type expansion valve. Note gasket location and low-pressure cutoff switch. (Chrysler Corporation)

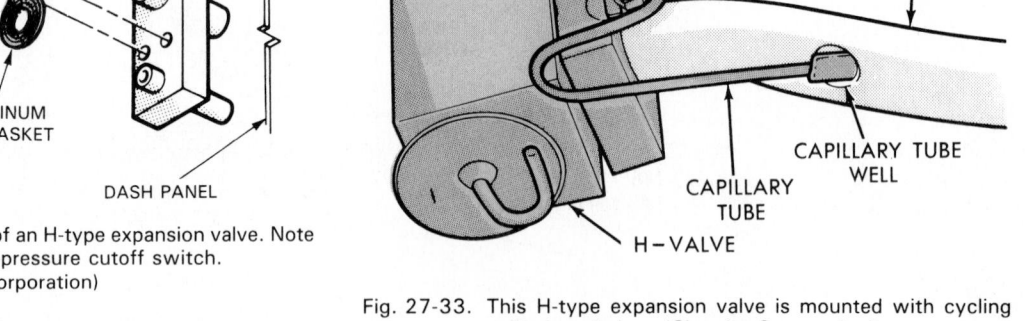

Fig. 27-33. This H-type expansion valve is mounted with cycling clutch switch. (Chrysler Corporation)

evaporator and is used in place of the thermostatic expansion valve. The orifice tube is the dividing line between the high- and low-pressure sections of the system. However, its operation does not depend on comparison of evaporator pressure and temperature. It is a fixed orifice and its flow rate is determined by the pressure difference across the orifice. Cycling the compressor on and off controls flow.

Fig. 27-1 illustrates a system using an orifice expansion tube. It is a common type system on new cars because of its simplicity.

Filter screens are located at both the inlet and outlet of the orifice tube to catch contaminants that may have entered the system. An orifice tube assembly is usually nonserviceable and must be replaced when defective.

27-18 SUCTION PRESSURE CONTROL VALVE

Some systems use suction pressure valves to maintain a certain pressure in the evaporator. This pressure is independent of the compressor low-side pressure and independent of the cooling demand. In most cases, a diaphragm or bellows in the valve responds only to the pressure in the evaporator.

When the evaporator pressure (for R-12) is above 29 to 31 psi (200 to 213 kPa), the valve opens. The valve closes if the pressure tends to be below these settings. Fig. 27-34 shows a cross section of a suction pressure valve with a manual pressure adjustment.

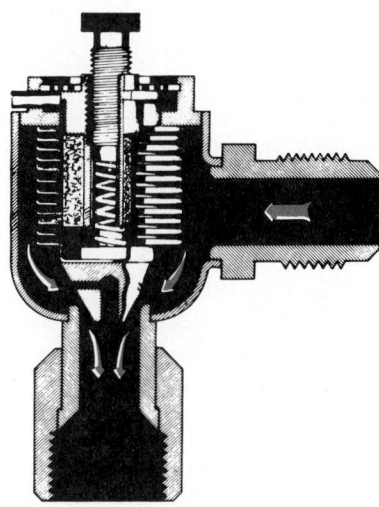

Fig. 27-34. Suction pressure regulator valve. Manual pressure adjustment controls evaporator temperature.

Several types of suction pressure valves have been used:
1. Suction throttling valves (STV).
2. Evaporator pressure regulators (EPR).
3. Pilot-operated absolute valves (POA).

Most of these valves are adjustable. Some have valve core access ports, which are used to connect service lines and gauge manifolds.

The main purpose of a suction pressure valve is to keep the evaporator above a freezing temperature. This prevents the moisture that condenses on the evaporator from freezing as the air flows through the evaporator core.

Some suction pressure valves maintain 28 psi (193 kPa) pressure in the evaporator until the air temperature becomes too cold. Then, a manual adjustment or a vacuum adjustment raises the pressure to about 30 psi (207 kPa).

Fig. 27-35 shows a vacuum control unit. At zero vacuum, the valve will operate at 3 psi (21 kPa) higher evaporator pressure. These valves usually have a service attachment to permit a pressure check of the evaporator. Notice that the vacuum attachment body can be adjusted by threading it in or out of the main body of the valve. The

adjustment and locknut are shown in Fig. 27-35.

The POA valve, developed by General Motors Corp., uses a sealed pressure element that maintains a constant pressure independent of the altitude of the car. See Fig. 27-36. This valve is equipped with an access valve service connection, an oil bleed connection (to remove oil from evaporator), and a thermostatic expansion valve equalizer connection. A cross section of this valve is in Fig. 27-37.

There are two basic combination metering devices built into a single container: The VIR (Valves-In-Receiver) and the

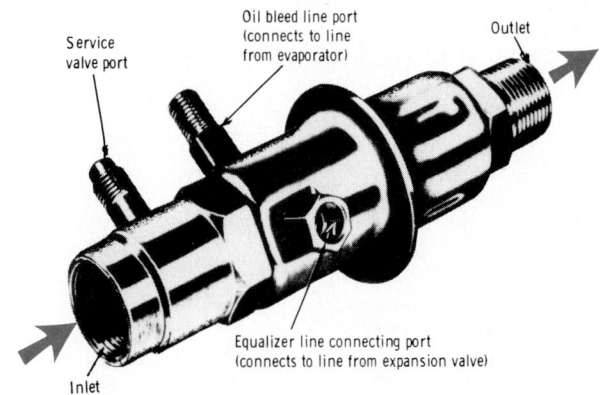

Fig. 27-36. Pilot-operated absolute suction throttling valve. Note flow direction. (Buick Motor Div., General Motors Corp.)

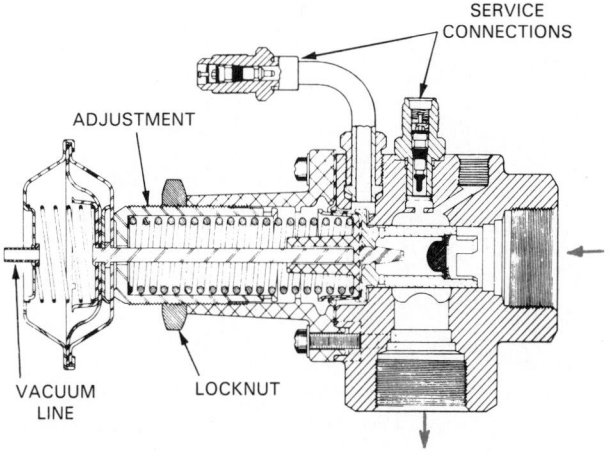

Fig. 27-35. Parts of vacuum-operated suction pressure regulator valve. Note the parts.

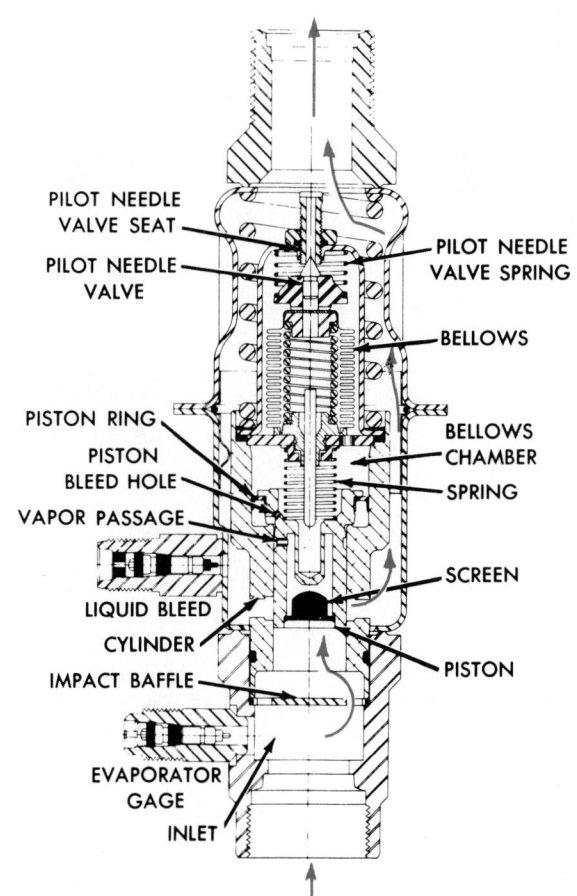

Fig. 27-37. Cross section of (POA) pilot-operated absolute valve. Note refrigerant flow through it. (Cadillac Motor Car Div., General Motors Corp.)

EEVIR (Evaporator Equalized Valves-In-Receiver). These units combine the POA valve, thermostatic expansion valve, receiver-dehydrator, and sight glass into a single unit.

27-19 SERVICE VALVES

A variety of service valves have been used on automotive air conditioning systems. Some are the standard front- and back-seat (one-way and two-way) service valves (see Chapters 2 and 14). Some are access valves (see Chapter 11).

All systems have some means by which high-pressure and low-pressure gauges can be connected to the system for pressure testing and to permit charging or adding oil to the system.

Fig. 27-38 shows a compressor fitting with an access valve connection.

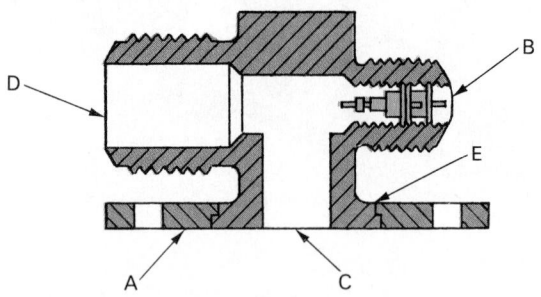

Fig. 27-38. A compressor mount equipped with Schrader valve attachment. A—Flange. B—Refrigerant service connection with Schrader valve. C—Opening to compressor. D—Refrigerant line opening. E—Swivel joint.

27-20 REFRIGERANT

The refrigerant which has been used in automobile air conditioning systems is R-12. The EPA ruling, that R-12 may not be vented after July, 1991, is resulting in the phasing out of its use in automobile air conditioning systems. Indications are that HFC-134a will be used in its place. See Chapters 9 and 28.

R-12 evaporates at 28 °F to 30 °F (-2 °C to -1 °C) in the evaporator, and condenses at about 120 °F (49 °C) in the condenser. The low-side evaporating pressure is about 28 psi (193 kPa). The condensing pressure varies from 130 psi (896 kPa) on cool days up to 200 psi (1378 kPa) on hot days.

27-21 OIL

Only a specially prepared oil should be used in the refrigerating system. This oil circulates throughout the system with the refrigerant. Most of it stays in the compressor. It must lubricate whether very cold or very hot. It must be dry and pure. Even a very small amount of moisture will freeze at the refrigerant metering device and may also form a sludge. The oil must be wax free.

Chapters 2 and 9 both describe refrigerant oil properties. Automotive systems usually use 500 viscosity oil.

27-22 AIR DISTRIBUTION

Air distribution for heating, ventilating, air conditioning, and defrost is accomplished through the use of ducts, vent doors, and louvers. Air is drawn into the vehicle's plenum by a blower through vents in the hood or at the base of the windshield. See Fig. 27-39. The air is then passed

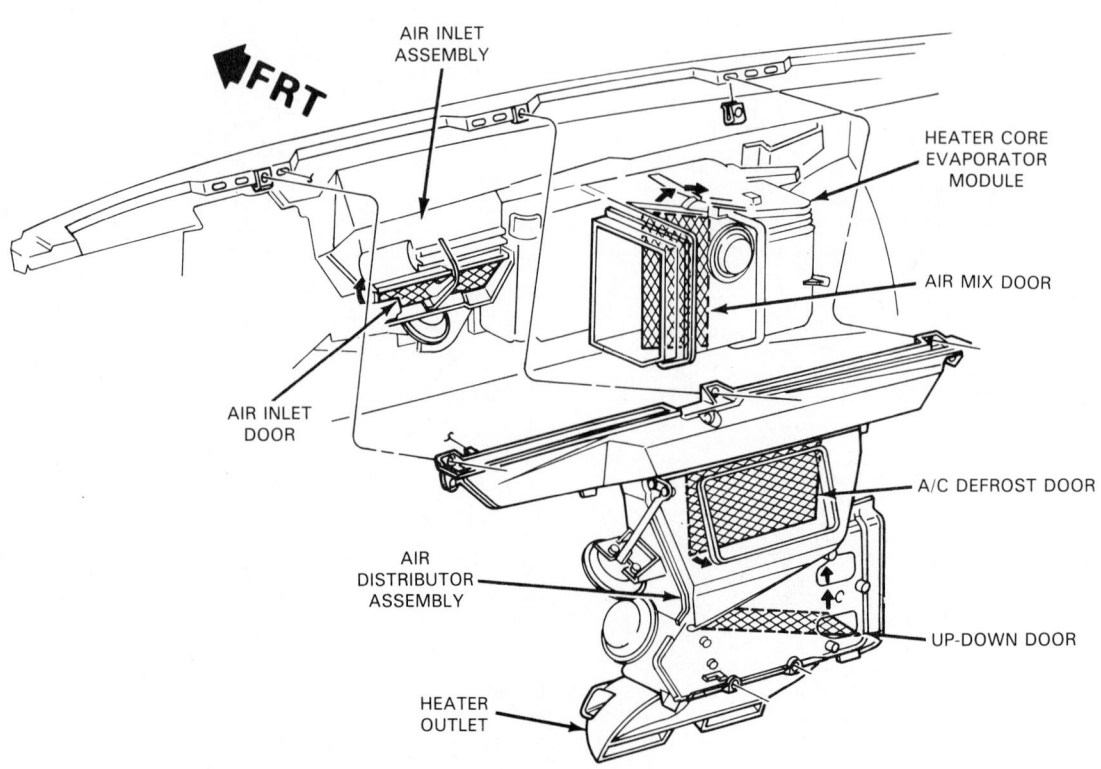

Fig. 27-39. Evaporator and plenum assembly with combined heater core and evaporator module. Note its location under the dashboard. (Cadillac Motor Car Div., General Motors Corp.)

through the heater core and/or the evaporator.

The air is directed by the specific setting on the control panel to heater or air conditioning ducts by vent doors operated electrically, mechanically, or by vacuum. Air conditioning ducts are usually mounted on the face of the dashboard and are often of the adjustable louver type. Defrost vents are located on the top of the dashboard and direct air up and over the windshield. Heater ducts usually empty below the dash, near the floor of the passenger compartment.

27-23 DUCTS

Heating and cooling ducts are made of metal or plastic. Fairly high air velocities are used, since the noise of air movement is not as critical as it is in an office or residence.

The duct system includes:
1. Fresh air inlet.
2. Return air inlet.
3. Evaporator housing.
4. Drain pan and drain connection.
5. Plenum chamber.
6. Conditioned air outlets:
   a. Defrost or deice (upper).
   b. Vent (center).
   c. Air conditioning (center).
   d. Heat (lower).
7. Dampers to change airflow:
   a. Manual.
   b. Power operated.
8. Grilles or louvers.

The heater core and evaporator are usually in series (regarding airflow), and the same duct system is used for both systems. The typical path of conditioned air is shown in Fig. 27-40.

Air dampers may be controlled by cables, electric motors, or vacuum-powered diaphragms. The diaphragm moves the damper in one direction; a spring moves the damper in the other direction. *These units are also called servo motors.*

Fig. 27-41 shows the application of vacuum-powered dampers. A vacuum-operated power servo motor is shown in Fig. 27-42. The damper diverter valve controls door movement.

27-24 INSULATION

Most car bodies are insulated. Fiberglass, glass wool, and various low K value, nonsettling, flexible insulations are used in automobiles.

The large amount of window area in an automobile allows considerable heat leakage and also a high sun load. However, the use of tinted glass reduces radiant heat load considerably. Light colored cars will absorb considerably less radiant heat than dark colored or black cars.

To avoid air conditioning the trunk space, insulation is usually placed over the back of the rear seat of the car.

The body of the car must be tightly sealed at all joints. Door gaskets must be in good condition. In addition to the usual water tightness test, the car body is also tested for air tightness.

27-25 ELECTRICAL CIRCUITS

Automotive air conditioning electrical systems are powered by the vehicle's 12 V (actually 12.6 V) dc bat-

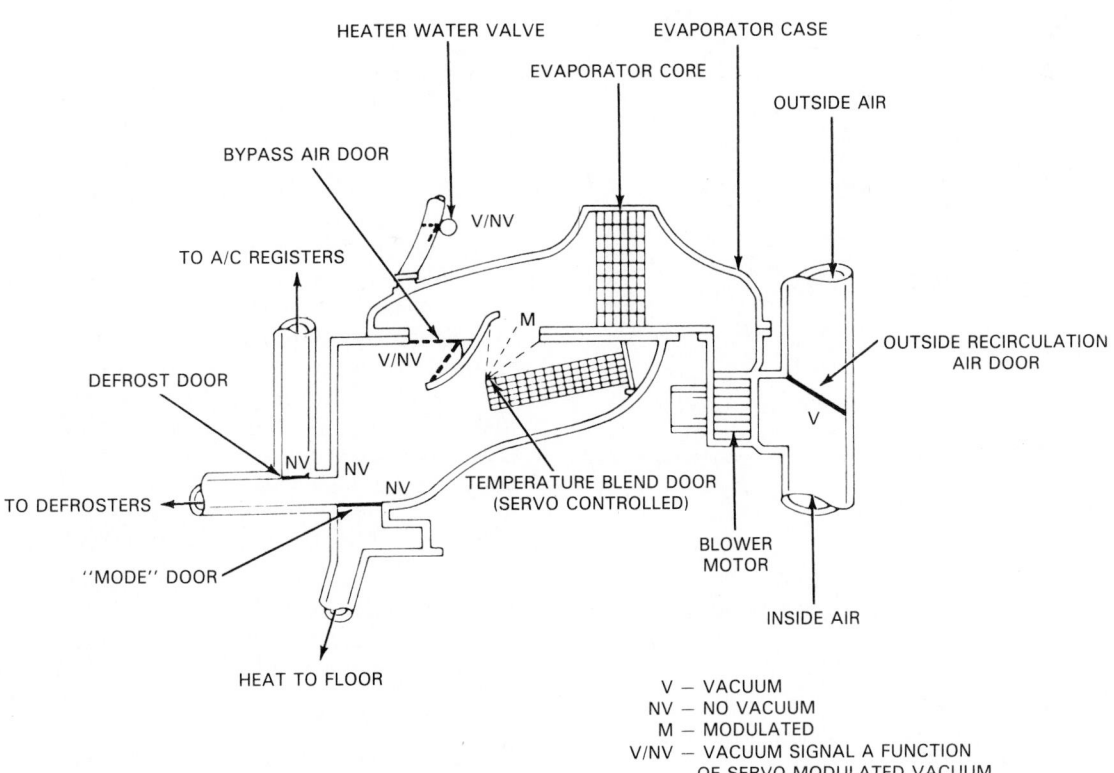

Fig. 27-40. A duct system and plenum chamber uses several doors to control direction of airflow in ducts.
(Ford Customer Service Div.)

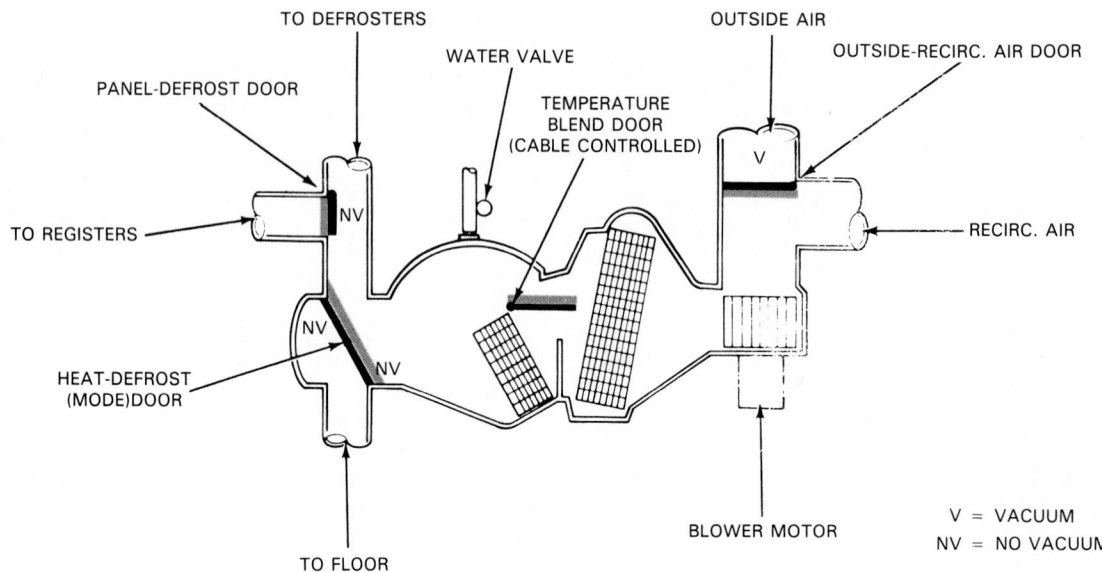

Fig. 27-41. Plenum chamber and duct system shows vacuum-operated doors (dampers).

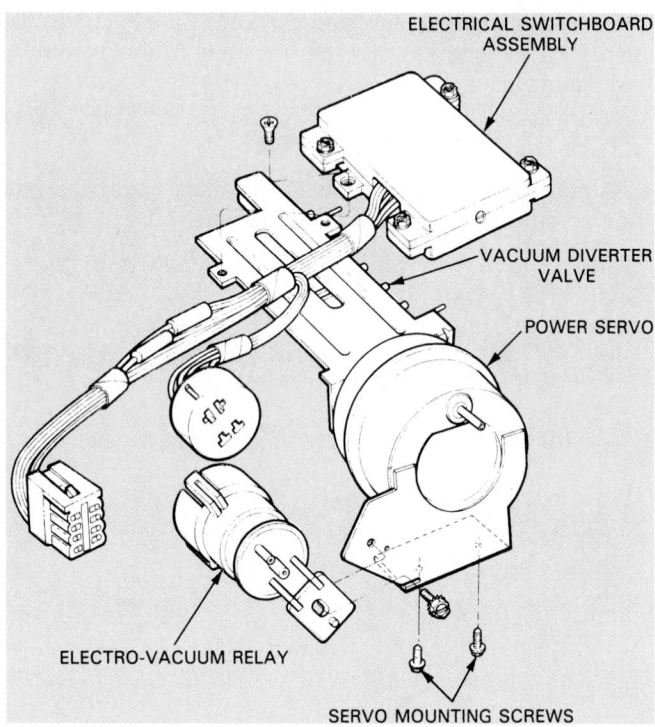

Fig. 27-42. A power servo unit. It controls the vacuum diverter valve.

tery or by alternator (generator) output (13 to 15 V). The basic electrical components of a modern system are:
1. Blower motor.
2. Vacuum-electrical servo motors.
3. Electromagnetic clutch.
4. Electronic sensors.
5. Electronic control unit.

Most automotive air conditioning wiring uses wires wrapped in convoluted tubing. The wires have pin connector terminals on their ends. Vacuum-actuated dampers are used on manual control systems. Electronic control systems may use a combination of a vacuum control system and electric motors for damper operation.

27-26 BLOWERS

Blowers used to circulate air in automobile air conditioning applications are a radial flow type (squirrel cage or centrifugal). The blowers are driven by dc motors, usually 12 V. Generally, the blowers are flexibly mounted to reduce noise. They are of several models, including single speed, two speed, and three speed.

The blower motors are either single shaft or double shaft, with sealed-in bearings that usually do not require oiling. If a motor develops bearing noise, the trouble may be worn bearings or too much end play. Worn bearings require motor replacement. Too much end play can sometimes be remedied by installing end-play washers. However, most units are sealed and cannot be repaired.

The motor's electrical load can be checked by inserting an ammeter in the wiring to the blower motor. If the ammeter reading is higher than specifications in the service manual, and if the motor is overheated, the windings may be shorted. Remove the motor. Check it again, and replace it if faulty. If the reading is lower than specifications, there may be a loose or dirty connection. Use a voltmeter to check for voltage drops. Correct the cause, when excessive voltage drop is located.

27-27 TYPES OF CONTROL SYSTEMS

There are two types of automotive air conditioning systems widely used today: electronic and manual. Both types use many of the same major components: compressor, evaporator, condenser, and heater core. The difference in the two systems is in how these devices are controlled.

The electronic control system uses a computer to monitor and control compressor cycling, blower speed, and damper positions. In the manual system, the user must operate switches and levers to control in-car temperature.

27-28 MANUAL CONTROL SYSTEMS

The manual control system regulates in-car temperature through the use of cables and switches. The cables are connected to the control panel and operate dampers. The switch controls blower speed.

Manual climate control systems require the driver to operate switches and levers on the control head to adjust for changes in the vehicle's inside temperature. See Fig. 27-43. If the inside of the car is too warm, the driver must move the blower switch to a higher speed and adjust the temperature control to a cooler position.

27-29 ELECTRONIC CONTROL SYSTEMS

Many recent automotive air conditioning, heating, ventilating, and defrost systems are controlled by a computer (electronic control circuit). Electronic control systems automatically adjust dampers, blower speed, and compressor cycling to maintain a desired temperature. These systems are called automatic temperature or climate control systems.

The driver simply inputs a desired temperature on the control panel and the system automatically adjusts numerous devices to obtain the desired temperature and airflow inside the vehicle, Fig. 27-44.

Several electronic control systems contain an outside temperature feature and self-diagnostic capability.

The primary control for the system is an interior temperature sensor, usually a thermistor. As passenger compartment temperature changes, the internal resistance of the sensor changes. The computer can use the resulting change in resistance and current flow to calculate whether more cooling or heating is needed. Input from this and other sensors allows the computer to produce outputs for automatically controlling the system.

27-30 ELECTRONIC CONTROL DIAGNOSTICS

Many electronic climate control systems have a self-diagnostic or a self-test capability. This feature helps the service technician locate trouble and faulty components. In a self-diagnostic system, the microprocessor (computer) sends fault information to the control panel. This trouble information may be obtained by touching a combination of buttons on the control panel or by using a jumper wire across two test terminals. When the technician has entered the diagnostic mode, a problem area is indicated by a flashing warning light or digital readout on the control panel. A chart in the service manual will explain what each number or code means.

The climate control panel may also serve as an information center. The control panel will display information such as blower voltage, blower door positions, coolant temperature, high- and low-side temperatures, and engine component status. By checking the control panel information and then using the manufacturer's trouble code charts, the service technician can quickly isolate the problem area and check the status of the vehicle's components. Refer to Fig. 27-45.

27-31 ELECTRONIC CONTROL MICROPROCESSOR

The climate control microprocessor or computer may be attached to the electronic control panel. It can also be located behind the glove box or under the dashboard. The climate control computer receives input from the vehicle sensors and the control panel key pad.

The microprocessor then uses this data to control the system. The microprocessor is usually a sealed unit with multiple pin wire connections. It may not be internally serviced and must be replaced if defective. Microprocessors are not interchangeable with different vehicles.

The primary information received by the microprocessor is a result of thermistor type sensors. Some common sensing devices are:
1. In-car temperature sensor.
2. Outside temperature sensor.
3. Engine coolant temperature sensor.
4. High-side temperature sensor.
5. Low-side temperature sensor.

The in-car temperature sensor measures the actual temperature of the vehicle's passenger compartment. This information is used in automatic climate control systems to cycle the compressor, and to control blower speed and

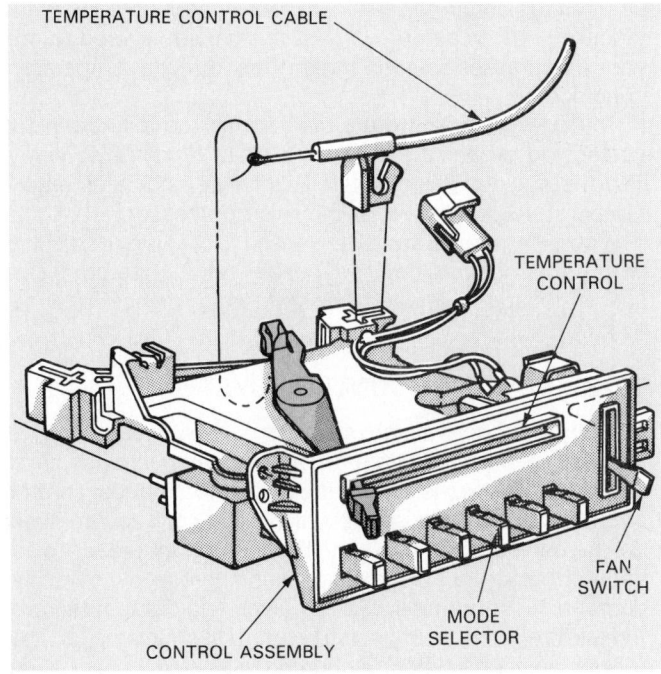

Fig. 27-43. A manual climate control panel. Note manual temperature control slide switch. (Chrysler Corporation)

Fig. 27-44. An electronic control panel. Note various temperature selectors. (Cadillac Motor Car Div., General Motors Corp.)

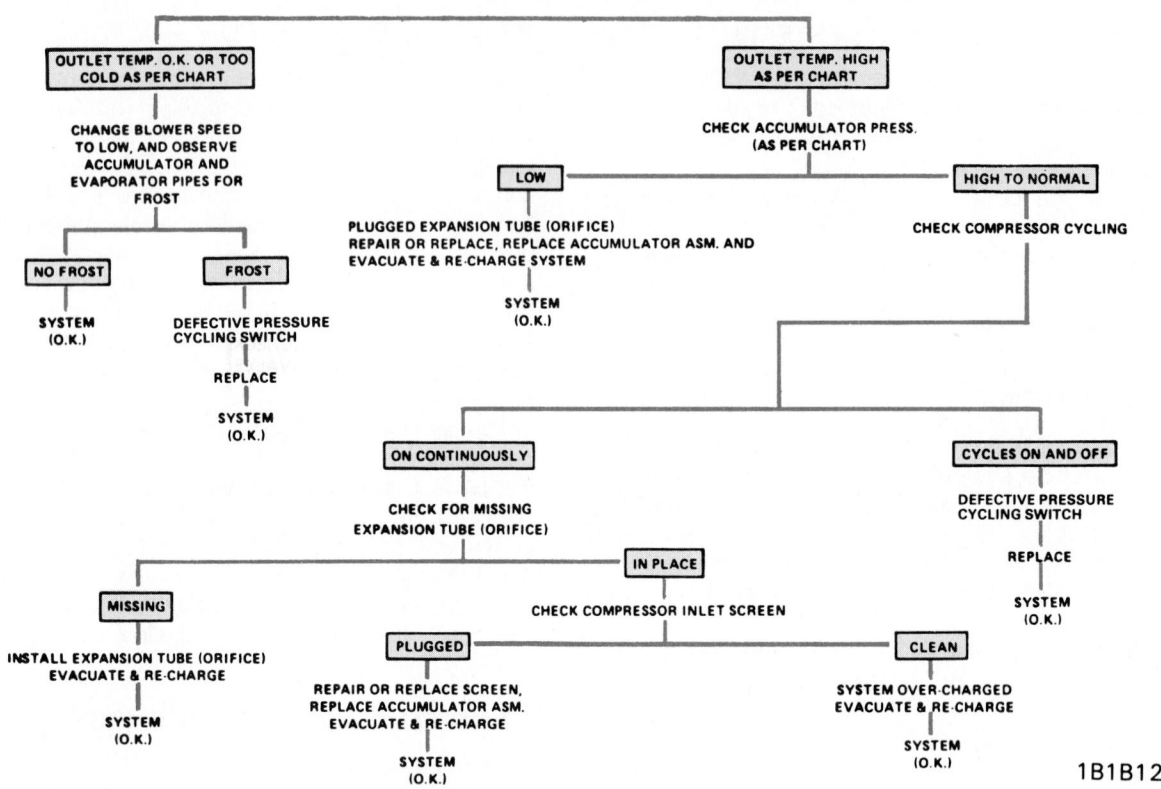

Fig. 27-45. A typical manufacturer's troubleshooting chart. Technicians use it to locate problems and service automobile air conditioners. (Buick Motor Div., General Motors Corp.)

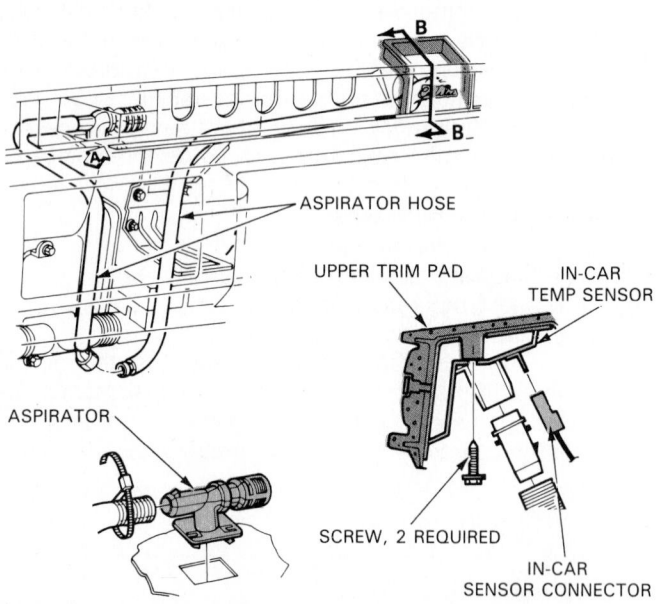

Fig. 27-46. In-car temperature sensor with aspirator. Sensor is usually located behind upper dashboard. (Cadillac Motor Car Div., General Motors Corp.)

vent door operation. The in-car temperature sensor is usually mounted behind the instrument panel.

A small amount of air is drawn into a plastic tubing passed over the in-car temperature sensor. The air movement is accomplished by the use of a connector called an

"aspirator." The aspirator is mounted on top of the heater and has a plastic hose connected to the in-car sensor. Aspiration involves forcing air across a tube which is open to the in-car sensor. This air movement draws a sample of air from the passenger compartment, Fig. 27-46.

The outside temperature sensor indicates the outside or ambient air temperature. It is usually mounted near the radiator grille. In some vehicles, this sensor is used to prevent compressor operation during low outside temperature conditions.

The coolant temperature sensor may be located in the coolant hose near the intake manifold. It may also screw into the engine water jacket. This sensor indicates engine temperature conditions to the microprocessor.

High-side temperature is sensed near the condenser before the metering device. The low-side temperature sensor is located between the metering device and the evaporator.

27-32 VACUUM CONTROL SYSTEMS

Many air conditioning systems use a vacuum power system (vacuum actuators) to operate dampers and valves. A vacuum system is shown in Fig. 27-47. Manual controls or automatic controls select which tubing or vacuum hoses will be subjected to vacuum. The diaphragms react to the vacuum and move against spring pressure to move dampers. This opens or closes valves and dampers to control airflow. The springs return the mechanisms to their former positions when vacuum is released.

The source for the vacuum power system is usually the engine intake manifold vacuum. A gasoline engine produces

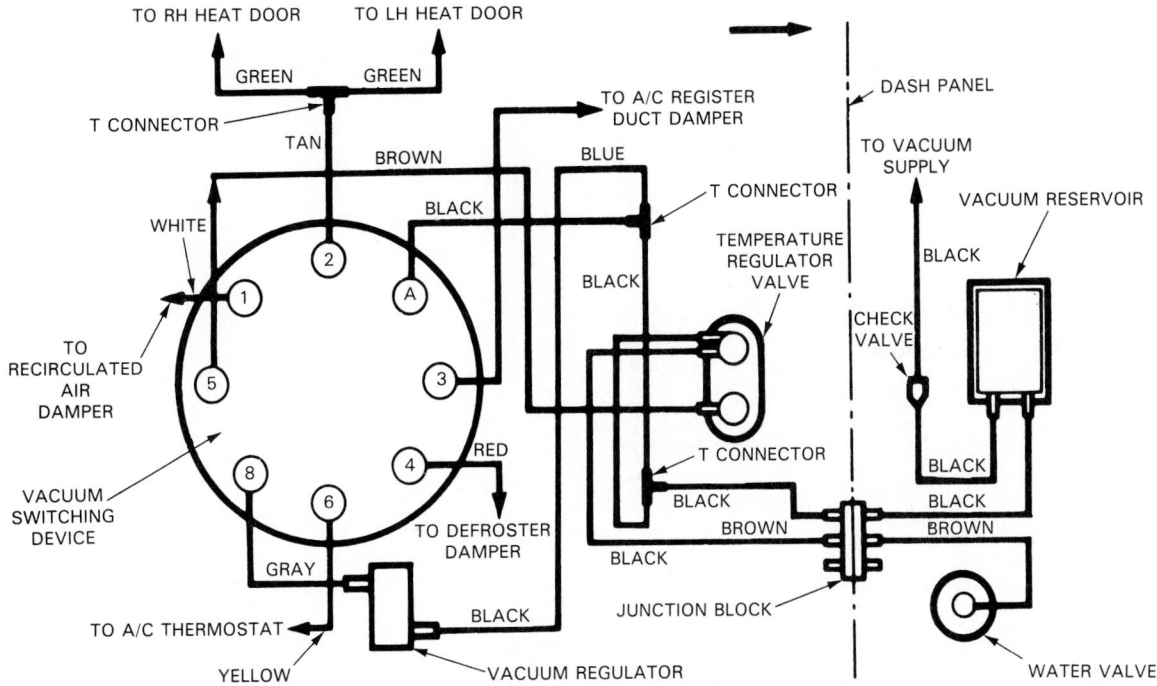

Fig. 27-47. Vacuum piping and vacuum actuators used in automotive air conditioning system. Vacuum is supplied by engine. (Ford Motor Co.)

from 5 to 20 in. Hg (127 to 508 mm Hg) depending upon engine speed and load. Most vacuum control systems will work with only 5 in. Hg (127 mm Hg). Some cars use an engine-driven or an electric vacuum pump as a source of power for vacuum-operated accessory units. Since diesel engines do not produce intake manifold vacuum, they must use a vacuum pump to operate any vacuum devices.

An electro-vacuum relay is shown in Fig. 27-48. Its solenoid coil operates a vacuum line valve and also controls blower operation. During cold engine operation, it is used to stop the blower by opening the contacts. The solenoid also closes the fresh air door by using the sealing ball to close the vacuum line. During maximum heat load, it closes the fresh air door to provide all recirculating air.

To check for leaks in the vacuum control system, connect a vacuum gauge into the system. Run the engine until the system has 16 in. Hg (406 mm Hg). Stop the engine. If the vacuum gauge starts to creep back toward zero (atmospheric), there may be a leak in the system or the valve in the line to the intake manifold is leaking.

To check actuator operation, install the vacuum gauge. Run the engine until the vacuum is 16 in. Hg (406 mm Hg). Stop the engine and watch the vacuum gauge as each actuator is operated. The vacuum should creep a little toward zero as each actuator is moved into operation. If the vacuum does not decrease:
1. That actuator has a pinched tube.
2. The line is plugged.
3. The damper is binding.

The dampers usually have an adjustment for full closing or opening. Check that they are set properly.

Vacuum control systems usually have a vacuum reservoir tank so that the dampers will keep working after shutting off the engine.

27-33 THERMOSTAT

Some systems use a thermostat to control compressor cycling. The thermostat is wired in series with the compressor clutch. The thermostat is usually located on the evaporator blower assembly. A refrigerant charged sensing bulb is attached to the inlet of the evaporator. The sensing bulb measures evaporator temperature and opens a normally closed set of contacts when a specified temperature is reached.

27-34 TRUCK/BUS CONDITIONING

The cabs of many trucks, buses, long distance hauling tractors (trucks), farm tractors, and earth movers are air conditioned.

Some truck systems use a remote condenser mounted on the roof of the cab. This type of installation removes the condenser from in front of the radiator, so that the radiator can operate at full efficiency. This is especially important during long pulls in low gear when the engine could overheat.

Some buses have combination roof-mounted air conditioning and heating systems. The system has an electronic panel in the evaporator that controls ventilation, heating, and air conditioning. The temperature controller is mounted on the dash. See Fig. 27-49.

These systems are similar to the automobile air conditioner and are installed and serviced in the same general way. Most of this equipment is installed after the cab has been made or as aftermarket equipment.

27-35 SERVICING AUTOMOBILE AIR CONDITIONERS

Servicing the automobile air conditioner is about the same as servicing standard air conditioning systems and commer-

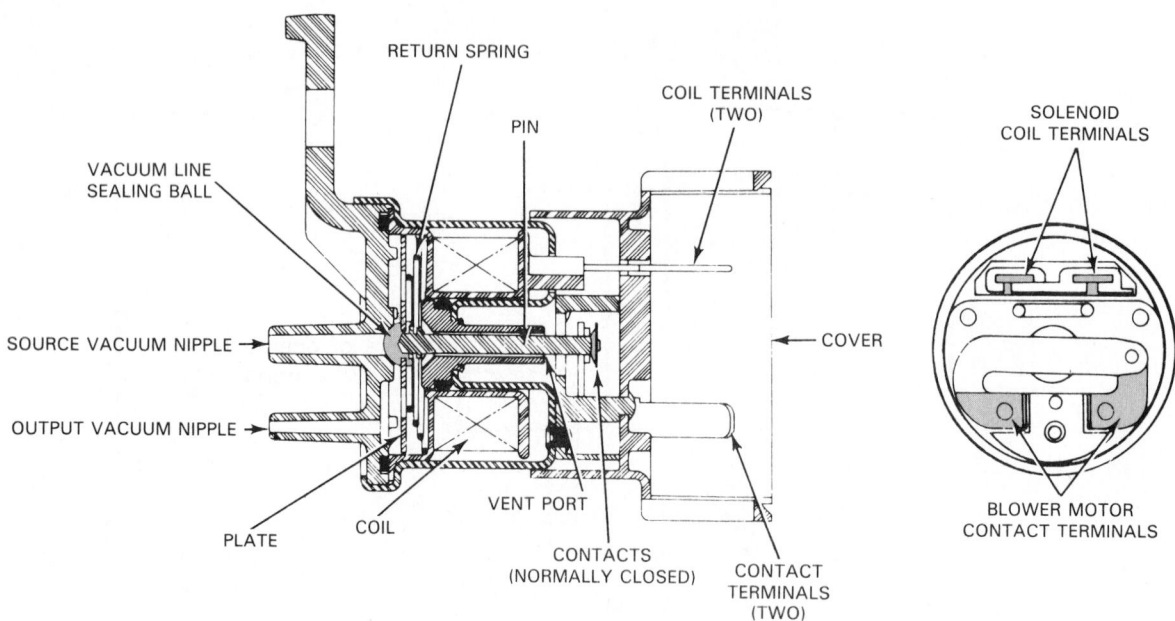

Fig. 27-48. An electro-vacuum relay. When solenoid coil is energized, it opens circuit to blower motor. It also closes vacuum line to fresh air door, causing door to close. (Ford Customer Service Div.)

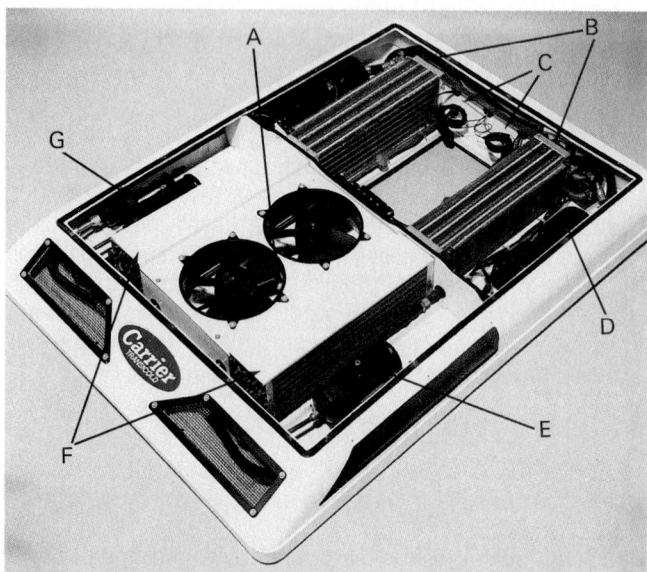

Fig. 27-49. Roof-mounted bus air conditioning and heating system. A—Condenser fans. B—Evaporator coils. C—Expansion valves. D—Evaporator fans. E—Receiver. F—Condenser coils. G—Filter-drier. (Carrier Transicold Division, Carrier Corporation)

cial systems. Study Chapters 11 and 14 carefully before trying to work on an automobile refrigerating system.

Servicing usually begins with a customer's complaint or during an annual check of the system. Common owners' complaints are:
1. No cooling.
2. Noise.
3. Intermittent cooling
4. Vibration.

There may be several causes for each complaint. Check the system thoroughly to find the cause.

The method of installing the gauge manifold is similar to the procedure described in Chapters 11 and 14. Shut off the engine when installing gauges to avoid injuries. Always clean the connections before removing any caps or plugs.

Fig. 27-50 illustrates the evacuation of an air conditioning system. This vacuum pump removes moisture from the refrigerant. All systems should be evacuated whenever they have been opened.

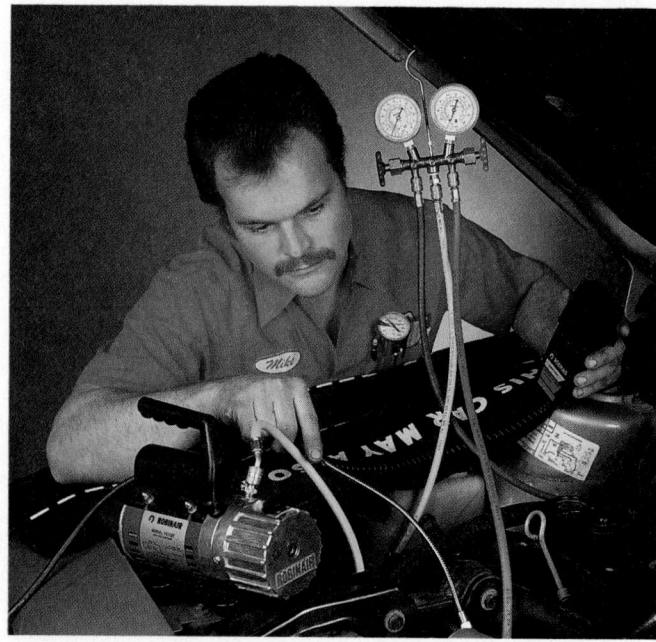

Fig. 27-50. Vacuum pump being used to remove air and moisture by reducing the pressure inside the air conditioning system. The lower pressure boils, or vaporizes, the moisture, and then exhausts it. By lowering system pressure, creating a vacuum, it "boils off" all the contaminating moisture. (Robinair Division, SPX Corporation)

For personal protection when removing access type valves or lines, wear gloves or use a pad made from several layers of cloth. Liquid refrigerants will freeze the skin or eyes. The gauge manifold and lines must be clean and dry, both inside and outside.

For personal safety, always wear goggles when working on the air conditioning system, when handling service cylinders, or disposable refrigerant containers. Avoid breathing any escaped refrigerant. Although the refrigerant is not toxic, the fact that it excludes oxygen makes it dangerous.

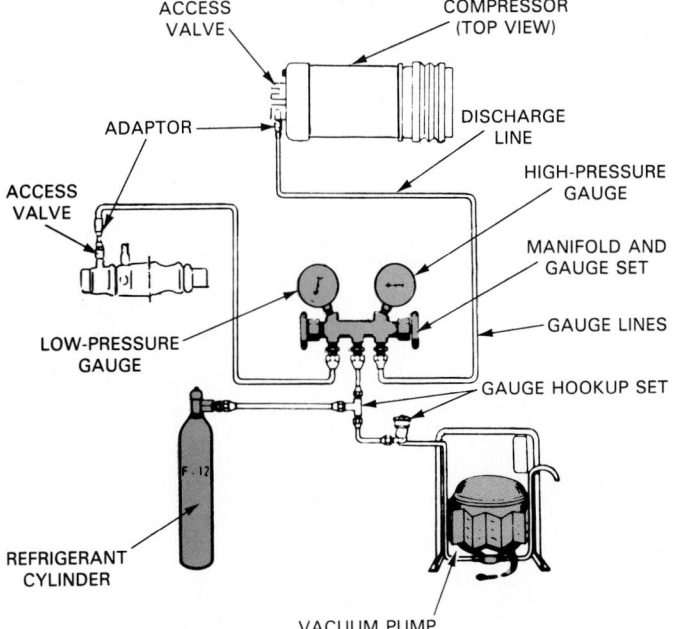

Fig. 27-51. Gauge and service manifold is installed in typical system. Both charging cylinder and vacuum pump are connected to middle opening of manifold.

Always attach a gauge manifold to the air conditioning system before service. See Fig. 27-51. Never use a manifold set that has been left open to the air until after the manifold and lines have been purged (cleaned) and dried.

Some compressors are fitted with gauge openings at both the suction service valve and the discharge service valve. The service valve stems on some compressors must be back-seated to seal the gauge openings. Some compressors use the valve core method with Schrader or Dill cores.

Before servicing an automobile air conditioning system, know what performance to expect from the system. Fig. 27-52 is a table of operating conditions of a system at various temperatures. Abnormal low-side and/or high-side pressures or noise will signal the need for repair.

Altitude affects vacuum actuators and some suction pressure controls, particularly if any part of the bellows or diaphragm movement is exposed to the pressure of the atmosphere. Fig. 27-53 shows the steps necessary to check a system, to charge the system with refrigerant, to purge the receiver, and to evacuate the system.

27-36 SERVICING BELTS

An engine having several belt-driven accessories may be equipped with three or four belts. Some late model engines only use one large, ribbed belt for all accessory units. Belt tension is adjusted by moving the generator, power steering pump, or the idler pulley.

Belts stretch in use. Therefore, they should be periodically checked for tightness and adjusted to specifications. A loose belt will soon fail. Pulleys can fail due to wear caused by belt slippage. The belt should be dry. Remove excessive oil on the pulleys. Replace any belts contaminated with oil. A shrill squeal when engine speed is increased indicates loose belts or glazed belt surfaces. A few drops of belt dressing on a glazed belt may stop the squeal temporarily. However, it is best to replace the belt.

Compressor belts with signs of oil contamination, cracks, or frayed edges should be replaced.

C.C.O.T. A/C SYSTEM DIAGNOSTIC PROCEDURE "INSUFFICIENT COOLING"

PERFORMANCE PRESSURE — TEMPERATURE DATA
VERIFY REFRIGERANT CHARGE

TEMPERATURE OF AIR ENTERING CONDENSER		70°F. (21°C.)	80°F. (27°C.)	90°F. (32°C.)	100°F. (38°C.)	110°F. (43°C.)
*COMPRESSOR OUT PRESSURE (BEFORE ORIFICE)	A SERIES	110-150 psi 758-1034 kPa	130-170 psi 896-1172 kPa	160-200 psi 1103-1379 kPa	195-235 psi 1344-1620 kPa	230-250 psi 1586-1724 kPa
	B SERIES	120-160 psi 827-1103 kPa	150-190 psi 1034-1310 kPa	185-225 psi 1276-1551 kPa	215-255 psi 1482-1758 kPa	270-310 psi 1861-2137 kPa
*EVAPORATOR PRESSURE (AT ACCUMULATOR)**	A SERIES	24-30 psi 165-207 kPa	24-30 psi 165-207 kPa	24-30 psi 165-207 kPa	24-30 psi 165-207 kPa	26-30 psi 179-207 kPa
	B	25-30 psi 173-207 kPa	25-30 psi 173-207 kPa	25-30 psi 173-207 kPa	26-32 psi 180-220 kPa	26-32 psi 180-220 kPa
*DISCHARGE AIR TEMPERATURE— LEFT CENTER OUTLET	A SERIES	34-38°F. 1.1-3.3°C.	34-38°F. 1.1-3.3°C.	35-39°F. 1.7-3.9°C.	38-42°F. 3.3-5.6°C.	40-44°F. 4.4-6.7°C.
	B	38-42°F. (3-6°C.)	38-42°F. (3-6°C.)	38-42°F. (3-6°C)	38-42°F. (3-6°)	42-48°F. (6-9°C.)

*JUST BEFORE COMPRESSOR DISENGAGES
**AT SEA LEVEL

Fig. 27-52. Performance chart for a system operation at various temperatures. Both conventional and metric pressures are shown. (Buick Motor Div., General Motors Corp.)

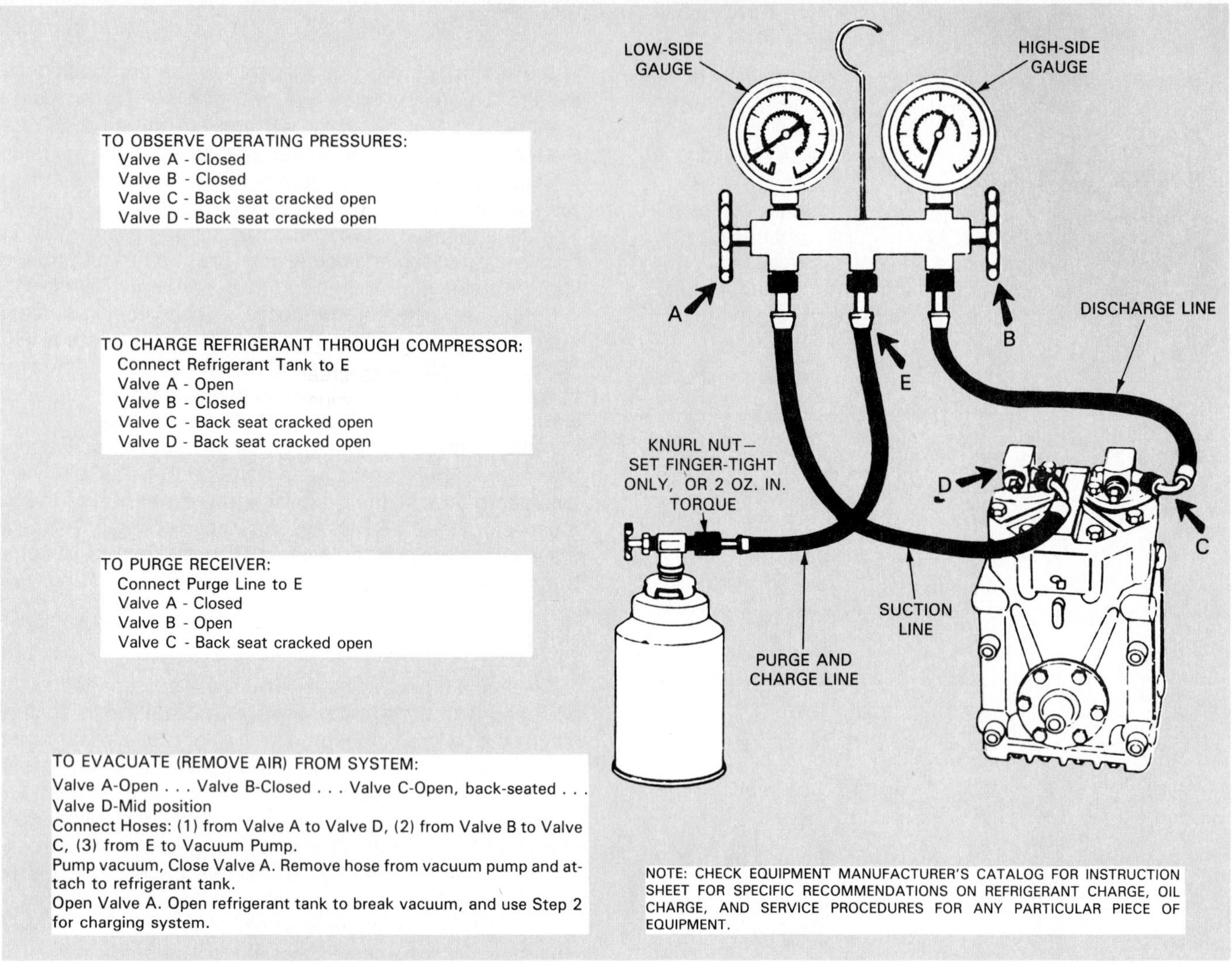

TO OBSERVE OPERATING PRESSURES:
Valve A - Closed
Valve B - Closed
Valve C - Back seat cracked open
Valve D - Back seat cracked open

TO CHARGE REFRIGERANT THROUGH COMPRESSOR:
Connect Refrigerant Tank to E
Valve A - Open
Valve B - Closed
Valve C - Back seat cracked open
Valve D - Back seat cracked open

TO PURGE RECEIVER:
Connect Purge Line to E
Valve A - Closed
Valve B - Open
Valve C - Back seat cracked open

TO EVACUATE (REMOVE AIR) FROM SYSTEM:
Valve A-Open . . . Valve B-Closed . . . Valve C-Open, back-seated . . .
Valve D-Mid position
Connect Hoses: (1) from Valve A to Valve D, (2) from Valve B to Valve C, (3) from E to Vacuum Pump.
Pump vacuum, Close Valve A. Remove hose from vacuum pump and attach to refrigerant tank.
Open Valve A. Open refrigerant tank to break vacuum, and use Step 2 for charging system.

LOW-SIDE GAUGE HIGH-SIDE GAUGE

A   E   B   DISCHARGE LINE

KNURL NUT— SET FINGER-TIGHT ONLY, OR 2 OZ. IN. TORQUE

D   C

SUCTION LINE

PURGE AND CHARGE LINE

NOTE: CHECK EQUIPMENT MANUFACTURER'S CATALOG FOR INSTRUCTION SHEET FOR SPECIFIC RECOMMENDATIONS ON REFRIGERANT CHARGE, OIL CHARGE, AND SERVICE PROCEDURES FOR ANY PARTICULAR PIECE OF EQUIPMENT.

Fig. 27-53. Procedures are given for observing operating pressures, methods of charging a system, purging a receiver, and evacuating air.

Always remember that a "short life" belt or a broken belt may be the result of an unusual overload (excessive pressures), pulley out of line, wrong type belt, engine oil leak, or incorrect tension. Determine the cause of the failure and remedy it.

Always loosen a belt before removing it. Forcing a belt over a pulley may injure and weaken the belt. This could also bend the pulleys.

27-37 SERVICING: TESTING FOR LEAKS

Check for refrigerant leaks by using a trace chemical, halide torch, electronic leak detector, foam leak detector (soap bubbles), or pressure rise method. These leak testing techniques are described in Chapters 11 and 14.

Some technicians will put into the system a refrigerant colored with reddish dye. Then, red discoloration on the metal surfaces will reveal the source of the leak.

Most frequently, leaks are found with the halide torch leak detector. It will locate a leak so small that it would amount to 1 lb. (0.45 kg) in about fourteen years. When using a halide torch, an exploring tube end sniffer is placed near the joint being checked. If there is a leak, some escaping refrigerant is drawn up the tube where it passes over a propane or acetylene heated copper element. If there is

refrigerant vapor in the air sample, the flame will turn green.

Danger! When R-12 is burned, very poisonous phosgene gas is produced. Avoid breathing fumes when leak testing an air conditioning system with a torch type tester.

The ultraviolet fluorescent leak detection system is used on residential, commercial, and automotive systems. It uses a high-intensity ultraviolet lamp, Fig. 27-54, a mist infuser tool, and specially formulated fluorescent additives to find the smallest possible leaks in the system. It is effective on leaks as small as 1/4 of an ounce per year. It can be used on any type of refrigerant. The technician inserts a pre-measured amount of fluorescent additive into the refrigerant or air conditioning system with the mist infuser. Then the lamp is used to pinpoint the leaks in the fitting, tubing, coils, or compressor, or wherever it may be. The additive remains in the system, allowing leak inspection with the ultraviolet lamp.

Leaks can be detected when a vacuum is being drawn on a system. With the vacuum pump running, shut off the vacuum valve on the manifold. If the vacuum gauge needle starts to creep back toward zero (atmospheric pressure), there is a leak in the system. The leak must be located and corrected before completing the vacuum operation for drying out the system.

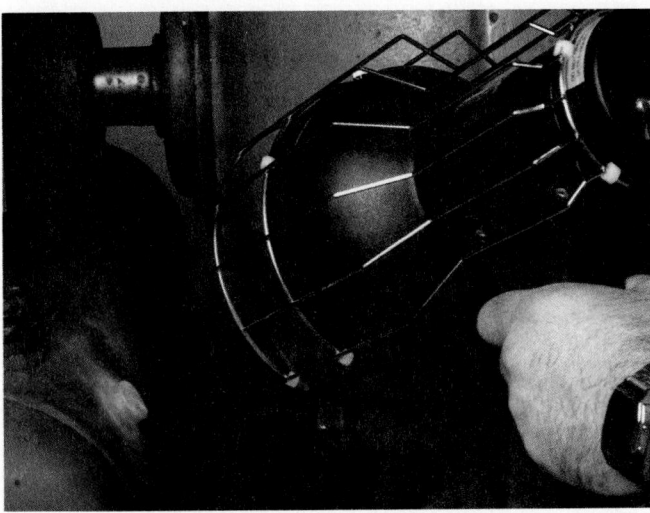

Fig. 27-54. Ultraviolet lamp locates refrigerant leak. (Spectronics Corporation)

Leak detection is done by pressurizing the system with R-12, then using the torch type leak detector, electronic leak detector, or foam leak detector (soap bubbles).

27-38 SERVICING: TESTING COMPRESSOR

The compressor must pump efficiently. If the capacity of the compressor decreases (worn valves, cylinders, or rings), maximum cooling will not be obtained.

A compressor should pump vacuum to 15 in. Hg (381 mm Hg) in a short time against normal head pressure. If this cannot be done, the pistons, rings, or intake valves are leaking (worn).

One method to test compressor seals is to allow the system to run. Stop the engine. Then turn the discharge service valve in all the way. If the head pressure drops, the exhaust valve in the compressor is leaking.

R-12 is used for testing for compressor leaks. Gauges are used to test the pumping capacity and valve condition of the compressor.

Note! Never run a compressor unless it has the correct amount of clean refrigerant oil in it.

Some compressors have a screen installed in the compressor body under the suction line service valve mounting. This screen removes foreign particles such as dirt, sand, and metal chips to prevent damage to the compressor. It should be removed and cleaned each time the refrigerant is removed from the system. If the screen is blocked (clogged), or almost blocked, the system will not refrigerate. In addition, the compressor crankcase would be under a vacuum, low-side pressure would be above normal, and high-side pressure would be below normal. Little or no refrigerant would circulate.

27-39 SERVICING: ADDING OIL TO SYSTEM

The compressor must have the correct amount and type of oil. For hot systems, 500 viscosity refrigerant oil is best. It must be clean and dry (no moisture content). Too much oil will cause oil pumping, reducing the efficiency of the system and possibly causing damage to the compressor valves. Too little oil will cause rapid wear of the compressor

bearings, pistons, rings, and valves. It will also cause scoring of the shaft seal. Therefore, it is important to check the oil level each time a unit is serviced.

To check system oil, install the gauge manifold. Purge the lines. Turn the suction service valve in all the way. Run the compressor until the compound (low-side) gauge reads zero psi. Then, turn the discharge service valve in all the way. Remove the oil level plug, and check the oil level. If the oil is too high, it will drain out. If too low, add oil by siphoning it into the compressor with the compressor crankcase at a partial vacuum. Use either a vacuum pump or the compressor to create this vacuum.

Some compressors must be removed from the system to check the oil level. Others require a hand-made dipstick. The wire dipstick is inserted through a bolt hole in the compressor crankcase to check the oil level. Check service manual procedures if in doubt.

27-40 SERVICING: CHARGING THE SYSTEM

Some automotive systems have a sight glass to aid in charging the system. See Fig. 27-55. When the system is operating normally, there will be no bubbles in the sight glass after the system has run for a few minutes. If the system is short of refrigerant, bubbles will appear regularly in the sight glass. A system without refrigerant, or with very little refrigerant, will not have enough liquid to form bubbles.

An overcharged unit may be detected by excessive head pressure, but this condition will not show in the sight glass. If high head pressure is shown on the high-side gauge, determine the correct system pressure and purge the excessive refrigerant.

The system is charged by means of the service manifold. Connect the manifold to the system. Then connect the charging cylinder to the manifold and purge the lines.

If only a small amount of refrigerant is to be added, charge through the low side with the cylinder upright. If the complete charge of the cylinder is to be put in the system, and if the system is under a vacuum, charge the refrigerant into the high side in liquid form. Do this by inverting the

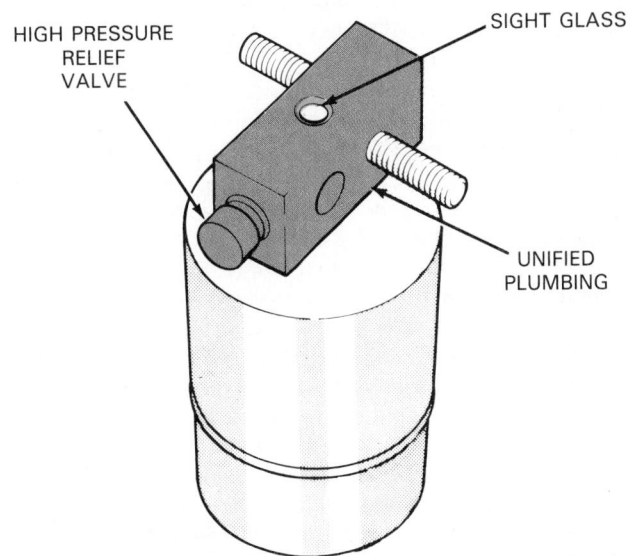

HIGH PRESSURE RELIEF VALVE

SIGHT GLASS

UNIFIED PLUMBING

Fig. 27-55. Filter-drier with sight glass for use as a refrigerant level indicator. Presence of bubbles always indicates a problem. (Chrysler Corp.)

cylinder and opening the service valves. Chapter 11 gives full instructions on the operation of the gauge manifold.

Fig. 27-56 illustrates a portable charging station. Note the gauge manifold, charging cylinder, and vacuum pump. The gauge manifold has three gauges. The two gauges back of the manifold handwheels are the suction gauge and the high-pressure gauge. The gauge on the right of the manifold is used during vacuum operations. The compressor is equipped with access valve service connections.

Fig. 27-57. Technician servicing an automobile air conditioning system. A—Vacuum-charging station. B—Refrigerant recovery and recycling system. (Robinair Division, SPX Corporation)

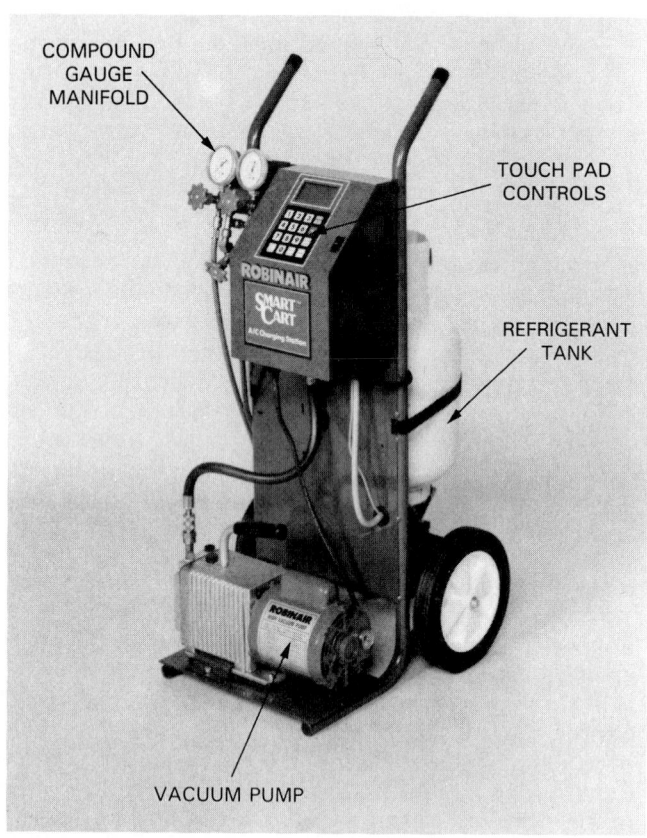

Fig. 27-56. Portable charging and discharging unit. Programming of unit for charging and evacuation is achieved by entering time needed for evacuation and amount of refrigerant to be charged. (Robinair Division, SPX Corporation)

Fig. 27-57 shows a technician using a touch pad microprocessor or programmer. It is a portable cart-style vacuum and charges automatically. The unit adds refrigerant to the system in 0.2-pound increments for partial charging or leak checking. Next to the service technician is a recovery and recycling unit. This unit is used to conform to the EPA Acts which require all R-12 refrigerant to be removed from a system without allowing it to enter the atmosphere and the recycling of the recovered refrigerant.

The units have separate controls for recovery and recycling procedures. To recover, a manifold gauge set is attached to the air conditioning/refrigeration system. The exhaust hose is connected to the recovery inlet. The manifold and tank valves are opened, and the recovery start button is pressed. The unit will run until recovery is complete, and will then automatically shut off. The unit can then be used as a recycling system. The unit is operated until the moisture

Fig. 27-58. A computerized air conditioning service station. A—Microprocessor key pad. B—Response indicator reading. C—Low-side gauge. D—High-side gauge. (Robinair Division, SPX Corporation)

has been removed, and the indicator light turns green. Refrigerant can then be reused.

Fig. 27-58 illustrates a microprocessor-controlled service station with vacuum pump and gauge manifold.

Some automotive air conditioning systems have a receiver for holding refrigerant. Therefore, the charge can vary somewhat and system operation will still be efficient. If the charge is 4 lb. (1.8 kg), the system may operate on as low as 2 1/2 lb. (1.1 kg) of refrigerant.

27-41 SERVICING: PERIODIC MAINTENANCE

The service technician should tell car owners to run the air conditioning a few minutes each month in fall, winter, and spring to keep compressor parts (especially the shaft seal) lubricated. 500 viscosity oil makes a good seal.

The owner should check the system each spring and fall as follows:
1. Condenser (clean fins and tubes of leaves, lint, and insects).
2. Refrigerant lines (check for signs of chafing or wear).
3. Belts (check for belt deterioration and proper adjustment).

The service technician should check the system each spring and fall or each 10,000 miles (16 100 km) as follows:
1. Clean all parts externally, including condenser.
2. Straighten fins on condenser.
3. Check refrigerant charge:
   a. Sight glass.
   b. Pressures in system.
4. Check oil level in compressor.
5. Check for leaks using a leak detector or color trace.
6. Make sure belts are in good condition and adjusted correctly.

When a vehicle equipped with an air conditioning system is stored for a long period, the air conditioner should be started very carefully. Sometimes the compressor binds during storage. It is best to raise the hood and watch the compressor and belt while turning on the air conditioner. If the belt starts slipping, stop the engine at once. This indicates that the compressor is turning with difficulty or is "frozen" (stuck). Try to "free" the compressor by slow and careful turning. If it will not turn, remove it at once for reconditioning.

27-42 REVIEW OF SAFETY

There are several very important safety precautions to be observed when working on automotive air conditioning systems.

The engine must sometimes be running to provide power for the air conditioner and air for the condenser. Therefore, connect the car's tail pipe to an air exhaust system when in an enclosed area. Do not touch the exhaust manifold or serious burns may result. Avoid putting tools, hands, or clothing in contact with the revolving fan. Moving belts are dangerous, too. Loose clothing presents a hazard around any moving parts.

It is best not to wear rings, a bracelet, or a wristwatch. These metal parts may short an electric circuit and burn the wearer, or may catch in a moving part and cause injury. One solution is to put a temporary shield over the fan. Usually, they are made of plastic and fasten to the radiator of the car.

Be careful about putting hands or tools near the spark plugs or plug wires. The electrical shock is not harmful itself, but the shock may cause you to jump against something, fall down, or jerk into moving parts of the engine. If possi-

ble, stop the engine before performing work on air conditioner parts under the hood. Always block the wheels of the car when running the engine.

Be sure to wear goggles and gloves when charging the system or when working on parts that contain refrigerant. R-12 boils at $-21.7\,°F$ ($-29\,°C$). When it is spilled on the skin or gets in the eye, it will cause freeze injury.

Keep a protective cap on the refrigerant cylinder when it is not in use. When using the cylinder, fasten it to a part of the vehicle or to some sturdy stand. Otherwise, it may fall and break the connections of the valve.

If it is necessary to heat a refrigerant cylinder, use warm water only. Never use a torch, electric heat, steam heat, stove, or radiator, because the cylinder may explode.

Avoid welding, brazing, or steam cleaning near the air conditioning system unless the refrigerant has been removed. Otherwise, the excessive pressures may damage the system and injure people in the area.

Breathing quantities of any refrigerant is harmful to a human or an animal. Ventilate the area to keep the vapor concentration to a minimum.

When discharging a system near an open flame, the refrigerant tends to break down and form toxic gases. These gases also tarnish metal and plated surfaces.

Most engine cooling systems are pressurized. If a pressure cap is removed when the engine is hot, hot water will erupt out of the radiator and can cause severe burns.

Blower fans can cause painful injuries to the hands. The sharp fins on the coils of the condenser and evaporator can cause deep cuts.

Manifold service lines must be kept clear of pulleys, belts, and fans.

27-43 TEST YOUR KNOWLEDGE

1. Of what material are air conditioning compressors usually made?
2. Name the three types of wobble plate compressors.
3. What will occur if the operating low-side pressure is too low?
4. What is a proper tension range for a new automotive compressor belt?
5. Name three common types of automotive metering devices.
6. At what location is the automotive type of air conditioner condenser usually mounted?
7. What type of liquid solution is used in the heater core to warm the passenger compartment?
8. What is the purpose of an accumulator?
9. How are air dampers usually operated?
10. What are the two most common types of automotive air conditioning control systems?
11. What type of control governs compressor cycling, blower speed, and damper positions in the electronic control system?
12. Where is the in-car temperature sensor usually located?
13. What is the purpose of the wide-open throttle switch?
14. What is the purpose of a POA (pilot-operated absolute) valve?
15. Where is the outside air temperature sensor usually located?
16. Name two types of compressor service valves.
17. What type of refrigerant is used in most automotive air conditioning systems, and what is a possible new

alternate refrigerant?

18. What are four common methods used to find a leak in the system?

19. What is the purpose of a sight glass?

20. What are the three major components of a portable charging station?

21. What items should the owner check each spring and fall in an automotive air conditioning system?

22. What viscosity oil is normally used in automotive air conditioning systems?

23. What is a sign of an overcharged system?

24. What is the effect of too little oil in a system?

25. What are two popular types of clutches that use an electromagnet?

26. Where should the thermostatic expansion valve (TEV) sensing bulb be located?

27. In a self-diagnostic system, where is the information from the microprocessor usually received?

28. What happens to refrigerant when it comes in contact with a flame?

29. Where is the system microprocessor or computer usually located?

30. How is in-car temperature measured with an electronic control system?

Chapter 28

REFRIGERANT RECOVERY—RECYCLING—RECLAIMING

After studying this chapter, the technician will be able to:
☐ Describe the effect of CFC refrigerants on the ozone layer in the atmosphere.
☐ Understand the Environmental Protection Agency (EPA) rules governing the phasing out of fully halogenated refrigerants (CFCs).
☐ Follow the EPA regulations regarding recycling of refrigerants.
☐ Discuss the proper procedures to recover, recycle, and reuse chlorofluorocarbon refrigerants (CFCs).
☐ Identify the various types of refrigerant recovery and recycling equipment and their use.
☐ Follow the procedures as set forth by the Department of Transportation regarding the transportation of refrigerant cylinders and drums.

28-1 CHLOROFLUOROCARBONS (CFCs) AND THE OZONE LAYER

The ozone layer is a fairly thin layer of the earth's upper atmosphere, approximately thirty-five miles above the ground. It is often called a screen or shield. The ozone layer has been credited with protecting the earth from the damaging ultraviolet rays of the sun. The ozone layer functions as a filter for the sun's ultraviolet rays and prevents human, plant, and sea life from the damaging effects, Fig. 28-1.

Fig. 28-1. The depletion of the ozone layer can cause problems for plant and sea life. (General Filters, Inc.)

The destruction of the ozone shield or cover by the release of chlorofluorocarbons (CFCs) into the atmosphere by some refrigerants is of concern in the use of these refrigerants.

Chlorofluorocarbons (CFCs) are a family of chemicals containing chlorine, fluorine, and carbon. CFCs R-12, R-11, and others are used as refrigerants and as blowing agents for the manufacture of insulation, packaging, etc. The stability of these compounds, coupled with their chlorine content, has caused the earth's protective ozone layer to be depleted.

This concern has prompted countries throughout the world to pass legislation preventing the use of chemicals that can affect the ozone layer. The United States Environmental Protection Agency (EPA) has enacted regulations which state that the following fully halogenated CFC refrigerants must be phased out by the turn of the century:

R-11 (trichloromonofluoromethane)
R-12 (dichlorodifluoromethane)
R-113 (trichlorotrifluoroethane)
R-114 (dichlorotetrafluoroethane)
R-115 (chloropentachloroethane)

Production phaseout of the above refrigerants is scheduled to be reduced from 1986 production levels as follows:

1991—15% reduction
1992—20%
1993—25%
1994—35%
1995—50%
1996—60%
1997—85%
1998—85%
1999—85%
2000—total phaseout

The EPA official regulations also indicate that recycling of refrigerants used in air conditioning in mobile vehicles, such as autos, trucks, etc., become mandatory in all service shops in July, 1992.

Schedule for HCFC phaseout is as follows:

2015—production freeze and use limitations.
2020—prohibited for new air conditioning and refrigeration use.
2030—total phaseout.

PHYSICAL DATA

	113	11	114	12	500	22	502	13	503	HCFC 141b	HCFC 123	HFC 134a
Chemical Formula.	C_2Cl_3F	CCl_3F	$C_2Cl_2F_4$	CCl_2F_2		$CHClF_2$		$CClF_3$		CCl_2FCH_3	$CHCl_2CF_3$	CF_3CH_2F
Molecular Weight.	187.4	137.4	170.9	120.9	99.3	86.5	111.6	104.5	87.5	116.95	152.91	102.03
Boiling Point @ 1 Atmos. (°F).	117.6	74.9	38.8	−21.6	−28.3	−41.4	−49.8	−114.6	−126.1	89.7	82.2	−15.08
Freezing Point @ 1 Atmos. (°F).	−31	−168	−137	−252	−254	−256	—	−294	—	−154.3	−160.6	−141.9
Critical Temperature, (°F).	417	388	294	234	222	205	180	84	67	410.4	363.2	214
Critical Pressure, (psia).	499	640	473	597	642	722	591	561	632	673.0	540.0	589.8
Saturated Liquid Density @ 86 °F*	96.8	91.4	89.8	80.8	71.2	73.0	74.5	82.4Φ	78.5Φ	76.31	90.41	74.17
Specific Heat of Liquid @ 86 °F (Btu/lb. °F).	0.22	0.21	0.24	0.24	0.30	0.31	0.30	0.24Φ	0.28Φ	0.35	0.21	0.36
Specific Heat of Vapor at constant pressure. (Cp), at 86 °F and 1 Atmos. (Btu/lb. °F).	0.15¹	0.14	0.17	0.15	0.18	0.20	0.17	0.13Φ	0.14Φ	0.17¹	0.17	0.21
Specific Heat Ratio of Vapor ($k=Cp/Cv$) at 86° F.	1.08¹	1.13	1.08	1.14	1.14	1.18	1.14	1.18Φ	1.21Φ	—	1.10	1.12
Flammability and Explosivity.	None	None	None	None	None	None	None	None	None	7.6–17.7†	None	None
Toxicity Rating**	4-5	6	6	6	5	5	5	6	6	Not Avail.	Not Avail.	Not Avail.

genetron 113
Trichlorotrifluoroethane
Used in low capacity centrifugal chiller packaged units. Operates with very low system pressures, high gas volumes.

genetron 11
Trichlorofluoromethane
A centrifugal refrigerant with low operating pressures. Gives higher capacity than Genetron 113. Is also used as a secondary coolant in low temperature systems. A popular choice for use in thermal insulation construction projects.

genetron 114
Dichlorotetrafluoroethane
Intermediate in pressure and displacement. Principally used with centrifugal compressors for higher capacities or for lower evaporator temperature process type applications. Also used in foam applications.

genetron 12
Dichlorodifluoromethane
A very versatile and widely used refrigerant. Common in reciprocating and rotary type equipment. For all types of applications, household to industrial. Also employed in some centrifugal designs and in several special applications such as sterilant gas, blowing agents and aerosols.

genetron 500
Azeotrope
An azeotrope of Genetron 12 which has slightly higher vapor pressures and provides higher capacities from the same compressor displacement.

genetron 22
Chlorodifluoromethane
As a refrigerant, operates with higher system pressures but offers low compressor displacement requirement. Popular in residential, commercial and industrial applications. Used as a blowing agent in aerosols and as an intermediate to produce fluoropolymers.

genetron 502
Azeotrope
An azeotrope of CFC-115 and HCFC-22 which is especially suited to low evaporation temperature applications. Handles high temperature lifts well and simultaneously provides capacity gains.

genetron 13
Chlorotrifluoromethane
A specialty low temperature refrigerant used in the low stage of cascade systems to provide evaporator temperatures in the range of −100 °F.

genetron 503
Azeotrope
An azeotrope of CFC-13 and HFC-23 which is used in the low stage of cascade type systems where it provides gains in compressor capacity and in low temperature capability.

HCFC 141b
Dichlorofluoroethane
HCFC-141b presents significant opportunities as a blowing agent alternative for use in rigid board, foam systems, flexible foam and other end-use applications.

HCFC 123
Dichlorotrifluoroethane
As a leading candidate in the next generation of blowing agents, HCFC-123 may offer effective solutions in such diverse applications as rigid applications and foam board and foam systems insulation. Also may be used in centrifugal refrigeration equipment and in specialized solvent applications.

HFC 134a
Tetrafluoroethane
A hydrofluorocarbon with an ozone depletion potential of zero. HFC-134a holds great promise as a CFC substitute for a wide range of air conditioning and refrigeration systems in residential, commercial, and industrial applications.

* (lbs./cu.ft.)
** (based on Underwriters' system)
Φ @ 0.2 Atmos. press.
¹ @ -30°
† Upper and lower vapor flammability (Vol. %)

DISCLAIMER
All statements, information, and data given herein are believed to be accurate and reliable but are presented without guaranty, warranty, or responsibility of any kind, express or implied. Statements or suggestions concerning possible use of our products are made without representation or warranty that any such use is free of patent infringement and are not recommendations to infringe any patent. The user should not assume that all safety measures are indicated, or that other measures may not be required.

Fig. 28-2. Physical properties of refrigerants. Note difference between 134a and 12. (Allied Signal, Inc.)

Penalties and fines for violating the above provisions can be rather severe. The EPA is authorized to seek legal action against any person violating these provisions.

Numerous studies have indicated that HCFC-22 and R-502 are considered less of a problem than fully halogenated chlorofluorocarbons. The latter are made of hydrocarbon molecules in which all the hydrogen atoms have been replaced by the halogen atoms "chlorine" and "fluorine." R-22 and R-502 are not fully halogenated. R-22 has two fluorine atoms, one chlorine, and a hydrogen atom—all connected to a carbon atom. These molecules will tend to break down in the lower atmosphere before getting to the stratosphere, where the chlorine can damage the ozone layer. Therefore, non-fully halogenated refrigerants like R-22 and R-502 cause much less ozone damage.

Research continues to determine how existing equipment may be modified to accept the refrigerants such as 134a and 123 that have little or no effect on the ozone. R-134a is currently being used in automotive air conditioning systems, and R-123 in new commercial applications.

The chart in Fig. 28-2 shows a comparison of the physical properties for present refrigerants, possible replacements, and their uses.

28-2 RECOVERY, RECYCLING, RECLAIMING OF REFRIGERANTS

Laws governing the release of chlorofluorocarbon refrigerants (CFCs) into the atmosphere have resulted in the development of procedures to recover, recycle, and reuse these refrigerants.

The industry has adopted specific definitions for these terms:

Recovery: To remove refrigerant in any condition from a system and store it in an external container without necessarily testing or processing it in any way.
Recycling: To clean refrigerant for reuse by oil separation and single or multiple passes through devices, such as replaceable core filter-driers, which reduce moisture, acidity, and matter. This term usually applies to procedures implemented at the field job site or at a local service shop.
Reclaim: To reprocess refrigerant to new product specifications by means which may include distillation. This will require chemical analysis of the refrigerant to determine that appropriate product specifications are met. This term usually refers to the use of processes or procedures available only at a reprocessing or manufacturing facility. This also includes on-site or local service shops that are equipped with highly technical equipment.

Many companies have developed the equipment necessary for the service technician to prevent the unnecessary release of chlorofluorocarbons into the atmosphere.

Refrigerant recovery management equipment, Fig. 28-3, can be divided into three categories:
1. Recovery: A unit that recovers or removes the refrigerant.
2. Recovery/Recycle: A unit that will recover and recycle the refrigerant.
3. Reclaim: A unit that will reclaim the refrigerant within the Environmental Protection Agency (EPA) standards.

Fig. 28-3. Refrigerant recovery management system. Unit A is a recovery/recycle unit which handles R-12, R-22, R-500, and R-502. Unit B is used for R-12 as a recovery/recycle unit. Unit C is a recovery unit for R-12. (Robinair Division, SPX Corporation)

28-3 REFRIGERANT RECOVERY EQUIPMENT

Recovery machines are available in various designs. The basic small units are intended for use with R-12 and act as recovery stations which are not vented to the atmosphere, Fig. 28-4.

The refrigerant is removed from the system in its present condition and stored in a disposable or transferable cylinder. The refrigerant then can be recycled at the service center or sent to a reclaiming station and reused at a later date.

The method of removing refrigerant by some small recovery equipment is the vapor recovery method. Using a refrigerant recovery device, the technician is capable of removing refrigerant from light commercial, automotive, residential, and appliance applications. During the recovery

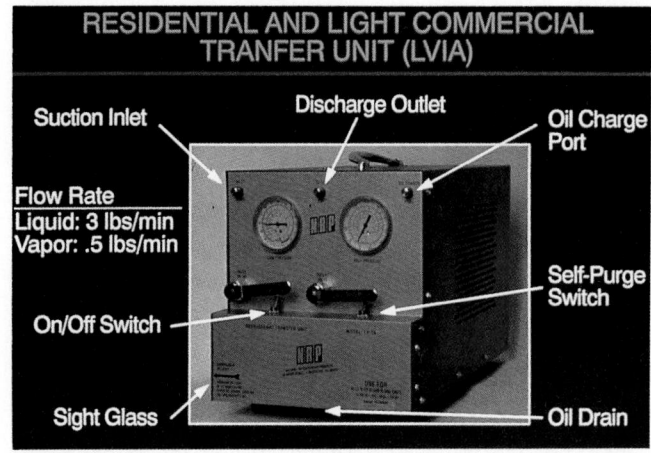

Fig. 28-4. Small recovery unit weighs approximately 45 lbs. and operates on 115 volts. (National Refrigeration Products)

process, the refrigerant is removed from the system by using the vapor within the system and the pumping power of a recovery machine, Fig. 28-5.

Recovery is similar to evacuating a system with the vacuum pump. Procedures vary with each manufacturer. Basically, the hose is connected from a low-side access port to the suction valve of the recovery unit. Once the exhaust hose is attached, the recovery device is turned on and the recovery begins. Some units have a signal device to indicate when the recovery is completed. This means that no more vapor is being processed by the recovery equipment. In some instances, the recovery device automatically closes off the vacuum system.

When the recovery is completed, the low-side isolation valve is shut off. The system should sit for at least five minutes. If the pressure rises to 10 psi or more, it may be an indication that there are pockets of cold liquid refrigerant throughout the system, and it may be necessary to start recovery again.

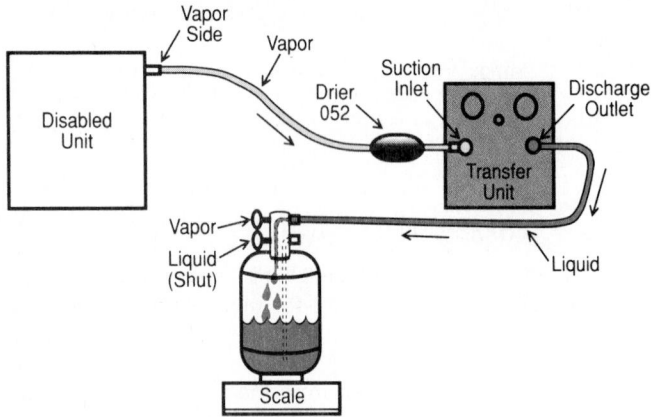

Fig. 28-5. The recovery of vapor refrigerant from a system. Note the direction of refrigerant flow from the disabled unit to the transfer unit.   (National Refrigeration Products)

Since it is much faster to recover liquid refrigerant rather than vapor refrigerant, the technician may prefer to use a machine that can remove the liquid refrigerant. Many machines are designed to accomplish this using standard refrigerant cylinders. Some small transfer units use special recovery cylinders that allow the technician to remove a liquid and vapor refrigerant.

Fig. 28-6 shows a procedure for removing the refrigerant using a liquid transfer concept. This type of recovery unit requires a cylinder with two ports.

The transfer unit pumps the refrigerant vapor from the top of the cylinder and pressurizes the air conditioning unit. The difference in pressure between the cylinder and the unit transfers the liquid refrigerant to the cylinder. Once the liquid has been removed, the remaining vapor is removed by changing the hook-up.

It is recommended that the compressor oil from the recovery unit be changed after recovery from a burned-out system or before recovery of a different refrigerant. It is also recommended the drier be replaced and the transfer unit and hoses be evacuated before transferring a different refrigerant.

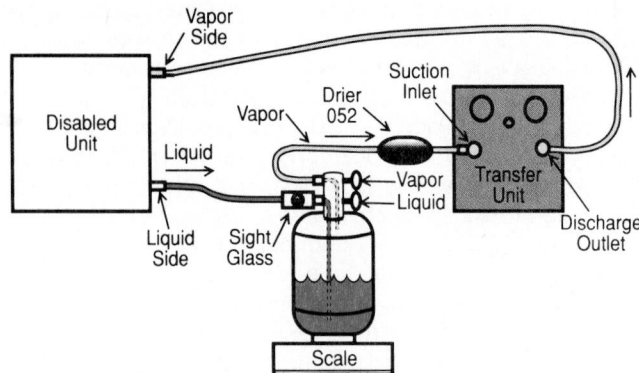

Fig. 28-6. The recovery of refrigerant from a system using a liquid transfer method. Note the direction of refrigerant flow from the disabled unit to the transfer unit.   (National Refrigeration Products)

The technician should make certain that the vessels being filled are not overfilled. Eighty percent capacity is normal. As the cylinder is filling, the pressure should be watched. If the recovery unit has a moisture indicator sight glass, any changes that may occur should be noted.

If the technician uses a system that only recovers the refrigerant, the recharging can be accomplished in many ways.

A computer charging or discharging system is shown in Fig. 28-7. This allows the technician to charge a system through the use of a computer scale. A predetermined weight is set on the keyboard. The maximum gross capacity is 110 pounds or 50 kilograms. The system is switchable from U.S. standard to metric. The hold key allows the technician to interrupt the charging or recovery cycle without losing the program.

Another method of recovery is through the use of an electronic refrigerant charger, Fig. 28-8. The unit is operated by rechargeable batteries and will charge R-12, R-22, R-500, R-502, and 134a. Refrigerant oil can also be added to the system by using the oil injector.

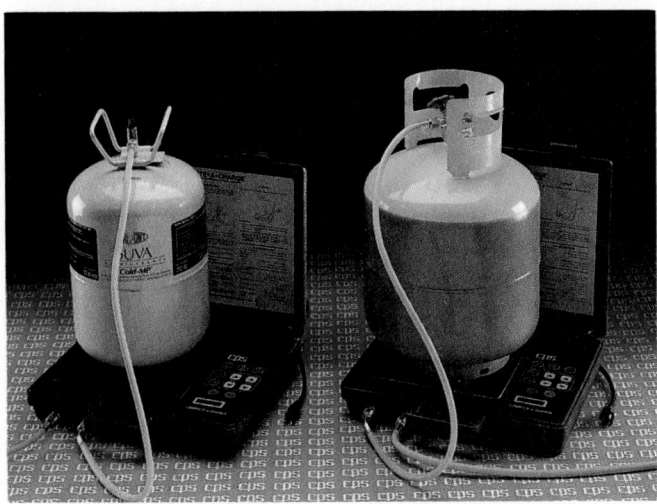

Fig. 28-7. The instrument on the left is set to automatically charge a system to a programmed amount. On the right, the instrument is set up to automatically remove a predetermined amount of refrigerant as set on the keyboard.   (CPS Products, Inc.)

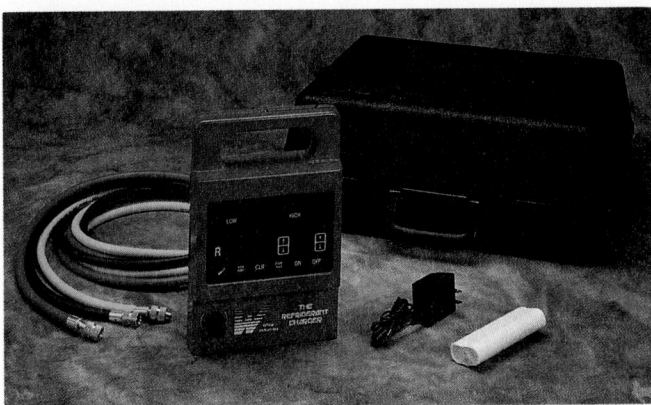

Fig. 28-8. Refrigerant charger will display pressure or corresponding temperature for refrigerant selected. (K-Whit Tools, Inc.)

28-4 REFRIGERANT RECYCLING EQUIPMENT

In the past, refrigerant was typically vented into the atmosphere. Using current technologies, this refrigerant can now be recovered and recycled. However, old or damaged chlorofluorocarbons cannot be reused simply by removal from a system and compressing to a vapor. The vapor, to be reused, must be clean. Recovery/recycling machines, Fig. 28-9, are designed to recover and clean the refrigerant at the job site or at a local service shop. Recycling as per-

Fig. 28-9. A combination R-12 reclamation and recharging system. Recovers refrigerant from automotive air conditioning system. It then purifies refrigerant to SAE standards. The refrigerant is then stored in heated cylinder and charged back to air conditioning system. (Ig-Lo, Inc.)

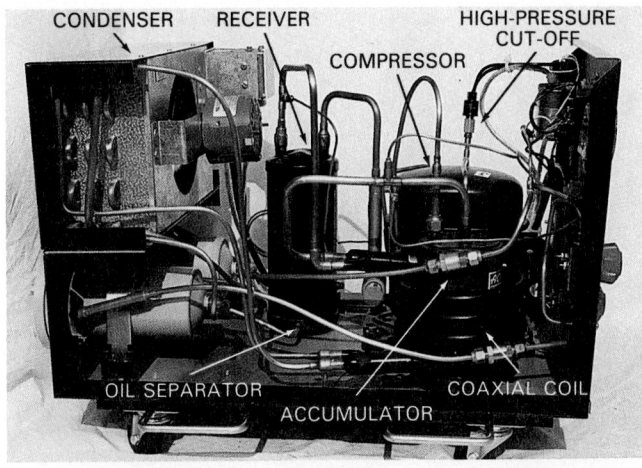

Fig. 28-10. Refrigerant recovery station mechanism. Note that unit has two filters.
(Thermal Engineering Company, Division of Seakay Co., Inc.)

formed by most of the machines on the market today reduces the contaminants through oil separation and filtration. This cleans the refrigerant, but not necessarily to the manufacturer's originial specifications of purity. Fig. 28-10 illustrates the operational mechanism of a system capable of handling R-12, R-22, R-500, and R-502.

Many of these units, known as refrigerant transfer units, are designed to pump down the system. This provides an on-site recycling machine that will return the recycled refrigerants to the same system. Some of the units have equipment to separate the oil and acid and measure the amount of oil in the vapor. The used refrigerant can be made reusable by the recycling machine, using replaceable core filter driers or other devices which reduce moisture, particles, acidity, etc. Oil separation of the used refrigerant is achieved by using single or multiple passes through the unit. The single-pass recycling machine processes refrigerant through a filter-drier and/or uses distillation. It makes only one trip from the recycling process through the machine and then into the storage cylinder. The multiple-pass machine recirculates refrigerant through the filter-drier many times. After a given period of time, or a given number of cycles, the refrigerant is transferred into the storage cylinder.

The following recovery/recycling guidelines are applicable for this type of equipment:

1. Properly maintain the recovery/recycling equipment as per manufacturer's guidelines, changing filters as recommended. Check the system and the recycling equipment for leaks.
2. The proper time to utilize the equipment and the recovery/recycling procedures is when, in the past, refrigerant would have been exhausted into the atmosphere.
3. Keep the refrigerant contained and air out. The procedure of connect-and-charge-and-disconnect is no longer needed. Most new units have shut-off valves that operate automatically as the hose is connected or disconnected.
4. Obtain an approximate appropriate vacuum and run a leak check. See Fig. 28-11. Repair all leaks.
5. Use basic principles of refrigerant flow and heat transfer to speed the recovery process. When transferring from

Fig. 28-11. Technician using ultrasonic leak detector.
(Amprobe Instrument)

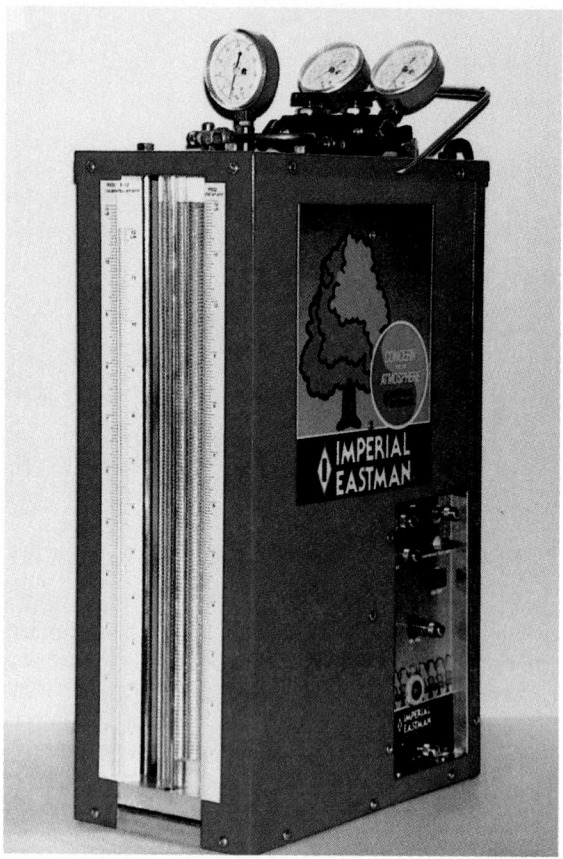

Fig. 28-12. Portable refrigerant recovery recycling equipment. Unit
has built-in holding tank.   (Imperial Eastman, Imperial Division)

one container to another, if possible transfer liquid from one tank to the other. This will allow transferring the entire liquid from one tank to the other without frosting the tank and limiting the flow rate from the room to the tank.

6. Always use appropriate refillable containers. Fill to the maximum of 80% of volume with liquid. Do not use disposable containers or those not approved.

7. Do not mix refrigerants. Mark containers. Thoroughly clean containers and all fittings upon completion.

Fig. 28-12 illustrates a refrigerant/recycle unit called a Refrigerant Management System. This unit weighs approximately fifty-nine pounds and has an internal storage capacity of 10 pounds. It is operated as a recovery unit and has a 1/2-hp compressor. Its recovery ability is approximately two pounds per minute for R-12, R-22, R-500, and R-502.

The unit has a low-pressure gauge, a high-pressure gauge, and a high-pressure condenser gauge. The front side of the unit has a refrigerant scale for various refrigerants and a refrigerant level indicator with a refrigerant column sight glass.

The lower front section has a low-side access port, high-side access port, vacuum discharge port, compressor power switch, and the recovery compressor oil fill. An oil sight glass indicator and a recovery compressor oil drain are located in the bottom section.

When hooking up the refrigerant management systems, it is recommended that high-side and low-side reclaiming be practiced. This procedure will avoid restrictions through the refrigerant control: expansion valve, cap tube, or orifice restrictor. If the technician were to recover from one side only, the result may be excessive recovery time or even incomplete refrigerant recovery. Therefore, the hoses are connected to the high and low side of the recovery system and then through the high and low side of the refrigeration system. Under no circumstances should liquid be removed from the system on a continual basis. The system is designed for vapor recovery. The initial recovery of high-side pressure "refrigerant" would be approximately 200 psig.

As the unit operates and vapor recovery is taking place, the point will be reached where liquid refrigerant appears in the column. The recovery is complete when the lowest vacuum on the low-side gauge is achieved, and the liquid refrigerant in the column stops rising.

This recovery is a procedure that should not be confused with the procedure used to evacuate a system by the use of a vacuum pump. This type of unit operates with an acceptable range of 15 to 20 inches of vacuum for removal of refrigerant.

A recovery recycling unit with an electronic charging scale designed for use in vehicular air conditioners is shown in Fig. 28-13. This allows the technician to recover, recycle, and recharge the system on site.

28-5 REFRIGERANT RECLAIM PROCEDURE

Reclaiming is defined as the reprocessing of a refrigerant to original production specifications as verified by chemical

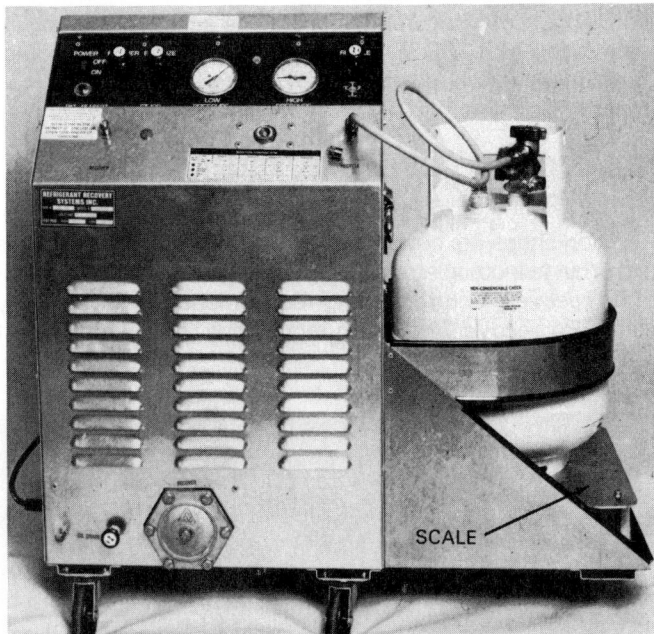

Fig. 28-13. A recovery system designed for use on automotive air conditioning system using R-12. Note built-in electronic charging scale. (Refrigerant Recovery Systems, Inc.)

analysis. In order to accomplish this, the machine must meet the SAE standards and remove 100% of the moisture and oil particulates. Many recovery/recycling machines cannot guarantee that the refrigerant will be returned to its original specifications.

An on-site recycling station must be able to provide separation of oil, acid, hard particle contaminants, moisture, and air in order to clean the used refrigerant.

A commercial unit is shown in Fig. 28-14. This type of unit is available for use with R-12, R-22, R-500, and R-502, and is designed for the continuous use required on a long run recovery recycling procedure.

This type of system can best be described as follows:
1. The refrigerant is accepted into the system as either vapor or liquid.
2. Refrigerant is violently boiled at high temperature under extreme high pressure.
3. Refrigerant then enters a large unique separator chamber where the velocity is radically reduced. This allows the vapor, at high temperature, to rise. During this phase, contaminants—copper chips, carbon, oil, acid, and all other contaminants—drop to the bottom of the separator to be removed during the "oil out" operation.
4. The distilled vapor passes to the air-cooled condenser and is converted to liquid.
5. The liquid passes into the on-board storage chamber(s). Within the chamber(s), an evaporator assembly lowers the liquid approximately 100°F (56°C) to a subcooled temperature of 38°F (3°C) to 40°F (4°C).
6. A replaceable filter-drier in this circuit removes the moisture while it continues the cleaning process to remove microscopic contaminants.
7. Chilling the refrigerant also facilitates the transfer to any external cylinders which are at room temperature.
Numerous refrigerant manufacturers and others have set up refrigerant recovery/reclaiming services that provide

refrigeration and air conditioning technicians a way to dispose of used refrigerant and obtain pure replacements as needed. The service technician must use Department of Transportation (DOT) approved returnable cylinders and proper tags, Fig. 28-15. Standard cylinders will hold approximately 100 pounds of used refrigerant and oil, although other containers range from 40 pounds to one ton. The technician uses a refrigerant transfer unit designed to pump down the system. See Fig. 28-16. This unit may also be used for charging a system.

The positive displaced compressed air machine removes both liquid and vapor. The refrigerant is reprocessed to designated purity specifications.

On large commercial installations, the service technician is provided with sample cylinders that are sent back to a reclaiming center. This is to obtain refrigerant analysis of contaminants prior to evacuation.

After approved for reclaiming, the refrigerant is removed. See Fig. 28-17. Technicians then take the refrigerant to the service center, where it is shipped back to the company and processed accordingly and returned for future sale as a used refrigerant. Reprocessing may be used for low-pressure refrigerants R-11 and R-113 and high-pressure refrigerants R-12, R-22, R-114, R-500, and R-502.

Company standards vary in regards to the type of vessel used to transport the refrigerant from the servicing area to the manufacturer. Some accept 55-gallon, 10-gallon minimal, etc. See Fig. 28-18. Each manufacturer has a procedure which must be followed, Fig. 28-19, and numerous documents are required by each company. See Fig. 28-20.

A reclamation company also provides a solution for the disposal of unwanted refrigerant.

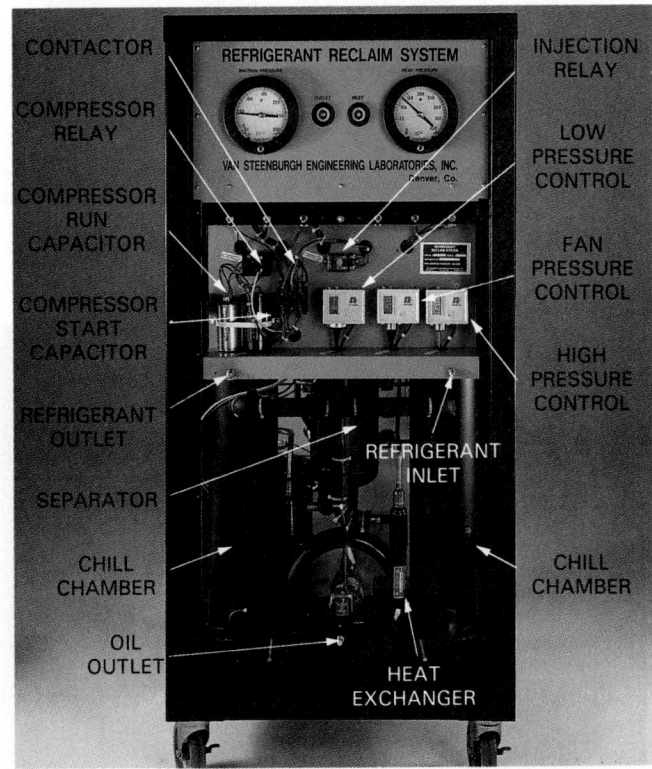

Fig. 28-14. Refrigerant reclaim system. Note location of inlet and outlet between suction pressure and head pressure gauges. (Van Steenburgh Engineering Laboratories, Inc.)

IMPORTANT SAFETY RECOMMENDATIONS

■ Only fill containers which are currently DOT approved for fluorocarbon refrigerants. Always inspect the container for pressure rating and latest hydrostatic test date. Be sure to thoroughly check each container for dents, gouges, bulges, cuts, or any other imperfections which may render it unsafe to hold refrigerant for storage or transportation.

■ DO NOT FILL A CONTAINER WHICH DOES NOT HAVE A CURRENT HYDROSTATIC DATE. DO NOT RISK FILLING A CONTAINER WHICH APPEARS UNSAFE OR MAY LEAK.

■ ALWAYS USE A SCALE WHEN FILLING ANY CONTAINER. DO NOT OVERFILL.

■ CAUTION: LIQUID REFRIGERANT CAN CAUSE FROSTBITE IF SKIN CONTACT OCCURS. ■ BE AWARE THAT THE REFRIGERANT/OIL BEING REMOVED FROM A SYSTEM MAY CONTAIN CONTAMINANTS WHICH MAY BE HARMFUL TO BREATHE OR CONTACT WITH THE SKIN. ■ ALWAYS PROVIDE FRESH AIR WHEN WORKING IN ENCLOSED AREAS. ■ AVOID BREATHING VAPORS. ■ ALWAYS WEAR SAFETY GLASSES AND GLOVES (cold resistant for pressurized refrigerants and rubber-type for R-11 or R-113. ■ AVOID CONTACT WITH CLOTHING.

■ BE SURE ALL CONNECTIONS ARE MADE TIGHT BEFORE TRANSFERRING REFRIGERANT INTO CONTAINERS. BE SURE ALL CLOSURES ARE MADE TIGHT ON THE CONTAINER IMMEDIATELY AFTER FILLING.

■ IT IS HIGHLY RECOMMENDED THAT YOU READ THE AIR-CONDITIONING AND REFRIGERATION INSTITUTE (ARI) "GUIDELINE K—GUIDELINE FOR RECOVERED FLUOROCARBON REFRIGERANTS."

■ READ AND FOLLOW ALL INSTRUCTIONS CAREFULLY.

Fig. 28-21. Standard safety recommendations to be followed when removing refrigerant from a system.   (National Refrigerants, Inc.)

may be covered are CFC storage and handling, transportation, recovery equipment and procedures, hazardous waste handling, storage, and disposal regulations. It is also essential that the service technician have a full understanding of the safety involved in handling and storage of refrigerant. See Fig. 28-21. Certification programs approved by the EPA are also being offered. The procedures for removal, basic field testing of refrigerant for purity, isolation of system components to prevent refrigerant venting, leak detection, isolation of leaks, and leak repairs are areas covered by most workshops, Fig. 28-22.

Fig. 28-22. Service technicians taking a motor vehicle refrigerant recycling and recovery examination.
(National Institute for Automotive Service Excellence [A.S.E.])

It is the technician's responsibility to follow **safe practice** procedures. This includes the replacement of both suction line and liquid line driers. If the system only has one, insert another in the opposite side. This will aid the refrigeration purification process.

Recycled refrigerants follow a standard that has been set by the Air-Conditioning and Refrigeration Institute (ARI), Standard ARI-700-88.

28-7 MOBILE AIR CONDITIONING

Clean Air Act Section 609 establishes the requirements in regard to mobile air conditioners. It states that, effective January 1, 1992, no person repairing or servicing motor vehicles for consideration may perform service on any motor vehicle air conditioner without properly using approved refrigerant recycling equipment, Fig. 28-23. Also, no person may perform such service unless that person has been properly trained and certified. The requirements do not become effective until January 1, 1993, for establishments which certify that they performed such service on fewer than 100 vehicles during 1990.

Rules governing stationary air conditioner and refrigeration equipment, the disposal of CFCs during service or repair, the certification of recovery/recycling equipment, certification of technicians, and record keeping are being developed further.

28-8 REVIEW OF SAFETY

The refrigerants identified for use in the refrigeration and air conditioning industry present no problems in normal use and handling. However, they should always be used in the proper manner to avoid potential hazards. Most refrigerants have low boiling points, thus protective clothing and eye protection should always be used to avoid frostbite. The higher boiling liquids can cause skin irritation. The refrigerant at atmospheric pressure vaporizes readily. If refrigerant contacts the skin, the latent heat of vaporization removes the heat from the skin and frostbite can occur. If refrigerant comes in contact with the eye, it may freeze the eye, causing blindness. Accidents involving refrigerants should be immediately referred to a doctor.

Fig. 28-23. Automotive service technician servicing automotive air conditioning system in keeping with the E.P.A. regulations. (National Institute for Automotive Service Excellence [A.S.E.])

The technician should make certain the refrigerant service cylinder is not filled completely with liquid refrigerant. Refrigerants should be stored only in cylinders marked on the label for that specific refrigerant.

Fluorocarbon vapors are heavier than air and tend to accumulate in low-lying areas. They replace the air in the room. The person breathing this will lose consciousness. Instruments that measure the percent of oxygen in the air are available and should be used. Air should contain a minimum of 19.1% oxygen. Exposure to refrigerants should be avoided. The inhalation of excessive amounts can lead to possible cardiac arrest and death.

28-9 TEST YOUR KNOWLEDGE

1. What is the ozone layer?
2. What is the ozone layer's major protection?
3. Identify the type of refrigerants that have the greatest negative effect on the ozone.
4. What is the EPA?
5. What is meant by the term *Recovery?*
6. What is meant by the term *Recycling?*
7. What is meant by the term *Reclaim?*
8. What is the primary method of removing refrigerant with the use of recovery equipment?
9. When recovery of the refrigerant has been completed, how can the system be checked to see if there are pockets of cold liquid refrigerant?
10. What is the maximum amount of liquid that a refrigerant cylinder or vessel should contain?
11. How are the contaminants in a refrigerant removed?
12. In the past, refrigerant was vented into the atmosphere. What is the proper procedure now?
13. What percent of the moisture and oil particulates must be removed from CFC refrigerants to meet the SAE standards?
14. How are unwanted refrigerants disposed?
15. What is the major function of an on-site recycling station?
16. In a system that contains CFCs, what driers should be replaced or added to aid in the purification process?
17. What type of identification must be on recovery cylinders and drums used for transporting fluorocarbon refrigerants?
18. Why is a small sample amount of refrigerant from a large commercial installation sent to a reclaiming center prior to removal?
19. What can happen if an excessive amount of refrigerant is inhaled while being removed from a system?
20. What happens if the refrigerant/oil being removed from a system comes in contact with the skin?

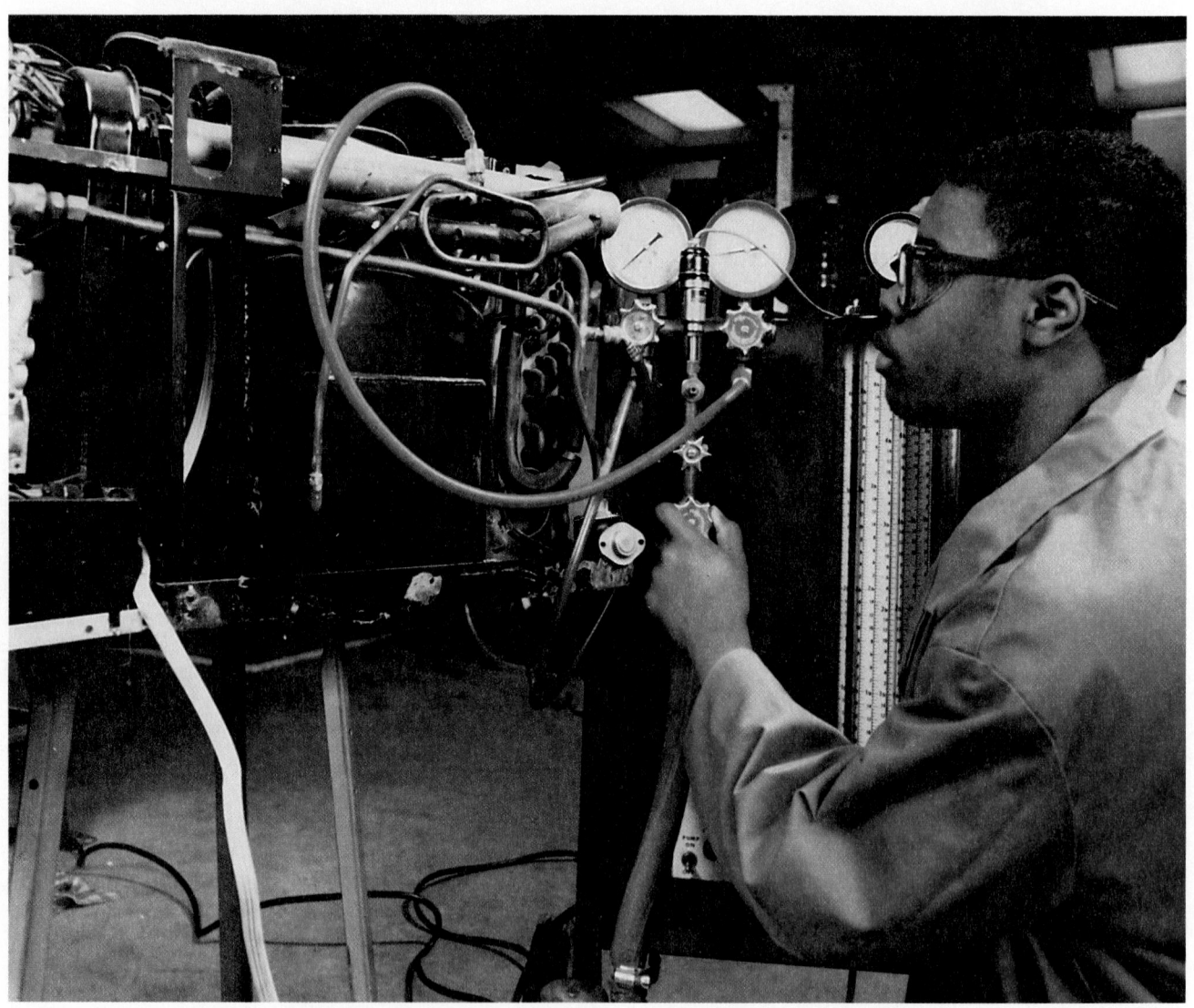

Technician using a combination vacuum pump and charging unit.   (Southeast Oakland Vocational Education Center)

Chapter 29

SERVICING AND TROUBLESHOOTING SIMPLIFIED

After studying this chapter, the technician will be able to:
☐ Explain the definition of servicing and troubleshooting.
☐ Follow a standard procedure in troubleshooting and servicing.
☐ Approach a problem in a logical and systematic sequence.
☐ Obtain from the owner and others the specific problems.
☐ By the use of a three-step procedure, determine the remedy for the problem.
☐ Demonstrate the proper servicing and troubleshooting procedures.
☐ Properly use the various types of manufacturer's troubleshooting charts.

29-1 SERVICING/TROUBLESHOOTING

There is a close relationship between the terms *troubleshooting* and *servicing.* Troubleshooting is often referred to as the analysis of the problem. This analysis is generally guided by a chart provided by the manufacturer of the equipment. Servicing is the manual labor needed to correct the problem that was identified in the troubleshooting sequence. See Fig. 29-1. Servicing is a procedure whereby the technician manually makes corrections to a refrigeration or air conditioning system to bring it back to the standards at which it was originally designed to operate.

Each refrigeration and air conditioning system has its own unique features and consequently its own problems. This chapter will indicate typical or common troubleshooting areas and the recommended procedure to follow when servicing such equipment.

29-2 TROUBLESHOOTING PROCEDURES

One of the key requirements for a service technician is to follow a standard procedure. An example is as follows:
1. Obtain from the owner a description of the problem.
2. From the problem identified by the owner, determine the possible cause of the problem.
3. Identify a specific remedy for the problem.

Using a standard procedure will save time, money, and frustration. By following the same procedure, the technician becomes more efficient as he/she uses repetitive skills.

It is important, whenever possible, to obtain from the building owner, or from the technician's own resources, the service manual or troubleshooting chart which is written by the company whose equipment is being serviced. There are

Fig. 29-1. Service technician charging a heat pump.
(Du Pont Company)

many different types of manuals in use. See Fig. 29-2. However, they utilize the same basic concepts.

Most charts have three basic columns. Normally, these column headings are:
1. Problem (Trouble, Complaint)
2. Possible Cause (Probable Cause, Have You Checked . . .)
3. Remedy (Repair, You May Need To . . .)
When using the troubleshooting charts, it is important that the technician follow the chart on this step-by-step basis.

Upon arrival at the location, the technician should become familiar with the system in question. The system should be

	Trouble	Probable Cause	Suggested Remedy
1	Complaint	Cause	Repair
2	Problem	Have You Checked	You May Need To
3			
4	**PROBLEM**	**POSSIBLE CAUSE**	**REMEDY**

PROBLEM	POSSIBLE CAUSE	REMEDY
F. Unit operates long or continuously.	1. Shortage of refrigerant. 2. Control contacts stuck or frozen closed. 3. Refrigerated space has excessive load or poor insulation. 4. System inadequate to handle load. 5. Evaporator coil iced. 6. Restriction in refrigeration system. 7. Dirty condenser. 8. Filter dirty.	1. Fix leak, add charge. 2. Clean contacts or replace control. 3. Determine fault and correct. 4. Replace with larger system. 5. Defrost. 6. Determine location and remove. 7. Clean condenser. 8. Clean or replace.

Fig. 29-2. Four common types of troubleshooting chart headings. Chart #4 shows the problem, possible cause, and remedy.

visually inspected and all components and wiring examined for any evidence of malfunction. The technician should then consider examination of the system's electrical wiring and component diagram. This will detail all the components the system contains.

When responding to a service call, the technician must always approach the problem in a logical and systematic sequence. There should be no attempt to make a quick decision that may only temporarily fix the problem. This may not uncover the root or cause of the immediate problem, and will often result in a callback for an identical service call. An example of this would be if the trouble is inadequate cooling and the technician simply adds refrigerant to a system low on refrigerant, without finding the actual leak. This would provide a temporary solution and an incorrect remedy. The technician might be required to return to the same job site in the very near future for the same problem.

29-3 OWNER'S DESCRIPTION OF PROBLEM

The first column of the troubleshooting chart normally lists the problem. This column would be the complaint given to the service technician by the owner—usually in general terms.

The technician begins troubleshooting by carefully listening to the owner's complaint. The complaint usually comes from an individual not familiar with the major operation of the air conditioning or refrigeration system. As a result, frequently the individual will use terms which are not the same as those which are used in the field. Therefore, the technician must listen carefully.

When analyzing the problem presented by the owner, as much information as possible should be obtained about how the system is operating and how it was operating before any malfunction.

The service technician should then proceed to obtain from others in the building any additional information available as to how the system has been functioning.

29-4 CHECKING POSSIBLE CAUSE

The next step would be for the technician to check the possible cause column and to analyze this listing in terms of the major components of the system. Each problem/complaint related to a malfunction in the system has a specific possible cause.

Fig. 29-2, Part 4, lists some of the possible causes which could be related to a problem in a self-contained commercial food storage unit. More extensive charts are shown later in this chapter.

As shown in Part 4 of Fig. 29-2, the owner indicated the major problem is a unit that is operating long or continuously. Possible causes could be shortage of refrigerant, control contacts stuck or frozen closed, or any of the eight items identified.

After a thorough investigation of the possible cause column, the technician proceeds to identify the part of the system listed as the possible cause of the problem or symptom. The technician should then be able to determine a specific cause or malfunction and to identify the specific faulty part.

29-5 SUGGESTED REMEDY

The final column on the troubleshooting chart may have a heading of Remedy. This is the third step when using troubleshooting charts. The technician will perform the appropriate task from this third column.

There are many steps to follow in repairing refrigeration or air conditioning equipment. Each part is checked in a step-by-step manner. The actual procedure will vary depending upon the specific remedy selected, the type of part or device being checked, and the specific system. The sequence for checking and repairing or replacing a part will vary. Procedures for an electrical device will be different from that of a mechanical device.

Basic service and safety procedures are always followed as the technician repairs the system.

Fig. 29-3. Technician using a combination volt-ammeter to check the electrical system of a combination heating and cooling system. (Southeast Oakland Vocational Education Center)

Servicing must always be accomplished through the use of the proper tools, gauges, electrical analyzing equipment, and other necessary equipment, Fig. 29-3.

29-6 TROUBLESHOOTING CHARTS

The use of a troubleshooting chart is relatively simple. The technician must understand it is a helpful map which leads from Step 1—Problem, to Step 2—Possible Cause, and Step 3—Remedy.

One must be very careful to utilize the specific troubleshooting chart from the manufacturer of the equipment being serviced. Any additional components added to the system must also be taken into consideration.

Troubleshooting charts vary, depending upon the purpose of the equipment and the particular manufacturer. Four different troubleshooting charts, covering domestic, commercial food storage, ice machines, and gas-fired furnaces, are illustrated here for study.

Fig. 29-4, Hermetic System Diagnosis Troubleshooting and Service Chart: This is written for use with hermetic refrigeration systems. Note the general introductory procedures and the electrical check of the system if the compressor will not run. The troubleshooting chart is broken down into three basic columns: *Complaint, Possible Cause, and Repair.*

Fig. 29-5, Domestic and Light Commercial Systems Troubleshooting and Service Chart: The introductory sec-

---

HERMETIC REFRIGERATION SYSTEM DIAGNOSIS

1. GENERAL
Each complaint is followed by probable causes and suggested repairs. To isolate the possible cause, proceed in a systematic manner to determine the faulty component. This guide does not cover all possible troubles and deficiencies that may occur under conditions of operation.

2. ELECTRICAL CHECK:
A—COMPRESSOR DOES NOT RUN—
1. Check power at outlet receptacle. Your compressor is designed to operate (see serial no. and data plate) on 115-60-1 with a voltage range of 126.5-103.5. Your 240/220-50-1 will operate within a range of 264-198.
2. Check thermostat for proper setting and continuity. Make sure control setting is not in an "Off" position. Continuity may be verified by following instructions on the individual compressor motor circuits.
3. Look for obvious loose or broken wiring.
4. Following systematically the instructions listed on the compressor motor circuitry, check the relay, overload, and, if employed, the start capacitor for continuity. Replace any components found faulty with the recommended service parts.

3. TROUBLESHOOTING AND SERVICE CHART

COMPLAINT	POSSIBLE CAUSE	REPAIR
A. Compressor will not start—no hum.	1. Line disconnect switch open. 2. Fuse removed or blown. 3. Overload protector tripped. 4. Control stuck in open position. 5. Control off due to cold location. 6. Wiring improper or loose	1. Close start of disconnect switch. 2. Replace fuse. 3. Refer to electrical diagram. 4. Repair or replace control. 5. Relocate control. 6. Check wiring against diagram.
B. Compressor will not start—hums.	1. Improperly wired. 2. Low voltage to unit. 3. Starting capacitor defective. 4. Relay failing to close. 5. Compressor motor has winding open or shorted. 6. Internal mechanical trouble in compressor.	1. Check wiring against diagram. 2. Determine reason and correct. 3. Determine reason and replace. 4. Determine reason and correct, replace if necessary. 5. Replace compressor. 6. Replace compressor.

Fig. 29-4. Hermetic refrigeration system diagnosis for a self-contained commercial food storage unit. (Silver King Division of Stevens Lee Company)

TROUBLESHOOTING AND SERVICE CHART

COMPLAINT	POSSIBLE CAUSE	REPAIR
C. Compressor will not start—hums but trips on overload protector	1. Improperly wired. 2. Low voltage to unit. 3. Relay failing to open. 4. Run capacitor defective. 5. Excessively high discharge pressure. 6. Compressor motor has a winding open or shorted. 7. Internal mechanical trouble in compressor (tight).	1. Check wiring against diagram. 2. Determine reason and correct. 3. Determine reason and correct, replace if necessary. 4. Determine reason and replace. 5. Check discharge shutoff, possible overcharge or insufficient cooling on condenser. 6. Replace compressor. 7. Replace compressor.
D. Compressor starts and runs, but short cycles on overload protector.	1. Additional current through overload protector. 2. Low voltage to unit. 3. Overload protector defective. 4. Run capacitor defective. 5. Excessive discharge pressure. 6. Suction pressure too high. 7. Compressor too hot—return gas hot. 8. Compressor motor has a winding shorted.	1. Check wiring diagram, check for added fan motors, pumps, etc., connected to wrong side of protector. 2. Determine reason and correct. 3. Check current, replace protector. 4. Determine reason and replace. 5. Check ventilation, restrictions in cooling medium, restrictions in refrigeration system. 6. Check for possibility of misapplication. Use stronger unit. 7. Check refrigerant charge (fix leak), add if necessary. 8. Replace compressor.

E. Unit runs ok, but short cycles.	1. Overload protector
	2. Thermostat.
	3. High pressure cut-out due to:
	  a. Insufficient air.
	  b. Overcharge.
	  c. Air in system.
	4. Low pressure cut-out due to:
	  a. Undercharge.
	  b. Restriction in expansion device.
		1. See D above.
		2. Differential set too close—widen.
		
		3a. Check air to condenser—correct.
		3b. Reduce refrigerant charge.
		3c. Purge.
		
		4a. Fix leak, add refrigerant.
		4b. Replace device.

F. Unit operates long or continuously.	1. Shortage of refrigerant. 2. Control contacts stuck or frozen closed. 3. Refrigerated space has excessive load or poor insulation. 4. System inadequate to handle load. 5. Evaporator coil iced. 6. Restriction in refrigeration system. 7. Dirty condenser. 8. Filter dirty.	1. Fix leak, add charge. 2. Clean contacts or replace control. 3. Determine fault and correct. 4. Replace with larger system. 5. Defrost. 6. Determine location and remove. 7. Clean condenser. 8. Clean or replace.

G. Start capacitor open, shorted, or blown.	1. Relay contacts not operating properly.
	2. Prolonged operation on start cycle due to:
	  a. Low voltage to unit.
	  b. Improper relay.
	  c. Starting load too high.
	
	3. Excessive short cycling.
	
	4. Improper capacitor.
		1. Check and replace.
		
		2a. Determine reason and correct.
		2b. Replace.
		2c. Correct by using pump down arrangement if necessary.
		3. Determine reason for short cycling (E above) and correct.
		4. Determine correct size and replace.

H. Relay defective or burned out.	1. Incorrect relay. 2. Incorrect mounting angle. 3. Line voltage too high or too low. 4. Excessive short cycling. 5. Relay being influenced by loose vibrating mounting. 6. Incorrect run capacitor.	1. Check and replace. 2. Remount relay in correct position. 3. Determine reason and correct. 4. Determine reason (see E) and correct. 5. Remount rigidly. 6. Replace with proper capacitor.
I. Space temperature too high.	1. Control setting too high. 2. Inadequate air circulation.	1. Reset control. 2. Improve air movement.
J. Suction line frosted or sweating.	1. Evaporator fan not running. 2. Overcharge of refrigerant.	1. Determine reason and correct. 2. Correct charge.
K. Liquid line frosted or sweating.	1. Restriction in dehydrator or strainer.	1. Replace part.
L. Unit noisy.	1. Loose parts or mounting. 2. Tubing rattle. 3. Bent fan blade causing vibration. 4. Fan motor bearings worn.	1. Find and tighten. 2. Reform to be free of contact. 3. Replace blade. 4. Replace motor.

Fig. 29-4. Continued.

DOMESTIC AND LIGHT COMMERCIAL SYSTEMS

The "Troubleshooting and Service Chart" is quite self-explanatory; however, a discussion of some of the complaints, possible causes, and repair solutions may be of some additional assistance.

COMPLAINT "A" is Compressor will not start. No hum. Possible causes are:

1. *Switch open.* Rather obvious, but maybe it would be wise to determine why or who opened it.
2. *Fuse removed or blown.* Again, was there a reason?
3. *Overload protector tripped.* Here, too, it is not a case of waiting until the overload resets, but rather to determine why.
4. *Control stuck in open position.* Faulty contactors may be a cause, although every effort is made to provide the best quality contactors. And heed this warning: Don't use the insulated end of a screwdriver to hold the contactor in. In doing so you run the risk of burning out a good compressor.

COMPLAINT "B" is Compressor will not start—hums and trips on overload.

1. & 2. have been discussed (see Complaint "A").
3. *Starting capacitor defective.* It says "determine reason." Possibly a start capacitor was installed which had too low a voltage rating.
4. *Relay failing to close.* Is the correct one being used? There seems to be a considerable tendency to substitute something other than the one specified—if it works leave it. We say, please don't do it.
5. *Compressor has a winding open or shorted.* The repair specified says simply "replace compressor." This means if the cause indicated has been proved—without doubt—conclusively—no question about it—only then replace the compressor. Remember that replacing the compressor is generally the most costly repair bill an owner can get. So be sure—first.
6. *Internal mechanical trouble in compressor.* If the serviceman has proved without question that none of the other possible causes are the reason for the condition, then and then only can it be mechanical trouble.

A "Troubleshooting Chart" of this kind is not the entire answer. There are probably a number of other reasons for the cause of each "complaint" listed, so keep in mind that application of knowledge gained through experience and common sense are as much a part of troubleshooting as the use of any chart.

COMPLAINT "C" is Compressor starts but stays on run winding. How do you know this condition is occurring? If the amps stay higher than normal. Or if you don't hear the changeover.

1. through 3. Covered previously (see Complaint "A" and "B").
4. *Run capacitor defective.* If the run capacitor is failed closed, there will be a period of time when the current and running sounds will seem to indicate the relay has not switched. In a relatively short period of time, the start winding will burn, so time is of the essence in this case.
5. *High head pressure.* Be sure to check all the things listed in the "repair" column.

COMPLAINT "D" is Compressor starts and runs, but short cycles on overload protector.

1. Mentioned before (see Complaint "A").
2. *Low voltage to unit (or unbalanced if three-phase).* In the matter of three-phase unbalance—this is an instance in which it is probably wise to call in the power company, or check with the building owner to determine what other equipment is on the source of power to cause the unbalance.
3. *Overload protector defective.* Sometimes difficult to determine. One good clue is how it looks—does it show to have been overheated?
6. *Suction pressure too high.* This will occur more often on refrigeration than air conditioning, especially on low temperature equipment.
7. *Compressor hot—insufficient gas cooling.* Usually a result of a low charge.

Most of the other complaints, causes, and repair suggestions are straightforward, and the best suggestion is to follow the chart.

WARNING: ELECTRICAL POWER MUST BE DISCONNECTED WHEN TERMINAL PROTECTIVE COVER IS NOT IN PLACE TO PROTECT AGAINST ELECTROCUTION OR VENTED TERMINAL.

TROUBLESHOOTING AND SERVICE CHART

	COMPLAINT	POSSIBLE CAUSE	REPAIR
A	Compressor will not start—no hum.	1. Line disconnect switch open. 2. Fuse removed or blown. 3. Overload protector tripped. 4. Control stuck in open position. 5. Control off due to cold location. 6. Wiring improper or loose.	1. Close start or disconnect switch. 2. Replace fuse. 3. Refer to electrical section. 4. Repair or replace control. 5. Relocate control. 6. Check wiring against diagram.
B	Compressor will not start—hums but trips on overload protector.	1. Improperly wired. 2. Low voltage to unit. 3. Starting capacitor defective. 4. Relay failing to close. 5. Compressor motor has a winding open or shorted. 6. Internal mechanical trouble in compressor. 7. Liquid refrigerant in compressor.	1. Check wiring against diagram. 2. Determine reason and correct. 3. Determine reason and replace. 4. Determine reason and correct, replace if necessary. 5. Replace compressor. 6. Replace compressor. 7. Add crankcase heater and/or accumulator.
C	Compressor starts, but does not switch off of start winding.	1. Improperly wired. 2. Low voltage to unit. 3. Relay failing to open. 4. Run capacitor defective. 5. Excessively high discharge pressure. 6. Compressor motor has a winding open or shorted. 7. Internal mechanical trouble in compressor (tight).	1. Check wiring against diagram. 2. Determine reason and correct. 3. Determine reason and correct, replace if necessary. 4. Determine reason and replace. 5. Check discharge shut-off valve, possible overcharge, or insufficient cooling on condenser. 6. Replace compressor. 7. Replace compressor.

Fig. 29-5. Domestic and light commercial systems troubleshooting and service chart.   (Tecumseh Products Company)

	COMPLAINT	POSSIBLE CAUSE	REPAIR
D	Compressor starts and runs, but short cycles on overload protector.	1. Additional current passing through overload protector. 2. Low voltage to unit (or unbalanced if three phase). 3. Overload protector defective. 4. Run capacitor defective. 5. Excessive discharge pressure. 6. Suction pressure too high. 7. Compressor too hot—return gas hot. 8. Compressor motor has a winding shorted.	1. Check wiring diagram. Check for added fan motors, pumps, etc., connected to wrong side of protector. 2. Determine reason and correct. 3. Check current, replace protector. 4. Determine reason and replace. 5. Check ventilation, restrictions in cooling medium, restrictions in refrigeration system. 6. Check for possibility of misapplication. Use stronger unit. 7. Check refrigerant charge (fix leak), add if necessary. 8. Replace compressor.

E	Unit runs OK, but short cycles on.	1. Overload protector.
		2. Thermostat.
		3. High-pressure cut-out due to:
		  a. Insufficient air or water supply.
		
		  b. Overcharge.
		  c. Air in system.
		4. Low-pressure cut-out due to:
		  a. Liquid line solenoid leaking.
		  b. Compressor valve leak.
		  c. Undercharge.
		  d. Restriction in expansion device.
			1. See D above.
			2. Differential set too close—widen.
			
			3a. Check air or water supply to condenser—correct.
			3b. Reduce refrigerant charge.
			3c. Purge.
			
			4a. Replace.
			4b. Replace.
			4c. Fix leak, add refrigerant.
			4d. Replace device.

F	Unit operates long or continuously.	1. Shortage of refrigerant. 2. Control contacts stuck or frozen closed. 3. Refrigerated or air conditioned space has excessive load or poor insulation. 4. System inadequate to handle load. 5. Evaporator coil iced. 6. Restriction in refrigeration system. 7. Dirty condenser. 8. Filter dirty.	1. Fix leak, add charge. 2. Clean contacts or replace control. 3. Determine fault and correct. 4. Replace with larger system. 5. Defrost. 6. Determine location and remove. 7. Clean condenser. 8. Clean or replace.

G	Start capacitor open, shorted, or blown.	1. Relay contacts not operating properly.
		2. Prolonged operation on start cycle due to:
		  a. Low voltage to unit.
		  b. Improper relay.
		  c. Starting load too high.
		
		3. Excessive short cycling.
		
		4. Improper capacitor.
			1. Clean contacts or replace relay if necessary.
			
			2a. Determine reason and correct.
			2b. Replace.
			2c. Correct by using pump down arrangement if necessary.
			3. Determine reason for short cycling (E above) and correct.
			4. Determine correct size and replace.

H	Run capacitor open, shorted, or blown.	1. Improper capacitor. 2. Excessively high line voltage (110% of rated-max.).	1. Determine correct size and replace. 2. Determine reason and correct.
I	Relay defective or burned out.	1. Incorrect relay. 2. Incorrect mounting angle. 3. Line voltage too high or too low. 4. Excessive short cycling. 5. Relay being influenced by loose vibrating mounting. 6. Incorrect run capacitor.	1. Check and replace. 2. Remount relay in correct position. 3. Determine reason and correct. 4. Determine reason (see E above) and correct. 5. Remount rigidly. 6. Replace with proper capacitor.
J	Space temperature too high.	1. Control setting too high. 2. Expansion valve too small. 3. Cooling coils too small. 4. Inadequate air circulation.	1. Reset control. 2. Use larger valve. 3. Add surface or replace. 4. Improve air movement.
K	Suction line frosted or sweating.	1. Expansion valve passing excess refrigerant or is oversized. 2. Expansion valve stuck open. 3. Evaporator fan not running. 4. Overcharge of refrigerant.	1. Readjust valve or replace with smaller valve. 2. Clean valve of foreign particles, replace if necessary. 3. Determine reason and correct. 4. Correct charge.
L	Liquid line frosted or sweating.	1. Restriction in dehydrator or strainer. 2. Liquid shut-off (king valve) partially closed.	1. Replace part. 2. Open valve fully.
M	Unit noisy.	1. Loose parts or mountings. 2. Tubing rattle. 3. Bent fan blade causing vibration. 4. Fan motor bearings worn.	1. Find and tighten. 2. Reform to be free of contact. 3. Replace blade. 4. Replace motor.

Fig. 29-5. Continued.

tion lists some of the common complaints, common causes, and repair. This is a written analysis of four common complaints—A, B, C, and D—pertaining to compressor malfunctions (shown in column form in the troubleshooting chart along with other frequent complaints).
Possible Cause, and *Repair.* At the bottom of each troubleshooting chart this company emphasizes precaution regarding electrical problems: "WARNING: ELECTRICAL POWER MUST BE DISCONNECTED WHEN TERMINAL PROTECTIVE COVER IS NOT IN PLACE TO PROTECT AGAINST ELECTROCUTION OR VENTED TERMINAL."

Fig. 29-6, Ice Flaker Machine Troubleshooting and Service Guide: This chart first gives a sequence of operation and a schematic drawing of the unit. It is essential to know how the machine operates prior to attempting to use the troubleshooting chart and servicing it. The troubleshooting chart is broken down into three basic columns: *Trouble, Cause,* and *Remedy.*

ICE FLAKER MACHINE
TROUBLESHOOTING AND SERVICE CHART

SEQUENCE OF OPERATION

1. At start-up all thermostats are warm, electric power flows from fuse thru Bin Thermostat directly to the compressor (through High- and Low-Pressure Cut-Outs as applicable) and through the warm side of the Gear Thermostat to the Gearmotor (and in parallel to the Green light).
   1a. Original pull down of Flaker may be slow, 5 to 10 minutes before ice is produced. During the last half of pull down the suction line at the compressor will be frosted, and the sight glass will show bubbles during pull down. After ice starts to be produced, frost will leave the compressor suction line and within 5 minutes the sight glass will show clear or only a small bubble in the top, which will disappear slowly.
2. As ice begins to form the suction line from the evaporator (location of Gear Thermostat cap tube) gets cold enough (below 32°) to trip Gear Thermostat to the cold side bypassing the Bin Thermostat (resetting for later Ice Unloading period) so power flows directly through the Gear Thermostat to the Gearmotor (and Green light); no change is apparent.
   2a. [On rare occasions with some particle-free waters, the water will subcool substantially below freezing, then suddenly crystallize and stall the Gearmotor (designed to withstand a stall). The evaporator will continue to get colder until the bottom (location of the Defrost Thermostat cap tube) gets below freezing, tripping the Defrost Thermostat cold, sending power to the Defrost Valve (and Red light) which sends hot refrigerant through the evaporator, releasing the Auger and Gearmotor until the bottom of the evaporator reaches about 45° tripping the Defrost Thermostat open and normal refrigeration restarts. Since subcooling with crystallization occurs only 2 or 3 times out of 100, it is unlikely to happen and very unlikely to repeat.]
3. During ice formation ice is drawn to the top of the evaporator, lowering the water level, opening the float valve. When water-flow is seen, it indicates ice formation even before ice appears in the bin coming from the ice tube. Water flows through the bottom of the auger and from a hole near the bottom flight into the ice chamber, cools to 32°, then freezes in thin layers on the wall of the evaporator cylinder where the auger flights remove it and spiral it upward. The ice thickens until it is a fairly dense mass as it rotates under the helical path of the deflector until it is forced into the discharge spout by the deflector. The ice then flows by gravity through the ice tube to give it velocity which spreads it in the bin.
4. When ice piles over about half the length of the 1/4" stainless tube which holds the Bin Thermostat cap tube, the thermostat opens shutting off power to the compressor. (Turning on the Yellow light which is connected in parallel with the thermostat). The Green light stays on as the Gear Thermostat is still cold supplying power from the fuse to the Gearmotor (and Green light) to Unload Ice from the evaporator until the freezing surface is above 32°. The Gearmotor must remain on a couple minutes to complete unloading and usually stays on 5 to 15 minutes until the suction line out of the evaporator warms to about 45°. Then the Gear Thermostat trips warm connecting the Gearmotor to the output side of the bin thermostat (resetting for the next start-up) shutting off the Gearmotor and Green light.
5. When the bin is full, only the Yellow light is illuminated indicating power is on and fuse okay. Only a trickle of power is used through Yellow neon-type light during flaker-off periods. Flaker may start-up while a small amount of ice is still on the 1/4" stainless steel tube. This is okay as the deflector can push ice through the ice tube creating some spread around the top of the bin until the stainless steel tube is fairly well covered again.

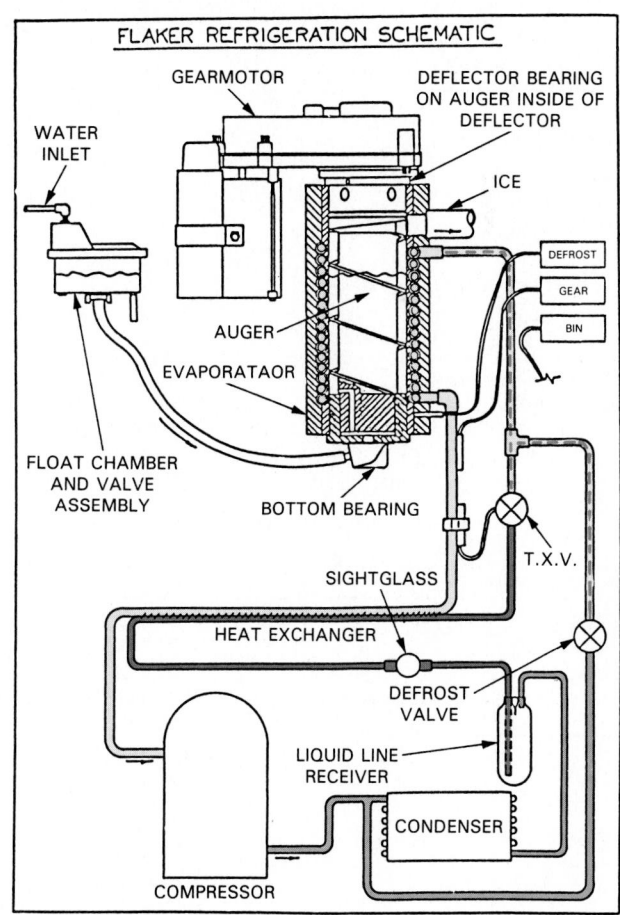

Fig. 29-6. Ice flaker machine service procedures and troubleshooting charts. (Kold-Draft Division, Uniflow Manufacturing Company)

TROUBLE, CAUSE, AND REMEDY		
TROUBLE	CAUSE	REMEDY
1. Flaker will not operate. No lights on in control box. 1a. Flaker will not operate. Yellow light on, but no ice on bin thermostat.	1. Line fuse blown. 2. Loose connection in control box or in power supply line. 1a. Bin control set too warm in a cold room between 45° and 55°. 2a. Room below 45°. 3a. Bin control has lost charge.	1. Check circuits for short or ground. Replace fuse. 2. Check for power supply at controls in control box. Check connections to bin thermostat. 1a. Set bin thermostat colder (cw) but recheck with ice to be sure it will shut off. 2a. Add heat to the room. 3a. Replace bin control.
2. Condensing fan and gearmotor operate (green light on) but not the compressor. 2a. Water-cooled flaker: gearmotor operates (green light on) but not the condensing fan or compressor.	1. Inoperative capacitors or relay. 2. Overload switch defective. 3. Loose connections or defective compressor. 1a. High-pressure cut-out open; inadequate water supply. 2a. Water supply okay, high pressure cut-out won't close with condenser cool.	1. Replace capacitors or relay. 2. Replace overload switch. 3. Check for power at compressor C-R terminals, C-S terminals. With power off, remove "C" connection, check ohms between C and R, also C and S. 1a. Check water supply and condenser water valve. 2a. Replace defective high pressure cut-out.
3. Compressor operating but fan off.	1. Circuit not complete. 2. Fan motor burned out.	1. Check circuit. 2. Replace motor.
4. Condenser fan operating but compressor unit operating intermittently.	1. Dirty Condenser Coil. 2. High or low voltage. 3. Excessive refrigerant.	1. Clean coil. 2. Correct to proper voltage within 10% of nameplate. 3. Remove some refrigerant, check sight glass.
5. Intermittent defrost, red light cycling on and off. Water level normal in float tank.	1. Water line elbow in bottom bearing in too far. 2. Defrost thermostat misadjusted, very cold supply water. 3. TXV too far open. 4. Deflector partially closed. 5. Gearmotor not running. 6. Gearmotor stalled with power on, bottom evaporator mounting screw too tight. 7. Gearmotor stalled.	1. Back out elbow 1 turn. 2. See service manual for defrost thermostat setting. Also see Abnormal Defrost Thermostat Setting in the manual. 3. Close TXV 1/4 turn, see service manual for proper setting. 4. See service manual for correct deflector setting. 5. Check power to gearmotor receptacle in bottom of control box. 6. Back off evaporator mounting screw 1 turn to see if gearmotor will operate after it cools down sufficiently for overload to cut back in. Also see #7. 7. Remove gearmotor and auger assembly and check for operation on workbench (take care not to damage auger flights). Remove bottom bearing, check for condition and check for snug but not a tight fit on bottom of auger. Check clearance between top of auger and bottom of deflector (see Flaker Mechanism Assem. Instr. #5).
6. Wet ice.	1. High water level. 2. Undercharge, bubbles going through sight glass. 3. Misadjusted TXV.	1. Lower water level. 2. Check for leaks, add R-12. 3. Adjust TXV (see "Expansion Valve Adjustment").
7. Ice too hard.	1. Low water level. 2. Deflector incorrectly adjusted. 3. TXV closed much too far. 4. Moisture in system and TXV partially frozen shut.	1. Raise water level. 2. SEE DEFLECTOR ADJUSTMENTS PROCEDURE. 3. See service manual for proper suction pressure and suction line temperature. 4. Dehydrate and recharge system.

Fig. 29-6. Continued.

TROUBLE, CAUSE, AND REMEDY		
TROUBLE	CAUSE	REMEDY
8. No ice with gearmotor, compressor and condenser fan operating. (Red light not on no power to defrost valve.)	1. Very low refrigerant. 2. Stuck defrost valve, defrost line warm, suction pressure above 20 pounds.	1. Repair leak, see service manual for proper charge. 2. Repair or replace defrost valve.
9. Flaker does not turn off.	1. Misadjusted bin thermostat. 2. Bin thermostat will not open when set warmest with ice on thermostat cap tube. 3. Mislocated bin thermostat cap tube.	1. Adjust bin thermostat (ccw), check with ice on coil. 2. Replace bin thermostat. 3. Check location of 1/4'' stainless cap tube holder parallel and below ice path coming from ice tube. Check thermostat cap tube located in the 1/4'' stainless tube.
10. Flaker cycles off and on.	1. Ice falling on bin thermostat capillary tube. 2. Ice tube not on outlet spout.	1. Relocate tube. 2. Remount ice tube and check for restriction.
11. Low production.	1. High head pressure. 2. Inadequate water supply. 3. TXV misadjusted. 4. High ambient. 5. Deflector closed. 6. Low head pressure.	1a. Clean condenser. Improve water supply to water-cooled condenser. Improve ventilation. 1b. Replace head pressure control valve on remote condenser models. 2. Check and clean filters. Raise water level. 3. Adjust TXV. 4. Decrease ambient to 90 °F.Max. 5. SEE DEFLECTOR ADJUSTMENTS PROCEDURE. 6a. Adjust water regulating valve. 6b. Add R-12 to remote condenser models. 6c. Refer to Remote Condenser section of Service Manual.
12. Flaker spouts coming off.	1. I.D. of discharge tubing too small. 2. Roll pins breaking on deflector. 3. Hose kinking.	1. Change to braided nylon tubing. 2. Readjust defrost thermostat. 3. Change to braided nylon tubing, reroute presently used tubing.

Fig. 29-6. Continued.

Fig. 29-7, Gas-Fired Forced-Air Furnace Servicing and Heating Analysis Guide: This chart uses a different format than the standard three-column chart. However, the three-step concept can still be utilized. The first step is the *Complaint,* which is divided into two categories—*No Heat* or *Unsatisfactory Heat.* The trouble is further identified by three conditions under ''No Heat'' and six under ''Unsatisfactory Heat.'' For ease of clarification, the complaint in this example was ''Unsatisfactory Heat.'' It was further indicated by the owner that there was too much heat. Following down the chart to the dots, the technician can identify Step No. 2, *Possible Cause,* which could be any of the four items indicated. The third step, *Test Method/Remedy,* indicates that the technician should see Service Procedure S-17. The technician would then perform the service procedures as indicated and, if all steps check out okay, would then proceed to check the other possible causes.

The simplified step-by-step procedure is followed in each of the above troubleshooting charts. Study these charts to become more efficient in your service ability.

29-7 MAINTENANCE SERVICE CONTRACTS

Society in general has become very familiar with the maintenance service contract concept. All new cars today are sold with an extended maintenance contract available to the owners. By this means, the purchaser can extend the service contract on the vehicle for a longer period of time. Maintenance agreements presently are utilized for many items: telephones, household appliances, television sets, VCRs, etc. This wide application has created a public awareness of the benefits of maintenance contracts.

In the air conditioning field, many companies offer this service to their customers. Service contracts vary depending on the particular company. Their service agreement may cover all repairs and replacements that may be needed in a given period of time. In Fig. 29-8, the agreement indicates specifically what is included in the tune-up and cleaning twice a year. The basic concepts of ''Tune-Up'' are identified—cleaning the evaporator coil, replacing filters, checking fan blades for tightness, etc. The agreement also

SERVICING
HEATING ANALYSIS GUIDE

POSSIBLE CAUSE	COMPLAINT									TEST METHOD-REMEDY	SEE SERVICE PROCEDURE
	NO HEAT			UNSATISFACTORY HEAT							
	SYSTEM WILL NOT START	BURNER WON'T IGNITE	BURNER IGNITES-LOCKS OUT	MAIN BURNER SHUTS OFF PRIOR TO T-STAT BEING SATISFIED	SHORT CYCLES	LONG CYCLES	SOOT AND/OR FUMES	TOO MUCH HEAT	NOT ENOUGH HEAT		
NO MAIN POWER	●									TEST VOLTAGE	S-1
FAULTY TRANSFORMER	●									TEST TRANSFORMER	S-4
FAULTY THERMOSTAT	●									TEST THERMOSTAT	S-3
FAULTY LIMIT SWITCH		●		●						TEST FAN AND LIMIT CONTROL	S-6
FAULTY FLAME SENSOR			●							TEST FLAME SENSOR	S-22
FAULTY IGNITION CONTROL MODULE		●	●							TEST IGNITION CONTROL MODULE	S-21
GAS SUPPLY VALVES IN OFF POSITION OR GAS VALVE OFF		●								TURN VALVES TO "ON" POSITION	
FAULTY INDUCED DRAFT BLOWER MOTOR	●									TEST MOTOR	S-8, S-9
OPEN DOOR SWITCH	●									CHECK DOOR CLOSED-TEST SWITCH	
FAULTY WIRING HARNESS		●								TEST WIRING	S-2
BROKEN OR SHORTED IGNITER		●								TEST IGNITER	S-20
FAULTY COMBUSTION RELAY	●			●						TEST RELAY	S-5
SENSOR NOT IN FLAME, LOW MICRO-AMPS			●							TEST FLAME SENSOR	S-22
FAULTY GAS VALVE OPR.		●								TEST GAS VALVE MAIN OPERATOR	S-10
OPEN AUX. LIMIT		●		●						PUSH MANUAL RESET	S-7
IMPROPER HEAT ANTICIPATOR SETTING				●	●	●				CHECK HEAT ANTICIPATOR SETTING	S-3B
IMPROPER AIR FLOW OR DISTRIBUTION				●	●			●	●	CHECK DUCT STATIC	S-17
CYCLING ON LIMIT					●				●	CHECK CONTROLS & TEMP. RISE	S-6, S-7, S-18
IMPROPER THERMOSTAT LOCATION					●	●		●	●	RELOCATE THERMOSTAT	
DELAYED IGNITION							●			TEST FOR DELAYED IGNITION	S-15
FLASHBACK							●			TEST FOR FLASHBACK	S-16
ORIFICE SIZE							●			CHECK ORIFICES	S-13
GAS PRESSURE		●					●		●	CHECK GAS PRESSURE	S-14
BURNED OUT HEAT EXHANGER, CHECK TEMPERATURE RISE							●			CHECK TEMP. RISE	S-18
THERMOSTAT OUT OF CALIBRATION								●	●	RECALIBRATE OR REPLACE	
STUCK GAS VALVE								●		REPLACE GAS VALVE	
FURNACE UNDERSIZED									●	REPLACE FURNACE WITH PROPER SIZE	
NO LOW VOLTAGE	●									CHECK WIRING	S-2
FAULTY PRESSURE SWITCH		●		●						TEST PRESSURE SW.	S-19
BLOCKED OR RESTRICTED FLUE·		●		●						CHECK FLUE	S-19
HIGH RESISTANCE GROUND			●							TEST GROUND	S-21
VENT SAFETY SWITCH OPEN	●									TEST CONTINUITY	S-23

Fig. 29-7. Heating analysis guide for gas-fired furnace. The following has been selected as an example: Step 1—The complaint was unsatisfactory heat—too much heat. Step 2—The possible causes are: —Improper air flow or distribution. —Improper thermostat location. —Thermostat out of calibration. —Stuck gas valve. Step 3—The possible remedies are: —Check duct stat (this information would be obtained from the Service Procedure S-17). —Relocate thermostat. —Recalibrate or replace thermostat. —Replace gas valve. (Amana Refrigeration, Inc.)

SERVICE PROCEDURE S-17

S-17 CHECKING DUCT STATIC

The maximum and minimum allowable external static pressures are found in the specification section. These tables also show the amount of air being delivered at a given static by a given motor speed or pulley adjustment.

The furnace motor cannot deliver proper air quantities (CFM) against statics other than those listed.

Too great of an external static pressure will result in insufficient air that can cause excessive temperature rise, resulting in limit tripping and etc. Whereas not enough static may result in motor overloading.

To determine proper air movement, proceed as follows:

1. With clean filters in the furnace, use a draft gauge (inclined manometer) to measure the static pressure of the return duct at the inlet of the furnace. (Negative Pressure.)
2. Measure the static pressure of the supply duct. (Positive Pressure.)
3. Add the two readings together for total external static pressure.

NOTE: Both reading may be taken simultaneously and read directly on the manometer if so desired. If an air conditioning coil or Electronic Air Cleaner is used in conjunction with the furnace, the readings must also include these components.

4. Consult proper tables for the quantity of air.

If the total external static pressure exceeds the minimum or maximum allowable statics, check for closed dampers, registers, undersized and/or oversized poorly laid out duct work.

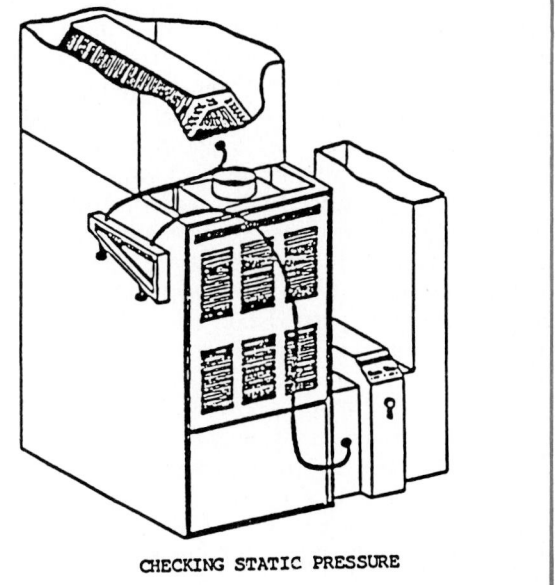

CHECKING STATIC PRESSURE

Fig. 29-7. Continued.

ENERGY SAVING AGREEMENT™

Precision Tune-Up and Professional Cleaning

This agreement provides you with a precision tune-up and professional cleaning twice a year.

TUNE-UP INCLUDES

Clean evaporator coil	Lubricate fan motor	Set manifold pressure
Monitor expansion valve	Check start and run capacitors	Measure air flow
Clean evaporator drain	Check start and run relays	Evaluate ductwork
Adjust bypass dampers	Monitor refrigerant level	Inspect valves
Clean or replace filters	Adjust operating pressures	Replace filters/nozzles
Tighten electrical connections	Measure voltage differences	Check flame baffle
Evaluate safety controls	Measure amperage draw	Clean combustion chamber
Measure temperature difference	Clean and adjust burners	Clean heat exchanger
Clean condenser coil	Evaluate vent system	Test thermocouple
Adjust thermostat calibration	Set burner adjustment	Apply protective coating to unit
Check fan blades for tightness	Measure gas input	

BENEFITS

Reduce Cooling and Heating Cost by $200	Inflation Proof
Improve Efficiency 21%	Priority Customer
Add 1 to 8 Years to Equipment Life	Agreement is Transferable
Improve Capacity 25%	24-Hour Emergency Service 365 days a year
10% Discount on Repair Service	

NAME __________

MAILING ADDRESS __________ LOCATION OF EQUIPMENT __________

PHONE __________ PHONE __________

EQUIPMENT	MAKE	WARRANTY DATE	MODEL NO.	SERIAL NO.

TERMS ☐ VISA ☐ MASTERCARD ☐ ONE YEAR ☐ TWO YEAR ☐ THREE YEAR TOTAL INVESTMENT $______

ACCOUNT NUMBER __________ EXPIRATION DATE __________

Air & Energy, Inc.
3500 East Bay Dr.
Holmes Beach, FL 34217
(813) 778-0773

SIGNATURE __________ DATE __________

CUSTOMER SIGNATURE __________ DATE __________

© 1987 John Young

NO-RISK MONEY-BACK GUARANTEE

Try this amazing service for three months. Your Air Conditioning and Heating system will work more efficiently. If you're not satisfied **for any reason** your entire investment will be refunded. **No Questions Asked.**

YOU CAN'T LOSE

Fig. 29-8. A twice-a-year cleaning and tune-up agreement. Note the numerous items that the technician performs. (Air & Energy, Inc.)

identifies what benefits may be derived from this contract by the owners: cooling and heating costs reduced by $200 per year, efficiency of units improved by 21%, a 24-hour emergency service 365 days a year. The agreement then proceeds to identify the specific equipment make, warranty date, model number, and serial number. Also note the no-risk three-month money-back guarantee, which helps encourage the home owner to sign this type of contract.

Another service agreement asks whether the customer can: inspect the air conditioning system, replace filters, clean coils, check motor and compressor, lubricate motor, calibrate thermostat, and other items. It then indicates that, if this is not possible by the owner, the company will do so and furnish all parts, labor, and material necessary to accomplish this, for a given cost.

29-8 TECHNICIAN APPEARANCE/CONDUCT

The technician's appearance and conduct contribute to the company image. A neat personal appearance will help to create the sense of confidence which is necessary in dealing with the customer, and will affect the customer's attitude toward the service performed.

Being on time on the job and displaying good work habits will create a desirable impression. Accurate and efficient work and, above all, a respect for the customer's property will build a sense of trust and satisfaction on the part of the customer regarding the technician. It is well for the service technician to be aware that positive remarks concerning the company will contribute to a good customer relationship. It is also important that the technician have respect for company vehicles and equipment, reflecting a concern toward the total operation of the company.

29-9 CUSTOMER RELATIONS

The role of the refrigeration, air conditioning, and service technician has changed tremendously in recent years. The service technician, in addition to being knowledgeable in refrigeration and air conditioning, is expected to meet the customer's needs and be knowledgeable regarding business operations and contractual agreements.

One of the keys to a successful business operation is customer relations. Customer relations are based on job performance and the activities of the technician as evaluated by the owners for whom the service is being provided. It is the responsibility of the technician to instill a sense of trust, value, and satisfaction regarding the work performed. As noted, this is achieved through not only the verbal interaction of the technician with the customer, but also in his/her general attitude and appearance, so that a positive impression is given to the customer.

It is essential that the technician *always* be courteous when dealing with customers. This is especially true when the customer feels that things are not being handled properly. The technician must always remain courteous, even when things are not going well. It is imperative that the technician also think in terms of treating the customer's needs as being an emergency situation—because, in the eyes and mind of the customer, it is an emergency for them.

29-10 ARRIVING ON THE JOB

When first appearing on the job, the technician should identify him/herself as well as the company, and the reason for the call should be stated. The customer should then be asked some specific questions: what has occurred, when it was first noticed, how many times it has occured, and

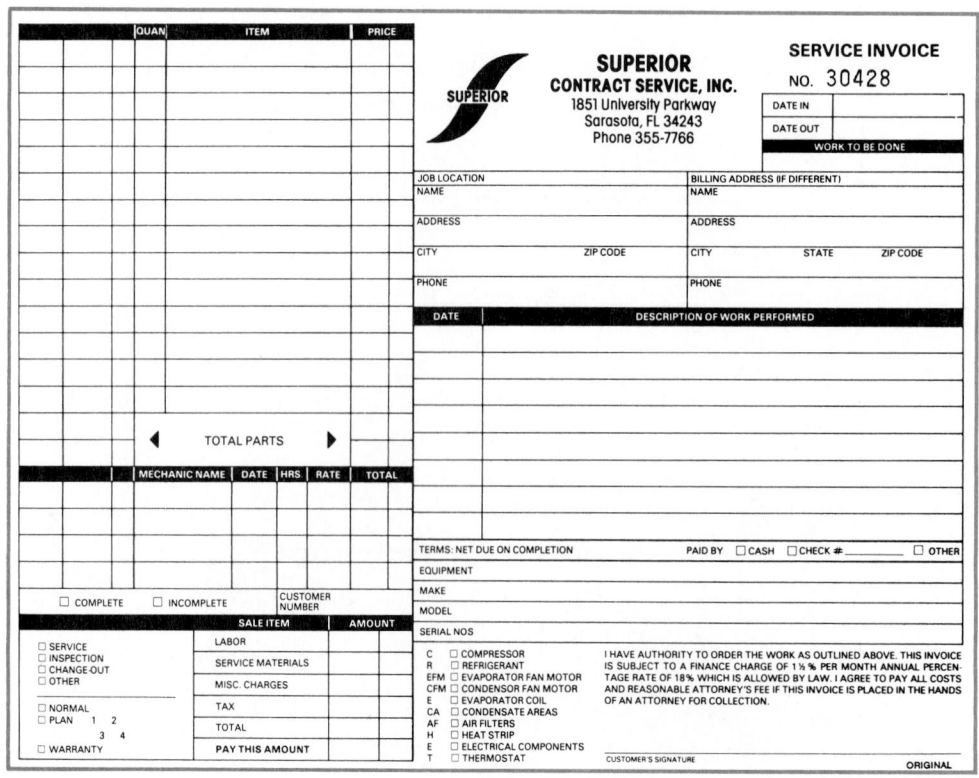

Fig. 29-9. Service invoice. Note the various categories: Item; Description of Work Performed; Equipment, Make, Model; type of payment, etc. (Superior Contract Service, Inc.)

other questions that are applicable to the particular situation. A polite attitude when asking these questions will help in obtaining the information needed to determine the problem and make the needed repair.

The technician should write on the service contract any information the customer volunteers concerning previous problems with the equipment. Also, any interest the customer has regarding add-on equipment or new contractual agreements, such as humidistats, etc., should be noted.

When the service operation has been completed, the technician should be certain to have the proper billing forms, Fig. 29-9, for the customer to sign. If applicable, indicate to the customer what can be done in the future so that this problem will not recur.

29-11 CONTRACTUAL AGREEMENT

Many contractual agreements are purchased during the initial service call and the form is filled out by the technician after speaking with the owner.

A contractual agreement is shown in Fig. 29-10. Note the technician is to indicate the equipment ages for air conditioning and heating equipment. The ages will determine the actual cost of the maintenance contract. For a system that is five years old, the contract charge is $115, versus $145 for one that is ten years old or more.

Contractual agreements vary depending upon the particular company. Some contracts cover domestic refrigerators, automatic ice makers, and other major appliances, while others cover commercial appliances. The sale of the contractual agreement gives the contracting company the opportunity to sell other additional services that will help to reduce the owner's cost of operating the unit. An example is to have a time delay installed in the system. The time delay would prevent the unit from starting immediately after it has stopped during power failures. Fast starting and stopping can cause damage to the units. Another add-on to the unit often sold is a humidistat. This provides for the operation of the air conditioning unit while the home is not occupied, based on the amount of humidity in the air. This type of service is often sold in southern states for owners who spend the summer up north and are not in the home to change the air conditioning setting based upon the humidity.

It is the responsibility of the technician who is making the service agreement to have a high level of loyalty with the customer. In developing the service agreement, the technician must be familiar with all the various types of service agreements and also understand the benefits that can be obtained from them for both the company and the owner. Service agreements are priced to cover all costs in-

PLAN A

UNIQUE AIR INC.
COMMERCIAL • RESIDENTIAL • LICENSED • INSURED
"Selling The Best—Servicing The Rest"
750 BELL ROAD
VENICE SARASOTA, FLORIDA 34240-9499 BRADENTON
484-7588 377-0153 756-3000

NAME
ADDRESS
CITY STATE ZIP
MAILING ADDRESS IF DIFFERENT FROM ABOVE
ADDRESS
CITY STATE ZIP

PRIVATE HOMES

THE FOLLOWING RATES APPLY FOR AIR CONDITIONING SYSTEMS WITH ELECTRIC HEAT AND ALSO HEAT PUMP SYSTEMS. 1 - 5 TONS.

EQUIPMENT AGES	1 - 5 YEARS	6 - 7 YEARS	8 - 9 YEARS	10 & OVER	AGE	PRICE
	115.00	125.00	135.00	145.00		$
Central Air	INCLUDED IN PLAN A —BASE PRICE					
Compressor						
Central Heat						
Humidistat						
Thermostat						
2 Preventive Maintenance Inspections						
OPTIONAL EQUIPMENT COVERAGE	1 - 5 YEARS	6 - 7 YEARS	8 - 9 YEARS	10 & OVER	AGE	PRICE
Electronic Air Cleaner	20.00	25.00	30.00	40.00		$.
Heat Recovery Unit	30.00	30.00	35.00	40.00		$.

OPTIONAL IN - HOUSE PLUMBING COVERAGE (LABOR ONLY)

Age of Residence	1 - 5 YEARS	6 - 7 YEARS	8 - 9 YEARS	10 & OVER	AGE	PRICE
Plumbing-Up to 2 Bathrooms	45.00	55.00	65.00	75.00		$.

Please add $10.00 for each additional bathroom.

OPTIONAL MAJOR APPLIANCE COVERAGE

EQUIPMENT AGES	1 - 5 YEARS	6 - 7 YEARS	8 - 9 YEARS	10 & OVER	AGE	PRICE
Refrigerator (including Compressor)	12.00	17.00	17.00	22.00		$.
With Cold Water/Ice Disp. in Door (Add)	12.00	12.00	17.00	17.00		$.
With Automatic Ice Maker add	12.00	12.00	17.00	17.00		$.
Oven and Range (including hood)	12.00	12.00	17.00	22.00		$.
With Self Cleaning Oven add	7.00	12.00	12.00	17.00		$.
Microwave (labor only)	12.00	17.00	22.00	27.00		$.
Water Heater (electric only)	7.00	12.00	17.00	17.00		$.
Dishwasher	12.00	12.00	17.00	22.00		$.
Trash Compactor	12.00	12.00	17.00	22.00		$.
Garbage Disposal	12.00	12.00	17.00	22.00		$.
Dryer	12.00	12.00	17.00	22.00		$.
Washer	17.00	22.00	27.00	32.00		$.

SEE CONTRACT FOR COMPLETE DETAILS

FOR GAS OR OIL FIRED SYSTEMS...PLEASE CALL OUR OFFICE FOR THE CONTRACT PRICE.

YOUR PLAN A CONTRACT WILL BECOME EFFECTIVE AFTER OUR INSPECTOR HAS FOUND YOUR EQUIPMENT TO BE IN GOOD WORKING ORDER.

AIR COND/HEATING EQUIP.	$.
IN - HOUSE PLUMBING	$.
TOTAL APPLIANCES	$.
TAX ON APPLIANCES 7%	$.
TOTAL CONTRACT PRICE	**$.**

Fig. 29-10. Contractual agreement. Note that the age of the equipment will determine the price. (Unique Air, Inc.)

volved in the necessary procedures and still produce a profit for the service company. Contracts are legal agreements that specify the terms and conditions of service.

29-12 INSPECTION REPORT

Many contractual agreements provide two inspections of the equipment per year. They will also provide for answering any service calls owners place to them. The inspection report provides a breakdown of procedures to be followed by the technician. Fig. 29-11 is an inspection report that is used by a typical company. For example, in the residential central air system, there are basically ten steps that the technician would follow:

1. A visual check is made of the coil, both front and back, and it is sprayed with self-rinsing coil cleaner.
2. The drain line is cut near the evaporator, and a wet-dry vacuum used to remove any water or particles of dirt. Then a piece of clear plastic tubing is replaced in the cut drain line.
3. The drain pan is vacuumed and algicide tablets placed in the pan.
4. The contact points are visually checked for excessive burn. Note: any contactors that have been used will show a noticeable marking.
5. A check is made for erosion at the coil. Plants, shrubs, branches, etc., should be clear of the unit, and there should be good airflow across the coil. If possible, the coil should be rinsed with clear water and cleaned.
6. The amp draw is read on each element, with the heat on.
7. Time delays that have been installed are to be disconnected and reconnected, and the starting time checked.
8. Temperature differential is checked, and should be 16 °F (−9 °C) to 20 °F (−7 °C).
9. A visual check is made of the ducts, for leaks.
10. Settings should be noted before inspection, so these devices can be reset when technician is finished.

Any unusual parts that will have to be replaced in the future should be indicated on the inspection report, such as belts, filters, etc., so that on the next service call the technician will not have to make a trip back to the company warehouse for parts. This becomes a record in the office and will make it easier for a technician to follow through in all future service calls to that particular site. By keeping a good maintenance record, the amount of call-backs and emergency service calls will be reduced, and this record should result in fewer problems with the equipment and therefore less cost in regards to general maintenance of the system.

29-13 REVIEW OF SAFETY

Check the company service manual for the proper wiring diagrams before servicing. Always make certain the main breaker is open and all power is off before attempting to service any refrigeration system. Follow the basic tests and procedures related to electrical leakage and current.

When a technician is working on the electrical components of a refrigeration or air conditioning system, the danger of electrical shock or damage to the instruments can be reduced by first checking the voltage of the system before beginning any electrical work.

The technician should never wear lose jewelry, rings, watches, or bracelets that can come in contact with the electrical terminals or wires.

SUPERIOR
CONTRACT SERVICE, INC.
P.O. BOX 2327
SARASOTA, FLORIDA 34230
SALES 355-7768
SERVICE 355-7766

INSPECTION REPORT 94771

BILLING LOCATION
NAME:
ADDRESS:
CITY/STATE:
TELEPHONE: WK
CONTACT

SERVICE LOCATION
NAME:
ADDRESS:
CITY/STATE:
TELEPHONE: WK
CONTACT
GEO CODE DEV
TENANT
AGENT
SPECIAL INST. ______

CUSTOMER NUMBER

CONDENSOR
MAKE
MODEL
SERIAL

EVAPORATOR
MAKE
MODEL
SERIAL

REMARKS: MECHANIC: ______ DATE ___/___/___

QTY.	DESCRIPTION	PRICE

INSPECTION CHARGE ______
MATERIALS CHARGE ______
TAX ______
TOTAL ______

OFFICE USE ONLY
☐ 1 ☐ 2 ☐ 3

INSPECTION REPORT:

CONDENSOR
HEAD PRESSURE ______
SUCTION PRESSURE ______
COMP. FLA ______
FAN FLA ______
LUBRICATE FAN MOTOR ______
CRANKCASE HEATER WARM ______
COIL CLEAN ______
CONTACTOR POINTS ______
ADDED FREON ______

EVAPORATOR
MOTOR HP ______
MOTOR FLA ______
LUBRICATE FAN MOTOR ______
COIL CLEAN ______
DRAIN LINE ______
DRAIN PAN ______

ELECTRIC HEAT
KW HEAT STRIP ______
1 ______
2 ______
3 ______
ELEMENTS ______

MISCELLANEOUS
UNIT CLEAN INSIDE & OUT ______
UNIT RUSTING ______
UNIT NEEDS RUST PROOFING ______
COMPRESSOR TIME DELAY ______
CALIBRATE THERMOSTAT ______
TEMPERATURE DIFF ______

ENERGY CONCERNS
DUCT SYSTEM ______
T - STAT ______
H - STAT ______

OIL HEAT
MAKE
MODEL
CHAMBER ______
RELAYS ______
CAD CELL ______
TRANSFORMER ______
DRAFT DAMPER ______
ADJUST FLAME ______

GAS HEAT
MAKE
MODEL
CLEAN PILOT LIGHT ______
CLEAN BURNER ______
CHECK THERMOCOUPLE ______
CHECK FAN CONTROLS ______
CHECK GAS VALVE ______
CHECK FLAME ______
CHECK LIMIT CONTROL ______

FILTERS

We, the undersigned, hereby acknowledge completion of the work described hereon and accept same and acknowledge receipt of a copy of this contract. It is agreed that payment will be made on demand after the above date of completion and that subsequent to that time we will be responsible for attorney's fees and costs incurred in collection.

CUSTOMER SIGNATURE ______

Fig. 29-11. Inspection report that is used when unit is checked by technician. (Superior Contract Service, Inc.)

When welding, soldering, or brazing, provide plenty of ventilation. Fluorocarbons are nonflammable, but—when mixed with flammable liquids or gas—can become flammable. The refrigeration system should be evacuated, purged, and left open prior to welding. This avoids a possible increase in hydrostatic pressure and the decomposition of the refrigerant.

Always properly evacuate and replace with fresh air any area where fluorocarbons have been emitted.

It is also important that proper shoes and clothing be worn so the technician will not slip on wet surfaces or have loose-fitting clothing become entangled in moving parts such as fans.

29-14 TEST YOUR KNOWLEDGE

1. Define troubleshooting.
2. Define servicing.
3. Where can the technician obtain the service manual or troubleshooting charts?
4. What is the first step in troubleshooting?
5. What is the second step in troubleshooting?
6. What is the third step in troubleshooting?
7. Why should the technician examine the system's electrical wiring and components diagram upon arrival at the job?
8. What may happen if the technician makes a quick decision in regards to solving the problem?
9. What does the first column in a troubleshooting chart list?
10. What does the second column in a troubleshooting chart list?
11. What does the third column in a troubleshooting chart list?
12. Where can the proper electrical wiring diagram be obtained?
13. Why is the procedure given in a troubleshooting chart a good method to use?
14. If the problem with a hermetic unit is that it operates long or continuously and the technician identifies the possible cause as a shortage of refrigerant, what will be the remedy?
15. How does the technician select the correct response from the possible cause column?
16. Why must there be adequate ventilation when welding or soldering?
17. What may happen if the technician is wearing lose jewelry or rings when on a service call?
18. When responding to a service call, in what way should the technician approach the problem?
19. If the owner of a commercial food storage unit complains that the liquid line is frosted or sweating, what is the possible cause, and what will be the remedy?
20. What should be done if, upon answering a service call, the technician finds a unit has a fuse removed or blown?
21. Why will maintenance agreements ensure that the equipment will last longer and prevent breakdowns?
22. Identify three important sections of a maintenance agreement.
23. Why are records maintained for services performed on the maintenance call and on regular service calls?

Chapter 30

TECHNICAL CHARACTERISTICS

CONTENTS

Kata thermometer, 1005
Infrared thermometer, 1005
Weights, specific heats of
  substances, 1006
Types of energy, 1006
Conventional energy
  equivalents, 1006
Metric energy equivalents, 1006
Conventional to metric linear
  equivalents, 1007
Area equivalents, 1007
Volume equivalents, 1007
Pressure equivalents, 1008
Velocity equivalents, 1008
Liquid measure equivalents, 1008
Weight equivalents, 1008
Flow equivalents, 1009
Temperature conversion table, 1009
Heat equivalents, 1010
Compression ratio, 1010
Pumping ratio, 1010
Laws of thermodynamics, 1010
Entropy, 1010
Theory of matter, 1011
Electron theory, 1011
Atom principles, 1011

Electron principles, 1012
Electrical units and symbols, 1013
Motor size calculations, mean
  effective pressure method, 1013
Lowering voltage saves power, 1014
Resistance of conductors, semicon-
  ductors, and nonconductors, 1014
Color code for resistors, 1014
Series resistance, 1014
Parallel resistance, 1014
Electrical Code, 1014
Air moisture holding properties, 1015
Moisture evaporation sources, 1015
Desiccants, 1015
Gas and vapor, 1016
Little-used refrigerants, 1016
Metric pressure-heat diagram, 1016
Moisture in liquid refrigerants, 1016
Dryness of refrigerants, 1019
Refrigerant oils, 1019
Eutectic compounds, 1019
Cryogenic temperatures, 1020
Brine freezing temperatures, 1020
Single energy source, 1021
Solvents and cleaning, 1021
Heating value of fuels, 1021

Twist drill sizes, 1022
Color code for piping, 1022
Food preservation by radiation
  treatment, 1022
Clean rooms, 1022
Solders and brazing metals, 1022
Heat conductivity, 1023
Galvanic action sequence, 1023
Electrolysis, 1023
Standard temperature, 1023
Standard pressure, 1023
Standard air, 1023
Refrigerant properties, 1024
Flare nut wrench sizes, 1025
Boyle's Law, 1025
Charles' Law, 1025
Gas Law, 1026
Adiabatic and isothermal expansion
  and contraction of gases, 1026
Root mean squared, 1026
Environmetal effects of CFCs and
  HCFCs, 1027
Psychrometric chart formula, 1027
SEER, 1027
Safety, 1028

Chapter 30

TECHNICAL CHARACTERISTICS

30-1 KATA THERMOMETER

A Kata thermometer is used to measure air currents in open spaces. It is an alcohol thermometer with a Fahrenheit scale etched in the glass. Two scales are available. One reads from 95 to 100 °F; the other from 125 to 130 °F.

To use the thermometer, heat it to the higher value in a hot water bath. Thoroughly dry the thermometer and suspend it in the air current. Then record the time it takes, in seconds, to cool to the lower reading on the F scale. Using this information, determine the air movement in feet per minute from a Kata thermometer table.

30-2 INFRARED THERMOMETER

Temperatures may be measured with an infrared thermometer, Fig. 30-1. This optical-electronic instrument gives almost instantaneous readings. It is aimed at the surface for which the temperature is to be determined. The temperature is read directly from the temperature scale.

Different materials and types of surfaces give off heat at different rates. It is, therefore, necessary to set the instrument according to the nature of the surface. The adjustment for this is called the emissivity adjustment. Some emissivity adjustments are given in Fig. 30-2.

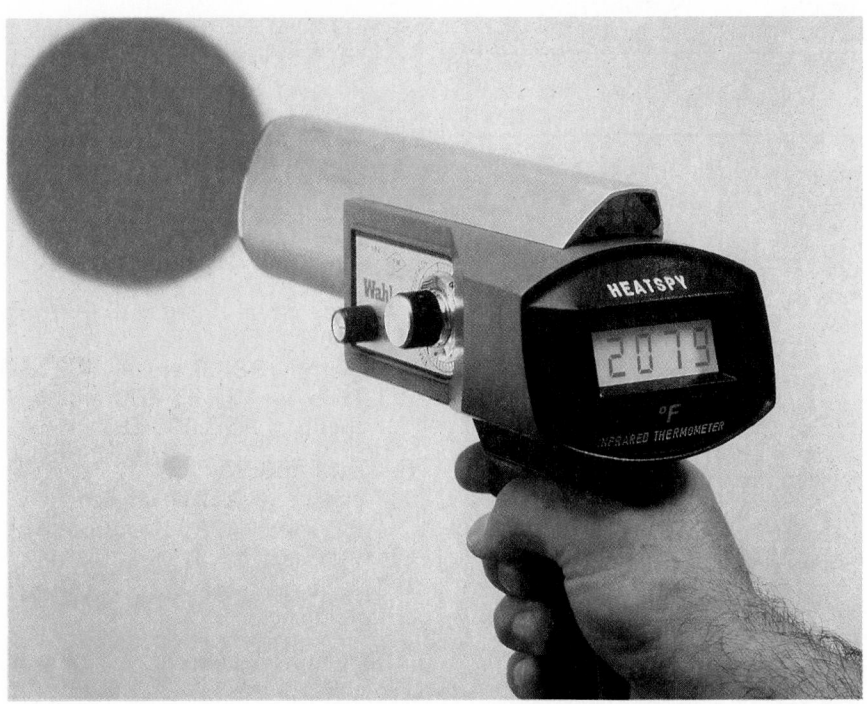

Fig. 30-1. An infrared thermometer. This instrument measures temperatures at distances of 1 to 40 feet from the source. Digital temperature readings from − 40 °F to 3100 °F are obtainable. Batteries are in handle.   (Wahl Instruments, Inc.)

INFRARED THERMOMETER ADJUSTMENT VALUES

MATERIAL		EMISSIVITY
ALUMINUM	BRIGHT	0.09
	ANODIZED	0.55
	OXIDIZED	0.2 to 0.3
BRASS	BRIGHT	0.03
	OXIDIZED	0.61
CHROMIUM	POLISHED	0.08
COPPER	BRIGHT	0.05
	OXIDIZED	0.78
IRON AND STEEL	POLISHED	0.55
	OXIDIZED	0.85
NICKEL	POLISHED	0.05
	OXIDIZED	0.95
ZINC	BRIGHT	0.23
	OXIDIZED	0.23
BRICK	BUILDING	0.45
PAINTS	WHITE	0.9
	BLACK	0.86
	OIL PAINTS (ALL COLORS)	0.92
ROOFING PAPER		0.91
RUBBER		0.94
SILICA		0.42 to 0.62
WATER		0.92

Fig. 30-2. Emissivity of various surfaces.

30-3 WEIGHTS AND SPECIFIC HEATS OF SUBSTANCES

MATERIAL	WEIGHT LBS./CU.FT.	SPECIFIC HEAT Btu/LB.
GASES		
Air (normal temp.)	.075	.24
METALS		
Aluminum	166.5	.214
Copper	552	.094
Iron	480	.118
Lead	710	.030
Mercury	847	.033
Steel	492	.117
Zinc	446	.096
LIQUIDS		
Alcohol	49.6	.60
Glycerine	83.6	.576
Oil	57.5	.400
Water	62.4	1.000
OTHERS		
Concrete	147	.19
Cork	15	.48
Glass	164	.199
Ice	57.5	.504
Masonry	112	.200
Paper	58	.324
Rubber	59	.48
Sand	100	.195
Stone	138-200	.20
Tar	75	.35
Wood, Oak	48	.57
Wood, Pine	38	.47

30-4 ENERGY

There are two kinds of energy—potential and kinetic.

Potential energy is like a body of water controlled by a dam. The water has a great potential for doing work. But until the water flows through a water wheel, no energy is produced. A mass at height H has PE = MgH.

Kinetic energy is the energy of a moving body. The formula is KE = $1/2\,MV^2$ = Foot Pounds. M = Weight, divided by 32 (the acceleration due to gravity). V = Velocity in feet per second (fps). In metric, KE = $1/2\,MV^2$ = Joules. M = Mass in kilograms. V = Velocity in meters per second (m/sec).

30-5 ENERGY EQUIVALENTS—CONVENTIONAL

1 Btu	= 778 ft. lb.
	= 252 gram-calories
	= 1054.8 joules
1 Horsepower	= 33,000 ft. lb./min.
	= 550 ft. lb./sec.
	= 746 watts
	= 2545.6 Btu/hr.
	= 42.42 Btu/min.
	= 1.014 hp (metric)
1 Horsepower hour	= 1 horsepower for 1 hr.
	= 1,980,000 ft. lb.
	= 746 watt hours
	= .746 kilowatt hours
	= 2545.6 Btu
1 Watt	= 3.414 368 Btu/hr.
1 Kilowatt	= 1000 watts
	= 1.34 horsepower
1 Kilowatt hour	= 1 kilowatt for 1 hr.
	= 1000 watt hours
ICE MELTING EFFECT (IME) 1 ton of Refrigeration	= 288,000 Btu/day
	= 12,000 Btu/hr.
	= 200 Btu/min.
	= 83.3 lb. of ice/hr.
	= 3.515 kW

30-6 ENERGY EQUIVALENTS—METRIC

ENERGY

1 dyne cm = 1 erg = 0.001 g·cm = 7.38×10^{-8} ft. lb.

1 g·cm = 980.6 ergs = 7.233×10^{-5} ft. lb.

1 ft. lb. = 13,557,300 ergs = 13,825.5 g·cm

1 therm = 100,000 Btu

RATE OF ENERGY

1 erg/sec. = 1 dyne cm/sec. = 7.38×10^{-8} ft. lb./sec.

1 g·cm/sec = 980.6 ergs/sec. = 7.24×10^{-5} ft. lb./sec.

1 ft. lb./sec. = 13,557,300 ergs/sec. = 13,800 g·cm/sec.

SI metric unit for energy is the joule. The erg and the dyne are part of the CGS metric system. CGS stands for centimeter-gram-second, a metric system that is widely used in all branches of science throughout the world before adoption of SI units.

30-7 LINEAR MEASURE EQUIVALENTS— US CONVENTIONAL—METRIC

1 inch		= 2.54 centimeters = 25.4 millimeters = 25 400 microns
1 foot	= 12 inches	= .304 meters = 30.48 centimeters
1 yard	= 3 feet	= .914 meters = 91.44 centimeters
1 micron		= .000 039 4 inches
1 millimeter	= 1000 microns	= .0394 inches
1 centimeter	= 10 millimeters	= .3937 inches
1 meter	= 100 centimeters	= 39.37 inches
1 kilometer	= 1000 meters	= .62137 miles

30-9 AREA EQUIVALENTS

1 sq. in.   = .0065 sq. meters (m²)
1 sq. ft. = 144 sq. in.   = .093 sq. meters (m²)
1 sq. yd. = 9 sq. ft.   = .836 sq. meters (m²)
1 sq. yd. = 1296 sq. in.

30-10 VOLUME EQUIVALENTS

1 cu. in.		= .016 liters = 16.39 cm³
1 cu. ft.	= 1728 cu. in. = 7.481 gal.	= 28.317 liters = .0283 m³ = 28 317.00 cm³
1 cu. yd.	= 27 cu. ft. = 46,656 cu. in.	
1 gal.	= .1337 cu. ft. = 231 cu. in.	= 3.79 liters = 3 785 cm³
1 cm³	= .0610237 cu. in.	
1 liter	= 61.03 cu. in. = .2642 gal.	= 1 000 cm³

30-8 FRACTIONS, DECIMALS, AND MILLIMETERS OF THE PARTS OF AN INCH

INCH	DECIMAL INCH	MILLIMETER	INCH	DECIMAL INCH	MILLIMETER
1/64	0.0156	0.3967	33/64	0.5162	13.0968
1/32	0.0312	0.7937	17/32	0.5312	13.4937
3/64	0.0468	1.1906	35/64	0.5468	13.8906
1/16	0.0625	1.5875	9/16	0.5625	14.2875
5/64	0.0781	1.9843	37/64	0.5781	14.6843
3/32	0.0937	2.3812	19/32	0.5937	15.0812
7/64	0.1093	2.7781	39/64	0.6093	15.4781
1/8	0.125	3.175	5/8	0.625	15.875
9/64	0.1406	3.5718	41/64	0.6406	16.2718
5/32	0.1562	3.9687	21/32	0.6562	16.6687
11/64	0.1718	4.3656	43/64	0.6718	17.0656
3/16	0.1875	4.7625	11/16	0.6875	17.4625
13/64	0.2031	5.1593	45/64	0.7031	17.8593
7/32	0.2187	5.5562	23/32	0.7187	18.2562
15/64	0.2343	5.9531	47/64	0.7343	18.6531
1/4	0.25	6.5	3/4	0.75	19.05
17/64	0.2656	6.7468	49/64	0.7656	19.4468
9/32	0.2812	7.1437	25/32	0.7812	19.8437
19/64	0.2968	7.5406	51/64	0.7968	20.2406
5/16	0.3125	7.9375	13/16	0.8125	20.6375
21/64	0.3281	8.3343	53/64	0.8281	21.0343
11/32	0.3437	8.7312	27/32	0.8437	21.4312
23/64	0.3593	9.1281	55/64	0.8593	21.8281
3/8	0.375	9.525	7/8	0.875	22.225
25/64	0.3906	9.9218	57/64	0.8906	22.6218
13/32	0.4062	10.3187	29/32	0.9062	23.0187
27/64	0.4218	10.7156	59/64	0.9218	23.4156
7/16	0.4375	11.1125	15/16	0.9375	23.8125
29/64	0.4531	11.5093	61/64	0.9531	24.2093
15/32	0.4687	11.9062	31/32	0.9687	24.6062
31/64	0.4843	12.3031	63/64	0.9843	25.0031
1/2	0.50	12.7	1	1.000	25.4

30-11 PRESSURE EQUIVALENTS

Fig. 30-3 shows how psia units are related to kPa units, and how psig units are related to kPa units.

1 psia	= 0.068 atmosphere = 144 lb./sq. ft. = .703 meters water = 2.036 in. of mercury = 70.3 cm water = 2.307 ft. of water = 51.7 mm Hg = 27.7 in. of water = 6.9 kPa
1 psig 0 psig	= 15.7 psia = 108 kPa = 14.7 psia = 101.3 kPa
1 oz./sq. in.	= .128 in. of mercury = 1.73 in. of water
1 in. of mercury	= .0334 atmosphere = 3.386 kPa = .491 psi = 25.4 mm Hg = 1.13 ft. of water = .3453 m water = 13.6 in. of water = 70.73 psf
1 ft. of water	= .0295 atmosphere = 2.985 kPa = .434 psi = 22.42 mm Hg = 62.43 lb./sq. ft. = .305 m water = .03 atmosphere = .883 in. of mercury (Hg)

1 atmosphere	= 29.92 in. of mercury = 101.28 kPa = 33.94 ft. of water = 760 mm Hg) = 14.696 psi = 10.33 m water = 2116.35 psf
1 psf	= .007 psi = 0.048 kPa = 4.725×10^{-4} = .359 mm Hg atmosphere = .01414 in. Hg = .0049 m water = .016 ft. water
1 kilogram/ sq. cm	= 14.22 psi = 10 meters of water = 2048.17 psf = .967 atmosphere = 97.98 kPa = 28.96 in. Hg = 32.8083 ft. water
1 meter of water	= 1.42 psi = 73.55 mm Hg = 204.8 psf = 9.78 kPa = .097 atmosphere = 2.896 in. Hg = 3.28 ft. water
1 mm Hg	= .019 psi = 0.133 kPa = 2.78 psf = .0136 m water = .001316 atmosphere = .039 in. Hg = .0446 ft. water

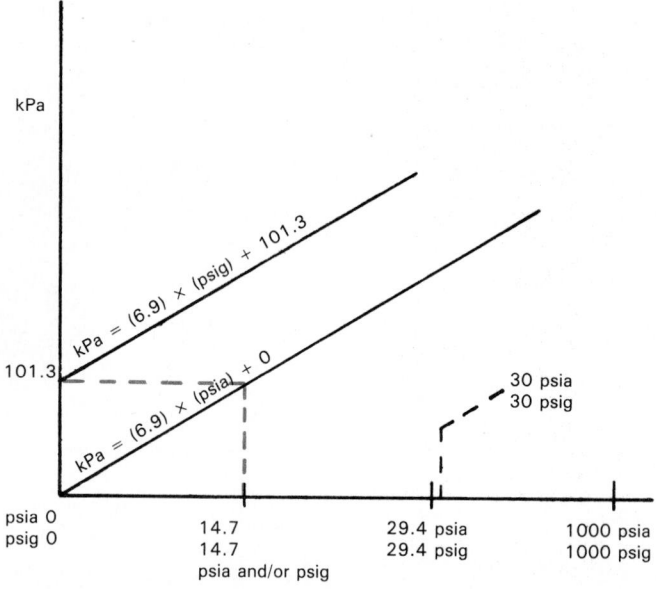

Fig. 30-3. The graph shows the following relationships: 1. Between psia and kPa (bottom graph line). 2. Between psig and kPa (top graph line). Most technicians prefer to use the lower graph line because it is simpler; it begins at zero. It is best to convert all psig readings to psia before trying to find the kPa reading.

30-13 LIQUID MEASURE EQUIVALENTS

Liquid Measure	U.S.	Metric
1 pint	= 16 ounces	= .473 liters
1 quart	= 2 pints	= .946 liters
1 quart	= 32 ounces	
1 gallon	= 4 quarts	= 3.785 liters
1 gallon	= 8 pints	
1 gallon	= 231 cubic inches	
1 cubic foot	= 7.48 gallons	
1 gallon	= 8.34 pounds of water	
1.136 quart		= 1 liter

30-14 WEIGHT EQUIVALENTS

AVOIRDUPOIS		
1 ounce	= 437 grains	= 28.35 grams = .028 kilogram
1 pound (lb.)	= 7000 grains	= .4536 kilograms = 453.6 grams
1 pound	= 16 ounces	= 453.6 grams
1 grain	= .000143 pounds	= .064 80 grams
1 ton	= 2000 pounds	= 907.2 kilograms
1 gram	= 15.43 grains = .03527 ounces = .002205 pounds	= .001 kilogram
1 kilogram	= 2.2 pounds	

30-12 VELOCITY EQUIVALENTS

1 mi./hr.	= 1.47 ft./sec. = .87 knots	= 1.61 km/hr. = .45 meters/sec.
1 ft./sec.	= .68 mi./hr. = 60 ft./min. = .59 knots	= 1.1 km/hr. = .305 meters/sec.

1 meter/sec.	= 3.28 ft./sec. = 2.24 mi./hr. = 1.94 knots	= 3.6 km/hr.
1 km/hr.	= .91 ft./sec. = .62 mi./hr. = .54 knots	= .28 meters/sec.

SPECIFIC WEIGHTS (DENSITY)

1 lb./cu. in	= 1728 lb./cu. ft.	= 27.68 g/cm³
		= 2.768 × 10⁷ g/m³
1 lb./cu. ft.	= 5.787 × 10⁻⁴	
	lb./cu. in.	= .016 g/cm³
1 gm/cm³	= 62.43 lb./cu. ft.	
1 kg/m³	= .06243 lb./cu. ft.	

30-15 FLOW EQUIVALENTS

1 cu. ft. per min.	= 7.481 gal./min.	= 28 317 cm³/min.
	= 449 gal./hr.	= 28.32 liters/min. (l/min.)
		= 1 700.00 l/hr.
1 cu. ft. per hour	= .0167 cu. ft./min.	= .472 l/min.
	= .1247 gal./min.	= 28.317 l/hr.
	= 7.481 gal./hr.	= 472 cm³/min.
1 gal. per min.	= .1337 cu. ft./min.	= 3.79 l/min.
	= 8.022 cu. ft./hr.	= 3785 cm³/min.
1 liter per min.	= .0353 cu. ft./min.	= 1 000 cm³/min.
	= 2.118 cu. ft./hr.	
	= .2642 gal./min.	
	= 15.852 gal./hr.	

30-16 TEMPERATURE CONVERSION TABLE

To use the following table, find the temperature to be converted in the center column. If converting to Celsius, read to the left. If converting to Fahrenheit, read to the right.

°C		°F	°C		°F
−273.15	−459.67		−56.7	−70	−94
−268	−450		−51.1	−60	−76
−262	−440		−45.6	−50	−58
−257	−430		−40.0	−40	−40
−251	−420		−34.4	−30	−22
−246	−410		−28.9	−20	−4
−240	−400		−23.3	−10	14
−234	−390				
−229	−380		−17.8	0	32
−223	−370				
−218	−360		−17.2	1	33.8
−212	−350		−16.7	2	35.6
−207	−340		−16.1	3	37.4
−201	−330		−15.6	4	39.2
−196	−320		−15.0	5	41.0
−190	−310		−14.4	6	42.8
−184	−300		−13.9	7	44.6
−179	−290		−13.3	8	46.4
−173	−280		−12.8	9	48.2
−169	−273	−459.4	−12.2	10	50.0
−168	−270	−454	−11.7	11	51.8
−162	−260	−436	−11.1	12	53.6
−157	−250	−418	−10.6	13	55.4
−151	−240	−400	−10.0	14	57.2
−146	−230	−382	−9.4	15	59.0
−140	−220	−364	−8.9	16	60.8
−134	−210	−346	−8.3	17	62.6
−129	−200	−328	−7.8	18	64.4
−123	−190	−310	−7.2	19	66.2
−118	−180	−292	−6.7	20	68.0
−112	−170	−274	−6.1	21	69.8
−107	−160	−256	−5.6	22	71.6
−101	−150	−238	−5.0	23	73.4
−95.6	−140	−220	−4.4	24	75.2
−90.0	−130	−202	−3.9	25	77.0
−84.4	−120	−184	−3.3	26	78.8
−78.9	−110	−166	−2.8	27	80.6
−73.3	−100	−148	−2.2	28	82.4
−67.8	−90	−130	−1.7	29	84.2
−62.2	−80	−112	−1.1	30	86.0

°C		°F	°C		°F
−0.6	31	87.8	43	110	230
0	32	89.6	49	120	248
0.6	33	91.4	54	130	266
1.1	34	93.2	60	140	284
1.7	35	95.0	66	150	302
2.2	36	96.8	71	160	320
2.8	37	98.6	77	170	338
3.3	38	100.4	82	180	356
3.9	39	102.2	88	190	374
4.4	40	104.0	93	200	392
5.0	41	105.8	99	210	410
5.6	42	107.6			
6.1	43	109.4	100	212	413
6.7	44	111.2			
7.2	45	113.0	104	220	428
7.8	46	114.8	110	230	446
8.3	47	116.6	116	240	464
8.9	48	118.4	121	250	482
9.4	49	120.2	127	260	500
10.0	50	122.0	132	270	518
10.6	51	123.8	138	280	536
11.1	52	125.6	143	290	554
11.7	53	127.4	149	300	572
12.2	54	129.2	154	310	590
12.8	55	131.0	160	320	608
13.3	56	132.8	166	330	626
13.9	57	134.6	171	340	644
14.4	58	136.4	177	350	662
15.0	59	138.2	182	360	680
15.6	60	140.0	188	370	698
16.1	61	141.8	193	380	716
16.7	62	143.6	199	390	734
17.2	63	145.4	204	400	752
17.8	64	147.2	210	410	770
18.3	65	149.0	216	420	788
18.9	66	150.8	221	430	806
19.4	67	152.6	227	440	824
20.0	68	154.4	232	450	842
20.6	69	156.2	238	460	860
21.1	70	158.0	243	470	878
21.7	71	159.8	249	480	896
22.2	72	161.6	254	490	914
22.8	73	163.4	260	500	932
23.3	74	165.2	266	510	950
23.9	75	167.0	271	520	968
24.4	76	168.8	277	530	986
25.0	77	170.6	282	540	1004
25.6	78	172.4	288	550	1022
26.1	79	174.2	293	560	1040
26.7	80	176.0	299	570	1058
27.2	81	177.8	304	580	1076
27.8	82	179.6	310	590	1094
28.3	83	181.4	316	600	1112
28.9	84	183.2	321	610	1130
29.4	85	185.0	327	620	1148
30.0	86	186.8	332	630	1166
30.6	87	188.6	338	640	1184
31.1	88	190.4	343	650	1202
31.7	89	192.2	349	660	1220
32.2	90	194.0	354	670	1238
32.8	91	195.8	360	680	1256
33.3	92	197.6	366	690	1274
33.9	93	199.4	371	700	1292
34.4	94	201.2	377	710	1310
35.0	95	203.0	382	720	1328
35.6	96	204.8	388	730	1346
36.1	97	206.6	393	740	1364
36.7	98	208.4	399	750	1382
37.2	99	210.2	404	760	1400
37.8	100	212.0	410	770	1418

°C		°F	°C		°F
416	780	1436	599	1110	2030
421	790	1454	604	1120	2048
427	800	1472	610	1130	2066
432	810	1490	616	1140	2084
438	820	1508	621	1150	2102
443	830	1526	627	1160	2120
449	840	1544	632	1170	2138
454	850	1562	638	1180	2156
460	860	1580	643	1190	2174
466	870	1598	649	1200	2192
471	880	1616	704	1300	2372
477	890	1634	760	1400	2552
482	900	1652	816	1500	2732
488	910	1670	871	1600	2912
493	920	1688	927	1700	3092
499	930	1703	982	1800	3272
504	940	1724	1038	1900	3452
510	950	1742	1093	2000	3632
516	960	1760	1149	2100	3812
521	970	1778	1204	2200	3992
527	980	1796	1260	2300	4172
532	990	1814	1316	2400	4352
538	1000	1832	1371	2500	4532
543	1010	1850	1427	2600	4712
549	1020	1868	1482	2700	4892
554	1030	1886	1538	2800	5072
560	1040	1904	1593	2900	5252
566	1050	1922	1649	3000	5432
571	1060	1940			
577	1070	1958			
582	1080	1976			
588	1090	1994			
593	1100	2012			

INTERPOLATION VALUES

°C		°F	°C		°F
0.56	1	1.8	3.33	6	10.8
1.11	2	3.6	3.89	7	12.6
1.67	3	5.4	4.44	8	14.4
2.22	4	7.2	5.00	9	16.2
2.78	5	9.0	5.56	10	18.0

Courtesy of Thermatron Corporation

30-17 HEAT EQUIVALENTS

```
1 Btu       = 252 calories
1 Btu       = 1054.4 J
1 kilocalorie = 1 kcal = 1000 calories
1 kcal      = 4.1840 kJ
1 kcal/kg   = 4.1840 kJ/kg
1 Btu/lb.   = 0.5556 kcal/kg
1 Btu/lb.   = (4.1840 kJ/kcal)/(1.8 °F/ °C)
            = 2.3244 kJ/kg
1 kcal/kg   = 1.8 Btu/lb.
1 Btu/hr.   = 0.2931 watts
```

The most correct definition of the kilocalorie is 1 kcal = 4.1840 kJ. This defines the unit called the *thermochemical calorie*. The thermochemical calorie is slightly more accurate than the unit used earlier in this book (1 kcal = 4.187 kJ).

Unit of heat per pound Celsius or Centigrade heat unit is the heat required to raise one pound of water 1 °C and equals 1.8 Btu.

1 Btu/hr./sq. ft./°F = 4.88 kgcal/hr./m²/ °C
1 kgcal/hr./m²/ °C = 0.205 Btu/hr./sq. ft./ °F

30-18 COMPRESSION RATIO

Compression ratio is the ratio of the total volume of the cylinder to the clearance space in a compressor, or that space which remains at the end of the compression cycle.

For Example: A compressor has a 2-in. bore and a 2-in. stroke and a clearance space of .010 in. The volume in the valve ports equals .05 cu. in. The piston displacement is:

$\pi r^2 \times S.$
$\pi = 3.1416$
$r = 1$
$S = 2$

The volume equals 3.1416 × 1 × 2 = 6.2832 cu. in.
The volume of the clearance space equals $\pi r^2 \times r \times S$
$\pi = 3.1416$
$r = 1$
$S = .010$

The volume equals 3.1416 × 1 × .010 = .031416 = .03 cu. in.
Total clearance space: .05 + .03 = .08 cu. in.
The total volume of the cylinder is:
6.2832 + .08 = 6.3632 = 6.36
The compression ratio is 6.36 ÷ .08 = 79.5:1

30-19 PUMPING RATIO

The pumping ratio is the ratio of the suction pressure to the condensing or head pressure expressed in absolute pressures. Absolute pressure equals gauge pressure plus 15. For example, if the low-side pressure of a system is 15 psi:

15 + 15 = 30 psia;

and the high-side pressure is 150 psi:

150 + 15 = 165 psia,

the pumping ratio equals $\frac{30}{165} = \frac{1}{5.5} = 5.5:1$.

The pumping ratio should not exceed a certain value for each particular refrigerant. If the pumping ratio is too high, the temperature of the high-pressure vapor going through the exhaust valve would overheat the mechanism and cause some of the oil in the exhaust pocket to become carbonized.

30-20 LAWS OF THERMODYNAMICS

FIRST LAW OF THERMODYNAMICS is a formula for the conversion of heat into work or work into heat. The formula is: 778 foot pounds of work is equivalent to the heat energy of 1 Btu.

SECOND LAW OF THERMODYNAMICS states that heat will only flow from a body at a certain temperature to another body which is at a lower temperature.

30-21 ENTROPY

Entropy is the heat available measured in Btu per pound degree change for a substance.

Entropy calculations are made from generally accepted temperature bases. For heating and steam power using water as the medium, the accepted base is 32 °F (0 °C). For domestic and most commercial refrigeration calculations, the base is −40 °F (−40 °C). For research and very low temperature work, a base of a lower temperature may be selected.

Entropy is generally used only in engineering calculations. Entropy tables have been worked out and are contained in most engineering handbooks.

30-22 THEORY OF MATTER

To understand electricity and electronics, one must know the properties of matter. All materials are made up of molecules. A molecule is the smallest part of a substance that has all the properties of that substance. All molecules are made up of atoms. The molecules may contain as few as one atom or up to hundreds of atoms.

Atoms are made mainly of electrons, neutrons, and protons. All the material in the universe is made from only about 110 different atoms. Each of these is different, depending on the number of electrons, protons, and neutrons in it.

All substances have mass, which is determined by the amount of matter contained in a substance. Mass of a body, then, is determined by dividing the weight of the body by the acceleration due to gravity, 32.2

The total of mass and energy is constant. Mass can be turned into energy. Energy can be turned into mass. The equation (A. Einstein's) is: $E = MC^2$. E = energy. M = mass. C = speed of light.

30-23 ELECTRON THEORY

Various principles of construction and operation of electrical equipment are based on theories which have come to be accepted principles of operation.

Scientific studies indicate that all substances are made up of molecules. Molecules are the smallest part of a substance that has all the properties of the substance. Molecules are made up of atoms. Atoms are made up of electrons, neutrons, and protons. Electrons are negative charges of electricity. Neutrons are neither negatively nor positively charged. Protons are positive charges of electricity.

The electrons revolve around the nucleus (center). The nucleus is made of protons and neutrons. This theory is illustrated in Fig. 30-4.The revolving electrons are of two types:

1. Free electrons.
2. Bound electrons.

If an atom has free electrons, the atom can transfer or conduct electrical energy. If an atom collects an electron charge for an instant, it is negatively charged. If the atom loses an electron, it becomes positively charged. The electrons travel from atoms having extra electrons to atoms having a lack of electrons.

30-24 THE ATOM

An elementary drawing of an atom, Fig. 30-4, looks like the solar system, with a sun and planets. The electrons revolve around the nucleus (center). The nucleus is made of protons and neutrons. Atoms also seem to have many other particles, but these rare particles are, at present, only of interest and use to research and nuclear scientists.

The electrons seem to be little clouds of energy rather than a small sphere. Two electrons opposite each other seem to be shaped like a dumbbell. Some of the smaller particles may be what scientists see in a cloud chamber as they do research.

The revolving electrons are attracted to the center (nucleus), yet their speed of travel keeps them a small distance from the nucleus by centrifugal force. The force trying to pull the electron into the center is called the negative electrical charge (negative sign is −). The nucleus attraction is called the positive electrical charge (positive sign is +). Based on this attraction, electricity is movement of the electrons (−) toward the nucleus (+).

The revolving electrons are of two types:

1. Free electrons.
2. Bound electrons.

If an atom has free electrons, the atoms can transfer or conduct electricity (electrons can move from one atom to another). See Fig. 30-5.

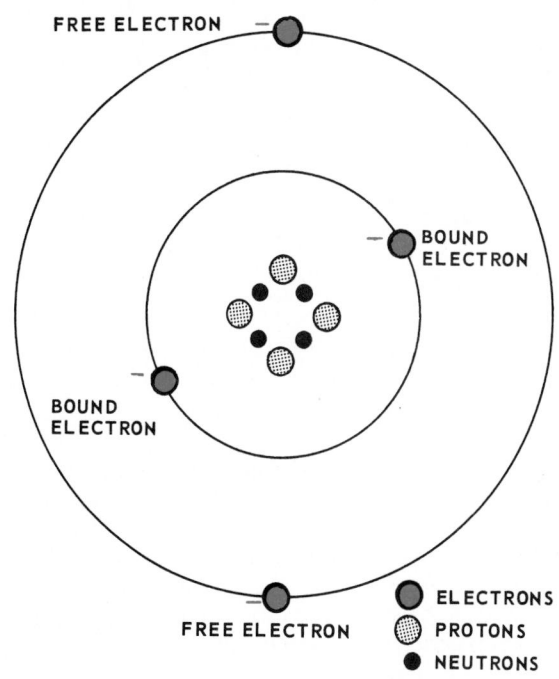

ELECTRONS •
PROTONS ○
NEUTRONS ◯

Fig. 30-4. Makeup of an atom, showing electrons, protons, neutrons, and their relative motion.

Fig. 30-5. An atom with two free electrons, two bound electrons, four protons, and four neutrons.

When an atom enters a chemical change, it does so in the form of a charged particle. These particles are called ions. Ions are of two types:
1. Negative.
2. Positive.

A negative ion is an atom that is negatively charged (has an extra electron). A positive ion is an atom that is positively charged (lacks an electron). Like charged ions repel. Unlike charged ions attract.

In the lighter elements, the atom contains one proton for each neutron. In the heavier elements, the atom contains more neutrons than protons. For example:

Helium = 2 protons, 2 neutrons
Boron = 5 protons, 5 neutrons
Mercury = 80 protons, 120 neutrons

A proton and a neutron are equal in mass. Either one is about 1845 times as great in mass as the electron mass. For example, if one can imagine that the outer electrons of a typical atom structure are in an orbit 200 yards in diameter, then the nucleus in the center would be 1/2 in. in diameter.

In most atoms, the number of electrons equals the number of protons in the atom. See Fig. 30-6. The negative charges equal the proton charges, and the resulting (net) charge of the atom is zero (0) if no chemical action is taking place or if no electrical flow is taking place.

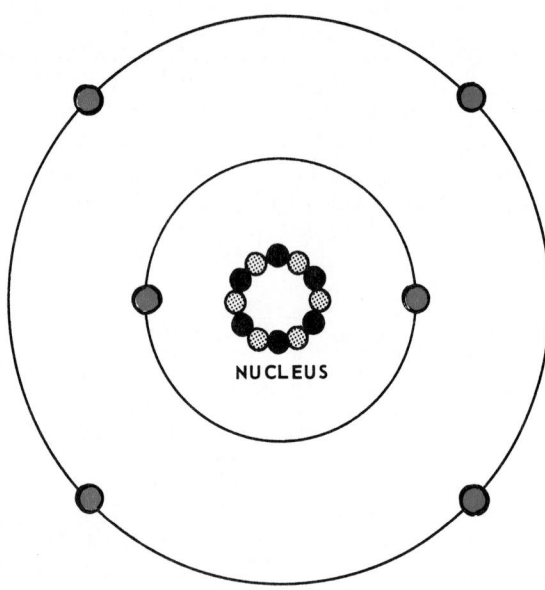

Fig. 30-6. A carbon atom with four free electrons, two bound electrons, six protons, and six neutrons.

If an external force is applied to an atom, one or more of the outermost electrons may be removed. These removable electrons are commonly called "free" electrons.

When the outer electrons of an atom interact with electrons of another atom, they combine to form a compound.

In some substances, each atom will retain all its electrons. In other substances, one or more outer atoms will be gained or lost as a result of the bonding.

The nucleus does not enter into chemical or electrical processes. To disrupt the nucleus, a vast amount of energy is needed, such as in an explosion of an atom bomb.

Two forces act on revolving electrons:
1. Centrifugal force.
2. Centripetal force.

Centrifugal force tends to make electrons move away from protons.

Centripetal force tends to make electrons move toward protons. (Electron has a negative charge; proton has a positive charge.)

The electron orbit is a balance of these two forces.

The electron has two energies:
1. Kinetic (motion).
2. Potential (position).

These two energies together determine the radius of the electron orbit. To stay in a certain orbit, the electron must neither gain nor lose energy.

30-25 THE ELECTRON

Electrons are negative charges of electricity. Fig. 30-7 shows an atom including its nucleus. Protons are positive charges of electricity. If an atom collects an extra free electron for a moment, it is negatively charged. If an atom loses a free electron, it is positively charged. The electrons travel from atoms with extra electrons to atoms with a lack of electrons.

Electricity is the flow of these electrons (or negative charges). The number of electrons that flow is extremely high. The term coulombs is used to represent the flow of 6.24×10^{18} or 6,240,000,000,000,000,000 electrons. One coulomb flowing for one second equals one ampere. The ampere is known as the quantity of electricity.

If there are too many protons, and electrons move to make up the difference, this movement or pressure to move

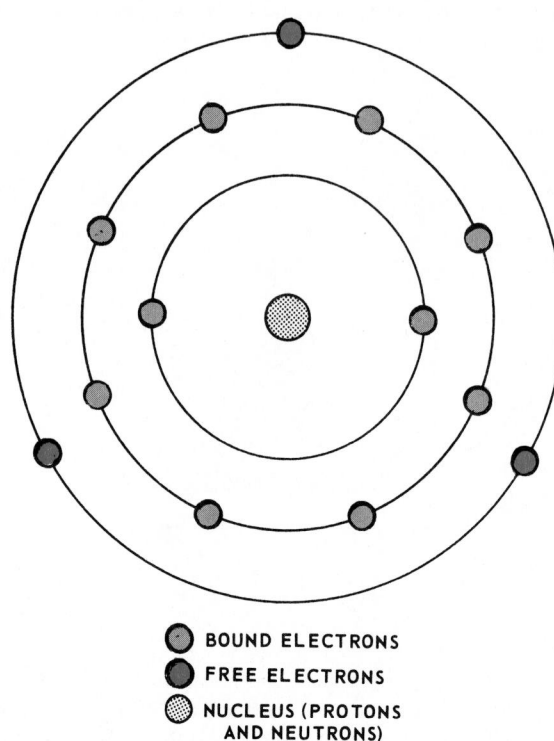

● BOUND ELECTRONS
● FREE ELECTRONS
◉ NUCLEUS (PROTONS AND NEUTRONS)

Fig. 30-7. An atom with three free electrons, ten bound electrons, and a nucleus of 13 protons and 13 neutrons.

is called electromotive force. Electric current consists of electrons traveling in a conductor, which may be made of a solid, a liquid, or a gas.

30-26 ELECTRICAL UNITS AND SYMBOLS

ELECTRICAL ENGINEERING UNITS AND CONSTANTS
As adopted by NBS

Symbols and Units

Quantity	Symbol	Unit	Symbol
charge	Q	coulomb	C
current	I	ampere	A
voltage, potential difference	V	volt	V
electromotive force	$\mathscr{E}$	volt	V
resistance	R	ohm	Ω
conductance	G	mho (siemens)	A/V, or mho (S)
reactance	X	ohm	Ω
susceptance	B	mho	A/V, or mho
impedance	Z	ohm	Ω
admittance	Y	mho	A/V, or mho
capacitance	C	farad	F
inductance	L	henry	H
energy, work	W	joule	J
power	P	watt	W
resistivity	ρ	ohm-meter	Ωm
conductivity	σ	mho per meter	mho/m
electric displacement	D	coulomb per sq. meter	C/m²
electric field strength	E	volt per meter	V/m
permittivity (absolute)	ϵ	farad per meter	F/m
relative permittivity	ϵ_r	(numeric)	
magnetic flux	Φ	weber	Wb
magnetomotive force	$\mathscr{F}$	ampere (ampere-turn)	A
reluctance	$\mathscr{R}$	ampere per weber	A/Wb
permeance	$\mathscr{P}$	weber per ampere	Wb/A
magnetic flux density	B	tesla	T
magnetic field strength	H	ampere per meter	A/m
permeability (absolute)	μ	henry per meter	H/m
relative permeability	μ_r	(numeric)	
length	l	meter	m
mass	m	kilogram	kg
time	t	second	s
frequency	f	hertz	Hz
angular frequency	ω	radian per second	rad/s
force	F	newton	N
pressure	p	newton per sq. meter	N/m²
temperature (absolute)	T	degree Kelvin	K
temperature (International)	t	degree Celsius	C

RECOMMENDED UNIT PREFIXES

Multiples and submultiples	Prefixes	Symbols	Pronunciation
10^{12}	tera	T	tĕr' á
10^9	giga	G	jī'gá
10^6	mega	M	mĕg' á
10^3	kilo	k	kĭl' ô
10^2	hecto	h	hĕk' tô
10	deka	da	dĕk' á
10^{-1}	deci	d	dĕs' ĭ
10^{-2}	centi	c	sĕn' tĭ
10^{-3}	milli	m	mĭl' ĭ
10^{-6}	micro	μ	mī' krô
10^{-9}	nano	n	năn' ô
10^{-12}	pico	p	pē' kô
10^{-15}	femto	f	fĕm' tô
10^{-18}	atto	a	ăt' tô

DEFINED VALUES AND CONVERSION FACTORS

Meter...................	1 650 763.73 wavelengths of the transition $2p_{10}$ – $5d_5$ in ⁸⁶Kr
Kilogram..............	mass of the international kilogram
Second................	1/31 556 925.974 7 of the tropical year 1900
Degree Kelvin.......	In the thermodynamic scale, 273.16 °K = triple point of water (fp, 273.15 °K=0 °C)
Unified atomic mass unit, u	1/12 the mass of an atom of the ¹²C nuclide
Standard acceleration of free fall	9.806 65 m s⁻², 980.665 cm s⁻²
Normal atmosphere...	101 325 N m⁻², 1 013 250 dyn cm⁻²
Thermochemical calorie...	4.1840 J, 4.1840×10⁷ erg
Int. Steam Table calorie	4.1868 J, 4.1868×10⁷ erg
Liter...................	0.001 000 028 m³, 1 000.028 cm³ (recommended by CIPM, 1950)
Inch...................	0.0254 m, 2.54 cm
Pound (avdp.).......	0.453 592 37 kg, 453.592 37 g

30-27 MOTOR SIZE CALCULATIONS— MEAN EFFECTIVE PRESSURE METHOD

Horsepower of a motor required to drive a refrigeration compressor can be calculated. The usual method uses the mean effective pressure (mep) in the cylinder. This is the pressure which is bearing down on the piston head and which must be overcome by the electric motor when driving the compressor. The mep is determined by a formula.

Basic variables of the formula are the low-side pressure, the high-side pressure, and the ratio of the specific heat at constant pressure to the specific heat at constant volume $\frac{C_p}{C_v}$ = K (for the kind of refrigerant). The formula follows:

$$mep = P_1 \times \frac{K}{K-1} \left[\left(\frac{P_2}{P_1}\right)^{\frac{K-1}{K}} - 1 \right]$$

P_1 = suction pressure, psia

P_2 = condenser pressure, psia

$$K = \frac{C_p}{C_v}$$

This value, when multiplied by the area of the piston, by the length of the stroke, and by the rpm, will give the foot-pounds per minute needed to drive the compressor.

$$\text{Ft. lb. per min.} = mep \times \frac{\pi D^2}{4} \times S \times N \times R$$

D = piston diameter

S = stroke in inches

N = number of cylinders

R = rpm

Convert the foot-pound per min. into hp by dividing by 33,000

$$hp = \frac{\text{ft. lb./min.}}{33,000}$$

The mep may be determined from the above or may be determined by obtaining the indicator card of the compressor being studied. Engineers' handbooks set forth the methods of using and obtaining indicator cards.

For example, if the indicator card shows a mep of 30 lb. per sq. in., the indicator hp necessary to drive a one cylinder compressor with a 2-in. bore and a 2-in. stroke, running at 200 rpm, would be as follows:

$$hp = \frac{30\ psi \times area \times stroke \times rpm}{33,000}$$

The dimensions have to be in foot-pounds = pressure × area × stroke. The resistance is the length of the compression stroke × rpm. The horsepower is:

$$\frac{30\ psi \times \frac{\pi D^2}{4} \times \frac{2}{12} \times 200}{33,000} =$$

$$\frac{30 \times \pi D^2 \times 2 \times 200}{4 \times 33,000 \times 12} = \frac{30 \times \pi 4 \times 2 \times 200}{4 \times 33,000 \times 12} =$$

$$\frac{\pi \times 200}{6600} = \frac{\pi}{33} = .0952\ or$$

approximately .10 hp.

This is the theoretical hp and neglects friction, oil pumping, starting load, and drive losses. Up to a 1-ton machine, one should double this hp. For example, 1/4 hp, calculated, will need a 1/2 hp motor. Up to 5 tons, this ratio gradually tapers off until adding 30 percent at 5 tons capacity will take into consideration the above losses. The reason for the decrease is that some of the losses remain constant, while others do not increase as rapidly in proportion to the increase in the size of the unit.

30-28 LOWERING VOLTAGE SAVES POWER

During times of power shortage, some utilities reduce the voltage supplied to the customer. This may result in some power saving. The largest saving will come from noninductive loads, such as lighting, resistance heating, and electric cooking. In these appliances, a reduction in voltage applied will cause some drop in the rate of current flow. Electrical power is equal to V × C − E × I equals watts—therefore, if both the voltage and current is reduced, the E × I will result in the reduction in the amount of power used.

In inductive loads there will not be a great saving. This means there will be very little saving in the operation of such appliances as refrigerators, air conditioners, and ventilating equipment. If the voltage is dropped, the current required will automatically increase. Therefore, power consumed, E × I equals W, will be affected very little. If "E" goes down and "I" goes up, power required will be affected very little. If motor speed is reduced, there will be some savings.

30-29 RESISTANCES OF CONDUCTORS, SEMI-CONDUCTORS, AND NONCONDUCTORS

The value, in ohms, of the resistance of conductors, semiconductors, and nonconductors (insulators) varies from very large numbers to very small numbers. Instead of working with long numbers and long decimal numbers, the usual practice is to use the significant number multiplied by a power of 10.

10^3 means $10 × 10 × 10 = 1000$
10^2 means $\quad\quad 10 × 10 = \quad 100$
10^1 means $\quad\quad\quad\quad\quad\quad\quad 10$
10^0 means $\quad\quad\quad\quad\quad\quad\quad\quad 1$

For decimals, a negative power of 10 is used (-10):

10^{-1} means $\quad\quad\quad\quad\quad\quad .1$
10^{-2} means $\quad\quad .1 × .1 = \quad .01$
10^{-3} means $.1 × .1 × .1 = .001$

The following table lists the approximate resistance of conductors, semiconductors, and nonconductors in ohms per centimeter of length:

For conductors, the resistance in ohms usually varies from:

$.000\ 001 = 10^{-6}$
$.000\ 01 = 10^{-5}$
$.000\ 1 = 10^{-4}$
$.001 = 10^{-3}$

For semiconductors, the resistance in ohms usually varies from:

$1 = 10^0$
$10 = 10^1$
$100 = 10^2$
$1000 = 10^3$
$10,000 = 10^4$
$100,000 = 10^5$
$1,000,000 = 10^6$

For nonconductors, the resistance in ohms usually varies from:

10^9 to 10^{18}

30-30 COLOR CODE FOR RESISTORS

The color bands on a resistor, Fig. 30-8, indicate its resistance value. The first band (A) is the first digit of the

COLOR CODE FOR RESISTORS

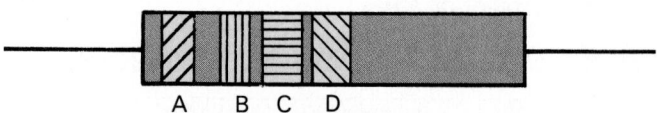

A — First Significant Figure
B — Second Significant Figure
C — Multiplier
D — Tolerance (%)

Color	Band A & B	Band C	Band D
Black	0	1	
Brown	1	10	
Red	2	100	
Orange	3	1,000	
Yellow	4	10,000	
Green	5	100,000	
Blue	6	1,000,000	
Violet	7		
Gray	8		
White	9		
Silver		.01	± 10
Gold		.1	± 5
No color			± 20

Fig. 30-8. Color coding indicating specific size of resistor.

resistance value. The second band (B) is the second digit of the resistance value. The third band (C) is the multiplier, and the fourth band (D) determines the tolerance.

The resistance of a resistor color-coded orange, blue, red, and gold would have a value of 36 × 100 = 3600 ohms. Since the fourth band is gold, the tolerance would be 5 percent. A tolerance of 5 percent means that the actual resistance can be between 3420 and 3780 ohms.

30-31 RESISTANCE IN SERIES

The total resistance of resistances in series is the sum of the separate resistances.

30-32 RESISTANCES IN PARALLEL

The total resistance of a parallel circuit is always less than any individual resistor or branch resistance. It can be computed in the following manner: Add the reciprocals of each resistance. The reciprocal of that sum is equal to the total resistance. Refer to Fig. 30-9.

30-33 ELECTRICAL CODE

The National (United States) Electrical Code and the Canadian Electrical Code each deal with the specifications having to do with refrigerating and air conditioning electrical circuits. In addition, most local communities have codes dealing with these circuits.

The refrigeration and air conditioning service technician should either have or have access to both the national codes and the local codes. In most cases, the refrigeration service technicians do not install electrical circuits. They should, however, be familiar with the codes to know whether the supply circuit is properly installed and safe to use.

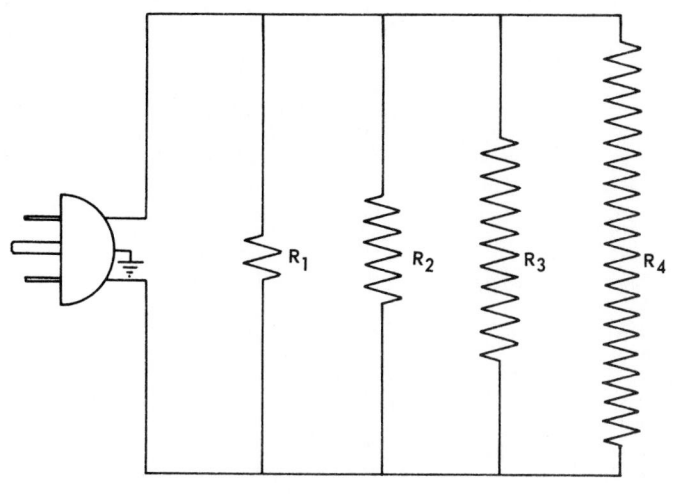

R_1 = 8 ohms
R_2 = 10 ohms
R_3 = 20 ohms
R_4 = 40 ohms
Total Resistance, R_T =

$$R_T = \cfrac{1}{\cfrac{1}{R_1} + \cfrac{1}{R_2} + \cfrac{1}{R_3} + \cfrac{1}{R_4}} =$$

$$\cfrac{1}{1/8 + 1/10 + 1/20 + 1/40} =$$

$$\cfrac{1}{5/40 + 4/40 + 2/40 + 1/40} = \frac{1}{12/40} = \frac{1}{.3} = 3.33 \text{ ohms}$$

R_T = 3.33 ohms

Fig. 30-9. Resistances in parallel.

30-34 AIR MOISTURE HOLDING PROPERTIES

Moisture holding capacity of air depends on its temperature. *The warmer the temperature, the greater the amount of moisture it will hold.* Fig. 30-10 shows a table of moisture holding properties or air. Fig. 30-11 shows the saturation pressure, the heat in the liquid, and the total heat after vaporization for mixtures of air and water vapor.

30-35 MOISTURE EVAPORATION SOURCES

OPERATION	POUNDS OF MOISTURE PER DAY
BATHING	.1 to 5
CLOTHES:	
Drying—average family—unvented	26.0
Washing—average family	4.0
COOKING:	
Breakfast	.9
Lunch	1.2
Dinner	2.7
DISHWASHING:	
Breakfast and Lunch	.2
Dinner	.7
HUMANS:	
Average	.4
MOPPING per 100 sq. ft.	3.0

30-36 DESICCANTS

Desiccants are used in driers installed in refrigerator liquid and/or suction lines to collect and remove moisture (water) from the system. Some common desiccants are:
Activated alumina.
Calcium sulfate.
Silica gel.
Alumina gel.
Molecular sieve.
Most driers will remove moisture, sediment, and acids from the circulating refrigerant.

Driers are usually placed in the liquid line close to the refrigerant control valve. They should always be placed in a cool place in the line, because heat may cause moisture to leave the desiccant and continue to circulate.

Liquid line driers are usually installed permanently and are only replaced when they lose their effectiveness.

Following a burnout, a large-capacity drier is usually placed in the suction line. Suction line driers should be removed as soon as the cleanup is completed.

Driers which have been used in a refrigerator system should not be reactivated by heating.

DEW POINT TEMPERATURE F°	POUNDS OF MOISTURE PER POUND OF DRY AIR	GRAINS OF MOISTURE PER POUND OF DRY AIR	PPM (WT.)
120	.08	570	80,000
110	.06	400	60,000
100	.044	300	42,000
90	.03	210	30,000
80	.022	150	22,000
70	.015	110	15,000
60	.011	76	11,000
50	.0075	53	7,500
40	.005	36	5,000
30	.0033	24	3,300
20	.002	15	2,000
10	.0014	9	1,400
0	.0008	5.5	800

Fig. 30-10. Moisture holding properties of air. (ASHRAE Handbook of Fundamentals)

Temp. F.	Sat. Press. in Hg.	ICE, WATER, WATER VAPOR Heat in Liquid Btu/lb.	Total Heat after Vapor-ization Btu/lb.	Vol. of Water Vapor cu.ft./lb.	DRY AIR Volume cu.ft./lb.	Specific Heat Btu/lb.	Amount of Water Vapor Saturate Grains
−40	3.790×10^{-3}	−177.1	1043.4	1.343×10^5	10.567	−9.61	.5508
−30	7.503×10^{-3}	−172.7	1047.8	7.441×10^4	10.820	−7.21	1.018
−20	1.259×10^{-2}	−168.2	1052.3	4.237×10^4	11.073	−4.81	1.830
−10	2.203×10^{-2}	−163.6	1056.7	2.475×10^4	11.326	−2.40	3.206
0	3.764×10^{-2}	−159.0	1061.1	1.481×10^4	11.579	0.00	5.480
10	6.286×10^{-2}	−154.2	1065.5	9060	11.832	2.40	9.161
20	.1027	−149.4	1069.9	5662	12.085	4.81	14.99
30	.1645	−144.4	1074.3	3608	12.338	7.21	24.07
32	.1803	−143.4	1075.2	3305	12.389	7.69	26.40
35	.2034	0.0	1076.5	2948	12.464	8.41	29.80
40	.2477	8.0	1078.7	2445	12.591	9.61	36.34
45	.3002	13.1	1080.9	2037	12.717	10.82	44.14
50	.3624	18.1	1083.1	1704	12.844	12.02	53.40
55	.4356	23.1	1085.2	1431	12.970	13.22	64.36
60	.5216	28.1	1087.4	1207	13.096	14.42	77.29
65	.6221	33.1	1089.6	1022	13.223	15.62	92.51
70	.7392	38.1	1091.8	868.0	13.349	16.83	110.4
72	.7911	40.1	1092.6	814.0	13.399	17.31	118.4
75	.8751	43.1	1093.9	740.0	13.475	18.03	131.3
80	1.0323	48.1	1096.1	633.0	13.602	19.23	155.8
85	1.2136	53.1	1098.3	543.3	13.738	20.43	184.4
90	1.4219	58.0	1100.4	467.9	13.854	21.64	217.6
95	1.6607	63.0	1102.6	404.2	13.981	22.84	256.4
100	1.9334	68.0	1104.7	350.2	14.107	24.04	301.5
120	3.4477	88.0	1113.3	203.2	14.612	28.85	569.0
140	5.8842	108.0	1121.7	123.0	15.117	33.67	1071.
160	9.6556	128.0	1129.9	77.27	15.622	38.48	2090.
180	15.295	148.0	1137.9	50.22	16.128	43.30	4598.
200	23.468	168.1	1145.8	33.64	16.632	48.12	16052.
212	29.921	180.1	1150.4	26.80	16.900	50.00	− − −

Fig. 30-11. Air-water vapor values. Properties of a mixture of air and water vapor for various temperatures from −40°F to 212°F (−40°C to 100°C) are shown. Note large volume occupied by 1 lb. of water vapor at the lower temperature.

30-37 GAS AND VAPOR

A true gas exists as a gas at standard temperatures and pressures.

Vapor is the gaseous state of a substance which is a liquid at standard atmospheric pressures and temperatures (such as water). However, vapor is also the term applied to refrigerants in the gaseous state inside the refrigerating system.

30-38 CHARACTERISTICS OF LITTLE-USED REFRIGERANTS

Characteristics of the most used refrigerants are described in Chapter 9.

Some refrigerants which were in common use at one time are listed in this paragraph. Their physical properties and use as refrigerants are explained in the ASHRAE Handbook of Fundamentals. These refrigerants are:

R-13	Monochlorotrifluoromethane	$CClF_3$
R-13 Bl	Monobromotrifluoromethane	$CBrF_3$
R-21	Dichloromonofluoromethane	$CHCl_2F$
R-30	Methylene Chloride	CH_2Cl_2
R-40	Methyl Chloride	CH_3Cl
R-113	Trichlorotrifluoroethane	CCl_2FCClF_2
R-114	Dichlorotetrafluoroethane	$CClF_2CClF_2$
R-160	Ethyl Chloride	C_2H_5Cl
R-290	Propane	C_3H_8

R-600	Butane	C_4H_{10}
R-717	Ammonia	NH_3
R-744	Carbon Dioxide	CO_2
R-764	Sulphur Dioxide	SO_2

30-39 METRIC PRESSURE-HEAT DIAGRAMS

With increased use of the SI (metric) system of units, metric pressure-heat diagrams for refrigerants are also being used more extensively.

The metric system pressure-heat diagram for R-12 is shown in Fig. 30-12. Fig. 30-13 shows the metric system pressure-heat diagram for R-22.

The standard refrigerating cycle of −15°C evaporating temperature and 30°C condensing temperature is shown on each diagram.

The US conventional unit pressure-heat diagrams for these two refrigerants are shown in Chapter 9.

30-40 MOISTURE IN LIQUID REFRIGERANTS

Moisture is soluble in most liquid refrigerants, and more of it may be safely present in some refrigerants than in others.

The amount of the allowable moisture varies with the temperature. Fig. 30-14 shows the maximum allowable moisture that may be present in six of the common refrigerants. In low-temperature applications, a very small amount of moisture may cause trouble.

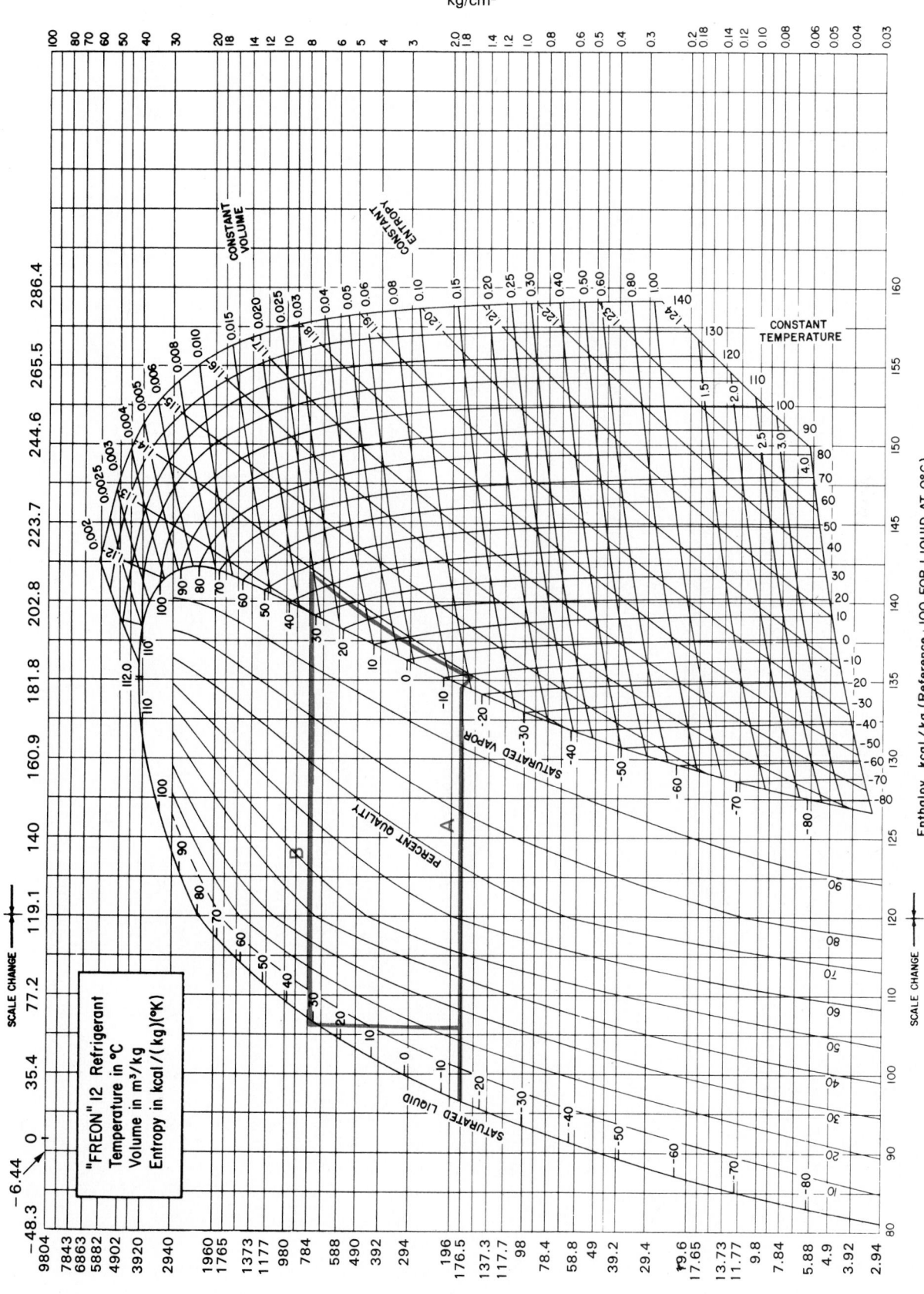

kg/cm²

Enthalpy, kJ/kg (Reference: O FOR LIQUID AT −40°C)

Enthalpy, kcal / kg (Reference : IOO FOR LIQUID AT 0°C)

"FREON" 12 Refrigerant
Temperature in °C
Volume in m³/kg
Entropy in kcal /(kg)(°K)

CONSTANT VOLUME

CONSTANT ENTROPY

CONSTANT TEMPERATURE

SATURATED VAPOR

PERCENT QUALITY

SATURATED LIQUID

SCALE CHANGE

PRESSURE kPa

Fig. 30-12. Pressure-heat diagram for R-12 expressed in metric units. The standard refrigerating cycle of evaporating temperature is shown at A and condensing temperature at B. (Adapted from Du Pont Company)

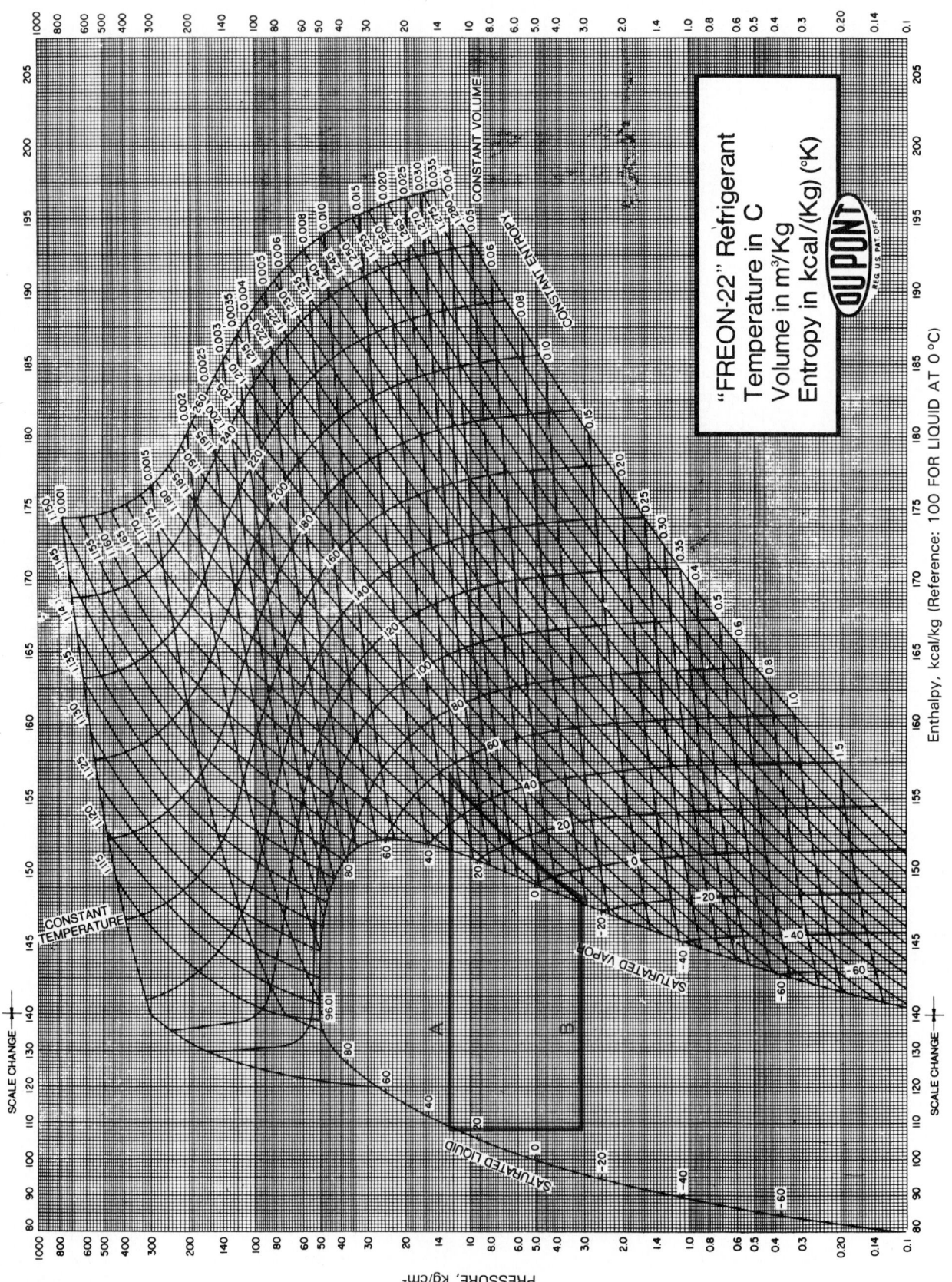

Fig. 30-13. Pressure-heat diagram for R-22 expressed in metric units. The standard refrigerating cycle of evaporating temperature is shown at A and condensing temperature is shown at B. (Du Pont Company)

Temp. °F	"Freon" 11	"Freon" 12	"Freon" 21	"Freon" 22	"Freon" 500	"Freon" 502
50	55	44	800	830	350	335
40	44	32	645	690	300	278
30	34	23.3	512	573	250	225
20	26	16.6	398	472	210	180
10	20	11.8	314	384	170	146
0	15	8.3	260	308	140	115
−10	11	5.7	205	244	110	90
−20	8	3.8	158	195	88	69
−30	6	2.5	118	152	67	53
−40	4	1.7	90	120	50	40
−50	3	1.1	66	91	40	30

SOLUBILITY OF WATER IN THE LIQUID PHASE OF THE "FREON" PRODUCTS (PARTS PER MILLION BY WEIGHT)

Fig. 30-14. Safe moisture content in six popular refrigerants.
(Du Pont Company)

30-41 DRYNESS OF REFRIGERANTS

Any gas holding water in the vapor state will, when cooled, reach a temperature at which free moisture (liquid water) will appear. The greater the concentration of water vapor in a gas, the higher the temperature at which water will condense out (free water).

This freeing of moisture is readily shown by frost on cold lines, beads of water on glasses of cold liquids, and, in the refrigeration system, by ice in the orifice of the expansion valve. Free water must be present in a refrigerating system before ice can be formed.

The highest temperature at which free water is liberated, on cooling, is called the dew point. There is only one dew point temperature for a given moisture concentration in a gas such as air or a refrigerant vapor. Knowing the dew point of a refrigerant determines the lowest temperature at which the refrigerant control will operate satisfactorily.

To prevent the formation of sludges and plating, the dew point should be 10 °F (6 °C) lower than the low-side refrigerant temperatures. To prevent refrigerant control freeze-up, the refrigerant should have a dew point of 5 °F (3 °C) lower than the low-side refrigerant temperature.

30-42 REFRIGERANT OILS

Requirements for a satisfactory refrigerant oil are rather severe. It must provide good lubricating qualities. It must be of the correct viscosity for the refrigerant and the machine in which it is being used. It must be free of moisture.

The viscosity of a refrigerant oil is measured using the Saybolt Universal Viscosity Test. A Saybolt Viscosimeter should be used. In this instrument, oil at 100 °F (38 °C) is allowed to flow through the standard Saybolt orifice. The time required in seconds for 60 cubic centimeters (cm³) of the oil to flow through this orifice is recorded.

The amount of moisture in refrigerant oil may be measured by its resistance to the flow of a current of electricity through it. This is known as its dielectric property.

Good refrigerant oil should have a minimum dielectric value of 25,000 V.

Another test which may be given to refrigerant oil is known as the floc test. This test applies to oils which are used with completely miscible (mixable) refrigerants, such as R-11, R-12, and R-22. The test is conducted by mixing 10 percent refrigerant with 90 percent oil and sealing it in a glass tube. Then it is cooled slowly until a flocculent (cloudy) precipitate of wax appears. The maximum temperature at which this occurs is recorded as the floc point.

Since the viscosity of oil changes with the temperature, oils at very low temperatures may not pour and may become a plastic solid. The temperature at which oil will just flow is called the pour point. This temperature is recorded either in Fahrenheit or Celsius as the pour point.

The flash point of an oil may be determined by slowly heating a quantity of the oil in an open container. Use a thermometer to determine the temperature of the oil. Bring a lighted candle or other flame to the surface of the oil as it is being heated. The temperature at which the vapors from the surface of the oil burn or flash is the flash point of the oil.

Refrigerant oils are sometimes given a very small amount of antifoam inhibitor to reduce foaming. Compressor parts are sometimes given a phosphating treatment to improve lubrication. Tricresyl phosphate has also been added to refrigerant oils to improve lubrication.

30-43 EUTECTIC

Proprietary compounds (made and sold by only one firm) are available that provide an active thermal energy source to maintain a specified temperature range. This is called thermal energy storage (TES). It is created by the heat of fusion or heat of crystallization.

These eutectic compounds can absorb or release heat with little or no change in temperature while they are changing from one physical state to another. The melting of ice is an example. In the process of cooling water, 1 Btu of heat is removed to lower the temperature of 1 pound of water 1 degree. This is true down to the freezing temperature. At the freezing temperature, 144 Btu per pound must be removed to freeze 1 pound of water. The temperature remains constant until the water is frozen.

A separate compound or mixture is needed for different temperature ranges. Fig. 30-15 is a table of some substances which may be used to maintain various temperatures.

Heat is absorbed or expelled by the eutectic as it changes its state. For low temperatures, the heat of fusion is used; for high temperatures, the heat of crystallization.

Eutectic compounds used in eutectic plates, as described in Chapter 12, contain other substances to prevent expansion on freezing. Freezing might otherwise rupture the containers. They also contain substances to reduce oxidation of the container materials. Useful life of the materials is thus extended.

30-44 CRYOGENIC TEMPERATURES

The cryogenic temperature range extends from −250.0 °F to Absolute 0 (−459.69 °F). The cryogenic ranges may also be expressed as follows:

−156.6 °C to −273.16 °C
116.5 K to 0 K
209.69 °F to 0 °R

F = Fahrenheit.
C = Celsius.
K = Kelvin.
R = Rankine.

Do not use low carbon steel for piping or pressure parts. The metal becomes very brittle when cold. Some nickel and/or some chromium improves the properties of steel at very low temperature.

30-45 BRINE FREEZING TEMPERATURES

Brines are water mixed with a substance which will go into solution with the water. The mixture will provide a fluid which can readily flow at temperatures below 32 °F (0 °C). See Fig. 30-16.

Brines are of several types: alcohol, salt, and glycol.
1. Alcohol brines are usually made of ethyl alcohol.
2. Salt brines are usually made of sodium chloride and/or calcium chloride. Eutectic point for sodium choride (common salt) solution is (−6 °F (−21 °C) and for calcium chloride, −60 °F (−51 °C).
3. Glycol brines with noncorrosive properties are usually made of glycerine, ethylene glycol, and/or propylene glycol.

A hydrometer can measure the density of brine solutions. The freezing temperature may be worked out from the hydrometer reading.

This is the same as measuring the freezing temperature of the cooling solution in an automobile radiator. If the hydrometer reading indicates that the freezing temperature of the solution is too high, it may be lowered by adding more of the antifreeze compound to the solution. Note that in the case of alcohol brine—since alcohol is lighter than water—the density decreases with the lowering of the

CHARACTERISTICS OF THERMAL ENERGY STORAGE COMPOUNDS			
TES MATERIAL	MELTING TEMPERATURE F.	MELTING TEMPERATURE C.	HEAT OF FUSION Btu/lb.
*TRANS TEMP 12	12	−11	115
N-DODECANE	10	−12	47
ETHYLENE GLYCOL	9	−13	63
*TRANS TEMP 27	27	−3	142
H₂O	32	0	144
*TRANS TEMP 40	40	4	75
N-TETRADECANE	42	6	98
*TRANS TEMP 65	65	18	80
N-HEXADECANE	62	17	102
ACETIC ACID	62	17	80
*TRANS TEMP 130	130	54	72
PARAFFIN WAX	130	54	63
TRISTEARIN	133	56	82
DRY ICE	−109	−78	241

*PROPRIETARY MATERIAL

Fig. 30-15. Characteristics of some eutectic thermal energy storage (TES) materials.

FREEZING TEMPERATURE °F AND °C							
BRINE SPECIFIC GRAVITY	20 (−7 °C)	10 (−12 °C)	0 (−18 °C)	−10 (−23 °C)	−20 (−29 °C)	−30 (−34 °C)	−40 (−40 °C)
ALCOHOL (FORMULA NO. 1) SPECIFIC GRAVITY AT 60 °F		.9691	.9592	.9486	.9345		
CALCIUM CHLORIDE SPECIFIC GRAVITY AT 60 °F	1.090	1.140	1.175	1.201	1.227	1.254	1.265
PERCENT OF CHEMICAL	10	17	20.5	23	25	27	28
SODIUM CHLORIDE USABLE ONLY DOWN TO 0 °F SPECIFIC GRAVITY AT 60 °F	1.072	1.118	1.158				
PERCENT OF CHEMICAL	10	16	21				
ETHYLENE GLYCOL SPECIFIC GRAVITY AT 60 °F	1.05	1.07	1.075	1.08	1.09	1.096	1.105
PERCENT OF CHEMICAL	32	40	43	45	50	53	57

Fig. 30-16. Table of properties of various brines.

freezing temperature. A hydrometer with an alcohol testing scale must be used.

30-46 TOTAL ENERGY—SINGLE ENERGY

In some buildings, only a single source of energy—such as natural gas—is supplied. All other needed energies are generated in the building from this gas. As an example, gas-burning engines may be used to drive electric generators. The waste heat from the cooling of the engines and the exhaust may be used either for heating water, space heating, or as heat for industrial processes. Also, the heat from lighting may be used in heating water or air.

30-47 SOLVENTS AND CLEANING

From use or repair operations, parts may become coated with lubricating oils, greases, oxides, dirt, metallic particles, or abrasives. Many methods may be used successfully to clean these parts. Lubricants or greases made from animal or vegetable oils or fats—such as tallow, lard oil, palm oil, olive oil—can usually be removed by saponification (making a soap of the oil or fat). Parts are soaked in an alkaline solution where the oils react with the alkali to form water soluble soap compounds.

Mineral oils which cannot be made into a soap solution—such as kerosene, machine oil, cylinder oil, and general lubricating oils—are usually cleaned by an emulsification process using soaps, wetting agents, and dispersing agents. (Emulsify means to make a liquid of solids by mixing them with a liquid. The solids do not dissolve but are suspended in the liquid and make it thick.)

Dirt, abrasives, metal dust, and inert materials are generally removed by one or both of these processes.

SOLVENT CLEANING is used for removing most of the oils from coated pieces. They are dipped (immersed) in a solvent such as mineral spirits. Solvent tanks should have safety lids and should be hooded and vented.

VAPOR CLEANING is also used to remove oils. Parts are held in a container where solvent vapors can condense on the parts. The condensed solvent washes away the oily coating, leaving surfaces dry and nearly clean.

Production degreasing machines have two or three compartments. The work is immersed in the first compartment containing a boiling solution of the solvent. It is then dipped into the second section, which contains clean cold solvent. Finally, it is hung in the third section, where only clean vapors condense on and wash over the work. The degreasing unit is selfpurifying. Oils and waste gather at the bottom of the third section.

Job shop cleaning uses the third section only. Solvents should be those approved by the Occupational Safety and Health Act (OSHA). *Venting is extremely important for safety to the operator.*

ALKALINE CLEANING baths are used primarily for the removal of oils, greases, solid particles of dirt, and metal particles, by immersing pieces in hot alkaline solutions. The chemicals saponify (make soap of) vegetable and animal oils and fats, emulsify mineral oils and greases, and suspend the solid material.

The combination of heat, active chemicals, and agitations (shaking) are important. Soap is either added or is formed by the saponification of vegetable or animal fats. Caustic soda, soda ash, and causticized soda are the cheapest and most direct methods of producing alkalinity in the bath. However, such materials, as a general rule, have less surface activity than more complex materials. Sodium metasilicate trisodium phosphate, and similar salts, are often used to obtain alkalinity in a solution.

A number of proprietary preparations are used for EMULSION CLEANING. They comprise an emulsification agent which disperses organic solvents in water solutions. Emulsifiable cleaners are miscible (mix) with oils and can be washed off with water; however a film of oil may remain on the work and make subsequent alkali cleaning necessary. Dragout costs are high.

Alkaline materials are used in ELECTROLYTIC CLEANING. The bath is maintained at as near boiling as possible without heavy tarnishing. The gas evolved tends to lift off the soil, making a clean surface for subsequent operations. The work to be cleaned is usually made the cathode. There are many formulas available for this work, but the one used depends upon the nature of the materials to be used and the degree of tarnish permissible. In many cases—particularly with carbon steel or cast iron—unusual results can be obtained by switching polarity several times during cleaning.

30-48 HEATING VALUE OF FUELS

Fuels commonly used for heating purposes are coal, oil, gas and wood. Burning these fuels in atmospheric air produces the heat needed. The heating value of coal is given in Fig. 30-17.

The American Society for Testing and Materials (ASTM) establishes the commercial grades of heating oil. Grade numbers are 1 through 6, although No. 3 and No. 4 are used very little. The grade of heating oil may be determined with a special heating oil hydrometer. Grade No. 1 has a gravity of between 45 and 38, as shown in Fig. 30-18. The heating value of oil is also shown in this figure.

FUEL	HEAT RELEASED — Btu Per Lb.
COAL:	
BITUMINOUS	12,000 TO 15,000
ANTHRACITE	13,000 TO 14,000

Fig. 30-17. Heating value of coal. As noted, the heating value of coal varies considerably depending upon the moisture, sulphur content, and other impurities.

COMMERCIAL GRADE NO.	Btu PER GALLON	GRAVITY RANGE DEG. ASTM	AVERAGE WEIGHT PER GALLONS IRS
1	137,000	45 — 38	6.8
2	140,000	40 — 30	7.1
3	140,000*		
4	141,000*	32 — 12	7.7
5	148,000	20 — 8	8.1
6	152,000	18 — 6	8.2
*NOT IN COMMON USE			

Fig. 30-18. Table of heating values of fuel oils. Numbers 5 and 6 are high viscosity oils and need preheating before use. ASTM means American Society for Testing and Materials.

FUEL	HEAT RELEASED – Btu PER CUBIC FOOT
GAS: NATURAL	1000 TO 1100*
MANUFACTURED	500 TO 600
LP (LIQUID PETROLEUM)	2500 TO 3200
*CHECK WITH THE LOCAL GAS COMPANY	

Fig. 30-19. Heating value of fuel gases.

	DECIMAL SIZES OF NUMBERED DRILLS						
No.	Size of Drill In Inches	No.	Size of Drill In Inches	No.	Size of Drill In Inches	No.	Size of Drill In Inches
1	.2280	21	.1590	41	.0960	61	.0390
2	.2210	22	.1570	42	.0935	62	.0380
3	.2130	23	.1540	43	.0890	63	.0370
4	.2090	24	.1520	44	.0860	64	.0360
5	.2055	25	.1495	45	.0820	65	.0350
6	.2040	26	.1470	46	.0810	66	.0330
7	.2010	27	.1440	47	.0785	67	.0320
8	.1990	28	.1405	48	.0760	68	.0310
9	.1960	29	.1360	49	.0730	69	.0292
10	.1935	30	.1285	50	.0700	70	.0280
11	.1910	31	.1200	51	.0670	71	.0260
12	.1890	32	.1160	52	.0635	72	.0250
13	.1850	33	.1130	53	.0595	73	.0240
14	.1820	34	.1110	54	.0550	74	.0225
15	.1800	35	.1100	55	.0520	75	.0210
16	.1770	36	.1065	56	.0465	76	.0200
17	.1730	37	.1040	57	.0430	77	.0180
18	.1695	38	.1015	58	.0420	78	.0160
19	.1660	39	.0995	59	.0410	79	.0145
20	.1610	40	.0980	60	.0400	80	.0135

Fig. 30-20. Numbered twist drill sizes.

The heating value of gas is indicted in Fig. 30-19. (The heating value of natural gas varies greatly, depending upon its composition.)

The heating value of wood is approximately 6200 Btu/lb.

30-49 TWIST DRILL SIZES

Sizes of twist drills may be given in fractions, decimals, numbers, or letters. Metric drill sizes are indicated in millimeters and decimals of a millimeter.

The service technician will use mostly fractional, numbered, and lettered drills. If it is necessary to do very much tapping and threading, drills will be used from each of these categories.

Fractional size drills start at 1/16 in. and increase by 1/64 in. through the required range. Decimal size drills are mostly used in manufacturing, hardly ever by the service technician. Number size drills are marked from 1 through 80. The range extends from .0135 in. up to less than 1/4 in. A table of numbered drills is shown in Fig. 30-20.

Letter size drills extend from A through Z, roughly from 1/4 in. through .413 in. Letter size twist drill sizes are shown in Fig. 30-21. Tap drill sizes are shown in Fig. 2-65.

30-50 COLOR CODE FOR PIPING

USE	COLOR
Fire protection equipment	Red
Safe material	Green (or, if needed, white, gray or aluminum)
Protective material	Bright blue
Extra valuable material	Deep purple
Dangerous material	Yellow or orange

Letter	Dia. In.	Letter	Dia. In.	Letter	Dia. In.
A	0.234	J	0.277	S	0.348
B	0.238	K	0.281	T	0.358
C	0.242	L	0.290	U	0.368
D	0.246	M	0.295	V	0.377
E	0.250	N	0.302	W	0.386
F	0.257	O	0.316	X	0.397
G	0.261	P	0.323	Y	0.404
H	0.266	Q	0.332	Z	0.413
I	0.272	R	0.339		

Fig. 30-21. Lettered twist drill sizes.

30-51 FOOD PRESERVATION BY RADIATION TREATMENT

Experiments have been conducted using atomic energy radiations to preserve food. Fresh foods have been put in sealed containers. These containers were exposed to a form of atomic energy radiation. It was found, upon inspection, that the food became sterile. It remained in its fresh state with no change in appearance, flavor, or food value as a result of being treated in this manner.

30-52 CLEAN ROOMS

In many laboratory situations, it is necessary to provide absolute control of the temperature, humidity, air cleanliness, and air chemistry. Such rooms are called clean rooms. Space industry is the primary user of clean room technology. Also some assembly operations must be conducted in clean rooms.

Dust particles too small to be seen may interfere with the operation of some very complicated mechanisms if allowed to enter the mechanisms during assembly. These devices are such that particles in the range of 20 to 50 microns will interfere with the operation. The use of clean rooms is increasing both in hospitals and industry.

30-53 SOLDERS AND BRAZING METALS

The following table lists some common solder alloys used in refrigeration work:

SOLDER	MELTING PT. °F	FLOW PT. °F	SHEAR STRENGTH psi
50-50 Tin-Lead	358	414	83.4
95-5 Tin-Antimony	450	465	327.0
Silver Solder 45 Ag., 15 Cu, 24 Cd, 16 Zn	1120	1145	8340
Phosphorous-Copper	1310	1650	8340

30-54 HEAT CONDUCTIVITY

HEAT CONDUCTIVITY OF MISCELLANEOUS SUBSTANCES		
MATERIAL	k*	R**
Air	.175	5.714
Concrete wall	8.00	.125
Glass	5.0	.2
Lead	243.0	.004
Vacuum, High	.004	250.

HEAT CONDUCTIVITY OF MISCELLANEOUS INSULATING MATERIALS			
MATERIAL	Density Lb./cu. ft.	k* Conductivity	R**
Cork, granulated	8.1	.34	2.941
Cork, granulated impregnated with pitch	17.79	0.428	2.336
Balsa	7.05	0.32	3.125
Felt	16.9	.25	4.0
Glass wool (curled pyrex)	4.0	0.29	3.448
Kapok	.87	.24	4.167
Mineral (slag) wool, loose packed	12.0	0.26	3.846
Polyurethane	1.5	0.16	6.25
Rock wool (fibrous rock, also felted)	6.0	0.26	3.846
Rubber, cellular	5.0	.37	2.703
Sawdust, pine	18.76	0.57	1.754
Straw fibers, pressed	8.67	0.32	3.125
Wood fibers (kingia australis)	8.4	0.33	3.03
Wool, pure	4.99	0.26	3.846

HEAT CONDUCTIVITY OF PROPRIETARY MATERIALS			
TRADE NAME	Density Lb./cu. ft.	k* Conductivity	R**
Armstrong's corkboard	7.3	0.285	3.509
Celotex	13.2	0.31	3.226
Dry-Zero	1.0	0.24	4.167
Nu Wood	15.0	0.32	3.125
United's 100 percent pure corkboard	9.0	0.27	3.704
U.S. mineral wool	12.0	0.26	3.846
Ferro Therm metal sheet (4 sheets)	4 oz./sq. ft./ sheet	0.226	4.425

Note: k* = Btu·in./hr./sq. ft./degree F.
       R** = Reciprocal of K.
       (American Society of Heating Refrigerating and Air Conditioning Engineers)

30-55 GALVANIC ACTION SEQUENCE

Certain materials produce electricity by chemical action. This is known as galvanic action. Some materials are much more active than others. Magnesium is one of the most active and platinum is one of the least active.

Galvanic action causes rapid deterioration at the place where galvanic action is occurring. To minimize galvanic action, joints where two unlike galvanic materials touch should be electrically insulated from one another.

Fig. 30-22 lists the galvanic action sequence for a series of common materials. The items at the top of the list are the most active and most likely to corrode. In galvanic action, these materials become an anode-positive. The items at the end of the list are least likely to corrode and in galvanic action these become a cathode-negative.

The corrosion of materials (usually metals) is often due to galvanic action.

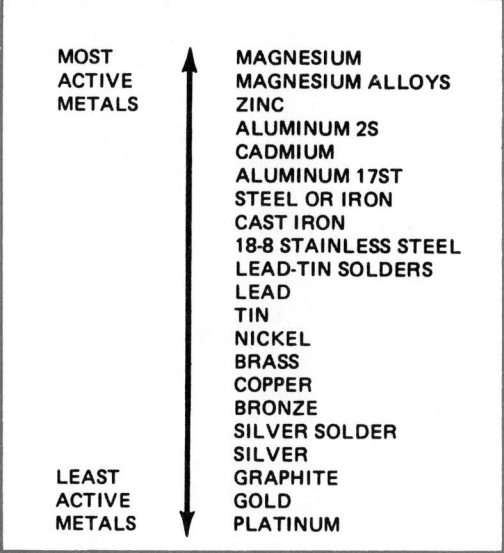

MOST ACTIVE METALS	MAGNESIUM
	MAGNESIUM ALLOYS
	ZINC
	ALUMINUM 2S
	CADMIUM
	ALUMINUM 17ST
	STEEL OR IRON
	CAST IRON
	18-8 STAINLESS STEEL
	LEAD-TIN SOLDERS
	LEAD
	TIN
	NICKEL
	BRASS
	COPPER
	BRONZE
	SILVER SOLDER
	SILVER
LEAST ACTIVE METALS	GRAPHITE
	GOLD
	PLATINUM

Fig. 30-22. Galvanic action sequence.

30-56 ELECTROLYSIS

Electrolysis is the movement of dc electricity through a substance which causes chemical change in the substance or its container. Electrolysis can be used for producing useful results such as storage battery operation and electroplating.

However, it can also cause severe damage. Whenever there are two electrical conductors and moisture, dc electricity will flow if the conductors are of different activity levels. As the dc electricity flow continues (electrons from − to +), one of the conductors will become coated with a new chemical and the other will have material removed (pits and holes). The pitted and "eaten away" areas are called corrosion areas. To avoid electrolytic corrosion, one must remove the moisture, seal off (insulate) the conductors, or use conductors of the same activity level.

30-57 STANDARD TEMPERATURE

The weight of a certain volume of air depends on its temperature. The standard temperature used for determining the weight of air is 32 °F (0 °C). This is the temperature of melting ice.

30-58 STANDARD PRESSURE

The weight of a certain volume of air depends on its pressure. The standard pressure for determining the weight of air is 29.92 in. Hg. or 760 mm Hg. This is the pressure of one atmosphere.

30-59 STANDARD TEMPERATURE AND PRESSURE

A sample of air at standard temperature and under standard pressure is known as a sample at STP conditions.

30-60 STANDARD AIR

Standard air and standard conditions are the same. Both conditions include pressure, temperature, and air density of 0.075 lb./cu. ft. (1.2007 kg/m³). These values are:

Refrigerant Number	Name	Chemical Symbol	Trade Name	Molecular Weight	Odor	Toxicity	Flammability	Pressure psia at 5°F	Pressure psia at 86°F	Latent Heat at 5°F	Sp. Heat of Liquid at 5°F	Critical Temperature of	Critical Pressure psia	Sp. Volume of Gas at 5°F	Density of Liquid in a 5°F #/cu. ft.	Cp/Cv Ratio	Sp. Heat of Vapor at 86°F
R-764	Sulphur Dioxide	SO_2		64.06	Pungent	High	Non	11.81	66.45	172.3	.34	314.8	1141.5	6.421	92	1.256 *	.34
R-40	Methyl Chloride	CH_3Cl		50.489	Sweet	Med.	Slight	20.89	95.53	180.6	.45	289.6	969.2	4.530	61	1.20	.4
R-717	Ammonia	NH_3		17.031	Pungent	High	Slight	34.27	169.2	565	1.10	271.4	1651	8.150	41.11	1.247	1.10
R-160	Ethyl Chloride	C_2H_5Cl	Alcozol	64.51	Etheral	Med.	Yes	4.65	27.10	177	.47	369	764	17.55	59.00	1.13	.42
R-12		CCl_2F_2		120.9	Sweet	Low	Non	26.61	107.9	68.2	.215	232.6	582.0	1.485	90	. . .	.243
R-13		$CClF_3$		104.46	Sweet	Low	Non	. . .	. . .	63.85 (-115°F)	.247 (-22°F)	84	561	.431 (-115°F)	77 (0°F)	1.172	. . .
R-744	Carbon Dioxide	CO_2		44.005	Non	Low	Non	334.4	1039.0	116	.5	87.8	1066.2	.2673	61.22 (0°F)	1.30 **	1.95
R-611	Methyl Formate	$C_2H_3O_2$	Carrene No. 1	60.04	Slight	. . .	Slight	1.96	13.69	236	.515	417	870	46.7	. . .	. . .	.515
R-30	Methylene Chloride	CH_2Cl_2		84.9	Sweet	. . .	Yes	1.17	10.6	162.1	.34	421	670	50.58	. . .	. . .	.34
R-21		$CHCl_2F$	Thermon	102.92	Sweet	Low	Non	5.5	30.5	105.5	.26	353.3	750.0	8.83	90.1	. . .	.26
R-22		$CHClF_2$		86.48	Sweet	Low	Non	43.02	174.5	93.43	11.97	204.8	716	1.246	83.34	. . .	.34
R-11		CCl_3F	Carrene No. 2	137.38	Sweet	Low	Non	29.31	18.28	84.0	.197	388.4	635.0	12.3	97.8	1.088	.20
R-114		$CClF_2CClF_2$		170.93	Sweet	Low	Non	.98	7.86	58.9	.238	294	474	.488	73.1	. . .	.160
R-113		$C_2Cl_3F_3$		187.4	Sweet	Low	Non	. . .	. . .	70.62	.199	417.4	495	27.04	103	. . .	.26
R-500		$CCl_2F_2 + CH_3CHF_2$		-99.29	Sweet	Low	Non	31.07	128.14	85.03	11.83	221.1	631	1.5227	80.10	4.61	. . .
R-502		$CHClF_2 + CClF_2CF_3$		111.64	Sweet	Low	Non	50.68	175.1	68.86	.027	194.1	618.7	.825	87.24	4.37	.07
R-290	Propane	C_3H_8		44.06	Sweet	Low	Yes	41.9	155	170.2	.56	302	661.5	2.48	34.33	. . .	.55
R-170	Ethane	C_2H_6		30.04	Sweet	Low	Yes	236.0	675.0	150.5	.66	90.1	730	.533	26.96	. . .	.83
R-600	Butane	C_4H_{10}		58.12	Sweet	Low	Yes	8.2	41.6	170.7	.51	308	529	9.98	38.41	. . .	.51
R-13B1	Kulene 131	CF_3Br			Etheral	Low	Non	77.93	261.8	44.88	.182	1535	587	.3854	112	. . .	.10
R-115	Monochloropenta-fluoroethane	$CClF_2CF_3$		154.48	Sweet	Low	Non	38	148.9	. . .	. . .	175.9	. . .	.82	. . .	3.76	. . .

* at 70°F   ** at 32°F   # at 70°F

Fig. 30-23. Table of properties of refrigerants.

US CONVENTIONAL SYSTEM	METRIC SYSTEM
Pressure = 29.92 in. Hg =	760 mm Hg
14.696 psia = 0 psi	101.28 kPa
Temperature = 69.8 °F	21 °C
Specific Volume = 13.33 cu.ft./lb.	0.833 m³/kg

30-61 REFRIGERANT PROPERTIES

Several refrigerant tables and pressure-heat enthalpy diagrams are given in this section. Most of the physical properties of refrigerants may be found in these tables and charts.

Chapter 9 describes the characteristics and uses of the more popular refrigerants in more detail.

Fig. 30-23 is a table showing the comparable properties of many of the refrigerants.

30-62 FLARE NUT WRENCH SIZES

A table of flare nut wrench sizes is shown in Fig. 30-24. Old and new wrench sizes across the flats of the flare nut are given. Tube size is based on outside diameter.

TUBE SIZE O.D.	WRENCH SIZE ACROSS FLATS	
	OLD	NEW
1/4 IN.	3/4 IN.	5/8 IN.
3/8 IN.	7/8 IN.	13/16 IN.
1/2 IN.	1 IN.	15/16 IN.

Fig. 30-24. A table of flare nut wrench sizes.

30-63 BOYLE'S LAW

Boyle's Law expresses a very interesting relationship between the pressure and volume of a gas. It is stated as follows:

"The volume of a gas varies inversely as the pressure, provided the temperature remains constant."

This means that if a quantity of gas has its pressure doubled, the volume becomes half of what it originally was. Or, if the volume is doubled, the gas has its pressure reduced by half. If a perfect gas is considered, Boyle's Law may be expressed as a formula:

Pressure × Volume = A constant number.

This being true, one can say that when either the pressure or the volume is changed, the corresponding pressure or volume is changed in the opposite direction. Therefore:

Old Pressure × Old Volume = the New Pressure × the New Volume

Expressed in letter form:

$$P_o \times V_o = P_n \times V_n$$

- P_o = Old Absolute Pressure
- P_n = New Absolute Pressure
- V_o = Old Volume
- V_n = New Volume

Note: This formula will be true only if the pressures are expressed as absolute pressures (psia).

Example: What will be the new volume of 5 cubic feet (cu. ft.) of gas at 20 psi if it is compressed to 60 psi, providing the temperature remains constant?

For calculation purposes, atmospheric pressure = 15 psi

$$P_o \times V_o = P_n \times V_n$$

- P_o = 20 psi (atmospheric pressure)
  = (20 + 15) = 35 psia (absolute pressure)
- P_n = 60 psi + atmospheric pressure =
  (60 + 15) = 75 psia

$P_o \times V_o = P_n \times V_n$, substituting the preceding values in this formula

$$35 \times 5 = 75 \times V_n \text{ or}$$

$$\frac{35 \times 5}{75} = V_n$$

$$V_n = \frac{35}{15}$$

V_n = 2.33 cu. ft. at 60 psi

In the metric system, volume is meters cubed (m³).

Pressure is in newtons per meter squared = N/m² = 1 pascal = 1 Pa.

Atmospheric pressure = 101 kPa

One bar = atmospheric pressure

1 bar = 100,000 Pa = 10⁵ Pa

Be careful to use the same volume units and pressure units in the same problem.

30-64 CHARLES'S LAW

Gases behave consistently with temperature changes. This is stated in Charles's Law:

"At a constant pressure the volume of a confined gas varies directly as the absolute temperature; and, at a constant volume, the pressure varies directly as the absolute temperature."

Absolute pressures and absolute temperatures must always be used in the equations. In the equation form:

AT CONSTANT VOLUME

The Old Absolute Pressure (P_o) × the New Absolute Temperature (T_n) = the New Absolute Pressure (P_n) × the Old Absolute Temperature (T_o).

$$P_o \times T_n = P_n \times T_o$$

Example: What is the pressure of a quantity of confined gas when raised to 60 °F (16 °C) if its original pressure was 35 pounds per square inch gauge (psig) and temperature 40 °F (4 °C), at a constant volume?

Solution: (Note that absolute temperatures and pressure must be used.)

$$P_o \times T_n = P_n \times T_o$$

- P_o = (35 + 15) = 50
- T_n = (60 + 460) = 520
- T_o = (40 + 460) = 500

$$P_n = \frac{P_o \times T_n}{T_o}$$

$$P_n = \frac{50 \times 520}{500} = 52 \text{ psia}$$

P_n = 52 − 15 = 37 psi

AT CONSTANT PRESSURE

The Old Volume (V_o) × the New Absolute Temperature (T_n) = the New Volume (V_n) × the Old Absolute Temperature (T_o).

$$V_o \times T_n = V_n \times T_o$$

Example: 5 cubic feet of gas at 37 °F is raised to 90 °F at constant pressure. What is the new volume?

Solution:

$$Vo \times Tn = Vn \times To$$
$$5 \times (90 + 460) = Vn \times (37 + 460)$$
$$5 \times 550 = Vn \times 497$$
$$\frac{5 \times 550}{497} = \frac{2750}{497} = Vn$$
$$Vn = 5.53 \text{ cubic feet}$$

In the metric system:
1. The volume will be in meters cubed (m^3).
2. The pressures will be in newtons per square meter (N/m^2) which is called pascal (Pa).
   $$1 \ N/m^2 = 1 \ Pa$$

30-65 GAS LAW

Boyle's Law and Charles's Law may be combined to solve true gas problems. The formula is:

$$\frac{Po \times Vo}{To} = \frac{Pn \times Vn}{Tn}$$

To = Old Absolute temperature
Tn = New Absolute temperature
Po, Vo, and To represent original conditions
Pn, Vn, and Tn represent new conditions

Absolute values for temperature and pressure must be used in this equation.

Example: If 2 cu. ft. of gas at 15 psi and 140 °F is stored in a 4 cu. ft. container at 30 psi, what is the new temperature?

Formula:

$$\frac{Po \times Vo}{To} = \frac{Pn \times Vn}{Tn}$$

Po = (15 + 15) = 30
Pn = (30 + 15) = 45
Vo = 2
Vn = 4
To = (140 + 460) = 600

$$Tn = \frac{Pn \times Vn \times To}{Po \times Vo}$$

$$Tn = \frac{45 \times 4 \times 600}{30 \times 2} = \frac{108,000}{60} = 1800 \ R. \ (F_A.)$$
$$= 1800 - 460 = 1340 \,°F$$

In the metric system the same formula is used but the units are different:
1. The volume is in cubic meters (m^3).
2. The pressure is in newtons per square meter — (N/m^2) or pascal = (Pa).

30-66 ADIABATIC EXPANSION AND CONTRACTION OF A GAS

When the term adiabatic is used, it refers to a natural process in a gas. The adiabatic process or property allows the gas to be expanded or contracted (compressed) without absorbing heat from outside or without transferring heat out of the gas.

Adiabatic expansion and compression of gases would occur if the gas were placed in a perfectly insulated cylinder with a frictionless piston so that heat could not enter the gas during expansion or escape during compression.

During adiabatic compression and expansion of ideal gases, the work performed (compression and expansion) is obtained FROM THE GAS. When work is done on a gas

as it is adiabatically compressed, the heat generated is not lost, but increases the temperature and therefore the pressure of the confined gas. During expansion, the pressure and temperature of the ideal gas decreases.

30-67 ISOTHERMAL EXPANSION AND CONTRACTION OF A GAS

An isothermal condition exists when the expansion or contraction of a gas occurs without a change in temperature (Boyle's Law, Para. 30-63). This condition can occur either during the expansion or the compression of a gas.

During expansion, the gas is cooled and heat needed to keep the gas at a constant temperature must come from an outside source. The heat so obtained must be exactly equal to that given up by the gas during its expansion. This is necessary to keep the temperature constant.

A similar condition must exist during the isothermal compression of a gas. Heat must be removed in an amount equal to the heat energy of compression in order to maintain a constant temperature. Regardless of whether the compressor is air-cooled or water-cooled, the heat removed must be equal to the heat input from the work of compressing the gas.

Gas compression and expansion operations cannot be performed without a change in the temperature of the confined gas. If the action results in raising the temperature of the confined gas, it is called EXOTHERMAL. If the action results in lowering the temperature of the confined gas, it is called ENDOTHERMAL.

30-68 ROOT MEAN SQUARED (rms)

The root mean squared (rms) values of voltage and current are used to calculate the power in a circuit when the current and voltage are not in phase. In this calculation a mean voltage and mean current are used.

To find the root mean squared value of a quantity which varies over a cycle, the following steps must be done:
1. Find the instantaneous values of the current and voltage and SQUARE them.
2. Find the MEAN over one cycle of the SQUARE of the instantaneous values.
3. Find the square ROOT of this value.

This gives the ROOT MEAN SQUARED value.

The calculation can be done two ways: with two currents in phase and with the currents not in phase.
1. If the currents are in phase and vary as a sine curve, as shown in Fig. 30-25, then the rms current (I_{rms}) is:

$$I_{rms} = I_{max} \times (\frac{\sqrt{2}}{2}) = I_{max} \times 0.707.$$

If the voltage varies as a sine curve, then the rms voltage is:

$$V_{rms} = V_{max} \times 0.707.$$

The average power is then:

$$P = I_{rms} \times V_{rms}$$
$$= (I_{max} \times 0.707) \times (V_{max} \times 0.707)$$
$$P = 0.5 \times I_{max} V_{max}.$$

2. If the currents are not in phase, but vary as sine curves, much of the previous calculation is still useful.

$$I_{rms} = I_{max} \times 0.707$$
$$V_{rms} = V_{max} \times 0.707$$

The average power is:

$$P = I_{rms} \times V_{rms} \times \cos \theta$$

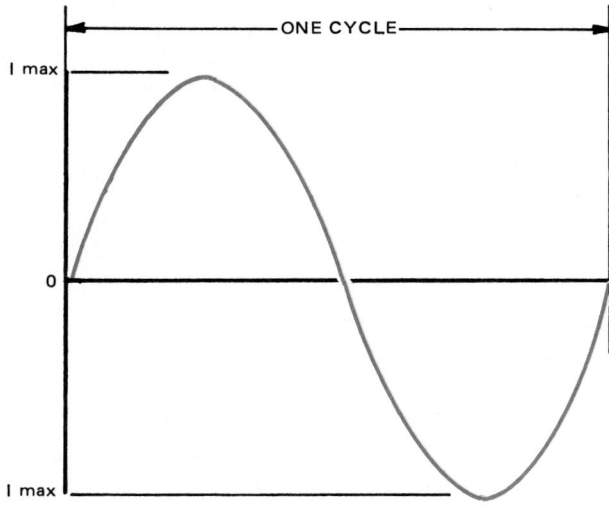

Fig. 30-25. Sine curve represents the current in a circuit.

The term "cos θ" is the cosine of the phase angle between the current and the voltage. The phase angle, θ, can be determined with an oscilloscope.

The dual-trace oscilloscope or oscillograph is best to use, but the phase angle can be determined on a single-trace oscilloscope. With a dual-trace oscilloscope, just note the amount of shift from right to left of the two traces. The shift can be from 0° to 90° (ignore the negative shifts like "–70°" for now). Two wave traces on top of each other give a 0° shift, two waves with a zero of one above a maximum of the other give 90°, and other amounts are in between as counted on the grid lines of the oscilloscope.

30-69 ENVIRONMENTAL EFFECTS OF CFCs AND HCFCs

Scientists have established that fully halogenated chlorofluorocarbons (CFCs), and hydrochlorofluorocarbons (HCFCs), have affected the ozone layer. This will cause changes in the types of refrigerants used, in recovery/recycling/reclaiming procedures for specific refrigerants, and in numerous related laws. See Chapters 9 and 28 regarding EPA (Environmental Protection Agency) rulings.

30-70 PSYCHROMETRIC CHART FORMULA

A psychrometric chart does not give the relative humidity for dry bulb temperatures below about 20 °F (-7 °C). A formula can be used. A formula is also useful if a chart is unavailable. Use the following formula:

$$RH = 100e^{(\frac{T_W - T_D}{28.116})} - 17.4935(T_D - T_W)e^{-\frac{T_D}{28.116}}$$

RH is the relative humidity in percent. T_D is the dry bulb temperature in °F. T_W is the wet bulb temperature in °F. The symbol e refers to the exponential function. The exponential function is taken to the power represented by the number above the symbol e. (The function e is the usual one used where 2.71828 is the value of e.)

The formula for grains of water per pound of dry air is as follows:

$$\frac{GRAIN}{LB.} = 0.09065RHe^{\frac{T_D}{28.116}}$$

30-71 SEASONAL ENERGY EFFICIENCY RATING (SEER)

The EER (energy efficiency rating) and the COP (coefficient of performance) are being combined into the SEER (seasonal energy efficiency rating). See Para. 15-63 and Para. 21-17. An average is obtained with the following formula:

$$SEER = 1.2 \frac{winter\ days}{summer\ days + winter\ days} EER$$
$$+ 0.8\frac{summer\ days}{summer\ days + winter\ days} (COP + 1) \cdot$$
$$3.414368$$

The EER is found in the usual ways as follows:

$$EER = \frac{Btu/hr.\ of\ heating}{electrical\ power\ in\ watts}$$

There is a better formula which uses all numbers given in Btu/hr. Use the following formula:

$$SEER_Q = 1.2 \frac{WD}{SD + WD} EER_Q$$
$$+ 0.8 \frac{SD}{SD + WD} (COP + 1)$$

The EER_Q is found as follows:

$$EER_Q = \frac{Btu/hr.\ of\ heating}{electrical\ power\ in\ watts} 0.293$$

For all formulas, the COP is as follows:

$$COP = \frac{Btu/hr.\ of\ cooling}{electrical\ power\ in\ watts} 0.293$$

30-72 SAFETY

A safety code for mechanical refrigeration is sponsored by the American Society of Heating, Refrigeration and Air Conditioning Engineers (ASHRAE). This code has been developed according to the rules and regulations of the American Standards Association. Copies of this safety code are available from the American Society of Heating, Refrigeration and Air Conditioning Engineers.

The Occupational Safety and Health Act provides many regulations and controls for the installation, operation, and service of refrigerating and air conditioning mechanisms.

The American Conference of Governmental Industrial Hygienists have published Threshold Limit Values for substances that may be toxic under certain conditions and length of exposure. The threshold values for some of these substances for an eight-hour exposure time are:

SUBSTANCE	ppm
Acetone	1000
Ammonia	50
Carbon dioxide	5000
Carbon monoxide	100
Carbon tetrachloride-Skin	10
Chlorine	1
Dichlorodifluoromethane	1000
Dichloromonofluoromethane	1000
Ethyl ether	400
Fluorine	0.1
Gasoline	500
Methyl acetylene	1000
Methyl alcohol (Methanol)	200
Methyl chloride	100
Ozone	0.1
Sulfur dioxide	5
Xylene (xylol)	200

Two service technicians repairing a rooftop air conditioning unit.   (Superior Contract Service, Inc.)

Chapter 31

CAREER OPPORTUNITIES IN REFRIGERATION AND AIR CONDITIONING

Few phases of modern living are untouched by modern refrigeration and air conditioning. Business, commercial operations, manufacturing processes, storage, and shipping are nearly all now carried on under controlled temperature conditions.

Skilled specialists are required to design, install, and maintain controlled environments in enclosed areas, from homes to space satellites. The use of computerized equipment has increased the need for total energy controlled facilities.

IMPORTANCE OF REFRIGERATION AND AIR CONDITIONING

The refrigeration and air conditioning industry helps make possible this system of living. Many fruits and vegetables are refrigerated immediately upon being harvested. The quality of such products is much better for this fact. Air conditioning has improved business and industrial efficiency, while adding to human comfort. More and more factories and heavy industries are being air conditioned. The present scale of farming is made possible, to a great extent, by the use of air conditioned tractor cabs and refrigerated harvesting equipment.

Cooling and freezing of meat and meat products makes possible their handling in a much more sanitary way than would be possible without mechanical refrigeration. Beverages, desserts, as well as staple foods, are all at least partially processed by refrigeration equipment.

THE CHALLENGE

Designing, manufacturing, selling, installing, and maintaining this equipment provides for many, many jobs that did not exist less than a generation ago. Since refrigeration is used in so many enterprises, it follows that anyone who has to work in these industries has to be familiar with the basic air conditioning and refrigeration processes.

Opportunities for employment in writing specifications for refrigeration and air conditioning equipment and selling this equipment have naturally grown with the industry.

All the careers are available to anyone interested, regardless of race, creed, color, or sex.

THE AIR CONDITIONING AND REFRIGERATION INDUSTRY

The air conditioning and refrigeration industry is usually divided into three industries:

1. Domestic.
2. Commercial.
3. Industrial.

The domestic field covers home refrigerators, freezers, and window air conditioners.

The commercial field includes all small automatic systems such as for stores, supermarkets, domestic central air conditioning, water coolers, beverage coolers, marine refrigeration and air conditioning, automotive air conditioning, and truck refrigeration and air conditioning systems.

The industrial field includes the large processing systems and air conditioning systems, packing plants, cold storage, and ice rinks. These systems require the attention of a refrigeration operating engineer.

CAREERS

Some of the employment opportunities include:
1. Jobs at various levels:
   a. Senior skilled.
   b. Skilled.
   c. Technicians.
   d. Supervisors.
   e. Professional personnel.
2. Various specialties (partial list):
   a. Engineers.
   b. Technicians.
   c. Test technicians.
   d. Sales engineers.
   e. Application engineers.
   f. Installers.
   g. Testers.
   h. Maintenance technicians.
   i. Service persons.
   j. Repair specialists.
      Helpers.
   k. Wholesalers.
      Sales engineers.
      Sales persons.
      Counter persons.
      Parts persons.
      Shipping and receiving persons.
   l. Operating engineers.
      Refrigeration.
      Industrial.
   m. Sheet metal experts.
      Helpers.

THE WORK THEY DO

As one might expect, the responsibilities of the person working in the air conditioning and heating industry vary greatly. So does the kind of work that person will do.

Consider the air conditioning, heating, and refrigeration technicians, for example. Working under the supervision of engineers, they help design, manufacture, sell, and service equipment. But often they specialize in one area, like research and development.

Those working in manufacturing of equipment may design and test or supervise production. They may also work as manufacturer's representatives or field sales persons. In such cases, responsibilities would include supplying contractors and engineering firms with data on installation, maintenance, operating costs, and performance specifications of equipment.

Some technicians are employed by contractors to help design and prepare installation instructions. Others work in customer relations and may be responsible for supervision of installation and maintenance of equipment.

Another group employed by the industry works on installation and service. They travel about in service trucks, servicing units in homes, offices, schools, and other buildings. This group includes:

1. Air conditioning and refrigeration technicians who install and repair units ranging in size from small window air conditioners up to large central systems. They must follow blueprints and specifications to install motors, compressors, evaporators, and other components. They will connect ducts, refrigerant lines, and piping as well as make power hookups. In event of breakdown, they find the cause and make repairs.
2. Furnace installers or heating equipment installers. They read blueprints and specifications and install oil, gas, and electric heating. Installation work includes placement of fuel supply lines, air ducts, pumps, and other parts of a heating system. After connecting the electrical wiring and controls, they check units for proper operation.
3. Oil heating system technicians who keep oil-fueled

A full mechanical air conditioning center being installed. A sales engineer was required to select and specify the proper equipment. Skilled workers install it. A service technician will be required to service and maintain the system.   (Tyler Refrigeration Corp.)

heating systems in good working order. Their work varies with the season. In fall and winter they service and adjust burners. During the summer they service the heating unit, replace oil and air filters, vacuum vents, ducts, and other parts of the system.
4. Gas heating system technicians or gas appliance service persons. Their duties are similar to those of the oil burner technician. They determine why a burner will not work and adjust or repair it.

Cooling and heating systems are sometimes installed or repaired by other types of technicians or tradespeople. For example, ductwork on a large heating or air conditioning job may be done by sheet metalworkers; electrical work by an electrician; and piping by pipe fitters. This is often the case on large installations where unions might be involved.

Installation and servicing of home air conditioners is one of the many careers open to a person with knowledge and hands-on training in refrigeration and air conditioning.
(Lennox International, Inc.

EDUCATION REQUIREMENTS

To qualify for employment, one should have good communication skills, a knowledge of practical mathematics, and some physics and chemistry.

One should also take at least a one-year training program in the theory and hands-on laboratory work on refrigeration, heating, and air conditioning systems.

Additional employment information is available from the nearest branch of the United States Employment Service and the local State Employment Service. Local schools or public libraries also offer reference materials such as the DICTIONARY OF OCCUPATIONAL TITLES and the OCCUPATIONAL OUTLOOK HANDBOOK.

EMPLOYMENT OPPORTUNITIES

EDUCATIONAL REQUIREMENTS	EDUCA-TION	CODE	JOB TITLES	AREAS OF EMPLOYMENT	
A GRADE 12 GRADUATE	D	1	• Outside Sales Engineer	**CONSULTING** 5, 7, 9, 10, 11, 14, 16	**CONSULTING** Companies devoted to designing and engineering the heating, cooling, ventilating, plumbing and electrical systems for buildings. Includes preparation of drawings, specifications, estimate of cost, supervision of installation and final approval.
	AC	2	• Outside Sales Tech. or Rep.		
	ACD	3	• Inside Sales Representative		
	AC	4	• Inside Sales Order Desk/Counter		
	ABCD	5	• Estimator		
	CD	6	• Sales Manager	**CONTRACTING** 5, 7, 8, 9, 10, 11, 12, 15, 17, 19, 20	**CONTRACTING** Companies that sell and install mechanical systems. Includes installation and fabrication of system components, preparation of drawings, estimation of costs, and supervision of installation according to specifications.
	CD	7	• Administration		
B APPRENTICESHIP VOCATIONAL/ TECHNICAL SCHOOL	CD	8	• Application Engineer/Tech.		
	AC	9	• Draftsman - Layout		
	D	10	• Consulting/Design Engineer	**SERVICING** 1, 2, 3, 4, 5, 6, 7, 8, 15, 17, 18, 19, 20, 21, 24	**SERVICING** Companies that repair and maintain mechanical systems. Includes repair and maintaining of system components, sale and installation of replacement components for economic and efficient system operation.
	CD	11	• Project Manager		
	ACD	12	• Purchasing		
	ABC	13	• Production		
	BCD	14	• Inspector		
C COMMUNITY COLLEGE	BC	15	• Job Foreman or Supervisor	**MANUFACTURING** 1, 2, 3, 4, 5, 6, 7, 8, 9, 10, 12, 13, 14, 17, 18, 19, 20, 21, 22, 23, 24, 25	**MANUFACTURING** Companies that purchase raw material or components and fabricate or assemble into equipment for sale. Includes sales and marketing, production, design and research development.
	ABCD	16	• Specification Writer		
	BC	17	• Field Service & Installation		
	ABC	18	• Shop Service and Repair		
	B	19	• Journeyman-Mechanic	**MERCHANDISING & SALES** 1, 2, 3, 4, 5, 6, 7, 8, 12, 25	**MERCHANDISING AND SALES** Companies which promote and sell equipment and components which have been manufactured by others. Includes sales promotion, advertising, warehousing and technical assistance.
D UNIVERSITY	AB	20	• Apprentice Mechanic		
	ABCD	21	• Service Manager		
	C	22	• Lab. Technician/Technologist		
	CD	23	• Research & Development		
	ABC	24	• Service Salesman	**GOVERNMENT & UTILITIES** 5, 7, 8, 9, 10, 11, 12, 14, 16, 19, 20, 22	**GOVERNMENT AND UTILITIES** Governments, utilities and other agencies provide their own consulting and servicing functions, set standards, test equipment and approve installations.
	AC	25	• Inventory Control		

AMERICAN SOCIETY OF HEATING, REFRIGERATING AND AIR-CONDITIONING ENGINEERS, INC.

Chapter 32
DICTIONARY OF TECHNICAL TERMS

A

ABSOLUTE HUMIDITY: Amount of moisture in the air, indicated in grains per cu. ft.

ABSOLUTE PRESSURE: Gauge pressure plus atmospheric pressure (14.7 lb. per sq. in.), equals absolute pressure.

ABSOLUTE TEMPERATURE: Temperature measured from absolute zero.

ABSOLUTE ZERO TEMPERATURE: Temperature at which all molecular motion ceases (−460 °F and −273 °C).

ABSORBENT: Substance with the ability to take up or absorb another substance.

ABSORPTION REFRIGERATOR: Refrigerator which creates low temperatures by using the cooling effect formed when a refrigerant is absorbed by chemical substance.

ACCESSIBLE HERMETIC: Assembly of motor and compressor inside a single bolted housing unit.

ACCUMULATOR: Storage tank which receives liquid refrigerant from evaporator and prevents it from flowing into suction line before vaporizing.

ACID CONDITION IN SYSTEM: Condition in which refrigerant or oil in system is mixed with fluids that are acid in nature.

ACR TUBING: Tubing used in air conditioning and refrigeration. Ends are sealed to keep tubing clean and dry.

ACTIVATED ALUMINA: Chemical which is a form of aluminum oxide. It is used as a drier or desiccant.

ACTIVATED CARBON: Specially processed carbon used as a filter-drier; commonly used to clean air.

ACTUATOR: That portion of a regulating valve which converts mechanical fluid, thermal energy, or electrical energy into mechanical motion to open or close the valve seats.

ADIABATIC COMPRESSION: Compressing refrigerant gas without removing or adding heat.

ADSORBENT: Substance with the property to hold molecules of fluids without causing a chemical or physical change.

AERATION: Act of combining substance with air.

AGITATOR: Device used to cause motion in confined fluid.

AIR, STANDARD: Air having a temperature of 68 °F (20 °C), a relative humidity of 36 percent, and pressure of 14.7 psia (101.3 kPa). The gas industry usually considers 60 °F (16 °C) as the temperature of standard air.

AIR BREAK: An inverted opening placed in the chimney of a gas furnace to prevent back pressure from outside wind from reaching the furnace flame or pilot.

AIR CLEANER: Device used for removal of airborne impurities.

AIR COIL: Coil on some types of heat pumps used either as an evaporator or a condenser.

AIR CONDITIONER: Device used to control temperature, humidity, cleanliness, and movement of air in conditioned space.

AIR CONDITIONING: Control of the temperature, humidity, air movement, and cleaning of air in a confined space.

AIR COOLER: Mechanism designed to lower temperature of air passing through it.

AIR CORE SOLENOID: Solenoid which has a hollow core instead of a solid core.

AIR DIFFUSER: Air distribution outlet or grille designed to direct airflow into desired patterns.

AIR GAP: The space between magnetic poles or between rotating and stationary assemblies in a motor or generator.

AIR HANDLER: Fan-blower, heat transfer coil, filter, and housing parts of a system.

AIR INFILTRATION: Leakage of air into rooms through cracks, windows, doors, and other openings.

AIR VENT: Valve, either manual or automatic, used to remove air from the highest point of a coil or piping assembly.

AIR WASHER: Device used to clean air while increasing or lowering its humidity.

AIR-COOLED CONDENSER: Heat of compression is transferred from condensing coils to surrounding air. This may be done either by convection or by a fan or blower.

AIR-SENSING THERMOSTAT: Thermostat unit in which sensing element is located in refrigerated space.

ALCOHOL BRINE: Water and alcohol solution which remains a liquid below 32 °F (0 °).

ALGAE: Low form of plant life, found floating free in water.

ALLEN WRENCH: Hexagonal (6-point) tip used to fit socket head screws or setscrews.

ALLEN-TYPE SCREW: Screw with recessed, hex-shaped head.

ALTERNATING CURRENT (ac): Electric current in which direction of flow alternates or reverses. In 60-cycle (Hertz) current, direction of flow reverses every 1/120th of a second.

AMBIENT TEMPERATURE: Temperature of fluid (usually air) which surrounds object on all sides.

AMERICAN STANDARD PIPE THREAD: Type of screw thread commonly used on pipe and fittings to assure a tight seal.

AMMETER: Electric meter, calibrated in amperes, used to measure current.

AMMONIA: Chemical combination of nitrogen and hydrogen (NH_3). Ammonia refrigerant is identified as R-117.

AMPERAGE: Electron or current flow of one coulomb per second past given point in circuit.

AMPERE: Unit of electric current equivalent to flow of one coulomb per second.

AMPERE TURNS: Term used to measure magnetic force. Represents product of amperes times number of turns in coil of electromagnet.

AMPLIFIER: Electrical device which increases electron flow in a circuit.

ANEMOMETER: Instrument for measuring the rate of airflow or motion.

ANGLE VALVE: Type of globe valve design, having pipe openings at right angles to each other. Usually, one opening is on the horizontal plane and one is on the vertical plane.

ANHYDROUS CALCIUM SULPHATE: Dry chemical made of calcium, sulphur, and oxygen ($CaSO_4$).

ANNEALING: Process of heat treating metal to get desired properties of softness and ductility (easily formed into new shape).

ANODE: Positive terminal of electrolytic cell.

ARCING, ELECTRICAL: Band of sparks formed when an electrical discharge from a conductor jumps to another conductor.

ARMATURE: Part of an electric motor, generator, or other device moved by magnetism.

A.S.A.: Formerly, abbreviation for American Standards Association. Now known as American National Standards Institute.

A.S.M.E. BOILER CODE: Standard specifications issued by the American Society of Mechanical Engineers for the construction of boilers.

ASPECT RATIO: Ratio of length to width of rectangular air grille or duct.

ASPIRATING PSYCHROMETER: Device which draws sample of air through it to measure humidity.

ASPIRATION: Movement produced in a fluid by suction.

A.S.T.M. STANDARDS: Standards issued by the American Society of Testing Materials.

ATMOSPHERIC DUST SPOT EFFICIENCY: Measurement of a device's ability to remove atmospheric air from test air.

Dictionary of Technical Terms

ATMOSPHERIC PRESSURE: Pressure that gases in air exert upon the earth; measured in pounds per square inch (kilopascals [kPa]).

ATOM: Smallest particle of element that can exist alone or in combination.

ATOMIZE: Process of changing a liquid to minute particles or a fine spray.

AUTOMATIC CONTROL: Valve action reached through self-operated or self-actuated means, not requiring manual adjustment.

AUTOMATIC DEFROST: System of removing ice and frost from evaporators automatically.

AUTOMATIC EXPANSION VALVE (AEV): Pressure-controlled valve which reduces high-pressure liquid refrigerant to low-pressure liquid refrigerant.

AUTOMATIC ICE CUBE MAKER: Refrigerating mechanism designed to automatically produce ice cubes in quantity.

AUTOTRANSFORMER: Transformer in which both primary and secondary coils have turns in common. Step-up or step-down of voltage is accomplished by taps on common winding.

AZEOTROPE: Having constant maximum and minimum boiling points.

AZEOTROPIC MIXTURE: Example of azeotropic mixture: refrigerant R-502 is mixture consisting of 48.8 percent refrigerant R-22 and 51.2 percent R-115. The refrigerants do not combine chemically, yet azeotropic mixture provides refrigerant characteristics desired.

B

BACK PRESSURE: Pressure in low side of refrigerating systems; also called suction pressure or low-side pressure.

BACK SEATING: Fluid opening/closing such as a gauge opening; to seat the joint where the valve stem goes through the valve body.

BAFFLE: Plate or vane used to direct or control movement of fluid or air within confined area.

BALL CHECK VALVE: Valve assembly (ball) which permits flow of fluid in one direction only.

BAR: Unit of pressure. One bar equals .9869 atmospheres.

BAROMETER: Instrument for measuring atmospheric pressure. It may be calibrated in pounds per square inch, in inches of mercury in a column, or millimeters.

BATH: Liquid solution used for cleaning, plating, or maintaining a specified temperature.

BATTERY: Electricity-producing cells which use interaction of metals and chemicals to create electrical current flow.

BAUDELOT COOLER: Heat exchanger in which water flows by gravity over the outside of the tubes or plates.

BEARING: Low friction device for supporting and aligning a moving part.

BELLOWS: Corrugated cylindrical container which moves as pressures change, or provides a seal during movement of parts.

BELLOWS SEAL: Method of sealing the valve stem. The ends of the sealing material are fastened to the bonnet and to the stem. Seal expands and contracts with the stem level.

BELT: A rubberlike, continuous loop placed between two or more pulleys to transfer rotary motion.

BENDING SPRING: Coil spring which is placed on inside or outside of tubing to keep it from collapsing while bending it.

BERNOULLI'S THEOREM: In stream of liquid, the sum of elevation head, pressure head, and velocity remains constant along any line of flow, provided no work is done by or upon liquid on course of its flow; decreases in proportion to energy lost in flow.

BIMETAL STRIP: Temperature regulating or indicating device which works on principle that two dissimilar metals with unequal expansion rates, welded together, will bend as temperatures change.

BLAST FREEZER: Low-temperature evaporator which uses a fan to force air rapidly over the evaporator surface.

BLEEDING: Slowly reducing the pressure of liquid or gas from a system or cylinder by slightly opening a valve.

BLEED-VALVE: Valve with small opening inside which permits a minimum fluid flow when valve is closed.

BOILER: Closed container in which a liquid may be heated and vaporized.

BOILER, HIGH-PRESSURE: Boiler furnishing steam at pressures of 15 pounds per square inch gauge or higher (205 kPa).

BOILER, HOT-WATER AND LOW-PRESSURE STEAM: A boiler furnishing hot water at pressures not more than 30 pounds per square inch gauge (308 kPa) or steam at pressures not more than 15 pounds per square inch gauge (205 kPa).

BOILER HORSEPOWER: Term now seldom used, meaning equivalent to a heating capacity of 33,475 Btu/hr. (9804 watts).

BOILING POINT: Boiling temperature of a liquid under a pressure of 14.7 psia (101.3 kPa).

BOILING TEMPERATURE: Temperature at which a fluid changes from a liquid to a gas.

BOOSTER: Common term applied to the use of a compressor when used as the first stage in the cascade refrigerating system.

BORE: Inside diameter of a cylinder.

BOURDON TUBE: Thin-walled tube of elastic metal flattened and bent into circular shape, which tends to straighten as pressure inside is increased. Used in pressure gauges.

BOYLE'S LAW: Law of physics: volume of a gas varies as pressure varies, if temperature remains the same. Example: If absolute pressure is doubled on quantity of gas, volume is reduced one half. If volume becomes doubled, gas has its pressure reduced by half.

BRAZING: Method of joining metals with nonferrous filler (without iron) using heat between 800 °F (427 °C) and melting point of base metals.

BREAKER STRIP: Strip of wood or plastic used to cover joint between outside case and inside liner of refrigerator.

BREECHING: Space in hot water or steam boilers between the end of the tubing and the jacket.

BRINE: Water saturated with a chemical such as salt.

BRITISH THERMAL UNIT (Btu): Quantity of heat required to raise temperature of one pound of water one degree Fahrenheit.

BULB, SENSITIVE: Part of sealed fluid device which reacts to temperature. Used to measure temperature or to control a mechanism.

BUNKER: Space where ice or cooling element is placed in commercial installations.

BURNER: Device in which burning of fuel takes place.

BUTANE: Liquid hydrocarbon (C_4H_{10}) commonly used as fuel for heating purposes.

BYPASS: Passage at one side of, or around, a regular passage.

C

CALCIUM SULFATE: Chemical compound ($CaSO_4$) which is used as a drying agent or desiccant in liquid line driers.

CALIBRATE: Position indicators to determine accurate measurements.

CALORIE: Two different calorie units are used by scientists. The calorie used by medical science is a small heat unit. It equals the heat required to raise the temperature of one gram of water one degree celsius (C). The calorie used by engineering science is a large heat unit. It is equal to the amount of heat required to raise the temperature of one kilogram of water one degree C. In the SI system it is recommended that the Joule unit of energy be used in place of the calorie.

CALORIMETER: Device used to measure quantities of heat or determine specific heats.

CAM: Mechanical component which is oblong, giving a reciprocating motion when rotated.

CAPACITANCE (C): Property of a nonconductor (condenser or capacitor) that permits storage of electrical energy in an electrostatic field.

CAPACITIVE REACTANCE: The opposition or resistance to an alternating current as a result of capacitance; expressed in ohms.

CAPACITOR: Type of electrical storage device used in starting and/or running circuits on many electric motors.

CAPACITOR-START MOTOR: Motor which has a capacitor in the starting circuit.

CAPACITY: Refrigeration rating system. Usually measured in Btu per hour or watts (metric).

CARBON DIOXIDE: Compound of carbon and oxygen (CO_2) which is sometimes used as a refrigerant. Refrigerant number is R-744.

CARBON FILTER: Air filter using activated carbon as air cleansing agent.

CARBON MONOXIDE: Colorless, odorless, and poisonous gas (CO) produced when carbon or carbonaceous fuels are burned with too little air.

CARBON TETRACHLORIDE: Colorless, nonflammable, and very toxic liquid used as a solvent. It must never be allowed to touch skin and fumes must not be inhaled.

CARRENE: Refrigerant in Group One (R-11). Chemical combination of carbon, chlorine, and fluorine.

CASCADE SYSTEMS: Arrangement in which two or more refrigerating systems are used in series; uses evaporator of one machine to cool condenser of other machine. Produces ultra-low temperatures.

CATHODE: Negative terminal of an electrical device. Electrons leave at this terminal.

CAVITATION: Localized gaseous condition that is found within a liquid stream.

CELSIUS TEMPERATURE SCALE: Temperature scale used in metric system. Freezing point of water is 0 °C, boiling point is 100 °C.

CENTIGRADE TEMPERATURE SCALE: See CELSIUS TEMPERATURE SCALE.

CENTIMETER: Metric unit of linear measurement which equals .3937 in.

CENTRALIZED COMPUTERIZED CONTROLLER: Energy control device,

centrally located, which makes control decisions based on operating data, programmed information, and stored data. Can be used to optimize energy consumption of many devices throughout a building.

CENTRAL STATION: Central location of condensing unit with either wet or air-cooled condenser. Evaporator located as needed and connected to the central condensing unit.

CENTRIFUGAL COMPRESSOR: Pump which compresses gaseous refrigerants by centrifugal force.

CERAMIC IGNITOR: Electric ignition system used in a water glycol solution, forced-air furnace. Electrically heated to create ignition of the gas-air mixture in the combustion chamber.

CHANGE OF STATE: Condition in which a substance changes from a solid to a liquid or a liquid to a gas due to addition of heat. Or, the reverse, in which a substance changes from a gas to a liquid, or a liquid to a solid, due to removal of heat.

CHARGE: Amount of refrigerant placed in a refrigerating unit.

CHARGING BOARD: Specially designed panel or cabinet fitted with gauges, valves, and refrigerant cylinders used for charging refrigerant and oil into refrigerating mechanisms.

CHARLES' LAW: Volume of a given mass of gas at a constant pressure varies according to its temperature.

CHECK VALVE: Device which permits fluid flow in one direction.

CHEMICAL REFRIGERATION: System of cooling using a disposable refrigerant. Also called an expendable refrigerant system.

CHILL FACTOR: Calculated number based on temperature and wind velocity.

CHILLER: Air conditioning system which circulates chilled water to various cooling coils in an installation.

CHIMNEY: Vertical shaft enclosing one or more flues for carrying flue gases to the outside atmosphere.

CHIMNEY CONNECTOR: Conduit (pipe) connecting the heating appliance (furnace) with the vertical flue.

CHIMNEY EFFECT: Tendency of air or gas to rise when heated.

CHIMNEY FLUE: Flue gas passageway in a chimney.

CHOKE TUBE: Throttling device used to maintain correct pressure difference between high-side and low-side in refrigerating mechanism. Capillary tubes are sometimes called choke tubes.

CIRCUIT: Tubing, piping, or electrical wire installation which permits flow to and from the energy source.

CIRCUIT, PARALLEL: Arrangement of electrical devices in which the current divides and travels through two or more paths and then returns through a common path.

CIRCUIT, PILOT: Secondary circuit used to control a main circuit or a device in the main circuit.

CIRCUIT, SERIES: Electrical wiring; electrical path (circuit) in which electricity to operate second lamp or device must pass through first; current flow travels, in turn, through all devices connected together.

CIRCUIT BREAKER: Safety device which automatically opens an electrical circuit if overloaded.

CLEARANCE POCKET COMPRESSOR: Small space in a cylinder from which compressed gas is not completely expelled. This space is called the compressor clearance space or pocket. For effective operation, compressors are designed to have as small a clearance space as possible.

CLIMATE CONTROL: A space in which an ideal climate is maintained by some devices.

CLOSED CIRCUIT: Electrical circuit in which electrons are flowing.

CLOSED CONTAINER: Container sealed by means of a lid or other device so that neither liquid nor vapor will escape from it at ordinary temperatures.

CLUTCH, MAGNETIC: Clutch built into automobile compressor flywheel, operated magnetically, which allows pulley to revolve without driving compressor when refrigerating effect is not required.

CODE INSTALLATION: Refrigeration or air conditioning installation which conforms to the local code and/or the national code for safe and efficient installations.

COEFFICIENT OF CONDUCTIVITY: Measure of the relative rate at which different materials conduct heat. Copper is a good conductor of heat and, therefore, has a high coefficient of conductivity.

COEFFICIENT OF EXPANSION: Increase in unit length, area, or volume for one degree rise in temperature.

COEFFICIENT OF PERFORMANCE (cop): Ratio of work performed or accomplished as compared to the energy used.

COGENERATION: Primary source of energy that is also used to produce a secondary source of energy. Example: The use of waste heat from an electrical energy generation system to heat a building.

CO₂ INDICATOR: Instrument used to indicate the percentage of carbon dioxide in stack gases.

COLD: The absence of heat; a temperature considerably below normal.

COLD JUNCTION: That part of a thermoelectric system which absorbs heat as the system operates.

COLD WALL: Refrigerator construction which has the inner lining of refrigerator serving as the cooling surface.

COLLECTOR: Semiconductor section of transistor, connected to the same polarity as the base.

COLLOIDS: Miniature cells peculiar to meats, fish, and poultry which, if disrupted, cause food to become rancid. Low temperatures minimize this action.

COMBUSTIBLE LIQUIDS: Liquid having a flash point at or above 140 °F (60 °C); known as Class 3 liquids.

COMFORT CHART: Chart used in air conditioning to show the dry bulb temperature, humidity, and air movement for human comfort conditions.

COMFORT COOLER: System used to reduce the temperature in the living space in homes. These systems are not complete air conditioners as they do not provide complete control of heating, humidifying, dehumidification, and air circulation.

COMFORT ZONE: Area on psychrometric chart which shows conditions of temperature, humidity, and sometimes air movement in which most people are comfortable.

COMMUTATOR: Part of rotor in electric motor which conveys electric current to rotor windings.

COMPOUND GAUGE: Instrument for measuring pressure both above and below atmospheric pressure.

COMPOUND REFRIGERATING SYSTEMS: System which has several compressors or compressor cylinders in series. The system is used to pump low-pressure vapors to condensing pressures.

COMPRESSION: Term used to denote increase of pressure on a fluid by using mechanical energy.

COMPRESSION GAUGE: Instrument used to measure positive pressures (pressures above atmospheric pressures) only. Gauge dial usually runs from 0 to 300 lb. per sq. in. gauge, (psig) (101.3 − 2 170 kPa).

COMPRESSION RATIO: Ratio of the volume of the clearance space to the total volume of the cylinder. In refrigeration it is also used as the ratio of the absolute low-side pressure to the absolute high-side pressure.

COMPRESSOR: Pump of a refrigerating mechanism which draws a low pressure on cooling side of refrigerant cycle and squeezes or compresses the gas into the high-pressure or condensing side of the cycle.

COMPRESSOR, EXTERNAL DRIVE: See COMPRESSOR, OPEN TYPE.

COMPRESSOR, HERMETIC: Compressor in which the driving motor is sealed in the same dome or housing as the compressor.

COMPRESSOR, MULTIPLE STAGE: Compressor having two or more compressive steps. Discharge from each step is the intake pressure of the next in series.

COMPRESSOR, OPEN TYPE: Compressor in which the crankshaft extends through the crankcase and is driven by an outside motor. Commonly called external drive compressor.

COMPRESSOR, RECIPROCATING: Compressor which uses a piston and cylinder mechanism to provide pumping action.

COMPRESSOR, ROTARY: Compressor which uses vanes, eccentric mechanisms, or other rotating devices to provide pumping action.

COMPRESSOR, SINGLE-STAGE: Compressor having only one compressive step between low-side pressure and high-side pressure.

COMPRESSOR DISPLACEMENT: Volume, in cubic inches, represented by the area of the compressor piston head or heads multiplied by the length of the stroke.

COMPRESSOR SEAL: Leakproof seal between crankshaft and compressor body in open type compressors.

COMPUTER: Series of electrical components which accepts inputs from an operator and controls outputs.

COMPUTER LANGUAGES: Specific wording or codes, such as BASIC, FORTRAN, and COBOL, which direct a computer to accept and store information and control outputs.

CONDENSATE: A fluid formed when a gas is cooled to its liquid state.

CONDENSATE PUMP: Device to remove water condensate that collects beneath an evaporator.

CONDENSATION: Liquid or droplets which form when a gas or vapor is cooled below its dew point.

CONDENSE: Action of changing a gas or vapor to a liquid.

CONDENSER: The part of refrigeration mechanism which receives hot, high-pressure refrigerant gas from compressor and cools gaseous refrigerant until it returns to its liquid state.

CONDENSER, AIR-COOLED: Heat exchanger which transfers heat to surrounding air.

CONDENSER, WATER-COOLED: Heat exchanger designed to transfer heat from hot gaseous refrigerant to water.

Dictionary of Technical Terms

CONDENSER COMB: Comb-like device, metal or plastic, used to straighten the metal fins on condensers or evaporators.

CONDENSER FAN: Forced air device used to move air through air-cooled condenser.

CONDENSING FURNACE: High efficiency, gas forced-air furnace that extracts the latent heat lost in conventional gas forced-air furnaces.

CONDENSING PRESSURE: Pressure inside a condenser at which refrigerant vapor gives up its latent heat of vaporization and becomes a liquid. This varies with the temperature.

CONDENSING TEMPERATURE: Temperature inside a condenser at which refrigerant vapor gives up its latent heat of vaporization and becomes a liquid. This varies with the pressure.

CONDENSING UNIT: Part of a refrigerating mechanism which pumps vaporized refrigerant from the evaporator, compresses it, liquefies it in the condenser, and returns it to the refrigerant control.

CONDENSING UNIT SERVICE VALVES: Shutoff valves mounted on condensing unit to enable service technicians to install and/or service unit.

CONDUCTION: The flow of heat between substances by molecular vibration.

CONDUCTIVITY: Ability of a substance to conduct or transmit heat and/or electricity.

CONDUCTOR: Substance or body capable of transmitting electricity or heat.

CONSOLE: A total unit or system of controls located in one area and enclosed. A window air conditioner is a console air conditioner.

CONSTRICTOR: Tube or orifice used to restrict flow of a gas or a liquid.

CONTAMINANT: Substance such as dirt, moisture, or other matter foreign to refrigerant or refrigerant oil in system.

CONTINUOUS CYCLE ABSORPTION SYSTEM: System which has a continuous flow of energy input.

CONTROL: Automatic or manual device used to stop, start, and/or regulate flow of gas, liquid, and/or electricity.

CONTROL, COMPRESSOR: See MOTOR CONTROL.

CONTROL, DEFROSTING: Device to automatically defrost evaporator. It may operate by means of a clock, door cycling mechanism, or during "off" portion of refrigerating cycle.

CONTROL, LOW-PRESSURE: Cycling device connected to low-pressure side of system.

CONTROL, MOTOR: Temperature or pressure-operated device used to control running of motor.

CONTROL, PRESSURE MOTOR: High- or low-pressure control connected into the electrical circuit and used to start and stop motor. It is activated by demand for refrigeration or for safety.

CONTROL, REFRIGERANT: Device used to regulate flow of liquid refrigerant into evaporator. Can be a capillary tube, expansion valves, or high-side and low-side float valves.

CONTROL, TEMPERATURE: Temperature-operated thermostatic device which automatically opens or closes a circuit.

CONTROL SYSTEM: All of the components required for the automatic control of a process variable.

CONTROL VALVE: Valve which regulates the flow or pressure of a medium which affects a controlled process. Control valves are operated by remote signals from independent devices using any of a number of control media such as pneumatic, electric, or electrohydraulic.

CONTROLLER: A group of controls and circuits used to accurately and automatically operate a device.

CONVECTION: Transfer of heat by means of movement or flow of a fluid or gas.

CONVECTION, FORCED: Transfer of heat resulting from forced movement of liquid or gas by means of a fan or pump.

CONVECTION, NATURAL: Circulation of a gas or liquid due to difference in density resulting from temperature differences.

CONVERSION FACTORS: Force and power may be expressed in more than one way. A horsepower is equivalent to 33,000 ft. lb. of work per minute, 746 watts, or 2546 Btu per hour. These values can be used for changing horsepower into foot pounds, British thermal units, or watts.

COOLER: Heat exchanger which removes heat from a substance.

COOLING TOWER: Device which cools by water evaporation in air. Water is cooled to wet bulb temperature of air.

COPPER PLATING: Abnormal condition developing in some units in which copper is electrolytically deposited on some compressor surfaces.

CORE, AIR: Coil of wire not having a metal core.

CORE, MAGNETIC: Magnetic center of a magnetic field.

CORROSION: Deterioration of materials from chemical action.

COULOMB: The quantity of electricity transferred by an electric current of one ampere in one second.

COUNTER EMF: Tendency for reverse electrical flow as magnetic field changes in an induction coil.

COUNTERFLOW: Flow in opposite direction.

COUPLINGS: Mechanical device joining refrigerant lines.

CRACKAGE: Joint in a structure which permits movement of a gas or vapor through it, even under a small pressure difference.

"CRACKING" A VALVE: Opening a valve a small amount.

CRANK THROW: Distance between center line of main bearing journal and center line of the crankpin or eccentric.

CRANKSHAFT SEAL: Leakproof joint between crankshaft and compressor body.

CRISPER: Drawer or compartment in refrigerator designed to provide high humidity along with low temperature to keep vegetables-especially leafy vegetables-cold and crisp.

CRITICAL PRESSURE: Compressed condition of refrigerant which gives liquid and gas the same properties.

CRITICAL TEMPERATURE: Temperature at which vapor and liquid have same properties.

CRITICAL VIBRATION: Vibration which is noticeable and harmful to structure.

CROSS CHARGED: Sealed container of two fluids which together create a desired pressure-temperature curve.

CRYOGENIC FLUID: Substance which exists as a liquid or gas at ultralow temperatures (−250°F or lower).

CRYOGENICS: Refrigeration which deals with producing temperatures of 250°F below zero and lower.

CURRENT: Transfer of electrical energy in a conductor by means of electrons changing position.

CURRENT RELAY: Device which opens or closes a circuit. It is made to act by a change of current flow in that circuit.

CUT-IN: The temperature value or the pressure value at which the control circuit closes.

CUT-OUT: Temperature value or pressure value at which the control circuit opens.

CYCLE: Series of events or operations which have tendency to repeat in the same order.

CYLINDER: 1—Device which converts fluid power into linear mechanical force and motion. This usually consists of movable elements such as a piston and piston rod, plunger or ram, operating within a cylindrical bore. 2—Closed container for fluids.

CYLINDER, REFRIGERANT: Cylinder in which refrigerant is stored and dispensed. Color code painted on cylinder indicates kind of refrigerant.

CYLINDER HEAD: Plate or cap which encloses compression end of compressor cylinder.

CYLINDRICAL COMMUTATOR: Commutator with contact surfaces parallel to the rotor shaft.

D

DALTON'S LAW: Vapor pressure created in a container by a mixture of gases is equal to sum of individual vapor pressures of the gases contained in mixture.

DAMPER: Device for controlling airflow.

DEAERATION: Act of separating air from substances.

DECIBEL (dB): Unit used for measuring relative loudness of sounds. One decibel is equal to approximate difference of loudness ordinarily detectable by human ear, the range of which is about 130 decibels on scale beginning with one for faintest audible sound.

DECK (COIL DECK): Insulated horizontal partition between refrigerated space and evaporator space.

DEFROST CYCLE: Refrigerating cycle in which evaporator frost and ice accumulation is melted.

DEFROST TIMER: Device, connected into electrical circuit, which shuts unit off long enough to permit ice and frost accumulation on evaporator to melt.

DEFROSTING: Process of removing frost accumulation from evaporators.

DEFROSTING TYPE EVAPORATOR: Evaporator operating at such temperatures that ice and frost on surface melts during off part of operating cycle.

DEGREASING: Solution or solvent used to remove oil or grease from refrigerator parts.

DEGREE-DAY: Unit that represents one degree of difference from inside temperature and the average outdoor temperature for one day; often used in estimating fuel requirements for a building.

DEHUMIDIFIER: Device used to remove moisture from air.

DEHYDRATED OIL: Lubricant which has had most of its water content removed (dry oil).

DEHYDRATOR: See DRIER.

DEHYDRATOR-RECEIVER: Small tank which serves as liquid refrigerant reservoir and which also contains a desiccant to remove moisture. Used on most automobile air conditioning installations.

DEICE CONTROL: Device for operating a refrigerating system in such a way as to provide melting of the accumulated ice and frost.

DELTA TRANSFORMER: Three-phase electrical transformer which has ends of each of three windings electrically connected to form a triangle.

DEMAND METER: Instrument which measures the kilowatt-hour usage of a circuit or group of circuits.

DENSITY: Closeness of texture or consistency of particles within a given substance. The weight per unit volume.

DEODORIZER: Device which absorbs or adsorbs various odors, usually by principle of absorption. Activated charcoal is commonly used.

DESICCANT: Substance used to collect and hold moisture in refrigerating system. A drying agent. Common desiccants are activated alumina and silica gel.

DESIGN PRESSURE: Highest or most severe pressure expected during operation. Sometimes used as the calculated operating pressure plus an allowance for safety.

DETECTOR, LEAK: Device used to detect and locate refrigerant leaks.

DEW: Condensed atmospheric moisture deposited in small drops on cool surfaces.

DEW POINT: Temperature at which vapor (at 100 percent humidity) begins to condense and deposit as liquid.

DIAC: A two-lead alternating current semiconductor that allows current to flow in both directions at a preset voltage.

DIAPHRAGM: Flexible material usually made of thin metal, rubber, or plastic.

DICHLORODIFLUOROMETHANE: Refrigerant commonly known as R-12.

DIE CASTING: Process of molding low-melting-temperature metals in accurately shaped metal molds.

DIELECTRIC FLUID: Fluid with high electrical resistance.

DIFFERENTIAL: The temperature or pressure difference between cut-in and cut-out temperature or pressure of a control.

DIODE: Two-element electron tube which will allow more electron flow in one direction in a circuit than in the other direction; tube which serves as a rectifier.

DIRECT CURRENT (dc): Electron flow which moves continuously in one direction in circuit.

DIRECT DIGITAL CONTROL (DDC): Use of digital computer to perform required automatic control operations in a total energy management system.

DIRECT EXPANSION EVAPORATOR: Evaporator using either an automatic expansion valve (AEV) or a thermostatic expansion valve (TEV) refrigerant control.

DISPLACEMENT, PISTON: Volume obtained by multiplying area of cylinder bore by length of piston stroke.

DISTILLING APPARATUS: Fluid-reclaiming device used to reclaim used refrigerants. Reclaiming is usually done by vaporizing and then recondensing refrigerant.

DISTRIBUTION CONTROLS: Systems which help evenly and efficiently transfer the heating or cooling medium to the area where it is needed.

DISTRICT HEATING AND COOLING: Use of a central utility system designed to provide heating and cooling to large residential and industrial areas.

DOME-HAT: Sealed metal container for the motor compressor of a refrigerating unit.

DOUBLE DUTY CASE: Commercial refrigerator in which a part of space is for refrigerated storage and part is equipped with glass windows for display purposes.

DOUBLE THICKNESS FLARE: Copper, aluminum, or steel tubing end which has been formed into two-wall thickness, 37 to 45 deg. bell mouth or flare.

DOWEL PIN: Accurately dimensioned pin pressed into one assembly part and slipped into another assembly part to insure accurate alignment.

DRAFT GAUGE: Instrument used to measure air movement by measuring air pressure differences.

DRAFT INDICATOR: Instrument used to indicate or measure chimney draft or combustion gas movement. Draft is measured in units of .1 in. of water column.

DRAFT REGULATOR: Device which maintains a desired draft in a combustion-heated appliance by automatically controlling the chimney draft to the desired value.

DRIER: Substance or device used to remove moisture from a refrigeration system.

DRIP PAN: Pan-shaped panel or trough used to collect condensate from evaporator.

DRY BULB: An instrument with a sensitive element to measure ambient air temperature.

DRY BULB TEMPERATURE: Air temperature as indicated by an ordinary thermometer.

DRY CAPACITOR CONDENSER: Electrical device made of dry metal and dry insulation; used to store electrons.

DRY CELL BATTERY: Electrical device used to provide dc electricity, having no liquid in the cells.

DRY ICE: Refrigerating substance made of solid carbon dioxide which changes directly from a solid to a gas (sublimates). Its subliming temperature is $-109\,°F\ (-78\,°C)$.

DRY SYSTEM: Refrigeration system which has the evaporator liquid refrigerant mainly in the atomized or droplet condition.

DUCT: Tube or channel through which air is conveyed or moved.

DYNAMOMETER: Device for measuring power output or power input of a mechanism.

E

ECCENTRIC: Circle or disk mounted off center on a shaft.

ECOLOGY: Science of life balance on earth.

EDDY CURRENTS: Induced currents flowing in a core.

EER: See ENERGY EFFICIENCY RATIO.

EFFECTIVE AREA: Actual flow area of an air inlet or outlet. Gross area minus area of vanes or grille bars.

EFFECTIVE TEMPERATURE: Overall effect on a human of air temperature, humidity, and air movement.

EFFICIENCY: Output of a device, system, or activity, divided by the input necessary to create the output. In a compressor the efficiency would be the work output, as measured by pressure change, divided by the energy input (usually electrical).

EJECTOR: Device which uses high fluid velocity, such as a venturi, to create low pressure or vacuum at its throat to draw in fluid from another source.

ELECTRIC DEFROSTING: Use of electric resistance heating coils to melt ice and frost off evaporators during defrosting.

ELECTRIC HEATING: System in which heat from electrical resistance units is used to heat the building.

ELECTRIC WATER VALVE: Solenoid type (electrically operated) valve used to turn water flow on and off.

ELECTRODEPOSITION PROCESS: Process in which metallic particles are applied to another metal surface through the use of an electric current.

ELECTROLYSIS: Movement of electricity through a substance which causes a chemical change in the substance or its container.

ELECTROLYTIC CONDENSER-CAPACITOR: Plate or surface capable of storing small electrical charges.

ELECTROMAGNET: Coil of wire wound around a soft iron core. When electric current flows through wire, the assembly becomes a magnet.

ELECTROMAGNETIC ENERGY: Energy which has both electrical and magnetic characteristics. Solar energy is electromagnetic.

ELECTROMOTIVE FORCE (EMF) VOLTAGE: Electrical force which causes current (free electrons) to flow or move in an electrical circuit. Unit of measurement is the volt.

ELECTRON: Elementary particle or portion of an atom which carries a negative charge.

ELECTRONIC CONTROL DIAGNOSTICS: Trouble codes which may be referenced on an automatic climate control system to diagnose problems.

ELECTRONIC LEAK DETECTOR: Electronic instrument which measures electronic flow across gas gap. Electronic flow changes indicate presence of refrigerant gas molecules.

ELECTRONIC RELAY: Electronic switch, such as a triac, which controls a power consuming device.

ELECTRONIC SIGHT GLASS: Device that sends an audible signal when system is low in refrigerant.

ELECTRONIC THERMISTOR: Electrical device that senses temperature change to control an output source; see THERMISTOR.

ELECTRONIC THERMOSTAT: Thermostat that uses electronic components to accomplish various sensing, switching, timing, staging, and display functions.

ELECTRONICS: Field of science dealing with electron devices and their uses.

ELECTROSTATIC FILTER: For cleaning air, a type of filter which gives dust particles an electric charge, This causes particles to be attracted to a plate so they can be removed from air.

EMITTER: The lead of a transistor shown using an arrow with a head on it.

END BELL: End structure of plate of electric motor which usually holds motor bearings.

END PLAY: Slight movement of shaft along its center line.

ENDOTHERMAL: Chemical reaction in which heat is absorbed.

ENERGY: Actual or potential ability to do work.

ENERGY AUDIT: Process of accurately determining the current energy consumption for a given area.

ENERGY CONSERVATION: Process, upon reviewing the calculations for determining head loads, of instituting changes that will result in energy savings.

ENERGY EFFICIENCY RATIO (EER): The ratio of the rated cooling capacity in Btu per hour divided by the amount of electrical power used in watts.

ENERGY MANAGEMENT CONTROL SYSTEM: Controllers used in a system which optimizes total energy usage in a building or residence.

ENERGY UTILIZATION INDEX (EUI): A number which is used to compare energy usages for different areas. It is calculated by dividing the energy consumption (in Btus) by the square footage of the conditioned area.

ENTHALPY: Total amount of heat in one pound of a substance calculated from accepted temperature base. Temperature of 32 °F (0 °C) is accepted base for water vapor calculation. For refrigerator calculations, accepted base is −40 °F (−40 °C).

ENTROPY: Mathematical factor used in engineering calculations. Energy in a system.

ENVIRONMENT: The surrounding conditions.

ENZYME: Complex organic substance, originating from living cells, that speeds up chemical changes in foods. Enzyme action is slowed by cooling.

EPOXY (RESINS): Synthetic plastic adhesive.

ETHANE (R-170): Refrigerant sometimes added to other refrigerants to improve oil circulation.

EUI: See ENERGY UTILIZATION INDEX.

EUTECTIC: That certain mixture of two substances providing lowest melting temperature of all the various mixes of the two substances.

EUTECTIC POINT: Freezing temperature for eutectic solutions.

EVACUATION: Removal of air (gas) and moisture from a refrigeration or air conditioning system.

EVAPORATION: Term applied to the changing of a liquid to a gas. Heat is absorbed in this process.

EVAPORATIVE CONDENSER: Device which uses open spray or spill water to cool a condenser. Evaporation of some of the water cools the condenser water and reduces water consumption.

EVAPORATOR: Part of a refrigerating mechanism in which the refrigerant vaporizes and absorbs heat.

EVAPORATOR, DRY TYPE: Evaporator in which the refrigerant is in the liquid droplet form.

EVAPORATOR, FLOODED: Evaporator containing liquid refrigerant at all times.

EVAPORATOR FAN: Fan which increases airflow over the heat exchange surface of evaporators.

EXFILTRATION: Slow flow of air from the building to the outdoors.

EXHAUST PORT: That opening which carries the fluid to the downstream pressure of a fluid system.

EXHAUST VALVE: A movable port which provides an outlet for the cylinder gases in a compressor or engine.

EXOTHERMAL: Chemical reaction in which heat is released.

EXPANSION JOINT: Device in piping designed to allow movement of the pipe caused by the pipe's expansion and contraction.

EXPANSION VALVE: Device in refrigerating system which reduces the pressure from the high side to the low side and is operated by pressure.

EXPENDABLE REFRIGERANT SYSTEM: System which discards the refrigerant after it has evaporated.

EXTERNAL DRIVE: Term used to indicate a compressor driven directly from the shaft or by a belt using an external motor. Compressor and motor are serviceable separately.

EXTERNAL EQUALIZER: Tube connected to low-pressure side of a thermostatic expansion valve diaphragm and to exit end of evaporator.

F

FAHRENHEIT SCALE: On a Fahrenheit thermometer, under standard atmospheric pressure, boiling point of water is 212 degrees and freezing point is 32 degrees above zero.

FAIL-SAFE CONTROL: Device which opens a circuit when sensing element loses its pressure.

FAN: Radial or axial flow device used for moving or producing flow of gases.

FARAD: Unit of electrical capacity; capacity of a condenser which, when charged with one coulomb of electricity, gives difference of potential of one volt.

FARADAY EXPERIMENT: Silver chloride absorbs ammonia when cool and releases it when heated. This is basis on which some absorption refrigerators operate.

FAST FOOD FREEZING: Method that uses liquid nitrogen or carbon dioxide to turn fresh food into long lasting frozen food. It is often referred to as cryogenic food freezing and freeze drying.

FEEDBACK: Information on current operation of a system or device used by the control system to modify future operation.

FEEDBACK CONTROL SYSTEM: Control system that is constantly correcting the condition. Also called a "closed loop system."

FEMALE THREAD: The internal thread on fittings, valves, machine bodies, and the like.

FIELD POLE: Part of stator of motor which concentrates magnetic field of field winding.

FILTER: Device for removing small foreign particles from a fluid.

FLAME TEST FOR LEAKS: Tool which is principally a torch. When a halogen mixture is fed to the flame, this flame will change color in the presence of heated copper.

FLAMMABLE LIQUIDS: Liquids having a flash point below 140 °F (60 °C) and a vapor pressure not exceeding 40 psia (276 kPa) at 100 °F (38 °C).

FLAPPER VALVE: Thin metal valve used in refrigeration compressors which allows gaseous refrigerants to flow in only one direction.

FLARE: An enlargement at the end of a piece of flexible tubing by which the tubing is connected to a fitting or another piece of tubing. This enlargement is made at about a 45 deg. angle. Fittings grip it firmly to make the joint leakproof and strong.

FLARE NUT: Fitting used to clamp tubing flare against another fitting.

FLASH GAS: Instantaneous evaporation of some liquid refrigerant in evaporator which cools remaining liquid refrigerant to desired evaporation temperature.

FLASH POINT: Temperature at which flammable liquid will give off sufficient vapor to support a flash flame but will not support continuous combustion.

FLASH WELD: Resistance type weld in which mating parts are brought together under considerable pressure while a heavy electrical current is passed through the joint to be welded.

FLEXIBLE DUCT: A duct that can be routed around obstacles by bending it gradually.

FLOAT VALVE: Type of valve which is operated by sphere or pan which floats on liquid surface and controls level of liquid.

FLOODED SYSTEM: Type of refrigerating system in which liquid refrigerant fills most of the evaporator.

FLOODED SYSTEM, HIGH-SIDE FLOAT: Refrigeration system which has a float operated by the level of the high-side liquid refrigerant.

FLOODED SYSTEM, LOW-SIDE FLOAT: Refrigerating system which has a low-side float refrigerant control.

FLOODING: Act of allowing a liquid to flow into a part of a system.

FLOW CHECK PISTON: Piston assembly, with an orifice in the center, which can operate as an expansion valve.

FLOW METER: Instrument used to measure velocity or volume of fluid movement.

FLUE: Gas or air passage which usually depends on natural convection to cause the combustion gases to flow through it. Forced convection may sometimes be used.

FLUID: Substance in either a liquid or gaseous state; substance containing particles which move and change position without separation of the mass.

FLUSH: Operation to remove any material or fluids from refrigeration system parts by purging them to the atmosphere using refrigerant or other fluids.

FLUX (BRAZING, SOLDERING): Substance applied to surfaces to be joined by brazing or soldering to keep oxides from forming and to produce joints.

FLUX, MAGNETIC: Lines of force of a magnet.

FOAM LEAK DETECTOR: System of soap bubbles or special foaming liquids brushed over joints and connections to locate leaks.

FOAMING: Formation of a foam in an oil-refrigerant mixture due to rapid evaporation of refrigerant dissolved in the oil. This is most likely to occur when the compressor starts and the pressure is suddenly reduced.

FOOT POUND: Unit of work. A foot pound is the amount of work done in lifting one pound one foot.

FORCE: Force is accumulated pressure and is expressed in pounds. If the pressure is 10 psi on a plate 10 in. sq., the force is 100 lb. If pressure is 10 kg/cm² on a plate 10 cm² in area, the force is 100 kg.

FORCED CONVECTION: Movement of fluid by mechanical force such as fans or pumps.

FORCE-FEED OILING: Lubrication system which uses a pump to force oil to surfaces of moving parts.

FREE WHEELING: Continued rotation of magnetic clutch on automotive compressor when clutch is disengaged.

FREEZE DRYING: Uses liquid nitrogen or carbon dioxide to turn fresh food into long lasting, frozen food. It is also referred to as fast food freezing and cryogenic food freezing.

FREEZER ALARM: A bell or buzzer used in many freezers which sounds an alarm when freezer temperature rises above safe limit.

FREEZER BURN: Condition applied to food which has not been properly wrapped and that has become hard, dry, and discolored.

FREEZE-UP: 1—Formation of ice in the refrigerant control device which may stop the flow of refrigerant into the evaporator. 2—Frost formation on an evaporator which may stop the airflow through the evaporator.

FREEZING: Change of state from liquid to solid.

FREEZING POINT: Temperature at which a liquid will solidify upon removal of heat. The freezing temperature for water is 32°F (0°C) at atmospheric pressure.

FREEZING POINT DEPRESSION: Temperature at which ice will form in solution of water and salt.

FREON: Trade name for a family of synthetic chemical refrigerants manufactured by the Du Pont Company.

FROST BACK: Condition in which liquid refrigerant flows from evaporator into suction line; usually indicated by sweating or frosting of the suction line.

FROST CONTROL, AUTOMATIC: Control which automatically cycles refrigerating system to remove frost formation on evaporator.

FROST CONTROL, MANUAL: Manual control used to change operation of refrigerating system to produce defrosting conditions.

FROST CONTROL, SEMIAUTOMATIC: Control which starts defrost part of a cycle manually and then returns system to normal operation automatically.

FROST FREE REFRIGERATOR: Refrigerated cabinet which operates with an automatic defrost during each cycle.

FROSTING TYPE EVAPORATOR: Refrigerating system which maintains the evaporator at frosting temperatures during all phases of cycle.

FROZEN: 1—Water in its solid state. 2—Seized (as in machine parts) due to lack of lubrication. The term "freeze-up" is often applied to this situation.

FUEL OIL: Kerosene or any hydrocarbon oil as specified by U.S. Department of Commerce Commercial Standard CS12 or ASTM D296, or the Canadian Government Specification Board, 3-GP-28, and having a flash point not less than 100°F (38°C).

FULL FLOATING: Mechanism construction in which a shaft is free to turn in all parts in which it is inserted.

FURNACE, CENTRAL WARM AIR: Self-contained appliance designed to supply heated air through ducts to spaces remote from or adjacent to the appliance location.

FUSE: Electrical safety device consisting of strip of fusible metal in circuit which melts when circuit is overloaded.

FUSIBLE PLUG: Plug or fitting made with a metal of a known low-melting temperature. Used as safety device to release pressures in case of fire.

G

GALVANIC ACTION: Wasting away of two unlike metals due to electrical current passing between them. The action is increased in the presence of moisture.

GAS: Vapor phase or state of a substance.

GAS, NONCONDENSABLE: Gas which will not form into a liquid under the operating pressure-temperature conditions.

GAS VALVE: Device in a pipeline for starting, stopping, or regulating flow of gas.

GASKET: Resilient (spongy) or flexible material used between mating surfaces of refrigerating unit parts, or on refrigerator doors, to give a leakproof seal.

GASKET, FOAM: Joint sealing material made of rubber or plastic foam strips.

GAUGE, COMPOUND: Instrument for measuring pressures both above and below atmospheric pressure.

GAUGE, HIGH-PRESSURE: Instrument for measuring pressures in range of 0 psig to 500 psig (101.3 kPa to 3 600 kPa).

GAUGE, LOW-PRESSURE: Instrument for measuring pressures in range of 0 psia to 50 psia (0 kPa to 350 kPa).

GAUGE MANIFOLD: Chamber device constructed to hold both compound and high-pressure gauges. Valves control flow of fluids through it.

GAUGE PORT: Opening or connection provided for a service technician to install a gauge.

GAUGE VACUUM: Instrument used to measure pressures below atmospheric pressure.

GLYCOL WATER SOLUTION FORCED-AIR FURNACE: Furnace with 50% glycol and 50% distilled water solution, which passes through tube-and-fin heat exchanger to distribute heat through furnace duct system.

GRAIN: Unit of weight and equal to 1/7000 lb. It is used to indicate the amount of moisture in the air.

GRILLE: Ornamental or louvered opening placed in a room at the end of an air passageway.

GROMMET: Plastic, metal, or rubber doughnut-shaped protectors which line holes where wires or tubing pass through panels.

GROUND, SHORT CIRCUIT: Fault in an electrical circuit allowing electricity to flow into the metal parts of a mechanism.

GROUND COIL: Heat exchanger buried in the ground. May be used either as an evaporator or as a condenser.

GROUND WIRE: Electrical wire which will safely conduct electricity from a structure into the ground.

H

HALIDE REFRIGERANTS: Family of refrigerants containing halogen chemicals.

HALIDE TORCH: Type of torch used to safely detect halogen refrigerant leaks in system.

HALOGENS: Substance containing fluorine, chlorine, bromine, and iodine.

HANGER: Device attached to walls or other structure for support of pipe lines.

HEAD: Pressure, usually expressed in feet of water, inches of mercury, or millimeters of mercury.

HEAD, STATIC: Pressure of fluid expressed in terms of height of column of the fluid, such as water or mercury.

HEAD, TOTAL STATIC: Static head from the surface of the supply source to the free discharge surface.

HEAD FRICTION: Head required to overcome friction of the interior surface of a conductor and between fluid particles in motion.

HEAD PRESSURE: Pressure which exists in condensing side of refrigerating system.

HEAD PRESSURE CONTROL: Pressure-operated control which opens electrical circuit if high-side pressure becomes too high.

HEAD PRESSURE SAFETY CUTOUT: Motor protection device wired in series with motor; will shut off the motor when excessive head pressures occur.

HEAD VELOCITY: Height of fluid equivalent to its velocity pressure in flowing fluid.

HEADER: Length of pipe or vessel, to which two or more pipe lines are joined, that carries fluid from a common source to various points of use.

HEAT: Form of energy which acts on substances to raise their temperature; energy associated with random motion of molecules.

HEAT EXCHANGER: Device used to transfer heat from a warm or hot surface to a cold or cooler surface. (Evaporators and condensers are heat exchangers.)

HEAT INTENSITY: Heat concentration in a substance as indicated by the temperature of the substance through use of a thermometer.

HEAT LAG: The time it takes for heat to travel through a substance heated on one side.

HEAT LEAKAGE: Flow of heat through a substance.

HEAT LOAD: Amount of heat, measured in Btu or watts, which is removed during a period of 24 hours.

HEAT OF COMPRESSION: Mechanical energy of pressure changed into energy of heat.

HEAT OF FUSION: Heat released from a substance to change it from a liquid state to a solid state. The heat of fusion of ice is 144 Btu per pound (335 kJ/kg).

HEAT OF RESPIRATION: Process by which oxygen and carbohydrates are assimilated by a substance; also when carbon dioxide and water are given off by a substance.

HEAT PIPE, GAS FORCED-AIR: High efficiency gas furnace that uses vertical liquid filled pipes. The pipes are heated by a burner at their base, and the liquid boils and vaporizes within the pipe. The furnace blower circulates air over the pipes for heating.

HEAT PUMP: Compression cycle system used to supply heat to a temperature-controlled space. Same system can also remove heat from the same space.

HEAT RECOVERY SYSTEM: Produces and stores hot water by transferring heat from condenser to cooler water.

HEAT SINK: Relatively cold surface capable of absorbing heat.

Dictionary of Technical Terms

HEAT TRANSFER: Movement of heat from one body or substance to another. Heat may be transferred by radiation, conduction, convection or a combination of these three methods.

HEAT TRANSFER MODULE: Primary system of heat transfer in a glycol water solution forced-air furnace. The heat transfer module contains the ignitor, burner, and primary solution circulating coil.

HEATING COIL: Heat transfer device consisting of a coil of piping, which releases heat.

HEATING CONTROL: Device which controls temperature of a heat transfer unit which releases heat.

HEATING VALUE: Amount of heat which may be obtained by burning a fuel. The heating value is usually expressed in Btu per lb., Btu per gal., or kJ/kg.

HERMETIC COMPRESSOR: Compressor which has the driving motor sealed inside the compressor housing. The motor operates in an atmosphere of the refrigerant.

HERMETIC MOTOR: Compressor drive motor sealed within same casing which contains compressor.

HERMETIC SYSTEM: Refrigeration system which has a compressor driven by a motor contained in compressor dome or housing.

HERTZ (Hz): Correct terminology for cycles per second.

Hg. (MERCURY): Heavy silver-white metallic element; only metal that is liquid at ordinary room temperature.

HIGH EFFICIENCY GAS FURNACE: Furnace which uses recycling of combustion gases or pulse combustion to obtain operating efficiencies from 85 to 95 percent.

HIGH SIDE: Parts of a refrigerating system which are under condensing or high-side pressure.

HIGH-PRESSURE CUT-OUT: Electrical control switch operated by the high-side pressure which automatically opens electrical circuit if too high pressure is reached.

HIGH-SIDE FLOAT: Refrigerant control mechanism which controls the level of the liquid refrigerant in the high-pressure side of mechanism.

HIGH-VACUUM PUMP: Mechanism which can create a vacuum in the 1000 to 1 micron range.

HOLLOW-TUBE GASKET: Sealing device made of rubber or plastic with tubular cross section.

HORSEPOWER: Unit of power equal to 33,000 ft. lb. of work per minute. One electrical horsepower equals 746 W.

HOT GAS BYPASS: Piping system in refrigerating unit which moves hot refrigerant gas from condenser into low-pressure side.

HOT GAS DEFROST: Defrosting system in which hot refrigerant gas from the high side is directed through evaporator for short period of time and at predetermined intervals, in order to remove frost from evaporator.

HOT JUNCTION: That part of thermoelectric circuit which releases heat.

HOT WATER HEATING SYSTEM: System in which water is circulated through heating coils.

HOT WIRE: 1—Resistance wire in an electrical relay which expands when heated and contracts when cooled. 2—Electrical lead which has a voltage difference between it and the ground.

HUMIDIFIER: Device used to add to and control humidity.

HUMIDISTAT: Electrical control which is operated by changing humidity.

HUMIDITY: Moisture. Dampness of air.

HYDRAULICS: Branch of physics having to do with the mechanical properties of water and other liquids in motion.

HYDROCARBONS: Organic compounds containing only hydrogen and carbon atoms in various combinations.

HYDROMETER: Floating instrument used to measure specific gravity of a liquid.

HYDRONIC: Heating system which circulates a heated fluid, usually water, through baseboard coils by means of a circulating pump which is controlled by a thermostat.

HYGROMETER: Instrument used to measure degree of moisture in the atmosphere.

HYGROSCOPIC: Ability of a substance to absorb and release moisture and change physical dimensions as its moisture content changes.

I

ICE CREAM CABINET: Commercial refrigerator which operates at approximately 0°F (−18°C); used for storage of ice cream.

ICE MELTING EQUIVALENT (IME—ICE MELTING EFFECT): Amount of heat absorbed by melting ice at 32°F (0°C) is 144 Btu per pound of ice or 288,000 Btu per ton.

IDENTIFICATION PLATE: Provides information such as manufacturer, part number, and specifications. Frequently mounted on the outside housing of motors and compressors.

IDLER: Pulley used on some belt drives to provide proper belt tension and to eliminate belt vibration.

IGNITION TRANSFORMER: Transformer designed to provide a high-voltage current. Used in many heating systems to ignite fuel; provides a spark gap.

IMPEDANCE: Opposition in an electrical circuit to the flow of an alternating current that is similar to the electrical resistance to a direct current.

IMPELLER: Rotating part of a pump.

INDUCED MAGNETISM: Ability of a magnetic field to produce magnetism in a metal.

INDUCTION MOTOR: An ac motor which operates on principle of rotating magnetic field. Rotor has no electrical connection, but receives electrical energy by transformer action from field windings.

INDUCTIVE REACTANCE: Electromagnetic induction in a circuit creates a counter or reverse (counter) emf (voltage) as the original current changes. It opposes the flow of alternating current.

INFILTRATION: Passage of outside air into building through doors, cracks around windows, etc.

INFRARED LAMP: Electrical device which emits infrared rays; invisible rays just beyond red in the visible spectrum.

INHIBITOR: Substance which prevents chemical reaction such as corrosion or oxidation.

INSTRUMENT: Used broadly to denote a device that has measuring, recording, indicating, and/or controlling abilities.

INSULATION, ELECTRIC: Substance which has almost no free electrons.

INSULATION, THERMAL: Material which is a poor conductor of heat; used to retard or slow down flow of heat through wall or partition.

INTEGRATED CIRCUIT: A circuit that incorporates multiple transistors and other semiconductors to a single circuit, sometimes called a "chip."

INTEGRATED CIRCUIT BOARD: Electronic circuit made from transistors, resistors, etc., all placed into a package referred to as a "chip," since all circuits are on one base of semiconductor material.

INTERMITTENT CYCLE: Cycle which repeats itself at varying time intervals.

INTERSTATE COMMERCE COMMISSION (ICC): Government body which controls the design and construction of pressure containers.

ION: Group of atoms or an atom electrically charged.

IR DROP: Electrical term indicating the loss in a circuit expressed in amperes times resistance (I x R) or voltage drop.

ISOTHERMAL: Changes of volume or pressure under conditions of constant temperature.

ISOTHERMAL EXPANSION AND CONTRACTION: Action which takes place without a temperature change.

J

JET COOLING SYSTEM: Jet pump is used to produce a vacuum so water or refrigerant may evaporate at relatively low temperatures. These systems usually require large condenser and have a low efficiency to remove heat.

JOINT: Connecting point as between two pipes.

JOULE: Metric unit of heat.

JOULE-THOMSON EFFECT: The change in the temperature of a gas on its expansion through a porous plug from a higher pressure to a lower pressure.

JOURNAL, CRANKSHAFT: Part of shaft which contacts the bearing on the large end of the piston rod.

JUNCTION BOX: Box or container housing group of electrical terminals.

K

KATA THERMOMETER: Large-bulb alcohol thermometer used to measure air speed or atmospheric conditions by means of cooling effect.

KELVIN SCALE (K): Thermometer scale on which unit of measurement equals the Celsius degree and according to which absolute zero is 0 degree, the equivalent of −273.16°C. Water freezes at 273.16°K and boils at 373.16°K.

KILOMETER: Metric unit of linear measurement = 1000 meters.

KILOPASCAL (kPa): See PASCAL.

KILOVOLT AMPERE (kVA): Unit of electrical flow equal to volts multiplied by amperes and divided by one thousand. Unit of power used when power circuit has power factor other than 1.kW = kVA x cosθ.

KILOWATT: Unit of electrical power, equal to 1000 watts.

KING VALVE: Liquid receiver service valve.

kVA: See KILOVOLT AMPERE.

L

LACQUER: Protective coating or finish which dries to form a film by evaporation of a volatile (easily goes from liquid to gas) constituent.

LADDER DIAGRAM: Electrical diagram that indicates order of electrical devices in a specific electrical circuit.

LAG: Delay in response.

LAMP, STERI: Lamp which has a high-intensity ultraviolet ray used to kill bacteria. Also used in food storage cabinets and in air ducts.

LAPPING: Smoothing a metal surface to high degree of refinement or accuracy using a fine abrasive.

LATENT HEAT: Heat energy absorbed in process of changing form of substance (melting, vaporization, fusion) without change in temperature or pressure.

LATENT HEAT OF CONDENSATION: Amount of heat released (lost) by a pound of a substance to change its state from a vapor (gas) to a liquid.

LATENT HEAT OF VAPORIZATION: Amount of heat required, per pound of substance, to change its state from a liquid to a vapor (gas).

LEAK DETECTOR: Device or instrument such as a halide torch, an electronic sniffer, or soap solution used to detect leaks.

LEGIONNAIRE'S DISEASE BACTERIUM (LDB): Is thought to be transmitted by airborne routes, possibly by open air cooling towers or evaporative condensers in commercial systems. Disease is named after an outbreak of illness at an American Legion convention in July, 1976.

LIMIT CONTROL: Control used to open or close electrical circuits as temperature or pressure limits are reached.

LIQUEFIED GASES: A gas below a certain temperature and above a certain pressure, that becomes liquid.

LIQUID: Substance whose molecules move freely among themselves, but do not tend to separate like those of gases.

LIQUID ABSORBENT: Chemical in liquid form which has the property to "take on" or absorb other fluids.

LIQUID DESUPERHEATER: Valve that permits small flow of refrigerant to enter low side of systems to cool suction gas.

LIQUID INDICATOR: Device located in liquid line which provides a glass window through which liquid flow may be watched.

LIQUID LINE: Tube which carries liquid refrigerant from the condenser or liquid receiver to the refrigerant control mechanism.

LIQUID NITROGEN: Nitrogen in liquid form used as a low-temperature refrigerant in expendable or chemical refrigerating systems.

LIQUID RECEIVER: Cylinder (container) connected to condenser outlet for storage of liquid refrigerant in a system.

LIQUID RECEIVER SERVICE VALVE: Two- or three-way manual valve located at the outlet of the receiver and used for installation and service purposes. It is sometimes called the king valve.

LIQUID-VAPOR VALVE REFRIGERANT CYLINDER: Dual hand valve on refrigerant cylinders which is used to release either gas or liquid refrigerant from the cylinder.

LIQUOR: Solution used in absorption refrigeration.

LITER: Metric unit of volume which equals 61.03 cu. in.

LITHIUM BROMIDE: Chemical commonly used as the absorbent in absorption cooling system. Water would then be the refrigerant.

LOCALIZED CONTROLLERS: Independent energy control device located near the system it is controlling.

LOW PRESSURE SAFETY CUTOUT: Motor protection device that senses low-side pressure. Control is wired in series with the motor and will shut off during periods of excessively low suction pressure.

LOW SIDE: That portion of a refrigerating system which is below evaporating pressure.

LOW-SIDE FLOAT VALVE: Refrigerant control valve operated by level of liquid refrigerant in low-pressure side of system.

LOW-SIDE PRESSURE: Pressure in cooling side of refrigerating cycle.

LOW-SIDE PRESSURE CONTROL: Device used to keep low-side evaporating pressure from dropping below certain pressure.

LOW-SIDE PRESSURE LIMITER: Motor protection device that senses low-side pressure. See LOW PRESSURE SAFETY CUTOUT.

LP FUEL: Liquefied petroleum used as a fuel gas.

M

MACHINE ROOM: Area where commercial and industrial refrigeration machinery—except evaporators—is located.

MAGNETIC CLUTCH: Device operated by magnetism to connect or disconnect a power drive.

MAGNETIC FIELD: Space in which magnetic lines of force exist.

MAGNETIC GASKET: Door-sealing material which keeps door tightly closed with small magnets inserted in gasket.

MAGNETISM: A field of force which causes a magnet to attract materials made of iron, nickel-cobalt or other ferrous material.

MALE THREAD: External thread on pipe, fittings, and valves for making connections that screw together.

MANIFOLD, SERVICE: Chamber equipped with gauges and manual valves, used by service technicians to service refrigerating systems.

MANOMETER: Instrument for measuring pressure of gases and vapors. Gas pressure is balanced against column of liquid, such as mercury, in U-shaped tube.

MASS: Quantity of matter held together so as to form one body.

MBH: Thousands of British Thermal Units (82 MBH = 82,000 Btu).

MEAN EFFECTIVE PRESSURE (mep): Average pressure on a surface when a changing pressure condition exists.

MECHANICAL CYCLE: Cycle which is a repetitive series of mechanical events.

MEGOHM: A unit of measure for electrical resistance. One megohm is equal to a million ohms.

MEGOHMMETER: Instrument for measuring extremely high resistances (in the millions of ohms ranges).

MELTING POINT: Temperature at atmospheric pressure at which a substance will melt.

MERCOID BULB: Electrical circuit switch which uses a small quantity of mercury in a sealed glass tube to make or break electrical contact with terminals within the tube.

MET: Term applied to the heat release from a human at rest. It equals 18.4 Btu/sq. ft./hr. (50 kcal/m²/hr or 58 W/m²).

METRIC SYSTEM: Decimal system of measuring.

MICRO: One millionth part of unit specified.

MICROFARAD: Unit of capacitor electrical capacity equal to 1/1,000,000 farad. Written mfd or μf.

MICROMETER: Precision measuring instrument used for making measurements accurate to .001 to .0001 in. or .001 cm.

MICRON: Unit of length in metric system; a thousandth part (1/1 000) of one millimeter.

MICRON GAUGE: Instrument for measuring vacuums very close to a perfect vacuum.

MICROPROCESSOR: Electrical component consisting of integrated circuits which may accept information, store it, and control an output device.

MILLI: Combining form denoting one thousandth (1/1 000); for example, millivolt means one thousandth of a volt.

MINIMUM STABLE SIGNAL (MSS): Correct setting for an expansion valve where it is utilizing the evaporator efficiently but remains free from "hunting."

MISCIBILITY: Substances that are capable of being mixed.

MODULATING: Type of device or control which tends to adjust by increments (minute changes) rather than by either "full on" or "full off" operation.

MODULATING REFRIGERATION CYCLE: Refrigerating system of variable capacity.

MOISTURE INDICATOR: Instrument used to measure moisture content of a refrigerant.

MOLECULE: Smallest portion of an element or compound that retains chemical identity with the substance in mass.

MOLLIER'S DIAGRAM: Graph of refrigerant pressure, heat, and temperature properties.

MONOCHLORODIFLUOROMETHANE: Refrigerant better known as Freon 22 or R-22. Chemical formula is $CHClF_2$. Cylinder color code is green.

MOTOR: Rotating machine that transforms fluid or electric energy into a mechanical motion.

MOTOR, CAPACITOR: Single-phase induction motor with an auxiliary starting winding connected in series with a condenser (capacitor) for better starting characteristics.

MOTOR, FOUR-POLE: 1800 rpm, 60 Hz electric motor (synchronous speed).

MOTOR, TWO-POLE: 3600 rpm, 60 Hz electric motor (synchronous speed).

MOTOR BURNOUT: Condition in which the insulation of an electric motor has deteriorated (become poor in quality) due to overheating.

MOTOR CONTROL: Device to start and/or stop a motor or hermetic motor compressor at certain temperature or pressure conditions.

MOTOR STARTER: High-capacity electric switches usually operated by electromagnets.

MSS POINT (MINIMUM STABLE SIGNAL): Best superheat setting which will provide constant or little temperature change at the thermostatic expansion valve temperature sensing element while the system is running.

MUFFLER, COMPRESSOR: Sound absorber chamber in refrigeration system. Used to reduce sound of gas pulsations.

MULLION: Stationary frame member between two doors.

MULLION HEATER: Electrical heating element mounted in the mullion. Used to keep mullion from sweating or frosting.

Dictionary of Technical Terms

MULTIPLE SYSTEM: Refrigerating mechanism in which several evaporators are connected to one condensing unit.

N

NATURAL CONVECTION: Movement of a fluid caused only by temperature differences (density changes).

NEEDLE POINT VALVE: Type of valve having a needle point plug and a small seat orifice for low-flow metering.

NEGATIVE TEMPERATURE COEFFICIENT THERMISTOR (NTC): Electronic thermistor which decreases in resistance as temperature increases.

NEOPRENE: Synthetic rubber which is resistant to hydrocarbon oil and gas.

NEUTRALIZER: Substance used to counteract acids in refrigeration system.

NEUTRON: That part of an atom core which has no electrical potential; electrically neutral.

NEWTON: Force exerted on an object that has mass of 1 kilogram and gravitational acceleration of 1 m/sec².

NITROGEN DIOXIDE: Mildly poisonous gas (NO_2) often found in smog or automobile exhaust fumes.

NO-FROST FREEZER: Low-temperature refrigerator cabinet in which no frost or ice collects on freezer surfaces or materials stored in cabinet.

NOISE DOSIMETER: Instrument used to measure sound in dBA.

NOMINAL SIZE TUBING: Tubing measurement which has an inside diameter the same as iron pipe of the same stated size.

NON-CODE INSTALLATION: Functional refrigerating system installed where there are no local, state, or national refrigeration codes in force.

NONCONDENSABLE GAS: Gas which does not change into a liquid at operating temperatures and pressures.

NONFERROUS: Group of metals and metal alloys which contain no iron.

NONFROSTING EVAPORATOR: Evaporator which never collects frost or ice on its surface.

NORMAL CHARGE: Thermal element charge which is part liquid and part gas under all operating conditions.

NORTH POLE, MAGNETIC: End of magnet out of which magnetic lines of force flow.

NTC: See NEGATIVE TEMPERATURE COEFFICIENT THERMISTOR.

O

OCTAVE: Frequency difference between harmonic vibrations; the doubling of the frequency of sound.

OCTYL ALCOHOL—ETHYL HEXANOL: Additive in absorption machines that reduces surface tension in absorber.

ODOR: That property of air contaminants that affect the sense of smell.

OFF CYCLE: Segment of refrigeration cycle when system is not operating.

OHM (R): Unit of measurement of electrical resistance. One ohm exists when one volt causes a flow of one ampere.

OHMMETER: Instrument for measuring resistance in ohms.

OHM'S LAW: Mathematical relationship between voltage, current, and resistance in an electric circuit; discovered by George Simon Ohm. It is stated as follows: voltage (E) equals amperes (I) times ohms (R); or $E = I \times R$.

OIL, REFRIGERATION: Specially prepared oil used in refrigerator mechanism which circulates, to some extent, with refrigerant.

OIL BINDING: Condition in which oil layer on top of refrigerant liquid may prevent it from evaporating at normal pressure-temperature.

OIL PRESSURE SAFETY CUTOUT: Motor protection device that senses oil pressure in the compressor. Is wired in series with the compressor and will shut it off during periods of low oil pressure.

OIL SEPARATOR: Device used to remove oil from gaseous refrigerant.

OPEN CIRCUIT: Interrupted electrical circuit which stops flow of electricity.

OPEN COMPRESSOR: Term used to indicate an external drive compressor. Not hermetic.

OPEN DISPLAY CASE: Commercial refrigerator designed to maintain its contents at refrigerating temperatures even though the contents are in an open case.

OPEN TYPE SYSTEM: Refrigerating system which uses a belt-driven or a coupling-driven compressor.

OPERATING PRESSURE: Actual pressure at which the system works under normal conditions. This pressure may be positive or negative (vacuum).

ORGANIC: Pertaining to or derived from living organisms.

ORIFICE: Accurate size opening for controlling fluid flow.

ORIFICE TUBE: Metering device consisting of a restricting tube with inlet and outlet screens.

O-RINGS: Sealing devices used between parts where there may be some motion.

OSCILLOSCOPE: Fluorescent-coated tube which visually shows an electrical wave.

OVERLOAD: Load greater than that for which system or mechanism was intended.

OVERLOAD PROTECTOR: Device, either temperature, pressure, or current operated, which will stop operation of unit if dangerous conditions arise.

OZONE: A form of oxygen, O_3, having three atoms to the molecule, usually produced by discharge of electricity through the air. The *ozone layer* is the outermost layer of the earth's atmosphere, which absorbs ultraviolet light from the sun and shields the lower layers and the earth from harmful rays. Holes caused by chlorine are occurring in the ozone layer. Chlorofluorocarbons (CFCs) contain chlorine and, when released to the atmosphere, cause deterioration of the ozone layer.

P

PACKAGE UNITS: Complete refrigerating system including compressor, condenser, and evaporator located in refrigerated space.

PACKING: Sealing device consisting of soft material or one or more mating soft elements. Reshaped by manually adjustable compression to obtain or maintain a leak-proof seal.

PARTIAL PRESSURES: Condition where two or more gases occupy a space and each one creates part of the total pressure.

PARTS PER MILLION (PPM): Unit of concentration of one element in another.

PASCAL (Pa): Also kilopascal (kPa); the force per unit area.

PASCAL'S LAW: Pressure imposed upon a fluid is transmitted equally in all directions.

PCB: See POLYCHLORINATED BIPHENYL.

PELTIER EFFECT: When direct current is passed through two adjacent metals, one junction will become cooler and the other will become warmer. This principle is the basis of thermoelectric refrigeration.

PERIMETER HOT GAS TUBE SYSTEM: System that has a tube located on the surface of the outer portion of the cabinet to prevent condensation from forming.

PERMANENT MAGNET: Material which has its molecules aligned and has its own magnetic field; bar of metal which has been permanently magnetized.

pH: Measurement of the free hydrogen ion concentration in an aqueous solution. A pH of 7 is neutral.

PHASE: Distinct functional operation during a cycle.

PHASE LOSS MONITOR: Motor protection device for polyphase motors that measures current flow to detect phase loss.

PHIAL: Term sometimes used to denote the sensing element on a thermostatic expansion valve.

PHOTOELECTRICITY: Physical action wherein an electrical flow is generated by light waves.

PHOTON: Particle of electromagnetic energy found in solar radiation.

PHOTOSTATICALLY: Method by which the molecular formation of an element changes due to light.

PHOTOVOLTAIC CELL: See SOLAR CELL.

PIEZOELECTRIC: Property of quartz crystal that causes it to vibrate when a high frequency (500 kHz or higher) voltage is applied. Concept is used to atomize water in a humidifier.

PISTON: Close-fitting part or plug which moves up and down in a cylinder.

PISTON DISPLACEMENT: Volume displaced by piston as it travels the full length of its stroke.

PITOT TUBE: Tube used to measure air velocities.

PLANCK'S CONSTANT: Constant value (6.626×10^{-34} watt seconds squared) which, when multiplied by the frequency of radiation, determines the amount of energy in a photon.

PLENUM CHAMBER: Chamber or container for moving air or other gas under a slight positive pressure.

POLYCHLORINATED BIPHENYL (PCB): Dielectric fluid used in capacitors and transformers that is very toxic. Use of PCB in transformers and capacitors is strictly regulated by the Environmental Protection Agency.

POLYPHASE MOTOR: Electrical motor designed to be used with a three- or four-phase electrical circuit.

POLYSTYRENE: Plastic used as an insulation in some refrigerated structures.

POLYURETHANE: Any synthetic rubber polymers produced from the polymerization of an HO and NCO group from two different compounds. Often used in insulation and molded products.

PONDED ROOF: Flat roof designed to hold a quantity of water which acts as a cooling device.

PORCELAIN: Ceramic china-like coating applied to steel surfaces.

PORTABLE SERVICE CYLINDER: Container used to store refrigerant. Two most common types are disposable and refillable.

POSITIVE TEMPERATURE COEFFICIENT THERMISTOR (PTC): Electronic thermistor which increases in resistance as temperature increases.

POTASSIUM PERMANGANATE: Used in carbon filters to help reduce odors.

POTENTIAL, ELECTRICAL: Electrical force which moves, or attempts to move, electrons along a conductor or resistance.

POTENTIAL RELAY: Electrical switch which opens on high voltage and closes on low voltage.

POTENTIOMETER: Instrument for measuring or controlling by sensing small changes in electrical resistance.

POUND FORCE: Force applied to a 1-lb. mass; has an acceleration of 32.173 ft./s².

POUR POINT: Lowest temperature at which a liquid will pour or flow.

POWER: 1—Time rate at which work is done or energy emitted. 2—Source or means of supplying energy.

POWER ELEMENT: Sensitive element of a temperature-operated control.

POWER FACTOR: Correction coefficient for the changing current and voltage values of ac power.

PPM: See PARTS PER MILLION.

PRECOOLER CONDENSER: Used to cool the refrigerant prior to entering main condenser.

PRESSURE: Energy impact on a unit area; force or thrust on a surface.

PRESSURE, ABSOLUTE: See ABSOLUTE PRESSURE.

PRESSURE, ATMOSPHERIC: See ATMOSPHERIC PRESSURE.

PRESSURE, BACK: See BACK PRESSURE.

PRESSURE, GAUGE: Reading in pounds per square inch (psi) above atmospheric pressure.

PRESSURE, HEAD: Force caused by the weight of a column or body of fluids. Expressed in feet, inches, or psi.

PRESSURE, OPERATING: Pressure at which a system is operating.

PRESSURE, SUCTION: Pressure in low-pressure side of a refrigerating system.

PRESSURE DROP: Pressure difference at two ends of a circuit, or part of a circuit; the two sides of a filter.

PRESSURE GAUGE: Instrument for measuring the pressure exerted by the contents on its container.

PRESSURE LIMITER: Device which remains closed until a certain pressure is reached, then opens and releases fluid to another part of system or breaks an electric circuit.

PRESSURE MOTOR CONTROL: Device which opens and closes an electrical circuit as pressures change.

PRESSURE REGULATOR, EVAPORATOR: Automatic pressure regulating valve mounted in suction line between evaporator outlet and compressor inlet. Purpose is to maintain a predetermined pressure and temperature in the evaporator.

PRESSURE SWITCH: Switch operated by a rise or drop in pressure.

PRESSURE WATER VALVE: Device used to control water flow. It is responsive to head pressure of refrigerating system.

PRESSURE-HEAT DIAGRAM: Graph of refrigerant pressure, heat and temperature properties. (Mollier's diagram).

PRESSURE-OPERATED ALTITUDE (POA) VALVE: Device which maintains a constant low-side pressure, independent of altitude of operation.

PRIMARY COIL: A tube-and-fin circular coil that contains a water glycol solution which surrounds the ignitor and burner. This coil is used in a water glycol gas forced-air furnace.

PRIMARY CONTROL: Device which directly controls operation of heating system.

PROCESS TUBE: Length of tubing fastened to hermetic unit dome, used for servicing unit.

PROPANE: Volatile hydrocarbon used as a fuel or as a refrigerant.

PROPORTIONAL: Being in the proper relative quantity or balance.

PROTECTOR, CIRCUIT: Electrical device which will open an electrical circuit if excessive electrical conditions occur.

PROTON: Particle of an atom with a positive charge.

PSI: Symbol or initials used to indicate pressure measured in pounds per square inch.

PSIA: Symbol or initials used to indicate pressure measured in pounds per square inch absolute. Absolute pressure equals gauge pressure plus atmospheric pressure.

PSIG: Symbol or initials used to indicate pressure in pounds per square inch gauge. The "g" indicates that it is gauge pressure and not absolute pressure.

PSYCHROMETER OR WET BULB HYGROMETER: Instrument for measuring the relative humidity of atmospheric air.

PSYCHROMETRIC CHART: Chart that shows relationship between the temperature, pressure, and moisture content of the air.

PSYCHROMETRIC MEASUREMENT: Measurement of temperature, pressure and humidity using a psychrometric chart.

PTC: See POSITIVE TEMPERATURE COEFFICIENT THERMISTOR.

PULLEY: Flat wheel with a "V" groove. When attached to a drive and drive members, the pulley provides a means for driving the compressor.

PULSE: Term referring to one cycle of ignition and combustion of a gas-air mixture in a pulse combustion furnace.

PULSE COMBUSTION PROCESS: Repeated ignition of a gas and air mixture in a high efficiency gas furnace.

PULSE FURNACE: Furnace which has a "tuned" (resonant) combustion chamber. Part of the energy normally lost through flue is returned to start next "pulse" of combustion.

PUMP: Any one of various machines which force gas or liquid into—or draw it out of—something as by suction or pressure.

PUMP, CENTRIFUGAL: Pump which produces fluid velocity and converts it to pressure head.

PUMP, FIXED DISPLACEMENT: A pump in which the displacement per cycle cannot be varied.

PUMP, RECIPROCATING SINGLE PISTON: A pump having a single reciprocating (moving up and down or back and forth) piston.

PUMP, SCREW: Pump having two interlocking screws rotating in a housing.

PUMP DOWN: The act of using a compressor or a pump to reduce the pressure in a container or a system.

PURGING: Releasing compressed gas to atmosphere through some part or parts for the purpose of removing contaminants from that part or parts.

PYROMETER: Instrument for measuring high temperatures.

Q

QUENCHING: Submerging hot solid object in cooling fluid.

QUICK-CONNECT COUPLING: A device which permits easy and fast connecting of two fluid lines.

R

R-11, TRICHLOROMONOFLUOROMETHANE: Low pressure, synthetic chemical refrigerant which is also used as a cleaning fluid.

R-12, DICHLORODIFLUOROMETHANE: Popular refrigerant known as Freon 12.

R-22, MONOCHLORODIFLUOROMETHANE: Low temperature refrigerant with boiling point of −41 °F (−40.5 °C) at atmospheric pressure.

R-113, TRICHLOROTRIFLUOROETHANE: Synthetic chemical refrigerant which is nontoxic and nonflammable.

R-160, ETHYL CHLORIDE: Toxic refrigerant now seldom used.

R-170, ETHANE: Low temperature application refrigerant.

R-290, PROPANE: Low temperature application refrigerant.

R-500: Refrigerant which is an azeotropic mixture of R-12 and R-152a.

R-502: Refrigerant which is azeotropic mixture of R-22 and R-115.

R-503: Refrigerant which is azeotropic mixture of R-23 and R-13.

R-504: Refrigerant which is azeotropic mixture of R-32 and R-115.

R-600, BUTANE: Low-temperature application refrigerant; also used as a fuel.

R-611, METHYL FORMATE: Low pressure refrigerant.

R-717, AMMONIA: Popular refrigerant for industrial refrigerating systems; also a popular absorption system refrigerant.

RADIAL COMMUTATOR: Electrical contact surface on a rotor which is perpendicular or at right angles to the shaft center line.

RADIANT HEATING: Heating system in which warm or hot surfaces are used to radiate heat into the space to be conditioned.

RADIATION: Transfer of heat by heat rays.

RAM AIR: Air forced through the condenser due to the rapid movement of the vehicle along the highway.

RANGE: Pressure or temperature settings of a control; change within limits.

RANKINE SCALE: Name given the absolute (Fahrenheit) scale. Zero (0 °R) on this scale is −460 °F.

REACTANCE: That part of the impedance of an alternating current circuit due to capacitance or inductance or both.

RECEIVER-DRIER: Cylinder (container) in a refrigerating system for storing liquid refrigerant and which also holds a quantity of desiccant.

RECEIVER HEATING ELEMENT: Electrical resistance mounted in or around liquid receiver. It is used to maintain head pressures when ambient temperature is low.

RECIPROCATING: Back and forth motion in a straight line.

RECORDING AMMETER: Electrical instrument which uses a pen to record amount of current flow on a moving paper chart.

RECORDING THERMOMETER: Temperature measuring instrument which has a pen marking a moving chart.

RECTIFIER, ELECTRIC: Electrical device for converting ac to dc.

RECUPERATIVE COIL: Secondary coil in glycol water forced-air furnace which extracts latent heat from combustion gases.

RECYCLING: Passing of flue gases from combustion in a furnace to a secondary heat exchanger to remove latent heat.

REED VALVE: Thin, flat, tempered steel plate fastened at one end.

REFRIGERANT: Substance used in refrigerating mechanism. It absorbs heat in evaporator by change of state from a liquid to a gas, and releases its heat in a condenser as the substance returns from the gaseous state back to a liquid state.

REFRIGERANT CHARGE: Quantity of refrigerant in a system.

REFRIGERANT CONTROL: Device which meters flow of refrigerant between two areas of a refrigerating system. It also maintains pressure difference between high-pressure and low-pressure side of the mechanical refrigerating system while unit is running.

REFRIGERANT DYE: Coloring agent that can be added to refrigerant to help locate leaks in a system.

REFRIGERANT RECLAIM: To reprocess refrigerant to new-product specifications. This will require chemical analysis, and usually refers to the processes available at a reprocessing or manufacturing facility.

REFRIGERANT RECOVERY: To remove refrigerant and place in a cylinder without necessarily testing it.

REFRIGERANT RECYCLING: To clean refrigerant for reuse by reducing moisture, acidity, and matter. Usually applies to procedures at job site or local service shop.

REGISTER: Combination grille and damper assembly covering an air opening or the end of an air duct.

RELATIVE HUMIDITY: Ratio of (difference between) amount of water vapor present in air to greatest amount possible at same temperature.

RELAY: An electromagnetic mechanism moved by a small electrical current in a control circuit. It operates a valve or switch in an operating circuit.

RELIEF VALVE: Safety device on a sealed system. It opens to release fluids before dangerous pressure is reached.

RELUCTANCE: A force working against the passage of magnetic lines of force (flux) through a magnetic substance.

REMOTE CONTROLLER: Energy control device capable of controlling multiple devices. It can be located away from the devices it is controlling.

REMOTE POWER ELEMENT CONTROL: Device with sensing element located apart from operating mechanism it controls.

REMOTE SYSTEM: Refrigerating system in which condensing unit is away from space to be cooled.

REPULSION-START INDUCTION MOTOR: An electric motor type which has an electrical winding on the rotor for starting purposes.

RESISTANCE: An opposition to flow or movement; a coefficient of friction.

RESISTANCE (R), ELECTRICAL: The difficulty electrons have moving through a conductor or substance.

RESISTOR: Electrical device that is a poor conductor of electricity and produces a given amount of resistance to current flow.

RESTRICTOR: A device for producing a deliberate pressure drop or resistance in a line by reducing the cross-sectional flow area.

RETROFIT: Term used in describing reworking an older installation to bring it up to date with modern equipment or to meet new code requirements.

REVERSE CYCLE DEFROST: Method of heating evaporator for defrosting. Valves move hot gas from compressor into evaporator.

REVERSING VALVE: Device used to reverse direction of the refrigerant flow, depending upon whether heating or cooling is desired.

REYNOLDS NUMBERS: A numerical ratio of the dynamic forces of mass flow to the sheer stress due to viscosity.

RINGLEMANN SCALE: Device for measuring smoke density.

RISER VALVE: Device used to manually control flow of refrigerant in vertical piping.

ROTARY BLADE (VANES) COMPRESSOR: Mechanism for pumping fluid by revolving blades inside cylindrical housing.

ROTARY COMPRESSOR: Mechanism which pumps fluid by using rotating motion.

ROTOR: Rotating or turning part of a mechanism.

RUNNING TIME: Amount of time a condensing unit is run per hour or per 24 hours.

RUNNING WINDING: Electrical winding of motor which has current flowing through it during normal operation of motor.

S

SADDLE VALVE (TAP-A-LINE): Valve body shaped so it may be silver brazed or clamped onto a refrigerant tubing surface.

SAFETY CAN: Approved container of not more than 5-gal. capacity. It has a spring-closing lid and spout cover. It is designed to relieve internal pressure safely when exposed to fire.

SAFETY CONTROL: Device to stop refrigerating unit if unsafe pressure, temperatures, and/or dangerous conditions are reached.

SAFETY MOTOR CONTROL: Electrical device used to open circuit to motor if temperature, pressure, and/or current flow exceed safe conditions.

SAFETY PLUG: Device which will release the contents of a container before rupture pressures are reached.

SAFETY VALVE: Self-operated, quick opening valve used for fast relief of excessive pressures.

SATURATION: Condition existing when substance contains all of another substance it can hold for that temperature and pressure.

SCALE FREE SYSTEM: Eliminates deposits in condensers by picking up electrical energy from water, allowing the deposits to be carried through the system and disposed.

SCAVENGER PUMP: Mechanism used to remove fluid from sump or container.

SCHRADER VALVE: Spring-loaded device which permits fluid flow in one direction when a center pin is depressed and in other direction when a pressure difference exists.

SCOTCH YOKE: Mechanism used to change reciprocating motion into rotary motion or vice versa. Used to connect crankshaft to piston in refrigeration compressor.

SCREW PUMP: Compressor constructed of two mated revolving screws.

SEAL, SHAFT: Device used to prevent leakage between shaft and housing.

SEAL LEAK: Escape of oil and/or refrigerant at the junction where a shaft enters a housing.

SEALED UNIT: See HERMETIC SYSTEM. Motor compressor assembly in which motor and compressor operate inside sealed housing.

SEASONAL ENERGY EFFICIENCY RATIO (SEER): A measure of cooling capacity.

SEAT: That portion of a valve mechanism against which the valve presses to effect shutoff.

SECOND LAW OF THERMODYNAMICS: Heat will flow only from material at higher temperature to material at lower temperature.

SECONDARY REFRIGERATING SYSTEM: Refrigerating system in which condenser is cooled by evaporator of another or primary refrigerating system.

SEEBECK EFFECT: When two different adjacent metals are heated, an electric current is generated between the metals.

SEER: See SEASONAL ENERGY EFFICIENCY RATIO.

SELECTIVE ABSORBER SURFACE: Surface used to increase the temperature of a solar collector.

SELF-INDUCTANCE: Magnetic field induced in conductor carrying the current.

SEMICONDUCTOR: A class of solids whose ability to conduct electricity lies between that of a conductor and that of an insulator.

SEMIHERMETIC COMPRESSOR: Hermetic compressor with service valves.

SENSIBLE HEAT: Heat which causes a change in temperature of a substance.

SENSOR: Material or device which goes through physical change or electronic characteristic change as surrounding conditions change.

SEPARATOR: Device to separate one substance from another.

SEPARATOR, OIL: Device to separate refrigerant oil from refrigerant gas and return the oil to compressor crankcase.

SEQUENCE CONTROLS: Group of devices which act in series (one after another) or in time order.

SERVICE VALVE: Manually operated valve mounted on refrigerating systems used for service operation.

SERVICEABLE HERMETIC: Hermetic unit housing containing motor and compressor assembly by use of bolts or cap screws.

SERVO: A servo motor supplies power to a servo-mechanism. A servo-mechanism is a low-power device (electrical, hydraulic, or pneumatic) used to put in operation and control a more complex or powerful mechanism.

SHADED-POLE MOTOR: Small ac motor designed to start under light loads.

SHELL-AND-TUBE FLOODED EVAPORATORS: Use water flow through tubes built into cylindrical evaporator or vice versa.

SHELL TYPE CONDENSER: Cylinder or receiver which contains condensing water coils or tubes.

SHORT CIRCUIT: Electrical condition where part of circuit touches another part of circuit and causes all or part of current to take wrong path.

SHORT CYCLING: Refrigerating system that starts and stops more frequently than it should.

SHROUD: Housing over condenser, evaporator, or fan.

SHUNT: Type of field coil with a specific resistance placed in parallel with an ammeter.

SI UNIT SYSTEM (Le Systeme International d'Unites): Metric system of measurement adopted by most technical industries throughout the world.

SIGHT GLASS: Glass tube or glass window in refrigerating mechanism. It shows amount of refrigerant or oil in system and indicates presence of gas bubbles in liquid line.

SILICA GEL: Absorbent chemical compound used as a drier. When heated, moisture is released and compound may be reused.

SILICON-CONTROLLED RECTIFIER (SCR): Electronic semiconductor which contains silicon. Controls current by timing pulses.

SILVER BRAZING: Brazing process in which brazing alloy contains some silver as part of joining alloy.

SINE WAVE, AC CURRENT: Wave form of single frequency alternating current; wave whose displacement is sine of angle proportional to time or distance.

SINGLE-PHASE MOTOR: Electric motor which operates on single-phase alternating current.

SINGLE-POLE, DOUBLE-THROW SWITCH, (SPDT): Electric switch with one blade and two contact points.

SINGLE-POLE, SINGLE-THROW SWITCH, (SPST): Electric switch with one blade and one contact point.

SINGLE-STAGE COMPRESSOR: Compressor having only one compressive step between inlet and outlet.

SKIN CONDENSER: Condenser using the outer surface of the cabinet as the heat radiating medium.

SLING PSYCHROMETER: Measuring device with wet and dry bulb thermometers. Moved rapidly in air, it measures relative humidity.

SLUG: 1—Unit of mass equal to the weight (in US units) of object divided by 32.2 (acceleration due to the force of gravity). 2—Detached mass of liquid or oil which causes an impact or hammer in a circulating system.

SLUGGING: Condition in which mass of liquid enters compressor causing hammering.

SMOKE TEST: Test made to determine completeness of combustion.

SOLAR CELL: Also known as a photovoltaic cell. Device which converts solar radiation directly to electricity.

SOLAR COLLECTOR: Device used to trap solar radiation, usually using an insulated black surface.

SOLAR ENERGY SYSTEMS: Systems used to collect, convert, and distribute solar energy in forms useful within a business or residence. A passive system uses no additional energy from other sources for the distribution of the solar generated heat. An active system may use blowers, supplemetary coils, etc.

SOLAR HEAT: Heat created by visible and invisible energy waves from the sun.

SOLDERING: Joining two metals by adhesion of a metal with a melting temperature of less than 800 °F (427 °C).

SOLENOID VALVE: Electromagnet with a moving core. It serves as a valve or operates a valve.

SOLID FUEL HEATING: The use of solid natural resources such as wood or coal to provide heat.

SOLID-STATE CONTROLLED ICE MAKER: Includes controls for ice maker that utilizes components made of semiconductors.

SOLID-STATE ELECTRONIC RELAYS: See ELECTRONIC RELAYS.

SOLUTION: A liquid which has another liquid or solid completely dissolved in it. A lithium bromide water solution, commonly used in absorption systems, is water with a quantity of lithium bromide dissolved in it. "Strong" and "weak" solutions are those with respectively high and low concentrations of another liquid or solid.

SONE: Calculated sound loudness rating.

SOUND TRACER: Instrument which helps locate sources of sound.

SOUTH POLE, MAGNETIC: That part of magnet into which magnetic flux lines flow.

SPECIFIC GRAVITY: Weight of a liquid compared to water, which is assigned value of 1.0.

SPECIFIC HEAT: Ratio of quantity of heat required to raise temperature of a body 1-deg. to that required to raise temperature of equal mass of water 1-deg.

SPECIFIC VOLUME: Volume per unit mass of a substance.

SPLASH SYSTEM, OILING: Method of lubricating moving parts by agitating or splashing oil in the crankcase.

SPLIT-PHASE MOTOR: Motor with two stator windings. Both windings are in use while starting. One is disconnected by centrifugal switch after motor attains speed. Motor then operates on other winding only.

SPLIT SYSTEM: Refrigeration or air conditioning installation which places condensing unit outside or away from evaporator. Also applicable to heat pump installations.

SPRAY COOLING: Method of refrigerating by spraying expendable refrigerant or by spraying refrigerated water.

SQUIRREL CAGE: Fan which has blades parallel to fan axis and moves air at right angles or perpendicular to fan axis.

STANDARD ATMOSPHERE: Condition when air is at 14.7 psia pressure, at 68 °F (20 °C) temperature and a relative humidity of 36 percent.

STANDARD CONDITIONS: Used as a basis for air conditioning calculations: temperature of 68 °F (20 °C), pressure of 29.92 inches of mercury (Hg.), and relative humidity of 30 percent.

STARTING RELAY: Electrical device which connects and/or disconnects starting winding of electric motor.

STARTING WINDING: Winding in electric motor used only briefly while motor is starting.

STATIONARY BLADE COMPRESSOR: Rotary pump which uses a non-rotating blade inside pump to separate intake chamber from exhaust chamber.

STATOR, MOTOR: Stationary part of electric motor.

STEAM: Water in vapor state.

STEAM HEATING: Heating system in which steam from a boiler is piped to radiators in space to be heated.

STEAM JET REFRIGERATION: Refrigerating system which uses a steam venturi to create high vacuum (low pressure) on a water container causing water to evaporate at low temperature.

STEAM TRAP: Automatic valve which traps air but allows condensate to pass while preventing passage of steam.

STETHOSCOPE: Instrument used to detect sounds and locate their origin.

STRAINER: Device such as a screen or filter used to retain solid particles while liquid passes through.

STRATIFICATION OF AIR: Condition in which there is little or no air movement in room; air lies in temperature layers.

STRIKE: Metal plate fastened to frame and into which the bolt of a latch or lock slides.

SUBCOOLING: Cooling of liquid refrigerant below its condensing temperature.

SUBLIMATION: Condition where a substance changes from a solid to a gas without becoming a liquid.

SUBSTANCE: Any form of matter or material.

SUCTION LINE: Tube or pipe used to carry refrigerant gas from evaporator to compressor.

SUCTION PRESSURE CONTROL VALVE: Device located in the suction line which maintains constant pressure in evaporator during running portion of cycle.

SUCTION SERVICE VALVE: Two-way, manually operated valve located at the inlet to compressor. It controls suction gas flow and is used to service unit.

SUCTION SIDE: Low-pressure side of the system extending from the refrigerant control through the evaporator to the inlet valve of the compressor.

SUPERHEAT: 1—Temperature of vapor above its boiling temperature as a liquid at that pressure. 2—The difference between the temperature at the evaporator outlet and the lower temperature of the refrigerant evaporating in the evaporator.

SUPERHEATER: Heat exchanger arranged to take heat from liquid going to evaporator and using it to superheat vapor leaving evaporator.

SURFACE PLATE: Tool with a very accurate flat surface. It is used for measuring purposes and for lapping flat surfaces.

SURGE: Regulating action of temperature or pressure before it reaches its final value or setting.

SURGE TANK: Container connected to the low-pressure side of a refrigerating system which increases gas volume and reduces rate of pressure change.

SWAGING: Enlarging one tube end so end of other tube of same size will fit within.

SWAMP COOLER: Evaporative type cooler in which air is drawn through porous mats soaked with water.

SWASH PLATE-WOBBLE PLATE: Device used to change rotary motion to reciprocating motion. Used in some refrigeration compressors.

SWEATING: 1—Condensation of moisture from air on cold surface. 2—Method of soldering in which the parts to be joined are first coated with a thin layer of solder.

SWEET WATER: Term sometimes used to describe tap water.

SYLPHON SEAL: Corrugated metal tubing used to hold seal ring and provide leakproof connection between seal ring and compressor body or shaft.

SYNTHETIC DUST WEIGHT ARRESTANCE: Measurement of filter's ability to remove synthetic dust from test air.

SYNTHETIC RUBBER, NEOPRENE: Soft resilient material made of a synthetic chemical compound.

Dictionary of Technical Terms

T

TAIL PIPE: Outlet pipe from the evaporator.

TANK, SUPPLY: Separate tank connected directly or by a pump to the oil-burning appliance.

TAP (SCREW THREAD): Tool used to cut internal threads.

TEMPERATURE: 1—Degree of hotness or coldness as measured by a thermometer. 2—Measurement of speed of motion of molecules.

TEMPERATURE-HUMIDITY INDEX: Actual temperature and humidity of air sample compared to air at standard conditions.

TEMPERATURE SENSING BULB: Bulb containing a volatile fluid and bellows or diaphragm. Temperature increase on the bulb causes the bellows or diaphragm to expand.

TEST LIGHT: Light provided with test leads. Used to test or probe electrical circuits to determine if they have electricity.

THERM: Quantity of heat equal to 100,000 Btu.

THERMAL RELAY (HOT WIRE RELAY): Heat-operated electrical control used to open or close a refrigeration system electrical circuit. This system uses a resistance wire to convert electrical energy into heat energy.

THERMISTOR: Basically a semiconductor which has electrical resistance that varies with temperature.

THERMOCOUPLE: Device which generates electricity, using principle that if two unlike metals are welded together and junction is heated, a voltage will develop across the open ends.

THERMOCOUPLE THERMOMETER: Electrical instrument using thermocouple as source of electrical flow, connected to milliammeter calibrated in temperature degrees.

THERMODISC DEFROST CONTROL: Electrical switch with bimetal disc controlled by temperature changes.

THERMODYNAMICS: Part of science which deals with the relationships between heat and mechanical action.

THERMOELECTRIC REFRIGERATION: Refrigerator mechanism that depends on Peltier effect. Direct current flowing through electrical junction between unlike metals provides heating or cooling effect depending on direction of current flow.

THERMOMETER: Device for measuring temperatures.

THERMOMODULE: Number of thermocouples used in parallel to achieve low temperatures.

THERMOPILE: Number of thermocouples used in series to create a higher voltage.

THERMOSTAT: Device which senses ambient temperature conditions and, in turn, acts to control a circuit.

THERMOSTATIC CONTROL: Device which operates system or part of system based on temperature change.

THERMOSTATIC EXPANSION VALVE (TEV): Control valve operated by temperature and pressure within evaporator. It controls flow of refrigerant. Control bulb is attached to outlet of evaporator.

THERMOSTATIC MOTOR CONTROL: Device used to control cycling of unit through use of control bulb. Bulb reacts to temperature changes.

THERMOSTATIC VALVE: Valve controlled by temperature change response elements.

THERMOSTATIC WATER VALVE: Valve used to control flow of water through system, actuated (made to work) by temperature difference. Used in units such as water-cooled compressor and/or condenser.

THREE-PHASE: Operating by means of combination of three alternating current circuits which differ in phase by one-third of a cycle.

THREE-WAY VALVE: Multi-orifice (opening) flow control valve with three fluid flow openings.

THROTTLING: Expansion of gas through orifice or controlled opening without gas performing any work as it expands.

TIMERS: Clock-operated mechanism used to control opening and closing of an electrical circuit.

TIMER-THERMOSTAT: Thermostat control which includes a clock mechanism. Unit automatically controls room temperature and changes temperature range depending on time of day.

TON OF REFRIGERATION: Refrigerating effect equal to the melting of 1 ton of ice in 24-hours. This may be expressed as follows: 288,000 Blu/24 hr., 12,000 Btu/1 hr., 200 Btu/min.

TON REFRIGERATION UNIT: Unit which removes same amount of heat in 24-hours as melting of 1 ton of ice.

TORQUE: Turning or twisting force.

TORQUE, FULL LOAD: Maximum torque delivered without overheating.

TORQUE, STALL: Torque developed when starting.

TORQUE, STARTING: Amount of torque available, when at 0 speed, to start and accelerate the load.

TORQUE WRENCHES: Wrench which may be used to measure torque or pressure applied to a nut or bolt.

TOTAL ENERGY MANAGEMENT: Conservation concept where a building is looked at in terms of its total energy usage, rather than analyzing the requirements of separate systems.

TOTAL HEAT: Sum of both the sensible and latent heat.

TRANSDUCER: Device turned on by change of power from one source for purpose of supplying power in another form to second system.

TRANSFORMER: Electromagnetic device which transfers electrical energy from primary circuit into variations of voltage in secondary circuit.

TRANSFORMER-RECTIFIER: Combination transformer and rectifier in which input ac current may be varied and then rectified into dc current.

TRANSISTOR: Electronic device commonly used for amplification. Similar in use to electron tube. Depends on conducting properties of semiconductors in which electrons moving in one direction are considered as leaving holes that serve as carriers of positive electricity in opposite direction.

TRANSMISSION: Heat loss or gain from a building through exterior components such as windows, walls, floors, etc.

TRIAC: Three-lead semiconductor that allows current flow in two directions when a preset voltage is applied at one of the leads.

TRICHLOROTRIFLUOROETHANE: Complete name of refrigerant R-113. Group 1 refrigerant in rather common use. Chemical compounds which make up this refrigerant are chlorine, fluorine, and ethane.

TRIPLE POINT: Pressure-temperature condition in which a substance is in equilibrium (balance) in solid, liquid, and vapor states.

TROPOSPHERE: Part of the atmosphere immediately above the earth's surface in which most weather disturbances occur.

TRUCK, REFRIGERATED: Commercial vehicle equipped to maintain below-ambient temperatures.

TUBE, CONSTRICTED: Tubing reduced in diameter.

TUBE-WITHIN-A-TUBE: Water-cooled condensing unit in which a small tube is placed inside large unit. Refrigerant passes through outer tube, water through the inner tube.

TUBING: Fluid-carrying pipe which has a thin wall.

TWO-TEMPERATURE VALVE: Pressure-opened valve used in suction line on multiple refrigerator installations which maintains evaporators in system at different temperatures.

TWO-WAY VALVE: Valve with one inlet port and one outlet port.

U

ULTRAVIOLET: Invisible radiation waves with frequencies shorter than wave lengths of visible light and longer than X ray.

UNITARY SYSTEM: A heating/cooling system factory assembled in one package and usually designed for conditioning one space or room.

UNIVERSAL MOTOR: Electric motor which will operate on either ac or dc.

URETHANE FOAM: Type of insulation which is foamed in between inner and outer walls of a container.

V

VACUUM: Pressure lower than atmospheric pressure.

VACUUM ACTIVATORS: Dampers and control valves used in automotive air conditioning system; controlled by the vacuum created by engine intake manifold vacuum.

VACUUM CONTROL SYSTEM: Intake manifold vacuum is used to operate dampers and controls in some automobile systems.

VACUUM PUMP: Special high efficiency device used for creating high vacuums for testing or drying purposes.

VALVE: Device used for controlling fluid flow.

VALVE, EXPANSION: Type of refrigerant control which maintains constant pressure in the low side of refrigerating mechanism. Valve is caused to operate by pressure in low or suction side. Often referred to as an automatic expansion valve or AEV.

VALVE, SERVICE: Device used to check pressures, service, and charge refrigerating systems.

VALVE, SOLENOID: Valve made to work by magnetic action through an electrically energized coil.

VALVE, SUCTION: Valve in refrigeration compressor which allows vaporized refrigerant to enter cylinder from suction line and prevents its return.

VALVE, WATER: In most water cooling units, a valve that provides a flow of water to cool the system while it is running.

VALVE PLATE: Part of compressor located between top of compressor body and head. It contains compressor valves and ports.

VAPOR: 1—vaporized refrigerant is preferred to the word gas. 2—A gas which is often found in its liquid state while in use.

VAPOR, SATURATED: Vapor condition which will result in condensation into droplets of liquid if vapor temperature is reduced.

VAPOR BARRIER: Thin plastic or metal foil sheet used in air-conditioned structures to prevent water vapor from penetrating insulating material.

VAPOR LOCK: Condition where liquid is trapped in line because of bend or improper installation. Such vapor prevents liquid flow.

VAPOR PRESSURE: Pressure imposed by either a vapor or gas.

VAPOR PRESSURE CURVE: Graphic presentation of various pressures produced by refrigerant under various temperatures.

VAPORIZATION: Change of liquid into a gaseous state.

VARIABLE AIR VOLUME (VAV) CONTROLLER: Device having electronic components used to regulate the volume of air in a distribution system.

VARIABLE PITCH PULLEY: Pulley which can be adjusted to provide different pulley drive ratios.

VAV: Variable air volume.

V-BELT: Type of belt commonly used in refrigeration work. Has a contact surface with a pulley which is in the shape of a V.

VELOCIMETER: Instrument that measures air speeds using a direct-reading air speed indicating scale.

VELOCITY: Quickness or rapidity of motion, swiftness, speed.

VENTILATION: Forced airflow, by design, between one area and another.

VIBRATION ARRESTORS: Soft or flexible substance or device which will reduce the transmission of a vibration.

VISCOSITY: Measurement of thickness of oil or its resistance to flow.

V MAX: Maximum (peak) voltage in alternating current cycle.

VOLTAGE: 1—Term used to indicate the electrical potential or electromotive force in an electrical circuit. 2—Voltage or electrical pressure which causes current to flow. 3—Electromotive force.

VOLTAGE CONTROL: Device used to provide some electrical circuits with uniform or constant voltage.

VOLTMETER: Instrument for measuring voltage in electrical circuit.

VOLUMETRIC EFFICIENCY: Term used to express the relationship between the actual performance of a compressor or of a vacuum pump and calculated performance of the pump based on its displacement.

VORTEX TUBE: Mechanism for cooling or refrigerating which accomplishes cooling effect by releasing compressed air through a specially designed tube.

VORTEX TUBE REFRIGERATION: Refrigerating or cooling device using principle of vortex tube, as in mining suits.

V RMS: Voltage root mean square; average voltage equal to the maximum voltage multiplied by a constant.

W

WALK-IN COOLER: Larger, commercially refrigerated space kept below room temperature. Often found in supermarkets or wholesale meat distribution centers.

WATER-COOLED CONDENSER: Condensing unit which is cooled through use of water flow.

WATER DEFROSTING: Use of water to melt ice and frost from evaporator during off-cycle.

WATER HAMMER: Noise generated by back pressure of water when a valve is closed.

WATT: Unit of electrical power.

WAX: Ingredient in many lubricating oils which may separate from the oil if cooled enough.

WET BULB: Device used in measurement of relative humidity. Evaporation of moisture lowers temperature of wet bulb compared to dry bulb temperature of same air sample.

WET BULB TEMPERATURE: Measure of the degree of moisture. It is the temperature of evaporation for an air sample.

WET CELL BATTERY: Cell or connected group of cells that converts chemical energy into electrical energy by reversible chemical reactions.

WET HEAT: Heating system using hot water (hydronic) heat or steam heat.

WHEATSTONE BRIDGE: Electronic circuit consisting of resistors and thermistor. A temperature change on thermistor causes the bridge to become unbalanced, which sends a signal to the output device.

WIDE OPEN THROTTLE SWITCH: Switch that disengages the automotive compressor circuit during periods of high acceleration.

WINDOW UNIT: Air conditioner which is placed in a window.

WOBBLE PLATE-SWASH PLATE: Type of compressor designed to compress gas, with piston motion parallel to crankshaft. Often used in automobile air conditioning system.

Z

ZERO ICE: Trade name for dry ice. See DRY ICE.

ZONE CONTROLS: Controls used to maintain each specific area or zone within a building at a desired condition. This is a type of distribution control often used in hydronic heating system.

ACKNOWLEDGMENTS

The production of a book of this nature would not be possible without the cooperation of the Refrigeration and Air Conditioning industry along with other associations, groups, and individuals. In preparing the manuscript for MODERN REFRIGERATION AND AIR CONDITIONING, the industry and related groups have been most cooperative. The authors acknowledge the cooperation of these people and associations with great appreciation:

A C & R Components, Inc.
Abacus International
Abbeon Cal, Inc.
Acme Electric Corp.
Action Instruments, Inc.
Addison Products Company
Aeroquip Corp.
Air Balance Inc.
Air Conditioning Contractors of America (ACCA)
Air and Energy, Inc.
Airserco Manufacturing Co., Inc.
Alco Controls Division, Emerson Electric Company
Allen Bradley Co., Presence Sensing Products, Fireye Division
Allied Signal, Inc.
Alnor Instrument Co.
Amana Refrigeration, Inc.
American Metal Products Co., Div. of MASCO Corp.
American Panel Corp.
American Seal-Tite
American Society of Heating, Refrigerating and Air Conditioning Engineers, Inc. (ASHRAE)
American Ultraviolet Co.
Amprobe Instrument
Anamet, Inc., Anaconda Metal Hose
Andersen Corp.
Anemostat Products Div.
Arnold Engineering Development Center, Arnold Air Force Base, TN
Auto Flo Co., Div. of MASCO Corp.
BDP Company
Bacharach, Inc.
Bally Engineered Structures, Inc.
Baltimore Aircoil Co., Inc.
Barber-Colman Co.
Beckett Corporation
R. W. Beckett Corp.
Bell and Gossett, ITT
Bernz-O-Matic Corp.
Bohn Refrigeration Products, a Unit of Heatcraft, Inc.
BOTSBALL, Howard Mfg. & Consulting, Inc.
Brasch Mfg. Co., Inc.
Bristol Babcock, Inc.
Buchbinder, Chicago, IL
Burnham Corporation
CPS Products, Inc.
Calgon Corp.
The Carlin Company
Carrier Corporation, Subsidiary of United Technologies Corporation
Carrier Transicold Division, Carrier Corporation
Chatleff Controls, Inc.
Chicago Valve Plate & Steel Co.

Chrysler Corporation
David Clark Co., Inc.
Cleveland Twist Drill Co.
The Coleman Co., Inc.
Compressed Air Magazine
Conservation Energy Systems
Control Systems International
The Cooper Group, Nicholson
Cooper Instrument Corporation
Copeland Corporation
DEC International, Inc.
Dairy Equipment Co.
Dale Electronics, Inc.
Danfoss, Inc.
Desa International
Detroit Edison Co.
Detroit Public Schools
The Dickson Company
The Henry G. Dietz Co., Inc.
DoALL Co.
Dodge Engineering
The Dometic Corporation
Dongan Electric Mfg. Co.
Du Pont Company
Dunham-Bush, Inc.
Dupont Energy Management Co., Inc.
Duro/Indestro, Duro Metal Products Co.
Dwyer Instruments, Inc.
Eaton Corp.
Ebco Manufacturing Company
Edison Electric Institute
Edwards Engineering Corp.
Eklind Tool Co.
Electrolux AB
Elkay Mfg. Co.
Emerson Electric Co.
Excelsior Mfg. & Supply Corp.
Farr Co.
Fedders North America
The Field Controls Co.
John Fluke Mfg. Co., Inc.
Ford Motor Co.
Frick Co.
Frigidaire Company
Fusite Division, Emerson Electric Co.
The Gates Rubber Co.
Gem Products, Inc.
General Electric Co.
General Filters, Inc.
General Motors Corporation
General Motors Corporation, Buick Motor Division
General Motors Corporation, Cadillac Motor Car Div.
General Motors Corporation, Harrison Radiator Div.
General Motors Corporation, Pontiac Motor Div.
Goodman Manufacturing Corporation

Goodway Tools Corp.
Goss Inc.
Grainger, Div. of W.W. Grainger, Inc.
Grasso, Inc.
Hackney Brothers Body, Wilson, NC
Harper-Wyman Co.
Hart & Cooley, Div. of Clevepak
Harvey-R.J. Instrument Corp.
Heliodyne, Inc.
Henry Valve Co.
Hobart Corp.
Home Planners, Inc.
Honeywell, Inc.
Howe Corp.
Hussmann Corporation
Hypertherm, Incorporated
ICI Americas, Inc.
ITT McDonnell & Miller
Ice-O-Matic
Ig-Lo, Inc.
Imperial Eastman, Imperial Division
Inter-City Products Corporation, Tempstar Heating and Cooling Products
Inter-City Products Corporation, ZoneAire
Jarrow Products, Inc.
Johnson Controls, Inc.
Kalglo Electronics Co., Inc.
Kinney Vacuum Co.
Klein Tools, Inc.
Knauf Fiber Glass
Kold-Draft Division, Uniflow Manufacturing Company
Koolatron Industries
Kramer Trenton Co.
K-Whit Tools, Inc.
Kysor/Warren Division of Kysor Industrial Corp.
Larchmont Engineering
Larkin, Inc.
Lennox International, Inc.
Lucas-Milhaupt, Inc., A Handy & Harman Company
Lutron Electronics Co., Inc.
Magnetek
Maid-O'-Mist Div.
Maneurop Inc.
The Marley Cooling Tower Company
Marsh Instrument Co.
Mast Development Co.
Maurey Mfg. Corp.
Maxitrol Co.
McQuay, SnyderGeneral Corp.
Michigan Farmer
Micro Switch, Div. of Honeywell, Inc.
Midco International, Inc.
Mitsubishi Electronics America, Inc.
Monarch Mfg. Works, Inc.

Mueller Brass Co.
Paul Mueller Company
Murray Corp.
Mycom Corp.
National Electrical Contractors
  Association
National Electrical Manufacturers
  Association (NEMA)
National Institute for Automotive Service
  Excellence (ASE)
National Refrigerant, Inc.
National Refrigeration Products
Owens-Corning Fiberglas Corp.
PSG Industries, Inc.
Packless Industries
Paragon Electric Co., Inc.
Parker Hannifin Corp., Fluidex Division
Parker Hannifin Corp., Refrigeration &
  Air Conditioning Div.
Parker Hannifin Corp., Refrigerating
  Specialties Div.
Parker Seal Co.
Peerless of America, Inc.
The Peerless Heater Company
Qualimetrics, Inc.
Ranco North America
Refrigerant Recovery Systems, Inc.
Refrigeration Research, Inc.
Refrigeration Service Engineers Society
  (RSES)
Revcor, Inc.
Rheem Mfg. Co., Air Conditioning Div.
Rheem Mfg. Co., Scientific Products Div.
Ridge Tool Company
Ritchie Engineering Company, Inc.
Robertshaw Controls Co.
Robinair Division, SPX Corporation
Rosemount Analytical, Inc.
Ross-Temp/IMI Cornelius Americas, Inc.
Rubatex Corp.
S-T Industries, Inc.
Sani-Serv

Scale Free Systems, Inc.
Schaefer Brush Mfg. Co., Inc.
Scotsman Ice Systems
Sealed Unit Parts Co., Inc.
Sensidyne, Inc.
SIBIR
Simpson Electric Company
A.O. Smith Co.
Snap-on Tools Corp.
SnyderGeneral Corp.
Solar Components Corporation
Solarex Corporation
Southeast Oakland Vocational Education
  Center
Spectronics Corporation
A.W. Sperry Instruments, Inc.
Sporlan Valve Co.
Sprague Electric Co.
Sprinkool Systems, Inc.
Stal Refrigeration AB
Standard Refrigeration Co.
Stevens Lee Company, Silver King
  Division
Suburban Manufacturing Company
Suntec Industries, Inc.
Superior Contract Service, Inc.
Superior Valve Co., Division of Amcast
  Industrial Corporation
Sweden-Alco Dispensing Systems, A
  Division of Alco Foodservice
  Equipment Co.
TIF Instruments, Inc.
TRW Greenfield Tap & Die Div.
TSI Incorporated
Taylor Freezer Co.
Taylor Scientific, Consumer
  Instruments, Sybron Corp.
Tectronic Products Company, Inc.
Tecumseh Products Company
Teletrol Systems, Inc.
Temprite Div. of Elkay Mfg. Co.
Texas Instruments, Inc.

Thermal Engineering Company
Thermatron Corporation
Therm-O-Disc, Incorporated
Tjernlund Products, Inc.
Torrington Research
Toyota Motor Corporation
The Trane Co.
Tranter, Inc., Edgefield Division
H.O. Trerice Co.
Trol-A-Temp, Division of Trolex
  Corporation
Tutco
Tyler Refrigeration Corp.
Union Carbide Corp., Linde Div.
Unique Air, Inc.
Utica Boilers, Inc.
Vaisala, Inc.
Van Steenburgh Engineering
  Laboratories, Inc.
Vickers, Inc.
Vilter Manufacturing Corporation
Virginia KMP Corp.
Vortec Corp.
WCI Freezer Division
Wahl Instruments, Inc.
Watsco Components, Inc.
Webster Heating and Speciality
  Products, Inc.
Weil-McLain, A Division of The Marley
  Company
Weksler Instruments Corp.
Whirlpool Corporation
White-Rodgers Div., Emerson Electric Co.
White-Rodgers Div., Emerson Electric Co.
  Electro-air/
Williamson Corporation
Wirsbo Company
Y/P Products, Inc.
York International Corp., Central
  Environmental Systems
Yukon Energy Corporation

The authors gratefully acknowledge the cooperation of the
following companies for the use of their equipment shown
on the cover:
  Amprobe Instrument
  Copeland Corporation
  Maneurop Inc.
  Robinair Division, SPX Corporation
  Simpson Electric Company
  TIF Instruments, Inc.

INDEX

A

Abbreviations, 37
Above-atmospheric-pressure
  element, 264
Abrasives, 68
  aluminum oxide, 68
  coarseness, 68
  emery, 68
  sandpaper, 68
  silicon carbide, 68
Absolute,
  pressure scales, 21
  pressure, 21-23
  temperature scales, 14
  temperatures, 13, 14
  zero, 13, 14
Absorbent, 93, 376
Absorber, 92
  heat, 34, 102
  selective, 867, 868
Absorbers, vibration, 468, 469
Absorption chemicals, 95, 627
Absorption cycle, 91, 93, 94, 612,
  689
  air conditioning, 628
  continuous, 91, 94, 612
  intermittent, 91-93, 613
  servicing, 630
Absorption generator, 91
Absorption mechanism, 91, 93, 95
  air conditioning, 622
  automatic defrost, 616
  burners, 611, 612
  gas controls, 619, 620
  gas supply, 618
  installing, 616, 630
  servicing, 626, 630
  thermostats, 616, 619
  wiring diagrams, 616
Absorption refrigerators, 91-95
  installing, 616, 630
  servicing, 626, 630
  thermostats, 616, 619
Absorption system, 91, 93, 611
  ammonia, 612
  cooling process, 614
  commercial, 627
  construction, 616, 625
  controls, 619, 620
  domestic, 614
  effectiveness, 630
  efficiency, 630
  Electrolux, 94, 616, 617, 619
  Faraday, 93, 95, 612
  hydrogen, 612
  installing, 616
  refrigerants, 613
  service items, 630
  Superfex, 613
  Trukold, 613

Absorption to emissions ratio, 867
Absorptivity, 867
Access port, 851, 852
Accessible hermetic units,
  401, 404
Accessories, domestic refrigerators,
  305
Accessory heater, 139
Accumulator, 102, 104, 310, 459,
  828, 959
Acetylene, 49
  air soldering, 48, 49
  torch, 48, 49
Acid,
  conditions in system, 300,
    380, 383
  test kit, 380
Acoustics, 782
ACR tubing, 39
Activated alumina, 376, 471, 1015
Activated carbon, 814
Adaptors,
  flaring tool, 45
  grounding, 203
  piercing valve, 364
  process tube, 341, 359, 360, 362
  Schrader core valve, 364, 365
  service valve wrench, 58
  valve, 342, 356, 359, 361, 362
Adding oil to system, 355, 372,
  373, 387, 551, 973
Additives, fuel oil, 736
Additives, refrigerating oil, 137
Adhesion, 47
Adhesive filters, 809
Adhesive kits, 52
Adiabatic compression, 31,
  589, 998
Adiabatic expansion, 1026
Adjustable pressure limiter, 154
Adjustable wrench, 54, 57
Adjusting,
  burner, 738
  controls, 256
  expansion valves, 142
  motor control, 256
  screw, 256
  thermostatic expansion valves,
    543, 545
  water valves, 541
Adjustments,
  altitude, 258, 259
  differential, 249, 254-258
  emissivity, 1005
  flame, 746
  motor controls, 251
  range, 249, 253, 254, 258
  refrigerant controls, 547
Adsorber, 152
Adsorption, 151, 152
  system, 611

AEV, (see Automatic expansion
  valve)
Agitator, 492
Air, 23, 36
  balance systems, 803
  break, 718
  characteristics, 803
  chemical reaction, 808
  circulation, 794, 799, 800
  circulation systems, 313, 856
  cleaner, 261, 807
  cleaning, 807
  coil, 834, 835
  compressed, 640
  condensation, 808
  conditioned, 779
  conditioner, energy efficiency
    ratio, 776
  conditioners, 687, 949
  contaminants, 667
  cooler, 761
  cooling, 762
  curtains, 482, 815
  defrosting, 423
  diffuser, 783
  dilution, 809
  distribution, 783-807, 964
  ducts, 783
  exfiltration, 780
  film, 927
  filtering, 808
  flow, 803
  forced, 794
  fumes, 807
  gravity, 794
  heat in, 779
  in systems, 370, 387, 510, 532,
    533, 592
  infiltration, 780
  inlet grilles, 795
  leakage, 925-933
  makeup, 662
  mists, 808
  moisture determination, 652
  moisture-holding properties, 1015
  movement, 105-107, 577,
    659, 795
  noise, 674
  outlet velocity, 795
  pollutants, 668
  pressure, 780
  properties of, 650, 779, 1015
  psychrometric chart, 656
  psychrometric properties of, 655
  purification, 668
  R-729, 296
  removing from system, 370,
    386, 509
  return, 803
  smoke, 808
  standard conditions of, 649, 1023

sterilization, 879
stratification of, 783
temperatures, 663
terminal velocity, 795
throw, 795
velocity measurement, 659
volumes, 819
 for cooling, 793, 937
 for heating, 793, 937
washer, 813
weight, 779
Air conditioning,
 absorption unit, 622, 628
 air filtering, 649
 air movement, 649, 659
 automobile, 11, 699, 949
  operation, 950
 bus, 969
 careers, 1001
 central plant, 846-851
 circuits, 879
 comfort coolers, 763
 complete systems, 823, 841
 console, 772
 controls, 259-261, 879, 880
 cooling units, 687, 761
 cycles, 85, 763, 825
 definition of, 649
 dehumidifiers, 694
 development of, 11
 education requirements, 1031
 equipment, 708, 762, 775
 fans, 804
 farm vehicles, 969
 fundamentals, 649
 heat loads, 925, 938, 939, 945
 heating equipment, 709
 humidifiers, 694, 708, 755,
  757, 758
 humidity control, 649, 655
 industry, 1030
 installation, 852
 instruments, 779, 815-818
 insulation, 944
 maintenance, 973
 mobile (CFC standards), 986
 mobile home, 758
 plant, central, 846
 plant, remote, 774
 polarized plug, 767
 principles, 761
 purpose of, 707
 ratings, 30
 recreational vehicle, 774
 refrigerant controls, 907
 safety devices, 880
 servicing, 767, 772
 split systems, 856, 914
 systems, 679, 707, 761,774
  cleaning, 779
  distributing, 779
 temperature, 663
 temperature control, 879, 880
 testing operation, 767
 thermometers, 670
 thermostats, 880-893
 through-the-wall, 687, 688, 758,
  759, 841
 timers, 890
 truck, 969

 unit, 708, 761, 763
 wall plate, 884
 window, 687, 763
 year-around, 779, 841
Air cooling evaporators, 104
Air-acetylene torch, 48, 49
Air-cooled condenser,
 capacities, 597
 head pressure, 297
Air-cooled condensers, 397, 398,
  406, 415-417, 533, 534, 548
 commercial, 107, 109, 415
 console, 772
 domestic, 107, 108
 finned-forced convection,
  107, 109
 finned-static, 107, 109
 outdoor, 406-411
Airflow, 803
 controls, 911
 frost-free refrigerator-freezer,
  323, 325
 indicators, 820
 refrigerator with frozen-food
  compartment, 312, 313
Air-to-air, 695, 834
Air-to-water, 839
Alarm system, temperature, 279
Alcohol, 13, 69
Alcohol brine, 1020
Alignment, belts, 241
Allen setscrews, 58, 71
Allen wrenches, 58
 L-shaped, 58
Alternating current, 169,170
 ammeter, 176
 characteristics, 170
 cycles, 183, 184
Altitude,
 adjustment, 258, 259
 effect on refrigeration
  temperatures, 258
Alumina, activated, 376,
  471, 1015
Aluminum,
 evaporators, repairing, 382
 oxide, 68
 tubing, 39, 52
Ambient compensator, 310, 318
Ambient temperature, 31
 correction, 882
American Society for Testing and
  Materials (ASTM), 1021
American Standard Safety Code for
  Mechanical Refrigeration, 283
American Standard Taper Pipe
  Thread (ASA), 61
Ammeter, 63, 176, 177, 341
 alternating current, 176
 direct current, 176
Ammonia (R-717), 40, 91-93, 292,
  293, 304
 absorption, 612
 open cycle refrigeration, 639
 properties of, 293
 testing for leaks, 294
Amount of refrigerant required, 301
Amperage, 172
 relay, 267
Ampere turns, 186

Amperes, 176, 182
 kilovolt, 207
 volt, 207
Amplifier, 194, 196-198, 880, 892
Analogy, water, 171, 172
Analyzer, motor compression, 387,
  394, 395
 stand, 387
Anemometer, 541, 659
Angle, 18
Angle deflector plates, 764, 765
Angular measurements, 18, 19
Annealed, 39
Annealing, 46
 tubing, 46
Anticipators, 250, 884, 889
 current, 884
Antifoam inhibitor, 1019
Antifreeze, 383
Anvil block, 52
Appliance hand truck, 342, 343
Appliances, cordless, 170
Application of latent heat, 27, 588
Arc, 18, 19
Arcing, 882
Area,
 cabinet, 571
 door, 933
 equivalents, 17, 1007
 evaporator, 581
 measurement, 17
 of a circle, 18, 19, 210
 wall, 931
 window, 932
Argon (R-740), 295
Arithmetic, basic, 15
Armature (see Rotor), 156,
  190, 191
 electric motor, 190, 191, 215
Arrestors, flashback, 49
Articulating connecting rod, 116
Artificial ice, 11, 75
ASA, 52, 61
Aspect ratio, duct, 792
Aspirating devices, 459
Aspirating psychrometer, 652
Aspiration, 497
Assembling,
 compressor, 386, 529
 hermetic unit, 387, 388
 refrigeration systems, 548
 swaged tubing, 49
Assembly devices, 72
Atmosphere, 14, 21, 22, 649
Atmospheric pressure, 19, 21-23
 gauges, 21
 standard, 14, 23
Atom, 13, 181, 1011
Atomizing, fuel oil, 737
Attachment, service valve, 359
Attachment plug configurations,
  210
Attic fan, 807
Attraction, magnetic, 184, 185
Attractive force, 19
A-type evaporator, 848, 849
Audit, energy, 919
Auto ranging voltmeters, 176
Auto transformer, 207
Automatic control systems, 879

Index

Automatic controls,
  air conditioning, 259, 879, 894
  defrosting, 271, 273, 274, 312,
    449
  electric, 247
  expansion valve, 79, 110, 112,
    141
  motor, 247, 249-253, 264, 265
  refrigerant, 141, 311, 313, 315,
    319, 320, 322, 325, 330, 333,
    907
Automatic defrosting, 311, 315,
  616
  controls, 247, 271-273
  electric heater system, 312
  hot gas system, 312, 316
  refrigerator-freezer, 311-315
  timer, 315, 317, 319
Automatic expansion valve,
  adjusting, 141
  design, 141, 142
  diagnosing troubles, 374
  principles, 141
Automatic expansion valves, 79,
  110, 141
  bellow type, 142, 143
  bleeder type, 144
  bypass type, 144
  diaphragm type, 143, 144
  refrigerant flow control, 110, 112
Automatic ice cube maker, 86,
  428-434, 490
  controls, 261
  external servicing, 350
  installing, 344
Automatic reset, 249, 251
Automobile air conditioning, 11,
  699, 949
  accumulator, 959
  air distribution, 964
  ammeter, 966
  automatic controls, 968
  belts, 957, 971
  blowers, 966
  bulb mounting, 961
  comfort cooling, 950, 952
  compressor seal, 957
  compressors, 954
  condenser, 959
  control switches, 953
  controls, 966
  cooling, 950, 964
  cooling capacity, 952
  crankshaft, 954, 957
  ducts, 965
  electrical circuits, 965
  electronic control systems, 967
  evaporators, 961
  expansion valves, 961
  farm vehicle, 969
  fastening, 960
  filters, 959
  flared fittings, 960
  heater core, 961
  heating, 961, 964
  hoses, 960
  insulation, 965
  leaks, 972
  lines, 960
  magnetic clutch, 952

  magnetic coils, 953
  maintenance, 973
  manual control systems, 967
  metering devices, 961
  muffler, 959
  oil, 964, 973
  operating conditions, 951
  O-ring, 960
  pressure, low-side, 951, 953
  pulley, 957
  receiver-drier, 959
  refrigerant, 301, 964
  refrigerant lines, 960
  refrigerant loss, 957
  rheostat, 950
  safety, 975
  Scotch yoke, 956
  service valves, 959, 971
  servicing, 969-974
  sight glass, 952, 960, 973
  storing refrigerant, 973
  suction pressure control
    valves, 962
  swash plate, 954
  teflon seal, 957
  thermostatic expansion valve, 961
  thermostats, 969
  truck air conditioning, 969
  two-cylinder compressor, 954
  vacuum actuators, 968
  vacuum control systems, 968, 969
  V-type compressor, 954
  wobble plate, 954
Autotransformer, 207
Auxiliary evaporators, 479
Average temperature-pressure
  conditions, refrigerator-freezer,
  345, 346
Avoirdupois, 1008
Awnings, 939
Axial flow, 804
Azeotropic mixture refrigerants, 289
  R-500, 288, 289
  R-502, 289-291
  R-503, 290-292

B

Bacteria, 11, 13, 808
Baffle, 577
Bakeries, refrigeration for, 492
Balance point, 840
Balancing the system, 803
Ball peen hammers, 59
Ball type valve, 148, 149
Bar, 21
Barograph, 672
Barometer, 21, 672
  mercury, 65
Baseboard heating, 749
Basement,
  fan, 807
  heat loss, 936
  specifications, 936
Basic arithmetic, 15
Basic current electricity, 169
Basic cycles, electricity, 169
Basic refrigeration systems, 75
Bath, ice and water, 545
Bath, mineral spirits, 384, 386
Battery, 169

  back up, 892
Batts, 944
Baudelot coolers, 585, 598
Bearings,
  compressor, 118, 527
  electric motors, 239
  lubrication, 238
  replacing, 527
  servicing, 238, 239
Beaufort scale, 664
Bellows, 115, 142, 143
Bellows operated low-pressure
  control, 264
Bellows operated thermometer, 63, 64
Bellows type AEV, 142, 143
Bellows type TEV, 145, 146
Below-atmospheric-pressure
  element, 264
Belts, 116
  alignment, 241
  automobile air conditioning,
    957, 971
  care of, 241
  motors, electric, 116, 240
  serpentine, 241
  sizes, 241
  tension, 241
  V-, 214, 240
Bender, tube, 42
  lever-type, 42
Bending
  springs, 42, 341
  tools, 42
  tubing, 42
Bernoulli's Equation, 650
Beverage coolers, 426-428, 584
  brine, 427
  sweet water, 427
Beverage cooling, 11, 426
Bimetal disc, 253
Bimetal motor controls, 251-253
  overload protection, 266
Bimetal strip, 63, 209
Bin level controls, 261
Binding, oil, 114, 161
Blade,
  compressor, 130
  construction, 129
  radial, 134, 135
  rotary, 127
  stationary, 127
Blankflange, 526
Blank-off plates, 852, 854
Bleeder type AEV, 144
Bleeder valve, 148, 149
Block-and-punch, 45
Blocks,
  anvil, 52
  flaring, 44
  lapping, 528
Blow holes, 386
Blowback, 737
Blow-by, 386
Blower control, 684
Blower plates, 633
Blower type evaporator, 423-426,
  475, 478, 495
Blowers, 966
Board, charging, 551
Bodies, truck, 495

Boiler (see Generator)
Boiler, 733 •
Boiling point, 14
Boiling pressure, 29
Boiling temperature, 14, 28
Bolted (accessible) motor
  compressor, 384
Bolted-on piercing valves, 363
Bolts,
  loosening, 56
  metric size, 56
Bonnet, 679
Boost-and-buck transformer, 206
Booster compressor, 127
Bore, 595
Borosilicate, 753
Bottle (see Cylinder)
Bottoming taps, 61
Bound electrons, 1011
Bourdon tube, 64
Box wrench, 54, 56
  double offset, 56
  offset, 56
  sizes, 59
  straight, 56
Boxes, refrigerator (see Cabinets)
Boyle's Law, 102, 154, 1025
Braces, hand, 60
Brass fittings, 41, 46
Brass tubing, 40
Brazed tube fittings, 47, 48
Brazed tubing connections, 43
Brazed-on piercing valves, 363, 364
Brazing, 49, 71, 517, 1022
  figure-eight motion, 51
  flux, 342
  joint cleaning, 47, 49, 369, 519
  joint designs, 50, 51
  oxyacetylene equipment, 49
  safety, 49
  silver, 47, 342, 519
  soldering torch, 341
  temperature, 51
Brazing filler metals, 47, 49
  copper alloy, 47, 49
  silver alloy, 47, 49
Breaker strip, 336, 337, 374
Breaker trim, 307
Breathing, 495
Brine, 23, 34, 1020
  alcohol, 1020
  calcium chloride, 34, 1020
  composition, 1020
  freezing temperatures, 1020
  glycol, 1020
  immersed evaporator, 34, 427
  sodium chloride, 34, 1020
  specific gravity, 23, 34, 1020
British thermal units, 13, 25, 565
Bromine, 283
Brushes, 69, 190, 191
  cleaning, 69
  file, 71
Btu (see British thermal unit)
Bubble solution, leak detector,
  366, 367
Building, heat sources, 943
Bulb, sensing (see Bulb, sensitive)
Bulb, sensitive, 111, 115, 144,
  145, 150-152, 251

mounting, 155
Bulb, temperature sensing (see Bulb,
  sensitive)
Bunker, 475
Burn, freezer, 305
Burned-out winding, 384
Burner,
  adjusting, 738
  installation, 743, 745
  pump, 738
  system parts, 720
Burners, 612
  Bunsen, 713
  electric ignition, 722
  gas, 679, 720
  gun type, 737
  kerosene, 612
  oil, 680, 681, 737
Burnouts, 371, 471, 472, 550
  clean up, 380
Bus air conditioning, 969
Butane (R-600), 294, 618
Butt welded, 40
Butter conditioner, 306, 335
Bypass, 97, 144, 436, 437
  automatic expansion valve, 144
  drain sump, 317
  hot gas, 400, 426, 427, 436,
  437
Bypass cycle, 590
Bypass systems, 96, 98, 426, 427

C

Cabinet,
  accessories, 336
  areas, 571
  construction, 305, 475, 565
  door construction, 475
  door removal, 395
  finishes, 306, 338, 475
  gaskets, 337
  hardware, 336, 337, 350
  ice accumulation, 335
  insulation, 306, 347
  leveling, 336, 344
  sizes, 477
  temperature, 297, 339
  thermometers, 338, 339
  volume, 572,
Cabinet humidifiers, 758
Cabinets, 306, 565
  care of, 335
  chest type freezer, 328-332
  commercial, 475, 565
  dairy, 480
  display, 399, 480
  domestic refrigerator, 305, 306
  domestic refrigerator-freezer, 305,
  306, 309, 312, 315, 319, 320,
  322, 324, 325
  florist, 478
  freezer, 328-334
  fresh food, 306
  frozen foods, 328-334, 481, 482
  grocery, 475
  ice cream, 484, 586
  reach-in, 475
  soda fountain, 484
  upright freezer, 333, 334
  walk-in, 476

Calcium chloride, 30, 34, 1020
  brine, 34
Calcium sulphate, 1015
Calculations,
  air conditioning, 776, 927
  commercial, 566
  duct, 799
  heat loads, 925
  of infiltration, 938
  wattage, 177
Calibration, 63-67
  cut-in, 254
  cutout, 254
  gauges, 65-67
  thermometer, 14
Caliper, micrometer, 67
Calorie, 25, 183
Cam action, ice maker, 263, 264
Cam gear mechanism, 326
Cap screw, 71
Capacitance, 170, 188, 189
Capacitor,
  construction, 189
  design, 189
  effect, 213, 221
  motor, 215
  running, 215
  servicing, 236, 237
  shorting, 351
  start, capacitor-run motor,
  213, 221
  start, motor, 221
  testing, 351, 555
Capacity,
  balancing, 576
  compressor, 594
  condenser, 576, 577, 597, 598
  condensing unit, 577
  evaporator, 576, 581, 584, 585
  liquid lines, 600
  refrigerant control, 608
  refrigerant line, 603
  specific heat, 26
  suction lines, 600, 603
  system, 594
  valves, thermostatic, 156, 159
  varying system, 85
Capillary tube, 81, 102, 111, 141,
  162, 163, 307, 313, 316, 342
  adjustable, 164
  applications, 164
  capacities, 164
  cleaner, 341, 382, 383
  cleaning, 355, 382, 383
  clogged, 376, 377
  controls, 111, 162, 374
  designs, 81, 111, 163
  diagnosing troubles, 374
  fittings, 164
  installation, 376
  normal charge, 376
  overcharged, 376
  principles of, 162
  refrigerant control, 81, 110, 111,
  162, 374
  replacement kit, 383
  replacing, 355
  replacing with automatic expansion
  valve, 143
  servicing, 374, 382

sizes, 164
sizing kit, 341, 383
strainer, 383
systems, 81, 110, 162, 374
Car, railway, refrigerated, 497
Carbon, 123, 152
Carbon dioxide (R-744), 23, 152,
  283, 294, 366, 633, 646, 650
solidified, 34, 90, 294
Carbon filter, 478, 814
Carbon tetrachloride warning, 70,
  384, 386
Cards, indicator, 1013
Care of,
  belts, 77, 241
  cabinets, 335
  gauges, 66, 507
  refrigerant cylinder, 72, 296, 370
  refrigerator or freezer, 335
  service valves, 58, 73, 370, 466,
  508, 509
Career opportunities, 1029
Carrene (see R-11)
Cascade systems, 84, 85, 127,
  590, 643
Case-hardened, 528
Cases (also see Cabinets)
  dairy, 480
  display, 480-482, 579
  double-duty, 479
  fast freezing, 484
  single-duty, 479
Casings, 134
Caustic solution, cleaning, 69
Ceilings, 933
  radiant, 753
Cell, solar, 196, 874-876
Celsius, 13-16
  absolute, 14
  comparison with Fahrenheit, 14
  conversion to Fahrenheit, 16
  conversion to Kelvin, 16
  scale, 14
  thermometer, 14
Celsius, Anders, 14
Center, magnetic, 242
Center punch, 70
Centigrade temperature scale, 14
Centimeter, 17, 18, 1007
  cubic, 18
  square, 17
Central air conditioning, 688, 846
  controls, 260, 261
  inspecting, 855
  installing, 852
  servicing, 855
Centralized control, 918
Centrifugal,
  compressor, 116, 133, 134, 520
  force, 116, 134
  units, 858
  switch, 215, 238
Centrifugal force, 134
CGC (see Combination gas control)
Chain wrench, 58
Chamber, environmental test, 100
Change of state, 23, 27
Change-of-state point, 27
Changing refrigerants, 301
Charcoal, 152

Charge-and-tap piercing valve, 364
Charges, negative, 169
Charges, positive, 169
Charging,
  board, 551
  cylinder, 359, 360, 371, 372
  line, 359
  network, R-C, 234
  station, 393, 394
Charging a system, 355, 356, 360,
  369-372, 393
  apparatus, 510
  automobile, 973
  checking, 513
  commercial, 510
  hermetic, 369
  high side, 510
  low side, 510
  portable charging cylinder,
  371, 372
  rebuilt, 393
Charles' Law, 102, 154, 1025
Charts,
  comfort, 663-667
  psychrometric, 656
  troubleshooting, 348, 349
  991, 996
Check valves, 128, 138, 139, 165,
  456, 458
  commercial, 456, 458
  domestic, 122, 127
Checking,
  airflow, 820
  compressor, 523
  door gaskets, 337
  for leaks, 366-369, 509
  oil charge, 551
  refrigerant charge, 513
  relays, 270
Chemical,
  dehumidifier, 776
  electrical generating, 169
  energy, 169
  refrigeration, 88, 295, 498, 633
  scavengers, 733
  stability, refrigeration oil, 300
Chest-type freezer, 326, 328-331
  lid counterbalancing, 326, 336
Chill factor, 664, 665
Chilled water, 858
  pump, 213
Chiller, water, commercial, 400,
  409
Chilling units (see Evaporator)
Chisels, cold, 70
Chloride, calcium, 30, 34
Chloride, sodium, 30, 34
Chlorine, 283
Chlorofluorocarbons (CFCs), 283,
  523, 977
  recovery, recycling, reclaiming,
  977-987
Chlorotetrafluoroethane, (see
  R-124), 301
Chlorotrifluoromethane (see R-13)
Chromate, 535
Circle,
  arc of, 18, 19
  area of, 18, 20, 210
  degrees of, 18, 19

Circuit,
  boards, 194, 197, 200, 326, 328
  breaker, 169, 209, 210, 227
  chips, integrated, 892
  connection, multiple, 352, 353
  continuity, 211
  fundamentals, 171
  interrupter, ground fault, 203
  protectors, 209, 210
  symbols, 172
  testing instruments, 211
  troubles, 170
Circuits, electrical, 169, 172, 182,
  204, 247, 275, 879, 965
  closed, 172
  continuous, 207
  control, 898
  Delta, 204, 205
  integrated, 194, 199
  motor, 205
  open, 172
  parallel, 182, 183, 247
  power, 204
  series, 182, 183, 247
  series-parallel, 183
  servicing, 552
  short, 172
  solar cell, 876
  star (Wye), 205
  thermocouple, 723
Circuits, water, 540
Circular charts, 815
Circular mils, 210
Circulation of air, 794
  systems, 856
Clampmeter, 176
Class 2 wiring, 209
Classification of refrigerants, 283
Claw hammers, 59
Clean Air Act, 301
Clean room, 813, 1022
Cleaning,
  air, 667, 779, 807
  alcohol, 69
  bath, 69
  brazed joint, 51
  cabinets, 335
  carbon tetrachloride warning, 70
  caustic solution, 69
  cloths, 342
  compressor, 380, 381, 526, 527
  condenser parts, 353
  domestic refrigerator or freezer,
  308, 335
  electrostatic, 810
  external mechanism, 350, 353
  facilities, 384
  filters, 809
  fluids, 69
  furnaces, 728
  gasoline, warning, 70
  methods, 69
  motors, 239
  oleum or mineral spirits, 69
  parts, 69
  R-11, 69
  refrigerators, 69
  solvents, 54, 69, 1021
  safety, 69, 70
  steam, 69

tanks, 384
vapor degreasing, 69
Clearance, compressor, 137
Clearance space, 119
Climate, outdoor and indoor, 663
Climatic control, 823, 841
system installation, 852
Clocks, defrosting, 277
Clogged capillary tube, 376, 377
Clogged screens, servicing, 142
Closed circuit, 172
Closed Delta transformer, 205, 208
Closed water system, solar,
869, 870
forced convection, 870
natural convection, 870
Closed-loop (feedback) control
systems, 249
Clutch,
armature, 953
field, 953
magnetic, 952, 953
plate, 953
Coal furnaces, 707
Coarseness, abrasive, 68
Code, electrical, 209, 1014
Code, piping, 1022
Code, size no., 46
Code installations, 518
requirements, 725
testing, 518
Coefficient of performance (COP),
596, 610, 827
Cogeneration, 755
Coil,
blower, 633
cooling, 36
finned, 416
heat pump, 827
air, 834
ground, 824, 835
inside/outside, 831, 832
water, 697, 834
heating, installation, 754
heating, service, 754
helical wound, 749
magnetic, 953
plate type, 953
recuperative, 714, 715
solenoid, 882
Cold, 13, 776
bar, 307
bath for thermostatic controls,
279, 558
chisels, 70
junction, 199
plate, 646
plate cooling, 633
reserve, water coolers, 261
storage temperatures, 568
traps, 393
Collapsible element, gas-charged,
154
Collector, solar,
angle, 869
covers, 868
heat insulation, 869
lens, 874
panel, 840
surface, 867

Collectors (solar),
concentrating, 876
flat plate, 867
liquid, 839
swimming pool, 872
vacuum tube, 868
Colloids, 305
Color code,
gauges, 65
ground wires, 203
pilot light, 722
piping, 1022
refrigerant cylinders, 297
refrigeration systems, 13
Column,
mercury, 21
water, 21, 661, 724
Combination furnace and cooling,
721
Combination gas control valve
(CGC), 901
Combination thermostats, 889
Combustion, pulse, 714
Combustion efficiency, 816
Comfort charts, 663-667
Comfort conditions, 664
Comfort cooler, 858, 907, 908
console type, 772
unit, 763
window unit, 763
Comfort cooling, 11, 762, 907,
908
automobile air conditioning,
950, 952
controls, 259-261
large, 774, 858
remote, 774
temperature range, 141
Comfort zone, 663
Comfort-Health Index (CHI), 666
Commercial,
absorption system, 627
applications of refrigeration, 490
cabinets, 475
calculations, 565, 573, 576
compressor, 405, 412, 520
compression control, 404,
414-416
condenser, 415-422
condensing units, servicing,
501, 520
construction of refrigeration
mechanisms, 397
controls, 443-453, 461
defrost systems, 423, 435-443
driers, 471, 472
equipment, installation of, 397
evaporators, 422-435
heat loads, 565, 573, 576
hermetic systems, 401
insulation, 504
liquid receivers, 422
low-pressure control, 443
mechanism, 397
motor controls, 213-215, 220,
443-447
motors, 111, 213-215, 220,
221, 223
refrigerant controls, 443
refrigeration, 11

screw type compressor, 132
systems, 95, 397
accessory devices, 95
application, 475
charging, 510
evaporator, 100
installing, 501
moisture in, 548
safety, 473
servicing, 519
water chiller, 400, 409
Commutator, 190, 191, 215
Compact refrigerator, 306
Comparison, refrigerant controls,
164, 165
Compass, magnetic, 184
Complete systems, 688, 823, 841
Composition gaskets, 529
Compound,
gauge, 21, 65, 341, 346, 347
refrigerating systems, 82, 83
refrigeration cycle, 82, 83
wound motors, 226
Compression, adiabatic, 31
Compression, heat of, 31
Compression cycle, 101, 102
operation, 102
parts, 102
pressure, 102, 105
rings, 119
system, 101
temperature, 102
typical systems, 102
Compression fittings, 52
Compression gauge, 64-66
Compression ratio, 138, 1010
Compression stroke, 101
Compression system, 100, 101
AEV, 79
capillary tube, 81
high-side float, 77
low-side float, 76
refrigerant controls, 141
safety, 139
TEV, 80
Compressor,
analyzer, 387, 394, 395
assembling, 386, 527, 529
bearings, 118, 119, 125, 238
burnouts, 371, 380
capacities, 594
checking, 374, 384-386, 523
checking leaks, 530
cleaning, 526, 527
connecting rods, 116, 120, 409
construction, 116, 412
cooling, 136
crankshafts, 116, 123, 400,
415
cylinders, 118, 119, 400, 401,
412-415
dehydrating, 387, 530
design, 116, 412, 594
dismantling, 379, 526
dome, 116, 386
drives, 123, 124
efficiency, 522, 596
electric tester, 394
evacuating, 387, 522, 532
flywheel, 123

gaskets, 122, 139, 530
housing, crankcase, 126
identification plate, 350
installing, 381
installing open type, 501, 531
liquid refrigerant in, 355
locating troubles in, 374
low-side pressure control
 valve, 461
lubrication, 137
muffler, 135, 136, 469
oil (see Refrigerant oil)
opening, 384
overhauling, 383, 527
pistons, 119
protection devices, 458
purpose of, 33, 101
rebuilding, 384-386
removing, 379, 380, 525
replacing, 355, 381, 382
rotor construction, 128, 130
seal, 123, 528, 957
service valve, 73, 105, 106, 356,
 507, 525
servicing, 379, 381, 532
(stuck), starting, 355
terminals, 353, 354
testing,
 automobile air conditioning, 973
 commercial, 523, 530, 532
 domestic, 370, 371, 375, 386,
  387
types, 116
unloader, 412, 414, 415
valves, 122, 461
 repair, 385, 529
volumetric efficiency, 137, 594
Compressors, 28, 31, 33, 76, 101,
 102, 106, 107, 412
automobile, 952-956
bolted (accessible), 384
booster, 127
centrifugal, 116, 133, 134, 520
commercial, 397, 520
commercial, hermetic, 401
double acting, 126
external drive, 77, 115, 123,
 400, 403, 412, 413, 523
heat pump, 695, 827
hermetic, 116, 412, 413
 reciprocating, 116, 384, 415
 repairing, 379, 381, 532
 rotary, 127, 385, 386, 415
modulating, 827
motor, 116
 faults, 374
 servicing, 242, 243, 355
multi-cylinder, reciprocating,
 118, 122
noisy, 524
open type, 116, 118, 123
parallel motor, 406
reciprocating, 116, 117, 125,
 126
rotary, 116, 127, 520
 hermetic, 127-130
 rotating vane, 127-129
 stationary blade, 127-130
rotating blade, 129
Scotch yoke, 956

screw type, 116, 130-133
 commercial, 132, 520
scroll, 134, 135
stationary blade type, 129
swash plate, 125, 126, 954
tandem motor, 406
three (multistage system),
 644, 645
two-cylinder, 118, 954
two-speed motor, 215
two-stage, 84, 590
variable displacement, 954, 955
welded, 384
wobble plate, 125, 954
Compressor-running tests,
 370, 371
Computers, 200
digital control, 917
languages, 201
microprocessors, 200
programming, 201
Concentrating collector, solar, 876
Condensate, 36
Condensate pump, 213, 425, 426
Condensation, 102
latent heat of, 27
Condensed, 102
Condenser,
cleaning, 353
coils, heat pump, 829, 833, 834
comb, 534
construction, 415-422
design, 306
fan, 213
fin, 416
head pressures, 297
installing, air-cooled, 533
installing, water-cooled, 537
removing, air-cooled, 533
removing, water-cooled, 533
repairing, 537
repairing, air-cooled, 533
replacing, 355
servicing, air-cooled, 533, 548
servicing, commercial, 519, 532
servicing, water-cooled, 534
standard temperatures, 283
troubles, 382
types, 107, 415-422
water pump, 213
Condensers 12, 32, 101-103,
 107, 170
air-cooled, 107, 108, 397, 398,
 415-417, 597
 finned-forced convection,
  107-109
 finned-static, 107, 109
automobile, 959
capacity of air-cooled, 597
capacity of water-cooled, 598
commercial, air-cooled, 107,
 415, 416
commercial, water-cooled,
 107, 417
domestic, 107, 108, 307,
 313, 316
evaporative, 76, 421, 422, 690
 servicing, 537
frozen food unit, 330, 333
oil-cooler, 329

outdoor air-cooled, 416
shell type, 417, 418
static, 107
tube-within-a-tube, 417-419
water-cooled, 417
 shell, 107, 108
 tube-within-a-tube, 107, 108
Condensing,
furnace, 714
pressure, 297, 592
temperature, 34, 298
unit,
 air-cooled, 406-411, 415, 416
 capacities, 577
 frozen foods, 330, 333
 installing, 501, 514
 location, 344
 service valves, 466
 servicing, 537
Conditioned area control, 879
Conditioners, unit air, 763
Conduction, 33, 34
Conductivity,
electricity, 169, 181
heat, 33, 34, 1023
of insulating materials, 566
Conductor, 169, 178, 180, 181,
 1014
Conductor of electricity, 34
Conductor of heat, 33, 34
Conical spiral, 52
Connecting a gauge manifold, 356
Connecting rods, 112, 116, 117
 articulating, 112
Connecting tubing, 43
Connections,
brazing, 43, 49
flared, 43, 57
soldering, 43, 47
Connector, multi-circuit, 353
Connector block, 315
Console air conditioners, 771, 772
installation, 772
servicing, 772
Constant pressure valve, 142, 454
Constants, electrical, 1013
Constrictor, tubing, 52
Construction, building, 934
Construction, commercial, 946
Construction, roof, 935
Consumption, energy, 916, 919
Consumption, water, 586, 598
Consumption, wattage, 236
Containerized refrigeration, 646
Contaminants, 376
air, 650, 667
micron size, 668
Continuity, 172, 555
Continuous circuit, 172
Continuous cycle, 94, 612
Continuous cycle, absorption,
 system, 91, 612, 613
controls, 612, 619
gas supply, 618
Contracts, service 561, 997, 1001
Control,
adjustments, 253-259
dampers, 313
direct digital, 917
heat flow, 34

humidity, 856
override, 892
panel, electrical, 169
point, 250
switches, 953
systems, 879
  diagnostics, 967
  electronic, 967
  fundamentals, 249
  manual, 967
  servicing, 922
temperature (see Temperature control)
valve, 641
Controllers, energy management, 879, 894, 917
  centralized computerized, 917, 918
  diagnostics, 893
  localized, 917
  programmable, 892, 917
  remote, 917
  thermostat 880
  variable air volume (VAV), 911
Controlling instrument action, 249-251
Controls, 141
  absorption, 612, 619
  air conditioning, 259-261, 879
  airflow, 911
  anti-icing, 909
  automatic, 110, 115, 141
  automatic defrost, 271, 272
  automatic expansion valve (AEV), 110, 112, 141-144
  automobile air conditioning, 966
  bellows operated, low-pressure, 264, 265
  blower, 684
  capillary tube (cap.), 110, 111, 141, 162-164
  central air conditioning, 260, 261
  circuit, 227, 247, 898
  combination gas, 901
  comfort cooling, 259-261, 907, 908
  commercial, 101, 257, 269, 443-448
  comparing, 164,165
  compression systems, 141
  compressor (see Motor control)
  condenser, 548
  conditioned area, 879, 898
  damper motor, 911, 914
  defrost timer, 277
  defrosting, 265-269, 311-317, 319
  defrosting clock, 277
  deicing, 276
  draft, 816
  dry system, 110
  duct, 914
  electric, 247
  electronic, 891
    cleaner, 910
  expansion valve, 141
  fan, 411, 910
  feedback, 897
  float, high-pressure, 77
  float, low-pressure, 76

flooded system, 110
flue draft, 720
fluid flow, 908
freezer, 258
frost control, 694
furnace, 880, 895, 896
gas, 612, 724, 879, 899
heat distribution, 911
heat pump, 260, 276, 828
heating, 260
hermetic, 249-253, 264, 265
high limit, 679, 683, 684
high pressure, 247, 264, 265, 446
high-side float (HSF), 110, 114, 115, 141, 161, 162
hot gas defrost, 274, 275
humidity, 260, 276, 277, 655, 909
hydraulic, 858
hydronic, 906
ice bank, 275
ice maker, 261-264, 448
icing, 907
infrared heat, 905
limit, 894, 896, 899, 917
low-side float (LSF), 110, 114, 141, 161, 162
low-side pressure, 461, 953
microprocessor, 132
modulating, 898, 901
motor (also see Controls, hermetic), 80, 115, 342, 443
  freezer, 309, 311, 313, 315, 318, 320, 325, 326, 328
  safety, 265, 266, 446
  servicing, 557
  temperature, 80, 339
oil furnace, 902
operating, 894, 896
pneumatic, 414, 911
pressure, 911
pressure, motor, 77, 264-266, 443, 444
pressure regulating, 454
primary, 894, 895, 899, 902, 917
refrigerant, 32, 101, 102, 110-115, 141, 164, 307, 310, 313, 316, 330, 332, 342, 374, 443, 543, 608, 828, 907
refrigerator, 258
relay, 894
safety, 279, 880, 896, 900, 923
safety, motor, 446, 447
semiautomatic defrost, 273, 274
sequential, 896, 917
solid state, 880
split system, 914
temperature, 663, 879
testing, 545, 546
thermal-electric expansion valve (THEXV), 141, 142, 153
thermostatic, expansion valve (TEV), 80, 110-113, 141, 144-152, 155, 156, 158, 159, 160
  motor, 251-253, 445
troubleshooting, 922
two-temperature valve, 454-456, 547

vacuum control system, 968
vapor pressure, 253
vending machine, 448
water cooler, 261
water level, 906
water valve, 463-466, 539, 540
zone, 914
Convection,
  forced, 34, 423
  natural, 34, 423
  of heat, 33, 34
Conventional measurements, 11
Conventional units, 501
Conversion, pressure, 298
Conversion, temperature, 1009
Coolers,
  beverage, 426-428, 489, 585
  comfort, 762, 763
  milk, 491
  swamp, 703
  unit, 761
  walk-in, 476
  water, 426, 487
Cooling,
  air volume, 762
  capacity, 30
  coil (see Evaporator), 36, 101
  controls, 260
  effect, 30, 32
  equipment, 762
  loads, 585, 586, 827, 926, 938
  systems, 761, 762
  thermostat, 889
  tower, 107, 109, 418, 420, 421, 537, 585, 693
  types,
    absorption system, 614
    cold plate, 633
    comfort, 762
    compression system, 101
    compressor, 136
    district systems, 863
    evaporative, 703, 774
    heat pump, 825
    ice cream, 586
    roof, 703
    solar energy, 874
    spray, 633
    steam jet, 642, 701
    vortex tube, 702
    water, 86, 585
Cooling unit (see Evaporator coils)
COP (see Coefficient of performance)
Copper alloy brazing filler metals, 47
Copper pipe,
  hard drawn, 39, 40
  nominal size, 39, 41
  sizes, 39
  soft, 39, 342
  thickness, 39
  tubing, 39, 40
  swaging, 52
Copper-and-phosphorus alloy wire, 342
Copper-to-steel joint, heating, 51
Cordless appliances, 170
Cords, power (see Power cords)
Core, magnetic, 186

Core valves, 364, 365
Corrosion, 40, 711, 732
CO² indicator, 816
Coulomb, 176, 188
Counter emf, 177, 193
Counterflow, 417-419
Couplings, 53, 382
  pressure, 164
  quick-connect, 851
  reducing, 53
  self-sealing, 851
  tube, 52
Cracking a valve, 58, 73, 507
Crank arrangements,
  piston-cylinder, 118
Crankcase, 116, 118, 119, 126
  compressor housing, 126
  heater, 122, 139
Crankshaft, 116, 118, 123
  crank throw type, 122
  eccentric type, 116, 121, 123,
    125, 400, 401
  Scotch yoke type, 124
Crankshaft journal, 120, 121
Crankshaft seal, 116, 123, 528
  construction, 124
  repair, 528
Crankthrow, 122, 125
Crimping, 71
Crispers, vegetable, 306
Critical pressure, 34
Critical temperature, 34
Critical vibration, 121, 123
Cross-charged, 144, 150
Cryogenic,
  boiling temperature, 35, 295
  fluids, 295
  refrigerants, 295
  refrigeration, 646
  temperature range, 295, 1020
Cryogenics, 14, 35, 295, 500,
  646, 1020
  safety, 295
Crystal, piezoelectric, 755
Cubic measurement, 17-19
Cupronickel coil, 835
Current, 115, 170, 176
  alternating, 169-171
  alternating cycles, 183, 184
  controls, 266-268, 271
  direct, 169, 170
  electric, 169, 170
  relay, 221, 266-268, 271
  starting, 217
  types, 170
Curves, temperature-pressure,
  282, 283
  use of, 283
Cut-in, 254-258
Cutout, 254-258
Cutter, tubing, 41
Cutting,
  capillary tubing, 51
  compounds, thread, 53
  thread, 61
    lubricant, 62
  tubing, 41
Cutting pliers, 59
  linesman's, 59
  diagonal, 59

Cycles, 101
  absorption, 91-94, 611, 689
  air conditioning, 688, 762,
    763, 825
  alternating current (Hz), 183, 184
  automobile air conditioning, 949
  bypass, 590
  commercial mechanical, 397
  compression, 101, 102
  continuous, 91, 611, 612
  cooling, 75, 101, 611
  cycling time, 395
  domestic,
    freezer, 330, 333
    fresh food refrigerator, 307, 308
    refrigerator-freezer, 309, 310,
      313, 317, 318, 320
  electricity, 169
  heat pump, 825
  heating, 710
  hermetic, 307, 310, 313, 317,
    318, 320, 330, 333
  intermittent, 91, 611, 613
  modulating, 85
  polyphase, 203
  pressure heat, 591, 592
  refrigerating for automobile, 950
  refrigeration, 101
  reverse, 825
  secondary refrigeration, 308, 317,
    318, 320, 330
  single-phase, 203
  Sterling, 646
  three-phase, 203
Cycling, short, 355, 406, 561
Cylinder,
  arrangement, 400, 415
  color code, 297
    R-11, 289
    R-12, 286
    R-22, 287
    R-134a, 303
    R-500, 289
    R-502, 290
    R-503, 292
  dimensions, 295, 296
  displacement, 595
  head, 121, 122, 530
  volume, 18
  charging, 295, 296, 359,
    360, 371
  compressor, 116, 118, 119
  disposable, 296, 342, 360
  horizontal, 634, 636
  low-side float, 161
  refrigerant, 295-297
  returnable, 295
  rotary, 129
  service, 295, 341, 360
  storage, 295
Cylindrical commutator, 215

D

Dairy cases, 479, 480
Dalton's Law, 36, 91
Damper, 796, 798
  fire, 798
  motor control, 911
D'Arsonval voltmeter, 174

dBA Scale, 674-676
DC (see Direct current)
DDC (see Direct digital control)
Deaeration, 732
Decimals, 15, 1007
Decimeter, 17
  cubic, 18
  square, 17
Deep vacuum, 387-389
Defrost control clock, 277
Defrost control timer, 277
Defrost controls,
  automatic, 271, 272
  comfort cooling, 258-261
  hot gas, 274, 275
  semiautomatic, 273, 274
Defrost cycle, 315, 316, 320,
  325, 333, 334, 435-443
Defrost system,
  clocks, 277
  cycles (see Defrost cycle)
  terminator, 315
  timers, 313, 315, 317, 319, 334,
    449, 451
Defrost systems,
  commercial, 423, 435-443
  domestic, 308, 312, 316
  electric heating, 98, 99, 312,
    441, 442
  hermetic, 308, 312, 316
  hot brine, 440
  hot fluid, 441
  hot gas, 97, 274, 312, 316,
    378, 436-450, 823
  nonfreezing solution, 440
  programmable, 453
  reverse cycle, 435, 442
  warm air, 443
  water, 440, 441
Defrosting,
  air, 423
  automatic, 271, 272, 312, 616
  clocks, 277
  controls (see Defrost control)
  evaporators, 435
  heat pump, 835
  heat sources, 568
  manual, 308
  solid state, 453
  type evaporator, 423
  valve, hot gas, 374
  warm air, 443
Degreasing, 527
  solvents, 384
  vapor, 69
Degree days, 667, 946
Degrees (see Temperature)
  absolute, 14
  Celsius, 14
  Centigrade (see Celsius)
  Fahrenheit, 14
  Kelvin, 14
  Rankine, 14
Degrees, electrical, 215
Degrees of,
  a circle, 18, 19
  an arc, 18, 19
Dehumidifiers, 276, 277, 694
Dehumidifying equipment, 276,
  277, 775

Dehumidifying systems, 761
Dehydrated oil, 119, 122, 123, 137
Dehydration, 478
  of system, 391
Dehydrator-receiver (see Drier-receiver)
Dehydrators (see Drier)
Deicing controls, 276
Delta, 15
  motor circuit, 224
  transformer, 204, 205
    closed, 205, 208
    open, 205
Demand meter, 206
Density, 23
  relative, 23
Deodorizers, 814
Department of Transportation (DOT), 295, 983, 984
Desert bag, 75, 76
Desiccants, 376, 654, 1015
Design,
  baffle, 577
  condensers, 106
  evaporator, 583
  furnace, 710
Design temperatures, 934
De-superheating, 415
Detector leak,
  366-369, 972
  sound, 344, 345
Determining,
  dryness of refrigerants, 1019
  heat leakage, 573
  usage load, 573
Development of refrigeration, 11
Devices, fastening, 71
Devices, gas control, 724
Devices, humidifying, 684
Dew point, 652, 909, 1019
  flue gas, 726
Diacs, 194
Diagnosing troubles (see Troubleshooting)
  external, 350
  internal, 355, 374
Diagnostic code, 323
Diagonal pliers, 59
Diagrams, enthalpy (see Enthalpy)
Diagrams, wiring, 174, 247
Dial-stem thermometers, 63
Diameter, 18
Diaphragm, 115, 142, 143, 145, 546
  automatic expansion valve, 143, 144
  thermostatic expansion valve, 145, 148, 149
Dichlorodifluoromethane, R-12, 284-286, 301, 303, 977
  characteristics, 285, 1024
Dichloromonofluoromethane, R-21, 1024
Dichlorotetrafluoromethane, R-114, 1024
Dichlorotrifluoroethane (CHCl$_2$CF$_3$), R-123, 301, 303, 527, 978, 979
Dictionary of Technical Terms, 1032-1046

Die, pipe, 52
Dielectric, 189
  electronic leak detector, 367-369
  of oil, 1019
  value, 1019
Dies, 62
  threading, 62
Diestock, 52, 62
  power driven, 53
  ratchet, 53
  standard, 52
Difference,
  pressure, 20
  temperature, 15, 34
    calculating, 15
Differential, 254-258
  adjustment mechanisms, 256
  adjustments, 249, 254-258
    cut-in type, 254
    cutout type, 254
    double type, 254
    thermostatic expansion valve, 543, 544
  amplifiers, 198
  controlling instrument action, 249
  types, 254
Diffusers, 796
Digital,
  computer control, 917
  control signals, 184
  manifold gauge, 66
  multimeter, 175
  thermometer, 63
  voltmeter, 175
Digits, 15
Dimensions, 17
  cylinders, 295, 296
Diodes, 181, 194
Direct current, 169, 170
  ammeter, 176
  motors, 185, 226
Direct digital control (DDC), 917
Direct drive socket wrench, 56
Direct expansion system, 541, 542
Discharge header safety spring, 122
Discharge line piping, 608
Discharge valves, 120, 122
Discharging refrigerant,
  commercial system, 509
  domestic system, 355, 393, 394
Discoloration, refrigerant oil, 300
Disk, valve, 123
Dismantling,
  compressor, 384, 525
  general instructions, 379, 521, 522
  hermetic units, 379
  refrigerating system, 379, 521, 522
  system, 379
Dispenser, milk, 492
Dispenser light, 323
Dispensing freezer, 486
Displacement, compressor, 137, 595
  variable, 954
Displacement, piston, 137, 595
Display case, 478, 579
  dairy products, 479, 480
  frozen foods, 481

open, 399, 480, 481
Disposable refrigerant cylinders, 296, 342, 360
  hand valve, 296
Dissipator, heat, 102
Distillate, fuel oil, 737
Distribution, air, 779
Distributor, refrigerant, 159-161
Distributor tube, 159, 161
District heating and cooling systems, 863
Diverter, 718
Dividing block, 128, 129
Dome hat, 116
Domestic,
  freezers, 168, 305, 326-335
  fresh food refrigerator, 305, 306-309
  hot water use, 870
  refrigeration, 305
    mechanical, 11
  refrigerator humidity, 305
  refrigerator temperature, 297, 298, 305, 339
  refrigerator-freezer, 305, 309-327
  system,
    charging, 371, 372, 393
    compression, 101
    compressors, 116
    condensers, 107, 108, 306
    discharging, 369, 383
    evaporators, 103, 306
    installing, 343
    motor controls, 115, 249-253, 264, 265, 352
    refrigerant controls, 81, 110, 111
  systems, absorption, 611, 614, 616
  water heating, solar,
    one tank system, 870
    two tank system, 870
Door alarm, 323
Door construction,
  gaskets, 337
  mechanisms, 336
  seal, 337
Door open monitor, 323
Door removal, 395
Doors,
  heat leakage, 933
  improperly hung (warped), 338
Double broach socket wrench, 56
Double offset box wrenches, 56
Double thickness flare, 44, 45
Double type differential adjustment, 254-256
Double-acting compressor, 126
Double-duty case, 479
Dowel pin, 386
Draft, 662, 782, 783, 816
  forced, 710, 718, 804
  induced, 710
Draft control, 717, 718, 816
  instruments, 816
Draft gauge, 662
Draft indicator, 662, 820
Draft measurement, 662
Drain heater, 834

Drain sump,
  bypass line, 317
  heater, 318
Drain system, chest-type freezer,
  330
Drier, filter-, 96, 101, 102, 342,
  365, 374, 471, 472, 959
  use of, 105, 106, 109, 110,
  376-378, 380, 471, 472
Drier coils, 315, 423
Drift punch, 70
Drill bits, 60, 61
Drill presses, 60
Drills, 60, 61
  electric, 60
  fractional-size, 60
  letter-size, 61
  number-size, 61
  tap, 61, 62
  twist, 60, 1022
Drink dispenser, 485, 486, 489,
  490
Drinking water cooler, 86, 87
Drip pan, 308
Drive pulley, 701
Driver, nut, 56
Drives, compressor motor, 123
Drop, pressure, 158, 365,
  541, 800
Drop, voltage, 171, 183, 184
Drums, refrigerant
  (see Cylinders)
Dry bulb temperature, 651
Dry bulb thermometers, 651
Dry capacitor, 216, 237
Dry cell battery, 169
Dry evaporators, 428, 541
  refrigerant controls, 543
Dry ice, 34, 90, 91
  refrigeration, 90, 91
  safety, 91
Dry system (see Domestic refrigeration), 103
  evaporator, 103
  removing, 541
Dryers (see Driers)
Drying, hermetic systems, 365,
  376-378, 380
  oven, 389
Dryness, determining, 387,
  470, 650
Dryness of refrigerants, 1019
Dual-pressure regulator, 154
Duckbill pliers, 60
Duct,
  calculations, 799
  construction, 788
  controls, 914
  elbows, 803
  heaters, 752
  installations, 750, 752
  insulation, 789
  joint riveter, 788
  maintenance, 804
  noise, 781
  perimeter, 789
  problems, 804
  resistance, 799
  sizes, 792
  stat, 857, 901

systems, 784, 857
  tapering, 791
  temperature, 793
  ventilators, 788
  volumes, 792
Ducts, 679, 783, 965
  air, 783
  automobile air conditioning, 699,
  950, 965
  loose, 344
  rectangular, 784
  return air, 803
  system, 784
  round, 790
Dust, 649, 668
Dyes, refrigerant (leak detector),
  366, 367
Dynamometer, 241

E

Early refrigeration, 11
Eccentric, 116, 123, 125,
  400, 401
Eccentric type crankshaft, 116,
  121
Ecology, 676
Edge connector, 326
Education requirements, 1031
Effect, cooling, 30, 32
Effect, ice melting, 1006
Effect of pressure, 14, 28, 29
Effective latent heat, 589
Efficiency,
  compressor, 595
  furnace, 713
  motor, 239, 553, 597
  volumetric, 137, 595
Ejector, 701
  nozzle, 701
Elbows, 54, 803
Electric,
  controls, 247
  current, 169, 170
  cycle, 171
  defrosting, 97-99, 311, 312, 441,
  442
  drills, 60
  furnaces, 750
  heater defrost system, 441, 442
  heaters, 139, 683, 697, 748
  heating, 683, 748
    appreciations, 748
    baseboard, 749
    building design, 749
    building insulation, 946
    elements, open wire, ribbon
    type, tubular cased, 749
    primary controls, 899
    principles, 748
    supplementary, 752, 873
  humidifiers, 757
  ignition, 900
  gun type, 742
  resistance heat,
    forced warm air, 683
    hydronic, 685
    room units, 686
    wires, 310
  sensors, 899
  supply, 201

thermostats, 882
  water valve, 463, 464, 538
Electric motor,
  applications, 213
  armatures (see Rotor)
  bearings, 213, 238
  belts, 240
  centrifugal switch, 215, 238
  cleaning, 239
  commutator, 215
  compressors, 116, 117
  connections, 219
  construction, 213, 217
  design, 213, 220
  external troubles, 243, 350
  fuses, 227
  grounding wire, 227, 232, 236
  horsepower, 227, 596
  lubrication, 238
  operation, 191, 214
  overhauling, 238, 242, 383, 553
  phase loss monitors, 225
  principles, 191, 214
  protection, 227
  radio interference, 235, 236
  relay, 214, 230, 266-271
  removing, 238, 242, 379, 553
  rotation, 217, 226, 244
  rotor, 130
  safety controls, 446, 447
  servicing, 236, 242, 352, 383,
  384, 552
  speed formula, 220
  standard data, 231
  starter, 266, 447
  starting and running windings, 217
  stator, 134
  synchronous speed, 216
  temperature, 230
  thermistor, 227, 230
  test stand, 241, 554
  transistor switching, 234
  troubles, external, 237, 243, 350,
  553
  types, 135, 213
  wattage, 217, 227
  windings, 217, 223
Electric motors, 213
  capacitor-start, induction run, 215
  commercial, 401-404
  conventional, 135
  direct current, 226
  elementary, 191
  fan, 233
  four-pole, 192, 193, 216
  hermetic, 135, 220
  induction, 214
  open type, 135, 213
  permanent split capacitor (PSC),
  223
  polyphase, 223
  repulsion-start, induction-run, 215
  single-phase, 193, 203
  split-phase, 214
  three-phase, 203, 223, 225
  two-pole, 192, 216
Electric refrigerators (see Domestic
  system)
Electrical,
  ammeter, 176, 177

batteries, 170
burnout, 135
capacitor, 170, 188-190
circuit,
  breakers, 169, 170, 209, 210
  protectors, 209
  symbols, 172
  testing instruments, 211
circuits, 169, 172, 182, 183,
  247-249, 640, 898, 965
  degrees, 215
  domestic freezer, 305, 328-331,
    333-335
  domestic refrigerator, 305, 308
  domestic refrigerator-freezer,
    310, 311, 314, 315, 318-320,
    321, 324-327, 552
  ice maker, 262-264
  safety review, 279, 280
codes, 209, 1014
condenser, 170
conductors, 169, 180, 181
connections, 210, 505
control panel, 169
controls, 115, 247
degrees, 215
dry cells, 169
efficiency, 201
energy, 24, 31, 169, 1006
external servicing (hermetic), 350
fuses, 169, 170
generator, 169, 188, 190
grounding, 202, 203, 279
horsepower, 177
instrument connecting and handling,
  180
instruments, 174, 176, 178, 179,
  236, 271, 277, 394, 552
internal servicing, 355
leads, 181
load, 177
magnetic fundamentals, 169
materials, 180
meter, 820
nonconductors (insulators),
  179, 181
ohmmeter, 179
potential (emf), 169, 190
power, 177, 201
power factor meter, 178
pressure, 169, 172, 174
relays, 208, 209
safety, 212
semiconductors, 179, 181
shunt, 180, 181
standard symbols, 174
storage battery, 169
supply, refrigerator-freezer, 342
switches, 210, 211
symbols, 172, 1013
tester, compressor, 394
three-wire receptacle, 202, 203
transformers, 204-206, 208
troubles, 211, 236, 242, 270
troubleshooting, 352
units, 1013
voltmeter, 171, 174, 175, 177
wattmeter, 177, 178
wiring, 169, 172, 181, 209, 210,
  247-249, 352

wiring diagrams, 174
Electrical-magnetic fundamentals,
  169
Electricity, 169
  alternating current (ac), 169-171
  ampere, 174, 176, 182
  basic, 169
  capacitance, 170, 188
  conductivity, 169, 181
  conductors of, 34
  coulombs, 176, 188
  current, 169, 170, 176
  cycles, 164, 183, 184
  direct current (dc), 169, 170
  electron, 169, 174, 176, 181
  electrostatic, 169
  farad, 170, 188
  from solar energy, 874
  generating, 169
  grounding, 172, 202
  Hertz (Hz), 169, 183, 184
  impedance, 178
  kilovolt, 174
  kilowatt, 177, 201
  megawatt, 201
  microfarad, 188
  microvolt, 174
  millivolt, 174
  ohm, 174, 179, 182
  phase, 177
  power factor, 177, 201, 202
  proton, 174
  reactance, 190
  resistance, 171, 178, 179, 182,
    183
  static, 169
  storage, 188
  types, 169
  voltage, 169, 174, 182
  voltage drop, 171, 183, 184
  watts, 177, 201
Electrodeposition, 868
Electrodynamic (dynamotor movement) voltmeter, 175
Electrolux system, 11, 94
Electrolysis, 724, 1023
Electroyte, 169
Electrolytic capacitor, 188
Electromagnetic,
  amplitude, 865
  field, 169, 186, 187
  inductance, 193
  induction, 188
  inductors, 193
  voltmeter types, 174,175
  wave, 865
  wavelengths, 865
Electromagnetism, 186, 188, 953
  left hand rule, 186, 188
Electromagnets, 186
  polarity, 188
  solenoid, 187
Electromechanical electrical
  generating, 169
Electrometer, 169
Electromotive force (emf), 174, 201,
  1013
  counter emf, 177, 193
Electron, 169, 1012
  flow, 169, 1012

holes, 90
theory, 180, 1011
Electronic,
  air cleaner (EAC), 808, 810
  circuitry voltmeter, 175
  cleaner control, 910
  control system, 879, 967
    automobile air conditioning, 967
    diagnostics, 967
    microprocessor, 967
  high-vacuum gauges, 392
  leak detector, 341, 366-369
  monitor consoles, domestic, 320,
    322, 325
  sight glass, 470
  sound tracer, 341, 344, 345
  thermostats, 851, 891
    programmable, 892
    solid state, 880
    variable speed motors, 234
Electronics, 169, 194, 1011
  pulse wave, 184
  solid state, 181
Electrons, 169, 174, 176, 181,
  186, 188
Electroscope, 169
Electrostatic,
  air cleaners, 261, 810
  electricity, 169
  filter, 810
  type voltmeter, 175
  water treatment, 538
Element, thermoelectric, 114
Elementary,
  absorption, 91, 611
  electric motor, 191
  mechanical refrigerator, 32, 33
  refrigerator, 32
Ell, street, 53
Emery, 68
EmF (see Electromotive force)
Emissivity adjustment, 1005
Emits, 867
Enamel, 338
End bell, 213, 239
End play, 213, 385
Endothermal, 1026
Energy, 24
  electrical, 24, 31, 169, 1006
  heat, 24, 31, 101, 1006
  kinetic, 24, 1006
  mechanical, 24, 31, 101, 1006
  potential, 24, 1006
  radiant, 865
  solar, 865
  stored, 24
  thermal, storage, 839, 859
  total, 862, 1021
Energy audit, 919
Energy consumption, 916, 919
Energy conversion factors, 24,
  31, 946
Energy Efficiency Ratio (EER), 231,
  609, 776
Energy management, 879
  control system, 879
    types, 917
      determination and usage, 919
      diagnostics and repair, 922
    energy audit, 919

energy consumption, 915, 919
energy utilization index, 919
heat loss and gain factors, 916
smart building; 915
total, 879, 914
Energy units, 31
Energy Utilization Index (EUI), 919
Engine-driven systems, 472
English micrometer, 67
English units (see U.S. conventional)
Enthalpy, 34, 35, 587
specific, 35
Enthalpy diagram,
R-12, 284, 285
R-22, 287, 288
R-134a, 302
R-500, 289
R-502, 291
R-503, 292
Entrophy, 1010
Environment, 301, 676
Environmental Protection Agency
(EPA), 301, 977
Environmental test chamber, 100
Enzymes, 305
Epoxy resins, 52, 53
EPR (see Evaporator pressure
regulator)
Equalizer tube, 158, 159
Equilibrium, state of,
refrigerants, 283
Equivalents,
area, 17, 1007
energy conversion, 24, 31, 1006
flow, 1009
force, 24
heat, 13, 25-27, 1010
latent heat, 27, 28
linear measure, 17, 1007
liquid measure, 1008
mass, 19
power, 24
pressure, 20, 21, 1008
refrigerative effect, 31
specific heat, 26
specific volume, 23
temperature, 14-16
velocity, 1008
volume, 18, 1007
weight, 19, 1008
Error, offset, controlling instrument
action, 250
Estimates, service, 562
Ethane (R-170), 294
Ethyl chloride (R-160), 1024
Ethyl hexanol, 631
Ethylene glycol, 869
EUI (see Energy Utilization Index)
Eutectic, 1019
plates, 423, 424, 497, 633
solution, 497, 585, 646
Evacuating, 74
compressor, 387, 509, 525
data, 371
equipment, 387, 510
evaporator, 387, 541
hermetic system, 371
liquid line, 509, 541, 542, 558
suction line, 548, 558
system, 371, 387-390, 509, 555

Evaporating,
cooling, 775
pressures, 28, 298
temperatures, 28, 297, 298
Evaporation, 27, 28, 75
source, 1015
Evaporative,
condenser, 76, 421, 422, 693
cooling, 703, 775
refrigeration, 75, 76
Evaporator,
area, 576, 581
capacities and calculations, 576,
581, 584, 585
code installation, 514, 515
coils, heat pump, 834, 835
construction, 103, 422, 580
defrosting, 308, 435
design, 103, 422, 580, 583
fan, 104, 312, 313, 423, 425
fins, 423, 424, 580
installing, 381, 503, 515, 577
location, 578
mounting, 435, 436, 503, 515
noncode installation, 501, 503
removing, 379, 541
repairing, 53, 382, 542
aluminum, 382
stainless steel, 382
replacing, 355
servicing, 347, 382, 542
size, 297
standard temperature, 283
system, 633
temperatures, 297, 298
types, 103, 423, 580
Evaporator coils (see Evaporators)
Evaporator pressure regulator
(EPR), 963
Evaporator repair kit, 54
Evaporators, 12, 36, 101, 103
air conditioning, 761
air-cooling, 104, 580
A-type, 848, 849
automobile air conditioning, 961
beverage, 426-428
blower type, 423, 425, 426
commercial, 104, 422-435, 578
defrost type, 311, 315, 319, 423
domestic, 103, 305, 307, 313,
316
dry-type, 103, 428
expansion, 541, 542
expendable refrigerant system, 633
fin-type, 103, 423, 424
flat-type, 848
flooded, 103, 838
forced circulation, 423, 424
forced convection, 105, 423
freezer, 306, 328-333, 423, 425
freezing dispenser, 434, 435
frosting, 347, 423, 826
ice cube maker, 428-434
immersed, 427
liquid cooling, 104, 426
multiple, 81, 82, 399
natural convection, 423
nonfrosting, 423, 424
plate-type, 424, 633
pressure-type, 427

shelf-type, 103, 104
shell-type, 103
slant-type, 848, 849
starved, 142
submerged, 426, 427
wall-type, 103, 105
Exchanger, heat, 80, 141, 443,
839
Exfiltration, 780, 937
Exhaust muffler, 136, 137
Exhaust parts, 127
Exhaust valve, 102, 120, 122
Exothermal, 1026
Expander, 52
Expansion valve,
adjustments, 142
construction, 110, 111, 142,143
installation, 543
operation, 110, 111
removing, 544
repairing, 545, 546
servicing, 544
systems, 79, 80, 141, 144
testing, 545, 546
Expansion valves, 79, 110-113,
141
automatic (AEV), 79, 110, 112,
141-144
automatic bypass, 144
automotive, 961
bellows type, 142
diaphragm type, 143
pressure controlled, 110, 112
pressure limited, 153, 154
temperature controlled, 111, 113,
144
thermal-electric (THEXV), 141,
152, 153
thermostatic (TEV), 80, 110-112,
141, 144-152, 155, 156,
158-160, 828, 829
Expendable refrigerant, 32,
294, 633
R-704, 294
R-728, 294
R-744, 294
refrigeration, 32, 88, 497, 633
Exponent, 15
Extension cord, insulated, 248
plug configurations, 248
External drive compressor,
115, 123
commercial, 400, 403, 412
four-cylinder V-type, 115
installing, 531, 554
motors, 134
servicing, 520, 523
system, 77
External equalizer, 158, 159
External motors, 214, 237
installing, 554
servicing, 552
External servicing, hermetic systems,
350
diagnosing troubles, 350
guide, 350
operations, 350
cabinet hardware, 350
cleaning, 350, 353
electrical, 350, 352

ice cube maker, 350
  noise, 350
Extrinsic semiconductors, 194

F

Fahrenheit,
  absolute scale, 14, 16
  conversion formulas, 14, 16
  temperature scale, 13, 14
  thermometer, 14
Fail-safe action, 264
Fan
  capacity, 806
  cleaning, 807
  motors, 233
  pressure, 806
  servicing, 242
  speed controls, 411
  speeds, 806
Fans, 104 (also see Blowers)
  air conditioning system, 804
  attic, 807
  axial flow, 805
  basement, 807
  radial flow, 804
Farad, 170, 188
Faraday, Michael, 11, 93, 95
Faraday experiment, 93, 95
Farm vehicle air conditioning, 969
Fast freezers, 484
Fast freezing, 13, 646
Fast freezing case, 484
Fasteners, spring, 71
Fastening devices, 71
Featheredging, 338
Feedback (closed-loop) control
  systems, 249
Feedback control, 897
  types, 898
Feeler gauge, 337
Feet of water column, 21
Female rotors, 130
Fiberglass insulation, 306
Field, electromagnetic, 169
Field, magnetic, 184, 186, 187
Field erected, 848
Field pole, 186, 193
Field serviceable hermetic unit,
  401, 404
Figure-eight motion, brazing, 51
File brushes, 71
File card, 71
Files, 70
  flat, hand, mill, 42, 70
Filler metals, brazing, 47, 49
  copper alloy, 47
  silver alloy, 47, 49
Filter,
  chambers, 740
  cleaning, 808-810
  efficiency, 670
  measuring, 670
  servicing, 809, 812
Filter-drier, 101, 102, 109, 110,
  300, 334, 342, 365, 376, 377,
  471, 472
  diagnosing trouble in, 374
  hermetic, 376
  installing, 355, 378, 380

liquid line, 109, 110
  low-side, 105, 106
  replacement cartridge, 472
  replacing, 355, 376-378
Filters, 81, 549, 808, 809, 814
  adhesive, 809
  air, 808
  automobile, 959
  carbon, 478, 814
  electrostatic, 810, 812
  loaded, 809
  oil saturated, 809
  throw-away, 809
  water, 463, 538
Filters and strainers, 740, 747
Fin and tube type evaporator, 103
Fin tube, 103, 597
Finishes, cabinet, repairing, 338
Finned-forced convection condenser,
  107, 108
Finned-static (natural convection)
  condenser, 107-109
Fins, 416, 423, 424, 542, 583
Fire dampers, 798
First Law of Thermodynamics, 26,
  1010
Fittings,
  brass, 46
  capillary tube, 164
  commercial, 515
  compression, 52
  flared, 39, 46, 52, 342
    code number, 46
    sizes, 416
  flexible hose, 40, 41, 47
  mechanical, screw-on, 52
  metric size, 47
  pipe, 52
    types, 53
  self-sealing coupling, 851
  soft soldering, 39
  soldered, 47, 342
  tubing, 44
    brazed, 47, 48
    metric size, 47
    soldered, 47
Fixed gas, 612
Fixture temperature, 297
Flake ice, 429, 430, 490
Flame,
  adjustment, 738, 746
  color, 722
  mirror, 743
  neutral, 49
  out, 743
  propagation, 726
  test, leaks, 367
Flammable refrigerants, 283
Flange,
  blank, 526
  mounting, 739
Flapper valve, 385
Flare,
  double thickness, 44, 45
  making a, 44
  single thickness, 43, 44
  testing a, 49
Flare nut, 44, 57
  wrench, 57, 1025
Flare plug, 542

Flared tubing connections, 43, 57
Flared tubing fittings, 39, 43, 46,
  52, 960
  code number, 46
  sizes, 46
Flaring block, 44
Flaring tool, 44
  adaptors, 45
Flash gas (vapor), 105, 141,
  148, 443
Flash point, 69, 384
  of oil, 1019
Flashback, 726
Flashback arrestors, 49
Flat plate solar collector, 867
Flat valve seat, 148
Flat-type evaporator, 848
Flexible insulation, 944
Flexible service and refrigerant line,
  358, 359
Flexible tubing (hose), 40
  fittings, 40, 41
Float, 107
  high-side, 77, 79, 110, 114, 115
  low-side, 77, 79, 110, 114
Float control,
  high-pressure side, 77
  low-pressure side, 76, 161, 427
Float switch, 447
Float valves, 161, 162
  high-pressure, 161
  low-pressure, 161
  operation of, 161, 162
  systems, 76, 161, 162
Floats, 161
  pan type, 162
Floc test, 1019
Floodback, 836
Flooded evaporator, 103, 836
Flooded system, 76, 77, 103, 110,
  114, 161, 162
  coil, 426, 427
  control, 76, 161, 162, 427
Flooding, 76, 77, 512
Florist cabinets, 478
  construction, 478
Flow,
  control, refrigerant, 110
  equivalents, 1009
  meter, 820
  of heat, 13, 34
Flue, 617, 710, 717
  absorption system, 617
Flue draft control, 720
Flue gas, 726
Fluid (see Liquids)
Fluid bulb, 722
Fluid solar system, 839
Fluorescent leak detection, 367, 972
Fluorine, 283
Fluoro refrigerants (danger), 1027
Flush, 383
Flux, 47, 51
  brazing, 342, 557
  magnetic, 184, 186, 188
  soldering, 47, 49, 342
Flywheel, 77, 116
  puller, 123
Foam leak detector, 366, 367
Foamed-in-place insulation, 475, 477

Fogs, 808
Food freezants, 294
Food preservation, 11, 13, 305
Foot, 17
  cubic, 18
  square, 17
Foot pounds, 24, 59
Force, 24
  attractive, 19
  centrifugal, 134, 1012
  centripetal, 134, 1012
  electromotive, 174
  gravitational, 19
  lines of-magnetic, 184, 186, 187
  pound, 24
Force feed systems, lubrication,
  137
Force fits, 71
Forced,
  circulation evaporator, 423, 424,
    761
  convection, 34, 107, 406, 423,
    569, 584, 752
  finned condenser, 107, 108
  draft, 718
  feed oiling, 137
  warm air furnace, 679, 680, 683
Formulas, temperature conversion,
  16
Fountains, soda, 484
Four cylinder V-type external drive
  compressors, 116
Four-pipe system, 857, 858
Four-pole electric motor, 192, 193
Four-way reversing solenoid valve,
  157
Four-way valve, 825
Fractional-inch wrenches, 56
Fractional-size twist drills, 60
Fractions, 1007
Frame, motor, 213
Free electrons, 188, 1011
Freezants, 294
Freeze drying, 495
Freezer,
  alarms, 331
  burns, 305
  cabinet, 305, 479, 481, 482
    gaskets, 336, 337
    hardware, 337
    thermometer, 338, 339
  condensing units, 328-333, 397
  controls, 258, 328-333
  cycling time, 395
  electrical circuit, 305
  evaporator, 328-333
  ice accumulation, 335
  insulation, 306
  mechanism, 305
  motor controls, 80
  motors, 213-215
  wax in system, 349, 365
Freezers, 305
  automatic defrost, 311-315
  care of, 335
  chest type, 326, 328-332
  domestic, 305
  dispensing, 486
  fast, 305, 495
  frost-free, 319-321

installing, 336, 342, 343
  electrical supply, 342
  grounding, 343
  leveling, 344
  locating position of, 342
locating trouble in, 347-349
manual defrost, 309-311
repairing finishes, 338
servicing, 335
shutting down, 395
side-by-side, 320-327
starting, 344
uncrating, 342
upright, 331, 333-335
Freezing, 11, 13, 305
  cycle, ice maker, 262, 263
  dispenser evaporator, 434, 435
  fast, 13
  food, industrial, 494
  ice cream, 586
  plant, 494
  point, 14
    depression, 30
  preservation radiation, 1022
  storage data, 298
  temperature, 13, 14, 29
  temperature, brine, 34, 1020
Freon jets, 642
Fresh food refrigerator,
  cabinet, 305, 306
  compact, 306
  electrical circuits, 308, 309
  mechanisms, 307
  temperature, 305
Fresh food storage, 305
  temperature ranges, 141, 305
Friction, 136, 169
Frost back, 142, 371
Frost control,
  automatic, 271, 449
  element, 696
  hot gas, 274
  semiautomatic, 273
Frost removal, 97
Frostbite, 303
Frost-free refrigerator-freezer, 104,
  272, 319-321
  cycle, 312
  side-by-side, 320-327
Frosting, suction line, 355
Frosting type evaporator, 423
Frozen food compartments, 305
Frozen foods (also see Freezer and
  Freezers),
  air curtains, 482
  display case, 481
  fish, 570
  freezer burn, 305
  fruits, 570
  industrial, 494
  locker plants, 494
  meat, 570
  poultry, 570
  preparation, 305
  refrigerants, 297, 298
  storage, 305
    cabinet, 482
    temperature range, 141
    time, 306
  temperature, 570

vegetables, 570
Fuel,
  gas, 720, 1021
  gas burner, 679
  oil, 736, 1021
  oil additives, 736
  oil atomizing, 737
  oil distillate, 737
Full floating piston pin, 119
Fundamentals, control systems, 249
Fundamentals of refrigeration, 11
Fur storage, 493
Furnace,
  construction, 710
  controls, 712, 880, 895, 896
  designs, 710
  heat pipe, 715
  inspection, 728
  puffback, 743
  venting, 717
Furnaces,
  condensing, 714
  electric, 750
  forced warm air, 679
  gas, 711
  glycol, 714
  gravity, warm air, 679
  high efficiency, 713
  hydronic, 728
  oil, 737
  pulse, 716
  tubular, 749
Fuse, electric motor, 182, 227
Fuse plug, 614
Fuses, 169, 209
Fuses and circuit breakers, 227
Fusible links, 842
Fusible plugs, 295, 467
Fusion, latent heat of, 27, 29
Fusion processes, 865

G

Gallon, 18
Galvanic action, 504, 1023
Galvanometric, 815
Gas, 20, 23, 36, 1016
  adiabatic expansion and contrac-
    tion, 1026
  burner, 720
  charged sensing element, 144,
    151, 152
  charged thermostatic expansion
    valve, 151, 152
    collapsible, 154
  constant, 35
  control, 619, 620, 724
    combination, 901
  cross charged sensing bulb, 144,
    151
    adsorption, 151
  definition of, 23
  equation, perfect, 35
  flash, 105, 141, 148
  flow, 712
  flue, 726
  fuel, 720
  furnaces, 711
    condensing, 714
    electronic controls, 879

fired radiant heat, 753
glycol, 714
heat pipe, 715
high efficiency, 713
inspection, 728
installation, 725
primary controls, 899
pulse, 716
servicing, 727
starting, 727
heating, supplementary, 873
heavy, 23
hot gas defrosting, 97, 274, 312, 316, 436-440
hydrogen, 646
ideal (perfect), 35
isothermal expansion and contraction, 1026
law, 1026
light, 23
natural, 639
noncondensible, 152, 154, 859
pressure, 20, 712
specific gravity, 23
supply, 618, 720
values (constants), 35
valve, 619
Gas and vapor, 36, 1016
saturated, 36
Gas pliers, 59
Gaskets, 71, 123, 139, 337
composition of, 72
compressor, 139
door, 337
Gasket-type piercing valve, 364
Gasoline, 70
Gauge manifold, 341, 356-360, 505
connecting, 356
construction, 357-360
flexible service and refrigeration line, 358, 359
types, 357-360
valves, 358
Gauge pressures, 20, 21
Gauges, 20, 21, 63-67
atmospheric, 21
calibration of, 66
care of, 66
color code for, 65
compound, 21, 65-67, 346, 347
construction of, 64-67
high-pressure, 65, 346, 347
high vacuum, 392
low pressure, 65
manifold for testing and servicing, 341, 357-360, 505
McLeod, 66
mercury, 21
metric, 20
pressure, 64-67
removing, 507
temperature, 63
tension, 241
use of, 505
vacuum, 65
Gaugler, R.S., 645
Gauss, 186
Gear compressor, 130
drive, 123
Generating electricity, 169

chemical, 169
electromagnetic, 169
Generator, absorption, 91
Generator, electrical, 169, 188, 190
armature, 190
brushes, 190, 191
commutators, 190, 191
Genetron 12/31, 286
Genetrons (see Refrigerants)
Gland, packing, 58, 73
Glass, sight, 82, 109, 469, 470, 505, 520
Glass panel heaters, 753
Glass stem thermometers, 63
Glycol, 625, 869
brine, 1020
furnace, 714
Goggles, safety, 61
Grain, 18, 650, 755
Gravitational force, 19
Gravity, 19
of brine, 23
of gases, 23
of water, 23
specific, 23
Gravity warm air furnace, 679
Grille, 796
plastic, 849
Grille controls, 796
Grinder, 383, 384
Grocery cabinets, 475
Grooves, oil, 119
Ground, electrical, 172
fault protector, 203
short circuit, 172
(three-wire) power cord, 248, 279
wire, 202
Ground coil, 697, 824, 835
Grounded spot, 384
Grounding, 202, 227, 248, 279, 343, 351
adaptor, 203
prong, 202, 203
Group One refrigerants, 283
pressure-temperature curves, 282, 283
R-11, 283, 287
R-12, 283-286
R-22, 283, 286, 287, 288
R-500, 283, 288, 289
R-502, 283, 290, 291
R-503, 283, 290-292
R-744, 283, 294
Group Two refrigerants, 283, 292
pressure-temperature curves, 292, 293
R-40, 292
R-717, 292-294
R-764, 292
Group Three refrigerants, 283, 294
R-170, 294
R-290, 294
R-600, 294
Gun, oil burner, 737
Gun type, 737
wiring diagram, 903

H

Hacksaws, 41, 71
Halide,

leak detector, 283, 367, 368
refrigerants, 283, 301
testing for leaks, 367
torch, 283, 289, 290, 366-368
Halogen compounds, 283
bromine, 283
chlorine, 283
fluorine, 283
iodine, 283
Halogen gas, 367
Halogenated refrigerants, 301, 977
Hammers, 59, 342
Hand braces, 60
Hand tools, 54
cold chisel, 70
files, 70, 71
hacksaws, 71
hammers, 59
mallet, 59
pliers, 59, 60
punches, 70
screwdrivers, 60
stamps, 60
wrenches, 54, 56-58
Hand truck, appliance, 342
Hand valves, 358, 466
disposable refrigerant cylinders, 296
needle valve, 341
piercing valve, 364
Hard hat, 846
Hard-drawn copper tubing, 39, 40
Harden, work, 46
Hardware, 336
Harvest cycle, ice maker, 262-264, 326
Hastelloy, 148
Head pressure, 137, 297, 406
control, 406
for common refrigerants, 541
high side, 297
safety cutout, 265
Heads, screw 71
Heat, 9
absorber, 102
absorbing capacity of ice, 27-30
anticipators, 884
area, 588
chart, 587, 588
comfort zone, 663
conduction, 33
conductivity, 34, 1023
conductors, 34
convection, 34
definition, 13
diagram, 286, 587
dissipator, 102
distribution controls, 911
energy, 24, 31, 169
enthalpy chart, 587
equivalents, 26, 1010
exchange, 673
exchanger, 141, 325, 326, 443, 592, 711
flow, 13, 34
gain, 916, 926, 938
in air, 779
insulation, 672
lag, 943
lamps, 543

leakage, 565, 566, 573, 926-933
  air conditioning, 926
  calculations, 565
  water cooling, 585
load, 565, 568, 573, 576, 925,
  926, 938, 939, 945
  calculations, 573, 576, 925, 938
  commercial, 565
  cooling, 926, 938
  heating, 925
  sun, 663, 939
  total, 573, 576
loss, 916, 925
  basement, 936
  unheated spaces, 937
measurement, 13
mechanical equivalent, 26
pipe, 637
pipe furnace, 715
radiation, 34
recovery systems, 491
reducing, 938
removing, 11, 12
sensible, 26, 27, 677
sensible, ratio, 656
sink, 672, 704
sources, buildings, 943
sources, defrosting, 568
storage, solar energy, 867, 868
  closed water system, 869
  rocks, 868
  warm air system, 869
sublimation, 90
superheat, 105, 145, 589
types,
  electric, 748
  latent, 27, 28, 589
  radiant, 704, 752
  residual, 885
  resistance, 748
  sensible, 26, 27
  specific, 26, 29, 589, 1006
  supplementary, 752, 873
unit, 25
Heat of,
  compression, 31
  fusion, 29
Heat pump,
  air coil, 834
  coils, 827
  compressor, 827
  condensers and evaporators, 834
  controls, 261, 276, 828
  cooling cycle, 696, 825
  cycles, 699, 825
  defrosting, 835
  exchanger, 592, 839
  ground coil, 824, 835
  heating cycle, 695, 696, 825
  Hi/Re/Li system, 838
  installation, 823, 836
  lake coil, 834
  motors, 827
  mounting, 834, 836
  operation, 695, 823
  performance, 826
  purpose, 695, 823
  reversing valves, 829
  servicing, 836
  systems, 825, 827

theory, 823
valves, 829
water heaters, 840
well water coil, 835
wiring diagrams, 829
Heat pumps, 101, 695, 697, 823,
  869, 873
  air coil, 834
  ground coil, 835
  lake coil, 834
  safety, 863
  solar, 839, 840, 869, 873
  through-the-wall, 836
  two-speed, 838
  well water coil, 835
Heat transfer, 33, 101
  conduction, 33
  control, 34
  convection, 34
  radiation, 33
  rate, 927
  regenerator, 646
  respiration, 568
  wire, 310
Heater core, 961
Heaters,
  accessory, 139
  automatic electric defrost, 315
  baseboard electric, 749
  crankcase, 119, 139
  drain, 834
  drain sump, 318
  duct, 752
  electrical, 310, 748, 835
  forced convection electric, 752
  glass panel, 753
  mullion, 315
  radiant electric, 753
  stile, 334, 335
  unit, 754
Heating,
  coil, 754
  controls, 260
  copper-to-steel joint, 51
  cycle, 756, 825
  equipment, 708, 709
  equipment for air conditioning, 707
  installation, 725, 733, 735, 745
  servicing, 727, 735, 746, 754
  thermostat, 885, 889
  types, 707
    district systems, 863
    electric, 697, 748
    gun type oil, 710
    radiant, 705
    solar energy, 866
    solid fuel, 747
    supplementary, 697, 752, 836
    underfloor hydronic system, 728
  value, fuel, 712, 1021
Heavy gas, 23
Helical rotors, 130-133
  female, 130
  male, 130
Helium, 646, 819
  liquid (R-704), 35, 294, 295
Helix design, 427
Hermetic refrigeration, 11
  adaptors, 359
  adding oil, 372, 393

assembling, 387, 388
centrifugal compressor, 134, 135
charging, 356, 359, 360, 369-372
  rebuilt, 393
circuit wiring, 308, 311, 314,
  319, 321, 324, 327, 331,
  333-335
cleaning, 353
compressor, 116, 117, 412, 413
condensers, 306, 307, 313, 316
controls, 230, 249-253, 264, 265
core valves, 364
cycles, 307, 308
defrosting systems, 308, 312,
  435-443
diagnosing electrical troubles, 350
dismantling, 379
driers and filters, 376
electrical circuit, 305, 308, 310,
  311, 314, 315, 318, 319-321,
  324-331, 333-335
evacuating, 355, 387
installations, 501, 514
installing, 341
leaks, repairing, 369
leaks, testing for, 366-369
mechanism, 401
motor controls, 115
motor terminals, 225
motors, 135, 220-225
  capacitor-start, capacitor-run, 213,
  214, 221, 222
  capacitor-start, induction-run,
  213-215, 221
  electrical characteristics of, 220
  hermetic split-phase, 221
  installing of, 557
  polyphase, 223
  polyphase three-phase, 225, 443
  removal of, 555
  repairing of, 243, 384, 557
  servicing, 243, 554-556
  starting a stuck motor, 243
  types of, 221
oil, 372, 1019
overhauling, 383
  equipment, 383
reciprocating compressor, 123
refrigerants, 281
removing the unit, 373
repairing compressor, 384-386,
  532
repairing motors, 384
repairing units, 532
rotary compressor, 128, 130,
  385, 386
safety, 395
screw type compressor, 130, 132
service valves and adaptors, 359,
  360, 361
servicing, 341, 532
  capillary tubes, 374
  check valves, 456
  compressor testing after overhaul,
  386, 530
  diagnosing electrical troubles,
  350, 352
  diagnosing mechanical troubles,
  355
  external, 350

guide, 350
internal, 350, 355, 365, 374
in-the-field service operations,
    347-349
mounting tubing service valves,
    359-361
overhauling sealed units, 350,
    383
process tubes, 362
rewelding compressor domes,
    386
sealed, 637
service valves and adaptors,
    359-361
tools and supplies, 341, 342
troubleshooting, 348, 349
use of gauge and service
manifold,
    356-360
use of high-vacuum pump,
    390-394
using Schrader valves, 364
systems, 306, 309, 311, 315,
    319, 326, 331, 387
testing rebuilt, 393
tubing, 369
types,
    commercial, 401
    domestic, 306
    serviceable, 127
units, 401
welding, 386
wiring (see Hermetic refrigeration,
    electrical circuits)
Hertz (Hz), ac cycles, 169, 171,
    183, 184
    sine wave, 184
Hex key, 58
Hex wrenches, 58
Hg (mercury), 21
High head pressure, 532
High limit control, 679, 682
High-efficiency furnaces, 713
    condensing, 714
    glycol, 714
    heat pipe, 715
    pulse, 716
    recycling, 713
High-pressure cutout, 557
High-pressure gauges, 65, 346,
    347
High-pressure safety control, 446
High-pressure side, 33, 101, 102
    float (HSF), 77, 114, 115, 141,
    161
        principles, 161, 162
        system, 77, 79
    flooded system, 114, 115
High-side pressure, 102, 297
High-side restriction, 355
High-side service valve, 106
High-vacuum gauges, 392
High-vacuum pump, 341, 355,
    390, 391
Hinges, 336
Hi/Re/Li system, 838, 839
Holding tanks, 693, 869
Holes, electron, 90
Horizontal type compressor, 400
Horsepower, 24, 177

electrical, 177
heat equivalents of, 31, 1006
hour, 1006
metric equivalent of, 24
Hose, flexible tubing, 40
Hose clamp fitting, 960
Hot brine defrost, 440
Hot gas,
    automatic defrost system, 312,
    316
    bypass, 400, 426, 427, 436,
    437, 548
    defrost, 97, 312, 316, 426, 427,
    436, 437, 450, 823
        controls, 274, 275
        troubles, 378
        valve, 374
    driers, 315
    mechanisms, 315
    troubles, 378
Hot junction, 199
Hot water heating, 728, 870
Hot wire, 172, 270, 271
Hot wire anemometer, 659
Housing, compressor, 126
Humidifier, 276, 277, 694, 755,
    757
    cabinet type, 758
    heat load, 945
    installing, 757
    servicing, 757
Humidifying equipment, 708
    operation, 707
    systems, 707
Humidistat, 261, 276, 909
Humidity, 36, 650
    definition of, 36
    domestic refrigerator, 305
    excessive, 756
    increasing, 756
    low, 652
    relative, 36, 650, 652, 658
Humidity control, 11, 841, 856
Humidity controls, 261, 276, 277,
    655, 909
Humidity measurement, 653
Humidity recorder, 655
Humidity water source, 756
Hunting, 141, 151, 156, 160, 161
Hydraulic controls, 412, 858, 859
Hydrochloric acid, 380
Hydrochlorofluorocarbons (HCFCs),
    301, 977, 979
Hydrofluoric acid, 380
Hydrogen (R-702), 23, 295, 612,
    646, 649, 650
    liquid, 35
Hydronic, 728
    furnace, 679, 681, 685, 707
    installation, 733
    radiator, 729, 731
    servicing, 734
    thermostats, 893
    underfloor heating system, 728
    zone, 729, 732
Hygrograph, 671
Hygrometer, 63, 651, 653
Hygroscopic,
    elements, 909
    substances, 654, 761

I
Ice, 11, 13, 20
    accumulation, 335, 347
    as a refrigerant, 75
    bank controls, 275
    controls, 261-264
    crystals, 13, 305
    cube bin, 86
    cube defrost cycle, 430, 431
    cube freezing cycles, 430
    cube maker, 86, 428-434, 449,
    450, 490, 538
        controls, 261-264, 448
        cycles, 263, 264
        external servicing, 350
        installing, 344
    dispenser, 322-328
    in cabinet insulation, 347
    level switch, 326
    maker, 86, 322-328, 428-434
        control, 448
        cycles, 326
        flake, 429, 430
        ice level switch, 326
        nugget, 429
        solid state controlled, 326, 328
    melting capacity, 30, 31
        effect, 1006
        equivalent, 29-31
        IME, 1006
        temperature, 13
    refrigerating effect, 29
    refrigeration, 75
    refrigerator, 75
    selection pad, 323
    specific heat capacity, 29
    types,
        artificial, 11, 75
        dry, 34, 90
        manufactured, 490
        natural, 11, 75
        zero, 34
Ice and salt mixture, 30, 75, 427
Ice cream,
    cabinets, 484
    freezing, 586
    makers, 486, 491
Icing control, 907
Ideal (perfect) gas, 35
Identification plate, motor com-
    pressor, 350
Identifying refrigerants, 281, 301
    by number, 281
    instrument, 301
Ignition method, 722
Ignition, electrical, 722, 742, 900
    transformer, 742
Immersed evaporator coil, 426
    brine, 34, 427
    sweet water, 34, 427
Immersion, 646
Impedance, 178
Impeller, 128, 134
Incandescent units, 283, 748
Inch, 17, 1007
    cubic, 18
    square, 17
Inch pound, 24, 59
Inches of mercury vacuum, 21
    conversion into psia, 21

Inches of water column, 21
Inclined-type compressor, 400
Increments, 63
Incubators, laboratory, refrigerated, 492
Index, Comfort-Health (CHI), 666
Indicator cards, 1013
Indicators, airflow, 820
Indicators, liquid, 82
Indicators, moisture, 300, 470, 471
Indoor climate, 663
  airflow, 820
Induced magnetism, 185, 186
Inductance, 193, 194
  mutual, 194
  self, 193
Induction, electromagnetic, 188
  capacitor, 188, 215
  motors, 188, 214
  repulsion-start, 213, 215
  split-phase, 215
Inductive loads, 177
Inductive reactance, 190
Inductors, 193
Industrial,
  applications, 493
  freezing of food, 494
  storing of foods, 493
Inefficient unit, 560
Inertia, 803
Infiltration, 780, 937
Infrared,
  heat controls, 905
  lamp, 753
  radiation, 14, 866
  thermometer, 1005
Inhibitor, 626
Injection, oil, 134
Inside coil, 831, 832
Inside diameter (ID), 164
Inspecting residential central air
  conditioning, 855
Inspection,
  periodic, 559
  report, 1002
  steam heating, 736
  warm air furnace, 728
Installation, 343, 501
  absorption refrigeration, 630
  air conditioning, 852
  code, 514, 518
  condensing unit, 501, 514
  domestic refrigerator, 343
  ducts, 789
  evaporator, 515, 577, 578
  gas burner, 725
  heat pump, 823, 836
  heating coil, 754
  hydronic system, 733
  noncode, 501
  oil burner, 745
  refrigerant lines, 504
  rooftop units, 841-843
  solar space heating, 870
  steam heating, 735
  testing,
    code, 518
    noncode, 501
  tubing, 503
  valves, manual, 466

window unit, 763, 764
Installing,
  absorption refrigerators, 630
  air-cooled condenser, 533
  capillary tube, 376
  commercial systems, 501
  console air conditioners, 772
  cooling unit (see Evaporator)
  direct expansion evaporators, 542
  domestic units, 341-343, 616, 630
  evaporators, 503
  expansion valve, 144, 543
  external drive compressor, 531
    motor, 554
  filter-drier, 355, 378, 380
  gauge manifold, 356
  gauges, 64-66
  hermetic motor compressor, 557
  humidifiers, 757
  ice cube maker, 344
  instruments, tools, supplies,
    341, 342
  motor compressor, 381
  oil tank, 743
  open type compressor (see external
    drive compressor)
  refrigerant lines, 515
  residential central air conditioning,
    852
  thermostats, 277, 278
  tubing, 39, 515
  two-temperature valves, 547
  water valves, 541
  water-cooled condenser, 417
Instructions to the owner, 335
Instruments, 63, 815
  air conditioning, 815
  air distribution, 815
  airflow measuring, 815, 820
  ammeter, 176, 236, 237
  anenometers, 541, 659, 815, 817
  barometer, 21, 672, 815
  circuit testing, 211
  combustion efficiency, 816
  connecting and handling of, 180
  draft control, 816
  electrical, 174, 176-179, 243,
    394, 395
    connecting and handling, 180
  gauges, 63-67
  hygrometer, 653
  installing and servicing, 341
  manometer, 22, 815, 817
  phase loss monitor, 225
  pitot tube, 660
  recording, 655, 815
  thermometer, 13, 14, 63
  velocimeter, 660
  voltmeter, 174, 236, 237
  water analysis, 820
  watt meter, 177, 236
Insulated tools, 212
Insulation,
  air conditioning, 944
  automobile, 965
  batt type, 944
  building, 926
  commercial, 565, 1023
  domestic, 306
  fiberglass, 306

freezer, 306
  heat, 34
  ice in the, 347
  motor, 220
  refrigerator, 306
  tubing, 504
  urethane, 306
Insulators (nonconductors), 34,
  179, 181
Intake,
  muffler, 136, 137
  ports, 127
  valves, 101, 122
Integrated circuits, 194, 199, 922
  boards, 194, 200, 922
  chips, 200, 892
Interference, radio, tv, 235, 236
Intermittent absorption system,
  91-93, 613
Intermittent cycle, 91, 613
Internal combustion engine, 646
Internal servicing, hermetic, 350
Internal threads, 61
Internal trouble, hermetic system,
  diagnosing, 355, 374
  electrical, 355
  lack of refrigerant, 355
  liquid refrigerant in compressor,
    355
  locating, 365
  restriction on high side, 355
  servicing operation, 355
  sweating or frosting suction line,
    355
Internal-type pipe wrench, 58
International Institute of Ammonia
  Refrigeration, 294
International system of units (see Le
  Systeme International d'Unites)
Intrinsic semiconductors, 194
Inverter, 196
Invoice, service, 1000
Iodine, 283
Ionizing air, 810
Ions, 1012
1 R drop, 183, 201
Iron vane voltmeter, 175
Isothermal expansion and contrac-
  tion, 1026

J

Jet nozzle, 701
Joint, repairing, 53
Joint, sweat, 47
Joint designs, brazing, 50, 51
Jones, Frederick McKinley, 11
Joule, 13, 24, 25
Journal, crankshaft, 120, 121
Junction,
  cold, 640
  hot, 640
Junction box, 248

K

"K" factor, 566, 927
Kata thermometer, 1005
Kelvin, 13-16, 26
  conversion to Celsius, 16
  conversion to Rankine, 16
  temperature scale, 14

Kelvinator, 11
Kerosene burners, 91, 92
Kerosene-fired refrigerators, 91,
  612, 613
Key, hex, 58
Key, torx, 58
Kilocalorie, 26
Kilogram, 19
Kilojoule, 25
Kilometer, 17, 1007
Kilopascal (kPa), 19
Kilovolt, 174
Kilovolt amperes (KVA), 207
Kilowatt, 24, 177, 201
  hour, 1006
Kinetic energy, 24, 1006
Klixon valve, 270

L

Laboratory, refrigerated
  incubators, 492
Lack of oil, 372
Lack of refrigerant (see Shortage of
  refrigerant)
Ladder diagrams, electrical circuit,
  248, 249
Lag, 885, 943
Lamps,
  heat, 543
  quartz, 753
  radiant heat, 753
  ultraviolet, 478, 814
Lapping,
  blocks, 385, 529
  compound, 385, 529
  tool, 385
Large capacity TEV, 158, 160
Larsen, J.M., 11
Latches, 336
  magnetic, 336
Latent heat, 27, 28, 587, 588
  applications, 27, 588
  effective, 587, 589
  equivalents, 27
  of condensation, 27, 28
  of fusion, 27-29
  of melting, 27-29
  of sublimation, 34
  of vaporization, 27, 28
  of refrigerants, 28, 588, 1025
  values, 281, 588, 1024
Law, Ohm's, 182
Law, Pascal's, 20
Law, Wiedmann-Frank, 34
Laws of refrigeration, 101
  Boyle's, 102, 154, 1025
  Charles's, 102, 154, 1025
  Dalton's, 36
  gas, 1026
  Lenz's, 191, 194
  Peltier, 88
Laws of thermodynamics, 13, 26,
  31, 101, 1010
Le Systeme international
  d'Unites, 11
Leads, electrical, 181
Leak detection,
  R-11, 287
  R-12, 285
  R-22, 287

R-500, 289
R-502, 290
R-503, 292
R-717, 294
Leak detector, 366-369, 509
  bubble solutions, 366, 367
  electronic, 341, 366-369, 519
  flame (halide torch), 283,
    366-368, 519
  refrigerant dye, 366, 367
  ultrasound, 344, 345
  ultraviolet fluorescent, 367, 972
Leakage, heat, 565, 566, 573, 926
Leaks,
  emergency measures, 362
  locating, 355, 366
  repairing, 355, 369
  testing for, 283, 366, 367,
    509, 972
Lefthand motor rule, 192
Lefthand rule, 186-188
Legionnaire's disease, 421
Length measurement, 17
Lens, solar collector, 874
Lenz, Emil, 199
Lenz's Law, 191, 194
Letter drill, 61, 1022
  sizes, 1022
Leveling refrigerator-freezer, 344
  devices, 336, 344
Lever type swaging tool, 52
Lid counterbalancing, 326, 336
Lift pump oil circuits, 842
Light,
  diffused, 809
  pilot, 722
  speed of, 1011
Light energy, 169
Light gas, 23
Light switches, 342
Lights, testing, 211
Limit controls, 249, 894, 896,
  899, 917
Limit switches, 888, 897
Limiters, pressure, 153, 279,
  408, 443
  adjustable, 154
Limitizer valve, 408
Line,
  liquid, 101, 102, 110
    capacities, 601
  refrigerant, 468, 601, 960
  suction, 101, 102, 104
    capacities, 603
Line tap piercing valve, 364
Line thermostats, 882
Line voltage transformer, 206
Linear measurement, 17
  equivalents, 17, 1007
  symbols, 17
Linesman's pliers, 59
Liquefication, 639
Liquid,
  absorbent, 611, 808
  charged power, 251
  charged sensing bulb, 144,
    150, 151
  collector, 839
  cooling, 426
    evaporators, 104, 426

cross charged sensing bulb, 144,
  150, 151
equivalents, 1008
floodback, 836
helium (R-704), 35, 294, 295
hydrogen, 35
impurities, 808
injection, 400
level indicator, 82, 371, 952, 960
line, 101, 102, 109, 558
  filter-drier, 105, 106, 471
  servicing, 558
measure equivalents, 1008
natural gas, 88
nitrogen (R-728), 35, 88, 294,
  295, 497, 633, 634, 646
nitrogen control valve, 634, 635
oxygen, 646
pressure, 20
receiver, 101, 102, 108, 110,
  422
  sizes, 601
refrigerant, 12, 28, 101
  moisture in, 1016, 1019
sensing element, 150
slugging, 828, 836
state, 20
Liquid crystal display (LCD), 65
Liquids, 23
  volatile, 645
Liquid-vapor valve, refrigerant
  cylinder, 296
Liter, 18, 1008
Lithium bromide, 95
Lithium bromide system, 627,
  628, 631
Lithium chloride, 95
Loads,
  cooling, 565, 827, 925,
    926, 938
  electrical, 177
  heat, 565, 925, 938, 939, 945
  inductive, 177
  noninductive, 177
Localized controller, 917
Locating trouble, 344, 345,
  347-349, 355, 365, 366, 374,
  519, 557
Location of,
  condensing unit, 501
  evaporator, 578
  refrigerator, 342, 501
  thermostat, 893
Locked rotor condition, 193
Locker plants, 494
Lockout relay, 894, 895
Loose baffles and ducts, 344
Loosening,
  bolts, 56
  nuts, 56
Louvers, 548
Low humidity, 652
Low voltage thermostat, 883
Low wax content, 300
Low-pressure gun type oil
  burner, 737
Low-pressure safety cutout,
  265, 266
Low-pressure side, 76, 101, 102,
  114, 141

Index

filter-drier, 105, 106
float (LSF) control, 110, 114, 141, 161, 162 •
  design, 161, 162
  principles, 161, 162
gauges, 65
pressure, 102, 297
  controls, 461, 962
  limiter, 279
service valve, 105, 106
systems, 76, 161
valve, 73, 161
Low-side float (LSF), 141
  cylinder, 161
  open pan (pan-type), 161
  sealed ball, 161
Low-side pressure, 102
  limiter, 279
  motor control, 110, 114
LP fuel, 612, 720
LRSV, 541
L-shaped Allen wrench, 58
Lubricant, thread cutting, 61
Lubricating oil, 106, 137
Lubrication, compressor, 137, 300, 372, 551, 1019
  centrifugal, 134, 135
  motor, 238, 553
  reciprocating, 137
  screw-type, 137
Lubrication methods, compressor,
  ball bearing, 238
  bushing, 238
  pressure (force feed) system, 137, 138
  ring, 238
  splash system, 137
  wick, 238

M

Machine screws, 61, 71
Magnet, 185-187
  permanent, 185, 252
Magnet pump, 622
Magnetic,
  attraction, 184, 185
  behavior fundamentals, 188, 189
  center, 242
  clutch, 952
  compass, 184
  field, 184, 186
  flux, 184, 185, 188
  gasket, 337
  latches, 336
  lines of force, 184, 186, 187, 253
  permeability, reluctance, 188, 189
  pole, 184, 185
  relay,
    amperage, 267
    current, 185, 221, 267, 268
    electronic, 269
    hot wire, 270, 271
    potential, 268, 269
    solid state, 269
    starter, 447, 448
    thermal, 269, 270
    voltage, 267
  repulsion, 184, 185
  slippage, 216

Magnetic-electrical fundamentals, 169
Magnetism, 169, 184
Magnets, electro-, 156, 157, 185-187
Maintenance, automobile air conditioning, 973
Makeup air, 662
Male rotors, 130
Mallets, 59
Manifold,
  gauge, 65, 341, 356-360, 505
  lines, 361-364
  service, 357, 358, 505
Manometer, 22, 671
  mercury, 65
Manual control systems, automobile air conditioning, 967
Manual defrost, 306-311
Manual valve, 73, 466
Marine refrigeration, 301, 499
Market, coolers (see Walk-in coolers), 476
Mass, 18
  equivalents, 19
Mass and weight, 18
Master switch, 247
Materials, refrigeration, 39
Matter, theory of, 1011
Maximum Operating Pressure (MOP), 154
  thermostatic charge, 154
  thermostatic expansion valves, 155
Maximum voltage (V_{max}), 171, 202
McLeod gauge, 65
Mean effective pressure, 596, 1013
Measurement, 17
  angular, 18, 19
  area, 17, 18
  draft, 662
  filter efficiencies, 670
  heat, 13, 14
  humidity, 650-655
  linear, 17, 979
  noise, 674
  pressure, 20-22
  temperature, 13, 14
  volume, 17, 18
Measuring,
  rules, 67
  tapes, 67
  tube, 371, 372
Meat, frozen, 305, 570
Mechanical,
  cycles, 397
  domestic refrigeration, 11
  energy, 31, 169
    conversion factors, 31
  equivalent of heat, 31
  fittings, screw-on, 52
  refrigeration, 11, 99
  refrigerator, operation, 11, 12, 32
Mechanism,
  absorption, 611
  air conditioning, 761, 779, 950
  automotive, 950
  commercial, 397

domestic,
  freezer, 305, 326, 328-330, 332, 333
  fresh food refrigerator, 307
  refrigerator-freezer, 309, 310, 313, 315, 319, 320, 322, 323, 325
  system, 341
hermetic refrigerator, 401
hot gas refrigerator, 315
refrigerator, 307
Megavolt, 555
Megawatt, 201
Melting,
  capacity, ice, 30, 31
  latent heat, 27
  point, 13
  temperature of ice, 13
MEP, 596, 1013
Mercury, 13
  barometer, 65
  bulb, 557
  column (Hg), 21
  conversion into psia, 21
  gauge, 21
  inches (of), 21
  manometer, 65
  switch, 251, 252
  tube, 557
Metal,
  cleaning, 527
  determining type, 369
Meter, 17, 1007
Meter, electrical 820
Metering devices, automobile air conditioning, 961
Metering type two-temperature valve, 454, 455
Methods of adding refrigerants, 369
Methyl chloride (R-40), 292
  pressure-temperature curves, 293
  properties of, 1016
Methyl formate, 1024
Methylene chloride (R-30), 1024
Metric,
  gauges, 20
  micrometers, 68
  system, 11 (also see Equivalents)
  units, 11, 17, 18
Metric size,
  bolts, 56
  nuts, 56
  pipe thread, 61
  tube fittings, 47
  tube, 40
Microampere, 180
Microfarad, 188
Micrometer, 67
  caliper, 67
  English or metric, 67, 68
Micron (μ), 17, 391, 866
Microorganism, 305
Microprocessor control, 132, 200, 892, 917, 967
Microvolt, 174
Mil, circular, 210
Mile, 17
Milk cooler, 491
Milk dispensers, 492
Mill file, 42

Millibar (mb), 21
Millimeter (mm), 17, 1007
Millivolt, 174, 884
Mineral oil, 137
Mineral spirits, 69, 384,386
Minimum Stable Signal (MSS), 150
Minutes,
  of a degree, 18
  seconds of a, 18
Mirror, inspection, 345
Mirrors, 874
Mixtures, ice and salt, 30,
  427, 1020
Mobile air conditioning, CFC
  standards, 986
Mobile home
  air conditioning, 758,
  refrigerators, 622
Modular refrigeration system, 489
Modulating, 85
  controls, 898
  cycle, 85
Module, 90, 640
Moisture,
  absorbent, 376
  in air, 36, 650, 1015
  in cabinet insulation, 347
  in compressor, 380
  in installation, 335
  in refrigerant, 300, 301
  in systems, 109, 365, 470,
  471, 548
  indicator, 301, 470, 471, 1019
  liquid refrigerants, 1016, 1019
  refrigerant oils, 1019
  removal, 365, 391, 548
  sources, 1015
  table of safe moisture
  contents, 301
Mold, 808
Molecular,
  motion, 102
  sieve, 471, 1015
  theory, 13, 102
  vibrations, 33
Molecules, 13, 23, 102
Monitor, odor, 669
Monitor, phase loss, 225
Monitor top, 11
Monochlorodifluoromethane (see
  R-22), 283, 286, 345
Monochloromonofluoromethane
  (R-31), 286
Motor (also see Electric motors),
  analyzer, 387, 394, 395
  applications, 213
  assembling, 386
  bearings, 239, 240, 527
  burnout, 370, 380, 471, 472
  centrifugal switch, 215, 238
  characteristics, 227
  circuits, 205
    Delta, 224
    Wye, 224
  cleaning, 239
  compressor fault, 374
  compressor, 116, 135, 311
    parallel, 405, 407
    tandem, 405, 406
  connections, 218

control relay, 309, 311, 330, 331
controls, 115, 249-253, 264,
  265, 342, 443, 557
  differential adjustment, 254-257
  overload protection, 265, 266
  pressure, 77, 115, 443-445
  range adjustment, 254
  safety, 265, 266, 446, 447
  temperature, 115, 230
  thermostatic, 77, 115, 251, 445
data, 231
design, 191, 213, 220
direction of rotation, 217
efficiency, 597
electrical testers, 394, 395
fan motors, 233
fans, 804
fuses, 227
horsepower, 24, 227
identification plate, 350
installing, 381, 557
insulation, 220
left-hand motor rule, 192
lubrication, 238
magnetic center, 242
MEP, 596, 1013
oiling, 238
opening, 384
overloads, 228, 229
phase loss monitors, 225
pressure control, 77, 78, 115,
  264-266, 443-445
principles, 213
protection, 228, 229
rebuilding, 384-386
removing, 379, 380, 553, 555
repair, 238, 242, 384
replacement terminals, 225
replacing, 381, 382
rotation, direction of, 217
rotor, 134, 135, 191, 213, 405
safety control, 228, 229,
  446, 447
servicing, 236, 238, 242, 383,
  384, 553, 554, 556, 557
size, 596
  heat input method, 596
  mean effective pressure
  method, 985
speed, 216
  formula, 220
starters, 443, 447, 448
starting relays, 214, 266-269
  checking and testing, 270, 271
starting stuck motor, 243, 355
stator, 134, 136, 191, 213, 405
stuck, 243, 355
synchronous speed, 216
temperatures, 230
terminal, 225, 353, 354
testing, 375, 386, 387,
  553, 554
testing stand, 241
theory, 191
thermistor, 227, 230
thermostatic motor controls, 77,
  252, 308, 445
troubles, 236, 238, 242,
  552-557
  locating, 374

types, 213
  ac, 196
  capacitor, 215, 222
  dc, 185, 226
  domestic, 135, 191
  elementary, 191
  external drive compressor, 135
  heat pump, 828
  hermetic, 116, 213, 220
  induction, 188, 214
  open type, 213, 554
  permanent split capacitor,
  223, 770
  polyphase, 223
  repulsion-start, induction-
  run, 215
  servo, 185
  shaded pole, 234
  single-phase, 213
  split-phase, 214, 221
  universal, 170, 226
  variable speed, 234
  winding, insulation, 217
Motorized water valves, 540
Mount, flange, 739
Mounting sensitive bulb, 155
Moving instructions, 501
Moving magnet voltmeter, 175
Moving plate voltmeter, 175
Moving vane voltmeter, 175
Mufflers, 135-137, 469, 959
Mullion heater, 315
Multi-cylinder compressors, 118
Multimeter, 175, 176, 820
Multiple evaporator systems, 81,
  82, 399
  belts, 123
  code installations, 514
  installations, 501, 517
  servicing dry coils, 541
Multiple thermocouple, 884
Multistage systems, 643
Multistage thermostats, 890
Mutual inductance, 194

N

National Board of Fire Underwriters,
  The, 283
National coarse (NC) thread, 52, 61
National Electrical Code, 209, 986
National Electrical Manufacturers
  Association (NEMA), 210
National fine (NF) thread, (SAE), 46,
  61
National pipe (NP) thread, 46, 52
National Refrigeration Safety Code,
  The, (NRSC), 283
National safety code (electrical),
  1014, 1027
Natural convection, 34, 749
  evaporators, 423
  finned-static condenser, 107, 108
Natural gas, 88, 618, 639, 720
Natural ice, 11, 7 5
Nautical mile, 17
Needle valve, 107, 114, 144,
  147, 341
Negative charges, 169
Negative temperature
  coefficient, 252

Neon (R-720), 295
Neutral flame, 49
Newton (N), 19, 20, 24
Nipple, 53
Nitrogen, 35, 39, 88, 294, 295,
  495, 633, 634, 646, 649,
  650, 819
  dry, 495
  liquid (R-728), 35, 88, 294, 295,
  495, 633, 634, 646
  use of, pressure testing, 366
No-frost refrigerator-freezer,
  319-327
  cycle, 312
Noise, 561, 673, 781
  acoustics, 782
  elimination, 344
  levels, 782
  locating, 344, 345
  measurement, 674
  reduction, 135
  reflectors, 781
  vibration, 781
Noise in,
  compressor, 135
  duct, 781, 803
  expansion valve, 544
  hermetic system, 556
  seal, 528
  valve, 385
Nominal size tubing, 39, 41
Noncandescent units, 748
Noncode installation, 501
Noncondensible gas, 152,
  154, 859
Nonconductors, 179, 181, 1014
Nonfreezing solution defrost
  system, 440
Nonfrosting evaporator, 423, 424
Nonfrosting system, 423
Noninductive loads, 177
Normal indicator, 323
North pole, magnetic, 184
Nozzle,
  ejector, 701
  jet, 701
NTC (see Negative temperature
  coefficient)
N-type semiconductors, 90
Nuclear power, 810
Nugget ice machine, 429
Numbered twist drills, 61
Numbers, rounding, 15
Nut driver, 56
Nut pliers, 60
Nuts, 71
  flare, 44, 57
    wrench, 57
  loosening, 56
  metric size, 56
Nylon flexible tubing fittings, 40, 41
Nylon yarn hose, 40

O

Occupational Safety and Health
  Administration (OSHA) standards,
  384, 473
Octane, 676
Octyl alcohol, 478, 631
Odor,

monitor, 669
  of refrigeration oil, 300
  removing, 478, 479, 814
Off cycle, 81
Off-on recorder, 341
Offset box wrenches, 56
Offset (error), controlling instrument
  action, 250
Offset screwdriver, 60
Ohm, 174, 179, 182
Ohmmeter, 63, 179, 341
  connections, 179
Ohm's Law, 182
Oil,
  acidity, 300
  adding to the system, 355, 372,
    373, 387, 551, 973
  additives, 137
  antifoam inhibitor, 1019
  binding, 114
  burner, 737, 739
    electric ignition, 742
    flame failure, 746
    forced warm air, 680
    gun-type, 737
    high-pressure, 737
    hydronic, 681
    installation, 743, 745
    low-pressure, 737
    nozzle, 737
    primary controls, 902
    pump, 739
    servicing, 746
    starting, 745
    vent plug, 745
  charging apparatus, 373
  control systems, 459
  cooler, 329, 330
  dielectric property, 991
  filter, 740
  flash point, 991
  floc point, 991
  furnaces, 737
  grades, 736
  grooves, 119
  heating, supplementary, 873
  injection, 134
  lack, 372
  level regulator, 459, 460
  level sight glass, 524
  lubricating, 106, 137
  moisture, 301
  operation, 106
  overcharge, 372
  pour point, 73, 300
  pressure safety control, 265,
    446, 447
  pump, 137
    gear type, 741
    rotary type, 739
    scavenger, 138
  purging, 379, 383
  refrigeration, 300
  removing, 383
  rings, 119
  sampler, 532
  separator, 85, 106, 107,
    459-461
  sight port, 391
  sludging, 300

tank installation, 743
  traps, 379, 607 (also see Oil
    separators)
  types,
    fuel, 736
    motor, 238
    refrigerant, 73, 137, 300, 342,
      964, 1019
    silicone, 538
  used in,
    automobile air conditioning, 964
    compressor, 137, 300, 372,
      551, 1019
    hermetic, 300, 372, 1019
    vacuum pumps, 391
  viscosity, 73, 137, 300,
    553, 1019
  wax content, 365
Oilless bushings, 239
Oleum, 69
On-and-off switch, 694
One-way service valve, 73
Opaque, 33
Open,
  circuit, 172
  cycle ammonia, 634
  cycle refrigeration, 85, 294, 634
  Delta transformer, 205
  display case, 480, 482
  end wrench, 54, 57
  external drive, 520
  frozen display case, 482
  motor,
    cleaning, 238
    compressor, 116
    lubrication, 238
  type system, 77, 116, 520
Open pan low-side float, 161, 162
Opening a motor compressor, 384
Operating controls, 249
Operating pressure, maximum
  (MOP), 154
Operation of compression
  cycle, 102
Operation of mechanical refrigerator,
  11, 32
Organism, micro-, 305
Orifice, 608
  sizes, 608
  tube, 961
O-rings, 41, 123, 138, 960
Orsat, 818
O-shaped wire terminal, 277, 278
Ounce, 18
Outdoor, air-cooled condenser,
  416, 548
  climate, 663
  condensing, unit, 406-411, 416
Out-of-phase, 202
Outside coil, 834, 835
Outside complete systems, 841
Over-capacity valve, 142
Overcharge of oil, 372
Overhauling,
  compressor, 526
  hermetic system, 350, 383
Overload, 229, 236
  motor, 229
  protection, 115, 229, 265, 266,
    331, 342, 351

relief valve, 467, 468
Override, 892
Overshoot, 884
Oxyacetylene, 49
  brazing equipment, 49
Oxygen, 295, 646, 649
  liquid (R-732), 295, 646
Ozone, 669
  layer, 301, 907

P

Packing gland, 58, 73
Pan type float control, 161, 162
Panel, solar, 869
Parallel circuits, 182, 247
Parallel motor compressors,
  405-407
Parallel-series circuits, 183
Partial pressure, 36
Partial vacuum, 22
Parts, cleaning, 69, 384, 527
Parts per million (ppm), 287, 300
Pascal (Pa), 19, 20
Pascal's Law, 20
Pasteurization, 305
Peen, 59
Peltier, Jean, 199
Peltier effect, 88
Pentafluoroethane (see R-125)
Percolation pump, 630
Perfect gas equation, 35
Perfect vacuum, 22, 391
Performance, compressor, 595
  coefficient, 596, 827
Perimeter drier, 310, 318
Perkins, Jacob, 11
Permanent magnet, 185, 252
  moving coil (D'Arsonval)
  voltmeter, 174
Permanent split capacitor motor
  (PSC), 223
Permeability, magnetic, 188, 189
Permeability-reluctance, 188
Petcocks, 521, 736
pH, 820
  factor, 537
Phase,
  electricity, 177
  single-, 201
  three-, 201
Phase loss monitor, 225
Phillips screwdriver, 60
Phosphorous-copper alloy wire, 342
Photocell, 181, 199, 900, 904
Photoconductors, 199
Photoelectric devices, 194, 199
Photoelectricity, 194
Photoemissive devices, 199
Photons, 865
Photosensitive detector, 904
Photovoltaic devices, 199
Photovoltaic solar cell, 874
Physical states of substances, 22
  gases, 23
  liquids, 23
  solids, 22
Physics of refrigeration, 11
Pi (π), 17
Piercing valve,
  adaptors, 364

gasket types, 364
Piercing valves, 342, 346, 347,
  356, 363, 364
  bolted-on, 363
  brazed-on, 363, 364
Piezoelectric crystal, 755
Pilot light, 722
  flame color, 722
Pilot-controlled TEV, 146
Pilot-operated absolute valve
  (POA), 963
Pilot-operated solenoid valve, 158
Pin punches, 70
Pinch-off tool, 362, 363, 379, 380
Pins, 71
Pins, piston, 119
Pipe,
  die, 52
  fittings, 52, 53
  sizes, 52
  threads, 52
    repairing, 53
    tap, 53
  wrench, 57, 58
Pipes,
  copper, 39, 342
  heat, 645
  polyvinyl chloride, 714, 716, 717
Pipe-within-a-pipe (see Tube-
  within-a-tube)
Piping,
  capacity, 600
  color code, 1022
  diagrams, 603
  heat loads, 565
  size, 600
Piston,
  cylinder, 117
  displacement, 137
  pins, 119
  rings, 116, 119
Piston-cylinder crank
  arrangement, 118.
Pistons,
  compressor, 116, 117
  Scotch yoke, 120
Pitot tube, 660, 815, 817
Plain tubing, 597
Plastic grilles, 849
Plastic tape, 342
Plastic tubing, 39, 40
Plate, condenser, 107, 108
Plates, eutectic, 423, 424,
  497, 633
Plates, evaporator, 423, 424, 633
Plates, valve, 122, 123
Plenum chamber, 950
Pliers, 59, 342
  cutting, 59
  duckbill, 60
  gas, 59
  nut, 60
  round nose, 60
  safety, 60
  slim nose, 60
Plug configurations, 210
Plug taps, 61
Plugs, fusible, 467, 614
Plumb, 236
Pneumatic control, 412, 414, 911

POA (see Pilot operated
  absolute valve)
Pocket thermometers, 63
Polarity of electromagnets, 188
Polarized circuit, 334
Polarized iron voltmeter, 175
Pole, magnetic, 184, 185
Pollen count, 670
Pollutants, 668
Polyethylene tubing, 40
Polyphase cycle, 203
Polyphase motor, 213, 223
Polypropylene rods, heat
  storage, 868
Polyurethane, 306, 477
Polyvinyl chloride (PVC) pipe, 714,
  716, 717
Ponded roof, 703, 944
Poppet valves, 122
Porcelain finishes, 338
Port,
  exhaust, 127
  intake, 127
  valve, 829
Portable,
  absorption refrigerators, 621
  charging cylinder, 371, 372
  thermometer, 14
  thermostat, 893
Positive charges, 169
Positive temperature coefficient,
  252, 266
Potential, electrical, 169, 174, 190
Potential energy, 24, 1006
Potential relay (magnetic), 266,
  268, 269, 271
Potentiometer, 14, 234, 902
Pound, 18
Pound force, 24
Pounds per square inch (see Psi,
  Psia, Psig)
Pour point, refrigeration oil, 73,
  300
Power, 24
  circuits, 204
  cords, three-wire (ground), 248
  element, 150, 151
  element mounting, 155
  factor, 177, 201, 202
    meter, 178
  loss, 183
  of ten, 15
  saver switch, 315, 326
  sources,
    electrical, 201
    nuclear, 810
ppm (see Parts per million)
Prefixes, unit, 15, 1013
Preparation of food for refrigerating,
  and freezing, 305
Preservation of food, 11, 13, 305
Presses, drill, 60
Pressure,
  control, 264-266
  definition, 19
  differences, 20
  disk, 738
  drop, 158, 593, 602, 799, 800
  effect on boiling point, 14, 28,
    29, 298

effect on freezing temperature, 14, 29
effect on latent heat, 27, 588
element,
  above atmospheric, 264
  below atmospheric, 264
equivalents, 19, 1008
gauges, 63-65, 341
limiters, 153, 279, 408
  adjustable, 154
lubrication, 137
measurement, 21, 22
motor control, 77, 115, 264-266, 397, 443-445
  commercial, 443
  domestic, 247
operated altitude valves (POA), 258, 259
recorder, 341
reducer valve, 454
regulating valves, 454, 455
regulator, 366, 367, 724
  dual, 154
  evaporator, 963
scales, 21, 394
sound, 674
system, lubrication, 137
testing for leaks, 366, 367
  safety, 366
type evaporator, 427
units, 19-21
valve, suction, 165
water valve, 463-465
Pressure of,
  condensation, 297
  gas, 20
  liquid, 20
  solid, 20
Pressure-heat,
  areas, 588
  cycles, 591
  diagrams, 587, 591
  enthalpy chart, 285-293, 587
  graphs, 587-594, 1016-1018
Pressure-limiting expansion valve, 153, 154
Pressures, 14, 19
  absolute, 21
  atmospheric, 19-21
  barometric, 21
  compression cycle, 102
  compressor valve, 122
  condensing, 297
  critical, 34
  electrical, 169, 172, 174
  evaporating, 297
  expansion valve, 110, 141
  gas, 712
  gauge, 21
  head (also see Head pressure), 137, 297
  high-side, 102, 297
  low-side, 102, 297
  maximum operating (MOP), 154
  mean effective (MEP), 596
  negative, 780
  partial, 36
  positive, 780
  refrigerant, 297
  standard, 1023

standard atmospheric, 14
suction, 76
total, 36
vapor, 32
velocity, 660
Pressurestats, 880, 898, 907
Pressure-temperature,
  charts, 298
  conditions, 345, 346
    in compression cycle, 102
  curves, 282, 283
  tables, 298
Pressure-time diagram, 164, 165
Prick punches, 70
Primary container, 634
Primary controls, 895, 917
  air, 794
  electric heat, 905
  gas, 899
  oil, 902
  power, 204
  programmable, 892, 917
Printed circuit boards, 200, 326
Process tube, 356, 359, 360, 362
  adaptor, 341, 359, 360, 362
  kit, 362
Processing ammonia vapor, 639
Processing plant, 494
Product storage, 570
Programmable controllers, 892, 917
Propane (R-290), 294, 622
Proportional controlling instrument action, 249-251
Propylene glycol, 869
Protector, circuit, 209, 210, 223
Protector, overload, 115, 229, 236, 265, 266, 342, 351
Proton, 174
PSC (permanent split capacitor), 770
Psi, 19, 21
Psia, 20, 21
Psig, 20, 21
Psychrometer, 651
  aspirating, 652
  recording sling, 651
Psychrometric,
  chart,
    comfort, 658
    formula, 1027
  measurement, 653, 656, 658
  properties of air, 655
PTC (see Positive temperature coefficient)
P-type semiconductors, 90
Puffback, 727, 743
Pull-down, 151
Puller, wheel, 526
Pulley, 77, 116, 123, 240, 957
  drive, 701
  variable pitch, 240
Pulsation, 737
Pulse, 714, 716, 717
Pulse furnace, 716
Pulse wave, 184
Pump,
  circulating, 728
  condensate, 425, 426
  heat, 101, 823

oil, 137
  scavenger, 138
oil burner, 738
  gun type, 739
  rotary, 390, 391
  vacuum, 387-394, 519
Pump down, 387-394, 441, 509
Pumping ratio, 1010
Punches, 70
  center, 70
  drift, 70
  pin, 70
  prick, 70
Punch-type swaging tool, 52
Purging, 74, 506
  oil, 379, 383
  refrigerant, 355, 370-372, 379, 506, 532, 533, 535
  service line, 341, 356, 357
PVC pipe (see Polychlorinated vinyl pipe)

Q

Quality, refrigerant, 588
Quantity, refrigerant, 302
Quart, 18
Quartz lamps, 753
Quick-connect coupling, 851
Quick-disconnect wire terminals, 277, 278

R

R factor, 926, 931
R-11 (trichloromonofluoromethane), 281, 283, 287, 289, 297, 299, 303, 367, 977
  applications of, 287
  cleaning use, 69
  cylinder color code, 289
  leak detection, 287
  properties of, 287
R-12 (dichlorodifluoromethane), 281, 283-286, 297, 299, 301, 303, 342, 345, 346, 367, 964, 977
  applications, 285
  cylinder color code, 286
  leak detection, 285
  pressure-heat (enthalpy) diagram, 284, 285
  properties, 284
R-13, 1016
R-13Bl, 1016
R-21, 1016
R-22, (monochlorodifluoromethane), 281, 283, 286-288, 297-299, 301, 342, 345, 367
  applications, 286
  cylinder color code, 287
  enthalpy diagram, 287, 288
  leak detection, 287
  properties, 286
R-23, 283
R-30, 1016
R-31 (monochloromonofluoro-methane), 286
R-40 (methyl chloride), 292, 1016
R-113 (trichlorotrifluoroethane), 297, 299, 977, 1016

R-114, 977, 1016
R-115, 283, 977
R-123 (dichlorotrifluoroethane), 301, 303, 527, 978, 979
R-124 (chlorotetrafluoroethane), 301
R-125 (pentafluoroethane), 301
R-134a (tetrafluoroethane), 301-303, 978, 979
R-160 (ethyl chloride), 1016
R-170 (ethane), 294
R-290 (propane), 294, 1016
R-500, 281, 283, 288, 289, 297-299, 367
  applications, 289
  cylinder color code, 289
  enthalpy diagram, 289
  leak detection, 289
  properties, 288
R-502, 281, 283, 290, 291, 297-299, 342, 367
  applications, 290
  cylinder color code, 290
  enthalpy diagram, 291
  leak detection, 290
  properties, 290
R-503, 281, 283, 290-292, 297, 299
  applications, 290
  cylinder color code, 292
  enthalpy diagram, 292
  leak detection, 292
  properties, 291
R-600 (butane), 294, 1016
R-702 (hydrogen), 295
R-704 (liquid helium), 294, 295
R-717 (ammonia) 40, 281, 292-294, 297, 298, 304, 1016
  applications, 292
  leak detection, 294
  properties, 293
  safety, 304
  safety code, 294
R-720, 295
R-728 (liquid nitrogen), 294, 295
R-729 (air), 295
R-732 (oxygen), 295
R-740 (argon), 295
R-744 (carbon dioxide), 283, 294, 1016
R-764 (sulphur dioxide), 292, 304, 1016
Radial blades, 134, 135
Radial-type compressor, 400, 415
Radiant energy, 865, 874
Radiant heating, 705, 752
  gas fired, 753
Radiant heating installation, 753
Radiant heating lamps, 753
Radiant heating source, 753
Radiation, 866, 867
  infrared, 14, 865, 866
  preservation of food by, 1022
  ultraviolet, 865, 866
Radiation of heat, 33, 34, 865
Radiator, 729-731, 735
Radiator of heat, 34
Radio interference, 236
Radiometer, 14
Radius, 17

Railway car construction, 497
Railway car refrigeration, 497
Range adjustment, 249, 253, 254, 258
  adjusting screws, 254
  controlling instrument action, 249
Rankine,
  conversion formulas, 16
  temperature scale, 14
Rapid pressure balancing (RPB), 149
Ratchet die stock, 53
Ratchet wrench, 58, 59
  reversible, 58
Ratings, refrigeration and air conditioning, 30
Ratio, compression, 137
R-C charging network, 234
Reach-in cabinet, 475
Reactance, 188, 189
  capacitive, 190
  inductive, 190
Reamer, 41, 239
  bearing, 239
Rebuilt systems, charging, 393
Rebuilt systems, testing, 393
Receivers, 537
  drier, 959
  liquid, 101, 102, 108, 110, 422
  sizes, 598
  repairing, 537
Receptacle,
  configurations, 210
  three-wire grounded, 202, 203
Reciprocating compressor, 116, 117, 119, 137, 138, 415, 520
  rebuilding of, 384
  swash plate, 125, 126
Reclaiming refrigerant, 977-987
  procedure, 982
  safety, 986
Recommended temperatures for commercial refrigerators, 568, 570
Recording,
  ammeter, 206
  instruments, 815
  pressure gauge, 341
  thermometer, 63, 381, 519
Recovery, refrigerant, 562, 977-987
  equipment, 979
  safety, 986
  system, 564
Recreational vehicle air conditioning, 774
Rectifier, electrical, 181, 194, 195
Rectifiers, silicon-controlled (SCR), 181, 194, 196
Recycling,
  energy, 713
  refrigerant, 562, 977-987
    equipment, 981
    safety, 986
RED system (G.E. data) 352-354
Reducing coupling, 53
Reed valve, 122
Reflector of heat, 34
Refractory cement liner, 737
Refractory material, 680

Refrigerant,
  amount required in system, 301
  applications, 298, 299
  changing, 301
  characteristics, 1016
  charge, 355, 369-372, 393, 394, 520
  checking, 301
  classification, 283
  color code, cylinders, 297
  container, 646
  controls (see Refrigerant controls)
  curves, use of, 283
  cylinders (see Refrigerant cylinders)
  discharging, 355, 393, 521
  distributor, 159-161
  dryness, 1019
  enthalpy, 34, 35
  flow control, 110-115
  head pressure, 541
  identification, 281, 301
    by number, 281
    instrument, 301
  lack (also see Shortage), 592
  latent heat, 27, 28
  leaks, 355, 362, 366-369
  line capacities,
    liquid, 603
    suction, 603
  lines, 109, 358, 359, 468, 600, 603, 960
    installing, 515
    servicing, 558
  loss, 957
  moisture, 300, 301, 365, 549, 1016
  molecules, 102
  numbers, 281
  oil, 73, 300, 342, 1019
  pipes and tubing, 47, 515, 600
  pressure-temperature curves, 282-283
    Group One refrigerants, 282, 283
    Group Two refrigerants, 292, 293
  pressure-temperature tables, 298
  pressures, 297
  properties, 281, 586, 1024
  quality, 588
  quantity, 301
  reclaiming, 977-987
    procedure, 982
    safety, 986
  recovery, 562, 977-987
    equipment, 564, 979
    safety, 986
  recycling, 562, 977-987
    equipment, 981
    safety, 986
  removing (purging), 355, 370-372, 379, 521
  requirements, 281
  safety, 303, 304
  saturated vapor, 589
  shortage, 355, 366, 369, 376, 377
  sightglass, 300, 301
  state of equilibrium, 283
  substituting, 301

superheating, 589
tables, use of, 298
transferring, 551
vapor, 28, 31, 104, 281, 1016
Refrigerant control,
  capacity, 608
  characteristics, 141
  comparing, 164, 165
  safety, 165
Refrigerant controls, 32, 110, 141,
  164, 374, 443, 543, 894
  automatic expansion valve (AEV),
    79, 141-144
  capillary tube, 81, 141,
    162-164, 374
  compression systems, 141
  dry evaporator, 543
  freezer, 328, 330, 333
  heat pump, 828
  high-side float, 77, 79, 162
  low-side float, 76, 77, 161, 162
  thermal-electric expansion valve
    (THEXV), 141, 152, 153
  thermostatic expansion valve
    (TEV), 80, 141, 144-152, 155,
    156
Refrigerant cylinder,
  color code, 297
  hand valve, 296
  safety regulations, 295, 993-996
  valves, 295, 296, 358
Refrigerant cylinders, 295-297, 550
  disposable, 296
  returnable service, 295, 296
  service, 295
  storage, 295, 296
Refrigerants (see R number), 12,
  31, 281-304, 977, 978
  air (see R-729)
  ammonia (see R-717)
  argon (see R-740)
  automobile air conditioning,
    301, 964
  azeotropic mixtures, 289
  butane (see R-600)
  carbon dioxide (see R-744)
  Carrene No. 2 (see R-11)
  chlorofluorocarbons (CFCs), 283,
    301, 977
  chlorotetrafluoroethane (see
    R-124)
  comparison of, 281
  cryogenic fluids, 295
  dichlorodifluoromethane (see R-12)
  dichlorotrifluoroethane (see R-123)
  ethane (see R-170)
  ethyl chloride (see R-160)
  expendable, 32, 88, 294, 633
  flammable, 283
  food freezants, 294
  gas, 281, 1016
  Group No. 1, 283
  Group No. 2, 283, 292
  Group No. 3, 283, 294
  halide, 283, 367
  halogenated, fully, 301, 977
  helium, liquid (see R-704)
  hydrochlorofluorocarbons (HCFCs),
    301, 977, 979
  hydrogen (see R-702)

liquid, 28, 31, 101
little used, 1016
methyl chloride (see R-40)
methyl formate (see R-611)
monochlorodifluoromethane
  (see R-22)
monochloromonofluoromethane
  (see R-31)
nitrogen, liquid (see R-728)
oxygen (see R-732)
pentafluoroethane (see R-125)
propane (see R-290)
safest, 283
sulphur dioxide (see R-764)
synthetic, 286
tetrafluoroethane (see R-134a)
toxic, 283
tricholoromonofluoromethane
  (see R-11)
water, 294
Refrigerating effect of ice, 29
  equivalents, 30
Refrigerating temperatures, 13
Refrigeration,
  basic systems, 75
  basis, mechanical, 11
  career opportunities, 1029
  compression cycle, 101, 102
  cycle, 101
    absorption, 611
    modulating, 85
    typical, 282
  development, 11
  electric heaters, 315
  fittings, 46, 47
  fundamentals, 11
  history, 11
  industry, 1030
  installation, 501
  laws, 101
  materials, 39
  operation, 11, 32, 33
  ratings, 30
  supplies, 68
    abrasives, 68
    brushes, 69
    cleaning solvents, 69
  systems,
    basic, 75
    cascade, 84
    commercial, 95, 96, 397
    compound, 82, 83
    continuous cycle absorption,
      91, 94
    expendable refrigerant, 88
    intermittent absorption, 91-93
    modular, 489
    multiple evaporator, 91, 92
  technician, educational
   requirements, 1031
  thermodynamics, 586
  ton (of), 30, 1006
  tools, 39
  troubleshooting, 348, 349,
   556, 592
  types,
    chemical, 88, 294
    commercial, 11
    cryogenic, 646
    domestic, 11

dry ice, 90, 91
early, 11
elementary, 32, 33
evaporative, 75, 76
expendable, 32, 633
hermetic, 11
ice, 75
marine, 301, 499
mechanical, 11, 32, 33, 101
open cycle, 88, 294, 633
railway car, 497
solid absorbent, 95
steam jet, 294
supermarket, 301, 475
thermoelectric, 88, 89, 640
truck, 495
unit, assembling, 387, 501
  supplies, 70
uses, 11
Refrigeration oil, 73, 137, 300
  acidic condition, 300
  additives, 137
  chemical stability, 300
  discoloration, 300
  odor, 300
  pour point, 300
  properties, 300
  thermal stability, 300
  viscosity, 300
  wax content, 300
Refrigerator,
  accessories, 305
  cabinet,
    construction, 305
    finishes, 338
    gaskets, 337
    hardware, 336, 337
    insulation, 306
    location, 342, 501
    temperatures, 297, 298, 305,
     339
    thermometers, 338, 339
  capillary tubes, 307, 313, 316
  care, 335
  charging, 369, 506
  cleaning, 335
  compressor, 307
  condenser, 306, 307, 313, 316
  controls, 258
  cycling, time, 395
  defrosting, 308, 312
  doors, 336-338
  electrical circuits, 305
  electrical supply, 342
  evaporators, 103, 305, 307,
   313, 316
  fresh food compartment, 306-308
  frozen food cabinets, 305, 479,
   480, 482, 483
  frozen food compartments, 309
  function, 12
  grounding, 343
  installing, 342, 343
  leveling, 344
  locating position, 342
  locating trouble, 347-349
  mechanisms, 307, 309, 310,
   313, 315, 319, 320, 322, 323,
   325
  shutting down, 395

starting, 344
temperatures, 297-299
uncrating, 342
wiring, 308, 311, 314, 319,
  321, 324, 327, 331, 333-335
Refrigerator-freezer,
automatic defrost, 311-319
frost-free, 319
manual defrost, 306-311
side-by-side, 320-327
Refrigerators,
automatic electric, 311
compact, 306
domestic, 305
elementary, 32, 33
frost-free, 104
ice, 75
manual defrost, 306
mechanical, 32, 33
  operation, 11
Registers, 796
Regular screwdriver, 60
Regulating valves, pressure, 620
Regulator, pressure, 366, 367
dual, 154
evaporator (EPR), 963
Regulators, gas pressure, 724
Relative density, 23
of gases, 23
Relative humidity, 36, 650
Relay, 208, 209, 266-271, 342,
  894
contactors, 894
lockout, 894, 895
servicing, 270, 271
tester, 270, 271
testing, 270, 271
types, 266
  current, magnetic, 267, 268
  electronic, 269
  motor starting, 266-271, 309,
    351
  potential, magnetic, 268, 269
  thermal, hot wire, 269
Relief valve, 467, 468
Reluctance, 188, 189
Remote,
air conditioner, 774
condensing unit, 476
controller, 917
system, 774
temperature sensing element, 264
Removing,
air from system, 388, 509, 534
air-cooled condenser, 533
breaker strips, 374
cabinet doors, 395
compressor, 525
electric motor, 553
evaporator unit, 542
expansion valve, 545
gauges, 507
heat from refrigerator, 11, 31, 32
hermetic motor compressors, 379,
  380, 557
moisture, 365
oil, 383
refrigerant, 379, 521
service valve, 525
system, 373, 521

water valve, 540
water-cooled condenser, 537
Repair kit, evaporator, 53
Repairing,
air-cooled condenser, 533
clogged TEV screen, 545
condensers and receivers, 537
control systems, 922
crankshaft seal, 528
electric motors, 553
electronic filters (see Servicing)
evaporators, 382
expansion valves, 546
finishes, 338
hardware, 336
hermetic motors, 384, 557
leak, 355, 362, 369
pipe threads, 53
refrigerating sytems, 341, 501
valve plate, 529
water valves, 541
Replacing,
capillary tube, 355
  kit, 383
compressor, 355, 381, 382
condenser, 355
evaporator, 355
filter-drier, 355, 376-378
valve reed, 385
Repulsion, magnetic, 184
Repulsion start-induction-run motors,
  213, 215
Resin, epoxy, 52, 53
Resistance, electrical, 171, 178,
  179, 182, 183, 1014
heaters, 315, 836
heating, 683-687, 748
in parallel, 1014
in series, 1014
Resistance, valve, elbows, tee, 604
Resistor-heating, 683, 748
Resistor-impedance, 178
Resistors, electrical, 178, 179
Restricted system, 374, 377
Restricted water flow, 538
Restriction, high-side, 355
Restrictor, 625
Retainers, valve, 122
Retarder, 65
Retrofit, 1043
Return air ducts, 803
Returnable service cylinders, 295,
  296
Reverse cycle, 823
defrost, 435, 442
Reversible ratchet, 58
Reversing valve, 829, 830, 831
Rheostat, 950
Riggers, 844
Ring wire terminal, 277, 278
Ringelmann scale, 819
Rings, 116, 119
O-, 123, 139
piston, 116, 119
  compression, 119
  oil, 119
Riser valve, 466, 467
Rivets, 71
Rocks, heat storage, 868, 869
Rods, connecting, 116, 120, 121

Roller, 128
Roof cooling, 703
Roof design, 935
Rooftop units, 841-846
Room, clean, 1022
Room dehumidifiers, 694
Room heating, electrical
  resistance, 686
Room humidifiers, 694
Root mean square (rms), 171,
  202, 1026
Rotary compressors, 116, 127,
  128, 415, 520
hermetic, 128, 130
rebuilding, 385, 386
rotating vane, 127, 128
stationary blade, 127-130
Rotary cylinder construction, 128
Rotary vacuum pump, 390
Rotary-type fuel oil pump, 739
Rotating anemometer, 659
Rotating vane compressor, 127,
  128
Rotation, motor, 218, 219
Rotor, 191
construction, 134, 136
Rotors, helical, 130-133
female, 130
male, 130
Rottenstone, 338
Round nose pliers, 60
Rounding to whole numbers, 15
RPB (see Rapid pressure balancing)
Rules, measuring, 67
Running winding, 215, 217
Rupture disk, 467, 468

S

Saddle, tap line, 363, 364
Saddle valve, 106
SAE, flared connection, 43-46
SAE, 1008 mild steel, 40
Safety, 37
absorption systems, 631
air conditioning, 677, 705, 776,
  821, 862, 947
air conditioning controls, 923
automobile air conditioning, 975
brazing, 49
cleaning solvents, 69
code for mechanical
  refrigeration, 1027
code, R-717 (ammonia), 294
cold chisels, 70
commercial systems, 473, 500,
  562, 610
compression systems, 139
control, 890, 896
cryogenic fluids, 295
cutout, 446
domestic refrigerator-freezer, 339
dry ice, 91
electrical, 212, 244, 279, 280
expendable refrigerant, 647
goggles, 61, 971
heat pumps, 863
heating systems, 758
hermetic systems, 395
limit controls, 249, 265,
  266, 279

motor, 395
motor control, 265, 446, 447
pliers, 60
plug, fusible, 467
pressure-testing for leaks, 366
punches, 70
refrigerant, 303
  controls, 165
  cylinders, 290, 296,
recovery, recycling,
  reclaiming, 986
release, 477, 478
relief valves, 467, 468
service valves, 73
servicing, 395, 562, 1002
solar systems, 876
standards, CFCs, 977, 984
thermometers, 63
tools, supplies, equipment, 74
toxic substances, threshold limit
  values, 1027
twist drills, 63
valve, 620
vents, 634
wrenches, 54, 57
Safety-stat, 679
Salt, 34
Salt and ice mixture, 30, 75,
  427, 1020
Salt spray test, 727
Sandpaper, 68, 342
Saturated vapor, 36, 589
Saturation, 651
Saw cutter, 41
Saybolt viscosimeter, 1019
Saybolt viscosity test, 1019
Scale, weighing, 18, 19, 371
Scales,
  absolute pressure, 21
  absolute temperature, 14
  Beaufort, 664
  dBA, 675
  pressure, 21
  temperature, 14, 81
Scavenger oil pump, 137
Schrader valve, 364, 550, 971
Scotch yoke, 124, 126, 956
  piston, 120
SCR, 181, 196, 880
Screens, 549
Screw heads, 71
Screw sizes, 71
Screw type compressors, 116,
  130-133
  commercial, 132, 520
Screwdrivers, 60, 342
Screw-on mechanical fittings, 52
Screws,
  cap, set, 71
  machine, 61, 71
  range adjusting, 254
  tap, 61
Scroll compressor, 134, 135
Sea level, 14
Seal,
  compressor, 123, 957
  crankshaft, 116, 123
    repairing, 528
  leaking, 524
  shaft, 123

Sealants, duct, 789
Sealed ball low-side float, 161
Sealed units (see Hermetic)
Sealing compounds, thread,
  53, 342
Seasonal energy efficiency ratio
  (SEER), 610, 1027
Seat, valve, 144, 148
Second Law of Thermodynamics,
  13, 31, 1010
Secondary, container, 634
Seconds of a minute, 18
Seebeck, Thomas, 199
Seebeck effect, 199
Selective absorber, 867
Selectrastat, 902
Self inductance, 194
Semiautomatic defrost controls,
  273, 274
Semiconductors, 90, 179, 181,
  193, 194, 1014
  extrinsic and intrinsic, 194
  N-type and P-type, 90
Semihermetic commercial
  systems, 397
Sensible heat, 26, 27, 677
Sensible heat ratio (SHR), 656
Sensing bulb, 80, 111, 140, 141,
  155, 251
  gas charged, 144, 151, 152
  gas cross charged, 144, 151, 152
  liquid charged, 144, 150, 151
  liquid cross charged, 144,
    150, 151
  mounting, 155
  temperature, 111
Sensing devices, 863
Sensing element, 150, 151, 264
Sensors, 194, 197
Separators, oil, 83, 85, 106, 107,
  459-461
Sequencing, 896
Sequential controls, 895, 917
Series circuit, 182, 183, 247
Series-parallel circuits, 183
Serpentine belt, 241
Service,
  contracts, 561, 997, 1001
  cylinders, 295, 341, 360
  equipment, 519
  estimates, 562
  inspection report, 1002
  instructions, 519
  invoice, 1000
  lines, 358
  manifold (see Gauges, manifold)
  notes, 559
  records, 1002
  technician, requirements, 341,
    1000, 1031
Service valve,
  attachment, 359, 363, 364
  wrench, 58
    adaptors, 58
Service valves, 73, 105, 135, 356,
  359-361, 404, 466, 507,
  525, 964
  high-side, 106
  low-side (suction), 105, 106
  one-way, 73

  safety, 73
  two-way, 73
Serviceable hermetic, 126,
  401, 404
Servicing, 989-1003
  absorption systems, 626, 630
  adding oil, 372
  air conditioners, 767, 772
  air conditioning systems, 855,
    856, 969
  air-cooled condenser, 532
  automobile, 969-974
  burnout, 380, 550
  cabinet insulation, moisture/
    ice, 347
  capacitor, 236, 237
  capillary tubes, 374, 376,
    382, 383
  commercial systems, 501,
    519, 559
  condensers, 382
  condensing units, 520, 532, 537
  console air conditioners, 772
  control system, 922
  cooling towers, 537
  direct expansion evaporators,
    541, 542
  dismantling system, 379
  domestic systems, 341
  electric motors, 236, 242,
    352, 552
  electrical troubleshooting,
    352, 552
  electronic filters, 812
  evaporative condensers, 537
  evaporator, iced, 347
  evaporators, 382, 541, 542
  external drive motors, 237, 541
  fan motors, 242
  filter and drier replacement, 376
  fuel oil burner, 746
  gas furnace, 727
  gauge manifold use, 356
  heat pump systems, 836
  heating coil, 754
  hermetic motors, 243, 374,
    384, 554
  hermetic systems, 341, 344-356,
    365-395, 532
    assembling, 387
    external, 350
    internal, 350, 355, 365, 374
    overhaul, 350, 383
    removing, 373
  hot gas bypass valves, 548
  hot gas defrost, 378
  humidifiers, 757
  hydronics system, 734
  ice makers, 538
  instruments, tools, supplies,
    341, 342
  liquid line, 558
  lithium bromide systems, 631
  motor,
    bearings, 239
    compressor, 243, 355, 379,
      381, 384-387
    controls, 557
    lubrication of, 238
    noise, locating and removing, 344

oil, removing, 383
oil burner, 746
open systems, 520
outdoor air-cooled condenser
 controls, 548
rebuilt system, 393
refrigerant,
 charging, 369-372
 leaks, 366-369
 moisture, 365
 removing, 379
 shortage, 366
relays, 270
restricted systems, 374
rooftop units, 846
safety, 395, 562, 1002
solenoid valves, 558
steam heating system, 735
stuck motor compressor,
 243, 355
suction line, 558
temperature-pressure conditions,
 determining, 345
thermoelectric units, 640
thermostat, 893
thermostatic expansion valves,
 544, 547
thermostats, 278
trouble signals, 346
two-temperature valve, 547
water valves, 538-540
water-cooled condenser, 534
wax clogging, 365
window units, 767
Servo motors, 185, 965
Set point level, controlling
 instrument action, 249, 250
Setscrews, Allen, 71
Shaded-pole motor, 234
Shadow indicating devices, 869
Shelf-type evaporator, 103
Shell and coil condenser, 417, 418
Shell and tube condenser, 417
Shell and tube flooded evaporator,
 85, 417
Shell-type condenser, 107,
 109, 417
Shell-type evaporator, 103
Shelving, 305, 306
Short circuit, 172, 384
Shortage of refrigerant, 355, 366,
 369, 376
Short-cycling, 355, 406, 561, 736
Shorting capacitors, 351
SHR (see Sensible heat ratio)
Shroud, 107
Shunt, 180, 181
Shutting down refrigerators and
 freezers, 395
SI (see Le Systeme International
 d'Unites)
Side, high-pressure, 101, 102
Side, low-pressure, 101, 102
Side-by-side arrangement, 118
Side-by-side refrigerator-freezer,
 320-327
Sight glass, 82, 109, 300, 301,
 342, 469, 470, 520, 960, 973
 electronic, 470
Sight port, oil, 391

Silica gel, 152, 376, 471, 1015
Silicates, synthetic, 376
Silicon carbide, 68
Silicon-controlled rectifier (SCR),
 181, 194, 196
Silver alloys, 49
Silver brazing, 49
 wire, 342
Silver chloride, 93
Sine wave, 184, 195, 204
Single energy, 1021
Single flare, 43, 44
Single-phase cycle, 203
Single-phase motor, 203, 213
Single-pole, double-throw switch,
 260, 880
Single-pole, single-throw switch,
 260, 880
Single-stage, two-stage pumps,
 390
Sink, heat, 704
Siphon action, 616
Sizes, pipe, 52
Slant type evaporators, 848, 849
Sliding port valve, 829
Slim nose pliers, 60
Slip joint combination pliers, 59
Slot blade screwdriver, 60
Sludging oil, 300
Slugging, 828, 836
Smoke, 808
 candle, 726, 820
 test, 819, 820
Snap-action,
 permanent magnet, 251
 switch, 115
 toggle mechanism, 246, 251,
 256, 257
 two-temperature valve, 455
Sniffer, 368
Snow making, 76, 499
Soap bubbles, 366, 367
Society of Automotive Engineers
 (SAE) standards, 39
Socket wrench, 54, 56, 58, 59
Soda fountain, 484, 485
Sodium chloride (NaCl), 30,
 34, 1020
Soft copper tubing, 39, 342
Soft ice cream maker, 486
Soft soldering fittings, 39
Solar, electric energy inverter, 196
Solar cell, 196, 874-876
 circuit, 875, 876
 construction, 874
 efficiency, 876
 performance, 876
 photovoltaic, 874, 875
Solar collector, 867, 868
 angle, 869
 cover, 868
 flat plate, 867
 heat insulating, 869
 lens, 874
 panel, 840, 869
 surface, 867
 swimming pool, 872
Solar energy, 865
 converting to electricity, 874
 definition, 865

domestic water heating, 870
 one tank system, 870
 two tank system, 870
electomagnetic, 865
flux, 866, 867
nature of, 865
radiant, 865, 874
shadow indicating devices, 869
space heating installations,
 870, 873
storage, 868, 869
systems,
 active, 866
 cooling, 874
 heating, 866
 passive, 866
 trapping principles, 867
Solar heat, 939
Solar heat pump combination,
 839, 840
Solar panel, 840, 869
Soldered tube fittings, 43, 47, 342
 soft soldering, 39
Soldering, 47, 71
 equipment, 341, 517
 flux, 342
 wire, 342
Solders, 47, 1022
Solenoid, 187, 894
 coil, 882
 plungers, 157
Solenoid valve, principles, 156
 servicing, 558
 stuck, 378
 types, 157, 158
Solenoid valves, 157
 reversing, 829
 two-temperature, 456, 457
Solid absorbent, 93, 95
Solid adsorbent, 611
Solid fuel heating, 747
Solid-state, 153, 181
 controlled ice maker, 326, 328
 defrost controls, 453
 diodes, 181, 194
 electronic relays, 266, 269
 expansion valve, 152, 153
 inverters, 196
 photocells, 181
 printed circuit board, 326, 328
 silicon controlled rectifiers (SCR),
 181, 194, 196
 temperature control, 251, 252
 thermistors, 197, 198, 251, 252
 transistors, 181, 194, 196
Solids, 22
 pressure of, 20
Solution, eutectic, 497, 585, 646
Solvents, cleaning, 54, 69,
 384, 1021
 safety, 69
Sones, 676
Sound pressure, 674
Sound tracer,
 electronic, 341,
 ultrasound, 344, 345
Source detector, ion, 367
South pole, magnetic, 184, 185
Space heating, solar,
 installations, 870

Spade wire terminals, 277, 278
Special refrigeration systems and
   applications, 633
Special tubing, 517
Specific enthalpy, 35
   of common refrigerants, 35
   of water, 35
Specific gravity, 23
   brine, 23
   gases, 23
   water, 23
Specific heat, 26, 589
   common substances, 1006
   equivalents, 26
   of water and ice, 29
Specific volume, 23
Speeds, motor, 216
   synchronous, 216
Spirit level, 771
Splash oiling, 137
Splash system, oil, 137
Split joint twist drills, 60
Split system, 856, 914
Split-phase motor, 193, 214,
   215, 221
Spray cooling, 634
Spray headers, 634
Sprays, water, 813
Spring bender, 42
Spring fasteners, 71
Spring loaded valve, 467, 468
Spring tube bending, 42
Square,
   centimeters, 17
   decimeter, 17
   foot, 17
   inch, 17
   meter, 17
   yard, 17
Square wave output, 196
Squirrel cage, 215
Stack thermometer, 816
Stainless steel evaporator,
   repairing, 382
Stainless steel tubing, 39, 40
   sizes, 40
Stamps, 60
Standard,
   atmosphere, 1023
   atmospheric pressure, 14
   conditions of air, 14, 1023
   pressure, 1023
   temperature, 283, 1023
Star (Wye) transformer, 204, 205
Starters, motor, 447, 448
   magnetic, 447, 448
Starting,
   dry system, 513
   gas furnace, 727
   oil burners, 745
   refrigerator-freezer, 344
   stuck compressors, 244, 355
   system, 510, 513
Starting current, 217
Starting kit, PSC motors, 771
Starting relays, 214, 266-271,
   351
Starting winding, 214, 217
Starve (starving) evaporator,
   142, 150

Stat, stack, 680
State, change of, 23, 27
State of equilibrium,
   refrigerants, 283
States of substance, 20, 22
Static, 107
Static electricity, 169
Static head, 602
Static pressure disk, 738
Stationary blade compressor,
   127-130
Stator, compressor, 134
   construction, 134, 136
Stator, motor, 191, 215
Statute mile, 17
Steam, 13
Steam heating, 734
   inspection, 736
   installation, 735
   servicing, 735
   testing, 735
   troubles, 735
Steam-jet cooling, 294, 642, 701
Steel tubing, 39, 40
Steel wool, 342
Stellite, 147
Step-down transformers, 204
Sterilization, air, 879
Sterling cycle, 646
Stethoscope, 344, 345
Stile heater, 334, 335
Storage,
   fresh foods, 305
      temperature ranges, 141
   frozen foods, 305, 481, 482
      temperature ranges, 141
   fur, 493
   produce, 570
   solar energy, 868, 869
Storage battery, 169
Storage cylinders, refrigerant, 294,
   295, 296
Storage, thermal energy, 859
   tank, 839
Stored energy, 24
Storing, cold, 775
Straddle-plugged-in units, 405
Straight blade screwdriver, 60
Straight box wrenches, 56
Straight shank twist drills, 60
Strainer (see Filter)
Strainer chamber, 740
Stratification of air, 673, 873
Street ell, 53
Striker plate, 337
Strip charts, 815
Strong solution, 631
Stubby screwdriver, 60
Stuck motor compressor, 243, 355
Sub-base, 884
Subcooling, 397, 400
Sublimation, 34, 90
   latent heat of, 34
Substances, states of, 20, 22
Substituting refrigerants, 301
Suction line, 101, 102, 104, 548
   accumulators, 459, 828
   filter-drier, 105, 106, 472
   frost back, 142, 355
   pressure drop, 602

   servicing, 558
   sweat back, 142, 355
Suction pressure valve, 165,
   462, 962
Suction service valve, 105,
   106, 405
Sulphur candle, 294, 630
Sulphur dioxide, (R-764), 292,
   304, 650
Sulphur spray vapor, 294
Summer cooling, 823
Sun (see Solar)
Sun effect, 926, 939
Sun heat load, 663, 939
Sun's spectrum, 865
Superfex, 613
Superheat, 105, 141, 145,
   150-152, 161, 587, 589
Superheated vapor, 587, 589
Superheating, 587, 589
Supermarket refrigeration, 301, 475
Supplemental resistance
   heaters, 835
Supplementary heat, 873
   electric, 752, 873
   oil and gas, 873
Supplies, refrigeration, 68
   abrasives, 68
   brushes, 69
   cleaning solvents, 69
   installation and servicing,
      341, 342
Surge tanks, 458
Surging, 160
Swaging, 47, 557
   copper tubing, 52
   tools, 52
Swamp cooler, 703
Swash plate compressor, 125, 126,
   954, 955
Sweat back, 142
Sweat joint, 47
Sweating, 477
   suction line, 355
Sweet water, 34, 427
   bath, 427
Swimming pool heating, 872
Swing temperature, 888
Swinging vane velocimeter, 660
Switch,
   centrifugal, 215, 238
   control, 953
   electrical, 210, 211
   light, 342
   limit, 888, 897
   master, 247
   mercury, 251, 252
   power saver, 315
   reversing, 640
   SPDT, 260, 880
   SPST, 260, 880
   thermal element, 896
   toggle, 115, 252, 256, 257
Symbols, 17, 37, 172, 1013
   circuit, 172
   standard, 174
Synchronous speed, 192
Synthetic silicates, 376
System capacity, 594
System lag, 885

Systeme International d'Unites (See Le Systeme International d'Unites)
Systems,
  absorption, 94, 611, 614, 628
  adding oil to, 372, 551
  adsorption, 611
  air circulation, 856
  air conditioning, 679, 707, 761, 774
  alarm, 331
  automatic, 879
  automobile, 949
  basic refrigeration, 75
  capillary tube, 81, 110, 147, 374, 382, 383
  cascade, 84, 590
  central, 846, 848, 849
  charging, 393, 510, 973
  check/reset, 323
  chilled water, 858, 859
  climatic control, 823, 841
  comfort cooling, 761, 762, 858
  commercial 95, 96, 397
    servicing, 501
  complete, 688, 823, 841
  compound, 82, 83
  compression, 101
  continuous absorption, 611-618
  control, 879-923
  defrost, 308, 312, 316, 435-443
  discharging, 122, 379, 383
  district heating and cooling, 862
  dry ice, 103
  duct, 752, 784, 857
  engine-driven, 472
  evacuating, 387, 509
  evaporator, 633
  expendable refrigerant, 88, 633
  external drive, 77, 78
  Faraday, 93, 612
  flooded, 103, 110, 114
  force feed, 137
  four-pipe system, 857
  heat pump, 827
  heat recovery, 491
  hermetic, 306, 309, 311, 315, 319, 326, 331
  high-side float, 102, 114
  Hi/Re/Li, 838
  hydronic, 728, 906, 913
  intermittent, 611, 613
  lithium bromide, 631
  low-side float, 102, 114
  mechanical refrigerating, 32, 33
  multiple, 81, 82, 399
  multistage, 643
  odor, 889
  outside, 841
  refrigerant recovery, 564
  restricted, 374
  solar energy, 865, 866, 874
  solar heat pump, 839, 840, 869, 873
  split system air conditioning, 856, 914
  spray, 633
  starting, 510, 513
  steam heating, 734
  steam-jet, 642
  testing, 393

  thermoelectric, 640
  through-the-wall, 841
  total energy, 859, 1021
    management, 914
  two-duct, 857
  two-pipe, 732
  unitary, 841
  vacuum control, 968
  warm air, 728

T

TAB, 803
Tables, refrigerants, 298
Tandem motor compressor, 405, 406
Tanks,
  holding, 693, 869
  oil installation, 743, 744
  refrigerant (see Cylinder)
  surge, 458
Tap drills, 61
  sizes, 62
Tap screws, 61
Tap valves, 135, 344
Tap water type cooler, 488
Tap wrench, 53
Tape, plastic, 342
Taper taps, 57
Tapes, measuring, 63
Taps, 61
  bottoming, 61
  plug, 61
  taper, 61
Technical characteristics, 1004-1028
Technical terms, 1032-1046
Teflon, 123
  seal, 957
TEM (see Total Energy Management)
Temperature,
  alarm system, 279
  Celsius, 14
  Centigrade, 14
  changes, 881
  chart recorder, 341, 345, 346, 393, 394
  charts, 297, 298
  comparison, 14, 15
  conditions, 102, 345, 346
  control, 251, 252, 649, 663
  control alarm system, 279
  conversion formulas, 16
  conversion table, 1009
  correction,
    ambient, 882
  curves, 282, 283
  differences, 15, 34
  droop, 888
  equation, 16
  Fahrenheit, 13, 14
  gauges, 63
  Kelvin, 14
  lag, system, 885
  measurement, 13, 14
  motor controls (thermostatic), 115
  of maximum density of water, 834
  of mixes, 486
  ranges, 141

  cryogenic, 295, 1020
  Rankine, 14
  scales, 14
  sensing bulb, 80, 111, 115, 142, 155, 251
  sensing element,
    remote, 264
  swing, 888
  symbols, 13, 14
  tables, 297, 298
Temperature-pressure,
  conditions, 345
  tables, 298
Temperatures, 13
  absolute, 14
  air, 650, 663
  air conditioning, 649
  ambient, 31
  boiling, 14, 28
  brine freezing, 1020
  cabinet, 297, 305, 339
  compression cycle, 102
  condensing, 34, 297, 298
  critical, 34
  cryogenic, 35, 1020
  design, 934
  dry bulb, 651
  effective, 664
  evaporating, 28, 297, 298
  freezing, 13, 29
  fresh food, 13
  frozen food, 570
  motor, 230
  refrigerating, 13
  refrigerator, 297-299
  room and duct, 793
  standard, 283, 1023
  storage case, 299
  wet bulb, 651
Ten, power of, 15
Tension gauge, 241
Terminals,
  attachment plug, 210
  motor, 225, 353, 354
  wire (also see type), 277, 278
TES, 1019
Test chamber, environmental, 100
Test cords, 352, 353
Test lights, 211, 341
Test stand, 241, 387
Tester,
  expansion valve, 545
  insulation, 181, 182
  leak, 509
Testing,
  capacitors, 351
  capillary tubes, 376, 377
  code installation, 518
  electric motors, 375, 386, 387, 553
  filter-driers, 376, 377
  for leaks, 283, 366, 367, 509, 533, 545, 972
  pressures, 366
  rebuilt system, 393
  stand, compressor, 241, 530
  thermostatic expansion valve, 545
  valves, 385
Tetrafluoroethane (CF_3CH_2F), R-134a, 301-303, 978, 979

TEV (see Thermostatic expansion valve)
Therm, 25
Thermal,
  cutouts, 254
  electric, 144, 153
  energy storage, 839, 859
  tank, 839
  insulation, 1023
  laws of refrigeration, 101
  relay, 266, 269, 270
  resistance, 926
  stability, refrigeration oil, 300
  unit, British (Btu), 13, 25
  unit, kilojoule, 25
Thermal-electric expansion valve (THEXV), solid state, 141
Thermistor, 14, 63, 153, 194, 197, 198, 230, 251, 252, 266
  circuit protection, 209
  thermometer, 63
Thermistor-type leak detector, 367
Thermocouple, 14, 194, 198, 199, 723
  gauges, 65
  thermometer, 63
Thermocouples, multiple, 884
Thermodynamics, 586, 1010
  First Law of, 26
  Second Law of 13, 31
Thermoelectric,
  compared to compressor system, 640
  devices, 194, 198
  element, 114
  expansion valve, 153
  module, 640
  operation, 88, 640
  power supply, 640
  refrigeration, 88, 640
  thermocouple, 63, 198, 199
Thermo-humidigraphs, 655, 671
Thermometer, 13, 14
  safety, 63
  scales, 14
  type thermostat, 882
Thermometers, 13, 63, 341, 670
  air conditioning, 670
  alcohol, 13
  cabinet, 338, 339
  comparison, 14
  dial stem, 63
  digital, 63
  dry bulb, 651
  glass stem, 14, 63
  infrared, 1005
  kata, 1005
  mercury, 13
  pocket, 63
  radiometer, 14
  recording, 63, 381, 519
  stack, 816
  thermistor, 63
  thermocouple, 63
  wet bulb, 651
Thermomodule, 640
Thermopane, 932
Thermopile, 198
Thermostat,
  anticipators, 884

controller, 880
droop, 888
guard, 893
heat pump, 851
installing, 277, 278
locations and mountings, 893
operation, 881
sensing bulb, 251
sensitive element, 881
servicing, 278, 893
temperature correction, 882
testing, 278
timer, 890
troubles, 278
types, 880
Thermostatic charge, MOP, 154
Thermostatic control, 115, 252, 253, 445
Thermostatic expansion valve (TEV),
  adjusting, 545
  capacities, 156, 159
  construction, 147
  control, 147
  cross charged power element, 151
  design, 147
  gas charged power element, 151
  liquid charged power element, 150
  MOP, 155
  normal charged power element, 150
  power element mounting, 155
  pressure, limiters, 153
  principles, 144, 145
  refrigerant flow control, 110-113
  servicing, 544-546
  superheat, 145, 150-152, 161
  temperature, 80
  testing and adjusting, 545
  types, 144, 145
Thermostatic expansion valves, 80, 82, 111-113, 141, 144, 152, 153, 560, 828, 961
  bellows type, 145, 146
  diaphragm type, 145, 148, 149
  large capacity, 160, 162
  pilot-controlled, 146
  special, 159
Thermostatic (temperature) motor controls, 77, 115, 251-253, 445
  bimetal, 251, 252
  sensing bulb, 251
  solid state, 251, 252
Thermostatic two-temperature valves, 456
Thermostatic water valve, 466
Thermostats, 33, 102, 879, 880
  air coil, 834
  automobile air conditioning, 969
  combination, 889
  comfort cooling, 259, 260
  cooling, 889
  electric heating, 882, 888, 905
  electronic, 891
  heating, 885, 889
  hydronic, 893
  line voltage, 882
  low voltage, 883
  millivolt, 884
  multistage, 890

portable, 893
rated, 886
refrigerator, 308
thermometer type, 882
unit, 686
vapor pressure type, 251-253
wall, 687
Thread,
  American standard taper pipe, 61
  metric size, 61
Thread cutting, 61
  compound, 53
  lubricant, 62
Threads,
  external, 62
  internal, 61
Threads, repairing, 53
Three physical states, 22
Three-phase cycle, 203, 204
Three-phase four-wire transformer, 208
Three-phase motor, 203, 208
Three-phase overload protection, 266
Three-way solenoid valve, 157, 158
Three-wire grounded,
  power cord, 248
  receptacle, 202, 203
Three-wire thermostat, 882
Through-the-wall air conditioners, 687, 688, 758, 759, 836, 841
Throwaway filters, 809
Timed "on-off" controlling instrument action, 249, 250
  anticipator, 250
Timers,
  defrost, 277, 315, 317, 319, 334, 449, 451
  thermostat, 890
Toggle switch mechanism, 115, 252, 256, 257
Ton, 18
"Ton" of refrigeration, 30
Tongs, ammeter, 176
Tool,
  block-and-punch, 45
  flaring, 44
Tools, 39
  drill presses, 60
  electric drills, 60
  hand, 54, 55
    cold chisels, 70
    files, 70
    hacksaws, 71
    hammers, 59, 342
      mallets, 59
    pliers, 50, 60, 342
    punches, 70
    screwdrivers, 60, 342
    stamps, 60
    wrenches, 54-58, 342
  hand braces, 60
  installation, 341
  insulated, 212
  lapping, 385
  pinch-off, 362, 363, 379, 380
  refrigeration, 39
  servicing, 341
  swaging, 52, 381

taps, 61
tubing, 342, 381
twist drills, 60
Torch, soldering-brazing, 341
Torque, 56, 58, 227
wrench, 55, 58, 59
Torr, 21, 390
Total energy, 862, 1021
Total Energy Management
(TEM), 914
smart building, 915
system determination and
usage, 919
system types and functions, 917
Total heat load, 573, 576
Total pressure, 36
Towers, cooling, 107, 109, 418,
420, 421, 537, 693
Toxic refrigerants, 283
Tracing water circuit troubles, 540
Transducer, 194, 198
Transfer of heat, 33, 34, 102
Transferring refrigerants, 550
Transformer,
characteristics, 205
circuits, 205, 206
principles, 204
Transformer-rectifier, 90
Transformers, 204, 205, 206, 208
auto-, 207
boost-and-buck voltage, 206
Delta, 204, 205, 208
line voltage, 205, 206
step-down, 204
three-phase, four-wire, 208
Wye (star), 205, 208
Transistor, 181, 194, 196, 197,
880, 892
Translucent, 33
Trapping principle, solar
energy, 867
Triacs, 194, 196, 880, 906
Trichloromonofluoromethane (R-11),
283, 287, 297, 299, 301,
303, 977
Trichlorotrifluoroethane (R-113),
297, 299, 303, 977, 1016
Triple evacuation, 388-390
Tripper, 890
Trouble signals, 346
Troubleshooting, 341, 342, 348,
349, 352, 382, 556, 559, 560,
592, 893, 922, 989-1003
chart, 348, 349, 991-996
electrical, 352
procedures, 989
Truck air conditioning 969
Truck refrigeration, 495
TruKold, 613
Tube constrictor, 52
Tubes,
bleeder, 757
Bourdon, 64
capillary, 81, 110, 141,
162-164, 342
distributor, 159, 161
equalizer, 158, 159
pitot, 660, 815, 817
process, 362
vortex, 640, 702

Tube-within-a-tube condenser, 107,
109, 417-419
Tubing, 39-41
ACR, 39
annealed, 39
annealing, 46
bending, 41, 503
brass, 40
brazing, 49
connecting, 43
coolant, 960
copper, hard, 40
copper, soft, 39, 342
cutting, 41
finned, 597
flaring, 43-45
flexible hose, 40
fittings, 40
hard drawn, 40
installing, 503
metric size, 40
nominal size, 41
plain, 597
plastic, 40
sizes, 40
polyethylene, 40
soldering, 47
stainless steel, 40
sizes, 40
steel, 39, 40
swaging, 52
Tubing coupling, 382
Tubing fittings, 46, 47
Tubing insulation, 504
Tubing sizes, 40, 41
OD, 40, 41
Tubing tools, 342, 381
Tubing winding, capillary, 163
Turbine flow meters, 820
Twist drill sizes, 1022
Two evaporating temperatures, 81
Two-cylinder compressors,
118, 954
Two-duct system, 857
Two-pipe system, 732
Two-pole electric motor, 192
Two-position "on-off" controlling
instrument action, 249, 250
Two-speed motor compressor, 116,
216, 838
Two-stage compressor, 590
Two-stage heating, 891
Two-stage oil pump, 739
Two-stage vacuum pump, 390,
391
Two-temperature valve, 81,
454-456, 547
servicing, 547
types, 454
Two-way service valves, 73
Two-way solenoid valve, 157
Types of compressors, 116

U

U factor, 567, 926
U.S. conventional units, 11, 37
Ultralow temperature, 643, 644
Ultrasonic sound detector, 344
Ultrasound, 344

Ultraviolet, 248, 814, 866
fluorescent leak detection,
367, 972
Uncrating cabinets, 342
Under-capacity valve, 142
Under-floor hydronic heating
system, 728
Undersize wire, 172
Underwriters Laboratory, 209
Unheated spaces, 937
Union, 53
Unit,
angular measurement, 18
dimensional measurement, 17, 18
heaters, 754
of energy, 24, 31
of force, 24
of heat, 25
of power, 24
of pressure, 19-21
of work, 24
prefixes, 15
electrical, 1013
temperature, 13
thermostat, 686
Unitary systems, 841, 846
Units,
air conditioning, 761
centrifugal, 858
condensing, 406-411, 415, 416
rooftop, 841-845
through-the-wall, 841
Units and symbols, electrical, 1013
Universal motors, 170, 226
Unloader, 138, 412, 414, 415
Upright freezer, 331, 333-335
Upright frozen food case, 481
Urethane insulation, 306, 475, 484
Usage load, 573
Uses of refrigeration, 11
Unshaped wire terminals, 277, 278

V

Vacuum, 21
deep, 387-389
inches of mercury, 21
partial, 22
perfect, 22, 391
Vacuum cleaner, hand, 341
Vacuum control systems, 968, 969
Vacuum gauges, 65, 392
Vacuum pump, 390, 519
oil, 391
portable, 391
rotary, 390
single- and two-stage, 390, 391
sizes of, 393
use of, 392
valve, 358
vapor (diffusion type), 390
Vacuum tube collector, 868
Values, interpolation, 1010
Valve,
adaptor, 342, 356, 359, 361,
362, 364
capacity, 142
disks, 123
kit, 362
plate, 122, 123, 529

retainers, 122
seat, 144, 148
testing, 385
Valves, 122
  absorption system, 620
  automatic, expansion (AEV), 79,
    110, 112, 141-144, 374
  ball type, 148, 149
  bellows-type expansion, 142
  bleed, 148, 149
  bolted-on, 363
  brazed on, 363, 364
  burred, 385
  bypass, 144
  check, 128, 138, 139, 165, 341,
    456, 458
  combination gas control, 901
  compressor, 122
    low-side pressure control,
      461, 962
  core, 342, 364, 365
    adaptor, 364
  cracking, 73
  diaphragm type, 143
  Dill core, 971
  discharge, 122
  electric water, 463, 464
  exhaust, 102, 122
  expansion, 141-144, 543, 961
    servicing, 544
  flapper, 385
  float, 161, 162
  four-way, 158, 825, 829, 830
  gas, 679
  gauge manifold, 358
  hand, 73, 358, 466
  heat pump reversing, 829
  high-side service, 106
  hot gas bypass, 374
    servicing, 548
  installation, manual, 466
  intake, 101, 122
  leaking, 385, 524
  limitizer, 408
  liquid vapor, 296
  low-side service, 105, 106
  manual, 73, 466
  needle, 107, 114, 115, 147, 341
  noisy, 385
  one-way, 73, 295, 829
  over-capacity, 142
  piercing, 342, 346, 347, 356,
    363, 364
  pilot-operated absolute (POA), 963
  poppet, 122
  port type, 829
  pressure controlled expansion,
    79, 141
  pressure regulating, 454, 620
  pressure-limiting expansion,
    153, 154
  reed, 122, 385
  refrigerant cylinder, 358
  refrigerant line, 518
  relief, 467, 468
  repairing compressor, 529
  reversing, 695, 829-831
  riser, 466, 467
  saddle, 106
  Schrader core, 364, 971

service, 73, 105, 135, 356,
  359-361, 466, 507, 525, 964
  cracking, 58
  one-way, 73
  safety, 73
  two-way, 73
  wrench, 58
snap-action type two-
  temperature, 455
solenoid, 157, 158, 829
  principles, 156, 157
  servicing, 378, 379, 558
  types, 157, 158
solid state, 153
suction pressure, 79, 141, 165,
  462, 962, 963
suction service, 105, 106, 405
suction throttling, 701, 962
tapping, 135, 344
thermal-electric expansion
  (THEXV), 141, 152, 153
thermostatic expansion (TEV), 80,
  110-113, 141, 145-152, 155,
  156, 158-160
three-way, 158, 518, 829
two-temperature, 81,
  454-456, 547
two-way, 73, 158, 518
under-capacity, 142
vacuum pump, 358
water, 463-467
weight, 78
Valves-in-Receiver (VIR), 963
Vanes, 127, 128, 385
  construction, 129
Vapor, 12, 13, 20, 36, 1016
  barrier, 659, 936, 944
  bubbles, 301
  charged, 151
    cross charged sensing bulb, 144,
      151, 152
    sensing bulb, 144, 151, 152
  definition (gas) 23, 36
  degreasing, 69
  lock, 163
  pressure, 32
    curve, 283
  pressure thermostat, 251
    construction, 252, 253
    volatile fluid, 251
  type,
    refrigerant, 12, 104
    saturated, 36, 589
    superheated, 150, 589
    water, 36
Vaporization, latent heat of, 27, 28
Vaporized refrigerant, 28, 31, 76
Vapor-type vacuum pump, 390
Variable Air Volume (VAV), 911
Variable controlling instrument
  action, 249, 250
Variable displacement compressors,
  954, 955, 956
Variable pitch pulley, 240
VAV (see Variable Air Volume)
V-belt, 240, 957
  drive, 123
Vegetable crisper, 306
Vegetables, fast freezing, 570
Velocimeter, 660

Velocity,
  gas, 603
  liquid, 601
Velocity equivalents, 1008
Velocity measurements, 659
Velocity pressure, 660
Velometer (see Velocimeter)
Vending machine, 491
  controls, 448
Vent pipe, 745
Vent plug, 745
Ventilation, 662
  requirements, 780
Venting, furnace, 717
Venting, job shop, 1021
Vents, 1021
  safety, 634
Vertical baffle, 477
Vertical cylinder compressor,
  400, 415
Vibration, 344, 539
  absorbers, 468, 469
  eliminator (damper), 95, 505
VIR (see Valves-in-Receiver)
Viscosity, refrigerant oil, 73, 137,
  300, 372, 1019
  Saybolt test, 1019
Vises, 60
Volatile fluid, vapor pressure
  thermostat, 251
Volt, 169, 174, 182
Volt amperes (VA), 207
Voltage, 169, 174
  boost-and-buck transformer, 206
  drop, 171, 183, 184
  fluctuation, 206
  lowering, 267, 986
  maximum (V_{max}), 171, 202
  relay, 221, 266
  root mean square, 202
Voltage-controlled relays, 266
Voltmeter, 63, 171, 174, 175,
  177, 341
  autoranging, 176
  digital (electronic circuitry), 175
  electromagnetic (types), 174, 175
Volume, 17, 18
  air, 819
  cabinet, 572
  constant, 1025
  specific, 23
Volume equivalents, 18, 23, 979
Volume measurement, 17, 18
Volume of a cylinder, 18
Volumetric efficiency,
  compressor, 137, 595
  check, 523, 524
Vortex tube, 640
  cooling, 702
  refrigeration, 640
V-type compressor, 116, 400-403,
  415, 954

W

Walk-in cabinets, 476
Walk-in coolers, 477
Wall,
  construction, 936
  electrical outlet, 343

heat leakage, 931
insulation, 944, 1023
thermostàt, 687
type evaporator, 103, 105
Warm air defrost, 443, 452
Warm air furnace, 679, 680, 683
Warm air heating, 728
inspection, 728
Warm air system, solar, 869, 870
Warm temperature monitor, 323
Warning light, 331
Water,
analysis instruments, 820
boiler water treatment, 733
boiling temperature, 14
chiller, commercial, 400
circuit troubles, tracing, 540
coil, 697, 834
column, 21
cooler, 86, 87, 487
cold reserve, 261
controls, 261
cooling loads, 585
defrost system, 440, 441
defrosting, 440, 441
electrostatic treatment, 538
enthalpy, 35
evaporative refrigeration, 75, 76
flow, restricted, 538
flow, too much, 539
freezing temperature, 14
heaters, heat pump, 840
heating, solar, 869, 870, 872
ice, 11, 13, 20, 29
latent heat, 27-30
level controls, 906
manometer, 22, 630, 721, 726
physical states, 20, 22
pipes and drains, 47
refrigerant, 294
specific gravity, 23
specific heat capacity, 29
sprays, 813
strainer, 463
sweet, 34
temperature of maximum
density, 834
valve, 463-465, 538-540
electric, 463, 464
pressure, 463-465
pressure setting, 465
thermostatic, 466
vapor, 20, 36
Wash electronic air cleaner, 808
Water cooled condenser, 417, 418,
534-537, 598
head pressure, 297
tube-in-a-tube and shell, 107, 109
Water-cooled tubing, 598
Water-cooled units, 137
Watt, 24, 177, 201
reading, 236
recorder, 341

Wattage, 177, 178, 201
calculation, 177
consumption, 236
Wattmeter, 177, 178
Wave, electromagnetic, 865
Wave, pulse, 184
Wave, sine, 183, 184
Wax, 73, 301, 365
Weak solution, 631
Weather, 663
Weighing scale, 18, 19
Weight and mass, 18
Weight equivalents, 19, 1008
Weight of air, 779
Weight of substances, 1006
Weight valve, 78
Weld grinding machine, 384
Welded motor compressor, 384
Welding, 71
compressor dome, 386
equipment, 517
Wet bulb, 651, 653
temperature, 653
Wet color, 300, 301
Wet roof cooling, 703
Wheatstone bridge, 252
Wick, 645
Wiedmann-Frank Law, 34
Wind chill, 664
Windings, 186
main, 216
running, 214, 216, 217
squirrel cage, 215
starting, 215-217
Windmills, 650
Window heat load, 938
Window unit, 687, 688, 763
installing, 764
servicing, 767
Windows, 932, 963
primary, 932
storm, 932
thermopane, 932
Winterizing air conditioning
unit, 771
Wire,
electrical, 169, 181
hot, 172
phosphorous-copper alloy, 342
silver brazing, 342
solder, 342
undersize, 172
Wire arcing, 882
Wire sizes, 210
Wire terminals, 277, 278
Wire ''walking,'' 882
Wirebrush, 69
Wire-static, condenser, 107, 108
Wiring, class 2, 209
Wiring diagrams, 174, 247
domestic freezer, 168, 331,
333-335

domestic fresh-food
refrigerator, 308
domestic refrigerator-freezer, 311,
314, 315, 319, 321, 324, 327
electrical circuit, 247, 248
ice maker, 262-264
system, 351
Wobble plate, 125
compressor, 125, 126, 954
Wool insulation, 1023
Work (W), 24
Work hardening, 39
Wrench adaptors, 58
Wrench safety, 54, 57
Wrenches, 54-59, 342
adjustable, 54, 57
Allen, 58
L-shaped, 58
box, 56, 57
chain, 58
combination, 57
flare nut, 57, 1025
hex, 58
internal type, 58
metric, 56
nut driver, 56
offset, 56, 57
open end, 54, 57
pipe, 57, 58
ratchet handle, 56, 58, 59
handle, 56
reversible, 58, 59
service valve, 58
adaptors, 58
socket, 54, 56
swivel handle, 56
T-handle, 56
torque, 55, 58, 59
W-type compressor, 400, 403, 415
Wye motor circuit, 224, 231
Wye (star) transformer, 204,
205, 208

X

X-type compressor, 415

Y

Yard, 17
cubic, 18
square, 17
Year-around air conditioner, 649
Yoke, Scotch, 120, 124, 126
Y-type compressor, 415

Z

Zero, absolute, 13
Zero ice, 34
Zero pressure, 20
Zone, comfort, 663
Zone control, 729, 914